BIOCHEMISTRY
The Chemical Reactions of Living Cells

BIOCHEMISTRY
The Chemical Reactions of Living Cells

David E. Metzler

Distinguished Professor Emeritus
Iowa State University

Volume 1

in association with
Carol M. Metzler

designed and illustrated by
David J. Sauke

HARCOURT
ACADEMIC
PRESS

San Diego London Boston
New York Sydney Tokyo Toronto

Second Edition

 **A discount coupon for Volume 2** can be found at the end of this volume.

Feedback. Please post typographical and scientific errors at **<http://www.harcourt-ap.com>**. Brief explanations of important alternative or controversial interpretations and important new information are also welcome.

Study question answers are posted at **<http://www.harcourt-ap.com>**.

Copyright © 2001, 1977 by HARCOURT/ACADEMIC PRESS ∞

Academic Press
A Harcourt Science and Technology Company
525 B Street, Suite 1900, San Diego, California 92101-4495, U.S.A.
http://www.academicpress.com

Academic Press
Harcourt Place, 32 Jamestown Road, London NW1 7BY, UK
http://www.academicpress.com

Harcourt/Academic Press
A Harcourt Science and Technology Company
200 Wheeler Road, Burlington, Massachusetts 01803
http://www.harcourt-ap.com

Library of Congress Catalog Card Number: 00-106082

International Standard Book Number: 0-12-492543-X (Set)
International Standard Book Number: 0-12-492540-5 (Volume 1)
International Standard Book Number: 0-12-492541-3 (Volume 2)

PRINTED IN THE UNITED STATES OF AMERICA
00 01 02 03 04 05 CO 9 8 7 6 5 4 3 2 1

A Brief
Table of Contents
Volume 1

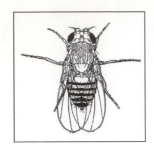

A Brief
Table of Contents
Volume 2

Volume 1

Chapter 1
The Scene of Action

Chapter 2
Amino Acids, Peptides, and Proteins

Chapter 3
Determining Structures and Analyzing Cells

Chapter 4
Sugars, Polysaccharides, and Glycoproteins

Chapter 5
The Nucleic Acids

Chapter 6
Thermodynamics and Biochemical Equilibria

Chapter 7
How Macromolecules Associate

Chapter 8
Lipids, Membranes, and Cell Coats

Chapter 9
Enzymes: The Catalysts of Cells

Chapter 10
An Introduction to Metabolism

Chapter 11
The Regulation of Enzymatic Activity and Metabolism

Chapter 12
Transferring Groups by Displacment Reactions

Chapter 13
Enolate Anions in Enzymatic Addition, Elimination, Isomerization, and Condensation Reactions

Chapter 14
Coenzymes: Nature's Special Reagents

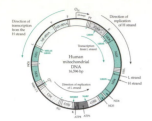

Human mitochondrial DNA 16,596-bp

Volume 2

Chapter 17
The Organization of Metabolism

Chapter 18
Electron Transport, Oxidative Phosphorylation, and Hydroxylation

Chapter 30
Chemical Communication between Cells

A. The Hormones
1. Receptors, Feedback Loops, and Cascades
2. The Vertebrate Hormones
3. Non-vertebrate Hormones
4. Plant Hormones

B. Neurochemistry
1. The Anatomy and Functions of Neurons
2. Organization of the Brain
3. Neuronal Pathways and Systems
4. The Propagation of Nerve Impulses
5. Ion Conducting Channels
6. Neuronal Metabolism
7. Synapses and Gap Junctions
8. Neurotransmitters
9. Neurotoxins
10. Mental Illness and Drugs
11. Addictive and Psychotropic Drugs
12. Odor, Taste, and Hearing
13. The Chemistry of Thinking
14. Circadian Cycles and Sleep

D. Development of Animal Embryos
1. Meiosis, Germ Cells, and Fertilization
2. Cell Axes and Patterns
3. Morphogens
4. Homeotic Genes
5. The Y Chromsome and the Sexes
6. Development of *Drosophila*
7. *Xenopus* and Limb Development

E. Mammalian Tissues
1. The Cardiovascular System
2. Blood Cells
3. Muscle; Cell Fusion
4. The Nervous System
5. The Endocrine System
6. Aging

F. The Development of Green Plants

G. Ecological Matters (Author's Personal Postscript)

Chapter 31
Biochemical Defense Mechanisms

A. Humoral Immunity
1. The Antibodies
2. Antigenicity
3. Triggering an Immune Response
4. Complement

B. Cell-mediated Immunity
1. Recognizing "Self": The Major Histocompatibility Complex
2. T-Cell Receptors
3. MHC Restriction

C. Genes for Immunoglobulins and T-Cell Receptors

D. Self Tolerance and Idiotypic Networks

E. Disorders of the Immune System

F. Other Defense Mechanisms

Chapter 32
Growth and Development

A. Growth and Differentiation of Bacteria
1. The Cell Cycle for *Escherichia coli*
2. Development of Bacterial Motility and Chemotaxis
3. Caulobacteria and Stem Cells
4. Spore Formation

B. Developmental Programs for Eukaryotes
1. The Cell Cycle
2. Transcriptional Controls
3. Programmed Alterations of the Genome
4. Programmed Cell Death
5. Cell-cell Signalling and Adhesion

C. Differentiation in Simple Eukaryotes
1. The Cellular Slime Molds
2. Yeast
3. The Hydra
4. Cell-constant Animals

Preface

It has been nearly 24 years since the first edition of this textbook appeared in 1977. During this period biochemistry and the related disciplines of biochemical genetics and molecular biology have blossomed. It is impossible to describe all the spectacular advances in knowledge in these areas in a single book. What I have tried to do is to provide a well-integrated, up-to-date and reliable text and reference book that covers the basic chemistry underlying biological phenomena and which conveys the excitement of present day biochemical studies. The book is written primarily for graduate students and for undergraduates with adequate training in chemistry. However, it contains enough introductory material that it should be useful in a variety of biochemistry courses as well as for self study.

The main text is focused on a clear presentation of biochemical principles. At the same time I have used boxes, as well as certain reference chapters, to add much additional material that I hope will be of interest to readers. The book has a high content of information but because it has been organized carefully I think that teachers will find it easy to use for biochemistry courses having a variety of organizational patterns. The book has a detailed index, tables of contents for individual chapters and many page cross-references to facilitate its use. The book is filled with information about the human body and about nature around us. Throughout the book the discoveries arising from application of the techniques of molecular biology, biochemical genetics and biophysical approaches have been introduced systematically. I hope that the result is not only a readable and reliable textbook but that it will serve also as a useful reference book for students as they continue their studies. The book contains over 17,000 references, more than 3000 of them to 1998-2001 literature. My goal in providing these references is to encourage students to pursue their own interests through study of current literature. Even though the references cited will not be "fresh" for long, they provide an introduction to many advanced or specialized topics. The papers cited have been written by many authors whose current publications can be sought for newer infor-mation. To facilitate use of the literature I have defined many terms and have tried to use currently accepted nomenclature while, at the same time, avoiding jargon and names that seem inappropriate or which I suspect will not stand the test of time.

The first edition contained 1050 pages of text in 16 chapters. I have doubled the number of chapters to allow for easier use and an improved organization. The first section, which now consists of 8 chapters, contains material on cell structure, molecular architecture, chemical properties of molecules, biochemical equilibria and thermodynamics. Chapter 1 provides quantitative information on sizes of cells and organelles, on the genetic complexity of organisms, and on the chemical composition of cells. Another part describes characteristics of important species considered in later chapters. It is designed to be helpful to students with a minimum of biological knowledge.

At the request of a number of teachers the new edition contains separate chapters on proteins, carbohydrates, nucleic acids, and lipids and membranes. Chapter 2, which describes protein structures and properties, also contains a brief review of structural principles, electrostatic interactions, hydrogen bonding and hydrophobic interactions, which are important for all classes of biopolymers. Chapter 3 is a condensed review of methods of separation and characterization of compounds and of modern methods of structure determination including X-ray diffraction and NMR techniques. The emphasis is on proteins. Chapter 4 deals with structures and reactions of carbohydrates and Chapter 5 with nucleic acids. In all cases a modern approach is followed. Short sections at the ends of these chapters provide additional information on methods.

The sixth chapter deals with thermodynamics and biochemical equilibria. A short review of basic ideas is followed by series of tabulated thermodynamic data. All of the Gibbs energy and enthalpy values cited later in the text are consistent with those in the tables. The SI unit of kilojoules per mole, which has become the accepted norm in the international biochemical literature, is used throughout. The change from kilocalories per mole is simple. An innovation, introduced in the first edition, is tabulation of Gibbs energies of combustion by NAD^+, whose use simplifies the arithmetic in evaluating Gibbs energy changes for metabolic processes. Thermodynamic quantities are given for individual ionic species, which facilitates approximate calculations. At the same time, the treatment of equilibrium constants will allow students to understand the complexities involved in a rigorous thermodynamic approach and also to understand the "Biochemical Standard State" proposed recently by R. A. Alberty and the International Union of Biochemistry. The section on biochemical equilibria in Chapter 6 describes the adenylate system as well as association and dissociation constants, acid–base chemistry of multiprotic compounds, and binding of metal ions. Chapter 7 deals with the binding of small molecules to macromolecules and with the structures of oligomeric enzymes, flagella, microtubules and viruses and hemoglobin. The allosteric effects so vital to these structures are also treated in depth. The chemical properties of lipids and the composition and properties of membranes and of the external cell coats that surround them are described in Chapter 8.

In the second major section of the book, which includes Chapters 9 through 16, the properties of enzymes, the kinetics of chemical reactions, and the specific mechanisms employed in enzymatic catalysis are discussed. Chapter 9 deals both with kinetics and with fundamental principles of enzymatic catalysis. Chapter 10 provides a brief introduction to metabolism, a background that will help students to appreciate the significance of the enzymes discussed in Chapters 12-16. The mechanisms by which enzymatic reactions are controlled within cells are considered in Chapter 11 as are important concepts about the regulation of metabolism. Chapters 12-16 describe specific groups of enzymes and coenzymes and the chemical mechanisms of their actions.

The third major section of the book treats reaction sequences found in metabolism. Following the approach introduced in the first edition, Chapter 17, which introduces volume 2, explores the logic of metabolic cycles and other pathways. Patterns are shown to arise as a natural result of the kind of chemistry required to obtain a necessary product from a given reactant. Many otherwise confusing aspects of metabolism are shown to be simple when it is understood that certain steps are needed to couple cleavage of ATP to biosynthesis. Chapter 18 deals with electron transport and ATP synthase, not only in mitochondria and chloroplasts but in bacterial membranes as well. Chapter 19, which considers the use of both "proticity" and ATP to drive the biological motors of muscles and microtubular transport follows logically. Chapters 20-22 are specialized chapters dealing with additional topics in the metabolism of carbohydrates and lipids. They may be used as a reference source.

A unique chapter, Light and Life (Chapter 23), covers photosynthesis, vision, and other biological responses to light, and also the fundamentals of light absorption spectra, circular dichroism, and fluorescence. Chapters 24 and 25 provide details of the biosynthesis and catabolism of an enormous number of nitrogenous compounds and aromatic substances. They can serve as a starting point for individual literature research projects by students.

Beginning the final section of the book, Chapter 26 provides an introduction to the biosynthesis of nucleic acids and proteins and also a succinct summary of methods used in the study of biochemical genetics. This is followed, in Chapter 27-29 with details of chromatin structure, the replication of DNA, transcription of genes, and synthesis of proteins. The final three chapters deal with cell communication and neurochemistry (Chapter 30), biological defense (Chapter 31) and differentiation and development (Chapter 32).

I acknowledge with thanks the many people who have helped me in the preparation of this text. Among them are my colleagues at Iowa State University and many scientists in other institutions. Special thanks are due to those who helped me to revise and rewrite the text and who entered the text and references into my computer. These include Wilma Holdren, Cathy Achenbach, Ann Kalvik, Nicole Long, Elitza Markova, and Kim McDermott, who has also managed all of the permission requests. I am extremely grateful to David Sauke who has been the artist and designer, and also to printing consultant Rob Louden. Carol Metzler looked after every reference citation to keep the bibliography as error free as possible, compiled the index and helped in numerous other ways. I am also grateful to my daughter Shanda, for her work on the project, and for her patience and friendly smiles of encouragement. Financial assistance was provided by my aunt Irene Whitaker.

David E. Metzler

Acknowledgments

I wish to express my appreciation to the following reviewers, each of whom read parts of the manuscript, for their generous assistance: Vernon Anderson, Case Western Reserve University; Amy Andreotti, Iowa State University; Jon B. Applequist, Iowa State University; Michael Blaber, Florida State University; Guillaume Chanfreau, University of California, Los Angeles; Eric E. Conn, University of California, Davis; H. Brian Dunford, University of Alberta; Herbert J. Fromm, Iowa State University; Michael Garrick, State University of New York at Buffalo; Donald Graves, Iowa State University; Mark R. Harpel, Dupont Pharmaceutical Company; Mark S. Hargrove, Iowa State University; Jack Horowitz, Iowa State University; Maja Irkovic-Jensen, University of Iowa; Nenad Kostić, Iowa State University; L. Andrew Lyon, Georgia Institute of Technology; John Macklin, University of Washington; Donald B. McCormick, Emory University; Ruby I. McDonald, Northwestern University; Alan Myers, Iowa State University; Andy Norris, Iowa State University; Bryce Plapp, University of Iowa; Daniel Purich, University of Florida College of Medicine; Dagmar Ringe, Brandeis University; John Robyt, Iowa State University; Philip Ryals, Mississippi State University; Robert Rucker, University of California, Davis; Louisa Tabatabai, National Animal Disease Center, ARS, USDA; Carl Tipton, Iowa State University; Joanne Yeh, Brown University.

Special thanks are due Jack Kirsch, University of California, Berkeley and to my colleagues at Iowa State University for providing many of the study questions.

Journal Acknowledgments

The cooperation of the publishers of journals in permitting inclusion of material in this book is gratefully acknowledged. The figures listed have been taken directly from the following journals or have been redrawn and adapted from published drawings. The literature citations appear in the chapter bibliographies.

Accounts of Chemical Research
Copyright by the American Chemical Society
Fig. 3-29B

Advances in Chemical Physics
Copyright by John Wiley and Sons, Inc.
Fig. 8-8B

Advances in Protein Chemistry
Copyright by Academic Press, London
Figs. 2-14, 2-16B, 2-22, 2-28

Annual Review of Biochemistry
Copyright by Annual Reviews
Fig. 7-25A, Fig. 8-27A,B, & Ch. 8 banner,
Fig. 8-29B

Annual Review of Biophysics and Biophysical Chemistry
Copyright by Annual Reviews
Fig. 13-8B, Fig. 13-9A & Ch. 13 banner (rt.)

Biochemical Journal
Copyright by the Biochemical Society
Fig. 3-9, Fig. 6-5

Biochemistry
Copyright by the American Chemical Society
Fig. 3-10, Fig. 3-26A,B, Fig. 3-28,
Fig. 3-29A, Fig. 4-21, Fig. 5-27,
Fig. 5-51 A,B,C, Fig 5-54C, Fig. 6-1,
Fig. 7-18A,B, Fig. 7-30C, Fig. 8-6B,
Fig. 8-9, Fig. 8-10A,B, Fig. in Box 8C,
Fig. 12-2, Fig. 12-5A,B, Fig. 12-7,
Fig. 12-12A,B, Fig. 12-19, Fig. 12-24,
Fig. 12-27A,B,C, Fig. 12-28A,B,C,
Fig. 12-28C, Fig. 12-32, Fig. 13-2A,
Fig. 13-4A, Fig. 13-6, Ch. 15 banner (rt.),
Fig. in Box 15C, Fig. 15-19, Fig. 15-23B,
Ch. 16 banner (left), Fig. 16-2A, Fig. 16-4,
Fig. 16-9, Fig. 16-20A, Fig. 16-24

Biochimica et Biophysica Acta
Copyright by Elsevier Science
Fig. 8-6A, Ch. 5 banner

Biophysical Journal
Copyright by the Biophysical Society
Fig. 5-14B, Fig. 8-7, Fig. 8-11,
Fig. 8-20A,B

Biopolymers
Copyright by John Wiley and Sons, Inc.
Fig. 5-11C

Bulletin of the Chemical Society of Japan
Copyright by The Chemical Society of Japan
Fig. 9-9A

Carbohydrate Research
Copyright by Elsevier Science
Fig. 4-19, Fig. 4-20

Cell
Copyright by Cell Press
Fig. 11-7B,C & Ch. 11 banner (rt.)

Chemical and Engineering News
Copyright by the American Chemical Society
Fig. in Box 8-G

Current Opinion in Structural Biology
Copyright by Elsevier Science
Fig. 7-7A

Current Topics in Membrane Transport
Copyright by Academic Press
Fig. 8-17

EMBO Journal
Copyright by the European Molecular Biology Organization
Fig. 5-15A,B, Fig. 8-20C, Fig. 16-20B,C

FASEB Journal
Copyright by The Federation of American Societies for Experimental Biology
Fig. 4-9, Fig. 8-33C, Fig. 11-5, Fig. 16-15A-D

Inorganica Chimica Acta
Copyright by Elsevier Science
Fig. 16-8C

Journal of the American Chemical Society
Copyright by The American Chemical Society
Fig. 5-14A, Fig. 5-25, Fig. 5-37, Fig. 8-22C,
Fig. 12-21, Fig. 12-22, Fig. 13-1B & Ch. 13 banner (left), Fig. 16-13, Fig. 16-21,
Fig. 16-26A,B

Journal of Biochemistry
Copyright by the Japanese Biochemical Society
Fig. 3-2, Fig. 15-11

Journal of Biological Chemistry
Copyright by the American Society for Biochemistry and Molecular Biology, Inc.
Fig. 2-21B, Fig. in Box 3C, Fig. 5-16A,B,C,
Fig. 5-48, Fig. 6-7, Fig. in Box 6D, Fig. 7-24B,
Fig. 8-14A-E, Fig. 8-16B, Fig. 8-19A,
Fig. 8-23, Fig. 8-24A,B, Fig. 8-28, Fig. 8-31,
Fig. 8-32A,B, Fig. 9-7A, Fig. 9-7B,
Fig. 12-20A-E, Fig. in Box 12C, Fig. 13-3,
Ch. 14 banner (rt.), Fig. 14-11, Fig. 15-9,
Fig.15-12, Fig. 15-14C,D, Ch. 16 banner (rt.),
Fig. 16-8A,B, Fig. 16-22A

Journal of Cell Biology
Copyright by The Rockefeller University Press
Fig. 1-15A

Journal of Molecular Biology
Copyright by Academic Press, London
Fig. 2-12, Fig. 2-13A, Fig. 2-16A, Fig. 2-30,
Fig. 3-19, Fig. 3-22, Fig. 3-25A,B, Fig 4-10,
Fig. 5-21A, Fig. 5-21B,C, Fig. 5-28B,
Fig. 5-32A,B, Fig. 5-36A,B, Fig 5-4,
Fig. 5-53A,B, Fig. 6-8, Fig. in Box 6E,
Ch. 7 banner (left), Fig. 7-5, Fig. 7-7B,C,
Fig. 7-8C, Fig. 7-12, Fig. 7-13A,B & Ch. 7 banner (cntr.), Fig. 7-14D, Fig. 7-20B,
Table 7-2, Fig. 7-28A,B, Fig. 7-29, Fig. 7-30B,
Fig. in Box 7A, (4,5,6), Fig. 8-19B, Fig. 9-14,
Fig. in Box 9F, Fig. in Box 10C, Fig. in Box 11A(B,C) Fig. 11-8B, Fig. 11-8C,
Fig. 12-3, Fig. 12-9, Fig. 12-30, Fig. 13-1A,
Fig. 13-7, Fig. 13-8A, Fig. 13-10A,B,
Fig. 13-11A, Fig. 15-10, Fig. 15-14A,B,
Fig. 15-23A & Ch. 15 banner (left),
Fig. 16-11, Fig. 16-16B

Microbiological Reviews
Copyright by the American Society for Microbiology
Fig. 4-5B

Nature (London)
Copyright by Macmillan Magazines Limited
Fig. 1-15E, Fig. in Box 2A, Fig. 2-21A,
Fig. 2-27, Fig. 3-18, Fig. 5-12B, Fig. 5-29,
Fig. 5-35A,B, Fig. 5-39A,B, Fig. 5-40, Fig. in Box 5D, Fig. 7-9B,C, Fig 7-25B, Fig. 7-26,
Fig. 7-34, Fig. 8-18B, Fig. 8-18C, Fig. 8-26A,B,
Fig. 11-7A, Fig. 11-8A, Fig. 11-12A,B,
Fig. 12-18, Fig. 12-28B, Fig. 16-16C

Nature New Biology
Copyright by Macmillan Magazines Limited
Fig. 5-9

Nature Structural Biology
Copyright by Macmillan Magazines Limited
Fig. 2-23D

Nucleic Acids Research
Copyright by Oxford University Press
Fig. 5-54A,B

Proceedings of the National Academy of Sciences, U. S. A.
Copyright by The National Academy of Sciences
Fig. 1-5, Fig 2-16C, Fig. 7-15A,B,
Fig. 7-30A, Fig. 8-18A, Ch. 12 banner,
Fig. 16-16A

Protein Science
Copyright by The Protein Society
Fig. in Box 11A (a) and Fig. Box 11A p. 13,
Fig. 11-6A,C & Ch. 11 banner (cntr.),
Fig. 12-6

Science
Copyright by the American Association for the Advancement of Science
Fig. 2-15, Fig. 2-17, Fig. 2-23B,C, Fig. 3-17,
Fig. 4-18A, Fig. 5-17, Fig. 5-18, Fig. 5-31A,B,
Fig. 5-38 A,B,C, Fig. 5-46 A,B, Fig. 7-24A,
Fig. 7-8B, Fig. in Box 7A (1,2) & Ch. 7 banner (rt.), Fig. in Box 7A (3), Fig. in Box 7A (7, 8, 9, 10), Fig. 8-16A,C, Fig. 8-21 A,B,C, Fig. 8-22A, Fig. 11-14, Fig. 12-26A,
Fig. 12-26B,C, Fig. 13-5, Fig. 14-2A,
Fig. 16-18A-E, Fig. 16-19, Fig. 16-25,
Fig. 16-31B,C

Scientific American
Copyright by Scientific American, Inc.
Fig. 7-8A

Starch/Stärke
Copyright by Wiley-VCH
Fig. 4-6

Structure
Copyright by Elsevier Science
Fig. 16-23

Trends in Biochemical Science
Copyright by Elsevier Science
Fig. 4-1, Fig. 5-52, Fig. 7-14A,B,C,
Fig. 7-19, Fig. 7-31A,B, Fig. 8-12A,B,
Fig. 12-10B, Fig. 16-3A, Fig. 16-10

Book Acknowledgments

The cooperation of the publishers of books in permitting inclusion of material in this book is gratefully acknowledged.

Academic Press
Reproduced by permission of the publisher.

Fig. 2-9 from *Conformation of Biopolymers*, Vol. 2, by V. Sasisekharian, A.V. Lakshminarayanan, and G.N. Ramachandran, © 1967

Fig. 3-3 from P. Flodin, and K. Aspberg, *Biol. Struct. Funct. Proc. IUB/IUBS Int. Sym.*, 1st ed., Vol. 1, © 1961

Fig. 4-14 from *Biogenesis of Plant Cell Wall Polysaccharides* by P. Albersheim, W.D. Bauer, K. Keestra, and K.W. Talmadge, (F. Loewus, ed.), © 1973

Fig. 5-45 from *The Biochemistry of Nucleic Acids* by J.N. Davidson, 7th ed., © 1972

Fig. 6-4 from *NMR in Biology* by P.J. Seeley, P.A. Sehr, D.G. Gadian, P.B. Garlick, and G.K. Radda, © 1977

Fig. 8-33A,B from *Extracellular Matrix Assembly and Structure* by Yurchenco, Birk, and Mecham, eds., ©1994

Fig. in Box 8F from *Black Skin Structure and Function* by W. Montagna, G. Prota, and J.A. Kenney, Jr., © 1993

American Society for Microbiology
Reproduced by permission of the publisher.

Box 7-C (Left) from *Bacteriophage T_4* by C.K Mathews, E.M. Kutter, G. Mosig, and P.B. Berget, eds., © 1983

Carnegie Institute
Reproduced by permission of the publisher.

Fig. 5-46 from *Carnegie Instit. Washington Yearbook*, 65th ed., by R.J. Britten and D.E. Kohne, © 1967

Chapman and Hall, Ltd.
Reproduced by permission of the publisher.

Table 8-2 from *An Introduction to the Chemistry and Biochemistry of Fatty Acids and Their Glycerides*, 2nd ed., by F.D. Gunstone, © 1967

Cold Spring Harbor Laboratory Press
Reproduced by permission of the publisher.

Fig. 3-16 from *Phage Display: A Laboratory Manual* by C.F. Barbas, D.B. Burton, J.K. Scott, and G.J. Silverman, © 2000

Cornell University Press
Reproduced by permission of the publisher.

Table 7-1 from *The Proton in Chemistry* by R.P. Bell, 2nd ed., © 1973

Elsevier Science Ltd.
Reproduced by permission of the publisher.

Fig. 4-16B from *Dynamics of Connective Tissue Macromolecules* by P.M.C. Burleigh and A.R. Poole, eds., © 1975

Fig. 7-27 from *Man's Hemoglobins*, by H. Lehmann and R.G. Huntsman, © 1966

Fig. 16-3B,C,D from *The Biochemistry and Physiology of Iron* by P.E. Bourne, P.M. Harrison, D.W. Rice, J.M.A. Smith, and R.F.D. Stansfield, (P. Saltman, J. Hegenauer, eds) © 1982

The McGraw-Hill Companies
Reproduced by permission of the publisher.

Fig. 8-31 from *The Metabolic and Molecular Bases of Inherited Disease*, 7th ed., Vol. 3, by P.H. Byers, (C.R. Scriver, A.L. Beaudet, W.S. Sly, D. Valle, eds), © 1995

John Wiley & Sons, Inc.
Reproduced by permission of the publisher.

Fig. 1-15C,D from *Cell Communication* by N.B. Gilula, © 1974

Table 6-10 from *Advanced Inorganic Chemistry*, 3rd ed., by F.A. Cotton and G. Wilkinson, © 1972

Fig. 8-8A from *Micelles, Monolayers, and Biomembranes* by M.N. Jones, and D. Chapman, © 1995

Fig. 8-34 from *Biological Mineralization* by J.A. Weatherell, and C. Robinson, (I. Zipkin, ed.), © 1973

Mosby
Reproduced by permission of the publisher.

Figs. 1-12A,C, 1-14A from *Biology of the Invertebrates* by C.P. Hickman, © 1973

Figs. 1-13B from *Integrated Principles of Zoology* by C.P. Hickman, © 1966

Oxford University Press
Reproduced by permission of the publisher.

Fig. 1-17 from *The Biological Chemistry of the Elements: The Inorganic Chemistry of Life* by J.J.R. Fraústo da Silva and R.J.P. Williams, © 1991

Portland Press Ltd.
Reproduced by permission of the publisher.

Fig. 5-10A,B,C from *Biochemical Nomenclature and Related Documents* by C. Liébecq, 2nd ed., © 1992

Saunders College Publishing
Reproduced by permission of the publisher.

Figs. 1-12B, 1-14C from *General Zoology*, 4th Ed., by C.A. Villee, W.F. Walker, and R.D. Barnes, © 1973

Waveland Press, Inc.
Reproduced by permission of the publisher.

Fig. 3-6 from *Biochemical Techniques, Theory and Practice*, by J.F. Robyt and B.J. White, © 1990

Williams and Wilkins Co.
Reproduced by permission of the publisher.

Fig. 1-15B from *Cellular Membranes and Tumor Cell Behavior, a collection of papers presented at the 28th annual Symposium on Fundamental Cancer Research*, by N.B. Gilula, © 1975

Many drawings in this book have been made using Dr. Per J. Kraulis' program MolScript. (Kraulis, P. J. (1991) *Journal of Applied Crystallography*, **24**, 946–950. MOLSCRIPT, a program to produce both detailed and schematic plots of protein structures.) These include Figures 2-6, 5-35B, 7-8C, 7-18, 7-20B, 7-30A, 8-19B, 8-20A,B,C, 8-24A, Box 11-A (fig. A), 12-3, 12-18, 15-19A,B, 18-10C, 19-15, 23-25, 23-30, 27-12A, and 28-6.

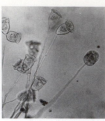

Left: Cells of the pathogenic O157:H7 strain of *Escherichia coli* attached to the surface epithelium of the cecum of a neonatal piglet. Electron-dense filaments (presumably polymerized actin) in the host cytoplasm can be seen subjacent to attached bacteria. The bacteria have effaced most micro-villi but some remain between the bacterial cells. Courtesy of Evelyn A. Dean-Nystrom, National Animal Disease Center, USDA, Agricultural Research Service, Ames, IA. Center: Many unicellular organisms such as these *Vorticella* inhabit wet and moist environments throughout the biosphere. Invertebrates have evolved as long as humans and have complex specializations such as the contractile stem of these protozoa. Courtesy of Ralph Buchsbaum. Right: Although 97% of animals are invertebrates, ~ 3% of the several million known species have backbones. Giraffe: © M. P. Kahl, Photo Researchers

Contents

The Scene of Action

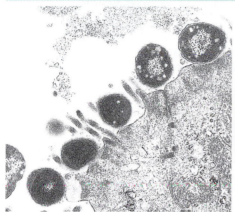 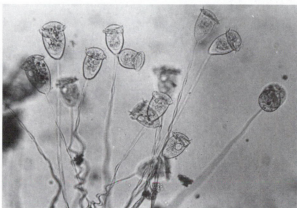 

This book is about the chemistry of living cells with special emphasis on the trillions of cells that make up your own body. Every aspect of life depends upon the chemical makeup of cells and on the chemical properties of the remarkable molecules found within the cells. The information presented here will give the reader a solid foundation for understanding not only the chemical basis of life but also the revolutionary developments in molecular biology, biochemical genetics, medicine, and agriculture which dominate today's scientific news and which will play an increasingly important role in our lives.

The first theme of the book is **biomolecular structure**. We'll look carefully at the complex structures of proteins, carbohydrates, RNA, DNA, and many other substances. We'll not only examine in-depth their molecular architecture but also study the chemical properties that make life possible.

A second theme is **metabolism**, the unceasing, complex network of thousands of chemical reactions by which cells grow and reproduce, take up foods and excrete wastes, move, and communicate with each other. Within cells we have a **steady state**, a condition in which the complex chemical constituents of cells are continuously being synthesized in one series of reactions and degraded in another. The result is a marvelous system of self-renewal or "turnover" of tissues. We'll examine the chemical reactions involved in these processes as well as the ways in which they are controlled. We will consider both the reaction sequences and the techniques such as cloning of genes, isotopic labeling, X-ray diffraction, and nuclear magnetic resonance spectroscopy, which are used today to study metabolism.

Human beings are surrounded by many other living creatures whose activities are important to us. Photosynthetic organisms obtain energy from sunlight and synthesize compounds that the human body requires but cannot make. Microorganisms cause decay of organic matter and convert it into forms usable by plants. This book deals with the chemical reactions occurring in all of these organisms. We'll look at strange and unusual reactions, along with those metabolic sequences common to most living things.

Each one of the thousands of chemical reactions of metabolism is catalyzed by an **enzyme**. Most of these enzymes are proteins, but others are made from **RNA** (ribonucleic acid). In both cases enzymes are very large molecules with precise three-dimensional structures. The study of the properties of enzymes and of **enzymatic catalysis** is a third theme of the book. Not only are the chemical mechanisms by which enzymes act of interest but also enzymes are often targets for useful drugs. Incorrectly formed enzymes can result in serious diseases.

The sequences of the amino acids in the chains from which proteins are constructed are encoded in the nucleotide sequences of **DNA** (deoxyribonucleic acid). The coding sequence for a protein in the DNA is found in the **structural gene** for that protein. The RNA enzymes are also encoded by DNA genes. A fourth major theme of the book deals with the nature of the **genetic code** used in DNA and with the processes by which cells read and interpret the code. It also includes study of the methods by which thousands of genes have been mapped to specific positions in chromosomes, isolated, cloned, and sequenced.

A large number of proteins present in the outer surfaces of cells serve as **receptors** that receive chemical messages and other signals from outside the cell. The receptors, which are sometimes enzymes, respond by generating internal signals that control metabolism and cell growth. Such **molecular signaling** is another major area of contemporary biochemistry.

Biologists have described over a million species, and several millions of others probably exist.[1] Many of these organisms have very specialized ways of life. However, they all have much chemistry in common. The same 20 amino acids can be isolated from proteins of plants, animals, and microorganisms. Formation of lactic acid in both bacteria and human muscle requires the same enzymes. Except for some small variations, the genetic code is universal—the same for all organisms. Thus, there is a unity of life and we can study metabolism as the entirety of chemical transformations going on in all living things. However, the differences among species are also impressive. Each species has its own gene for almost every protein.

When the enzyme that catalyzes a particular metabolic reaction is isolated from a number of different organisms, it is usually found to have similar properties and a similar mechanism of catalysis, regardless of the source. However, the *exact* sequence of amino acids in the enzyme will be almost unique to the organism that produced it. When the three-dimensional structures are compared it is found that differences between species often affect only the peripheral parts of an enzyme molecule. The interior structure of the protein, including the catalytic machinery, is highly conserved. However, the surface regions, which often interact with other macromolecules, vary greatly. Such interactions help to control metabolism and may account for many differences in the metabolism among living beings.

Variations in protein structures are not limited to differences between species. Individuals differ from one another. Serious genetic diseases sometimes result from the replacement of a single amino acid unit in a protein by a different amino acid. Genetic deviations from the "normal" structure of a protein result from **mutations**. Many mutations, whether they occurred initially in our own cells or in those of our ancestors, are detrimental.

However, such mutations also account for variation among individuals of a species and allow for evolution. The chemical nature and consequences of mutations and their significance to health, medicine, and agriculture are dealt with throughout the book. We now have reliable methods for inducing in the laboratory mutations at any specific place in a protein sequence and also for synthesizing new DNA sequences. These techniques of **genetic engineering** have given biochemists the ability to modify protein structures freely, to create entirely new proteins, and to provide a basis for the rapidly developing field of **genetic therapy**.

It should be clear from this introduction that **biochemistry** deals with virtually every aspect of life. The distinguishing feature of the science is that it approaches biological questions in terms of the underlying chemistry. The term **molecular biology** is often regarded as synonymous with biochemistry. However, some scientists use it to imply a more

biological approach. These molecular biologists also emphasize structure and function but may have a goal of understanding biological relationships more than chemical details. **Biophysics**, a closely related science, encompasses the application of physical and mathematical tools to the study of life.

A. The Simplest Living Things

The simplest organisms are the **bacteria**.[2–5] Their cells are called **prokaryotic** (or procaryotic) because no membrane-enclosed nucleus is present. Cells of all other organisms contain nuclei separated from the cytoplasm by membranes. They are called **eukaryotic**. While viruses (Chapter 5) are sometimes regarded as living beings, these amazing parasitic objects are not complete organisms and have little or no metabolism of their own. The smallest bacteria are the **mycoplasmas**.[6–8] They do not have the rigid cell wall characteristic of most bacteria. For this reason they are easily deformed and often pass through filters designed to stop bacteria. They are nutritionally fussy and are usually, if not always, parasitic. Some live harmlessly in mucous membranes of people, but others cause diseases.

BOX 1-A ABOUT MEASUREMENTS

In 1960 the International General Conference on Weights and Measures adopted an improved form of the metric system, **The International System of Units** (SI). The units of mass, length, and time are the kilogram (kg), meter (m), and second (s). The following prefixes are used for fractions and multiples:

10^{-18}, *atto* (a)	10^{-6}, *micro* (μ)	10^{9}, *giga* (G)
10^{-15}, *femto* (f)	10^{-3}, *milli* (m)	10^{12}, *tera* (T)
10^{-12}, *pico* (p)	10^{3}, *kilo* (k)	10^{15}, *peta* (P)
10^{-9}, *nano* (n)	10^{6}, *mega* (M)	10^{18}, *exa* (E)

There is an inconsistency in that the prefixes are applied to the gram (g) rather than to the basic unit, the kilogram.

SI units have been used throughout the book whenever possible. There are no feet, microns, miles, or tons. Molecular dimensions are given uniformly in nanometers rather than in angstrom units (Å; 1Å = 0.1 nm). Likewise the calorie and kilocalorie have been replaced by the SI unit of energy, the **joule** (J; 1 calorie = 4.184 J).

Throughout the book frequent use is made of the following symbols:

, "appr oximately equal to"
~, "approximately" or "about"

For example, *Mycoplasma pneumoniae* is responsible for primary atypical pneumonia.

Cells of mycoplasmas sometimes grow as filaments but are often spherical and as small as 0.3 micrometer (μm) in diameter. Their outer surface consists of a thin **cell membrane** about 8 nanometers (nm) thick. This membrane encloses the **cytoplasm**, a fluid material containing many dissolved substances as well as submicroscopic particles. At the center of each cell is a single, highly folded molecule of DNA, which constitutes the bacterial chromosome. Besides the DNA there may be, in a small spherical mycoplasma, about 1000 particles ~20 nm in diameter, the **ribosomes**. These ribosomes are the centers of protein synthesis. Included in the cytoplasm are many different kinds of proteins, but there is room for a total of only about 50,000 protein molecules. Several types of RNA as well as many smaller molecules are also present. Although we don't know what minimum quantities of proteins, DNA, and other materials are needed to make a living cell, it is clear that they must all fit into the tiny cell of the mycoplasma.

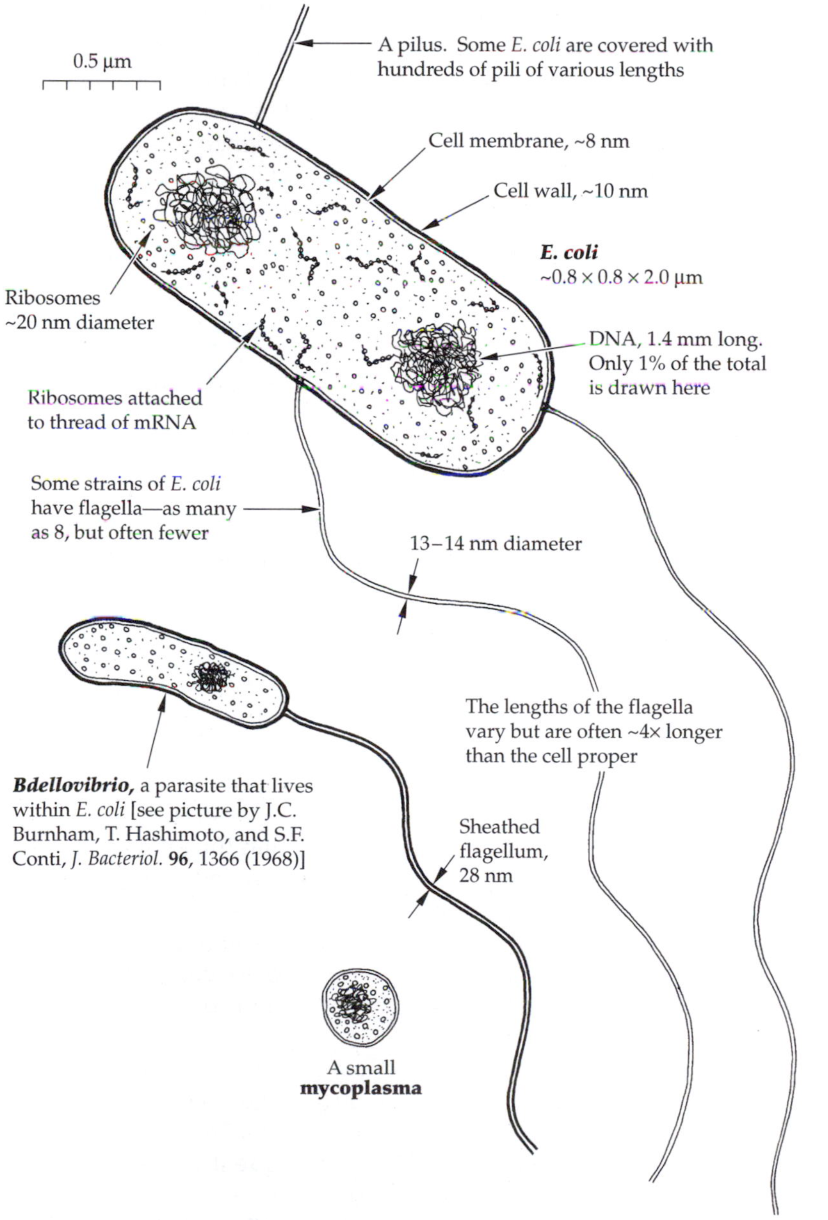

1. *Escherichia coli*

The biochemist's best friend is *Escherichia coli*, an ordinarily harmless inhabitant of our intestinal tract. This bacterium is easy to grow in the laboratory and has become the best understood organism at the molecular level.[4,9] It may be regarded as a typical true bacterium or **eubacterium**. The cell of *E. coli* (Figs. 1-1, 1-2) is a rod ~2 μm long and 0.8 μm in diameter with a volume of ~1 μm^3 and a density of ~1.1 g/cm^3. The mass is ~1 x 10^{-12} g, i.e., 1 picogram (pg) or ~0.7 x 10^{12} daltons (Da) (see Box 1-B).[4] It is about 100 times bigger than the smallest mycoplasma but the internal structure, as revealed by the electron microscope, resembles that of a mycoplasma.

Each cell of *E. coli* contains from one to four identical DNA molecules, depending upon how fast the cell is growing, and ~15,000–30,000 ribosomes. Other particles that are sometimes seen within bacteria include food stores such as fat droplets and granules (Fig. 1-3). The granules often consist of **poly-β-hydroxybutyric acid**[10] accounting for up to 25% of the weight of *Bacillus megaterium*. **Polymetaphosphate**, a highly polymerized phosphoric acid, is sometimes stored in "metachromatic granules." In addition, there may be droplets of a separate aqueous phase, known as **vacuoles**.

2. The Bacterial Genome

The genetic instructions for a cell are found in the **DNA molecules**. All DNA is derived from four different kinds of monomers, which we call **nucleotides**. DNA molecules are double-stranded: two polymer chains are coiled together, their nucleotide units being associated as **nucleotide pairs** (see Fig. 5-7). The genetic messages in the DNA are in the form of

0.5 μm

A pilus. Some *E. coli* are covered with hundreds of pili of various lengths

Cell membrane, ~8 nm

Cell wall, ~10 nm

E. coli
~0.8 × 0.8 × 2.0 μm

Ribosomes ~20 nm diameter

Ribosomes attached to thread of mRNA

DNA, 1.4 mm long. Only 1% of the total is drawn here

Some strains of *E. coli* have flagella—as many as 8, but often fewer

13–14 nm diameter

The lengths of the flagella vary but are often ~4× longer than the cell proper

Bdellovibrio, a parasite that lives within *E. coli* [see picture by J.C. Burnham, T. Hashimoto, and S.F. Conti, *J. Bacteriol.* **96**, 1366 (1968)]

Sheathed flagellum, 28 nm

A small **mycoplasma**

Figure 1-1 *Escherichia coli* and some smaller bacteria.

sequences of nucleotides. These sequences usually consist of a series of code "words" or **codons**. Each codon is composed of three successive nucleotides and specifies which one of the 20 different kinds of amino acids will be used at a particular location in a protein. The sequence of codons in the DNA tells a cell how to order the amino acids for construction of its many different proteins.

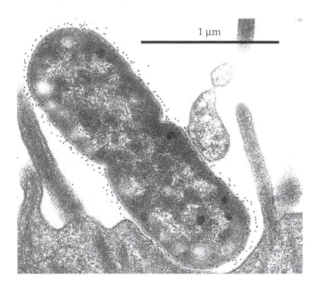

Figure 1-2 Transmission electron micrograph of a dividing cell of *Escherichia coli* O157:H7 attached to the intestinal epithelium of a neonatal calf. These bacteria, which are able to efface the intestinal microvilli, form characteristic attachments, and secrete shiga toxins, are responsible for ~73,000 illnesses and 60 deaths per year in the U. S.[10a] After embedding, the glutaraldehyde-fixed tissue section was immunostained with goat anti-O157 IgG followed by protein A conjugated to 10-nm gold particles. These are seen around the periphery of the cell bound to the O-antigen (see Fig. 8-28). Notice the two microvilli of the epithelium. Courtesy of Evelyn A. Dean-Nystrom, National Animal Disease Center, USDA, Agricultural Research Service, Ames, IA.

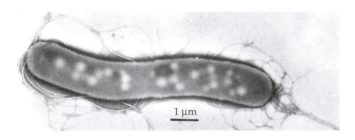

Figure 1-3 A cell of a *Spirillum* negatively stained with phosphotungstic acid. Note the tufts of flagella at the ends, the rough appearance of the outer surface, the dark granules of poly-β-hydroxybutyric acid and the light-colored granules of unknown nature. Courtesy of F. D. Williams, Gail E. VanderMolen, and C. F. Amstein.

Assume that a typical protein molecule consists of a folded chain of 400 amino acids. Its structural gene will therefore be a sequence of 1200 nucleotide pairs. Allowing a few more nucleotides to form spacer regions between genes we can take ~1300 as the number of nucleotide pairs in a typical bacterial gene. However, some genes may be longer and some may be much shorter. The **genome** is the quantity of DNA that carries a complete set of genetic instructions for an organism. In bacteria, the genome is a single chromosome consisting of one double-stranded DNA molecule. *Mycoplasma genitalium* is the smallest organism for which the DNA sequence is known.[11] Its genome is a double-helical DNA circle of 580,070 nucleotide pairs and appears to contain about 480 genes (an average of ~1200 nucleotides per gene).

The average mass of a nucleotide pair (as the disodium salt) is 664 Da. It follows that the DNA of *M. genitalium* has a mass of ~385 x 10^6 Da. The relative molecular mass (M_r) is 0.385 x 10^9 (See Box 1-B for definitions of dalton and M_r). The DNA of *E. coli* is about seven times larger with a mass of ~2.7 x 10^9 Da. It contains ~4.2 x 10^6 nucleotide pairs and encodes over 4000 different proteins (see Table 1-3).

Each nucleotide pair contributes 0.34 nm to the length of the DNA molecule; thus, the total length of DNA of an *E. coli* chromosome is 1.4 mm. This is about 700 times the length of the cell which contains it. Clearly, the molecules of DNA are highly folded, a fact that accounts for their appearance in the electron microscope as dense aggregates called **nucleoids**, which occupy about one-fifth of the cell volume (Fig. 1-4).

BOX 1-B RELATIVE MOLECULAR MASS, M_r, AND DALTONS

Atomic and molecular masses are assigned relative to the mass of the carbon isotope, ^{12}C, whose atomic weight is defined as exactly 12. The actual mass of a single atom of ^{12}C is defined as 12 **daltons**, one dalton being 1.661 x 10^{-24} g. The mass of a molecule can be given in daltons **(Da)** or kilodaltons **(kDa)**. This molecular mass in daltons is numerically equivalent to the relative molecular mass **(M_r)** or molecular weight **(MW)**[a] and also to the molar mass **(g/mol)**. However, it is not correct to use the dalton for the unitless quantity M_r. Masses of structures such as chromosomes, ribosomes, mitochondria, viruses, and whole cells as well as macromolecules can be given in daltons.[b]

---

[a] The Union of Pure and Applied Chemistry renamed molecular weight as **relative molecular mass** with the symbol M_r; M_r = MW.
[b] J. T. Edsall (1970) *Nature* (*London*) **228**, 888.

Each bacterial nucleoid contains a complete set of genetic "blueprints" and functions independently. Each nucleoid is **haploid**, meaning that it contains only a single complete set of genes. In addition to their chromosome, bacteria often contain smaller DNA molecules known as **plasmids**. These plasmids also carry genetic information that may be useful to bacteria. For example, they often encode proteins that confer resistance to antibiotics. The ability to acquire new genes from plasmids is one mechanism that allows bacteria to adapt readily to new environments.[12] Plasmids are also used in the laboratory in the cloning of genes and in genetic engineering (Chapter 26).

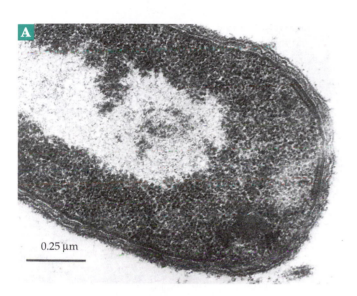

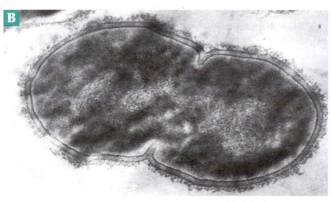

Figure 1-4 (A) Thin (~60 nm) section of an aquatic gram-negative bacterium, *Aquaspirillum fasciculus*. Note the light-colored DNA, the dark ribosomes, the double membrane characteristics of gram-negative bacteria (Chapter 8, Section E), and the cell wall. In addition, an internal "polar membrane" is seen at the end. It may be involved in some way in the action of the flagella. (B) A thin section of dividing cell of *Streptococcus*, a gram-positive organism. Note the DNA (light-stranded material). A portion of a mesosome is seen in the center and septum can be seen forming between the cells. Micrographs courtesy of F. D. Williams, Gail E. VanderMolen, and C. F. Amstein.

3. Ribonucleic Acids (RNA) and the Transcription and Translation of Genetic Information

The genetic information in the DNA is not utilized directly by the protein-synthesizing machinery of cells. Instead, molecules of ribonucleic acid (RNA) are synthesized according to the instructions encoded in the DNA, a process called **transcription**. Although they differ from DNA significantly in their structure, these RNA molecules carry the same coded information as is found in a length of DNA that contains one or a few genes. If DNA is regarded as the "master blueprint" of the cell, molecules of RNA are "secondary blueprints." This concept is embodied in the name **messenger RNA** (mRNA) which is applied to a small, short-lived fraction of RNA that carries information specifying amino acid sequences of proteins. Each molecule of mRNA carries the genetic message from one or more genes to the ribosomes where the proteins are made.

Ribosomes are extraordinarily complex little protein-synthesizing machines. Each ribosome of *E. coli* has a mass of 2.7×10^6 Da and contains 65% of a stable **ribosomal RNA** and ~35% protein. About 50 different kinds of protein molecules are present as parts of the ribosomal structure. Working together with a variety of **transfer RNA** molecules and enzymes, the ribosomes are able to read the genetic messages from mRNA and to accurately assemble any kind of protein molecule that a gene may specify. This process is called **translation** of the genetic message.

4. Membranes and Cell Walls

Like the mycoplasma, the *E. coli* cell is bounded by an 8-nm membrane which consists of ~50% protein and 50% lipid. When "stained" (e.g., with permanganate) for electron microscopy, this single membrane appears as two very thin (2.0 nm) dark lines separated by an unstained center band (~3.5 nm) (Fig. 1-4; see also Fig. 8-4). Single membranes of approximately the same thickness and staining behavior occur in all cells, both of bacteria and of eukaryotes.

A cell membrane is much more than just a sack. It serves to control the passage of small molecules into and out of the cell. Its outer surface carries receptors for recognition of various materials. The inside surface of bacterial membranes contains enzymes that catalyze most of the oxidative metabolism of the cells. Bacterial cell membranes are sometimes folded inward to form internal structures involved in photosynthesis or other specialized reactions of metabolism such as oxidation of ammonia to nitrate.[2] In *E. coli* replication of DNA seems to occur on certain parts of the membrane surface, probably under the control of membrane-bound enzymes. The formation of the new membrane which divides multiplying cells proceeds synchronously with the synthesis of DNA.

A characteristic of true bacteria (**eubacteria**) is a rigid **cell wall** which surrounds the cell membrane. The 40-nm-thick wall of *E. coli* is a complex, layered structure five times thicker than the cell membrane. Its chemical makeup is considered in Chapter 8. One of the layers is often referred to as the **outer membrane**. In some bacteria the wall may be as much as 80 nm thick and may be further surrounded by a thick **capsule** or **glycocalyx** (slime layer).[13] The main function of the wall seems to be to prevent osmotic swelling and bursting of the bacterial cell when the surrounding medium is hypotonic.

If the osmotic pressure of the medium is not too low, bacterial cell walls can sometimes be dissolved, leaving living cells bounded only by their membranes. Such **protoplasts** can be produced by action of the enzyme lysozyme on gram-positive bacteria such as *Bacillus megaterium*. Treatment of cells of gram-negative bacteria with penicillin (Box 20-G) produces **spheroplasts**, cells with partially disrupted walls. Spheroplasts and protoplasts are useful in biochemical studies because substances enter cells more readily when the cell wall is absent. Strains of bacteria lacking rigid walls are known as **L forms**.

5. Flagella and Pili

Many bacteria swim at speeds of 20–60 μm/s, ten or more body lengths per second! Very thin thread-like **flagella** of diameter 13–20 nm coiled into a helical form are rotated by the world's smallest "electric motors" to provide the motion.[14] While some bacteria have a single flagellum, the corkscrew-like *Spirillum* (Fig. 1-3) synchronously moves tufts of flagella at both ends. Some strains of *E. coli* have no flagella, but others contain as many as eight flagella per cell distributed over the surface. The flagella stream out behind in a bundle when the bacterium swims. The flagella of the helical **spirochetes** are located inside the outer membrane.[15,16]

In addition to flagella, extremely thin, long, straight filaments known as **pili** or **fimbriae** (Fig. 1-2) project from the surfaces of many bacteria.[14] The "sex pili" (F pili and I pili) of *E. coli* have a specific role in sexual conjugation. The similar but more numerous common pili or fimbriae range in thickness from 3 to 25 nm and in length from 0.2 to 2 μm. Pili are involved in adhesion of bacteria to surrounding materials or to other bacteria and facilitate bacterial infections.[17–19] A typical *E. coli* cell has 100–300 pili.[5]

6. Classification and Evolution of Bacteria

Bacteria vary greatly in their chemistry and metabolism, and it is difficult to classify them in a rational way. In higher organisms species are often defined as forms that cannot interbreed and produce fertile offspring, but such a criterion is meaningless for bacteria whose reproduction is largely asexual and which are able readily to accept "visiting genes" from other bacteria.[12] The classification into species and genera is therefore somewhat arbitrary. A currently used scheme (Table 1-1)[20] classifies the prokaryotes into 35 groups on the basis of many characteristics including shape, staining behavior, and chemical activities. Table 1-1 also includes genus names of most of the bacteria discussed in this book.

Bacteria may have the shape of spheres or straight or curved rods. Some, such as the **actinomycetes**, grow in a branching filamentous form. Words used to describe bacteria often refer to these shapes: a **coccus** is a sphere, a **bacillus** a rod, and a **vibrio** a curved rod with a flagellum at one end. A **spirillum** is screw-shaped. These same words are frequently used to name particular genera or families. Other names are derived from some chemical activity of the bacterium being described.

The **gram stain** provides an important criterion of classification that depends upon differences in the structure of the cell wall (see Chapter 20). Bacterial cells are described as **gram-positive** or **gram-negative** according to their ability to retain the basic dye crystal violet as an iodine complex. This difference distinguishes two of four large categories of bacteria.[20] Most actinomycetes, the spore-forming bacilli, and most cocci are gram-positive, while *E. coli*, other enterobacteria, and pseudomonads are gram-negative. A third category consists of eubacteria that lack cell walls, e.g. the mycoplasma.

Comparisons of amino acid sequences of proteins and the nucleotide sequences of DNA and RNA have provided a new approach to classification of bacteria.[21–28] Although the origins of life are obscure, we can easily observe that the genome changes with time through mutation and through the enzyme-catalyzed process of **genetic recombination**. The latter gives rise to the deletion of some nucleotides and the insertion of others into a DNA chain. When we examine sequences of closely related species, such as *E. coli* and *Salmonella typhimurium*, we find that the sequences are very similar. However, they differ greatly from those of many other bacteria. Consider the 23S ribosomal RNA, a molecule found in the ribosomes of all bacteria. It contains ~3300 nucleotides in a single highly folded chain. The basic structure is highly conserved but between any two species of bacteria there are many nucleotide substitutions caused by mutations as well as deletions and insertions. By asking what is the minimum number of mutations that could have converted one 23S RNA into another and by assuming a more or less constant rate of mutation over millions of years it is possible to construct a **phylogenetic tree** such as that shown in Fig. 1-5.

One conclusion from these comparisons is that the methane-producing bacteria, the **methanogens**,[24] are only distantly related to most other bacteria. Methano-gens together with the cell wall-less *Thermoplasma*,[28] some salt-loving **halobacteria**, and some **thermophilic** (heat-loving) sulfur bacteria form a fourth major category. They are often regarded as a separate kingdom, the **archaeobacteria**,[25] which together with the kingdom of the eubacteria form the superkingdom prokaryota. Certain archaeobacteria have biochemical characteristics resembling those of eukaryotes and

TABLE 1-1
A Systematic Classification Scheme for Bacteria[a,b]

Kingdom Procaryotae

The bacteria are classified according to the following 35 groups. Within these groups many genera are classified into subgroups or families. A few genera, most of which are mentioned elsewhere in this book, are listed here by name. Members of a single subgroup are placed together and are separated by semicolons from members of other subgroups.

1. The spirochetes (long bacteria, up to 500 μm, that are propelled by the action of filaments wrapped around the cell between the membrane and wall). *Borrelia* (*B. burgdorferi*, Lyme disease), *Leptospira*, *Treponema* (*T. pallidum*, syphilis)
2. Aerobic spiral and curved motile gram-negative bacteria. *Bdellovibrio*, *Campylobacter* (*C. jejuni*, diarrhea), *Helicobacter* (*H. pylori*, gastric ulcers), *Spirillum*
3. Nonmotile gram-negative curved bacteria
4. Gram-negative, aerobic rods and cocci. *Acetobacter*, *Agrobacterium*, *Azotobacter*, *Brucella* (*B. abortus*, brucellosis), *Flavobacterium*, *Gluconobacter*, *Legionella* (*L. pneumophila*, Legionnaire's disease), *Methylomonas*, *Neisseria* (*N. gonorrhea*, gonorrhea), *Pseudomonas*, *Rhizobium*, *Thermus*, *Xanthomonas*, *Rochalimaea* (*R. henselae*, cat scratch disease)
5. Gram-negative, facultatively anaerobic rods. *Enterobacter*,[c] *Proteus*, *Yersinia* (*Y. pestis*, plague), *Escherichia*, *Klebsiella*, *Salmonella* (*S. typhi*, typhoid fever), *Serratia*, *Shigella* (*S. dysenteriae*, bacterial dysentery), *Haemophilus*; *Vibrio* (*V. cholerae*, Asiatic cholera); *Zymomonas*
6. Gram-negative, anaerobic bacteria. *Butyrivibrio*
7. Dissimilatory sulfate- or sulfur-reducing bacteria. *Desulfovibrio*
8. Anaerobic gram-negative cocci. *Veillonella*
9. The rickettsias (parasitic bacteria with exacting nutritional requirements and small genome sizes) and chlamydias. *Chlamydia* (*C. trachomatis*, trachoma), *Rickettsia* (*R. rickettsii*, Rocky Mountain spotted fever)
10. Anoxygenic photosynthetic bacteria. Green sulfur bacteria. *Chlorobium*, *Prosthecochloris*; purple nonsulfur bacteria: *Rhodopseudomonas*, *Rhodospirillum*; purple sulfur bacteria: *Chromatium*, *Thiospirillum*

11. Oxygenic photosynthetic bacteria. Cyanobacteria (blue-green algae): *Synechocystis*; *Anabaena*, *Nostoc*; *Oscillatoria*
12. Aerobic, chemolithotrophic bacteria. Colorless sulfur bacteria: *Thiobacillus*; iron or manganese-oxidizing bacteria, magnetotactic bacteria; nitrifying bacteria: *Nitrobacter*, *Nitrosomonas*
13. Budding and/or appendaged bacteria. *Caulobacter*
14. Sheathed bacteria
15. Nonphotosynthetic, nonfruiting gliding bacteria. *Beggiatoa* (a filamentous bacterium containing sulfur granules)
16. Fruiting, gliding bacteria. *Myxococcus*
17. Gram-positive cocci. *Leuconostoc*, *Micrococcus*, *Peptococcus*, *Staphylococcus* (*S. aureus*, boils, infections), *Streptococcus* (*S. pyogenes*, scarlet fever, throat infections, *S. pneumoniae*, pneumonia)
18. Endospore-forming gram-positive rods and cocci. Aerobic: *Bacillus* (*B. anthracis*, anthrax), anaerobic: *Clostridium* (*C. tetani*, tetanus; *C. botulinum*, botulism)
19. Regular nonsporing gram-positive rods. *Lactobacillus*
20. Irregular nonsporing gram-positive rods. *Actinomyces*, *Bifidobacterium*, *Corynebacterium* (*C. diphtheriae*, diphtheria), *Propionibacterium*
21. Mycobacteria. *Mycobacterium* (*M. tuberculosis*, tuberculosis; *M. leprae*, leprosy)

22-29. Actinomycetes
30. Mycoplasmas. *Acholeplasma*, *Mycoplasma*
31. Methanogens. *Methanobacterium*; *Methanosarcina*; *Methanospirillum*
32. Archaeal sulfate reducers
33. Halobacteria. *Halobacterium*
34. Cell wall-less archaeobacteria. *Thermoplasma*
35. Very thermophilic S^0-Metabolizers. *Sulfolobus*; *Thermococcus*

---

[a] From Bergey's Manual of Systematic Bacteriology, 9th ed. J. G. Holt, N. R. Krieg, P. H. A. Sneath, J. T. Staley and S. T. Williams, Eds. (1994) Williams and Wilkins, Baltimore, Maryland. For another recent list see http://www.ncbi.nlm.nih.gov/

[b] The human diseases caused by some species are also listed.

[c] Formerly *Aerobacter*.

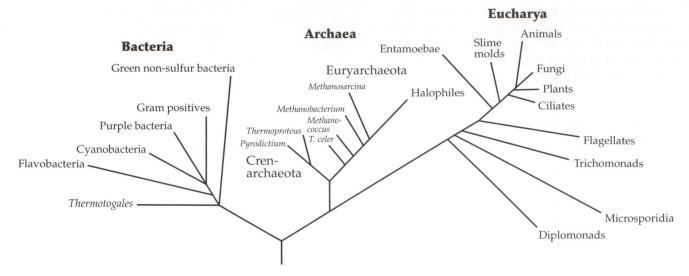

Figure 1-5 Universal phylogenetic tree. From Wheelis *et al.*[29]

some biologists therefore classify them as **archaea** and rank their kingdom as equal to that of the bacteria and the eukaryotes (Fig. 1-5).[27,29,30,30a,30b] Others disagree.[31] In Table 1-1, the archaeobacteria are found in groups 31–35. Most bacteria are very small in size but there are species large enough to be confused with eukaryotic protozoa. The record for bacteria seems to be held by *Epulopiscium fishelsoni*, a parasite of the surgeonfish intestinal tract. A single cell measured > 600 μm by 80 μm diameter, over 10^6 times larger in volume than a cell of *E. coli*.[32] The organism is a gram-positive bacterium as judged by analysis of its cloned ribosomal RNA genes.

7. Nutrition and Growth of Bacteria

Autotrophic (self-nourishing) bacteria can synthesize all of their organic cell constituents from carbon dioxide, water, and inorganic forms of nitrogen and sulfur. The **photoautotrophs** extract their energy from sunlight, while the **chemoautotrophs** obtain energy from inorganic chemical reactions. For example, the hydrogen bacteria oxidize H_2 to H_2O and sulfur bacteria oxidize H_2S to H_2SO_4. Like the fungi and animals, most bacteria are **chemoheterotrophic**; they obtain energy from the breakdown of organic compounds. Some of these heterotrophic bacteria are **anaerobes** which live without O_2. Many of them metabolize complex organic substances such as sugars in the absence of oxygen, a process called **fermentation**. Others oxidize organic compounds with an inorganic oxidant such as nitrite or sulfate. Members of the genus *Clostridium* are poisoned by oxygen and are known as **obligate anaerobes**. Others, including *E. coli*, are **facultative anaerobes**, able to grow either

in the presence or in the absence of oxygen. **Obligate aerobes** depend for energy upon combustion of organic compounds with oxygen.

One of the largest groups of strictly aerobic heterotrophic bacteria, the pseudomonads (*Pseudomonas* and related genera), are of interest to biochemists because of their ability to oxidize organic compounds, such as alkanes, aromatic hydrocarbons, and steroids, which are not attacked by most other bacteria. Often, the number of oxidative reactions used by any one species of bacteria is limited. For example, the acetic acid bacteria that live in wine and beer obtain all of their energy by oxidation of ethanol to acetic acid:

$$CH_3CH_2OH + O_2 \rightarrow CH_3COOH + H_2O$$

Bacteria can grow incredibly fast. Under some conditions, it takes a bacterial cell only 10–20 min to double its size and to divide to form two cells.[4] An animal cell may take 24 h for the same process. Equally impressive are the rates at which bacteria transform their foods into other materials. One factor contributing to the high rate of bacterial metabolism may be the large surface to volume ratio. For a small spherical bacterium (coccus) of diameter 0.5 μm, the ratio of the surface area to the volume is 12×10^6 m^{-1}, while for an ameba of diameter 150 μm the ratio is only 4×10^4 m^{-1} (the ameba can increase this by sticking out some pseudopods). Thimann[33] estimated that for a 90-kg human, the ratio is only 30 m^{-1}.

When food is limited, some bacteria such as the *Bacillus* form **spores**. These are compact little cells that form inside the vegetative cell and are therefore called **endospores**. They sometimes have only 1/10 the volume of the parent cell. Their water content is very low, their metabolic rate is near zero, and they are extremely resistant to heat and further desiccation. Under suitable conditions, the spores can "germinate" and renew their vegetative growth. Spore formation is one of several examples of the development of specialized cells or **differentiation** among prokaryotes.

8. Photosynthetic and Nitrogen-Fixing Prokaryotes

It is likely that the earth was once a completely anaerobic place containing water, ammonia, methane, formaldehyde, and more complicated organic compounds. Perhaps the first forms of life, which may have originated about 3.5×10^9 years ago, resembled present-day anaerobic bacteria. The purple and green photosynthetic bacteria may be related to organisms that developed at a second stage of evolution: those able to capture energy from sunlight. Most of these gram-negative photosynthetic bacteria are strict anaerobes. None can make oxygen as do higher plants. Rather, the hydrogen needed to carry out the reduction of carbon dioxide in the photosynthetic process is obtained by the splitting of inorganic compounds, such as H_2S, thiosulfate, or H_2, or is taken from organic compounds. Today, photosynthetic bacteria are found principally in sulfur springs and in deep lakes, but at one time they were probably far more abundant and the only photosynthetic organisms on earth.

Before organisms could produce oxygen a second complete photosynthetic system, which could cleave H_2O to O_2, had to be developed. The simplest oxygen-

BOX 1-C IN THE BEGINNING

No one knows how life began. Theories ranging from the biblical accounts to recent ideas about the role of RNA are plentiful but largely unsatisfying. In the 1800s the great physical chemist Arrhenius was among scientists that preferred the idea held by some scientists today that a "seed" came from outer space. Until recently the only concrete data came from fossils. Making use of a variety of isotopic dating methods it can be concluded that cyanobacteria were present 2.2×10^9 years ago and eukaryotes 1.4×10^9 years ago. About 0.5×10^9 years ago the "Cambrian explosion" led to the appearance of virtually all known animal phyla. Many of these then became extinct about 0.2×10^9 years ago.

New insights published in 1859[a] were provided by Charles Darwin. However, his ideas were only put into a context of biochemical data after 1950 when sequencing of proteins and later nucleic acids began. From an astonishingly large library of sequence data available now we can draw one firm conclusion: *Evolution can be observed;*[b] *it does involve mutation of DNA.* Comparisons of sequences among many species allow evolutionary relationships to be proposed.[c-e] In general these are very similar to those deduced from the fossil record. They support the idea that evolution occurs by natural selection and that duplication of genes and movements of large pieces of DNA within the genome have occurred often. As many as 900 "ancient conserved regions" of DNA in the *E. coli* genome corresponding to those in human, nematode, and yeast DNA are thought to date back perhaps 3.5×10^9 years.[f] However, nobody has explained how life evolved before there was DNA.

One of the first scientists to devote his career to biochemical evolution was I. V. Oparin,[g] who published a book on the "origin of life" in 1924. Oparin and J. B. S. Haldane, independently, proposed that early life was anaerobic and that energy was provided by fermentation. In 1951 Stanley Miller built an apparatus that circulated CH_4, NH_3, H_2O, and H_2, compounds thought to be present in a primitive atmosphere, past an electric discharge. He found glycine, alanine, β-alanine, and other amino acids among the products formed.[h] Schrödinger pointed out that a flux of energy through a system will tend to organize the system. The solar energy passing through the biosphere induces atmospheric circulation and patterns of weather and ocean currents.[i,j] Perhaps in the primordeal oceans organic compounds arose from the action of light and lightning discharges. These compounds became catalysts for other reactions which eventually evolved into a rudimentary cell-less metabolism. It is a large jump from this to a cell! Among other problems is the lack of any explanation for the development of individual cells or of their genomes. However, because it helps to correlate much information *we will always take an evolutionary approach in this book* and will discuss the "beginnings" a little more in later chapters.

[a] Maynard-Smith, J. (1982) *Nature (London)* **296**, 599–601
[b] Lenski, R. E., and Travisano, M. (1994) *Proc. Natl. Acad. Sci. U.S.A.* **91**, 6808–6814
[c] Wilson, A. C. (1985) *Sci. Am.* **253**(Oct), 164–173
[d] Eigen, M., Gardiner, W., Schuster, P., and Winkler-Oswatitsch, R. (1981) *Sci. Am.* **244**(Apr), 88–118
[e] Doolittle, R. F. (1992) *Protein Sci.* **1**, 191–200
[f] Green, P., Lipman, D., Hillier, L., Waterston, R., States, D., and Claverie, J.-M. (1993) *Science* **259**, 1711–1716
[g] Broda, E. (1980) *Trends Biochem. Sci.* **5**, IV–V
[h] Miller, S. L. (1953) *Science* **117**, 528–529
[i] Mason, S. (1993) *Trends Biochem. Sci.* **18**, 230–231
[j] Welch, G. R. (1996) *Trends Biochem. Sci.* **21**, 452

producing creatures existing today are the **cyano-bacteria**,[34] also known as blue-green algae. Many cyanobacteria are unicellular, but others such as *Oscillatoria*, a slimy "plant" that often coats the inside walls of household aquaria, consist of long filaments about 6 μm in diameter (see Fig. 1-11). All cyanobacteria contain two groups of pigments not found in other prokaryotes: **chlorophyll a** and **β-carotene**, pigments that are also found in the chloroplasts of true algae and in higher plants. A recently discovered group of bacteria, the **prochlorophytes**, are even closer to chloroplasts in their pigment composition.[35]

In addition to pigmented cells, some cyanobacteria contain paler cells known as **heterocysts**. They have a specialized function of fixing molecular nitrogen. The

development of the ability to convert N_2 into organic nitrogen compounds represents another important evolutionary step. Because they can both fix nitrogen and carry out photosynthesis, the blue-green algae have the simplest nutritional requirements of any organisms. They need only N_2, CO_2, water, light, and minerals for growth.

Evolution of the photosynthetic cleavage of water to oxygen was doubtless a major event with far-reaching consequences. Biologists generally believe that as oxygen accumulated in the earth's atmosphere, the obligate anaerobes, which are poisoned by oxygen, became limited to strictly anaerobic environments. Meanwhile, a new group of bacteria, the **aerobes**, appeared with mechanisms for detoxifying oxygen

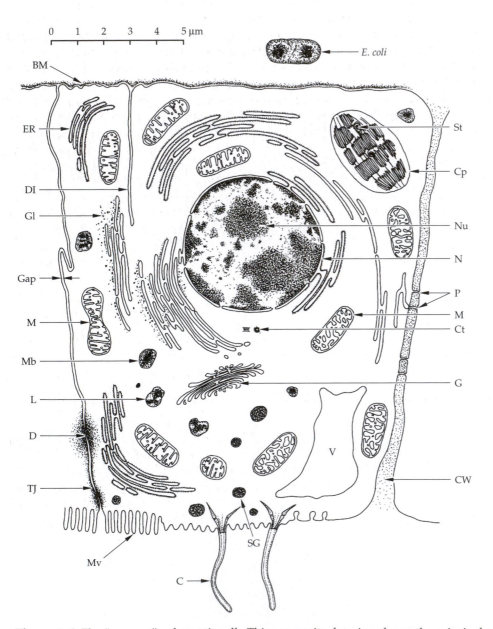

Abbreviations:

BM, basement membrane
ER, rough endoplasmic reticulum (with ribosomes attached; smooth ER is depicted nearer the nucleus and on the right side of the cell.)
DI, deep indentation of plasma membrane
GI, glycogen granules
Gap, space ~10-20 nm thick between adjacent cells
M, mitochondrion
Mb, microbody
L, lysosome
D, desmosome
TJ, tight junction
Mv, microvilli
C, cillium
SG, secretion granule
V, vacuole
Nu, nucleolus
G, Golgi apparatus
CW, cell wall (of a plant)
Ct, centrioles
P, plasmodesmata
N, nucleus
Cp, chloroplast
St, starch granule

Figure 1-6 The "average" eukaryotic cell. This composite drawing shows the principal organelles of both animal and plant cells approximately to the correct scale. (Adapted from a drawing by Michael Metzler.)

and for using oxygen to oxidize complex organic compounds to obtain energy.

B. Eukaryotic Cells

Cells of the **eukaryotes** contain true nuclei and are much larger and more complex internally than are those of prokaryotes. The nucleus of a cell contains most of its DNA and is separated from the cytoplasm by membranes. Within the cytoplasm are various **organelles** with characteristic structures. These include **mitochondria**, **lysosomes**, **peroxisomes**, and **centrioles**. Eukaryotic cells come in so many sizes and shapes and with so many specialized features that it is impossible to say what is typical. Nevertheless, Fig. 1-6 is an attempt to portray some sort of "average" cell, partly plant and partly animal.

As can be seen from Table 1-2, which lists the diameters and volumes of several roughly spherical cells, there is a great variation in size. However, a diameter of 10–20 μm may be regarded as typical for both plants and animals. For growth of a large cell such as the ovum, many adjacent cells assist in synthesis of foodstuffs which are transferred to the developing egg cell. Plant cells are often large but usually 90% or more of the cell is filled with a **vacuole** or **tonoplast**,[36] which is drawn unrealistically small in Fig. 1-6. The metabolically active protoplasm of plant cells often lies in a thin layer at their peripheries.

Many cells are far from spherical; for example, human red blood cells are discs 8 x 8 x 1 to 2 μm with a volume of 80 μm³. Plant fiber cells may be several millimeters in length. Nerve cells of animals have long extensions, the **axons**, which in the human sometimes attain a length of a meter. Muscle cells fuse to give very long multinucleate fibers.

1. The Nucleus

In a typical animal cell the nucleus has a diameter of ~5 μm and a volume of 65 μm³. Except at the time of cell division, it is densely and almost uniformly packed with DNA. The amount of DNA present is larger than that in bacteria as is indicated in Table 1-3. Yeast contains about three times as much genetic matter as *E. coli* and a human being or a mouse about 700 times as much. However, genes are sometimes duplicated in higher organisms and large amounts of **repetitive DNA** of uncertain significance are often present. Some amphibians have 25 times *more* DNA per cell than do humans. The fruit fly *Drosophila* contains about 13,600 functioning genes and a human being perhaps 50,000.[37]

Because of its acidic character, DNA is stained by basic dyes. Long before the days of modern biochemistry, the name **chromatin** was given to the material in the nucleus that was colored by basic dyes. At the time of cell division, the chromatin is consolidated into distinct **chromosomes** which contain, in addition to 15% DNA, about 10% RNA and 75% protein.

Nearly all of the RNA of the cell is synthesized (transcribed) in the nucleus, according to the instructions encoded in the DNA. Some of the RNA then moves out of the nucleus into the cytoplasm where it functions in protein synthesis and in some other ways. Many eukaryotic genes consist of several sequences that may be separated in the DNA of a chromosome by **intervening sequences** of hundreds or thousands of base pairs. The long RNA transcripts made from these **split genes** must be cut and **spliced** in the nucleus to form the correct messenger RNA molecules which are then sent out to the ribosomes in the cytoplasm.

Each cell nucleus contains one or more dense **nucleoli**, regions that are rich in RNA and may contain 10–20% of the total RNA of cells. Nucleoli are sites of synthesis and of temporary storage of ribosomal RNA, which is needed for assembly of ribosomes. The **nuclear envelope** is a pair of membranes, usually a few tens of nanometers apart, that surround the nucleus. The two membranes of the pair separate off a thin **perinuclear space** (Fig. 1-7). The membranes contain "pores" ~130 nm in diameter with a complex structure (see Fig. 27-8).[38,39] There is a central channel ~42 nm in diameter, which provides a route for controlled passage of RNA and other large molecules from the nucleus into the cytoplasm and also from the cytoplasm to the nucleus. Smaller ~10 nm channels allow passive diffusion of ions and small molecules.

TABLE 1-2
Approximate Sizes of Some Cells

Cell	Diameter (μm)	Approximate volume (μm³)
E. coli	1	1.0
Small thymus cell	6	120
Liver cell	20	4,000
Human ovum (mature)	120	500,000
Hen's egg (white excluded)	20,000	4×10^{12}
Yeast cell	10	500
Onion root (meristematic cell)	17	2,600
Parenchyma cell of a fruit	1,000	1×10^8

TABLE 1-3
Haploid Genome Sizes for Several Organisms

Organism; see footnotes for sequence information	Millions of nucleotide base pairs (Mb)	Number of chromosomes (haploid)	Estimated number of genes
Mycoplasma genitalium[a–c]	0.580	1	482
Rickettsia prowazekii[d]	1.11	1	834
Haemophilus influenzae[c,e–g]	1.83	1	1,709
Methanococcus janaschii (an archaeon)[h]	1.66	1	1,738
Bacillus subtilis[i]	4.16	1	
Escherichia coli[j,k]	4.64	1	4,288
Myxococcus xanthis[f]	9.2	1	
Synechocystis sp. (a cyanobacterium)[l]	3.57	1	3,169
Saccharomyces cerevisiae (a yeast)[k,m–p,t]	13.5	17	6,241
Giardia lamblia (a protozoan)[q]	12		
Plasmodium falciparum (malaria parasite)[v]	25–30	14	
Dictyostelium discoideum (a slime mold)[r]	34		
Caenorhabditis elegans (a nematode)[s,t]	97	6	18,424
Sea Urchin	900		
Drosophila melanogaster (fruit fly)[u,t]	180	4	13,601
Danio rerio (zebrafish)[w]	1,700	2.5	
Fugu rubripes (pufferfish)[x]	400		60,000
S. African lungfish[y]	102,000	19	
Mus musculus (mouse)[z]	~3,000	20	80,000
Bos (cow)[aa]	~3,000	30	80,000
Homo sapiens (human)[bb,cc]	~3,000	23	50,000– 150,000
Arabidopsis thaliana (green plant)[dd,ee]	115.4	5	25,498
Rice[ff–hh]	450	12	
Maize[ii] or Wheat[ff]	~2,700		
Lily[jj]	>100,000		

2. The Plasma Membrane

The thin (8 nm) outer cell membrane or "plasma-lemma" (Fig. 1-7) controls the flow of materials into and out of cells, conducts impulses in nerve cells and along muscle fibrils, and participates in chemical communication with other cells. Deep infoldings of the outer membrane sometimes run into the cytoplasm. An example, is the "T system" of tubules which functions in excitation of muscle contraction (Figs. 19-7, 19-21). Surfaces of cells designated to secrete materials or to absorb substances from the surrounding fluid, such as the cells lining kidney tubules and pancreatic secretory cells, are often covered with very fine projections or **microvilli** which greatly increase the surface area. In other cases projections from one cell interdigitate with those of an adjacent cell to give more intimate contact.

3. Vacuoles, Endocytosis, and Lysosomes

Cells often contain vacuoles or smaller vesicles that are separated from the cytosol by a *single* membrane. Their content is often quite

[a] Fraser, C. M., and 28 other authors (1995) *Science* **270**, 397–403

[b] Goffeau, A. (1995) *Science* **270**, 445–446

[c] Brosius, J., Robison, K., Gilbert, W., Church, G. M., and Venter, J. C. (1996) *Science* **271**, 1302–1304

[d] Andersson, S. G. E., Zomorodipour, A., Andersson, J. O., Sicheritz-Pontén, T., Alsmark, U. C. M., Podowski, R. M., Näslund, A. K., Eriksson, A.-S., Winkler, H. H., and Kurland, C. G. (1998) *Nature (London)* **396**, 133–140

[e] Fleischmann, R. D., and 39 other authors (1995) *Science* **269**, 496–512

[f] He, Q., Chen, H., Kupsa, A., Cheng, Y., Kaiser, D., and Shimkets, L. J. (1994) *Proc. Natl. Acad. Sci. U.S.A.* **91**, 9584–9587

[g] Mrázek, J., and Karlin, S. (1996) *Trends Biochem. Sci.* **21**, 201–202

[h] Bult, C. J., and 39 other authors; corresponding author Venter, J. C. (1996) *Science* **273**, 1058–1073

[i] Azevedo, V., Alvarez, E., Zumstein, E., Damiani, G., Sgaramella, V., Ehrlich, S. D., and Serror, P. (1993) *Proc. Natl. Acad. Sci. U.S.A.* **90**, 6047–6051

[j] Blattner, F. R. and 16 other authors (1997) *Science* **277**, 1453–1462

[k] Winzeler, E. A., and 51 other authors (1999) *Science* **285**, 901–906

[l] Kaneko, T., and 23 other authors (1996) *DNA Res.* **3**, 109–136. See also http://www.kazusa.or.jp/cyano

[m] Dujon, B., and 107 other authors (1994) *Nature (London)* **369**, 371–378

[n] Taguchi, T., Seko, A., Kitajima, K., Muto, Y., Inoue, S., Khoo, K.-H., Morris, H. R., Dell, A., and Inoue, Y. (1994) *J. Biol. Chem.* **269**, 8762–8771

[o] Johnston, M., and 34 other authors. (1994) *Science* **265**, 2077–2082

[p] Williams, N. (1995) *Science* **268**, 1560–1561

[q] Sogin, M. L., Gunderson, J. H., Elwood, H. J., Alonso, R. A., and Peattie, D. A. (1989) *Science* **243**, 75–77

[r] Loomis, W. F., and Insall, R. H. (1999) *Nature (London)* **401**, 440–441

[s] The *C. elegans* Sequencing Consortium (1998) *Science* **282**, 2012–2018 (See this article for list of authors.)

[t] Rubin, G. M., and 54 other authors. (2000) *Science* **287**, 2204–2215

[u] Adams, M. D., and 194 other authors. (2000) *Science* **287**, 2185–2195

[v] Su, X.-z, Ferdig, M. T., Huang, Y., Huynh, C. Q., Liu, A., You, J., Wootton, J. C., and Wellems, T. E. (1999) *Science* **286**, 1351–1353

[w] Postlethwait, J. H., Johnson, S. L., Midson, C. N., Talbot, W. S., Gates, M., Ballinger, E. W., Africa, D., Andrews, R., Carl, T., Eisen, J. S., Horne, S., Kimmel, C. B., Hutchinson, M., Johnson, M., and Rodriguez, A. (1994) *Science* **264**, 699–703

[x] Fishman, M. C. (1999) *Proc. Natl. Acad. Sci. U.S.A.* **96**, 10554–10556

[y] Kornberg, A., and Baker, T. A. (1992) *DNA Replication*, 2nd ed., pp. 19–21, Freeman, New York

[z] Dietrich, W. F., Copeland, N. G., Gilbert, D. J., Miller, J. C., Jenkins, N. A., and Lander, E. S. (1994) *Proc. Natl. Acad. Sci. U.S.A.* **92**, 10849–10853

[aa] Anonymous (1994) *Nature (London)* **368**, 167

[bb] Schuler, G. D. and 102 other authors (1996) *Science* **274**, 540–546

[cc] Koonin, S. E. (1998) *Science* **279**, 36–37

[dd] Olson, M. V. (1993) *Proc. Natl. Acad. Sci. U.S.A.* **90**, 4338–4344

[ee] The Arabidopsis Genome Initiative, (2000) *Nature* **408**, 796–815

[ff] Stevens, J. E. (1994) *Science* **266**, 1186–1187

[gg] Shimamoto, K. (1995) *Science* **270**, 1772–1773

[hh] Singh, K., Ishii, T., Parco, A., Huang, N., Brar, D. S., and Khush, G. S. (1996) *Proc. Natl. Acad. Sci. U.S.A.* **93**, 6163–6168

[ii] Carels, N., Barakat, A., and Gernardi, G. (1995) *Proc. Natl. Acad. Sci. U.S.A.* **92**, 11057–11060

[jj] Alberts, B., Bray, D., Lewis, J., Raff, M., Roberts, K., and Watson, J. D. (1994) *Molecular Biology of the Cell*, 3rd ed., Garland, New York

acidic.[40] Small vesicles sometimes bud inward from the plasma membrane in a process called **endocytosis**. In this manner the cell may engulf particles (**phago-cytosis**) or droplets of the external medium (**pinocytosis**). The resulting endocytotic vesicles or **endosomes** often fuse with **lysosomes**, which are small acidified vesicles containing a battery of enzymes powerful enough to digest almost anything in the cell. In cells that engulf bits of food (e.g., ameba) lysosomes provide the digestive enzymes. Lysosomes also take up and digest denatured or damaged proteins and may digest "worn out" or excess cell parts including mitochondria. Lysosomes are vital components of cells,[41] and several serious human diseases result from a lack of specific lysosomal enzymes.

4. The Endoplasmic Reticulum and Golgi Membranes

Although cytoplasm is fluid and in some organisms can undergo rapid streaming, the electron microscope

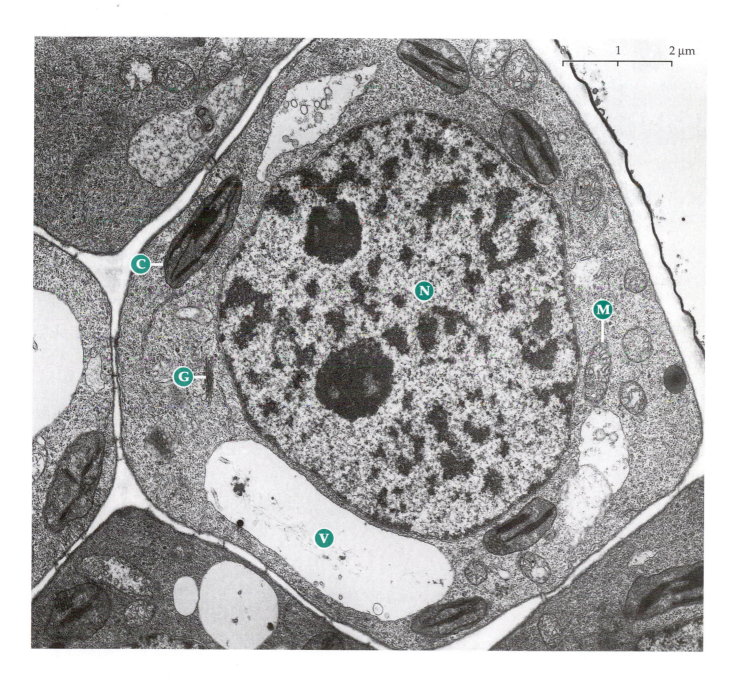

Figure 1-7 Electron micrograph of a thin section of a young epidermal cell of a sunflower. The tissue was fixed and stained with uranyl acetate and lead citrate. Clearly visible are the nucleus (N), mitochondria (M), chloroplasts (C), a Golgi body dictyosome (G), endoplasmic reticulum, vacuole (V), cell wall, plasmodesmata, and cuticle (upper right, thin dark layer). Micrograph courtesy of H. T. Horner.

has revealed that within the liquid portion, the **cytosol**, there is a complex network of membranes known as the **endoplasmic reticulum** (ER). The membranes of the ER form tubes, vesicles, and flattened sacs called **cisternae**. The intracisternal spaces appear to connect with the perinuclear space and to a series of 3–12 flattened, slightly curved disk-shaped membranes known as the **Golgi apparatus** (Figs. 1-7, 20-8).[42,43] This organelle was first reported by Camillo Golgi in 1898.[44] Its existence was long doubted, but it is known now to play a vital role in metabolism.

The ER, the Golgi membranes, and secretion granules apparently represent an organized system for synthesis of secreted protein and formation of new membranes. Parts of the ER, the **rough endoplasmic reticulum** are lined with many ribosomes of 21–25 nm diameter. While resembling those of bacteria, these eukaryotic ribosomes are about 50% heavier (4×10^6 Da). The **smooth endoplasmic reticulum** lacks ribosomes but proteins made in the rough ER may be modified in the smooth ER, e.g., by addition of carbohydrate chains. Small membrane vesicles break off from the smooth ER and pass to the Golgi membranes which lie close to the smooth ER on the side toward the center of the cell. Here additional modification reactions occur (Chapter 20). At the outer edges the membranes of the Golgi apparatus pinch off to form vacuoles which are often densely packed with enzymes or other proteins. These **secretion granules** move to the surface and are released from the cell. In this process of **exocytosis** the membranes surrounding the granules fuse with the outer cell membrane. The rough ER appears to contribute membrane material to the smooth ER and Golgi apparatus, while material from Golgi membranes can become incorporated into the outer cell membrane and into lysosomes. Outer mitochondrial membranes and membranes around vacuoles in plant cells may also be derived directly from the ER. Outer membrane materials are probably "recycled" by endocytosis.

The term **microsome**, frequently met in the biochemical literature, refers to small particles of 50–150 nm diameter which are mostly fragments of the ER together with some material from the plasma membrane. Microsomes are formed when cells are ground or homogenized. Upon centrifugation of the disrupted cells, nuclei and other large fragments sediment first, then the mitochondria. At very high speeds (e.g., at 100,000 times the force of gravity) the microsomes, whose masses are 10^8–10^9 Da, settle. With the electron microscope we see that in the microsomes the membrane fragments have closed to give small sacs to the outside of which the ribosomes still cling:

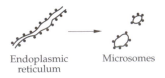

Endoplasmic   Microsomes
reticulum

5. Mitochondria, Plastids, and Peroxisomes

Mitochondria, complex bodies about the size of bacteria and bounded on the outside by a double membrane (Figs. 1-6 and 1-7), are present in all eukaryotic cells that use oxygen for respiration. The numbers per cell appear to vary from the *one* for certain tiny trypanosomes to as many as 3×10^5 in some oocytes. Liver cells often contain more than 1000 mitochondria apiece.[45] Study of ultrathin serial sections of a single yeast cell by electron microscopy has shown that under some growth conditions all of the yeast mitochondria are interconnected.[46] More recent evidence from new imaging procedures, e.g. using the green fluorescent protein (Box 23-A)[46a, 46b] also supports the idea that mitochondria are interconnected in a reticulum that can become fragmented under some conditions. The inner membrane of a mitochondrion is often highly folded to form the **cristae** (crests). The outer membrane is porous to small molecules but the passage of substances into and out of the inner space of the mitochondrion, known as the **matrix,** is tightly controlled by the inner membrane. Although some of the oxidative chemical activites of the cells are located in the ER and in peroxisomes, the major energy-yielding reactions for aerobic organisms are found in the mitochondria, which are also the principal site of utilization of oxygen. Within each mitochondrion is a small circular molecule of DNA whose genes encode only a few of the many proteins needed in this organelle. Also present within mitochondria are ribosomes of a size similar to those of bacteria and smaller than those lining the rough ER.

Plastids are organelles of plant cells that serve a variety of purposes.[45] Most important are the **chloroplasts**, the chlorophyll-containing sites of photosynthesis. Like mitochondria they contain folded internal membranes (see Fig. 23-19) and several small molecules of DNA.

Fragile organelles, the **peroxisomes** or **microbodies,** occur in many cells.[47–50] In green leaves they may occur in numbers up to one-third those of mitochondria. Peroxisomes are often about the size of mitochondria but have only a single membrane and do not contain DNA. They often contain an apparently crystalline "core." The single membrane of peroxisomes is porous to small molecules such as sucrose. This permits these organelles to be separated from mitochondria by centrifugation in a sucrose gradient where the microbodies assume a density of about 1.25 g/cm^3 compared to 1.19 for the impervious mitochondria.

Peroxisomes are rich in enzymes that produce and decompose hydrogen peroxide. They often make a major contribution to the oxidative metabolism of cells. In germinating oilseeds **glyoxysomes**, a type of peroxisome, contain enzymes that catalyze reactions of the biosynthetic "glyoxylate pathway" of metabolism.[51] Organelles that resemble peroxisomes in appearance but which are functionally more closely related to mitochondria are the **hydrogenosomes** of anaerobic protozoa.[52] As the name suggests, these organelles are the site of formation of molecular hydrogen, a common product of anaerobic metabolism.

6. Centrioles, Cilia, Flagella, and Microtubules

Many cells contain **centrioles**,[53] little cylinders about 0.15 μm in diameter and 0.5 μm long, which are *not* enclosed by membranes. Each centriole contains a series of fine **microtubules** of 25 nm diameter. A pair of centrioles are present near the nucleus in most animal cells and play an important role in cell division. Together with surrounding materials they form the **centrosome**. However, centrioles have never been observed in plant cells.

Related in structure to centrioles are the long **flagella** and shorter **cilia** (the two words are virtually synonymous) which are commonly present as organelles of locomotion in eukaryotic cells. Stationary cells of our own bodies also often have cilia. For example, there are 10^9 cilia/cm^2 in bronchial epithelium.[54] Modified flagella form the receptors of light in our eyes and of taste in our tongues. Flagella and cilia have a diameter of about 0.2 μm and a characteristic internal structure. Eleven hollow microtubules of ~24 nm diameter are usually arranged in a "9 + 2" pattern with nine pairs of fused tubules surrounding a pair of single tubules (Figs. 1-8 and 19-23). Each microtubule

resembles a bacterial flagellum in appearance, but there are distinct and significant chemical differences. The **basal body** of the flagellum, the **kinetosome** (Fig. 1-8), resembles a centriole in structure, dimensions, and mode of replication. Recently a small 6–9 megabase pair DNA has been found in basal bodies of the protozoan *Chlamydomonas*.[55,56]

Microtubules similar to those found in flagella are also present in the cytoplasm. Together with thinner **microfilaments** of several kinds they form an internal **cytoskeleton** that provides rigidity to cells.[58,59] Microtubules also form the "spindle" of dividing cells. In nerve axons (Chapter 30) the microtubules run parallel to the length of the axons and are part of a mechanical transport system for cell constituents.

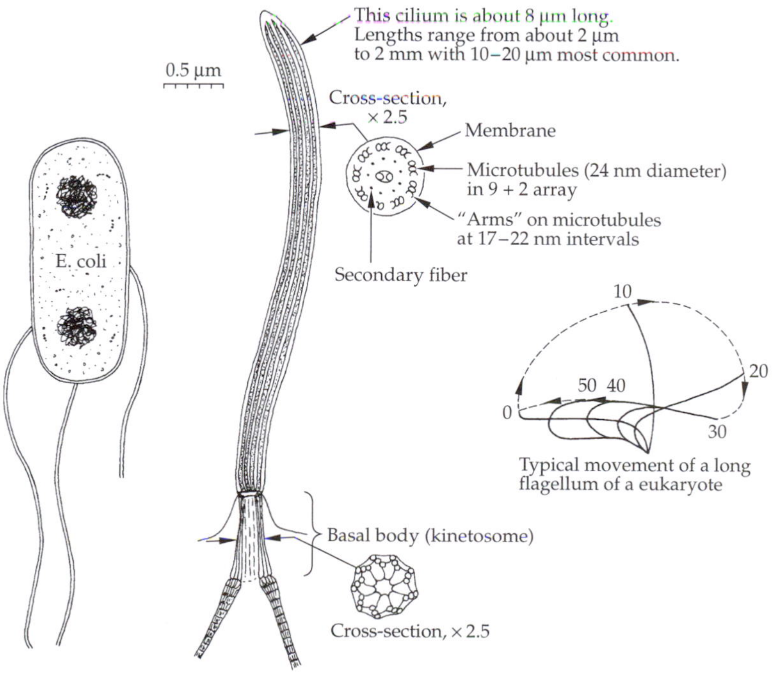

7. Cell Coats, Walls, and Shells

Like bacteria, most cells of higher plants and animals are surrounded by extracellular materials. Plants have rigid walls rich in cellulose and other carbohydrate polymers. Outside surfaces of plant cells are covered with a **cuticle** containing layers of a polyester called **cutin** and of wax (Fig. 1-6). Surfaces of animal cells are usually lined with carbohydrate molecules which are attached to specific surface proteins to form **glycoproteins**. Spaces between cells are filled with such "cementing substances" as **pectins** in plants and **hyaluronic acid** in animals. Insoluble proteins such as **collagen** and **elastin** surround connective tissue cells. Cells that lie on a surface (epithelial and endothelial cells) are often lined on one side with a thin, collagen-containing **basement membrane** (Figs. 1-6 and 8-31). Inorganic deposits such as calcium phosphate (in bone), calcium carbonate (eggshells and spicules of sponges), and silicon dioxide (shells of diatoms) are laid down, often by cooperative action of several or many cells.

C. Inheritance, Metabolic Variation, and Evolution of Eukaryotes

The striking differences between eukaryotic and prokaryotic cells have led to many speculations about the evolutionary relationship of these two great classes of living organisms. A popular theory is that mitochondria, which are characteristic of most eukaryotes, arose from aerobic bacteria. After cyanobacteria had developed and oxygen

Figure 1-8 Structure of cilia and flagella of eukaryotes. After P. Satir.[57]

BOX 1-D INHERITED METABOLIC DISEASES

In 1908 Archibald Garrod[a,b] proposed that **cystinuria** (Chapter 8) and several other defects in amino acid and sugar metabolism were "inborn errors of metabolism", i.e. inherited diseases. Since that time the number of recognized genetic defects of human metabolism has increased at an accelerating rate to ~4000.[c–e] Hundreds of other genetic problems have also been identified. For over 800 of these the defective gene has been mapped to a specific chromosome.[f] An example is **sickle cell anemia** (Box 7-B) in which a defective hemoglobin differs from the normal protein at one position in one of its constituent polypeptide chains. Many other defects involve loss of activity of some important enzyme.

Most genetic diseases are rare, affecting about one person in 10,000. However, **cystic fibrosis** affects one in 2500. There are so many metabolic diseases that over 0.5% of all persons born may develop one. Many die at an early age. A much greater number (>5%) develop such conditions as diabetes and mental illness which are, in part, of genetic origin. Since new mutations are always arising, genetic diseases present a problem of continuing significance.

At what rate do new mutations appear? From the haploid DNA content (Table 1-2) we can estimate that the total coding capacity of the DNA in a human cell exceeds two million genes (actually two million *pairs* of genes in diploid cells). However, only a fraction of the DNA codes for proteins. There are perhaps 50,000 pairs of structural genes in human DNA. The *easily detectable* rate of mutation in bacteria is about 10^{-6} per gene, or 10^{-9} per base per replication.[g] As a result of sophisticated "proofreading" and repair systems, it may be as low as 10^{-10} per base in humans.[h] Thus, in the replication of the 3×10^9 base pairs in diploid human chromosomes we might anticipate about one mistake per cell division. Only about 1/50 of these would be in structural genes and potentially harmful. Thus, if there are 10^{16} division cycles in a normal life span[h] each parent may pass on to future generations about 2 mutations in protein sequences. The ~10^{14} body cells (somatic cells) also undergo mutations which may lead to cancer and to other problems of aging. Most mutations may be harmless or nearly so and a few may be beneficial.

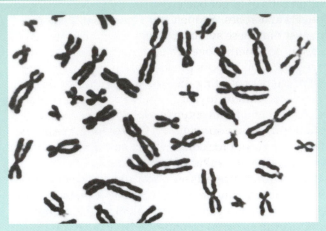

Photomicrograph of human male metaphase chromosomes. © Photo Researchers

However, many are damaging and some are *lethal*. If a mutation is lethal, a homozygote will not survive and will be lost in an early (and usually undetected) spontaneous abortion. Healthy individuals carry as many as ten lethal recessive mutations as well as at least 3–5 autosomal recessive mutations of a seriously harmful type. Harmful dominant mutations are also frequent in the population. These include an elevated lipoprotein content of the blood and an elevated cholesterol level which are linked to early heart disease.

Biochemical disorders are also important because of the light they shed on metabolic processes. No other species is observed as carefully as *Homo sapiens*. As a consequence frequent reference will be made to genetic diseases throughout the book. A goal is to find ways to prevent or ameliorate the effects of these disorders. For example, in the treatment of **phenylketonuria** (Chapter 25) or of **galactosemia** (Chapter 20), a change in the diet can prevent irreversible damage to the brain, the organ most frequently affected by many of these diseases. Injection of a missing enzyme is giving life to victims of Gaucher's disease (Chapter 20). In many other cases no satisfactory therapy is presently available, but the possibilities of finding some way to supply missing enzymes or to carry out "genetic surgery" are among the most exciting developments of contemporary medical biochemistry (Chapter 26).

[a] Garrod, A. E. (1909) *Inborn Errors of Metabolism*, Oxford, London

[b] Bearn, A. G. (1993) *Archibald Garrod and the Individuality of Man*, Oxford, New York

[c] Scriver, C. R., Beaudet, A. L., Sly, W. S., and Valle, D., eds. (1995) *The Metabolic and Molecular Bases of Inherited Disease*, 7th ed., McGraw-Hill, New York

[d] Davies, K. E. (1992) *Molecular Basis of Inherited Disease*, Oxford, New York

[e] McKusick, V. A. (1994) *Mendelian Inheritance in Man*, 11th ed., Johns Hopkins Univ. Press, Baltimore, Maryland

[f] McKusick, V. A., Amberger, J., and Steinberg, J. (1994) *J. NIH Res.* **6**, 115–134

[g] Watson, J. D. (1976) *Molecular Biology of the Gene*, 3rd ed., Benjamin, Menlo Park, California (p. 254)

[h] Koshland, D. E., Jr. (1994) *Science* **266**, 1925

had become abundant, a **symbiotic** relationship could have arisen in which small aerobic bacteria lived within cells of larger bacteria that had previously been obligate anaerobes. Sequence similarities of proteins suggest that these symbionts may have been related to present-day methanogens[60] and thermophilic sulfur bacteria.[61] The aerobes presumably used up any oxygen present, protecting the surrounding anaerobic organisms from its toxicity. The relationship became permanent and led eventually to the mitochondria-containing eukaryotic cell.[62–65] Further symbiosis with cyanobacteria or prochlorophytes could have led to the chloroplasts of the eukaryotic plants.

A fact that supports such ideas is the existence among present-day organisms of many endosymbiotic relationships. For example, the green paramecium (*Paramecium bursaria*) contains, within its cytoplasm, an alga (*Chlorella*), a common green plant that is quite capable of living on its own. Perhaps by accident it took up residence within the paramecium.[62] Some dinoflagellates (Fig. 1-9) contain endosymbiotic cyanobacteria[66] and recently a ciliate that contains endosymbiotic purple photosynthetic bacteria has been discovered.[67] These bacteria do not produce O_2 but utilize products of the host ciliates' metabolism such as acetate, lactate, and H_2 as electron donors for photosyntheses. They also utilize O_2 for respiration and may protect their hosts from the toxicity of O_2, just as may have happened in the distant past. According to this theory the symbionts would eventually have lost their photosynthetic ability and have become mitochondria. The relationship of mitochondria to bacteria is also supported by many biochemical similarities.

Fossils of bacteria and blue-green algae have been obtained from rocks whose age, as determined by geochemical dating, is more than 3×10^9 years.[68,69] However the first eukaryotic cells may have appeared about 1×10^9 years ago[70] and started to evolve into the more than one million species that now exist.[1,71,72]

1. A Changing Genome

How is it possible for the genome of an organism to increase in size as it evolved from a lower form to a higher one? Simple mutations that cause alterations in protein sequences could lead to changes in form and behavior of the organisms but could not, by themselves, account for the increase in genetic material that accompanied evolution. As a result of new techniques of genetic mapping and determining the sequence of nucleotides in DNA we are rapidly acquiring a detailed knowledge of the organization of the genome. It has been found that genes are often present as duplicate but not entirely identical copies. This suggests that there are mechanisms by which cells can acquire extra copies of one or more genes. Indeed it seems probable

that at some time in the past the entire genome of bacteria was doubled and that it was later doubled again.[73] Evidence for this is that the masses of bacterial chromosomes group around values of 0.5, 1.4, and 2.7×10^9 Da. Genes can also be duplicated during the process of genetic recombination, which is discussed in Chapter 27. In addition, the size of the genome may have increased by incorporation of genetic material from extrachromosomal plasmids.

A possible advantage to a cell possessing an extra copy of a gene is that the cell would survive even when mutations rendered unusable the protein encoded by one of the copies. As long as one of the genes remained "good," the organism could grow and reproduce. The extra, mutated gene could be carried for many generations. As long as it produced only harmless, nonfunctioning proteins there might be little selection pressure to eliminate it and it might undergo repeated mutations. After many mutations and many generations later, the protein for which it coded could prove useful to the cell in some new way.

An example of evolution via gene duplication is provided by the oxygen-carrying proteins of blood. It appears that about a billion years ago, the gene for an ancestral **globin**, the protein of hemoglobin, was doubled. One gene evolved into that of present-day **globins** and the other into the gene of the muscle protein **myoglobin**. Still later, the globin gene again doubled leading to the present-day α and β chains of hemoglobin (Chapter 7). These are two distinctly different but related protein subunits whose genes are not even on the same chromosome. To complicate the picture further, most human beings have two or more copies of their α chain gene[74] as well as genes for fetal and embryonic forms of hemoglobin. However, some populations have lost one or more α chain genes. Thus, the genome changes in many details, even today.

2. Genetic Recombination, Sex, and Chromosomes

Bacteria usually reproduce by simple fission. The single DNA molecule of the chromosome is duplicated and the bacterium divides, each daughter cell receiving an identical chromosome. However, genetic recombination, which is accomplished in several ways by bacteria (Chapter 27), provides a deliberate process for mixing of genes. This process has been most fully developed in eukaryotic organisms that undergo sexual reproduction. The growth of a multicelled individual begins with the fusion of two haploid **gametes**, an egg and a spermatozoon. Each gamete carries a complete set of genetic instructions, and after the nuclei fuse the fertilized egg or **zygote** is **diploid**. Each diploid cell contains *two* complete sets of genetic blueprints of quite different origin. Even if a gene from one parent is defective, the chances are that the gene from the other parent will be good. Sexual reproduction and the associated genetic recombination also provide a means for mixing of genes.

When eukaryotic cells prepare to divide in the process called **mitosis** (Fig. 26-11), the DNA molecules of the nucleus, which become spread out through a large volume, coil and fold. Together with proteins and other molecules they form the compact bodies known as chromosomes. Some organisms, such as *Ascaris* (a roundworm), have only two chromosomes, a **homologous pair**, one inherited from the father and one from the mother. Both chromosomes divide in every mitotic cell division so that every cell of the organism has the homologous pair. Higher organisms usually have a larger number of chromosomes. Thus, humans have 23 homologous pairs. The mouse has 20, the toad 11, onions 8, mosquitos 3, and *Drosophila* 4. Human chromosomes vary in size but are usually 4–6 μm long and ~1 μm in diameter.

By the successive divisions of mitosis, a single fertilized eukaryotic egg cell can grow to an adult. Less than 50 successive mitotic divisions will produce the ~10^{14} cells of a human. However, formation of gametes, which are haploid, requires the special process of **meiosis** (Fig. 26-12), by which the number of chromosomes is divided in half. During meiosis one chromosome of each of the homologous pairs of the diploid cell is passed to each of the gametes that are formed. In an organism such as *Ascaris*, which contains only a single pair of chromosomes, a gamete receives either the chromosome of maternal origin or that of paternal origin but not both. In organisms that have several pairs of chromosomes, one chromosome of each pair is passed to the gamete in a random fashion during meiosis. Most gametes receive some chromosomes of maternal and some of paternal origin. An important feature of meiosis is the genetic recombination that occurs during **crossing-over**. In this process, the strands of DNA are cut and genetic material is exchanged between the chromosomes of maternal and paternal origin. Thus, crossing-over breaks the **linkage** between genes and provides for greater variability in the offspring than would otherwise be possible. Each of us receives half of our genes from our mother and half from our father, but some of these genes have been inherited from each grandparent on both sides of the family, some from each great-grandparent, etc.

Many genes are passed down through many generations without substantial change, but others are evidently designed to be scrambled readily within somatic cells. Cell surface proteins[75] and antibody molecules are among the proteins whose genes undergo alteration during growth and differentiation of the tissues of the body (Chapter 32).

3. Haploid and Diploid Phases

In human beings and other higher animals, meiosis leads directly to formation of the gametes, the egg and sperm cells. These fuse to form a diploid nucleus and the adult develops by repeated mitosis of the diploid cells. While meiosis also occurs in the life cycle of all eukaryotic creatures, it is not always at a point corresponding to that in the human life cycle. Thus, the cells of many protozoa and of fungi are ordinarily haploid. When two haploid nuclei fuse to form a diploid cell, meiosis quickly occurs to produce haploid individuals again. Among lower plants and animals there is often an alternation of haploid and diploid phases of the life cycle. For example, gametes of ferns fall to the ground and germinate to form a low-growing green mosslike haploid or **gametophyte** form. The latter produces motile haploid gametes which fuse to a diploid zygote that grows into the larger and more obvious **sporophyte** form of the fern.

It is presumably the ability to survive as a heterozygote, even with one or more highly deleterious mutations, that has led to the dominance of the diploid phase in higher plants and animals.[76] However, to the biochemical geneticist organisms with a haploid phase offer experimental advantages because recessive mutants can be detected readily.

D. Survey of the Protists

Unicellular eukaryotes have traditionally been grouped together with multicellular organisms in which all cells have similar functions, with little or no differentiation into tissues, as the kingdom **Protista**.[77,78] The fungi may also be included or may be regarded as a separate kingdom.[79] With present-day emphasis on DNA sequence comparisons the traditional classification is changing, however.[26]

1. Protozoa

Among the best known of the animal-like protista is the **ameba** (subphylum Sarcodina or Rhizopoda). The most striking feature of the ameba (Fig. 1-9) is its method of locomotion, which involves the transformation of cytoplasm from a liquid state to a semisolid gel. As the ameba moves, the cytoplasm at the rear liquifies and flows to the front and into the extending pseudopodia where it solidifies along the edges. The ameba poses several important biochemical questions: What chemistry underlies the reversible change from liquid to solid cytoplasm? How can the cell membranes break and reform so quickly when an ameba engulfs food particles?[80]

Relatives of the ameba include the **Radiolaria**,

marine organisms of remarkable symmetry with complex internal skeletons containing the carbohydrate polymer chitin together with silica (SiO_2) or strontium sulfate. The **Foraminifera** deposit external shells of calcium carbonate or silicon dioxide. Over 20,000 species are known and now as in the distant past their minute shells fall to the bottom of the ocean and form limestone deposits.

Tiny ameboid parasites of the subphylum **Sporozoa** attack members of all other animal phyla. Several genera of **Coccidia** parasitize rabbits and poultry causing enormous damage. Humans are often the victims of species of the genus *Plasmodium* (Fig. 1-9) which invade red blood cells and other tissues to cause

malaria, one of our most serious ailments on a worldwide basis.[81–84] Throughout history malaria has probably killed more persons than any other disease. *Toxoplasma gondii* is another parasite which, in its haploid phase, is found throughout the world in wild animals and in humans. Although its presence usually elicits no symptoms, it sometimes causes blindness and mental retardation in children and can be fatal to persons with AIDS. Its sexual cycle occurs exclusively in cats.[85,86]

Another subphylum of protozoa, the Mastigophora, are propelled by a small number of flagella and are intermediate between animals and the algae. One of these is *Euglena viridis*, a small freshwater organism with a long flagellum in front, a flexible tapered body, green chloroplasts, and a light-sensitive "eyespot" which it apparently uses to keep itself in the sunshine (Fig. 1-9). Euglena is also able to live as a typical animal if there is no light. Treatment with streptomycin (Box 20-B) causes *Euglena* to lose its chloroplasts and to become an animal permanently. The **dinoflagellates** (Fig. 1-9), some colorless and some green, occur in great numbers among the plankton of the sea. *Giardia lamblia* is a troublesome intestinal parasite.

The **hemoflagellates** are responsible for some of our most terrible diseases. Trypanosomes (genus *Trypanosoma*) invade the cells of the nervous system causing African sleeping sickness. Mutating their surface proteins frequently by gene-scrambling mechanisms, these and other parasites are able to evade the immune response of the host.[87,88] For the same reason it is difficult to prepare vaccines against them. Other flagellates live in a symbiotic relationship within the alimentary canals of termites (Fig. 1-9) and roaches. Termites depend upon bacteria that live within the cells of these symbiotic protozoans to provide the essential enzymes needed to digest the cellulose in wood.

Members of the subphylum Ciliophora, structurally the most complex of the protozoa, are covered with a large number of cilia which beat together in an organized pattern.[89] The following question immediately comes to mind: How

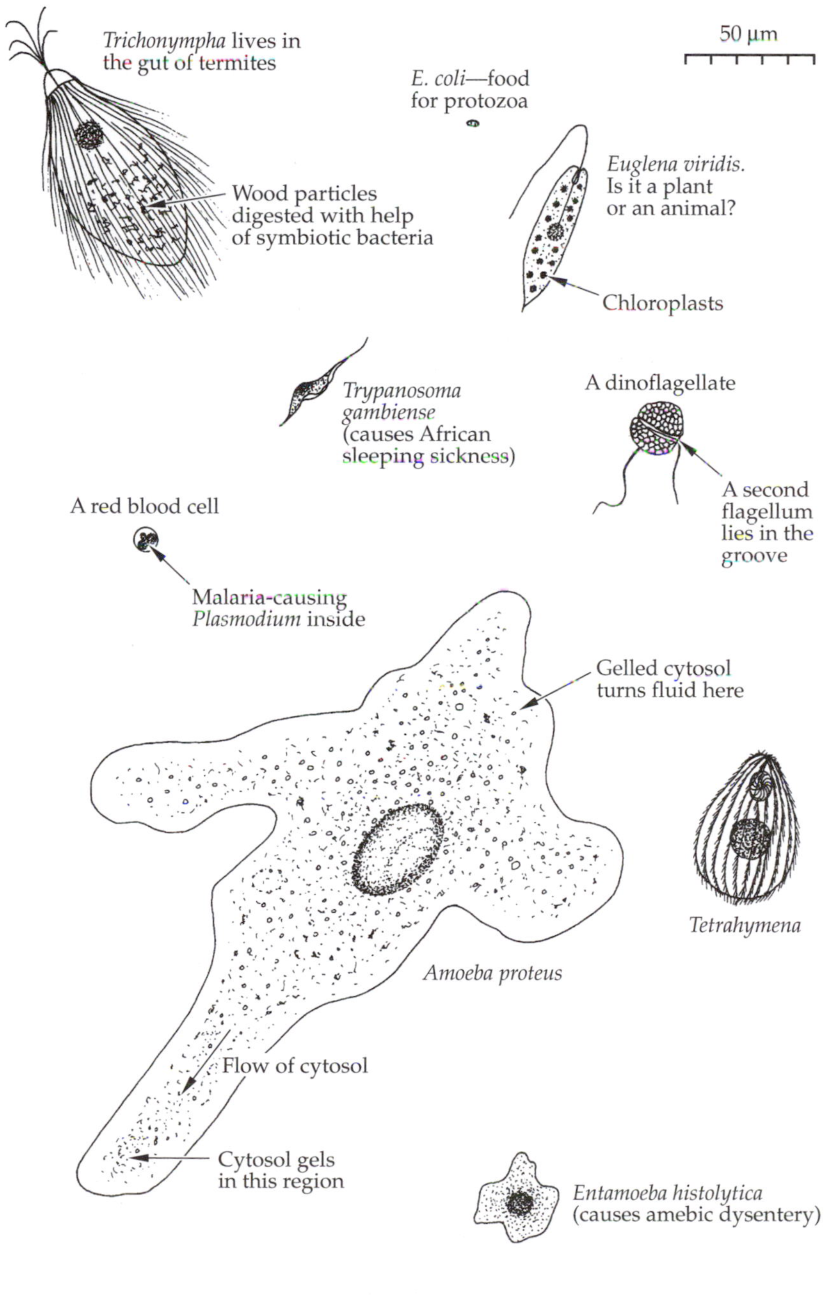

Trichonympha lives in the gut of termites

E. coli—food for protozoa

50 μm

Euglena viridis. Is it a plant or an animal?

Chloroplasts

Wood particles digested with help of symbiotic bacteria

Trypanosoma gambiense (causes African sleeping sickness)

A dinoflagellate

A second flagellum lies in the groove

A red blood cell

Malaria-causing *Plasmodium* inside

Gelled cytosol turns fluid here

Tetrahymena

Amoeba proteus

Flow of cytosol

Cytosol gels in this region

Entamoeba histolytica (causes amebic dysentery)

Figure 1-9 A few well-known protists.

are the cilia able to communicate with each other to provide this organized pattern? Two ciliates that are often studied by biochemists are *Tetrahymena* (Fig. 1-9), one of the simplest, and *Paramecium*, one of the more complex.

The **Myxomycetes** or "slime molds" are more closely related to protozoa than to fungi.[90] Members of the family Acrasieae, the best studied member of which is *Dictyostelium discoideum*, start life as small amebas. After a time, when the food supply runs low, some of the amebas begin to secrete pulses of a chemical attractant **cyclic AMP**. Neighboring amebas respond to the pulses of cyclic AMP by emitting their own pulses about 15 s later, then moving toward the original source.[91,92] The ultimate effect is to cause the amebas to stream to centers where they aggregate and form fungus-like fruiting bodies. Asexual spores are formed and the life cycle begins again. Other Myxomycetes grow as a multinucleate (diploid) **plasmodium** containing millions of nuclei but no individual cell membranes. *Physarum polycephalum*, a species whose plasmodium may spread to a diameter of 30 cm, has become popular with biochemists. The 800,000 nuclei per square millimeter all divide synchronously.

2. Fungi

Lacking photosynthetic ability, living most often in soil but sometimes in water, the fungi are represented by almost half as many species (~10^5) as are the vascular plants.[93] The distinguishing characteristics of fungi are the lack of chlorophyll and growth as a series of many branched tubules (usually 6–8 μm diameter), the **hyphae**, which constitute the **mycelium**. The hyphae are not made up of separate cells but contain a mass of protoplasm with many nuclei. Only occasional septa divide the tubules. Most fungi are saprophytic, living on decaying plants or animal tissues. However, others are parasites that produce serious and difficult-to-treat infections in humans. An important medical problem is the lack of adequate antibiotics for treating fungal infections (mycoses).[94–96] On the other hand, fungi produce important antibiotics such as **penicillin**. Still others form some of the most powerful toxins known!

The lower fungi or **Phycomycetes** include simple aquatic molds and mildew organisms. Higher fungi are classified as **Ascomycetes** or **Basidiomycetes** according to the manner in which the sexual spores are born. In the Ascomycetes these spores are produced in a small sac called an **ascus** (Fig. 1-10). Each ascus contains four or eight spores in a row, a set of four representing the results of a single pair of meiotic divisions. A subsequent mitotic division will give eight spores. This is one of the features that has made *Neurospora crassa* (Fig. 1-10) a favorite subject for genetic

studies.[97] The ascospores can be dissected out in order from the ascus and cultivated separately to observe the results of crossing-over during meiosis.

Neurospora also reproduces via haploid spores called **conidia**. The haploid mycelia exist as two mating types and conidia or mycelia from one type can fertilize cells in a special body (the protoperithecium) of the other type to form zygotes. The latter immediately undergo meiosis and mitosis to form the eight ascospores. Among other Ascomycetes are the highly prized edible truffles and morels. However, most mushrooms and puffballs are fruiting bodies of Basidiomycetes. Other Basidiomycetes include the **rusts**, which cause enormous damage to wheat and other grain crops.

Yeasts are fungi adapted to life in an environment of high sugar content and which usually remain unicellular and reproduce by budding (Fig. 1-10). Occasionally the haploid cells fuse in pairs to form diploid cells and sexual spores. Some yeasts are related to the Ascomycetes, others to Basidiomycetes. *Saccharomyces cerevisiae*, the organism of both baker's and brewer's yeast, is an Ascomycete. It can grow indefinitely in either the haploid or diploid phase. The genetics and biochemistry of this yeast have been studied extensively.[98–102] The genome is relatively small with 13.5×10^6 base pairs in 17 chromosomes. The sequence of the 315,000 base pairs of chromosome III was determined in 1992[101,102] and the sequence of the entire genome is now known.[103]

Fungi often grow in symbiotic association with other organisms. Of special importance are the **mycorrhizae** (fungus roots) formed by colonization of fine roots by beneficial soil fungi. Almost all plants of economic importance form mycorrihizae.[104]

3. Algae

Algae are chlorophyll-containing eukaryotic organisms which may be either unicellular or colonial.[105] The colonial forms are usually organized as long filaments, either straight or branched, but in some cases as blades resembling leaves. However, there is little differentiation among cells. The gold-brown, brown, and red algae contain special pigments in addition to the chlorophylls.

The euglenids (**Euglenophyta**) and dinoflagellates (**Pyrrophyta**), discussed in the protozoa section, can equally well be regarded as algae. The bright green **Chlorophyta**, unicellular or filamentous algae, are definitely plants, however. Of biochemical interest is *Chlamydomonas*, a rather animal-like creature with two flagella and a carotenoid-containing eyespot or **stigma** (Fig. 1-11). *Chlamydomonas* contains a single chloroplast. The "pyrenoid", a center for the synthesis of starch, lies, along with the eyespot, within the chloroplast. The organism is haploid with "plus" and "minus" strains and motile gametes. Zygotes immediately undergo meiosis to form haploid spores. With a well-established genetic map, *Chlamydomonas* is another important organism for studies of biochemical genetics.[106]

The filamentous *Ulothrix* shows its relationship to the animals through formation of asexual spores with four flagella and biflagellate gametes. Only the zygote is diploid. On the other hand, the incomparably beautiful *Spirogyra* (Fig. 1-11) has no motile cells. The ameboid male gamete flows through a tube formed between the two mating cells, a behavior suggesting a relationship to higher green plants.

Some unicellular algae grow to a remarkable size. One of these is *Acetabularia* (Fig. 1-11), which lives in the warm waters of the Mediterranean and other tropical seas. The cell contains a single nucleus which lies in the base or rhizoid portion. In the mature alga, whose life cycle in the laboratory is 6 months, a cap of characteristic form develops. When cap development is complete, the nucleus divides into about 10^4 secondary nuclei which migrate up the stalk and out into the rays of the cap where they form cysts. After the cap decays and the cysts are released, meiosis occurs and the flagellated gametes fuse in pairs to form zygotes which again grow into diploid algae. Because of its large size and the location of the nucleus in the base, the cells can be cut and grafted. Nuclei can be removed or transplanted and growth and development can be studied in the presence or absence of a nucleus.[107–110] The green algae **Volvox** live in wheel-like colonies of up to several thousand cells and are useful for biochemical studies of differentiation.[111]

Look through the microscope at almost any sample of algae from a pond or aquarium and you will see little boatlike **diatoms** slowly gliding through the water. The most prominent members of the division Chrysophyta, diatoms are characterized by their external "shells" of silicon dioxide. Large and ancient deposits of diatomaceous earth contain these durable silica skeletons which are finely marked, often with beautiful patterns (Fig. 1-11). The slow motion of diatoms is accomplished by streaming of protoplasm through a groove on the surface of the cell. Diatoms are an important part of marine plankton, and it is estimated that three-fourths of the organic material of the world is produced by diatoms and dinoflagellates. Like the brown algae, Chrysophyta contain the pigment **fucoxanthin**.

Other groups of algae are the brown and red marine algae or seaweed. The former (**Phaeophyta**) include the giant kelps from which the polysaccharide **algin** is obtained. The **Rhodophyta** are delicately branched plants containing the red pigment **phycoerythrin**. The polysaccharides, **agar** and **carrageenin**,

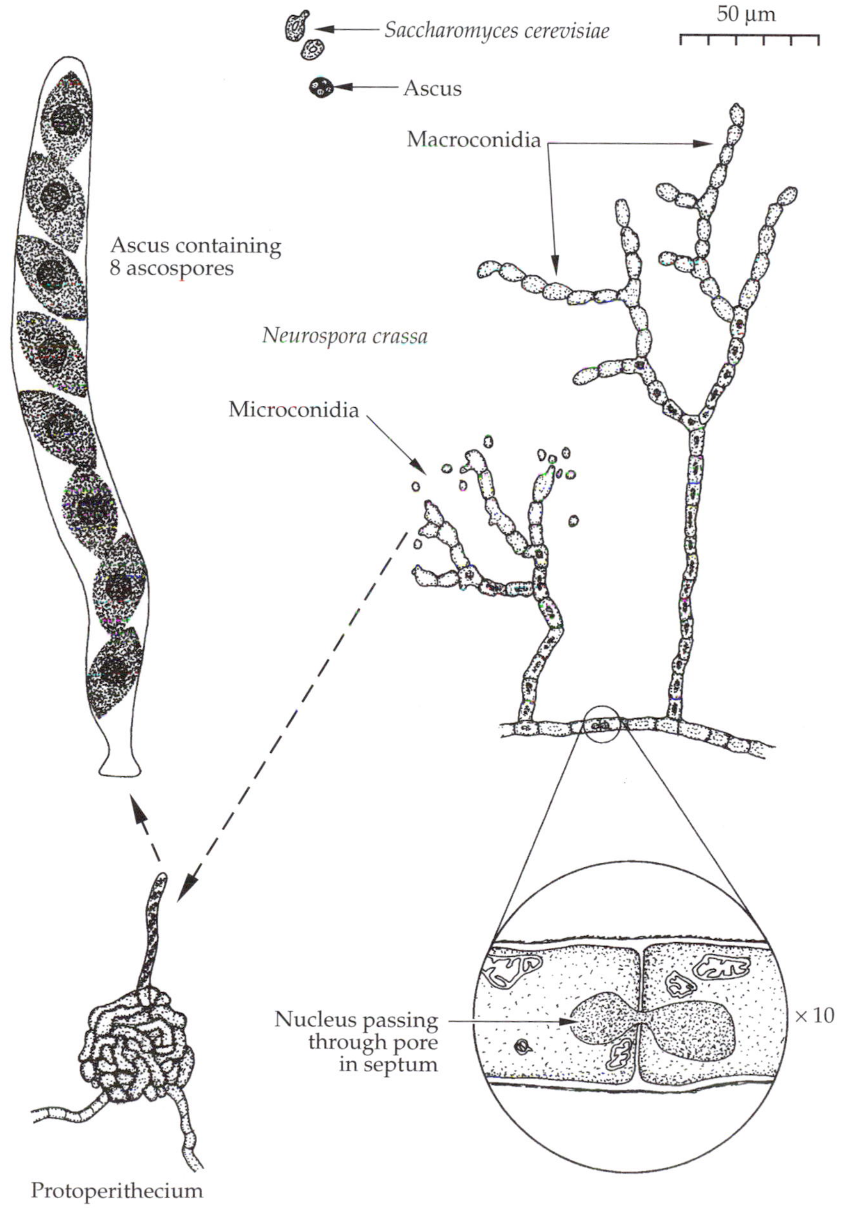

Figure 1-10 Two frequently studied fungi. Top (including ascus): the yeast *Saccharomyces cerevisiae*. Below: *Neurospora crassa* showing various stages. After J. Webster.[93]

a popular additive to chocolate drinks and other foods, come from red algae.

Symbiotic associations of fungi with either true algae or with cyanobacteria are known as **lichens**. Over 15,000 varieties of lichens grow on rocks and in other dry and often cold places. While the algae appear to benefit little from the association, the fungi penetrate the algae cells and derive nutrients from them.[112] Although either of the two partners in a lichen can be cultured separately, the combination of the two is capable of producing special pigments and phenolic substances known as **depsides** which are not formed by either partner alone.

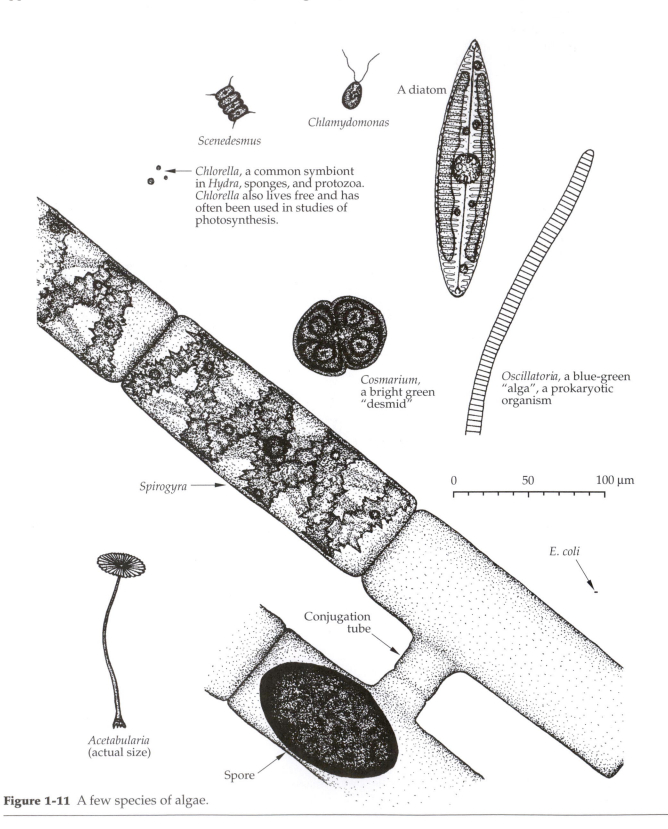

Figure 1-11 A few species of algae.

E. The Variety of Animal Forms

In this section, we will consider only a few biochemical and other aspects of multicellular animals or **Metazoa**. The sudden appearance of a large number of Metazoans about 0.5×10^9 years ago[113,114] may have been an outcome of the appearance of split genes (see Section B, 1). As a result of gene duplication the coding pieces of split genes, the **exons**, could be moved to new locations in a chromosome where they could have become fused with other pieces of DNA to form entirely new genes.[115]

1. Major Groups of Multicellular Animals

The simplest metazoa are tiny symbiotic worms of the phylum (or subkingdom) **Mesozoa**, which live in the kidneys of deep sea-dwelling cephalopods (octopi and squid). Each worm is made up of only 25 cells in a single layer enclosing one or a small number of elongated axial cells (Fig. 1-12). Mesozoa have been regarded as parasitic, but they appear to facilitate excretion of NH_3 by the host through acidification of the urine.[118,119]

Porifera or sponges are the most primitive of multicelled animals.[120] They lack distinct tissues but contain several specialized types of cells. The body is formed by stationary cells that pump water through the pores to bring food to the sponge. Within the body **amebocytes** work in groups to form the **spicules** of calcium carbonate, silicon dioxide, or the protein **spongin** (Fig. 1-12). Sponges appear to lack a nervous system.

Individuals of the next most complex major phylum, **Cnidaria** (formerly Coelenterata), are radially symmetric with two distinct cell layers, the **endoderm** and **ectoderm**. Many species exist both as a polyp or **hydra** form (Fig. 1-13) and as a **medusa** or jellyfish. The jellyfish apparently has no brain but the ways in which its neurons interconnect in a primitive radial net are of interest. The Cnidaria have a very simple body form with remarkable regenerative powers. The freshwater hydra, a creature about 1 cm long (Fig. 1-13), contains a total of $\sim 10^5$ cells. A complete hydra can be regenerated from a small piece of tissue if the latter contains some of both the inner and the outer cell layers.[121,122]

The body of flatworms (phylum **Platyhelminthes**) consists of two external cell layers (endoderm and ectoderm) with a third layer between. A distinct excretory system is present. In addition to a nerve net resembling that of the Cnidaria, there are a cerebral ganglion and distinct eyes. One large group of flatworms, the **planarians** (typically about 15 mm in length, Fig. 1-14), inhabit freshwater streams. They are said to be the simplest creatures in which *behavior* can be studied.

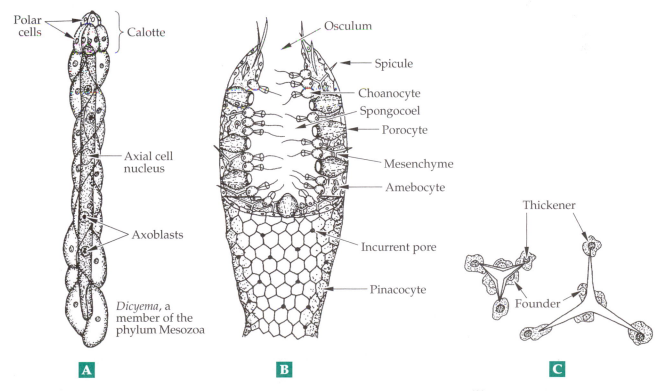

Figure 1-12 Some lower forms of Metazoa. (A) Mesozoa (25 cells). After C. P. Hickman.[116] (B) A small asconoid sponge. After C. A. Villee, W. F. Walker, Jr., and R. D. Barnes.[117] (C) Ameboid cells of a sponge forming spicules. After Hickman.

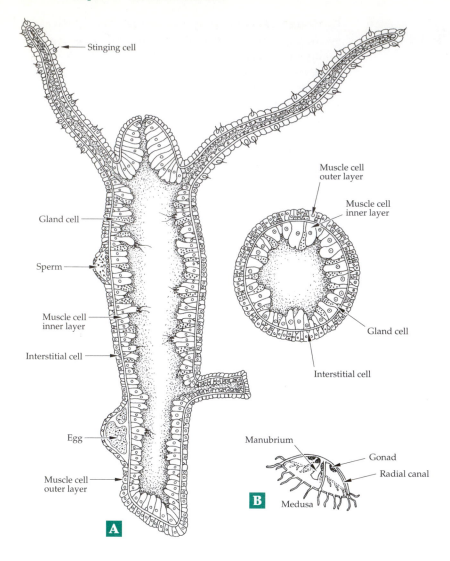

Figure 1-13 (A) Hydra. After Loomis.[123] (B) The medusa stage of *Obelia*, a hydroid coelenterate.[123a]

Many parasitic flatworms (tapeworms and flukes) attack higher organisms.[124] Among them are the **Schistosoma**, tiny worms that are transmitted to humans through snails and which attack the blood vessels. The resulting **schistosomiasis** is one of the most widespread debilitating diseases on earth today, affecting 200 million people or more.[125,126]

The roundworms (**Nematoda**)[127–129] have, in addition to the **enteron** (alimentary tract), a separate body cavity. Free-living nematodes abound in water and soil but many species are parasitic. They do enormous damage to plants and to some animal species. Trichina, hookworms[129a], and filaria worms attack humans. However, in the laboratory the 1-mm-long, 810-cell nematode, *Caenorhabditis elegans* (Fig. 1-14) has become an important animal. In 1963 Sydney Brenner launched what has become a worldwide effort to make this tiny worm the equivalent in the animal kingdom of

E. coli in the bacterial world.[129] The 10^8 nucleotides in the worm's six chromosomes contain ~13,600 genes. *C. elegans* has become an important animal in which to study differentiation. Already the exact lineage of every cell has been traced, as has every connection among the 302 neurons in the animal's nervous system. The related **rotifers**,[130] with whirling "wheels" of cilia on their heads (Fig. 1-14) and transparent bodies, are a delight to the microscopist. Like nematodes, they are "cell constant" organisms. The total number of cells in the body is constant as is that in almost every part of every organ. Part of the developmental plan of such organisms is a "programmed cell death" (Chapter 32).

The **Annelida** (segmented worms)[131] are believed to be evolutionary antecedents of the arthropods. Present-day members include earthworms, leeches, and ~10^5 species of marine **polychaetes**. Annelids have a true body cavity separate from the alimentary canal and lined by a peritoneum. They have a well-developed circulatory system and their blood usually contains a type of hemoglobin.

About 10^6 species of **arthropods** (80% of all known animals) have been described. Most are very small.[72] These creatures, which have a segmented exoskeleton of **chitin** and other materials, include the horseshoe crabs, the Arachnida (scorpions, spiders, and mites), the Crustacea, Myriopoda (centipedes and millipedes), and the Insecta. Important biochemical problems are associated with the development and use of insecticides and with our understanding of the metamorphosis that occurs during the growth of arthropods.[132] The fruit fly *Drosophila melanogaster* has provided much of our basic knowledge of genetics and continues to be the major species in which development is studied.[133–134a]

Among the molluscs (phylum **Mollusca**) the squids and octopuses have generated the most interest among biochemists. The neurons of squid contain giant axons, the study of which has led to much of our knowledge of nerve conduction. Octopuses show signs of intelligence not observed in other invertebrates whose nervous reactions seem to be entirely "preprogrammed." The brains of some snails contain only 10^4 neurons, some of which are unusually large. The

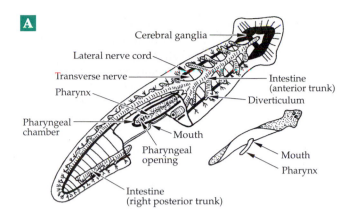

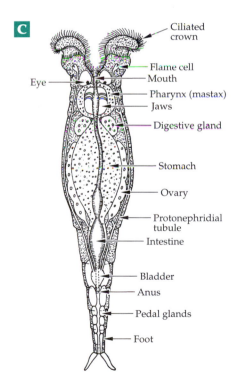

Figure 1-14 (A) A planarian, length 15 mm. After Hickman.[116] Diagram of digestive and nervous sytems; cutaway section shows ventral mouth. Small drawing shows pharynx extended through ventral mouth. (B) The nematode *Caenorhabditis elegans. Ascaris* is very similar in appearance. From Buchsbaum.[77] (C) A rotifer, *Philodina* (~10^3 cells). After C. A. Villee *et al.*[117]

Echinodermata or spiny-skinned animals (starfish, sea urchins, and sea cucumbers) are regarded as a highly advanced phylum. Their embryological development has been studied intensively.

The phylum **Chordata**, to which we ourselves belong, includes not only the vertebrates but also more primitive marine animals that have a spinal cord. Among these primitive species, which may be related to early ancestral forms, are the **tunicates** or sea squirts. They have a very high concentration of vanadium in their blood.

2. Cell Types and Tissues

Isolated animal cells in tissue culture, no matter how highly differentiated, tend to revert quickly to one of three basic types known as **epitheliocytes**, **mechanocytes**, and **amebocytes**. Epitheliocytes are closely adherent cells derived from epithelial tissues and thought to be related in their origins to the two surface layers of the embryonic blastula. Mechanocytes, often called **fibroblasts** or **fibrocytes**, are derived from muscle, supporting, or connective tissue. Like the amebocytes, they arise from embryonic mesenchymal tissue cells that have migrated inward from the lower side of the blastula (Chapter 32). **Neurons**, **neuroglia**, and **lymphocytes** are additional distinct cell types.

BOX 1-E   ERRORS, MISCONCEPTIONS, AND SPECULATION

Warning: Not everything in this book is true. Despite all efforts to get it right, there are unintentional errors and misinterpretations of experimental results. Indeed, the history of biochemistry is replete with accounts of experimental findings that were interpreted incorrectly. Yet, the ideas expressed often stimulated others to develop a more correct picture later. The same is true today. Students should be critical, should look at experimental details, and *consult original literature as much as possible.*

Progress in science depends both upon careful observations and measurements and upon imaginative interpretations of unexpected findings. Speculative ideas, a number of which are mentioned in this book, provide an important stimulus in science. They should neither be ignored nor accepted as facts. I have tried to write in such a way that established facts will not often be confused with speculation.

Tissues. Cells aggregate to form four major kinds of tissue. Epithelial tissues line the primary surfaces of the body: the skin, the digestive tract, urogenital tract, and glands. External skin is composed of flat platelike squamous epithelial cells whereas internal surfaces are often formed by columnar epithelial cells. Glands (sweat, oil, mammary, and internal secretory) as well as the sensory organs of the tongue, nose, and ear are all composed of epithelial cells. Epithelial cells are among the most highly polarized of cells. One side of each cell faces the outside, either air or water, while the other side is often directly against a basement membrane.

Supporting and connective tissues include the fatty **adipose tissue** as well as **cartilage** and **bone**. Both of the latter contain large amounts of intercellular material or **ground substance** consisting largely of complex polymers. Embryonic fibroblasts differentiate into white fibers, which produce collagen, and yellow fibers, which form elastin. The fibrils of both of these proteins are assembled in the intercellular space where they are embedded in the ground substance. **Osteoblasts** form bone by deposition of calcium phosphate in $3-7$ μm thick layers within a ground substance that contains special proteins.

A third tissue is **muscle**, which is classified into three types: **striated** (voluntary skeletal muscle), **cardiac** (involuntary striated muscle), and **smooth** (involuntary) muscle. There are two major groups of cells in **nervous tissue**, the fourth tissue type. **Neurons** are the actual conducting cells whose cell membranes carry nerve impulses. Several kinds of **glial cells** lie between and around the neurons.

Blood cells. Blood and the linings of blood vessels may be regarded as a fifth tissue type.[135,135a] The human body contains 5×10^9 **erythrocytes** or red blood cells per ml, a total of 2.5×10^{13} cells in the five liters of blood present in the body. Erythrocytes are rapidly synthesized in the bone marrow. The nucleus is destroyed, leaving a cell almost completely filled with hemoglobin. With an average lifetime of 125 days, human red blood cells are destroyed by leukocytes in the spleen and liver.

The white blood cells or **leukocytes** are nearly a thousandfold less numerous than red cells. About 7×10^6 cells are present per ml of blood. There are three types of leukocytes: **lymphocytes** (~26% of the total), **monocytes** (~7% of the total), and **polymorphonuclear leukocytes** or **granulocytes** (~70% of the total). Lymphocytes are about the same size as erythrocytes and are made in lymphatic tissue. Individual lymphocytes may survive for as long as ten years. They function in antibody formation and are responsible for maintenance of long-term immunity.

Monocytes, two times larger, are active in ingesting bacteria. These cells stay in the blood only a short time before they migrate into the tissues where they become **macrophages**,[136] relatively fixed phagocytic cells. Macrophages not only phagocytize and kill invading bacteria, protozoa, and fungi but also destroy cancer cells. They also destroy damaged cells and cellular debris as part of the normal turnover of tissues. They play an essential role in the immune system by "processing" antigens and in releasing stimulatory proteins.

Granulocytes of diameter $9-12$ μm are formed in the red bone marrow. Three types are distinguished by staining: **neutrophils**, **eosinophils**, and **basophils**. Neutrophils are the most numerous phagocytic cells of our blood and provide the first line of defense against bacterial infections. The functions of eosinophils and basophils are less well understood. The number of eosinophils rises during attacks of hay fever and asthma and under the influence of some parasites, while the basophil count is increased greatly in leukemia and also by inflammatory diseases. Granules containing histamine, heparin, and leukotrienes are present in the basophils. Blood **platelets** or **thrombocytes** are tiny ($2-3$ μm diameter) cell-like bodies essential for rapid coagulation of blood. They are formed by fragmentation of the cytoplasm of bone marrow **megakaryocytes**. One mature megakaryocyte may contribute 3000 platelets to the $1-3 \times 10^8$ per ml present in whole blood.

Cell culture. Laboratory growth of isolated animal cells has become very important in biochemistry.[137] Sometimes it is necessary to have many cells with as nearly as possible identical genetic makeup. Such bacterial cells are obtained by plating out the bacteria and selecting a small colony that has grown from a single cell to propagate a "pure strain." Similarly, single eukaryotic cells may be selected for tissue culture and give rise to a **clone** of cells which remains genetically identical until altered by mutations.

The culture of embryonic fibroblasts is used to obtain enough cells to perform prenatal diagnosis of inherited metabolic diseases (Box 1-D). Tissue culture is easiest with embryonic or cancer cells, but many other tissues can be propagated. However, the cells that grow best and which can be propogated indefinitely are not entirely normal; the well-known **HeLa** strain of human cancer cells which was widely grown for many years throughout the world contains $70-80$ chromosomes per cell compared with the normal 46.

3. Communication

Plants are able to maintain their form because the cells are surrounded by thick walls that cement the cells together. However, animal cells lack rigid walls and must be held together by specialized contacts.[138,139] Contacts between cells of both plants and animals are

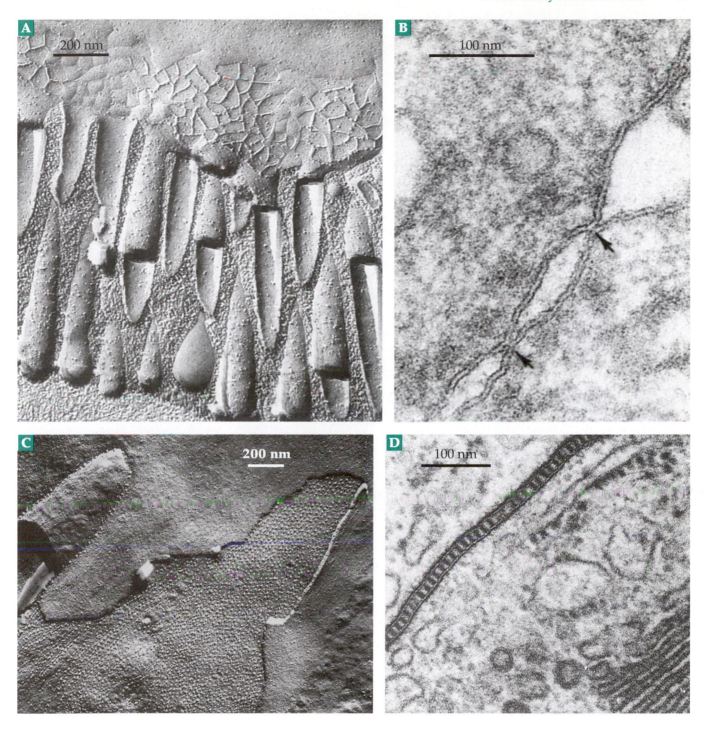

Figure 1-15 Electron micrographs of cell junctions of three types. (A) Freeze-fractured zona occludens (occlusion zone) between epithelial cells of the rat small intestine. The tight junctions are repesented as a meshwork of ridges (in the P or protoplasmic fracture face) or grooves (in the E fracture face which looks toward the extracellular space). These represent the actual sites of membrane fusion. Microvilli are seen in the lower part of the photograph. From D. S. Friend and N. B. Gilula.[141] (B) Thin cross section of tight junction between mouse hepatocytes. The arrows indicate points of membrane fusion. From Gilula.[142] Copyright 1975 by The Williams & Wilkins Co., Baltimore. (C) A freeze-fractured septate junction from ciliated epithelium of a mollusc. This type of junction forms a belt around the cells. Fracture face P (central depressed area) contains parallel rows of membrane particles that correspond to the arrangmenet of the intracellular septa seen in thin sections. The surrounding fracture face E contains a complementary set of grooves. Particles in nonjunctional membrane regions (upper right corner) are randomly arranged. (D) Thin section of a septate junction of the type shown in (C). The plasma membranes of the two cells are joined by a periodic arrangement of electron-dense bars or septa, which are present within the intercellular space. Note the Golgi membranes in the lower right part of the photograph. (C) and (D) are from N. B. Gilula[143]

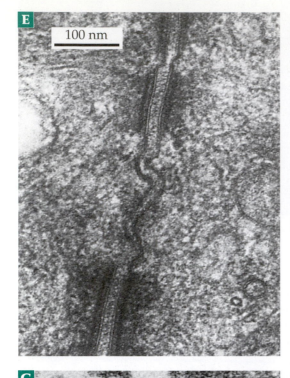

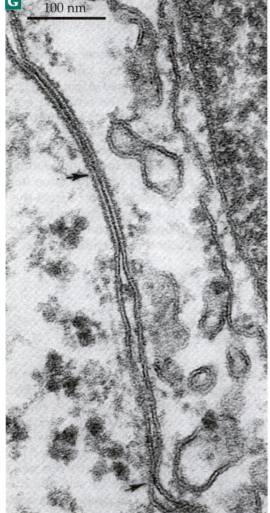

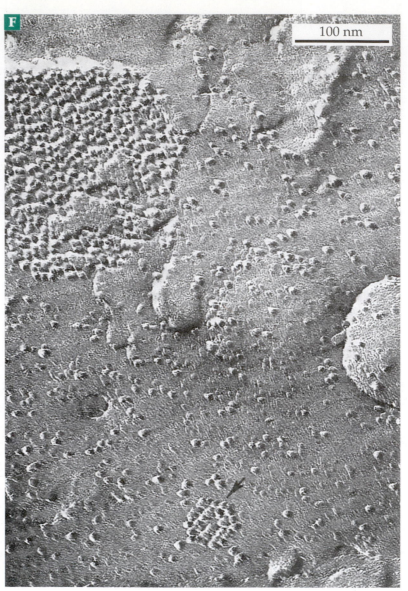

Figure 1-15 (continued) (E) Desmosomes (macula adherens) in rat intestinal epithelium. Features include a wide (25–35 nm) intercellular space containing dense material, two parallel cell membranes, a dense plaque associated with the cytoplasmic surface and cytoplasmic tono-filaments that converge on the dense plaque. From Gilula.[143] (F) Freeze-fractured surface through gap junctions between communicating cells in culture. Both a large junction and a smaller one below (arrow) can be seen. (G) Gap junctions in thin section. (F) and (G) are from N. B. Gilula.[144]

important for a second reason: Cells must communicate, one with another.

   Cell contacts and junctions. Many epithelial cells, e.g., those lining the border of kidney tubules and secretory glands, form **tight junctions** with adjacent cells. Electron microscopy shows that in these junctions the outer portions of the membranes actually fuse in some places (Fig. 1-15). One way in which this has been demonstrated is to freeze a tissue rapidly and to fracture it in the frozen state within a vacuum chamber. The fractured tissue is kept at about −100°C in a vacuum for a short time while water molecules evaporate from the fractured surfaces. A thin plastic replica is then made of the etched surfaces, which sometimes pass through tight junctions revealing details of their structure (Fig. 1-15, A, E). Study of electron micrographs of such surfaces shows that some cells are completely surrounded by belts of tight junctions, sometimes referred to as **occlusion zones** or **terminal bars**. Tight junctions between endothelial cells of blood capillaries in the brain prevent free diffusion of compounds from the blood stream into brain cells and form the **blood–brain barrier**.[140] Tight junctions between neurons and adjacent cells surround the **nodes of Ranvier** (Chapter 30).

Contacts of another type, known as **septate desmosomes** or **adhesion discs**, form a belt around the cells of invertebrate epithelia. In these contacts a space of ~18 nm between adjacent cell membranes is bridged in a number of places by thin walls. Behind the desmosomes the membrane is often backed up at these points by an electron-dense region to which are attached many fine microfilaments of ~6−10 nm diameter (Fig. 1-15, D).

One method of communication between cells is by passage of chemical substances through special junctions which, because of their appearance in electron micrographs of thin sections (Fig. 1-15, G) are known as **gap junctions**.[139,145,146] Gap junctions may cover substantial areas of the cell interface. In cross section, a thin 3−4 nm gap between the adjacent cell membranes is bridged by a lattice-like structure, which may appear in freeze-fractured surfaces as a hexagonal array of particles (Fig. 1-15, F, lower junction). These particles or **connexons** are each thought to be composed of six protein subunits. A central channel in the connexon is able to pass molecules of molecular mass up to about 500 Da.[147,148] Small molecules may be able to pass freely from one cell to another through the gap junction. Because of their low electrical resistance, gap junctions allow "electrical coupling" of cells. Such junctions form the **electrotonic synapses** that link some neurons to other excitable cells. Heart cells are all electrically coupled through gap junctions.[149]

Another type of communicating junction is also found in **synapses** of the nervous system. At these specialized contacts a nerve impulse transmitted along the membrane of one neuron triggers the release of a **neurotransmitter**, a chemical substance that passes across the gap between cells of the synapse and initiates a nerve impulse in the second neuron (Chapter 30).

Cell recognition. Cells of higher organisms are able to recognize other cells as identical, as belonging to another tissue, or as being "foreign." This ability is developed most highly by cells of the immune system but is possessed to some extent by others. For example, cells of sponges can be separated by partial digestion of the protein "cement" that holds them together. When dissociated cells from orange sponges were mixed with those from yellow sponges, the cells clumped together to reform small sponges.[150,151] Furthermore, orange cells stuck to orange cells and yellow to yellow cells. Similar results have been obtained using a mixture of cultivated liver, kidney, and embryonic brain cells. When a wound heals, epithelial cells grow and move across the wound surface but they stop when they meet. Cells in tissue culture and growing on a glass surface experience this same **contact inhibition**[152] and spread to form a unicellular layer. Cancer cells in culture do not stop but climb one on top of the other, apparently lacking proper recognition and communication. Many chemical signals appear to pass between cells. An important goal of contemporary biochemistry is to understand how cells recognize each other and respond to signals that they receive.

F. Higher Plants and Plant Tissues

Botanists recognize two divisions of higher plants. The **Bryophyta** or moss plants consist of the Musci (mosses) and Hepaticae (liverworts). These plants grow predominantly on land and are characterized by swimming sperm cells and a dominant gametophyte (haploid) phase. **Tracheophyta**, or vascular plants, contain conducting tissues. About 2×10^5 species are known. The ferns (class Filicineae, formerly Pteridophyta) are characterized by a dominant diploid plant and alternation with a haploid phase. Seed plants are represented by two classes: **Gymnosperms** (cone-bearing trees) and **Angiosperms**, the true flowering plants.

Genetically the simplest of the angiosperms is the little weed *Arabidopsis thaliana*, whose generation time is as short as five weeks. Its five chromosomes contain only 10^8 base pairs in all, the smallest known genome among angiosperms[153] and one whose complete nucleotide sequence is being determined. Its biochemistry, physiology, and developmental biology are under intensive study. It may become the "fruit fly" of the plant kingdom.

There are several kinds of plant tissues. Undifferentiated, embryonic cells found in rapidly growing regions of shoots and roots form the **meristematic tissue**. By differentiation, the latter yields the simple tissues, the parenchyma, collenchyma, and sclerenchyma. **Parenchyma** cells are among the most abundant and least specialized in plants. They give rise through further differentiation to the **cambium layer**, the growing layer of roots and stems. They also make up the pith or pulp in the center of stems and roots, where they serve as food storage cells.

The **collenchyma**, present in herbs, is composed of elongated supporting cells and the **sclerenchyma** of woody plants is made up of supporting cells with hard lignified cell walls and a low water content. This tissue includes **fiber cells**, which may be extremely long; e.g., pine stems contain fiber cells of 40 μm diameter and 4 mm long.

Two complex tissues, the **xylem** and **phloem**, provide the conducting network or "circulatory system" of plants. In the xylem or woody tissue, most of the cells are dead and the thick-walled tubes (**tracheids**) serve to transport water and dissolved minerals from the roots to the stems and leaves. The phloem cells provide the principal means of downward conduction of foods from the leaves. Phloem cells are joined end to end by **sieve plates**, so-called because they are perforated by numerous minute pores through which cytoplasm of adjoining sieve cells appears to be connected by strands 5–9 μm in diameter.[154] Mature sieve cells have no nuclei, but each sieve cell is paired with a nucleated "companion" cell.

Epidermal tissue of plants consists of flat cells, usually containing no chloroplasts, with a thick outer wall covered by a heavy waxy **cuticle** about 2 μm thick. Only a few specialized cells are found in the epidermis. Among them are the paired **guard cells** that surround the small openings known as **stomata** on the undersurfaces of leaves and control transpiration of water. Specialized cells in the root epidermis form **root hairs**, long extensions (~1 mm) of diameter 5–17 μm. Each hair is a single cell with the nucleus located near the tip.

Figure 1-16 shows a section from a stem of a typical angiosperm. Note the thin cambium layer between the phloem and the xylem. Its cells continuously

undergo differentiation to form new layers of xylem increasing the woody part of the stem. New phloem cells are also formed, and as the stem expands all of the tissues external to the cambium are renewed and the older cells are converted into bark.

Plant **seeds** consist of three distinct portions. The **embryo** develops from a zygote formed by fusion of a sperm nucleus originating from the pollen and an egg cell. The fertilized egg is surrounded in the gymnosperms by a nutritive layer or **endosperm** which is **haploid** and is derived from the same gametophyte tissue that produced the egg. In angiosperms *two* sperm nuclei form; one of these fertilizes the egg, while the other fuses with *two* haploid **polar nuclei** derived from the female gametophyte. (The polar nuclei are formed by the same mitotic divisions that formed the egg.) From this develops a *3n* **triploid** endosperm.

G. The Chemical Composition of Cells

Water is the major component of living cells, but the amount varies greatly. Thus, the pig embryo is 97% water; at birth a new-born pig is only 89% water. A lean 45-kg pig may contain 67% water but a very fat 135-kg animal only 40% water. Similar variations are encountered with other constituents.

The water content of a tissue is often determined by thoroughly drying a weighed sample of tissue at low temperature in vacuum and then weighing it a second time. The solid material can then be extracted with a solvent that will dissolve out the fatty compounds. These are referred to collectively as **lipids**. After evaporation of the solvent the lipid residue may be weighed. By this procedure a young leafy vegetable might be found to contain 2–5% lipid on a dry weight

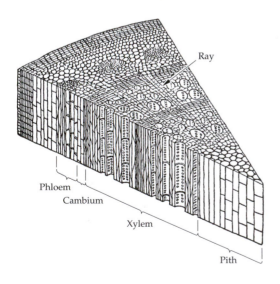
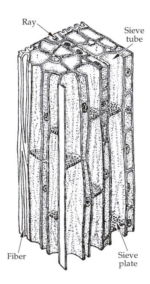
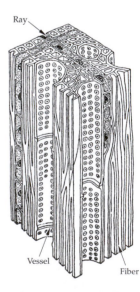

Figure 1-16 Section of the stem of an angiosperm. Enlarged sections showing tubes of the phloem (left) and xylem (right). From S. Biddulph and O. Biddulph.[155] Drawn by Bunji Tagawa.

basis. Even very lean meats contain 10–30% lipid.

The residue remaining after removal of the lipid consists predominately of three groups of compounds: **proteins**, **nucleic acids**, and **carbohydrates**. Most of the nitrogen present in tissues is found in the proteins and the protein content is sometimes estimated by determining the percentage of nitrogen and multiplying by 6.25. In a young green plant, 20–30% of the dry matter may be protein, while in very lean meat it may reach 50–70%.

TABLE 1-4
Approximate Composition of Metabolically Active Cells and Tissues[a]

Component	E. coli[b] (%)	Green plant (spinach, Spinacia oleracea)[c]	Rat liver[d] (%)
H_2O	70	93	69
Protein	15	2.3	21
Amino acids	0.4		
DNA	1		0.2
RNA	6		1.0
Nucleotides	0.4		
Carbohydrates	3	3.2	
Cellulose		0.6	
Glycogen			3.8
Lipids	2	0.3	6
Phospholipids			3.1
Neutral lipids			1.6
Sterols			0.3
Other small molecules	0.2		
Inorganic ions	1	1.5	
K^+			0.4

	Equivalents per liter in rat liver
Amino acid residues	2.1
Nucleotide units	0.03
Glycogen (glucose units)	0.22
K^+	0.1

[a] Data were not readily available for spaces left blank
[b] From J. D. Watson (1976) *Molecular Biology of the Gene*, 3rd ed., p. 69, Benjamin, New York The amounts of amino acids, nucleotides, carbohydrates, and lipids include precursors present in the cell.
[c] From B. T. Burton (1976) *Human Nutrition*, 3rd ed., McGraw-Hill, New York (p. 505)
[d] From C. Long, ed., (1961) *Biochemists' Handbook*, pp. 677–679, Van Nostrand-Reinhold, Princeton, New Jersey

A dried tissue sample may be burned at a high temperature to an **ash**, which commonly amounts to 3–10% and is higher in specialized tissues such as bone. It is a measure of the inorganic constituents of tissues.

The carbohydrate content can be estimated by the difference of the sum of lipid, protein, and ash from 100%. It amounts to 50–60% in young green plants and only 2–10% in typical animal tissues. In exceptional cases the carbohydrate content of animal tissues may be higher; the glycogen content of oysters is 28%.

The amount of nucleic acid in tissues varies from 0.1% in yeast and 0.5–1% in muscle and in bacteria to 15–40% in thymus gland and sperm cells. In these latter materials of high nucleic acid content it is clear that multiplication of % N by 6.25 is not a valid measure of protein content. For diploid cells of the body the DNA content per cell is nearly constant.

Table 1-4 compares the composition of a bacterium, of a green plant, and of an active animal tissue (rat liver). Although the solid matter of cells consists principally of C, H, O, N, S, and P, many other chemical elements are also present. Among the cations, Na^+, K^+, Ca^{2+}, and Mg^{2+} are found in relatively large amounts. Thus, the body of a 70 kg person contains 1050 g Ca (mostly in the bones), 245 g K, 105 g Na, and 35 g Mg. Iron (3 g), zinc (2.3 g), and rubidium (1.2 g) are the next most abundant. Of these iron and zinc are essential to life but rubidium is probably not. It is evidently taken up by the body together with potassium.

The other metallic elements in the human body amount to less than 1 g each, but at least seven of them play essential roles. They include copper (100 mg), manganese (20 mg), and cobalt (~5 mg). Others, such as chromium (<6 mg), tin, and vanadium, have only recently been shown essential for higher animals.[156,157] Nickel, lead, and others may perhaps be needed.

Nonmetallic elements predominating in the ash are phosphorus (700 g in the human body), sulfur (175 g), and chlorine (105 g). Not only are these three elements essential to all living cells but also selenium, fluorine, silicon (Box 4-B), iodine, and boron are needed by higher animals and boron by plants (Fig. 1-17). Iodine deficiency may affect one billion human beings and may cause 20 million cases per year of **cretinism**, or less severe brain damage.[158]

What is the likelihood that other elements will be found essential? Consider a human red blood cell, an object of volume ~80 μm^3 and containing about 3×10^8 protein molecules (mostly hemoglobin). About 7×10^5 atoms of the "trace metal" copper and 10^5 atoms of the nutritionally essential tin are present in a single red cell. Also present are 2×10^4 atoms of silver, a toxic metal. Its concentration, over 10^{-7} M, is sufficient that it could have an essential catalytic function. However, we know of none and it may simply have gotten into our bodies from handling money, jewelry, and other

BOX 1-F ABOUT THE REFERENCES

The lists of references at the ends of chapters are provided to encourage readers to look at original research articles. The lists are neither complete nor critically selected, but they do increase the information given in this book many-fold. I apologize for the important papers omitted. However, the references that are here will help a student to get started in reading the literature. Each reference contains other references and names of persons active in the field. By searching recent journal indices or a computer database it is easy to find additional articles by the same authors or on the same subject.

Look at the various types of scientific articles including reviews, preliminary reports and full research papers. Be sure to examine those in the *primary source journals* which publish detailed research results. These articles have always been sent to referees, active scientists, who check to see that the experiments are described accurately, that the authors have cited relevant literature, and that the conclusions are logical. Some journals, e.g.,

Biochem. Biophys. Res. Comm. and *FEBS Letts.* are dedicated to rapid publication of short reports but are also refereed. Other journals provide mostly reviews or a mixture of reviews. Periodical review series, such as *Advances in Nucleic Acid Chemistry and Related Topics*, often appear annually. Every student who intends to become a professional biochemist should consider purchasing the *Annual Review of Biochemistry* each year. This indispensible source of current information on most aspects of biochemistry is available to students at a very low price.

Many journal papers are difficult to read. To start, pick papers that have an understandable introduction. Choose reviews that are short, such as those in *Trends in Biochemical Sciences*. Then go on to the more comprehensive ones. Never sit back and hope that your computer will automatically fetch just what you need! Many journals carry papers of biochemical importance. Those specializing in biochemistry include the following:

Full Title	Abbreviation
Advances in Carbohydrate Chemistry and Biochemistry[a]	Adv. Carbohydr. Chem. Biochem.
Advances in Protein Chemistry[a]	Adv. Protein Chem.
Analytical Biochemistry	Anal. Biochem.
Annual Review of Biophysics and Biomolecular Structure[a]	Ann. Rev. Biophys. Biomolec. Struct.
Annual Review of Biochemistry[a]	Ann. Rev. Biochem.
Archives of Biochemistry and Biophysics	Arch. Biochem. Biophys.
Biochemical and Biophysical Research Communications	Biochem. Biophys. Res. Commun.
Biochemical Journal	Biochem. J.
Biochemistry	
Biochimica et Biophysica Acta	Biochim. Biophys. Acta
Bioorganic Chemistry	Bioorg. Chem.
Carbohydrate Research	Carbohydr. Res.
EMBO Journal[b]	EMBO J.
European Journal of Biochemistry	Eur. J. Biochem.
FASEB Journal[c]	FASEB J.
Journal of Bacteriology	J. Bacteriol.
Journal of Biochemistry	J. Biochem.
Journal of Biological Chemistry	J. Biol. Chem.
Journal of Lipid Research	J. Lipid Res.
Journal of Molecular Biology	J. Mol. Biol.
Journal of the American Chemical Society	J. Am. Chem. Soc.
Journal of Theoretical Biology	J. Theor. Biol.
Methods in Enzymology[a]	Methods Enzymol.
Nature	
Nucleic Acids Research	Nucleic Acids Res.
Proceedings of the National Academy of Sciences, USA	Proc. Natl. Acad. Sci. U.S.A.
Science	
Structure	
Trends in Biochemical Sciences	Trends Biochem. Sci. or TIBS

[a] These are not journals but series of review and reference books. There are many other series of "Advances in..." and "Annual Reviews of ..." that are not listed here.
[b] EMBO–European Molecular Biology Organization
[c] FASEB–Federation of American Societies for Experimental Biology

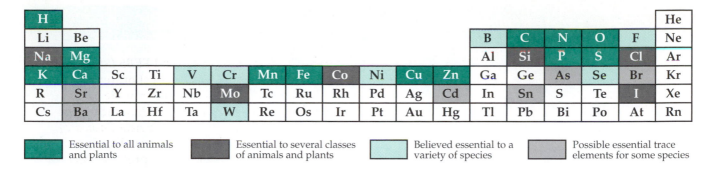

H																	He
Li	Be											B	C	N	O	F	Ne
Na	Mg											Al	Si	P	S	Cl	Ar
K	Ca	Sc	Ti	V	Cr	Mn	Fe	Co	Ni	Cu	Zn	Ga	Ge	As	Se	Br	Kr
R	Sr	Y	Zr	Nb	Mo	Tc	Ru	Rh	Pd	Ag	Cd	In	Sn	S	Te	I	Xe
Cs	Ba	La	Hf	Ta	W	Re	Os	Ir	Pt	Au	Hg	Tl	Pb	Bi	Po	At	Rn

Essential to all animals and plants   Essential to several classes of animals and plants   Believed essential to a variety of species   Possible essential trace elements for some species

Figure 1-17 Elements known to be essential to living things (after da Silva and Williams[157]). Essential elements are enclosed within shaded boxes. The 11 elements–C, H, O, N, S, P, Na, K, Mg, Ca, and Cl–make up 99.9% of the mass of a human being. An additional 13 are known to be essential for higher animals in trace amounts. Boron is essential to higher plants but apparently not to animals, microorganisms, or algae.

silver objects. The red blood cell also contains boron and aluminum (3×10^5 atoms each), arsenic (7×10^5 atoms), lead (7×10^4 atoms), and nickel (2×10^4 atoms). Of the elements (uranium and below) in the periodic table, only four (Ac, Po, Pa, and Ra) are present, on the average, in quantities less than one atom per cell.[156]

Of the apparently nonessential elements, several, e.g., Cs, Rb, Sr, and Ni (possibly essential) are not toxic at low concentrations. Others, such as Sb, As, Ba, Be, Cd, Pb, Hg, Ag, Tl, and Th, are highly toxic.

The ionic compositions of tissues and of body fluids vary substantially. Blood of marine organisms is similar to that of seawater in its content of Na^+, Cl^-, Ca^{2+}, and Mg^{2+}. Blood of freshwater and terrestrial organisms contains about ten times less Na^+ and Cl^- and several times less Ca^{2+} and Mg^{2+} than is present in seawater, but it is nevertheless relatively rich in these ions.

In general, cells are rich in K^+ and Mg^{2+}, the K^+ predominating by far, and are poor in Na^+ and Ca^{2+}. Chloride is the principal inorganic anion, but organic carboxylate and phosphate groups contribute most of the negative charges (Table 1-4), many of which are fixed to proteins or other macromolecules. Ling estimated that cells typically contain about 1.66 M of amino acid residues in their proteins. Of these residues, 10% have negatively charged side chains and 8% positively charged. The difference is a net negative charge amounting to 33 mM within cells.[159]

References

1. May, R. M. (1992) *Sci. Am.* **267**(Oct), 42–48
2. Stanier, R. Y., Ingraham, J. L., Wheelis, M. L., and Painter, P. R. (1986) *The Microbial World*, 5th ed., Prentice-Hall, Englewood Cliffs, New Jersey
3. Brock, T. D., Smith, D. W., and Madigan, M. T. (1988) *Biology of Microorganisms*, 5th ed., Prentice-Hall, Englewood Cliffs, New Jersey
4. Neidhardt, F. C., ed. (1987) *Escherichia coli and Salmonella typhimurium*, Am. Soc. for Microbiology, Washington, D.C.
5. Neidhardt, F. C., Ingraham, J. L., and Schaechter, M. (1990) *Physiology of the Bacterial Cell*, Sinauer, Sunderland, Mass.
6. Maniloff, J., and Morowitz, H. J. (1972) *Bacteriol. Rev.* **36**, 263–290
7. Razin, S. (1978) *Microbiol. Rev.* **42**, 414–470
8. Barile, M. F., Razin, S., Tully, J. G., and Whitcomb, R. F., eds. (1979) *The Mycoplasmas*, Vol. I, Academic Press, New York
9. Watson, J. D. (1976) *Molecular Biology of the Gene*, 3rd ed., Benjamin, Menlo Park, California (p. 61)
10. Horowitz, D. M., and Sanders, J. K. M. (1994) *J. Am. Chem. Soc.* **116**, 2695–2702
10a. Gansheroff, L. J., and O'Brien, A. D. (2000) *Proc. Natl. Acad. Sci. U.S.A.* **97**, 2959–2961
11. Fraser, C. M., Gocayne, J. D., White, O., Adams, M. D., Clayton, R. A., Fleischmann, R. D., Bult, C. J., Kerlavage, A. R., Sutton, G., Kelley, J. M., Fritchman, J. L., Weidman, J. F., Small, K. V., Sandusky, M., Fuhrmann, J., Nguyen, D., Utterback, T. R., Saudek, D. M., Phillips, C. A., Merrick, J. M., Tomb, J.-F., Dougherty, B. A., Bott, K. F., Hu, P.-C., Lucier, T. S., Peterson, S. N., Smith, H. O., Hutchison, C. A., III, and Venter, J. C. (1995) *Science* **270**, 397–403
12. Sonea, S. (1988) *Nature (London)* **331**, 216
13. Costerton, J. W., Geesey, G. G., and Cheng, K.-J. (1978) *Sci. Am.* **238**(Jan), 86–95
14. Macnab, R. M. (1987) in *Escherichia coli and Salmonella typhimurium* (Niedhardt, F. C., ed), pp. 70–83, Am. Soc. for Microbiology, Washington, D.C.
15. Margulis, L., Ashen, J. B., Sonea, S., Solé, M., and Guerrero, R. (1993) *Proc. Natl. Acad. Sci. U.S.A.* **90**, 6966–6970
16. Goldstein, S. F., Charon, N. W., and Kreiling, J. A. (1994) *Proc. Natl. Acad. Sci. U.S.A.* **91**, 3433–3437
17. Levine, M. M. (1985) *N. Engl. J. Med.* **313**, 445–447
18. Kuehn, M. J., Heuser, J., Normark, S., and Hultgren, S. J. (1992) *Nature (London)* **356**, 252–255
19. Alper, J. (1993) *Science* **262**, 1817
20. Holt, J. G., Krieg, N. R., Sneath, P. H. A., Staley, J. T., and Williams, S. T., eds. (1994) *Bergey's Manual of Determinative Bacteriology*, 9th ed., Williams & Wilkins, Baltimore, Maryland
21. Doolittle, R. F. (1992) *Protein Sci.* **1**, 191–200
22. Schwartz, R. M., and Dayhoff, M. O. (1978) *Science* **199**, 395–403
23. Cavalier-Smith, T. (1986) *Nature (London)* **324**, 416–417
24. Jones, W. J., Nagle, D. P., Jr., and Whitman, W. B. (1987) *Microbiol. Rev.* **51**, 135–177
25. Woese, C. R. (1987) *Microbiol. Rev.* **51**, 221–271
26. Sogin, M. L., Gunderson, J. H., Elwood, H. J., Alonso, R. A., and Peattie, D. A. (1989) *Science* **243**, 75–77
27. Garrett, R. A., Dalgaard, J., Larson, N., Kjems, J., and Mankin, A. S. (1991) *Trends Biochem. Sci.* **16**, 22–26
28. Searcy, D. G. (1982) *Trends Biochem. Sci.* **7**, 183–185

29. Wheelis, M. L., Kandler, O., and Woese, C. R. (1992) *Proc. Natl. Acad. Sci. U.S.A.* **89**, 2930–2934
30. Woese, C. R., Kandler, O., and Wheelis, M. L. (1990) *Proc. Natl. Acad. Sci. U.S.A.* **87**, 4576–4579
30a. Doolittle, W. F. (1999) *Science* **284**, 2124–2128
30b. Pennisi, E. (1999) *Science* **284**, 1305–1307
31. Cavalier-Smith, T. (1992) *Nature (London)* **356**, 570
32. Angert, E. R., Clements, K. D., and Pace, N. R. (1993) *Nature (London)* **362**, 239–241
33. Thimann, K. V. (1963) *The Life of Bacteria*, 2nd ed., Macmillan, New York
34. Stanier, R. Y., and Cohen-Bazire, G. (1977) *Ann. Rev. Microbiol.* **31**, 225–274
35. Morden, C. W., and Golden, S. S. (1989) *Nature (London)* **337**, 382–385
36. Marme, D., Marre, E., and Hertel, R., eds. (1982) *Plasmalemma and Tonoplast: Their Functions in the Plant Cell*, Elsevier, Amsterdam
37. Collins, F. S. (1995) *Proc. Natl. Acad. Sci. U.S.A.* **92**, 10821–10823
38. Hinshaw, J. E., Carraghen, B. O., and Milligan, R. A. (1992) *Cell* **69**, 1133–1141
39. Dingwall, C., and Lasky, R. (1992) *Science* **258**, 942–947
40. Nolta, K. V., Padh, H., and Steck, T. L. (1991) *J. Biol. Chem.* **266**, 18318–18323
41. de Duve, C. (1975) *Science* **189**, 186–194
42. Griffiths, G., and Simons, K. (1986) *Science* **234**, 438–443
43. Hartley, S. M. (1992) *Trends Biochem. Sci.* **17**, 325–327
44. Golgi, C. (1898) *Arch. Ital. Biol.* **30**, 60 and 278
45. Kirk, J. T. O., and Tilney-Bassett, R. A. E. (1978) *The Plastids: Their Chemistry, Structure, Growth, and Inheritance*, 2nd ed., Elsevier-North Holland Biomedical Press, Amsterdam
46. Hoffmann, H. P., and Avers, C. J. (1973) *Science* **181**, 749–751
46a. Capaldi, R. A. (2000) *Trends Biochem. Sci.* **25**, 212–214
46b. Rutter, G. A., and Rizzuto, R. (2000) *Trends Biochem. Sci.* **25**, 215–221
47. Fukui, Y., and Tanaka, A. (1979) *Trends Biochem. Sci.* **4**, 246–249
48. Kindl, H., and Lazarow, P. B., eds. (1982) *Peroxisomes and Glyoxysomes*, Vol. 386, Annals of the New York Academy of Sciences, New York
49. de Duve, C. (1983) *Sci. Am.* **248**(May), 74–84
50. Masters, C., and Crane, D. (1995) *The Peroxisome: A Vital Organelle*, Cambridge Univ. Press, London
51. Lord, J. M., and Roberts, L. M. (1980) *Trends Biochem. Sci.* **5**, 271–274
52. Cerkasov, J., Cerkasovova, A., Kulda, J., and Vilhelmova, D. (1978) *J. Biol. Chem.* **253**, 1207–1214
53. Glover, D. M., Gonzalez, C., and Raff, J. W. (1993) *Sci. Am.* **268**(Jun), 62–68
54. Olmsted, J. B., and Borisy, G. G. (1973) *Ann. Rev. Biochem.* **42**, 507–540
55. Hall, J. L., Ramanis, Z., and Luck, D. J. L. (1989) *Cell* **59**, 121–132
56. Hyams, J. S. (1989) *Nature (London)* **341**, 485–486
57. Satir, P. (1961) *Sci. Am.* **204**(Feb), 108–116
58. Lazarides, E. (1980) *Nature (London)* **283**, 244–256
59. Amos, L. A., and Amos, W. B. (1991) *Molecules of the Cytoskeleton*, Guilford, New York
60. Iwabe, N., Kuma, K., Hasegawa, M., Osawa, S., and Miyata, T. (1989) *Proc. Natl. Acad. Sci. U.S.A.* **86**, 9355–9359
61. Lake, J. A. (1991) *Trends Biochem. Sci.* **16**, 46–50

62. Margulis, L. (1981) *Symbiosis in Cell Evolution*, Freeman, San Francisco, California
63. Hurst, L. D. (1994) *Nature (London)* **369**, 451
64. Gupta, R. S., and Golding, G. B. (1996) *Trends Biochem. Sci.* **21**, 166–171
65. de Duve, C. (1996) *Sci. Am.* **274** (Apr), 50–57
66. Markert, C. L., Shaklee, J. B., and Whitt, G. S. (1975) *Science* **189**, 102–114
67. Fenchel, T., and Bernard, C. (1993) *Nature (London)* **362**, 300
68. Schopf, J. W. (1993) *Science* **260**, 640–646
69. Horgan, J. (1993) *Sci. Am.* **269**(Aug), 24
70. Doolittle, W. F. (1980) *Trends Biochem. Sci.* **5**, 146–149
71. May, R. M. (1986) *Nature (London)* **324**, 514–515
72. Seger, J. (1989) *Nature (London)* **337**, 305–306
73. Riley, M., and Anilionis, A. (1978) *Ann. Rev. Microbiol.* **32**, 519–560
74. Goossens, M., Dozy, A. M., Embury, S. H., Zachariades, Z., Hadjiminas, M. G., Stamatoyannopoulos, G., and Kan, Y. W. (1980) *Proc. Natl. Acad. Sci. U.S.A.* **77**, 518–521
75. Williams, R. O., Young, J. R., and Majiwa, P. A. O. (1979) *Nature (London)* **282**, 847–849
76. Kondrashov, A. S. (1988) *Nature (London)* **336**, 435–440
77. Buchsbaum, R., Buchsbaum, M., Pearse, J., and Pearse, V. (1987) *Animals Without Backbones*, 3rd ed., Univ. Chicago Press, Chicago, Illinois
78. Margulis, L., and Schwartz, K. V. (1999) *Five Kingdoms: An Illustrated Guide to the Phyla of Life on Earth*, 3rd ed., Freeman, New York
79. Laver, W. G., Air, G. M., Dopheide, T. A., and Ward, C. W. (1980) *Nature (London)* **283**, 454–457
80. McKanna, J. A. (1973) *Science* **179**, 88–90
81. Miller, L. H. (1992) *Science* **257**, 36–37
82. Wyler, D. J. (1993) *N. Engl. J. Med.* **329**, 31–37
83. Miller, L. H., Good, M. F., and Milon, G. (1994) *Science* **264**, 1878–1883
84. Nussenzweig, R. S., and Long, C. A. (1994) *Science* **265**, 1381–1382
85. Sibley, L. D., and Boothroyd, J. C. (1992) *Nature (London)* **359**, 82–85
86. Soldati, D., and Boothroyd, J. C. (1993) *Science* **260**, 349–352
87. Donelson, J. E., and Turner, M. J. (1985) *Sci. Am.* **252**(Feb), 44–51
88. Blum, M. L., Down, J. A., Gurnett, A. M., Carrington, M., Turner, M. J., and Wiley, D. C. (1993) *Nature (London)* **362**, 603–609
89. Gall, J. G., ed. (1986) *The Molecular Biology of Ciliated Protozoa*, Academic Press, Orlando, Florida
90. Olive, L. S. (1975) *The Mycetozoans*, Academic Press, New York
91. Kimmel, A. R., ed. (1989) *Molecular Biology of Dictyostelium Development*, Liss, New York
92. Traynor, D., Kessin, R. H., and Williams, J. G. (1992) *Proc. Natl. Acad. Sci. U.S.A.* **89**, 8303–8307
93. Webster, J. (1980) *Introduction to Fungi*, 2nd ed., Cambridge Univ. Press, New York
94. Davies, D. A. L., and Pope, A. M. S. (1978) *Nature (London)* **273**, 235–236
95. Georgopapadakou, N. H., and Walsh, T. J. (1994) *Science* **264**, 371–373
96. Sternberg, S. (1994) *Science* **266**, 1632–1634
97. Davis, R. H., and DeSerres, F. J. (1970) *Meth. Enzymol.* **17A**, 79–148
98. Botstein, D., and Fink, G. R. (1988) *Science* **240**, 1439–1443
99. Pringle, J. R., Broach, J. R., and Jones, E. W., eds. (1991–1994) *The Molecular and Cellular Biology of the Yeast Saccharomyces*, Cold Springs Harbor Laboratory Press, Cold Spring Harbor, New York (3 Vols.)

References

100. Wheals, A. E., Rose, A. H., and Harrison, S. J., eds. (1995) *The Yeasts*, 2nd ed., Vol. 6, Academic Press, San Diego, California

101. Gimeno, C. J., Ljungdahl, P. O., Styles, C. A., and Fink, G. R. (1992) *Cell* **68**, 1077–1090

102. Oliver, S. G., and 146 other authors (1992) *Nature (London)* **357**, 38–46

103. Williams, N. (1995) *Science* **268**, 1560–1561

104. Ruehle, J. L., and Marx, D. H. (1979) *Science* **206**, 419–422

105. Canter-Lund, H., and Lund, J. W. G. (1995) *Freshwater Algae: Their Microscopic World Explored*, Biopress, Bristol, England

106. Harris, E. H. (1989) *The Chlamydomonas Sourcebook*, Academic Press, San Diego, California

107. Brachet, J. L. A. (1965) *Endeavour* **24**, 155–161

108. Bonotto, S., Kefeli, V., and Puiseux-Dao, S. (1979) *Developmental Biology of Acetabularia*, Elsevier-North Holland, Amsterdam

109. Berger, S., and Kaever, M. J. (1992) *Dasycladales: An Illustrated Monograph of a Fascinating Algal Order*, Oxford Univ. Press, New York

110. Menzel, D., and Elsner-Menzel, C. (1990) *Protoplasm* **157**, 52–63

111. Jaenicke, L. (1982) *Trends Biochem. Sci.* **7**, 61–64

112. Ahmadjian, V. (1962) in *Physiology and Biochemistry of Algae*, Vol. 2 (Lewin, R. A., ed), pp. 817–822, Academic Press, New York

113. Morris, S. C. (1993) *Nature (London)* **361**, 219–221

114. Levinton, J. S. (1992) *Sci. Am.* **267**(Nov), 84–91

115. Keese, P. K., and Gibbs, A. (1992) *Proc. Natl. Acad. Sci. U.S.A.* **89**, 9489–9493

116. Hickman, C. P. (1973) *Biology of the Invertebrates*, Mosby, St. Louis, Missouri

117. Villee, C. A., Walker, W. F., Jr., and Barnes, R. D. (1973) *General Zoology*, Saunders, Philadelphia, Pennsylvania

118. Lapan, E. A., and Morowitz, H. (1972) *Sci. Am.* **227**(Dec), 94–101

119. Lapan, E. A. (1975) *Comp. Biochem. Physiol.* **52A**, 651–657

120. Bergquist, P. R. (1978) *Sponges*, Hutchinson, London

121. Grimmelikhuijzen, C. J. P., and Schaller, H. C. (1979) *Trends Biochem. Sci.* **4**, 265–267

122. Lenhoff, H. M., and Lenhoff, S. G. (1988) *Sci. Am.* **258**(Apr), 108–113

123. Loomis, W. F. (1959) *Sci. Am.* **200**(Apr), 150

123a. Hickman, C. P. *Integrated Principles of Zoology* (1966) Mosby, St. Louis

124. Trager, W. (1986) *Living Together: The Biology of Animal Parasitism*, Plenum, New York

125. Cherfas, J. (1991) *Science* **251**, 630–631

126. Kolberg, R. (1994) *Science* **264**, 1859–1861

127. Wood, W. B., ed. (1988) *The Nematode Caenorhabditis elegans*, Cold Spring Harbor Lab. Press, Cold Spring Harbor, New York

128. Kenyon, C. (1988) *Science* **240**, 1448–1453

129. Roberts, L. (1990) *Science* **248**, 1310–1313

129a. Hotez, P. J., and Pritchard, D. I. (1995) *Sci. Am.* **272**(Jun), 68–74

130. Dumont, H. J., and Green, J., eds. (1980) *Rotatoria*, p. 42, Junk, The Hague

131. Mill, P. J., ed. (1978) *Physiology of Annelids*, Academic Press, New York

132. Florkin, M., and Scheer, B. T., eds. (1970, 1971) *Chemical Zoology*, Vol. 5 and 6, Academic Press, New York

133. Goldstein, L. S. B., and Fyrberg, E. A., eds. (1994) *Drosophila melanogaster: Practical Uses in Cell and Molecular Biology*, Academic Press, San Diego, California

134. Lawrence, P. (1992) *The Making of a Fly: The Genetics of Animal Design*, Blackwell, Oxford

134a. Rubin, G. M., and Lewis, E. B. (2000) *Science* **287**, 2216–2218

135. Maddy, A. H. (1992) *Trends Biochem. Sci.* **17**, 125–126

135a. Harris, J. R., ed. (1991) *Blood Cell Biochemistry*, Plenum, New York (Three volumes)

136. Johnston, R. B., Jr. (1988) *N. Engl. J. Med.* **318**, 747–754

137. Freshney, R. I. (1994) *Culture of Animal Cells*, 3rd ed., Wiley-Liss, New York

138. Feldman, J., Gilula, N. B., and Pitts, J. D., eds. (1978) *Intercellular Junctions and Synapses*, Halsted Press, New York

139. Staehelin, L. A., and Hull, B. E. (1978) *Sci. Am.* **238**(May), 141–153

140. Bradbury, M. (1979) *The Concept of a Blood-Brain Barrier*, Wiley, Chichester, New York

141. Friend, D. S., and Gilula, N. B. (1972) *J. Cell Biol.* **53**, 771

142. Gilula, N. B. (1975) *Cellular Membranes and Tumor Cells*, Williams & Wilkins Co., Baltimore, Maryland (p. 221)

143. Gilula, N. B. (1974) in *Cell Communication* (Cox, R. P., ed), pp. 1–29, Wiley, New York

144. Gilula, N. B. (1972) *Nature (London)* **235**, 262–265

145. Zimmer, D. B., Green, C. R., Evans, W. H., and Gilula, N. B. (1987) *J. Biol. Chem.* **262**, 7751–7763

146. Hertzberg, E. L., and Johnson, R. G., eds. (1988) *Gap Junctions*, Liss, New York

147. Robinson, S. R., Hampson, E. C. G. M., Munro, M. N., and Vaney, D. I. (1993) *Science* **262**, 1072–1074

148. Stauffer, P. L., Zhao, H., Luby-Phelps, K., Moss, R. L., Star, R. A., and Muallem, S. (1993) *J. Biol. Chem.* **268**, 19769–19775

149. Veenstra, R. D., and DeHaan, R. L. (1986) *Science* **233**, 272–274

150. Moscona, A. A. (1961) *Sci. Am.* **205**(Sep), 143–162

151. Alberts, B., Bray, D., Lewis, J., Raff, M., Roberts, K., and Watson, J. D. (1983) *Molecular Biology of the Cell*, pp. 679–680, Garland Publ., New York

152. Heaysman, J. E. M. (1978) *Intl. Rev. Cytol.* **55**, 49–66

153. Palca, J. (1989) *Science* **245**, 131

154. Lloyd, A. M., Barnason, A. R., Rogers, S. G., Byrne, M. C., Fraley, R. T., and Horsch, R. B. (1986) *Science* **234**, 464–466

155. Biddulph, S., and Biddulph, O. (1959) *Sci. Am.* **200**(Feb), 44–49

156. Bowen, H. J. M. (1966) *Trace Elements in Biochemistry*, Academic Press, New York

157. Fraústo da Silva, J. J. R., and Williams, R. J. P. (1991) *The Biological Chemistry of the Elements: The Inorganic Chemistry of Life*, Clarendon Press, Oxford

158. Hetzel, B. S. (1994) *N. Engl. J. Med.* **331**, 1770–1771

159. Ling, G. N. (1962) *A Physical Theory of the Living State*, Ginn (Blaisdell), Boston, Massachusetts

Study Questions

1. Describe the principal structural or organizational differences between prokaryotic and eukaryotic cells.

2. Describe two or more principal functions of proteins within cells, one function of DNA, two or more functions of RNA, and one function of lipids.

3. Compare the chemical makeup of ribosomes, of cell membranes, and of bacterial flagella.

4. Assume the following dimensions: Mycoplasma, sphere, 0.33 μm diameter; *E. coli*, cylinder, 0.8 μm diameter x 2 μm; liver cell, sphere 20 μm; root hair, cylinder, 10 μm diameter x 1 mm.

   a. Calculate for each cell the total volume, the mass in grams and in daltons (assume a specific gravity of 1.0).

   b. Assume that bacterial ribosomes are approximately spherical with a diameter of 23 nm. What is their volume? If the mass of a bacterial ribosome is 2.7×10^6 daltons, what is its apparent density (divide mass by volume)? Experimentally the buoyant density of bacterial ribosomes in a cesium chloride gradient (Chapter 5) is about 1.6 g/cm^3. How can this difference be explained? If eukaryotic ribosomes are 1.17 times larger than bacterial ribosomes in linear dimensions, what is the volume of a eukaryotic ribosome?

   c. What fraction of volume of *E. coli* consists of cell wall, of plasma membrane, of ribosomes (assume 15,000 are present)? If a cell of *E. coli* is 80% water, what fraction by weight of the total solids consists of ribosomes? Of DNA (assuming 2 chromosomes per cell)?

   d. What fraction by volume of a liver cell is composed of ribosomes, of plasma membrane, of mitochondria (assume 1000 mitochondria)? What fraction is accounted for by the nucleus?

5. a. What is the molar concentration of an enzyme of which only one molecule is present in an *E. coli* cell?

   b. Assume that the concentration of K^+ within an *E. coli* cell is 150 mM. Calculate the number of K^+ ions in a single cell.

   c. If the pH inside the cell is 7.0, how many H^+ ions are present?

6. If chromosomes (and chromatin) are 15% DNA, what will be the mass of 23 pairs of chromosomes in a human diploid cell? If the nucleus has a diameter of 5 mm and a density of 1.1 g/cm^3, what fraction by weight of the nucleus is chromatin?

7. Compare the surface to volume ratios for an *E. coli* cell, a liver cell, the nucleus of a eukaryotic cell, a root hair. If a cell of 20 μm diameter is 20% covered with microvilli of 0.1 μm diameter and 1 μm length centered on a 0.2 μm spacing, how much will the surface/volume ratio be increased?

8. It has been shown that the code for specifying a particular amino acid in a protein is determined by a sequence of three nucleotides (a codon) in a DNA chain. There are four different kinds of nucleotide units in DNA. How many different codons exist? Note that this is larger than the number of different amino acids (20) that are incorporated into proteins plus the three stop (termination) codons (see Tables 5-5 and 5-6 for a list of codons).

9. State two similarities and two differences between cyanobacteria (blue-green algae) and green algae.

10. Compare the sizes and structures of bacterial and eukaryotic flagella.

11. How much larger in volume is a typical eukaryotic cell compared to a bacterium?

12. Compare the structure and properties of mitochondria, chloroplasts, and peroxisomes.

13. What are the possible origins of mitochondria and chloroplasts? What evidence can you cite to support your answer?

14. How many different kinds of polymers, e.g. proteins, RNA, that are present in or around living cells, can you name? Can you name some sub-groups in any of your categories?

15. Compare the composition of these three, especially with respect to C, H, O, N, S, P, Fe, Cu, Al, and Si:

   a. The earth's crust

   b. Ocean water

   c. Cytoplasm

A spider's orb-web is formed by extrusion of a concentrated protein solution and stretching of the resulting fiber. The cross-strands, which are stronger than steel, resemble silkworm silk. The molecules contain microcrystalline β sheet domains that are rich in Gly-Ala repeats as well as polyalanine segments. The capture spiral is formed from much more elastic molecules that contain many β-turn-forming sequences. These assume a spring-like β spiral. See Box 2-B.

© John Gerlach/DRK PHOTO

Contents

Amino Acids, Peptides, and Proteins

Thousands of different proteins make up a very large fraction of the "machinery" of a cell. Protein molecules catalyze chemical reactions, carry smaller molecules through membranes, sense the presence of hormones, and cause muscle fibers to move. Proteins serve as structural materials within cells and between cells. Proteins of blood transport oxygen to the tissues, carry hormones between cells, attack invading bacteria, and serve in many other ways. No matter what biological process we consider, we find that a group of special proteins is required.

The amino acid units that make up a protein molecule are joined together in a precise sequence when the protein is made on a ribosome. The chain is then folded, often into a very compact form. Sometimes the chain is then cut in specific places. Pieces may be discarded and parts may be added. A metal ion, a coenzyme derived from a vitamin, or even a single methyl group may be attached to form the biologically active protein. The final product is a complex and sophisticated machine, often with moving parts, that is exquisitely designed for its particular role.

The biological functioning of a protein is determined both by the properties of the chemical groups in the amino acids that are joined to form the protein chain and by the way the chain is folded. The ways in which the different parts of the protein interact with each other and with other molecules are equally important. These interactions play a major role in determining the folding pattern and also provide much of the basis for the biological functioning of proteins. Similar considerations apply also to carbohydrates, nucleic acids, and other biopolymers. For these reasons it is appropriate to review some fundamentals of molecular structure and geometry.

A. Structural Principles for Small Molecules

Stable organic molecules are held together by covalent bonds which are usually very strong. The standard Gibbs energies of formation $(\Delta G_f°)$[†] of many covalent single bonds are of the order of -400 kJ/mol (96 kcal/mol). The bonds have definite directions, which are measured by **bond angles** and definite **bond lengths**.

1. Bond Angles

Because of the tetrahedral arrangement of the four bonds around single-bonded carbon atoms and most phosphorus atoms, all six of the bond angles about the central atom have nearly the same tetrahedral angle of 109.5°.

Bond angles within chains of carbon atoms in organic compounds vary only slightly from this, and even atoms that are attached to fewer than four groups usually have similar angles; for example, the H–O–H angle in a water molecule is 105°, and the H–N–H angles of ammonia are 107°. In ethers the C–O–C angle is 111°. However, bond angles of only 101° are present in H_2O_2 and of 92° in H_2S and PH_3.

[†]See Chapter 6 for a review of thermodynamics

The presence of **double bonds** leads to **planarity** and to compounds with bond angles of 120°, the internal angle in a hexagon. The planar geometry imposed upon an atom by a double bond is often transmitted to an adjacent nitrogen or oxygen atom as a result of **resonance** (Section 6). For example, the amide groups that form the peptide linkage in proteins (see Fig. 2-5) are nearly planar and the angles all fall within four degrees of 120°.

2. Bond Lengths

Chemists describe bond lengths as the distances between the nuclei of bonded atoms. The C–C single bond has a length of 0.154 nm (1.54 Å). The C–O bond is ~0.01 nm shorter (0.143 nm), and the typical C–H bond has a length of ~0.109 nm. The C–N bond distance is halfway between that for C–C and C–O (0.149 nm). Other lengths, such as that of O–H, can be estimated from the covalent radii given in Table 2-1.

The length of a double bond between any two atoms (e.g., C=C) is almost exactly 0.020 nm less than that for a single bond between the same atoms. If there is resonance, hence only partial double bond character, the shortening is less. For example, the length of the C–C bond in benzene is 0.140 nm; the C–O distances in the carboxylate anion are 0.126 nm.

Carboxylate ion

Using simple geometry, it is easy to calculate overall lengths of molecules; here are two distances worth remembering:

|← 0.254 →| nm

Distance between alternate C atoms in fully extended hydrocarbon chain

0.31 nm

Distance across a benzene ring

In the preceding simplified structural formula for benzene the six hydrogen atoms have been omitted. Resonance between the two possible arrangements of the three double bonds[1] is indicated by the circle. Chemical shorthand of the following type is used throughout the book. Carbon atoms may be represented by an angle or the end of a line, but other atoms will always be shown.

$P \equiv$ —$\overset{\overset{O}{\|}}{\underset{\underset{O^-}{|}}{P}}$—OH = phospho group

3. Contact Distances

Covalent bond distances and angles tell us how the atomic nuclei are arranged in space but they do not tell us anything about the outside surfaces of molecules. The distance from the center of an atom to the point at which it contacts an adjacent atom in a packed structure such as a crystal (Fig. 2-1) is known as the **van der Waals radius**. The ways in which biological molecules fit together are determined largely by the van der Waals contact radii. These, too, are listed in Table 2-1. In every case they are approximately equal to the *covalent radius plus 0.08 nm*. Van der Waals radii

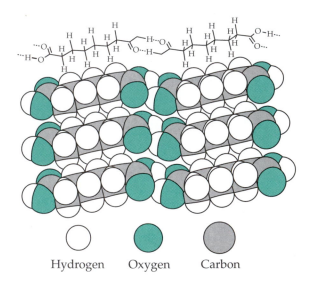

Hydrogen   Oxygen   Carbon

Figure 2-1 Packing of molecules of suberic acid HOOC–$(CH_2)_6$–COOH in a crystal lattice as determined by neutron diffraction.[2] Notice the pairs of hydrogen bonds that join the carboxyl groups at the ends of the molecules and also the close contact of hydrogen atoms between the chains. Only the positions of the hydrogen nuclei were determined; the van der Waals radii have been drawn around them. However, the radii were originally determined from X-ray and neutron diffraction data obtained from many different crystalline compounds.

are not as constant as covalent radii because atoms can be "squeezed" a little, but only enough to decrease the contact radii by 0.005–0.01 nm. The radii of space-filling molecular models are usually made a little smaller than the actual scaled van der Waals radii to permit easier assembly.

4. Asymmetry: Right-Handed and Left-Handed Molecules

The left hand looks much like the right hand, but they are different. One is the mirror image of the other. A practical difference is that your right hand will not fit into a left-handed glove. Despite our daily acquaintance with "handedness" it may seem difficult to explain in words how a right and a left hand differ. However, since most biochemical compounds are asymmetric,[3] it is important to be able to visualize these molecules in three dimensions and to draw their structures on paper. One of the best ways of learning to do this is to study molecular models. You may learn the most by making your own models (see Appendix).

Whenever four different groups are bonded to a central carbon atom, the molecule is asymmetric and the four groups can be arranged in two different **configurations**. Consider alanine, one of the alpha (α)-amino acids from which proteins are built.

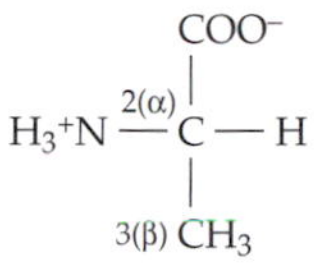

α–Alanine
According to the Fisher projection formula it is L-alanine.

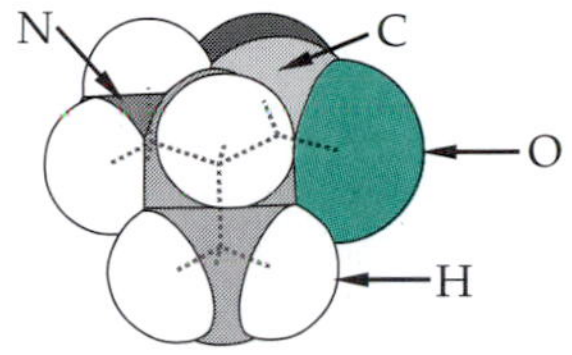

It is called an α-amino acid because the amino group is attached to the α (or number 2) carbon atom. To indicate its three-dimensional structure on a flat piece of paper, the bonds that project out of the plane of the paper and up toward the reader are often drawn as elongated triangles, while bonds that lie behind the plane of the paper are shown as dashed lines. The isomer of alanine having the configuration about the α-carbon atom shown in the following structural formulas is called S-alanine or L-alanine. The isomer which is a mirror image of S-alanine is R-alanine or D-alanine. Pairs of R and S compounds (see Section B for definitions) are known as **enantiomorphic** forms or **enantiomers**.

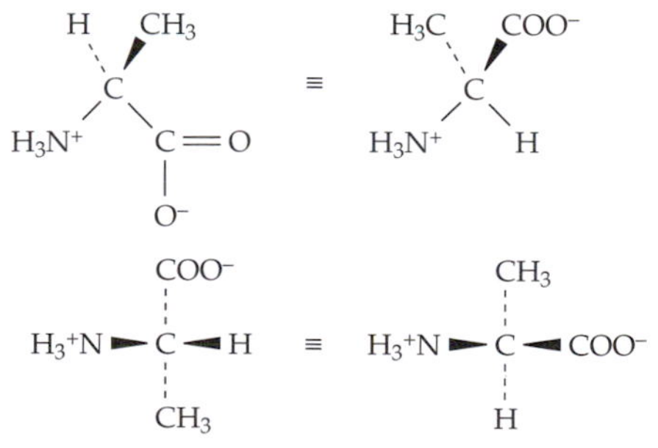

S-Alanine (L-Alanine)

Notice that in the foregoing drawings, the carboxyl group ($-C\!\!\stackrel{O}{\diagdown}_{OH}$), abbreviated –COOH, is shown as having lost a proton to form the carboxylate ion –COO⁻. Likewise, the amino group (–NH$_2$) has gained a proton to form the –NH$_3^+$ ion. The resulting **dipolar ionic** or **zwitterionic** structure is the one that actually exists for amino acids both in solution and in crystals.

TABLE 2-1
The Sizes of Some Atoms[a–c]

Element	Covalent radius (nm)[a,b,d]	van der Waals radii (nm)[d]	
		a Sideways contact[c]	*b* Polar contact[c]
H	0.030	0.12	
F	0.064	0.138	0.130
C	0.077	0.16	
N	0.070	0.160	0.160
O	0.066	0.154	0.154
Cl	0.099	0.178	0.158
Si	0.117		
P	0.110	0.19	
S	0.104	0.203	0.160
Br	0.114	0.184	0.154
I	0.133	0.213	0.176
Se		0.215	0.170
"Radius" of methyl group		0.20	
Half-thickness of aromatic molecules		0.170	

[a] From Pauling, L. (1960) *The Nature of the Chemical Bond*, 3rd ed., Cornell Univ. Press, Ithaca, New York (pp. 224–227 and 260).
[b] Covalent radii for two atoms can be summed to give the interatomic distance. The van der Waals radii determine how closely molecules can pack. The closest observed contacts between atoms in macromolecules are approximately 0.02 nm less than the sum of the van der Waals radii. From Sasisekharian, V., Lakshminarayanan, A. V., and Ramachandran, G. N. (1967) in *Conformation of Biopolymers*, Vol. 2 (Ramachandran, G. N., ed), Academic Press, New York, (p. 641)
[c] Nyburg, S. C and Faerman, C. H. (1985) *Acta Crystal.* **B41**, 274–279 Shapes of many atomic surfaces are elliptical. The major radius *a* applies to sideways contacts and the minor radius *b* to "polar" contacts along a covalent bond axis. Distances are for atoms singly bonded to C and may differ slightly if bonds are to other atoms.
[d] For distances in Å multiply by 10.

The D- and L- families of amino acids.

The amino acids of which proteins are composed are related to L-alanine but have various side chains (R groups) in place of the methyl group of alanine. In the preceding section the structure of L-alanine was given in four different ways. To recognize them all as the same structure, we can turn them in space to an orientation in which the carboxyl group is at the top, the side chain (–CH₃) is down, and both project behind the paper. The amino group and hydrogen atom will then project upward from the paper at the sides as shown below. According to a convention introduced at the beginning of this century by Emil Fischer, *an amino acid is L if, when oriented in this manner, the amino group lies to the left and D if it lies to the right.*

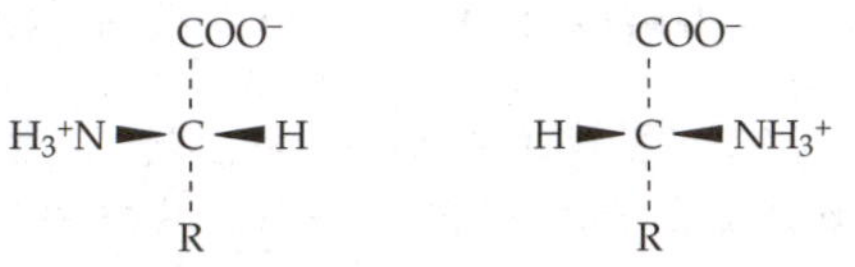

L-Amino acid   D-Amino acid

Fischer further proposed that the amino acid in this orientation could be projected onto the paper and drawn with ordinary lines for all the bonds. This gives the previously shown **Fischer projection formula** of L-alanine.

Although the D and L system of designating configuration is old it is still widely used. Remember that D and L refer to the absolute configurations about a selected reference atom in the molecule; for an amino acid this is the number 2 or α-carbon. A quantity that is related to the asymmetry of molecules is the experimentally measurable **optical rotation** (Chapter 23). The sign of the optical rotation (+ or –) is sometimes given together with the name of a compound, e.g., D(+)-glucose. The older designations *d* (dextro) and *l* (levo) indicated + and –, respectively. However, compounds with the D configuration may have either + or – optical rotation.

In older literature optical isomerism of the type represented by D and L pairs was usually discussed in terms of "asymmetric carbon atoms" or "asymmetric centers." Now the terms **chiral** (pronounced *ki-ral*) **molecules**, chiral centers, and **chirality** (Greek: "handedness") are preferred.

The RS notation for configuration.

This notation, devised by Cahn, Ingold, and Prelog, provides an unambiguous way of specifying configuration at any chiral center.[4,5] It is especially useful for classes of compounds for which no well-established DL system is available. The groups or atoms surrounding the central carbon atom, or other central atom, are ranked according to a **priority sequence**. The priority of a group is determined by a number of sequence rules, the first of which is (1) *Higher atomic number precedes*

lower. In the following illustration, the priorities of the groups in D-alanine are indicated by the letters $a > b > c > d$. The highest priority (a) is assigned to the NH₂ groups which contain nitrogen bonded to the central atom. To establish the configuration, the observer views the molecule down the axis connecting the central atom to the group having the lowest priority, i.e., to group d. Viewed in this way, the sequence of groups a, b, and c can either be that of a right-handed turn (clockwise) as shown in the drawing or that of a left-handed turn (counterclockwise).

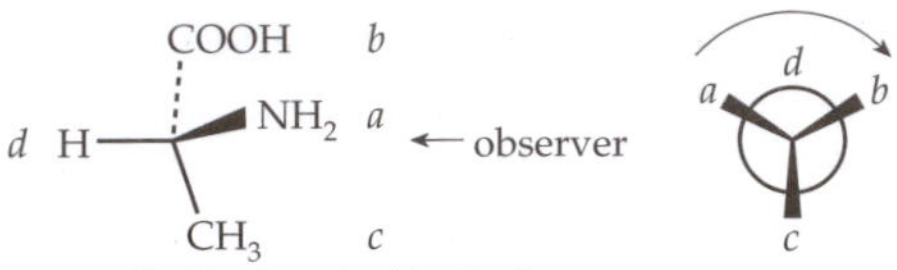

R-Alanine (D-Alanine)

The view down the axis and toward the group of lowest priority (d), which lies behind the page. The right-handed turn indicates the configuration R (rectus = right); the opposite configuration is S (sinister = left).

To establish the priority sequence of groups first look at the atoms that are bonded directly to the central atom, arranging them in order of decreasing atomic number. Then if necessary, move outward to the next set of atoms, again comparing atomic numbers. In the case of alanine, groups b and c must be ordered in this way because they both contain carbon directly bonded to the central atom. When double bonds are present at one of the atoms being examined, e.g., the carboxyl group in alanine, imagine that **phantom atoms** that replicate the real ones are present at the ends of the bonds:

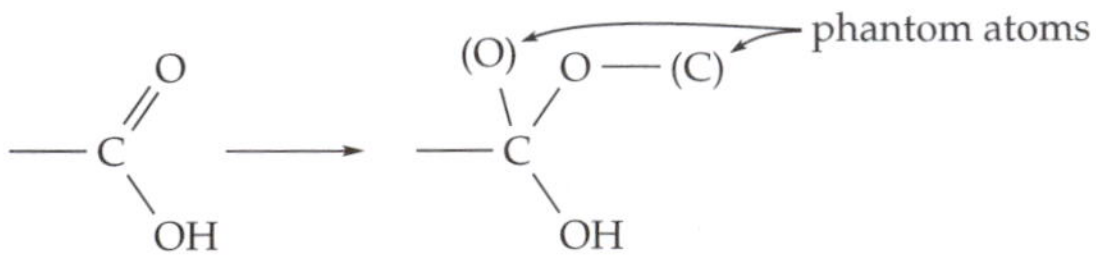

These phantom atoms fill out the valences of the atoms involved in the multiple bonds and are considered to have zero atomic number and zero mass. They are not considered in establishing priorities.

If the first rule and the expansion of multiple bonds are not sufficient to establish the priority, use these additional rules: (2) *Higher atomic mass precedes lower.* (3) When a double bond is present Z precedes E (see Geometrical isomers). For ring systems a *cis* arrangement of the highest priority substituents precedes *trans*. (4) When a pair of chiral centers is present *R,R* or *S,S* precedes *R,S* or *S,R*. (5) *An R chiral center precedes S.* For further details see Eliel *et al.*[5] and Bentley.[6] The following groups are ordered in terms of *decreasing priority*[6]: SH > OR > OH > NH–COCH₃ > NH₂ > COOR > COOH > CHO > CH₂OH > C₆H₅ > CH₃ > ³H > ²H > H.

Although the *RS* system is unambiguous, closely

related compounds that belong to the same configurational family in the DL system may have opposite configurations in the RS system. Thus, L-cysteine (side chain –CH$_2$SH) has the R configuration. This is one of the reasons that the DL system is still used for amino acids and sugars.

Diastereoisomers. Whereas compounds with one chiral center exist as an enantiomorphic pair, molecules with two or more chiral centers also exist as diastereoisomers (diastereomers). These are pairs of isomers with an opposite configuration at one or more of the chiral centers, but which are not complete mirror images of each other. An example is L-threonine which has the 2S, 3R configuration. The diastereoisomer with the 2S, 3S configuration is known as L-allo-threonine. L-isoleucine, whose side chain is –CH(CH$_3$) CH$_2$CH$_3$, has the 2S, 3R configuration. It can be called 2(S)-amino-3(R)-methyl-valeric acid but the simpler name L-isoleucine implies the correct configuration at both chiral centers.

Sometimes the subscript s or g is added to a D or L prefix to indicate whether the chirality of a compound is being related to that of serine, the traditional configurational standard for amino acids, or to that of glyceraldehyde. In the latter case the sugar convention (Chapter 4) is followed. In this convention the configurations of the chiral centers furthest from C1 are compared. Ordinary threonine is L$_s$- or D$_g$-threonine. The configuration of dextrorotatory (+)-tartaric acid can be described as 2R, 3R, or as D$_s$, or as L$_g$.

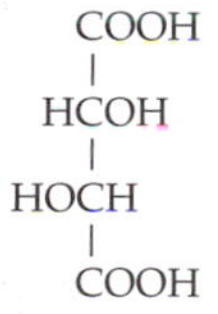

$$
\begin{array}{c}
\text{COOH} \\
|\\
\text{HCOH} \\
|\\
\text{HOCH} \\
|\\
\text{COOH}
\end{array}
$$

2R, 3R-tartaric acid

Biochemical reactions are usually stereospecific and a given enzyme will catalyze reactions of molecules of only a single configuration. A related fact is that proteins ordinarily consist entirely of amino acids of the L series.

Geometrical isomers. The RS system also gives an unambiguous designation of geometrical isomers containing a double bond.[5,7] At each end of the bond, select the group of highest priority. If these two groups lie on the same side of the double bond the configuration is **Z** (from the German **zusammen**, "together"); if on opposite sides **E** (**entgegen**, "opposite").

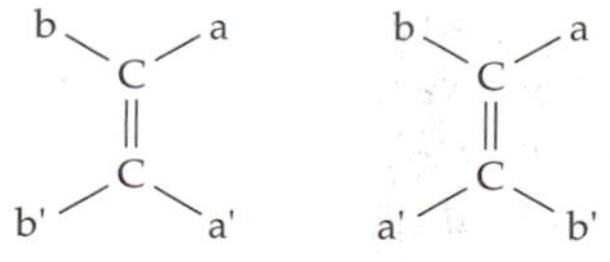

a and a' higher priority than b or b'

Configurations of amide or ester linkages may also be specified in this manner. This is possible because the C–N bond of an amide has partial double-bond character, as to a lesser extent does the C–O bond to the bridge oxygen in an ester. In this case, assign the lowest priority to the unshared electron pair on the ester bridge oxygen.

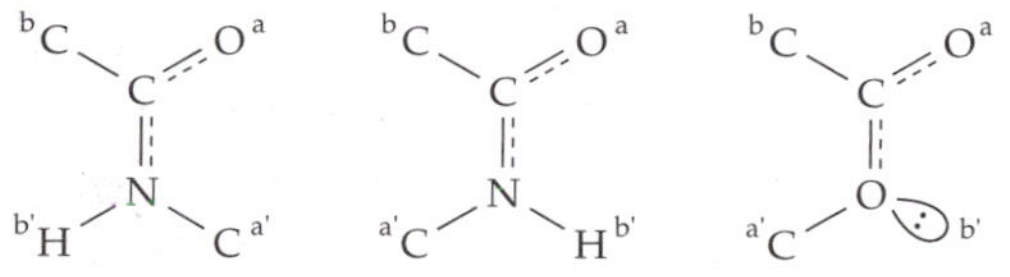

Z or trans   E or cis

An amide of the Z configuration is ordinarily referred to as trans in protein chemistry because the main chain atoms are trans.

5. Conformations: The Shapes That Molecules Can Assume

As important to biochemists as configurations, the stable arrangements of bonded atoms, are **conformations**, the various orientations of groups that are caused by rotation about single bonds.[5,8] In many molecules such rotation occurs rapidly and freely at ordinary temperatures. We can think of a –CH$_3$ group as a kind of erratic windmill, turning in one direction, then another. However, even the simplest molecules have preferred conformations, and in more complex structures rotation is usually very restricted.

Consider a molecule in which groups A and B are joined by two CH$_2$ (methylene) groups. If A and B are pulled as far apart as possible, the molecule is in its fully extended **anti** or **staggered** conformation:

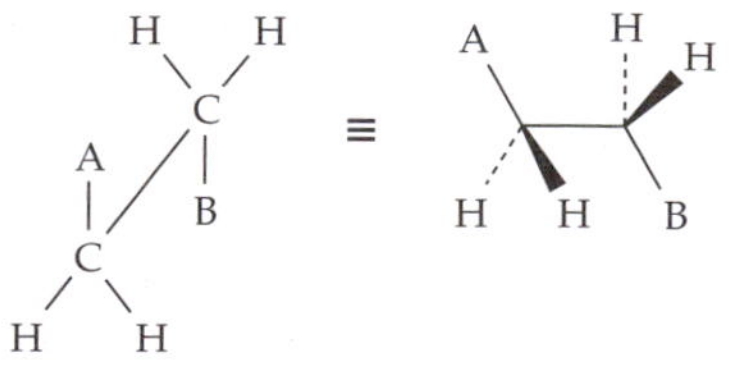

The antiperiplanar conformation

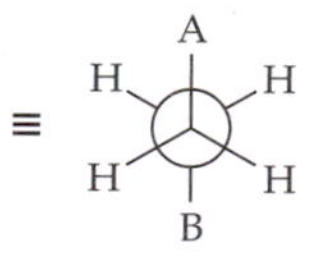

View down the axis joining the C atoms (Newton projection)

Groups A and B are said to be **antiperiplanar** (ap) in this conformation. Not only are A and B as far apart as possible but also all of the hydrogen atoms are at their maximum distances one from the other. This can be seen by viewing the molecule down the axis joining the carbon atoms (Newman projection). Rotation of the second carbon atom 180° around the single bond yields the **eclipsed** conformation in which groups A and B are synperiplanar.

Eclipsed or synperiplanar conformation

If A and B are large bulky groups they will bump together, attainment of the eclipsed conformation will be almost impossible, and rotation will be severely restricted. Even if A and B are hydrogen atoms (ethane), there will be a rotational barrier in the eclipsed conformation which amounts to ~12 kJ (3 kcal) per mole because of the crowding of the hydrogen atoms as they pass each other.[5,9] This can be appreciated readily by examination of space-filling molecular models.

If groups A and B are methyl groups (butane), the steric hindrance between A and B leads to a rotational barrier of ~25 kJ (6 kcal) per mole. The consequence of this simple fact is that in fatty acids and related substances and in polyethylene the chains of CH_2 groups tend to assume fully extended zigzag conformations.

In addition to this extended conformation there are two **gauche** (**skewed** or synclinal) conformations which are only slightly less stable than the staggered conformation and in which A and B interfere only if they are very bulky. In one of the two gauche conformations B lies to the right of A and in the other to the left of A when viewed down the axis.

One of the two gauche or synclinal conformations

These two conformations are related to right-handed and left-handed screws, respectively. The threads on an ordinary right-handed household screw, when viewed down the axis from either end, move backward from left to right in the same fashion as do the groups A and B in the illustration. The angle ϕ is the **torsion angle** and is positive for right-handed conformations. Gauche conformations are important in many biological molecules; for example, the sugar alcohol **ribitol** stacks in crystals in a "sickle" conformation,[6] in which the chain starts out (at the left) in the zigzag arrangement but shifts to a gauche conformation around the fourth carbon atom, thereby

minimizing steric interference between the OH groups on the second and fourth carbons.

Ribitol in sickle conformation

The complete series of possible conformations is shown in Fig. 2-2.

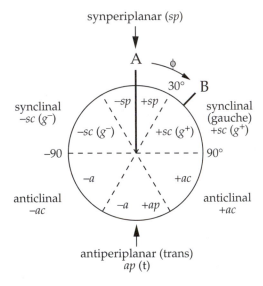

Figure 2-2 Description of conformations about a single bond in the terminology of Klyne and Prelog[10,11] using the Newman projection. Group A is on the front atom at the top: the conformation is given for each possible position of group B on the other atom.

In the chain of methylene units, the hydrogen atoms on alternate carbon atoms of the fully extended chain barely touch (Fig. 2-1) but larger atoms cannot be accommodated. Thus, when fluorine atoms of van der Waals radius 0.135 nm replace the hydrogen atoms of radius 0.12 nm, a fully extended chain is no longer possible. For this reason the torsion angle in polyfluoroethylene is changed from the 180° of polyethylene to 166°, enough to relieve the congestion but not enough to cause severe eclipsing of the fluorines on adjacent carbons. The resulting **helical structure** is reminiscent of those occurring in proteins and other biopolymers. We see that helix formation can be a natural result of steric hindrance between groups of atoms.

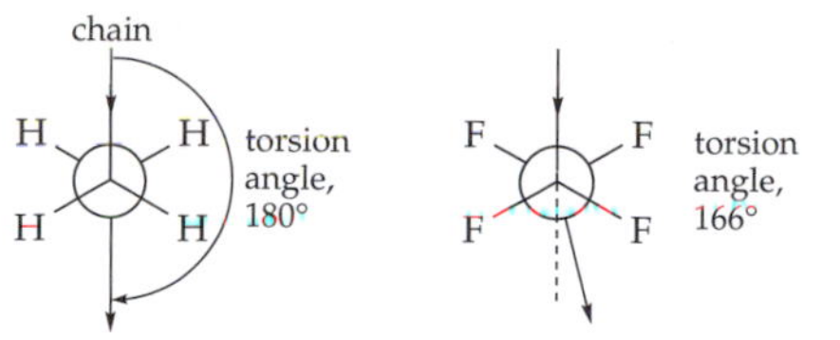

Polyethylene   Polyfluoroethylene (Teflon)

Conformations of ring-containing molecules are dealt with in Chapter 4.

6. Tautomerism and Resonance

Many simple organic compounds exist as mixtures of two or more rapidly interconvertible isomers or **tautomeric forms**. Tautomers can sometimes be separated one from the other at low temperatures where the rate of interconversion is low. The classic example is the **oxo-enol** (or keto-enol) equilibrium (Eq. 2-1).

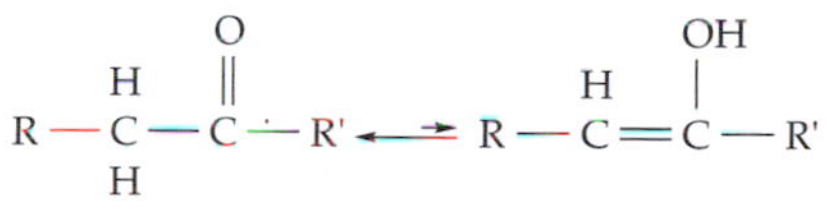

Oxo form   Enol   (2-1)

Although usually less stable than the oxo (keto) form, the enol is present in a small amount. It is formed readily from the oxo tautomer by virtue of the fact that hydrogen atoms attached to carbon atoms that are immediately adjacent to carbonyl (C=O) groups are remarkably acidic. Easy dissociation of a proton is a prerequisite for tautomerism. Since most hydrogen atoms bound to carbon atoms do not dissociate readily, tautomerism is unusual unless a carbonyl or other "activating group" is present.

Since protons bound to oxygen and nitrogen atoms usually *do* dissociate readily, tautomerism also exists in amides and in ring systems containing O and N (Eqs. 2-2 to 2-5).

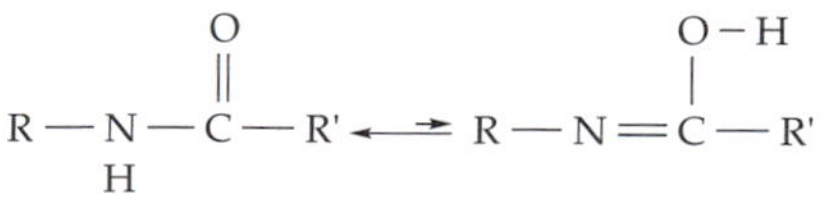

A   B   (2-2)

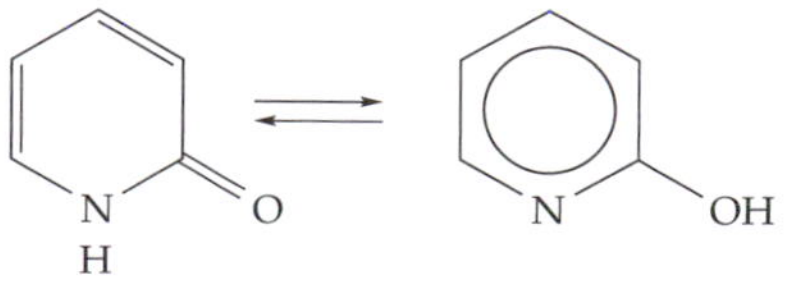

2-Pyridone   2-Hydroxypyridine   (2-3)

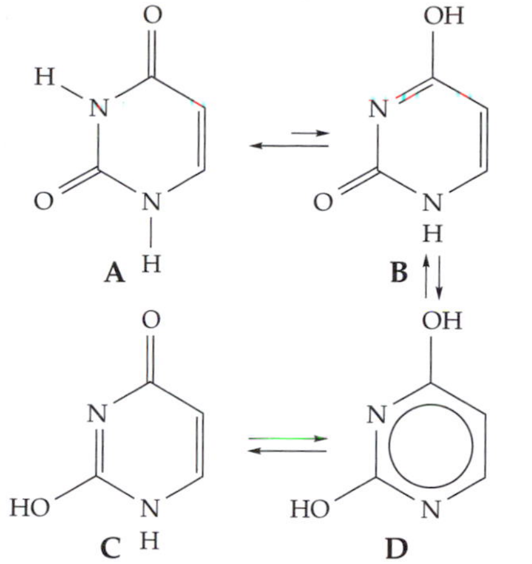

(2-4)

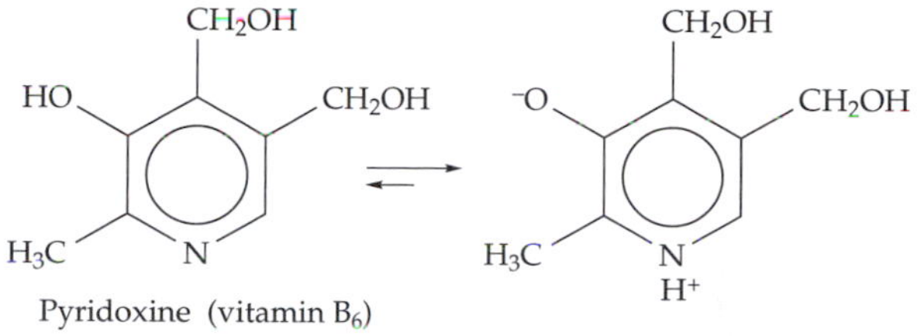

Pyridoxine (vitamin B_6)

A   B   (2-5)

The tautomerism in Eq. 2-2 is the counterpart of that in the oxo-enol transformation. However, the equilibrium constant for aqueous conditions favors form A very strongly. 2-Pyridone is tautomerized to 2-hydroxypyridine (Eq. 2-3) to a greater extent. Pyrimidines (Eq. 2-4) and purines can form a variety of tautomers. The existence of form D of Eq. 2-4 is the basis for referring to uracil as dihydroxypyrimidine. However, the di-oxo tautomer A predominates. Pyridoxine (vitamin B_6) exists in water largely as the dipolar ionic tautomer B (Eq. 2-5) but in methanol as the uncharged tautomer A. In a pair of tautomers, a hydrogen atom always moves from one position to another and the lengths and bond character of these bonds also change.

The equilibrium constant for a tautomeric interconversion is simply the ratio of the mole fractions of the two forms; for example, the ratio of enol to oxo forms of acetone[12] in water at 25°C is 6.0 x 10^{-9}, while that for isobutyraldehyde is 1.3 x 10^{-4}. The ratio of 2-hydroxypyridine to 2-pyridone is about 10^{-3} in water but increases to 0.6 in a hydrocarbon solvent and to 2.5 in the vapor phase.[13,14] The ratio of dipolar ion to uncharged pyridoxine (Eq. 2-5) is ~4 at 25°C in water.[15] The ratios of tautomers B, C, and D to the tautomer A of uracil (Eq. 2-4) are small, but it is difficult to measure them quantitatively.[16] These tautomeric ratios are defined for given overall states of protonation (see Eq. 6-82). The constants are independent of pH but will change if the overall state of protonation of the molecule is changed. They may also be altered by changes in temperature or solvent or by binding to a protein or other molecule.

It is important to distinguish tautomerism from **resonance**, a term used to indicate that the properties of a given molecule cannot be represented by a *single* valence structure but can be represented as a hybrid of two or more structures in which all the nuclei remain in the same places. Only bonding electrons move to convert one resonance form into another. Examples are the **enolate anion**, which can be thought of as a hybrid of structures A and B, and the amide linkage, which can be represented by a similar pair of resonance forms.

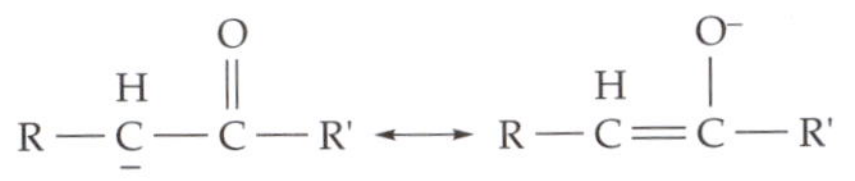

A   **B**

Enolate anion

A double-headed arrow is often used to indicate that two structures drawn are resonance structures rather than tautomers or other separable isomers.

Although they are distinctly different, tautomerism and resonance are related. Thus, the acidity of carbon-bound hydrogens in ketones, which allows formation of enol tautomers, results from the fact that the enolate anion produced by dissociation of one of these hydrogens is stabilized by resonance. Similarly, tautomerism in the imidazole group of the amino acid histidine is related to resonance in the imidazolium cation. Because of this resonance, if a proton approaches structure A of Eq. 2-6 and becomes attached to the left-hand nitrogen atom (N^δ), the positive charge in the resulting intermediate is distributed over both nitrogen atoms. This makes the proton on N^ϵ acidic, permitting it to dissociate to tautomer B.

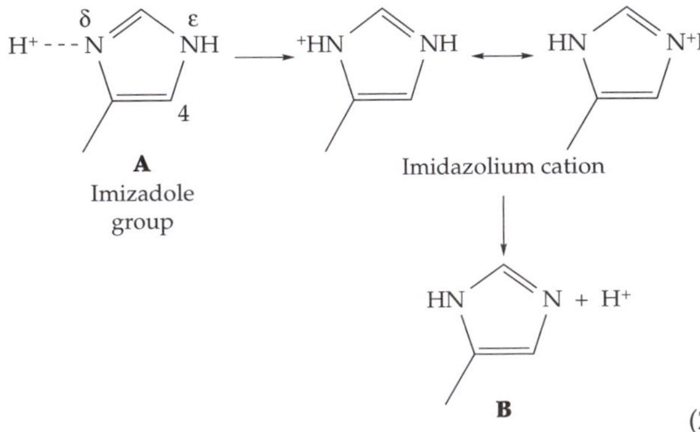

A
Imizadole
group

Imidazolium cation

(2-6)

Note: The nitrogen atom designated N^δ (or ND1)[11] in Eq. 2-6 may also be called N1 or N^π (*pros*-N). Likewise, N^ϵ (NE2) may be designated N3 or N^τ (*tele*- N). Since

N^σ has sometimes also been called N3, it is best not to use the numerical designations for the nitrogen atoms.

The tautomeric ratio of B to A for histidine in water (Eq. 2-6) has been estimated, using ^{15}N- and ^{13}C-NMR, as 5.0 when the α-amino group is protonated and as 2.5 when at high pH it is unprotonated.[17] This tautomerism of the imidazole group is probably important to the function of many enzymes and other proteins; for example, if N^ϵ of structure A (Eq. 2-6) is embedded in a protein, a proton approaching from the outside can induce the tautomerism shown with the release of a proton in the interior of the protein, perhaps at the active site of an enzyme. The form protonated on N^δ (B of Eq. 2-6), which is the minor form in solution, predominates in some positions within proteins.[18]

B. Forces between Molecules and between Chemical Groups

The structure of living cells depends very much on the covalent bonds within individual molecules and on covalent crosslinks that sometimes form between molecules. However, weaker forces acting between molecules and between different parts of the same molecule are responsible for many of the most important properties of biochemical substances. These are described as **van der Waals forces**, **electrostatic forces**, **hydrogen bonds**, and **hydrophobic interactions**. In the discussion that follows the thermodynamic quantities ΔH, ΔS, and ΔG will be used. If necessary, please see Chapter 6 for definitions and a brief review.

1. Van der Waals Forces

All atoms have a weak tendency to stick together, and because of this even helium liquifies at a low enough temperature. This is a result of the van der Waals or "London dispersion forces" which act strongly only at a very short distance. These forces arise from electrostatic attraction between the positively charged nucleus of one atom and the negatively charged electrons of the other.[19–21] Because nuclei are screened by the electron clouds surrounding them, the force is weak. The energy (enthalpy) of binding of one methylene ($-CH_2-$) unit into a monomolecular layer of a fatty acid is about $-\Delta H° = 1.7$ kJ/mol.[22] Although this is a small quantity, when summed over the 16 or more carbon atoms of a typical fatty acid the binding energy is substantial. When a methylene group is completely surrounded in a crystalline hydrocarbon, its van der Waals energy, as estimated from the heat of sublimation, is 8.4 kJ/mol; that of H_2O in liquid water at the melting point of ice is 15 kJ/mol.[22]

While van der Waals forces between individual atoms act over very short distances, they can be felt at surprisingly great distances when exerted by large molecules or molecular aggregates.[23] Forces between very smooth surfaces have been measured experimentally at distances as great as 10 nm and even to 300 nm.[23a] However, these "long-range van der Waals forces" probably depend upon layers of oriented water molecules on the plates[23] (see also Section 5).

2. Attraction between Charged Groups (Salt Linkages)

Fixed positive and negative charges attract each other strongly. Consider a carboxylate ion in contact with $-NH_3^+$ or with an ion of calcium:

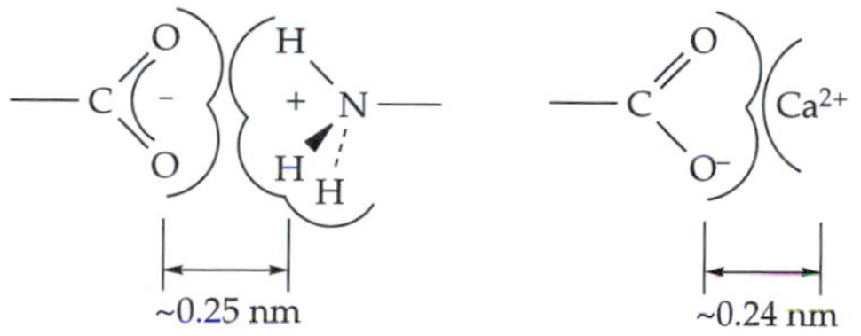

From the van der Waals radii of Table 2-1 and the ionic crystal radius of Ca^{2+} of 0.10 nm, we can estimate an approximate distance between the centers of positive and negative charge of 0.25 nm in both cases. It is of interest to apply Coulomb's law to compute the force F between two charged particles which are almost in contact. Let us choose a distance of 0.40 nm (4.0 Å) and apply Eq. 2-7.

$$F = 8.9875 \times 10^9 \times \frac{qq'}{r^2\varepsilon} \text{ newtons}$$

$$(2\text{-}7)$$

In this equation r is the distance in meters, q and q' are the charges in coulombs (one electronic charge = 1.6021×10^{-19} coulombs), ε is the dielectric constant, and F is the force in newtons (N). The force per mole is NF where N is Avogadro's number.

An uncertainty in this kind of calculation is in the dielectric constant ε, which is 1.0 for a vacuum, about 2 for hydrocarbons, and 78.5 for water at 25°C. If ε is taken as 2, the force for $r = 0.40$ nm is 4.3×10^{14} N/mol. The force would be twice as great for the $Ca^{2+} - COO^-$ case. To move two single charges further apart by just 0.01 nm would require 4.3 kJ/mol, a substantial amount of energy. However, if the dielectric constant were that of water, this would be reduced almost 40-fold and the electrostatic force would not be highly significant in binding. It is extremely difficult to assign a dielectric constant for use in the interior of proteins.[23b] For charges spaced far apart within proteins the effective

dielectric constant is usually as high as 30–60.[24] For closely spaced charges in hydrophobic niches it may be as low as 2–4.[25-27]

A calculation that is often made is the work required to remove completely two charges from a given distance apart (e.g., 0.40 nm) to an infinite distance (Eq. 2-8).

$$W \text{ (kJ mol}^{-1}) = 8.9875 \times 10^6 \times \frac{qq'}{r\varepsilon} \text{ N}$$

$$= \frac{138.9}{\varepsilon r \text{ (in nm)}}$$

$$(2\text{-}8)$$

If $\varepsilon = 2$, this amounts to 174 kJ/mol for single charges at a distance of 0.40 nm; 69 kJ/mol at 1 nm; and only 6.9 kJ/mol at 10 nm, the distance across a cell membrane. We see that very large forces exist between closely spaced charges.

Electrostatic forces are of great significance in interactions between molecules and in the induction of changes in conformations of molecules. For example, attraction between $-COO^-$ and $-NH_3^+$ groups occurs in interactions between proteins. Calcium ions often interact with carboxylate groups, the doubly charged Ca^{2+} bridging between two carboxylate or other polar groups. This occurs in carbohydrates such as agarose, converting solutions of these molecules into rigid gels (Chapter 4).

Individual macromolecules as well as cell surfaces usually carry a net negative charge at neutral pH. This causes the surfaces to repel each other. However, at a certain distance of separation the van der Waals attractive forces will balance the electrostatic repulsion.[21] Protruding hydrophobic groups may then interact and may "tether" bacteria or other particles at a fixed distance, often ~5 nm, from a cell surface.[28]

3. Hydrogen Bonds

One of the most important weak interactions between biologically important molecules is the hydrogen bond (H-bond).[29,30,30a] These "bonds" are the result of electrostatic attraction caused by the uneven distribution of electrons within covalent bonds. For example, the bonding electron pairs of the H–O bonds of water molecules are attracted more tightly to the oxygen atoms than to the hydrogen atoms. A small net positive charge is left on the hydrogen and a small net negative charge on the oxygen. Such **polarization** of the water molecules can be indicated in the following way:

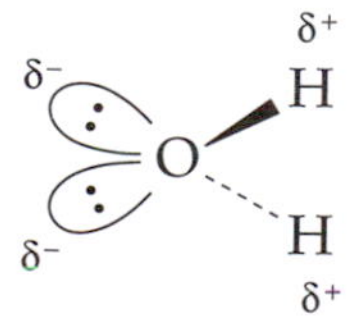

Here the δ^+ and δ^- indicate a fraction of a full charge present on the hydrogen atoms and on the nonbonded electron pairs of the oxygen atom, respectively. Molecules such as H_2O, with strongly polarized bonds, are referred to as **polar molecules** and functional groups with such bonds as **polar groups**. They are to be contrasted with such nonpolar groups as the $-CH_3$ group in which the electrons in the bonds are nearly equally shared by carbon and hydrogen.

A hydrogen bond is formed when the positively charged end of one of the **dipoles** (polarized bonds) is attracted to the negative end of another dipole. Water molecules tend to hydrogen bond strongly one to another; each oxygen atom can be hydrogen-bonded to two other molecules and each hydrogen to another water molecule. Thus, every water molecule can have up to four hydrogen-bonded neighbors.

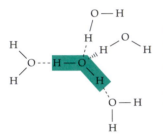

A water molecule hydrogen bonded to four other water molecules; note the tetrahedral arrangement of bonds around the central oxygen.

Many groups in proteins, carbohydrates, and nucleic acids form hydrogen bonds to one another and to surrounding water molecules. For example, an imidazole group of a protein can bond to an OH group of an amino acid side chain or of water in the following ways:

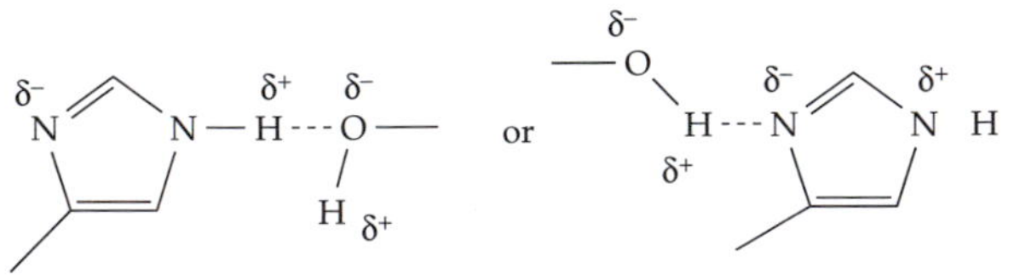

Remember that hydrogen bonds are always formed between pairs of groups, with one of them, often C=O or C=N-, containing the negative end of a dipole and the other providing the proton. The proton acceptor group, often OH or NH and occasionally SH, and even CH in certain structures,[31,31a,31b] donates an unshared pair of electrons. Dashed arrows are sometimes drawn from the hydrogen atom to the electron donor atom to indicate the direction of a hydrogen bond. Do not confuse these arrows with the curved arrows that indicate flow of electrons in organic reactions.

The strength of hydrogen bonds, as measured by the bond energy, varies over the range 10–40 kJ/mol. The stronger the hydrogen bond the shorter its length. Because hydrogen atoms can usually not be seen in

X-ray structures of macromolecules, the lengths of hydrogen bonds are often measured between the surrounding heavy atoms:

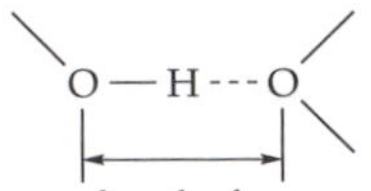

length of a
hydrogen bond

A typical —OH---O hydrogen bond will have a length of about 0.31 nm; a very strong hydrogen bond may be less than 0.28 nm in length, while weak hydrogen bonds will approach 0.36 nm, which is the sum of the van der Waals contact distances plus the O–H bond length. Beyond this distance a hydrogen bond cannot be distinguished easily from a van der Waals contact.

Hydrogen bonds are strongest when the hydrogen atom and the two heavy atoms to which it is bonded are in a straight line. For this reason hydrogen bonds tend to be linear. However, *the dipoles forming the hydrogen bond do not have to be colinear for strong hydrogen bonding*:[32,32a] There is some preference for hydrogen bonding to occur in the direction of an unshared electron pair on the oxygen or nitrogen atom.[33–35]

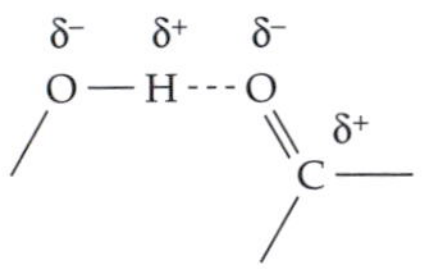

A linear O–H- - -O hydrogen bond with
dipoles at an angle one to another.

Both ammonia, NH_3, and the $-NH_2$ groups of proteins are good electron donors for hydrogen bond formation. However, the hydrogen atoms of uncharged $-NH_2$ groups tend to be poor proton donors for H-bonds.[36] Do hydrogen bonds have some covalent character? The answer is controversial.[36a,36b,36c]

Hydrogen bonding is important both to the internal structure of biological macromolecules and in interactions between molecules. Hydrogen bonding often provides the specificity necessary to bring surfaces together in a complementary way. Thus, the location of hydrogen-bond forming groups in surfaces between molecules is important in ensuring an exact alignment of the surfaces.[37] The hydrogen bonds do not always have to be strong. For example, Fersht and coworkers, who compared a variety of mutants of an enzyme of known three-dimensional structure, found that deletion of a side chain that formed a good hydrogen bond to the substrate weakened the binding energy by only 2–6 kJ/mol. However, loss of a hydrogen bond to a charged group in the substrate caused a loss of 15–20 kJ/mol of binding energy.[37] Study of mutant proteins created by genetic engineering is now an important tool for experimentally investigating the biological roles of hydrogen bonding.[37–39]

4. The Structure and Properties of Water

Water is the major constituent of cells and a remarkable solvent whose chemical and physical properties affect almost every aspect of life. Many of these properties are a direct reflection of the fact that most water molecules are in contact with their neighbors entirely through hydrogen bonds.[40–48] Water is the only known substance for which this is true.

In ordinary ice all of the water molecules are connected by hydrogen bonds, six molecules forming a hexagonal ring resembling that of cyclohexane. The structure is extended in all directions by the formation of additional hydrogen bonds to adjacent molecules (Fig. 2-3). As can be seen in this drawing, the molecules in ice assume various orientations in the hexagonal array, and frequently rotate to form their hydrogen bonds in different ways. This randomness remains as the temperature is lowered, and ice is one of few substances with a residual entropy at absolute zero.[49,50] Ice is unusual also in that the molecules do not assume closest packing in the crystal but form an open structure. The hole through the middle of the hexagon and on through the hexagons lying below it is ~0.06 nm in diameter.

The short hydrogen-bond length (averaging 0.276 nm) in ice indicates of strong bonding. The heat of sublimation ($\Delta H°$) of ice is −48.6 kJ/mol. If the van der Waals dispersion energy of −15 kJ/mol is subtracted from this, the difference of −34 kJ/mol can be attributed entirely to the hydrogen bonds—two for each molecule. Their average energy is 17 kJ/mol apiece. However, some of the hydrogen bonds are stronger and others weaker than the average.[51]

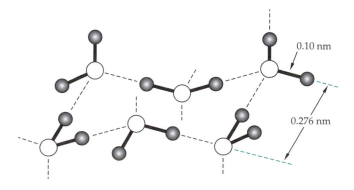

Figure 2-3 Six water molecules in the lattice of an ice crystal. The hydrogen bonds, which connect protons with electron pairs of adjacent molecules, are shown as dashed lines.

In a gaseous water dimer the hydrogen bond is linear, a fact that suggests some covalent character.[52]

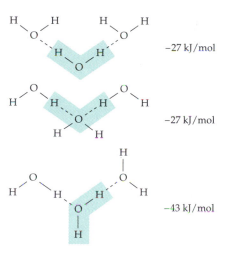

Its length is distinctly greater than that in ice. This is one of a number of pieces of evidence suggesting cooperativity in formation of chains of hydrogen bonds.[40,53,54] Consider the following three trimers for which theoretical calculations have predicted the indicated hydrogen bond energies.[40] In the first case the central water molecule *donates* two protons for hydrogen-bond formation; in the second it *accepts* the protons. In the third case it is both an electron acceptor and a donor. The OH dipoles are oriented "head to tail" and the hydrogen bonds are stronger than in the other

cases. Long chains of similarly oriented hydrogen bonds exist in ice and this may account for the short hydrogen bond lengths. Closed rings of hydrogen bonds oriented to give a maximum cooperative effect also exist in liquid water clusters[55] and within proteins, carbohydrates, and nucleic acids.[53,54]

The nature of liquid water is still incompletely understood,[46–48,56] but we know that water contains ice-like clusters of molecules that are continually breaking up and reforming in what has been called a "flickering cluster" structure. Judging by the infrared spectrum of water, about 10% of the hydrogen bonds are broken when ice melts.[41] A similar conclusion can be drawn from the fact that the heat of melting of ice is −5.9 kJ/mol. It has been estimated that at 0°C the average cluster contains about 500 water molecules.[41] At 50°C there are over 100 and at the boiling point about 40. Although most molecules in liquid water are present in these clusters, the hydrogen bonds are rapidly broken and reformed in new ways, with the average lifetime of a given hydrogen bond being ~10^{-12} s.

5. Hydration of Polar Molecules and Ions

Water molecules are able to hydrogen bond not only to each other but also to polar groups of dissolved compounds. Thus, every group that is capable of forming a hydrogen bond to another organic group is also able to form hydrogen bonds of a somewhat similar strength with water. For this reason, hydrogen bonding is usually not a significant force in holding small molecules together in aqueous solutions. Polar molecules that stick together through hydrogen bonding when dissolved in a nonpolar solvent often do not associate in water. How then can biochemists assert that hydrogen bonding is so important in biochemistry? Part of the answer is that proteins and nucleic acids can be either properly folded with hydrogen bonds formed internally or denatured with hydrogen bonds from those same groups to water. The Gibbs energy change between these two states is small.

Every ion in an aqueous solution is surrounded by a shell of oriented water molecules held by the attraction of the water dipoles to the charged ion. The hydration of ions has a strong influence on all aspects of electrostatic interactions and plays a dominant role in determining such matters as the strength of acids and bases, the Gibbs energy of hydrolysis of ATP, and the strength of bonding of metal ions to negatively charged groups. For example, the previously considered interaction between carboxylate and calcium ions would be much weaker if both ions retained their hydration shells.

Consider the following example. $\Delta G°$ for dissociation of acetic acid in water is +27.2 kJ/mol (Table 6-5). The enthalpy change $\Delta H°$ for this process is almost zero (–0.1 kJ/mol) and $\Delta S°$ is consequently –91.6 J K^{-1}. This large entropy decrease reflects the increased amount of water that is immobilized in the hydration spheres of the H^+ and acetate$^-$ ions formed in the dissociation reaction. In contrast, dissociation of NH_4^+ to NH_3 and H^+ converts one positive ion to another. $\Delta H°$ is large (+52.5 kJ/mol) but the entropy change $\Delta S°$ is small (–2.0 J K^{-1}, Table 6-5).

Although effects of hydration are important in almost all biochemical equilibria, they are difficult to assess quantitatively. It is hard to know how many molecules of water are freed or immobilized in a given reaction. Charged groups in proteins are often hydrated. However, if they are buried in the interior of the protein, they may be solvated by polarizable protein side chain groups such as –OH or by backbone or side chain amide groups.[57,58]

6. Hydrophobic Interactions

Fats, hydrocarbons, and other materials whose molecules consist largely of nonpolar groups have a low solubility in water and a high solubility in nonpolar solvents. Similarly, the long alkyl groups of fatty acid esters aggregate within membranes and nonpolar side chains of proteins are often packed together in the centers of protein molecules. Because it is as if the nonpolar groups "fear" water, this is known as the **hydrophobic** effect.[59–70] The terms hydrophobic forces, hydrophobic interactions, and hydrophobic bonding have also been used. However, the latter term can be misleading because the hydrophobic effect arises not out of any special attraction between nonpolar groups but primarily from the strong internal cohesion of the hydrogen-bonded water structure.

How strongly do nonpolar groups "attract" each other in water? A partial answer can be obtained by measuring the standard Gibbs energy ΔG of transfer of a hydrocarbon molecule from a dilute aqueous solution into a dilute solution in another hydrocarbon. By studying a series of alkanes, Abraham[62] calculated the Gibbs energy change per CH_2 unit (Eq. 2-9) as: $\Delta G° = –3.8$ kJ/mol.

$$-CH_2-(\text{water}) \longrightarrow -CH_2-(\text{hexane})$$
$$\Delta G° = -3.8 \text{ kJ/mol} \qquad (2\text{-}9)$$

This equation is a quantitative statement of the fact that the CH_2 group prefers to be in a nonpolar environment than to be surrounded by water. A similar Gibbs energy change would be expected to accompany the bringing together of a methylene unit from a small molecule and a hydrophobic surface on a protein molecule. However, in the latter case the accompanying decrease in entropy would make $\Delta G°$ less negative.

What causes the decrease in Gibbs energy when nonpolar groups associate in water? Jencks[60] suggested that we think of the transfer of a nonpolar molecule from a nonpolar solvent into water in two steps: (1) Create a cavity in the water of about the right size to accommodate the molecule. Since many hydrogen bonds will be broken, the Gibbs energy of cavity formation will be high. It will be principally an enthalpy (ΔH) effect. (2) Allow the water molecules in the solvent to make changes in their orientations to accommodate the nonpolar molecule that has been placed in the cavity. The water molecules can move to give good van der Waals contacts and also reorient themselves to give the maximum number of hydrogen bonds. Since hydrogen bonds can be formed in many different ways in water, there may be as many or even more hydrogen bonds after the reorientation than before. This will be true especially at low temperature where most water exists as large icelike clusters. For dissolved hydrocarbons, the enthalpy of formation of the new hydrogen bonds often almost exactly balances the enthalpy of creation of the cavity initially so that ΔH for the overall process (transfer from inert solvent into water) is small. For the opposite transfer (from water to hydrocarbon hexane; Eq. 2-9) $\Delta H°$ is usually a small positive number for aliphatic hydrocarbons and is nearly zero for aromatic hydrocarbons.

Since $\Delta G° = \Delta H° - T\Delta S°$ (see Chapter 6), it follows that the negative value of $\Delta G°$ for hydrophobic interactions must result from a positive entropy change, which may arise from the restricted mobility of water molecules that surround dissolved hydrophobic groups. When two hydrophobic groups come together to form a "hydrophobic bond," water molecules are freed from the structured region around the hydrophobic surfaces and the entropy increases. The $\Delta S°$ for Eq. 2-9 is about 12 J deg^{-1} mol^{-1}. Attempts have been made to relate this value directly to the increased number of orientations possible for a water molecule when it is freed from the structured region.[64] However, interpretation of the hydrophobic effect is complex and controversial.[65–71a]

The formation constant K_f for hydrophobic associations often increases with increasing temperature. This is in contrast to the behavior of K_f for many association reactions that involve polar molecules and for which $\Delta H°$ is often strongly negative (heat is released). An example of the latter is the protonation of ammonia in an aqueous solution (Eq. 2-10).

$$NH_3 + H^+ \rightarrow NH_4^+$$
$$\Delta H° = -52.5 \text{ kJ/mol} \qquad (2\text{-}10)$$

Since $R\ln K_f = -\Delta G°/T = -\Delta H°/T + \Delta S°$, K_f decreases with increasing temperature if $\Delta H°$ is negative. Because for a hydrophobic interaction with a positive value of $\Delta H°$ K_f increases with increasing temperature, an increase in stability at higher temperatures is sometimes used as a criterion for hydrophobic bonding. However, this criterion does not always hold. For example, base stacking interactions in polynucleotides (Chapter 5), whose strength does not increase with increasing temperature, are still thought to be hydrophobic.

The water molecules that are in immediate contact with dissolved nonpolar groups are partially oriented. They form a cagelike structure around each hydrophobic group. When particles surrounded by such hydration layers are 1–2 nm apart, they sometimes experience either a fairly strong repulsion or an enhanced attraction caused by these hydration layers.[21,64–66,72] Direct experimental measurements have shown that these effects extend to distances of 10 nm[21,63] and can account for the previously mentioned long-range van der Waals forces.

Various efforts have been made to develop scales of **hydrophobicity** that can be used to predict the probability of finding a given amino acid side chain buried within a protein or in a surface facing water.[59,73] A new approach has been provided by the study of mutant proteins. For example, deletion of a single $-CH_2-$ group from an interior hydrophobic region of a protein was observed to decrease the stability of the protein by 4.6 kJ/mol.[74]

C. Amino Acids and Peptides

Twenty α-amino acids are the monomers from which proteins are made. These amino acids share with other biochemical monomers a property essential to their role in polymer formation: *They contain at least two different chemical groups able to react with each other to form a covalent linkage.* In the amino acids these are the protonated amino (NH_3^+) and carboxylate (COO^-) groups. The characteristic linkage in the protein polymer is the **peptide** (amide) linkage whose formation can be imagined to occur by the splitting out of water between the carboxyl of one amino acid and the amino group of another (Eq. 2-11).

This equation is not intended to imply a mechanism for peptide synthesis. The equilibrium position for this reaction in an aqueous solution favors the free amino acids rather than the peptide. Therefore, both biological and laboratory syntheses of peptides usually do not involve a simple splitting out of water. Since the dipeptide of Eq. 2-11 still contains reactive carboxyl and amino groups, other amino acid units can be joined by additional peptide linkages to form **polypeptides**. These range from short-chain **oligomers** to polymers of from ~50 to several thousand amino acid units, the proteins.[75–77]

α-Amino acids

A peptide linkage in a dipeptide

$$(2\text{-}11)$$

1. Properties of α-Amino Acids

The amino acids have in common a dipolar ionic structure and a chiral center. They are differentiated, one from another, by the structures of their **side chain groups**, designated R in the foregoing formulas. These groups are of varying size and chemical structure. The side chain groups fill much of the space in the interior of a protein molecule and also protrude from the external surfaces of the protein where they determine many of the chemical and physical properties of the molecule.

Table 2-2 shows the structures of the side chains

of the amino acids commonly found in proteins. The complete structure is given for proline. Both the three-letter abbreviations and one-letter abbreviations used in describing sequences of amino acids in proteins are also given in this table. Amino acids of groups **a**–**c** of Table 2-2 plus phenylalanine and methionine are sometimes grouped together as *nonpolar*. They tend to be found in a hydrophobic environment on the inside of a protein molecule. Groups **f** and **i** contain *polar, charged* side chains which usually protrude into the water surrounding the protein. The rest are classified as *polar* but *noncharged*.

To get acquainted with amino acid structures, learn first those of **glycine, alanine, serine, aspartic acid**, and **glutamic acid**. The structures of many other amino acids can be related to that of alanine (R=CH$_3$) by replacement of a β hydrogen by another

group. Metabolic interrelationships will make it easier to learn structures of the rest of the amino acids later.

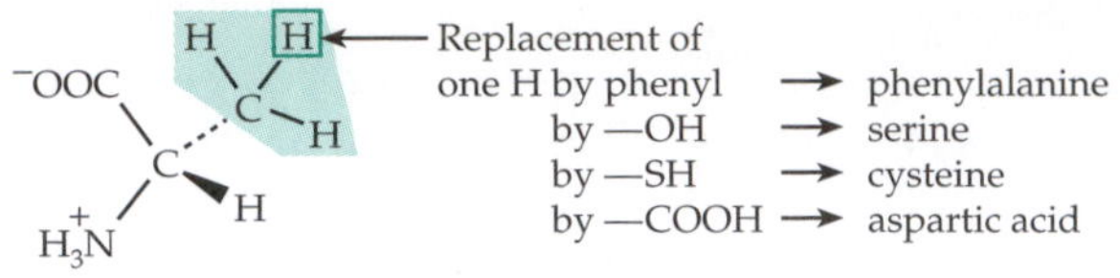

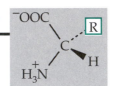

L-Alanine

Replacement of
one H by phenyl → phenylalanine
by —OH → serine
by —SH → cysteine
by —COOH → aspartic acid

Since the –COOH groups of glutamic and aspartic acids are completely dissociated to –COO$^-$ at neutral pH, it is customary in the biochemical literature to refer to these amino acids as **glutamate** and **aspartate** without reference to the nature of the cation or cations present as counter ions. Such "-ate" endings are also used for most other acids (e.g., malate, oxaloacetate,

TABLE 2-2
Structure and Chemical Properies of Side Chain Groups (R) of Amino Acids

a. Glycine "side chain" = —H (Gly, G)[a]   Strictly a link in the peptide chain, glycine provides a minimum of steric hindrance to rotation and to placement of adjacent groups.

b. Amino acids with alkyl groups as side chains

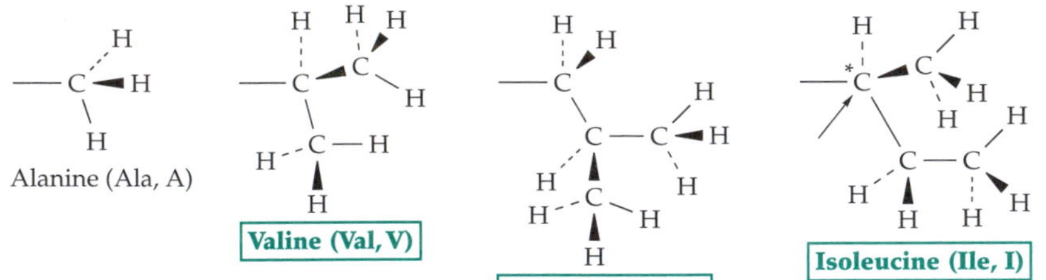

Alanine (Ala, A) Valine (Val, V) Leucine (Leu, L) Isoleucine (Ile, I)

These bulky groups of distinctive shapes participate in hydrophobic interactions in protein interiors and in forming binding sites of specific shapes.

c. The imino acid proline. Because the side chain is fused to the α-amino group, the entire structure, not just the side chain, is shown.

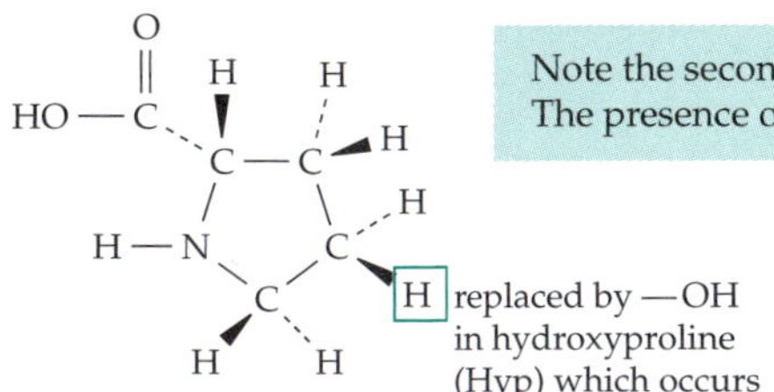

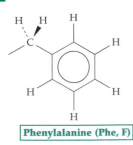

Proline (Pro, P)

Note the secondary amino group and the relatively rigid conformation. The presence of proline strongly influences the folding of protein chains.

H replaced by —OH in hydroxyproline (Hyp) which occurs in collagen

d. Aromatic amino acids

Phenylalanine (Phe, F) Tyrosine (Tyr, Y) Tryptophan (Trp, W)

Acidic protein, pK$_a$ ~9.5–10.9. H donor in hydrogen bonds

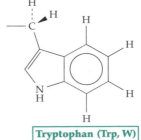

Tyrosine, phenylalanine, and tryptophan can form hydrophobic bonds and may be especially effective in bonding to other flat molecules.

TABLE 2-2
(continued)

e. Amino acid alcohols

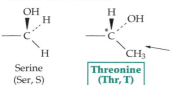

Serine
(Ser, S)

**Threonine
(Thr, T)**

Note the second chiral center.
The L-amino acid with the
opposite configuration in the
side chain is L-*allo*-threonine.

The —OH group is very weakly acidic (pK_a ~13.6). It can form
esters with phosphoric acid or organic acids and is a site of
attachment of sugar rings in glycoproteins. Hydroxyl groups
of serine are found at the active centers of some enzymes.

f. Acidic amino acids

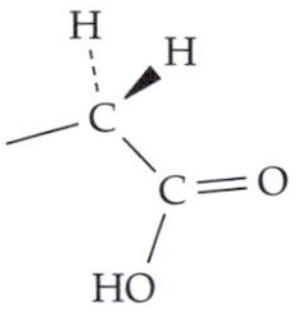

Aspartic acid (Asp, D) 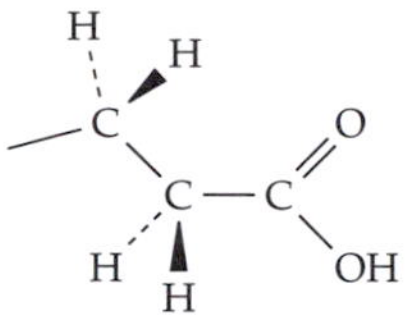   Glutamic acid (Glu, E)

Carboxyl groups of these side chains are dissociated
at neutral pH (pK_a values are 4.3–4.7) and provide
anionic (–) groups on the surfaces of proteins.

g. Amides of aspartic acid and glutamic acid

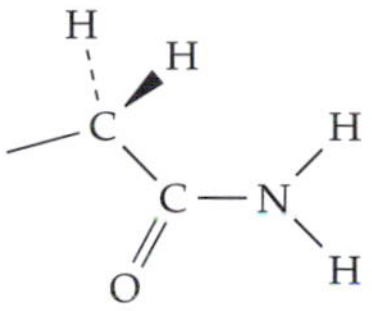

Asparagine (Asn, N)

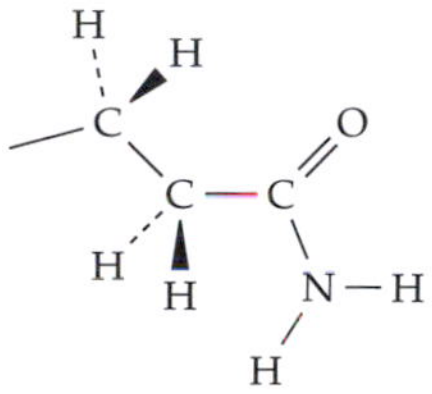

Glutamine (Gln, Q)

The amide group is not acidic but is polar
and participates in hydrogen bonding.

If it is uncertain whether a position in a
protein is occupied by aspartic acid or
asparagine, it may be designated Asx or
B. If glutamic or glutamine, it may be
designated as Glx or Z.

h. Sulfur-containing amino acids

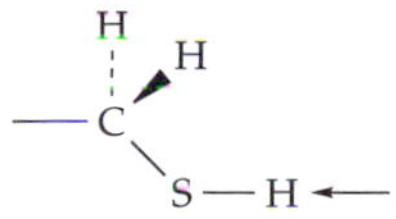

Cysteine (Cys, C)

S — H ← A weak acid,
pK$_a$ ~8.5

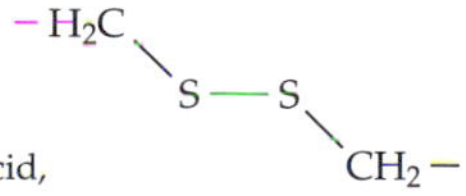

Two cysteine SH groups can be oxidatively joined
to form a disulfide bridge in the "double-headed"
amino acid cystine.

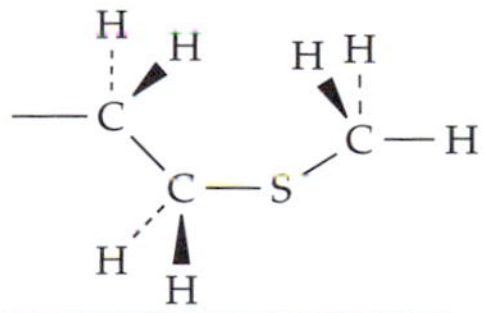

Methionine (Met, M)

Key

☐ **Essential in the human diet**

⬚ **Essential if phenylalanine or cysteine is inadequate**

▒ **Essential for rapid growth**

[a] The three-letter and one-letter abbreviations used for the
amino acid residues in peptides and proteins are given in
parentheses. B, J, U, X, and Z can be used to indicate
modified or unusual amino acids.

i. Basic amino acids

This basic site accepts a proton to
form a conjugate acid of pK_a ~6.4–
7.0 and carrying a positive charge.

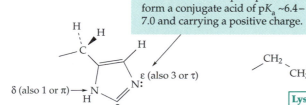

δ (also 1 or π) → N
H
2

ε (also 3 or τ)

Histidine (His, H)

The **imidazole** groups in histidine side
chains are parts of the active sites of
many enzymes. Like other basic
groups in proteins they also may bind
metal ions.

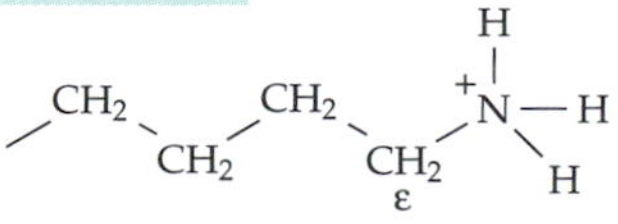

Lysine (Lys, K)

A flexible side arm with a
potentially reactive amino
group at the end. The
high pK$_a$ of ~10.5 means
that lysine side chains are
ordinarily protonated in
neutral solutions.

Guanidinium ion

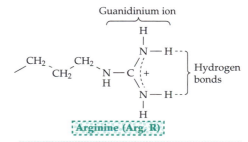

Hydrogen
bonds

Arginine (Arg, R)

The guanidinium group has a high pK_a of
over 12 and remains protonated under
most circumstances. It is stabilized by
resonance as indicated by the dashed line.
Guanidinium groups are often sites for
binding of phosphate or carboxylate
groups by pairs of hydrogen bonds.

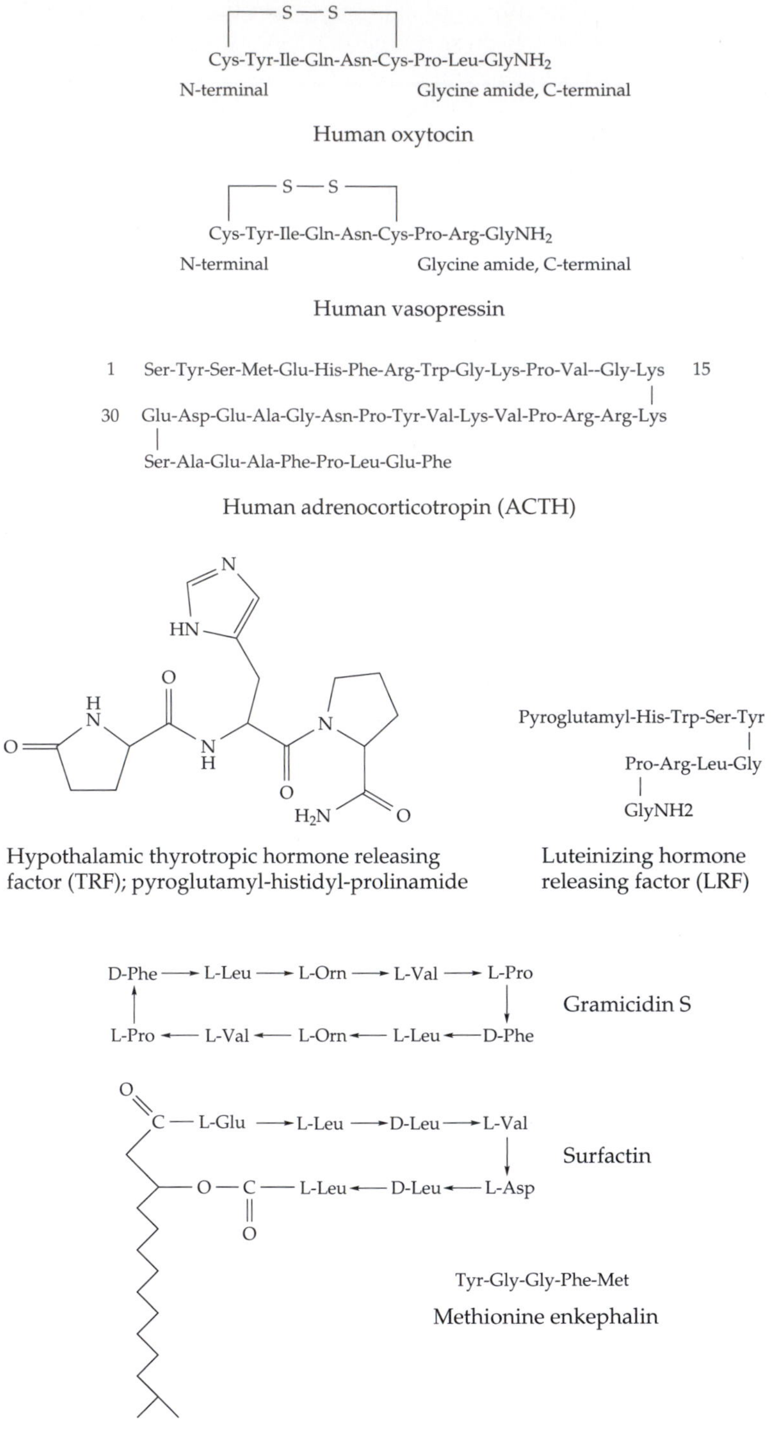

Cys-Tyr-Ile-Gln-Asn-Cys-Pro-Leu-GlyNH₂
N-terminal   Glycine amide, C-terminal

Human oxytocin

Cys-Tyr-Ile-Gln-Asn-Cys-Pro-Arg-GlyNH₂
N-terminal   Glycine amide, C-terminal

Human vasopressin

1   Ser-Tyr-Ser-Met-Glu-His-Phe-Arg-Trp-Gly-Lys-Pro-Val--Gly-Lys   15

30   Glu-Asp-Glu-Ala-Gly-Asn-Pro-Tyr-Val-Lys-Val-Pro-Arg-Arg-Lys

Ser-Ala-Glu-Ala-Phe-Pro-Leu-Glu-Phe

Human adrenocorticotropin (ACTH)

Hypothalamic thyrotropic hormone releasing
factor (TRF); pyroglutamyl-histidyl-prolinamide

Pyroglutamyl-His-Trp-Ser-Tyr
Pro-Arg-Leu-Gly
GlyNH2

Luteinizing hormone
releasing factor (LRF)

D-Phe ⟶ L-Leu ⟶ L-Orn ⟶ L-Val ⟶ L-Pro
L-Pro ⟵ L-Val ⟵ L-Orn ⟵ L-Leu ⟵ D-Phe

Gramicidin S

Surfactin

Tyr-Gly-Gly-Phe-Met

Methionine enkephalin

Figure 2-4 Structures of some naturally occurring peptides. Oxytocin and vasopressin are hormones of the neurohypophysis (posterior lobe of the pituitary gland). Adrenocorticotropin is a hormone of the adenohypophysis (anterior pituitary). Hormones of the adenohypophysis are released under the influence of releasing factors (regulatory factors) produced in the neighboring hypothalamus (a portion of the brain) in response to neural stimulation. Structures of two releasing factors are shown. Note that the γ-carboxyl groups of the N-terminal glutamine residues have reacted, with loss of NH₃, with the neighboring terminal –NH₂ groups to form cyclic amide (pyroglutamyl) groups.[78] Gramicidin S is an antibiotic made by *Bacillus brevis*, and surfactin is a depsipeptide (containing an ester linkage), a surface active antibiotic of *Bacillus subtilis*. Methionine enkephalin is a brain peptide with opiate-like activity.[79]

phosphate, and adenylate) and in names of enzymes (e.g., lactate dehydrogenase).

During the formation of polypeptides, the α-amino and carboxyl groups of the amino acids are converted into the relatively unreactive and uncharged amide (peptide) groups except at the two chain termini. In many cases the terminal amino and carboxyl groups are also converted within cells into uncharged groups (Chapter 10). Immediately after the protein is synthesized its terminal carboxyl group is often converted into an amide. The N terminus may be acetylated or cyclized to a pyroglutamyl group. Sometimes a cyclic peptide is formed (Fig. 2-4).

The properties of polypeptides and proteins are determined to a large extent by the chemistry of the side chain groups, which may be summarized briefly as follows. Glycine in a peptide permits a maximum of conformational mobility. The nine relatively nonpolar amino acids–alanine, valine, leucine, isoleucine, proline, methionine, phenylalanine, tyrosine, and tryptophan–serve as building blocks of characteristic shape. Tyrosine and tryptophan also participate in hydrogen bonding and in aromatic: aromatic interactions within proteins.

Much of the chemistry of proteins involves the side chain functional groups –OH, –SH, –COO⁻, –NH₃⁺, and imidazole (Eq. 2-6) and the guanidinium group of arginine. The side chains of asparagine and glutamine both contain the amide group $CONH_2$, which is relatively inert chemically but which can undergo hydrogen-bonding interactions. The amide linkages of the polypeptide backbone must also be regarded as important functional groups. Most polar groups are found on the outside surfaces of proteins where they can react chemically in various ways. When inside proteins they form H-bonds to the peptide backbone and to other polar groups.

2. Acidic and Basic Side Chains

The side chains of aspartic and glutamic acids carry negatively charged carboxylate groups at pH 7 while those of lysine and arginine carry the positively charged $-NH_3^+$ and guanidinium ions, respectively.

Guanidinium group

At pH 7 the weakly basic imidazole group of histidine may be partially protonated. Both the –SH group of cysteine and the phenolic –OH of tyrosine are weakly acidic and will dissociate and thereby acquire negative charges at a sufficiently high pH.

The number of positive and negative charges on a protein at any pH can be estimated approximately from the acid dissociation constants (usually given as pK_a values) for the amino acid side chains. These are given in Table 2-2. However, pK_a values of buried groups are often greatly shifted from these, especially if they associate as **ion pairs**. In addition, many proteins have free amino and carboxyl-terminal groups at the opposite ends of the peptide chain. These also participate in acid–base reactions with approximately the following pK_a values.

terminal —COOH, $pK_a = 3.6$–3.7
terminal —NH_3^+, $pK_a = 7.5$–7.9

The acid–base properties of an amino acid or of a protein are described by titration curves of the type shown in Figs. 3-1 and 3-2. In these curves the number of equivalents of acid or base that have reacted with an amino acid or protein that was initially at neutral pH are plotted against pH. The net negative or positive electrical charge on the molecule can be read directly from the curves. Both the net electrical charges and the distribution of positively and negatively charged groups are often of crucial importance to the functioning of a protein.

Additional aspects of the acid–base chemistry of amino acids and proteins are considered in Chapter 3, Section A and Chapter 6, Section E. The student may find it appropriate to study these sections at this time and to work the associated study problems.

3. The Peptide Unit

The very ability of a protein to exist as a complex three-dimensional structure depends upon the properties of the amide linkages between the amino acid units. Many of these properties follow from the fact that an amide can be viewed as a resonance hybrid of the following structures. Because of the partial double-bond character, the C–N bond is shorter than that of a normal single bond and the C=O bond is lengthened.

The observed lengths in nanometers determined by X-ray diffraction measurements are given in Fig. 2-5 (top). The partial double-bond character of the C–N bond has important consequences. The peptide unit is nearly planar as is indicated by the dashed parallelogram in Fig. 2-5.

However, the bonds around the nitrogen retain some pyramidal character (Fig. 2-5, bottom). Even more important is the fact that there is flexibility. As a result, the torsion angle ω may vary over a range of $\pm 15°$ or even more from that in the planar state.[80,81,81a]

The resonance stabilization of the amide linkage is thought to be about 85 kJ/mol. Rotation around the C–N bond through 90° would be expected to require

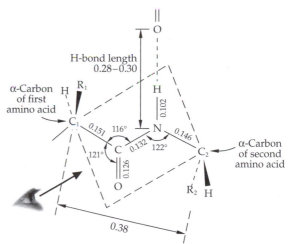

Distance (nm) between successive α-carbon atoms in a protein.

View down C—N axis
(as indicated by heavy arrow above)

This angle, ω, may be as great as $\pm 15°$

Figure 2-5 Dimensions of the peptide linkage. Interatomic distances in nm, including the hydrogen bond length to an adjacent peptide linkage, are indicated. The atoms enclosed by the dotted lines all lie *approximately* in a plane. However, as indicated in the lower drawing, the nitrogen atom tends to retain some pyramidal character.

about this much energy. This fact immediately suggests a way in which proteins may sometimes be able to store energy—by having one or more peptide units twisted out of complete planarity.[69]

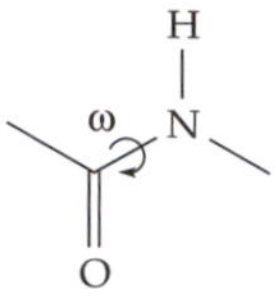

An important effect of the resonance of the amide linkage is that the oxygen atom acquires some negative charge and the NH group some positive charge. Some of the positive charge is usually depicted as residing on the nitrogen, but some is found on the hydrogen atom. The latter can be pictured as arising from a contribution of a fourth resonance form that contains no bond to hydrogen.

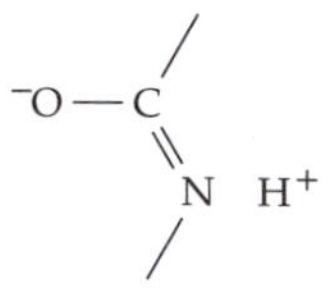

Nevertheless, this picture is inadequate. Various evidence indicates that the nitrogen actually carries a net negative charge.[81a]

The positive and negative ends of the dipoles in the amide group tend to associate to form strong hydrogen bonds. These hydrogen bonds together with the connecting amide linkages can form chains that may run for considerable distances through proteins. The tendency for cooperativity in hydrogen bond formation may impart unusual stability to these chains. As with individual amide linkages, these chains of hydrogen-bonded amides can also be thought of as resonance hybrids:

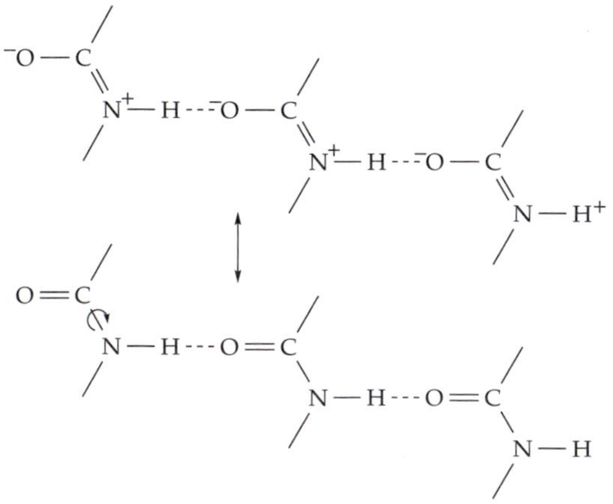

The two structures pictured are extreme forms, the true structure being something in between. In the lower form, rotation about the C–N bond would be permitted but then the charge separation present in

the upper structure would no longer exist. Thus, the hydrogen bonds would be weakened. We can conclude that if an amide linkage in such a chain becomes twisted, the hydrogen bonds that it forms will be weakened. If there is cooperativity, the hydrogen bonds will all be strongest when there is good planarity in all of the amides in the chain.

Amides have very weak basic properties and protonation is possible either on the oxygen (A) or on the nitrogen (B).[82,83]

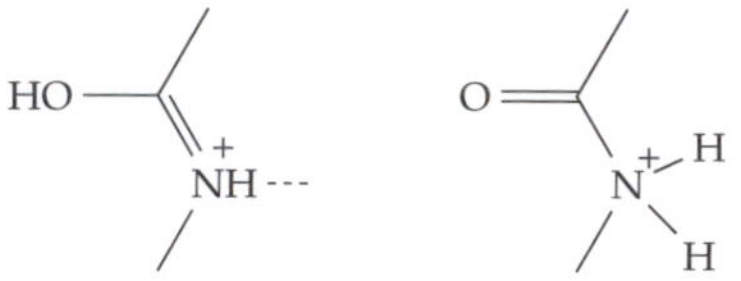

The pK_a values for such protonation are usually less than zero, but it is possible that a correctly placed acidic group in a protein could protonate either oxygen or nitrogen transiently during the action of a protein. Protonation on oxygen would strengthen hydrogen bonds from the nitrogen whereas protonation on nitrogen would weaken hydrogen bonds to oxygen and might permit rotation. The amide group has a permanent dipole moment of 3.63 debyes oriented as follows:

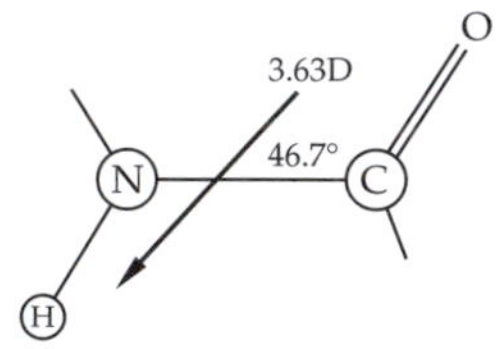

Here the arrow points toward the *positive* end of the dipole.

While the *trans* peptide linkage shown in Fig. 2-4 is usual, the following *cis* peptide linkage, which is ~8 kJ/mol less stable than the *trans* linkage, also occurs in proteins quite often. The nitrogen atom is usually but not always from proline.[81,84]

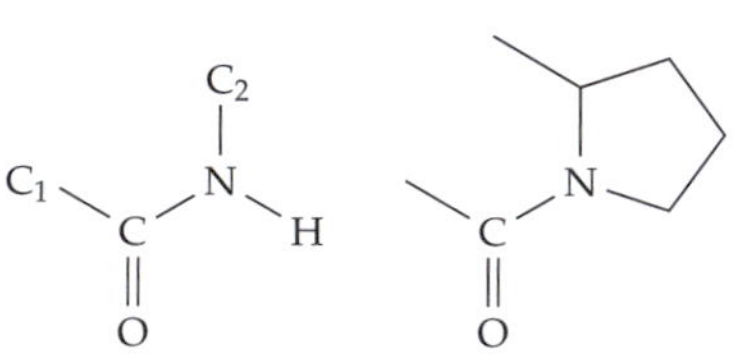

cis Peptide linkages

4. Polypeptides

The chain formed by polymerization of amino acid molecules provides the **primary structure** of a protein. Together with any covalent crosslinkages and other modifications, this may also be called the

covalent structure of the protein. Each monomer unit in the chain is known as an amino acid **residue**. This term acknowledges the fact that each amino acid has lost one molecule of H_2O during polymerization. To be more precise, the number of water molecules lost is *one less* than the number of residues. Peptides are named according to the amino acid residues present and beginning with the one bearing the terminal amino group. Thus, L-alanyl-L-valyl-L-methionine has the following structure:

Like amino acids, this tripeptide is a dipolar ion. The same structure can be abbreviated Ala-Val-Met or, using one-letter abbreviations, AVM. It is customary in describing amino acid sequences to place the amino-terminal (N-terminal) residue at the left end and the carboxyl-terminal (C-terminal) residue at the right end. Residues are numbered sequentially with the N-terminal residue as 1. An example is shown in Fig. 2-6.

The sequence of amino acid units in a protein is always specified by a gene. The sequence determines how the polypeptide chain folds and how the folded protein functions. For this reason much effort has gone into "sequencing," the determination of the precise order of amino acid residues in a protein. Sequences of several hundreds of thousands of proteins and smaller peptides have been established and the number doubles each year.[75,87,88,88a] Most of these

A
```
     1              10                  20                  30                  40
     A P P S V F A E V P Q A Q P V L V F K L I A D F R E D P D P R K V N L G V G A Y
              50                  60                  70                  80
     R T D D C Q P W V L P V V R K V E Q R I A N N S S L N H E Y L P I L G L A E F R
              90                  100                 110                 120
     T C A S R L A L G D D S P A L Q E K R V G G V Q S L G G T G A L R I G A E F L A
              130                 140                 150                 160
     R W Y N G T N N K D T P V Y V S S P T W E N H D G V F T T A G F K D I R S Y R Y
              170                 180                 190                 200
     W D T E K R G L D L Q G F L S D L E N A P E F S I F V L H A C A H N P T G T D P
              210                 220                 230                 240
     T P E Q W K Q I A S V M K R R F L F P F F D S A Y Q G F A S G N L E K D A W A I
              250                 260                 270                 280
     R Y F V S E G F E L F C A Q S F S K N F G L Y N E R V G N L T V V A K E P D S I
              290                 300                 310                 320
     L R V L S Q M Q K I V R V T W S N P P A Q G A R I V A R T L S P D E L F H E W T
              330                 340                 350                 360
     G N V K T M A D R I L S M R S E L R A R L E A L K T P G T W N H I T D Q I G M F
              370                 380                 390                 400
     S F T G L N P K Q V E Y L I N E K H I Y L L P S G R I N M C G L T T K N L D Y V
              410
     A T S I H E A V T K I Q
```

B

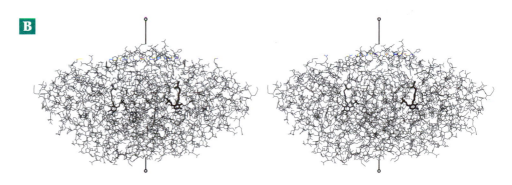

Figure 2-6 (A) The complete amino acid sequence of the cytoplasmic enzyme aspartate aminotransferase from pig heart. The peptide has the composition Lys_{19}, His_8, Arg_{26}, $(Asp + Asn)_{42}$, Ser_{26}, Thr_{26}, $(Glu + Gln)_{41}$, Pro_{24}, Gly_{28}, Ala_{32}, Cys_5, Val_{29}, Met_6, Ile_{19}, Leu_{38}, Tyr_{12}, Phe_{23}, Trp_9. The molecular mass is 46.344 kDa and the complete enzyme is a 93.147-kDa dimer containing two molecules of the bound coenzyme pyridoxal phosphate attached to lysine-258 (enclosed in box).[85,86] (B) A stereoscopic view of a complete enzyme molecule which contains two identical subunits with the foregoing sequence. Coordinates from Arthur Arnone. In this "wire model" all the positions of all of the nearly 7000 atoms that are heavier than hydrogen are shown. The > 8000 hydrogen atoms have been omitted. The view is into the active site of the subunit on the right. The pyridoxal phosphate and the lysine residue to which it is attached are shown with heavy lines. The active site of the subunit to the left opens to the back side as viewed here. The drawing may be observed best with a magnifying viewer available from Abrams Instrument Corp., Lansing, Michigan or Luminos Photo Corp., Yonkers, New York. However, with a little practice, it is possible to obtain a stereoscopic view unaided. Hold the book with good illumination about 20–30 cm from your eyes. Allow your eyes to relax as if viewing a distant object. Of the four images that are visible, the two in the center can be fused to form the stereoscopic picture. Drawings by program MolScript (Kraulis, 1991).

BOX 2-A PROTEINS OF BLOOD PLASMA

Among the most studied of all proteins are those present in blood plasma.[a-f] Their ready availability and the clinical significance of their study led to the early development of electrophoretic separations. Electrophoresis at a pH of 8.6 (in barbital buffer) indicated six main components. The major and one of the fastest moving proteins is **serum albumin**. Trailing behind it are the α_1-, α_2-, and β-**globulins**, **fibrinogen**, and γ-**globulins**. Each of these bands consists of several proteins and two-dimensional separation by electrophoresis and isoelectric focusing (Chapter 3) reveals over 30 different proteins.[e] Many of these contain varying numbers of attached carbohydrate units and appear as families of spots.

Fractionation of large quantities of plasma together with immunochemical assays has led to identification of over 200 different proteins. Sixty or more are enzymes, some in very small quantitites which may have leaked from body cells. Normally plasma contains 5.7 – 8.0 g of total protein per 100 ml (~1 mM). Albumin accounts for 3.5 – 4.5 g/100 ml. An individual's liver synthesizes about 12 g each day. Next most abundant are the **immunoglobulins**. One of these (IgG or γ-globulin) is present to the extent of 1.2–1.8 g/100 ml. Also present in amounts greater than 200 mg per 100 ml are α- and β-**lipoproteins**, the α_1 **antitrypsin**, α_2-**macroglobulin**, **haptoglobin**, **transferrin**, and fibrinogen.

Plasma proteins have many functions. One of them, fullfilled principally by serum albumin, is to impart enough osmotic pressure to plasma to match that of the cytoplasm of cells. The heart-shaped human serum albumin consists of a single 65 kDa chain of 585 amino acid residues coiled into 28 helices.[g] Three homologous repeat units or domains each contain six disulfide bridges, suggesting that gene duplication occurred twice during the evolution of serum albumins. The relatively low molecular mass and high density of negative charges on the surface make serum albumin well adapted for the role of maintaining osmotic pressure. However, serum albumin is not essential to life.

Over 50 mutant forms have been found and at least 30 persons have been found with no serum albumin in their blood.[h,i] These analbuminemic individuals are healthy and have increased concentrations of other plasma proteins.

A second major function of plasma proteins is transport. Serum albumin binds to and carries many sparingly soluble metabolic products, including fatty acids, tryptophan, cysteine, steroids, thyroid hormones, Ca^{2+}, Cu^{2+}, Zn^{2+}, other metal ions, bilirubin, and various drugs. There are also many more specialized transporter proteins. **Transferrin** carries iron and **ceruloplasmin** (an α_2 globulin) transports copper. **Transcortin** carries corticosteroids and progesterone, while another protein carries sex hormones. **Retinol-binding protein** carries vitamin A and **cobalamin-binding proteins** vitamin B_{12}. **Hemopexin** carries heme to the liver, where the iron can be recovered.[j] Haptoglobin binds hemoglobin released from broken red cells and also assists in the recycling of the iron in the heme.[k] **Lipoproteins** (see Table 21-1) carry phospholipids, neutral lipids, and cholesterol esters. Most of the mass of these substances is lipid.

Immunoglobulins, α_1-trypsin inhibitor and α_2-**macroglobulin**,[k] ten or more blood clotting factors; and proteins of the **complement system** all have protective functions that are discussed elsewhere in this book. Hormones, many of them proteins, are present in the blood as they are carried to their target tissues. Many serum proteins have unknown or poorly understood functions. Among these are the **acute phase proteins**, whose concentrations rise in response to inflammation or other injury.

[a] Allison, A. C., ed. (1974) *Structure and Function of Plasma Proteins*, Vol. 1, Plenum, New York

[b] Allison, A. C., ed. (1976) *Structure and Function of Plasma Proteins*, Vol. 2, Plenum, New York

[c] Putnam, F. W., ed. (1975) *The Plasma Proteins*, 2nd ed., Vol. 1 and 2, Academic Press, New York

[d] Blomback, B., and Hanson, L. A., eds. (1979) *Plasma Proteins*, John Wiley, Chichester, Oklahoma

[e] Geisow, M. J., and Gordon, A. H. (1978) *Trends Biochem. Sci.* **3**, 169–171

[f] Smith, E. L., Hill, R. L., Lehman, I. R., Lefkowitz, R. J., Handler, P., and White, A. (1983) in *Principles of Biochemistry, Mammalian Biochemistry*, 7th ed., McGraw-Hill, New York, pp. 3–37

[g] He, X. M., and Carter, D. C. (1992) *Nature* **358**, 209–215

[h] Peters, T., Jr. (1996) *All About Albumin: Biochemistry, Genetics and Medical Applications*, Academic Press, San Diego, California

[i] Madison, J., Galliano, M., Watkins, S., Minchiotti, L., Porta, F., Rossi, A., and Putnam, F. W. (1994) *Proc. Natl. Acad. Sci: USA* **91**, 6476–6480

[j] Satoh, T., Satoh, H., Iwahara, S.-I., Hrkal, Z., Peyton, D. H., and Muller-Eberhard, U. (1994) *Proc. Natl. Acad. Sci: U.S.A.* **91**, 8423–8427

[k] Feldman, S. R., Gorias, S. L., and Pizzo, S. V. (1985) *Proc. Natl. Acad. Sci: U.S.A.* **82**, 5700–5704

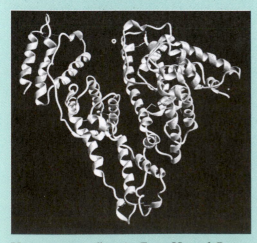

Human serum albumin. From He and Carter.[g]

have been deduced from the sequences of nucleotides in DNA. Sequences of some small peptide hormones and antibiotics are shown in Fig. 2-4 and that of a 412-residue protein in Fig. 2-6. The molecular mass of a protein can be estimated from the chain length by assuming that each residue adds 100–115 Da.

The amino acid composition varies greatly among proteins. A typical globular protein contains all or most of the 20 amino acids. The majority are often present in roughly similar amounts but His, Cys, Met, Tyr, and Trp tend to be less abundant than the others. Specialized proteins sometimes have unusual amino acid compositions. For example, collagen of connective tissue contains 33 mole% glycine and 21% of proline + hydroxyproline residues; the major proteins of saliva contain 22% of glutamate + glutamine and 20–45% proline.[89] Cell walls of plants contain both high proline and high glycine polypeptides. One from petunias is 67% glycine.[90] Silk fibroin contains 45% glycine and 29% alanine. A DNA repair protein of yeast has 13 consecutive aspartate residues.[91] The tough eggshell (chorion) of the domesticated silkmoth *Bombyx mori* contains proteins with ~30% cysteine.[92] Many proteins consist, in part, of repeated short sequences. For example, the malaria-causing *Plasmodium falciparum* in its sporozoite stage is coated with a protein that contains 37 repeats of the sequence NANP interspersed with 4 repeats of NDVP.[93] These two sequences have been indicated with single-letter abbreviations for the amino acids.

With a large number of protein and DNA sequences available, it has become worthwhile to compare sequences of the same protein in different species or of different proteins within the same or different species. Computer programs make it possible to recognize **similarities** and **homologies** between sequences even when deletions and insertions have occurred.[88,88a,94–97] The term homology has the precise biological definition *"having a common evolutionary origin,"* but it is often used to describe any close similarity in sequence.[98] Among a pair of homologous proteins, a change at a given point in a sequence may be either **conservative**, meaning that a residue of similar character (large, small, positively charged, nonpolar, etc.) has been substituted, or it may be **nonconservative**.

D. The Architecture of Folded Proteins

All proteins are made in the same way but as the growing peptide chains peel off from the ribosome, each of the thousands of different proteins in a living cell folds into its own special **tertiary structure**.[88a] The number of possible conformations of a protein chain is enormous. Consider a 300-residue polypeptide which could stretch in fully extended form for ~100 nm. If the chain were folded back on itself about 13

times it could form a 7-nm square sheet about 0.5 nm thick. The same polypeptide could form a thin helical rod 45 nm long and ~1.1 nm thick. If it had the right amino acid sequence it could be joined by two other similar chains to form a collagen-type triple helix of 87 nm length and about 1.5 nm diameter (Fig. 2-7).

The highly folded **globular proteins** vary considerably in the tightness of packing and the amount of internal water of hydration.[99,100] However, a density of ~1.4 g cm^{-3} is typical. With an average mass per residue of 115 Da our 300-residue polypeptide would have a mass of 34.5 kDa or 5.74×10^{-20} g and a volume of 41 nm^3. This might be approximated by a cube 3.45 nm in width, a "brick" of dimensions $1.8 \times 3.6 \times 6.3$ nm, or a sphere of diameter 4.3 nm. Although protein molecules are usually very irregular in shape,[101] for purposes of calculation idealized ellipsoid and rod shapes are often assumed (Fig. 2-7).

It is informative to compare these dimensions with those of the smallest structures visible in cells; for example, a bacterial flagellum is ~13 nm in diameter and a cell membrane ~8–10 nm in thickness. Bricks of the size of the 300-residue polypeptide could be used to assemble a bacterial flagellum or a eukaryotic microtubule. Helical polypeptides may extend through cell membranes and project on both sides, while a globular protein of the same chain length may be almost completely embedded in the membrane.

1. Conformations of Polypeptide Chains

To understand how a polypeptide chain folds we need to look carefully at the possible conformations of the peptide units. Since each peptide unit is nearly planar, we can think of a polypeptide as a chain of flat units fastened together as in Fig. 2-8. Every peptide unit is connected to the next by the α-carbon of an amino acid. This carbon provides two single bonds to the chain and rotation can occur about both of them (except in the cyclic amino acid proline). To specify the conformation of an amino acid unit in a polypeptide chain, we must describe the torsion angles about both of these single bonds.[11,76,102] These angles are indicated by the symbols φ **(phi)** and ψ **(psi)** and are assigned the value 180° for the fully extended chain as shown in Fig. 2-8. Each angle is taken as zero for the impossible conformation in which the two chain ends are in the eclipsed conformation. By the same token, the torsion angle ω **(omega)** around the C–N bond of the amide is 0° for a planar *cis* peptide linkage and 180° for the usual *trans* linkage.

Since both φ and ψ can vary for each residue in a protein, there are a large number of possible conformations. However, many are excluded because they bring certain atoms into collision. This fact can be established readily by study of molecular models.

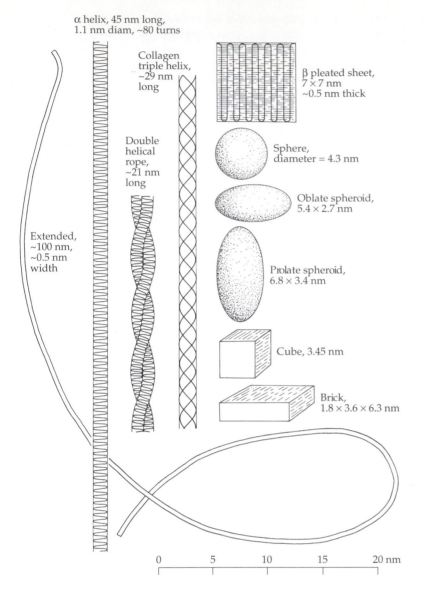

Figure 2-7 Some idealized shapes that a 34.5 kDa protein molecule of 300 amino acids might assume.

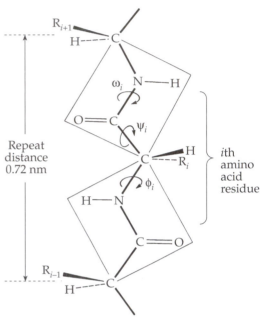

Figure 2-8 Two peptide units in the completely extended β conformation. The torsion angles ϕ_i, ψ_i, and ω_i are defined as 0° when the main chain atoms assume the *cis* or eclipsed conformation. The angles in the completely extended chain are all 180°. The distance from one α carbon atom (C_α) to the next in a peptide chain is always 0.38 nm, no matter how the chain is folded.

Using a computer, it is possible to study the whole range of possible combinations of ϕ and ψ. This has been done for the peptide linkage by Ramachandran. The results are often presented as plots of ϕ vs ψ (Ramachandran plots or **conformational maps**)[102a,103,103a] in which possible combinations of the two angles are indicated by blocked out areas. The original Ramachandran plots were made by representing the atoms as hard spheres of appropriate van der Waals radii, but the version shown in Fig. 2-9 was calculated using a complex potential energy function to represent the van der Waals attraction and the repulsion from close contact.[103] This map was calculated for poly-L-alanine but it would be very similar for most amino acids.

Notice the two areas connected by a higher energy bridge on the left side of Fig. 2-9. The upper area

contains the pairs of torsion angles for the extended **β structures** as well as for **collagen**. The lower area contains allowed conformations for the *right-handed helices.* As can be seen from Fig. 2-10, most of the observed conformations of peptide units in a real protein fall into these regions. Glycyl residues are an exception. Since glycine has no β-carbon atom, the conformations are less restricted. Out of nearly 1900 non-glycine residues in well-determined protein structures, 66 were found in disallowed areas of the Ramachandran diagram.[104a] These were often accommodated by local distortions in bond angles. The positions at which such steric strain occurs are often in regions concerned with function.[104b] One residue, which lies in a disallowed region in Fig. 2-10 is asparagine 297 of aspartate aminotransferase (Fig. 2-6). It is located

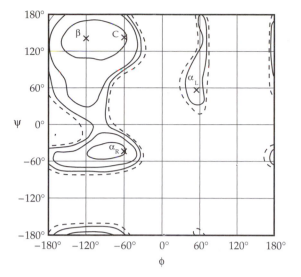

Figure 2-9 Potential energy distribution in the $\phi-\psi$ plane for a pair of peptide units with alanyl residues calculated using potential parameters of Scheraga and Flory. Contours are drawn at intervals of 1 kcal (4.184 kJ) per mol going down from 0 kcal per mol. The zero contour is dashed. From Ramachandran *et al.*[104] The points marked x are for the four ideal structures: twisted β structure (β), collagen (C), right-handed α helix (α_R), and the less favored left-handed α helix (α_L).

adjacent to the coenzyme site. The possible conformations of proline residues are limited. The angle ϕ is always $-60 \pm 20°$, while ψ for the residue adjacent to the proline N can be either ~150° or ~ $-30°$.[105] Typical ϕ, ψ angles for some regular peptide structures are given in Table 2-3.

Conformations of side chain groups are designated by a series of torsion angles designated χ_i.[11] Within proteins there are strong preferences for certain χ_i/χ_2 pairs.[107] Torsion angles within proline rings are considered in Chapter 5,A,6.

2. The Extended Chain β Structures

As was first pointed out by Pauling and Corey,[108,109] an important structural principle is that within proteins the maximum possible number of hydrogen bonds involving the C=O and N–H groups of the peptide chain should be formed. One simple way to do this is to line up fully extended chains ($\phi = \psi = 180°$) and to form hydrogen bonds between them. Such a structure

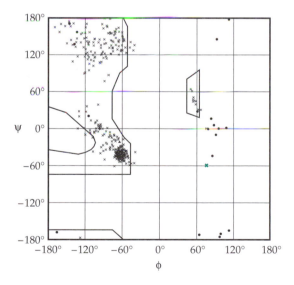

Figure 2-10 Ramachandran plot for cytosolic aspartate aminotransferase. The angles ϕ and ψ were determined experimentally from X-ray diffraction data at 0.16 nm resolution and model building. The majority of conformations are those of α helices or of β structure. Glycine residues are indicated by filled circles, while all other residues are denoted by an "x". One of these (green) lies quite far from an allowed area and must give rise to localized strain.[104b] Extreme lower limit "allowed" regions by the hard-sphere criteria are shown in outline. From coordinates of Arthur Arnone *et al.* (unpublished).[106]

TABLE 2-3
Approximate Torsion Angles for Some Regular Peptide Structures[a]

Structure			ϕ (deg.)	ψ (deg.)
Hypothetical fully extended polyglycine chain[b]			-180	$+180$
β-Poly(L-alanine) in antiparallel-chain pleated sheet			-139	$+135$
Parallel-chain pleated sheet			-119	$+113$
Twisted β strand			-120	$+140$
Polyglycine II			-80	$+150$
Poly(L-proline) II			-78	$+149$
Collagen[c]			-60 ± 15	-140 ± 15
Right-handed α helix[d]			-57	-47
Left-handed α helix			$+57$	$+47$
β Bends:	Type I,	residue 2	-60	-30
		residue 3	-90	0
	Type II,	residue 2	-60	120
		residue 3	80	0
	Type III,	residue 2 & 3	-60	-30
β Bulges[e]:	"Classical" β bulge residue 1		-100	-45
	G1 bulge	residue 1	85	0

[a] From Liébecq, C., ed. (1992) *Biochemical Nomenclature*, Portland Press, London and Chapel Hill (for the International Union of Biochemistry and Molecular Biology).
[b] Torsion angles for the fully extended chain can be designated either $+180°$ or $-180°$, the two being equivalent. They are given as $\phi = -180°$ and $\psi = +180°$ to facilitate comparison with the other structures.
[c] Ramachandran, G. N. (1967) *Treatise on Collagen*, Vol. 1, p. 124, Academic Press, New York
[d] Both ϕ and ψ are quite variable but $\phi + \psi = -104°$.
[e] For residues other than these indicated the torsion angles are about those of typical β structures.

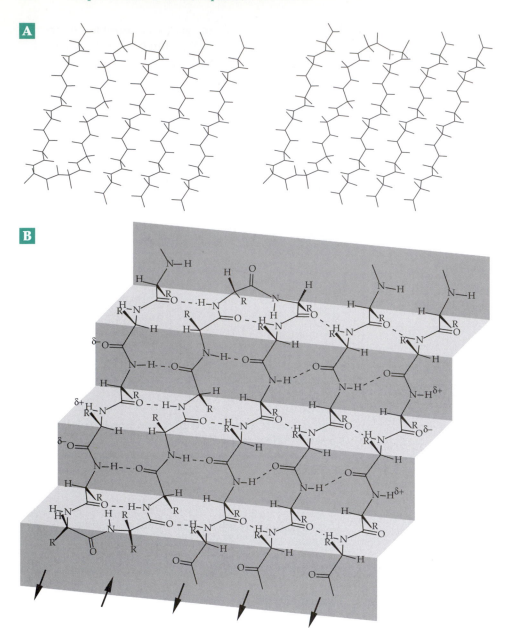

Figure 2-11 The extended chain β pleated sheet structures. (A) Stereoscopic drawing without atomic symbols. (B) Drawing with atomic symbols. At the left is the *antiparallel* structure. The 0.70 nm spacing is slightly decreased from the fully extended length. The amino acid side chains (R) extend alternately above and below the plane of the "accordion pleated" sheet. The pairs of linear hydrogen bonds between the chains impart great strength to the structure. The chain can fold back on itself using a "β turn" perpendicular to the plane of the pleated sheet. The *parallel* chain structure (right side) is similar but with a less favorable hydrogen bonding arrangement. Arrows indicate chain directions.

exists for polyglycine and resembles that in Fig. 2-11. Notice that on the left side of this figure, the adjacent chains run in opposite directions; hence, the term **antiparallel β structure**. The antiparallel arrangement not only gives the best hydrogen bond formation between chains but also permits a single chain to fold back on itself giving a compact hairpin loop.

Pleated sheets. While a fully extended polyglycine chain is possible, the side chains of other amino

acids cannot be accommodated without some distortion of the structure. Thus, the peptide chains in silk fibroin have a repeat distance of 0.70 nm compared with the 0.72 nm for the fully extended chain (Fig. 2-8). Pauling and Corey[108] showed that this shortening of the chain could result from rotation of angle ϕ by ~40° (to −140°) and rotation of ψ in the opposite direction by ~45° (to +135°) to give a slightly puckered chain. The resulting multichain structure (shown in Fig. 2-11) is known as a **pleated sheet**. As in this figure, both

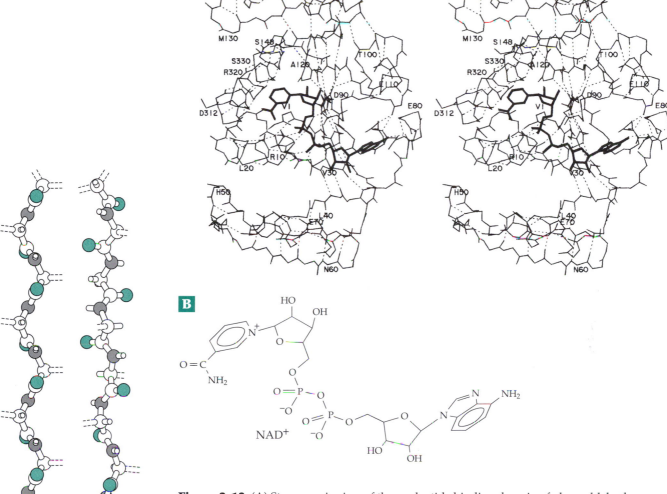

A

B

NAD⁺

Figure 2-12 Straight (left) and twisted (right) peptide chains in extended β conformations. From Chothia.[110]

Figure 2-13 (A) Stereoscopic view of the nucleotide binding domain of glyceraldehyde phosphate dehydrogenase. The enzyme is from *Bacillus stearothermophilus* but is homologous to the enzyme from animal sources. Residues are numbered 0–148. In this wire model all of the main chain C, O, and N atoms are shown but side chains have been omitted. The large central twisted β sheet, with strands roughly perpendicular to the page, is seen clearly; hydrogen bonds are indicated by dashed lines. Helices are visible on both sides of the sheet. The coenzyme NAD⁺ is bound at the end of the β sheet toward the viewer. Note that the two phosphate groups in the center of the NAD⁺ are H-bonded to the N terminus of the helix beginning with R10. From Skarzynski *et al.*[111a] (B) Structural formula for NAD⁺.

parallel and antiparallel strands are often present in a single β sheet within a protein.

The β structure is one of the most important **secondary structures** in proteins. It occurs in about 80% of the soluble globular proteins whose structures have been determined. In many cases almost the entire protein is made up of β structure. Single strands of extended polypeptide chain are sometimes present within globular proteins but more often a chain folds back on itself to form a hairpin loop. A second fold may be added to form an antiparallel "β meander"[102] and additional folds to form β sheets. Beta structures are found in silk fibers (Box 2-B) as well as in soluble proteins.

Twisted sheets. X-ray diffraction studies have shown that β pleated sheets are usually not flat but are twisted. In a twisted sheet the individual polypeptide chains make a shallow left-handed helix. However, when successive carbonyl groups are viewed along the direction of the chain, a **right-handed twist** is seen (Fig. 2-12).[110] Such twisted β sheets are often found in the globular proteins. An example (Fig. 2-13) is the "nucleotide-binding" domain of a dehydrogenase enzyme. The twist of the sheet is seen clearly in this stereoscopic view. When such chains are associated into β sheets, whether parallel or antiparallel, and are viewed in a direction perpendicular to the chains and looking down the edge of the sheet, a left-handed "propeller" is seen. Such a propeller is visible in the drawing of carboxypeptidase A (Fig. 2-14).

The cause of the twist in β sheets appears to lie in noncovalent interactions between hydrogen atoms on the β-carbon atoms of side chains and the peptide backbone atoms. For side chains of most L- amino acids these interactions provide a small tendency

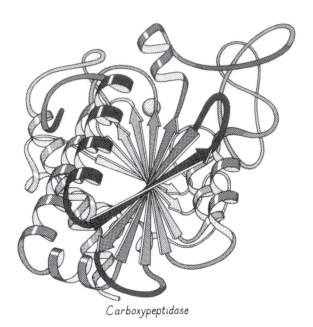

Carboxypeptidase

Figure 2-14 A "ribbon" drawing of the 307- residue protein-hydrolyzing enzyme carboxypeptidase A. In this type of drawing wide ribbons are used to show β strands and helical turns while narrower ribbons are used for bends and loops of the peptide chains. The direction from the N terminus to C terminus is indicated by the arrowheads on the β strands. No individual atoms are shown and side chains are omitted. Courtesy of Jane Richardson.[117]

towards the observed right-handed twist.[111] Nonplanarity in the amide groups (Fig. 2-5) may also contribute. Interstrand interactions seem to be important.[111b]

Properties of β sheets. In antiparallel β sheets, nonpolar residues are often present on one side of the sheet and polar residues on the other. The nonpolar side may be buried in the protein, perhaps backed up against another β sheet of similar structure as in the carbohydrate-binding **lectin** shown in Fig 2-15 to give a **β sandwich**. To accommodate this packing arrangement, nonpolar and polar residues tend to alternate in the amino acid sequence. The facing nonpolar surfaces of the two sheets are essentially smooth. The β strands in one sheet lie at an angle of ~30° to those in the other sheet. The twist of the strands allows pairs of nonpolar side chains from the two sheets to maintain good van der Waals contact with each other for a considerable distance along the strands. Beta sheets of silks (Box 2-B) and of immunoglobulin domains (Fig 2-16B) are also thought to be associated in a back-to-back fashion. *Parallel* β structures have been found only when there are five or more strands, some of which may be antiparallel to the others. The parallel β structure is apparently less stable than the antiparallel structure. Parallel strands are usually buried in the protein, being surrounded by either other extended strands or helices.

Chains of hydrogen bonds and amide linkages pass across β sheets perpendicular to the chain direction. There are partial positive and negative charges along the outside edges of the sheet where these hydrogen bond chains terminate (Fig. 2-11). The polarity of these chains alternates and the positive end of one peptide bond is relatively near to the negative end of the next along the edges of the sheets. These "unsatisfied ends" of hydrogen bond chains are often sites of

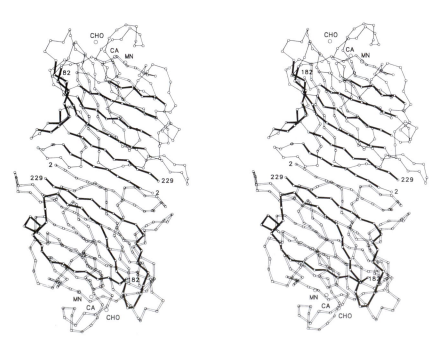

Figure 2-15 A stereoscopic **alpha-carbon plot** showing the three-dimensional structure of favin, a sugar-binding lectin from the broad bean (*Vicia faba*). In this plot only the α-carbon atoms are shown at the vertices. The planar peptide units are represented as straight line segments. Side chains are not shown. The protein consists of two identical subunits, each composed of a 20-kDa α chain and a 20-kDa β chain. The view is down the twofold rotational axis of the molecule. In the upper subunit the residues involved in the front β sheet are connected by double lines, while those in the back sheet are connected by heavy solid lines. In the lower subunit the α chain is emphasized. Notice how the back β *sheet* (not the chain) is continuous between the two subunits. Sites for bound Mn²⁺ (MN), Ca²⁺ (CA), and sugar (CHO) are marked by larger circles. From Reeke and Becker.[112]

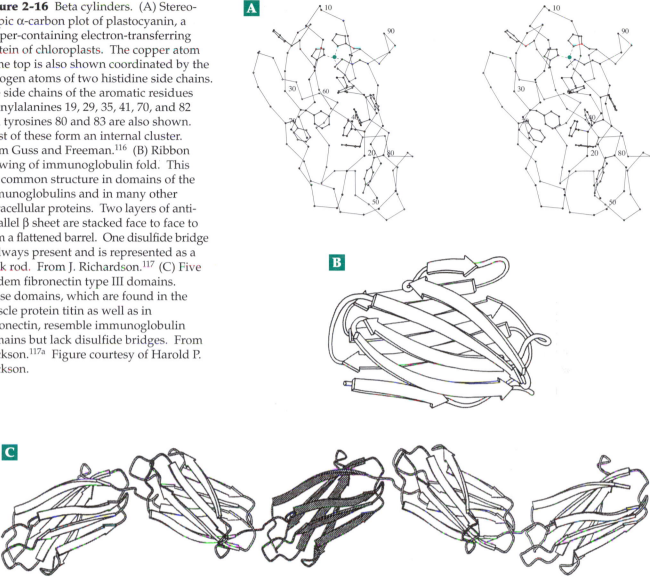

Figure 2-16 Beta cylinders. (A) Stereoscopic α-carbon plot of plastocyanin, a copper-containing electron-transferring protein of chloroplasts. The copper atom at the top is also shown coordinated by the nitrogen atoms of two histidine side chains. The side chains of the aromatic residues phenylalanines 19, 29, 35, 41, 70, and 82 and tyrosines 80 and 83 are also shown. Most of these form an internal cluster. From Guss and Freeman.[116] (B) Ribbon drawing of immunoglobulin fold. This is a common structure in domains of the immunoglobulins and in many other extracellular proteins. Two layers of antiparallel β sheet are stacked face to face to form a flattened barrel. One disulfide bridge is always present and is represented as a thick rod. From J. Richardson.[117] (C) Five tandem fibronectin type III domains. These domains, which are found in the muscle protein titin as well as in fibronectin, resemble immunoglobulin domains but lack disulfide bridges. From Erickson.[117a] Figure courtesy of Harold P. Erickson.

interaction of polar groups from side chains. Thus, OH groups of serine or threonine residues, amide groups of asparagine and glutamine residues, etc., often fold back and hydrogen-bond to these ends.

Edges of β sheets can also serve as binding sites for other polar molecules. For example, substrates bind to an edge of a β sheet in the active sites of trypsin and other proteases (Chapter 12). Some proteins, e.g. the lectin shown in Fig. 2-15, form dimers by joining identical edges of a β sheet in antiparallel orientation.[112]

Cylinders and barrels. The twisted β sheets of proteins are often curved to form structures known as **β cylinders** or **β barrels** (Fig. 2-16).[113,114] Simple cylinders formed by parallel β strands form the backbones of the electron transport protein plastocyanin, the enzyme superoxide dismutase, the oxygen carrier

hemerythin (Fig. 16-20), transporter proteins that carry hydrophobic ligands,[115] and the immunoglobulins in which each domain contains a β barrel.

The eight-stranded β cylinder of plastocyanin (Fig. 2-16A) is somewhat flattened and can also be regarded as a β sandwich.[116,118] However, the β barrel of triose phosphate isomerase (see Fig. 2-28) is surrounded by eight α helices which provide additional stability and a high symmetry. Bacterial outer membranes contain pores created by very large β cylinders within proteins called **porin**s.[119,120] The one shown in Fig. 8-20 has 16 strands.

A single β strand can also be wound into a cylinder with the hydrogen bonds running parallel to the helix axis. A right-handed **parallel β helix** of this type has been found in the bacterial enzyme **pectate lyase**.[121,122] The polypeptide chains of the 353-residue protein contain seven complete turns of about 22

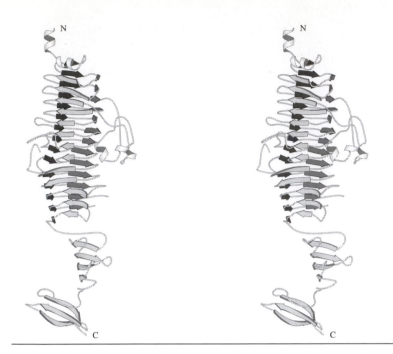

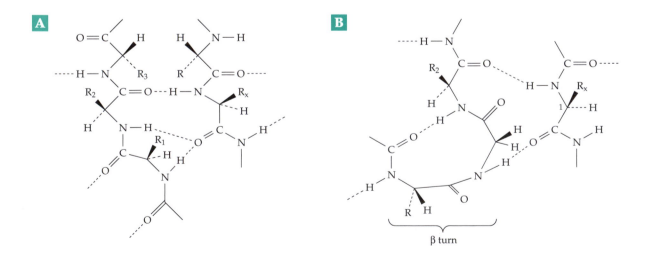

Figure 2-17 Wire model of the tailspike protein of bacteriophage P22 of *Salmonella*. Three of these fish-shaped molecules associate as a trimer to form the spike. From Steinbacher *et al.*[123]

residues each which form a cylindrical parallel β sheet. The cylinder is folded inward into roughly an L-shape with packed side chains filling the remaining inside space. The tailspikes of bacteriophage P22, a virus of *Salmonella,* are formed from 666-residue protein subunits which contain 13 turns of parallel β helix (Fig. 2-17).[123]

The tendency for a β sheet to fold into a cylinder is encouraged in antiparallel β structures by the existence of a common irregularity called the β **bulge**.[124,125] As illustrated in Fig. 2-18, a β bulge contains an extra residue inserted into one of the chains. In the second type of bulge shown in Fig. 2-18, the extra residue is glycine with torsion angles of about $\phi = 85°$, $\psi = 0°$, which are possible only for glycine. In the two β cylinders of trypsin, chymotrypsin, and elastase (Fig. 12-9), there are seven β bulges.

Beta structures are found in many small peptides. The hormone oxytocin (Fig. 2-4), the antibiotics gramicidin S (Fig. 2-4) and valinomycin (Fig. 8-22), and the mushroom peptide antamanide (Box 28-B) are among these. The cyclic structures of these compounds favor formation of antiparallel β strands with sharp turns at the ends. Polypeptide antibiotics that have alternating

Figure 2-18 Typical β bulges in antiparallel pleated sheets. The residues R_1, R_2, and R_x identify the bulges. (A) A "classic" β bulge, in which ϕ_1 and ψ_1 are nearly those of an α helix while other torsion angles are approximately those of regular β structures. (B) The G1 bulge in which the first residue is glycine with $\phi_1 = 85°$, $\psi_1 = 0°$. It is attached to a type II β turn of which the glycine (labeled 1) is the third residue.

sequences of D and L residues can be coiled into a β helix[126,127] or a pair of polypeptide chains arranged in an antiparallel fashion can form a double stranded β helix with an 0.3 to 0.4-nm hydrophilic pore through the center. These peptide antibiotics appear to exert their antibacterial action by creating pores through cell membranes and allowing ions to pass through without control.

Beta propellers. Another major folding pattern is a circular array of four to eight "blades" that form a propeller-like structure. Each blade is a small, roughly triangular four-stranded antiparallel β sheet (See Figs. 11-7 and 15-23). Sequences that fold into these blades can often, but not always, be recognized as **WD repeats**. These are typically 44- to 60-residue sequences that have the sequence GH (Gly-His) about 11–24 residues from the N terminus and WD (Trp-Asp) at the C terminus.[127a,127b] This repeat sequence encodes the

---

BOX 2-B SILKS

The green lacewing fly *Chrysopa flava* lays its eggs on 1 cm silk stalks glued to the undersides of leaves. It has been proposed, as for other silks, that the peptide chains in the stalks are aligned perpendicular to the long dimension of the fiber and are folded back on themselves many times to form a β sheet with only ~8 residues between folds.[a] The chains of **silk fibroin**, the major protein of silkworm silk, contain 50 repeats of the sequence:[b]

GAGAGSGAAG(SGAGAG)$_8$Y

All of the alanine and serine side chains of this sequence presumably protrude on one side of a β sheet while the other side has only the hydrogen atoms of the glycine. This permits an efficient stacking of the sheets with interdigitation of the side chains of the alanine and serine residues of adjacent sheets.

However, DNA sequences of silk genes have revealed a greater complexity. The silk fibroin sequence suggests that at bends between β strands there are often S – S bridges and the crystalline β sheet domains are interspersed with 100–200 residue segments of amorphous protein[b–d] whose coiled chains can be stretched greatly. The protein from *Bombyx mori* consists of 350-kDa heavy chains linked by a disulfide bond to 25-kDa light chains.[e] A silkworm **housing silk** contains polyalanine sequences, e.g. (Ala)$_{10-14}$. These evidently form α-helical regions.[f] Silk molecules are synthesized and stored in glands as a concentrated solution of apparently globular molecules. The silk is extruded through a spinneret, whose diameter is ~10 μm, after which the silkworm stretches the silk, causing it to form a stiff fiber. Apparently the stretching of the folded polypeptide chains permits the β-sheet-forming

Courtesy of Ralph Buchsbaum

sequences to find each other and form oriented crystalline regions.[d]

Spiders may produce silk from as many as seven different types of glands. Dragline silk, by which a spider itself may descend, is stronger than steel but because of the coiled amorphous regions may be stretched by 35%.[g−1] The most elastic silks are those of the catching spirals of orb-webs.[i,j,l] They can be stretched 200% and contain a variety of repeated sequences, including GPCC(X)$_n$. The latter is similar to sequences in elastin (Section 4), and is able to form type II β bends, which may have proline in position 2 and must have glycine in position 3 (Fig. 2-24). A chain of repeated β bend motifs can form a flexible spring, a β spiral as proposed for elastin. Genes for spider silk have been cloned and are being used to engineer new proteins with commercial uses, e.g., to help anchor cells in regenerating body tissues.[d,h,j]

^{a} Geddes, A. J., Parker, K. D., Atkins, E. D. T., and Beighton, E. (1968) *J. Mol. Biol.* **32**, 343–368

^{b} Garel, J.-P. (1982) *Trends Biochem. Sci.* **7**, 105–108

^{c} Vollrath, R. (1992) *Sci. Am.* **266**(Mar), 70–76

^{d} Calvert, P. (1998) *Nature (London)* **393**, 309–311

^{e} Mori, K., Tanaka, K., Kikuchi, Y., Waga, M., Waga, S., and Mizuno, S. (1995) *J. Mol. Biol.* **251**, 217–228

^{f} van Beek, J. D., Beaulieu, L., Schäfer, H., Demura, M., Asakura, T., and Meier, B. H. (2000) *Nature (London)* **405**, 1077–1079

^{g} Simmons, A. H., Michal, C. A., and Jelinski, L. W. (1996) *Science* **271**, 84–87

^{h} Hinman, M. B., and Lewis, R. V. (1992) *J. Biol. Chem.* **267**, 19320–19324

^{i} Hayashi, C. Y., and Lewis, R. V. (1998) *J. Mol. Biol.* **275**, 773–784

^{j} Guerette, P. A., Ginzinger, D. G., Weber, B. H. F., and Gosline, J. M. (1996) *Science* **272**, 112–115

^{k} Spek, E. J., Wu, H.-C., and Kallenbach, N. R. (1997) *J. Am. Chem. Soc.* **119**, 5053–5054

^{l} Hayashi, C. Y., and Lewis, R. V. (2000) *Science* **287**, 1477–1479

fourth β strand of one blade followed by the first three strands of the next blade. This overlap snaps the blades together to form the propeller. At least one tight turn is present in each blade. A seven-bladed propeller is present in the β subunits of the regulatory GTP-hydrolyzing **G proteins**, which couple extracellular signals to intracellular enzymes and ion channels (Fig. 11-7). A similar β propeller is present in **clathrin**, which forms cage-like enclosures around endocytic vesicles (Chapter 8).[127c] Six-bladed propellers are predicted to occur in many extracellular proteins.[127d] A β propeller binds the coenzyme PQQ in bacterial dehydrogenases (Fig. 15-23).

3. Helices

The **alpha helix** represents the second major structural element of soluble proteins[108,128] and is also found in many fibrous proteins, including those of muscle and hair. In the α helix both φ and ψ are about −50 to −60° and except at the N-terminal helix end, *each NH is hydrogen bonded to the fourth C=O further down the chain.* All of the N–H and C=O groups of the peptide linkages lie roughly parallel to the helix axis; the N–H groups point toward the N terminus of the chain and the carbonyl groups toward the C terminus (Fig. 2-19A).

The number of amino acid units per turn of the helix is ~3.6, with five turns of the helix containing 18 residues. The **pitch** (repeat distance) of the helix, which can be determined experimentally from X-ray diffraction data, is 0.54 nm. Polar coordinates for the α helix have been tabulated.[130] With L-amino acids, the right-handed helix, is more stable than the left-handed helix which has so far not been found in proteins. Frequently, however, a few residues have the φ, ψ angles of this helix. The φ, ψ angles of the α helix are given in Table 2-3 as −57°, −47°, but are much more variable in real helices. In erythrocruorin, for which an accurate structure determination has been made,[131]

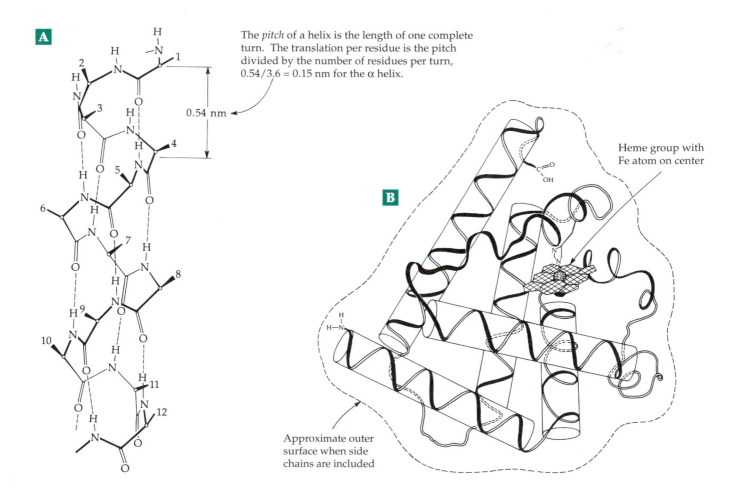

The *pitch* of a helix is the length of one complete turn. The translation per residue is the pitch divided by the number of residues per turn, 0.54/3.6 = 0.15 nm for the α helix.

0.54 nm

Heme group with Fe atom on center

Approximate outer surface when side chains are included

Figure 2-19 The α helix. (A) The right-handed α helix with vertical hydrogen bonds indicated by dotted lines. The positions of the amino acid side chains are indicated by the numbers. (B) The conformation of the peptide backbone of myoglobin.[129] Five long α helices are indicated as rods. Several shorter helices can also be seen. The overall size of the molecule is approximately 4.4 × 4.4 × 2.5 nm.

the average values are $\phi = -64°$, $\psi = -40°$. More important is the observation that $\phi + \psi = 104°$ within about $\pm 10°$ for most residues in helices of this and other proteins.[132,133] The deviation from the ideal dimensions results in part from the hydrogen bonding of water molecules or polar protein side chains to the carbonyl oxygen atoms in the helix.[131]

Helix formation is spontaneous for peptides as short as 13 residues in water.[134,135] Although the difference in thermodynamic stability between an α helix and an unfolded **"random coil"** conformation is small, poly-L-alanine peptides form helices in water. Glycine destabilizes helices, presumably because of the increased entropy of the unfolded chain which results from the wider range of the conformational angles ϕ and ψ for glycyl residues. Proline destabilizes helices even more because its restricted ϕ and ψ angles cause the helix to be kinked.[136,137] However, prolyl residues are often present at ends of helices. Other amino acids all fit into helices but may stabilize or destabilize the helix depending upon immediately neighboring groups.[138–140] Bulky side chain groups with a low dielectric constant stabilize helices by strengthening the hydrogen bonds within the helices.[141]

Helices of smaller and larger diameter than that of the α helix are possible. The most important is the **3_{10} helix** (or 3.0_{10} helix), which has exactly three residues per turn.[140,142–144] Each NH forms a hydrogen bond to the third C=O on down the chain; thus, the 3_{10} helix is tighter than the α helix. Although long 3_{10} helices are seldom found in proteins, a single turn of this tighter helix frequently occurs as a "defect" at the end of an α helix. A polymer of α-aminoisobutyric acid forms long 3_{10} helices because the α-dimethyl side chains constrain ϕ and ψ to appropriate values.[145] Short helical peptide

chains in water may exist as a mixture of α and 3_{10} forms in equilibrium.[146] The π **helix**, with 4.4 residues per turn is of a larger diameter than the α helix and has only been found in proteins as a single turn, usually at a C terminus.[132,132a]

Properties of helices. The dipoles of the backbone amide linkages of an α helix are all oriented in the same direction. The positive end of each dipole is associated through hydrogen bonding with the negative end of another. This leaves three partial positive charges at the N terminus of the helix and three partial

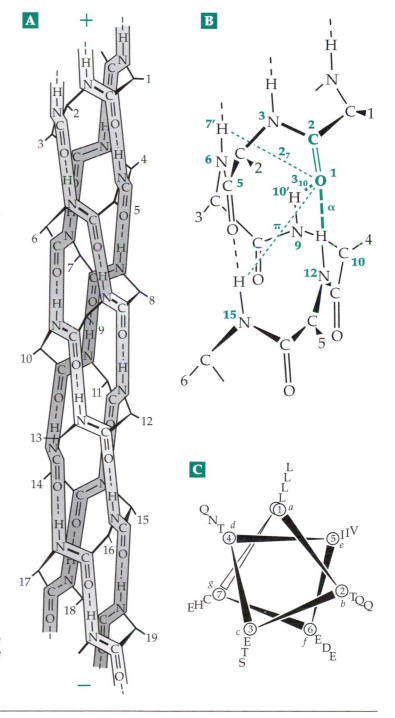

Figure 2-20 (A) The α helix showing the three chains of interconnected amide groups and hydrogen bonds with partial net positive and negative charges at the ends. These chains run across the turns of the helical polypeptide backbone. (B) Scheme illustrating hydrogen-bonding pattern for 2_7 ribbon, 3_{10} helix, α helix, and π helix. (C) The α helix represented as a helical wheel. Imagine viewing the helix from the N-terminal end of the segment with the lines corresponding to the backbone of the peptide. Residues, which are designated by single letters, are spaced 100° apart since there are 3.6 residues/turn. The peptide shown is a 22-residue sequence from a "leucine-zipper" domain of a protein that participates in gene regulation. From Fathallah-Shaykh *et al.*[145a]

negative charges at the C terminus and creates a "macro-dipole." It has been estimated that each end of the helix carries one-half an elementary unit of charge.[147–148b] However, this may be an overestimate. The partial charges are connected by three chains of hydrogen bonds that run across the turns of the helix as is indicated in Fig. 2-20A. These chains are polarized and also possess a large **hyperpolarizability**,[149] i.e., the polarization of the helix increases more rapidly than in direct proportion to an applied electrical field.

In a 3_{10} helix, there are just two chains of hydrogen bonds across the polypeptide chain and running the length of the helix. The characteristic hydrogen bonding of the α helix and that of the 3_{10} helix are often both present within a helix. Irregularities with 3_{10} type hydrogen bonds, arising from interactions with amino acid side chains or with solvent molecules, may cause a helix to be kinked or curved.[132] Side chains of polar residues, including those of Asn, Asp, Thr, and Ser (and less often Glu, Gln, or His), frequently fold back and hydrogen bond to the NH and CO groups that carry the partial charges at the helix ends (Fig. 2-20). The side chain of the third residue in the helix may also hydrogen bond to the NH of the first residue.[133,150–157] The hydrogen bonding of a negatively charged side chain group to the N-terminal end of the helix or of a positively charged group to the C-terminal end provides an **N-cap** or a **C-cap** which helps to stabilize the helix by strengthening its hydrogen bonds.[158] However, the most frequent residue at the C-terminal end is glycine.[151,155,159] Helices often point toward active sites of enzymes and interactions of the helix dipoles with substrates undergoing reaction may be important to the mechanism of action of these catalysts.[148,149,160]

Stacking of helices in proteins. Many proteins are made up almost entirely of α helices. One of these, **myoglobin**, was the first protein for which the complete three-dimensional structure was worked out by X-ray crystallography.[129] Myoglobin is a small oxygen-carrying protein of muscle. Its structure is closely related to that of hemoglobin of blood. Its 153 amino acid residues are arranged in eight different α-helical segments containing from 7 to 26 residues each. These rodlike helices are stacked together in an irregular fashion as shown in Fig. 2-19. Serum albumin (Box 2-A) has 28 helices organized into three homologous domains. In contrast, the filamentous bacterial viruses have protein coats made up of small subunits, each coiled as a single α helix. These are packed in a regular manner to form the rodlike virus coats (Fig. 7-7).

Coiled coils. In a large family of proteins, two right-handed α helices are coiled around each other in a left-handed **superhelix** (Fig. 2-21).[161–167] This **coiled coil** structure was first suggested by Francis Crick[166]

to account for the fact that the X-ray diffraction pattern of the **keratins** of skin and hair indicated a pitch of 0.51 nm rather than the 0.54 nm expected for an α helix.

A

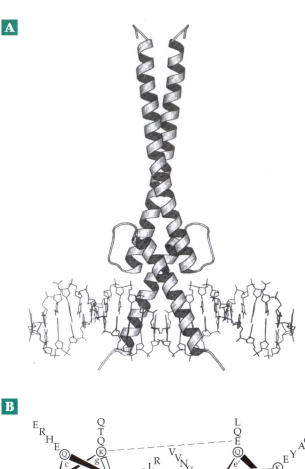

B

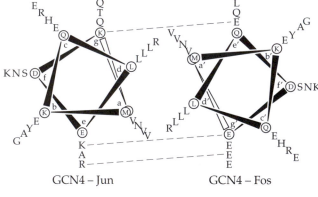

GCN4 – Jun   GCN4 – Fos

Figure 2-21 (A) Ribbon drawing of the transcription factor called Max in a complex with DNA. The C termini of the peptide chains are at the top.[169] Courtesy of S. K. Burley. (B) Helical wheel representation of residues 2–31 of the coiled coil portion of the leucine zipper (residues 249–281) of the related transcription factor GCN4 from yeast. The view is from the N terminus and the residues in the first two turns are circled. Heptad positions are labeled a–g. Leucine side chains at positions d interact with residues d' and e' of the second subunit which is parallel to the first. However, several residues were altered to give a coiled coil that mimics the structure of the well-known heterodimeric oncoproteins Fos and Jun (see Chapter 11). This dimer is stabilized by ion pairs which are connected by dashed lines. See John et al.[172]

Crick suggested that two supercoiled α helices inclined at an angle of 20° to each other would produce the apparent shortening in the helix pitch and would also permit nonpolar side chains from one strand to fit into gaps in the surface of the adjacent strand, a **knobs-into-holes** bonding arrangement. Helical strands tend to coil into ropes because a favorable interstrand contact can be repeated along the length of the strands only if the strands coil about each other. A coiled coil can often be recognized by **heptad repeats**:

$$(\mathbf{a}\text{-}b\text{-}c\text{-}\mathbf{d}\text{-}e\text{-}f\text{-}\mathbf{g})_n$$
$$1\ 2\ 3\ 4\ 5\ 6\ 7$$

Here, residues a, d, and g (1, 4, and 7) often carry non-polar side chains. These come together in the coiled coil as is illustrated in the **helical wheel** representations in Fig. 2-21B and provide a longitudinal hydrophobic strip along the helix.[167] Charged groups are often present in other locations and in such a way as to provide electrostatic stabilization through interactions between residues within a single α helix[170-171a] or between the pair of helices. The latter type of interaction also determines whether the coiled-coil consists of a parallel or antiparallel pair and whether the two helices are of identical or of differing amino acid sequence.[171b]

Recent attention has been focused on a DNA-binding structure called the **leucine zipper**. A pair of parallel α helices are joined as a coiled coil at one end but flare out at the other end to bind to DNA. In the yeast transcription factor GCN4, whose three-dimensional structure has been determined to high resolution (Fig. 2-21B),[172] the d position of the coiled coil is occupied by leucine and the e and g positions are often occupied by charged groups that form stabilizing ion pairs. Residues at positions b, c, and f are generally on the outside and exposed to solvent.[168,171,173] The coiled coil flares out at the C-terminal ends and carries DNA-binding groups. The structure of a related transcription factor is shown in Fig. 2-21A.[169]

The muscle proteins **myosin**[174] and **tropomyosin** also both consist of pairs of identical chains oriented in the same direction. The two 284-residue tropomyosin chains each contain 40 heptads and are linked by a single disulfide bridge. X-ray crystallographic studies[175,176] and electron microscopy[177] show that the molecule is a rod of 2.0 nm diameter and 41 nm length, the dimensions expected for the coiled coil. However, as with other "regular" protein structures, there are some irregularities. Myosin chains (Chapter 19) contain 156 heptads.

Another group of proteins have parallel coiled coils flanked by nonhelical domains in subunits that associate as filaments. These include the **keratins** of skin as well as the intermediate fiber proteins of the cytoskeleton (Chapter 7).[164,178] Natural coiled coils often have a parallel orientation, but synthetic peptides have been designed to form antiparallel coiled coils.[179,179a]

Helix bundles. A third peptide chain can be added to a coiled coil to form a triple-stranded bundle.[180–183] An example is the glycoprotein **laminin** found in basement membranes. It consists of three peptide chains which, for ~600 residues at their C-terminal ends, form a three-stranded coil with heptad repeats.[182,184] Numerous proteins are folded into four helical segments that associate as **four-helix bundles** (Fig. 2-22).[185–188] These include electron carriers, hormones, and structural proteins. The four-helix bundle not only is a simple packing arrangement, but also allows interactions between the + and − ends of the macro-dipoles of the helices.

Membranes contain many largely α-helical proteins. Cell surface receptors often appear to have one, two, or several membrane-spanning helices (see Chapter 8). The single peptide chain of the bacterial light-operated ion pump **bacteriorhodopsin** (Fig. 23-45) folds back upon itself to form seven helical rods just long enough to span the bacterial membrane in which it functions.[189] Photosynthetic reaction centers contain an α helix bundle which is formed from two different protein subunits (Fig. 23-31).[190] A recently discovered α,α **barrel** contains 12 helices. Six parallel helices form an inner barrel and 6 helices antiparallel to the first 6 form an outer layer (see Fig. 2-29).[191–193]

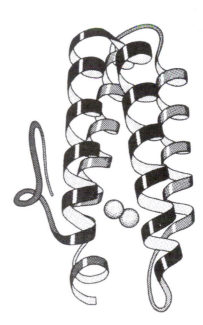

Figure 2-22 Ribbon drawing of an up-and-down four-helix bundle in **myohemerythrin**. The two spheres represent the two iron atoms which carry an O_2 molecule. They are coordinated by histidine and aspartate side chains. Courtesy of J. Richardson.[117]

4. Polyglycine II and Collagen

In a second form of polyglycine each amino acid residue is rotated 120° from the preceding one about a 3-fold screw axis as is shown in the end view of Fig. 2-23A. The angle ψ is about 150° while ϕ is about –80° for each residue. The distance along the axis is 0.31 nm/residue and the repeat distance is 0.93 nm. The molecules can coil into either a right-handed or a left-handed helix. In this structure, the N–H and C=O groups protrude perpendicular to the axis of the helix and, as in the β structure, can form H-bonds with adjacent chains.

Poly (L-proline) assumes a similar helical structure. However, the presence of the bulky side chain groups induces a *left-handed* helical twist. **Collagen**, the principal protein of connective tissue, basement membranes, and other structures, is the most abundant protein in the animal body. Its fundamental unit of structure is a triple helix of overall dimensions 1.4 x 300 nm which resembles the structure of polyglycine II but contains only three chains.[195] The left-handed helices of the individual chains are further wound into a right-handed superhelix (Fig. 2-23B,C). Collagen contains 33% glycine and 21% (proline + hydroxyprolines). The reason for the high glycine content is that bulky side chain groups cannot fit inside the triple helix. A hydrogen-bonded cylinder of hydration surrounds each triple helix and is anchored by hydrogen bonds to the –OH groups of the hydroxyproline residues.[196] Sequences of several collagen chains have been established. One of these contains ~1050 residues, ~1020 of which consist of ~340 triplets of sequence GlyXY where Y is usually proline and X is often hydroxyproline (Hyp).[197,198] The commonest triplet is Gly-Hyp-Pro.

There are several types of collagen. In one type, two identical chains of one kind are coiled together with a third dissimilar chain to form the triple helix. Several of these triple helices associate to form 8-nm microfibrils (Fig 2-23D).[199] Once synthesized, collagen is extensively modified and crosslinked. See Chapter 8.

Collagen-like triple helices also occur within other proteins. One of these is protein C1q, a component of the **complement system** of blood (Chapter 31). This protein interacts with antibodies to trigger a major aspect of the immune response. C1q has six subunits, each made up of three different polypeptide chains of about 200 residues apiece. Beginning a few residues from the N termini, there are over 80 residues in each chain with collagen-like sequences. The three chains apparently form a triple helix within each subunit. However, the C-terminal portions are globular in nature.[200] Collagen-like tails also are present on some forms of the enzyme acetylcholinesterase (see Chapter 12C,10). The **extensins** of plant cell walls contain 4-hydroxyproline and evidently have a structure

related to that of collagen.[201,202] Shorter 4- to 8-residue segments of left-handed polyproline helix are found in many proteins.[203]

5. Turns and Bends

To form a globular protein, a polypeptide chain must repeatedly fold back on itself. The turns or bends by which this is accomplished can be regarded as a third major secondary structural element in proteins. Turns often have precise structures, a few of which are illustrated in Fig. 2-24. As components of the loops of polypeptide chains in active sites, turns have a special importance for the functioning of enzymes and other proteins. In addition, tight turns are often sites for modification of proteins after their initial synthesis (Section F).

The **β turn** (Fig. 2-24) is often found in hairpin or reverse turns at the edges of β sheets (Fig. 2-11) and at other locations.[204–212] If all four residues that contribute to β bends are counted, they constitute about one-third of the amino acid residues in most proteins.[124] In many β turns, the C=O of the first residue hydrogen bonds to the NH of the fourth residue. This hydrogen bond may be part of the hydrogen bond network of a β pleated sheet. The peptide unit between α-carbon atoms 2 and 3 of the turn is perpendicular to the sheet. There are two possibilities for the orientation of this peptide unit. In a *type I turn*, the C=O is down when the turn is viewed as in Fig. 2-24, while the side chains of residues 2 and 3 point upwards or outward on the opposite side of the bend. In a *type II turn*, the C=O is up and the NH down. Residue 3 is always glycine in a type II turn because the side chain would collide with the C=O group if any other amino acid were present. As is seen in Fig. 2-24, a *trans*-proline can fit at position 2 in a type II turn[206] as well as in type I turns. A *cis*-proline residue can fit at position 2 or position 3 in a type I β turn.[137,213]

The *type III β turn* is similar to a type I turn but has the ϕ, ψ angles of a 3_{10} helix and the two chains emerging from the turn are not as nearly parallel as they are in type I turns. Beta turns of the less common types I', II', and III' have a left-handed twist. As can be seen in Fig. 2-24, this permits a better match to the twist of strands in a β sheet. Unless glycine is present, these bends are less stable because of steric hindrance.[214,215] Polar side chain groups such as those of aspartate or asparagine often form hydrogen bonds to the central peptide units of β turns.[214]

The tight γ turn[215] and the proline-containing β turn shown in Fig. 2-24 are thought to be major components of the secondary structure of **elastin**.[216–218] This stretchable polymer, which consists largely of nonpolar amino acids, is the most abundant protein of the elastic fibers of skin, lungs, and arteries. The

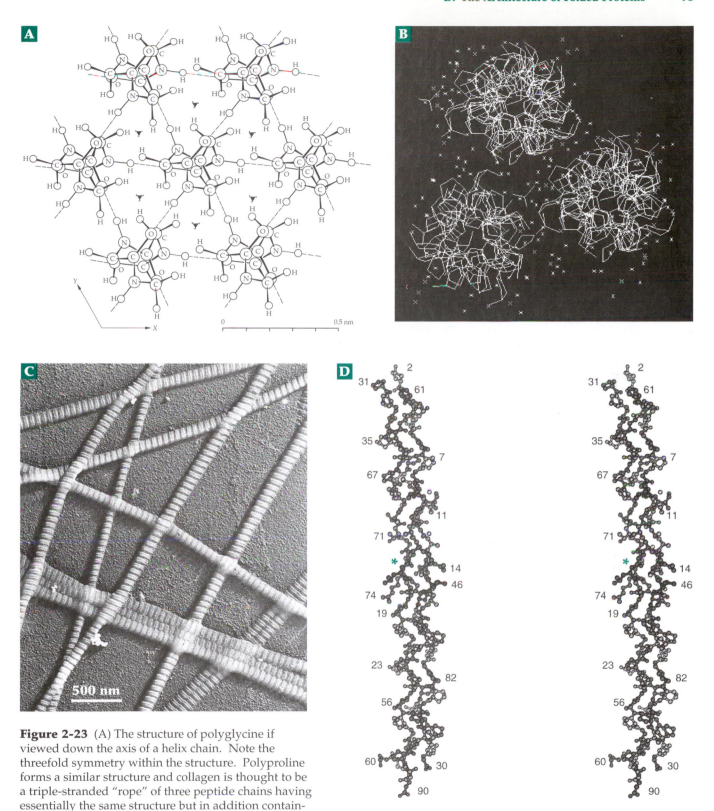

Figure 2-23 (A) The structure of polyglycine if viewed down the axis of a helix chain. Note the threefold symmetry within the structure. Polyproline forms a similar structure and collagen is thought to be a triple-stranded "rope" of three peptide chains having essentially the same structure but in addition containing a right-handed supercoil. This is illustrated by the crystal structure of a collagen-like peptide shown in B and C. (B) View similar to that in (A), but showing how three chains form the triple-stranded ropes separated by a cylinder of hydration. The structure is shown as a wire model. The x's are water molecules. They form an extensive network of H-bonds to one another and to peptide groups. From Bella et al.[194] C) Electron micrograph of collagen fibrils, each of which consists of many triple-helical units. Deposited from suspension and shadow cast with chromium. Courtesy of Jerome Gross, M.D. (D) Stereoscopic view of a collagen-like model peptide. Each of the three parallel 30-residue chains contains a (Pro-Hyp-Gly)$_3$ "cap" at each end and the 12-residue sequence Ile-Thr-Gly-Ala-Arg-Gly-Leu-Ala-Gly-Pro-Hyp-Gly in the center. This sequence is found in human type III collagen and includes a site (green asterisk) of known mutations (see Chapter 8). From Kramer et al.[194a] Courtesy of Helen Berman.

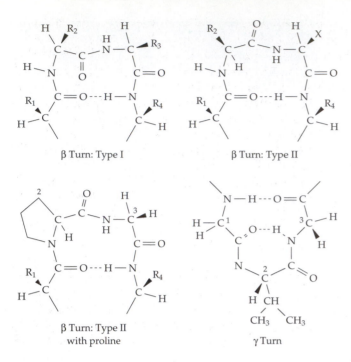

β Turn: Type I

β Turn: Type II

β Turn: Type II
with proline

γ Turn

Figure 2-24 Tight turns found in polypeptide chains. Two types of β turn are shown. A third variant, the type III or 3_{10} turn resembles the type I turn but has the φ, ψ angles of a 3_{10} helix. Type II β turns containing proline and tighter γ turns are thought to be major structural components of elastin. Another β turn, lacking the hydrogen bond has a *cis*-proline residue at position 3.

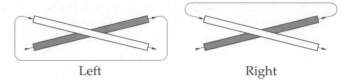

Left   Right

Figure 2-25 Right- and left-handed crossover connections in proteins. These connections are nearly always right-handed. The broad arrows represent β strands. The crossover often contains a helix. Units of two adjacent β strands (βαβ units) with an α helix between are found frequently in globular proteins.

6. Domains, Subunits, and Interfaces

Many proteins are organized into tightly folded globular "domains" consisting of 50–150 amino acid residues.[117,222–227] Smaller proteins may have 2 or 3 domains but large proteins may have more. For example, the immunoglobulin "heavy chains" (Chapter 31) have four or five domains similar to that shown in Fig. 2-16B. The enormous 3000-kDa muscle protein **titin** contains 260 domains, many of which are of the immunoglobulin typeFig. 2-16,C.[228,117a] In most proteins the polypeptide chain folds to form one domain, then passes through a "hinge" to the next. In others, the C terminus or the N terminus of the polypeptide folds back across two or more domains as a kind of "strap" that helps to hold the protein together. Even when a protein contains only one domain, it often consists of two distinct lobes with a cleft between them. Many proteins, e.g. hemoglobin (Fig. 7-25), consist of subunits about the size of the globular domains in larger proteins.

Structural domains of proteins are sometimes encoded by a single coding segment of DNA i.e., by a single exon in a split gene. Domains of this type may have served as **evolutionarily mobile modules** that have spread to new proteins and multiplied during evolution. For example, the immunoglobulin structural domain is found not only in antibodies but also in a variety of cell surface proteins.[229–232]

Whether we deal with domains connected by a flexible hinge or with subunits, there are **interfaces** between the different parts of the protein. These interfaces are often formed largely of nonpolar groups. However, they frequently contain a small number of hydrogen bonds that bridge between one domain and another or between one subunit and another. In the case of hemoglobin, important changes occur in this hydrogen bonding and movement occurs along one of the interfaces between two subunits. Likewise the active sites of enzymes are often located at interfaces between domains. During catalysis, movement and reorganization of the hydrogen bonds and side chain packing in the interfaces may take place.[232a]

70-kDa chains are thought to have 70- to 75-residue regions in which the polypeptide folds back on itself repeatedly with a large number of bends in a broad left-handed **β spiral**.[216,218a] The consensus sequence VPGVP, which tends to form a type II β turn with proline in position 2 (Fig. 2-24) is present in long tandem repeats e.g., (VPGVG)$_{11}$. These extensible regions alternate with short α helices which are crosslinked to other chains. Similar structures are present in silks (Box 2-B) and in proteins of wheat gluten.[217a,218b]

Besides hairpin turns and broader U-turns, a protein chain may turn out and fold back to reenter a β sheet from the opposite side. Such **crossover connections**, which are necessarily quite long, often contain helices. Like turns, crossover connections have a handedness and are nearly always right-handed (Fig. 2-25).[117,219] Most proteins also contain poorly organized loops on their surfaces. Despite their random appearance, these loops may be critical for the functioning of a protein.[220] In spite of the complexity of the folding patterns, peptide chains are rarely found to be knotted.[221]

7. Packing of Side Chains

In Fig. 2-19B myoglobin is pictured as a cluster of α-helical rods surrounding the heme core. This picture is incomplete because the space between the rods and inside the molecule is tightly packed with amino acid side chains almost all of which are hydrophobic. The same is true for the β barrels of Fig. 2-16, which are filled largely with nonpolar side chains. As the structures of more and more proteins have been determined, a consistent pattern has emerged. Within the interior of proteins the side chain groups are packed together remarkably well.[99,100,222,233–236] Although occasional holes are present, they are often filled by water molecules.[99,237a] The **packing density**, the volume enclosed by the van der Waals envelope divided by the total volume, is ~0.75 for the interior of the lysozyme and ribonuclease molecules compared with the theoretical value of 0.74 for close-packed spheres. However, regions with many hydrogen bonds may be less tightly packed.[238]

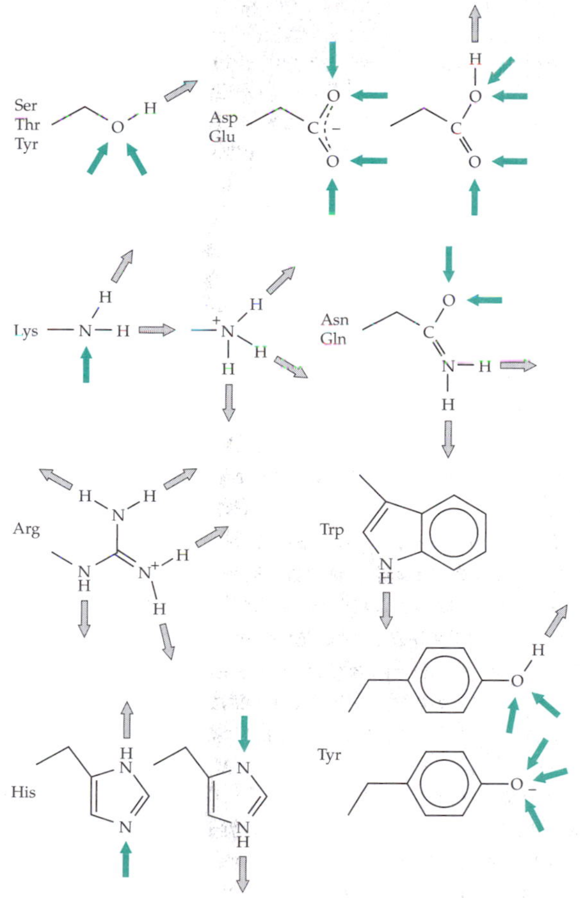

The interiors of proteins often contain large numbers of aromatic side chains which are frequently associated as pairs or as **aromatic clusters**.[239,240] For example, see the structure of plastocyanin in Fig. 2-16A. Rings may lie perpendicular one to another or be "stacked" face to face but offset. Oxygen atoms often lie in contact with the edges of aromatic rings.[241] The planar guanidinium groups of arginine side chains often stack against aromatic rings and amide groups may sometimes do the same.[242] It has been suggested that both aromatic: aromatic and aromatic:oxygen interactions may be associated with additional stabilization of the protein by ~4- to 8- kJ/mol. Tyrosine side chain –OH groups often stabilize ends of β strands by forming H-bonds to backbone atoms.[243]

Most polar groups are on the surfaces of proteins, and those that are not are almost always hydrogen bonded to other groups in the interior.[244] While most nonpolar groups are inside proteins, they are also present in the outer surfaces where they are often clustered into **hydrophobic regions** or "**patches**." The latter may be sites of interaction with other proteins or with lipid portions of membranes.

8. The Network of Internal Hydrogen Bonds

The fact that nonpolar groups tend to be buried in the interior of proteins suggests that the inside of a protein might be a flexible blob of oily material. In fact, the nonpolar groups tend to be densely packed and aromatic rings often impart considerable rigidity to the hydrophobic cores. Buried polar groups, which are invariably hydrogen bonded to other side chain groups, to amide groups of the polypeptide backbone,[244–247] or to buried water molecules, form a well-defined internal network. When a series of closely related proteins, for example, those having the same function in several different species, are compared, the hydrogen bonded network is often nearly invariant. This suggests a functional significance. The hydrogen bonding possibilities for some of the side chain groups in proteins are indicated in Fig. 2-26.

Charged side chains are sometimes buried in the interior of proteins, usually together with an ion of the opposite charge to form a **hydrogen-bonded ion pair**.[248,248a,248b] However, there is sometimes a single charge that is "neutralized" by the interaction with dipoles of polar groups.[217] Sometimes an undissociated carboxyl group is hydrogen bonded to a carboxylate ion.[249]

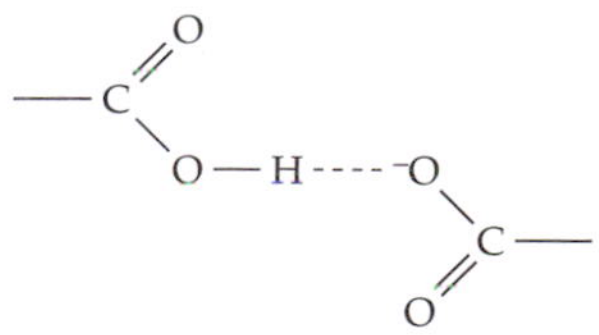

Figure 2-26 Some of the possibilities for hydrogen bonding of side chain groups in proteins. Oxygen atoms can and frequently do form up to three hydrogen bonds at once. Open arrows point *from* H-atoms and *toward* electron donor pairs.

A buried carboxyl group of this type will display a pK_a value far higher than the normal value.

Arginine side chains are large and able to form multiple hydrogen bonds[247,250] as well as salt linkages[251] to different parts of a folded peptide. Cations of arginine and lysine are never buried unless in ion pairs. Protons from strong acids have long been known to bind to centers of benzene rings and water can form weak hydrogen bonds to the centers of aromatic rings.[252,253,253a] Such bonds also occur within proteins and often involve the binding of guanidinium groups or of inorganic cations to indole rings of tryptophans.[253b,253c] Protonated imidazole groups may also bind to aromatic rings.[253d]

$$H_2O - benzene \quad hydrogen\ bond$$

Another important aspect of the structure of proteins is the presence of **hydrogen-bonded water molecules** in pockets and cracks. These molecules, as well as a much larger number of water molecules bound at the outer surface, are clearly visible from X-ray studies. They often occur singly, bonded to the ends of amide groups, especially to the carbonyl ends. Internal bends of the peptide chain are almost always hydrated.[254] These water molecules often make two or more hydrogen bonds to different parts of the protein or to other water molecules. Clusters of water molecules,[49,255,255a] sometimes in the form of pentagonal rings,[256] are often present. NMR spectroscopy has shown that water molecules bound to protein *surfaces* exchange rapidly with the bulk water in which the protein is dissolved; the "residence time" on the protein is typically less than a nanosecond. *Interior* water molecules have much longer residence times of 10^{-2} to 10^{-8} s for a small protein.[257] They may be regarded as part of the protein structure.

E. Folding Patterns and Protein Families

Proteins are folded in many ways. We have already considered several simple patterns: the antiparallel β cylinder (Fig. 2-16), the 2-helix coiled coil (Fig. 2-21) and the 3- and 4-helix bundles (Fig. 2-22). Another simple motif that has been found repeatedly is the **helix–turn –helix** or **helix–loop–helix** in which two helices at variable angles, one to another and with a turn or short loop between them, form a structural unit. DNA-binding **repressors** and **transcription factors** (see Fig. 2-21 and also Chapter 5) often contain this motif as do many Ca^{2+}-binding proteins. Proteins containing 3–6 helical segments, often fold into a roughly polyhedral shape.[258,259] An example is myoglobin (Fig. 2-19B).

1. Complex Folding Patterns

Proteins often contain elements of both α and β structure. One of the first of the complex folding motifs to be recognized was a nucleotide-binding domain identified by Rossmann and associates.[260–262] This Rossman fold contains six parallel β strands which alternate with six helices. The result of the parallel β structure is that the helices are also parallel and that their amino-terminal ends, which carry partial positive charges, are aligned in approximately the same directions. The positive end of one of the helices lies behind the negatively charged phosphate groups[148] which characteristically bind at the edge of the sheet containing the C termini of the β strands. This can be seen in Fig. 2-13, which shows binding of the coenzyme NAD to the nucleotide binding domain of **glyceraldehyde phosphate dehydrogenase**. Similar nucleotide-binding domains are found in many other dehydrogenases whose members constitute a **protein family**.

Figure 2-27 depicts **topology diagrams** for the Rossman fold and for two related families of proteins. These families bind the nucleotides called GTP and ATP, respectively. Both are structural relatives of NAD. A major part of the structure of all of the proteins in these families consist of β–α units, each one containing a β strand followed by a helix. They are

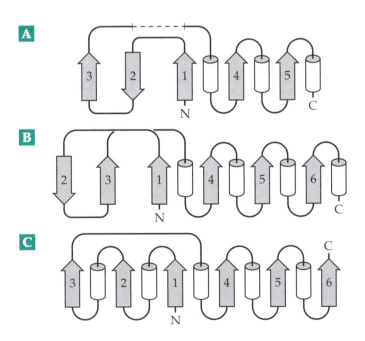

Figure 2-27 Topologies of the folds of three families of nucleotide binding α/β proteins. Cylinders represent α helices and arrows β strands. (A) The ATPase fold for the clathrin-uncoating ATPase; (B) The G-protein fold that binds GTP and is found in *ras* proteins; (C) The Rossmann fold that binds NAD in several dehydrogenases. From Brändén.[262]

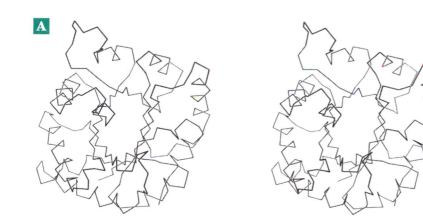
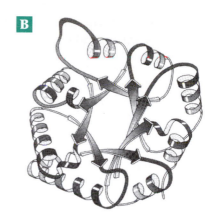

Figure 2-28 The eight-fold α/β barrel structure of triose phosphate isomerase. From Richardson. (A) Stereoscopic view. (B) Ribbon drawing. Courtesy of Jane Richardson.[117]

classified as α/β proteins. The Rossman fold is composed of six β–α units. Recently a ribonuclease inhibitor protein with 15 consecutive β–α units has been characterized.[263] Each β–α unit contains several residues of leucine. This **leucine-rich repeat** occurs in many other proteins as well.[264,264a,264b]

The **α/β barrel** shown in Fig. 2-28 consists of 8 consecutive β–α units in a symmetric array.[265,266] By 1995 over 40 of these barrels had been identified in a diverse group of enzymes. One bifunctional enzyme contains two α/β barrels. Although the nature of the reaction catalyzed varies, the active site is always found in the center of the barrel at the C-terminal ends of the 8 parallel β strands and therefore between the N termini of the surrounding helices. The enzyme sequences show no homology and frequent occurrence of the 8-stranded barrel may reflect the fact that it is a natural packing arrangement of β–α units. However, a 10-stranded barrel of this type has also been found.[267]

This barrel can be compared with that of the 12-helix α,α barrel of a fungal glucoamylase whose structure is shown in Fig. 2-29. Numerous more complex folding patterns have been discovered. They have been classified by Jane Richardson.[117,122] Many of the proteins described by these folding patterns can be grouped into families and "superfamilies."[227] Chothia suggested that there may be about 1,000 families in nature;[268] over 700, with over 360 distinct folds have been identified.[268a]

2. Symmetry

A sometimes puzzling feature of protein structure is the widespread occurrence of an approximate two-fold axis of symmetry. This often arises as a natural result of association of a pair of irregular subunits (Chapter 7). The association is such that rotation

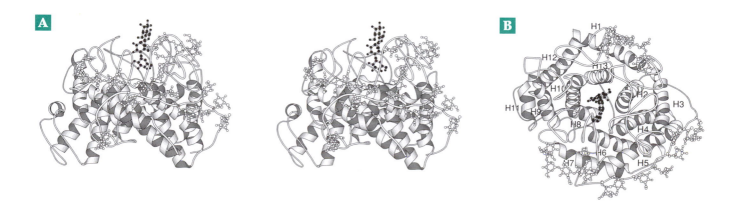

Figure 2-29 Structure of the α,α barrel of a fungal enzyme glucoamylase. (A) side view (stereoscopic); (B) top view. The active site, which cleaves glucose units from the ends of starch chains, is in the depression in the center of the barrel. Here it is occupied by an inhibitor. See Aleshin et al.[192] Courtesy of Alexander Aleshin.

about the twofold axis will cause the two subunits to exchange positions and to remain in an identical chemical environment. Approximate symmetry is often observed also *within* single peptide chains. For example, in the Rossman fold (Figs. 2-13, 2-27), an approximate twofold axis passes between the center strands of the β sheet residues and relates the two flanking helices, which begin with residues R10 and T100, respectively. The bound NAD^+ also possesses an approximate twofold axis, but it is not quite symmetrically placed at the end of the β sheet. Both phospho groups are seen to interact with the N terminus of the helix beginning at residue 10. The small bacterial protein ferredoxin (Fig. 16-16B) contains two iron-sulfur clusters related by an approximate 2-fold axis. The two β cylinders of elastase (Fig. 12-9) as well as the two sides of the flattened β barrel of copper–zinc superoxide dismutase[269] are approximately related by twofold axes. The enzyme thiosulfate: cyanide sulfurtransferase (Eq. 24-46) is remarkably symmetric but the active site is located in just one half. The widespread existence of this approximate symmetry suggests a biological significance that remains to be discovered.

3. Effects of Sequence on Folding

Studies of synthetic polypeptides as well as examination of known protein structures reveal that some amino acids, e.g., Glu, Ala, Leu, tend to promote α helix formation. Others, such as Tyr, Val, and Ile, are more often present in β structure, while Gly, Pro, and Asn are likely to be found in bends.[270,270a,270b] The frequencies with which particular amino acids appear in helices, β structure, or turns were first compiled by

TABLE 2-4
Classification of Protein Residues According to Their Tendencies to Form α Helix, β Structure, and β Turns[a]

Amino acid	P_α	Helix-forming tendency	P_β	β structure-forming tendency	P_t
Glu⁻	1.51	++	0.37	br+	0.44
Met	1.45	++	1.05	+	0.67
Ala	1.42	++	0.83	i	0.57
Leu	1.21	++	1.30	+	0.53
Lys+	1.16	+	0.74	br	1.01
Phe	1.13	+	1.38	+	0.71
Gln	1.11	+	1.10	+	0.56
Trp	1.08	+	1.37	+	1.11
Ile	1.08	+	1.60	++	0.58
Val	1.06	+	1.70	++	0.30
Asp⁻	1.01	w	0.54	br+	1.26
His⁺	1.00	w	0.87	i	0.69
Arg⁺	0.98	i	0.93	i	1.00
Thr	0.83	i	1.19	+	1.00
Ser	0.77	i	0.75	br	1.56
Cys	0.70	i	1.19	+	1.17
Tyr	0.69	br	1.47	++	1.25
Asn	0.67	br	0.89	i	1.68
Pro	0.57	br+	0.55	br+	1.54
Gly	0.57	br+	0.75	br	1.68

++ = strong former	i = indifferent
+ = former	br = breaker
w = weak former	br+ = strong breaker

[a] The conformational parameters P_α, P_β, and P_t (β turn) are the frequencies of finding a particular amino acid in an α helix, β structure, or β turn (in 29 proteins of known structure) divided by the average frequency of residues in those regions. Residues are arranged in order of decreasing tendency toward helix formation. From Chou, P. V. and Fasman, G. D (1974) *Biochemistry* **13**, 222–245.

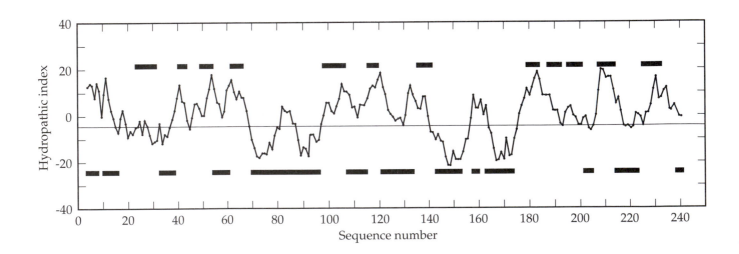

Figure 2-30 Plot of hydropathy index versus sequence number for bovine chymotrypsinogen. The indices for individual residues have been averaged nine at a time. The solid bars at the top of the plot mark interior regions as determined by crystallography. The solid bars below the plot indicate regions that are on the outside of the molecule. From Kyte and Doolittle.[280]

Chou and Fasman[270] (Table 2-4). These frequencies also differ significantly between parallel and antiparallel β sheets. Some of the preferences are readily understandable. For example, glycine is too flexible to stabilize a helix, but it can occur in helices between residues that are better helix formers. Because the peptide linkage to its nitrogen lacks an NH group, proline can fit into only one of three positions at the N terminus of a helix. For the strong helix formers, the β-CH_2 groups fit into the helix well and provide stabilization through van der Waals interactions. However, the β-methyl groups of valine and isoleucine cause crowding. For these residues, a β structure is more favorable than a helix.[271] Side chains of Asp, Asn, Ser, Thr, and Cys can hydrogen bond to backbone amide groups and can either stabilize or destabilize a helix or β sheet depending upon their location.

It appears that the folding pattern of a peptide is encoded in the sequence itself.[272-274] Thus, when several residues that favor helix formation are clustered together, a helix may form. Chou and Fasman suggested that when four helix formers out of six residues are clustered, nucleation of a helix takes place. The helix can then be elongated in both directions until terminated by a proline or other "helix breaker"; additional folding can then occur. If three out of five β formers are clustered, a β strand may form. If random folding brings two or more of these strands together they may associate to form the nucleus for a β sheet. Some success has been achieved using this approach in predicting whether a given residue will be found in a helix, a β strand or a loop.[275] However, prediction of complete folding patterns is much more difficult. Many new approaches are being explored.[275-279d] The problem is an important one. Although three-dimensional structures have been determined for thousands of proteins, sequences are known for hundreds of thousands. The number is growing rapidly. Being able to predict correctly a three-dimensional protein structure will be a major scientific accomplishment with many practical consequences.

A way of examining the entire sequence for polar or nonpolar character of the side chains was introduced by Kyte and Doolittle.[280,281] A **hydropathy index** based on the polarity of the side chains of a given residue and of its neighbors in the sequence is plotted against residue number (Fig. 2-30). Helices are often found to have an **amphipathic** character, hydrophobic on one side and hydrophilic on the other. Such helices can be characterized by a plot such as that in Fig. 2-30. The hydrophobic side of an amphipathic helix can pack against a hydrophobic core of a protein, can lie against a membrane, or can be aligned with other helices to form coiled coils (Fig. 2-21) or to give a hydrophilic channel with an outer hydrophobic surface that can fit into a cell membrane.

F. Chemical Modification and Crosslinking

Some newly synthesized proteins, upon folding, are ready to "go to work" immediately but others must be modified. Pieces are frequently cut from the ends or out of the center of a folded peptide chain. Sometimes the amino and carboxyl termini are converted to nonionic groups, e.g.,

$$—NH_3^+ \longrightarrow —NH-\overset{\overset{\displaystyle O}{\|}}{C}—R \quad R=CH_3 \text{ or long alkyl} \qquad (2\text{-}12)$$

$$—\overset{\overset{\displaystyle O}{\|}}{C}—\underset{H}{N}—CH_2–COO^- \longrightarrow \longrightarrow —\overset{\overset{\displaystyle O}{\|}}{C}—NH_2 \qquad (2\text{-}13)$$

Side chains may be modified in a very large number of different ways.[282-284] These include acetylation and other kinds of acylation (Eq. 2-12),[285-287] **methylation** (Eqs. 2-14, 2-15), **phosphorylation** (Eq. 2-16), **phosphoadenylation**,[288] formation of **sulfate esters** (Eq. 2-17),[289,290] and hydrolysis (Eqs. 2-18, 2-19).[291] In at least a few proteins some L-amino acid residues are converted to D-.[292-294]

$$—COO^- \longrightarrow —C\overset{\displaystyle O}{\underset{\displaystyle OCH_3}{\diagup}} \qquad (2\text{-}14)$$

$$\text{~~~~}NH_3^+ \longrightarrow \text{~~~~}{}^+N\overset{\displaystyle CH_3}{\underset{\displaystyle CH_3}{-CH_3}} \qquad (2\text{-}15)$$

$$—CH_2OH \longrightarrow —CH_2—O—\overset{\overset{\displaystyle O}{\|}}{\underset{\underset{\displaystyle O^-}{|}}{P}}—O^- \qquad (2\text{-}16)$$

$$—CH_2\!\!-\!\!\bigcirc\!\!-\!\!OH \longrightarrow —CH_2\!\!-\!\!\bigcirc\!\!-\!\!O—SO_3^- \qquad (2\text{-}17)$$

$$\underset{\substack{\text{Asn, Gln}\\\text{side chain}}}{—\overset{\overset{\displaystyle O}{\|}}{C}\underset{NH_2}{\diagdown}} + H_2O \longrightarrow —\overset{\overset{\displaystyle O}{\|}}{C}\underset{O^-}{\diagdown} + NH_4^+ \qquad (2\text{-}18)$$

$$\underset{H}{—N}—\overset{\overset{\displaystyle NH^+}{\|}}{C}\underset{NH_2}{\diagdown} + H_2O \longrightarrow \underset{H}{—N}—\overset{\overset{\displaystyle O}{\|}}{C}\underset{NH_2}{\diagdown} + NH_4^+$$

$$\text{Arginine} \qquad\qquad\qquad \text{Citrulline} \qquad (2\text{-}19)$$

Usually, only one or a small number of side chains of a protein is modified. However, glycoproteins may contain many different attached sugar molecules. An example is the glucoamylase shown in Fig. 2-29. In several proteins involved in blood clotting (Fig. 12-17), as many as 10 glutamic acid side chains are carboxylated (Eq. 2-20).

$$— CH_2 — CH_2 — COO^- \longrightarrow — CH_2 — C\overset{H}{\underset{COO^-}{\diagdown}}COO^- \qquad (2\text{-}20)$$

The 216-residue hen egg yolk storage protein **phosvitin** contains 123 serine residues, most of which have been phosphorylated (Eq. 2-16).[295] A basic protein of the myelin sheath of neurons contains as many as 6 specific residues of **citrulline** (Eq. 2-19).[296] An adhesive protein from the foot of a marine mollusk contains ~80 repeated sequences containing hydroxy-proline 2,3-dihydroxyproline and **3,4-dihydroxyphenylalanine** (Dopa).[297,298]

Some modification reactions alter the electrical charge on a side chain and, as a consequence, can affect the ways in which the protein interacts with other molecules. Negative charges added in formation of clusters of γ-carboxyglutamates (Eq. 2-20) create strong **calcium ion-binding centers** in the modified proteins. Acylation of N termini or of Ser, Thr, or Cys side chains by long-chain fatty acids provides hydrophobic "tails" able to anchor proteins to membrane surfaces.[285,286] Addition of polyprenyl groups to cysteine side chains near the C-termini has a similar effect.[299,300] These and more complex membrane anchors are considered in Chapter 8. While modified amino acids can be found at many places in a protein, they are often located at turns. For example, serine or threonine residues in turns are often phosphorylated or glycosylated. Modification of proteins is dealt with further in Chapter 10 and at other points in the book. Glycosylation of proteins is considered in Chapter 20.

In addition to deliberate enzyme-catalyzed processes, there are nonenzymatic processes that alter proteins. These include the degradative reactions described in Section 5 and also *reversible* reactions that may be physiologically important. For example, the N-terminal amino groups of peptides, and other amino groups of low pK_a can form **carbamates** with bicarbonate (Eq. 2-21).[301–303] This provides an important mechanism of carbon dioxide transport in red blood cells (Chapter 7) and a way by which CO_2 pressure can control some metabolic processes.

$$— NH_3^+ + CO_2 \rightleftharpoons — NH — C\overset{O}{\underset{OH}{\diagdown}} + H^+ \qquad (2\text{-}21)$$

1. Disulfide Crosslinkages

The linking together of two different parts of a peptide chain or of two different peptide chains is extremely important to living beings.[304,305] One of the most widespread of crosslinkages is the **disulfide bridge**. It forms spontaneously when two –SH groups of cysteine side chains are close together and are oxidized by O_2 or some other reagent (Eq. 2-22).

$$2 — CH_2 — SH \longrightarrow — H_2C\overset{CH_2 —}{\underset{S — S}{\diagdown\diagup}} \qquad (2\text{-}22)$$

The disulfide group is inherently chiral, with the preferred torsion angle about the S–S bond being ~90°. Both right- and left-handed disulfide bridges occur in proteins.[305,306]

Two S atoms viewed down S—S axis

Disulfide linkages are frequently present in proteins that are secreted from cells but are less common in enzymes that stay within cells. Perhaps because the latter are in a protected environment, the additional stabilization provided by disulfide bridges is not needed. Disulfide bridges are not only found within single-subunit proteins but they also link different peptide chains. For example, the four chains of each immunoglobulin molecule (Chapter 31) are joined by disulfide bridges and each domain within the chains is stabilized by a disulfide bridge. One of the most highly crosslinked proteins known occurs in the keratin matrix of hair. Breakage of the –S–S– linkages of this protein is an essential step in the chemical "permanent wave" process. A thiol compound is used to reductively cleave the crosslinks and after the hair is reset new crosslinks are formed by air oxidation.

2. Other Crosslinkages

Another common crosslink is an amide formed between the γ-carboxyl group of a glutamic acid side chain and an amino group from a lysine residue.[307] This **isopeptide** linkage is formed from a residue of gluta-mine through the action of the enzyme **transglutaminase** (Eq. 2-23). Isopeptide crosslinks are found in hair, skin, connective tissue, and blood clots.

$$(2\text{-}23)$$

Occasionally an isopeptide linkage joins amino groups of lysine side chains to the C-terminal carboxyl groups of other peptide chains to give branched chains, e.g. see **ubiquitin** (Box 10-C). Elaborate crosslinks derived from lysine are found in collagen and elastin (Chapter 8). Dityrosine linkages formed by oxidative joining of the aromatic rings of tyrosine are found in insect cuticle and in the plant cell wall extensins[202] (Chapters 20, 25).

G. Dynamic Properties of Proteins

Sometimes using energy from the cleavage of chemical bonds, sometimes depending only upon energy provided by the random bombardment by solvent molecules, proteins perform their specific functions with amazing speed. A question that has long intrigued biochemists is "to what extent do proteins stretch or flex, unfold and refold, or undergo other conformational movements during their action?" To answer this and related questions, many techniques are being applied to study the dynamic properties of proteins.[308,309]

1. Motion of Backbone and Side Chains

Even in the *crystalline state* there is evidence of movement. In the images constructed from X-ray or neutron diffraction experiments side chains on the surfaces of protein molecules are often not clearly visible because of rapid rotational movement. Some segments of the polypeptide chain may be missing from the image. However, side chain groups within the core of a domain are usually seen clearly. They probably move only in discrete steps. However, they may sometimes shift rapidly between different conformations, all of which maintain a close-packed interior.[310–312]

Studies of nuclear magnetic resonance spectra (Chapter 3) and of polarization of fluorescence (Chapter 23), have shown that there is rapid though restricted rotational movement of side chains of proteins *in solution*. Even buried phenylalanine and tyrosine side chains often rotate rapidly whereas movement of the

bulkier tryptophan rings is more limited. Peptide NH protons in unfolded polypeptide chains undergo rapid acid or base catalyzed exchange with H_2O.[313] However, in globular proteins, the rate of exchange of buried NH protons is often orders of magnitude slower.[314–316] For most proteins there appears to be little tendency to unfold completely and then refold; the major domains hold together tightly. However, there may be local unfolding, for example, of a helix at the surface of a protein, which will allow more rapid exchange. Using NMR techniques (Chapter 3), rates of exchange of all of the individual peptide NH protons within small proteins can be measured.[315] Cracks may open up in proteins. This is suggested by the fact that O_2, I_2, and certain other small molecules are apparently able to penetrate the protein freely and to quench the fluorescence of buried tryptophan side chains (Chapter 23).

Since packing density tends to be lower at active sites than in the bulk of the protein, it is probable that more conformational alterations occur near active sites than elsewhere.[317] Lumry and Rosenberg[318] suggested that the "defects" of poor packing and poor hydrogen bonding of some regions of a folded peptide chain provide a store of potential energy that can be important to the functioning of a protein. Even in a very well-packed protein domain there are defects. Some atoms are compressed by the folding of the peptide chain and are closer together by over 0.04 nm than predicted by the van der Waals radii. These packing defects have been estimated to destabilize the protein by as much as 250 kJ/mol.[319] Details of structural heterogeneity within several proteins for which very precise structural data are available have been described.[311]

2. Conformational Changes

We have seen that some polypeptides assume an extended β conformation while others form helices. In some cases, the same protein can do both. For example, hair can be stretched greatly, the α helices of the keratin molecules uncoiling and assuming a β conformation with hydrogen bonds *between* chains instead of *within* a single chain. Thus, a polymer may have more than one conformational state in which the folding and hydrogen bonding are different.[320,321] With soluble proteins more than one folded conformation is possible with different sets of hydrogen bonds and internal hydrophobic interactions. Some of the conformations of a globular protein are more stable than others, and a protein will ordinarily assume one of the energetically most favorable conformations. However, there may be other conformations of almost equal energy.

A large body of evidence suggests that many proteins do exist in two or more different but well-defined conformational states and that the ability of a protein to undergo easy conversion from one to another is of great biological significance. In some cases such as that of hemoglobin (Chapter 7), there are changes in the interactions between subunits. Alterations in the hinge regions between domains of immunoglobulins have been seen. In many enzymes, including the dehydrogenases (Chapter 15), kinases (Chapter 12), and aspartate aminotransferase (Fig. 2-6), a cleft between two domains appears to open and close.[322] In other cases, more subtle alterations in conformational state, involving mainly changes in the internal hydrogen bonded network together with small localized changes in chain folding, have been observed.

Conformational alterations in proteins are probably facilitated by the fact that some hydrogen-bonded groups are found within the hydrophobic interior. *All of the buried hydrogen atoms suitable for hydrogen bond formation are ordinarily hydrogen bonded to an electron donor group. However, because oxygen atoms in proteins each have two unshared electron pairs, there are, in general, more electron donor groups than there are hydrogen atoms to which they can bind. This sets the stage for a competition between electronegative centers for particular proton suitable for hydrogen bonding and provides a molecular basis for the easy triggering of conformational changes.*[323]

3. Denaturation and Refolding

An extreme conformational alteration is the **denaturation** of proteins, which may be caused by heating or by treatment with reagents such as strong acids and bases, **urea**, **guanidinium chloride**, and **sodium dodecyl sulfate** (SDS).

Urea

Guanidinium chloride

$$CH_3 - (CH_2)_{11} - OSO_3^- \ Na^+$$

Sodium dodecyl sulfate (SDS)

Denaturation leads to unfolding of a protein to a more random conformation. In the denatured state the amide groups of the peptide chain form hydrogen bonds with surrounding water molecules and with denaturants such as urea or the guanidinium ion rather than with each other.[324,325] Denaturants also diminish the strength of the hydrophobic interactions that promote folding.[326] Characteristic biochemical activities are lost and physical properties such as sedimentation constant, viscosity, and light absorption are altered. The ease of denaturation of proteins and the

fact that denaturation is sometimes reversible show that the energy differences between the folded conformations and the open random coil conformation are usually not great.[274,327,328] However, it has been difficult to establish the amount of stabilization of a folded polypeptide chain provided by buried hydrogen bonds[328a,328b,328c] or the role of cooperatively formed hydrogen-bonded chains[328c,328d] or clusters.[328c,328d,328e]

Complete denaturation of a protein was generally regarded as an irreversible process prior to 1956 when Anfinsen showed that denatured ribonuclease (Chapter 12) could refold spontaneously.[273] This 124-residue protein contains four disulfide (–S–S–) bridges and thus is tied firmly together. When these bridges are broken reductively in the presence of a denaturing agent, the enzyme becomes inactive. Anfinsen found that upon reoxidation under appropriate conditions, complete activity reappeared. The molecules had folded spontaneously into the correct conformation, the one in which the correct *one* of 105 (7 x 5 x 3 x 1) possible pairings of the eight –SH groups present needed to reform the four disulfide bridges had taken place. This observation has had an important influence on thinking about protein synthesis and folding of polypeptide chains into biologically active molecules.

A puzzling problem was posed by Levinthal many years ago.[329] We usually assume that the peptide chain folds into one of the most stable conformations possible. However, proteins fold very rapidly. Even today, no computer would be able, in our lifetime, to find by systematic examination the thermodynamically most stable conformation.[328] It would likewise be impossible for a folding protein to "try out" more than a tiny fraction of all possible conformations. Yet folded and unfolded proteins often appear to be in a thermodynamic equilibrium! Experimental results indicate that denatured proteins are frequently in equilibrium with a **compact denatured state** or "molten globule" in which hydrophobic groups have become clustered and some secondary structures exists.[330–336] From this state the polypeptide may rearrange more slowly through other **folding intermediates** to the final "native conformation."[336a,336b]

It is generally assumed that within cells the folding of the peptide chain commences while the chain is still being synthesized on a ribosome. The growing chain probably folds rapidly in a random way until it finds one or more stable conformations which serve as folding intermediates for slower conversion to the finished protein.[337,338] However, any process within a cell is affected by the complex intracellular environment.[339] Folding can be catalyzed or inhibited by proteins known as **molecular chaperones**. Folding may sometimes require isomerization of one or a few proline residues from *trans* to *cis*.[340,341] For example, during the refolding of ribonuclease the isomerization of Pro 93 appears to be a rate-limiting step.[342,343] Such isomerizations may be assisted by **peptidyl-prolyl-(cis–trans) isomerases** (Box 9-F). Disulfide linkages are sometimes formed incorrectly.[344,345] A **protein-disulfide isomerase** catalyzes cleavage and reformation of these bridges, allowing the protein to find the most stable crosslinking arrangement. The actions of these enzymes and of molecular chaperones are considered further in Chapters 10 and 12.

4. Effects of pH and Solvent

Because polypeptide chains contain many acidic and basic amino acid side chains, the properties of proteins are greatly influenced by pH. At low pH carboxylates, $-S^-$, and imidazole groups accept protons causing the overall net charge on the macromolecule to be strongly positive. At high pH protons are lost and the protein becomes negatively charged. Electrostatic repulsion between like charges may cause proteins to denature at low or high pH. More stable proteins may be very soluble at low or high pH. Proteins often have a minimum solubility and a maximum stability near the **isoelectric point,** the pH at which the net charge is zero.[336,346,347] Activities of enzymes, abilities to bind specifically to other proteins, and

BOX 2-C THE NOBEL PRIZES

Many young scientists dream of one day winning a Nobel Prize. Although denounced by some, the much sought and highly publicized award has, since 1901, been given to an outstanding group of scientists. Many of these have made major contributions to biochemistry or to techniques important to biochemists. Here is a partial list.

Year	Name	Prize[a]	Discovery or subject of study
1901	Wilhelm C. Röntgen	Physics	Discovery of X-rays
1902	Emil H. Fischer	Chemistry	Synthesis of sugars and purines
1903	Svante A. Arrhenius	Chemistry	Electrolytic dissociation; a founder of physical chemistry
1903	Antoine Henri Becquerel; Marie S. Curie, and Pierre Curie	Physics	Discovery and study of radioactivity
1906	Camillo Golgi and S. Ramon y Cajal	Physiology/Medicine	Discovery of Golgi apparatus
1907	Edward Buchner	Chemistry	Biochemistry, cell-free fermentation
1910	Albrecht Kossel	Physiology/Medicine	Isolation of nucleic acid bases
1914	Max von Laue	Physics	Discovery of X-ray diffraction by crystals
1915	Richard M. Willstätter	Chemistry	Plant pigments, especially chlorophyll
1915	William H. and William L. Bragg	Physics	Analysis of crystal structure by X-rays
1919	Jules Bordet	Physiology/Medicine	Discovery of blood complement; complement fixation test
1920	Walther H. Nernst	Chemistry	Thermochemistry
1922	Archibald V. Hill and Otto F. Meyerhof	Physiology/Medicine	Chemistry of muscle contraction
1923	Frederick G. Banting and John J. R. Macleod	Physiology/Medicine	Discovery of insulin and treatment of diabetes
1926	Theodor Svedburg	Chemistry	Study of high M_r compounds, development of ultracentrifuge
1927	Heinrich O. Wieland	Chemistry	Bile acids
1928	Adolf O. R. Windaus	Chemistry	Sterols and vitamins
1929	Frederick G. Hopkins and Christiaan Eijkman	Physiology/Medicine	Discovery of vitamins, tryptophan, vitamin B_1
1929	Arthur Harden and Hans A. S. von Euler-Chelpin	Chemistry	Fermenting enzymes, fermentation of sugars
1930	Karl Landsteiner	Physiology/Medicine	Blood groups A, B, O, Rh
1930	Hans Fischer	Chemistry	Structures and chemistry of porphyrins, chlorophyll
1931	Otto H. Warburg	Physiology/Medicine	Respiratory enzymes
1933	Thomas H. Morgan	Physiology/Medicine	Chromosome theory and chromosome maps
1934	George R. Minot, William P. Murphy, and George H. Whipple	Physiology/Medicine	Treatment of pernicious anemia
1936	Henry H. Dale and Otto Loewi	Physiology/Medicine	Acetylcholine release at nerve endings
1937	Albert von Szent-Györgyi	Physiology/Medicine	Vitamin C
1937	Walter N. Haworth and Paul Karrer	Chemistry	Carbohydrate structures, structures of carotenoids, flavins, vitamin A
1938	Richard Kuhn	Chemistry	Carotenoids and vitamins
1939	Gerhard Domagk	Physiology/Medicine	Prontosil, first antibacterial sulfa drug
1939	Adolf F. J. Butenandt and L. Ruzicka	Chemistry	Isolation and study of sex hormones, study of polymethylenes, terpenes
1943	E. A. Doisy and Carl Henrik Dam	Physiology/Medicine	Isolation and study of vitamin K
1945	Alexander Fleming, Ernst B. Chain, and Howard W. Florey	Physiology/Medicine	Discovery and structure of penicillin
1945	A. J. Virtanen	Chemistry	Nutritional chemistry
1946	James B. Sumner, J. H. Northrop, and W. M. Stanley	Chemistry	Crystallization of enzymes and virus proteins
1947	Carl F. Cori and Gerty T. Cori	Physiology/Medicine	Glycogen metabolism, the Cori cycle
1947	B. A. Houssay and Robert Robinson	Chemistry	Investigation of plant alkaloids
1948	Arne W. K. Tiselius	Chemistry	Electrophoresis, study of serum proteins
1950	Phillip S. Hench, Edward C. Kendall, and Tadeus Reichstein	Physiology/Medicine	ACTH
1952	Selman A. Waksman	Physiology/Medicine	Discovery of streptomycin
1952	A. J. P. Martin and Richard L. M. Synge	Chemistry	Paper chromatography
1953	Fritz A. Lipmann and Hans A. Krebs	Physiology/Medicine	Coenzyme A, citric acid cycle
1954	Linus C. Pauling[b]	Chemistry	The nature of the chemical bond
1955	A. H. T. Theorell	Physiology/Medicine	Oxidative enzymes
1955	Vincent du Vigneaud	Chemistry	Synthesis of biotin and oxytocin
1957	Daniel Bovet	Physiology/Medicine	First antihistamine drug
1957	Alexander R. Todd	Chemistry	Work on nucleotides coenzymes
1958	Joshua Lederberg, George W. Beadle, and Edward L. Tatum	Physiology/Medicine	One gene-one enzyme hypothesis from genetic studies of neurospora
1958	Frederick Sanger[b]	Chemistry	Protein sequencing, insulin

numerous physiological processes are controlled by effects of pH on proteins.

The solvent for most proteins in nature is water. However, many enzymes function well in organic solvents if they retain only a small amount of essential structural and catalytic water.[348]

5. Irreversible Damage to Proteins

Every protein in an organism has its own characteristic lifetime. No sooner is it synthesized than degradation begins. Enzyme-catalyzed cleavage of the peptide linkages leads to turnover of proteins but before this occurs a number of spontaneous damaging reactions may alter the protein. Prevalent among these is **deamidation** of asparaginyl residues to aspartyl or

BOX 2-C THE NOBEL PRIZES (continued)

Year	Name	Prize[a]	Discovery or subject of study
1959	Arthur Kornberg and Severo Ochoa	Physiology/Medicine	Enzymatic synthesis of DNA
1960	F. M. Burnet and P. Medawar	Physiology/Medicine	Immunological tolerance in animals
1961	Melvin Calvin	Chemistry	Photosynthesis using $^{14}CO_2$
1962	Francis H. C. Crick, James D. Watson, and Maurice H. F. Wilkins	Physiology/Medicine	Molecular structure of DNA
1962	John C. Kendrew and Max F. Perutz	Chemistry	Structures of protein by X-ray diffraction
1962	Linus C. Pauling[b]	Peace	Ending atmospheric testing of nuclear weapons
1964	Konrod E. Bloch and Feodor Lynen	Physiology/Medicine	Pathways of cholesterol biosynthesis
1964	Dorothy M. Crowfoot Hodgkin	Chemistry	X-ray structures, vitamin B_{12}
1965	Andre Lwolf, Jacques Monod, and Francois Jacob	Physiology/Medicine	Messenger RNA, regulation of transcription
1967	Ragnar Granit, Haldan Keffer Hartline, and George Wald	Physiology/Medicine	Chemistry of vision
1967	Manfred Eigen and Norrish Porter	Chemistry	Study of high-speed chemical reactions
1968	Robert W. Holley	Physiology/Medicine	RNA sequence
1968	H. Gobind Khorana and Marshall W. Nirenberg	Physiology/Medicine	The genetic code
1969	Max Delbrück, Alfred D. Hershey, and Salvador E. Luria	Physiology/Medicine	Replication and genetic structures of viruses
1970	Julius Axelrod, Bernard Katz, and Ulf von Euler	Physiology/Medicine	Transmission of nerve impulses, noradrenaline
1970	Luis F. Leloir	Chemistry	Role of sugar nucleotides in biosynthesis of carbohydrates
1971	Earl W. Sutherland, Jr.	Physiology/Medicine	Cyclic AMP
1972	Gerald M. Edelman and Rodney R. Porter	Physiology/Medicine	Structure of antibodies
1972	Christian B. Anfinsen, Sanford Moore, and William H. Stein	Chemistry	Ribonuclease, structure and activity
1974	Albert Claude, Christian R. de Duve, and George E. Palade	Physiology/Medicine	Cell structure
1975	David Baltimore, Renato Dulbecco, and Howard M. Temin	Physiology/Medicine	Reverse transscriptase
1975	John W. Cornforth and Vladimir Prelog	Chemistry	Stereochemistry of organic molecules and of enzymatic reactions
1977	Rosalyn S. Yalow, Roger C. L. Guillemin, and Andrew V. Schally	Physiology/Medicine	Radioimmunoassay thyrotropin-releasing hormone
1978	Daniel Nathans, Werner Arber, and Hamilton O. Smith	Physiology/Medicine	Restriction enzymes
1978	Peter Mitchell	Chemistry	Biological energy transfer
1980	Paul Berg, Walter Gilbert, and Frederick Sanger[b]	Chemistry	Recombinant DNA, methods of sequence determination for DNA
1982	Sune K. Bergstrom, Bergt I. Samuelsson, and John R. Vane	Physiology/Medicine	Isolation and study of prostaglandins
1982	Aaron Klug	Chemistry	Development of crystallographic electron microscopy
1983	Barbara McClintock	Physiology/Medicine	Gene transposition
1984	Niels K. Jerne, Georges J. F. Köhler, and Cesar Milstein	Physiology/Medicine	Cellular basis of immunology
1984	R. Bruce Merrifield	Chemistry	Solid-phase synthesis of peptides
1985	Joseph L. Goldstein and Michael S. Brown	Physiology/Medicine	Control of cholesterol metabolism
1986	Rita Levi-Montalcini and Stanley Cohen	Physiology/Medicine	Nerve growth factor
1987	Susumu Tonegawa	Physiology/Medicine	Genetics of antibody formation
1988	Johan Diesenhofer, Robert Huber, and Hartmut Michel	Chemistry	Three-dimensional structure of a photosynthetic reaction center
1988	Gertrude Elion, George Hitchings, and James Black	Physiology/Medicine	Principles of drug treatment and design of many important drugs
1989	J. Michael Bishop, Harold E. Varmus, and Joseph E. Murray	Physiology/Medicine	Origin of retroviral oncogenes
1989	Sidney Altman and Thomas R. Cech	Chemistry	Catalytic RNA
1991	Erwin Neher and Bert Sakmann	Physiology/Medicine	Functioning of single ion channels in cells
1991	Richard R. Ernst	Chemistry	High-resolution NMR
1992	Edmond H. Fischer and Edwin G. Krebs	Physiology/Medicine	Reversible protein phosphorylation in biological regulation
1992	Rudolph A. Marcus	Chemistry	Theory of electron transfer reactions
1993	Richard J. Roberts and Phillip A. Sharp	Physiology/Medicine	Discovery of split genes
1993	Michael Smith and Kary B. Mullis	Chemistry	Oligonucleotide-directed mutagenesis and polymerase chain reaction
1994	Alfred Gilman and Martin Rodbell	Physiology/Medicine	"G-proteins"
1995	Edwin B. Lewis, Christiane Nusslein-Volhard, and Eric Wieschaus	Physiology/Medicine	Homeotic mutations in *Drosophila*
1995	Paul Crutzen, Sherwood Rowland, and Mario Molina	Chemistry	Damage to the stratospheric ozone layer
1996	Peter C. Doherty and Rolf M. Zinkernagel	Physiology/Medicine	Specificity of cell-mediated immune response
1997	Paul D. Boyer and John E. Walker	Chemistry	Mechanism of ATP synthesis
1997	Jens C. Skou	Chemistry	Discovery of Na^+, K^+-ATPase
1997	Stanley B. Prusiner	Physiology/Medicine	Discovery of prions
1998	Robert F. Furchgott, Louis J. Ignarro, and Ferid Murad	Physiology/Medicine	Nitric oxide as a signaling molecule
1999	Günter Blobel	Physiology/Medicine	Intrinsic signals that govern transport and localization of proteins
2000	Arvid Carlsson	Physiology/Medicine	Identification of dopamine as signaling molecule in brain
	Paul Greengard	Physiology/Medicine	Discovery of the dopamine signaling cascade
	Eric R. Kandel	Physiology/Medicine	Molecular basis of learning

[a] Prizes are given in Physics, Chemistry, Physiology or Medicine, Literature and Peace.
[b] Sanger and Pauling each have been awarded two Nobel Prizes.

isoaspartyl groups (Eq. 2-24).[349,350] Aspartyl residues can undergo the same type of cyclization at low pH. The sequence Asn-Gly is especially susceptible to rearrangement according to Eq. 2-24. However, the peptide torsion angles may be more important in determining whether deamidation occurs. The intermediate succinimide may be racemized easily at the chiral center marked by the asterisk in Eq. 2-24. Thus, the Asn-Gly sequence represents a weak linkage, which is nevertheless present in many proteins.

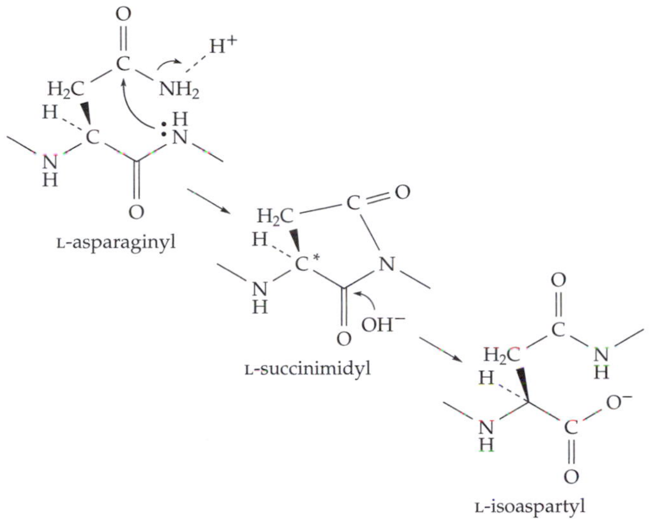

L-asparaginyl

L-succinimidyl

L-isoaspartyl

(2-24)

Cystine residues in disulfide bridges are also not completely stable but undergo β elimination[305,351] (Chapter 13) at slightly alkaline pH values according to Eq. 2-25. The free thiol group formed (HSR) is still attached to the protein but may sometimes be in a position to enter into thiol-disulfide exchange reactions with other S–S bridges causing further degradation. Methionine side chains in proteins can be oxidized to sulfoxides: $-CH_2-CH_2-\overset{O}{\underset{\|}{S}}-CH_3$ and hydroxyl groups can be introduced into aromatic rings by oxidation. Hydrogen peroxide and other oxidants may be responsible for such oxidation within cells (see Chapter 18). The very long-lived proteins of the lens of the eye are especially susceptible to deamidation, racemization, oxidation, and accumulation of covalently attached blue fluorescent compounds.[352–354]

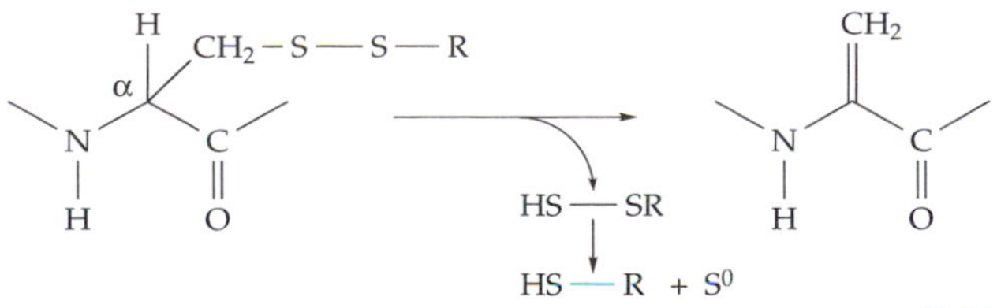

(2-25)

H. Design and Engineering of Proteins

Methods of chemical synthesis of polypeptides and of cloning and mutating genes now allow us to alter peptide sequences at will and to design completely new proteins.[355–356b] The methods are discussed in Chapters 3, 5, and 26. The following are examples.

Peptides that form α helices that associate as coiled coils,[357] or as three- or four-helix tetrameric bundles[179a,358–360] or amphipathic helices that associate with lipid bilayers have been made.[355,361] More difficult has been the design of proteins that form β sheets.[362–364a] Many efforts are being made to understand protein stability[365–367] by systematic substitutions of one residue for another. Addition of new disulfide linkages at positions selected by study of three-dimensional structures sometimes stabilizes enzymes.[368–371] On the other hand elimination of unnecessary cysteine residues can enhance stability by preventing β elimination[351] and replacement of asparagine by threonine can improve the thermostability of enzymes by preventing deamidation.[372,373]

Artificial mutants of subtilisin[373] and of many other enzymes are helping us to understand the mechanisms of catalysis. Artificially prepared temperature-sensitive mutants (Chapter 26) and naturally occurring mutants are providing new insight into pathways of folding of proteins whose three-dimensional structures are known. For these studies it is necessary to follow a rational strategy in deciding which of the astronomical number of possible mutant forms may be of interest. For example, residues within active sites may be changed. Introduction of mutations at random locations can also be useful in finding regions of interest.[374] When mutant proteins can be crystallized, an exact understanding of the effect of the mutation on the structure is possible. Sometimes two or more mutations may have to be introduced to obtain the desired modification.

It is also possible to incorporate a range of "unnatural" amino acids in specific sites in a polypeptide sequence and to observe resulting effects on a protein's properties.[375] A quite different approach is to design polypeptides that mimic a natural peptide but consist of D-amino acids. The peptide chain is *reversed*, i.e., the N terminus becomes the C terminus and every peptide linkage is also reversed. The amino acid side chains preserve their relationships one to another and the backbone atoms tend to preserve their hydrogen bonding pattern. Peptides made in this way tend to be resistant to cleavage by enzymes. Some may be useful as drugs.[376,377]

References

1. Glendening, E. D., Faust, R., Streitwieser, A., Vollhardt, K. P. C., and Weinhold, F. (1993) *J. Am. Chem. Soc.* **115**, 10952–10957
2. Gao, Q., Weber, H.-P., Craven, B. M., and McMullan, R. K. (1994) *Acta Crystallogr.* **B50**, 695–703
3. Hegstrom, R. A., and Kondepudi, D. K. (1990) *Sci. Am.* **262**(Jan), 108–115
4. Cahn, R. S., Ingold, C., and Prelog, V. (1966) *Angew. Chem. Int. Ed. Engl.* **5**, 385–415
5. Eliel, E. L., Wilen, S. H., and Mander, L. N. (1994) *Stereochemistry of Organic Compounds*, Wiley, New York
6. Bentley, R. (1969) *Molecular Asymmetry in Biology*, Vol. 1, Academic Press, New York (pp. 49–56)
7. Blackwood, J. E., Gladys, C. L., Loening, K. L., Petrarca, A. E., and Rush, J. E. (1968) *J. Am. Chem. Soc.* **90**, 509–510
8. Hanson, K. R. (1966) *J. Am. Chem. Soc.* **88**, 2731–2742
9. Lambert, J. B. (1970) *Sci. Am.* **222**(Jan), 58–70
10. Klyne, W., and Prelog, V. (1960) *Experientia* **16**, 521–523
11. Liébecq, C., ed. (1992) *Biochemical Nomenclature*, Portland Press, London and Chapel Hill, North Carolina (for the International Union of Biochemistry and Molecular Biology)
12. Chiang, Y., Kresge, A. J., and Tang, Y. S. (1984) *J. Am. Chem. Soc.* **106**, 460–462
13. Beak, P., Fry, F. S., Jr., Lee, J., and Steele, F. (1976) *J. Am. Chem. Soc.* **98**, 171–179
14. Beak, P. (1977) *Acc. Chem. Res.* **10**, 186–192
15. Metzler, D. E., Harris, C. M., Johnson, R. J., Siano, D. B., and Thomson, J. A. (1973) *Biochemistry* **12**, 5377–5392
16. Katritzky, A. R., and Karelson, M. (1991) *J. Am. Chem. Soc.* **113**, 1561–1566
17. Blomberg, F., Maurer, W., and Rüterjans, H. (1977) *J. Am. Chem. Soc.* **99**, 8149–8159
18. Walters, D. E., and Allerhand, A. (1980) *J. Biol. Chem.* **255**, 6200–6204
19. Fersht, A. (1977) *Enzyme Structure and Mechanism*, Freeman, San Francisco, California
20. Gabler, R. (1978) *Electrical Interactions in Molecular Biophysics*, Academic Press, New York (p. 188)
21. Israelachvili, J. N., and McGuiggan, P. M. (1988) *Science* **241**, 795–800
22. Salem, L. (1962) *Can. J. Biochem. Physiol.* **40**, 1287–1298
23. Tsao, Y.-H., Fennell Evans, D., and Wennerström, H. (1993) *Science* **262**, 547–550
23a. Kurihara, K., and Kunitake, T. (1992) *J. Am. Chem. Soc.* **114**, 10927–10933
23b. Yang, L., Valdeavella, C. V., Blatt, H. D., and Pettitt, B. M. (1996) *Biophys. J.* **71**, 3022–3029
24. Warshel, A. (1987) *Nature (London)* **330**, 15–16
25. Honig, B., and Nicholls, A. (1995) *Science* **268**, 1144–1149
26. Meot-Ner, M. M. (1988) *J. Am. Chem. Soc.* **110**, 3075–3080
27. Loewenthal, R., Sancho, J., Reinikainen, T., and Fersht, A. R. (1993) *J. Mol. Biol.* **232**, 574–583
28. Gristina, A. G., Oga, M., Webb, L. X., and Hobgood, C. D. (1985) *Science* **228**, 990–993
29. Jeffrey, G. A., and Saenger, W. (1991) *Hydrogen Bonding in Biological Structures*, Springer-Verlag, Berlin
30. Gilli, P., Bertolasi, V., Ferretti, V., and Gastone, G. (1994) *J. Am. Chem. Soc.* **116**, 909–915
30a. Jeffrey, G. A. (1997) *An Introduction to Hydrogen Bonding*, Oxford Univ. Press, New York
31. Derewenda, Z. S., Derewenda, U., and Kobos, P. M. (1994) *J. Mol. Biol.* **241**, 83–93
31a. Wahl, M. C., and Sundaralingam, M. (1997) *Trends Biochem. Sci.* **22**, 97–102

31b. Bella, J., and Berman, H. M. (1996) *J. Mol. Biol.* **264**, 734–742
32. Schulz, G. E., and Schirmer, R. H. (1979) *Principles of Protein Structure*, Springer, New York (p. 36)
32a. Platts, J. A., Howard, S. T., and Bracke, B. R. F. (1996) *J. Am. Chem. Soc.* **118**, 2726–2733
33. Taylor, R., Kennard, O., and Versichel, W. (1983) *J. Am. Chem. Soc.* **105**, 5761–5766
34. Murray-Rust, P., and Glusker, J. P. (1984) *J. Am. Chem. Soc.* **106**, 1018–1025
35. Görbitz, C. H., and Etter, M. C. (1992) *J. Am. Chem. Soc.* **114**, 627–631
36. Nelson, D. D., Jr., Fraser, G. T., and Klemperer, W. (1987) *Science* **238**, 1670–1674
36a. Ghanty, T. K., Staroverov, V. N., Koren, P. R., and Davidson, E. R. (2000) *J. Am. Chem. Soc.* **122**, 1210–1214
36b. Cornilescu, G., Ramirez, B. E., Frank, M. K., Clore, G. M., Gronenborn, A. M., and Bax, A. (1999) *J. Am. Chem. Soc.* **121**, 6275–6279
36c. Benedict, H., Shenderovich, I. G., Malkina, O. L., Malkin, V. G., Denisov, G. S., Golubev, N. S., and Limbach, H.-H. (2000) *J. Am. Chem. Soc.* **122**, 1979–1988
37. Fersht, A. R., Shi, J., Knill-Jones, J., Lowe, D. M., Witkinson, A. J., Blow, D. M., Brick, P., Carter, P., Woye, M. M. Y., and Winter, G. (1985) *Nature (London)* **314**, 235–238
38. Eriksson, A. E., Baase, W. A., Zhang, X.-J., Heinz, D. W., Blaber, M., Baldwin, E. P., and Matthews, B. W. (1992) *Science* **255**, 178–183
38a. Takano, K., Yamagata, Y., Kubota, M., Funahashi, J., Fujii, S., and Yutani, K. (1999) *Biochemistry* **38**, 6623–6629
39. Lim, W. A., Hodel, A., Sauer, R. T., and Richards, F. M. (1994) *Proc. Natl. Acad. Sci. U.S.A.* **91**, 423–427
40. Schuster, P., Zundel, G., and Sandorfy, C. (1976) *The Hydrogen Bond*, North Holland Publ. Co., Amsterdam (3 vols.)
41. Luck, W. A. P., and Kleeberg, H. (1978) in *Photosynthetic Oxygen Evolution* (Metzner, H., ed), pp. 1–29, Academic Press, London
42. Edsall, J. T. (1983) *Trends Biochem. Sci.* **8**, 29–31
43. Grunwald, E. (1986) *J. Am. Chem. Soc.* **108**, 5719–5726
44. Westhof, E. (1993) *Water and Biological Macromolecules*, CRC Press, Boca Raton, Florida
45. Kusalik, P. G., and Svishchev, I. M. (1994) *Science* **265**, 1219–1221
46. Libnau, F. O., Toft, J., Christy, A. A., and Kvalheim, O. M. (1994) *J. Am. Chem. Soc.* **116**, 8311–8316
47. Benson, S. W., and Siebert, E. D. (1992) *J. Am. Chem. Soc.* **114**, 4269–4276
48. Liu, K., Cruzan, J. D., and Saykally, R. J. (1996) *Science* **271**, 929–933
49. Pauling, L. (1935) *J. Am. Chem. Soc.* **57**, 2680
50. Wittebort, R. J., Usha, M. G., Ruben, D. J., Wemmer, D. E., and Pines, A. (1988) *J. Am. Chem. Soc.* **110**, 5668–5671
51. Li, J., and Ross, D. K. (1993) *Nature (London)* **365**, 327–329
52. Dyke, T. R., Mack, K. M., and Muenter, J. S. (1977) *J. Chem. Phys.* **66**, 498–510
53. Saenger, W. (1979) *Nature (London)* **279, 280**, 343–344, 848
54. Saenger, W. (1987) *Ann. Rev. Biophys. Biophys. Chem.* **16**, 93–114
55. Pribble, R. N., and Zwier, T. S. (1994) *Science* **265**, 75–79
56. Gregory, J. K., Clary, D. C., Liu, K., Brown, M. G., and Saykally, R. J. (1997) *Science* **275**, 814–817
57. Quiocho, F. A., Sack, J. S., and Vyas, N. K. (1987) *Nature (London)* **329**, 561–564
58. Pflugrath, J. W., and Quiocho, F. A. (1988) *J. Mol. Biol.* **200**, 163–180

59. Kauzmann, W. (1959) *Adv. Prot. Chem.* **14**, 1–63
60. Jencks, W. P. (1987) *Catalysis in Chemistry and Enzymology*, Dover, Mineola, New York (p. 393)
61. Tanford, C. (1980) *The Hydrophobic Effect: Formation of Micelles and Biological Membranes*, 2nd ed., Wiley, New York
62. Abraham, M. H. (1980) *J. Am. Chem. Soc.* **102**, 5910–5912
63. Ben-Náim, A. (1980) *Hydrophobic Interactions*, Plenum, New York
64. Murphy, K. P., Privalov, P. L., and Gill, S. J. (1990) *Science* **247**, 559–561
65. Muller, N. (1992) *Trends Biochem. Sci.* **17**, 459–463
66. Spolar, R. S., Livingstone, J. R., and Record, J., MT. (1992) *Biochemistry* **31**, 3947–3955
67. Stites, W. E., Meeker, A. K., and Shortle, D. (1994) *J. Mol. Biol.* **235**, 27–32
68. Hecht, D., Tadesse, L., and Walters, L. (1993) *J. Am. Chem. Soc.* **115**, 3336–3337
69. Gill, S. J., and Wadsö, I. (1976) *Proc. Natl. Acad. Sci. U.S.A.* **73**, 2955–2958
70. Sharp, K. A., Nicholls, A., Fine, R. F., and Honig, B. (1991) *Science* **252**, 106–109
71. Van Oss, C. J. (1994) *Interfacial Forces in Aqueous Media*, Dekker, New York
71a. Hummer, G., Garde, S., García, A. E., Pohorille, A., and Pratt, L. R. (1996) *Proc. Natl. Acad. Sci. U.S.A.* **93**, 8951–8955
72. Israelachvili, J., and Wennerström, H. (1996) *Nature (London)* **379**, 219–225
73. Radzicka, A., and Wolfenden, R. (1988) *Biochemistry* **27**, 1664–1670
74. Kellis, J. T., Jr., Nyberg, K., Sali, D., and Fersht, A. R. (1988) *Nature (London)* **333**, 784–786
75. Sanger, F. (1988) *Ann. Rev. Biochem.* **57**, 1–28
76. Creighton, T. E. (1993) *Proteins, Structures and Molecular Principles*, 2nd ed., Freeman, New York
77. Kyte, J. (1995) *Structure in Protein Chemistry*, Garland Publ., New York
78. Guillemin, R., and Burgus, R. (1972) *Sci. Am.* **227**(Nov), 24–33
79. Hochster, R. M., Kates, M., and Quastel, J. H., eds. (1973) *Metabolic Inhibitors*, Vol. 3, Academic Press, New York (p. 312)
80. Ramachandran, G. N., and Kolaskar, A. S. (1973) *Biochim. Biophys. Acta.* **303**, 385–388
81. MacArthur, M. W., and Thornton, J. M. (1996) *J. Mol. Biol.* **264**, 1180-1195
81a. Milner-White, E. J. (1997) *Protein Sci.* **6**, 2477–2482
82. Moffat, J. B. (1973) *J. Theor. Biol.* **40**, 247–258
83. Fersht, A. R. (1971) *J. Am. Chem. Soc.* **93**, 3504–3515
84. Karle, I. L. (1974) *Biochemistry* **13**, 2155–2162
85. Ovchinnikov, Y. A. (1973) *FEBS Lett.* **29**, 31–33
86. Doonan, S. (1974) *FEBS Lett.* **38**, 229–233
87. Bhown, A. S. (1990) *Handbook of Proteins*, A & M Publ., Birmingham, Alabama
88. Boguski, M., and McEntyre, J. (1994) *Trends Biochem. Sci.* **19**, 71
88a. Branden, C., and Tooze, J. (1991) *Introduction to Protein Structure*, Garland Publ., New York
89. Clements, S., Mehansho, H., and Carlson, D. M. (1975) *J. Biol. Chem.* **260**, 13471–13477
90. Condit, C. M., and Meagher, R. B. (1986) *Nature (London)* **323**, 178–181
91. Reynolds, P., Weber, S., and Prakash, L. (1985) *Proc. Natl. Acad. Sci. U.S.A.* **82**, 168–172
92. Rodakis, G. C., and Kafatos, F. C. (1982) *Proc. Natl. Acad. Sci. U.S.A.* **79**, 3551–3555
93. Young, J. F., Hockmeyer, W. T., Gross, M., Ballou, W. R., Wirtz, R. A., Trosper, J. H., Beaudoin, R. L., Hollingdale, M. R., Miller, L. H., Diggs, C. L., and Rosenberg, M. (1985) *Science* **228**, 958–962

94. Doolittle, R. F. (1987) *Of URFS and ORFS: A Primer on How to Analyze Derived Amino Acid Sequences*, Univ. Science Books, Mill Valley, California

95. Doolittle, R. F. (1989) *Trends Biochem. Sci.* **14**, 244–245

96. Pearson, W. R., and Lipman, D. J. (1988) *Proc. Natl. Acad. Sci. U.S.A.* **85**, 2444–2448

97. Holm, L., and Sander, C. (1993) *J. Mol. Biol.* **233**, 123–138

98. Lewin, R. (1987) *Science* **237**, 1570

99. Fleming, P. J., and Richards, F. M. (2000) *J. Mol. Biol.* **299**, 487–498

100. Gekko, K., and Hasegawa, Y. (1986) *Biochemistry* **25**, 6563–6571

101. Goodsell, D. S., and Olson, A. J. (1993) *Trends Biochem. Sci.* **18**, 65–68

102. Schultz, G. E., and Schirmer, R. H. (1979) *Principles of Protein Structure*, Springer-Verlag, New York

102a. Ramachandran, G. N., Ramakrishnam, C., and Sasisekharan, V. (1963) *J. Mol. Biol.*, **7**, 95–99

103. Ramachandran, G. N., Venkatachalam, C. M., and Krimm, S. (1966) *Biophys. J.* **6**, 849–872

103a. Karplus, P. A. (1996) *Protein Sci.* **5**, 1406–1420

104. Sasiskharian, V., Lakshminarayanan, A. V., and Ramachandran, G. N. (1967) in *Conformation of Biopolymers*, Vol. 2 (Ramachandran, G. N., ed), p. 641–654, Academic Press, New York

104a. Gunasekaran, K., Ramakrishnan, C., and Balaram, P. (1996) *J. Mol. Biol.* **264**, 191–198

104b. Herzberg, O., and Moult, J. (1991) *Proteins* **11**, 223–229

105. Milner-White, E. J., Bell, L. H., and Maccallum, P. H. (1992) *J. Mol. Biol.* **228**, 725–734

106. Rhee, S. (1990) *High resolution X-ray diffraction studies of cytosolic aspartate aminotransferase* [Ph.D. Dissertation], University of Iowa, Iowa City, Iowa

107. Dunbrack, R. L., Jr., and Karplus, M. (1993) *J. Mol. Biol.* **230**, 543–574

108. Pauling, L., and Corey, R. B. (1951) *Proc. Natl. Acad. Sci. U.S.A.* **37**, 729–740

109. Rich, A. (1994) *Nature (London)* **371**, 285

110. Chothia, C. (1973) *J. Mol. Biol.* **75**, 295–302

111. Chou, K.-C., Némethy, G., and Scheraga, H. A. (1983) *Biochemistry* **22**, 6213–6221

111a. Skarzynski, T., Moody, P. C. E., and Wonacott, A. J. (1987) *J. Mol. Biol.* **193**, 171–187

111b. Wang, L., O'Connell, T., Tropsha, A., and Hermans, J. (1996) *J. Mol. Biol.* **262**, 283–293

112. Reeke, G. N., Jr., and Becker, J. W. (1986) *Science* **234**, 1108–1111

113. Murzin, A. G., Lesk, A. M., and Chothia, C. (1994) *J. Mol. Biol.* **236**, 1369–1381

114. Murzin, A. G., Lesk, A. M., and Chothia, C. (1994) *J. Mol. Biol.* **236**, 1382–1400

115. LaLonde, J. M., Bernlohr, D. A., and Ranaszak, L. J. (1994) *FASEB J.* **8**, 1240–1241

116. Guss, J. M., and Freeman, H. C. (1983) *J. Mol. Biol.* **169**, 521–563

117. Richardson, J. S. (1981) *Adv. Prot. Chem.* **34**, 167–339

117a. Erickson, H. P. (1994) *Proc. Natl. Acad. Sci. U.S.A.* **91**, 10114–10118

118. Redinbo, M. R., Cascio, D., Choukair, M. K., Rice, D., Merchant, S., and Yeates, T. O. (1993) *Biochemistry* **32**, 10560–10567

119. Weiss, M. S., and Schulz, G. E. (1993) *J. Mol. Biol.* **231**, 817–824

120. Kreusch, A., and Schulz, G. E. (1994) *J. Mol. Biol.* **243**, 891–905

121. Yoder, M. D., and Jurnak, F. (1995) *FASEB J.* **9**, 335–342

122. Richardson, J. S. (1994) *FASEB J.* **8**, 1237–1239

123. Steinbacher, S., Seckler, R., Miller, S., Steipe, B., Huber, R., and Reinemer, P. (1994) *Science* **265**, 383–385

124. Richardson, J. S., Getzoff, E. D., and Richardson, D. C. (1978) *Proc. Natl. Acad. Sci. U.S.A.* **75**, 2574–2578

125. Chan, A. W. E., Hutchinson, E. G., Harris, D., and Thornton, J. M. (1993) *Protein Sci.* **2**, 1574–1590

126. Urry, D. W. (1972) *Proc. Natl. Acad. Sci. U.S.A.* **69**, 1610–1614

127. Nicholson, L. K., and Cross, T. A. (1989) *Biochemistry* **28**, 9379–9385

127a. Smith, T. F., Gaitatzes, C., Saxena, K., and Neer, E. J. (1999) *Trends Biochem. Sci.* **24**, 181–185

127b. Garcia-Higuera, I., Fenoglio, J., Li, Y., Lewis, C., Panchenko, M. P., Reiner, O., Smith, T. F., and Neer, E. J. (1996) *Biochemistry* **35**, 13985–13994

127c. ter Naar, E., Harrison, S. C., and Kirchhausen, T. (2000) *Proc. Natl. Acad. Sci. U.S.A.* **97**, 1096–1100

127d. Springer, T. A. (1998) *J. Mol. Biol.* **283**, 837–862

128. Pauling, L. (1993) *Protein Sci.* **2**, 1060–1063

129. Kendrew, J. C. (1969) *Sci. Am.* **205**(Dec), 96–110

130. Arnott, S., and Dover, S. D. (1967) *J. Mol. Biol.* **30**, 209–212

131. Steigemann, W., and Weber, E. (1979) *J. Mol. Biol.* **127**, 309–338

132. Barlow, D. J., and Thornton, J. M. (1988) *J. Mol. Biol.* **201**, 601–619

132a. Rajashankar, K. R., and Ramakumar, S. (1996) *Protein Sci.* **5**, 932–946

133. Harper, E. T., and Rose, G. D. (1993) *Biochemistry* **32**, 7605–7609

134. Chakrabartty, A., Schellman, J. A., and Baldwin, R. L. (1991) *Nature (London)* **351**, 586–588

135. Creamer, T. P., and Rose, G. D. (1995) *Protein Sci. U.S.A.* **4**, 1305–1314

136. Dempsey, C. E. (1992) *Biochemistry* **31**, 4705–4712

137. Ramachandran, G. N., and Mitra, A. K. (1976) *J. Mol. Biol.* **107**, 85–92

138. Blaber, M., Zhang, X.-j, and Matthews, B. W. (1993) *Science* **260**, 1637–1639

139. Chakrabartty, A., Kortemme, T., and Baldwin, R. L. (1994) *Protein Sci.* **3**, 843–852

140. Toniolo, C., and Benedetti, E. (1991) *Trends Biochem. Sci.* **16**, 350–353

141. Avbelj, F., and Moult, J. (1995) *Biochemistry* **34**, 755–764

142. Karpen, M. E., De Haseth, P. L., and Neet, K. E. (1992) *Protein Sci.* **1**, 1333–1342

143. Fiori, W. R., Miick, S. M., and Millhauser, G. L. (1993) *Biochemistry* **32**, 11957–11962

144. Smythe, M. L., Huston, S. E., and Marshall, G. R. (1993) *J. Am. Chem. Soc.* **115**, 11594–11595

145. Bavosa, A., Benedetti, E., Di Blasio, B., Pavone, V., Pedone, C., Toniolo, C., and Bonora, G. M. (1986) *Proc. Natl. Acad. Sci. U.S.A.* **83**, 1988–1992

145a. Fathallah-Shaykh, H., Wolf, S., Wong, E., Posner, J. B., and Furneaux, H. M. (1991) *Proc. Natl. Acad. Sci. U.S.A* **88**, 3451–3454

146. Miick, S. M., Martinez, G. V., Flori, W. R., Todd, A. P., and Millhauser, G. L. (1992) *Nature (London)* **359**, 653–655

147. Wada, A. (1976) *Adv. Biophys.* **9**, 1–63

148. Hol, W. G. J., van Duijnen, P. T., and Berendsen, H. J. C. (1978) *Nature (London)* **273**, 443–446

148a. Hol, W. G. J., Halie, L. M., and Sander, C. (1981) *Nature (London)* **294**, 532–536

148b. Lockhart, D. J., and Kim, P. S. (1992) *Science* **257**, 947–951

149. Levine, B. F., and Bethea, C. G. (1976) *J. Chem. Phys.* **65**, 1989–1993

150. Gray, T. M., and Matthews, B. W. (1984) *J. Mol. Biol.* **175**, 75–81

151. Richardson, J. S., and Richardson, D. C. (1988) *Science* **240**, 1648–1652

152. Nicholson, H., Anderson, D. E., Dao-pin, S., and Matthews, B. W. (1991) *Biochemistry* **30**, 9816–9828

153. Åqvist, J., Luecke, H., Quiocho, F. A., and Warshel, A. (1991) *Proc. Natl. Acad. Sci. U.S.A.* **88**, 2026–2030

154. Lyu, P. C., Wemmer, D. E., Zhou, H. X., Pinker, R. J., and Kallenbach, N. R. (1993) *Biochemistry* **32**, 421–425

155. Viguera, A. R., and Serrano, L. (1995) *J. Mol. Biol.* **251**, 150–160

156. Doig, A. J., Chakrabartty, A., Klingler, T. M., and Baldwin, R. L. (1994) *Biochemistry* **33**, 3396–3403

157. Doig, A. J., Macarthur, M. W., Stapley, B. J., and Thornton, J. M. (1997) *Protein Sci.* **6**, 147–155

158. Sali, D., Bycroft, M., and Fersht, A. R. (1988) *Nature (London)* **335**, 740–743

159. Aurora, R., and Rose, G. D. (1998) *Protein Sci.* **7**, 21–38

160. Metzler, D. E. (1979) *Adv. Enzymol.* **50**, 1–40

161. Lupas, A. (1996) *Trends Biochem. Sci.* **21**, 375–382

161. Lupas, A. (1996) *Trends Biochem. Sci.* **21**, 375–382

162. Kohn, W. D., Mant, C. T., and Hodges, R. S. (1997) *J. Biol. Chem.* **272**, 2583–2586

163. Fraser, R. D. B., MacRae, T. P., Parry, D. A. D., and Suzuki, E. (1986) *Proc. Natl. Acad. Sci. U.S.A.* **83**, 1179–1183

164. Johnson, L. D., Idler, W. W., Zhou, X.-M., Roop, D. R., and Steinert, P. M. (1985) *Proc. Natl. Acad. Sci. U.S.A.* **82**, 1896–1900

165. Cooper, D., and Sun, T.-T. (1986) *J. Biol. Chem.* **261**, 4646–4654

166. Crick, F. H. C. (1953) *Acta Crystallogr.* **6**, 689–697

167. Vazquez, S. R., Kuo, D. Z., Bositis, C. M., Hardy, L. W., Lew, R. A., and Humphreys, R. E. (1992) *J. Biol. Chem.* **267**, 7406–7410

168. O'Shea, E. K., Llemm, J. D., Kim, P. S., and Alber, T. (1991) *Science* **254**, 539–544

169. Ferré-D'Amaré, A. R., Prendergast, G. C., Ziff, E. B., and Burley, S. K. (1993) *Nature (London)* **363**, 38–44

170. Lyu, P. C., Marky, L. A., and Kallenbach, N. R. (1989) *J. Am. Chem. Soc.* **111**, 2733–2734

171. Graddis, T. J., Myszka, D. G., and Chaiken, I. M. (1993) *Biochemistry* **32**, 12664–12671

171a. Fairman, R., Chao, H.-G., Lavoie, T. B., Villafranca, J. J., Matsueda, G. R., and Novotny, J. (1996) *Biochemistry* **35**, 2824–2829

171b. Kohn, W. D., Kay, C. M., and Hodges, R. S. (1998) *J. Mol. Biol.* **283**, 993–1012

172. John, M., Briand, J.-P., Granger-Schnarr, M., and Schnarr, M. (1994) *J. Biol. Chem.* **269**, 16247–16253

173. Zhou, N. E., Kay, C. M., and Hodges, R. S. (1992) *Biochemistry* **31**, 5739–5746

174. Cohen, C., and Parry, D. A. D. (1986) *Trends Biochem. Sci.* **11**, 245–248

175. Phillips, G. N., Jr., Fillers, J. P., and Cohen, C. (1986) *J. Mol. Biol.* **192**, 111–131

176. Mo, J., Holtzer, M. E., and Holtzer, A. (1993) *Protein Sci.* **2**, 128–130

177. Xie, X., Rao, S., Walian, P., Hatch, V., Phillips, G. N., Jr., and Cohen, C. (1994) *J. Mol. Biol.* **236**, 1212–1226

178. Eckert, R. L. (1988) *Proc. Natl. Acad. Sci. U.S.A.* **85**, 1114–1118

179. Gernert, K. M., Surles, M. C., Labean, T. H., Richardson, J. S., and Richardson, D. C. (1995) *Protein Sci.* **4**, 2252–2260

179a. Monera, O. D., Zhou, N. E., Lavigne, P., Kay, C. M., and Hodges, R. S. (1996) *J. Biol. Chem.* **271**, 3995–4001

180. Lovejoy, B., Choe, S., Cascio, D., McRorie, D. K., DeGrado, W. F., and Eisenberg, D. (1993) *Science* **259**, 1288–1293

References

181. Harbury, P. B., Kim, P. S., and Alber, T. (1994) *Nature (London)* **371**, 80–83
182. Efimov, V. P., Nepluev, I. V., Sobolev, B. N., Zurabishvili, T. G., Schulthess, T., Lustig, A., Engel, J., Haener, M., Aebi, U., Venyaminov, S. Y., Potekhin, S. A., and Mesyanzhinov, V. V. (1994) *J. Mol. Biol.* **242**, 470–486
183. Lieberman, M., Tabet, M., and Sasaki, T. (1994) *J. Am. Chem. Soc.* **116**, 5035–5044
184. Utani, A., Nomizu, M., Timpl, R., Roller, P. P., and Yamada, Y. (1994) *J. Biol. Chem.* **269**, 19167–19175
185. Robinson, C. R., and Sligar, S. G. (1993) *Protein Sci.* **2**, 826–837
186. Steif, C., Weber, P., Hinz, H.-J., Flossdorf, J., Cesareni, G., and Kokkinidis, M. (1993) *Biochemistry* **32**, 3867–3876
187. Redfield, C., Boyd, J., Smith, L. J., Smith, R. A. G., and Dobson, C. M. (1992) *Biochemistry* **31**, 10431–10437
188. Harris, N. L., Presnell, S. R., and Cohen, F. E. (1994) *J. Mol. Biol.* **236**, 1356–1368
189. Chou, K.-C., Carlacci, L., Maggiora, G. M., Parodi, L. A., and Schulz, M. W. (1992) *Protein Sci.* **1**, 810–827
190. Rees, D. C., Komiya, H., Yeates, T. O., Allen, J. P., and Feher, G. (1989) *Ann. Rev. Biochem.* **58**, 607–633
191. Aleshin, A. E., Hoffman, C., Firsov, L. M., and Honzatko, R. B. (1994) *J. Mol. Biol.* **238**, 575–591
192. Aleshin, A. E., Firsov, L. M., and Honzatko, R. B. (1994) *J. Biol. Chem.* **269**, 15631–15639
193. Harris, E. M. S., Aleshin, A. E., Firsov, L. M., and Honzatko, R. B. (1993) *Biochemistry* **32**, 1618–1626
194. Bella, J., Eaton, M., Brodsky, B., and Berman, H. M. (1994) *Science* **266**, 75–81
194a. Kramer, R. Z., Bella, J., Mayville, P., Brodsky, B., and Berman, H. M. (1999) *Nature Struct. Biol.* **6**, 454–457
195. Brodsky, B., and Shah, N. K. (1995) *FASEB J.* **9**, 1537–1546
196. Redeker, V., Levilliers, N., Schmitter, J.-M., Le Caer, J.-P., Rossier, J., Adoutte, A., and Bré, M.-H. (1994) *Science* **266**, 1688–1691
197. Miller, A. (1982) *Trends Biochem. Sci.* **7**, 13–18
198. Eyre, D. R. (1980) *Science* **207**, 1315–1322
199. Parry, D. A. D., and Craig, A. S. (1979) *Nature (London)* **282**, 213–215
200. Porter, R. R., and Reid, K. B. M. (1979) *Adv. Prot. Chem.* **33**, 1–71
201. Ashford, D., and Neuberger, A. (1980) *Trends Biochem. Sci.* **5**, 245–248
202. Chen, J., and Varner, J. E. (1985) *Proc. Natl. Acad. Sci. U.S.A.* **82**, 4399–4403
203. Adzhubei, A. A., and Sternberg, M. J. E. (1994) *Protein Sci.* **3**, 2395–2410
204. Richardson, J. S. (1985) *Nature (London)* **316**, 102–103
205. Hutchinson, E. G., and Thornton, J. M. (1994) *Protein Sci.* **3**, 2207–2216
206. Venkatachalam, C. M. (1968) *Biopolymers* **6**, 1425–1436
207. Perczel, A., Foxman, B. M., and Fasman, G. D. (1992) *Proc. Natl. Acad. Sci. U.S.A.* **89**, 8210–8214
208. Yan, Y., Tropsha, A., Hermans, J., and Erickson, B. W. (1993) *Proc. Natl. Acad. Sci. U.S.A.* **90**, 7898–7902
209. Perczel, A., McAllister, M. A., Császár, P., and Csizmadia, I. G. (1993) *J. Am. Chem. Soc.* **115**, 4849–4858
210. Mattos, C., Petsko, G. A., and Karplus, M. (1994) *J. Mol. Biol.* **238**, 733–747
211. Hynes, T. R., Hodel, A., and Fox, R. O. (1994) *Biochemistry* **33**, 5021–5030
212. Haque, T. S., Little, J. C., and Gellman, S. H. (1994) *J. Am. Chem. Soc.* **116**, 4105–4106

213. Gierasch, L. M., Deber, C. M., Madison, V., Niu, C.-H., and Blout, E. R. (1981) *Biochemistry* **20**, 4730–4738
214. Rees, D. C., Lewis, M., and Lipscomb, W. N. (1983) *J. Mol. Biol.* **168**, 367–387
215. Sapse, A.-M., Mallah-Levy, L., Daniels, S. B., and Erickson, B. W. (1987) *J. Am. Chem. Soc.* **109**, 3526–3529
216. Urry, D. W. (1993) *Angew. Chem. Int. Ed. Engl.* **32**, 819–841
217. Sandberg, L. B., Soskel, N. T., and Leslie, J. G. (1981) *N. Engl. J. Med.* **304**, 566–579
217a. Tatham, A. S., and Shewry, P. R. (2000) *Trends Biochem. Sci.* **25**, 567–571
218. Gotte, L., Volpin, D., Horne, R. W., and Mammi, M. (1976) *Micron* **7**, 95–102
218a. Urry, D., Luan, C.-H., and Peng, S. (1995) *Ciba. Found. Symp.* **192**, 4–30
218b. van Dijk, A. A., de Boef, E., Bekkers, A., van Wijk, L. L., van Swieten, E., Hamer, R. J., and Robillard, G. T. (1997) *Protein Sci.* **6**, 649–656
219. Sternberg, M. J. E., and Thornton, J. M. (1976) *J. Mol. Biol.* **105**, 367–382
220. Leszczýnski, J. F., and Rose, G. D. (1986) *Science* **234**, 849–855
221. Liang, C., and Mislow, K. (1994) *J. Am. Chem. Soc.* **116**, 11189–11190
222. Chothia, C. (1984) *Ann. Rev. Biochem.* **53**, 537–572
223. Resnick, D., Pearson, A., and Krieger, M. (1994) *Trends Biochem. Sci.* **19**, 5–8
224. Shoyab, M., Plowman, G. D., McDonald, V. L., Bradley, J. G., and Todaro, G. J. (1989) *Science* **243**, 1074–1076
225. Doolittle, R. F. (1995) *Ann. Rev. Biochem.* **64**, 287–314
226. Berman, A. L., Kolker, E., and Trifanov, E. N. (1994) *Proc. Natl. Acad. Sci. U.S.A.* **91**, 4044–4047
227. Orengo, C. A., Jones, D. T., and Thornton, J. M. (1994) *Nature (London)* **372**, 631–634
228. Pan, K.-M., Damodaran, S., and Greaser, M. L. (1994) *Biochemistry* **33**, 8255–8261
229. Doolittle, R. F., and Bork, P. (1993) *Sci. Am.* **269**(Oct), 50–56
230. Baron, M., Norman, D. G., and Campbell, I. D. (1991) *Trends Biochem. Sci.* **16**, 13–17
231. Mulichak, A. M., Tulinsky, A., and Ravichandran, K. G. (1991) *Biochemistry* **30**, 10576–10588
232. Bork, P., Holm, L., and Sander, C. (1994) *J. Mol. Biol.* **242**, 309–320
232a. Bogan, A. A., and Thorn, K. S. (1998) *J. Mol. Biol.* **280**, 1–9
233. Tsai, J., Taylor, R., Chothia, C., and Gerstein, M. (1999) *J. Mol. Biol.* **290**, 253–266
233a. Gerstein, M., and Chothia, C. (1996) *Proc. Natl. Acad. Sci. U.S.A.* **93**, 10167–10172
234. Varadarajan, R., and Richards, F. M. (1992) *Biochemistry* **31**, 12315–12327
235. Allewell, N. (1987) *Trends Biochem. Sci.* **12**, 417–418
236. Lim, W. A., Farruggio, D. C., and Sauer, R. T. (1992) *Biochemistry* **31**, 4324–4333
237. Hubbard, S. J., and Argos, P. (1994) *Protein Sci.* **3**, 2194–2206
237a. Hubbard, S. J., and Argos, P. (1996) *J. Mol. Biol.* **261**, 289–300
238. Hunt, N. G., Gregoret, L. M., and Cohen, F. E. (1994) *J. Mol. Biol.* **241**, 214–225
239. Hunter, C. A., and Sanders, K. M. (1990) *J. Am. Chem. Soc.* **112**, 5525–5534
240. Burley, S. K., and Petsko, G. A. (1985) *Science* **229**, 23–28
241. Thomas, K. A., Smith, G. M., Thomas, T. B., and Feldmann, R. J. (1982) *Proc. Natl. Acad. Sci. U.S.A.* **79**, 4843–4847
242. Flocco, M. M., and Mowbray, S. L. (1994) *J. Mol. Biol.* **235**, 709–717

243. Hemmingsen, J. M., Gernert, K. M., Richardson, J. S., and Richardson, D. C. (1994) *Protein Sci.* **3**, 1927–1937
244. Teeter, M. M. (1991) *Ann. Rev. Biophys. Biophys. Chem.* **20**, 577–600
245. McDonald, I. K., and Thornton, J. M. (1994) *J. Mol. Biol.* **238**, 777–793
246. Bordo, D., and Argos, P. (1994) *J. Mol. Biol.* **243**, 504–519
247. Shimoni, L., and Glusker, J. P. (1995) *Protein Sci.* **4**, 65–74
248. Dao-pin, S., Anderson, D. E., Baase, W. A., Dahlquist, F. W., and Matthews, B. W. (1991) *Biochemistry* **30**, 11521–11529
248a. Kumar, S., and Nussinov, R. (1999) *J. Mol. Biol.* **293**, 1241–1255
248b. Strop, P., and Mayo, S. L. (2000) *Biochemistry* **39**, 1251–1255
249. Sawyer, L., and James, M. N. G. (1982) *Nature (London)* **295**, 79–80
250. Borders, C. L., Jr., Broadwater, J. A., Bekeny, P. A., Salmon, J. E., Lee, A. S., Eldridge, A. M., and Pett, V. B. (1994) *Protein Sci.* **3**, 541–548
251. Mrabet, N. T., Van den Broeck, A., Van den brande, I., Stanssens, P., Laroche, Y., Lambeir, A.-M., Matthiissens, G., Jenkins, J., Chiadmi, M., van Tilbeurgh, H., Rey, F., Janin, J., Quax, W. J., Lasters, I., De Maeyer, M., and Wodak, S. J. (1992) *Biochemistry* **31**, 2239–2253
252. Dougherty, D. A. (1996) *Science* **271**, 163–168
253. Mitchell, J. B. O., Nandi, C. L., McDonald, I. K., Thornton, J. M., and Price, S. (1994) *J. Mol. Biol.* **239**, 315–331
253a. De Wall, S. L., Meadows, E. S., Barbour, L. J., and Gokel, G. W. (1999) *J. Am. Chem. Soc.* **121**, 5613–5614
253b. Gallivan, J. P., and Dougherty, D. A. (1999) *Proc. Natl. Acad. Sci. U.S.A.* **96**, 9459–9464
253c. Wouters, J. (1998) *Protein Sci.* **7**, 2472–2475
253d. Fernández-Recio, J., Romero, A., and Sancho, J. (1999) *J. Mol. Biol.* **290**, 319–330
254. Rose, G. D., Young, W. B., and Gierasch, L. M. (1983) *Nature (London)* **304**, 654–657
255. Watenpaugh, K. D., Margulis, T. N., Sieker, L. C., and Jensen, L. H. (1978) *J. Mol. Biol.* **122**, 175–190
255a. Nakasako, M. (1999) *J. Mol. Biol.* **289**, 547–564
256. Teeter, M. M. (1984) *Proc. Natl. Acad. Sci. U.S.A.* **81**, 6014–6018
257. Otting, G., Liepinsh, E., and Wüthrich, K. (1991) *Science* **254**, 974–980
258. Murzin, A. G., and Finkelstein, A. V. (1988) *J. Mol. Biol.* **204**, 749–770
259. Chothia, C. (1989) *Nature (London)* **337**, 204–205
260. Buehner, M., Ford, G. C., Moras, D., Olsen, K. W., and Rossmann, M. G. (1974) *J. Mol. Biol.* **90**, 25–49
261. Rossmann, M. G., Liljas, A., Branden, C.-I., and Banaszak, L. J. (1975) *The Enzymes*, 3rd ed., Vol. 11, Academic Press, New York (p. 61)
262. Brändén, C. (1990) *Nature (London)* **346**, 607–608
263. Kobe, B., and Deisenhofer, J. (1995) *Nature (London)* **374**, 183–186
264. Kobe, B., and Deisenhofer, J. (1994) *Trends Biochem. Sci.* **19**, 415–421
264a. Kajava, A. V. (1998) *J. Mol. Biol.* **277**, 519–527
264b. Kobe, B., and Kajava, A. V. (2000) *Trends Biochem. Sci.* **25**, 509–515
265. Farber, G. K., and Petsko, G. A. (1990) *Trends Biochem. Sci.* **15**, 228–234
266. Reardon, D., and Farber, G. K. (1995) *FASEB J.* **9**, 497–503
267. Uhlin, U., and Eklund, H. (1994) *Nature (London)* **370**, 533–539
268. Chothia, C. (1992) *Nature (London)* **357**, 543–544
268a. Zhang, C., and DeLisi, C. (1998) *J. Mol. Biol.* **284**, 1301–1305
269. McLachlan, A. D. (1980) *Nature (London)* **285**, 267–268

References

270. Chou, P. Y., and Fasman, G. D. (1974) *Biochemistry* **13**, 222–245

270a. Myers, J. K., Pace, C. N., and Scholtz, J. M. (1997) *Biochemistry* **36**, 10923–10929

270b. Spek, E. J., Olson, C. A., Shi, Z., and Kallenbach, N. R. (1999) *J. Am. Chem. Soc.* **121**, 5571–5572

271. Minor, D. L., Jr., and Kim, P. S. (1994) *Nature (London)* **367**, 660–663

272. Richards, F. M. (1991) *Sci. Am.* **264**(Jan), 54–63

273. Anfinsen, C. B. (1973) *Science* **181**, 223–230

274. Dill, K. A., Bromberg, S., Yue, K., Fiebig, K. M., Yee, D. P., Thomas, P. D., and Chan, H. S. (1995) *Protein Sci.* **4**, 561–602

275. Rost, B., Schneider, R., and Sander, C. (1993) *Trends Biochem. Sci.* **18**, 120–123

276. Wako, H., and Blundell, T. L. (1994) *J. Mol. Biol.* **238**, 693–708

277. Yi, T.-M., and Lander, E. S. (1993) *J. Mol. Biol.* **232**, 1117–1129

278. Monge, A., Friesner, R. A., and Honig, B. (1994) *Proc. Natl. Acad. Sci. U.S.A.* **91**, 5027–5029

279. Smith, C. K., Withka, J. M., and Regan, L. (1994) *Biochemistry* **33**, 5510–5517

279a. Lacroix, E., Viguera, A. R., and Serrano, L. (1998) *J. Mol. Biol.* **284**, 173–191

279b. Gotoh, O. (1996) *J. Mol. Biol.* **264**, 823–838

279c. Dandekar, T., and Argos, P. (1996) *J. Mol. Biol.* **256**, 645–660

279d. Jones, D. T. (1999) *J. Mol. Biol.* **292**, 195–202

280. Kyte, J., and Doolittle, R. F. (1982) *J. Mol. Biol.* **157**, 105–132

281. Reyes, V. E., Phillips, L., Humphreys, R. E., and Lew, R. A. (1989) *J. Biol. Chem.* **264**, 12854–12858

282. Wold, F. (1986) *Trends Biochem. Sci.* **11**, 58–59

283. Graves, D. J., Martin, B. L., and Wang, J. H. (1994) *Co-and post-translational modification of proteins*, Oxford Univ. Press, New York

283a. Angeletti, R. H., ed. (1998) *Protein, Analysis and Design*, Academic Press, San Diego, California

284. Yan, S. C. B., Grinnell, B. W., and Wold, F. (1989) *Trends Biochem. Sci.* **14**, 264–268

285. Berthiaume, L., Deichaite, I., Peseckis, S., and Resh, M. D. (1994) *J. Biol. Chem.* **269**, 6498–6505

286. McIlhinney, R. A. J. (1990) *Trends Biochem. Sci.* **15**, 387–391

287. Violand, B. N., Schlittler, M. R., Lawson, C. Q., Kane, J. F., Siegel, N. R., Smith, C. E., Kolodziej, E. W., and Duffin, K. L. (1994) *Protein Sci.* **3**, 1089–1097

288. Hilz, H., Fanick, N., and Klapproth, K. (1986) *Proc. Natl. Acad. Sci. U.S.A.* **83**, 6267–6271

289. Hortin, G., Fok, K. F., Toren, P. C., and Strauss, A. W. (1987) *J. Biol. Chem.* **262**, 3082–3085

290. Bateman, A., Solomon, S., and Bennett, H. P. J. (1990) *J. Biol. Chem.* **265**, 22130–22136

291. Takahara, H., Tsuchida, M., Kusubata, M., Akutsu, K., Tagami, S., and Sugawara, K. (1989) *J. Biol. Chem.* **264**, 13361–13368

292. Kreil, G. (1994) *J. Biol. Chem.* **269**, 10967–10970

293. Mor, A., Amiche, M., and Nicolas, P. (1992) *Trends Biochem. Sci.* **17**, 481–485

294. Kreil, G. (1994) *Science* **266**, 996–997

295. Prescott, B., Renugopalakrishnan, V., Glimcher, M. J., Bhushan, A., and Thomas, G. J., Jr. (1986) *Biochemistry* **25**, 2792–2798

296. Wood, D. D., and Moscarello, M. A. (1989) *J. Biol. Chem.* **264**, 5121–5127

297. Waite, J. H., Housley, T. J., and Tanzer, M. L. (1985) *Biochemistry* **24**, 5010–5014

298. Taylor, S. W., Waite, J. H., Ross, M. M., Shabanowitz, J., and Hunt, D. F. (1994) *J. Am. Chem. Soc.* **116**, 10803–10804

299. Clarke, S. (1992) *Ann. Rev. Biochem.* **61**, 355–386

300. Marshall, C. J. (1993) *Science* **259**, 1865–1866

301. Mroz, E. A. (1989) *Science* **243**, 1615

302. Rothgeb, T. M., England, R. D., Jones, B. N., and Gurd, R. S. (1978) *Biochemistry* **17**, 4564–4571

303. Lorimer, G. H. (1983) *Trends Biochem. Sci.* **8**, 65–68

304. Friedman, M. (1977) *Protein Crosslinking*, Vol. 2, Plenum, New York

305. Torchinsky, Y. M. (1981) *Sulfur in Proteins*, Pergamon, Oxford

306. Harrison, P. M., and Sternberg, M. J. E. (1994) *J. Mol. Biol.* **244**, 448–463

307. Folk, J. E., and Finlayson, J. S. (1977) *Adv. Prot. Chem.* **31**, 1–133

308. Gurd, F. R. N., and Rothgeb, T. M. (1979) *Adv. Prot. Chem.* **33**, 73–165

309. Karplus, M., and McCammon, J. A. (1983) *Ann. Rev. Biochem.* **53**, 263–300

310. Caspar, D. L. D., Clarage, J., Salunke, D. M., and Clarage, M. (1988) *Nature (London)* **332**, 659–662

311. Smith, J. L., Hendrickson, W. A., Honzatko, R. B., and Sheriff, S. (1986) *Biochemistry* **25**, 5018–5027

312. Frauenfelder, H., Sligar, S. G., and Wolynes, P. G. (1991) *Science* **254**, 1598–1603

313. Buck, M., Radford, S. E., and Dobson, C. M. (1994) *J. Mol. Biol.* **237**, 247–254

314. Kim, K.-S., Fuchs, J. A., and Woodward, C. K. (1993) *Biochemistry* **32**, 9600–9608

315. Rohl, C. A., and Baldwin, R. L. (1994) *Biochemistry* **33**, 7760–7767

316. Spyracopoulos, L., and O'Neil, J. D. J. (1994) *J. Am. Chem. Soc.* **116**, 1395–1402

317. Huber, R. (1979) *Trends Biochem. Sci.* **4**, 271–276

318. Lumry, R., and Rosenberg, A. (1974) *Colloques Internationaux du C.N.R.S.* **246**, 53

319. Mao, B., Pear, M. R., McCammon, J. A., and Northrup, S. H. (1981) *J. Mol. Biol.* **151**, 199–202

320. Brack, A., and Spach, G. (1981) *J. Am. Chem. Soc.* **103**, 6319–6323

321. Frauenfelder, H., Parak, F., and Young, R. D. (1988) *Ann. Rev. Biophys. Biophys. Chem.* **17**, 451–479

322. Gerstein, M., Lesk, A. M., and Chothia, C. (1994) *Biochemistry* **33**, 6739–6749

323. Kretsinger, R. H., and Nockolds, C. E. (1973) *J. Biol. Chem.* **248**, 3313–3326

324. Liepinsh, E., and Otting, G. (1994) *J. Am. Chem. Soc.* **116**, 9670–9674

325. Scholtz, J. M., Barrick, D., York, E. J., Stewart, J. M., and Baldwin, R. L. (1995) *Proc. Natl. Acad. Sci. U.S.A.* **92**, 185–189

326. Kamoun, P. P. (1988) *Trends Biochem. Sci.* **13**, 424–425

327. Dill, K. A., and Shortle, D. (1991) *Ann. Rev. Biochem.* **60**, 795–825

328. Baker, D., and Agard, D. A. (1994) *Biochemistry* **33**, 7505–7509

328a. Sharp, K. A., and Englander, S. W. (1994) *Trends Biochem. Sci.* **19**, 526–529

328b. Pace, C. N., Shirley, B. A., McNutt, M., and Gajiwala, K. (1996) *FASEB J.* **10**, 75–83

328c. Sippl, M. J. (1996) *J. Mol. Biol.* **260**, 644–648

328d. Meyer, E. (1992) *Protein Sci.* **1**, 1543–1562

328e. Hill, R. B., Hong, J.-K., and DeGrado, W. F. (2000) *J. Am. Chem. Soc.* **122**, 746–747

329. Levinthal, C. (1968) *J. Chim. Phys.-Chim. Biol.* **65**, 44–45

330. Lattman, E. E., Fiebig, K. M., and Dill, K. A. (1994) *Biochemistry* **33**, 6158–6166

331. Mark, A. E., and van Gunsteren, W. F. (1992) *Biochemistry* **31**, 7745–7748

332. Peng, Z., and Kim, P. S. (1994) *Biochemistry* **33**, 2136–2141

333. Hagihara, Y., Tan, Y., and Goto, Y. (1994) *J. Mol. Biol.* **237**, 336–348

334. Yu, Y., Makhatadze, G. I., Pace, C. N., and Privalov, P. L. (1994) *Biochemistry* **33**, 3312–3319

335. Dobson, C. M., Evans, P. A., and Radford, S. E. (1994) *Trends Biochem. Sci.* **19**, 31–37

336. Yang, A.-S., and Honig, B. (1994) *J. Mol. Biol.* **237**, 602–614

336a. Baker, D. (2000) *Nature (London)* **405**, 39–42

336b. Dinner, A. R., Sali, A., Smith, L. J., Dobson, C. M., and Karplus, M. (2000) *Trends Biochem. Sci.* **25**, 331–339

337. Udgaonkar, J. B., and Baldwin, R. L. (1988) *Nature (London)* **335**, 694–699

338. Roder, H., Elöve, G. A., and Englander, S. W. (1988) *Nature (London)* **335**, 700–704

339. Gething, M.-J., and Sambrook, J. (1992) *Science* **355**, 33–45

340. Odefey, C., Mayr, L. M., and Schmid, F. X. (1995) *J. Mol. Biol.* **245**, 69–78

341. Chazin, W. J., Kördel, J., Drakenberg, T., Thulin, E., Brodin, P., Grundström, T., and Forsén, S. (1989) *Proc. Natl. Acad. Sci. U.S.A.* **86**, 2195–2198

342. Lin, L.-N., and Brandts, J. F. (1987) *Biochemistry* **26**, 3537–3543

343. Herning, T., Yutani, K., Taniyama, Y., and Kikuchi, M. (1991) *Biochemistry* **30**, 9882–9891

344. Chatrenet, B., and Chang, J.-Y. (1993) *J. Biol. Chem.* **268**, 20988–20996

345. Talluri, S., Rothwarf, D. M., and Scheraga, H. A. (1994) *Biochemistry* **33**, 10437–10449

346. Pace, C. N. (1990) *Trends Biochem. Sci.* **15**, 14–17

347. Yang, A.-S., and Honig, B. (1993) *J. Mol. Biol.* **231**, 459–474

348. Hartsough, D. S., and Merz, K. M., Jr. (1993) *J. Am. Chem. Soc.* **115**, 6529–6537

349. Aswad, D. W., and Johnson, B. A. (1987) *Trends Biochem. Sci.* **12**, 155–158

350. Brennan, T. V., and Clarke, S. (1993) *Protein Sci.* **2**, 331–338

351. Volkin, D. B., and Klibanov, A. M. (1987) *J. Biol. Chem.* **262**, 2945–2950

352. Wells-Knecht, M. C., Huggins, T. G., Dyer, D. G., Thorpe, S. R., and Baynes, J. W. (1993) *J. Biol. Chem.* **268**, 12348–12352

353. Luthra, M., Ranganathan, D., Ranganathan, S., and Balasubramanian, D. (1994) *J. Biol. Chem.* **269**, 22678–22682

354. Lubec, G., Weninger, M., and Anderson, S. R. (1994) *FASEB J.* **8**, 1166–1169

355. Kaiser, E. T. (1987) *Trends Biochem. Sci.* **12**, 305–309

355a. Beasley, J. R., and Hecht, M. H. (1997) *J. Biol. Chem.* **272**, 2031–2034

356. Oxender, D. E., and Fox, C. F., eds. (1987) *Protein Engineering*, Liss, New York

356a. Dahiyat, B. I., and Mayo, S. L. (1997) *Science* **278**, 82–87

356b. DeGrado, W. F., Summa, C. M., Pavone, V., Nastri, F., and Lombardi, A. (1999) *Ann. Rev. Biochem.* **68**, 779–819

357. Bryson, J. W., Betz, S. F., Lu, H. S., Suich, D. J., Zhou, H. X., O'Neil, K. T., and DeGrado, W. F. (1995) *Science* **270**, 935–941

358. Nautiyal, S., Woolfson, D. N., King, D. S., and Alber, T. (1995) *Biochemistry* **34**, 11645–11651

359. Betz, S. F., Liebman, P. A., and DeGrado, W. F. (1997) *Biochemistry* **36**, 2450–2458

360. Handel, T. M., Williams, S. A., and DeGrado, W. F. (1993) *Science* **261**, 879–885

361. Struthers, M. D., Cheng, R. P., and Imperiali, B. (1996) *Science* **271**, 342–345

362. Yan, Y., and Erickson, B. W. (1994) *Protein Sci.* **3**, 1069–1073

363. Quinn, T. P., Tweedy, N. B., Williams, R. W., Richardson, J. S., and Richardson, D. C. (1994) *Proc. Natl. Acad. Sci. U.S.A.* **91**, 8747–8751

364. Hecht, M. H. (1994) *Proc. Natl. Acad. Sci. U.S.A.* **91**, 8729–8730

364a. Kortemme, T., Ramírez-Alvarado, M., and Serrano, L. (1998) *Science* **281**, 253–256

References

365. Kellis, J. T., Jr., Nyberg, K., and Fersht, A. R. (1989) *Biochemistry* **28**, 4914–4922

366. Goldenberg, D. P. (1988) *Ann. Rev. Biophys. Biophys. Chem.* **17**, 481–507

367. Alber, T. (1989) *Ann. Rev. Biochem.* **58**, 765–798

368. Katz, B. A., and Kossiakoff, A. (1986) *J. Biol. Chem.* **261**, 15480–15485

369. Pantoliano, M. W., Ladner, R. C., Bryan, P. N., Rollence, M. L., Wood, J. F., and Poulos, T. L. (1987) *Biochemistry* **26**, 2077–2082

370. Wetzel, R. (1987) *Trends Biochem. Sci.* **12**, 478–482

371. Mitchinson, C., and Wells, J. A. (1989) *Biochemistry* **28**, 4807–4815

372. Ahern, T. J., Casal, J. I., Petsko, G. A., and Klibanov, A. M. (1987) *Proc. Natl. Acad. Sci. U.S.A.* **84**, 675–679

373. Estell, D. A., Graycar, T. P., Miller, J. V., Powers, D. B., Burnier, J. P., Ng, P. G., and Wells, J. A. (1986) *Science* **233**, 659–663

374. Loeb, D. D., Swanstrom, R., Everitt, L., Manchester, M., Stamper, S. E., and Hutchison, C. A., III. (1989) *Nature (London)* **340**, 397–440

375. Mendel, D., Ellman, J. A., Chang, Z., Veenstra, D. L., Kollman, P. A., and Schultz, P. G. (1992) *Science* **266**, 1798–1802

376. Brady, L., and Dodson, G. (1994) *Nature (London)* **368**, 692–694

377. Liu, N., Deillon, C., Klauser, S., Gutte, B., and Thomas, R. M. (1998) *Protein Sci.* **7**, 1214–1220

Study Questions

1. Name all of the isometric tripeptides which could be formed from one molecule each of tyrosine, alanine, and valine.

2. What functional groups are found in protein side chains? Of what importance to protein structure and function are (a) hydrophobic groups, (b) acidic and basic groups, (c) sulfyhydryl groups?

3. If placed in water and adjusted to a pH of 7, will the following migrate toward the anode or the cathode if placed in an electrical field? (a) Aspartic acid, (b) alanine, (c) tyrosine, (d) lysine, (e) arginine, (f) glutamine.

4. Draw the following hydrogen-bonded structures:
   (a) A dimer of acetic acid.
   (b) A tyrosine–carboxylate bond in the interior of a protein.
   (c) A phosphate–guanidinium ion pair in an enzyme–substrate complex.

5. Contrast the properties of the amino acids with those of the saturated fatty acids with respect to solubility in water and in ether and to physical state.

6. Describe in as much detail as you can the characteristic properties of (a) β sheets, (b) α helices, (c) turns in peptide chains, and (d) collagen.

7. Predict whether the following peptide segments will be likely to exist as an α helix or as part of a β structure within a protein:
   (a) Poly-L-leucine
   (b) Poly-L-valine
   (c) Pro-Glu-Met-Val-Phe-Asp-Ile
   (d) Pro-Glu-Ala-Leu-Phe-Ala-Ala

8. Describe three ways in which a side chain of a serine residue can fold back and hydrogen bond to a C=O or N–H group of the backbone and two ways by which an asparagine side chain can do the same. There are yet other possibilities.

9. Compare structural features and properties of the following proteins: silk fibroin, α-keratin, collagen, and bovine serum albumin.

10. In what way do the solubilities of proteins usually vary with pH? Why?

11. Compare the following: the diameters of (a) a carbon atom in an organic molecule (b) a bacterial cell, e.g. of *E. coli* (c) a human red blood cell (d) a ribosome (e) the length of a peptide unit in an extended polypeptide chain (f) the length of the carbon atom chain in an 18-carbon fatty acid.

12. Compare: (a) the length of a peptide unit (residue) in a polypeptide in an extented (β) conformation. (b) the length by which an α helix is extended by the addition of one amino acid unit (c) the length of one turn of an α helix. (d) the diameter of an α helix (both using atom centers in the backbones and using van der Waals radii) for a poly-L-alanine helix.

13. What are: albumins, globulins, protamines, scleroproteins, glycoproteins, lipoproteins?

14. Where are the following found and what are their functions? Gamma globulin, hemocyanin, pepsin, glucagon, ferritin, phosphorylase.

15. List the nutritionally essential amino acids for human beings. Compare these needs with those of other species, including lactic acid bacteria, malaria parasites, green plants, etc.

16. Define: chiral, enantomer, diastereomer, epimer, anomer (see Chapter 4), prochiral (see Chapter 9). What is meant by the statement that biochemical reactions are stereochemically specific? Why is such stereospecificity to be expected in organisms (which are constructed from asymmetric units)? See Chapter 9 for further discussion.

17. What are disulfide bridges and of what significance are they in protein structure?

18. What is meant by "denaturation" of a protein? Mention several ways in which denaturation can be brought about. How is denaturation explained in terms of structure?

19. A chain of L-amino acids can form either a right-handed or a left-handed helix. From the Ramachandran diagram in Fig. 2-9, can you say anything about the relative stabilities of right and left-handed helices? What do you predict for polyglycine?

20. What similarities and differences would you predict for two proteins of identical amino acid sequence but one made from all L-amino acids and the other from all D-amino acids?

Study Questions

21. Complete the following peptide structure for L-seryl-L-valyl-L-asparaginyl—etc. Extend the chain in the C-terminal direction to form a **beta turn** with the chain coming back to form a **beta sheet**. Add a third segment of peptide *parallel* to the folded back chain to form a 3-stranded beta sheet. Indicate all hydrogen bonds correctly. Draw the side chains of the seryl and asparaginyl residues in positions 1 and 3 so that they form proper hydrogen bonds to groups in the peptide backbone.

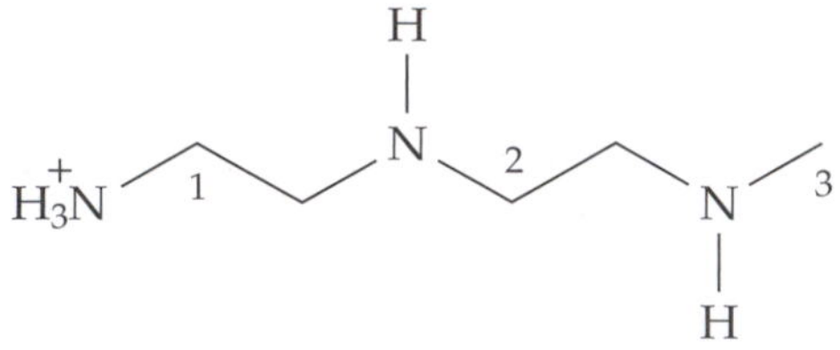

22. Complete the following structure to form a short segment of alpha helix. Extend the chain by at least three residues. Form all hydrogen bonds correctly. Add at least three side chains with the correct chirality at the alpha carbon positions. Add one electrically charged side chain and show how it interacts with the peptide backbone at either the C- or N-terminus to help *stabilize* the helix.

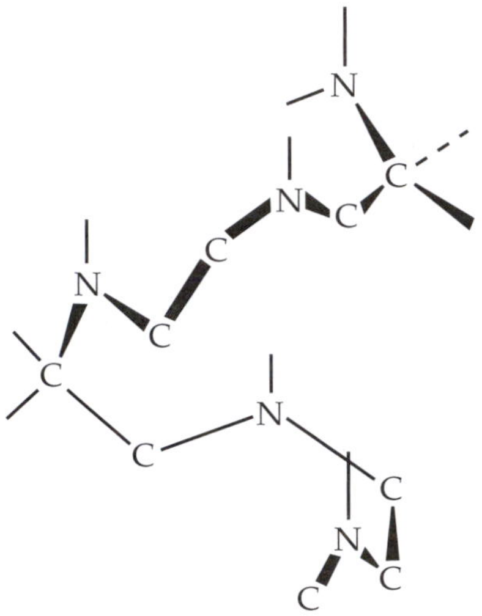

23. What is the relationship between Ångstrom units (Å) and nanometers (nm)? Give the indicated distance in Å or nm.

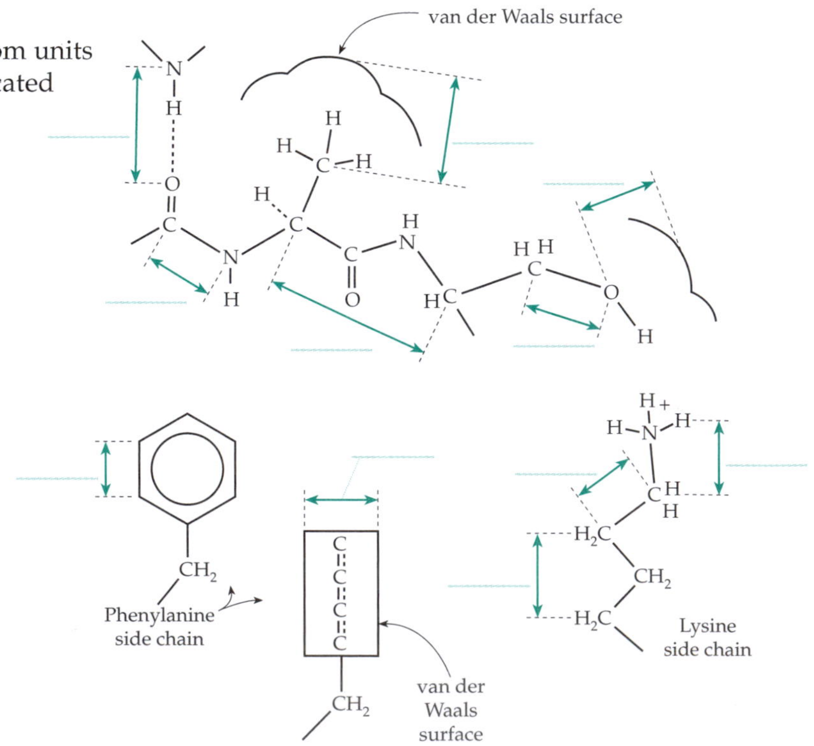

Study Questions

24. Draw three residues of a polypeptide chain constructed of L-amino acids using the top template of the two below. Now, using the second template, whose polypeptide chain begins at the right, use the same three residues, *numbering from the C-terminus* using D-amino acids. This is know as a *retro*-inverso polypeptide.

Compare the outer surface of of the standard polypeptide and its *retro*-inverso analog. How would your answer be affected by the presence of threonine or isoleucine in the peptides?

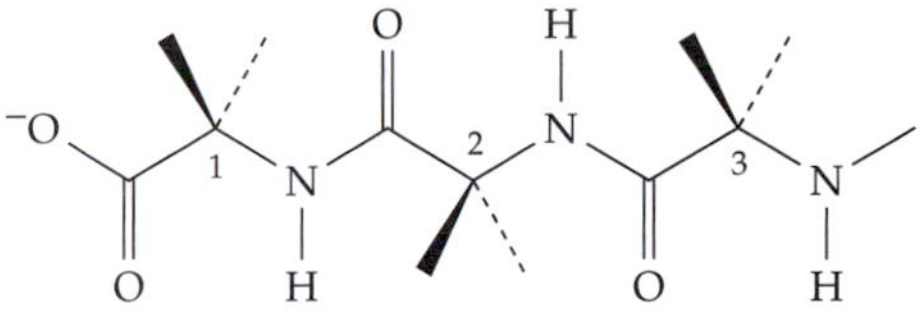

If the two polypeptides were folded into a hairpin loop of β structure how would the exterior surfaces compare? How would the hydrogen bonding compare?

What possible value can you imagine for *retro*-inverso polypeptides in design of drugs? See Brady and Dodson, *Nature* **368**, 692–694 (1994) or Guichard *et al., Proc. Natl. Acad. Sci. USA* **91**, 9765–9769 (1994)

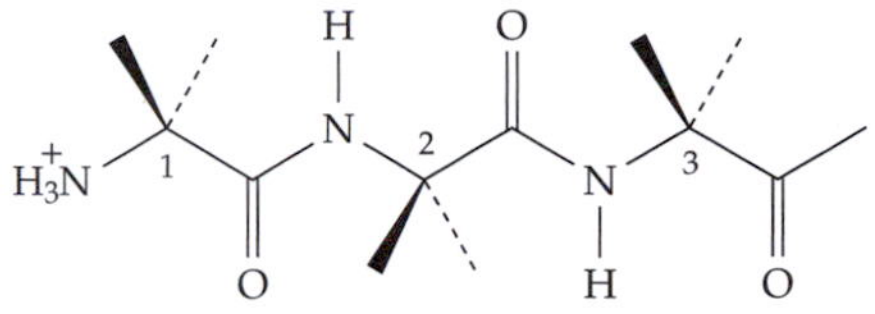

Polypeptide

Retro-inverso polypeptide

A Laue X-ray diffraction pattern from a protein crystal. A stationary crystal is irradiated with very intense white, multiwavelength X rays from a synchrotron source. The diffraction pattern is rich in information. A single 0.1 ms X-ray pulse may provide a pattern with enough information to determine a three-dimensional structure. The pattern consists of thousands of diffraction spots arranged on intersecting rings. The coordinates of the diffraction spots together with their measured intensities provide the necessary information for structure determination. Courtesy of Louise Johnson.

Contents

Determining Structures and Analyzing Cells *3*

How have chemists deduced the thousands of structural formulas that we write for the substances found in nature? The answer is far too complex to give here in detail. However, the separation of compounds, the analysis of mixtures, and the unraveling of structures remain essential parts of biochemistry. A "minireview" of methods, with emphasis on proteins, follows. Additional procedures having to do primarily with carbohydrates, nucleic acids, or lipids are given in Chapters 4, 5, and 8, respectively.

A. Understanding pH and Electrical Charges on Macromolecules

Proteins, nucleic acids, and carbohydrates all contain acidic or basic functional groups. The strengths of the acidic groups vary over a broad range from that of the strongly acidic phosphate and sulfate esters to that of the very weakly acidic alcoholic – OH group. The net electrical charge, as well as the spatial distribution of the charged groups, affects the properties of a macromolecule greatly. Therefore, it will be worthwhile for us to consider some aspects of acid–base chemistry before discussing other topics.

1. Strengths of Acids and Bases: the pK$_a$'s

The strength of an acid is usually described by the acid dissociation constant K_a

$$HA = A^- + H^+ \qquad (3\text{-}1)$$

$$K_a = [A^-][H^+] / [HA] \qquad (3\text{-}2)$$

For strong acids K_a is high and for weak acids it is low. Since the values of K_a vary by many orders of magnitude it is customary to use as a measure of the acid strength pK_a. This is the negative logarithm of K_a (pK_a = $-\log K_a$) . For very strong acids pK_a is less than zero, while very weak acids have pK_a values as high as 15 or more.

In the biochemical literature *the strength of a base is nearly always given by the pK$_a$ of the conjugate acid.* Thus, A^- in Eq. 3-1 is a base and HA its conjugate acid. The base could equally well be uncharged A and its conjugate acid HA^+. In both cases Eq. 3-2 would hold. This defines the strength of both the acid HA and the base A^-. It follows that strong bases have weak conjugate acids with high pK_a values, while weak bases have strong conjugate acids with low pK_a values.

For a compound containing several acidic groups we define a series of consecutive dissociation constants K_{1a}, K_{2a}, K_{3a}, etc. For the sake of simplicity we will omit the a's and call these K_1, K_2, K_3, $\cdots$.

$$H_3A \xrightarrow{K_1} H_2A \xrightarrow{K_2} HA \xrightarrow{K_3} A \qquad (3\text{-}3)$$

If there are n consecutive dissociation constants there will be $n + 1$ ionic species H_3A, H_2A, etc. Notice that H_3A could be a neutral molecule with H_2A, HA, and A bearing changes of -1, -2, and -3, respectively. Alternatively, H_3A might carry 1, 2, or 3 positive changes. In every case the mathematical expressions will be the same. For this reason the charges have been deliberately omitted from Eq. 3-3 and others that follow. Each constant in Eq. 3-3 is defined as in Eq. 3-2, i.e., $K_2 = [HA][H^+] / [H_2A]$, etc. Keep in mind that there are significant differences between the *apparent equilibrium constants* (concentration equilibrium constants) that we ordinarily use and thermodynamic equilibrium constants that are obtained by extrapolation to zero ionic strength.[1] A related complication is the uncertainty associated with the measurement of pH. This is often considered a measurement of hydrogen ion activity but this is not a correct statement. (The matter is considered briefly in Chapter 6). However, for all practical purposes, in the range of about pH 4–10 the pH can be equated with $-\log [H^+]$.

Often only one of the ionic forms of Eq. 3-3 will be important in a biochemical reaction. A particular ionic species may be the substrate for an enzyme. Likewise, an enzyme–substrate complex in only a certain state of protonation may react to given products. In these cases, and whenever pH affects an equilibrium, it is useful to relate the concentration $[A_i]$ of a given ionic form of a compound to the total of all ionic forms $[A]_t$ using Eqs. 3-4 and 3-5.

$$[A_i] = [A]_t / F_i \qquad (3\text{-}4)$$

$$[A]_t = [A] + [HA] + [H_2A] + \cdots [H_nA] \qquad (3\text{-}5)$$

For Eq. 3-4, $A_1 = H_nA$, $A_2 = H_{n-1}A$, etc. and F_1, F_2, etc. are the *Michaelis pH functions*,[2,3] which were proposed by L. Michaelis in 1914. For the case represented by Eq. 3-3 there are four ionic species and therefore four Michaelis pH functions which have the following form (Eq. 3-6). Here, K_1, K_2, etc. are the usual consecutive acid dissociation constants.

$$F_1 = 1 + K_1/[H^+] + K_1K_2/[H^+]^2 + K_1K_2K_3/[H^+]^3$$

$$F_2 = [H^+]/K_1 + 1 + K_2/[H^+] + K_2K_3/[H^+]^2$$

$$F_3 = [H^+]^2/K_1K_2 + [H^+]/K_2 + 1 + K_3/[H^+]$$

$$F_4 = [H^+]^3/K_1K_2K_3 + [H^+]^2/K_2K_3 + [H^+]/K_3 + 1$$

$$(3\text{-}6)$$

If there are only three ionic forms the first three of these equations will apply if the final term is dropped from each. The student should be able to verify these equations and to write the appropriate pH functions for other cases. Since these relationships are met so often in biochemistry it is worthwhile to program a computer to evaluate the Michaelis pH functions and to apply them as needed. From Eq. 3-4 it can be seen that *the reciprocal of the Michaelis pH function for a given ionic form represents the fraction of the total compound in that form* and that the sum of these reciprocals for all the ionic forms is equal to one. Examples of the use of the Michaelis pH functions in this book are given in Eq. 6-50, which relates the Gibbs energy of hydrolysis of ATP to the pH, and in Eqs. 9-55 to 9-57, which de-

scribe the pH dependence of enzymatic action.

Notice that in Eq. 3-3 single arrows have been used rather than the pairs ($\rightleftarrows$) that are often employed to indicate reversible equilibria. This is done so that *the direction of the arrow indicates whether we are using a dissociation constant or an association constant.* The use of single arrows in this manner to indicate how the equilibrium constants are to be written is a good practice when dealing with complex equilibria.

2. Titration Curves

When a neutral amino acid is titrated with acid, the carboxylate groups become protonated and acid is taken up. Likewise titration with base removes protons from the protonated amino groups and base is taken up. If we plot the number of equivalents of acid or base that have reacted with the neutral amino acid versus pH, a titration curve such as that shown in Fig. 3-1 is generated. The curve for histidine contains three steps; the first corresponds to the titration of the carboxylate group with acid, the second to the titration of the protonated imidazole of the side chain, and the third to the titration of the protonated amino group with base. Each step is characterized by a midpoint that is equal to the pK_a value for the group being titrated. The ends of the curve at low and high pH, which are drawn with a dashed line, are obtained only after corrections have been applied to the data. If only the equivalents of acid or base *added* rather than the number *reacted* are plotted, we obtain the solid line shown in Fig. 3-1. We see that at the low pH end there is no distinct end point. As we add more acid to try to complete the titration, the correction that must be applied to the data becomes increasingly greater. The difference between the dashed and solid lines reflects the fact that at the low pH end much of the acid added is used to simply lower the pH. Therefore, we have a large free $[H^+]$. Similarly, at the high pH end we have a high free $[OH^-]$.

The exact shapes of the ends of the titration curve depend heavily on the total concentration. Likewise, the magnitude of the correction required to obtain a plot of equivalents of acid or base reacted varies with the concentration and is smaller the higher the concentration of the substance being titrated (see problems 2 and 3 at the end of this chapter). An important rule in doing acid-base titrations, especially when very small samples are available, is to use the highest possible concentration of sample in the smallest possible volume and to titrate with relatively concentrated acid or base. Because of the difficulty of adequately correcting titration curves at low pH it is hard to estimate the pK_a values of the carboxyl groups of amino acids accurately from titration. An additional experimental difficulty exists at the high pH end because of the tendency for basic

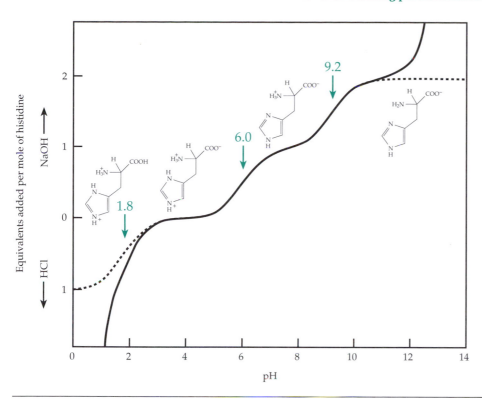

Figure 3-1 Titration curve for histidine. The solid line represents the uncorrected titration curve for 3 mM histidine monohydrochloride titrated with 0.2 M HCL to lower pH or with 0.2 M NaOH to higher pH assuming pK_a values of 1.82, 6.00, and 9.17. The dashed line represents the corrected curve showing the number of protons bound or lost per mole of histidine monohydrochloride.

solutions to absorb CO_2 from the air. Sometimes formaldehyde is added to shift the apparent pK_a of the amino groups to lower values and make the titration more satisfactory.

Despite the difficulties, real proteins can be titrated successfully to estimate the numbers of carboxyl, histidine, tyrosine, and other groups.[4] An example is shown in Fig. 3-2. The experimental data have been fitted with a theoretical curve based on the pK_a's of the carboxyl, histidine, tyrosine, and amino groups as determined by NMR measurements.[5] Account has been taken of the effect of the electrical field created by the many charged groups in distorting the curve from that obtained by summing the theoretical titration curves of the component groups. However, when there are multiple acid–base groups that are close together in a protein, a more complex situation involving tautomerism arises. This is discussed in Chapter 6.

Titration curves based on plots of light absorption versus pH or of NMR chemical shifts versus pH (see Fig. 3-29) are often useful. They have the important advantage that no special correction for free acid or base is needed at low or high pH.

3. Buffers

A mixture of a weak acid HA and of its **conjugate base** A constitutes a buffer which resists changes in pH. This can be seen most readily by taking logarithms of both sides of Eq. 3-2. By replacing log K with $-pK_a$

and log [H^+] with $-pH$ and rearranging we obtain Eq. 3-7 (the Henderson–Hasselbalch equation). It is sometimes useful to rewrite this as Eq. 3-8, where α is the fraction of the acid that is dissociated at a given pH.

$$pH - pK_a = \log([A]/[HA]) \qquad (3\text{-}7)$$

$$pH - pK_a = \log[\alpha/(1-\alpha)];$$

$$\alpha = [A] / [HA] + [A] = \frac{10^{(pH - pK_a)}}{1 + 10^{(pH - pK_a)}} \qquad (3\text{-}8)$$

Logarithms to the base 10 are used in both equations. These equations are useful in preparing buffers and in thinking about what fraction of a substance exists in a given ionic form at a particular value of pH. From Eq. 3-7 it is easy to show that when the pH is near the pK_a relatively large amounts of acid or base must be added to change the pH if the concentrations of the buffer pair A and HA are high.

Buffers are often added to maintain a constant pH in biochemical research[6] and naturally occurring buffer systems within body fluids and cells are very important (Box 6-A). Among the most important natural buffers are the proteins themselves, with the imidazole groups of histidine side chains providing much of the buffering capacity of cells around pH 7 (Figs. 3-1 and 3-2). Table 3-1 lists some useful biochemical buffers and their pK_a values. Here are a few

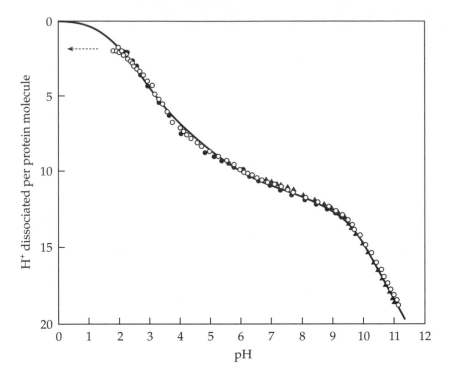

Figure 3-2 Acid–base titration curve for hen lysozyme at 0.1 ionic strength and 25°C. ○, initial titration from the pH attained after dialysis; ●, back titration after exposure to pH 1.8; ▲, back titration after exposure to pH 11.1. The solid curve was constructed on the basis of "intrinsic" pK_a values based on NMR data. From Kuramitsu and Hamaguchi[5]

which can be removed by vacuum evaporation or lyophilization.[8]

Before structural work can begin, pure substances must be separated from the complex mixtures in which they occur in cells and tissues.[4,13–24] Often, a substance must be isolated from a tissue in which it is present in a very low concentration. After it is isolated in pure form, if it is a large molecule, it must often be cut up into smaller pieces which are separated, purified, and identified. Accurate quantitative analysis is required to determine the ratios of these fragments. Considerable ingenuity may then have to be exercised in putting the pieces of the jigsaw puzzle back together to determine the structure of the "native" molecule. Many books, a few of which are cited here,[4,13–23,25–45] provide instructions. There are also journals and other periodicals dedicated to biochemical methods.[46–53]

practical hints about buffer preparation. Buffers containing monovalent ions tend to change pH with dilution less than do those with multivalent ions such as HPO_4^{2-} or $HP_2O_7^{3-}$. The pK_a values of carboxylic acids and of phosphoric acid, or of its organic derivatives, change very little with change in temperature. The pH of a buffer prepared with such components is nearly independent of temperature (Table 3-1). However, the pK_a of the $-NH_3^+$ group changes greatly with temperature. Buffer composition can be calculated readily from Eq. 3-8 and pK_a values from Table 3-1. It is convenient to keep in the laboratory standardized (to ~1% error) 1 M HCl and 1 M NaOH for use in buffer preparation. Compositions calculated from Eq. 3-8 will usually yield buffers of pH very close to those expected. Final adjustment with HCl may be needed if the pH is more than one unit away from a pK_a value. When two buffering materials are present, the composition should be calculated independently for each. The measurement of pH should always be done with great care because it is easy to make errors. Everything depends upon the reliability of the standard buffers used to calibrate the pH meter.[7] Often, especially during isolation of small compounds, it is desirable to work in the neutral pH region with volatile buffers, e.g., trimethylamine and CO_2 or ammonium bicarbonate,

1. Fractionation of Cells and Tissues

A fresh tissue or a paste of packed cells of a microorganism, usually collected by centrifugation, may be the starting material.[23,54–57] Tissue is often ground in a kitchen-type blender or, for gentler treatment, in a special **homogenizer**. The popular Potter–Elvehjem homogenizer is a small apparatus in which a glass or plastic pestle rotates inside a tight-fitting mortar tube (see standard laboratory equipment catalogs for pictures). Microbial cells are frequently broken with supersonic oscillation (**sonication**) or in special pressure cells. It is important to pay attention to the pH, buffer composition, and, if subcellular organelles are to be separated, the osmotic pressure. To preserve the integrity of organelles, 0.25 M sucrose is frequently used as the suspending medium, and $MgCl_2$ as well as a metal complexing agent such as ethylenediaminetetraacetate (EDTA) (Table 6-10) may be added. Soluble enzymes are often extracted without addition of sucrose, but reducing compounds such as glutathione (Box 11-B), mercaptoethanol, or dithiothreitol (Eq. 3-23) may be added. The crude **homogenate** may be strained and is usually centrifuged briefly to remove cell fragments and other "debris." Large-scale purification of

TABLE 3-1
Practical pK_a Values for Some Useful Buffer Compounds at 25°C and Ionic Strength 0.1[a]

Compound	pK_a	Grams per mole	d(pK_a)/dT	Charge on conjugate base
Citric acid (pK_1)		192		−1
Formic acid	3.7	0	−1	
Citric acid (pK_2)	4.45	192	−0.0016	−2
Acetic acid	4.64	60	0.0002	−1
Succinic acid (pK_2)	5.28	118	0	−2
Citric acid (pK_3)	5.80	192	0	−3
3,3-Dimethylglutaric acid	5.98	160	0.006	−2
Piperazine (pK_1)	(5.68) 6.02	86		0
Cacodylic acid (dimethylarsinic acid)	6.1	138		−1
MES[b]	6.1	195	−0.011	
BIS-TRIS[b]	6.41	209	−.017	
Carbonic acid (pK_1)	6.4[c]			
Pyrophosphoric acid (pK_3)	6.76	178		-2
Phosphoric acid (pK_2)	6.84	98	−.0028	−2
PIPES[b]	6.90	353	−.0085	−2
Imidazole	7.07	68	−.020	0
BES[b]	7.06	213	−.016	−1
Diethylmalonic acid	7.2	136		
TES[b]	7.37	279	−.020	−1
HEPES[b]	7.46	238	−.014	−1
N-Ethylmorpholine	7.79	115	−.022	0
Triethanolamine	7.88	149	−.020	0
TRICINE[b]	8.02	178	−.021	−1
TRIS[b]	8.16	121	−.031	0
Glycylglycine	8.23	132	−.028	−1
BICINE[b]	8.26	163	−.018	−1
4-Phenolsulfonic acid	8.70	174	−.013	−2
Diethanolamine	9.00	105	−.024	0
Ammonia	9.2		−0.031	0
Boric acid	9.2		−0.008	mixed
Pyrophosphoric acid (pK_4)	9.41	178		-3
Ethanolamine	9.62	61	−.029	0
Glycine (pK_2)	9.8	75	−0.025	−1
Piperazine (pK_2)	9.82	86		0
Carbonic acid (pK_2)	10.0		−0.009	−2
Piperidine	11.1	85		0

[a] Based on compilation by Ellis and Morrison[9] with additional data from Good et al.[10,11] and Dawson et al.[12] The Good buffers have dipolar ionic constituents. Since no form is without electrically charged groups they are unlikely to enter and disrupt cells.

[b] Abbreviations used:

BES	N,N-Bis(2-hydroxyethyl)-2-aminoethanesulfonic acid
BICINE	N,N-Bis(2-hydroxyethyl)glycine
BIS-TRIS	Bis(2-hydroxyethyl)iminotris(hydroxymethyl)methane
HEPES	N-2-Hydroxyethylpiperazine-N'-2-ethanesulfonic acid
MES	2-(N-Morpholino)ethanesulfonic acid
PIPES	Piperazine-N,N'-bis(2-ethanesulfonic acid)
TES	N-Tris(hydroxymethyl)methyl-2-aminoethanesulfonic acid
TRICINE	N-Tris(hydroxymethyl)methylglycine
TRIS	Tris (hydroxymethyl) aminomethane

[c] For CO_2 (solid) + $H_2O \rightarrow H^+ + HCO_3^-$, apparent p$K_a$.

proteins is often initiated with such a crude homogenate.

Cell organelles are also often separated by centrifugation. In one procedure a homogenate in 0.25 M sucrose (**isotonic** with most cells) is centrifuged for 10 min at a field of 600–1000 times the force of gravity (600–1000 g) to sediment nuclei and whole cells. The supernatant fluid is then centrifuged another 10 min at ~10,000 g to sediment mitochondria and lysosomes. Finally, centrifugation at ~100,000 g for about an hour yields a pellet of microsomes (p. 14),[58] which contains both membrane fragments and ribosomes. Each of the separated components can be resuspended and recentrifuged to obtain cleaner preparations of the organelles. The sedimented particles can often be solubilized by chemical treatment, for example, by the addition of either ionic or nonionic detergents. Membrane proteins can be isolated following solubilization in this way (Chapter 8). The soluble supernatant fluid remaining after the highest speed centrifugation provides the starting material for isolation of soluble enzymes and many small molecules.

2. Separations Based on Molecular Size, Shape, and Density

The simplest way to separate very large dissolved molecules is to let the small ones pass through a suitable sieve which may be a membrane with holes or a bed of gel particles. If the size of the particles approaches that of the holes in the sieve, the rate of passage will depend upon shape as well as size.

Dialysis, ultrafiltration, and perfusion chromatography. In dialysis[59] and ultrafiltration,[60] a thin membrane, e.g., made of cellulose acetate (cellophane) and containing holes 1–10 nm in diameter (typically 5 nm), is used as a semipermeable barrier. Small molecules pass through but large ones are retained. Dialysis depends upon diffusion and can be hastened by adequate stirring. Ultrafiltration requires a pressure difference across the membrane. The more sophisticated procedures of gel filtration and perfusion chromatography were introduced in 1959.[61–65] A column is packed with material such as the crosslinked dextran Sephadex, polyacrylamide gels (such as the Bio-Gel P Series), or agarose gels (e.g., Bio-Gels A and Sepharoses). These come in the form of soft beads, the interior network of which is a three-dimensional network of polymer strands (Fig. 4-10). Recently porous beads of hard crosslinked polystyrene, glass, or various other silicate materials have been employed.[66–68] The interstices between strands, whose size depends upon the degree of crosslinking introduced chemically into the gel, are small enough to exclude large molecules but to admit smaller ones. If a mixture of materials of different molecular size is passed through such a column the

smaller molecules are retarded because of diffusion into the gel, while the larger molecules pass through unretarded (Fig. 3-3). Sephadex G-25 excludes all but salts and compounds no larger than a simple sugar ring. Sephadex G-200, which is much less crosslinked, permits separation of macromolecules in the range of 5–200 kDa. As is explained in Section B, gel filtration also provides an important way of estimating M_r for proteins and other macromolecules.

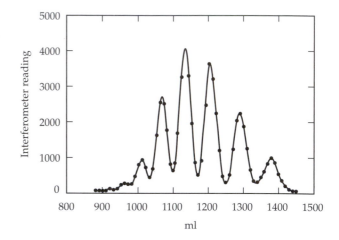

Figure 3-3 Separation of oligosaccharides by gel filtration. The sugars dissolved in distilled water were passed through a column of Sephadex G-25. The peaks contain (right to left) glucose, cellobiose, cellotriose, etc. From Flodin and Aspberg.[64]

Centrifugation. Centrifuges of many sizes and speeds are used in the laboratory to remove debris as well as to collect precipitated proteins and other materials at various steps in a purification scheme. The most remarkable are the **ultracentrifuges** which produce forces greater than 4×10^5 times that of gravity. They can be used both for separation of molecules and for determination of M_r (see Section C).

When macromolecules in a solution are subjected to an ultracentrifugal field they are accelerated rapidly to a constant velocity of sedimentation. This is expressed as a **sedimentation constant s**, which is the rate (cm/s) per unit of centrifugal force. The unit of s is the second but it is customary to give it in Svedberg units, S ($1S = 10^{-13}s$). Sizes of particles are often cited by their S values. The sedimentation constant is affected by the sizes, shapes, and densities of the particles as is discussed further in Section C. If carried out at constant velocity an equilibrium will eventually be attained in which sedimentation is just balanced by diffusion and a smooth concentration gradient forms from the top to the bottom of the centrifuge cell or tube. Concentration gradients can also be formed by centrifuging a concentrated solution of small molecules.[13,69,70] Such a concentration gradient is also a **density gradient** which can be made very steep. For example, a gradient with over a 10% increase in density from top to bottom can be created using 6 M **cesium chloride** (CsCl) and is widely used in DNA separations. If DNA is added prior to centrifugation it will come to rest in a narrow band or bands determined by the **buoyant densities** of the species of DNA present (see Chapter 5). After centrifugation, which is usually done in a plastic tube, a hypodermic needle is inserted through the bottom of the tube and the contents are pumped or allowed to flow by gravity into a fraction collector.

Another type of gradient centrifugation (**zone centrifugation**) utilizes a preformed gradient to stabilize bands of cell fragments, organelles, or macromolecules as they sediment.[71–74] For example, RNA can be separated into several fractions of differing sedimentation constants in a centrifuge tube that contains **sucrose** ranging in concentration from 25% at the bottom to 5% at the top. This is prepared by a special mixing device or "gradient maker" prior to centrifugation. The solution of RNA is carefully layered on the top, the tube is centrifuged at a high speed for several hours, and the different RNA fractions separate into slowly sedimenting sharp bands. A 20–60% gradient of sucrose or glycerol may be used in a similar way to separate organelles.[58]

3. Separations Based on Solubility

Some fibrous proteins are almost insoluble in water and everything else can be dissolved away. More often soluble proteins are precipitated from aqueous solutions by adjustment of the pH or by addition of large amounts of salts or of organic solvents. The solubility of any molecule is determined both by the forces that hold the molecules together in the solid state and by interactions with solvent molecules and with salts or other solutes that may be present.[58a]

Proteins usually have many positively and negatively charged groups on their surfaces. If either a positive or a negative charge predominates at a given pH the protein particles will tend to repel each other and to remain in solution. However, near the isoelectric point (pI), the pH at which the net charge is zero (see Section 7), the solubility will usually be at a minimum. The pH of a tissue extract may be adjusted carefully to the pI of a desired protein. Any protein that precipitates can be collected by centrifugation and redissolved to give a solution enriched in the protein sought. Some proteins, such as those classified (by an old system) as **globulins**, are insoluble in water but are readily **"salted in"** by addition of low concentrations (e.g., up to 0.1 M) of salts. Low concentrations of salts increase the solubility of most proteins because

the salt ions interact with the charged groups on the protein surfaces and interfere with strong electrostatic forces that are often involved in binding protein molecules together in the solid state. Some salts, including $CaCl_2$ and NaSCN, which bind to proteins, are especially effective in salting in.[75,76] Addition of *high* concentrations of salt causes precipitation of most proteins from aqueous solutions. The most effective and most widely used materials for this **"salting out"** of proteins are $(NH_4)_2SO_4$ and Na_2SO_4. Because the salt ions interact so strongly with water, the protein molecules interact less with water and more with each other. A similar intramolecular effect may cause the stabilization of proteins of halophilic bacteria by 1–4 M KCl.[77]

Different proteins precipitate at different concentrations of an added salt. Hence, a fraction of proteins precipitating between two different concentrations of salt can be selected for further purification (Fig. 3-4). Protein concentrations can be estimated as described in Box 3-A.

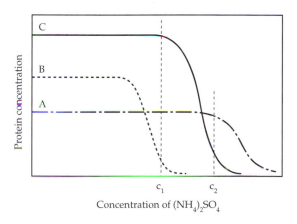

Concentration of $(NH_4)_2SO_4$

Figure 3-4 Hypothetical behavior of a solution containing three proteins, A, B, and C, upon ammonium sulfate fractionation. The concentration of protein remaining in the solution is plotted against ammonium sulfate concentration (usually expressed as % saturation). Addition of ammonium sulfate to concentration c_1 will precipitate largely protein B, which can be removed by centrifugation. Addition of additional salt to c_2 will precipitate largely protein C, while A remains in solution.

Precipitation methods are popular first steps in purification of proteins because they can be carried out on a large "batch" scale. The amounts of ammonium sulfate used are often expressed as percentage saturation, i.e., as a percentage of the amount required to saturate the solution (4.1 M at 25°). Convenient tables are available[12,78] that permit one to weigh out the correct amount of solid ammonium sulfate to give a desired percentage saturation or to go from one percentage saturation to a higher one.

Proteins are often stabilized by low concentrations of simple alcohols or ketones[76] and by higher concentrations of polyhydroxy alcohols, such as glycerol[77] and sucrose,[78] and also by certain inert, synthetic polymers such as **polyethyleneglycol** (PEG).[79] The latter is a widely used precipitant. The polyhydroxy-alcohols and PEG are all hydrated but tend not to interact strongly with the protein molecules. On the other hand, simple alcohols may denature proteins by their interaction with nonpolar regions.[77]

4. Separation by Partition

Many of the most important separation methods are based on repeated equilibration of a material between two separate phases, at least one of which is usually liquid. Small molecules may be separated by **countercurrent distribution** in which a material is repeatedly partitioned between two immiscible liquid phases, one more polar than the other. New portions of both liquids are moved by a machine in a "countercurrent manner" between the equilibration steps or are moved continuously through coiled tubes.[80–82,82a]

A similar result is accomplished by using as one phase a solid powder or fine "beads" packed in a vertical column or spread in a thin layer on a plate of glass. The methods are usually referred to as **chromatography**, a term proposed by Tswett to describe separation of materials by color. In 1903 Tswett passed solutions of plant leaf pigments (chlorophylls and carotenes) in nonpolar solvents such as hexane through columns of alumina and of various other adsorbents and observed separation of colored bands which moved down the column as more solvent was passed through. Individual

---

BOX 3-A QUANTITATIVE ESTIMATION OF PROTEIN CONCENTRATIONS

Biochemists often need to estimate the content of protein in a sample. For example, in devising a purification procedure for an enzyme it is customary to estimate the number of units of enzyme activity (as defined in Chapter 9) per milligram of protein (U/ mg). As progress is made in the purification this ratio increases. It becomes constant with respect to additional purification attempts, when a homogeneous enzyme is obtained.

One of the most widely used and most sensitive protein assays (for 0.1–1 mg/ml of protein) is the colorimetric procedure of **Lowry**.[a–c] It makes use of a phosphomolybdic–phosphotungstic acid reagent (the Folin–Ciocalteu reagent) which is reduced by proteins in the presence of alkaline Cu^{2+} to characteristic "blue oxides" whose color can be monitored at 750 nm. Much of the color comes from the reducing action of tyrosine and tryptophan. The color yield varies greatly from protein to protein and users may have trouble with reproducibility. A related method utilizes **bicinchoninic acid** which forms a purple color (measured at 362 nm) with the Cu^{+1} that is formed by reduction of alkaline Cu^{2+} by the protein.[d–f] This reagent is easier to use than that of the Lowry procedure and gives stable and reproducible readings.

A third widely used procedure, introduced by **Bradford**[g] and modified by others, measures the binding of the dye Coomassie brilliant blue whose peak absorption shifts from 465 nm to 595 nm upon binding. The change occurs within two minutes and is stable. However, the color yield varies from one protein to another.

Less sensitive but very simple and precise is measurement of the light absorption around 280 nm. This is discussed in the main text in Section D,6. For a typical protein an absorbance of 1.0 at 280 nm corresponds to a protein concentration of 1 mg/ml.[h] The very old **biuret method** is also useful for samples containing 1–10 mg/ml of protein. The violet color that arises upon addition of copper sulfate to an alkaline solution of a peptide or protein is recorded at 540–560 nm.[h] The color is especially intense for longer polypeptides. The name of the method arises from the fact that biuret gives a similar color[i] (see also Eq. 6-85).

Biuret

---

[a] Lowry, O. H., Rosebrough, N. J., Farr, A. L., and Randall, R. J. (1951) *J. Biol. Chem.* **193**, 265–275
[b] Peterson, G. L. (1979) *Anal. Biochem.* **100**, 201–220
[c] Larson, E., Howlett, B., and Jagendorf, A. (1986) *Anal. Biochem.* **155**, 243–248
[d] Smith, P. K., Krohn, R. I., Hermanson, G. T., Mallia, A. K., Gartner, F. H., Provenzano, M. D., Fujimoto, E. K., Goeke, N. M., Olson, B. J., and Kenk, D. C. (1985) *Anal. Biochem.* **150**, 76–85
[e] Davis, L. C., and Rodke, G. A. (1987) *Anal. Biochem.* **161**, 152–156
[f] Hill, H. D., and Straka, J. G. (1988) *Anal. Biochem.* **170**, 203–208
[g] Bradford, M. M. (1976) *Anal. Biochem.* **72**, 248–254
[h] Fruton, J. S., and Simmonds, S. (1958) *General Biochemistry*, 2nd ed., Wiley, New York (p. 130)
[i] Layne, E. (1957) *Methods Enzymol.* **III**, 450–451

pure pigments could be **eluted** from the column by continued passage of solvent. This important method is called **adsorption chromatography**. It is assumed that the pigments are absorbed on the surface rather than being disssolved in the solid material. A related method is **hydrophobic interaction chromatography** of proteins.[83–86] The packing material is similar to that described in Section 6 (Affinity Chromatography) but bears long-chain alkyl groups that can interact with the hydrophobic patches on surfaces of proteins. A very different adsorbent that is very useful in separation of proteins is carefully prepared microcrystalline **hydroxylapatite**.[87,88] It presumably functions in part by ion exchange.

Column packing materials such as **silica gel** contain a large amount of water, and separation involves partition between an immobilized aqueous phase in the gel and a mobile, often organic, solvent flowing through the column. Usually materials elute sooner when they are more soluble in the mobile phase than in the aqueous phase. These methods are closely related to perfusion chromatography, which is described in Section 2.

Aromatic amino acids, lipids, and many other materials can be separated on **reversed-phase** columns in which nonpolar groups, usually long-chain alkyl groups, are covalently attached to silica gel, alumina, or other inert materials.[66,80] The mobile phase is a more polar solvent, often aqueous, and gradually made less polar by addition of an organic solvent. In reversed-phase chromatography more polar compounds migrate faster through the system than do nonpolar materials, which experience hydrophobic interaction with the solid matrix.

Many traditional chromatographic methods including reversed-phase chromatography have been adapted for use in automatic systems which employ columns of very finely divided solid materials such as silica, alumina, or ion exchange materials coated onto fine glass beads.[89–91] These **high-performance liquid chromatographic** (**HPLC**) systems often utilize pressures as high as 300 atmospheres. Separations are often sharper and faster than with other chromatographic methods.[92–97] Reversed-phase columns in which the solid matrix may carry long (e.g. C_{18}) hydrocarbon chains have been especially popular for separation both of peptides and of proteins. Proteins may also be separated by gel filtration, ion exchange, or other procedures with HPLC equipment.

A sheet of high-quality filter paper containing adsorbed water serves as the stationary phase in **paper chromatography**. However, **thin-layer chromatography**, which employs a layer of silica gel or other material spread on a glass or plastic plate, has often supplanted paper chromatography because of its rapidity and sharp separations (Fig. 3-5).[16,96a,98–100] An approach that requires no stationary phase at all is

field flow fractionation.[101] Here a suitable external field (e.g., electrical, magnetic, and centrifugal) or a thermal gradient is imposed on the particles flowing through a narrow channel.

For volatile materials **vapor phase chromatography** (gas chromatography) permits equilibration between the gas phase and immobilized liquids at relatively high temperatures. The formation of volatile derivatives, e.g., methyl esters or trimethylsilyl derivatives of sugars, extends the usefulness of the method.[103,104] A method which makes use of neither a gas nor a liquid as the mobile phase is **supercritical fluid chromatography**.[105] A gas above but close to its critical pressure and temperature serves as the solvent. The technique has advantages of high resolution, low temperatures, and ease of recovery of products. Carbon dioxide, N_2O, and xenon are suitable solvents.

5. Ion Exchange Chromatography

Separation of molecules that contain electrically charged groups is often accomplished best by ion exchange chromatography.[105a] The technique depends upon interactions between the charged groups of the molecules being separated and fixed ionic groups on

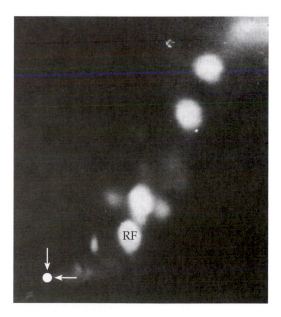

Figure 3-5 Photograph of a two-dimensional thin layer (silica gel) chromatogram of a mixture of flavins formed by irradiation of ~10 µg of the vitamin riboflavin. The photograph was made by the fluorescence of the compounds under ultraviolet light. Some riboflavin (RF) remains. The arrows indicate the location of the sample spot before chromatography. Chromatography solvents: a mixture of acetic acid, 2-butanone, methanol, and benzene in one direction and n-butanol, acetic acid, and water in the other. See Treadwell et al.[102]

an immobile matrix. Separation depends upon small differences in pK_a values and net charges and upon varying interactions of nonpolar parts of the molecules being separated with the matrix. Since changes in pH can affect both the charges on the molecules being separated and those of the ion exchange material, the affinities of the molecules being separated are strongly dependent on pH. For example, proteins and most amino acids are held tightly by cation exchangers at low pH but not at all at high pH.

Aqueous solutions are usually employed and the columns are packed with beads of **ion exchange resins**, porous materials containing bound ionic groups such as $-SO_3^-$, $-COO^-$, $-NH_3^+$, or quaternary nitrogen atoms. Synthetic resins based on a cross-linked polystyrene are usually employed for separation of small molecules. For larger molecules chemical derivatives of cellulose or of crosslinked dextrans (Sephadex), agarose, or polyacrylamide are more appropriate. Positively charged ions, such as amino acids in a low pH solution, are placed on a cation exchange resin such as Dowex 50, which contains dissociated sulfonic acid groups as well as counter ions such as Na^+, K^+, or H^+. The adsorbed amino acids are usually eluted with buffers of increasing pH containing sodium or lithium ions. The procedure,

which was developed by Moore and Stein,[106–109] is widely used for automatic quantitative analysis of amino acid mixtures obtained by hydrolysis of a protein or peptide (Fig. 3-6).[110–113]

Ion exchange chromatography of proteins and peptides is often done with such ion exchange materials as carboxymethyl-Sephadex and phosphocellulose, which carry negatively charged side chains or diethyl-aminoethylcellulose (DEAE-cellulose), which carries positively charged amino groups.[114] These materials do not denature proteins or entrap them and have a large enough surface area to provide a reasonable absorptive capacity. The mobile phase is usually buffered. For anion exchangers the pH should be above ~4.4 to keep most carboxylate side chains on the proteins ionized. The pH may be increased to pH 7–10, where most histidine imidazolium ions have dissociated, increasing the mobility of many proteins. For cation exchange the pH is usually buffered below pH 6 or 7 (Fig. 3-6).[115,116]

6. Affinity Chromatography

In this technique the chromatographic absorbent is designed to make use of specific biochemical inter-

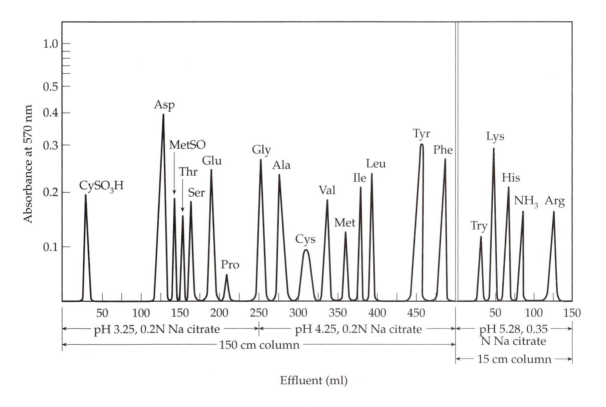

Figure 3-6 Separation of amino acids by cation-exchange chromatography on a sulfonated polystyrene resin in the Na^+ form by the method of Moore and Stein.[110] The amino acids were detected by reaction with ninhydrin (Box 3-C); areas under the peaks are proportional to the amounts. Two buffers of successively higher pH are used to elute the amino acids from one column, while a still higher pH buffer is used to separate basic amino acids on a shorter column. From Robyt and White.[13]

actions to "hook" selectively a particular macromolecule or group of macromolecules.[117–122] Affinity chromatography is used in many ways, including the purification of enzymes, antibodies, and other proteins that bind tightly to specific small molecules.

Because of their open gel structure (Fig. 4-10) agarose derivatives in bead form provide a good solid support matrix. The hydroxyl groups of the agarose are often linked to amino compounds. In one widely used procedure[119] the agarose is treated with cyanogen bromide (Br – C ≡N) in base to "activate" the carbohydrate (Eq. 3-9, step *a*). Then the amino compound is added (step *b*). The isourea product shown is the major one with agarose gels but dextran-based matrices tend to react by steps *c−e*.

Absorbents containing a large variety of R groups of shapes specifically designed to bind to the desired proteins can be made in this way. If the coupling is done with a diamine [R = (CH$_2$)$_n$—NH$_2$], the resulting ω-aminoalkyl agarose can be coupled with other compounds by reaction with **carbodiimide** (Eq. 3-10). For reaction in nonaqueous medium dicyclohexylcarbodiimide (R″ = cyclohexyl) is often used, but for linking groups to agarose a water-soluble reagent such as 1-ethyl-3-(3-dimethylaminopropyl) carbodiimide is recommended.[120] Carbodiimides are widely used for forming amide or phosphodiester linkages in the laboratory. The formation of an amide with a side chain of ω-aminoalkyl agarose can be pictured as in Eq. 3-10.

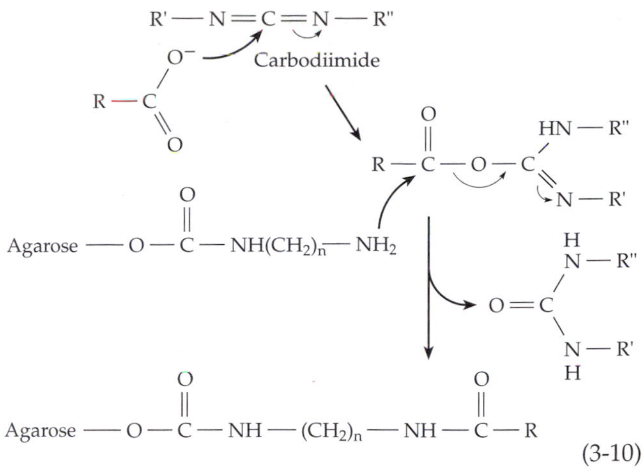

Many other means of preparation of adsorbents for affinity chromatography are also available.[121,122] For example, 1,1'-carbonyl-diimidazole can be used to couple a diamine to the matrix (Fig. 3-11). This reagent has the advantage that it does not depend upon the relatively unstable isourea linkages formed by Eq. 3-9 to hold the specific affinity ligands.[121]

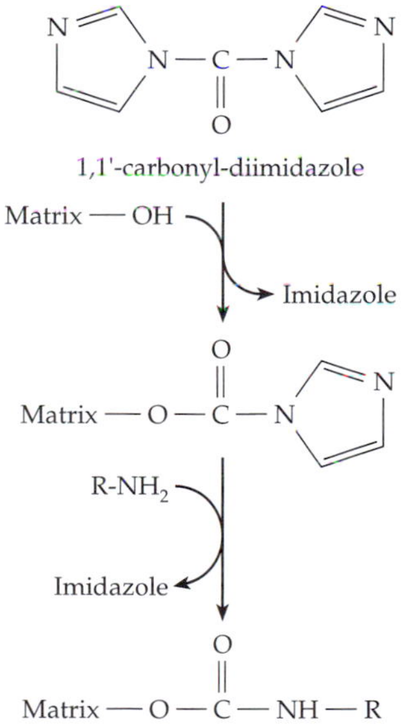

(3-10)

1,1'-carbonyl-diimidazole

(3-11)

An example of a successful application of affinity chromatography is the isolation of the enzyme cytidine deaminase from cells of *E. coli*. Cytidine was linked covalently via long **spacer** arms to the agarose beads as in the following diagram:

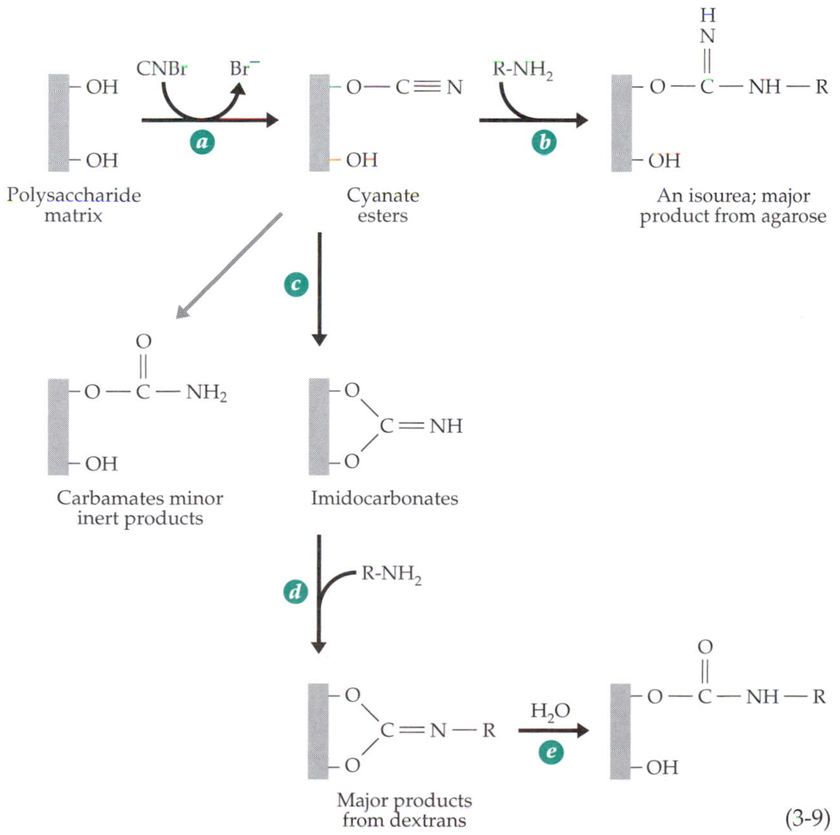

(3-9)

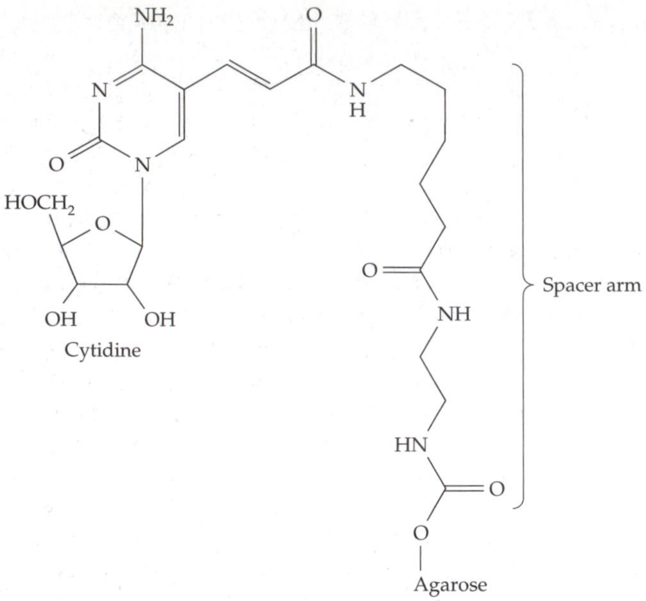

Cytidine

Spacer arm

Agarose

A cell extract was subjected to ammonium sulfate fractionation and the dialyzed protein was then poured through the affinity column which held the cytidine deaminase molecules because of their affinity for the cytidine structures that were bound to the agarose. The protein was eluted with a borate buffer; the borate formed complexes with the adjacent hydroxyl groups of the cytidine and thereby released the protein. After passage of the protein through an additional column of DEAE-Sephadex the deaminase had been purified 1700-fold compared to the crude extract.[123]

Another technique is to engineer genes to place a polyhistidine "tag" at the C terminus of a protein chain. Commercial cloning vehicles and kits are available for this purpose.[124] The protein produced when the engineered gene is expressed can be captured by the affinity of the polyhistidine tag for Cu^{2+}, Ni^{2+}, Co^{2+}, or Zn^{2+} held in chelated form on an affinity column.[124–127]

A surprising discovery was that certain dyes, for example Cibachron blue, when covalently coupled to a suitable matrix, often bind quite specifically to proteins that have a nonpolar binding pocket near a positive charge.[128] This includes many enzymes that act on nucleotides.

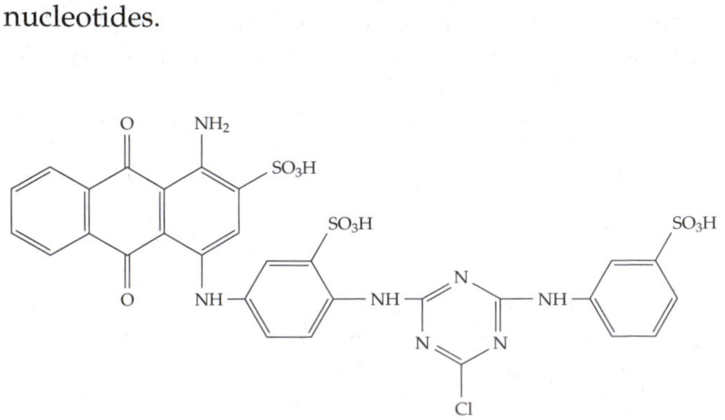

Cibacron blue F3GA

7. Electrophoresis and Isoelectric Focusing

The methods considered in this section make use of movement of molecules in an electrical field. Separation depends directly upon differences in the net charge carried by molecules at a fixed pH. The net charge for compounds containing various combinations of acidic and basic groups can be estimated by considering the pK_a of each group and the extent to which that group is dissociated at the selected pH using Eqs. 3-3 to 3-5. At some pH, the **isoelectric point** (pI), a molecule will carry no net charge and will be immobile in an electric field. At any other pH it will move toward the anode (+) or cathode (–). The pH at which the protein carries no net charge in the complete absence of added electrolytes is called the **isoionic point**.[129]

Electrophoresis, the process of separating molecules, and even intact cells[130] (Box 3-B), by migration in an electrical field, is conducted in many ways.[28,131–140] In **zone electrophoresis**, a tiny sample of protein solution, e.g., of blood serum, is placed in a thin line on a piece of paper or cellulose acetate. The sheet is moistened with a buffer and electrical current is passed through it. An applied voltage of a few hundred volts across a 20-cm strip suffices to separate serum proteins in an hour. To hasten the process and to prevent diffusion of low-molecular-weight materials, a higher voltage may be used. Two to three thousand volts may be applied to a sample cooled by water-chilled plates. Large-scale electrophoretic separations may be conducted in beds of starch or of other gels.

One of the most popular and sensitive methods for separation of proteins is electrophoresis in a column filled with **polyacrylamide** or **agarose gel** or on a thin layer of gel on a plate. The method depends upon both electrical charge and molecular size and has been referred to as **electrophoretic molecular sieving**.[28,133,135–137,141–143] Polyacrylamide gel electrophoresis is often carried out in the presence of ~1% of the denaturing detergent sodium docecyl sulfate which coats the polypeptide chain rather evenly. This method, which is often referred to as **SDS–PAGE**, has the advantage of breaking up complex proteins composed of more than one subunit and sorting the resultant monomeric polypeptide chains according to molecular mass (see Box 3-C). A disulfide-reducing reagent (see Eq. 3-23) such as ~1% 2-mercaptoethanol is usually present but may be omitted to permit detection of crosslinked peptides.

Capillary electrophoresis is increasingly popular[144–149a] and can be used to separate attomole amounts.[150] It can be used not only for separation of proteins but also for rapid estimation of the net charge on a protein.[151] The separation is conducted in tubes with internal diameters as small as 10–15 μm and as short as 1 cm. Multiple channels cut into a glass chip

BOX 3-B SORTING AND ANALYZING SINGLE CELLS

It is often important to examine and analyze individual cells.[a] For example, large numbers of single blood cells can be tested for the presence of specific antigenic determinants that arise by mutation. This permits assessment of the frequency of these mutations.[b] The complex chemical processing of neuropeptides can be studied on the contents of a single neuron (Chapter 30) using mass spectrometry.[c]

Several methods for separating cells have been devised. These include electrophoresis[d] or use of magnetic microspheres.[b] Micromanipulation can sometimes be used to select single cells for analysis. The most impressive technique is **flow cytometry**,[e,f] which is used daily on human blood samples in clinical laboratories. A suspension of cells is passed at a high rate of flow through a narrow capillary of ~0.2 mM diameter. The sample stream, which is surrounded by a larger "sheath" stream, has a

smaller diameter of ~20 μm. One or more laser beams are used to record information about each cell over a period of a few microseconds as the cells pass by at a rate of as much as 10^5 cells / s.

Flow cytometers developed from simpler cell counters, but now they are used to record cell size (from light scattering), optical absorbance, fluorescence, and phosphorescence. The optical properties are often enhanced by staining. The use of two dyes that fluoresce at different wavelengths permits the construction of two-dimensional plots as in the accompanying figure.

Capillary electrophoresis is one of the techniques able to separate constituents of single cells and is illustrated in the second figure.

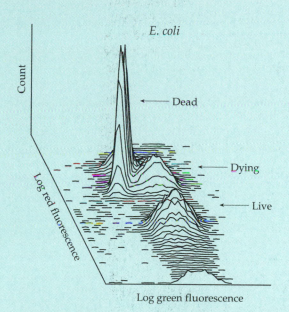

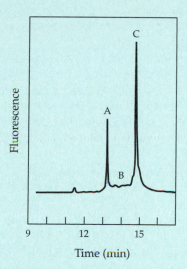

Electropherogram of major proteins from a single erythrocyte. Peaks A, B, and C are carbonic anhydrase (~7 amol), methemoglobin (~ 5 amol), and hemoglobin A_0 (~ 450 amol), as identified from migration times relative to standards.[a] Courtesy of Edward S. Yeung.

Flow cytometric histogram of fluorescently labeled live and dead *E. coli* bacteria. The dye kit (BacLightTM) that was used stains membrane-compromised dead bacteria with a red fluorescing dye and live bacteria with a green fluorescing dye. Cells were analyzed on an EPICS XL cytometer (Coulter Corporation). By integrating the area under each population it was possible to discern the percentage of dead (54), dying (16), and live (30) bacteria within a mixed population. The use of fluorescent dyes such as these has proven useful for studying various mechanisms employed by the food industry for killing microorganisms in food products and for studying a variety of bacteria derived from seawater, soil, plant materials, and laboratory-grown cultures. Courtesy of Kristi Harkins.

[a] Yeung, E. S. (1994) *Acc. Chem. Res.* **27**, 409–414

[b] Jovin, T. M., and Arndt-Jovin, D. J. (1980) *Trends Biochem. Sci.* **5**, 214–219

[c] Li, K. W., Hoek, R. M., Smith, F., Jiménez, C. R., van der Schors, R. C., van Veelen, P. A., Chen, S., van der Greef, J., Parish, D. C., Benjamin, P. R., and Geraerts, W. P. M. (1994) *J. Biol. Chem.* **269**, 30288–30292

[d] Bauer, J. (1994) *Cell Electrophoresis*, CRC Press, Boca Raton, Florida

[e] Shapiro, H. M. (1995) *Practical Flow Cytometry*, 3rd ed., Wiley-Liss, New York

[f] Darzynkiewicz, Z., Robinson, J. P., and Crissman, H. A., eds. (1994) *Flow Cytometry*, 2nd ed., Academic Press, San Diego

can be used.[152]

Whereas in conventional zone electrophoresis most of the electrical current is carried by the buffer, in **isotachophoresis**[153,154] the ions being separated carry most of the current. In **isoelectric focusing**,[28,155–157] a pH gradient is developed electrochemically in a vertical column or on a thin horizontal plate between an anode and a cathode. The pH gradient in a column is stabilized by the presence of a density gradient, often formed with sucrose, and the apparatus is maintained at a very constant temperature. Proteins within the column migrate in one direction or the other until they reach the pH of the isoelectric point where they carry no net charge and are "focused" into a narrow band. As little as 0.01 pH unit may separate two adjacent protein bands which are located at positions corresponding to their isoelectric points. A newer development is the use of very narrow pH gradients that are immobilized on a polyacrylamide matrix.[158–160] With this technique some hemoglobin mutants differing only in substitution of one neutral amino acid for another have been separated.[161] Special techniques are needed for highly basic proteins.[162]

A two-dimensional method in which proteins are separated by isoelectric focusing (preferably with an immobilized pH gradient) in the first dimension and by SDS-gel electrophoresis in the second has become a popular and spectacularly successful method for studying complex mixtures of proteins (Box 3-C).[163–166] Over 2000 proteins can be separated on a single plate. A similar procedure but without SDS can be used to examine undenatured proteins.[167–169] Computer-assisted methods are being developed to catalog the thousands of proteins being identified in this way[170–172] and also to allow rapid identification of spots by mass spectrometry. The technique can be applied to intact proteins in subpicomole quantities, even in whole cell lysates,[150,173,173a] or an enzyme such as trypsin can be used to cut the proteins into pieces on the gel plate and the mixtures of peptides can be analyzed by mass spectrometry.[174–176] Capillary electrophoresis or capillary isoelectric focusing can be applied before samples are sent to the mass spectrometer.

C. Determining the Relative Molecular Mass, M_r

The evaluation of M_r is often of critical importance. Minimum values of M_r can often be computed from the content of a minor constituent, e.g., the tryptophan of a protein or the iron of hemoglobin. However, physicochemical techniques provide the basis for most measurements.[177] Observations of osmotic pressure or light scattering can also be used and provide determinations of M_r that are simple in principle, but which have pitfalls.[178]

1. Ultracentrifugation

Some of the most reliable methods for determining M_r depend upon **analytical ultracentrifuges**. These instruments, capable of generating a centrifugal field as much as 4×10^5 times that of gravity, were developed in the 1920s and 1930s by T. Svedburg and associates in Uppsala, Sweden.[179–181] Driven by oil turbines, the instruments were expensive and difficult to use, but by 1948 a reliable electrically driven machine, the Beckman Model E ultracentrifuge, came into widespread use. It has had a major impact on our understanding of proteins, on methods of purification of proteins, and on our understanding of interactions of protein molecules with each other and with small molecules.[182,183] Nevertheless, it was not until 1990 that a truly "user-friendly" analytical untracentrifuge became available.[180,184–186] The Beckman Model XL-A centrifuge has a very small rotor driven by an air-cooled induction motor and is computer controlled. Data are recorded automatically in digital form and computer programs are available to carry out the necessary computations. The instrument can record ultraviolet-visible spectra at multiple radial positions in the sample cell (Fig. 3-7).

A straightforward determination of M_r is obtained by centrifuging until an equilibrium distribution of the molecules of a protein or other macromolecular material is obtained and by recording the variation in concentration from the center to the periphery of the centrifuge cell[177,183,185,187–190] (see also Section A,2). Using short cells, this **sedimentation equilibrium** can be attained in 1–5 hours instead of the 1–2 days needed with older instruments. For a single component system the concentration distribution at equilibrium is given by Eq. 3-12.

$$c(r) = c(a) \exp [M (1 - \bar{v}\rho) \omega^2 (r^2 - a^2) / 2RT] \quad (3\text{-}12)$$

Here $c(r)$ is the concentration c at the radial position r (measured from the centrifuge axis), a is the radial distance of the meniscus, M is the molecular mass in daltons, and $\bar{v}$ is the partial specific volume in ml/gram. For most proteins $\bar{v}$ varies from 0.69–0.75. It is the reciprocal of the density of the particle. Rho (ρ) is the density (g/ml) of the solvent. A plot of log $c(r)$ against r^2 is a straight line of slope $M (1 - \bar{v}\rho) / 2RT$. The computer can also accommodate mixtures of proteins of differing molecular masses, interacting mixtures, etc.[185,191]

Sedimentation velocity. The relative molecular mass M_r can also be measured from observation of the velocity of movement of the boundary (or boundaries for multicomponent systems) between solution and solvent from which the macromolecules have sedi-

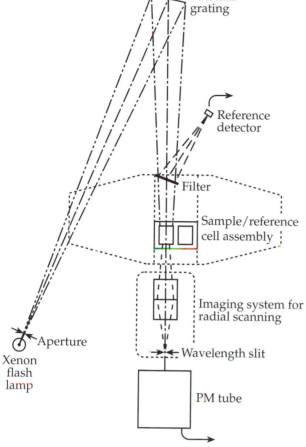

Figure 3-7 The scanning absorption optical system of the Beckman Optima™ XL-A ultracentrifuge. Courtesy of Beckman Coulter.

using Eq. 3-14.

$$D(\text{cm}^2 \text{ s}^{-1}) = k_B T/f = RT/Nf \qquad (3\text{-}14)$$

By combining Eqs. 3-13 and 3-14 we obtain the Svedberg equation:

$$M = \frac{RTs}{D(1-\bar{v}\rho)} \qquad (3\text{-}15)$$

Here R is in the cgs (cm-gram-second) unit of 8.31 x 10^7 erg mol^{-1} deg^{-1}. Using this equation the relative molecular mass M_r, which is numerically the same as M, can be evaluated from the sedimentation constant s. Since s, D, and $\bar{v}$ must all be measured with care, the method is demanding. It is often necessary to measure s and D at several concentrations and to extrapolate to infinite dilution. It is also customary to correct the data to give the values $s°_{20,w}$ and $D°_{20,w}$ expected at 20°C in pure water at infinite dilution.

To the extent that we can regard protein molecules as spherical we can substitute for f in Eq. 3-13 the frictional coefficient of a sphere:

$$f_{\text{sphere}} = 6\pi\eta r_h \qquad (3\text{-}16)$$

Here r_h is the hydrated radius or **Stokes radius** of the protein. On this assumption s will be expected to increase with the relative molecular mass approximately as $M_r^{2/3}$. A plot of log s against log M_r should be a straight line. Figure 3-8 shows such a plot for a number of proteins. The plots for nucleic acids, which can often be approximated as rods rather than spheres, fall on a different line from those of proteins. Furthermore, the sedimentation constant falls off more rapidly with increasing molecular mass than it should for spheres.

From analysis of a variety of well-characterized proteins, Squire and Himmel[192] observed that if proteins are assumed to contain 0.53 g H_2O per gram of protein and to have a mean value for $\bar{v}$ of 0.730 g / cm^3 the value of M_r can be predicted by Eq. 3-17 with the standard deviation indicated. Here, S is the sedimentation constant in Svedberg units (10^{-13}s).

$$M_r = 6850 \text{ S}^{3/2}) 0.090 M_r \qquad (3\text{-}17)$$

For proteins with various values of $\bar{v}$, Eq. 3-18 applies.

$$M_r = 922 [\text{S} / (1-\bar{v}\rho)]^{3/2}) \pm 0.066 M_r \qquad (3\text{-}18)$$

mented. This boundary, which can be visualized by optical methods, is quite sharp initially, but it broadens with time, because of diffusion, as the macromolecules sediment.

A molecule in a centrifuge is acted upon not only by the applied centrifugal force but also by an opposing **buoyant force** that depends upon the difference in density of the sedimenting particles and the solvent and by a frictional drag, which is proportional to a **frictional coefficient** f. Setting the sum of these forces to zero for the hydrodynamic steady state yields Eq. 3-13, which defines the **sedimentation constant** s.

$$s = \frac{v}{\omega^2 r} = \frac{M(1-\bar{v}\rho)}{Nf} \qquad (3\text{-}13)$$

Here f is the frictional coefficient which is difficult to predict or to measure but is often assumed to be the same as the frictional coefficient that affects **diffusion**. It can be obtained from the diffusion coefficient D

BOX 3-C ISOTOPES IN BIOCHEMICAL INVESTIGATIONS

Both stable[a] and radioactive[b–e] isotopes are widely used in chemical and biological investigations. The study of metabolism was revolutionized by the introduction of isotopic tracers. In one of the first biological experiments with the stable isotope ^{15}N (detected by mass spectrometry), Schoenheimer and associates in 1937 established the previously unsuspected turnover of protein in living tissues (Chapter 24, Section B). In 1937 Ruben et al. reported the uptake of radioactive $^{11}CO_2$ by plants.[f] A few years later Calvin and associates first traced the pathway of carbon in photosynthesis using the much longer lived $^{14}CO_2$ (Box 17-F). Wood and Werkman, in 1941, employed the stable isotope ^{13}C in studies of bacterial and mammalian metabolism (Box 17-C). The radioactive ^{32}P and ^{35}S have served to elucidate the metabolism of phosphorus and sulfur. Titrium (^{3}H) has been used to label many organic substances including thymine, which has been used extensively in the study of nucleic acid metabolism. Radioactive isotopes provide the basis for sensitive analytical procedures such as **radio-immunoassays** of minute quantities of hormones (Box 31-D). Through **radioautography** these isotopes facilitate numerous analytical procedures (see accompanying photo) and have provided the basis for important end-group methods used in sequence determination of polynucleotides (Eq. 5-24).

Several isotopes used in biochemistry are listed in the following table. For each radioactive isotope, the half-life is given, as is the type of particle emitted, and the energy of the particle. Gamma rays, such as those given off in decay of ^{125}I or ^{131}I, are very penetrating and easy to count precisely, as is the energetic β radiation from ^{32}P. On the other hand, ^{3}H (tritium) is relatively difficult to detect[g] but its weak β particle, which can travel only a short distance through a sample, makes it uniquely suitable for radioautography on a microscopic scale. Positrons ($β^{+}$) travel some distance, e.g., up to a few millimeters in the case of ^{13}N. They are then destroyed by reacting with an electron to produce a pair of γ rays of energy 0.511 MeV, equal to the sum of the rest masses of an electron plus a positron. The half-life (Eq. 9-4) determines the isotopic abundance needed to achieve a given radiation rate, a practical matter in providing a sufficient rate of decay to permit counting with an acceptably low statistical error. Even very short-lived isotopes such as ^{13}N, have proved useful as tracers.[h] The amount of an isotope giving 3.7×10^{10} disintegrations per second (this is 1 g of pure radium, 0.3 mg of ^{3}H, or 0.22 g of ^{14}C) is known as the **curie** (Ci). One millicurie (mCi) provides 2.22×10^{9} disintegrations / min

Isotope	Half-life	Maximum energy of radiation (MeV) β	γ
^{2}H (deuterium)	Stable		
^{3}H (tritium)	12.26 years	0.018	
^{11}C	20.4 min	0.511	($β^{+}$)
^{13}C	Stable		
^{14}C	5730 years	0.156	
^{13}N	9.96 min	1.2 0.511	($β^{+}$)
^{15}N	Stable		
^{15}O	20.4 min	0.511	($β^{+}$)
^{18}O	Stable		
^{18}F	110 min	0.511	($β^{+}$)
^{22}Na	2.6 years	0.55 1.28	
^{32}P	14.3 days	1.71	
^{35}S	87.2 days	0.167	
^{36}Cl	3×10^{5} years	0.716	
^{40}K	1.3×10^{9} years	1.4 1.5	
^{45}Ca	165 days	0.26	
^{59}Fe	45 days	0.46 1.1	
^{65}Zn	250 days	0.32 1.14	
^{90}Sr	29 years	0.54	
^{125}I	60 days	0.036	
^{131}I	8.06 days	0.61 0.36	

(dpm). Radiolabeled substances ordinarily contain only a small fraction of the unstable isotope together with a larger number of unlabeled molecules. Compounds are usually sold in millicurie or microcurie quantities and with a stated specific activity as mCi $mmol^{-1}$. For example, a compound labeled at a single position with ^{3}H and having a specific activity of 50 mCi $mmol^{-1}$ would contain about 0.17% ^{3}H at that position.

Because of the development of new NMR techniques and improvements in mass spectrometry stable isotopes, such as ^{2}H, ^{13}C, ^{19}F, and ^{31}P, are being used more frequently to study metabolism.[j] Carbon-13 containing compounds can fulfill many

BOX 3-C ISOTOPES IN BIOCHEMICAL INVESTIGATIONS (continued)

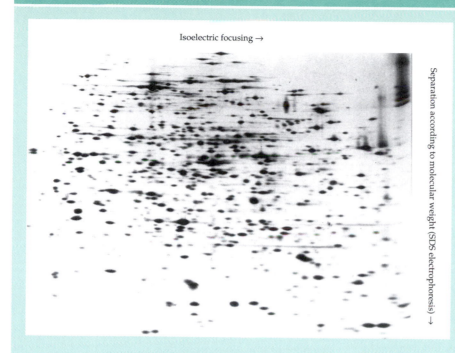

Isoelectric focusing →

Separation according to molecular weight (SDS electrophoresis) →

Radioautogram showing the separation of proteins of *E. coli* labeled with ^{14}C amino acids. From O'Farrell.[i] Twenty-five µl of sample containing 180,000 cpm and ~10 µg of protein were subjected to isoelectric focusing in a 2.5 x 130-mm tube containing polyacrylamide gel to separate proteins according to isoelectric point. The gel was then extruded from the column and was placed on one edge of a slab of polyacrylamide gel. Then SDS electrophoresis in the second dimension separated the proteins according to size. Over 1000 spots could be seen in the original radioautogram, which was obtained by placing a piece of photographic film over the gel slab and exposing it to the radiation for 875 hours. For details see O'Farrell.[i]

of the same tracer functions as [14]C. Even the radioactive ^{3}H nucleus can be utilized for *in vivo* NMR.[k,l] Although radioisotope labeling is very sensitive, it gives little information unless compounds are isolated and laboriously degraded to determine the positions of the labels. NMR spectroscopy is less sensitive but can give direct chemical information about the positions of ^{13}C in compounds within living cells. A compound containing only ^{13}C in one or in many positions can be safely administered to human individuals as well as to other organisms and spectra of products that arise can be observed. High-resolution deuterium NMR spectroscopy has been used to follow ^{2}H incorporated at C-1, C-2, or C-6 positions in glucose.[m]

As a result of metabolic reactions an isotope may appear at more than one position in a product, yielding two or more isotope isomers or **isotopomers**. These are seen individually by NMR spectroscopy and the concentration and isotope labeling patterns of the labeled compounds can be followed over a period of time. The use of this **isotopomer analysis** in studies of the citric acid cycle is illustrated in Box 17-C and its use in studies of glucose metabolism is considered in Chapter 17, Section L.

A change in isotopic mass, especially from ^{1}H to ^{2}H or ^{3}H, often produces a strong effect on reaction rates and the study of **kinetic isotope effects** has provided many insights into the mechanisms of enzymatically catalyzed reactions. Isotopes have permitted a detailed understanding of the stereo-

chemistry of enzymatic reactions, an impressive example being the synthesis and use of chiral acetate (Chapter 13)[n] and chiral phosphate groups (Chapter 12). Specific isotopic properties provide the basis for NMR (Section G).

[a] Matwiyoff, N. A., and Ott, D. G. (1973) *Science* **181**, 1125–1132
[b] Wang, C. H., Willis, D. L., and Loveland, W. D. (1975) *Radiotracer Methodology in the Biological, Environmental, and Physical Sciences*, Prentice-Hall, Englewood Cliffs, New Jersey
[c] Wang, Y., ed. (1969) *Handbook of Radioactive Nuclides*, CRC Press, Cleveland, Ohio
[d] Thornburn, C. C. (1972) *Isotopes and Radiation in Biology*, Butterworth, London
[e] Slater, R. J., ed. (1990) *Radioisotopes in Biology: A Practical Approach*, IRL Press, Oxford
[f] Ruben, S., Hassid, W. Z., and Kamen, M. D. (1939) *J. Am. Chem. Soc.* **61**, 661–663
[g] Bransome, J., ed. (1970) *Liquid Scintillation Counting*, Grune & Stratton, New York
[h] Cooper, A. J. L. (1985) *Adv. Enzymol.* **57**, 251–356
[i] O'Farrell, P. H. (1975) *J. Biol. Chem.* **250**, 4007–4021
[j] Wolfe, R. R. (1992) *Radioactive and Stable Isotope Tracers in Biomedicine*, Wiley, New York
[k] Newmark, R. D., Un, S., Williams, P. G., Carson, P. J., Morimoto, H., and Klein, M. P. (1990) *Proc. Natl. Acad. Sci. U.S.A.* **87**, 583–587
[l] Bergerat, A., Guschlbauer, W., and Fazakerley, G. V. (1991) *Proc. Natl. Acad. Sci. U.S.A.* **88**, 6394–6397
[m] Aguayo, J. B., Gamcsik, M. P., and Dick, J. D. (1988) *J. Biol. Chem.* **263**, 19552–19557
[n] Cornforth, J. W. (1976) *Science* **193**, 121–125

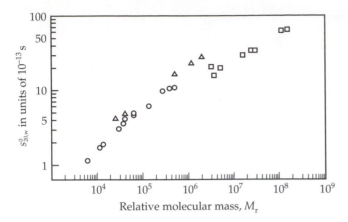

Figure 3-8 Plots of the logarithm of the sedimentation constant s against the logarithm of the molecular weight for a series of proteins and nucleic acids: ($\bigcirc$) globular proteins, ($\triangle$) RNA, and ($\square$) DNA. Proteins include a lipase (milk), cytochrome c, ribonuclease (pancreatic), lysozyme (egg white), follicle-stimulating hormone, bacterial proteases, human hemoglobin, prothrombin (bovine), malate dehydrogenase, γ-globulin (horse), tryptophanase (*E. coli*), glutamate dehydrogenase (chicken), and cytochrome a. Double-stranded DNA molecules are those of bacteriophage ϕX174 (replicative form), T_7, λ_{b2}, T_2, and T_4, and that of a papilloma virus. The RNA molecules are tRNA, rRNA, and mRNA of *E. coli*, and that of turnip yellow mosaic virus.[190a,191a]

2. Gel Filtration and Gel Electrophoresis

Several newer methods of molecular mass determination were developed in the 1960s–1980s. One is gel filtration. A column of gel beads such as Sephadex is prepared carefully and is calibrated by passing a series of protein solutions through it. The volume V_e at which a protein peak emerges from the column can be expressed as the sum of two terms (Eq. 3-19) in

$$V_e = V_o + \sigma V_i \qquad (3\text{-}19)$$

which V_o is the **void volume**, i.e., the elution volume that is observed for very large particles that are completely excluded from the gel, and V_i is the internal volume within the beads of gel. The value of σ is inversely related to the diffusion constant D, which for a spherical particle D is related by Eq. 3-20 to the Stokes radius r_h. This equation comes directly from Eqs. 3-16 and 3-14.

$$r_h = \frac{k_B T}{6\pi\eta D} \qquad (3\text{-}20)$$

This suggests a proportionality between σ and the molecular radius. In fact, Eq. 3-21, in which a and b are constants provides a fairly good approximation for σ and V_e is correlated approximately with log M_r as shown in Fig. 3-9.[193,194]

$$\sigma = a \log r_h + b \qquad (3\text{-}21)$$

A series of reference proteins of known molecular masses are used to calibrate the column and M_r for an unknown protein is estimated from its position on the graph.[195,196] Another modification of the method depends upon chromatography in a high concentration of the denaturing salt guanidinium chloride. The assumption is made that proteins are denatured into random coil conformations in this solvent.[196]

Probably the most widely used method for determining the molecular mass of protein subunits is gel electrophoresis in the presence of the denaturing detergent sodium dodecyl sulfate (SDS). The protein molecules are not only denatured but also all appear to become more or less evenly coated with detergent.[197] The resulting rodlike molecules usually show a uniform dependence of electrophoretic mobility on molecular mass (plotted as log M_r). An example is shown in Fig. 3-10. Again, the molecular mass of the protein under investigation is estimated by comparison of its rate of migration with that of a series of marker proteins.[195,198]

3. Mass Spectrometry

Mass spectrometry has played a role in biochemistry since the early 1940s when it was introduced for use in following isotopic labels during metabolism.[199–200c] However, it was not until the 1990s that suitable commercial instruments were developed to permit mass spectrometry using two new methods of ionization. The techniques are called **matrix-assisted laser desorption / ionization time-of-flight (MALDI-TOF)** and **electrospray ionization (ESI)** mass spectrometry.

In the MALDI technique a pulsed laser beam strikes a solid sample and heats, vaporizes, and ionizes compounds with little decomposition.[201–209] Proteins or other biopolymers are mixed with a "matrix" that absorbs the heat of the laser beam. The protein sample together with the matrix is dried. Most proteins form crystals and the laser beam is directed toward individual protein crystals or aggregates. Various materials are used for the matrix. Compounds as simple as glycerol, succinic acid, or urea can be used with an infrared laser. For proteins an ultraviolet nitrogen laser tuned to 337 nm is usually employed with an ultraviolet light-absorbing matrix such as hydroxybenzoic acid, 2,5-dihydroxybenzoic acid, α-hydroxy-

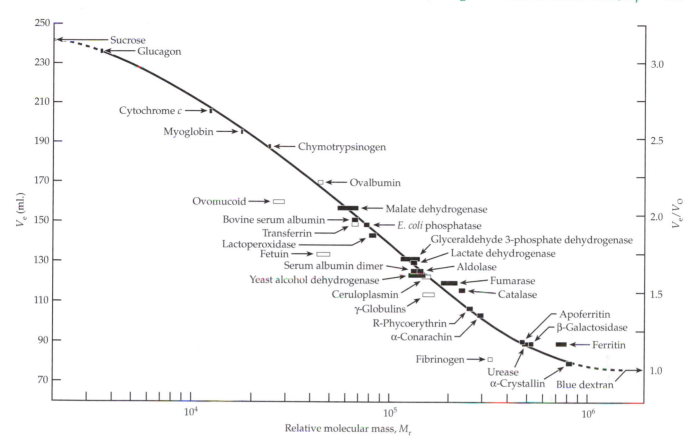

Figure 3-9 Elution volume of various proteins on a column of Sephadex G-200 as a function of molecular mass. The right-hand vertical axis shows the ratio of the elution volumes to that of blue dextran, a high-molecular-mass polysaccharide that is excluded from the internal volume. After Andrews.[193]

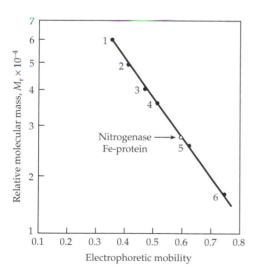

Figure 3-10 Estimation of the molecular mass of the polypeptide chain of the nitrogenase Fe-protein using SDS-polyacrylamide electrophoresis; from a set of four standard curves. The marker proteins are (1) catalase, (2) fumarase, (3) aldolase, (4) glyceraldehyde-phosphate dehydrogenase, (5) α-chymotrypsinogen A, and (6) myoglobin. (o) indicates position of azoferredoxin. From Nakos and Mortenson.[195]

cinnamic acid, or sinapinic acid (Chapter 25). The matrix ionizes, desorbs from the surface, and transfers energy to the crystalline protein, causing it to ionize and desorb from the surface. Oligosaccharides and oligonucleotides can be ionized in a similar way.

MALDI spectra are relatively simple (Fig. 3-11), often containing a single major peak corresponding to the singly charged molecular ion [M + H]$^+$ of mass $m + 1$ and perhaps a doubly charged molecular ion [M + 2H]$^{2+}$ of mass $(m + 2)/2$. For oligomeric proteins the major peak is often that of the monomer with weaker peaks for oligomers. The instrument can also be adjusted to generate negative ions whose detection is useful for study of phosphorylated peptides, many oligosaccharides, and oligonucleotides. With a TOF spectrometer there is no upper limit to the mass range and masses of over 100 kDa can be measured to about ±0.1%. Femtomole quantities can be detected.

The MALDI method is especially useful for complex mixtures of peptides and can be utilized in peptide sequencing. The technique is also appropriate for studying mixtures of glycoproteins. Negative-ion MALDI can be applied to oligonucleotide mixtures. Further improvements in resolution in both MALDI and ESI methods are anticipated as a result of developement of Fourier transform mass spectrometers.[202]

In ESI mass spectrometry[201,203–205,210–213] the sample, dissolved in an appropriate solvent (usually a 50:50 mixture of methanol and water for proteins), is infused directly into the ionization chamber of the spectrometer through a fused silica capillary. At the end of the capillary the solution is subjected to electrical stress created by a voltage difference of about 5 kV between the electrospray needle and the sampling orifice (the counter-electrode). The process results in the formation of singly and / or multiply charged molecular ions which are guided into the analyzer for mass analysis. For proteins every arginine, lysine, and histidine may bind a hydrogen ion to form a variety of positive ions. A 100-kDa protein may easily bind 100 protons bringing

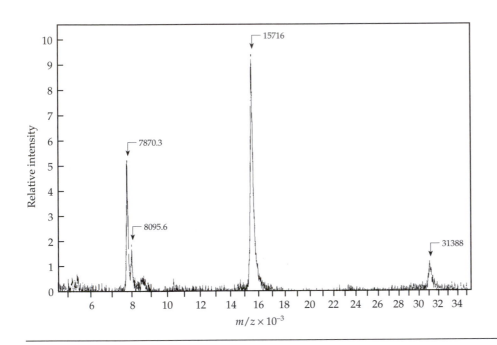

Figure 3-11 Matrix-assisted laser desorption / ionization time-of-flight (MALDI-TOF) mass spectrum of bovine erythrocyte Cu-Zn superoxide dismutase averaged over ten shots with background smoothing. One-half μl of solution containing 10 pmol of the enzyme in 5 mM ammonium bicarbonate was mixed with 0.5 μl of 50 mM α-cyanohydroxycinnamic acid dissolved in 30% (v / v) of acetonitrile-0.1% (v / v) of trifluoroacetic acid. The mixture was dried at 37° C before analysis. The spectrum shows a dimer of molecular mass of 31,388 Da, singly charged and doubly charged molecular ions at 15,716, and 7870 Da, respectively. The unidentified ion at mass 8095.6 may represent an adduct of the matrix with the doubly charged molecular ion. Courtesy of Louisa Tabatabai.

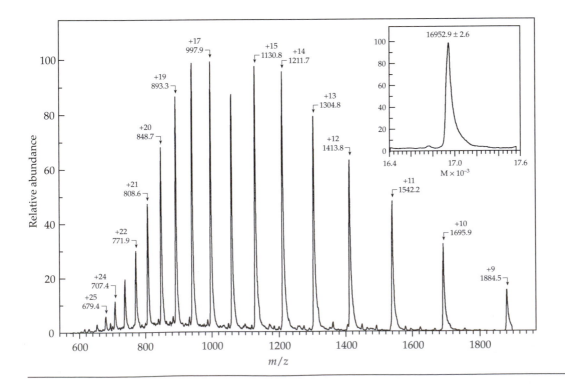

Figure 3-12 Positive ion electrospray mass spectrum of horse apomyoglobin (M_r 16,950.4). The net charge on each ion as well as the mass to charge ratio m/z is indicated at the top of each peak. The inset shows a computer "deconvolution" of the spectrum with the calculated value of molecular mass. Courtesy of Kamel Harrata.

the m/z ratio for the fully charged protein to 1000, well below the maximum m/z ratio of ~ 2400 for a typical quadrupole mass spectrometer. Since not all basic groups are protonated the spectra consist of families of peaks of differing m/z (Fig. 3-12). For a single pure protein the molecular mass can be calculated from the ratio of m/z values from any pair of adjacent peaks. It is better, especially if there is a mixture of proteins, to use a computer to extract M_r.[214] The accuracy can be quite high, typically $\pm 0.01\%$: one mass unit in 10,000.

Some complexity arises from the fact that each carbon atom in a protein contains about 1% of ^{13}C. This means that for a protein of mass > 10 kDa there will be a confusing array of peaks in the mass spectrum and it may be difficult to pick out the relatively minor "monoisotopic" peak that arises from molecules containing only ^{12}C, ^1H, ^{14}N, ^{16}O, and ^{32}S. In fact, the peak representing the most abundant mass will be a few mass units higher than the monoisotopic peak.[215,216] (see Study Question 15). New computer programs have been devised to assist in the analysis. Use of ^{13}C **and ^{15}N-depleted nutrients** also extends the applicability of mass spectrometry.[217]

Electrospray mass spectrometry utilizes a soft ionization technique at nearly atmospheric pressure. As a result, intact molecular ions are formed in high yield. The instrument can be interfaced readily to HPLC or capillary electrophoresis columns and sub-femtomole amounts of proteins can be detected. A disadvantage is that salt concentrations must be kept low ($< mM$) and that the protein tends to bind Na^+, K^+, and anions that may confuse interpretation of spectra.

D. Determining Amino Acid Composition and Sequence

When they are first isolated, proteins are usually characterized by M_r, isoelectric point, and other easily measured properties. Among these is the amino acid composition[112] which can be determined by completely hydrolyzing the protein to the free amino acids. Later, it is important to establish the primary structure or amino acid sequence.[51] This has been accomplished traditionally by cutting the peptide chain into smaller pieces that can be characterized easily. However, most protein sequences are now deduced initially from the corresponding DNA sequences, but further chemical characterization is often needed.

1. Cleavage of Disulfide Bridges

Before a polypeptide chain can be degraded it is usually necessary to break any disulfide bridges.[218-220] For some proteins such as the keratins of hair, these linkages must be broken even to get the protein into solution. Oxidation with performic acid (Eq. 3-22) has been used on ribonuclease but is not often employed because the performic acid also oxidizes tryptophan residues.

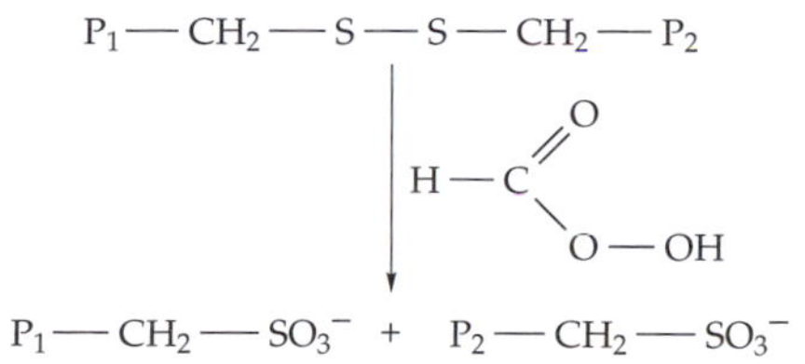

$$P_1 - CH_2 - S - S - CH_2 - P_2$$
$$\downarrow H-C{<}^O_{O-OH}$$
$$P_1 - CH_2 - SO_3^- \ + \ P_2 - CH_2 - SO_3^- \qquad (3\text{-}22)$$

Reduction by **dithiothreitol** or **dithioerythritol** (Eq. 3-23) is usually successful.

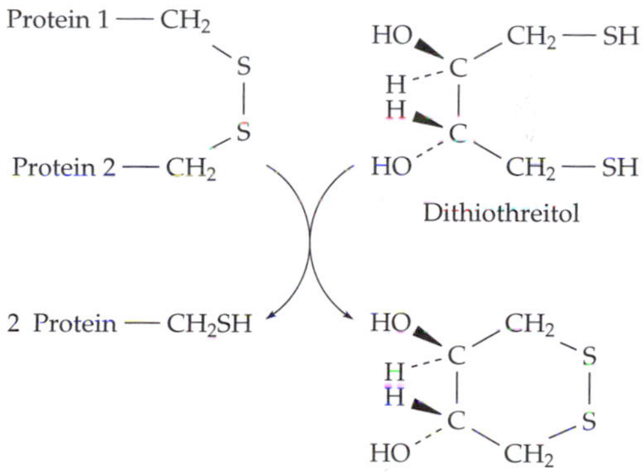

$$(3\text{-}23)$$

Upon oxidation these dithiols cyclize to form stable disulfides, driving the reaction to completion. These same compounds are also widely used to protect SH groups in enzymes against accidental oxidation by oxygen and to dissolve highly crosslinked insoluble proteins. Mercaptoethanol, $HS-CH_2-CH_2-OH$, may be used for the same purposes but requires higher concentrations and has a disagreeable odor.

Once cleaved, disulfide bridges may be prevented from reforming by conversion of the resulting thiol groups to stable derivatives, e.g., with iodoacetate and iodacetamide (Eq. 3-24).

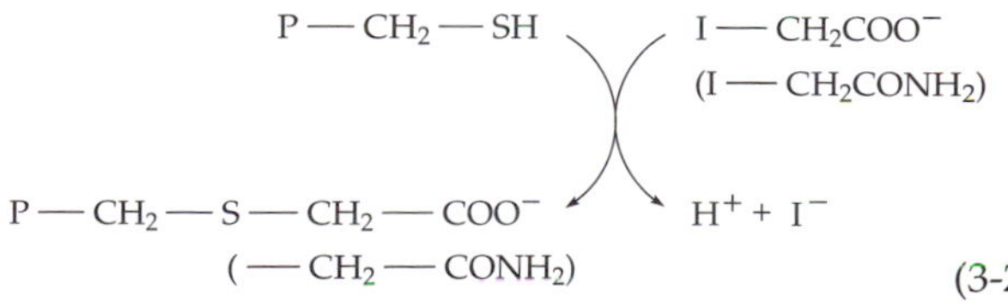

$$(3\text{-}24)$$

A better reagent is acrylonitrile (Eq. 3-25).

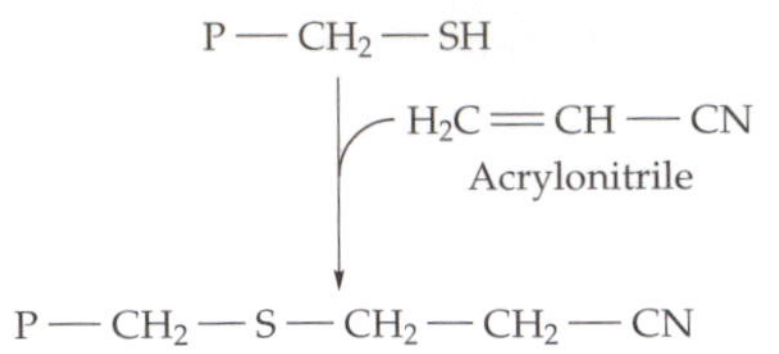

$$P — CH_2 — SH$$

$$H_2C = CH — CN$$
Acrylonitrile

$$P — CH_2 — S — CH_2 — CH_2 — CN \qquad (3\text{-}25)$$

In modern sequencing methods vinylpyridine, which reacts in a similar way, is often used. It can be detected during amino acid analysis or sequencing after derivatization with phenylisothiocyanate (Eq. 3-30).

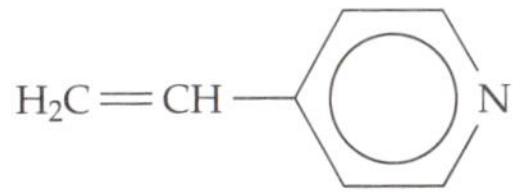

$$H_2C = CH —$$ ⟨pyridine ring⟩ $$N$$

Vinylpyridine

2. Hydrolysis and Other Chain Cleavage Reactions

Most biopolymers are inherently unstable with respect to cleavage to monomer units by reaction with water. Hydrolysis can be catalyzed by protons, by hydroxyl ions, or by the protein-hydrolyzing enzymes that are discussed in Chapter 12. Complete hydrolysis of proteins is usually accomplished by heating under nitrogen with 6 M HCl at 150°C for 65 min. Some amino acids, especially tryptophan, are destroyed and the amides in the side chains of asparagine and glutamine are converted to the free acids. Some peptide linkages such as Val-Val are very resistant and tend to be incompletely hydrolyzed. No procedure has been found which gives the ideal complete hydrolysis. Use of 4 M methanesulfonic acid containing 3-(2-aminoethyl)indole instead of 6 M HCl gives less decomposition of tryptophan.[221] Base-catalyzed hydrolysis of proteins also gives good yields of tryptophan but causes extensive racemization of amino acids.[222]

Complete enzymatic digestion of proteins can be accomplished with a mixture of enzymes including proteases produced by fungi (Pronase). However, the enzymes attack each other, making quantitative analysis difficult. The problem can be circumvented by immobilizing the hydrolytic enzymes in a column of agarose gel. The protein to be hydrolyzed is passed through the gel and the constituent amino acids emerge from the bottom of the column.[223,224]

Selective enzymatic hydrolysis. The traditional strategy in sequence determination is to cut protein chains into smaller pieces which can be separated by chromatography or electrophoresis and sequenced individually. Enzymatic cleavage is especially useful because of its specificity. **Trypsin**, a so-called **endo-**

peptidase, cleaves peptide chains at a rapid rate only if the carbonyl group of the amide linkage cleaved is contributed by one of the basic amino acids lysine, arginine, or aminoethylcysteine (see Fig. 12-10). If the protein is treated with maleic anhydride (Eq. 3-26),

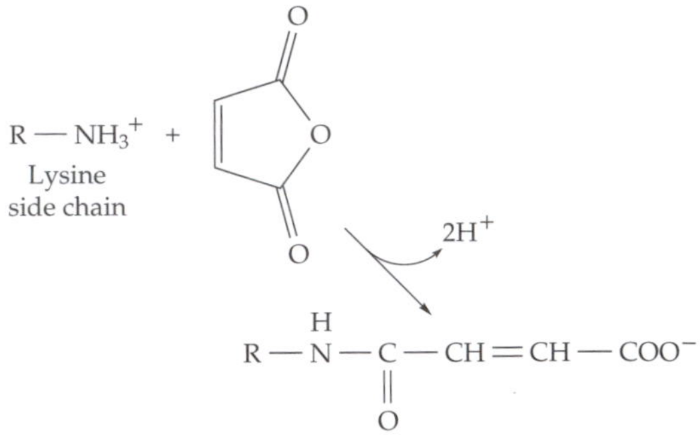

$$R — NH_3^+ \; +$$
Lysine side chain

⟨maleic anhydride structure⟩

$$2H^+$$

$$R — \overset{H}{\underset{\underset{O}{\|}}{N} — C} — CH = CH — COO^- \qquad (3\text{-}26)$$

BOX 3-D   THE PROTEOME

The ability to separate rapidly and detect minute amounts of proteins has spawned a new concept: The **proteome** is envisioned as a record of all proteins being actively synthesized by a cell – or of all genes being actively "expressed."[a–c] The concept evolved from efforts to automate the cataloging of spots on two-dimensional gels such as that shown in Box 3-B where each spot represents a single protein. The ability to unambiguously identify the spots by mass spectroscopy has brought new optimism to the attempt to use gels automatically to analyze all of the proteins formed by a cell.[a,c] The methods are potentially very important to developing new diagnostic procedures for human medicine. Watching changes in the proteome, including posttranslational modifications in proteins, as cells develop and grow will provide new insights into biochemical regulation.

A similar concept is that of a **complete transcriptional map**, a record of all of the different RNA molecules being synthesized by a cell.[d,e] These include many different mRNAs, each of which may give rise to more than one protein, as well as many RNAs with other functions.

[a] Kahn, P. (1995) *Science* **270**, 369–370
[b] Swinbanks, D. (1995) *Nature (London)* **378**, 653
[c] Shevchenko, A., Jensen, O. N., Podtelejnikov, A. V., Sagliocco, F., Wilm, M., Vorm, O., Mortensen, P., Shevchenko. A, Boucherie, H., and Mann, M. (1996) *Proc. Natl. Acad. Sci. U.S.A.* **93**, 14440–14445
[d] Richard, G.-F., Fairhead, C., and Dujon, B. (1997) *J. Mol. Biol.* **268**, 303–321
[e] Schena, M., Shalon, D., Heller, R., Chai, A., Brown, P. O., and Davis, R. W. (1996) *Proc. Natl. Acad. Sci. U.S.A.* **93**, 10614–10619

the lysine residues are protected and trypsin will cleave only at the Arg-X positions. If the resultant peptides are separated and held at pH 3.5 overnight, the blocking groups are hydrolyzed off and a second trypsin treatment can be used to cleave at the Lys-X positions.[225] The number of cleavage sites for trypsin can be increased by converting an –SH group to a positively charged one by aminoethylation (Eq. 3-27).[226] The reaction can be accomplished either with ethyleneimine (caution: carcinogen) as shown in this equation or by bromoethylamine, which eliminates Br^- to form ethyleneimine.

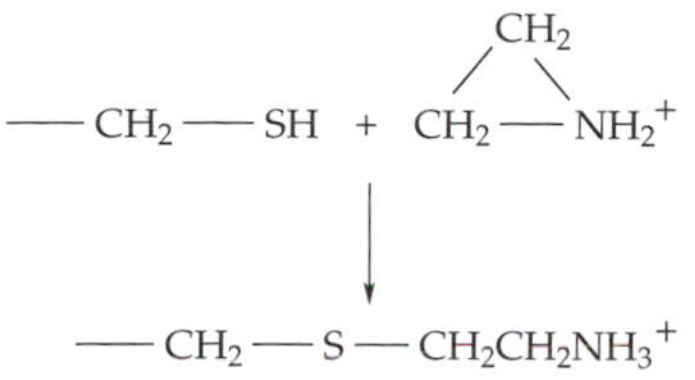

$$—CH_2—SH \; + \; CH_2 \overset{\displaystyle CH_2}{\diagup} NH_2^+$$

$$—CH_2—S—CH_2CH_2NH_3^+ \qquad (3\text{-}27)$$

Because of its specificity for basic residues, trypsin converts a protein into a relatively small number of **tryptic peptides** which may be separated and characterized. Trypsin acts primarily on denatured proteins, and to obtain good results the disulfide bridges must be broken first. **Chymotrypsin** is less specific than trypsin and **pepsin** is even less specific (Table 3-2). Nevertheless, they can be used to cut a peptide chain into smaller fragments whose sequences can be determined. To establish the complete amino acid sequence

for a protein, "overlapping" peptide fragments must be found that contain sequences from ends of two different tryptic fragments. In this way the tryptic peptides can be placed in the order in which they occurred in the native protein. This tedious procedure is rarely used today. Peptide sequencing is still important but is usually coordinated with gene sequencing, X-ray structure determination, or mass spectroscopy which minimize the need for overlapping fragments.

While trypsin cuts the peptide linkages Lys-X and Arg-X, a fungal protease cleaves only X-Lys.[227] A protease from the submaxillary glands of mice cleaves only Arg-X,[228] one from *Staphylococcus* specifically at Glu-X,[229,230] and one from kidneys at Pro-X.[231]

Several enzymes catalyze stepwise removal of amino acids from one or the other end of a peptide chain. **Carboxypeptidases**[232] remove amino acids from the carboxyl-terminal end, while **aminopeptidases** attack the opposite end. Using chromatographic methods, the amino acids released by these enzymes may be examined at various times and some idea of the sequence of amino acids at the chain ends may be obtained. A **dipeptidyl aminopeptidase** from bovine spleen cuts dipeptides one at a time from the amino terminus of a chain. These can be converted to volatile trimethylsilyl derivatives and identified by mass spectrometry.[233] If the chain is shortened by one residue using the Edman degradation (Section 3) and the dipeptidyl aminopeptidase is again used, a different set of dipeptides that overlaps the first will be obtained and a sequence can be deduced. Carboxypeptidase Y can be used with MALDI mass spectrometry to deduce the C-terminal amino acid sequence for a peptide. However, Ile and Leu cannot be distinquished.

Nonenzymatic cleavages. Of the various nonenzymatic methods that have been proposed, one has been outstandingly useful. Cyanogen bromide, $N \equiv C – Br$, cleaves peptide chains adjacent to methionine residues. The sulfur of methionine displaces the bromide ion (Eq. 3-28) and because of a favorable spatial relationship, the resulting sulfonium compound undergoes C–S bond cleavage through participation of the adjacent peptide group (Eq. 3-28, step *b*). The C=N of the product is then hydrolyzed with cleavage of the peptide chain in step *c*.

The linkage Asp-Gly can often be cleaved specifically by treatment with hydroxylamine at high pH.[195] Procedures for specific cleavage of tryptophanyl bonds have been de-

TABLE 3-2
Specificities of Commonly Used Protein-Hydrolyzing Enzymes

Trypsin	Lys-X, Arg-X	X not Pro
Chymotrypsin		
rapidly:	Phe-X, Tyr-X, Trp-X	X not Pro
slowly:	Y-X	X not Pro
	Y=Leu, Asn, Gln, His, Met, Ser, Thr	
Staphylococcus aureus		
protease V-8	Glu-X	X not Pro
Clostripain	Arg-X	
Pepsin		
preferentially:	X-Phe-X, X-Tyr-X, X-Leu-X	
less so:	X-Ala-X	
Thermolysin		
rapidly:	X-Y	
	Y=Ile, Leu, Val, Ala, Phe, Met	
slowly:	X-Y	
	Y=Tyr, Gly, Thr, Ser	

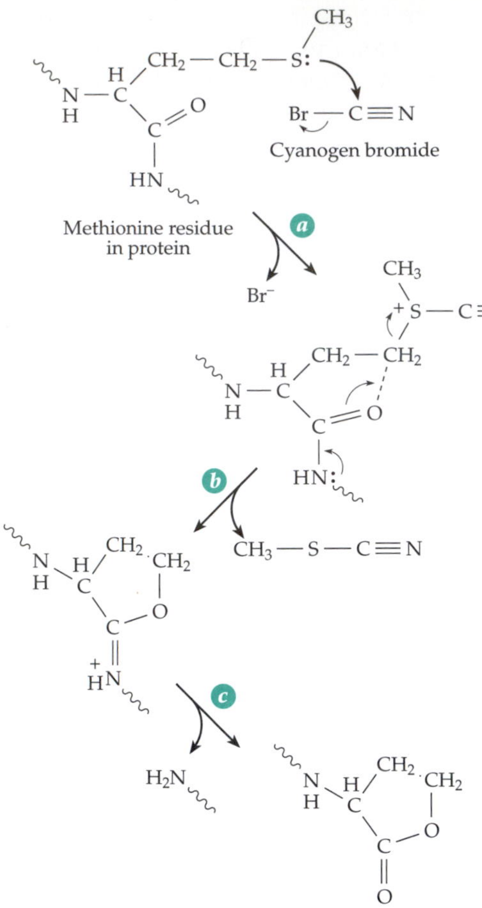

Methionine residue in protein

Cyanogen bromide

Peptidyl homoserine lactone   (3-28)

vised.[234] The Asp-Pro linkage is susceptible to cleavage by trifluoroacetic acid, which is used in the automated Edman degradation employed in peptide sequencing (Eq. 3-30). Cleavage by trifluoroacetic acid can be used to generate peptides for subsequent sequence determination.

Separating the peptides. A procedure that has been very important in the development of protein chemistry is **peptide mapping** or "fingerprinting." The procedure begins with cleavage of the disulfide linkages, denaturation, and digestion with trypsin or some other protease. The sizes and amino acid compositions of the resulting series of peptides are characteristic of the protein under study. The mixture of peptides is placed on a thin layer plate and subjected to chromatography in one direction, then to electrophoresis in the other direction, with the peptides separating into a characteristic pattern or fingerprint. Fingerprinting has been especially useful in searching for small differences in protein structure, for example, between genetic variants of the same protein (Fig. 7-27). Currently, the peptides are usually separated on ion exchange or gel filtration columns, by reversed-phase HPLC, or by capillary electrophoresis and are then often passed, in subpicomole amounts, into a mass spectrometer.[176,235,236]

3. Determining Amino Acid Sequence

The covalent structure of insulin was established by Frederick Sanger in 1953 after a 10-year effort. This was the first protein sequence determination.[237,238] Sanger used partial hydrolysis of peptide chains whose amino groups had been labeled by reaction with 2,4-dinitrofluorobenzene[239] to form shorter end-labeled fragments. These were analyzed for their amino acid composition and labeled and hydrolyzed again as necessary. Many peptides had to be analyzed to deduce the sequence of the 21-residue and 30-residue chains that are joined by disulfide linkages in insulin.[237,238]

The Sanger method is mainly of historic interest, although end-labeling may still be used for various purposes. A more sensitive labeling reagent than was used by Sanger is **dansyl chloride**. It reacts to form a sulfonamide linkage that is stable to acid hydrolysis and is brilliantly fluorescent (Eq. 3-29). The related reagent dimethylaminoazobenzene-4'-sulfonyl chloride gives highly colored derivatives easily seen on thin-layer chromatography plates.[240]

The Edman degradation. One of the most important reagents for sequence analysis is **phenylisothiocyanate,** whose use was developed by P. Edman.[241–243] This reagent also reacts with the N-terminal amino group of peptides (Eq. 3-30, step *a*). The resulting adduct undergoes cyclization with cleavage of the peptide linkage (Eq. 3-30, step *b*) under acidic conditions. After rearrangement (step *c*) the resulting **phenylthiohydantoin** of the N-terminal amino acid can be identified. The procedure can then be repeated on the shortened peptide chain to identify the amino acid residue in the second position. With careful work the Edman degradation can be carried down the chain for several tens of residues.

Ingenious **protein sequenators** have been devised to carry out the Edman degradation automatically.[242,244–246] Each released phenylthiohydantoin is then identified by HPLC or other techniques. Commercial sequenators have often required 5–20 nmol of peptide but new microsequenators can be used with amounts as low as 5–10 picomoles or less.[247,248]

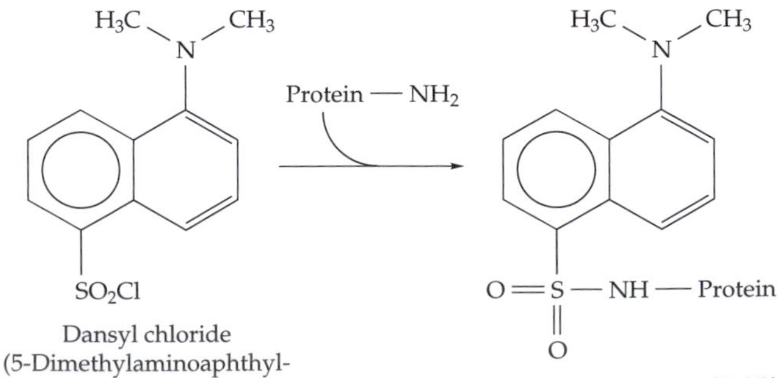

Dansyl chloride
(5-Dimethylaminoaphthyl-sulfonyl chloride)   (3-29)

Microsequencers permit sequence analysis on minute amounts of protein. Microsequencing can be used in conjunction with two-dimensional electrophoretic separations of proteins such as that shown in Box 3-C. The proteins in the polyacrylamide gel are electrophoretically transferred onto a porous sheet (membrane) of an inert material such as polyvinyl difluoride.[249–251] After staining, a selected spot is cut out and placed into the sequencer. To avoid the problems associated with blocked N termini, the protein may be treated with proteases on the membrane and the resulting peptide fragments may then be separated on a narrow-bore HPLC column and sequenced.[240]

Because many proteins are modified at the N terminus, blocking application of the Edman degradation, it would be useful to have a similar method for sequencing from the C terminus. It has been difficult to devise a suitable strategy, but there has been some success.[252–254]

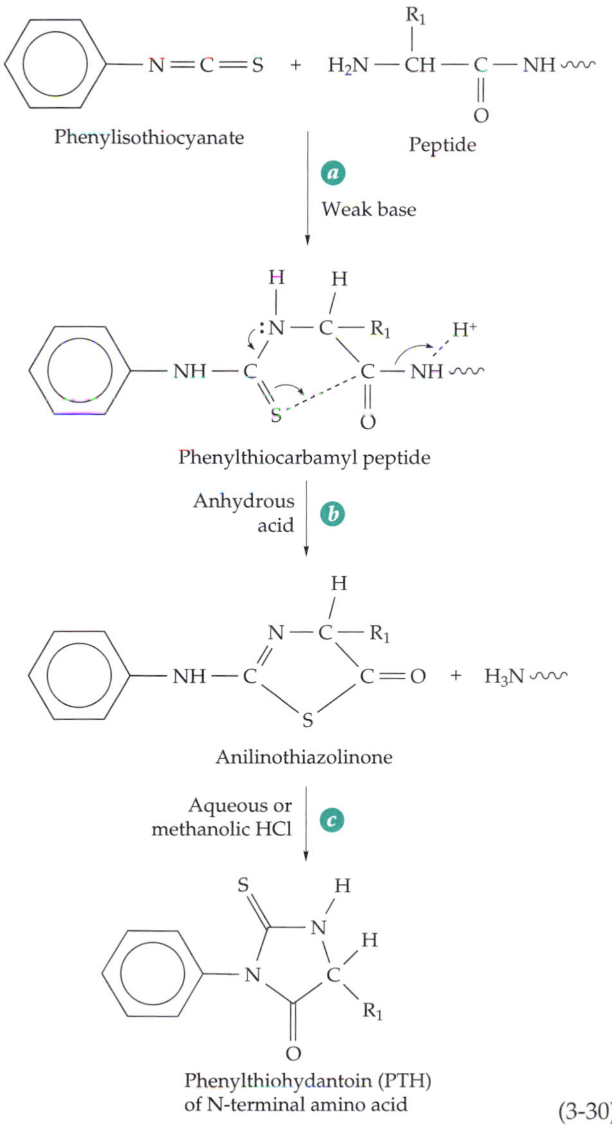

Phenylisothiocyanate   Peptide

a Weak base

Phenylthiocarbamyl peptide

Anhydrous acid **b**

Anilinothiazolinone

Aqueous or methanolic HCl **c**

Phenylthiohydantoin (PTH) of N-terminal amino acid   (3-30)

Protein sequences from the genes. Complete sequences of large numbers of genes have been determined and the corresponding sequences of proteins can be read directly from those of the corresponding genes. One method for sequencing a gene is to isolate a specific messenger RNA, which does not contain intervening sequences. A DNA copy (cDNA) is made from the mRNA and is used to ascertain the sequence of the encoded protein. The genomic DNA is also often sequenced. Introns are recognized by the nucleotide sequences at their ends and the correct amino acid sequence for the encoded protein is deduced.

In many instances, however, a gene can be identified only after part of the protein, often an N-terminal portion has been sequenced. This knowledge permits synthesis of an **oligonucleotide probe** that can be used to locate the gene (Chapter 5). Nucleotide sequences can be verified by comparison with sequences of tryptic or other fragments of a protein. Similarly, protein sequences are often checked by sequencing the corresponding genes as well as by study of X-ray structures. Substantial numbers of errors are made in sequencing of both DNA and protein so that checking is important.

Mass spectrometry in sequencing. Proteins can also be sequenced by mass spectrometry or by a combination of Edman degradation and mass spectrometry.[213,255,255a] Until recently the peptides had to be converted to volatile derivatives by extensive methylation and acetylation or by other procedures. However, newer ionization methods including MALDI (Fig. 3-11) and ESI (Fig. 3-12) have made it possible to obtain mass spectra on unmodified peptides. In one procedure a nonspecific protease cleaves a peptide chain into a mixture of small oligopeptides which are separated by HPLC into 20–40 fractions, each of which may contain 10–15 peptides but which can be sent directly into the ionization chamber of the mass spectrometer.[256] Peptides can be generated from a protein using immobilized enzymes, separated on a chromatographic column, and introduced sequentially into the mass spectrometer. Examination of peptide mixtures by mass spectrometry provides a way of verifying sequences deduced from DNA sequencing.[257] Mass spectrometry is also used widely to study covalently modified proteins.[173,216,257a,257b] As a rule, these cannot be recognized from gene sequences.

4. Locating Disulfide Bridges

A final step in sequencing is often the location of S–S bridges. The reduced and alkylated protein can be cleaved enzymatically (e.g., with elastase, pepsin, or thermolysin) to relatively small fragments, each of which contains no more than one modified cysteine.

The same enzymatic cleavage can then be applied to the unreduced enzyme. Pairs of peptide fragments remain linked by the S–S bridges. These crosslinked pairs can be separated, the disulfide bridges cleaved, and the resulting peptides identified, each as one of the already sequenced fragments. Mass spectrometry provides a rapid method for their identification.[258]

Another elegant way of locating S–S bridges employs **diagonal electrophoresis**. Electrophoresis of the digest containing the crosslinked pairs is conducted in one direction on a sheet of filter paper. Then the paper is exposed to performic acid vapor to cleave the bridges according to Eq. 3-22 and electrophoresis is conducted in the second direction and the paper is sprayed with ninhydrin. The spots falling off the diagonal are those that participated in S–S bridge formation. They can be associated in pairs from their positions on the paper and can be identified with peptides characterized during standard sequencing procedures.[259]

Diagonal electrophoresis and its relative diagonal chromatography are useful for other purposes as well. After electrophoresis or chromatography is conducted in one direction, the paper or thin-layer plate may be sprayed with a reagent that will react with some components or may be irradiated with light before the separation is repeated in the second direction (Fig. 3-5).[102,260]

5. Detecting Products

Important to almost all biochemical activity is the ability to detect, and to measure quantitatively, tiny amounts of specific compounds. "Color reagents," which develop characteristic colors with specific compounds, are especially popular. For example, **ninhydrin** (Box 3-E) can be used as a "spray reagent" to detect a small fraction of a micromole of an amino acid or peptide in a spot on a chromatogram. It can also be used for a quantitative determination, the color being developed in a solution. More sensitive than absorption of light (color) is fluorescence. **Fluorescamine** (Eq. 3-31) reacts with any primary amine to form a highly fluorescent product. As little as 50 pmol of amino acid can be determined quantitatively.[261] A yet more sensitive fluorogenic reagent for detection of amino acids, peptides, and amines of all types is o-phthaldialdehyde.[262,263]

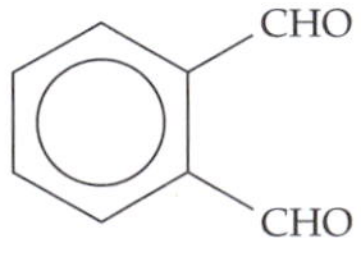

o-Phthaldialdehyde

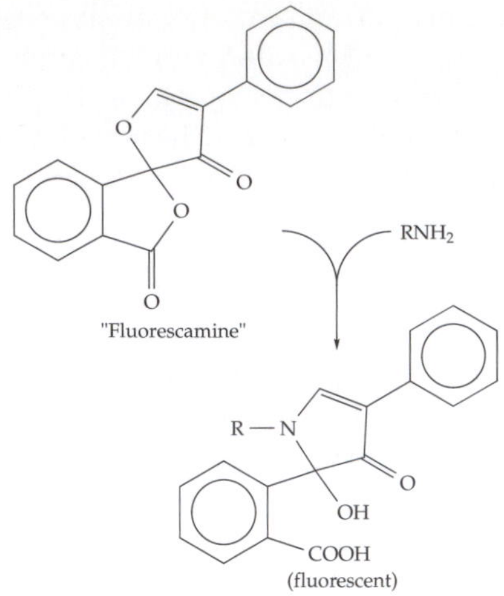

"Fluorescamine"

(fluorescent)   (3-31)

Reaction with naphthalene 2,3-dicarboxaldehyde (Eq. 3-32) increases the limit of detection 100-fold or more.[264]

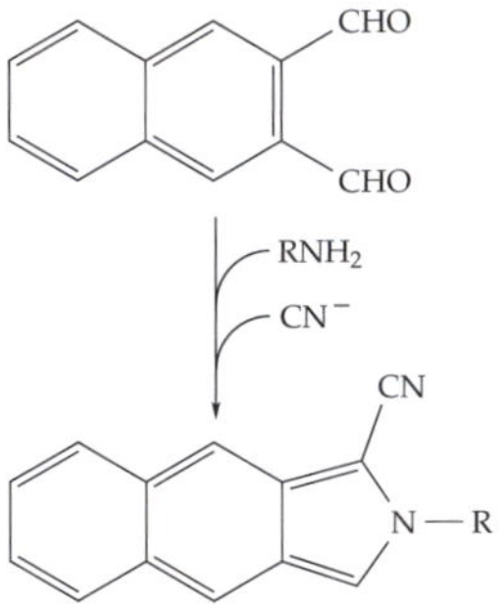

(3-32)

Detection of proteins on thin-layer plates, gel slabs, or membranes is often accomplished by staining with a dye,[265–267] the most widely used being Coomassie brilliant blue.[268] Various silver-containing stains may also be used. After separation of a protein mixture by electrophoresis and transfer to an inert membrane,

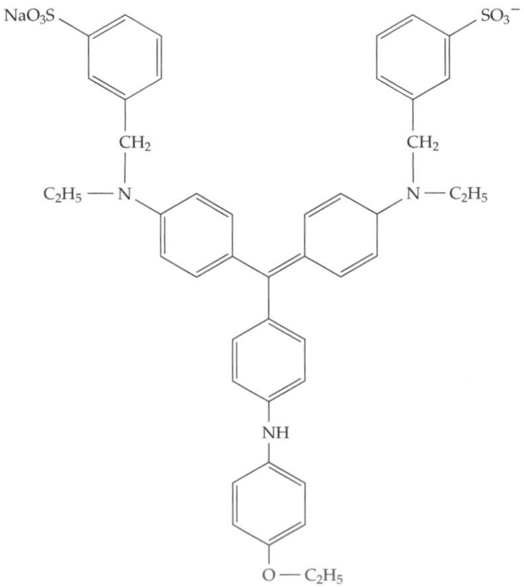

Coomassie brilliant blue

BOX 3-E NINHYDRIN

Ninhydrin (1,2,3-indantrione monohydrate) forms Schiff bases (ketimines) with amino acids. These react in ways similar to those of Schiff bases of pyridoxal phosphate (Chapter 14). Decarboxylation of the ketimines followed by hydrolysis of the resulting aldimines yields an intermediate amine that can couple with a second molecule of ninhydrin to form a characteristic purple color.[a,b] First reported by Ruhemann in 1910, the intermediate amine can also be hydrolyzed to free ammonia. Therefore, to ensure maximum color yield ninhydrin solutions for quantitative analysis usually contain reduced ninhydrin, which can react with free NH_3 and ninhydrin to form Ruhemann's purple (see scheme). The reaction has been widely used in chromatography and in quantitative amino acid analysis and also as a convenient spray reagent for paper and thin-layer chromatography. While α-amino acids react most readily, primary amines and peptides also form Ruhemann's purple. In these cases a proton rather than CO_2 is lost from the ketimine. When pyridoxamine (Chapter 14) on chromatograms reacts, a bright orange product, presumably the aldimine, appears. Secondary amines, such as proline, give a yellow color.

Both the ninhydrin reaction and pyridoxal phosphate-catalyzed decarboxylation of amino acids (Chapter 14) are examples of the **Strecker degradation**. Strecker reported in 1862 that alloxan causes the decarboxylation of alanine to acetaldehyde, CO_2, and ammonia.[c]

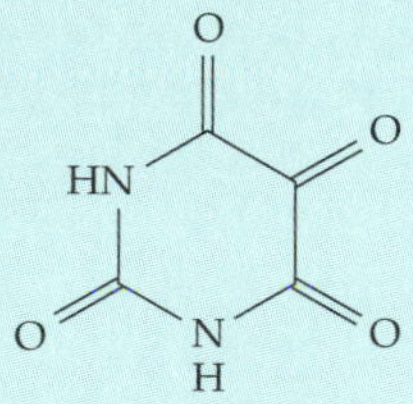

Alloxan

Many other carboxyl compounds, e.g., those of the general structure

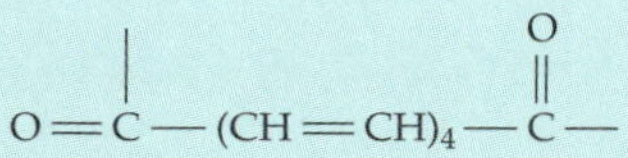

$$O{=}\overset{|}{C}{-}(CH{=}CH)_4{-}\overset{O}{\overset{\|}{C}}{-}$$

and p-nitrosalicylaldehyde also cause the Strecker degradation.[d]

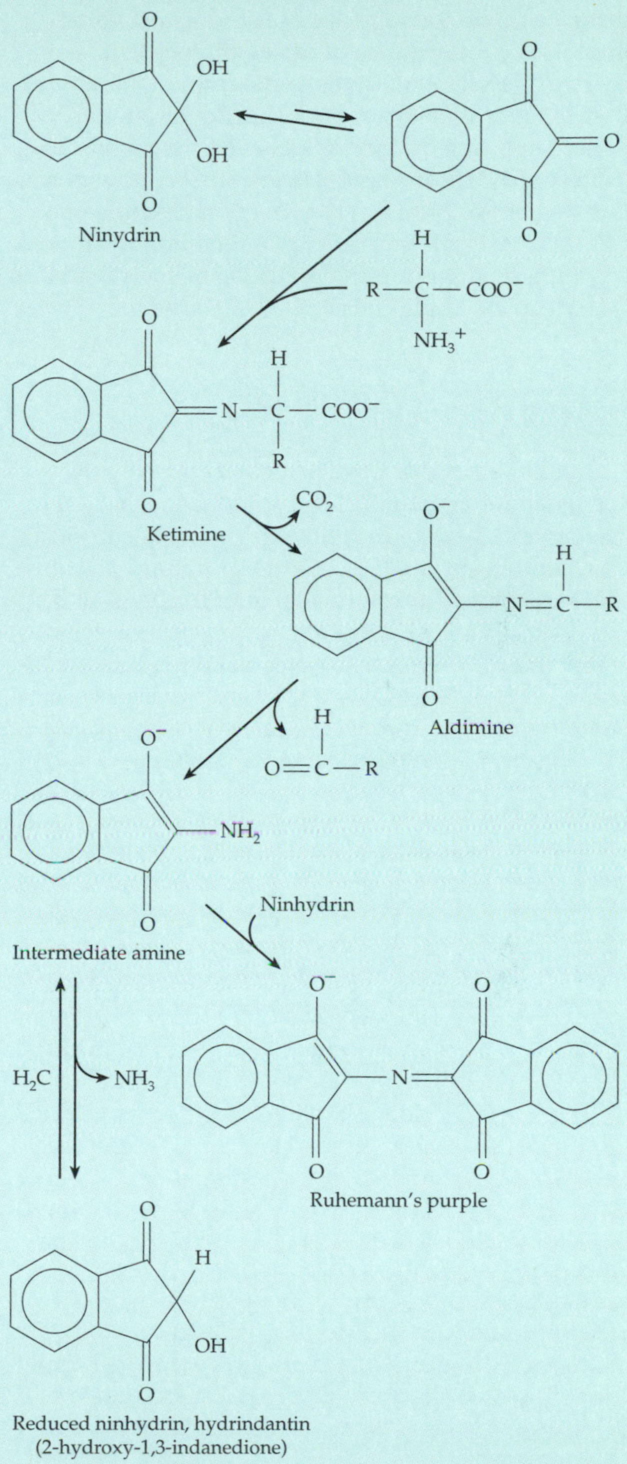

Ninydrin

Ketimine

Aldimine

Intermediate amine

Ruhemann's purple

Reduced ninhydrin, hydrindantin
(2-hydroxy-1,3-indanedione)

[a] Wigfield, D. C., and Croteau, S. M. (1980) *Biochem. Edu.* **8**, 26–27
[b] Friedman, M., and Williams, L. D. (1974) *Bioorg. Chem.* **3**, 267–280
[c] Strecker, A. (1862) *Annalen* **123**, 363–365
[d] Schonberg, A., and Moubacher, R. (1952) *Chem. Rev.* **50**, 261–277

the resulting protein "blots" can be stained with specific antibodies.[269,270] Flame ionization detectors can measure as little as a few picomoles of almost any substance leaving a vapor-phase chromatographic column. The importance of developing new, more sensitive analytical methods by which the quantity of material investigated can be scaled down can hardly be overemphasized. Increasingly sensitive methods of detection, including mass spectrometry, now permit measurement of fmol (10^{-15} mol) quantities in some cases. With this ability the output of neurotransmitters from a single neuron in the brain can be measured and the contents of single cells can be analyzed.

6. Absorption of Light

Side chains of the three aromatic amino acids phenylalanine, tyrosine, and tryptophan absorb ultraviolet light in the 240- to 300-nm region, while histidine and cystine absorb to a lesser extent. Figure 3-13 shows the absorption spectrum of a "reference compound" for tyrosine. There are three major absorption bands, the first one at 275 nm being a contributor to the well-

known 280-nm absorption band of proteins. There is a much stronger absorption band at about 240 nm. Sensitive methods for estimating protein concentration depend upon the measurement of this absorption together with that from other side chains at around 280 or 230 nm.[13,271–274] There is an even stronger absorption band at 192 nm. However, at these wavelengths even air absorbs light and experimental difficulties are extreme. At 280 nm, and even more at 230 nm, it is easy to contaminate samples with traces of light-absorbing material invisible to the eye. Therefore, most estimations of protein concentration from light absorption depend upon the 280-nm band.

Figure 3-14 shows the spectra of *N*-acetyl ethyl esters of all three of the aromatic amino acids and of cystine. To a first approximation, the absorption spectra of proteins can be regarded as a summation of the spectra of the component amino acids. However, the absorption bands of some residues, particularly of tyrosine and tryptophan, are shifted to longer wavelengths than those of the reference compounds in water. This is presumably a result of being located within nonpolar regions of the protein. Notice that the spectra for tyrosine, phenylalanine, and cystine in Fig.

BOX 3-F BIOSENSORS AND ELECTRONIC NOSES

A new approach to detection of molecules of biological interest is the development of biosensors. These are small devices that detect the binding of specific molecules to a **receptor** which is in intimate contact with a specially prepared surface that serves as a **transducer**. The receptor might be a layer of enzyme, antibody, hormone receptor, lectin, or oligonucleotide. Binding of substrate, antigen, hormone, sugar, or complementary polynucleotide strand, respectively, induces a response consisting of some kind of electrical or optical signal.[a–d] If the sensor is constructed on a semiconductor chip changes in an imposed potential difference may be detected.[a,e] However, changes in optical properties are more often observed. Fluorescence of dyes incorporated into the tranducing layer may be induced by binding of a molecule to a protein that undergoes an allosteric modification (see Chapter 9).[f] Many biosensors measure **surface plasmon resonance**, a change in the evanescent wave that develops in a surface when a light beam at the angle of total reflectance strikes the surface. This induces a change in the dielectric constant which can be measured.[f–m] Biosensors are used to estimate binding constants and also rate constants. However, read the article by Schuck and Milton[m] for tests of the validity of kinetic data. Biosensors can serve as "electronic noses." One possible application is in

the analysis of compounds in human breath as an aid to medical diagnosis. Over 400 volatile organic compounds have been identified in breath using gas chromatography and mass spectrometry.[n]

[a] Briggs, J. (1987) *Nature (London)* **329**, 565–566

[b] Zurer, P. (1997) *Chem. Eng. News* **September 15**, 7

[c] Kress-Rogers, E., ed. (1997) *Handbook of Biosensors and Electronic Noses: Medicine, Food, and the Environment*, CRC Press, Boca Raton, Florida

[d] Cunningham, A. J. (1998) *Introduction to Bioanalytical Sensors*, Wiley, New York

[e] McConnell, H. M., Owicki, J. C., Parce, J. W., Miller, D. L., Baxter, G. T., Wada, H. G., and Pitchford, S. (1992) *Science* **257**, 1906–1912

[f] Marvin, J. S., Corcoran, E. E., Hattangadi, N. A., Zhang, J. V., Gere, S. A., and Hellinga, H. W. (1997) *Proc. Natl. Acad. Sci. U.S.A.* **94**, 4366–4371

[g] Raether, H. (1988) *Surface Plasmons, Springer Tracts in Modern Physics*, Vol. 111, Springer-Verlag, Berlin

[h] Peterlinz, K. A., Georgiadis, R. M., Herne, T. M., and Tarlov, M. J. (1997) *J. Am. Chem. Soc.* **119**, 3401–3402

[i] Hendrix, M., Priestley, E. S., Joyce, G. F., and Wong, C.-H. (1997) *J. Am. Chem. Soc.* **119**, 3641–3648

[j] Salamon, Z., Brown, M. F., and Tollin, G. (1999) *Trends Biochem. Sci.* **24**, 213–219

[k] Chao, H., Houston, M. E., Jr., Grothe, S., Kay, C. M., O'Connor-McCourt, M., Irvin, R. T., and Hodges, R. S. (1996) *Biochemistry* **35**, 12175–12185

[l] McNally, A. J., Mattsson, L., and Jordan, F. (1995) *J. Biol. Chem.* **270**, 19744–19751

[m] Schuck, P., and Minton, A. P. (1996) *Trends Biochem. Sci.* **21**, 458–460

[o] Phillips, M. (1992) *Sci. Am.* **267**(July), 74–79

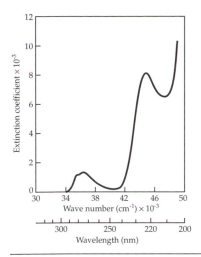

Figure 3-13 The absorption spectrum of *N*-acetyltyrosine ethyl ester in an aqueous phosphate buffer of pH 6.8. Absorbance (as molar extinction coefficient, Eq. 23-5) is plotted against increasing energy of light quanta in units of wave number. The more commonly used wavelength scale is also given. Spectra are most often presented with the low wavelength side to the left. In the convention adopted here the energy of a quantum increases to the right. There are three $\pi-\pi^*$ electronic transitions that give rise to absorption bands of increasing intensity. The third $\pi-\pi^*$ transition of the aromatic ring is at ~ 52,000 cm^{-1} (192 nm) and reaches a molar extinction coefficient of ~ 40,000. The $n-\pi^*$ and $\pi-\pi^*$ transitions of the amide group in this compound also contribute to the high energy end of the spectrum (see Chapter 23 for further discussion).

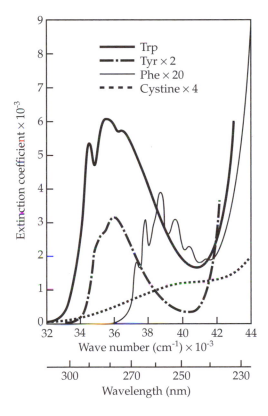

Figure 3-14 The spectra of the first electronic transitions of the *N*-acetyl derivatives of the ethyl esters of phenylalanine, tyrosine, and tryptophan together with that of the dimethyl ester of cystine in methanol at 25°C. The spectra for the Tyr, Phe, and cystine derivatives have been multiplied by the factors given on the graph.[272]

3-14 have been multiplied by factors of 2 to 20. It is evident that if all of the light-absorbing side chains were present in equal numbers tryptophan would dominate the absorption band and that phenylalanine would contribute little except some small wiggles. The molar extinction coefficient ε can be estimated

from the numbers of residues of each type per molecule as follows:[274]

$$\varepsilon_{280}\ (M^{-1}cm^{-1}) = 5500\ (\text{no. Trp}) + 1490\ (\text{no. Tyr}) +$$
$$125\ (\text{no. cystine}) \qquad (3\text{-}33)$$

For proteins of unknown composition, a useful approximation is that a solution containing 1 mg / ml of protein has an absorbance at 280 nm of about 1.0.

E. Quantitative Determinations and Modification Reactions of Side Chain Groups

The functional groups present in the side chains of proteins include $-NH_2$, $-SH$, $S-S$, $-OH$, $-COO^-$, the imidazole group of histidine, the guanidine group of arginine, the phenolic group of tyrosine, the indole ring of tryptophan, and the $-S-CH_3$ group of methionine. These are able to enter into a great variety of chemical reactions, most of which make use of the nucleophilic properties of these groups. The reactions are most often those of nucleophilic **addition** or nucleophilic **displacement**. The basic chemistry of these reactions often parallels biochemical reactions that are discussed in Chapters 12 and 13. In many instances, the reactions are nonspecific; amino, thiol, and hydroxyl groups may all react with the same reagent. The usefulness of the reactions depends to a large extent on the discovery of conditions under which there is some selectivity. It is also important that the reactions be complete. Only a few reactions will be considered here; these and others have been reviewed by Glazer *et al.*[225,275]

1. Reactions of Amino Groups

The numerous amino groups of lysine residues and of the N termini of peptide chains usually protrude into the aqueous surroundings of a protein. Chemical modification can be done in such a way as to preserve the net positive charge which amino groups carry at most pH values, to eliminate the positive charge leaving a neutral side chain, or to alter the charge to a negative value. Alterations of these charges can greatly affect interactions of the protein molecules with each other and with other substances.

Amino groups react reversibly with carbonyl compounds to form Schiff bases.[275–277] Reduction of the latter by sodium borohydride or sodium cyanoborohydride causes an irreversible change (Eq. 3-34, steps *a* and *b*). Cyanoborohydride is specific for Schiff bases and does not reduce the carbonyl compound. However, side products may cause problems.[277,278] Depending upon which carbonyl compound is used, the net positive charge on the amino group may be retained or it may be replaced with a different charge by this "reductive alkylation" sequence. Formaldehyde will react according to Eq. 3-34 in two steps to give a dimethyl amino group with no change of net charge.[279] Pyridoxal phosphate (Chapter 14) is converted by Eq. 3-34 into a fluorescent label. With a limited amount of pyridoxal phosphate only one or a few lysine residues may be labeled, often at active centers of enzymes. Schiff bases formed from glyceraldehyde in Eq. 3-34 can undergo the Amadori rearrangement (Eq. 4-8) to form stable products which, however, can be reconverted to the original amino groups upon acid hydrolysis. The borohydride reduction product of Eq. 3-34 with glyceraldehyde can be reconverted to the original amine by periodate oxidation (Eq. 4-11).[277]

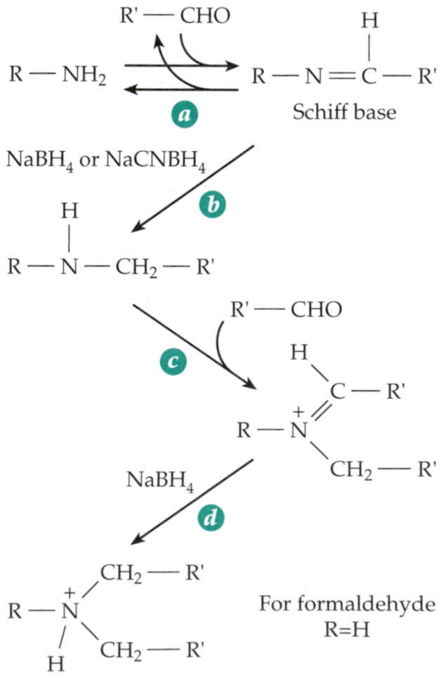

Another addition reaction of amino groups is **carbamoylation** with sodium cyanate (Eq. 3-35). A displacement reaction by an amino group on an acid

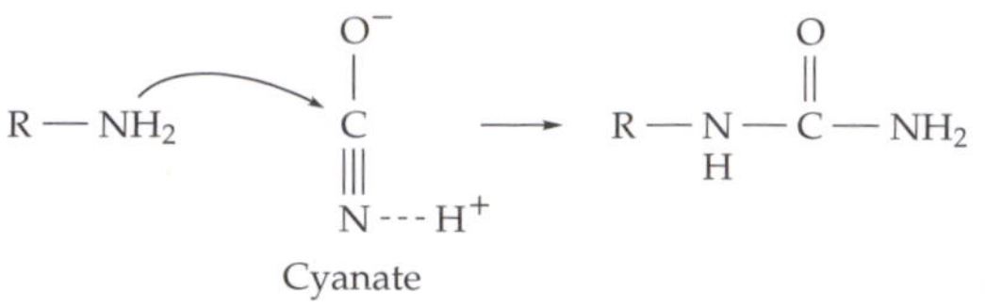

$$\text{Cyanate} \tag{3-35}$$

anhydride such as acetic anhydride (Eq. 3-36) leads to **acylation**, a nonspecific reaction which is also undergone by thiol, hydroxyl, and other groups. When acetic anhydride is used, the net positive charge of an

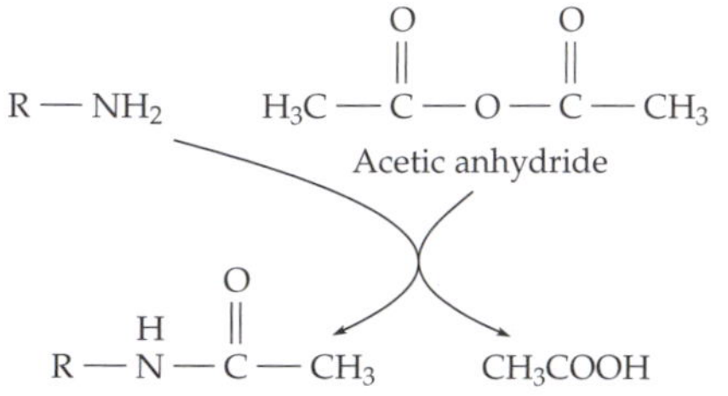

$$\tag{3-36}$$

amino group is lost. However, the product obtained with succinic anhydride or maleic anhydride (Eq. 3-26) carries a negative charge. In the latter case, the modification can readily be reversed by altering the pH.

Both **amidination** (Eq. 3-37) and **guanidination** (Eq. 3-38) lead to retention of the positive charge.

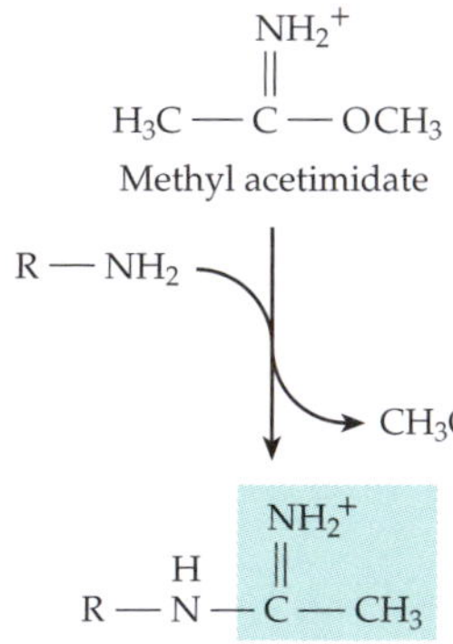

$$\tag{3-34}$$

$$\tag{3-37}$$

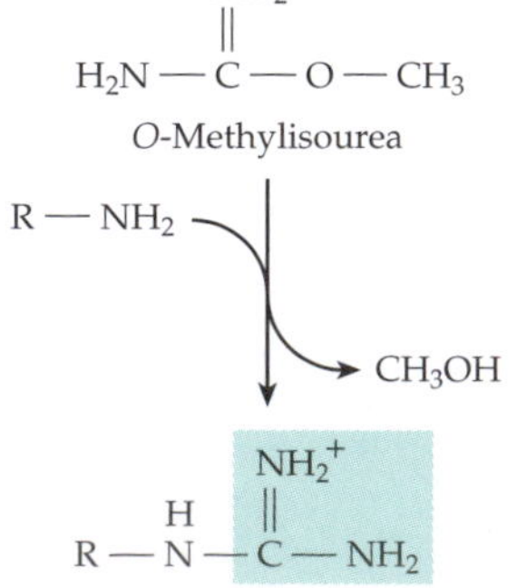

$$\tag{3-38}$$

Bifunctional imidoesters such as **dimethylsuberimidate** may be used to establish whether or not two different proteins or subunits are close together in a complex or in a supramolecular structure such as a membrane or ribosome.

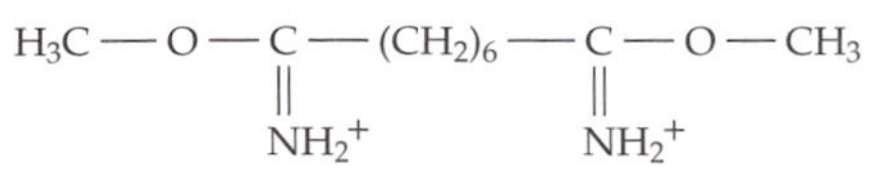

$$H_3C-O-C-(CH_2)_6-C-O-CH_3$$
$$\quad\quad\;\; \overset{\parallel}{NH_2^+} \quad\quad\quad\quad \overset{\parallel}{NH_2^+}$$

Dimethylsuberimidate

Another useful reaction of amino side chains is that with dansyl chloride (Eq. 3-29). Many lysine derivatives can be determined quantitatively by amino acid analysis.[280]

2. Reactions of SH Groups

In addition to the alkylation with iodoacetate (Eq. 3-24), sulfhydryl groups can react with N-ethylmaleimide (Eq. 3-39).[281] This reaction blocks the SH groups irreversibly and has often been used in attempts to establish whether or not a thiol group plays a role in the functioning of a protein. Loss of function in the

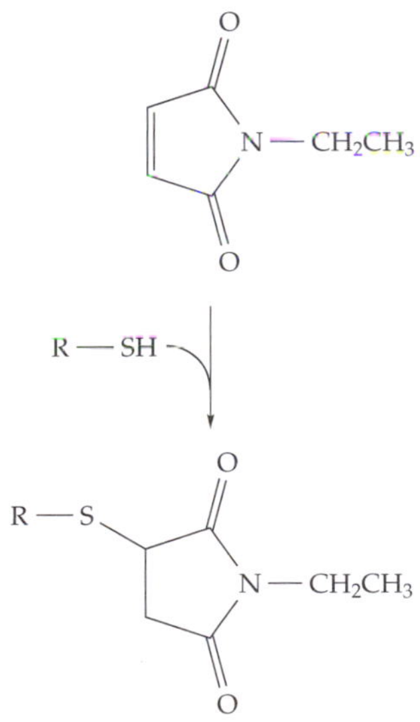

(3-39)

presence of this "sulfhydryl reagent" may mean that an SH group has an essential role or it could be a result of the bulk of the group added. The N-ethylmaleimide group is large and could prevent proper contact between an enzyme and substrate or between two proteins. To avoid the possible effect of excessive bulk, it is useful to convert the SH to the small thiocyanate group – SCN (Eq. 3-40).[282]

Ellman's reagent 5,5'-dithiobis(2-nitrobenzoic acid; DTNB) reacts quantitatively with –SH groups (Eq. 3-41) to form mixed disulfides with release of a thiolate anion that absorbs light at 412 nm with a molar extinction

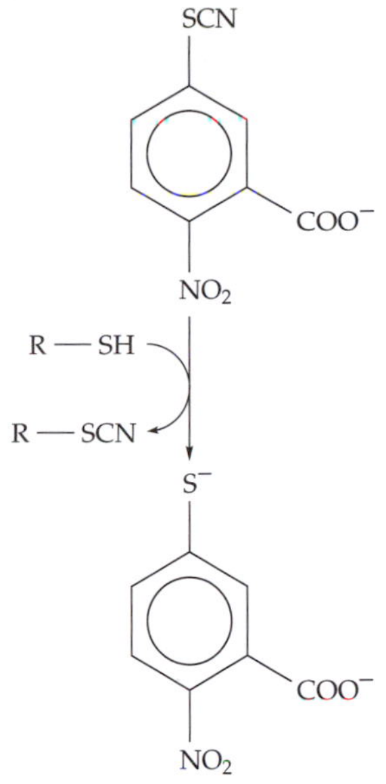

(3-40)

coefficient of 14,150 M^{-1} cm^{-1}.[283] While DTNB has been widely used to determine the content of –SH groups in proteins, there are some disadvantages.

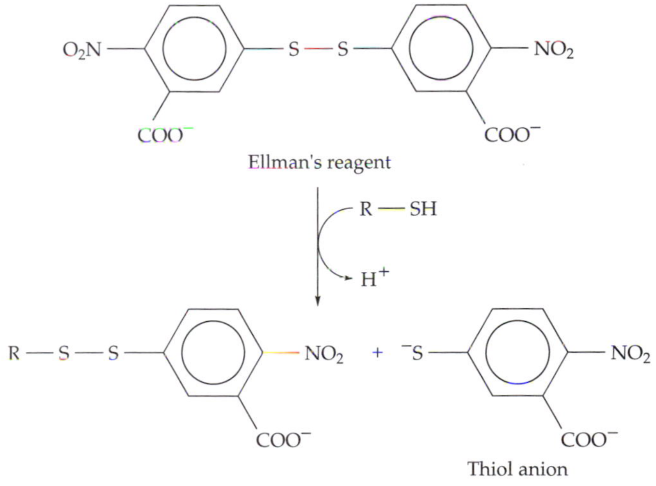

(3-41)

Pyridyldisulfides such as 2-pyridyldisulfide or the isomeric 4-pyridyldisulfide react more completely and with greater selectivity.[281]

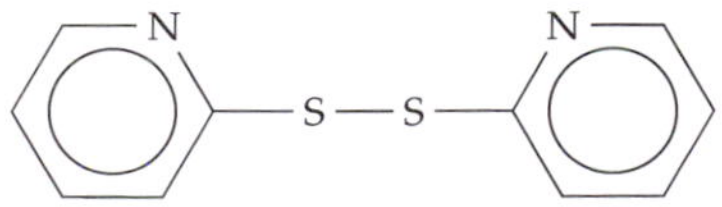

Thiol groups have a high affinity for mercury ions including organic mercury derivatives, which are widely used in the determination of protein structures by X-ray crystallography (Section F). Titration of SH groups in proteins is often accomplished with p-mercuribenzoate (Eq. 3-42). The reaction may be followed spectrophotometrically at 250 – 255 nm, a region in which the mercaptide product absorbs strongly.

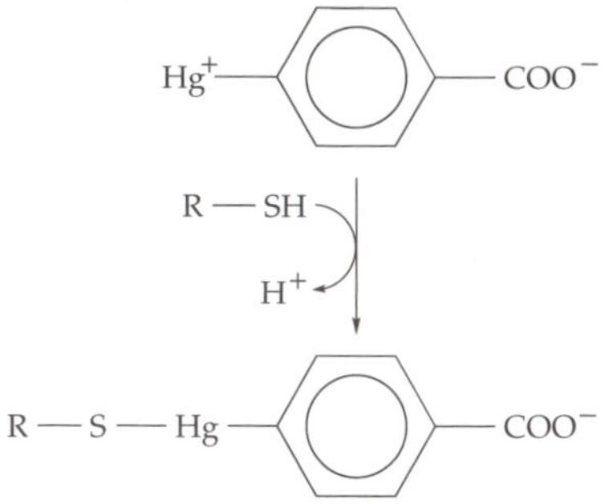

(3-42)

3. Reactions of Other Side Chains

There are no highly selective reactions for –OH, –COO$^-$, or imidazole groups. However, some hydroxyl groups in active sites of enzymes are unusually reactive in nucleophilic addition or displacement and can be modified by acylation, phosphorylation, or in other ways. Carboxyl groups, which are exceedingly numerous on protein surfaces, can be modified by treating with a water-soluble carbodiimide (Eq. 3-10) in the presence of a high concentration of an amine such as the ethyl ester of glycine. The imidazole groups of residues of histidine can often be selectively destroyed by dye-sensitized photooxidation (Ch. 12, Section D,5) or can be acylated with ethoxyformic anhydride.

Compounds with two adjacent carbonyl groups such as 1,2-cyclohexanedione (Eq. 3-43) react selectively with guanidinium groups from arginine residues in proteins. Under certain conditions the product indicated in Eq. 3-43 predominates. Related reagents are derived from camphorquinone.[284]

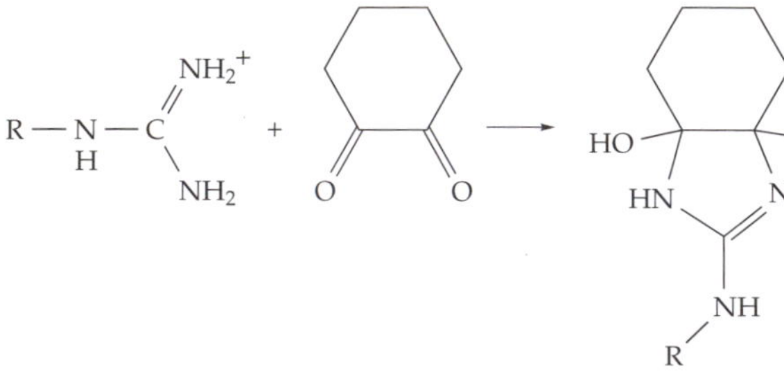

(3-43)

The phenolic group of tyrosine undergoes iodination (Eq. 3-44), acylation, coupling with diazonium compounds, and other reactions.

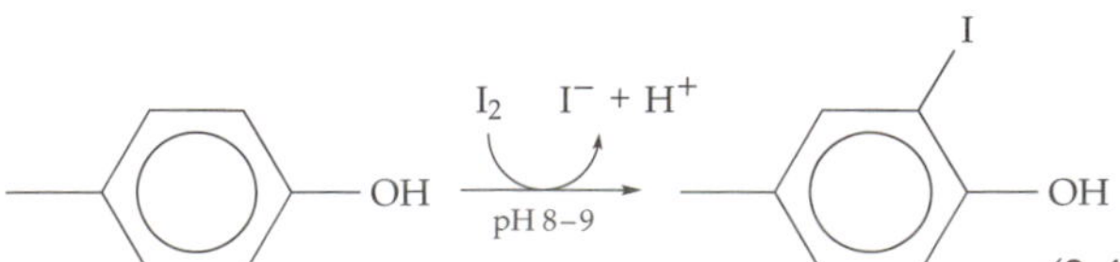

(3-44)

The following sparingly soluble chloroamide together with I$^-$ will also iodinate tyrosine and can be used to incorporate radiolabeled iodine into proteins.[285,286]

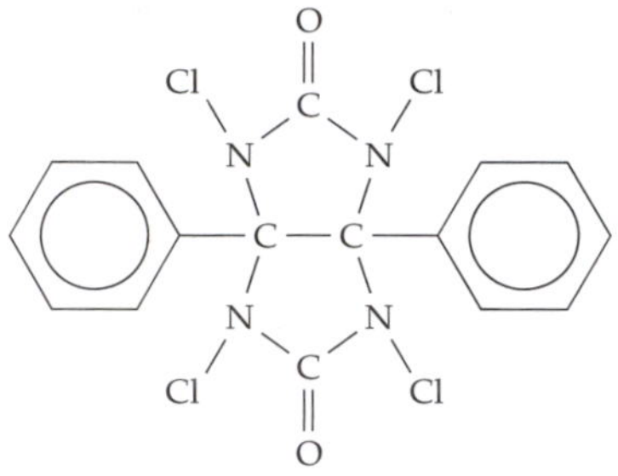

1,3,4,6-Tetrachloro-3α,6α-diphenylglycouril

Tetranitromethane reacts slowly with tyrosyl groups to form 3-nitrotyrosyl groups (Eq. 3-45). The by-product **nitroform** is intensely yellow with $\varepsilon_{350} = 14,400$. The reagent also oxidizes SH groups and reacts with other anionic groups.

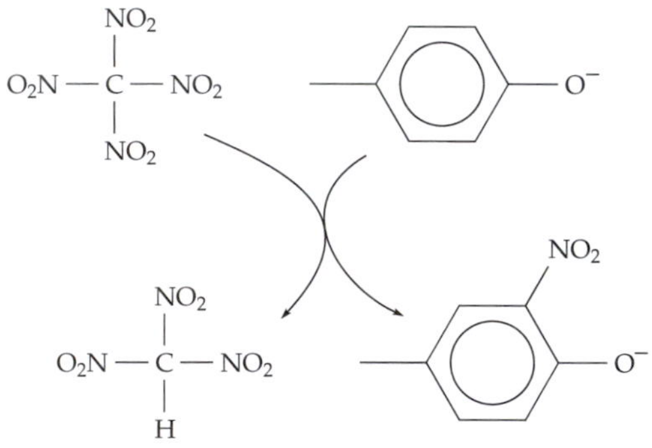

(3-45)

Koshland devised the following reagent for the indole rings of tryptophan residues (Eq. 3-46).

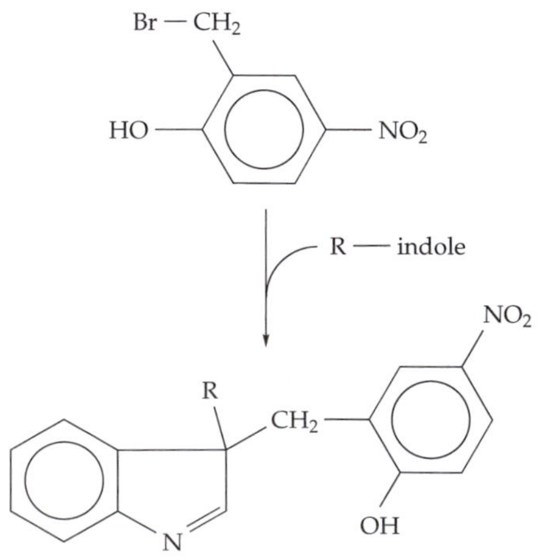

(3-46)

Imidazole, lysine amino groups, and tyrosine hydroxyl groups react with **diethylpyrocarbonate**

(Eq. 3-47) at low enough pH (below 6) that the reaction becomes quite selective for histidine.[70,287] Reactivity with this reagent is often used as an indication of histidine in a protein.[288–290]. The reaction may be monitored by observation of NMR resonances of imidazole rings.[288,290]

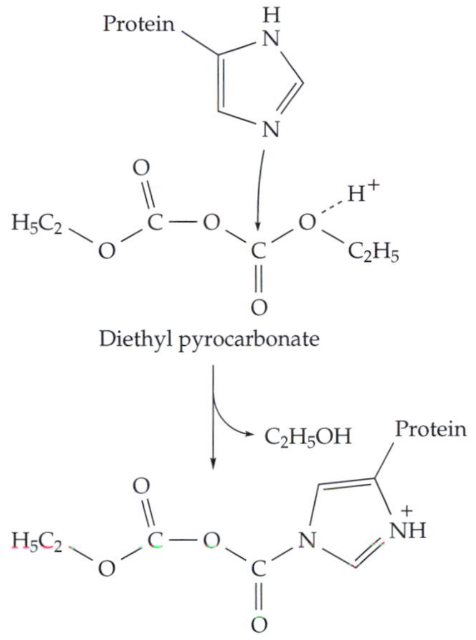

Diethyl pyrocarbonate

(3-47)

The thioether side chains of methionine units in proteins can be oxidized with hydrogen peroxide to the corresponding sulfones (Eq. 3-48). They can also be alkylated, e.g., by CH_3I to form $R—S^+(CH_3)_2$.

$$R—S—CH_3 + 2H_2O_2 \longrightarrow R—\overset{\overset{\textstyle O}{\|}}{\underset{\underset{\textstyle O}{\|}}{S}}—CH_3 + 2H_2O$$

(3-48)

4. Affinity Labeling

To identify groups that are part of or very near to the active site of a protein, reagents can be designed that carry a reactive chemical group into the active site.[291] The related **photoaffinity labeling**[292,293] is also widely used (see also Chapter 23).

F. Synthesis of Peptides

The synthesis of peptides of known sequence in the laboratory is extremely important to biochemical research. For example, we might want to know how the effects of a peptide hormone are altered by replacement of one amino acid in a particular position by another. The synthetic methods must be precise[294–298] and because there are so many steps the yield should be 98% or better for every step. Even so, it is still impractical to synthesize very large peptides. Those that have been made, such as the hormone insulin and the enzyme ribonuclease, have been obtained in low yields and have been difficult to purify.[298] It is usually more practical to obtain large peptides from natural sources. It is often practical to clone a suitable piece of DNA in a bacterial plasmid and to set up biological production of the desired peptide (Chapter 26). On the other hand, for smaller peptides, laboratory synthesis is feasible. Even for large peptides it is useful because it permits incorporation of unnatural amino acids as well as isotopic labels.

The general procedure for making a peptide in the laboratory is to "block" the amino group of what will become the N-terminal amino acid with a group that can be removed later. The subsequent amino acid units "activated" at their carboxyl end are then attached one by one. The chemical activation is often accomplished by conversion of the carboxyl group of the amino acid to an anhydride. At the end of the synthesis, the blocking group must be removed from the N terminus and also from various side chain groups such as those of cysteine and lysine residues. In many respects, this procedure is analogous to the biological synthesis of proteins whose basic chemistry is discussed in Chapter 29.

1. Solid-Phase Peptide Synthesis

Modern methods of peptide synthesis began with the solid-phase method introduced by Merrifield[299] in 1962 (Fig. 3-15). To begin the synthesis a suitably protected amino acid is covalently linked to a polystyrene bead. The blocking *t*-butoxycarbonyl (Boc) group is removed as isobutene by an elimination reaction to give a bound amino acid with a free amino group. This can then be coupled to a second amino acid with a blocked amino group using dicyclohexylcarbodiimide (Eq. 3-10). The removal of the blocking group and addition of a new amino acid residue can then be repeated as often as desired. The completed peptide is removed from the polystyrene by action of a strong acid such as HF.

Advantages of the Merrifield procedure are that the peptide is held tightly and can be washed thoroughly at each step. Problems arise from repeated use of trifluoroacetic acid and the need to use HF or other strong acid to cleave the peptide from the matrix and also to remove blocking benzyl groups that must be present on many side chain groups. Newer variations of the procedure include a more labile linkage to a polyamide type of polymer and use of blocked amino acids.[297,300–301a] These "active esters" will spontaneously condense with the free amino group of the growing peptide and with suitable catalysis will eliminate

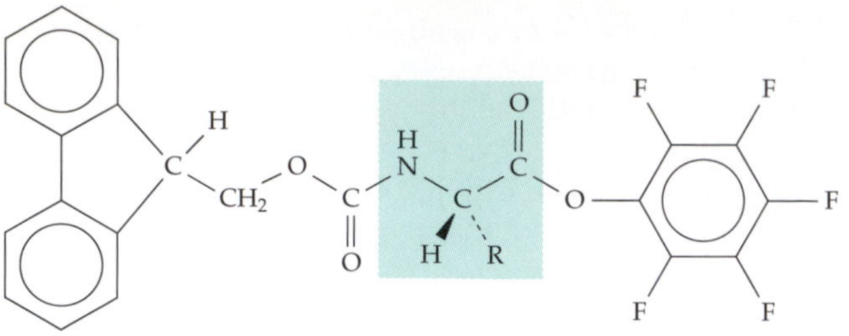

Blocked amino acid

pentafluorophenol. The fluorenylmethoxycarbonyl (Fmoc) blocking group is removed under mildly basic conditions. The whole procedure has been automated in commercially available equipment.

Smaller peptides may be joined to form longer ones.[298] Also useful is enzymatic synthesis. Protein-hydrolyzing enzymes under appropriate conditions will form peptide linkages, for example, joining together oligopeptides.[302,303] Other new methods have been devised to join unprotected peptides.[304,305]

Semisynthetic approaches can also be used to place unnatural amino acids into biologically synthesized proteins through the use of suppressor transfer RNAs (Chapter 29).[306]

2. Combinatorial Libraries

Many chemists devote all of their efforts to the synthesis of new compounds, including polypeptides, that might be useful as drugs. Traditionally, this has involved the tedious preparation of a large number of compounds of related structure which can be checked individually using various biochemical or biological tests. In recent years a new approach using "combinatorial chemistry" has become very popular and is continually being adapted for new purposes.[307–310] There are several approaches to creating a combinatorial library.

In "split synthesis" procedures a solid-phase

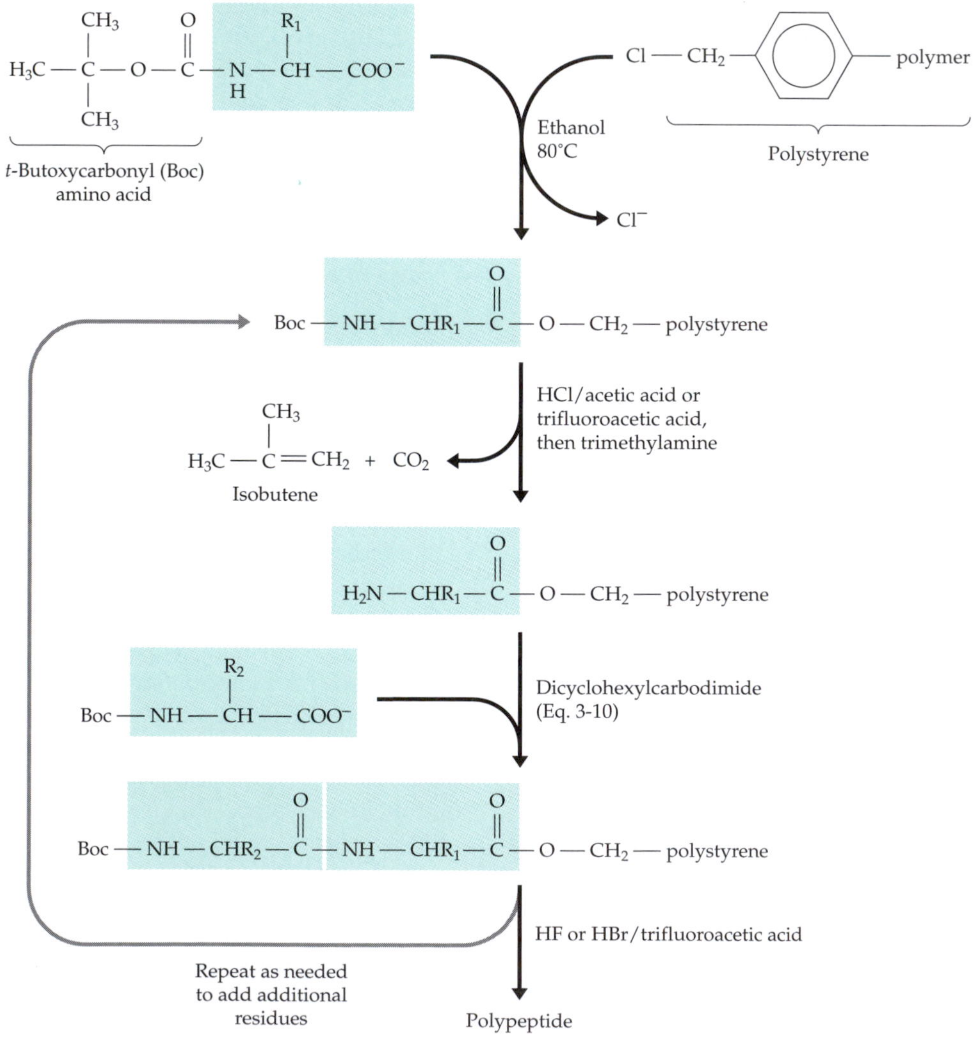

Figure 3-15 Procedure for solid-phase peptide synthesis devised by Merrifield.[299]

synthesis is conducted on beads. For example, a family of peptides, each with the same C terminus, can be started on a large number of beads. After the first amino acid residue is attached the beads are divided into up to 20 equal portions and different amino acids are added to each portion. The beads can then be mixed and again subdivided. The third residue will again contain many different amino acids attached to each of the different amino acids in the second position. By repeating the procedure again, perhaps for many steps, a "library" of random peptides with each bead carrying a single compound will be formed.

To test whether a polypeptide or other compound carried on a given bead has a derived biological activity, such as the ability to inhibit a certain enzyme, various assays that require only one bead can be devised. However, if a particular bead carries a compound of interest, how can it be identified? The bead carries only a small amount of compound but it may be possible using microsequencing procedures to identify it. An alternative procedure is to use an encoding method to identify the beads.

An alternative to the "one bead–one peptide" approach is to incorporate random sequences of a DNA segment into a gene that can be used to "display" the corresponding peptide sequence. This is illustrated in Fig. 3-16. A protein segment (which may be a random sequence) can be displayed either on the major coat proteins along the shaft or on the minor coat proteins at the end of the bacterial virus fd (see Fig. 7-7). In the case of random insertions, each virus particle may display a different sequence (as many as 10^8). Peptides may be selected by binding to a desired receptor or monoclonal antibody and the DNA encapsulated in the virus particle can be used to produce more peptides for identification purposes.[311–316] Many other ingenious systems for constructing and testing libraries of peptides[317–320] and other molecules[319,321,322] are being devised. One of these involves a photolithography procedure for immobilizing macromolecules in a regular addressable array, e.g., in a 0.5-mm checkerboard pattern, on a flat surface.[323,324]

G. Microscopy

The light microscope[325] was developed around 1600 but serious studies of cell structure (histology) did not begin until the 1820s. By 1890, microscope lenses had reached a high state of perfection[326] but the attainable resolution was limited by the wavelength of light. For 450 nm blue light the limit is about 300 nm and for ultraviolet light, viewed indirectly, about 200 nm.[325,327] By the 1940s the electron microscope with its far superior resolving power had overshadowed the light microscope.

For both light and electron microscopy, the preparation of thin sections of cells is a very important technique. Only with very thin sections is the image sufficiently focused. However, **confocal scanning optical microscopy**, invented in the 1950s but not used commercially until much later,[328,329] provided an alternative solution to the focusing problem. A conical beam of light focused to a point is scanned across the sample and the transmitted light (or light emitted by fluorescence) passes through a small "pinhole" aperture located in the primary image plane to a photomultiplier tube where its intensity is recorded. The illuminating beam is moved to scan the entire field sequentially. A series of pinholes in a spinning disk may accomplish the same result. The focal plane can

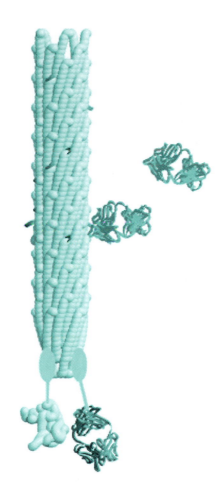

Figure 3-16 Model of bacteriophage fd engineered to display peptides as inserts in the coat proteins of the virus. The native virus structure is shown in gray; proteins not present in the native virus are shown black or green. Inserted near the N-termini of some major coat proteins is a 6-residue peptide. To one of these peptides a specific Fab antibody fragment (green) has bound from solution, and a second Fab is shown nearby. The N-terminal region of a minor coat protein at the end of the virion has been engineered to display a (different) Fab fragment. Steric constraints are less stringent for inserts in the minor proteins, but fewer copies per virion are possible. Reprinted with permission from Barbas, *et al.*[313a]

be varied so that an image of a thick object such as a cell can be optically sectioned into layers of less than 1 μm thickness. Stereoscopic pairs can also be generated (Fig. 3-17).[330]

A newer development in confocal microscopy is the use of two-photon and three-photon excitation of the fluorescent molecules that occur naturally within cells using short pulses of short-wavelength high-energy laser light.[331] Distribution of such compounds as NADH,[332] DNA, and the neurotransmitter seroto-nin[333] can be observed without damaging cells. Individual storage granules, each containing ~5 x 10^8 molecules of serotonin in a concentration of ~50 mM, can be seen.[334] Another new instrument, the **near-field scanning optical microscope** (NSOM),[335,336] is a lensless instrument in which the illuminating beam passes through a very small (e.g., 100 nm diameter) hole in a probe that is scanned in front of the sample. It may extend the limit of optical microscopy to ~1/50 the wavelength of the light.

Since it first became commercially available in 1939, the electron microscope has become one of the most important tools of cell biology.[337,338] The practical resolution is about 0.4 nm, but recent developments in scanning electron microscopy have resulted in resolution of 0.14 nm.[339] Of major importance was the development around 1950 of microtomes and knives capable of cutting thin (20–200 nm) sections of tissues embedded in plastic.[340] A bacterium such as *E. coli* can be sliced into as many as 10 thin longitudinal slices (see Fig. 1-4) and a eukaryotic cell of 10 μm diameter into 100 slices. Serial sections can be examined to determine three-dimensional structures. For some results see Bubel.[341]

If a slice of fresh (frozen) tissue is examined directly, little is seen because most of the atoms found in cells are of low atomic mass and scatter electrons weakly and uniformly. Therefore, thin sections must be "stained' with atoms of high atomic mass, e.g., by treatment with potassium permanganate or osmium tetroxide. Tissues must also be "fixed" to prevent disruption of cell structures during the process of

removal of water and embedding in plastic. Fixatives such as formaldehyde react with amino groups and other groups of proteins and nucleic acids. Some proteins are precipitated in place and digestive enzymes that otherwise would destroy much of the fine structure of the cell are inactivated. Glutaraldehyde (a five-carbon dialdehyde) is widely used to fix and crosslink protein molecules in the tissue. The methods continue to be improved.[342]

Small particles, including macromolecules, may be "shadowed." Chromium or platinum can be evaporated in a vacuum from an angle onto the surface of the specimen, Individual DNA molecules can be "seen" in this way.[343] In fact, only the "shadows" are seen and they are 2–3 times wider than the DNA molecules. In the **negative contrast** method a thin layer of a solution containing the molecules to be examined, together with an electron-dense material such as 1% sodium phosphotungstate, is spread on a thin carbon support film. Upon drying, a uniform electron-dense layer is formed. Where the protein molecules lie, the phosphotungstate is excluded, giving an image of the protein molecule.

Surfaces of cells, slices, or intact bacteria can be coated with a deposit of platinum or carbon. The coating, when removed, provides a "negative" **replica** which can be examined in the microscope. Alternatively, a thin plastic replica can be made and can be shadowed to reveal topography. In "freeze fracturing" and "freeze etching," fresh tissue, which may contain glycerol to prevent formation of large ice crystals, is frozen rapidly. Such frozen cells can often be revived; hence, they may be regarded as still alive until the moment that they are sliced! The frozen tissue is placed in a vacuum chamber within which it is sliced or fractured with a cold knife. If desired, the sample can be kept in the vacuum chamber at about –100°C for a short time, during which some water molecules evaporate from the surface. The resultant etching reveals a fine structure of cell organelles and membranes in sharp relief. After etching, a suitable replica is made and examined (Fig. 1-15A and E). Fracturing tends to take place through lipid portions of cell membranes.

Small viruses, bacterial flagella, ribosomes, and even molecules can be seen by electron microscopy. However, to obtain a clear image in three dimensions requires a computer-based technique of **image reconstruction** or **electron microscope tomography**, which was developed initially by Aaron Klug and associates.[344–349] A sample is mounted on a goniometer, a device that allows an object to be tilted at exact angles. Electron

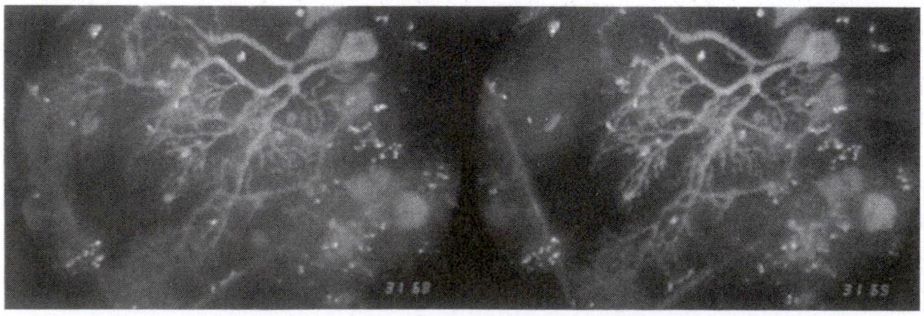

Figure 3-17 Confocal micrograph showing a forty-micrometer stereo slice in a 90-μm thick section of mouse cerebellum.[330] Courtesy of A. Boyde.

micrographs are prepared with the sample untilted and tilted in several directions at various angles, e.g., up to 90° in 10° increments. The micrographs are digitized and a computer is used to reconstruct a three-dimensional image.

In **electron crystallography**[350] micrographs of two-dimensional crystalline arrays of molecules or larger particles are prepared. A Fourier transform of the micrograph gives a diffraction pattern which can be treated in a manner similar to that usual for X-ray diffraction to give a three-dimensional image. An important milestone in use of this technique was the determination of the structure of bacteriorhodopsin (Fig. 23-45) at 0.3-nm resolution. Bear in mind that X-ray crystallography can also be viewed as a form of microscopy.

Invention of the **scanning tunneling microscope** (STM) by Binnig and Roher[351 – 353] initiated a new revolution in microscopy. The STM and similar scanned probe microscopes examine surfaces by moving a fine probe mounted in a piezoelectric x,y,z-scanner[354] across the surface to be examined. The tiny tungsten probe of the STM is so fine that its tip may consist of a single atom. When a small voltage is applied a minuscule quantum mechanical tunnelling current flows across the small gap between the probe and the surface and a high-resolution image, sometimes at atomic resolution, is created from the recorded variation in current.[354-359] The STM theoretically responds only to surfaces that conduct electrons, but nonconducting samples have been imaged at high humidity; presumably by conductance of electrons or ions through the surface water layer. The success of the STM spurred the development of many other types of scanned probe microscopes. Among these, the **atomic force microscope** (AFM; Fig. 3-18) has been especially useful for biological materials, including proteins and nucleic acids. The AFM moves a fine-tipped stylus directly across the sample surface or, alternatively, vibrates the probe above the surface. The small up-and-down movements of the stylus are recorded[359–366] and thereby create a topographic or force-field map of the sample. AFM images contain three-dimensional information and can be used to view individual molecules (Fig. 3-19).[367,367a] **Chemical force microscopy** is sensitive to adhesion and friction as a function of the interaction between defined chemical groups on the tip and sample.[368]

An emerging field is force spectroscopy, in which the AFM measures interaction forces between and within individual molecules.[369 – 371a] Under development are NMR microscopes (Section I). There is continual effort to see small objects more directly and more clearly!

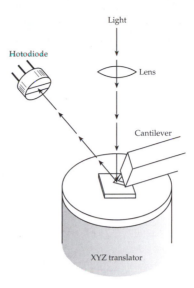

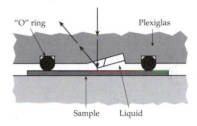

Figure 3-18 Schematic diagram of the atomic force microscope.[360] Courtesy of Paul Hansma.

Figure 3-19 AMF images of cholera toxin[364] (see also Box 11-A). Courtesy of Z. Shao.

H. X-ray and Neutron Diffraction

One of the most important techniques by which we have learned bond lengths and angles and precise structures of small molecules is X-ray diffraction. Today, this technique, which involves measurements of the scattering of X-rays by crystalline arrays of molecules, is being used with spectacular success to study macromolecules of biochemical and medical importance.[372–378]

X-rays were described by Röntgen in 1896 but there was uncertainty as to their wave nature.[274] It was not until 1912 that the wavelengths of X-rays had been measured and it was recognized that they were appropriate for the use of X-rays in structure determination. Consider the fact that with a conventional light microscope we cannot distinguish two small objects that are much closer together than the distance represented by the wavelength of the light with which we observe them. This is about 460 nm for blue light. By comparison the wavelength of the copper K_α radiation, which is used in protein crystallography, is 0.1542 nm entirely appropriate for seeing the individual atoms of which matter is composed. Recognizing this, W. L. Bragg in Cambridge, England, in 1913 used X-ray diffraction to establish the structures of NaCl, KCl, and KBr in the crystalline state. The science of X-ray crystallography had been founded.[379]

In 1926, James Summer crystallized the enzyme urease (Chapter 16) and crystallization of other enzymes soon followed.[380] In 1934, J. B. Bernal brought back to Cambridge from Uppsala, Sweden, some crystals of pepsin almost 2 mm long that had been grown in T. Svedberg's laboratory. Bernal and Crowfoot showed that these delicate crystals, which contained almost 50% water, gave a sharp diffraction pattern when they were protected by enclosure in a narrow capillary tube containing some of the mother liquor from which the crystals had been grown.[381,382] After this, diffraction patterns were obtained for many protein crystals, and in 1937 Max Perutz chose for his thesis work at Cambridge the X-ray crys-

tallography of hemoglobin. The project seemed hopeless at times, but in 1968, 31 years later, Perutz had determined the structure of hemoglobin.[383–386]

If X-rays could be focused easily, could one build an X-ray microscope that would permit the immediate viewing of molecular structures? X-ray holograms at the molecular level have been obtained.[387,388] However, currently the only practical X-ray microscope for protein structures involves the measurement of diffraction patterns created by the scattering of X-rays (or of neutrons) from the crystalline lattice. The details of this procedure can be found in other sources.[372–375,378,389–390] It is sufficient to point out here that a pattern of many

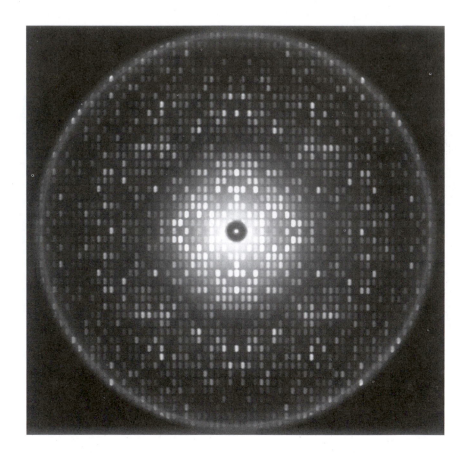

Figure 3-20 An X-ray diffraction photograph such as was used to determine the structure of hemoglobin. This precession photograph was obtained from a crystal of human deoxyhemoglobin by rotating the crystal along two different axes in a defined manner before a narrow X-ray beam. The film also was moved synchronously. The periodicity observed in the photograph is a result of a diffraction phenomenon arising from the periodic arrangement of atoms in the crystal. The distances of the spots from the origin (center) are inversely related to the distances between planes of atoms in the crystal. In this photograph (which shows only two dimensions of the three-dimensional diffraction pattern) the spots at the periphery represent a spacing of 0.28 nm. If the intensities of the spots are measured, and if the phases of the harmonic functions required for the Fourier synthesis can be assigned correctly, the structure can be deduced to a resolution of 0.28 nm from a set of patterns of this type. In the case of human deoxyhemoglobin, a complete set of data would consist of about 16,000 spots. With modern equipment the resolution limit can be extended to better than 0.15 nm (or more than 100,000 unique spots). Photograph courtesy of Arthur Arnone.

spots, such as that in Fig. 3-20, is obtained. In a typical determination of a protein structure from 10,000 to several times that number of spots must be measured. The needed information is contained in the coordinates of the spots (i.e., in the angles through which the scattering occurs) and in the intensities of the spots. The structure is obtained by a Fourier synthesis in which a large number of mathematical functions in the form of three-dimensional sinusoidal waves are summed. The wavelengths of these functions are determined by the positions of the spots in the diffraction pattern. The amplitudes of the waves are related to the intensities of the spots.[374] However, the waves are not all in phase and the phases must be learned in some other way. This presents a difficult problem. Mathematical methods have been devised that automatically determine the phases for small molecules and usually allow the structures to be established quickly. For proteins it is more difficult.

In 1954, Perutz introduced the **isomorphous replacement method** for determining phases. In this procedure a heavy metal, such as mercury or platinum, is introduced at one or more locations in the protein molecule. A favorite procedure is to use mercury derivatives that combine with SH groups. The resulting heavy metal-containing crystals must be isomorphous with the native, i.e., the molecules must be packed the same and the dimensions of the crystal lattice must be the same. However, the presence of the heavy metal alters the intensities of the spots in the diffraction pattern and from these changes in intensity the phases can be determined. Besides the solution to the phase problem, another development that was absolutely essential was the construction of large and fast computers. It would have been impossible for Perutz to determine the structure of hemoglobin in 1937, even if he had already known how to use heavy metals to determine phases.

The first protein structure to be learned was that of myoglobin, which was established by Kendrew et al. in 1960.[391–393] That of the enzyme lysozyme was deduced by Blake et al. in 1965.[394] Since then, new structures have appeared at an accelerating rate so that today we know the detailed architecture of over 6000 different proteins[395] with about 300 distinctly different folding patterns.[396] New structures are being determined at the rate of about one per day. X-ray diffraction has also been very important to the study of naturally or artifically oriented fibrous proteins[397] and provided the first experimental indications of the β structure of proteins.

Suppose that you have isolated a new protein. How can you learn its three-dimensional structure? The first step is crystallization of the protein, something that a biochemist may be able to do. Crystallization is done in many ways, often by the slow diffusion of one solution into another or by the slow removal of solvent through controlled evaporation.[374,398] Ammonium sulfate and polyethylene glycol are two commonly used precipitants. The presence of the neutral detergent β-octyl glucoside improves some crystals.[399] Droplets of protein solution mixed with the precipitant are often suspended on microscope cover glasses in small transparent wells or are placed in depression plates within closed plastic boxes.[374,400,401] In either case, a reservoir of a solution with a higher concentration of precipitant is present in the same compartment. Water evaporates from the samples into the larger reservoir, concentrating the protein and causing its crystallization. Crystals grow slightly better in a spacecraft than on Earth.[402,403] Some proteins, notably myosin from muscle, crystallize well only after reductive methylation of all lysine side chains to dimethyllysine with formaldehyde and sodium borohydride (Eq. 3-34).[278]

The next step is for a protein crystallographer to mount a small perfect crystal in a closed silica capillary tube and to use an X-ray camera to record diffraction patterns such as that in Fig. 3-20. These patterns indicate how perfectly the crystal is formed and how well it diffracts X-rays. The patterns are also used to calculate the dimensions of the unit cell and to assign the crystal to one of the seven **crystal systems** and one of the 65 enantiomorphic **space groups**. This provides important information about the relationship of one molecule to another within the unit cell of the crystal. The unit cell (Fig. 3-21) is a parallelopiped

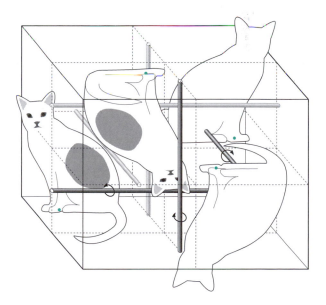

Figure 3-21 Diagram showing an asymmetric unit as it might appear in a unit cell of space group P2₁2₁2₁. This unit cell has three pairs of nonintersecting twofold **screw axes** which are marked by the shaded rods. These are designated by arrows and the symbol. Two asymmetric units are related one to another by rotation around a twofold axis together with translation by one-half the dimension of the unit cell.

whose sides are parallel to crystallographic axes and which, by translation in three dimensions, gives rise to the entire crystal. The unit cell must contain at least one **asymmetric unit**, the smallest unit of structure that lacks any element of crystal symmetry. The asymmetric unit may consist of one or a small number of protein subunits.

In crystals of the triclinic system all of the asymmetric units are aligned in the same manner and there are no axes of symmetry. Monoclinic crystals have a single **twofold** or **dyad axis** (see Fig. 7-11). The unit cell might contain two molecules related one to another by the dyad or one dimeric molecule with the dyad located as in Fig. 7-11C. Orthorhombic crystals have three mutually perpendicular twofold axes and the unit cell is a rectangular solid. Trigonal, tetragonal, and hexagonal crystal systems have three, four and sixfold axes of symmetry, respectively, while the cubic crystal contains four threefold axes along with diagonals of the cube as well as two-fold axes passing through the faces (see Fig. 15-14). Within each crystal system there are several space groups. An example is the orthorhombic space group $P2_12_12_1$, which is often met with small organic molecules and proteins. In this space group the unit cell contains three mutually perpendicular but nonintersecting twofold screw axes (Fig. 3-21). The position of one molecule in the unit cell is related to the next by both a 180° degree rotation about the twofold axes and a translation of one-half the length of the unit cell. There is such a screw axis for each of the three directions.[372]

The third step in the structure determination is collection of the X-ray diffraction data. This may be done with a **diffractometer** in which a narrow collimated pencil source of X-rays is aimed at the crystal and the intensities and positions of the diffracted beams are measured automatically. The computer-controlled diffractometer is able to measure the angles to within less than one-hundredth of a degree. If sufficient time is allowed, very weak spots can be counted. Today, diffractometers are more likely to be used for preliminary measurements, while the major data collection is done with an **area detector**, an

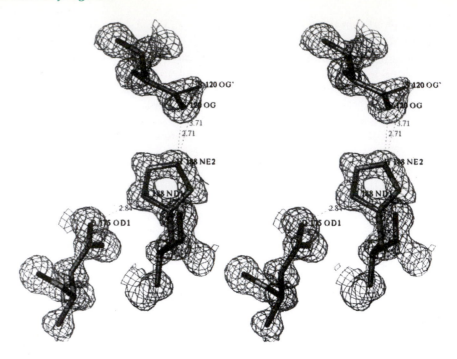

Figure 3-22 Stereoscopic view of a section of the structure of cutinase from the fungus *Fusarium solani* determined to a resolution of 0.10 nm. The three amino acid residues shown are serine 120 (top), histidine 188, and aspartate 175 (lower left). The structure is presented as a contour map with a "wire mesh" drawn at a "cutoff" level of density equal to 1σ above the average, where σ is the root mean square density of the entire map. The side chains of these three residues constitute the "catalytic triad" in the active site of this enzyme (see Chapter 12). At this resolution more than one conformation of a group may often be seen. For example, the gamma oxygen (OG) of S120 is seen in two positions, the major one being toward His 188. When the map is drawn with a lower contour level the N-H proton on His 188 that is hydrogen bonded to Asp 175 can also be seen.[410] Courtesy of Christian Cambillau.

instrument that collects many reflections simultaneously. There are some difficulties. For example, during the long periods of irradiation needed to measure the weak spots with a diffractometer the protein crystals decompose and must be replaced frequently. Data collection may last for months. The newest methods utilize more powerful X-ray sources and often **synchrotron radiation**, which delivers very short and extremely intense pulses of X-rays and allows data to be collected on very small well-formed crystals.[404]

The fourth step is the preparation of isomorphous crystals of heavy metal-containing derivatives. The heavy metal may be allowed to react with the protein before crystallization or may be diffused into preformed crystals. A variety of both cationic and anionic metal complexes, even large $Ta_6Br_{12}^{2+}$ tantalum clusters, have been used.[405] Two or more different heavy metal derivatives are often required for calculation of the phases. The heavy metal atoms must be present at only a very small number of locations in the unit cell.

An entire data set must be collected for each of these derivatives. The evaluation of the phases from these data is a complex mathematical process which usually involves the calculation first of a "difference Patterson projection."[406] This is derived by Fourier transformation of the differences between the scattering intensities from the native and heavy atom-containing crystals. The Patterson map is used to locate the coordinates of the heavy metal atoms which are then refined and used to compute the phases for the native protein.

Alternative methods of solving the phase problem are also used now. When a transition metal such as Fe, Co, or Ni is present in the protein, anomolous scattering of X-rays at several wavelengths (from synchrotron radiation) can be used to obtain phases. Many protein structures have been obtained using this multiple wavelength anomalous diffraction (MAD phasing) method.[404,407,408] Selenocysteine is often incorporated into a protein that may be produced in

bacteria using recombinant DNA procedures. Crystals are prepared both with protein enriched in Se and without enrichment. In some ways it is better to incorporate tellurium (^{127}Te) in telluromethionine.[409]

In the fifth step of an X-ray structure determination the **electron density map** is calculated using the intensities and phase information. This map can be thought of as a true three-dimensional image of the molecule revealed by the X-ray microscope. It is usually displayed as a stereoscopic view on a computer graphics system (Fig. 3-22). It is also often prepared in the form of a series of transparencies mounted on plastic sheets. Each sheet represents a layer, perhaps 0.1 nm thick, with contour lines representing different levels of electron density.

Using the electron density map a three-dimensional model of the protein can be built. For years, the customary procedure was to construct a model at a scale of 2 cm = 0.1 nm using an **optical comparator**, but

Figure 3-23 (A) Stereoscopic α-carbon plot of the cystolic aspartate aminotransferase dimer viewed down its dyad symmetry axis. Bold lines are used for one subunit (subunit 1) and dashed lines for subunit 2. The coenzyme pyridoxal 5'-phosphate (Fig. 3-24) is seen most clearly in subunit 2 (center left). (B) Thirteen sections, spaced 0.1 nm apart, of the 2-methylaspartate difference electron density map superimposed on the α-carbon plot shown in (A). The map is contoured in increments of ± 2σ (the zero level omitted), where σ = root mean square density of the entire difference map. Positive difference density is shown as solid contours and negative difference density as dashed contours. The alternating series of negative and positive difference density features in the small domain of subunit 1 (lower right) show that the binding of L-2-methylaspartate between the two domains of this subunit induces a right-to-left movement of the small domain. (Continues)

crystallographers now use computer graphics systems. The three-dimensional image of the electron density map and a computer-generated atomic model are superimposed on the computer screen. An example is shown in Fig. 3-22. When the superposition has been completed, the coordinates of all atoms are present in the memory of the computer.

The final step in the structure determination is **refinement** using various mathematical methods. From the coordinates of the atoms in the model the expected diffraction pattern is computed and is compared with that actually observed. The differences between predicted and observed density are squared and summed. The sum of the squares constitutes an error function which is then minimized by moving the various atoms in the model short distances while keeping bond lengths, angles, and van der Waals distances within acceptable limits and recalculating the error function. This complex "refinement" procedure must be repeated literally hundreds of thousands of times with every part of the structure being varied. Structures at very high resolution (0.07–0.1 nm) may reveal multiple positions for hydrogen-bonded side chains[410a,b] as well as hydrogen atoms (Fig. 3-22) and even bonding electrons.[410b]

Once the three-dimensional structure of the pro-

tein is known, further experiments are usually done using the X-ray diffraction technique. Since protein crystals contain channels of solvent between the packed molecules, it is usually possible to diffuse small molecules into the crystal. These may be substrates, inhibitors, or allosteric effectors. Diffraction data are collected after diffusion of the small molecules into the crystal and a **difference electron density map** may be calculated. This may show exactly where those molecules were bound. An example is shown in Fig. 3-23. Sometimes difference maps not only show the binding to an enzyme of a substrate or other small molecule but also reveal conformational changes in proteins. Such is the case for Fig. 3-23, which shows the binding of α-methylaspartate, an inhibitor that behaves initially like a substrate and goes part way through the reaction sequence for aspartate aminotransferase until further reaction is blocked by the methyl group. This difference map also shows that the part of the protein to the left of the binding site in the figure has moved.[411] Using the X-ray data it was possible subsequently to deduce the nature of the conformational change.

Examination of the effect of temperature on the diffraction pattern of a protein can give direct information about the mobility of different parts of the mole-

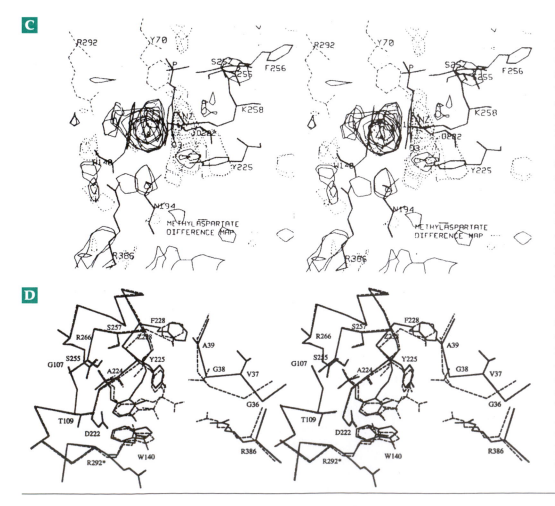

Figure 3-23 (Cont.) (C) Nine sections, spaced 0.1 nm apart, of a part of the 2-methylaspartate difference map superimposed on the atomic model shown in (A) and (D). The coenzyme is shown as the internal aldimine with Lys 258 (see Fig. 14-6, 14-10). The positive and negative contours on the two sides of the coenzyme ring indicate that the coenzyme tilts over to form the external aldimine when substrates react.[413] (D) Superimposed structure of the active site of the enzyme in its free form as in (A) (bold lines) and the refined structure of the α-methylaspartate complex, (dashed lines).[411] This illustrates the tilting of the coenzyme ring, which is also shown in Eq. 14-39 and Fig. 14-10. Courtesy of Arthur Arnone and Sangkee Rhee.

cules.[412] This complements information obtainable in solution from NMR spectra.

Because hydrogen atoms contain only one electron, and therefore scatter X-rays very weakly, they are usually not seen at all in X-ray structures of proteins. However, neutrons are scattered strongly by hydrogen atoms and **neutron diffraction** is a useful tool in protein structure determination.[414,415] It has been used to locate tightly bonded protons that do not exchange with ^2H_2O as well as bound water (^2H_2O).

The development of synchrotron radiation as an X-ray source[404,416–418] has permitted accumulation of data for electron density difference maps in less than 1 s and it is expected that such data can eventually be acquired in ~1 ps.[419–421] If a suitable photochemical reaction can be initiated by a picosecond laser flash, a substrate within a crystalline enzyme can be watched as it goes through its catalytic cycle. An example is the release of inorganic phosphate ions from a "caged phosphate" (Eq. 3-49) and study of the reaction of the released phosphate with glycogen phosphorylase (Chapter 12).[422,423]

"Caged phosphate"   (3-49)

However, caged substrates usually must diffuse some distance before reacting, so very rapid events cannot be studied. An alternative approach is to diffuse substrates into crystals at a low temperature at which reaction is extremely slow but a substrate may become seated in an active site ready to react. In favorable cases such "frozen" Michaelis complexes may be heated by a short laser pulse to a temperature at which the reaction is faster and the steps in the reaction may be observed by X-ray diffraction.[424,425]

I. Nuclear Magnetic Resonance (NMR)

Organic chemists and biochemists alike have long relied on NMR spectroscopy to assist in identification and determination of structures of small compounds. Most students have some familiarity with this technique and practical information is available in many places.[426–428] Measurements can be done on solids, liquids, or gases but are most often done on solutions held in special narrow NMR tubes. Volumes of samples are typically 0.5 ml (for a 1–5-mM solution containing 1–5 μmol of protein) but less for small molecules and

higher concentrations. Newer techniques allow use of a volume as small as 5 nl and containing <0.1 nmol of sample.[429] For many years progress in biochemical application of NMR was slow, but a dramatic increase in the power of the spectrometers, driven in great measure by the revolution in computer technology, has made NMR spectroscopy a major force in the determination of structures and functions of proteins and nucleic acids.[430,431]

1. Basic Principles of NMR Spectroscopy

The basis of NMR spectroscopy lies in the absorption of electromagnetic radiation at radiofrequencies by atomic nuclei.[426,427,432–437] All nuclei with odd mass numbers (e.g., ^1H, ^{13}C, ^{15}N, ^{17}O, ^{19}F, and ^{31}P), as well as those with an even mass number but an odd atomic number, have magnetic properties. Absorption of a quantum of energy $E = h\nu$ occurs only when the nuclei are in the strong magnetic field of the NMR spectrometer and when the frequency ν of the applied electromagnetic radiation is appropriate for "resonance" with the nucleus being observed. In the widely used "500-megahertz" NMR spectrometers the liquid helium-cooled superconducting electromagnet has a field strength of 11.75 tesla (T). In this field a proton resonates at ~500 megahertz (MHz) and nuclei of ^{31}P, ^{13}C, and ^{15}N at ~202, 125, and 50 MHz, respectively. At 500 MHz the energy of a quantum is only $E = 3.3 \times 10^{-33} \times 10^8$ J $= 0.2$ J mol^{-1}, more than four orders of magnitude less than the average energy of thermal motion of molecules (3.7 kJ mol^{-1}). Thus, the spin transitions induced in the NMR spectrometer have no significant effect on the chemical properties of molecules.

The resonance frequency ν at which absorption occurs in the spectrometer is given by Eq. 3-50, where H_o is the strength of the external magnetic field, μ is the magnetic moment of the nucleus being investigated, and h is Planck's constant. The basis for NMR spectroscopy lies in the fact that nuclei in different positions in a molecule resonates at slightly different frequencies. In a protein each one of the hundreds or thousands of protons resonate at its own frequency. With older NMR instruments a spectrum at the constant magnetic field H_o can be obtained by varying the frequency and observing the values at which absorption occurs, much as is done for ultraviolet, visible, and infrared spectra (Chapter 23). With newer pulsed NMR spectrometers the measurement is done differently but the spectra

$$\nu = \mu H_o / h \qquad (3-50)$$

look the same. The higher the magnetic field, the greater the variation in resonance frequency and the higher the sensitivity. The most powerful commercial NMR spectrometers currently available operate at about 750 MHz for ^{1}H and a few higher frequency instruments have been built.

The **proton NMR** spectrum of the coenzyme pyridoxal phosphate in ^{2}H$_2$O is shown in Fig. 3-24 as obtained with a 60-MHz spectrometer. Four things can be measured from such a spectrum: (1) the **intensity** (area under the band). In a proton NMR spectrum, areas are usually proportional to the numbers of equivalent protons giving rise to absorption bands; (2) the **chemical shift**, the difference in frequency between the peak observed for a given proton and a peak of some standard reference compound. In Fig. 3-24 the reference peak is at the right edge; (3) the **width** at half-height (in hertz), a quantity that can provide information about molecular motion and about chemical exchange; and (4) **coupling constants** which measure interactions between nearby magnetic nuclei. These are extremely important to the determination of structures of both small and large molecules.

With a magnetic field of H$_o$ = 11.75 T and a 500-MHz oscillator, the positions of proton resonances in organic compounds are spread over a range of ~10,000 Hz. This is 20 parts per million (ppm) relative to 500 MHz. Positions of individual resonances are usually given in ppm and are always measured in terms of a shift from the resonance position of some standard substance. For protons this is most often **tetramethylsilane (TMS)**, an inert substance that can be added directly to the sample in its glass tube. Biochemists

often use ^{2}H$_2$O as solvent and the water-soluble sodium 3-trimethylsilyl 1-propane sulfonate (**DSS** or Tier's salt) as a standard. Its position is insignificantly different from that of TMS. For NMR spectra measured in ^{2}H$_2$O, the "pD" of the medium is sometimes indicated. It has often been taken as the pH meter reading plus 0.4. However, because of uncertainty about the meaning of pD, most workers cite the apparent pH measured with a glass electrode and standardized against aqueous buffers.[438] It is important to describe how the measurement was made when publishing results.

Band widths. The narrowness of a band in an NMR spectrum is limited by the **Heisenberg uncertainty principle**, which states that $\Delta E \times \Delta t = h/2\pi$, where h is Planck's constant, ΔE is the uncertainty in the energy, and Δt is the lifetime of the magnetically excited state. Since $E = h\nu$ for electromagnetic radiation, E is directly proportional to the width of the absorption band (customarily measured at one-half its full height). The magnetic nucleus is well shielded from external influences and the lifetime of its excited state tends to be long. Hence, $\Delta \nu$ is small, often amounting to less than 0.2 Hz. This fact is very favorable for the success of high-resolution proton magnetic resonance. However, bands are often much broader for large macromolecules.

The chemical shift. In a molecule such as TMS, the electrons surrounding the nuclei "shield" the nucleus so that it does not experience the full external magnetic field. For this reason, absorption occurs at a high frequency (high energy). Protons that are bound

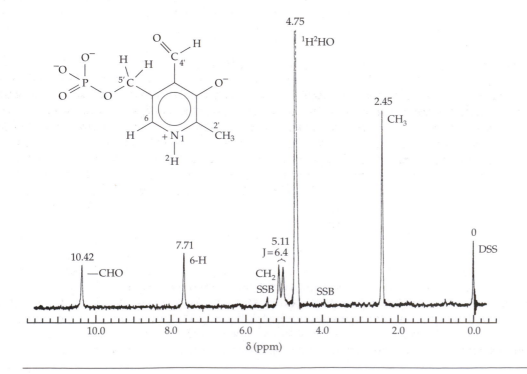

Figure 3-24 The 60-MHz proton magnetic resonance spectrum of pyridoxal 5'-phosphate at neutral pH (apparent pH = 6.65). The internal standard is DSS. Chemical shifts in parts per million are indicated beside the peaks. Spectrum courtesy of John Likos.

to an atom deficient in electrons (because of attachment to electron withdrawing atoms or groups) are **deshielded**. The greater the deshielding, the further **downfield** from the TMS position is the NMR peak.

The magnitude of this chemical shift may be stated in hertz, but it is most often expressed in ppm as δ (Eq. 3-51). The value of δ is the shift in frequency relative to frequency of the oscillator in parts per million and is independent of the field strength. It still depends upon use of a particular reference standard which must be stated when a δ value is given. In the spectrum shown in Fig. 3-24, the methyl protons appear 2.45 ppm below the DSS peak but still at a relatively high field. Characteristic chemical shift ranges for other protons (Table 3-3) extend to ~20 ppm.

$$\delta \, (\text{ppm}) = \frac{\Delta \nu \, (\text{Hz}) \times 10^6}{\nu \, (\text{Hz}) \text{ of oscillator}} \qquad (3\text{-}51)$$

Aromatic rings lead to strong deshielding of attached protons because of a **ring current** induced in the circulating π electrons. Thus, in Fig. 3-24 the peaks of the methylene protons which are adjacent to the aromatic ring occur at 5.10 and 5.12 ppm. The 6-H, which is bound directly to the ring, is more strongly deshielded and appears at 7.71 ppm. The hydrogen of the aldehyde groups is deshielded as a result of a similar "diamagnetic electronic circulation" in the carbonyl group. Its peak is even further downfield at 10.4 ppm. Ring current and other effects on chemical shifts are important in NMR spectroscopy of proteins. Aromatic proton resonances sometimes stand out because they have been shifted far downfield. Ring current effects on chemical shifts can be predicted quite accurately if three-dimensional structures are known.[439] Additionally, computer programs are available for predicting them.[440-443]

Hydrogen bonding has a very large effect on the chemical shift of protons. The resonance of a strongly hydrogen-bonded proton is usually shifted downfield from its position in non-hydrogen-bonding media. This is especially true for hydrogen bonds to charged groups, e.g., the NH of a histidine or tryptophan side chain hydrogen bonded to a carboxylate group, a situation often met in the active sites of proteins. The chemical shift of ^{13}C in the carboxyl group is also affected.[444,445] While **ring current shifts** can be predicted quite well, it is much more difficult to predict the total chemical shift.[446,447]

Scalar coupling (J coupling). The energy of the spin transition of a hydrogen nucleus is strongly influenced by the local presence of other magnetic nuclei, e.g., other protons that are covalently attached to the same or an adjacent atom. These neighboring protons can be in either of the two spin states, a fact that results

in easily measured differences in the energy of the NMR transition under consideration. This spin–spin interaction (coupling) leads to a splitting of NMR bands of protons into two or more closely spaced bands. The ethyl group often appears in NMR spectra as a "quartet" of four evenly spaced peaks that arise from the CH_2 group and a triplet of peaks arising from the CH_3 protons. Protons attached to the same carbon (**geminal** protons), and in similar environments, do not ordinarily split each other's peaks, while the protons on the neighboring carbon do.

The **coupling constant** J is the difference in hertz between the successive peaks in a multiplet. It is a field-independent quantity and the same no matter what the frequency of the spectrometer. In Fig. 3-24 the peak of the methylene protons is split by $^1H-^{31}P$ coupling, with a value of $J \sim 6.4$ Hz. While spin–spin coupling is most pronounced when magnetic nuclei are close together in a structure, the effect can sometimes be transmitted through up to five covalent bonds. The technique of **double irradiation** or **spin decoupling** can be used to detect spin coupling. The sample is irradiated at the resonance frequency of one of the nuclei involved in the coupling, while the spectrum is observed in the frequency region of the other nucleus of the coupled pair. Under these conditions the multiplet collapses into a singlet and the mutual coupling of the two nuclei is established. The coupling can be seen directly in appropriate two-dimensional NMR spectra.

The coupling constant between two **vicinal protons** which are attached to adjacent carbon atoms (or other atoms) depends upon the torsion angle.

The **Karplus equation** (Eq. 3-52) relates J to the torsion angle φ' (so labeled to distinguish it from peptide torsion angle φ; Fig. 2-8).

$$J_{H,H'} \quad A \, \cos^2 \phi' + B \cos \phi' + C \qquad (3\text{-}52)$$

This equation was predicted on theoretical grounds, but the constants, A, B, and C are empirical.[450] Other forms of the equation, some of them simplified, have also been proposed.[451] The Karplus relationship is often used to estimate time-averaged torsion angles in peptides. For a $C_\alpha H-NH$ torsion angle the parameters A, B and C of Eq. 3-52 are ~ 6.4, 1.4, and 1.9, respectively. For a $C_\alpha H - C_\beta H$ they are ~ 9.5, 1.6, and

1.8, respectively.[452] Coupling constants between ^{1}H and ^{13}C or ^{15}N are of importance in determination of three-dimensional structure of proteins.[453]

The nuclear Overhauser effect (NOE) is the result of transfer of magnetization from one nucleus to a nearby nucleus directly through space rather than via *J*-coupling.[427,454] This was observed first as a result of irradiation of a resonance in a one-dimensional spectrum resulting in an increased intensity of the resonance of the nearby nucleus. Magnetization transfer can occur from a given nucleus to one or more nearby nuclei. Each such transfer that is detected is usually referred to simply as an NOE. For an NOE to be observable the two nuclei must be very close together, <0.5 nm (5 Å). The strength of the magnetization transfer falls off approximately as the sixth power of the interatomic distance. Consider two hydrogen atoms, both tightly hydrogen-bonded to an intervening oxygen atom, e.g., of a carboxylate or phosphate group. The expected H–H distance would be ~0.3 nm and a strong NOE between them would be anticipated. Two nonbonded hydrogen atoms (e.g., on methyl groups of amino acid side chains) can be as close together as 0.24 nm at van der Waals contact and could show a very strong NOE. However, contact is rarely this close in proteins unless in a hydrogen bond.

2. Nuclei Other than Hydrogen

Deuterium (²H). The natural abundance is very low so that use of [2]H-labeled compounds is practical for study of metabolism, e.g., for following an ^{2}H label in glucose into products of fermentation[455] or in mammalian blood flow.[456] Deuterium NMR has been used extensively to study lipid bilayers (Chapter 8).

Carbon - 13. Use of ^{13}C in NMR developed slowly because of the low natural abundance of this isotope. Another complication was the occurrence of [13]C–[1]H coupling involving the many protons normally present in organic compounds. The latter problem was solved by the development of **wide-band proton decoupling** (noise decoupling). With a natural abundance of only 1.1%, ^{13}C is rarely present in a molecule at adjacent positions. Thus, [13]C–[13]C coupling does not introduce complexities and in a noise-decoupled natural abundance spectrum each carbon atom gives rise to a single peak. Even so, ^{13}C NMR spectroscopy was not practical until pulsed Fourier transform (FT) spectrometers were developed (Section 2).

Chemical shifts in ^{13}C spectra are often 200 ppm or more downfield relative to TMS. The effects of substituents attached to a carbon atom are often additive when two or more substituents are attached to the same atom.[457,458] It is often necessary (but costly) to prepare compounds enriched in ^{13}C beyond the natural abundance. For proteins this may be done by growing an organism on a medium containing [^{13}C]glucose or a single amino acid enriched in [13]C.[459,460] Using metabolites enriched in ^{13}C, it is also possible to observe metabolism of living tissues directly. For example, glycogen synthesis from ^{13}C-containing glucose has been observed in a human leg muscle using a wide-bore magnet and surface coils for transmitting and receiving.[461] This topic is discussed further in Box 17-C.

Nitrogen - 15. Despite difficulties associated with low natural abundance (0.37%) and low sensitivity, ^{15}N NMR is practical and with isotopically enriched samples has become very important. Proteins with a high content of ^{15}N can be produced easily and inexpensively from cloned genes in bacterial plasmids. For example, cells of *E. coli* can be grown on a minimal medium containing [^{15}N] NH₄Cl. Since ^{13}C can also be introduced in a similar way it is possible to incorporate both isomers simultaneously. Production of uniformly labeled protein containing ^{15}N and / or ^{13}C provides the basis for multidimensional isotope-edited spectra necessary for protein structure determination (next section) and for study of tautomerization of histidine rings (Eq. 2-6).[460,462–464] [15]N chemical shifts of groups in proteins are spread over a broad range (Table 3-3).[465]

Phosphorus - 31. NMR spectroscopy using ^{31}P, the ordinary isotope of phosphorus, also has many uses.[466] Application of ^{31}P NMR to living tissues has been extraordinarily informative[467] and is dealt with in Chapter 6. The many phosphorus nuclei in nucleotides, coenzymes, and phosphorylated metabolites and proteins are all suitable objects of investigation by NMR techniques.

Fluorine - 19. Although not abundant in nature, ^{19}F gives an easily detected NMR signal and can be incorporated in place of hydrogen atoms into many biochemical compounds including proteins.[468,469] In one study genetic methods were used to place 3-fluorotyrosine separately into eight positions in the lac repressor protein (Chapter 29).[470] Measurements of ^{19}F NMR spectra were used to study domain movement. Active site groups of enzymes can be modified to incorporate [19]F.[471] Binding of fluorinated substrates can be studied.[472] Nontoxic ^{19}F-containing compounds are useful as intracellular pH indicators, the NMR spectrometer serving as the pH meter (Box 6-A).[473,474] An atom of fluorine attached to an aromatic ring is highly and predictibly sensitive to inductive effects of substituents in the para position to the fluorine.[475] For example, fluorine at the 6-position in pyridoxal phosphate (Fig. 14-4) can be observed in enzymes and reports changes in coenzyme structure.[476,477]

Some other nuclei. Here are a few reported uses of NMR on other nuclei. ^{3}He, binding into little cavities in fullerenes;[478] ^{11}B, binding of boronic acids to active sites;[479] ^{23}Na, measurement of intracellular [Na^+];[480–482] ^{35}Cl and ^{37}Cl, binding to serum albumin;[483] ^{113}Cd,

reporter that can replace Zn^{2+} (no magnetic moment) in active sites of many enzymes and in nonenzymatic systems as well;[484] Tl, replacing K^+ in enzyme binding sites;[485] ^{17}O, study of dynamics of protein hydration;[486] and ^{77}Se, observation of acetylchymotrypsin intermediate.[487]

TABLE 3-3
Approximate Chemical Shift Ranges in ^{1}H- and in ^{15}N- NMR Spectra

Group	^{1}H chemical shift (ppm from TMS)	^{15}N chemical shift (ppm from liquid NH_3)
–CH_3	0–4.0	
–CH_2–	1.1–4.4	
–CH	2.4–5	
Peptide αH random coil[a]	3.9–5.0	
–OH[b]	~5–6	
>C=C<H	5–8	
–NH_2		31–37
NH		
Peptide	7–12	103–142
Aromatic H	7–9	
Imidazole		
$C^{ε1}$–H	~7.7	
$C^{δ2}$–H	~7.0	
N–H	~10	165–180
Imidazolium		
$C^{ε1}$–H	~8.7	
$C^{δ2}$–H	~7.4	
N–H	10–18	
–CHO Aldehyde	9.4–10.4	
–COOH	11.3–12.2	
Indole NH	~10	130–145
Guanidinium		
–$N^{η}H_2$		69–77
–$N^{ε}$ H		31–37

a See Wishart *et al.*[448]
b See Linderstrøm-Lang[449]

3. Fourier Transform Spectrometers and Two-Dimensional NMR

Although NMR spectroscopy was widely used by the 1950s it was revolutionized by two developments, pioneered by Richard Ernst in the mid 1960s and 1970s.[488] The first of these was pulsed Fourier transform (FT) spectroscopy, which permits rapid accumulation of high-resolution spectra. In an FT NMR spectrometer a strong pulse of radiofrequency (RF) radiation is delivered to the sample over a period of a few microseconds and its effects are observed at all frequencies simultaneously. Although 1–2 or more seconds must be allowed before the next pulse is delivered, one complete NMR spectrum is obtained with each pulse. Often the results of hundreds, thousands, or even hundreds of thousands of pulses are added to provide greater sensitivity. With good temperature control this may be accomplished over periods of minutes to days.[489,490]

Free induction decay. The strong exciting RF pulse is delivered with an orientation at right angles to that of the static field H_o of the magnet and whose direction defines the z axis. As a result of this pulse, the magnetization of a nucleus is tilted away from the z axis and *precesses around the z axis* at its resonance (Larmor) frequency, which is ~500 MHz for ^{1}H in a 500-MHz spectrometer. The frequency that is measured is actually a difference from the "carrier frequency" of the RF pulse. The precessing magnetization of the nuclei has a component in the *xy* plane which induces an electrical signal in the coil of the NMR probe which defines the *y* axis. This signal, which contains the Larmor frequencies of all of the nuclei of a given element, is recorded as a function of time over a period of a few seconds. The signal decays away exponentially. However, this curve of **free induction decay** (FID) is not smooth but contains within it all of the Larmor frequencies. If enough points (perhaps 500 in a 2-s acquisition) are recorded and stored in the computer's memory, Fourier transformation of the data will produce the frequency-dependent NMR spectrum.[489,490]

Relaxation times T_1 and T_2. When a very strong pulse of electromagnetic radiation is applied in the NMR spectrometer, virtually all of the nuclei are placed in the magnetically excited state. If another pulse were applied immediately, little energy would be absorbed because the system is **saturated**. In the older "continuous-wave" NMR spectrometers, the energy is always kept small so that little saturation occurs. However, in FT NMR instruments, the strong pulses lead to a high degree of saturation. Application of repeated pulses would produce no useful information were it not for the fact that the excited nuclei soon relax back to their equilibrium energy distribution. Relaxation occurs through interactions of the nuclei with fluctuating magnetic fields in the environment. For organic molecules in solution the fluctuations that are most often effective in bringing about relaxation are the result of moving electrical dipoles in the immediate vicinity. Even so, relaxation of protons in water requires seconds.

Relaxation of nuclear magnetic states is characterized by two relaxation times. The longitudinal or **spin-lattice relaxation time** T_1 measures the rate of relaxation of the net magnetic vector of the nuclei in the direction of H_0. The transverse or **spin–spin relaxation time** T_2 measures the relaxation in the xy plane perpendicular to the direction of H_0. The two relaxation times can be measured independently. In general, $T_2 < T_1$. For solids, T_2 is quite short ($\sim 10^{-5}$ s) whereas relaxation times of seconds are observed in solutions. This lengthening of the lifetime of the excited state in going from solid to liquid leads to a narrowing of absorption lines and explains why NMR bands in liquids are often narrow. However, an increase in viscosity or a loss of fluidity in a membrane leads to broadening.

How can T_2 and T_1 be measured? T_2 for fluids can often be estimated from the width of the band Δv at half-height (Eq. 3-53). However, pulsed NMR methods are usually employed, the measurement of T_1 being especially easy.

$$T_2 \approx 1/\pi\,\Delta v \qquad (3\text{-}53)$$

An attempt is often made to relate T_1 and T_2 to the molecular dynamics of a system. For this purpose a relationship is sought between T_1 or T_2 and the **correlation time** τ_c of the nuclei under investigation. The correlation time is the time constant for exponential decay of the fluctuations in the medium that are responsible for relaxation of the magnetism of the nuclei. In general, $1/\tau_c$ can be thought of as a rate constant made up of the sum of all the rate constants for various independent processes that lead to relaxation. One of the most important of these ($1/\tau_1$) is for molecular tumbling.

$$1/\tau_1 \approx (3k_BT)/4\pi\eta r^3 \qquad (3\text{-}54)$$

This equation is closely related to that of rotational diffusion (Eq. 9-35). Another term is the reciprocal of the **residence time** τ_m, the mean time that a pair of

dipoles are close enough together to lead to relaxation.

In the usual solvents at room temperature, τ_c is of the order of 10^{-12} to 10^{-10} s. Thus, relaxation rates in solutions are considerably faster than the frequencies of radiation absorbed in the NMR spectrometer ($\sim 10^8$ s^{-1}). Relaxation is relatively ineffective and T_1 and T_2 are usually large and equal. Bands remain sharp. As the correlation time increases (as happens, for example, if the viscosity is increased), T_1 and T_2 decrease with T_1 reaching a minimum when $\tau_c \sim v$, the frequency of the absorbed radiation. Lines are broadened and hyperfine lines (from coupling between nuclei) cannot be resolved. As τ_c is increased further, T_2 reaches a constant low value, while T_1 rises again. NMR measurements can be made in the region where τ_c exceeds v, a circumstance that is favored by the use of high-frequency spectrometers. On the other hand, in fluids it is more customary to work in the range of "extreme motional narrowing" at low values of τ_c. Both T_1 and T_2 rise as the mobility of the molecules increases.

A limitation of use of NMR measurements of proteins comes from the increase in tumbling time with increasing size of the molecules. Since $1/\tau_r$ is often the most important term in the relaxation rate constant, only small proteins of mass <20 kDa give very sharp bands. Nevertheless, usable spectra are often obtainable on proteins ten times this size.

A practical problem in ^{13}C NMR arises from slow relaxation (long T_1). Partial saturation is attained and signal intensities are reduced for those carbon atoms for which relaxation is especially ineffective. Relaxation times can be measured separately for each carbon atom in a molecule and can yield a wealth of information about the **segmental motion** of groups within a molecule. Although the relationships between relaxation times and molecular motion are complex, they are often relatively simple for ^{13}C. Carbon atoms are usually surrounded by attached hydrogen atoms, and dipole–dipole interactions with these hydrogen atoms cause most of the nuclear relaxation. For a carbon atom attached to N equivalent protons in a molecule undergoing rapid tumbling, Eq. 3-55 holds. Where $h = h/2\pi$ and γ_c and γ_H are the magnetogyric ratios of carbon and hydrogen nuclei.

$$1/T_1 \approx \frac{Nh^2\gamma_c{}^2\gamma_H{}^2\tau_{eff}}{r^6} \qquad (3\text{-}55)$$

This equation permits a calculation of an effective correlation time τ_{eff} for each carbon atom.[491] T_1 and T_2 can also be evaluated for individual ^{15}N or ^{13}C nuclei in labeled proteins.[491]

Two-dimensional and multidimensional NMR spectra. Proteins have such complex NMR spectra that, except for small regions at the upfield and downfield ends (see Fig. 3-26A), it is impossible to interpret one-dimensional spectra. A solution to this problem came from the development by Jeener, Ernst, and Freeman of methods of displaying NMR spectra in two dimensions.[428,490,492,493] All two-dimensional and multidimensional NMR methods make use of one basic procedure: After the initial RF pulse, a second or a series of subsequent RF pulses are introduced. This is done before the nuclei have had time to relax completely. The time t_1 from the initial pulse to the second pulse is called the **evolution period** and allows accumulation of information about NOEs or *J*- coupling, whether homonuclear (e.g. ^1H–^1H or ^{13}C–^{13}C) or heteronuclear (e.g., ^1H–^{13}C or ^1H–^{15}N).

Two-dimensional spectra usually require hours or days of acquisition because separate FIDs are collected

for a series of many different values of t_1. This provides a second timescale for the experiment. The second pulse is often used to rotate the directions of magnetization of the nuclei that are being observed from the xy plane into the yz plane but with the z component at 180° to the H° vector. One very important type of two-dimensional plot is the **NOESY** (NOE spectroscopy) spectrum, which detects NOEs between all excited nuclei that are close enough together (Fig. 3-25B). The transfer of magnetization occurs during a **mixing time** Δ, which follows the second pulse. For a protein Δ may be 25–300 ms. A third 90° pulse returns the z components of the magnetization to be parallel with the y axis and the FID is collected over the period t_2. The amplitude of the resonances detected is modulated by the frequencies that existed during the evolution

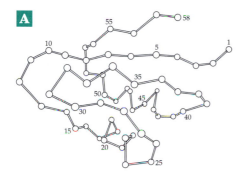

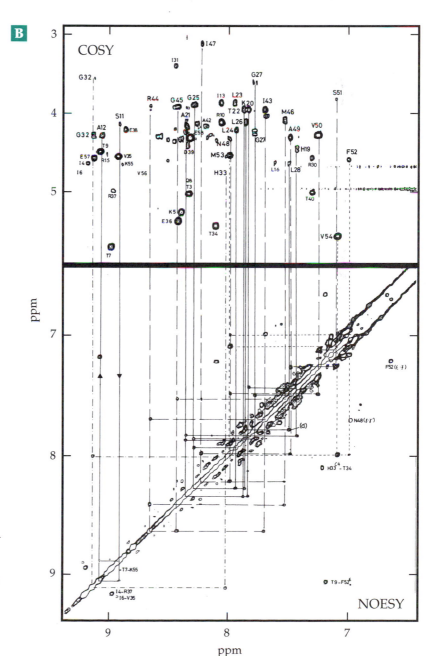

Figure 3-25 (A) Alpha-carbon plot of the structure of ribosomal protein L30 from *E. coli* as deduced by NMR spectroscopy and model building. (B) Combined COSY-NOESY diagram for ribosomal protein L30 used for elucidation of d_{NN} connectivities (see Fig. 3-27). The upper part of the diagram represents the fingerprint region of a COSY spectrum recorded for the protein dissolved in H_2O. The sequential assignments of the crosspeaks is indicated. The lower part of the diagram is part of a NOESY spectrum in H_2O. The d_{NN} "walks" are indicated by : (→—) S11-A12; (—) H19 to L26; (- - - -) I43 to S51; (- - - -) G32 to H33; (——) F52 to V54. From van de Ven and Hilbers.[494]

period. After Fourier analysis[436,490] a two-dimensional plot with two frequency axes is generated and is usually displayed as a contour plot. Along the diagonal of Fig. 3-25B are peaks representing the one-dimensional spectrum. All of the peaks off of the diagonal are NOEs which can be related back to the peaks on the diagonal as shown by the horizontal and vertical lines.

A second important two-dimensional method is **correlation spectroscopy (COSY)**, in which the pairs of off-diagonal peaks result from spin–spin coupling. A related method called **TOCSY** provides correlations that extend through more than three bonds. The COSY plot in Fig. 3-26B is for the synthetic cyclic decapeptide *cyclo*-(Δ³-Pro–D-*p*-Cl-Phe–D-Trp–Ser–Tyr–D-Trp–N-Me-Leu–Arg–Pro–β-Ala). It was obtained in six hours on a 500-MHZ instrument. The region marked I reveals couplings of protons on α-carbons to those on adjacent β carbons within the same residue ($J_{\alpha\beta}$; Fig. 3-27) and other couplings within the side chain. Each amino acid has a characteristic pattern. From careful study of this region it is possible to correlate each α-H resonance with a particular amino acid side chain. However, some residues are difficult to distinguish, e.g., His, Trp, Phe, and Tyr have similar $J_{\alpha\beta}$ values. Region II of Fig. 3-26B reveals connectivities of α and β hydrogens to N-H protons in the 7– 9 ppm region. Each α-H is coupled to the N-H of the same residue ($J_{\alpha H}$, Fig. 3-27). A section of a COSY plot is also shown in Fig. 3-25B and indicates how resonances can be related to those in the NOESY plot made on the same sample.

Of great importance in the determination of protein structures is the use of ¹⁵N- or ¹³C-enriched samples to obtain **isotope-edited** spectra. For example in **HSQC** or in **¹⁵N-multiple quantum cohenence (HMQC)** spectra we see only NH protons in a plot of ¹H chemical shift in one dimension versus the ¹⁵N chemical shift of the attached

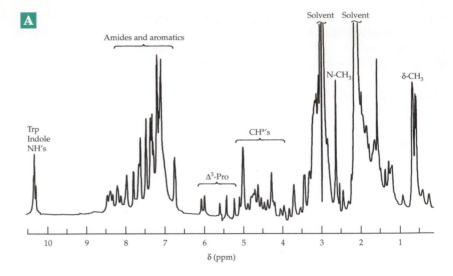

A

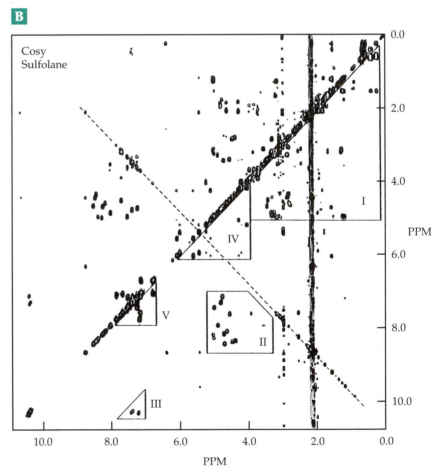

B

Figure 3-26 Proton NMR spectra of a cyclic decapeptide analog of the gonadotropin-releasing hormone in the solvent sulfolane at 500 MHz. (A) One-dimensional spectrum. This figure also illustrates upfield methyl group regions, α-hydrogen, amide, aromatic, and downfield (10–20 ppm) regions. The indole NH resonances are shifted downfield by the ring current of the indole. (B) COSY spectrum plotted as a contour map. The outlined areas represent five unique *J*-coupled regions; area I, the Cᵅ to Cᵝ to Cᵞ etc., side chain connectivities; area II, the NH to Cᵅ connectivities; area III, the indole NH of tryptophan; area IV, the connectivities of the Δ³-Prol residue; and area V, the aromatic resonances. Peaks that appear on the solid vertical line and the dashed diagonal lines are artifacts. From Baniak *et al.*[495]

nitrogen atom in the other (Fig. 3-28). Furthermore, as shown in this figure, particular types of NH bonds (peptide, –NH_2, imidazole, indole, amide, and guanidinium) appear in different regions. There is only one peptide NH per residue and, for a small protein, each may be separately visible.

4. Three-Dimensional Structures and Dynamics of Proteins

The first three-dimensional structure of a small protein was determined solely from NMR measurements in 1984. To date hundreds of "NMR structures" have been deduced. For small proteins of $M_r < 10,000$ two-dimensional COSY and NOESY spectra can suffice. For larger proteins use of ^{15}N- and/or ^{13}C-enriched proteins is essential.[428,430,496] This permits generation of a third and even a fourth frequency axis and three- and four dimensional NMR. For example, using various complex pulse sequences an ^{15}N-correlated ^{1}H – ^{2}H NOESY spectrum can be generated. This shows directly which NOEs arise from NH protons with resonances in the amide region of an HSQC or HMQC spectrum (Fig. 3-28). With both ^{15}N and ^{13}C present, many additional coupling patterns and J values can be observed.[497–502] It is also possible to measure NOEs from atoms in a protein to those of a relatively weakly bound ligand such as a coenzyme or substrate analog and to determine the conformation of the bound ligand from this **transferred NOE**.[503–505]

Assignment of resonances. After acquisition of the necessary data, which may require 10–15 mg of protein, the observed resonances must be assigned to specific amino acid residues in the peptide chain. The connectivities of the individual CH and NH groups that have been identified in the COSY spectrum and information about the relationship of one atom to another atom nearby in space are required. The closest neighbors to either αH or peptide NH protons are often protons in a neighboring residue ($d_{\alpha N}$, d_{NN}, Fig. 3-27). As was pointed out in the preceding section, NOE correlations obtained from plots such as that in Fig. 3-25B provide much of the information needed to establish which resonances belong to each residue in a known sequence.[494] The three-dimensional structure of the ribosomal protein L30 of *E. coli.* (Fig. 3-25A) was deduced entirely by NMR spectroscopy. A downfield part of the NOESY and COSY plots used is shown in Fig. 3-25B. This figure also shows how cross-peaks in the NOESY spectrum were correlated with identified COSY peaks. It is helpful initially to hunt for unique dipeptides that can be identified in the NOESY spectrum. Ambiguities that arise can be resolved by use of various additional techniques.

NOEs and distance constraints. NOESY plots also contain the essential information needed to determine which side chains *distant* in the sequence are close together in space. A NOE observed for a pair of nuclei falls off as the inverse sixth power of the distance between them. For this reason, NOEs are observed only for pairs of atoms closer than about 0.4 nm. It is possible, in principle, to calculate distances between nuclei from the NOE intensity, but this is not accurate. Often, the NOE cross-peaks are grouped into three categories that correspond to maximum possible distances of 0.25, 0.30, and 0.40 nm. These can be related for the most part to intraresidue (e.g., $d_{N\alpha'}$ of Fig. 3-27) sequential (e.g., $d_{\alpha N'}$ and $d_{NN'}$, Fig. 3-27) and long range backbone–backbone distances. These values constitute a series of **distance constraints** which are applied while making an automated computer search for a folding pattern that will meet these constraints and at the same time have acceptable torsion angles and good side chain packing throughout. This process makes use of distance geometry algorithms and other methods.[430,506,507] An early success was the solution of a 75-residue amylase inhibitor independently by crystallographers[508] and NMR spectroscopists.[509] The NMR structure was based on 401 NOE distance constraints, 168 distance constraints imposed by hydrogen bonds and 50 torsion angles deduced from J values. Recently, refinement of NMR structures has been done as in X-ray crystallography.[507] Some structures have been refined using both NMR and X-ray data.[442]

The spectra in Figs. 3-25, 3-26 and 3-28 are for relatively small proteins. Spectra of larger proteins are more complex and lines are broader. Many techniques are used to simplify spectra. The NMR spectrum of a protein is simplified considerably if the protein is denatured by heating, and ^{1}H NMR spectra of "random coil" proteins can be predicted well from tables of standard chemical shifts for the individual amino acids.[510] Many amide NH protons exchange with solvent rapidly, making it easier to assign the remaining peaks. However, nearly all NH peaks will be seen in an HMQC or HSQC (Fig. 3-28) spectrum. Partial, or even complete, substitution of deuterium for hydrogen will also simplify spectra.[511–513] Microorganisms that will grow in a medium rich in D_2O can be used as sources of partially deuterated proteins. Because the remaining protons usually have ^{2}H rather than ^{1}H as a neighbor, dipolar line broadening is reduced and sharper resonances are observed.[514] Substitution of ^{15}N for ^{14}N in the backbone amide groups can also yield spectra with narrower lines.[515] **Isotope-edited** NMR spectra allow simplification of complex two-dimensional spectra by observation of only those protons attached to an isotopically labeled nucleus, e.g., ^{13}C or ^{15}N.[516–518] Measurement of ^{15}NH-C$_{\alpha H}$ J couplings facilitates structure determinations.[519]

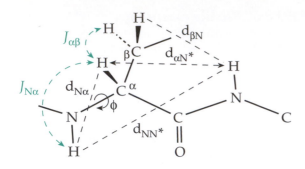

Figure 3-27 Illustration of some distances (d_{NN}, $d_{\alpha N}$ and $d_{\beta N}$) obtained from NOESY spectra and some coupling constants ($J_{\alpha\beta}$ and $J_{N\alpha}$ obtained from ^{1}H COSY spectra or J-resolved spectra. The coupling constants and proton chemical shifts provide a "fingerprint" for each residue and $J_{N\alpha}$ may also provide an estimated value for torsion angle f. The distances establish residue-to-residue connectivities as well as distance constraints that may permit a calculation of three-dimensional structure. Additional coupling constants can be measured for ^{15}N- or ^{13}C- enriched proteins. Coupling from β-hydrogens to other side chain hydrogens provides "fingerprint" information about individual residues.

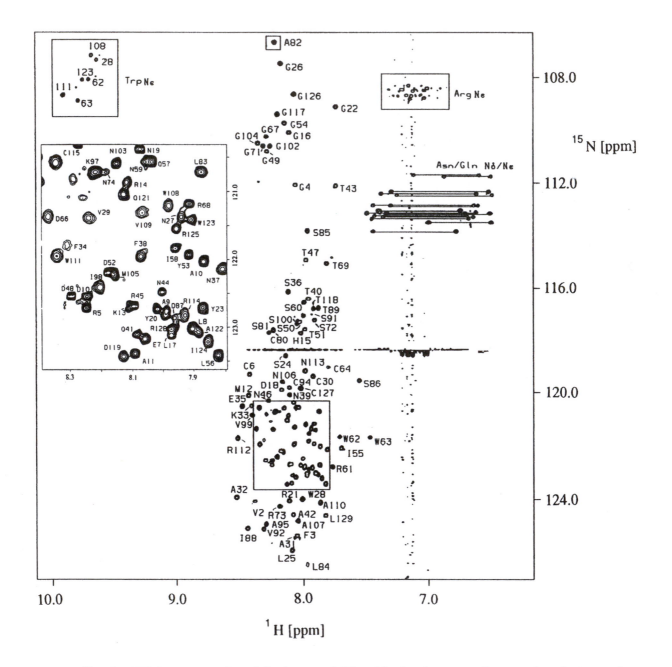

Figure 3-28 A ^{15}N – ^{1}H HSQC spectrum of partially denatured 129-residue hen lysozyme. Boxes enclose the tryptophan indole region (upper left), the arginine side chain N$^\varepsilon$ region (upper left), and a portion of the amide NH region (lower center and enlarged in the insert). Resonances of pairs of hydrogen atoms in side chain (Asn and Gln) amide groups are indicated by horizontal lines. From Buck *et al.*[524]

Many aspects of the dynamics of the action of enzymes of this size can be studied by NMR spectroscopy. New approaches allow study of proteins up to ~30 kDa in size.[519a] Computer programs can analyze data and make automatic assignments of resonances.[520,521]

Also useful are techniques for measuring NMR spectra on solids, including microcrystalline proteins[522] (see also Box 9-C). In some cases, e.g., for heme proteins,[522,523] the crystals can be oriented in a magnetic field permitting measurements of NMR spectra with more than one orientation of the crystals. This potentially affords more information than the usual techniques. For example, growth of bacteria in ^2H_2O media containing ^1H-containing pyruvate yields proteins with almost complete deuteration in the C-α an C-β positions but with highly protonated methyl groups. This gives rise to good $– CH_3$ to $– CH_3$ NOEs and provides other advantages in both NMR spectroscopy and mass spectrometry.[512] Isotopically enriched amino acids, e.g., ^{13}C-enriched leucine,[525] or isoleucine containing ^{15}N, ^{13}C, and ^2H as well[526] can be fed to growing bacteria. The use of paramagnetic shifts by ions such as Gd^{3+} may be helpful.[527] Very large shifts are sometimes induced in heme proteins by the Fe^{3+} of the heme that is embedded within the protein.[528]

5. Other Information from NMR Spectra

NMR titrations. How do pH changes affect NMR resonances? Resonances of ^1H nuclei close enough to a proton with a pK_a in the pH region under study will experience a shift, which may be either upfield or downfield when the proton dissociates. The resonance of ^{13}C in a carboxyl group will shift downfield when the proton on the carboxyl group dissociates. If the proton that dissociates is tightly hydrogen bonded in a protein, its rate of dissociation may be *slow* compared to the NMR frequency used. If so, the original peak will decrease in value as the pH is raised and a new peak will appear at a position characteristic of the dissociated form. However, protons attached to N or O are usually in *rapid* exchange with the solvent. In this case, the NMR resonance of the nucleus being observed will move continuously from one chemical shift value at low pH to a different one at high pH. In an intermediate case the resonance will shift and broaden.

Both the $C^{δ2}$-H and $C^{ε1}$-H protons of histidine can often be seen in proteins (Fig. 3-29A).[529 – 531] As is shown in Fig. 3-29 A and B, their chemical shifts are strongly dependent upon the state of protonation of the ring nitrogen atoms. At low pH, the positive charge that is shared by the two NH groups attracts electrons away from both CH positions, causing deshielding of the CH protons. The effect is greater for the $C^{ε1}$-H than for $C^{δ2}$-H protons (Table 3-3). If the groups being titrated do not interact strongly with other nearby basic or

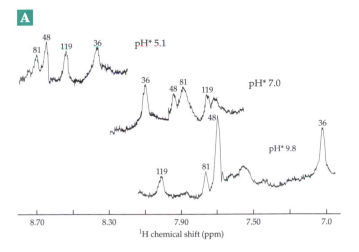

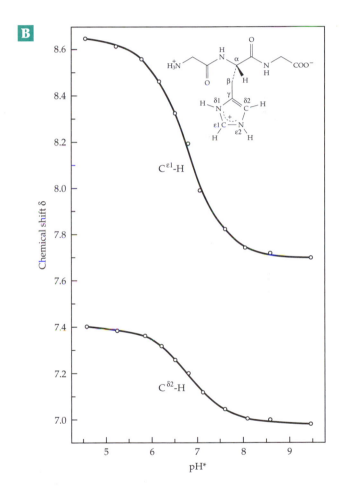

Figure 3-29 (A) ^1H NMR spectra of human myoglobin in D_2O showing the C-H^{ε1} resonances of the imidazole rings of histidines 36, 48, 81, and 119 at three values of the apparent pH (pH*). From Bothelho and Gurd.[529] (B) ^1H NMR titration curves for the histidine $C^{ε1}$ and C-^{δ2} ring protons of glycyl-L-histidylglycine. Data obtained at 100 MHz using 0.1 M tripeptide in D_2O containing 0.3 M NaCl. The pH values (labeled pH*) are uncorrected glass electrode pH meter readings of D_2O solutions using an electrode standardized with normal H_2O buffers. Chemical shifts are downfield from TMS. Courtesy of J. L. Markley.[438]

acidic groups, pK_a values can be estimated for individual histidines.[530,532] With ^{15}N-containing proteins the tautomeric states ($NH^{\epsilon 2}$ vs $NH^{\delta 1}$) of the imidazole rings can also be deduced.[533] Observation of imidazole rings shows that in proteins pK_a values of imidazolium groups are sometimes less than five and sometimes greater than ten.[463,534]

Observing exchangable protons. ^1H spectra of proteins are often recorded in D_2O because of interference from the very strong absorption of H_2O at approximately 4.8 ppm. However, resonances of some rapidly exchanging protons are lost. Special pulse sequences, as well as improvements in spectrometer design, allow many of these resonances to be seen in H_2O.[535–537] At the far downfield end ($\delta > 10$) of ^1H NMR spectra there are often weak peaks arising from NH protons of imidazole or indole side chains that can be observed in H_2O.[463,538–540] These resonances are often shifted 2–5 ppm downfield from the positions given in Table 3-3. The ^1H resonance for a carboxyl (– COOH) proton is shifted to 20 ppm or more in spectra of very strongly hydrogen-bonded anionic complexes such as the malonate dianion.[541–544] (see Chapter 9, Section D,4). The strongest hydrogen bonds cause the greatest downfield shifts because the negative charge pulls the ^1H proton away from the electrons of the atom to which it is attached, deshielding the proton.[545]

A good example is provided by the imidazole NH proton of the active site of trypsin and related serine proteases (Chapter 12), which is seen at ~16 ppm. Another example is provided by aspartate aminotransferase, whose ^1H NMR spectrum in a dilute aqueous phosphate buffer is shown in Fig. 3-30. The peak labeled A is the resonance of the NH proton on the ring of the pyridoxal phosphate coenzyme (marked in Fig. 14-6) and peak B belongs to an adjacent imidazole group of histidine 143. Both of these protons move with pH changes around a pK_a of ~ 6.2 which is associated with the Schiff base proton 6–8 nm away.[534,546–551] Peak A moves upfield 2.0 ppm and peak B downfield 1.0 ppm when the pH is raised around this pK_a. These hydrogen-bonded protons act as sensitive "reporters" of the electronic environment of the active site. Many proteins contain carboxylate or phosphate groups in their active sites and observation of NMR resonances of protons hydrogen bonded to them or to groups in substrates or inhibitors may be a useful technique for study of many enzymes and other proteins.

Exchange rates of amide protons. The NH protons of the peptide backbone can be observed in H_2O in the 6–11 ppm region. If the spectrum is recorded for a sample in D_2O, many of these resonances disappear gradually as the protons of the peptide units exchange with the deuterium ions of the medium. A study of the observed exchange rates can shed light on

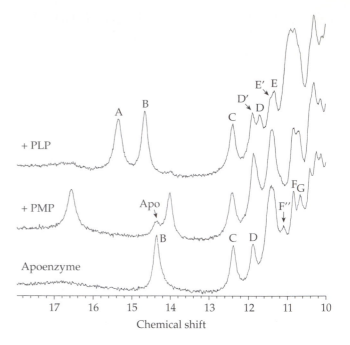

Figure 3-30 Spectra of the pyridoxal phosphate (PLP), pyridoxamine phosphate (PMP) and apoenzyme forms of pig cytosolic aspartate aminotransferase at pH 8.3, 21°C. Some excess apoenzyme is present in the sample of the PMP form. Spectra were recorded at 500 MHz. Chemical shift values are in parts per million relative to that of H_2O taken as 4.80 ppm at 22°C. Peak A is from a proton on the ring nitrogen of PLP or PMP, peaks B and D are from imidazole NH groups of histidines 143 and 189 (see Fig. 14-6), and peaks C and D′ are from amide NH groups hydrogen bonded to carboxyl groups.

the dynamics of protein molecules in solution. Exchange of amide protons in proteins with ^2H or ^3H has been studied on a relatively slow timescale by such techniques as observation of infrared vibrations of the amide group (see Chapter 23).[552,553] However, the development of two-dimensional NMR dramatically improved the ability to study proton exchange.[552,554–556] The measurements may involve use of quenching by rapid solvent change,[556] special pulse sequences,[555] and study of T_1 relaxation rates (seconds timescale). Recently, electrospray mass spectrometry has also been exploited.[557,558] Exchange patterns for large proteins may be followed,[559] and proteolytic fragmentation into short peptides may be used after various lengths of exchange time to investigate the exchange in specific regions of a protein.[216,560] Unfolding of a protein under denaturing conditions can be studied, as can refolding of a completely denatured protein.[561]

Amide hydrogen exchange is usually discussed in terms of a model proposed by Linderstrøm-Lang in 1955.[554,556] He suggested that portions of protein molecules unfold sporadically to allow rapid exchange which can be catalyzed by H^+, HO^-, or other acids or bases. The pH dependence of exchange for a given amide NH can be described as follows:

$$k_{ex} = k_{H} [H^+] + k_{OH} [OH^-] + k_{w} \qquad (3\text{-}56)$$

where k_H, k_{OH}, and k_w are rate constants for acid-catalyzed, base-catalyzed, and the very slow water-catalyzed exchange.[554,556,562] While many amide protons exchange rapidly, hydrogen-bonded NH protons in well-packed hydrophobic core regions exchange slowly,[554,563] sometimes remaining in the protein for years in a D_2O solution. Some unusually stable small proteins such as the seed protein **crambin** show little exchange. The $C^{\varepsilon 1}$ protons of histidine imidazoles also exchange slowly with D_2O from the medium. The exchange rates[529,564] are rapid at higher temperatures, with an average half-life of about 11 min at 65°C. Different residues may exchange at different rates. Binding of substrates or inhibitors can stabilize the protein, slowing all exchange rates of both amide and imidazole groups.

Solid-state NMR and other topics. Little can be said about these topics, but NMR measurement on solid crystalline materials is now practiced and is providing a wealth of information. It is being applied more often to biochemically related problems.[465,542,565–567]

NMR spectroscopy of nucleic acids is discussed briefly in Chapter 5. An important medical application of NMR is in **imaging**, a topic dealt with in Box 30-A.

J. The Protein Data Bank, Three-Dimensional Structures, and Computation

The x,y,z coordinates of all atoms in published, refined three-dimensional structures have been deposited in the Protein Data Bank (Table 3-4).[568–571] Many other related databases are available,[572] e.g., covering molecular modeling,[573] gene sequences, proteome data,[574] and much, much more. A good way to keep up to date is to read the "computer corner" in *Trends in Biochemical Sciences (TIBS)*. Most databases can be reached on the World Wide Web.[572] A selected list is

TABLE 3-4
Selected World Wide Web Servers Related to Protein Structures and Sequences[a]

Question / area	Tool	Access / URL
Database search by comparison of 3D structures	Dali server	*http://www.embl-heidelberg.de/dali/dali.html*
Structural classification of proteins	SCOP	*http://www.bio.cam.ac.uk/scop/*
Summary and analysis of PDP structures	PDBSum	*http://www.biochem.ucl.ac.uk/bsm/pdbsum*
Retrieve 3D coordinates	Protein Data Bank	*http://www.rcsb.org/pdb/*
SWISS-PROT sequence database, Swissmodel homology modeling, etc.	ExPASy	*http://expasy.hcuge,ch/*
Molecular graphics viewer for PCs and workstations	RasMol	*http://www.bernstein-plus-sons.com/software/rasmol*
World Wide Web -Entrez and Molecular Modeling Database access		*http://www.ncbi.nlm.nih.gov*
Protein Science Kinemages	MAGE and PREKIN	*http://www.prosci.uci.edu/kinemages/KinemageIndex.html*
Pedro's Biomolecular Research Tools		*http://www.fmi.ch/biology/research_tools.html*
Predict secondary structure from sequence	Predict Protein server	*http://www.sander.embl-heidelberg.de*
Browse databanks in molecular biology	SRS server	*http://www.embl-heidelberg/srs/srsc*
The Human Genome Database		*http://gdbwww.gdb.org/*
Image Library of Biological Macromolecules		*http://www.imb-jena.de/IMAGE.html*
Bacterial Nomenclature		*http://www.gbf-braunschweig.de/DSMZ/bactnom/bactnam.htm*

[a] From Holm and Sander,[570] Hogue *et al.*,[573] Laskowski *et al.*,[580] and Walsh *et al.*[572]

given in Table 3-4. The widely used viewer called RasMol can be used with your PC or MacIntosh computer[575–576a] or UNIX workstation[569,575]. Another way to view macromolecules is to use the *Protein Science* Kinemages ("kinetic images") using the program MAGE.[577,578]

Although it is mentioned in a few places, this book does not begin to describe the rapid growth of computation in biochemistry, biophysics, and biology in general. Very fast methods of protein structure determination are being developed.[579,582] One of the major goals in current computation is to predict folding patterns of proteins from their sequences.[582,583] By

comparing sequences we can often guess an approximate structure but accurate predictions are still not possible. Having a structure, we would like to predict properties and reactivities and to be able to guess how two more macromolecules interact to form macromolecular complexes. We would like to understand the complex chain of nonpolar and electrostatic interactions that underlie the fundamental properties of catalysis, movement, and responsiveness of organisms. Many computers and ingeneous minds are working to help us match theory with reality in these areas. Read the current journals!

References

1. Strang, R. (1981) *Trends Biochem. Sci.* **6**, VII–VIII
2. Michaelis, L. (1922) *Die Wasserstoffionenkonzentration*, p. 48, Springer-Verlag, Berlin
3. Dixon, M., and Webb, E. C., eds. (1979) *Enzymes*, 3rd ed., pp.138–163, Academic Press, New York
4. Dryer, R. L., and Lata, G. F. (1989) *Experimental Biochemistry*, Oxford Univ. Press, New York
5. Kuramitsu, S., and Hamaguchi, K. (1980) *J. Biochem.* **87**, 1215–1219
6. Beynon, R. J., and Easterby, J. S. (1996) *Buffer Solutions: The Basics*, IRL Press, Oxford
7. Bates, R. G. (1964) *Determination of pH*, Wiley, New York
8. Stoll, V. S., and Blanchard, J. S. (1990) in *Guide to Protein Purification* (Deutscher, M. P., ed), pp. 24–38, Academic Press, San Diego, California
9. Ellis, K. J., and Morrison, J. F. (1982) *Methods Enzymol.* **87**, 405–426
10. Good, N. E., Winget, G. D., Winter, W., Connolly, T. N., Izawa, S., and Singh, R. M. M. (1966) *Biochemistry* **5**, 467–477
11. Ferguson, W. J., Braunschweiger, K. I., Braunschweiger, W. R., Smith, J. R., McCormick, J. J., Wasmann, C. C., Jarvis, N. P., Bell, D. H., and Good, N. E. (1980) *Anal. Biochem.* **104**, 300–310
12. Dawson, R. M. C., Elliot, D. C., Elliot, W. H., and Jones, K. M. (1986) *Data for Biochemical Research*, 3rd ed., Oxford Science Pub., London
13. Robyt, J. F., and White, B. J. (1990) *Biochemical Techniques, Theory and Practice*, 2nd ed., Waveland Press, Inc., Prospect Height, Ill
14. Cooper, T. G. (1977) *The Tools of Biochemistry*, Wiley, New York
15. Alexander, R. R., and Griffiths, J. M. (1992) *Basic Biochemical Methods*, 2nd ed., Wiley, New York
16. Wilson, K., and Walker, J., eds. (1994) *Principles and Techniques of Practical Biochemistry*, 4th ed., Cambridge Univ. Press, New York
17. Boyer, R. F. (1992) *Modern Experimental Biochemistry*, 2nd ed., Benjamin-Cummings Publ., Redwood City, California
18. le Maire, M., Chabaud, R., and Herve, G. (1991) *Laboratory Guide to Biochemistry, Enzymology, and Protein Physical Chemistry*, Plenum, New York
19. Fini, C., Floridi, A., and Finelli, V. N., eds. (1989) *Laboratory Methodology in Biochemistry*, CRC Press, Boca Raton, Florida
20. Scopes, R. K. (1993) *Protein Purification: Principles and Practice*, 3rd ed., Springer, New York
21. Harris, E. L. V., and Angal, S. (1990) *Protein Purification: A Practical Approach*, IRL Press, Oxford, 2 volumes

22. Deutscher, M. P., ed. (1990) *Guide to Protein Purification: Methods in Enzymology*, Vol. 182, Academic Press, San Diego, California
23. Celis, J. E., ed. (1994) *Cell Biology: A Laboratory Handbook*, Academic Press, New York (3 volumes)
24. Fasman, G. D., ed. (1989) *CRC Practical Handbook of Biochemistry and Molecular Biology*, CRC Press, Boca Raton, Florida
25. Rosenberg, I. M. (1996) *Protein Analysis and Purification*, Birkhäuser, Cambridge, Massachusetts
26. Walker, J. M., ed. (1996) *The Protein Protocols Handbook*, Humana Press, Totowa, New Jersey
27. Marshak, D. R., Kadonaga, J. T., Burgess, R. R., Knuth, M. W., Brennan, W. A., Jr., and Lin, S.-H. (1996) *Strategies for Protein Purification and Characterization*, Cold Spring Harbor Lab. Press, Cold Spring Harbor, New York
28. Creighton, T. E., ed. (1989) *Protein Structure: A Practical Approach*, IRL Press, Oxford
29. Marshak, D. R., Kadonaga, J. T., Burgess, R. R., Knuth, M. W., Brennan, W. A., and Lin, S.-H. (1995) *Strategies for Protein Purification and Characterization*, Cold Spring Harbor Lab. Press, Cold Spring Harbor, New York
30. Crabb, J. W., ed. (1995) *Techniques in Protein Chemistry*, Vol. VI, Academic Press, San Diego, California
31. Walker, J. M., ed. (1994) *Basic Protein and Peptide Protocols*, Humana Press, Totowa, New Jersey
32. Bollag, D. M., and Edelstein, S. J. (1990) *Protein Methods*, Wiley-Liss, New York
33. Creighton, T. E., ed. (1989) *Protein Structure and Protein Function: A Practical Approach*, IRL Press, Oxford
34. Copeland, R. A. (1994) *Methods for Protein Analysis. A Practical Guide to Laboratory Protocols*, Chapman and Hall, New York
35. Matsuidaira, P. T., ed. (1995) *A Practical Guide to Protein and Peptide Purification for Microsequencing*, Academic Press, San Diego, CA
36. Hugli, T. E., ed. (1989) *Techniques in Protein Chemistry*, Academic Press, New York
37. Means, G. E., and Feeney, R. E. (1973) *Chemical Modification of Proteins*, Holden-Day, San Francisco
38. Keleti, G., and Lederer, W. H. (1974) *Handbook of Micromethods for the Biological Sciences*, Van Nostrand-Reinhold, New York
39. Switzer, R. L., and Garrity, L. F. (1999) *Experimental Biochemistry:Theory and Exercises in Fundamental Methods*, 3rd ed., Freeman, New York
40. Schulz, G. E., and Schirmer, R. H. (1979) *Principles of Protein Structure*, Springer-Verlag, New York

41. Scopes, R. K. (1994) *Protein Purification. Principles of Protein Structure*, Springer-Verlag, New York
42. Villafranca, J., ed. (1990) *Current Research in Protein Chemistry: Techniques, Structure, and Function*, Academic Press, New York
43. Coligan, J., Dunn, B., Ploegh, H., Speicher, D., and Wingfield, P., eds. (1995) *Current Protocols in Protein Science*, Wiley, New York
44. Matsudaira, P. T., ed. (1989) *A Practical Guide to Protein and Peptide Purification for Microsequencing*, Academic Press, San Diego, California
45. Doonan, S., ed. (1996) *Protein Purification Protocols*, Humana Press, Totowa, New Jersey
46. Little, P., and Hughes, S., eds. (1992) *Technique*, Vol. 4, Academic Press, London
47. *Methods: A Companion to Methods in Enzymology*, 1990–present, Academic Press, San Diego, California
48. Wankat, P. C., Van Oss, C. J., and Henry, J. D., Jr., eds. (1992) *Separation and Purification Methods*, Dekker, New York
49. *protein expression and purification*, (1990 – present), Academic Press, San Diego, California
50. Coligan, J., Dunn, B., Ploegh, H., Speicher, D., and Wingfield, P., eds. (1996–present) *Current Protocols in Protein Science*, Wiley, New York
51. Hirs, C. H. W., and Timasheff, S. N., eds. (1983) *Methods in Enzymology, Enzyme Structure*, Vol. 91, Academic Press, New York
52. Doolittle, R. F., ed. (1990) *Methods in Enzymology, Molecular Evolution: Computer Analysis of Protein and Nucleic Acid*, Vol. 183, Academic Press, New York
53. Bollag, D. E., Rozycki, M. D., and Edelstein, S. J., eds. (1996) *Protein Methods*, 2nd ed., Wiley, New York
54. Birnie, G. D., and Fox, S. M., eds. (1969) *Subcellular Components*, Butterworth, London
55. Estabrook, R. W., and Pullman, M. E., eds. (1967) *Methods in Enzymology*, Vol. 10, Academic Press, New York
56. Kaback, H. R. (1971) in *Methods in Enzymology*, Vol. 22 (Jakoby, W. B., ed), pp. 99–120, Academic Press, New York
57. Dingle, J. T. (1972) *Lysosomes. A Laboratory Handbook*, North-Holland Publ., Amsterdam
58. Rasmussen, N. (1996) *Trends Biochem. Sci.* **21**, 319–321
58a. Jenkins, W. T. (1998) *Protein Sci.* **7**, 376–382
59. McPhie, P. (1971) in *Methods in Enzymology*, Vol. 22 (Jakoby, W. B., ed), pp. 23–32, Academic Press, New York
60. Blatt, W. F. (1971) in *Methods in Enzymology*, Vol. 22 (Jakoby, W. B., ed), pp. 39–49, Academic Press, New York
61. Porath, J., and Flodin, P. (1959) *Nature (London)* **183**, 1657–1659

References

62. Porath, J. (1978) *Trends Biochem. Sci.* **3**, N100
63. Lathe, G. H. (1978) *Trends Biochem. Sci.* **3**, N99–N100
64. Flodin, P., and Aspberg, K. (1961) *Biol. Struct. Funct. Proc. IUB/IUBS Int. Sym*, 1st ed., Vol. 1, pp. 345-349 (p. 346)
65. Weir, M., and Sparks, J. (1987) *Biochem. J.* **245**, 85
66. Regnier, F. E. (1991) *Nature (London)* **350**, 634–635
67. Afeyan, N. B., Gordon, N. F., and Mazsaroff, I. (1990) *J. Chromatogr.* **519**, 1–29
67a. Wu, C.-s. (1999) *Column Handbook for Size Exclusion Chromatography*, Academic Press, San Diego, California
68. Liapis, A. I., and McCoy, M. A. (1992) *J. Chromatogr.* **599**, 87–104
69. Cantor, C. R., and Schimmel, P. R. (1980) *Biophysical Chemistry, Part II*, Freeman, San Francisco, California (pp. 549–685)
70. Kyte, J. (1995) *Structure in Protein Chemistry*, Garland Publ., New York
71. Brakke, M. K. (1953) *Arch. Biochem. Biophys.* **45**, 275–290
72. Britten, R. J., and Roberts, R. B. (1960) *Science* **131**, 32–33
73. Martin, R. G., and Ames, B. N. (1961) *J. Biol. Chem.* **236**, 1372–1379
74. Meselson, M., Stahl, F. W., and Vinograd, J. (1957) *Proc. Natl. Acad. Sci. U.S.A.* **43**, 581–588
75. da Silva, J. R. R. F. (1978) in *New Trends in Bio-organic Chemistry* (Williams, R. J. P., and da Silva, J. R. R. F., eds), Academic Press, London
76. Arakawa, T., and Timasheff, S. N. (1982) *Biochemistry* **21**, 6545–6552
77. Rao, J. K. M., and Argos, P. (1981) *Biochemistry* **20**, 6536–6543
78. Green, A. A., and Hughes, W. L. (1955) in *Methods in Enzymology*, Vol. 1 (Colowick, S. P., and Kaplan, N. O., eds), pp. 67–90, Academic Press, New York
79. Dawson, R. M. C., Elliott, D. C., Elliott, W. H., and Jones, K. M. (1969) *Data for Biochemical Research*, 2nd ed., Clarendon Press, Oxford
80. Ito, Y. (1982) *Trends Biochem. Sci.* **7**, 47–50
81. Ito, Y. (1987) *Nature (London)* **326**, 419–420
82. Mandava, N. B., and Ito, Y., eds. (1988) *Countercurrent Chromatography*, Dekker, New York
82a. Menet, J.-M., and Thiébaut, D., eds. (1999) *Countercurrent Chromatography*, Dekker, New York
83. Er-el, Z., Zaidenzaig, Y., and Shaltiel, S. (1972) *Biochem. Biophys. Res. Commun.* **49**, 383–390
84. Arakawa, T., and Timasheff, S. N. (1982) *Biochemistry* **21**, 6536–6544
85. Atha, D. H., and Ingham, K. C. (1981) *J. Biol. Chem.* **256**, 12108–12117
86. Kenney, A., and Fowell, S., eds. (1992) *Practical Protein Chromatography*, Humana Press, Totowa, New Jersey
87. Tiselius, A., Hjerten, S., and Levin, O. (1956) *Arch. Biochem. Biophys.* **65**, 132
88. Wolf, W. J., and Sly, D. A. (1964) *J. Chromatogr.* **15**, 247–250
89. Regnier, F. E. (1987) *Science* **238**, 319–323
90. Hutchens, T. W., and Porath, J. (1987) *Biochemistry* **26**, 7199–7204
91. Molnár, I., Horváth, C., and Jatlow, P. (1978) *Chromatographia* **11**, 260–265
92. Hamilton, R. J., and Sewell, P. A. (1982) *Introduction to High Performance Liquid Chromatography*, 2nd ed., Chapman and Hall, London
93. Berridge, J. C. (1985) *Techniques for the Automated Optimization of HPLC Separations*, Wiley, New York
94. Simpson, C. F. (1982) *Techniques in Liquid Chromatography*, Wiley, New York
95. Mant, C. T., and Hodges, R. S. (1991) *High-Performance Liquid Chromatography of Peptides and Proteins: Separation, Analysis, and Conformation*, CRC Press, Boca Raton, Florida
96. Gooding, K. M., and Regnier, F. E., eds. (1990) *HPLC of Biological Macromolecules*, Dekker, New York
96a. Beesley, T. E., Buglio, B., and Scott, R. P. W. (2000) *Quantitative Chromatographic Analysis*, Dekker, New York
97. Oliver, R. W. A., ed. (1989) *HPLC of Macromolecules: A Practical Approach*, IRL Press, Oxford
98. Fried, B., and Sherma, J., (1999) *Thin-Layer Chromatography*, 4th ed., Dekker, New York
99. Zlatkis, A., and Kaiser, R. E. (1977) *High Performance Thin-Layer Chromatography*, Elsevier Scientific-North Holland, Amsterdam
100. Sherma, J., and Fried, B., eds. (1996) *Practical Thin-Layer Chromatography: A Multidisciplinary Approach*, CRC Press, Boca Raton, Florida
101. Liu, M.-K., Li, P., and Giddings, J. C. (1993) *Protein Sci.* **2**, 1520–1531
102. Treadwell, G. E., Cairns, W. L., and Metzler, D. E. (1968) *J. Chromatogr.* **35**, 376–388
103. Novotny, M. V. (1989) *Science* **264**, 51–57
104. Schomvurg, G. (1990) *Gas Chromatography: A Practical Course*, VCH Publ., Cambridge, UK
105. Lee, M. L., and Markides, K. E. (1987) *Science* **235**, 1342–1347
105a. Muraviev, D., Gorshkov, V., and Warshawsky, A. (1999) *Ion Exchange*, Dekker, New York
106. Moore, S., and Stein, W. H. (1949) *Cold Spring Harbor Symp. Quant. Biol.* **14**, 179–190
107. Moore, S., and Stein, W. H. (1949) *J. Biol. Chem.* **178**, 53–77
108. Manning, J. M. (1993) *Protein Sci.* **2**, 1188–1191
109. Svasti, J. (1980) *Trends Biochem. Sci.* **5**, 8–9
110. Moore, S., and Stein, W. H. (1951) *J. Biol. Chem.* **192**, 663–681
111. Moore, S., and Stein, W. H. (1954) *J. Biol. Chem.* **211**, 893–906
112. Moore, S., and Stein, W. H. (1963) in *Methods in Enzymology*, Vol. 6 (Colowick, S. P., and Kaplan, N. O., eds), pp. 819–831, Academic Press, New York
113. Yamamoto, S., Nakanishi, K., and Matsuno, R. (1988) *Ion Exchange Chromatography of Proteins*, Dekker, New York
114. Peterson, E. A. (1970) *Cellulosic Ion Exchangers*, Elsevier-North Holland, Amsterdam
115. Nölting, B., Golbik, R., Neira, J. L., Soler-Gonzalez, A. S., Schreiber, G., and Fersht, A. R. (1997) *Proc. Natl. Acad. Sci. U.S.A.* **94**, 826–830
116. Lundell, N., and Markides, K. (1992) *Chromatographia* **34**, 369–375
117. Dean, P. D. G., Johnson, W. S., and Middle, F. A., eds. (1985) *Affinity Chromatography, A Practical Approach*, IRL Press, Oxford
118. Parikh, I., and Cuatrecasas, P. (1985) *Chem. Eng. News* **63** (Aug 26), 17–32
119. Porath, J. (1968) *Nature (London)* **218**, 834–838
120. Inman, J. K. (1974) in *Methods in Enzymology*, Vol. 34 (Jakoby, W. B., and Wilchek, M., eds), pp. 30–58, Academic Press, New York
121. Bethell, G. S., Ayers, J. S., Hearn, M. T. W., and Hancock, W. S. (1981) *J. Chromatogr.* **219**, 361–372
122. Wilchek, M., and Miron, T. (1987) *Biochemistry* **26**, 2155–2161
123. Ashley, G. W., and Bartlett, P. A. (1984) *J. Biol. Chem.* **259**, 13615–13620
124. Hengen, P. N. (1995) *Trends Biochem. Sci.* **20**, 285–286
125. Loo, T. W., and Clarke, D. M. (1995) *J. Biol. Chem.* **270**, 21449–21452
126. Janknecht, R., de Martynoff, G., Lou, J., Hipskind, R. A., Nordheim, A., and Stunnenberg, H. G. (1991) *Proc. Natl. Acad. Sci. U.S.A.* **88**, 8972–8976
127. Hochuli, E., Dobeli, H., and Schacher, A. (1987) *J. Chromatogr.* **411**, 177–184
128. Turner, A. J. (1981) *Trends Biochem. Sci.* **6**, 171–173
129. Vesterberg, O. (1971) in *Methods in Enzymology*, Vol. 22 (Jakoby, W. B., ed), pp. 389–412, Academic Press, New York
130. Bauer, J. (1994) *Cell Electrophoresis*, CRC Press, Boca Raton, Florida
131. Allison, S. A., and Tran, V. T. (1995) *Biophys. J.* **68**, 2261–2270
132. Shuster, L. (1971) in *Methods in Enzymology*, Vol. 22 (Jakoby, W. B., ed), pp. 412–433, Academic Press, New York
133. Chrambach, A., and Rodbard, D. (1971) *Science* **172**, 440–451
134. Andrews, A. T. (1986) *Electrophoresis: Theory, Techniques and Biochemical and Clinical Applications*, 2nd ed., Clarendon-Oxford Univ. Press, Oxford
135. Gordon, A. H. (1980) *Electrophoresis of Proteins in Polyacrylamide and Starch Gels*, Revised ed., Elsevier-North Holland, Amsterdam
136. Chrambach, A., Dunn, M. J., and Radola, B. J., eds. (1987) *Advances in Electrophoresis*, Vol. I, VCH Publ., Cambridge, UK
137. Serwer, P., Easom, R. A., Hayes, S. J., and Olson, M. S. (1989) *Trends Biochem. Sci.* **14**, 4–7
138. *electrophoresis*, (1995–present), VCH Publ., New York
139. *advances in electrophoresis*, (1987–present), VCH Publ., Cambridge
140. Hawcroft, D. M. (1997) *Electrophoresis: The Basics*, IRL Press, Oxford
141. Hengen, P. N. (1995) *Trends Biochem. Sci.* **20**, 202–203
142. Garfin, D. E. (1990) *Methods Enzymol.* **182**, 425–441
143. Hames, B. D., and Rickwood, D., eds. (1990) *Gel Electrophoresis of Proteins: A Practical Approach*, 2nd ed., IRL Press, Oxford
144. Righetti, P. G., ed. (1996) *Capillary Electrophoresis in Analytical Biotechnology*, CRC Press, Boca Raton, Florida
145. Camilleri, P., ed. (1993) *Capillary Electrophoresis: Theory and Practice*, CRC Press, Inc., Boca Raton
146. Altria, K. D., ed. (1996) *Capillary Electrophoresis Guidebook*, Humana Press, Totowa, New Jersey
147. Landers, J. P., ed. (1994) *Handbook of Capillary Electrophoresis*, CRC Press, Boca Raton, Florida
148. Landers, J. P. (1993) *Trends Biochem. Sci.* **18**, 409–414
149. Wehr, T., Rodríguez-Díaz, R., and Zhu, M. (1998) *Capillary Electrophoresis of Proteins*, Dekker, New York
149a. Yao, S., Anex, D. S., Caldwell, W. B., Arnold, D. W., Smith, K. B., and Schultz, P. G. (1999) *Proc. Natl. Acad. Sci. U.S.A.* **96**, 5372–5377
150. Valaskovic, G. A., Kelleher, N. L., and McLafferty, F. W. (1996) *Science* **273**, 1199–1202
151. Gao, J., Gomez, F. A., Härter, R., and Whitesides, G. M. (1994) *Proc. Natl. Acad. Sci. U.S.A.* **91**, 12027–12030
152. Harrison, D. J., Fluri, K., Seiler, K., Fan, Z., Effenhauser, C. S., and Manz, A. (1993) *Science* **261**, 895–897
153. Radola, B. J., and Graesslin, D., eds. (1977) *Electrofocusing and Isotachophoresis*, de Gruyter, Berlin
154. Bocek, P., Dem, M., Gebauer, P., and Dolnik, V. (1987) *Analytical Isotachophoresis*, VCH Publ., Cambridge, UK

References

155. Righetti, P. G. (1983) *Isoelectric Focusing: Theory, Methodology and Applications*, Elsevier - North Holland Biomedical Press, Amsterdam

156. Catsimpoolas, N., and Drysdale, J., eds. (1977) *Biological and Biomedical Applications of Isoelectric Focusing*, Plenum, New York

157. Egen, N. B., Bliss, M., Mayersohn, M., Owens, S. M., Arnold, L., and Bier, M. (1988) *Anal. Biochem.* **172**, 488–494

158. Bjellqvist, B., Ek, K., Righetti, P. G., Gianazza, E., Görg, A., Westermeier, R., and Postel, W. (1982) *J. Biochem. Biophys. Methods* **6**, 317–339

159. Righetti, P. G., Gianazza, E., and Gelfi, C. (1988) *Trends Biochem. Sci.* **13**, 335–338

160. Righetti, P. G. (1990) *Immobilized pH Gradients: Theory and Methodology*, Elsevier, Amsterdam

161. Whitney, J. B., III, Cobb, R. R., Popp, R. A., and O'Rourke, T. W. (1985) *Proc. Natl. Acad. Sci. U.S.A.* **82**, 7646–7650

162. Madden, M. S. (1995) *Anal. Biochem.* **229**, 203–206

163. Anderson, N. (1979) *Nature (London)* **278**, 122–123

164. Wade, N. (1981) *Science* **211**, 33–35

165. Henslee, J. G., and Srere, P. A. (1979) *J. Biol. Chem.* **254**, 5488–5491

166. Hanash, S. M., and Strahler, J. R. (1989) *Nature (London)* **337**, 485–486

167. Neukirchen, R. O., Schlosshauer, B., Baars, S., Jackle, H., and Schwarz, U. (1982) *J. Biol. Chem.* **257**, 15229–15234

168. Racine, R. R., and Langley, C. H. (1980) *Nature (London)* **283**, 855–857

169. Celis, J. E., and Bravo, R., eds. (1984) *Two-Dimensional Gel Electrophoresis of Proteins: Methods and Applications*, Academic Press, New York

170. Celis, J. E., and Bravo, R. (1981) *Trends Biochem. Sci.* **6**, 197–201

171. Garrels, J. I. (1989) *J. Biol. Chem.* **264**, 5269–5282

172. Krauss, M. R., Collins, P. J., and Blose, S. H. (1989) *Nature (London)* **337**, 669–670

173. Wood, T. D., Chen, L. H., Kelleher, N. L., Little, D. P., Kenyon, G. L., and McLafferty, F. W. (1995) *Biochemistry* **34**, 16251–16254

173a. McLafferty, F. W., Fridriksson, E. K., Horn, D. M., Lewis, M. A., and Zubarev, R. A. (1999) *Science* **284**, 1289–1290

174. Shevchenko, A., Jensen, O. N., Podtelejnikov, A. V., Sagliocco, F., Wilm, M., Vorm, O., Mortensen, P., Shevchenko.A, Boucherie, H., and Mann, M. (1996) *Proc. Natl. Acad. Sci. U.S.A.* **93**, 14440–14445

175. Wilm, M., Shevchenko, A., Houthaeve, T., Breit, S., Schweigerer, L., Fotsis, T., and Mann, M. (1996) *Nature (London)* **379**, 466–469

176. Hess, D., Covey, T. C., Winz, R., Brownsey, R. W., and Aebersold, R. (1993) *Protein Sci.* **2**, 1342–1351

177. Van Holde, K. E. (1985) *Physical Biochemistry*, 2nd ed., Prentice-Hall, Englewood Cliffs, New Jersey

178. Cantor, C. R., and Schimmel, P. R. (1980) *Biophysical Chemistry*, Freeman, San Francisco, California

179. Svedburg, T., and Pederson, K. O. (1940) *The Ultracentrifuge*, Oxford Univ. Press, London

180. Schachman, H. K. (1989) *Nature (London)* **341**, 259–260

181. Schachman, H. K. (1992) in *Analytical Ultracentrifugation in Biochemistry and Polymer Science* (Harding, S. E., Rowe, A. J., and Horton, J. C., eds), pp. 3–15, Royal Society for Chemistry, Cambridge, UK

182. Schachman, H. K. (1959) *Ultracentrifugation in Biochemistry*, Academic Press, New York

183. Rivas, G., and Minton, A. P. (1993) *Trends Biochem. Sci.* **18**, 284–287

184. Harding, S. E., Rowe, A. J., and Horton, J. C., eds. (1992) *Analytical Ultracentrifugation in Biochemistry and Polymer Science*, CRC Press, Boca Raton, Florida

185. Hansen, J. C., Lebowitz, J., and Demeler, B. (1994) *Biochemistry* **33**, 13155–13163

186. Ralston, G. (1993) *Introduction to Analytical Ultracentrifugation*, Beckman Instruments, Fullerton, California

187. Stafford, W. F., and Schuster, T. M. (1995) in *Introduction to Biophysical Methods for Protein and Nucleic Acid Research* (Glasel, J. A., and Deutscher, M. P., eds), pp. 111–145, Academic Press, San Diego

188. Bowen, T. J. (1970) *An Introduction to Ultracentrifugation*, Wiley (Interscience), New York

189. Bothwell, M. A., Howlett, G. J., and Schachman, H. K. (1978) *J. Biol. Chem.* **253**, 2073–2076

190. Pollet, R. J., Haase, B. A., and Standaert, M. L. (1979) *J. Biol. Chem.* **254**, 30–33

190a. *CRC Handbook of Biochemistry*, (1968) Chem. Rubber Publ. Co., Cleveland, Ohio

191. Jacobsen, M. P., Wills, P. R., and Winzor, D. J. (1996) *Biochemistry* **35**, 13173–13179

191a. Colowick, S. P., and Kaplan, N. O., eds., (1968) *Methods Enzymol.* **12B**, 388–389

192. Squire, P. G., and Himmel, M. E. (1979) *Arch. Biochem. Biophys.* **196**, 165–177

193. Andrews, P. (1965) *Biochem. J.* **96**, 595–606

194. Cantor, C. R., and Schimmel, P. R. (1980) *Biophysical Chemistry*, Freeman, San Fransico, California (pp. 670–684)

195. Nakos, G., and Mortenson, L. (1971) *Biochemistry* **10**, 455–458

196. Mann, K. G., and Fish, W. W. (1972) *Methods Enzymol.* **26C**, 28–42

197. Igou, D. K., Lo, J. T., Clark, D. S., Mattice, W. L., and Younathan, E. S. (1974) *Biochem. Biophys. Res. Commun.* **60**, 140–145

198. Weber, K., Pringle, J. R., and Osborn, M. (1972) *Methods Enzymol.* **26C**, 3–29

199. Wood, H. G., Werkman, C. H., Hemingway, A., and Nier, A. O. (1941) *J. Biol. Chem.* **139**, 377–381

200. Biemann, K. (1995) *Protein Sci.* **4**, 1920–1927

200a. Niessen, W. M. A. (1998) *Liquid Chromatography - Mass Spectrometry*, 2nd ed., Dekker, New York

200b. Chapman, J. R., ed. (2000) *Mass Spectrometry of Proteins and Peptides*, Hamana Press, Totowa, New Jersey

200c. Burlingame, A. L., Carr, S. A., and Baldwin, M. A., eds. (2000) *Mass Spectrometry in Biology and Medicine*, Hamana Press, Totowa, New Jersey

201. Senko, M. W., and McLafferty, F. W. (1994) *Annu. Rev. Biophys. Biomol. Struct.* **23**, 763–785

202. Li, Y., Hunter, R. L., and McIver, R. T., Jr. (1994) *Nature (London)* **370**, 393–395

203. Siuzdak, G. (1994) *Proc. Natl. Acad. Sci. U.S.A.* **91**, 11290–11297

204. Caprioli, R. M., and Suter, M. J.-F. (1995) in *Introduction to Biophysical Methods for Protein and Nucleic Acid Research* (Glasel, J. A., and Deutscher, M. P., eds), pp. 147-204, Academic Press, San Diego

205. Burlingame, A. L., and Carr, S. A., eds. (1996) *Mass Spectrometry in the Biological Sciences*, Humana Press, Totowa, New Jersey

206. Karas, M., and Hillenkamp, F. (1988) *Anal. Chem.* **60**, 2299–2303

207. Tanaka, K., Waki, H., Ido, Y., Akita, S., Yoshida, Y., and Yoshida, T. (1988) *Rapid Commun. Mass Spectrom.* **2**, 151–156

208. Beavis, R. C., and Chait, B. T. (1989) *Rapid Commun. Mass Spectrom.* **3**, 233–237

209. Hillenkamp, F., Karas, M., Beavis, R. C., and Chait, B. T. (1991) *Anal. Chem.* **63**, 1193A–1198A

210. Smith, R. D., Loo, J. A., Edmonds, C. G., Barinaga, C. J., and Udseth, H. R. (1990) *Anal. Chem.* **62**, 882–899

211. Fenn, J. B., Mann, M., Meng, C. K., Wong, S. F., and Whitehouse, C. M. (1990) *Mass Spectrom. Rev.* **9**, 37

212. Dole, M., Mack, L. L., Hines, R. L., Mobley, R. C., Ferguson, L. D., and Alice, M. B. (1968) *J. Chem. Phys.* **49**, 2240

213. Mann, M., and Wilm, M. (1995) *Trends Biochem. Sci.* **20**, 219–224

214. Mann, M., Meng, C. K., and Fenn, J. B. (1989) *Anal. Chem.* **61**, 1702–1708

215. Covey, T. R., Bonner, R. F., Shushan, B. I., and Henion, J. (1988) *Rapid Commun. Mass Spectrom.* **2**, 249–256

216. Miranker, A., Kruppa, G. H., Robinson, C. V., Aplin, R. T., and Dobson, C. M. (1996) *J. Am. Chem. Soc.* **118**, 7402–7403

217. Marshall, A. G., Senko, M. W., Li, W., Li, M., Dillon, S., Guan, S., and Logan, T. M. (1997) *J. Am. Chem. Soc.* **119**, 433–434

218. Torchinsky, Y. M. (1981) *Sulfur in Proteins*, Pergamon, Oxford

219. Friedman, M. (1973) *The Chemistry and Biochemistry of the Sulfhydryl Group in Amino Acids, Peptides and Proteins*, Pergamon, Oxford

220. Jocelyn, P. C. (1972) *Biochemistry of the SH Group*, Academic Press, New York

221. Simpson, R. J., Neuberger, M. R., and Liu, T.-Y. (1976) *J. Biol. Chem.* **251**, 1936–1940

222. Hugli, T. E., and Moore, S. (1972) *J. Biol. Chem.* **247**, 2828–2834

223. Chin, C. C. Q., and Wold, F. (1974) *Anal. Biochem.* **61**, 379–391

224. Royer, G. P., Schwartz, W. E., and Liberatore, F. A. (1977) *Methods Enzymol.* **47E**, 40–45

225. Glazer, A. N. (1976) in *The Proteins*, 3rd ed., Vol. II (Neurath, H., and Hill, R. L., eds), pp. 1–103, Academic Press, New York

226. Plapp, B. V., Raftery, M. A., and Cole, R. D. (1967) *J. Biol. Chem.* **242**, 265–270

227. Doonan, S., and Fahmy, H. M. A. (1975) *Eur. J. Biochem.* **56**, 421–426

228. Mitchell, W. M. (1977) *Methods Enzymol.* **47E**, 165–169

229. Gadasi, H., Maruta, H., Collins, J. H., and Korn, E. D. (1979) *J. Biol. Chem.* **254**, 3631–3636

230. Drapeau, G. R. (1977) *Methods Enzymol.* **47E**, 189–191

231. Koida, M., and Walter, R. (1976) *J. Biol. Chem.* **251**, 7593–7599

232. Hayashi, R. (1977) *Methods Enzymol.* **47E**, 84–93

233. Krutzsch, H. C., and Pisano, J. J. (1977) *Methods Enzymol.* **47E**, 391–404

234. Mahoney, W. C., and Hermodson, M. A. (1979) *Biochemistry* **18**, 3810–3814

235. Hancock, W. S., ed. (1996) *New Methods in Peptide Mapping for the Characterization of Proteins*, CRC Press, Boca Raton, Florida

236. Lundblad, R. L. (1995) *Techniques in Protein Modification*, CRC Press, Boca Raton, Florida

237. Sanger, F. (1949) *Biochem. J.* **45**, 563–574

238. Sanger, F. (1988) *Ann. Rev. Biochem.* **57**, 1–28

239. Ruthen, R. (1993) *Sci. Am.* **269**(Oct), 30–31

240. Aebersold, R. H., Leavitt, J., Saavedra, R. A., Hood, L. E., and Kent, S. B. H. (1987) *Proc. Natl. Acad. Sci. U.S.A.* **84**, 6970–6974

241. Edman, P. (1956) *Acta Chem. Scand.* **10**, 761–768

242. Edman, P., and Begg, C. (1967) *Eur. J. Biochem.* **1**, 80–91

243. Niall, H. D. (1977) *Nature (London)* **268**, 279–280

244. Schlesinger, D. H., ed. (1988) *Macromolecular Sequencing and Synthesis*, Liss, New York

245. Shively, J. E., ed. (1986) *Methods of Protein Microcharacterization*, Humana Press, Clifton, New Jersey

References

246. Bhown, A. S. (1988) *Protein/Peptide Sequence Analysis: Current Methodoligies*, CRC Press, Boca Raton, Florida

247. Hunkapiller, M. W., and Hood, L. E. (1983) *Science* **219**, 650–659

248. Totty, N. F., Waterfield, M. D., and Hsuan, J. J. (1992) *Protein Sci.* **1**, 1215–1224

249. Gershoni, J. M. (1985) *Trends Biochem. Sci.* **10**, 103–106

250. Moos, M., Jr., Nguyen, N. Y., and Liu, T.-Y. (1988) *J. Biol. Chem.* **263**, 6005–6008

251. Baldo, B. A., and Tovey, E. R., eds. (1989) *Protein Blotting*, S. Karger, Farmington, CT

252. Wellner, D., Panneerselvam, C., and Horecker, B. L. (1990) *Proc. Natl. Acad. Sci. U.S.A.* **87**, 1947–1949

253. Bailey, J. M., Nikfarjam, F., Shenoy, N. R., and Shively, J. E. (1992) *Protein Sci.* **1**, 1622–1633

254. Bailey, J. M., Shenoy, N. R., Ronk, M., and Shively, J. E. (1992) *Protein Sci.* **1**, 68–80

255. Johnson, R. S., and Walsh, K. A. (1992) *Protein Sci.* **1**, 1083–1091

255a. Keough, T., Youngquist, R. S., and Lacey, M. P. (1999) *Proc. Natl. Acad. Sci. U.S.A.* **96**, 7131–7136

256. Hunt, D. F., Shabanowitz, J., Yates, J. R., III, Zhu, N.-Z., Russell, D. H., and Castro, M. E. (1987) *Proc. Natl. Acad. Sci. U.S.A.* **84**, 620–623

257. Takao, T., Hitouji, T., Shimonishi, Y., Tanabe, T., Inouye, S., and Inouye, M. (1984) *J. Biol. Chem.* **259**, 6105–6109

257a. Jaffe, H., Veeranna, and Pant, H. C. (1998) *Biochemistry* **37**, 16211–16224

257b. Whitelegge, J. P., Gundersen, C. B., and Faull, K. F. (1998) *Protein Sci.* **7**, 1423–1430

258. Yazdanparast, R., Andrews, P. C., Smith, D. L., and Dixon, J. E. (1987) *J. Biol. Chem.* **262**, 2507–2513

259. Brown, J. R., and Hartley, B. S. (1966) *Biochem. J.* **101**, 214–228

260. Svobodova, S., Hais, I. M., and Kostir, J. V. (1953) *Chem. Listy* **47**, 205

261. Chen, R. F., Smith, P. D., and Maly, M. (1978) *Arch. Biochem. Biophys.* **189**, 241–250

262. Lee, K. S., and Drescher, D. G. (1979) *J. Biol. Chem.* **254**, 6248–6251

263. Chen, R. F., Scott, C., and Trepman, E. (1979) *Biochim. Biophys. Acta.* **576**, 440–455

264. Lunte, S. M. et al. (1989) *Anal. Biochem.* **178**, 202–207

265. Sharma, Y., Rao, C. M., Rao, S. C., Krishna, A. G., Somasundaram, T., and Balasubramanian, D. (1989) *J. Biol. Chem.* **264**, 20923–20927

266. Wilson, C. M. (1983) *Methods Enzymol.* **91**, 236–247

267. Merril, C. R. (1990) *Methods Enzymol.* **182**, 477–488

268. Finn, F. M., and Hofmann, K. (1976) in *The Proteins*, 3rd ed., Vol. II (Neurath, H., and Hill, R. L., eds), pp. 105–253, Academic Press, New York

269. Burnette, W. N. (1981) *Anal. Biochem.* **112**, 195–203

270. Karey, K. P., and Sirbasku, D. A. (1989) *Anal. Biochem.* **178**, 255–259

271. Strickland, E. H., Billups, C., and Kay, E. (1972) *Biochemistry* **11**, 3657–3662

272. Metzler, D. E., Harris, C., Yang, I.-Y., Siano, D., and Thomson, J. A. (1972) *Biochem. Biophys. Res. Commun.* **46**, 1588–1597

273. Morton, R. A. (1975) *Biochemical Spectroscopy*, Wiley, New York

274. Pace, C. N., Vajdos, F., Fee, L., Grimsley, G., and Gray, T. (1995) *Protein Sci.* **4**, 2411–2423

275. Glazer, A. N., Delange, R. J., and Sigman, D. S. (1976) *Chemical Modification of Proteins: Selected Methods and Analytical Procedures*, Elsevier, New York

276. Feeney, R. E., Blankenhorn, G., and Dixon, H. B. F. (1975) *Advances in Protein Chemistry*, Vol. 29, Academic Press, New York (pp. 135–203)

277. Acharya, A. S., and Manjula, B. N. (1987) *Biochemistry* **26**, 3524–3530

278. Rypniewski, W. R., Holden, H. M., and Rayment, I. (1993) *Biochemistry* **32**, 9851–9858

279. Jentoft, N., and Dearborn, D. G. (1979) *J. Biol. Chem.* **254**, 4359–4365

280. Hennecke, M., and Plapp, B. N. (1984) *Anal. Biochem.* **136**, 110–118

281. Brocklehurst, K. (1979) *Int. J. Biochem* **10**, 259–274

282. Degani, Y., and Patchornik, A. (1974) *Biochemistry* **13**, 1–11

283. Riddles, P. W., Blakeley, R. L., and Zerner, B. (1983) *Methods Enzymol.* **91**, 49–60

284. Pande, C. S., Pelzig, M., and Glass, J. D. (1980) *Proc. Natl. Acad. Sci. U.S.A.* **77**, 895–899

285. Fraker, P. J., and Speck, J. C., Jr. (1978) *Biochem. Biophys. Res. Commun.* **80**, 849–857

286. Tolan, D. R., Lambert, J. M., Boileau, G., Fanning, T. G., Kenny, J. W., Vassos, A., and Traut, R. R. (1980) *Anal. Biochem.* **103**, 101–109

287. Miles, E. W. (1977) *Methods Enzymol.* **47**, 431–442

288. Kawata, Y., Sakiyama, F., Hayashi, F., and Kyogoku, Y. (1990) *Eur. J. Biochem.* **187**, 255–262

289. Secundo, F., Carrea, G., D'Arrigo, P., and Servi, S. (1996) *Biochemistry* **35**, 9631–9636

290. Miura, S., Tomita, S., and Ichikawa, Y. (1991) *J. Biol. Chem.* **266**, 19212–19216

291. Colman, R. F. (1983) *Ann. Rev. Biochem.* **52**, 67–91

292. Chowdhry, V., and Westheimer, F. H. (1979) *Ann. Rev. Biochem.* **48**, 293–325

293. Jelenc, P. C., Cantor, C. R., and Simon, S. R. (1978) *Proc. Natl. Acad. Sci. U.S.A.* **75**, 3564–3568

294. Simmonds, R. J. (1992) *Chemistry of Biomolecules*, CRC Press, Boca Raton, Florida

295. Bodanszky, M., Bodanszky, A., and Trost, B. M., eds. (1994) *The Practice of Peptide Synthesis*, 2nd ed., Springer-Verlag, New York

296. Lloyd-Williams, P., Albericio, F., and Giralt, E. (1997) *Chemical Approaches to the Synthesis of Peptides and Proteins*, CRC Press, Boca Raton, Florida

297. Kent, S. B. H. (1988) *Ann. Rev. Biochem.* **57**, 957–989

298. Kaiser, E. T., Mihara, H., Laforet, G. A., Kelly, J. W., Walters, L., Findeis, M. A., and Sasaki, T. (1989) *Science* **243**, 187–192

299. Merrifield, B. (1986) *Science* **232**, 341–347

300. Atherton, E., Logan, C. J., and Sheppard, R. C. (1981) *J. Chem. Soc., Perkin, I.*, 538–546

301. Atherton, E., and Sheppard, R. C. (1985) *J. Chem. Soc. Chem. Commun.*, 165–166

301a. Miller, S. C., and Scanlan, T. S. (1998) *J. Am. Chem. Soc.* **120**, 2690–2691

302. You-shang, Z. (1983) *Trends Biochem. Sci.* **8**, 16–17

303. Kullman, W. (1987) *Enzymatic Peptide Synthesis*, CRC Press, Boca Raon, Florida

304. Liu, C.-F., and Tam, J. P. (1994) *J. Am. Chem. Soc.* **116**, 4149–4153

305. Tam, J. P., Lu, Y.-A., Liu, C.-F., and Shao, J. (1995) *Proc. Natl. Acad. Sci. U.S.A.* **92**, 12485–12489

306. Noren, C. J., Anthony-Cahill, S. J., Griffith, M. C., and Schultz, P. G. (1989) *Science* **244**, 182–188

307. Borman, S. (1996) *Chem. Eng. News* **74** (Feb 12), 29–35

308. Youngquist, R. S., Fuentes, G. R., Lacey, M. P., and Keough, T. (1995) *J. Am. Chem. Soc.* **117**, 3900–3906

309. Cortese, R., ed. (1996) *Combinatorial Libraries: Synthesis, Screening and Application Potential*, Walter de Gruyter & Co., New York

310. Ellman, J., Stoddard, B., and Wells, J. (1997) *Proc. Natl. Acad. Sci. U.S.A.* **94**, 2779–2782

311. Kishchenko, G., Batliwala, H., and Makowski, L. (1994) *J. Mol. Biol.* **241**, 208–213

312. Scott, J. K. (1992) *Trends Biochem. Sci.* **17**, 241–245

312a. Barbas, C. F., Burton, D. B., Scott, J. K., and Silverman, G. J. (2000) *Phage Display: A Laboratory Manual* Cold Spring Harbor Press, Cold Spring Harbor, New York

313. Marvin, D. A., (1998) *Current Opinions in Structural Biology* **8**, 150–158

313a. Barbas, C. F., Burton, D. B., Scott, J. K., and Silverman, G. J. (2000) *Phage Display: A Laboratory Manual* Cold Spring Harbor Lab. Press, Cold Spring Harbor, New York

314. McConnell, S. J., and Hoess, R. H. (1995) *J. Mol. Biol.* **250**, 460–470

315. Katz, B. A. (1995) *Biochemistry* **34**, 15421–15429

316. Allen, J. B., Walberg, M. W., Edwards, M. C., and Elledge, S. J. (1995) *Trends Biochem. Sci.* **20**, 511–516

317. Gates, C. M., Stemmer, W. P. C., Kaptein, R., and Schatz, P. J. (1996) *J. Mol. Biol.* **255**, 373–386

318. Bastos, M., Maeji, N. J., and Abeles, R. H. (1995) *Proc. Natl. Acad. Sci. U.S.A.* **92**, 6738–6742

319. Han, H., Wolfe, M. M., Brenner, S., and Janda, K. D. (1995) *Proc. Natl. Acad. Sci. U.S.A.* **92**, 6419–6423

320. Chu, Y.-H., Dunayevskiy, Y. M., Kirby, D. P., Vouros, P., and Karger, B. L. (1996) *J. Am. Chem. Soc.* **118**, 7827–7835

321. Déprez, B., Williard, X., Bourel, L., Coste, H., Hyafil, F., and Tartar, A. (1995) *J. Am. Chem. Soc.* **117**, 5405–5406

322. Kim, R. M., Manna, M., Hutchins, S. M., Griffin, P. R., Yates, N. A., Bernick, A. M., and Chapman, K. T. (1996) *Proc. Natl. Acad. Sci. U.S.A.* **93**, 10012–10017

323. Sundberg, S. A., Barrett, R. W., Pirrung, M., Lu, A. L., Kiangsoontra, B., and Holmes, C. P. (1995) *J. Am. Chem. Soc.* **117**, 12050–12057

324. Baum, R. M. (1991) *Chem. Eng. News* **69**, 21–22

325. Slayter, E. M., and Slayter, H. S. (1992) *Light and Electron Microscopy*, Cambridge Univ. Press, New York

326. Bracegirdle, B. (1989) *Trends Biochem. Sci.* **14**, 464–468

327. Matthews, C. K., and van Holde, K. E. (1996) *Biochemistry*, 2nd ed., Benjamin, Menlo Park, California

328. Lichtman, J. W. (1994) *Sci. Am.* **271**(Aug), 40–45

329. Shotton, D., and White, N. (1989) *Trends Biochem. Sci.* **14**, 435–439

330. Boyde, A. (1985) *Science* **230**, 1270–1276

331. Gura, T. (1997) *Science* **276**, 1988–1990

332. Bennett, B. D., Jetton, T. L., Ying, G., Magnuson, M. A., and Piston, D. W. (1996) *J. Biol. Chem.* **271**, 3647–3651

333. Pennisi, E. (1997) *Science* **275**, 480–481

334. Maiti, S., Shear, J. B., Williams, R. M., Zipfel, W. R., and Webb, W. W. (1997) *Science* **275**, 530–532

335. Pool, R. (1988) *Science* **241**, 25–26

336. Betzig, E., and Chichester, R. J. (1993) *Science* **262**, 1422–1425

337. Dykstra, M. J. (1992) *Biological Electron Microscopy*, Plenum, New York

338. Hand, A. R. (1995) in *Introduction to Biophysical Methods for Protein and Nucleic Acid Research* (Glasel, J. A., and Deutscher, M. P., eds), pp. 205-260, Academic Press, San Diego

339. Nellist, P. D., McCallum, B. C., and Rodenburg, J. M. (1995) *Nature (London)* **374**, 630–632

References

340. Schliwa, M. (1997) *Nature (London)* **387**, 764
341. Bubel, A., ed. (1989) *Microstructure and Function of Cells*, 1st ed., Ellis Harwood, New York
342. Penman, S. (1995) *Proc. Natl. Acad. Sci. U.S.A.* **92**, 5251–5257
343. Morel, G., ed. (1995) *Visualization of Nucleic Acids*, CRC Press, Boca Raton, Florida
344. Caspar, D. L. D., and DeRosier, D. J. (1982) *Science* **218**, 653–655
345. Frank, J., ed. (1992) *Electron Tomography: Three-Dimensional Imaging with the Transmission Electron Microscope*, Plenum, New York
346. Skoglund, U., and Daneholt, B. (1986) *Trends Biochem. Sci.* **11**, 499–503
347. Lambert, O., Boisset, N., Taveau, J.-C., and Lamy, J. N. (1994) *J. Mol. Biol.* **244**, 640–647
348. Taveau, J.-C., Boisset, N., Lamy, J., Lambert, O., and Lamy, J. N. (1997) *J. Mol. Biol.* **266**, 1002–1015
349. De Rosier, D. J. (1997) *Nature (London)* **386**, 26–27
350. Dorset, D. L. (1997) *Proc. Natl. Acad. Sci. U.S.A.* **94**, 1791–1794
351. Binnig, G., and Rohrer, H. (1987) *Rev. Mod. Phys.* **59**, 615–625
352. Binnig, G., and Rohrer, H. (1985) *Sci. Am.* **253**, 50–56
353. Wiesendanger, R. (1997) *Proc. Natl. Acad. Sci. U.S.A.* **94**, 12749–12750
354. Wickramasinghe, H. K. (1989) *Sci. Am.* **261**(Oct), 98–105
355. Hansma, P. K., Elings, V. B., Marti, O., and Bracker, C. E. (1988) *Science* **242**, 209–216
356. Pool, R. (1990) *Science* **247**, 634–636
357. Edstrom, R. D., Yang, X., Lee, G., and Evans, D. F. (1990) *FASEB J.* **4**, 3144–3151
358. Engel, A. (1991) *Ann. Rev. Biophys. Biophys. Chem.* **20**, 79–108
359. Amato, I. (1997) *Science* **276**, 1982–1985
360. Prater, C. B., Butt, H. J., and Hansma, P. K. (1990) *Nature (London)* **345**, 839–840
361. Hansma, H. G., and Hoh, J. H. (1994) *Annu. Rev. Biophys. Biomol. Struct.* **23**, 115–139
362. Radmacher, M., Tillmann, R. W., Fritz, M., and Gaub, H. E. (1992) *Science* **257**, 1900–1905
363. Karrasch, S., Hegerl, R., Hoh, J. H., Baumeister, W., and Engel, A. (1994) *Proc. Natl. Acad. Sci. U.S.A.* **91**, 836–838
364. Mou, J., Yang, J., and Shao, Z. (1995) *J. Mol. Biol.* **248**, 507–512
365. Yip, C. M., and Ward, M. D. (1996) *Biophys. J.* **71**, 1071–1078
366. Mou, J., Sheng, S., Ho, R., and Shao, Z. (1996) *Biophys. J.* **71**, 2213–2221
367. Kuznetsov, YuG., Malkin, A. J., Land, T. A., DeYoreo, J. J., Barba, A. P., Konnert, J., and McPherson, A. (1997) *Biophys. J.* **72**, 2357–2364
367a. Czajkowsky, D. M., and Shao, Z. (1998) *FEBS Letters* **430**, 51–54
368. Noy, A., Frisbie, C. D., Rozsnyai, L. F., Wrighton, M. S., and Lieber, C. M. (1995) *J. Am. Chem. Soc.* **117**, 7943–7951
369. Boland, T., and Ratner, B. D. (1995) *Proc. Natl. Acad. Sci. U.S.A.* **92**, 5297–5301
370. Miles, M. (1997) *Science* **277**, 1845–1847
371. Hansma, H. G., Laney, D. E., Bezanilla, M., Sinsheimer, R. L., and Hansma, P. K. (1995) *Biophys. J.* **68**, 1672–1677
371a. Fisher, T. E., Oberhauser, A. F., Carrion-Vazquez, M., Marszalek, P. E., and Fernandez, J. M. (1999) *Trends Biochem. Sci.* **24**, 379–384
372. Blundell, T. L., and Johnson, L. N. (1976) *Protein Crystallography*, Academic Press, New York
373. Cantor, C. R., and Schimmel, P. R. (1980) *Biophysical Chemistry*, Freeman, San Francisco, California (pp. 687–811)

374. McPherson, A. (1982) *Preparation and Analysis of Protein Crystals*, Wiley, New York
375. Wyckoff, H. W., Hirs, C. H. W., and Timasheff, S. N., eds. (1985) *Methods in Enzymology*, Vol. 114,115, Academic Press, Orlando, Florida
376. Allewell, N. M., and Trikha, J. (1995) in *Introduction to Biophysical Methods for Protein and Nucleic Acid Research* (Glasel, J. A., and Deutscher, M. P., eds), pp. 381–431, Academic Press, San Diego
377. Verlinde, C. L. M. J., Merritt, E. A., Van Den Akker, F., Kim, H., Feil, I., Delboni, L. F., Mande, S. C., Sarfaty, S., Petra, P. H., and Hol, W. G. J. (1994) *Protein Sci.* **3**, 1670–1686
378. Woolfson, N. M. (1997) *An Introduction to X-ray Crystallography*, 2nd ed., Cambridge Univ. Press, New York
379. Thomas, J. M. (1993) *Nature (London)* **364**, 478–482
380. Dounce, A. L., and Allen, P. Z. (1988) *Trends Biochem. Sci.* **13**, 317–320
381. Bernal, J. D., and Crowfoot, D. (1934) *Nature (London)* **133**, 794–795
382. Glusker, J. P. (1994) *Protein Sci.* **3**, 2465-2469
383. Perutz, M. F., Muirhead, H., Cox, J. M., Goaman, L. G. G., Mathews, F. S., McGandy, E. L., and Webb, L. E. (1968) *Nature (London)* **219**, 29–32
384. Perutz, M. (1992) *Protein Structure: New Approaches to Disease and Therapy*, W.H. Freeman, San Fransico and New York
385. Eisenberg, D. (1994) *Protein Sci.* **3**, 1625–1628
386. Perutz, M. F. (1978) *Sci. Am.* **239**(Dec), 92–125
387. Tegze, M., and Faigel, G. (1996) *Nature (London)* **380**, 49–51
388. Fadley, C. S., and Len, P. M. (1996) *Nature (London)* **380**, 27–28
389. Rhodes, G. (1993) *Crystallography Made Crystal Clear*, Academic Press, New York
389a. Glusker, J. P., and Trueblood, K. N. (1985) *Crystal Structure Analysis A Primer*, 2nd ed., Oxford Univ. Press, Oxford
389b. Drenth, J. (1994) *Principles of Protein X-ray Crystallography*, Springer-Verlag, New York
389c. McRee, D. E. (1999) *Practical Protein Crystallography*, Academic Press, San Diego, California
390. Sweet, R. M. (1985) *Methods Enzymol.* **114**, 19–46
391. Kendrew, J. C. (1963) *Science* **139**, 1259–1266
392. Kendrew, J. C., Dickerson, R. E., Strandberg, B. E., Hart, R. G., Davies, D. R., Phillips, D. C., and Shore, V. C. (1960) *Nature (London)* **185**, 422–427
393. Rossman, M. G. (1994) *Protein Sci.* **3**, 1731–1733
394. Blake, C. C. F., Koenig, D. F., Mair, G. A., North, A. C. T., Phillips, D. C., and Sarma, V. R. (1965) *Nature (London)* **206**, 757–761
395. Laskowski, R. A., Hutchinson, E. G., Michie, A. D., Wallace, A. C., Jones, M. L., and Thornton, J. M. (1997) *Trends Biochem. Sci.* **22**, 488–490
396. Gerstein, M., and Levitt, M. (1997) *Proc. Natl. Acad. Sci. U.S.A.* **94**, 11911–11916
397. Marvin, D. A., and Nave, C. (1982) in *Structural Molecular Biology* (Davies, D. B., Saenger, W., and Danyluk, S. S., eds), pp. 3–44, Plenum, New York
398. Ducruix, A., and Giegé, R., eds. (1999) *Crystallization of Nucleic Acids and Proteins*, 2nd ed., IRL Press, Oxford
399. McPherson, A., Koszelak, S., Axelrod, H., Day, J., Williams, R., Robinson, L., McGrath, M., and Cascio, D. (1986) *J. Biol. Chem.* **261**, 1969–1975
400. McPherson, A., Jr. (1976) *J. Biol. Chem.* **251**, 6300–6303
401. Ray, W. J., Jr., and Puvathingal, J. M. (1986) *J. Biol. Chem.* **261**, 11544–11549

402. Day, J., and McPherson, A. (1992) *Protein Sci.* **1**, 1254–1268
403. Koszelak, S., Day, J., Leja, C., Cudney, R., and McPherson, A. (1995) *Biophys. J.* **69**, 13–19
404. Service, R. F. (1999) *Science* **285**, 1342–1346
405. Knäblein, J., Neuefeind, T., Schneider, F., Bergner, A., Messerschmidt, A., Löwe, J., Steipe, B., and Huber, R. (1997) *J. Mol. Biol.* **270**, 1–7
406. Glusker, J. P. (1984) *Trends Biochem. Sci.* **9**, 328–330
407. Guss, J. M., Merritt, E. A., Phizackerley, R. P., Hedman, B., Murata, M., Hodgson, K. O., and Freeman, H. C. (1988) *Science* **241**, 806–811
408. Walter, R. L., Ealick, S. E., Friedman, A. M., Blake, R. C., II, Proctor, P., and Shoham, M. (1996) *J. Mol. Biol.* **263**, 730–751
409. Budisa, N., Karnbrock, W., Steinbacher, S., Humm, A., Prade, L., Neuefeind, T., Moroder, L., and Huber, R. (1997) *J. Mol. Biol.* **270**, 616–623
410. Longhi, S., Czjzek, M., Lamzin, V., Nicolas, A., and Cambillau, C. (1997) *J. Mol. Biol.* **268**, 779–799
410a. Yamano, A., Heo, N.-H., and Teeter, M. M. (1997) *J. Biol. Chem.* **272**, 9597–9600
410b. Lamzin, V. S., Morris, R. J., Dauter, Z., Wilson, K. S., and Teeter, M. M. (1999) *J. Biol. Chem.* **274**, 20753–20755
411. Rhee, S., Silva, M. M., Hyde, C. C., Rogers, P. H., Metzler, C. M., Metzler, D. E., and Arnone, A. (1997) *J. Biol. Chem.* **272**, 17293–17302
412. Frauenfelder, H., Petsko, G. A., and Tsernoglou, D. (1979) *Nature (London)* **280**, 558–563
413. Arnone, A., Rogers, P. H., Hyde, C. C., Briley, P. D., Metzler, C. M., and Metzler, D. E. (1985) in *Transaminases* (Christen, P., and Metzler, D. E., eds), pp. 138–155, Wiley, New York
414. Wlodawer, A. (1982) *Prog. Biophys. and Mol. Biol.* **40**, 115–159
415. Kossiakoff, A. A. (1985) *Ann. Rev. Biochem.* **54**, 1195–1227
416. Butler, D. (1994) *Nature (London)* **371**, 469
417. Hellemans, A. (1997) *Science* **277**, 1214–1215
418. Glanz, J. (1995) *Science* **267**, 1904–1906
419. Pool, R. (1988) *Science* **241**, 295
420. Eisenberger, P., and Suckewer, S. (1996) *Science* **274**, 201–202
421. Neutze, R., and Hajdu, J. (1997) *Proc. Natl. Acad. Sci. U.S.A.* **94**, 5651–5655
422. Hajdu, J., and Johnson, L. N. (1990) *Biochemistry* **29**, 1669–1678
423. Duke, E. M. H., Wakatsuki, S., Hadfield, A., and Johnson, L. N. (1994) *Protein Sci.* **3**, 1178–1196
424. Jacoby, M. (1997) *Chem. Eng. News* **Dec 8**, 5
425. Rischel, C., Rousse, A., Uschmann, I., Albouy, P.-A., Geindre, J.-P., Audebert, P., Gauthier, J.-C., Förster, E., Martin, J.-L., and Antonetti, A. (1997) *Nature (London)* **390**, 490–492
426. Rattle, H. (1995) *An NMR Primer for Life Scientists*, Partnership Press, Tuckerton, NJ
427. Roberts, G. C. K., ed. (1993) *NMR of Macromolecules. A Practical Approach*, IRL Press, Oxford
428. Evans, J. N. S. (1995) *Biomolecular NMR Spectroscopy*, Oxford University Press, New York
429. Olson, D. L., Peck, T. L., Webb, A. G., Magin, R. L., and Sweedler, J. V. (1995) *Science* **270**, 1967–1969
430. Wuthrich, K. (1986) *NMR of Proteins and Nucleic Acids*, Wiley, New York
431. Nageswara Rao, B. D., and Kemple, M. D., eds. (1996) *NMR as a structural tool for macromolecules: Current status and future directions*, Plenum, New York
432. Bovey, F. A. (1969) *Nuclear Magnetic Resonance Spectroscopy*, Academic Press, New York

References

433. Sanders, J. K. M., and Hunter, B. K. (1993) *Modern NMR Spectroscopy*, 2nd ed., Oxford Univ. Press, London

434. Dybowski, C., and Lichter, R. L., eds. (1987) *NMR Spectroscopy Techniques*, Vol. 5, Dekker, New York

435. Bangerter, B. W. (1995) in *Introduction to Biophysical Methods for Protein and Nucleic Acid Research* (Glasel, J. A., and Deutscher, M. P., eds), pp. 317-379, Academic Press, San Diego

436. Freeman, R. (1997) *A Handbook of Nuclear Magnetic Resonance*, 2nd ed., Addison Wesley Longman Limited, Essex, England

437. Emsley, J. W., ed. (1996) *Encyclopedia of NMR*, John Wiley & Sons, Ltd., Sussex, England

438. Markley, J. L. (1975) *Acc. Chem. Res.* **8**, 70–80

439. Perkins, S. J. (1982) in *Biological Magnetic Resonance*, Vol. 4 (Berliner, L. J., and Reuben, J., eds), pp. 193-334, Plenum Press, New York and London

440. Osapay, K., and Case, D. A. (1991) *J. Am. Chem. Soc.* **113**, 9436–9444

441. Szilágyi, L. (1995) *Prog. Nucl. Magn. Reson. Spectrosc.* **27**, Part 4, 326–443

442. Williamson, M. P., Kikuchi, J., and Asakura, T. (1995) *J. Mol. Biol.* **247**, 541–546

443. Sitkoff, D., and Case, D. A. (1997) *J. Am. Chem. Soc.* **119**, 12262–12273

444. Gu, Z., Zambrano, R., and McDermott, A. (1994) *J. Am. Chem. Soc.* **116**, 6368–6372

445. Gu, Z., Ridenour, C. F., Bronnimann, C. E., Iwashita, T., and McDermott, A. (1996) *J. Am. Chem. Soc.* **118**, 822–829

446. de Dios, A. C., Pearson, J. G., and Oldfield, E. (1993) *Science* **260**, 1491–1496

447. Sulzbach, H. M., Schleyer, P. V. R., and Schaefer, H. F., III (1994) *J. Am. Chem. Soc.* **116**, 3967–3972

448. Wishart, D. S., Sykes, B. D., and Richards, F. M. (1992) *Biochemistry* **31**, 1647–1651

449. Linderstrøm-Lang, K. (1955) *Chemical Society Special Publication*, No. 2, 1–20

450. Barfield, M., and Karplus, M. (1969) *J. Am. Chem. Soc.* **91**, 1–16

451. Karplus, M. (1963) *J. Am. Chem. Soc.* **85**, 2870–2871

452. Bax, A. (1989) *Ann. Rev. Biochem.* **58**, 223–256

453. Wang, A. C., and Bax, A. (1996) *J. Am. Chem. Soc.* **118**, 2483–2494

454. Stonehouse, J., Adell, P., Keeler, J., and Shaka, A. J. (1994) *J. Am. Chem. Soc.* **116**, 6037–6038

455. Aguayo, J. B., Gamcsik, M. P., and Dick, J. D. (1988) *J. Biol. Chem.* **263**, 19552–19557

456. Ackerman, J. J. H., Ewy, C. S., Kim, S.-G., and Shalwitz, R. A. (1987) *Ann. N.Y. Acad. Sci.* **508**, 89–98

457. Breitmaier, E., and Voelter, W. (1987) *Carbon-13 NMR Spectroscopy*, 3rd ed., VCH Publishers, New York

458. Beckmann, N. (1995) *Carbon-13 NMR Spectroscopy of Biological Systems*, Academic Press, New York

459. Gronenborn, A. M., Clore, G. M., Schmeissner, U., and Wingfield, P. T. (1986) *Eur. J. Biochem.* **161**, 37–43

460. Clore, G. M., Bax, A., Driscoll, P. C., Wingfield, P. T., and Gronenborn, A. M. (1990) *Biochemistry* **29**, 8172–8184

461. Jue, T., Rothman, D. L., Shulman, G. I., Tavitian, B. A., DeFronzo, R. A., and Shulman, R. G. (1989) *Proc. Natl. Acad. Sci. U.S.A.* **86**, 4489–4491

462. Markley, J. L. (1989) *Methods Enzymol.* **176**, 12–64

463. Pelton, J. G., Torchia, D. A., Meadow, N. D., and Roseman, S. (1993) *Protein Sci.* **2**, 543–558

464. Rajagopal, P., Waygood, E. B., and Klevit, R. E. (1994) *Biochemistry* **33**, 15271–15282

465. Solum, M. S., Altmann, K. L., Strohmeier, M., Berges, D. A., Zhang, Y., Facelli, J. C., Pugmire, R. J., and Grant, D. M. (1997) *J. Am. Chem. Soc.* **119**, 9804–9809

466. Verkade, J. G., and Quin, L. D., eds. (1987) *Phosphorus-31 NMR Spectroscopy in Stereochemical Analysis*, VCH Publishers, Deerfield Beach, Florida

467. Ugurbil, K., Kingsley-Hickman, P. B., Sako, E. Y., Zimmer, S., Mohanakrishnan, P., Robitaille, P. M. L., Thoma, W. J., Johnson, A., Foker, J. E., From, A. H. L. (1987) *Ann. N.Y. Acad. Sci.* **508**, 265–286

468. Feeney, J., McCormick, J. E., Bauer, C. J., Birdsall, B., Moody, C. M., Starkmann, B. A., Young, D. W., Francis, P., Havlin, R. H., Arnold, W. D., and Oldfield, E. (1996) *J. Am. Chem. Soc.* **118**, 8700–8706

469. Lau, E. Y., and Gerig, J. T. (1997) *Biophys. J.* **73**, 1579–1592

470. Jarema, M. A. C., Lu, P., and Miller, J. H. (1981) *Proc. Natl. Acad. Sci. U.S.A.* **78**, 2707–2711

471. Slebe, J. C., and Martinez-Carrion, M. (1990) *J. Biol. Chem.* **253**, 2093–2097

472. Briley, P. A., Eisenthal, R., Harrison, R., and Smith, G. D. (1990) *Biochem. J.* **163**, 325–331

473. Roberts, J. K. M., Wade-Jardetzky, N., and Jardetzky, O. (1981) *Biochemistry* **20**, 5389–5394

474. Deutsch, C. J., and Taylor, J. S. (1987) *Ann. N.Y. Acad. Sci.* **508**, 33

475. Gutowsky, H. S., McCall, D. W., McGarvey, B. R., and Meyer, L. H. (1952) *J. Am. Chem. Soc.* **74**, 4809–4817

476. Chang, Y. C., and Graves, D. J. (1985) *J. Biol. Chem.* **260**, 2709–2714

477. Scott, R. D., Chang, Y.-C., Graves, D. J., and Metzler, D. E. (1985) *Biochemistry* **24**, 7668–7681

478. Saunders, M., Jiménez-Vázquez, H. A., Cross, R. J., Mroczkowski, S., Freedberg, D. I., and Anet, F. A. L. (1994) *Nature (London)* **367**, 256–257

479. Adebodun, F., and Jordan, F. (1988) *J. Am. Chem. Soc.* **110**, 309–310

480. Springer, C. S., Jr. (1987) *Ann. Rev. Biophys. Biophys. Chem.* **16**, 375–399

481. Wittenberg, B. A., and Gupta, R. K. (1985) *J. Biol. Chem.* **260**, 2031–2034

482. Jelicks, L. A., and Gupta, R. K. (1989) *J. Biol. Chem.* **264**, 15230–15235

483. Price, W. S., Ge, N.-H., Hong, L.-Z., and Hwang, L.-P. (1993) *J. Am. Chem. Soc.* **115**, 1095–1105

484. Wang, S. M., and Gilpin, R. K. (1983) *Anal. Chem.* **55**, 493–497

485. Markham, G. D. (1986) *J. Biol. Chem.* **261**, 1507–1509

486. Denisov, V. P., and Halle, B. (1995) *J. Mol. Biol.* **245**, 682–687

487. Mullen, G. P., Dunlap, R. B., and Odom, J. D. (1986) *Biochemistry* **25**, 5625–5632

488. Aldhous, P. (1991) *Nature (London)* **353**, 689

489. Becker, E. D., and Farrar, T. C. (1972) *Science* **178**, 361–368

490. Bax, A., and Lerner, L. (1986) *Science* **232**, 960–967

491. Yamazaki, T., Muhandiram, R., and Kay, L. E. (1994) *J. Am. Chem. Soc.* **116**, 8266–8278

492. Farrar, T. C., and Becker, E. D. (1971) *Pulse and Fourier Transform NMR*, Academic Press, New York

493. Wüthrich, K. (1989) *Science* **243**, 45–50

494. van de Ven, F. J. M., and Hilbers, C. W. (1986) *J. Mol. Biol.* **192**, p. 419-441

495. Baniak, E. L., Rivier, J. E., Struthers, R. S., Hagler, A. T., and Gierasch, L. M. (1987) *Biochemistry* **26**, 2642–2656

496. Wüthrich, K. (1990) *J. Biol. Chem.* **265**, 22059–22062

497. Bax, A., and Grzesiek, S. (1993) in *NMR of Proteins* (Clore, G. M., and Gronenborn, A. M., eds), pp. 33–52, CRC Press, Boca Raton, Florida

498. Clore, G. M., and Gronenborn, A. M. (1994) *Protein Sci.* **3**, 372–390

499. Hoitnik, C. W. G., Driscoll, P. C., Hill, H. A. O., and Canters, G. W. (1994) *Biochemistry* **33**, 3560–3571

500. Yamazaki, T., Tochio, H., Furui, J., Aimoto, S., and Kyogoku, Y. (1997) *J. Am. Chem. Soc.* **119**, 872–880

501. Yamazaki, T., Lee, W., Arrowsmith, C. H., Muhandiram, D. R., and Kay, L. E. (1994) *J. Am. Chem. Soc.* **116**, 11655–11666

502. Hu, J.-S., and Bax, A. (1997) *J. Am. Chem. Soc.* **119**, 6360–6368

503. Song, S., Velde, D. V., Gunn, C. W., and Himes, R. H. (1994) *Biochemistry* **33**, 693–698

504. Barsukov, I. L., Lian, L.-Y., Ellis, J., Sze, K.-H., Shaw, W. V., and Roberts, G. C. K. (1996) *J. Mol. Biol.* **262**, 543–558

505. Wu, W.-J., Anderson, V. E., Raleigh, D. P., and Tonge, P. J. (1997) *Biochemistry* **36**, 2211–2220

506. Metzler, W. J., Hare, D. R., and Pardi, A. (1989) *Biochemistry* **28**, 7045–7052

507. McDowell, L. M., Lee, M., McKay, R. A., Anderson, K. S., and Schaefer, J. (1996) *Biochemistry* **35**, 3328–3334

508. Pflugrath, J. W., Wiegand, G., Huber, R., and Vértesy, L. (1986) *J. Mol. Biol.* **189**, 383–386

509. Kline, A. D., Braun, W., and Wüthrich, K. (1986) *J. Mol. Biol.* **189**, 377–382

510. McDonald, C. C., and Phillips, W. D. (1969) *J. Am. Chem. Soc.* **91**, 1513–1521

511. Nietlispach, D., Clowes, R. T., Broadhurst, R. W., Ito, Y., Keeler, J., Kelly, M., Ashurst, J., Oschkinat, H., Domaille, P. J., and Laue, E. D. (1996) *J. Am. Chem. Soc.* **118**, 407–415

512. Rosen, M. K., Gardner, K. H., Willis, R. C., Parris, W. E., Pawson, T., and Kay, L. E. (1996) *J. Mol. Biol.* **263**, 627–636

513. Venters, R. A., Farmer, B. T., II, Fierke, C. A., and Spicer, L. D. (1996) *J. Mol. Biol.* **264**, 1101–1116

514. LeMaster, D. M., and Richards, F. M. (1988) *Biochemistry* **27**, 142–150

515. Bax, A., Kay, L. E., Sparks, S. W., and Torchia, D. A. (1989) *J. Am. Chem. Soc.* **111**, 408–409

516. Torchia, D. A., Sparks, S. W., and Bax, A. (1989) *Biochemistry* **28**, 5509–5524

517. Lowry, D. F., Redfield, A. G., McIntosh, L. P., and Dahlquist, F. W. (1988) *J. Am. Chem. Soc.* **110**, 6885–6886

518. Stockman, B. J., Reily, M. D., Westler, W. M., Ulrich, E. L., and Markley, J. L. (1989) *Biochemistry* **28**, 230–236

519. Qian, H., Mayo, S. L., and Morton, A. (1994) *Biochemistry* **33**, 8167–8171

519a. Riek, R., Pervushin, K., and Wüthrich, K. (2000) *Trends Biochem. Sci.* **25**, 462–468

520. Günter, P., Braun, W., Billeter, M., and Wüthrich, K. (1989) *J. Am. Chem. Soc.* **111**, 3997–4004

521. Oh, B.-H., Westler, W. M., and Markley, J. L. (1989) *J. Am. Chem. Soc.* **111**, 3083–3085

522. Cross, T. A., and Opella, S. J. (1983) *J. Am. Chem. Soc.* **105**, 306–308

523. Rothgeb, T. M., and Oldfield, E. (1981) *J. Biol. Chem.* **256**, 1432–1446

524. Buck, M., Schwalbe, H., and Dobson, C. M. (1995) *Biochemistry* **34**, 13219–13232

525. Nicholson, L. K., Kay, L. E., Baldisseri, D. M., Arango, J., Young, P. E., Bax, A., and Torchia, D. A. (1992) *Biochemistry* **31**, 5253–5263

526. Gardner, K. H., and Kay, L. E. (1997) *J. Am. Chem. Soc.* **119**, 7599–7600

527. Lenkinski, R. E., Dallas, J. L., and Glickson, J. D. (1979) *J. Am. Chem. Soc.* **101**, 3071–3077

References

528. Carver, J. A., and Bradbury, J. H. (1984) *Biochemistry* **23**, 4890–4905
529. Bothelho, L. H., and Gurd, F. R. N. (1978) *Biochemistry* **17**, 5188–5196
530. Bothelho, L. H., Friend, S. H., Matthew, J. B., Lehman, L. D., Hanania, G. I. H., and Gurd, F. R. N. (1978) *Biochemistry* **17**, 5197–5205
531. Morino, Y., Yamasaki, M., Tanase, S., Nagashima, F., Akasaka, K., Imoto, T., and Miyazawa, T. (1984) *J. Biol. Chem.* **259**, 3877–3882
532. Cocco, M. J., Kao, Y.-H., Phillips, A. T., and Lecomte, J. T. J. (1992) *Biochemistry* **31**, 6481–6491
533. Bachovin, W. W., Wong, W. Y. L., Farr-Jones, S., Shenvi, A. B., and Kettner, C. A. (1988) *Biochemistry* **27**, 7689–7697
534. Mollova, E. T., Metzler, D. E., Kintanar, A., Kagamiyama, H., Hayashi, H., Hirotsu, K., and Miyahara, I. (1997) *Biochemistry* **36**, 615–625
535. Guéron, M., Plateau, P., and Decorps, M. (1991) *Progr. in NMR Spectr.* **23**, 135–209
536. Plateau, P., and Guéron, M. (1982) *J. Am. Chem. Soc.* **104**, 7310–7311
537. Sklenar, V., and Bax, A. (1987) *J. of Magnetic Resonance* **74**, 469–479
538. Bullitt, E., and Makowski, L. (1995) *Nature (London)* **373**, 164–167
539. Griffin, J. H., Cohen, J. S., and Schechter, A. N. (1973) *Biochemistry* **12**, 2096–2099
540. Tong, H., and Davis, L. (1995) *Biochemistry* **34**, 3362–3367
541. Gunnarsson, G., Wennerström, H., Egan, W., and Forsén, S. (1976) *Chem. Phys. Lett.* **38**, 96–99
542. McDermott, A., and Ridenour, C. F. (1996) in *Encyclopedia of NMR* (Emsley, J. W., ed), pp. 3820–3824, J. Wiley & Sons, Ltd., Sussex, UK
543. Zhao, Q., Abeygunawardana, C., Gittis, A. G., and Mildvan, A. S. (1997) *Biochemistry* **36**, 14616–14626
544. Perrin, C. L., and Nielson, J. B. (1997) *J. Am. Chem. Soc.* **119**, 12734–12741
545. Tjandra, N., and Bax, A. (1997) *J. Am. Chem. Soc.* **119**, 8076–8082
546. Kintanar, A., Metzler, C. M., Metzler, D. E., and Scott, R. D. (1991) *J. Biol. Chem.* **266**, 17222–17229
547. Metzler, C. M., Metzler, D. E., Kintanar, A., Scott, R. D., and Marceau, M. (1991) *Biochem. Biophys. Res. Commun.* **178**, 385–392
548. Firsov, L. M., Neustroev, K. N., Aleshin, A. E., Metzler, C. M., Metzler, D. E., Scott, R. D., Stoffer, B., Christensen, T., and Svensson, B. (1994) *Eur. J. Biochem.* **223**, 293–302
549. Metzler, D. E., Metzler, C. M., Mollova, E. T., Scott, R. D., Tanase, S., Kogo, K., Higaki, T., and Morino, Y. (1994) *J. Biol. Chem.* **269**, 28017–28026
550. Metzler, D. E., Metzler, C. M., Scott, R. D., Mollova, E. T., Kagamiyama, H., Yano, T., Kuramitsu, S., Hayashi, H., Hirotsu, K., and Miyahara, I. (1994) *J. Biol. Chem.* **269**, 28027–28033
551. Metzler, D. E. (1997) *Methods Enzymol.* **280**, 30–40
552. Englander, S. W., Mayne, L., Bai, Y., and Sosnick, T. R. (1997) *Protein Sci.* **6**, 1101–1109
553. Kossiakoff, A. A. (1982) *Nature (London)* **296**, 713–721
554. Rohl, C. A., and Baldwin, R. L. (1994) *Biochemistry* **33**, 7760–7767
555. Gemmecker, G., Jahnke, W., and Kessler, H. (1993) *J. Am. Chem. Soc.* **115**, 11620–11621
556. Zhang, Y.-Z., Paterson, Y., and Roder, H. (1995) *Protein Sci.* **4**, 804–814
557. Zhang, Z., and Smith, D. L. (1993) *Protein Sci.* **2**, 522–531
558. Wagner, D. S., Melton, L. G., Yan, Y., Erickson, B. W., and Anderegg, R. J. (1994) *Protein Sci.* **3**, 1305–1314
559. Zhang, Z., Post, C. B., and Smith, D. L. (1996) *Biochemistry* **35**, 779–791
560. Zappacosta, F., Pessi, A., Bianchi, E., Venturini, S., Sollazzo, M., Tramontano, A., Marino, G., and Pucci, P. (1996) *Protein Sci.* **5**, 802–813
561. Kim, K.-S., Fuchs, J. A., and Woodward, C. K. (1993) *Biochemistry* **32**, 9600–9608
562. Eriksson, M. A. L., Härd, T., and Nilsson, L. (1995) *Biophys. J.* **69**, 329–339
563. Spyracopoulos, L., and O'Neil, J. D. J. (1994) *J. Am. Chem. Soc.* **116**, 1395–1402
564. Markley, J. L. (1975) *Biochemistry* **14**, 3546–3554
565. Schmidt-Rohr, K., and Spiess, H. W. (1994) *Multidimensional Solid-State NMR and Polymers*, Academic Press, San Diego, California
566. Schmidt-Rohr, K. (1996) *J. Am. Chem. Soc.* **118**, 7601–7603
567. Wu, C. H., Ramamoorthy, A., Gierasch, L. M., and Opella, S. J. (1995) *J. Am. Chem. Soc.* **117**, 6148–6149
568. Meyer, E. F. (1997) *Protein Sci.* **6**, 1591–1597
569. Stampf, D. R., Felder, C. E., and Sussman, J. L. (1995) *Nature (London)* **374**, 572–573
570. Holm, L., and Sander, C. (1995) *Trends Biochem. Sci.* **20**, 478–480
571. Murzin, A. G., Brenner, S. E., Hubbard, T., and Chothia, C. (1995) *J. Mol. Biol.* **247**, 536–540
572. Walsh, L., Newell, M., Ruczaj, H., Reid, T., and Desmond, P., eds. (1997) *Trends Guide to the Internet*, Elsevier, Cambridge, UK
573. Hogue, C. W. V., Ohkawa, H., and Bryant, S. H. (1996) *Trends Biochem. Sci.* **21**, 226–229
574. Wilkins, M. R., Hochstrasser, D. F., Sanchez, J.-C., Bairoch, A., and Appel, R. D. (1996) *Trends Biochem. Sci.* **21**, 496–497
575. Peitsch, M. C., Wells, T. N. C., Stampf, D. R., and Sussman, J. L. (1995) *Trends Biochem. Sci.* **20**, 82–84
576. Sayle, R. A., and Milner-White, E. J. (1995) *Trends Biochem. Sci.* **20**, 374–376
576a. Bernstein, H. J. (2000) *Trends Biochem. Sci.* **25**, 453–454
577. Sokolik, C. W. (1995) *Trends Biochem. Sci.* **20**, 122–124
578. Hunt, T. (1996) *Trends Biochem. Sci.* **21**, 74
579. Service, R. F. (2000) *Science* **287**, 1954–1956
580. Laskowski, R. A., Hutchinson E. G., Michie, A. D., Wallace, A. C., Jones, M. L., and Thornton, J. M. (1997) *Trends Biochem. Sci* **22**, 488–490
581. Cohn, E. J., and Edsall, J. T. (1943) *Proteins, Amino Acids and Peptides as Ions and Dipolar Ions*, Van Nostrand-Reinhold, Princeton, New Jersey (pp. 90–93)
582. Abbott, A. (2000) *Nature (London)* **408**, 130–132
583. Simmerling, C., Lee, M. R., Ortiz, A. R., Kolinski, A., Skolnick, J., and Kollman, P. A. (2000) *J. Am. Chem. Soc.* **122**, 8392–8402

Study Questions

Flask No.	Mol HCl	Mol NaOH	pH
1	0.010		1.71
2	0.009		1.85
3	0.006		2.25
4	0.002		2.94
5		0.002	9.00
6		0.004	9.37
7		0.005	9.60

1. a. From the expression for the dissociation constant of an acid, HA (Eq. 3-1), derive the logarithmic form $pH - pK_a = \log[A^-]/[HA] = \log_{10} \alpha/(1-\alpha)$ (Eq. 3-11), where α is the fraction of the acid in the ionized form.

   b. The apparent dissociation constant K_a for the $H_2PO_4^-$ ion at 25°C and 0.5 M total phosphate concentration is 1.380×10^{-7} M. What will be the pH of a solution 0.025 M in KH_2PO_4 and 0.25 M in Na_2HPO_4? This is a National Bureau of Standards buffer (see Bates[7]).

   c. Suppose that you wanted to prepare a buffer of pH = 7.00 at 25°C from anhydrous KH_2PO_4 ($M_r = 136.09$) and Na_2HPO_4 ($M_r = 141.98$). If you placed 3.40 g of KH_2PO_4 in a 1 liter volumetric flask, how much anhydrous Na_2HPO_4 would you have to weigh and add before making to volume to obtain the desired pH? If you wanted to have the correct pH to ± 0.01 unit, how accurately would you have to weigh your salts? NOTE: It is quicker to prepare a buffer of precise pH this way than it is to titrate a portion of buffer acid to the desired pH with sodium hydroxide.

2. The apparent pK_a for 0.1 M formic acid is 3.70 at 25°C.

   a. Concentrated HCl was added to a liter of 0.1 M sodium formate until a pH of 1.9 was attained. Calculate the concentration of formate ion and that of unionized formic acid.

   b. Calculate the hydrogen ion concentration.

   c. How many equivalents of HCl had to be added in part a to bring the pH to 1.9?

3. Exactly 0.01-mol portions of glycine were placed in several 100-ml volumetric flasks. The following exact amounts of HCl or NaOH were added to the flasks, the solutions were made to volume with water, mixed, and the pH measured. Calculate the pK_a values for the carboxyl and amino groups from the following, making as many independent calculations of each pKa as the data permit. At low pH values *you must correct for the free hydrogen ion concentration* (see question 2c).[581]

4. Using the pK_a values from problem 3, construct the theoretical titration curve showing the equivalents of H^+ or OH^- reacting with 1 mol of glycine as a function of pH. Note that the shape of this curve is independent of the pK_a. Sketch similar curves for glutamic acid (pK_a's equal 2.19, 4.25, and 9.67), histidine (pK_a's equal 1.82, 6.00, and 9.17) and lysine (pK_a's equal 2.18, 8.95, and 10.53).

   Compare your plot for glycine with a plot of 1 M acid or base added to 0.01 mol of glycine in 100 ml of water. You may also compare your curves with those for glycine published in other textbooks.

5. Make a table of characteristic pK_a values for acidic and basic groups in proteins. Which of these groups contribute most significantly to the titration curves of proteins?

6. If placed in water and adjusted to a pH of 7, will the following migrate toward the anode or the cathode if placed in an electrical field? (a) Aspartic acid, (b) alanine, (c) tyrosine, (d) lysine, (e) arginine, and (f) glutamine

7. The tripeptide L-Ala – L-His – L-Gln had the following pK_a values: 3.0 (α-COOH), 9.1 (α-NH_3^-), and 6.7 (imidazolium).

   a. What is the isoelectric pH (pI) of the peptide, i.e., the pH at which it will carry no net charge? Hint, the pI for amino acids is usually given approximately as the arithmetic mean of two pK_a values.[581]

   b. Draw the structures of the ionic forms of the peptide that occur at pH 5 and at pH 9. At each pH compute the fraction of the peptide in each ionic form.

Study Questions

8. a. Write the structure for glycyl-L-tryptophanyl-L-prolyl-L-seryl-L-lysine.

    b. What amino acids could be isolated from it following acid hydrolysis?

    c. Following alkaline hydrolysis?

    d. After nitrous acid treatment followed by acid hydrolysis

    e. In an electrolytic cell at pH 7.0 would the peptide migrate toward the cathode or toward the anode? What is the approximate isoelectric point of the peptide?

    f. If a solution of this peptide were adjusted to pH 7, and then titrated with sodium hydroxide in the presence of 10% formaldehyde, how many equivalents of base would be required per mole of peptide to raise the pH to 10?

9. A peptide was shown to contain only L-lysine and L-methionine. Titration of the peptide showed 3 free amino groups for each free carboxyl group present, and each amino group liberated 1 mole of N_2 when the peptide was treated with HNO_2 in the Van Slyke apparatus. When the deaminated peptide was hydrolyzed completely in acid and the hydrolyzate again treated with HNO_2, the same amount of N_2 is liberated as that derived from the intact peptide. A sample of the original peptide was treated with excess dinitrofluorobenzene to give a dinitrophenyl (DNP) peptide, which was shown spectrophotometrically to contain three DNP groups per free carboxyl group. When this DNP-peptide was completely hydrolyzed, the following products were found: a colorless compound containing S (A_1) and a yellow compound containing S (A_3). Partial hydrolysis of the DNP-peptide yields A_1, A_2, A_3, plus 4 additional yellow compounds, B_1, B_2, B_3, and B_4. On complete hydrolysis, B_1 yields A_1, A_2, and A_3; B_2 yields A_1 and A_2; B_3 yields A_1 and A_3; and B_4 yields A_3 only.

    What is the most probable structure of the original peptide?

10. (a) Explain two advantages of the isotope dilution method of analysis. (b) From the following data, calculate the amount of cyclic AMP (cAMP) present per ml of human gluteus maximus muscle cells. Cells were treated with ^{32}P-enriched cAMP, S.A. = 50 μCi / μmol for 0.2 h (all cAMP was taken up by cells), cells were homogenized, and the soluble cAMP was isolated and purified. The specific activity of the isolated cAMP = 10 μCi / μmol. The total amount of cAMP added was 1.0×10^{-7} mol per ml of cells.

11. The figure in Box 3-C shows a high-resolution separation of the soluble proteins of *E. coli*. The investigator labeled the proteins with ^{14}C-containing amino acids.

    a. How would you carry out the labeling experiment?

    b. What other isotope(s) could be used to label proteins? What chemical form(s) would you use? What limitations might there be?

    c. What soluble components of an *E. coli* cell sonicate might interfere with the two-dimensional separation, and how could they be removed?

    d. What technique(s) other than radioactive labeling could be used for locating proteins?

    e. If *all* the soluble proteins of *E. coli* were detected, about how many separate proteins would you expect to see?

    f. Indicate two or more properties of the resolution technique which are most significant in making it applicable to a system containing a very large number of proteins.

12. ^{35}S is a beta emitter, with no gamma or other type of radiation. It has the following properties: $t_{1/2} = 86.7$ days, $\varepsilon_{max} = 0.168$ MeV.

    a. Write the equation for the radiochemical decomposition of ^{35}S.

    b. Discuss the advantages and limitation of the use of ^{35}S as an isotopic tracer.

13. Given a quantity of agarose, cyanogen bromide, 6-aminohexanoic acid, and any other necessary reagents, show the chemical reactions for making an affinity column bearing the amino acid tryptophan bound through its amino group.

14. Imagine that you have isolated a 55-kDa protein and that you have a cDNA (Chapter 5) copied from its structural gene. The cDNA sequence indicates that the protein contains four cysteine residues. How would you determine the location of the two disulfide bridges present in the oxidized form of the protein? What uncertainties might you encounter?

15. From the m/z ratios shown in Fig. 3-12 calculate the molecular mass of the protein (apomyoglobin, which lacks the bound heme of myoglobin). First show that for two peaks differing by one unit in the charge z, the mass m_2 of the larger fragment is related to the molecular mass M of the uncharged protein, n_2 (numerically equal to z_2 in this case), the number of charged ions (in this case H^+) bound and to the mass X of the bound ion (in this case $X = 2$) by the equation:

$$m_2 = (M + n_2X) / n_2$$

Show also that the mass of the smaller fragment is:

$$m_1 = [M + (n_2 + 1) X] / (n_2 + 1)$$

From consecutive pairs of peaks compute values of n (Do these agree with those printed in Fig. 3-12?) and of M, which is numerically equal to M_r. Is your value of M_r *average* or *monoisotopic*? See Covey et al.[215]

16. The following values of M_r were measured for a protein:

In dilute buffer at pH 7: 300,000

In 6 M guanidium chloride: 100,000

In 6 M guanidinium chloride and 100 mM β-mercaptoethanol: 50,000

Treatment with dansyl chloride followed by hydrolysis yielded dansyl-Ala and dansyl-Ser. Describe the quaternary structure of the protein.

Each cotton fiber is a single cell seed hair, ~30 mm in length. The dry fiber is ~95% cellulose, which constitutes the secondary cell wall (See Fig. 20-4,D) and is also present in the primary cell wall. The fibers, which are ~30 nm in length, consist of many parallel chains (Fig. 4-5) ~5 μm in length, each containing ~10,000 glucose residues. Van der Waals forces, together with one hydrogen bond per glucose, contribute to the stability of the tightly packed fiber. Cotton is one of the major agricultural crops, ~87 million bales, each ~220 kg, being produced annually world-wide. Photo and information from A. D. French and M. A. Godshall, Southern Regional Research Center, USDA, New Orleans, LA.

Contents

Sugars, Polysaccharides, and Glycoproteins

4

We are all familiar with sugars, important components of our diet which are present in fruits, honey, table sugar, and syrups. Within our bodies the simple sugar D-glucose is an essential source of energy. It is present in blood at a relatively constant concentration of 5.5 mM and is carried to all tissues. A **polysaccharide** called **glycogen**, a polymeric form of glucose, provides a reserve of readily available energy within our cells. **Starches** and other polysaccharides store energy within plants. Polysaccharides also have major *structural* functions in nature. **Cellulose**, another polymer of glucose, forms the fibers of cotton, plant cell walls, and wood. Both the tough exoskeletons of anthropods and the cell walls of fungi depend for their strength on the nitrogen-containing polysaccharide **chitin**. Polysaccharides that carry many negative charges, such as **hyaluronan,** form a protective layer between animal cells, while **pectins** play a similar role in plants.

A third function of sugar residues (**glycosyl groups**) is in biological recognition and communication. The outer surfaces of cells are nearly covered by covalently attached **oligosaccharides**, small polymeric arrays of sugar rings. Some are attached to side chain groups of proteins and others to lipids to form **glycoproteins** and **glycolipids,** respectively. Because of the variety of different sugars, the various ways in which they can be linked, and their ability to form oriented hydrogen bonds, these oligosaccharides provide much of the chemical code for identifying cells. This coding enables cells to attach to each other in correct ways during development of a multicelled organism. It helps to activate our immune system to attack parasites and also helps bacteria to attack us! We are only beginning to understand the numerous critical functions of the glycosyl groups of cell surfaces.

A. Structures and Properties of Simple Sugars

The simple sugars, or **monosaccharides**, are polyhydroxyaldehydes (**aldoses**) or polyhydroxyketones (**ketoses**).[1-5] All have the composition $(CH_2O)_n$, hence the family name **carbohydrate**. A typical sugar, and the one with the widest distribution in nature, is glucose.

D-Mannose (Man) has the opposite configuration at C-2

D-Galactose (Gal) has the opposite configuration at C-4

D-Glucose (Glc) showing numbering of atoms

Sugars are named as D or L according to the configuration about this carbon atom, one removed from the terminal position

L-Glucose

The carbonyl group of this and other sugars is highly reactive and a characteristic reaction is **addition** of electron-rich groups such as –OH. If a sugar chain is long enough (4–6 carbon atoms) one of the hydroxyl groups of the same molecule can add to the carbonyl group to form a cyclic **hemiacetal** or ring form, which reaches an equilibrium with the free aldehyde or ketone form (Eq. 4-1). The six-membered rings formed in this way (**pyranose** rings) are especially stable, but five-membered **furanose** rings also exist in many carbohydrates.

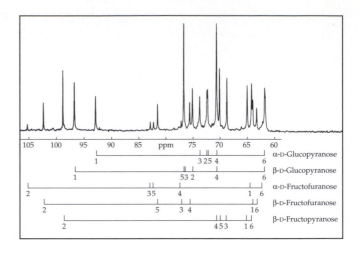

(4-1)

Figure 4-1 Natural abundance ^{13}C-NMR spectrum of honey showing content of α and β pyranose ring forms of glucose and both pyranose and furanose ring forms of fructose. Open chain fructose, with a peak at 214 ppm, was present in a trace amount. From Prince *et al.*[6]

The ring forms of the sugars are the monomers used by cells to form polysaccharides. Indeed, it is the natural tendency of 5- and 6-carbon sugars (**pentoses** and **hexoses**) to cyclize that permits formations of stable sugar polymers from the reactive and unstable monomers. When a sugar cyclizes a new chiral center is formed at the **anomeric carbon atom**, the atom that was present in the original carbonyl group. The two configurations about this carbon atom are designated α and β as indicated in Eq. 4-1. In an equilibrium mixture, *ring forms of most sugars predominate over open chains.*[6–13] Thus, at 25°C in water glucose reaches an

OH
|
H — C — OH
|
R

Covalent hydrate
of R—CHO

equilibrium with ~0.001% free aldehyde, 0.004% covalent hydrate of the aldehyde (see also Eq. 13-1), 39% α-pyranose form, 61% β-pyranose form, and 0.15% each of the much less stable α and β furanose forms.[9,12] The ketose sugar **fructose** in solution exists as ~73% β-pyranose, 22% β-furanose, 5% α-furanose, and 0.5% open chain form (Fig. 4-1).[6,10,13] Although polysaccharides are composed almost exclusively of sugar residues in ring forms, the open chain forms are sometimes metabolic intermediates.

CH₂OH
|
C = O
|
H — O — C — H
|
H — C — O — H
|
H — C — O — H
|
CH₂OH

D-Fructose (Fru), a ketose that is a close structural and metabolic relative of D-glucose. It occurs in honey and fruit juices in free form, in the disaccharide sucrose (table sugar) as a 5-membered furanose ring, and in other oligosaccharides and polysaccharides.

Because of their many polar hydroxyl groups, most sugars are very soluble in water. However, hydrogen bonds between molecules stabilize sugar crystals making them insoluble in nonpolar solvents. Intermolecular hydrogen bonds between chains of sugar rings in cellulose account for much of the strength and insolubility of these polysaccharides.

1. The Variety of Monosaccharides

Sugars contain several chiral centers and the various diastereomers are given different names. The commonly occurring sugars **D-glucose**, **D-mannose**, and **D-galactose** are just three of the 16 diastereomeric aldohexoses. The Fischer projection formulas for the entire family of eight D-aldoses with 3–6 carbon atoms are given in Fig. 4-2. Several of these occur only rarely in nature.

Monosaccharides are classified as D or L according to the configuration at the chiral center farthest from the carbonyl group. If the –OH group attached to this carbon atom lies to the right when the sugar is oriented according to the Fischer convention, the sugar belongs to the D family. The simplest of all the chiral sugars is glyceraldehyde. The family of aldoses in Fig. 4-2 can be thought of as derived from D-glyceraldehyde by upward extension of the chain. Besides D-glucose, D-mannose, and D-galactose, the most abundant naturally occurring sugar shown in Fig. 4-2 is **D-ribose**, a major component of RNA. Another abundant aldose is **D-xylose**, a constituent of the polysaccharides of wood.

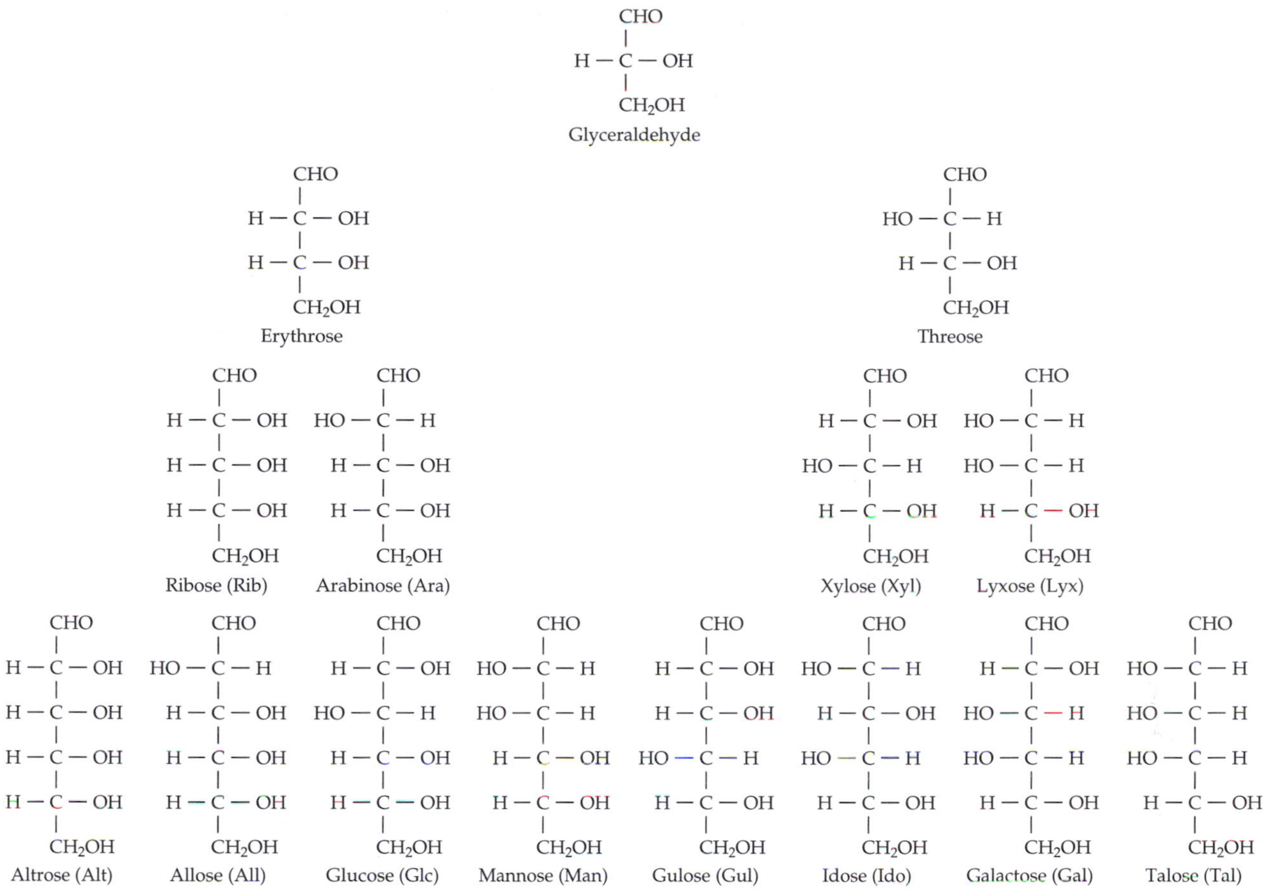

$$
\begin{array}{c}
\text{CHO} \\
| \\
\text{H} - \text{C} - \text{OH} \\
| \\
\text{CH}_2\text{OH}
\end{array}
$$

Glyceraldehyde

Erythrose

Threose

Ribose (Rib)   Arabinose (Ara)   Xylose (Xyl)   Lyxose (Lyx)

Altrose (Alt)   Allose (All)   Glucose (Glc)   Mannose (Man)   Gulose (Gul)   Idose (Ido)   Galactose (Gal)   Talose (Tal)

Figure 4-2 Formulas for the D-aldoses. Prefixes derived from the names of these aldoses are used in describing various other sugars including ketoses. The prefixes include *erythro*, *threo*, *arabino*, *ribo*, *galacto*, *manno*, etc., D-Fructose can be described in this manner as D-*arabino*-hexulose. The prefixes refer to the configurations of a series of consecutive but not necessarily contiguous chiral centers.[2] Thus, 3-*arabino*-hexulose is an isomer of fructose with the carbonyl group at the 3 position. The prefix *deoxy*, which means "lacking oxygen," is often used to designate a modified sugar in which an –OH has been replaced by –H, e.g., 2-deoxyribose.

For each D sugar there is an L sugar which is the complete mirror image or enantiomer of the D form. Although **L-arabinose** occurs widely in plants, and some derivatives of L sugars are present in glycoproteins, most naturally occurring sugars have the D configuration. A pair of sugars, such as glucose and galactose, differing in configuration at only one of the chiral centers are known as **epimers**. The D-ketose sugars with the carbonyl group in the 2 position (Fig. 4-3) are also abundant in nature, often occurring in the form of phosphate esters as intermediates in metabolism.

Ways of indicating configuration. The Fischer projection formulas used in Figs. 4-2 and 4-3 are convenient in relating the sugar structures by their individual carbon configurations to each other, but they give an unrealistic three-dimensional picture. According to the Fischer convention each carbon atom must be viewed with both vertical bonds projecting behind the atom viewed. In fact, the molecule cannot assume such a conformation because the chain folds back on itself, bringing many atoms into collision.

Ring forms of sugars are also often drawn according to the Fischer convention; making use of elongated bent lines to represent ordinary simple bonds:

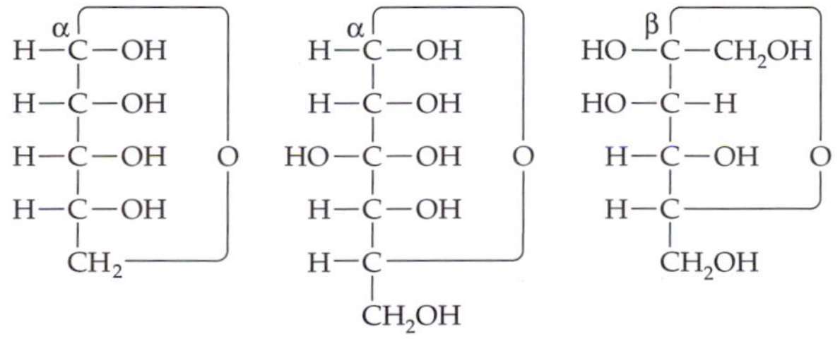

α-D-Ribopyranose   α-D-Glucopyranose   β-D-Fructofuranose

In this representation the hydroxyl at the anomeric

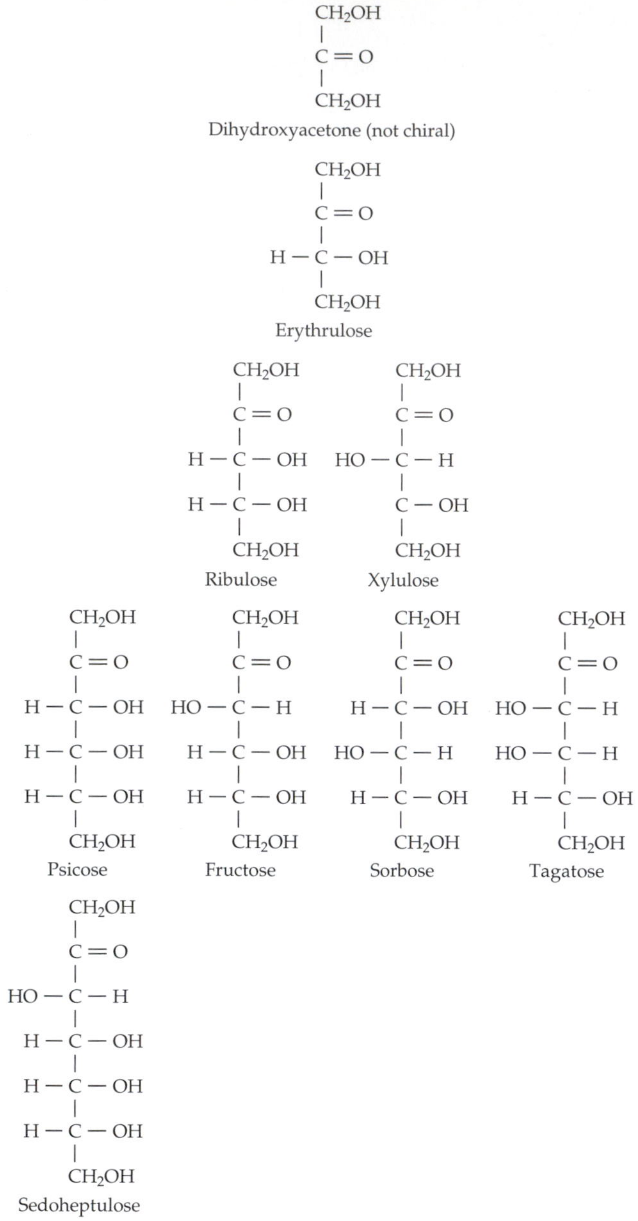

CH₂OH
|
C=O
|
CH₂OH
Dihydroxyacetone (not chiral)

CH₂OH
|
C=O
|
H—C—OH
|
CH₂OH
Erythrulose

CH₂OH .... CH₂OH
| .... |
C=O .... C=O
| .... |
H—C—OH ...... HO—C—H
| .... |
H—C—OH ...... C—OH
| .... |
CH₂OH .... CH₂OH
Ribulose ...... Xylulose

CH₂OH ..... CH₂OH ..... CH₂OH ..... CH₂OH
| .... | .... | .... |
C=O .... C=O .... C=O .... C=O
| .... | .... | .... |
H—C—OH .. HO—C—H .. H—C—OH .. HO—C—H
| .... | .... | .... |
H—C—OH .. H—C—OH .. HO—C—H .. HO—C—H
| .... | .... | .... |
H—C—OH .. H—C—OH .. H—C—OH .. H—C—OH
| .... | .... | .... |
CH₂OH ..... CH₂OH ..... CH₂OH ..... CH₂OH
Psicose .... Fructose ... Sorbose .... Tagatose

CH₂OH
|
C=O
|
HO—C—H
|
H—C—OH
|
H—C—OH
|
H—C—OH
|
CH₂OH
Sedoheptulose

Figure 4-3 Formulas for the open forms of the D-ketoses.

carbon atom is always on the right side of the molecules for α forms in the D series of sugars and at the left side for the β forms. For the L sugars the opposite is true. For example, since α-D-glucopyranose and α-L-glucopyranose are enantiomers, their Fischer formulas must be mirror images.

Simplified sugar rings are often drawn with **Haworth structural formulas**. The lower edge of the ring, which may be shown as a heavy line, is thought of as projecting out toward the reader and the other edge as projecting behind the plane of the paper.

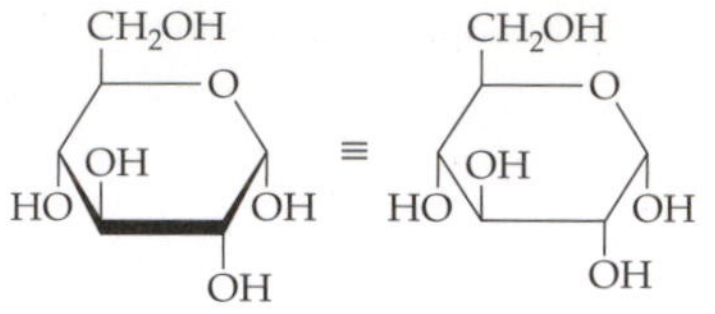

α-D-Glucopyranose
(Haworth structural
formula)

Haworth structures are easy to draw and unambiguous in depicting configurations,[14] but they also do not show the spatial relationships of groups attached to other rings correctly. For this reason conformational formulas of the type described in Section 2 and shown in Fig. 4-4 are used most often in this book.

Natural derivatives of sugars. The aldehyde group of an aldose can be oxidized readily to a carboxyl group to form an **aldonic acid**. Among the several aldonic acids that occur naturally is 6-phosphogluconic acid, which is pictured here as the 6-phosphogluconate ion:

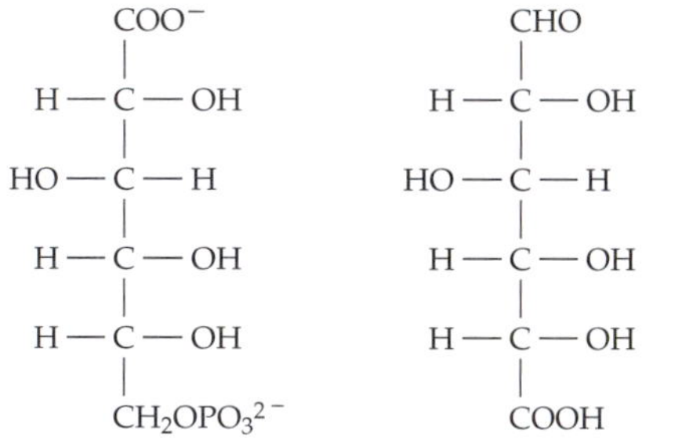

COO⁻ .... CHO
| .... |
H—C—OH .... H—C—OH
| .... |
HO—C—H .... HO—C—H
| .... |
H—C—OH .... H—C—OH
| .... |
H—C—OH .... H—C—OH
| .... |
CH₂OPO₃²⁻ .... COOH
6-Phosphogluconate ion   Glucuronic acid (GlcA)

The enzymatically formed **uronic acids** have –COOH in the terminal position. This is position 6 in **glucuronic acid**, whose structure is given here and as a ring in Fig. 4-4. Sugar chains with –COOH at both ends are called **aldaric acids**, e.g., glucaric acid. The –OH group in the 2 position of glucose may be replaced by –NH₂ to form 2-amino-2- deoxyglucose, commonly called **glucosamine** (GlcN), or by –NH–CO–CH₃ to form **N-acetylglucosamine** (GlcNAc). Similar derivatives of other sugars exist in nature. In many poly-saccharides, sulfate groups are attached in ester linkage to the sugar units. The sulfo (–SO₃⁻) sugar **6-sulfo-α-D-quinovose** (Fig. 4-4) is found in lipids of photosynthetic membranes.[15,16]

The sugar alcohols, in which the carbonyl group has been reduced to –OH, also occur in nature. For example, D-**glucitol** (D-sorbitol), the sugar alcohol obtained by reducing either D-glucose or L-sorbose (Eq. 4-2), is a major product of photosynthesis and widely distributed in bacteria and throughout the eukaryotic kingdom. It is present in large amounts in berries of the mountain ash and in many other fruits. It exists in a high concentration in human semen and

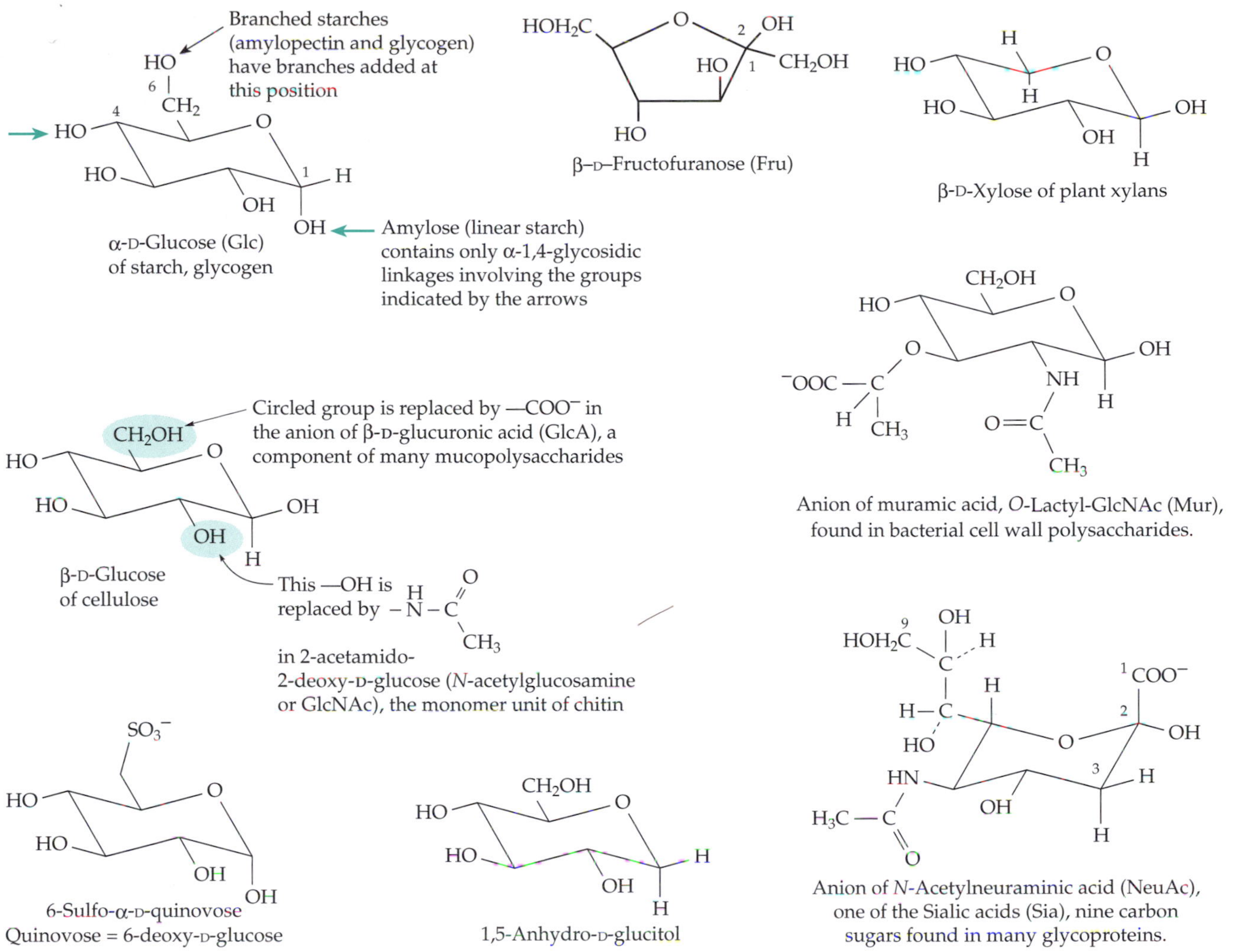

α-D-Glucose (Glc) of starch, glycogen

Branched starches (amylopectin and glycogen) have branches added at this position

Amylose (linear starch) contains only α-1,4-glycosidic linkages involving the groups indicated by the arrows

β–D–Fructofuranose (Fru)

β-D-Xylose of plant xylans

Circled group is replaced by —COO⁻ in the anion of β-D-glucuronic acid (GlcA), a component of many mucopolysaccharides

β-D-Glucose of cellulose

This —OH is replaced by $-N-C$ in 2-acetamido-2-deoxy-D-glucose (N-acetylglucosamine or GlcNAc), the monomer unit of chitin

Anion of muramic acid, O-Lactyl-GlcNAc (Mur), found in bacterial cell wall polysaccharides.

6-Sulfo-α-D-quinovose
Quinovose = 6-deoxy-D-glucose

1,5-Anhydro-D-glucitol

Anion of N-Acetylneuraminic acid (NeuAc), one of the Sialic acids (Sia), nine carbon sugars found in many glycoproteins.

Figure 4-4 Some simple sugars and sugar derivatives in ring forms. Most of these are present in polysaccharides.

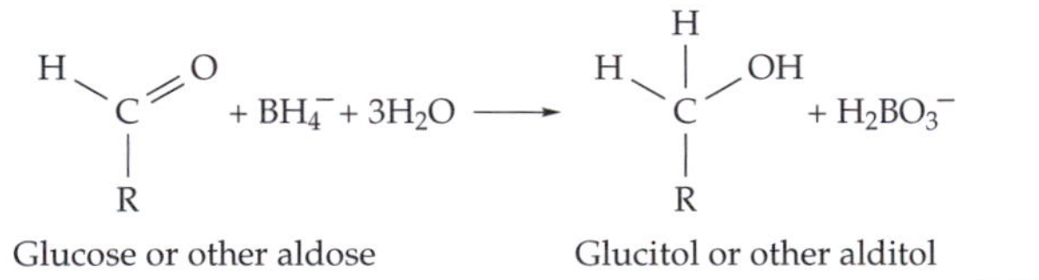

Glucose or other aldose

Glucitol or other alditol

(4-2)

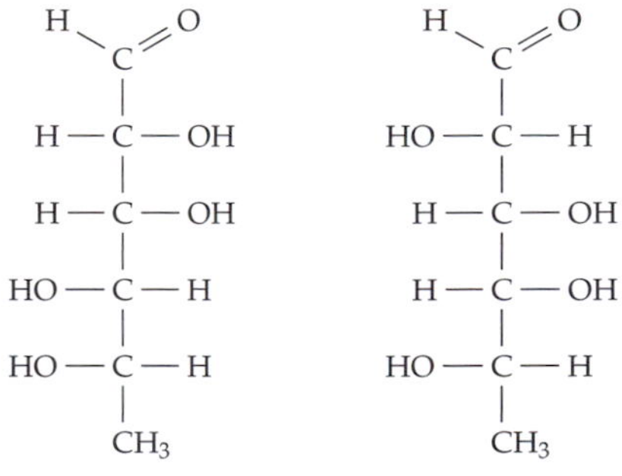

L-Rhamnose (Rha)

L-Fucose (Fuc)

accumulates in lenses of diabetics. D-Glucitol and other sugar alcohols arise in some fungi during metabolism of the corresponding sugars.[17] Mannitol, another product of photosynthesis, is also present in many organisms.[18] Another polyol that is found in human blood in significant concentrations (0.06–0.25 mM) is 1,5-anhydro-D-glucitol (Fig. 4-4). It is largely of dietary origin.[19]

Two common 6-deoxy sugars which lack the hydroxyl group at C-6 are **rhamnose** and **fucose**. Both are of the "unnatural" L configuration but are derived metabolically from D-glucose and D-mannose, respectively.

Vitamin C (ascorbic acid, Box 18-D) is another important sugar derivative. **Neuraminic acid** is a 9-carbon sugar made by transferring a 3-carbon piece onto a hexosamine. Its N-acetyl and N-glycolyl derivatives are called **sialic acids**. Their names may be abbreviated NeuAc and NeuGl, respectively, or simply as Sia (see also Fig. 4-4).

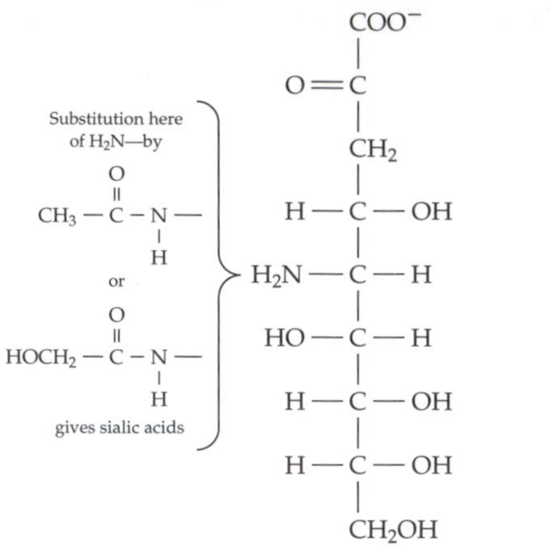

Neuraminic acid anion (Neu)

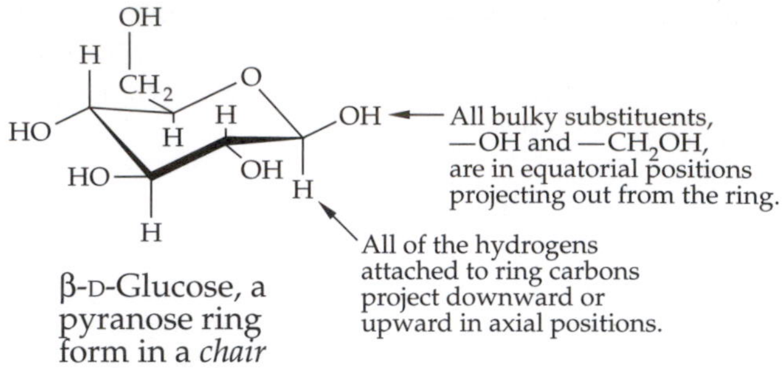

The sialic acids are prominent constituents of the glycoproteins of cell surfaces. More than 30 modified forms, for example, with added methyl or acetyl groups, are known.[20–24]

2. Conformations of Five- and Six-Membered Ring Forms

Single-bonded six-membered molecular rings, such as those in cyclohexane and in sugars, most often assume a **chair (C)** conformation. For sugars in pyranose ring forms, there are two possible chair conformations,[4,25,26] which are designated $^4C_1$(or C1) and $^1C_4$ (or 1C). The superscript and subscript numbers on the designations indicate which atoms are above and

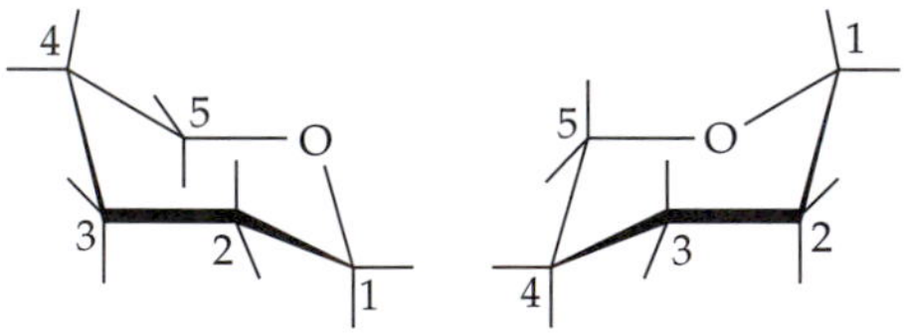

$^4C_1$ (or C1) $^1C_4$ (or 1C)

below the plane of the other four ring atoms. *These conformers are not easily interconvertible unless the ring is opened by a chemical reaction*, e.g., that of Eq. 4-1. However, manipulation with an atomic force microscope has shown that stretching of single polysaccharide chains can cause interconversion of the two chair conformations of pyranose rings, which are separated by an energy barrier of ~46 kJ/mol.[26a]

Most sugars occur in the chair conformation that places the largest number of substituents in equatorial positions and is therefore most stable thermodynamically. For D-aldoses this is usually the $^4C_1$ conformation:

β-D-Glucose, a pyranose ring form in a *chair* conformation.

← All bulky substituents, —OH and —CH₂OH, are in equatorial positions projecting out from the ring.

All of the hydrogens attached to ring carbons project downward or upward in axial positions.

For L-aldoses it is the $^1C_4$ form. There are exceptions to this rule. For example, α-D-idopyranose as well as iduronate rings assume the $^1C_4$ conformation because this conformation places the maximum number of bulky groups in equatorial positions.[27,28] It is noteworthy that electronegative substituents on the anomeric carbon atom of a sugar ring often prefer an axial orientation. This **anomeric effect** is also reflected in a shortening by about 0.01 nm of hydrogen bonds involving the hydrogen atom of the anomeric axial OH group. These effects can be explained partially as a result of coulombic repulsion of the two C–O dipoles and of resonance of the following type:[29–31]

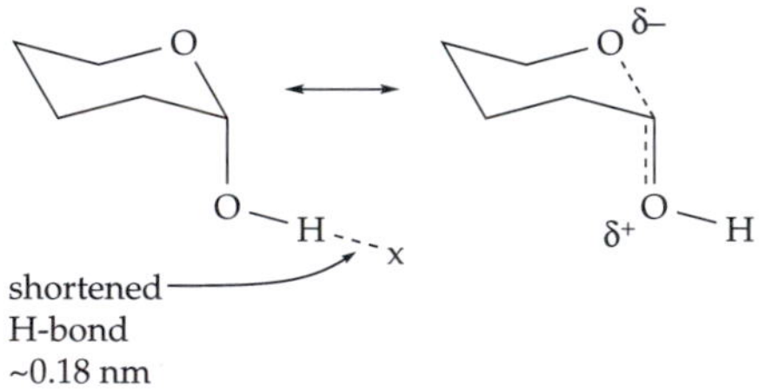

shortened H-bond ~0.18 nm

In addition to the chair conformations of six-membered rings the less stable **boat (B)** conformations are also possible. The six boat forms are smoothly interconvertible through intermediate **twist (T)** forms, which are also called **skew (S)** forms.[3,5] Since the

Boat   Skew or twist

internal angle in a pentagon is 108° (close to the tetrahedral angle) we might anticipate a nearly planar five-membered ring. However, eclipsing of hydrogen atoms on adjacent carbons prevents formation of such a flat structure. One of the atoms may be buckled out of the plane of the other four about 0.05 nm, into an **envelope conformation** (e.g., ^1E), or only three atoms may be in a plane, as in a twist conformation.

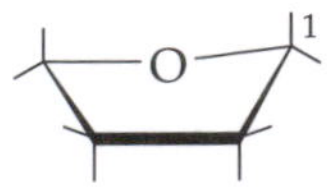

The ^1E envelope conformation

Any one of the five atoms of the ring can be either above or below the plane defined by the other four in the envelope conformation. The energy barriers separating them are very low, and in cyclopentane or in proline all of the envelope conformations are freely interconvertible through intermediate skew forms.[32] Furanose sugar rings are very flexible but the presence of the bulky substituents reduces the number of possible conformations.[33-36a] See Chapter 5 for further discussion.

3. Characteristic Reactions of Monosaccharides

The aldehyde group of aldoses can either be oxidized or reduced and ketoses can be reduced. The best laboratory reagent for reduction is **sodium borohydride** which acts rapidly in neutral aqueous solutions (see Eq. 4-2). Since both NaB^3H_4 and NaB^2H_4 are available, radioactive or heavy isotope labels can be introduced in this way. The aldehyde groups can be oxidized by a variety of agents to the corresponding aldonic acids, a fact that accounts for the reducing properties of these sugars. In alkaline solution aldoses reduce Cu^{2+} ions to cuprous oxide (Eq. 4-3), silver ions

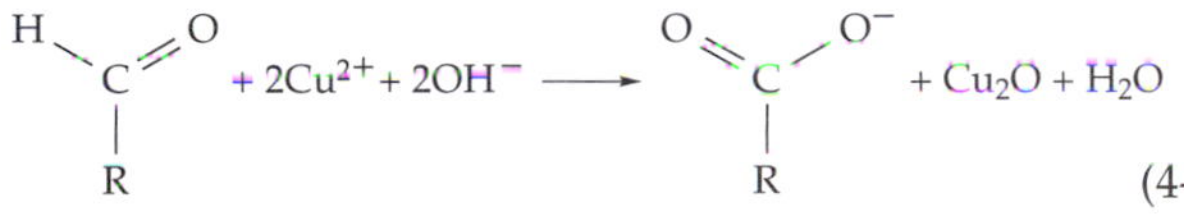

$$\underset{\substack{|\\R}}{\overset{\substack{H\\ \diagdown\\C\\ \diagup\diagup\\O}}{}} + 2Cu^{2+} + 2OH^- \longrightarrow \underset{\substack{|\\R}}{\overset{\substack{O\\ \diagdown\diagup\diagup\\C\\ \diagdown\\O^-}}{}} + Cu_2O + H_2O \qquad (4\text{-}3)$$

to the free metal, or hexacyanoferrate (III) to hexacyanoferrate (II). These reactions provide the basis for sensitive analytical procedures. Even though the aldoses tend to exist largely as hemiacetals (Eq. 4-1) the reducing property is strongly evident. Oxidation by metal-containing reagents is usually via the free aldehyde, but oxidation by hypobromite BrO^- (Br_2 in alkaline solution) yields the lactone, as does enzymatic oxidation (Eq. 15-10).

Sugars are unstable in acid. Boiling with concentrated HCl or H_2SO_4 converts pentoses to furfural (Eq. 4-4a) and hexoses to hydroxymethylfurfural (Eq. 4-4b).

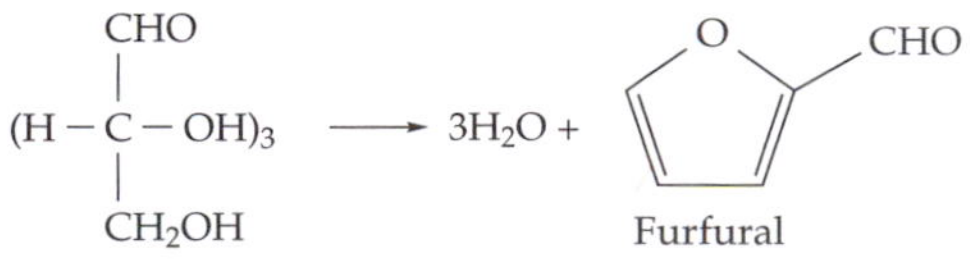

$$\underset{\substack{|\\CH_2OH}}{\overset{\substack{CHO\\|}}{(H-C-OH)_3}} \longrightarrow 3H_2O + \text{[furan ring]} CHO$$

Furfural (4-4a)

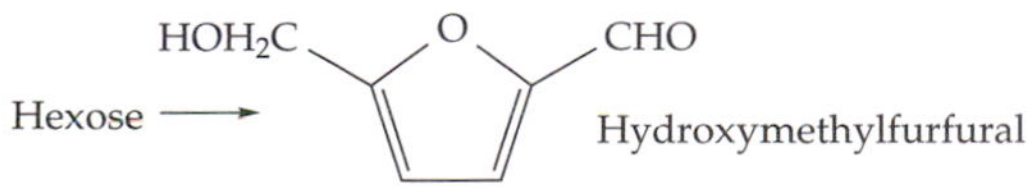

$$\text{Hexose} \longrightarrow HOH_2C\text{[furan ring]}CHO$$

Hydroxymethylfurfural (4-4b)

The aldehydes produced in these reactions can be condensed with various phenols or quinones to give colored products useful in quantitative estimation of sugar content or in visualizing sugars on thin-layer chromatographic plates. Phenol and sulfuric acid yield a product whose absorbance at 470 nm can be used as a measure of the total carbohydrate content of most samples. Resorcinol (1,3-dihydroxybenzene) in 3 M HCl (Seliwanoff's reagent) gives a red precipitate with ketoses. Orcinol (5-methylresorcinol) reacts rapidly with pentoses and with ribonucleosides and ribonucleotides. Since 2-deoxy sugars react slowly this can be used as a test for RNA. Diphenylamine with H_2SO_4 gives a blue-green color specifically with 2-deoxy sugars and can be used to test for DNA if the sample is first hydrolyzed.[37]

B. Glycosides, Oligosaccharides, Glycosylamines, and Glycation

The hydroxyl group on the anomeric carbon atom of the ring forms of sugars is reactive and can be replaced by another nucleophilic group such as $-O-R$ from an alcohol. The product is a glycoside (Eq. 4-5). The reaction with methanol occurs readily with acid catalysis under dehydrating conditions, e.g., in 100% methanol.

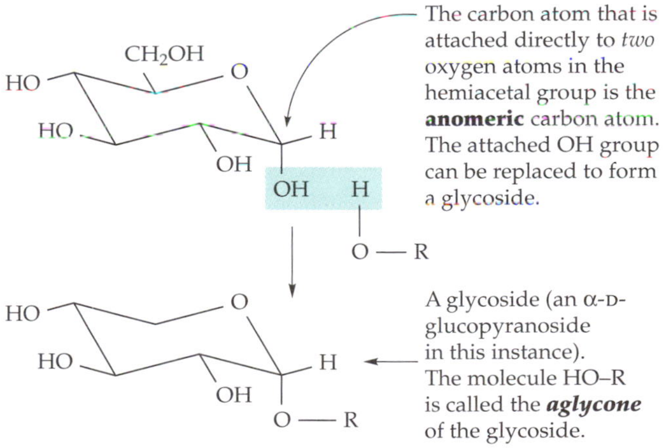

The carbon atom that is attached directly to *two* oxygen atoms in the hemiacetal group is the **anomeric** carbon atom. The attached OH group can be replaced to form a glycoside.

A glycoside (an α-D-glucopyranoside in this instance). The molecule HO–R is called the **aglycone** of the glycoside.

(4-5)

The alcohol in this equation can be a simple one such as methanol or it can be any of the –OH groups of another sugar molecule. For example, two molecules of α-D-glucopyranose can be joined, in an indirect synthesis, to form **maltose** (Eq. 4-6). Maltose is formed by the hydrolysis of starch and is otherwise not found in nature. There are only three abundant naturally occurring **disaccharides** important to the metabolism of plants and animals.[38] They are **lactose** (milk), **sucrose** (green plants), and **trehalose** (fungi and insects).

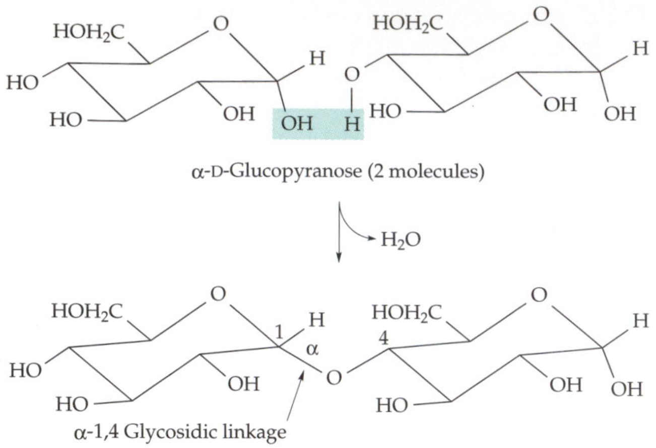

α-D-Glucopyranose (2 molecules)

→ H_2O

α-1,4 Glycosidic linkage

Maltose, a disaccharide   (4-6)

Disaccharides are linked by **glycosidic** (acetal) linkages. The symbol α-1,4, used in Eq. 4-6, refers to the fact that in maltose the glycosidic linkage connects carbon atom 1 (the anomeric carbon atom) of one ring with C-4 of the other and that the configuration about the anomeric carbon atom is α. While the α and β ring forms of free sugars can usually undergo ready inter-conversion, the configuration at the anomeric carbon atom is "frozen" when a glycosidic linkage is formed. To describe such a linkage, we must state this configu-ration together with the positions joined in the two rings (see Eq. 4-6). Lactose, whose structure follows, can be described as a disaccharide containing one galactose unit in a β-pyranose ring form and whose anomeric carbon atom (C-1) is joined to the 4 position of glucose, giving a β-1,4 linkage:

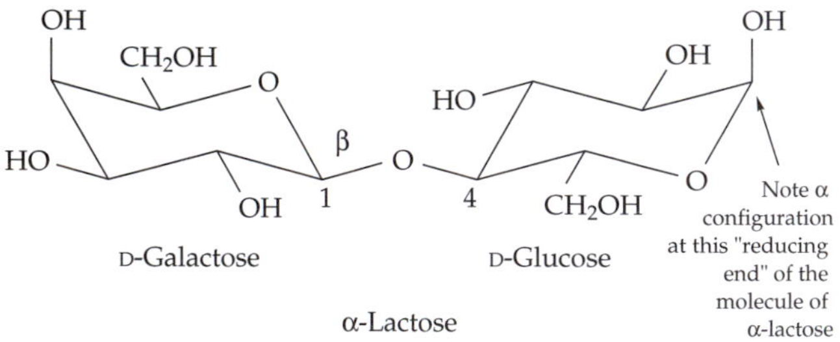

α-Lactose

The systematic name for α-lactose, O-β-D-galacto-pyranosyl-(1→4)-α-D-glucopyranose, provides a com-plete description of the stereochemistry, ring sizes, and mode of linkages. This name may be abbreviated β-D-Galp-(1→4)-α-D-Glcp. Since pyranose rings are so common and since most natural sugars belong to the D family, the designations D and p are often omitted. It may be assumed that in this book sugars are always D unless they are specifically designated as L. When linkages remain uncertain abbreviated formulas are given. Because the glucose ring in lactose is free to open to an aldehyde and to equilibrate (in solution) with other ring forms, the name lactose does not imply

a fixed ring structure for the glucose half. Thus, the name lactose can be abbreviated as βGal-(1→4)-Glc or, more succinctly, as Galβ1→4Glc. However, in crystal-line form the sugar exists either as α-lactose or β-lactose. The latter is more soluble and sweeter than α-lactose, which sometimes crystallizes in ice cream upon pro-longed storage and produces a "sandy" texture. An isomer of lactose, Galβ1→6Glc or **allolactose**, is an important inducer of transcription in cells of *E. coli* (Chapter 28).

Notice that in the drawing of the lactose structure the glucose ring has been "flipped over" with respect to the orientation of the galactose ring, a consequence of the presence of the β-1,4 linkage. For maltose, where the linkage is α-1,4, the two rings are usually drawn with the same orientation (Eq. 4-6). Maltose can be described as α-D-Glcp-(1→4)-D-Glcp or more simply as Glcα1→4Glc.

In sucrose and in α,α-trehalose the reducing groups of two rings are joined. Each of these sugars exists in a single form. Sucrose serves as the major transport sugar in green plants, while trehalose plays a similar role in insects, as does D-glucose in our blood.

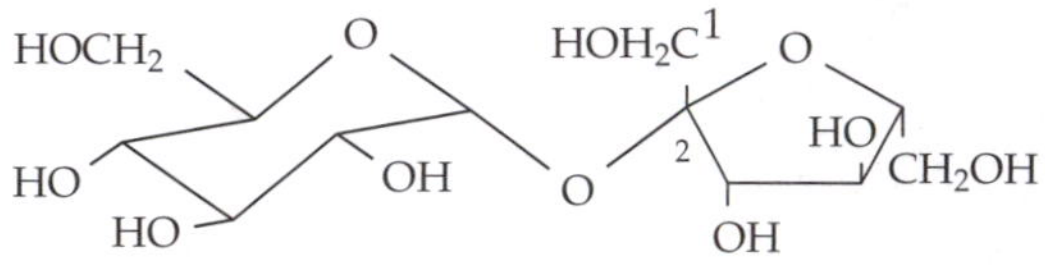

Sucrose: Glcpα1→β2Fruf

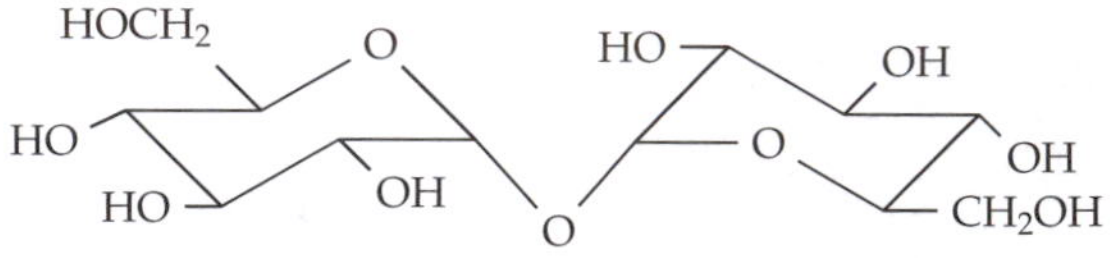

α, α-Trehalose: Glcpα1→α1Glcp

Trehalose, or "mushroom sugar," is found not only in fungi but also in many other organisms.[39-41]. It serves as the primary transport sugar in the hemolymph of insects and also acts as an "antifreeze" in many species. It accounts for up to 20% of the dry weight of **anhy-drobiotic organisms**, which can survive complete dehydration. These include spores of some fungi, yeast cells, macrocysts of *Dictyostelium*, brine shrimp cysts (dried gastrulas of *Artemia salina*), some nema-todes, and the resurrection plant. These organisms can remain for years in a dehydrated state. Hydrogen bonding between the trehalose and phosphatidylcholine may stabilize the dry cell membranes.[18,40,41] Although they can be desiccated, fungal spores remain dormant even when considerable water is present. One of the first detectable changes when the spores germinate is a rapid increase in the activity of the enzyme **trehalase** which hydrolyzes trehalose to glucose.[42,43]

Disaccharides, as well as higher oligosaccharides and polysaccharides, are thermodynamically unstable with respect to hydrolysis, for example, for lactose in aqueous solution:

$$\text{Lactose} + H_2O \rightarrow \text{D-glucose} + \text{D-galactose}$$
$$\Delta G° \approx -8.7 \pm 0.2 \text{ kJ mol}^{-1} \text{ at } 25°C \qquad (4\text{-}7)$$

The corresponding equilibrium constant K = [D-glucose] [D-galactose]/[lactose] $\approx$ 34 M. For other oligosaccharides K varies from 17 to 500 M.[44,45] However, sucrose is far less stable with $K = 4.4 \times 10^4$ M and $\Delta G° = -26.5 \pm 0.3$ kJ mol^{-1} for its hydrolysis.[46] Sucrose is also less stable kinetically and undergoes rapid acid-catalyzed hydrolysis. This fact is exploited when vinegar is used to convert sucrose to the less crystallizable mixture of glucose and fructose during candy-making. In marked contrast, trehalose is extremely resistant to acid-catalyzed hydrolysis.

The joining of additional sugar rings through glycosidic linkages to a disaccharide leads to the formation of **oligosaccharides**, which contain a few residues, and to **polysaccharides**, which contain many residues. Among the well-known oligosaccharides are the substituted sucroses **raffinose**, Gal$p$(1→6) Glc$p$(1→2) Fru$f$, and **stachyose**, Gal$p$(1→6) Gal$p$(1→6) Glc$p$(1→2) Fru$f$. Both sugars are found in many legumes and other green plants in which they are formed by attachment of the galactose rings to sucrose. Oligosaccharides have many functions. For example, gram-negative bacteria often synthesize oligosaccharides of 6−12 glucose units in β-1,2 linkage joined to sn-1-phosphoglyceryl groups. They are found in the periplasmic space between the inner and outer cell membranes and may serve to control osmotic pressure.[47] Oligosaccharides of 10−14 α-1,4-linked D-galacturonic acid residues serve as signals of cell wall damage to plants and trigger defensive reactions against bacteria in plants.[48]

Just as alcohols can be linked to sugars by glycoside formation (Eq. 4-5), amines can react similarly to give **glycosylamines** (N-glycosides). In this instance it is usually the free aldehyde that reacts via formation of a Schiff base (Eq. 4-8a). The latter can cyclize (Eq. 4-8b) to the glycosylamine with either the α or β configuration. Another important reaction of Schiff bases of sugars is the **Amadori rearrangement** (Eq. 4-8c,d), which produces a secondary ketoamine with a 1-oxo-2-deoxy structure.

This can cyclize as in step e of Eq. 4-8. In addition, epimerization at position 2 occurs through the reversal of step d. The overall reaction of Eq. 4-8 is often called **glycation** to distinguish it from **glycosylation,** the transfer of a glycosyl group. The Amadori rearrangement is important in nitrogen metabolism and in nonenzymatic reactions of sugars. For example, small

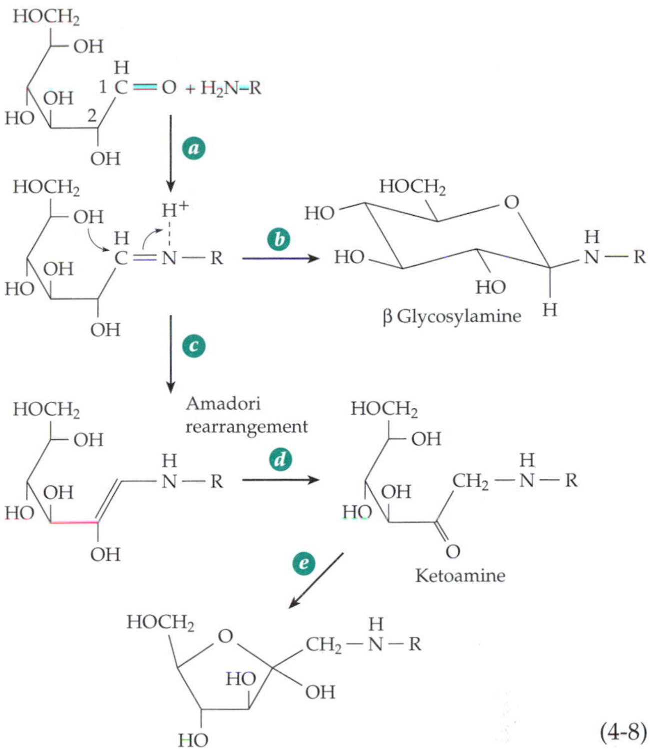

$$(4\text{-}8)$$

amounts of glycosylated hemoglobin and other proteins modified by glycation of the protein amino groups are normally present in the blood.[49-51] People with diabetes, who have a high concentration of blood glucose, have increased amounts of glycated protein. High concentration of either glucose or fructose[41,52] may cause serious problems. For example, modification of the protein **crystallins** of the lens of the eye may lead to cataracts. Similar problems with galactose may accompany galactosemia (Chapter 20).[53] Ketoamines formed by glycation of proteins may undergo crosslinking reactions with side chains of other protein molecules and this may be one cause of aging.[42,43] Other reactions of the ketoamines lead to formation of fluorescent and colored products.[41,43,50,54−56] Oxidation products of sugars also participate in these reactions,[57] and nucleic acid bases also react.[56]

C. Polysaccharides (Glycans)

Polymers of sugars are present in all cells and serve a variety of functions.[58-60] The simple sugars commonly used in the assembly of polysaccharides include D-glucose, D-mannose, D-galactose, D-fructose, D-xylose, L-arabinose, related uronic acids, and amino sugars (Fig. 4-4). These monomer units can be put together in many ways, either as **homopolysaccharides** containing a single kind of monomer or as **heteropolysaccharides** containing two or more different monomers. Because there are many sugars and many ways in which they can be linked, there is a bewildering

TABLE 4-1
Some of the Many Polysaccharides Found in Nature

Name	Source	Monomer	Main linkage	Branch linkages
Starch	Green plants			
Amylose		D-Glucose	α1,4	
Amylopectin		D-Glucose	α1,4	α1,6
Glycogen	Animals, bacteria	D-Glucose	α1,4	α1,6
Cellulose	Green plants, some bacteria	D-Glucose	β1,4	
Dextrans	Some bacteria	D-Glucose	α1,6	α1,3
Pullulan	Yeast	D-Glucose	α1,6 + α1,4	
Callose	Green plants	D-Glucose	β1,3	
Yeast glucan	Yeast	D-Glucose	β1,3	
Schizophyllan, curdlan, paramylon		D-Glucose	β1,3	β1,6 on every third residue
Mannans	Algae	D-Mannose	1,4	
	Yeast	D-Mannose	α1,6	
Xylans	Green plants	D-Xylose	β1,3	
	Brown seaweed			
Inulin	Some plant tubers	D-Fructose	β2,6	
Chitin	Fungi, anthropods	N-acetyl-D-Glucosamine	β1,4	

Alternating polysaccharides

Name	Source	Monomer	Main linkage	Branch linkages
Hyaluronan	Animal connective tissue	Glucuronic acid + N-acetylglucosamine	β1,4	
Chondroitin sulfate		D-Glucosamine N-acetyl-D-Galactosamine	β1,3 + β1,4	
Dermatan sulfate		α-L-Iduronate + N-acetyl-D-Galactosamine	β1,3 + β1,4	
Pectin	Higher plants	D-Galactunonate + others	β1,4 + others	
Alginate	Seaweed	D-Mannuronate + L-Guluronate	β1,4 + α1,4	
Agar-agar	Red seaweed	Galactose	β1,4 and α1,3	
Carageenan	Red seaweed	Galactose-4-sulfate + 3,6-anhydro-D-Galactose-2-sulfate	β1,4 + α1,3	
Murein	Bacterial cell wall	N-acetyl-D-Glucosamine + N-acetyl-D-Muramic acid	β1,4	

variety of different polysaccharides. Their chains can be linear or helical. The most numerous functional groups present are the free hydroxyl groups, some of which may form additional glycosidic linkages to produce **branched chains**. Polysaccharides may also contain –COOH, –NH$_2$, –NHCOCH$_3$, and other groups. After polymerization, hydroxyl groups are sometimes methylated or converted to sulfate esters or to ketals formed with pyruvic acid. Structural characteristics of some of the major polysaccharides are listed in Table 4-1.

1. Conformations of Polysaccharide Chains

Despite the variety of different monomer units

and kinds of linkage present, the conformational possibilities for carbohydrate chains are limited. The sugar ring is a rigid unit and the connection of one unit to the next can be specified by means of two torsion angles ϕ and ψ just as with peptides.[61–63] To specify a torsion angle four atoms must be selected – the two at the ends of the bond about which rotation is being considered and two others. There is more than one way to define the zero angle for ϕ and ψ. As illustrated in the drawing on p. 172, ϕ may be taken as the H1–C1–O–C4' dihedral angle and ψ as the H4'–C4'–O–C1 angle. The zero angle is when H1 and H4' are eclipsed. A related alternative is to take ϕ and ψ as 0° when the two midplanes of the sugar rings are coplanar.

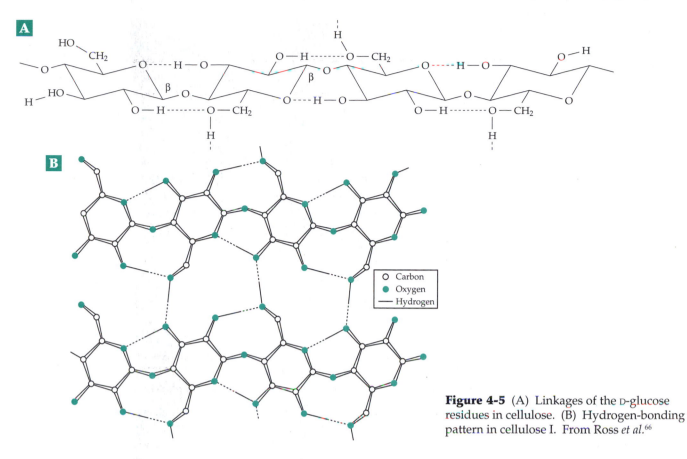

Figure 4-5 (A) Linkages of the D-glucose residues in cellulose. (B) Hydrogen-bonding pattern in cellulose I. From Ross *et al.*[66]

BOX 4-A CYCLODEXTRINS

An enzyme produced by some bacteria of the genus *Bacillus* cuts chains of amylose and converts them into tiny rings consisting of six (α-cyclodextrin), seven (β-cyclodextrin), eight (γ-cyclodextrin), or more glucose units. The cyclodextrins, which were first isolated by F. Schardinger in 1903, have intrigued carbohydrate chemists for many years by their unusual properties.[a] They are surprisingly resistant to acid hydrolysis and to attack by amylases and cannot be fermented by yeast.

The torus-like cyclodextrin molecules have an outer polar surface and an inner nonpolar surface.

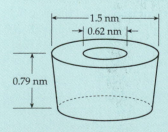

The small hydrophobic cavities within the cyclodextrins have diameters of 0.50, 0.62, and 0.79 nm, respectively, for the α, β, and γ dextrins.[b] Cavities are potential binding sites for a great variety of both inorganic and organic molecules.[a–g] Complexes of simple alcohols, polyiodides,[c] ferrocene,[d] and many other compounds have been observed. Cyclodextrins can be used to "encapsulate" food additives[b] and their complexes may be useful in separation of enantiomers of drugs.[f,g] They can be chemically modified by adding catalytic groups to serve as enzyme models[h,i] or as color-change indicators sensitive to binding of organic molecules.[j] They can be linked together to form molecular **nanotubes**. Polymer chains can even be threaded through the tubes.[k] What practical applications may yet come from this?

[a] French, D. (1957) *Adv. Carbohydr. Chem. Biochem.* **12**, 189–260
[b] Korpela, T., Mattsson, P., Hellman, J., Paavilainen, S., and Mäkelä, M. (1988–89) *Food Biotechnology* **2**, 199–210
[c] Noltemeyer, M., and Saenger, W. (1980) *J. Am. Chem. Soc.* **102**, 2710–2722
[d] Menger, F. M., and Sherrod, M. J. (1988) *J. Am. Chem. Soc.* **110**, 8606–8611
[e] Hamilton, J. A., and Chen, L. (1988) *J. Am. Chem. Soc.* **110**, 4379–4391
[f] Armstrong, D. W., Ward, T. J., Armstrong, R. D., and Beesley, T. E. (1986) *Science* **232**, 1132–1135
[g] Lipkowitz, K. B., Raghothama, S., and Yang, J. (1992) *J. Am. Chem. Soc.* **114**, 1154–1162
[h] Anslyn, E., and Breslow, R. (1989) *J. Am. Chem. Soc.* **111**, 8931–8932
[i] Granados, A., and de Rossi, R. H. (1995) *J. Am. Chem. Soc.* **117**, 3690–3696
[j] Ueno, A., Kuwabara, T., Nakamura, A., and Toda, F. (1992) *Nature* **356**, 136–137
[k] Harada, A., Kamachi, J. L., and Kamachi, M. (1993) *Nature* **364**, 516–518

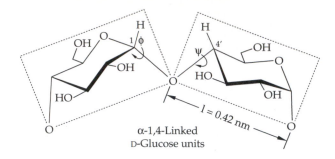

α-1,4-Linked
D-Glucose units

The IUB has proposed another convention.[64]

2. The Glucans

From glucose alone various organisms synthesize a whole series of polymeric glucans with quite different properties. Of these, **cellulose,** an unbranched β-1,4-linked polyglucose (Fig. 4-5A), is probably the most abundant. It is the primary structural polysaccharide of the cell walls of most green plants.[65] For the whole Earth, plants produce ~10^{14} kg of cellulose per year.

A systematic examination of the possible values for φ and ψ shows that for cellulose these angles are constrained to an extremely narrow range which places the monomer units in an almost completely extended conformation.[62] Each glucose unit is flipped over 180° from the previous one. The polymer has a twofold screw axis and there is a slight zigzag in the plane of the rings.[66–68] Remember that in the chair form of glucose all of the –OH groups lie in equatorial positions and are able to form hydrogen bonds with neighboring chains. This feature, together with the rigidity of conformation imposed by the β configuration of the monomer units, doubtless accounts for the ability of cellulose to form strong fibers.

It has been impossible to obtain large single crystals of cellulose. However, from 60 to 90% of native cellulose is thought to be clustered to form the needle-like crystalline microfibrils. These microcrystals of **cellulose I** can be separated from other plant materials by prolonged boiling with dilute NaOH and HCl. Their structure has been established by electron diffraction[67] and by comparison with high-resolution X-ray diffraction structure of the tetrasaccharide β-D-cellotetraose.[66,69] Two closely similar parallel-chained, hydrogen-bonded structures appear to be present. One is shown in Fig. 4-5B. The hydrogen bonds and van der Waals forces bind the chains into sheets which are stacked to form fibers. A typical fiber of plant cellulose has a diameter of 3.5–4 nm and contains 30–40 parallel chains, each made up of 2000–10,000 glucose units. The chain ends probably overlap to form essentially endless fibers that can extend for great distances through the cell wall. They interact with other polysaccharides as is illustrated in Fig. 4-14. A single cotton

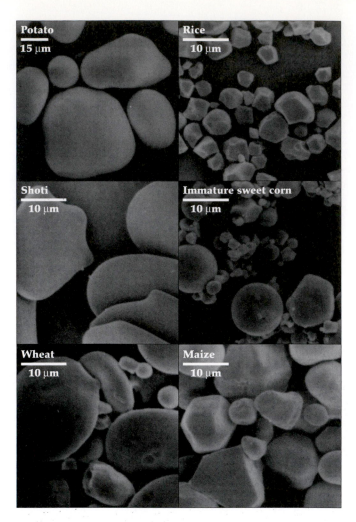

Figure 4-6 Scanning electron micrographs of starch granules. Magnification 2500× and 3000×. From Jane et al.[73]

fiber may be 2–3 cm in length.

Cotton thread treated with concentrated NaOH shrinks and has an increased luster. The resulting "Mercerized" cellulose has changed into other crystalline forms. The major one is **cellulose II**, in which the chains in the sheets are antiparallel.[69a] Cellulose II may also occur to some extent in nature. Many other modified celluloses, e.g., **methylcellulose**, in which some –OH groups have been converted to methyl ethers[70] are important commercial products.

Starch, another of the most abundant polymers of glucose, is stored by most green plants in a semi-crystalline form in numerous small granules. These granules, which are usually formed within colorless membrane-bounded plastids, have characteristic shapes and appearances (Fig. 4-6) that vary from plant to plant. One component of starch, **amylose**, is a linear polymer of many α-D-glucopyranose units in 1,4 linkage (Fig. 4-7) as in maltose. Starch granules always contain a second kind of molecule known as **amylopectin**.[58]

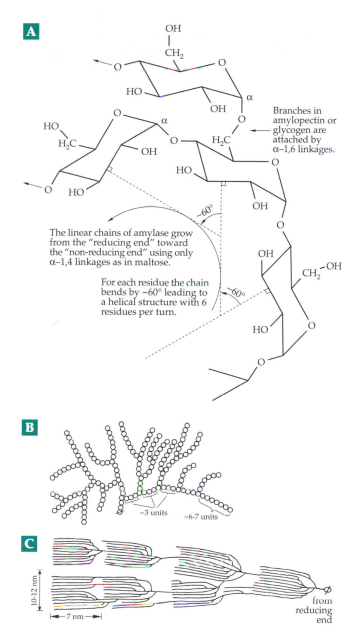

The linear chains of amylase grow from the "reducing end" toward the "non-reducing end" using only α–1,4 linkages as in maltose.

For each residue the chain bends by ~60° leading to a helical structure with 6 residues per turn.

Branches in amylopectin or glycogen are attached by α–1,6 linkages.

~3 units   ~6-7 units

10-12 nm

7 nm

from reducing end

Figure 4-7 (A) Linkages of the glucose residues in starches and in glycogen. (B) Schematic diagram of the glycogen molecule as proposed originally by K. H. Meyer.[74] The circles represent glucose residues which are connected by α-1,4 linkages and, at the branch points, by α-1,6 linkages. The symbol φ designates the reducing group. From D. French.[75] (C) Proposed broomlike clusters in amylopectin. After D. French.[71]

Both amylopectin and **glycogen** (animal starch) consist of highly branched bushlike molecules. Branches are attached to α-1,4-linked chains through α-1,6 linkages (Fig. 4-7). When glucose is being stored as amylopectin or glycogen the many "nonreducing" ends appear to grow like branches on a bush, but when energy is needed the tips are eaten back by enzymatic action (Fig. 11-2). There is only one **reducing**

end of a starch molecule where one might expect to find a hemiacetal ring in equilibrium with the free aldehyde. However, this end may be attached to a protein.

Starch granules show a characteristic pattern of growth rings of 0.3- to 0.4-μm thickness with thinner dense layers about 7 nm apart. Study of this layered structure suggests that the individual amylopectin molecules have branches close together in broomlike clusters (Fig. 4-7C).[61,71,72] The amylopectin chains are 120–400 nm long and their relative molecular masses may reach 15–30 million. In addition, starch granules usually contain molecules of the straight-chain amylose, each containing several hundred glucose units and having molecular masses of ~100 kDa. Most starches contain 20–21% amylose but there are special varieties of plants that produce starch with 50–70% amylose. On the other hand, the "waxy" varieties of maize form only amylopectin and lack amylose.

In both starch and glycogen the glucose units of the main chains are linked with α-1,4 linkages. An extended conformation is not possible and the chains tend to undergo helical coiling. One of the first helical structures of a biopolymer to be discovered (in 1943)[76,77] was the left-handed helix of amylose wound around molecules of pentaiodide (I_5^-) in the well-known blue starch–iodine complex[78] (Fig. 4-8). The helix contains six residues per turn, with a pitch of 0.8 nm and a diameter of nearly 14 nm. Amylose forms complexes of similar structure with many other small molecules.[79]

Another more tightly coiled double-helical form of amylose has been proposed.[80] Each chain would contain six glucose units per turn and the two chains could be arranged in either parallel or antiparallel directions. The average amylose molecule contains 2000 glucose units and could be stretched to a slender chain over 1 μm long, longer than the crystalline regions observed in starch granules. Thus, the chains within the granules would have to fold back on themselves, possibly in hairpin fashion:

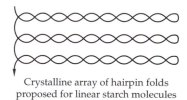

Crystalline array of hairpin folds proposed for linear starch molecules

X-ray diffraction studies support the double-helical structure but suggest a *parallel* orientation of the amylose chains.[81] Since amylose has not been obtained as single crystals the diffraction data do not give a definitive answer. However, if double helices are formed by adjacent branches in amylopectin and glycogen the two strands would be parallel. Starch granules also contain amorphous starch which appears to contain single helices, possibly wrapped around lipid materials.[82]

Glycogen is stored in the cytoplasm of animal cells and to some extent in the lysosomes as enormous 100- to 200-MDa particles. These appear in the electron microscope as aggregates of smaller particles of molecular masses up to 20–40 MDa. A laminated internal structure with surface bumps is suggested by STM microscopy (Fig. 4-9). Biosynthesis of glycogen may be initiated by a 37-kDa protein called **glycogenin**, which remains covalently attached to the reducing end of the glycogen (Chapter 20).[83] Despite the huge molecular masses of glycogen particles, both ^{1}H and ^{13}C NMR resonances are sharp, indicating a high degree of mobility of the glycosyl units.[84]

Beta-1,3-linked glucans occur widely in nature. When a new green plant cell is formed the first poly-

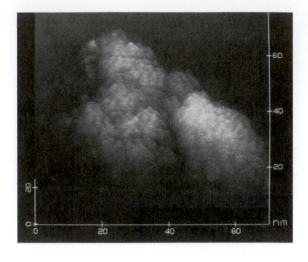

Figure 4-9 A scanning tunneling microscopic (STM) image of three glycogen molecules. The corrugated surface suggests a laminated structure. These molecules have been purified from tissues by treatment with strong alkali, which breaks the larger aggregates into particles of $M_r = 1-10 \times 10^6$ and diameter 25–30 nm. Courtesy of Fennell Evans.[93]

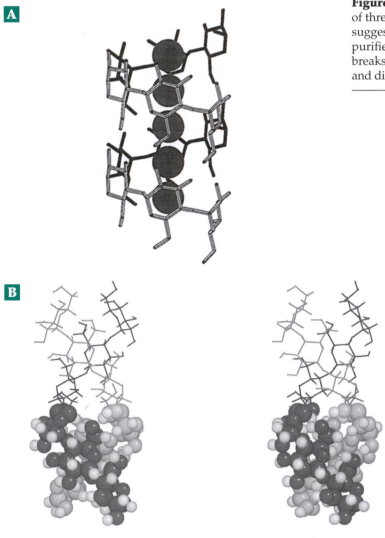

Figure 4-8 (A) Structure of the helical complex of amylose with I_3^- or I_5^-. The iodide complex is located in the interior of the helix having six glucose residues per turn. (B) Model of a parallel-stranded double helix. There are six glucose units per turn of each strand. The repeat period measured from the model is 0.35 nm per glucose unit. Courtesy of Alfred French.

saccharide to be synthesized is not cellulose but the β-1,3-linked glucose polymer **callose**. Cellulose appears later. Callose is also produced in some specialized plant tissues, such as pollen tubes,[85] and is formed in massive amounts at the site of wounds or of attack by pathogens. The major structural component of the yeast cell wall is a β-1,3-linked glucan with some β-1,6 branches.[86] **Schizophyllan** is a β-1,3-linked glucan with a β-1,6-linked glucosyl group attached to every third residue. A glucan from the coleoptiles of oats contains 30% β-1,3 linkages in a linear chain that otherwise has the structure of cellulose.[87] Other β-1,3-linked glucans serve as energy storage molecules in lower plants and in fungi. Among these are β-1,3-linked glucans such as **paramylon**,[88,89] which is stored by the euglena. A similar polysaccharide, **curdlan**, is formed by certain bacteria. [90]

Some other bacteria, e.g., *Leuconostoc mesenteroides*, make 1,6-linked poly-D-glucose or **dextrans**.[91,92] These always contain some α-1,3-linked branches and may also have α-1,4 and α-1,2 linkages, the structures varying from species to species. Dextrans formed by bacteria such as *Streptococcus mutans* growing on the surfaces of teeth are an important component of dental plaque. Bacterial

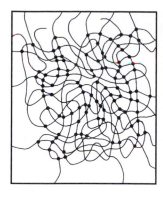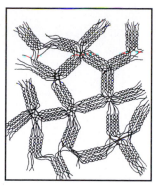

Figure 4-10 A schematic representation of the gel networks of Sephadex (left) and agarose (right). Note that the aggregates in agarose gels may actually contain 10–10⁴ helices rather than the smaller numbers shown here. From Arnott *et al.*[95]

dextrans are also produced commercially and are chemically crosslinked to form gels (Sephadex) which are widely used in biochemical separation procedures (Figs. 3-1 and 4-10). A yeast polysaccharide **pullulan** is a regular polymer of maltotriose units with α-1,4 linkages joined in a single chain by α-1,6 linkages.[94] One glucan, called **alternan** is a linear polymer with alternate α-1,6 and 1,3 linkages.

3. Other Homopolysaccharides

Cell walls of yeasts contain **mannans** in which the main α-1,6-linked chain carries short branches of one to three mannose units joined in α-1,2, α-1,3, and sometimes α-1,6 linkages.[96] These are covalently linked to proteins.[97] A β-1,4-linked mannan forms microfibrils in the cell wall of some algae such as *Acetabularia* (Fig. 1-11) which do not make cellulose.[98,99] The cell walls of some seaweeds contain a β-1,3-linked **xylan** instead of cellulose. This polysaccharide forms a three-stranded right-handed triple helix.[99,100] Even though xylose is a five-carbon sugar, the polymer contains the thermodynamically more stable six-membered pyranose rings. On the other hand, fructose, a 6-carbon sugar, is present as five-membered furanose rings in **inulin**, the storage polysaccharide of the Jerusalem artichoke and other Compositae, and also in sweet potatoes. The difference has to do with biosynthetic pathways. Furanose rings arise both in inulin and in sucrose because the biosynthesis occurs via the 6-phosphate ester of fructose, making it impossible for the phosphate derivative to form a 6-membered ring.

The major structural polysaccharide in the exoskeletons of arthropods and of other lower animal forms is **chitin**, a linear β-1,4-linked polymer of *N*-

acetylglucosamine whose structure resembles that of cellulose. In β chitin the individual parallel chains are linked by hydrogen bonds to form sheets in which parallel chains are held together by NH --- O = C hydrogen bonds between the carboxamide groups. In the more abundant α chitin the chains in alternate sheets have opposite orientations,[101,102] possibly a result of hairpin folds in the strands. Native chitin exists as microfibrils of 7.25 nm diameter. These contain a 2.8-nm core consisting of 15–30 chitin chains surrounded by a sheath of 27-kDa protein subunits. The microfibrils pack in a hexagonal array, but the structure is not completely regular. Several proteins are present; some of the glucosamine units of the polysaccharide are not acetylated and the chitin core is often calcified.[103] The commercial product **chitosan** is a product of alkaline deacetylation of chitin but it also occurs naturally in some fungi.[102] Chitin is also present in cell walls of yeasts and other fungi. It is covalently bonded to a β-1,3-linked glycan which may, in turn, be linked to a mannoprotein (see Section D,2).[97]

4. Heteropolysaccharides of the Animal Body

Many polysaccharides contain repeating units consisting of more than one different kind of monomer.[104,105] Some of these are composed of two sugars in a simple alternating sequence. Examples are **hyaluronan** (hyaluronic acid) and the **chondroitin**, **dermatan**, **keratan**, and **heparan sulfates**. They are important components of the "ground substance" or intracellular cement of connective tissue in animals. Hyaluronan,[106,107] which is abundant in synovial fluid and the vitreous humor of the eye, is a repeating polymer of glucuronic acid and *N*-acetylglucosamine with the structure shown in Fig. 4-11 and M_r of several million. The chondroitin sulfates and dermatan sulfate are similar polymers but with substitution by *N*-acetylgalactosamine and α-L-iduronic acid, respectively, and with sulfate ester groups in the positions indicated in Fig. 4-11.

Hyaluronan solutions are remarkably viscous and at a concentration of only 0.1% can have over 80% of the typical viscosity of biological fluids. This property may result from the presence of hydrogen bonds between the carboxylate, carboxamide, and hydroxyl groups of adjacent sugar residues as in Fig. 4-12. The hydrogen bonds stiffen the chain to give a slender rod. The tetrasaccharide shown in the figure is the repeating unit in a threefold helix.[109] However, the charged molecules do not associate to form strong fibrils like those of cellulose. The chain can be bent easily with breakage of the H-bonds at various positions to give a random coil structure.

While hyaluronan is not covalently attached to proteins, it is usually anchored to cell surfaces and to

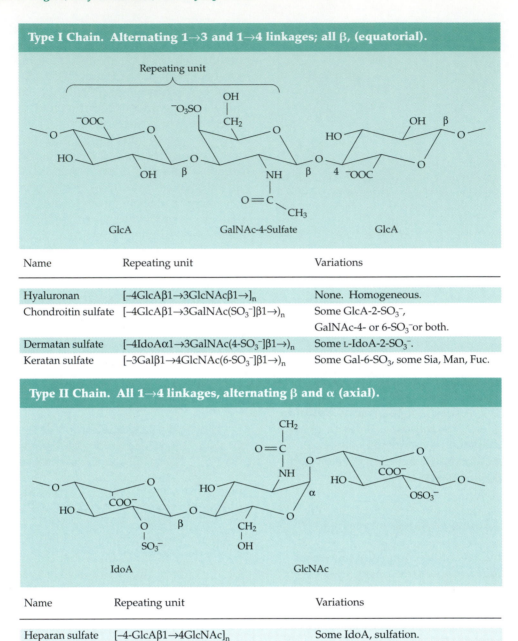

Type I Chain. Alternating 1→3 and 1→4 linkages; all β, (equatorial).

Name	Repeating unit	Variations
Hyaluronan	[–4GlcAβ1→3GlcNAcβ1→]$_n$	None. Homogeneous.
Chondroitin sulfate	[–4GlcAβ1→3GalNAc(SO$_3^-$)β1→)$_n$	Some GlcA-2-SO$_3^-$, GalNAc-4- or 6-SO$_3^-$or both.
Dermatan sulfate	[–4IdoAα1→3GalNAc(4-SO$_3^-$)β1→)$_n$	Some L-IdoA-2-SO$_3^-$.
Keratan sulfate	[–3Galβ1→4GlcNAc(6-SO$_3^-$)β1→)$_n$	Some Gal-6-SO$_3$, some Sia, Man, Fuc.

Type II Chain. All 1→4 linkages, alternating β and α (axial).

Name	Repeating unit	Variations
Heparan sulfate	[–4-GlcAβ1→4GlcNAc]$_n$	Some IdoA, sulfation.
Heparin	[–4-L-IdoA(SO$_3^-$)α1→4GlcN(SO$_3^-$)α1→]$_n$	Some IdoA-2-SO$_3^-$, GlcNAc. 24 different disaccharides possible.

Figure 4-11 The repeating disaccharide units of hyaluronan and other glycosaminoglycans. See Fransson[108] and Hardingham and Fosang.[107]

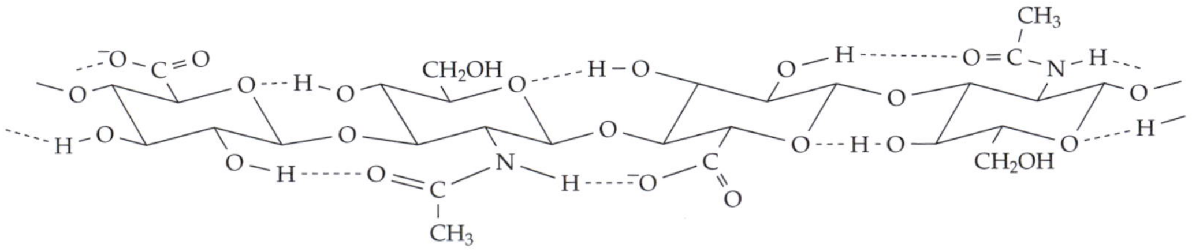

Figure 4-12 Proposed hydrogen-bonding scheme for the "native" conformation of hyaluronan. See Morris *et al.*[109]

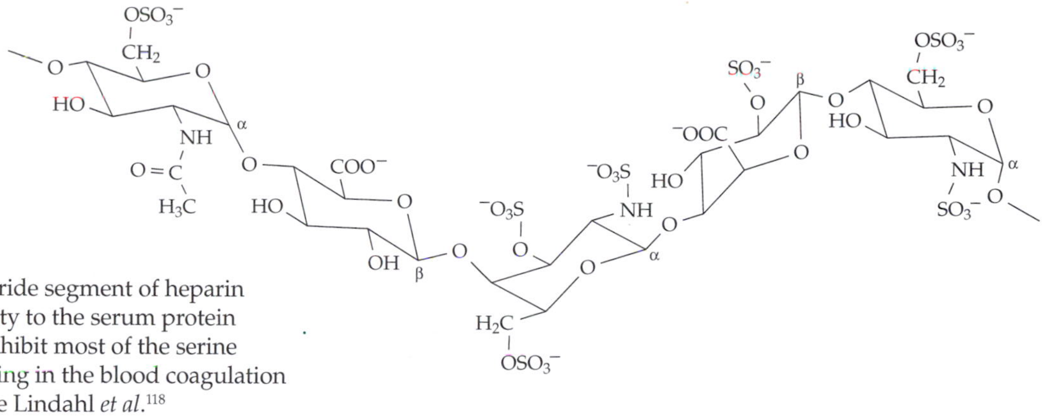

Figure 4-13 A pentasaccharide segment of heparin which binds with high affinity to the serum protein antithrombin causing it to inhibit most of the serine protease enzymes participating in the blood coagulation process (see Chapter 12). See Lindahl *et al.*[118]

protein "receptors" within the extracellular matrix. In contrast, chondroitin, dermatan, keratan, and heparan sulfates are attached covalently to the proteins at the reducing ends of the polymer chains (see Section D). The attached polymers undergo enzyme-catalyzed chemical alteration. In dermatan most of the glucuronate residues found in chondroitin have been epimerized to iduronate and sulfate groups in ester linkages have been added. Chondroitin sulfate is especially abundant in cartilage; dermatan sulfate is concentrated in skin. Heparan sulfates are more heterogeneous than the other polymers of this group and have been described as "the most complex polysaccharides on the surface of mammalian cells."[110]

In the mast cells of lungs, liver, and other cells heparan sulfate chains are modified further and released into the bloodstream as **heparin**, a material with important anticoagulant properties. Both the amino groups and the 6-hydroxyls of the glucosamine residues of heparin carry sulfate groups. In some units D-glucuronic acid or glucuronate-2-sulfate[111] is present in α-1,4 linkage, but more often L-iduronic acid-2-sulfate is the first unit in the disaccharide.[112–114] The iduronate ring appears to have unusually high conformational flexibility which may influence the biological activity of heparin and related polysaccharides.[115] Because of its anticoagulant property, heparin is an important drug for prevention of blood clot formation.[116] In the United States, in 1976 six metric tons of heparin were used to treat 10 million patients.[117] The anticoagulant activity resides in large part in a nonrepeating pentasaccharide (Fig. 4-13).[112,118] This portion of the heparin, especially if part of a larger octasaccharide,[113] binds to several proteins including the enzyme inhibitors **heparin cofactor II** and **plasma antithrombin III** (Chapter 12). Heparin greatly accelerates the rate at which these proteins bind and inactivate blood clotting factors. See also Section D,1. Lower invertebrates, as well as marine brown algae, contain heavily sulfated fucans which are largely 1,3-linked.[119]

5. Plant Heteropolysaccharides

Fibers of cellulose, which run like rods through the amorphous matrix of plant cell walls, appear to be coated with a monolayer of **hemicelluloses**. Predominant among the latter is a **xyloglucan**, which has the basic cellulose structure but with α-1,6-linked xylose units attached to three-fourths of the glucose residues.[104,120,121] L-Fucose may also be present in trisaccharide side chains: L-Fucα1→2Galβ1→2Xylα1→.[122] **Pectins** of higher plants contain β-1,4-linked polygalacturonates interrupted by occasional 1,2-linked L-rhamnose residues. Some of the carboxyl groups of these **rhamnogalacturonan** chains are methylated.[123] **Arabinans** and **galactans** are also present in pectin. A possible arrangement of cellulose fibers, hemicelluloses, and pectic materials in a cell wall has been proposed (Fig. 4-14).

Agarose, an alternating carbohydrate polymer consisting of ~120-kDa chains, is the principal component of agar and the compound that accounts for most of the gelling properties of that remarkable substance. A solid agar gel containing 99.5% water can be formed. Agarose molecules form left-handed double helices with a threefold screw axis, a pitch of 1.90 nm, and a central cavity containing water molecules.[124,125]

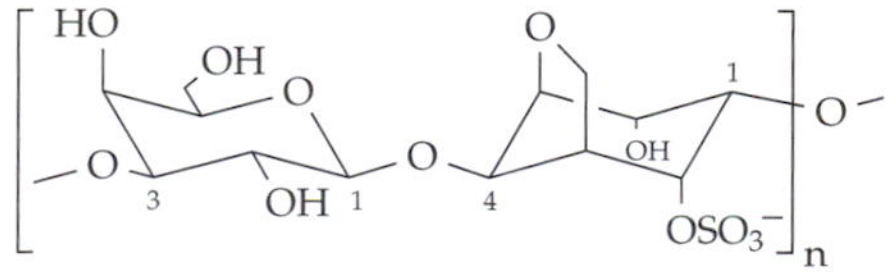

Agarose:
[-3-D-Galβ1→4-(3,6-anhydro)-L-Galα1→]n

A similar structure has been established for the gel-forming **carrageenans** from red seaweed. The X-ray data suggest that three of the disaccharide units form one turn of a right-handed helix with a pitch of 2.6 nm. A second chain with a parallel orientation, but displaced by half a turn, wraps around the first helix.[124] Such double-helical regions provide "tie points" for the formation of gels (Fig. 4-10).[104,127,128]

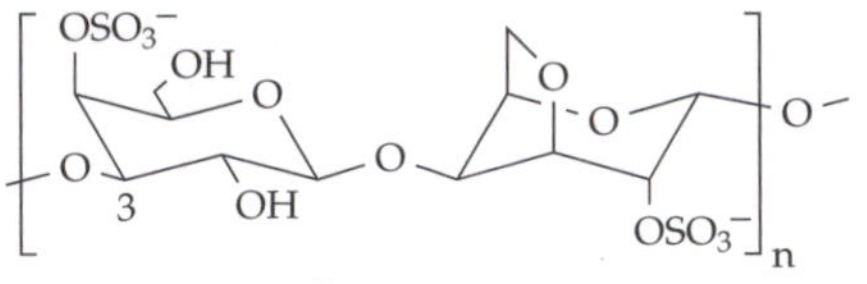

ι-Carrageenan:
[3Gal (4 sulfate) β1→4 (3,6-anhydro) Gal (2-sulfate) α1→]n

Sulfate groups protrude from the structure in

pairs and provide binding sites for calcium ions, which stabilize the gel. The presence of occasional extra sulfate groups in these polymers causes kinks in the chains because the derivatized pyranose rings reverse their conformation to the other chair form. This prevents the entire polysaccharide chain from assuming a regular helical structure.[100,104]

Alginates, found in cell walls of some marine algae and also formed by certain bacteria, consist in part of a linear β-1,4-linked polymer of D-mannuronate with a cellulose-like structure. Alginates also contain

BOX 4-B SILICON: AN ESSENTIAL TRACE ELEMENT

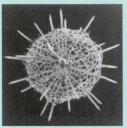

Silica skeleton of a radiolarian[e]

No one can doubt that diatoms, which make their skeletons from SiO_2, have an active metabolism of silicon. They can accumulate as much as 0.7 μM soluble silicon, possibly attached to proteins.[a-d]

The radiolaria and sponges often accumulate silicon and limpets make opal base plates for their teeth. Silicon may account for as much as 4% of the solids of certain grasses. Although silicon is usually not considered an essential nutrient for all plants, there is much evidence that it is essential to some and that it is often beneficial.[f] Silicon is found in soil primarily as silicic acid, H_4SiO_4, whose concentration ranges from 0.1–0.6 mM. Most of the silicon taken up by plants is deposited within cells, in cell walls, between cells, or in external layers as hydrated SiO_2. Presumably the organic components of the plant control the deposition. The SiO_2 in some plants takes the form of sharp particles which may have a defensive function. They abrade the enamel surfaces of the teeth of herbivores and can cause other illnesses.[g]

Silicon is essential for growth and development of higher animals,[h-l] and it has been suggested that humans may require 5–20 mg per day.[m] In the chick, silicon is found in active calcification sites of young bone.[i] Silicon-deficient animals have poorly calcified bones and also an elevated aluminum content in their brains.[m] Silicon is present in low amounts in the internal organs of mammals but makes up ~0.01% of the skin, cartilage, and ligaments, in which it is apparently bound to proteoglycans such as chondroitin-4-sulfate, dermatan sulfate, and heparan sulfate (Fig. 4-11).[m,n] These polymers contain ~0.04% silicon or one atom of silicon per 130–280 repeating units of the polysaccharides. Plant pectins contain about five times this amount. The silicon is appar-

ently bound tightly in *ether* linkage. Perhaps orthosilicic acid, $Si(OH)_4$, reacts with hydroxyl groups of the carbohydrates to form bridges between two chains as follows:

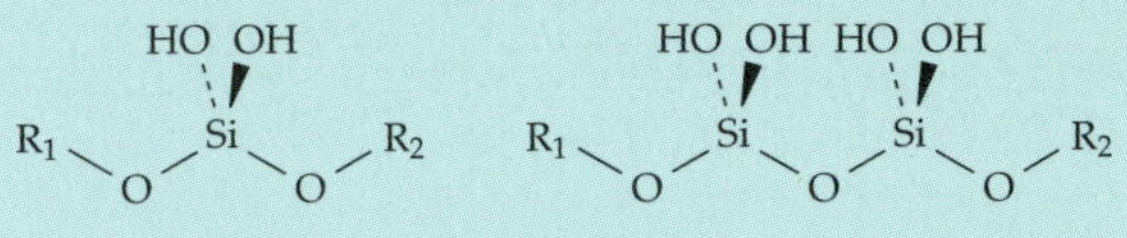

In each of these formulas additional free OH groups are available on the silicon so that it is possible to crosslink more than two polysaccharide chains. Silicon may function as a biological crosslinking agent in connective tissue. **Silaffins**, small polypeptides containing polyamine side chains of modified lysine residues, apparently initiate silica formation from silicic acid in diatoms.[o]

[a] Robinson, D. H., and Sullivan, C. W. (1987) *Trends Biochem. Sci.* **12**, 151–154
[b] Round, F. E. (1981) in *Silicon and Siliceous Structures in Biological Systems* (Simpson, T. L., and Volcani, B. E., eds), pp. 97–128, Springer, New York
[c] Evered, D., and O'Connor, M. (1986) *Silicon Biochemistry*, Wiley, New York
[d] Round, F. E., Crawford, R. M., and Mann, D. G. (1990) *The Diatoms*, Cambridge Univ. Press, Cambridge UK
[e] Buchsbaum, R., Buchsbaum, M., Pearse, J., and Pearse, V. (1987) *Animals Without Backbones*, 3rd ed., Univ. Chicago Press, Chicago
[f] Epstein, E. (1994) *Proc. Natl. Acad. Sci. U.S.A.* **91**, 11–17
[g] McNaughton, S. J., and Tarrants, J. L. (1983) *Proc. Natl. Acad. Sci. U.S.A.* **80**, 790–791
[h] Schwarz, K. (1970) in *Trace Element Metabolism in Animals* (Mills, F., ed), pp. 25–38, Livingstone, Edinburgh, UK
[i] Schwarz, K., and Milne, D. B. (1972) *Nature* **239**, 333–334
[j] Carlisle, E. M. (1972) *Science* **278**, 619–621
[k] Hoekstra, W. H., Suttie, J. W., Ganther, H. E., and Mertz, W., eds. (1974) *Trace Element Metabolism in Animals-2*, University Park Press, Baltimore, Maryland
[l] Carlisle, E. M. (1988) *Science Total Environment* **73**, 95–106
[m] Nielsen, F. H. (1991) *FASEB J.* **5**, 2661–2667
[n] Schwarz, K. (1973) *Proc. Natl. Acad. Sci. U.S.A.* **70**, 1608–1612
[o] Kröger, N., Deutzmann, R., and Sumper, M. (1999) *Science* **286**, 1129–1132

α-L-guluronate, sometimes in homopolymeric "blocks" and sometimes alternating with the mannuronate residues. Groups of adjacent guluronate units are thought to impart calcium-binding properties to alginates.[129]

Polysaccharides with calcium-binding sites may also serve to initiate deposition of calcium carbonate. For example, the unicellular alga *Pleurochrysis carterae* contains an unusual polysaccharide with the following highly negatively charged repeating unit:

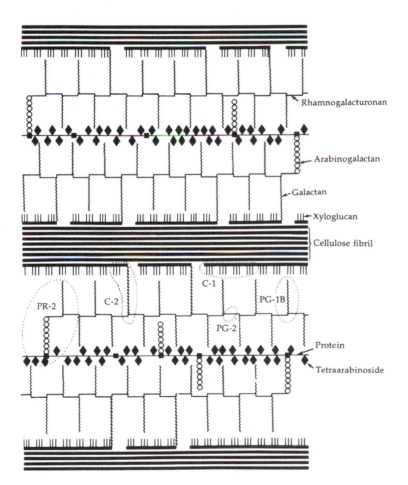

[-4GlcAβ1→2-*meso*-tartrate-3→1-glyoxylate-1→]ₙ

Meso-tartrate is joined to glucuronic acid in glycosidic linkage and by acetal formation to the aldehyde **glyoxylic acid**, HOOC—CHO, which is also joined in an ether linkage to the next repeating unit.[130] Similar **open acetal linkages** join monosaccharide units in some bacterial lipopolysaccharides and may occur more widely.[130a]

Bacteria form and secrete a variety of heteropolysaccharides, several of which are of commercial value because of their useful gelling properties. **Xanthan gum** (formed by *Xanthomonas campestris*) has the basic cellulose structure but every second glucose residue carries an α-1,3-linked trisaccharide consisting of 6-*O*-acetylmannose, glucuronic acid, and mannose in the following repeating unit:[131,132]

$$(-4\text{Glc}\beta1\text{-}4\text{Glc}\beta1\text{-})_n$$
$$3$$
$$\uparrow$$
$$6\text{-}O\text{-Acetyl-Man}\alpha1$$
$$2$$
$$\uparrow$$
$$\text{Man}\beta1\text{-}4\text{GlcA}\beta1$$

The polymer is further modified by reaction of about half of the mannosyl residues with pyruvate to form ketals (Eq. 4-9).

Mannosyl residue

(4-9)

Acetan of *Acetobacter xylinum* has pentasaccharide side chains that contain L-rhamnose.[122] A helical structure for the strands has been observed by atomic force microscopy.

Figure 4-14 Tentative structure of the walls of suspension-cultured sycamore cells. The wall components are in approximately proper proportions but the distance between cellulose elementary fibrils is expanded to allow room to present the interconnecting structure. There are probably between 10 and 100 cellulose elementary fibrils across a single primary cell wall. From Albersheim *et al.*[126]

6. Polysaccharides of Bacterial Surfaces

The complex structure of bacterial cell walls is discussed in Chapter 8. However, it is appropriate to mention a few bacterial polysaccharides here. The innermost layer of bacterial cell walls is a porous network of a highly crosslinked material known as **peptidoglycan** or **murein** (see Fig. 8-29). The backbone of the peptidoglycan is a β-1,4-linked alternating polymer of *N*-acetyl-D-glucosamine and *N*-acetyl-D-muramic acid. Alternate units of the resulting chitin-like molecule carry unusual peptides that are attached to the lactyl groups of the *N*-acetyl-muramic acid units and crosslink the polysaccharide chains. In *E. coli* and other gram-negative bacteria the peptidoglycan forms a thin (2 nm) continuous network around the cell. This "baglike molecule" protects the organism from osmotic stress. In addition, gram-negative bacteria have an outer membrane and on its outer surface a complex lipopolysaccharide. The projecting ends of the lipopolysaccharide molecules consist of long carbohydrate chains with repeating units that have antigenic properties and are called **O antigens**. Specific antibodies can be prepared against these polysaccharides, and so varied are the structures that 1000 different "serotypes" of *Salmonella* are known. These are classified into 17 principal groups. For example, group E3 contains the following repeating unit, where *n* may be ~50 on the average. Rha = L-rhamnose.

$$\begin{array}{c} \text{Glc}\alpha 1 \\ \downarrow \\ 4 \\ (\text{Gal}\beta 1 \rightarrow 6\text{Man}1 \rightarrow 4\text{Rha} \rightarrow)_n \end{array}$$

Polysaccharides of groups A, B, and D contain the repeating unit

$$\begin{array}{c} \text{X} \\ \downarrow \\ (\text{Man} \rightarrow \text{Rha} \rightarrow \text{Gal} \rightarrow)_n \end{array}$$

where X is a 3,6-dideoxyhexose: **paratose** in type A, **abequose** in type B, and **tyvelose** in type D (Fig. 4-15). The existence of the many serotypes depends on the variety of components, on the many types of linkage (α and β, 1→2, 1→3, 1→4, and 1→6) in the repeating units, and on further structural variations at the chain ends.

   At the inner end of the O antigen is a shorter polysaccharide "core" whose structure is less varied than that of the outer ends but which is remarkable in containing two sugars found only in bacterial cell walls: a seven-carbon heptose and an eight carbon α-oxo sugar acid, **ketodeoxyoctonate** (KDO). The structures are given in Fig. 4-15 and the arrangement of these sugars in the *Salmonella* lipopolysaccharide is shown in Fig. 8-30. That figure also shows the manner in which the oligosaccharide that bears the O antigen is attached to a lipid anchoring group that is embedded in the outer membrane of the bacteria.

   A great variety of polysaccharides are present in the outer layers of other types of bacteria. For example, the mycobacteria have an alternating 5- and 6-linked β-D-Gal*f* polymer attached to their peptidoglycan.

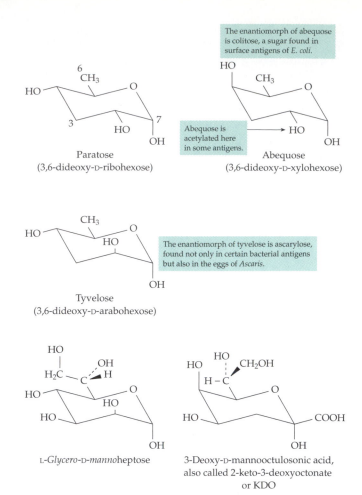

Figure 4-15 Structures of special sugars found in the "antigens" of the outer cell walls of gram-negative bacteria.

Attached to this galactan are branched penta-D-arabinose units:

$$(\text{Ara}f\beta 1 \rightarrow 2\text{Ara}f\alpha 1)_2 \rightarrow 3,5\text{Ara}f\alpha—$$

These are further modified by esterification with mycolic acids.[133] Information on some other cell wall components is given in Chapter 20.

D. Glycoproteins and Proteoglycans

   Many proteins, including almost all of those that are secreted from cells and many that are components of cell surfaces, carry covalently attached oligosaccharides.[134–139] These **glycoproteins** may carry just one or a few, often highly branched, oligosaccharide chains. For example, ribonuclease B has a structure identical to that of ribonuclease A (Fig. 12-25) except for the presence of an oligosaccharide on asparagine 34.[140] In other instances proteins carry a large number of such chains and the carbohydrate may account for over half of the mass of a glycoprotein. Most carbohydrate chains are attached either as O-glycosides with the hydroxyl groups of the side chains of serine, threonine, or other hydroxyamino acid residues or as N-glycosyl groups through linkage to the amide groups of asparagine side chains. Both types of linkage may be present in a single protein. Here are some examples.

Xylβ1→O-Ser(Thr)	Proteoglycans of connective tissue; thyroglobulins
$\begin{array}{c}\varepsilon\\ \rightarrow N\!-\!Lys\\ H\end{array}$	Some dermatan sulfates
Galβ1→O-Hydroxylysine (Hydroxyproline)	Collagen, extension
L-Araα1→O-4-Hydroxy-proline	Plants
GalNAcα1→O-Ser(Thr)	Many glycoproteins
GlcNAcα1→O-Ser(Thr)	Glycoproteins of cytoplasmic surfaces
GlcNAcβ1→NH-βCH$_2$-Asn	Many glycoproteins Some dermatan and heparan sulfates

Linkage of a glycosyl group to a carbon atom of an indole ring of tryptophan has also been demonstrated.[141]

1. O-Linked Glycosyl Groups

In the O-linked glycoproteins the sugar that is attached directly to the protein is usually either xylose, galactose, or N-acetylgalactosamine, all in the pyranose ring form. Xylose is found only in the intercellular **proteoglycans** which carry the chondroitin, dermatan, and related sulfated polysaccharide chains of connective tissues.[141a,b] Since amino sugars are a major constituent, proteoglycans are often called **glycosaminoglycans**. Chondroitin, dermatan, and heparan sulfates are all attached to "core" proteins by the same linkage, which is illustrated here for chondroitin

Repeating unit of chondroitin as in Fig. 4-11	Terminal unit

$$(\rightarrow4GlcAβ1\rightarrow3GalNAβ1)_n\rightarrow4GlcAβ1$$
$$\downarrow$$
$$3Galβ1$$
$$\downarrow$$
$$3Galβ1$$
$$\downarrow$$
$$4Xylβ1\rightarrow O\text{-Ser}$$

Some IdoA may be present in the terminal unit of dermatan sulfate.[142] Keratan sulfate has its own core proteins[143] and has different terminal units including the following:[107]

$$Keratan\rightarrow Gal\rightarrow GlcNAc$$
$$\searrow$$
$$GalNAc\rightarrow O\text{-Ser}$$
$$\nearrow$$
$$Sia\rightarrow Gal$$

The large proteoglycan of human cartilage is built upon the 246 kDa protein **aggrecan**. In the central half of the peptide chain are many Ser-Gly sequences to which about one hundred 10- to 25-kDa chondroitin sulfate chains are attached. About 30 keratan sulfate chains as well as other oligosaccharide groups are also present. These proteoglycan subunits are joined with the aid of 44- to 49-kDa **link protein** to molecules of hyaluronan[144–146] (Fig. 4-16). Several types and sizes of proteoglycan are known.[143,145–147] Dermatan sulfates may be linked to these through either serine or asparagine, depending upon the tissue. The polysaccharide chains of the proteoglycans also bind to collagen fibrils to form a "fiber-reinforced composite material" between cells. Chondroitin and heparan sulfates may be attached at different Ser-Gly sites in a single peptide chain.[148] Degradation of heparan proteoglycans may lead to the shorter free carbohydrate chains found in the circulating heparin. Commercial heparin preparations used as anticoagulants are produced by oxidative destruction of the attached proteins.[113]

A quite different situation holds for **collagen** in which β-galactosyl units and glucosyl-β-galactosyl disaccharide units are attached to side chains of hydroxylysine formed by postsynthetic modification of the original procollagen chain.

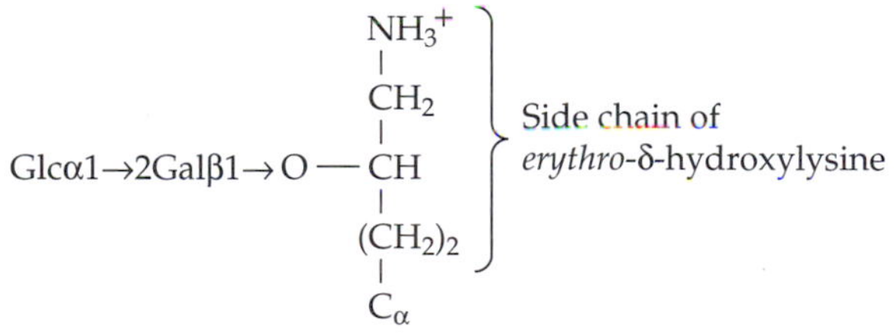

$$Glcα1\rightarrow2Galβ1\rightarrow O\!-\!\!\begin{array}{c}NH_3^+\\ |\\ CH_2\\ |\\ CH\\ |\\ (CH_2)_2\\ |\\ C_\alpha\end{array}\quad\begin{array}{l}\text{Side chain of}\\ erythro\text{-δ-hydroxylysine}\end{array}$$

A great deal of variation in the amount of glycosylation is observed from one species to another. The human α-1(II) chains of collagen usually carry four disaccharides and four monosaccharide units. In the related collagen-like **extensins**, which are found in plant cell walls, the hydroxyproline (Hyp) side chains are O-glycosylated, largely by short oligosaccharides of arabinose in furanose ring form,[150–152] e.g., Ara$f$β1→2Ara$f$β1→2Ara$f$β1→4Hyp. There are as many as 25 repeats of Ser-Pro-Pro-Pro-Pro encoded in an extensin gene. Most of the prolines are hydroxylated and glycosylated. The presence of two or more contiguous proline residues seems to be the signal for the hydroxylation reaction to take place.[152]

Another distinct family of O-linked glycoproteins are the **mucins**, which are present in saliva and other

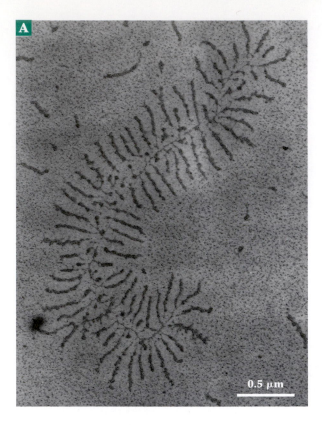

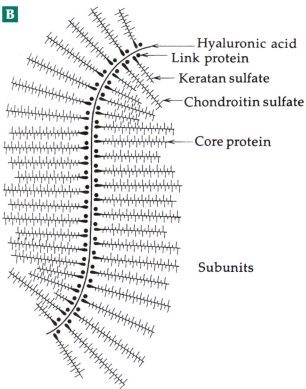

Figure 4-16 (A) Dark field electron micrograph of a proteo-glycan aggregate from bovine articular cartilage (from bearing surfaces of joints). Courtesy of Joseph A. Buckwalter. The filamentous backbone consists of hyaluronic acid, as in (B). The proteoglycan subunits extend from the backbone. From Rosenberg.[149]

mucous secretions. The polypeptides consist largely of serine, threonine, proline, and glycine, with up to one-third of the residues being Ser + Thr. Mucins may contain 70 to 85% carbohydrate linked through *N*-acetylgalactosamine residues to the serine and threonine hydroxyls.[153–160a] Salivary mucins of cows, sheep, and dogs contain largely the following disac-charide, which may contain either *N*-acetyl- or *N*-glycolyl-neuraminic acid:

$$\text{NeuNAc}\alpha2\rightarrow6\text{GlcNAc}\beta\rightarrow\text{Ser(Thr)}$$

Up to 800 disaccharides are present on a single large protein which is composed of disulfide-linked subunits. The many negatively charged sialic acid (*N*-glycolyl-neuraminic acid) groups are thought to cause expan-sion and rigidity which increase the viscosity of the protein. Some other mucins contain predominantly large oligosaccharides, which often carry **blood group determinants**.[161,162]

The ABO(H) family of blood group determinants (Box 4-C) are oligosaccharide groups assembled from D-galactose, *N*-acetyl-D-galactosamine, and L-fucose. They are carried on the nonreducing ends of *O*-linked oligosaccharides which may be attached to cell surface proteins, to mucins, or to the sphingolipid known as ceramide (Chapter 8). Attachment to proteins is usually via GalNAc. An example of a blood type A determi-nant on an *O*-linked oligosaccharide is the following, where the linkage between the galactose and *N*-acetyl-glucosamine may be either β-1,3 or β-1,4:

$$\text{GalNAc}\alpha1\rightarrow3\text{Gal}\beta1\rightarrow3(4)\text{GlcNAc}\beta1\rightarrow3\text{GalNAc}\rightarrow O\text{-Ser(Thr)}$$
$$\underset{\underset{\text{Fuc}\alpha1}{\uparrow}}{2}$$

2. Asparagine-Linked Oligosaccharides

In many glycoproteins oligosaccharides are linked through *N*-acetylglucosamine to the side chain of asparagine.[134,163,164]

This structure also illustrates one of the hydrogen-bonding possibilities through which the sugar can interact with the protein. Since the site of glycosylation is often at β bends in the surface of the protein, the amide groups of the N-acetylglucosamine may alternatively hydrogen bond to amide groups of the peptide.[165] This asparagine linkage is very common. For example, it is present in 97% of glycoproteins of blood plasma[166] and it is also predominent in the glycoproteins of tissue surfaces.[167] There are numerous structures for asparagine-linked oligosaccharides but many contain the following core to which additional glycosyl groups may be attached:

$$Y_3\alpha1$$
$$\downarrow$$
$$6$$
$$Y_2\alpha1\rightarrow3Man\alpha1 \qquad\qquad \pm Fuc\alpha1$$
$$\searrow 6 \qquad\qquad\qquad \downarrow 6$$
$$Man\beta1\rightarrow4GlcNA\beta1\rightarrow4GlcNA\beta1\rightarrow Asn$$
$$\nearrow 3$$
$$Y_1\alpha1\rightarrow2Man\alpha1$$

Here the ±Fuc indicates that this residue is present only on some of the chains.

Notice the three mannose residues on the left side. In the **high mannose type** oligosaccharide, Y_1, Y_2, and Y_3 are additional mannose units. In many instances $Y_1 = Y_2 = Y_3 = Man\alpha\rightarrow Man$. These "extra" mannose units are put onto the oligosaccharide during the original biosynthesis and before it is attached to the protein (Chapter 20). Some of the mannose units may then be removed during the "processing" of the oligo-

saccharide in the endoplasmic reticulum and residues of glucosamine, galactose, and sialic acid (Sia) may be added. Thus, Y_1 and Y_2 in the foregoing structure often become

$$Sia\alpha2\rightarrow3(6)Gal\beta1\rightarrow4GlcNAc\beta1\rightarrow$$

Here, the sialic acid may be linked either 2,3 or 2,6 and the GlcNAc either 1,2 or 1,3. Both Y groups may consist of trisaccharides of this type in "biantennary" oligosaccharides and a third trisaccharide (Y_3) may be added to form a "triantennary" molecule. An additional N-acetylglucosamine is often linked by β-1,4 linkage to the central mannose of the core and fucosyl residues in α-1,6 linkage are often linked to the N-acetylglucosamine next to the asparagine.[168] The Galβ1→4GlcNAc disaccharide unit in the above Y group is also called **N-acetyllactosamine** because of its relationship to lactose. It is often repeated in long Y groups, e.g., as $(\rightarrow3Gal\beta1\rightarrow4Glc-NAc\beta1—)_n$, the oligosaccharides being called poly-N-acetyllactosamino-glycans. These structures are also principal carriers of the ABO blood type determinants on erythrocyte surfaces.[169]

Many N-linked oligosaccharides are highly branched. For example, in ovomucoid, a protease inhibitor of hen eggs, "pentaantennary" oligosaccharides have two and three N-acetylglucosamine rings, respectively, attached to the terminal mannose units of the oligosaccharide core in 1,2, 1,4 and 1,6 linkages. Another large

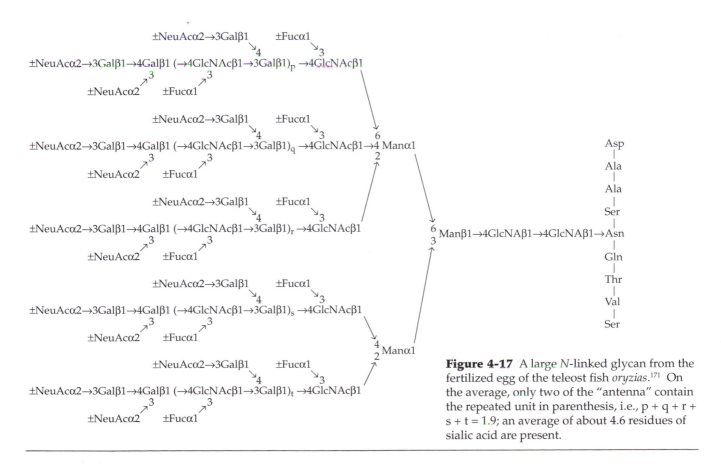

Figure 4-17 A large N-linked glycan from the fertilized egg of the teleost fish *oryzias*.[171] On the average, only two of the "antenna" contain the repeated unit in parenthesis, i.e., p + q + r + s + t = 1.9; an average of about 4.6 residues of sialic acid are present.

BOX 4-C THE BLOOD GROUP DETERMINANTS

The role of carbohydrates in biological communication is well illustrated by the human blood types.[a,b] According to the ABO system first described by Landsteiner in 1900, individuals are classified into types A, B, AB, and O. Blood of individuals of the same type can be mixed without clumping of cells, but serum from a type O individual contains antibodies that agglutinate erythrocytes of persons of types A and B. Serum of persons of type B causes type A cells to clump and vice versa. Individuals of none of the four types have antibodies against type O erythrocytes. For this reason, persons with type O blood are sometimes inaccurately described as "universal donors."

The ABO blood types are determined by specific **blood group determinants** which are attached to the nonreducing ends of *O*-linked oligosaccharides of surface glycoproteins, mucins, glycolipids, and, to a lesser extent, *N*-linked oligosaccharides. The blood group determinants are found on erythrocytes and all endothelial cells of the body. In about 80% of the population they are also present on glycoproteins of the saliva and other secretions.

The minimal determinant structures, attached to "carrier" R, are as follows:

$$\begin{array}{l}
\text{GalNAc}\alpha1\rightarrow3\ \text{Gal}\beta2\rightarrow\text{R} \qquad \text{A determinant}\\
\phantom{\text{GalNAc}\alpha1\rightarrow3\ \text{Gal}}2\\
\phantom{\text{GalNAc}\alpha1\rightarrow3\ \text{Gal}}\uparrow\\
\phantom{\text{GalNAc}\alpha1\rightarrow3\ \text{Gal}}\text{Fuc}\alpha1
\end{array}$$

$$\begin{array}{l}
\text{Gal}\alpha1\rightarrow3\ \text{Gal}\beta1\rightarrow\text{R} \qquad \text{B determinant}\\
\phantom{\text{Gal}\alpha1\rightarrow3\ \text{Gal}}2\\
\phantom{\text{Gal}\alpha1\rightarrow3\ \text{Gal}}\uparrow\\
\phantom{\text{Gal}\alpha1\rightarrow3\ \text{Gal}}\text{Fuc}\alpha1
\end{array}$$

$$\begin{array}{l}
\text{Gal}\beta1\rightarrow\text{R} \qquad \text{H(O) determinant}\\
2\\
\uparrow\\
\text{Fuc}\alpha1
\end{array}$$

Here R refers to a "carrier oligosaccharide" which can be as simple as

→3(4)GlcNAcβ1→3Galβ1→4Glcβ1—O—Ser (Thr, ceramide)

or may be a complex lactosaminoglycan or an oligosaccharide such as that in Fig. 4-17. The minimal determinants can be linked to the carrier oligosaccharide by either a β-1,3 (type I chain) or β-1,4 (type II chain) glycosidic bond.

The genetic basis for the ABO blood groups is well understood. There are three **alleles**, variants of a gene, that encodes a **glycosyltransferase**. In A type individuals, this enzyme transfers *N*-acetylgalactosamine from a carrier molecule, called UDP

(Chapter 17), onto the terminal positions of the H(O) determinant. The enzyme specified by the *B* allele transfers galactose. The two enzymes differ in only four amino acid residues but the result is an altered substrate specificity.[a,c,d] The *O* allele produces inactive enzyme as a result of a single base deletion in the gene.[a] The H gene has been identified as that of an α-1,2 fucosyltransferase that transfers α-L-fucose from the carrier GDP to the galactose unit in the foregoing structure.[e] Persons with an inactive *H* transferase gene may have the rare type I, which results from addition of glucosamine branches to the repetitive H antigen structures of polylactosaminoglycans.[f] More often though, the H antigen

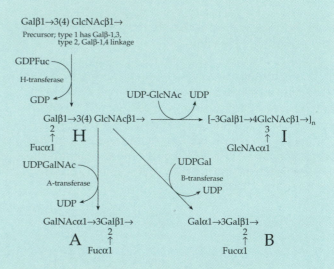

is acted upon by a different fucosyltransferase encoded by the *Le* gene, which determines the **Lewis blood group**. This enzyme places α-1,4-linked L-fucose onto the H antigen to give the *Le*[b] antigen. The same fucosyltransferase (which is

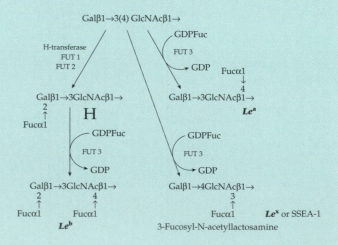

BOX 4-C Continued

designated FUT 3) acting on type 2 precursor chains forms the Le^a antigen. The same transferase adds a β-1,3-linked fucose to type 2 chains to give the Le^x antigen, which is also called SSEA-1.[g,h] Lactosamine-type precursors carrying α-2,3-linked sialic acid at their nonreducing ends on surfaces of granulocytes, monocytes, and natural killer cells can be converted to sialylated Le^x (s-Le^x).[i,j]

$$NeuAc\alpha2 \rightarrow 3Gal\beta1 \rightarrow 4GlcNAc \rightarrow$$
$$\underset{\uparrow}{\overset{2}{}}\qquad \textbf{\textit{s-Le}}^{\textbf{x}}$$
$$Fuc\alpha1$$

Individuals with active gene Se (for secretion) secrete glycoproteins bearing the blood group substance into saliva and other body fluids. The Se gene is another H-transferase present in epithelial cells and salivary glands. In individuals with active Se genes (~80% of most populations) most soluble H antigens are converted to Le^b oligosaccharides.[k]

There are at least 12 other well established human blood groups, several of which involve oligosaccharides attached to glycoproteins or glycolipids. The MN antigens consist of a sequence of amino acids near the N terminus of the protein glycophorin (Chapter 8) with attached sialic acid-containing O-linked oligosaccharides,[l] e.g., as in the following structure:

$$NeuAc\alpha2$$
$$\downarrow$$
$$6$$
$$NeuAc\alpha2 \rightarrow 3Gal\beta1 \rightarrow 3GalNAc \rightarrow O—Ser \ (Thr)$$
$$\underset{\uparrow}{\overset{4}{}}$$
$$GalNAc\beta1 \ \} \ \text{in Cad antigen}$$

The Cad antigen, also found on glycophorin as well as on gangliosides, has an additional β-1,4-linked GalNAc at the indicated position.[m] The Kell blood group antigens are carried on a 93 kDa glycoprotein of erythrocyte surfaces.[n] The P blood group depends upon surface carbohydrates such as the tetrasaccharide part of the ganglioside called globoside; see Fig. 20-9. Its characteristic antigenic activities are destroyed by treatment with dilute periodate.

Other antigens, including those of the Rh and LW groups, are represented by exposed parts of proteins on the erythrocyte surfaces.[o]

[a] Yamamoto, F., Clausen, H., White, T., Marken, J., and Hakomori, S. (1990) *Nature* **345**, 229–233

[b] Frevert, J., and Ballou, C. E. (1985) *Biochemistry* **24**, 753–759

[c] Feizi, T. (1990) *Trends Biochem. Sci.* **15**, 330–331

[d] Yamamoto, F., and Hakomori, S. (1990) *J. Biol. Chem.* **265**, 19257–19262

[e] Larsen, R. D., Ernst, L. K., Nair, R. P., and Lowe, J. B. (1990) *Proc. Natl. Acad. Sci. U.S.A.* **87**, 6674–6678

[f] van den Eijnden, D., Koenderman, A., and Schiphorst, W. (1988) *J. Biol. Chem.* **263**, 12461–12471

[g] Mollicone, R., Reguigne, I., Kelly, R. J., Fletcher, A., Watt, J., Chatfield, S., Aziz, A., Cameron, H. S., Weston, B. W., Lowe, J. B., and Oriol, R. (1994) *J. Biol. Chem.* **269**, 20987–20994

[h] Nishihara, S., Narimatsu, H., Iwasaki, H., Yazawa, S., Akamatsu, S., Ando, T., Seno, T., and Ikuyo, N. (1994) *J. Biol. Chem.* **269**, 29271–29278

[i] Natsuka, S., Gersten, K. M., Zenita, K., Kannagi, R., and Lowe, J. B. (1994) *J. Biol. Chem.* **269**, 16789–16794

[j] Murray, B. W., Wittmann, V., Burkart, M. D., Hung, S.-C., and Wong, C.-H. (1997) *Biochemistry* **36**, 823–831

[k] Kelly, R. J., Rouquier, S., Giorgi, D., Lennon, G. G., and Lowe, J. B. (1995) *J. Biol. Chem.* **270**, 4640–4649

[l] Adamany, A., Blumenfeld, O., Sabo, B., and McCreary, J. (1983) *J. Biol. Chem.* **258**, 11537–11545

[m] Gillard, B., Blanchard, D., Bouhours, J.-F., Cartron, J.-P., van Kuik, J. A., Kamerling, J., Vliegenthart, J., and Marcus, D. (1988) *Biochemistry* **27**, 4601–4606

[n] Redman, C., Avellino, G., Pfeffer, S., Mukherjee, T., Nichols, M., Rubinstein, P., and Marsh, W. L. (1986) *J. Biol. Chem.* **261**, 9521–9525

[o] Bloy, C., Hermand, P., Blanchard, D., Cherif-Zahar, B., Goossens, D., and Cartron, J.-P. (1990) *J. Biol. Chem.* **265**, 21482–21487

N-linked glycan is pictured in Fig. 4-17. Like many others, it has a number of sialic acid residues at the nonreducing ends and also contains N-acetyllactosamine units. The major component of the cell walls of yeast (*S. cerevisiae*) is a mannoprotein that carries long N-linked oligosaccharides with highly branched outer chains of over 100 mannose residues[170] (see also Section C,3).

The special importance of sialic acids in glycoproteins may lie in the negative electrical charges and the resultant Ca^{2+}-binding properties that they impart to cell surfaces. Some glycoproteins carry sulfate or phosphate groups that have similar effects. The glycoproteins of the slime mold *Dictyostelium* contain both mannose-6-P and mannose-6-sulfate residues.[163] Oligosaccharides on the protein subunits of flagella of halobacteria contain sulfate esters of glucuronic acid.[172] Some N-linked oligosaccharides carry chains of keratan sulfate in place of the sialic acid in the Y_1 and Y_2 groups of the oligosaccharide.[173]

3. Glycoproteins in Biological Recognition

The clusters of sugar rings that form the oligosaccharides on glycoproteins play a vital role in many aspects of biological recognition.[139,174–176] A good example is provided by the human blood groups whose characteristics are determined largely by oligosaccharides (Box 4-C). The adhesion of viruses, bacteria, and eukaryotic parasites to cell surfaces and of one cell to another in multicellular organisms is also dependent on carbohydrates. Recently, it has become clear that the oligosaccharides of cell surfaces change during growth and development and provide an important mechanism by which cells can recognize each other and respond. Why are carbohydrates used for this purpose? It has been pointed out by Sharon and Lis[175] that four different nucleotides can make only 24 distinct tetranucleotides but that four different monosaccharides can make 35,560 unique tetrasaccharides.

Lectins and other carbohydrate-binding proteins.

Much of biological carbohydrate-dependent recognition is a result of interaction of individual glycosyl groups or of oligosaccharides on a glycoprotein with a second protein. In some cases that protein is referred to as a **receptor**; in other cases the glycosyl groups may be called the receptor. Carbohydrate-binding proteins include antibodies, enzymes, and carriers that help sugars to cross cell membranes. In addition, there is a large group of carbohydrate binding proteins called **lectins** (from the latin *lectus*; to select).[24,174,177,178]

The first lectins discovered were proteins of plant seeds with specific sugar-binding properties and the ability to agglutinate erythrocytes. **Ricin**, a very toxic protein from castor beans, was isolated in 1888.[178,179] Perhaps the best known lectin is **concanavalin A**, a protein crystallized by Sumner in 1919.[180] Concanavalin A makes up 2–3% of the protein of the jack bean. It is one of a family of legume lectins that resemble favin, whose structure is shown in Fig. 2-15.[177,181,182] Many lectins, including ricin, have quite different three-dimensional structures but share the common characteristic of having a selective binding site for one or more glycosyl rings. Concanavalin A binds to α-D-mannopyranose or α-D-glucopyranose with unmodified hydroxyl groups at C-3, C-4, and C-6.[183,184] Tighter binding is observed if additional mannose residues are present in an oligosaccharide.[178,183] The protein also has specific binding sites for Ca^{2+} and for a transition metal ion such as Mn^{2+}. Soybean lectin binds D-*N*-acetylgalactosamine and D-galactose units, while wheat lectin is specific for D-*N*-acetylglucosamine.

Animal cells also produce lectins.[24,105,185–193] The amebas of the cellular slime mold *Dictyostelium* synthesize a classical lectin called **discoidin I** that binds GalNAc or Gal. It is absent from cells until they are ready to differentiate into an aggregating form (Chapter 1). Then it is produced in abundance.[178] Discoidin I has a second binding site specific for the peptide sequence Arg-Gly-Asp (RGD) which is known to be involved in cell adhesion and which binds to such surface proteins as fibronectin and laminin (Chapter 8). Many animal tissues contain soluble lactose-binding lectins known as **galectins** or S-Lac lectins. The best known member is a dimer of 14-kDa subunits;[187,188] many other related lectins have been found.[189,192] Another family are Ca^{2+}-dependent or C-type lectins which are specific for mannose, L-fucose, or other sugars.[105,193] **Lectin domains** are being discovered in many proteins.

Carbohydrate-binding sites.

The structures of the sites that recognize and bind sugar rings in lectins, enzymes, transporter proteins, and other carbohydrate-binding proteins vary greatly, as does the tightness of binding. However, there are certain common features: Sugar rings are bound by hydrogen bonds, which are often numerous. An example is the galactose chemoreceptor protein from *E. coli*. It binds both α and β anomers of either D-glucose or D-galactose and is utilized by the bacteria in searching for food (see Chapter 19). The structure of D-glucose bound to this protein is shown in Fig. 4-18. Notice the many hydrogen bonds. Two of the –OH groups have the maximum of three hydrogen bonds apiece. There are three negatively charged aspartate side chains and one positively charged guanidinium group. These provide strong ion–dipole interactions which add strength to the bonds. The presence of ionized groups in varying numbers and constellations is another common feature of protein–carbohydrate interactions. A third common feature is the presence of aromatic rings, which often lie against one face of the sugar. The stereoscopic drawing of Fig. 4-18A shows an indole ring of a tryptophan residue in front and a phenylalanine side chain behind the sugar.[194] Sugars bind to lectins,[181,184,195] to enzymes (Chapter 12), and to antibodies[196] through similar interactions.

Binding of viruses and bacteria to cells.

The cholera toxin and a related toxic protein from *E. coli* bind to Sia→Gal groups attached to glycolipids (gangliosides, Chapter 8) of erythrocytes and other cells.[197] The influenza virus gains access to our body cells by first binding through a viral surface **hemagglutinin**. This is a protein that binds specifically to NeuAcα2→6Gal or NeuAcα2→3Gal of cell surface oligosaccharides.[174,175,198] Removal of sialic acid from erythrocyte surfaces abolishes the ability of the influenza and some other viruses to bind. It seems somewhat surprising that a second surface protein on the influenza virus is a **neuraminidase** (or **sialidase**) which catalyzes the removal of sialic acid from cell surface proteins, destroying unoccupied virus receptors.[199,200] This may facilitate movement of a virus particle through the mucin layer surrounding a cell. Bacteria and other invading parasites also produce neuraminidases.[201] *Trypanosamas cruzi*, the causative agent of chagas disease, employs a **transsialidase** to transfer sialic acid from a Galβ on the host cell onto a protein on the parasite surface. This is essential for successful invasion of the host.[202]

The adherence of cells of *E. coli* to mannosyl units of cell surface proteins may initiate the infections that sometimes occur with this bacterium.[174,203] However, cells of *E. coli* from strains that cause urinary infections bind to Galα1→4Gal on glycolipids that carry the blood group P antigens (Box 4C).[174,204] Neuropathogenic strains of *E. coli* or of *Neisseria meningitidis*, which may cause neonatal meningitis, bind to α-2,8-polysialic acid chains on nerve cells.[135,205,206] *Helicobacter pylori*, the stomach ulcer bacterium, binds to the human Lewis[b] blood group antigen (Box 4-C).[207] *Entamoeba histolytica*, which causes amebic dystentery, binds to Gal and GalNAc-containing oligosaccharides such as GalNAcα1→3Galα.[208]

Aggregation and adherence of cells.

Differences among cell surfaces are fundamental to the formation of multicellular organisms and to many physiological processes. Proteins, carbohydrates, and lipids all contribute material to exposed cell surfaces. The adhesion of one cell to another is mediated by a group of **adhesion proteins** together with oligosaccharide groups and sometimes polysaccharides. Families of adhesion proteins include **integrins** and **cadhedrins** (discussed in Chapter 8), various members of the **immunoglobulin superfamily**, the **cell differentiation antigens**, (often designated CD44, etc.), the C-type lectins known as **selectins**, and proteoglycans.

Here are two of many known examples of specific cell–cell adhesion. The species-specific reaggregation of dissociated cells of marine sponges (Chapter 1) depends upon a 20-kDa proteoglycan of unique structure[209–211] together with a cell surface receptor protein and calcium ions. The recognition of egg cell surfaces by sperm[212–214] is species specific and depends upon interaction of sperm receptors with *O*-linked oligosaccharides of the extracellular coat of the ovum.

Growth and differentiation. The exact structures of the oligosaccharides and polysaccharides of cell surfaces vary not only with cell and tissue type but also with the position of a cell and with time. Actions of numerous glycosyl transferases alter these saccharide groups as an organism grows and develops. Other enzymes alter them by hydrolytic removal of sugars, by isomerizaion, oxidation, and addition of other components such as phospho, sulfo, and acetal groups. For example, the presence of the H-antigen determinant, whose structure is shown in Box 4-C, is strictly regulated, both temporally and spatially, during vertebrate development.[215,216] The relative amounts of the H determinant vs Le[x], sLe[x] (Box 4-C)

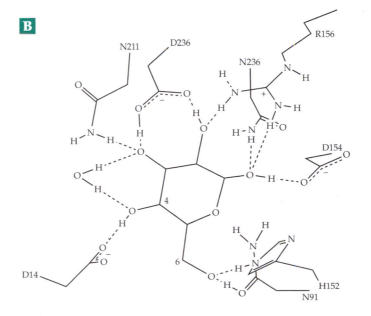

Figure 4-18 (A) Stereoscopic view of the interactions between the *E. coli* galactose chemoreceptor protein and a bound molecule of D-glucose. (B) Schematic drawing showing many of these interactions and the state of ionization deduced for the aspartate and arginine side chains. From Vyas *et al*.[194]

and other surface groupings is controlled by fucosyl-transferases, sialyltransferases, etc. Human cancers often accumulate large amounts of fucosylated glyco-proteins and glycolipids carrying Le[a], Le[x], and sialyl-Le[x] antigens[175,216,217] and sialomucins.[218] Glycoproteins help to control many metabolic processes. For example, a protein **calnexin** (a chaperonin, Chapter 10) helps glycoproteins to fold correctly.[219] On the other hand, removal of terminal sialic acid residues from blood plasma proteins leads to rapid removal of the proteins from circulation and to catabolism by liver cells.[175] This process depends upon a receptor protein specific for oligosaccharides with terminal galactosyl residues (Chapter 20).

Recognition and adhesion by leukocytes. A group of three calcium-dependent lectins known as **selectins** bind the sialyl Lewis x (sLe[x]) antigen and play important roles in adhesion to cells of the vascular endothelium and leukocytes[175,220–224] and also to platelets. Although all of the selectins bind the sLe[x] antigen, the binding is weak and these multidomain proteins may simultaneously bind to other ligands such as heparan sulfate.[225] The leukocyte L-selectin (CD62L) binds very tightly to the 3'-sialyl-6'-sulfo-Le[x] determinant[220,226] which occurs on mucin-like glyco-proteins. The interaction with P selectin helps leuko-cytes to bind to surfaces on endothelial cells in lymph nodes and sites of chronic or acute inflammation. In a similar way E selectin is synthesized in vascular endo-thelial cells that have been transiently "activated" by cytokines in response to injury and other inflammatory stimuli. The E selectin binds the sulfated sLe[x] antigen on surfaces of neutrophils, monocytes, eosinophils, and basophils.[220,222,227–230] P selectin is stored in secre-tory granules of platelets and endothelial cells and is released to the cell membrane upon activation by thrombin.[220]

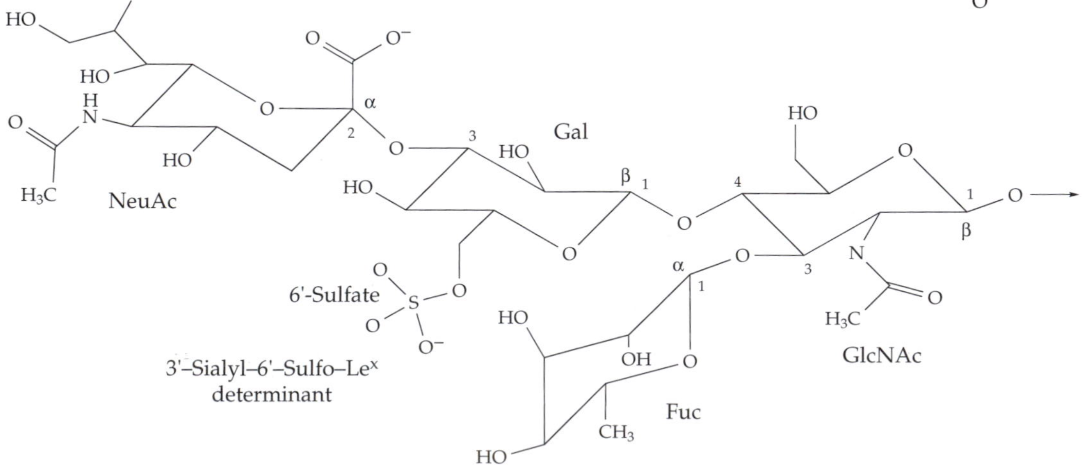

E. Some Special Methods

Small monosaccharides and oligosaccharides can be separated readily from polymeric constituents and can be purified further by chromatographic procedures including gel filtration as is illustrated in Fig. 3-3. However, polysaccharides and complex oligosaccha-rides are harder to purify. A few of these, such as cellulose and glycogen, are sufficiently stable that other materials can be dissolved away from them by prolonged boiling in strongly basic solutions. Complex carbohydrates are usually cut into smaller oligosaccha-rides or glycopeptides. These may be separated by HPLC, capillary electrophoresis, or thin-layer chroma-tography[134,231–235] (Fig. 4-19) or by chromatography on immobilized lectins.[234] Quantities of less than 25 pico-moles can be separated by use of mass spectrometry with liquid chromatography.[236,237] High resolution Fourier transform mass spectroscopy is very useful in the study of posttranslational glycosylation of proteins.[237a,b]

1. Release of Oligosaccharides from Glycoproteins

The *O*-linked oligosaccharides of glycoproteins or glycolipids can be split off from the proteins by β elimination (see also Chapter 13):

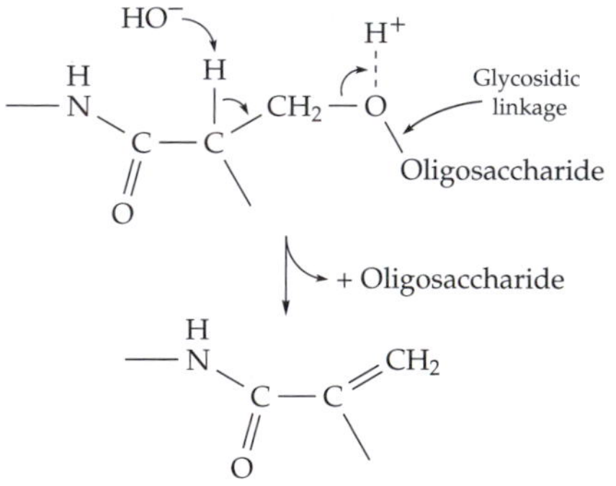

(4-10)

6'-Sulfate

3'–Sialyl–6'–Sulfo–Le[x] determinant

NeuAc Gal GlcNAc Fuc

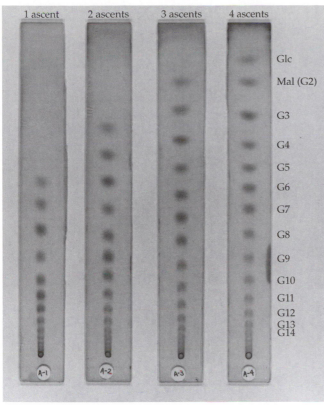

Solvent: MeCN/EtOAc/PrOH-1/H₂O
             85   20   50   50

Figure 4-19 Ascending thin-layer chromatography of a mixture of maltooligosaccharides obtained by the hydrolysis of linear starch. G2 (maltose), G3 (maltotriose), G4 (malto-tetraose), G5 (maltopentaose) – G14 represent oligosaccharides with the indicated number of glucose residues, all in α-1,4 linkage. In the multiple ascent technique the chromatographic solvent, whose composition (by volume) is indicated, is allowed to ascend the thin-layer plate repeatedly with the plate allowed to dry between ascents. The separation of the higher oligosaccharides is distinctly improved by a larger number of ascents. Photograph courtesy of John Robyt.

Treatment with 0.1–0.5 M NaOH for several hours will completely liberate the oligosaccharides, whose released carbonyl groups may then be reduced with NaBH₄, NaB²H₄, or ³H-labeled borohydride to form stable sugar alcohols (Eq. 4-2).[238] Asparagine-linked oligosaccharides are often recovered as glycopeptides prepared by complete proteolytic digestion of the denatured glycoprotein. These can be separated by high voltage electrophoresis in borate buffers.[239] The oligosaccharides can be released from the glycopeptides by enzymes such as endo-N-acetylglucosamini-dase[240] or glycopeptidyl amylase of almonds.[241,242] These release the oligosaccharide as a 1-amino derivative.

When a dry glycoprotein is heated at 105° with anhydrous hydrazine for 8–12 h all of the asparagine-

linked oligosaccharide chains are released (Eq. 4-11).[239,243–245] In addition to the glycosylamine product of step *a*, the corresponding 1–OH and –NH–NH₂ compounds are also formed. These are all converted to the stable acetyl derivative (step *b*), after which the oligosaccharides may be characterized by mass spectrometry and high-field NMR spectoscopy.[234,246–248] Crystallization of oligosaccharides in complex mixtures is difficult to impossible.

2. Hydrolysis

Most glycosidic linkages are hydrolyzed readily by heating with 1 N mineral acids. The mechanism of the hydrolytic reaction is similar to that employed by the enzyme lysozyme (Chapter 12). Some linkages are unusually sensitive to acid and a few are very resistant. Thus, a variety of conditions may be applied for partial acid hydrolysis as an aid to characterization.[249] Acetolysis, cleavage by acetic anhydride, is also of value.[250] A battery of hydrolytic enzymes specific for sugars that are joined in a given type of glycosidic linkage are available.[251] These are useful in determining sequences of oligosaccharides released from glycoproteins. Radioactive tracer techniques can also be applied.[252]

3. Methylation

An important general method in characterization of carbohydrates is the classical **exhaustive methylation** (permethylation). Repeated treatment with a methylating agent such as methyl iodide converts all free OH groups to OCH₃ groups. Then, complete acid

$$(4\text{-}11)$$

hydrolysis, followed by separation of the methylated sugars and their quantitative determination, reveals the relative amounts of **end units** (containing four methoxyl groups), straight **chain units** (containing three methoxyl groups), and **branch points** (containing two methoxyl groups). In addition, the structure of the methylated derivatives provides information on the positions of the linkages in the sugar rings.

After a methylated polysaccharide is subjected to partial hydrolysis, the newly exposed hydroxyl groups (or those created by borohydride reduction of carbonyl groups) can be labeled by ethylation or propylation. One procedure for sequencing complex carbohydrates makes use of high-resolution reversed-phase liquid chromatography to separate the many alkylated oligosaccharides produced by methylation followed by partial acid hydrolysis, reduction, and ethylation.[233] Reductive cleavage of the glycosidic linkages in methylated polysaccharides allows uniquivocal determination of ring size.[253] Branch points may be located by methanolysis of the permethylated polysaccharide followed by conversion of free –OH groups to *p*-bromobenzoate esters. The latter are separated and the circular dichroism (Chapter 23) is measured. Mass spectrometry has also been applied successfully.[246] A simple procedure that can be conducted in any laboratory using thin-layer chromatography is illustrated in Fig. 4-20.[254]

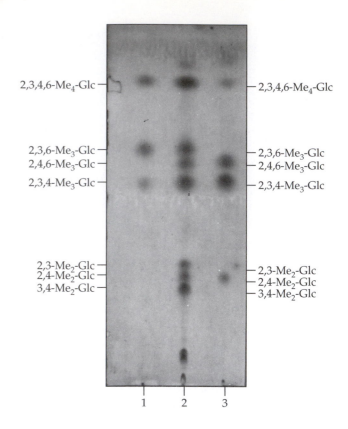

Figure 4-20 Separation of a mixture of *O*-methylated glucoses by ascending thin-layer chromatography. Whatman K6 TLC plates were used with two ascents of the solvent acetonitrile/chloroform/methanol in the ratio 3/9/2, V/V/V. Courtesy of John Robyt.

4. Periodate Oxidation (Smith Degradation)

One of the most important reagents in investigations of carbohydrate structure is periodic acid (or sodium periodate).[255] This reagent oxidatively cleaves C–C bonds bearing adjacent OH or NH_2 groups to form dialdehydes (Eq. 4-12). The method is quantitative. After some hours of reaction, excess periodate not consumed in the oxidation can be determined. If three consecutive carbon atoms bear hydroxyl or amino groups, formic acid is liberated from the central atom and can also be measured quantitatively. After destruction of excess periodate the dialdehyde can be reduced by addition of solid sodium borohydride to form stable CH_2OH groups. Following mild acid hydrolysis to split the acyclic acetal linkages, the fragments can be separated and identified. The sequence of reactions is known as the Smith degradation. If sodium borotritide (NaB^3H_4) is used for the reduction the fragments will be radioactive and can be located on chromatograms by fluorography. Periodate oxidation can also be used to alter all surface oligosaccharides.[99] Removal of *O*-linked oligosaccharides from glycoproteins can be accomplished by periodate oxidation and alkaline β elimination from the dialdehyde product of Eq. 4-12 if OR′ is part of a serine or other hydroxyamino acid side chain in a protein.[256]

5. Nuclear Magnetic Resonance

As in other areas of biochemistry, NMR has become an extremely important tool in carbohydrate research. The mixtures of anomers of various ring

$$(4\text{-}12)$$

BOX 4-D ANTIFREEZE AND ICE-NUCLEATION PROTEINS

Fish living in Arctic and Antarctic waters may encounter temperatures as low as −1.9°C. The freezing point depression provided by dissolved salts and proteins in the blood is insufficient to protect the fish from freezing. As winter approaches, they synthesize and accumulate in their blood serum a series of eight or more special antifreeze proteins.[a–d] One type of antifreeze glycoprotein from winter flounder contains the following unit repeated 17–50 times.

$$(-Ala-Ala-Thr-)_n$$
$$|$$
$$GalNAc\alpha1 \rightarrow O$$
$$3$$
$$\uparrow$$
$$Gal\beta1$$

Destruction of the galactosyl residues by oxidation with periodate, acetylation of the free hydroxyl groups of the oligosaccharides, or their removal by β elimination all lead to loss of antifreeze activity.

The same fish contain a second series of alanine-rich antifreeze polypeptides that are *not* glycosylated but which exist as amphipathic helices. One of these (Type I) contains ~40 residues in a single helix.[e–i] A third family of antifreeze proteins (Type II), found in the sea raven are globular proteins, rich in cysteine and β structures. They are members of the lectin family.[d,j] A fourth type (Type III) found in the sea pout and some other fishes are 62- to 66-residue globular proteins containing an orthogonal β sandwich structure.[k,l] Messenger RNA molecules coding for the antifreeze proteins are found in the livers of flounder in the winter but are absent in the summer.[m]

Antifreeze proteins, that are 3 – 4 times as effective as those in fish, have been isolated from some insects and other arthropods.[m,n,o] They help beetle larvae to overwinter.[m] The insect proteins have a parallel β helix structure resembling that in Fig. 2-17 and stabilized by S — S bridges.[o,p] Some plants also synthesize antifreeze proteins.[n,q,r] One of these, isolated from carrots, is a member of the leucine-rich-repeat family.[q]

How do antifreeze proteins work? The major effect is to greatly slow the freezing rather than to decrease the freezing point. The proteins apparently accomplish this by binding to the surfaces of small ice crystals and preventing their growth.[d,f,h,i,k,l,s,t] This provides the fish with enough time for the blood to pass back into the liver, in which a high enough temperature is maintained to melt any microcrystals before the blood again circulates through the colder tissues. Some of the proteins have clusters of polar side chains that bind to specific faces of the ice crystals and inhibit growth.[s]

A few fishes tolerate a high internal osmotic pressure and accumulate glycerol in their blood up to a concentration of 0.4 M.[t] Insects may accumulate up to 3 M glycerol and some species utilize various other cryoprotectants, such as mannitol, sorbitol, erythritol, threitol, trehalose, glucose, fructose, proline, and alanine.[u] Some amphibians and reptiles can survive freezing and recover fully. For the most studied wood frog, rapid freezing is fatal, but slow freezing leaves the frog, whose heart ceases to function, with a 200-fold increased glucose concentration and a decreased water content in its organs. It resumes normal activities within 14–24 h of thawing.[t]

Having an effect opposite to that of the antifreeze proteins are surface proteins of some bacteria of the genera *Pseudomonas*, *Erwinia*, and *Xanthomonas*. These proteins provide nuclei for growth of ice crystals from supercooled water.[v,w]

a Feeney, R. E., Burcham, T. S., and Yeh, Y. (1986) *Ann. Rev. Biophys. Biophys. Chem.* **15**, 59–78
b Eastman, J. T., and DeVries, A. L. (1986) *Sci. Am.* **255** (Nov), 106–114
c Davies, P. L., and Hew, C. L. (1990) *FASEB J.* **4**, 2460–2468
d Gronwald, W., Loewen, M. C., Lix, B., Daugulis, A. J., Sönnichsen, F. D., Davies, P. L., and Sykes, B. D. (1998) *Biochemistry* **37**, 4712–4721
e Chakrabartty, A., Ananthanarayanan, V. S., and Hew, C. L. (1989) *J. Biol. Chem.* **264**, 11307–11312
f Yang, D. S. C., Sax, M., Chakrabartty, A., and Hew, C. L. (1988) *Nature (London)* **333**, 232–237
g Madura, J. D., Wierzbicki, A., Harrington, J. P., Maughon, R. H., Raymond, J. A., and Sikes, C. S. (1994) *J. Am. Chem. Soc.* **116**, 417–418
h Sicheri, F., and Yang, D. S. C. (1995) *Nature (London)* **375**, 427–431
i Wierzbicki, A., Taylor, M. S., Knight, C. A., Madura, J. D., Harrington, J. P., and Sikes, C. S. (1996) *Biophys. J.* **71**, 8–18
j Ng, N. F. L., and Hew, C. L. (1992) *J. Biol. Chem.* **267**, 16069–16075
k Chao, H., Sönnichsen, F. D., DeLuca, C. I., Sykes, B. D., and Davies, P. L. (1994) *Protein Sci.* **3**, 1760–1769
l Jia, Z., DeLuca, C. I., Chao, H., and Davies, P. L. (1996) *Nature (London)* **384**, 285–288
m Gourlie, B., Lin, Y., Price, J., DeVries, A. L., Powers, D., and Huang, R. C. C. (1984) *J. Biol. Chem.* **259**, 14960–14965
n Li, N., Chibber, B. A. K., Castellino, F. J., and Duman, J. G. (1998) *Biochemistry* **37**, 6343–6350
o Graether, S. P., Kuiper, M. J., Gagné, S. M., Walker, V. K., Jia, Z., Sykes, B. D., and Davies, P. L. (2000) *Nature (London)* **406**, 325– 328
p Liou, Y.-C., Tocilj, A., Davies, P. L., and Jia, Z. (2000) *Nature (London)* **406**, 322–324
q Worrall, D., Elias, L., Ashford, D., Smallwood, M., Sidebottom, C., Lillford, P., Telford, J., Holt, C., and Bowles, D. (1998) *Science* **282**, 115–117
r Sidebottom, C., Buckley, S., Pudney, P., Twigg, S., Jarman, C., Holt, C., Telford, J., McArthur, A., Worrall, D., Hubbard, R., and Lillford, P. (2000) *Nature (London)* **406**, 256
s Knight, C. A., (2000) *Nature (London)* **406**, 249–251
t Costanzo, J. P., Lee, R. E., Jr., DeVries, A. L., Wang, T., and Layne, J. R., Jr. (1995) *FASEB J.* **9**, 351–352
u Storey, K. B., and Storey, J. M. (1990) *Sci. Am.* **263** (Dec), 92–97
v Wolber, P., and Warren, G. (1989) *Trends Biochem. Sci.* **14**, 179–182
w Gurian-Sherman, D., and Lindow, S. E. (1993) *FASEB J.* **7**, 1338–1343

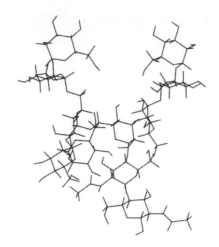

Figure 4-21 Stereoscopic view of an N-linked oligosaccharide whose structure has been deduced by two-dimensional NMR spectroscopy and energy calculations. This is one of a range of allowed conformations. From Homans *et al.*[257]

sizes (Eq. 4-1) can be analyzed with proton or ^{13}C NMR.[10,257] A variety of newer NMR techniques, some of which have been described in Chapter 3, have been applied.[28,258–267] A problem with NMR spectroscopy of these compounds has been the lack of the large number of nuclear Overhauser enhancements (NOEs) that can be observed.[268,269] Nevertheless, when combined with energy calculations it is possible to use

NMR measurements to deduce conformations of sugar rings, three-dimensional structures, and the degree of conformational flexibility in various parts of N-linked oligosaccharides (Fig. 4-21). Measurement of C–O–C –C spin-coupling constants is also of value.[270] Use of multidimensional NMR has permitted analysis of mixtures of cellulose oligosaccharides.[269]

References

1. Pigman, W. W., and Horton, D., eds. (1972) *The Chemistry of Carbohydrates*, 2nd ed., Academic Press, New York
2. Guthrie, R. D. (1974) *Guthrie and Honeyman's Introduction to Carbohydrate Chemistry*, 4th ed., Oxford Univ. Press (Clarendon), London and New York (p. 17)
3. Binkley, R. W. (1988) *Modern Carbohydrate Chemistry*, Dekker, New York
4. Shalenberger, R. S. (1982) *Advanced Sugar Chemistry: Principles of Sugar Stereochemistry*, AVI Publ. Co., Westport, Connecticut
5. El-Khadem, H. S. (1988) *Carbohydrate Chemistry*, Academic Press, San Diego, California
6. Prince, R. C., Gunson, D. E., Leigh, J. S., and McDonald, G. G. (1982) *Trends Biochem. Sci.* **7**, 239–240
7. Wertz, P. W., Garver, J. C., and Anderson, L. (1981) *J. Am. Chem. Soc.* **103**, 3916–3922
8. Ha, S., Gao, J., Tidor, B., Brady, J. W., and Karplus, M. (1991) *J. Am. Chem. Soc.* **113**, 1553–1557
9. Angyal, S. J. (1979) in *Asymmetry of Carbohydrates* (Harmon, R. E., ed), p. 15, Decker, New York
10. Gray, G. R. (1976) *Acc. Chem. Res.* **9**, 418
11. Pierce, J., Serianni, A. S., and Barker, R. (1985) *J. Am. Chem. Soc.* **107**, 2448–2456
12. Maple, S. R., and Allerhand, A. (1987) *J. Am. Chem. Soc.* **109**, 3168–3169
13. Goux, W. J. (1985) *J. Am. Chem. Soc.* **107**, 4320–4327
14. Barker, S. A., and Baggett, N. (1978) *Trends Biochem. Sci.* **3**, 140–141
15. Benson, A. A. (1963) *Adv. Lipid Res.* **1**, 387–394

16. Benning, C., Beatty, J. T., Prince, R. C., and Somerville, C. R. (1993) *Proc. Natl. Acad. Sci. U.S.A.* **90**, 1561–1565
17. Dijkema, C., Kester, H. C. M., and Visser, J. (1985) *Proc. Natl. Acad. Sci. U.S.A.* **82**, 14–18
18. Tarczynski, M. C., Jensen, R. G., and Bohnert, H. J. (1992) *Proc. Natl. Acad. Sci. U.S.A.* **89**, 2600–2604
19. Yamanouchi, T., Tachibana, Y., Sekino, N., Akanuma, H., Akaoka, I., and Miyashita, H. (1994) *J. Biol. Chem.* **269**, 9664–9668
20. Nadano, D., Iwasaki, M., Endo, S., Kitajima, K., Inoue, S., and Inoue, Y. (1986) *J. Biol. Chem.* **261**, 11550–11557
21. Schauer, R. (1985) *Trends Biochem. Sci.* **10**, 357–360
22. Faillard, H. (1989) *Trends Biochem. Sci.* **14**, 237–241
23. Roth, J., Kempf, A., Reuter, G., Schauer, R., and Gehring, W. J. (1992) *Science* **256**, 673–675
24. Powell, L. D., and Varki, A. (1995) *J. Biol. Chem.* **270**, 14243–14246
25. Stoddart, J. F. (1971) *Stereochemistry of Carbohydrates*, Wiley-Interscience, New York
26. Brady, J. W. (1986) *J. Am. Chem. Soc.* **108**, 8152–8160
26a. Marszalek, P. E., Pang, Y.-P., Li, H., Yazal, J. E., Oberhauser, A. F., and Fernandez, J. M. (1999) *Proc. Natl. Acad. Sci. U.S.A.* **96**, 7894–7898
27. Venkataraman, G., Sasisekharan, V., Cooney, C. L., Langer, R., and Sasisekharan, R. (1994) *Proc. Natl. Acad. Sci. U.S.A.* **91**, 6171–6175
28. Hajduk, P. J., Horita, D. A., and Lerner, L. E. (1993) *J. Am. Chem. Soc.* **115**, 9196–9201

29. Wolfe, S., Whangbo, M.-H., and Mitchell, D. J. (1979) *Carbohydrate Res.* **69**, 1–26
30. Kirby, A. J. (1983) *The Anomeric Effect and Related Stereoelectronic Effects at Oxygen*, Springer, Berlin
31. Perrin, C. L., Armstrong, K. B., and Fabian, M. A. (1994) *J. Am. Chem. Soc.* **116**, 715–722
32. Cui, W., Li, F., and Allinger, N. L. (1993) *J. Am. Chem. Soc.* **115**, 2943–2951
33. Westhof, E., and Sundaralingam, M. (1980) *J. Am. Chem. Soc.* **102**, 1493–1500
33a. French, A. D., Dowd, M. K., and Reilly, P. J. (1997) *J. Mol. Struct. (Theochem)*, 395–396
34. Gabb, H. A., and Harvey, S. C. (1993) *J. Am. Chem. Soc.* **115**, 4218–4227
35. Plavec, J., Tong, W., and Chattopadhyaya, J. (1993) *J. Am. Chem. Soc.* **115**, 9734–9746
36. Ellervik, U., and Magnusson, G. (1994) *J. Am. Chem. Soc.* **116**, 2340–2347
36a. French, A. D., and Murphy, V. G. (1973) *Carbohydr. Res.* **27**, 391–406
37. Robyt, J. F., and White, B. J. (1996) *Biochemical Techniques, Theory and Practice*, 2nd ed., Waveland Press, Prospect Heights, Illinois
38. Lipták, A., Fügedi, P., Szurmai, Z., and Harangi, J. (1990) *CRC Handbook of Oligosaccharides. Vol. I: Disaccharides*, CRC Press, Boca Raton, Florida
39. Cerami, A. (1986) *Trends Biochem. Sci.* **11**, 311–314
40. Watkins, N. G., Neglia-Fisher, C. I., Dyer, D. G., Thorpe, S. R., and Baynes, J. W. (1987) *J. Biol. Chem.* **262**, 7207–7212
41. Súarez, G., Rajaram, R., Oronsky, A. L., and Gawinowicz, M. A. (1989) *J. Biol. Chem.* **264**, 3674–3679

References

42. Takata, K., Horiuchi, S., Araki, N., Shiga, M., Saitoh, M., and Morino, Y. (1988) *J. Biol. Chem.* **263**, 14819–14825

43. Hayase, F., Nagaraj, R. H., Miyata, S., Njoroge, F. G., and Monnier, V. M. (1989) *J. Biol. Chem.* **264**, 3758–3764

44. Goldberg, R. N., and Tewari, Y. B. (1989) *J. Biol. Chem.* **264**, 9897–9900

45. Tewari, Y. B., and Goldberg, R. N. (1989) *J. Biol. Chem.* **264**, 3966–3971

46. Goldberg, R. N., Tewari, Y. B., and Ahluwalia, J. C. (1989) *J. Biol. Chem.* **264**, 9901–9904

47. Weissborn, A. C., Rumley, M. K., and Kennedy, E. P. (1991) *J. Biol. Chem.* **266**, 8062–8067

48. Reymond, P., Grünberger, S., Paul, K., Müller, M., and Farmer, E. E. (1995) *Proc. Natl. Acad. Sci. U.S.A.* **92**, 4145–4149

49. Nacharaju, P., and Acharya, A. S. (1992) *Biochemistry* **31**, 12673–12679

50. Baynes, J. W., and Monnier, V. M., eds. (1989) *the Maillard Reaction in Aging, Diabetes and Nutrition*, Liss, New York

51. Larsen, M. L., Hørder, M., and Mogensen, E. F. (1990) *N. Engl. J. Med.* **323**, 1021–1025

52. Luthra, M., and Balasubramanian, D. (1993) *J. Biol. Chem.* **268**, 18119–18127

53. Urbanowski, J. C., Cohenford, M. A., and Dain, J. A. (1982) *J. Biol. Chem.* **257**, 111–115

54. Grandhee, S. K., and Monnier, V. M. (1991) *J. Biol. Chem.* **266**, 11649–11653

55. Brownlee, M., Cerami, A., and Vlassara, H. (1988) *N. Engl. J. Med.* **318**, 1315–1320

56. Papoulis, A., Al-Abed, Y., and Bucala, R. (1995) *Biochemistry* **34**, 648–655

57. Wells-Knecht, K. J., Zyzak, D. V., Litchfield, J. E., Thorpe, S. R., and Baynes, J. W. (1995) *Biochemistry* **34**, 3702–3709

58. French, D. (1975) in *International Review of Biochemistry*, Vol. 5 (Whelan, W. J., ed), p. 269, Univ. Park Press, Baltimore, Maryland (Series One)

59. Atkins, E. D. T., ed. (1986) *Polysaccharides, Topics in Structure and Morphology*, VCH Publisher, New York

60. Whistler, R. L., and BeMiller, J. N., eds. (1993) *Industrial Gums*, 3rd ed., Academic Press, San Diego, California

61. Rao, V. S. R., and Sathyanarayana, B. K. (1972) *Biopolymers* **11**, 1379–1394

62. French, A. D., and Dowd, M. K. (1993) in *Cellulostics: Chemical,Biochemical and Materials Aspects* (Kennedy, J. F., Phillips, G. O., and Williams, P. A., eds), pp. 51–56, Ellis Harwood, New York

63. Duda, C. A., and Stevens, E. S. (1993) *J. Am. Chem. Soc.* **115**, 8487–8488

64. Liébecq, C., ed. (1992) *Biochemical Nomenclature*, Portland Press, London and Chapel Hill, North Carolina (for the International Union of Biochemistry and Molecular Biology)

65. Preston, R. D. (1986) *Trends Biochem. Sci.* **11**, 377–380

66. Ross, P., Mayer, R., and Benziman, M. (1991) *Microbiol. Rev.* **55**, 35–58

67. Sugiyama, J., Vuong, R., and Chanzy, H. (1991) *Macromolecules* **24**, 4168–4175

68. Dudley, R. L., Fyfe, C. A., Stephenson, P. J., Deslandes, Y., Hamer, G. K., and Marchessault, R. H. (1983) *J. Am. Chem. Soc.* **105**, 2469–2472

69. Gebler, K., Kraub, N., Steiner, T., Betzel, C., Sandmann, C., and Saenger, W. (1994) *Science* **266**, 1027–1029

69a. Langan, P., Nishiyama, Y., and Chanzy, H. (1999) *J. Am. Chem. Soc.* **121**, 9940–9946

70. Grover, J. A. (1993) in *Industrial Gums*, 3rd ed. (Whistler, R. L., and BeMiller, J. N., eds), pp. 475–504, Academic Press, San Diego, California

71. French, D. (1973) *J. Animal Sci.* **37**, 1048–1061

72. Maddelein, M. L., Libessart, N., Bellanger, F., Delrue, B., D'Hulst, C., Van den Koornhuyse, N., Fontaine, T., Wieruszeski, J. M., Decq, A., and Ball, S. (1994) *J. Biol. Chem.* **269**, 25150–25157

73. Jane, J., Kasemsuwan, T., Leas, S., Zobel, H., and Robyt, J. F. (1994) *Starch/Stärke* **46**, 121–129

74. Meyer, K. H. (1943) *Adv. Enzymol.* **3**, 109–136

75. French, D. (1969) in *Symposium on Foods: Carbohydrates and their Roles* (Schultz, H. W., ed), pp. 26–54, AVI Publ. Co., Westport, Connecticut

76. Rundle, R. E., and French, D. (1943) *J. Am. Chem. Soc.* **65**, 1707–1710

77. Hybl, A., Rundle, R. E., and Williams, D. E. (1965) *J. Am. Chem. Soc.* **87**, 2779–2788

78. Teitelbaum, R. C., Ruby, S. L., and Tobin, J. M. (1980) *J. Am. Chem. Soc.* **102**, 3322–3328

79. Winter, W. T., and Sarko, A. (1974) *Biopolymers* **13**, 1461–1482

80. Kainuma, K., and French, D. (1972) *Biopolymers* **11**, 2241–2250

81. Wu, H.-C., H, and Sarko, A. (1978) *Carbohydr. Res.* **61**, 7–25

82. Gidley, M. J., and Bociek, S. M. (1988) *J. Am. Chem. Soc.* **110**, 3820–3829

83. Lomako, J., Lomako, W. M., Whelan, W. J., Dombro, R. S., Neary, J. T., and Norenberg, M. D. (1993) *FASEB J.* **7**, 1386–1393

84. Chen, W., Avison, M. J., Zhu, X. H., and Shulman, R. G. (1993) *Biochemistry* **32**, 11483–11487

85. Ohana, P., Delmer, D. P., Steffens, J. C., Matthews, D. E., Mayer, R., and Benziman, M. (1991) *J. Biol. Chem.* **266**, 13742–13744

86. Mol, P. C., Park, H., Mullins, J. T., and Cabib, E. (1994) *J. Biol. Chem.* **269**, 31267–31274

87. Yamamoto, R., and Nevins, D. J. (1978) *Carbohydrate Res.* **67**, 275–280

88. Marchessault, R. H., and Deslandes, Y. (1979) *Carbohydrate Res.* **75**, 231–242

89. Kang, M. S., and Cabib, E. (1986) *Proc. Natl. Acad. Sci. U.S.A.* **83**, 5808–5812

90. Harada, T., Terasaki, M., and Harada, A. (1993) in *Industrial Gums*, 3rd ed. (Whistler, R. L., and BeMiller, J. N., eds), pp. 427–445, Academic Press, San Diego, California

91. Walker, G. J. (1978) in *International Review of Biochemistry*, Vol. 16 (Manners, D. J., ed), p. 75, Univ. Park Press, Baltimore, Maryland

92. de Belder, A. N. (1993) in *Industrial Gums*, 3rd ed. (Whistler, R. L., and BeMiller, J. N., eds), pp. 399–425, Academic Press, San Diego, California

93. Yang, X., Miller, M. A., Yang, R., Evans, D. F., and Edstrom, R. D. (1990) *FASEB J.* **4**, 3140–3143

94. Tsujisaka, Y., and Mitsuhashi, M. (1993) in *Industrial Gums*, 3rd ed. (Whistler, R. L., and BeMiller, J. N., eds), pp. 447–460, Academic Press, San Diego, California

95. Arnott, S., Fulmer, A., Scott, W. E., Dea, I. C. M., Moorehouse, R., and Rees, D. A. (1974) *J. Mol. Biol.* **90**, 269–284

96. Shibata, N., Ikuta, K., Imai, T., Satoh, Y., Satoh, R., Suzuki, A., Kojima, C., Kobayashi, H., Hisamichi, K., and Suzuki, S. (1995) *J. Biol. Chem.* **270**, 1113–1122

97. Kollár, R., Petráková, E., Ashwell, G., Robbins, P. W., and Cabib, E. (1995) *J. Biol. Chem.* **270**, 1170–1178

98. Preston, R. D. (1968) *Sci. Am.* **218**(Jun), 102–108

99. Turvey, J. R. (1978) in *International Review of Biochemistry*, Vol. 16 (Manners, D. J., ed), Univ. Park Press, Baltimore, Maryland

100. Kirkwood, S. (1974) *Ann. Rev. Biochem.* **43**, 401–417

101. Minke, R., and Blackwell, J. (1978) *J. Mol. Biol.* **120**, 167–181

102. Roberts, G. A. F. (1992) *Chitin Chemistry*, MacMillan, London

103. Blackwell, J., and Weih, M. A. (1980) *J. Mol. Biol.* **137**, 49–60

104. Rees, D. A., and Welsh, E. J. (1977) *Angew. Chem. Int. Ed. Engl.* **16**, 214–224

105. Lee, Y. C. (1992) *FASEB J.* **6**, 3193–3200

106. Laurent, T. C., and Fraser, J. R. E. (1992) *FASEB J.* **6**, 2397–2404

107. Hardingham, T. E., and Fosang, A. J. (1992) *FASEB J.* **6**, 861–870

108. Fransson, L. (1987) *Trends Biochem. Sci.* **12**, 406–411

109. Morris, E. R., Rees, D. A., and Welsh, E. J. (1980) *J. Mol. Biol.* **138**, 383–400

110. David, G. (1993) *FASEB J.* **7**, 1023–1030

111. Yamada, S., Murakami, T., Tsuda, H., Yoshida, K., and Sugahara, K. (1995) *J. Biol. Chem.* **270**, 8696–8705

112. Lane, D. A., and Björk, I., eds. (1992) *Heparin And Related Polysaccharides*, Plenum, New York

113. Edens, R. E., Fromm, J. R., Linhardt, R. J., and Weiler, J. M. (1995) *Biochemistry* **34**, 2400–2407

114. Linhardt, R. J., Ampofo, S. A., Fareed, J., Hoppensteadt, D., Mulliken, J. B., and Folkman, J. (1992) *Biochemistry* **31**, 12441–12445

115. Casu, B., Petitou, M., Provasoli, M., and Sinaÿ, P. (1988) *Trends Biochem. Sci.* **13**, 221–225

116. Oates, J. A., and Wood, A. J. J. (1991) *N. Engl. J. Med.* **324**, 1565–1574

117. Rosenberg, R. D., and Lam, L. (1979) *Proc. Natl. Acad. Sci. U.S.A.* **76**, 1218–1222

118. Lindahl, U., Kusche-Gullberg, M., and Kjelln, L. (1998) *J. Biol. Chem.* **273**, 24979–24982

119. Mulloy, B. A., Ribeiro, A.-C., Alves, A.-P., Vieira, R. P., and Mourao, P. A. S. (1994) *J. Biol. Chem.* **269**, 22113–22123

120. Gibeaut, D., and Carpita, N. C. (1994) *FASEB J.* **8**, 904–915

121. Hayashi, T., and Matsuda, K. (1981) *J. Biol. Chem.* **256**, 11117–11122

122. Millane, R. P. (1992) in *Frontiers in Carbohydrate Research-2* (Chandrasekaran, R., ed), pp. 168–190, Elsevier, London

123. Rolin, C. (1993) in *Industrial Gums*, 3rd ed. (Whistler, R. L., and BeMiller, J. N., eds), pp. 257–293, Academic Press, San Diego, California

124. Arnott, S., Scott, W. E., Rees, D. A., and McNab, C. G. A. (1974) *J. Mol. Biol.* **90**, 253–267

125. Selby, H. H., and Whistler, R. L. (1993) in *Industrial Gums*, 3rd ed. (Whistler, R. L., and BeMiller, J. N., eds), pp. 87–103, Academic Press, San Diego, California

126. Albersheim, P., Bauer, W. D., Keestra, K., Talmadge, K. W. (1973) *Biogenesis of Plant Cell Wall Polysaccharides*, (Loewus, F., ed.), Academic Press, New York

127. Rees, D. A. (1972) *Biochem. J.* **126**, 257–273

128. Therkelsen, G. H. (1993) in *Industrial Gums*, 3rd ed. (Whistler, R. L., and BeMiller, J. N., eds), pp. 145–180, Academic Press, San Diego, California

129. Clare, K. (1993) in *Industrial Gums*, 3rd ed. (Whistler, R. L., and BeMiller, J. N., eds), pp. 105–143, Academic Press, San Diego, California

130. Marsh, M. E., Chang, D. C., and King, G. C. (1992) *J. Biol. Chem.* **267**, 20507–20512

130a. Vinogradov, E., and Bock, K. (1999) *Angew. Chem. Ed. Engl.* **38**, 671–674

131. Sutherland, I. W. (1979) *Trends Biochem. Sci.* **4**, 55–59

References

132. Kang, K. S., and Pettitt, D. J. (1993) in *Industrial Gums*, 3rd ed. (Whistler, R. L., and BeMiller, J. N., eds), pp. 341–397, Academic Press, San Diego, California

133. Wolucka, B. A., McNeil, M. R., de Hoffmann, E., Chonjnacki, T., and Brennan, P. J. (1994) *J. Biol. Chem.* **269**, 23328–23335

134. Kobata, A. (1984) in *Biology of Carbohydrates*, Vol. 2 (Ginsburg, V., and Robbins, P. W., eds), pp. 87–161, Wiley, New York

135. Lennarz, W. J., ed. (1980) *The Biochemistry of Glycoproteins and Proteoglycans*, Plenum, New York

136. Montreul, J. (1982) in *Comprehensive Biochemistry*, 1st ed., Vol. 19B, Part II (Florkin, M., and Stotz, E. H., eds), pp. 1–188, Elsevier, Amsterdam

137. Rademacher, T. W., Parekh, R. B., and Dwek, R. A. (1988) *Ann. Rev. Biochem.* **57**, 785–838

138. Roberts, D. D., and Mecham, R. P., eds. (1993) *Cell Surface and Extracellular Glycoconjugates: Structure and Function*, Academic Press, San Diego, California

138a. Varki, A., Cummings, R., Esko, J., Freeze, H., Hart, G., and Marth, J., eds (1999) *Essentials of Glycobiology*, Cold Spring Harbor Lab. Press, Cold Spring Harbor, New York

139. Fukuda, M., and Hindsgaul, O., eds (1994) *Molecular Glycobiology*, Oxford Publishing, Oxford

140. Williams, R. L., Greene, S. M., and McPherson, A. (1987) *J. Biol. Chem.* **262**, 16020–16031

141. Hofsteenge, J., Müller, D. R., de Beer, T., Löffler, A., Richter, W. J., and Vliegenthart, J. F. G. (1994) *Biochemistry* **33**, 13524–13530

141a. Iozzo, R. V. (1998) *Ann. Rev. Biochem.* **67**, 609–652

141b. Bernfield, M., Götte, M., Park, P. W., Reizes, O., Fitzgerald, M. L., Lincecum, J., and Zako, M. (1999) *Ann. Rev. Biochem.* **68**, 729–777

142. Sugahara, K., Ohkita, Y., Shibata, Y., Yoshida, K., and Ikegami, A. (1995) *J. Biol. Chem.* **270**, 7204–7212

143. Funderburgh, J. L., Funderburgh, M. L., Brown, S., Vergnes, J.-P., Hassell, J. R., Mann, M. M., and Conrad, G. W. (1993) *J. Biol. Chem.* **268**, 11874–11880

144. Neame, P. J., Christner, J. E., and Baker, J. R. (1987) *J. Biol. Chem.* **262**, 17768–17778

145. Scott, J. E. (1987) *Trends Biochem. Sci.* **12**, 318–321

146. Rosenberg, L., Tang, L.-H., Pal, S., Johnson, T. L., and Choi, H. U. (1988) *J. Biol. Chem.* **263**, 18071–18077

147. Choi, H. U., Johnson, T. L., Pal, S., Tang, L.-H., Rosenberg, L., and Neame, P. J. (1989) *J. Biol. Chem.* **264**, 2876–2884

148. Kokenyesi, R., and Bernfield, M. (1994) *J. Biol. Chem.* **269**, 12304–12309

149. Rosenberg, L. (1975) in *Dynamics of Connective Tissue Macromolecules* (Burleigh, P. M. C., and Poole, A. R., eds), p. 107, North-Holland Publ., Amsterdam

150. Chen, J., and Varner, J. E. (1985) *Proc. Natl. Acad. Sci. U.S.A.* **82**, 4399–4403

151. Kieliszewski, M., deZacks, R., Leykam, J. F., and Lamport, D. T. A. (1992) *Plant Physiol.* **98**, 919–926

152. Kieliszewski, M. J., O'Neill, M., Leykam, J., and Orlando, R. (1995) *J. Biol. Chem.* **270**, 2541–2549

153. Eckhardt, A. E., Timpte, C. S., Abernethy, J. L., Toumadje, A., Johnson, W. C., and Hill, R. L. (1987) *J. Biol. Chem.* **262**, 11339–11344

154. Timpte, C. S., Eckhardt, A. E., Abernethy, J. L., and Hill, R. L. (1988) *J. Biol. Chem.* **263**, 1081–1088

155. Gupta, R., and Jentoft, N. (1989) *Biochemistry* **28**, 6114–6121

156. Carter, S. R., Slomiany, A., Gwozdzinski, K., Liau, Y. H., and Slomiany, B. L. (1988) *J. Biol. Chem.* **263**, 11977–11984

157. Jentoft, N. (1990) *Trends Biochem. Sci.* **15**, 291–294

158. Butenhof, K. J., and Gerken, T. A. (1993) *Biochemistry* **32**, 2650–2663

159. Toribara, N. W., Roberton, A. M., Ho, S. B., Kuo, W.-L., Gum, E., Hicks, J. W., Gum, J. R., Byrd, J. C., Siddiki, B., and Kim, Y. S. (1993) *J. Biol. Chem.* **268**, 5879–5885

160. Sadler, J. E. (1984) in "*Biology of Carbohydrates*", Vol. 2 (Ginsburg, V., and Robbins, P. W., eds), pp. 199–288, Wiley, New York

160a. Perez-Vilar, J., and Hill, R. L. (1999) *J. Biol. Chem.* **274**, 31751–31754

161. Fukuda, M., Carlsson, S. R., Klock, J. C., and Dell, A. (1986) *J. Biol. Chem.* **261**, 12796–12806

162. Gerken, T. A., and Jentoft, N. (1987) *Biochemistry* **26**, 4689–4699

163. Freeze, H. H., and Wolgast, D. (1986) *J. Biol. Chem.* **261**, 127–134

164. Carver, J. P., and Brisson, J.-R. (1984) in *Biology of Carbohydrates*, Vol. 2 (Ginsburg, V., and Robbins, P. W., eds), pp. 289–329, Wiley, New York

165. Bush, C. A., Duben, A., and Ralapati, S. (1980) *Biochemistry* **19**, 501–504

166. Finne, J., and Krusius, T. (1979) *Eur. J. Biochem.* **102**, 583–588

167. Järnefelt, J., Finne, J., Krusius, T., and Rauvala, H. (1978) *Trends Biochem. Sci.* **3**, 110–114

168. Kornfeld, R., and Kornfeld, S. (1980) in *The Biochemistry of Glycoproteins and Proteoglycans* (Lennarz, W. J., ed), Plenum, New York

169. Fukuda, M., Dell, A., and Fukuda, M. N. (1984) *J. Biol. Chem.* **259**, 4782–4791

170. Hernández, L. M., Olivero, I., Alvarado, E., and Larriba, G. (1992) *Biochemistry* **31**, 9823–9831

171. Taguchi, T., Seko, A., Kitajima, K., Muto, Y., Inoue, S., Khoo, K.-H., Morris, H. R., Dell, A., and Inoue, Y. (1994) *J. Biol. Chem.* **269**, 8762–8771

172. Wieland, F., Paul, G., and Sumper, M. (1985) *J. Biol. Chem.* **260**, 15180–15185

173. Funderburgh, J. L., Funderburgh, M. L., Mann, M. M., and Conrad, G. W. (1991) *J. Biol. Chem.* **266**, 14226–14231

174. Sharon, N., and Lis, H. (1989) *Science* **246**, 227–229

175. Sharon, N., and Lis, H. (1993) *Sci. Am.* **268**(Jan), 82–89

176. Opdenakker, G., Rudd, P. M., Ponting, C. P., and Dwek, R. A. (1993) *FASEB J.* **7**, 1330–1337

177. Sharon, N. (1993) *Trends Biochem. Sci.* **18**, 221–226

178. Barondes, S. H. (1988) *Trends Biochem. Sci.* **13**, 480–482

179. Sphyris, N., Lord, J. M., Wales, R., and Roberts, L. M. (1995) *J. Biol. Chem.* **270**, 20292–20297

180. Sumner, J. B., and Howell, S. F. (1936) *J. Bacteriol.* **32**, 227–237

181. Bourne, Y., Rougés, P., and Cambillau, C. (1990) *J. Biol. Chem.* **265**, 18161–18165

182. Sharon, N., and Lis, H. (1990) *FASEB J.* **4**, 3198–3208

183. Mandal, D. K., and Brewer, C. F. (1993) *Biochemistry* **32**, 5116–5120

184. Derewenda, Z., Yariv, J., Helliwell, J. R., Kalb, A. J., Dodson, E. J., Papiz, M. Z., Wan, T., and Campbell, J. (1989) *EMBO J.* **8**, 2189–2193

185. Drickamer, K. (1988) *J. Biol. Chem.* **263**, 9557–9560

186. Krusius, T., Gehlsen, K. R., and Ruoslahti, E. (1987) *J. Biol. Chem.* **262**, 13120–13125

187. Gitt, M. A., and Barondes, S. H. (1991) *Biochemistry* **30**, 82–89

188. Barondes, S. H., Cooper, D. N. W., Gitt, M. A., and Leffler, H. (1994) *J. Biol. Chem.* **269**, 20807–20810

188a. Varela, P. F., Solís, D., Díaz-Maurino, T., Kaltner, H., Gabius, H.-J., and Romero, A. (1999) *J. Mol. Biol.* **294**, 537–549

189. Chiu, M. L., Parry, D. A. D., Feldman, S. R., Klapper, D. G., and O'Keefe, E. J. (1994) *J. Biol. Chem.* **269**, 31770–31776

190. Madsen, P., Rasmussen, H. H., Flint, T., Gromov, P., Kruse, T. A., Honoré, B., Vorum, H., and Celis, J. E. (1995) *J. Biol. Chem.* **270**, 5823–5829

191. Cho, M., and Cummings, R. D. (1995) *J. Biol. Chem.* **270**, 5207–5212

192. Gitt, M. A., Wiser, M. F., Leffler, H., Herrmann, J., Xia, Y.-R., Massa, S. M., Cooper, D. N. W., Lusis, A. J., and Barondes, S. H. (1995) *J. Biol. Chem.* **270**, 5032–5038

193. Weis, W. I., Kahn, R., Fourme, R., Drickamer, K., and Hendrickson, W. A. (1991) *Science* **254**, 1608–1615

194. Vyas, N. K., Vyas, M. N., and Quiocho, F. A. (1988) *Science* **242**, 1290–1295

195. Shaanan, B., Lis, H., and Sharon, N. (1991) *Science* **254**, 862–865

196. Cygler, M., Rose, D. R., and Bundle, D. R. (1991) *Science* **253**, 442–445

197. Ångström, J., Teneberg, S., and Karlsson, K.-A. (1994) *Proc. Natl. Acad. Sci. U.S.A.* **91**, 11859–11863

198. Sauter, N. K., Hanson, J. E., Glick, G. D., Brown, J. H., Crowther, R. L., Park, S.-J., Skehel, J. J., and Wiley, D. C. (1992) *Biochemistry* **31**, 9609–9621

199. Janakiraman, M. N., White, C. L., Laver, W. G., Air, G. M., and Luo, M. (1994) *Biochemistry* **33**, 8172–8179

200. Bossart-Whitaker, P., Carson, M., Babu, Y. S., Smith, C. D., Laver, W. G., and Air, G. M. (1993) *J. Mol. Biol.* **232**, 1069–1083

201. Crennell, S. J., Garman, E. F., Laver, E. G., Vimr, E. R., and Taylor, G. L. (1993) *Proc. Natl. Acad. Sci. U.S.A.* **90**, 9852–9856

202. Colli, W. (1993) *FASEB J.* **7**, 1257–1264

203. Ofek, I., Beachey, E. H., and Sharon, N. (1978) *Trends Biochem. Sci.* **3**, 159–160

204. Lindberg, F., Lund, B., and Normark, S. (1986) *Proc. Natl. Acad. Sci. U.S.A.* **83**, 1891–1895

205. Cho, J.-W., and Troy, F. A. (1994) *Proc. Natl. Acad. Sci. U.S.A.* **91**, 11427–11431

206. Baumann, H., Brisson, J., Michon, F., Pon, R., and Jennings, H. J. (1993) *Biochemistry* **32**, 4007–4013

207. Falk, P. G., Bry, L., Holgersson, J., and Gordon, J. I. (1995) *Proc. Natl. Acad. Sci. U.S.A.* **92**, 1515–1519

208. Adler, P., Wood, S. J., Lee, Y. C., Lee, R. T., Petri, W. A., and Schnaar, R. L. (1995) *J. Biol. Chem.* **270**, 5164–5171

209. Misevic, G. N., and Burger, M. M. (1990) *J. Biol. Chem.* **265**, 20577–20584

210. Spillmann, D., Hård, K., Thomas-Oates, J., Vliegenthart, J. F. G., Misevic, G., Burger, M. M., and Finne, J. (1993) *J. Biol. Chem.* **268**, 13378–13387

211. Spillmann, D., Thomas-Oates, J. E., van Kuik, J. A., Vliegenthart, J. F. G., Misevic, G., Burger, M. M., and Finne, J. (1995) *J. Biol. Chem.* **270**, 5089–5097

212. Rosati, F., and De Santis, R. (1980) *Nature (London)* **283**, 762–764

213. Seppo, A., Penttilä, L., Niemelä, R., Maaheimo, H., and Renkonen, O. (1995) *Biochemistry* **34**, 4655–4661

214. Litscher, E. S., Juntunen, K., Seppo, A., Penttilä, L., Niemelä, R., Renkonen, O., and Wassarman, P. M. (1995) *Biochemistry* **34**, 4662–4669

215. Hitoshi, S., Kusunoki, S., Kanazawa, I., and Tsuji, S. (1995) *J. Biol. Chem.* **270**, 8844–8850

216. de Vries, T., Srnka, C. A., Palcic, M. M., Swiedler, S. J., van der Eijnden, D. H., and Macher, B. A. (1995) *J. Biol. Chem.* **270**, 8712–8722

217. Stroud, M. R., Levery, S. B., Mårtensson, S., Salyan, M. E. K., Clausen, H., and Hakomori, S. (1994) *Biochemistry* **33**, 10672–10680

218. Hull, S. R., Sugarman, E. D., Spielman, J., and Carraway, K. L. (1991) *J. Biol. Chem.* **266**, 13580–13586

219. Ware, F. E., Vassilakos, A., Peterson, P. A., Jackson, M. R., Lehrman, M. A., and Williams, D. B. (1995) *J. Biol. Chem.* **270**, 4697–4704

220. Leppänen, A., Mehta, P., Ouyang, Y.-B., Ju, T., Helin, J., Moore, K. L., van Die, I., Canfield, W. M., McEver, R. P., and Cummings, R. D. (1999) *J. Biol. Chem.* **274**, 24838–24848

221. Chandrasekaran, E. V., Jain, R. K., Larsen, R. D., Wlasichuk, K., and Matta, K. L. (1995) *Biochemistry* **34**, 2925–2936

222. Kogan, T. P., Revelle, B. M., Tapp, S., Scott, D., and Beck, P. J. (1995) *J. Biol. Chem.* **270**, 14047–14055

223. Waddell, T. K., Fialkow, L., Chan, C. K., Kishimoto, T. K., and Downey, G. P. (1995) *J. Biol. Chem.* **270**, 15403–15411

224. Bertozzi, C. R., Fukuda, S., and Rosen, S. D. (1995) *Biochemistry* **34**, 14271–14277

225. Norgard-Sumnicht, K., and Varki, A. (1995) *J. Biol. Chem.* **270**, 12012–12024

226. Hemmerich, S., Leffler, H., and Rosen, S. D. (1995) *J. Biol. Chem.* **270**, 12035–12047

227. Yuen, C.-T., Bezouska, K., O'Brien, J., Stoll, M., Lemoine, R., Lubineau, A., Kiso, M., Hasegawa, A., Bockovich, N. J., Nicolaou, K. C., and Feizi, T. (1994) *J. Biol. Chem.* **269**, 1595–1598

228. Patel, T. P., Goelz, S. E., Lobb, R. R., and Parekh, R. B. (1994) *Biochemistry* **33**, 14815–14824

229. Graves, B. J., Crowther, R. L., Chandran, C., Rumberger, J. M., Li, S., Huang, K.-S., Presky, D. H., Familletti, P. C., Wolitzky, B. A., and Burns, D. K. (1994) *Nature (London)* **367**, 532–538

230. Lasky, L. A. (1995) *Ann. Rev. Biochem.* **64**, 113–139

231. Ginsburg, V. E. (1982) *Meth. Enzymol.* **83**, all

232. Churms, S. C. (1990) *CRC Handbook of Chromatography Carbohydrates*, Vol. 2, CRC Press, Boca Raton, Florida

233. Valent, B. S., Darvill, A. G., McNeil, M., Robertsen, B. K., and Albersheim, P. (1980) *Carbohydrate Res.* **79**, 165–192

234. Chaplin, M. F., and Kennedy, J. F., eds. (1994) *Carbohydrate Analysis A Practical Approach*, Oxford University, Oxford

235. Robyt, J. F., and Mukerjea, R. (1994) *Carbohydr. Res.* **251**, 187–202

236. Green, E. D., and Baenziger, J. U. (1989) *Trends Biochem. Sci.* **14**, 168–172

237. Carr, S. A., Huddleston, M. J., and Bean, M. F. (1993) *Protein Sci.* **2**, 183–196

237a. Fridriksson, E. K., Beavil, A., Holowka, D., Gould, H. J., Baird, B., and McLafferty, F. W. (2000) *Biochemistry* **39**, 3369–3376

237b. Burlingame, A. L., Carr, S. A., and Baldwin, M. A., eds. (2000) *Mass Spectrometry in Biology and Medicine*, Hamana Press, Totowa, New Jersey

238. Downs, F., and Pigman, W. (1969) *Biochemistry* **8**, 1760–1766

239. Narasimhan, S., Harpaz, N., Longmore, G., Carver, J. P., Grey, A. A., and Schacter, H. (1980) *J. Biol. Chem.* **255**, 4876–4884

240. Cohen, R. E., and Ballou, C. E. (1980) *Biochemistry* **19**, 4345–4358

241. Plummer, T. H., Jr., and Tarentino, A. L. (1981) *J. Biol. Chem.* **256**, 10243–10246

242. Risley, J. M., and Van Etten, R. L. (1985) *J. Biol. Chem.* **260**, 15488–15494

243. Takasaki, S., Mizuochi, T., and Kobata, A. (1982) *Meth. Enzymol.* **83**, 263–268

244. Parente, J. P., Wieruszeski, J.-M., Strecker, G., Montreuil, J., Fournet, B., van Halbeek, H., Dorland, L., and Vliegenthart, J. F. G. (1982) *J. Biol. Chem.* **257**, 13173–13176

245. Patel, T., Bruce, J., Merry, A., Bigge, C., Wormald, M., Jaques, A., and Parekh, R. (1993) *Biochemistry* **32**, 679–693

246. Sweeley, C. C., and Nunez, H. A. (1985) *Ann. Rev. Biochem.* **54**, 765–801

247. Dua, V. K., Roa, B. N. N., Wu, S.-S., Dube, V. E., and Bush, C. A. (1986) *J. Biol. Chem.* **261**, 1599–1608

248. Tsai, P.-K., Dell, A., and Ballou, C. E. (1986) *Proc. Natl. Acad. Sci. U.S.A.* **83**, 4119–4123

249. Lindberg, B., Lönngren, L., and Svensson, S. (1975) *Adv. Carbohyd. Chem. Biochem.* **31**, 185–240

250. Guthrie, R. D., and McCarthy, J. F. (1967) *Adv. Carbohydr. Chem. Biochem.* **22**, 11–23

251. Edge, C., Parekh, R., Rademacher, T., Wormald, M., and Dwek, R. (1992) *Nature (London)* **385**, 693–694

252. Varki, A. (1991) *FASEB J.* **5**, 226–235

253. Rolf, D., and Gray, G. R. (1982) *J. Am. Chem. Soc.* **104**, 3539–3541

254. Mukerjea, R., Kim, D., and Robyt, J. F. (1996) *Carbohydrate Res.* **292**, 11–20

255. Bobbitt, J. M. (1956) *Adv. Carbohydr. Chem. Biochem.* **11**, 1–41

256. Rich, A., Nordheim, A., and Wang, A. H.-J. (1984) *Ann. Rev. Biochem.* **53**, 791–846

257. Homans, S. W., Dwek, R. A., and Rademacher, T. W. (1987) *Biochemistry* **26**, 6571–6578

258. Homans, S. W., Pastore, A., Dwek, R. A., and Rademacher, T. W. (1987) *Biochemistry* **26**, 6649–6655

259. Cumming, D. A., Shah, R. N., Krepinsky, J. J., Grey, A. A., and Carver, J. P. (1987) *Biochemistry* **26**, 6655–6663

260. Reuben, J. (1985) *J. Am. Chem. Soc.* **107**, 1747–1755

261. Cumming, D. A., Dime, D. S., Grey, A. A., Krepinsky, J. J., and Carver, J. P. (1986) *J. Biol. Chem.* **261**, 3208–3213

262. Vuister, G. W., de Waard, P., Boelens, R., Vliegenthart, J. F. G., and Kaptein, R. (1989) *J. Am. Chem. Soc.* **111**, 772–774

263. Gronenborn, A. M., and Clore, G. M. (1989) *Biochemistry* **28**, 5978–5984

264. Abeygunawardana, C., and Bush, C. A. (1991) *Biochemistry* **30**, 8568–8577

265. Rivière, M., and Puzo, G. (1992) *Biochemistry* **31**, 3575–3580

266. van Duynhoven, J. P. M., Goudriaan, J., Hilbers, C. W., and Wijmenga, S. S. (1992) *J. Am. Chem. Soc.* **114**, 10055–10056

267. Wu, J., Bondo, P. B., Vuorinen, T., and Serianni, A. S. (1992) *J. Am. Chem. Soc.* **114**, 3499–3505

268. Homans, S. W. (1990) *Biochemistry* **29**, 9110–9118

269. Bose, B., Zhao, S., Stenutz, R., Cloran, F., Bondo, P. B., Bondo, G., Hertz, B., Carmichael, I., and Serianni, A. S. (1998) *J. Am. Chem. Soc.* **120**, 11158–11173

270. Flugge, L. A., Blank, J. T., and Petillo, P. A. (1999) *J. Am. Chem. Soc.* **121**, 7228–7238

Study Questions

1. A nonreducing disaccharide gives an octamethyl derivative with dimethyl sulfate and alkali. On acid hydrolysis, this derivative yields 1 mol of 2,3,4,6-tetramethyl-D-glucose and 1 mol of 2,3,4,6-tetramethyl-D-galactose. The disaccharide is hydrolyzed rapidly by either maltase or lactase (a β-galactosidase).

   Give an adequately descriptive name of the disaccharide, and draw its Haworth projection formula.

2. An aldopentose (A) of the D-configuration on oxidation with concentrated nitric acid gives a 2,3,4-trihydroxypentanedioic acid (a trihydroxyglutaric acid) (B) which is optically inactive. (A) on addition of HCN, hydrolysis, lactonization, and reduction gives two stereoisomeric aldohexoses (C) and (D). (D) on oxidation affords a 2,3,4,5-tetrahydroxy-hexanedioic acid (a saccharic acid) (E) which is optically inactive. Give structures of compounds (A)-(E).

3. What products are formed when periodic acid reacts with sorbitol?

4. A 10.0 g sample of glycogen gave 6.0 millimol of 2,3-di-O-methylglucose on methylation and acid hydrolysis.

   a. What percent of the glucose residues in glycogen have chains substituted at the α–1→6 position?

   b. What is the average number of glucose residues per chain?

   c. How many millimols of 2,3,6-tri-O-methyl-glucose were formed?

   d. If the molecular weight of the polysaccharide is 2×10^6, how many glucose residues does it contain?

   e. How many nonreducing ends are there per molecule or equivalently how many chains are there per molecule?

5. D-Mannitol is a symmetric molecule, yet it is optically active. Explain.

6. When D-glucose is treated with acidic methanol, the first products which can be isolated are mainly methyl furanosides, but after extensive reaction the furanosides disappear and methyl glucopyranosides accumulate. Why?

7. Write the structural formulas for (a) 1,6 anhydro β-D-glucopyranose; (b) 1,6 anhydro β-D-altrose. Compound (b) is many times more stable than compound (a). Explain this on stereochemical grounds.

8. The disaccharide *nigerose* is α-D-Glu*p*-(1→3)-D-glu. Write out its structure. How would you prove this structure using methylation, periodate oxidation, and other methods.

9. Inositol is 1,2,3,4,5,6-hexahydroxycyclohexane. Draw configurational formulas for all possible stereoisomers and indicate which would be expected to be optically active.

10. Why do you suppose that the major form of D-fructose in solution is the pyranose form but D-fructose in sucrose is in the furanose form?

11. D-Xylose is an easily prepared sugar, potentially available in enormous quantity. What is a common source? How can it be obtained from this source?

12. How can xylitol be obtained from xylose? Discuss the stereochemical properties of xylitol.

13. The enzyme xylose isomerase is important industrially. Why?

14. Glucose reacts non-enzymatically with amino acids and proteins, including hemoglobin, egg-white proteins and serum albumin. For example, if glucose is not removed prior to drying, dried egg whites slowly turn brown and develop off-flavors and odors. What do you propose as the most likely first step in the non-enzymatic glucose-protein chemical reaction? How can the first product transform spontaneously into a ketose derivative?

15. What characteristics would you expect in a binding site for a sugar ring in an enzyme, lectin, or other proteins?

16. Using structural formulas, descirbe the two major types of linkage of carbohydrate chains or clusters to proteins to form glycoproteins or proteoglycans.

17. What products would you expect from cellulose as a result of methylation analysis? Periodate oxidation? The Smith degradation? The action of an alpha amylase?

18. When glycoproteins are treated with alkaline brohydride, amino acid analysis often indicates a decrease in the amount of serine and a corresponding increase in the amount of alanine, or a decrease in threonine with the appearance of α-aminobutyric acid. Explain.

DNA spreading from the broken head of a bacteriophage T2 phage. This classic electron micrograph, published by A. K. Kleinschmidt and coworkers in 1962 (*Biochem. Biophys. Acta.* **61**, 857–864, 1962) was prepared by spreading the phage particles suspended in a protein–salt solution as a mixed monolayer on a water-air interface. The resultant osmotic shock burst the head and confined the DNA as a single thread near the phage ghost. After transfer to a suitable carbon surface, removal of water, and shadowing with platinum, the micrograph was obtained. Courtesy of Albrecht K. Kleinschmidt

Contents

The Nucleic Acids

5

The phosphorus- and nitrogen-containing materials that came to be known as nucleic acids were first isolated from cells around 1870 by Friedrich Miescher but were long regarded as something of a curiosity.[1] Nevertheless, the structures of the monomer units, the **nucleotides**, were established by 1909 and the correct **polynucleotide** structure of the chains of DNA and RNA was proposed by Levene and Tipson in 1935.[2,3]

The nucleotides are made up of three parts:

1. One of the **pyrimidine** or **purine "bases"** uracil, cytosine, adenine, or guanine (Fig. 5-1). All four of these bases are present in RNA, while DNA contains thymine instead of uracil. Atoms in the bases are numbered 1–6 or 1–9.
2. A **sugar**, either D-ribose or D-2-deoxyribose. Carbon atoms in sugars are numbered 1'–5.'
3. **Phosphoric acid**

Although the biological synthesis is indirect, we can imagine that nucleotides are formed from these parts by elimination of two molecules of water as indicated in Eq. 5-1. In nucleic acids the nucleotides are combined through phosphodiester linkages between the 5'-hydroxyl of the sugar in one nucleotide and the 3'-hydroxyl of another. Again, we can imagine that these linkages were formed by the elimination of water (Eq. 5-2). The structures of a pair of short polynucleotide strands in DNA are shown in Fig. 5-2. That of a segment of double-helical DNA is shown in Fig. 5-3 and that of a transfer RNA in Figs. 5-30 and 5-31.

Despite the fact that Levene had deduced the correct structure for polynucleotides, he was thrown off the trail of a deeper understanding by the roughly equal amounts of the four bases found in either DNA or RNA. He assumed that nucleic acids must be regular repeating polymers for which there was no obvious biological function. It was not until 1944 that there

Figure 5-1 Structures of the major pyrimidine and purine bases of DNA and RNA.

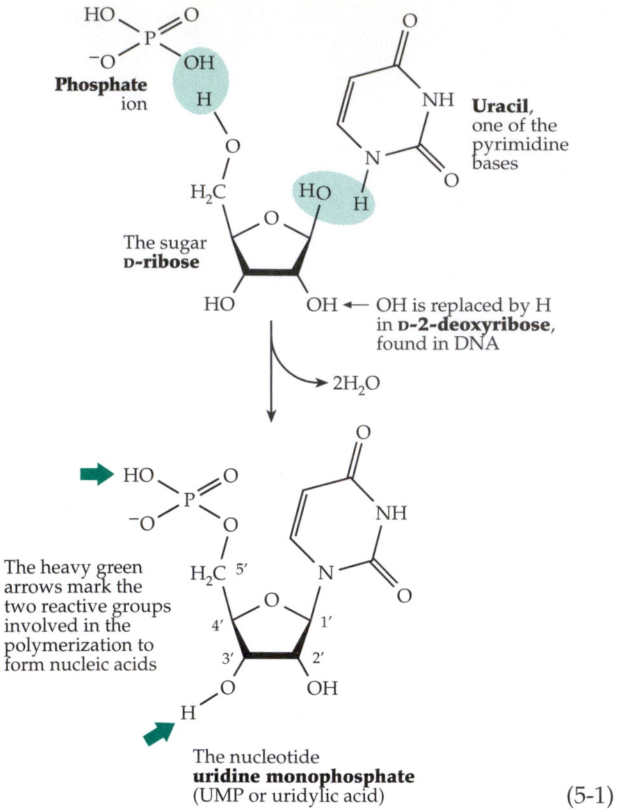

The sugar **D-ribose**

← OH is replaced by H in **D-2-deoxyribose**, found in DNA

Phosphate ion

Uracil, one of the pyrimidine bases

The heavy green arrows mark the two reactive groups involved in the polymerization to form nucleic acids

The nucleotide **uridine monophosphate** (UMP or uridylic acid)

$$(5\text{-}1)$$

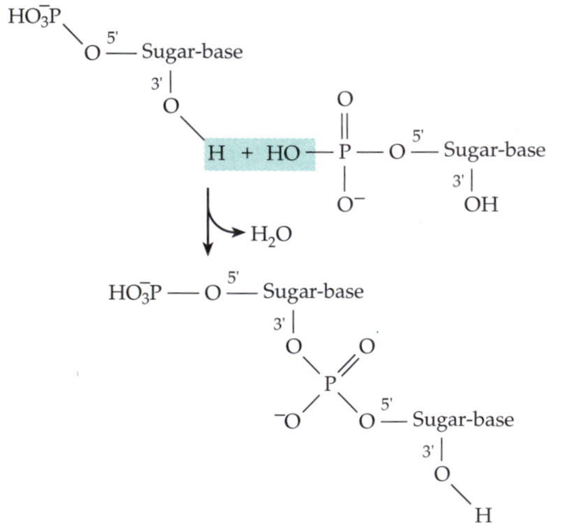

$$(5\text{-}2)$$

artificially formed DNA fibers. An additional key piece of information was the discovery by Erwin Chargaff that *in all double-stranded DNA the content of adenine equals that of thymine and the content of guanine equals that of cytosine.*

The most significant feature of the proposed structure was the pairing of bases in the center through hydrogen bonding. The pairs and triplets of hydrogen bonds (Fig. 5-2) could form in the manner shown only if adenine (A) was paired with thymine (T) and cytosine (C) with guanine (G) at every point in the entire DNA structure. Thus, *the nucleotide sequence in one chain is complementary to but not identical to that in the other chain.* It was apparent almost immediately that the sequence of bases in a DNA chain must convey the encoded genetic information. The complementarity of the two strands suggested a simple mechanism for replication of genes during cell divisions. The two strands could separate and a complementary strand could be synthesized along each strand to give two molecules of the DNA, one for each of the two cells.

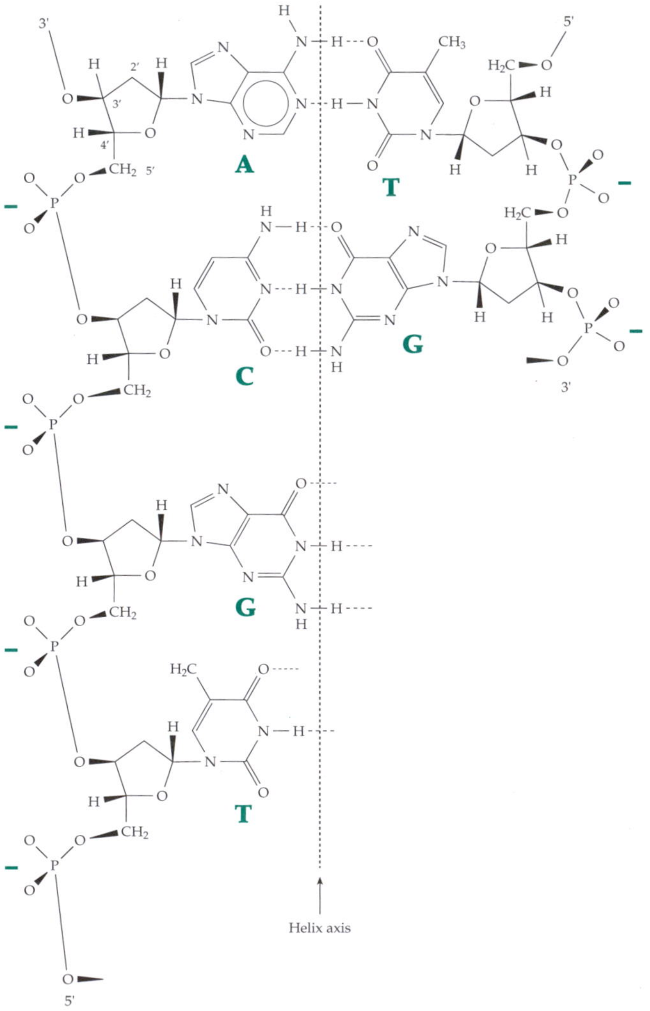

Figure 5-2 A distorted (flattened) view of the Watson–Crick structure of DNA showing the hydrogen-bonded base pairs.

was concrete evidence that DNA carried genetic information (see Chapter 26). However, it was James Watson and Francis Crick's recognition of the double-helical structure of DNA[4-9] in 1953 and the mechanism of replication that this structure implied that captured the imagination of biologists and chemists alike and paved the way for the present-day explosion of knowledge of DNA, RNA, and of the encoded proteins.

Watson and Crick proposed that DNA is a double helix of two antiparallel polynucleotide chains (Figs. 5-2 and 5-3). The structure was deduced from model building together with knowledge of the X-ray diffraction data of Maurice F. Wilkins and Rosalind Franklin[9a] on

The essential correctness of the concept has been proved.

Two extremely important developments came in the 1970s: (1) Methods were found for cutting and rejoining DNA fragments and for **cloning** them in bacteria and (2) ways were devised for rapid determination of **nucleotide sequences**. The application of these techniques is now providing startling advances in biology and medicine. In 1970 we knew virtually nothing about the sequences of nucleotides in genes but today, we know the sequences for many thousands of genes of all types. By the 1980s sequences had been established for hemoglobin,[11] γ-globulins,[12] collagen,[13] and for many enzymes. An example is the gene sequence for mitochondrial aspartate aminotransferase (Fig. 5-4).[14] Its coding regions consist of 1299 pairs of mononucleotides, the bases being paired as in Fig. 5-2 in double helical form. Complete DNA sequences are known for numerous viruses including the 9740-base pair (bp) DNA provirus form of the RNA virus that causes AIDS[15] and for bacterial viruses such as T7 bacteriophage (39,936 bp).[16] Also determined in the 1980s were sequences of human mitochondrial DNA (16,598 bp),[17] and of chloroplast DNA from the tobacco plant (155,844 bp).[18]

In 1995 the first complete sequences of bacterial genomes were obtained (Table 1-3).[19,20] These were followed by sequences of many other bacterial genomes,[21] including the 4.2×10^6 bp *E. coli* genome (Table 1-3). Sequencing of the 16 chromosomes of the 12.07 Mbp genome of yeast[22,23] containing ~6300 genes was completed in 1996[24] and that of the 97 Mbp genome of the nematode *Caenorhabditis elegans* in 1998. By 2000 the sequence of the 180 Mbp genome of the fruit fly *Drosophila melanogaster* was largely completed, and most sequences of the 3×10^9 bp human genome were known.[24a]

Cloning of DNA has not only provided an essential step in sequence determination but also has given birth to a new industry devoted to producing proteins from genes cloned in bacteria, yeast, or other cells. Human insulin and the antiviral protein interferon were two of the first proteins produced in this way. Methods now in use permit us to introduce at will alterations at any point in a DNA sequence. We are able to locate and study the genes responsible for many genetic defects.

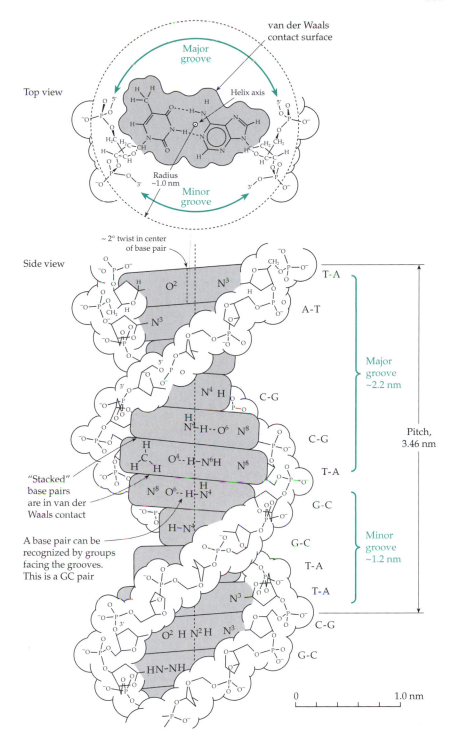

Figure 5-3 The double-helical structure of DNA. The structure shown is that of the B form and is based on coordinates of Arnott and Hukins.[10] The major and minor grooves, discussed on p. 213, are marked.

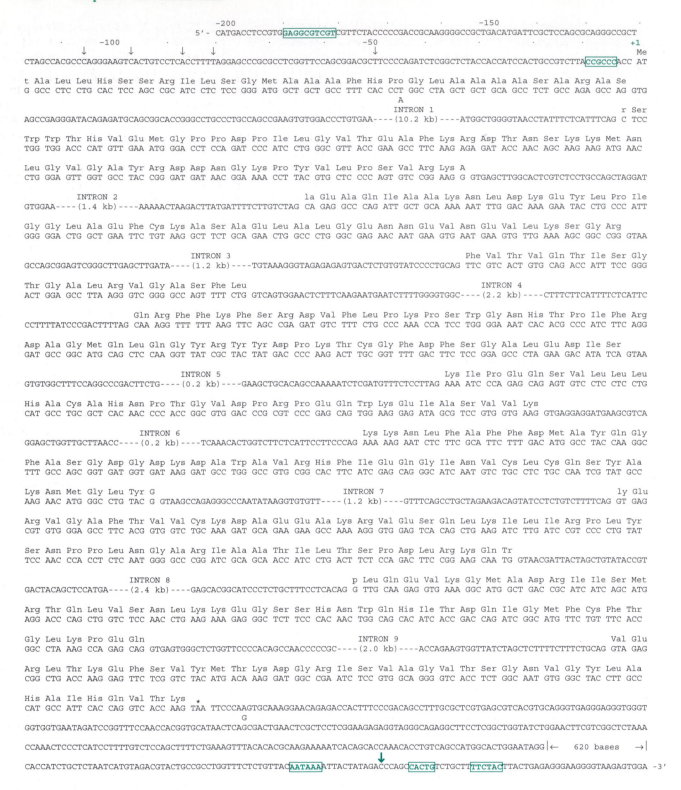

Figure 5-4 The nucleotide sequence of the gene for mitochondrial aspartate aminotransferase from the mouse. From Tsuzuki *et al.*[14] The gene encodes a 433-residue protein requiring 1299 nucleotide pairs (1.3 kb). However, 29 residues are cut off from the N terminus to form the mature mitochondrial protein. In addition, the gene is split by nine introns which vary in length from 0.2–10.2 kb. The sequences at the ends of the introns are shown. There is also an "upstream" region (of which 200 nucleotides are shown) at the 5'-end of the gene. It contains two binding sites for **transcription factor Sp1** (boxed). At the 3' end the 993 additional nucleotides contain signals (boxed) surrounding the **polyadenylation site** (green arrow) for 3' processing and termination of transcription. The mature messenger RNA is about 2400 nucleotides (2.4 kb) in length but the gene, with introns, occupies about 25 kb. The +1 marks the position of the first nucleotide of the initiation codon ATG (encoding methionine) and the asterisk (*) the termination codon TAA. These are AUG and UAA in the mRNA. Other codons are indicated by amino acid abbreviations.

The first successes in using virus-like particles to carry new pieces of DNA into the cells to help correct these defects have been reported. Our ability to breed new varieties of plants and microorganisms has been enormously enhanced. We can foresee the production of artificially designed enzymes to conduct many industrial chemical processes. These are among the many reasons for the excitement today in the fields of nucleic acid chemistry and molecular genetics.

A. Structure and Chemistry of Nucleotides

1. Names and Abbreviations

The purine and pyrimidine ring compounds found in nucleic acids are known as "bases," even though some of them have almost no basic character. **Nucleosides** are the N-glycosyl derivatives of the bases with ribose or 2-deoxyribose. The **nucleotides** are phosphate esters of nucleosides. Similar names are applied to related compounds such as adenosine triphosphate (ATP) that are not present in DNA or RNA. The names of the principal nucleotides from which the nucleic acids are formed are given in Table 5-1. The novice may find these confusing! Even worse than the names in the table is **hypoxanthine** (Hyp), which is derived from adenine by replacement of its $-NH_2$ group with $-OH$ and tautomerization:

The nucleoside formed from hypoxanthine and ribose is known as **inosine** (Ino or I) and the corresponding nucleotide as **inosinic acid**. Further substitution at C-2 of $-H$ by $-OH$ and tautomerization yields **xanthine** (Xan). Its nucleoside is xanthosine (Xao, X). A similar hydroxylation at C-7 converts xanthine to **uric acid**, an important human urinary excretion product derived from nucleic acid bases.

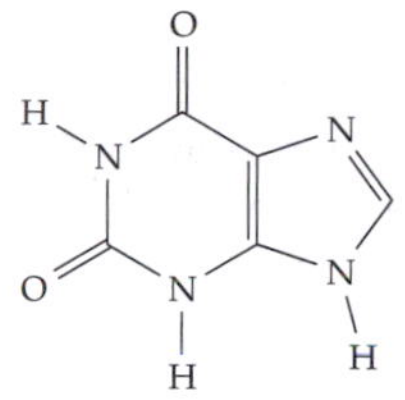

Xanthine 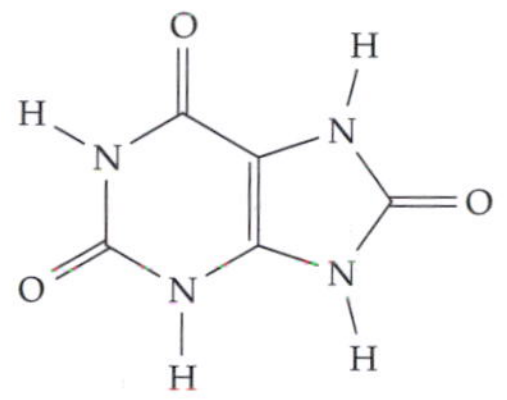   Uric acid

TABLE 5-1
Names of Pyrimidine and Purine Bases, Nucleosides, and 5'-Nucleotides[a]

A   Nucleotide units of RNA (abbreviations in parentheses)[b]

Base:	Uracil (Ura)	Cytosine (Cyt)	Adenine (Ade)	Guanine (Gua)
Nucleoside:	Uridine (Urd or U)	Cytidine (Cyd or C)	Adenosine (Ado or A)	Guanosine (Guo or G)
5'-Nucleotide:	Uridine 5'-phosphate or 5'-uridylic acid (Urd-5'-P or UMP)	Cytidine 5'-phosphate or 5'-cytidylic acid (Cyd-5'-P or CMP)	Adenosine 5'-phosphate or 5'-adenylic acid (Ado-5'-P or AMP)	Guanosine 5'-phosphate or 5'-guanylic acid (Guo-5'-P or GMP)

B   Nucleotide units of DNA

These contain 2-deoxyribose and the nucleosides and nucleotides are called deoxyadenosine (dAdo or dA), deoxyadenosine 5'-phosphate (dAMP), etc.

DNA contains thymine (Thy) rather than uracil. The deoxyribose derivatives are thymidine (dThd or dT) and thymidine 5'-phosphate. The ribose derivatives of thymine are the nucleoside ribosylthymidine (Thd) and ribosylthymidine 5'-phosphate (Thd-5'-P).

C   Abbreviations used in describing polynucleotide sequences

U,T,C,A,G	Uracil, thymine, cytosine, adenine, guanine
Y or Pyr	Pyrimidine (T or C)
R or Pur	Purine (A or G)
M	Amino base (A or C)
K	Keto base (G or T)
S	Strongly pairing (G or C)
W	Weakly pairing (A or T)
H	Not G (any other base)
B	Not A
V	Not T or U
D	Not C
N	Any base

[a] From "Biochemical Nomenclature" Liébecq, C., ed.[25]

[b] Isomers of the 5'-nucleotides, in which the phosphate is attached to the oxygen on C-3', are the 3'-nucleotides. Care must be taken to avoid ambiguity. The simple abbreviations UMP, CMP, AMP, and GMP always refer to the 5'-nucleotides.

Nucleic acid structures are abbreviated in several ways. For example, the sugar rings may be portrayed by vertical lines. The abbreviations A, C, U, T, and G for the individual bases or Pu (purine) and Py (pyrimidine) are placed at the upper ends of the lines, and slanted lines with P in the centers represent the 3'–5' phosphodiester linkages in a polynucleotide.

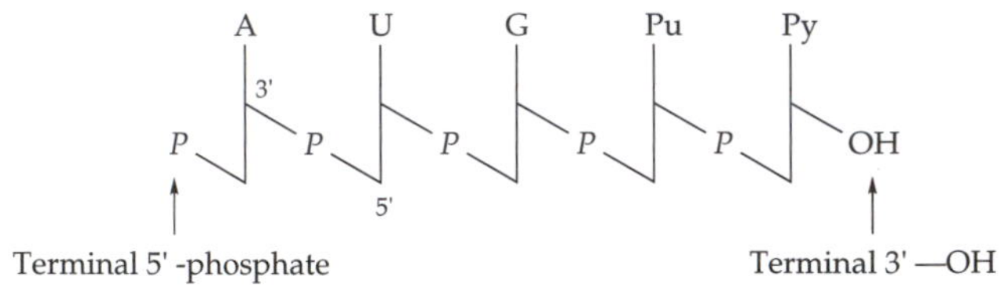

Terminal 5'-phosphate   Terminal 3'—OH

The same structure can be further abbreviated

	5' end	pApUpGpRpY	3' end
or	5' end	A U G R Y	3' end

Here purine is abbreviated R and pyrimidine Y. By convention the 5' end of a polynucleotide is ordinarily placed to the left in these formulas. Lengths of nucleic acid chains are usually given as a number of bases or kilobases (kb). For double-stranded DNA (dsDNA) the length is given as base pairs (bp), kilobase pairs (kbp), or megabase pairs (Mbp). However, in most places, including this book, the abbreviations kb and Mb will be used for a length of DNA whether single or double stranded.

For double-stranded DNA, one strand, usually the **coding strand,** from which the amino acid sequence can be read using the code in Tables 5-4 or 5-5 (Section C), has the 5' end at the left while the complementary strand has the 3' end at the left, e.g.,

Ala	Tyr	Gly	Pro	Phe	Cys	Gln	Termination
5' G C C	T A T	G G A	C C T	T T C	T G C	C A G	T G A 3'
3' C G G	A T A	C C T	G G A	A A G	A C G	G T C	A C T 5'

2. Acid–Base Chemistry and Tautomerism

The ionized phosphate groups of the polymer "backbone" give nucleic acid molecules a high negative charge. For this reason DNA in cells is usually associated with basic proteins such as the **histones** or **protamines** (in spermatozoa), with polycations of amines such as **spermidine** ($H_3^+NCH_2CH_2CH_2CH_2$ $NH_2^+CH_2CH_2CH_2NH_3^+$), or with alkaline earth cations such as Mg^{2+}. If the pH of a solution containing double-helical DNA is either lowered to ~3 or raised to ~12, the two strands unravel and can be separated.

Over the entire range of pH the alternating sugar-phosphate backbone of the polymeric chains remains negatively charged. However, depending on the pH, the bases can be protonated or deprotonated with a resultant breaking of the hydrogen bonds that hold the pairs of bases together.[26,27]

Pyrimidines and purines, which contain the $-NH_2$ group, are weakly basic. The cationic protonated conjugate acid forms of cytidine, adenosine, and guanosine have pK_a values of 4.2, 3.5, and 2.7, respectively. Similar values are observed for the 5'-nucleotides. In these compounds it is not the $-NH_2$ group that binds the proton but an adjacent nitrogen atom in the ring (Eq. 5-3).

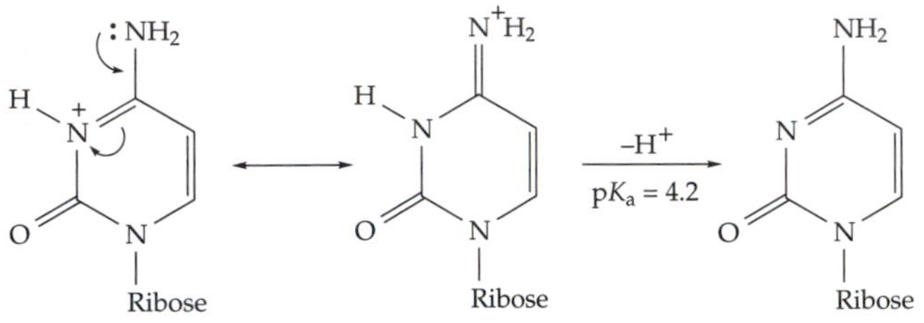

$$(5-3)$$

We can understand this if we recognize that the bases have substantial aromatic character.[26,28] In aniline (aminobenzene) electrons are withdrawn from the amino group into the aromatic ring with a strong decrease in basicity of this $-NH_2$ group (the pK_a is 4.6). Similarly, electrons are withdrawn from the NH_2 groups of cytosine, adenosine, and guanine into the pyrimidine and purine rings as is indicated by the small curved arrows on the left-hand structure of Eq. 5-3. The effect is even stronger than in aniline, largely because of the presence of the nitrogen atoms in the rings. In cytosine it is primarily N-3 that serves as the electron acceptor. As a consequence this nitrogen becomes more basic than the $-NH_2$ group and is the major site of protonation. However, as is indicated in Eq. 5-3, the positive charge on the cation is shared by resonance with the exocyclic amino group.

Adenosine is similar to cytosine in its acid–base chemistry; N-1, adjacent to the $-NH_2$ group, is the principal site of protonation. A tautomer of the cation protonated at N-3 is formed in smaller amounts. Guanosine is electronically more complex, being protonated mainly at N-7 and to a lesser extent at N-3[29]. This can be understood in terms of electronic interaction with the adjacent oxygen as indicated in the resonance structure to the right in the following diagram:

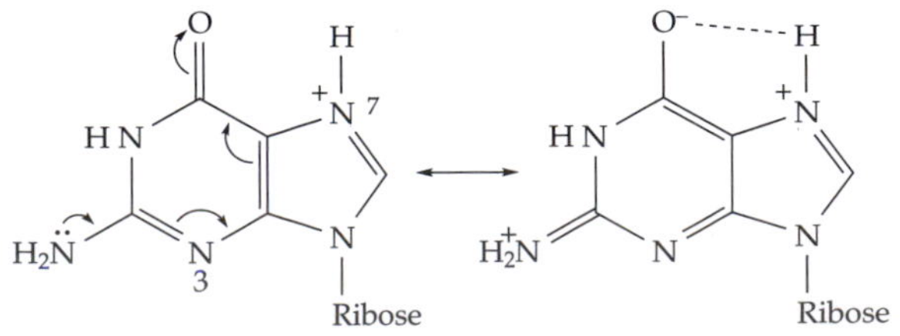

Under basic conditions the proton on N-3 of uridine or thymine or on N-1 of guanosine can dissociate with a pK_a of ~9.2. These "bases" are actually weak acids!

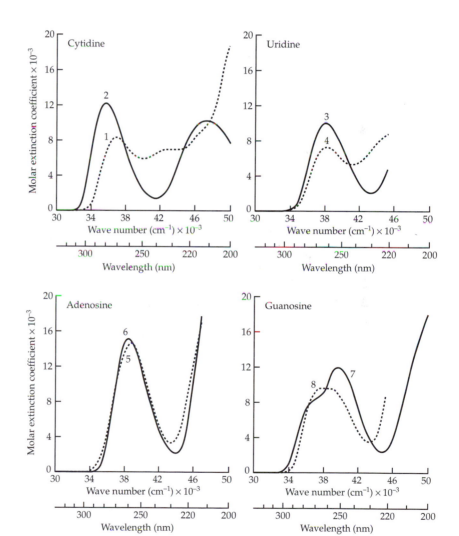

(5-4)

Look at the "Kekule" resonance structure shown in the center of Eq. 5-4. The two negative charges flanking the dissociable proton are sufficient to hold it firmly to the ring at low pH. The proton is half dissociated only when the pH is raised to 9.2. At still higher values of pH (pK_a ~12.4) a proton dissociates from a ribose hydroxyl group.

The tautomerism of pyrimidines was discussed in Chapter 2, Section A,6. The tautomeric forms shown in Fig. 5-1 predominate. However, it is possible that minor tautomers such as the following are sometimes preferentially bound into active sites of enzymes where the dielectric

A minor tautomer of uridine

constant may be low and where the geometrical arrangement of amino acid functional groups may favor protonation on oxygen rather than nitrogen.[30,31]

3. Absorption of Ultraviolet Light

Nucleic acids strongly absorb ultraviolet light of wavelengths below about 300 nm, with an absorption maximum at ~260 nm and a stronger one below 200 nm. This property is

biologically important because of the resultant induction of mutations, a natural result of exposure to sunlight. In the laboratory the same property is useful in identification and quantitative analysis of nucleotides. The ultraviolet spectra of four nucleosides are shown in Fig. 5-5. Notice the changes that accompany protonation or deprotonation of the ring. The absorption bands, which are related to those of benzene[26,32] (see Chapter 23), provide another indication of the partial aromatic character of the bases. Chapter 23 also provides information on photochemical reactions of the pyrimidines.

Figure 5-5 Near ultraviolet absorption spectra of cytidine, uridine, adenosine, and guanosine. 1. Monoprotonated form of cytidine (for which $pK_a = 4.2$). 2. Neutral form (pH ~ 7) of cytidine. 3. Neutral form of uridine (for which $pK_a = 9.2$). 4. Monoanionic form of uridine. 5. Monoprotonated form of adenosine ($pK_a = 3.5$). 6. Neutral form of adenosine. 7. Neutral form of guanosine ($pK_a = 9.2$). 8. Monoanion of guanosine.

BOX 5-A THE ALKALI METAL IONS

The many negative charges along a nucleic acid backbone interact with all of the cations in a cell. This box discusses some of these ions with emphasis on the group IA metal ions. Although sodium and potassium occur in similar amounts in the crust of the earth, living cells all accumulate potassium ions almost to the exclusion of sodium.[a-c] Sodium ions may be required only by certain marine organisms and by multicellular animals that regulate their internal body fluids. Most nonmarine plants have no demonstrable need for sodium.

The tendency to accumulate K^+ is even more remarkable since seawater is ~0.46 M in Na^+ and only 0.01 M in K^+. Other alkali metals occur in even smaller amounts, e.g., 0.026 mM Li^+, 0.001 mM Rb^+, and a trace of Cs^+. Soil water is ~0.1 mM in K^+ and 0.65 mM in Na^+. Again, strong discrimination in favor of potassium is observed in uptake by plants.

Intracellular concentrations of K^+ range from 200 mM in *E. coli* and 150 mM in mammalian muscle to ~30 mM in freshwater invertebrates such as clams, hydra, and some protozoa. While K^+ cannot be replaced by Na^+, a partial replacement by Rb^+ and to a lesser extent by Cs^+ is usually possible. In many microorganisms rubidium can almost completely replace potassium, and even a rat can survive for a *short* time with almost complete substitution of K^+ by Rb^+. Protons replace most K^+ in brown algae.[d] The human nutritional requirement for potassium is high, amounting to ~2 g/day. Present populations may suffer a chronic deficiency of potassium as a result of food processing and boiling of vegetables.[e]

Sodium is also essential to higher animals, and rats die on a sodium-free diet. The sodium content of cells varies among species, but it is usually no more than 0.1 –0.2 times that of K^+. A measurement of the $[Na^+]$ within heart cells gave a concentration of ~9 mM, which was increased by a factor of ~2.5 in a low Ca^{2+} insulin-containing medium.[f] In this measurement the NMR resonance of the abundant external Na^+ was shifted by use of a paramagnetic reagent (e.g., a dysprosium (III) complex), that remained outside the cell. The signal from the internal Na^+ was then seen clearly. In blood, the relationship between Na^+ and K^+ concentrations is reversed from that within cells. Human plasma is 0.15 M in Na^+ and 0.005 M in K^+. Curiously, the taste for salt in the diet appears to be largely an acquired one.[g]

It is not immediately obvious why K^+ is the preferred counterion within tissues, but a fundamental reason may lie in the differences in hydration between Na^+ and K^+ (Chapter 6). On the other hand, the relationship of these ions to the excitability of membranes (Chapter 30) may be of paramount importance, even in bacteria. The concentration differences in the two ions across membranes represent a readily available source of Gibbs energy for a variety of membrane-associated activities. Cells actively pump Na^+ out and K^+ into cells (Chapter 8).

Many intracellular enzymes require K^+ for activity.[b,c] These include those promoting phosphorylation of carboxyl groups or enolate anions and elimination reactions yielding enols as well as some enzymes dependent upon the coenzyme pyridoxal phosphate.[h-j] In all of these enzymes NH_4^+, Rb^+, or Tl^+ can usually replace K^+. This permits study of the binding site for Tl^+ by the very sensitive ^{205}Tl NMR spectroscopy.[k] The discovery that K^+ is preferentially bound in some tetraplex DNA structures (see Fig. 5-8) further emphasizes the significant difference in biological properties of the alkali metal ions. Various synthetic macrocyclic compounds are also able to selectively bind specific alkali metal ions.[l]

The following tabulation gives concentrations not only of K^+ and Na^+ but also of the other principal ionic constituents in human blood plasma and within cells of skeletal muscle.[m] Units are mmol/kg H_2O.

Ion	Blood plasma	Skeletal muscle (intracellular)
Na^+	150	14
K^+	5	150
Mg^{2+}	0.9	8
Ca^{2+}	2.5	1
Cl^-	105	16
HCO_3^-	27	10
Proteins⁻	17*	50*
Other anions†	6	146

* Milliequivalents/kg H_2O
† Phosphates and other nonprotein anions.

[a] Kernan, R. P. (1965) *Cell K*, Butterworth, London

[b] Suelter, C. H. (1974) in *Metal Ions in Biological Systems*, Vol. 3 (Sigel, H., ed), pp. 201–251, Dekker, New York

[c] Suelter, C. H. (1970) *Science* **168**, 789–795

[d] Steinbach, H. B. (1962) *Comp. Biochem. Physiol.* **4**, 677–720

[e] Weber, C. E. (1970) *J. Theor. Biol.* **29**, 327–328

[f] Wittenberg, B. A., and Gupta, R. K. (1985) *J. Biol. Chem.* **260**, 2031–2034

[g] Kaunitz, H. (1956) *Nature (London)* **178**, 1141–1144

[h] Toney, M. D., Hohenester, E., Cowan, S. W., and Jansonius, J. N. (1993) *Science* **261**, 756–759

[i] Antson, A. A., Demidkina, T. V., Gollnick, P., Dauter, Z., Von Tersch, R. L., Long, J., Berezhnoy, S. N., Phillips, R. S., Harutyunyan, E. H., and Wilson, K. S. (1993) *Biochemistry* **32**, 4195–4206

[j] Metzler, C. M., Viswanath, R., and Metzler, D. E. (1991) *J. Biol. Chem.* **266**, 9374–9381

[k] Markham, G. D. (1986) *J. Biol. Chem.* **261**, 1507–1509

[l] Christensen, J. J., Hill, J. O., and Izatt, R. M. (1971) *Science* **174**, 459–467

[m] Composite data from Muntwyler, E. (1968) *Water and Electrolyte Metabolism and Acid–Base Balance*, p. 14. Mosby, St. Louis, Missouri; White, A., Handler, P., and Smith, E. L. (1973) *Principles of Biochemistry*, 5th ed., p. 802. McGraw-Hill, New York; Long, C. (1961) *Biochemist's Handbook*, p. 670. Van Nostrand, Princeton, New Jersey. Reported ranges for some constituents are very wide.

4. Chemical Reactions of the Bases

The purines and pyrimidines are relatively stable compounds with considerable aromatic character. Nevertheless, they react with many different reagents and, under some relatively mild conditions, can be completely degraded to smaller molecules. The chemistry of these reactions is complex and is made more so by the fact that a reaction at one site on the ring may enhance the reactivity at other sites. The reactions of nucleic acids are largely the same as those of the individual nucleosides or nucleotides, the rates of reaction are often influenced by the position in the polynucleotide chain and by whether the nucleic acid is single or double stranded. The reactions of nucleosides and nucleotides are best understood in terms of the electronic properties of the various positions in the bases.[26,33] Most of the chemical reactions are nucleophilic addition or displacement reactions of types that are discussed in Chapters 12 and 13.

Positions 2, 4, and 6 of pyrimidine bases are deficient in electrons and are therefore able to react with nucleophilic reagents. The 6 position is especially reactive toward additions, while the 2 position is the least reactive. The corresponding electron-deficient positions in the purine bases are 2, 6, and 8. These positions, which are marked by asterisks on the following structures, have electrophilic character in all of the commonly occurring pyrimidines and purines.

All of the oxygen and nitrogen atoms in the pyrimidines, as well as the 5 position of the ring, have nucleophilic character and can therefore react with electrophilic centers of various reagents. A number of specific reactions that have been found useful to biochemists are described in Section H,3.

5. Base Pairs, Triplets, and Quartets

The purine and pyrimidine bases are the "side chains" of the nucleic acids. The polar groups that are

$$>C=O, \quad >NH, \text{ and } -NH_2$$

present in the bases can form hydrogen bonds to other nucleic acid chains, e.g., in the base pairs of the DNA

double helix, and also to proteins. Figure 5-6 shows the shapes and the hydrogen bonding groups available in the bases. The number of both electron donor groups and proton donors available for hydrogen bonding is large and more than one mode of base pairing is possible.

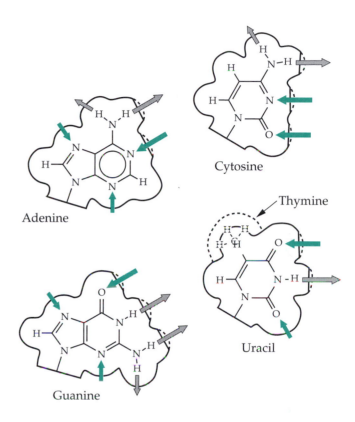

Figure 5-6 Outlines of the purine and pyrimidine bases of nucleic acids showing van der Waals contact surfaces and some of the possible directions in which hydrogen bonds may be formed. Large arrows indicate the hydrogen bonds present in the Watson–Crick base pairs. Smaller arrows indicate other hydrogen bonding possibilities. The directions of the green arrows are from a suitable hydrogen atom in the base toward an electron pair that serves as a hydrogen acceptor. This direction is *opposite* to that in the first edition of this book to reflect current usage.

The base pairs proposed by Watson and Crick are shown in Fig. 5-2 and again in Fig. 5-7. While X-ray diffraction studies indicate that it is these pairs that usually exist in DNA, other possibilities must be considered. For example, Hoogsteen proposed an alternative A-T pairing using the 6-NH_2 and N-7 of adenine.[34] Here the distance spanned by the base pair, between the C-1' sugar carbons, is 0.88 nm, less than the 1.08 nm of the Watson–Crick pairs. Duplexes of certain substituted poly (A) and poly (U) chains contain only Hoogsteen base pairs[35] and numerous X-ray structure determinations have established that Hoogsteen pairs

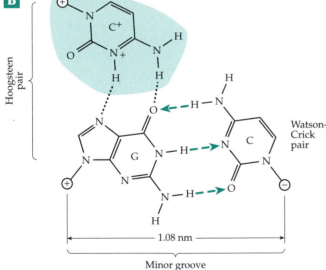

Hoogsteen A*T base pair

0.88 nm

do occur in true nucleic acids.[36] Figure 5-7 shows the structures of two Hoogsteen pairs. In each case the purine component has formed a Watson–Crick pair with a third base to give a **base triplet**. The first of these triplets may be designated T*A•T, where * represents the Hoogsteen hydrogen bonding and • represents the Watson–Crick bonding. The second triplet in Fig. 5-7, C⁺*G•C, can form only with the N-1 protonated form (low pH form) of cytosine.

Reversed Watson–Crick or Hoogsteen AU or AT pairs are formed if the 2-carbonyl rather than the 4-carbonyl of the U or T makes a hydrogen bond with the amino group of adenine.[27] Because of the resulting arrangement of the ribose rings the base pairs cannot fit into the ordinary Watson–Crick double-stranded DNA structure. A reversed Watson–Crick pair can also be formed between G and C but with only two hydrogen bonds, while a reversed Hoogsteen pair can form only if a minor tautomer of cytosine is used.

Hoogsteen pairs were first observed in nature in transfer RNA molecules (Fig. 5-31). These molecules contain mostly Watson–Crick base pairs but there are also two reversed Hoogsteen pairs. One of them, between U8 and A14, is invariant in all tRNAs studied. Hoogsteen pairing also occurs in four-stranded DNA, which has important biological functions. A **G quartet** from a DNA tetraplex held together by Hoogsteen base pairs is shown in Fig. 5-8.

Figure 5-7 Two base triplets that form in triple-stranded DNA and involve both Watson–Crick and Hoogsteen base pairing. (A) The triplet T*A•T, where the T (marked T*) of the third strand is hydrogen bonded as a Hoogsteen pair (·······) to an adenine of a Watson–Crick AT pair (whose hydrogen bonds are indicated (‒ ‒ ‒). (B) The triplet C⁺*G•C, where C⁺ is cytosine in its N-1 protonated (low pH) form. The Watson–Crick strands are antiparallel, as indicated by the ⊕ and ⊖ signs. The third strand may have either orientation, but when it contains largely pyrimidines it is parallel to a purine-rich strand. An example is shown in Fig. 5-24.

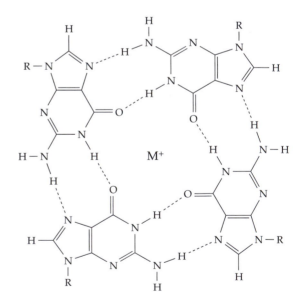

Figure 5-8 A guanine quartet held together by Hoogsteen base pairing. This structure is found in the telomeres at the ends of linear chromosomes. Four segments of DNA, each of which may be part of a single folded strand, (see p. 227) give structures in which four or more of these G quartets are stacked one above the other. Monovalent ions, usually K⁺ or Na⁺, are bound in the center, although not always in the plane of the bases. See Fang and Cech[38] and Gellert *et al.*[39]

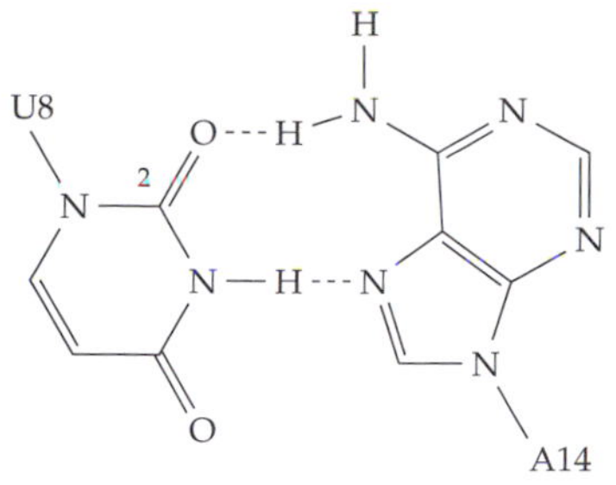

U8

A reversed
Hoogsteen base
pair present in
tRNA molecules

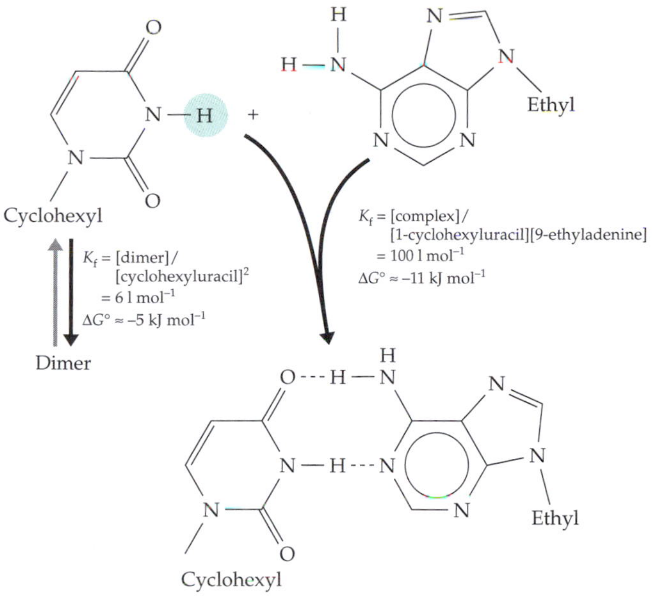

A14

K_f = [dimer]/
[cyclohexyluracil]^2
= 6 l mol^{-1}
$\Delta G° \approx -5$ kJ mol^{-1}

Dimer

K_f = [complex]/
[1-cyclohexyluracil][9-ethyladenine]
= 100 l mol^{-1}
$\Delta G° \approx -11$ kJ mol^{-1}

Ethyl

Cyclohexyl

Ethyl

Cyclohexyl

Complex (base pair)

$$(5\text{-}5)$$

Another pairing that occurs in tRNAs allows guanine to pair with uracil, e.g., G4 with U69. This was originally proposed to account for codon–anticodon interactions betweentRNA molecules and messenger RNA (Chapter 29). It is commonly called **wobble pairing** because the uracil must wobble away from its orientation in the normal Watson–Crick pair.[27,37]

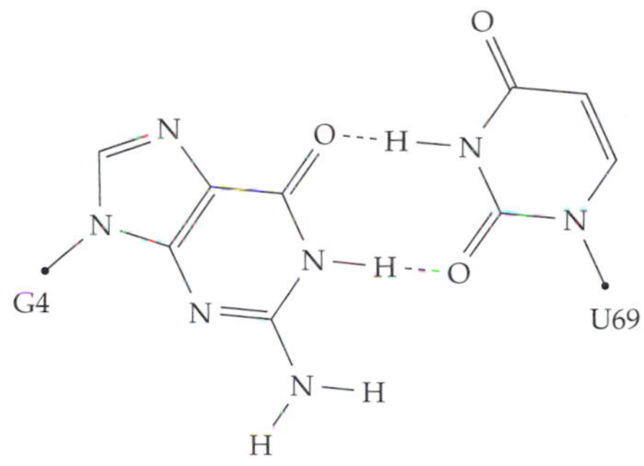

G4

U69

A "wobble" base pair

Strengths of base pairs. How strong are the bonds between pairs of bases in DNA? The question is hard to answer because of the strong interaction of the molecules with polar solvents through hydrogen bonding and hydrophobic effects. Some insight has come from studies of the association of bases in nonpolar solvents. Thus, 1-cyclohexyluracil forms a dimer involving either hydrogen bonding or stacking, but the association is weak with the Gibbs energy of formation $\Delta G_f° = 5$ kJ mol^{-1}. When the same compound was mixed with 9-ethyladenine, a base-paired complex formed between the two compounds with a formation constant over ten-fold greater than that for the dimer (Eq. 5-5).[40] When the circled hydrogen atom in Eq. 5-5 was replaced by $-CH_3$, which blocked the pairing, K_f fell below 1 kJ/mol ($\Delta G° > 0$). The difference in ΔG_f in the two cases was only 7 kJ mol^{-1}. Many other estimates have been made.[40a,40b] The small energies summed over the many base pairs present in the DNA molecule help provide stability to the structure.

Stacking of bases. The purines and pyrimidines of nucleic acids, as well as many other compounds with flat ring structures and containing both polar and nonpolar regions, are sparingly soluble in either water or organic solvents. Molecules of these substances prefer neither type of solvent but adhere tightly to each other in solid crystals. Both experimental measurements and theoretical computations[41] suggest that hydrogen bonding is the predominant force in the pairing of bases in a vacuum or in nonpolar solvents. However, in water stacking becomes important.[42] In a fully extended polynucleotide chain consecutive bases are 0.7 nm apart, twice the van der Waals thickness of a pyrimidine or purine ring, but in double-helical DNA of the B type (Figs. 5-3, 5-12) the distance between consecutive base pairs is only 0.34 nm. They are touching.

One effect of stacking is a decrease in the expected intensity of light absorption. The molar extinction coefficient of a solution of double-helical DNA or RNA is always less by up to 20–30% than that predicted from the spectra of the individual nucleosides (Fig. 5-5). This **hypochromic effect** is considered further in Chapter 23.

Because both hydrogen bonding and stacking are involved, the thermodynamics of base pairing in nucleic acids is complicated.[43] The hydrophobic parts of exposed bases tend to induce an ordering of the surrounding water molecules and therefore a decrease in their entropy. However, hydrogen bonding of the polar groups of the bases to the solvent causes a decrease in water structure. This is greater than the increase in structure around the hydrophobic regions and the stacking of bases leads to a net decrease in entropy. The entropy change ΔS for addition of a base pair to the end of a double-stranded RNA helix in a hairpin loop such as that displayed in Fig. 5-9 ranges from −0.05 to −0.15 kJ/degree per base pair.[44,45]

The enthalpy change ΔH tends to be small and positive for association of alkyl groups in water and nearly zero for association of aromatic hydrocarbons

(Chapter 2). However, ΔH is distinctly negative for association of heterocyclic bases. This has also been attributed to a decrease in the ordering of solvent around the bases as a result of exclusion of water. Attraction or repulsion of partial charges on the polar groups comprising the purine and pyrimidine bases may also be an important factor.[43,46–48] For addition of a base pair to an RNA helix, the change in enthalpy, ΔH, varies from about –24 to –60 kJ/mol.[44,45]

Since $\Delta G = \Delta H - T\Delta S$, the net result is a negative value of ΔG, a "hydrophobic effect" that favors association of bases. Substantial efforts have been made to estimate quantitatively the Gibbs energies of formation of helical regions of RNA molecules in hairpin stem-loops such as that of Fig. 5-9.[44,45,49–51] Table 5-2 shows the observed increments in $\Delta G_f°$ of such a helix upon addition of one base pair at the end of an existing helix. Addition of an AU pair supplies only –4 to –5 kJ

to $\Delta G_f°$. The exact amount depends upon whether an A or a U is at the 5' end in the existing helix. If an AU pair is added to the helix terminating in CG or GC, about –9 kJ/mol is added. Larger increases in $-\Delta G_f°$ result from addition of GC pairs, which contain three hydrogen bonds between the bases versus the two in AU pairs.[27,41,52]

UG pairs provide a very small amount of stabilization to an RNA double helix, while the presence of unpaired bases has a destabilizing effect. The most stable hairpin loops contain four or five bases. Depending upon whether the loop is "closed" by CG or AU, the helix is destabilized by 20–30 kJ/mol. "Bulge loops," which protrude from one side of a helix, have a smaller destabilizing effect. An example of the way in which Table 5-2 can be used to estimate the energies of formation of a loop in a straight-chain RNA is illustrated in Fig. 5-9. Similar analysis of base pairing in DNA can also be done.[53–55]

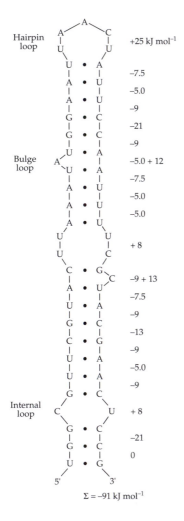

Figure 5-9 The contribution of base-paired regions and loops to the Gibbs energy of a possible secondary structure for a 55 base fragment from R17 virus. The stem-loop structure shown here is part of a larger one considered by Tinoco *et al.*[58]

TABLE 5-2
Gibbs Energies of Formation $\Delta G_f°$ at 25°C for Addition of One Base Pair to an Existing RNA Helix[a,b]

Base pair at end of existing helix	Base pair added	$\Delta G_f°$ (kcal mol^{-1}±10%)	$\Delta G_f°$ (kJ mol^{-1}±10%)
A • U	A • U	–1.2	–5.0
U[c] • A	A • U	–1.8	–7.5
C • G or G[d] • C	A • U	–2.2	–9
C • G	G • C	–3.2	–13
G • C	C • G or G • C	–5.0	–21
G • U	U • G	–0.3	–1

Hairpin loops of 4 or 5 bases		
Closed by GC	+5	+21
Closed by AU	+7	+29
Bulge loops		
1 base	+3	+12
4–7 bases	+5	+21

[a] Table modified from that of Tinoco *et. al.*[58]

[b] All base pairs in the table are oriented as follows: 5'—A→3'
                                                        3'←U—5'

[c] $\Delta G_f°$ is the same for U added to an A end.
     A   U

[d] $\Delta G_f°$ is the same for C or G added to an A end.
     G   C   U

Tautomerism and base pairing. Tautomerism has an interesting relationship to the formation of the pairs and triplets of hydrogen bonds in DNA or RNA. Each base exists predominately as one preferred

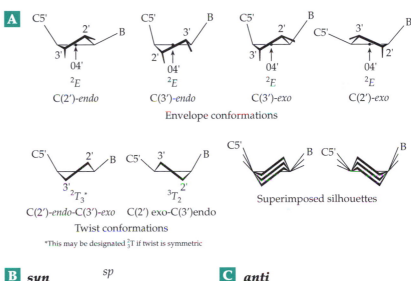

A   B   (5-6)

tautomer, but at any moment a very small fraction is present as less stable tautomers. Watson and Crick suggested that this fact may be responsible for the occurrence of some mutations.[56,57] Thus, tautomer B of Eq. 5-6 would not be able to pair with thymine, its proper pairing partner, but could pair with cytosine.

Similarly, tautomer B of uracil in Eq. 2-4 could pair with guanine instead of its proper partner adenine. If a similar event occurred to an AT pair during gene replication an incorrect copy of the gene, differing in a single "code letter," would be formed. However, because the tautomerism is affected so strongly by the environment (Chapter 2), the extent to which it may cause mispairing while replication enzymes act is uncertain.[59] Both bases in a pair could be tautomerized by synchronous transfer of protons in two parallel hydrogen bonds. However, theoretical calculations predict a high energy barrier to this process.[60] Proton transfer, which can also be induced by light, has been studied on a femtosecond scale.[61] After

photochemical transfer of one proton a second is transferred within a few picoseconds.

6. Conformations of Nucleotides

The furanose ring of ribose or deoxyribose is flexible and can be interconverted smoothly among an infinite number of envelope (E) and skew or twist (T) conformations. See Chapter 4, Section A,2. However, there are limits set by steric and anomeric effects.[27,62–64] Conformations are often described as in Fig. 5-10 by stating which atom in an envelope conformation lies mostly out of the plane of the other four atoms. If this atom lies above the ring, i.e., toward the base, the ring

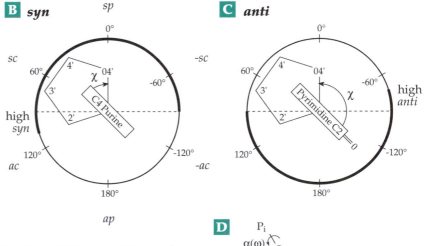

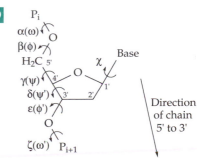

Figure 5-10 Conformational properties of nucleosides. (A) Representations of several conformations of a ribose or deoxyribose ring in a nucleoside. (B) View down the N–C axis joining a purine base to the sugar in a nucleoside in a *syn* conformation. The atom marked C4 is in the six-membered ring, which protrudes further over the sugar ring. The angle χ is measured from 0° to ±180°. *Syn* conformations are those for which χ falls in the region of the heavy semicircle. (C) An *anti* conformation of a pyrimidine nucleoside. Other *anti* conformations fall in the region of the heavy semicircle. (D) Labeling of the conformational angles of the main chain in a polynucleotide.[25,27,37,37a] See also Fig. 2-2.

conformation is known as **endo**; when below the ring, it is known as **exo**. The C(2')-*endo* (^2E) and C(3')-*endo* (^3E) conformations are most commonly approximated in nucleotides and nucleic acids.[65] The C(3')-*exo* conformation is designated E_3, the twist conformation C(2')-*endo*-C(3')-*exo* as $^2T_3$, etc. A conformation can be specified more precisely by the five torsion angles ν_0 to ν_4 (which have also been designated τ_0 to τ_4). All of the envelope and twist conformers can be interconverted readily. The interconversions can be imagined to occur in a systematic way by **pseudorotation**, a rotation of the pucker around the sugar ring. In a commonly used convention a pseudorotation phase angle P ranges from $0°$ to $360°$ as the pucker moves *twice* around the ribose

ring to restore the original conformation. P is taken as $0°$ for the symmetric $^3T_2$ twist, $-18°$ for E_2, $+18°$ for ^3E, $+54°$

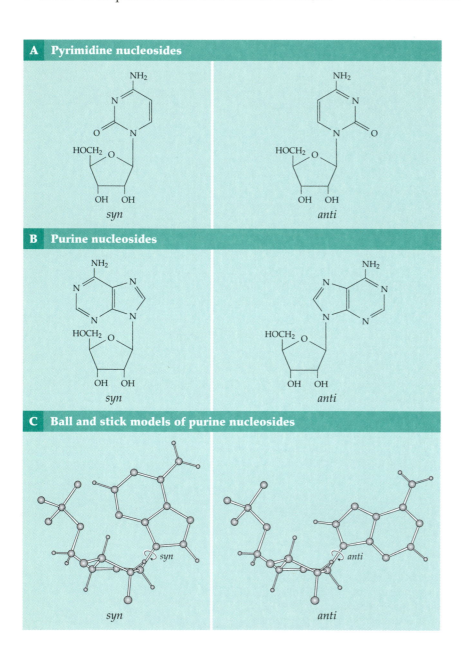

for ^4E, $+198°$ for E_3, etc. The five torsion angles ν_j, for j = 0 – 4 as defined in the foregoing diagram, are related to P as follows:

$$\nu_j = \nu_{max} \cdot \cos[P + j \cdot \phi] \quad (5\text{-}7)$$

where $\phi = 720°/5 = 144°$. The maximum torsion angle, ν_{max}, is about $40°$. See Saenger[27] for details.

Conformational alterations of ribose and deoxyribose rings can occur within polynucleotides and are of biochemical importance. An interesting consequence of changes in ring conformation is that the distance between the C(5') and N atoms attached to the sugar ring of a nucleoside may vary by as much as 0.05 nm (see Fig. 5-10A). The orientation of a base with respect to the sugar is specified by the angle χ. In one convention (Fig. 5-10B,C) its zero value is taken as that in which the N(1)–C(2) bond of a pyrimidine or the N(9)–C(4) bond of a purine is *cis* to the C(1')–O(4') bond, [sometimes called the C(1')–O(1') bond]. However, other definitions have been used.[27] Typical values of χ for nucleotides and nucleosides vary between $-75°$ and $-165°$. In these **anti** conformations the CO and NH groups in the 2 and 3 positions of the pyrimidine ring (or in positions 1, 2, and 6 of the purine ring) are *away* from the sugar ring, while in the **syn** conformations they lie over the ring. *Anti* conformations are more often present than *syn* in nucleic acids. A different view of *syn* and *anti* pairs of nucleotides is shown in Fig. 5-11.

An additional five torsion angles are needed to specify the backbone conformation of a polynucleotide. According to the convention adopted by the IUB the six angles are desig-

Figure 5-11 Views of a nucleoside in *syn* and *anti* conformations.[66] (C) Courtesy of Dr. Muttaiya Sundaralingam.

nated α, β, γ, δ, ε, and ζ, as is indicated in Fig. 5-10D.[25,27] In an older but much used convention,[66] starting at any phosphorus atom the angles ω, ϕ, and ψ specify the next three torsion angles as one moves toward the 3' end of the chain, while ω', ϕ', and ψ' specify the angles lying toward the 5' end (these are shown in parentheses in Fig. 5-10D).   Notice that $\delta(\psi') = \nu_3 + 120°$.

B. Double Helices

1. The B Form of DNA

This form of DNA, whose structure is depicted in Figs. 5-3 and 5-12, is stable at high humidity and is thought to approximate that of most DNA in cells.[27,37,67] If we look directly at the axis of the double helix and perpendicular to one of the base pairs, and ignore the fact that the base pair is asymmetric, we see that the nucleotide unit in one chain is related to the nucleotide unit lying across from it in the opposite plane by a two-fold axis of rotation (**dyad axis**). This symmetry element, which arises from the antiparallel arrangement of the chains, makes the DNA molecules from the outside look nearly identical whether viewed from one end or the other – and whether viewed as a model by the human eye or through contact with an enzyme which might act on the molecule. Actually, *the two chains are not identical*, and the genetic information can be read off from the functional groups exposed in two **grooves** in the surface of the helix (Figs. 5-3 and 5-7). The broader groove in the B form, which is referred to as the **major groove**, is about 0.85 nm deep and 1.1–1.2 nm wide when allowance is made for the van der Waals radii of the atoms. In some other forms of DNA the major groove is narrow, but it can always be identified by the larger of the arcs that can be drawn between the two N–C bonds of the nucleosidic linkages in a nucleotide pair. The **minor groove** or narrow groove, which is defined by the smaller arc between the two N–C bonds, is ~ 0.75 nm deep and 0.6 nm wide in B-DNA.

The diameter of the double helix of B-DNA, measured between phosphorus atoms, is just 2.0 nm. The rise per turn, the **pitch**, is 3.4 nm. There are about ten base pairs per turn (9.7 and 10.6 in two different crystal forms).[68,69] Thus, the rise per base pair is 0.34 nm, just the van der Waals thickness of an aromatic ring (Table 2-1). It is clear that the bases are stacked in the center of the helix. A 1000-bp (1-kb) gene would be a segment of DNA rod about 340 nm long, about 1/40 the length of the molecule in the electron micrograph of Fig. 5-13.

As is appropriate for the cell's master blueprint, DNA in the double helix is stable. Factors contributing to this stability are (1) the pairs and triplets of hydrogen bonds between the bases; (2) the van der Waals

attraction between the flat bases which stack together; (3) the fact that on the outside of the molecule are many oxygen atoms, some negatively charged, which are able to form strong hydrogen bonds with water, with small ions, or with proteins that surround the DNA; and (4) the ability to form superhelices (see Section C,3). Nevertheless, the long DNA chains present in our chromosomes are frequently broken and an elaborate system of repair enzymes is needed to preserve the reliability of this master code for the cell.[73,74]

Today it is generally accepted that most DNA exists in nature as a double helix resembling the B form. However, doubts were expressed as recently as 1980. The reason for the uncertainty lay in the fact that X-ray data from the stretched "paracrystalline" DNA fibers used in earlier studies are much less precise than those from true crystals.[10] The X-ray data for B-DNA fibers could also be interpreted in terms of an alternative "side-by-side" structure which would permit easier separation of strands during replication.[75] However, numerous high-resolution X-ray structure determinations on single crystals of synthetic DNA fragments have confirmed the double-helical structure and the presence of Watson–Crick base pairs.[68,69,76–84] Similarly, fragments related to the double helices of RNA have been crystallized and the structures determined to atomic resolution.[85–87] The right-handed helical structure of DNA in solution has also been confirmed by independent methods based on electron microscopy,[88] scanning probe microscopy,[89,90] fluorescence resonance energy transfer,[91] and computation.[92]

Dickerson and associates discovered[78,79] that B-DNA has a "spine" of water in the minor groove of regions rich in A-T base pairs. Two water molecules per base pair are hydrogen bonded to form a long chain. Half of the water molecules in the chain also form hydrogen bonds to oxygen and nitrogen atoms that are exposed in the minor groove while the others hydrogen-bond to a second chain of water molecules, forming a ribbon of hydration (Fig. 5-14A).[81,93–95] Additional water binds to polar groups in the major groove. The hydration pattern is largely *local*, i.e., each base has characteristic hydration sites (Fig. 5-14B).[94] Some of the hydration sites may be occupied partially by the nomovalent ions, Na^+ or K^+, depending upon the medium. Bound divalent metal ions are also sometimes seen in X-ray structures.[81,83a] However, most cations are thought to remain mobile.

2. Other Double-Helical Forms of DNA

The B form of fibrous DNA is stable under conditions of high (~ 93%) humidity but at 75% humidity it is converted into **A-DNA** in which the base pairs are inclined to the helix axis by about 13° and in which the ribose rings are primarily C3'-*endo* rather than C2'-*endo*.

B-form

A-form

Z-form

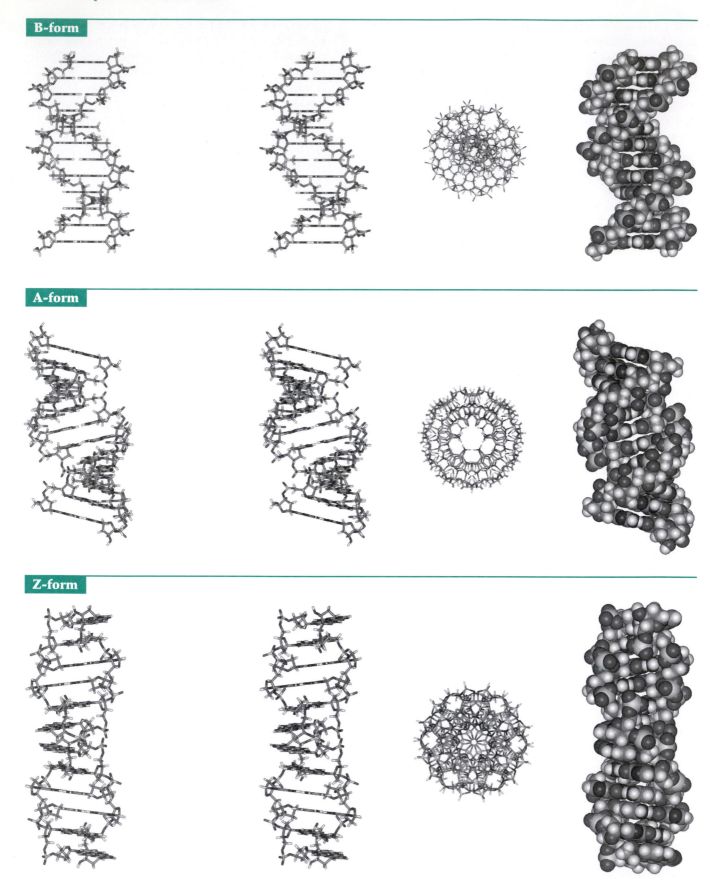

Figure 5-12 (Left) Stereoscopic skeleton models of the B, A, and Z forms of double helical DNA. See Schlick.[70] (Center) End views (Right) Space-filling models of the same three DNA forms: B, A, and Z. Courtesy of Tamar Schlick.

There are 10.9 base pairs per turn[93,96–98] (Fig 5-12; Table 5-3). In the A form the major groove is very deep (~ 1.35 nm) and narrower (~ 0.27 nm) than in the B form and extensively hydrated.[99] The minor groove is wider (~ 1.1 nm) and shallower (~ 0.28 nm) than in the B form. The crystal structure for a form intermediate between A and B has also been reported.[99a] Paracrystalline forms of DNA known as B', C, and D have also been observed[100,101] and others have been proposed.[102,103]

The most interesting additional form, **Z-DNA**, was discovered by X-ray studies of the alternating oligodeoxyribonucleotides d(CpGp-CpGpCpG) and d(CpGpCpG).[104] The helix of Z-DNA is *left-handed*.[71,72,105–112] The repeating unit consists of two Watson–Crick base pairs, the backbone following a zigzag pattern (Fig. 5-12). There are 12 base pairs per turn. The cytosine groups have the usual *anti* conformation but the guanosine groups are *syn*. Some of the ribose rings have the C2'-*endo* conformation characteristic of B-DNA but some are C3'-*endo*. The alternating CpG sequence is not essential for Z-DNA formation. However, alternating *syn* and *anti* conformations are important and purines assume the *syn* conformation more readily than do pyrimidines. The major groove of Z-DNA is shallow (Fig. 5-12), allowing it to accomodate bulky substituents at C8 of purines or C5 of pyrimidines. Such substituents favor Z-DNA.

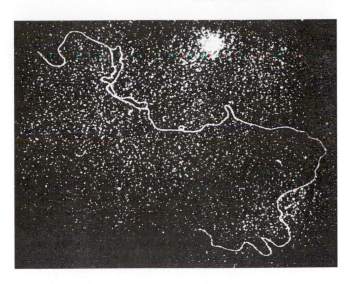

Figure 5-13 Electron micrograph of a DNA molecule (from a bacterial virus bacteriophage T7) undergoing replication. The viral DNA is a long (~ 14 μm) duplex rod containing about 40,000 base pairs. In this view of a replicating molecule an internal "eye" in which DNA has been duplicated is present. The DNA synthesis was initiated at a special site (origin) about 17% of the total length from one end of the duplex. The DNA was stained with uranyl acetate and viewed by dark field electron microscopy. Micrograph courtesy J. Wolfson and D. Dressler.

Figure 5-14 (A) Stereoscopic drawing showing two layers of water molecules that form a "spine" or "ribbon" of hydration in the minor groove of B-DNA. The inner layer is shown as larger filled circles; water molecules of the outer layer are depicted with smaller dots and are numbered. Hydrogen bonds are shown as dashed lines. (B) Electron density map. (A) and (B) from Tereshko *et al.*[95] (C) Stereoscopic representation of the superimposed electron densities of 101 water molecules observed to hydrate 14 guanine rings found in 14 B-DNA molecules for which high-resolution X-ray structures were available. Positions of 101 water molecules within 0.34 nm from any atom of the 42 guanines are plotted. From Schneider and Berman.[94]

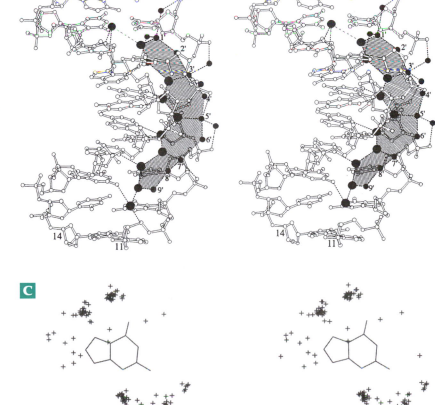

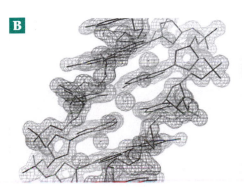

The B form is the most hydrated and most stable form of DNA under conditions of high humidity but even in solution it can be converted to A-DNA and Z-DNA by a high concentration of NaCl.[113] This is presumably because the salt dehydrates the DNA. Saenger and coworkers pointed out that the oxygen atoms of successive phosphate groups in the polynucleotide backbone of B-DNA are at least 0.66 nm apart, too far apart to be bridged by a water molecule.[114] However, the phosphates are individually hydrated. On the other hand, in A-DNA and Z-DNA the oxygen atoms on successive phosphates are as close as 0.53 and 0.44 nm, respectively. This allows one H_2O molecule to bridge between two phosphates, stabilizing these forms in environments of low humidity. This may be one factor that affects the B to A and Z transitions.[115–117] In the narrow Z-DNA helix repulsion between negative charges on the phospho groups of opposite strands is strong. By shielding these charges high salt concentrations also help to stabilize Z-DNA.

Since the Z form of DNA is favored in regions rich in G-C pairs,[118,119] it is reasonable to expect that it may occur in nature. Antibodies have been prepared which bind specifically to regions of DNA in the Z form and have been used to identify many such regions.[111,120] Genetic studies in *E. coli* have also provided strong evidence that left-handed DNA sequences are formed *in vivo* in that bacterium.[121,122] Z-DNA-forming sequences have also been found in the *Halobacterium* genome.[119] Segments of Z-DNA may occur in control regions called **enhancers** (Chapter 28)[121–124] and Z-DNA may also form behind RNA polymerase molecules that are moving along a gene while synthesizing mRNA. The associated negative supercoiling of the DNA could cause it to assume the Z form. This region of Z-DNA can, in turn, be a site for interaction with specific proteins.

In the Watson–Crick structure the two strands are *antiparallel*, an essential for replication. However, stable segments of double-stranded DNA with *parallel* strands can also be formed and may occur in specialized regions of the genome.[125–128]

3. The Conformational Flexibility of Double Helices

Local variations in the sequence of nucleotides affect the conformation of a DNA molecule and it is clear that the helix is not uniformly coiled throughout the entire length.[80,103,124,129–134] While most helix segments probably have a right-handed twist others may be left-handed. Most DNA is probably in the B form but there are segments in the A form. These may arise from formation of hybrid duplexes with RNA, which assume the A conformation and are also favored by certain base sequences. Rules for predicting the DNA conformation from the nucleotide sequence have been proposed.[116,135] In the simplest case[135] we consider each pair of adjacent nucleotides in the double helix. There are 16 possible pairs in one chain. These can be designated as in the following examples: (AA,TT), (CG,CG), and (AG,CT). The first two letters within the parentheses represent the sequence (from 5' to 3') in one chain while the second pair of letters represent the sequence (again from 5' to 3') in the complementary chain. The rules state that (AA,TT) or (TT,AA) repeated in a sequence will stabilize the B form of DNA. Repetition of (CC,GG) or (GG,CC) will favor conversion to the A form. Repetitions of (CG,CG) favor the Z form, especially if an alternating sequence of purines and pyrimidines is present throughout the (G + C)-rich region.

Even within the regular B structure the

TABLE 5-3
Helix Parameters for Three Types of DNA[a]

Parameter	Form of DNA or RNA (A-form)		
	B	A	Z
Helical twist, degrees	28–42	16–44	GC −51 ± 2
mean	36 ± 4	33 ± 6	CG −8.5 ± 1
Base pairs per turn	10.0 (9.7–10.6)	11-12	12
Helix rise per base pair, nm	0.34 ± .04	0.29 ± .04	GC 0.35 ± .02 CG 0.41 ± .02
Base inclination, degrees	−2.0 ± 5	13 ± 2	8.8 ± .7
Propeller twist, degrees	12 ± 5	15 ± 6	4.4 ± 3
Base roll degrees	−1.0 ± 5	6 ± 5	3.4 ± 2
Predominant conformation of deoxyribose	C2'-endo	C3'-endo	C C2'-endo G C3'-endo
Depth of grooves (nm) Major	0.85	1.35	very shallow
Minor	~0.75	~0.28	very deep
Width of grooves (nm) Major	1.1–1.2	0.27	broad
Minor	0.6	~1.1	narrow

[a] See R. E. Dickerson[78,78a]; based on single-crystal X-ray analysis.

torsion angles χ and δ (Fig. 5-10) are variable. Their changes are highly correlated.[136,136a] As χ ranges from −140° to −90°, ζ varies between 80° and 160°. Other pairs of torsion angles are also correlated. Besides allowing for changes between B, A, and Z conformations, this flexibility of the DNA helix together with cooperativity with adjacent base pairs may allow transmission of conformational effects for some distances along a DNA helix.[137,138] Supercoiling, discussed in Section C,3, also affects the helical conformation.

DNA can be stretched into yet another form or forms. Application of a force of 65 – 70 piconewtons (pN) stretches B-DNA by 70%. This "overstretched DNA" may also be important biologically.[130,139–141a]

Rotational and translational movements of bases.

In considering the conformational flexibility of a polynucleotide it is useful to define the parameters associated with movement of base pairs or individual bases. The possible movements are indicated in Fig. 5-15 and in Table 5-3. A **long axis** is established for each base pair in a structure for which X-ray data are available. Making use of these axes, we can move along the helix and measure the **angle of twist** from one base pair to the next.[142–145] While the mean value

of the twist for B-DNA is 36°, values of 28–42° have been observed in structures of oligonucleotides determined by X-ray analysis. The **base tilt**, which is nearly zero for B-DNA, is less variable. Within each base pair there is also a **propeller twist** around the long axis.[146] It averages about 12° for B-DNA. In addition, the whole base pair may **roll**, i.e., be rotated around its long axis by several degrees. These motions relieve steric interferences in the center of the helix, for example, those that would arise between purines in adjacent base pairs if the same helix parameters were imposed on each base pair.[142,147] A more detailed analysis reveals the large range of motions, mostly small, which are described in Fig. 5-15.[145,148] The mathematics needed to deal with computer-based modeling of polynucleotide structures has been developed.[80,149,150] Its application showed that there is a strong correlation between twist and roll. Typical average values of base tilt, propeller twist, and roll are given in Table 5-3 for B-, A-, and Z-DNA.

Bends and bulges.

If we overlook the ridges and grooves on their surfaces the DNA structures shown in Figs. 5-3 and 5-12 are straight rods. However, real DNA rods are crooked and may contain distinct bends. The

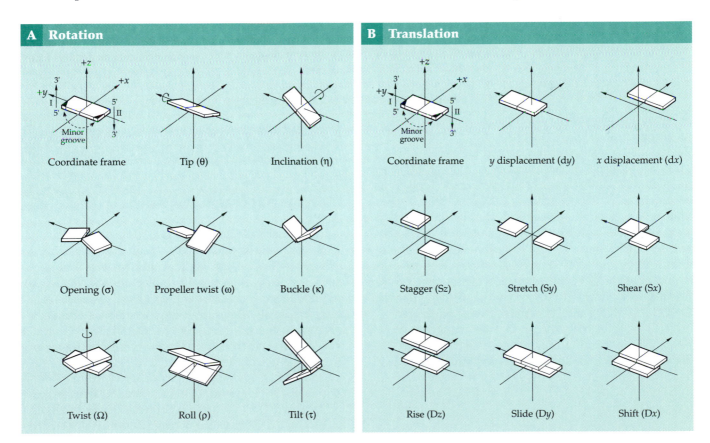

Figure 5-15 (A) Drawing illustrating various rotational movements of bases in polynucleotides. *Upper two rows*: rotations of two bases of a pair. *Lower row*: rotations involving two successive base pairs. (B) Translational movements. *Upper two rows*: involving two bases of a pair. *Lower row*: two successive base pairs. From Diekmann.[145]

existence of **bent DNA** in nature was first discovered from study of fragments enzymatically cut from kinetoplast DNA of tropical parasites.[151,152] The bent fragments moved anomolously slowly during electrophoresis in polyacrylamide gels of small pore size. Evidently, the bent shape impedes movement through the pores. Since the initial discovery, regions of bent DNA have been found in origins of replication[152–154] and other specialized sites. Bending of DNA because of local sequence variations may also be important to the positioning of nucleosomes (Fig. 27-3) on DNA.[155]

Several causes for DNA bending have been identified.[124,138,142,156–164] For example, a sharp 26° bend is expected at the junction between B- and A-DNA segments.[158] The presence of a thymine photodimer (Eq. 23-26) may cause a 30° bend.[165] Some naturally occurring bent DNAs contain repeated (A+T)-rich sequences. This suggested that the sequence (AA,TT), containing adjacent thymine methyl groups, can be thought of as a wedge, consisting of both roll and tilt components. Such a wedge repeated at intervals averaging 10.5 bp along the helix might cause the helix to bend.[129,144] However, computer modeling experiments suggest that thymine methyl groups do not distort the DNA helix. The bending may be a result of the greater tendency for AT pairs to roll and to enhance the van der Waals and electrostatic attractions between the 2'-O of cytosine and the 2'-NH$_2$ of guanine in the CG base pairs of bent DNA.[157,166] Bending can occur in both (A+T)- and (G+C)-rich regions and is often induced by binding to proteins or protein complexes that act in replication, transcription, and recombination.[167]

Various errors are made during replication or recombination of DNA. Incorporation of an incorrect nucleotide will cause a mismatched base pair in which proper hydrogen bonds cannot be formed. Most of the mispaired bases that result from mistakes in replication are removed by repair processes (Chapter 27). Those that remain can often assume alternative pairings that distort the helix only slightly. For example, a G-T wobble pair fits readily into an oligodeoxyribonucleotide double helix. Even a bulky G-A base pair causes little perturbation of the helix.[168] Incorporation of an extra nucleotide into one strand of the DNA will create a bulge in the helix.[169] NMR spectroscopic studies on bulged oligonucleotides have shown that the extra base can be stacked into the helix causing a sharp bend.[170,171] However, some oligonucleotides with mismatched bases crystallize as straight helices with the extra nucleotide looped out.[171–173] Mismatched base pairs tend to destabilize helices.[174]

Interactions with ions. Because each linking phospho group carries a negative charge (two charges per base pair, Fig. 5-2) the behavior of polynucleotides is strongly affected by cations of all kinds. The predominant small counterions within cells are K$^+$ and

Mg^{2+}. They are attracted to the negative charges on the polynucleotide backbone and, although they remain mobile, they tend to occupy a restricted volume.[175–177] Some may bind in well-defined locations as in Fig. 5-8. Because of the presence of these positive ions the interactions of nucleic acids with cationic groups of proteins are strongly affected by the salt concentration.

Organic cations compete with the simple counterions K$^+$ and Mg^{2+}. Among these, the polyamines are predominant.[178] Crystal structures have revealed that spermine binds across the deep grooves of tRNA and the major groove of B-DNA. Spermine also binds into the deep groove of A-DNA interacting by hydrogen bonding with bases in GTG sequence in both strands.[179] It binds tightly to CG-rich sequences in the minor groove of Z-DNA, where it tightens the structure and shortens the helix.[110] At higher concentrations of DNA, as occur in cell nuclei, spermidine induces the conversion of the DNA into liquid crystalline phases.[180]

Heavier metal ions and metal complexes can find sites on nitrogen atoms of the nucleic acid bases. Examples are the platinum complex **cisplatin** and the DNA-cleaving antibiotic **neocarzinostatin** (Box 5-B). Can metals interact with the π electrons of stacked DNA bases? A surprising result has been reported for intercalating complexes of ruthenium (Ru) and rhodium (Rh). Apparent transfer of electrons between Ru (II) and Rh (III) over distances in excess of 4.0 nm, presumably through the stacked bases, has been observed,[181] as has electron transfer from other ions.[181a] Stacked bases are apparently semiconductors.[182]

C. The Topology and Dynamics of Nucleic Acids

1. Rings, Catenanes, and Knots

While a DNA molecule may exist as a straight rod, the two ends are often covalently joined. Thus, the chromosomes of *E. coli* and of other bacteria are single closed circles. Circular DNA molecules are also found in mitochondria, chloroplasts, and many viruses. Further complexity arises from the fact that the circles of DNA are sometimes interlocked in chainlike fashion (**catenated**). An unusual example of this phenomenon is the presence of thousands of small catenated DNA circles in the single mitochondrion of a trypanosome (Fig. 5-16).[183] Sometimes circular DNA is **knotted** as in Fig. 5-17.[184–186] Knots and catenanes often appear as intermediate forms during replication and recombination, especially involving circular DNA.[187,188]

Methods have been devised for synthesis of even very complex DNA knots.[185,186] Let's look briefly at the topology of knots. The three simple knots shown here have a chirality beyond that of the nucleotide

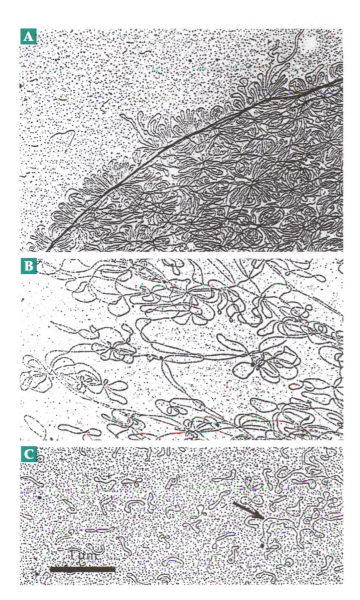

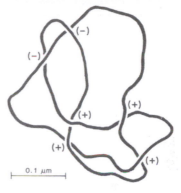

Figure 5-17 Electron micrograph of a six-noded knot made by the Tn3 resolvase which is involved in movement of the Tn3 transposon (Chapter 27) from one location to another within the genome. Putative six-noded knot DNA was isolated by electroelution from an agarose gel. The knots, which are nicked in one strand, were denatured to allow the nicked strand to slide away and leave a ssDNA knot. This was coated with *E. coli* recA protein (Fig. 27-24) to greatly thicken the strand and to permit the sign of each node (designated in the tracing) to be seen. From Wasserman *et al.*[184]

Figure 5-16 (A) Electron micrograph of the network of catenated DNA circles in the mitochondrion of the trypanosome *Crithidia fasciculata*. (B) and (C) The same network after treatment with a **topoisomerase** from bacteriophage T4 that catalyzes a decatenation to form individual covalently closed circles (Chapter 27). Five times as much enzyme was added in (C) as in (B). Two sizes of circles are present. Most are "minicircles", each containing about 2300 bp but a smaller number of larger ~35-kb "maxicircles" are also present. One of these is marked by the arrow. From Marini, Miller, and Englund.[183]

The node is negative if the crossing is like that in a left-handed supercoil and positive if like that in a right-handed supercoil:[191,192]

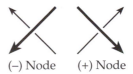

(−) Node   (+) Node

units.[189,190] This can be expressed by indicating the sign of each node in the knot:

Trefoil knot (−)   Figure-8 knot   Trefoil knot (+)

2. Supercoiled DNA

Double-stranded DNA in solutions of low salt content usually assume the B-DNA conformation with 10.4–10.5 base pairs per turn. If the two ends are joined the resulting covalently closed circular DNA will be "relaxed." However, there are topoisomerases that act on this form of DNA by cutting both strands, holding the ends, and twisting the two chains (Chapter 27).

The energy source for the process is provided by cleavage of ATP. **DNA gyrase** untwists relaxed circular dsDNA one turn at a time and reseals the cut ends. A **reverse gyrase** from certain bacteria twists the relaxed DNA more tightly. In both cases the change causes the DNA to form **superhelical turns**.[67,193–197] These may be either **solenoidal** or **plectonemically interwound** (as a twisted thread; Fig. 5-18).

The geometric and topological properties of closed supercoiled DNA molecules may be described by three quantities: The **linking number** (*Lk* also called the winding number, α), the **twist** (*Tw*), and the **writhe** (*Wr*). If a segment of double helical DNA were laid on a flat surface and the ends were joined to form a relaxed circle both *Lk* and *Tw* would equal the number of helical turns in the DNA. The writhe *Wr* would be zero. The linking number is a topological property. It has an integral value which is unchanged if the DNA molecule is distorted. It can be changed only for DNA with open ends. When the ends are joined the linking (winding) number is constant unless one or both chains are cleaved. However, twist and writhe are geometric properties, which can change according to Eq. 5-8 while the *Lk* remains constant. The twist is related to the number of helical turns while the writhe is related to the number of superhelical turns.[27,67,198–200]

$$Lk = Tw + Wr \qquad (5\text{-}8)$$

However, *Tw* and *Wr* do not usually have integral values. It is hard to define the relaxed state for a circular DNA. For example, changing the ionic composition will alter *Tw* and *Wr* according to equation 5-8. Both *Lk* and *Tw* are taken as positive for a right-handed **toroidal** (solenoidal) supercoil or a left-handed interwound twist.

For relaxed B-DNA, *Lk* is equal to the total number of base pairs in the circle divided by 10.5, *Wr* = 0, and *Tw* = *Lk*. Both *Lk* and *Tw* are positive in the right-handed B- and A-DNA forms, but *Tw* is negative for Z-DNA. In closed circular DNA the value of *Wr* is usually negative, the secondary structure being a fully formed Watson–Crick helix but with right-handed interwound superhelical turns or left-handed toroidal superhelical turns. The helix is said to be **underwound** (*Lk* < *Tw*).

Some of the topological properties of double-stranded DNA can be demonstrated by twisting together two pieces of flexible rubber tubing whose ends can be joined with short rods to form a closed circle.[196] Twist the tubing in a right-handed fashion as tightly as possible without causing supercoiling (*Lk* = *Tw*; *Wr* = 0). If the ends are now joined the circle will be relaxed. Now twist one turn tighter before joining the ends. A right-handed toroidal supercoil will be formed (Δ*Wr* = 1; Δ*Lk* = 1 + *Tw*; *Tw* is the same as before). If twisting is continued until several supercoils appear before the ends are joined, two interconvertible forms result. One form has right-handed toroidal supercoils and the other left-handed interwound supercoils as in Fig. 5-19. If relaxed circular DNA is unwound by one turn by cutting and resealing one chain, a single right-handed inter-wound supercoil will be formed (Δ*Wr* = – 1, *Lk* = *Tw* – 1). On the other hand, if the two chains in a closed relaxed helix are pried apart, as happens during intercalation (Section 3), *Lk* must remain constant, *Tw* will decrease, and *Wr* will increase with appearance of left-handed supercoiling.

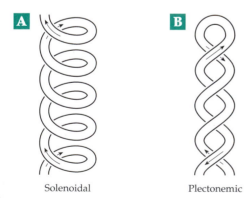

Solenoidal   Plectonemic

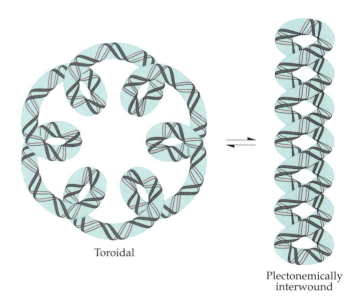

Toroidal

Plectonemically interwound

Figure 5-18 Two forms of supercoiling of a DNA duplex: (A) Solenoidal. (B) Plectonemically interwound. These are both negatively supercoiled, as can be deduced from the arrows at the nodes, but the solenoid is left-handed and the plectonemic form right-handed. From Wasserman and Cozzarelli.[187]

Figure 5-19 Topological equivalence of toroidal (solenoidal) and plectonemically interwound forms of a circular DNA. These two forms have a constant value of the linking number *Lk* (or α), the twist *Tw*, and writhing number *Wr*.

During replication of DNA (Chapter 27) pairing of some bases associated with the replication apparatus is prevented. Upon release of the constraint the newly replicated DNA forms base pairs and becomes supercoiled.

Since Lk is constant in circularly closed DNA a change of one turn of B-DNA ($Tw = 1$) into a turn of Z-DNA ($Tw = -1$) will cause the writhing number Wr to change by -2. Conversely, if the writhing number is forced to change by -2, a turn of Z-DNA may develop somewhere in a suitable (G + C)-rich region of the DNA.

Some of the information about supercoiled DNA can be summarized as follows:

Relaxed circular DNA	$Lk = Tw, Wr = 0$
Underwound circular DNA	$Lk < Tw$; Wr is negative; right-handed interwound supercoils or left-handed toroidal supercoils may be present. Alternatively $Lk = Tw$ and left-handed Z-DNA regions may appear.
DNA with intercalated molecules	DNA is partially untwisted; Tw is lowered; Wr is increased with decrease in number of negative supercoils. As Tw approaches Lk the DNA becomes relaxed.

The **superhelix density** of a DNA molecule is often expressed as $\sigma = Wr / Tw \approx$ number of superhelical turns per 10 bp.[31,198,199a,199b,201] In most naturally occurring circular DNA molecules σ is negative, a typical value being -0.05 (~5 negative superhelical turns per 1000 bp). The presence of superhelices in circular DNA molecules can be recognized readily by its effect upon the sedimentation constant of the DNA. Naturally occurring supercoiled DNA from polyoma virus sediments rapidly but after nicking of one of the strands of the double helix by brief exposure to a DNA-hydrolyzing enzyme the resulting relaxed form of the molecule sediments more slowly. Supercoiling lowers the viscosity of solutions of DNA and increases the electrophoretic mobility (Fig. 5-20) and may also be recognized by electron microscopy.

Naturally occurring or artificially prepared supercoiled DNA molecules can often be separated by electrophoresis into about ten forms, each differing from the other by one supercoiled turn and by $\Delta Lk = \pm 1$ (Fig. 5-20). The relative amounts of these **topological isomers** form an approximately Gaussian distribution. The isomers apparently arise as a result of thermal fluctuations in the degree of supercoiling at the time that the circles were enzymatically closed.[202]

Why is DNA in cells supercoiled? One effect of supercoiling is to contract the very long, slender double helices into more compact forms. In eukaryotic cells much of the DNA exists in **nucleosomes**. Each bead-like nucleosome consists of a core of eight subunits of proteins called **histones** around which an ~140 bp length of DNA is coiled into two negative, left-handed toroidal superhelical turns (Fig. 5-21).[203,204] There is some spacer DNA between nucleosomes. Otherwise a nucleosome providing two superhelical turns per 200 bp would produce a superhelix density of -0.1, twice that observed. A detailed analysis of the geometric properties of DNA wrapped around nucleosomes or other protein particles has been developed.[199–199c] Nucleosome formation is thought to protect DNA and to keep it in a more compact state than when it is fully active and involved in transcription. Nucleosomes, discussed further in Chapter 27, are also important in the regulation of transcription. Even though nucleosomes as such are absent, bacterial DNA also has a superhelix density of -0.05, apparently a result of interaction with other proteins such as the histonelike HU.[205] Interaction with smaller molecules can also affect supercoiling.[206]

The presence of naturally supercoiled DNA and a variety of topoisomerases suggests that the control of DNA supercoiling is biologically important. In fact,

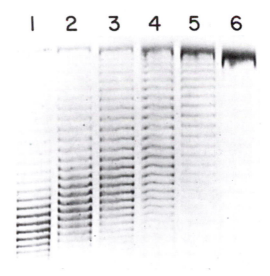

Figure 5-20 Electrophoresis of DNA from the SV40 virus with varying numbers of superhelical turns. Molecules of the native DNA (lane 1) move rapidly toward the anode as a series of bands, each differing from its neighbors by one superhelical turn. The average number of superhelical turns is about 25. Incubation with a topoisomerase from human cells causes a stepwise removal of the superhelical turns by a cutting and resealing of one DNA strand. The DNA incubated with this enzyme at 0°C for periods of 1, 3, 6, 10, and 30 min (lanes 2–6) is gradually converted to a form with an average of zero supercoils. For details see Keller.[202,212] Electrophoresis was carried out in an 0.5% agarose –1.9% polyacrylamide slab gel (17 x 18 x 0.3 cm). The bands of DNA were visualized by staining with the fluorescent intercalating dye ethidium bromide. Photograph courtesy of Walter Keller.

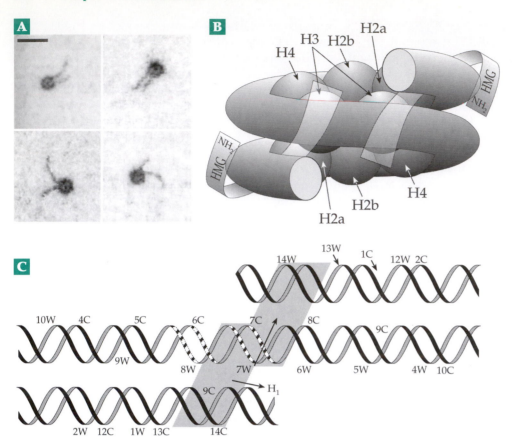

Figure 5-21 Nucleosomes. (A) Electron micrographs of individual nucleosomes reconstituted from 256-bp DNA fragments and separated proteins. From Hamiche et al.[213] Courtesy of Ariel Prunell. (B) Model of a nucleosome core. The 1.75-turn (145-bp) DNA superhelix winds around the histone octomer which consists of two subunits apiece of histones H2A, H2B, H3, and H4. In addition, two elongated molecules of proteins HMG-14 or HMG-17 are indicated (see also Chapter 27). (C) Schematic radial projection of the double-helical DNA showing areas protected from cleavage by hydroxyl radicals (see Fig. 5-50) by the bound proteins. The shaded areas are those protected by HMGs. The zigzag lines near the dyad axis indicate the most prominent regions of protection. (B) and (C) are from Alfonso et al.[214]

topoisomerases play essential roles in both replication of DNA and in transcription of genes. Supercoiling requires energy and topoisomerases that induce super-coiling must provide energy, e.g., by cleavage of ATP. Conversely, supercoiled DNA can be a source of energy for biological processes. It has been estimated that the ΔH, ΔS, and ΔG per mole for formation of a single superhelical turn are 35.9 kJ, 68 J/ °K, and 14.6 kJ, respectively.[207] A reduction in superhelix density is accompanied by a decrease in Gibbs energy and can therefore be coupled to other processes that have positive values of ΔG. An example is conversion of (G + C)-rich regions into the Z form of DNA, which is favored by negative supercoiling.[207,208] Supercoiling affects binding of various proteins to DNA[197,209,210] as well as intercalation (discussed in the next section) and formation of cruciform structures (Section D,3).

The folding of DNA into compact forms, such as that in chromosomes, is also influenced by supercoiling. In the absence of nucleosomes supercoiled DNA may assume the plectonemic form (Figs. 5-18, 5-19). Segments of the resulting rods may aggregate side-by-side in a liquid crystalline state.[200,211] The relatively high cation concentration within cells favors this trans-formation.[215–217] Polyamines such as spermidine are especially effective in promoting aggregation of DNA, in formation of Z-DNA,[110] and possibly in facilitating

cooperative processes that require that two DNA molecules interact with each other.[218,219]

3. Intercalation

Flat, aromatic, hydrophobic rings are often able to insert themselves between the base pairs of a DNA duplex. Such **intercalation** is observed for many anti-biotics, drugs, dyes, and environmental pollutants. Among them are proflavine, ethidium bromide, actino-mycin (Box 28-A), hycanthone (Fig. 5-22), and dauno-mycin (Fig. 5-23). Hycanthone, employed in the treat-ment of schistosomiasis, is one of the most widely used drugs in the world. Since intercalating agents can be mutagenic, such drugs are not without their hazards.

Intercalation is often used to estimate the amount of negative supercoiling of DNA molecules. Varying amounts of the intercalating agent are added, and the sedimentation constant or other hydrodynamic prop-erty of the DNA is observed. As increasing intercala-tion occurs, the secondary turns of DNA are unwound (the value of Tw in Eq. 5-8 decreases). Each intercalated ring causes an unwinding of the helix of ~26°. Since for a closed covalent duplex the value of Lk in Eq. 5-8 is constant, the decrease in Tw caused by increased intercalation leads to an increase in the value of Wr, which is usually negative for natural DNA. When sufficient intercalation has occurred to raise *Wr* to zero, a minimum sedimentation rate is observed. Addition of further intercalating agent causes a positive supercoiling.

When the "replicative form" of DNA of the virus φX174 (Chapter 27), a small circular molecule containing ~5000 bp, was treated with proflavine,[221] the binding of 0.06 mol of proflavine per mole of nucleotides reduced *Wr* to zero. From this it was estimated that σ = 0.055, corresponding to –27 superhelical turns at 25°, pH 6.8, ionic strength ~0.2. Changes in temperature, pH, and ionic environment strongly influence supercoiling. In general σ becomes less negative by ~3.3 x 10⁻⁴ per degree of temperature increase.[222] For example, the observed value[221] of σ for φX174 DNA was –0.059 at 15°C and –0.040 (–20 superhelical turns) at 75°C at an ionic strength of ~0.2.

The exact ways in which intercalating substances can fit between the base pairs of nucleic acids are being revealed by X-ray diffraction studies of complexes with nucleosides, dinucleotides, and other oligonucleotides.[220,223–225] The structure of a complex in which daunomycin is intercalated between two GC base pairs in DNA is shown as in Fig. 5-23. Daunomycin and other related anthracycline antibiotics also have an amino sugar ring that binds into the minor groove of the DNA, providing both electrostatic stabilization and hydrogen bonding. Substituents on this aliphatic ring also hydrogen bond to DNA bases.[226]

Does intercalation of flat molecules into nucleic acid chains have a biochemical function? Aromatic rings of amino acid side chains in proteins designed to interact with nucleic acids may sometimes intercalate into nucleic acid helices serving a kind of "bookmark" function.[227] Changes in superhelix density caused by such intercalation may be important in the orderly handling of DNA by enzymes within cells.

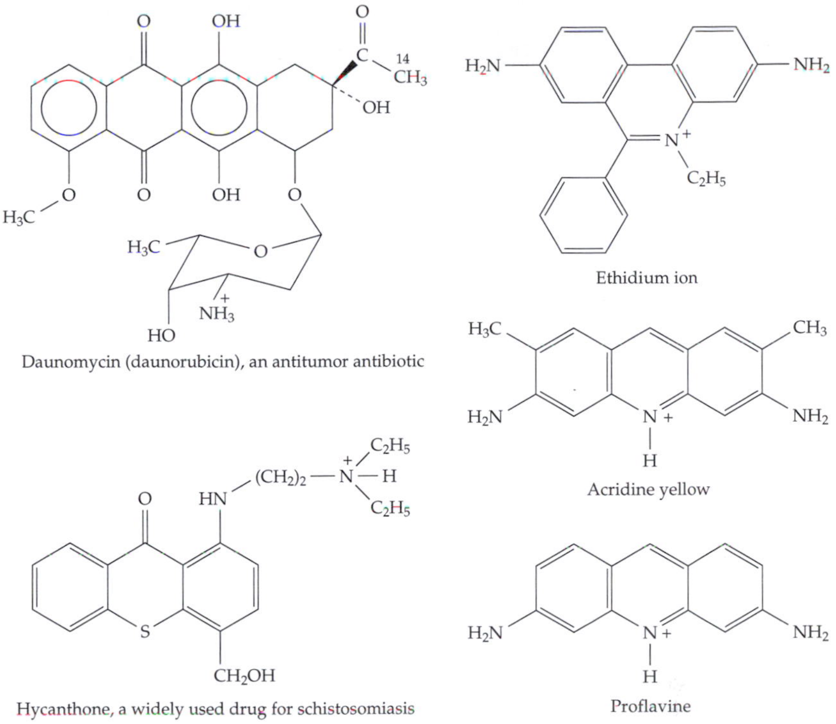

Figure 5-22 Structures of some substances that tend to "intercalate" into DNA structures. See also Fig. 5-23.

Daunomycin (daunorubicin), an antitumor antibiotic

Hycanthone, a widely used drug for schistosomiasis

Ethidium ion

Acridine yellow

Proflavine

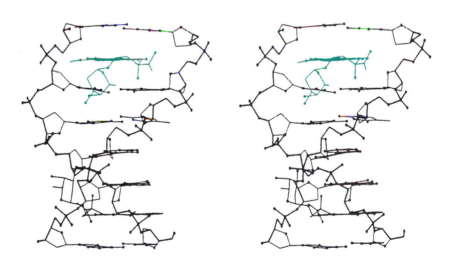

Figure 5-23 Stereoscopic drawing showing a molecule of daunomycin (Fig. 5-22) intercalated between two base pairs in a molecule of double-helical DNA, d(CGTACG). Nitrogen and oxygen atoms are shown as dots. From Quigley *et al.*[220] Both daunomycin and adriamycin (doxorubicin; 14-hydroxy-daunomycin) are important but seriously toxic anticancer drugs.

BOX 5-B ANTITUMOR DNA DRUGS

Chemotherapy of cancer at present involves simultaneous use of two or more drugs. For example, antifolates (Chapter 15) or nucleoside analogs such as 5-fluorouridine may be used together with a drug that binds directly to DNA and inhibits the replication of cancer cells. In 1963, it was discovered accidentally that platinum ions released from supposedly inert platinum electrodes inhibited the division of *E. coli* cells. This led Rosenberg and associates to test platinum compounds against animal cells.[a] Among many compounds tested *cis*-dichlorodiammineplatinum(II) (*cis*-DDP or **cisplatin**) emerged as an important anticancer drug that is especially effective against testicular and ovarian cancers.[b–d] The *trans* isomer, however, is inactive.

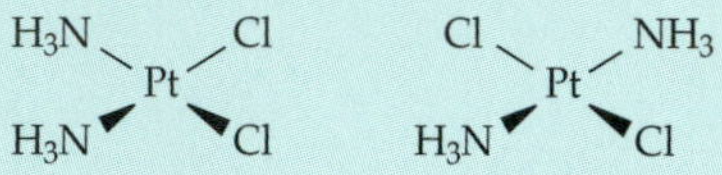

cis-DDP   *trans*-DDP
(Cisplatin)   (inactive)

cis-DDP binds to adjacent deoxyguanosines within one strand of ds or ssDNA. The stacking of adjacent bases is disrupted as the platinum binds to N-7 nitrogen atoms of the two guanine rings by replacement of the two chloride ions. The product has the following structure,[c,e] with the Pt lying in the minor groove of dsDNA:

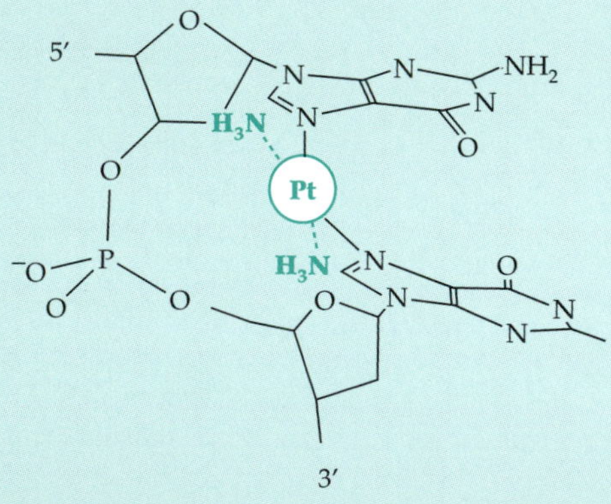

The two guanines are no longer stacked, but the structure is impossible for the *trans* isomer. The cisplatin adducts appear to prevent proper DNA repair and to induce programmed cell death (apoptosis).[f]

Since the 1950s, using a different approach, the U.S. National Cancer Institute, as well as agencies in other countries, has sought to find natural anticancer compounds in plants, fungi, microorganisms, and marine invertebrates.[g] Among these are many antibiotics that intercalate into DNA helices, e.g.,

daunomycin (Figs. 5-22, 5-23), **menogaril,**[h] **triostin A,**[i] and the antitrypanosomal drug **berenil.**[j] Some of these are also alkylating agents that contain double bonds to which such groups as the 2-NH_2 of guanine may add:

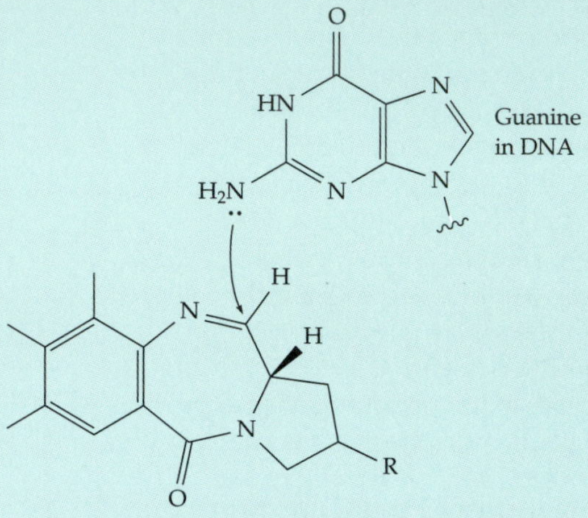

Tomaymycin and related antibiotics[k]

Diol epoxides and cyclic imines such as **mitomycin** also form adducts specifically with guanine 2-NH_2 groups:[l] **Neocarzinostatin** is an antitumor protein with a nonprotein "chromophore." After intercalation and binding into the minor groove of bulged DNA, it undergoes "activation" by addition of a thiol group. The enediyne structure undergoes rearrangement with formation of a reactive diradical that attacks the DNA.[m] A family of related antitumor enediynes has also been discovered.[n]

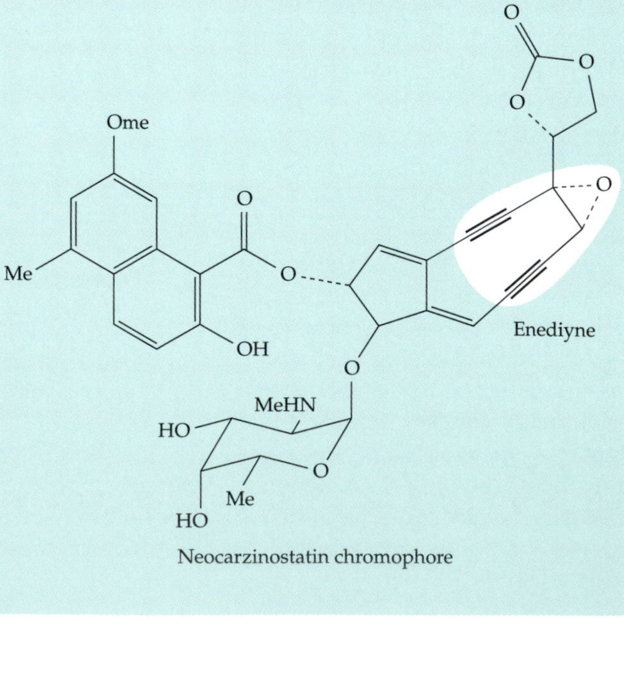

Neocarzinostatin chromophore

BOX 5-B   (continued)

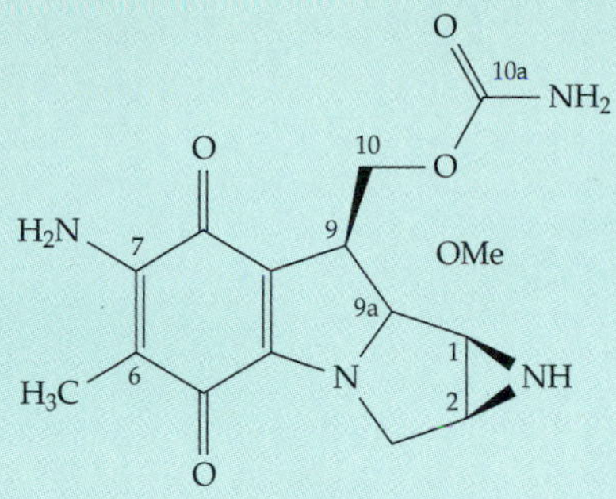

Mitomycin

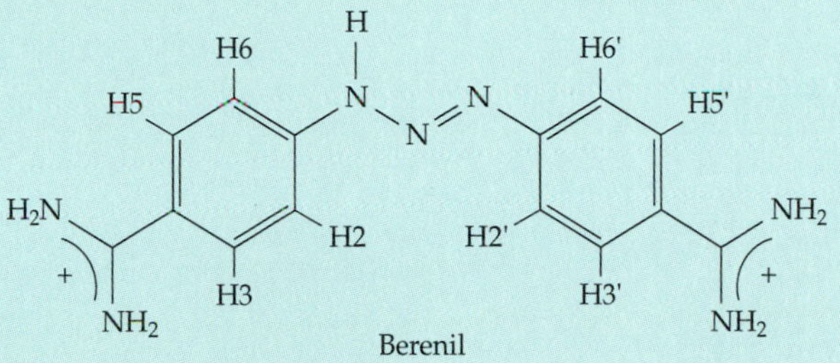

Berenil

The antibiotic **bleomycin**, which also binds in the minor groove of B-DNA with some specificity for G-C Sites, forms an iron (II) complex. It can be

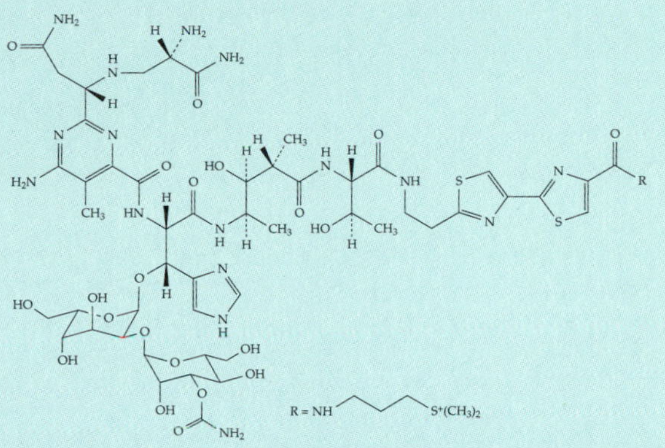

Bleomycin A₂

R = NH

Another group of drugs occupy extended binding sites in the minor grooves of DNA double helices, often with specificity for a particular base sequence. Examples are the antitrypanosomal drug **berenil**,[j] toxic *Streptomyces* antibiotics **netropsin, distamycin**, and related synthetic compounds.[o–q] Netropsin lies within the minor groove in regions with two or more consecutive AT pairs, displacing the spine of hydration as shown in the following stereoscopic drawing. Binding depends upon both electrostatic interactions and formation of specific hydrogen bonds involving the amide groups of the antibiotics.

oxygenated to form an Fe(II)–O₂ complex (see Chapter 16) which cleaves the DNA chain.[s,t] Synthetic compounds that do the same thing have been made by connecting an EDTA–iron, or other iron chelate complex covalently to the DNA-binding compound.[n,u–x] A goal is to direct drugs to selected target sites in DNA and to induce bond cleavage at those sites in a manner analogous to that observed with restriction endonucleases. Most current chemotherapeutic agents are very toxic. Present research is designed to target these drugs more precisely to specific DNA sequences and to identify target sequences peculiar to cancers. See also **Designed third strands** in main text.

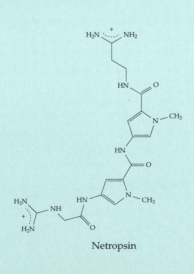

Netropsin

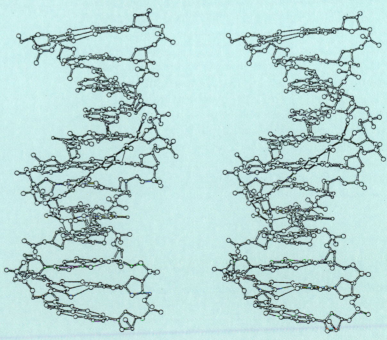

Netropsin lying in the minor groove of B-DNA hydrogen-bonded to bases in the central ATAT tetranucleotide. From Coll *et al.*[r]

BOX 5-B Continued

a Rosenberg, B., Van Camp, L., and Krigas, T. (1965) *Nature (London)* **205**, 698–699
b Zamble, D. B., and Lippard, S. J. (1995) *Trends Biochem. Sci.* **20**, 435–439
c Pilch, D. S., Dunham, S. U., Jamieson, E. R., Lippard, S. J., and Breslauer, K. J. (2000) *J. Mol. Biol.* **296**, 803–812
d Temple, M. D., McFadyen, W. D., Holmes, R. J., Denny, W. A., and Murray, V. (2000) *Biochemistry* **39**, 5593–5599
e Huang, H., Zhu, L., Reid, B. R., Drobny, G. P., and Hopkins, P. B. (1995) *Science* **270**, 1842–1845
f Zlatanova, J., Yaneva, J., and Leuba, S. H. (1998) *FASEB J.* **12**, 791–799
g Booth, W. (1987) *Science* **237**, 969–970
h Chen, H., and Patel, D. J. (1995) *J. Am. Chem. Soc.* **117**, 5901–5913
i Wang, A. H.-J., Ughetto, G., Quigley, G. J., Hakoshima, T., van der Marel, G. A., van Boom, J. H., and Rich, A. (1984) *Science* **225**, 1115–1121
j Pilch, D. S., Kirolos, M. A., Liu, X., Plum, G. E., and Breslauer, K. J. (1995) *Biochemistry* **34**, 9962–9976
k Barkley, M. D., Cheatham, S., Thurston, D. E., and Hurley, L. H. (1986) *Biochemistry* **25**, 3021–3031
l Kumar, G. S., Lipman, R., Cummings, J., and Tomasz, M. (1997) *Biochemistry* **36**, 14128–14136

m Yang, C. F., Stassinopoulos, A., and Goldberg, I. H. (1995) *Biochemistry* **34**, 2267–2275
n Nicolaou, K. C., Dai, W.-M., Tsay, S.-C., Estevez, V. A., and Wrasidlo, W. (1992) *Science* **256**, 1172–1178
o Goodsell, D. S., Ng, H. L., Kopka, M. L., Lown, J. W., and Dickerson, R. E. (1995) *Biochemistry* **34**, 16654–16661
p Rentzeperis, D., Marky, L. A., Dwyer, T. J., Geierstanger, B. H., Pelton, J. G., and Wemmer, D. E. (1995) *Biochemistry* **34**, 2937–2945
q Tanious, F. A., Ding, D., Patrick, D. A., Tidwell, R. R., and Wilson, W. D. (1997) *Biochemistry* **36**, 15315–15325
r Coll, M., Aymami, J., van der Marel, G. A., van Boom, J. H., Rich, A., and Wang, A. H.-J. (1989) *Biochemistry* **28**, 310–320
s Burger, R. M., Drlica, K., and Birdsall, B. (1994) *J. Biol. Chem.* **269**, 25978–25985
t Kane, S. A., Hecht, S. M., Sun, J.-S., Garestier, T., and Hélène, C. (1995) *Biochemistry* **34**, 16715–16724
u Veal, J. M., and Rill, R. L. (1988) *Biochemistry* **27**, 1822–1827
v Campisi, D., Morii, T., and Barton, J. K. (1994) *Biochemistry* **33**, 4130–4139
w Mack, D. P., and Dervan, P. B. (1992) *Biochemistry* **31**, 9399–9405
x Han, H., Schepartz, A., Pellegrini, M., and Dervan, P. B. (1994) *Biochemistry* **33**, 9831–9844

4. Polynucleotides with Three or Four Strands

Some nucleotide sequences in DNA favor the formation of a regular triple-helical (**triplex**) structure. This is possible when there are long stretches of adjacent pyrimidines having any sequence of C and T in one strand of double-helical DNA. The other strand of the DNA will contain the correct purines for formation of Watson–Crick base pairs. With such a structure it is always possible to add a third strand of a polypyrimidine using Hoogsteen base pairing. Triads formed in this way contain either two T's and one A or two C's and one G, as is shown in Fig. 5-7. In the latter case one of the C's must be protonated to allow formation of the pair of hydrogen bonds.[227a] These triplex structures can be formed only for stretches of DNA containing all pyrimidines in one strand (a **homopyrimidine** strand) and all purines in the other (a **homopurine** strand). A poly (AAU) triplex can also be formed, using the hydrogen bond pattern of the first triplet in Fig. 5-7. Triple-stranded synthetic polynucleotides of these types were prepared by Felsenfeld and others as early as 1957.[228–230] The third strand in a triplex can be either parallel or antiparallel to the homopurine strand. The third strand is either homopurine or a mixture of purine and thymine. The triplets are G•C*C, A*A•T, or T*A•T and are formed by Watson–Crick (•) and **reversed Hoogsteen** (*) pairing.[231] Recently, there has been a renewed interest in DNA triplets because of their occurrence in natural DNA, their possible importance in genetic recombination, and the potential for design of powerful inhibitors of replication and transcription that function via triplex formation.[35,232–238]

H-DNA. Strands of DNA contain many homopyrimidine "tracts" consisting of repeated sequences of pyrimidines, e.g., $d(T-C)_n$, which may be abbreviated more simply as $(TC)_n$. In this example the complementary strand would contain the two purines in the repeated sequence $(GA)_n$. At low pH, where protonation of the cytosine rings occurs, or in negatively supercoiled DNA, the two strands of the repeating sequence may separate, with the $(TC)_n$ strand folding back to form a triple helix in which the base triplets have the hydrogen bonding pattern of Fig. 5-7. The resulting structure, which is shown in Fig. 5-24, is known as **H-DNA**.[239–241] A variety of related "nodule" and looped forms of DNA can also be formed.[35,242]

R-DNA. A different type of triplex DNA may be formed during genetic recombination. A Watson–Crick duplex is brought together by one or more proteins with a single strand that is, for at least a considerable distance, an exact copy of one of the strands of the duplex. It is within such a triplex that cutting of a strand of the duplex takes place to initiate recombination (see Chapter 27 for a detailed discussion). Can a triplex structure containing two identical chains be formed? Possible base triplets include the following C•G*G triplet which occurs in a crystalline oligonucleotide structure.[243] This triplet contains a variation on Hoogsteen pairing and is related to the first triplet in Fig. 5-7. The other triplets needed for the proposed **R-DNA** recombination intermediates are G•C*C, T•A*A, and A•T*T. While there is keen interest in R-DNA[35,244–247] the formation of a stable intermediate triplex is still uncertain.[35,247]

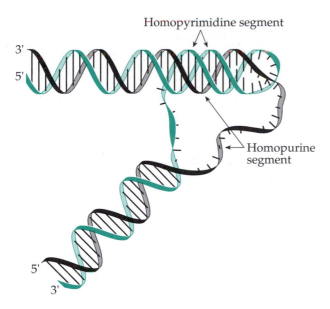

Homopyrimidine segment

Homopurine segment

3'
5'

5'
3'

Figure 5-24 Proposed structure of H-DNA which can be formed when a homopyrimidine segment and its complementary homopurine segment separate as a result of protonation of the cytosine rings or of negative supercoiling stress.[239,240] The triple-helical portion contains base triplets of the kind shown in Fig. 5-7.

are synthetic phosphoramidates in which the phosphodiester linkage between nucleosides is replaced by $3'\text{-}NH\text{-}PO_3^-\text{-}O\text{-}5'$. The resulting oligonucleotide is resistant to digestion by phosphodiesterases present in cells. Its geometry favors binding to A-DNA or RNA.[251] Another oligonucleotide mimic, dubbed PNA (peptide nucleic acid), contains monomer units of the following type[252–253b] and is able to form a triple helix:

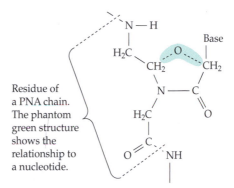

Residue of a PNA chain. The phantom green structure shows the relationship to a nucleotide.

Yet another approach is to synthesize a "hairpin polyamide" that contains pyrrole, hydroxypyrrole, and imidazole groups in a sequence that favors tight binding to a specific dsDNA sequence.[254] Some possible geometries for triplex DNA are illustrated in Fig. 5-25. These are based on computer-assisted modeling.[238]

Tetraplex (quadruplex) structures. The ends of linear chromosomes, the **telomeres**, have unusual nucleotide sequences repeated hundreds or thousands of times.[255–258] There is usually a **guanine-rich strand** running 5' to 3' toward the end of the chromosome and consisting of sequences such as TTAGGG in vertebrates, TTTTGGGG in the ciliate *Oxytricha*, and TG_{1-3} in *Saccharomyces cerevisiae*. The complementary strand is cytosine rich. The guanine-rich strand is longer than its complement, "overhanging" by about 2 repeat units. The significance for the replication of chromosomes is discussed in Chapter 27.

The thing that has attracted most attention to telomeric DNA is the unusual structure of the G-rich strand that was signaled by the first NMR studies.[259] Subsequent investigation[260,261] revealed the presence of G quartets (Fig. 5-8) which are apparently stacked in folding patterns such as the following:[255,262–268]

Designed third strands. If triplex DNA segments can form naturally it should be possible to design oligonucleotides that will bind into the major groove of a DNA duplex to form a triplex at a specific location or locations in the genome. Dervan and associates are studying this approach systematically.[248–250] Such oligonucleotides may be modified chemically to provide stronger binding to targeted locations and may prove to be useful therapeutic agents. Among these

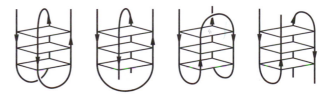

These structures are stabilized by the presence of univalent cations, K^+ being more effective than Na^+,

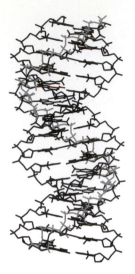

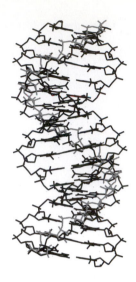

Figure 5-25 Two superposed stereoscopic diagrams illustrating the lowest energy triple helical conformations of one pyrimidine, one purine, and one Hoogsteen DNA strand (in black). Superposed is an RNA purine strand in gray. See Srinivasan and Olson.[238] Courtesy of Wilma K. Olson.

Li[+], Rb[+], or Cs[+].[260,269–271] A protein isolated from *Tetrahymena* binds specifically to the G4-DNA regions.[272] The presence of 5-methylcytosine (m[5]c) also stabilizes quadruplex structures.[273] The oligoribonucleotide UG$_4$U forms G$_4$ quartets so stable that it takes days at 40°C in D$_2$O for the hydrogen-bonded NH protons forming the quartet to be exchanged for [2]H.[274]

Tetraplex structures have also been observed for G-rich repeating sequences associated with the human **fragile X syndrome**.[275,276] This is the most common cause of inherited mental retardation and appears to arise as a result of the presence of an excessive number of repeats of the trinucleotide sequence (CGG)$_n$. For normal persons n = 60 or less; for healthy carriers n may be as high as 200 but for sick individuals it may be much higher.[275] The structure in solution, as determined by NMR spectroscopy, is shown in Fig. 5-26. Another variant of four-stranded DNA, which arises from cytosine-rich DNA, contains C•CH[+] pairs such as the following at low pH.[277–279]

Hemiprotonated d(CH[+]•C)

In sequences such as d(TC$_5$) and d(C$_3$T) these C•CH[+] pairs are intercalated as is shown in Fig. 5-27. This intercalated DNA (**I-DNA**) may provide an alternative conformation for some telomeric sequences.[281,282] An I-DNA motif has also been identified in oligonucleotides from the DNA of human centromeres.[283] The seemingly unusual forms of DNA described in this section may represent only a fraction of the naturally occurring DNA structures of biological significance.

5. Junctions

Special structural features may be found at junctions between different types of DNA, e.g., between A-DNA and B-DNA.[284–286] However, the most interesting junctions are *branched*.[287–290] For example, Fig. 5-28 shows a four-way junction in which all of the bases form Watson–Crick pairs. This junction is better known as a **Holliday junction** because it was proposed by Holliday in 1964 as an intermediate in genetic recombination.[291] As shown at the top of Fig. 5-28A the junction is formed from *two homologous DNA duplexes*. These are identical except for the boxed and shaded base pairs. The ends of the first duplex are marked I and II and those of the second III and IV. The Holliday junction appears to arise by cleavage of one strand of each duplex with rejoining of the strands as indicated by the green arrows. Rotation gives the untwisted Holliday junction structure

Figure 5-26 Structure of a G•C∗G•C∗ tetrad present in a quadruplex structure formed by the oligonucleotide d(GCGCTTTGCGC) in Na[+]-containing solution. See Kettani *et al.*[276]

shown. The postulated three-dimensional structure of the junction is indicated in Fig. 5-28B.[292–294]

An important characteristic of Holliday junctions formed from homologous duplexes is that they can move by a process called **branch migration**.[295] Because of the twofold symmetry of the branched structure the hydrogen bonds of one base pair can be broken while those of a new base pair are formed, the branch moving as shown in Fig. 5-28. Notice that, in this example, the nonhomologous (boxed) base pairs TA and GC have become *mispaired* as TG and AC after branch migration. More significantly, the junction may be cut by a **resolvase** at the points marked

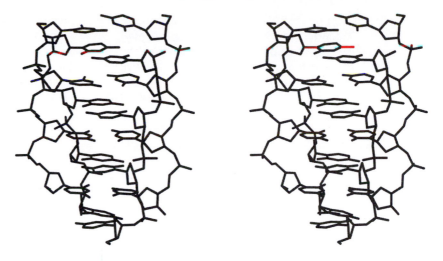

Figure 5-27 Stereoscopic view of a four-stranded intercalated DNA or I-DNA formed from $d(C_4)$. Two parallel duplexes with $C \cdot CH^+$ pairs are intercalated into each other. From Chen *et al.*[280]

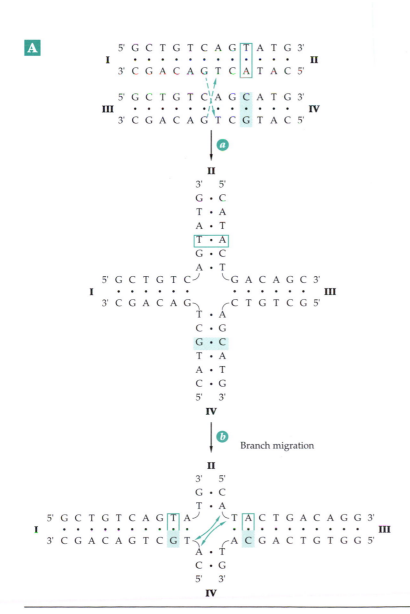

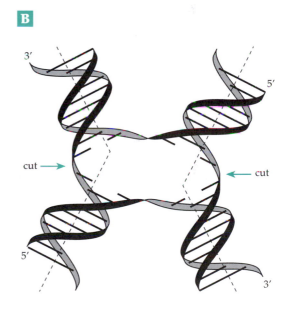

Figure 5-28 (A) Abbreviated reaction sequence for formation of a four-way Holliday junction between two homologous DNA duplexes. In step *a* strands are cut and rejoined with movement of the strands to a roughly antiparallel orientation. The resulting structure is thought to resemble that shown below the four-stranded representation. In step *b* branch migration takes place, separating the nonhomologous base pairs TA and CG and causing mismatched pairs which will be subject to repair. (B) Proposed three-dimensional structure (after drawing by Bennett and West).[292]

by the green double-headed arrow. If the strand break is then resealed, and the DNA strands are replicated, the boxed base pair in duplex I–II will have been transferred into a strand replicated from III–IV and genetic recombination will have been accomplished. Branch migration can occur over much longer distances than are indicated in this figure, so the alteration transferred may be far from the site of the initial cleavage and whole genes or groups of genes can be transferred. Recombination is considered in more detail in Chapter 27.

The Holliday junctions formed during recombination are mobile, but synthetic **immobile Holliday junctions** can be synthesized by using nonhomologous base sequences or by locking the junctions.[288] This has permitted careful physical study of these and other more elaborate synthetic junctions. With suitable choices of base sequences for the oligonucleotides from which they are made, such junctions will assemble spontaneously. Double-stranded DNA with immobile junctions is a very suitable construction material on a "nanochemical" scale. It has been assembled into knots, rings, cubes, and more complex polyhedra.[296–297b]

D. Ribonucleic Acids (RNA)

The best known forms of RNA are: (1) the long chains of messenger RNA (mRNA), which carry genetic messages copied from DNA to the ribosomes where proteins are made; (2) the much shorter transfer RNAs (tRNAs) which participate in reading the genetic code, correctly placing each amino acid in its sequence in the proteins; and (3) ribosomal RNAs (rRNA), which provide both structural material and a catalytic center for peptide bond formation. In addition there are numerous small RNAs that function in the splicing and editing[298 – 300] of mRNA, processing of tRNA precursors, methylation of ribosomal RNA,[301] transfer of proteins across membranes, and replication of DNA.[302] The genomes of many viruses consist of RNA. There are doubtless additional as yet undiscovered types of RNA.

Unlike DNA, which exists largely as double helices, the single chains of RNA can fold into complex forms containing many bulges and loops of the sort depicted in Fig. 5-9.[300,303,304] These loops are closed by double-stranded **stems** which have the A conformation. The B conformation is impossible because of the presence of the 2'-hydroxyl groups on the ribose rings in RNA. Even one ribonucleotide in a 10-nucleotide oligomer prevents formation of the B structure.[55,83] The 2'-OH groups not only keep the RNA in the A form but also engage in hydrogen-bond formation. Hydrogen bonds may form between the 2'-OH and the oxygen atom in the next ribose ring in the 3' → 5' direction. The – OH groups also hydrogen bond to water molecules which form a network within the minor groove.[305,306]

These bound water molecules, in turn, can bond to associated protein and to other atoms of the complex loops found in RNA molecules. The 2'-OH groups also act as ligands for divalent metal ions in some tRNAs and in some RNA catalytic sites. Transient **hybrid DNA–RNA double helices** also exist within cells and they too usually have the overall shape of A-DNA.[55,307–309] However, the minor groove is intermediate in width between that expected for the A and B forms.

1. RNA Loops and Turns

Like polypeptides, polynucleotide chains have preferred ways of bending or turning. The loops at the ends of the hairpin turns of RNA molecules sometimes consist of a trinucleotide such as UUU,[304] but are usually larger. In tRNA there are typically seven bases that do not participate in regular Watson–Crick pairing (see Figs. 5-30, 5-31). The tetranucleotide **UUCG** is frequently present, and the sequence 5'-GGAC**UUCG**GUCC forms an unusually stable hairpin.[85,310] Other tetranucleotides, such as UGAA,[311] CCCG (also found in DNA loops),[312] GCAA,[313] and GAAA,[314] occur often. The latter are members of a larger group of loop structures with the consensus sequence GNRA, where N is any nucleotide and R is a purine.[315,315a] These sequences are very common in highly folded structures of ribosomal RNAs. Until recently high-resolution X-ray structures were available for only a few tRNAs and ribozymes. To help remedy this deficiency the structures of a great variety of oligonucleotide stem/loop (hairpin) structures are being determined, most by NMR spectroscopy.[304,316,317] The sharp turns in the loops involve mostly rotation about the two torsion angles around the phosphorus atom of the third, from the 5' end, of the seven nucleotides. Base bulges on stems (Fig. 5-9) not only introduce kinks and bends in RNA stems[169] but also provide well-defined hydrogen-bonded binding sites for proteins. Numerous branched three-way and more complex junctions provide other important motifs in folded RNA.[318]

Among the new RNA structures are those of RNA–antisense RNA pairs in "kissing" hairpin complexes.[87,319] Another interesting complex folding pattern in RNA is the **pseudoknot**, a structural feature that has been identified in many RNA sequences.[85,320–327] A pseudoknot can be formed if nucleotide sequences favorable to formation of two short RNA stems are overlapped as shown in Fig. 5-29. After stem 2 in this drawing is formed (step a) additional base pairing can lead to formation of stem 1 (step b). The base pairs of the two stems can stack coaxially to form the pseudoknot (step c).

For the formation of base-paired stems the RNA must contain antiparallel sequences that allow Watson

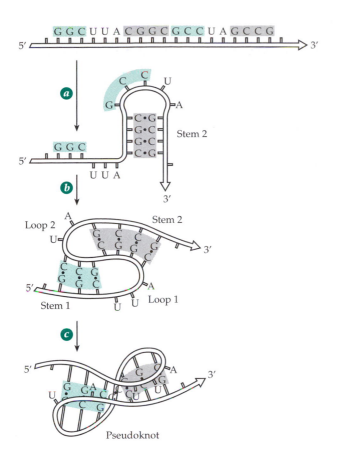

Figure 5-29 Formation of a pseudoknot in an RNA chain. After Puglisi *et al.*[325]

–Crick or wobble (GU) base pairing in the stems. This requires certain relationships in the sequences of the DNA in the genes that encode these molecules as discussed in Section E,3. Because of the base-pairing requirements, some of the bases in the stems protrude as bulges and fail to form pairs. Often, it is possible to find more than one reasonable structure, each having some bases unpaired.[328]

2. Transfer RNAs

In all tRNAs the bases can be paired to form "cloverleaf" structures with three hairpin loops and sometimes a fourth as is indicated in Fig. 5-30.[329–331] This structure can be folded into the L-shape shown in Fig. 5-31. The structure of a phenylalanine-carrying tRNA of yeast, the first tRNA whose structure was determined to atomic resolution by X-ray diffraction, is shown.[170,332–334] An aspartic acid-specific tRNA from yeast,[335] and an *E. coli* chain-initiating tRNA, which places *N*-formyl-methionine into the N-terminal position of proteins,[336,337] have similar structures. These molecules are irregular bodies as complex in conformation as globular proteins. Numerous NMR studies show that the basic

structure is conserved in all tRNAs. However, animal mitochondrial tRNAs often lack some of the usual stem-loop "arms" as well as the invariant nucleotides in the dihydrouridine and TψC loops (Fig. 5-30).[307,338,339] At the bottom of the structure as shown in Figs. 5-30 and 5-31 is the **anticodon**, a triplet of bases having the correct structures to permit pairing with the three bases of the codon specifying a particular amino acid (see Table 5-5), in this case phenylalanine.

While tRNAs consist largely of loops and stems containing Watson–Crick base pairs, they also contain Hoogsteen pairs, wobble pairs, and triplets such as the following.

The first of these contains a Watson–Crick A•U pair with a second A bound to form a reversed Hoogsteen A∗A pair. The second contains a G•C Watson–Crick pair with a Hoogsteen N-7-guanine∗G (m^7G∗G) pair. Among the complex base associations present in tRNAs are some that also involve hydrogen bonding to the 2'-hydroxyl groups of ribose rings and to at least one of the phosphate groups. There are over 100 internal hydrogen bonds, a large proportion of which are relatively invariant among the known tRNAs.

3. Properties and Structures of Some Other RNA Molecules

A few specialized RNA molecules are listed in

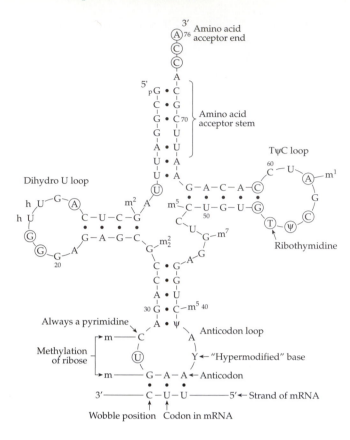

Figure 5-30 Schematic cloverleaf structure of a phenylalanine-specific transfer RNA (tRNA[Phe]) of yeast. The dots represent pairs or triplets of hydrogen bonds. Nucleosides common to almost all tRNA molecules are circled. Other features common to most tRNA molecules are also marked. The manner in which the anticodon may be matched to a codon of mRNA is indicated at the bottom.

TABLE 5-4
Some Specialized RNA Molecules

Kind of RNA	Number of nucleotides
Transfer RNAs (Figs. 5-30, 5-31)	60 – 85
Ribosomal RNAs	
5S	~ 120
5.8S (rat)	158
16S (*E. coli*; Fig. 5-32)	1542
18S (rat)	1874
23S (*E. coli*)	2904
28S (rat)	4718
Telomerase guide RNA[a-d]	159 (*Tetrahymena*)
M1 RNA of Ribonuclease P[e,f]	350 – 410
	377 (*E. coli*)
Tetrahymena Intron ribozyme (Fig. 12-26)[g]	413
Viroid hammerhead ribozyme (Fig. 12-27)[h]	~ 55
Signal recognition particle 7S RNA[i]	295 (human)
Small nuclear RNAs[j,k]	65 – 1200
RNA-editing guide RNA[l]	~ 60
Thermotolerance factor (G8 RNA)[m]	~ 300
Viroid RNA (Fig. 28-19)	240 – 380
Virus MS2 genome (Chapter 29)	3569

[a] Greider, C. W., and Blackburn, E. H. (1989) *Nature (London)* **337**, 331–337

[b] Bhattacharyya, A., and Blackburn, E. H. (1994) *EMBO J.* **13**, 5721–5731

[c] Singer, M. S., and Gottschling, D. E. (1994) *Science* **266**, 404–409

[d] Feng, J., Funk, W. D., Wang, S.-S., Weinrich, S. L., Avilion, A. A., Chiu, C.-P., Adams, R. R., Chang, E., Allsopp, R. C., Yu, J., Le, S., West, M. D., Harley, C. B., Andrews, W. H., Greider, C. W., and Villeponteau, B. (1995) *Science* **269**, 1236–1241

[e] Stark, B. C., Kole, R., Bowman, E. J., and Altman, S. (1978) *Proc. Natl. Acad. Sci. U.S.A.* **75**, 3717–3721

[f] Mattsson, J. G., Svärd, S. G., and Kirsebom, L. A. (1994) *J. Mol. Biol.* **241**, 1–6

[g] Cech, T. R. (1987) *Science* **236**, 1532–1539

[h] Hertel, K. J., Herschlag, D., and Uhlenbeck, O. C. (1994) *Biochemistry* **33**, 3374–3385

[i] Li, W.-Y., Reddy, R., Henning, D., Epstein, P., and Busch, H. (1982) *J. Biol. Chem.* **257**, 5136–5142

[j] Maxwell, E. S., and Fournier, M. J. (1995) *Ann. Rev. Biochem.* **35**, 897–934

[k] Nicoloso, M., Qu, L.-H., Michot, B., and Bachellerie, J.-P. (1996) *J. Mol. Biol.* **260**, 178–195

[l] Kable, M. L., Seiwert, S. D., Heidmann, S., and Stuart, K. (1996) *Science* **273**, 1189–1195

[m] Fung, P. A., Gaertig, J., Gorovsky, M. A., and Hallberg, R. L. (1995) *Science* **268**, 1036–1039

Table 5-4. There are many others.[302] In addition, there are thousands of different mRNAs within a cell. The most abundant RNA molecules in cells are those of the ribosomes. Ribosomes consist of two elaborate RNA-protein subunits, a large subunit with sedimentation constant ~ 30S in bacteria and ~ 40S in eukaryotes. The small subunit contains 16S or 18S RNA and the large subunit 23S or 28S as well as smaller 5S and 5.8S RNAs (Table 5-4; Table 28-1). A proposed three-dimensional structure[341,342] of a bacterial 16S ribosomal RNA and the corresponding ribosomal subunit with its 21 proteins are shown in Fig. 5-32. It might seem impossible that the folding pattern of the RNA was deduced correctly before an X-ray structure was available. However, a **phylogenetic approach**, the comparison of nucleotide sequences among several species, suggested that the stem structures of rRNAs are highly conserved (see discussion in Chapter 29). This fact, together with a variety of other chemical

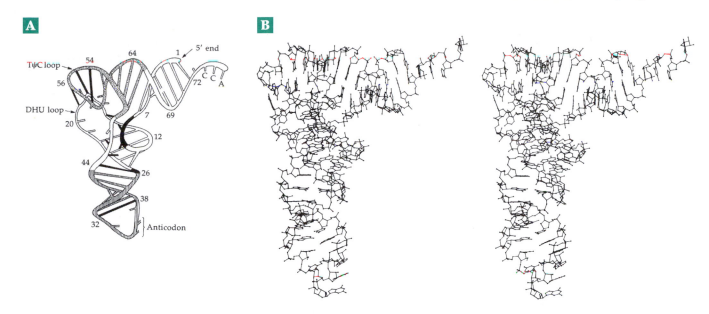

Figure 5-31 The three-dimensional structure of a phenylalanine-specific transfer RNA of yeast. (A) Perspective diagram of folding of polynucleotide chain. The ribose phosphate backbone is drawn as a continuous cylinder with bars to indicate hydrogen-bonded base pairs. The positions of single bases are indicated by rods which are intentionally shortened. The TψC arm is heavily stippled, and the anticodon arm is marked by vertical lines. Tertiary structure interactions are illustrated by black rods. Redrawn from Quigley and Rich.[340] (B) Stereoscopic view of the structure of yeast tRNA[Phe] as revealed by X-ray crystallography. The acceptor stem with the protruding ACCA sequence at the 3' end is to the right. The anticodon GAA is at the bottom right side of the drawing. The guanine ring is clearly visible at the very bottom. The middle adenine of the anticodon is seen exactly edge-on as is the "hypermodified" base Y (see Fig. 5-33) which lies just above the anticodon. Its side chain is visible at the back of the drawing. Preceding the anticodon on the left (5' side) are two unpaired bases (C and U). The 2'-hydroxyl groups of the cytosine and of the guanosine in the anticodon are methylated. Moving far up the anticodon stem one can see two groups of base triplets which utilize both Watson–Crick and Hoogsteen types of base pairing. Drawing courtesy of Alexander Rich.

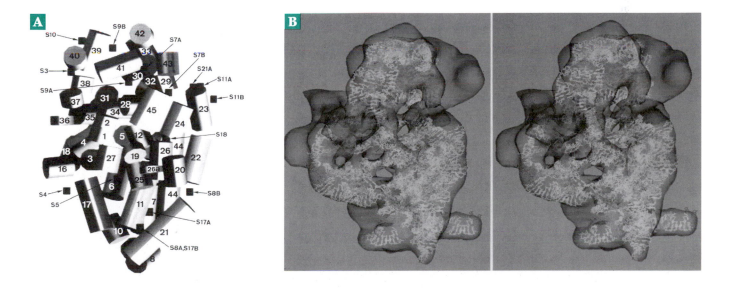

Figure 5-32 (A) A three-dimensional computer graphics model proposed by Brimacombe et al.[341] for the single chain of E. coli 16S ribosomal RNA. The helices are depicted as cylinders, which are all connected. The small dark squares denote the positions of artificially formed RNA–protein crosslinks, marked with the appropriate protein number. For proteins exhibiting more than one crosslink site (e.g., S17), the sites are denoted A or B, in each case A being the site nearer to the 5' terminus of the 16S RNA. (B) Stereoscopic view of tentative atomic model of 16S RNA in the 30S ribosomal subunit. The viewing direction is different from that in (A). From Mueller and Brimacombe.[342] Courtesy of Richard Brimacombe.

evidence,[341,343,344] allowed the prediction of the structure shown as well as many characteristics of the rRNA of the large subunit.[345] Culminating decades of effort[346-350] (Chapter 29), complete structures of bacterial ribosomes were established by 2000[342,351-354] and the peptidyl transferase center was identified as a ribozyme[355-357] (discussed in Chapter 12).

4. Modified Nucleotides

The picture of DNA or RNA as chains of only four kinds of nucleotides is not quite accurate. DNA contains

Uridine

Guanosine

Nucleoside Y

N^6-(Δ^2-Isopentenyl)adenosine

(Abbrev: iA) found at 3' end of anticodons ending in A (codons starting with U)

a cytokinin

Nucleoside Q (Queosine)

found in wobbling positions of anticodons

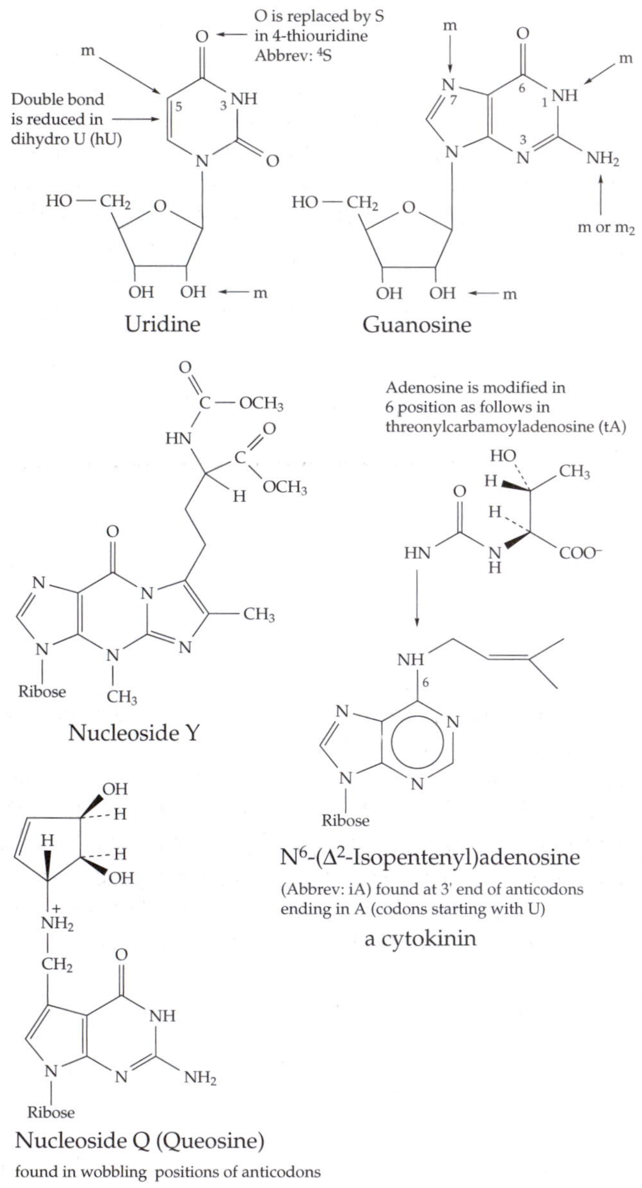

Figure 5-33 Structures of some nucleosides containing modified bases and found in tRNA molecules. Positions where methylation may occur are designated m. Some other abbreviations are e (ethyl), ac (acetyl), and f (formyl).[25]

a significant number of methylated bases among which are **5-methylcytosine** (5mC or m^{5}C) and 6-methyladenine (6mA). The former is regularly present in the nuclear DNA of higher animals and plants. In mammals from two to seven percent of the cytosine is methylated. It is likely that these methylated bases mark special points in the genetic blueprint. As is discussed in Chapter 27, methylation, which is accomplished after the synthesis of the polynucleotide, may block the expression of certain genes.[358] This appears to happen when one of the two X-chromosomes becomes inactivated in cells of females. Methylation may also be used to **imprint** certain genes, i.e. to mark them as coming from a specific one of the two parents.[359]

Another function of methylation is to protect DNA from attack by enzymes (restriction endonucleases) formed in response to invading viruses (Chapter 26). Some viruses, notably the bacteriophage of the T-even series that attack *E. coli* (Box 7-C), have developed their own protective devices. They contain **5-hydroxymethylcytosine** (HOMeC) in place of cytosine. The extra hydroxyl groups provided in this fashion often carry one or two glucose units in glycosidic linkage.[360] A bacteriophage attacking *Bacillus subtilis* substitutes hydroxymethyluracil for uracil and 5-dihydroxypentyluracil for thymine, and phage W14 of *Pseudomonas acidovorans* substitutes the 5-methyl of thymine with $-CH_2-NH-(CH_2)_4-NH_3^+$.[360,361]

The modifications carried out on RNA molecules are more varied and more extensive than those of DNA. Sixty or more modification reactions are known for tRNA, with the number and the extent of modification depending upon the species. Structures of some of the modified bases are indicated in Fig. 5-33. Uridine can be methylated either on the base or on the 2'-hydroxyl of the sugar. Methylation at the 5 position of uridine yields **ribothymidine**. Cytidine can be modified in the same positions. Reduction of the 5,6 double bond of uridine gives **dihydrouridine** (hU). Replacement of the oxygen at position 4 by sulfur gives **4-thiouridine** (⁴sU). Positions in the guanosine structure that can be methylated are also indicated in Fig. 5-33. The symbol m is commonly used to designate methylation in nucleic acid bases; m$_2$ indicates dimethylation, e.g., 6,6-dimethyladenine is abbreviated m$_2^6$A.

A remarkable transformation is that of uridine into pseudouridine (ψ).

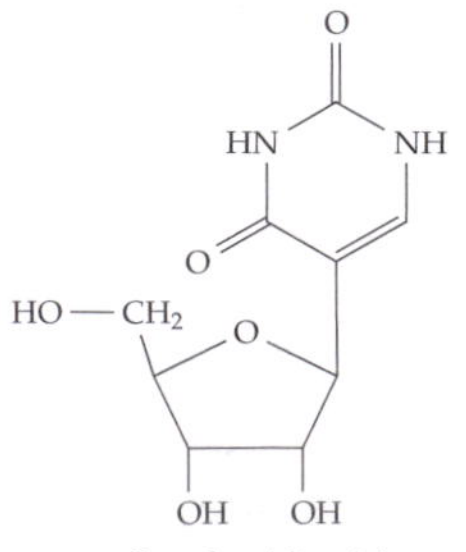

Pseudouridine (ψ)

BOX 5-C THE RNA WORLD

The discovery in the 1980s that RNA molecules often have catalytic properties and may serve as true enzymes (ribozymes; Chapter 12) stimulated new thinking about evolution. Although RNA catalysts are not as fast as the best enzymes they are able to catalyze a wide variety of different reactions. Could it be that in the early evolution of organisms RNA provided both the genetic material and catalysts? The "RNA world" would have been independent of both DNA and protein.[a,b] Later DNA could have been developed as a more stable coding molecule and proteins could have evolved as more efficient catalysts. Plausible reactions by which both cytosine and uracil could have arisen in drying ponds on early Earth have been demonstrated.[c]

A major objection to the RNA world is the lack of stability of ribose and the inability to demonstrate the nonenzymatic synthesis of ribose in significant amounts. Even if ribose were present, it would be largely in the pyranose ring forms. Initial formation of the 5-phosphate would be required to allow formation of a nucleotide with a furanose ring. These and other obstacles to the RNA world have led to the suggestion that some *other* genetic material preceded RNA and DNA.[d,e] One possibility is a peptide-like RNA analog.[e] A simple coding system could also have been used, e.g. one based on only two bases, such as C and G, instead of four.[f,g]

Perhaps it is more probable that formation of proteins *and* the present coding system evolved simultaneously? The major metabolic cycles (Chapter 10) could also have developed at the same time. RNAs could not have been the *first catalysts*. Hydrogen ions, hydroxyl ions, ammonium, cyanide, and other simple ions as well as amines and peptides could all have played a role in prebiotic chemistry. Another speculation suggests an "iron-sulfur world" in which organic materials would be formed on mineral surfaces through reactions involving reduction of bicarbonate by iron sulfide and H_2S.[h]

If the RNA world did exist, has it left us with any real clues? Benner *et al.* suggest that modern metabolism is a palimpsest of the RNA world, a parchment that has been inscribed two or more times, with previous texts imperfectly erased and therefore still partially legible.[i] If so, can we find a way to read the text of the ancient RNAs?

---

[a] Gesteland, R. F., Cech, T. R., and Atkins, J. F., eds. (1999) *The RNA World*, 2nd ed. Cold Spring Harbor Lab. Press, Cold Spring Harbor, New York

[b] Orgel, L. E. (1994) *Sci. Am.* **271** (Oct), 77 – 83

[c] Robertson, M. P., and Miller, S. L. (1995) *Nature (London)* **375**, 772 – 774

[d] Larralde, R., Robertson, M. P., and Miller, S. L. (1995) *Proc. Natl. Acad. Sci. U.S.A.* **92**, 8158 – 8160

[e] Böhler, C., Nielsen, P. E., and Orgel, L. E. (1995) *Nature (London)* **376**, 578 – 581

[f] Sievers, D., and von Kiedrowski, G. (1994) *Nature (London)* **369**, 221 – 224

[g] Piccirilli, J. A. (1995) *Nature (London)* **376**, 548 – 549

[h] Maden, E. H. (1995) *Trends Biochem. Sci.* **20**, 337 – 341

[i] Benner, S. A., Ellington, A. D., and Tauer, A. (1989) *Proc. Natl. Acad. Sci. U.S.A.* **86**, 7054 – 7058

Pseudouridine is formed by enzymatic rearrangement of uridine in the original transcript (Eq. 28-3). It can form a base pair with adenine in the same manner as does uracil. Pseudouridine is found not only in tRNA but also in several places in both large and small ribosomal RNA subunits. For example, it is present at position 516 in the *E. coli* 16S RNA,[364] at a specific position in the 23S RNA, and at many more locations in eukaryotic rRNA.

The bases called "Y" and "Q" are highly modified guanines (Fig. 5-33). Q is found at the 5' end of some anticodons in the "wobble position" (see Fig. 5-30). Two **hypermodified** adenosines are also shown in Fig. 5-33. The N^6-isopentenyladenosine is found at the 3' end of the anticodons that pair with codons starting with U. This compound is also a plant hormone, a **cytokinin** (Chapter 30). Another highly modified purine, threonylcarbamoyladenine, occurs adjacent to the end of anticodons pairing with codons starting with A. The function of these hypermodified bases is uncertain, but they appear to promote proper binding to ribosomes. The modifications are often not absolutely essential for function.

Another source of modified bases in both DNA and RNA is spontaneous or "accidental" alteration. Nucleic acids encounter many highly reactive and mutagenic materials including hydroxyl radicals, formed from O_2, and are able to convert guanine rings into 7,8-dihydro-8-oxoguanine.[362] Other reactive and carcinogenic compounds can form adducts with nucleic acid bases.[363] See Eq. 5-18 and also Chapter 27.

5. RNA Aptamers

Ellington and Szostak[365] synthesized a random "pool" of ~10^{15} different oligodeoxyribonucleotides, each ~100 nucleotides in length. They "amplified" these using the polymerase chain reaction (PCR; Section H,6) and prepared a mixture of the corresponding RNAs by *in vitro* transcription. From the ~10^{13} different sequences still present they selected individual oligonucleotides by affinity chromatography on columns that contained well-defined immobilized ligands such as organic dyes. They called the selected RNAs **aptamers**. Their approach is being used to find RNA sequences that bind to such ligands as ATP, FMN,[366] the bronchodilator theophylline,[367] aminoglycoside antibiotics,[368] arginine,[369] etc.[370] Many of the selected aptamers bind their ligands very tightly and studying them may shed light on interactions of RNA with proteins and on the catalytic activities of RNA, which are discussed in Chapter 12.

E. The Genetic Code

The general nature of the genetic code was suggested by the structure of DNA. Both DNA and proteins are linear polymers. Thus, it was logical to suppose that the sequence of the bases in DNA codes for the sequence of amino acids. There are only four bases in DNA but 20 different amino acids in proteins at the time of their synthesis. It is obvious that each amino acid must be specified by some combination of more than one base. While 16 pairs of bases are possible, this is still too few to specify 20 different amino acids. Therefore, it appeared that at least a triplet group of three nucleotides would be required to code for one amino acid.[371] Sixty-four (4^3) such triplet **codons** exist, as is indicated in Tables 5-5 and 5-6.

Simplicity argues that the genetic blueprint specifying amino acid sequences in proteins should consist of consecutive, nonoverlapping triplets. This assumption turned out to be correct, as is illustrated by the DNA sequence for a gene shown in Fig. 5-5. In addition to the codons that determine the sequence of amino acids in the protein, there are **stop codons** that tell the ribosomal machinery when to terminate the polypeptide chain. One methionine codon serves as an **initiation codon** that marks the beginning of a polypeptide sequence. One of the valine codons sometimes functions in the same way.

How does a cell read the code? This question is dealt with in detail in Chapters 28 and 29. A key step is the positioning of each amino acid on the ribosome in proper sequence. This is accomplished by the pairing of codons of messenger RNA with the anticodons of the appropriate transfer RNA molecules as is indicated at the bottom of Fig. 5-30. Each tRNA carries the appropriate "activated" amino acid at its 3' end ready to be inserted into the growing peptide.

1. The "Reading Frames"

It is immediately obvious that there are three ways of reading the genetic code in mRNA depending upon which nucleotide is used to start each codon. For example, in the following mRNA sequence either codons GCA, CAG, or AGC could be selected as first.

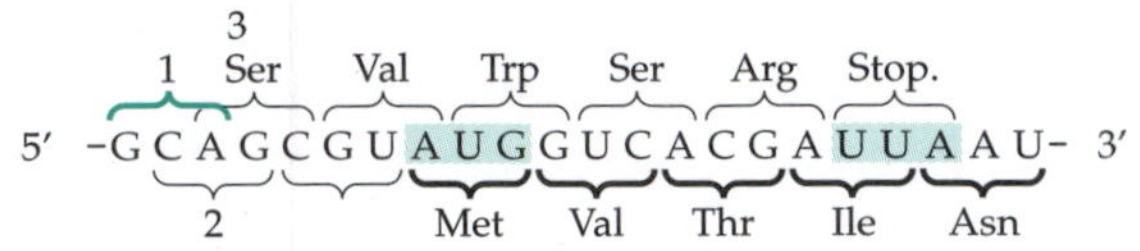

These codons define the three reading frames or phases in which the code may be read. Here, the term *frame*

TABLE 5-5
The Genetic Code[a]

Amino acid	Codons	Total number of codons
Alanine	GCX	4
Arginine	CGX, AGA, AGG	6
Asparagine	AAU, AAC	2
Aspartic acid	GAU, GAC	2
Cysteine	UGU, UGC	2
Glutamic acid	GAA, GAG	2
Glutamine	CAA, CAG	2
Glycine	GGX	4
Histidine	CAU, CAG	2
Isoleucine	AAU, AUC, AUA	3
Leucine	UUA, UUG, CUX	6
Lysine	AAA, AUG	2
Methionine		
(also initiation codon)	AUG	1
Phenylalanine	UUU, UUC	2
Proline	CCX	4
Serine	UCX, AGU, AGC	6
Threonine	ACX	4
Tryptophan	UGG	1
Tyrosine	UAU, UAC	2
Valine		
(GUG is sometimes		
an initiation codon)	GUX	4
Termination	UAA (*ochre*)	
	UAG (*amber*)	
	UGA	3
Total		64

[a] The codons for each amino acid are given in terms of the sequence of bases in messenger RNA. From left to right, the sequence is from the 5' end to the 3' end. The symbol X stands for any one of the four RNA bases. Thus, each codon symbol containing X represents a group of four codons.

does not designate a single codon, although frame does designate a single exposure in a motion picture film. **Reading frame** designates which of the three possible sets of codons we are using. In the foregoing sequence the codons in reading frames 2 and 3 are labeled. Reading frame 2 contains the initiation codon AUG (shaded) which could mark the beginning of an encoded protein sequence. Reading frame 3 contains a termination (stop) codon which, when the mRNA transcript is read by ribosomes, will terminate polypeptide synthesis. It may be in the position shown but not have any real function. However, it could represent the end of the coding sequence that is marked if genes for the two proteins overlap a little at the ends, a situation that actually occurs in nature.

A reading frame in a specified part of a DNA sequence is said to be **"open"** if there is an initiation codon preceded by suitable regulatory signals (an **operator** region). This means that it *could* encode a protein. The reading frame of a sequence is open until the next termination codon. Recently another usage has appeared. Many writers refer to an **open reading frame** as a segment of DNA in which any one of the

three reading frames is open. Another complexity in the reading of genetic messages arises because splicing of RNA may sometimes cause a shift in the reading frame. For example, a mRNA being transcribed from the sequence in reading frame 2 in the foregoing example may skip over a nucleotide part of the time to form an RNA in which the first part is encoded by reading frame 2 and the second by reading frame 3 (a + 1 frameshift). Alternatively, a nucleotide could be read twice with a − 1 frameshift with the sequence of reading frame 1 for the latter part of the mRNA. Frameshifts can also occur during protein synthesis as the mRNA is being read.

In the present example we have examined the sequence in mRNA. In the DNA there are two strands. One is the **coding strand** (also called the nontranscribing or nontranscribed strand), which has a sequence that corresponds to that in the mRNA and the one that is given in Fig. 5-4. The second antiparallel and complementary strand can be called the **template strand** or the noncoding, transcribing, or transcribed strand.[372] The mRNA that is formed is sometimes referred to as a **sense strand**. The complementary mRNA, which corresponds in sequence to the noncoding strand of DNA, is usually called **antisense RNA**.

TABLE 5-6
The Sixty-Four Codons of the Genetic Code

5'–OH Terminal base	Middle base				3'–OH Terminal base
	U(T)	C	A	G	
U(T)	Phe	Ser	Tyr	Cys	U(T)
	Phe	Ser	Tyr	Cys	C
	Leu	Ser	Term[c]	Term[d]	A
	Leu	Ser	Term	Trp	G
C	Leu	Pro	His	Arg	U(T)
	Leu	Pro	His	Arg	C
	Leu[a]	Pro	Gln	Arg	A
	Leu	Pro	Gln	Arg	G
A	Ile	Thr	Asn	Ser	U(T)
	Ile	Thr	Asn	Ser	C
	Ile	Thr	Lys	Arg	A
	Met[b]	Thr	Lys	Arg	G
G	Val	Ala	Asp	Gly	U(T)
	Val	Ala	Asp	Gly	C
	Val	Ala	Glu	Gly	A
	Val[b]	Ala	Glu	Gly	G

[a] The codon CUA (CTA) encodes threonine and the codon AUA (ATA) methionine in mammalian mitochondria.
[b] Initiation codons. The methionine codon AUG is the most common starting point for translation of a genetic message but GUG can also serve. In such cases it codes for methionine rather than valine.
[c] The "termination codon" UAA (TAA) encodes glutamine in *Tetrahymena*.
[d] The termination codon UGA (TGA) encodes tryptophan in mitochondria and selenocysteine in some contexts in nuclear genes.

2. Variations

Is the genetic code "universal" or does it vary from one organism to another? Studies with bacteria, viruses, and higher organisms including humans have convinced us that the code is basically the same for all organisms. However, there are some variations. For example, in mitochondria of both humans and yeast the codon TGA is not a termination codon but represents tryptophan. In mammalian mitochondria CTA represents threonine rather than leucine, and ATA encodes methionine instead of isoleucine. These differences in the code are related to the fact that mitochondria contain their own piece of DNA. It encodes not only several proteins but also tRNA molecules whose anticodon structures are altered to accommodate the changed meanings of the codons of the mitochondrial DNA and mRNA.[373,374]

Variations in the code for cytoplasmic proteins have been found. In *Tetrahymena* and other ciliates the codon TAA represents glutamine rather than being a termination codon.[375] A few proteins, including some in the human body, contain **selenocysteine**, the selenium-containing analog of cysteine. Selenocysteine is encoded by termination codon TGA. See Chapter 29 for details. However, even though TGA is occasionally used in this way, it serves as a termination codon for most proteins within the same cells.[376] Thus, the *context* in which the codon TGA occurs determines how it is read by the ribosomal machinery.

3. Palindromes and Other Hidden Messages

The sentence "Madam, I'm Adam" reads the same either forward or backward. Such sentences, known as **palindromes**,[377] are infrequent in the English language. However, most DNA contains many palindromes, sequences of base pairs that read the same in forward and reverse directions. Consider, for example, the gene that specifies the sequence of nucleotides for the tRNA molecules of Fig. 5-30. It is a double-stranded DNA segment in which one strand has a sequence identical to that of Fig. 5-30 except for the substitution of T for U and for ψ (pseudouridine) and for the lack of methylation and other base modifications. The second strand is the exact complement. Figure 5-34 shows the part of this gene (residues 49–76) that corresponds to the 3' end of the tRNA molecule. This DNA segment could exist in a second conformation having a loop on each side of the molecule (Fig. 5-34). The stems of the two loops in this **cruciform** conformation are identical and symmetrically disposed around the center of the molecule. If we overlook the seven nucleotides in the center of the loop, the message in the stems reads the same in both directions, as is indicated by the green arrows.

Palindromes are often imperfect as is the one shown in Fig. 5-34. Here the two stems in the cruciform structure are related by an exact twofold rotational symmetry but the loops at the ends of the stems are not. Unpaired bases may bulge at various points in double-stranded stems of longer palindromes. These imperfect palindromes in the DNA are responsible for much of the tertiary structure of the various kinds of RNA. The tertiary structure, in turn, often determines the interaction of the RNA with enzymes and other proteins.

Special properties may be observed for palindromes containing homopurine tracts in one strand, and therefore homopyrimidine tracts in the other. If two identical palindromes of this type occur close together it is possible that the pyrimidine-containing strand of one can join with a hairpin

loop of the other to form a triplex base structure. A related triplex structure may be formed when inverted repeat sequences occur within a homopyrimidine tract in one chain of DNA (Fig. 5-34B).[378] These have been called **mirror-repeats** or **H-palindromes** to distinguish them from true palindromes.[237,379] Each base triplex structure contains one set of Watson–Crick hydrogen bonds and one set of Hoogsteen hydrogen bonds. The triplexes of H-DNA are all either TAT or CGC[+], where one C is protonated (see also Fig. 5-24).

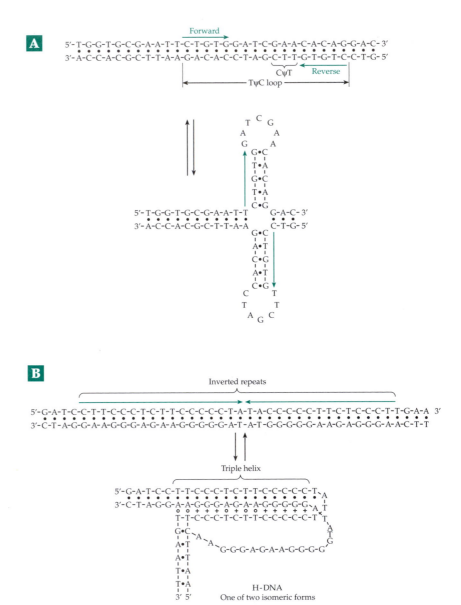

Figure 5-34 (A) Two conformations of a segment of the yeast phenylalanine tRNA gene. The segment shown codes for the 3' end of the tRNA molecule shown in Fig. 5-30, including the TψC loop. (B) Formation of H-DNA (Fig. 5-24) proposed for a sequence in plasmid pGG32. The major element of the structure is the triplex, which is formed from the Watson–Crick duplex (•) associated with the homopyrimidine loop through Hoogsteen base pairing (○, +). One of the two possible "isomeric" forms is shown. See Mirkin *et al.*[378]

Segments of H-DNA might block transcription, and proteins that bind to the triplex H-DNA may be involved in transcriptional regulation.[379]

Until rather recently there had been little to indicate that DNA actually assumes cruciform conformations in cells. However, strong experimental evidence suggests that some cruciform structures do form naturally.[380] Their formation from palindromic DNA [like the formation of Z-DNA from (G + C)-rich sequences] is a way of relieving torsional strain induced by supercoiling. Whether or not cruciform structures occur frequently within cells, there is no doubt that palindromic sequences are of great importance in the interaction of nucleic acids with symmetric dimeric and tetrameric protein molecules such as the gene repressor protein shown in Fig. 5-35.[381-383]

DNA contains numerous other protein binding sites which are not palindromes but whose sequences represent additional encoded information. The RNA transcripts likewise contain sequences that direct the catalytic machinery involved in splicing, that bind to ribosomal proteins, that control rates of transcription, and that cause termination of transcription.

4. The Base Composition of DNA and RNA

The nucleotide composition of DNA is surprisingly variable. The sum of the percentage cytosine plus the percentage guanine (C + G) for bacteria varies from 22 to 74%. That for *E. coli* is 51.7%. Among eukaryotic organisms, the range is somewhat narrower (28 to 58%; for humans, 39.7%). The fact that bacterial DNA molecules are more varied than those of higher organisms is not surprising. The prokaryotes have evolved for just as many more million of years as have we. Perhaps because of their simpler structure and rapid rate of division, nature has done more experimentation with genetic changes in bacteria than in people.

Comparisons of the C + G content of organisms have been used as a basis for establishing genetic relatedness.[384] However, since thymine is especially susceptible to photochemical alteration by ultraviolet light, bacteria with a high (C + G) content may have evolved in environments subject to strong sunlight or high temperatures, whereas those with a low (C + G) content have developed in more protected locations.[385,386]

F. Interaction of Nucleic Acids with Proteins

Most functions of DNA and RNA are dependent on proteins. Cells contain a vast array of polypeptides that bind to nucleic acids in many specific ways.[387-390] We have seen (Fig. 5-21) that the histones hold supercoiled DNA in the form of nucleosomes[391] in eukaryotic nuclei and similar proteins bind to and protect the double-stranded DNA of bacteria.[205,392] Sperm cells "package" a large amount of DNA into a small space with the help of small arginine-rich proteins called **protamines**.[393] Cells always have some single-stranded DNA segments as well as single-stranded DNA binding (**SSB**) proteins. Among the latter are proteins from *E. coli*[394,395] and from viruses.[396-399] Specialized proteins bind to DNA sequences in telomeres[38,272] and centromeres.[400] A large number of proteins interact with RNA in ribosomes, spliceosomes, and other complexes.

A host of enzymes, which are described elsewhere in the book, act on DNA and RNA. They include hydrolytic nucleases, methyltransferases, polymerases, topoisomerases, and enzymes involved in repair of damaged DNA and in modifications of either DNA or RNA. While most of these enzymes are apparently proteins, a surprising number are **ribozymes**, which consist of RNA or are RNA–protein complexes in which the RNA has catalytic activity.

1. The Helix–Turn–Helix Motif

Much current interest in DNA–protein interactions is focused on regulatory processes. In prokaryotes the initiation of transcription of a large fraction of the genes is blocked by the binding of proteins known as **repressors**. While their structures are varied, one large group of repressors have DNA-binding domains with a similar helix–turn–helix architectural motif. They bind with high affinity to specific control regions which contain palindromic DNA sequences such as the following one, which defines a binding site for the *E. coli* **trp** (tryptophan) **repressor**:

$$5'\text{-}T\text{-}G\text{-}T\text{-}A\text{-}C\text{-}T\text{-}A\text{-}G\text{-}T\text{-}T\text{-}A\text{-}A\text{-}C\text{-}T\text{-}A\text{-}G\text{-}T\text{-}A\text{-}C\text{-}3'$$
$$3'\text{-}C\text{-}A\text{-}T\text{-}G\text{-}A\text{-}T\text{-}C\text{-}A\text{-}A\text{-}T\text{-}T\text{-}G\text{-}A\text{-}T\text{-}C\text{-}A\text{-}T\text{-}G\text{-}T\text{-}5'$$

These repressor proteins form dimers joined through a rigid central domain with a pair of arms, each containing two helices that form the helix–turn–helix motif. One helix of each pair fits into the major groove of the DNA helix and forms a "reading head" that carries a specific arrangement of amino acid side chains able to locate the symmetric nucleotide sequences flanking the twofold axis of the palindrome. In the case of the *trp* repressor, these are the ACTAGT hexanucleotides (marked by the arrows in the preceding structure). The interaction is depicted in Fig. 5-35.[401-405] The repressor has a high affinity for the DNA only if one molecule of L-tryptophan is bound to each subunit at a specific site near the DNA helix. Binding of the tryptophan causes a conformational change, which is pictured in Fig. 5-35. If tryptophan is absent the repressor binds to the palindromic DNA only weakly and transcription of genes needed for tryptophan biosynthesis occurs freely. However, if tryptophan accumulates within the cell it binds to the repressor molecules causing them to bind firmly to DNA and prevent transcription. There are at least three of these palindromic sequences in *E. coli*, each one regulating a set of genes (**operons**) involved in tryptophan synthesis. In contrast to the effect of tryptophan, the tryptophan analog indole-3-propionate, which lacks the amino group of tryptophan, *derepresses* the same operons. It also binds to the *trp* repressor, but with the indole ring flipped over by 180° so that its carboxylate group contacts the phosphate groups of the DNA repelling them through both electrostatic and steric effects.[406,407] The picture in Fig. 5-35 is an oversimplification. In fact, at some binding sites (operator sites) more than one dimeric repressor binds in a tandem fashion.[405]

The helix–turn–helix motif is also found in many other proteins. One of these is the bacterial **lac** (lactose) **repressor** which controls the *lac* operon and for which

A

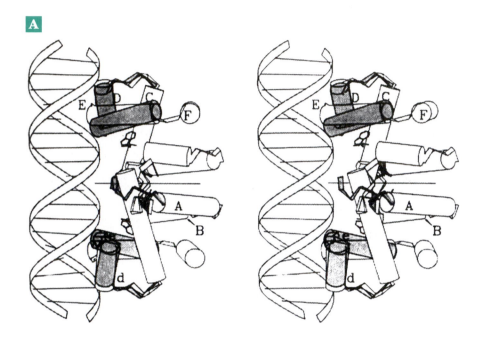

B

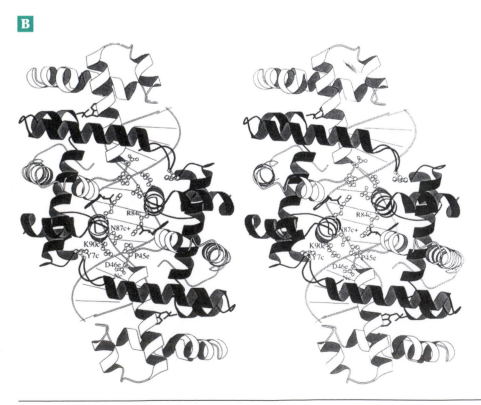

Figure 5-35 Stereoscopic drawings illustrating the binding of a dimeric molecule of the Trp repressor protein to a palindromic sequence in DNA. (A) Schematic view showing structures of the aporepressor (partly shaded gray) and the holorepressor with bound tryptophan (unshaded) are superimposed. Cylinders represent the α helices in (B). From Zhang et al.[402] (B) MolScript ribbon diagram with a few side chains that interact with the DNA shown. Two tandemly bound dimeric repressor molecules are shown. Two bound molecules of tryptophan are visible in each dimer. The DNA is drawn as a double helix with lines representing the base pairs. From Lawson and Carey.[405]

the terms operator, promoter, repressor, and operon were first introduced (Chapter 28).[408–410] Some bacterial viruses, such as **phage lambda (λ)** of *E. coli,* encode repressors that allow the virus to reside in the bacteria without immediately destroying them. The λ repressor and other closely related proteins also utilize the helix –turn–helix motif[411–415] as do some proteins that *activate* transcription (see Chapter 28).

2. Other DNA-Binding Motifs

Nobody knows how many different DNA-binding structures may be discovered. However, most of those that are designed to recognize specific DNA sequences have some part that fits into the major groove of B-DNA.[416] Usually the DNA structure must be in a specific form: B-, A- or Z-, but it may sometimes be bent or distorted. The DNA recognition motifs in the proteins may consist of helices, β strands, or loops.

Leucine zipper proteins. Several transcription factors have the leucine zipper structure, which has been described in Chapter 2 and was illustrated there by the structure of transcription factor Max (Fig. 2-21).[417] Related structures include those of the transcription

activators c-Jun and c-Fos[418] and of the yeast transcription factor GCN4.[419,420] The latter is a dimeric protein in which the C-terminal halves of the two monomers form the helices of the leucine zipper. The helices, which are continuous for over 60 residues, fan out to interact with the DNA double helix as in Fig. 2-21. A 19-residue basic domain of each protein helix crosses the major groove of the DNA with side chain groups interacting with the DNA as is illustrated in Fig. 5-36.

Zinc fingers. Another large group of transcription factors contain a bound zinc ion in a "finger" motif. The Zn^{2+} is held by two cysteine $–S^-$ groups present on a loop of β structure and by two imidazole groups present on an α helix (Fig. 5-37).[421–424] Amino acid sequences found in many proteins that regulate transcription tend to form zinc fingers and to interact with DNA. These were recognized first in the transcription factor TFIIIA from *Xenopus laevis* in 1985.[421,425,426] This protein contains a zinc finger motif repeated nine times. X-ray structures, one of which is shown in Fig. 5-38, are known for proteins with three[423] and five[427] zinc fingers. (See Chapter 28.)

Beta ribbons. An antiparallel double-stranded β ribbon can fit into the major groove of DNA and form

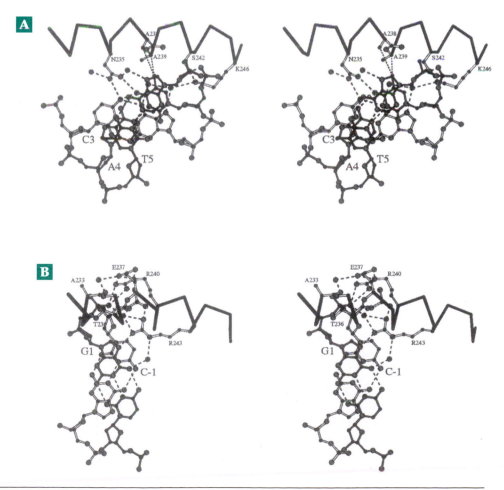

Figure 5-36 Stereoscopic diagrams showing some of the interactions between an N-terminal helical domain of the yeast transcription factor GCN4-bZIP, a leucine zipper protein, and a specific palindromic DNA binding site:

```
   –5       –1 1       5
5' –A T G A C G T C A T
```

(The bases are numbered outward from the central C and G.) The small solid spheres are water molecules. Notice the water mediated interactions of the basic arginine and lysine side chains with the nucleic acid bases and also the interaction of R240 and R243 (in B) with a backbone phosphate. The overall structure of the protein is similar to that of another leucine zipper shown in Fig. 2-21. From Keller *et al.*[419] Drawings courtesy of Timothy J. Richmond.

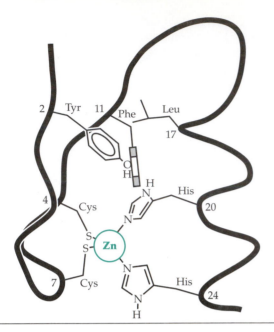

Figure 5-37 Three-dimensional structure of a zinc finger. This is formed by the binding of Zn^{2+} to the following sequence in a protein:

$$\phi \; X \; C \; X_{2-4} \; C \; X \; X \; X \; F \; X_5 \; L \; X \; X \; H \; X_{3,4} \; H$$

$$\underset{\beta}{\xrightarrow{\hspace{1cm}}} \text{turn} \underset{\beta}{\xrightarrow{\hspace{1cm}}} \text{turn} \underset{\alpha \text{ helix}}{\xrightarrow{\hspace{1cm}}}$$

Here ϕ is a hydrophobic amino acid and X may be any amino acid. After Krizek *et al.*[428]

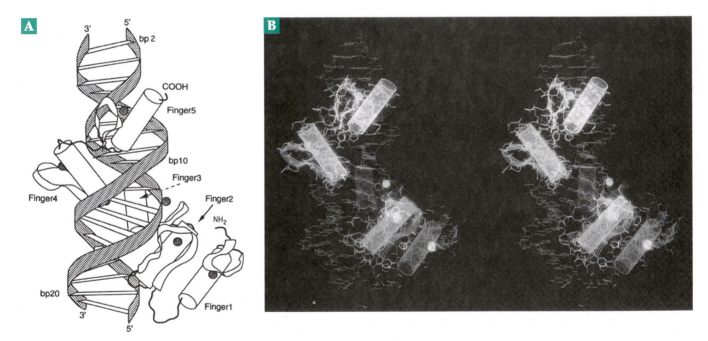

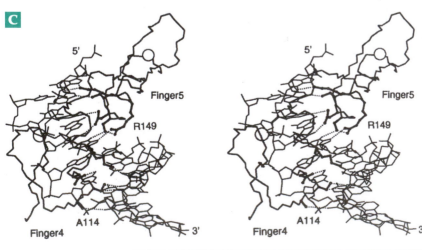

Figure 5-38 (A) Sketch of a complex of a five-zinc finger protein bound to a high-affinity site in DNA. This protein is a human oncogene called GL1. Its gene is amplified in many tumors, but it is similar to many other transcription factors. The cylinders and ribbons highlight the α helices and β sheets. Cobalt ions, which replaced the Zn^{2+} in the crystals, are shown as spheres. (B) Stereoscopic view of the complex in a similar orientation. (C) Stereoscopic view emphasizing the contacts of side chains from fingers 5 and 6 with the DNA. The view is similar to that in (A) and (B) but the structure has been tilted back. Drawings courtesy of Nikola Pavletich.[427]

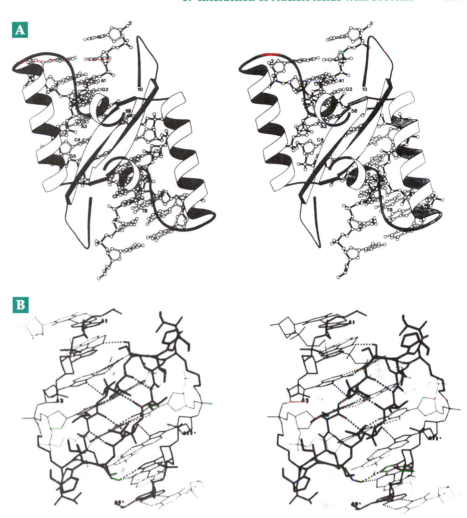

Figure 5-39 (A) Stereoscopic ribbon representation of the *E. coli* methionine repressor–operator complex. Two subunits form a dimer with a double-stranded antiparallel β ribbon that fits into the major groove of the DNA in the B form. One strand is shaded more darkly than the other. (B) View of the β ribbon and its interactions with the DNA. Notice the direct hydrogen bonds from amino acid side chains of lysine and threonine residues to bases in the specific palindromic sequence AGACGTCT. From Somers and Phillips.[429] Drawings courtesy of S. E. V. Phillips.

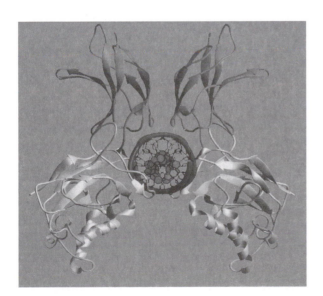

Figure 5-40 Structure of a protein known as transcription factor NF-κB bound to its DNA target. Each subunit of the dimeric protein contains two β barrel domains. The loops at the ends of the barrels interact with the DNA in the center. From Müller *et al.*[433] Courtesy of Stephen C. Harrison.

hydrogen bonds and other interactions with the DNA. This is the basis for specific recognition of the operator sequence for the *E. coli* **met** (methionine) **repressor** as is illustrated in Fig. 5-39.[429] The same binding motif is employed by some other repressors[430,431] and also by the abundant bacterial DNA-binding protein **HU**.[432] It has also been utilized in designing the previously mentioned hairpin polyamide DNA-binding compounds.

The winged helix family. A group of large protein transcription factors contain an N-terminal DNA-binding domain with the striking winged helix motif shown in Fig. 5-40.[433–436] It occurs in proteins from a wide range of organisms from yeast to human.

3. RNA-Binding Proteins

Among the many proteins that bind to RNA molecules[437–439] are the aminoacyl-tRNA synthetases, a variety of other well known enzymes,[440] the ribosomal proteins discussed in Chapter 29, and various proteins with dual functions of catalysis and regulation of translation. A widely found RNA-binding **ribonucleoprotein domain** (RNP domain), occurs in hundreds of proteins including many of the RNA processing proteins considered in Chapter 28.[437,441,442] This 70–90 residue αβ module binds an RNA strand against a β sheet surface. Another ~ 70-residue protein motif binds ds RNA.[443]

G. Viruses and Plasmids

Attacking every living thing from the smallest mycoplasma to human beings the nucleoprotein particles known as viruses have no metabolism of their own. However, they "come alive" when the nucleic acid that they contain enters a living cell. Viruses are significant to us not only because of the serious disease problems that result from their activities but also as tools in the study of molecular biology. A mature virus particle or **virion** consists of one or more nucleic acid molecules in a protein coat or **capsid**, usually of helical or icosahedral form. The capsid is made up of morphological subunits called **capsomers**. They can sometimes be seen clearly with the electron microscope. The capsomers in turn are usually composed of a number of smaller protein subunits. Some of the larger viruses are surrounded by membranous envelopes. Others, such as the T-even **bacteriophage** which attack *E. coli*, have extraordinarily complex structures (Box 7-C).

Most viruses contain a genome of either double-stranded DNA or single-stranded RNA, but some small viruses have single-stranded DNA and others have double-stranded RNA. The number of nucleotides in a virus genome may vary from a few thousand to several hundred thousand and the number of genes from 3 to over 200. Sometimes the nucleic acid molecules within the virion are circular, but in other cases they are linear. Table 5-7 lists a few of the known types of viruses as well as some individual viruses.[72,444] The size of the genome, in kilobases (kb), or kilobase pairs (kbp) is indicated. The number of genes in a virus is often somewhat more than one per kbp of DNA. A vast amount of information is available about these infectious particles. Only a few reference sources are cited here.[72,444–446] The architectures of some helical and icosahedral protein coats that surround genomic DNA or RNA are described further in Chapter 7.

1. Viruses with Single-Stranded DNA

The very small **helical bacteriophages** of the Ff family, such as fd, f1, and M13 (see Fig. 7-7) resemble thin bacterial pili but each virus particle contains a molecule of single-stranded circular ~6400 nucleotide DNA of M_r ~2 x 10^6 which encodes only ten proteins.[447–449] Bacteriophage φX174 (Fig. 5-41), an **icosahedral** DNA-containing virus 25 nm in diameter, is only three times as thick as the thinnest cell membrane. Its DNA contains just 5386 nucleotides.[450–452] The similar bacteriophage G4 contains 5577.[453,454]

It is remarkable that such tiny viruses are able to seize control of the metabolic machinery of the cell and turn it all in the direction of synthesis of more virus particles. There are only 11 known proteins encoded by the genes of φX or G4. Three genes encode the three kinds of protein subunits of the virus coat. Sixty copies of each are needed, as are 12 copies of a "pilot" protein.[451] Eight of the genes are spaced closely together, occupying most of the DNA. The other three genes are embedded within some of the first eight but in different reading frames. In one short region of the G4 chromosome the same nucleotides are part of three different genes, using all three possible reading frames. In addition to their own genes, these small viruses make use of many components of the cell that they infect. A large group of animal viruses, the **parvoviruses**, are similar in size and architecture to bacteriophage φX174.

Canine parvovirus, first identified in 1978, is now endemic.[455,456] Childhood **fifth disease** is also caused by a parvovirus.[456,457] When these single-stranded DNA viruses infect cells a double-stranded **replicative form** of DNA arises by synthesis of the complementary **negative strand** alongside the original **positive** DNA strand. Many copies of the replicative form are then synthesized. The negative strands of the replicative forms serve as templates for synthesis of numerous new positive strands that are incorporated into the progeny viruses. The whole process may take only 20 minutes. Some parvoviruses are unable to reproduce unless the cell is also infected by a larger adenovirus.

While most plant viruses contain dsRNA, the **geminiviruses**,[458] which cause a number of plant diseases, contain single-stranded DNA. The virus particle consists of a fused pair of incomplete icosahedra, evidently containing a single ~2500-bp DNA strand. Replication may require coinfection with two virus particles of differing sequence.

2. Viruses with Double-Stranded DNA

One of the smallest viruses containing double-stranded DNA is the 3180-nucleotide human **hepatitis B virus**. It infects millions of people throughout the world causing chronic hepatitis and often liver cancer.[459–461] The circular DNA is surrounded by an envelope consisting of proteins, carbohydrate, and a lipid bilayer. Many icosahedral viruses also contain dsDNA. Among them are the **papovaviruses**, some of which cause warts and others malignant tumors. Much studied by biochemists is the 5386-nucleotide **simian virus 40 (SV40)**, a monkey virus capable of inducing tumors in other species.[462,463] Closely related

TABLE 5-7
Characteristics of Some Individual Viruses and Groups of Viruses

Type of genome[a] and group or individual virus name	Shape[b]	Diameter[c] (nm)	Masses in daltons x 10⁻⁶		Thousands of bases or base pairs (kilobases)
			Total	DNA or RNA	
DNA, single-stranded					
Bacteriophages *fd, f1, M13*	H	~6 x ~880 (length)	17.6	2.1	6.4
Bacteriophages φ174, G4	I	25	6.2	1.8	5.4–5.6
Parvoviruses	I	18–25		1.8	5.5
Geminiviruses	I (fused pairs)				
DNA, double-stranded					
Hepatitis B virus	I (enveloped)			2.1	3.18
Papoviruses	I	45–55			
SV40 (monkey)	I		17.6	3.5	5.22
Polyoma (mouse)	I	45	23.6	3.3	4.96
BK virus (human)	I			3.4	4.96
Papilloma (human wart)	I	56		5.3	8.0
Bacteriophage φ29	T			12	19.3
Adenoviruses	I	70		20–30	30–45
Bacteriophage Mu				25	38
Bacteriophage T7	T(short)				39.9
Bacteriophage P22	T			28.5	43.2
Bacteriophage λ	T			32	48.6
T-even bacteriophage	T	100 x 80 (head)	215	130	166
Baculoviruses of insects	I	70–130			
Herpesviruses	I	100			
core		78	~1000	80–120	80–140
envelope		150–200			
Pox viruses, e.g.,					
Smallpox, vaccinia (cowpox)	C	160 x 250	~4000	150–240	240–300
Cauliflower mosaic virus	I				8.0
RNA, single-stranded					
Unsheathed					
Potato spindle tuber viroid				0.116	0.30
Hepatitis delta virus					1.7
Sheathed, *plus-strand*					
Tobacco necrosis satellite	I	18	1.7	0.4	1.20
Small bacteriophages					
R17, MS2, Qβ	I	23–26	3.6–4.0	1.2–1.5	3.5–4.5
Picornaviruses	I		8.4	2.6	7.9
Polioviruses	I	27	6.4	2	6.1
Rhinoviruses	I	27–30	7–8	2.2–2.8	6.7–8.5
Turnip yellow mosaic virus	I	28	5.0–6.0	2.0	6.1
Tobacco mosaic virus	H	18 x 300	40	2.2	6.7
Togaviruses	I	20–40		4	11
Negative-strand viruses					
Influenza virus	I	80–100	200	2.0	6.1
Bullet-shaped viruses					
Rhabdoviruses	C	20 x 130			
Retroviruses	I	80–100		7–10	20–30
RNA, double-stranded					
Reoviruses	I	55–60		11–12	16–18
Mobillivirus	I (enveloped)	38			
Rotavirus	Wheel-shaped	70			
Leishmania RNA virus					5.28

[a] Complete nucleotide sequences are known for most of these viruses.
[b] Shapes are indicated as I, icosahedral; H, helical; T, a tailed phage; C, complex.
[c] The second dimension given for some helical and complex viruses is the length (nm).

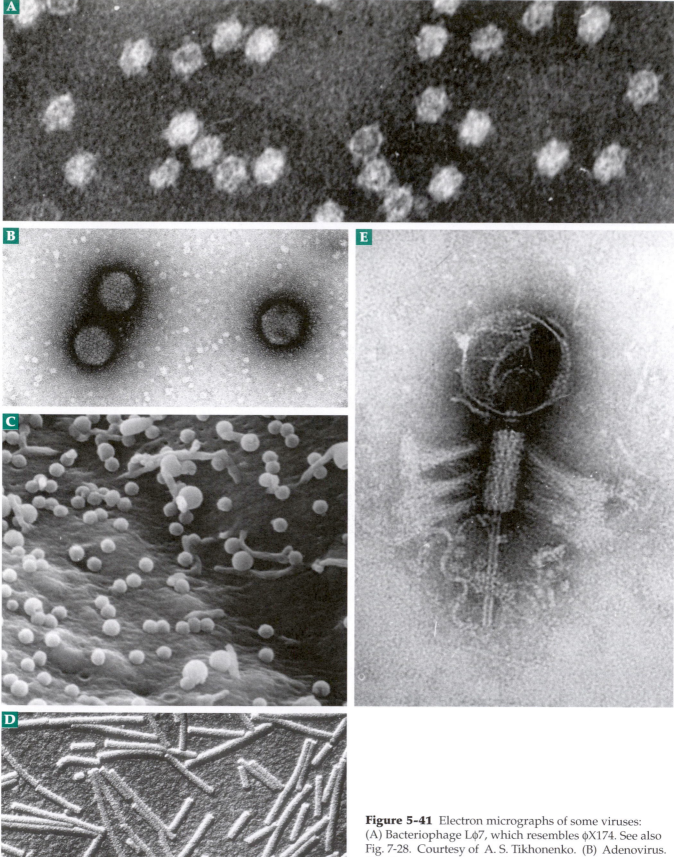

Figure 5-41 Electron micrographs of some viruses: (A) Bacteriophage Lφ7, which resembles φX174. See also Fig. 7-28. Courtesy of A. S. Tikhonenko. (B) Adenovirus. (C) Cytomegalovirus particles being released from a fibroblast. (B) and (C) courtesy of T. Moninger. (D) Tobacco mosaic virus. © Omikron, Photo Researchers. (E) A tailed bacteriophage. Courtesy of A. S. Tikhonenko.

is the **polyomavirus** of the mouse[464] and the **BK virus** of humans,[465] also suspected of causing cancer. The **papillomaviruses** cause warts and perhaps cancer[466] and the larger (70 nm diameter) **adenoviruses** (Fig. 5-41),[467,468] cause respiratory infections. Some 32 types infect humans. Many of the details of eukaryotic transcription were first studied using the adenovirus.[469] The very large **herpesviruses** are enveloped by a lipid-containing membrane.[470–472] Among them are herpes simplex viruses, which infect human mucous membranes and varicella-zoster virus, which causes chicken pox and shingles. Other herpesviruses include cytomegaloviruses (Fig. 5-41), another common human pathogen, and the Epstein–Barr virus, which causes mononucleosis and is suspected of causing cancer.[473] Another herpesvirus has been associated with multiple sclerosis.[474] The very large icosahedral **baculoviruses** cause polyhedroses in insects. One that infects the fly *Tipula* measures 130 nm in diameter. **Poxviruses** are also large and complex.[475] The tailed bacteriophages (Box 7-D) range in size from the small ~29-kDa phage P22 to the very complex T-even phage. Some plant viruses also contain dsDNA. One of these, the **cauliflower mosaic virus**,[476,477] is transmitted by aphids. It has proved useful as a gene-transfer vehicle for genetic engineering.

3. Viruses Containing RNA

Several plant diseases including the **potato spindle tuber disease** are caused by **viroids**, molecules of single-stranded RNA only 240–380 nucleotides in length with a folded structure.[478–481] Such an RNA could code for a protein containing only about 100 amino acids. However, the known 359-nucleotide sequence of the potato spindle tuber virus contains no AUG initiation codon. It seems impossible that the virus carries any gene for a protein. Conserved features of viroid sequences suggest a close relationship to the intervening DNA sequences known as type I introns (see Fig 28-19). Whatever its genetic message, a plant viroid causes the plant cell to replicate many copies of the viroid molecule, which may then be transmitted to other plants by aphids, or on the surface of tools, by humans. A larger 1678-nucleotide viroid-like RNA (**hepatitis delta virus**) has been identified in human patients with severe chronic hepatitis and who were also infected with hepatitis B virus. The ssRNA in this virus does encode at least one protein.[482–484]

One of the smallest of the encapsulated RNA-containing viruses is the **satellite tobacco necrosis virus**. It replicates only when the plant is also infected with the larger tobacco necrosis virus. The satellite virus, whose three-dimensional structure is known from X-ray diffraction studies,[485] contains a 1200-nucleotide strand of RNA which encodes a 195-residue protein.

Sixty copies of the latter are assembled around the RNA in an icosahedral array (Fig. 7-14) to form the virion. The structure of the similar satellite tobacco mosaic virus has also been described in detail.[486,487]

RNAs of the small bacteriophages f2, R17, MS2 and Qβ contain 3500 – 4500 nucleotides and are enclosed in icosahedral shells made up of 90 identical subunits.[488–490] Initially only three genes were evident but a fourth small gene in a different reading frame has since been found in at least one of them.[491] The RNA molecules in these and other **positive-strand** viruses serve as mRNA within the host cells. Another group of small RNA viruses are the **picornaviruses** (picoRNA, meaning very little RNA). Many of these icosahedral viruses of 15–30 nm diameter attack humans. Among them are the **enteroviruses** including the **polioviruses**,[492,493] the **hepatitis A virus**,[494] the **coxsackieviruses**, and some of the **echoviruses**. A second class of picornaviruses include **rhinoviruses** which cause the common cold. More than 100 types are known.[495–497] The **foot–and–mouth disease virus**,[498–500] which attacks the cloven-footed animals, and the **Mengo virus**,[501] which can cause a fatal encephalitis in mice, are also picornaviruses. The three-dimensional structures of polio virus, human rhinoviruses, Mengo virus, and many other eicosahedral viruses are known. Their architecture is discussed in Chapter 7.

The **togaviruses**, which are a little larger than the picornaviruses, have an icosahedral core surrounded by a lipid membrane. Yellow fever and rubella (German measles) are both caused by togaviruses. Other togaviruses, such as **Sindbis virus**[502] and **Semliki Forest virus**,[503] have become important in biological research.

A large number of icosahedral RNA viruses of diameter 28–30 nm (Fig. 7-14) attack plants, causing diseases such as tomato bushy stunt,[504] southern bean mosaic,[505] or turnip yellow mosaic. Best known of the helical RNA viruses is the **tobacco mosaic virus** (Figs. 5-41, 7-8).[506–507a] Its genome contains 6395 nucleotides as linear ssRNA. Many strains are known. Related viruses cause cucumber green mottle[508] and other plant diseases.

Large viruses of 80 – 100 nm diameter bearing 8–10 spikes at the vertices of the icosahedra cause influenza,[509,510] mumps, measles, and related diseases. The internal structure must be complex. Only 1% of the virus is RNA, and that consists of several relatively small pieces. These are **negative strand viruses** whose RNA is of the opposite polarity to the mRNA. The latter must be formed by transcription from the negative strand. The viruses carry their own RNA polymerase for this purpose. Of even more complex structure are the bullet-shaped **rhabdoviruses** which cause rabies and vesicular stomatitis.[511] The diameter of these viruses is 65–90 nm and the length 120–500 nm. The internal structure includes a helical arrangement of nucleoprotein. They also are negative strand viruses.

Among the **retroviruses**[512-514] are types B and C **oncoviruses** which induce malignant tumors in mammals and birds[515] and the **human immunodeficiency virus** (HIV), the apparent causative agent of AIDS. Their RNA functions in a surprising way. Each virion contains a **reverse transcriptase**, an enzyme that transcribes copies of circular dsDNA copies from the one or two mRNA-like molecules that make up the virus genome. Following action of the reverse transcriptase, one of the transcribed DNA circles becomes covalently spliced into the host's own cellular DNA. There it remains permanently as a **provirus**. RNA molecules transcribed from the provirus serve as mRNA for virus-encoded proteins and also as the genomes for new virus particles.

Double-stranded RNA is unusual in nature but constitutes the genome of the **reoviruses**.[516] The RNA of these viruses fragments into segments upon infection. One member of the group is thought to be the cause of acute diarrhea of infants.[517]

4. Viruses without Nucleic Acid?

The cause of the slow, fatal neurological disease of sheep known as **scrapie** has been a mystery for many years. Similar human diseases include **kuru** and **Creutzfeldt-Jakob** disease.[518-520] Scrapie can be transmitted by injection and this has permitted isolation of the apparent infective agent, a 27- to 30-kDa hydrophobic protein particle[521] which is devoid of DNA or RNA. Prusiner[521] suggested the name **prion** (proteinaceous infectious particle) for the scrapie agent. However, mRNA for the prion is present in normal as well as infected brains, and protein produced in mouse cells from cloned prion genes did not cause scrapie infections. Therefore, there was doubt about the causative agent for the disease. The prion concept is now generally accepted and is considered further in Chapter 29. There are still some who are looking for a nucleic acid component.[519,522-524]

5. Life Cycles

Viruses have many modes of life. They enter cells in various ways. Some enter through coated pits from which they are taken into lysosomes via endocytosis. Others are literally injected into the cells (See Box 7-C). Within cells some viruses are assembled in the nucleus, some in the cytoplasm, and some in membranes. The typical life cycle of a virus leads to rapid formation of large numbers of progeny. Within 20 minutes after entrance into a bacterial cell, a bacteriophage can induce the formation of 100–200 new bacteriophage particles. One of the bacteriophage genes encodes a protein that is also synthesized by the host and which induces lysis of the cell membrane and destruction of the cell. Many animal viruses destroy cells in a similar fashion.

Temperate bacteriophage, the best known being phage λ, have a very different life cycle. Their DNA usually becomes integrated at a specific point into the genome of the bacterium (Chapter 27). Only rarely is an infected cell lysed. The retroviruses that attack mammals and birds have a similar characteristic. Their DNA is also integrated into the host genome. Some viruses that usually produce lysis of cells, e.g., SV40, adenoviruses, herpes viruses, and hepatitis B virus, can occasionally be integrated into the DNA of the host. If such integration occurs in the middle of a gene, that gene will be mutated. This is one way in which such viruses may induce cancers.

One of the most important results of integration of viral DNA into the host genome is that the integrated genes are replicated as part of the genome and are transmitted from one generation to the next. Among these are the cancer-causing **viral oncogenes** (*v-onc*), which are discussed in Box 11-D and in Chapter 11, Section H. While viruses are important causes of cancer in some animals, relatively few human cancers are thought to result directly from the action of viruses. However, the Epstein–Barr virus, which causes mononucleosis, can sometimes be integrated into epithelial cells of nasal regions and can evidently cause cancer. The same virus appears to be responsible for Burkitt's lymphoma, a common cancer in certain areas in Africa.[525]

6. Plasmids and Transposable Genetic Elements

In addition to their chromosomal DNA, bacteria often carry extra small pieces of DNA as permanent parts of their genome. These **plasmids** (sometimes called **episomes**), which are about the size of the DNA of viruses, replicate independently of the host chromosomes. Each bacterial cell usually contains more than a single copy of the plasmid. For example, the "colicinogenic" plasmid **ColE1**, that infects *E. coli* is a circular piece of DNA of molecular mass 4.2 x 10^6 Da. Over 20 copies are normally found per cell but in the presence of a suitable concentration of the drug chloramphenicol the number may rise to 1000–2000.

Plasmids carry a variety of genes which are often useful to bacteria. Some proteins encoded by plasmid genes confer drug resistance to a bacterium. Some are antibiotics. For example, a protein encoded by a gene in plasmid ColE1 is toxic to other strains of *E. coli*. Some plasmids carry genes for enzymes needed for the oxidation of hydrocarbons. Some plasmids contain genes for the **restriction endonucleases** which have become essential to present-day molecular biology and genetic engineering (Section H, 2).

As with some viruses, the DNA of many plasmids can become integrated into the genome of the host. An example is provided by the large 62-kDa plasmids known as **sex factors**. They contain genes encoding the protein subunits of the sex pili (Chapter 7) and can become integrated into the bacterial chromosome. Bacteria containing integrated sex factors are "male" and are able to transfer genes not only of the sex factor but also of virtually the entire bacterial genome into other susceptible bacterial cells. This provides bacteria with the means for sexual reproduction. The transfer of DNA between the bacteria may occur via the sex pili (see Chapter 26). In this respect the sex factors are similar to viruses such as M13 that also appear to gain entrance to bacteria via sex pili.[72]

Integrated viruses are also related to **transposable genetic elements** (transposons). These are segments of DNA that allow genes to move from place to place within the chromosomes (Chapter 27).

H. Methods of Study

Many of the methods discussed in Chapter 3 are directly applicable to nucleic acids. A few additional methods will be considered in this section.[525a]

1. Isolation and Separation of Nucleic Acids

RNA is often extracted from lysed cells or tissues, separated ribosomes, mitochondria, plastids, or nuclei by warming with aqueous phenol and a detergent such as sodium dodecyl sulfate (SDS). Proteins are denatured by this treatment and are dissolved by the phenol, while RNA remains in the lighter aqueous layer. Depending on the conditions DNA may either remain in the aqueous layer or be removed.[526–528] Various precipitation and extraction procedures may be used to separate the RNA in the aqueous layer from polysaccharides, from DNA (if present), and from their components.[526,528,529] DNA may be extracted from cells or nuclei as a nucleic acid–protein complex using 1 M NaCl. The protein can then be denatured with an organic solvent, by detergents, or by phenol. It is desirable to digest proteins away with a nonspecific protein-hydrolyzing enzyme such as proteinase K.[528] After removal of proteins DNA is often precipitated with cold ethanol.

During isolation of RNA, bentonite (a type of clay) or other inhibitors of ribonuclease are often added. For the same reason, chelating agents that complex metal ions needed for the action of deoxyribonucleases are used to protect DNA. Care is necessary to avoid

shearing of the very long, narrow strands of DNA. Even rapid pipetting of solutions will cause such breakage.

Extracted RNA molecules may be separated from each other by centrifugation in a sucrose gradient (Chapter 3).[528] Fragments of DNA are purified in the same way or by equilibrium centrifugation in **CsCl gradients**.[37] Concentration gradients in the dense salt solution are stable, and the sharpness of banding of particles is ensured by use of a high centrifugal field. Single-stranded DNA may be separated from double-stranded DNA, and DNAs of differing G + C content can be separated. The latter separation is based on differences in buoyant densities ρ in CsCl which are approximately shown in Eq. 5-9.

$$\rho = 1.660 + 0.098 \text{ (mole fraction C + G)} \qquad (5-9)$$

One of the most important methods for separating either RNA or DNA mixtures is zone electrophoresis through polyacrylamide or agarose gels. The separated bands may be visualized by scanning in ultraviolet light or by fluorescence of intercalated dyes such as ethidium bromide (Figs. 5-20, 5-22). This method is being displaced to some extent by HPLC using DEAE type ion exchange columns[530,531] for small lengths of DNA including plasmids. The procedure called **pulsed field electrophoresis** makes it possible to isolate very large pieces of DNA, up to several million base pairs in length.[532–534] The separation is carried out in agarose gels, through which the long DNA rods must move in a snakelike fashion.[534] The current is delivered in a pulse and then, after a period of a second to several minutes, a second pulse in a different direction, usually at 90° to the first. The procedure is repeated many times. The size of the DNA seems to affect the time required to reorient the molecules and to start moving in the second direction. Intact DNA from small chromosomes can be separated (Fig. 5-42). To prevent breakage of the DNA by shearing, intact cells are suspended in liquid agarose and allowed to gel into a block about 2 x 5 x 10 mm in size. The block is treated with enzymes and detergents to lyse the cells and to remove all protein and RNA.[534] The block, containing the residual DNA molecules, is then embedded in the electrophoresis gel. Other methods of DNA separation include chromatography on hydroxyl-apatite and gel filtration.

2. Hydrolysis and Analysis

Both DNA and RNA are easily broken down by acid-catalyzed hydrolysis. Thus, heating at 100°C for one hour in 12 M $HClO_4$ is sufficient to hydrolyze nucleic acids to their constituent bases. However, for analysis of RNA it is better to heat in 1 N HCl for 1 h at

100°C. The products are adenine, guanine, cytidine-5'-phosphate, and uridine-5'-phosphate.[535] As is suggested by this distribution of products, the glycosylamine linkages to purines are more labile than those to pyrimidines. The linkages are also less stable in DNA than in RNA. A procedure based on these differences and useful in sequencing by the Maxam-Gilbert method, is to leave DNA overnight in the cold at pH 2 to cleave off all of the purine bases. The resulting polymer is known as an **apurinic** acid.

In alkaline solutions RNA is hydrolyzed to a mixture of 2'- and 3'-nucleotides.

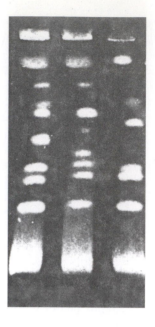

Figure 5-42 Intact DNA from the chromosomes of three strains of the malaria parasite *Plasmodium falciparum*, ranging from 750 Kb to 5 Mb, separated by pulsed-field gel electrophoresis. Courtesy of C. Smith and T. E. Wellems. Reproduced by permission of Amersham Pharmacia Biotech Inc.

The mechanism involves participation of the free 2'-OH of the ribose groups and formation of cyclic 2', 3'-phosphates and is similar to that of pancreatic ribonuclease (Chapter 12). Because deoxyribose lacks the free 2'-OH, the phosphodiester linkages in DNA are quite stable in base.

Hydrolytic cleavages of nucleic acids by the enzymes known as **nucleases** are of great practical value. Pancreatic ribonuclease, an **endonuclease**, cuts a chain adjacent to a pyrimidine in nearly random fashion, leaving phospho groups attached to the 3' position in the nucleotide products (Fig. 5-43). **Exonucleases** cleave from the ends of chains. For example, the phosphodiesterase of snake venom cleaves from the 3' end, which must have a free 3'-OH group, to give 5'-nucleotides. On the other hand, the phosphodiesterase from spleen has the opposite polarity, cleaving chains from the 5' end to give 3'-nucleotides. Similar variations in specificity are found among enzymes that cleave DNA. For example, pancreatic DNAase I, which cleaves preferentially between adjacent purines and pyrimidines, yields 5' mononucleotides whereas DNAase II gives 3'-mononucleotides. Various hydrolytic cleavage reactions of polynucleotides are summarized in Fig. 5-43.

The most striking specificity in DNA hydrolysis is displayed by the **restriction endonucleases** which are discussed further in Chapter 26. These fussy catalysts cleave only at points within or close to a defined sequence of several nucleotides in double-stranded DNA. For example, the enzyme EcoR I cuts only at the following palindromic sequence:

$$\downarrow$$
5' – G A A T T C – 3'
• • • • • •
3' – C T T A A G – 5'
$$\uparrow$$

The cuts in the two strands are made at the points indicated by the arrows. This one endonuclease will cut almost any DNA into long pieces averaging about 5000 base pairs each. These pieces can in turn be cleaved by other restriction endonucleases to form smaller fragments. Since there are about 2400 of these enzymes known, with 188 different specificities,[536] it is possible to cut any piece of DNA down to a size of 100–500 base pairs, ideal for sequencing.[537–539] Each fragment has known sequences at the two ends. Some restriction enzymes cleave outside their specific recognition sequence (see Table 26-2). Some recognize 16-nucleotide palindromes and cut at rare sites.

It is sometimes desirable to cut a large DNA molecule at only a few points. One approach is to protect most sites of a restriction enzyme's action by methylating them (see Chapter 26) while protecting the desired cleavage site, for example by a repressor protein[540] or by a PNA molecule (p. 227) of specific sequence designed to "clamp" the site chosen for protection.[541] Ribozymes (Chapter 12) have been engineered to be as specific or more specific than endonucleases.[542,543] Other new approaches are being developed.[544]

The base composition of either RNA or DNA can be determined after hydrolysis catalyzed by 98% formic acid at 175°C for 30 min or by 12 M perchloric acid at 100°C for 1 h.[545] The bases can then be separated by ion exchange chromatography on a sulfonated polystyrene resin. RNA can be hydrolyzed to a mixture of nucleoside 2'- and 3'-phosphates by 0.3 M NaOH at

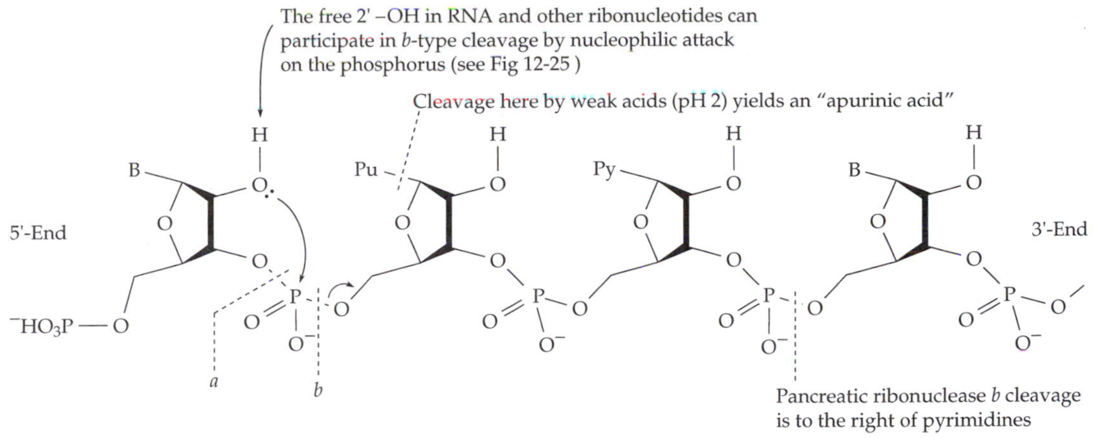

The free 2' –OH in RNA and other ribonucleotides can participate in *b*-type cleavage by nucleophilic attack on the phosphorus (see Fig 12-25)

Cleavage here by weak acids (pH 2) yields an "apurinic acid"

Pancreatic ribonuclease *b* cleavage is to the right of pyrimidines

A. Cleavage at point *a* is catalyzed
   1. Throughout the molecule by endonucleases
      Pancreatic deoxyribonuclease I
   2. Only at the 3' end by exonucleases
      Venom diesterase, nonspecific, attacks DNA and RNA. A free 3'-OH is essential

B. Cleavage at point *b* is catalyzed
   1. Randomly throughout the molecule by endonucleases and by bases (nonenzymatically)
      Pancreatic ribonuclease cleaves only to the right of a pyrimidine-containing nucleotide
      Ribonuclease T1 of *Aspergillus oryzae* cleaves to the right of a guanine-containing residue (3'-guanylate) in ssRNA
      Ribonuclease T2 of *Aspergillus oryzae* cleaves to the right of an adenine-containing residue (3'-adenylate) in ssRNA
      Pancreatic deoxyribonuclease (DNase) II
      Micrococcal DNase
   2. Only at the 5' end by exonucleases
      Bovine spleen phosphodiesterase hydrolyzes both polyribo- and polydeoxyribonucleotides

Figure 5-43 Some hydrolytic cleavage reactions of polynucleotides. Reactions of both RNA and DNA are included.

37°C for 16 h and DNA can be hydrolyzed to nucleotides enzymatically. The negatively charged nucleotides can then be separated by ion exchange chromatography on a quaternary base-type resin (Chapter 3). Periodate cleavage (Eq. 4-12) and reduction of the resulting dialdehydes by [^{3}H]NaBH$_4$ to trialcohols allows introduction of a radioactive label (Fig. 5-44). Alternatively, the dialdehydes can be reductively alkylated by an amine plus NaCNBH$_3$.[546]

The total content of RNA + DNA in tissues may be estimated from the phosphorus content or by color reactions of the sugars.[37,545] These reactions depend upon dehydration to furfural or deoxyfurfural by concentrated sulfuric acid or HCl (Eq. 4-4). Furfural formed from RNA reacts with orcinol (3,5-dihydroxytoluene) and ferric chloride to produce a green color useful in colorimetric estimation of RNA. A similar reaction of DNA with diphenylamine yields a blue color.

Quantitative determination of over 90 free nucleotide compounds found within cells can be accomplished by thin layer chromatographic procedures on as few as 10^6 bacterial cells (~2 µg) labeled by growth in a ^{32}P$_i$-containing medium.[547]

3. Characteristic Reactions of the Bases and Backbone

Reactions of nucleophiles. A number of nucleophilic reagents add reversibly at the 6 position of pyrimidines. Thus, bisulfite adds to uridine (Eq. 5-10).[528] Hydroxylamine (HONH$_2$) adds in a similar fashion to give a compound with –HNOH in the 6 positions.[528] Sodium borohydride (NaBH$_4$), which can be viewed as a donor of a hydride ion (H$^-$), reduces uridine to the 5,6-dihydro derivative. This presumably occurs by attack of the hydride ion at position 6 in a manner analogous to the reaction of bisulfite in Eq. 5-10.

Cytidine reacts in the same way, but the bisulfite addition compound is unstable. These C5–C6 adducts of cytidine all have a greatly enhanced reactivity at C4,

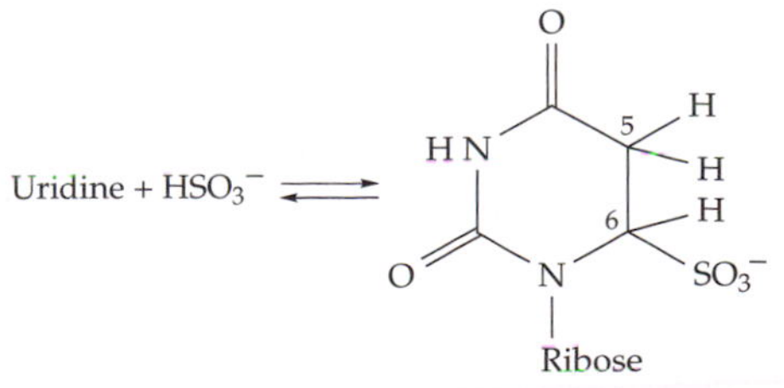

Uridine + HSO$_3^-$ ⇌

(5-10)

presumably because of the lessened aromatic character of the ring. Cytidine is slowly deaminated by base, presumably as a result of attack by hydroxyl ion on the electrophilic center at C4 and subsequent elimination of NH_3 (Eq. 5-11). The reaction is catalyzed by buffer

salts and by bisulfite and hydroxylamine. Catalysis probably occurs, at least in part, as a result of addition of these nucleophiles at the 6 position to form compounds with increased nucleophilic reactivity at C4.[548] Hydroxylamine and methoxyamine (NH_2OCH_3) participate in reactions parallel to that of the hydroxyl ion in Eq. 5-11. Products contain $-NHOH$ or $-NH-OCH_3$ in place of $-NH_2$ but tautomerize to the more stable forms shown in Eq. 5-12. Similar substitution reactions occur with other amines.[33] The C6 adduct with hydrazine can undergo ring cleavage (Eq. 5-13). The initial product then undergoes β elimination, leaving ribosylurea or deoxyribosylurea. The same reaction can be carried out on intact strands of DNA and is widely used in determination of nucleotide sequences. The conversion of 5-hydroxymethylcytosine to the

(5-11)

(5-12)

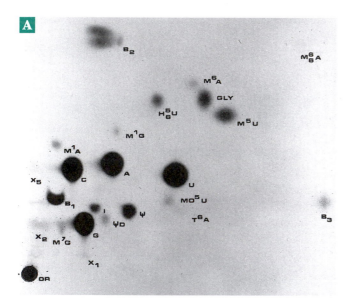

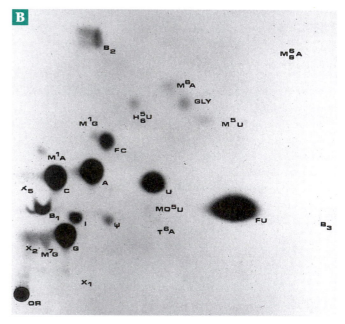

Figure 5-44 Fluorographs of 2'3'-[³H] nucleoside trialcohols from *Bacillus subtilis* grown in the absence (A) and presence (B) of 5-fluorouracil. About 2.3 nmol of nucleosides from each sample was chromatographed and exposed to X-ray film for 90 h. at −80 °C. Or (origin) B_1, B_2, and B_3 contain unidentified material present in a reaction mixture lacking RNA. Abbreviations used: FU, 5-fluorouridine; FC, 5-fluorocytidine; U, uridine; C, cytidine; G, guanosine; A, adenosine; I, inosine; m'G, 1-methylguanosine; m^7G, 7-methylguanosine; m'A, 1-methyladenosine; m^6A, 6-methyladenosine; m^6_6A, 6,6-dimethyladenosine; t^6A, N-[9-(β-D-ribofuranosyl) purin-6-yl carbamoyl] threonine; H^5_6U, 5,6-dihydrouridine; ψ, pseudouridine; $ψ_D$, decomposition product of ψ; m^5U, 5-methyluridine (ribosylthymine); mo^5U, 5-methoxyuridine; N', a nucleoside trialcohol obtained by reduction of a nucleoside dialdehyde with [³H]NaBH₄; FU-5 and FU-20 samples correspond to tRNAs from cells grown at that final concentration of 5-fluorouracil in μg / ml. Courtesy of Ivan Kaiser.

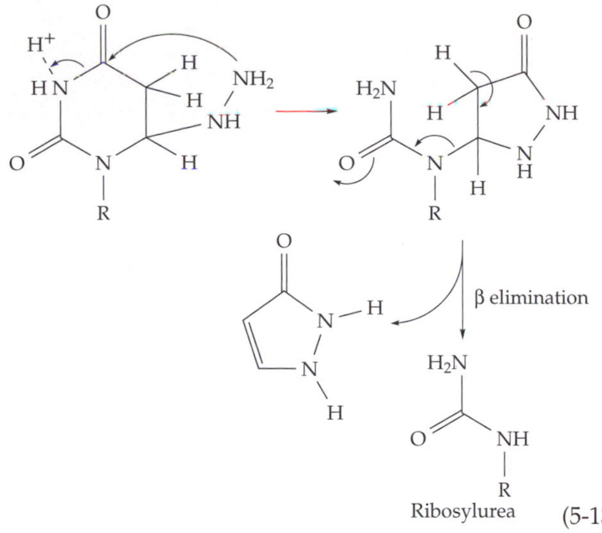

(5-13) Ribosylurea

β elimination

5-methylenesulfonate (5-CH$_2$SO$_3^-$) by reaction with bisulfite should also be mentioned.[549] This is a nucleophilic displacement on the electron-deficient methylene group of this base.

Reactions with electrophilic reagents. Reactions of nucleic acids with the simplest electrophile, the proton, have been considered in Section A2. Somewhat similar are the reactions by which metal ions bind at many sites on both the bases and the phosphate groups of the backbone.[550]

An important reaction is the deamination of amines by dilute **nitrous acid**. This reagent, by a complex mechanism, converts the amino groups of cytidine, adenosine, and guanosine to hydroxyl groups; hydroxy compounds tautomerize to the corresponding amides (Eq. 5-14).

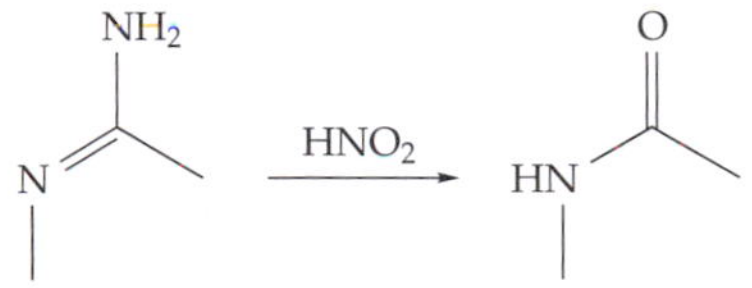

$$\text{(5-14)}$$

Cytidine reacts more rapidly than does adenosine which in turn reacts more rapidly than guanosine. The reaction converts cytosine into uracil and adenine into hypoxanthine. The changes are mutagenic because during replication the modified bases of the DNA pair differently than do the original bases. Guanine is converted to xanthine but this is not likely to be highly mutagenic. Nitrous acid can also convert uridine to 5-nitrouridine.

The amino groups of the bases react reversibly with aldehydes but to a lesser extent than do the more strongly basic amino groups of the amino acids. Formaldehyde forms adducts containing either one or two molecules of the aldehyde (Eq. 5-15).

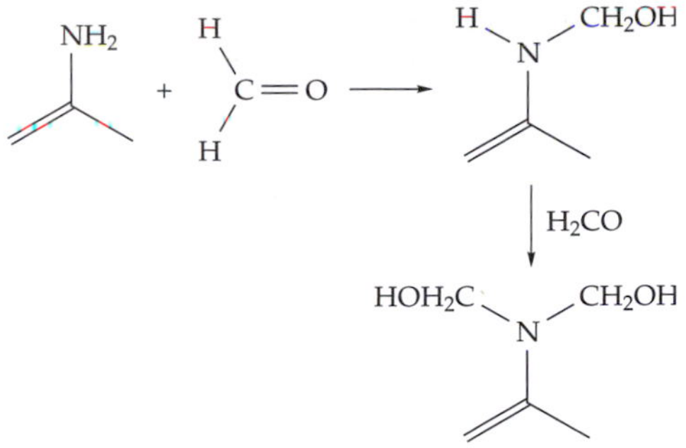

$$\text{(5-15)}$$

These are reversible reactions. A more nearly irreversible crosslinking can occur by elimination of water between one of these products and a nucleophilic group in another base. A dicarbonyl reagent that is widely used because of its specificity toward guanine is **kethoxal** (Eq. 5-16).

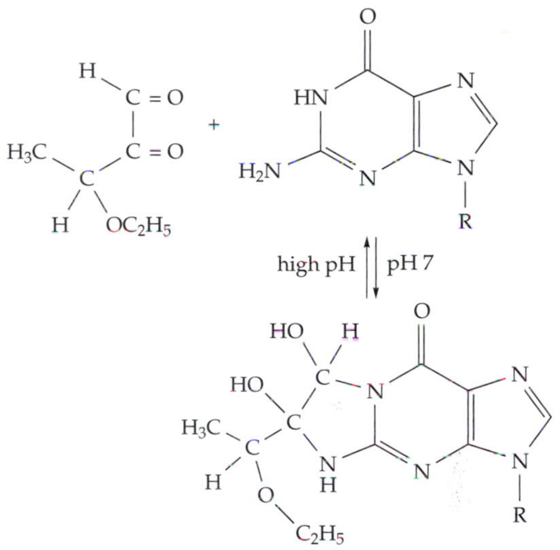

$$\text{(5-16)}$$

Formation of the cyclic product is a consequence of the presence of the adjacent amino and NH groups in the guanine ring.

Pyrimidines undergo **halogenation** at position 5 (Eq. 5-17), while guanine reacts at position 8. Adenine is quite unreactive. Elemental halogens or a variety of other halogenating reagents may be used. Of special

$$\text{Br}_2 + \text{cytidine} \rightarrow \text{5-Br-cytidine} + \text{H}^+ + \text{Br}^- \tag{5-17}$$

value is iodination with ^{131}I or ^{125}I, by which a high level of radioactivity may be introduced into nucleic acids.

Alkylation reactions are not only of use in structural studies but also provide the basis for the action of a large class of mutagenic compounds.[528] Treatment of a nucleoside, nucleotide, or nucleic acid with an alkyl iodide or a dialkylsulfate converts residues of guanosine to an N^7-alkyl-guanosine (Eq. 5-18).

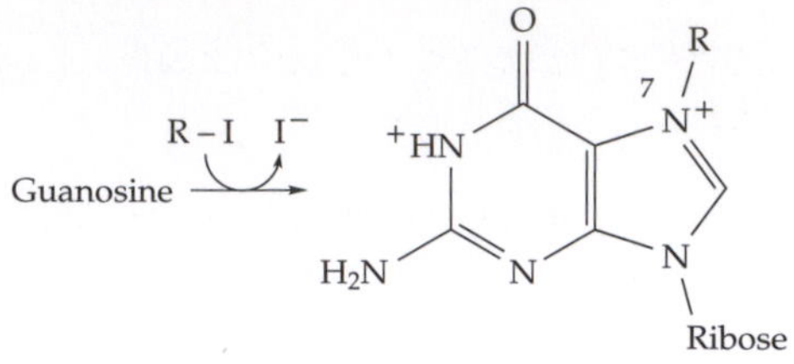

$$\text{Guanosine} \xrightarrow{\text{R}-\text{I} \quad \text{I}^-} \quad (5\text{-}18)$$

Reaction occurs at other nitrogen atoms as well as the oxygen atom of the base and of the ribose ring to a lesser extent. Adenine is alkylated preferentially at N-1 and cytosine and thymine at the corresponding position (N-3) almost exclusively. Uridine and thymidine react very slowly. Adenine is also alkylated at N-3, N-7 and at the exocyclic N-6.

Other alkylating reagents include the powerful mutagens dialkylnitrosamines and alkylnitrosoureas.

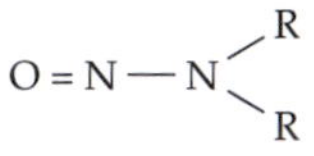

$$O = N - N \begin{smallmatrix} R \\ \\ R \end{smallmatrix}$$

Dialkylnitrosamines

Epoxides alkylate by a nucleophilic displacement reaction that opens the ring (Eq. 5-19).

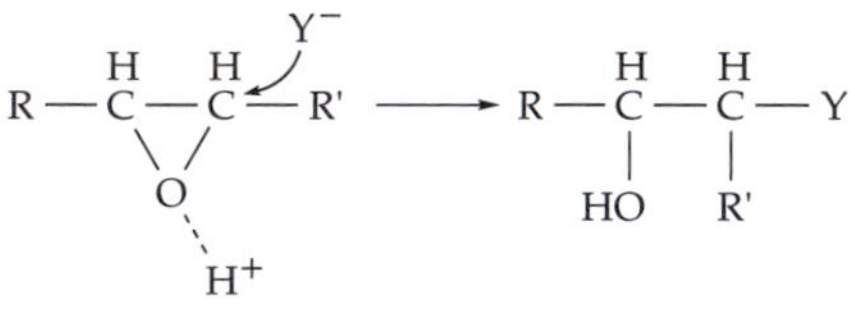

$$(5\text{-}19)$$

The nitrogen and sulfur mustards undergo internal ring closure to an iminium ion (Eq. 5-20) which can then open by attack of a nucleophilic atom of the nucleic acid.

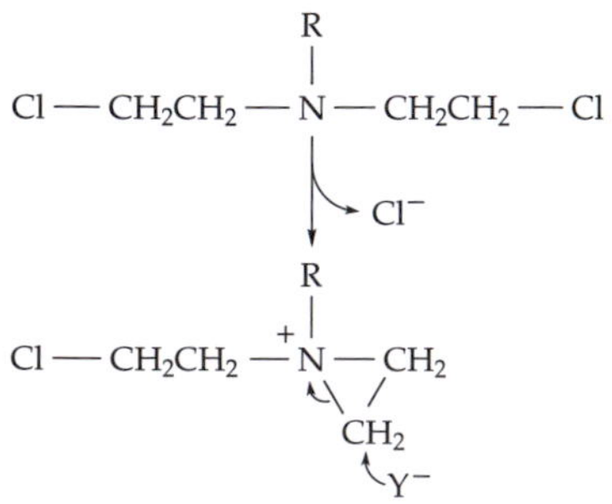

$$\text{Cl} - CH_2CH_2 - \overset{R}{\underset{|}{N}} - CH_2CH_2 - Cl \qquad \text{A nitrogen mustard gas}$$

$$\downarrow Cl^-$$

$$(5\text{-}20)$$

Other alkylating agents react through nucleophilic addition to a carbon–carbon double bond. Thus, acrylonitrile reacts with the nitrogen or oxygen atoms of nucleic acids in the same manner as does the SH group in Eq. 3-25. The water-soluble carbodiimides react as in the first step of Eq. 3-10 to form adducts of the following type:

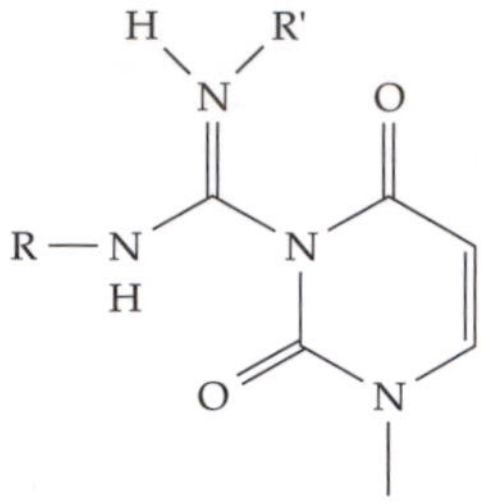

There are many other alkylating agents which often display widely varying reactivity and specificity toward particular nucleic acid bases and particular nucleotide sequences.

A striking effect of alkylation of guanine in nucleic acids is the labilization of the *N*-glycosyl linkage to the ribose or deoxyribose. This effect can be understood in terms of the induction by resonance of a partial positive charge on the nitrogen of the glycosyl linkage.

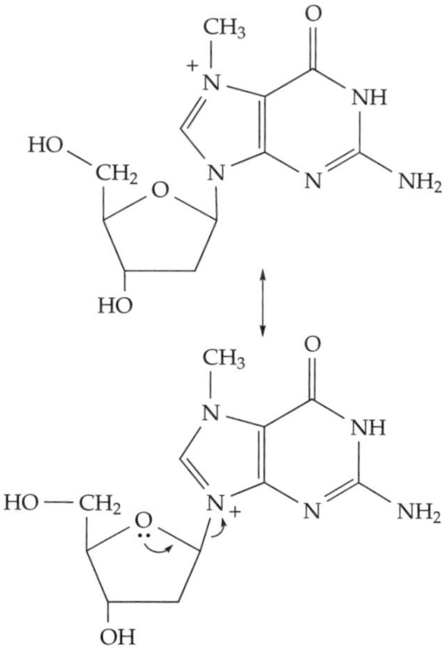

As is indicated by the small arrows on the right-hand structure, the positive charge assists in an elimination reaction that produces an oxycarbocation. The latter can then react with a hydroxyl ion from water.

Reactions causing cleavage of the sugar-phosphate backbone. Treatment of DNA with 16–18 M hydrazine (Eq. 5-13) leads to the destruction of the pyrimidine rings. The reaction can be made somewhat specific for cytosine by carrying it out in the presence of a high concentration of chloride.[551] The remaining polymer, an apyrimidinic acid, contains residues of ribosylurea. These undergo an amino-catalyzed displacement and a β elimination sequence that cleaves the polynucleotide chain (Eq. 5-21). Hydration of the aldehyde (Eq. 13-1) and several

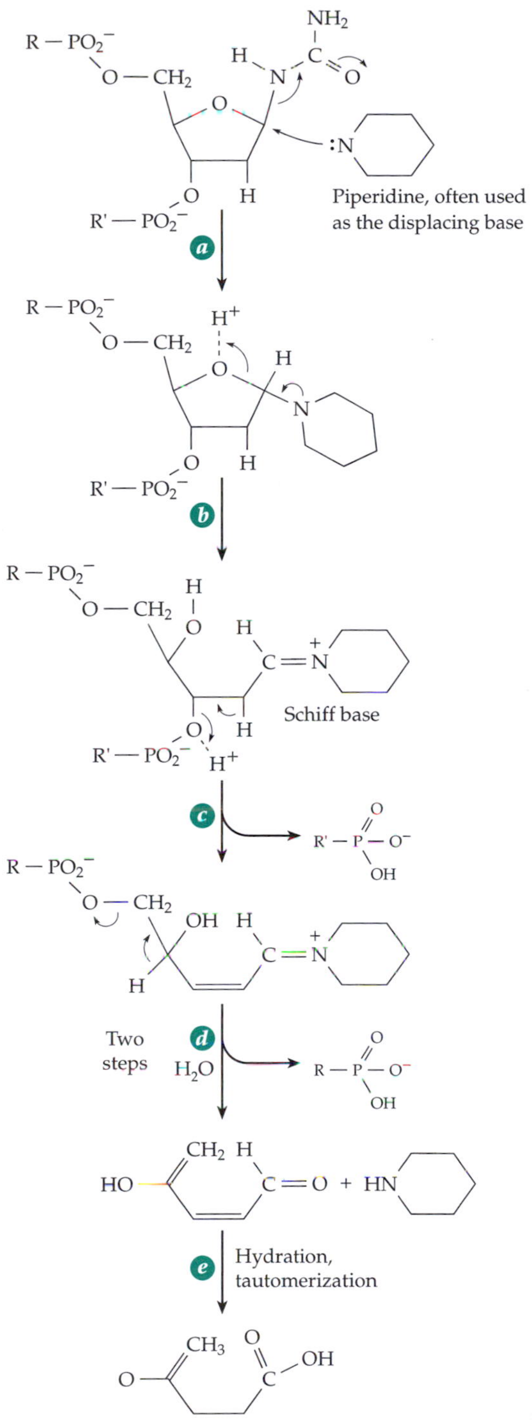

R — PO₂⁻
O — CH₂

NH₂
H—N—C=O

:N (piperidine)

O
R' — PO₂⁻   H   Piperidine, often used
as the displacing base

a

R — PO₂⁻
O — CH₂   H⁺
O   H
N (piperidine)
O   H
R' — PO₂⁻

b

R — PO₂⁻
O — CH₂   H
O   H
C=N⁺ (piperidine)
Schiff base
O   H
R' — PO₂⁻   H⁺

c

→ R'—P—O⁻ (O, OH)

R — PO₂⁻
O — CH₂   OH   H
C=N⁺ (piperidine)
H

Two steps   **d**   H₂O   → R—P—O⁻ (O, OH)

CH₂   H
HO — ( ) — C=O + HN (piperidine)

e   Hydration, tautomerization

CH₃   O
O — ( ) — C—OH

$$(5\text{-}21)$$

tautomerization steps are involved in step *e* of this equation. This reaction is very useful in sequence determination (Section 6). Notice that "tracts" of purine nucleotides remain intact after this treatment. A similar base-catalyzed reaction sequence can be used to displace N⁷-methylguanine and to cleave the polynucleotide. Ethylnitrosourea, in its reaction with purines, is useful as a structural probe of RNA.

4. Melting, Hybridization, and Polynucleotide Probes

Like proteins, nucleic acids can undergo denaturation. The strands of the double helix of DNA are separated and the double-stranded regions of RNA molecules "melt." Denaturation can be accomplished by addition of acids, bases, and alcohols or by removal of stabilizing counter ions such as Mg^{2+}. The product is a random coil and denaturation can be described as a helix → coil transition. Denaturation of nucleic acids by heat, like that of proteins, is cooperative (Chapter 7, Section A,3) and can be described by a characteristic **melting temperature**.

A plot of the optical absorbance at 260 nm (the wavelength of maximum light absorption by nucleic acids) versus temperature is known as a **melting curve** (Fig. 5-45). The absorbance is lower, by up to 40%, for native than for denatured nucleic acids. This **hypochromic effect** (Chapter 23) is a result of the interaction between the closely stacked bases in the helices of the native molecules. The melting temperature T_m is taken as the midpoint of the increase in absorbance (Fig. 5-45). As the percentage of G + C increases, the nucleic acid becomes more stable toward denaturation because of the three hydrogen bonds in each GC pair. T_m increases almost linearly with increases in the G + C content. In the "standard" citrate buffer (0.15 M NaCl + 0.015 M sodium citrate, pH 7.0) Eq. 5-22 holds. The exact numerical relationship depends strongly upon the ionic composition and pH of the medium.[37,72,552,553]

$$\%\,(G + C) = 2.44\,(T_m - 69.3); \; T_m \text{ in } °C \qquad (5\text{-}22)$$

The curves in Fig. 5-45 appear simple, but using newer apparatus and plotting the first derivative of the melting curve yields a complex pattern that depends on the sequence of bases.[555]

Complete denaturation of DNA leads to separation of the two complementary strands. If a solution of denatured DNA is cooled quickly, the denatured strands remain separated. However, if the temperature is held for some time just below T_m (a process known as **annealing**), the native double-stranded structure can be reformed. An important tool for studying DNA has been the measurement of the **kinetics of reassociation** of separated strands of relatively short DNA fragments.[72,556,557]

Because it depends upon the concentration of two separated strands, reassociation obeys second-order kinetics (Chapter 9) and Eq. 5-23, which is readily derived by integrating Eq. 9-8 for [A] = [B] = C from time 0 to *t*:

$$C / C_0 = 1 / (1 + k\,C_0 t) \qquad (5\text{-}23)$$

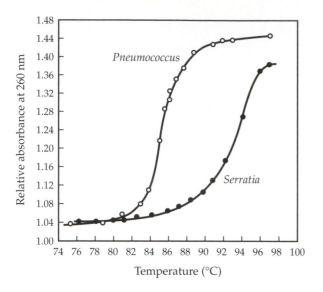

Figure 5-45 A melting curve for DNA molecules from two different sources. From Davidson.[554]

drastically different between two species do not (Fig. 5-46). One way to do such an experiment is to immobilize the long-chain denatured DNA from the one organism by embedding it in an agar gel[558] or by absorbing it onto a nitrocellulose filter.[559,560] The DNA fragments from the second organism are passed through a column containing "beads" of the DNA-containing agar or through the filter with adsorbed DNA. Pairing of fragments with complementary sequences occurs

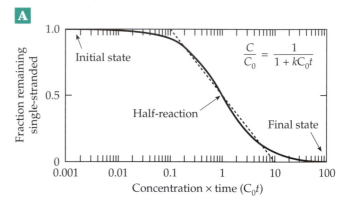

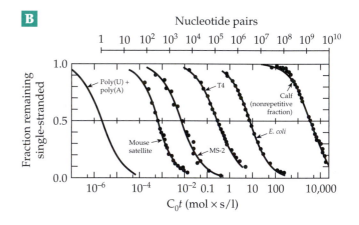

Figure 5-46 Reassociation curves for DNA from Britten and Kohne.[556,561] (A) Time course of an ideal, second-order reaction to illustrate the features of the log $C_0 t$ plot. The equation represents the fraction of DNA which remains single-stranded at any time after the initiation of the reaction. For this example, k is taken to be 1.0, and the fraction remaining single-stranded is plotted against the product of total concentration and time on a logarithmic scale. (B) Reassociation of double-stranded nucleic acids from various sources. The genome size is indicated by the arrows near the upper nomographic scale. Over a factor of 10^9, this value is proportional to the $C_0 t$ required for half-reaction. The DNA was sheared, and the other nucleic acids are reported to have approximately the same fragment size (about 400 nucleotides, single-stranded). Correction has been made to give the rate that would be observed at 0.18 M sodium-ion concentration. The temperature in each case was optimal, i.e., ~30°C below the melting temperature T_m. The extent of reassociation was established by measuring optical rotation (calf thymus DNA), ribonuclease resistance (MS-2), or hypochromicity.

The initial concentration of denatured DNA, C_0, is related in this way to the concentration C of DNA remaining dissociated at time t. A plot of the fraction of molecules remaining single-stranded versus the logarithm of $C_0 t$ (Fig. 5-46A) is a convenient way of displaying data. As indicated in Fig. 5-46B, the value of $C_0 t$ increases in direct proportion to the length of the DNA chain in the genome, but it is very much decreased if the sequence of bases is highly repetitive [poly(T) and poly(A)]. The slope of the plot at the midpoint gives an indication of the heterogeneity of the DNA fragments in a solution.

Denatured DNA fragments can sometimes reassociate with DNA from a different source to form **hybrid duplexes**. Such double helices, in which one strand comes from one strain of an organism and the other strand from a genetic variant of the same organism or from a different species, are known as **heteroduplex**. Some mutations consist of **deletions** or **additions** of one or a substantial number of bases to a DNA chain. Heteroduplexes prepared from DNA of such mutants hybridized with that from a nonmutant strain have normal hydrogen-bonded Watson–Crick base pairs for the most part. However, they may have single-stranded loops in regions where long deletions or additions prevent complementary base pairing.

Hybridization measurements have been used in many studies of **homology** of nucleic acids from different species. A nucleic acid is cut (e.g., by sonic oscillation) into pieces of moderate length (~ 1000 nucleotides) and is denatured. The denatured DNA fragments are mixed with denatured DNA of another species. Nucleotide sequences that are closely similar between species tend to hybridize, whereas sequences that are

and such paired fragments are retained while strands that do not pair pass on through the column (or filter).

Both DNA hybrids and **DNA–RNA hybrid duplexes** are very important to present day genetic research.[560,562] Molecules of mRNA that represent transcripts of a particular gene will hybridize only with one of the two separated strands of DNA for that gene.

A major use of hybridization is to locate a gene or other DNA or RNA sequence by means of a **synthetic probe**.[563] This is a small piece of DNA or RNA which is labeled in some way, e.g., with a radioisotope such as ^3H, ^{32}P, or ^{125}I. Alternatively, the probe may carry a highly fluorescent dye or a "tag" that can be recognized by a specific antibody.[564,565] An example of the latter is the use of the vitamin biotin and the specific binding protein **avidin** (see Box 14-B).[566] Related procedures employ labeling with **digoxigenin** and often employ chemiluminescent detection.[567–569] Several methods for preparation of probes are in use. Some are enzymatic but the direct chemical synthesis of oligonucleotide probes is probably used the most.

One of the first methods devised for making a highly radioactive DNA probe is called **nick translation**.[570] A piece of dsDNA, e.g., a "restriction" fragment cut from a larger piece of DNA by restriction endonucleases, is selected. A small amount of pancreatic DNase I is added. It creates "nicks" in which one strand has been cut and some nucleotides have been removed leaving a gap. Now the DNA is incubated with DNA polymerase I (pol I) and a mixture of the four mononucleotide triphosphates, the precursors of biological synthesis of DNA (Chapter 27). Usually a ^{32}P or ^{35}S label with high radioactivity is present in one of the nucleotide triphosphates as indicated in the following structure. The polymerase fills the gap, adding nucleotides to the exposed 3' end of the nicked chain. *E. coli* pol I has a second enzymatic activity which allows it to digest a polynucleotide chain from

The four nucleotide triphosphate precursors of DNA

the 5' end. Thus, as synthesis proceeds at the 3' side the nick is "translated" as shown in (Eq. 5-24).

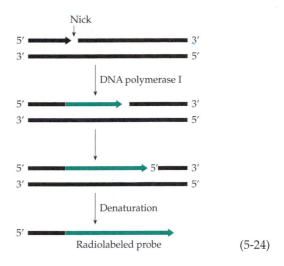

(5-24)

Probes may also consist of DNA copied from mRNA. This is known as **cDNA** and is also widely used to determine indirectly the sequences of mRNA molecules. Messenger RNA may be isolated from the total cellular RNA by affinity chromatography on bound poly (dT) or poly (U). These materials selectively hold RNA with the poly (A) tails characteristic of most eukaryotic mRNA (see Chapter 28). Another source of mRNA is polyribosomes (polysomes), which are "reading" mRNA and actively making proteins. An antibody to the protein for which mRNA is desired will often bind to the protein chains being synthesized and precipitate the polysomes. The mRNA can be recovered and used as a template for cDNA.

Synthesis of cDNA, usually in radiolabeled form is accomplished with **reverse transcriptase**, the enzyme from retroviruses that synthesize a DNA–RNA hybrid from ssRNA.[570–572] A short oligo (dT) primer is usually hybridized to the 3' poly (A) tail to initiate synthesis. Reverse transcriptase also has ribonuclease (RNase H) activity and will digest away the RNA. If desired, synthesis of the second strand can be carried out by a DNA polymerase to give a complete DNA duplex. Many gene sequences have been deduced from cDNA copies.

Often the most practical approach to obtaining a DNA probe is synthesis of a mixture of short oligonucleotides, often in radioactive form as described in the next section. The "redundancy" in the genetic code, i.e., the existence of two or more codons for most amino acids, presents a problem in designing an oligonucleotide probe based on amino acid sequence information. Examination of Table 5-5 suggests part of the solution. Whereas only Met and Trp have single unambiguous codons nine amino acids have only two codons each. We should try to find an amino sequence that contains Met and Trp and as many of the nine others as possible. We should avoid sequences that contain Ser, Leu, or Arg because each has six codons. We can then make a mixture of oligonucleotides, using the various coding possibilities. Mixtures of as many as 1024 (2^{10}) oligonucleotides have been used. One of these may bind tightly and specifically to the desired DNA segment. Instead of such a complex mixture it may be more useful to incorporate a modified base at the most ambiguous positions. For example, inosine, which occurs in the wobble position in anticodons (Fig. 5-30), can pair with A, C, or T.[573] Substitution of 2-aminoadenosine can cause a probe to bind more tightly because a third hydrogen bond will be present in each AT pair.[574]

Another important procedure is labeling ends of polynucleotides. Most often the 5' end is labeled with a radioisotope or by covalent attachment of a fluorescent dye. For example, a **polynucleotide kinase** can be used to transfer a radioactive γ-phospho group from ATP to the 5' end of a polynucleotide that has a free 5'-OH group.

5. Synthesis of Oligonucleotides and Polynucleotides

Efficient solid-phase methods of synthesis analogous to those for polypeptides (Fig. 3-15) have been devised. The pioneering work was done by H. G. Khorana, who made the first synthetic gene[575] and later synthesized a gene for the visual pigment rhodopsin (Chapter 23). Several synthetic approaches have been developed.[576] Currently the most popular method involves the use of phosphite esters. Most nucleophilic groups of the monomers are derivatized with removable blocking groups. For example, N-4 of cytosine and N-6 of adenosine may carry benzoyl groups. The 5'–OH of each nucleotide is blocked by a di-p-anisylphenylphenylmethyl (also called dimethoxytrityl, DMTO) group. The 3'–OH is converted to one of a number of activated derivatives such as the following N, N-diisopropylamino phosphines.[575,577–579]

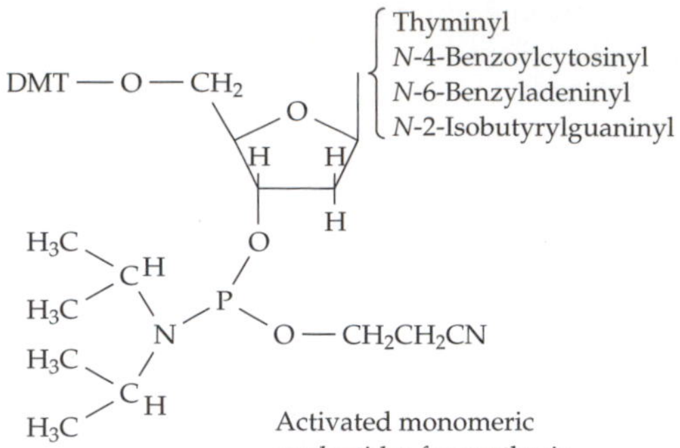

Solid-phase synthesis is usually done on a silica support with a covalently attached succinamide as shown in Eq. 5-25. The first nucleotide at the 3' end of the chain to be synthesized is attached by an ester linkage to the bound succinamide (step a, Eq. 5-25). The 5'-protecting group is removed in step b and the 5'-OH reacts with the activated phosphine of the second nucleotide (step c, Eq. 5-25). Steps b and c are then repeated as often as necessary to complete the chain. The finished polynucleotide can be removed from the solid support, the cyanoethyl groups removed

from the phosphorus atoms by β elimination and all of the other blocking groups removed by treatment with concentrated NH_3. The whole procedure has been automated.[580–582]

If a large piece of DNA is needed several oligonucleotides can be joined end to end enzymatically (Eq.

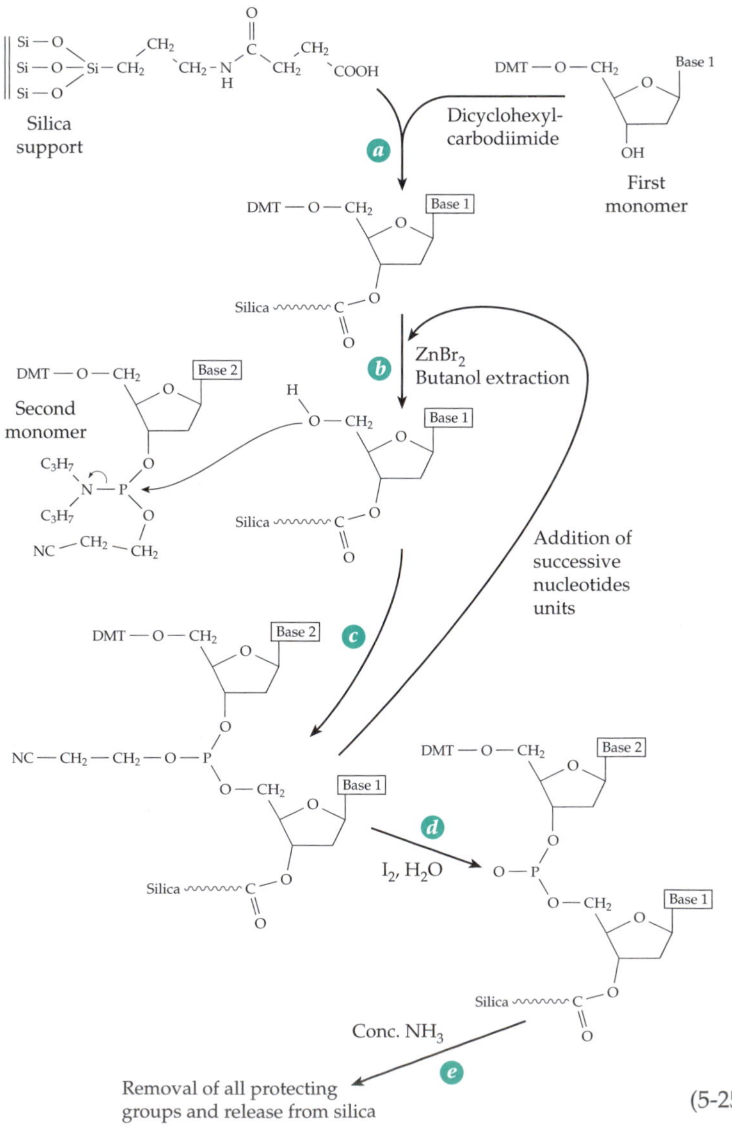

Activated monomeric nucleotides for synthesis of polynucleotides

Removal of all protecting groups and release from silica

(5-25)

BOX 5-D DNA FINGERPRINTING

Jeffreys *et al.*[a-c] digested human DNA to completion with *Hin* f I and *Sau*3A restriction endonucleases. Certain fragments, which originated from "minisatellite" bands of repetitive DNA showed a very high degree of **polymorphism** among the population. Many different fragments sharing these repeated sequences were formed in the restriction digest. If a suitable labeled probe was used, it hybridized with as many as 80 different bands.[a,d-f] The resulting pattern appeared, like a fingerprint, to be different for every individual, as is shown in the accompanying photo. Unlike a fingerprint the DNA pattern also contains information that often allows deductions about parentage.

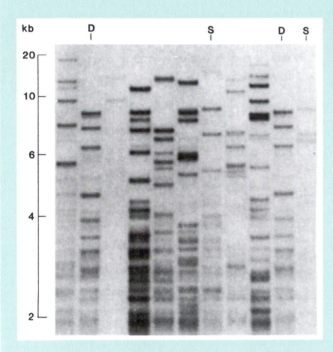

DNA "fingerprints" made from one or two drops of blood from ten different individuals. DNA was isolated, digested to completion with restriction endonuclease *Hin* f I, and subjected to electrophoresis in a 20-cm-long agarose gel until all DNA fragments smaller than 1.5 kb in length had passed off the gel. The DNA was then transferred to a nitrocellulose filter by Southern's method and hybridized with a ^{32}P-labeled single-stranded DNA probe prepared from cloned human minisatellite DNA. The probe used had the "consensus" composition (AGAGGTGGGCAG-GTGG). Within the 29 tandem repeats in this 0.46-kb probe there are various sequences close to the one shown. Filters were then autoradiographed for four days. Two duplicate samples (marked D) were taken from the same individual and two others (marked S) from two sisters. A number of bands in common are evident. From Jeffreys *et al.*[c]

The technique has come into widespread use in forensic analysis, with DNA typing being possible from a single hair.[g] In a famous early case it was used to allow an immigrant child to be reunited with his mother[h] and it is being used regularly to protect innocent persons accused of rape or murder.[i] It is also widely used to provide evidence of guilt. However, the very small chance of a close match between unrelated persons prevents the use of DNA typing alone as proof of guilt. Because DNA samples are often "amplified" by PCR (Section H,6), there is also a possibility of contamination and forensic use of DNA typing is still controversial.[i-m]

However, DNA typing continues to be improved[f,n,o] and to be applied in a great variety of ways. For example, the skeletal remains of a murder victim were identified by DNA fingerprints after being buried for eight years.[p] DNA typing is also used to study mating habits of birds,[q] the genetic variability of populations of whales sampled by biopsy,[r] etc.

[a] Jeffreys, A. J., Wilson, V., and Thein, S. L. (1985) *Nature (London)* **314**, 67–73

[b] Lewin, R. (1986) *Science* **233**, 521–522

[c] Jeffreys, A. J., Wilson, V., and Thein, S. L. (1985) *Nature (London)* **316**, 76–79

[d] Vassart, G., Georges, M., Monsieur, R., Brocas, H., Lequarre, A. S., and Christophe, D. (1987) *Science* **235**, 683–684

[e] Huang, L.-S., and Breslow, J. L. (1987) *J. Biol. Chem.* **262**, 8952–8955

[f] Kirby, L. T. (1990) *DNA Fingerprinting*, Stockton Press, New York

[g] Higuchi, R., von Beroldingen, C. H., Sensabaugh, G. F., and Erlich, H. A. (1988) *Nature (London)* **332**, 543–546

[h] Jeffreys, A. J., Brookfield, J. F. Y., and Semeonoff, R. (1985) *Nature (London)* **317**, 818 – 819

[i] Balding, D. J., and Donnelly, P. (1994) *Nature (London)* **368**, 285–286

[j] Neufeld, P. J., and Colman, N. (1990) *Sci. Am.* **262**(May), 46–53

[k] Lewontin, R. C., and Hartl, D. L. (1991) *Science* **254**, 1745–1750

[l] Lander, E. S., and Budowle, B. (1994) *Nature (London)* **371**, 735–738

[m] Lewontin, R. C. (1994) *Nature (London)* **372**, 398

[n] Uitterlinden, A. G., Slagboom, P. E., Knook, D. L., and Vijg, J. (1989) *Proc. Natl. Acad. Sci. U.S.A.* **86**, 2742–2746

[o] Jeffreys, A. J., MacLeod, A., Tamaki, K., Neil, D. L., and Monckton, D. G. (1991) *Nature (London)* **354**, 204–209

[p] Hagelberg, E., Gray, I. C., and Jeffreys, A. J. (1991) *Nature (London)* **352**, 427 – 429

[q] Burke, T., and Bruford, M. W. (1987) *Nature (London)* **327**, 149–152

[r] Hoelzel, A. R., and Amos, W. (1988) *Nature (London)* **333**, 305

27-5). For example, a functional 17-bp gene for the 53-residue human epidermal growth factor was synthesized by joining ten oligonucleotides of lengths 11–59 bp.[575] DNA is often synthesized enzymatically using methods described in Chapter 26. Cloned sequences of synthetic DNA can also be transcribed to produce **polyribonucleotides** of any desired sequence.[583] New nonenzymatic methods for RNA synthesis have also been devised.[583–587]

6. The Polymerase Chain Reaction (PCR)

This important technique was first described in 1971–1974 by Khorana and associates[588,589] but was not used until it was rediscovered in 1983 by Mullis.[590–593] It was quickly developed[591,592,594–596] and has played a major role in biochemistry ever since. It continues to be applied in numerous ways.[589,597–599a]

The PCR technique provides a way of "amplifying" a small number of DNA molecules, i.e., to produce many copies. This is often done by cloning but PCR offers a quick and easy way to obtain millions of copies of a desired relatively short segment of DNA. Standard PCR can be used for up to about 5000-nucleotide pieces. More recently modified procedures have allowed 35-kb segments to be amplified.[600]

The basic PCR procedure is initiated by hybridizing two oligonucleotide primers onto opposite strands of denatured DNA, one at each end of the section chosen for amplification (Fig. 5-47). A DNA polymerase is then used to convert each of the separated strands into a duplex. The mixture of products is then heated to denature the two new duplexes. After cooling, the primers, which are present in great excess, hybridize to all four strands. In a second cycle of polymerase action these are all converted to duplexes, etc. After 20 cycles millions of copies will be made. At first, copies with tails extending beyond the limits specified by the oligonucleotide primers will be formed. However, it is easy to see that after a few cycles, most molecules will be of just the desired length. A heat-stable polymerase from *Thermus aquaticus* (*Taq* polymerase) is used so that the enzyme is not denatured by the repeated cycles of heating and cooling, which are conducted automatically by a simple apparatus.

The polymerase chain reaction is being used to speed up prenatal diagnosis of genetic diseases, to detect viral infections, for tissue typing needed for organ transplantation, in forensic procedures, and in the study of the DNA of ancient tissues such as those of frozen wooly mammoths.[601–603] If suitable restriction enzyme sites are present in the primers, the amplified DNA can be cloned readily.[604] The 3.3×10^{-9} fmol of a DNA sequence found in a diploid chromosome pair in a single cell can be amplified in 50 cycles to 5–500 fmol, enough to study by hybridization with radioactive

probes.[605] A large sample of a few pg of DNA can be amplified in 20 cycles to micrograms. By placing sequencing primers within the amplified segments, it is possible to generate DNA that can be sequenced directly using the dideoxy sequencing technique (Chapter 5) without cloning.[606] The PCR technique has also been used to amplify cDNA molecules formed from RNA transcripts present in very low abundance. One of the problems with the PCR is that priming may occur by DNA fragments other than the added primers. Contamination must be scrupulously avoided. Another problem is that errors are introduced into DNA during amplification by PCR. Perhaps 1 in 200 of the copies will contain an incorrect base.[607] If such a molecule is cloned the error will be perpetuated. Good practice requires that more than one clone is selected and sequenced to allow such errors to be avoided.

7. Sequence Determination

Satisfactory (but slow) methods for determining sequences of RNA molecules have been known for over 30 years. The procedures are somewhat parallel to those used in sequencing proteins. However, no similar method could be devised for DNA. Little progress was made until rather recently when new approaches led to extremely rapid procedures for sequencing DNA. As a consequence, it is now much easier to learn the sequences of genes than it is to sequence the proteins which they encode!

Preparing the DNA. The first step is to obtain a sample of enough identical DNA molecules to permit sequence analysis. This in itself may be a complex undertaking. Perhaps we want to know the sequence of one particular gene in the 3,500,000 kilobase pairs of DNA present in a single human cell. How can this gene be found and the DNA be obtained for analysis? Three techniques have been essential: cutting the DNA with restriction endonucleases, hybridization, and cloning. More recently PCR and related methods[608] have simplified the sample preparation. DNA can often be amplified using primers that contain sequences that will later serve as **sequencing primers**.

Restriction maps and Southern blots. Although it doesn't require the synthesis of a primer, the Maxam–Gilbert procedure usually demands that a "restriction map" of the DNA be prepared to help keep track of the fragments being sequenced.[609] See also Chapter 26. Figure 5-48[610–612] shows the restriction map of the mitochondrial DNA gene, *oxi3* from yeast. This gene, which encodes one of the subunits of cytochrome oxidase (Chapter 18), consists of 9979 base pairs. It was cloned in a suitable plasmid after which the restriction map (Fig. 5-48) was prepared by cutting with 18 different restriction enzymes.[611] The protein subunit contains 510 residues and therefore requires a coding capacity in the DNA of 1530 base pairs. This is only 16% of the total length of the gene, the majority of whose DNA is found in the three large introns.

One way to select a desired segment of DNA from a digest of chromosomal DNA is to sort out the "restriction fragments" by gel electrophoresis.[613] The DNA from the gel can be transferred to a nitrocellulose sheet while retaining the separation pattern using a method devised by Southern.[560,614,615] In this **Southern blot** technique, solvent flows from a pool beneath the gel up through the gel and the nitrocellulose sheet into paper towels. The DNA is trapped on the nitrocellulose in the same pattern observed in the electropherogram. A suitably labeled probe such as cDNA with

incorporated ^{32}P is flowed repeatedly across the nitrocellulose sheet under conditions that favor formation of hybrids. Only the DNA complementary to the cDNA probe will retain the label. This DNA can then be located with the help of a autoradiogram. It is important that single-stranded DNA be used. If double-stranded restriction fragments are separated on the electropherogram they must be denatured while in place in the gel before hybridization is attempted.

Once the desired piece of DNA has been identified it is usually necessary to increase its amount. The conventional approach is to incorporate the DNA fragment into a plasmid and clone by the methods described in Chapter 26. The selected DNA can usually be cut cleanly from the plasmid used for cloning with the same restriction endonuclease originally used in

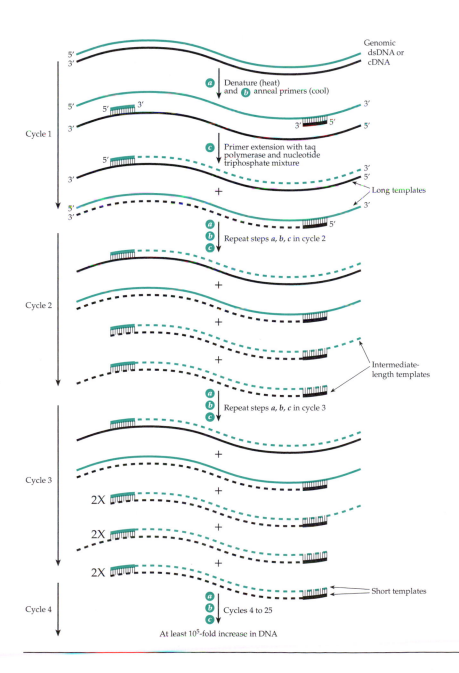

Figure 5-47 Amplification of DNA using the polymerase chain reaction (PCR). Double-stranded DNA is denatured by heating to 90–99° C (step a) and oligonucleotide primers complementary to short 12–18 nucleotide sequences at the two ends of the piece of DNA to be amplified are annealed to the separated strands by cooling to 40–75° C (step b). The two DNA strands serve as templates for synthesis of new complementary strands using a heat-stable DNA polymerase and a mixture of the four nucleotide triphosphates. Nucleotide units are added to the 3' ends of the primers, with the new chains growing in the 5' → 3' direction (step c). Steps a, b, and c are then repeated as many as 30 times using a thermal cycler device that periodically raises and lowers the temperature with a cycle time of a few minutes. The polymerase is unharmed by the heating and is reused in each cycle. An excess of the primer and of the nucleotide triphosphates sufficient for all of the cycles is present initially. In the early cycles new long and intermediate length templates are created. However, the number of short templates increases exponentially and the final product consists predominately of the short selected DNA segment (short templates).

fragmenting the DNA. An alternative procedure is to clone a mixture of DNA fragments and then sort colonies of bacteria containing the cloned fragments using DNA–RNA hybridization.[616] Bacteria from the selected colonies are then propagated to produce large amounts of the plasmid DNA. Alternatively, PCR can be used directly on the selected DNA fragment. This is more often the preferred choice.[604,606]

The Sanger dideoxy method. The rapid sequencing methods all depend upon the fact that single-stranded DNA fragments under denaturing conditions migrate on electrophoresis in polyacrylamide gels strictly according to their length. Thus, if a mixture contains all lengths of radiolabeled polynucleotides from very short oligonucleotides to fragments containing 200 or 300 bases, the polynucleotides will all appear, one above the other, as a series of bands that can be visualized by radioautography. The first of these methods was published by Sanger and Coulson[617] in 1975 and was followed in 1977 by the method which is now used.[618–621]

A sample of double-stranded DNA is denatured. One of the resulting single strands is used as a template to direct the synthesis of a complementary strand of radioactive DNA using a suitable DNA polymerase. The "Klenow fragment" of *E. coli*, DNA polymerase I, reverse transcriptase from a retrovirus, bacteriophage T7 DNA polymerase, *Taq* polymerase, and specially engineered enzymes produced from cloned genes have all been used.

Before the sequencing begins it is necessary to prepare a short **primer** that is complementary to a sequence at one end of the DNA strand to be sequenced. This may be prepared enzymatically,[622,623] or by non-enzymatic synthesis. The short primer is annealed to the end of the DNA and the resulting molecule is incubated with a DNA polymerase and a mixture of the four mononucleotide triphosphates, one of which is radiolabeled in this position. Four reaction mixtures are prepared. Each mixture contains all four nucleoside triphosphates and also one of four different **chain-terminating inhibitors**, the most popular of which are the 2', 3'–dideoxyribonucleoside triphosphates:

2',3'-Dideoxyribonucleotide triphosphates

These inhibitors are added in a ratio of about 100:1 with the natural substrates. In this ratio they are incorporated into the growing DNA chain about once

in 200 times on the average. However, in the various growing DNA chains they are incorporated at different points ranging from the very first nucleotide to the last. Since the incorporated dideoxy monomer lacks the 3-hydroxyl group needed for polymer formation, chain growth is terminated abruptly. Synthesized polynucleotides are denatured and subjected to electrophoresis in four adjacent lanes. The resulting patterns contain bands corresponding to all of the successive oligonucleotides but not all in the same lane. A given lane will contain only the bands of the oligonucleotides terminated by the particular inhibitor used. The other bands will be found in the other three lanes. Each band will have been terminated by the inhibitor employed in that lane. The nucleotide sequence can be read directly from the banding pattern as is shown in Fig. 5-49. Arabinosyl nucleotide triphosphates have also been used as chain-terminating inhibitors. A sequence determined by the Sanger method is usually checked by also sequencing the complementary strand.

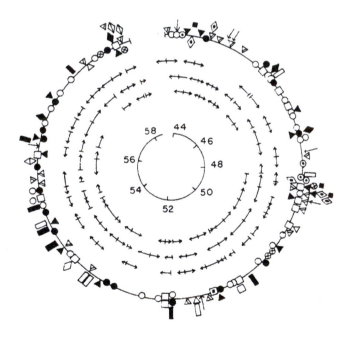

Figure 5-48 A physical map of the *oxi3* locus of yeast mitochondrial DNA. The restriction fragments used for DNA sequencing are indicated by the *arrows*. The extent to which the sequences were read is represented by the lengths of the arrows. The map units are shown in the *inner circle*. The following symbols, together with the names of the restriction enzymes (Chapter 26), are used for the restriction sites:

▲ *Hinf* I	○ *Alu* I	⊙ *Hha* I
△ *Hpa* II	◈ *Pvu* II	▲ *Rsa* I
▯ *Hae* III	◈ *Hinc* II	▮ *Hph* I
◆ *Taq* I	⟷ *Hind* III	⊡ *Blg* II
● *Mbo* I	⊗ *Eco* RI	◇ *Bam* HI
□ *Mbo* II	↓ *Eco* RII	■ *Hpa* I

From Bonitz *et al.*[611]

A major factor in the success of the dideoxy sequencing method has been the development of cloning techniques that provide ssDNA in a form ready for use. In any cloning procedure the DNA that is to be sequenced has been covalently ligated to the end of a DNA strand of the **cloning vehicle**, a modified plasmid or virus (Chapter 26). Sequencing is often done on DNA cloned in a modified ssDNA bacterial virus such as M13. Although dsDNA is ligated to the ds replicating form of viral DNA, the virus particles produced when the virus is propagated in *E. coli* cells are single stranded. Cleavage of the viral DNA with its incorporated "passenger DNA" with an appropriate restriction endonuclease releases the passenger DNA (to be sequenced) with a short piece of DNA from the cloning vehicle attached at the 3' end. Since the sequence of this small piece of the cloning vehicle is known, a suitable primer of length ~12–18 nucleotides can be synthesized (or purchased) and annealed to the DNA. This serves as the primer and allows the sequence to be read from the 5' end to the 3' end of the synthesized complementary strand. Double-stranded DNA attached to vehicles such as the pUC plasmids can also be sequenced directly if the DNA is denatured by alkali treatment. After neutralization and precipitation an appropriate primer is annealed to one or the other of the two strands.[624]

Since its introduction, many modifications and improvements have been developed. The sequencing gels have been improved. The use of ^{35}S labeling has given sharper autoradiographs.[625] Alternatively, a silver stain can be used with unlabeled primers.[626] GC-rich DNA sequences are often difficult to sequence, probably because even in the denaturing polyacrylamide gels used for sequencing they tend to form hairpin loops, perhaps as a result of formation of Hoogsteen base pairs (Fig. 5-7). Formation of these loops results in uneven spacing between the adjacent bands in the sequencing gel, so-called "compression artifacts." Use of a 7-deaza-dGTP in place of dGTP in the sequencing reaction ameliorates this problem.[627] Sensitivity can be improved by use of "cycle sequencing" in which a heat-stable polymerase such as *Taq* polymerase is used, and after heating the same template DNA is used repeatedly to give a higher yield of labeled fragments.[628,629]

About 200 – 400 bases can be successfully sequenced manually in a single run. By prolonging the time of electrophoresis in a second run, the sequence can be extended considerably. By using the just obtained sequence information, it is possible to select a new start point 200 or more nucleotides further along the template chain and to synthesize an oligonucleotide primer to anneal to the template at this point. In this way it is possible to "walk" along the template adding additional sequences at each step. Another procedure is to delete by mutation various segments of the cloned DNA above the sequence that binds the primer. This

Figure 5-49 A DNA sequencing gel obtained using a segment of DNA from salmon sperm selected by suitable oligonucleotide primers, amplified by PCR, and sequenced with a ^{35}S label in the primer. Four samples were used, one with each of the four dideoxy chain terminators (A, G, C, T, A, C, G, T from left to right). After electrophoresis the shorter fragments are at the lower end of the gel. The sequence of the strand complementary to the template strand whose sequence is being determined is read from the bottom of the gel. Here it starts CTATGATAC. Reproduced by permission of Amersham Pharmacia Biotech, Limited.

permits analysis of the whole cloned fragment via an overlapping set of sequences from the deletion mutants.[630,631]

Many DNA sequences continue to be determined manually by the well-developed long gel procedures as illustrated in Fig. 5-49. However, sequencing whole genomes has depended upon the development of high-speed automated procedures.[632] Instead of radiolabeling, fluorescent dyes may be joined to the primer to allow detection of the chain fragments produced during sequencing. Automatic sequencers use four dyes that fluoresce with different colors.[633,634] A different dye is used for each of the reaction mixtures. Then the four samples are mixed together and the DNA fragments are subjected to electrophoresis in a single lane. A laser beam excites the fluorescence, scanning several lanes with different samples as the electrophoresis progresses. A photomultiplier tube records the fluorescence intensity of each band through a series of four filters in a rotating wheel. This allows automatic recognition of the four different colors of fluorescence and therefore of the nucleic acid base present in each position in the sequence. Improved strategies for "primer walking"[635] and for "shotgun sequencing"[19,632,636] have been devised. In the shotgun strategy, whole bacterial genomes have been cut by restriction enzymes into large numbers of overlapping fragments which have been separated and sequenced.[637] A computer program is used to analyze and assemble the sequences into a complete genomic sequence. An example is provided by the genome of the *Methanococcus jannaschii*. Its large circular chromosome contains 1,664,976 bp and there are two additional pieces to the genome, one containing 58,407 bp and the other 16,550 bp. The sequences were deduced from 36,718 individual sequencing runs on high-speed automatic sequencers. For each run, on average, 481 bp could be read.[636] To sequence the human genome faster methods are needed.[638] Capillary electrophoresis with a single laser beam scanning the output of 24 capillaries has been demonstrated.[639,640] Extremely rapid sequencing of oligonucleotides up to 100 bp in length can be accomplished by mass spectrometry.[641] This may be an important technique for diagnosis of genetic defects (Chapter 26).

The method of Maxam and Gilbert.

The nonenzymatic method devised by Maxam and Gilbert[642–644] can be used to sequence either ss or dsDNA. Before the sequencing is begun, a radioactive label is incorporated, usually at the 5' end. This is often done by cleaving off any phosphate groups present on the 5' end with alkaline phosphatase and then transferring a new radioactive phospho group with the assistance of the enzyme polynucleotide kinase and radioactive γ-^{32}p-labeled ATP. If dsDNA is used the strands are separated so that each has a label only at one end.

The key step in sequencing by the Maxam–Gilbert procedure is to cleave chemically the DNA at random locations using reagents that have some specificity for particular bases. The cleavage process involves three distinct steps: (1) chemical modification, as specific as is possible for the chosen base; (2) displacement of the modified base from the sugar; and (3) elimination and chain cleavage using amine catalysis (Eq. 5-21). Two consecutive steps can often be combined. There are several versions of the method; one involves dimethyl sulfate as the specific reagent for guanine. It forms N^7-methylguanosine (see Eq. 5-18) which upon heating with the strong base piperidine at 90°C undergoes addition of hydroxyl ion with ring opening and displacement of the modified base according to Eq. 5-26. The product, a glycosylamine of piperidine, is in equilibrium with a Schiff base which can undergo chain cleavage as in Eq. 5-21.

A second sample of DNA is treated with a piperidine-formate buffer of pH 2 in the cold. This promotes the acid depurination of both guanine and adenosine. A third sample is treated with hydrazine, with both cytidine and thymidine being cleaved to ribosylurea according to Eq. 5-13. Again, this is followed by displacement and β elimination (Eq. 5-21) catalyzed by piperidine. The fourth sample is also treated with hydrazine but in the presence of a high concentration of NaCl which inhibits the reaction with thymidine, by lowering the pK_a of the thymine, and allows the cleavage to be more nearly specific for cytidine. Each of the reactions is conducted in such a way that on the average only one cleavage event occurs per molecule of DNA. Since the cleavages occur at many different points, a family of nested radioactively labeled oligonucleotides, one from each original molecule, is produced. When these are sorted by polyacrylamide gel electrophoresis, the pattern of the oligonucleotides in the four adjacent channels allows the nucleotide sequence to be read directly from the autoradiogram.

The Maxam–Gilbert method doesn't require synthesis of a primer and it sometimes works well for sequences that are difficult to obtain with the Sanger–Coulson procedure. The two methods may both be used to provide additional certainty about a sequence. The Maxam–Gilbert method is very convenient for sequencing small oligonucleotides which often react poorly with the polymerase used for the chain termination method. The method usually requires that a restriction map be prepared.

Sequencing RNA.

The first known RNA sequence, that of an alanine tRNA, was determined by Holley and associates in 1965. The RNA was subjected to partial hydrolysis with pancreatic ribonuclease and ribonuclease T_1 (Fig. 5-43). The small oligonucleotide fragments were separated by ion exchange chromatography under denaturing conditions (7 M urea) and were then characterized individually.[645] The availability of additional enzymes such as ribonuclease U$_2$, the *B cereus* ribonuclease and ribonuclease Phy M of *Physarum* (Fig. 5-43) and the use of radiolabeling and of two-dimensional fingerprinting of digests have made the procedures more versatile.[646] A valine tRNA was sequenced independently by Bayev[647] and Campbell.[648]

Since the development of the rapid methods for sequencing DNA, many mRNA sequences have been determined by using reverse transcriptase to make a cDNA strand complementary to the RNA. The cDNA is then sequenced.[649] Rapid sequencing methods parallel to those used for DNA have also been devised.[650–652]

Nearest neighbor analysis. A technique developed by Kornberg and associates before the availability of sequencing methods is the **nearest neighbor sequence analysis**. Using a single radioactive ^{32}P-containing nucleoside triphosphate together with the three other unlabeled nucleoside triphosphates, a primer chain of DNA is elongated from the 3' end along a ssDNA template chain using a DNA polymerase. The incorporation of ^{32}P from the α position of the nucleotide triphosphate occurs in the bridge phosphates that connect the nucleotide originally carrying the ^{32}P to the 3' position of the neighboring nucleotide. Cleavage of the ^{32}P-containing product of the reaction with a mixture of micrococcal DNase and spleen phosphodiesterase, which catalyze b-type cleavage (Fig. 5-43), gives fragments in which the ^{32}P will now be attached to what was the 5' nearest neighbor to the radioactive nucleotide in the DNA.[72,653] Measurement of the radioactivity in each of the 3' nucleotides of thymine, cytosine, adenine, and guanine gives the frequencies of the adjacent pairs, TA, CA, AA, and GA. Using the other radioactive nucleoside triphosphates one at a time in separate experiments, all of the nearest neighbor frequencies can be obtained. From such an experiment it was possible to deduce that the strands in the double helix were oriented in an antiparallel fashion, as predicted by Watson and Crick. If the strands had been parallel, different nearest neighbor frequencies would have been observed.

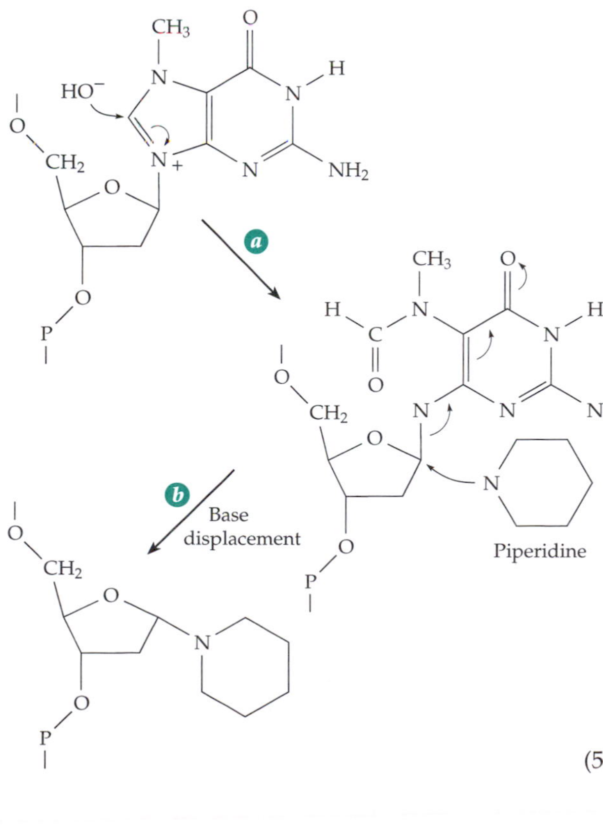

Understanding sequences. Sequences of over 20 million nucleotides from hundreds of organisms had been determined by 1988 and the number is doubling each 2–3 years.[654] Sequencing the human genome has required rates of millions of bases per day. With the massive amount of data already available it has become of great interest to compare sequences of genes, whether they encode similar or dissimilar proteins, to make comparisons between species, and to search for sequences that bind specific proteins or that encode particular regulatory signals. Relationships of common evolutionary origin or homology as well as other sequence similarities are often sought.

To handle the mass of existing data, powerful computer programs have been developed and various graphical procedures have also been developed to help the human mind comprehend the results.[654,655] One important problem is to define and locate what are called **consensus sequences**. The problem is best illustrated by examples.[654] The cleavage site for the *Eco*RI restriction endonuclease is **GAATTC**. There is no ambiguity. In a DNA of random sequence this would be expected to occur by chance in about $(1/4)^6$ nucleotides (4 kb). On the other hand, the *Hin*II restriction endonuclease cleaves within the consensus sequence **GTYRAC** where Y = C or T and R = A or G. It would be expected to occur by chance in about $1/4^5$ nucleotides. Many binding sites for RNA polymerase, the so called **promoters** (Chapter 28) contain the consensus sequence **TAtAaT**, at position -10, ahead of the 5' end of the sequence that is transcribed into mRNA. The lower case t and a used here imply that other nucleotides may often replace T or A at these positions. There are many promoters and over 70% of those described have this consensus sequence. All have the less restricted sequence **TAxxxT**, where x may be any nucleotide. Our definition of consensus sequence is somewhat arbitrary. Now consider the problem of locating a −35 site whose consensus sequence is **TTGACA** but which may, for different genes, be shifted backward or forward by a nucleotide or two. This is a consensus sequence. Therefore, in many cases one or more substitutions in the sequence will have been made. The result is that the sequence of nucleotides in

(5-26)

which the consensus sequence is to be found is likely to appear entirely random. Sophisticated computer programs are helpful in locating it.[654]

8. Protein–DNA Interactions

The most detailed information about interactions of proteins with DNA is coming from X-ray crystallographic studies. Examples are seen in Figs. 5-35 to 5-40. Several other methods have also been very useful. Much has been learned from the effects of mutations in DNA-binding proteins or in regions of DNA to which a protein binds. Binding of proteins to DNA can also be recognized by its effects on the mobility of DNA during gel electrophoresis.[656,657] Chemical[658] or laser-induced crosslinking can show that within a complex a specific residue in a protein is adjacent to a certain sequence in the DNA.

The technique of **protection mapping** or **"footprinting"** is widely used to determine which nucleotides in a sequence are covered by a bound protein.[659] A reagent which attacks and cleaves DNA nearly randomly is used. DNase I was first introduced for this purpose[659] and has been used widely. An important finding is that certain sites that are readily cleaved (**hypersensitive sites**) are frequently located in chromatin undergoing transcription. Footprinting has also been accomplished with other nucleases, with dimethylsulfate (which acts on A and C), with carbodiimides (which act on U and G), and with the Maxam–Gilbert guanine-specific cleavage (Eq. 5-26). One of the most popular methods employs cleavage by hydroxyl radicals.[660–663] **Photofootprinting** depends upon decreased or increased sensitivity to ultraviolet light at sites bound by proteins.[214,662,664] In footprinting experiments the DNA to be studied is radioactively labeled at one end of one strand. In the absence of the protecting protein, denaturation and electrophoresis of the cleaved fragments yields a nearly random "ladder" of DNA fragments. In the presence of the binding protein some cleavage products will be missing from the ladder. The bound protein leaves a "footprint" (Fig. 5-50A,C).

Related methods are being applied to the determination of the secondary structure of RNA molecules[665,666] and to the study of interactions with proteins. For example, treatment with dimethyl sulfate under appropriate conditions methylates bases that are not paired, giving largely 1-methyladenosine and 3-methylcytidine.[667]

9. Nuclear Magnetic Resonance

Much of the initial effort to study polynucleotides by NMR spectroscopy was directed toward transfer RNAs, only a few of which have been crystallized in a form suitable for X-ray diffraction. Study of the other tRNAs by NMR techniques has established that all of the tRNAs have a similar architecture and that the structures observed in the crystals are preserved in solution.[668] Figure 5-51 shows the low-field end of the NMR spectrum of a valine-specific tRNA from *E. coli*. The spectrum is run in H_2O rather than D_2O so that exchangeable hydrogens in the hydrogen bonds of the Watson–Crick base pairs can be observed.[669] The protons giving rise to the downfield resonances are primarily those attached to nitrogen atoms of the rings and in hydrogen-bonded positions. These protons are shielded by adjacent electron-donating groups and by their attachment to the semiaromatic rings of the bases. The NMR signals are further shifted downfield to varying degrees depending upon whether or not the proton being observed is attached to a base that is stacked with other bases. The size of the shift also depends upon which neighboring bases are present. The proton on N-3 of AU base pairs is deshielded more than the proton on the N-1 of GC base pairs. Therefore, the AU protons appear further downfield than the GC protons. The stronger ring current in A than in C enhances this separation.

All of the resonances in Fig. 5-51 have been assigned to particular bases. This was done in part by varying the temperature, changing the magnesium ion concentration, and predicting shifts caused by ring currents in adjacent bases making use of the X-ray crystal structures. NMR spectra of hairpin helical fragments also provided essential information. From integration of the areas under the peaks it was concluded that the 20 resonances seen below −11 ppm represent 27 protons. Twenty of these are in Watson-Crick base pairs and correspond to those expected from the X-ray structure. Six more belong to protons involved in tertiary interactions, such as base pair triplets or non-Watson–Crick pairs. One of these, labeled "G" in the figure, is in the dihydrouridine stem and involves the ring proton of N-1 of m^7G46 which is hydrogen bonded to N-7 of G22 in the major groove of the RNA. G22 is located at the beginning of the "extra loop."

Measurements of the NOE of nearby protons in both small RNA molecules[668,669] and DNA oligonucleotides[670,671] provided much additional information. Figures 5-51B and C show NOESY spectra of the tRNA^{Val} and the way in which weak cross-peaks between the H-bonded imino protons in adjacent base pairs (Fig. 5-51B) can be used to establish connectivities.[669] Beginning with resonance B, it is possible to establish the sequence of the NH groups giving rise to these resonances as OBUGJNT. Using other data as well, it was concluded that these represent the seven base pairs of the acceptor stem (see Fig. 5-30), with resonance C representing the first GC pair, resonance B the second, etc. Resonance A was identified as coming from the base triplet containing a Hoogsteen

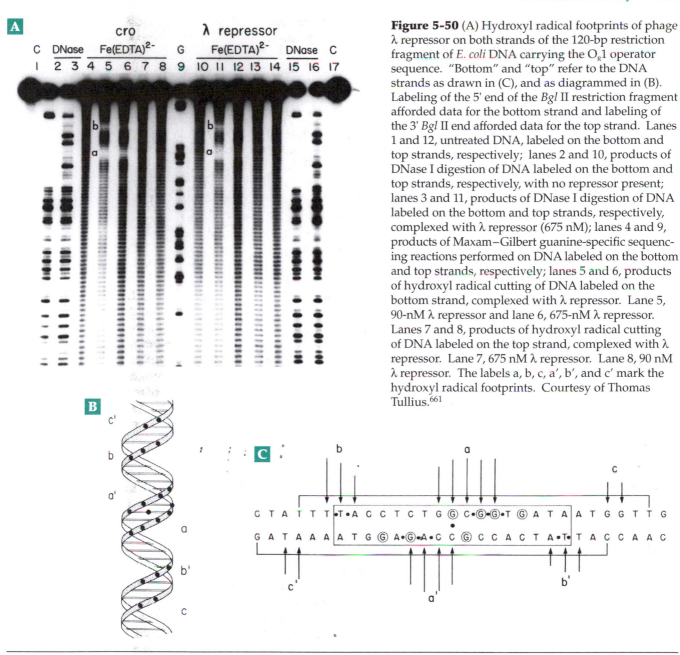

Figure 5-50 (A) Hydroxyl radical footprints of phage λ repressor on both strands of the 120-bp restriction fragment of *E. coli* DNA carrying the O_R1 operator sequence. "Bottom" and "top" refer to the DNA strands as drawn in (C), and as diagrammed in (B). Labeling of the 5' end of the *Bgl* II restriction fragment afforded data for the bottom strand and labeling of the 3' *Bgl* II end afforded data for the top strand. Lanes 1 and 12, untreated DNA, labeled on the bottom and top strands, respectively; lanes 2 and 10, products of DNase I digestion of DNA labeled on the bottom and top strands, respectively, with no repressor present; lanes 3 and 11, products of DNase I digestion of DNA labeled on the bottom and top strands, respectively, complexed with λ repressor (675 nM); lanes 4 and 9, products of Maxam–Gilbert guanine-specific sequencing reactions performed on DNA labeled on the bottom and top strands, respectively; lanes 5 and 6, products of hydroxyl radical cutting of DNA labeled on the bottom strand, complexed with λ repressor. Lane 5, 90-nM λ repressor and lane 6, 675-nM λ repressor. Lanes 7 and 8, products of hydroxyl radical cutting of DNA labeled on the top strand, complexed with λ repressor. Lane 7, 675 nM λ repressor. Lane 8, 90 nM λ repressor. The labels a, b, c, a', b', and c' mark the hydroxyl radical footprints. Courtesy of Thomas Tullius.[661]

base pair of 4-thiouracil at position 8 with A14 (see Fig. 5-7). Its connectivity to the sequence KCEO is also outlined in Fig. 5-51B. However, it could not be established without additional data which also helped to identify the sequence O-K as residues 10–13 of the dihydrouracil stem. Peak O is a multiproton peak representing not only GC 10 but also UA 7. In general, the GC protons are at the higher field side of the spectrum and the AU protons at the lower side. However, AU 7 is shifted to an anomolously high position. Cross-peaks between imino protons of uracil and the nearby C2 protons of adenine in Watson–Crick AU base pairs or C8 protons of Hoogsteen AU pairs can be observed in the 6.5 – 9 ppm region as shown in Fig. 5-35B. This region also contains information about other protons bound to the nuclei acid bases.

Similar techniques are being used for the study of DNA.[672] The presence of a second hydrogen in the 2' position of the deoxyribose rings of DNA adds several H-H distances (Fig. 5-52) that can be measured in addition to those seen in RNAs. Characteristic differences are seen in the NOESY plots of A, B, and Z forms of DNA.[670,671,673,674] Although detailed structural information has been obtained for short segments of DNA, spectra of larger oligonucleotides are impossible to analyze with two-dimensional methods because of extensive overlap of resonances.[665] The difficulty is already apparent in the 17 base pair DNA segment for which a one-dimensional spectrum as well as COSY and NOESY spectra are shown in Fig. 5-53.

Some help with the complexity can be obtained by incorporation of ^{13}C-enriched methyl groups into

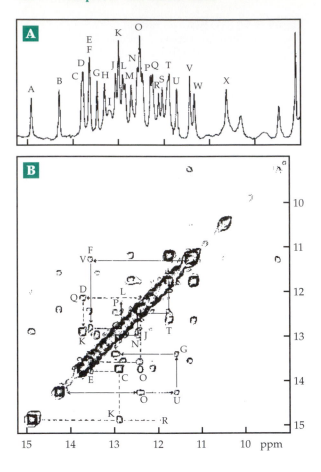

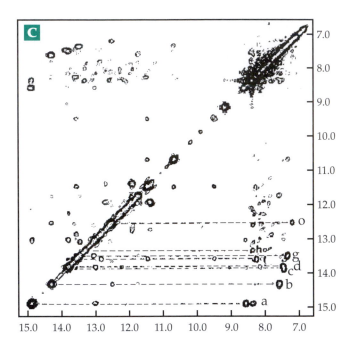

Figure 5-51 (A) The low-field region of the one-dimensional ^{1}H NMR spectrum of *E. coli* tRNA$_1^{Val}$ at 27°C in H$_2$O. Resonances are identified by letters A – X. (B) NOESY spectrum of the same tRNA under similar conditions showing the imino-imino NOEs. In the lower right sector the connectivity traces of the acceptor helix and dihydrouridine helix are shown as solid and dotted lines, respectively. In the NOESY sample the two protons in peak EF are partially resolved whereas the two protons in peak T have coalesced. (C) NOESY spectrum of *E. coli* tRNA$_1^{Val}$ at 32°C showing the imino and aromatic proton regions. AU-type imino protons have been connected horizontally by a dotted line to the cross-peak of their proximal C2-H or C8-H in the 7 to 9 ppm region, which has been labeled with the corresponding lower-case letter. From Hare *et al.*[669] Courtesy of Brian Reid.

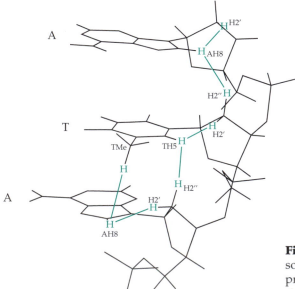

Figure 5-52 Segment of a DNA chain in the B conformation illustrating some intrachain NOEs that may be observed. The close juxtaposition of proton pairs is provided by the H2'-*endo* (^{2}E) conformation of the sugar rings with *anti* base conformation (Fig. 5-11). After Cohen.[675]

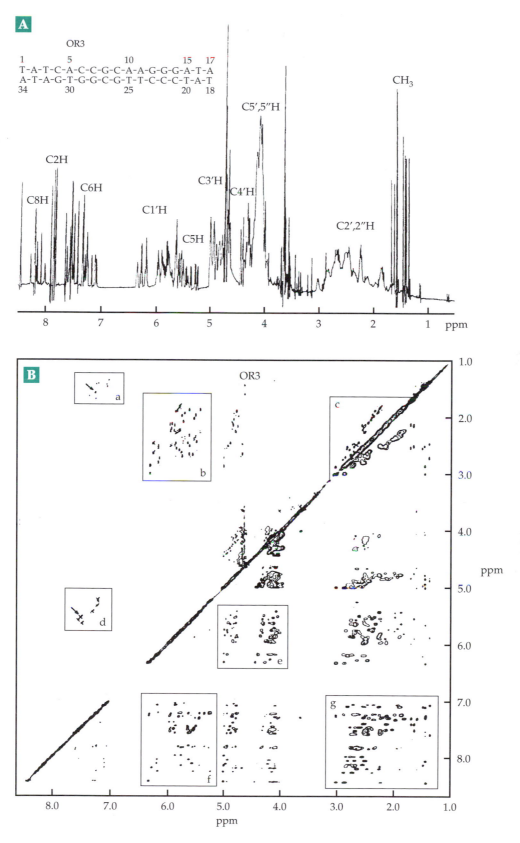

Figure 5-53 (A) ^{1}H NMR spectrum of a 17 base-pair DNA segment from the operator sequence OR3 from bacteriophage λ in D$_2$O at 37°C. (B) Combined COSY above the diagonal and NOESY (below the diagonal) spectra. C5H and C6H J coupling is established from cross-peaks in box d for cytosines and in box a for thymines. Two unresolved cross-peaks give rise to the more intense spots marked by arrows. Box b contains cross-peaks from scalar coupling of the two H2' protons to the H1' protons of the deoxyribose rings. Most of the aromatic proton resonances could be assigned using the NOE cross-peaks in box f. For further details see Wemmer *et al.*[676] See also Bax and Lerner.[672] Courtesy of B. Reid.

polynucleotides. One way to do this is to grow yeast on a medium containing methionine with ^{13}C in its methyl group. The most intense peaks in ^{13}C NMR spectra of tRNA molecules from this yeast will represent methyl groups in modified bases in the tRNA. Incorporated ^{13}C also permits study of internal motion within tRNA or other oligonucleotides by NMR methods.[677] However, as with protein NMR spectroscopy the major recent advances have come from systematic incorporation of both ^{13}C and ^{15}N into nucleic acids and the development of three- and four-dimensional NMR methods.[678–684b] Also important are methods for replacing some hydrogen atoms with deuterium to simplify spectra.[685,686]

The very sensitive ^{19}F nucleus can be introduced into tRNAs by incorporation of 5-fluorouracil in place of uracil[687] (Fig. 5-54A,B). Phosphorus 31 NMR spectra (Fig. 5-54C) can provide information about conformations of the chain.[688]

A few recent NMR investigations of polynucleotides include studies of triple-helical DNA,[689] Holliday junctions,[290] double-stranded oligonucleotides containing adducts of carcinogens,[690,691] of hairpin loops with sheared A•A and G•G pairs,[692] and of proton exchange in both imino and amino groups.[693]

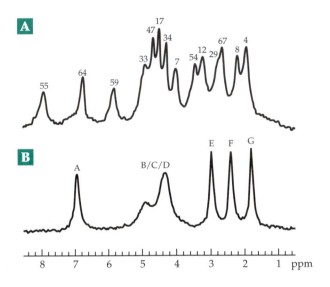

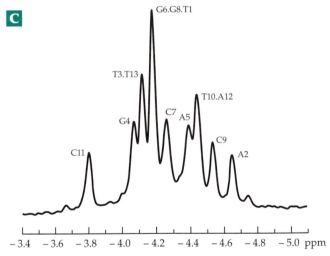

Figure 5-54 (A) An ^{19}F NMR spectrum of the 76-residue *E. coli* tRNA^{Val} containing 5-fluorouracil in 14 positions. Recorded at 47°C. The numbers above the resonances indicate the position in the sequence. (The sequence is not identical to that for the yeast tRNA shown in Fig. 5-30.) Modified from Chu *et al.*[694] Courtesy of Jack Horowitz. (B) A similar spectrum for a 35-residue "minihelix" that contains the acceptor stem of the tRNA^{Val} and seven fluorouracils. The broad peaks B, D, and E are shifted far upfield by reaction with bisulfite (Eq. 5-11) suggesting that they are not hydrogen bonded and are present in the loop of the stem–loop structure. Peaks A, E, F, and G correspond to resonances 64, 7, 67, and 4, respectively, in (A) and represent fluorouracil in the stem structure. From Chu *et al.*[694] Courtesy of Jack Horowitz. (C) A ^{31}P NMR spectrum of a synthetic 14 base-pair DNA segment related to the *E. coli lac* operator. The palindromic sequence is TCTGAGCGCTCAGA. The numbers refer to the positions from the 5' end. From Schroeder *et al.*[688]

References

1. Portugal, F. H., and Cohen, J. S. (1977) *A Century of DNA: A History of the Discovery of the Structure and Function of the Genetic Substance*, MIT Press, Cambridge, Massachusetts
2. Levene, P. A., and Tipson, R. S. (1935) *J. Biol. Chem.* **109**, 623–630
3. Schlenk, F. (1988) *Trends Biochem. Sci.* **13**, 67–69
4. Watson, J. D., and Crick, F. H. C. (1953) *Nature (London)* **171**, 737–738
5. Watson, J. D. (1968) *The Double Helix*, Atheneum, New York
6. Crick, F. (1988) *What a Mad Pursuit: A Personal View of Scientific Discovery*, Basic Books, New York

7. Olby, R. (1974) *The Path to the Double Helix*, Macmillan, London
8. Wilson, H. R. (1988) *Trends Biochem. Sci.* **13**, 275–278
9. Chambers, D. A., Reid, K. B. M., and Cohen, R. L. (1994) *FASEB J.* **8**, 1219–1226
9a. Piper, A. (1998) *Trends Biochem. Sci.* **23**, 151–154
10. Arnott, S., and Hukins, D. W. L. (1973) *J. Mol. Biol.* **81**, 93–105
11. Proudfoot, N. J., Shander, M. H. M., Manley, J. L., Gefter, M. L., and Maniatis, T. (1980) *Science* **209**, 1329–1336

12. Tucker, P. W., Liu, C.-P., Mushinski, J. F., and Blattner, F. R. (1980) *Science* **209**, 1353–1360
13. Yamada, Y., Mudryj, M., Sullivan, M., and de Crombrugghe, B. (1983) *J. Biol. Chem.* **258**, 2758–2761
14. Tsuzuki, T., Obaru, K., Setoyama, C., and Shimada, K. (1987) *J. Mol. Biol.* **198**, 21–31
15. Ratner, L., and 18 other authors (1985) *Nature (London)* **313**, 277–284
16. Dunn, J. J., and Studier, F. W. (1983) *J. Mol. Biol.* **166**, 477–535
17. Peden, K. W. C. (1983) *Gene* **22**, 277–280

References

18. Anderson, S., Bankier, A. T., Barrell, B. G., deBruijn, M. H. L., Coulson, A. R., Drouin, J., Eperon, I. C., Nierlich, D. P., Roe, B. A., Sanger, F., Schreier, P. H., Smith, A. J. H., Staden, R., and Young, I. G. (1981) *Nature (London)* **290**, 457–470

19. Fraser, C. M., and 28 other authors (1995) *Science* **270**, 397–403

20. Fleischmann, R. D., and 39 other authors (1995) *Science* **269**, 496–512

20a. Doolittle, R. F. (1998) *Nature (London)* **392**, 339–342

21. Nowak, R. (1995) *Science* **267**, 172–174

22. Dujon, B., and 107 other authors (1994) *Nature (London)* **369**, 371–378

23. Johnston, M., and 34 other authors (1994) *Science* **265**, 2077–2082

24. Goffeau, A., Barrell, B. G., Bussey, H., Davis, R. W., Dujon, B., Feldmann, H., Galibert, F., Hoheisel, J. D., Jacq, C., Johnston, M., Louis, E. J., Mewes, H. W., Murakami, Y., Philippsen, P., Tettelin, H., and Oliver, S. G. (1996) *Science* **274**, 546–557

24a. Macilwain, C. (2000) *Nature (London)* **405**, 983–984

25. Liébecq, C., ed. (1992) *Biochemical Nomenclature*, Portland Press, London and Chapel Hill, North Carolina (for the International Union of Biochemistry and Molecular Biology)

26. Shabarova, Z., and Bogdanov, A. (1994) *Advanced Organic Chemistry of Nucleic Acids*, VCH Publ., Weinheim

27. Saenger, W. (1984) *Principles of Nucleic Acid Structure*, Springer, New York

28. Hurst, D. T. (1980) *An Introduction to the Chemistry and Biochemistry of Pyrimidines, Purines and Pteridines*, Wiley, New York

29. Markowski, V., Sullivan, G. R., and Roberts, J. D. (1977) *J. Am. Chem. Soc.* **99**, 714–718

30. Katritzky, A. R., and Karelson, M. (1991) *J. Am. Chem. Soc.* **113**, 1561–1566

31. Szczesniak, M., Leszczynski, J., and Person, W. B. (1992) *J. Am. Chem. Soc.* **114**, 2731–2733

32. Scanlan, M. J., and Hillier, I. H. (1984) *J. Am. Chem. Soc.* **106**, 3737–3745

33. Brown, D. M. (1974) in *Basic Principles in Nucleic Acid Chemistry*, Vol. II (Tso, P. O. P., ed), pp. 2–90, Academic Press, New York

34. Hoogsteen, K. (1963) *Acta Crystallogr.* **16**, 907–916

35. Frank-Kamenetskii, M. D., and Mirkin, S. M. (1995) *Ann. Rev. Biochem.* **64**, 65–95

36. Portugal, J. (1989) *Trends Biochem. Sci.* **14**, 127–130

37. Adams, R. L. P., Knowler, J. T., and Leader, D. P. (1992) *The Biochemistry of the Nucleic Acids*, 11th ed., Chapman & Hall, London

37a. Saenger, W. (1973) *Angew. Chem. Int. Ed. Engl.* **12**, 591–601

38. Fang, G., and Cech, T. R. (1993) *Biochemistry* **32**, 11646–11657

39. Gellert, M., Lipsett, M. N., and Davies, D. R. (1962) *Proc. Natl. Acad. Sci. U.S.A.* **48**, 2013–2018

40. Kennell, D. E. (1971) *Prog. Nucleic Acid Res. Mol. Biol.* **11**, 259–301

40a. Stofer, E., Chipot, C., and Lavery, R. (1999) *J. Am. Chem. Soc.* **121**, 9503–9508

40b. Peyret, N., Seneviratne, P. A., Allawi, H. T., and SantaLucia, J., Jr. (1999) *Biochemistry* **38**, 3468–3477

41. Pranata, J., Wierschke, S. G., and Jorgensen, W. L. (1991) *J. Am. Chem. Soc.* **113**, 2810–2819

42. Dang, L. X., and Kollman, P. A. (1990) *J. Am. Chem. Soc.* **112**, 503–507

43. Newcomb, L. F., and Gellman, S. H. (1994) *J. Am. Chem. Soc.* **116**, 4993–4994

44. Borer, P. N., Dengler, B., Tinoco, I., Jr., and Uhlenbeck, O. C. (1974) *J. Mol. Biol.* **86**, 843–853

45. Freier, S. M., Kierzek, R., Jaeger, J. A., Sugimoto, N., Caruthers, M. H., Neilson, T., and Turner, D. H. (1986) *Proc. Natl. Acad. Sci. U.S.A.* **83**, 9373–9377

46. Sarai, A., Mazur, J., Nussinov, R., and Jernigan, R. L. (1988) *Biochemistry* **27**, 8498–8502

47. Hunter, C. A. (1993) *J. Mol. Biol.* **230**, 1025–1054

48. Friedman, R. A., and Honig, B. (1995) *Biophys. J.* **69**, 1528–1535

49. DePrisco Albergo, D., Marky, L. A., Breslauer, K. J., and Turner, D. H. (1981) *Biochemistry* **20**, 1409–1413

50. Privalov, P. L., and Filimonov, V. V. (1978) *J. Mol. Biol.* **122**, 447–464

51. Privalov, P. L., and Filimonov, V. V. (1978) *J. Mol. Biol.* **122**, 465–470

52. Petersen, S. B., and Led, J. J. (1981) *J. Am. Chem. Soc.* **103**, 5308–5313

53. Petruska, J., and Goodman, M. F. (1995) *J. Biol. Chem.* **270**, 746–750

54. SantaLucia, J., Jr., Allawi, H. T., and Seneviratne, P. A. (1996) *Biochemistry* **35**, 3555–3562

55. Egli, M., Usman, N., and Rich, A. (1993) *Biochemistry* **32**, 3221–3237

56. Watson, J. D., and Crick, F. H. C. (1953) *Nature (London)* **171**, 964–967

57. Topal, M. D., and Fresco, J. R. (1976) *Nature (London)* **263**, 285–289

58. Tinoco, I., Jr., Borer, P. N., Dengler, B., Levine, M. D., Uhlenbeck, O. C., Crothers, D. M., and Gralla, J. (1973) *Nature New Biol.* **246**, 40–41

59. Fazakerley, G. V., Gdaniec, Z., and Sowers, L. C. (1993) *J. Mol. Biol.* **230**, 6–10

60. Florián, J., Hrouda, V., and Hobza, P. (1994) *J. Am. Chem. Soc.* **116**, 1457–1460

61. Douhal, A., Kim, S. K., and Zewail, A. H. (1995) *Nature (London)* **378**, 260–263

62. Gabb, H. A., and Harvey, S. C. (1993) *J. Am. Chem. Soc.* **115**, 4218–4227

63. Ellervik, U., and Magnusson, G. (1994) *J. Am. Chem. Soc.* **116**, 2340–2347

64. Thibaudeau, C., Plavec, J., and Chattopadhyaya, J. (1994) *J. Am. Chem. Soc.* **116**, 8033–8037

65. Santos, R. A., Tang, P., and Harbison, G. S. (1989) *Biochemistry* **28**, 9372–9378

66. Sundaralingam, M. (1969) *Biopolymers* **7**, 821–860

67. Vologodskii, A. (1992) *Topology and Physics of Circular DNA*, CRC Press, Boca Raton, Florida

68. Baikalov, I., Grzeskowiak, K., Yanagi, K., Quintana, J., and Dickerson, R. E. (1993) *J. Mol. Biol.* **231**, 768–784

69. Lipanov, A., Kopka, M. L., Kaczor-Grzeskowiak, M., Quintana, J., and Dickerson, R. E. (1993) *Biochemistry* **32**, 1373–1389

70. Schlick, T., Hingerty, B. E., Peskin, C. S., Overton, M. L., and Broyde, S. (1991) *Theoretical Biochemistry and Molecular Physics*, Vol. 1 (Beveridge, D. L. and Lavery, R., eds), pp. 39–58, Adenine Press, Guilderland, New York

71. Wang, A. H.-J., Quigley, G. J., Kolpak, F. J., Crawford, J. L., van Boom, J. H., van der Marel, G., and Rich, A. (1979) *Nature (London)* **282**, 680–686

72. Kornberg, A., and Baker, T. A. (1992) *DNA Replication*, 2nd ed., Freeman, New York

73. Kaufmann, W. K., and Paules, R. S. (1996) *FASEB J.* **10**, 238–247

74. Lindahl, T. (1993) *Nature (London)* **362**, 709–715

75. Rodley, G. A., and Bates, R. H. T. (1980) *Trends Biochem. Sci.* **5**, 231–233

76. Kennard, O., and Salisbury, S. A. (1993) *J. Biol. Chem.* **268**, 10701–10704

77. Drew, H. R., and Dickerson, R. E. (1981) *J. Mol. Biol.* **151**, 535–556

78. Dickerson, R. E. (1983) *Sci. Am.* **249**(Dec), 94–111

78a. Dickerson, R. E. (1997) in *International Tables for Macromolecular Crystallography* (Rossmann, M. G., and Arnold, E., eds), (Chapter 23, Part c)

79. Shui, X., McFail-Isom, L., Hu, G. G., and Williams, L. D. (1998) *Biochemistry* **37**, 8341–8355

80. Gorin, A. A., Zhurkin, V. B., and Olson, W. K. (1995) *J. Mol. Biol.* **247**, 34–48

81. Tereshko, V., Minasov, G., and Egli, M. (1999) *J. Am. Chem. Soc.* **121**, 470–471

82. Yuan, H., Quintana, J., and Dickerson, R. E. (1992) *Biochemistry* **31**, 8009–8021

83. Ban, C., Ramakrishnan, B., and Sundaralingam, M. (1994) *J. Mol. Biol.* **236**, 275–285

83a. Kielkopf, C. L., Ding, S., Kuhn, P., and Rees, D. C. (2000) *J. Mol. Biol.* **296**, 787–801

84. Johansson, E., Parkinson, G., and Neidle, S. (2000) *J. Mol. Biol.* **300**, 551–561

85. Shen, L. X., Cai, Z., and Tinoco, I., Jr. (1995) *FASEB J.* **9**, 1023–1033

86. Rosenberg, J. M., Seeman, N. C., Day, R. O., and Rich, A. (1976) *J. Mol. Biol.* **104**, 145–167

87. Marino, J. P., Gregorian, R. S., Jr., Csankovszki, G., and Crothers, D. M. (1995) *Science* **268**, 1448–1454

88. Iwamoto, S., and Hsu, M.-T. (1983) *Nature (London)* **305**, 70–72

89. Morel, G., ed. (1995) *Visualization of Nucleic Acids*, CRC Press, Boca Raton, Florida

90. Arscott, P. G., Lee, G., Bloomfield, V. A., and Evans, D. F. (1989) *Nature (London)* **339**, 484–486

91. Jares-Erijman, E. A., and Jovin, T. M. (1996) *J. Mol. Biol.* **257**, 597–617

92. Duan, Y., Wilkosz, P., Crowley, M., and Rosenberg, J. M. (1997) *J. Mol. Biol.* **272**, 553–572

93. Auffinger, P., and Westhof, E. (2000) *J. Mol. Biol.* **300**, 1113–1131

93a. Soler-López, M., Malinina, L., Liu, J., Huynh-Dinh, T., and Subirana, J. A. (1999) *J. Biol. Chem.* **274**, 23683–23686

94. Schneider, B., and Berman, H. M. (1995) *Biophys. J.* **69**, 2661–2669

95. Tereshko, V., Minasov, G., and Egli, M. (1999) *J. Am. Chem. Soc.* **121**, 3590–3595

96. Franklin, R. E., and Gosling, R. G. (1953) *Nature (London)* **172**, 156–157

97. Ramakrishnan, B., and Sundaralingam, M. (1995) *Biophys. J.* **69**, 553–558

98. Ramakrishnan, B., and Sundaralingam, M. (1993) *J. Mol. Biol.* **231**, 431–444

99. Eisenstein, M., and Shakked, Z. (1995) *J. Mol. Biol.* **248**, 662–678

99a. Ng, H.-L., Kopka, M. L., and Dickerson, R. E. (2000) *Proc. Natl. Acad. Sci. U.S.A.* **97**, 2035–2039

100. Selsing, E., Wells, R. D., Alden, C. J., and Arnott, S. (1979) *J. Biol. Chem.* **254**, 5417–5422

101. Mahendrasingam, A., Forsyth, V. T., Hussain, R., Greenall, R. J., Pigram, W. J., and Fuller, W. (1986) *Science* **233**, 195–197

102. Hopkins, R. C. (1981) *Science* **211**, 289–291

103. Gupta, G., Bansal, M., and Sasisekharan, V. (1980) *Proc. Natl. Acad. Sci. U.S.A.* **77**, 6486–6490

104. Arnott, S., Chandrasekaran, R., Birdsall, D. L., Leslie, A. G. W., and Ratliff, R. L. (1980) *Nature (London)* **283**, 743–745

105. Herbert, A., and Rich, A. (1996) *J. Biol. Chem.* **271**, 11595–11598

106. Gessner, R. V., Frederick, C. A., Quigley, G. J., Rich, A., and Wang, A. H.-J. (1989) *J. Biol. Chem.* **264**, 7921–7935

107. Zhou, G., and Ho, P. S. (1990) *Biochemistry* **29**, 7229–7236

References

108. Parkinson, G. N., Arvanitis, G. M., Lessinger, L., Ginell, S. L., Jones, R., Gaffney, B., and Berman, H. M. (1995) *Biochemistry* **34**, 15487–15495

109. Sadasivan, C., and Gautham, N. (1995) *J. Mol. Biol.* **248**, 918– 930

110. Bancroft, D., Williams, L. D., Rich, A., and Egli, M. (1994) *Biochemistry* **33**, 1073–1086

111. Rich, A., Nordheim, A., and Wang, A. H.-J. (1984) *Ann. Rev. Biochem.* **53**, 791–846

112. Moore, M. H., Van Meervelt, L., Salisbury, S. A., Lin, P. K. T., and Brown, D. M. (1995) *J. Mol. Biol.* **251**, 665–673

113. Wang, Y., Thomas, G. A., and Peticolas, W. L. (1987) *Biochemistry* **26**, 5178–5186

114. Saenger, W., Hunter, W. N., and Kennard, O. (1986) *Nature (London)* **324**, 385–388

115. Frank-Kamenetskii, M. (1986) *Nature (London)* **324**, 305

116. Basham, B., Schroth, G. P., and Ho, P. S. (1995) *Proc. Natl. Acad. Sci. U.S.A.* **92**, 6464–6468

117. Misra, V. K., and Honig, B. (1996) *Biochemistry* **35**, 1115–1124

118. McLean, M. J., Blaho, J. A., Kilpatrick, M. W., and Wells, R. D. (1986) *Proc. Natl. Acad. Sci. U.S.A.* **83**, 5884–5888

119. Kim, J.-m, and DasSarma, S. (1996) *J. Biol. Chem.* **271**, 19724–19731

120. Nordheim, A., Pardue, M. L., Lafer, E. M., Möller, A., Stollar, B. D., and Rich, A. (1981) *Nature (London)* **294**, 417–422

121. Jaworski, A., Hsieh, W.-T., Blaho, J. A., Larson, J. E., and Wells, R. D. (1987) *Science* **238**, 773–777

122. Zacharias, W., O'Conner, T. R., and Larson, J. E. (1988) *Biochemistry* **27**, 2970–2978

123. Nordheim, A., and Rich, A. (1983) *Nature (London)* **303**, 674–679

124. Wells, R. D. (1988) *J. Biol. Chem.* **263**, 1095–1098

125. Rentzeperis, D., Kupke, D. W., and Marky, L. A. (1994) *Biochemistry* **33**, 9588–9591

126. Liu, K., Miles, H. T., Frazier, J., and Sasisekharan, V. (1993) *Biochemistry* **32**, 11802–11809

127. Zhou, N., Germann, M. W., van de Sande, J. H., Pattabiraman, N., and Vogel, H. J. (1993) *Biochemistry* **32**, 646–656

128. Otto, C., Thomas, G. A., Rippe, K., Jovin, T. M., and Peticolas, W. L. (1991) *Biochemistry* **30**, 3062–3069

129. Trifonov, E. N. (1991) *Trends Biochem. Sci.* **16**, 467–470

130. Rich, A. (1998) *Proc. Natl. Acad. Sci. U.S.A.* **95**, 13999–14000

131. Hageman, P. J. (1988) *Ann. Rev. Biophys. Biophys. Chem.* **17**, 265–286

132. Poncin, M., Hartmann, B., and Lavery, R. (1992) *J. Mol. Biol.* **226**, 775–794

133. MacKerell, A. D., Jr., Wiórkiewicz-Kuczera, J., and Karplus, M. (1995) *J. Am. Chem. Soc.* **117**, 11946–11975

134. Kosikov, K. M., Gorin, A. A., Zhurkin, V. B., and Olson, W. K. (1999) *J. Mol. Biol.* **289**, 1301–1326

135. Peticolas, W. L., Wang, Y., and Thomas, G. A. (1988) *Proc. Natl. Acad. Sci. U.S.A.* **85**, 2579–2583

136. Dickerson, R. E., and Drew, H. R. (1981) *Proc. Natl. Acad. Sci. U.S.A.* **78**, 7318–7322

136a. Packer, M. J., and Hunter, C. A. (1998) *J. Mol. Biol.* **280**, 407–420

137. Hogan, M., Dattagupta, N., and Crothers, D. M. (1979) *Nature (London)* **278**, 521–524

138. Mitra, C. K., Sarma, M. H., and Sarma, R. H. (1981) *J. Am. Chem. Soc.* **103**, 6727–6737

139. Cluzel, P., Lebrun, A., Heller, C., Lavery, R., Viovy, J.-L., Chatenay, D., and Caron, F. (1996) *Science* **271**, 792–794

140. Smith, S. B., Cui, Y., and Bustamante, C. (1996) *Science* **271**, 795–799

141. Strick, T. R., Allemand, J. F., Bensimon, D., Bensimon, A., and Croquette, V. (1996) *Science* **271**, 1835–1837

141a. Allemand, J. F., Bensimon, D., Lavery, R., and Croquette, V. (1998) *Proc. Natl. Acad. Sci. U.S.A.* **95**, 14152–14157

142. Dickerson, R. E. (1983) *J. Mol. Biol.* **166**, 419–441

143. Tung, C.-S., and Harvey, S. C. (1986) *J. Biol. Chem.* **261**, 3700–3709

144. Ulanovsky, L. E., and Trifonov, E. N. (1987) *Nature (London)* **326**, 720–722

145. Diekmann, S. (1989) *EMBO J.* **8**, 1–4

146. El Hassan, M. A., and Calladine, C. R. (1996) *J. Mol. Biol.* **259**, 95–103

147. Calladine, C. R. (1982) *J. Mol. Biol.* **161**, 343–352

148. Hunter, C. A., and Lu, X.-J. (1997) *J. Mol. Biol.* **265**, 603–619

149. Babcock, M. S., Pednault, E. P. D., and Olson, W. K. (1994) *J. Mol. Biol.* **273**, 125–156

150. Lu, X.-J., El Hassan, M. A., and Hunter, C. A. (1997) *J. Mol. Biol.* **273**, 668–680

151. Kitchin, P. A., Klein, V. A., Ryan, K. A., Gann, K. L., Rauch, C. A., Kang, D. S., Wells, R. D., and England, P. T. (1986) *J. Biol. Chem.* **261**, 11302–11309

152. Koo, H.-S., Wu, H.-M., and Crothers, D. M. (1986) *Nature (London)* **320**, 501–506

153. Zahn, K., and Blattner, F. B. (1987) *Science* **236**, 416–422

154. Snyder, M., Buchman, A. R., and Davis, R. W. (1986) *Nature (London)* **324**, 87–89

155. Travers, A. A. (1987) *Trends Biochem. Sci.* **12**, 108–112

156. Grzeskowiak, K., Goodsell, D. S., Kaczor-Grzeskowiak, M., Cascio, D., and Dickerson, R. E. (1993) *Biochemistry* **32**, 8923–8931

157. Allewell, N. (1988) *Trends Biochem. Sci.* **13**, 193–195

158. Haran, T. E., Kahn, J. D., and Crothers, D. M. (1994) *J. Mol. Biol.* **244**, 135–143

159. Strauss, J. K., and Maher, L. J., III. (1994) *Science* **266**, 1829–1834

160. Olson, W. K., Marky, N. L., Jernigan, R. L., and Zhurkin, V. B. (1993) *J. Mol. Biol.* **232**, 530–554

161. Price, M. A., and Tullius, T. D. (1993) *Biochemistry* **32**, 127–136

162. DiGabriele, A. D., and Steitz, T. A. (1993) *J. Mol. Biol.* **231**, 1024–1039

163. Chuprina, V. P., Fedoroff, O. Y., and Reid, B. R. (1991) *Biochemistry* **30**, 561–568

164. Crothers, D. M. (1994) *Science* **266**, 1819–1820

165. Husain, I., Griffith, J., and Sancar, A. (1988) *Proc. Natl. Acad. Sci. U.S.A.* **85**, 2558–2562

166. Srinivasan, A. R., Torres, R., Clark, W., and Olson, W. K. (1987) *J. Biomol. Struct. Dyn.* **5**, 459–496

167. Lutter, L. C., Halvorson, H. R., and Calladine, C. R. (1996) *J. Mol. Biol.* **261**, 620–633

168. Privé, G. G., Heinemann, V., Chandrasegaran, S., Kan, L.-S., Kopka, M. L., and Dickerson, R. E. (1987) *Science* **238**, 498–504

169. Lilley, D. M. J. (1995) *Proc. Natl. Acad. Sci. U.S.A.* **92**, 7140–7142

170. Woodson, S. A., and Crothers, D. M. (1988) *Biochemistry* **27**, 3130–3141

171. Miller, M., Harrison, R. W., Wlodawer, A., Appella, E., and Sussman, J. L. (1988) *Nature (London)* **334**, 85–86

172. Hunter, W. N., Brown, T., Kneale, G., Anand, N. N., Rabinovich, D., and Kennard, O. (1987) *J. Biol. Chem.* **262**, 9962–9970

173. Lane, A., Martin, S. R., Ebel, S., and Brown, T. (1992) *Biochemistry* **31**, 12087–12095

174. Li, Y., and Agrawal, S. (1995) *Biochemistry* **34**, 10056–10062

175. Manning, G. S. (1978) *Q. Rev. Biophys.* **11**, 179–246

176. Mascotti, D. P., and Lohman, T. M. (1990) *Proc. Natl. Acad. Sci. U.S.A.* **87**, 3142–3146

177. Li, A. Z., Huang, H., Re, X., Qi, L. J., and Marx, K. A. (1998) *Biophys. J.* **74**, 964–973

178. Frydman, L., Rossomando, P. C., Frydman, V., Fernandez, C. O., Frydman, B., and Samejima, K. (1992) *Proc. Natl. Acad. Sci. U.S.A.* **89**, 9186–9190

179. Jain, S., Zon, G., and Sundaralingam, M. (1989) *Biochemistry* **28**, 2360–2364

180. Pelta, J., Durand, D., Doucet, J., and Livolant, F. (1996) *Biophys. J.* **71**, 48–63

181. Rajski, S. R., Kumar, S., Roberts, R. J., and Barton, J. K. (1999) *J. Am. Chem. Soc.* **121**, 5615–5616

181a. Aich, P., Labiuk, S. L., Tari, L. W., Dalbaere, L. J. T., Roesler, W. J., Falk, K. J., Steer, R. P., and Lee, J. S. (1999) *J. Mol. Biol.* **294**, 477–485

182. Fink, H.-W., and Schönenberger, C. (1999) *Nature (London)* **398**, 407–410

183. Marini, J. C., Miller, K. G., and Englund, P. T. (1980) *J. Biol. Chem.* **255**, 4976–4979

184. Wasserman, S. A., Dungan, J. M., and Cozzarelli, N. R. (1985) *Science* **229**, 171–174

185. Du, S. M., Wang, H., Tse-Dinh, Y.-C., and Seeman, N. C. (1995) *Biochemistry* **34**, 673–682

186. Du, S. M., Stollar, B. D., and Seeman, N. C. (1995) *J. Am. Chem. Soc.* **117**, 1194–1200

187. Wasserman, S. A., and Cozzarelli, N. R. (1986) *Science* **232**, 951–960

188. White, J. H., and Cozzarelli, N. R. (1984) *Proc. Natl. Acad. Sci. U.S.A.* **81**, 3322–3326

189. Cipra, B. (1992) *Science* **255**, 403

190. Adams, C. C. (1994) *The Knot Book*, Freeman, New York

191. Krasnow, M. A., Stasiak, A., Spengler, S. J., Dean, F., Koller, T., and Cozzarelli, N. R. (1983) *Nature (London)* **304**, 559–560

192. Du, S. M., and Seeman, N. C. (1992) *J. Am. Chem. Soc.* **114**, 9652–9655

193. Wang, J. C. (1980) *Trends Biochem. Sci.* **5**, 219–221

194. Cantor, C. R., and Schimmel, P. R. (1980) *Biophysical Chemistry, Part 3*, Freeman, San Francisco, California (pp. 1265–1290)

195. Lebowitz, J. (1990) *Trends Biochem. Sci.* **15**, 202–207

196. Bates, A. D., and Maxwell, A. (1993) *DNA Topology*, Oxford Univ. Press, London

197. Yang, Y., Westcott, T. P., Pedersen, S. C., Tobias, I., and Olson, W. K. (1995) *Trends Biochem. Sci.* **20**, 313–319

198. Vinograd, J., Lebowitz, J., and Watson, R. (1968) *J. Mol. Biol.* **33**, 173–197

199. White, J. H., Cozzarelli, N. R., and Bauer, W. R. (1988) *Science* **241**, 323–327

199a. Vologodskii, A. V., and Cozzarelli, N. R. (1994) *Annu. Rev. Biophys. Biomol. Struct.* **23**, 609–643

199b. Cozzarelli, N. R., Boles, T. C., and White, J. H. (1990) in *DNA Topology and its Biological Effects*, pp. 139–184, Cold Spring Harbor Lab. Press, Cold Spring Harbor, New York

199c. Podtelezhnikov, A. A., Cozzarelli, N. R., and Vologodskii, A. V. (1999) *Proc. Natl. Acad. Sci. U.S.A.* **96**, 12974–12979

200. Reich, Z., Levin-Zaidman, S., Gutman, S. B., Arad, T., and Minsky, A. (1994) *Biochemistry* **33**, 14177–14184

201. Eriksson, M. A. L., Härd, T., and Nilsson, L. (1995) *Biophys. J.* **68**, 402–426

202. Keller, W. (1975) *Proc. Natl. Acad. Sci. U.S.A.* **72**, 2550–2554

203. Kornberg, R. D., and Klug, A. (1981) *Sci. Am.* **244**(Feb), 52–64

204. Wang, B.-C., Rose, J., Arents, G., and Moudrianakis, E. N. (1994) *J. Mol. Biol.* **236**, 179–188

References

205. Malik, M., Bensaid, A., Rouviere-Yaniv, J., and Drlica, K. (1996) *J. Mol. Biol.* **256**, 66–76

206. Serrano, M., Salas, M., and Hermoso, J. M. (1993) *Trends Biochem. Sci.* **18**, 202–206

207. Seidl, A., and Hinz, H.-J. (1984) *Proc. Natl. Acad. Sci. U.S.A.* **81**, 1312–1316

208. Rahmouni, A. R., and Wells, R. D. (1989) *Science* **246**, 358–363

209. Spolar, R. S., and Record, M. T., Jr. (1994) *Science* **263**, 777–784

210. White, J. H., Gallo, R. M., and Bauer, W. R. (1992) *Trends Biochem. Sci.* **17**, 7–12

211. van Holde, K., and Zlatanova, J. (1995) *J. Biol. Chem.* **270**, 8373–8376

212. Keller, W. (1975) *Proc. Natl. Acad. Sci. U.S.A.* **72**, 4876–4880

213. Hamiche, A., Schultz, P., Ramakrishnan, V., Oudet, P., and Prunell, A. (1996) *J. Mol. Biol.* **257**, 30–42

214. Alfonso, P. J., Crippa, M. P., Hayes, J. J., and Bustin, M. (1994) *J. Mol. Biol.* **236**, 189–198

215. Bednar, J., Furrer, P., Stasiak, A., Dubochet, J., Egelman, E. H., and Bates, A. D. (1994) *J. Mol. Biol.* **235**, 825–847

216. Kumar, A. (1995) *Biochemistry* **34**, 12921–12925

217. Stigter, D. (1995) *Biophys. J.* **69**, 380–388

218. Baeza, I., Gariglio, P., Rangel, L. M., Chavez, P., Cervantes, L., Arguello, C., Wong, C., and Montañez, C. (1987) *Biochemistry* **26**, 6387–6392

219. Schmid, M. B. (1988) *Trends Biochem. Sci.* **13**, 131–135

220. Quigley, G. J., Wang, A. H.-J., Ughetto, G., van der Marel, G., van Boom, J. H., and Rich, A. (1980) *Proc. Natl. Acad. Sci. U.S.A.* **77**, 7204–7208

221. Campbell, A. M., and Jolly, D. J. (1973) *Biochem. J.* **133**, 209–226

222. Depew, R. E., and Wang, J. C. (1975) *Proc. Natl. Acad. Sci. U.S.A.* **72**, 4275–4279

223. Westof, E., and Sundaralingam, M. (1980) *Proc. Natl. Acad. Sci. U.S.A.* **77**, 1852–1856

224. Sobell, H. M. (1974) *Sci. Am.* **231**(Aug), 82–91

225. Chen, H., and Patel, D. J. (1995) *J. Am. Chem. Soc.* **117**, 5901–5913

226. Chaires, J. B., Satyanarayana, S., Suh, D., Fokt, I., Przewloka, T., and Priebe, W. (1996) *Biochemistry* **35**, 2047–2053

227. Rajeswari, M. R., Bose, H. S., Kukreti, S., Gupta, A., Chauhan, V. S., and Roy, K. B. (1992) *Biochemistry* **31**, 6237–6241

227a. Leitner, D., Schröder, W., and Weisz, K. (2000) *Biochemistry* **39**, 5886–5892

228. Felsenfeld, G., Davies, D., and Rich, A. (1957) *J. Am. Chem. Soc.* **79**, 2023–2024

229. Felsenfeld, G., and Miles, H. T. (1967) *Ann. Rev. Biochem.* **36**, 407–448

230. Arnott, S., and Selsing, E. (1974) *J. Mol. Biol.* **88**, 509–521

231. Wang, E., Koshlap, K. M., Gillespie, P., Dervan, P. B., and Feigon, J. (1996) *J. Mol. Biol.* **257**, 1052–1069

232. Radhakrishnan, I., and Patel, D. J. (1994) *Biochemistry* **33**, 11405–11416

233. Radhakrishnan, I., and Patel, D. J. (1994) *J. Mol. Biol.* **241**, 600–619

234. Rhee, S., Han, Z.-j, Liu, K., Miles, H. T., and Davies, D. R. (1999) *Biochemistry* **38**, 16810–16815

235. Roberts, R. W., and Crothers, D. M. (1996) *Proc. Natl. Acad. Sci. U.S.A.* **93**, 4320–4325

236. van Dongen, M. J. P., Heus, H. A., Wymenga, S. S., van der Marel, G. A., van Boom, J. H., and Hilbers, C. W. (1996) *Biochemistry* **35**, 1733–1739

237. Klysik, J. (1995) *J. Mol. Biol.* **245**, 499–507

238. Srinivasan, A. R., and Olson, W. K. (1998) *J. Am. Chem. Soc.* **120**, 484–491

239. Hanvey, J. C., Shimizu, M., and Wells, R. D. (1988) *Proc. Natl. Acad. Sci. U.S.A.* **85**, 6292–6296

240. Htun, H., and Dahlberg, J. E. (1989) *Science* **243**, 1571–1576

241. Mirkin, S. M., and Frank-Kamenetskii, M. D. (1994) *Annu. Rev. Biophys. Biomol. Struct.* **23**, 541–576

242. Singh, S., Patel, P. K., and Hosur, R. V. (1997) *Biochemistry* **36**, 13214–13222

243. Van Meervelt, L., Vlieghe, D., Dautant, A., Gallois, B., Précigoux, G., and Kennard, O. (1995) *Nature (London)* **374**, 742–744

244. Reddy, G., Jwang, B., Rao, B. J., and Radding, C. M. (1994) *Biochemistry* **33**, 11486–11492

245. Kim, M. G., Zhurkin, V. B., Jernigan, R. L., and Camerini-Otero, R. D. (1995) *J. Mol. Biol.* **247**, 874–889

246. Cox, M. M. (1995) *J. Biol. Chem.* **270**, 26021–26024

247. Jain, S. K., Cox, M. M., and Inman, R. B. (1995) *J. Biol. Chem.* **270**, 4943–4949

248. Best, G. C., and Dervan, P. B. (1995) *J. Am. Chem. Soc.* **117**, 1187–1193

249. Greenberg, W. A., and Dervan, P. B. (1995) *J. Am. Chem. Soc.* **117**, 5016–5022

250. Hashem, G. M., Pham, L. P., Vaughan, M. R., and Gray, D. M. (1998) *Biochemistry* **37**, 61–72

251. Gryaznov, S. M., Lloyd, D. H., Chen, J.-K., Schultz, R. G., DeDionisio, L. A., Ratmeyer, L., and Wilson, W. D. (1995) *Proc. Natl. Acad. Sci. U.S.A.* **92**, 5798–5802

252. Betts, L., Josey, J. A., Veal, J. M., and Jordan, S. R. (1995) *Science* **270**, 1838–1841

253. Tomac, S., Sarkar, M., Ratilainen, T., Wittung, P., Nielsen, P. E., Nordén, B., and Gräslund, A. (1996) *J. Am. Chem. Soc.* **118**, 5544–5552

253a. Ray, A., and Nordén, B. (2000) *FASEB J.* **14**, 1041–1060

253b. Izvolsky, K. I., Demidov, V. V., Nielsen, P. E., and Frank-Kamenetskii, M. D. (2000) *Biochemistry* **39**, 10908–10913

254. White, S., Szewczyk, J. W., Turner, J. M., Baird, E. E., and Dervan, P. B. (1998) *Nature (London)* **391**, 468–471

255. Marsh, T. C., and Henderson, E. (1994) *Biochemistry* **33**, 10718–10724

256. Blackburn, E. H. (1991) *Nature (London)* **350**, 569–573

257. Blackburn, E. H. (1991) *Trends Biochem. Sci.* **16**, 378–381

258. Moyzis, R. K. (1991) *Sci. Am.* **265**(Aug), 48–55

259. Henderson, E., Hardin, C. C., Wolk, S. K., Tinoco, I., Jr., and Blackburn, E. (1987) *Cell* **51**, 899–908

260. Laughlan, G., Murchie, A. I. H., Norman, D. G., Moore, M. H., Moody, P. C. E., Lilley, D. M. J., and Luisi, B. (1994) *Science* **265**, 520–527

261. Kang, C. H., Zhang, X., Ratliff, R., Moyzis, R., and Rich, A. (1992) *Nature (London)* **356**, 126–131

262. Smith, F. W., and Feigon, J. (1993) *Biochemistry* **32**, 8682–8692

263. Balagurumoorthy, P., and Brahmachari, S. K. (1994) *J. Biol. Chem.* **269**, 21858–21869

264. Williamson, J. R. (1994) *Annu. Rev. Biophys. Biomol. Struct.* **23**, 703–730

265. Frank-Kamenetskii, M. (1992) *Nature (London)* **356**, 105

266. Wang, K. Y., Swaminathan, S., and Bolton, P. H. (1994) *Biochemistry* **33**, 7517–7527

267. Wang, Y., and Patel, D. J. (1995) *J. Mol. Biol.* **251**, 76–94

268. Laporte, L., and Thomas, G. J., Jr. (1998) *Biochemistry* **37**, 1327–1335

269. Ross, W. S., and Hardin, C. C. (1994) *J. Am. Chem. Soc.* **116**, 6070–6080

270. Sen, D., and Gilbert, W. (1990) *Nature (London)* **344**, 410–414

271. Miura, T., Benevides, J. M., and Thomas, G. J., Jr. (1995) *J. Mol. Biol.* **248**, 233–238

272. Schierer, T., and Henderson, E. (1994) *Biochemistry* **33**, 2240–2246

273. Hardin, C. C., Corregan, M., Brown, B. A., II, and Frederick, L. N. (1993) *Biochemistry* **32**, 5870–5880

274. Cheong, C., and Moore, P. B. (1992) *Biochemistry* **31**, 8406–8414

275. Chen, F.-M. (1995) *J. Biol. Chem.* **270**, 23090–23096

276. Kettani, A., Kumar, R. A., and Patel, D. J. (1995) *J. Mol. Biol.* **254**, 638–656

277. Kang, C. H., Berger, I., Lockshin, C., Ratliff, R., Moyzis, R., and Rich, A. (1994) *Proc. Natl. Acad. Sci. U.S.A.* **91**, 11636–11640

278. Gehring, K., Leroy, J.-L., and Guéron, M. (1993) *Nature (London)* **363**, 561–565

279. Lacroix, L., Mergny, J.-L., Leroy, J.-L., and Hélène, C. (1996) *Biochemistry* **35**, 8715–8722

280. Chen, L., Cai, L., Zhang, X., and Rich, A. (1994) *Biochemistry* **33**, 13540–13546

281. Kang, C. H., Berger, I., Lockshin, C., Ratliff, R., Moyzis, R., and Rich, A. (1995) *Proc. Natl. Acad. Sci. U.S.A.* **92**, 3874–3878

282. Benevides, J. M., Kang, C., and Thomas, G. J., Jr. (1996) *Biochemistry* **35**, 5747–5755

283. Gallego, J., Chou, S.-H., and Reid, B. R. (1997) *J. Mol. Biol.* **273**, 840–856

284. Selsing, E., Wells, R. D., Alden, C. J., and Arnott, S. (1979) *J. Biol. Chem.* **254**, 5417–5422

285. Nishizaki, T., Iwai, S., Ohkubo, T., Kojima, C., Nakamura, H., Kyogoku, Y., and Ohtsuka, E. (1996) *Biochemistry* **35**, 4016–4025

286. Salazar, M., Champoux, J. J., and Reid, B. R. (1993) *Biochemistry* **32**, 739–744

287. Seeman, N. C., and Kallenbach, N. R. (1994) *Annu. Rev. Biophys. Biomol. Struct.* **23**, 53–86

288. Zhang, S., Fu, T.-J., and Seeman, N. C. (1993) *Biochemistry* **32**, 8062–8067

289. Pikkemaat, J. A., van den Elst, H., van Boom, J. H., and Altona, C. (1994) *Biochemistry* **33**, 14896–14907

290. Carlström, G., and Chazin, W. J. (1996) *Biochemistry* **35**, 3534–3544

291. Holliday, R. (1964) *Genet. Res. Camb.* **5**, 282–304

292. Bennett, R. J., and West, S. C. (1995) *J. Mol. Biol.* **252**, 213–226

293. Grainger, R. J., Murchie, A. I. H., and Lilley, D. M. J. (1998) *Biochemistry* **37**, 23–32

294. Lilley, D. M. J. (1997) *Proc. Natl. Acad. Sci. U.S.A.* **94**, 9513–9515

295. Panyutin, I. G., Biswas, I., and Hsieh, P. (1995) *EMBO J.* **14**, 1819–1826

296. Chen, J., and Seeman, N. C. (1991) *Nature (London)* **350**, 631–633

297. Zhang, Y., and Seeman, N. C. (1994) *J. Am. Chem. Soc.* **116**, 1661–1669

297a. Liu, F., Sha, R., and Seeman, N. C. (1999) *J. Am. Chem. Soc.* **121**, 917–922

297b. Fahlman, R. P., and Sen, D. (1999) *J. Am. Chem. Soc.* **121**, 11079–11085

298. Feagin, J. E. (1990) *J. Biol. Chem.* **265**, 19373–19376

299. Stuart, K. (1991) *Trends Biochem. Sci.* **16**, 68–72

300. Simons, R. W., and Grunberg-Manago, M., eds. (1997) *RNA Structure and Function*, Cold Spring Harbor Lab. Press, Cold Spring Harbor, New York

301. Nicoloso, M., Qu, L.-H., Michot, B., and Bachellerie, J.-P. (1996) *J. Mol. Biol.* **260**, 178–195

302. Gold, L., Polisky, B., Uhlenbeck, O., and Yarus, M. (1995) *Ann. Rev. Biochem.* **64**, 763–797

303. Draper, D. E. (1996) *Trends Biochem. Sci.* **21**, 145–149

304. Sich, C., Ohlenschläger, O., Ramachandran, R., Görlach, M., and Brown, L. R. (1997) *Biochemistry* **36**, 13989–14002

References

305. Egli, M., Portmann, S., and Usman, N. (1996) *Biochemistry* **35**, 8489–8494

306. Auffinger, P., and Westhof, E. (1997) *J. Mol. Biol.* **274**, 54–63

307. Salazar, M., Fedoroff, O. Y., Miller, J. M., Ribeiro, N. S., and Reid, B. R. (1993) *Biochemistry* **32**, 4207–4215

308. Fedoroff, O. Y., Salazar, M., and Reid, B. R. (1993) *J. Mol. Biol.* **233**, 509–523

309. Ratmeyer, L., Vinayak, R., Zhong, Y. Y., Zon, G., and Wilson, W. D. (1994) *Biochemistry* **33**, 5298–5304

310. Varani, G., Cheong, C., and Tinoco, I., Jr. (1991) *Biochemistry* **30**, 3280–3289

311. Butcher, S. E., Dieckmann, T., and Feigon, J. (1997) *J. Mol. Biol.* **268**, 348–358

312. van Dongen, M. J. P., Wijmenga, S. S., van der Marel, G. A., van Boom, J. H., and Hilbers, C. W. (1996) *J. Mol. Biol.* **263**, 715–729

313. SantaLucia, J., Jr., Kierzek, R., and Turner, D. H. (1992) *Science* **256**, 217–219

314. Murphy, F. L., and Cech, T. R. (1994) *J. Mol. Biol.* **236**, 49–63

315. Jucker, F. M., Heus, H. A., Yip, P. F., Moors, E. H. M., and Pardi, A. (1996) *J. Mol. Biol.* **264**, 968–980

315a. Gutell, R. R., Cannone, J. J., Konings, D., and Gautheret, D. (2000) *J. Mol. Biol.* **300**, 791–803

316. Fountain, M. A., Serra, M. J., Krugh, T. R., and Turner, D. H. (1996) *Biochemistry* **35**, 6539–6548

317. Huang, S., Wang, Y.-X., and Draper, D. E. (1996) *J. Mol. Biol.* **258**, 308–321

318. Nowakowski, J., and Tinoco, I., Jr. (1996) *Biochemistry* **35**, 2577–2585

319. Chang, K.-Y., and Tinoco, I., Jr. (1997) *J. Mol. Biol.* **269**, 52–66

320. Rosendahl, G., Hansen, L. H., and Douthwaite, S. (1995) *J. Mol. Biol.* **249**, 59–68

321. Pleij, C. W. A. (1990) *Trends Biochem. Sci.* **15**, 143–147

322. ten Dam, E., Pleij, K., and Draper, D. (1992) *Biochemistry* **31**, 11665–11676

323. Shen, L. X., and Tinoco, I., Jr. (1995) *J. Mol. Biol.* **247**, 963–978

324. Pinard, R., Payant, C., and Brakier-Gingras, L. (1995) *Biochemistry* **34**, 9611–9616

325. Puglisi, J. D., Wyatt, J. R., and Tinoco, I., Jr. (1988) *Nature (London)* **331**, 283–286

326. Qiu, H., Kaluarachchi, K., Du, Z., Hoffman, D. W., and Giedroc, D. P. (1996) *Biochemistry* **35**, 4176–4186

326a. Kolk, M. H., van der Graaf, M., Wijmenga, S. S., Pleij, C. W. A., Heus, H. A., and Hilbers, C. W. (1998) *Science* **280**, 434–438

327. Shi, P.-Y., Brinton, M. A., Veal, J. M., Zhong, Y. Y., and Wilson, W. D. (1996) *Biochemistry* **35**, 4222–4230

328. Hubbard, J. M., and Hearst, J. E. (1991) *Biochemistry* **30**, 5458–5465

329. Soll, D., and Abelson, J., eds. (1979) *Transfer RNA*, Cold Spring Harbor Lab. Press, Cold Spring Harbor, New York

330. Zimmermann, R. A. (1996) *Science* **271**, 1240–1241

331. Steinberg, S., Gautheret, D., and Cedergren, R. (1994) *J. Mol. Biol.* **236**, 982–989

332. Holbrook, S. R., Sussman, J. L., Warrant, R. W., and Kim, S.-H. (1978) *J. Mol. Biol.* **123**, 631–660

333. Rich, A., and Kim, S. H. (1978) *Sci. Am.* **238** (Jan), 52–62

334. Rich, A. (1978) *Trends Biochem. Sci.* **3**, 34–37

335. Moras, D., Comarmond, M. B., Fischer, J., Weiss, R., Thierry, J. C., Ebel, J. P., and Giegé, R. (1980) *Nature (London)* **288**, 669–674

336. Woo, N. H., Roe, B. A., and Rich, A. (1980) *Nature (London)* **286**, 346–351

337. Wakao, H., Romby, P., Westhof, E., Laalami, S., Grunberg-Manago, M., Ebel, J.-P., Ehresmann, C., and Ehresmann, B. (1989) *J. Biol. Chem.* **264**, 20363–20371

338. Hou, Y.-M., and Schimmel, P. (1992) *Biochemistry* **31**, 4157–4160

339. Leehey, M. A., Squassoni, C. A., Friederich, M. W., Mills, J. B., and Hagerman, P. J. (1995) *Biochemistry* **34**, 16235–16239

340. Quigley, G. J., and Rich, A. (1976) *Science* **194**, 796–806

341. Brimacombe, R., Atmadja, J., Stiege, W., and Schüler, D. (1988) *J. Mol. Biol.* **199**, 115–136

342. Mueller, F., and Brimacombe, R. (1997) *J. Mol. Biol.* **271**, 524–544

343. Turner, D. H., and Sugimoto, N. (1988) *Ann. Rev. Biophys. Biophys. Chem.* **17**, 167–192

344. Brimacombe, R. (1988) *Biochemistry* **27**, 4207–4213

345. Schnare, M. N., Damberger, S. H., Gray, M. W., and Gutell, R. R. (1996) *J. Mol. Biol.* **256**, 701–719

346. Yonath, A. (1984) *Trends Biochem. Sci.* **9**, 227–230

347. Pennisi, E. (1999) *Science* **285**, 2048–2051

348. Svergun, D. I., Koch, M. H. J., Pedersen, J. S., and Serdyuk, I. N. (1994) *J. Mol. Biol.* **240**, 78–86

349. Vladimirov, S. N., Druzina, Z., Wang, R., and Cooperman, B. S. (2000) *Biochemistry* **39**, 183–193

350. Gregory, S. T., and Dahlberg, A. E. (1999) *J. Mol. Biol.* **285**, 1475–1483

351. Ban, N., Nissen, P., Hansen, J., Moore, P. B., and Steitz, T. A. (2000) *Science* **289**, 905–920

352. Cate, J. H., Yusupov, M. M., Yusupova, G. Z., Earnest, T. N., and Noller, H. F. (1999) *Science* **285**, 2095–2104

353. Culver, G. M., Cate, J. H., Yusupova, G. Z., Yusupov, M. M., and Noller, H. F. (1999) *Science* **285**, 2133–2135

354. Carter, A. P., Clemons, W. M., Brodersen, D. E., Morgan-Warren, R. J., Wimberly, B. T., and Ramakrishnan, V. (2000) *Nature (London)* **407**, 340–348

355. Nissen, P., Hansen, J., Ban, N., Moore, P. B., and Steitz, T. A. (2000) *Science* **289**, 920–930

356. Muth, G. W., Ortoleva-Donnelly, L., and Strobel, S. A. (2000) *Science* **289**, 947–950

357. Cech, T. R. (2000) *Science* **289**, 878–879

358. Pennisi, E. (1996) *Science* **273**, 574–575

359. O'Gara, M., Klimasauskas, S., Roberts, R. J., and Cheng, X. (1996) *J. Mol. Biol.* **261**, 634–645

360. Gommers-Ampt, J. H., and Borst, P. (1995) *FASEB J.* **9**, 1034–1042

361. Maltman, K. L., Neuhard, J., and Warren, R. A. J. (1981) *Biochemistry* **20**, 3586–3591

362. Lipscomb, L. A., Peek, M. E., Morningstar, M. L., Verghis, S. M., Miller, E. M., Rich, A., Essigmann, J. M., and Williams, L. D. (1995) *Proc. Natl. Acad. Sci. U.S.A.* **92**, 719–723

363. Leonard, G. A., McAuley-Hecht, K. E., Gibson, N. J., Brown, T., Watson, W. P., and Hunter, W. N. (1994) *Biochemistry* **33**, 4755–4761

364. Wrzesinski, J., Bakin, A., Nurse, K., Lane, B. G., and Ofengand, J. (1995) *Biochemistry* **34**, 8904–8913

365. Ellington, A. D., and Szostak, J. W. (1990) *Nature (London)* **346**, 818–822

366. Fan, P., Suri, A. K., Fiala, R., Live, D., and Patel, D. J. (1996) *J. Mol. Biol.* **258**, 480–500

367. Jenison, R. D., Gill, S. C., Pardi, A., and Polisky, B. (1994) *Science* **263**, 1425–1429

368. Hamasaki, K., Killian, J., Cho, J., and Rando, R. R. (1998) *Biochemistry* **37**, 656–663

369. Connell, G. J., Illangesekare, M., and Yarus, M. (1993) *Biochemistry* **32**, 5497–5502

370. Patel, D. J., Suri, A. K., Jiang, F., Jiang, L., Fan, P., Kumar, R. A., and Nonin, S. (1997) *J. Mol. Biol.* **272**, 645–664

371. Crick, F. H. C. (1966) *Cold Spring Harbor Symp. Quant. Biol.* **31**, 3–9

372. Cornish-Bowden, A. (1996) *Trends Biochem. Sci.* **21**, 155

373. Hall, B. D. (1979) *Nature (London)* **282**, 129–130

374. Barrell, B. G., Bankier, A. T., and Drouin, J. (1979) *Nature (London)* **282**, 189–194

375. Horowitz, S., and Gorovsky, M. A. (1985) *Proc. Natl. Acad. Sci. U.S.A.* **82**, 2452–2455

376. Chambers, I., and Harrison, P. R. (1987) *Trends Biochem. Sci.* **12**, 255–256

377. Gardner, M. (1970) *Sci. Am.* **223**(Aug), 110–112

378. Mirkin, S. M., Lyamichev, V. I., Drushlyak, K. N., Dobrynin, V. N., Filippov, S. A., and Frank-Kamenetskii, M. D. (1987) *Nature (London)* **330**, 495–497

379. Guieysse, A.-L., Praseuth, D., and Hélène, C. (1997) *J. Mol. Biol.* **267**, 289–298

380. Panayotatos, N., and Fontaine, A. (1987) *J. Biol. Chem.* **262**, 11364–11368

381. Pabo, C. O., and Sauer, R. T. (1984) *Ann. Rev. Biochem.* **53**, 293–321

382. Berg, O. G., and von Hippel, P. H. (1988) *Trends Biochem. Sci.* **13**, 207–211

383. Marmorstein, R. Q., Joachimiak, A., Sprinzl, M., and Sigler, P. B. (1987) *J. Biol. Chem.* **262**, 4922–4927

384. Muto, A., and Osawa, S. (1987) *Proc. Natl. Acad. Sci. U.S.A.* **84**, 166–169

385. Singer, C. E., and Ames, B. N. (1979) *Science* **170**, 822–826

386. Singer, C. E., and Ames, B. N. (1972) *Science* **175**, 1391

387. Harrison, S. C. (1991) *Nature (London)* **353**, 715–719

388. Churchill, M. E. A., and Travers, A. A. (1991) *Trends Biochem. Sci.* **16**, 92–97

389. Travers, A. (1993) *DNA-Protein Interactions*, Chapman & Hall, New York

390. Jones, S., van Heyningen, P., Berman, H. M., and Thornton, J. M. (1999) *J. Mol. Biol.* **287**, 877–896

391. Arents, G., and Moudrianakis, E. N. (1995) *Proc. Natl. Acad. Sci. U.S.A.* **92**, 11170–11174

392. Starich, M. R., Sandman, K., Reeve, J. N., and Summers, M. F. (1996) *J. Mol. Biol.* **255**, 187–203

393. Hud, N. V., Milanovich, F. P., and Balhorn, R. (1994) *Biochemistry* **33**, 7528–7535

394. Lohman, T. M., and Ferrari, M. E. (1994) *Ann. Rev. Biochem.* **63**, 527–570

395. Overman, L. B., and Lohman, T. M. (1994) *J. Mol. Biol.* **236**, 165–178

396. Folmer, R. H. A., Nilges, M., Folkers, P. J. M., Konings, R. N. H., and Hilbers, C. W. (1994) *J. Mol. Biol.* **240**, 341–357

397. Olah, G. A., Gray, D. M., Gray, C. W., Kergil, D. L., Sosnick, T. R., Mark, B. L., Vaughan, M. R., and Trewhella, J. (1995) *J. Mol. Biol.* **249**, 576–594

398. Kanellopoulos, P. N., Tsernoglou, D., van der Vliet, P. C., and Tucker, P. A. (1996) *J. Mol. Biol.* **257**, 1–8

399. Bogdarina, I., Fox, D. G., and Kneale, G. G. (1998) *J. Mol. Biol.* **275**, 443–452

400. Pluta, A. F., Mackay, A. M., Ainsztein, A. M., Goldberg, I. G., and Earnshaw, W. C. (1995) *Science* **270**, 1591–1594

401. Schevitz, R. W., Otwinowski, Z., Joachimiak, A., Lawson, C. L., and Sigler, P. B. (1985) *Nature (London)* **317**, 782–786

402. Zhang, R.-G., Joachimiak, A., Lawson, C. L., Schevitz, R. W., Otwinowski, Z., and Sigler, P. B. (1987) *Nature (London)* **327**, 591–597

References

403. Otwinowski, Z., Schevitz, R. W., Zhang, R.-G., Lawson, C. L., Joachimiak, A., Marmorstein, R. Q., Luisi, B. F., and Sigler, P. B. (1988) *Nature (London)* **335**, 321–329

404. Shakked, Z., Guzikevich–Guerstein, G., Frolow, F., Rabinovich, D., Joachimiak, A., and Sigler, P. B. (1994) *Nature (London)* **368**, 469–473

405. Lawson, C. L., and Carey, J. (1993) *Nature (London)* **366**, 178–182

406. Lawson, C. L., and Sigler, P. B. (1988) *Nature (London)* **333**, 869–871

407. Guenot, J., Fletterick, R. J., and Kollman, P. A. (1994) *Protein Sci.* **3**, 1276–1285

408. Rastinejad, F., Artz, P., and Lu, P. (1993) *J. Mol. Biol.* **233**, 389–399

409. Markiewicz, P., Kleina, L. G., Cruz, C., Ehret, S., and Miller, J. H. (1994) *J. Mol. Biol.* **240**, 421–433

410. Lewis, M., Chang, G., Horton, N. C., Kercher, M. A., Pace, H. C., Schumacher, M. A., Brennan, R. G., and Lu, P. (1996) *Science* **271**, 1247–1254

411. Brennan, R. G., and Matthews, B. W. (1989) *J. Biol. Chem.* **264**, 1903–1906

412. Jordan, S. R., and Pabo, C. O. (1988) *Science* **242**, 893–899

413. Wolberger, C., Dong, Y., Ptashne, M., and Harrison, S. C. (1988) *Nature (London)* **335**, 789–795

414. Aggarwal, A. K., Rodgers, D. W., Drottar, M., Ptashne, M., and Harrison, S. C. (1988) *Science* **242**, 899–907

415. Bell, A. C., and Koudelka, G. B. (1993) *J. Mol. Biol.* **234**, 542–553

416. Mandel-Gutfreund, Y., Schueler, O., and Margalit, H. (1995) *J. Mol. Biol.* **253**, 370–382

417. Ferré-D'Amaré, A. R., Prendergast, G. C., Ziff, E. B., and Burley, S. K. (1993) *Nature (London)* **363**, 38–44

418. Junius, F. K., O'Donoghue, S. I., Nilges, M., Weiss, A. S., and King, G. F. (1996) *J. Biol. Chem.* **271**, 13663–13667

419. Keller, W., König, P., and Richmond, T. J. (1995) *J. Mol. Biol.* **254**, 657–667

420. Harbury, P. B., Kim, P. S., and Alber, T. (1994) *Nature (London)* **371**, 80–83

421. Klug, A., and Schwabe, J. W. R. (1995) *FASEB J.* **9**, 597–604

422. Rhodes, D., and Klug, A. (1993) *Sci. Am.* **268** (Feb), 56–65

423. Pavletich, N. P., and Pabo, C. O. (1991) *Science* **252**, 809–817

424. Mackay, J. P., and Crossley, M. (1998) *Trends Biochem. Sci.* **23**, 1–4

425. Miller, J., McLachlan, A. D., and Klug, A. (1985) *EMBO J.* **4**, 1609–1614

426. Pieler, T., and Theunissen, O. (1993) *Trends Biochem. Sci.* **18**, 226–230

427. Pavletich, N. P., and Pabo, C. O. (1993) *Science* **261**, 1701–1707

428. Krizek, B. A., Amann, B. T., Kilfoil, V. J., Merkle, D. L., and Berg, J. M. (1991) *J. Am. Chem. Soc.* **113**, 4518–4523

429. Somers, W. S., and Phillips, S. E. V. (1992) *Nature (London)* **359**, 387–393

430. Phillips, S. E. V. (1994) *Annu. Rev. Biophys. Biomol. Struct.* **23**, 671–701

431. Breg, J. N., van Opheusden, H. J., Burgering, M. J. M., Boelens, R., and Kaptein, R. (1990) *Nature (London)* **346**, 586–589

432. Bonnefoy, E., Takahashi, M., and Yaniv, J. R. (1994) *J. Mol. Biol.* **242**, 116–129

433. Müller, C. W., Rey, F. A., Sodeoka, M., Verdine, G. L., and Harrison, S. C. (1995) *Nature (London)* **373**, 311–317

434. Ghosh, G., Van Duyne, G., Ghosh, S., and Sigler, P. B. (1995) *Nature (London)* **373**, 303–310

435. Baltimore, D., and Beg, A. A. (1995) *Nature (London)* **373**, 287–288

436. Kaufmann, E., Müller, D., and Knöchel, W. (1995) *J. Mol. Biol.* **248**, 239–254

437. Burd, C. G., and Dreyfuss, G. (1994) *Science* **265**, 615–621

438. Nagai, K., and Mattaj, I. W., eds. (1994) *RNA-Protein Interactions*, IRL Press, Oxford

439. Draper, D. E. (1995) *Ann. Rev. Biochem.* **64**, 593–620

440. Hentze, M. W. (1994) *Trends Biochem. Sci.* **19**, 101–103

441. Allain, F. H.-T., Gubser, C. C., Howe, P. W. A., Nagai, K., Neuhaus, D., and Varani, G. (1996) *Nature (London)* **380**, 646–650

442. Kenan, D. J., Query, C. C., and Keene, J. D. (1991) *Trends Biochem. Sci.* **16**, 214–220

443. Fierro-Monti, I., and Mathews, M. B. (2000) *Trends Biochem. Sci.* **25**, 241–246

444. Voyles, B. A. (1993) *The Biology of Viruses*, Mosby, St. Louis, Missouri

445. Fields, B. N., and Knipe, D. M., eds. (1990) *Fields Virology*, 2nd ed., Raven Press, New York

446. Cann, A. (1993) *Principles of Molecular Virology*, Academic Press, San Diego, California

447. Endemann, H., and Model, P. (1995) *J. Mol. Biol.* **250**, 496–506

448. Williams, K. A., Glibowicka, M., Li, Z., Li, H., Khan, A. R., Chen, Y. M. Y., Wang, J., Marvin, D. A., and Deber, C. M. (1995) *J. Mol. Biol.* **252**, 6–14

449. Wen, Z. Q., Overman, S. A., and Thomas, G. J., Jr. (1997) *Biochemistry* **36**, 7810–7820

450. Sanger, F., Air, G. M., Barrell, B. G., Brown, N. L., Coulson, A. R., Fiddes, J. C., Hutchison, C. A., III, Slocombe, P. M., and Smith, M. (1977) *Nature (London)* **265**, 687–695

451. McKenna, R., Ilag, L. L., and Rossmann, M. G. (1994) *J. Mol. Biol.* **237**, 517–543

452. McKenna, R., Xia, D., Willingmann, P., Ilag, L. L., Krishnaswamy, S., Rossmann, M. G., Olson, N. H., Baker, T. S., and Incardona, N. L. (1992) *Nature (London)* **355**, 137–143

453. Godson, G. N. (1978) *Trends Biochem. Sci.* **3**, 249–253

454. McKenna, R., Bowman, B. R., Ilag, L. L., Rossmann, M. G., and Fane, B. A. (1996) *J. Mol. Biol.* **256**, 736–750

455. Wu, H., and Rossmann, M. G. (1993) *J. Mol. Biol.* **233**, 231–244

456. Tsao, J., Chapman, M. S., Agbandje, M., Keller, W., Smith, K., Wu, H., Luo, M., Smith, T. J., Rossmann, M. G., Compans, R. W., and Parrish, C. R. (1991) *Science* **251**, 1456–1464

457. Chipman, P. R., Agbandje-McKenna, M., Kajigaya, S., Brown, K. E., Young, N. S., Baker, T. S., and Rossmann, M. G. (1996) *Proc. Natl. Acad. Sci. U.S.A.* **93**, 7502–7506

458. Howarth, A. J., and Goodman, R. M. (1982) *Trends Biochem. Sci.* **7**, 180–182

459. Tiollais, P., Pourcel, C., and Dejean, A. (1985) *Nature (London)* **317**, 489–495

460. Tiollais, P., and Buendia, M.-A. (1991) *Sci. Am.* **264**(Apr), 116–123

461. Böttcher, B., Wynne, S. A., and Crowther, R. A. (1997) *Nature (London)* **386**, 88–91

462. Reddy, V. B., Thimmappaya, B., Dhar, R., Subramanian, K. N., Zain, B. S., Pan, J., Ghosh, P. K., Celma, M. L., and Weissman, S. M. (1978) *Science* **200**, 494–502

463. Liddington, R. C., Yan, Y., Moulai, J., Sahli, R., Benjamin, T. L., and Harrison, S. C. (1991) *Nature (London)* **354**, 278–284

464. Griffith, J. P., Griffith, D. L., Rayment, I., Murakami, W. T., and Caspar, D. L. D. (1992) *Nature (London)* **355**, 652–654

465. Yang, R. C. A., and Wu, R. (1979) *Science* **206**, 456–462

466. Howley, P. M. (1986) *N. Engl. J. Med.* **315**, 1089–1090

467. Hess, M., Cuzange, A., Ruigrok, R. W. H., Chroboczek, J., and Jacrot, B. (1995) *J. Mol. Biol.* **252**, 379–385

468. Athappilly, F. K., Murali, R., Rux, J. J., Cai, Z., and Burnett, R. M. (1994) *J. Mol. Biol.* **242**, 430–455

469. Witkowski, J. A. (1988) *Trends Biochem. Sci.* **13**, 110–113

470. Sugden, B. (1991) *Trends Biochem. Sci.* **16**, 45–46

471. Zhou, Z. H., Prasad, B. V. V., Jakana, J., Rixon, F. J., and Chiu, W. (1994) *J. Mol. Biol.* **242**, 456–469

472. Trus, B. L., Booy, F. P., Newcomb, W. W., Brown, J. C., Homa, F. L., Thomsen, D. R., and Steven, A. C. (1996) *J. Mol. Biol.* **263**, 447–462

473. Henle, W., Henle, G., and Lennette, E. T. (1974) *Sci. Am.* **241**(Jul), 48–59

474. Berardelli, P. (1997) *Science* **278**, 1710

475. Senkevich, T. G., Bugert, J. J., Sisler, J. R., Koonin, E. V., Darai, G., and Moss, B. (1996) *Science* **273**, 813–816

476. Hohn, T., Hohn, B., and Pfeiffer, P. (1985) *Trends Biochem. Sci.* **10**, 205–209

477. Rosa, P., Mantovani, S., Rosboch, R., and Huttner, W. B. (1992) *J. Biol. Chem.* **267**, 12227–12232

478. Diener, T. O. (1991) *FASEB J.* **5**, 2808–2813

479. Diener, T. V. (1984) *Trends Biochem. Sci.* **9**, 133–136

480. Diener, T. O. (1979) *Viroids and Viroid Diseases*, Wiley, New York

481. Martínez-Soriano, J. P., Galindo-Alonso, J., Maroon, C. J. M., Yucel, I., Smith, D. R., and Diener, T. O. (1996) *Proc. Natl. Acad. Sci. U.S.A.* **93**, 9397–9401

482. Polson, A. G., Bass, B. L., and Casey, J. L. (1996) *Nature (London)* **380**, 454–456

483. Wang, K.-S., Choo, Q.-L., Weiner, A. J., Ou, J.-H., Najarian, R. C., Thayer, R. M., Mullenbach, G. T., Denniston, K. J., Gerin, J. L., and Houghton, M. (1986) *Nature (London)* **323**, 508–514

484. Lai, M. M. C. (1995) *Ann. Rev. Biochem.* **64**, 259–286

485. Unge, T., Liljas, L., Strandberg, B., Vaara, I., Kannan, K. K., Fridborg, K., Nordman, C. E., and Lentz, P. J., Jr. (1980) *Nature (London)* **285**, 373–377

486. Larson, S. B., Koszelak, S., Day, J., Greenwood, A., Dodds, J. A., and McPherson, A. (1993) *Nature (London)* **361**, 179–182

487. Larson, S. B., Koszelak, S., Day, J., Greenwood, A., Dodds, J. A., and McPherson, A. (1993) *J. Mol. Biol.* **231**, 375–391

488. Golmohammadi, R., Valegård, K., Fridborg, K., and Liljas, L. (1993) *J. Mol. Biol.* **234**, 620–639

489. LeCuyer, K. A., Behlen, L. S., and Uhlenbeck, O. C. (1995) *Biochemistry* **34**, 10600–10606

490. Stonehouse, N. J., Valegård, K., Golmohammadi, R., van den Worm, S., Walton, C., Stockley, P. G., and Liljas, L. (1996) *J. Mol. Biol.* **256**, 330–339

491. Beremarand, M. N., and Blumenthal, T. (1979) *Cell* **18**, 257–266

492. Hogle, J. M., Chow, M., and Filman, D. J. (1985) *Science* **229**, 1358–1365

493. Hogle, J. M., Chow, M., and Filman, D. J. (1987) *Sci. Am.* **256**(Mar), 42–49

494. Najaran, R., Caput, D., Gee, W., Potter, S. J., Renard, A., Merryweather, J., Van Nest, G., and Dina, D. (1985) *Proc. Natl. Acad. Sci. U.S.A.* **82**, 2627–2631

495. Rossmann, M. G., Arnold, E., Griffith, J. P., Kamer, G., Luo, M., Smith, T. J., Vriend, G., Rueckert, R. R., Sherry, B., McKinlay, M. A., Diana, G., and Otto, M. (1987) *Trends Biochem. Sci.* **12**, 313–318

References

496. Olson, N. H., Kolatkar, P. R., Oliveira, M. A., Cheng, R. H., Greve, J. M., McClelland, A., Baker, T. S., and Rossmann, M. G. (1993) *Proc. Natl. Acad. Sci. U.S.A.* **90**, 507–511

497. Hadfield, A. T., Oliveira, M. A., Kim, K. H., Minor, I., Kremer, M. J., Heinz, B. A., Shepard, D., Pevear, D. C., Rueckert, R. R., and Rossmann, M. G. (1995) *J. Mol. Biol.* **253**, 61–73

498. Küpper, H., Keller, W., Kurz, C., Forss, S., Schaller, H., Franze, R., Stohmaier, K., Marquardt, O., Zaslavsky, V. G., and Hofschneider, P. H. (1981) *Nature (London)* **289**, 555–559

499. Brown, F. (1981) *Trends Biochem. Sci.* **6**, 325–327

500. Acharya, R., Fry, E., Stuart, D., Fox, G., Rowlands, D., and Brown, F. (1989) *Nature (London)* **337**, 709–716

501. Luo, M., Vriend, G., Kamer, G., Minor, I., Arnold, E., Rossmann, M. G., Boege, U., Scraba, D. G., Duke, G. M., and Palmenberg, A. C. (1987) *Science* **235**, 182–191

502. Choi, H.-K., Tong, L., Minor, W., Dumas, P., Boege, U., Rossmann, M. G., and Wengler, G. (1991) *Nature (London)* **354**, 37–43

503. Simons, K., Garoff, H., and Helenius, A. (1982) *Sci. Am.* **246**(Feb), 58–66

504. Harrison, S. C., Olson, A. J., Schutt, C. E., Winkler, F. K., and Bricogne, G. (1978) *Nature (London)* **276**, 368–373

505. Erickson, J. W., Silva, A. M., Murthy, M. R. N., Fita, I., and Rossmann, M. G. (1985) *Science* **229**, 625–629

506. Bloomer, A. C., Champness, J. N., Bricogne, G., Staden, R., and Klug, A. (1978) *Nature (London)* **276**, 362–368

507. Holmes, K. C. (1980) *Trends Biochem. Sci.* **5**, 4–7

507a. Lauffer, M. A. (1984) *Trends Biochem. Sci.* **9**, 369–371

508. Wang, H., and Stubbs, G. (1994) *J. Mol. Biol.* **239**, 371–384

509. Kaplan, M. M., and Webster, R. G. (1977) *Sci. Am.* **237**(Dec), 88–106

510. von Itzstein, M., and 17 other authors. (1993) *Nature (London)* **363**, 418–423

511. Clarke, D. K., Duarte, E. A., Elena, S. F., Moya, A., Domingo, E., and Holland, J. (1994) *Proc. Natl. Acad. Sci. U.S.A.* **91**, 4821–4824

512. Varmus, H. (1988) *Science* **240**, 1427–1434

513. Darlix, J.-L., Lapadat-Tapolsky, M., de Rocquigny, H., and Roques, B. P. (1995) *J. Mol. Biol.* **254**, 523–537

514. Löwer, R., Löwer, J., and Kurth, R. (1996) *Proc. Natl. Acad. Sci. U.S.A.* **93**, 5177–5184

515. Stephenson, J. R., ed. (1980) *Molecular Biology of RNA Tumor Viruses*, Academic Press, New York

516. Compans, R. W., and Bishop, D. H. L., eds. (1983) *Double-Stranded RNA Viruses*, Elsevier, Biomedical, New York

517. Thornton, A., and Zuckerman, A. J. (1975) *Nature (London)* **254**, 557–558

518. Baldwin, M. A., Cohen, F. E., and Prusiner, S. B. (1995) *J. Biol. Chem.* **270**, 19197–19200

519. Manuelidis, L., Sklaviadis, T., Akowitz, A., and Fritch, W. (1995) *Proc. Natl. Acad. Sci. U.S.A.* **92**, 5124–5128

520. Prusiner, S. B. (1996) *Trends Biochem. Sci.* **21**, 482–487

521. Prusiner, S. B. (1998) *Proc. Natl. Acad. Sci. U.S.A.* **95**, 13363–13383

522. Weissmann, C. (1999) *J. Biol. Chem.* **274**, 3–6

523. Chesebro, B. (1998) *Science* **286**, 660–662

524. Balter, M. (1999) *Science* **286**, 660–662

525. Temin, H. M. (1978) *Trends Biochem. Sci.* **3**, N80–N83

525a. Sambrook, J., and Russell, D. (2000) *Molecular Cloning: A Laboratory Manual*, 3rd ed., Cold Spring Harbor Lab. Press, Plainview, New York

526. Grossman, L., and Moldave, K., eds. (1967) *Methods in Enzymology*, Vol. 12A, Academic Press, New York (pp. 531–708)

527. Robyt, J. F., and White, B. J. (1996) *Biochemical Techniques, Theory and Practice*, 2nd ed., Waveland Press, Prospect Heights, Illinois

528. Blackburn, G. M., and Gait, M. J., eds. (1996) *Nucleic Acids in Chemistry and Biology*, 2nd ed., Oxford Univ. Press, Oxford

529. Rajeswari, M. R., Montenay-Garestier, T., and Helene, C. (1987) *Biochemistry* **26**, 6825–6831

530. Hecker, R., Colpan, M., and Riesner, D. (1985) *Journal of Chromatography* **326**, 251–261

531. Edwardson, P. A. D., Atkinson, T., Lowe, C. R., and Small, D. A. P. (1986) *Anal. Biochem.* **152**, 215–220

532. Lawrance, S. K., Smith, C. L., Srivastava, R., Cantor, C. R., and Weissman, S. M. (1987) *Science* **235**, 1387–1390

533. Cantor, C. R., Smith, C. L., and Matthew, M. K. (1988) *Ann. Rev. Biophys. Biophys. Chem.* **17**, 287–304

534. Smith, C. L., and Cantor, C. R. (1987) *Trends Biochem. Sci.* **12**, 284–287

535. Kochetkov, N. K., and Budovskii, E. I. (1972) *Organic Chemistry of Nucleic Acids*, Plenum, New York (Part A, pp. 137–147)

536. Bozic, D., Grazulis, S., Siksnys, V., and Huber, R. (1996) *J. Mol. Biol.* **255**, 176–186

537. Smith, H. O. (1979) *Science* **205**, 455–462

538. Air, G. M. (1979) *CRC Crit. Revs. Biochem.* **6**, 1–33

539. Roberts, R. J. (1980) *Methods Enzymol.* **65**, 1–15

540. Koob, M., and Szybalski, W. (1990) *Science* **250**, 271–273

541. Veselkov, A. G., Demidov, V. V., Frank-Kamenetskii, M. D., and Nielsen, P. E. (1996) *Nature (London)* **379**, 214

542. Zaug, A. J., Been, M. D., and Cech, T. R. (1986) *Nature (London)* **324**, 429–433

543. Uhlenbeck, O. C. (1987) *Nature (London)* **328**, 596–600

544. Landgraf, R., Chen, C.-hB., and Sigman, D. S. (1994) *Biochemistry* **33**, 10607–10615

545. Adams, R. L. P., Burdon, R. H., Campbell, A. M., and Smellie, R. M. S. (1976) *Davidson's The Biochemistry of the Nucleic Acids*, 8th ed., Academic Press, New York (pp. 50–82)

546. Rayford, R., Anthony, D. D., Jr., O'Neill, R. E., Jr., and Merrick, W. C. (1985) *J. Biol. Chem.* **260**, 15708–15713

547. Bochner, B. R., and Ames, B. N. (1982) *J. Biol. Chem.* **257**, 9759–9769

548. Chen, H., and Shaw, B. R. (1994) *Biochemistry* **33**, 4121–4129

549. Hayatsu, H., and Shiragami, M. (1979) *Biochemistry* **18**, 632–647

550. Spiro, T. G., ed. (1980) *Nucleic Acid-Metal Ion Interactions*, Wiley, New York

551. Dische, Z. (1955) *Nucleic Acids* **1**, 755

552. Marmur, J., and Doty, P. (1962) *J. Mol. Biol.* **5**, 109–118

553. Korolev, N., Lyubartsev, A. P., and Norden-skiöld, L. (1998) *Biophys. J.* **75**, 3041–3056

554. Davidson, J. N. (1972) *The Biochemistry of Nucleic Acids*, 7th ed., Academic Press, New York (p. 148)

555. Wada, A., Yabuki, S., and Husimi, Y. (1980) *Crit. Revs. Biochem.* **9**, 87–144

556. Britten, R. J., and Kohne, D. E. (1968) *Science* **161**, 529–540

557. Wilson, D. A., and Thomas, C. A., Jr. (1974) *J. Mol. Biol.* **84**, 115–144

558. Bendich, A. J., and Bolton, E. T. (1968) in *Methods in Enzymology*, Vol. 12B (Grossman, L., and Moldave, K., eds), pp. 635–640, Academic Press, New York

559. Gillespie, D. (1968) in *Methods in Enzymology*, Vol. 12B (Grossman, L., and Moldave, K., eds), pp. 641–668, Academic Press, New York

560. Hall, B. D., Haarr, L., and Kleppe, K. (1980) *Trends Biochem. Sci.* **5**, 254–256

561. Britten, R. J., and Kohne, D. E. (1967) *Carnegie Instit. Washington Yearbook*, 65th ed., Carnegie Instit., Washington, D. C. (pp. 78–106)

562. Marmur, J. (1994) *Trends Biochem. Sci.* **19**, 343–346

563. Szabo, P., and Ward, D. C. (1982) *Trends Biochem. Sci.* **7**, 425–427

564. Kessler, C., ed. (1992) *Nonradioactive Labeling and Detection of Biomolecules*, Springer-Verlag, New York

565. Leitch, A. R., Chwarzacher, T., Jackson, D., and Leitch, I. J. (1994) *In Situ Hybridization: a Practical Guide*, Bios, Oxford

566. Kumar, A., Tchen, P., Roullet, F., and Cohen, J. (1988) *Anal. Biochem.* **169**, 376–382

567. Isaac, P. G., ed. (1994) *Protocols for Nucleic Acid Analysis by Nonradioactive Probes*, Humana Press, Totowa, New Jersey

568. Gillevet, P. M. (1990) *Nature (London)* **348**, 657–658

569. Negro, F., Pacchioni, D., Shimizu, Y., Miller, R. H., Bussolati, G., Purcell, R. H., and Bonino, F. (1992) *Proc. Natl. Acad. Sci. U.S.A.* **89**, 2247–2251

570. Maniatis, T., Fritsch, E. F., and Sambrook, J. (1982) *Molecular Cloning. A Laboratory Manual*, Cold Spring Harbor Lab. Press, Cold Spring Harbor, New York

571. Gubler, U., and Hoffman, B. J. (1983) *Gene* **25**, 263–269

572. Watson, C. J., and Jackson, J. F. (1985) in *DNA Cloning Vol. I: A Practical Approach* (Glover, D. M., ed), pp. 79–100, IRL Press, Washington, DC

573. Ohtsuka, E., Matsuki, S., Ikehara, M., Takahashi, Y., and Matsubara, K. (1985) *J. Biol. Chem.* **260**, 2605–2608

574. Warner, B. D., Warner, M. E., Karus, G. A., Ku, L., Brown-Shimer, S., and Urdea, M. S. (1983) *DNA* **4**, 401–411

575. Sekiya, T., Takeya, T., Brown, E. L., Belagaje, R., Contreras, R., Fritz, H.-J., Gait, M. J., Lees, R. G., Ryan, M. J., Khorana, H. G., and Norris, K. E. (1979) *J. Biol. Chem.* **254**, 5787–5801

576. Wada, T., Sato, Y., Honda, F., Kawahara, S.-i, and Sekine, M. (1997) *J. Am. Chem. Soc.* **119**, 12710–12721

577. Matteucci, M. D., and Caruthers, M. H. (1981) *J. Am. Chem. Soc.* **103**, 3185–3191

578. Beaucage, S. L., and Caruthers, M. H. (1981) *Tetrahedron Letters* **22**, 1859–1862

579. Air, G. M. (1979) *CRC Crit. Revs. Biochem.* **6**, 1–33

580. Alvarado-Urbina, G., Sathe, G. M., Liu, W.-C., Gillen, M. F., Duck, P. D., Bender, R., and Ogilvie, K. K. (1981) *Science* **214**, 270–274

581. Caruthers, M. H. (1985) *Science* **230**, 281–285

582. Lashkari, D. A., Hunicke-Smith, S. P., Norgren, R. M., Davis, R. W., and Brennan, T. (1995) *Proc. Natl. Acad. Sci. U.S.A.* **92**, 7912–7915

583. Usman, N., Ogilvie, K. K., Jiang, M.-Y., and Cedergren, R. J. (1987) *J. Am. Chem. Soc.* **109**, 7845–7854

584. Ebe, K., Schöed, M., Rossi, J. J., and Wallace, R. B. (1987) *DNA* **6**, 497–504

585. Usman, N., and Cedergren, R. (1992) *Trends Biochem. Sci.* **17**, 334–339

586. Eaton, B. E., and Pieken, W. A. (1995) *Ann. Rev. Biochem.* **64**, 837–863

587. Rohatgi, R., Bartel, D. P., and Szostak, J. W. (1996) *J. Am. Chem. Soc.* **118**, 3340–3344

588. Kleppe, K., Ohtsuka, E., Kleppe, R., and Khorana, H. G. (1971) *J. Mol. Biol.* **56**, 341–361

589. Panet, A., and Khorana, H. G. (1974) *J. Biol. Chem.* **249**, 5213–5221

590. Saiki, R. K., Scharf, S., Faloona, F., Mullis, K. B., Horn, G. T., Erlich, H. A., and Arnheim, N. (1985) *Science* **230**, 1350–1354

591. Mullis, K. B. (1990) *Sci. Am.* **262**(Apr), 56–65

592. Rabinow, P. (1996) *Making PCR: A Story of Biotechnology*, Univ. Chicago Press, Chicago, Illinois

593. Mullis, K., Ferré, F., and Gibbs, R., eds. (1994) *The Polymerase Chain Reaction*, Birkhäuser, Boston, Massachusetts

594. Saiki, R. K., Gelfand, D. H., Stoffel, S., Scharf, S. J., Higuchi, R., Horn, G. T., Mullis, K. B., and Erlich, H. A. (1988) *Science* **239**, 487–491

595. Erlich, H. A., Gelfand, D. H., and Saiki, R. K. (1988) *Nature (London)* **331**, 461–462

596. Appenzeller, T. (1990) *Science* **247**, 1030–1032

597. Newton, C., and Graham, A. (1997) *PCR*, 2nd ed., Springer-Verlag, New York

598. Bloch, W. (1991) *Biochemistry* **30**, 2735–2747

599. White, B. A., ed. (1993) *PCR Protocols*, Humana Press, Totowa, New Jersey

599a. Innis, M. A., Gelfand, D. H., and Sninsky, J. J., eds. (1999) *PCR Applications*, Academic Press, San Diego, California

600. Cohen, J. (1994) *Science* **263**, 1564–1565

601. Pääbo, S. (1993) *Sci. Am.* **269**(Nov), 86–92

602. Pääbo, S., Higuchi, R. G., and Wilson, A. C. (1989) *J. Biol. Chem.* **264**, 9709–9712

603. Höss, M., Pääbo, S., and Vereshchagin, N. K. (1994) *Nature (London)* **370**, 333

604. Scharf, S. J., Horn, G. T., and Erlich, H. A. (1986) *Science* **233**, 1076–1078

605. Li, H., Gyllensten, U. B., Cui, X., Saiki, R. K., Erlich, H. A., and Arnheim, N. (1988) *Nature (London)* **335**, 414–417

606. Engelke, D. R., Hoener, P. A., and Collins, F. S. (1988) *Proc. Natl. Acad. Sci. U.S.A.* **85**, 544–548

607. Karlovsky, P. (1990) *Trends Biochem. Sci.* **15**, 419

608. Hengen, P. N. (1995) *Trends Biochem. Sci.* **20**, 372–373

609. Wong, G. K.-S., Yu, J., Thayer, E. C., and Olson, M. V. (1997) *Proc. Natl. Acad. Sci. U.S.A.* **94**, 5225–5230

610. Bernardi, G. (1978) *Nature (London)* **276**, 558–559

611. Bonitz, S. G., Coruzzi, G., Thalenfeld, B. E., Tzagoloff, A., and Macino, G. (1980) *J. Biol. Chem.* **255**, 11927–11941

612. Martin, N. C., Miller, D. L., and Donelson, J. E. (1979) *J. Biol. Chem.* **254**, 11729–11734

613. Fischer, S. G., and Lerman, L. S. (1979) *Methods Enzymol.* **68**, 183–191

614. Southern, E. M. (1975) *J. Mol. Biol.* **98**, 503–517

615. Southern, E. (1979) *Methods Enzymol.* **68**, 152–176

616. Grunstein, M., and Wallis, J. (1979) *Methods Enzymol.* **68**, 379–389

617. Sanger, F., and Coulson, A. R. (1975) *J. Mol. Biol.* **94**, 441–448

618. Sanger, F., Nicklen, S., and Coulson, A. R. (1977) *Proc. Natl. Acad. Sci. U.S.A.* **74**, 5463–5467

619. Alphey, L. (1997) *DNA Sequencing*, Springer-Verlag, New York

620. Ansorge, W., Voss, H., and Zimmerman, J., eds. (1997) *DNA Sequencing Strategies - Automated and Advanced Approaches*, Wiley, New York

621. Griffin, H. G., and Griffin, A. M., eds. (1993) *DNA Sequencing Protocols*, Humana Press, Totowa, New Jersey

622. Smith, A. J. H. (1980) *Methods Enzymol.* **65**, 560–580

623. Gillam, S., and Smith, M. (1980) *Methods Enzymol.* **65**, 687–701

624. Chen, E. Y., and Seeburg, P. H. (1985) *DNA* **4**, 165–170

625. Ornstein, D., Moen, P. T., and Kashdan, M. A. (1985) *DNA* **4**, 94

626. Bassam, B. J., Caetano-Anolles, G., and Gresshoff, P. M. (1991) *Anal. Biochem.* **196**, 80–83

627. Mizusawa, S., Nishimura, S., and Seela, F. (1986) *Nucleic Acids Res.* **14**, 1319–1324

628. Gerken, T. A., Gupta, R., and Jentoft, N. (1992) *Biochemistry* **31**, 639–648

629. Hengen, P. N. (1996) *Trends Biochem. Sci.* **21**, 33–34

630. Poncz, M., Solowiejczyk, D., Ballantine, M., Schwartz, E., and Surrey, S. (1982) *Proc. Natl. Acad. Sci. U.S.A.* **79**, 4298–4302

631. Henikoff, S. (1984) *Gene* **28**, 351–359

632. Adams, M. D., Kerlavage, A. R., Kelley, J. M., Gocayne, J. D., Fields, C., Fraser, C. M., and Venter, J. C. (1994) *Nature (London)* **368**, 474–475

633. Smith, L. M., Sanders, J. Z., Kaiser, R. J., Hughes, P., Dodd, C., Connell, C. R., Heiner, C., Kent, S. B. H., and Hood, L. E. (1986) *Nature (London)* **321**, 674–679

634. Ju, J., Ruan, C., Fuller, C. W., Glazer, A. N., and Mathies, R. A. (1995) *Proc. Natl. Acad. Sci. U.S.A.* **92**, 4347–4351

635. Fu, D.-J., Broude, N. E., Köster, H., Smith, C. L., and Cantor, C. R. (1995) *Proc. Natl. Acad. Sci. U.S.A.* **92**, 10162–10166

636. Bult, C. J., and 39 other authors. (1996) *Science* **273**, 1058–1073

637. Sutcliffe, J. G. (1995) *Trends Biochem. Sci.* **20**, 87–90

638. Rowen, L., Mahairas, G., and Hood, L. (1997) *Science* **278**, 605–607

639. Mathies, R. A., Huang, X. C., and Quesada, M. A. (1992) *Anal. Chem.* **64**, 2149–2154

640. Yeung, E. S., and Li, Q. (1998) in *High Performance Capillary Electrophoresis* (Khaldi, M. G., ed) Wiley, New York

641. Little, D. P., Aaserud, D. J., Valaskovic, G. A., and McLafferty, F. W. (1996) *J. Am. Chem. Soc.* **118**, 9352–9359

642. Maxam, A. M., and Gilbert, W. (1977) *Proc. Natl. Acad. Sci. U.S.A.* **74**, 560–564

643. Gilbert, W. (1981) *Science* **214**, 1305–1312

644. Maxam, A. M., and Gilbert, W. (1980) *Methods Enzymol.* **65**, 499–560

645. Brownlee, G. G. (1972) in *Determination of Sequences in RNA* (Work, T. S., and Work, E., eds), North-Holland Publ., Amsterdam

646. Pilly, D., Niemeyer, A., Schmidt, M., and Bargetzi, J. P. (1978) *J. Biol. Chem.* **253**, 437–445

647. Bayev, A. A. (1995) *Comprehensive Biochemistry* **38**, 439–479

648. Campbell, P. N. (1995) *Trends Biochem. Sci.* **20**, 259–260

649. Myers, T. W., and Gelfand, D. H. (1991) *Biochemistry* **30**, 7661–7666

650. Simoncsits, A., Brownlee, G. G., Brown, R. S., Rubin, J. R., and Guilley, H. (1977) *Nature (London)* **269**, 833–836

651. Stanley, J., and Vassilenko, S. (1978) *Nature (London)* **274**, 87–89

652. Kitamura, N., and Wimmer, E. (1980) *Proc. Natl. Acad. Sci. U.S.A.* **77**, 3196–3200

653. Josse, J., Kaiser, A. D., and Kornberg, A. (1961) *J. Biol. Chem.* **236**, 864–875

654. Stormo, G. D. (1988) *Ann. Rev. Biophys. Biophys. Chem.* **17**, 241–263

655. Goad, W. B. (1986) *Ann. Rev. Biophys. Biophys. Chem.* **15**, 79–95

656. Garner, M. M., and Revzin, A. (1986) *Trends Biochem. Sci.* **11**, 395–396

657. Berger, R., Duncan, M. R., and Berman, B. (1993) *BioTechniques* **15**, 650–652

658. Welsh, J., and Cantor, C. R. (1984) *Trends Biochem. Sci.* **9**, 505–508

659. Galas, D., and Schmitz, A. (1978) *Nucleic Acids Res.* **5**, 3157–3170

660. Tullius, T. D. (1987) *Trends Biochem. Sci.* **12**, 297–300

661. Tullius, T. D., and Dombroski, B. A. (1986) *Proc. Natl. Acad. Sci. U.S.A.* **83**, 5469–5473

662. Prigodich, R. V., and Martin, C. T. (1990) *Biochemistry* **29**, 8017–8019

663. Flaus, A., Luger, K., Tan, S., and Richmond, T. J. (1996) *Proc. Natl. Acad. Sci. U.S.A.* **93**, 1370–1375

664. Tornaletti, S., and Pfeifer, G. P. (1995) *J. Mol. Biol.* **249**, 714–728

664a. Chaulk, S. G., Pezacki, J. P., and MacMillan, A. M. (2000) *Biochemistry* **39**, 10448–10453

665. Jaeger, J. A., SantaLucia, J., Jr., and Tinoco, I., Jr. (1993) *Ann. Rev. Biochem.* **62**, 255–287

666. Hüttenhofer, A., and Noller, H. F. (1994) *EMBO J.* **13**, 3892–3901

667. Laughrea, M., and Tam, J. (1992) *Biochemistry* **31**, 12035–12041

668. Reid, B. R. (1981) *Ann. Rev. Biochem.* **50**, 969–996

669. Hare, D. R., Ribeiro, N. S., Wemmer, D. E., and Reid, B. R. (1985) *Biochemistry* **24**, 4300–4306

670. Ikuta, S., Chattopadhyaya, R., Ito, H., Dickerson, R. E., and Kearns, D. R. (1986) *Biochemistry* **25**, 4840–4849

671. Lefèvre, J.-F., Lane, A. N., and Jardetzky, O. (1987) *Biochemistry* **26**, 5076–5090

672. Bax, A., and Lerner, L. (1986) *Science* **232**, 960–967

673. Patel, D. J., Shapiro, L., and Hare, D. (1987) *Ann. Rev. Biophys. Biophys. Chem.* **16**, 423–454

674. Flynn, P. F., Kintanar, A., Reid, B. R., and Drobny, G. (1988) *Biochemistry* **27**, 1191–1197

675. Cohen, J. S. (1987) *Trends Biochem. Sci.* **12**, 133–135

676. Wemmer, D. E., Chou, S.-H., and Reid, B. R. (1984) *J. Mol. Biol.* **180**, 41–60

677. Schmidt, P. G., Sierzputowska-Gracz, H., and Agris, P. F. (1987) *Biochemistry* **26**, 8529–8534

678. Farmer, B. T., Jr., and Müller, L. (1993) *J. Am. Chem. Soc.* **115**, 11040–11041

679. Nikonowicz, E. P., and Pardi, A. (1992) *Nature (London)* **355**, 184–186

680. Tate, S.-i, Ono, A., and Kainosho, M. (1994) *J. Am. Chem. Soc.* **116**, 5977–5978

681. Marino, J. P., Prestegard, J. H., and Crothers, D. M. (1994) *J. Am. Chem. Soc.* **116**, 2205–2206

682. Moore, P. B. (1995) *Acc. Chem. Res.* **28**, 251–256

683. Pardi, A. (1995) *Nature Struct. Biol.* **1**, 846–849

684. Mer, G., and Chazin, W. J. (1998) *J. Am. Chem. Soc.* **120**, 607–608

684a. Nikonowicz, E. P., Michnicka, M., and DeJong, E. (1998) *J. Am. Chem. Soc.* **120**, 3813–3814

684b. Liu, A., Majumdar, A., Hu, W., Kettani, A., Skripkin, E., and Patel, D. J. (2000) *J. Am. Chem. Soc.* **122**, 3206–3210

685. Tolbert, T. J., and Williamson, J. R. (1996) *J. Am. Chem. Soc.* **118**, 7929–7940

686. Xu, J., Lapham, J., and Crothers, D. M. (1996) *Proc. Natl. Acad. Sci. U.S.A.* **93**, 44–48

687. Horowitz, J., Ofengand, J., Daniel, W. E. Jr., and Cohn, M. (1977) *J. Biol. Chem.* **252**, 4418–4420

688. Schroeder, S. A., Roongta, V., Fu, J. M., Jones, C. R., and Gorenstein, D. G. (1989) *Biochemistry* **28**, 8292–8303

689. Radhakrishnan, I., Patel, D. J., and Gao, X. (1992) *Biochemistry* **31**, 2514–2523

690. Schurter, E. J., Sayer, J. M., Oh-hara, T., Yeh, H. J. C., Yagi, H., Luxon, B. A., Jerina, D. M., and Gorenstein, D. G. (1995) *Biochemistry* **34**, 9009–9020

691. Mao, H., Deng, Z., Wang, F., Harris, T. M., and Stone, M. P. (1998) *Biochemistry* **37**, 4374–4387

692. Chou, S.-H., Zhu, L., Gao, Z., Cheng, J.-W., and Reid, B. R. (1996) *J. Mol. Biol.* **264**, 981–1001

693. Kettani, A., Guéron, M., and Leroy, J.-L. (1997) *J. Am. Chem. Soc.* **119**, 1108–1115

694. Chu, Wen-C., Liu, J. C.-H., and Horowitz, J. (1997) *Nucleic Acids Res.* **25**, 3944–3949

Study Questions

1. Describe the typical distribution pattern of RNA and DNA in bacterial cells and in eukaryotic cells.

2. Draw the structures of the Watson–Crick base pairs guanine–cytosine (GC) and adenine–thymine (AT). Also draw the GU pair, which is not a Watson–Crick pair.

3. Draw the tautomeric structures possible for the cation formed by protonation of 9-methyladenine.

4. What unusual base pairs could arise from a minor tautomer of cytosine or from a minor tautomer of guanine?

5. The minor imino tautomeric form of adenosine occurs infrequently in DNA. Can this cause mutations? Explain; draw structures to illustrate your answer.

6. Will the substitution of hypoxanthine for adenine in DNA result in mutation? Explain.

7. Why is the methylation of DNA to form O^6-methyl guanine mutagenic?

8. Draw the structure of a dinucleotide that might be obtained by the partial hydrolysis of RNA. Indicate the following:
   a) The 5' end
   b) The 3' end
   c) The torsion angle χ
   d) The point of cleavage by pancreatic ribonuclease
   e) The point of cleavage by periodic acid
   f) Two points at which the structure might be methylated by modifying enzymes acting on a polynucleotide

9. Draw the structure of guanosine-5'-phosphate in such a way that the configurations of the sugar ring and of the glycosidic linkage are clearly indicated. State whether you have drawn a *syn* or an *anti* conformer. Circle the most acidic proton in the guanine ring and indicate its approximate pK_a. Which is the most basic center? What is the approximate pK_a of the conjugate acid?

10. What are the chemical functional groups in DNA? In RNA?

11. Electrophoresis of a mixture of the dinucleotides ApC and ApU at pH 3.5 separates two components. Identify these and explain the order of migration. Be as quantitative as possible.

12. Draw the structure of the predominant form of pGpC as it occurs at pH 3.5.

13. Why is DNA denatured at pH 11?

14. Draw a schematic representation of the polynucleotide portion of a DNA molecule and of an RNA molecule and indicate positions of cleavage by the following treatments:
   a) Mild HCl
   b) More vigorous HCl
   c) Mild NaOH
   d) More vigorous NaOH
   e) Pancreatic RNase
   f) Pancreatic DNase
   g) Splenic DNase
   h) Splenic phosphodiesterase
   i) Snake venom phophodiesterase
   j) Dnase from *Micrococcus*

15. A sample of DNA from a virus was hydrolyzed by acid and was found to have the following base composition (in mol%): adenine, 30; thymine, 39; guanine, 18; cytosine, 13. This differs from that of most DNA preparations. Offer a possible explanation. Sketch the expected temperature-absorbance profile of this DNA. Do you expect much hyperchromicity? Explain your answer.

16. Adenine is found to constitute 16.3% of the nucleic acid bases in a sample of bacterial DNA. What are the percentages of the other three bases?

17. For the following DNA sequence

   3'-CGATACGGCTATGCCATAGGC-5'

   write a) the sequence of the complementary DNA strand;
   b) the sequence of the corresponding segment of mRNA formed using the DNA segment above as the template;
   c) the amino acid sequence encoded by this segment.

18. What is the molecular mass of a segment of B-DNA that encodes a 386-residue protein? What is the length in nm? in Å? Do not make allowance for introns.

19. Complete the following table:

Name	Monomer	Linkage	Range of molecular masses
Protein			
Polysaccharide			
Nucleic acid			
Teichoic acid			
Poly-β-hydroxybutyrate			

20. What is meant by the T_m of a DNA sample? How does T_m vary with base composition and what is the explanation of this?

21. Isolated "naked" bacterial DNA, from which proteins have been removed, is supercoiled. DNA in the bacterial chromosome is also supercoiled. When naked DNA is nicked, its supercoiling is abolished. In contrast nicking the chromosomal DNA does not abolish its supercoiling. Explain.

22. A closed circular duplex DNA has a 90 base-pair segment of alternating G and C residues. Upon transfer to a solution containing a high salt concentration, this segment undergoes a transition from the B conformation to the Z conformation.
    a) Explain why the high salt concentration induces a B→Z transformation.
    b) What changes would you expect in (1) the linking number Lk, (2) the writhe Wr, and (3) the twist Tw of the DNA as a result of this transition.

23. Name two or more characteristics of a DNA sequence or of its environment that will favor conversion of B-DNA into Z-DNA.

24. Suppose one double helical turn of a superhelical DNA molecule changes from a B conformation to the Z conformation. Calculate the approximate changes in (1) the linking ΔLk, (2) the writhe ΔWr, and (3) the twist ΔTw of the DNA as a result of this transition. Show your calculations and explain your answers. For this problem assume that the B form of DNA has 10.4 bp per turn. Why is the B→Z transition favored in naturally occurring supercoiled DNA?

25. A circular DNA plasmid of length 1144 bp is supercoiled with a twist (Tw) of 110. Assume that the DNA has 10.4 bp per turn in its relaxed state.
    a) What is the linking number Lk and the writhe Wr in the plasmid?
    b) Is the plasmid negatively or positively supercoiled?
    c) Ethidium bromide is an intercalating agent that inserts between the stacked base pairs, separating the stacks and causing local unwinding that decreases the value of Tw. What effect would ethidium bromide have on the migration rate of the plasmid during electrophoresis?
    d) If part of the plasmid were to undergo a transition from B-DNA to Z-DNA, what would be the effect on Lk, Tw, and Wr?

26. You have been given a sample of nucleic acid, describe two ways you could determine whether it is RNA or DNA and two ways to determine whether it is single- or double-stranded.

27. What conclusions can you draw about the nature of the protein binding site on DNA from the observation that methylation of cytosine residues in the protein-recognition sequence inhibits protein binding?

28. The anticodon loop of one of the tRNA Gly molecules from *E. coli* is as follows. Identify the anticodon, reading from 3' to 5'. This tRNA recognizes two different Gly codons. What are they? Write them from 5' to 3'.

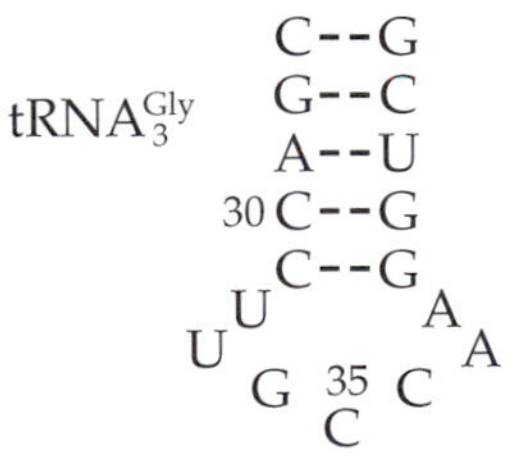

$$\text{tRNA}^{\text{Gly}}_3 \quad \begin{array}{c} \text{C--G} \\ \text{G--C} \\ \text{A--U} \\ 30\,\text{C--G} \\ \text{C--G} \\ {}_{\text{U}}{}^{\text{U}}\!\!\quad\!\!{}^{\text{A}}_{\text{A}} \\ \text{G} \;\; 35\; \text{C} \\ \text{C} \end{array}$$

The complete tRNA contains 75 nucleotides. Sketch the rest of the molecule in the cloverleaf representation. Label the 5' and 3' ends and the dihydrouridine and TψC loops. What are the last three nucleotides at the 3' end?

29. Are viruses alive? Explain your answer.

Ice and water are in equilibrium at 0°C and atmospheric pressure. When ice melts under these conditions the heat Q absorbed from the surroundings is the **enthalpy change**, ΔH, which equals 6.008 kJ mol^{-1}. For a reversible reaction the **entropy change**, ΔS, equals $\Delta H/T$ and the entropy increases when the ice melts by 6.008×10^3 J / 273.16 = 22.0 JK^{-1}. Along the frozen edges of the river the ice and water are not in a true equilibrium but in a steady state. It is an **open system** in which water flows and energy is exchanged with the surroundings. Similarly, the cells of *Vorticella* represent open systems through which water and nutrients flow, and energy is exchanged with the surrounding water. Many chemical reactions within the cells are near equilibrium while the organism maintains its own structure. Photos: Frosty River, Banff Natl. Park, Alberta © Stephen J. Kraseman; *Vorticella* courtesy of Ralph Buchsbaum.

Contents

Thermodynamics and Biochemical Equilibria

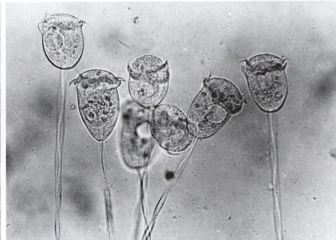

6

We all know from experience the importance of energy to life. We know that we must eat and that hard work not only tires us but also makes us hungry. Our bodies generate heat, an observation that led Lavoisier around 1780 to the conclusion that respiration represented a slow combustion of foods within the body. It soon became clear that respiration must provide the energy for both the mechanical work done by muscles and the chemical synthesis of body constituents. All organisms require energy and the ways in which living things obtain and utilize energy is a major theme of biology.

The discovery of the first and second laws of thermodynamics permitted the development of precise, quantitative relationships between heat, energy, and work. It also allowed **chemical equilibria** to be understood. Modern biochemical literature abounds with references to the thermodynamic quantities **energy** E, **enthalpy** H, **entropy** S, and **Gibbs energy** (also called free energy) G. The purposes of this chapter are: (1) to provide a short review of thermodynamic equations, (2) to provide tables of thermodynamic quantities for biochemical substances and to explain the use of these data in the consideration of equilibria in biochemical systems, and (3) to introduce the **adenylate system**, which consists of adenosine triphosphate (ATP), adenosine diphosphate (ADP), adenosine mono-phosphate (AMP), and inorganic phosphate (P_i). This system plays a central role in energy metabolism, and (4) to provide a quantitative understanding of the effects of pH and of metal ions on biochemical equilibria. *Many readers will want to go directly to Section D, which deals with the adenylate system and its significance for life.*

A. Thermodynamics

Thermodynamics is concerned with the quantitative description of heat and energy changes and of chemical equilibria.[1–10] Knowledge of *changes* in thermodynamic quantities, such as ΔH and ΔS, enables us to predict the equilibrium positions in reactions and whether or not under given circumstances a reaction will or will not take place. Furthermore, the consideration of thermodynamic quantities provides insight into the nature of forces responsible for bonding between molecules, enzymatic catalysis, functioning of DNA and RNA, and many other phenomena.

It is important to realize that while thermodynamic information will tell us whether or not a reaction can take place it says nothing about the rate of the reaction. It will not even say whether a reaction will proceed at all within a given period of time. This has led to the occasional assertion that thermodynamics is not relevant to biochemistry. This is certainly not true; it is important to understand energy relationships in biochemical reactions. At the same time, one should avoid the trap of assuming that thermodynamic calculations appropriate for *equilibrium* situations can always be applied directly to the *steady state* found in a living cell.

Thermodynamics is an exact science and its laws deal with measurable quantities whose values are determined only by the **state** of the **system** under consideration. For example, the system might be the solution in a flask resting in a thermostated bath. To specify its state we would have to say whether it is pure solid, liquid, or gas, or a solution of specified composition and give the temperature and pressure. The flask, the bath, and everything else would be called **surroundings** or **environment**. The system plus surroundings is sometimes referred to as **the universe**.

1. The First Law of Thermodynamics

The first law of thermodynamics asserts the conservation of energy and also the equivalence of **work** and **heat**. Work and heat are both regarded as energy in transit. Heat may be absorbed by a system from the surroundings or evolved by a system and absorbed in the surroundings. Work can be done *by* a system on the surroundings or it can be done *on* a system. The first law postulates that there is an **internal energy** E (also designated U), which is dependent only on the present state of the system and in no way is dependent upon the history of the system. The first law states that *E can be changed only by the flow of energy as heat or by work*. In other words, energy can neither be created nor destroyed.

In mathematical form, the first law is given as follows:

$$\Delta E = E \text{ (products)} - E \text{ (reactants)} = Q - W \quad (6\text{-}1)$$

Here Q is the heat absorbed by the system from the surroundings and W is the work done by the system on the surroundings. Energy, heat, and work are all measured in the same units. Chemists have traditionally used the **calorie** (cal) or **kilocalorie** (kcal) but are switching to the SI unit, the **joule** (Table 6-1).[10a] Work done by the system may be **mechanical** (e.g., by changing the volume of the surroundings), **electrical** (e.g., by charging of a battery), or **chemical** (e.g., by effecting the synthesis of a polypeptide from amino acids).

2. Enthalpy Changes and Thermochemistry

We are most often interested in the *changes* in the thermodynamic functions when a chemical reaction takes place; for example, the heat absorbed by the system within a **bomb calorimeter** where the volume stays constant (Q_v) is a direct measure of the change in E:

$$Q_V = \Delta E$$

$$(6\text{-}2)$$

To measure ΔE for combustion of a biochemical compound, the substance may be placed in a bomb together with gaseous oxygen and the mixture ignited within the calorimeter by an electric spark. In this case, heat will be evolved from the bomb and will pass into the surroundings. Q_v and ΔE will be negative. The bomb calorimeter is designed to measure Q_v and thereby to give us a way of determining ΔE for reactions.

Processes at constant pressure. Chemical and biochemical reactions are much more likely to be conducted at constant pressure (usually 1 atm) than they are at constant volume. For this reason, chemists tend to use the **enthalpy** H more often than the internal energy E.

$$H = E + PV \quad (6\text{-}3)$$

It follows from Eq. 6-3 that *if the pressure is constant*, ΔH_p is equal to $\Delta E_p + P \, \Delta V$. Since in a process at constant pressure, $P \, \Delta V$ is exactly the pressure–volume work done on the surroundings, the heat absorbed at constant pressure (Q_p) is a measure of ΔH_p.

$$Q_P = \Delta E_P + P \, \Delta V = \Delta H_P \quad (6\text{-}4)$$

Since enthalpy changes can be obtained directly from measurement of heat absorption at constant pressure, even small values of ΔH for chemical and biochemical reactions can be measured using a microcalorimeter.[11,12] Using the technique of pulsed acoustic calorimetry, changes during biochemical processes can be followed on a timescale of fractions of a millisecond. An example is the laser-induced dissociation of a carbon monoxide–myoglobin complex.[13]

The term *enthalpy* was coined to distinguish H from E, but we sometimes tend to be careless about language and many discussions of energy in the literature are in fact about enthalpy. The difference is often not significant because if the pressure–volume work is negligible, E and H are the same.

Enthalpies of combustion and physiological fuel values. The **heat of combustion** ($-\Delta H_c$) of an organic substance is usually determined from ΔE_c, which is measured in a bomb calorimeter. Since ΔE_v and ΔE_p are nearly identical, it follows that $\Delta H_p = \Delta E_v + P \, \Delta V$. Here ΔV is the volume change which would have occurred if the reaction were carried out at constant pressure P; thus, ΔH_p can be estimated by calculation. Since ΔH is desired for combustion to carbon dioxide, water, elemental nitrogen (N_2), and sulfur, correction must be made for the amounts of the latter elements converted into oxides. By these procedures, it has been possible to obtain highly accurate values of ΔH_c both for biochemical compounds and for mixed foodstuffs. In nutrition, $-\Delta H_c$ is sometimes referred to as the **gross energy**. Values are usually expressed in kilocalories (kcal) by chemists but often as **Cal** (with a capital C) in the nutritional literature.

Caloric values of foods (physiological fuel values) are enthalpies of combustion but with an opposite sign, ($-\Delta H_c$), and corrected for energy lost in urine (e.g., as urea) and feces. While enthalpies of combustion of foods are all negative, the caloric values are given as positive numbers. Caloric values for proteins are

TABLE 6-1
Units of Energy and Work and the Values of Some Physical Constants

The joule, SI unit of energy
$$1 J = 1 \text{ kg m}^2 \text{ s}^{-2}$$
$$= 1 \text{ N m (newton meter)}$$
$$= 1 \text{ W s (watt second)}$$
$$= 1 \text{ C V (coulomb volt)}$$

Thermochemical calorie
$$1 \text{ cal} = 4.184 J$$

Large calorie
$$1 \text{ Cal} = 1 \text{ kcal} = 4.184 \text{ kJ}$$

Work required to raise 1 kg 1 m on earth
(at sea level) $= 9.807 J$

Gibbs energy of hydrolysis of 1 mole of ATP
at pH 7, millimolar concentrations $=$
$-12.48 \text{ kcal} = -52.2 \text{ kJ}$

Work required to concentrate 1 mole of a substance
1000-fold, e.g., from 10^{-6} to 10^{-3} M
$= 4.09 \text{ kcal} = 17.1 \text{ kJ}$

Avogadro's number, the number of particles in a mole
$N = 6.0220 \times 10^{23}$

Faraday $1 F = 96,485 \text{ C mol}^{-1}$ (coulombs per mole)

Coulomb $1 C = 1 \text{ A s (ampere second)}$
$= 6.241 \times 10^{18}$ electronic charges

The Boltzmann constant
$k_B = 1.3807 \times 10^{-23} \text{ J deg}^{-1}$

The gas constant, $R = N k_B$
$R = 8.3144 \text{ J deg}^{-1} \text{ mol}^{-1}$
$= 1.9872 \text{ cal deg}^{-1} \text{ mol}^{-1}$
$= 0.08206 \text{ 1 atm deg}^{-1} \text{ mol}^{-1}$
and at 25°C $RT = 2.479 \text{ kJ mol}^{-1}$

The unit of temperature is the kelvin (K); 0°C $=$
273.16 K

$\ln x = 2.3026 \log x$

One atmosphere (atm) $= 101.325$ kilopascals (kPa)

calculated for the conversion of the nitrogen to urea, the major nitrogenous excretion product in mammals, rather than to elemental nitrogen. Typical values are shown in Table 6-2.

TABLE 6-2
Caloric Values of Food Components

Component	Caloric values per gram	
Carbohydrates	4.1 kcal	17 kJ
Pure glucose	3.75 kcal	15.7 kJ
Lipids	9.3 kcal	39 kJ
Proteins[a]	4.1 kcal	17 kJ

[a] Nitrogen excreted as urea.

From a thermochemical viewpoint, can a human or animal be regarded as just a catalyst for the combustion of foodstuffs? To answer this question, large calorimeters were constructed into which an animal or a human being was placed. If, while in the calorimeter, the subject neither gained nor lost weight, the heat evolved should have been just equal to $-\Delta H$ for combustion of the food consumed to CO_2, water, and urea. That this prediction was verifed experimentally does not seem surprising, but at the time that the experiments were first done in the early years of the century there may have been those who doubted that the first law of thermodynamics applied to mammals.

In practice, animal calorimetry is quite complicated because of the inherent difficulty of accurate heat measurements, uncertainties about the amount of food stored, and the necessity of corrections for ΔH_c of the waste products. However, the measurement of energy metabolism has been of considerable importance in nutrition and medicine. Indirect methods of calorimetery have been developed for use in measuring the **basal metabolic rate** of humans. For a good discussion see White *et al.*[14]

The basal metabolic rate is the rate of heat evolution in the resting, postabsorptive state, in which the subject has not eaten recently. In this condition, stored foods provide the energy and are oxidized at a relatively constant rate. The basal metabolic rate tends to be *proportional to the surface area;* which can be approximated (in units of m^{-2}) as 1/60 [height (cm) × mass (kg)].[15] For a young adult female, the basal metabolic rate is typically ~154 kJ h^{-1} m^{-2} and for a young adult male ~172 kJ h^{-1} m^{-2}. This is ~320 – 360 kJ h^{-1} for a 70-kg person. Note that 360 kJ h^{-1} is the same as the power output of a 100-watt lightbulb. While there is considerable variation among individuals, basal metabolic rates far below or above normal may indicate a pathological condition such as an insufficiency or oversupply of the thyroid hormone thyroxine. Metabolic rates fall somewhat below the basal value during sleep and are much higher than basal during hard exercise. A human may attain rates as high as 2500 kJ (600 kcal) per hour. At a basal rate of 320 kJ (76 kcal) per hour, a person requires 7680 kJ (1835 kcal) each 24 h to supply his or her basal needs, plus additional energy during periods of muscular exercise. Routine light exercise as in the office or during housework increases metabolism to about double the basal rate. Although the caloric values in Table 6-2 are reliable for prediction of metabolic energy needs, they must be adjusted to predict the efficiency of utilization for growth. In one study[16] a group of rats deposited 28% of the available energy from sucrose as body protein and fats, but fats were deposited with an efficiency of 36%.

3. The Second Law of Thermodynamics

Why does heat flow from a warm body into a cold one? Why doesn't it ever flow in the reverse direction? We can see that differences in temperature control the direction of flow of heat, but this observation raises still another question: What *is* temperature? Reflection on these questions, and on the interconversion of heat and work, led to the discovery of the second law of thermodynamics and to the definition of a new thermodynamic function, the **entropy S**.

Consider the melting of ice. This is a phase transition that usually takes place at constant temperature and pressure. At a temperature just above 0°C ice melts completely, but at a temperature just below 0°C it does not melt at all. At 0°C we have an equilibrium. In the language of thermodynamics, the melting of ice at 0°C is a **reversible reaction**. What criterion could be used to predict this behavior for water? For many familiar phenomena, e.g., combustion, a spontaneous reaction is accompanied by the evolution of a large amount of heat, i.e., ΔH is negative. However, when ice melts it *absorbs* heat. The ΔH of fusion amounts to 6.008 kJ mol^{-1} at 0°C and is nearly the same just below 0°C, where the ice does not melt, and just above 0°C, where the ice melts completely. In the latter case, the melting of ice is a spontaneous reaction for which ΔH is positive. It is clear from such facts that *the sign of the enthalpy change does not serve as a criterion of spontaneity*.

A correct understanding of the ice–water transition came when it was recognized that when ice melts not only does H increase by 6.008 kJ mol^{-1}, as the molecules acquire additional internal energy of translation, vibration, and rotation, but also *the molecules become more disordered*. Although historically entropy was introduced in a different context, it is now recognized to be a measure of "microscopic disorder." When ice melts, the entropy S increases because the structure becomes less ordered.

The second law of thermodynamics is stated in many different ways, but the usual mathematical formulation asserts that for the universe (or for an isolated system)

$$\Delta S \text{ (system + surroundings)} = 0$$
for reversible processes

$$\Delta S > 0$$
for real (nonreversible) processes   (6-5)

The second law is sometimes stated in another way: *The entropy of the universe always increases.*

The second law also defines both S and the thermodynamic temperature scale as follows:

$$dS_{\text{reversible}} = q/T   \quad (6\text{-}6)$$

Here q is an infinitesimal quantity of heat absorbed from the surroundings by the system and T is measured in kelvins (K). For a *reversible phase transition* such as the melting of ice at constant pressure and temperature, the change in entropy of the H_2O is just $\Delta H/T$.

$$\Delta S)_{P,T,\text{ reversible}} = Q/T = \Delta H/T   \quad (6\text{-}7)$$

Entropy is measured in units of joules per kelvin (or °C) or calories per K, the latter sometimes being abbreviated as e. u. (entropy units). Since the melting of ice at 0°C is a reversible process, the second law asserts that the entropy of the surroundings decreases by the same amount that the entropy of the water increases. The value of $T \Delta S$ is numerically equal to the heat of fusion, 6.008 kJ mol^{-1} in the case of water at 0°C. Thus, the entropy increase in the ice as it melts at 0°C is 6.008 x 10^3 J/273.16 K = 22.0 J K^{-1}.

The thermodynamic temperature. The definition of thermodynamic temperature in kelvins (Eq. 6-8) also follows from Eq. 6-6. See textbooks of thermodynamics for further treatment.

$$T = (\partial E/\partial S)_V = (\partial H/\partial S)_P   \quad (6\text{-}8)$$

The entropy of a substance can be given a precise mathematical formulation involving the degree of molecular disorder (Eq. 6-9).

$$S = k_B \ln \Omega   \quad (6\text{-}9)$$

Here k_B is the **Boltzmann constant** (see Table 6-1) and Ω is given precisely as the number of microscopic states (different arrangements of the particles) of the system corresponding to a given macroscopic state, i.e., to a given temperature, pressure, and quantity. It increases as volume or temperature is increased and in going from solid to liquid to gaseous states. Equation 6-9 is not part of classical thermodynamics (which deals only with macroscopic systems, i.e., with large collections of molecules). However, using the methods of statistical thermodynamics,[17] this equation can be used to predict the entropies of gases.

The *racemization of an amino acid* provides a biochemical example that can be related directly to Eq. 6-9. A solution of an L-amino acid will be efficiently changed into the racemic mixture of 50% D and 50% L by the action of an enzyme (a **racemase**) with no uptake or evolution of heat. Thus, $\Delta H = 0$ and the only change is an entropy change. Let us designate Ω for the pure isomer as Ω'. Since there are just two choices of configuration for each of the N molecules in 1 mole of the racemate we see that for the racemate

$$\Omega = 2^N \Omega'   \quad (6\text{-}10)$$

Applying Eq. 6-9 we calculate ΔS as follows:

$$\Delta S = k_B (\ln 2^N + \ln \Omega') - k_B \ln \Omega'$$
$$= N k_B \ln 2 = R \ln 2 = 5.76 \text{ J K}^{-1} \text{ mol}^{-1}$$
$$(6\text{-}11)$$

Entropies from measurement of heat capacities.

It follows from Eq. 6-9 that $S = 0$ when $T = 0$ for a perfect crystalline substance in which no molecular disorder exists. The *third law of thermodynamics* asserts that as the thermodynamic temperature T approaches 0 K the entropy S also approaches zero for perfect crystalline substances. From this it follows that at any temperature above 0 K, the entropy is given by Eq. 6-12.

$$S = \int_0^T C_p \, d \ln T \qquad (6\text{-}12)$$

In this equation C_p is the heat capacity at constant pressure:

$$C_P = (\partial H / \partial T)_P \qquad (6\text{-}13)$$

If C_p is measured at a series of low temperatures down to near zero K, Eq. 6-12 can be used to evaluate the absolute entropy S. If phase transitions occur as the temperature is raised, entropy increments given by Eq. 6-7 must be added to the value of S given by Eq. 6-12. For a few compounds, such as water (Chapter 2),

TABLE 6-3
Entropies of Selected Substances[a]

Substance	State[b]	Entropy S	
		cal K^{-1} mol^{-1}	J K^{-1} mol^{-1}
C (diamond)	s	0.55	2.3
C (graphite)	s	1.36	5.7
Cu	s	8.0	33
Na	s	12.2	51
H_2O (ice)	s	9.8	41
H_2O	l	16.7	70
H_2O	g (1 atm)	45.1	189
He	g	30.1	126
H_2	g	31.2	131
N_2	g	45.8	192
CO_2	g	51.1	214
Benzene	g	64.3	269
Cyclohexane	g	71.3	298

[a] All values are given in entropy units (e.u.) of calories per Kelvin per mole and in joules per Kelvin per mole at 25°C (298.16 K).
[b] Here s stands for solid, l for liquid, and g for gaseous.

molecular disorder is present in the crystalline state even at 0 K. For these substances a term representing the entropy at 0 K must be added to Eq. 6-12.

The entropies of a few substances are given in Table 6-3. Notice how the entropy increases with increasing complexity of structure, with transitions from solid to liquid to gas, and with decreasing hardness of solid substances.

Measurements of C_p versus temperature for solutions of macromolecules or for biological membranes (Fig. 8-9) over a narrower temperature range are also of interest. These can be obtained with a **differential scanning calorimeter**[18,19] or by an indirect procedure.[20] Denaturation of polymers or phase changes in membranes may be observed. Larger values of C_p are observed for open, denatured, or random-coil structures that are usually present at higher temperatures than for tightly folded molecules.

4. A Criterion of Spontaneity: The Gibbs Energy

We have seen that while many spontaneous processes, e.g., combustion of organic compounds, are accompanied by liberation of heat (negative ΔH), others are accompanied by absorption of heat from the surroundings (positive ΔH). An example of the latter is the melting of ice at a temperature just above 0°C, during which there is a large increase in the entropy of the water. As we have seen, at 0°C at equilibrium $T \Delta S$ is just equal to $-\Delta H$ (Eq. 6-7).

The recognition that $\Delta H - T \Delta S = 0$ for a system at equilibrium led J. W. Gibbs to realize that the proper thermodynamic function for determining the spontaneity of a reaction is what is now known as the **Gibbs energy** or Gibbs function **G** (Eq. 6-14).

$$G = H - TS \qquad (6\text{-}14)$$

In the older literature the Gibbs energy was usually called the **free energy** or Gibbs free energy and was often given the symbol F. For a process at constant temperature and pressure the change in G is given by Eq. 6-15 in which all quantities refer to the system.

$$\Delta G_{T,P} = \Delta H - T \Delta S \qquad (6\text{-}15)$$

For a reversible (equilibrium) process doing only pressure–volume work:

$$\Delta G)_{T,\text{reversible}} = \Delta H - T \Delta S = 0 \qquad (6\text{-}16)$$

It can also be shown readily that ΔG is *negative for any spontaneous* (irreversible) *process*. Such a process is called **exergonic**. Likewise, if ΔG is positive, a given reaction will *not* proceed spontaneously and is called **endergonic**. The magnitude of the decrease in the Gibbs energy ($-\Delta G$) is a direct measure of the maximum work which could be obtained from a given chemical reaction if that reaction could be coupled in some fashion reversibly to a system able to do work. It represents the maximum amount of electrical work that could be extracted or the maximum amount of muscular work or osmotic work obtainable from a reaction in a biological system. In any real system, the amount of work obtainable is necessarily less than $-\Delta G$ because real processes are irreversible, i.e., entropy is created.

Returning to the older assumption that the magnitude of ΔH might be an index of work obtainable, we note that $T\,\Delta S$ amounts to only a few kilojoules for most reactions. Therefore, if ΔH is large, as in the combustion of foodstuffs, it is not greatly different from ΔG for the same process. Therefore, we can justify use of the caloric value of a food as an approximate measure of the work obtainable from its metabolism in the body.

5. Practical Thermochemistry

For thermodynamic data to be useful in chemical calculations, we must agree upon **standard states** for elements and compounds. If we wish to talk about the change in the Gibbs energy that occurs when one or more pure compounds are converted to other pure substances, we must agree upon a state (crystalline, liquid, gaseous, or in solution) and upon a pressure (especially when gases are involved) at which the data apply. The standard pressure is usually 1 atm. Standard states of the elements are the pure *crystalline, solid*, or *gaseous materials*, e.g., C (graphite), S (crystalline, rhombic), P (crystalline, white), and N_2, O_2, and H_2 (gaseous). It is also essential to specify the temperature. Thermodynamic data are most often given for 25°C, but there is a standard state for each substance at each temperature.

It is usually impractical to measure the values of G or H, but ΔG and ΔH for a chemical reaction can be evaluated. Changes in Gibbs energy can be calculated from tables of ΔG for formation of compounds from the elements (Eq. 6-17). These values of ΔG_f can be obtained experimentally by measuring ΔH of combustion for the compound of interest and for H_2, elemental carbon, and other elements present in the compound and also by obtaining entropies from heat capacity measurements. Many other tabulated ΔG_f values have been obtained indirectly utilizing data from equilibrium constants. The resulting **standard Gibbs energies of formation** are given the symbol $\Delta G_f°$. *The values of $\Delta G_f°$ for the elements in their standard states are all exactly zero.*

Summing changes in Gibbs energy. A convenient feature of thermodynamic calculations is that if two or more chemical equations are summed, ΔG for the resulting overall equation is just the sum of the ΔG's for the individual equations as illustrated in Eqs. 6-17 to 6-20. The same applies for ΔH and ΔS.

$$CH_3COOH\ (l) \longrightarrow 2\,C + 2\,H_2 + O_2$$
$$-\Delta G_f° \text{ for acetic acid} = +396.4 \text{ kJ mol}^{-1} \quad (6\text{-}17)$$

$$2\,O_2 + 2\,C \longrightarrow 2\,CO_2$$
$$2\,\Delta G_f° \text{ for } CO_2 = -788.8 \text{ kJ mol}^{-1} \quad (6\text{-}18)$$

$$O_2 + 2\,H_2 \longrightarrow 2\,H_2O$$
$$2\,\Delta G_f° \text{ for } H_2O = -474.4 \text{ kJ mol}^{-1} \quad (6\text{-}19)$$

$$\overline{CH_3COOH\ (l) + 2\,O_2 \longrightarrow 2\,CO_2 + 2\,H_2O\ (l)}$$
$$\Delta G_c° \text{ for acetic acid} = -866.8 \text{ kJ mol}^{-1} \quad (6\text{-}20)$$

In this example an equation for the decomposition of acetic acid into its elements (Eq. 6-17) has been summed with Eqs. 6-18 and 6-19, which represent the formation of the proper number of molecules of CO_2 and H_2O from the elements. The sum of the three equations gives the equation for the combustion of acetic acid to CO_2 and water, and the sum of the ΔG values for the three equations gives ΔG for combustion of acetic acid. The resulting value of ΔG is for combustion of pure liquid acetic acid by oxygen at 1 atm to give CO_2 at 1 atm and pure liquid water, all reactants and products being in their standard states.

The process described in the preceding paragraph is represented by Eq. 6-21, which is a general equation for calculation of $\Delta G°$ for any reaction from $\Delta G_f°$ of products and reactants.

$$\Delta G° = \Sigma\Delta G_f° \text{(products)} - \Sigma\Delta G_f° \text{(reactants)}$$
$$(6\text{-}21)$$

How does the change in Gibbs energy vary if we go from the standard state of a compound to some other state? Consider a change of pressure in a gas. It is easy to show (see any thermodynamics text) that

$$(\partial G/\partial P)_T = V \quad (6\text{-}22)$$

Using Eq. 6-22 together with the perfect gas law, we obtain the relationship (Eq. 6-23) between the Gibbs energy $\overline{G}$ of one mole of a substance at pressure P and the standard Gibbs energy $\overline{G}°$ at pressure $P°$.

$$\overline{G} - \overline{G}° = RT \ln \frac{P}{P°} = RT\ \Delta \ln P \quad (6\text{-}23)$$

Here the bar over the symbol G indicates that the Gibbs energy is for one mole of substance. Since $P°$ is by definition 1 atm, the Gibbs energy change per mole upon changing the pressure from $P°$ to P is just $RT \ln P$.

Reactions in solution. It is customary in books on thermodynamics to develop most of the important thermodynamic equations as applied to a perfect gas, but we will move at this point to a consideration of biochemical substances in solution. Biochemists are usually interested in the behavior of substances dissolved in relatively dilute aqueous solutions but also in cytoplasm, in which some concentrations may be very high. Sometimes the interest may be in nonaqueous solutions. In any case, it is necessary to establish a standard state for the solute. The standard state of a substance in aqueous solution is customarily taken as a strictly hypothetical one **molal** solution (one mole of solute per kilogram of water) *whose properties are those of a solute at infinite dilution*. An equation exactly analogous to Eq. 6-23 can be written relating the Gibbs energy of one mole of dissolved solute $\overline{G}_i$ to the Gibbs energy $\overline{G}_i°$ in the hypothetical standard state of unit activity and to the **activity** a_i of the solute (Eq. 6-24).

$$\overline{G}_i = \overline{G}_i° + RT \ln a_i \qquad (6\text{-}24)$$

Here the subscript i designates a particular component in a solution which also contains solvent and, perhaps, other components. To be precise, $\overline{G}_i$ is a *partial molar Gibbs energy*, i.e., the changes in total Gibbs energy of a very large volume of solution when one mole of the component is added.

From Eq. 6-24 it follows that the *Gibbs energy change for dilution* from one activity a_1 to another a_2 is:

$$\Delta\overline{G} \text{ (dilution from } a_1 \text{ to } a_2) = RT \ln (a_2/a_1) \qquad (6\text{-}25)$$

Equation 6-24 and the equations that follow from it apply to molal activities. However, the concentration can be substituted for activity in very dilute solution where the behavior of the dissolved molecules approximates that of the hypothetical ideal solution for which the standard state is defined. For any real solution, the activity can be expressed as the product of an activity coefficient and the concentration (Eq. 6-26).

$$a = \gamma c \qquad (6\text{-}26)$$

where a = activity, γ = activity coefficient, and c = molal concentration. Thus, to use the tabulations of thermodynamic functions for substances in solution to predict behavior in other than very dilute solution, we must multiply the concentration of every component by the appropriate activity coefficient. For the approximate

calculations which are often of interest to biochemists, it is customary to equate concentration with activity. Furthermore, in dilute solutions, the more usual **molar** concentrations (moles per liter) are nearly equal to molal concentrations. The same equations can be used with *mole fractions* rather than molal concentrations.

$\Delta G°$ and the equilibrium constant. Consider the following generalized chemical equation (Eq. 6-27) for reaction of a moles of A with b moles of B to give products C and D, etc.

$$a A + b B + \cdots = c C + d D \cdots \qquad (6\text{-}27)$$

The standard Gibbs energy change $\Delta G°$ for the process is given by Eq. 6-28:

$$\Delta G° = c\,\overline{G}°(C) + d\,\overline{G}°(D) + \cdots$$
$$- a\,\overline{G}°(A) - b\,\overline{G}°(B) \cdots \qquad (6\text{-}28)$$

The symbol $G°$ (A) designates the Gibbs energy of A, etc. The value of ΔG for any desired concentrations of reactants or products can be related to this $\Delta G°$ by applying to each component Eq. 6-24 with the following result:

$$\Delta G = \Delta G° + RT \ln \frac{a_C^c a_D^d \cdots}{a_A^a a_B^b \cdots} \qquad (6\text{-}29)$$

Here a_C represents the activity of component C, etc. This useful equation permits us to calculate ΔG for the low concentrations usually found in biochemical systems. These are more often in the millimolar range or less rather than approaching the hypothetical 1 M of the standard state. Often concentrations are substituted in Eq. 6-29 for activities:

$$\Delta G \approx \Delta G° + RT \ln \frac{[C]^c[D]^d}{[A]^a[B]^b} \qquad (6\text{-}30)$$

Equation 6-29 is used in another way by noting that $\Delta G = 0$ when a system is at equilibrium and that at equilibrium the product $a_C^c a_D^d \ldots / a_A^a a_B^b \ldots$ is just the equilibrium constant K. It follows that

$$\Delta G° = - RT \ln K = - 2.303RT \log K$$
$$= - 19.145T \log K \text{ J mol}^{-1}$$
$$= - 5.708 \log K \text{ kJ mol}^{-1} \text{ at } 25°C$$
$$= - 1.364 \log K \text{ kcal mol}^{-1} \text{ at } 25°C \qquad (6\text{-}31)$$

Although the units of $\Delta G°$ are kJ mol^{-1}, the Gibbs energy change in Eq. 6-31 is that for the reactions of a moles of

A, b moles of B, etc., as in the equation used to define K (Eq. 6-27 in this instance). It is also important to realize that the log term in Eq. 6-31 must be unitless. Although we usually write $\ln K$ or $\log K$, K here represents K/Q, where $Q = 1$ because it has the same form as K but with all components in their standard states. Since the units of K and Q are the same, $\log K$ is unitless. Similar considerations apply to expressions of K in exponential form.

Activity coefficients and concentration equilibrium constants.

Strictly speaking, Eq. 6-31 applies only to thermodynamic equilibrium constants – that is, to constants that employ activities rather than concentrations. The experimental determination of such constants requires measurements of the apparent equilibrium constant or **concentration equilibrium constant**[21] K_c at a series of different concentrations and extrapolation to infinite dilution (Eq. 6-32).

K_c = concentration equilibrium constant

$$= \frac{[C]^c[D]^d}{[A]^a[B]^b} \text{ at equilibrium}$$

(6-32)

Extrapolation of K_c to infinite dilution to give K is usually easy because the activity coefficients of most ionic substances vary in a regular manner with **ionic strength** and follow the **Debye–Hückel** equation (Eq. 6-33) in very dilute solutions (ionic strength < 0.01).

$$\log \gamma = -0.509 z_1 z_2 \sqrt{\mu}$$

(6-33)

The integers z_1 and z_2 are the numbers of charges (valences) for the cation and anion of the salt. The ionic strength (μ, or I) is evaluated as follows:

$$\mu = \tfrac{1}{2} \Sigma_i c_i z_i^2$$

(6-34)

Here c_i are the molar concentrations of the ions. The summation is carried out over all the ions present. The activity coefficient γ (Eq. 6-33) is the mean activity coefficient for both the cation and anion.

Equation 6-33 suggests that extrapolation of equilibrium constants to infinite dilution is done appropriately by plotting $\log K_c$ vs $\sqrt{\mu}$. For example, Fig. 6-1 shows plots of pK'_a for dissociation of $H_2PO_4^-$, AMP^-, and ADP^{2-}, and ATP^{3-} vs $\sqrt{\mu}$. The variation of pK'_a with $\sqrt{\mu}$ at low concentrations (Eq. 6-35) is derived by application of the Debye-Hückel equation (Eq. 6-33):

$$pK_a' = pK_a - 0.509 (z_A^2 - z_{HA}^2) \sqrt{\mu}$$

(6-35)

Straight lines of slope $-0.509 (z_A^2 - z_{HA}^2)$ are expected. The observed (negative) slopes (Fig. 6-1) are ~1.5 for $H_2PO_4^-$ and AMP^-, ~2.5 for ADP^{2-}, and ~3.5 for ATP^{3-}. The data over the entire range of ionic strength are fitted by empirical relationships of the type of Eq. 6-35a:

$$pK_a' = pK_a - a\sqrt{\mu} + b\mu \quad \text{for } \mu < 0.2$$

(6-35a)

in which a and b are empirically determined constants. For example, for $H_2PO_4^-$ $a = 1.52$ and $b = 1.96$. The value of pK_a found was 7.18, about 0.22 greater than the value at $\mu = 0.2$, an ionic strength more commonly used in the laboratory and close to that found in tissues. Note that the difference between the extrapolated pK_a for ATP^{3-} of 7.68 and the observed value of ~ 7.04 at $\mu = 0.2$ is even greater. Serious errors can be introduced into calculations by using extrapolated values for K for solutions of appreciable ionic strength. The errors will be maximal for ions of high charge type such as ATP^{3-} and ATP^{4-}.

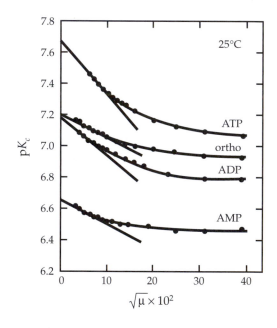

Figure 6-1 The apparent pK_a values for secondary ionizations of AMP, ADP, ATP, and H_3PO_4 (abbreviated ortho) plotted against $\sqrt{\mu}$. Temperature: 25°C. From R. C. Phillips *et al.*[22]

Another problem with equilibrium constants for reactions that use or produce hydrogen ions is that there is no rigorous relationship between pH and a_{H^+} or [H^+]. Indeed, the concept of an activity of a single ion has little meaning in thermodynamics. Nevertheless, in the pH range of interest to biochemists, results that are very close to those obtained by more rigorous methods are achieved by assuming that the pH meter responds to hydrogen ion activity. The almost universal practice of biochemists is to assume the pH meter reading obtained with a glass electrode equal to $-\log a_{H^+}$ and to substitute the value of a_{H^+} so obtained for $[H^+]$ in defining the concentration equilibrium constant, K_c. It is also customary in most branches of chemistry to use values of equilibrium constants and of Gibbs energy which have *not* been extrapolated to $\mu = 0$. Thus most values of K and ΔG, including many of those in this book, are actually of K_c and ΔG_c. An international Commission on Biothermodynamics[21] recommended that values of K_c be measured with the lowest effective buffer concentration and that the ionic strength be brought to 0.1 with KCl.

Changes in equilibria with temperature. At constant pressure ΔG varies with absolute temperature as follows:

$$\frac{d(\Delta G/T)}{dT} = -\Delta H/T^2 \tag{6-36}$$

The corresponding variation in K is described by the **van't Hoff equation**:

$$\frac{d \ln K}{dT} = \frac{\Delta H^\circ}{RT^2}$$

or

$$\frac{d \ln K}{d(1/T)} = \frac{-\Delta H^\circ}{R}$$

or

$$\Delta H^\circ(\text{kJ mol}^{-1}) = \frac{-0.01914 \ d\log_{10} K}{d(1/T)} \tag{6-37}$$

If ΔH° can be assumed constant over the temperature range of an experiment, a plot of $\ln K$ vs $1/T$ provides a convenient estimate of ΔH° (or $\Delta H'$ if $\ln K'$ is plotted). The slope of the line will be $-\Delta H^\circ/R$. Since ΔG° can be calculated from K, the method also permits evaluation of ΔS° using Eq. 6-15. However, unless great care is taken the method is of low accuracy[23] and it is preferable to establish ΔH by direct calorimetry. Also, especially for proteins, the assumption that ΔH° is constant over a significant temperature range may be erroneous.

From observations at only two temperatures, T_1, and T_2, Eq. 6-37 becomes

$$\Delta H^\circ = R \ln (K_2/K_1)[T_1 T_2/(T_2 - T_1)] \text{ kJ mol}^{-1} \tag{6-37a}$$

6. Thermodynamics and Life Processes

Can thermodynamics be applied to living organisms? Classical thermodynamics deals with equilibria, but living beings are never in equilibrium. The laws of thermodynamics are usually described as statistical laws. How can such laws apply to living things, some of which contain all of their genetic information in a single molecule of DNA?[24,25] The ideal, reversible reactions of classical thermodynamics occur at infinitesimal speeds. How can thermodynamics be applied to the very rapid chemical reactions that take place in organisms? One answer is that thermodynamics can be used to decide *whether or not a reaction is possible* under given conditions. Thus, if we know the steady-state concentrations of reactants and products within a cell, we can state whether a reaction will or will not tend to go in a given direction.

We may still ask whether there are generalities comparable to the laws of thermodynamics that apply to the kind of steady state or "dynamic equilibrium" that exists in organisms. Lars Onsager showed that such relationships can be found for conditions that are *near equilibrium*. Ilya Prigogine and associates extended Onsager's findings and showed that under conditions that are *far from equilibrium* the system tends to become unstable and to spontaneously develop new structures, which Prigogine calls **dissipative structures**.[26-28] Vortices in flowing water, tornados and hurricanes are examples of dissipative structures. The maintenance of dissipative structures depends upon a flow of energy and matter through the system. The flow of energy is sufficient to "organize" a system. Thus, if a flask of water is placed on a hot plate, a cycle is established. The water moves via cyclic convection currents that develop as a result of the flow of energy through the system. Morowitz reckoned that the $6 \pm 3 \times 10^{18}$ kJ/year of solar energy that falls on the earth supplies the organizing principle for life.[28] Just as it drives the great cycles within the atmosphere and within the seas, it gives rise to the branching and interconnecting cycles of metabolism. This idea may even make the spontaneous development of the organized systems that we call life from inanimate precursors through evolution seem a little more understandable.

A characteristic of this nonequilibrium or irreversible thermodynamics is that time is explicitly introduced. Furthermore, open systems, in which materials and energy flow into and out of the system, are considered. Clearly, a living organism is an open system not a closed one of classical thermodynamics. Because of the flow of materials concentration gradients are set up and transport phenomena often become of primary importance. Articles and books that provide an introduction to nonequilibrium thermodynamics and to the literature in the field include the following.[10,26,28-34] Whether these methods can be applied in a practical way to metabolic systems has been debated.[35,36]

TABLE 6-4
Gibbs Energies of Formation and of Oxidation at 25°C for Compounds of Biochemical Interest[a,b]

Compound	Formula	$\Delta G_f°$ (kJ mol^{-1})	$\Delta G_c°$ (kJ mol^{-1})	For oxidation by NAD$^+$ $\Delta G°_{ox}$ (kJ mol^{-1})	$\Delta G°_{ox}$ (pH 7) (kJ mol^{-1})	Number of electrons
Acetaldehyde	C_2H_4O	−139.7	−1123.5	171.5	−28.3	10
Acetic acid	$C_2H_4O_2$	−369.4	−866.8	169.2	9.3	8
Acetate$^-$		−369.2	−894.0	142.0	22.1	8
Acetyl-CoA		−374.1*	−889.1*	146.9*	−13.0*	8
Acetyl-P		−1218.4	−901.7	134.3	−25.6	8
Acetylene[c]	C_2H_2	209.2	−1235.2	59.8	−140.0	10
Acetoacetate$^-$	$C_4H_5O_3^-$	−493.7	−1795.4	276.5	−3.2	16
Acetone	C_3H_6O	−161.2	−1733.6	338.4	18.7	16
cis-Aconitate^{3-}	$C_6H_3O_6^{3-}$	−920.9	−2157.0	173.9	−65.8	18
L-Alanine	$C_3H_7O_2N$	−371.3	−1642.0	300.4	0.8	15
L-Asparagine	$C_4H_8O_3N_2$	−526.6	−1999.7	331.2	−28.4	18
L-Aspartate$^-$	$C_4H_6O_4N^-$	−700.7	−1707.0	235.4	−24.3	15
n-Butanol	$C_4H_{10}O$	−171.8	−2591.7	516.2	36.7	24
n-Butyric acid	$C_4H_8O_2$	−380.2	−2146.1	443.8	44.2	20
n-Butyrate$^-$	$C_4H_7O_2^-$	−352.6	−2173.7	416.2	56.6	20
Butyryl-CoA		−357.5*	−2168.8*	421.1*	21.5*	20
Caproate$^-$	$C_6H_{11}O_2^-$	−329.7	−3459.7	684.1	84.7	32
CO_2 (g)		−394.4	0.0	0.0	0.0	0
CO_2 (aq)		−386.2	−8.2	−8.2	−8.2	0
HCO_3^-		−587.1	−44.5	−44.5	−4.6	0
CO (g)		−137.3	−257.1	1.9	−38.1	2
Citrate^{3-}	$C_6H_5O_7^{3-}$	−1166.6	−2148.4	182.5	−57.3	18
Creatine	$C_4H_9O_2N_3$	−264.3	−2380.6	338.8	−80.8	21
Creatinine	$C_4H_7ON_3$	−28.9	−2378.8	340.6	−79.0	21
Crotonate$^-$	$C_4H_5O_2^-$	−275.7	−2013.4	317.5	−2.1	18
Cysteine	$C_3H_7O_2NS$	−339.8	−2178.3	541.1	121.5	21
Cystine	$C_6H_{12}O_4N_2S_2$	−665.3	−4133.8	1046.0	246.9	40
Dihydroxyacetone-P[d]		−1293.2	−1458.4	95.5	−144.2	12
Erythrose 4-P[d]		−1439.1	−1944.1	127.8	−191.9	16
Ethanol	C_2H_6O	−181.5	−1318.8	235.1	−4.6	12
Ethylene (g)[c]	C_2H_4O	68.1	−1331.3	222.7	−17.1	12
Formaldehyde	CH_2O	−130.5	−501.0	16.9	−63.0	4
Formic acid	CH_2O_2	−356.1	−275.5	−16.5	−56.5	2
Formate$^-$	CHO_2^-	−350.6	−281.0	−22.0	−22.0	2
Fructose	$C_6H_{12}O_6$	−915.4	−2874.1	233.8	−245.7	24
Fructose 6-P[d]		−1758.3	−2888.1	219.8	−259.7	24
Fructose di-P[d]		−2600.8	−2902.5	205.4	−274.1	24
Fumaric acid	$C_4H_4O_4$	−647.1	−1404.8	149.2	−90.6	12
Fumarate$^-$	$C_4H_3O_4^-$	−604.2	−1447.7	106.2	−93.6	12
α-D-Galactose	$C_6H_{12}O_6$	−923.5	−2865.9	242.0	−237.5	24
α-D-Glucose	$C_6H_{12}O_6$	−917.2	−2872.2	235.6	−243.8	24
Glucose 6-P		−1760.3	−2886.0	221.8	−257.6	24
L-Glutamate$^-$	$C_5H_8O_4N^-$	−696.8	−2342.5	376.9	−2.7	21
L-Glutamine	$C_5H_{10}O_3N_2$	−524.8	−2633.1	474.8	−4.7	24
3-P glycerate$^-$[d]		−1515.7	−1235.9	59.0	−100.8	10
2-P glycerate$^-$[d]		−1509.9	−1241.8	53.2	−106.6	10
Glyceraldehyde 3-P[d]		−1285.6	−1466.0	87.9	−151.8	12
Glycerol	$C_3H_8O_3$	−488.5	−1643.4	169.5	−110.2	14
Glycerol-P		−1336.2	−1652.6	160.3	−119.3	14
Glycine	$C_2H_5O_2N$	−373.5	−1008.3	157.2	−22.6	9
Glycogen	$C_6H_{10}O_5$	−665.3	−2887.0	220.9	−258.6	24
Glycolate$^-$	$C_2H_3O_3^-$	−523.4	−739.7	37.2	−42.7	6
Glyoxylate$^-$	$C_2HO_3^-$	−461.1	−564.9	−46.9	−86.9	4
H_2O (l)		−237.2	0.0	0.0	0.0	0
OH$^-$		−157.3	−79.9	−79.9	−39.9	0
H$^+$		0.0	0.0	0.0		0
H_2 (g)		0.0	−237.2	21.8	−18.2	2
H_2O_2		−136.8	−100.4	−359.4	−319.4	−2
H_2S		−27.4	−714.6	321.3	161.5	8
HS$^-$		12.6	−754.5	281.4	161.5	8
β-Hydroxybutyric acid	$C_4H_8O_3$	−531.4	−1994.9	336.0	−23.6	18
β-Hydroxybutyrate$^-$	$C_4H_7O_3^-$	−506.3	−2020.0	310.9	−8.8	18
Hydroxypyruvate	$C_3H_4O_4$	−615.9	−1041.6	−5.7	−165.5	8

TABLE 6-4
(continued)

Compound	Formula	ΔG_f° (kJ mol^{-1})	ΔG_c° (kJ mol^{-1})	For oxidation by NAD$^+$ ΔG°_{ox} (kJ mol^{-1})	$\Delta G'_{ox}$ (pH 7) (kJ mol^{-1})	Number of electrons
Hypoxanthine	C_5H_6O	89.5	−2773.0	334.8	−144.6	24
Isocitrate^{3-}	$C_6H_5O_7^{3-}$	−1160.0	−2155.1	175.8	−63.9	18
α-Ketoglutarate^{2-}	$C_5H_4O_5^{2-}$	−798.0	−1885.5	186.4	−53.3	16
Lactate$^-$	$C_3H_5O_3^-$	−516.6	−1378.1	175.9	−23.9	12
α-Lactose	$C_{12}H_{22}O_{11}$	−1515.2	−5826.5	389.3	−569.7	48
L-Leucine	$C_6H_{13}O_2N$	−356.3	−3551.7	721.6	62.3	33
Mannitol	$C_6H_{14}O_6$	−942.6	−3084.0	282.8	−236.6	26
Malate^{2-}	$C_4H_4O_5^{2-}$	−845.1	−1444.0	109.9	−49.9	12
Methane (g)	CH_4	−50.8	−818.0	218.0	58.2	8
Methanol	CH_4O	−175.2	−693.5	83.4	−36.4	6
NH_4^+		−79.5	−276.3	112.2	12.3	3
NO_2^-		−34.5	−84.1	−472.6	−372.7	−3
NO (g)		86.7	−86.7	−345.7	−305.7	−2
NO_3^-		−110.5	−8.1	−655.6	−515.7	−5
Oxalate^{2-}	$C_2O_4^{2-}$	−674.9	−351.1	−92.1	−52.1	2
Oxaloacetate^{2-}	$C_4H_2O_5^{2-}$	−797.2	−1254.7	40.2	−79.7	10
H_3PO_4 (aq)[e]		−1147.3	0.0	0.0	0.0	0
$H_2PO_4^-$ (aq)[e]		−1135.1	−12.1	−12.1	27.8	0
HPO_4^{2-} (aq)[e]		−1094.1	−53.1	−53.1	26.8	0
n-Propanol	C_3H_8O	−175.8	−1956.1	374.8	15.2	18
Isopropanol	C_3H_8O	−185.9	−1946.0	384.9	25.3	18
Propionate$^-$	$C_3H_5O_2^-$	−360.0	−1534.7	278.2	38.5	14
Pyruvate$^-$	$C_3H_3O_3^-$	−474.5	−1183.1	111.9	−47.9	10
Phosphoenolpyruvate^{3-}		−1269.5	−1245.0	50.0	−109.8	10
Ribose 5-P[d]		−1599.9	−2414.9	175.0	−224.6	20
Ribulose 5-P[d]		−1597.6	−2417.1	172.8	−226.8	20
Sedoheptulose 7-P[d]		−1913.3	−3364.6	261.2	−298.2	28
Sedoheptulose di-P[d]		−2755.8	−3379.0	246.9	−312.5	28
Sorbitol	$C_6H_{14}O_6$	−942.7	−3083.9	282.9	−236.5	26
Succinate^{2-}	$C_4H_4O_4^{2-}$	−690.2	−1598.9	214.1	14.3	14
Succinyl-CoA		−686.7*	−1602.4*	210.6*	−29.2*	14
Sucrose	$C_{12}H_{22}O_{11}$	−1551.8	−5789.9	425.9	−533.1	48
SO_4^{2-}		−742.0	0.0	0.0	79.9	0
SO_3^{2-}		−497.1	−244.9	14.1	54.0	2
$S_2O_3^{2-}$		−513.4	−733.4	302.5	222.6	8
L-Threonine	$C_4H_9O_3N$	−514.6	−2130.3	330.1	−49.5	19
L-Tyrosine	$C_9H_{11}O_3N$	−387.2	−4466.8	842.5	23.4	41
Urea	CH_4ON_2	−203.8	−664.9	112.0	−7.8	6
Uric acid	$C_5H_4O_3N_4$	−356.9	−2089.4	241.5	−118.1	18
L-Valine	$C_5H_{11}O_2N$	−360.0	−2916.5	579.9	40.5	27
Xanthine	$C_5H_5O_2N_4$	−139.3	−2425.6	293.8	−125.7	21
D-Xylulose	$C_5H_{10}O_5$	−748.1	−2409.8	180.1	−219.5	20

[a] The quantities tabulated are ΔG_f°, the standard free energy of formation from the elements; ΔG_c°, the standard free energy of combustion; ΔG°_{ox}, the standard free energy of oxidation by NAD$^+$ to products NADH + H$^+$, CO_2, H_2O, N_2, HPO_4^{2-}, and SO_4^{2-}; $\Delta G'_{ox}$ (pH 7), the apparent standard free energy change at pH 7. All values are in kJ mol^{-1} at 25°C in aqueous solution unless indicated otherwise. If a compound is designated (g) the values are for the gaseous phase at 1 atm pressure. The number of electrons involved in complete oxidation to CO_2, H_2O, N_2, and H_2SO_4 is given in the final column. If this number is negative, the compound must be reduced to obtain the products, e.g., $2\,NO_3^- + 10\,e^- + 12\,H^+ \rightarrow N_2 + 6\,H_2O$. The data for phosphate esters refer to the compounds with completely dissociated phosphate groups ($-O-PO_3^{2-}$). The values of ΔG_f° for many of these compounds were calculated as ΔG_f° (nonphosphorylated compound) − ΔG° for hydrolysis (to HPO_4^{2-}, Table 6-6) −ΔG_f° for H_2O (one molecule for each phosphate ester formed) + ΔG_f° for HPO_4^{2-} (from this table). Data from Bassman and Krause[d] were used directly. For acyl-CoA derivatives CoA (−SH) is treated as an "element," i.e., the values of ΔG_f° given and designated with an asterisk (*) are for formation from the elements plus free CoA. The values of ΔG_c° and ΔG_{ox} are for oxidation to the usual products plus CoA. Values of ΔG° of hydrolysis (Table 6-6) were used in computing ΔG_f° for each of these compounds from that of the corresponding alcohol or carboxylate anion. Another source containing an extensive table of Gibbs energy values is Wilhoit, R. C. (1969) in *Biochemical Microcalorimetry* (Brown, H. D. ed.), pp. 305–317. Academic Press, New York

[b] The major source is Long, C., ed. (1961) *Biochemists Handbook*, pp. 90–92. Van Nostrand, Reinhold, Princeton, New Jersey. Most of the values in this collection are from Burton, K. (1957) *Ergeb. Physiol., Biol. Chem. Exp. Pharmakol.* **49**, 275–298

[c] From Stull, D.R., Westrum, E. F., Jr., and Sinke, G. C. (1969) *The Chemical Thermodynamics of Organic Compounds*. Wiley, New York

[d] Bassham, J. A. and Krause, G. H. (1969) *Biochim. Biophys. Acta.* **189**, 207–221

[e] Van Wazer, J. R. (1958) *Phosphorus and Its Compounds*, Vol. I, p. 889. Wiley (Interscience), New York

B. Tables of $\Delta G°$ Values for Biochemical Compounds

1. Gibbs Energies of Formation

Table 6-4 gives, in the first column, standard values of Gibbs energies of formation from the elements $\Delta G_f°$ for a variety of pure solids, gases, and liquids as well as values for substances in solution at the hypothetical 1 M activity. As an example, consider the value of $\Delta G_f°$ for pure liquid acetic acid, -389.1 kJ mol^{-1}. The equation for its formation from the elements is:

$$2\,C\,(s) + 2\,H_2\,(g, 1\,atm)$$
$$+ O_2\,(g, 1\,atm) \longrightarrow C_2H_4O_2\,(l)$$
$$\Delta G_f° = -389.1\,\text{kJ mol}^{-1} \tag{6-38}$$

To obtain the Gibbs energy of formation in aqueous solution, we must have solubility data as well as activity coefficients of acetic acid at various concentrations. From these data the change in Gibbs energy for solution of the liquid acetic acid in water to give aqueous acetic acid in the hypothetical 1 molal standard state (Eq. 6-39) can be obtained.

$$\text{Acetic acid (l)} \longrightarrow \text{acetic acid (aq)}$$
$$\Delta G = -7.3\,\text{kJ mol}^{-1} \tag{6-39}$$

Summing Eqs. 6-38 and 6-39 we obtain:

$$2\,C + 2\,H_2 + O_2 \longrightarrow \text{acetic acid (aq)}$$
$$\Delta G_f° = -396.4\,\text{kJ mol}^{-1} \tag{6-40}$$

In many computations it is convenient to have ΔG values for single ions, e.g., for acetate$^-$. We can obtain $\Delta G_f°$ of acetate$^-$ (aq) from that of acetic acid (aq) by making use of $\Delta G°$ of dissociation (Eq. 6-41).

$$\text{Acetic acid} \longrightarrow H^+ + \text{acetate}^-$$
$$\Delta G° = -5.708 \log K_a = 5.708\,\text{p}K_a$$
$$= +27.2\,\text{kJ mol}^{-1}\,\text{at}\,25°C \tag{6-41}$$

By convention *we define the Gibbs energy of formation of H^+ as zero.* Then, by summing Eqs. 6-40 and 6-41 we obtain $\Delta G_f°$ of acetate$^-$ = -369.2 kJ mol^{-1}.

2. Gibbs Energies of Dissociation of Protons

Table 6-5 gives thermodynamic dissociation constants and values of $\Delta G°$ and $\Delta H°$ for a number of acids of interest in biochemistry. Some of these values were used in obtaining the values of $\Delta G_f°$ for the ions of Table 6-4. The data of Table 6-5 can also be used in evaluation of Gibbs energy changes for reactions of ionic forms not given in Table 6-4.

3. Group Transfer Potentials

Recall that the equilibria for reactions by which monomers are linked to form biopolymers (whether amides, esters, phosphodiesters, or glycosides) usually favor *hydrolysis* rather than formation (condensation). The equilibrium positions depend on the exact structures. Some linkages are formed easily if monomer concentrations are high enough, but others are never formed in significant concentrations. Likewise, hydrolysis may be partial at equilibrium or it may be 99.9% or more complete.

Let us compare the hydrolysis of the two organic phosphates **adenosine triphosphate** (ATP) and **glucose 6-phosphate** (Eqs. 6-42 and 6-43).

$$HATP^{3-} + H_2O \longrightarrow HADP^{2-} + H_2PO_4^-$$
$$\Delta G° \approx -32.9\,\text{kJ mol}^{-1}$$
$$\Delta H° \approx -22.6\,\text{kJ mol}^{-1}\,\text{at}\,25°C,\ \mu = 0.25 \tag{6-42}$$

$$\alpha\text{-D-Glucose 6-phosphate}^- + H_2O \longrightarrow$$
$$\text{glucose} + H_2PO_4^-$$
$$\Delta G° = -16.4\,\text{kJ mol}^{-1}\,\text{at}\,25°C \tag{6-43}$$

The decrease in Gibbs energy upon hydrolysis is twice as large for ATP as it is for glucose 6-phosphate. Glucose phosphate is thermodynamically more stable than ATP. It would be easier to form than would ATP by a reversal of the hydrolysis reaction and also easier to form biosynthetically. From the Gibbs energies of hydrolysis it follows that a **phospho group** could be transferred spontaneously from ATP to glucose in the presence of a suitable catalyst but not vice versa.

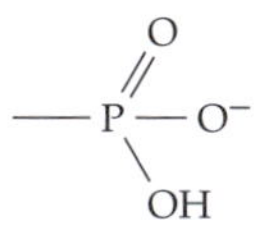

$$\begin{array}{c} O \\ \| \\ -P-O^- \\ | \\ OH \end{array}$$

Phospho (phosphoryl) group

Because it reflects quantitatively the thermodynamic tendency for a group to be transferred to another

TABLE 6-5
Values of pK_a, $\Delta G°$, and $\Delta H°$ for Ionization of Acids at 25°C [a,b]

Acid	pK_a	$\Delta G°$ (kJ mol^{-1})	$\Delta H°$ (kJ mol^{-1})
Formic acid	3.75	21.4	0.04
Acetic acid	4.76	27.2	−0.1
Propionic acid	4.87	27.8	−0.6
Lactic acid	3.97 (35°C)	23.4	2.2
Pyruvic acid	2.49	14.2	12.1
NH_4^+	9.25	52.8	52.2
$CH_3NH_4^+$	10.59	60.4	55.4
Alanine			
–COOH	2.35	13.4	3.1
–NH_3^+	9.83	56.1	45.4
β-Alanine			
–COOH	3.55	20.3	4.5
–NH_3^+ (apparent)	10.19	58.2	
L-Alanyl-L-alanine	3.34	19.1	−0.5
Aspartic acid			
–COOH	2.05	11.7	7.7
–COOH	3.87	22.1	4.0
–NH_3^+	10.60	60.5	38.8
H_2CO_3, pK_1	6.35[c]	36.2	9.4
pK_2	10.33	59.0	15.1
H_3PO_4, pK_1	2.12	12.1	−7.9
pK_2	7.18[d]	41.0	3.8
(apparent)	6.78[e]	38.7	3.3
pK_3	12.40	70.8	17.6
Glycerol 1-phosphate, pK_2	6.66	38.0	−3.1
Glucose 6-phosphate	6.50	37.1	−1.8
Pyrophosphoric acid, $H_4P_2O_7$			
pK_3	6.7	38.1	−1.3
(apparent)	6.12[e]	34.9	0.5
pK_4	9.4	53.6	−7.1
(apparent)	8.95[e]	51.2	1.7
Adenosine	3.5	20.1	13.0
AMP			
pK_1 (ring, apparent)	3.74[e]	21.3	4.2
pK_2 (phosphate)	6.67[d]	38.1	3.6
(apparent)	6.45[e]	36.8	3.6
ADP			
pK_2 (ring, apparent)	3.93[e]	22.4	4.2
pK_3 (diphosphate)	7.20[d]	41.1	−5.7
(apparent)	6.83[f]	39.0	−5.7
ATP			
pK_3 (ring, apparent)	4.06	23.2	0
pK_4 (triphosphate)	7.68[d]	43.8	−7.0
(apparent)	7.06[d]	40.2	−7.0
Pyridine	5.17	29.5	20.1
Phenol	9.98	56.9	23.6

[a] These are thermodynamic values (infinite dilution) except for those labeled apparent. The latter apply at an ionic strength of 0.2–0.25.

[b] Most data are from Jencks, W. P. and Regenstein, J. (1976) in *Handbook of Biochemistry and Molecular Biology*, 3rd ed., Vol. I (Fasman, G. D. ed.), pp. 305–351, CRC Press, Cleveland, Ohio.

[c] Here, pK_1 is for $K_1 = [H^+][HCO_3^-] / [CO_2] + [H_2CO_3]$. From Forster, R. E., Edsall, J. T., Otis, A. B., and Roughton, F. J. W., eds. (1969) *NASA Spec. Publ.* 188.

[d] From Phillips, R. C., George, P., and Rutman, R. J. (1963) *Biochemistry* 2, 501–508.

[e] From Alberty, R. A. (1969) *J. Biol. Chem.* 244, 3290–3302.

[f] Values used by Alberty, R. A. (1972) *Horizons of Bioenergetics*, pp. 135–147, Academic Press, New York, calculated for 0.2 ionic strength from equations of Phillips, R. C., George, P., and Rutman, R. J. (1966) *J. Am. Chem. Soc.* 88, 2631–2640.

nucleophile (see Chapter 12), *the Gibbs energy decrease (−ΔG°) upon hydrolysis* is sometimes called the **group transfer potential**. During the hydrolysis of ATP (Eq. 6-42) the phospho group of ATP is transferred to a hydroxyl ion from water with a value $\Delta G° = -32.9$ kJ mol^{-1}. The group transfer potential of this phospho group is 32.9 kJ/mol and that of the phospho group of glucose 6-phosphate is only 27.6 kJ mol^{-1}. While the choice of water as the reference nucleophile for expression of the group transfer potential is somewhat arbitrary, it is customary. Transfer of groups is important in energy metabolism and in biosynthesis of polymers. Gibbs energies of hydrolysis are given in Table 6-6 for several compounds.

4. "Constants" That Vary with pH and Magnesium Ion Concentrations

Equation 6-42 is written for hydrolysis of HATP^{3-} to HADP^{2-} + $H_2PO_4^-$, a stoichiometry that applies well in the pH range around 6. However, at a pH above ~7 most of the ATP is in the form ATP^{4-} and is cleaved to HPO$_4^{2-}$ according to Eq. 6-44.

$$ATP^{4-} + H_2O \rightarrow ADP^{3-} + HPO_4^{2-} + H^+$$
$$\Delta G° \approx +5.4 \text{ kJ mol}^{-1} \text{ at 25°C, } \mu = 0.25$$
$$\Delta H° \approx -19.7 \text{ kJ mol}^{-1}$$

(6-44)

The value of $\Delta G° = +5.4$ kJ mol^{-1} for this reaction is hardly the large negative number expected for a highly spontaneous reaction. What is the matter? The problem is that H$^+$ is produced and that the standard state of H$^+$ is 1 M, not 10^{-7} M. Because of this, biochemists often prefer to use another kind of **apparent dissociation constant** and an **apparent ΔG** such that the standard state of H$^+$ is taken as that of the pH at which the experiments were done, usually pH 7. The symbol K' has often be used and is used in this book to represent the following pH-dependent equilibrium constant (Eq. 6-45) which will be a constant only at a single pH.

$$K' = \frac{[ADP^{3-}][HPO_4^{2-}]}{[ATP^{4-}]} \quad (6\text{-}45)$$

If one proton is produced in the reaction as in Eq. 6-44, the following relationship will hold.

$$\Delta G' = \Delta G° - 5.708 \times pH \text{ kJ mol}^{-1} \text{ at } 25°C \quad (6\text{-}46)$$

Note that $\Delta G' = -RT \ln K'$ and that $[H^+]$ does not appear in the expression for K' given by Eq. 6-45. From the value $\Delta G° = +5.4$ kJ mol^{-1} and applying Eq.

TABLE 6-6
Gibbs Energies of Hydrolysis at 25°C (in kJ mol^{-1})[a]

Compound	Products	$\Delta G°$	$\Delta G'$(pH 7)	$\Delta H°$
ATP^{4-}	$ADP^{3-} + HPO_4^{2-} + H^+$	5.41[b]	−34.54	−19.71
ATP^{4-}	$AMP^{2-} + HP_2O_7^{3-} + H^+$	2.54[c]	−37.4	−19.0
$MgATP^{2-}$	$MgADP^- + HPO_4^{2-} + H^+$	16.0[d]	−24.0	−14.2
ADP^{3-}	$AMP^{2-} + HPO_4^{2-} + H^+$	3.67[c]	−36.3	−13.5
AMP^{2-}	Adenosine + HPO_4^{2-}	−9.6[e]	−9.6	0
ATP^{4-}	Adenosine + $HP_3O_{10}^{4-}$	−36.0[e]	−36.0	−7.9
$HP_2O_7^{3-}$	$2 HPO_4^{2-} + H^+$	6.54[c]	−33.4	−12.6
Acetyl phosphate^{2-}	Acetate$^-$ + HPO_4^{2-} + H^+	−7.7[f]	−47.7	
1,3-Diphosphoglycerate^{4-}	3-Phosphoglycerate^{3-} + HPO_4^{2-} + H^+	−14.5[g]	−54.5	
Phosphoenolpyruvate^{3-}	Pyruvate$^-$ + HPO_4^{2-}	−61.9	−61.9	−25.1
Carbamoyl phosphate^{2-}	$H_2N-\overset{\overset{\text{O}}{\|\|}}{C}-O^- + HPO_4^{2-} + H^+$	−11.5	−51.5	
Creatine phosphate$^-$	Creatine$^+$ + HPO_4^{2-}	−43.1	−43.1	
Phosphoarginine$^-$	Arginine + HPO_4^{2-}		−38.1[h]	(Mg^{2+} present)
Glycerol phosphate^{2-}	Glycerol + HPO_4^{2-}	−9.2	−9.2	
α-D-Glucose 6-phosphate^{2-}	α-D-Glucose + HPO_4^{2-}	−13.8	−13.8	−2.5
Glucose 1-phosphate^{2-}	Glucose + HPO_4^{2-}	−20.9	−20.9	
Maltose (or glycogen)	2-Glucose	−16.7	−16.7	
Sucrose	Glucose + fructose	−29.3	−29.3	
UDP glucose^{2-}	Glucose + UDP^{3-} + H^+	9.4	−30.5	
N^{10}-Formyltetrahydrofolic acid	Formate$^-$ + H^+ + tetrahydrofolic acid	14.1	−25.9[h]	
Acetic anhydride	2-Acetate$^-$ + 2H^+	31.1	−48.9	
Acetyl-CoA	Acetate$^-$ + H^+ + CoA	4.9	−35.1[i]	
Succinyl-CoA$^-$	Succinate^{2-} + H^+ + CoA	−3.5	−43.5[j]	
Ethyl acetate	Ethanol + acetate$^-$ + H^+	20.2	−19.7	
Asparagine	Aspartate$^-$ + NH_4^+	−15.1	−15.1	
Glycine ethyl ester$^+$ (39°C)	Glycine + ethanol + H^+	4.9	−35.1	
Valyl-tRNA$^+$	Valine + tRNA + H^+	4.9	−35.1	

[a] Unless indicated otherwise, the values are based on tables from Jencks, W. P. (1976) *Handbook of Biochemistry and Molecular Biology*, 3rd ed., Vol I (Fasman,G.D. ed.), pp. 296–304. CRC Press, Cleveland, Ohio. For a reaction producing one proton at pH 7 $\Delta G' = -39.96$ kJ mol^{-1}.

[b] Guynn, R.W. and Veech, R.L. (1973) *J. Biol. Chem.* **248**, 6966–6972.

[c] Based on +11.80 kcal mol^{-1} for hydrolysis to $P_2O_7^{4-}$ plus $\Delta G°$ of dissociation of $HP_2O_7^{3-}$ as quoted by Alberty, R. A. (1969) *J.Biol.Chem.* **244**, 3290–3324. However, 1.017 kcal mol^{-1} was added to the value of 11.80 to make it consistent with that for hydrolysis of ATP to ADP. Reevaluation by Frey and Arabshahi (1995) *Biochemistry* **34**, 11307–11310, indicates that $\Delta G'$ (pH 7) for hydrolysis of ATP to AMP + PP$_i$ is ~ 10 kJ mol^{-1} more negative than is shown here.

[d] Alberty, R.A. (1972) *Horizons of Bioenergetics*, Academic Press, New York, pp. 135–147.

[e] George, P. , Witonsky, R.J., Trachtman, M., Wu, C., Dorwart, W., Richman, L., Richman, W., Shurayh, F., and Lentz, B. (1970) *Biochim. Biophys. Acta.* **223**, 1–15.

[f] Based on $\Delta G^{\ddagger} = 3.0$ kcal mol^{-1} for acetyl phosphate + CoA → acetyl-CoA + P$_i$ from Stadtman, E. R. (1973) *The Enzymes*, (Boyer, P.D., ed.), 3rd ed., Vol. 8. pp. 1–49. Academic Press, New York, together with $\Delta G'$ (pH 7) for hydrolysis of acetyl-CoA.

[g] Estimated from $\Delta G°$ for ATP hydrolysis + $\Delta G' = -19.9$ kJ mol^{-1} for the 3-phosphoglycerate kinase reaction: Burton, K. and Krebs, H. A. (1953) *Biophys. J.* **54**, 94–107; and (1955) *Biophys. J.* **59**, 44–46.

[h] Estimated from data at pH 7.7 or 8.0 (tables of Jencks).

[i] Guynn, R. W., Gelberg, H. J., and Veech, R. L. (1973) *J. Biol. Chem.* **248**, 6957–6965 found $\Delta G° = -35.75$ kJ mol^{-1} at 38°C. Without data on ΔH, correction to 25°C is difficult. Burton, K., (1955) *Biophys. J.* **59**, 44–46, gave $\Delta G'$ (pH 7) for ATP^{4-} + acetate$^-$ + CoA → ADP^{3-} + acetyl-CoA + HPO_4^{2-} as approximately zero at 25°C. Guynn *et al.* found −0.56 kJ mol^{-1} at 38°C. This same value (−0.56 kJ) at 25°C was assumed to obtain the figure given here. This is equivalent to assuming $\Delta H°$ of hydrolysis as almost the same for ATP and acetyl-CoA, an unsupported assumption.

[j] Assumed 2 kcal mol^{-1} more negative than that of acetyl-CoA as in tables of Jencks.

6-46, we obtain for the hydrolysis of ATP at 25°C, $\mu = 0.25$:

$$\Delta G' \text{ (pH 7)} = -34.5 \text{ kJ mol}^{-1}$$
$$= -8.26 \text{ kcal mol}^{-1} \qquad (6\text{-}47)$$

An additional set of standard states is frequently met in the biochemical literature. An equilibrium constant, designated in this book as $K^{\dagger}$, is used to relate the *total concentrations of all ionic forms* of the components present at the pH of the experiment. Thus,

$$K^{\dagger} = \frac{[\text{ADP, all forms}][\text{phosphate, all forms}]}{[\text{ATP, all forms}]} \qquad (6\text{-}48)$$

What is the pH within a cell? Is it constant or does it vary with physiological conditions? Do all cells operate at similar pH? The answers to these important questions have been sought using a variety of techniques.[a] Tiny microelectrodes with tips only 1 μm in diameter have been inserted into cells. Indicator dyes have been diffused into cells and either light absorption or fluorescence[b] measured. The distribution of a suitable radiolabeled weak acid or weak base, such as [14C]methylamine, that is able to permeate cells can be used to calculate the difference between internal and external pH.[c] The activity of the enzyme carbonic anhydrase, which is very pH sensitive, can be used to monitor the pH of mitochrondia.[d] Since 1973 NMR methods have been popular.[e-1] The chemical shifts of 31P in inorganic phosphate ($pK_a = 6.9$), ATP, glucose 6-phosphate, and of other metabolites of 13C in citrate and bicarbonate,[l] give direct estimates of pH. However, caution must be exercised if internal pH values good to ± 0.1 unit are to be obtained.[g] More sensitive measurements over the pH range 1.3 – 9.1 can be made by diffusing one of a series of aminophosphonates into cells and measuring the 31P chemical shift.[h] Fluorinated probes such as dimethylfluoroalanine ($pK_a = 7.3$) are also useful because of their low toxicity and high sensitivity of the 19F NMR signal. These alanine derivatives can be diffused into cells as their methyl esters, which are rapidly cleaved within cells, allowing the fluorinated amino acids to accumulate.[i,m]

The pH within cells appears to be tightly controlled although small variations are sometimes observed. Red blood cells, thymocytes, liver, skeletal muscles, and intact hearts all maintain a pH in the range 7.0 – 7.3.[b,h,i] However, the pH can fall to 6.2 within 13 minutes of oxygen deprivation (ischemia) and to 6.1 after exhaustive exercise.[n,o] The 31P NMR technique permits the monitoring of pH as well as the state of the adenylate system (Section D) in human limbs suffering from circulatory insufficiency.[c]

In higher plants the pH of cytoplasm is 7.4–7.5 but vacuoles are acidic with a pH of 4.5–6.[l] The cytoplasm of maize root tips has a pH of 7.1 but the vacuoles are at a pH of 5.5.[f] The pH of granules of the chromaffin cells of the adrenal cortex, which accumulate high concentrations of ATP and catecholamines, is also low, ~ 5.7.[p] The bacterium *Streptococcus faecalis* maintains a higher internal pH of ~ 8.0, even when the pH of the medium varies from 6.5 to 8.0[q] while *E. coli* operates at pH 7.6, the extremes of variation being 7.4 – 7.8 when the external pH changes from 5.5–9.[r]

Changes of internal pH during developmental events such as fertilization of sea urchin eggs (+0.3 unit) have been recorded. However, the significance of pH changes in metabolic regulation remains uncertain.[c]

[a] Kotyk, A., and Slavik, J. (1989) *Intracellular pH and its Measurement*, CRC Press, Boca Raton, Florida
[b] Rogers, J., Hesketh, T. R., Smith, G. A., and Metcalfe, J. C. (1983) *J. Biol. Chem.* **258**, 5994–5997
[c] Nuccitelli, R., and Deamer, D. W., eds. (1982) *Intracellular pH: Its Measurement, Regulation and Utilization in Cellular Functions*, Liss, New York
[d] Dodgson, S. J., Forster, R. E., II, and Storey, B. T. (1982) *J. Biol. Chem.* **257**, 1705–1711
[e] Moon, R. B., and Richards, J. H. (1973) *J. Biol. Chem.* **248**, 7276–7278
[f] Roberts, J. K. M., Ray, P. M., Wade-Jardetsky, N., and Jardetsky, O. (1980) *Nature (London)* **283**, 870–872
[g] Avison, M. J., Hetherington, H. P., and Shulman, R. G. (1986) *Ann. Rev. Biophys. Biophys. Chem.* **15**, 377–402
[h] Pietri, S., Miollan, M., Martel, S., Le Moigne, F., Blaive, B., and Culcasi, M. (2000) *J. Biol. Chem.* **275**, 19505–19512
[i] Taylor, J. S., and Deutsch, C. (1983) *Biophys. J.* **43**, 261–267
[j] Bailey, I. A., Williams, S. R., Radda, G. K., and Gadian, D. G. (1981) *Biochem. J.* **196**, 171–178
[k] Barton, J. K., Den Hollander, J. A., Lee, T. M., MacLaughlin, A., and Shulman, R. G. (1980) *Proc. Natl. Acad. Sci. U.S.A.* **77**, 2470–2473
[l] Gout, E., Bligny, R., and Douce, R. (1992) *J. Biol. Chem.* **267**, 13903–13909
[m] Deutsch, C. J., and Taylor, J. S. (1987) *Ann. N.Y. Acad. Sci.* **508**, 33
[n] Garlick, P. B., Radda, G. K., and Seeley, P. J. (1979) *Biochem. J.* **184**, 547–554
[o] Pan, J. W., Hamm, J. R., Rothman, D. L., and Shulman, R. G. (1988) *Proc. Natl. Acad. Sci. U.S.A.* **85**, 7836–7839
[p] Pollard, H. B., Shindo, H., Creutz, C. E., Pazoles, C. J., and Cohen, J. S. (1979) *J. Biol. Chem.* **254**, 1170–1177
[q] Kobayashi, H., Murakami, N., and Unemoto, T. (1982) *J. Biol. Chem.* **257**, 13246–13252
[r] Slonczewski, J. L., Rosen, B. P., Alger, J. R., and MacNab, R. M. (1981) *Proc. Natl. Acad. Sci. U.S.A.* **78**, 6271–6275

and

$$\Delta G^{\dagger} = -RT \ln K^{\dagger} \qquad (6\text{-}49)$$

The Gibbs energy change $\Delta G^{\dagger}$ can be related to $\Delta G'$ by considering the relationship of $K^{\dagger}$ to K'. For ATP hydrolysis in the pH range of 2–10, $K^{\dagger}$ is given by Eq. 6-50.

$$K^{\dagger} = \frac{K'\left(1 + \dfrac{[H^+]}{K_{HADP^{2-}}} + \dfrac{[H^+]^2}{K_{HADP^{2-}}K_{H_2ADP^-}}\right)\left(1 + \dfrac{[H^+]}{K_{HPO_4^{2-}}}\right)}{\left(1 + \dfrac{[H^+]}{K_{HATP^{3-}}} + \dfrac{[H^+]^2}{K_{HATP^{3-}}K_{H_2ATP^{2-}}}\right)} \qquad (6\text{-}50)$$

In this equation K_{HADP2-}, etc., are consecutive dissociation constants as given in Table 6-4. The expressions in parentheses are the **Michaelis pH functions**, which were considered in Chapter 3 (Eqs. 3-4 to 3-6). In Eq. 6-50 they relate the total concentration of each component to the concentration of the most highly dissociated form. Thus, for the pH range 2–10

$$[P_i]_{total} = [HPO_4^{2-}](1 + [H^+] / K_{H_2PO_4^-}) \qquad (6\text{-}51)$$

Using apparent pK_a values ($\mu = 0.2$) for $H_2PO_4^-$, $HADP^{2-}$, and $HATP^{3-}$ of 6.78, 6.83, and 7.06 (Table 6-5) and taking $\Delta G'$ at pH 7 as -34.5 kJ mol^{-1}, we compute $\Delta G^{\dagger} = -35.0$ kJ mol^{-1} at pH 7. The difference between $\Delta G'$ and $\Delta G^{\dagger}$ in this case is small, but it would be larger if the ionic forms in Eq. 6-44 were not the ones predominating at pH 7.

To obtain the Gibbs energy change for a reaction under *other than standard conditions*, Eq. 6-29 must be applied. Thus, at pH 7 and 0.01 M activities of ADP^{3-}, ATP^{4-}, and HPO_4^{2-}, ΔG for hydrolysis of ATP according to Eq. 6-44 is $-34.5 - (2 \times 5.71) = -45.9$ kJ mol^{-1} = -11.0 kcal mol^{-1}. We see that at concentrations existing in cells (usually in the millimolar range) ATP has a substantially higher group transfer potential than under standard conditions.

The reader should bear in mind that there is no accepted standard usage of K' and that $K^{\dagger}$ is just for this book! An international committee[21] has recommended that K' be used with the same meaning as $K^{\dagger}$ in this book and more changes may be coming (see the next section).

To obtain ΔG at a temperature other than 25°C, we must know ΔH for the reaction. Using Eq. 6-37a it is easy to show that ΔG at temperature T_2 is related to that at T_1 as follows:

$$\Delta G_2 \approx \frac{T_2 \Delta G_1 - (T_2 - T_1) \Delta H}{T_1} \qquad (6\text{-}52)$$

The enthalpy of hydrolysis of ATP according to Eq. 6-44 is approximately -19.7 kJ mol^{-1}. Using this value and applying Eq. 6-52 we can calculate that $\Delta G'$ (pH 7) for the hydrolysis of ATP at 38°C is -35.2 kJ mol^{-1}. The value of $\Delta G'$ (25°C, pH 7), used in obtaining this answer, is -35.54 kJ mol^{-1} which was computed from the value of $\Delta G'$ (pH 7), 38°C of 35.19 kJ mol^{-1} reported by Guynn and Veech[37] using Eq. 6-52. Bear in mind that all of the foregoing Gibbs energy changes are *apparent* values applying to solutions of ionic strength ~0.25.

Both ADP and ATP as well as inorganic pyrophosphate form complexes with metal ions. Since the magnesium complexes are often the predominant forms of ADP and ATP under physiological conditions, we must consider the following Gibbs energy changes. These are apparent values for $\mu = 0.2$ at 25°C.

$$ATP^{4-} + Mg^{2+} \rightarrow Mg\,ATP^{2-}$$
$$\Delta G° = -26.3; \Delta H° = 13.8 \text{ kJ mol}^{-1} \qquad (6\text{-}53)$$

$$ADP^{3-} + Mg^{2+} \rightarrow Mg\,ADP^{-}$$
$$\Delta G° = -19.8; \Delta H° = 15.1 \text{ kJ mol}^{-1} \qquad (6\text{-}54)$$

Combining the apparent $\Delta G°$ values for Eqs. 6-47, 6-53, and 6-54, we obtain

$$MgATP^{2-} + H_2O \rightarrow MgADP^- + HPO_2^{2-} + H^+$$
$$\Delta G' \text{ (pH 7)} = -28.0 \text{ kJ mol}^{-1} \qquad (6\text{-}55)$$

The stoichiometry of Eq. 6-55 never holds exactly. Some of the Mg^{2+} dissociates from the $MgADP^-$; both protons and Mg^{2+} bind to $H_2PO_4^-$; and $HATP^{3-}$ and $HADP^{2-}$ are present and bind Mg^{2+} weakly.[38] Thus, the observed value of $\Delta G^{\dagger}$ varies with both pH and magnesium concentration as well as with changes in ionic strength. Tables and graphs showing the apparent value of $\Delta G^{\dagger}$ under various conditions have been prepared by Alberty[38] and by Phillips et al.[39] An example is shown in Fig. 6-2. From this graph we find that $\Delta G^{\dagger}$ for hydrolysis of ATP at pH 7, 25°C, $\mu = 0.2$, and 1 mM Mg^{2+} (a relatively high intracellular concentration[40]) is -30.35 kJ mol^{-1} (-7.25 kcal mol^{-1}).

Figure 6-2 was drawn using the equations of Alberty, but the value of $\Delta G'$ (pH 7) of hydrolysis of ATP = 34.54 kJ mol^{-1} at $[Mg^{2+}] = 0$ based on results of Guynn and Veech[37] was used. The pK_a values and formation constants of Mg^{2+} complexes were those of Alberty.[38] Note that George et al.[40] provided formation constants of these complexes at infinite dilution where the values of $\Delta G°$ of formation are considerably more negative than those given in Eqs. 6-53 to 6-55.

From the foregoing considerations we see that complexing with Mg^{2+} somewhat decreases the group transfer potential of the phospho group of ATP. Furthermore, changes in the concentration of free Mg^{2+} with time and between different regions of a cell may

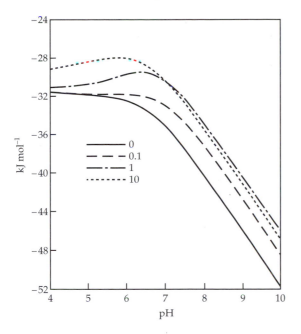

Figure 6-2 Plots of the apparent Gibbs energy $\Delta G^{\dagger}$ for hydrolysis of ATP as a function of pH at a series of different concentrations of free magnesium ions. Millimolar $[Mg^{2+}]$ is indicated by the numbers by the curves. Computer-drawn graphs courtesy of Carol M. Metzler.

have significant effects.[41] While Mg^{2+} is a principal cation in tissues, it is by no means the only one. Thus, Ca^{2+}, Mn^{2+}, and even K^+ will affect equilibria involving polyphosphates such as ATP.

It is not easy to measure the group transfer potential of ATP and published values vary greatly. George et al.[40] reported $\Delta G°$ for Eq. 6-42 as -39.9 kJ mol^{-1} at 25°C, $\mu = 0.2$, and as -41.25 kJ mol^{-1} at infinite dilution. They regarded these values as good to ± 4 kJ mol^{-1}. However, most other estimates have been at least 4 kJ mol^{-1} less negative.[42–44] The self-consistent set of thermodynamic data used throughout this book are based in part on the value of ΔG for hydrolysis of ATP obtained by Guynn and Veech.[37]

5. A New Standard for Biochemical Thermodynamics?

Because of the complexities of the equilibria involved in biochemical reactions Alberty and others[45–47] and a Panel on Biochemical Thermodynamics[48] have suggested that tables of thermodynamic properties for biochemical use be tabulated for the following conditions: $T = 298.15$ K, $P = 1$ bar (0.1 MP$_a$ $\equiv$ 0.987 atm), pH = 7, pMg = 3 ($[Mg^{2+}] = 10^{-3}$ M), and ionic strength I (μ in this chapter) = 0.25 M. There appears to be both advantages and disadvantages to this. The proponents suggest that the symbols $\Delta G'$, K', etc. be used for this

new standard. To change all of the numbers in this book, which uses a self-consistent set of thermodynamic quantities, is impractical. However, it is worthwhile to compare the value of Alberty and Goldberg for $\Delta G'°$ for hydrolysis of ATP to ADP + P$_i$ under the proposed biochemical standard conditions with other values given in this chapter, all at 25°C.

$\Delta G'° = -32.49$ kJ mol^{-1} (new proposed biochemical standard)
$\Delta G^{\dagger}$ (pH 7) -35.0 kJ mol^{-1} (Eq. 6-50)
$\Delta G'$ (pH 7) -34.5 kJ mol^{-1} (for ATP^{4-} $\rightarrow$ ADP^{3-} + HPO4^{2-}; Eq. 6-47)
$\Delta G'$ (pH 7) -28.0 kJ mol^{-1} (for MgATP^{2-} $\rightarrow$ MgADP$^-$ + HPO4^{2-}; Eq. 6-55)

6. Bond Energies and Approximate Methods for Estimation of Thermodynamic Data

For approximate estimation of enthalpy changes during reactions, use can be made of empirical bond energies (Table 6-7) which represent the approximate enthalpy changes ($-\Delta H°$) for formation of compounds in a gaseous state from atoms in the gas phase. Other more comprehensive methods of approximation have been developed.[49,50]

7. Gibbs Energies of Combustion by O$_2$ and by NAD$^+$

Since oxidation processes are so important in the metabolism of aerobic organisms, it is often convenient to discuss Gibbs energies of combustion. These are easily derived from the Gibbs energies of formation. For example, ΔG_c for acetate ion (aqueous) may be obtained as follows:

$Acetate^- + H^+ \rightarrow 2\ C + 2\ H_2 + O_2$
$\Delta G° = +369.2$ kJ mol^{-1}
$\qquad = -\Delta G_f°$ of acetate (from Table 6-3)

$2\ H_2 + O_2 \rightarrow 2\ H_2O$
$\Delta G° = 2 \times -237.2 = -474.4$ kJ mol^{-1}
$\qquad = 2 \times \Delta G_f°$ of H_2O $\qquad$ (6-56)

$2\ C + 2\ O_2 \rightarrow 2\ CO_2$
$\Delta G° = 2 \times -394.4 = -788.8$ kJ mol^{-1}
$\qquad = 2 \times \Delta G_f°$ of CO_2

$Acetate^-$ (1 M) $+ H^+$ (1 M) $+ 2\ O_2$ (1 atm) $\rightarrow$
$\qquad 2\ CO_2$ (1 atm) $+ H_2O$ (l)
$\Delta G° = -894.0$ kJ mol^{-1}
$\qquad = \Delta G_c°$ of acetate$^-$

TABLE 6-7
Empirical Bond Energies and Resonance Energies[a]

Energy values for single bonds (kJ mol^{-1})		Energy values for multiple bonds (kJ mol^{-1})		Empirical resonance energy values (kJ mol^{-1})	
H–H	436	C=C	615	Benzene	155
C–C	346	N=N	418	Naphthalene	314
Si–Si	177	O=O	402 ('Δ state)		
N–N	161	C=N	615	Styrene	155 + 21
O–O	139	C=O	686 (formaldehyde)	Phenol	155 + 29
S–S	213		715 (aldehydes)		
C–H	413		728 (ketones)	Benzaldehyde	155 + 17
Si–H	295	C=S	477	Pyridine	180
N–H	391	C≡C	812	Pyrrole	130
P–H	320	N≡N	946 (N$_2$)	Indole	226
O–H	463	C≡N	866 (HCN)		
S–H	339		891 (nitriles)	$-C{\underset{OH}{\overset{O}{\diagup\!\!\!\diagup}}}$	117
C–Si	290				
C–N	292			$-C{\underset{OR}{\overset{O}{\diagup\!\!\!\diagup}}}$	100
C–O	351				
C–S	259			$-C{\underset{NH_2}{\overset{O}{\diagup\!\!\!\diagup}}}$	88
C–F	441				
C–Cl	328			Urea	155
C–Br	276			CO	439
C–I	240			CO$_2$	151
Si–O	369				

[a] From Pauling, L. (1960) *The Nature of the Chemical Bond*, 3rd ed., pp. 85, 189, and 195–198, Cornell Univ. Press, Ithaca, New York

Table 6-4 lists values of $\Delta G_c{}^\circ$ as well as those of $\Delta G_f{}^\circ$. Besides CO_2 and H_2O, the other products assumed in calculating the values of $\Delta G_c{}^\circ$ in Table 6-4 are N_2 (1atm), H_3PO_4 (1 M), and H_2SO_4 (1 M).

Much of the oxidation occurring in cells is carried out by the biological oxidizing agent **nicotinamide adenine dinucleotide** (**NAD$^+$**) or by the closely related NADP$^+$ (Chapter 10). It is convenient to tabulate values of ΔG° for complete oxidation of compounds to CO_2 using NAD$^+$ rather than O_2 as the oxidizing agent. These values, designated $\Delta G^\circ{}_{ox}$ and $\Delta G'{}_{ox}$ (pH 7), are also given in Table 6-3. Notice that these values are relatively small, corresponding to the fact that little energy is made available to cells by oxidation with NAD$^+$; for example (Eq. 6-57):

Acetate$^-$ + 2 H$_2$O + 4 NAD$^+$ → 2 CO$_2$ + 4 NADH + 3 H$^+$ (10^{-7} M)
$\Delta G'{}_{ox}$ (pH 7) = +22.1 kJ mol^{-1} (6-57)

When the reduced NAD$^+$ (designated NADH) so formed is reoxidized in mitochondria (Eq. 6-58), a large amount of energy is made available to cells.

4 NADH + 4 H$^+$ + 2 O$_2$ → 4 H$_2$O + 4 NAD$^+$
$\Delta G'{}_{ox}$ (pH 7) = −876.1 kJ mol^{-1} (6-58)

The sum of Eqs. 6-57 and 6-58 is the equation for combustion of acetate by O_2 (Eq. 6-59) and the two ΔG values sum to ΔG_c for acetate$^-$.

Acetate$^-$ + H$^+$ + 2 O$_2$ →
2 CO$_2$ + 2 H$_2$O
$\Delta G_c{}^\circ$ = −894.0 kJ mol^{-1}
$\Delta G_c'$ (pH 7) = −854.0 kJ mol^{-1} (6-59)

The values of ΔG_{ox} (Table 6-4) not only give an immediate indication of the relative amounts of energy available from oxidation of substrate with NAD$^+$ but also are very convenient in evaluating ΔG for fermentation reactions. For example, consider the fermentation of glucose to ethanol (Eq. 6-60):

α-D-Glucose → 2 CO$_2$ + 2 ethanol (6-60)

The Gibbs energy change $\Delta G'$ (pH 7) for fermentation of glucose to ethanol and CO$_2$ can be written immediately from the data of Table 6-4 (Eq. 6-61).

$\Delta G'$ (pH 7) = −243.8 − 2(−4.6)
= −234.6 kJ mol^{-1} (6-61)

The values of ΔG_{ox} for H_2O, CO_2, and H$^+$ are always zero and need not be considered. The same computation can be made using the awkwardly large values of $\Delta G_c{}^\circ$ or using values of $\Delta G_f{}^\circ$ The latter are also large, and CO_2 and water must be considered in the equations. Table 6-4 can be used to obtain values of ΔG° for many metabolic reactions considered later in the book. Data from any column in the table may be used for this purpose, but for simplicity try the values in the $\Delta G'{}_{ox}$ column.

Note, however, that for reactions involving oxidation by O_2 or by any oxidant, other than NAD$^+$, not appearing in Table 6-4, the following two-step procedure is necessary if the ΔG_{ox} values are used. From the ΔG_{ox} values compute ΔG° or $\Delta G'$ for the reaction under consideration using NAD$^+$ as the oxidant. Then add to this the Gibbs energy of oxidation by O_2 (or other oxidant) of the appropriate number of moles of NADH formed. The latter is given for O_2 in Table 6-8 and can also be evaluated for a number of other oxidants, such as Fe^{3+} and cytochrome *c*, from the data in Table 6-8.

BOX 6-B MAGNESIUM

After potassium ion Mg^{2+} is the most abundant cation present in tissues[a,b]. The average adult ingests ~10–12 mmol of magnesium ion daily (~1/4 g). Of this, about one-third is absorbed from the digestive tract. An equivalent amount is excreted in the urine to maintain homeostasis. Sixty percent of the magnesium in the body is found in the bones. The total Mg^{2+} concentration in serum is ~ 1 mM, while various tissues contain as much as 5 – 17 mM.[c] The concentration of *free* Mg^{2+} is difficult to measure. A variety of measurements using ^{31}P NMR spectroscopy of ATP,[d–f] ^{19}F NMR of added fluorine-containing chelators,[g] fluorescent chelators,[h] ion selective electrodes,[c] and other indirect procedures[g,i,j] indicate that the free magnesium concentration is almost the same in extracellular fluids, cytosol, and mitochondria.[h] Most values have been about 0.5 mM, somewhat less in erythrocytes. However, the most recent estimates indicate an intracellular $[Mg^{2+}]$ of 0.8–1.1 mM.[c,f]

The additional Mg^{2+} present in cells is bound to proteins, nuclei acids, and soluble compounds such as ATP, ADP, citrate,[k] and other phosphate- and carboxylate-containing substances. The binding is reversible and equilibration is usually rapid. It has been suggested that $[Mg^{2+}]$, like $[H^+]$, remains relatively constant within cells and that these two ions are in free equilibrium in the blood serum.[l] Nevertheless, there are instances in which at least temporary alterations in concentrations of both free Mg^{2+} and H^+ occur.[m] During rapid catabolism of carbohydrates the formation of lactic acid by glycolysis leads to acidification of muscle cells, the pH falling from 7.3 to as low as 6.3. This drop in pH causes a large decrease in the extent of binding of Mg^{2+} to molecules such as ATP and to a transient increase in $[Mg^{2+}]$. Similarly, the release of bisphosphoglycerate from hemoglobin upon oxygenation leads to a decreased concentration of free Mg^{2+} as the latter coordinates with bisphosphoglycerate.[n] The 0.25 mM $[Mg^{2+}]$ of aerobic red blood cells rises to 0.67 mM under anaerobic conditions.[d] Such changes in the free Mg^{2+} concentrations will affect many equilibrium[o] and may be of significance in metabolic regulation.

The magnesium ion has a smaller radius than Ca^{2+}, a fact that may account for its more ready entry into cells. Mg^{2+} can often be replaced by Mn^{2+} with full activity for enzymes that require it. On the other hand, high concentrations of Ca^{2+} are often antagonistic to Mg^{2+}. This antagonism is clearly seen in the effect of the two ions on irritability of protozoa.[p] Both deficiency of magnesium and excess of calcium in the surrounding medium cause in-

creased irritability. Excess magnesium leads to anesthesia. The Mg^{2+} concentration is high in hibernating animals.

Over 300 enzymes are dependent upon Mg^{2+}, the largest single group being the phosphotransferases, for which MgATP complex may be regarded as the substrate.[q] Magnesium has a special role in photosynthesis as a component of chlorophyll.

One of the most toxic metals is beryllium. It has been suggested that Be^{2+} competes with Mg^{2+} at many enzyme sites, including those of phosphoglucomutase and of phosphatases. However, as pointed out by Petsko,[r] because of its small size beryllium tends not to form Be^{2+} ions but, rather, covalent complexes such as $BeF_3 \cdot OH_2^-$, BeF_4^{2-}, and $MgADP \cdot BF_2(OH)$. In the latter complex beryllium is covalently linked to the oxygen atom of the β phospho group of ADP to give an analog of MgATP, which inhibits many enzymes. See for example, Fig. 19-16A, in which $MgADP \cdot BeF_x$ occupies the active site of myosin.

[a] Cowan, J. A., ed. (1995) *The Biological Chemistry of Magnesium*, VCH Publ., New York

[b] Strata, P., and Carbone, E., eds. (1991) *Mg^{2+} and Excitable Membranes*, Springer-Verlag, Berlin and New York

[c] McGuigan, J. A. S., Blatter, L. A., and Buri, A. (1991) in *Mg^{2+} and Excitable Membranes* (Strata, P., and Carbone, E., eds), Springer-Verlag, Berlin and New York

[d] Gupta, R. K., and Moore, R. D. (1980) *J. Biol. Chem.* **255**, 3987–3993

[e] Garfinkel, L., and Garfinkel, D. (1984) *Biochemistry* **23**, 3547–3552

[f] Clarke, K., Kashiwaya, Y., King, M. T., Gates, D., Keon, C. A., Cross, H. R., Radda, G. K., and Veech, R. L. (1996) *J. Biol. Chem.* **271**, 21142–21150

[g] Levy, L. A., Murphy, E., Raju, B., and London, R. E. (1988) *Biochemistry* **27**, 4041–4048

[h] Jung, D. W., Apel, L., and Brierley, G. P. (1990) *Biochemistry* **29**, 4121–4128

[i] Corkey, B. E., Duszynski, J., Rich, T. L., Matschinsky, B., and Williamson, J. R. (1986) *J. Biol. Chem.* **261**, 2567–2574

[j] Magneson, G. R., Puvathingal, J. M., and Ray, W. J., Jr. (1987) *J. Biol. Chem.* **262**, 11140–11148

[k] Kwack, H., and Veech, R. L. (1992) *Curr. Top. Cell. Regul.* **33**, 185–207

[l] Veloso, D., Guynn, R. W., Oskarrson, M., and Veech, R. L. (1973) *J. Biol. Chem.* **248**, 4811–4819

[m] Purich, D. L., and Fromm, H. J. (1972) *Curr. Top. Cell. Regul.* **6**, 131–167

[n] Bunn, H. F., Ransil, B. J., and Chao, A. (1971) *J. Biol. Chem.* **246**, 5273–5279

[o] Cornell, N. W. (1979) *J. Biol. Chem.* **254**, 6522–6527

[p] Meli, J., and Bygrave, F. L. (1972) *Biochem. J.* **128**, 415–420

[q] Vink, R., McIntosh, T. K., and Faden, A. I. (1991) in *Mg^{2+} and Excitable Membranes* (Strata, P., and Carbone, E., eds), Springer-Verlag, Berlin and New York

[r] Petsko, G. A. (2000) *Proc. Natl. Acad. Sci. U.S.A.* **97**, 538–540

C. Electrode Potentials and Gibbs Energy Changes for Oxidation–Reduction Reactions

We live under a blanket of the powerful oxidant O_2. By cell respiration oxygen is reduced to H_2O, which is a very poor reductant. Toward the other end of the scale of oxidizing strength lies the very weak oxidant H^+, which some bacteria are able to convert to the strong reductant H_2. The O_2–H_2O and H^+–H_2 couples define two biologically important oxidation–reduction (**redox**) systems. Lying between these two systems are a host of other pairs of metabolically important substances engaged in oxidation–reduction reactions within cells.

There are two common methods for expressing the oxidizing or reducing powers of redox couples in a quantitative way. On the one hand, we can list values of ΔG for oxidation of the reduced form of a couple to the oxidized form by O_2. A compound with a large value of $-\Delta G$ for this oxidation will be a good reductant. An example is H_2 for which ΔG of combustion at pH 7 (Table 6-4) is -237 kJ / mol. Poor reductants such as Fe^{2+} are characterized by small values of ΔG of oxidation (-8.5 kJ mol^{-1} for 2 $Fe^{2+} \rightarrow$ 2 Fe^{3+}). The Gibbs energies of oxidation of biological hydrogen carriers, discussed in Chapter 15, for the most part fall between those of H_2 and Fe^{2+}.

A second way of expressing the same information is to give **electrode potentials** (Table 6-8). Electrode potentials are also important in that their direct measurement sometimes provides an experimental approach to the study of oxidation–reduction reactions within cells. To measure an electrode potential it must be possible to reduce the oxidant of the couple by flow of electrons (Eq. 6-62) from an electrode surface, often of specially prepared platinum.

$$A + 2\,H^+ + 2\,e^- \rightarrow AH_2 \qquad (6\text{-}62)$$

Equation 6-62 represents a reversible reaction taking place at a single electrode. A complete electrochemical cell has two electrodes and the reaction occurring is the sum of two half-reactions. The electrode potential of a given half-reaction is obtained from the measured electromotive force of a complete cell in which one half-reaction is that of a standard **reference electrode** of known potential. Figure 6-3 indicates schematically an experimental setup for measurement of an electrode potential. The standard hydrogen electrode consists of platinum over which is bubbled hydrogen gas at one atmosphere pressure. The electrode is immersed in a solution containing hydrogen ions at unit activity (a_{H^+} = 1). The potential of such an electrode is conventionally taken as zero. In practice it is more likely that the reference electrode will be a calomel electrode or some other electrode that has been established experimentally as reliable and whose potential is accurately known.

The standard electrode is connected to the experimental electrode compartment by an electrolyte-filled bridge. In the experimental compartment the reaction represented by Eq. 6-62 occurs at the surface of another electrode (often platinum). The voltage difference between the two electrodes is measured with a potentiometer. The difference between the observed voltage and that of the reference electrode gives the electrode potential of the couple under investigation. It is important that the electrode reaction under study be strictly reversible. When the electromotive force (emf) of the experimental cell is balanced with the potentiometer against an external voltage source, no current flows through the cell. However, for a reversible reaction a slight change in the applied voltage will lead to current flow. The flow will be in either of the two directions, depending upon whether the applied voltage is raised or lowered.

Not all redox couples are reversible. This is especially true of organic compounds; for example, it is not possible to determine readily the electrode potential for an aldehyde–alcohol couple. In some cases, e.g., with enzymes, a readily reducible dye with a potential similar to that of the couple being measured can be added. A list of suitable dyes has been described by Dutton.[51] If the dye is able to rapidly exchange electrons with the couple being studied, it is still possible to measure the electrode potential directly. In many cases electrode potentials appearing in tables have been calculated from Gibbs energy data. The student should be able to calculate many of the potentials in Table 6-8 from Gibbs energy data from Table 6-4. If A, H^+, and AH_2 are all present at unit activity in the experimental cell, the observed potential for the half-reaction will be the **standard electrode potential** $E°$. If the emf of the hypothetical cell with the standard hydrogen electrode is positive when electron flow is in the direction indicated by the arrow in Fig. 6-3, the potential of the couple A / AH_2 is also taken as positive (and is often called a **reduction potential**). This is the convention used in establishing Table 6-8, but potentials of exactly the opposite sign (oxidation potentials) are used by some chemists. To avoid confusion in reading it is best to be familiar with values of one or two potentials such as those of the O_2^-–H_2O and NAD^+–NADH couples.

When electrons flow in the external circuit the maximum amount of work that can be done per mole of electrochemical reaction ($-\Delta G$) is given by Eq. 6-63

$$\Delta G = nEF = nE \times 96.487\ \text{kJ mol}^{-1}\ \text{V}^{-1}$$
$$= -nE \times 23.061\ \text{kcal mol}^{-1}\ \text{V}^{-1} \qquad (6\text{-}63)$$

where F equals the number of coulombs per mole of electrons (Avogadro's number multiplied by the charge on the electron = 96,487 coulombs). E is measured in volts and represents the difference of the electrode potentials of the two half-cells. In the case of a cell using the standard hydrogen electrode, E is the electrode potential of the experimental couple. The number of moles of electrons transferred in the reaction equation (n) is usually 1 or 2 in biochemical reactions (2 for Eq. 6-62).

Since the reactants and products need not be at unit activity, we must define the observed electrode potential E as a function of $E°$ and the activities (concentrations) of A, AH_2, and H^+ (Eq. 6-64).

$$E = E° + \frac{RT}{nF} \ln \frac{[A][H^+]^2}{[AH_2]}$$

If $n = 2$

$$E = E° + 0.0296 \log \frac{[A][H^+]^2}{[AH_2]} \quad \text{volts at 25°C}$$

$$(6\text{-}64)$$

In the biochemical literature values of the apparent standard electrode potential at pH 7 ($E°'$) are usually tabulated instead of values of $E°$ (Table 6-8, second column). Note that $E°'$ (pH 7) for the hydrogen electrode is not zero, as it is at pH 0, but −0.414 V. Values of E are related to $E°'$ by Eq. 6-64 with $[H^+]^2$

TABLE 6-8
Reduction Potentials of Some Biologically Important Systems[a,b]

Half-reaction	$E°$ (V)	$E°'$ (pH 7) (V)	$-\Delta G'$ (pH 7) (kJ mol^{-1}) for oxidation by O_2 (per 2 electrons)
$O_2 + 4 H^+ + 4 e^- \rightarrow 2 H_2O$	+1.229	+0.815	0.0
$Fe^{3+} + e^- \rightarrow Fe^{2+}$	0.771	0.771	8.5
$NO_3^- + 2 H^+ + 2 e^- \rightarrow NO_2^- + H_2O$	0.42	176.0	
Cytochrome f (Fe^{3+}) + $e^- \rightarrow$ cytochrome f (Fe^{2+})		0.365	86.8
Fe (CN)$_6^{3-}$ (ferricyanide) + $e^- \rightarrow$ Fe (CN)$_6^{4-}$		0.36	87.8
$O_2 + 2 H^+ + 2 e^- \rightarrow H_2O_2$	0.709	0.295	100.3
Cytochrome a (Fe^{3+}) + $e^- \rightarrow$ Cytochrome a (Fe^{2+})		0.29	101.3
p-Quinone + 2 H$^+$ + 2 $e^- \rightarrow$ hydroquinone	0.699	0.285	102.3
Cytochrome c (Fe^{3+}) + $e^- \rightarrow$ cytochrome c (Fe^{2+})		0.254	108.3
Adrenodoxin (Fe^{3+}) + $e^- \rightarrow$ adrenodoxin (Fe^{2+})		0.15	128.3
Cytochrome b_2 (Fe^{3+}) + $e^- \rightarrow$ cytochrome b_2 (Fe^{2+})		0.12	134.1
Ubiquinone + 2 H$^+$ + 2 $e^- \rightarrow$ ubiquinone H$_2$		0.10	138.0
Cytochrome b (Fe^{3+}) + $e^- \rightarrow$ cytochrome b (Fe^{2+})		0.075	142.8
Dehydroascorbic acid + 2 H$^+$ + 2 $e^- \rightarrow$ ascorbic acid		0.058	146.1
Fumarate^{2-} + 2 H$^+$ + 2 $e^- \rightarrow$ succinate^{2-}		0.031	151.3
Methylene blue + 2 H$^+$ + 2 $e^- \rightarrow$ leucomethylene blue (colorless)		0.011	155.2
Crotonyl-CoA + 2 H$^+$ + 2 $e^- \rightarrow$ butyryl-CoA		−0.015	160.2
Glutathione + 2 H$^+$ + 2 $e^- \rightarrow$ 2-reduced glutathione		~ −0.10	176.6
Oxaloacetate^{2-} + 2 H$^+$ + 2 $e^- \rightarrow$ malate^{2-}		−0.166	189.3
Pyruvate$^-$ + 2 H$^+$ + 2 $e^- \rightarrow$ lactate^{1-}		−0.185	193.0
Acetaldehyde + 2 H$^+$ + 2 $e^- \rightarrow$ ethanol		−0.197	195.3
Riboflavin + 2 H$^+$ + 2 $e^- \rightarrow$ dihydroriboflavin		−0.208	197.4
Acetoacetyl-CoA + 2 H$^+$ + 2 $e^- \rightarrow$ β-hydroxybutyryl-CoA		−0.238 (38°C)	203.2
$S + 2 H^+ + 2 e^- \rightarrow H_2S$	0.14	−0.274	210.2
Lipoic acid + 2 H$^+$ + 2 $e^- \rightarrow$ dihydrolipoic acid		−0.29	213.2
$NAD^+ + H^+ + 2 e^- \rightarrow NADH$	−0.113	−0.32	219.0
$NADP^+ + H^+ + 2 e^- \rightarrow NADPH$		−0.324	219.8
Ferredoxin (Fe^{3+}) + $e^- \rightarrow$ ferredoxin (Fe^{2+}) (Clostridia)		−0.413	237.0
$2 H^+ + 2 e^- \rightarrow H_2$	0	−0.414	237.2
$CO_2 + H^+ + 2 e^- \rightarrow$ formate$^-$		−0.42 (30°C)	238.3
Ferredoxin (Fe^{3+}) + $e^- \rightarrow$ ferredoxin (Fe^{2+}) (spinach)		−0.432	240.6

[a] A compound with a more positive potential will oxidize the reduced form of a substance of lower potential with a standard free energy change $\Delta G° = -nF \Delta E° = -n \Delta E° \times 96.49$ kJ mol^{-1} where n is the number of electrons transferred from reductant to oxidant. The temperature is 25°C unless otherwise indicated. $E°$ refers to a standard state in which the hydrogen ion activity = 1; $E°'$ refers to a standard state of pH 7, but in which all other activites are unity.

[b] The major source is Loach, P. A. (1976) in *Handbook of Biochemistry and Molecular Biology* 3rd ed. Vol. I (Fasman, G. D. ed.), pp. 122–130, CRC Press, Cleveland, Ohio.

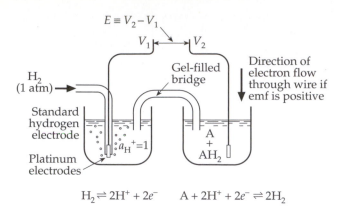

$$E \equiv V_2 - V_1$$

$$H_2 \rightleftharpoons 2H^+ + 2e^- \qquad A + 2H^+ + 2e^- \rightleftharpoons 2H_2$$

Figure 6-3 Device for measurement of electrode potentials. The electrode reactions are indicated below each half-cell. The maximum electrical work that can be done by such a cell on its surroundings is $-\Delta G = nEF$, where $E = V_2 - V_1$ as measured by a potentiometer. If A is reduced to AH_2 by H_2, electrons will flow through an external circuit as indicated. A will be reduced in the right-hand cell. H_2 will be oxidized to H^+ in the left-hand cell. Protons will flow through the gel bridge from left to right as one of the current carriers in the internal circuit.

deleted from the numerator (since the term in log $[H^+]$ is contained in $E°'$). On the scale of $E°'$ (pH 7) the potential of the oxygen–water couple is 0.815 V, while that of the NAD$^+$–NADH couple is –0.32 V.

D. The Adenylate System

Adenosine triphosphate
ATP^{4-}

Hydrolysis here yields AMP + pyrophosphate (PP_i)

Hydrolysis here yields ADP + inorganic phosphate (P_i)

Of central importance to the energy metabolism of all cells is the **adenylate system** which consists of adenosine 5'-triphosphate (**ATP**), adenosine 5'-diphosphate (**ADP**), and adenosine 5'-monophosphate (**AMP**) together with **inorganic phosphate** (P_i), **pyrophosphate** (PP_i), and **magnesium** ions. Remember that P_i refers to the mixture of ionic forms of phosphoric acid present under experimental conditions. Between pH 4 and pH 10 this will be mainly $H_2PO_4^-$

($pK_a = 6.8$) and HPO_2^{2-}. Likewise, the symbols AMP, ADP, and ATP refer to mixtures of ionic forms and PP_i refers to a mixture of the ions of pyrophosphoric acid. Above pH 4.4 only $H_2P_2O_7^{2-}$ ($pK_a = 6.1$), $HP_2O_7^{3-}$ ($pK_a = 9.0$), and $P_2O_7^{4-}$ contribute appreciably to PP_i.

1. Storage and Utilization of Energy

ATP is a thermodynamically unstable molecule with respect to hydrolysis to either ADP or AMP as is indicated in the foregoing diagram. The standard Gibbs energy of hydrolysis, $\Delta G'$, of ATP^{4-} to ADP^{3-} + HPO_4^{2-} at pH 7 is -34.5 kJ mol^{-1} and that of hydrolysis of ADP^{3-} to AMP^{2-} + HPO_4^{2-} is -36.3 kJ mol^{-1} at 25°C (Table 6-6). The exact value of these changes in Gibbs energy depends on pH and on the concentration of Mg^{2+} as is detailed in Sections B, 4, 5. Rates of reaction of components may also depend upon metal ions. Magnesium ion is especially important and complexes such as $MgATP^{2-}$ are regarded as the true substrates for many ATP-utilizing enzymes.

The large *Gibbs energy decreases upon hydrolysis (high group transfer potentials)* enable cells to use ATP and ADP as stores of readily available energy. Energy from the adenylate system is used for many purposes including biosynthesis, transport of ions and molecules across membranes, and for doing mechanical work. The mechanisms by which this energy is utilized are considered later (see Chapters 10, 12, and 17–19). The first step most often involves transfer of either the terminal γ-phospho group of ATP to a site on a different molecule or transfer of the entire AMP portion of the molecule onto another group. Thus the products of cleavage of ATP in these energy-utilizing processes may be either ADP + P_i or AMP + PP_i. In the latter case the pyrophosphate is usually hydrolyzed rapidly to two molecules of P_i by pyrophosphatases. This process, too, serves an essential function in the adenylate system, because it removes a product of the initial ATP cleavage reaction, shifting the overall equilibrium of the reaction sequence in the direction of the products.

The adenylate system provides the major store of rapidly available energy but the whole family of **nucleoside triphosphates** that are related to ATP in structure have similar functions. These include guanosine triphosphate (GTP), uridine triphosphate (UTP), cytidine triphosphate (CTP), and deoxyribonucleotide triphosphates (Table 5-1). These compounds are formed by successive transfer to the nucleoside monophosphate (GMP, UMP, CMP, dGMP, etc.) from two different molecules of ATP of two phospho groups. The resulting compounds are used to provide energy for a variety of specific biosynthetic processes, including synthesis of RNA and DNA. Inorganic **polyphosphates**, linear polymers of orthophosphate (P_i), are present in nearly all living forms.[52-54] Like ATP, they can also store energy and in some organisms substitute for ATP in certain enzymatic reactions.

2. Synthesis of ATP

The **phosphate anhydride** (pyrophosphate) linkages of ATP are generated by the joining of ADP and inorganic phosphate by means of special **phosphorylation** reactions. The most important of the latter occur in the photosynthetic membranes of chloroplasts (**photosynthetic phosphorylation**) and in oxygen-utilizing membranes of bacteria and of mitochondria (**oxidative phosphorylation**). Conversion of AMP to ADP is accomplished by transfer of the terminal phospho group from an ATP molecule to AMP in a reaction catalyzed by the extremely active enzyme **adenylate kinase** (Chapter 12) which is found in all cells. The following equations indicate how one of the special phosphorylation reactions must be used twice for the conversion of one molecule of AMP into one molecule of ATP.

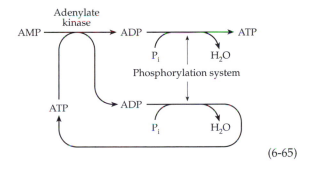

$$(6\text{-}65)$$

Various measures of the phosphorylating potential of the adenylate system within cells have been proposed. One measure is the product $[ATP] / [ADP][P_i]$, which will be called the **phosphorylation state ratio** or R_p in this book. It is also sometimes called the "phosphorylation potential." This ratio directly affects the Gibbs energy of hydrolysis of ATP, as is shown by Eq. 6-29. The value of R_p may be as high as 10^4 to 10^5 M^{-1} within cells[55] adding -22.8 kJ mol^{-1} to ΔG of hydrolysis of ATP. Another quantity that is sometimes cited is the "energy charge," the mole fraction of adenylic acid "charged" by conversion to ATP. ADP is regarded as "half-charged."[56,57]

$$\text{Energy charge} = \frac{[ATP] + \frac{1}{2}[ADP]}{[ATP] + [ADP] + [AMP]}$$

$$(6\text{-}66)$$

The energy charge varies from 0 if only AMP is present to 1.0 if all of the AMP is converted to ATP.

Measurements on a variety of cells and tissues show that the energy charge is usually between 0.75 and 0.90. Although it is easy to calculate its numerical value, the energy charge cannot be used in chemical equations and the idea that the energy charge of a cell plays a key role in regulation of metabolism has been challenged.[58]

3. Creatine Phosphate, an Energy Buffer

Although ATP provides the immediate source of energy for operating muscle, its concentration is only about 5 mM. However, muscle contains, in addition, a **phosphagen**, an N-phospho derivative of a guanidinium compound. In mammalian muscle, the phosphagen is **creatine phosphate**. Related compounds including **arginine phosphate** serve in various invertebrates. The group transfer potential ($-\Delta G^{\circ\prime}$ of hydrolysis) for creatine phosphate is 43.1 kJ mol^{-1}. Thus, the transfer of a phospho group from creatine phosphate to ADP to form ATP (Eq. 6-67) is spontaneous with $\Delta G' = -8.6$ kJ mol^{-1}. Recent values under a variety of conditions have been reported by Taugue and Dobson.[59] Creatine phosphate, which is present at a concentration of 20 mM, provides a reserve of high-energy phospho groups and keeps the adenylate system of muscle buffered at a high value of R_p.

$$\Delta G' = -8.6 \text{ kJ mol}^{-1}$$

$$(6\text{-}67)$$

4. Phosphorus-31 NMR

A spectacular development is the ability to observe components of the adenylate system as well as phosphocreatine and other phosphate esters in living cells by ^{31}P NMR.[60–61a] Spectra can be recorded on suspensions of cells[60] or of mitochondria,[62] on individual excised muscles (Fig. 6-4),[63,64] or on perfused organs[65,66] including beating rat hearts and surgically exposed animal organs.[61,61a] Metabolism can be observed in human erythrocytes[67] and even in human limbs, liver, hearts, and brain.[68–70] The method can play a valuable role in understanding human diseases.[63,67,70] As is seen in Fig. 6-4, each phosphorus atom of ATP gives a distinct resonance. The area of the P_γ peak provides a direct measure of the ATP concentration, and phosphocreatine (creatine-*P*) and P_i can be estimated in a similar way. Knowing the concentrations of ATP, P_i, and creatine-*P*, the amount of ADP present, usually too low to be estimated by NMR, may be calculated.[61] Barnacle muscle contains, instead of phosphocreatine, a high concentration of phosphoarginine which has also been measured by NMR.[71]

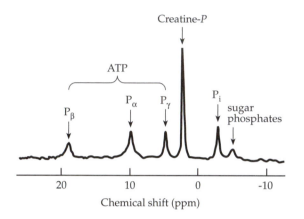

Figure 6-4 Phosphorus-31 NMR spectrum of an excised rat muscle (*vastus lateralis*) in Ringer solution at 15°C. The spectrum represents the accumulation of 400 scans. From P. J. Seeley *et al.*[64]

Usually no ADP can be seen in NMR spectra, although there may be about 0.5 mM ADP according to chemical analysis of rapidly frozen tissues. It has been concluded that most ADP is bound to proteins. In muscle the proteins myosin and actin (Chapter 20) hold most of the ADP leaving only about 0.02 mM free.[62,72] The same conclusion has been reached on the basis of other evidence.[55]

If a perfused heart in an NMR spectrometer is deprived of oxygen, the level of phosphocreatine drops rapidly and that of P_i increases while that of ATP remains relatively constant until the phosphocreatine is gone. Then it too falls and becomes undetectable after 17 min.[72] Similar changes occur when a tourniquet is placed on a human arm in the NMR spectrometer. Study of the rate and extent of recovery of the adenylate system when oxygen is readmitted is helping to provide a better means of protecting kidneys and other organs during transplantation operations. Diagnosis of circulatory ailments in human limbs by ^{31}P NMR may soon be routine. Use of radiofrequency coils placed on the body surface allows monitoring of the adenylate system in the heart, brain, and other tissues deep within the body.[69,73] Changes of concentration in the adenylate system and of phosphocreatine can be monitored very rapidly and evenly throughout the cardiac heartbeat cycle.[74]

E. Complex Biochemical Equilibria

The binding of small molecules to larger ones is basic to most biological phenomena. Substrates bind to enzymes and hormones bind to receptors. Metal ions bind to ATP, to other small molecules, and to metalloproteins. Hydrogen ions bind to amino acids, peptides, nucleotides, and most macromolecules. In this section we will consider ways of describing mathematically the equilibria involved.

The strength of bonding between two particles can be expressed as a **formation constant** (or **association constant**) K_f. Consider the binding of a molecule X to a second molecule P, which may be a protein or some other macromolecule. If there is on the surface of P only one single binding site for X, the process can be described by Eq. 6-68 and the equilibrim constant K_f by Eq. 6-69.

$$X + P \rightarrow PX \qquad (6\text{-}68)$$

$$K_f = [PX] / [P][X] \qquad (6\text{-}69)$$

The units for K_f are liters per mole (M^{-1}). The constant K_f is a direct measure of the strength of the binding: The higher the constant, the stronger the interaction. This fact can be expressed in an alternative way by giving the standard Gibbs energy change ($\Delta G°$) for the reaction (Eq. 6-70). The more negative $\Delta G_f°$, the stronger the binding.

$$\begin{aligned} \Delta G_f° &= -RT \ln K_f = -2.303 RT \log K_f \\ &= -5.708 \log K_f \text{ kJ mol}^{-1} \text{ at } 25°C \end{aligned} \qquad (6\text{-}70)$$

To avoid confusion, it is important to realize that the frequently used **dissociation constants** (K_d) are reciprocal association constants (Eq. 6-71). The use of

$$K_d = 1 / K_f \qquad (6\text{-}71)$$

association constants and of dissociation constants is firmly entrenched in different parts of the chemical literature; be sure to keep them straight. Dissociation constants are customarily used to describe acid–base chemistry, while formation constants are more often employed to describe complexes with metal ions or associations of macromolecules (Section C). However, both types of equilibria can be described using either formation constants or dissociation constants.

Logarithms of formation constants, which are proportional to the Gibbs energies of association, are often tabulated. The logarithms of formation constants and pK_a values of dissociation constants are identical (Eq. 6-72) and are a measure of the standard

$$\log K_f = -\log K_d = pK_d \qquad (6\text{-}72)$$

Gibbs energy decrease in the association reaction. The difference in $\Delta G°$ corresponding to a change of one unit in log K_f or pK_d is -5.7 kJ mol^{-1}, a handy number to remember.

1. Effects of pH on Equilibria

Compounds that contain several acidic or basic groups can exist in a number of different ionic forms, H_nA, $H_{n-1}A$, H_2A, HA, A, etc., as is indicated in the following equation, which is identical in form to Eq. 3-3.

$$H_nA \xrightarrow{K_1} H_{n-1}A \longrightarrow \cdots H_2A \xrightarrow{K_{n+1}} HA \xrightarrow{K_n} A$$

(6-73)

The dissociation constants K_1 ---- K_n for a multiprotic acid H_nA are defined as stepwise or **macroscopic constants** (also called molecular constants). For some compounds, e.g. alanine, the pK_a values are far apart (pK_1 and pK_2 are 2.4 and 9.8, respectively). The macroscopic constants can be assigned specifically, K_1 to the carboxyl group and K_2 to the protonated amino group. At the isoelectric pH of 6.1 the alanine exists almost entirely as the dipolar ion. However, for compounds in which the macroscopic pK_a values are closer together, they cannot be assigned to specific groups. We will consider some specific examples in the next section.

2. Microscopic Dissociation Constants and Tautomerization

A microscopic constant applies to a single site. Consider the dissociation of a simple carboxylic acid:

$$R—COOH \longrightarrow R—COO^- + H^+ \qquad (6-74)$$

The dissociation constant is about 1.7×10^{-5} and $pK_a = 4.8$. Since there is only one proton, the observed dissociation constant is also the microscopic dissociation constant. Now consider the cation of pyridoxine which has two dissociable protons bound to distinctly different sites, the phenolic oxygen and the ring nitrogen.

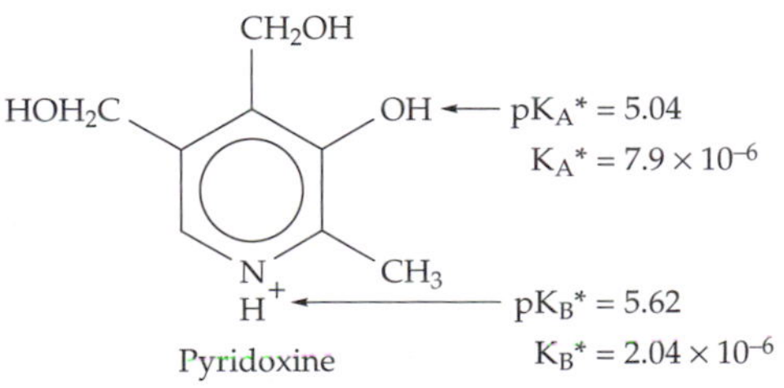

Pyridoxine
(vitamin B$_6$)

$pK_A^* = 5.04$
$K_A^* = 7.9 \times 10^{-6}$

$pK_B^* = 5.62$
$K_B^* = 2.04 \times 10^{-6}$

Either of the two protons might dissociate first as the pH is raised (Eq. 6-75). However, the two microscopic dissociation constants pK_A^* and pK_B^* are distinctly different. The result is that at 25°C in the neutral (monoprotonated) form 80% of the molecules carry a proton on the N, while the other 20% are protonated on the less basic – O$^-$. Notice that the subscripts a and b used in this discussion do *not* refer to acidic and basic but to the individual dissociation steps shown in Eq. 6-75. Microscopic constants will always be indicated with asterisks in this discussion.

The two monoprotonated forms of pyridoxine are the tautomeric pair shown in Eq. 6-75 and whose concentrations are related by the **tautomeric ratio**, $R =$ [neutral form]/[dipolar ion], a *pH-independent equilibrium constant* with a value of $0.204/0.796 = 0.26$ at 25°C.[75] Evaluation of microscopic constants for dissociation of protons from compounds containing non-identical groups depends upon measurement of the tautomeric ratio, or ratios if more than two binding sites are present. In the case of pyridoxine, a spectrophotometric method was used to estimate R.

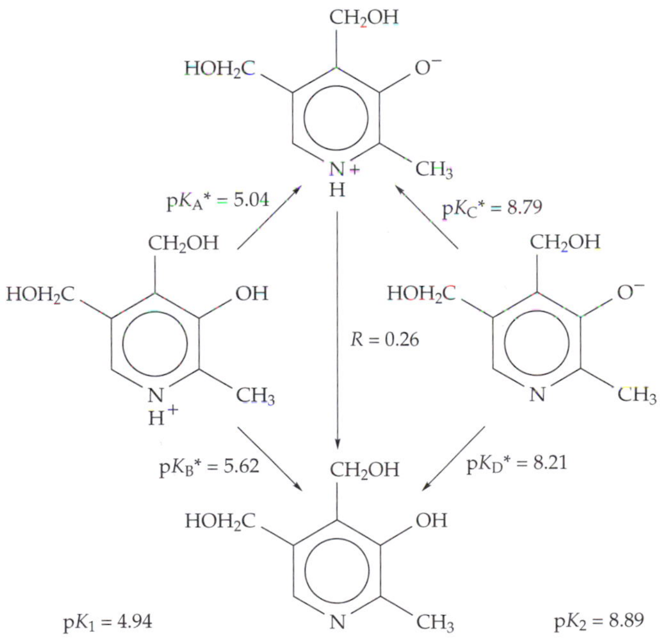

$pK_A^* = 5.04$ $R = 0.26$ $pK_C^* = 8.79$

$pK_B^* = 5.62$ $pK_D^* = 8.21$

$pK_1 = 4.94$ $pK_2 = 8.89$

(6-75)

To calculate microscopic constants from stepwise constants and tautomeric ratios, consider Eq. 6-76 in which $[HP]_A$ and $[HP]_B$ are the concentrations of the two tautomers and K_1 is the first **stoichiometric** or **macroscopic** dissociation constant for the diprotonated species H_2P.

$$K_1 = \frac{([HP]_A + [HP]_B)[H^+]}{[H_2P]} = K_A^* + K_B^*$$

$pK_1 = 4.94$; $K_1 = 1.15 \times 10^{-5} = 9.1 \times 10^{-6} + 2.4 \times 10^{-6}$

(6-76)

We see that K_1 is just the sum of the two microscopic constants K_A^* and K_B^* for dissociation of H_2P to the pair of tautomers HP(A) and HP(B). In a similar fashion it can be shown that the second stoichiometric constant K_2 is related to the microscopic constants K_C^* and K_D^* for dissociation of HP(A) and HP(B) to form P (Eq. 6-77).

$$1 / K_2 = 1 / K_C^* + 1 / K_D^* \qquad (6-77)$$

Since the tautomeric ratio R equals $[HP]_B / [HP]_A$, Eqs. 6-76 and 6-77 can be rearranged to Eqs. 6-78 to 6-81. These allow the evaluation of all of the microscopic constants from the two stoichiometric constants K_1 and K_2 plus the tautomeric ratio R.

$$pK_A^* = pK_1 + \log (1 + R) \qquad (6-78)$$

$$pK_B^* = pK_A^* - \log R \qquad (6-79)$$

$$pK_C^* = pK_2 - \log (1 + R) \qquad (6-80)$$

$$pK_D^* = pK_C^* + \log R \qquad (6-81)$$

For pyridoxine pK_1 and pK_2 were determined spectrophotometrically as 4.94 and 8.89. These values, together with that of R given above, were used to estimate the microscopic constants that are given in Eq. 6-74.[75] Notice that the microscopic constants of Eq. 6-74 are not all independent; if any three of the five equilibrium constants are known the other two can be calculated readily. In describing and measuring such equilibria it is desirable to select one pathway of dissociation, e.g., $H_2P \rightarrow HP(A) \rightarrow P$, and to relate the species HP(B) to it via the pH-independent constant R.

Often more complex situations arise in which additional tautomers or other forms arise via pH-independent reactions. These can all be related back to the reference ionic species by additional ratios R, which may describe equilibria for tautomerization, hydration, isomerization, etc. (Eq. 6-82).[76] In the case illustrated, only one of the ratios, namely R_2 or R_3, is likely to be a tautomerization constant because, as a rule, H_2P and P will not have tautomers. Equations analogous to Eqs. 6-76 to 6-82 can be written easily to derive K_C^*, K_D^* and any other microscopic constants desired from the stoichiometric constants plus the ratios R_1 to R_4. While it is easy to describe tautomerism by equations such as Eqs. 6-76 and 6-82 *it is often difficult*

to measure the tautomeric ratios R.[77] In favorable cases measurements of spectra of one kind or another allow their evaluation. However, because tautomerism may be extremely rapid, NMR spectra will often show only one peak for a proton present in a mixture of tautomers.

As was pointed out in Chapter 2, tautomerization ratios are often affected strongly by changes in solvent. Tautomerism among monoprotonated forms of cysteine, glutathione, and histidine (Eq. 2-6) has received considerable attention by biochemists[77-79] as has tautomerism in binding of protons and other small ligands to proteins.[80,81] For cysteine,[78,79] for which the following species coexist in the alkaline pH range, the distribution of the various ionic species including the two tautomers is shown in Fig. 6-5. A similar situation is met in the small protein thioredoxin (Box 15-C) which has a buried aspartate carboxylate that interacts with a nearby cysteine –SH group,[82] in papain where –SH and imidazole groups interact (Fig. 12-15),[83] and in carbohydrases where two or more carboxyl and carboxylate side chains interact (Chapter 12).[84]

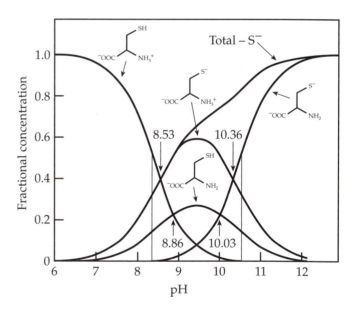

Figure 6-5 Distribution of various ionic species of cysteine as a function of pH. The function in each ionic and tautomeric form is plotted. Microscopic pK_a values are given numerically and macroscopic pK_a's are indicated by the vertical lines. From Dixon and Tipton.[79]

$$
\begin{array}{ccccc}
H_2P(A) & \xrightarrow{\ K_a^*\ } & HP(A) & \xrightarrow{\ K_b^*\ } & P(A) \\
\downarrow R_1 & & R_2 \swarrow \ \searrow R_3 & & \downarrow R_4 \\
H_2P(B) & & HP(B) \quad HP(C) & & P(B)
\end{array}
$$

$$(6-82)$$

Because it is usually difficult to measure tautomeric ratios, dissociation constants are frequently assumed identical to those of compounds in which one of the acidic groups has been modified by methylation or esterification or in some other manner to prevent dissociation of a proton. For example, the cation of 1-methyluracil can be dissociated to the two tautomers shown in Eq. 6-83.

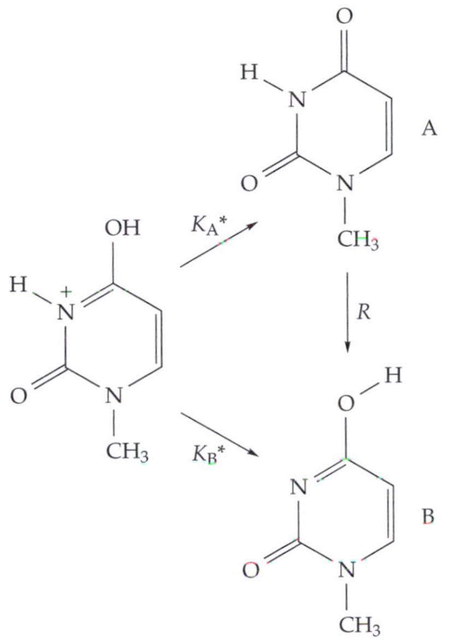

$$(6-83)$$

The apparent pK_a values were estimated for dissociation of the following two dimethylated cations which resemble that of Eq. 6-83.

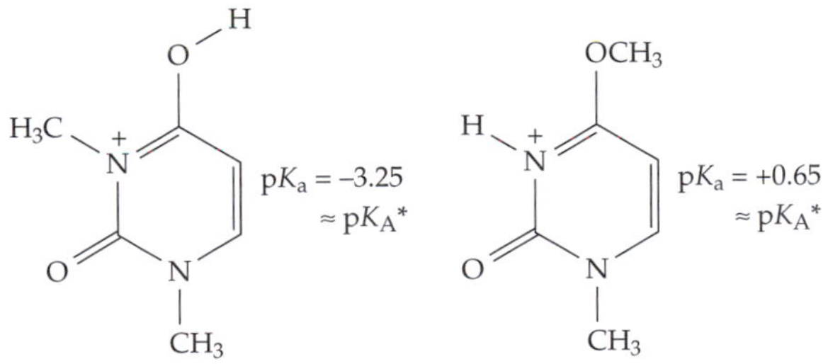

It can reasonably be assumed that these pK_a values approximate pK_A^* and pK_B^* as indicated. Thus, applying Eq. 6-85, $\log R \approx -3.25 - 0.65 + 3.90$. This result indicates that tautomer A of Eq. 6-89 is overwhelmingly predominant in water.[85,86] It also suggests that, within experimental error, the pK_a for 1-methyluracil will equal pK_A^*, namely, -3.25. In fact, it is close to this (-3.40).[85]

3. The Binding of Metal Ions

Equilibria in the formation of complex ions with metals are treated exactly as is the binding of small molecules and ions to macromolecules.[87–89] Stepwise constants are defined for the formation of complexes containing one, two, or more ligands L bound to a central metal ion M. The binding constants K_f's are usually referred to as β's as in Eq. 6-84.

$$\beta_1 = K_{f1} \quad \beta_n = K_{f1} K_{f2} \dots K_{fn} \quad (6-84)$$

Many important questions can be asked about the binding of metal ions within living cells. For example, What fraction of a given metal ion is free and what fraction is bound to organic molecules? To what ligands is a metal bound? Since many metal ions are toxic in excess, it is clear that homeostatic mechanisms must exist. How do such mechanisms sense the free metal ion activity within cells? How does the body get rid of unwanted metal ions? Answers to all these questions depend upon the quantitative differences in the binding of metal ions to the variety of potential binding sites found within a cell.

Table 6-9 gives formation constants for 1:1 complexes of several metal ions and a number of inorganic as well as organic ligands.[89] Only the values of $\log K_1$ are given when a series of stepwise constants have been established. However, in many cases two or more ligands can bind to the same metal ion. Thus for cupric ion and ammonia there are four constants.

$$Cu^{2+} + NH_3 \quad \log K_i = 4.0, 3.3, 2.7, 2.0$$

They are all separated by more than the "statistical distance," which in this case is less than the 0.6 logarithmic units for two equivalent binding sites (see Chapter 7, Section A,2). Thus, the second ligand binds less tightly than the first and **anticooperativity** in binding of successive ligands is observed. Most metal ions will also bind two or three successive amino acids. In the case of copper, whose preferred coordination number is four, two ligands may be bound. Again, a distinct anticooperativity is evident in the spread of the two constants.

$$Cu^{2+} + alanine \quad \log K_i = 8.1, 6.8$$

Factors affecting the strength of binding of a metal in a complex. More basic ligands tend to bind metal ions more tightly just as they do protons. However, the strength of bonding to metal ions to a group is more nearly proportional to the **nucleophilic character** (Chapter 12) which is only partly determined by basicity to protons.

BOX 6-C   LINEAR GIBBS ENERGY RELATIONSHIPS

Organic functional groups exert characteristic electronic effects upon other groups to which they are attached. The quantitative expression of such effects can sometimes be correlated by linear Gibbs energy relationships. The best known of these is the **Hammett equation**, which deals with the transmission of electronic effects across a benzene or other aromatic ring. Consider the acid dissociation constants of three classes of compounds:

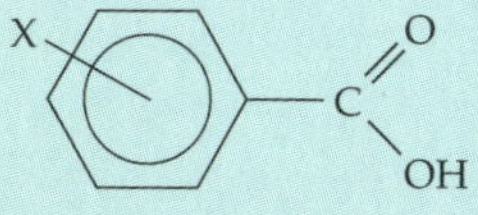

Benzoic acids 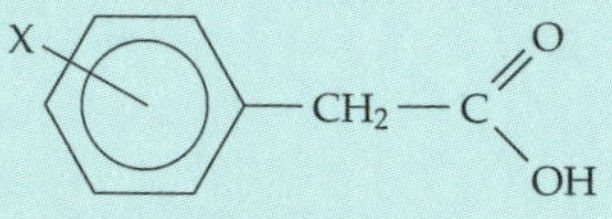   Phenylacetic

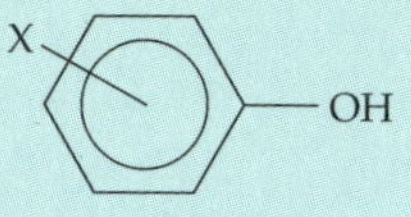

Phenol

The values of pK_a given in the following table have been observed for the parent compounds and for the *meta*-chloro- and the *meta*-nitro-substituted compounds.

Values of pK_a for Unsubstituted and Substituted Benzoic Acids, Phenylacetic Acids, and Phenols

Meta substituent	Benzoic acids*	Phenylacetic acids	Phenols
— H (parent compound)	4.202	4.31	9.92
— Cl	3.837	4.13	9.02
— NO₂	3.460	3.97	8.39

* For an independent set of pK_a values for substituted benzoic acids see Bolton and Fleming.[a] The values of σ calculated from them are slightly different from those given here.

Since pK_a is the negative logarithm of a dissociation constant, it follows from Eq. 6-30 that values of pK_a are directly proportional to values of $\Delta G°$ for dissociation of protons. In the Hammett treatment differences in pK_a values, rather than differences in $\Delta G°$, are considered. When a hydrogen atom in the meta position of benzoic acid is replaced by the electron-withdrawing Cl or NO_2, pK_a is lowered, i.e., the basicity of the conjugate base of the acid is decreased. The decrease in pK_a amounts to −0.365 for *m*-chlorobenzoic acid and −0.742 for *m*-nitrobenzoic acid. The Hammett treatment asserts that these changes in pK_a are a measure of the electron-withdrawing power of the meta substituent.[b] Thus, the nitro group is about twice as strong as the chloro group in this respect. The numerical values of these

changes in the dissociation constant of benzoic acid define the **substituent constants** σ, which are used in the Hammett equation and are given in the following table. In this equation we use the symbol pK_0 to represent the pK_a of the unsubstituted parent compound and pK to represent the pK_a of the substituted molecule.

For substituted benzoic acids: $pK_0 - pK = \sigma$

The decreases in the pK of phenylacetic acid occasioned by replacement of the *meta*-hydrogen with Cl or NO_2 are −0.18 and −0.34, respectively, substantially less than for the benzoic acids. On the other hand, for the phenols the differences amount to −0.90 and −1.53, much *greater* than for the benzoic acids. The Hammett equation asserts that for reactions such as the dissociation of protons from phenylacetic acids or from phenols, the changes in ΔG occasioned by meta substitutions are proportional to the σ values, i.e., to the changes in ΔG for the standard reaction – dissociation of a proton from benzoic acid.[b-d]

Substituent Constants for Use in the Hammett and Taft Equations[b-d]

Substituent	σ_m	σ_p	σ_p^-	σ_p^+	σ^* (Taft)
−O⁻	−0.71	−0.52			
−NH₂	−0.16	−0.66		−1.11	0.62
−OH	0.121	−0.37		−0.85	1.34
−OCH₃	0.115	−0.27		−0.78	1.81
−CH₃	−0.069	−0.17		−0.31	0.00
−NH−COCH₃	0.21	−0.01		−0.25	
−H	0	0		0	0.49
−CH₂OH	0.08	0.08			0.56
−COO⁻	−0.10	0.00		0.11	−1.06
−SO₃	−0.05	0.09		0.12	
−SH	0.25	0.15		0.019	1.68
−CH₂Cl		0.18			1.05
−CONH₂	0.28	0.36			
−F	0.337	0.06	0.02	−0.07	
−I	0.352	0.18		0.135	
−Cl	0.373	0.227		0.114	
−CHO	0.36	0.22	0.37		
−COCH₃	0.376	0.502	0.85		
−COOH	0.37	0.45		0.42	2.08
−COOCH₃	0.39	0.31		0.49	
−SO₂NH₂	0.55	0.62		0.61	
−CN	0.56	0.66	0.89	0.66	
−C≡CH					2.18
−CF₃	0.43	0.54		0.61	2.61
−CCl₃					2.65
−NO₂	0.710	0.778	1.25	0.790	4.0
−NH₃⁺	1.13	1.70			3.76
N (pyridine)	0.73	0.83			
NH⁺ (pyridine)	2.18	2.42		4.0	

BOX 6-C   (continued)

$$\log (K/K_0) = pK_0 - pK = \rho\sigma$$

The Hammett equation

The proportionality constant ρ, which also appears in the equation, is a measure of the *sensitivity* of the reaction to the presence of electron-withdrawing or electron-donating substituents in the ring. For benzoic acid, ρ is taken as 1.00. Using the data from the accompanying table together with many other data, an average value of $\rho = 0.49$ has been found for phenylacetic acids. Likewise, $\rho = 2.23$ for phenols, and $\rho = 5.7$ for dissociation of protons from substituted pyridinium ions. The sensitivity to substituent changes is highest (highest ρ) in the latter case where proton dissociation is directly from an atom in the ring and is lowest when the basic center is removed farthest from the ring (phenylacetic acid).

Through knowledge of the value of ρ for a given reaction, it is possible to predict the effect of a substituent on pK_a using the tabulated values of σ. In many cases the effects are additive for multisubstituted compounds.

$$pK_a = pK_0 - \rho\Sigma\sigma$$

Since substituents in o, m, and p positions have quantitatively different influences, different substituent constants σ are defined for each position. Moreover, since special complications arise from ortho interactions, it is customary to tabulate σ values only for meta and para positions. These are designated σ_m and σ_p. Apparent σ values for *ortho* substituents are also available.[d] An additional complication is that certain reactions are unusually sensitive to para substituents that are able to interact by resonance directly across the ring. An example is the acid dissociation of phenols. While σ_p for the nitro group is ordinarily 0.778, a correct prediction of the effect of the *p*-nitro group on dissociation of phenol is given only if σ_p is taken as 1.25. This higher σ value is designated σ_p^-. The resonance in the phenolate anion giving rise to this enhanced effect of the nitro group may be indicated as follows:

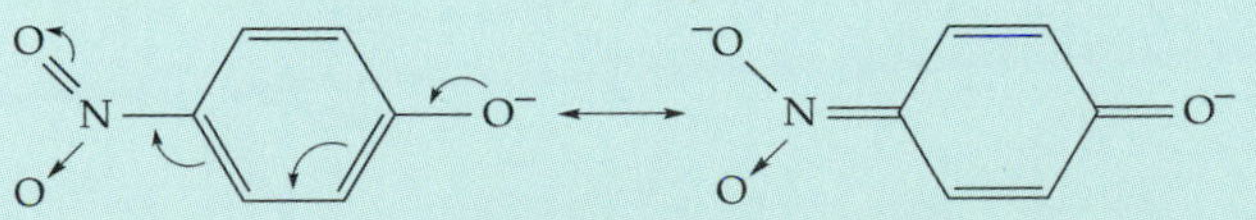

(Here the curved arrows represent the direction of flow of electrons needed to convert from the one resonance structure to the other.) For similar reasons, some reactions require the use of special σ_p^+ constants for strongly electron-donating substitu-

ents such as OH which are able to interact across the ring by resonance. Thus σ_p for the OH group is -0.37, while σ_p^+ is -0.85.[e] For the methoxyl group ($-OCH_3$) $\sigma_p = -0.27$, $\sigma_p^+ = -0.78$, and $\sigma_m = +0.12$.

The use of different kinds of substituent constants complicates the application of the Hammett equation and over 20 different sets of σ values have been proposed. A simplification is the representation of substituent constants as linear combinations of two terms, one representing "field" or "inductive" effects and the other resonance effects.[e,f]

Many other linear Gibbs energy relationships have been proposed; for example, the acid strengths of aliphatic compounds can be correlated using the "Taft polar substituent constants" σ^*.

$$\log (K/K_0) = \rho^* \sigma^*$$

For example, the following give good approximations of pK_a values.[d]

for R—COOH $pK_a = 4.66 - 1.62\,\sigma^*$
for R—CH_2—COOH $pK_a = 5.16 - 0.73\,\sigma^*$

While the examples chosen here concern only dissociation of protons, the Hammett equation has a much broader application. Equilibria for other types of reactions can be treated. Furthermore, since rates of reactions are related to Gibbs energies of activation, many rate constants can be correlated. For these purposes the Hammett equation can be written in the more general form in which k may be either an equilibrium constant or a rate constant.[b] The subscript j denotes the reaction under consideration and i the substituent influencing the reaction.

$$\log k_{ij} - \log k_{0j} = \rho_i \sigma_j$$

An example of a linear Gibbs energy relationship that is widely used in discussing mechanisms of enzymatic reactions is the Brönsted plot (Eqs. 9-90 and 9-91).

[a] Bolton, P. D., Fleming, K. A., and Hall, F. M. (1972) *J. Am. Chem. Soc.* **94**, 1033–1034
[b] Hammett, L. P. (1970) *Physical Organic Chemistry*, 2nd ed., p. 356, McGraw-Hill, New York
[c] Wells, P. R. (1963) *Chem. Rev.* **63**, 171–219
[d] Barlin, G. B., and Perrin, D. D. (1966) *Q. Rev., Chem. Soc.* **20**, 75–101
[e] Swain, C. G., and Lupton, E. C., Jr. (1968) *J. Am. Chem. Soc.* **90**, 4328–4337
[f] Hansen, L. D., and Hepler, L. G. (1972) *Can. J. Chem.* **50**, 1030–1035

TABLE 6-9
Logarithms of Binding Constants for Some 1:1 Metal Complexes at 25°C[a]

Ligand	H[+]	Mg[2+]	Ca[2+]	Mn[2+]	Cu[2+]	Zn[2+]
Hydroxide, OH[-]	14.0	2.5	1.4		6.5	4.4
Acetate[-]	4.7	~0.65	0.5	~1.0	2.0	1.5
Lactate[-]	3.8	~1.0	~1.2	1.3	3.0	2.2
Succinate[2-]	5.2	1.2	1.2		3.3	2
NH$_3$	9.3	~0	−0.2	0.8	4.0	2.4
Ethylenediamine	10.2	0.4		2.8	10.8	6.0
Glycine[-]	9.6	2.2	1.4*	2.8	8.2	5.0
Glycine amide	8.1			~1.5	5.3	3.3
Alanine[-]	9.7	2.0*	1.2*	3.0*	8.1	4.6
Aspartate[2-]	9.6	2.4	1.6		8.6	5.8
Glycylglycine[-]	8.1		1.2*	2.2*	5.5	3.4
Pyridine	5.2			0.1	2.5	1
Imidazole	7.5			1.6	4.6	2.6
Histidine	9.1			3.3	10.2	6.6
Adenine	9.8				8.9	6.4
Citrate[3-]	5.6	3.2	4.8	3.5	~4	4.7
EDTA[4- b]	10.2	8.8	10.6	13.8	18.7	16.4
EGTA[4- c]	9.4	5.3	10.9	12.2	17.6	12.6
ATP[4-]	6.5	4.2	3.8	4.8	6.1	4.9

[a] All values are for log K_1. Included is the highest pK_a for protons. Data for amino acids are from Martell, A.E. and Smith, R.M. (1974, 1975) *Critical Stability Constants*, 3 vols., Plenum, New York. Others are from Sillén, L.G. and Martell, A.E. (1964) *Stability Constants of Metal–Ion Complexes*, Spec. Publ. No. 17, Chemical Society, London. Most constants for amino acids are for ionic strength 0.1. Some (designated by asterisks) are for zero ionic strength. The values shown for other ligands have been selected from a large number reported without examination of the original literature.

[b] Ethylenediaminetetraacetic acid (EDTA), a chelating agent widely used by biochemists for

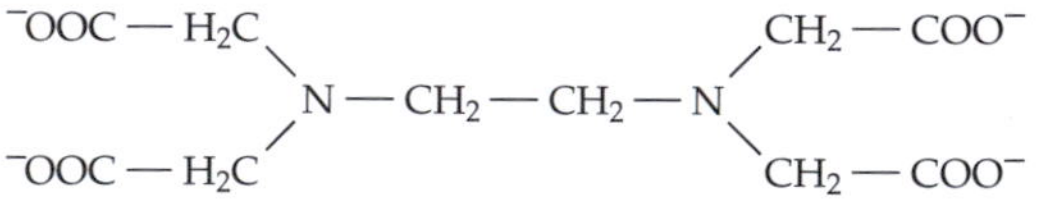

$$^-OOC-H_2C \diagdown \qquad CH_2-COO^- \diagup$$
$$N-CH_2-CH_2-N$$
$$^-OOC-H_2C \diagup \qquad CH_2-COO^- \diagdown$$

preventing unwanted reactions of metal ions. The high formation constants ensure that most metal ions remain bound to the EDTA.

[c] EGTA is similar to EDTA but has the group –CH$_2$–CH$_2$–O–CH$_2$–CH$_2$–O–CH$_2$–CH$_2$– joining the two nitrogen atoms in place of –CH$_2$–CH$_2$– of EDTA. Note that EGTA has a higher selectivity for Ca[2+] compared to Mg[2+] than does EDTA.

One of the most important factors in determining the affinity of organic molecules for metal ions is the **chelate effect**. The term chelate (pronounced "keel-ate") is from a Greek word meaning crab's claw. It refers to the greatly enhanced binding of metal ions resulting from the presence of two or more complexing groups in the same organic molecule. The chelate effect has been exploited by nature in the design of many important metal-binding molecules, including porphyrins (Fig. 16-5), chlorophyll (Fig. 23-20), the siderophores (Fig. 16-1), and metal-binding proteins. Structures of two **chelate complexes** are shown here. Notice from Table 6-9 that many simple compounds, such as the α-amino acids and citric acid, often form strong chelate complexes with metal ions.

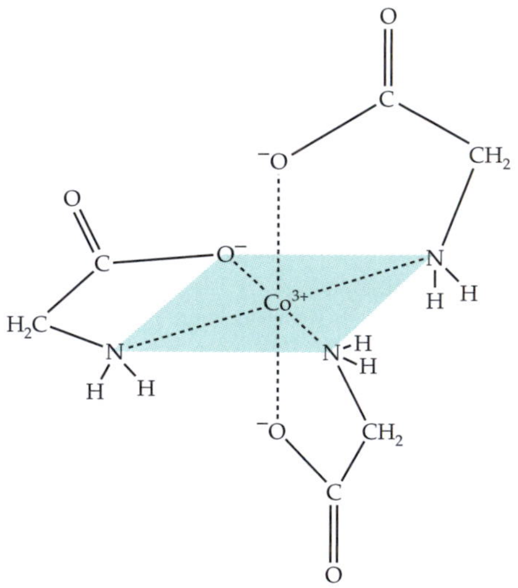

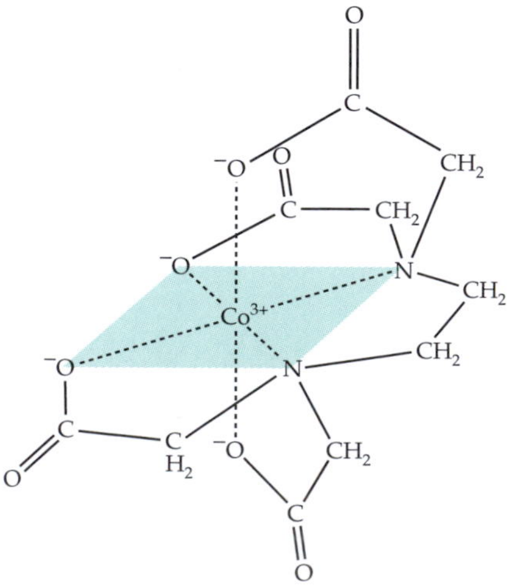

The pH of the medium always has a strong effect on metal binding. Competition with protons means that metal complexes tend to be of weak stability at low pH. Anions of carboxylic acids are completely protonated below a pH of ~4 and a metal can combine only by displacing a proton. However, at pH 7 or higher, there is no competition from protons. On the other hand, in the case of ethylenediamine, whose pK_a values are 10.2 and 7.5 (Table 6-9), protons are very strong competitors at pH 7, even with a strongly complexing metal ion such as Cu[2+]. At high pH there may be competition between the ligand and hydroxyl ion. At pH 7 about one-half of Cu[2+] dissolved in water is complexed as CuOH[+]. Aluminum forms soluble complexes Al(OH)[2+], Al(OH)$_2$[+], and Al(OH)$_4$[+] (Box 12-F).

How properties of the metal ion affect chelation.
The **charge**, the **ionic radius** (Table 6-10), the **degree of hydration**, and the **geometry of orbitals** used in covalent bonding between metal and chelating groups all affect the formation constants of a complex. Multi-charged ions form stronger complexes than do mono-valent ions, which have a lower charge density.

Among ions of a given charge type (e.g., Na^+ vs K^+; Mg^{2+} vs Ca^{2+}), the smaller ions are more strongly hydrated than are the larger ions in which the charge is dispersed over a greater surface area. Most cations, except for the large ones, have a primary hydration sphere containing about six molecules of water. Four molecules of water can be placed around the ion in one plane as shown in the following drawing of water molecules coordinated to Mg^{2+}.

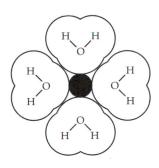

One additional water molecule can be bound above and another below to provide six molecules in an

array of **octahedral** geometry. An alternative tetra-hedral arrangement of four molecules of water around the ion has been suggested for Li^+ and Na^+.[90] In either case additional solvent molecules are held in a looser secondary sphere. For instance, electrochemical trans-ference experiments indicate a total of ~16 molecules of water around Na^+ and ~10 around K^+.

To form a chelate complex a metal ion must usually lose most of its hydration sphere. For this reason, the larger, less hydrated metal ions often bind more strongly than do the smaller, more hydrated ones. For example, Ca^{2+} binds to EDTA more tightly than does Mg^{2+} (Table 6-9). However, the reverse order may be observed with negatively charged ligands such as OH^- in which the charge in highly concentrated. The same is true for ATP^{4-}, which binds Mg^{2+} more tightly than Ca^{2+} (Table 6-9; Section B,5).

Differences in both the charge density and the hydration of ions determine the **Hofmeister series** (lyotropic series).[91]

$$Ba^{2+} > Ca^{2+} > Mg^{2+} > Li^+ > Na^+ > K^+ > Cs^+ > NH_4^+$$

This was originally defined as the order of effectiveness in precipitating colloids or protein molecules. The ions to the left are less hydrated than those to the right. A similar series can be defined for anions. The following, a well-known sequence of the stabilities of complexes of metals of the first transition series, applies to many different types of complexes including those of the α-amino acids, as is shown in Fig. 6-11.

$$Mn^{2+} < Fe^{2+} < Co^{2+} < Ni^{2+} < Cu^{2+} > Zn^{2+}$$

Simple electrostatic theory based upon differences in the ionization potentials or electronegativity of the ions would predict a gradual monotonic increase in chelate stability from manganese to zinc. In fact, with nitrogen-containing ligands Cu^{2+} usually forms by far the strongest complexes. Cobalt, nickel, and iron ions also show an enhanced tendency to bind to nitrogen-containing ligands. The explanation is thought to lie in the ability of the transition metals to supply *d* orbitals which can participate in covalent bond formation by accepting electrons from the ligands. Iron, copper, and cobalt are often located in the centers of nitrogen-containing structures such as the heme of our blood (iron, Fig. 16-5) and vitamin B_{12} (Box 16-B).

Metal binding sites in cells. Functional groups that participate in metal binding include negatively charged carboxylate $-COO^-$, thiolate $-S^-$, phenolate $-O^-$, and phosphate$^-$ as well as uncharged amino, imida-zole, $-OH$, and the polarizable carbonyl groups of peptide and amide side chains. To which of these ligands will specific ions tend to bind? The alkali metal ions Na^+ and K^+ are mostly free within cells

TABLE 6-10
Ionic Radii in Nanometers for Some Metallic and Nonmetallic Ions[a]

				Mn^{2+}	0.080				
Li^+	0.060			Fe^{2+}	0.076			H^-	0.21
Na^+	0.095			Co^{2+}	0.074			F^-	0.136
K^+	0.133			Ni^{2+}	0.069			Br^-	0.195
Rb^+	0.148			Cu^{2+}	0.072[b]			I^-	0.216
				Zn^{2+}	0.074				
				Cd^{2+}	0.097				
Be^{2+}	0.031								
Mg^{2+}	0.065			Al^{3+}	0.050				
Ca^{2+}	0.099			Fe^{3+}	0.064				
Sr^{2+}	0.113			Mo^{4+}	0.070				
Ba^{2+}	0.135			Mo^{6+}	0.062				

[a] Radii are calculated according to the method of Pauling and are taken from Cotton, F.A. and Wilkinson, G. (1972) *Advanced Inorganic Chemistry*, 3rd ed. Wiley (Interscience), New York .

[b] From Ahrens, L. H. as given by Sienko, M. J. and Plane, R. A. (1963) *Physical Inorganic Chemistry*, pp. 68–69. Benjamin, New York.

(Box 5-A) but are in part bound to defined sites in proteins. Both Ca^{2+} and Mg^{2+} tend to remain partially free and complexed with the numerous phosphate and carboxylate ions present in cells. However, they may find very specific tight binding sites such as that of Mg^{2+} in chlorophyll (Fig. 23-20). The heavier metal ions, including those of zinc, copper, iron, and other transition metals, usually bind to nitrogen or sulfur atoms.[92] For example, small peptides react with Cu^{2+} to form chelate complexes in which the peptide carbonyl oxygen binds to the metal (Eq. 6-85, step *a*).[93,94] By losing a proton the peptide NH can sometimes also function as a metal ligand (Eq. 6-85, step *b*).

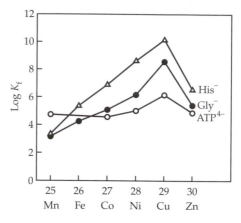

Figure 6-6 Logarithms of formation constants of metal complexes of histidine, glycine, and ATP plotted against the atomic number of elements from manganese to zinc.

In many **metalloproteins** the metals are found in **prosthetic groups** such as the porphyrin of heme proteins and the molybdopterin of molybdenum containing enzymes (Fig. 16-31). Very often clusters of carboxylate, imidazole, and other groups are used to create binding sites. In carboxypeptidase A (Fig. 12-16) two imidazole groups and one carboxylate of a glutamate site chain hold Zn^{2+}. In carbonic anhydrases, three imidazoles hold a zinc ion (Fig. 13-1), while in one kind of superoxide dismutase both Zn^{2+} and Cu^{2+} are bound in adjacent locations with six imidazoles and one carboxylate group participating in the binding. In contrast, Zn^{2+} and Cd^{2+} are bound in metallothioneins by clusters of thiol groups from cysteine side chains (Box 6-E). While Fe^{2+} is bound by from four to six nitrogen atoms in heme proteins, it is also found attached to oxygen atoms of tyrosine side chains, as is shown for a transferrin in Fig. 16-2. One imidazole, one carboxylate group, and a bicarbonate ion also bind to the iron. This site is also quite satisfactory for Al^{3+}

(Box 12-F), which tends to occupy a fraction of the transferrin sites in blood, although it binds much more weakly than does Fe. Several other binding sites for transition metals are pictured in Chapter 16.

Calcium-binding proteins. Much information about the functions of transition metal ions is given in Chapter 16. Here and in Box 6-D we consider calcium ions which, because of their broad distribution and range of functions, will be discussed in virtually every chapter of this book. The concentration of Ca^{2+} varies greatly in different parts of a cell and also with time (Box 6-D). Many **calcium-binding proteins** participate in mediating the physiological effects of these changes and also in buffering the calcium ion concentration. These include a large family of **helix–loop–helix** or **EF-hand proteins**.[95–97] The first of these to be discovered was **parvalbumin**, a 108-residue protein from carp muscle.[96,98-99a] The structure of the pair of metal-binding sites of the protein is shown in Fig. 6-7.

Each consists of two helices that are almost perpendicular and connected by a loop that forms the Ca^{2+}-binding site. In the site at the left side in Fig. 6-7A the Ca^{2+} is bound by four carboxylate groups from aspartate and glutamate side chains, a hydroxyl group of serine, and the residue 57 carbonyl oxygen of the peptide backbone.[99b] The same peptide group is hydrogen bonded to a carbonyl group of another segment of peptide chain near the second Ca^{2+} site (to the right in Fig. 6-7A). This site contains four carboxylate ions, one of which coordinates the Ca^{2+} with both oxygen atoms, and another peptide carbonyl group. By attaching itself to several different side chain groups, the metal ion can induce a substantial change in conformation from that present in the calcium-free protein.[97]

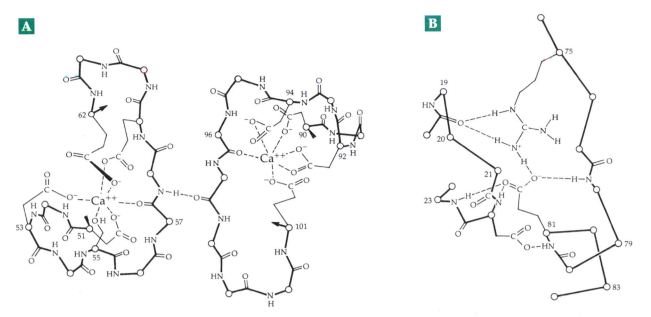

Figure 6-7 (A) Part of the 108-residue peptide chain of the calcium-binding protein of carp muscle. The two calcium-binding loops are shown together with a hydrogen bond between them. (B) A view of the intricate network of hydrogen bonds linking two segments of the peptide chain in the interior of the molecule. Note especially the bonding of the guanidinium group from arginine-75 to the carboxylate of glutamic acid 81 and to the peptide carbonyl of residue 18. Note that the carboxylate group also interacts with two different peptide NH groups. From Kretsinger and Nockolds.[98]

Kretzinger, who discovered the structure of parvalbumin, named the Ca^{2+}-binding helix–loop–helix motif an **EF-hand** because it is formed by helices E and F and resembles a hand with pointer finger extended along the E helix and the thumb in the direction of the F helix, the flexed middle finger forming the Ca^{2+}-binding loop. Helices C and D also form a hand. The resulting pair of hands can be visualized better at the top of the calmodulin structure in Fig. 4-8. The EF-hand motif has been identified in over 1000 proteins.[100]

Parvalbumins, which are also found in other vertebrates, are high-affinity Ca^{2+}-buffers.[99] Additional calcium buffers with EF-hand structures are the vitamin D-induced **calbindins**. One 9-kDa calbindin is found in mammalian intestinal tissue and in skin. It has two helix–loop–helix Ca^{2+}-binding sites of differing affinity[101,102] that presumably function in the absorption of calcium. A 28-kDa vitamin D-dependent protein from chicken intestine contains six similar Ca^{2+}-binding loops.[97,103]

Another group of calcium-buffering and storage proteins with remarkable Ca^{2+}-binding properties are the 40- to 45-kDa **calsequestrins**, which are found in the lumen of the ER (sarcoplasmic reticulum) of skeletal muscle. Calsequestrins are not typical EF-hand proteins but have a high content of glutamate and aspartate. They bind ~50 Ca^{2+} per molecule of protein with K_d ~ 1 mM.[104,105] Similar proteins called **calreticulins** are found in most non-muscle cells.[106,107]

A very large number of EF-hand proteins have signaling functions. The best known of these is the 148-residue **calmodulin**, which regulates many enzymes and cellular processes (Box 6-D; also Chapter 11).[108,109] The protein, which is present in all eukaryotes, has a conserved sequence that forms two pairs of helix–loop–helix Ca^{2+}-binding sites that are separated by a long helix (Fig. 6-8).[108,110,111] Two of the sites bind Ca^{2+} tightly and cooperatively,[110,112] with K_d values in the micromolar range. Calmodulin from almost all species contains the modified amino acid ε-N-trimethyllysine at position 115. However, octopus calmodulin has ordinary lysine at this position and seems to function well.[108] Calmodulin's controlling functions result from Ca^{2+}-induced conformational changes that modify its affinity for other proteins whose activity may be increased or decreased by bound calmodulin.[109,113] A protein with a similar dumbell shape and structure is **troponin C** of skeletal muscles.[114,115] Troponin C binds to a complex of proteins that assemble on the thin actin filaments of muscle fibers and control contraction in response to changes in the calcium ion concentration (Chapter 19).[116] Other proteins that contain EF-hand motifs and are therefore responsive to Ca^{2+} include **spectrin** of cell membranes,[117] **clathrin light chains** from coated vesicles,[118,119] the extracellular **osteonectin** of bones and teeth,[120] and a birch pollen antigen.[121] Another group of 17 or more small **S100 EF-hand** proteins play a variety of other roles.[122–123a] One of these, which has a high affinity for Zn^{2+}, has been named **psoriasin** because of its 5-fold or greater

BOX 6-D CALCIUM

The essentiality of calcium ions to living things was recognized in the last century by S. Ringer, who showed that ~1 µM Ca^{2+} was needed to maintain the beat in a perfused frog heart. Later, calcium was shown essential for repair of ruptures in the cell membrane of the protozoan *Stentor* and for the motion of amebas. The animals quickly died in its absence. The role in the frog heart was traced to transmission of the nerve impulse from the nerve to the heart muscle. More recently it has been recognized that Ca^{2+} is required for blood clotting and is involved in triggering many responses by cells. Calcium ions are also an integral part of many enzymes and have structural roles in proteins, carbohydrate gels, and biological membranes.

Like Na^+, the calcium ion is actively excluded from cells. Indeed, 99% of the calcium in the human body is present in the bones.[a-d] The blood serum concentration of Ca^{2+} is ~3 mM, of which ~1.5 mM is free. The rest is chelated by proteins, carbohydrates, and other materials. Within cells the concentration of free Ca^{2+} is < 1 µM and typically ~0.05–0.2 µM for unexcited cells.[d-f] However, the total intracellular Ca^{2+} is considerably higher and may be in excess of 1 mM. Approximate total concentrations are: red blood cells, 20 µM; liver, 1.6 mM; and heart, 4 mM. A gradient in $[Ca^{2+}]$ of 10^3 or more is maintained across membranes by the calcium ion pump (Chapter 8). The action of this pump is counteracted by a very slow diffusion of the external Ca^{2+} back through the membrane via an Na^+–Ca^{2+} exchange into the cells.[g]

Free Ca^{2+} lacks spectroscopic properties suitable for its direct measurement at the low concentrations present in cells. However, it can be measured indirectly by the use of chelating agents that are relatively specific for Ca^{2+} and which have a measurable property that changes upon calcium-binding. The fluorescent photoprotein **aequorin** (Chapter 23), which may be injected into cells or synthesized within cells from transferred genes is often employed.[h,i] Various synthetic Ca^{2+} indicators also fluoresce brilliantly upon chelation.[j-n] Others contain fluorine or a suitably placed atom of ^{13}C which changes its NMR chemical shift upon chelation with Ca^{2+}.[o] These compounds may be carried into cells in the form of esters which pass through membranes but are then hydrolyzed leaving the indicators trapped in the cytoplasm.[j] Chelate compounds that bind Ca^{2+} within cells and release it rapidly upon irradiation with ultraviolet light have also been developed.[p,q]

Consistent with their role in signaling, calcium ions are unevenly distributed within cells. Mitochondria, endoplasmic reticulum, Golgi, and nuclei may all take up calcium ions. Cytoplasmic Ca^{2+}

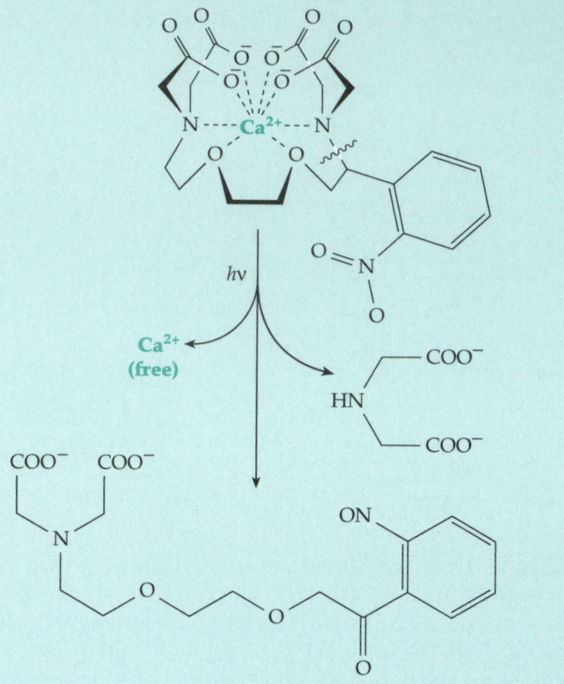

Calcium chelate compound that releases free calcium ion within cells upon irradiation with ultraviolet light.[p,v]

may sometimes be sequestered in microvesicles (**calciosomes**) or in intracellular granules[c,r,s] in which the $[Ca^{2+}]$ is ~0.5–1 mM but may reach 5–10 mM.[f] Many regulatory mechanisms exist. For example, **calcitonin** and **parathyroid hormone** interact with **vitamin D** and its metabolites in the small intestine, bones, and kidneys to control the deposition of calcium in bones, a topic considered in Box 22-C.

A characteristic function of Ca^{2+} in living things is **activation** of various metabolic processes. This occurs when a sudden change in permeability of the plasma membrane or in the membranes of the endoplasmic reticulum (ER) allows Ca^{2+} to diffuse into the cytoplasm. For example, during the contraction of muscle, the Ca^{2+} concentration rises from ~0.1 to ~10 µM as a result of release from storage in the calciosomes of the ER. The calcium ions bind to **troponin C** initiating muscle contraction (Chapter 19).[t] The ER membranes of muscle are rich in a Ca^{2+} pump protein[u,v] (Fig. 8-26), and in a series of calcium-binding proteins such as **calsequestrin** (see text).[f] Their combined action soon restores the $[Ca^{2+}]$ to the original low value.

Skeletal muscle is activated by nerve impulses which induce Ca^{2+} release through the action of a **voltage sensor**, a protein also known as the **dihydropyridine receptor**,[w,x] together with a **calcium release channel** known as the **ryanodine**

BOX 6-D   (continued)

receptor.[w,y–aa] Inositol trisphosphate (Fig. 11-9), cyclic ADP-ribose, as well as pH changes[bb] are involved in controlling these channels,[cc,dd] a topic discussed in Chapter 11, E. The release of Ca^{2+} can be visualized using high-speed digital imaging microscopy and fluorescent $[Ca^{2+}]$ indicators.[ee,ff] Release of Ca^{2+} stored in sea urchin eggs is induced by cyclic ADP-ribose and by nicotinic acid adenine dinucleotide phosphate (NAADP$^+$, Eq. 15-16).[dd,gg]

In a similar manner, when a nerve impulse reaches a neuron ending (synapse) calcium ions are released into the cytoplasm and provide the trigger that causes stored neurotransmitters to be dumped into the narrow synaptic cleft that separates the endings of two communicating neurons.[y,hh] The released neurotransmitter initiates an impulse in the "postsynaptic neuron" usually again with an inflow of Ca^{2+} (see Chapter 30).[e] Hormones and various other compounds often stimulate the flow of calcium ions into cells.[ii] There Ca^{2+} regulates enzymes[jj,kk] (Chapter 11), microtubules, clathrin of coated vesicles (Chapter 8), K^+ channels in nerve membranes, and events within mitochondria,[ll,mm] in the ER,[op] and in the nucleus.[nn,oo] A substantial fraction of these responses are mediated by **calcium-binding regulatory proteins**. Among the most prominent of these is **calmodulin** (Fig. 6-8), which, upon binding of Ca^{2+} activates a host of metabolic processes[pp] as indicated in the following scheme, which is modified from that of Cheung.[qq]

In many cases metabolic control by Ca^{2+} is modulated by phosphorylation and dephosphorylation

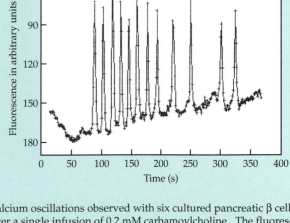

Calcium oscillations observed with six cultured pancreatic β cells after a single infusion of 0.2 mM carbamoylcholine. The fluorescence intensity of the Ca^{2+} indicator dye fura 2, with excitation at 380 nm, was recorded versus time. From Pretki *et al.*[ss]

or other covalent modification of proteins. Such modification may alter the affinity for Ca^{2+}, allowing the latter to either bind and induce a conformational change or remain unbound,[rr] without a change in $[Ca^{2+}]$. Nevertheless, the free Ca^{2+} concentration within cells can change greatly and very rapidly. For example, the oscillatory change in intracellular $[Ca^{2+}]$ shown above was observed in pancreatic insulin-secreting β cells responding to stimulation by the agonist carbamoylcholine. The free $[Ca^{2+}]$ was evaluated from fluorescence measurements using the Ca^{2+} indicator dye fura 2 (From Prentki *et al*[ss]). Oscillations in $[Ca^{2+}]$ have been observed under many circumstances.[ff,oo,tt–ww] These are of particular interest because of the possible relationship to the initiation of oscillatory nerve conduction (Chapter 30).[uu] Released Ca^{2+} often appears to move across cells in waves and sometimes to be released as "puffs" or "sparks."[ee,ff,oo]

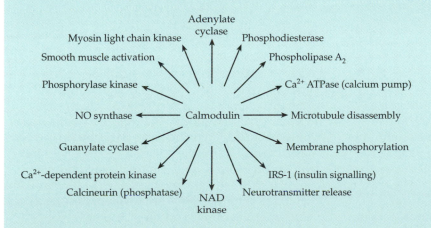

Adenylate cyclase
Myosin light chain kinase
Phosphodiesterase
Smooth muscle activation
Phospholipase A$_2$
Phosphorylase kinase
Ca^{2+} ATPase (calcium pump)
NO synthase — Calmodulin — Microtubule disassembly
Guanylate cyclase
Membrane phosphorylation
Ca^{2+}-dependent protein kinase
IRS-1 (insulin signalling)
Calcineurin (phosphatase)
NAD kinase
Neurotransmitter release

---

[a] Bianchi, C. P. (1968) *Cell Ca*$^{++}$, Appleton, New York

[b] Means, A. R., ed. (1994) *Calcium Regulation of Cellular Function*, Lippincott-Raven, Hagerstown, Maryland

[c] Volpe, P., Krause, K.-H., Hashimoto, S., Zorzato, F., Pozzan, T., Meldolesi, J., and Lew, D. P. (1988) *Proc. Natl. Acad. Sci. U.S.A.* **85**, 1091–1095

[d] Carafoli, E. (1987) *Ann. Rev. Biochem.* **56**, 395–433

[e] Ghosh, A., and Greenberg, M. E. (1995) *Science* **268**, 239–247

[f] Meldolesi, J., and Pozzan, T. (1998) *Trends Biochem. Sci.* **23**, 10–14

[g] Khananshvili, D. (1991) *J. Biol. Chem.* **266**, 13764–13769

(continued)

BOX 6-D (continued)

[h] Rizzuto, R., Simpson, A. W. M., Brini, M., and Pozzan, T. (1992) *Nature (London)* **358**, 325–327

[i] Knight, M. R., Campbell, A. K., Smith, S. M., and Trewavas, A. J. (1991) *Nature (London)* **352**, 524–526

[j] Tsien, R. Y., Pozzan, T., and Rink, T. J. (1984) *Trends Biochem. Sci.* **9**, 263–266

[k] Davies, P. L., and Hew, C. L. (1990) *FASEB J.* **4**, 2460–2468

[l] Ganz, M. B., Rasmussen, J., Bollag, W. B., and Rasmussen, H. (1990) *FASEB J.* **4**, 1638–1644

[m] Lloyd, Q. P., Kuhn, M. A., and Gay, C. V. (1995) *J. Biol. Chem.* **270**, 22445–22451

[n] Etter, E., Minta, A., Poenie, M., and Fay, F. (1996) *Proc. Natl. Acad. Sci. U.S.A.* **93**, 5368–5373

[o] Robitaille, P.-M. L., and Jiang, Z. (1992) *Biochemistry* **31**, 12585–12591

[p] Ellis-Davies, G. C. R., and Kaplan, J. H. (1994) *Proc. Natl. Acad. Sci. U.S.A.* **91**, 187–191

[q] Ellis-Davies, G. C. R., Kaplan, J. H., and Barsotti, R. J. (1996) *Biophys. J.* **70**, 1006–1016

[r] Burgoyne, R. D., and Cheek, T. R. (1991) *Trends Biochem. Sci.* **16**, 319–320

[s] Golovina, V. A., and Blaustein, M. P. (1997) *Science* **275**, 1643–1648

[t] Gagné, S. M., Li, M. X., and Sykes, B. D. (1997) *Biochemistry* **36**, 4386–4392

[u] MacLennan, D. H., Rice, W. J., and Green, N. M. (1997) *J. Biol. Chem.* **272**, 28815–28818

[v] Yonekura, K., Stokes, D. L., Sasabe, H., and Toyoshima, C. (1997) *Biophys. J.* **72**, 997–1005

[w] Yano, M., El-Hayek, R., and Ikemoto, N. (1995) *J. Biol. Chem.* **270**, 3017–3021

[x] Mitterdorfer, J., Sinnegger, M. J., Grabner, M., Striessnig, J., and Glossmann, H. (1995) *Biochemistry* **34**, 9350–9355

[y] McPherson, P. S., and Campbell, K. P. (1993) *J. Biol. Chem.* **268**, 13765–13768

[z] Sakube, Y., Ando, H., and Kagawa, H. (1997) *J. Mol. Biol.* **267**, 849–864

[aa] Zorzato, F., Fujii, J., Otsu, K., Phillips, M., Green, N. M., Lai, F. A., Meissner, G., and MacLennan, D. H. (1990) *J. Biol. Chem.* **265**, 2244–2256

[bb] Donoso, P., Beltrán, M., and Hidalgo, C. (1996) *Biochemistry* **35**, 13419–13425

[cc] Lee, M. G., Xu, X., Zeng, W., Diaz, J., Wojcikiewicz, R. J. H., Kuo, T. H., Wuytack, F., Racymaekers, L., and Muallem, S. (1997) *J. Biol. Chem.* **272**, 15765–15770

[dd] Graeff, R. M., Franco, L., De Flora, A., and Lee, H. C. (1998) *J. Biol. Chem.* **273**, 118–125

[ee] Isenberg, G., Etter, E. F., Wendt-Gallitelli, M.-F., Schiefer, A., Carrington, W. A., Tuft, R. A., and Fay, F. S. (1996) *Proc. Natl. Acad. Sci. U.S.A.* **93**, 5413–5418

[ff] Horne, J. H., and Meyer, T. (1997) *Science* **276**, 1690–1693

[gg] Lee, H. C., and Aarhus, R. (1997) *J. Biol. Chem.* **272**, 20378–20383

[hh] Miller, R. J. (1992) *J. Biol. Chem.* **267**, 1403–1406

[ii] Quitterer, U., Schröder, C., Müller-Esterl, W., and Rehm, H. (1995) *J. Biol. Chem.* **270**, 1992–1999

[jj] Suzuki, K. (1987) *Trends Biochem. Sci.* **12**, 103–105

[kk] Martin, B. L., and Graves, D. J. (1986) *J. Biol. Chem.* **261**, 14545–14550

[ll] Moudy, A. M., Handran, S. D., Goldberg, M. P., Ruffin, N., Karl, I., Kranz-Eble, P., DeVivo, D. C., and Rothman, S. M. (1995) *Proc. Natl. Acad. Sci. U.S.A.* **92**, 729–733

[mm] Kasumi, T., Tsumuraya, Y., Brewer, C. F., Kersters-Hilderson, H., Claeyssens, M., and Hehre, E. J. (1987) *Biochemistry* **26**, 3010–3016

[nn] Santella, L., and Carafoli, E. (1997) *FASEB J.* **11**, 1091–1109

[oo] Lipp, P., Thomas, D., Berridge, M. J., and Bootman, M. D. (1997) *EMBO J.* **16**, 7166–7173

[op] Corbett, E. F., and Michalak, M. (2000) *Trends Biochem. Sci.* **25**, 307–311

[pp] Ikura, M., Clore, G. M., Gronenborn, A. M., Zhu, G., Klee, C. B., and Bax, A. (1992) *Science* **256**, 632–638

[qq] Cheung, W. Y. (1980) *Science* **207**, 19–27

[rr] Geisow, M. J. (1987) *Trends Biochem. Sci.* **12**, 120–121

[ss] Prentki, M., Glennon, M. C., Thomas, A. P., Morris, R. L., Matschinsky, F. M., and Corkey, B. E. (1988) *J. Biol. Chem.* **263**, 11044–11047

[tt] Tang, Y., and Othmer, H. G. (1995) *Proc. Natl. Acad. Sci. U.S.A.* **92**, 7869–7873

[uu] Putney, J. W., Jr. (1998) *Science* **279**, 191–192

[vv] Berridge, M. J. (1997) *Nature (London)* **386**, 759–760

[ww] Hajnóczky, G., and Thomas, A. P. (1997) *EMBO J.* **16**, 3533–3543

increase above the normal level in keratinocytes of patients with the skin disease psoriasis.[123b]

Calcium ions are often involved in holding negatively charged groups together, for example, in the binding of proteins to phospholipid membranes. Among these membrane-associated proteins are the vitamin K-dependent proteins, all of which contain several residues of the chelating amino acid γ-carboxyglutamate (Chapter 15) at their calcium-binding sites.[124,125] Many of these are involved in the clotting of blood (Chapter 12). Some of the membrane-binding proteins called **annexins** (Chapter 8) are also Ca^{2+}-dependent ion channels.[126,127] Other Ca^{2+}-requiring lipid-binding proteins include the lipocortins, calpactins, and calelectrins.[128] Cadherins bind Ca^{2+} and help provide cohesion between cells.[129] Many proteins contain bound Ca^{2+} in precisely defined sites where it plays a structural role. These include the galactose-binding protein of bacterial transport and chemotaxis (Fig. 4-18)[130] and α-lactalbumin of milk. Although α-lactalbumin does not contain the helix–loop–helix pattern, its Ca^{2+}-binding site does consist of three carboxylate groups and two backbone carbonyl oxygen atoms.[131] The α-amylases, thermolysin, and staphylococcal nuclease (Chapter 12), and the lectin favin (Fig. 2-15) all contain bound Ca^{2+}. Calcium ions also bind to anionic groups in carbohydrates, e.g., to the sulfate groups in carageenin gels (Chapter 4) where they provide structural stability.

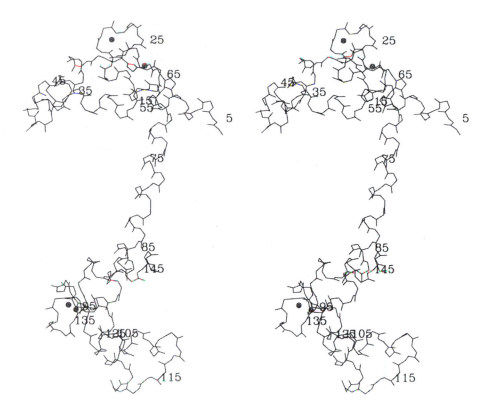

Figure 6-8 Stereoscopic backbone trace of a 148-residue recombinant calmodulin. The two helix–turn–helix (EF-hand) loops and their bound Ca²⁺ (as concentric circles) are at the top and the two near the C terminus are at the bottom. The long central helix, seen in this crystal structure, may undergo conformational changes during the functioning of this Ca²⁺-sensing molecule.[132] From Chattopadhyaya *et al.*[111] Courtesy of F. A. Quiocho.

BOX 6-E   METALLOTHIONEINS

If animals ingest excessive amounts of Zn(II), Cd (II), Hg(II), or Cu(I) their livers and kidneys accumulate these metals as complexes of proteins called metallothioneins.[a–e] In mammals at least three related genes encode these metal-binding proteins. The best known, metallothionein II, has a highly conserved 61-residue sequence containing 20 cysteine residues and no aromatic residues. The protein is organized into two domains, each able to bind a cluster of metal ions via thiolate side chains. The three-dimensional structure of rat liver metallothionein containing five Cd²⁺ and two Zn²⁺ ions is shown in the accompanying stereoscopic diagram.[f] The 61 alpha carbons, the beta carbon and sulfur atoms (green) of cysteine residues and the bound metal ions are indicated. The N-terminal domain (residues 1–29) contains one Cd²⁺ and two Zn² and nine cysteine sulfurs which bind the metal ions. Three of the sulfur atoms form bridges between pairs of metals. The second cluster contains four Cd²⁺ held by 11 cysteine sulfur atoms,

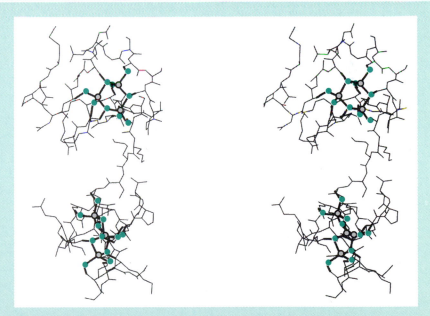

Metallothionein containing bound Cd²⁺ and Zn²⁺. From Robbins *et al.*[f]

five of which bridge between pairs of metals. All of the metal ions are tetrahedrally coordinated. The polypeptide chains of metallothioneins consist predominantly of beta turns.[g] Important techniques in the study of these proteins include ¹¹³Cd NMR,[h,i]

BOX 6-E   METALLOTHIONEINS (continued)

spectroscopic methods,[j,k] and X-ray absorption.[l]

Transcription of metallothionein genes is induced by metal ions, and toxic metals such as Cd and Hg accumulate as metallothionein complexes, suggesting that one function is to protect against metal toxicity.[m] However, synthesis is also induced by glucocorticoid hormones,[n] and accumulation of a high concentration of copper and zinc in fetal metallothionein suggests a role in storage of these essential metals.[c] Another metal that binds to metallothioneins is Au(I),[o] which is widely used in thiolate salts as a chemotherapeutic agent for rheumatoid arthritis.

Metallothioneins are also found in insects, lower invertebrates, and even in bacteria.[i,p] Nevertheless, there are other metal-binding proteins.[q,r] For example, albacore tuna contain a 66-kDa glycoprotein that contains eight mole percent histidine and binds three Zn^{2+}, each by a cluster of three His.[s] Plants and fungi contain **phytochelins**, peptides consisting largely of repeated γ-glutamylcysteine units.[a,t,u] These appear to protect plants against toxicity of cadmium in the same manner as do the metallothioneins in our bodies.

[a] Mehra, R. K., Garey, J. R., Butt, T. R., Gray, W. R., and Winge, D. R. (1989) *J. Biol. Chem.* **264**, 19747–19753

[b] Kägi, J. H. R., and Kojima, Y., eds. (1987) *Metallothionein II*, Birkhäuser, Basel, Berlin

[c] Fischer, E. H., and Davie, E. W. (1998) *Proc. Natl. Acad. Sci. U.S.A.* **95**, 3333–3334

[d] Kägi, J. H. R., and Schäffer, A. (1988) *Biochemistry* **27**, 8509–8515

[e] Hamer, D. H. (1986) *Ann. Rev. Biochem.* **55**, 913–951

[f] Robbins, A. H., McRee, D. E., Williamson, M., Collett, S. A., Xuong, N. H., Furey, W. F., Wang, B. C., and Stout, C. D. (1991) *J. Mol. Biol.* **221**, 1269–1293

[g] Pande, J., Pande, C., Gilg, D., Vasák, M., Callender, R., and Kägi, J. H. R. (1986) *Biochemistry* **25**, 5526–5532

[h] Cismowski, M. J., Narula, S. S., Armitage, I. M., Chernaik, M. L., and Huang, P. C. (1991) *J. Biol. Chem.* **266**, 24390–24397

[i] Vasák, M., Hawkes, G. E., Nicholson, J. K., and Sadler, P. J. (1985) *Biochemistry* **24**, 740–747

[j] Willner, H., Vasák, M., and Kägi, J. H. R. (1987) *Biochemistry* **26**, 6287–6292

[k] Lu, W., and Stillman, M. J. (1993) *J. Am. Chem. Soc.* **115**, 3291–3299

[l] George, G. N., Byrd, J., and Winge, D. R. (1988) *J. Biol. Chem.* **263**, 8199–8203

[m] Sadhu, C., and Gehamu, L. (1988) *J. Biol. Chem.* **263**, 2679–2684

[n] Karin, M., Haslinger, A., Holtgreve, H., Richards, R. I., Krauter, P., Westphal, H. M., and Beato, M. (1984) *Nature (London)* **308**, 513–519

[o] Laib, J. E., Shaw, C. F., III, Petering, D. H., Eidsness, M. K., Elder, R. C., and Garvey, J. S. (1985) *Biochemistry* **24**, 1977–1986

[p] Freedman, J. H., Slice, L. W., Dixon, D., Fire, A., and Rubin, C. S. (1993) *J. Biol. Chem.* **268**, 2554–2564

[q] Waalkes, M. P., and Peratoni, A. (1986) *J. Biol. Chem.* **261**, 13097–13103

[r] Willuhn, J., Schmitt-Wrede, H. P., Greven, H., and Wunderlich, F. (1994) *J. Biol. Chem.* **269**, 24688–24691

[s] Dyke, B., Hegenauer, J., Saltman, P., and Laurs, R. M. (1987) *Biochemistry* **26**, 3228–3234

[t] Reese, R. N., and Winge, D. R. (1988) *J. Biol. Chem.* **263**, 12832–12835

[u] Clemens, S., Kim, E. J., Neumann, D., and Schroeder, J. I. (1999) *EMBO J.* **18**, 3325–3333

1. Lewis, G. N., and Randall, M. (1961) *Thermodynamics and the Free Energy of Chemical Substances*, 2nd ed., McGraw-Hill, New York
2. Mahan, B. H. (1963) *Elementary Chemical Thermodynamics*, Benjamin, New York
3. Everdell, M. H. (1965) *Introduction to Chemical Thermodynamics*, Norton, New York
4. Dickerson, R. E. (1969) *Molecular Thermodynamics*, Benjamin, New York
5. Jones, M. N., ed. (1988) *Biochemical Thermodynamics*, 2nd ed., Elsevier, Amsterdam
6. Hinz, H.-J., ed. (1986) *Thermodynamic Data for Biochemistry and Biotechnology*, Springer-Verlag, Berlin
7. Gutfreund, H., and Edsall, J. T. (1983) *Biothermo-dynamics*, Wiley, New York
8. Harold, F. M. (1986) *The Vital Force: A Study of Bioenergetics*, Freeman, San Francisco, California
9. Garby, L., and Larsen, P. S. (1995) *Bioenergetics; Its Thermodynamic Foundations*, Cambridge Univ. Press, London and New York
10. Jou, D., and Llebot, J. E., eds. (1990) *Introduction to the Thermodynamics of Biological Processes*, Prentice Hall, Englewood Cliffs, New Jersey
10a. Wadsö, I. (1985) *Eur. J. Biochem.* **153**, 429–434
11. Sturtevant, J. M. (1971) in *Techniques of Chemistry*, Vol. I, Part V (Weissberger, A., ed), pp. 347–425, Wiley, New York
12. Brown, H. D., ed. (1969) *Biochemical Microcalorimetry*, Academic Press, New York
13. Peters, K. S., and Snyder, G. J. (1988) *Science* **241**, 1053–1057
14. White, A., Handler, P., and Smith, E. L. (1968) *Principles of Biochemistry*, 4th ed., McGraw-Hill, New York (pp. 291–301)
15. Mosteller, R. D. (1988) *N. Engl. J. Med.* **318**, 1130
16. Donata, K., and Hegsted, D. M. (1985) *Proc. Natl. Acad. Sci. U.S.A.* **82**, 4866–4870
17. Ben-Naim, A. (1992) *Statistical Thermodynamics for Chemists and Biochemists*, Plenum, New York
18. Donovan, J. W. (1984) *Trends Biochem. Sci.* **9**, 340–344
19. Tanaka, A., Flanagan, J., and Sturtevant, J. M. (1993) *Protein Sci.* **2**, 567–576
20. Pace, C. N., and Laurents, D. V. (1989) *Biochemistry* **28**, 2520–2525
21. Interunion Commission on Biothermodynamics. (1976) *J. Biol. Chem.* **251**, 6879–6885
22. Phillips, R. C., George, P., and Rutman, R. J. (1963) *Biochemistry* **2**, 501–508
23. Liu, Y., and Sturtevant, J. M. (1995) *Protein Sci.* **4**, 2559–2561
24. Halling, P. J. (1989) *Trends Biochem. Sci.* **14**, 317–318
25. Örstan, A. (1990) *Trends Biochem. Sci.* **15**, 137–138
26. Nicolis, G., and Prigogine, I. (1977) *Self-organization in Non-equilibrium Systems: From Dissipative Structure to Order Through Fluctuations*, Wiley-Interscience, New York
27. Procaccia, I., and Ross, J. (1977) *Science* **198**, 716–717
28. Morowitz, H. J. (1968) *Energy Flow in Biology*, Academic Press, New York
29. Caplan, S. R., and Essig, A. (1983) *Bioenergetics and Linear Nonequilibrium Thermodynamics/ The Steady State*, Harvard Univ. Press, Cambridge, Massachusetts
30. Peacocke, A. R. (1983) *An Introduction to the Physical Chemistry of Biological Organization*, Clarendon Press, Oxford
31. Blumenfeld, L. A. (1983) *Physics of Bioenergetic Processes*, Springer-Verlag, Berlin

32. Katchalsky, A., and Curran, P. F. (1965) *Non-equilibrium Thermodynamics in Biophysics*, Harvard Univ. Press, Cambridge, Massachusetts
33. Prigogine, I. (1967) *Introduction to Thermodynamics of Irreversible Processes*, 3rd ed., Wiley, New York
34. Coveney, P. V. (1988) *Nature (London)* **333**, 409–415
35. Wilson, D. F. (1982) *Trends Biochem. Sci.* **7**, 275–278
36. Westerhoff, H. V. (1982) *Trends Biochem. Sci.* **7**, 275–279
37. Guynn, R. W., and Veech, R. L. (1973) *J. Biol. Chem.* **248**, 6966–6972
38. Alberty, R. A. (1969) *J. Biol. Chem.* **244**, 3290–3302
39. Phillips, R. C., George, P., and Rutman, R. J. (1969) *J. Biol. Chem.* **244**, 3330–3342
40. George, P., Phillips, R. C., and Rutman, R. J. (1963) *Biochemistry* **2**, 508–512
41. Purich, D. L., and Fromm, H. J. (1972) *Curr. Top. Cell. Regul.* **6**, 131–167
42. Jencks, W. P. (1968) in *Handbook of Biochemistry* (Sober, H. A., ed), pp. J-148, CRC, Cleveland, Ohio
43. Bassham, J. A., and Krause, G. H. (1969) *Biochim. Biophys. Acta.* **189**, 207–221
44. Rosing, J., and Slater, E. C. (1972) *Biochim. Biophys. Acta.* **267**, 275–290
45. Alberty, R. A. (1992) *Biophysical Chem.* **42**, 117–131
46. Alberty, R. A., and Cornish-Bowden, A. (1993) *Trends Biochem. Sci.* **18**, 288–291
47. Alberty, R. A., and Goldberg, R. N. (1992) *Biochemistry* **31**, 10610–10615
48. Alberty, R. A. (1994) *Pure & Appl. Chem.* **66**, 1641–1666
49. Stokes, G. B. (1988) *Trends Biochem. Sci.* **13**, 422–424
50. Mavrovouniotis, M. L. (1991) *J. Biol. Chem.* **266**, 14440–14445
51. Dutton, P. L. (1971) *Biochim. Biophys. Acta.* **226**, 63–80
52. Pepin, C. A., and Wood, H. G. (1986) *J. Biol. Chem.* **261**, 4476–4480
53. Wood, H. G., and Clark, J. E. (1988) *Ann. Rev. Biochem.* **57**, 235–260
54. Crooke, E., Akiyama, M., Rao, N. N., and Kornberg, A. (1994) *J. Biol. Chem.* **269**, 6290–6295
55. Veech, R. L., Lawson, J. W. R., Cornell, N. W., and Krebs, H. A. (1979) *J. Biol. Chem.* **254**, 6538–6547
56. Swedes, J. S., Sedo, R. J., and Atkinson, D. E. (1975) *J. Biol. Chem.* **250**, 6930–6938
57. Chapman, A. G., Fall, L., and Atkinson, D. E. (1971) *J. Bacteriol.* **108**, 1072–1086
58. Purich, D. L., and Fromm, H. J. (1973) *J. Biol. Chem.* **248**, 461–466
59. Teague, W. E., Jr., and Dobson, G. P. (1992) *J. Biol. Chem.* **267**, 14084–14093
60. Cohen, S. M., Ogawa, S., Rottenberg, H., Glynn, P., Yamane, T., Brown, T. R., and Shulman, R. G. (1978) *Nature (London)* **273**, 554–556
60a. Ugurbil, K., Kingsley-Hickman, P. B., Sako, E. Y., Zimmer, S., Mohanakrishnan, P., Robitaille, P. M. L., Thoma, W. J., Johnson, A., Foker, J. E., and From, A. H. L. (1987) *Ann. N.Y. Acad. Sci.* **508**, 265–286
61. Brosnan, M. J., Chen, L., Van Dyke, T. A., and Koretsky, A. P. (1990) *J. Biol. Chem.* **265**, 20849–20855
61a. Saupe, K. W., Spindler, M., Hopkins, J. C. A., Shen, W., and Ingwall, J. S. (2000) *J. Biol. Chem.* **275**, 19742–19746

62. Hutson, S. M., Williams, G. D., Berkich, D. A., LaNoue, K. F., and Briggs, R. W. (1992) *Biochemistry* **31**, 1322–1330
63. Burt, C. T., Glonek, T., and Bárány, M. (1977) *Science* **195**, 145–149
64. Seeley, P. J., Sehr, P. A., Gadian, D. G., Garlick, P. B., and Radda, G. K. (1977) in *NMR in Biology* (Dwek, R. A., Campbell, I. D., Richards, R. E., and Williams, R. J. P., eds), pp. 247–275, Academic Press, London
65. Masson, S., and Quistorff, B. (1992) *Biochemistry* **31**, 7488–7493
66. Jeffrey, F. M., Storey, C. J., Nunnally, R. L., and Malloy, C. R. (1989) *Biochemistry* **28**, 5323–5326
67. Kagimoto, T., Higaki, T., Nagata, K., Morino, Y., and Takatsuki, K. (1989) *NMR in Biomedicine* **2**, 93–97
68. Park, J. H., Brown, R. L., Park, C. R., Cohn, M., and Chance, B. (1988) *Proc. Natl. Acad. Sci. U.S.A.* **85**, 8780–8784
69. Radda, G. K. (1992) *FASEB J.* **6**, 3032–3038
70. Radda, G. K. (1986) *Science* **233**, 640–645
71. Burt, C. T. (1985) *Trends Biochem. Sci.* **10**, 404–406
72. Grove, T. H., Ackerman, J. J. H., Radda, G. K., and Bore, P. J. (1980) *Proc. Natl. Acad. Sci. U.S.A.* **77**, 299–302
73. Bottomley, P. A. (1985) *Science* **229**, 769–772
74. Fossel, E. T., Morgan, H. E., and Ingwall, J. S. (1980) *Proc. Natl. Acad. Sci. U.S.A.* **77**, 3654–3658
75. Metzler, D. E., Harris, C. M., Johnson, R. J., Siano, D. B., and Thomson, J. A. (1973) *Biochemistry* **12**, 5377–5392
76. Johnson, R. J., and Metzler, D. E. (1970) *Methods Enzymol.* **18**, 433–471
77. Dixon, H. B. F. (1992) *Essays in Biochemistry* **27**, 161–176
78. Vander Jagt, D. L., Hansen, L. D., Lewis, E. A., and Han, L. B. (1972) *Arch. Biochem. Biophys.* **153**, 55–61
79. Dixon, H. B. F., and Tipton, K. F. (1973) *Biochem. J.* **133**, 837–842
80. Klotz, I. M., and Hunston, D. L. (1975) *J. Biol. Chem.* **250**, 3001–3009
81. Weber, G. (1975) *Adv. Prot. Chem.* **29**, 1–83
82. Chivers, P. T., Prehoda, K. E., Volkman, B. F., Kim, B.-M., Markley, J. L., and Raines, R. T. (1997) *Biochemistry* **36**, 14985–14991
83. Johnson, F. A., Lewis, S. D., and Shafer, J. A. (1981) *Biochemistry* **20**, 44–48
84. McIntosh, L. P., Hand, G., Johnson, P. E., Joshi, M. D., Körner, M., Plesniak, L. A., Ziser, L., Wakarchuk, W. W., and Withers, S. G. (1996) *Biochemistry* **35**, 9958–9966
85. Katritzky, A. R., and Waring, A. (1962) *J. Chem. Soc.*, 1540–1548
86. Katritzky, A. R., and Karelson, M. (1991) *J. Am. Chem. Soc.* **113**, 1561–1566
87. Bjerrum, J. (1941) *Metal Amine Formation in Aqueous Solution*, Haase & Son, Copenhagen
88. Eichhorn, G. L., ed. (1973) *Inorganic Biochemistry*, 2 vols, Elsevier, Amsterdam
89. Martell, A. E., and Smith, R. M. (1975) *Critical Stability Constants*, Plenum, New York
90. Michaelian, K. H., and Moskovits, M. (1978) *Nature (London)* **273**, 135–136
91. Lewsin, S. (1974) *Displacement of Water and its Control in Biochemical Reactions*, Academic Press, London
92. Karlin, S., Zhu, Z.-Y., and Karlin, K. D. (1997) *Proc. Natl. Acad. Sci. U.S.A.* **94**, 14225–14230
93. Freeman, H. C., Healy, M. J., and Scudder, M. L. (1977) *J. Biol. Chem.* **252**, 8840–8847
94. Torrado, A., Walkup, G. K., and Imperiali, B. (1998) *J. Am. Chem. Soc.* **120**, 609–610
95. Nakayama, S., and Kretsinger, R. H. (1994) *Annu. Rev. Biophys. Biomol. Struct.* **23**, 473–507

References

96. Kretsinger, R. H. (1976) *Ann. Rev. Biochem.* **45**, 239–266

96a. Pawlowski, K., Bierzynski, A., and Godzik, A. (1996) *J. Mol. Biol.* **258**, 349–366

97. Ikura, M. (1996) *Trends Biochem. Sci.* **21**, 14–17

98. Kretsinger, R. H., and Nockolds, C. E. (1973) *J. Biol. Chem.* **248**, 3313–3326

99. McPhalen, C. A., Sielecki, A. R., Santarsiero, B. D., and James, M. N. G. (1994) *J. Mol. Biol.* **235**, 718–732

99a. Declercq, J.-P., Evrard, C., Lamzin, V., and Parello, J. (1999) *Protein Sci.* **8**, 2194–2204

99b. Biekofsky, R. R., Martin, S. R., Browne, J. P., Bayley, P. M., and Feeney, J. (1998) *Biochemistry* **37**, 7617–7629

100. Drake, S. K., Zimmer, M. A., Miller, C. L., and Falke, J. J. (1997) *Biochemistry* **36**, 9917–9926

101. Akke, M., Forsén, S., and Chazin, W. J. (1995) *J. Mol. Biol.* **252**, 102–121

102. Denisov, V. P., and Halle, B. (1995) *J. Am. Chem. Soc.* **117**, 8456–8465

103. Heizmann, C. W., and Hunziker, W. (1991) *Trends Biochem. Sci.* **16**, 98–103

104. Ikemoto, N., Ronjat, M., Mészáros, L. G., and Koshita, M. (1989) *Biochemistry* **28**, 6764–6771

105. Meldolesi, J., and Pozzan, T. (1998) *Trends Biochem. Sci.* **23**, 10–14

106. Randolph, J. T., McClure, K. F., and Danishefsky, S. J. (1995) *J. Am. Chem. Soc.* **117**, 5712–5719

107. Fliegel, L., Burns, K., MacLennan, D. H., Reithmeier, R. A. F., and Michalak, M. (1989) *J. Biol. Chem.* **264**, 21522–21528

108. Cohen, P., and Klee, C. B., eds. (1988) *Calmodulin*, Elsevier, Amsterdam

109. James, P., Vorherr, T., and Carafoli, E. (1995) *Trends Biochem. Sci.* **20**, 38–42

110. Lafitte, D., Capony, J. P., Grassy, G., Haiech, J., and Calas, B. (1995) *Biochemistry* **34**, 13825–13832

111. Chattopadhyaya, R., Meador, W. E., Means, A. R., and Quiocho, F. A. (1992) *J. Mol. Biol.* **228**, 1177–1192

112. Pedigo, S., and Shea, M. A. (1995) *Biochemistry* **34**, 10676–10689

113. da Silva, A. C. R., and Reinach, F. C. (1991) *Trends Biochem. Sci.* **16**, 53–57

114. Gagné, S. M., Li, M. X., and Sykes, B. D. (1997) *Biochemistry* **36**, 4386–4392

115. Trigo-Gonzalez, G., Awang, G., Racher, K., Neden, K., and Borgford, T. (1993) *Biochemistry* **32**, 9826–9831

116. Spyracopoulos, L., Li, M. X., Sia, S. K., Gagné, S. M., Chandra, M., Solaro, R. J., and Sykes, B. D. (1997) *Biochemistry* **36**, 12138–12146

117. Travé, G., Lacombe, P.-J., Pfuhl, M., Saraste, M., and Pastore, A. (1995) *EMBO J.* **14**, 4922–4931

118. Näthke, I., Hill, B. L., Parham, P., and Brodsky, F. M. (1990) *J. Biol. Chem.* **265**, 18621–18627

119. Pley, U. M., Hill, B. L., Alibert, C., Brodsky, F. M., and Parham, P. (1995) *J. Biol. Chem.* **270**, 2395–2402

120. Engel, J., Taylor, W., Paulsson, M., Sage, H., and Hogan, B. (1987) *Biochemistry* **26**, 6958–6965

121. Engel, E., Richter, K., Obermeyer, G., Briza, P., Kungl, A. J., Simon, B., Auer, M., Ebner, C., Rheinberger, H.-J., Breitenbach, M., and Ferreira, F. (1997) *J. Biol. Chem.* **272**, 28630–28637

122. Schäfer, B. W., and Heizmann, C. W. (1996) *Trends Biochem. Sci.* **21**, 134–140

123. Drohat, A. C., Baldisseri, D. M., Rustandi, R. R., and Weber, D. J. (1998) *Biochemistry* **37**, 2729–2740

123a. Gribenko, A. V., and Makhatadze, G. I. (1998) *J. Mol. Biol.* **283**, 679–694

123b. Brodersen, D. E., Nyborg, J., and Kjeldgaard, M. (1999) *Biochemistry* **38**, 1695–1704

124. Christiansen, W. T., Tulinsky, A., and Castellino, F. J. (1994) *Biochemistry* **33**, 14993–15000

125. Schwalbe, R. A., Ryan, J., Stern, D. M., Kisiel, W., Dahlbäck, B., and Nelsestuen, G. L. (1989) *J. Biol. Chem.* **264**, 20288–20296

126. Demange, P., Voges, D., Benz, J., Liemann, S., Göttig, P., Berendes, R., Burger, A., and Huber, R. (1994) *Trends Biochem. Sci.* **19**, 272–276

127. Saurel, O., Cézanne, L., Milon, A., Tocanne, J.-F., and Demange, P. (1998) *Biochemistry* **37**, 1403–1410

128. Moss, S. E., and Crumpton, M. J. (1990) *Trends Biochem. Sci.* **15**, 11–12

129. Shapiro, L., Fannon, A. M., Kwong, P. D., Thompson, A., Lehmann, M. S., Grübel, G., Legrand, J.-F., Als-Nielsen, J., Colman, D. R., and Hendrickson, W. A. (1995) *Nature (London)* **374**, 327–336

130. Vyas, N. K., Vyas, M. N., and Quiocho, F. A. (1988) *Science* **242**, 1290–1295

131. Stuart, D. I., Acharya, K. R., Walker, N. P. C., Smith, S. G., Lewis, M., and Phillips, D. C. (1986) *Nature (London)* **324**, 84–87

132. Williams, R. J. P. (1991) *Trends Biochem. Sci.* **16**, 206

Study Questions

1. a) From $\Delta G°$ for hydrolysis of sucrose (Table 6-6) calculate the equilibrium constant

$$K = [glucose][fructose] / [sucrose]$$

   at 25°C. Call this hydrolysis reaction 1.
   b) Is the sucrose in a 1 M solution stable? Explain.
   c) If acid is added to a 1 M sucrose solution to catalyze its hydrolysis, what will be the final sucrose concentration at equilibrium? (Assume that concentrations equal activities for the purpose of these calculations.)
   d) Reaction 2 is the hydrolysis of α-D-glucose 1-phosphate to glucose and inorganic phosphate (P_i). Using $\Delta G°$ for this reaction (Table 6-6) calculate the equilibrium constant.
   e) Sucrose phosphorylase from the bacterium *Pseudomonas saccharophila* catalyzes the following reaction (reaction 3):

$$\text{Sucrose} + P_1 \rightarrow$$
$$\text{α-D-glucose 1-phosphate + fructose}$$

   Calculate the equilibrium constant and the standard Gibbs energy change at 25°C for reaction 3 from the equilibrium constants obtained above for reactions 1 and 2. Show that $\Delta G°$ for reaction 3 = $\Delta G°$ of reaction 1 - $\Delta G°$ of reaction 2.
   f) Could the bacterium carry out reaction 3 in the following two consecutive steps? Explain.

$$\text{Sucrose} \rightarrow \text{glucose + fructose}$$
$$\text{Glucose} + P_i \rightarrow \text{glucose 1-phosphate}$$

2. For each of the following reactions, state whether the equilibrium constant will be between 0.1 and 10 (i.e., about one), greater than 100, or less that 0.01. Assume that the pH is constant at 7.0.
   a) $2\,ADP^{3-} \rightarrow ATP^{4-} + AMP^{2-}$
   b) $ATP^{3-} + \text{glucose} \rightarrow \text{glucose 6-phosphate}^{2-} + ADP^{2-} + H^+$
   c) $ADP^{2-} + HPO_4^{2-} + H^+ \rightarrow ATP^{3-}$
   d) Glucose 6-phosphate$^{2-} \rightarrow$ fructose 6-phosphate^{2-}
   e) Phosphoenolpyruvate^{3-} + glucose $\rightarrow$ glucose 1-phosphate^{2-} + pyruvate$^-$

3. The combustion of 1 mol of solid urea to liquid water and gaseous carbon dioxide and nitrogen (N_2) in a bomb colorimeter at 25°C (constant volume) liberated 666 kJ of heat energy. Calculate ΔH, the change in heat content (enthalpy), for this reaction.

4. Using data of Table 6-4 calculate $\Delta G'$ (pH 7) for the following reactions:
   a) Glucose $\rightarrow$ 2 lactate$^-$ + 2 H$^+$
   b) $2\,NH_4^+ + HCO_3^- \rightarrow \text{urea} + 2\,H_2O + H^+$
   c) 2-oxoglutarate^{2-} + 1/2 O_2 + CoA + H$^+ \rightarrow$ succinyl-CoA + H_2O + CO_2

5. What is the ionic strength of a 0.2 M solution of NaCl? of 0.2 M Na_2SO_4?

6. The [ATP]/[ADP] ratio in an actively respiring yeast cell is about 10. What would be the intracellular [3-phosphoglycerate]/[1,3-bisphosphoglycerate] ratio have to be to make the phosphoglycerate kinase reaction (Fig. 9-7, reaction 7) proceed toward 1,3-bisphosphoglycerate synthesis at 25°C, pH 7?

7. a) Using data from Table 6-8 determine the equilibrium constant for the reaction between malate and methylene blue, assuming all reactants present initially at the same concentration. Indicate clearly the direction of the reaction for which the Gibbs energy change is written.
   b) Calculate the percentage of the reduced (leuco) form of methylene blue present at pH 7 and 25°C in a system for which the measured electrode potential is 0.065 V.

8. NAD$^+$ is a coenzyme for both pyruvate dehydrogenase and ethanol dehydrogenase. Using the values of E_o' from Table 6-8 calculate the Gibbs energy change and the equilibrium constant for the reaction.

$$\text{Lactate}^- + \text{acetaldehyde} \rightarrow \text{pyruvate}^- + \text{ethanol}$$

9. Consider the oxidation of acetate at 25°C:

$$CH_3COO^- \,(0.1\,M) + 2\,O_2\,(g, 0.2\,atm) +$$
$$H^+ \,(10^{-7}\,M) \rightarrow 2\,H_2O\,(l) + 2\,CO_2\,(g)$$

   a) What is the equilibrium pressure of CO_2 if the reaction is not coupled to any other reaction?
   b) What is the equilibrium pressure of CO_2 if the reaction is coupled to the formation of 0.01 M ATP from 0.02 M ADP and 0.01 M HPO_4^{2-} in the citric acid cycle?
   c) What do the above calculations tell you about the prospects of gaining 100% efficiency of energy storage in ATP from the citric acid cycle?
   d) If the actual pressure of CO_2 is 0.01 atm, what is the efficiency of energy storage under the conditions in (b)?

Study Questions

10. The equilibrium constant for the following reaction, which is catalyzed by creatine kinase, has been determined by chemical analysis. The data are given below. [S. A. Kuby and E. A. Noltman, in *The Enzymes*, 2nd ed. (P. D. Boyer, H. Lardy, and K. Myrbäck, eds), Vol. VI, pp. 515–602. Academic Press, New York, 1962]

$$ATP^{4-} + creatine \rightarrow ADP^{3-} + creatine\ phosphate^{2-} + H^+$$

t (°C)	K
20	6.30×10^{-9}
30	5.71×10^{-9}
38	5.47×10^{-9}

  a) What are $\Delta G°$, $\Delta H°$, and $\Delta S°$ for the reaction at 25°C?
  b) What are $\Delta G'$, $\Delta H'$, and $\Delta S'$ (pH 7) for the reaction at 25°C?
  c) What are $\Delta G'$, $\Delta H'$, and $\Delta S'$ (pH 7) for the hydrolysis of creatine phosphate at 25°C?

11. The following reaction was carried out in a calorimeter at 25°C in 0.1 M phosphate buffer at pH 7.4 in the presence of a particulate suspension containing the mitochondrial electron transport system [M. Poe, H. Gutfreund, and R. W. Estabrook, *ABB* **122**, 204–211 (1967)]:

$$NADH + H^+ + 1/2\ O_2\ (aq., sat.) \rightarrow NAD^+ + H_2O$$

The oxygen consumption was monitored continuously with an oxygen electrode. The temperature was monitored simultaneously with a thermocouple immersed in the solution. At the start of the reaction 96 μmol NADH was added to 29.0 ml buffer containing O_2. A nearly zero-order reaction was observed with the rate of O_2 consumption of 6.87 μmol/min and the rate of temperature rise of 0.01171 K/min. The heat capacity of the calorimeter and contents was 254.6 J/K. What is ΔH for the above reaction? NOTE: The H^+ is supplied by the phosphate buffer, which has a ΔH of dissociation of 5.4 kJ mol^{-1}.

12. Enthalpy and Gibbs energy changes for the following reaction at 25°C are given in Table 6-6.

$$ATP^{4-}\ (1\ M) + H_2O \rightarrow$$
$$ADP^{3-}\ (1\ M) + H^+\ (10^{-7}\ M) + HPO_4^{2-}\ (1\ M)$$

  a) How much heat is evolved at constant temperature and pressure if the reaction takes place in a test tube without doing any work other than $p\ \Delta V$ work?
  b) How much heat is evolved or absorbed by the foregoing reaction if it is coupled with 100% efficiency to an endergonic reaction?
  c) What efficiency of coupling to an endergonic reaction is required in order that the foregoing reaction neither evolve nor absorb heat?

13. Microorganisms use a great variety of fermentation reactions for obtaining energy. Could the following reaction be used for such a purpose? Explain the reasons for your answer.

$$C_4H_6O_4\ (succinic\ acid) + H_2O \rightarrow$$
$$C_3H_8O_3\ (glycerol) + CO_2$$

14. a) Calculate the work done in kJ and in kcal by a 70 kg person in climbing up stairs three stories (13 m).
  b) Calculate how much ATP (in mmol and in grams) will be needed for the climb if muscles can use the ATP with 50% efficiency and if the phosphorylation state $R_p = [ATP] / [ADP] [P_i]$ is 10^3 M^{-1}. The standard value of $\Delta G°'$ (pH 7) for hydrolysis of MgATP to MgADP and P_i may be taken as −31 kJ/mol at 37°C.
  c) How much of this ATP could be provided by transfer of phospho groups from the stored creatine phosphate (Cr-P) to ADP in muscle? Assume that [Cr-P] = 20 mM and may fall to 10 mM during the climb. $\Delta G°'$ (pH 7) for hydrolysis of creatine phosphate is about −43 kJ/mol. If the creatine kinase reaction attains equilibrium what will the value of R_p be for the adenylate system?

15. Acid–base titration gave the following three pK_a values for cysteine: 1.70, 8.36, and 10.53. Spectrophotometric data allowed H. B. F. Dixon and K. F Tipton (1973, *Biochem J.* **133**, 837–842) to estimate the following ratio of tautomeric species at pH 9.4 [$^-OOC – C(NH_3^+)– CH_2S^-$] / [$^-OOC – C(NH_2) – CH_2SH$] = 2.12, as is also shown in Fig. 6-5. Verify and assign the four microscopic pK_a values.

16. The coenzyme pyridoxal phosphate (PLP; structure on p. 740) has pK_a values of 3.62, 6.10, and 8.33, as determined by acid–base titration or by spectrophotometric titration. The pK_a of 6.10 belongs primarily to the phosphate group and is nearly independent of the others. However, the pK_a values of 3.62 and 8.33 are shared by the protonated ring nitrogen and the phenolic –OH group. PLP exists as an equilibrium mixture of aldehyde together with its covalent hydrate (see Eq. 13-1). The equilibrium constants for hydrate formation (K_h = [hydrate] / [aldehyde]) are independent of pH but differ for each ionization state of the ring. Consider only the equilibria in the pH range 7–12.

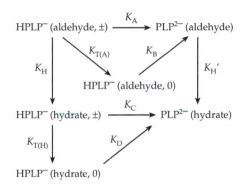

The UV-visible absorption spectrum of the monoprotonated form of PLP was divided mathematically into individual bands for the aldehyde with dipolar ionic ring (±), the aldehyde tautomer with an uncharged ring (0) and the hydrate of the dipolar ion. The following fractions were estimated (Harris *et al.*, 1976, *Biochem. Biophys. Acta.* **521**, 181–194.

Aldehyde (±)	52%
Aldehyde (0)	28%
Hydrate (±)	16%
Hydrate (0)	4%, estimated indirectly

From 300 MHz ^1H NMR spectra areas for the following hydrogen atoms were estimated by Robitaille.

	pH 7	pH 12
4'-H aldehyde	.86	.900
4'-H hydrate	.14	.012
6-H, aldehyde	.96	1.059
6-H, hydrate	.17	.014
2'-CH_3, aldehyde	2.49	2.96
2'-CH_3, hydrate	.51	.038

Evaluate the tautomerization constants $K_{T(A)}$ and $K_{T(H)}$, the hydration constants K_H and $K_{H'}$, and the microscopic pK_a values pK_A, pK_B, pK_C, and pK_D in the foregoing scheme. Assume that the hydration ratios K_h and $K_{h'}$ are the same in H_2O (spectrophotometric data) and in D_2O (NMR data).

Earthworm hemoglobin   Ferritin shell

Some ways in which protein subunits associate. (Left) The 3.66 MDa hemoglobin of the earthworm *Lumbricus terrestris* contains 144 globin subunits organized as 12 cylindrical disulfide-linked dodecamers. Two 6-dodecamer layers, each a ring with 6-fold cyclic symmetry, lie back-to-back. This reconstructed particle also contains three types of linker proteins in the center region. From Lamy *et al.* (2000) *J. Mol. Biol.* **298**, 638. (Center) The iron storage protein ferritin is formed from 24 19- to 21-kDa 4-helix-bundle subunits with cubic symmetry. As many as 4500 atoms of iron, as hydrated iron oxide, may be stored in the internal cavity. See Fig. 7-13. From Trikha *et al.* (1995) *J. Mol. Biol.* **248**, 954. Courtesy of Elizabeth Theil. (Right) The 2 MDa molecular chaperone GroEL consists of two back-to-back 7-subunit rings, each subunit formed from domains E, I and A. A 7-subunit cap of the smaller GroES may cover either end to form a compartment in which polypeptides fold. See Box 7-A.

Contents

How Macromolecules Associate

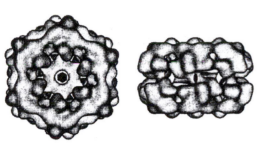

7

Earthworm hemoglobin

Ferritin shell

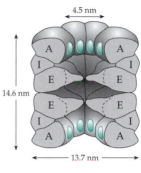

4.5 nm

14.6 nm

13.7 nm

GroEL Chaperonin
◖ Hydrophobic binding patches

The complicated shapes and internal structures of cells are determined to a large extent by the way in which proteins and other macromolecules are bonded one to another. In addition, intimate association of macromolecules is essential to such biological processes as the motion of flagella, the contraction of muscle, the action of antibodies, the transmission of nerve impulses, the replication of DNA, and the synthesis of proteins. Equally important is the binding of small molecules to large ones. In this chapter we will first examine methods of measuring binding with an emphasis on protons and small molecules. Then we will consider the ways in which macromolecules stick together as well as the role of conformational changes within macromolecules.

A. Describing Binding Equilibria

In previous discussions of pH we have dealt with dissociation constants, but in this section we will use formation constants K_f, where $K_f = 1/K_d$. Measurement of the strength of association of molecules is an everyday aspect of modern biochemical research. It may be important to know how strongly a hormone binds to a receptor in a cell membrane or how well a feedback inhibitor binds to an enzyme to determine whether the interaction is significant physiologically. The binding of O_2 to hemoglobin and other oxygen carriers is vitally important, but the description of these oxygenation reactions is mathematically complex. This is especially so because we must consider effects of pH changes and of changing concentrations of allosteric effectors on the binding equilibria.

In considering such equilibria we must first examine the individual interactions of different domains of a protein, one with another. These can be described by association constants or, alternatively, by the Gibbs energy changes for the association reaction.[1,2] The average kinetic energy of motion of a molecule in solution is about $3/2k_BT$, where k_BT is Boltzmann's constant. For one mole the kinetic energy is $3/2RT$ or 3.7 kJ (0.89 kcal) mol^{-1} at 25°C. Thus, if $K_f = 10$ M^{-1} ($\Delta G° = -5.7$ kJ mol^{-1} or -1.36 kcal mol^{-1}) the binding energy is only slightly in excess of the thermal energy of the molecules and the complex is weakly bound. In this instance, if X and P are both present in 10^{-4} molar concentrations (typical enough for biochemical systems), only 0.1% of the molecules will exist as the complex ([complex] = $K_f[X][P]$). If the formation constant is higher by a factor of 1000, i.e., $K_f = 10^4$ M^{-1} ($\Delta G° = -22.8$ kJ mol^{-1}), 38% of the molecules will exist as the complex; while if $K_f = 10^7$ M^{-1} (extremely strong binding, $\Delta G° = -40$ kJ or -9.55 kcal mol^{-1}), 97% of the molecules will be complexed.

1. Analyzing Data

The extent of binding of a molecule X to another molecule P (Eqs. 7-1, 7-2) is measured by varying the concentrations of X and P and observing changes in the concentration of the complex [PX]. The first

$$X + P \rightleftharpoons PX \qquad (7\text{-}1)$$

$$K_f = [PX]/[P][X]) \qquad (7\text{-}2)$$

prerequisite is to find a measurable property that is different for the complex than for either of the free components. For example, the complex may be colored and the components colorless. More commonly, the complex simply has a different light absorbance (A) at a certain wavelength than do the components. Likewise, the circular dichroism or the chemical shift of a peak in the NMR spectrum may change. If P is an enzyme, only the complex PX will undergo decomposition to products. Sometimes (but not always) the rate of breakdown of PX (the enzyme–substrate complex) to form products will be relatively slow compared to the rate at which the equilibrium between X, P, and PX is established. In this case the concentration of complex PX will be proportional to the observed rate of formation of product.

Whatever change of property is measured, its value will increase with increasing concentrations of X if the total concentration of the macromolecule P is kept constant. In the usual experimental design, the molar concentration of P is small and it is possible to increase the concentration of X to quite large values. When this is done, it is usually observed that at high enough values of [X] almost all of the P is converted to PX, and the change being measured (e.g., ΔA for increased light absorption) no longer increases. This effect is known as **saturation** and is observed in most binding studies and also in many physiological phenomena.

The property being measured (ΔA) reaches a maximum value ΔA_{max} at saturation and when all of compound P has been converted to PX. The ratio of [PX] to the total concentration of all forms of P present $[P]_t$ is known as the **saturation fraction** and is often given the symbol Y. If P has more than one binding site for X, Y is defined as the fraction of the total binding sites occupied. If n is the number of sites per molecule, the total number of sites is $n[P]$. The value of Y is often taken as $\Delta A / \Delta A_{max}$, an equality that holds for multisite macromolecules only if the change in A is the same for each successive molecule of X added. This is not always true, but when it is Eq. 7-3 is followed.

$$\sum_i i \frac{[PX_i]}{n[P]_t} = Y = \frac{\Delta A}{\Delta A_{max}} \tag{7-3}$$

Here i represents the number of ligands X bound to P and may vary from 0 to n. When $n = 1$ the saturation fraction Y and ΔA are related to the concentration of free unbound X and the formation constant as follows:

$$Y = \frac{K_f[X]}{1 + K_f[X]} \qquad \Delta A = \frac{\Delta A_{max} K_f[X]}{1 + K_f[X]} \tag{7-4}$$

A plot of Y or ΔA against [X] is shown in Fig. 7-1. This kind of plot is sometimes called an **adsorption isotherm** because it describes binding only at a constant temperature. Notice, from both Fig. 7-1 and Eq. 7-4,

that Y reaches a value of 0.5 when [X] is just equal to $1/K_f$ (or to K_d). Note also that as [X] increases saturation is reached slowly and that even at the point representing the highest concentration of X ($8/K_f$ in Fig. 7-1) saturation is less than 90%. Since in the usual experimental situation, we do not know Y but only ΔA, it is difficult to estimate the limiting value ΔA_{max} from a plot of this type unless K_f is very high. However, we need to know ΔA_{max} to evaluate K_f. For this reason, plots like that of Fig. 7-1 are seldom used, this one being included mainly to illustrate a point of nomenclature. The curve shown in Fig. 7-1 is a rectangular hyperbola, and the type of saturation curve shown is frequently referred to as **hyperbolic**. This is in contrast to certain other binding curves (Section 3) which, when plotted in this way, are **sigmoidal** (S-shaped).

A better type of plot is often that of Y against log [X] (Fig. 7-2). It has the following features. (1) The curve is symmetric about the midpoint at log [X] = log K_f. (2) No matter how high or low the concentration range used in the experiments, it is easy to choose a scale that puts all the points on the same sheet of paper. (3) Spacing between points tends to be more uniform than in a plot against [X]; e.g., compare Figs. 7-1 and 7-2 for which the experimental points represent the same data and for which values of [X] for successive points are each twofold greater than the preceding one. (4) The same logarithmic scale can be used for all compounds, no matter how strong or weak the binding, and the same shape curve is obtained for all 1:1 complexes. The midpoint slope, $dY/d\log$ [X], is 0.576; the change in log [X] in going from 10 to 90% saturation is 1.81. The curve is familiar to most chemists because it is frequently used for pH titration curves in which pH substitutes for $-\log$ [X]. To represent a complex with tighter binding, the curve is simply moved to the left, and for weaker binding, it is moved to the right.

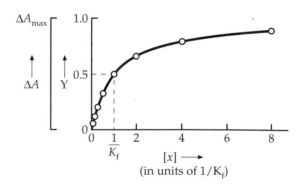

Figure 7-1 An adsorption isotherm, a plot of the saturation fraction Y or of some change in a measured property ΔA vs [X], the concentration of a substance that binds reversibly to a macromolecule. The curve is hyperbolic and [X] = $1/K_f$ when Y = 0.5.

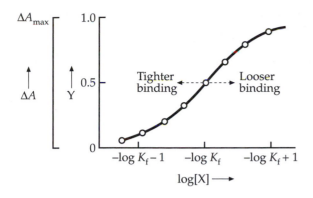

Figure 7-2 A saturation curve plotted on a logarithmic scale for [X]. The data points are the same as those used in Fig. 7-1.

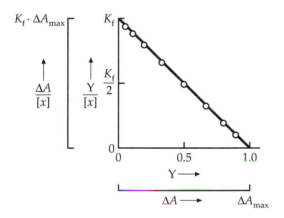

Figure 7-3 A Scatchard plot of the same data shown in Figs. 7-1 and 7-2. This is the best of the linear plots for studying binding.

Saturation data are often plotted in yet another form known as the **Scatchard plot** (Fig. 7-3). The value of $\Delta A/[X]$ (or of $Y/[X]$) is plotted against ΔA (or Y) and a straight line is fitted to the points, preferably using the "method of least squares." The intercept on the x axis and the slope of the fitted line give values of $\Delta A_{max}/K_f$ and K_f, respectively, as indicated by Eq. 7-5, which follows directly from Eq. 7-4.

$$Y/[X] = K_f - YK_f$$
$$\Delta A/[X] = \Delta A_{max} K_f - \Delta A K_f \qquad (7\text{-}5)$$

The Scatchard plot is the best of the various linear transformations of the saturation equation and is preferred to "double reciprocal plots" analogous to that shown in Fig. 9-3.

Scatchard's original equation was formulated to deal with the binding of two or more ligands to a single macromolecule.[3–5] If we let $[X]_b$ represent the concentration of bound X and $[P]_t$ the total molar concentration of the protein or other macromolecule and if there is only one binding site on the protein, $[X]_b/[P]_t$ will equal Y. However, if there are n independent binding sites that have the same binding constant K_f Eq. 7-5a will hold.

$$\frac{[X]_b}{[P]_t} \cdot \frac{1}{[X]} = nK_f - \frac{[X]_b}{[P]_t} K_f \qquad (7\text{-}5a)$$

If $[X]_b/[P]_t[X]$ is plotted against $[X]_b/[P]_t$ the resulting linear plot will have an intercept of K_f on the y axis and n on the x axis. Thus, n is directly apparent, whereas in Eq. 7-5 it is incorporated into Y. A problem arises if, as discussed in the next section, the multiple binding sites are not independent but interact. Curved Scatchard plots result and attempts to extract more than one binding constant can lead to very large errors. Before measuring saturation curves, the student should read additional articles or books on the subject.[2,6–10]

2. Multiple Binding Sites on a Single Molecule

A macromolecule may often be able to bind several molecules of a second compound X. Consider the case in which the macromolecule P binds successively one molecule of X, then a second, and a third, up to a total of n. We define the stepwise formation constants, $K_1, K_2, \ldots, K_n$, as follows:

$$P + X \xrightarrow{K_1} PX$$
$$PX + X \xrightarrow{K_2} PX_2 \ldots$$
$$PX_{n\text{-}1} + X \xrightarrow{K_n} PX_n \qquad (7\text{-}6)$$

Remember that these are reversible reactions even though unidirectional arrows are used. The general expression for the ith stepwise formation constant is given by Eq. 7-7.

$$K_i = \frac{[PX_i]}{[PX_{i-1}][X]} \qquad (7\text{-}7)$$

Remember that Y is the fraction of total binding sites saturated. The number of moles of X bound per mole of P is nY and is obtained by summing the concentrations $[PX] + 2[PX_2] + \ldots$ and dividing the sum of all the forms of P:

For two binding sites $(n = 2)$

$$2Y = \frac{[PX] + 2[PX_2]}{[P] + [PX] + [PX_2]} \qquad (7\text{-}8)$$

For the general case

$$nY = \left(\sum_{i=1}^{n} i\,[PX_i] \right) \bigg/ \left([P] + \sum_{i=1}^{n} [PX_i] \right) \qquad (7\text{-}9)$$

The summations are over all of the integral values of i from 1 to n. Now, by expressing each concentration, $[PX_i]$, in terms of the concentrations $[X]$ and $[P]$ of *free* X and P, together with the stepwise formation constants, we obtain Eq. 7-10.

For $n = 2$

$$2Y = \frac{K_1[X] + 2K_1K_2[X]^2}{1 + K_1[X] + K_1K_2[X]^2} \qquad (7\text{-}10)$$

A similar equation can be written for the general case. Note that the concentration of P does not appear in Eq. 7-10 and that Y is a function only of $[X]$ and the stepwise formation constants. Such equations define the isotherms for binding of two or more molecules of X to P. From an experimental plot of Y (or of ΔA) vs $[X]$ or log $[X]$, it is possible in favorable cases to determine the stepwise constants K_1, K_2, ..., K_n. However, this becomes quite complicated. To simplify Eq. 7-10 and the corresponding equation for the general case, we can group the constants together and designate the products of constants (K_1, K_1K_2, $K_1K_2K_3$, etc.) as ψ_1, ψ_2, . . ., ψ_n. Our equations are now as follows:

For $n = 2$

$$2Y = \frac{\psi_1[X] + 2\psi_2[X]^2}{1 + \psi_1[X] + \psi_2[X]^2} \qquad (7\text{-}11)$$

For the general case

$$nY = \sum_{i=1}^{n} i\psi_i[X]^i \bigg/ \left(1 + \sum_{i=1}^{n} \psi_i[X]^i \right) \qquad (7\text{-}12)$$

From experimental data, it is usually easiest to first determine the ψ's (there are n of them), and then to calculate from the ψ's the stepwise constants. For example:

$$K_1 = \psi_1 \qquad K_2 = \psi_2/K_1 \text{, etc.} \qquad (7\text{-}13)$$

While Eq. 7-12, known as the **Adair equation**,[11] might seem to provide a complete description of the binding process, it usually does not. In many cases, there is more than one kind of binding site on a macromolecule and Eq. 7-12 tells us nothing about the distribution of the ligand X among different sites in complex PX. To consider this problem we must examine the *microscopic binding constants*.

Microscopic binding constants and statistical effects. As discussed in Chapter 6, Section E,2, microscopic binding constants represent the constants for binding to specific individual sites. Now, consider a straight-chain dicarboxylic acid which has two *identical* binding sites for protons. If the chain connecting the two carboxylate anions is long enough, the carboxylate groups will be far enough apart that they do not influence each other through electrostatic interaction.

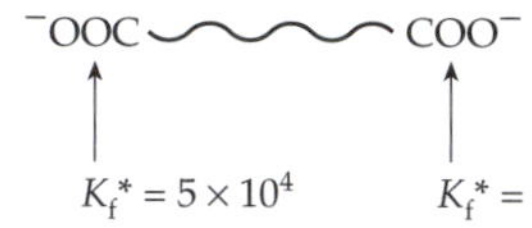

Each group will have a microscopic binding constant (K_f^*) of 5×10^4 M^{-1}. The constant K_f^* can also be called an **intrinsic binding constant**, because it is characteristic of a carboxylate group that is free of interactions with other groups. Intuition tells us (correctly) that, in its binding of protons, a solution of this dicarboxylic acid dianion will behave exactly like a solution of the monovalent anion $R-COO^-$ at twice the concentration. A single intrinsic binding constant suffices to describe both binding sites. It may seem surprising then that the **stepwise formation constants** (also called **stoichiometric** or **macroscopic** formation constants) K_1 and K_2 differ: $K_1 = 10 \times 10^4$ M^{-1} and $K_2 = 2.5 \times 10^4$ M^{-1}. This fact reflects the so-called statistical effect. Either of the two carboxylate groups in the molecule can bind a proton in the first step to give two indistinguishable molecules, PH:

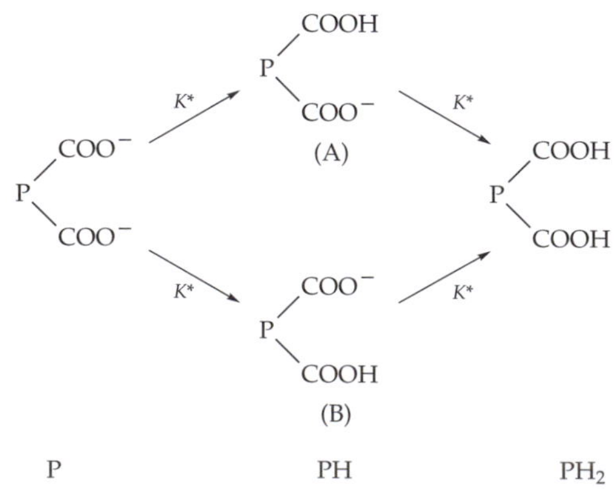

(7-14)

If we label the two forms of PH as A and B (Eq. 7-14) and consider that each one of them is independently in equilibrium with P through formation constant K_f^*, we obtain Eq. 7-15 (which may be compared with Eqs. 6-75 and 6-76, which are written for *dissociation* constants).

$$K_1 = \frac{[PH]_A + [PH]_B}{[P][H^+]} = 2K^* \quad \text{and} \quad K_2 = K^*/2 \tag{7-15}$$

This result is related to probability and arises for the same reason that if you reach into a barrel containing 50% white balls and 50% black balls, you will pull out one of each just twice as often as you will pull out a pair of white or a pair of black. In the general case of n equivalent binding sites, the microscopic formation constants K_i^* are related[12,13] to the stepwise constants K_i as follows:

$$K_i = \frac{(n + 1 - i)}{i} K_i^* \tag{7-16}$$

It is also easy to show,[14] using Eqs. 7-12 and 7-16 that for n *completely equivalent and independent* binding sites Eqs. 7-17 and 7-18 hold:

$$Y = \frac{K^*[X](1 + K^*[X])^{n-1}}{(1 + K^*[X])^n} \tag{7-17}$$

or

$$Y = \frac{K^*[X]}{1 + K^*[X]} \tag{7-18}$$

In this case the microscopic association constants are all identical and represent a single **intrinsic** constant applicable to all of the sites. In fact, Eq. 7-18 is identical to that for association of a single proton (or other ligand) with a single binding site, satisfying our intuitive notion that a set of n completely independent binding sites should behave just like a solution of an n-fold more concentrated compound with a single binding site. Thus, our arithmetic has led us to a conclusion that was already obvious. However, it is rarely true that binding sites on a single macromolecule are completely independent; there is almost always *interaction* between them, and the equations that we have derived for evaluation of stepwise and intrinsic constants cannot be applied without modification.

Electrostatic repulsion: anticooperativity.

As we have seen, a hypothetical acid with an infinite distance between the carboxylate groups and log $K_f^* = 4.8$ would have two macroscopic binding constants

TABLE 7-1

Binding Constants of Protons to Dianions of Dicarboxylic Acids[a]

Acid dianion	No. of CH_2 groups	log K_1 (pK_2)	log K_2 (association) (pK_1) (dissociation)
Hypothetical dianion with log $K^* = 4.8$	∞	5.1	4.5
Azelaic	7	5.41	4.55
Adipic	4	5.41	4.42
Succinic	2	5.48	4.19
Malonic	1	5.69	2.83

[a] From R. P. Bell, (1973) *The Proton in Chemistry*, 2nd ed., p. 96. Cornell Univ. Press, Ithaca, New York

separated by the statistical distance (log 4 = 0.6). Compare these values with the observed binding constants for protons with the dianions of acids containing 7, 4, 2, and 1 CH_2 groups given in Table 7-1. For the longest chain (that of azelaic acid) the log K_f values are not very different from those of the hypothetical long-chain acid. However, as the groups come closer together, the first binding constant is increased markedly because of the additional electrostatic attraction and the second is decreased. The spread between the two log K_f values increases from 0.6 to as much as 2.9 as a result of interaction between the binding sites.

In malonic and succinic acids the first proton bound can be shared by both carboxyl groups through formation of a hydrogen bond. (See discussion in Chapter 9, Section D.) Additional factors operate in oxalic acid where the carboxyl groups are connected directly and for which pK_a values (pK_a = log K_f) are 4.19 and 1.23. In all of these examples the binding of the first proton makes it harder to bind a second proton. Such negative interaction or **anticooperativity** between binding sites is very common and always leads to a spread of the formation constants and a broadening of the curve of Y vs log [X]. This is shown graphically on the right side of Fig. 7-4 where the binding curves for protons with acetate ion and with succinic acid dianion are compared. Notice that binding of protons *increases* as log [H$^+$] increases, giving the curves an unfamiliar appearance when compared with the more familiar curve of *dissociation vs* pH.

Can we predict the pK_a values in Table 7-1? With an appropriate dielectric constant chosen Eq. 2-8 can be applied. The difference between the two successive log K_f values reduced by 0.6 (the statistical factor) is a

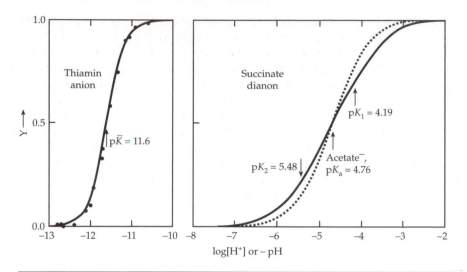

Figure 7-4 Binding of protons to the thiamin anion, the succinate dianion, and the acetate anion. Acetate (dashed line) binds a single proton with a normal width binding curve. Succinate dianion binds two protons with anticooperativity, hence a broadening of the curve. The thiamin anion (yellow form, see Eq. 7-19) binds two protons with complete cooperativity and a steep binding curve.

measure of the electrostatic effect. For malonic acid this ΔpK_a is 2.25 and for succinic acid it is 0.69 (Table 7-1). In 1923, N. Bjerrum proposed that the value of ΔpK_a could be equated directly with the work needed to bring the two negative charges together to a distance representing the charge separation in the malonate dianion.

Thus, applying Eq. 6-31, $\Delta G = 5.708 \, \Delta pK_a$ kJ mol^{-1} = 12.84 kJ mol^{-1} for malonate. Equating this with W in Eq. 2-8 and assuming a dielectric constant of 78.5 (that of water), the distance of charge separation r is calculated to be 0.138 nm. This is much too small. The computation was improved by Westheimer and Kirkwood, who assumed a dielectric constant of 2.0 *within* the molecule. By approximating the molecule as an ellipsoid of revolution, they were able to make reasonably accurate calculations of electrostatic effects on pK_a values.[15] Thus, for malonic acid Westheimer and Shookhoff[16] predicted $r = 0.41$ nm for malonic acid dianion. Recently more sophisticated calculations[17] have been used to predict pK_a values for the compounds in Table 7-1 and others.[18]

Electrostatic theory has also been used successfully to interpret titration curves of proteins in which the net negative or positive charge distributed over the surface of the protein varies continuously from high pH to low as more protons are added.[19]

Electrostatic effects can be transmitted extremely effectively through aromatic ring systems, a fact that explains some of the significance of heterocyclic aromatic systems in biochemical molecules. Consider the

microscopic binding constant of the phenolate anion of pyridoxine as influenced by the state of protonation of the ring nitrogen. These are shown in Eq. 6-75, where $pK_a^* = 4.94$ and $pK_d^* = 8.20$ define the binding constants for protonation of a phenolate ion when the ring nitrogen is protonated or unprotonated, respectively. We see that $\Delta pK_a = 3.26$, even greater than that of malonic acid.

3. Cooperative Processes

Can it ever happen that interaction between groups leads to a *decrease* from the statistical separation between values of the stepwise constants instead of to an increase? At first glance, the answer seems to be no. A decreased separation would imply that the intrinsic binding constant for the second proton bound is higher than that for the first, but common sense tells us that the first proton ought to bind at the site with the highest binding constant. However, look at the experimental binding curve of protons with the anion of thiamin shown in Fig. 7-4. Instead of being broadened from the curve of acetate, it is just half as wide. The explanation depends upon some rather amusing chemistry of thiamine. Under suitable conditions, this vitamin can be crystallized as a yellow sodium salt, the structure of whose anion is shown in Eq. 7-19. Weak binding of a proton to one of the nitrogens as shown in Eq. 7-19 creates an electron deficiency at the adjacent carbon and the $-S^-$ anion adds to the C=N group, closing the ring to an unstable tricyclic form of thiamin. This tricyclic form can be observed in methanol and can be crystallized. It is unstable in water because the central ring can open, with the electrons flowing as indicated by the small arrows to create a strongly basic site on the same nitrogen. A second proton combines at this basic nitrogen with a high binding constant to form a cation.

The key to the reversed order of strength of the binding

constants lies in the molecular rearrangements intervening between the two binding steps.[20,21] In this particular case, we cannot measure the successive binding constants K_1 and K_2 directly from the titration curve because K_2 is almost two orders of magnitude larger than K_1. Consequently, the binding curve shown in Fig. 7-4 is (within the experimental error) twice as steep at the center (the slope is 2 x 0.576) as that for acetate ion and is accurately represented in Eq. 7-20. Comparison of Eq. 7-20 with Eq. 7-10 shows how the latter has been simplified because no significant concentration of the form PX is present in the cooperative case.

$$Y = \frac{\bar{K}^2[X]^2}{1 + \bar{K}^2[X]^2} \qquad (7\text{-}20)$$

This is only part of the story about the acid–base chemistry of thiamin. For the rest, see Chapter 14, Section D,1.

The binding of protons by the thiamin anion is an example of a **cooperative process**, so named because binding of the first proton makes binding of the second easier. Although relatively rare among small molecules, cooperative processes are very common and important in biochemistry.[22,23] A cooperative binding curve is sometimes referred to as **sigmoidal** because the plot of Y against [X] (the binding isotherm) is S shaped. The maximum possible cooperativity is observed when the binding of the first ligand enhances the affinity of all other sites so much that no species other than P and PX_n are present in significant concentration.

It is easy to show that, for n binding sites with such *completely cooperative binding*, the saturation fraction is:

$$Y = \frac{\bar{K}^n[X]^n}{1 + \bar{K}^n[X]^n} \qquad (7\text{-}21)$$

where $\bar{K} = (K_1 \ldots K_n)^{1/n}$. The midpoint slope in the binding curve (Y vs log [X]) is $0.576n$ and the change, $\Delta\log[X]$, between Y = 0.1 and Y = 0.9 is $1.81/n$.

Equation 7-21 can be rewritten as

$$Y/(1 - Y) = \bar{K}^n[X]^n \qquad (7\text{-}22)$$

Taking logarithms (Eq. 7-22)

$$\log[Y/(1 - Y)] = n \log \bar{K} + n \log[X] \qquad (7\text{-}23)$$

A plot of $\log[Y/(1 - Y)]$ vs $\log[X]$ is known as a **Hill plot**. According to Eq. 7-22, it is linear with a slope of n. Remember that this equation was derived for an ideal case of completely cooperative binding at n sites.[24] However, Hill plots are often used to plot experimental data for systems in which cooperativity is incomplete. Thus, the experimentally measured slope of a Hill plot (n_{Hill}) is not an integer and is usually less than n, the number of binding sites. A comparison of n_{Hill} with n is often used as a measure of the degree of cooperativity: $n_{Hill}/n = 1.00$ for complete cooperativity but is less than one if cooperativity, is incomplete. An example of a very high degree of cooperativity is provided by the hexameric enzyme glutamate dehydrogenase, whose saturation curve for substrate displays n_{Hill} approaching six.[25] It is not necessary to make a Hill plot to get n_{Hill}. From the usual binding curve of Y (or ΔA) vs log [X] the midpoint slope can be measured with satisfactory precision. Alternatively, the difference, $\Delta\log[X]$, between 0.1 and 0.9 saturation can be evaluated and n_{Hill} calculated from Eq. 7-24.

CH₂CH₂OH structure — Yellow anion of thiamin (vitamin B₁)

$\xrightarrow{\text{H}^+\ \text{weak binding}}$

A strongly basic site is created here by ring opening

Cation (colorless) $\xleftarrow{\text{H}^+\ \text{strong binding}}$ "Tricyclic" unstable form

$$(7\text{-}19)$$

$$n_{Hill} = \frac{\text{midpoint slope}}{0.576} = \frac{1.81}{\Delta \log[X]} \qquad (7\text{-}24)$$

Binding curves sometimes show more than one step; in such cases Hill plots are not linear and no simple measure of cooperativity can be defined.

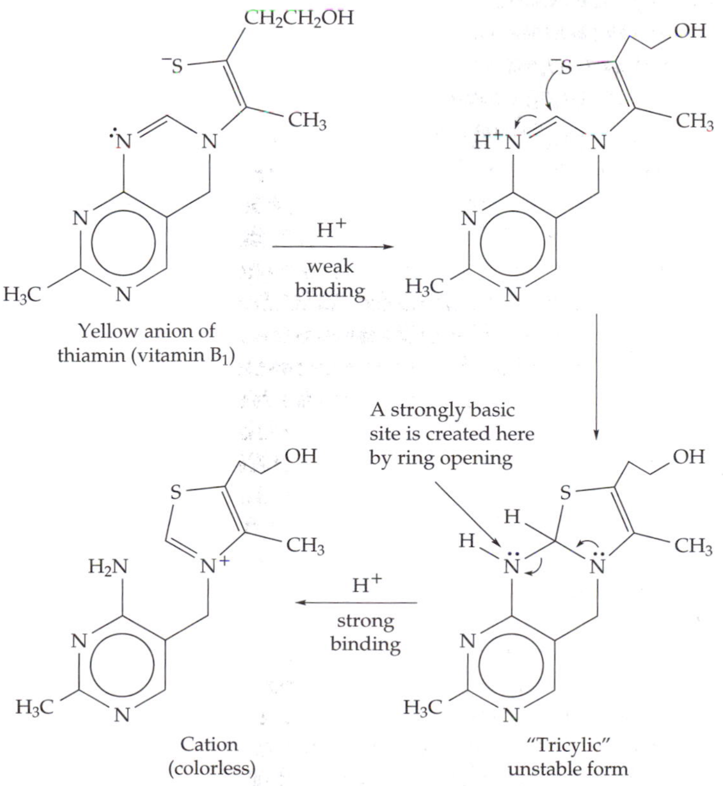

A second example of cooperativity is provided by the reversible denaturation of coiled peptide chains. Some proteins can be brought to a pH of 4 by addition of acid but without protonation of buried groups with intrinsic pK_a values greater than four. When a little more acid is added, some less basic group is protonated, permitting the protein to unfold and to expose the more basic hidden groups. Thus, cooperative proton binding is observed. As in the case of thiamin the cooperativity depends upon the occurrence of a conformational change in the molecule linked to protonation of a particular group.

The reversible transformation between an α helix and a random coil conformation is also cooperative. In this case, once a helix is started, additional turns form rapidly and the molecule is completely converted into the helix. Likewise, once it unfolds it tends to unfold completely. Melting of DNA (Chapter 5) or, indeed, of any crystal is cooperative.[8] The stacking of nucleotides alongside a template polynucleotide can also be cooperative. For example, the binding of an adenylate residue to two strands of polyuridylic acid leads to cooperative formation of a triple-helical complex (Chapter 5, Section C,4). Here the stacking interactions make helix growth energetically easier than initiation of new helical regions.[26]

B. Complementarity and the Packing of Macromolecules

Because the forces acting between them are weak, two molecules will cling together tightly only if there is a close fit between their surfaces. For a firm bond to be formed many atoms must be in contact and the two molecular surfaces must be *complementary* one to the other. If a "knob," such as a –CH_3 group, is present on one surface, there must be an appropriate hollow in the complementary surface. A positive charge in one surface is likely to be opposite a negative charge in the other. A proton donor group can form a hydrogen bond only if it is opposite a group with unshared electrons; nonpolar (hydrophobic) groups must be opposite each other if hydrophobic interaction is to occur. An important principle is that *two molecules with complementary surfaces tend to join together and interact, whereas molecules without complementary surfaces do not interact*. Watson called this "selective stickiness."[27] Selective stickiness permits the **self-assembly** of biological macromolecules having surfaces of complementary shape into fibers, tubes, membranes, and polyhedra. It also provides the means for specific pairing of purine and pyrimidine bases during the replication of DNA and during the synthesis of RNA and of proteins.

Complementarity of surfaces is equally important to the chemical reactions of cells. Each of these reactions is catalyzed by an enzyme, which contains reactive chemical groupings in the right places and in the right orientations to interact with and promote a chemical change in another molecule, the **substrate**. Specific catalysis is one of the most basic characteristics of living things. Enzymatic catalysis provides the basis not only for the reactions of metabolism but also for the movement of muscle fibers, the flowing of the cytoplasm in the ameba, and virtually all other biological responses. To understand these phenomena requires an examination of the structures of the macromolecules involved and of the ways in which they can fit together.

Just as the amino acids, sugars, and nucleotides are the building blocks for formation of proteins, polysaccharides, and nucleic acids, these three kinds of macromolecule are the units from which larger subcellular structures are assembled. Fibers, microtubules, virus "coats," and small symmetric groups of **subunits** in **oligomeric proteins** all result from the packing of macromolecules in well-defined ways, something that is often called **quaternary structure**.

1. Rings and Helices

Consider first the aggregation of identical protein subunits. While many protein molecules are nearly spherical, they are nevertheless asymmetric. In the drawings that follow the asymmetry is exaggerated, but the principles illustrated are valid. One easily observed lesson from nature is that even though living things are made up of asymmetric materials, a great deal of symmetry is evident.[28] At the molecular level the symmetry of crystalline arrays of atoms or molecules is described mathematically by the elements of symmetry present in **space groups** (p. 133). There are 230 of these but only 65 accommodate asymmetric objects (Chapter 3).[29] Two of the natural ways for identical asymmetric subunits to interact lead to rings and helices, respectively.

Molecules with cyclic symmetry. Consider a subunit (**protomer**) of the shape shown in Eq. 7-25 and containing a region *a* that is complementary to the surface *j* on another part of the same molecule. Two such protomers will tend to stick together to form a dimer, region *a* of one protomer sticking to region *j* of the other. The dimer will still contain a free region *a* at one end and a region *j* at the other which are not involved in bonding. Other protomers can stick to these free ends. In some instances long chains can be formed. However, if the geometry is just right, a third subunit can fit in to form a closed ring (a trimer). Depending on the geometry of the subunits, the ring can be even smaller (a dimer) or it can be larger (a tetramer, pentamer, etc.). The bonding involved is between two different regions (a and j) of a subunit and is sometimes described as **heterologous**.[30] To obtain a closed ring of subunits, the angle between the bonding groups a and j must be correct or the ring cannot be completed.

A ring formed using exclusively heterologous interactions possesses **cyclic symmetry**. The trimer in Eq. 7-25 has a **threefold axis**: Each subunit can be superimposed on the next by rotation through $360°/3$. The oligomer is said to have C_3 symmetry. Many real proteins, including all of those with 3, 5, or another uneven number of identical protomers, appear to be formed of subunits arranged with cyclic symmetry. An example is the cholera toxin from *Vibrio cholerae*, which forms a pentamer with an outer ring of subunits with C_5 symmetry (Fig. 7-5).

Now consider the quantitative aspects of heterologous interactions with ring formation. Let K_f be the formation constant and $\Delta G°$ the Gibbs energy change for the reaction of the j end of protomer P with the a end of a second protomer to form the dimer P_2 (Eq. 7-25).

In the second step (Eq. 7-25) a third protomer combines. It forms *two* new aj interactions. If we assume for this step that $\Delta G_f°$ is $2\,\Delta G°$, K_f will be equal to K^2. The overall association constant for formation of a trimer from three protomers will be given by Eq. 7-26.

$$3\,P \to P_3$$
$$K_f = K^3$$
$$\Delta G_f° = 3\,\Delta G° \qquad (7\text{-}26)$$

This will be true only if $\Delta G°$ for formation of both new aj bonds in the trimers is exactly the same as that for formation of the aj bond in the dimer. The reader may wish to criticize this assumption[30a,b,c] and to suggest conditions that might lead to overestimation or underestimation of K_f for the trimer as calculated previously.

Now consider a hypothetical example: Protomer P is continuously synthesized by a cell. At the same time some subunits are degraded to a nonaggregating form via a second metabolic reaction. The two reactions are balanced so that [P] is always present at a steady state

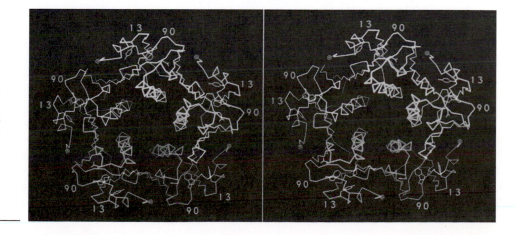

Step 1
$$K_f = K \quad \Delta G_f° = \Delta G°$$

Step 2
$$K_f \approx K^2 \quad \Delta G_f° \approx 2\Delta G°$$

$$(7\text{-}25)$$

value of 10^{-5} M. Suppose that a value for a single aj interaction of $K = 10^4$ (and $\Delta G° = -22.8$ kJ mol^{-1}) governs aggregation to form dimers and trimeric rings. What concentration of dimers and of trimer rings will be present in the cell in equilibrium with the 10^{-5} concentration of P. Using Eq. 7-25 we see that the concentration of dimers [P_2] is $10^4 \times (10^{-5})^2 = 10^{-6}$ M. (Note that the amount of material in this concentration of dimer is equivalent to 2×10^{-6} M of the monomer units.) The concentration of rings [P_3] is $(10^4)^3 \times (10^{-5})^3 = 10^{-3}$ M, equivalent to 3×10^{-3} M of the monomer units. Thus, of the *total* P present in the cell ($10^{-5} + 0.2 \times 10^{-5} + 300 \times 10^{-5}$ M), 99.6% is associated with trimers, 0.33% is still monomers, and only 0.07% exists as dimers. Thus, the formation of two heterologous bonds simultaneously to complete a ring imparts a high degree of cooperativity to the association reaction of Eq. 7-25. We will find in a cell mostly either rings or monomer but little dimer.

Figure 7-5 Stereoscopic view of the B$_5$ pentamer of cholera toxin B. The pentamer, known as **choleragenoid**, has a central hole of ~ 1.5 nm diameter into which a helix from the A subunit is inserted. As viewed here, the front surface of the pentamer has binding sites for the oligosaccharide chains of ganglioside GM$_1$, which serves as the toxin receptor. The back side binds the A subunit. See also Box 11-A. From Zhang *et al.*[31]

Now consider what will happen to the little rings within the cell if the process that removes P to a non-associating form suddenly becomes more active so that [P] falls to 10^{-6} M. If K is still 10^4, what will be the percentages of P, P_2, and P_3 at equilibrium? Here we note a characteristic of cooperative processes: A higher than first power dependence on a concentration.

Helical structures. If the angle at the interface *aj* is slightly different, instead of a closed ring, we obtain a helix as shown in Fig. 7-6A. The helix may have an integral number of subunits per turn or it may have a nonintegral number, as in the figure. The same type of heterologous interaction *aj* is involved in joining each subunit to the preceding one, but in addition other interactions occur. If the surfaces involved in these additional interactions are complementary and the geometry is correct, groups from two different parts of the molecule (e.g., *b* and *k*) may fit together to form another heterologous bond. Still a third heterologous interaction *cl* may be formed between two other parts of the subunit surfaces. If interactions *aj*, *bk*, and *cl* are strong (i.e., if the surfaces are highly complementary over large areas), extremely strong microtubular structures may be formed, such as those in the flagella of eukaryotic organisms (Fig. 1-8). If the interactions are weaker, labile microfilaments and microtubules, such as are often observed to form and dissociate within cells, may arise.

The geometry of subunits within a helix is often advantageously displayed by imagining that the surface of the structure can be unfolded to give a radial projection (Fig. 7-6B). Here subunits corresponding to those in the helix in Fig. 7-6A are laid out on a plane obtained by slitting the cylinder representing the surface of the helix and laying it out flat. In the example shown, the number of subunits per turn is about 4.8 but it can be an integral number. The interactions *bk* between subunits along the direction of the fiber axis may sometimes be stronger than those (*aj*) between adjacent subunits around the spiral. In such cases the microtubule becomes frayed at the ends through breaking of the *aj* interactions. This phenomenon can be observed under the electron microscope for the microtubules from flagella of eukaryotic organisms. Figures 7-7 to 7-10 show four helical structures from the molecular domain. They are a filamentous bacteriophage, a plant virus, a bacterial pilus, and an actin microfibril. Each is composed largely of a single kind of protomer. A larger and more complex helical structure, the microtubule, is shown in Fig. 7-34.

Filamentous bacteriophages. Bacteriophages of the Ff family include the fd, f1, and M13 strains.[31a,32–36] Phage M13 is widely used in cloning genes and for many other purposes (Chapter 26). The genome is a circular, single-stranded DNA of ~6400 nucleotides

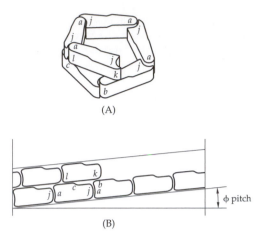

(A)

(B)

Figure 7-6 (A) Heterologous bonding of subunits to form a helix. (B) Radial projection of subunits arranged as in helix A. Different bonding regions of the subunit are designated *a*, *b*, *c*, *j*, *k*, and *l*.

which is held in an elongated double-stranded form by a helical sheath of about 2700 subunits of a 50-residue protein. The rod is about 6 nm in diameter and 880 nm long and it is capped by two specialized proteins at one end and a different pair of proteins at the other end. The five coat proteins are encoded by five of the 11 genes present in these little viruses.[37,38]

Each coat subunit in the Ff viruses is coiled into an α-helical rod of 7 nm length. These are arranged in the virus in a right-handed helical pattern with a pitch of 1.5 nm and with 4.4 subunits per turn (Fig. 7-7). The protein rods are inclined to the helix axis and extend inward. This arrangement permits a "knobs-in-holes" hydrophobic bonding between subunits. The helix of pitch 1.5 nm is the **primary** or **one-start helix**. However, in every regular helical structure we can also trace a two-start helix, a three-start helix, etc. In this instance the five-start helix is easiest to see.

The protein coat of these viruses provides an elongated cylindrical cavity to protect the circular, single-stranded DNA molecule that is the genome. Although there are two antiparallel strands of DNA, a regular base-paired structure is impossible and the DNA is probably not present in a highly ordered form.[38a] There are about 2.4 nucleotides in the DNA per protein subunit. However, there are related viruses with ratios as low as one nucleotide per protein subunit and containing more highly extended DNA.[34,39]

A rod-shaped plant virus. The tobacco mosaic virus (Figs. 5-41, 7-8) is a 300-nm-long rod constructed from 2140 identical wedge-shaped subunits whose detailed molecular structure is known.[40] Each 158-residue subunit contains five helices and a small β sheet. A single strand of RNA containing 6395 nucleotides (~ 3 per protein subunit) lies coiled in a groove where it interacts with side chains from two of the helices (Fig. 7-8B).[41–44a] The virus is assembled by the binding of a region of the RNA 800 – 1000 nucleotides from the 3' end to a two-turn helix of subunits that appears to form spontaneously. Additional subunits then add at each end, binding to the RNA as well as the adjacent protein subunits.[42,44,45] A relative with a very similar structure is cucumber green mottle mosaic virus.[46]

Bacterial pili. The adhesion pili, or fimbriae,[48] of bacteria are also helical arrays of subunits. The P pili of *E. coli* are encoded by a cluster of 11 genes in the *pap* (pilus associated with pyelonephritis) cluster. They are needed to allow the bacteria to colonize the human urinary tract. The bulk of the ~1-μm-long pilus is made up of about 1000 subunits of a 185-residue protein encoded by gene *PapA*. They form a right-handed helix of ~7 nm diameter with 3.28 subunits per turn and a pitch of ~2.5 nm (Fig. 7-9A).[49–51] The rod is anchored to the bacterial outer membrane by a protein encoded by gene *PapH*, while subunits encoded by *PapE* and *PapF* fasten the adhesin protein (PapG) to the tips of the pili.[49,52,53,53a] The adhesin binds to the Galα1→4Gal ends of glycolipids in

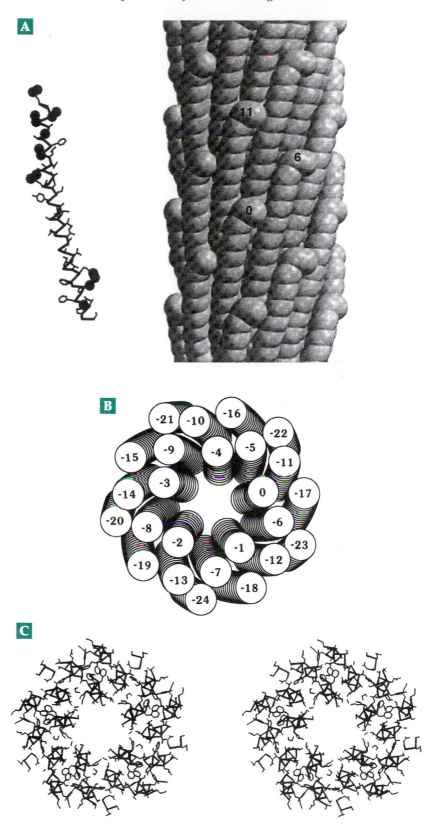

Figure 7-7 Structure of the virus fd protein sheath. (A) Left. A single coat subunit, with its N terminus towards the top, as if moved from the left side of the sheath. The dark circles represent charged atoms of Asp, Glu, and Lys side chains. The backbone of the protein is a Cα diagram. The positively charged atoms near the C terminus line the inner surface of the sheath neutralizing the negative charge of the DNA core. Right. Each subunit is represented by a helical tube through successive Cα atoms. Three nearest neighbors, indexed as 0, 6, and 11, are indicated. The axial slab shown represents ~1% of the total length of the virion. From Marvin.[31a] (B) A 2.0 nm section through the virus coat with the helices shown as curved cylinders. The view is down the axis from the N-terminal ends of the rods. The rods extend upward and outward. The rods with indices 0 to −4 start at the same level, forming a five-start helical array. The rods with more negative indices start at lower levels and are therefore further out when they are cut in this section. (C) The same view but with "wire models" of the atomic structure of the rods. From Marvin *et al.*[32]

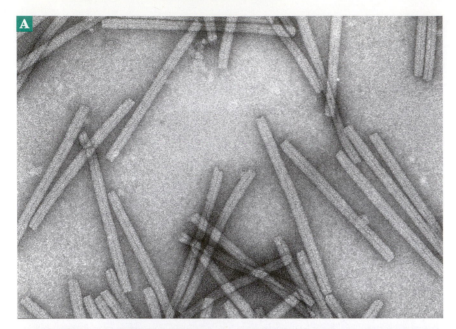

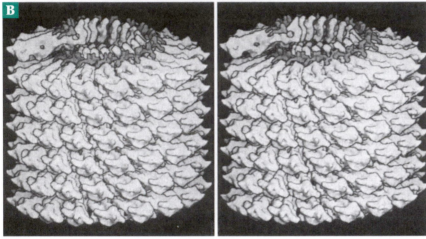

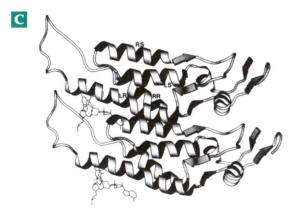

Figure 7-8 (A) Electron micrograph of the rod-shaped particles of tobacco mosaic virus. © Omikron, Photo Researchers. See also Butler and Klug.[42] (B) A stereoscopic computer graphics image of a segment of the 300 nm long tobacco mosaic virus. The diameter of the rod is 18 nm, the pitch of the helix is 2.3 nm, and there are 16 1\3 subunits per turn. The coat is formed from ~2140 identical 17.5-kDa subunits. The 6395-nucleotide genomic RNA is represented by the dark chain exposed at the top of the segment. The resolution is 0.4 nm. From Namba, Caspar, and Stubbs.[47] (C) A MolScript ribbon drawing of two stacked subunits. From Wang and Stubbs.[46]

the kidney.[51,54] The PapE, F, and G subunits form a thin ~2.5-nm-thick by 15 nm "fibrillum" which is attached by an adapter protein encoded by gene *PapK*. A special chaperonin (*PapD* gene) is also required for pilus assembly in *E. coli*[53a] as well as other bacteria.[55] Another *E. coli* pilus adheres to mannose oligosaccharides.[53b]

Similar pili of *Neisseria gonnorheae* are used by that bacterium. The three-dimensional structure of the 158-residue pilin subunit is that of a globular subunit with an 8.5 nm α-helical spine at one end.[56,57] A proposed model of the intact pilus shaft is shown in Fig. 7-9B,C. Notice the similarity of the packing of the α-helical spines in the center to the packing arrangement in the bacteriophage coat in Fig. 7-7. Similar features may be present in the P pilus rod shown in Fig. 7-9A. However, there is uncertainty about the packing arrangement. The *E. coli* type 1 pilus subunits contain immunoglobulin folds that are completed by donation of an N-terminal strand from a neighboring subunit.[53a] In thin fimbriae of *Salmonella* extended, parallel β helices may be formed (see Fig. 2-17)[53e,53f]

Other types of pili are also well-known.[53c,d] F pili or conjugative pili are essential for sexual transfer of DNA between bacterial cells (Chapter 26). F+ strains of *E. coli* form hollow pili of 8.5 nm diameter with a 2.0-nm central hole.[58,59] Their 90-residue subunits apparently form rotationally symmetric pentamers which stack to form the pili.[59] These pili are essential to establishing the initial contact between conjugating bacterial cells.

The thin filaments of muscle.
An essential component of skeletal muscle (discussed further in Chapter 19) is filamentous **actin** (F-actin). It is composed of 375-residue globular subunits of a single type and with a highly conserved sequence.[60,61] It is found not only in muscle but also in other cells where it is a component of the cytoskeleton. The actin microfilament has the geometry of a left-

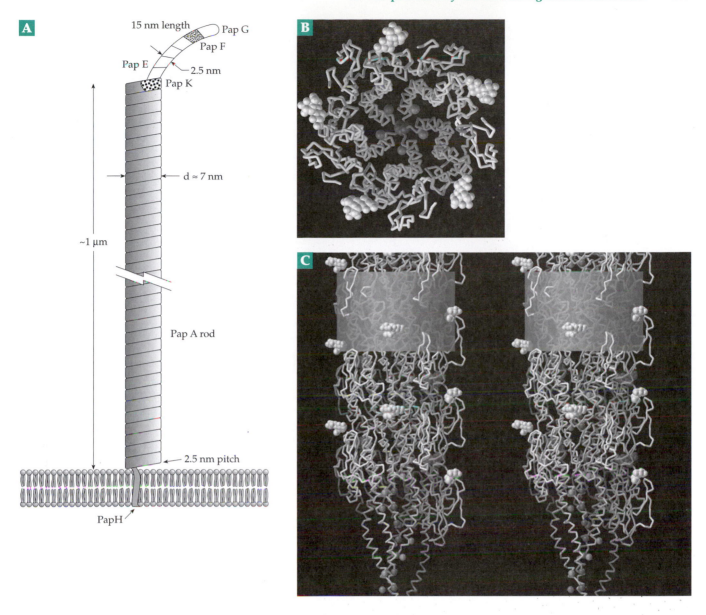

Figure 7-9 (A) Schematic diagram of a bacterial P pilus. The ~1-μm-long helical rod is anchored to the outer cell membrane by protein Pap H. The adhesion Pap G binds to galactosyl glycolipids of the host. (B, C) The structure of pilus fiber from *Neisseria gonnorrheae* modeled from the atomic structure of the 54-residue pilin subunit. The exact structure of the fiber is uncertain, but the model generated here by trying various possible helical packings matches the dimensions obtained from fiber diffraction patterns and electron microscopic images. (B) Cross section. (C) Stereoscopic view. The experimental dimensions of 4.1-nm pitch and 6.0-nm diameters are shown by the "transparent" ring in (B). From Parge *et al.*[57]

handed one-start or **primary helix** with a pitch of only 0.54 nm and with approximately two subunits per turn (Fig. 7-10).[62,63]
It can also be described as a right-handed two-start helix in which *two chains* of subunits coil around one another with a long pitch (Fig. 7-10).

2. Oligomers with Twofold (Dyad) Axes

Paired interactions. If two subunits are held together with interactions *aj* and are related by a twofold axis of rotation as shown in Fig. 7-11, we obtain an **isologous dimer**. Each point such as *a* in one subunit is related to the same point in the other subunit by reflection through the axis of rotation. In the center, along the twofold axis, points *c* and *c'* are *directly opposite the same points* in the other subunit. Figure 7-11

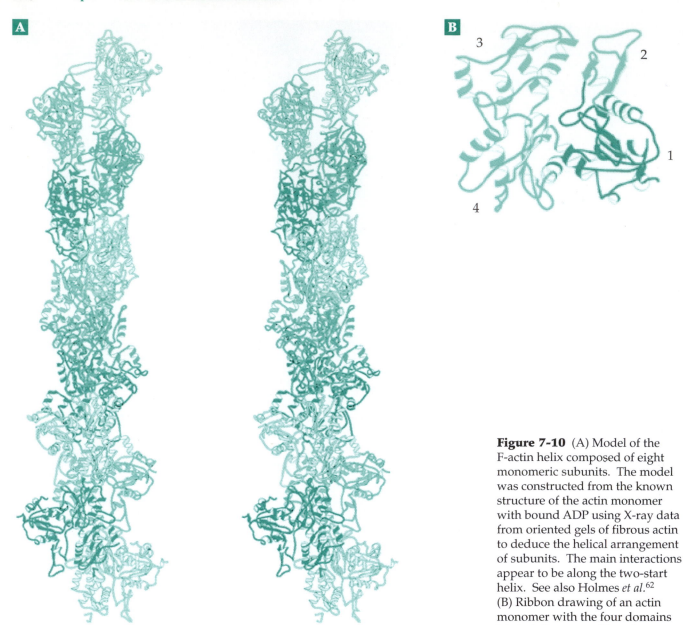

Figure 7-10 (A) Model of the F-actin helix composed of eight monomeric subunits. The model was constructed from the known structure of the actin monomer with bound ADP using X-ray data from oriented gels of fibrous actin to deduce the helical arrangement of subunits. The main interactions appear to be along the two-start helix. See also Holmes *et al.*[62] (B) Ribbon drawing of an actin monomer with the four domains labeled. Courtesy of Ivan Rayment.

is drawn with a hole in the center so that groups *c* and *c'* do not actually touch, and it is the paired interactions such as *aj* of groups not adjacent to the axis that contribute most to the bonding. However, a real protein dimer may or may not have such a hole. The pair of identical interactions in an isologous dimer may be referred to as a single **isologous bond**. Such a bond always contains the paired interactions between complementary groups (*aj*) and has pairs of identical groups along the axis. However, because they are identical those groups usually cannot interact in a specific complementary manner.

Isologous bonding is very important in oligomeric enzymes, and it has been suggested that isologous interactions evolved early. Initially there may not have been much complementarity in the bonding but two

"hydrophobic spots" on the surface of the subunits came together in a nonspecific association.[64] Later in evolution the more specific paired interactions could have been added.

Dihedral symmetry. Isologous dimers can serve as subunits in the formation of larger closed oligomers and helices; for example, an isologous pair of the sort shown in Fig. 7-11A can be flipped over onto the top of another similar pair as shown in Figs. 7-11B and 7-11C. Again, if the proper complementary surfaces exist, bonds can form as shown (*bk* in Fig. 7-11B and *bk* and *cl* in Fig. 7-11C). Both structures in Figs. 7-11B and 7-11C possess dihedral (D_2) symmetry.[65] In addition to the twofold axis of rotation lying perpendicular to the two rings, there are two other twofold axes of rotation as

BOX 7-A LIFE AND DEATH FOR PROTEINS: CHAPERONINS AND PROTEASOMES

In 1968, a tiny cylindrical particle, which appeared to be a stack of 11-nm rings, was observed by electron microscopy of an extract of erythrocytes.[a,b] Later, a similar particle was found in both the nucleus and the cytoplasm of other cells of many organisms. The particles were soon recognized as a new type of protein-hydrolyzing enzyme, a large 700-kDa particle consisting of 20–30 subunits of several different types which came to be known as the **multicatalytic protease** or **20S proteasome**.[c,d] Electron microscopy and X-ray diffraction showed that the particle is formed from four stacked rings, each of which consists of seven subunits whose molecular masses range from 21 to 31 kDa.[e–j]

Proteasomes are strikingly similar in architecture, though not in peptide sequences, to another particle found in both bacteria and eukaryotes: a molecular "chaperone" or **chaperonin**. The chaperonins, of which there are several types, protect proteins while they fold or undergo translocation within cells.[k] One of the best studied members is the *E. coli* protein **GroEL**, which is also composed of double rings of 14 subunits with seven-fold rotational symmetry and with two of these assemblies associated back-to-back with dihedral symmetry.[l] The dimensions of GroEL and 20S proteasomes are nearly the same. However, GroEL has only two rings of ~60-kDa subunits, more than twice the size of proteosomal subunits. The accompanying sketch illustrates this fact and also the basic structural similarity of 20S proteosomes with GroEL. The αβ pairs of the proteosome, correspond to single subunits of the chaperonin, but these subunits have three distinct domains–apical, intermediate, and equatorial

(labeled A, I, and E , respectively, in the drawing).[m] After a protein, whether correctly, incorrectly, or only partially folded, enters a cavity in GroEL, a second protein **GroES** of smaller size (~10 kDa) but with seven-fold symmetry binds to one end of the chaperonin.[p] Seven molecules of ATP also bind to sites on the GroEL ring to which GroES binds (the *cis* ring). The binding of the ATP and GroES evidently induces a major conformational change in the GroEL subunits[l,m,n,q] which causes the binding cavity to expand to over twice the original volume. This change (see drawing) also causes hydrophobic surfaces of the cavity to become buried and hydrophilic side chains to be exposed. The cavity surface was initially largely hydrophobic and able to bind many proteins nonspecifically, but upon expansion it becomes hydrophilic and less likely to bind. This releases the encased protein to complete its folding or to partially unfold and refold without interference from other proteins.

While a protein is adjusting its folding in the *cis* compartment another protein molecule may become trapped in the *trans* compartment. After some time the bound ATP molecules are hydrolyzed. As in the contraction of muscle, which is discussed in Chapter 19, the loss of inorganic phosphate (P_i) and ADP from the active site can be accompanied by movement. In the chaperonin this involves a conformational switch so that the ES heptamer is released and the conformation of the *trans* ring of EL is switched to that of the initial *cis* ring and vice versa. The new *cis* ring is ready to receive an ES cap and the new *trans* ring can release the folded protein.[r] A variety of experimental approaches are being used in an

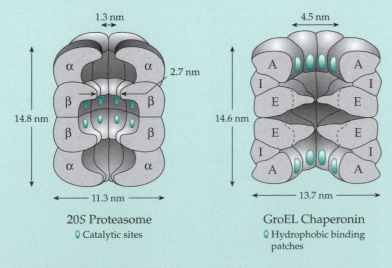

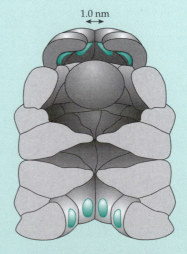

20S Proteasome
◊ Catalytic sites

GroEL Chaperonin
◊ Hydrophobic binding patches

From Weissman, Sigler and Horwich[g]. Illustrations by K. Sutliff.
Right: From Mayhew and Hartl[o]

Expanded GroEL-GroES complex with enclosed folding polypeptide[o]

BOX 7-A LIFE AND DEATH FOR PROTEINS: CHAPERONINS AND PROTEASOMES (continued)

effort to further understand the action of GroEL.[s–x] The chaperonin may function repeatedly before a protein becomes properly folded.[t]

While chaperonins assist proteins to fold correctly proteasomes destroy unfolded chains by partial hydrolysis, cutting the chains into a random assortment of pieces from 3 to 30 residues in length with an average length of ~ 8 residues.[y] Proteasomes destroy not only unfolded and improperly folded proteins but also proteins marked for destruction by the ubiquitin system described in Box 10-C. It has been hard to locate true proteosomes in most bacteria. However, they do contain protease particles with similar characteristics[z–bb] and archaeons, such as *Thermoplasma acidophilum*, have proteasomes similar to those of eukaryotes.[cc]

The *Thermoplasma* proteasome contains only two kinds of subunits, α and β, which have similar amino acid sequences. These form α_7 and β_7 rings which associate in $\alpha_7\beta_7$ pairs with two of these double rings stacked back-to-back with dihedral D7 symmetry: $\alpha_7\beta_7\beta_7\alpha_7$. The crystal structure has been determined for this 20S proteasome from *T. acidophilum*[g,dd] and for the corresponding proteasome from yeast (*Saccharomyces cerevisiae*).[f,ee] The accompanying drawings illustrate top and side views of the *T. acidophilum* proteasome. The particle contains three internal cavities. The outer two are formed between the α_7 and β_7 rings and the inner is formed between the two β_7 rings. A channel only 1.3 nm in diameter permits the entrance of peptide chains into the compartments.

The active sites of the enzymes[ff] are located in the β subunits in the central cavity.[dd] While the yeast and human proteasomes are similar to those of *Thermoplasma*, the β subunits consist of seven different protein-hydrolyzing enzymes whose catalytic activities and mechanisms are considered in Chapter 12. There are also seven different α subunits, all of whose sequences are known.[i,gg] To make the story more complex, additional subunits, some of which catalyze ATP hydrolysis, form a 600- to 700-kDa cap which adds to one or both ends of a 20S proteosome to give a larger **26S proteasome**.[b,d,e,w] These larger proteasomes carry out an ATP-dependent cleavage of proteins selected for degradation by the ubiquitin system (Box 10-C; Chapter 12). Some of the short peptide segments formed by proteasomes may leave cells and participate in intercellular communication. For example, pieces of antigenic peptides are used by cells of the immune system for "antigen presentation" (Chapter 31),[hh] an important process by which the immune system recognizes which cells are "self" and which are foreign or malignant and must be killed.

The structure of the caps on the 26S proteasome ends is complex. At least 20 different regulatory subunits have been identified.[ii,jj]

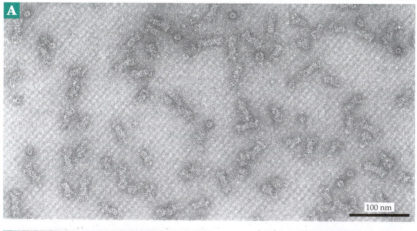

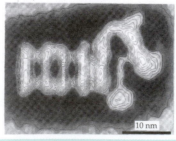

(A) Electron micrograph of 26S proteasomes from *Xenopus* oocytes negatively stained with 2% uranyl acetate. (B) Image of the 26S proteasome (left) and a 20S proteasome with only one end cap. These views were obtained by correlation averaging of 527 individual images of the 26S proteasome and 395 images of the single-ended form. From Peters *et al.*[e] Courtesy of Wolfgang Baumeister.

BOX 7-A (continued)

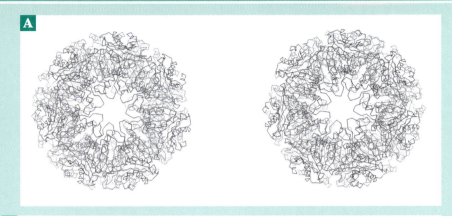

(A) Top view of the 20S proteasome as an α-carbon plot showing the seven-fold symmetry. The α subunits are in front of the β subunits. (B) Side view showing the proteasome cut open along its seven-fold axis. From Löwe et al.[dd] Courtesy of Robert Huber.

[a] Harris, J. R. (1968) *Biochim. Biophys. Acta.* **150**, 534–537

[b] Peters, J.-M. (1994) *Trends Biochem. Sci.* **19**, 377–382

[c] Bosch, G., Baumeister, W., and Essen, L.-O. (2000) *J. Mol. Biol.* **301**, 19–25

[d] Ferrell, K., Wilkinson, C. R. M., Dubiel, W., and Gordon, C. (2000) *Trends Biochem. Sci.* **25**, 83–88

[e] Peters, J.-M., Cejka, Z., Harris, J. R., Kleinschmidt, J. A., and Baumeister, W. (1993) *J. Mol. Biol.* **234**, 932–937

[f] Groll, M., Ditzel, L., Löwe, J., Stock, D., Bochtler, M., Bartunik, H. D., and Huber, R. (1997) *Nature (London)* **386**, 463–471

[g] Weissman, J. S., Sigler, P. B., and Horwich, A. L. (1995) *Science* **268**, 523–524

[h] DeMartino, G. N., and Slaughter, C. A. (1999) *J. Biol. Chem.* **274**, 22123–22126

[i] Gerards, W. L. H., de Jong, W. W., Bloemendal, H., and Boelens, W. (1998) *J. Mol. Biol.* **275**, 113–121

[j] Schmidtke, G., Schmidt, M., and Kloetzel, P.-M. (1997) *J. Mol. Biol.* **268**, 95–106

[k] Netzer, W. J., and Hartl, F. U. (1998) *Trends Biochem. Sci.* **23**, 68–73

[l] Sigler, P. B., Xu, Z., Rye, H. S., Burston, S. G., Fenton, W. A., and Horwich, A. L. (1998) *Ann. Rev. Biochem.* **67**, 581–608

[m] Kawata, Y., Kawagoe, M., Hongo, K., Miyazaki, T., Higurashi, T., Mizobata, T., and Nagai, J. (1999) *Biochemistry* **38**, 15731–15740

[n] Betancourt, M. R., and Thirumalai, D. (1999) *J. Mol. Biol.* **287**, 627–644

[o] Mayhew, M., and Hartl, F. U. (1996) *Science* **271**, 161–162

[p] Hunt, J. F., Weaver, A. J., Landry, S. J., Gierasch, L., and Deisenhofer, J. (1996) *Nature (London)* **379**, 37–45

[q] Cliff, M. J., Kad, N. M., Hay, N., Lund, P. A., Webb, M. R., Burston, S. G., and Clarke, A. R. (1999) *J. Mol. Biol.* **293**, 667–684

[r] Rye, H. S., Burston, S. G., Fenton, W. A., Beechem, J. M., Xu, Z., Sigler, P. B., and Horwich, A. L. (1997) *Nature (London)* **388**, 792–798

[s] Chatellier, J., Buckle, A. M., and Fersht, A. R. (1999) *J. Mol. Biol.* **292**, 163–172

[t] Nieba-Axmann, S. E., Ottiger, M., Wüthrich, K., and Plückthun, A. (1997) *J. Mol. Biol.* **271**, 803–818

[u] Gervasoni, P., Gehrig, P., and Plückthun, A. (1998) *J. Mol. Biol.* **275**, 663–675

[v] Torella, C., Mattingly, J. R., Jr., Artigues, A., Iriarte, A., and Martinez-Carrion, M. (1998) *J. Biol. Chem.* **273**, 3915–3925

[w] Horwich, A. L., Weber-Ban, E. U., and Finley, D. (1999) *Proc. Natl. Acad. Sci. U.S.A.* **96**, 11033–11040

[x] Buckle, A. M., Zahn, R., and Fersht, A. R. (1997) *Proc. Natl. Acad. Sci. U.S.A.* **94**, 3571–3575

[y] Kisselev, A. F., Akopian, T. N., and Goldberg, A. L. (1998) *J. Biol. Chem.* **273**, 1982–1989

[z] Kessel, M., Maurizi, M. R., Kim, B., Kocsis, E., Trus, B. L., Singh, S. K., and Steven, A. C. (1995) *J. Mol. Biol.* **250**, 587–594

[aa] Shin, D. H., Lee, C. S., Chung, C. H., and Suh, S. W. (1996) *J. Mol. Biol.* **262**, 71–76

[bb] Tamura, T., Tamura, N., Cejka, Z., Hegerl, R., Lottspeich, F., and Baumeister, W. (1996) *Science* **274**, 1385–1389

[cc] Maupin-Furlow, J. A., and Ferry, J. G. (1995) *J. Biol. Chem.* **270**, 28617–28622

[dd] Löwe, J., Stock, D., Jap, B., Zwickl, P., Baumeister, W., and Huber, R. (1995) *Science* **268**, 533–539

[ee] Stuart, D. I., and Jones, E. Y. (1997) *Nature (London)* **386**, 437–438

[ff] Voges, D., Zwickl, P., and Baumeister, W. (1999) *Ann. Rev. Biochem.* **68**, 1015–1068

[gg] Schmidt, M., and Kloetzel, P.-M. (1997) *FASEB J.* **11**, 1235–1243

[hh] Goldberg, A. L., and Rock, K. L. (1992) *Nature (London)* **357**, 375–379

[ii] Adams, G. M., Falke, S., Goldberg, A. L., Slaughter, C. A., DeMartino, G. N., and Gogol, E. P. (1997) *J. Mol. Biol.* **273**, 646–657

[jj] Knowlton, J. R., Johnston, S. C., Whitby, F. G., Realini, C., Zhang, Z., Rechsteiner, M., and Hill, C. P. (1997) *Nature (London)* **390**, 639–643

indicated in the drawings. Again, the interactions are paired and isologous; of many possible contacts two *bk* interactions and two *cl* interactions are marked for each pair of subunits in Fig. 7-11. There are a total of six pairs of these interactions, one between each combination of two subunits. This may be a little more difficult to see in Fig. 7-11B than in Fig. 7-11C because in the former the subunits are arranged in a more or less square configuration. Nevertheless, a pair of interactions between the left-hand subunit in the top ring and the subunit in the lower ring at the right does exist, even if it is only electrostatic and at a distance. An example of a tetrameric enzyme with perfect dihedral symmetry of the type shown in Fig. 7-11B is **lactate dehydrogenase** (Chapter 15). The plant agglutinin **concanavalin A** has a quaternary structure resembling that in Fig. 7-11C.

Square arrays of four subunits can be formed using either heterologous or isologous interactions. Both types of bonding can occur in larger aggregates. For example, two trimers such as that shown in Eq. 7-25 can associate to a hexamer having dihedral (D_3) symmetry; a heterologous "square tetramer" can dimerize to give a dihedral (D_4) octamer.[65] The enzymes **ornithine decarboxylase** (Fig. 7-12)[66] and **glutamine synthetase** (Chapter 24)[67] each consist of double rings of six subunits each. The upper ring is flipped over onto the lower giving dihedral symmetry (D_6) with one 6-fold axis and six 2-fold axes at right angles to it.

Oligomers with cubic symmetry (polyhedra).

Symmetrical arrangements containing more than one axis of rotation of order higher than 2-fold are said to have cubic symmetry. The **tetrahedron** is the simplest example. It contains four 3-fold axes which pass through the vertices and the centers of the faces and three 2-fold axes which pass through the midpoints of the six edges. *Since protein subunits are always asymmetric, a tetrameric protein cannot possess cubic symmetry.* As we have already seen, tetrameric enzymes have dihedral symmetry. However, a heterologous trimer with 3-fold symmetry can form a face of a tetrahedron containing a total of 12 asymmetric subunits. Twenty-four subunits can interact to form a **cube**. Three 4-fold axes pass through the centers of the faces, four 3-fold axes pass through the vertices, and six 2-fold axes pass through the edges (see Figs. 7-13 and 16-3).

The largest structure of cubic symmetry that can

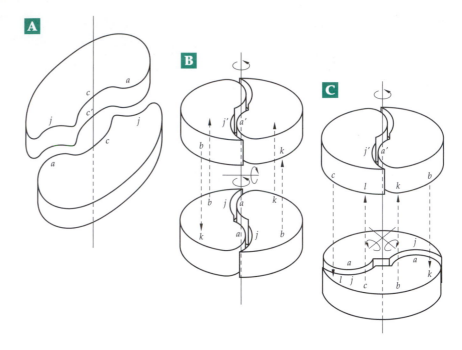

Figure 7-11 (A) Isologous bonding between pairs of subunits; (B) an "isologous square" arrangement of subunits; (C) an apparently "tetrahedral" arrangement of subunits. Note the three twofold axes.

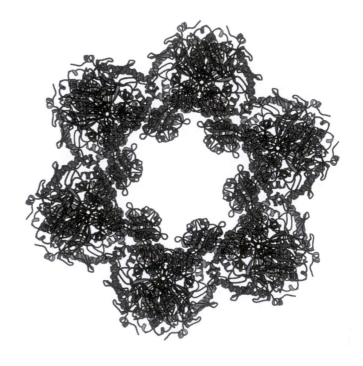

Figure 7-12 A ribbon representation of the ornithine decarboxylase dodecamer. Six dimers of the 730-residue subunits are related by C_6 crystallographic symmetry. MolScript drawing from Momany *et al.*[66] Courtesy of Marvin Hackert.

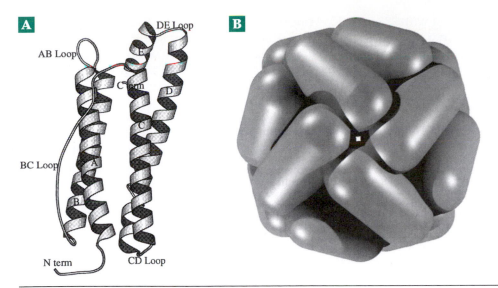

Figure 7-13 (A) MolScript ribbon drawing of a subunit of the iron oxide storage proteins L-ferritin from amphibian red cells. This 4-helix bundle is represented by cylinders of 1.3 nm diameter in the oligomer. (B) Helices A and C of the monomer are on the outer surfaces of the oligomer and helices B and D are on the inner surface. The oligomer consists of a shell of 24 subunits and is viewed down a 4-fold axis illustrating its 423 (cubic) symmetry. The molecule is illustrated further in Fig. 16-3. From Trikha *et al.*[74] Courtesy of Elizabeth Theil.

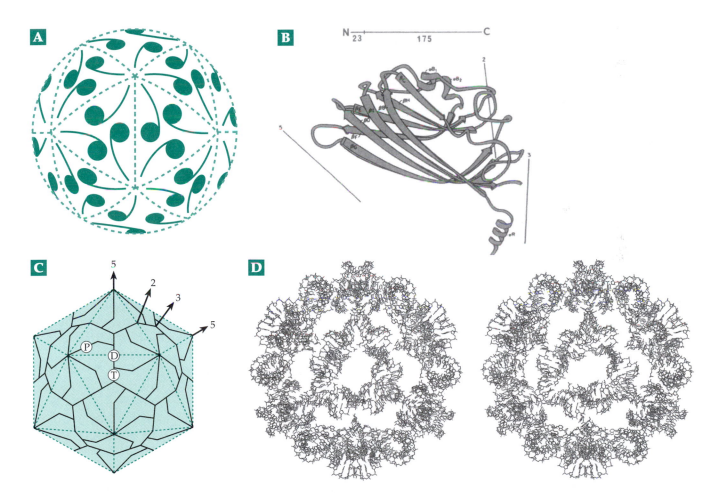

Figure 7-14 (A) Schematic drawing illustrating an icosahedrally symmetric structure with sixty identical asymmetric subunits all in equivalent positions. The 5-fold axes are located at the vertices of the icosohedron and the 2-fold and 3-fold axes can readily be seen. (B) Ribbon drawing of the 195-residue polypeptide chain of the coat subunit of satellite tobacco necrosis virus. The protein folds into an inwardly projecting N-terminal segment and a "β-jellyroll" domain. The packing of this subunit in the virus particle is shown schematically in (C). The symmetry axes drawn next to the subunit diagram (B) can be used to position it in the structure. Contacts between subunits are labeled D, T, and P ('dimer', 'trimer', 'pentamer'). Diagrams courtesy of Drs. Strandberg, Liljas, and Harrison.[68] (D) The distribution of RNA helical segments in a hemisphere of a virion of a similar small virus, the satellite tobacco mosaic virus. The virion viewed down a 3-fold axis from the virus exterior. The helical axes of the RNA segments are along icosahedral edges. From Larson *et al.*[75]

be made is the **icosahedron**, a regular solid with 20 triangular faces. Sixty subunits, or some multiple of 60, are required and at each vertex they form a heterologous pentamer. As with the tetrahedron, each face contains a heterologous trimer, while isologous bonds across the edges form dimers (Fig. 7-14C). Many viruses consist of roughly spherical protein shells (coats) containing DNA or RNA inside.[68–70] As with the filamentous viruses, the protein coats consist of many identical subunits, a fact that can be rationalized in terms of economy from the genetic viewpoint. *Only one gene is needed to specify the structure of a large number of subunits.*[70,71] Under the electron microscope the viruses often have an icosahedral appearance (Figs. 5-41A, 7-14), and chemical studies show that the number of the most abundant subunits is usually a multiple of 60. An example is the tiny **satellite tobacco necrosis virus**,[72] diameter ~18 nm, whose coat contains just 60 subunits of a 195-residue protein. Its genome is a 1239-nucleotide molecule of RNA. The structure of the coat has been determined to 0.25-nm resolution.[73]

Many virus coats have 180 subunits or a number that is some other multiple of 60. However, in these coats the subunits cannot all be in identical environments. Two cases may be distinguished. If all of the subunits have identical amino acid sequences they probably exist in more than one distinct conformation that permit them to pack efficiently. (Next section) Alternatively, two or more subunits of differing sequence and structure may associate to form 60 larger subunits that do pack with icosahedral symmetry. For example, the polioviruses (diameter 25 nm) contain three major coat proteins (α, β, and γ or VP1, VP2, and VP3). These are formed by cleavage of a large precursor protein into at least four pieces.[76,77] The three largest pieces of ~33-, 30-, and 25-kDa mass (306, 272, and 238 residues, respectively) aggregate as $(\alpha\beta\gamma)_{60}$. Sixty copies of a fourth subunit of 60 residues are found within the shell.

Related picorna viruses such as human rhinoviruses (Fig. 7-15),[69,78,79] foot-and-mouth disease virus, parvovirus,[80] and Mengo virus[81] have similar architectures. The small (diameter 25 nm) single-stranded DNA bacteriophages such as φX174 also have three different coat proteins, one of which forms small hollow spikes at the vertices of the icosahedral shell (Fig. 5-41A).[82]

Asymmetry and quasi-equivalence in oligomers. It is natural to think about association of subunits in symmetric ways. Consequently the observation of square, pentagonal, and hexagonal arrangements of subunits directly with the electron microscope led to a ready acceptance of the idea that protomers tend to associate symmetrically. However, consider the predicament of the two molecules shown in Fig. 7-16. They might get together to form an isologous dimer if it were not for the fact that their "noses" are in the way. Despite the obvious steric hindrance, an isologous dimer can be formed in this case if one subunit is able to undergo a small change in conformation (Fig. 7-16). In the resulting dimer the two subunits are only **quasi-equivalent**.

Unsymmetrical dimerization of proteins appears to be a common phenomenon that is often observed in protein crystals. For example, the

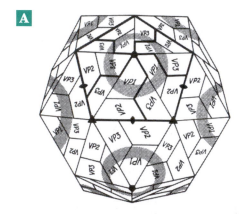

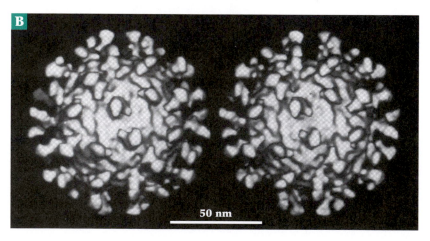

Figure 7-15 (A) Schematic diagram of the icosahedral shell of a human rhinovirus showing the arrangement of the three subunits VP1, VP2, and VP3, each present as 60 copies. (B) Stereoscopic view of an image of the virus "decorated" by the binding of two immunoglobulinlike domains of the intercellular adhesion molecule ICAM-1, a natural receptor for the virus. Part of this receptor binds into a groove or "canyon," which in marked in (A) by the dark bands. From Olson *et al.*[78] Courtesy of Michael Rossmann.

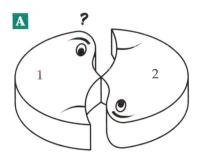 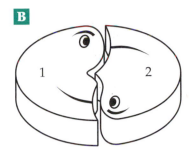

Figure 7-16 Nonsymmetric bonding in a dimer. (A) Two molecules which cannot dimerize because of a bad fit at the center. (B) A solution: Molecule 1 has refolded its peptide chain a little, changing shape enough to fit to molecule 2.

enzymes malic dehydrogenase and glyceraldehyde phosphate dehydrogenase (Chapter 15) are both tetramers of approximate dihedral symmetry but X-ray crystallography revealed distinct asymmetries[83,84] which include a weaker binding of the coenzyme NAD^+ in one subunit. This may simply reflect differences in environment within the crystal lattice. However, negative cooperativity in coenzyme binds has also been revealed by kinetic experiments.[85]

The polypeptide hormone **insulin** is a small protein made up of two chains (designated A and B) which are held together by disulfide bridges (Fig. 7-17A). Figure 7-17B is a sketch of the structure as revealed by X-ray crystallography,[86,87] with only the backbone of the peptide chains and a few side chains shown. In the drawing, the B chain lies behind the A chain. Beginning with the N-terminal Phe 1 of the B chain the peptide backbone makes a broad curve, and then falls into an helix of three turns lying more or less in the center of the molecule. After a sharp turn, it continues upward on the left side of the drawing in a nearly completely extended β structure. The A chain has an overall U shape with two roughly helical portions. The U shape is partly maintained by a disulfide bridge running between two parts of the A chain. Two disulfide bridges hold the A and B chains together, and hydrophobic bonding of internal side chain groups helps to stabilize the molecule.

Insulin in solution dimerizes readily, the subunits occupying quasi-equivalent positions. Figure 7-17C shows some details of the bonding between the subunits in the insulin dimer with a view from the outside of the molecule down the 2-fold axis (marked by the X in the center of the Phe 25 ring in the right-hand chain) through the dimer. The C-terminal ends of the B chains are seen in an extended conformation. The two antiparallel chains form a β structure with two pairs of hydrogen bonds. If there were perfect isologous bonding, the two pairs would be entirely equivalent and symmetrically related one to the other. A straight line drawn from a position in one chain and passing through the twofold axis (X) would also pass approximately through the corresponding position in the other chain. However, there are many deviations from perfect symmetry, the most striking of which is at the center

where the Phe 25 from the right-hand chain projects upward and to the left. If the symmetry were perfect the corresponding side chain from the left-hand chain would project upward and to the right and the two phenylalanines would collide, exactly as do the "noses" in Fig. 7-16. In insulin one phenylalanine side chain has been flipped back out of the way.

Under proper conditions, three insulin dimers associate to form a hexamer of approximate dihedral (D_3) symmetry that is stabilized by the presence of two zinc ions. Figure 7-17D is a crude sketch of the hexamer showing the three dimers, the 3-fold axis of symmetry, and the two pseudo 2-fold axes, one passing between the two subunits of the dimer and the other between two adjacent dimers. Figure 7-18 is a stereoscopic ribbon diagram of the atomic structure, with the A chains omitted, as obtained by X-ray diffraction.[87] The structure has also been obtained by NMR spectroscopy.[88] Note that each of the two zinc atoms lies on the threefold axis and is bound by three imidazole rings from histidines B-10. The significance of the zinc binding is uncertain but these hexamers readily form rhombohedral crystals, even within the pancreatic cells that synthesize insulin. The structure illustrates a feature that is common to many oligomers of circular or dihedral symmetry. A central "channel" is often quite open and protruding side chain groups, such as the imidazole groups in insulin, form handy nests into which ions or molecules regulating activity of proteins can fit. Conformational differences in insulin are induced by the binding of phenol. In Fig. 7-18A the C-terminal ends of the chains are extended but in the phenol complex (B) they have coiled to extend the α helices.

Quasi-equivalence in virus coats. A large number of icosahedral viruses have coats consisting of 180 identical subunits. For example, the small RNA-containing bacteriophage MS2 consists of an eicosahedral shell of 180 copies of a 129-residue protein that encloses one molecule of a 3569-residue RNA.[89] The virus also contains a single molecule of a 44-kDa protein, the A protein, which binds the virus to a bacterial pilus to initiate infection. Related bacteriophages GA, fr, f2, and Qβ[90,91] have a similar architecture. Many RNA-containing viruses of plants also have 180 subunits in their coats.[68] Much studied are the **tomato bushy stunt virus** (diameter ~33 nm, 40-kDa subunits),[68] and the related **southern bean mosaic virus**.[92] The human **wart virus** (diameter ~56 nm) contains 420 subunits, seven times the number in a regular icosahedron. **Adenoviruses** (diameter ~100 nm) have 1500 subunits, 25 times more than the 60 in a regular icosahedron.[93,94] Caspar and Klug[95] proposed a theory of quasi-equivalence of subunits

according to which the distances between the centers of subunits are preserved in a family of **icosadeltahedra** containing subunits in multiples of 20. However, the angles must vary somewhat from those in a regular icosahedron (compare with geodesic shells in which the angles are constant but the distances are not all the same). The resulting polyhedra contain hexamers as well as pentamers at vertices; for example, the shells of the 180-subunit viruses contain clusters of subunits forming 12 pentamers and 20 hexamers. There are

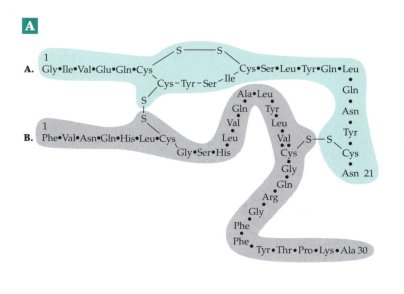

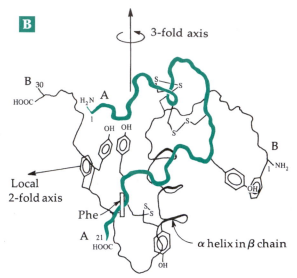

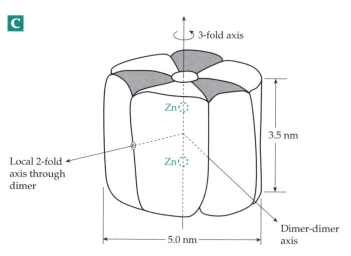

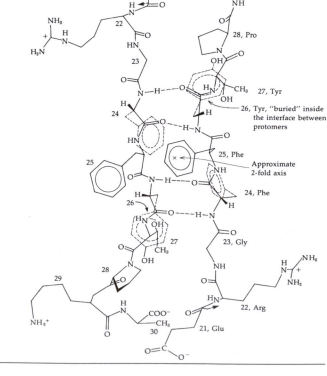

Figure 7-17 The structure of insulin. (A) The amino acid sequence of the A and B chains linked by disulfide bridges. (B) Sketch showing the backbone structure of the insulin molecule as revealed by X-ray analysis. The A and B chains have been labeled. Positions and orientations of aromatic side chains are also shown. (C) View of the paired N-terminal ends of the B chains in the insulin dimer. View is approximately down the pseudo-twofold axis toward the center of the hexamer. (D) Schematic drawing showing packing of six insulin molecules in the zinc-stabilized hexamer.

also 60 trimers (on the faces) and 90 dimers (across the edges) (Fig. 7-19). Such structures can be formed only for certain values of T where the number of subunits is $60T$ and there are 12 pentamers (pentons) and 10 (T-1) hexamers hexons. T can assume values of $h^2 + hk + k^2$ where h and k may be positive integers or zero. Some allowed T values are 1, 3, 4, 7, 9, 13, 25.[68,70,96,97]

The subunits in virus coats with T greater than one are not all in equivalent positions. For example, the three subunits labeled A, B, and C in Fig. 7-19 are each slightly differently positioned with respect to neighboring subunits. Since virus coats are usually tightly packed the subunits must assume more than one conformation. One kind of conformational change that allows quasi-equivalence of sub-units is observed in the tomato bushy stunt virus. Two structural domains are connected by a hinge which allows an outer protruding domain to move slightly to pre-serve good isologous interactions with a corresponding domain in another subunit.[68]

The southern bean mosaic virus has an eight-stranded anti-parallel β-barrel structure closely similar to that of the major domain of the bushy stunt viruses but lacking the second hinged domain. The problem of quasi-equivalence is resolved by the presence of an N-terminal extension that binds onto a subunit across the quasi-six-fold axis to give a set of three subunits (labeled C in Fig. 7-19) that associate with true three-fold symmetry and another set (B) with a slightly differ-ent conformation fitting between them.[68,92] The subunits A, which have a third conformation, fit to-gether around the five-fold axis in true cyclic symmetry.

A surprising finding is that the polyoma virus coat, which was expected to contain 420 (7 x 60) subunits, apparently contains only 360. The result is that the hexavalent morphological unit is a pentamer and that quasi-eqivalence appears

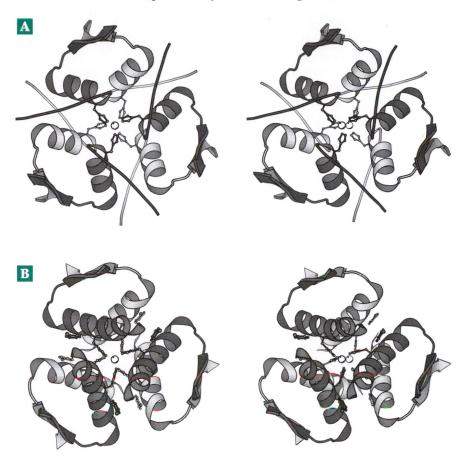

Figure 7-18 Stereoscopic MolScript ribbon drawings of the B chains (A chains omitted) of (A) hexameric 2-zinc pig insulin. (B) A phenol complex of the same protein. Within each dimer the B chains are shaded differently. The Zn^{2+} ions are represented by white spheres and the coordinating histidine side chains are shown. Six noncovalently bound phenol molecules can be seen, as can several conformational differences. From Whittingham *et al.*[87] Courtesy of Peter C. E. Moody.

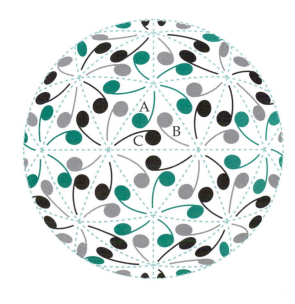

Figure 7-19 Schematic icosahedrally symmetric structure with 180 sub-units. The quasi-equivalent units A, B, and C are necessarily somewhat differently positioned with respect to their neighbors and must therefore assume different conformations in order to fit together tightly. From Harrison.[68]

to be violated.[98,99] Flexible arms tie the pentamers together.

Quasi-equivalence of subunits also provides the supercoil in bacterial flagella (Chapter 19) and accounts for some interesting aspects of the structure of tobacco mosaic virus. The protein subunits of the virus can exist either as a helix with 16.3 subunits per turn (Fig. 7-8) or as a flat ring of 17 subunits.[100] A very small conformational difference is involved. These rings dimerize but do not form larger aggregates. What is surprising is that the dimeric rings do not have dihedral symmetry, all of the subunits in the dimeric disk being oriented in one direction but with two different conformations. The disk may serve as an intermediate in virus assembly. The inner portions of the quasi-equivalent disk subunits have a jawlike appearance as if awaiting the incorporation of RNA. As the RNA becomes bound, the disks could dislocate to a "lock-washer" conformation to initiate and to propagate growth of the helical virus particle.[100,44a] However, there is uncertainty about this interpretation.[45,101]

Some enzymes, such as yeast hexokinase and creatine kinase (Chapter 12), associate in extremely asymmetric ways.[102] A dimer is formed by means of heterologous interactions but steric hindrance prevents the unsatisfied sets of interacting groups from joining with additional monomers to form higher polymers. As Galloway pointed out, many biological structures are not completely ordered but nevertheless possess well-defined and functionally important local relationships.[103]

Regulatory subunits and multienzyme complexes.

Proteins are often organized into large complexes, sometimes for the purpose of regulating metabolism. An example is **aspartate carbamoyltransferase** which catalyzes the first step in the synthesis of the pyrimidine rings of DNA and RNA (Chapter 25). The 310-kDa enzyme from *E. coli* can be dissociated into two 100-kDa trimers, referred to as **catalytic subunits**, and three 34-kDa dimers, the **regulatory subunits** which alter their conformations in response to changes in the ATP, UTP, and CTP concentrations.[104–107] The molecule is roughly triangular in shape[47,108] with a thickness of 9.2 nm and a length of the triangular side of 10.5 nm (Fig. 7-20). The symmetry is 3:2, i.e., it is dihedral with one 3-fold axis of rotation and three 2-fold axes. The two trimers of catalytic subunits lie face-to-face with the dimeric regulatory subunits fitting between them into the grooves around the edges of the trimers (Fig. 7-20). The dimers are not aligned exactly parallel with the 3-fold axis, but to avoid eclipsing, the upper half of the array is rotated around the 3-fold axis with respect to the lower half. In the center is an aqueous cavity of dimensions ~2.5 x 5.0 x 5.0 nm. The active sites of the enzyme are inside this cavity which is reached through six ~1.5-nm opening around the sides.

Many other oligomeric enzymes and other complex assemblies of more than one kind of protein subunit are known. For example, the **2-oxoacid dehydrogenases** are huge 2000- to 4000-kDa complexes containing three different proteins with different enzymatic activities in a cubic array (Fig. 15-14). The filaments of striated muscle (Chapter 19), antibodies and complement of blood (Chapter 31), and the tailed bacteriophages (Box 7-C) all have complex molecular architectures.

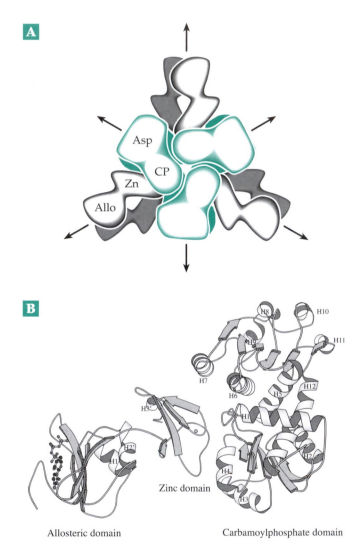

Figure 7-20 (A) Subunit assembly of two C_3 catalytic trimers (green) and three R_2 regulatory dimers around the periphery in aspartate carbamoyltransferase. After Krause *et al*.[109] Courtesy of William N. Lipscomb. The aspartate- and carbamoylphosphate-binding domains of the catalytic subunits are labeled Asp and CP, respectivley, while the zinc and allosteric domains of the regulatory subunits are labeled Allo and Zn, respectively. (B) Ribbon drawing of a single pair of regulatory (left) and catalytic (right) subunits with the structural domains labeled. MolScript drawing from Thomas *et al*.[110]

A substrate will bind better to some conformations of a protein than it will to others. This simple fact, together with the tendency for protein monomers to associate into clusters, allows for cooperative changes in conformation within oligomeric proteins. These changes provide the basis for important aspects of the regulation of enzymes and of metabolism. They impart cooperativity to the binding of small molecules such as that of oxygen to hemoglobin and of substrates and regulating molecules to enzymes. Many of the most fundamental and seemingly mysterious properties of living things are linked directly to cooperative changes within the fibrils, membranes, and other structures of the cell.

In 1965 a simple, appealing mathematical description of cooperative phenomena was suggested by Monod, Wyman, and Changeux[30,110a,110b] and focused new attention on the phenomenon. They suggested that conformational changes in protein subunits, which could be associated with altered binding characteristics, occur cooperatively within an oligomer. For example, binding of phenol to hexameric 2-zinc insulin (Fig. 7-18) could induce all six individual subunits to change their conformation together, preserving the D_6 symmetry of the complex. (In fact, it is more complex than this.[111]) The four subunits of hemoglobin could likewise change their conformation and affinity for O_2 synchronously. This is very nearly true and is of major physiological significance.

Consider an equilibrium (Eqs. 7-27 and 7-28) between protein molecules in two different conformations A and B (T and R in the MWC terminology) and containing a single binding site for molecule X. In the Monod–Wyman–Changeux (MWC) model the conformations are designated T (tense) and R (relaxed) but in the interest of providing a more general treatment the terminology used in this book is that of Koshland *et al.*[13,112–115]

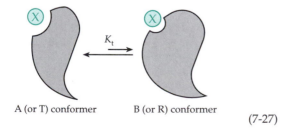

A (or T) conformer   B (or R) conformer

$$(7\text{-}27)$$

$$K_t = [B] / [A] \qquad (7\text{-}28)$$

If the equilibrium constant K_t is approximately 1, the two **conformers** have equal energies, but if $K_t < 1$, A is more stable than B.

Assume that conformer B binds X more strongly than does conformer A (as is suggested by the shapes of the binding sites in Eq. 7-27). The intrinsic binding constants to the A and B conformers K_{AX} and K_{BX} (or K_T and K_R) are defined by Eq. 7-29:

$$K_{AX} = [AX] / [A][X]$$
$$K_{BX} = [BX] / [B][X]) \qquad (7\text{-}29)$$

The entire set of equilibria for this system are shown in Eq. 7-30. Note that the constant relating

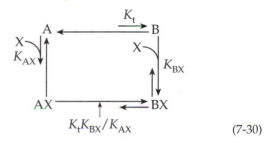

$$(7\text{-}30)$$

BX to AX is not independent of the other three constants but is given by the expression $K_t K_{BX} / K_{AX}$. Now consider the following situation. Suppose that A predominates in the absence of X but that X binds more tightly to B than to A. There will be largely either free A or BX in the equilibrium mixture with smaller amounts of AX and B. An interesting kinetic question arises. By which of the two possible pathways from A to BX (Eq. 7-30) will the reaction take place? The first possibility, assumed in the MWC model, is that X binds only to preformed B, which is present in a small amount in equilibrium with A. The second possibility is that X can bind to A but that AX is then rapidly converted to BX. We could say that X *induces a conformational change* that leads to a better fit. This is the basis for the **induced fit** theory of Koshland. Bear in mind that the equilibrium constants can give us the equilibrium concentrations of all four forms in Eq. 7-30. However, rates of reaction are often important in metabolism and we cannot say *a priori* which of the two pathways will be followed.

If K_{BX} / K_{AX} is very large, an insignificant amount of AX will be present at equilibrium. In such a case there is no way experimentally to determine K_{AX}. The two constants K_t and K_{BX} are sufficient to describe the *equilibria* but an induced fit mechanism may still hold.

Now consider the association of A and B to form oligomers in which the intrinsic binding constants K_{AX} and K_{BX} have the same values as in the monomers. Since more enzymes apparently exist as isologous dimers than as any other oligomeric form,[116] it is appropriate to consider the behavior of such dimers in some detail. Monod *et al.* emphasized that both conformers A and B (T and R) can associate to form isologous dimers in which symmetry is preserved (Eq. 7-31).

K = 1/L (in MWC terminology)

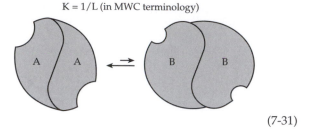

(7-31)

On the other hand, association of B and A would lead to an unsymmetric dimer in which bonding between subunits might be poor:

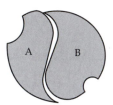

Mixed AB dimer
which associates weakly

In the MWC treatment, the assumption is made that the mixed dimer AB can be neglected entirely. However, a more general treatment requires that we consider all dimeric forms. The formation constants K_{AA}, K_{BB}, and K_{AB} are defined as follows[13,113] (Eqs. 7-32 to 7-34; note the statistical factor of 2 relating K_{AB} to the association constant K_f):

$$2A \rightleftarrows A_2 \quad K_{AA} = \frac{[A_2]}{[A]^2} \tag{7-32}$$

$$2B \rightleftarrows B_2 \quad K_{BB} = \frac{[B_2]}{[B]^2} = \frac{[B_2]}{K_t^2[A]^2} \tag{7-33}$$

$$A + B \rightleftarrows AB \quad K_f = 2K_{AB} = \frac{[AB]}{[A][B]} = \frac{[AB]}{K_t[A]^2} \tag{7-34}$$

2. Binding Equilibria for a Dimerizing Protein

All of the equilibria of Eqs. 7-28 through 7-34 involved in formation of dimers A_2, AB, and B_2 and in the binding of one or two molecules of X per dimer are depicted in Fig. 7-21. Above each arrow the microscopic

constant associated with that step is shown multiplied by an appropriate statistical factor. The fractional saturation Y is given by Eq. 7-35. Each of the nine terms in the numerator gives the concentration of

2Y (based on dimer) =

$$\frac{\begin{array}{c} [AX] + [BX] + [A_2X] + 2[A_2X_2] + [ABX] \\ + [AXB] + 2[ABX_2] + [B_2X] + 2[B_2X_2] \end{array}}{\begin{array}{c} \frac{1}{2}([A] + [AX] + [B] + [BX]) + [A_2] + [A_2X] \\ + [A_2X_2] + [AB] + [ABX] + [AXB] \\ + [ABX_2] + [B_2] + [B_2X] + [B_2X_2] \end{array}}$$

(7-35)

bound X represented by one of the nine forms containing X in Fig. 7-21. The 14 terms in the denominator represent the concentration of protein in each form including those containing no bound X. Protein concentrations are given in terms of the molecular mass of the dimer; hence, some of the terms in the denominator are multiplied by 1/2.

All of the terms in both the numerator and the denominator of Eq. 7-35 can be related back to [X], using the microscopic constants from Fig. 7-21 to give an equation (comparable to Eq. 7-8) which presents Y in terms of [X], K_{AX} and K_{BX}, K_t, and the interaction constants K_{AA}, K_{AB}, and K_{BB}. Since the equation is too complex to grasp immediately, let us consider several specific cases in which it can be simplified.

The Monod-Wyman-Changeux (MWC) model.
If both K_{AA} and K_{BB} are large enough, there will be no dissociation into monomers. The transition between conformation A and conformation B can occur cooperatively within the dimer or higher oligomer, and the mathematical relationships shown in Fig. 7-21 are still appropriate. One further restriction is needed to describe the MWC model. Only symmetric dimers are allowed. That is, K_{AA} and $K_{BB} >> K_{AB}$ (see Eq. 7-31), and only those equilibria indicated with green arrows in Fig. 7-21 need be considered.[30] In the absence of ligand X, the ratio $[B_2]/[A_2]$ is a constant, 1/L in the MWC terminology (Eq. 7-36; see also Eq. 7-31).

$$\frac{[B_2]}{[A_2]} = \frac{1}{L} = \frac{K_{BB}}{K_{AA}} K_t^2 \tag{7-36}$$

Both of the association constants K_{AA} and K_{BB} and the transformation constant K_t affect the position of the equilibrium. Thus, a low ratio of $[B_2]$ to $[A_2]$ could result if K_{BB} and K_{AA} were similar but K_t was small. If K_t were ~1 a low ratio could still arise because $K_{AA} >$

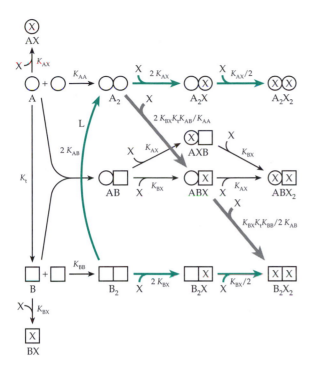

Figure 7-21 Possible forms of a dimerizing protein existing in two conformations with a single binding site per protomer for X. Green arrows indicate equilibria considered by MWC. Solid arrows indicate equilibria considered by Koshland *et al.*[13,114] Heavy gray arrows are for the simplest induced fit model with no dissociation of the dimer. Note that all equilibria are regarded as reversible (despite the unidirectional arrows). K_{AX} and K_{BX} are assumed the same for subunits in monomeric and dimeric forms.

For an oligomer with n subunits Monod *et al.* assumed that all sites in either conformer are independent and equivalent. The equation for Y (based on Eq. 7-17) is

$$Y = \frac{L \cdot K_{AX}[X](1 + K_{AX}[X])^{n-1} + K_{BX}[X](1 + K_{BX}[X])^{n-1}}{L(1 + K_{AX}[X])^{n} + (1 + K_{BX}[X])^{n}}$$

(7-39)

We assume initially that B_2 binds X more strongly than does A_2. Hence, if the equilibrium in Eq. 7-36 favors B_2 strongly (L is small), the addition of X to the system will not shift the equilibrium between the two conformations and binding will be noncooperative (Eq. 7-39 will reduce to Eq. 7-18). However, if the equilibrium favors A_2 (L is large), addition of X will shift the equilibrium in favor of B_2 (which binds X more tightly). Furthermore, since the expression for Y (Eq. 7-39) contains a term in $K^2_{BX}[X]^2$ in the numerator, binding will tend to be cooperative. In the extreme case that L is large and $K_{AX} \sim 0$, most of the terms in Eq. 7-39 drop out and it approaches the equation previously given for completely cooperative binding (Eq. 7-21) with $K = K_{BX}^2 L$. With other values of K_{AX}, K_{BX}, and L incomplete cooperativity is observed.[112]

Further development of the MWC theory as it applies to enzyme kinetics is given in Chapter 9, Section B.

The induced fit model. In this model, only A_2, ABX, and B_2X_2 are considered (heavy arrows in Fig. 7-15).[13,114] The expression for 2Y is:

$$2\,Y = \frac{[ABX] + 2\,[B_2X_2]}{[A_2] + [ABX] + [B_2X_2]}$$

$$= \frac{2\,K_{BX}K_t\dfrac{K_{AB}}{K_{AA}}[X] + 2\,(K_{BX}K_t)^2\dfrac{K_{BB}}{K_{AA}}[X]^2}{1 + 2\,K_{BX}K_t\dfrac{K_{AB}}{K_{AA}}[X] + (K_{BX}K_t)^2\dfrac{K_{BB}}{K_{AA}}[X]^2}$$

(7-40)

The constants used here are defined by Eqs. 7-8 through 7-10 and differ from those of Koshland, who sometimes arbitrarily set $K_{AA} = 1$ and redefined K_{BB} as an *interaction constant* equal to K_{BB} / K_{AA}. Although this simplifies the algebra it is appropriate only for completely associated systems and might prove confusing.

When K_{AB} is small (no "mixed" dimer) Eq. 7-16 also simplifies to Eq. 7-45 for completely cooperative binding with the value K given by Eq. 7-17. On the other hand, if K_{AB} is large compared to K_{AA} and K_{BB}, anticooperativity (negative cooperativity) will be observed. The saturation curve will contain two separate steps just as in the binding of protons by succinate dianion (Fig. 7-5).

K_{BB}, i.e., because the subunits are associated more tightly in A_2 than in B_2. For this case Eq. 7-35 simplifies to Eq. 7-37.

$$2Y = \frac{[A_2X] + 2\,[A_2X_2] + [B_2X] + 2\,[B_2X_2]}{[A_2] + [A_2X] + [A_2X_2] + [B_2] + [B_2X] + [B_2X_2]}$$

$$= \frac{\begin{aligned}&2K_{AA}K_{AX}[X] + 2K_{AA}K_{AX}^2[X]^2 \\ &\quad + 2K_{BB}K_{BX}K_t^2[X] + 2K_{BB}K_{BX}K_t^2[X]^2\end{aligned}}{\begin{aligned}&K_{AA} + 2K_{AA}K_{AX}[X] + K_{AA}K_{AX}^2[X]^2 + K_{BB}K_t^2 \\ &\quad + 2K_{BB}K_{BX}K_t^2[X] + K_{BB}K_{BX}^2K_t^2[X]^2\end{aligned}}$$

(7-37)

Substituting from Eq. 7-36 into Eq. 7-37 we obtain (Eq. 7-38):

Y (for dimer)

$$= \frac{L \cdot K_{AX}[X](1 + K_{AX}[X]) + K_{BX}[X](1 + K_{BX}[X])}{L(1 + K_{AX}[X])^2 + (1 + K_{BX}[X])^2}$$

(7-38)

$$\overline{K} = K_{BX}^2 K_t^2 \frac{K_{BB}}{K_{AA}} = \frac{K_{BX}^2}{L} \qquad (7\text{-}41)$$

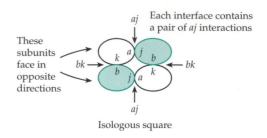

Isologous square

One conformational state dissociated. It may happen that A_2 is a dimer but that B_2 dissociates into monomers because K_{BB} is very small. In such a case binding of X leads to dissociation of the protein. A well-known example is provided by hemoglobin of the lamprey which is a dimer and which dissociates to a monomer upon binding of oxygen.[117] Equation 7-11 simplifies to Eq. 7-42. The reader may wish to consider whether the weakly cooperative binding of oxygen by lamprey hemoglobin is predicted by this equation.

$$2Y = \frac{[BX]}{[A_2] + \frac{1}{2}[BX]} \qquad (7\text{-}42)$$

Look again at the expression for L, the constant determining the relative amounts of a protein in conformations A and B in the absence of ligand. From Eq. 7-36 we see that a large value of L (conformer A favored) can result either because K_t is very small or because $K_{BB} << K_{AA}$. Thus, if $K_t \sim 1$ and L is large, the subunits must associate much more weakly in B_2 than in A_2 and the chances are that binding of X will dissociate the molecule as in the case of lamprey hemoglobin. On the other hand, if K_t is very small, implying that the molecule is held in conformation A because of some intrinsically more stable folding pattern in that conformation, K_{BB} might exceed K_{AA} very much; if K_{AA} were low enough A_2 could be completely dissociated. Binding of ligand would lead to association and to cooperative binding. This can be verified by writing down the appropriate terms from Eq. 7-35.

3. Higher Oligomers

Mathematical treatment of binding curves for oligomers containing more than two subunits is complex, but the algebra is straightforward. A computer can be programmed to do necessary calculations. Avoid picking an equation from the literature and assuming that it will be satisfactory. Consider the two tetrameric structures in Fig. 7-22. In the isologous square (also shown in Fig. 7-11) separate contributions to the free energy of binding can be assigned to the individual pairs of interactions aj and bk.

Thus, following Cornish-Bowden and Koshland[114] for assembly of the tetramer (Eq. 7-43):

$$\Delta G_f = 2\,\Delta G_{ajAA} + 2\,\Delta G_{bkAA}$$

$$K_f = K_{ajAA}^2 K_{bkAA}^2 \qquad (7\text{-}43)$$

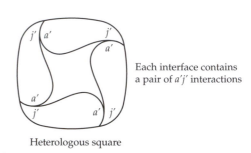

Heterologous square

Figure 7-22   Comparison of the interactions in isologous (dihedral) and heterologous (cyclic) square configurations of subunits.

Since Gibbs energies are additive, the formation constant will be the product of formation constants representing the individual interactions; thus, K_{ajAA} represents the formation constant of a dimer in which only the aj pair of bonds is formed.

In the isologous tetrahedron (Fig. 7-5) the third set of paired interactions cl must be taken into account. (However, the third interaction constant will not be an association constant of the type represented by K_{ajAA} and K_{bkAA} but a dimensionless number.) On the other hand, the heterologous square has only a single interaction constant.

Now consider the binding of one molecule of X to the isologous tetramer with a conformational change in one subunit (Eq. 7-44). We see that one pair of aj interactions and one pair of bk interactions have been

$$A_4 + X \longrightarrow \quad A_3BX \qquad (7\text{-}44)$$

altered. The equilibrium constant for the binding of X to the tetramer will be (Eq. 7-45) in which the 4 is a

$$K = 4\,\frac{K_{ajAB}K_{bkAB}}{K_{ajAA}K_{bkAA}}\,K_{BX}K_t \qquad (7\text{-}45)$$

statistical factor arising from the fact that there are four different ways in which to form A_3BX. When a second molecule of X is added three geometrical arrangements are possible:

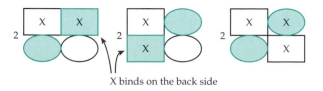

X binds on the back side

Each one can be formed in two ways. It is a simple matter to write down the microscopic constants for addition of the second molecule of X as the sum of three terms. Because values of the constants for *aj* and *bk* differ, it will be clear that the three ways of adding the second molecule of X are not equally probable. Thus, the oligomer will show preferred orders of "loading" with ligand X.

Two different geometries for the heterologous tetramer are possible in form $A_2B_2X_2$. Again, the different arrangements need not be equally probable and the relative distribution of each will be determined by the specific values of the interaction constants. In the heterologous tetramer A_4 only one type of an interaction is present between subunits. However, as soon as a single molecule of X is bound and one subunit of conformation B is present, two kinds of *aj* interactions exist. (One in which group *a* is present in conformation A and the other in which it is present in conformation B.) Since these interactions always occur in equal numbers they can be lumped together.

While the foregoing may seem like an unnecessarily long exercise, it should provide a basic approach which can be applied to specific problems. However, remember that mathematical models require simplification. Real proteins often have more than two stable conformations.[117] The entire outside surface of a protein is made up of potential binding sites for a number of different molecules, both small and large. Filling of almost any of these sites can affect the functioning of a protein.

D. The Oxygen-Carrying Proteins

1. Myoglobin and Hemoglobin

The most studied example of a conformational change in a multisubunit protein induced by binding of a small molecule is provided by the cooperative binding of oxygen to hemoglobin.[118–120] Mammalian hemoglobin is an $\alpha_2\beta_2$ tetramer of ~ 16-kDa subunits, each containing 140 – 150 residues. Within each subunit the peptide chain folds in a characteristic largely α-helical pattern around a single large flat iron-containing

ring structure called **heme** (Fig. 7-23). The folding is essentially the same in all hemoglobins, both in the α and β subunits and in the monomeric muscle oxygen storage protein, myoglobin. Amino acid residues are customarily designated by their position in one of the eight helices A–H. The imidazole group of histidine F-8 is coordinated with the iron in the center of the heme on the "proximal" side. The other side of the iron atom (the "distal" side) is the site of binding of a single molecule of O_2.

Although the folding of the peptide chain is almost the same in both subunits, and almost identical to that of myoglobin,[118,121,122] there are numerous differences in the amino acid sequence. If it were not for these differences, hemoglobin would be a highly symmetric molecule with the bonding pattern indicated in Fig. 7-5 with three 2-fold axes of rotation. In fact, hemoglobin has one true axis of rotation and two pseudo-twofold axes. There are two sets of true isologous interactions (those between the two α subunits and between the two β subunits) and two pairs of unsymmetrical interactions (between α and β subunits). The nearly symmetric orientation of different portions of the peptide backbone is clearly seen in the beautiful drawings of Geis.[119]

The contact region involved in one pair of interactions in hemoglobin ($\alpha_1\beta_1$) is more extensive than the other. There is close contact between 34 different amino acid side chains and 110 atoms lie within 0.4 nm of each other.[118] Hydrophobic bonding is the principal force holding the two subunits together, and only a few reciprocal contacts of the type found in a true isologous bond remain. The second contact designated $\alpha_1\beta_2$ involves only 19 residues and a total of 80 atoms. Because this interaction is weaker, hemoglobin dissociates relatively easily into αβ dimers held together by the $\alpha_1\beta_1$ contacts and motion occurs along the $\alpha_1\beta_2$ contacts during oxygenation. The truly isologous interactions (i.e., αα and ββ) are weak because the identical protomers hardly touch each other.

The binding of oxygen. Curves of percentage oxygenation (Y) vs the partial pressure of O_2 are given in Fig. 7-24 and illustrate the high degree of cooperativity. Depending upon conditions, values of n_{Hill} (Eq. 7-24) may be as high as 3. As a result of this cooperativity the hemoglobin, in the capillaries of the lungs at a partial pressure of oxygen of ~ 100 mm of mercury, is nearly saturated with oxygen. However, when the red cells pass through the capillaries of tissues in which oxygen is utilized the partial pressure of oxygen falls to about 5 mm of mercury. The cooperativity means that the oxygen is more completely "unloaded" than it would be if all four heme groups acted independently.

Deoxyhemoglobin has a low affinity for O_2, but the observed cooperativity in binding implies that in the fully oxygenated state the O_2 is held with a high affinity. The monomeric myoglobin also has a high affinity for oxygen, as does the abnormal **hemoglobin H**, which is made up of four β subunits. The latter also completely lacks cooperativity in binding.[123] These results can be interpreted according to the MWC model to indicate that deoxyhemoglobin exists in the T (A) conformation, whereas oxyhemoglobin is in the R (B) conformation. Myoglobin stays in the R conformation in *both* states of oxygenation as do the separated α and β chains of hemoglobin. The subunits of hemoglobin H also appear to be frozen in the R conformation, even though the quaternary structure is similar to that of deoxyhemoglobin.[123,124]

Oxygenation curves of hemoglobin are often fitted with the Adair equation (Eq. 7-12). Thus, at pH 7.4 under the conditions given in Table 7-2, Imai[125] found for the successive formation constants $K_1 = 0.004$, $K_2 = 0.009$, $K_3 = 0.002$ and $K_4 = 0.95$ in units of mm Hg^{-1}. From the definition of a formation constant the oxygen

pressure P_{O_2} required for 50% oxygenation in the first step will be at $P_{O_2} = 1/K_f$ or log P_{O_2} = log (1/.004) = 2.4. This is a high oxygen pressure, far to the right side of the oxygenation curve in Fig. 7-24A. However, log K_4 is about 0.02, well to the left on the oxygenation curve. From these formation constants we can say that after three of the subunits have become oxygenated the affinity of the remaining subunit has increased about 300-fold when the concentration of the effector 2,3-bisphosphoglycerate is present at the normal physiological concentraton (see Section 4).[125–127]

However, we must ask what uncertainties are present in the data used to obtain these constants. To extract four successive binding constants from a curve like that in Fig. 7-24A is extremely difficult.[129,130] This fact has encouraged the widespread use of the simpler MWC model.[30a,127a] When the same data were treated by Imai[131] using the MWC model it was found that $L = 2.8 \times 10^6$ and $c = K_{f\,(T)}/K_{f\,(R)} = 0.0038$. Changes in

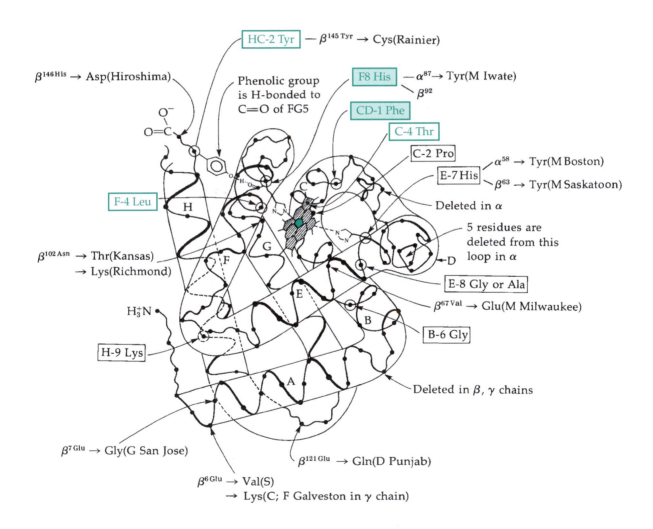

Figure 7-23 Folding pattern of the hemoglobin monomers. The pattern shown is for the β chain of human hemoglobin. Some of the differences between this and the α chain and myoglobin are indicated. Evolutionarily conserved residues are indicated by boxes, ☐ highly conserved, ▨ invariant. Other markings show substitutions observed in some abnormal human hemoglobins. Conserved residues are numbered according to their location in one of the helices A–H, while mutant hemoglobins are indicated by the position of the substitution in the entire α and β chain.

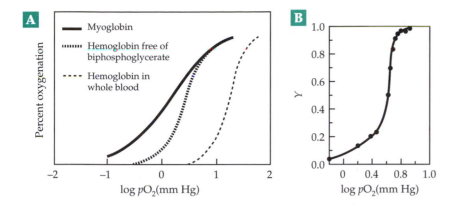

Figure 7-24 Cooperative binding of oxygen by hemoglobins. (A) Binding curve for myoglobin (noncooperative) and for hemoglobin in the absence and presence (in whole blood) of 2,3-bisphosphoglycerate. Oxygen affinity is decreased by bisphosphoglycerate. (B) Saturation curve for hemoglobin (erythrocruorin) of *Arenicola*, a spiny annelid worm. The molecule contains 192 subunits and 96 hemes. It shows very strong cooperativity with n_{Hill} ~ 6. From (A) Benesch and Benesch,[123] and (B) Waxman.[128]

enthalpy, entropy, and Gibbs energy are given in Table 7-2. Kinetic data[131a] as well as O_2-binding measurements with single crystals[132] are partially consistent with the MWC model.[133] However, the discovery of a third quaternary structure of hemoglobin, similar to the R state but distinct from it,[134–136a] emphasizes the complexity of this allosteric molecule.

Hemoglobin tetramers tend to stay tightly associated but some dissociation of oxyhemoglobin into dimers does occur ($K_f = 7 \times 10^5$ M).[137,138] Deoxyhemoglobin is about 40,000 times more tightly associated. All of the equilibria involved are strongly affected by pH and by the presence of salts such as NaCl.[126,139] This is in part a reflection of the strong role of ionic interactions in holding together the subunits in the T state as is discussed in the following sections.[127]

Structural changes accompanying oxygen binding. Perutz and associates, using X-ray crystallography, found *small* but real differences in the conformation of the subunits of deoxy- and oxy- hemoglobin.[140,141] More striking is the fact that upon oxygenation, both α and β subunits undergo substantial amounts of *rotation*, the net result being that the hemes of the two β subunits move about 0.07 nm closer together in the oxy form than in the deoxy form. Within the $\alpha_1\beta_1$ contacts (Fig. 7-25) little change is seen. On the other hand, contact $\alpha_1\beta_2$, the "allosteric interface," is altered drastically. As Perutz expressed it, there is a "jump in the dovetailing" of the CD region of the subunit relative to the FG region of the β subunit. The hydrogen-bonding pattern is also changed. A major difference is seen in the hydrogen-bonded salt bridges present at the ends of the molecules of deoxyhemoglobin. The $-NH_3^+$ group of Lys H-10 in each α subunit is hydrogen bonded to the carboxyl group of the C-terminal arginine of the opposite α chain. The guanidinium group of each C-terminal arginine is hydrogen bonded to the carboxyl group of Asp H-9 in the opposite α chain. It is also hydrogen bonded to an inorganic anion (phosphate or Cl⁻), which in turn is hydrogen-bonded to the α amino group of Val 1 of the opposite α chain[142] forming a pair of isologous interactions.

At the other end of the molecule, the C-terminal group of His 146 of each β chain binds to the amino group of Lys C-6 of the α chain, while the imidazole side chain binds to Asp FG-1 of the same β chain (Fig. 7-25). These salt bridges appear to provide extra stability to

TABLE 7-2
Thermodynamic Functions for Oxygenation of Hemoglobin[a]

Reaction	ΔH (kJ mol⁻¹)	ΔS (J°K⁻¹ mol⁻¹)	ΔG (kJ mol⁻¹)	K_f
$T \rightarrow T(O_2)_4$	-51 ± 1	-154 ± 4	-5.0 ± 1.7	7.5
$R \rightarrow R(O_2)_4$	-62 ± 2	-146 ± 6	-19 ± 2	2.0×10^3
$T \rightarrow R$ (unoxygenated)	-70 ± 7	-111 ± 25	-37 ± 10	3.6×10^{-7}
$T(O_2)_4 \rightarrow R(O_2)_4$ (oxygenated)			-19	2.1×10^3

Parameters for MWC model

$L = (3.6 \times 10^{-7})^{-1} = 2.8 \times 10^6$

$c = K_{f(T)} / K_{f(R)} = 0.0038$

[a] From Imai, K. (1979) *J. Mol. Biol.* **133**, 233–247. The measurements were made at pH 7.4 in the presence of 0.1 M chloride ion and 2 mM 2,3-bisphosphoglycerate to mimic physiological conditions. The values of ΔH, ΔS, and ΔG given are per mole of heme, i.e., per monomer unit. They must be multiplied by 4 to correspond to the reactions as shown for the tetramer.

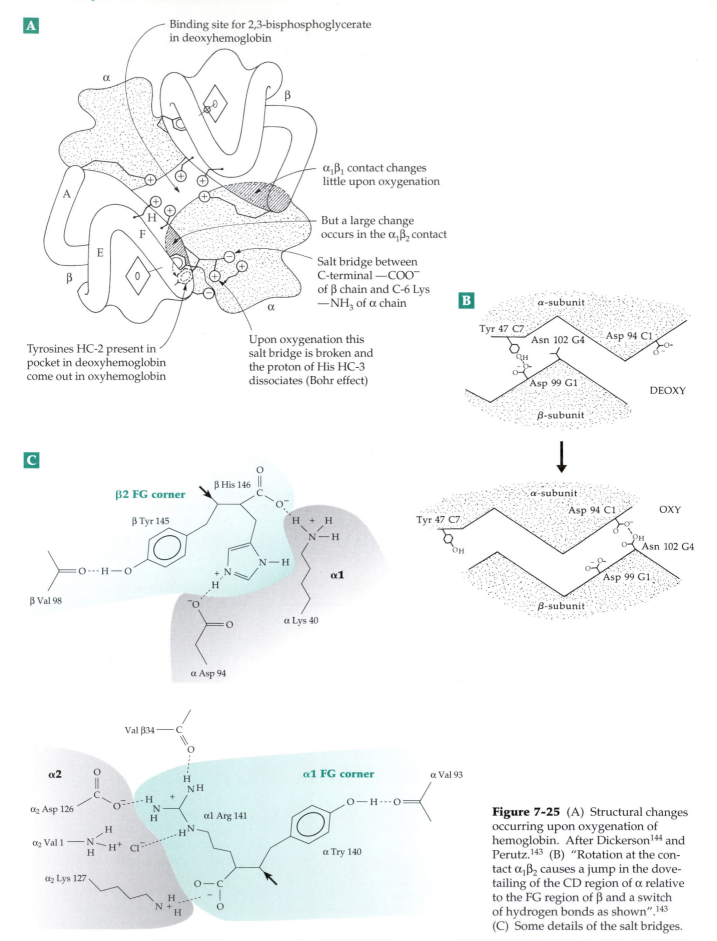

Figure 7-25 (A) Structural changes occurring upon oxygenation of hemoglobin. After Dickerson[144] and Perutz.[143] (B) "Rotation at the contact $\alpha_1\beta_2$ causes a jump in the dovetailing of the CD region of α relative to the FG region of β and a switch of hydrogen bonds as shown".[143] (C) Some details of the salt bridges.

deoxyhemoglobin and account for the high value of the constant L. In deoxyhemoglobin the side chain of the highly conserved Tyr HC-2 lies tucked into a pocket between the H and F helices and is hydrogen bonded to the main chain carbonyl of residue FG-5 (Figs. 7-23 and 7-25). Upon oxygenation this tyrosine in each subunit is released from its pocket; the salt bridges at the ends of the molecules are broken and the subunit shifts into the new bonding pattern characteristic of oxyhemoglobin.[143,144] Cooperativity in O_2 binding is absent or greatly decreased in mutant hemoglobins with substitutions in the residues involved in these salt bridges[145-147] or in residues lying in the $\alpha_1\beta_2$ interface.[148,149]

How does the binding of O_2 to the iron of heme trigger the conformational change in hemoglobin? An enormous amount of effort by many people has been expended in trying to answer this question. As is pointed out in Chapter 16, the iron atom in deoxyhemoglobin lies a little outside the plane of the heme rings. When oxygenation occurs the iron atom moves toward the oxygen and into the plane of the heme.[150,151] This movement probably amounts to only about 0.05 nm. Nevertheless, this small displacement evidently induces the other structural changes that are observed. The iron pulls the side chain of histidine F-8 with it and moves helix F which is also hydrogen bonded to this imidazole ring. Because of the tight packing of the various groups this motion cannot occur freely but is accompanied by a movement of the F helix by 0.1 nm across the heme plane. These movements may induce additional structural changes in the irregularly folded FG corners that allow the subunits to shift to the new stable position of the R state. All four subunits appear to change conformation together. This conformational change must also cause the affinity for oxygen of any unoxygenated subunits to rise dramatically, presumably by shifting the iron atoms into the planes of the heme rings. This ensures the cooperative loading of the protein by O_2.

While there is no doubt that the iron atom moves upon oxygenation, it is not obvious that this will lead to the observed cooperativity. Oxygenated heme has some of the characteristics of an Fe(III)–peroxide anion complex.[152] The iron atom acquires an increased positive charge upon oxygenation by donating an electron for bond formation.

$$Fe^{2+} \text{ (deoxy)} + O_2 \longrightarrow Fe^{3+} - O_2^- \qquad (7\text{-}46)$$

This may transmit an electronic effect through either His F-8 or the heme ring to the nearby $\alpha_1\beta_2$ interface and also affect the subunit interactions. The reaction of various heme proteins with oxygen is discussed further in Chapter 16.

The Bohr effect and allosteric regulators. The breaking of the salt bridges at the ends of the hemoglobin molecule upon oxygenation has another important result. The pK_a values of the N-terminal valines of the α subunits and of His HC-3 of the β subunits are abnormally high in the deoxy form because they are tied up in the salt bridges. In the oxy form in which the groups are free, the pK_a values are lower. If hemoglobin is held at a constant pH of 7, these protons dissociate upon oxygenation. This **Bohr effect**, described in 1904,[153-156] is important because acidification of hemoglobin stabilizes the deoxy form. In capillaries in which oxygen pressure is low and in which carbon dioxide and lactic acid may have accumulated, the lowering of the pH causes oxyhemoglobin to release oxygen more efficiently. These effects are also strongly dependent on the presence of chloride ions.[139,142,143,157,158]

Just as the conformational equilibria in hemoglobin can be shifted by attachment of oxygen to the heme groups, so the binding of certain other molecules at different sites can also affect the conformation. Such compounds are called **allosteric effectors** or **regulators** because they bind at a site other than the "active site." They are considered in more detail in Chapter 9. An important allosteric effector for human hemoglobin is **2,3-bisphosphoglycerate**, a compound found in human red blood cells in a high concentration approximately equimolar with that of hemoglobin. One molecule of bisphosphoglycerate binds to a hemoglobin tetramer in the deoxy form with $K_f = 1.4 \times 10^5$ but has only half this affinity for oxyhemoglobin.[159] X-ray crystallography shows that bisphosphoglycerate binds between the two β chains of deoxyhemoglobin directly on the twofold axis (Fig. 7-26).[159] Because of the presence

2,3-Bisphosphoglycerate

of 2,3-bisphosphoglycerate in erythrocytes the affinity of oxygen for hemoglobin in whole blood is less than that for isolated hemoglobin[160,161] (Fig. 7-24). This is important because it allows a larger fraction of the oxygen carried to be unloaded from red corpuscles in body tissues. The bisphosphoglycerate level of red cells varies with physiological conditions, e.g., people living at higher elevations have a higher concentration.[161] It has been suggested that artificial manipulation of the level of this regulatory substance in erythrocytes may be of clinical usefulness for disorders in oxygen transport.

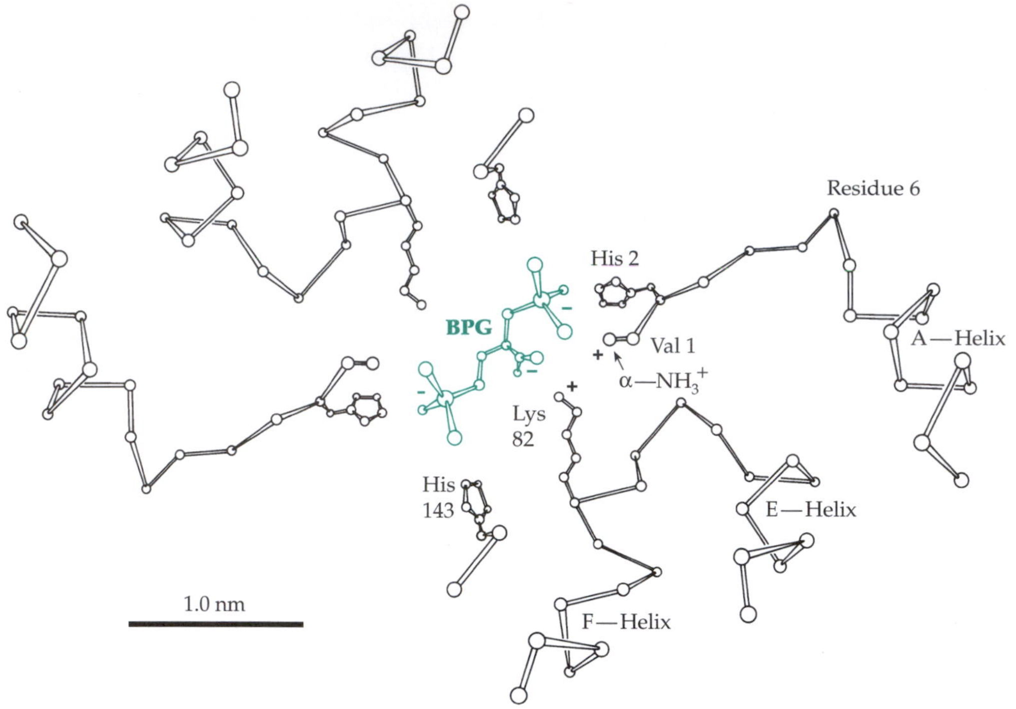

Figure 7-26 The allosteric effects of 2,3-bisphosphoglycerate (BPG) bound to the β chains of human deoxyhemoglobin. The phosphate groups of the BPG form salt bridges with valines 1 and histidines 2 and 143 of both β chains and with lysine 82 of one chain. This binding pulls the A helix and residue 6 toward the E helix. From Arnone.[159]

Not all species contain 2,3-bisphosphoglycerate in their erythrocytes. In birds and turtles its function appears to be served by inositol pentaphosphate. In crocodiles the site between the two β chains that binds organic phosphates in other species has been modified so that it binds bicarbonate ion, HCO_3^-, specifically. This ion, which accumulates in tissues as crocodiles lie under water, acts as an allosteric regulator in these animals.[162,163] It allows the animals to more completely utilize the O_2 from the hemoglobin and to remain under water longer. The Bohr effect, which was considered in the preceding section, can be viewed as resulting from the action of **protons** as allosteric effectors that bind to the amino and imidazole groups of the salt linkages. **Carbon dioxide** also acts as a physiological effector in mammalian blood by combining reversibly with NH_2-terminal groups of the α and β subunits to form **carbamino** ($-NH-COO^-$) groups (Eq. 7-47).[119,164] It is the N-terminal amino groups rather than lysyl side chain groups that undergo this reaction. Because of their relatively low pK_a values there is a significant

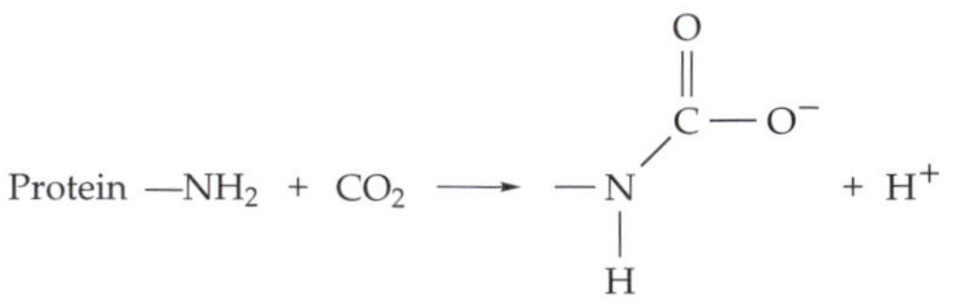

$$\text{Protein} -NH_2 + CO_2 \longrightarrow \begin{array}{c} O \\ \| \\ C-O^- \\ | \\ -N \\ | \\ H \end{array} + H^+ \qquad (7\text{-}47)$$

fraction of unprotonated $-NH_2$ groups at the pH of blood. The affinity for CO_2 is highest in deoxygenated hemoglobin. Consequently, unloading of O_2 is facilitated in the CO_2-rich respiring tissues. Hemoglobin carries a significant fraction of CO_2 to the lungs, and there the oxygenation of hemoglobin facilitates the dissociation of CO_2 from the carbamino groups. Hemoglobin is also one of the major pH buffers of blood.

Carbon monoxide, cyanide, and nitric oxide.

A danger to hemoglobin and other heme proteins is posed by competing ligands such as CO, CN^-, and NO. All of these are present within organisms and both CO and NO act as hormones. Hemoglobin and myoglobin are partially protected from carbon monoxide by the design of the binding site for O_2. The distal imidazole of histidine E7 hydrogen bonds to O_2 but not to the nonpolar CO. The site also accommodates the geometry of the bound O_2 better than that of CO.[165] Bound CO can be released from hemes by the action of light. Using X-ray diffraction[166,166a] and X-ray absorption measurements[167] at cryogenic temperatures, it has been possible to observe the motions of both the released CO and the heme in myoglobin, motions which may shed light on the normal oxygen transport cycle. Cooperativity in the binding of CO to hemoglobins has been studied extensively,[158,168,169] as has binding to model heme compounds.[170] Cyanide ions bind most tightly and also cooperatively[171–173] to the oxidized Fe^{3+} form, which is called **methemoglobin**.

Nitric oxide is a reactive, paramagnetic gaseous free radical which is formed in the human body and in other organisms by an enzymatic oxidation of L-arginine (Eq. 18-65). Since about 1980, NO has been recognized as a hormone with a broad range of effects

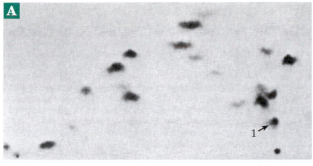

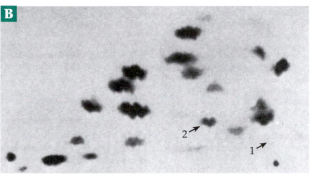

Figure 7-27 "Fingerprints" of human hemoglobins. The denatured hemoglobin was digested with trypsin and the 28 resulting peptides were separated on a sheet of paper by electrophoresis in one direction (horizontal in the figures; anode to the left) and by chromatography in the other direction (vertical in the figure). The peptides were visualized by spraying with ninhydrin or with specific reagents for histidine or tyrosine residues. Since trypsin cuts only next to lysine, which occurs infrequently, the petide pattern provides a fingerprint, characteristic for any pure protein. (A) The fingerprint of normal adult hemoglobin A. (B) Fingerprint of hemoglobin S (sickle cell hemoglobin). One histidine-containing peptide (1) is missing and a new one (2) is present. This altered peptide contains the first eight residues of the N-terminal chain of the subunit of the protein. From H. Lehmann and R. G. Huntsman, *Man's Haemoglobin*.[187]

(Chapters 11, 18). It binds to the iron of heme groups in either the Fe^{2+} or Fe^{3+} form and also reacts with thiol groups of proteins and small molecules to form *S*-nitrosothiols (R–S–N=O).[174–176] It reacts with the heme iron of myoglobin and hemoglobin and, by transfer of one electron, can oxidize the iron of hemoglobin to the Fe^{3+} methemoglobin with formation of the nitroxyl ion NO^-.[177,178] This reaction may be a major cause of methemoglobin formation.

One of the major effects of NO is to induce the relaxation of smooth muscle of blood vessels, an important factor in the regulation of blood pressure. Hemoglobin can carry NO both on its heme and on the thiol group of cysteine β93.[174,175] The affinity for NO is high in the T state and low in the R state. This allows hemoglobin to carry NO from the lungs to tissues, where it can be released and participate in the regulation of blood pressure.[174,179]

A cytoplasmic hemoglobin of the clam *Lucina pectinata* has evolved to carry oxygen to symbiotic chemoautotrophic bacteria located within cells of the host's gills. It is also readily oxidized to the Fe^{3+} methemoglobin form which binds **sulfide ions** extremely tightly[180–182] and is thought to transport sulfide to the bacteria.

2. Abnormal Human Hemoglobins

Many alterations in the structure of hemoglobin have arisen by mutations in the human population. It is estimated that one person in 20 carries a mutation that will cause a hemoglobin disorder in a homozy-gote.[183] There are also many unrecognized and harmless substitutions of one amino acid for another. However, substitutions near the heme group often adversely affect the binding of oxygen and substitutions in the interfaces between subunits may decrease the cooperative interaction between subunits.[184] One of the most common and serious abnormal hemoglobins is **hemoglobin S**, which is present in individuals suffering from **sickle cell disease** (see Box 7-B). In Hb S, glutamic acid 6 of the β chain is replaced by valine. Replacement of the same amino acid by lysine leads to **Hb C**[185] and is associated with a mild disease condition . A few of the many other substitutions that have been studied are indicated in Fig. 7-23. The locations of the defects in the hemoglobin structure have been established with the aid of protein "fingerprinting" (Fig. 7-27).

A group of serious defects are represented by the **hemoglobins M**. Only heterozygotic individuals survive. Their blood is dark because in Hb M the iron in half of the subunits is oxidized irreversibly to the ferric state. The resulting methemoglobin is present in normal blood to the extent of about 1%. While normal methemoglobin is reduced by a **methemoglobin reductase** system (Box 15-H), methemoglobins M cannot be reduced. All of the five hemoglobins M result from substitutions near the heme group. In four of them, one of the heme-linked histidines (either F-8 or E-7) of either the α or the β subunits is substituted by tyrosine. In the fifth, valine 67 of the β chains is substituted by glutamate. The two hemoglobins M that carry substitutions in the α subunits (M_{Boston} and M_{Iwate}) are frozen in the T (deoxy) conformation and therefore have low oxygen affinities and lack cooperativity.

BOX 7-B SICKLE CELL DISEASE, MALARIA, AND BLOOD SUBSTITUTES

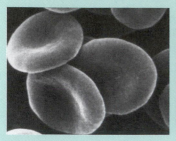 

Left: Normal erythrocytes, © Biophoto Associates, Photo Researchers.
Right: Sickled erythrocytes, © Nigel Calder.

Many persons, especially if they are of west African descent, suffer from the crippling and often lethal sickle cell disease.[a,b] In 1949, Pauling, Itano, and associates discovered that hemoglobin from such individuals migrated unusually rapidly upon electrophoresis.[c] Later, Ingram devised the method of protein fingerprinting illustrated in Fig. 7-27 and applied it to hemoglobin.[d] He split the hemoglobin molecule into 15 tryptic peptides which he separated by electrophoresis and chromatography. From these experiments the abnormality in sickle cell hemoglobin (hemoglobin S; Hb S) was located at position 6 in the β chain (see Fig. 7-23). The glutamic acid present in this position in hemoglobin A was replaced by valine in Hb S. This was the first instance in which a genetic disease was traced directly to the presence of a single amino acid substitution in a specific protein. The DNA of the normal gene for the β globin chain has since been sequenced and found to have the glutamic acid codon GAG at position 6. A single base change to GTG (see Table 5-5) causes the sickle cell mutation. Persons homozygous for this altered gene have sickle cell disease, while the much more numerous heterozygotes have, at most, minor problems.

When HbS is deoxygenated it tends to "crystallize" in red blood cells, which contain 33% by weight hemoglobin. The crystallization (actually gel formation) distorts the cells into a sickle shape and these distorted corpuscles are easily destroyed, leading to anemia. The introduction of the hydrophobic valine residue in Hb S at position 6 near the end of the molecule helps form a new bonding domain by which the hemoglobin tetramers associate to form long semicrystalline microfilamentous arrays.[b,e–g]

Why is there such a high incidence of the sickle cell gene, estimated to be present in three million Americans? The occurrence and spread of the gene in Africa was apparently the result of a balance between its harmful effects and a beneficial effect under circumstances existing there. The malaria parasite, the greatest killer of all time, lives in red blood cells during part of its life cycle (see Fig. 1-9).

Red cells that contain Hb S as well as Hb A are apparently less suitable than cells containing only Hb A for growth of the malaria organism. Thus, heterozygotic carriers of the sickle cell gene survived epidemics of malaria but at the price of seeing one-fourth of their offspring die of sickle cell disease.

What is the outlook for the many (50,000 in the United States alone) sufferers of sickle cell disease today? Careful medical care, including blood transfusion, can prolong life greatly[h] and intense efforts are under way to find drugs that will prevent Hb S from crystallizing.[i] The problem arises from a hydrophobic interaction of valine B6 with phenylalanine B85 and leucine B88 of another molecule in the filaments of Hb S. The latter two residues are on the outside surface of helix F (see Fig. 7-23). It is difficult to modify one of these residues chemically but various alterations at the nearby N-termini of the β chains do inhibit sickling. Cyanate does so by specifically carbamoylating these amino groups. However, although it was tested in humans, cyanate is too toxic for use.[j] Another approach employs an aldehyde that will

form Schiff bases (Eq. 13-4)[k] with the same amino groups.[k,l] A third approach is to use an acylating reagent. For example, methylacetyl phosphate[m] acetylates the same β Lys 82 amino groups that react with bisphosphoglycerate and with cyanate.

Aspirin (2-acetoxybenzoic acid) is also a mild acetylating reagent and "two-headed" aspirins such

BOX 7-B   (continued)

as the following react specifically to crosslink the hemoglobin β chains.

These compounds bind into the bisphosphoglycerate binding site (Fig. 7-26) and prevent the chains from spreading apart as far as they normally do in the deoxy (T) state. Since it is only the latter that crystallizes, the compounds have a powerful antisickling action.[n,o] Various other crosslinking reagents have been developed and more than one could be used together.[p–r] New drugs that serve as allosteric modifiers in the same fashion as bisphosphoglycerate may also be useful.[s]

A fouth approach to treatment of sickle cell disease is gene therapy. This might allow patients to produce, in addition to Hb S, an engineered hemoglobin with compensating mutations that would mix with the Hb S and prevent gelling.[t,u,v] This is impractical at present but there is another approach. Persons with sickle cell disease sometimes also have the disorder of hereditary persistence of fetal hemoglobin. They continue to make Hb F into adulthood. Great amelioration of sickle cell disease is observed in patients with 20–25% Hb F.[t] Hydroxyurea stimulates a greater production of Hb F and in patients with hereditary persistence of Hb F hydroxyurea may raise its level in erythocytes to ~50% of the total hemoglobin.[w]

Crosslinking of the *alpha* chains of normal deoxyhemoglobin through lysines 99 yields a hemoglobin with normal oxygen-binding behavior and an increased stability.[o] It makes a practical emergency blood substitute, whereas unmodified hemoglobin is unsatisfactory.[v] Unless encapsulated in an erythrocyte the hemoglobin tetramers tend to dissociate to dimers, losing cooperativity and escaping through kidneys. Suitable crosslinking helps to solve this problem.[x,y,z] Both α and β chains can be produced from cloned genes and reassembled to form hemoglobin.[aa,bb] This will probably allow genetic engineering to form more stable but suitably cooperative hemoglobins that can be used to avoid hazards of transmission of viruses by transfusion.

[a] Weatherall, D. J., Clegg, J. B., Higgs, D. R., and Wood, W. G. (1995) in *The Metabolic and Molecular Bases of Inherited Disease*, 7th ed., Vol. 1 (Scriver, C. R., Beaudet, A. L., Sly, W. S., and Valle, D., eds), pp. 3417–3484, McGraw-Hill, New York

[b] Harrington, D. J., Adachi, K., and Royer, W. E., Jr. (1998) *J. Biol. Chem.* **273**, 32690–32696

[c] Strasser, B. J. (1999) *Science* **286**, 1488–1490

[d] Ingram, V. M. (1957) *Nature (London)* **180**, 326–328

[e] Cretegny, I., and Edelstein, S. J. (1993) *J. Mol. Biol.* **230**, 733–738

[f] Padlan, E. A., and Love, W. E. (1985) *J. Biol. Chem.* **260**, 8272–8279

[g] Cao, Z., and Ferrone, F. A. (1996) *J. Mol. Biol.* **256**, 219–222

[h] Acquaye, C., Wilchek, M., and Gorecki, M. (1981) *Trends Biochem. Sci.* **6**, 146–149

[i] Klotz, I. M., Haney, D. N., and King, L. C. (1981) *Science* **213**, 724–731

[j] Harkness, D. R. (1976) *Trends Biochem. Sci.* **1**, 73–

[k] Acharya, A. S., Sussman, L. G., and Manning, J. M. (1983) *J. Biol. Chem.* **258**, 2296–2302

[l] San George, R. C., and Hoberman, H. D. (1986) *J. Biol. Chem.* **261**, 6811–6821

[m] Ueno, H., Pospischil, M. A., and Manning, J. M. (1989) *J. Biol. Chem.* **264**, 12344–12351

[n] Walder, J. A., Walder, R. Y., and Arnone, A. (1980) *J. Mol. Biol.* **141**, 195–216

[o] Chatterjee, R., Welty, E. V., Walder, R. Y., Pruitt, S. L., Rogers, P. H., Arnone, A., and Walder, J. A. (1986) *J. Biol. Chem.* **261**, 9929–9937

[p] Benesch, R. E., and Kwong, S. (1988) *Biochem. Biophys. Res. Commun.* **156**, 9–14

[q] Kluger, R., Wodzinska, J., Jones, R. T., Head, C., Fujita, T. S., and Shih, D. T. (1992) *Biochemistry* **31**, 7551–7559

[r] Jones, R. T., Shih, D. T., Fujita, T. S., Song, Y., Xiao, H., Head, C., and Kluger, R. (1996) *J. Biol. Chem.* **271**, 675–680

[s] Abraham, D. J., Wireko, F. C., Randad, R. S., Poyart, C., Kister, J., Bohn, B., Liard, J.-F., and Kunert, M. P. (1992) *Biochemistry* **31**, 9141–9149

[t] McCune, S. L., Reilly, M. P., Chomo, M. J., Asakura, T., and Townes, T. M. (1994) *Proc. Natl. Acad. Sci. U.S.A.* **91**, 9852–9856

[u] Cole-Strauss, A., Yoon, K., Xiang, Y., Byrne, B. C., Rice, M. C., Gryn, J., Holloman, W. K., and Kmiec, E. B. (1996) *Science* **273**, 1386–1388

[v] May, C., Rivella, S., Callegari, J., Heller, G., Gaensler, K. M. L., Luzzatto, L., and Sadelain, M. (2000) *Nature (London)* **406**, 82–86

[w] Eaton, W. A., and Hofrichter, J. (1995) *Science* **268**, 1142–1143

[x] Snyder, S. R., Welty, E. V., Walder, R. Y., Williams, L. A., and Walder, J. A. (1987) *Proc. Natl. Acad. Sci. U.S.A.* **84**, 7280–7284

[y] Dick, L. A., Heibel, G., Moore, E. G., and Spiro, T. G. (1999) *Biochemistry* **38**, 6406–6410

[z] Manjula, B. N., Malavalli, A., Smith, P. K., Chan, N.-L., Arnone, A., Friedman, J. M., and Acharya, A. S. (2000) *J. Biol. Chem.* **275**, 5527–5534

[aa] Yamaguchi, T., Pang, J., Reddy, K. S., Witkowska, H. E., Surrey, S., and Adachi, K. (1996) *J. Biol. Chem.* **271**, 26677–26683

[bb] Jeong, S. T., Ho, N. T., Hendrich, M. P., and Ho, C. (1999) *Biochemistry* **38**, 13433–13442

In hemoglobins Rainier and Nancy the usually invariant C-terminal tyrosine 145 of the β chains is substituted by cysteine and by aspartate, respectively. Oxygen affinity is high and cooperativity is lacking.[186] Hemoglobin Kansas, in which the β^{102} asparagine is substituted by threonine, also lacks cooperativity and has a very low oxygen affinity, while hemoglobin Richmond, in which the same amino acid is substituted by lysine, functions normally. In hemoglobin Creteil the β89 serine is replaced by asparagine with the result that the adjacent C-terminal peptide carrying tyrosine 145 becomes disordered. In hemoglobin Hiroshima the C-terminal histidine in the β chain is replaced by aspartic acid. This histidine is one that donates a Bohr proton and in the mutant hemoglobin the oxygen affinity is increased 3-fold and the Bohr effect is halved.[188] In hemoglobin Suresnes the C-terminal arginines 141 of the α chains are replaced by histidine with loss of one of the anion-binding sites mentioned in Section 3.[189]

3. Comparative Biochemistry of Hemoglobin

Even within human beings there are several hemoglobins. In addition to myoglobin, a brain protein neuroglobin,[189a] and adult hemoglobin A (**Hb A**, $\alpha_2\beta_2$), there is a minor hemoglobin A_2 ($\alpha_2\delta_2$). Prior to birth the blood contains **fetal hemoglobin**, also called hemoglobin F (**Hb F**, $\alpha_2\gamma_2$). In the presence of 2,3-bisphosphoglycerate Hb F has a 6-fold higher oxygen affinity than Hb A as befits its role in obtaining oxygen from the mother's blood.[190–192] Hemoglobin F disappears a few months after birth and is replaced by Hb A. Each of the hemoglobins differs from the others in amino acid sequence.

In other species the amino acid composition of hemoglobins varies more, as do the interactions between subunits. Hemoglobins and myoglobins are found throughout the animal kingdom and even in plants.[192a] The **leghemoglobins**[193–195] are formed in root nodules of legumes and are involved in nitrogen fixation by symbiotic bacteria. Other hemoglobins apparently function in the roots of plants.[196–197] Hemoglobins or myoglobins are found in some cyanobacteria[198] and in many other bacteria.[197,199] The globin fold of the polypeptide is recognizable in all of these.[200]

The quaternary structure of hemoglobin also varies. Myoglobin is a monomer, as is the leghemoglobin. Hemoglobin of the sea lamprey dissociates to monomers upon oxygenation.[201,201a] The clam *Scapharca inaequivalvis* has a dimeric hemoglobin that binds O_2 cooperatively even though the interactions between subunits are very different from those in mammalian hemoglobins.[202–204] Hemoglobin of the nematode *Ascaris* is an octamer.[205,206] It has puzzling properties, including a very high affinity for O_2 and a slow dissociation rate. The distal His E-7 is replaced by glutamine, which has a hydrogen-bonding ability closely similar to that of histidine.

Earthworms,[207] polychaete worms,[208,209] and leeches[210] have enormous hemoglobin molecules consisting of as many as 144 globin chains arranged into 12 dodecamers and held together by 36–42 linker chains. These hemoglobins are often called **erythrocruorins**. In a few families of polychaetes chloroheme (Fig. 16-5) substitutes for heme and the proteins are called **chlorocruorins**.[208]

What is common to all of the hemoglobins? The same folding pattern of the peptide chain is always present. The protein is always wrapped around the heme group in an identical or very similar manner. In spite of this striking conservation of overall structure, when animal hemoglobins are compared, *there are only ten residues that are highly conserved*. They are indicated in Fig. 7-23 by the boxes. The two glycines (or alanine) at B-6 and E-8 are conserved because the close contact between the B and E helices does not permit a larger side chain. Proline C-2 helps the molecule turn a corner. Four of the other conserved residues are directly associated with the heme group. Histidine E-7 and His F-8 are the "heme-linked" histidines. Tyrosine HC-2, as previously mentioned, plays a role in the cooperativity of oxygen binding. Only Lys H-9 is on the outside of the molecule. The reasons for its conservation are unclear.[211] When sequences from a broader range of organisms were determined five residues (see Fig. 7-23) were found to be highly conserved; *only two are completely conserved*. These are His F-8 and Phe CD-1, which binds the heme noncovalently.[212] Hemoglobins are not the only biological oxygen carriers. The two-iron **hemerythrins** (Fig. 16-20) are used by a few phyla of marine invertebrates, while the copper containing **hemocyanins** (Chapter 16, Section D,4) are used by many molluscs and arthropods.

E. Self-Assembly of Macromolecular Structures

While it is easy to visualize the assembly of oligomeric proteins, it is not as easy to imagine how complex objects such as eukaryotic cilia (Fig. 1-8) or the sarcomeres of muscle (Fig. 19-6) are formed. However, study of the assembly of bacteriophage particles and other small biological objects has led to the concepts of **self-assembly** and **assembly pathways**, concepts that are now applied to every aspect of the architecture of cells.

1. Bacteriophages

A remarkable example of self-assembly is that of the T-even phage (Box 7-C).[213–215] From genetic analysis (Chapter 26) at least 22 genes are known to be required for formation of the heads, 21 genes for the tails, and 7 genes for the tail fibers. Many of these genes encode

BOX 7-C THE T-EVEN BACTERIOPHAGES

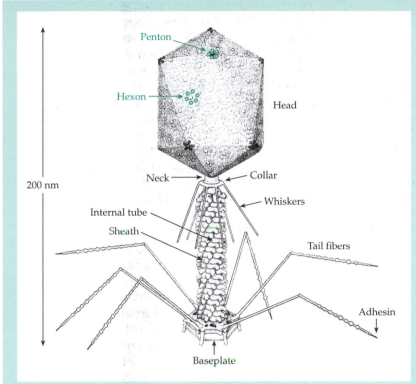

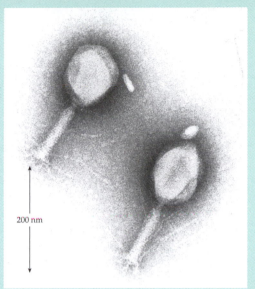

Bacteriophage T4. × 240,000
Micrograph courtesy of Tom Moninger.

Drawing courtesy of F. Eiserling and the
American Society for Microbiology.[a]

Among the most remarkable objects made visible by the electron microscope are the T-even bacteriophage (T2, T4, and T6) which attack *E. coli*.[a–e] While it is not often evident how a virus gains access to a cell, these "molecular syringes" literally inject their DNA through a hole dissolved in the cell wall of the host bacterium. The viruses, of length ~200 nm and mass ~225 x 10^6 Da each, contain 130 x 10^6 Da of DNA in a 100 x 70 nm head of elongated icosahedral shape. The head surface appears to be formed from ~840 copies of a 45-kDa protein known as gp23 (gene product 23; it is encoded by gene 23; see map in Fig. 26-2). These protein molecules are arranged as 140 hexamers (hexons) and together with ~55 copies of protein gp24, arranged as 11 pentamers (pentons), make up the bulk of the shell.[f] The head also contains at least nine other proteins, including three internal, basic proteins that enter the bacterium along with the DNA. Additional proteins form the **neck**, **collar**, and **whiskers**. The phage **tail**, which fastens to the collar via a **connector** protein,[g] contains an **internal tube** with a 2.5-nm hole, barely wide enough to accommodate the flow of the DNA molecule into the bacterium. The tube is made up of 144 subunits of gp19. The 8 x 10^6-Da **sheath** that surrounds the tail tube is made up of 144 subunits of gp18, each of mass 55 kDa, arranged in the form of 24 rings of six subunits each.[h,i] The sheath has contractile properties. After

the virus has become properly attached to the host it shortens from ~80 to ~30 nm, forcing the inner tube through a hole etched in the wall of the bacterium. At the end of the tail is a **baseplate**, a hexagonal structure bearing six short **pins**, each a trimer of a 55 kDa zinc metalloprotein.[j] One of the ten proteins known to be present in the baseplate is the enzyme T4 lysozyme (Chapter 12). The baseplate also contains six molecules of the coenzyme 7,8-dihydropteroyl-hexaglutamate (Chapter 15, Section D).

Six elongated, "jointed" **tail fibers** are attached to the baseplate. The proximal segment of each fiber is a trimer of the 1140-kDa protein gp34. A globular domain attaches it to the baseplate. The distal segment is composed of three subunits of the 109-kDa gp37, three subunits of the 23-kDa gp36, and a single copy of the 30-kDa gp35.[k] The C-terminal ~140 residues of the 1026-residue gp37 are the specific **adhesin** that binds to a lipopolysaccharide of *E. coli* type B cells or to the outer membrane protein OmpC (Chapter 8).[l] Among the smaller molecules present in the virus are the polyamines **putrescine** and **spermidine** (Chapter 24), which neutralize about 30% of the basic groups of the DNA.

How is infection by a T-even virus initiated? Binding of the tail fibers to specific receptor sites on the bacterial surface triggers a sequence of conformational changes in the fibers, baseplate, and

BOX 7-C THE T-EVEN BACTERIOPHAGES (continued)

sheath. The lysozyme is released from the baseplate and etches a hole in the bacterial cell wall. Contraction of the sheath is initiated at the baseplate and continues to the upper end of the sheath. The tail tube is forced into the bacterium and the DNA rapidly flows through the narrow hole into the host cell.

During contraction the subunits of the sheath undergo a remarkable rearrangement into a structure containing 12 larger rings of 12 subunits each.[h] Thus, a kind of mutual "intercalation" of subunits occurs. In its unidirectional and irreversible nature the shortening of the phage tail differs from the contraction of muscle. The protein subunits of the sheath seem to be in an unstable high energy state when the tail sheath of the phage is assembled. The stored energy remains available for later contraction.

[a] Mathews, C. K., Kutter, E. M., Mosig, G., and Berget, P. B., eds. (1983) *Bacteriophage T₄*, Am. Soc. Microbiology, Washington, D.C.
[b] Wood, W. B., and Edgar, R. S. (1967) *Sci. Am.* **217**(Jul), 60–74
[c] Mathews, C. K. (1971) *Bacteriophage Biochemistry*, Van Nostrand-Reinhold, Princeton, New Jersey
[d] Voyles, B. A. (1993) *The Biology of Viruses*, Mosby, St. Louis, Missouri
[e] Tikhonenko, A. S. (1970) *Ultrastructure of Bacterial Viruses*, Plenum, New York
[f] Branton, D., and Klug, A. (1975) *J. Mol. Biol.* **92**, 559–565
[g] Cerritelli, M. E., and Studier, F. W. (1996) *J. Mol. Biol.* **258**, 299–307
[h] Moody, M. F., and Makowski, L. (1981) *J. Mol. Biol.* **150**, 217–244
[i] Müller, D. J., Engel, A., Carrascosa, J. L., and Vélez, M. (1997) *EMBO J.* **16**, 2547–2553
[j] Zorzopulos, J., and Kozloff, L. M. (1978) *J. Biol. Chem.* **253**, 5543–5547
[k] Cerritelli, M. E., Wall, J. S., Simon, M. N., Conway, J. F., and Steven, A. C. (1996) *J. Mol. Biol.* **260**, 767–780
[l] Tétart, F., Repoila, F., Monod, C., and Krisch, H. M. (1996) *J. Mol. Biol.* **258**, 726–731

sequences of proteins that are incorporated into the mature virus, but several specify enzymes needed in the assembly process. Several mutant strains of the viruses are able to promote synthesis of all but one of the structural proteins of the virion. Proteins accumulating within these defective bacterial hosts have no tendency to aggregate spontaneously. However, when the missing protein (synthesized by bacteria infected with another strain of virus) is added complete virus particles are formed rapidly. Investigations resulting from this and other observations have led to the conclusion that during assembly *each different protein is added to the growing aggregate in a strictly specified sequence or assembly pathway*. The addition of each protein creates a binding site for the next protein. In some cases the protein that binds is an enzyme that cuts off a piece from the growing assembly of subunits and thereby creates a site for the next protein to bind.

Before considering this complex process further, let's look at the assembly of simpler filamentous bacteriophages (Fig. 7-7) and bacterial pili (Fig. 7-9). The filamentous bacteriophages are put together from hydrophobic protein subunits and DNA. After their synthesis the protein subunits are stored within the cytoplasmic membrane of the infected bacteria.[216,217] These small, largely α-helical rods can easily fit within the membrane and remain there until a DNA molecule also enters the membrane.[218,219] Two additional proteins (gene I proteins) also enter the membrane. One has a 348-residue length, while the second is a 107-residue protein formed by translational initiation at a later point in the DNA sequence of the gene. These proteins help to create an assembly site at a place where

the inner and outer membrane of the host bacterium are close together.[220] It isn't clear how the process is initiated, but it is likely that each subunit of the viral coat contains a nucleotide binding site that interacts with the DNA. Adjacent sides of the subunits are hydrophobic and interact with other subunits to spontaneously coat the DNA. As the rod is assembled, the hydrophobic groups become "buried." It is postulated that the remaining side chain groups on the outer surface of the virus are hydrophilic and that the formation of this hydrophilic rod provides a driving force for automatic extrusion of the phage from the membrane.[216]

Bacterial pili appear to be extruded in a similar manner. They arise rapidly and may possibly be retracted again into the bacterial membrane. The P pilus in Fig. 7-9A is made up of subunits PapA, G, F, E, and K which must be assembled in the correct sequence. A chaperonin PapD is also required as is an "**usher protein**," PapC,[50] and also the disulfide exchange protein DsbA (Chapter 10). DsbA helps PapD to form the correct disulfide bridges as it folds and PapD binds and protects the various pilus subunits as they accumulate in the periplasmic space of the host. The usher protein displaces the chaperonin PapD and "escorts" the subunits into the membrane where the extrusion occurs.[50,55]

Because eicosahedra are regular geometric solids and the faces can be made up of hexons and pentons of identical subunits, it might seem that self-assembly of eicosahedral viruses would occur easily. However, the subunits usually must be able to assume three or more different conformations and the shells can easily be assembled incorrectly. Several stcategies are employed to avert this problem.[221,222] Some viruses assemble an empty shell into which the DNA flows, but many others first form an internal **scaffolding** or **assembly core** around which the shell is assembled. An external scaffold may also be needed.[223] The RNA virus MS2 forms its T=3 capsid by using the RNA molecule as the assembly core.[224] Other viruses may have one or more core proteins which dissociate from the completed shell or are removed by the action of

proteases. This is a feature of the small φX phage (Fig. 7-28),[225,226,230] the tailed phages,[227,228] and double-stranded RNA viruses including human reoviruses.[229] Bacteriophage PRD1, another virus of *E. coli* and *Salmonella typhimurium*, has a membrane inside the capsid apparently playing a role in assembly.[232]

A general concept that seems to hold in all cases is one of "**local rule**." Several conformers of a virus subunit may equilibrate within a cell. However, *they*

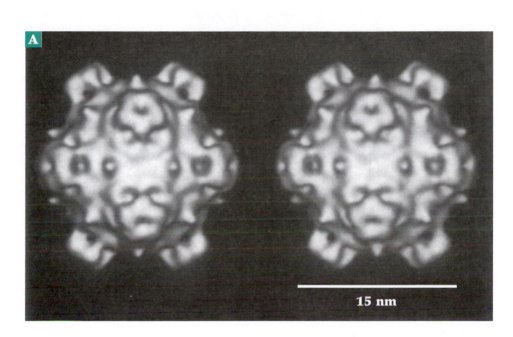

Figure 7-28 (A) Stereoscopic view of the φX174 114 S mature virion viewed down a twofold axis after a cryoelectron microscopy reconstruction. From McKenna *et al.*[225] (B) Morphogenesis of φX174 (based on a report of Hayashi *et al.*[231]). Proteins A and C are required for DNA synthesis. Drawing from McKenna *et al.*[225] Courtesy of Michael G. Rossmann.

15 nm

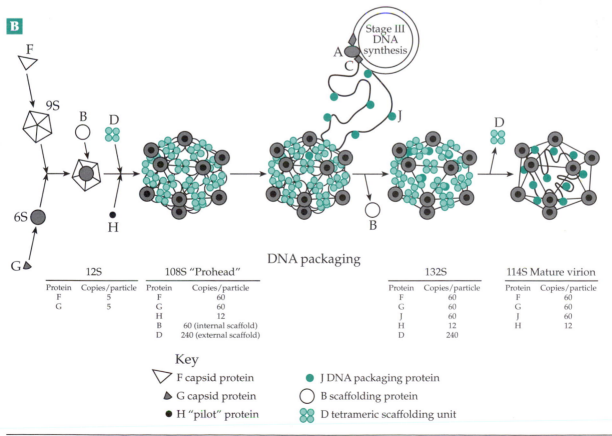

DNA packaging

12S			108S "Prohead"	
Protein	Copies/particle		Protein	Copies/particle
F	5		F	60
G	5		G	60
			H	12
			B	60 (internal scaffold)
			D	240 (external scaffold)

132S		114S Mature virion	
Protein	Copies/particle	Protein	Copies/particle
F	60	F	60
G	60	G	60
J	60	J	60
H	12	H	12
D	240		

Key

▽ F capsid protein   ● J DNA packaging protein

▲ G capsid protein   ○ B scaffolding protein

● H "pilot" protein   ✪ D tetrameric scaffolding unit

can associate only through surfaces that are complementary. A conformer that allows pentons to form cannot assemble into a hexon and only certain combinations of other conformers can give rise to hexons, etc.[233] If there is only one conformation and the shape is right a T=1 shell will be formed. If there are three conformers a T= 3 shell may arise. Another generalization is that in most cases the **procapsid** or **prohead** that is formed initially is fragile. Subunits may still be undergoing conformational changes. However, a final conformational alteration, which may include chain cleavage by a protease, usually occurs. This often expands the overall dimensions of the capsid and creates new intersubunit interactions which greatly strengthen the mature capsid.[234] Scaffolding proteins are then removed and DNA or RNA enters the capsid, again in a precise sequence. There are many variations and the detail that is known about virus assembly is far too great to describe here.

Figure 7-29 illustrates the assembly pathway for the very small φX174, a T=1 virus. The major capsid protein F is a 426-residue eight-stranded β-barrel. The 175-residue G protein forms pentameric spikes while 60 copies of the internal scaffolding protein and 240 copies of the external scaffolding protein D and 12 copies of the pilot protein are required to form the prohead. The single-stranded DNA enters along with 60 copies of a DNA packaging protein J.

Assembly of the tailed bacteriophages (Box 7-C) is even more complex. The genome of the viruses is large. The 166-kb circular dsDNA of phage T4 contains ~250 genes, many of which encode proteins of the virion or enzymes or chaperonins needed in assembly. The assembly pathway for the bacteriophage heads requires at least 22 gene products.[235] Seven of these form the assembly core which serves as a scaffolding around which the 840 copies of gp23 and 55 copies of gp24 (see Box 7-C) are added to give the elongated icosahedral prohead I. Most of the internal proteins are then dissolved by proteases, one of which is the phage-specified gp21. A protease also cuts a piece from each molecule of gp23 to form the smaller gp23*, the major protein of the mature **prohead II**. This cleavage also triggers the conformational change leading to head expansion. The empty proheads are now filled with DNA in a process which is assisted by another series of catalytic proteins.

The T4 phage tail is assembled in a separate sequence. Six copies of each of three different proteins form a "hub" with hexagonal symmetry (Fig. 7-29). In another assembly sequence, seven different proteins form wedge-shaped pieces, six of which are then joined to the hub to form the hexagonal baseplate. Two more proteins then add to the surface of the base plate and activate it for the growth of the internal tail tube. Only after assembly of the internal tube begins does the sheath

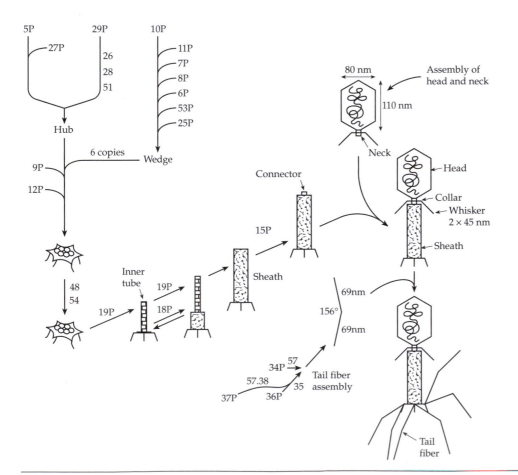

Figure 7-29 Assembly sequence for bacteriophage T4 with details for the tail. The numbers refer to the genes in the T4 chromosome map (Fig. 26-2). A "P" after the number indicates that the protein gene product is incorporated into the phage tail. Other numbers indicate gene products that are thought to have essential catalytic functions in the assembly process. Adapted from King and Mykolajewycz[236] and Kikuchi and King.[214]

begin to grow, and only when both of these tubular structures have reached the correct length, is a cap protein placed on top. The DNA-filled head is then attached by a special connector and only then do the tail fibers, which have been assembled separately, join at the opposite end.

How can each step in this complex assembly process set the stage for the next step? Apparently the structure of each newly synthesized protein monomer is stable only until a specific interaction with another protein takes place. The binding energy of this interaction is sufficient to induce a conformational alteration that affects a distant part of the protein surface and generates complementarity toward a binding site on the next protein that is to be added. Every one of the baseplate proteins must have such self-activating properties! Sometimes proteolytic cleavage of a subunit is required. If it occurs at an appropriate point in the sequence it provides thermodynamic drive for the assembly process.

The induction of a change in one protein by interaction with another protein is a phenomenon that is met also in the construction of microtubules, ribosomes, cilia, and myofibrillar assemblies of muscle. It is basic to the assembly of the many labile but equally real cascade systems of protein–protein interactions such as that involved in the clotting of blood (Chapter 12) and signaling at membrane surfaces.

2. "Kringles" and Other Recognition Domains and Motifs

The assemble of either transient or long lasting complexes of proteins is often dependent upon the presence of conserved structural domains of 30–100 residues. A similar domain may occur in many different proteins and often two or more times within a single protein. The sequences within such a domain are homologous, allowing it to be recognized from protein or gene sequences alone.[237,238] Domains are often named after the protein in which they were first discovered. For example, **EGF-like domains** resemble the 53-residue epidermal growth factor. **SH2** and **SH3** domains are src-homology domains, named after **Src** (c-src), the protein encoded by the *src* protooncogene

(Table 11-3).[239] The SH2 and SH3 domains are found near the N terminus of this 60-kDa protein. They are also found in many other proteins. An adapter protein called **Grb2**, important in cell signaling, consists of nothing but one SH2 domain and two SH3 domains (Figs. 11-13, 11-14).[240] The SH2 domains bind to phosphotyrosyl side chains of various proteins, while SH3 domains bind to a polyproline motif. Another phosphotyrosyl binding domain, the plekstrin homology or **PH** domain, is named for the protein in which it was discovered. **Kringle** and **apple** describe the appearances of the folded proteins in those domains. Structural domains often function to hold two proteins together or to help anchor them at a membrane surface by binding to specific protein groups, such as phosphotyrosyl, or calcium ions. Table 7-3 lists a few well-known folding domains and Fig. 7-30 shows three-dimensional structures of two of them.

Recognition domains often function transiently. For example, SH2 domains are often found in proteins that interact with phosphotyrosyl groups of "activated" cell surface receptors. The receptors become activated

TABLE 7-3
A Few Well-Known Structural Domains

Name	Length in amino acid residues	Specific ligands
EGF-like[241–244]	~45	Ca^{2+}
SH2[238,245–247] Structure[248–253]	~100	Phosphotyrosine
SH3[239,246,254] Structure[255–257]	~60	Polyproline, PXXP
PTB[238,258,259]		Proline-rich sequence
PH (plekstrin homology)[260,261] Structure[262,263]		Phosphotyrosine
PDZ[264–266]	80–100	C-terminal XS / TXV–COO⁻
Immunoglobulin repeat (Fig. 2-16)[267]	~100	
Kringles, blood clotting proteins[268–270]	80-85	Calcium binding
Apple, Blood clotting Factor X[271]	90	Calcium binding
WW (Trp–Trp)[272]	~38	Proline
Serine protease[273]		
P (Trefoil)[274]	~50	
TPR (Tetratrico peptide repeat)[247,275,276]		
ZBD (Zinc-binding domains) Zinc finger (Fig. 5-37)[277] Others[277–280]		

by conformational alteration resulting from the binding. The src protein is a tyrosine kinase and, when activated, uses ATP to phosphorylate tyrosyl groups of other proteins, and using its SH2 domains it will bind to such groups forming and passing an intracellular message to them.[249]

F. The Cytoskeleton

The cytoplasm of eukaryotic cells contains a complex network of slender rods and filaments that serve as a kind of internal skeleton. The properties of this **cytoskeleton** affect the shape and mechanical properties of cells. For example, the cytoskeleton is responsible

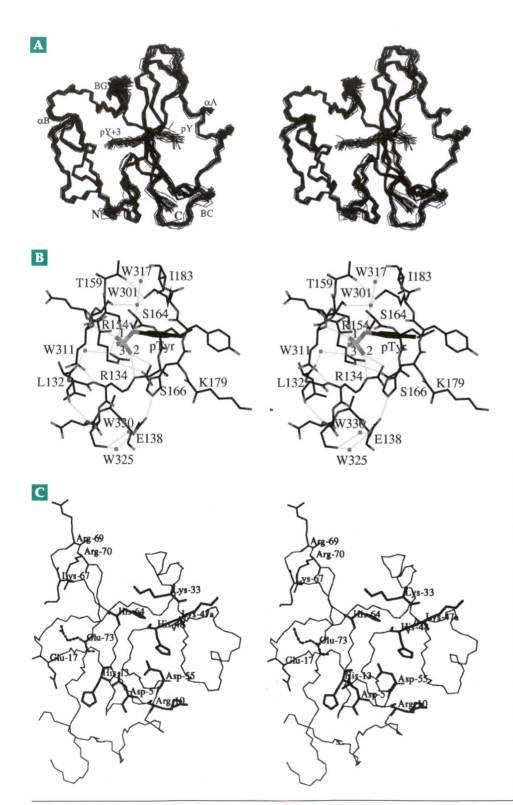

Figure 7-30 (A) Stereoscopic MolScript view showing 30 super-posed solution structures of the SH2/phosphopeptide complex from protein Shc calculated from NMR data. The N and C termini of the protein as well as the phospho-tyrosine (pY) and (pY + 3) residues of the phosphopeptide are indicated. From Zhou *et al*.[252] Courtesy of Stephen W. Fesik. (B) View of the phosphotyrosine side chain of the peptide pYEEI bound to a high-affinity SH2 domain from the human *src* tyrosine kinase called p56[lck]. The phosphate group forms a series of hydrogen bonds with groups in the protein and with water molecules (small dots) and an ion pair with the guanidinium group of R134. From Tong *et al*.[248] Courtesy of Liang Tong. (C) Structure of kringle 2 from human tissue plasminogen activator (see Chapter 12). From de Vos *et al*.[269] Courtesy of Abraham M. de Vos.

for the biconcave disc shape of erythrocytes and for the ameba's ability to rapidly interconvert gel-like and fluid regions of the cytoplasm.[281–283]

Three principal components of the cytoskeleton are **microfilaments** of ~6 nm diameter, **microtubules** of 23–25 nm diameter, and **intermediate filaments** of ~10 nm diameter. A large number of associated proteins provide for interconnections, for assembly, and for disassembly of the cytoskeleton. Other proteins act as **motors** that provide motion. One of these motors is present in **myosin** of muscle. This protein is not only the motor for muscular work but also forms **thick filaments** of 12–16 nm diameter, which are a major structural component of muscle (see Fig. 19-6).

1. Intermediate Filaments

In most cells the intermediate filaments provide the scaffolding for the cytoskeleton[284–286] They may account for only 1% of the protein in a cell but provide up to 85% of the protein in the tough outer layers of skin. Intermediate filament proteins are encoded by over 50 human genes[286] which specify proteins of various sizes, structures, and properties. However, all of them have central 300- to 330-residue α-helical regions through which the molecules associate in parallel pairs to form coiled-coil rods with globular domains at the ends (Fig. 7-31). Some of these proteins, such as the **keratin** of skin, are insoluble. Others, including the nuclear **lamins** (Chapter 27)[287] and **vimentin**,[288–289a] dissociate and reform filaments reversibly.

Vimentin is found in most cells and predominates in fibroblasts and other cells of mesenchymal origin. **Desmin** (55-kDa monomer) is found in both smooth and skeletal muscle.[289b,290,290a] In the latter, it apparently ties the contractile myofibrils to the rest of the cytoskeletal network and the individual myofibrils to each other at Z disc (see Fig. 19-6). The **glial filaments** from the astroglial cells of the brain are composed mainly of a single type of 55-kDa subunits of the **glial fibrillar acidic protein** but the **neurofilaments** of mammalian neurons are composed of three distinct subunits of 68-, 150-, and 200-kDa mass.[291–293] The larger subunits have C-terminal tails that are not required for filament formation but which can be phosphorylated and form bridges to neighboring neurofilaments and other cytoskeletal components and organelles. Keratin filaments, which eventually nearly fill the highly differentiated epidermal cells, are also made up of several different subunits.[294] Extensions of the keratin chains are rich in cysteine side chains which form disulfide crosslinkages to adjacent molecules to provide a network that can be dehydrated to form hair and the tough outer layers of skin.[286] Elastin-associated microfibrils are important constituents of elastic tissues of blood vessels, lungs, and skin.[295]

A common architecture of intermediate filaments is a staggered head-to-tail and side-by-side association of pairs of the coiled-coil dimers into 2- to 3-nm protofilaments and further association of about eight protofilaments to form the 10-nm intermediate filaments.[286,290,296]

2. Microfilaments

The most abundant microfilaments are composed of fibrous actin (F-actin; Fig. 7-10). The **thin filaments** of F-actin are also one of the two major components of the contractile fibers of skeletal muscle. There is actually a group of closely related actins encoded by a multigene family. At least four vertebrate actins are specific to various types of muscle, while two (β- and γ-actins) are cytosolic.[298,299] Actins are present in all animal cells and also in fungi and plants as part of the cytoskeleton. The microfilaments can associate to

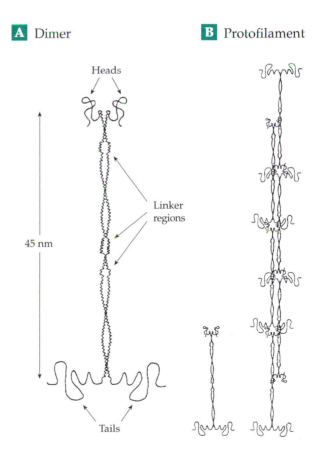

A Dimer   **B** Protofilament

Heads

Linker regions

45 nm

Tails

Figure 7-31 A model for the structure of keratin microfibrils of intermediate filaments. (A) A coiled-coil dimer, 45-nm in length. The helical segments of the rod domains are interrupted by three linker regions. The conformations of the head and tail domains are unknown but are thought to be flexible. (B) Probable organization of a protofilament, involving staggered antiparallel rows of dimers. From Jeffrey A. Cohlberg[297]

form larger arrays and actin often exists as thicker "cables," some of which form the **stress fibers** seen in cultured cells adhering to a glass surface. In the red blood cells the **spectrin–actin** meshwork (Fig. 8-14), which lies directly beneath the plasma membrane, together with the proteins that anchor it to the membrane, form the cytoskeleton.[284,300,301] Its mechanical properties appear to be responsible for the biconcave disc shape of the cell.

The **acrosomal process** of some invertebrate sperm cells is an actin cable that sometimes forms almost instantaneously by polymerization of the actin monomers and shoots out to penetrate the outer layers of the egg during fertilization (Chapter 32). The **stereocilia**, the "hairs" of the hair cells in the inner ear, contain bundles of actin filaments.[302] Motion of the stereocilia caused by sound produces changes in the membrane potential of the cells initiating a nerve impulse. In certain lizards each hair cell contains about 75 stereocilia of lengths up to 30 μm and diameter 0.8 μm and containing more than 3000 actin filaments in a semicrystalline array. Microvilli (Fig. 1-6) contain longitudinal arrays of actin filaments.

In every instance, groups of microfilaments are held together by other proteins. Stress fibers of higher eukaryotes contain the "muscle proteins" **tropomyosin**, **α-actinin**, and myosin (Chapter 19), although the latter is usually not in fibrillar form. **Filamin** (250-kDa) and a 235-kDa protein are associated with actin in platelets.[303] The high-molecular-weight **synemin** crosslinks vimentin and desmin filaments,[304] while the smaller, highly polar **filaggrin** provides a matrix around the keratin filaments in the external layers of the skin.[305]

Postsynthetic modifications of cytoskeletal microfilaments can also occur. For example, epidermal keratin has been found to contain lanthionine, (γ-glutamyllysine) and lysinoalanine, both presumably arising from crosslinkages.[306]

3. Microtubules

A prominent component of cytoplasm consists of microtubules which appear under the electron microscope to have a diameter of 24 ± 2 nm and a 13 - to 15-nm hollow core.[307–310] However, the true diameter of a hydrated microtubule is about 30 nm and the microtubule may be further surrounded by a 5–20 nm low density layer of associated proteins. Microtubules are present in the most striking form in the flagella and cilia of eukaryotic cells (Fig. 1-8). The **stable microtubules** of cilia are integral components of the machinery causing their motion (Chapter 19). **Labile microtubules**, which form and then disappear, are often found in cytoplasm in which motion is taking place, for example, in the pseudopodia of the ameba. The mitotic spindle

consists largely of microtubules which function in the movement of chromosomes in a dividing cell (Box 7-D and Chapter 26).

Microtubules in the long axons of nerve cells function as "rails" for the "fast transport" of proteins and other materials from the cell body down the axons. In fact, microtubules appear to be present throughout the cytoplasm of virtually all eukaryotic cells (Fig. 7-32) and also in spirochetes.[311] Motion in microtubular systems depends upon motor proteins such as **kinesin**, which moves bound materials toward what is known as the "negative" end of the microtubule,[312] **dyneins** which move toward the positive end.[310] These motor proteins are driven by the Gibbs energy of hydrolysis of ATP or GTP and in this respect, as well as in some structural details (Chapter 19), resemble the muscle protein myosin. Dynein is present in the arms of the microtubules of cilia (Fig. 1-8) whose motion results from the sliding of the microtubules driven by the action of this protein (Chapter 19).

Microtubules are assembled from ~ 55-kDa **tubulins**, which are mixed dimers of α subunits (450 residues) and β subunits (445 residues) with 40% sequence identity. The αβ dimers, whose structure is shown in

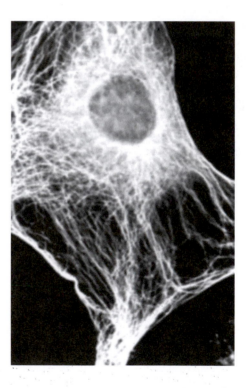

Figure 7-32 Micrograph of a mouse embryo fibroblast was obtained using indirect immunofluorescence techniques.[313] The cells were fixed with formaldehyde, dehydrated, and treated with antibodies (formed in a rabbit) to microtubule protein. The cells were then treated with fluorescent goat antibodies to rabbit γ-globulins (see Chapter 31) and the photograph was taken by fluorescent light emission. Courtesy of Klaus Weber.

BOX 7-D MITOSIS, TETRAPLOID PLANTS, AND ANTICANCER DRUGS

Microtubules in cells undergoing mitosis are the target of several important drugs. One of these is the alkaloid **colchicine** which is produced by various members of the lily family and has been used since ancient Egyptian times for the alleviation of the symptoms of gout.[a,b]

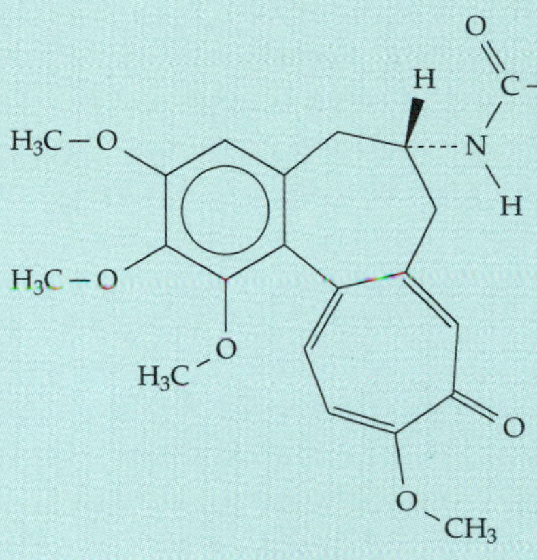

in treatment of breast, ovarian, and other cancers.[f] Binding sites for the compound have been located in β tubulin subunits (Fig. 7-33).[h,i] Attempts are being made to develop "taxoids" and other drugs more effective than taxol against cancer cells.[j,k]

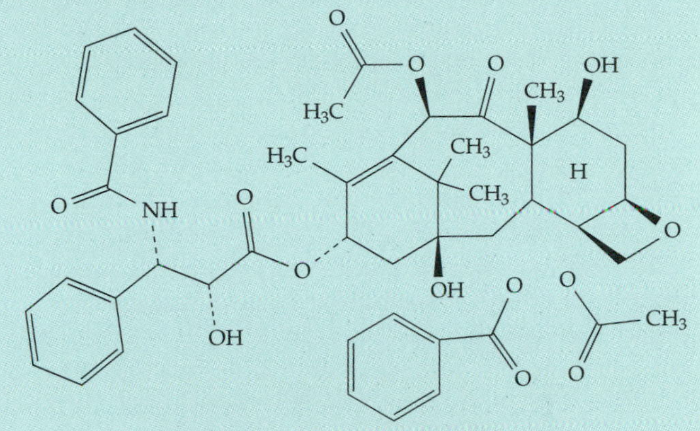

Colchicine

Taxol (paclitaxel)

This compound, with its tropolone ring system, binds specifically and tightly and prevents assembly of microtubules, including those of the mitotic spindle.[b] Colchicine forms a complex with soluble tubulin,[c] perhaps a dimeric αβ complex of the two subunits.[d] Dividing cells treated with colchicine appear to be blocked at metaphase (Chapter 26) and daughter cells with a high degree of polyploidy are formed. This has led to the widespread use of colchicine in inducing formation of tetraploid varieties of flowering plants. Similar effects upon microtubules are produced by the antitumor agents **vincristine** and **vinblastine**, alkaloids formed by the common plant *Vinca* (periwinkle), and also by a variety of other drugs.[e]

The more recently discovered Taxol (paclitaxel) was extracted from the bark of the Pacific yew.[f] It *stabilizes* microtubules, inhibiting their disassembly.[g] Taxol also blocks mitosis and causes the cells which fail to complete mitosis to die. Taxol has been synthesized[f] and is a promising drug that is being used

Laser scanning confocal micrograph of chromosomes at metaphase. Courtesy of Tom Moninger

Another group of drugs that bind to microtubules are **benzimidazole** and related compounds. These have been used widely to treat infection by parasitic nematodes in both humans and animals. Unfortunately resistance has developed rapidly. In a nematode that infects sheep a single tyrosine to phenylalanine mutation at position 200 in the β-tubulin subunit confers resistance.[l]

[a] Margulis, T. N. (1974) *J. Am. Chem. Soc.* **96**, 899–902
[b] Chakrabarti, G., Sengupta, S., and Bhattacharyya, B. (1996) *J. Biol. Chem.* **271**, 2897–2901
[c] Panda, D., Daijo, J. E., Jordan, M. A., and Wilson, L. (1995) *Biochemistry* **34**, 9921–9929
[d] Shearwin, K. E., and Timasheff, S. N. (1994) *Biochemistry* **33**, 894–901
[e] Hastie, S. B., Williams, R. C., Jr., Puett, D., and Macdonald, T. L. (1989) *J. Biol. Chem.* **264**, 6682–6688
[f] Nicolaou, K. C., Nantermet, P. G., Ueno, H., Guy, R. K., Couladouros, E. A., and Sorensen, E. J. (1995) *J. Am. Chem. Soc.* **117**, 624–633

[g] Derry, W. B., Wilson, L., and Jordan, M. A. (1995) *Biochemistry* **34**, 2203–2211
[h] Rao, S., He, L., Chakravarty, S., Ojima, I., Orr, G. A., and Horwitz, S. B. (1999) *J. Biol. Chem.* **274**, 37990–37994
[i] Makowski, L. (1995) *Nature (London)* **375**, 361–362
[j] Nicolaou, K. C., Guy, R. K., and Potier, P. (1996) *Sci. Am.* **274**(Jun), 94–98
[k] Kowalski, R. J., Giannakakou, P., and Hamel, E. (1997) *J. Biol. Chem.* **272**, 2534–2541
[l] Kwa, M. S. G., Veenstra, J. G., Van Dijk, M., and Roos, M. H. (1995) *J. Mol. Biol.* **246**, 500–510

Fig. 7-33[314–316] are thought to be packed into an imperfect helix as indicated in Fig. 7-34. The structure can also be regarded as an array of longitudinal protofilaments. Naturally formed microtubules usually have precisely 13 protofilaments and a discontinuity in the helical stacking of subunits as shown in Fig. 7-34. When grown in a laboratory the microtubules usually have 14 protofilaments[317] and rarely 10 or 16 protofilaments with regular helical packing.[318] Microtubules of some moths and also of male germ cells of *Drosophila* have 16-protofilament microtubules without a discontinuity, an architecture that is specified by the geometry of a specific β-tubulin isoform.[319]

Each tubulin dimer binds one molecule of GTP strongly in the α subunit and a second molecule of GTP or GDP more loosely in the β subunit. In this respect, tubulin resembles actin, whose subunits are about the same size. However, there is little sequence similarity. Labile microtubules of cytoplasm can be formed or disassembled very rapidly. GTP is essential for the fast growth of these microtubules and is hydrolyzed to GDP in the process.[320] However, nonhydrolyzable analogs of GTP, such as the one containing the linkage P–CH$_2$–P between the terminal and central phosphorus atoms of the GTP, also support polymerization.[321] Since microtubules have a distinct **polarity**, the two ends have different tubulin surfaces exposed, and polymerization and depolymerization can occur at different rates at the two ends. As a consequence, microtubules often grow at one end and disassemble at the other. Such **"treadmilling"** may be important in movement of chromosomes in neuronal migration[322] and in fast axonal transport of macromolecules (Chapter 30).[323] During mitosis the minus ends of the microtubules are believed to be tightly anchored at the centrosome while subunit exchanges can occur at the plus ends[323,324] where the β subunits are exposed. Using a phage display system (see Fig. 3-16) it could be shown that the N termini of the α subunits are exposed at the minus ends.[325] Kinesin can bind to the β subunits all along the microtubule.[326] Microtubules are formed by growth from microtubule nucleation sites in **microtubule organizing centers** found in centrosomes, spindle poles, and other locations.[327] Several proteins, including **γ-tubulin**, are required.[317,328,329] A proposed assembly pathway is illustrated in Fig. 7-34.

Isolated microtubules always contain small amounts of larger ~300-kDa **microtubule-associated proteins** (MAPS).[330] These elongated molecules may in part lie in the grooves between the tubulin subunits and in part be extended outward to form a low-density layer around the tubule.[283,309] Nerve cells that contain stable microtubules have associated stabilizing proteins.[331] A family of proteins formed by differential splicing of mRNA are known as **tau**. The tau proteins are prominent components of the cytoskeleton of neurons. They not only interact with microtubules but also undergo reversible phosphorylation. Hyperphosphorylated tau is the primary component of the paired helical filaments found in the brains of persons with Alzheimer disease.[330]

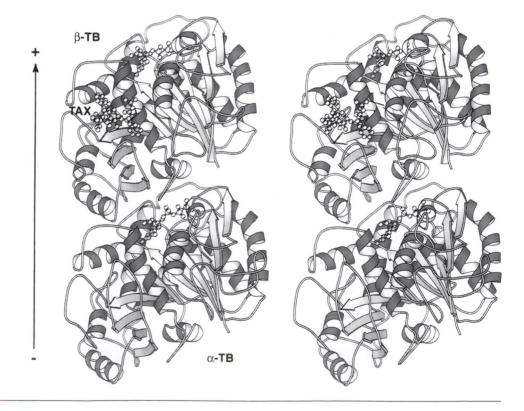

Figure 7-33 Stereoscopic ribbon diagram of the tubulin dimer with α-tubulin with bound GTP at the top and β-tubulin with bound GDP at the bottom. The β-tubulin subunit also contains a bound molecule of taxotere (see Box 7-D) which is labeled TAX. This model is based upon electron crystallography of zinc-induced tubulin sheets at 0.37-nm resolution and is thought to approximate closely the packing of the tubulin monomers in microtubules.[315] The arrow at the left points toward the plus end of the microtubule. Courtesy of Kenneth H. Downing.

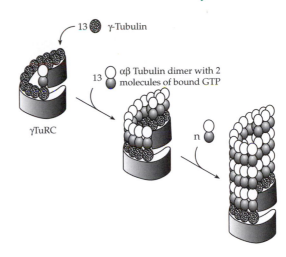

Figure 7-34 Growth of a microtubule from a γ-tubulin ring complex (γTURC). The helical γ-tubulin rings are formed in the microtubule organizing centers which, in animal cells, are the centrosomes. Thirteen γ-tubulin subunits are shown in a hypothetical array formed together with a base of other molecules of unknown structure. The microtubule grows by addition of successive layers of α/β-tubulin dimers, each a split ring of 13 dimers with the β-tubulin subunits toward the base, the **negative end**, and the α-tubulin subunits toward the growing **positive end**. After Zheng *et al.*[317]

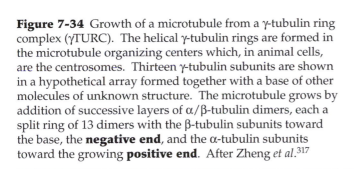

References

1. Williams, D. H., Searle, M. S., Mackay, J. P., Gerhard, U., and Maplestone, R. A. (1993) *Proc. Natl. Acad. Sci. U.S.A.* **90**, 1172–1178
2. Weber, G. (1992) *Protein Interactions*, Chapman and Hall, London
3. Scatchard, G. (1949) *Ann. N.Y. Acad. Sci.* **51**, 660–672
4. Zierler, K. (1989) *Trends Biochem. Sci.* **14**, 314–317
5. Munson, P. J., Rodbard, D., and Klotz, I. M. (1983) *Science* **220**, 979–981
6. Deranleau, D. A. (1969) *J. Am. Chem. Soc.* **91**, 4044–4049
7. Dowd, J. E., and Riggs, D. S. (1965) *J. Biol. Chem.* **240**, 863–869
8. Feldman, H. A. (1983) *J. Biol. Chem.* **258**, 12865–12867
8a. Dunford, H. B. (1984) *J. Chem. Educ.* **61**, 129–132
8b. Klotz, I. M. (1997) *Ligand-Receptor Energetics*, Wiley, New York
9. Conners, K. A. (1987) *Binding Constants*, Wiley, New York
10. Motulsky, H. J., and Ransnas, L. A. (1987) *FASEB J.* **1**, 365–374
11. Schejter, A., and Margoliash, E. (1985) *Trends Biochem. Sci.* **10**, 490–492
12. Steinhardt, J., and Reynolds, J. A. (1969) *Multiple Equilibria in Proteins*, Academic Press, New York
13. Koshland, D. E., Jr. (1970) in *The Enzymes*, 3rd ed., Vol. 1 (Boyer, P. D., ed), pp. 341–396, Academic Press, New York
14. Tanford, C. (1961) *Physical Chemistry of Macromolecules*, Wiley, New York (pp. 533–534)
15. Edsall, J. T., and Wyman, J. (1958) *Biophysical Chemistry*, Academic Press, New York
16. Westheimer, F. H., and Shookhoff, M. W. (1939) *J. Am. Chem. Soc.* **61**, 555–560
17. Honig, B., and Nicholls, A. (1995) *Science* **268**, 1144–1149
18. Rajasekaran, E., Jayaram, B., and Honig, B. (1994) *J. Am. Chem. Soc.* **116**, 8238–8240
19. Tanford, C. (1961) *Physical Chemistry of Macromolecules*, Wiley, New York (Chapter 8)
20. Maier, G. D., and Metzler, D. E. (1957) *J. Am. Chem. Soc.* **79**, 4386–4391
21. Metzler, D. E. (1960) in *The Enzymes*, 2nd ed., Vol. 2 (Boyer, P. D., Lardy, H., and Myrback, K., eds), pp. 295–337, Academic Press, New York
22. Applequist, J. (1977) *J. Chem. Educ.* **54**, 417–419
23. Hill, T. (1962) *Introduction to Statistical Thermodynamics*, Addison–Wesley, Reading, Massachusetts (pp. 235–241)

24. Weiss, J. N. (1997) *FASEB J.* **11**, 835–841
25. Wang, X.-G., and Engel, P. C. (1995) *Biochemistry* **34**, 11417–11422
26. Damle, V. N. (1972) *Biopolymers* **11**, 1789–816
27. Watson, J. D. (1976) *Molecular Biology of the Gene*, 3rd ed., Benjamin, Menlo Park, California (p. 98)
28. Bernal, J. D. (1967) *J. Mol. Biol.* **24**, 379–390
29. Blundell, T. L., and Johnson, L. N. (1976) *Protein Crystallography*, Academic Press, New York
30. Monod, J., Wyman, J., and Changeux, J. D. (1965) *J. Mol. Biol.* **12**, 88–118
30a. Dill, K. A. (1997) *J. Biol. Chem.* **272**, 701–704
30b. Boresch, S., and Karplus, M. (1995) *J. Mol. Biol.* **254**, 801–807
30c. Tamura, A., and Privalov, P. L. (1997) *J. Mol. Biol.* **273**, 1048–1060
31. Zhang, R.-G., Westbrook, M. L., Westbrook, E. M., Scott, D. L., Otwinowski, Z., Maulik, P. R., Reed, R. A., and Shipley, G. G. (1995) *J. Mol. Biol.* **251**, 550–562
31a. Marvin, D. A. (1998) *Current Opinion in Structural Biology* **8**, 150–158
32. Marvin, D. A., Hale, R. D., Nave, C., and Citterich, M. H. (1994) *J. Mol. Biol.* **235**, 260–286
33. Glucksman, M. J., Bhattacharjee, S., and Makowski, L. (1992) *J. Mol. Biol.* **226**, 455–470
34. Day, L. A., Marzec, C. J., Reisberg, S. A., and Casadevall, A. (1988) *Ann. Rev. Biophys. Biophys. Chem.* **17**, 509–539
35. Williams, K. A., Glibowicka, M., Li, Z., Li, H., Khan, A. R., Chen, Y. M. Y., Wang, J., Marvin, D. A., and Deber, C. M. (1995) *J. Mol. Biol.* **252**, 6–14
36. Overman, S. A., Tsuboi, M., and Thomas, G. J., Jr. (1996) *J. Mol. Biol.* **259**, 331–336
37. Makowski, L. (1992) *J. Mol. Biol.* **228**, 885–892
38. Rakonjac, J., and Model, P. (1998) *J. Mol. Biol.* **282**, 25–41
38a. Welsh, L. C., Marvin, D. A., and Perham, R. N. (1998) *J. Mol. Biol.* **284**, 1265–1271
39. Liu, D. J., and Day, L. A. (1994) *Science* **265**, 671–674
40. Bhyravbhatla, B., Watowich, S. J., and Caspar, D. L. D. (1998) *Biophys. J.* **74**, 604–615
41. Fraenkel-Conrat, H. (1964) *Sci. Am.* **211**(Oct), 47–54
42. Butler, P. J. G., and Klug, A. (1978) *Sci. Am.* **239**(Nov), 62–69
43. Bloomer, A. C., Champness, J. N., Bricogne, G., Staden, R., and Klug, A. (1978) *Nature (London)* **276**, 362–368
44. Holmes, K. C. (1980) *Trends Biochem. Sci.* **5**, 4–7

44a. Eisenstein, M., Shariv, I., Koren, G., Friesem, A. A., and Katchalski-Katzir, E. (1997) *J. Mol. Biol.* **266**, 135–143
45. Raghavendra, K., Kelly, J. A., Khairallah, L., and Schuster, T. M. (1988) *Biochemistry* **27**, 7583–7588
46. Wang, H., and Stubbs, G. (1994) *J. Mol. Biol.* **239**, 371–384
47. Namba, K., Casper, D. L. D., and Stubbs, G. J. (1985) *Science* **227**, 773–776
48. Eisenstein, B. I. (1987) Escherichia coli *and* Salmonella typhimurium, Am. Soc. Microbiology, Washington, D.C., FC Niedhardt, ed. (pp. 84–90)
49. Bullitt, E., and Makowski, L. (1995) *Nature (London)* **373**, 164–167
50. Bullitt, E., Jones, C. H., Striker, R., Soto, G., Jacob-Dubuisson, F., Pinkner, J., Wick, M. J., Makowski, L., and Hultgren, S. J. (1996) *Proc. Natl. Acad. Sci. U.S.A.* **93**, 12890–12895
51. Bullitt, E., and Makowski, L. (1998) *Biophys. J.* **74**, 623–632
52. Gong, M., and Makowski, L. (1992) *J. Mol. Biol.* **228**, 735–742
53. Kuehn, M. J., Ogg, D. J., Kihlberg, J., Slonim, L. N., Flemmer, K., Bergfors, T., and Hultgren, S. J. (1993) *Science* **262**, 1234–1241
53a. Sauer, F. G., Fütterer, K., Pinkner, J. S., Dodson, K. W., Hultgren, S. J., and Waksman, G. (1999) *Science* **285**, 1058–1061
53b. Choudhury, D., Thompson, A., Stojanoff, V., Langermann, S., Pinkner, J., Hultgren, S. J., and Knight, S. D. (1999) *Science* **285**, 1061–1065
53c. Skerker, J. M., and Shapiro, L. (2000) *EMBO J.* **19**, 3223–3234
53d. Hazes, B., Sastry, P. A., Hayakawa, K., Read, R. J., and Irvin, R. T. (2000) *J. Mol. Biol.* **299**, 1005–1017
53e. Collinson, S. K., Parker, J. M. R., Hodges, R. S., and Kay, W. W. (1999) *J. Mol. Biol.* **290**, 741–756
53f. Murphy, F. V., IV, Sweet, R. M., and Churchill, M. E. A. (1999) *EMBO J.* **18**, 6610–6618
54. Hanson, M. S., and Brinton, C. C., Jr. (1988) *Nature (London)* **332**, 265–268
55. St. Geme, J. W., III, Pinkner, J. S., Krasan, G. P., Heuser, J., and Bullitt, E. (1996) *Proc. Natl. Acad. Sci. U.S.A.* **93**, 11913–11918
56. Parge, H. E., Bernstein, S. L., Deal, C. D., McRee, D. E., Christensen, D., Capozza, M. A., Kays, B. W., Fieser, T. M., Draper, D., So, M., Getzoff, E. D., and Tainer, J. A. (1990) *J. Biol. Chem.* **265**, 2278–2285

References

57. Parge, H. E., Forest, K. T., Hickey, M. J., Christensen, D. A., Getzoff, E. D., and Tainer, J. A. (1995) *Nature (London)* **378**, 32–38

58. Folkhard, W., Leonard, K. R., Malsey, S., Marvin, D. A., Dubochet, J., Engel, A., Achtman, M., and Helmuth, R. (1979) *J. Mol. Biol.* **130**, 145–160

59. Paiva, W. D., Grossman, T., and Silverman, P. M. (1992) *J. Biol. Chem.* **267**, 26191–26197

60. Rubenstein, P. A. (1990) *BioEssays* **12**, 309–315

61. Venkatesh, B., Tay, B. H., Elgar, G., and Brenner, S. (1996) *J. Mol. Biol.* **259**, 655–665

62. Holmes, K. C., Popp, D., Gebhard, W., and Kabsch, W. (1990) *Nature (London)* **347**, 44–49

63. Lorenz, M., Poole, K. J. V., Popp, D., Rosenbaum, G., and Holmes, K. C. (1995) *J. Mol. Biol.* **246**, 108–119

64. Valentine, R. C. (1969) in *Symmetry and Function of the Biological Systems at the Macromolecular Level* (Engstrom, A., and Strandberg, B., eds), p. 165, Wiley, New York (11th Nobel Symp.)

65. Klotz, I. M., Darnall, D. W., and Langerman, N. R. (1975) in *The Proteins*, 3rd ed., Vol. 1 (Neurath, H., and Hill, R. L., eds), p. 293, Academic Press, New York

66. Momany, C., Ernst, S., Ghosh, R., Chang, N.-L., and Hackert, M. L. (1995) *J. Mol. Biol.* **252**, 643–655

67. Almassy, R. J., Janson, C. A., Hamlin, R., Xuong, N.-H., and Eisenberg, D. (1986) *Nature (London)* **323**, 304–309

68. Harrison, S. C. (1984) *Trends Biochem. Sci.* **9**, 345–351

69. Rossmann, M. G., and Johnson, J. E. (1989) *Ann. Rev. Biochem.* **58**, 533–573

70. Johnson, J. E., and Speir, J. A. (1997) *J. Mol. Biol.* **269**, 665–675

71. Watson, J. D. (1976) *Molecular Biology of the Gene*, 3rd ed., Benjamin, Menlo Park, California (p. 107)

72. Unge, T., Liljas, L., Strandberg, B., Vaara, I., Kannan, K. K., Fridborg, K., Nordman, C. E., and Lentz, P. J., Jr. (1980) *Nature (London)* **285**, 373–377

73. Küpper, H., Keller, W., Kurz, C., Forss, S., Schaller, H., Franze, R., Stohmaier, K., Marquardt, O., Zaslavsky, V. G., and Hofschneider, P. H. (1981) *Nature (London)* **289**, 555–559

74. Trikha, J., Theil, E. C., and Allewell, N. M. (1995) *J. Mol. Biol.* **248**, 949–967

75. Larson, S. B., Day, J., Greenwood, A., and McPherson, A. (1998) *J. Mol. Biol.* **277**, 37–59

76. Hogle, J. M., Chow, M., and Filman, D. J. (1985) *Science* **229**, 1358–1365

77. Hogle, J. M., Chow, M., and Filman, D. J. (1987) *Sci. Am.* **256** (Mar), 42–49

78. Olson, N. H., Kolatkar, P. R., Oliveira, M. A., Cheng, R. H., Greve, J. M., McClelland, A., Baker, T. S., and Rossmann, M. G. (1993) *Proc. Natl. Acad. Sci. U.S.A.* **90**, 507–511

79. Hadfield, A. T., Oliveira, M. A., Kim, K. H., Minor, I., Kremer, M. J., Heinz, B. A., Shepard, D., Pevear, D. C., Rueckert, R. R., and Rossmann, M. G. (1995) *J. Mol. Biol.* **253**, 61–73

80. Wu, H., and Rossmann, M. G. (1993) *J. Mol. Biol.* **233**, 231–244

81. Luo, M., Vriend, G., Kamer, G., Minor, I., Arnold, E., Rossmann, M. G., Boege, U., Scraba, D. G., Duke, G. M., and Palmenberg, A. C. (1987) *Science* **235**, 182–191

82. Tikhonenko, A. S. (1970) *Ultrastructure of Bacterial Viruses*, Plenum, New York

83. Birktoft, J. J., Rhodes, G., and Banaszak, L. J. (1989) *Biochemistry* **28**, 6065–6081

84. Skarzynski, T., Moody, P. C. E., and Wonacott, A. J. (1987) *J. Mol. Biol.* **193**, 171–187

85. Zimmerle, C. T., and Alter, G. M. (1993) *Biochemistry* **32**, 12743–12748

86. Hodgkin, D. C. (1975) *Nature (London)* **255**, 103

87. Whittingham, J. L., Chaudhuri, S., Dodson, E. J., Moody, P. C. E., and Dodson, G. G. (1995) *Biochemistry* **34**, 15553–15563

88. Chang, X., Jorgensen, A. M. M., Bardrum, P., and Led, J. J. (1997) *Biochemistry* **36**, 9409–9422

89. Golmohammadi, R., Valegård, K., Fridborg, K., and Liljas, L. (1993) *J. Mol. Biol.* **234**, 620–639

90. Liljas, L., Fridborg, K., Valegård, K., Bundule, M., and Pumpens, P. (1994) *J. Mol. Biol.* **244**, 279–290

91. Tars, K., Bundule, M., Fridborg, K., and Lijas, L. (1997) *J. Mol. Biol.* **271**, 759–773

92. Abad-Zapatero, C., Abdel-Meguid, S. S., Johnson, J. E., Leslie, A. G. W., Rayment, I., Rossmann, M. G., Suck, D., and Tsukihara, T. (1980) *Nature (London)* **286**, 33–39

93. Athappilly, F. K., Murali, R., Rux, J. J., Cai, Z., and Burnett, R. M. (1994) *J. Mol. Biol.* **242**, 430–455

94. Stewart, P. L., Chiu, C. Y., Huang, S., Muir, T., Zhao, Y., Chait, B., Mathias, P., and Nemerow, G. R. (1997) *EMBO J.* **16**, 1189–1198

95. Caspar, D. L. D., and Klug, A. (1962) *Cold Spring Harbor Symposia on Quantitative Biology* **27**, 1–24

96. Johnson, J. E. (1996) *Proc. Natl. Acad. Sci. U.S.A.* **93**, 27–33

97. Steven, A. C., Trus, B. L., Booy, F. P., Cheng, N., Zlotnick, A., Caston, J. R., and Conway, J. F. (1997) *FASEB J.* **11**, 733–742

98. Rayment, I., Baker, T. S., Caspar, D. L. D., and Murakami, W. T. (1982) *Nature (London)* **295**, 110–115

99. Liddington, R. C., Yan, Y., Moulai, J., Sahli, R., Benjamin, T. L., and Harrison, S. C. (1991) *Nature (London)* **354**, 278–284

100. Champness, J. N., Bloomer, A. C., Bricogne, G., Butler, P. J. G., and Klug, A. (1976) *Nature (London)* **259**, 20–24

101. Fukuda, M., and Okada, Y. (1987) *Proc. Natl. Acad. Sci. U.S.A.* **84**, 4035–4038

102. McDonald, R. C., Engelman, D. M., and Steita, T. A. (1979) *J. Biol. Chem.* **254**, 2942–2943

103. Galloway, J. (1984) *Trends Biochem. Sci.* **9**, 233–238

104. Kantrowitz, E. R., and Lipscomb, W. N. (1990) *Trends Biochem. Sci.* **15**, 53–59

105. Zhou, B.-B., and Schachman, H. K. (1993) *Protein Sci.* **2**, 103–112

106. Xi, X. G., De Staercke, C., Van Vliet, F., Triniolles, F., Jacobs, A., Stas, P. P., Ladjimi, M. M., Simon, V., Cunin, R., and Hervé, G. (1994) *J. Mol. Biol.* **242**, 139–149

107. Fetler, L., Vachette, P., Hervé, G., and Ladjimi, M. M. (1995) *Biochemistry* **34**, 15654–15660

108. Honzatko, R. B., Crawford, J. L., Monaco, H. L., Ladner, J. E., Edwards, B. F. D., Evans, D. R., Warren, S. G., Wiley, D. C., Ladner, R. C., and Lipscomb, W. N. (1982) *J. Mol. Biol.* **160**, 219–263

109. Krause, K. L., Volz, K. W., and Lipscomb, W. N. (1985) *Proc. Natl. Acad. Sci. U.S.A.* **82**, 1643–1647

110. Thomas, A., Field, M. J., and Perahia, D. (1996) *J. Mol. Biol.* **261**, 490–506

110a. Brunori, M. (1999) *Trends Biochem. Sci.* **24**, 158–161

110b. Di Cera, E., ed. (1998) *Advances in Protein Chemistry, Linkage Thermodynamics of Macromolecular Interactions*, Vol. 51, Academic Press, San Diego, California

111. Bloom, C. R., Choi, W. E., Brzovic, P. S., Ha, J. J., Huang, S.-T., Kaarsholm, N. C., and Dunn, M. F. (1995) *J. Mol. Biol.* **245**, 324–330

112. Hammes, G. G., and Wu, C.-W. (1971) *Science* **172**, 1205–1211

113. Koshland, D. E., Jr., Némethy, G., and Filmer, D. (1966) *Biochemistry* **5**, 365–385

114. Cornish-Bowden, A. J., and Koshland, D. E., Jr. (1970) *J. Biol. Chem.* **245**, 6241–6250

115. Cornish-Bowden, A. J., and Koshland, D. E., Jr. (1971) *J. Biol. Chem.* **246**, 3092–3102

116. Darnall, D. W., and Klotz, I. M. (1972) *Arch. Biochem. Biophys.* **149**, 1–14

117. Doci, Y., Sugita, Y., and Yoneyama, Y. (1973) *J. Biol. Chem.* **248**, 2354–2363

118. Perutz, M. F., Muirhead, H., Cox, J. M., and Goaman, L. C. G. (1968) *Nature (London)* **219**, 131–139

119. Dickerson, R. E., and Geis, I. (1983) *Hemoglobin: Structure, Function, Evolution and Pathology*, Benjamin / Cummings Publ., Redwood City, California

120. Antonini, E., and Brunori, M. (1971) *Hemoglobin and Myoglobin in Their Reactions with Ligands*, North-Hollard, Publ., Amsterdam

121. Kendrew, J. C. (1961) *Sci. Am.* **205** (Dec), 96–110

122. Fermi, G., Perutz, M. F., Shaanan, B., and Fourme, R. (1984) *J. Mol. Biol.* **175**, 159–174

123. Benesch, R., and Benesch, R. E. (1974) *Science* **185**, 905–908

124. Borgstahl, G. E. O., Rogers, P. H., and Arnone, A. (1994) *J. Mol. Biol.* **236**, 831–843

125. Imai, K. (1979) *J. Mol. Biol.* **133**, 233–247

126. Amiconi, G., Antonini, E., Brunori, M., Wyman, J., and Zolla, L. (1981) *J. Mol. Biol.* **152**, 111–129

127. Flanagan, M. A., Ackers, G. K., Matthew, J. B., Hanania, G. I. H., and Gurd, F. R. N. (1981) *Biochemistry* **20**, 7439–7449

127a. Shibayama, N. (1999) *J. Mol. Biol.* **285**, 1383–1388

128. Waxman, L. (1971) *J. Biol. Chem.* **246**, 7318–7327

129. Doyle, M. L., Di Cera, E., and Gill, S. J. (1988) *Biochemistry* **27**, 820–824

130. Edelstein, S. J. (1996) *J. Mol. Biol.* **257**, 737–744

131. Imai, K. (1983) *J. Mol. Biol.* **167**, 741–749

131a. Gibson, Q. H. (1999) *Biochemistry* **38**, 5191–5199

132. Paoli, M., Liddington, R., Tame, J., Wilkinson, A., and Dodson, G. (1996) *J. Mol. Biol.* **256**, 775–792

133. Henry, E. R., Jones, C. M., Hofrichter, J., and Eaton, W. A. (1997) *Biochemistry* **36**, 6511–6528

134. Silva, M. M., Rogers, P. H., and Arnone, A. (1992) *J. Biol. Chem.* **267**, 17248–17256

135. Jayaraman, V., and Spiro, T. G. (1995) *Biochemistry* **34**, 4511–4515

136. Schumacher, M. A., Zheleznova, E. E., Poundstone, K. S., Kluger, R., Jones, R. T., and Brennan, R. G. (1997) *Proc. Natl. Acad. Sci. U.S.A.* **94**, 7841–7844

136a. Tame, J. R. H. (1999) *Trends Biochem. Sci.* **24**, 372–377

137. Manning, L. R., Jenkins, W. T., Hess, J. R., Vandegriff, K., Winslow, R. M., and Manning, J. M. (1996) *Protein Sci.* **5**, 775–781

138. Holt, J. M., and Ackers, G. K. (1995) *FASEB J.* **9**, 210–218

139. Perutz, M. F., Shih, D. T.-b, and Williamson, D. (1994) *J. Mol. Biol.* **239**, 555–560

140. Perutz, M. F., Muirhead, H., Cox, J. M., Goaman, L. C. G., Mathews, F. S., McGandy, L. E., and Webb, L. E. (1968) *Nature (London)* **219**, 29–32

141. Perutz, M. (1990) *Mechanisms of Cooperativity and Allosteric Regulation in Proteins*, Cambridge Univ. Press, London

142. O'Donnell, S., Mandaro, R., Schuster, T. M., and Arnone, A. (1979) *J. Biol. Chem.* **254**, 12204–12208

143. Perutz, M. F. (1970) *Nature (London)* **228**, 726–739

144. Dickerson, R. E. (1972) *Ann. Rev. Biochem.* **41**, 815–842

References

145. Ishimori, K., Hashimoto, M., Imai, K., Fushitani, K., Miyazaki, G., Morimoto, H., Wada, Y., and Morishima, I. (1994) *Biochemistry* **33**, 2546–2553

146. Chao, H., Sönnichsen, F. D., DeLuca, C. I., Sykes, B. D., and Davies, P. L. (1994) *Protein Sci.* **3**, 1760–1769

147. Kim, H.-W., Shen, T.-J., Ho, N. T., Zou, M., Tam, M. F., and Ho, C. (1996) *Biochemistry* **35**, 6620–6627

148. Kavanaugh, J. S., Weydert, J. A., Rogers, P. H., and Arnone, A. (1998) *Biochemistry* **37**, 4358–4373

149. Kiger, L., Klinger, A. L., Kwiatkowski, L. D., De Young, A., Doyle, M. L., Holt, J. M., Noble, R. W., and Ackers, G. K. (1998) *Biochemistry* **37**, 4336–4345

150. Hoard, J. L., and Scheidt, W. R. (1973) *Proc. Natl. Acad. Sci. U.S.A.* **70**, 3919–3922

151. Fermi, G., Perutz, M. F., and Shulman, R. G. (1987) *Proc. Natl. Acad. Sci. U.S.A.* **84**, 6167–6168

152. De Baere, I., Perutz, M. F., Kiger, L., Marden, M. C., and Poyart, C. (1994) *Proc. Natl. Acad. Sci. U.S.A.* **91**, 1594–1597

153. Bohr, C., Hasselbalch, K., and Krogh, A. (1904) *Skand. Arch. Physiol.* **16**, 402–

154. Wyman, J., Jr. (1948) *Adv. Prot. Chem.* **4**, 407–531

155. Benesch, R. E., and Benesch, R. (1974) *Adv. Prot. Chem.* **28**, 211–237

156. Fang, T.-Y., Zou, M., Simplaceanu, V., Ho, N. T., and Ho, C. (1999) *Biochemistry* **38**, 13423–13432

157. Fronticelli, C., Pechik, I., Brinigar, W. S., Kowalczyk, J., and Gilliland, G. L. (1994) *J. Biol. Chem.* **269**, 23965–23969

158. Perrella, M., Ripamonti, M., and Caccia, S. (1998) *Biochemistry* **37**, 2017–2028

159. Arnone, A. (1972) *Nature (London)* **237**, 146–149

160. Benesch, R., and Benesch, R. E. (1969) *Nature (London)* **221**, 618–622

161. Brewer, G. J., and Eaton, J. W. (1971) *Science* **171**, 1205–1211

162. Perutz, M. F., Bauer, C., Gros, G., Leclercq, F., Vandecasserie, C., Schnek, A. G., Braunitzer, G., Friday, A. E., and Joysey, K. A. (1981) *Nature (London)* **291**, 682–684

163. Komiyama, N. H., Miyazaki, G., Tame, J., and Nagai, K. (1995) *Nature (London)* **373**, 244–246

164. Fantl, W. J., Di Donato, A., Manning, J. M., Rogers, P. H., and Arnone, A. (1987) *J. Biol. Chem.* **87**, 12700–12713

165. Sage, J. T., and Jee, W. (1997) *J. Mol. Biol.* **274**, 21–26

166. Teng, T.-Y., Srajer, V., and Moffat, K. (1997) *Biochemistry* **36**, 12087–12100

166a. Ostermann, A., Waschipky, R., Parak, F. G., and Nienhaus, G. U. (2000) *Nature (London)* **404**, 205–208

167. Chance, M. R., Miller, L. M., Fischetti, R. F., Scheuring, E., Huang, W.-X., Sclavi, B., Hai, Y., and Sullivan, M. (1996) *Biochemistry* **35**, 9014–9023

168. Ösapay, K., Theriault, Y., Wright, P. E., and Case, D. A. (1994) *J. Mol. Biol.* **244**, 183–197

169. Martin, K. D., and Parkhurst, L. J. (1990) *Biochemistry* **29**, 5718–5726

170. Tetreau, C., Lavalette, D., Momenteau, M., Fischer, J., and Weiss, R. (1994) *J. Am. Chem. Soc.* **116**, 11840–11848

171. Huang, Y., and Ackers, G. K. (1995) *Biochemistry* **34**, 6316–6327

172. Paoli, M., Dodson, G., Liddington, R. C., and Wilkinson, A. J. (1997) *J. Mol. Biol.* **271**, 161–167

173. Daugherty, M. A., Shea, M. A., and Ackers, G. K. (1994) *Biochemistry* **33**, 10345–10357

174. Stamler, J. S., Jia, L., Eu, J. P., McMahon, T. J., Demchenko, I. T., Bonaventura, J., Gernert, K., and Piantadosi, C. A. (1997) *Science* **276**, 2034–2037

175. Upmacis, R. K., Hajjar, D. P., Chait, B. T., and Mirza, U. A. (1997) *J. Am. Chem. Soc.* **119**, 10424–10429

176. Miller, L. M., Pedraza, A. J., and Chance, M. R. (1997) *Biochemistry* **36**, 12199–12207

177. Gow, A. J., and Stamler, J. S. (1998) *Nature (London)* **391**, 169–173

178. Lutter, L. C., Halvorson, H. R., and Calladine, C. R. (1996) *J. Mol. Biol.* **261**, 620–633

179. Jia, L., Bonaventura, C., Bonaventura, J., and Stamler, J. S. (1996) *Nature (London)* **380**, 221–226

180. Rizzi, M., Wittenberg, J. B., Coda, A., Fasano, M., Ascenzi, P., and Bolognesi, M. (1994) *J. Mol. Biol.* **244**, 86–99

181. Rizzi, M., Wittenberg, J. B., Coda, A., Ascenzi, P., and Bolognesi, M. (1996) *J. Mol. Biol.* **258**, 1–5

182. Nguyen, B. D., Zhao, X., Vyas, K., La Mar, G. N., Lile, R. A., Brucker, E. A., Phillips, G. N., Jr., Olson, J. S., and Wittenberg, J. B. (1998) *J. Biol. Chem.* **273**, 9517–9526

183. Weatherall, D. J., Clegg, J. B., Higgs, D. R., and Wood, W. G. (1995) in *The Metabolic and Molecular Bases of Inherited Disease*, 7th ed., Vol. 1 (Scriver, C. R., Beaudet, A. L., Sly, W. S., and Valle, D., eds), pp. 3417–3484, McGraw-Hill, New York

184. Bank, A., Mears, J. G., and Ramirez, F. (1980) *Science* **207**, 486–493

185. Hirsch, R. E., Juszczak, L. J., Fataliev, N. A., Friedman, J. M., and Nagel, R. L. (1999) *J. Biol. Chem.* **274**, 13777–13782

186. Arnone, A., Thillet, J., and Rosa, J. (1981) *J. Biol. Chem.* **256**, 8545–8552

187. Lehmann, H., and Huntsman, R. G. (1966) *Man's Haemoglobins*, North-Holland Publ. Co., Amsterdam (p. 246)

188. Olson, J. S., Gibson, Q. H., Nagel, R. L., and Hamilton, H. B. (1972) *J. Biol. Chem.* **247**, 7485–7493

189. Poyart, C., Bursaux, E., Arnone, A., Bonaventura, J., and Bonaventura, C. (1980) *J. Biol. Chem.* **255**, 9465–9473

189a. Burmester, T., Weich, B., Reinhardt, S., and Hankeln, T. (2000) *Nature* **407**, 520–523

190. Tomita, S. (1981) *J. Biol. Chem.* **256**, 9495–9500

191. Giardina, B., Scatena, R., Clementi, M. E., Cerroni, L., Nuutinen, M., Brix, O., Sletten, S. N., Castagnola, M., and Condò, S. G. (1993) *J. Mol. Biol.* **229**, 512–516

192. Clementi, M. E., Scatena, R., Mordente, A., Condò, S. G., Castagnola, M., and Giardina, B. (1996) *J. Mol. Biol.* **255**, 229–234

192a. Burr, A. H. J., Hunt, P., Wagar, D. R., Dewilde, S., Blaxter, M. L., Vanfleteren, J. R., and Moens, L. (2000) *J. Biol. Chem.* **275**, 4810–4815

193. Lee, H. C., Wittenberg, J. B., and Peisach, J. (1993) *Biochemistry* **32**, 11500–11506

194. Harutyunyan, E. H., Safonova, T. N., Kuranova, I. P., Popov, A. N., Teplyakov, A. V., Obmolova, G. V., Rusakov, A. A., Vainshtein, B. K., Dodson, G. G., Wilson, J. C., and Perutz, M. F. (1995) *J. Mol. Biol.* **251**, 104–115

195. Hargrove, M. S., Barry, J. K., Brucker, E. A., Berry, M. B., Phillips, G. N., Jr., Olson, J. S., Arredondo-Peter, R., Dean, J. M., Klucas, R. V., and Sarath, G. (1997) *J. Mol. Biol.* **266**, 1032–1042

196. Trevaskis, B., Watts, R. A., Andersson, C. R., Llewellyn, D. J., Hargrove, M. S., Olson, J. S., Dennis, E. S., and Peacock, W. J. (1997) *Proc. Natl. Acad. Sci. U.S.A.* **94**, 12230–12234

196a. Hargrove, M. S., Brucker, E. A., Stec, B., Sarath, G., Arredondo-Peter, R., Klucas, R. V., Olson, J. S., and Phillips, G. N., Jr. (2000) *Structure* **8**, 1005–1014

197. Hardison, R. C. (1996) *Proc. Natl. Acad. Sci. U.S.A.* **93**, 5675–5679

198. Potts, M., Angeloni, S. V., Ebel, R. E., and Bassam, D. (1992) *Science* **256**, 1690–1692

199. Yeh, S.-R., Couture, M., Ouellet, Y., Guertin, M., and Rousseau, D. L. (2000) *J. Biol. Chem.* **275**, 1679–1684

200. Aronson, H.-E. G., Royer, W. E., Jr., and Hendrickson, W. A. (1994) *Protein Sci.* **3**, 1706–1711

201. Honzatko, R. B., Hendrickson, W. A., and Love, W. E. (1985) *J. Mol. Biol.* **184**, 147–164

201a. Qiu, Y., Maillett, D. H., Knapp, J., Olson, J. S., and Riggs, A. F. (2000) *J. Biol. Chem.* **275**, 13517–13528

202. Condon, P. J., and Royer, W. E. J. (1994) *J. Biol. Chem.* **269**, 25259–25267

203. Royer, W. E., Jr., Heard, K. S., Harrington, D. J., and Chiancone, E. (1995) *J. Mol. Biol.* **253**, 168–186

204. Mozzarelli, A., Bettati, S., Rivetti, C., Rossi, G. L., Colotti, G., and Chiancone, E. (1996) *J. Biol. Chem.* **271**, 3627–3632

205. Yang, J., Kloek, A. P., Goldberg, D. E., and Mathews, F. S. (1995) *Proc. Natl. Acad. Sci. U.S.A.* **92**, 4224–4228

206. Huang, S., Huang, J., Kloek, A. P., Goldberg, D. E., and Friedman, J. M. (1996) *J. Biol. Chem.* **271**, 958–962

207. Lamy, J., Kuchumov, A., Taveau, J.-C., Vinogradov, S. N., and Lamy, J. N. (2000) *J. Mol. Biol.* **298**, 633–647

208. de Haas, F., Taveau, J.-C., Boisset, N., Lambert, O., Vinogradov, S. N., and Lamy, J. N. (1996) *J. Mol. Biol.* **255**, 140–153

209. de Haas, F., Zal, F., You, V., Lallier, F., Toulmond, A., and Lamy, J. N. (1996) *J. Mol. Biol.* **264**, 111–120

210. de Haas, F., Boisset, N., Taveau, J.-C., Lambert, O., Vinogradov, S. N., and Lamy, J. N. (1996) *Biophys. J.* **70**, 1973–1984

211. Zuckerkandl, E. (1965) *Sci. Am.* **212**(May), 110–118

212. Perutz, M. F. (1986) *Nature (London)* **322**, 405

213. Wood, W. B., and Edgar, R. S. (1967) *Sci. Am.* **217**(Jul), 60–74

214. Kikuchi, Y., and King, J. (1975) *J. Mol. Biol.* **99**, 645–672

215. Casjens, S., and King, J. (1975) *Ann. Rev. Biochem.* **44**, 555–611

216. Pollard, T. D., and Craig, S. W. (1982) *Trends Biochem. Sci.* **7**, 55–58

217. Ito, K., Date, T., and Wickner, W. (1980) *J. Biol. Chem.* **255**, 2123–2130

218. McDonnell, P. A., Shon, K., Kim, Y., and Opella, S. J. (1993) *J. Mol. Biol.* **233**, 447–463

219. Sanders, J. C., Haris, P. I., Chapman, D., Otto, C., and Hemminga, M. A. (1993) *Biochemistry* **32**, 12446–12454

220. Guy-Caffey, J. K., and Webster, R. E. (1993) *J. Biol. Chem.* **268**, 5496–5503

221. Duda, R. L., Martincic, K., and Hendrix, R. W. (1995) *J. Mol. Biol.* **247**, 636–647

222. Xie, Z., and Hendrix, R. W. (1995) *J. Mol. Biol.* **253**, 74–85

223. Marvik, C. J., Dokland, T., Nokling, R. H., Jacobsen, E., Larsen, T., and Lindqvist, B. H. (1995) *J. Mol. Biol.* **251**, 59–75

224. LeCuyer, K. A., Behlen, L. S., and Uhlenbeck, O. C. (1995) *Biochemistry* **34**, 10600–10606

225. McKenna, R., Ilag, L. L., and Rossmann, M. G. (1994) *J. Mol. Biol.* **237**, 517–543

226. Dokland, T., McKenna, R., Ilag, L. L., Bowman, B. R., Incardona, N. L., Fane, B. A., and Rossmann, M. G. (1997) *Nature (London)* **389**, 308–313

227. Thuman-Commike, P. A., Greene, B., Jakana, J., Prasad, B. V. V., King, J., Prevelige, P. E., Jr., and Chiu, W. (1996) *J. Mol. Biol.* **260**, 85–98

228. Parker, M. H., Stafford, W. F., III, and Prevelige, P. E., Jr. (1997) *J. Mol. Biol.* **268**, 655–665

References

229. Butcher, S. J., Dokland, T., Ojala, P. M., Bamford, D. H., and Fuller, S. D. (1997) *EMBO J.* **16**, 4477–4487

230. Olson, N. H., Baker, T. S., Willingmann, P., and Incardona, N. L. (1992) *J. Structural Biol.* **108**, 168–175

231. Hayashi, M., Aoyama, A., Richardson, D. L., Jr., and Hayashi, M. N. (1988) in *The Bacteriophages* (Calendar, R., ed), pp. 1–71, Plenum, New York

232. Butcher, S. J., Bamford, D. H., and Fuller, S. D. (1995) *EMBO J.* **14**, 6078–6086

233. Berger, B., Shor, P. W., Tucker-Kellogg, L., and King, J. (1994) *Proc. Natl. Acad. Sci. U.S.A.* **91**, 7732–7736

234. Steven, A. C., Greenstone, H. L., Booy, F. P., Black, L. W., and Ross, P. D. (1992) *J. Biol. Chem.* **228**, 870–884

235. van Driel, R. (1980) *J. Mol. Biol.* **138**, 27–42

236. King, J., and Mykolajewycz, N. (1973) *J. Mol. Biol.* **75**, 339–358

237. Feller, S. M., Ren, R., Hanafusa, H., and Baltimore, D. (1994) *Trends Biochem. Sci.* **19**, 453–458

238. Pawson, T. (1995) *Nature (London)* **373**, 573–580

239. Koch, C. A., Anderson, D., Moran, M. F., Ellis, C., and Pawson, T. (1991) *Science* **252**, 668–674

240. Maignan, S., Guilloteau, J.-P., Fromage, N., Arnoux, B., Becquart, J., and Ducruix, A. (1995) *Science* **268**, 291–293

241. Ullner, M., Selander, M., Persson, E., Stenflo, J., Drakenberg, T., and Teleman, O. (1992) *Biochemistry* **31**, 5974–5983

242. Meininger, D. P., Hunter, M. J., and Komives, E. A. (1995) *Protein Sci.* **4**, 1683–1695

243. Bersch, B., Hernandez, J.-F., Marion, D., and Arlaud, G. J. (1998) *Biochemistry* **37**, 1204–1214

244. Rand, M. D., Lindblom, A., Carlson, J., Villoutreix, B. O., and Stenflo, J. (1997) *Protein Sci.* **6**, 2059–2071

245. Müller, K., Gombert, F. O., Manning, U., Grossmüller, F., Graff, P., Zaegel, H., Zuber, J. F., Freuler, F., Tschopp, C., and Baumann, G. (1996) *J. Biol. Chem.* **271**, 16500–16505

246. Fry, M. J., Panayotou, G., Booker, G. W., and Waterfield, M. D. (1993) *Protein Sci.* **2**, 1785–1797

247. Malek, S. N., Yang, C. H., Earnshaw, W. C., Kozak, C. A., and Desiderio, S. (1996) *J. Biol. Chem.* **271**, 6952–6962

248. Tong, L., Warren, T. C., King, J., Betageri, R., Rose, J., and Jakes, S. (1996) *J. Mol. Biol.* **256**, 601–610

249. Xu, R. X., Word, J. M., Davis, D. G., Rink, M. J., Willard, D. H., Jr., and Gampe, R. T., Jr. (1995) *Biochemistry* **34**, 2107–2121

250. Mikol, V., Baumann, G., Zurini, M. G. M., and Hommel, U. (1995) *J. Mol. Biol.* **254**, 86–95

251. Pascal, S. M., Yamazaki, T., Singer, A. U., Kay, L. E., and Forman-Kay, J. D. (1995) *Biochemistry* **34**, 11353–11362

252. Zhou, M.-M., Meadows, R. P., Logan, T. M., Yoon, H. S., Wade, W. S., Ravichandran, K. S., Burakoff, S. J., and Fesik, S. W. (1995) *Proc. Natl. Acad. Sci. U.S.A.* **92**, 7784–7788

253. Metzler, W. J., Leiting, B., Pryor, K., Mueller, L., and Farmer, B. T., II. (1996) *Biochemistry* **35**, 6201–6211

254. Viguera, A. R., Arrondo, J. L. R., Musacchio, A., Saraste, M., and Serrano, L. (1994) *Biochemistry* **33**, 10925–10933

255. Liang, J., Chen, J. K., Schreiber, S. L., and Clardy, J. (1996) *J. Mol. Biol.* **257**, 632–643

256. Lim, W. A., Richards, F. M., and Fox, R. O. (1994) *Nature (London)* **372**, 375–379

257. Guruprasad, L., Dhanaraj, V., Timm, D., Blundell, T. L., Gout, I., and Waterfield, M. D. (1995) *J. Mol. Biol.* **248**, 856–866

258. van der Geer, P., and Pawson, T. (1995) *Trends Biochem. Sci.* **20**, 277–280

259. Zhou, M.-M., Ravichandran, K. S., Olejniczak, E. T., Petros, A. M., Meadows, R. P., Sattler, M., Harlan, J. E., Wade, W. S., Burakoff, S. J., and Fesik, S. W. (1995) *Nature (London)* **378**, 584–592

260. Rameh, L. E., Arvidsson, A.-k, Carraway, K. L., III, Couvillon, A. D., Rathbun, G., Crompton, A., VanRenterghem, B., Czech, M. P., Ravichandran, K. S., Burakoff, S. J., Wang, D.-S., Chen, C.-S., and Cantley, L. C. (1997) *J. Biol. Chem.* **272**, 22059–22066

261. Hemmings, B. A. (1997) *Science* **275**, 1899

262. Blomberg, N., Baraldi, E., Nilges, M., and Saraste, M. (1999) *Trends Biochem. Sci.* **24**, 441–445

263. Koshiba, S., Kigawa, T., Kim, J.-H., Shirouzu, M., Bowtell, D., and Yokoyama, S. (1997) *J. Mol. Biol.* **269**, 579–591

264. Saras, J., and Heldin, C.-H. (1996) *Trends Biochem. Sci.* **21**, 455–458

265. Cabral, J. H. M., Petosa, C., Sutcliffe, M. J., Raza, S., Byron, O., Poy, F., Marfatia, S. M., Chishti, A. H., and Liddington, R. C. (1996) *Nature (London)* **382**, 649–652

266. Ponting, C. P. (1997) *Protein Sci.* **6**, 464–468

267. Resnick, D., Pearson, A., and Krieger, M. (1994) *Trends Biochem. Sci.* **19**, 5–8

268. Chang, Y., Mochalkin, I., McCance, S. G., Cheng, B., Tulinsky, A., and Castellino, F. J. (1998) *Biochemistry* **37**, 3258–3271

269. de Vos, A. M., Ultsch, M. H., Kelley, R. F., Padmanabhan, K., Tulinsky, A., Westbrook, M. L., and Kossiakoff, A. A. (1992) *Biochemistry* **31**, 270–279

270. De Serrano, V. S., and Castellino, F. J. (1992) *Biochemistry* **31**, 11698–11706

271. Sun, M.-F., Zhao, M., and Gailani, D. (1999) *J. Biol. Chem.* **274**, 36373–36377

272. Macias, M. J., Hyvönen, M., Baraldi, E., Schultz, J., Sudol, M., Saraste, M., and Oschkinat, H. (1996) *Nature (London)* **382**, 646–649

273. Bogusky, M. J., Dobson, C. M., and Smith, R. A. G. (1989) *Biochemistry* **28**, 6728–6735

274. Hoffmann, W., and Hauser, F. (1993) *Trends Biochem. Sci.* **18**, 239–243

275. Lamb, J. R., Tugendreich, S., and Hieter, P. (1995) *Trends Biochem. Sci.* **20**, 257–259

276. Das, A. K., Cohen, P. T. W., and Barford, D. (1998) *EMBO J.* **17**, 1192–1199

277. Barlow, P. N., Luisi, B., Milner, A., Elliott, M., and Everett, R. (1994) *J. Mol. Biol.* **237**, 201–211

278. Lichtarge, O., Yamamoto, K. R., and Cohen, F. E. (1997) *J. Mol. Biol.* **274**, 325–337

279. Saurin, A. J., Borden, K. L. B., Boddy, M. N., and Freemont, P. S. (1996) *Trends Biochem. Sci.* **21**, 208–214

280. Pérez-Alvarado, G. C., Kosa, J. L., Louis, H. A., Beckerle, M. C., Winge, D. R., and Summers, M. F. (1996) *J. Mol. Biol.* **257**, 153–174

281. Porter, K. R., and Tucker, J. B. (1981) *Sci. Am.* **244**(Mar), 57–67

282. Weber, K., and Osborn, M. (1985) *Sci. Am.* **253**(Oct), 110–120

283. Amos, L. A., and Amos, B. W. (1991) *Molecules of the Cytoskeleton*, The Guilford Press, New York

284. Lazarides, E. (1980) *Nature (London)* **283**, 249–256

285. Fuchs, E., and Weber, K. (1994) *Ann. Rev. Biochem.* **63**, 345–382

286. Fuchs, E., and Cleveland, D. W. (1998) *Science* **279**, 514–519

287. Kaufman, S. H. (1989) *J. Biol. Chem.* **264**, 13946–13955

288. Ip, W., Hartzer, M. K., Pang, Y.-Y. S., and Robson, R. M. (1985) *J. Mol. Biol.* **183**, 365–375

289. Herrmann, H., Häner, M., Brettel, M., Müller, S. A., Goldie, K. N., Fedtke, B., Lustig, A., Franke, W. W., and Aebi, U. (1996) *J. Mol. Biol.* **264**, 933–953

289a. Herrmann, H., Strelkov, S. V., Feja, B., Rogers, K. R., Brettel, M., Lustig, A., Häner, M., Parry, D. A. D., Steinert, P. M., Burkhard, P., and Aebi, U. (2000) *J. Mol. Biol.* **298**, 817–832

289b. Heimburg, T., Schuenemann, J., Weber, K., and Geisler, N. (1996) *Biochemistry* **35**, 1375–1382

290. Geisler, N., Schünemann, J., and Weber, K. (1992) *Eur. J. Biochem.* **206**, 841–852

290a. Bellin, R. M., Sernett, S. W., Becker, B., Ip, W., Huiatt, T. W., and Robson, R. M. (1999) *J. Biol. Chem.* **274**, 29493–29499

291. Leterrier, J. F., Käs, J., Hartwig, J., Vegners, R., and Janmey, P. A. (1996) *J. Biol. Chem.* **271**, 15687–15694

292. Betts, J. C., Blackstock, W. P., Ward, M. A., and Anderton, B. H. (1997) *J. Biol. Chem.* **272**, 12922–12927

293. Carter, J., Gragerov, A., Konvicka, K., Elder, G., Weinstein, H., and Lazzarini, R. A. (1998) *J. Biol. Chem.* **273**, 5101–5108

294. Osborn, M., and Weber, K. (1986) *Trends Biochem. Sci.* **11**, 469–472

295. Gibson, M. A., Kumaratilake, J. S., and Cleary, E. G. (1989) *J. Biol. Chem.* **264**, 4590–4598

296. Steinert, P. M. (1991) *J. Structural Biol.* **107**, 157–174

297. Cohlberg, J. A. (1993) *Trends Biochem. Sci.* **18**, 360–362

298. Nakajima-Iijima, S., Hamada, H., Reddy, P., and Kakunaga, T. (1985) *Proc. Natl. Acad. Sci. U.S.A.* **82**, 6133–6137

299. Orlova, A., Chen, X., Rubenstein, P. A., and Egelman, E. H. (1997) *J. Mol. Biol.* **271**, 235–243

300. Fulton, A. B. (1984) *The Cytoskeleton: Cellular Architecture and Choreography*, Chapman and Hall, New York

301. Bershadsky, A. D., and Vasiliev, J. M. (1988) *Cytoskeleton*, Plenum, New York

302. DeRosier, D. J., Tilney, L. G., and Egelman, E. (1980) *Nature (London)* **287**, 291–296

303. Collier, N. C., and Wang, K. (1982) *J. Biol. Chem.* **257**, 6937–943

304. Moon, R. T., and Lazarides, E. (1983) *Proc. Natl. Acad. Sci. U.S.A.* **80**, 5494–5499

305. Meek, R. L., Lonsdale-Eccles, J. D., and Dale, B. A. (1983) *Biochemistry* **22**, 4867–4877

306. Steinert, P. M., and Idler, W. W. (1979) *Biochemistry* **18**, 5664–5669

307. Dustin, P. (1980) *Sci. Am.* **243**(Aug.), 67–76

308. Mandelkow, E., and Mandelkow, E.-M. (1994) *Current Opinion in Structural Biology* **4**, 171–179

309. Hyams, J. S., and Lloyd, C. W., eds. (1994) *Microtubules*, Wiley-Liss, New York

310. Sosa, H., and Milligan, R. A. (1996) *J. Mol. Biol.* **260**, 743–755

311. Margulis, L., To, L., and Chase, D. (1978) *Science* **200**, 1118–1124

312. Thormählen, M., Marx, A., Müller, S. A., Song, Y.-H., Mandelkow, E.-M., Aebi, U., and Mandelkow, E. (1998) *J. Mol. Biol.* **275**, 795–809

313. Weber, K., Pollack, R., and Bibring, T. (1975) *Proc. Natl. Acad. Sci. U.S.A.* **72**, 459–463

314. Wolf, S. G., Nogales, E., Kikkawa, M., Gratzinger, D., Hirokawa, N., and Downing, K. H. (1996) *J. Mol. Biol.* **262**, 485–501

315. Nogales, E., Wolf, S. G., and Downing, K. H. (1998) *Nature (London)* **391**, 199–203

316. Pennisi, E. (1998) *Science* **279**, 176–177

317. Zheng, Y., Wong, M. L., Alberts, B., and Mitchison, T. (1995) *Nature (London)* **378**, 578–583

318. Hackney, D. D. (1995) *Nature (London)* **376**, 215–216

References

319. Raff, E. C., Fackenthal, J. D., Hutchens, J. A., Hoyle, H. D., and Turner, F. R. (1997) *Science* **275**, 70–73

320. Angelastro, J. M., and Purich, D. L. (1992) *J. Biol. Chem.* **267**, 25685–25689

321. Terry, B. J., and Purich, D. L. (1980) *J. Biol. Chem.* **255**, 10532–10536

322. Rakic, P., Knyihar-Csillik, E., and Csillik, B. (1996) *Proc. Natl. Acad. Sci. U.S.A.* **93**, 9218–9222

323. Rodionov, V. I., and Borisy, G. G. (1997) *Science* **275**, 215–218

324. Hoenger, A., and Milligan, R. A. (1996) *J. Mol. Biol.* **263**, 114–119

325. Fan, J., Griffiths, A. D., Lockhart, A., Cross, R. A., and Amos, L. A. (1996) *J. Mol. Biol.* **259**, 325–330

326. Hirose, K., Fan, J., and Amos, L. A. (1995) *J. Mol. Biol.* **251**, 329–333

327. Knop, M., and Schiebel, E. (1997) *EMBO J.* **16**, 6985–6995

328. Moritz, M., Braunfeld, M. B., Sedat, J. W., Alberts, B., and Agard, D. A. (1995) *Nature (London)* **378**, 638–640

329. Berridge, M. J. (1990) *J. Biol. Chem.* **265**, 9583–9586

330. Arnold, C. S., Johnson, G. V. W., Cole, R. N., Dong, D. L.-Y., Lee, M., and Hart, G. W. (1996) *J. Biol. Chem.* **271**, 28741–28744

331. Bosc, C., Cronk, J. D., Pirollet, F., Watterson, D. M., Haiech, J., Job, D., and Margolis, R. L. (1996) *Proc. Natl. Acad. Sci. U.S.A.* **93**, 2125–2130

Study Questions

1. Rewrite Equations 6-75 through 6-77 in terms of dissociation constants. These may be labeled K_1, K_2, K_i, etc., as is conventional, but you may prefer to use K_{1d}, K_{2d}, K_{id}, etc., to avoid confusion.

2. A molecule has two identical binding sites for a ligand X. The Gibbs energy of interaction between ligands bound to the same molecule, ε, is defined as the change in Gibbs energy of binding of the ligand to the molecule that results from the prior binding of a ligand at the adjacent site. If the saturation fraction is Y, show from the equation for the binding isotherm that the following equation holds when $Y = 1/2$:

$$dY/d \ln[X] = \frac{1}{2(1 + e^{\varepsilon/RT})}$$

3. The hydrogen ion binding curve for succinate is shown is Fig. 7-4. From the curve estimate ε and the microscopic association constants.

4. A linear chain molecule has a very large number of identical binding sites for a ligand X. The Gibbs energy of interaction between ligands bound to adjacent sites is ε. Interactions between non-nearest neighbors are considered negligible. If the binding constant for a site adjacent to unoccupied sites is K_r, the binding isotherm is given by

$$Y = \frac{1}{2} + \frac{K[X]e^{-\varepsilon/RT} - 1}{2\{K[X]e^{-\varepsilon/RT} - 1)^2 + 4K[X]\}^{1/2}}$$

[Applequist, J. (1977) *J. Chem. Ed.* **54**, 417]. Show from the equation for the binding isotherm that the following equation holds at $Y = 1/2$;

$$dY/d \ln [X] = 1/4 e^{\varepsilon/2RT}$$

5. The binding of adenosine to polyribouridylic acid [poly(U)] has been studied by the method of equilibrium dialysis [Huang and Ts'o (1966) *J. Mol. Biol.* **16**, 523]. The table below gives the fraction of poly(U) sites occupied, Y at various molar concentrations of free adenosine [A] at 5°C. Assuming that the nearest-neighbor interaction model is correct, determine the intrinsic association constant for the binding of adenosine to poly U and the free energy of interaction of adjacent bound adenosines. Do the bound molecules attract or repel each other?

$[A] \times 10^3$	Y	$[A] \times 10^3$	Y
0.51	0	3.07	0.72
2.10	0	4.00	0.92
2.70	0.15	6.50	0.93
2.96	0.36	8.50	0.93
3.01	0.52	10.00	1.00

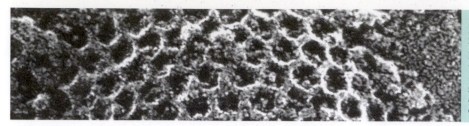

Portion of an endocytic vacuole forming in the plasma membrane of a cultured fibroblast. The view is from the inside of the cell and shows a large clathrin cage assembling to form a coated vesicle. The overall diameter of the vacuole is ~0.2 μm. Clathrin cages vary in size and in the number of faces but are typically ~0.1 μm in diameter (see Fig. 8-27 and associated references). Courtesy of Barbara Pearse.

Contents

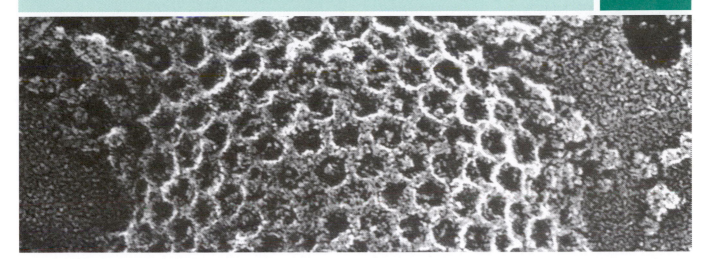

Lipids, Membranes, and Cell Coats

8

The boundary between a living cell and its surroundings is the incredibly thin (7–10 nm) plasma membrane. This vital partition, which controls the flow of materials into and out of a cell and which senses and controls the response of cells to hormones and other external signals, consists largely of **phospholipids** together with embedded proteins. The nonpolar chains of the phospholipids stick together to form a double molecular layer or **bilayer** which provides the basic structure of almost all biological membranes.

Phospholipids, together with other natural materials that have a high solubility in apolar solvents or are structurally related to compounds with such solubility properties, are classified as **lipids**. The most abundant lipids are the fats, compounds that are stored by animals and by many plants as an energy reserve (Fig. 8-1). Other lipids form the outer cuticle of plants and yet others serve as protective coatings on feathers and hair. Vitamins A, D, K, and E and ubiquinone are all lipids as are a variety of hormones and such light-absorbing plant pigments as the chlorophylls and carotenoids. Many of these compounds are dissolved in or partially embedded in the plasma membrane of bacteria or in the mitochondrial and chloroplast membranes of higher organisms. Membranes serve many purposes. The most obvious is to divide space into compartments. Thus, the plasma membrane forms cell boundaries and mitochondrial membranes separate the enzymes and metabolites of mitochondria from those of the cytosol. Membranes are semipermeable and regulate the penetration into cells and organelles of both ionic and nonionic substances. Many of these materials are brought into the cell against a concentration gradient. Hence, osmotic work must be done in a process known as **active transport**. Many enzymes,

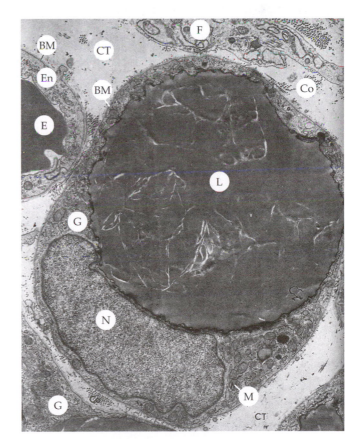

Figure 8-1 Electron micrograph of a thin section of a fat storage cell or adipocyte. L, the single large fat droplet; N, nucleus; M, mitochondria; En, endothelium of a capillary containing an erythrocyte (E); CT, connective tissue ground substance which contains collagen fibers (Co) and fibroblasts (F). The basement membranes (BM) surrounding the endothelium and the fat cell are also marked. From Porter and Bonneville.[6] Courtesy of Mary Bonneville.

including those responsible for most of the oxidative metabolism of cells, are found in membranes of bacteria and of mitochondria. Within the chloroplasts of green leaves, highly folded membranes containing chlorophyll absorb energy from the sunlight. Thin membranes contain the photoreceptor proteins that function in vision. Electrical impulses are transmitted along the membranes of nerve cells.

The outer surfaces of membranes are designed to interact with the cell's external world. Special receptors sense the presence of hormones. Binding proteins await the arrival of needed nutrients and help to bring them into cells. Highly individual arrangements of protein and of the carbohydrate "fuzz" of glycoproteins and sphingolipids screen the outer surface, helping to prevent attack by foreign bacteria, viruses, and toxins.

A. Lipid Structures

Unlike proteins, polysaccharides, and nucleic acids, most lipids are not polymers. However, they are made by linking together smaller molecules.[1-5] Among the "building blocks" of lipids are **fatty acids, glycerol, phosphoric acid**, and **sugars**. Many lipids have both polar and nonpolar regions. This gives them an **amphipathic** character, i.e., a tendency toward both hydrophobic and hydrophilic behavior, and accounts for their tendency to aggregate into membranous structures. Notice that in the following structure the carbon atoms of glycerol have been numbered 1 to 3. Although glycerol is *not* chiral, the positioning of the two hydroxymethyl groups is not equivalent. If the *priority*, used in the *RS* system (page 42) for the C-1 group is raised (e.g., by ester formation) to be higher than that for the C-3 group, the molecule would have the *S* configuration. The C-1 group is said to be pro-*S* and the C-3 group pro-*R*. According to the stereo-chemical numbering (*sn*) system, which is discussed further in Chapter 9, the carbon in the pro-*S* position is numbered 1.

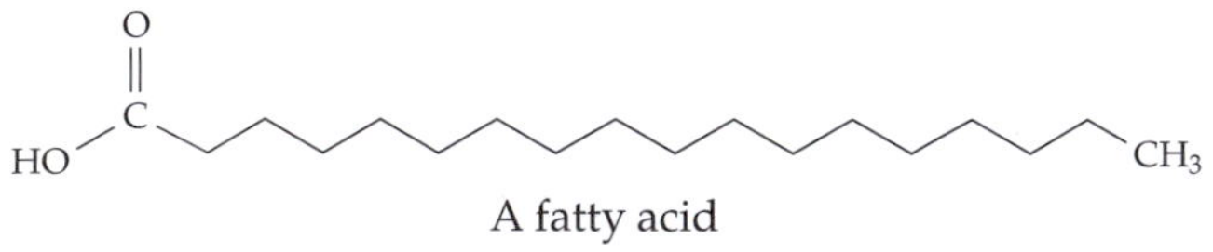

A fatty acid

Stearic acid, 18 carbon atoms, major
component of animal triacylglycerols

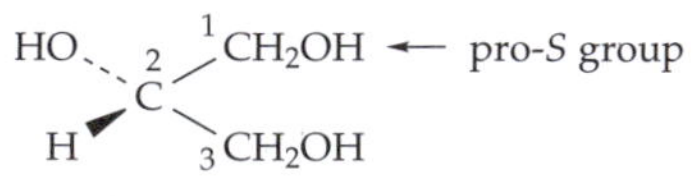

Glycerol

Labeled by the stereochemical
numbering (*sn*) system

1. Fatty Acids, Fatty Alcohols, and Hydrocarbons

The names and structures of some fatty acids are summarized in Table 8-1. Notice that these acids have straight carbon chains and may contain one or more double bonds. Except for the smallest members of the series, which are soluble in water, fatty acids are strongly hydrophobic. However, they are all acids with pK_a values in water of ~4.8. To the extent that free fatty acids occur in nature, they are likely to be found in interfaces between lipid and water with the carboxyl groups dissociated and protruding into the water. However, most naturally occurring fatty acids

TABLE 8-1
Some Important Fatty Acids

No. carbon atoms	Systematic name	Common name	Abbreviation[a]	Common name of acyl group
Saturated fatty acids				
1	Methanoic	Formic		Formyl
2	Ethanoic	Acetic		Acetyl
3	Propanoic	Propionic		Propionyl
4	Butanoic	Butyric	4:0	Butyryl
12	Dodecanoic	Lauric	12:0	Lauroyl[b]
14	Tetradecanoic	Myristic	14:0	Myristoyl
16	Hexadecanoic	Palmitic	16:0	Palmitoyl
18	Octadecanoic	Stearic	18:0	Stearoyl
20	Eicosanoic	Arachidic	20:0	
22	Docosanoic	Behenic	22:0	
24	Tetracosanoic	Lignoceric	24:0	
Unsaturated fatty acids[c]				
4		Crotonic	4:1(2*t*)	Crotonoyl
16		Palmitoleic	16:1(9*c*)	
18		Oleic	18:1(9*c*)	Oleoyl
18		Vaccenic	18:1(11*c*)	
18		Linoleic	18:2(9*c*,12*c*)	
18		Conjugated linoleic	18:2(9*c*,11*t*)	
18		Linolenic	18:3(9*c*,12*c*,15*c*)	
20		Arachidonic	20:4(5*c*,8*c*,11*c*,14*c*)	

[a] The number of carbon atoms is given first, then the number of double bonds. The positions of the lowest numbered carbon of each double bond and whether the configuration is *cis* (*c*) or *trans* (*t*) are indicated in parentheses.

[b] Official IUPAC names of these and other acyl groups have been designated by the Commission of the Nomenclature of Organic Chemistry in *Pure and Applied Chemistry* **10**, 111–125 (1965). In a number of cases IUPAC inserted an *o* in the traditional name, e.g., palmityl became palmitoyl and crotonyl became crotonoyl. However, acetyl was not changed. In many cases the systematic names, e.g., hexadecanoyl (from hexadecanoic acid), are preferable and IUPAC–IUB recommends that alkyl radicals always be designated by systematic names, e.g., hexadecyl, *not* palmityl alcohol. The older use of palmityl for both acyl and alkyl radicals was one reason for IUPAC's adoption of new names for acyl radicals.

[c] Systematic names are not often used because of their complexity, e.g., linolenic acid is *cis,cis,cis*-9,12,15-octadecatrienoic acid.

are esterified or combined via amide linkages in complex lipids. For example, ordinary fats are largely the fatty acid esters of glycerol called **triacylglycerols** (triglycerides).

There is a seemingly endless variety of fatty acids, but only a few of them predominate in any single organism. Most fatty acid chains contain an even number of carbon atoms. In higher plants the C_{16} **palmitic acid** and the C_{18} unsaturated **oleic** and **linoleic acids** predominate. The C_{18} saturated **stearic acid** is almost absent from plants and C_{20} to C_{24} acids are rarely present except in the outer cuticle of leaves. Certain plants contain unusual fatty acids which may be characteristic of a taxonomic group. For example, the Compositae (daisy family) contain acetylenic fatty acids and the castor bean contains the hydroxy fatty acid **ricinoleic acid**.

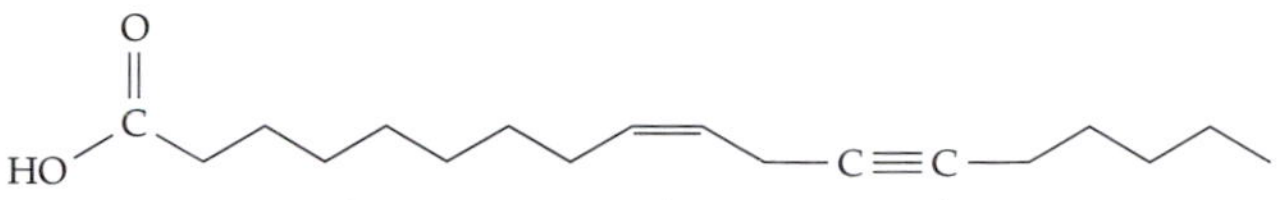

Crepenynic acid 18:2 (9c, 12a)
Accounts for 60% of the fatty acids in seeds of
Crepis foetida, a member of the compositae family

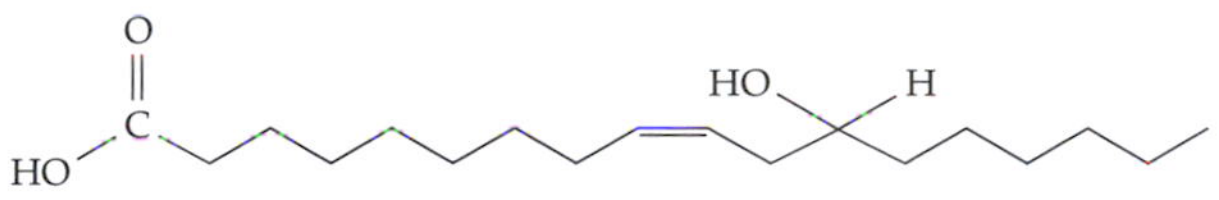

Ricinoleic acid (12-hydroxyoleic acid)
Accounts for up to 90% of the fatty acids of *Ricinus communis* (castor bean)

Like plants, animals contain palmitic and oleic acids. In addition, large amounts of stearic acid and small amounts of the C_{20}, C_{22}, and C_{24} acids are also present. Phospholipids of photoreceptor membranes of the retina contain fatty acid chains as long as C_{36}.[7] The variety of fatty acids found in animals is greater than in a given plant species. A large fraction of the fatty acids present in most higher organisms are *unsaturated* and contain strictly *cis* double bonds. Table 8-2 shows the fatty acid composition of some typical triacylglycerol mixtures.

Bacteria usually lack polyunsaturated fatty acids but often contain branched fatty acids, cyclopropane-containing acids, hydroxy fatty acids, and unesterified fatty acids. Mycobacteria, including the human pathogen *Mycobacterium tuberculosis*, contain **mycolic acids**. In these compounds the complex grouping R contains a variety of functional groups including $-OH$, $-OCH_3$,

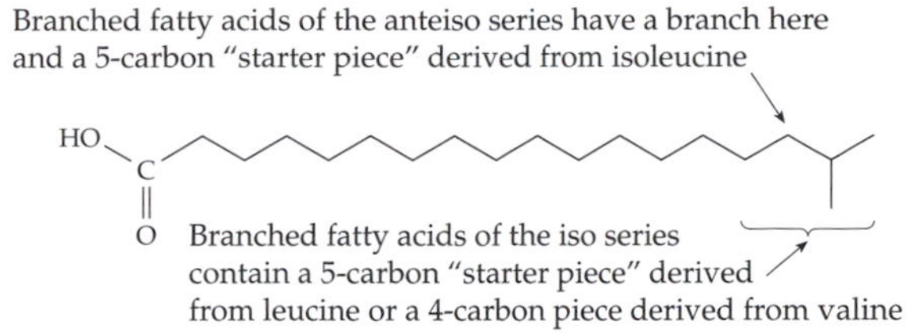

Branched fatty acids of the anteiso series have a branch here and a 5-carbon "starter piece" derived from isoleucine

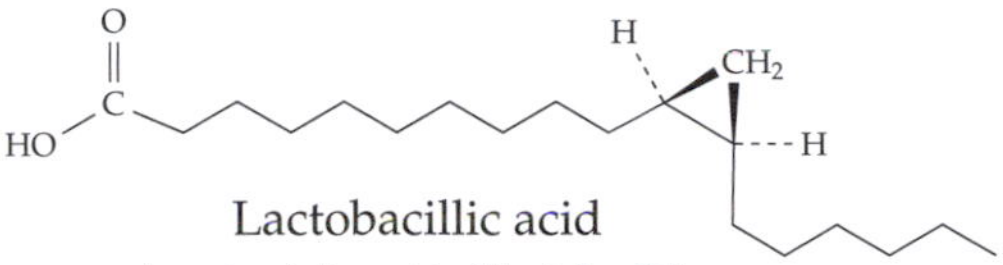

Branched fatty acids of the iso series contain a 5-carbon "starter piece" derived from leucine or a 4-carbon piece derived from valine

Branched fatty acids

Lactobacillic acid
A major fatty acid of lactobacilli

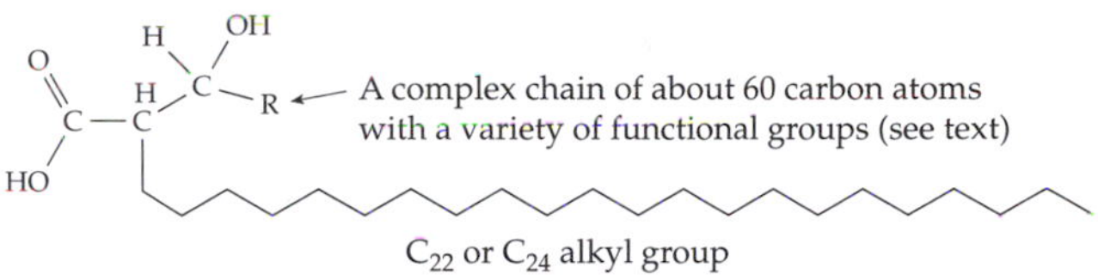

A complex chain of about 60 carbon atoms with a variety of functional groups (see text)

C_{22} or C_{24} alkyl group
A mycolic acid

$C=O$, $-COOH$, cyclopropane rings, methyl branches, and $C=C$ bonds. Each species of *Mycobacterium* contains about two dozen different mycolic acids[9,10] as well as other complex C_{30}–C_{56} fatty acids (see Eq. 21-5).[11]

Certain polyunsaturated fatty acids are essential in the human diet (see Box 21-B). One of these, **arachidonic acid** (which may be formed from dietary linoleic acid), serves as a precursor for the formation of the hormones known as **prostaglandins** and a series of related **prostanoids**. Lipids of animal origin also

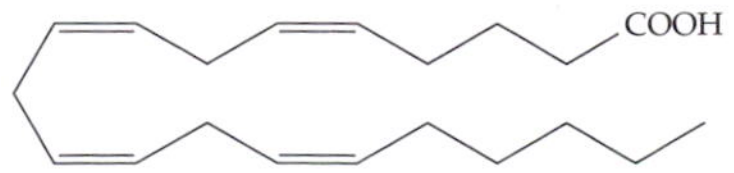

Arachidonic acid
A nutritionally essential fatty acid (Box 21-A)
is shown here in a folded conformation

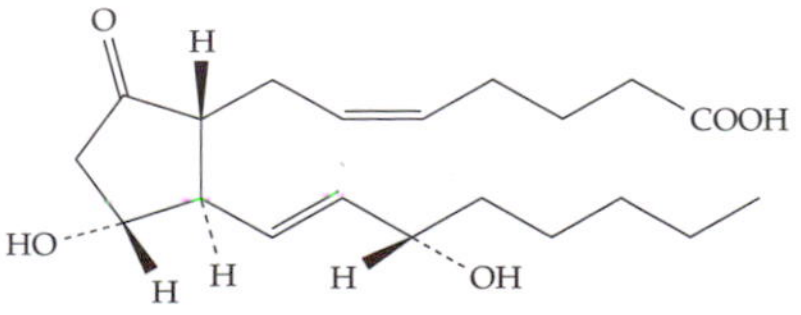

Prostaglandin PGE$_2$
One of a family of hormones, this one is
synthesized in tissues from arachidonic acid

TABLE 8-2
Fatty Acid Composition (in %) of Some Typical Fats and Oils[a]

Fats and oils	No. of carbon atoms and (following colon) the number of double bonds					
	14	16	18	16:1	18:1	18:2
Human depot fat	3	23	4	8	45	10
Beef tallow	4	30	25	5	36	1
Corn oil		13	2		31	54
Lard	1	28	15	3	42	9

[a] From Gunstone.[8]

contain unusual unsaturated fatty acids. Among them, **conjugated linoleic acids** are receiving attention for their possible cancer-preventive action. The predominant form in meats, dairy products, and the human body is the C_{18} 9-*cis*, 11-*trans* isomer whose two double bonds are conjugated.[11a]

Other lipid components include the **fatty alcohols** which are formed by reduction of the acids. These are esterified with fatty acids to form **waxes**. Both fatty alcohols and free fatty acids occur in waxes together with the esterified forms. These mixtures are found on exterior surfaces of plants and animals. Plants and, to a limited extent, animals are able to decarboxylate fatty acids in a multistep process to **alkanes** and these too are important constituents of some waxes. Small amounts of fatty acid amides such as *cis*-9,10-octadecenoamide are present in low concentrations in

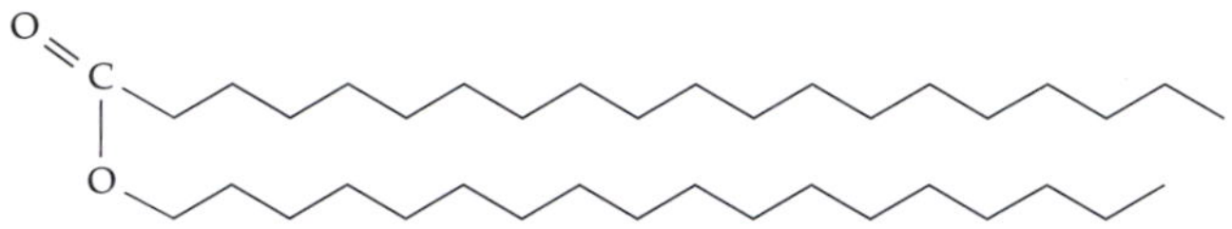

A wax
An ester of a fatty acid and a fatty alcohol

the cerebrospinal fluid of cats and rats as well as humans. This compound accumulates in cats that are deprived of sleep. When the synthetic compound was injected into rats they fell into apparently normal sleep.[12]

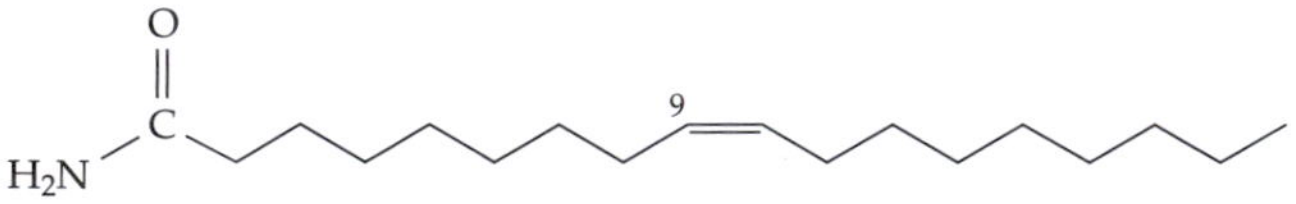

cis-9,10-Octadecenoamide (Oleic acid amide)

Insects make unsaturated as well as saturated hydrocarbons. The former as well as long-chain alcohols and their esters often form the volatile **pheromones** with which insects communicate. Thus, the female pink bollworm attracts a male with a sex pheromone consisting of a mixture of the *cis,cis* and *cis,trans* isomers of 7,11-hexadecadienyl acetate,[13] and European corn borer males are attracted across the cornfields of Iowa by *cis*-11-tetradecenyl acetate.[14] Addition of a little of the *trans* isomer makes the latter sex attractant much more powerful. Since more than one species uses the same attractant, it is possible that the males can distinguish between different ratios of isomers or of mixtures of closely related substances.

2. Acylglycerols, Ether Lipids, and Waxes

The components of complex lipids are linked in a variety of ways. Often, glycerol acts as the central unit,

e.g., combining in ester linkage with three fatty acids to form **triacylglycerols** (triglycerides), the common fats of adipose tissues and plant oils. Diacyl- and monoacyl-glycerols (diglycerides, and monoglycerides) are

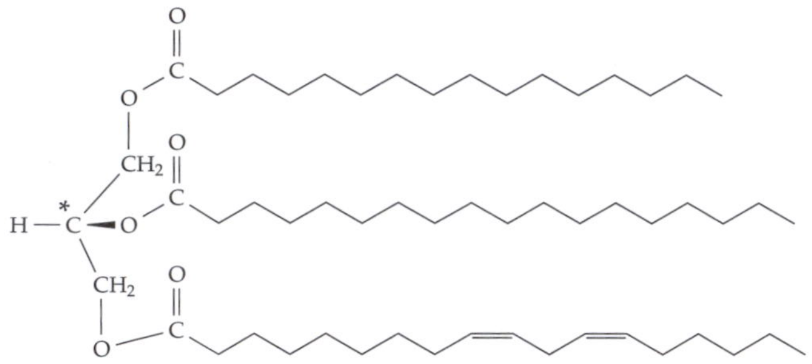

A triacylglycerol (fat)
Notice the chiral center, designated by the asterisk

present to a lesser extent. In addition, small amounts of **alkyl ethers** or **alkenyl ethers** are often present in isolated lipids. They are especially abundant in fish liver oils.

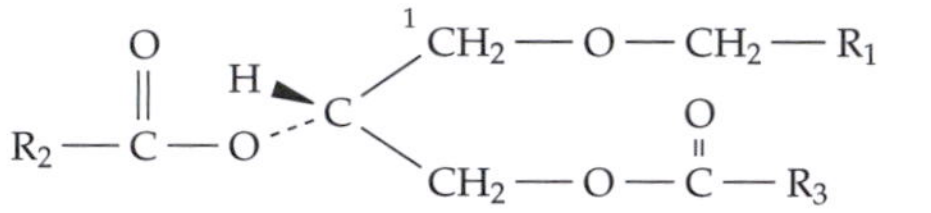

1-Alkyl-2,3-diacyl-*sn*-glycerol, an alkyl ether lipid

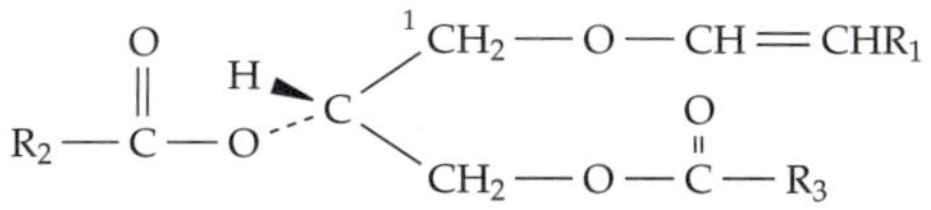

1-Alkenyl-2,3-diacyl-*sn*-glycerol, an alkenyl ether

These ether lipids are all chiral molecules with an *R* configuration but are derivatives of the nonchiral glycerol. The carbon atoms of glycerol are numbered using the stereochemical system which is described on p. 470. The ether linkage is to the *sn*-1 carbon atom. Most phospholipids are derivatives of the *sn*-3 phosphate ester of glycerol.

Triacylglycerols and the ether lipids described in the previous section are classified as **neutral lipids**. Other neutral lipids are alcohols, waxes, aldehydes, and hydrocarbons derived from fatty acids. These sometimes have specific biological functions. For example, fatty aldehydes are important in the bioluminescence of bacteria (Eq. 23-47).

3. Phospholipids

As major constituents of biological membranes, phospholipids play a key role in all living cells. The two principal groups of phospholipids are the

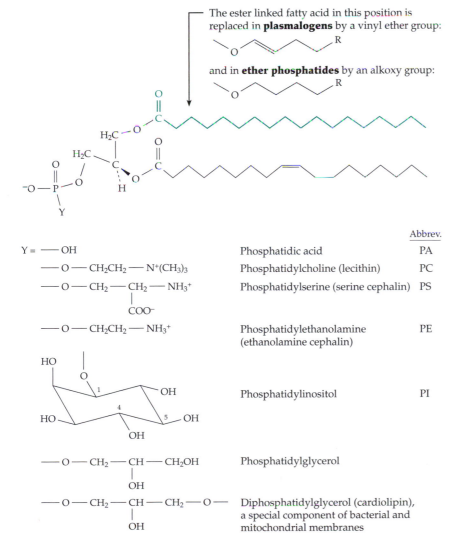

Figure 8-2 Structures of some phosphatides (glycerophospholipids).

glycerophospholipids (glycero-phosphatides; Fig. 8-2) which contain the alcohol glycerol and the sphingophospholipids which contain the alcohol sphingosine (Section 5). The glycerophospholipids can be thought of as arising from the building blocks glycerol, fatty acids, the dihydrogen phosphate ion $H_2PO_4^-$, and the appropriate alcohol by removal of four molecules of water (Eq. 8-1). They are derivatives of sn-glycerol-3-phosphate. Esterification of this alcohol with two fatty acids gives a **phosphatidic acid** (Fig. 8-2). Formation of a phosphate diester linkage to one of the alcohols

sn-Glycerol 3-phosphate
(L-α-glycerol phosphate)

choline, serine, or ethanolamine yields a glycerophospholipid. The resulting three groups of phospholipids are called **phosphatidylcholine** (lecithin), **phosphatidylserine**, and **phosphatidylethanolamine**, respectively (Fig. 8-2).

The phosphate and choline, ethanolamine, or serine portions of the phosphatide are electrically charged and provide a polar "head" for the molecule. In all three cases the positively charged group is able to fold back and form an ion pair with the negatively charged phosphate group. However, the methyl groups surrounding the nitrogen in phosphatidylcholine prevent a very close approach and with phosphatidylserine the adjacent carboxylate group weakens this electrostatic interaction. Unlike the triacylglycerols, most of which are liquid at body temperature, phospholipids are solid at this temperature. This property, like the ionic properties of the phosphatides, is doubtless related to their suitability for functioning in biological membranes.

Lecithins and related phospholipids usually contain a saturated fatty acid in the C-1 position but an unsaturated acid, which may contain from one to four double bonds, at C-2. Arachidonic acid is often present here. Hydrolysis of the ester linkage at C-2 yields a 1-acyl-3-phosphoglycerol, better known as a **lysophosphatidylcholine**. The name comes from the powerful detergent action of these substances which leads to lysis of cells. Some snake venoms contain phospholipases that form lysophosphatidylcholine. Lysophosphatidic acid (1-acyl-glycerol-3-phosphate) is both an intermediate in phospholipid biosynthesis (Chapter 21) and also a signaling molecule released into the bloodstream by activated platelets.[15]

Another group of phosphatides contain the hexahydroxycyclohexane known as **inositol** (Fig. 8-2, see also Chapter 21).[16] Phosphatidylinositol, as well as smaller amounts of phosphatides derived from phosphate esters of inositol are present in membranes of all eukaryotes and have a specific role in regulating responses of cells to hormones and other external agents. See Chapter 11 for details. Phosphatidylinositol also forms part of "anchors" used to hold certain proteins onto membrane surfaces (see Fig. 8-13).

Bacteria and plants often make the anionic **phosphatidyglycerol** in which the second glycerol is esterified at its *sn*-1 position with the phosphate. Bacteria, as well as mitochondria, contain diphosphatidylglycerol (**cardiolipin**) in which phosphatidyl groups are attached at both the 1 and 3 positions of glycerol (Fig. 8-2). Ether phospholipids, analogous to the ether lipids described in Section 2, are also widely distributed. The alkenyl ether analogs of phosphatidylcholine (Fig. 8-2) are called **plasmalogens**.[17] In neutrophils the 1-*O*-alkyl ethers contain the major share of the cell's arachidonic acid, which is esterified in the 2 position.[18,19]

In halophilic (salt loving), thermophilic, and methanogenic bacteria, most of the lipids present are either

---

BOX 8-A THE PLATELET-ACTIVATING FACTOR

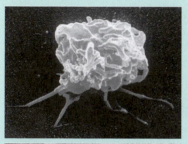

Scanning EM of activated blood platelets. © Quest, Photo Researchers

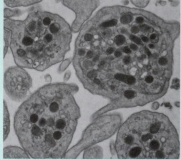

Transmission EM of thin section through activated blood platelets. © Photo Researchers

Part of the immune response consists of the release from stimulated neutrophils, macrophages, and other cells of a **platelet activating factor (PAF)**, a material that "activates" blood platelets. Activated platelets aggregate, a response that initiates clot formation. They may also be lysed and release stored substances that include platelet-derived growth factors (see Chapter 11) and fibrin stabilizing factor, a proenzyme that is converted to the protein-crosslinking enzyme transglutaminase (Eq. 2-17). See Fig. 12-17. However, the principal interest in the platelet-activating factor has arisen from its powerful effect in inducing **inflammation**

in surrounding tissues.

PAF has been identified as the following simple ether phospholipid:[a-d]

Platelet-activating factor (PAF)
1-*O*-alkyl-2-acetyl-*sn*-glycerophosphocholine

A remarkably potent compound, its effects on platelets are observed at concentrations of 10^{-11} to 10^{-10} M. Both lyso-PAF and the glycero-1- phosphocholine enantiomer are inactive. Specific receptors for this factor are evidently present on platelet surfaces.[e-g]

One effect of PAF on platelets is to induce a rapid (5–10s) cleavage of phosphatidylinositol 4,5-bisphosphate by phospholipase C to give **diacylglycerol** and **inositol 1,4,5-trisphosphate**. The subsequent effects of these two substances in causing a rapid influx of Ca^{2+} and in inducing a series of secondary responses are outlined in Fig. 11-9. Among these responses are the release of the materials stored in the platelet's granules. PAF also appears to inhibit adenylate cyclase[b] and causes vasodilation, a property not expected for a compound that stimulates clotting. Receptors for PAF are also present in the brain, where this phospholipid may function in regulation of development.[h]

phospholipids and glycolipids containing the C_{20} isoprenoid **phytanyl** group or the C_{40} **diphytanyl** group[20-25] (see also Section 4), related isoprenoid alcohols, or long-chain 1,2-diols.[25] An example of a diphytanylglycerophospholipid is the following:

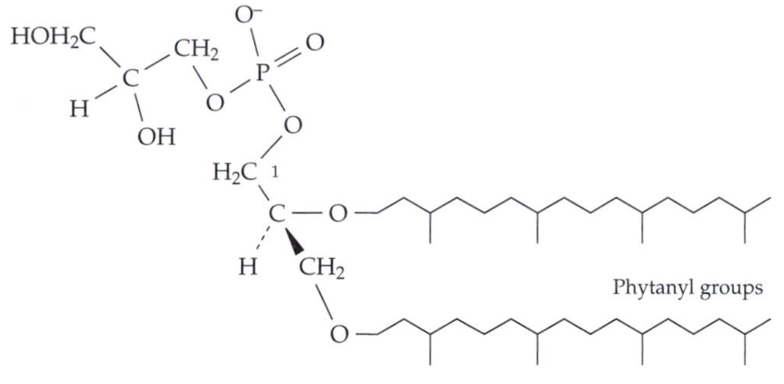

A phospholipid derived from 2,3-di-O-phytanyl-sn-glycerol

Many other phospholipids are present in small amounts or in a limited number of species. These include **phosphonolipids**, which contain a C–P bond and are abundant in ciliate protozoa such as *Tetrahymena* and in some other invertebrates.[26] Phosphonoethylamine replaces phosphoethanolamine in these lipids. A consequence of this structural alteration is a high

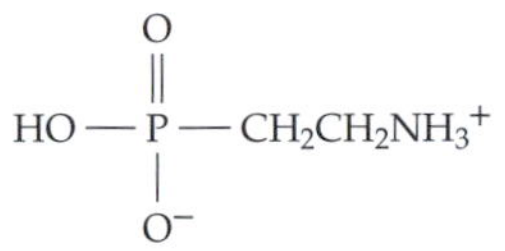

Phosphonoethylamine

degree of resistance to the action of the enzyme phospholipase C. The phosphonolipids of the external membrane of *Tetrahymena* are also ether lipids with an alkoxy group in the *sn*-1 position. This makes them

---

BOX 8-A   (continued)

Stimulated platelets release arachidonic acid rapidly from their phospholipids, apparently as a result of activation of phospholipase A_2. The released arachidonate can in turn be metabolized to endoperoxides and thromboxane A_2 (Chapter 21). These compounds are also potent activators of platelets and cause a self-activating or **autocrine** effect.[i,j] While PAF has a beneficial function, it can under some conditions contribute in a dangerous way to inflammation and to allergic responses including **anaphylaxis**,[j] **asthma**[g] and **cold-induced urticaria**.[k] Although the effect of PAF is separate from those of histamine and of leukotrienes, these agents may act cooperatively to induce inflammation.[l]

The biosynthesis of PAF is discussed in Chapter 21, Section C. One pathway is deacylation at the glycerol C-2 of a longer chain alkyl phospholipid followed by acetylation at the same position.[m,n]

Platelets can inactivate PAF by the inverse sequence: deacetylation followed by acylation with arachidonic acid. PAF can also be hydrolyzed by a PAF acetylhydrolase, a phospholipase.[o] Absence of this enzyme in the brain may be related to a human brain malformation.[h] Some tissues may contain a lipid that inhibits binding of PAF.[p] The following compound from a Chinese herb binds to PAF receptors and may be the forerunner of useful drugs.[q]

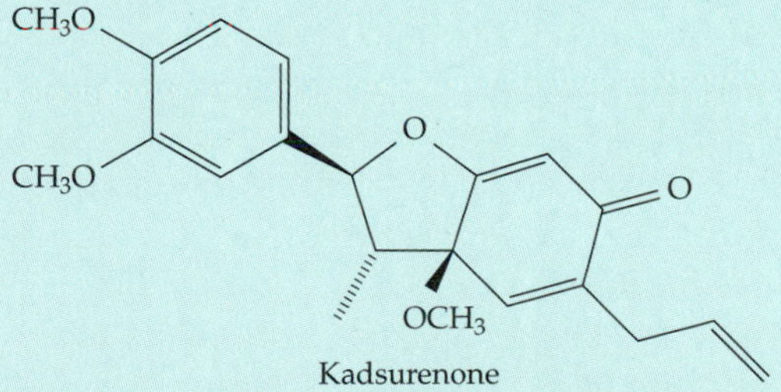

Kadsurenone

[a] Cusack, N. J. (1980) *Nature (London)* **285**, 193
[b] Hanahan, D. J. (1986) *Ann. Rev. Biochem.* **55**, 483–509
[c] Prescott, S. M., Zimmerman, G. A., and McIntyre, T. M. (1990) *J. Biol. Chem.* **265**, 17381–17384
[d] Winslow, C. M., and Lee, M. L., eds. (1987) *New Horizons in Platelet Activating Factor Research*, Wiley, New York
[e] Hwang, S.-B., Lam, M.-H., and Pong, S.-S. (1986) *J. Biol. Chem.* **261**, 532–537
[f] Ishii, I., Izumi, T., Tsukamoto, H., Umeyama, H., Ui, M., and Shimizu, T. (1997) *J. Biol. Chem.* **272**, 7846–7854
[g] Bazan, N. G. (1995) *Nature (London)* **374**, 501–502
[h] Hattori, M., Adachi, H., Aoki, J., Tsujimoto, M., Arai, H., and Inoue, K. (1995) *J. Biol. Chem.* **270**, 31345–31352
[i] Bussolino, F., Sironi, M., Bocchietto, E., and Mantovani, A. (1992) *J. Biol. Chem.* **267**, 14598–14603
[j] Darius, H., Lefer, D. J., Smith, J. B., and Leefer, A. M. (1986) *Science* **232**, 58–60

[k] Grandel, K. E., Farr, R. S., Wanderer, A. A., Eisenstadt, T. C., and Wasserman, S. I. (1985) *N. Engl. J. Med.* **313**, 405–409
[l] Tomeo, A. C., Egan, R. W., and Durán, W. N. (1991) *FASEB J.* **5**, 2850–2855
[m] Billah, M. M., Eckel, S., Myers, R. F., and Siegel, M. I. (1986) *J. Biol. Chem.* **261**, 5824–5831
[n] Lee, T.-c, Ou, M.-c, Shinozaki, K., Malone, B., and Snyder, F. (1996) *J. Biol. Chem.* **271**, 209–217
[o] Stafforini, D. M., McIntyre, T. M., Zimmerman, G. A., and Prescott, S. M. (1997) *J. Biol. Chem.* **272**, 17895–17898
[p] Miwa, M., Hill, C., Kumar, R., Sugatan, J., Olson, M. S., and Hanahan, D. J. (1987) *J. Biol. Chem.* **262**, 527–530
[q] Hwang, S.-B., Lam, M.-H., Biftu, T., Beattie, T. R., and Shen, T.-Y. (1985) *J. Biol. Chem.* **260**, 15639–15645

BOX 8-B DIPALMITOYLPHOSPHATIDYLCHOLINE AND THE SURFACTANT SYSTEM OF THE LUNGS

When air is exhaled the small alveoli of the lungs could collapse if it were not for the surface active material (surfactant) present in the fluid that coats the lungs.[a–e] In fact, the lack of adequate surfactant is the cause of **respiratory distress syndrome**, a major cause of death among premature infants and a disease that may develop in acute form in adults. The surfactant material forms a thin film of high fluidity at the air–liquid interface and lowers the surface tension from the 72 mN/m of pure water to <10 mN/m.[f,g] (Pay attention to the definition of surface tension.[h]) About 65% by weight of the surfactant is lecithin, mostly dipalmitoylphosphatidylcholine (see Fig. 8-4), a phospholipid resistant to attack by oxygen. Phosphatidylglycerol, in an unusually high concentration, accounts for ~ 12% of the human surfactant. Other phospholipids, plasmalogen,[i] cholesterol, proteins, and calcium ions are also present.

The surfactant contains four unique proteins, designated SP-A, SP-B, SP-C, and SP-D.[c,e] The major protein (SP-A) is a sialic acid-rich glycoprotein derived from a 26-kDa peptide, which contains a short collagen-like domain.[e,j,k] Like collagen, this domain contains glycosylated hydroxyproline. The C-terminal domain is a Ca^{2+}-dependent C-type lectin (Chapter 4), while the N-terminal domain is involved in oligomer formation through disulfide bridges. The overall structure is similar to that of the complement protein C1q (Chapter 31).[e,j] Protein D is also collagen-like[l] but evidently plays a very different functional role than SP-A. The latter associates with the major surfactant lipids but SP-D does not. It does bind phosphatidylinositol[m] and glucosylceramide, lipids that are present in small amounts. Perhaps SP-D helps to remove these polar lipids which might interfere with surfactant action.[e] Both proteins A and D may also have functions in the immune system.[l]

Proteins SP-B and SP-C are small extremely hydrophobic polypeptides consisting of 79 and 35 amino acid residues, respectively.[n,o] Aliphatic branched amino acids constitute 23 of the 35 residues of the C-terminal part of protein C, which is also palmitoylated on two cysteine residues. SP-B is formed from a large 381-residue precursor. The mature protein contains seven cysteines and disulfide bridges. Both proteins have major effects on the properties of the surfactant mixture. They promote rapid reorganization of lipid layers, an important consideration for the functioning of the surfactant. Infants lacking SP-B suffer severe respiratory failure with high mortality.[e]

The properties of the surfactant allow for rapid formation of a large area of lipid monolayer–air interface. The low surface tension and the ability to rapidly spread the mixture of lipids and proteins are essential.[o] At the end of the expiration stage of breathing the surfactant is present in the interface as a strong, tightly packed monolayer whose properties reflect the rigidity of the dipalmitoylphosphatidylcholine. The excess surfactant in the alveolar fluid forms liposome-like bilayer structures and also associates with proteins and calcium ions to form a lattice-like material called **tubular myelin**. Lipid in this form must be transferred rapidly into the air–liquid interface during inspiration.[p] The ability of phospholipids to pass through a hexagonal phase (Fig. 8-12) may also be important for this transition.[q,r]

One enzyme present in the surfactant fluid is an acid phosphatase able to hydrolyze phosphatidylglycerol phosphate, perhaps functioning in the final step of biosynthesis of the phosphatidylglycerol present in the surfactant.[c] Study of the action of the natural lung surfactant has led to development of artificial surfactant mixtures that are being used effectively to save many lives.[d]

[a] Rooney, S. A. (1979) *Trends Biochem. Sci.* **4**, 189–191

[b] Persson, A., Chang, D., Rust, K., Moxley, M., Longmore, W., and Crouch, E. (1989) *Biochemistry* **28**, 6361–6367

[c] Rooney, S. A., Young, S. L., and Mendelson, C. R. (1994) *FASEB J.* **8**, 957–967

[d] Jobe, A. H. (1993) *N. Engl. J. Med.* **328**, 861–868

[e] Kuroki, Y., and Voelkers, D. R. (1994) *J. Biol. Chem.* **269**, 25943–25946

[f] Shiffer, K., Hawgood, S., Haagsman, H. P., Benson, B., Clements, J. A., and Goerke, J. (1993) *Biochemistry* **32**, 590–597

[g] Pastrana-Rios, B., Flach, C. R., Brauner, J. W., Mautone, A. J., and Mendelsohn, R. (1994) *Biochemistry* **33**, 5121–5127

[h] Bangham, A. D. (1992) *Nature (London)* **359**, 110

[i] Rana, F. R., Harwood, J. S., Mautone, A. J., and Dluhy, R. A. (1993) *Biochemistry* **32**, 27–31

[j] McCormack, F. X., Calvert, H. M., Watson, P. A., Smith, D. L., Mason, R. J., and Voelker, D. R. (1994) *J. Biol. Chem.* **269**, 5833–5841

[k] Taneva, S., McEachren, T., Stewart, J., and Keough, K. M. W. (1995) *Biochemistry* **34**, 10279–10289

[l] Crouch, E., Persson, A., Chang, D., and Heuser, J. (1994) *J. Biol. Chem.* **269**, 17311–17319

[m] Ogasawara, Y., Kuroki, Y., and Akino, T. (1992) *J. Biol. Chem.* **267**, 21244–21249

[n] Korimilli, A., Gonzales, L. W., and Guttentag, S. H. (2000) *J. Biol. Chem.* **275**, 8672–8679

[o] Pastrana, B., Mautone, A. J., and Mendelsohn, R. (1991) *Biochemistry* **30**, 10058–10064

[p] Lipp, M. M., Lee, K. Y. C., Zasadzinski, J. A., and Waring, A. J. (1996) *Science* **273**, 1196–1199

[q] Perkins, W. R., Dause, R. B., Parente, R. A., Minchey, S. R., Neuman, K. C., Gruner, S. M., Taraschi, T. F., and Janoff, A. S. (1996) *Science* **273**, 330–332

[r] Discher, B. M., Maloney, K. M., Grainger, D. W., Sousa, C. A., and Hall, S. B. (1999) *Biochemistry* **38**, 374–383

resistant to phospholipase A_1 as well. These two properties appear to protect the naked cell membranes of the protozoa from their own phospholipases which may be secreted into the environment.[26] Marine algae form an arsenic-containing phospholipid *O*-phosphatidyltrimethylarsonium lactic acid.[27]

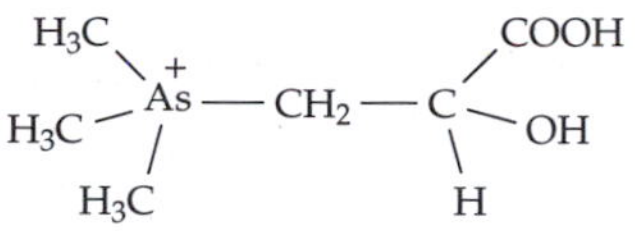

Trimethylarsonium lactic acid

They apparently do this as part of a scheme for detoxifying arsenate taken up with phosphate from the phosphate-poor ocean water.

4. Glycolipids

The polar heads of the **glycoglycerolipids** lack phospho groups but contain *sugars* in glycosidic linkage.[28] Large amounts of the galactolipids shown in the following structure are found in chloroplasts.[29] The monogalactosyl diacylglycerol is said to be the most abundant polar lipid in nature.

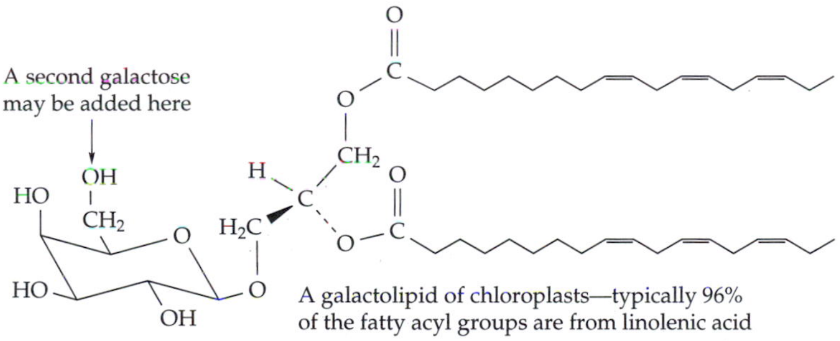

A second galactose may be added here

A galactolipid of chloroplasts—typically 96% of the fatty acyl groups are from linolenic acid

Chloroplasts also contain the following **sulfolipid**, an anionic sulfonate.

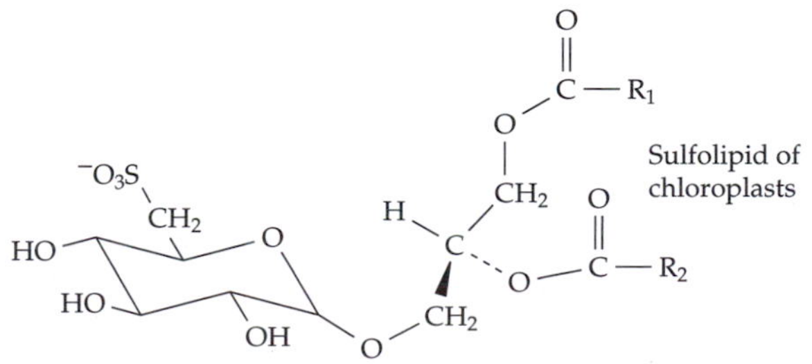

Sulfolipid of chloroplasts

6-Sulfo-6-deoxy-α-D glucopyranosyl diglyceride

Marine algae as well as aquatic higher plants accumulate **arsenophospholipids**.[30]

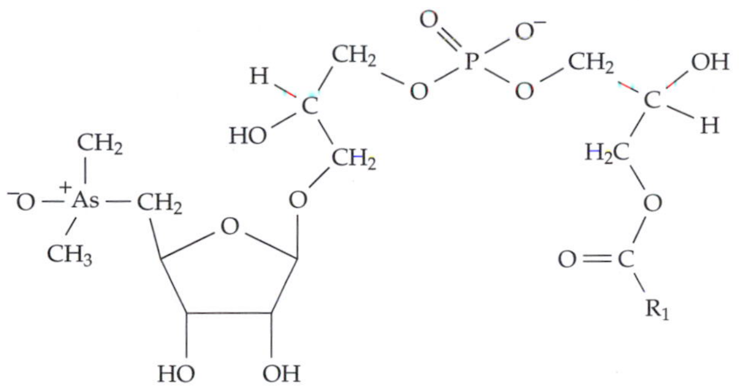

Arsenophospholipid of aquatic plants

The plasma membrane of mammalian male germ cells contains the following **sulfogalactosylglycerolipid**. It is found only in spermatozoa and testes, in which it accounts for 5–8% of total lipid, and in the brain, in which it accounts for only 0.2% of total lipid.[31]

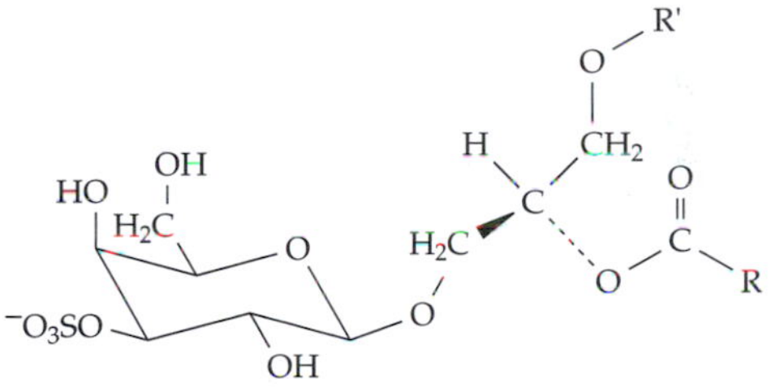

Sulfogalactosylglycerolipid, a sulfate ester

A variety of acylated glucolipids and phosphoglucolipids, including monoglucosyl and diglucosyl diacylglycerols, have been identified in membranes of the cell-wall-less bacterium *Acholeplasma laidlawii*.[32,32a] The following glycolipid from the methanogen *Methanosarcina*[33] is identical to the core structure of eukaryotic glycosylphosphatidylinositol membrane protein anchors (Fig. 8-13).

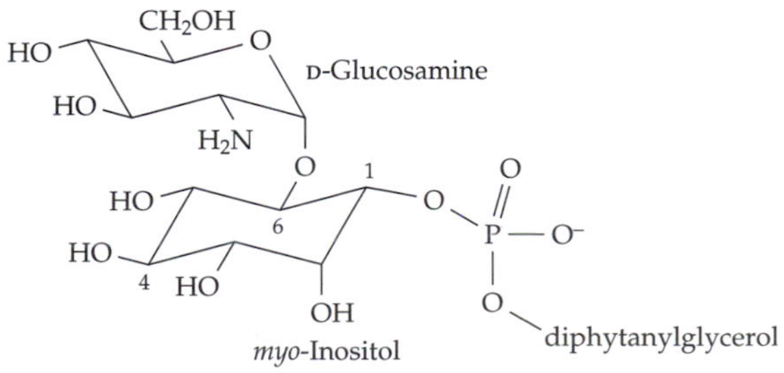

In addition to the previously mentioned phytanyl ether phospholipids, methanogens contain **diphytanyl tetraether lipids** that are both glycerophospholipids and glycolipids.

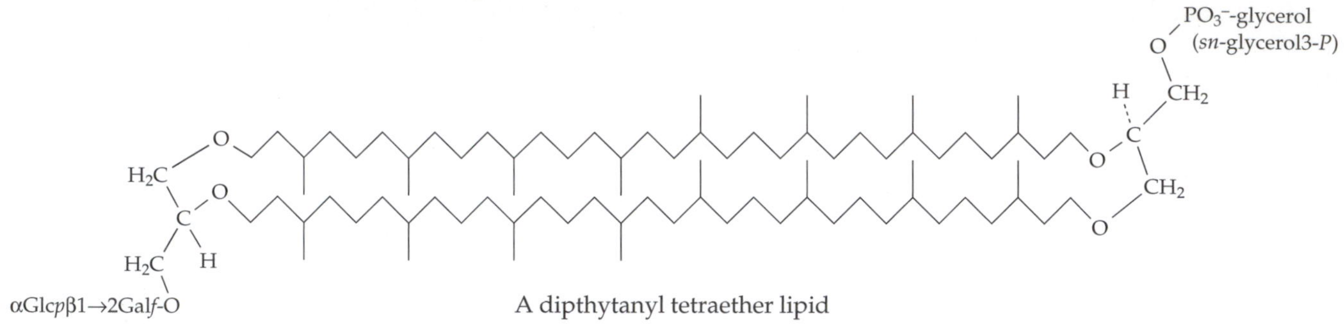

A dipthytanyl tetraether lipid
From a methanogenic bacterium.[21]

αGlcpβ1→2Galf-O

5. Sphingolipids

The backbone of the sphingolipids is the basic alcohol **sphingosine** (sphingenine) or a related long chain base. At least 60 such bases have been identified.[34,35] They vary in chain length from C_{14} to C_{22} and include members of the branched iso and anteiso series. Up to two double bonds may be present. Sphingosine contains 18 carbon atoms and is formed from palmitic acid and serine (Fig. 21-6). An intermediate in the formation of sphingosine is the saturated **sphinganine** (dihydrosphingosine), which is also a common component of animal sphingolipids. Hydroxylation of sphinganine to **phytosphingosine** occurs in both plants and animals, especially within glycolipids. The name comes from the fact that phytosphingosine was first discovered in plants.

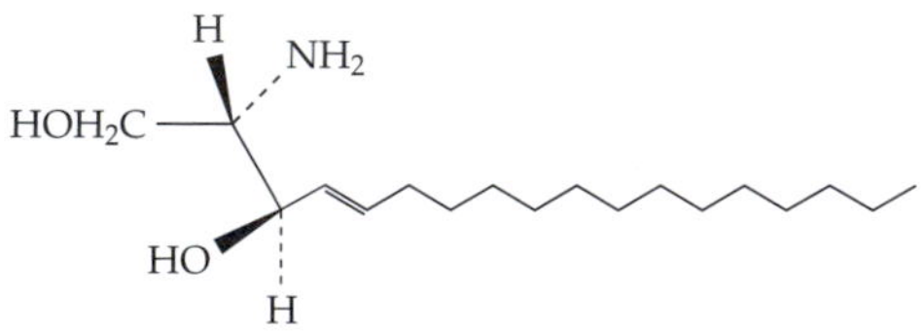

Sphingosine (sphingenine)

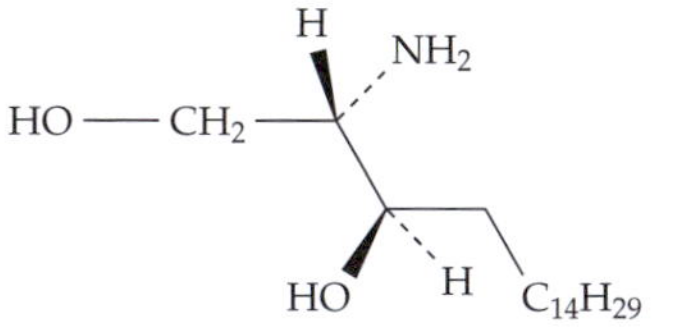

Sphinganine (dihydrosphingosine)

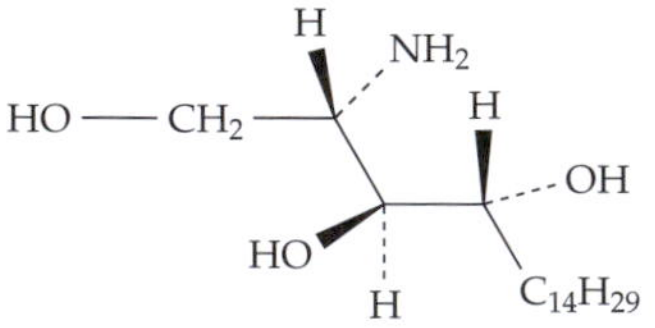

Phytosphingosine

Sphingosine-containing lipids are classified as **sphingophospholipids** (sphingomyelins) and **sphingoglycolipids**. In both cases the sphingosine is combined in amide linkage with a fatty acid to form a **ceramide** (Fig. 8-3) which still contains a free hydroxyl group able to combine with another component. In the sphingomyelins, which were first isolated from human brain by Thudicum in 1884, the additional component is usually phosphocholine (Fig. 8-3). Ceramide aminoethylphosphonates and related glycolipids occur in some invertebrates.[36,37]

The **cerebrosides** are glycosides of ceramide containing galactose or glucose. They are found in relatively large amounts in the brain where monogalactosylceramide predominates. Cerebrosides also occur in other animal tissues and to a lesser extent in plants. Many glycosphingolipids contain oligosaccharides of various sizes. When the oligosaccharide contains one or more residues of sialic acid the compound is known as a **ganglioside**.[38] Sialic acids are never found in plants. However, plants and fungi contain **phytoglycolipids**, which resemble gangliosides. Some contain inositol phosphates as well as sugars.[39–41] The structures of gangliosides may be very complex. Like glycoproteins, these substances are often located at the outer surface of cells where they may act as receptors for toxins, viruses, and hormones (see Section D).[42] Some of them carry attached blood group antigens (Box 4-C).[43,44] The sulfate esters of cerebrosides, known as **sulfatides** (Fig. 8-3), are also important components of membranes.[45]

Gliding bacteria such as *Capnocytophaga* contain sulfonolipids:[46]

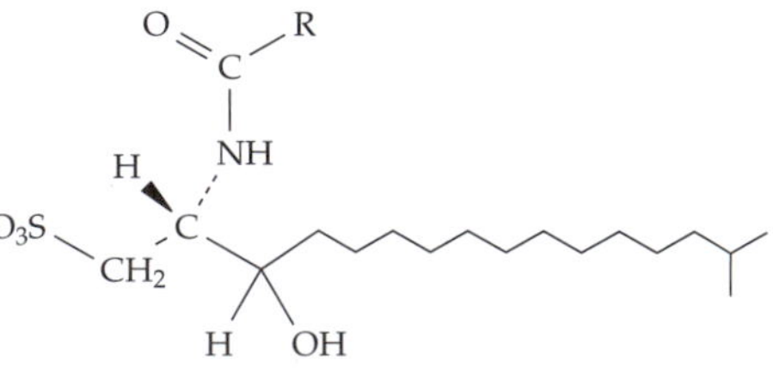

N-acylcapnine

A 1-deoxy-1-sulfonate analog of ceramides from gliding bacteria. R is a long-chain alkyl group.

These are analogs of the ceramides. The sulfono-lipids seem to be necessary for the gliding movement of these bacteria across solid surfaces.[47]

6. Sterols and Other Isoprenoid Lipids

A large group of isoprenoid lipids, including **sterols**, **terpenes**, and **carotenoid compounds**, are often present in membranes or in extracted lipids. Among these are the fat-soluble vitamins A, D, E, and K and the high polymers **rubber** and **gutta-percha**.

The phytyl group of chlorophyll (Fig. 23-30), the phytanyl and diphytanyl groups of the lipids of methanogens, and the side chain of the pigment heme *a* (Fig. 16-5) are all related and are all derived from the precursor **prenyl pyrophosphate** (isopentenyl diphosphate). The whole group of compounds are often referred to as **polyprenyl** or by the older designation **isoprenoid**. The major discussion of polyprenyl compounds is found in Chapter 22 but the role of cholesterol and related compounds in membranes and the function of polyprenyl groups as membrane anchors for some proteins will be considered in this chapter.

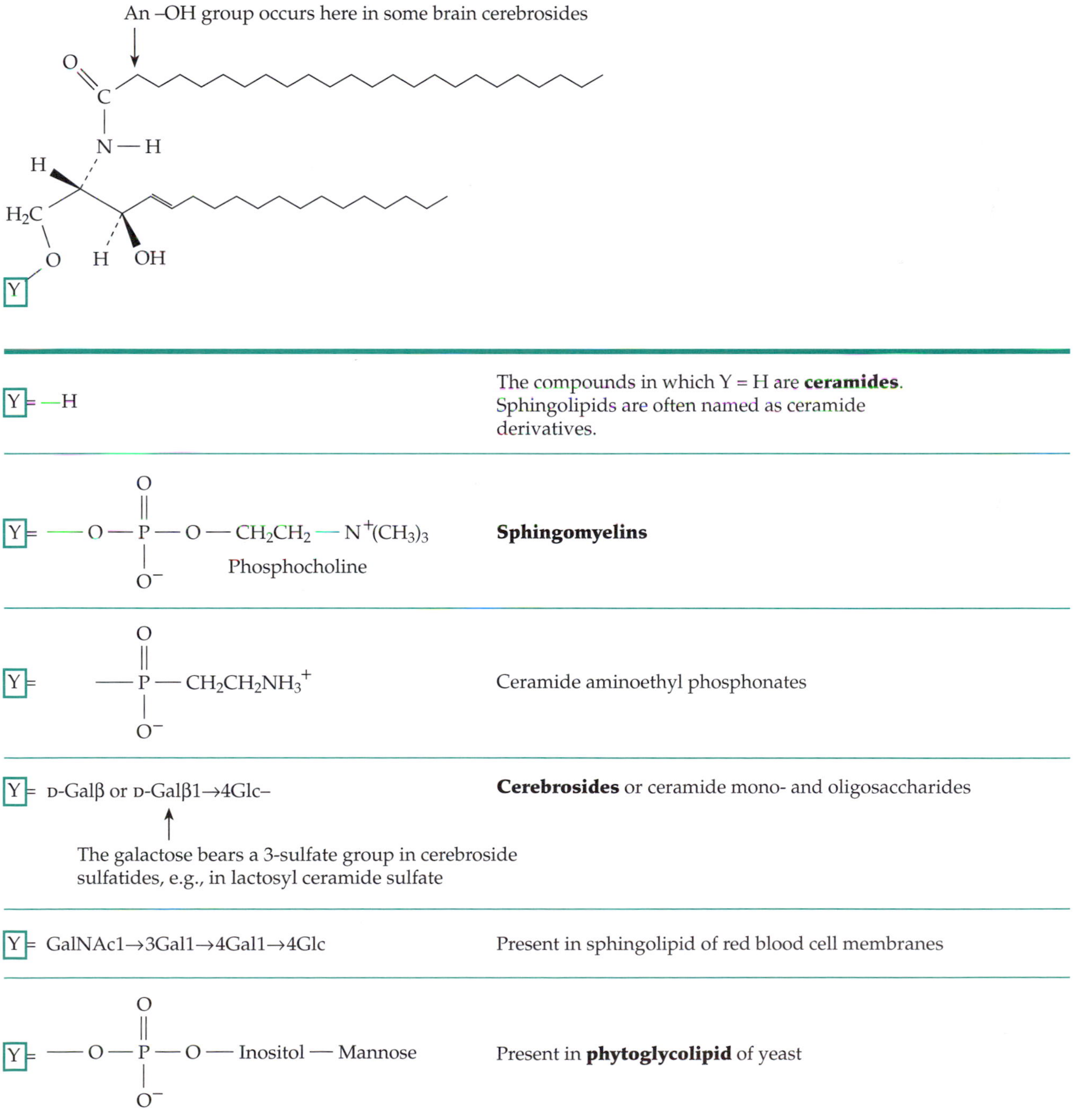

An –OH group occurs here in some brain cerebrosides

Y = —H

The compounds in which Y = H are **ceramides**. Sphingolipids are often named as ceramide derivatives.

$$Y = -O-\overset{\overset{\displaystyle O}{\|}}{\underset{\underset{\displaystyle O^-}{|}}{P}}-O-CH_2CH_2-N^+(CH_3)_3$$

Phosphocholine

Sphingomyelins

$$Y = -\overset{\overset{\displaystyle O}{\|}}{\underset{\underset{\displaystyle O^-}{|}}{P}}-CH_2CH_2NH_3^+$$

Ceramide aminoethyl phosphonates

Y = D-Galβ or D-Galβ1→4Glc–

Cerebrosides or ceramide mono- and oligosaccharides

The galactose bears a 3-sulfate group in cerebroside sulfatides, e.g., in lactosyl ceramide sulfate

Y = GalNAc1→3Gal1→4Gal1→4Glc

Present in sphingolipid of red blood cell membranes

$$Y = -O-\overset{\overset{\displaystyle O}{\|}}{\underset{\underset{\displaystyle O^-}{|}}{P}}-O-Inositol-Mannose$$

Present in **phytoglycolipid** of yeast

Figure 8-3 Structures of some sphingolipids.

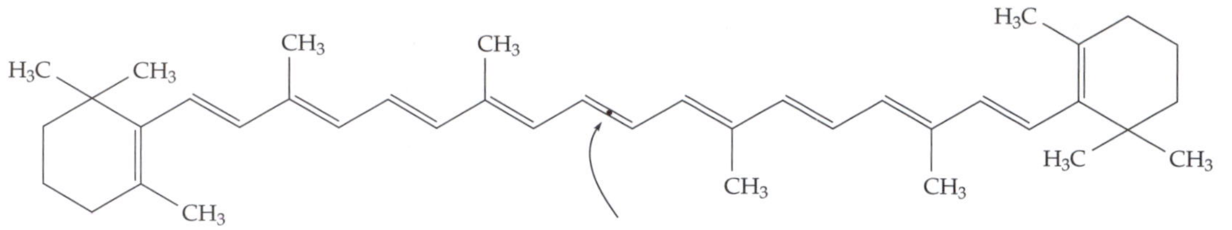

β-Carotene, a yellow pigment of carrots and other plants; converted in the human body to vitamin A.

Notice the center of symmetry in this 40-carbon molecule derived from 8 prenyl units

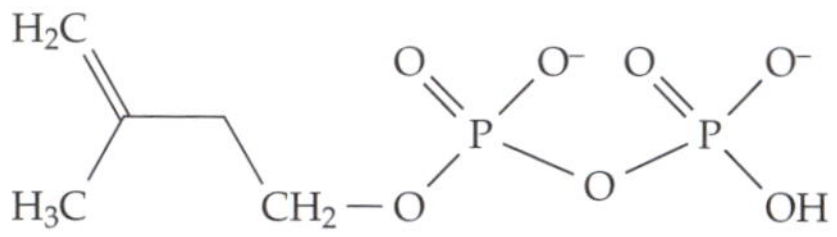

Prenyl diphosphate (pyrophosphate) (isopentenyl pyrophosphate), precursor of steroid, carotenoid, and other isoprenoid substances

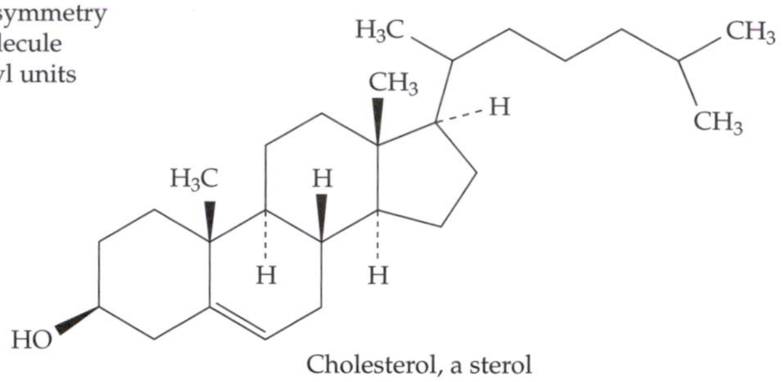

Cholesterol, a sterol

B. Membranes

In Chapter 7 we examined ways in which protein subunits can be stacked to form helices and closed oligomers. Another important arrangement of cell constituents is that of flat sheets or membranes.[48–54] Chemists, physicists, and biologists have mounted a sustained effort to understand these thin but remarkably tough outer surfaces of cells. However, consider the fact that a 7- to10-nm-thick plasma membrane of a cell of 10-μm diameter has less than 1/1000 the thickness of the cell and occupies only 0.5% of the total volume. The technical difficulties in studying such a membrane are great and are compounded by the fact that a cell contains more than one kind of membrane.

Membranes from many sources have been studied. One of these is **myelin**, the multilayered insulation that surrounds the axons of many nerve cells.[55–57] Myelin is derived from the plasma membrane of **Schwann cells** which lie adjacent to many neurons. Schwann cells literally wrap themselves around neuronal axons. Their cytoplasm is squeezed out leaving little but tightly packed membrane layers. Myelin membranes are the most stable known and also have the highest lipid content (80%). Another readily available experimental material is the plasma membrane of the human red bood cell, which can be prepared by osmotic rupture of the cells. The remaining **erythrocyte ghosts** contain ~1% of the dry matter of the cell and may have been studied more than any other membrane. A much investigated specialized membrane is the outer portion of the visual receptor cells known as **rods** (Chapter 23), which contains a closely packed and regular array of flat discs, each one consisting of a pair of membranes. Both membranes and cell walls of many kinds of bacteria have also been investigated.

1. The Structure of Membranes

Membranes consist largely of protein and lipid. The ratio (by weight) of protein to lipid varies from 0.25 in myelin to ~3.0 in bacterial membranes. In membranes of erythrocytes it is about 1.2 and a ratio of about 1.0 may be regarded as typical for animal cells. Small amounts of carbohydrates (<5%) are present, as are traces of RNA (<0.1%).

In 1926, Gorter and Grendel calculated that the erythrocyte ghost contained just enough lipid to form a 3.0- to 4.0-nm-thick layer around the cell. Apparently they reached this correct conclusion only because their measurements of pressures of surface films contained compensating errors.[58] Nevertheless, this information, together with the known propensity of lipids to aggregate in **micelles** in which the hydrocarbon "tails" clustered together and the polar "heads" protruded into the surrounding water,[59] led Danielli and Davsen in 1935 to propose the **lipid bilayer** structure for membranes.[60] Its essential features are indicated in Fig. 8-4. Hydrophobic bonding holds the extended hydrocarbon chains together, while the polar groups of the phospholipid molecules may interact with proteins on the sides of the bilayer. The original proposal assumed an extended β structure for the proteins, which would allow them to coat the bilayer uniformly on both sides. However, this is not correct. Proteins are sometimes embedded in the bilayer, sometimes protrude through it, and sometimes are attached on one surface, most often the cytoplasmic surface of the plasma membrane. These concepts were brought together by Singer and Nicolson in 1972 in the **fluid mosaic model** of membrane structure[61] (Fig. 8-5). The lipid bilayer still provides the basic structure upon which the complex membranes of living organisms are assembled.[52,54,62–65] The term "fluid" refers to the fact

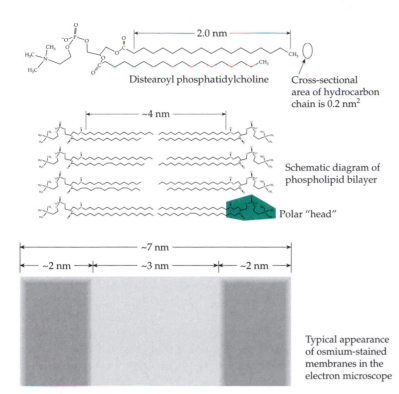

Distearoyl phosphatidylcholine

Cross-sectional area of hydrocarbon chain is 0.2 nm²

Schematic diagram of phospholipid bilayer

Polar "head"

Typical appearance of osmium-stained membranes in the electron microscope

Figure 8-4 Bimolecular lipid layers and membranes. (Top) A molecule of phosphatidylcholine. (Center) Lipid bilayer structure. (Bottom) Bilayer structure as seen by the electron microscope with osmium tetroxide staining.

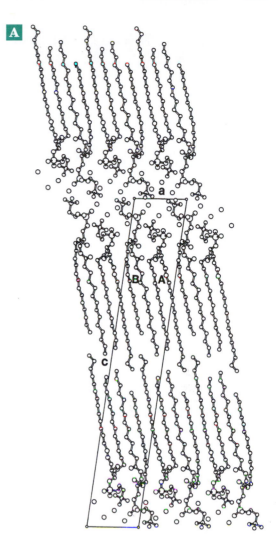

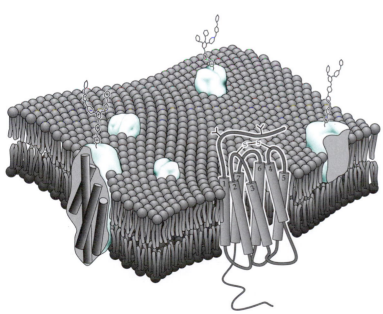

Figure 8-5 The fluid mosaic model of Singer and Nicolson.[61] Some integral membrane proteins, which are shown as irregular solids, are dissolved in the bilayer. Transmembrane proteins protrude from both sides. One of these is pictured as a seven-helix protein, a common type of receptor for hormones and for light absorption by visual pigments. Other proteins adhere to either the outer or the inner surface. Many membrane proteins carry complex oligosaccharide groups which protrude from the outer surface (Chapter 4). A few of these are indicated here as chains of sugar rings.

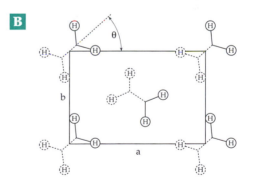

Figure 8-6 (A) Molecular packing of 2,3-dimyristoyl-D-glycero-1-phosphocholine dihydrate. The two molecules in the asymmetric unit are labeled 1 and 2. The position of the water molecules is indicated either by W1–W4 or by small open circles. Hydrogen bonds are represented by dotted lines. From Pascher *et al.*[66] (B) Two-dimensional "orthorhombic" arrangement of hydrocarbon chains in a crystalline alkane. The *a–b* plane corresponds to the plane of the bilayer surface; the long axes of the acyl chains project from the page. From Cameron *et al.*[67]

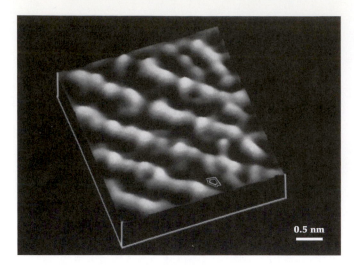

0.5 nm

Figure 8-7 Atomic force microscope image of a dimyristoyl-phosphatidylethanolamine bilayer deposited by the Langmuir–Blodgett technique (see Fig. 8-8) at a specific molecular area of 0.41 nm² and a surface pressure of 40 mN/m on a freshly cleaved mica substrate. The images were taken under water. The long, uniformly spaced rows are roughly 0.7–0.9 nm in spacing. The modulation along the rows, with rounded bright spots roughly every 0.5 nm, corresponds to the individual headgroups of the phosphatidylethanolamine molecules. The lattice spacing is identical to that measured by X-ray diffraction at the air–water interface. The area per molecule in the AFM image is ~0.4 nm². From Zasadzinski et al.[68]

that at temperatures suitable for growth and metabolism the hydrocarbon chains are not rigidly packed in the center of the bilayer but are "molten" (see Section 2). However, at a low enough temperature they become rigid and pack together in a manner similar to that of the chains in crystals of the phosphatidylcholine shown in Fig. 8-6A. These crystals consist of stacked bilayers of thickness 5.5 nm.[66,69] The scanning tunneling and atomic force microscopes have provided direct views of a similar arrangement of side chains in a monomolecular fatty acid layer[68,70,71] (Fig. 8-7). Measurements on multilamellar vesicles of dipalmitoylphosphatidylcholine give bilayer thicknessess from 5.4 nm for dehydrated vesicles to 6.7 nm for the biologically relevant fully hydrated bilayers.[72]

Lipids of membranes. Approximately 1500 different lipids have been identified in the myelin of the central nervous system of humans. About 30 of these are present in substantial amounts.[73] The distribution of the different lipids varies markedly between membranes from different sources (Table 8-3) making generalization difficult. However, phospholipids are apparently always present and, except in chloroplasts, make up from 40% to over 90% of the total lipid (Table 8-3).

Five kinds of phospholipid predominate: phosphatidylcholine, phosphatidylethanolamine, phosphatidylserine, phosphatidylglycerols, and sphingomyelin. Usually there are also small amounts of phosphatidylinositol. The major phospholipid in animal cells is phosphatidylcholine, but in bacteria it is phosphatidylethanolamine. The phospholipids of *E. coli* consist of 80% phosphatidylethanolamine, 15% phosphatidylglycerol, and 5% diphosphatidylglycerol (cardiolipin). Significant amounts of cardiolipin are found only in bacteria and in the inner membrane of mitochondria. Sphingomyelin is almost absent from mitochondria, endoplasmic reticulum, or nuclear membranes.

Glycolipids are important constituents of the plasma membranes, of the endoplasmic reticulum, and of chloroplasts. The cerebrosides and their sulfate esters, the sulfatides, are especially abundant in myelin. In plant membranes, the predominant lipids are the galactosyl diglycerides.[29,74] The previously described ether phospholipids (archaebacteria), ceramide aminoethylphosphonate (invertebrates), and sulfolipid (chloroplasts) are also important membrane components.

Cholesterol makes up 17% of myelin and is present in plasma membranes. However, it usually does not occur in bacteria and is present only in trace amounts in mitochondria. Related sterols are present in plant membranes. Esters of sterols occur as transport forms but are not found in membranes. Membrane bilayers, likewise, contain little or no triacylglycerols, the latter being found largely as droplets in the cytoplasm.

Quantitatively minor membrane components with important biological functions include **ubiquinone**, which is present in the inner mitochondrial membrane, and the **tocopherols**. Plant chloroplast membranes contain chlorophyll, carotenes, and other lipid-soluble pigments.

Liquid crystals, liposomes, and artificial membranes. Phospholipids dissolve in water to form true solutions only at very low concentrations (~10^{-10} M for distearoyl phosphatidylcholine). At higher concentrations they exist in **liquid crystalline phases** in which the molecules are partially oriented. Phosphatidylcholines (lecithins) exist almost exclusively in a **lamellar** (smectic) phase in which the molecules form bilayers. In a warm phosphatidylcholine–water mixture containing at least 30% water by weight the phospholipid forms **multilamellar vesicles**, one lipid bilayer surrounding another in an "onion skin" structure. When such vesicles are subjected to ultrasonic vibration they break up, forming some very small vesicles of diameter down to 25 nm which are surrounded by a single bilayer. These unilamellar vesicles are often used for study of the properties of bilayers. Vesicles of both types are often called **liposomes**.[75–77]

When liposomes are stained with osmium tetroxide or potassium permanganate, embedded, and sectioned for electron microscopy, their membranes show a characteristic three layered structure similar to that observed for biological membranes. Two darkly stained lines ~2–2.5 nm thick are separated by a clear space ~2.5–3.5 nm wide in the center. Both myelin and the retinal rod outer segments show closely spaced pairs of such membranes with a combined width of 18 nm. These results seemed to support the original Davsen–Danielli model. However, many questions must be raised about the interpretation of these results. Why does OsO_4 stain only the outer protein layer when it is known to react also with double bonds of hydrocarbon side chains of lipids to form osmate esters which are readily reduced to a diol and osmium?[77a-c,78] Membranes from which most of the lipid has been extracted still stain with OsO_4 to give three-layered electron micrographs. Perhaps little can be concluded from the three-layered appearance. We have learned that it is difficult to determine even the thickness, let alone the complete structure of an object that is only 6–10 nm thick.

Strong support for the lipid bilayer model comes from the preparation of another type of artificial mem-

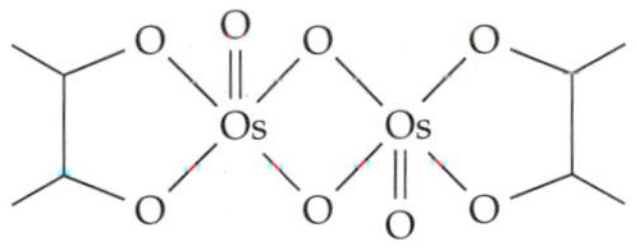

An osmate ester formed
from two unsaturated groups

brane which can be made from a solution of phosphatidylcholine or of a mixture of phospholipids plus cholesterol in a hydrocarbon solvent. A droplet of solution is placed on a small orifice in a plastic sheet, separating two compartments filled with an aqueous medium (Fig. 8-8). The solution in the orifice quickly drains, just as does a soap bubble, and the resulting film eventually becomes so thin that the bright colors disappear and a **"black membrane"** is formed. Similar membranes, but without a residual content of hydrocarbon solvent, have been created by apposition of two lipid monolayers formed at an air–water interface.[79,80] The thickness of such artificial membranes is thought to be only 6–9 nm. Resilient and self-sealing, the membranes can be stained with OsO_4 to give a typical three-layered pattern.

TABLE 8-3
Estimated Chemical Compositions of Some Membranes

Compound	Percentage of total dry weight of membrane[a]					
	Myelin (bovine)	Retinal rod	Plasma membrane (human erythrocyte)	Mitochondrial membranes	E. coli[b,c,d] (inner and outer membranes)	Chloroplasts[e]
Protein	22	59	60	76	75	48
Total lipid	78	41	40	24	25	52[f]
Phosphatidylcholine	7.5	13	6.9	8.8		
Phosphatidylethanolamine	11.7	6.5	6.5	8.4	18	
Phosphatidylserine	7.1	2.5	3.1			
Phosphatidylinositol	0.6	0.4	0.3	0.75		
Phosphatidylglycerol					4	
Cardiolipin[g]		0.4		4.3	3	
Sphingomyelin	6.4	0.5	6.5			
Glycolipid	22.0	9.5	Trace	Trace		23
Cholesterol	17.0	2.0	9.2	0.24		
Total phospholipid	33	27	24	22.5	25	4.7
Phospholipid as a percentage of total lipid	42	66	60	94	>90%	9

[a] Dewey, M. M. and Barr, L. (1970) *Curr. Top. Membr. Transp.* **1**, 6.
[b] Kaback, H. R. (1970) *Curr. Top. Membr. Transp.* **1**, 35–99.
[c] Mizushima, S. and Yamada, H. (1975) *Biochim. Biophys. Acta.* **375**, 44–53 .
[d] Yamato, I. Anraku, Y. and Hirosawa, H. (1975) *J. Biochem. (Tokyo)* **77**, 705–718. These investigators found 67% protein, 21% lipids, 10% carbohydrate, and 2% RNA.
[e] Lichtenthaler, H. K. and Park, R. B. (1963) *Nature (London)* **198**, 1070–1072.
[f] About 14% is accounted for by chlorophyll, carotenoids, and quinones.[e]
[g] Diphosphatidylglycerol (Fig. 8-2).

The study of monolayers formed on a water surface has also provided important information. A thin film of an amphiphilic (containing both polar and nonpolar groups) compound such as a fatty acid is prepared. This is done by depositing a small quantity of the compound dissolved in a volatile solvent on a clean aqueous surface between the barriers of a **Langmuir trough** (Fig. 8-8).[81,82] The difference in surface tension (π) across the barriers is measured with a suitable device[81] for different areas of the monolayer, i.e., for different positions of the moveable barrier. The value of π is low for expanded monolayers and falls to nearly zero when the surface is no longer completely covered. The pressure reaches a plateau when a compact monolayer is formed, after which it rises again (Fig. 8-8B). At very high values of π the monolayer collapses (buckles). Both the cross-sectional area per molecule in the monolayer and the collapse pressure can be determined. For typical fatty acids, regardless of chain length, the area covered is only ~ 0.2 nm^2 per molecule indicating that the fatty acid chains are stacked vertically to the surface in the monolayer. The collapse

pressure is higher for longer molecules as a result of the greater number of van der Waals interaction between the chains. **Langmuir–Blodgett layers** are prepared by transferring one or more monolayers onto a smooth solid surface (Fig. 8-7).[82,83]

Physical properties of membrane lipids.
A completely extended C$_{18}$ fatty acid chain as shown in Fig. 8-4 has a length of ~ 2.0 nm and occupies, either in crystals or in monolayers, when viewed "end-on," an area of ~ 0.2 nm^2. The hydrocarbon layer in a lipid bilayer containing such chains would have a thickness of about 4.0 nm; that determined by X-ray diffraction for myelin is ~ 3.5 nm. However, for artificial black membranes the thickness of the hydrocarbon layer can be as little as 3.1 nm when all solvent is removed.[84] These and many other results[85] indicate that the hydrocarbon chains are to some extent folded and that the membrane is expanded over that expected according to the simplest model.

Structure determinations on crystalline alkanes confirm that the chains exist in a completely extended conformation and that adjacent chains often pack together in the orthorhombic arrangement shown in Fig. 8-6B. As the temperature of such crystals is raised a series of solid–solid phase transitions is observed below the melting point of the crystals.[86] These can be detected by changes in the infrared absorption spectrum or by small amounts of heat absorption revealed by **differential scanning calorimetry** (Fig. 8-9). Each new phase permits a greater degree of mobility for the hydrocarbon chains. Thus, at a high enough temperature but below the melting point, the chains are able to rotate freely about their own axes in a so-called **hexagonal phase**. Now the chains are packed in a hexagonal array instead of the orthorhombic array of Fig. 8-6B. At intermediate temperatures, some of the chains may assume nonplanar conformations and changes in the tilt of the hydrocarbon chains (Fig. 8-6) may occur.

Similar phase transitions are observed for bilayers.[88–90] For dipalmitoyl phosphatidylcholine the first detectable **subtransition**[91] is centered at a temperature T_s of 18°C. The second, known as the **pretransition**, occurs at 35°C (T_p). The structure below T_s may be described as rigid or crystalline and that above T_s as a **gel** in which the hydrocarbon side chains twist and turn much more freely but in which the orthorhombic packing is maintained.[86] Above T_p the head groups become disordered. Although the orthorhombic packing of the tails may be maintained, there are several distinct phases,[92,93] including one or more in which the gel is thought to assume a structure analogous to that in the hexagonal phase of hydrocarbons. At 41°C the **main transition** occurs.

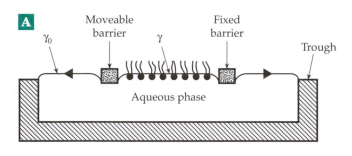

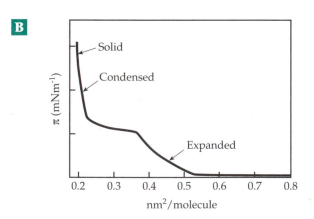

Figure 8-8 (A) The Langmuir–Adam film balance. Tension on the moveable barrier is recorded for different areas of the surface between the barriers. This gives the surface pressure π, which is the difference between the surface tension (γ_0) of a clean aqueous surface and that of a spread monolayer (γ): $\pi = \gamma_0 - \gamma$. Courtesy of Jones and Chapman.[81] (B) Surface pressure (π)–area per molecule isotherm for a typical fatty acid (e.g., pentadecanoic acid C$_{14}$H$_{29}$CO$_2$H) at the aqueous–air interface. From Knobler.[81a]

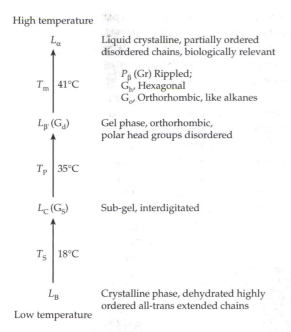

High temperature

L_α — Liquid crystalline, partially ordered disordered chains, biologically relevant

P_β (Gr) Rippled;
$G_{h'}$ Hexagonal
$G_{o'}$ Orthorhombic, like alkanes

T_m | 41°C

$L_{\beta'}$ (G_d) — Gel phase, orthorhombic, polar head groups disordered

T_p | 35°C

L_C (G_s) — Sub-gel, interdigitated

T_s | 18°C

L_B — Crystalline phase, dehydrated highly ordered all-trans extended chains

Low temperature

This is a sharper transition with a well-defined melting temperature designated T_m. Above T_m the lipid is in the lamellar liquid crystalline or L_α state. The bilayer continues to hold together, but the fatty acid chains have melted and are now free to rotate and undergo twisting movements more freely than at lower temperatures (Fig. 8-11). The main transition is highly, but not completely, cooperative. Thus, the melting of the membrane occurs over a range of several degrees. The presence in biological membranes of a variety of different components containing a variety of fatty acid chains leads to a broadening of the melting range.

The behavior of bilayers is strongly influenced by the lipid composition. Phospholipids containing saturated, long-chain fatty acids have high transition temperatures. The presence of unsaturated fatty acyl groups with *cis* double bonds in membrane lipids encourages folding of the hydrocarbon chains and lowers T_m. Even a single double bond lowers T_m, the decrease being greatest when the double bond is near the center of the chain.[94–96] While T_m for dipalmitoyl phosphatidylcholine is 41°C, that of 1,2-dipalmitoyl phosphatidyl-*sn*-glycerol, which lacks the phosphocholine head group, is 70°C. This falls to 11.6°C for the polyunsaturated 1-stereoyl-2-linoleoyl-*sn*-glycerol, whose melting curve is shown in Fig. 8-9.[87] This lipid also shows a complex phase behavior and a melting point for the stable, crystalline β' phase higher than that of the α phase.

Inclusion of other molecules of irregular shape within membranes also lowers T_m. However, a molecule of cholesterol can pack into a bilayer with a cross-sectional area of 0.39 nm², just equal to that of two hydrocarbon chains.[49] It tends to harden membranes above T_m but increases mobility of hydrocarbon chains below T_m.[97–100] A complex of cholesterol and phosphatidylcholine may form a separate phase within the membrane.[101,102] The ether-linked plasmalogens may account for over 30% of the phosphoglycerides of the white matter of the brain and of heart and ether linked phospholipids are the major lipids of many anaerobic bacteria.[103] Their T_m values are a few degrees higher than those of the corresponding acyl phospholipids.[104]

Between the pretransition temperature and T_m solid and liquid regions may coexist within a bilayer.[101] The term **lateral phase separation** has been applied to this phenomenon.[105,106] Since changes in the equilibrium between solid and liquid can be induced readily, e.g., by changes in the ionic environment surrounding the bilayer, lateral phase separation may be of significance in such phenomena as nerve conduction.[107]

The phase transitions in bilayers can be recognized in many ways. Differential scanning colorimetry has already been mentioned. Another approach is to measure the spacing between molecules by X-ray diffraction. The cross-sectional area occupied by a phospholipid in a bilayer is always greater than the 0.40 nm² expected for closest packing of a pair of extended hydrocarbon chains.[39,85]

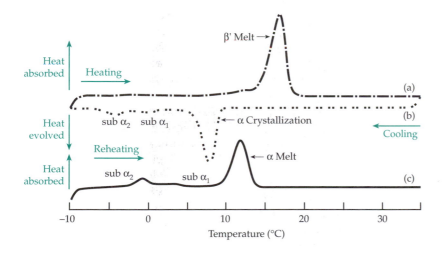

Figure 8-9 Differential scanning calorimetric curves for 1-stearoyl-2-linoleoyl-*sn*-glycerol. (A) Crystals of the compound grown from a hexane solution were heated from –10° to 35°C at a rate of 5°C per minute and the heat absorbed by the sample was recorded. (B) The molten lipid was cooled from 35° to –10°C at a rate of 5° per minute and the heat evolved was recorded as the lipid crystallized in the α phase and was then transformed through two sub-α phases. (C) The solid was reheated. From Di and Small.[87] Courtesy of Donald M. Small.

Below T_m the spacing between chains is about 0.42 nm corresponding to close packing of the fatty acid chains in a hexagonal array with an area per phospholipid of 0.41 nm^2. As the temperature is raised above T_m the spacing increases[85] to give an average area per phospholipid of 0.64–0.73 nm^2. Another technique (Box 8-C) is to study a **spin label** by EPR while yet another is to observe the fluorescence of a **polarity-dependent fluorescence probe** such as N-phenylnaphthylamine or other fluorescent probes[108] (see Chapter 23). The compound is incorporated into the membrane and undergoes changes in the intensity of its fluorescence when the state of the membrane is altered.

A variety of NMR techniques are being applied[109–113] both to liposomes and to natural membranes.[111,114] Incorporation of ^{13}C or ^{2}H into various positions in the hydrocarbon chains has allowed measurements of the relative degree of mobility of the chains at different depths in the bilayer (Fig. 8-10).[109,115–117] The results are in agreement with statistical mechanical predictions that configurational freedom increases with depth toward the midplane of the bilayer. Separation of a

bilayer into two or more phases can be observed using ^{2}H- or ^{31}P- NMR.[118–120] The orientation and dynamic behavior of various head groups has been explored,[110,121] as have effects of mixing into the bilayer other lipids such as glycosphingolipids[122] and cholesterol.[123,124] Crystalline phospholipids are being investigated by solid-state NMR.[125]

Fourier transform infrared spectroscopy[126,127] also provides information about conformation of both hydrocarbon chains and head groups. EPR spectroscopy (Box 8-C) with doxyl probes on carbon atoms at different depths within the bilayer has also been employed.[128]

In recent years **molecular dynamics simulations** have been used to predict behavior of membranes. As is indicated in Fig. 8-11, the molten interior of the liquid crystalline L_α state is portrayed clearly.[129–131] In the gel state the hydrocarbon chains maintain a closer packing and undergo coordinated movement.[88] It is difficult to know how realistic the simulations are. To calibrate the method efforts are made to correctly predict a series of known properties such as density and area per lipid (0.61 nm).[130]

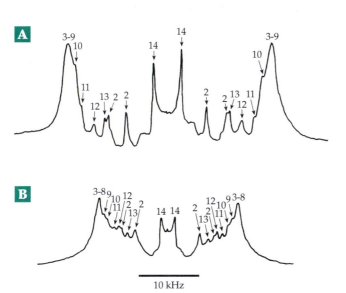

Figure 8-10 ^{2}H NMR spectra of dimyristoyl phosphatidylcholine-d$_{27}$/water in lamellar phases at 40°C. One chain of the phosphatidylcholine is fully deuterated, containing 27 atoms of ^{2}H. The mole ratios of water to lipid were 5.0 in (A) and 25.0 in (B). The average interfacial areas per alkyl chain as measured by X-ray diffraction were 0.252 nm^2 for (A) and 0.313 nm^2 for (B). ^{2}H NMR spectra are presented as "powder patterns" because the lipid molecules are randomly oriented in the magnetic field of the spectrometer as if in a powder. This gives rise to pairs of peaks symmetrically located on both sides of the origin. The separation distances are a measure of the quadrupole splitting of the NMR absorption line caused by the ^{2}H nucleus. The various splittings of the resonances of the 13 –CH$_2$– and one –CH$_3$ groups reflect differences in mobility.[109] The peaks have been assigned tentatively as indicated. From Boden, Jones, and Sixl.[115] Courtesy of N. Boden.

Functions of phospholipid head groups. The dipolar ionic head groups of phosphatidylcholine and phosphatidylethanolamine occupy about the same cross-sectional area as the two hydrocarbon tails. Thus, they are in rather close contact with each other. In crystals chains of hydrogen-bonded atoms may be formed. In phosphatidylethanolamine the phosphate and –NH$_3^+$ ions may alternate in these chains.[132]

In phosphatidylcholine, in which the nitrogen is surrounded by methyl groups and cannot form this kind of chain, water molecules bridge between the phosphates but the positive charges still interact with the adjacent negative charges.

The chains of hydrogen bonds between the head groups of phosphatidylethanolamine help to stabilize the bilayer and are apparently responsible for the elevation of T_m by 10–30° above that observed for phosphatidylcholine.[132] In contrast, the negatively charged carboxyl groups of phosphatidylserine make the membrane less stable. The melting point is increased if the pH is lowered, protonating these groups. Their presence also makes the membrane sensitive to the concentration of cations.[133] The same is true of phosphatidylglycerol, whose head group contains a negatively charged phosphate without an attached counterion. Addition of calcium ions increases T_m greatly and causes either phosphatidyglycerol or phosphatidylserine to form a separate phase with a more crystalline-like packing of the hydrocarbon side chains.[134] Hydrogen bonding between head groups also occurs with glycolipids.[135]

Non-bilayer structures of phospholipids.

Under appropriate conditions some aqueous phospholipids can exist in non-bilayer phases, a fact that may be of considerable biological importance.[119,136,137] In the presence of Ca^{2+} some pure phospholipids can be converted to the **inverted hexagonal** or H_{II} phase (Fig. 8-12).[136,138–140] In this phase the phospholipid heads are clustered together in cylindrical "inverted" micelles which pack in a hexagonal array. The ease with which this transition can occur is increased by the presence of small amounts of diacylglycerols or lysolecithins.[141] Some lipids, such as the galactosyldiacylglycerol of chloroplasts, do not form bilayers but prefer the hexagonal phase structure.[29,32a] This is

thought to be a result of very high curvature of a bilayer that arises from the sizes and packing of their head groups. Another phase, even though it is liquid, has a three-dimensional cubic symmetry.[142–144a] It apparently consists of a complex arrangement of polyhedral bilayer surfaces with interpenetrating water channels between them.[143]

Membrane fluidity and life.

In agreement with the known behavior of bilayers, the lipids of most membranes in all organisms are partially liquid at those temperatures suitable for life. Organisms have developed at least three distinct means of ensuring that membrane lipids remain liquid.[145] (1) In our bodies (as well as in *E. coli*) the unsaturated fatty acids that are present lower the melting point. Mutants of *E. coli* that are unable to synthesize unsaturated fatty acids cannot live unless these materials are supplied in the medium.[146] (2) In *Bacillus subtilis*, which contains no unsaturated fatty acids when grown at 37°C, and in other gram-positive bacteria, more than 70% of membrane

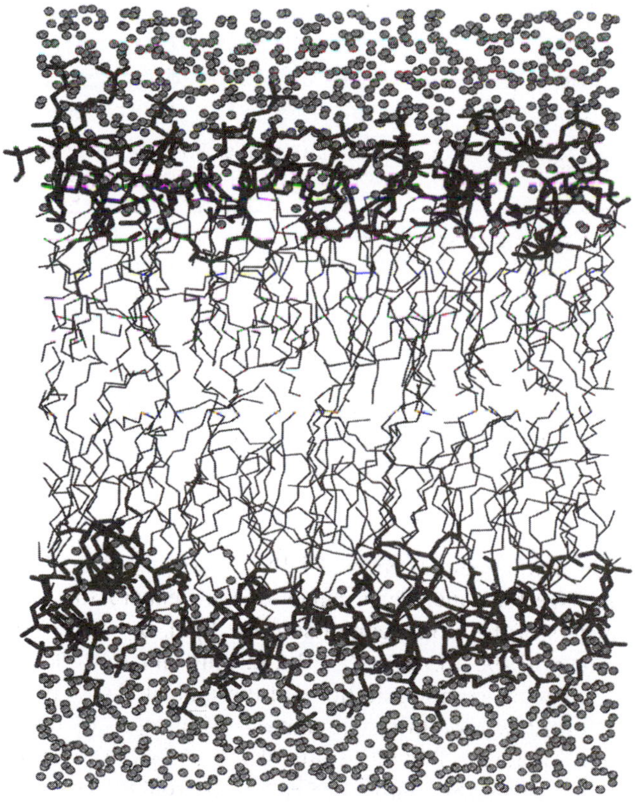

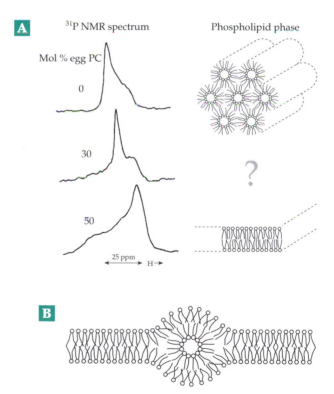

Figure 8-11 Results of simulated motion in a lipid bilayer consisting of 64 molecules of dipalmitoylphosphatidylcholine and 23 water molecules per lipid at a pressure of 2 atm and 50°C. The view is that observed after 500 ps of simulation. Bold lines represent the head group and glycerol parts of the structures and the thin lines the hydrocarbon chains. The gray spheres represent water molecules. From Berger, Edholm, and Jähnig.[130] Courtesy of Dr. Olle Edholm.

Figure 8-12 (A) ^{31}P NMR spectra of different phospholipid phases. Hydrated soya phosphatidylethanolamine adopts the hexagonal H_{II} phase at 30°C. In the presence of 50 mol% of egg phosphatidylcholine only the bilayer phase is observed. At intermediate (30%) phosphatidylcholine concentrations an isotropic component appears in the spectrum. (B) Inverted micelles proposed to explain "lipidic particles" seen in freeze fracture micrographs of bilayer mixture of phospholipids, e.g., of phosphatidylethanolanine + phosphatidylcholine + cholesterol. From de Kruijft *et al.*[119] Courtesy of B. de Kruijft.

BOX 8-C ELECTRON PARAMAGNETIC RESONANCE (EPR) SPECTRA AND "SPIN LABELS"

Unpaired electrons have magnetic moments and are therefore suitable objects for magnetic resonance spectroscopy. The technique is similar to NMR spectroscopy, but microwave frequencies of ~10^{10} Hz are employed, the energies being ~100 times greater than those used in NMR.[a,b] Unpaired electrons are found in organic free radicals and in certain transition metal ions, both of which are important to many enzymatic processes. Furthermore, **spin labels** in the form of stable organic free radicals, can be attached to macromolecules at many different points.[a,c–e] Coupling of such artificially introduced unpaired electrons with the magnetic moments of other unpaired electrons or of magnetic nuclei can often be observed by EPR techniques.

The conditions for absorption of energy in the EPR spectrometer are given by the equation

$$h\nu = g\beta H_o$$

which is identical in form to that for NMR spectroscopy. Here H_o is the external magnetic field strength and β is a constant called the Bohr magneton. The value of g, the **spectroscopic splitting factor**, is one of the major characteristics needed to describe an EPR spectrum. The value of g is exactly 2.000 for a free electron but may be somewhat different in radicals and substantially different in transition metals. One factor that causes g values to vary with environment is **spin-orbit coupling** which arises because the p and d orbitals of atoms have directional character. For the same reason g sometimes has three discrete values for the three different directions (g value anisotropy). The g value parallel to the direction of H_o ($g_{||}$) often differs from that in the perpendicular direction ($g_\perp$). Both values can be ascertained experimentally.

A second feature of an EPR spectrum is **hyperfine structure** which results from coupling of the magnetic moment of the unpaired electron with nuclear spins. The coupling is analogous to the spin–spin coupling of NMR (Chapter 3). The hyperfine splitting constant A, like the coupling constant J of NMR spectroscopy, is given in Hertz. Splitting may be caused by a magnetic atomic nucleus about which the electron is moving or by some adjacent nucleus or other unpaired electron. Sometimes important chemical conclusions can be drawn from the presence or absence of splitting. Thus, the EPR spectrum of a metal ion in a complex will be split by nuclei in the ligand only if covalent bonding takes place.

It is customary in EPR spectroscopy to plot the first derivative of the absorption rather than the

absorption itself. Thus, for the paramagnetic nitroxide 2,2,6,6-tetramethylpiperidine-1-oxyl the EPR spectrum consists of three equally spaced bands whose peaks are marked at the points where the steep

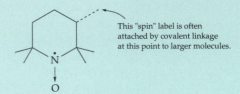

This "spin" label is often attached by covalent linkage at this point to larger molecules.

2,2,6,6-Tetramethylpiperidine-1-oxyl

lines in the middle of the first derivative plots cross the horizontal axis.

Coupling with the ^{14}N nuclear spin causes splitting into three lines as shown in the accompanying figure.

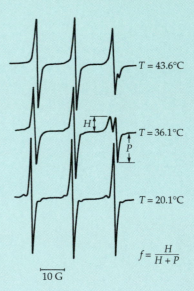

$$f = \frac{H}{H+P}$$

10 G

EPR spectrum of tetramethylpiperidine-1-oxyl dissolved in an aqueous dispersion of phospholipids. (Top) above the main bilayer transition temperature T_t; (center) between T_t and pretransition temperature; (bottom) below pretransition temperature. From Shimshick and McConnell.[f]

This nitroxide is more soluble in liquid regions of bilayers than it is in solid regions. As bilayers are warmed in the EPR spectrometer, the solubility of this spin-labeled compound in the lipid can be followed (see figure). The lower of the three spectra approximates that of the spin label in water alone, while the others are composite spectra for which part of the spin label has dissolved in the phospholipid bilayers.

BOX 8-C (continued)

Since frequencies for EPR spectroscopy are ~100 times higher than those for NMR spectroscopy, correlation times (Chapter 3) must be less than ~10^{-9} s if sharp spectra are to be obtained. Sharp bands may sometimes be obtained for solutions, but samples are often frozen to eliminate molecular motion; spectra are taken at very low temperatures. For spin labels in lipid bilayers, both the bandwidth and shape are sensitively dependent upon molecular motion, which may be either random or restricted. Computer simulations are often used to match observed band shapes under varying conditions with those predicted by theories of motional broadening of lines. Among the many spin-labeled compounds that have been incorporated into lipid bilayers are the following:

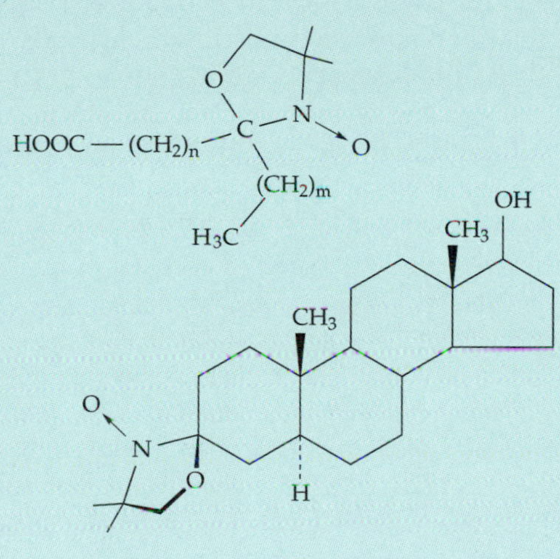

Much of the interpretation of the observed changes in EPR spectra of spin labels is empirical. For example, the spectra in the accompanying figure can be interpreted to indicate that the spin label dissolves in the lipid to a greater extent at higher temperatures. The ratio f (defined in the figure) is an empirical quantity whose change can be monitored as a function of temperature. Plots of f vs T have been used to identify transition and pretransition temperatures in bilayers.[f]

EPR spectroscopy is used widely in the study of proteins and of lipid–protein interactions.[c] It has often been used to estimate distances between spin labels and bound paramagnetic metal ions.[g] A high-resolution EPR technique that detects NMR transitions by a simultaneously irradiated EPR transition is known as electron-nuclear double resonance (ENDOR).[h]

[a] Berliner, L. J., and Reuben, J., eds. (1989) *Spinlabeling. Theory and Applications*, Vol. 8, Plenum, New York ((Biological Magnetic Resonance Series)

[b] Cantor, C. R., and Schimmel, P. R. (1980) *Biophysical Chemistry*, Freeman, San Francisco, California (pp. 525–536, 1352–1362)

[c] Marsh, D. (1983) *Trends Biochem. Sci.* **8**, 330–333

[d] Esmann, M., Hideg, K., and Marsh, D. (1988) *Biochemistry* **27**, 3913–3917

[e] Millhauser, G. L. (1992) *Trends Biochem. Sci.* **17**, 448–452

[f] Shimshick, E. J., and McConnell, H. M. (1973) *Biochemistry* **12**, 2351–2360

[g] Voss, J., Salwinski, L., Kaback, H. R., and Hubbell, W. L. (1995) *Proc. Natl. Acad. Sci. U.S.A.* **92**, 12295–12299

[h] Lubitz, W., and Babcock, G. T. (1987) *Trends Biochem. Sci.* **12**, 96–100

fatty acids contain methyl branches (Chapter 21)[147,148] which can decrease the melting point and increase the monolayer surface area by a factor of as much as 1.5. (3) Yet another mechanism for lowering the melting point of fats is the incorporation of **cyclopropane-containing fatty acids** (Chapter 21).

On the other hand, as we have already seen, cholesterol tends to reduce the mobility of molecules in membranes and causes phospholipid molecules to occupy a smaller area than they would otherwise. Myelin is especially rich in long-chain sphingolipids and cholesterol, both of which tend to stabilize artificial bilayers. Within our bodies, the bilayers of myelin tend to be almost solid. Bilayers of some gram-positive bacteria growing at elevated temperatures are stiffened by biosynthesis of bifunctional fatty acids with co-valently joined "tails" that link the opposite sides of a bilayer.[149]

Why must membrane lipids be mobile? One reason is probably to be found in the participation of

membranes in many vital transport processes. Biological membranes have a relatively high permeability to neutral molecules (including H_2O),[64,150] and it has been suggested that above T_m fatty acid chains are free to rotate by 120° around single bonds from trans to gauche conformations. When such rotation occurs about adjacent, or nearly adjacent single bonds, **kinks** are formed. If a kink originates near the bilayer surface, as will usually be the case, a small molecule may jump into the void created. Since the kink can easily migrate through the bilayer, a small molecule may be carried through with it.[151,152] The same factors may assist larger protein molecules which function in membrane transport. They probably also account for the substantial degree of **hydration** of bilayers which involves both the polar head groups and water diffusing through the nonpolar interior.[153]

Not only can molecules diffuse through membranes but also membrane lipids and proteins can move with respect to neighboring molecules. The rates of **lateral**

diffusion of lipids in bilayers and of antigenic proteins on cell surfaces are rapid.[80,154] If diffusion of phospholipids is assumed to occur by a pairwise exchange of neighboring molecules, the frequency of such exchanges can be estimated[155] as ~$10^7 s^{-1}$. However proteins may meet many obstacles to free diffusion.[156] Lateral diffusion is often measured by the technique of **fluorescence recovery after photobleaching**. One small spot in a bilayer that contains a dye attached to a lipid or a protein is bleached by a laser beam. Lateral diffusion of nearby unbleached molecules into the bleached spot can then be observed.[80] Lateral diffusion can also be observed by NMR spectroscopy[157] and by single-particle tracking.[158,159] In addition to diffusion there may often be a flow of membrane constituents in directions dictated by metabolism.[160] Although lateral diffusion is fast a "flip-flop" transfer of a phospholipid from one side of the bilayer to the other may require many seconds.[161] However, a sudden increase in the calcium ion concentration, an important intramolecular signal (Chapter 11), activates a "scramblase" protein which promotes a rapid transbilayer movement of phospholipids.[162]

Electrical properties of membranes. Biological membranes serve as barriers to the passage of ions and polar molecules, a fact that is reflected in their high electrical resistance and capacitance. The electrical resistance is usually 10^3 ohms cm^{-2}, while the capacitance is 0.5–1.5 microfarad (μF) cm^{-2}. The corresponding values for artificial membranes are ~10^7 ohms cm^{-2} and 0.6 – 0.9 μF cm^{-2}. The lower resistance of biological membranes must result from the presence of proteins and other ion-carrying substances or of pores in the membranes. The capacitance values for the two types of membrane are very close to those expected for a bilayer with a thickness of ~2.5 nm and a dielectric constant of 2.[54,84,163] The electrical potential gradient is steep.

Outer cell surfaces usually carry a net negative charge, the result of phosphate groups of phospholipids, of carboxylate groups on proteins, and of sialic acids attached to glycoproteins. This negatively charged surface layer attracts ions of the opposite charge (counterions), including protons, and repels those of the same charge. The result is development of a diffuse **electrical double layer** consisting of the fixed negative charges on the surface and a positive **ionic atmosphere** extending into the solution for a distance that depends upon the ionic strength.[164–166] This ionic atmosphere is analogous to that postulated by the Debye–Hückel theory (Chapter 6). At the physiological ionic strength of 0.145 M the thickness of the double layer, taken as the distance at which the electrical potential falls to a certain fraction of that at the cell surface,[164,165] is about 0.8 nm. However, the double-layer thickness increases to about three times this

value at an ionic strength of 10^{-3} M and to still greater distances at lower ionic strengths.

The net surface charge of a cell and the associated electrical double layer are important in interactions between cells and may influence the development of extracellular structure such as basement membranes. The net negative charge on cells also gives rise to an experimentally measurable electrophoretic mobility.

A characteristic of living cells is the maintenance of **ionic gradients** across the plasma membrane. Thus almost all cells accumulate K$^+$, even "pumping" it from very dilute external solutions. Cells also exclude sodium, pumping it out from the cytoplasm by mechanisms considered in Section C,2. If a microelectrode is inserted through a cell membrane and the potential difference is measured between the inside and outside of the cell, a **resting potential** which, in nerve cells, may be as high as 90 mV is observed. The origin of the potential appears to lie in the concentration differences of ions. From the value of ΔG for dilution of an ion (Eq. 6-25) and the relationship between ΔG and electrode potential (Eq. 6-63), the Nernst equation (Eq. 8-2) can be derived. According to this equation, which applies to a single ion for which the membrane is permeable,

$$E_\mathrm{m} = \frac{RT}{nF} \ln\left(\frac{c_1}{c_2}\right) = \frac{0.059}{n} \log\left(\frac{c_1}{c_2}\right) \quad \text{at } 25°C \tag{8-2}$$

a 10-fold concentration difference across the membrane for a monovalent ion ($n = 1$) would lead to a 59-mV membrane potential, E_m. Since membranes are relatively impermeable to sodium ions, it is generally conceded that for many membranes the origin of the membrane potential lies mainly with the potassium ion concentration difference which is maintained by the Na$^+$, K$^+$-ATPase (Section C). A more complete equation takes account of K$^+$, Na$^+$, and Cl$^-$ together with their respective permeabilities.[167–169] Note also that Eq. 6-64 is also often called the Nernst equation.[170]

Protons are also pumped across cytoplasmic and inner mitochondrial membranes, a topic of Chapter 18. The flow of protons from inside to outside also contributes to the membrane potential. The positive charges of H$^+$, K$^+$, and other cations associated with the external membrane surface are balanced by the negative charges of protein molecules as well as Cl$^-$ and phosphate anions that are in or near to the inner surface of the membranes.

Another possibility for proton flow has intrigued biophysicists for years. Membranes often display a substantial electrical conductivity in a lateral direction along the membrane surface.[171–173] Electrical conduction may involve movement of protons along hydrogen-bonded lines, e.g., involving ethanolamine head groups or phosphate groups and bridging water as previously discussed (see also Eq. 9-96). Alternatively, conduction may depend upon membrane-associated proteins.[174] This lateral proton conduction may be important to many proton-driven membrane processess, such as rotation of bacterial flagella, ATP synthesis, and pumping of ions (Chapter 18).

The two sides of a membrane.
Many observations indicate great differences between the inside and outside of the membranes that surround cells.[51,175,176] Bretscher and Raff[62] observed that, among the phospholipids of the erythrocyte membrane, phosphatidylcholine predominates in many mammals but is replaced by sphingomyelin in ruminants. Sheep erythrocytes are resistant to cobra venom phospholipase A, which is known to remove the fatty acid from the central position on the glycerol of phosphatidylcholine, causing lysis of the cells. The resistance of sheep erythrocytes suggested that the sphingomyelin is on the outside of the membrane while the phosphatidylethanolamine and other phospholipids are inside. By inference, phosphatidylcholine is also largely on the outside of plasma membranes. Supporting this conclusion is the observation that most of the reactive amino groups of phosphatidylethanolamine and phosphatidylserine are found on the inside (cytoplasmic) surfaces.[177] Since the total content of phosphatidylcholine and sphingomyelin often exceeds that of phosphatidylethanolamine and phosphatidylserine, the bilayer would be incomplete on the inside of the membrane were it not for the presence of proteins, which contribute more to the inside than to the outside surface.

Glycolipids are usually on the *outside* of plasma membranes with the attached sugar chains projecting into the surrounding water. An important generalization is that *sugar groups attached either to lipids or to proteins tend to be on outer cell surfaces or on materials that are being exported from cells.* An exception is found in the abundant galactolipids of chloroplasts.

2. Membrane Proteins

The many proteins present within or attached to membranes have a variety of functions. Some are obviously structural, tying other proteins to a membrane or providing a base for projecting fimbriae, flagella, and other appendages. Some proteins of the outer surface act as anchoring points for macromolecules that lie between cells. The inner surfaces of membranes are attached to the cytoskeleton (Chapter 7). Many of the proteins embedded in membranes control the passage of materials across membranes. Others serve as receptors that sense the presence of specific compounds or of light. Membranes may also contain foreign proteins such as subunits of virus coats. Proteins that are deeply embedded in membranes are referred to as **intrinsic** or **integral membrane proteins**. Proteins that are more loosely associated with the membrane, principally at the inner surface, are called **peripheral**.[61]

Integral membrane proteins.
Membrane proteins are hard to crystallize[178] and precise structures are known for only a few of them.[179–181] A large fraction of all of the integral membrane proteins contain one or more **membrane-spanning helices** with loops of peptide chain between them. Folded domains in the cytoplasm or on the external membrane surface may also be present. The best-known structure of a transmembrane protein is that of the 248-residue bacteriorhodopsin. It consists of seven helical segments that span the plasma membrane (Fig. 23-45) and serves as a light-activated proton pump. Other proteins with similar structures act as hormone receptors in eukaryotic membranes. A seven-helix protein embedded in a membrane is depicted in Fig. 8-5 and also, in more detail, in Fig. 11-6.

The most hydrophobic integral membrane proteins can be extracted into organic solvents such as mixtures of chloroform and methanol. One such **proteolipid protein**, the 23.5-kDa lipophilin, accounts for over half the protein of myelin.[57,182] The purified protein from rat brain contains 66% of nonpolar amino acids and six molecules of covalently bound palmitic acid and other fatty acids per peptide chain in thioester linkage to cysteine side chains. This protein evidently has four transmembrane helical segments with the six fatty acid chains incorporated into the membrane bilayer. It also has cytoplasmic and extracellular loops, one of which binds inositol hexakisphosphate (Ins P-6). (Fig. 11-9).[183] The myelin proteolipid is an essential component of the myelin sheath and defects in this protein are associated with some demyelinating diseases[57] which are discussed in Chapter 30.

There are many known topologies for helix-bundle membrane proteins with the number of membrane-spanning helices ranging from 1 to 14 or more e.g., see Fig. 8-23.[184–186] A topology can often be predicted using suitable computer programs.[187–191] A first step is to identify all sequences of 20 or more residues that could form a helix sufficiently hydrophobic to allow good nonpolar interactions with the bilayer core. Sequences that can form **amphipathic** (amphiphilic) helices must also be considered because two or more of these can pack together in a membrane with their hydrophilic sides together, sometimes forming pores (see Section C,1).[192]

Predictions of membrane protein structure are in part based on the **positive-inside rule** which states that positively charged lysine and arginine residues will not pass through a membrane but will remain on the negatively charged cytoplasmic surface. Often, N-terminal parts of a protein will pass through a membrane and will contain glutamate or aspartate residues whose side chains carry negative charges. These tend to remain outside of the membrane where they are attracted to the positive charges on the membrane outer surface. Positively charged residues may interact with phosphate groups of phospholipids and tyrosine and tryptophan side chains may interact with the carbonyl groups of the ester linkages.[192a,b]

After probable transmembrane helical regions have been identified a residue-by-residue attempt can be made to identify cytosolic and extracellular loops. Chemical reactivities of the naturally occurring side chains can also be examined. Residues within the helix bundle will be protected. Mutations can be prepared systematically by "**scanning mutagenesis**." For example, presumed extracellular loops in the erythrocyte band 3 protein (next section) have been mutated by introduction of N-glycosylation acceptor sites (Asn-X-Ser / Thr). If the sequence is at least 12–14 residues away from transmembrane sequences it will probably be glycosylated in a suitable laboratory test system.[193] Within a transmembrane domain amino acid residues can be systematically replaced with alanine ("alanine scanning mutagenesis") or with cysteine and effects on the protein can be observed. Substitution with cysteine allows another possibility: Two cysteines on adjacent transmembrane helices may be linked as disulfides if they are close enough.[194,195] Discovery of such neighboring pairs can be very valuable in attempting to establish relationships of one helix to another. A cloned gene can also be split into two pieces prior to crosslinking of cysteine side chains. This may facilitate mapping of tertiary interactions within transmembrane proteins.[196] Molecular dynamics methods for modeling helix bundle proteins are also being developed.[197]

Not all integral membrane proteins have a helix bundle structure. Some of the simplest transmembrane proteins are the subunits of bacterial viruses and pili (Figs. 7-7 and 7-9) The 7-nm α-helical rods of phage M13 have a 20-residue hydrophobic section (residues 25-46) which is preceded by a negatively charged sequence that appears to form a short *amphipathic* helix that lies on the external surface of the membrane and helps to anchor the subunit. A positively charged cluster at the opposite end of the hydrophobic helical region remains in the cytoplasm.[180,186,198] The **porins**, which form pores in outer membranes of bacteria and ribosomes, are large 16-strand β cylinders (Section C,1).[179,180] It has been suggested that this different basic architecture may be related to the fact that the porin polypeptides must be exported through the inner membrane before being refolded in the periplasm.[199] Keep in mind that a β strand of nine residues can span the 33-nm bilayer core just as well as can a 22-residue helix. Beta structures as well as shorter helices may well be present in proteins that also contain membrane-spanning helices.

A third important structural pattern involves extensive use of amphipathic helices that lie partially embedded in a membrane surface. For example, the blood lipoproteins are lipid particles partially coated by amphipathic helices (Chapter 21).[200,201]

Anchors for proteins. Proteins with membrane-spanning sequences are usually anchored into the membranes with the help of many polar side chains, often including glycosylated residues, in the ends and loops that protrude on the two sides of the membranes. Other proteins are anchored by insertion into membranes of nonpolar groups. These include fatty acyl groups at N termini, fatty acid ester groups on serine, threonine, or cysteine side chains, and polyprenyl groups attached to cysteine side chains as thioethers.[202] Many membrane-anchored proteins carry the saturated 14-carbon **myristoyl** group in amide linkage with an N-terminal glycine of the protein.[203–206] The fatty acid chain is added at the time of the protein synthesis, i.e., **cotranslationally**. It can intercalate into the membrane bilayer but provides a relatively weak anchor.[207] The 16-carbon **palmitoyl** group is typically added to a cysteine thiol in recognition sequences at various positions in a protein.[182,202,208,209] The modification occurs after protein synthesis, i.e., **posttranslationally**. Because thioesters are relatively unstable, palmitoylation is regarded as a reversible modification.[210]

A third lipid anchor is provided by the polyprenyl **farnesyl** (15-carbon) and **geranylgeranyl** (20-carbon) groups in thioether linkage to cysteine residues. These must be present in specific recognition sequences at the C termini of proteins, most often with the sequence CAAX.[211–215] The prenylation (also called isoprenylation) reaction is followed by proteolytic removal of the last three residues (AAX) and methylation of the new C-terminal carboxyl group as is discussed in Chapter 11, Section D,3. See also Chapter 22, Section A,4.

A lipoprotein present in the periplasmic space of *E. coli* is anchored to the outer bacterial membrane by a triacylated modified N-terminal cysteine containing a glyceryl group in thioether linkage as shown in the following structure (see also Section E,1).

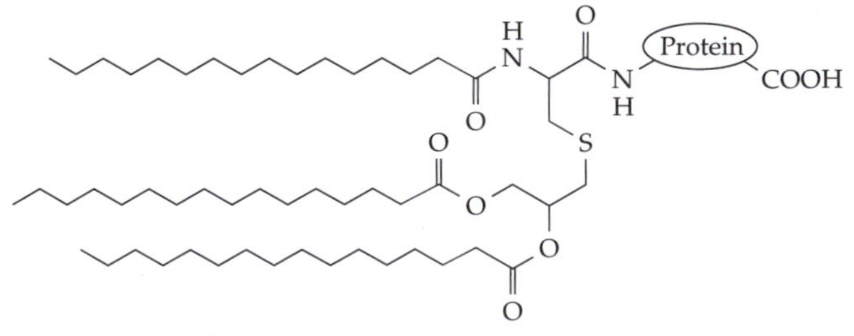

A related anchor that uses diphytanylglycerylation is found in certain proteins of archaeobacteria.[216]

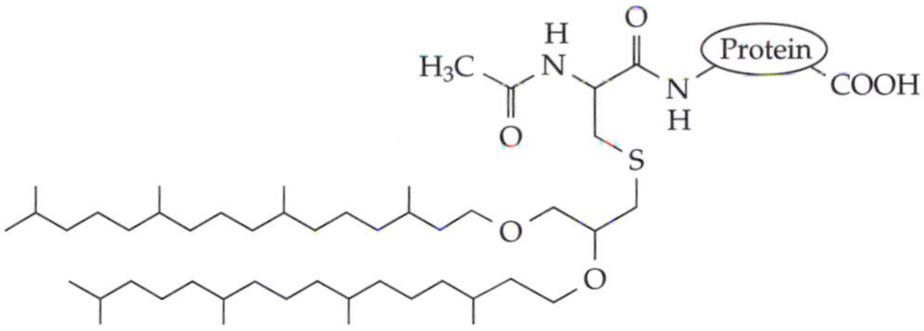

A series of **glycosylphosphatidylinositol** (GPI) anchors that are covalently linked to a variety of proteins utilize diacylglycerol or alkylacylglycerol for attachment to a bilayer. The proteins are joined through their C-terminal carboxyl groups to the diacylglycerol by a chain of covalently linked ethanolamine, phosphate, mannose, glucosamine, and *myo*-inositol as shown in Fig. 8-13.[217–223] The proteins are linked to the diacylglycerol through the conserved structure: H_2N-protein→ ethanolamine→P→6Manα1→2Manα1→ 6Manα1→4GlcNAc→Ins→diacylglycerol.

The structure in Fig. 8-13, which anchors the small Thy-1 antigen to surfaces of rat thymocytes,[224] contains additional mannose, *N*-acetylgalactosamine, and ethanolamine phosphate. These groups may be missing or substituted by other groups in other anchors.[225] For example, that of human erythrocyte acetylcholinesterase lacks the extra mannose and GalNAc of the Thy-1 anchor but contains a palmitoyl group attached to an oxygen atom of the inositol. This provides an additional hydrocarbon tail that can enter the bilayer.[218] Other PI-anchored proteins include enzymes, such as alkaline phosphatase and lipoprotein lipase, and surface proteins of the parasites *Trypanosoma*,[225]*Leishmania*, and *Toxoplasma*.[220] Some adhesion molecules and a variety of other outer surface proteins are similarly attached to membranes. Analysis of the genome of *Caenorhabditis elegans* suggests that the nematode contains over 40 GPI-tailed proteins and perhaps more than 120.[225a]

Analyzing erythrocyte membranes.

The proteins of red blood cell membranes were among the first to be studied. Because membrane proteins are present in small amounts and tend to be hard to dissolve without denaturation, they have been difficult to study.

They usually do not dissolve readily in water, but red cell membranes can be almost completely solubilized in water using a 5×10^{-3} M solution of the chelating agent EDTA (Table 6-9) or by 0.1 M tetramethyl ammonium bromide.[226] These observations suggested that ionic linkages between proteins, or between proteins and phospholipids, are important to membrane stability. Nonionic detergents such as those of the Triton X series or β-octylglucopyranoside also solubilize most membrane proteins,[178,227 – 229] whereas ionic detergents such as sodium dodecyl sulfate (SDS) often cause unfolding and denaturation of peptide chains.

Gel electrophoresis of plasma membrane proteins in SDS solution yields ~10 prominent bands and at

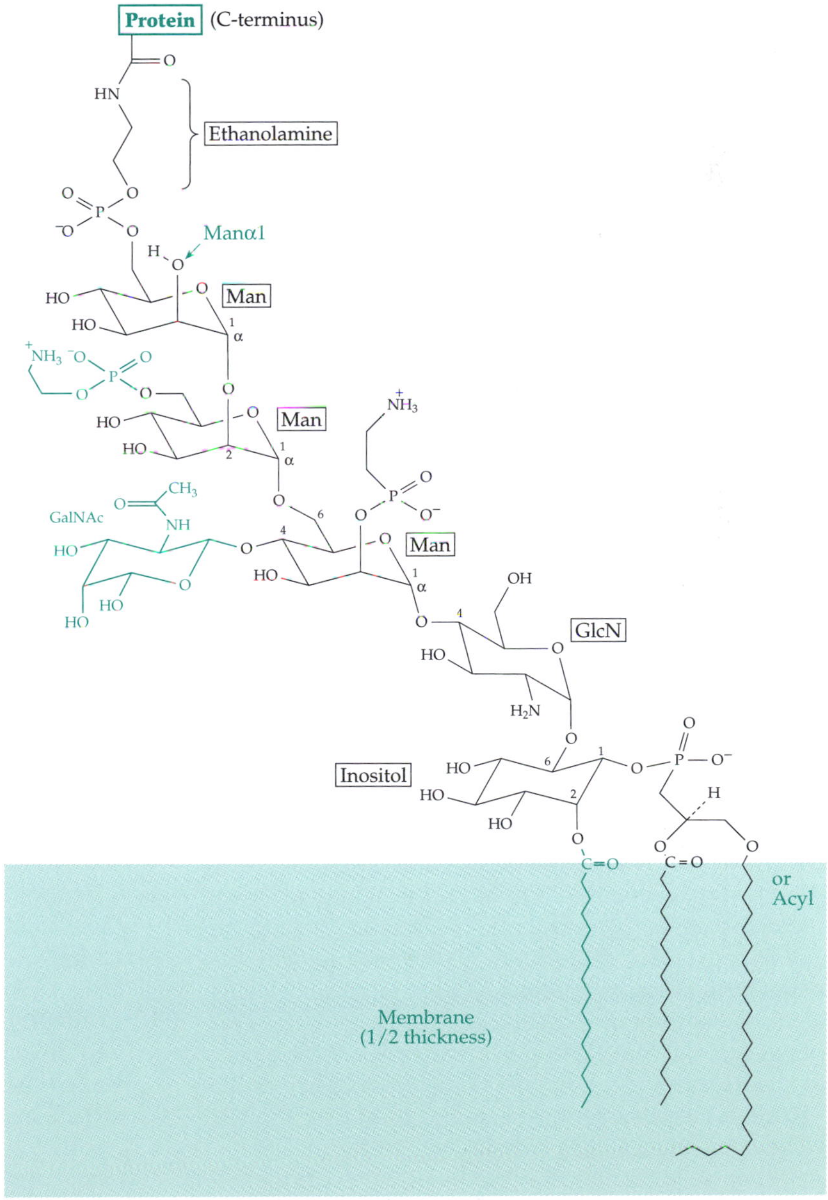

Figure 8-13 Structure of glycosylphosphatidylinositol (also called phosphatidylinositol-glycan) membrane anchors. The core structure is shown in black. The green parts are found in the Thy-1 protein and / or in other anchors.

Polyoxyethylene p-t-octyl phenol: the Triton X series of detergents.
n = 9–10 for Triton X-100 and 7–8 for Triton X-114.

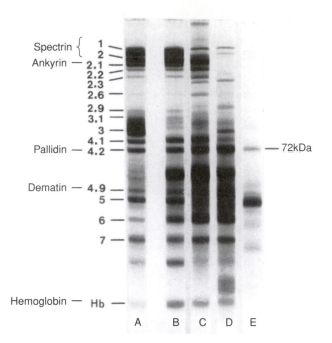

Figure 8-14 SDS-polyacrylamide gel electrophoresis of human erythrocyte ghosts. (A) From untreated cells. (B) From cells digested externally with chymotrypsin. (C) Inside-out vesicles prepared from cells pretreated with chymotrypsin. (D) The same inside-out vesicles after further treatment with chymotrypsin. (E) Polypeptides released by the chymotryptic treatment of the inside-out vesicles. The peptides are numbered according to the system of Steck[232]; Hb, hemoglobin. From Luna et al.[233]

least 30 less intense bands ranging in molecular mass from 10 to 360 kDa.[230,231] Proteins of erythrocyte ghosts, and the usual system for numbering them, are shown in Fig. 8-14. Some very important proteins known to be present in this membrane, such as $(Na^+ + K^+)$-activated ATPase (Section C,2), are found in such low quantities (e.g., a few hundred molecules in a single red blood cell)[62] that they do not show up in electropherograms. Mitochondrial membranes appear to be more complex than plasma membranes, but myelin has a somewhat simpler composition.

Glycoproteins. Many of the integral proteins of membranes are glycoproteins.[234–236] These may sometimes be recognized in electropherograms because they are stained by the periodic acid-Schiff (PAS) procedure. At least 20 glycoproteins are present in erythrocyte ghosts, and glycoproteins appear to be prominent protein components of the plasma membranes of all eukaryotes and of primitive archaebacteria such as *Halobacterium*.

The most abundant glycoprotein of red blood cell membranes is the 95-kDa PAS-reactive **band 3 protein** (Fig. 8-14) which makes up ~25% of the total membrane protein.[236–240] A variety of asparagine-linked oligosaccharides based on the core hexasaccharide structure shown in Chapter 4, Section D,2 are present and apparently project into the surrounding medium. Electron micrographs of freeze-fractured surfaces through the membrane bilayer (Fig. 8-15) show ~4200 particles of 8 nm diameter per square micrometer, randomly distributed and apparently embedded in the membrane. These probably represent dimers of the glycoprotein. The amino acid sequence of the 911-residue protein suggests 13 membrane-spanning helices in the 550-residue C-terminal domain and that the 41-kDa N-terminal domain projects inward into the cytoplasm.[236,240] Among the first 31 residues are 16 of aspartate or glutamate. These provide a highly negatively charged tail that is able to interact electrostatically with other proteins.[236] Among these are components of the cytoskeleton, which appears to be anchored to the membrane via the band 3 protein. The band 3 protein is also a substrate for transglutaminase (Eq. 2-23) which creates covalent crosslinks to other proteins.[237] Another major function of the band 3 glycoprotein is to form channels for the transport of anions (Section C,2).

Another integral glycoprotein of erythrocytes, the 31-kDa **glycophorin A** (PAS-1),[235,241–244] is ~60% by weight carbohydrate. Its 131-residue chain has a single ~23-residue membrane-spanning helix. The first 50 residues at the N terminus, which project from the outside surface of the membrane, include many serines and threonines. Their side chains carry 15 O-linked tetrasaccharides and one complex N-linked oligosaccharide. There are a total of ~160 sugar residues per peptide chain, largely *N*-acetylgalactosamine, galactose, and sialic acid. Some of the oligosaccharides contain MNO blood group determinants[234,241] (Box 4-C). Because of a high content of sialic acid, glycophorin also carries a large negative charge. The 35-residue C-terminal domain is hydrophilic and rich in proline, glutamate, and aspartate. It probably extends into the cytoplasm and may bind calcium ions or interact with $-NH_3^+$ groups on phospholipid heads.

If all the sugar residues of the glycophorin molecules in an erythrocyte were spread over the surface of the cell they could cover approximately one-fifth of its surface in a loose network. However, it is more likely that they form bushy projections of a more localized sort. These oligosaccharides not only act as immunological determinants but also serve as receptors for influenza viruses. Other glycoproteins related to glycophorin A occur in smaller amounts.[244]

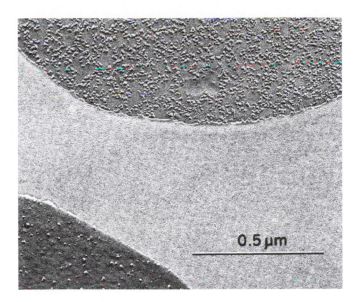

Figure 8-15 Freeze-fractured membranes of two erythrocyte "ghosts." The upper fracture face (PF) shows the interior of the membrane "half" closest to the cytoplasm. The smooth region is lipid and contains numerous particles. The lower face, the extracellular half (EF), possesses fewer particles. The space between the two is nonetched ice. See Figs. 1-4 and 1-15 for electron micrographs of sections through biological membranes. Courtesy of Knute A. Fisher.

Connections to the cytoskeleton.

About one-third of the protein of the red blood cell membrane is accounted for by a pair of larger hydrophobic peptides called **spectrin** with molecular masses of 280 kDa (α chain) and 246 kDa (β chain).[245–250a] These are found in bands 1 and 2 of Fig. 8-14. The spectrin monomers consist largely of 106- to 119- residue repeat sequences each of which folds into a short triple-helical bundle (Fig. 8-16B). The beaded-chain monomers associate readily to $\alpha\beta$ dimers, long ~100 nm thin flexible rods which associate further to $(\alpha\beta)_2$ tetramers. The latter, in turn, bind to monomers or to small oligomers of actin. In red blood cells the actin crosslinks the spectrin tetramers into a two-dimensional "fishnet" (Fig. 8-16A), the ~85,000 spectrin tetramers uniformly covering the entire inner surface (130 μm^2) of the erythrocyte.[245,251] The inner location of spectrin was established by the fact that chemical treatments[252] that covalently label groups on proteins exposed on the outer surface of erythrocytes (e.g., iodination with lactoperoxidase; Chapter 16) did not label spectrin.[257]

Spectrin also binds to one domain of another large 215-kDa peripheral protein called **ankyrin** (band 2.1, in Fig. 8-14) which anchors the spectrin network to the membrane.[258,259] Ankyrin is actually a multigene family of related proteins that are present in many metazoan tissues.[258,260] These are modular proteins with separate binding domains for spectrin and band 3 protein. The latter domain contains 24 **ankyrin repeats**, 33-residue modules, also found in a variety of other proteins.[258,259]

Ankyrin binds firmly to the band 3 glycoprotein which is embedded in the membrane. The 78 kDa band 4.1 protein is another major component of the erythrocyte **membrane skeleton**.[261,262] Protein 4.1 binds both to ankyrin and also to **glycophorin C**, providing another anchor to an integral membrane protein (Fig. 8-16C). Spectrin:actin:protein 4.1 in a 1:2:1 ratio are the major components of the membrane skeleton.[262] Other less abundant proteins include **adducin**, protein 4.2 (**pallidin**, which interacts with band 3 protein),[263] protein 4.9 (**dematin**, an actin bundling protein),[264] and the muscle proteins tropomyosin and tropomodulin (Chapter 19). While spectrin and ankyrin of erythrocyte membranes have been studied most intensively, related proteins occur in other cells.[265] Spectrin of brain and other tissues is also known as **fodrin**.[266] **Dystrophin** and **α-actinin**, actin-crosslinking proteins of muscle, are also members of the spectrin superfamily.[248] Protein 4.1 also occurs in various organisms,[267] in various tissues, and in various locations in cells.[268]

What is the function of the membrane skeleton? There is a group of hereditary diseases including **spherocytosis** in which erythrocytes do not maintain their biconcave disc shape but become spherical or have other abnormal shapes and are extremely fragile.[269–272] Causes of spherocytosis include defective formation of spectrin tetramers and defective association of spectrin with ankyrin or the band 4.1 protein.[265,273] Thus, the principal functions of these proteins in erythrocytes may be to strengthen the membrane and to preserve the characteristic shape of erythrocytes during their 120-day lifetime in the bloodstream. In other cells the spectrins are able to interact with microtubules, which are absent from erythrocytes, and to microtubule-associated proteins of the cytoskeleton (Chapter 7, Section F).[270] In nerve terminals a protein similar to erythrocyte protein 4.1 may be involved in transmitter release.[274] The cytoskeleton is also actively involved in transmembrane signaling.

Integrins and focal adhesions.

Mature erythrocytes have no nucleus and lack the microtubules and actin filaments that span other cells. In nonerythroid cells the major connections of the cytoskeleton to the membrane are through large $\alpha\beta$ heterodimeric membrane-spanning proteins called integrins (Fig. 8-17). These proteins, as well as the ends of the actin filaments, tend to be concentrated in regions long observed and described by microscopists as focal adhesions.[275–279] These are also sites of interaction with the external proteins that form the **extracellular matrix (ECM)**. There are at least 16 different integrin

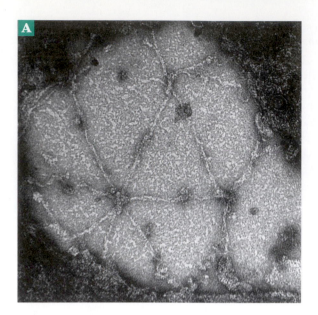

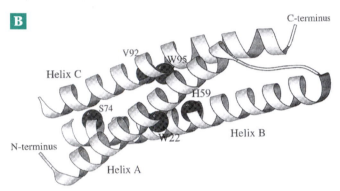

Figure 8-16 The erythrocyte membrane skeleton. (A) Electron micrograph showing a region of the membrane skeleton (negatively stained, X 200,000) and artificially spread to a surface area nine to ten times as great as in the native membrane. Spreading makes it possible to obtain clear images of the skeleton whose protein components are so densely packed and so subject to thermal flexing on the native, unspread membrane that it is difficult to visualize the individual molecules and the remarkably regular way that they are connected. The predominantly hexagonal and pentagonal network is composed of spectrin tetramers cross-linked by junctions containing actin oligomers and band 4.1 protein. Band 4.9 protein and tropomyosin are probably also bound to the oligomers, whose length (13 actin monomers long) corresponds to the length of a tropomyosin molecule. From Byers and Branton.[253,254] Courtesy of Daniel Branton. (B) Proposed triple α-helical structure of a single spectrin repeat unit.[255] Courtesy of Ruby I. MacDonald and Alfonso Mondragón. Each α spectrin chain consists of 20 and each β spectrin chain of 17 such repeats, which have only partially conserved sequences. The α and β chains are thought to associate in a side-by-side fashion and the αβ heterodimers in an end-to-end fashion to give tetramers. These are the rod-like structures seen in (A) and (C). (C) Cross section of the unspread membrane and cytoskeleton as pictured by Luna and Hitt.[256] Major interactions among components of the cytoskeleton are shown. Apparent sizes of the protein subunits, based on migration positions in SDS-polyacrylamide gel electrophoresis are: spectrin (260 and 225 kDa), adducin (105 and 100 kDa), band 3 (90 to 100 kDa), protein 4.1 (78 kDa), protein 4.2 (pallidin, 72 kDa), dematin (protein 4.9, 48 kDa), glycophorin C (~25 kDa), actin (43 kDa), and tropomyosin (29 and 27 kDa). Courtesy of Elizabeth J. Luna.

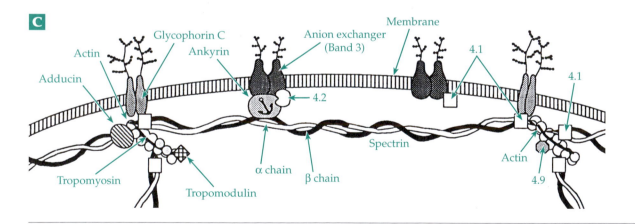

α subunits, whose external domains consist of up to 1114 residues, at least 8 β subunits with external domains of up to 678 residues, and at least 22 distinctly different αβ heterodimers.[280–282] The N-terminal part of each α subunit contains seven repeats of ~60 residues each, probably arranged as a β-propeller (see Fig. 11-7D or 15-23).[281] The integrins are structurally complex. Some contain a nucleotide-binding domain (see Figs. 2-13 and 2-27C).[283] Integrins have differing and quite exacting specificities toward the proteins of the external matrix to which they bind. They span the cell membrane and appear to be actively involved in communication between the cytoskeleton and external proteins.[275,277,284] They are often described as **receptors** for the proteins that bind to them.

Other cell adhesion molecules. Long before the discovery of integrins another class of transmembrane adhesion molecules were recognized as members of the immunoglobulin family. These were called cell

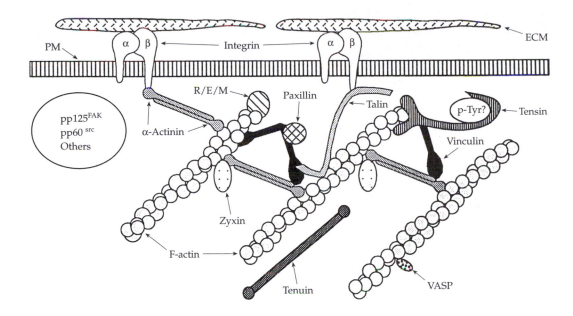

Figure 8-17 Working model of the protein–protein interactions in focal adhesions determined by *in vitro* binding experiments and immunolocalization. In addition, several interactions are of relatively low affinity in solution but may be enhanced at the membrane surface. Abbreviations are: ECM, extracellular matrix; PM, plasma membrane; p-Tyr-?, unknown phosphotyrosine-containing protein; R/E/M, member of the radixin/ezrin/moesin family; VASP, vasodilator-stimulated phosphoprotein. Diagram is modified from Simon *et al.*[285]

adhesion molecules (**CAMs**) by Edelman.[286,287] These molecules go by names such as intercellular adhesion molecule-1 (ICAM-1), neural cell adhesion molecule (NCAM),[288,289] liver cell adhesion molecule (LCAM), vascular cell adhesion molecule (VCAM), and platelet endothelial cell adhesion molecule (PECAM).[290,291] Many of these proteins, including the **T-cell antigen CD2**,[291a] were first recognized on leukocyte surfaces as differentiation antigens and are often designated by the "Cluster of Differentiation" names (Chapter 31). For example, ICAM-1 is also called CD54 and PECAM CD31.[290,292] The genes for these proteins are often expressed differentially in various tissues and the mRNA molecules formed undergo alternative splicing.[287,289] The extracellular domains of these adhesion molecules consist largely of Ig domains and most have a single transmembrane helix and a small cytoplasmic C-terminal domain. Some of the alternatively spliced forms are attached to the membrane by PGI tails.[287,289] ICAM-1 (Fig. 8-18) has five Ig domains and VCAM-1 has seven,[293] but some CAMs have only two.[294] The CAMs are glycoproteins, often with large *N*-linked oligosaccharides attached. The widely distributed N-CAM contains long $\alpha2\rightarrow8$ linked polysialic acid chains on two of the three *N*-glycosylation sites in the fifth Ig domain.[289] The CAMs are often referred to as **receptors**. Their ligands include surface proteins such as fibronectin (next section) but also the integrins, which are also called receptors. Integrins are **coreceptors** for

receptors of the Ig superfamily. Each of the coreceptors in a pair has binding sites for other ligands as well (Fig. 8-18). The CAMs and many other adhesion molecules are most abundantly expressed in embryonic tissues in which cells often migrate to new locations and for which the communication with neighboring cells is especially active. Many adhesion molecules bind only weakly and reversibly to their ligands, allowing cells to move.

The **cadherins** are calcium-dependent adhesion proteins that mediate direct cell–cell interactions.[295,296] The external parts of the cadherins also have repeated structural domains with the Ig fold.[297-298b] They have high affinity for each other, allowing cadherins from two different cells to interact and tie the cells together with a zipper-like interaction that is stabilized by the bound Ca^{2+} ions,[297,300] and may be relatively long-lived. The gene for cadherin E is often mutated in breast cancers and may be an important **tumor suppressor gene** (Box 11-D).[301]

Peripheral proteins of the outer membrane surface. Many integral membrane glycoproteins have their sugar-bearing portions exposed on the outer surface of the plasma membrane. Among these are receptors, ion pumps, and biochemical markers of individuality. In addition to these proteins, which are actually embedded in the bilayer, there are external peripheral proteins. One of the best known of these is

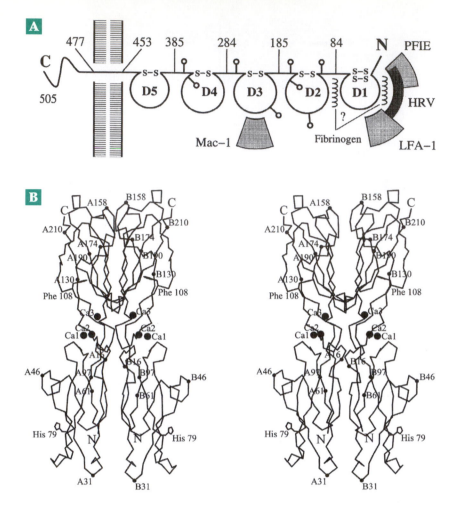

Figure 8-18 (A) Diagram of the ICAM-1 molecule. The structures labeled D1–D5 are the Ig domains. The glycosylation sites are labeled with small lollipops and approximate sites for binding of the α chain (CD11a) of the integrin called LFA-1, for the macrophage antigen Mac-1, for fibrinogen, for a ligand from erythrocytes infected by the malaria organism *Plasmodium falciparum* (PFIE), and for human rhinoviruses (HRV) are labeled. The binding sites are indicated schematically but each one is a complex interacting surface complementary to its ligand. From Bella *et al.*[299] Courtesy of Michael Rossman. (B) Structure of two N-terminal domains of E-cadherin. The molecules of the dimer are related by a noncrystallographic twofold symmetry axis running vertically in the plane of the page. Clusters of three calcium ions are bound in the linker regions, connecting the N- and C-terminal domains of each molecule and, in this view, are separated by the twofold axis. The N- and C-terminal domains are composed of seven-stranded β-barrels showing the same topology and similar three-dimensional structures. From Nagar *et al.*[296]

fibronectin (from *fibra*, "fiber", and *nectare*, "to bind").[302–309] This very large 470-kDa glycoprotein is a disulfide-linked dimer. Appearing under the electron microscope as having two 60-nm arms,[305] fibronectin molecules join together and surround animal cells, anchoring other proteins and carbohydrates of the ECM to the cells. Fibronectin binds tightly to several different cell surface integrins and also to collagen and to glycosaminoglycans of the matrix. It binds to the blood-clotting protein fibrinogen, to actin, and also to staphylococci and other bacteria. Each peptide chain of the fibronectin dimer is organized as several domains. Fibrin and staphylococci bind to the N-termi-

nal domain. A second domain binds collagen and a domain near the C-terminus binds heparin. Several different integrins of cell surfaces bind to the region of the 8th, 9th, and 10th type III repeats.[308] The specificity of the binding depends to a large extent on the presence of the specific tripeptide sequence Arg-Gly-Asp (**RGD**) in a type III repeat (Fig. 8-19B). Peptide sequences such as the PHSRN (Fig. 8-19B) and others[310] also participate in binding to specific integrins. Initial interactions are noncovalent but fibrin and collagen gradually become covalently attached to fibronectin through isopeptide linkages formed by transglutaminase (Eq. 2-23). This enzyme is abundant

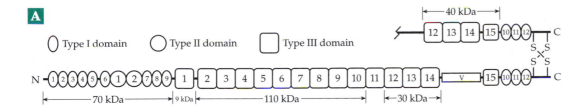

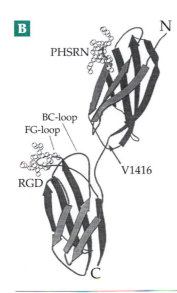

Figure 8-19 (A) Schematic diagram of a human fibronectin molecule showing one complete ~ 250-kDa chain consisting largely of 29 type I, II, or III fibronectin domains with a 12-kDa type III connecting segment or "V-region." The second chain, for which only the C-terminal portion is shown, is identical to the first except for the absence of the V-region as a result of alternative splicing of the mRNA. The two chains are joined in antiparallel fashion by a pair of –S–S– bridges near the C terminus.[313] The kDa sizes refer to fragments generated by the action of thermolysin. After Ingham *et al.*[306] (B) MolScript ribbon drawing of the structure of the ninth and tenth type III repeats in human fibronectin. The biologically active peptide segments PHSRN and RGD are represented in a stick and ball mode. The arrow marks valine 1416, next to which a polyglycine linker of various lengths has been inserted experimentally in order to study interactions between modules.[308] From Spitzfaden *et al.*[314]

in many developing tissues but often disappears as the tissues mature.[311,312] Fibronectin was formerly known as the "large external transformation-sensitive protein," a name derived from the fact that its quantity is greatly reduced in many virus-transformed cells. This might explain the loss of adhesiveness and of "contact inhibition" observed for cancer cells (Box 11-D).

There are 20 isoforms of human fibronectin. These arise by alternative splicing of the primary gene transcript. Their formation is developmentally regulated.[308,315] One of the isoforms is present as a soluble protein (cold-insoluble globulin) in blood plasma.[302] The fibronectins belong to a larger family of **cytoadhesins**, among which are the blood-clotting proteins **fibrinogen** and **von Willebrand factor** (Chapter 12), and also **thrombospondin**, **vitronectin**,[316] **tenascin**,[317–320] **laminin**,[321,322] **osteopontin**,[323] and **collagens**. These proteins all have a modular construction with repeated domains, often of several different types.[324] For example, tenascin contains 14 1/2 EGF domains (Table 7-3), 16 fibronectin type III repeats, and a C-terminal segment homologous to fibrinogens.[325] Laminin (see Fig. 8-33) also has EGF-like modules. Most of these proteins also have in common the presence of the RGD sequence, which binds the cytoadhesins to those integrins and other molecules that carry RGD receptor sites.[304,326–328] The extracellular matrix is discussed further in Section E,2.

3. Enzymes and Membrane Metabolism

Many of the proteins of membranes are enzymes. For example, the entire electron transport system of mitochondria (Chapter 18) is embedded in membranes and a number of highly lipid-soluble enzymes have been isolated. Examples are **phosphatidylserine decarboxylase**, which converts phosphatidylserine to phosphatidylethanolamine in biosynthesis of the latter, and **isoprenoid alcohol phosphokinase**, which participates in bacterial cell wall synthesis (Chapter 20). A number of **ectoenzymes** are present predominantly on the outsides of cell membranes.[329] Enzymes such as phospholipases (Chapter 12), which are present on membrane surfaces, often are relatively inactive when removed from the lipid environment but are active in the presence of phospholipid bilayers.[330,331] The distribution of lipid chain lengths as well as the cholesterol content of the membrane can affect enzymatic activities.[332]

Why are membranes so important to cells? Besides their obvious importance in enclosing and defining the limits of cells, membranes are the result of a natural aggregation of amphipathic molecules. They also represent a natural arrangement for boundaries between different aqueous phases within a cell. In addition, membranes are the "habitat" for many relatively nonpolar molecules formed by metabolism. These include proteins with hydrophobic surfaces and those with membrane anchors. The semiliquid interior of the membrane permits distortion of the bilayer and the addition or subtraction of proteins (and low-molecular-mass materials) in response to metabolic processes in the adjacent cytoplasm.

The principal factor providing stability to macromolecules and membranes is the hydrophobic nature of reduced organic compounds. This characteristic leads to the separation of lipids, proteins, and other molecules from the aqueous cytoplasm into oligomeric aggregates and membranes. However, the best catalysts, including most enzymes, are soluble in water. Thus, membranes represent thin regions of relative stability adjacent to aqueous regions in which chemical reactions occur readily and which tend to contain the more polar, the smaller, and the more water-soluble materials. The stability of membrane surfaces provides a means of bringing together reactants and of promoting complex sequences of biochemical reactions. For example, membranes contain both oxidative enzymes and reactive, dissolved quinones. The membrane–cytoplasmic interfaces may often be the metabolically most active regions of cells.

Despite their stability, membrane components have a metabolism of their own which is related to the high concentrations of oxidizing enzymes located in or on membranes. Oxidative reactions provide a mechanism for modification of hydrophobic membrane constituents. For example, sterols, prostaglandins, and other regulatory molecules are initially synthesized as hydrophobic chains attached to water-soluble "head groups" (Chapter 21). The hydrophobic products of these synthetic reactions tend to be deposited in membranes. However, attack by oxygen leads to introduction of hydroxyl groups and to a gradual increase in water solubility. As the hydrophilic nature of the compound is increased through successive enzymatic hydroxylation reactions, the hydrophobic membrane constituents eventually redissolve in the water and are completely metabolized. Another process that actively degrades membrane lipids is attack by hydrolytic enzymes such as the phospholipases.

C. The Transport of Molecules through Membranes

Small neutral molecules, such as water or ethanol, can penetrate membranes by **simple diffusion**.[64,150] The rate is determined by the solubility of a substance in the membrane, by its diffusion coefficient (see Eq. 9-24) in the membrane, and by the difference in its concentration between the outside and the inside of the cell. This concentration difference is commonly referred to as the **concentration gradient** across the membrane. The ease of diffusion through a membrane is described quantitatively by a **permeability**

coefficient P which is related to the diffusion coefficient D (Eq. 8-3).

$$J = -D_m K c / \Delta x = -P \Delta c \qquad (8-3)$$

Here J is the **flux** of molecules across the membrane, i.e., the number of molecules crossing one cm^2 per second. D_m is the diffusion coefficient in the bilayer, while K is a **partition coefficient**, the ratio of the concentration of the diffusing solute in the bilayer to that in water. The concentration gradient of the solute across the membrane is Δc, while Δx is the membrane thickness in centimeters. The permeability coefficient P for H_2O through biological membranes[333] is about $1 - 10 \ \mu m \ s^{-1}$. For H^+ and OH^- P is $0.1 \ \mu m \ s^{-1}$. For halide ions diffusing across liposome bilayers P ranges from 10^{-5} to $10^{-3} \ \mu m \ s^{-1}$,[334] fast enough to be of some biological significance. However, for most other ions P is less than $10^{-6} \ \mu m \ s^{-1}$. Because of their high lipid content membranes are quite permeable to nonpolar materials. For example, anesthetics usually have a high solubility in lipids, enabling them to penetrate nerve membranes.

1. Facilitated Diffusion and Active Transport

While simple diffusion may account for the entrance of water, carbon dioxide, oxygen, and anesthetic molecules into cells, movement of most substances is facilitated by protein **channels** and **transporters**.[335] Genes of 76 families of such proteins have been located in the genomes of 18 prokaryotes.[335a] Some of these provide for **facilitated diffusion**.[169,335] Like simple diffusion, it depends upon a concentration gradient and molecules always flow from a higher to a lower concentration. A distinguishing feature of facilitated diffusion is a **saturation effect**, i.e., a tendency to reach a maximum rate of flow through the membrane as the concentration of the diffusing substance, on the high concentration side, is increased. In this characteristic it is similar to enzymatic action (Chapter 9).

In **active transport** a material is carried across a membrane against a concentration gradient, i.e., from a lower concentration to a higher concentration. This process necessarily has a positive Gibbs energy change (as given by Eq. 6-25) of approximately $5.71 \log c_2 / c_1$ kJ mol^{-1}, where c_2 and c_1 are the higher and lower concentrations, respectively. The transport process must be coupled with a spontaneous exergonic reaction. In **primary active transport** there is a direct coupling to a reaction such as the hydrolysis of ATP to "pump" the solute across the membrane. **Secondary active transport** utilizes the energy of an **electrochemical gradient** established for a second solute; that is, a second solute is pumped against a concentration gradient and the first solute is then allowed to cross the membrane through an exchange process with the second solute (**antiport** or **exchange diffusion**).

Alternatively, both the first and the second solutes may pass through the membrane bound to the same carrier (**cotransport** or **symport**). Another form of active transport is **group translocation**, a process in which the substance to be transported undergoes covalent modification, e.g., by phosphorylation. The modified product enters the cell and within the cell may be converted back to the unmodified substance. Transport processes, whether facilitated or active, often require the participation of more than one membrane protein. Sometimes the name **permease** is used to describe the protein complexes utilized.

Like facilitated diffusion, active transport depends upon conformational changes in carrier or pore proteins, the equilibrium between the two conformations depending upon the coupled energy-yielding process. Thus, if ATP provides the energy a phosphorylated carrier will probably have a different conformation than the unphosphorylated protein. A carrier with Na^+ bound at one site may have a different affinity for glucose than the same carrier lacking Na^+. A hypothetical example of the kind of cycle that can function in active transport is provided by the picture of the "sodium pump" given in Fig. 8-25.

2. Pores, Channels, and Carriers

To accommodate the rapid diffusion that is often needed to supply food, water, and inorganic ions to cells, membranes contain a variety of small pores and channels. The pores may be nonspecific or they may be selective for anions or cations or for some other chemical characteristics. They may be permanently open or sometimes closed and referred to as **gated**. The gating may be controlled by the membrane electrical potential, by a hormone, by the specific ligand, or by other means. Some pores may be small enough to allow only small molecules such as H_2O to pass through. Others may be large enough to allow for nonspecific simple diffusion of molecules of low molecular mass. Structures are known for only a few.

Large pores tend to be nonspecific, but when the solute approaches the pore diameter in size the specificity increases. Furthermore, diffusion of ions through pores is influenced strongly by any charged groups in or near the pore. Thus, a cation will not enter a pore containing a net positive charge in its surface. Any electrical potential difference across the membrane, resulting from accumulation of excess negative ions within the cell, will also affect the diffusion of ions.[336,337]

Porins. The outer membranes of gram-negative bacteria contain several 34- to 38-kDa proteins known as porins. They form a large number of trimeric pores which allow molecules and ions with molecular masses <600 Da to enter. However, even small proteins are excluded. This appears to be a means for protecting the bacteria against enzymes such as lysozyme (Chapter 12). Four distinct porins are among the most abundant proteins of the outer membrane of *E. coli*.[338] They are designated according to the names of their genes. **OmpF** (a nonspecific, open channel) and OmpC (**osmoporin**) are encoded by *OmpF* (outer membrane protein F) and *OmpC* genes, respectively. **Maltoporin** (LamB) is selective for maltodextrins but also allows other small molecules and ions to pass.[338–341a] Its name comes from its original discovery as a receptor for bacteriophage lambda. PhoE (**phosphoporin**) is a porin with a preference for anions such as sugar phosphates while OmpF prefers cations.[342] Porin **FepA** is ligand gated, opening to take up the chelated iron from ferric enterochelin (Chapter 16). OmpF, OmpC, and PhoE all have 16-stranded β-barrel structures (Fig. 8-20).[199,343] Maltoporin forms a quite similar 18-stranded barrel. Similar porins are present in many bacteria.[199,344]

The porin monomers associate to form trimeric channels as is shown in Fig. 8-20B. They all have a central water-filled, elliptical channel that is constricted in the center to an "eye" ~0.8 x 1.1 nm in size. In this restriction zone the channel is lined with polar residues that provide the substrate discrimination and gating. For example, in OmpF and PhoE there are many positively and negatively charged side chains that form the edge of the eye (Fig. 8-20C). The electrostatic potential difference across the outer membrane is small, but apparently determines whether the porins are in an open or a closed state.[344a] The voltage difference has opposite effects on OmpF and PhoE, apparently as a result of the differing distribution of charged groups.[342,345,346] A key role in determining the voltage dependence may be played by Lys 18 (Fig. 8-20C).[342] Polyamines (Chapter 24), which are present in the outer membrane, induce closing of porin channels by binding to specific aspartate and tyrosine side chains.[347]

The most *abundant* protein in the *E. coli* outer membrane is OmpA. It appears to form a transmembrane helical bundle. Although it is regarded primarily as a structural protein it too acts, in monomeric form, as an inefficient diffusion pore.[350] Mitochondrial outer membranes contain nonspecific pores (mitochondrial porins) that allow passage of sucrose and other saccharides of molecular mass up to 2 to 8 kDa.[351,352] Similar pore-forming proteins have been found in plant peroxisomes.[353]

Aquaporins. Many biological membranes are not sufficiently permeable to water to allow for rapid osmotic flow. For example, the kidney membranes in portions of Henle's loop have permeabilities as high as 2500 μm s^{-1} (compared to 10–20 μm s^{-1} in a 1:1 cholesterol / phospholipid bilayer).[354-355a] This high permeability is provided by aquaporin-1 (**AQP-1**), formerly called CHIP (channel-forming integral membrane protein) or AQP-CHIP. Aquaporin-1 was first identified in erythrocyte membranes and is present in many tissues. The 28-kDa subunits of the protein form six-helix bundles, each with a pore in the center. These are associated as tetramers in the membrane.[356-358]

Other aquaporins with related sequences occur broadly. There are at least ten in mammals.[359,359a] Plants, which must accommodate to heavy loss of water in hot dry weather, have aquaporins in both plasma membranes and tonoplasts.[360] Bacteria also have aquaporins.[356,361] A defect in aquaporin-2 of the kidney collecting duct leads to **nephrogenic diabetes insipidus**, in which the kidneys fail to concentrate urine in response to secretion of the hormone vasopressin.[355a,362,363]

Ion channels. Most organisms contain a large number of ion channels. One of these, which plays a key role in nerve conduction, is the **voltage-gated K⁺ channel**. It is closed most of the time but opens when a nerve impulse arrives, dropping the membrane potential from its resting 50 - to 70-mV (negative inside) value to below zero. This voltage change opens the channel, allowing a very rapid outflow of K⁺ ions.[364-367] The channels then close spontaneously. There are many different K⁺ channels but most have a similar architecture.[368-370] The three-dimensional structure has been determined for the membrane-spanning part of the K⁺ channel of *Streptomyces lividans* (Fig. 8-21). The funnel-shaped tetrameric molecule has a narrow **conduction channel** which contains the **selectivity filter**.

The conduction channel is lined largely with hydrophobic groups. The selectivity filter, which discriminates between K⁺ and Na⁺, is a short (~1.2-nm-long) narrow (~1.0-nm-diameter) portion of the

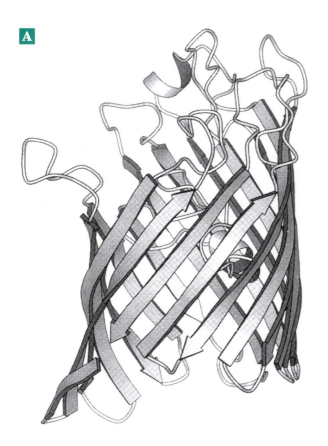

A

B

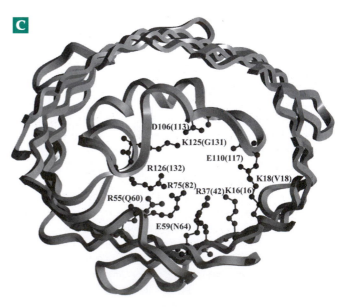

C

Figure 8-20 MolScript ribbon drawings of the OmpF porin of *E. coli.* (A) View of the 340-residue monomer. (B) View of the trimer looking down the threefold axis. From Watanabe *et al.*[348] From atomic coordinates of Cowan *et al.*[349] (C) Molecular model of the constriction zone of the PhoE porin. Locations of key residues are shown, with positions of homologous residues in OmpF given in parentheses. Extracellular loops have been omitted. Constructed from coordinates of Cowan *et al.*[349] by Samartzidou and Delcour.[342] Courtesy of Anne Delcour.

pore that is aligned roughly with the center of the bilayer. It is formed by four extended peptide chains, one from each subunit, each having the "**signature sequence**" for K^+ channels, TVGYG. In this sequence the peptide carbonyl groups all point into the channel. The consecutive groups of four carbonyls along the channel form binding sites for K^+, whose ionic diameter (Table 6-10) is 0.27 nm. The K^+ must lose its hydration sphere to fit into the 1.0-nm channel. The site of strongest binding, occupied by K^+ in Fig. 8-21C, lies just at the C-terminal ends of four helices, and the partial negative charges of the helix dipoles probably contribute to the binding. The pore may constrict to strengthen the bonds. When a second K^+ ion enters the channel it appears to bind ~0.75 nm from the central K^+, repelling it and weakening its interaction with the filter, allowing it to pass through the pore. Rb^+ (0.30 nm diameter) and Cs (0.34 nm diameter) also pass through.

How are the smaller Na^+ (0.19 nm diameter) and Li^+ (0.12 nm diameter) excluded from the pore? The pore is too small for the hydrated ions and perhaps too large to bind the dehydrated ion well enough to let it escape from its hydration sphere. Four negatively charged side chains in the cytoplasmic mouth of the pore presumably discourage anions from entering. At the other end, the extracellular entryway is a site that can be blocked specifically by 35- to 40-residue scorpion toxins.[368]

The *S. lividans* K^+ channel is *not* voltage gated. Voltage-gating mechanisms must be learned from study of other channels! The voltage-dependent K^+ channels from the rat have an $\alpha_4\beta_4$ composition. The α and β subunits coassemble in the endoplasmic reticulum and remain as a permanent complex.[370a–c] After insertion into the plasma membrane the α subunits form a channel as in Fig. 8-21. However, an additional intracellular domain of

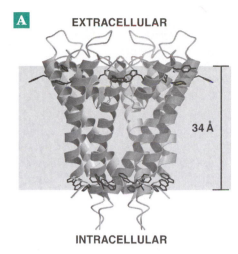

EXTRACELLULAR

34 Å

INTRACELLULAR

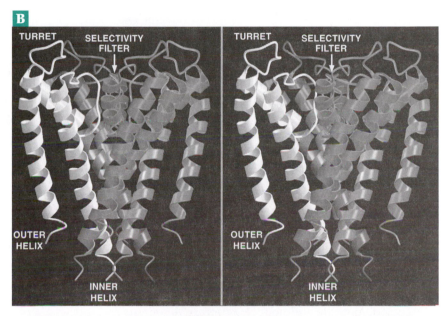

TURRET   SELECTIVITY FILTER   TURRET   SELECTIVITY FILTER

OUTER HELIX   OUTER HELIX

INNER HELIX   INNER HELIX

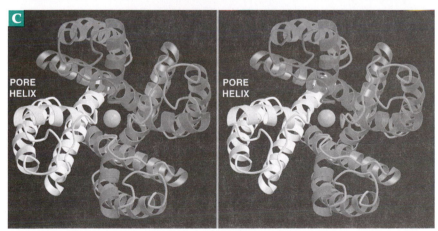

PORE HELIX   PORE HELIX

Figure 8-21 Views of the tetrameric K^+ channel from *Streptococcus lividans*. (A) Ribbon representation as an integral membrane protein. Aromatic amino acids on the membrane-facing surface are also shown. (B) Stereoscopic view. (C) Stereoscopic view perpendicular to that in (B) with a K^+ ion in the center. From Doyle *et al.*[366]

the α subunits, together with the β_4 complex, provide an elaborate and as yet poorly understood internal structure. The β subunits are oxidoreductases containing bound NADH, whose function is also uncertain.

There are many more pores and carriers of various kinds in biological membranes. Some, like the *S. lividans* K$^+$ channel, facilitate diffusion of a single ion or compound. They are **uniporters**. Others promote cotransport in which an ion such as H$^+$ or Na$^+$ also passes through the carrier. In some cases a pore is formed by a single molecule. Other pores follow a twofold, threefold, or fourfold axis (Fig. 8-21) of an oligomeric protein.[371] Two different conformations for the protein, one in which a "gate" opens to one side of the membrane and one in which it opens to the other, are usually involved. Interconversion between the two conformations is spontaneous but the equilibrium between them may be influenced by the binding of the solute, by the membrane potential, or by binding of inhibitors or activators. The latter may bind differently to the parts of the carrier (pore) protein exposed on the two sides of the membrane.

Channel-forming toxins and antibiotics.

Some of the bacterial toxins known as **colicins** (Box 8-D) kill susceptible bacteria by creating pores that allow K$^+$ to leak out of the cells. One part of the **complement system** of blood (Chapter 31) uses specific proteins to literally punch holes in foreign cell membranes. **Mellitin**, a 26-residue peptide of bee venom,[372,373] as well as other hemolytic toxins and antibiotic peptides of insects, amphibians, and mammals (Chapter 31) form amphipathic helices which associate to form voltage-dependent anion-selective channels in membranes.[374–377]

The polypeptide antibiotic **gramicidin A** consists of 15 nonpolar residues of alternating D- and L-configuration; two molecules can form a channel with a right-handed β helix structure.[378,379] The central 0.48-nm hole in the β helix is large enough to allow unhydrated cations such as Na$^+$ or K$^+$ to pass through. The same peptide, under other conditions, forms left-handed helical channels whose structures are known.[379] Aggregates of β_{10} helices may form channels between helices of **suzukacillin**.[380] For this antibiotic as well as for **alamethicin** (Chapter 30) the conductance depends upon the membrane potential, a characteristic shared with the pores of nerve membranes. A variety of synthetic channel-forming peptides have been prepared. Some of these form α-helical bundles with a central pore. A five helix bundle of this type[381] has a structure reminiscent of the acetylcholine receptor channel of neurons.[382]

Ionophores and other mobile carriers.

Facilitated diffusion of a molecule or ion is sometimes accomplished by binding to a **mobile carrier**. An example is the diffusion of a complex of K$^+$ with the low-molecular-mass lipid-soluble carrier, or ionophore **valinomycin** (Fig. 8-22). The K$^+$–valinomycin complex diffuses the short distance to the other side of the membrane and discharges the bound ion. If the rates of binding to a carrier and of release from the carrier are greater than those of the diffusion process, Michaelis–Menten kinetics are observed. The maximum velocity $V_{\max}$ and Michaelis constant K_m can be defined as in Eq. 9-15 for enzymatic catalysis.

Valinomycin is a **depsipeptide** which contains ester linkages as well as amide linkages. The antibiotic is made up of D- and L-valine, L-lactic acid, and D-hydroxyisovaleric acid. When incorporated into an artificial membrane bathed in a K$^+$-containing medium, valinomycin increases the conductance greatly and when it is added to a suspension of *Streptococcus faecalis* cells the high ratio of [K$^+$]$_i$ / [K$^+$]$_o$ falls rapidly.[383] The loss of K$^+$ from cells probably explains the antibiotic activity. However, under suitable conditions, with a high external [K$^+$], the bacteria will continue to grow and reproduce in the presence of the antibiotic.[384]

Uncomplexed valinomycin has a more extended conformation than it does in the potassium complex.[385,386] The conformational change results in the breaking of a pair of hydrogen bonds and formation of new hydrogen bonds as the molecule folds around the potassium ion. Valinomycin facilitates potassium transport in a passive manner. However, there are cyclic changes between two conformations as the carrier complexes with ions, diffuses across the membrane, and releases ions on the other side. The rate of transport is rapid, with each valinomycin molecule being able to carry ~10^4 potassium ions per second across a membrane. Thus, a very small amount of this ionophore is sufficient to alter the permeability and the conductance of a membrane.

Because the stability constant of its complex with potassium is much greater than that with sodium, valinomycin is a relatively specific potassium ionophore. In contrast, the mushroom peptide **antamanide** has a binding cavity of a different geometry and shows a strong preference for *sodium* ions.[388,390] The structure of the Na$^+$–antamanide complex is also shown in Fig. 8-22B. The *Streptomyces* polyether antibiotic **monensin** (Fig. 8-22D),[389,391] a popular additive to animal feeds, is also an ionophore. However, its mode of action, which involves disruption of Golgi functions, is uncertain.[392]

Anions of lipid-soluble phenols such as 2,4-dinitrophenol can serve as effective carriers of *protons* (Chapter 18). However, proteins usually serve as the natural carriers, both of protons and of other ions. A protein is sometimes pictured as rotating to present the solute-binding surface first to one side, then to the other side of a membrane. However, gated pores or channels are probable for most biological transport.

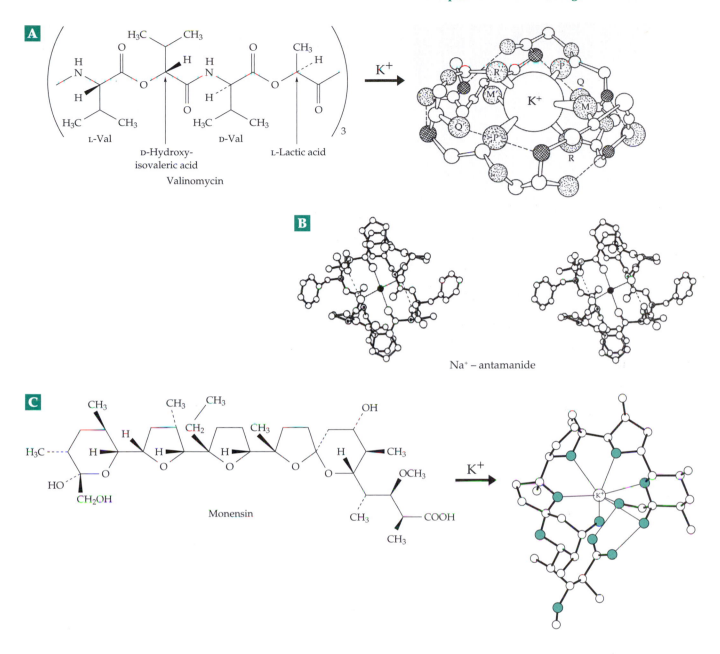

Figure 8-22 (A) Uncomplexed valinomycin and its complex with K^+ (from Duax *et al.*[387]). (B) Stereodiagram of the Na–[Phe^4, Val^6]antamanide complex. A molecule of C_2H_5OH, which forms the fifth ligand to the Na^+, is omitted for clarity. From Karle *et al.*[388] (C) Monensin and its complex with K^+. From Pangborn *et al.*[389]

3. The 12-Helix Major Facilitator Superfamily

A large family of transmembrane facilitators from bacteria and eukaryotes appear to consist largely of 12 transmembrane helices with intervening cytosolic and extracellular loops. Some of these transporters facilitate simple uniport diffusion, but others participate in active transport of the symport or antiport type.[393,394] Several hundred members of the family are known.[395]

Entrance of sugars into cells. It is important that sugars be able to enter cells rapidly. However, the

permeability coefficient P for D-glucose across a lipid bilayer is only 10^{-6} to 10^{-5} µm s^{-1}. For an intact erythrocyte, P is much greater: ~1 µm s^{-1}. This is the result of facilitated diffusion by a transport protein with a high specificity for hexose and pentose sugars having a pyranose ring in a C1 chain conformation.[396] This human erythrocyte glucose transporter, now known as **GLUT1**, has a K_m of 1.6 mM for D-glucose but of >3 M for L-glucose. It is a 55-kDa intrinsic membrane glycoprotein migrating in band 4.5 of Fig. 8-14. From the sequence of its cloned gene the unglycosylated carrier was deduced to be a 54-kDa peptide of 492 residues.

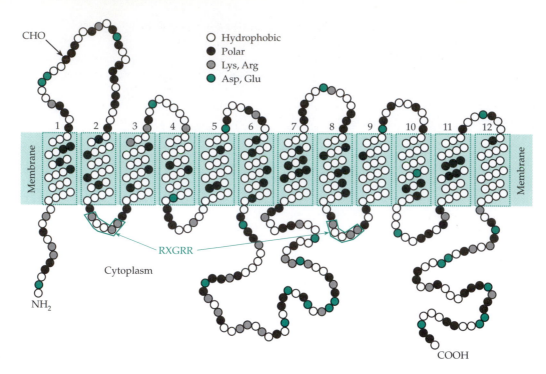

Figure 8-23 Predicted topology of the human glucose transporter GLUT1. The 12 predicted helices are numbered and the single external N-linked oligosaccharide is marked CHO. The sequence RXGRR (marked) is found in many 12-helix transporters. Its occurrence in these two positions suggests that the transporters may have evolved by duplication of a 6-helix motif. However, the human glucose transporters otherwise show no sequence similarity to other 12-helix transporters. After Bell *et al.*[398] See also Muekler.[400]

The glucose transporter is specifically inhibited by the fungal metabolite **cytochalasin B** which can also be used for photoaffinity labeling of the transporter.[397]

Cytochalasin B

There are at least six closely related facilitative sugar transporters in the human body.[398–400] GLUT1 is found not only in red cells but also in brain[401] and other tissues. GLUT2 is the principal liver transporter and GLUT3 is found along with GLUT1 and GLUT5 in the brain.[401] GLUT5 is primarily a fructose transporter.[402,403] The latter is present in spermatozoa, in which fructose transport is especially important (Box 20-A), and also in the small intestine. Both the GLUT1 and GLUT5 genes are overexpressed in breast cancer.[403]

GLUT6 is an unexpressed pseudogene (see Chapter 27) and GLUT7 has been found only in liver microsomes. GLUT4 of skeletal and cardiac muscle and adipose tissue has received a great deal of attention because of its response to insulin[400,404-405a] and to exercise[406] (see Chapter 17).

All of the GLUT family appear to be 12-helix transmembrane-regulated, gated-pore proteins[407,408] with relatively short cytosolic N- and C-terminal ends. The proposed topology of GLUT1, which has been supported by much experimental data,[400,406] is shown in Fig. 8-23. However, the helices are thought to be bundled with the glucose channel centered within a helix bundle. The structure is unknown, but it could resemble the α,α-barrel of glucoamylase (Fig. 2-29), which has 12 helices, a glucose-binding site in the center, and a structure that could easily be modified to form a central pore.

Dehydroascorbate, the oxidized form of vitamin C (Box 18-D) is also transported into cells by GLUT1 and GLUT3.[409] A related transporter carries L-fucose into mammalian cells.[410] Another facilitates the uptake of galactose in yeast.[411]

Cotransport of sugars and other nutrients with H⁺ or Na⁺. Epithelial cells of the small intestine or of kidney tubules must take up glucose at low concentrations and discharge it into the bloodstream at a higher concentration.[412] This active transport is accomplished by cotransport of glucose with Na^+ in a 1:2 ratio, with the sodium ion concentration gradient across the membrane providing a usable source of energy amounting to 5.8 kJ / equivalent of Na^+.[413] The 662-residue human, sodium-dependent transporter **SGLT1** may have 14 transmembrane helices, 5 of which have been proposed to provide the sugar pathway.[414]

Cotransport with Na^+ is also observed for transport of many other sugars, amino acids, neurotransmitters, and cofactors.[415] A confusing variety of transporter molecules have been identified and are now being classified into families based on gene sequences as well as function.[416,417] Transporters from intestinal mucosal cells, kidney membranes, and synaptic endings of neurons have been studied most and have been the source for many of the cloned genes.[418] The Na^+-dependent transporters of neutral amino acids from these tissues have long been classified as system A, system B, and system ASC (alanine, serine, cysteine) according to substrate specificities.[417,419-420a] Both cationic amino acids and cystine are taken up by kidney tubules and intestinal epithelial cells by another Na^+-dependent transporter which is defective in human **cystinuria**, a common metabolic genetic problem.[421,422]

The brain contains several transporters specialized for rapid uptake of neurotransmitters glutamate and aspartate,[423-425] glycine,[426] γ-aminobutyrate (Gaba),[427,428] and catecholamines and also for taurine, L-proline,[429] serotonin,[430] and other substances. Many of these are not only Na^+ dependent but also require cotransport of Cl^-.[428-430] There are several different glutamate transporter genes with specialized distribution in the brain and other tissues.[423,424]

Cotransport of sugars with H^+ is especially common in bacteria[431] but also occurs in eukaryotes. For example, the alga **chlorella** employs hexose / H^+ symporters.[432] The most investigated H^+ cotransporter is probably the **lactose** (*lac*) **permease** from *E. coli* which enables *E. coli* to take up lactose and other β-galactosides from very dilute solutions.[433-436] From a variety of measurements it has been possible to propose a stacking arrangement for the 12 helices[434] and to identify groups that are essential for function of the 417-residue protein. Glutamates as well as an arginine, and a histidine, all in transmembrane regions, are essential and may be involved in the gating and transport functions.[436] The 469-residue **melibiose permease** of *E. coli* transports α-D-galactopyranosides, including meliobiose (Gal*p*α1 → 6 Glc) and raffinose (Gal*p*α1 → 6 Glc*p*α1 → 2 Fru*f*).[437,438] This permease will couple sugar uptake to either the H^+ or Na^+ gradient (or to a gradient of Li^+ but *not* of K^+).[437] Cells of *E. coli* also contain H^+ symporters for D-galactose, D-xylose, and L-

arabinose that are homologous to the mammalian GLUT proteins.[439]

A series of specific H^+-linked cotransporters are found in the brush border membranes of small intestine and in kidney epithelial cells.[440,441] In green plants H^+-linked cotransport of amino acids is used in the distribution of amino acids synthesized in the roots and leaves to other parts of the plants.[442,443]

Antiporter or ion exchange transporters are also common. For example, *E. coli* uses a metal ion–tetracycline / H^+ transporter to carry the antibiotic tetracycline out of cells. This protein, when present, provides a high level of antibiotic resistance to the bacteria.[444]

4. Active Transport Systems

Both bacteria and eukaryotes also possess complex active transport systems for uptake of sugars, amino acids, and other nutrients and for pumping out toxic xenobiotics. One of the most important groups are the high affinity **ABC** (ATP-binding cassette) **transporters**, also called **traffic ATPases**.[445-448a] The *E. coli* genome contains genes for 80 ABC transporters, 54 of which had been identified before the genome sequence was completed.[447] In the human body an ABC transporter enables eukaryotic cells to pump out a large number of different drugs and other foreign compounds. This **multidrug resistance protein** (or P-glycoprotein) not only protects cells but also can seriously interfere with drug treatment. For example, cancer cells with increased amounts of this transporter often arise during chemotherapy. The transporter protein is a single 1280-residue, 170-kDa chain which probably has 12 transmembrane helices and two ATP-binding domains.[448,449] Other human ABC transporters include the antigen processing transporter **TAP** (Chapter 31), the 1480-residue anion transporter **CFTR**, which is defective in cystic fibroses (Box 26-C),[450] the erythrocyte glutathione-conjugate exporter (Box 11-B), and a long-chain fatty acid transporter of peroxisomes. ABC transporters usually consist of four domains. Two are hydrophobic intrinsic membrane domains, each with six membrane-spanning helices and two are peripheral membrane ATP binding domains. All four domains may be in a single peptide chain, as in CFTR, or they may be separate smaller proteins as in bacterial periplasmic permeases.[431,446]

Periplasmic permeases. Gram-negative bacteria contain numerous ABC transporters with components located on the periplasmic surfaces of their plasma membranes. Many of these can be dissociated from the surfaces by **osmotic shock**, i.e., by sudden changes in the osmotic pressure of the medium.[451,452] For example, cells of *E. coli* suspended in 0.5 M sucrose, treated with 10^{-4} M EDTA for 10 min, and then diluted with cold water release ~50 **binding proteins** that hold sugars, amino acids, ions, and other substances tightly with $K_f = 10^6–10^8$ M^{-1}.[453,454] An example is the **L-arabinose binding protein** of *E. coli*, a 306-residue peptide organized into two large α/β units with a deep cleft between them. The sugar is bound into this cleft[455,456] by an extensive network of hydrogen-bonding interactions between polar groups in the sugar and in the protein similar to the network that binds either D-galactose (Fig. 4-18) or D-glucose to another of the periplasmic binding proteins.[457,458] Proteins with the same general architecture bind D-ribose, L-arabinose,[458] and maltodextrins.[459] The α1,4-linked maltodextrins, as well as maltose, are major nutrients for *E. coli* and enter the periplasmic space via the previously men-

tioned LamB porin. Other periplasmic proteins bind histidine,[460] basic amino acids,[461] branched chain amino acids, only leucine, oligopeptides, polyamines,[462] and the tetrahedral anions phosphate[454] and sulfate.[463] The sulfate^{2-} anion is held to its binding protein by seven hydrogen bonds–one from a serine –OH, one from an indole ring NH, and five from peptide NH groups, three of which are at the positive ends of α helices. No permanently charged groups nor cations nor water molecules come into contact with the SO_4^{2-}.[463] The phosphate-binding protein also binds a tetrahedral dianion HPO_4^{2-} but it doesn't bind sulfate. It forms numerous hydrogen bonds with its ligand and also an ion pair with a guanidinium group. An aspartate carboxylate hydrogen bonds to the –OH

BOX 8-D COLICINS: ANTIBIOTIC PROTEINS

Certain strains of *E. coli* and related bacteria synthesize proteins known as colicins that kill cells of other susceptible strains.[a–c] Three kinds of colicins, each encoded in its own small DNA plasmid (colicinogenic factor), are known. **Colicin E3** is a 58-kDa ribonuclease (RNase) that attacks 26S ribosomal RNA of susceptible bacteria;[d] **colicin E2** is a deoxyribonuclease (DNase) that cleaves the bacterial chromosome.[e] **Colicin E1** and its relatives, colicins A, B and Ia, Ib and N, attack the bacterial inner membranes and form lethal pores which allow K$^+$ and other ions to flow out of the cell. The presence of a single channel will kill the bacterium. The effect is similar to that of valinomycin.[f,g] The small colicin V is an 88-residue peptide antibiotic that is secreted by a dedicated ABC export system.[h] Colicins are members of a larger group of **bacteriocins**. One of these proteins, megacin Cx from *Bacillus megaterium*, kills bacteria of sensitive strains by blocking protein synthesis.[i]

The channel-forming colicins bind to bacterial surface molecules that serve as their receptors. For example, colicin N binds to the abundant *E. coli* surface protein OmpF. Interaction with a complex of membrane proteins known as tol Q, R, A, and B then leads to translocation across both outer and inner membranes and refolding of the colicin in a pore-forming conformation.[g] This mechanism is also used by colicins A, E, and K. The single-stranded bacteriophages M13, fd, and f1 "parasitize" the same transport system. Colicins B, D, I, and M enter bacteria with the aid of a second transport system consisting of proteins TonB, ExbB, and ExbD which participate in uptake of chelated iron (Chapter 16) and of vitamin B$_{12}$. This system is also parasitized by bacteriophages T1, T5, and φ80.[j]

The N-terminal portion of the 522-residue polypeptide chain of colicin E1 appears to be required for transport into the membrane and the central part for binding to the receptor; the channel-forming property is characteristic of the C-terminal region.[k] A similar organization has been established for the smaller colicin N:translocation domain, (residues 1–66), receptor domain, (residues 67–182), and pore-forming domain (residues 183–387).

The colicin E1 plasmid is a 4.43 MDa circular double stranded DNA molecule consisting of 6646 base pairs.[l] Only one site is susceptible to cleavage by the restriction endonuclease ECoR1 (Chapter 26) This feature has led to its widespread use in cloning of genes.

[a] Luria, S. E. (1975) *Sci. Am.* **233**(Dec), 30–37
[b] Parker, M. W., Tucker, A. D., Tsernoglou, D., and Pattus, F. (1990) *Trends Biochem. Sci.* **15**, 126–129
[c] Parker, M. W., and Pattus, F. (1993) *Trends Biochem. Sci.* **18**, 391–395
[d] Escuyer, V., Boquet, P., Perrin, D., Montecucco, C., and Mock, M. (1986) *J. Biol. Chem.* **261**, 10891–10898
[e] Dvhsllrt, K., and Nomura, M. (1976) *Proc. Natl. Acad. Sci. U.S.A.* **73**, 3989–3993
[f] Wiener, M., Freymann, D., Ghosh, P., and Stroud, R. M. (1997) *Nature (London)* **385**, 461–464
[g] Evans, L. J. A., Labeit, S., Cooper, A., Bond, L. H., and Lakey, J. H. (1996) *Biochemistry* **35**, 15143–15148
[h] Fath, M. J., Zhang, L. H., Rush, J., and Kolter, R. (1994) *Biochemistry* **33**, 6911–6917
[i] Brusilow, W. S. A., and Nelson, D. L. (1981) *J. Biol. Chem.* **256**, 159–164
[j] Derouiche, R., Bénédetti, H., Lazzaroni, J.-C., Lazdunski, C., and Lloubès, R. (1995) *J. Biol. Chem.* **270**, 11078–11084
[k] Griko, Y. V., Zakharov, S. D., and Cramer, W. A. (2000) *J. Mol. Biol.* **302**, 941–953
[l] Chan, P. T., Ohmori, H., Tomizawa, J.-I., and Lebowitz, J. (1985) *J. Biol. Chem.* **260**, 8925–8935

of the phosphate. Since the sulfate ion lacks the necessary H to form this bond it is excluded from the binding site.[454]

All of these binding proteins have similar architectures. The ligands fit into a groove between two domains as is seen in Fig. 8-24 for the histidine binding protein. The histidine is held by formation of carboxylate–arginine and $-NH_3^+$–aspartate ion pairs and an additional hydrogen bond to the imidazole. The proteins are able to bend in the hinge region between the two domains to give a better fit to their ligands.

The periplasmic binding proteins function together with the other subunits of the ABC transporter system. One of the best understood systems is encoded by the histidine transport operon of *Salmonella typhimurium*.[445,453] There are four genes: *hisJ* (encoding the histidine-binding protein), *hisQ*, *hisM*, and *hisP*. The Q and M proteins are hydrophic integral membrane proteins which interact with two copies of the P protein, which contains the ATP-binding motif, to form a His QMP_2 membrane complex. The soluble HisJ transfers its bound histidine to this complex, which with the hydrolysis of ATP supplying the driving force transfers the histidine across the membrane, presumably via a channel.[445] In *E. coli* the gene *malE* encodes the periplasmic maltose binding protein, while *malF* and *malG* are genes for the integral plasma membrane components. The *malK* gene encodes an ATP-binding protein homologous to the *hisP* protein. As with the histidine permease, a multiprotein complex $MalFGK_2$ is formed. It accepts maltose or a maltodextrin and ATP is hydrolyzed to drive the transport.[459] *MalT* encodes a positive regulatory protein that stimulates transcription of the mal genes.[464] Membrane

transport systems are made more complex by the fact that several of the binding proteins serve also as receptors for stimulation of **chemotaxis** (Chapter 19).

The bacterial phosphotransferase system.

A third system for uptake of sugars is utilized by *E. coli* and by many other bacteria. This phosphoenolpyruvate-dependent phosphotransferase system converts glucose, mannose, fructose, other sugars, or mannitol into their 6-phosphate esters, at the same time transporting the latter across the membrane (a group translocation).[431,465] Four proteins form a cascade (Eq. 8-4). Phosphoenolpyruvate, an intermediate in sugar metabolism whose high energy of hydrolysis provides the

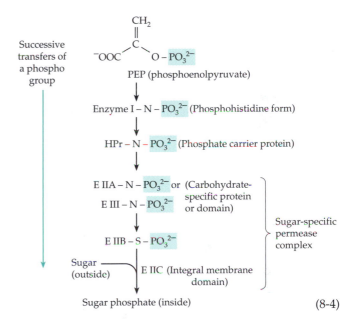

Successive transfers of a phospho group

PEP (phosphoenolpyruvate)

Enzyme I – N – PO_3^{2-} (Phosphohistidine form)

HPr – N – PO_3^{2-} (Phosphate carrier protein)

E IIA – N – PO_3^{2-} or   (Carbohydrate-
E III – N – PO_3^{2-} specific protein
                              or domain)

E IIB – S – PO_3^{2-}

Sugar (outside)

E IIC (Integral membrane domain)

Sugar phosphate (inside)

Sugar-specific permease complex

(8-4)

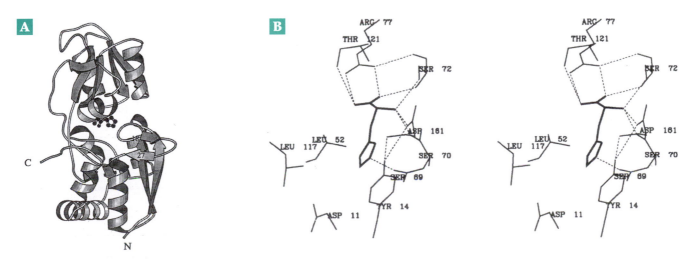

Figure 8-24 (A) MolScript ribbon drawing of the periplasmic histidine-binding protein HisJ, a component of an ABC transporter system of *Salmonella*. The bound L-histidine is shown as a ball-and-stick model. (B) Stereoscopic view of the histidine-binding site showing hydrogen-bonding interactions of protein side chains with the histidine. From Oh *et al*.[460] Courtesy of Giovanna Ferro-Luzzi Ames.

driving force, phosphorylates $N^{\varepsilon 2}$ of a histidine side chain on **enzyme I**, a large 64-kDa soluble membrane-associated protein.[466-467a]

The phospho group is then transferred sequentially to the small 88-residue dimeric phosphate carrier protein **HPr**, to the carbohydrate-specific membrane proteins IIA and IIB, and to the sugar being transported. A histidine side chain at position 15 of HPr is phosphorylated by PEP to form $N^{\delta 1}$-phosphohistidine.[468] The $NH^{\varepsilon 2}$ of the same histidine forms a hydrogen bond to C-terminal glutamate 85, which is also hydrogen bonded to an arginine side chain. This chain of interacting groups may function in the phospho transfer reactions.[469] Both enzyme I and HPr function in the transport of many ligands but there are specific **enzymes II** (EIIs) for each sugar or other ligand.[470] In *E. coli* there are at least 13 different PTS transporters.[471] A complete EII usually consists of three domains – EIIA (or EIII), EIIB, and EIIC. In some cases all three domains are in a single polypeptide chain, but in other cases they are individual proteins or some combination of individual and bifunctional proteins. The 637-residue mannitol-specific EII contains all three domains. Domains A and B are cytoplasmic while the N-terminal segments form the integral membrane C domain. The two phosphorylation sites are His 554 (EIIA) and Cys 384 (EIIB).[472] The glucose-specific EII from *E. coli* consists of two subunits, IIA and IICB.[471,473–475] The mannose transporter has three subunits representing domains IIAB, IIC, and an additional integral membrane subunit IID.[476] In addition to their direct transport functions, components of the PTS system play regulatory roles in chemotaxis, transcription, and control of other transporters.[477,478]

5. Transport of Ions

Cell membranes are impermeable to most ions. Only a small number of ions can enter cells readily and these usually do so with the assistance of protein channels or pores. The principal anion of plasma (Box 5A) is Cl^-, which passes through membranes readily by virtue of the presence of channel-forming proteins. Chloride ions are often distributed across membranes passively according to Eq. 8-5, which describes the **Donnan equilibrium**.[167,479,480]

$$[K^+]_i[Cl^-]_i = [K^+]_o[Cl^-]_o \qquad (8\text{-}5)$$

Here the subscripts i and o refer to the inside and the outside of the cell, respectively. The potassium ion concentration within a cell is maintained at a high value by the operation of the $Na^+ + K^+$ pump and by the presence of nondiffusible anions within the cell. According to Eq. 8-5, the internal chloride concentration must be low, with the product of $[K]_i[Cl]_i$ equaling

that of the low exterior $[K^+]$ and high exterior $[Cl^-]$. The internal $[Na^+]$ and $[Ca^{2+}]$ are both low, while the internal $[K^+]$ is high. These differences are also linked to the membrane potential (Eq. 8-2), which is ordinarily expressed as a negative voltage of the interior of a cell, mitochondrion, plastid, etc. with respect to a reference electrode in the external medium.

The maintenance of both the membrane potential and the steep gradients of ionic concentrations is essential to cells, both as a means of coupling metabolic energy to transport and other processes and for electrical signaling. The effects are most pronounced for mitochondrial membranes for which E_m may attain –140 to –170 mV and for plasma membranes of excitable cells such as neurons (E_m = –70 to –90 mV). For liver and kidney cells E_m of plasma membranes may be approximately – 35 mV and for erythrocytes only – 9 mV.[480]

In excitable cells electrical impulses are initiated by opening or closing ion channels. They are propagated along an axon by a complex sequence of opening and closing of voltaged-gated channels,[481] a process that is described in Chapter 30. All cells appear to also contain ATP-driven ion pumps as well as simple channels, cotransporter proteins, and ion exchangers.

Anions. Cell membranes have long been known to be relatively permeable to Cl^- and other small anions. However, the molecular basis of this permeability is quite complex.[482] Voltage-gated selective anion channels, often called **chloride channels**, are important in electrically excitable membranes, where they ensure a high resting chloride conductance and stability.[483] The gene for one of these channel proteins was first cloned from the electric ray *Torpedo*[484,485] and is designated *Clc*-0. Similar channels have been found in organisms ranging from bacteria to yeast, green plants, and vertebrate animals. The yeast genome contains just one *Clc* gene[486] but mammals have at least nine. One of these, *Clc*-2, is defective in **myotonia congenita**, a human disease of impaired muscle relaxation; and in similar diseases of mice, goats, and horses.[483,487] A mutation in the chloride channel *Clc*-5 causes kidney problems, including proteinuria, hypercalciuria, and kidney stones.[483]

Erythrocyte membranes permit rapid transport of anions to allow for the exchange of Cl^- within the red cells for HCO_3^- generated by tissue metabolism. The HCO_3^- binds to deoxyhemoglobin (Eq. 7-47) and is carried to the lungs, where it is released upon oxygenation of the hemoglobin. Then the reverse exchange of internal HCO_3^- for Cl^- occurs. This electroneutral ion exchange is mediated by the **band 3 protein** (Fig. 8-14) which contains a channel that allows anions but not cations to pass.[238,239,488] The band 3 protein, also known as **AE1** (anion exchanger 1), is found principally in red blood cells but is also present in kidney tubules.[489]

Related proteins occur in other tissues.[488] The 911-residue band 3 protein consists of two distinct parts of nearly equal size. The N-terminal portion is attached to the membrane skeleton (Fig. 8-16). The C-terminal part, which is embedded in the membrane, is thought to form 14 transmembrane helices and to contain the ion exchange channel or channels.[489a] As previously mentioned, defects in the N-terminal portion cause spherocytosis. The mutation Arg 589 His in the C-terminal half causes **renal tubular acidosis** in which the kidneys do not adequately remove acids from the body.[238,489] Band 3 proteins can also exchange phosphate, sulfate, and phosphoenolpyruvate for Cl^- or bicarbonate.

Another chloride channel, which is regulated by cyclic AMP (Chapter 11), functions in secretory epithelia. Its regulation is faulty in **cystic fibrosis** (Box 26-A), one of the most common human genetic defects, especially among persons of European descent.[490] As was previously mentioned, this **cystic fibrosis transmembrane conductance regulator** (CFTR) is a member of the ABC superfamily of transporters. The large 1480-residue protein apparently has two 6-helix membrane-spanning domains, two cytoplasmic nucleotide-binding domains,[491] and another large cytoplasmic regulatory domain.

In addition to AE1 (band 3 protein), kidneys depend upon other modes of reabsorption of HCO_3^- from the proximal tubules. These include a Na^+/HCO_3^- cotransporter, which seems to be related to the AE family of ion exchangers. However, it transfers three HCO_3^- ions per Na^+ and is therefore highly **electrogenic**.[492,493] Transporters for phosphate, sulfate, and small organic anions are found in many organisms. Bacterial periplasmic transporters have already been described. Plants employ $H^+/phosphate$,[494] $H^+/sulfate$,[495] and $H^+/nitrate$[496] cotransporters. Phosphate transporters are probably essential to all organisms. One of the best known is the mitochondrial P_i/H^+ cotransporter which carries phosphate ions originating from hydrolysis of ATP to ADP + P_i back into the mitochondria.[497,498] See also Table 18-8. A human Na^+/P_i cotransporter in the kidney is also essential. An X-linked trait leading to inadequate synthesis of the transporter causes hypophosphatemic vitamin D-resistant **rickets**.[499]

Monocarboxylates such as lactate and pyruvate enter animal cells with the aid of monocarboxylate/H^+ cotransporters of low specificity.[500] A $Cl^-/oxalate$ transporter is one of several ion exchange proteins in the kidney.[501] Transport systems for ADP, phosphate, dicarboxylates, and other anions are very active in mitochondrial membranes (Chapter 18).

Cation channels. When a nerve impulse passes along an axon gated pores or channels specifically permeable to Na^+ and K^+ open for short periods of time as a result of changes in membrane potential induced by the advancing wave of the action potential (Chapter 30). We have already considered the structure of a potassium ion channel (Fig. 8-21). However, there are at least 30 types of K^+ channels that can be distinguished.[502] Many of these appear to have similar channel structures but to serve a variety of purposes. While the K^+ channels of neurons are voltage gated many others are controlled by hormones, neurotransmitters, or mechanical stimuli.[503] One of the most investigated K^+ channels, known as K_{ATP}, sets the resting potential in the insulin-secreting β cells of the pancreas by facilitating a flow of K^+ *into* cells. Such channels, which help equilibrate intracellular and extracellular K^+ at near equilibrium, are found in cells with low negative values of E_m. They are called *inwardly rectifying*. When the internal glucose concentration in the β cells rises it initiates a complex signaling sequence involving blockage of the K_{ATP} channels by ATP, opening of voltage-sensitive Ca^{2+} channels, and insulin secretion.[504–506] The K_{ATP} channel is also blocked by **sulfonylureas**. As a result, these compounds induce insulin release and are useful in treatment of diabetes meltitus (Box 17-G). When its gene was cloned the **sulfonylurea receptor** was found to be a transmembrane protein of the ABC transporter family.[505] The potassium channel protein is another subunit of the transporter. Similar K_{ATP} channels are found in the kidneys[507] and also in embryonic cells in which they may participate in regulation of the cell cycle.[508]

Voltage-regulated **sodium channels** are the major participants in propagation of nerve impulses. The large 260-kDa α subunit of the sodium channel of nerve membranes contains four homologous repeat sequences, each of which may form transmembrane helices and also contain a loop that may participate in forming a pore similar to the K^+ pore of Fig. 8-21.[509–510a] However, the structure is uncertain.[511] The channel complex also contains 36- or 33-kDa $β_1$ and $β_2$ subunits that appear to be members of the Ig superfamily.

Epithelial cells contain a quite different Na^+ channel that participates in reabsorption of urinary Na^+ and in control of blood pressure (Box 22-D). The channel consists of at least three structurally similar subunits, each with two predicted transmembrane helices and a large extracellular domain.[512] These channels are blocked specifically by the diuretic compound **amiloride**.[513]

Amiloride, a diuretic

The best known sodium channel is, in fact, a general cation channel which is part of the "nicotinic" **acetylcholine receptor**. Acetylcholine is the principal excitatory transmitter in the peripheral nervous system and upon release from synaptic endings or neuromuscular junctions occupies receptors on the "postsynaptic" membranes of one or more adjacent neurons. When acetylcholine binds in the ion pore, the acetylcholine receptor opens and cations flow out, depolarizing the membrane. Under favorable circumstances, this initiates a nerve impulse (action potential) in the postsynaptic neuron. The receptor gene was first cloned from *Torpedo* and the receptor protein has been studied extensively.[382] The 290-kDa protein consists of five similar-sized subunits with an $\alpha_2\beta\gamma$ structure. These form a nearly symmetric fivefold oligomer with a pore in the center. Images of both the open and the closed states, obtained by electron microscopy at a resolution of 0.9 nm, suggest that the inner pore of ~2.6 nm diameter is formed by five α helices, one from each subunit. It is open when acetylcholine binds to the two α subunits and closes to a much smaller diameter when the acetylcholine leaves (and is destroyed by hydrolysis).

Calcium channels are a third major group of cation-selective channels.[514] As pointed out in Box 6-D, calcium ions are involved in a very wide range of signaling functions. These are discussed in several places in this book. Several of these functions depend upon voltage-gated Ca^{2+} channels. Muscle is rich in **L-type** or DHP-sensitive channels (Box 6-D) which play a role in transmission of nerve impulses to muscles by allowing rapid flow of calcium ions into cells from outside.[515]

The structure appears to be homologous to that of voltage-gated Na^+ channels with a large 170-kDa subunit with a fourfold repeat plus smaller subunits. The **ryanodine receptors** of muscle control the release of Ca^{2+} from stores in the endoplasmic reticulum[516,517] (see also Chapter 19). These receptors are ligand gated, being activated by cyclic ADP-ribose (Chapter 11). Two additional types of voltage-sensitive Ca^{2+} channels, **N** and **P**, are found in the central nervous system.[514,518] Another ligand gated calcium channel has been found in endothelial cells.[519] It is activated by **sphingosylphosphocholine**[519] rather than by cAMP-ribose or inositol triphosphate.[520] A sodium / calcium ion exchanger and cotransporter[521] utilizes the Na^+ gradient to exchange three external Na^+ ions for one internal Ca^{2+}.

Many aspects of calcium function are poorly understood. Among these is the role of a group of proteins known as **annexins** (formerly lipocortins, calpactins, endonexins, etc.).[522] The ten or more members of the annexin family[523–527] share the property of binding to phospholipid membranes in the presence of Ca^{2+}. One of the several proposed functions of annex-

ins is formation of Ca^{2+} channels, a function that is suggested by the modular three-dimensional structures.[525,526] Other suggested functions include roles in membrane fusion, exocytosis, and adhesion.

Active transport of cations. Most organisms take up ions from their surroundings by active transport. Green plants extract essential nutrients from the extremely dilute solutions in contact with their roots. Microorganisms such as yeast and bacteria have the same ability, and specific concentrating systems for many ions such as K^+, Ca^{2+}, sulfate, and phosphate have been identified. The skin of a frog can take up Na^+ from a 10^{-5} M solution of NaCl and extrude it into the internal fluids whose Na^+ concentration may be greater than 0.1 M. Ions can also be concentrated from internal fluids and excreted at higher concentrations. Some seabirds and marine animals rid their bodies of excess salt by secretion from salt glands. The lining of the human stomach is able to concentrate hydrogen ions in gastric juice to ~0.16 M.

Organelles within cells have their own ion-concentrating mechanisms. Thus, mitochondria can concentrate K^+, Ca^{2+}, Mg^{2+}, and other divalent metal ions as well as dicarboxylic acids (Chapter 18). The entrance and exit of many substances from mitochondria appear to occur by exchange diffusion, i.e., by secondary active transport. Such ion exchange processes may also occur in other membranes.

ATP-driven ion pumps. Within virtually all cells the sodium concentration is relatively low, while that of potassium is high (Box 5-A). One theory[528] regards the cytoplasm as analogous to an ion exchange resin with fixed charges in a lattice. Highly crosslinked ion exchange resins exhibit specificity toward binding of certain ions; e.g., sulfonic acid resins tend to bind K^+ preferentially, while phosphonic acid resins tend to bind Na^+. Do proteins also prefer K^+ to Na^+?

In contrast to the ion exchange theory, much evidence indicates that cells have an active **ion pump** that removes Na^+ from cells and introduces K^+. For example, the cytoplasm of the giant axons of nerves of squid can be squeezed out and replaced by ionic solutions. Erythrocyte ghosts can be allowed to reseal with various materials inside. Ion transport into or out of cells has been demonstrated with such preparations and also with intact cells of many types. Such transport is blocked by such inhibitors as cyanide ion, which prevents nearly all oxidative metabolism. However, the cyanide block can be relieved by introduction into the cells of ATP and other phosphate compounds of high group-transfer potential.

Uptake of K^+ by cells and extrusion of Na^+ from cells are also specifically blocked by "cardiac glycosides" such as **ouabain** (Fig. 22-12). Ouabain labeled with ^3H binds to the outer surface of cells, and from this binding it was estimated that erythrocytes possess 100–200 ion pumping sites per cell (~ 1 site / μm^2).[529] For the HeLa cell (a widely studied strain of human cancer cells) 10^5 to 10^6 sites / cell (~ 10 / μm^2) were found. Further experiments showed that in the presence of Na^+ within the cell and K^+ on the outside of the cell, ATP is hydrolyzed. The rate of hydrolysis was directly related to the concentrations of these two ions and to the number of ouabain binding sites and also required the presence of Mg^{2+}. These observations led to the concept of an **($Na^+ + K^+$)- activated ATPase** (often abbreviated **Na^+, K^+-ATPase**) as synonymous with the membrane-bound ion pump. Within the cell Na^+ must be located on one side of the membrane and K^+ on the other to activate this enzyme. However, the purified enzyme would be expected to hydrolyze ATP in the test tube in the presence of $Na^+ + K^+ + Mg^{2+}$. Such a protein was isolated from several sources and has been studied intensively. It is an $\alpha\beta$ mixed dimer with molecular masses of ~ 113 kDa for the α chains and ~ 55 kDa for the glycoprotein β chains.[530,531] The proteins may associate to $\alpha_2\beta_2$ tetramers in membranes. The genes for various isoforms of the proteins from several sources have been cloned and sequenced.[531,532] The large α subunit may span the bilayer of the membrane as many as ten times; the glycoprotein β subunit is thought to be largely on the outer surface and may have only one membrane-spanning helix.[533,534] A small 68-residue γ subunit copurifies with the pump protein. It may be involved in control of the ATPase, which has complex regulatory properties.[535,536] Sulfatides (ceramide galactose-3-sulfate) may also play a role in the enzymatic activity.[535]

The sodium–potassium pump displays a curious stoichiometry. *Three sodium ions are pumped from the inside and two potassium ions from the outside of a cell for each molecule of ATP cleaved.* Thus, an excess of positive ions is pumped out with the result that a negative charge develops inside the cell and a positive charge accumulates on the outside. This action of the Na^+, K^+-ATPase is the primary source of the membrane potential for most eukaryotic cells and is said to be **electrogenic**. Because the cell membrane is somewhat permeable to K^+, outward diffusion of K^+ through the "leaky" membrane along its concentration gradient helps to maintain the membrane potential as does inward leakage of Cl^-. At the same time, Na^+ diffuses inward, aided by the membrane potential. Even though the permeability of Na^+ is low, a steady state is reached at which the rate of passive inward diffusion of cations just balances the membrane potential set up by the active transport.

The energy for transport of Na^+ and K^+ by the ion pump is supplied by ATP. The Na^+,K^+-ATPase does not merely catalyze the hydrolysis of ATP but also couples its cleavage to the pumping of the ions. The pumping of sodium and potassium ions is one of the most important energy-requiring activities of cells. It is said to account for 23% of the ATP utilization in a resting human. *Thus, it constitutes an important fraction of the basal metabolic activity.*

The Na^+,K^+- ATPase is one of a family of over 50 ion pumps that are characterized by transfer of a phospho group from ATP to an aspartate side chain carboxylate in the invariant sequence DKTG to give an intermediate phosphoenzyme + ADP.[530,537]

Phosphoenzyme intermediate of Na^+,K^+- ATPase

These **P-type ATPases** are characterized by phosphoenzyme intermediates, by a conserved consensus sequence, and through inhibition by vanadate ion.[537–539] The structures are poorly known. Some consist of single chains (perhaps dimerized) and some have more than one chain. However, the major subunit always appears to have about ten transmembrane helices with a large ~ 430-residue cytoplasmic domain between the fourth and fifth helices. This domain contains the ATP binding site and the phosphoaspartyl group of the phosphoenzyme.[534] This is Asp 369 for the Na^+,K^+- ATPase.

In addition to the Na^+,K^+- ATPases there is a very active **Ca^{2+}-ATPase** which transports two Ca^{2+} from the inside of cells to the outside while returning two H^+ from outside per ATP.[540–543a] This is the primary transporter by which cells maintain a low internal $[Ca^{2+}]$. During its action it becomes phosphorylated on Asp 351. However, in neurons, in which the membrane potential is maintained at a high negative value by the sodium pump, an Na^+ / Ca^{2+} ion exchange plays an even more important role.[540]

Other P-type ATPases include the gastric H^+, K^+- ATPase, which acidifies the stomach and has a high degree of sequence homology with the Na^+, K^+- ATPase.[544] Secretion of HCl into the stomach apparently involves diffusion of K^+ together with Cl^- from the bloodstream through the cells lining the stomach. The K^+ is then pumped back into these cells in exchange for H^+ by H^+,K^+-ATPase.[545] The chloride channel may be in the same protein as the ($K^+ + H^+$) pump.[546] The kidney is the principle acid excretory organ of the body and as such also contains proton pumps. An electrogenic H^+- ATPase pumps H^+ alone outward through the plasma membranes of fungi and of green plants.[547,548] The resulting proton gradient may be used to provide energy for transport of other materials into cells. A group of metal ion P-type transporters carry copper and other nutrient ions into cells and extrude Cd^{2+} and other toxic ions.[539] Some **alkaliphilic bacteria** pump Na^+ to create a sodium ion gradient.[539a] All of these ion pumping systems require MgATP as the source of energy and function via phosphoenzyme intermediates. The ionic gradients generated can be used to move other ions or nonionic compounds into or out of cells by exchange or cotransport processes. For example, internal H^+ may be exchanged for external Na^+ in an exchanger-mediated process that assists in control of cytoplasmic pH.[549,549a] The reverse process in *E. coli*[549b] and many other bacteria provides the principal mechanism by which those cells export Na^+. This exchange is driven by the electrochemical gradient of the H^+ ion created by oxidative phosphorylation (Chapter 18). The Na^+ ion, in turn, can be used by bacterial cells to drive other uptake processes, e.g., sugar or amino acid-Na^+ cotransport.

What is the mechanism by which ATPase transporters function? We still do not know.[550] The pumping cycles for the Na^+,K^+- ATPase and the Ca^{2+}- ATPase are similar although different in details. The ATPases are reversible and with suitable ionic gradients will work as ATP synthases.[551] A strictly hypothetical model for the Na^+, K^+-ATPase is shown in Fig. 8-25. There are at least two conformations of the ion pump proteins.[552,552a] In one conformation the protein binds three sodium ions tightly, while in the other conformation it binds two potassium ions. The ATP operates the "motor" that carries out the conformational changes. In Fig. 8-25 the ion pump, in conformation A, is shown embedded in a membrane. In the center, perhaps between three or more transmembrane helices, there is a narrow cavity, perhaps resembling that of the K^+ channel (Fig. 8-21), into which chelating groups (e.g., C=O groups of the peptide chain) protrude. These groups form the three binding sites for the 0.19-nm-diameter Na^+ ion. The spontaneous binding of the sodium ions triggers a phosphorylation reaction by which a phospho group from the MgATP^{2-} complex is transferred to the side chain carboxyl of the active site aspartate. This phosphorylation in turn triggers a change to the second conformation in which the channel to the outside is open and that to the inside is closed. At the same time the affinity

for Na^+ is decreased over 100-fold and the sodium ions dissociate on the outside. The affinity for Na^+ may decrease because the diameter of the pore is increased to accommodate the larger 0.27-nm diameter of K^+ ions, perhaps by a twisting motion of the peptide chains that form the channel.

The next step is loading with two K^+ ions. The affinity for K^+ in the second conformation is high. The return to conformation 1 with release of K^+ to the inside is triggered by hydrolytic removal of the phospho group as inorganic phosphate (P_i). It may seem surprising that a channel could be opened and closed so readily with synchronous changes in the number and specificity of ion binding sites. However, recall the type of structural alteration occurring upon oxygenation of hemoglobin (Fig. 7-25). Rotation of the hemoglobin subunits with respect to one another causes small changes in the geometrical relationships of groups protruding into the central channel. This strongly

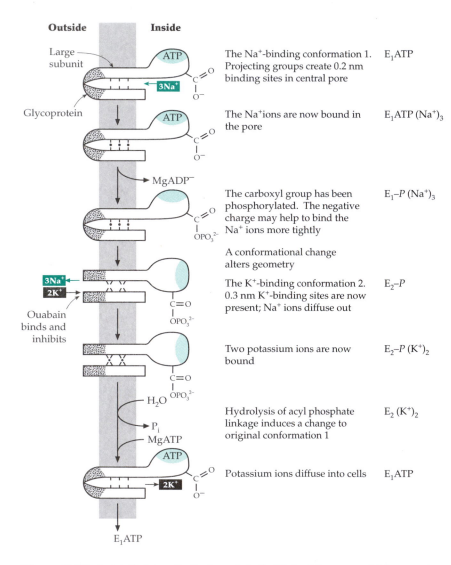

Figure 8-25 A strictly hypothetical model of a $Na^+ + K^+$ pump which operates by ATP-driven opening and closing of a channel at opposite ends and with alternate tight binding of Na^+ and K^+.

affects the binding of 2,3-bisphosphoglycerate. Very small movements could open up the Na^+ binding groups and create new binding sites for the larger K^+ ion, using in part the same chelating groups.

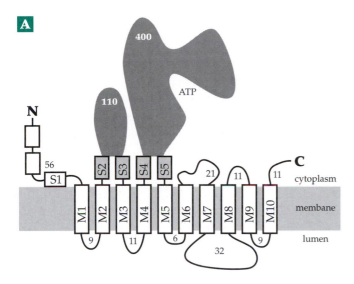

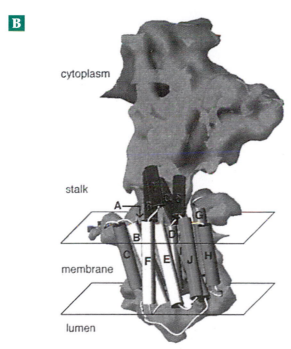

Figure 8-26 The structure of the 994-residue Ca^{2+}-ATPase of the endoplasmic reticulum of rabbit muscle at 0.8-nm resolution. (A) Predicted topology diagram organized to correspond to the electron density map prepared by electron crystallography of frozen-hydrated tubular crystals. The number of amino acid residues in each connecting loop is marked. (B) The electron density map with the predicted structure embedded. The relationships of the helices in (B) to those in (A) are not unambiguous. The helices marked B, D, E, and F in (B) may form the Ca^{2+} channel. The large cytoplasmic loops, which are black in (A), were not fitted. From Zhang et al.[553] Courtesy of David L. Stokes.

Images of both the Ca^{2+}-ATPase (Fig. 8-26)[553] and the H^+-ATPase of *Neurospora* plasma membranes[548] at 0.8 nm resolution reveal similar transmembrane regions and large cytoplasmic domains which are somewhat differently organized. The picture in Fig. 8-26 has been greatly clarified by determination of the structure by X-ray diffraction to a resolution of 0.26 nm.[553a,b,c] Two calcium-binding sites have been located in the transmembrane domain between the helices marked M4, M5, M6, and M8. The Ca^{2+} ions are apparently coordinated by side chains of Asp, Glu, Gln, and Thr. There are three cytosolic domains. The site of phosphorylation, Asp 351, lies within a large ~ 27 kDa **P** (phosphorylation) domain adjacent to the membrane. The ATP is held by a nucleotide-binding **N** domain which must at some point in the cycle move close to Asp 351 for phosphorylation to occur. The third cytosolic (**A**, actuator) domain is thought to be involved in control of the conformational alterations. The nucleotide-binding domain lacks the "P loop" characteristic of many ATPases and GTPases (see p. 648), but is homologous in its sequence to L-2-haloacid dehalogenase (Eq. 12-2)

Two other types of proton-pumping ATPases are considered in Chapter 18. One is the mitochondrial F_1F_0 ATPase, which ordinarily operates in the reverse direction as the body's principal **ATP synthase**. The other type, which in some ways resembles the mitochondrial F_1F_0 ATPase, is the **vacuolar ATPase** (V-ATPase). These are true proton pumps which acidify vacuoles of plants and also lysosomes and phagocytic vacuoles.[554,555] They are also considered in Chapter 18.

6. Exocytosis, Endocytosis, and the Flow of Membrane Constituents

Observation of cells under the microscope with time-lapse photography reveals that the plasma membrane as well as the mitochondria and other organelles are in a constant state of motion. Mitochondria twist and turn and the surface membrane undulates continuously. Vesicles empty their contents to the outside of the cells, while materials are taken into cells through endocytosis. In addition, chemical evidence indicates a directed flow of the materials of which membranes are constructed from the endoplasmic reticulum (ER) to the Golgi vesicles, excretion granules, and plasma membrane (see Fig. 10-8). Along this route new materials are inserted from the cytoplasmic side of the membrane, while enzymes within the vesicles add glycosyl units and make other modifications. The plasma membrane surface area grows quite rapidly. In secretory cells fusion of secretion granules with the plasma membrane also adds additional material to the membrane.

Counterbalancing this expansion of the plasma

membrane is active endocytosis of fluids and of solid materials from outside the cell. This not only brings new materials into the cell but also accomplishes removal of material from the plasma membrane and partial recycling of its components. One form of endocytosis is seen with the ameba.[555a] The cytoplasm flows around a smaller organism or other particle of food enclosing it in an internal membrane-bound compartment (**endocytic vacuole** or **endosome**). This vacuole then fuses with lysosomes which supply the necessary enzymes to digest the food. In a similar way phagocytic cells of our bodies engulf micro-organisms or other particles to form **phagosomes** which, over a period of more than 24 hours, undergo extensive biochemical changes.[556] They acquire digestive enzymes, vacuolar ATPase,[557] other proteins needed to kill bacteria, and other parasites.

Uptake of smaller particles including protein molecules occurs by **micropinocytosis**, a process that can be seen only by electron microscopy. This often takes place via **coated pits**, indentations of ~0.3 μm diameter underlain by a thickened membrane.[558–559b] The pit membrane is also coated with protein molecules and appears to have many short bristles or spikes protruding into the cytoplasm (Fig. 8-27A). After endocytosis the coated pits become **coated vesicles** of 0.15–0.25 nm (Fig. 8-27B). Within a few seconds, however, these vesicles lose their coat and become endosomes.

The major protein making up the coat is the 180-kDa **clathrin**, but smaller 33- to 36-kDa peptides of several types also contribute.[561] The coat forms a "basket" with pentagonal and hexagonal faces surrounding the lipid bilayer of the vesicle. At each vertex of the basket is a "triskelion," a trimer of clathrin together with an equal number of the smaller chains. The smallest baskets consist of 12 pentagons plus 4, 8 or more hexagons, a relationship that allows formation of a variety of larger baskets.[562,563] Additional 50- and 100-kDa accessory proteins form a shell around the clathrin cage. That clathrin is essential for normal cell growth has been established by the observation that deletion of its structural gene from yeast is lethal.[564] The addition of more trimer units from a reserve of soluble clathrin in the cytoplasm allows the vesicles to develop and break off from the membrane. From studies with inhibitors it is evident that metabolic energy is required to drive the process.

Other vesicles are surrounded by nonclathrin membrane coats. Some of these originate from **caveolae** (little caves), which act in endocytosis, exocytosis, and transmembrane signaling.[564a,b,c] A **coatomer** complex of eight subunits with molecular masses from 20 to 60 kDa coats vesicles involved in transport between compartments of the Golgi.[565–567]

What is inside a coated vesicle? Cells take up a variety of peptide hormones and proteins. This usual-ly occurs with the aid of specific receptor proteins located in or on the outside of the plasma membrane. Some of these, e.g., receptors for the low-density lipo-protein of plasma (Chapter 22), are clustered in coated pits. Other receptors, such as those for insulin or epidermal growth factor, are spread more evenly across the membrane but collect in coated pits when the hormone binds. Endocytosis provides a means for the cell to take up and in some cases destroy the hormone or the receptor or both.

Transmembrane proteins, including hormone receptors, are incorporated into coated vesicles with the help of **clathrin adapter proteins** (APs). These

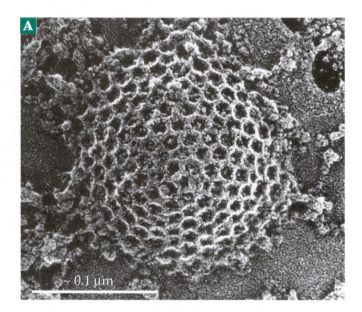

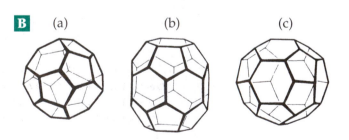

Figure 8-27 (A) Region of a coated membrane from fibroblasts at an intermediate stage of the budding process, demonstrated by deep etching and rotary replication (by J. E. Heuser[559b]). From Pearse and Bretscher.[560] Courtesy of Barbara Pearse. (B) Three structures identified among the smallest coated vesicles. Structure (*a*) contains 12 pentagons and four hexagons, the latter lying at the vertices of a tetrahedron; (*b*) has a barrel shape built of 12 pentagons and eight hexagons; structure (*c*) also has twelve pentagons and eight hexagons, but the latter are arranged in two arcs of four, related in the same way as the two parts of a tennis ball. Larger coats seem to be constructed on similar principles, with the addition of further hexagons. From Pearse and Bretscher.[560]

complex oligomeric proteins bind to recognition or "sorting" sequences such as dileucine on YXXφ (Y = Tyr, X = any amino acid, φ = bulky hydrophobic).[567a-d] The adapter proteins also bind to clathrin, the N-terminal β-propeller domain associating with the sequence LφXφD / E of some AP adapters.[567c] Completion of a coated vesicle requires membrane fusion as the vesicle is pinched off from the membrane surface. A GTPase (Chapter 11) called **dynamin** as well as another protein **endophilin I** are required. Endophilin I is an acyltransferase able to transfer the fatty acyl group of arachidonoyl-coenzyme A to lysophosphatidic acid in the cytosolic surface of the membrane. This may change the curvature of the membrane, assisting in vesicle formation.[567e]

Once inside a cell the vesicles lose their coats to become endosomes which may then fuse with lysosomes or with Golgi membranes. The removal of a clathrin coat requires ATP as well as the chaperonin Hsp 70 (Chapter 10) and a coat protein called **auxilin**.[568] Triskelion is distorted and displaced from the clathrin cage. The interior of the newly formed endosome is quickly acidified by the action of a proton pump in the vesicle walls.[554,569] This sometimes leads to dissociation of enclosed receptors from their ligands and permits recycling of receptors and lipids of the vesicle membranes to the cell surface. This is the case for the low-density lipoprotein receptor.[570,571]

Exocytosis, by which the content of a secretion vesicle is released to the outside of a cell, is just as important as endocytosis. The process is sometimes very specialized. For example, the release of a nematocyst from *Hydra* (Fig. 1-13) can occur in about 3 ms.[572] Exocytosis involves **fusion** of membranes,[562,573] a process also occurring during the movement of endosomes along the endocytic pathway,[574] during vesicular transport between Golgi compartments, and in many other biological processes. Exocytosis is often triggered by the binding of Ca^{2+} to specific proteins of the vesicle wall and of the cytoskeleton (Chapter 7).[575] The fusion of membranes at several stages in the vesicle-mediated transport of materials between Golgi compartments requires a specific protein known as the *N*-ethylmaleimide-sensitive fusion protein (**NSF**).[576,577] A host of other specialized proteins are also involved[562,578] and are discussed in Chapter 10. See also Chapters 20 and 29.

D. Communication

External coats and cell walls help to control the access of materials to a cell. However, it is the outer surface of the plasma membrane that makes the cell's first contact with nutrients, hormones, and other important chemicals. The membrane must often not only detect these materials but also send signals to the interior of the cell and sometimes to adjacent cells. These signals may be about changes in pH or nutrient concentration or the presence of hormones, neurotransmitters, or harmful materials. For these reasons cell membranes contain many embedded receptors and signaling complexes. These are discussed in Chapter 11 and later sections of the book.

The plasma membrane also contains many "markers" of the individuality of the species or of an individual. These are chemical groupings that, in higher animals, can be recognized by the immune system as "self" rather than as a foreign invader. These surface markers may also be used by parasites as camouflage to evade the immune response of the host. Such chemical groupings on cell surfaces are often described as **antigenic determinants** and the molecules that carry them as **antigens**. Each antigenic determinant elicits the production of antibodies that will bind specifically to it. Over 250 different antigenic groups have already been described for the surface of the red blood cell. They determine the blood type. Groups on the surfaces of other cells determine whether a transplanted tissue will be rejected. Various proteins from plant and other sources act as **agglutinins** by binding to surface groups much as do antibodies. Viruses that attack cells may also become adsorbed onto specific surface molecules, which act as receptors.

Immunoglobulins can also be receptors. For example, molecules of IgE bound to basophils and the related **mast cells** of tissues serve as receptors for allergens. Binding of an allergen to the IgE molecules stimulates the release of granules containing histamine and other substances (Chapter 31).

E. The Extracellular Matrix and Cell Walls

The surroundings of cells are extremely complex and vary from one organism to another and from one tissue to another. The principal function of cell walls and other surface coats is to protect cells against attack by organisms and against physical disruption.

1. The Structure of Bacterial Cell Walls

The plasma membrane of bacterial cells, other than the wall-less mycoplasmas and some archaebacteria, is surrounded by a multilayered wall which may be separated from the membrane by a thin **periplasm** (or periplasmic space). This can be seen most clearly in suitably prepared thin sections of cells of *E. coli* or other gram-negative bacteria as a relatively empty space of 11- to 25-nm thickness (Fig. 8-28).[579-581] The volume of this space (which may be filled with gelled material) depends upon the osmotic pressure of the medium. In *E. coli* it contains 20–40% of the total cell water. In gram-negative bacteria the innermost layer of the walls lies within the periplasm (Fig. 8-28). It is a porous network of a highly crosslinked material known as **peptidoglycan** or **murein**. The backbone of the peptidoglycan is a β-1,4-linked polymer of alternating N-acetylglucosamine and N-acetylmuramic acid residues. Alternate units of the resulting chitin-like molecule carry unusual peptides attached to the lactyl groups of the N-acetylmuramic acid units (Fig. 8-29). The peptide side chains are crosslinked as indicated in the figure. The peptides vary considerably in structure and crosslinkages.[582,583] In *E. coli* and other gram-negative bacteria the peptidoglycan forms a thin (2-nm) continuous network around the cell, but in gram-positive bacteria the highly crosslinked peptidoglycan forms a layer as much as 10 nm thick.[584] The peptidoglycan layer is surrounded by other layers whose structures vary from one organism to another, with the outermost antigenic layers being the most variable.

The outer membrane of gram-negative bacteria.

Outside the murein layer of *E. coli* and other gram-negative organisms is a phospholipid-containing **outer membrane** which has the thickness and something of the structure of a typical biological membrane (Fig. 8-28).[585,586] This membrane is attached to the peptidoglycan layer with the aid of a small hydrophobic 58-residue **lipoprotein** whose N terminus contains a **glycerylcysteine** which carries three fatty acids connected by ester and amide linkages:

Glycerylcysteine at N-terminus of lipoprotein and linked to 3 acyl groups

The fatty acid chains are evidently embedded in the outer membrane as an anchor. About one-third of the lipoprotein molecules are attached covalently to the peptidoglycan through an amide linkage between the side chain amino group of the C-terminal lysine of the protein and a diaminopimelic acid residue of the peptidoglycan (Fig. 8-29). Thus, the protein replaces one of the terminal D-alanine residues of about one in ten of the murein peptides. There are $\sim 2.5 \times 10^5$ molecules of the bound form of the lipoprotein per cell spread over a surface area of peptidoglycan of $\sim 3 \ \mu m^2$. They appear to be associated as trimers located primarily in the periplasmic space.[589]

About 10^5 copies per cell of the previously mentioned (Section C,2) larger 325-residue structural protein, **OmpA protein**,[350] after its gene symbol *ompA*

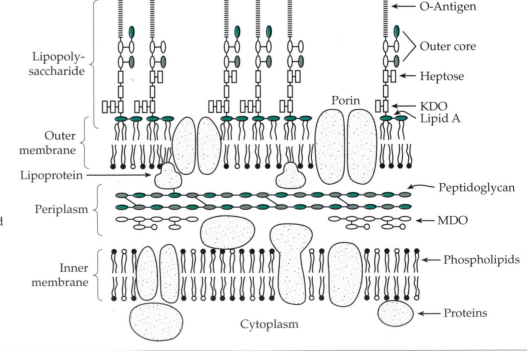

Figure 8-28 Schematic molecular structure of the *E. coli* envelope. Sugar residues are represented by ovals and rectangles. Circles represent polar head groups of phospholipids. MDO, membrane-derived oligosaccharides; KDO, 3-deoxy-*manno*-octulosonic acid (structures for KDO and heptose are in Fig. 4-15). From Raetz and Dowhan.[587]

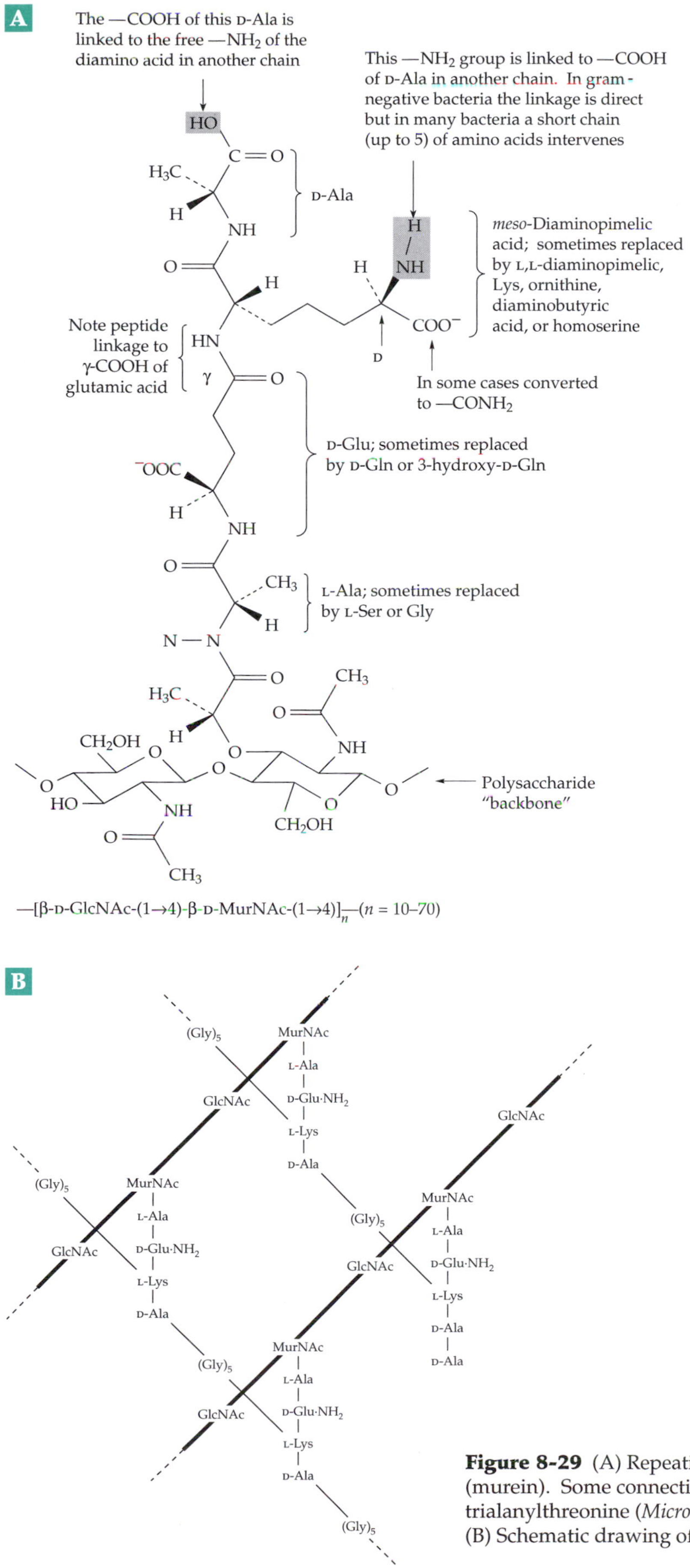

A The —COOH of this D-Ala is linked to the free —NH$_2$ of the diamino acid in another chain

This —NH$_2$ group is linked to —COOH of D-Ala in another chain. In gram-negative bacteria the linkage is direct but in many bacteria a short chain (up to 5) of amino acids intervenes

D-Ala

meso-Diaminopimelic acid; sometimes replaced by L,L-diaminopimelic, Lys, ornithine, diaminobutyric acid, or homoserine

Note peptide linkage to γ-COOH of glutamic acid

In some cases converted to —CONH$_2$

D-Glu; sometimes replaced by D-Gln or 3-hydroxy-D-Gln

L-Ala; sometimes replaced by L-Ser or Gly

Polysaccharide "backbone"

—[β-D-GlcNAc-(1→4)-β-D-MurNAc-(1→4)]$_n$—(n = 10–70)

B

(Gly)$_5$ MurNAc
            |
            L-Ala
            |
GlcNAc   D-Glu·NH$_2$
            |
            L-Lys
            |
            D-Ala

(Gly)$_5$ MurNAc
            |
            L-Ala
            |
GlcNAc   D-Glu·NH$_2$
            |
            L-Lys
            |
            D-Ala

GlcNAc

(Gly)$_5$ MurNAc
            |
            L-Ala
GlcNAc |
          D-Glu·NH$_2$
            |
            L-Lys
            |
            D-Ala
            |
            D-Ala

MurNAc
|
L-Ala
|
GlcNAc   D-Glu·NH$_2$
            |
            L-Lys
            |
            D-Ala

(Gly)$_5$

(Gly)$_5$

Figure 8-29 (A) Repeating unit of structure of a bacterial peptidoglycan (murein). Some connecting bridges are pentaglycine (*Staphylococcus aureus*), trialanylthreonine (*Micrococcus roseum*), and polyserine (*S. epidermis*). (B) Schematic drawing of the peptidoglycan of *S. aureus*. From Osborn.[588]

(outer membrane protein A), are also present. The outer membrane contains almost the same number of molecules of two porins. Together with phospholipids or with the lipopolysaccharide discussed in the next paragraphs, the porins and OmpA protein associate in hexagonal arrays which provide the basic framework structure of the outer membrane.[585] Two of the outer membrane proteins have been shown to contain some of the modified lysine α-aminoadipic acid 5-semialdehyde (**allysine**).[590] The aldehyde groups of allysine are able to form crosslinks to other proteins as has been well established for collagen and elastin in the human body.

A characteristic feature of the outer surface of gram-negative bacteria is a **lipopolysaccharide**[586] that is anchored in the outer membrane. It was discussed briefly in Chapter 4 where the structures of the repeating oligosaccharides known as O-antigens are given. Figures 8-28 and 8-30 show the manner in which the oligosaccharide bearing the O antigen is attached to a lipophilic anchoring group that is embedded in the outer membrane of the bacteria.[591] The anchor, which is called **lipid A**, is a β-1,6-linked disaccharide of *N*-acetylglucosamine. It is also linked both to phosphate groups and to the fatty acyl groups that fit into the lipid bilayer of the membrane. In *E. coli* and *Salmonella typhimurium* four molecules of **3-D-hydroxymyristic** acid are joined by ester and amide linkages to the two GlcN units (Fig. 8-30). Other fatty acids, including lauric, myristic, and palmitic acids, are esterified to the hydroxyl groups of two or three residues of hydroxymyristic acid (Fig. 8-30). Lipid A from other gram-negative bacteria is similar but with variations in the fatty acyl group composition and linkages to the carbohydrate.[592–594]

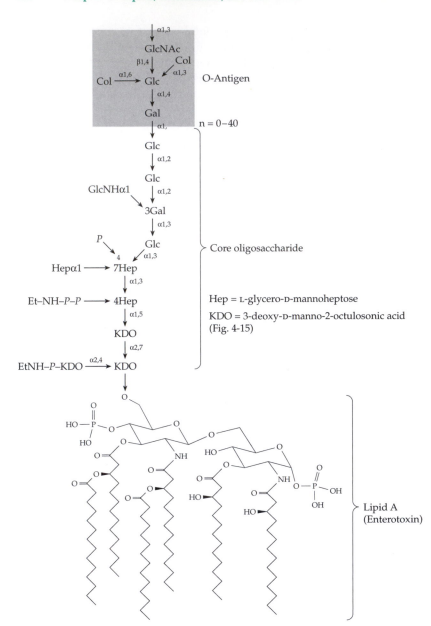

Hep = L-glycero-D-mannoheptose
KDO = 3-deoxy-D-manno-2-octulosonic acid
(Fig. 4-15)

Lipid A
(Enterotoxin)

Figure 8-30 Structures of the lipopolysaccharides of the outer membrane of *E. coli* and *S. typhimurium* including the bilayer anchor lipid A. For structures of L-glycero-D-mannoheptulose, KDO, and colitose, see Fig. 4-15.

In *Bacillus subtilus* D-alanine is attached to one of the green oxygen atoms in at least half of the units

Ribitolteichoic acids

In *B. subtilus* β-D-glucose is attached here

Glycerolteichoic acids

In *Lactobacillus arabinosus* D-alanine in ester linkage occupies this position but is replaced by β-D-glucose on about one unit in nine

Many differences are also found in the "core" oligosaccharide (Fig. 8-30) and in the O-antigens.[594–596]

Why do cells of *Salmonella* have a thousand distinguishable surface antigens, many based on differences in the O-antigens? The ends of these carbohydrate clusters are the groups to which the antibodies in animals clamp themselves if the bacteria enter the bloodstream. Mutants known as R forms (because of the growth as rough colonies on agar plates) completely lack the outer O-antigen. The R mutants of *Salmonella* are nonpathogenic, whereas the smooth strains with intact O-antigen often cause illness. Perhaps, if the O-antigen has the right cluster of sugar rings at the ends, the host organism does not recognize it as dangerous. This is only part of a continuous battle between the immune system of the body and camouflaged surfaces of attacking pathogens.

The lipopolysaccharides of bacterial surfaces have been identified as the "**endotoxins**" that cause many of the worst effects of gram-negative bacterial infections and that are often lethal.[597] Although the name endotoxin came from the assumption that there was an internal heat-resistant toxin that was released from the bacteria, the toxin is part of the bacterial cell wall. While the O-antigens provide for innumological recognition, it is the *lipid A* part that is responsible for the toxicity and unusually strong fever-inducing properties of the lipopolysaccharide.[592–594,597]

Teichoic and teichuronic acids. The cell walls of gram-positive bacteria are composed of a thick peptidoglycan layer which also contains proteins and additional polymers known as **teichoic acids** and **teichuronic acids**. In some species these account for 50% of the dry weight of the cells.[598,599] Teichoic acids are high polymers of the following general types: They are often attached through phosphodiester linkages to N-acetylglucosamine or a disaccharide which, in turn, is attached to muramic acid residues of the peptidoglycan. Since the teichoic acid is uniformly distributed in its attachment to the peptidoglycan, it must be intimately associated with peptidoglycan throughout the wall.

Teichoic acids are often covalently attached to glycolipids which are part of the plasma membrane. For example, the glycerolteichoic acid of *Streptococcus faecium* contains about 28 monomer units of glycerol phosphate, approximately 60% of which carry residues of kojibiose (Gluα1 → 2Glu) as a phosphatidylkojibiosyl diacylglycerol membrane anchor.[600] Teichuronic acids contain uronic acids:

$$[\rightarrow 4(N\text{-acetyl})\text{-Man}A\beta 1 \rightarrow 6\text{Glc}\alpha 1\text{-}]_n$$

From *Micrococcus luteus*[601]

$$\left[\begin{array}{c} \rightarrow 4\text{Glc}\beta 1 \rightarrow 3\text{Rha}\alpha 1 \rightarrow 4\text{Rha}\alpha 1\text{-} \\ 3 \\ \uparrow \\ \text{Glc}A\beta 1 \end{array} \right]_n$$

From *Bacillus megaterium*[602]

Nearly 800 of the foregoing tetrasaccharide units are joined to give a large, densely packed, ~500-kDa polymer. The reducing end of the polymer is covalently attached to the peptidoglycan layer.

Mycobacteria. Most mycobacterial species seem to be harmless saprophytic soil bacteria of the gram-positive group. However, tuberculosis (caused by *Mycobacterium tuberculosis*) may infect one-third of the inhabitants of the earth[603] and kills about three million people per year.[604] Leprosy, caused by *M. leprae*, affects 10 million or more.[605] Most antibiotics are not effective against mycobacteria because of the unusual nature of their multilayered cell walls. The muramic acid of their peptidoglycans contains glycolyl groups instead of acetyl groups, perhaps providing extra hydrogen bonding within the wall.[604] Most characteristically, they are rich in lipid materials. Among these are a unique polysaccharide, a highly branched **arabinogalactan**, that is covalently attached to muramic acid residues of the peptidoglycan. Clusters of arabinofuranosyl units at the nonreducing ends of the chains are esterified with **mycolic acids** (Section A,1).[10,604,605] Mycobacteria also contain **lipoarabinomannans** that act as major antigens.[603,606] Some mycobacterial species synthesize small glycopeptidolipids that may disrupt cell membranes of hosts.[607] Mycobacteria don't produce toxins of usual types but cause slow, long-lasting infections.

Other bacterial coats. Archaebacteria not only have unusual plasma membranes that contain phytanyl and diphytanyl groups (Section A,3)[608] but also have special surface layers (**S-layers**) that may consist of many copies of a single protein that is anchored in the cell membrane.[609] The surface protein of the hypothermic *Staphylothermus marius* consists of a complex structure formed from a tetramer of 92-kDa rods with an equal number of 85-kDa "arms."[610,611] S-layers are often formed not only by archaebacteria but also by eubacteria of several types and with quite varied structures.[612–614] While many bacteria carry adhesins on pili, in others these adhesive proteins are also components of surface layers.[615] Additional sheaths, capsules, or slime layers, often composed of dextrans (Chapter 4) and other carbohydrates, surround some bacteria.

2. The Surroundings of Animal Cells

Cells in the external epithelial layers are always surrounded by a protective covering of some kind. Our own skin is made up of specialized cells which become filled with microfibrils of keratin as they move outward and become the relatively dry nonliving external surface (Box 8-F). Internal epithelial cells secrete protein and carbohydrate materials that form a thin **basement membrane** around the exposed parts of the cells. The **connective tissue** that lies between organs and which also includes tendons, cartilage, and bone consists of a relatively small number of cells surrounded by a "matrix" consisting of the protein fibers collagen and elastin in a "ground substance" rich in proteoglycans (Chapter 4).[616–618] In bone, the calcium phosphate is deposited within this matrix.

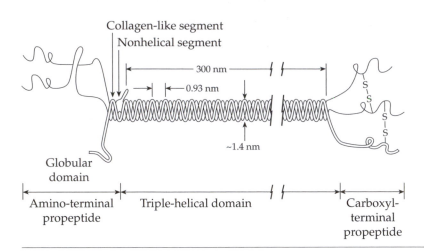

Collagen-like segment
Nonhelical segment

300 nm

0.93 nm

~1.4 nm

Globular domain

Amino-terminal propeptide   Triple-helical domain   Carboxyl-terminal propeptide

Figure 8-31 Schematic representation of type I procollagen. The molecule is composed of two identical proα1 chains (solid lines) and one proα2 chain (dashed line). In addition to the central triple-helical region that gives rise to the collagen molecule, as portrayed in Fig. 2-23(C), the precursor contains amino- and carboxyl-terminal non-triple-helical domains. The amino-terminal domain is composed of a presumably globular region, a short collagen-like segment, and a non-triple-helical region in which cleavages by the amino-terminal protease occur. Interchain disulfide bonds are limited to the carboxyl-terminal domain. The short telopeptides at the ends of collagen α chains represent the residual sequences of the linkage regions between the collagen helix and the terminal domains. After Prockop[628] and Byers.[625]

The collagens. The most abundant proteins in the body are the collagens,[619–624] a family of closely related materials that account for 20% of the total protein in higher animals. Collagens make up much of the organic mass of skin, tendons, blood vessels, bone, the cornea, vitreous humor of the eye, and basement membranes. They are found in every metazoan phylum studied. There are at least 16 types in the human body.[625] Type I collagen, which accounts for 90% of the total, is the major form occurring in skin, tendon, and bone. It is synthesized by the fibroblasts and is excreted into the extracellular space where it is polymerized into a durable long-lived material. Collagen II is found exclusively in cartilage and the vitreous humor of the eyes. Form III is located in blood vessels and intestines and is prominent in embryonic tissues. Collagen IV is the major form found in basement membranes. Some other collagens and characteristic features are listed in Table 8-4.

All collagens contain the triple-helical structure shown in Fig. 2-23. Collagen I consists of two chains of one kind (α1) and one of another (α2), while most other collagens have three identical chains. In these chains over 1000 residues have the characteristic GlyXY sequence.[624,626] At each end of the rodlike molecules short segments of the peptides fold into globular domains. For type I collagen 16 residues at the N termini and 25 at the C termini form these domains. Collagens are synthesized as intracellular precursors known as **procollagens**. The three chains of procollagens are much longer than in the mature proteins and have larger non-triple-helical ends (Fig. 8-31). The C-terminal extensions are crosslinked by S–S bridges.[625,627] Synthesis of the collagen chains requires at least six minutes after which they are released into the cisternal space of the ER, associate, and become crosslinked (Fig. 8-32). This crosslinking

ensures that the three chains remain in proper register in the triple helix while the procollagen is converted into collagen, a process involving additional crosslinking.

Before this "maturation" can occur there must be other modifications to procollagen. These begin while the peptide chains are still attached to ribosomes of the rough ER. Hydroxylases (Chapter 18) localized in the membranous vesicles of the ER convert some of the proline and lysine residues of the procollagen chains into **4-hydroxyproline**[625,629,630] and **hydroxylysine** (Eqs. 8-6 and 8-7). Lesser amounts of 3-hydroxyproline are formed. About 100 molecules of 4-hydroxyproline and 50 of 5-hydroxylysine are created in each α1 chain.

Proline residue   4-Hydroxyproline residue

(8-6)

Lysine side chain

(8-7)

Galactosyl units are then transferred onto some of the hydroxyl groups of the hydroxylysine side chains, and glucosyl groups are transferred onto some of the galactosyl groups. This glycosylation may prevent incorrect association of the procollagen molecules.

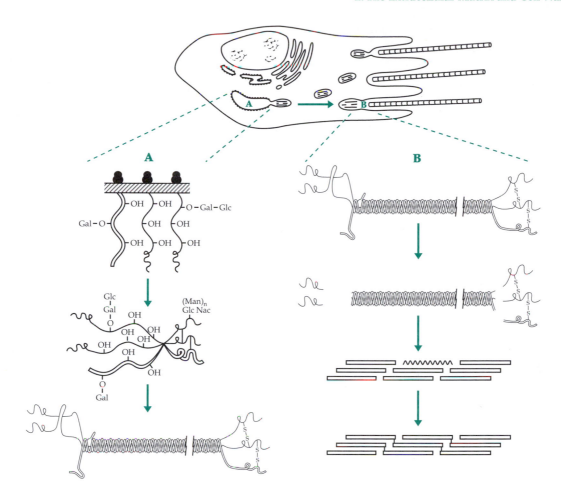

Figure 8-32 Scheme summarizing the biosynthesis of a fibrillar collagen by a fibroblast. (A) Assembly of pro-α chains in cisternae of the rough endoplasmic reticulum; posttranslational hydroxylations and glycosylations; association and disulfide bonding of C-propeptides; and zipper-like folding of the triple helix by nucleated growth. (B) Proteolytic processing of procollagen to collagen; self-assembly of fibrils by nucleated growth; and covalent crosslinking of the fibrils. The collagen molecule is first shown as a triple helix and then either as a *wavy line* to depict a molecule assembling on the surface of a fibril or as a *rectangle* to depict the quarter-staggered assay of monomers in a fibril. The proteolytic processing of procollagen and assembly of fibrils may occur within crypts of fibrils as shown here or perhaps at some distance from the cell. After D. J. Prockop.[628]

Within the extracellular space two **procollagen peptidases** act to cleave a 35-kDa peptide from the C terminus[631] and a 20-kDa peptide from the N-terminal end of each of the three chains of the secreted procollagen. The amino acid composition of the peptides removed is quite unlike that of the remaining collagen monomer (also called tropocollagen) which contains one-third glycine and much proline.

The three-stranded monomers of collagens I–III are rods of dimensions ~1.4 x 300 nm (Fig. 8-23). When they reach their final location they associate and become crosslinked to form strong fibers with diameters ranging from 8 nm to 0.5 m. Tendons tend to contain large fibrils, while those in bone are small. The smallest 8-nm fibrils must contain about 20 triple helices in a single cross section but the successive monomers are staggered by 6.4–6.7 nm (234 residues for type I collagen) in such a way that 3-nm gaps are left between the ends (Fig. 8-32). These gaps give rise to the characteristic banding pattern seen in the fibrils in Fig. 2-23D. Finer analysis of the bands together with the known sequences shows that the bands also reflect the locations of residues with charged side chains.[632] These ionized groups are thought to provide electrostatic stabilization to the fibrils.[633] The exact packing of the collagen molecules into sheetlike or microfibrillar crystalline arrays is still uncertain.[634] However, there is agreement on the staggered arrangement[634a] and upon the fact that precisely formed crosslinks with neighboring rods prevent the gaps from weakening the fibrils.

Crosslinking of collagen is initiated by oxidation of some of the lysyl and hydroxylysyl side chains from amino groups to aldehyde groups under the action of a copper-containing oxidase (Eq. 8-8, Chapter 18). The aldehyde groups enter into a variety of reactions that lead to crosslinking of the collagen monomers and to

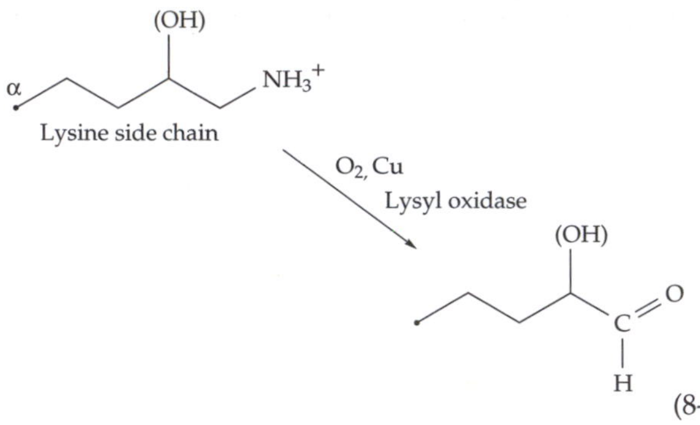

(8-8)

the formation of insoluble fibers.[635] One reaction is an aldol condensation followed by elimination of water (Eq. 8-9, step *a*). If one of the two aldehydes involved in the condensation is derived from hydroxylysine and the other from lysine, two isomeric condensation products are formed. The aldol condensation product can react further: An imidazole group from a histidine side chain can add to the carbon–carbon double bond and (either before or after this reaction) another lysine side chain can form a Schiff base with the free aldehyde. The results of these two processes are summarized in Eq. 8-9, step *b*. The final product **histidinohydroxymerodesmosine** links four different side chain groups. In other instances, simple Schiff bases (**aldimines**) are formed between aldehyde and ε-amino groups. If there is an adjacent hydroxyl group these can isomerize to ketoamines (Eq. 8-10). Two residues of hydroxylysine, one of which is glycosylated as shown in Eq. 8-10, are often involved. Borohydride reduction (Eq. 4-2) and hydrolysis leads to isolation of **dihydroxylysinonorleucine**, the predominant product of such treatment of bone or cartilage.

Crosslinkages reducible with borohydride are characteristic of newly formed collagen but these disappear and are replaced by more stable crosslinks as collagen matures. For example, a **3-hydroxypyridine** that joins three triple helices may be formed from the reaction of two ketoamine groupings (with elimination of one glycosylated hydroxylysine residue).[636–639] Similar chemistry can also produce pyrrole crosslinks.[639]

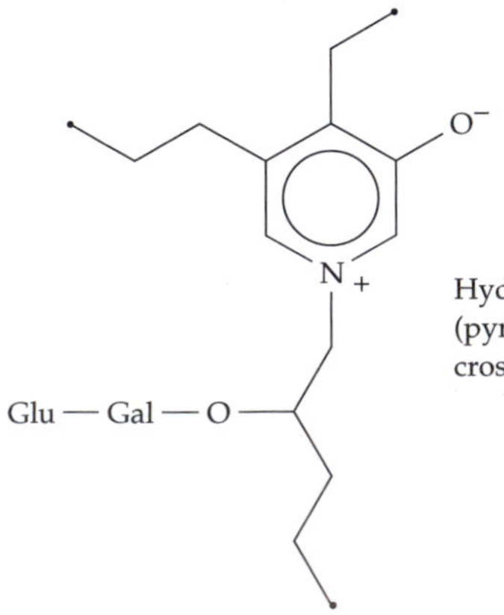

Hydroxypyridine
(pyridinoline)
crosslink

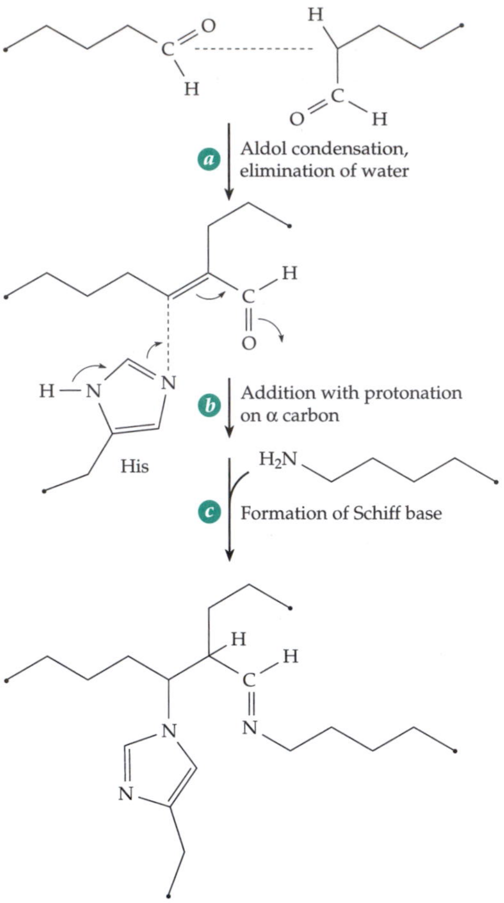

(8-9)

Histidinohydroxymerodesmosine

The crosslinkages in collagen are not located at random but are found in certain positions, often toward the ends of the collagen monomers. Thus, histidine residues are found only at positions 89, 929, and 1034 in the α1(I) peptide chains. Residue 9 in the N-terminal globular portion of one chain is linked to residue 946 of another while residue 103 is linked to 1047 in the globular C-terminal peptide.[640] The variety and number of crosslinkages vary among different species. As collagen ages through a lifetime, glycation (Eq. 4-8) leads to crosslinkages in which two ketoamines are formed via glycation cyclize.[641]

Collagens I, II, and III form fibrils with similar structures. However, other collagens are longer or shorter and aggregate in different ways. A pepsin-resistant part of the collagen V molecules resembles collagens I, II, and III but it may contain an additional non-collagenous segment at the N terminus. Collagens V and XI are quantitatively minor components of the extracellular matrix but are thought to provide a core around which the fibrils of collagens I and III may form.[642] Types XII, XIV, IX, and XVI collagens contain interruptions in the helix which create bends, flexible sites, and sites of increased proteolytic susceptibility. They may link the fibrils to other components of the surrounding matrix.[643]

TABLE 8-4
Types of Vertebrate Collagen

Type number	Location	Characteristics	Gene location: human chromosome
Forming quarter-staggered fibrils			
I[a,b]	Skin, bone, tendon, dentin	Most abundant, banded quarter-staggered fibrils	7, 17
III[a,c]	Skin, blood vessels	Abundant, small banded fibrils	2
V[a,d]	Most interstitial tissues; cartilage, bone	Abundant, small fibrils	2
Predominant in cartilage and bone			
II[a]	Hyaline cartilage, vitreous humor	Very abundant, small banded fibrils	12
XI[a]	Hyaline cartilage		
With interrupted triple helices			
XII[a,e]	Embryonic tendon, periodental ligaments	Fibril associated	
XIV[a,e]	Fetal skin, tendon	Fibril associated	
IX[a,e]	Cartilage, vitreous humor	Minor, contains attached glycosamino-glycan, fibril associated	
XVI[f]	Cartilage	Fibril associated	
Forming sheets and networks			
IV[a,g]	All basement membranes	Nonfibrillar network	13
X[a]	Mineralizing cartilage, growth plate	Short chain	
VIII[a,h]	Endothelial cells; Descemet's membrane of the cornea	Small helices linked in hexagonal arrays	
Forming beaded filaments			
VI[a]	Most interstitial tissues, intervertebral discs	Beaded microfilaments	
Forming anchoring fibrils			
VII[a,i]	Basement membranes	Long-chain, antiparallel dimers, anchoring fibrils	

[a] Martin, G. R., Timpl, R., Müller, P. K., and Kühn, K. (1985) *Trends Biochem. Sci.* **10**, 285-287
van der Rest, M., and Garrone, R. (1991) *FASEB J.* **5**, 2814-2823
Prockop, D. J., and Kivirikko, K. I. (1995) *Ann. Rev. Biochem.* **64**, 403-434
[b] Prockop, D. J. (1990) *J. Biol. Chem.* **265**, 15349-15352
[c] Nah, H.-D., Niu, Z., and Adams, S. L. (1994) *J. Biol. Chem.* **269**, 16443-16448
[d] Myers, J. C., Loidl, H. R., Stolle, C. A., and Seyer, J. M. (1985) *J. Biol. Chem.* **260**, 5533-5541

[e] Shaw, L. M., and Olsen, B. R. (1991) *Trends Biochem. Sci.* **16**, 191-194
[f] Myers, J. C., Yang, H., D'Ippolito, J. A., Presente, A., Miller, M. K., and Dion, A. S. (1994) *J. Biol. Chem.* **269**, 18549-18557
[g] Hudson, B. G., Reeders, S. T., and Tryggvason, K. (1993) *J. Biol. Chem.* **268**, 26033-26036
[h] Benya, P. D., and Radilla, S. R. (1986) *J. Biol. Chem.* **261**, 4160-4169
[i] Lunstrum, G. P., Sakai, L. Y., Keene, D. R., Morris, N. P., and Burgeson, R. E. (1986) *J. Biol. Chem.* **261**, 9042-9048

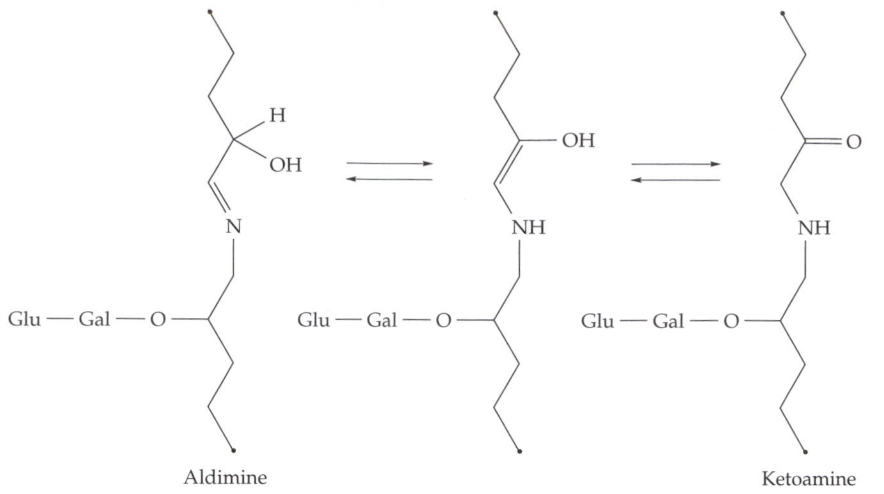

Aldimine

Ketoamine

(8-10)

The basement membrane collagen IV forms a non-fibrillar network.[644,645] The 400-nm-long molecules aggregate via their identical ends. Four molecules are held together through their triple-helical N termini, while the C-terminal globular domains connect pairs of molecules. Type IV collagen is also a proteoglycan with a glycoaminoglycan chain attached to one of its nonhelical domains. It may become covalently attached to type II collagen via a hydroxypyridine linkage.[646] Dimers of the microfibrillar collagen VI are formed from 105-nm-long monomers by antiparallel and staggered alignment, with the 75-nm overlapping helical segments twisting around each other to form coiled dimers. After a symmetrical association of dimers to tetramers, fibrillar structures are formed by end-to-end aggregation. The connection of monomers to dimers, tetramers, and polymers occurs by disulfide bonds between triple-helical segments and globular domains. The 450-nm-long collagen VII molecules associate into antiparallel dimeric structures which show a 60-nm overlap and which subsequently assemble laterally with their ends in register.[622]

At least 32 genes encode the α peptide chains of vertebrate collagens.[647,648] These chains are assembled into the 19 known types of collagen. Alternative splicing of the mRNAs provides additional isoforms.[649] The collagen α2(I) gene from both the chick and human DNA is ~38 kb in length and consists of 52 exons separated by introns ranging in length from 80 to 2000 base pairs.[647,650] At least nine of the exons that encode the triple-helical regions have exactly 54 bp. All are multiples of 9 bp, i.e., the length needed to encode one Gly-X-Y triplet (Chapter 2, Section D,4). The significance of these observations is unclear but the presence of so many introns does suggest ways in which collagen sequences could have been transferred into the genes for such proteins as acetylcholineesterase and the C1q component of complement.[647] The human α1(I) collagen gene also consists of 51 segments but they lie

within a shorter 18-kb length of DNA.[651] A collagen gene from *Drosophila* is much less fragmented.

Collagens are found in all metazoan organisms.[652] Invertebrate collagens play a variety of specialized roles.[653] For example, minicollagens from *Hydra* strengthen the walls of their nematocysts.[572]

Elastic fibers. The elastic properties of lung, skin, and large blood vessels are provided by elastic fibers in the extracellular matrix.[654] The fibers consist of amorphous material together with the insoluble protein **elastin**, which is rich in glycine, proline, and hydrophobic amino acids. Its special structure (Chapter 2) provides elasticity to the fibers. A 72-kDa precursor **tropoelastin** is secreted into the extracellular space where it is acted upon by lysyl oxidase (Eq. 8-7) and crosslinked into a rubber-like network.[654] Remarkable crosslinkages are formed. Three aldehyde groups derived by oxidation of lysine side chains combine with one lysine amino group through aldol condensations, dehydration, and oxidation to form residues of **desmosine** and **isodesmosine**.

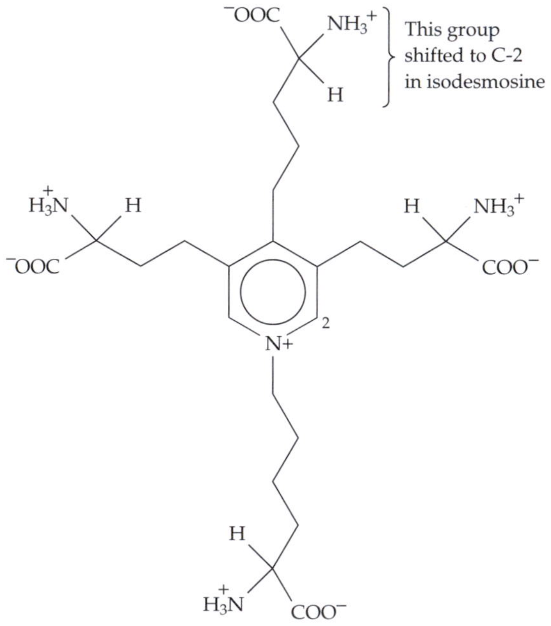

Desmosine

Cartilage and basement membranes. Tendons consist largely of collagen, but in most tissues the collagen fibrils are embedded in a matrix of proteoglycans and various other proteins.[655–657] Both the core proteins of the proteoglycans and the attached

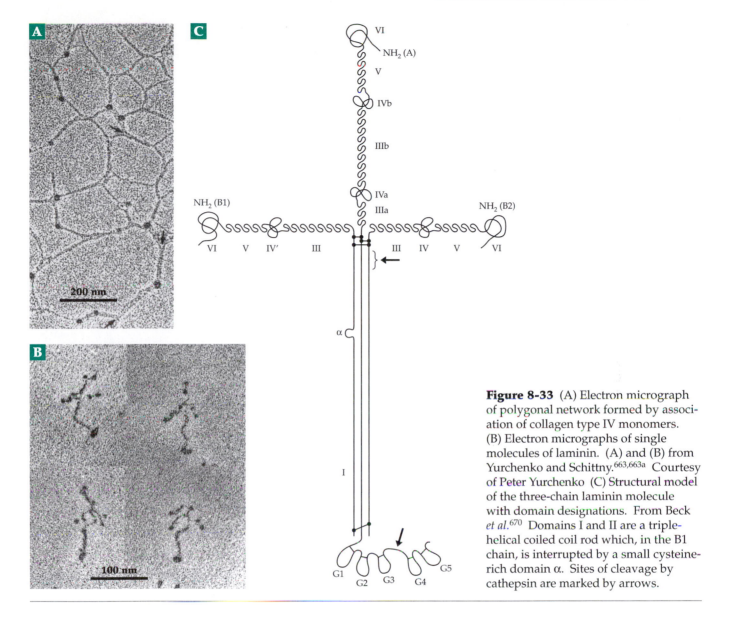

Figure 8-33 (A) Electron micrograph of polygonal network formed by association of collagen type IV monomers. (B) Electron micrographs of single molecules of laminin. (A) and (B) from Yurchenko and Schittny.[663,663a] Courtesy of Peter Yurchenko (C) Structural model of the three-chain laminin molecule with domain designations. From Beck *et al.*[670] Domains I and II are a triple-helical coiled coil rod which, in the B1 chain, is interrupted by a small cysteine-rich domain α. Sites of cleavage by cathepsin are marked by arrows.

polysaccharides interact with collagen fibrils and with fibronectin and other (previously discussed) cell surface proteins. The cartilage matrix[658,659] consists largely of proteoglycans (Fig. 4-16)[660] and of several difficult to study, insoluble proteins. One of these is the 148-kDa **cartilage matrix protein**, which yields 52-kDa subunits upon reduction.[661,662] It interacts with both proteoglycans and collagen and may help to integrate the cartilage matrix.

Basement membranes (Fig. 1-6)[663] function in part as an exoskeleton that helps keep cells positioned. However, the thick basement membranes of the capillary walls of the glomeruli of the kidney provide the ultrafilters that prevent most proteins from entering the urine. Basement membranes contain large amounts of collagen IV, which forms a polygonal network (Fig. 8-33A). A second macromolecular network is formed by the very large 950-kDa cross-shaped multisubunit protein called **laminin** (Fig.

8-33B).[664–666] Laminin is one of a series of extracellular proteins which appear to have arisen by shuffling of structural modules during evolution.[667] It contains sites for binding to heparin, to integrins,[668] to the heparin sulfate proteoglycan **agrin** (see also Fig. 4-11), and to the 150-kDa sulfated glycoprotein **nidogen** (entactin).[669] At least seven isoforms of laminin, with varying tissue distributions, are formed, in part as a result of alternative splicing of the mRNA transcripts.[666] Laminin is rich in EGF-like modules.[665] An X-ray structure of three of them shows that they form a continuous rod of complex structure.[321] A smaller 100 kDa basement membrane **fibulin** also contains multiple EGF-like repeats.[671] As with other extracellular structures, crosslinking of laminin and other components of basement membranes by transglutaminase provides additional stability.[672]

BOX 8-E GENETIC DEFECTS OF COLLAGEN STRUCTURE

Any major protein of the body is likely to be associated with a number of genetic problems. In the case of collagen, the possibility for harmful mutations is enhanced by the existence of a large number of genes that encode the more than 16 types of collagen which are expressed differently in different tissues. There are known human disorders resulting from defects in synthesis, secretion, or structure of types I, II, III, IV, and VII collagens. Other defects involve lysyl hydroxylase and procollagen N-proteinase.[a–d]

In the severe lethal form of **osteogenesis imperfecta** (brittle bone syndrome) the victims' collagen I may contain an α1 chain lacking as many as 100 residues or a shortened α2 chain. In other cases a cysteine, arginine, or other amino acid has been substituted for glycine in the triple-helical region of an α1 chain.[e,f] The cloning of collagen genes has permitted an exact description and precise location of the defects in these genes. Alterations toward the N-terminus of the α1 chain or in the α2 chain often cause a milder type of osteogenesis imperfecta.[g] Some patients have deletions in the pro-α2 chains of collagen I causing the chain to be incorporated into the collagen without removal of the N-terminal or C-terminal peptide to give a protein with poor stability.[h] In other cases amino acid substitutions in the α2 chain cause formation of chains with excessive posttranslational modification. Sometimes the α2 chain is not incorporated into the triple helix and the collagen I formed contains three α1 chains.

Another well established abnormality of collagen is found in cattle suffering from **dermatosparaxis**, a disease in which the skin is extremely brittle. The collagen chains are disorganized and have poor fiber-forming properties. The procollagen peptidase that cleaves a peptide from the N termini of the chains of procollagen is apparently defective. A similar human collagen disease is the **Ehlers–Danlos syndrome**, which in some instances is accompanied with recurrent joint dislocations and curvature of the spine. At least ten different types of the disorder are known.[d] The procollagen peptidase is sometimes lacking.[i] In other cases a person synthesizes an abnormal pro-α2 chain that is resistant to cleavage by the peptidase because of deletion of the normal cleavage site. In others collagen is formed in only small amounts or is degraded rapidly. Some individuals lack lysyl hydroxylase and others have a defect in mRNA splicing which causes loss of an exon from the mRNA and synthesis of shortened pro-α2 chains.[j]

Somewhat similar symptoms are observed in **lathyrism**, a disease which arises when animals ingest seeds of *Lathyris odoratus*, the common sweetpea. Since lathyris peas form part of the diet of some peoples, the condition is also known in humans and often causes curvature of the spine and rupture of the aorta.

The biochemical problem has been traced to the presence in the seeds of **β-cyanoalanine** and of its decarboxylation product **β-aminopropionitrile**.

$$N \equiv C - CH_2CH_2 - NH_3^+$$

Although the mode of action is not certain, this compound is an inhibitor of lysyl oxidase essential to the crosslinking of both collagen and elastin. A hereditary defect with a similar effect in the mouse involves a defect in lysyl oxidase.[k,l]

Collagen defects account for a variety of other skeletal problems[m] including some cases of the common **osteoarthritis**.[n] Mice lacking the α1 chain of collagen IX develop a degenerative joint disease resembling human osteoarthritis.[o] An inherited defect in the basement membrane collagen IV is responsible for the inherited **Alport disease** in which kidney filtration is defective.[p,q] Similar symptoms are observed with the acute autoimmune disease **Goodpasture syndrome** and in **diabetic nephropathy**.[p]

[a] Prockop, D. J. (1990) *J. Biol. Chem.* **265**, 15349–15352

[b] Kuivaniemi, H., Tromp, G., and Prockop, D. J. (1991) *FASEB J.* **5**, 2052–2060

[c] Prockop, D. J., and Kivirikko, K. I. (1995) *Ann. Rev. Biochem.* **64**, 403–434

[d] Byers, P. H. (1995) in *The Metabolic and Molecular Bases of Inherited Disease*, 7th ed., Vol. 3 (Scriver, C. R., Beaudet, A. L., Sly, W. S., and Valle, D., eds), pp. 4029–4077, McGraw-Hill, New York

[e] Cohen-Solal, L., Zylberberg, L., Sangalli, A., Gomez Lira, M., and Mottes, M. (1994) *J. Biol. Chem.* **269**, 14751–14758

[f] Lightfoot, S. J., Atkinson, M. S., Murphy, G., Byers, P. H., and Kadler, K. E. (1994) *J. Biol. Chem.* **269**, 30352–30357

[g] Marini, J. C., Lewis, M. B., Wang, Q., Chen, K. J., and Orrison, B. M. (1993) *J. Biol. Chem.* **268**, 2667–2673

[h] Mundlos, S., Chan, D., Weng, Y. M., Sillence, D. O., Cole, W. G., and Bateman, J. F. (1996) *J. Biol. Chem.* **271**, 21068–21074

[i] Holmes, D. F., Watson, R. B., Steinmann, B., and Kadler, K. E. (1993) *J. Biol. Chem.* **268**, 15758–15765

[j] Weil, D., D'Alessio, M., Ramirez, F., Steinmann, B., Wirtz, M. K., Glanville, R. W., and Hollister, D. W. (1989) *J. Biol. Chem.* **264**, 16804–16809

[k] Pope, F. M., Martin, G. R., Lichtenstein, J. R., Penttinen, R., Gerson, B., Rowe, D. W., and McKusick, V. A. (1975) *Proc. Natl. Acad. Sci. U.S.A.* **72**, 1314–1316

[l] Rowe, D. W., McGoodwin, E. B., Martin, G. R., and Grahn, D. (1977) *J. Biol. Chem.* **252**, 939–942

[m] Freisinger, P., Ala-Kokko, L., LeGuellec, D., Franc, S., Bouvier, R., Ritvaniemi, P., Prockop, D. J., and Bonaventure, J. (1994) *J. Biol. Chem.* **269**, 13663–13669

[n] Ala-Kokko, L., Baldwin, C. T., Moskowitz, R. W., and Prockop, D. J. (1990) *Proc. Natl. Acad. Sci. U.S.A.* **87**, 6565–6568

[o] Fässler, R., Schnegelsberg, P. N. J., Dausman, J., Shinya, T., Muragaki, Y., McCarthy, M. T., Olsen, B. R., and Jaenisch, R. (1994) *Proc. Natl. Acad. Sci. U.S.A.* **91**, 5070–5074

[p] Zhou, J., Hertz, J. M., Leinonen, A., and Tryggvason, K. (1992) *J. Biol. Chem.* **267**, 12475–12481

[q] Gunwar, S., Ballester, F., Noelken, M. E., Sado, Y., Ninomiya, Y., and Hudson, B. G. (1998) *J. Biol. Chem.* **273**, 8767–8775

BOX 8-F SKIN

Mammalian skin must be tough, water-resistant, self-renewing, and rapidly healing. The outer layers of cells or **epidermis** consist principally of **keratinocytes**, epithelial cells specialized for formation of keratin (Fig. 7-31). In the inner layer of the epidermis the **basal stem cells** divide, providing a constant outward flow of cells which become progressively flattened, dehydrated, and filled with keratin fibrils.[a] The outer layers contain only dead cells which are finally sloughed or abraded from the surface. Human epidermis is completely renewed in about 28 days!

About 25 different human genes encode the keratins of skin and other soft tissues. Others specify the keratins of hair and nails.[b] Both of these hard tissues as well as claws, hoofs, beaks, horns, scales, quills, and feathers are largely keratin. However, there are additional constituents. During the final stages of keratinocyte differentiation a 15-nm-thick crosslinked sheath of protein, the **cornified cell envelope** (CE), forms beneath the plasma membrane.[c] Crosslinkages between keratin and other proteins are formed by the action of transglutaminases.[d–f] A specialized protein **involucrin**, which contains glutamine-rich repeating sequences, provides many of the side chain amide groups for the crosslinking reaction (Eq. 2-23).[g] **Loricin**, a protein containing glycine-rich flexible loops,[h] is also a major partner in these cross-linkages.[c,h] The histidine-rich **filaggrin** undergoes complex processing before binding to keratin fibrils to provide another form of crosslinkage.[i,j] Small proline-rich proteins, desmosonal proteins, and others are also present in the CE.[c]

As the final outer **stratum corneum** is formed the phospholipid bilayer deteriorates and intercellular lipid layers are formed.[k,l] These contain principally ceramides, cholesterol, and free fatty acids. Some sphingolipids are covalently attached to proteins.[a]

Also present in the epidermis are embedded macrophage-like Langerhans cells as well as pigmented **melanocytes**, cells with highly branched dendrites, which lie just above the basal stem cell layer. Each melanocyte contains hundreds of pigmented organelles called **melanosomes**. They contain not only the black or reddish **melanin** pigments but also the enzymes needed to form them (Chapter 25).[n,o]

The dendrites of a melanocyte contact about 36 keratinocytes and are able to transfer melanosomes to these adjacent cells. The numbers and sizes of the melanosomes as well as melanin structure determine differences in skin color.[m] Similar cells in amphibians, the **melanophores**, also contain light receptors.[p] Their melanosomes are not transferred to other cells but may be either clustered near the center of the cell or dispersed. The location can be changed quickly by transport of the melanosomes along a network of microtubules allowing the animals to change in response to changes in light color.[q] Various stimuli, including ultraviolet irradiation of melanocytes, cause increased synthesis of melanin with a resultant tanning[o] and added protection against sunburn.

Beneath the basement membrane of the epidermis is the **dermis**, a thick, tough, collagen-rich connective tissue. Blood vessels and nerve endings are found in this layer, as are roots of hairs and oil and sweat glands.[r]

Skin suffers from a variety of ailments including serious hereditary diseases.[a] One group of **keratinization disorders**, known as **ichthyoses**, are characterized by thickened, scaly skin. Some hereditary ichthyoses result from defects in type II

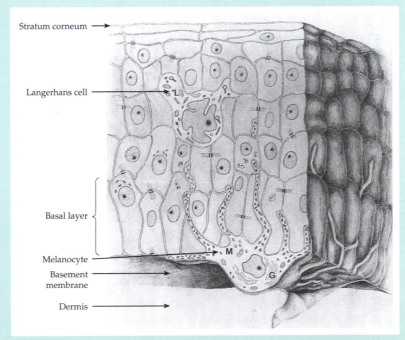

Diagram of a dendritic melanocyte surrounded by satellite keratinocytes. The Golgi area (G), where the melanosomes are synthesized, is shown around the nucleus. The other branched cell, higher in the epidermis, is a Langerhans cell with its tennis racquet-shaped granules. Courtesy of Dr. W. Quevedo, Jr. From Montagna *et al.*[m]

BOX 8-F SKIN (continued)

keratin.[b] **Lamellar ichthyosis** reflects defects in the crosslinking enzyme transglutaminase.[e,f] **Epidermolysis bullosa** is a heterogeneous group of disorders characterized by easy formation of blisters. One form has been traced to a defect in the anchoring fibrils of type VII collagen, which tie cells of the basal layer to the basement membrane.[s,t] Others are defects in keratins of the basal or intermediate layers.[a] Yet others involve the lipid metabolism of skin, e.g. a steroid sulfatase deficiency.[a]

The most frequent skin disorder, which affects about 2% of the world's population is **psoriasis**. The thickened, scaly patches can cover much of the skin and become disabling. The inflammation and excessive epidermal growth are usually a T-cell mediated immunologic response to antigenic stimuli.[p,u,v] However, there is a hereditary form.[u] Other common skin disorders include actinic keratosis induced by light and cancer.

[a] Roop, D. (1995) *Science* **267**, 474–475

[b] Takahashi, K., Paladini, R. D., and Coulombe, P. A. (1995) *J. Biol. Chem.* **270**, 18581–18592

[c] Robinson, N. A., Lapic, S., Welter, J. F., and Eckert, R. L. (1997) *J. Biol. Chem.* **272**, 12035–12046

[d] Kim, S.-Y., Chung, S.-I., and Steinert, P. M. (1995) *J. Biol. Chem.* **270**, 18026–18035

[e] Huber, M., Rettler, I., Bernasconi, K., Frenk, E., Lavrijsen, S. P. M., Ponec, M., Bon, A., Lautenschlager, S., Schorderet, D. F., and Hohl, D. (1995) *Science* **267**, 525–528

[f] Candi, E., Melino, G., Lahm, A., Ceci, R., Rossi, A., Kim, I. G., Ciani, B., and Steinert, P. M. (1998) *J. Biol. Chem.* **273**, 13693–13702

[g] Yaffe, M. B., Beegen, H., and Eckert, R. L. (1992) *J. Biol. Chem.* **267**, 12233–12238

[h] Hohl, D., Mehrel, T., Lichti, U., Turner, M. L., Roop, D. R., and Steinert, P. M. (1991) *J. Biol. Chem.* **266**, 6626–6636

[i] Resing, K. A., Walsh, K. A., Haugen-Scofield, J., and Dale, B. A. (1989) *J. Biol. Chem.* **264**, 1837–1845

[j] Mack, J. W., Steven, A. C., and Steinert, P. M. (1993) *J. Mol. Biol.* **232**, 50–66

[k] ten Grotenhuis, E., Demel, R. A., Ponec, M., Boer, D. R., van Miltenburg, J. C., and Bouwstra, J. A. (1996) *Biophys. J.* **71**, 1389–1399

[l] Bouwstra, J. A., Thewalt, J., Gooris, G. S., and Kitson, N. (1997) *Biochemistry* **36**, 7717–7725

[m] Montagna, W., Prota, G., and Kenney, J. A., Jr. (1993) *Black Skin Structure and Function*, Academic Press, San Diego, California

[n] Potterf, S. B., Muller, J., Bernardini, I., Tietze, F., Kobayashi, T., Hearing, V. J., and Gahl, W. A. (1996) *J. Biol. Chem.* **271**, 4002–4008

[o] Roméro-Graillet, C., Aberdam, E., Biagoli, N., Massabni, W., Ortonne, J.-P., and Ballotti, R. (1996) *J. Biol. Chem.* **271**, 28052–28056

[p] Greaves, M. W., and Weinstein, G. D. (1995) *N. Engl. J. Med.* **332**, 581–588

[q] Rogers, S. L., Tint, I. S., Fanapour, P. C., and Gelfand, V. I. (1997) *Proc. Natl. Acad. Sci. U.S.A.* **94**, 3720–3725

[r] Martin, P. (1997) *Science* **276**, 75–81

[s] Byers, P. H. (1995) in *The Metabolic and Molecular Bases of Inherited Disease*, 7th ed., Vol. 3 (Scriver, C. R., Beaudet, A. L., Sly, W. S., and Valle, D., eds), pp. 4029–4077, McGraw-Hill, New York

[t] Christiano, A. M., Ryynänen, M., and Uitto, J. (1994) *Proc. Natl. Acad. Sci. U.S.A.* **91**, 3549–3553

[u] Tomfohrde, J., Silverman, A., Barnes, R., Fernandez-Vina, M. A., Young, M., Lory, D., Morris, L., Wuepper, K. D., Stastny, P., Menter, A., and Bowcock, A. (1994) *Science* **264**, 1141–1145

[v] Boehncke, W.-H., Dressel, D., Zollner, T. M., and Kaufmann, R. (1996) *Nature (London)* **379**, 777

Fibrillin and Marfan's syndrome. Most connective tissues contain insoluble, beaded microfibrils 10–12 nm in diameter. A component of some of these microfibrils, which are often found in elastic tissue, was purified from media used to culture human fibroblasts in 1986. This protein, called fibrillin, is a single-chain 350-kDa glycoprotein which contains ~14% cysteine.[673–674] Using a DNA probe based on the partially cloned fibrillin gene, the location of the gene was established on the long arm of chromosome 15 at a site previously identified as that of a gene defective in Marfan's syndrome. This disorder often causes dislocation of lenses of the eyes and aortic aneurysm as well as elongated limbs and fingers. Point mutations in the fibrillin gene have been identified in both Marfan's patients and family members that carry the defective gene.[675,676]

The cuticles of invertebrates. The tough elastic cuticle of the nematode *Caenorhabditis* is largely collagen. However, the molecules are smaller than in vertebrates and there are ~100 different genes whose transcription gives rise to a large variety of similar proteins.[652] Cuticles of some annelids have unusually long collagens.[653,677] In contrast, the epithelial cells of insects and other arthropods secrete chitin which serves as the framework for development of a thick and often hard cuticle or exoskeleton. The cuticle also contains a variety of proteins.[678,679] In some instances mineralization by calcium carbonate occurs. During the later phases of the cuticle development extensive crosslinking of the proteins takes place. This is largely by reactions between modified aromatic side chains and resembles the chemistry of formation of melanin and lignin (Chapter 25).[680]

Bones, teeth, and shells. Living organisms are able to induce the formation of over 60 inorganic compounds.[681] Most of these are formed by animals. Two forms of calcium carbonate, **calcite** and **aragonite**, predominate.[682] These minerals form shells, exoskeleton bones, bones, teeth, and other specialized structures.

While some organisms promote mineral deposition completely outside of their cell coats, the mineralization is usually controlled by the proteins and polysaccharides lying around and between cells.[683–687]

Bone deposition begins in the proteoglycan and collagen II matrix of cartilage. Later these polymers are largely replaced by collagen I which accounts for 90% of the organic material in mature bone. Bone collagen has a distinctive pattern of crosslinking.[639,688,689] Some borohydride-reducible crosslinks remain throughout adult life.[639] However, complex hydroxypyridine and pyrrole linkages are more characteristic of bone collagen. Embedded in the bone matrix are spidery cells, the **osteocytes**. Among these are **osteoblasts**,[689a] which secrete the collagen and other proteins that promote the laying down of calcium phosphate. Also present are large multinucleate **osteoclasts**[689b] which dissolve bone and reabsorb calcium and phosphate. Cells of both types remain active in mature bone, which is both a structural material and a store of calcium and phosphorus.

How do osteoblasts induce calcium phosphate deposition? The first crystals laid down within the cartilage matrix are of carbonate apatite only 2–3 nm thick and tens of nanometers in length. They are intimately associated with the collagen and other components of the matrix[690] and may be formed within matrix vesicles.[691] From observations with large biologically formed calcium carbonate crystals it is known that proteins or other organic initiators of crystallization can be found embedded in mature crystals. If the same is true in bone what are the initiators? The answer is uncertain but it is known that without osteoblasts the partially mineralized cartilage will not become bone.[692]

A key to the development of osteoblasts appears to be an **osteoblast-specific transcription factor OS/2** or **Cbfa1**.[693–695] Mutations in human Cbfa1 are linked to a series of skeletal defects.[696] A unique change accompanying conversion of a precursor cell into an osteoblast is the formation of a 49-residue γ-carboxyglutamate (Gla)-containing protein called **osteocalcin**.[697] (See also Box 15-F). It is the most abundant noncollagenous protein of bone. Its three Gla residues doubtless help to bind calcium ions and osteocalcin may be an initiator of crystallization. Osteocalcin has also been found in fish scales.[698] Also present in bone is a 74-residue **matrix Gla protein** which has 5 Gla residues.[699]

Other phosphoproteins, glycoproteins, and proteoglycans may also be required for mineralization. For example, the 286-residue glycoprotein **osteonectin**[323,700,701] accounts for 3% of total bone protein. It contains two Ca^{2+}-binding motifs and inhibits growth of hydroxylapatite crystals, but its role in bone development is not clear. One of the phosphoproteins of developing bone has been identified as

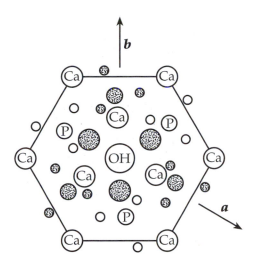

Figure 8-34 An end-on view of a crystallite of hydroxylapatite. The shaded atoms of Ca, P, and O represent an underlying layer. The OH⁻ groups form a longitudinal H-bonded array in the center. From J. A. Weatherell and C. Robinson,[705] p. 55. A small amount of Mg^{2+} is present in place of Ca^{2+} and a very small fraction of the OH⁻ is replaced by fluoride ion which has a bone strengthening effect.

the 24 kDa propeptide cut from the N-terminal end of the α1 chain of type I procollagen.[702] Another bone protein, the sialic acid-containing **osteopontin** has the cell-binding sequence GRGDS identical to that in fibronectin.[703,704] Since it also binds to hydroxylapatite this protein may form a bridge between cell surfaces and the mineral matrix.

The mineral phase of bone is largely hydroxylapatite, $Ca_{10}(PO_4)_6(OH)_2$ (Fig. 8-34), which is essentially in chemical equilibrium with the calcium and phosphate ions present in the blood serum. Thus, bone cells can easily promote either the deposition or dissolution of the mineral phase by localized change in pH, concentrations of Ca^{2+} or HPO_4^{2-}, or of chelating compounds such as ATP or inorganic pyrophosphate. Small 100 nm vesicles rich in acidic phospholipids and containing both Ca^{2+} and the enzymes alkaline phosphatase and pyrophosphatase, may play an essential role in calcification. Perhaps they release the calcium and enzymes that generate inorganic phosphate.[691]

Bone is noted for its continual "remodeling" caused by the action of both osteoblasts and osteoclasts.[704a,b] The latter are large multinuclear cells derived from the same precursors as give rise to macrophages.[689b,704] Osteoclasts generate a sealed, acidic compartment on a bone surface, using a vacuolar ATPase to pump protons from the cytosol of the osteoclast, using H_2CO_3 as a source of protons.[706] The HCO_3^- then leaves cells in exchange for Cl⁻ which then also enters the acidified compartment. Not only is the calcium phosphate dissolved but lysomal enzymes digest the organic materials. Each osteoclast creates a long tun-

BOX 8-G THE BIOCHEMISTRY OF TEETH

Like other bones, mammalian teeth are composed largely of collagen and hydroxylapatite. The much studied rat incisor is 65% mineral; about 85% of the organic material is type I collagen. A tooth consists of three mineralized tissues together with the internal blood vessels and nerves of the pulp.[a] The outer layer of **enamel** is formed by secretion from a thin layer of epithelial cells, the **ameloblasts**. The enamel matrix is devoid of collagen but contains two characteristic groups of proteins, the **amelogenins** and the **enamelins**.

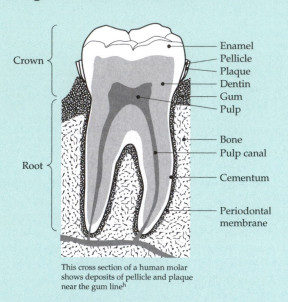

This cross section of a human molar shows deposits of pellicle and plaque near the gum line[h]

The amelogenins are hydrophobic but are also rich in proline, histidine, and glutamic acid.[a,b] They account for 90% of the matrix proteins but are replaced by the initially less abundant enamelins[c] in fully mineralized teeth.[a,d] Another protein present in developing enamel is **ameloblastin**, which appears to be unique to amyloblasts.[a] As mineralization of the enamel progresses the matrix is lost and 98% of the enamel is hydroxylapatite.[a]

Dentin is of epithelial-mesenchymal origin. The **odontoblasts** secrete an extracellular matrix that is rich in collagen I and which also contains all of the other major bone proteins as well as a **dentin sialoprotein**[e] and poorly characterized phosphate carriers, the **phosphophoryns**.

The third hard tissue of teeth is **cementum**, which binds teeth to the periodontal ligaments. The ligaments contain fibrillar collagen which inserts into the cementum.[f,g]

Except for the common cold, tooth decay (**caries**) is the most prevalent disease in the United States.[h,i] Caries is initiated by attack of acids produced by bacterial fermentations on the enamel. The saliva contains calcium and phosphate and is supersaturated with respect to these ions. As a result the enamel surface is continuously recalcified. The 43-residue salivary protein **statherin** retards precipitation of calcium phosphates from the saliva, preventing excessive build-up of calcified deposits on the teeth.[j] Other proline-rich proteins may play a role in recalcification repair.[k]

A freshly cleaned tooth surface quickly becomes coated with a thin **pellicle** of salivary proteins. This provides a surface for growth of **dental plaque**, which contains many bacteria and adhesive polysaccharides such as dextrans.[l] The latter are generated from dietary sucrose by such bacteria as *Streptococcus mutans*. (Chapter 20) and others.[m] Many factors affect the probability of tooth decay. A high sucrose diet promotes decay.[n] While most people have some trouble with tooth decay, 1 or 2 per thousand remain totally free of caries and seem to be immune. Many factors must affect resistance to caries. For example, individuals vary in the kinds and numbers of bacteria present on teeth and in the structure of tooth enamel.[o] Addition of fluoride ion to water supplies at a level of 1ppm (0.05 mM) is generally believed to reduce the incidence of tooth decay. However, caries has been declining in many developed countries at rates that are the same for water with or without fluoride.[p,q] If teeth escape caries **periodontal disease**, caused by bacteria, is often a major problem for older people.[r]

[a] Krebsbach, P. H., Lee, S. K., Matsuki, Y., Kozak, C. A., Yamada, K. M., and Yamada, Y. (1996) *J. Biol. Chem.* **271**, 4431–4435
[b] Renugopalakrishnan, V., Strawich, E. S., Horowitz, P. M., and Glimcher, M. J. (1986) *Biochemistry* **25**, 4879–4887
[c] Deutsch, D., Palmon, A., Fisher, L. W., Kolodny, N., Termine, J. D., and Young, M. F. (1991) *J. Biol. Chem.* **266**, 16021–16028
[d] Robinson, C., Weatherell, J. A., and Höhling, H. J. (1983) *Trends Biochem. Sci.* **8**, 284–287
[e] Ritchie, H. H., Hou, H., Veis, A., and Butler, W. T. (1994) *J. Biol. Chem.* **269**, 3698–3702
[f] Arzate, H., Olson, S. W., Page, R. C., Gown, A. M., and Narayanan, A. S. (1992) *FASEB J.* **6**, 2990–2995
[g] Yamauchi, M., Katz, E. P., and Mechanic, G. L. (1986) *Biochemistry* **25**, 4907–4913
[h] Sanders, H. J. (1980) *Chem. Eng. News* **(Feb. 25) 58**(8) 30–39
[i] Shaw, J. H. (1987) *N. Engl. J. Med.* **317**, 996–1004
[j] Schlesinger, D. H., and Hoy, D. I. (1977) *J. Biol. Chem.* **252**, 1689–1695
[k] Clements, S., Mehansho, H., and Carlson, D. M. (1985) *J. Biol. Chem.* **260**, 13471–13477
[l] Kolenbrander, P. E., Ganeshkumar, N., Cassels, F. J., and Hughes, C. V. (1993) *FASEB J.* **7**, 406–408
[m] Abeygunawardana, C., and Bush, C. A. (1990) *Biochemistry* **29**, 234–248
[n] Newbrun, E. (1982) *Science* **217**, 418–423
[o] Cevc, G., Cevc, P., Schara, M., and Skaleric, U. (1980) *Nature (London)* **286**, 425–426
[p] Diesendorf, M. (1986) *Nature (London)* **322**, 125–129
[q] Hileman, B. (1988) *Chem. Eng. News* **Aug. 1**, 26–42
[r] Williams, R. C. (1990) *N. Engl. J. Med.* **322**, 373–382

nel into the interior of the bone. However, the tunnel is soon lined with new osteoblasts and new bone is laid down.[707] In **osteoporosis**, a very common disease of older persons, the rate of resorption of bone by osteoclasts exceeds that of bone formation by osteoblasts. Women typically lose 50% and men ~30% of their calcium phosphate from vertebrae and ends of long bones as they age.[708] In osteopetrosis (marble bone disease) excessive calcification of bone and other tissues occurs as a result of a deficiency of carbonic anhydrase.[709] See also Chapter 13. A disease of progressive calcification of soft tissues[710] results from an excess of **bone morphogenic factor-4**, one of a group of protein factors acting on bone development (Chapter 32).[631] **Paget's disease** is another disorder of bone remodeling that leads to overproduction of bone of poor quality.[711] As discussed in Box 22-C, the circulating level of Ca^{2+} as well as the cellular uptake of this ion are controlled by vitamin D and its metabolites and by calcitonin and the parathyroid hormones.

Lack of a pyrophosphate ion pump in cartilage cells may cause a deficit in pyrophosphate in the surroundings of the forming bone. In mice with a defect in this pump bony spurs, similar to those in human osteoarthritis, are formed.[711a] Over half of the world's population of persons over 65 years of age are afflicted by arthritis, over 100 types being known.[711b]

Calcium carbonate in several different crystalline forms arises biologically.[705,712] Sometimes the mineral is deposited within vesicles inside the cell. Thus some species of the protozoan *Hymenomonas* use Golgi vesicles as sites for construction of segments of plantlike cell walls complete with crystalline calcium carbonate plates attached to the outer surface.[713] These wall segments are then transported to their final locations. The intricately sculptured spicules of sea urchins and sponges are single crystals of calcium carbonate which have grown within intracellular vesicles.[687,714] Diatoms also form their shells of nearly pure, hydrated SiO_2 (See Box 4-B) entirely within membrane-bound vesicles.[715] On the other hand, shells of molluscs are usually deposited outside the cells of the organism but again under the influence of a protein matrix.[716,717] The animals apparently actively pump bicarbonate outward where it reacts with Ca^{2+} (Eq. 8-11).[718]

$$Ca^{2+} + HCO_3^- \rightarrow CaCO_3 + H^+ \qquad (8\text{-}11)$$

The protons released then react with more bicarbonate to form carbonic acid which is converted to CO_2 and water (Eq. 8-12) by the enzyme carbonic anhydrase.

$$H_2CO_3 \xrightarrow{\text{Carbonic anhydrase}} CO_2 + H_2O \qquad (8\text{-}12)$$

A problem that arises within organisms is avoiding

crystallization under supersaturating conditions. For example, normal urine is supersaturated with calcium oxalate. To prevent formation of renal calculi (stones)[719] an inhibitory glycoprotein is present and slows the formation and growth of crystals.[720] Under some disease conditions calcium carbonate stones may form in pancreatic ducts. A 17 kDa lectinlike glycoprotein called **lithostatine** has been proposed to inhibit stone formation by binding to certain planes on $CaCO_3$ microcrystals just as antifreeze proteins (Box 4-D) inhibit ice formation.[721] However, this proposed function for lithostatine is doubtful.[722,723] Pathological deposits of crystalline calcium pyrophosphate and basic calcium phosphates are sometimes present in joints,[724] even in Neanderthal skeletons.[725]

3. Cell Walls of Fungi and Green Plants

The cell walls of yeasts and other fungi are made up largely of glucans, chitin, and a mannan-protein complex. Yeasts contain predominately glucans but chitin is the major polysaccharide in many other fungi. The most abundant glucan is a $\beta1,3$ linked polymer with about 3% $\beta1,6$ linkages to the branches and a molecular mass of ~ 240 kDa. In addition there is about 15% of a highly branched 1,6 linked glucan containing 1,3 linked branches. At the outer surface of the wall are mannoproteins[726–728] which carry small serine- or threonine-linked oligomannans as well as large highly branched mannans linked to asparagine through the usual *N*-acetylglucosamine-containing core structure. Some of these highly branched mannan chains serve as species-specific antigens.[727] Like those of the bacterial and animal cell surfaces, the antigens vary in structure, a fact with important medical implications. Curiously, many fungi have surface fimbriae which are composed of **collagen**, which is usually regarded as exclusively an animal protein.[729]

Cell walls of higher plants (Fig. 1-7, 4-14) are composed largely of polysaccharides. They are discussed briefly in Chapter 20.

References

1. Christie, W. W. (1973) *Lipid Analysis; Isolation, Preparation, Identification and Structural Analysis of Lipids*, Pergamon, Oxford
2. Hitchcock, C., and Nichols, B. W. (1971) *Plant Lipid Biochemistry*, Academic Press, New York
3. Kates, M. (1972) in *Laboratory Techniques in Biochemistry and Molecular Biology*, Vol. 3, Part II (Work, T. S., and Work, E., eds), North-Holland Publ., Amsterdam
4. Weete, J. W. (1980) *Lipid Biochemistry of Fungi and Other Organisms*, Plenum, New York
5. Gurr, M. I., and James, A. T. (1975) *Lipid Biochemistry, an Introduction*, 2nd ed., Cornell Univ. Press, Ithaca, New York
6. Porter, K. R., and Bonneville, M. A. (1973) *Fine Structure of Cells and Tissues*, 4th ed., Lea & Febiger, Philadelphia, Pennsylvania
7. Aveldaño, M. I., and Sprecher, H. (1987) *J. Biol. Chem.* 262, 1180–1186
8. Gunstone, F. D. (1967) *An Introduction to the Chemistry and Biochemistry of Fatty Acids and Their Glycerides*, 2nd ed., Chapman and Hall, London (Chapter 6)
9. Steck, P. A., Schwartz, B. A., Rosendahl, M. S., and Gray, G. R. (1978) *J. Biol. Chem.* 253, 5625–5629
10. McNeil, M., Daffe, M., and Brennan, P. J. (1991) *J. Biol. Chem.* 266, 13217–13223
11. Qureshi, N., Takayama, K., and Schnoes, H. K. (1980) *J. Biol. Chem.* 255, 182–189
11a. Adlof, R. O., Duval, S., and Emken, E. A. (2000) *Lipids* 35, 131–135
12. Cravatt, B. F., Prospero-Garcia, O., Siuzdak, G., Gilula, N. B., Henriksen, S. J., Boger, D. L., and Lerner, R. A. (1995) *Science* 268, 1506–1509
13. Hummel, H. E., Gaston, L. K., Shorey, H. H., Kaae, R. S., Byrne, K. J., and Silverstein, R. M. (1973) *Science* 181, 873–874
14. Klun, J. A., Chapman, O. L., Mattes, K. C., Wojtkowski, P. W., Beroza, M., and Sonnet, P. E. (1973) *Science* 181, 661–663
15. Moolenaar, W. H. (1995) *J. Biol. Chem.* 270, 12949–12952
16. Wells, W. W., and Eisenberg, F., Jr., eds. (1978) *Cyclitols and Phosphoinositides*, Academic Press, New York
17. DeBuch, H. (1978) *Trends Biochem. Sci.* 3, 44–45
18. Chilton, F. H., and Connell, T. R. (1988) *J. Biol. Chem.* 263, 5260–5265
19. Ford, D. A., and Gross, R. W. (1994) *Biochemistry* 33, 1216–1222
20. Friedman, S. M., ed. (1978) *Biochemistry of Thermophily*, Academic Press, New York
21. Kushwaha, S. C., Kates, M., Sprott, G. D., and Smith, I. C. P. (1981) *Science* 211, 1163–1164
22. Luzzati, V., Gambacorta, A., DeRosa, M., and Gulik, A. (1987) *Ann. Rev. Biophys. Biophys. Chem.* 16, 25–47
23. Goldfine, H., and Langworthy, T. A. (1988) *Trends Biochem. Sci.* 13, 217–221
24. Ferrante, G., Ekiel, I., and Sprott, G. D. (1986) *J. Biol. Chem.* 261, 17062–17066
25. Pond, J. L., Langworthy, T. A., and Holzer, G. (1986) *Science* 231, 1134–1136
26. Florin-Christensen, J., Florin-Christensen, M., Knudson, J., and Rasmussen, L. (1986) *Trends Biochem. Sci.* 11, 354–355
27. Cooney, R. V., Mumma, R. O., and Benson, A. A. (1978) *Proc. Natl. Acad. Sci. U.S.A.* 75, 4262–4264
28. Renou, J.-P., Giziewicz, J. B., Smith, I. C. P., and Jarrell, H. C. (1989) *Biochemistry* 28, 1804–1814
29. Gounaris, K., and Barber, J. (1983) *Trends Biochem. Sci.* 8, 378–381
30. Knowles, F. C., and Benson, A. A. (1983) *Trends Biochem. Sci.* 8, 178–180

31. Tupper, S., Wong, P. T. T., Kates, M., and Tanphaichitr, N. (1994) *Biochemistry* 33, 13250–13258
32. Andersson, A.-S., Rilfors, L., Bergqvist, M., Persson, S., and Lindblom, G. (1996) *Biochemistry* 35, 11119–11130
32a. Vikström, S., Li, L., and Wieslander, Å. (2000) *J. Biol. Chem.* 275, 9296–9302
33. Nishihara, M., Utagawa, M., Akutsu, H., and Koga, Y. (1992) *J. Biol. Chem.* 267, 12432–12435
34. Powles, T. J., Easty, D. M., Easty, G. C., Bondy, P. K., and Munro-Neville, A. (1973) *Nature (London), New Biol.* 245, 83
35. Lynch, D. V. (1993) in *Lipid Metabolism in Plants* (Moore, T. S., Jr., ed), pp. 285–308, CRC Press, Boca Raton, Florida
36. Araki, S., Satake, M., Ando, S., Hayashi, A., and Fujii, N. (1986) *J. Biol. Chem.* 261, 5138–5144
37. Araki, S., Abe, S., Odani, S., Ando, S., Fujii, N., and Satake, M. (1987) *J. Biol. Chem.* 262, 14141–14145
38. Song, W., and Rintoul, D. A. (1989) *Biochemistry* 28, 4194–4200
39. Hsieh, T. C.-Y., Lester, R. L., and Laine, R. A. (1981) *J. Biol. Chem.* 256, 7747–7755
40. Barr, K., and Lester, R. L. (1984) *Biochemistry* 23, 5581–5588
41. Dickson, R. C., Nagiec, E. E., Wells, G. B., Nagiec, M. M., and Lester, R. L. (1997) *J. Biol. Chem.* 272, 29620–29625
42. Yamakawa, T., and Nagai, Y. (1978) *Trends Biochem. Sci.* 3, 128–131
43. Hansson, G. C. (1983) *J. Biol. Chem.* 258, 9612–9615
44. Clausen, H., Holmes, E., and Hakomori, S.-I. (1986) *J. Biol. Chem.* 261, 1388–1392
45. Farooqui, A. A. (1981) *Adv. Lipid Res.* 18, 159–202
46. Godchaux, W., III, and Leadbetter, E. R. (1984) *J. Biol. Chem.* 259, 2982–2990
47. Abbanat, D. R., Leadbetter, E. R., Godchaux, W., III, and Escher, A. (1986) *Nature (London)* 324, 367–369
48. Jones, M. N., and Chapman, D., eds. (1994) *Micelles, Monolayers and Biomembranes*, Wiley, New York
49. Jain, M. (1988) *Introduction to Biological Membranes*, 2nd ed., Wiley, New York
50. Lofan, R., and Nicolson, G. L. (1981) in *Advanced Cell Biology* (Schwartz, L. M., and Azar, M. M., eds), Van Nostrand Reinhold Co., New York
51. Storch, J., and Kleinfeld, A. M. (1985) *Trends Biochem. Sci.* 10, 418–421
52. Robertson, R. N. (1983) *The Lively Membranes*, Cambridge Univ. Press, Cambridge
53. Starzak, M. (1984) *The Physical Chemistry of Membranes*, Academic Press, San Diego, California
54. Gennis, R. B. (1989) *Biomembranes: Molecular Structure and Function*, Springer-Verlag, New York
55. Morell, P., and Norton, W. T. (1980) *Sci. Am.* 242(May), 88–116
56. Martenson, R. E., ed. (1992) *Myelin: Biology and Chemistry*, CRC Press, Boca Raton, Florida
57. Weimbs, T., and Stoffel, W. (1994) *Biochemistry* 33, 10408–10415
58. Korn, E. D. (1966) *Science* 153, 1491–1498
59. Tanford, C. (1980) *The Hydrophobic Effect: Formation of Micelles and Biological Membranes*, 2nd ed., Wiley, New York
60. Danielli, J. F., and Davsen, H. (1985) *J. Cell. Comp. Physiol.* 5, 495–508
61. Singer, S. J., and Nicolson, G. L. (1972) *Science* 175, 720–731
62. Bretscher, M. S., and Raff, M. C. (1975) *Nature (London)* 258, 43–49

63. Quinn, P. J., and Cherry, R. J., eds. (1992) *Structural and Dynamic Properties of Lipids and Membranes*, Portland Press, London
64. Disalvo, E. A., and Simon, S. A., eds. (1995) *Permeability and Stability of Lipid Bilayers*, CRC Press, Boca Raton, Florida
65. Jacobson, K., Sheets, E. D., and Simson, R. (1995) *Science* 268, 1441–1442
66. Pascher, I., Lundmark, M., Nyholm, P.-G., and Sundell, S. (1992) *Biochim. Biophys. Acta* 1113, 339–373
67. Cameron, D. G., Gudgin, E. F., and Mantsch, H. H. (1981) *Biochemistry* 20, 4496–4500
68. Zasadzinski, J. A. N., Helm, C. A., Longo, M. L., Weisenhorn, A. L., Gould, S. A. C., and Hansma, P. K. (1991) *Biophys. J.* 59, 755–760
69. Wiener, M. C., and White, S. H. (1992) *Biophys. J.* 61, 428–433
70. Smith, D. P. E., Bryant, A., Quate, C. F., Rabe, J. P., Gerber, C. H., and Swalen, J. D. (1987) *Proc. Natl. Acad. Sci. U.S.A.* 84, 969–972
71. Yang, J., Tamm, L. K., Tillack, T. W., and Shao, Z. (1993) *J. Mol. Biol.* 229, 286–290
72. Nagle, J. F., Zhang, R., Tristram-Nagle, S., Sun, W., Petrache, H. I., and Suter, R. M. (1996) *Biophys. J.* 70, 1419–1431
73. Mueller, P., Rudin, D. O., Tien, H. T., and Wescott, W. C. (1962) *Nature (London)* 194, 979–980
74. Howard, K. P., and Prestegard, J. H. (1995) *J. Am. Chem. Soc.* 117, 5031–5040
75. Ostro, M. J. (1987) *Sci. Am.* 256(Jan), 103–111
76. Philippot, J. R., and Schuber, F., eds. (1995) *Liposomes as Tools in Basic Research and Industry*, CRC Press, Boca Raton, Florida
77. Harwood, J. L. (1992) *Trends Biochem. Sci.* 17, 203–204
77a. Schröder, M. (1980) *Chem. Rev.* 80, 187–213
77b. Mohapatra, S. K., and Behrman, E. J. (1982) *J. Inorg. Biochem* 16, 85–89
77c. Deetz, J. S., and Behrman, E. J. (1981) *Intl. J. Peptide Prot. Res.* 17, 495–500
78. Norrby, P.-O., Becker, H., and Sharpless, K. B. (1996) *J. Am. Chem. Soc.* 118, 35–42
79. Montal, M., and Mueller, P. (1972) *Proc. Natl. Acad. Sci. U.S.A.* 69, 3561–3566
80. Ladha, S., Mackie, A. R., Harvey, L. J., Clark, D. C., Lea, E. J. A., Brullemans, M., and Duclohier, H. (1996) *Biophys. J.* 71, 1364–1373
81. Jones, M. N., and Chapman, D. (1995) *Micelles, Monolayers, and Biomembranes*, Wiley-Liss, New York
81a. Knobler, C. M. (1990) *Adv. Chem. Phys.* 77, 397–449
82. Bryce, M. R., and Petty, M. C. (1995) *Nature (London)* 374, 771–776
83. Zasadzinski, J. A., Viswanathan, R., Madsen, L., Garnaes, J., and Schwartz, D. K. (1994) *Science* 263, 1726–1733
84. Fettiplace, R., Andrews, D. M., and Haydon, D. A. (1971) *J. Membr. Biol.* 5, 277–296
85. White, S. W., and King, G. I. (1985) *Proc. Natl. Acad. Sci. U.S.A.* 82, 6532–6536
86. Snyder, R. G., Maroncelli, M., Qi, S. P., and Strauss, H. L. (1981) *Science* 214, 188–190
87. Di, L., and Small, D. M. (1995) *Biochemistry* 34, 16672–16677
88. Tu, K., Tobias, D. J., Blasie, J. K., and Klein, M. L. (1996) *Biophys. J.* 70, 595–608
89. Katsaras, J., Raghunathan, V. A., Dufourc, E. J., and Dufourcq, J. (1995) *Biochemistry* 34, 4684–4688
90. Snyder, R. G., Liang, G. L., Strauss, H. L., and Mendelsohn, R. (1996) *Biophys. J.* 71, 3186–3198
91. Chen, S. C., Sturtevant, J. M., and Gaffney, B. J. (1980) *Proc. Natl. Acad. Sci. U.S.A.* 77, 5060–5063

References

92. Sun, W.-J., Tristram-Nagle, S., Suter, R. M., and Nagle, J. F. (1996) *Proc. Natl. Acad. Sci. U.S.A.* **93**, 7008–7012

93. Mou, J., Yang, J., and Shao, Z. (1994) *Biochemistry* **33**, 4439–4443

94. Wang, Z.-q, Lin, H.-n, Li, S., and Huang, C.-h. (1995) *J. Biol. Chem.* **270**, 2014–2023

95. Wang, G., Lin, H.-n, Li, S., and Huang, C.-h. (1995) *J. Biol. Chem.* **270**, 22738–22746

96. Wang, Z.-q, Lin, H.-n, Li, S., and Huang, C.-h. (1994) *J. Biol. Chem.* **269**, 23491–23499

97. Dufourc, E. J., Parish, E. J., Chitrakorn, S., and Smith, I. C. P. (1984) *Biochemistry* **23**, 6062–6071

98. Maulik, P. R., and Shipley, G. G. (1996) *Biochemistry* **35**, 8025–8034

99. McMullen, T. P. W., and McElhaney, R. N. (1997) *Biochemistry* **36**, 4979–4986

100. Finegold, L., ed. (1993) *Cholesterol in Membrane Models*, CRC Press, Boca Raton, Florida

101. Recktenwald, D. J., and McConnell, H. M. (1981) *Biochemistry* **20**, 4505–4510

102. Needham, D., McIntosh, T. J., and Evans, E. E. (1988) *Biochemistry* **27**, 4668–4673

103. Goldfine, H., Johnston, N. C., and Phillips, M. C. (1981) *Biochemistry* **20**, 2908–2916

104. Lewis, R. V. A. H., Pohle, W., and McElhaney, R. N. (1996) *Biophys. J.* **70**, 2736–2746

105. Shimshick, E. J., and McConnell, H. M. (1973) *Biochemistry* **12**, 2351–2360

106. Almeida, P. F. F., Vaz, W. L. C., and Thompson, T. E. (1992) *Biochemistry* **31**, 7198–7210

107. Träuble, H., and Eibl, H. (1974) *Proc. Natl. Acad. Sci. U.S.A.* **71**, 214–219

108. Basánez, G., Nieva, J. L., Rivas, E., Alonso, A., and Goni, F. M. (1996) *Biophys. J.* **70**, 2299–2306

109. Smith, R. L., and Oldfield, E. (1984) *Science* **225**, 280–288

109a. Smith, I. C. P. (1983) in *NMR of Newly Accessible Nuclei*, Vol. 2 (Laslo, P., ed), pp. 1–26, Academic Press, New York

110. Marassi, F. M., and Macdonald, P. M. (1992) *Biochemistry* **31**, 10031–10036

111. Oldfield, E., Bowers, J. L., and Forbes, J. (1987) *Biochemistry* **26**, 6919–6923

112. Bruzik, K. S., and Nyholm, P.-G. (1997) *Biochemistry* **36**, 566–575

113. Hong, M., Schmidt-Rohr, K., and Nanz, D. (1995) *Biophys. J.* **69**, 1939–1950

114. Thurmond, R. L., Niemi, A. R., Lindblom, G., Wieslander, Å., and Rilfors, L. (1994) *Biochemistry* **33**, 13178–13188

115. Boden, N., Jones, S. A., and Sixl, F. (1991) *Biochemistry* **30**, 2146–2155

116. Davis, J. H. (1993) in *Cholesterol in Membrane Models* (Finegold, L., ed), pp. 67–157, CRC Press, Boca Raton, Florida

117. Baenziger, J. E., Jarrell, H. C., and Smith, I. C. P. (1992) *Biochemistry* **31**, 3377–3385

118. De Boeck, H., and Zidovetzki, R. (1992) *Biochemistry* **31**, 623–630

119. de Kruijft, B., Cullis, P. R., and Verkleij, A. J. (1980) *Trends Biochem. Sci.* **5**, 79–81

120. Peng, X., and Jonas, J. (1992) *Biochemistry* **31**, 6383–6390

121. Hong, M., Schmidt-Rohr, K., and Zimmermann, H. (1996) *Biochemistry* **35**, 8335–8341

122. Lu, D., Singh, D., Morrow, M. R., and Grant, C. W. M. (1993) *Biochemistry* **32**, 290–297

123. Salmon, A., and Hamilton, J. A. (1995) *Biochemistry* **34**, 16065–16073

124. Ruocco, M. J., Siminovitch, D. J., Long, J. R., Das Gupta, S. K., and Griffin, R. G. (1996) *Biophys. J.* **71**, 1776–1788

125. Bruzik, K. S., and Harwood, J. S. (1997) *J. Am. Chem. Soc.* **119**, 6629–6637

126. Casal, H. L., and McElhaney, R. N. (1990) *Biochemistry* **29**, 5423–5427

127. Hübner, W., Mantsch, H. H., Paltauf, F., and Hauser, H. (1994) *Biochemistry* **33**, 320–326

128. Léger, C. L., Daveloose, D., Christon, R., and Viret, J. (1990) *Biochemistry* **29**, 7269–7275

129. Jakobsson, E. (1997) *Trends Biochem. Sci.* **22**, 339–344

130. Berger, O., Edholm, O., and Jähnig, F. (1997) *Biophys. J.* **72**, 2002–2013

131. Chiu, S.-W., Clark, M., Balaji, V., Subramaniam, S., Scott, H. L., and Jakobsson, E. (1995) *Biophys. J.* **69**, 1230–1245

132. Browning, J. L. (1981) *Biochemistry* **20**, 7144–7151

133. Cevc, G., Watts, A., and Marsh, D. (1981) *Biochemistry* **20**, 4955–4965

134. Harlos, K., and Eibl, H. (1980) *Biochemistry* **19**, 895–899

135. Koynova, R. D., Kuttenreich, H. L., Tenchov, B. G., and Hinz, H.-J. (1988) *Biochemistry* **27**, 4612–4619

136. Hayakawa, E., Naganuma, M., Mukasa, K., Shimozawa, T., and Araiso, T. (1998) *Biophys. J.* **74**, 892–898

137. Katsaras, J., Jeffrey, K. R., Yang, D. S.-C., and Epand, R. M. (1993) *Biochemistry* **32**, 10700–10707

138. Srisiri, W., Sisson, T. M., O'Brien, D. F., McGrath, K. M., Han, Y., and Gruner, S. M. (1997) *J. Am. Chem. Soc.* **119**, 4866–4873

139. Siegel, D. P., and Epand, R. M. (1997) *Biophys. J.* **73**, 3089–3111

140. Thurmond, R. L., Lindblom, G., and Brown, M. F. (1993) *Biochemistry* **32**, 5394–5410

141. Siegel, D. P., Banschbach, J., and Yeagle, P. L. (1989) *Biochemistry* **28**, 5010–5019

142. Siegel, D. P., and Banschbach, J. L. (1990) *Biochemistry* **29**, 5975–5981

143. Landau, E. M., and Rosenbusch, J. P. (1996) *Proc. Natl. Acad. Sci. U.S.A.* **93**, 14532–14535

144. Delacroix, H., Gulik-Krzywicki, T., and Seddon, J. M. (1996) *J. Mol. Biol.* **258**, 88–103

144a. Tenchov, B., Koynova, R., and Rapp, G. (1998) *Biophys. J.* **75**, 853–866

145. Russell, N. J. (1984) *Trends Biochem. Sci.* **9**, 108–112

146. deMendoza, D., and Cronan, J. E., Jr. (1983) *Trends Biochem. Sci.* **8**, 49–52

147. Mantsch, H. H., Madec, C., Lewis, R. N. A. H., and McElhaney, R. N. (1987) *Biochemistry* **26**, 4045–4049

148. Bechtel, D. B., Mueller, D. D., Whaley, T. W., and Bulla, L. A., Jr. (1985) *J. Biol. Chem.* **260**, 9784–9792

149. Bérubé, L. R., and Hollingsworth, R. I. (1995) *Biochemistry* **34**, 12005–12011

150. Huster, D., Jin, A. J., Arnold, K., and Gawrisch, K. (1997) *Biophys. J.* **73**, 855–864

151. Träuble, H. (1971) *J. Membr. Biol.* **4**, 193–208

152. Wilson, M. A., and Pohorille, A. (1996) *J. Am. Chem. Soc.* **118**, 6580–6587

153. Ho, C., and Stubbs, C. D. (1997) *Biochemistry* **36**, 10630–10637

154. Bretscher, M. S. (1980) *Trends Biochem. Sci.* **5**, VI–VII

155. Devaux, P., and McConnell, H. M. (1972) *J. Am. Chem. Soc.* **94**, 4475–4481

156. Kell, D. B. (1984) *Trends Biochem. Sci.* **9**, 86–88

157. Picard, F., Paquet, M.-J., Dufourc, E. J., and Auger, M. (1998) *Biophys. J.* **74**, 857–868

158. Schmidt, T., Schütz, G. J., Baumgartner, W., Gruber, H. J., and Schindler, H. (1996) *Proc. Natl. Acad. Sci. U.S.A.* **93**, 2926–2929

159. Sheets, E. D., Lee, G. M., Simson, R., and Jacobson, K. (1997) *Biochemistry* **36**, 12449–12458

160. Hopkins, C. R. (1992) *Trends Biochem. Sci.* **17**, 27–32

161. Kleinfeld, A. M., Chu, P., and Storch, J. (1997) *Biochemistry* **36**, 5702–5711

162. Zhao, J., Zhou, Q., Wiedmer, T., and Sims, P. J. (1998) *Biochemistry* **37**, 6361–6366

163. Wolff, D., Canessa-Fischer, M., Vargas, F., and Diaz, G. (1971) *J. Membr. Biol.* **6**, 304–314

164. Sherbet, G. V. (1978) *The Biophysical Characterisation of the Cell Surface*, Academic Press, London

165. Jones, M. N. (1975) *Biological Interfaces*, Elsevier, Amsterdam

166. McLaughlin, S. (1989) *Ann. Rev. Biophys. Biophys. Chem.* **18**, 113–136

167. Aidley, D. J. (1971) *The Physiology of Excitable Cells*, Cambridge Univ. Press, London and New York

168. Nystrom, R. A. (1973) *Membrane Physiology*, Prentice-Hall, Englewood Cliffs, New Jersey

169. Nicholls, D. G., and Ferguson, S. J. (1992) *Bioenergetics 2*, Academic Press, London

170. Wood, P. M. (1985) *Trends Biochem. Sci.* **10**, 106–107

171. Teissié, J., Prats, M., LeMassu, A., Stewart, L. C., and Kates, M. (1990) *Biochemistry* **29**, 59–65

172. Teissié, J., Gabriel, B., and Prats, M. (1993) *Trends Biochem. Sci.* **18**, 243–246

173. Heberle, J., Riesle, J., Thiedemann, G., Oesterhelt, D., and Dencher, N. A. (1994) *Nature (London)* **370**, 379–382

174. Gabriel, B., and Teissié, J. (1996) *Proc. Natl. Acad. Sci. U.S.A.* **93**, 14521–14525

175. Devaux, P. F. (1991) *Biochemistry* **30**, 1163–1173

176. Victorov, A. V., Janes, N., Taraschi, T. F., and Hoek, J. B. (1997) *Biophys. J.* **72**, 2588–2598

177. Shiffer, K. A., Rood, L., Emerson, R. K., and Kuypers, F. A. (1998) *Biochemistry* **37**, 3449–3458

178. Michel, H. (1983) *Trends Biochem. Sci.* **8**, 56–59

179. Cowan, S. W., and Rosenbusch, J. P. (1994) *Science* **264**, 914–916

180. von Heijne, G. (1994) *Annu. Rev. Biophys. Biomol. Struct.* **23**, 167–192

181. Bloom, M. (1995) *Biophys. J.* **69**, 1631–1632

182. Weimbs, T., and Stoffel, W. (1992) *Biochemistry* **31**, 12289–12296

183. Yamaguchi, Y., Ikenaka, K., Niinobe, M., Yamada, H., and Mikoshiba, K. (1996) *J. Biol. Chem.* **271**, 27838–27846

184. Jennings, M. L. (1989) *Ann. Rev. Biochem.* **58**, 999–1027

185. Bowie, J. U. (1997) *J. Mol. Biol.* **272**, 780–789

186. Gafvelin, G., Sakaguchi, M., Andersson, H., and von Heijne, G. (1997) *J. Biol. Chem.* **272**, 6119–6127

187. Jähnig, F. (1990) *Trends Biochem. Sci.* **15**, 93–95

188. Jones, D. T., Taylor, W. R., and Thornton, J. M. (1994) *Biochemistry* **33**, 3038–3049

189. Persson, B., and Argos, P. (1996) *Protein Sci.* **5**, 363–371

190. Fasman, G. D., and Gilbert, W. A. (1990) *Trends Biochem. Sci.* **15**, 89–92

191. Milpetz, F., Argos, P., and Persson, B. (1995) *Trends Biochem. Sci.* **20**, 204–205

192. Shai, Y. (1995) *Trends Biochem. Sci.* **20**, 460–464

192a. Killian, J. A., and von Heijne, G. (2000) *Trends Biochem. Sci.* **25**, 429–434

192b. Lew, S., Ren, J., and London, E. (2000) *Biochemistry* **39**, 9632–9640

193. Popov, M., Tam, L. Y., Li, J., and Reithmeier, R. A. F. (1997) *J. Biol. Chem.* **272**, 18325–18332

194. Lee, G. F., Burrows, G. G., Lebert, M. R., Dutton, D. P., and Hazelbauer, G. L. (1994) *J. Biol. Chem.* **269**, 29920–29927

195. Lee, G. F., Dutton, D. P., and Hazelbauer, G. L. (1995) *Proc. Natl. Acad. Sci. U.S.A.* **92**, 5416–5420

196. Yu, H., Kono, M., McKee, T. D., and Oprian, D. D. (1995) *Biochemistry* **34**, 14963–14969

197. Sansom, M. S. P., Son, H. S., Sankararamakrishnan, R., Kerr, I. D., and Breed, J. (1995) *Biophys. J.* **68**, 1295–1310

References

198. Stopar, D., Jansen, K. A. J., Páli, T., Marsh, D., and Hemminga, M. A. (1997) *Biochemistry* **36**, 8261–8268

199. Nikaido, H. (1994) *J. Biol. Chem.* **269**, 3905–3908

200. Mishra, V. K., and Palgunachari, M. N. (1996) *Biochemistry* **35**, 11210–11220

201. Mishra, V. K., Palgunachari, M. N., Lund-Katz, S., Phillips, M. C., Segrest, J. P., and Anantharamaiah, G. M. (1995) *J. Biol. Chem.* **270**, 1602–1611

202. Casey, P. J. (1995) *Science* **268**, 221–225

203. McIlhinney, R. A. J. (1990) *Trends Biochem. Sci.* **15**, 387–391

204. Buss, J. E., Mumby, S. M., Casey, P. J., Gilman, A. G., and Sefton, B. M. (1987) *Proc. Natl. Acad. Sci. U.S.A.* **84**, 7493–7497

205. Johnson, D. R., Bhatnagar, R. S., Knoll, L. J., and Gordon, J. I. (1994) *Ann. Rev. Biochem.* **63**, 869–914

206. Peseckis, S. M., and Resh, M. D. (1994) *J. Biol. Chem.* **269**, 30888–30892

207. McLaughlin, S., and Aderem, A. (1995) *Trends Biochem. Sci.* **20**, 272–276

208. Milligan, G., Parenti, M., and Magee, A. I. (1995) *Trends Biochem. Sci.* **20**, 181–186

209. Wilson, P. T., and Bourne, H. R. (1995) *J. Biol. Chem.* **270**, 9667–9675

210. Duncan, J. A., and Gilman, A. G. (1996) *J. Biol. Chem.* **271**, 23594–23600

211. Marshall, C. J. (1993) *Science* **259**, 1865–1866

212. Glomset, J. A., Gelb, M. H., and Farnsworth, C. C. (1990) *Trends Biochem. Sci.* **15**, 139–142

213. Casey, P. J., and Seabra, M. C. (1996) *J. Biol. Chem.* **271**, 5289–5292

214. Shipton, C. A., Parmryd, I., Swiezewska, E., Andersson, B., and Dallner, G. (1995) *J. Biol. Chem.* **270**, 566–572

215. Wedegaertner, P. B., Wilson, P. T., and Bourne, H. R. (1995) *J. Biol. Chem.* **270**, 503–506

216. Kikuchi, A., Sagami, H., and Ogura, K. (1999) *J. Biol. Chem.* **274**, 18011–18016

217. Turner, A. J. (1994) in *Essays in Biochemistry*, Vol. 28 (Tipton, K. F., ed), pp. 113–128, Portland Press, London

218. Low, M. G. (1987) *Biochem. J.* **244**, 1–13

219. Lisanti, M. P., and Rodriguez-Boulan, E. (1990) *Trends Biochem. Sci.* **15**, 113–118

220. Thomas, J. R., Dwek, R. A., and Rademacher, T. W. (1990) *Biochemistry* **29**, 5413–5422

221. Brewis, I. A., Ferguson, M. A. J., Mehlert, A., Turner, A. J., and Hooper, N. M. (1995) *J. Biol. Chem.* **270**, 22946–22956

222. Previato, J. O., Jones, C., Xavier, M. T., Wait, R., Travassos, L. R., Parodi, A. J., and Mendonça-Previato, L. (1995) *J. Biol. Chem.* **270, No 13**, 7241–7250

223. Heinz, D. W., Ryan, M., Smith, M. P., Weaver, L. H., Keana, J. F. W., and Griffith, O. H. (1996) *Biochemistry* **35**, 9496–9504

224. Williams, A. F., and Gagnon, J. (1982) *Science* **216**, 696–703

225. Mehlert, A., Richardson, J. M., and Ferguson, M. A. J. (1998) *J. Mol. Biol.* **277**, 379–392

225a. Eisenhaber, B., Bork, P., Yuan, Y., Löffler, G., and Eisenhaber, F. (2000) *Trends Biochem. Sci.* **25**, 340–341

226. Reynolds, J. A., and Trayer, H. (1971) *J. Biol. Chem.* **246**, 7337–7342

227. Pryde, J. G. (1986) *Trends Biochem. Sci.* **11**, 160–163

228. Bennett, V. (1985) *Ann. Rev. Biochem.* **54**, 273–304

229. Justice, J. M., Murtagh, J. J., Jr., Moss, J., and Vaughan, M. (1995) *J. Biol. Chem.* **270**, 17970–17976

230. Neville, D. M., Jr., and Glossmann, H. (1971) *J. Biol. Chem.* **246**, 6335–6338

231. Inaba, M., and Maede, Y. (1988) *J. Biol. Chem.* **263**, 17763–17771

232. Steck, T. L. (1974) *J. Cell Biol.* **62**, 1–19

233. Luna, E. J., Kidd, G. H., and Branton, D. (1979) *J. Biol. Chem.* **254**, 2526–2532

234. Viitala, J., and Järnefelt, J. (1985) *Trends Biochem. Sci.* **10**, 392–395

235. Marchesi, V. T., Furthmayr, H., and Tomita, M. (1976) *Ann. Rev. Biochem.* **45**, 667–698

236. Jay, D., and Cantley, L. (1986) *Ann. Rev. Biochem.* **55**, 511–538

237. Murthy, S. N. P., Wilson, J., Zhang, Y., and Lorand, L. (1994) *J. Biol. Chem.* **269**, 22907–22911

238. Tanner, M. J. A. (1996) *Nature (London)* **382**, 209–210

239. Müller-Berger, S., Karbach, D., König, J., Lepke, S., Wood, P. G., Appelhans, H., and Passow, H. (1995) *Biochemistry* **34**, 9315–9324

240. Erickson, H. K. (1997) *Biochemistry* **36**, 9958–9967

241. Fukuda, M., Lauffenburger, M., Sasaki, H., Rogers, M. E., and Dell, A. (1987) *J. Biol. Chem.* **262**, 11952–11957

242. MacKenzie, K. R., Prestegard, J. H., and Engelman, D. M. (1997) *Science* **276**, 131–133

243. Challou, N., Goormaghtigh, E., Cabiaux, V., Conrath, K., and Ruysschaert, J.-M. (1994) *Biochemistry* **33**, 6902–6910

244. Huang, C.-H., Reid, M., Daniels, G., and Blumenfeld, O. O. (1993) *J. Biol. Chem.* **268**, 25902–25908

245. Marchesi, V. T. (1985) *Ann. Rev. Cell Biol.* **1**, 531–561

246. Fujita, T., Ralston, G. B., and Morris, M. B. (1998) *Biochemistry* **37**, 264–271

247. Yan, Y., Winograd, E., Viel, A., Cronin, T., Harrison, S. C., and Branton, D. (1993) *Science* **262**, 2027–2030

248. Cherry, L., Menhart, N., and Fung, L. W.-M. (1999) *J. Biol. Chem.* **274**, 2077–2084

249. Begg, G. E., Harper, S. L., Morris, M. B., and Speicher, D. W. (2000) *J. Biol. Chem.* **275**, 3279–3287

250. Fujita, T., Ralston, G. B., and Morris, M. B. (1998) *Biochemistry* **37**, 272–280

250a. Rief, M., Pascual, J., Saraste, M., and Gaub, H. E. (1999) *J. Mol. Biol.* **286**, 553–561

251. Fowler, V. M. (1986) *Nature (London)* **322**, 777–778

252. Carraway, K. L. (1975) *Biochim. Biophys. Acta.* **415**, 379–410

253. Byers, T. J., and Branton, D. (1985) *Proc. Natl. Acad. Sci. U.S.A.* **82**, 6153–6157

254. Elgsaeter, A., Stokke, B. T., Mikkelsen, A., and Branton, D. (1986) *Science* **234**, 1217–1223

255. Pantazatos, D. P., and MacDonald, R. I. (1997) *J. Biol. Chem.* **272**, 21052–21059

256. Luna, E. J., and Hitt, A. L. (1992) *Science* **258**, 955–964

257. Reichstein, E., and Blostein, R. (1975) *J. Biol. Chem.* **250**, 6256–6263

258. Bennett, V. (1992) *J. Biol. Chem.* **267**, 8703–8706

259. Michaely, P., and Bennett, V. (1993) *J. Biol. Chem.* **268**, 22703–22709

260. Kordeli, E., Lambert, S., and Bennett, V. (1995) *J. Biol. Chem.* **270**, 2352–2359

261. An, X.-L., Takakuwa, Y., Nunomura, W., Manno, S., and Mohandas, N. (1996) *J. Biol. Chem.* **271**, 33187–33191

262. Workman, R. F., and Low, P. S. (1998) *J. Biol. Chem.* **273**, 6171–6176

263. Malik, S., Sami, M., and Watts, A. (1993) *Biochemistry* **32**, 10078–10084

264. Azim, A. C., Knoll, J. H. M., Beggs, A. H., and Chishti, A. H. (1995) *J. Biol. Chem.* **270**, 17407–17413

265. Wolfe, L. C., John, K. M., Falcone, J. C., Byrne, A. M., and Lux, S. E. (1982) *N. Engl. J. Med.* **307**, 1367–1374

266. Diakowski, W., and Sikorski, A. F. (1995) *Biochemistry* **34**, 13252–13258

267. Marfatia, S. M., Lue, R. A., Branton, D., and Chishti, A. H. (1995) *J. Biol. Chem.* **270**, 715–719

268. Krauss, S. W., Chasis, J. A., Rogers, C., Mohandas, N., Krockmalnic, G., and Penman, S. (1997) *Proc. Natl. Acad. Sci. U.S.A.* **94**, 7297–7302

269. Khodadad, J. K., Waugh, R. E., Podolski, J. L., Josephs, R., and Steck, T. L. (1996) *Biophys. J.* **70**, 1036–1044

270. Hansen, J. C., Skalak, R., Chien, S., and Hoger, A. (1996) *Biophys. J.* **70**, 146–166

271. Rice-Evans, C. A., and Dunn, M. J. (1982) *Trends Biochem. Sci.* **7**, 282–286

272. Mohandas, N., and Evans, E. (1994) *Annu. Rev. Biophys. Biomol. Struct.* **23**, 787–818

273. Agre, P., Casella, J. F., Zinkham, W. H., McMillan, C., and Bennett, V. (1985) *Nature (London)* **314**, 380–383

274. Baines, A. J., and Bennett, V. (1985) *Nature (London)* **315**, 410–413

275. Clark, E. A., and Brugge, J. S. (1995) *Science* **268**, 233–239

276. Dejana, E., Corada, M., and Lampugnani, M. G. (1995) *FASEB J.* **9**, 910–918

277. Malik, R. K., and Parsons, J. T. (1996) *J. Biol. Chem.* **271**, 29785–29791

278. Lo, S. H., An, Q., Bao, S., Wong, W.-K., Liu, Y., Janmey, P. A., Hartwig, J. H., and Chen, L. B. (1994) *J. Biol. Chem.* **269**, 22310–22319

279. Horwitz, A. F. (1997) *Sci. Am.* **276**(May), 68–75

280. de Pereda, J. M., Wiche, G., and Liddington, R. C. (1999) *EMBO J.* **18**, 4087–4095

281. Springer, T. A. (1997) *Proc. Natl. Acad. Sci. U.S.A.* **94**, 65–72

282. Bazzoni, G., and Hemler, M. E. (1998) *Trends Biochem. Sci.* **23**, 30–34

283. Emsley, J., King, S. L., Bergelson, J. M., and Liddington, R. C. (1997) *J. Biol. Chem.* **272**, 28512–28517

284. Luscinskas, F. W., and Lawler, J. (1994) *FASEB J.* **8**, 929–938

285. Simon, K. O., Otey, C. A., Pavelko, F. M., and Burridge, K. (1991) *Curr. Top. Membr. Transp.* **38**, 57–64

286. Edelman, G. M. (1983) *Science* **219**, 450–457

287. Becker, J. W., Erickson, H. P., Hoffman, S., Cunningham, B. A., and Edelman, G. M. (1989) *Proc. Natl. Acad. Sci. U.S.A.* **86**, 1088–1092

288. Cunningham, B. A., Hemperly, J. J., Murray, B. A., Prediger, E. A., Brackenbury, R., and Edelman, G. M. (1987) *Science* **236**, 799–806

289. Kudo, M., Kitajima, K., Inoue, S., Shiokawa, K., Morris, H. R., Dell, A., and Inoue, Y. (1996) *J. Biol. Chem.* **271**, 32667–32677

290. Newman, P. J., Berndt, M. C., Gorski, J., White, G. C., II, Lyman, S., Paddock, C., and Muller, W. A. (1990) *Science* **247**, 1219–1222

291. DeLisser, H. M., Yan, H. C., Newman, P. J., Muller, W. A., Buck, C. A., and Albelda, S. M. (1993) *J. Biol. Chem.* **268**, 16037–16046

291a. Chen, H. A., Pfuhl, M., McAlister, M. S. B., and Driscoll, P. C. (2000) *Biochemistry* **39**, 6814–6824

292. Casasnovas, J. M., Stehle, T., Liu, J.-h, Wang, J.-h, and Springer, T. A. (1998) *Proc. Natl. Acad. Sci. U.S.A.* **95**, 4134–4139

293. Wang, J.-H., Pepinsky, R. B., Stehle, T., Liu, J.-H., Karpusas, M., Browning, B., and Osborn, L. (1995) *Proc. Natl. Acad. Sci. U.S.A.* **92**, 5714–5718

294. Springer, T. A. (1990) *Nature (London)* **346**, 425–434

References

295. Wagner, G. (1995) *Science* **267**, 342

296. Nagar, B., Overduin, M., Ikura, M., and Rini, J. M. (1996) *Nature (London)* **380**, 360–364

297. Shapiro, L., Fannon, A. M., Kwong, P. D., Thompson, A., Lehmann, M. S., Grübel, G., Legrand, J.-F., Als-Nielsen, J., Colman, D. R., and Hendrickson, W. A. (1995) *Nature (London)* **374**, 327–336

298. Overduin, M., Harvey, T. S., Bagby, S., Tong, K. I., Yau, P., Takeichi, M., and Ikura, M. (1995) *Science* **267**, 386–389

298a. Nollet, F., Kools, P., and van Roy, F. (2000) *J. Mol. Biol.* **299**, 551–572

298b. Pertz, O., Bozic, D., Koch, A. W., Fauser, C., Brancaccio, A., and Engel, J. (1999) *EMBO J.* **18**, 1738–1747

299. Bella, J., Kolatkar, P. R., Marlor, C. W., Greve, J. M., and Rossmann, M. G. (1998) *Proc. Natl. Acad. Sci. U.S.A.* **95**, 4140–4145

300. Patel, D. J., and Gumbiner, B. M. (1995) *Nature (London)* **374**, 306–307

301. Berx, G., Cleton-Jansen, A.-M., Nollet, F., de Leeuw, W. J. F., van de Vijver, M. J., Cornelisse, C., and van Roy, F. (1995) *EMBO J.* **14**, 6107–6115

302. Vartio, T., and Vaheri, A. (1983) *Trends Biochem. Sci.* **8**, 442–444

303. Hynes, R. O. (1986) *Sci. Am.* **254**(Jun), 42–51

304. Ruoslahti, E. (1988) *Ann. Rev. Biochem.* **57**, 375–413

305. Engel, J., Odermatt, E., Engel, A., Madri, J. A., Furthmayr, H., Rohde, H., and Timpl, R. (1981) *J. Mol. Biol.* **150**, 97–120

306. Ingham, K. C., Brew, S. A., Huff, S., and Litvinovich, S. V. (1997) *J. Biol. Chem.* **272**, 1718–1724

307. Johnson, K. J., Sage, H., Briscoe, G., and Erickson, H. P. (1999) *J. Biol. Chem.* **274**, 15473–15479

308. Grant, R. P., Spitzfaden, C., Altroff, H., Campbell, I. D., and Mardon, H. J. (1997) *J. Biol. Chem.* **272**, 6159–6166

309. Hynes, R. O. (1990) *Fibronectins*, Springer-Verlag, New York

310. Moyano, J. V., Carnemolla, B., Domínguez-Jiménez, C., García-Gila, M., Albar, J. P., Sánchez-Aparicio, P., Leprini, A., Querzé, G., Zardi, L., and Garcia-Pardo, A. (1997) *J. Biol. Chem.* **272**, 24832–24836

311. Greenberg, C. S., Birckbichler, P. J., and Rice, R. H. (1991) *FASEB J.* **5**, 3071–3077

312. Jeong, J.-M., Murthy, S. N. P., Radek, J. T., and Lorand, L. (1995) *J. Biol. Chem.* **270**, 5654–5658

313. An, S. S. A., Jiménez-Barbero, J., Petersen, T. E., and Llinás, M. (1992) *Biochemistry* **31**, 9927–9933

314. Spitzfaden, C., Grant, R. P., Mardon, H. J., and Campbell, L. D. (1997) *J. Mol. Biol.* **265**, 565–579

315. Kornblihtt, A. R., Pesce, C. G., Alonso, C. R., Cramer, P., Srebrow, A., Werbajh, S., and Muro, A. F. (1996) *FASEB J.* **10**, 248–257

316. Yoneda, A., Ogawa, H., Kojima, K., and Matsumoto, I. (1998) *Biochemistry* **37**, 6351–6360

317. Fischer, D., Chiquet-Ehrismann, R., Bernasconi, C., and Chiquet, M. (1995) *J. Biol. Chem.* **270**, 3378–3384

318. Yokosaki, Y., Matsuura, N., Higashiyama, S., Murakami, I., Obara, M., Yamakido, M., Shigeto, N., Chen, J., and Sheppard, D. (1998) *J. Biol. Chem.* **273**, 11423–11428

319. Denda, S., Müller, U., Crossin, K. L., Erickson, H. P., and Reichardt, L. F. (1998) *Biochemistry* **37**, 5464–5474

320. Oberhauser, A. F., Marszalek, P. E., Erickson, H. P., and Fernandez, J. M. (1998) *Nature (London)* **393**, 181–185

321. Stetefeld, J., Mayer, U., Timpl, R., and Huber, R. (1996) *J. Mol. Biol.* **257**, 644–657

322. Colognato, H., MacCarrick, M., O'Rear, J. J., and Yurchenco, P. D. (1997) *J. Biol. Chem.* **272**, 29330–29336

323. Lane, T. F., and Sage, E. H. (1994) *FASEB J.* **8**, 163–173

324. Venstrom, K. A., and Reichardt, L. F. (1993) *FASEB J.* **7**, 996–1003

325. Gherzi, R., Carnemolla, B., Siri, A., Ponassi, M., Balza, E., and Zardi, L. (1995) *J. Biol. Chem.* **270**, 3429–3434

326. Ruoslahti, E., and Pierschbacher, M. D. (1987) *Science* **238**, 491–497

327. D'Souza, S. E., Ginsberg, M. H., and Plow, E. F. (1991) *Trends Biochem. Sci.* **16**, 246–250

328. Leahy, D. J., Hendrickson, W. A., Aukhil, I., and Erickson, H. P. (1992) *Science* **258**, 987–991

329. Stanley, K. K., Newby, A. C., and Luzio, J. P. (1982) *Trends Biochem. Sci.* **7**, 145–147

330. Roberts, M. F. (1996) *FASEB J.* **10**, 1159–1172

331. Han, S. K., Yoon, E. T., Scott, D. L., Sigler, P. B., and Cho, W. (1997) *J. Biol. Chem.* **272**, 3573–3582

332. Caruthers, A., and Melchior, D. L. (1986) *Trends Biochem. Sci.* **11**, 331–335

333. Gutknecht, J. (1987) *Proc. Natl. Acad. Sci. U.S.A.* **84**, 6443–6446

334. Paula, S., Volkov, A. G., and Deamer, D. W. (1998) *Biophys. J.* **74**, 319–327

335. Stein, W. D. (1990) *Channels, Carriers and Pumps: an Introduction to Membrane Transport*, Academic Press, San Diego, California

335a. Paulsen, I. T., Nguyen, L., Sliwinski, M. K., Rabus, R., and Saier, M. H., Jr. (2000) *J. Mol. Biol.* **301**, 75–100

336. Fox, R. O., Jr., and Richards, F. M. (1982) *Nature (London)* **300**, 325–330

337. Kempf, C., Klausner, R. D., Weinstein, J. N., Van Renswoude, J., Pincus, M., and Blumenthal, R. (1982) *J. Biol. Chem.* **257**, 2469–2476

338. Klebba, P. E., Hofnung, M., and Charbit, A. (1994) *EMBO J.* **13**, 4670–4675

339. Schirmer, T., Keller, T. A., Wang, Y.-F., and Rosenbusch, J. P. (1995) *Science* **267**, 512–514

340. Meyer, J. E. W., Hofnung, M., and Schulz, G. E. (1997) *J. Mol. Biol.* **266**, 761–775

341. Jordy, M., Andersen, C., Schülein, K., Ferenci, T., and Benz, R. (1996) *J. Mol. Biol.* **259**, 666–678

341a. Dumas, F., Koebnik, R., Winterhalter, M., and Van Gelder, P. (2000) *J. Biol. Chem.* **275**, 19747–19751

342. Samartzidou, H., and Delcour, A. H. (1998) *EMBO J.* **17**, 93–100

343. Pauptit, R. A., Schirmer, T., Jansonius, J. N., Rosenbusch, J. P., Parker, M. W., Tucker, A. D., Tsernoglou, D., Weiss, M. S., and Schulz, G. E. (1991) *J. Structural Biol.* **107**, 136–145

344. Kreusch, A., Neubüser, A., Schiltz, E., Weckesser, J., and Schulz, G. E. (1994) *Protein Sci.* **3**, 58–63

344a. Müller, D. J., and Engel, A. (1999) *J. Mol. Biol.* **285**, 1347–1351

345. Bainbridge, G., Mobasheri, H., Armstrong, G. A., Lea, E. J. A., and Lakey, J. H. (1998) *J. Mol. Biol.* **275**, 171–176

346. Van Gelder, P., Saint, N., Phale, P., Eppens, E. F., Prilipov, A., van Boxtel, R., Rosenbusch, J. P., and Tommassen, J. (1997) *J. Mol. Biol.* **269**, 468–472

347. Iyer, R., Wu, Z., Woster, P. M., and Delcour, A. H. (2000) *J. Mol. Biol.* **297**, 933–945

348. Watanabe, M., Rosenbusch, J., Schirmer, T., and Karplus, M. (1997) *Biophys. J.* **72**, 2094–2102

349. Cowan, S. W., Schirmer, T., Rummel, G., Steiert, M., Ghosh, R., Pauptit, R. A., Jansonius, J. N., and Rosenbusch, J. P. (1992) *Nature (London)* **358**, 727–733

350. Stathopoulos, C. (1996) *Protein Sci.* **5**, 170–173

351. Mannella, C. A. (1992) *Trends Biochem. Sci.* **17**, 315–320

352. Li, A. Z., Huang, H., Re, X., Qi, L. J., and Marx, K. A. (1998) *Biophys. J.* **74**, 964–973

353. Reumann, S., Maier, E., Benz, R., and Heldt, H. W. (1995) *J. Biol. Chem.* **270**, 17559–17565

354. Chrispeels, M. J., and Agre, P. (1994) *Trends Biochem. Sci.* **19**, 421–425

355. Knepper, M. A. (1994) *Proc. Natl. Acad. Sci. U.S.A.* **91**, 6255–6258

355a. Borgnia, M., Nielsen, S., Engel, A., and Agre, P. (1999) *Ann. Rev. Biochem.* **68**, 425–458

356. Ringler, P., Borgnia, M. J., Stahlberg, H., Maloney, P. C., Agre, P., and Engel, A. (1999) *J. Mol. Biol.* **291**, 1181–1190

357. de Groot, B. L., Heymann, J. B., Engel, A., Mitsuoka, K., Fujiyoshi, Y., and Grubmüller, H. (2000) *J. Mol. Biol.* **300**, 987–994

358. Cheng, A., van Hoek, A. N., Yeager, M., Verkman, A. S., and Mitra, A. K. (1997) *Nature (London)* **387**, 627–630

359. Ma, T., Song, Y., Gillespie, A., Carlson, E. J., Epstein, C. J., and Verkman, A. S. (1999) *J. Biol. Chem.* **274**, 20071–20074

359a. Zeuthen, T., and Klaerke, D. A. (1999) *J. Biol. Chem.* **274**, 21631–21636

360. Barone, L. M., Shih, C., and Wasserman, B. P. (1997) *J. Biol. Chem.* **272**, 30672–30677

361. Calamita, G., Bishai, W. R., Preston, G. M., Guggino, W. B., and Agre, P. (1995) *J. Biol. Chem.* **270**, 29063–29066

362. Deen, P. M. T., Verdijk, M. A. J., Knoers, N. V. A. M., Wieringa, B., Monnens, L. A. H., van OS, C. H., and van Oost, B. A. (1994) *Science* **264**, 92–95

363. Bai, L., Fushimi, K., Sasaki, S., and Marumo, F. (1996) *J. Biol. Chem.* **271**, 5171–5176

364. Jan, L. Y., and Jan, Y. N. (1994) *Nature (London)* **371**, 119–122

365. Armstrong, C. (1998) *Science* **280**, 56–57

366. Doyle, D. A., Cabral, J. M., Pfuetzner, R. A., Kuo, A., Gulbis, J. M., Cohen, S. L., Chait, B. T., and MacKinnon, R. (1998) *Science* **280**, 69–77

367. Heginbothan, L., Odessey, E., and Miller, C. (1997) *Biochemistry* **36**, 10335–10342

368. MacKinnon, R., Cohen, S. L., Kuo, A., Lee, A., and Chait, B. T. (1998) *Science* **280**, 106–109

369. Aidley, D. J., and Stanfield, P. R. (1996) *Ion Channels, Molecules in Action*, Cambridge Univ. Press, New York

370. Kreusch, A., Pfaffinger, P. J., Stevens, C. F., and Choe, S. (1998) *Nature (London)* **392**, 945–948

370a. Gulbis, J. M., Zhou, M., Mann, S., and MacKinnon, R. (2000) *Science* **289**, 123–127

370b. Kobertz, W. R., Williams, C., and Miller, C. (2000) *Biochemistry* **39**, 10347–10352

370c. Roux, B., Bernèche, S., and Im, W. (2000) *Biochemistry* **39**,

371. Klingenberg, M. (1981) *Nature (London)* **290**, 449–454

372. Prince, R. C., Gunson, D. E., and Scarpa, A. (1985) *Trends Biochem. Sci.* **10**, 99

373. Smith, R., Separovic, F., Milne, T. J., Whittaker, A., Bennett, F. M., Cornell, B. A., and Makriyannis, A. (1994) *J. Mol. Biol.* **241**, 456–466

374. Dathe, M., Schümann, M., Wieprecht, T., Winkler, A., Beyermann, M., Krause, E., Matsuzaki, K., Murase, O., and Bienert, M. (1996) *Biochemistry* **35**, 12612–12622

375. Kiyota, T., Lee, S., and Sugihara, G. (1996) *Biochemistry* **35**, 13196–13204

376. Monette, M., and Lafleur, M. (1996) *Biophys. J.* **70**, 2195–2202

377. Ojcius, D. M., and Young, J. D.-E. (1991) *Trends Biochem. Sci.* **16**, 225–229

378. Langs, D. A. (1988) *Science* **241**, 188–191

References

379. Doyle, D. A., and Wallace, B. A. (1997) *J. Mol. Biol.* **266**, 963–977

380. Iqbal, M., and Balaram, P. (1981) *Biochemistry* **20**, 7278–7284

381. Malashkevich, V. N., Kammerer, R. A., Efimov, V. P., Schulthess, T., and Engel, J. (1996) *Science* **274**, 761–765

382. Unwin, N. (1995) *Nature (London)* **373**, 37–43

383. Harrold, F. M., and Baarda, J. R. (1967) *J. Bacteriol.* **94**, 53–60

384. Harold, F. M., and Van Brunt, J. (1977) *Science* **197**, 372–373

385. Pressman, B. C. (1976) *Ann. Rev. Biochem.* **45**, 501–530

386. Hamilton, J. A., Sabesan, M. N., and Steinrauf, L. K. (1981) *J. Am. Chem. Soc.* **103**, 5880–5885

387. Duax, W. L., Hauptman, M., Weeks, C. M., and Norton, D. A. (1972) *Science* **176**, 911–914

388. Karle, I. L., Karle, J., Wieland, T., Burgemeister, W., Faulstich, H., and Witkop, B. (1973) *Proc. Natl. Acad. Sci. U.S.A.* **70**, 1836–1840

389. Pangborn, W., Duax, W., and Langs, D. (1987) *J. Am. Chem. Soc.* **109**, 2163–2165

390. Karle, I. L. (1985) *Proc. Natl. Acad. Sci. U.S.A.* **82**, 7155–7159

391. Inabayashi, M., Miyauchi, S., Kamo, N., and Jin, T. (1995) *Biochemistry* **34**, 3455–3460

392. Ledger, P. W., and Tanzer, M. L. (1984) *Trends Biochem. Sci.* **9**, 313–314

393. Marger, M. D., and Saier, M. H., Jr. (1993) *Trends Biochem. Sci.* **18**, 13–20

394. Goswitz, V. C., and Brooker, R. J. (1995) *Protein Sci.* **4**, 534–537

395. Seidel, H. M., and Knowles, J. R. (1994) *Biochemistry* **33**, 5641–5646

396. Cloherty, E. K., Sulzman, L. A., Zottola, R. J., and Carruthers, A. (1995) *Biochemistry* **34**, 15395–15406

397. Shanahan, M. F., and D'Artel-Ellis, J. (1984) *J. Biol. Chem.* **259**, 13878–13884

398. Bell, G. I., Burant, C. F., Takeda, J., and Gould, G. W. (1993) *J. Biol. Chem.* **268**, 19161–19164

399. Silverman, M. (1991) *Ann. Rev. Biochem.* **60**, 757–794

400. Mueckler, M. (1994) *Eur. J. Biochem.* **219**, 713–725

401. Maher, F., Vannucci, S. J., and Simpson, I. A. (1994) *FASEB J.* **8**, 1003–1011

402. Burant, C. F., Takeda, J., Brot-Laroche, E., Bell, G. I., and Davidson, N. O. (1992) *J. Biol. Chem.* **267**, 14523–14526

403. Zamora-León, S. P., Golde, D. W., Concha, I. I., Rivas, C. I., Delgado-López, F., Baselga, J., Nualart, F., and Vera, J. C. (1996) *Proc. Natl. Acad. Sci. U.S.A.* **93**, 1847–1852

404. Coderre, L., Kandror, K. V., Vallega, G., and Pilch, P. F. (1995) *J. Biol. Chem.* **270**, 27584–27588

405. Fischer, Y., Thomas, J., Sevilla, L., Munoz, P., Becker, C., Holman, G., Kozka, I. J., Palacín, M., Testar, X., Kammermeier, H., and Zorzano, A. (1997) *J. Biol. Chem.* **272**, 7085–7092

405a. Shepherd, P. R., and Kahn, B. B. (1999) *N. Engl. J. Med.* **341**, 248–257

406. Mueckler, M., and Makepeace, C. (1997) *J. Biol. Chem.* **272**, 30141–30146

407. Cloherty, E. K., Diamond, D. L., Heard, K. S., and Carruthers, A. (1996) *Biochemistry* **35**, 13231–13239

408. Seatter, M. J., De La Rue, S. A., Porter, L. M., and Gould, G. W. (1998) *Biochemistry* **37**, 1322–1326

409. Rumsey, S. C., Kwon, O., Xu, G. W., Burant, C. F., Simpson, I., and Levine, M. (1997) *J. Biol. Chem.* **272**, 18982–18989

410. Wiese, T. J., Dunlap, J. A., and Yorek, M. A. (1994) *J. Biol. Chem.* **269**, 22705–22711

411. Kasahara, M., Shimoda, E., and Maeda, M. (1997) *J. Biol. Chem.* **272**, 16721–16724

412. Turk, E., Kerner, C. J., Lostao, M. P., and Wright, E. M. (1996) *J. Biol. Chem.* **271**, 1925–1934

413. Chen, X.-Z., Coady, M. J., Jackson, F., Berteloot, A., and Lapointe, J.-Y. (1995) *Biophys. J.* **69**, 2405–2414

414. Panayotova-Heiermann, M., Eskandari, S., Turk, E., Zampighi, G. A., and Wright, E. M. (1997) *J. Biol. Chem.* **272**, 20324–20327

415. Prasad, P. D., Wang, H., Kekuda, R., Fujita, T., Fei, Y.-J., Devoe, L. D., Leibach, F. H., and Ganapathy, V. (1998) *J. Biol. Chem.* **273**, 7501–7506

416. Malandro, M. S., and Kilberg, M. S. (1996) *Ann. Rev. Biochem.* **65**, 305–336

417. Utsunomiya-Tate, N., Endou, H., and Kanai, Y. (1996) *J. Biol. Chem.* **271**, 14883–14890

418. Kong, C.-T., Yet, S.-F., and Lever, J. E. (1993) *J. Biol. Chem.* **268**, 1509–1512

419. Su, T.-Z., Wang, M., Syu, L.-J., Saltiel, A. R., and Oxender, D. L. (1998) *J. Biol. Chem.* **273**, 3173–3179

420. Yao, D., Mackenzie, B., Ming, H., Varoqui, H., Zhu, H., Hediger, M. A., and Erickson, J. D. (2000) *J. Biol. Chem.* **275**, 22790–22797

420a. Reimer, R. J., Chaudhry, F. A., Gray, A. T., and Edwards, R. H. (2000) *Proc. Natl. Acad. Sci. U.S.A.* **97**, 7715–7720

421. Mosckovitz, R., Udenfriend, S., Felix, A., Heimer, E., and Tate, S. S. (1994) *FASEB J.* **8**, 1069–1074

422. Chillarón, J., Estévez, R., Mora, C., Wagner, C. A., Suessbrich, H., Lang, F., Gelpí, J. L., Testar, X., Busch, A. E., Zorzano, A., and Palacín, M. (1996) *J. Biol. Chem.* **271**, 17761–17770

423. Trotti, D., Rossi, D., Gjesdal, O., Levy, L. M., Racagni, G., Danbolt, N. C., and Volterra, A. (1996) *J. Biol. Chem.* **271**, 5976–5979

424. Peghini, P., Janzen, J., and Stoffel, W. (1997) *EMBO J.* **16**, 3822–3832

425. Lebrun, B., Sakaitani, M., Shimamoto, K., Yasuda-Kamatani, Y., and Nakajima, T. (1997) *J. Biol. Chem.* **272**, 20336–20339

426. Ponce, J., Biton, B., Benavides, J., Avenet, P., and Aragón, C. (2000) *J. Biol. Chem.* **275**, 13856–13862

427. Pantanowitz, S., Bendahan, A., and Kanner, B. I. (1993) *J. Biol. Chem.* **268**, 3222–3225

428. Bennett, E. R., and Kanner, B. I. (1997) *J. Biol. Chem.* **272**, 1203–1210

429. Velaz-Faircloth, M., Guadano-Ferraz, A., Henzi, V. A., and Fremeau, R. T., Jr. (1995) *J. Biol. Chem.* **270**, 15755–15761

430. Chen, J.-G., Liu-Chen, S., and Rudnick, G. (1998) *J. Biol. Chem.* **273**, 12675–12681

431. Nikaido, H., and Saier, M. H., Jr. (1992) *Science* **258**, 936–942

432. Will, A., Grassl, R., Erdmenger, J., Caspari, T., and Tanner, W. (1998) *J. Biol. Chem.* **273**, 11456–11462

433. Kaback, H. R., Bibi, E., and Roepe, P. D. (1990) *Trends Biochem. Sci.* **15**, 309–314

434. Frillingos, S., Wu, J., Venkatesan, P., and Kaback, H. R. (1997) *Biochemistry* **36**, 6408–6414

435. Green, A. L., Anderson, E. J., and Brooker, R. J. (2000) *J. Biol. Chem.* **275**, 23240–23246

436. Zhao, M., Zen, K.-C., Hubbell, W. L., and Kaback, H. R. (1999) *Biochemistry* **38**, 7407–7412

437. Zani, M. L., Pourcher, T., and Leblanc, G. (1994) *J. Biol. Chem.* **269**, 24883–24889

438. Pourcher, T., Bibi, E., Kaback, H. R., and Leblanc, G. (1996) *Biochemistry* **35**, 4161–4168

439. McDonald, T. P., Walmsley, A. R., and Henderson, P. J. F. (1997) *J. Biol. Chem.* **272**, 15189–15199

440. Fei, Y.-J., Liu, W., Prasad, P. D., Kekuda, R., Oblak, T. G., Ganapathy, V., and Leibach, F. H. (1997) *Biochemistry* **36**, 452–460

441. Brandsch, M., Thunecke, F., Küllertz, G., Schutkowski, M., Fischer, G., and Neubert, K. (1998) *J. Biol. Chem.* **273**, 3861–3864

442. Boorer, K. J., Frommer, W. B., Bush, D. R., Kreman, M., Loo, D. D. F., and Wright, E. M. (1996) *J. Biol. Chem.* **271**, 2213–2220

443. Boorer, K. J., and Fischer, W.-N. (1997) *J. Biol. Chem.* **272**, 13040–13046

444. Someya, Y., Niwa, A., Sawai, T., and Yamaguchi, A. (1995) *Biochemistry* **34**, 7–12

445. Nikaido, K., Liu, P.-Q., and Ames, G. F.-L. (1997) *J. Biol. Chem.* **272**, 27745–27752

446. Létoffé, S., Delepelaire, P., and Wandersman, C. (1996) *EMBO J.* **15**, 5804–5811

447. Blattner, F. R., Plunkett, G., III, Bloch, C. A., Perna, N. T., Burland, V., Riley, M., Collado-Vides, J., Glasner, J. D., Rode, C. K., Mayhew, G. F., Gregor, J., Davis, N. W., Kirkpatrick, H. A., Goeden, M. A., Rose, D. J., Mau, B., and Shao, Y. (1997) *Science* **277**, 1453–1462

448. Decottignies, A., Lambert, L., Catty, P., Degand, H., Epping, E. A., Moye-Rowley, W. S., Balzi, E., and Goffeau, A. (1995) *J. Biol. Chem.* **270**, 18150–18157

448a. Driessen, A. J. M., Rosen, B. P., and Konings, W. N. (2000) *Trends Biochem. Sci.* **25**, 397–401

449. Sharom, F. J., DiDiodato, G., Yu, X., and Ashbourne, K. J. D. (1995) *J. Biol. Chem.* **270**, 10334–10341

450. Ko, Y. H., and Pedersen, P. L. (1995) *J. Biol. Chem.* **270**, 22093–22096

451. Nossal, N. G., and Heppel, L. A. (1966) *J. Biol. Chem.* **241**, 3055–3062

452. Ames, G. F.-L. (1986) *Ann. Rev. Biochem.* **55**, 397–425

453. Ames, G. F.-L., and Higgins, C. F. (1983) *Trends Biochem. Sci.* **8**, 97–100

454. Yao, N., Ledvina, P. S., Choudhary, A., and Quiocho, F. A. (1996) *Biochemistry* **35**, 2079–2085

455. Quiocho, F. A., and Vyas, N. K. (1984) *Nature (London)* **310**, 381–386

456. Gilliland, G. L., and Quiocho, F. A. (1981) *J. Mol. Biol.* **146**, 341–362

457. Vyas, N. K., Vyas, M. N., and Quiocho, F. A. (1988) *Science* **242**, 1290–1295

458. Vyas, N. K., Vyas, M. N., and Quiocho, F. A. (1991) *J. Biol. Chem.* **266**, 5226–5237

459. Shilton, B. H., Shuman, H. A., and Mowbray, S. L. (1996) *J. Mol. Biol.* **264**, 364–376

460. Oh, B.-H., Kang, C.-H., De Bondt, H., Kim, S.-H., Nikaido, K., Joshi, A. K., and Ames, G. F.-L. (1994) *J. Biol. Chem.* **269**, 4135–4143

461. Oh, B.-H., Ames, G. F.-L., and Kim, S.-H. (1994) *J. Biol. Chem.* **269**, 26323–26330

462. Sugiyama, S., Vassylyev, D. G., Matsushima, M., Kashiwagi, K., Igarashi, K., and Morikawa, K. (1996) *J. Biol. Chem.* **271**, 9519–9525

463. Pflugrath, J. W., and Quiocho, F. A. (1988) *J. Mol. Biol.* **200**, 163–180

464. Manson, M. D., Boos, W., Bassford, P. J., Jr., and Rasmussen, B. A. (1985) *J. Biol. Chem.* **260**, 9727–9733

465. Reizer, J., Michotey, V., Reizer, A., and Saier, M. H., Jr. (1994) *Protein Sci.* **3**, 440–450

466. Garrett, D. S., Seok, Y.-J., Liao, D.-I., Peterkofsky, A., Gronenborn, A. M., and Clore, G. M. (1997) *Biochemistry* **36**, 2517–2530

467. Chauvin, F., Brand, L., and Roseman, S. (1994) *J. Biol. Chem.* **269**, 20263–20269

467a. Zhu, P.-P., Szczepanowski, R. H., Nosworthy, N. J., Ginsburg, A., and Peterkofsky, A. (1999) *Biochemistry* **38**, 15470–15479

468. Jones, B. E., Rajagopal, P., and Klevit, R. E. (1997) *Protein Sci.* **6**, 2107–2119

469. El-Kabbani, O. A. L., Waygood, E. B., and Delbaera, L. T. J. (1987) *J. Biol. Chem.* **262**, 12926–12929

References

470. Reizer, J., Hoischen, C., Reizer, A., Pham, T. N., and Saier, M. H., Jr. (1993) *Protein Sci.* **2**, 506–521

471. Eberstadt, M., Grdadolnik, S. G., Gemmecker, G., Kessler, H., Buhr, A., and Erni, B. (1996) *Biochemistry* **35**, 11286–11292

472. Van Dijk, A. A., Scheek, R. M., Dijkstra, K., Wolters, G. K., and Robillard, G. T. (1992) *Biochemistry* **31**, 9063–9072

473. Gemmecker, G., Eberstadt, M., Buhr, A., Lanz, R., Golic Grdadolnik, S., Kessler, H., and Erni, B. (1997) *Biochemistry* **36**, 7408–7417

474. Pelton, J. G., Torchia, D. A., Remington, S. J., Murphy, K. P., Meadow, N. D., and Roseman, S. (1996) *J. Biol. Chem.* **271**, 33446–33456

475. Lanz, R., and Erni, B. (1998) *J. Biol. Chem.* **273**, 12239–12243

476. Nunn, R. S., Markovic-Housley, Z., Génovésio-Taverne, J.-C., Flükiger, K., Rizkallah, P. J., Jansonius, J. N., Schirmer, T., and Erni, B. (1996) *J. Mol. Biol.* **259**, 502–511

477. Roseman, S., and Meadow, N. D. (1990) *J. Biol. Chem.* **265**, 2993–2996

478. Hurley, J. H., Faber, H. R., Worthylake, D., Meadow, N. D., Roseman, S., Pettigrew, D. W., and Remington, S. J. (1993) *Science* **259**, 673–677

479. Cantor, C. R., and Schimmel, P. R. (1980) *Biophysical Chemistry*, Freeman, San Francisco, California

480. Masuda, T., Dobson, G. P., and Veech, R. L. (1990) *J. Biol. Chem.* **265**, 20321–20334

481. Catterall, W. A. (1995) *Ann. Rev. Biochem.* **64**, 493–531

482. Valverde, M. A., Hardy, S. P., and Sepúlveda, F. V. (1995) *FASEB J.* **9**, 509–515

483. Schmidt-Rose, T., and Jentsch, T. J. (1997) *Proc. Natl. Acad. Sci. U.S.A.* **94**, 7633–7638

484. Jentsch, T. J., Steinmeyer, K., and Schwarz, G. (1990) *Nature (London)* **348**, 510–514

485. Middleton, R. E., Pheasant, D. J., and Miller, C. (1994) *Biochemistry* **33**, 13189–13198

486. Schwappach, B., Stobrawa, S., Hechenberger, M., Steinmeyer, K., and Jentsch, T. J. (1998) *J. Biol. Chem.* **273**, 15110–15118

487. Gronemeier, M., Condie, A., Prosser, J., Steinmeyer, K., Jentsch, T. J., and Jockusch, H. (1994) *J. Biol. Chem.* **269**, 5963–5967

488. Pucéat, M., Korichneva, I., Cassoly, R., and Vassort, G. (1995) *J. Biol. Chem.* **270**, 1315–1322

489. Jarolim, P., Shayakul, C., Prabakaran, D., Jiang, L., Stuart-Tilley, A., Rubin, H. L., Simova, S., Zavadil, J., Herrin, J. T., Brouillette, J., Somers, M. J. G., Seemanova, E., Brugnara, C., Guay-Woodford, L. M., and Alper, S. L. (1998) *J. Biol. Chem.* **273**, 6380–6388

489a. Fujinaga, J., Tang, X.-B., and Casey, J. R. (1999) *J. Biol. Chem.* **274**, 6626–6633

490. Welsh, M. J., and Smith, A. E. (1995) *Sci. Am.* **273**(Dec), 52–59

491. Seibert, F. S., Linsdell, P., Loo, T. W., Hanrahan, J. W., Riordan, J. R., and Clarke, D. M. (1996) *J. Biol. Chem.* **271**, 27493–27499

492. Romero, M. F., Hediger, M. A., Boulpaep, E. L., and Boron, W. F. (1997) *Nature (London)* **387**, 409–413

493. Burnham, C. E., Amlal, H., Wang, Z., Shull, G. E., and Soleimani, M. (1997) *J. Biol. Chem.* **272**, 19111–19114

494. Muchhal, U. S., Pardo, J. M., and Raghothama, K. G. (1996) *Proc. Natl. Acad. Sci. U.S.A.* **93**, 10519–10523

495. Smith, F. W., Ealing, P. M., Hawkesford, M. J., and Clarkson, D. T. (1995) *Proc. Natl. Acad. Sci. U.S.A.* **92**, 9373–9377

496. Zhou, J.-J., Theodoulou, F. L., Muldin, I., Ingemarsson, B., and Miller, A. J. (1998) *J. Biol. Chem.* **273**, 12017–12023

497. Schroers, A., Burkovski, A., Wohlrab, H., and Krämer, R. (1998) *J. Biol. Chem.* **273**, 14269–14276

498. Phelps, A., Briggs, C., Mincone, L., and Wohlrab, H. (1996) *Biochemistry* **35**, 10757–10762

499. Collins, J. F., and Ghishan, F. K. (1996) *FASEB J.* **10**, 751–759

500. Garcia, C. K., Brown, M. S., Pathak, R. K., and Goldstein, J. L. (1995) *J. Biol. Chem.* **270**, 1843–1849

501. Kuo, S.-M., and Aronson, P. S. (1996) *J. Biol. Chem.* **271**, 15491–15497

502. Yellen, G. (1987) *Ann. Rev. Biophys. Biophys. Chem.* **16**, 227–246

503. Miller, C. (1991) *Science* **252**, 1092–1096

504. Philipson, L. H., and Steiner, D. F. (1995) *Science* **268**, 372–373

505. Inagaki, N., Gonoi, T., Clement, J. P., IV, Namba, N., Inazawa, J., Gonzalez, G., Aguilar-Bryan, L., Seino, S., and Bryan, J. (1995) *Science* **270**, 1166–1170

506. Bränström, R., Efendic, S., Berggren, P.-O., and Larsson, O. (1998) *J. Biol. Chem.* **273**, 14113–14118

507. Ruknudin, A., Schulze, D. H., Sullivan, S. K., Lederer, W. J., and Welling, P. A. (1998) *J. Biol. Chem.* **273**, 14165–14171

508. Day, M. L., Johnson, M. H., and Cook, D. I. (1998) *EMBO J.* **17**, 1952–1960

509. McCormick, K. A., Isom, L. L., Ragsdale, D., Smith, D., Scheuer, T., and Catterall, W. A. (1998) *J. Biol. Chem.* **273**, 3954–3962

510. Favre, I., Moczydlowski, E., and Schild, L. (1996) *Biophys. J.* **71**, 3110–3125

510a. Lipkind, G. M., and Fozzard, H. A. (2000) *Biochemistry* **39**, 8161–8170

511. Pérez-García, M. T., Chiamvimonvat, N., Marban, E., and Tomaselli, G. F. (1996) *Proc. Natl. Acad. Sci. U.S.A.* **93**, 300–304

512. Kosari, F., Sheng, S., Li, J., Mak, D.-O. D., Foskett, J. K., and Kleyman, T. R. (1998) *J. Biol. Chem.* **273**, 13469–13474

513. Garty, H. (1994) *FASEB J.* **8**, 522–528

514. Miller, R. J. (1992) *J. Biol. Chem.* **267**, 1403–1406

515. Mitterdorfer, J., Sinnegger, M. J., Grabner, M., Striessnig, J., and Glossmann, H. (1995) *Biochemistry* **34**, 9350–9355

516. McPherson, P. S., and Campbell, K. P. (1993) *J. Biol. Chem.* **268**, 13765–13768

517. Jeyakumar, L. H., Copello, J. A., O'Malley, A. M., Wu, G.-M., Grassucci, R., Wagenknecht, T., and Fleischer, S. (1998) *J. Biol. Chem.* **273**, 16011–16020

518. Kraus, R. L., Sinnegger, M. J., Glossmann, H., Hering, S., and Striessnig, J. (1998) *J. Biol. Chem.* **273**, 5586–5590

519. Kim, S., Lakhani, V., Costa, D. J., Sharara, A. I., Fitz, J. G., Huang, L.-W., Peters, K. G., and Kindman, L. A. (1995) *J. Biol. Chem.* **270**, 5266–5269

520. Mezna, M., and Michelangeli, F. (1996) *J. Biol. Chem.* **271**, 31818–31823

521. Nicoll, D. A., Hryshko, L. V., Matsuoka, S., Frank, J. S., and Philipson, K. D. (1996) *J. Biol. Chem.* **271**, 13385–13391

522. Crumpton, M. J., and Dedman, J. R. (1990) *Nature (London)* **345**, 212

523. Swairjo, M. A., and Seaton, B. A. (1994) *Annu. Rev. Biophys. Biomol. Struct.* **23**, 193–213

524. Benz, J., Bergner, A., Hofmann, A., Demange, P., Göttig, P., Liemann, S., Huber, R., and Voges, D. (1996) *J. Mol. Biol.* **260**, 638–643

525. Burger, A., Berendes, R., Liemann, S., Benz, J., Hofmann, A., Göttig, P., Huber, R., Gerke, V., Thiel, C., Römisch, J., and Weber, K. (1996) *J. Mol. Biol.* **257**, 839–847

526. Demange, P., Voges, D., Benz, J., Liemann, S., Göttig, P., Berendes, R., Burger, A., and Huber, R. (1994) *Trends Biochem. Sci.* **19**, 272–276

527. Campos, B., Mo, Y. D., Mealy, T. R., Li, C. W., Swairjo, M. A., Balch, C., Head, J. F., Retzinger, G., Dedman, J. R., and Seaton, B. A. (1998) *Biochemistry* **37**, 8004–8010

528. Ling, G. N. (1984) *In Search of the Physical Basis of Life*, Plenum, New York

529. Baker, P. F., and Willis, J. S. (1970) *Nature (London)* **226**, 521–523

530. Kyte, J. (1972) *J. Biol. Chem.* **247**, 7642–7649

531. Lingrel, J. B., and Kuntzweiler, T. (1994) *J. Biol. Chem.* **269**, 19659–19662

532. Lutsenko, S., Daoud, S., and Kaplan, J. H. (1997) *J. Biol. Chem.* **272**, 5249–5255

533. Scheiner-Bobis, G., and Schreiber, S. (1999) *Biochemistry* **38**, 9198–9208

534. Gatto, C., Wang, A. X., and Kaplan, J. H. (1998) *J. Biol. Chem.* **273**, 10578–10585

535. Rossier, B. C., Geering, K., and Kraehenbuhl, J. P. (1987) *Trends Biochem. Sci.* **12**, 483–487

536. Minor, N. T., Sha, Q., Nichols, C. G., and Mercer, R. W. (1998) *Proc. Natl. Acad. Sci. U.S.A.* **95**, 6521–6525

537. Lutsenko, S., and Kaplan, J. H. (1995) *Biochemistry* **34**, 15607–15613

538. Nelson, N., and Taiz, L. (1989) *Trends Biochem. Sci.* **14**, 113–116

539. Solioz, M., and Vulpe, C. (1996) *Trends Biochem. Sci.* **21**, 237–241

539a. Ueno, S., Kaieda, N., Koyama, N. (2000) *J. Biol. Chem.* **275**, 14537–14540

540. Carafoli, E. (1994) *FASEB J.* **8**, 993–1002

541. MacLennan, D. H., Rice, W. J., and Green, N. M. (1997) *J. Biol. Chem.* **272**, 28815–28818

542. Myung, J., and Jencks, W. P. (1995) *Biochemistry* **34**, 3077–3083

543. Dode, L., De Greef, C., Mountian, I., Attard, M., Town, M. M., Casteels, R., and Wuytack, F. (1998) *J. Biol. Chem.* **273**, 13982–13994

543a. Menguy, T., Corre, F., Bouneau, L., Deschamps, S., Vuust Moller, J., Champeil, P., le Maire, M., and Falson, P. (1998) *J. Biol. Chem.* **273**, 20134–20143

544. Melle-Milovanovic, D., Milovanovic, M., Nagpal, S., Sachs, G., and Shin, J. M. (1998) *J. Biol. Chem.* **273**, 11075–11081

545. Rulli, S. J., Horiba, M. N., Skripnikova, E., and Rabon, E. C. (1999) *J. Biol. Chem.* **274**, 15245–15250

546. Asano, S., Inoie, M., and Takeguchi, N. (1987) *J. Biol. Chem.* **262**, 13263–13268

547. Seto-Young, D., Hall, M. J., Na, S., Haber, J. E., and Perlin, D. S. (1996) *J. Biol. Chem.* **271**, 581–587

548. Auer, M., Scarborough, G. A., and Kühlbrandt, W. (1998) *Nature (London)* **392**, 840–843

549. Orlowski, J., and Grinstein, S. (1997) *J. Biol. Chem.* **272**, 22373–22376

549. Orlowski, J. and Grinstein, S. (1997) *J. Biol. Chem.* **272**, 22373–22376

549a. Wakabayashi, S., Pang, T., Su, X., and Shigekawa, M. (2000) *J. Biol. Chem.* **275**, 7942–7949

549b. Williams, K. A. (2000) *Nature (London)* **403**, 112–115

550. Kasho, V. N., Stengelin, M., Smirnova, I. N., and Faller, L. D. (1997) *Biochemistry* **36**, 8045–8052

551. Campos, M., and Beaugé, L. (1997) *Biochemistry* **36**, 14228–14237

552. Vilsen, B. (1997) *Biochemistry* **36**, 13312–13324

552a. Boxenbaum, N., Daly, S. E., Javaid, Z. Z., Lane, L. K., and Blostein, R. (1998) *J. Biol. Chem.* **273**, 23086–23092

553. Zhang, P., Toyoshima, C., Yonekura, K., Green, N. M., and Stokes, D. L. (1998) *Nature (London)* **392**, 835–839

553a. Toyoshima, C., Nakasako, M., Nomura, H., and Ogawa, H. (2000) *Nature (London)* **405**, 647–655

References

553b. MacLennan, D. H., and Green, N. M. (2000) *Nature (London)* **405**, 633–634

553c. Zhang, Z., Lewis, D., Strock, C., and Inesi, G. (2000) *Biochemistry* **39**, 8758–8767

554. Crider, B. P., Andersen, P., White, A. E., Zhou, Z., Li, X., Mattsson, J. P., Lundberg, L., Keeling, D. J., Xie, X.-S., Stone, D. K., and Peng, S.-B. (1997) *J. Biol. Chem.* **272**, 10721–10728

555. Nanda, A., Brumell, J. H., Nordström, T., Kjeldsen, L., Sengelov, H., Borregaard, N., Rotstein, O. D., and Grinstein, S. (1996) *J. Biol. Chem.* **271**, 15963–15970

555a. Kelly, R. B. (1999) *Trends Biochem. Sci.* **24**, M29–M33

556. Desjardins, M., Celis, J. E., van Meer, G., Dieplinger, H., Jahraus, A., Griffiths, G., and Huber, L. A. (1994) *J. Biol. Chem.* **269**, 32194–32200

557. Peng, S.-B., Crider, B. P., Tsai, S. J., Xie, X.-S., and Stone, D. K. (1996) *J. Biol. Chem.* **271**, 3324–3327

558. Pearse, B. M. F., and Crowther, R. A. (1987) *Ann. Rev. Biophys. Biophys. Chem.* **16**, 49–68

559. Schmid, S. L., and Damke, H. (1995) *FASEB J.* **9**, 1445–1453

559a. Marsh, M., and McMahon, H. T. (1999) *Science* **285**, 215–220

559b. Heuser, J. (1981) *Trends Biochem. Sci.* **6**, 64–68

560. Pearse, B. M. F., and Bretscher, M. S. (1981) *Ann. Rev. Biochem.* **50**, 85–101

561. Pishvaee, B., Munn, A., and Payne, G. (1997) *EMBO J.* **16**, 2227–2239

562. Rothman, J. E. (1996) *Protein Sci.* **5**, 185–194

563. Brodsky, F. M., Hill, B. L., Acton, S. L., Näthke, I., Wong, D. H., Ponnambalam, S., and Parham, P. (1991) *Trends Biochem. Sci.* **16**, 208–213

564. Lemmon, S. K., and Jones, E. W. (1987) *Science* **238**, 504–509

564a. Anderson, R. G. W. (1998) *Ann. Rev. Biochem.* **67**, 199–225

564b. Shin, J.-S., Gao, Z., and Abraham, S. N. (2000) *Science* **289**, 785–788

564c. Mulvey, M. A., and Hultgren, S. J. (2000) *Science* **289**, 732–733

565. Lowe, M., and Kreis, T. E. (1995) *J. Biol. Chem.* **270**, 31364–31371

566. Schekman, R., and Orci, L. (1996) *Science* **271**, 1526–1533

566a. Eugster, A., Frigerio, G., Dale, M., and Duden, R. (2000) *EMBO J.* **19**, 3905–3917

567. Bloomer, A. C., Champness, J. N., Bricogne, G., Staden, R., and Klug, A. (1978) *Nature (London)* **276**, 362–368

567a. Rapoport, I., Chen, Y. C., Cupers, P., Shoelson, S. E., and Kirchhausen, T. (1998) *EMBO J.* **17**, 2148–2155

567b. Huang, K. M., D'Hondt, K., Riezman, H., and Lemmon, S. K. (1999) *EMBO J.* **18**, 3897–3908

567c. Owen, D. J., Vallis, Y., Pearse, B. M. F., McMahon, H. T., and Evans, P. R. (2000) *EMBO J.* **19**, 4216–4227

567d. Wendland, B., Steece, K. E., and Emr, S. D. (1999) *EMBO J.* **18**, 4383–4393

567e. Scales, S. J., and Scheller, R. H. (1999) *Nature (London)* **401**, 123–124

568. Ungewickell, E., Ungewickell, H., Holstein, S. E. H., Lindner, R., Prasad, K., Barouch, W., Martin, B., Greene, L. E., and Eisenberg, E. (1995) *Nature (London)* **378**, 632–635

569. Mellman, I., Fuchs, R., and Helenius, A. (1986) *Ann. Rev. Biochem.* **55**, 663–700

570. Brown, M. S., and Goldstein, J. L. (1974) *Sci. Am.* **251**(Nov), 58–66

571. Brown, M. S., and Goldstein, J. L. (1986) *Science* **232**, 34–47

572. Holstein, T. W., Benoit, M., Herder, G., Wanner, G., David, C. N., and Gaub, H. E. (1994) *Science* **265**, 402–404

573. Ferro-Novick, S., and Jahn, R. (1994) *Nature (London)* **370**, 191–193

574. Mayorga, L. S., Berón, W., Sarrouf, M. N., Colombo, M. I., Creutz, C., and Stahl, P. D. (1994) *J. Biol. Chem.* **269**, 30927–30934

575. Burgoyne, R. D. (1988) *Nature (London)* **331**, 20

576. Wilson, D. W., Wilcox, C. A., Flynn, G. C., Chen, E., Kuang, W.-J., Henzel, W. J., Block, M. R., Ullrich, A., and Rothman, J. E. (1989) *Nature (London)* **339**, 355–359

577. Colombo, M. I., Taddese, M., Whiteheart, S. W., and Stahl, P. D. (1996) *J. Biol. Chem.* **271**, 18810–18816

578. Goda, Y. (1997) *Proc. Natl. Acad. Sci. U.S.A.* **94**, 769–772

579. Lugtenberg, B. (1981) *Trends Biochem. Sci.* **6**, 262–266

580. Graham, L. L., Beveridge, T. J., and Nanninga, N. (1991) *Trends Biochem. Sci.* **16**, 328–329

581. Oliver, D. B. (1987) in *Escherichia coli and Salmonella typhi-murium* (Neidhardt, F. C., ed), pp. 56–70, Am. Soc. for Micro-biology, Washington, DC

582. Glauner, B., Höltje, J.-V., and Schwarz, U. (1988) *J. Biol. Chem.* **263**, 10088–10095

583. Park, J. T. (1987) in *Escherichia coli and Salmonella typhimurium* (Neidhardt, F. C., ed), pp. 23–32, Am. Soc. for Microbiology, Washington, DC

584. Scherrer, R. (1984) *Trends Biochem. Sci.* **9**, 242–245

585. Nikaido, H., and Vaara, M. (1987) in *Escherichia coli and Salmonella typhimurium* (Neidhardt, F. C., ed), pp. 7–22, Am. Soc. for Microbiology, Washington, DC

586. Hancock, R. E. W., Karunaratne, D. N., and Bernegger-Egli, C. (1994) in *Bacterial Cell Wall (New Comprehensive Biochemistry)*, Vol. 27 (Ghuysen, J.-M., and Hakenbeck, R., eds), pp. 263–279, Elsevier, Amsterdam

587. Raetz, C. R. H., and Dowhan, W. (1990) *J. Biol. Chem.* **265**, 1235–1238

588. Osborn, M. J. (1969) *Ann. Rev. Biochem.* **38**, 501–538

589. Choi, D.-S., Yamada, H., Mizuno, T., and Mizushima, S. (1986) *J. Biol. Chem.* **261**, 8953–8957

590. Diedrich, D. L., and Schnaitman, C. A. (1978) *Proc. Natl. Acad. Sci. U.S.A.* **75**, 3708–3712

591. Raetz, C. R. H., Ulevitch, R. J., Wright, S. D., Sibley, C. H., Ding, A., and Nathan, C. F. (1991) *FASEB J.* **5**, 2652–2660

592. Hollingsworth, R. I., and Carlson, R. W. (1989) *J. Biol. Chem.* **264**, 9300–9303

593. Ogawa, T. (1994) *Eur. J. Biochem.* **219**, 737–742

594. Rietschel, E. T., Kirikae, T., Schade, F. U., Mamat, U., Schmidt, G., Loppnow, H., Ulmer, A. J., Zähringer, U., Seydel, U., Di Padova, F., Schreier, M., and Brade, H. (1994) *FASEB J.* **8**, 217–225

595. Masoud, H., Moxon, E. R., Martin, A., Krajcarski, D., and Richards, J. C. (1997) *Biochemistry* **36**, 2091–2103

596. Aspinall, G. O., and Monteiro, M. A. (1996) *Biochemistry* **35**, 2498–2504

597. Rietschel, E. T., and Brade, H. (1992) *Sci. Am.* **267**(Aug), 54–61

598. Costerton, J. W., Irwin, R. T., and Cheng, K. J. (1981) *Crit. Rev. Microbiol* **8**, 303–338

599. Fischer, W. (1994) in *Bacterial Cell Wall (New Comprehensive Biochemistry)*, Vol. 27 (Ghuysen, J.-M., and Hakenbeck, R., eds), pp. 199–215, Elsevier, Amsterdam

600. Ganfield, M.-C. W., and Pieringer, R. A. (1980) *J. Biol. Chem.* **255**, 5164–5169

601. Johnson, S. D., Lacher, K. P., and Anderson, J. S. (1981) *Biochemistry* **20**, 4781–4785

602. Ivatt, R. J., and Gilvarg, C. (1979) *J. Biol. Chem.* **254**, 2759–2765

603. Venisse, A., Rivière, M., Vercauteren, J., and Puzo, G. (1995) *J. Biol. Chem.* **270**, 15012–15021

604. Brennan, P. J., and Nikaido, H. (1995) *Ann. Rev. Biochem.* **64**, 29–63

605. Liu, J., Barry, C. E., III, Besra, G. S., and Nikaido, H. (1996) *J. Biol. Chem.* **271**, 29545–29551

606. Mikusová, K., Mikus, M., Besra, G. S., Hancock, I., and Brennan, P. J. (1996) *J. Biol. Chem.* **271**, 7820–7828

607. Lopez-Marin, L. M., Quesada, D., Lakhdar-Ghazal, F., Tocanne, J.-F., and Lanéelle, G. (1994) *Biochemistry* **33**, 7056–7061

608. Sprott, G. D., Ekiel, I., and Dicaire, C. (1990) *J. Biol. Chem.* **265**, 13735–13740

609. Peters, J., Nitsch, M., Kühlmorgen, B., Golbik, R., Lupas, A., Kellermann, J., Engelhardt, H., Pfander, J.-P., Müller, S., Goldie, K., Engel, A., Stetter, K.-O., and Baumeister, W. (1995) *J. Mol. Biol.* **245**, 385–401

610. Thomas, S. R., and Trust, T. J. (1995) *J. Mol. Biol.* **245**, 568–581

611. Peters, J., Baumeister, W., and Lupas, A. (1996) *J. Mol. Biol.* **257**, 1031–1041

612. Fischetti, V. A. (1991) *Sci. Am.* **264**(Jun), 58–65

613. Goward, C. R., Scawen, M. D., Murphy, J. P., and Atkinson, T. (1993) *Trends Biochem. Sci.* **18**, 136–139

614. Schneewind, O., Fowler, A., and Faull, K. F. (1995) *Science* **268**, 103–106

615. Knörle, R., and Hübner, W. (1995) *Biochemistry* **34**, 10970–10975

616. Shaper, N. L., Hollis, G. F., Douglas, J. G., Kirsch, I. R., and Shaper, J. H. (1988) *J. Biol. Chem.* **263**, 10420–10428

617. Lin, C. Q., and Bissell, M. J. (1993) *FASEB J.* **7**, 737–743

618. Har-El, R., and Tanzer, M. L. (1993) *FASEB J.* **7**, 1115–1123

619. Mayne, R., and Burgeson, R. E., eds. (1987) *Structure and Function of Collagen Types*, Academic Press, New York, Orlando, San Diego

620. Nimni, M. E., ed. (1988) *Collagen: Biochemistry, Biomechanics, Biotechnology*, CRC Press, Boca Raton, Florida

621. Ramachandran, G. N. (1988) *Intl. J. Peptide Prot. Res.* **31**, 1–16

622. Martin, G. R., Timpl, R., Müller, P. K., and Kühn, K. (1985) *Trends Biochem. Sci.* **10**, 285–287

623. Prockop, D. J., and Kivirikko, K. I. (1995) *Ann. Rev. Biochem.* **64**, 403–434

624. van der Rest, M., and Garrone, R. (1991) *FASEB J.* **5**, 2814–2823

625. Byers, P. H. (1995) in *The Metabolic and Molecular Bases of Inherited Disease*, 7th ed., Vol. 3 (Scriver, C. R., Beaudet, A. L., Sly, W. S., and Valle, D., eds), pp. 4029–4077, McGraw-Hill, New York

626. Brodsky, B., and Shah, N. K. (1995) *FASEB J.* **9**, 1537–1546

627. Koivu, J., and Myllylä, R. (1987) *J. Biol. Chem.* **262**, 6159–6164

628. Prockop, D. J. (1990) *J. Biol. Chem.* **265**, 15349–15352

629. Annunen, P., Autio-Harmainen, H., and Kivirikko, K. I. (1998) *J. Biol. Chem.* **273**, 5989–5992

630. Reiser, K., McCormick, R. J., and Rucker, R. B. (1992) *FASEB J.* **6**, 2439–2449

631. Li, S.-W., Sieron, A. L., Fertala, A., Hojima, Y., Arnold, W. V., and Prockop, D. J. (1996) *Proc. Natl. Acad. Sci. U.S.A.* **93**, 5127–5130

632. Meek, K. M., Chapman, J. A., and Hardcastle, R. A. (1979) *J. Biol. Chem.* **254**, 10710–10714

633. Chan, V. C., Ramshaw, J. A. M., Kirkpatrick, A., Beck, K., and Brodsky, B. (1997) *J. Biol. Chem.* **272**, 31441–31446

References

634. Wess, T. J., Hammersley, A. P., Wess, L., and Miller, A. (1998) *J. Mol. Biol.* **275**, 255–267

634a. Kramer, R. Z., Venugopal, M. G., Bella, J., Mayville, P., Brodsky, B., and Berman, H. M. (2000) *J. Mol. Biol.* **301**, 1191–1205

635. Eyre, D. R., Paz, M. A., and Gallop, P. M. (1984) *Ann. Rev. Biochem.* **53**, 717–748

636. Miller, A. (1982) *Trends Biochem. Sci.* **7**, 13–18

637. Eyre, D. R. (1980) *Science* **207**, 1315–1322

638. Wu, J.-J., and Eyre, D. R. (1984) *Biochemistry* **23**, 1850–1857

639. Hanson, D. A., and Eyre, D. R. (1996) *J. Biol. Chem.* **271**, 26508–26516

640. Kiss, I., Deák, F., Holloway, R. G., Jr., Delius, H., Mebust, K. A., Frinberger, E., Argraves, W. W., Tsonis, P. A., Winterbottom, N., and Goetink, P. F. (1989) *J. Biol. Chem.* **264**, 8126–8134

641. Tanaka, S., Avigad, G., Eikenberry, E. F., and Brodsky, B. (1988) *J. Biol. Chem.* **263**, 17650–17657

642. Kleman, J.-P., Aeschlimann, D., Paulsson, M., and van der Rest, M. (1995) *Biochemistry* **34**, 13768–13775

643. Mazzorana, M., Gruffat, H., Sergeant, A., and van der Rest, M. (1993) *J. Biol. Chem.* **268**, 3029–3032

644. Martin, G. R., Timpl, R., Miller, P. K., and Kuhn, K. (1985) *Trends Biochem. Sci.* **10**, 285–287

645. Gunwar, S., Ballester, F., Noelken, M. E., Sado, Y., Ninomiya, Y., and Hudson, B. G. (1998) *J. Biol. Chem.* **273**, 8767–8775

646. van der Rest, M., and Mayne, R. (1988) *J. Biol. Chem.* **263**, 1615–1618

647. de Crombrugghe, B., and Pastan, I. (1982) *Trends Biochem. Sci.* **7**, 11–13

648. Myers, J. C., Yang, H., D'Ippolito, J. A., Presente, A., Miller, M. K., and Dion, A. S. (1994) *J. Biol. Chem.* **269**, 18549–18557

649. Lui, V. C. H., Ng, L. J., Sat, E. W. Y., Nicholls, J., and Cheah, K. S. E. (1996) *J. Biol. Chem.* **271**, 16945–16951

650. de Wet, W., Bernard, M., Benson-Chanda, V., Chu, M.-L., Dickson, L., Weil, D., and Ramirez, F. (1987) *J. Biol. Chem.* **262**, 16032–16036

651. Chu, M.-L., de Wet, W., Bernhard, M., and Ramirez, F. (1985) *J. Biol. Chem.* **260**, 2315–2370

652. Kramer, J. M. (1994) *FASEB J.* **8**, 329–336

653. Engel, J. (1997) *Science* **277**, 1785–1786

654. Rosenbloom, J., Abrams, W. R., and Mecham, R. (1993) *FASEB J.* **7**, 1208–1218

655. Hukins, D. W. L. (1984) *Connective Tissue Matrix*, MacMillan, New York

656. Fransson, L.-Å. (1987) *Trends Biochem. Sci.* **12**, 406–411

657. Burg, M. A., Tillet, E., Timpl, R., and Stallcup, W. B. (1996) *J. Biol. Chem.* **271**, 26110–26116

658. Caplan, A. I. (1984) *Sci. Am.* **251**(Oct), 84–94

659. Heinegård, D., and Oldberg, Å. (1989) *FASEB J.* **3**, 2042–2051

660. Cheng, F., Heinegård, D., Fransson, L.-Å., Bayliss, M., Bielicki, J., Hopwood, J., and Yoshida, K. (1996) *J. Biol. Chem.* **271**, 28572–28580

661. Hauser, N., and Paulsson, M. (1994) *J. Biol. Chem.* **269**, 25747–25753

662. Beck, K., Gambee, J. E., Bohan, C. A., and Bächinger, H. P. (1996) *J. Mol. Biol.* **256**, 909–923

663. Yurchenco, P. D., and Schittny, J. C. (1990) *FASEB J.* **4**, 1577–1590

663a. Yurchenco, P. D., Birk, D. E., and Mecham, R. P., eds. (1994) *Extracellular Matrix Assembly and Structure*, Academic Press, San Diego, California

664. Utani, A., Nomizu, M., Timpl, R., Roller, P. P., and Yamada, Y. (1994) *J. Biol. Chem.* **269**, 19167–19175

665. Pöschl, E., Mayer, U., Stetefeld, J., Baumgartner, R., Holak, T. A., Huber, R., and Timpl, R. (1996) *EMBO J.* **15**, 5154–5159

666. Durkin, M. E., Gautam, M., Loechel, F., Sanes, J. R., Merlie, J. P., Albrechtsen, R., and Wewer, U. M. (1996) *J. Biol. Chem.* **271**, 13407–13416

667. Beckmann, G., Hanke, J., Bork, P., and Reich, J. G. (1998) *J. Mol. Biol.* **275**, 725–730

668. Colognato-Pyke, H., O'Rear, J. J., Yamada, Y., Carbonetto, S., Cheng, Y.-S., and Yurchenco, P. D. (1995) *J. Biol. Chem.* **270**, 9398–9406

669. Kramer, J. M., Cox, G. N., and Hirsch, D. (1985) *J. Biol. Chem.* **260**, 1945–1951

670. Beck, K., Hunter, I., and Engel, J. (1990) *FASEB J.* **4**, 148–160

671. Sasaki, T., Kostka, G., Göhring, W., Wiedemann, H., Mann, K., Chu, M.-L., and Timpl, R. (1995) *J. Mol. Biol.* **245**, 241–250

672. Aeschlimann, D., and Paulsson, M. (1991) *J. Biol. Chem.* **266**, 15308–15317

673. Reinhardt, D. P., Ono, R. N., and Sakai, L. Y. (1997) *J. Biol. Chem.* **272**, 1231–1236

673a. Kettle, S., Yuan, X., Grundy, G., Knott, V., Downing, A. K., and Handford, P. A. (1999) *J. Mol. Biol.* **285**, 1277–1287

673b. Ritty, T. M., Broekelmann, T., Tisdale, C., Milewicz, D. M., and Mecham, R. P. (1999) *J. Biol. Chem.* **274**, 8933–8940

674. Reinhardt, D. P., Keene, D. R., Corson, G. M., Pöschl, E., Bächinger, H. P., Gambee, J. E., and Sakai, L. Y. (1996) *J. Mol. Biol.* **258**, 104–116

675. McKusick, V. A. (1991) *Nature (London)* **352**, 279–281

676. Francke, U., and Furthmayr, H. (1994) *N. Engl. J. Med.* **330**, 1384–1385

677. Mann, K., Mechling, D. E., Bächinger, H. P., Eckerskorn, C., Gaill, F., and Timpl, R. (1996) *J. Mol. Biol.* **261**, 255–266

678. Durkin, M. E., Carlin, B. E., Vergnes, J., Bartos, B., and Merlie, J. (1987) *Proc. Natl. Acad. Sci. U.S.A.* **84**, 1570–1574

679. Richards, A. G. (1978) in *Biochemistry of Insects* (Rockstein, M., ed), Academic Press, New York (p. 205)

680. Schaefer, J., Kramer, K. J., Garbow, J. R., Jacob, G. S., Stejskal, E. O., Hopkins, T. L., and Speirs, R. D. (1987) *Science* **235**, 1200–1204

681. Simkiss, K., and Wilbur, K. M. (1989) *Biomineralization*, Academic Press, San Diego, California

682. DeOliveira, D. B., and Laursen, R. A. (1997) *J. Am. Chem. Soc.* **119**, 10627–10631

683. Weiner, S., and Addadi, L. (1991) *Trends Biochem. Sci.* **16**, 252–256

684. Albeck, S., Aizenberg, J., Addadi, L., and Weiner, S. (1993) *J. Am. Chem. Soc.* **115**, 11691–11697

685. Bonucci, E., ed. (1992) *Calcification in Biological Systems*, CRC Press, Boca Raton, Florida

686. Mann, S., Archibald, D. D., Didymus, J. M., Douglas, T., Heywood, B. R., Meldrum, F. C., and Reeves, N. J. (1993) *Science* **261**, 1286–1292

687. Aizenberg, J., Hanson, J., Ilan, M., Leiserowitz, L., Koetzle, T. F., Addadi, L., and Weiner, S. (1995) *FASEB J.* **9**, 262–268

688. Otsubo, K., Katz, E. P., Mechanic, G. L., and Yamauchi, M. (1992) *Biochemistry* **31**, 396–402

689. Fledelius, C., Johnsen, A. H., Cloos, P. A. C., Bonde, M., and Qvist, P. (1997) *J. Biol. Chem.* **272**, 9755–9763

689a. Ducy, P., Schinke, T., and Karsenty, G. (2000) *Science* **289**, 1501–1504

689b. Teitelbaum, S. L. (2000) *Science* **289**, 1504–1508

690. Weiner, S., and Traub, W. (1992) *FASEB J.* **6**, 879–885

691. McLean, F. M., Keller, P. J., Genge, B. R., Walters, S. A., and Wuthier, R. E. (1987) *J. Biol. Chem.* **262**, 10481–10488

692. Komori, T., Yagi, H., Nomura, S., Yamaguchi, A., Sasaki, K., Deguchi, K., Shimizu, Y., Bronson, R. T., Gao, Y.-H., Inada, M., Sato, M., Okamoto, R., Kitamura, Y., Yoshiki, S., and Kishimoto, T. (1997) *Cell* **89**, 755–764

693. Ducy, P., Zhang, R., Geoffroy, V., Ridall, A. L., and Karsenty, G. (1997) *Cell* **89**, 747–754

694. Dickman, S. (1997) *Science* **276**, 1502

695. Rodan, G. A., and Harada, S.-i. (1997) *Cell* **89**, 677–680

696. Mundlos, S., Otto, F., Mundlos, C., Mulliken, J. B., Aylsworth, A. S., Albright, S., Lindhout, D., Cole, W. G., Henn, W., Knoll, J. H. M., Owen, M. J., Mertelsmann, R., Zabel, B. U., and Olsen, B. R. (1997) *Cell* **89**, 773–779

697. Price, P. A., and Williamson, M. K. (1985) *J. Biol. Chem.* **260**, 14971–14975

698. Nishimoto, S. K., Araki, N., Robinson, F. D., and Waite, J. H. (1992) *J. Biol. Chem.* **267**, 11600–11605

699. Price, P. A., Rice, J. S., and Williamson, M. K. (1994) *Protein Sci.* **3**, 822–830

700. Hohenester, E., Maurer, P., and Timpl, R. (1997) *EMBO J.* **16**, 3778–3786

701. Kelm, R. J., Jr., Swords, N. A., Orfeo, T., and Mann, K. G. (1994) *J. Biol. Chem.* **269**, 30147–30153

702. Fisher, L. W., Robey, P. G., Tuross, N., Otsuka, A. S., Tepen, D. A., Esch, F. S., Shimasaki, S., and Termine, J. D. (1987) *J. Biol. Chem.* **262**, 13457–13463

703. Denhardt, D. T., and Guo, X. (1993) *FASEB J.* **7**, 1475–1482

704. Shanmugam, V., Chackalaparampil, I., Kundu, G. C., Mukherjee, A. B., and Mukherjee, B. B. (1997) *Biochemistry* **36**, 5729–5738

704a. Rodan, G. A., and Martin, T. J. (2000) *Science* **289**, 1508–1514

704b. Service, R. F. (2000) *Science* **289**, 1498–1500

705. Zipkin, I., ed. (1973) *Biological Mineralization*, Wiley, New York

706. Schlesinger, P. H., Blair, II. C., Teitelbaum, S. L., and Edwards, J. C. (1997) *J. Biol. Chem.* **272**, 18636–18643

707. Alberts, B., Bray, D., Lewis, J., Raff, M., Roberts, K., and Watson, J. D. (1994) *Molecular Biology of the Cell*, 3rd ed., Garland, New York (p. 1182–1186)

708. Riggs, B. L., and Melton, L. J. (1992) *N. Engl. J. Med.* **327**, 620–627

709. Sly, W. S., and Hu, P. Y. (1995) in *The Metabolic and Molecular Bases of Inherited Disease*, 7th ed., Vol. 3 (Scriver, C. R., Beaudet, A. L., Sly, W. S., and Valle, D., eds), pp. 4113–4124, McGraw-Hill, New York

710. Roush, W. (1996) *Science* **273**, 1170

711. Delmas, P. D., and Meunier, P. J. (1997) *N. Engl. J. Med.* **336**, 558–566

711a. Ho, A. M., Johnson, M. D., and Kingsley, D. M. (2000) *Science* **289**, 265–270

711b. Hagmann, M. (2000) *Science* **289**, 225–226

712. Hohling, H. J., Barckhaus, R. H., and Kreftin, E. R. (1980) *Trends Biochem. Sci.* **5**, 8–11

713. Outka, D. E., and Williams, D. C. (1971) *Journal of Protozoology* **18**, 285–297

714. Inoue, S., and Okazaki, K. (1977) *Sci. Am.* **236**(Apr), 83–92

715. Volcani, B. E. (1981) in *Silicon and Silicious Structures in Biological Systems* (Simpson, T. L., and Volcani, B. E., eds), Springer-Verlag, New York (p. 157)

716. Mann, S. (1988) *Nature (London)* **332**, 119–124

717. Calvert, P. (1988) *Nature (London)* **334**, 651–652

718. Watanabe, N., and Wilbur, K. M., eds. (1976) *The Mechanisms of Mineralization in Invertebrates and Plants*, Univ. N. Carolina Press, Columbia, South Carolina

References

719. Lemann, J., Jr. (1993) *N. Engl. J. Med.* **328**, 880–881

720. Nakagawa, Y., Abram, V., Kézdy, F. J., Kaiser, E. T., and Coe, F. L. (1983) *J. Biol. Chem.* **258**, 12594–12600

721. Bertrand, J. A., Pignol, D., Bernard, J.-P., Verdier, J.-M., Dagorn, J.-C., and Fontecilla-Camps, J. C. (1996) *EMBO J.* **15**, 2678–2684

722. Bimmler, D., Graf, R., Scheele, G. A., and Frick, T. W. (1997) *J. Biol. Chem.* **272**, 3073–3082

723. De Reggi, M., Gharib, B., Patard, L., and Stoven, V. (1998) *J. Biol. Chem.* **273**, 4967–4971

724. Cheung, H. S., Kurup, I. V., Sallis, J. D., and Ryan, L. M. (1996) *J. Biol. Chem.* **271**, 28082–28085

725. Rothschild, B. M., and Thillaud, P. L. (1991) *Nature (London)* **349**, 288

726. Cabib, E., Roberts, R., and Bowers, B. (1982) *Ann. Rev. Biochem.* **51**, 763–793

727. Ballou, C. E., and Raschke, W. C. (1974) *Science* **184**, 127–134

728. Frevert, J., and Ballou, C. E. (1985) *Biochemistry* **24**, 753–759

729. Celerin, M., Ray, J. M., Schisler, N. J., Day, A. W., Stetler-Stevenson, W. G., and Laudenbach, D. E. (1996) *EMBO J.* **15**, 4445–4453

Study Questions

1. Compare the chemical makeup of the extracellar "coat" or "matrix" materials secreted by the following cells: bacteria, fibroblasts, osteoblasts, plant cells, fungi.

2. The iodine number of a compound is defined as the number of grams of I_2 absorbed (through addition to C=C bonds to give a diiodo derivative) per 100 g of fat. NOTE: Iodine monochloride or iodine monobromide are the usual halogenating reagents but the iodine number is expressed in terms of grams of I_2. The saponification number is the number of milligrams of KOH needed to completely saponify (hydrolyze and neutralize the resulting fatty acids) 1 g of fat. A pure triglyceride has a saponification number of 198 and an iodine number of 59.7.

   a) What is the relative molecular mass?
   b) What is the average chain length of the fatty acids?
   c) What is the number of double bonds in the molecule?

3. Spermaceti (a wax from the head of the sperm whale) resembles high molecular mass hydrocarbons in physical properties and inertness toward Br_2 / $CHCl_3$ and $KMnO_4$; on qualitative analysis, it gives positive tests only for carbon and hydrogen. However, its IR spectrum shows the presence of an ester linkage, and quantitative analysis gives the empirical formula $C_{16}H_{32}O$. A solution of the wax in alcoholic KOH is refluxed for a long time. Titration of an aliquot shows that one equivalent of base is consumed for every 475 g of wax. Water and ether are added to the cooled reflux mixture, and the aqueous and etheral layers are separated. Acidification of the aqueous layer yields a solid A with a neutralization equivalent of 260 ± 5. Evaporation of the ether layer gives solid B, which could not be titrated. Reduction of either spermaceti or A by lithium aluminum hydride gave B as the only product.

   What is the likely structure of spermaceti?

4. Stearic acid (1.16 g) was dissolved in 100 ml of ethanol. A 10 µl portion of the resulting solution was pipetted onto a clean surface of a dilute HCl solution (in a shallow tray) where it spread to form a monolayer of stearic acid. The layer was compressed (by moving a Teflon barrier across the tray) until the surface pressure π started to rise sharply and reached ~20 dyn / cm. Note that $\pi = \gamma_o - \gamma$ where γ is the measured surface tension with the film present and γ_o is the higher surface tension of water alone. The compressed film occupied a 20 x 24 cm area. Calculate the cross-sectional area of an alkyl chain in stearic acid. [See J. B. Davenport, *in Biochemistry and Methodology of Lipids* (A. R. Johnson and J. B. Davenport, eds.), pp. 47-83. Wiley-Intersience, New York, 1971; and M. C. Phillips, *in Progress in Surface and Membrane Science* (J. F. Danielli, D. M. Rosenberg, and D. A. Cadenhead, eds.) Vol. 5, pp. 139-221. Academic Press, New York, 1972.]

5. In 1925, E. Gorter and F. Grendel (*J. Exp. Med.* **41**, 439) reported measurements in which they extracted lipid from red blood cell membranes with acetone, spread the lipids as a monolayer, and measured the area of the compressed monolayer. They then estimated the surface area of an erythrocyte and calculated that the ratio of the lipids (as a monolayer) to the surface area of the red blood cell was 1.9–2.0. More modern experiments gave the following: each erythrocyte membrane contains 4.5 x 10^{-16} mol of phospholipid and 3.1 x 10^{-16} mol of cholesterol.

   a) If the cross-sectional areas of phospholipid and cholesterol molecules in a membrane are taken as 0.70 and 0.38 nm^2, respectively, what surface area would be occupied in a monolayer?
   b) If the measured surface area of an erythrocyte is 167 μm^2, what is the ratio of the area calculated in (a) to the area of the cell surface?
   c) How might you explain the difference between this answer and that of Gorter and Grendel? See E. D. Korn (1966) *Science* **153**, 1491–1498.

Study Questions

6. The following experimental observations are related to biological membrane structure and function. Discuss the implications of each observation with respect to membrane structure.

   a) Many macrocyclic antibiotics (nonactin, valinomycin, and others) form 1: 1 complexes with alkali metal cations in a highly selective manner. The complexes are readily soluble in nonpolar organic solvents. These antibiotics increase the electrical conductance and permeability to alkali metal cations of synthetic phospholipid membranes. Valinomycin increases the electrical conductance of thylakoid membranes of chloroplasts in the presence of K^+ but not in the presence of Na^+; it also uncouples oxidative phosphorylation in mitochondria. (See Chapter 18.)

   b) Treatment of intact chloroplasts with a galactolipase releases glactose from galactosyl diglycerides. Treatment of red blood cell ghosts with phopholipase C releases about 75% of the lipid P in water-soluble form. In neither case is the structural integrity of the membrane destroyed.

   c) Using sphingomyelin as a hapten, antibodies specific for this lipid can be produced. When red blood cell ghosts are exposed to these antibodies, it can be shown that the antibodies react, but only on one side of the membrane.

7. Describe the structure of biological membranes and the characteristic functions of lipid-, protein-, and carbohydrate-containing components. Describe the differences between inner and outer membrane surfaces.

8. Compare the distribution of triglycerides, phosphatidylcholine, phosphatidylethanolamine, sphingomyelins, glycolipids, and cholesterol within cells. Consider differences between the two sides of membranes.

9. Consider the chemistry underlying the labeling of cell surfaces with each of the following:
   a) Lactoperoxidase,
   b) galactose oxidase,
   c) formylmethionylsulfone methyl phosphate,
   d) the diazonium salt of diiodosulfanilic acid,
   e) fluorescent antibodies,
   f) antibodies conjugated with ferritin.

   Write the equations for the chemical reactions involved. State what surface groups will be labeled by each reagent. List special advantages of each of these reagents.

10. Which would be the more effective detergent in the pH range 2 to 3, sodium lauryl sulfonate or sodium laurate? Why?

11. Suppose that a cell contains 10 mM Na^+ and 100 mM K^+ and that it is bathed in extracellular fluid containing 100 mM Na^+ and 5 mM K^+. How much energy will be required to transport three equivalents of Na^+ out and two equivalents of K^+ in? Compare this with $\Delta G'$ for hydrolysis of ATP at pH 7. Assume that the membrane is permeable to Cl^-.

12. An *E. coli* cell is said to contain about 10^5 molecules of an envelope protein of MW = 36,500. If the latter is spherical and the spheres are closely packed in a hexogonal lattice, how much of the surface area of the bacterium would be covered? What would the diameter of the protein be? What spacing would be required if 10^5 molecules covered the surface completely? Suggest a shape for the protein molecule that is consistent with the requirement.

Random Bi Bi Mechanism

Bimolecular reactions of two molecules, A and B, to give two products, P and Q, are catalyzed by many enzymes. For some enzymes the substrates A and B bind into the active site in an ordered sequence while for others, binding may be in a random order. The scheme shown here is described as random Bi Bi in a classification introduced by Cleland. Eighteen rate constants, some second order and some first order, describe the reversible system. Determination of these kinetic parameters is often accomplished using a series of double reciprocal plots (Lineweaver-Burk plots), such as those at the right.

Contents

Enzymes: The Catalysts of Cells

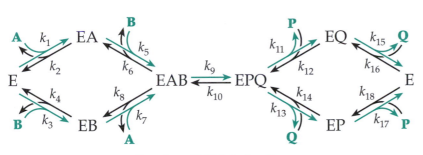

9

Random Bi Bi Mechanism

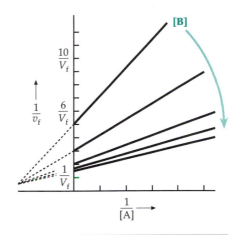

Most of the machinery of living cells is made of enzymes. Thousands of them have been extracted from cells and have been purified and crystallized. Many others are recognized only by their catalytic action and have not yet been isolated in pure form. Most enzymes are soluble globular proteins but an increasing number of RNA molecules are also being recognized as enzymes. Many structural proteins of the cell also act as catalysts. For example, the muscle proteins actin and myosin together catalyze the hydrolysis of ATP and link the hydrolysis to movement (Chapter 19). Catalysis is one of the most fundamental characteristics of life.

How do we recognize that a protein or RNA molecule *is* an enzyme? The answer is that *enzymes are recognized primarily by their ability to catalyze a chemical reaction.* For this reason, an everyday operation for many biochemists is the measurement of the catalytic activity of enzymes. Only by measuring the rates of the catalyzed reactions carefully and quantitatively has it been possible to isolate and purify these remarkable molecules.

Since the beginning of biochemical investigation enzymes have held a special fascination for chemists and biologists. How can these easily destroyed substances catalyze reactions with such speed and without formation of significant quantities of side products? Some enzymes increase the velocity of a single chemical reaction of a specific compound by a factor of as much as 10^{10}. How can a protein do this? In this chapter we'll consider both ways of measuring enzymatic activity and basic mechanisms of catalysis.

A. Information from Kinetics

The quantitative study of catalysis by enzymes, i.e., the study of enzyme kinetics, is a highly developed branch of biochemistry. It is one of our most important means of learning about the mechanisms of catalysis at the active sites of enzymes.[1-13a] By determining **rate constants *k*** under a variety of conditions we can learn just how fast an enzyme can act, how tightly it binds its substrates to form the enzyme-substrate (ES) complexes essential to catalysis, how specific it is with respect to substrate structure, and how it is affected by compounds that inhibit or activate the catalysis.

1. Measuring the Speed of an Enzymatic Reaction

A major goal in kinetic studies is to establish a **rate equation** which describes the velocity of a reaction in terms of **kinetic constants** and other experimentally measurable parameters. To measure the velocity of any chemical reaction precisely we must start the reaction at a definite time by rapidly mixing together the two or more reactants. Then, while keeping the mixture at an accurately constant temperature and pH, we must measure the concentration of a reactant or product after a fixed time interval, or at various times. No end of ingenuity has gone into devising ways of doing this for particular enzymes. Whatever the procedure, the information we must obtain is the **rate** at which some concentration changes with time. We can then construct a **progress curve** showing the decrease in concentration [S] of the reactant (**substrate**) or the increase in the concentration of **product** [P] with time (Fig. 9-1).

BOX 9-A A HISTORICAL NOTE ON ENZYMES

While the earliest physiologists postulated a "vital force" to explain the chemical reactions of cells, the existence of biochemical substances promoting reactions outside of the body was recognized at least by the early 1600s. However, the role of yeast in fermentation was still unknown and it was thought that both alcoholic fermentation and animal digestion were caused by unknown substances called **ferments**. In 1752, Reamur demonstrated the solvent power of the gastric juice of birds and by 1783 Spallanzani had extended the studies to humans and other species.[a] In 1836 Schwann isolated the enzyme **pepsin**[b] from gastric juice.

In the same year Jacob Berzelius introduced the concept of catalysis, which he developed as a result of studies of the effects of acids and bases in promoting the hydrolysis of starch and of the effects of metals on the decomposition of hydrogen peroxide. Berzelius proposed the term **catalyst** from the Greek "katalysis," meaning "dissolution." Although he had been concerned primarily with inorganic catalysts, Berzelius recognized that a natural catalyst,

an amylase that causes the hydrolysis of starch, had already been isolated from germinating barley in 1833 and that "in living plants and animals thousands of catalytic processes take place."[c] Some chemists and physiologists accepted Berzelius' concept of biochemical catalysis immediately, but many did not. The matter was complicated by Pasteur's discovery that yeast cells were the causative agent of alcoholic fermentation. Only after Edward Buchner's reports in 1897 that a juice formed by grinding yeast with sand and filtering could still ferment sugar (see Chapters 15 and 17) was the reality of enzymes in metabolism[d] generally accepted. The word **enzyme** (from the Greek "in yeast") was introduced earlier by F. W. Kuhne, professor of physiology in Heidelberg and a person who had accepted Berzelius' concept.[c]

[a] Richmond, C. (1986) *Trends Biochem. Sci.* **11**, 528–530
[b] Schwann, T. (1836) *Arch. Anat. Physiol.*, 90–138
[c] Hoffmann-Ostenhof, O. (1978) *Trends Biochem. Sci.* **3**, 186–187
[d] Buchner, E. (1897) *Ber. Deut. Chem. Ges.* **30**, 117–124

The velocity v of an enzymatic reaction is defined as the rate at which a substrate disappears or at which a product is formed, the two being identical:

$$v = -d[S]/dt = d[P]/dt \qquad (9\text{-}1)$$

Under the steady-state conditions that usually apply (see p. 449), *the rate of increase of product will be the same as the rate of decrease of substrate*. The units of velocity are moles per liter per second ($M\ s^{-1}$) or more traditionally in enzymology moles per liter per minute. We are interested in the **instantaneous velocity,** which at

any time is given by the slope of the progress curve (Fig. 9-1). We usually want to measure the velocity immediately after the reaction is started to avoid the decrease in rate that comes from the depletion of substrate or accumulation of products. In some cases, as in the example given in Fig. 9-1, this is difficult to do with accuracy.

In some chemical reactions, which involve first-order processes, a logarithmic plot of the progress curve (log[S] vs t) gives a straight line so that the initial slope need not be determined. However, most **enzyme assays** give progress curves that remain nearly linear for only short periods of time. In these cases we need very sensitive methods for detecting products. These may involve colorimetric or fluorimetric measurements or the use of radioactive substrates. One of the most sensitive approaches is to arrange the assay so that a product of the reaction serves as a catalyst for another enzymatic process, thus amplifying the amount of final product to be measured.[14,15] Another approach is to measure velocities of reactions in a very small volume, e.g., 200 nanoliters, and to continuously separate and measure products using capillary electrophoresis and fluorescence detection.[16]

If the progress curve is not a straight line at the beginning of the reaction (0 time) and if the amount of compound that reacts in a fixed

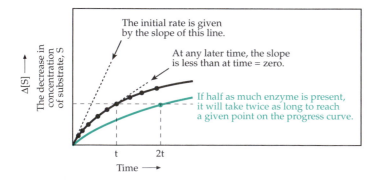

Figure 9-1 The progress curve for an enzymatic reaction in which the substrate S is converted into products.

interval of time is taken as the rate, an erroneous answer will be obtained. Sometimes an **integrated rate expression** that describes the time course of product formation can be used (see Eq. 9-22). Even when the progress curve is nonlinear as shown in Fig. 9-1, it is possible to estimate relative rates by noting that in most cases if one unit of enzyme yields a certain amount of product in time t_1, the same amount of product will be formed by n units of enzyme in time t_1 / n.

First-order reactions. In many chemical reactions the rate of decrease of the concentration of a given reactant [A] is found experimentally to be directly proportional to the concentration of that reactant at any given time:

$$v = -d[A] / dt = k[A] \qquad (9\text{-}2)$$

Such a reaction is described as **first order** and the proportionality constant **k** is known as the **rate constant.** Such first-order kinetics is observed for **unimolecular processes** in which a molecule of A is converted into product P in a given time interval with a probability that does not depend on interaction with another molecule. An example is radioactive decay. Enzyme–substrate complexes often react by unimolecular processes. In other cases, a reaction is **pseudo-first order**; compound A actually reacts with a second molecule such as water, which is present in such excess that its concentration does not change during the experiment. Consequently, the velocity is apparently proportional only to [A].

A first-order rate constant k has units of s^{-1}. When [A] = 1, $v = k$. Thus, k is a measure of the speed in mol $l^{-1} s^{-1}$ of the reaction of a substance at unit activity. As a first-order reaction proceeds, [A] decreases and at time t is given by any one of the following three equivalent expressions (Eq. 9-3). These are obtained by integration of Eq. 9-2 in which t_0 is the time at which the reaction was started.

$$[A] = A_0 e^{-kt}$$
$$\ln([A_0] / [A]) = kt$$
$$\log[A_0] - \log[A] = kt / 2.303 \qquad (9\text{-}3)$$

Equation 9-3 is the equation of exponential decay. A characteristic of exponential decay is that [A] is halved in a time that is independent of concentration. The **half-life** is $t_{1/2}$:

$$t_{1/2} = \ln 2 / k = 0.693 / k \qquad (9\text{-}4)$$

The **relaxation time** τ for A is defined by Eq. 9-5 and represents the time required for the concentration [A] to fall to $1 / e$ (or ~0.37) of its initial value.

$$\tau = 1 / k = t_{1/2} / \ln 2 \qquad (9\text{-}5)$$

Turnover numbers and units of activity. When an enzyme is catalyzing product formation at the maximum possible rate, V_{max}, we can usually assume that the active site of every molecule of enzyme contains a substrate in the form of an intermediate enzyme–substrate complex ES and that this complex is being converted to products according to Eq. 9-6:

$$d[P] / dt = V_{max} = k[ES] = k[E]_t$$
$$\text{or} \qquad V_{max} = k_{cat}[E]_t \qquad (9\text{-}6)$$

Here E_t is the total enzyme, namely, the free enzyme E plus enzyme–substrate complex ES. The equation holds only at **substrate saturation,** that is, when the substrate concentration is high enough that essentially all of the enzyme has been converted into the intermediate ES. The process is first order in enzyme but is zero order in substrate. The rate constant k is a measure of the speed at which the enzyme operates. When the concentration $[E]_t$ is given in moles per liter of *active sites* (actual molar concentration multiplied by the number of active sites per mole) the constant k is known as the **turnover number**, the **molecular activity**, or **k_{cat}**. The symbol k_{cat} is also used in place of k in Eq. 9-6 for complex rate expressions in which k_{cat} cannot represent a single rate constant but is an algebraic expression that contains a number of different constants.

Turnover numbers can be measured only when the concentration of the enzyme is known. Partly for this reason the activity of an enzyme is usually given as **specific activity,** the *units of activity per milligram of protein.* One **international unit** is the amount of enzyme that produces 1 μmol of product per minute under standard (usually optimal) conditions. The International Union of Biochemistry[17] has recommended a larger unit, the **katal** (kat), the amount of enzyme that converts one mol s^{-1} of substrate to product.

1 kat = 6 x 10^7 international units
1 international unit = 16.67 nkat (nanokatals)

If the enzyme is pure and is saturated with substrate under the standard assay conditions, the following relationships hold.

Turnover No. = katals/mol of active sites
= [nkat/mg] x M_r x $10^{-6}/n$
= [international units/mg] x M_r x $10^{-3}/60n$

Here M_r is the relative molecular mass of the enzyme and n is the number of active sites per molecule. Since the activity of an enzyme is dependent on both temperature and pH, these variables must be specified. Turnover numbers of enzymes vary from <1 to ~ 10^6 s^{-1}. Trypsin, chymotrypsin, and many intracellular enzymes have turnover numbers of ~10^2 s^{-1}. The fastest enzymes, which include **catalase** (Chapter 16), **carbonic anhydrase** (Chapter 13), and **Δ^5-3-oxosteroid isomerase** (Chapter 13), have maximum turnover numbers of 2×10^5 s^{-1} or more. Compare these reaction rates with those of a typical organic synthesis in the laboratory. A reaction mixture must often be heated for hours ($k < 10^{-3}$ s^{-1}). Many enzymes accelerate rates by factors of greater than 10^6 over those observed in the absence of an enzyme at a comparable temperature and pH. Enzymes often bring two or more substrates together, binding them at a specific location in their active sites. Because of this, rapid reactions can be catalyzed even when the reactants are present in low concentrations.

Second-order reactions. For a chemical reaction to occur between two molecules, A and B (Eq. 9-7), they must meet and collide.

$$A + B \xrightarrow{k_2} P \qquad (9\text{-}7)$$

The velocity of such a **second-order** process is characterized by a **bimolecular rate constant** k_2 and is proportional to the product of the concentrations of A *and* B:

$$v = k_2[A][B] \qquad (9\text{-}8)$$

The units of k_2 are M^{-1} s^{-1}. If [B] is present at unit activity, the rate is $k_2[A]$, a quantity with units of s^{-1}. We can see that the bimolecular, or second-order, rate constant for reaction of A with B may be compared with first-order constants when the second reactant B is present at unit activity. In many real situations, reactant B is present in large excess and in a virtually constant concentration. The reaction is pseudo-first order and the experimentally observed rate constant $k_2[B]$ is an *apparent first-order rate constant*. The bimolecular rate constant k_2 can be obtained by dividing the apparent constant by [B].

Reversible chemical reactions. In any reversible process, we must consider rate constants for both the forward and the reverse reactions. At equilibrium a reaction proceeds in the forward direction at exactly the same velocity as in the reverse reaction so that no change occurs. For this reason there is always a relationship between the equilibrium constant and the rate constants. For Eq. 9-9, k_1 is the bimolecular rate constant

$$A + B \underset{k_2}{\overset{k_1}{\rightleftharpoons}} P$$

$$\text{Equilibrium constant} = K = \frac{[P]}{[A][B]} = \frac{k_1}{k_2} \qquad (9\text{-}9)$$

for the forward reaction and k_2 the unimolecular rate constant for the reverse reaction. The equilibrium constant K can easily be shown, from Eq. 9-2 and 9-8, to equal k_1/k_2 for the reaction of Eq. 9-9.

The student should be aware that in kinetic equations rate constants are usually numbered consecutively via subscripts and that the subscripts do not imply anything about the molecularity. The system which is used here employs odd-numbered constants for steps in the forward direction and even-numbered constants for steps in the reverse direction. However, *many authors number the steps in the forward direction consecutively and those in the reverse direction with corresponding negative subscripts.*

What relationships exist between experimentally observable rates and k_1 and k_2 for a reversible reaction? Consider first the simplest case (Eq. 9-10):

$$A \underset{k_2}{\overset{k_1}{\rightleftharpoons}} B \qquad (9\text{-}10)$$

If pure A is placed in a solution, its concentration will decrease until it reaches an equilibrium with the B which has been formed. It is easy to show that in this case [A] does not decay exponentially but $[A] - [A]_{equil}$ does. If $\log([A] - [A]_{equil})$ is plotted against time a first-order rate constant k, characteristic of *the rate of approach to equilibrium*, will be obtained. Its relationship to k_1 and k_2 is given by Eq. 9-11.

$$k = k_1 + k_2 \qquad (9\text{-}11)$$

The relaxation time τ for approach to equilibrium can be expressed as follows:

$$\tau^{-1} = k = \tau_1^{-1} + \tau_2^{-1} = k_1 + k_2 \qquad (9\text{-}12)$$

and for the more complex case of Eq. 9-9 as

$$k = \tau^{-1} = k_1([A]_e + [B]_e) + k_2 \qquad (9\text{-}13)$$

where $[A]_e$ and $[B]_e$ are the equilibrium concentrations of A and B.[18]

2. Formation and Reaction of Enzyme–Substrate Complexes

An abundance of evidence indicates that the first step in enzymatic catalysis is the combining of the enzyme and substrate reversibly to form a complex, ES (Eq. 9-14). Formation of the complex is normally reversible. ES can either break up to form enzyme and substrate again or it can undergo conversion to a product or products, often by a unimolecular process. Three rate constants are needed to describe this system for a reaction that is irreversible overall. A complete description of the kinetic behavior is fairly involved.

$$E + S \underset{k_2}{\overset{k_1}{\rightleftharpoons}} ES \xrightarrow{k_3} E + P \qquad (9\text{-}14)$$

For example, the kinetics may be different within cells, where molar concentrations of enzymes often exceed those of substrate, than in the laboratory. In most laboratory experiments the enzyme is present at an extremely low concentration (e.g., 10^{-8} M) while the substrate is present in large excess. Under these circumstances the **steady-state approximation** can be used. For this approximation *the rate of formation of ES from free enzyme and substrate is assumed to be exactly balanced by the rate of conversion of ES on to P.* That is, for a relatively short time during the duration of the experimental measurement of velocity, the concentration of ES remains essentially constant. To be more precise, the steady-state criterion is met if the absolute rate of change of a concentration of a transient intermediate is very small compared to that of the reactants and products.[19]

The **Michaelis–Menten equation** (Eq. 9-15) describes the initial reaction rate of a single substrate with an enzyme under steady-state conditions.

$$v = \frac{V_{max}}{1 + K_m/[S]} = \frac{V_{max}[S]}{K_m + [S]}$$

$$\text{where } K_m = (k_2 + k_3)/k_1 \qquad (9\text{-}15)$$

This can be rearranged as follows:

$$\frac{V_{max}}{v} = 1 + K_m/[S] \qquad (9\text{-}16)$$

Equation 9-15 provides a relationship between the velocity observed at a particular substrate concentration and the maximum velocity that would be achieved at infinite substrate concentration. The quantities V_{max} and K_m are often referred to as the **kinetic parameters** of an enzyme and their determination is an important part of the characterization of an enzyme. Equation 9-15 can be derived by setting the rate of formation of the ES complex ($k1[E][S]$) in the steady state equal to its rate of breakdown, ($[k_2 + k_3][ES]$). Rearranging and substituting K_m, as defined in Eq. 9-15, we obtain Eq. 9-17.

$$[E][S] = \frac{(k_2 + k_3)}{k_1}[ES] = K_m[ES] \qquad (9\text{-}17)$$

Using this equation, together with a mass balance relationship ($[E] = [E]_t - [ES]$), we can solve for $[ES]/[E]_t$, the fraction of enzyme combined as enzyme–substrate complex (Eq. 9-18).

$$\frac{[ES]}{[E]_t} = \frac{[S]}{K_m + [S]} \qquad (9\text{-}18)$$

The maximum velocity, $V_{max} = k_3[E]_t$, is attained only when all of the enzyme is converted into ES. Under other conditions $v = k_3[ES]$ and Eq. 9-19 holds.

$$[ES] / [E]_t = v / V_{max} \qquad (9\text{-}19)$$

Substituting from Eq. 9-19 into Eq. 9-18 gives the Michaelis–Menten equation (Eq. 9-15).

In many cases, the rate at which ES is converted back to free E and S is much greater than the rate of conversion of ES to products ($k_2 \gg k_3$). In such cases K_m equals k_2/k_1, the dissociation constant for breakdown of ES to free enzyme and substrate (sometimes called K_s). Thus, K_m *sometimes has a close inverse relationship to the strength of binding of substrate to enzyme.* In such cases, $1/K_m$ is a measure of the affinity of the substrate for its binding site on the enzyme and for a series of different substrates acted on by the same enzyme. The more tightly bound substrates have the lower values of K_m. But beware! The condition that k_3 is negligible compared to k_2 may be met with some (poorer) substrates but may not always be met by others. From Eq. 9-15 we see that K_m is always greater than or equal to K_s, but it may be less than K_s for more complex mechanisms.

Figure 9-2 shows a plot of velocity against substrate concentration as given by Eq. 9-15. The position of V_{max} on the ordinate is marked, but it should be clear that the experimental velocity (v) can never attain V_{max} unless [S] is very high relative to K_m. The value of v approaches V_{max} asymptotically. Since K_m is defined as the value of [S] at which $v = V_{max}/2$, its value can be estimated from Fig. 9-2. However, K_m cannot be determined reliably because of the difficulty of establishing the value of V_{max} from a plot of this type. Notice that the curve of Figure 9-2 is identical in form to the saturation curve for reversible binding shown in Fig. 7-1.

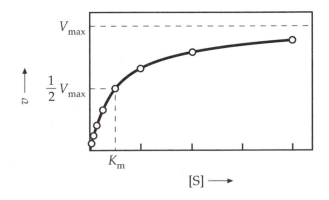

Figure 9-2 Plot of observed velocity v vs substrate concentration [S] for an enzyme-catalyzed reaction.

Linear forms for rate equations. To obtain K_m and V_{max} from experimental rate data, Eq. 9-15 can be transformed by algebraic rearrangement into one of several linear forms. The popular **double-reciprocal** or **Lineweaver–Burk** plot of $1/v$ against $1/[S]$ (Fig. 9-3) is described by Eq. 9-20. The values of K_m/V_{max} and $1/V_{max}$ can be evaluated from the slope and intercept, respectively, of this straight line plot.

$$\frac{1}{v} = \frac{1}{V_{max}} + \frac{K_m}{V_{max}}\frac{1}{[S]} \qquad (9\text{-}20)$$

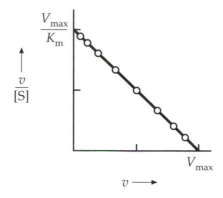

Figure 9-4 The Eadie–Hofstee plot of $v/[S]$ against v. The slope is $-1/K_m$; the intercept on the vertical axis is V_{max}/K_m and that on the horizontal axis is V_{max}.

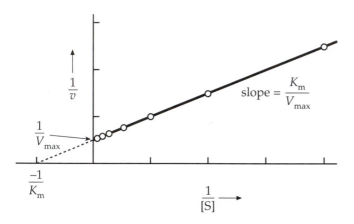

Figure 9-3 Double-reciprocal or Lineweaver–Burk plot of $1/v$ vs $1/[S]$. The intercept on the vertical axis gives $1/V_{max}$ and the slope gives K_m/V_{max}. The intercept on the horizontal axis equals $-1/K_m$.

Another linear plot, the **Eadie–Hofstee plot**, is that of $v/[S]$ vs v (Fig. 9-4). It is related to the Scatchard plot (Fig. 7-3) and is fitted by Eq. 9-21.

$$\frac{v}{[S]} = \frac{V_{max}}{K_m} - v\frac{1}{K_m} \qquad (9\text{-}21)$$

While the point for $[S] = 0$ and $v = 0$ cannot be plotted, the ratio $v/[S]$ approaches V_{max}/K_m as v approaches zero. Notice the distribution of the points in Fig. 9-4. Substrate concentrations were chosen such that the increase in velocity from point to point is more or less constant, a desirable experimental situation. The points on the Eadie–Hofstee plot are also nearly evenly distributed, but those of the Lineweaver–Burk plot are compressed at one end. (However, if the substrate concentrations for successive points are selected in the ratios 1, 1/3, 1/5, 1/7, and 1/9, the spacing will be uniform on the Lineweaver–Burk plot.) A second advantage of the Eadie–Hofstee plot is that the entire range of possible substrate concentration from near zero to infinity can be fitted onto a single plot.

As pointed out at the beginning of Section A, depletion of substrate with time lowers $[S]$ from its initial value to some extent. This can have an especially adverse effect on points obtained at low $[S]$, e.g., points on the right side of Fig. 9-3 or on the left side of Fig. 9-4. Equations 9-20 and 9-21 will be more precise if the average substrate concentration over the time period of the assay, rather than that at zero time,[20] is used.

Nonlinear equations and integrated rate equations. In discussions of the *control* of metabolism through regulation of enzymatic activity, it is often better to plot v against log $[S]$ as in Fig. 9-5. This plot also has the virture that the entire range of attainable substrate concentrations can be plotted on one piece of paper if the point for $[S] = 0$ (at minus infinity) is omitted. The same scale can be used for all enzymes. The plot is S-shaped, both for simple cases that are represented by Eq. 9-15 and for enzymes that bind substrate cooperatively (Section B,5). Thus, the classification or "hyperbolic" vs "sigmoidal" is lost. However, the degree of cooperativity can be directly measured from the midpoint slope of the curve. A disadvantage is that it is awkward to measure V_{max} from a plot of this type, and it may be preferable to obtain it from a linear plot (Figs. 9-3 or 9-4). Alternatively, computer-assisted methods can be used to obtain both K_m and V_{max} and to fit a curve to the experimental points as in Fig. 9-5.

An attractive alternative to the use of initial velocities and linear plots is to measure the kinetic parameters V_{max} and K_m using points all along the progress curve (Fig. 9-1). Various procedures for doing this have been devised.[21,22] For example, the integrated form of Eq. 9-15 can be given as Eq. 9-22 or 9-23.

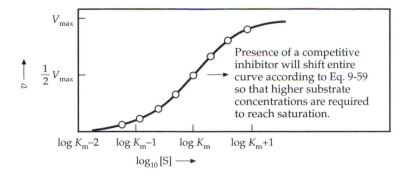

Presence of a competitive inhibitor will shift entire curve according to Eq. 9-59 so that higher substrate concentrations are required to reach saturation.

Figure 9-5 Plot of v against log [S] for an enzyme-catalyzed reaction.

$$V_{max}t = [P] + K_m \ln \frac{[S]_0}{[S]_0 - [P]} \qquad (9\text{-}22)$$

$$V_{max} t = [S]_0 - [S] + K_m \ln ([S]_0 / [S]) \qquad (9\text{-}23)$$

Here $[S]_0$ is the initial substrate concentration at $t = 0$ and [S] and [P] are the substrate and product concentrations at any later time. In one type of linear plot,[21] based on Eq. 9-23, $([S]_0 - [S]) / t$ is plotted against $1/t$ $\ln ([S]_0 / [S])$, the slope of the plot being $-K_m$ and the intercepts being V_{max} and V_{max} / K_m. Computer programs are available for analysis of enzymatic progress curves, even for complex mechanisms.[23,24]

Kinetics with high enzyme concentrations. Laboratory studies of the kinetics of purified enzymes are usually conducted with enzyme concentrations of 10^{-7} to 10^{-10} M, but within cells enzyme concentrations are probably often in the range 10^{-6} to 10^{-5} M,[25] which may be higher than the concentrations of the substrates upon which the enzymes act. Be cautious in drawing conclusions about kinetics under such circumstances! Methods have been devised for handling kinetic data when the concentration of enzyme is greater than K_m, a condition that can lead to intolerably high errors if the usual equations are applied.[26,27]

3. Diffusion and the Rate of Encounter of an Enzyme with Substrate

What determines the value of k_1 in Eq. 9-14? This rate constant represents the process by which the substrate and enzyme find each other, become mutually oriented, and bind to form ES. If orientation and binding are rapid enough, the rate will be determined by the speed with which the molecules can come together by diffusion. Large molecules in solution are free to travel for only a tiny fraction of their diameter as a result of their frequent collision with solvent molecules.

The result is visible in the **Brownian movement** of microscopic particles suspended in a fluid. If an individual particle is followed, it is seen to undergo a "random walk," moving in first one direction then another. Albert Einstein showed that if the distances transversed by such particles in a given time Δt are measured, the mean square of these Δx values Δ^2 can be related by Eq. 9-24 to the diffusion constant D (which is usually given in units of cm^2s^{-1}).

$$\Delta^2 = 2D \, \Delta t \qquad (9\text{-}24)$$

For molecules, the Brownian movement cannot be observed directly but the diffusion constant can be measured, for example, by observing the rate of spreading of a boundary between two different concentrations of the substance.[28] The diffusion constant for $^1H^2HO$ (HDO) in H_2O at 25°C is 2.27×10^{-5} cm^2 s^{-1}, and the values for the ions K^+ and Cl^- are about the same.[29] For many small molecules D is approximately 10^{-5} cm^2 s^{-1}, and the value decreases as the size of the molecule increases. Thus, for the 13.7-kDa ribonuclease, $D = 1.1 \times 10^{-6}$ cm^2 s^{-1}, and for the 500-kDa myosin, D is $\sim 1 \times 10^{-7}$ cm^2 s^{-1}. For the spherical particle the **Stokes–Einstein equation** (Eq. 9-25) can be used to relate the diffusion coefficient to the radius, the coefficient of viscosity η, the Boltzmann constant k_B, and the temperature (Kelvin) T.

$$D = \frac{k_B T}{6\pi\eta r} \qquad (9\text{-}25)$$

To estimate the rate constant for a reaction that is controlled strictly by the frequency of collisions of particles, we must ask how many times per second one of a number n of particles will be hit by another of the particles as a result of Brownian movement. The problem was analyzed in 1917 by Smoluchowski,[30,31] who considered the rate at which a particle B diffuses toward a second particle A and disappears when the two collide. Using Fick's law of diffusion, he concluded that the number of **encounters** per milliliter per second was given by Eq. 9-26.

Number of encounters/ml s^{-1} =
$$4 \pi (D_A + D_B) (r_A + r_B) n_A n_B \qquad (9\text{-}26)$$

Here D_A and D_B are the two diffusion coefficients, r_A and r_B are the radii (in Å) of the two particles, and n_A and n_B are the numbers of particles per milliliter. The number of encounters per liter per second is $(4\pi / 1000) (D_A + D_B)$ $(r_A + r_B) N^2 [A][B]$, where N is Avogadro's number. Dividing this frequency by N gives the rate of collision v in M s^{-1}, a velocity that can be equated (Eq. 9-27) with $k_D [A] [B]$, where k_D is a second-order rate constant which determines the **diffusion controlled limit**, that is, the maximum possible rate of a reaction.

$$v = \frac{4\pi}{1000}(D_A + D_B)(r_A + r_B)[A][B]N = k_D[A][B] \tag{9-27}$$

This equation can be rearranged to Eq. 9-28.

$$k_D = \frac{4\pi N}{1000}(D_A + D_B)(r_A + r_B)M^{-1}\,s^{-1} \tag{9-28}$$

While this equation is thought to overestimate the diffusion-limited rate constant slightly, it is a good approximation. If the diffusing particles are approximately spherical, diffusion constants D_A and D_B can be calculated from Eq. 9-25, and Eq. 9-28 becomes Eq. 9-29.

$$k_D = \frac{2RT}{3000\eta}\left(2 + \frac{r_A}{r_B} + \frac{r_B}{r_A}\right) \tag{9-29}$$

The value of k_D does not vary greatly as the ratio of radii r_A / r_B is changed, and in most cases it may be assumed that $r_A / r_B \approx 1$, in which case the equation simplifies (Eq. 9-30).

$$k_D \approx \frac{8RT}{3000\eta}M^{-1}\,s^{-1} \tag{9-30}$$

For water at 25°C, the coefficient of viscosity η is ~0.01 poise (1 poise = 10^{-5} newton cm^{-2}) which leads, according to Eq. 9-30, to $k_D \sim 0.7 \times 10^{10}\,M^{-1}\,s^{-1}$. It was shown by Debye[32] that this rate constant must be multiplied by a correction factor when charged particles rather than uncharged spheres diffuse together. This factor may be of the order of 5–10 for a substrate and enzyme carrying two or three charges and may act to either increase or decrease reaction rates. Since the viscosity of cytoplasm is 0.02–0.03 poise,[33] k_D will be reduced correspondingly.

A simple alternative derivation of Eq. 9-30 was developed by Dexter French,[34] who also provided the author with most of this discussion of encounter theory. Consider a small element of volume ΔV swept out by a particle as it moves through the solution for a distance equal to its own radius. This element of volume will equal πr^3:

$$\Delta V = \pi r^2 \times r = \pi r^3 \tag{9-31}$$

It will be swept out in a time Δt which can be calculated from Eq. 9-24 as $r^2 / 2D$. Substituting the value of D given by Eq. 9-25, we obtain for Δt:

$$\Delta t = \frac{3\pi\eta r^3}{k_B T} \tag{9-32}$$

Division of Eq. 9-31 by Eq. 9-32 gives the volume swept out per second by one particle:

$$\Delta V / \Delta t = k_B T / 3\eta \;\; cm^3\,s^{-1} \tag{9-33}$$

Since the collision radius for two particles of equal size is two times the particle radius, the effective volume swept out will be four times that given by Eq. 9-33. Since both particles are diffusing, the effective diffusion constant will be twice that used in obtaining Eq. 9-28. Thus, the effective volume swept out by the particle in a second will be eight times that given by Eq. 9-33. The volume swept out by one mole of particles is equal to k_D (recall that the second-order rate constant has dimensions of liter $mol^{-1}\,s^{-1}$). Thus, when converted to a moles per liter basis and multiplied by 8, Eq. 9-33 should (and does) become identical with the Smoluchowski equation (Eq. 9-30).

The volume given by Eq. 9-33 is about 1.4×10^{-11} cm^3, which could be represented approximately by a cube 2.4 μm on a side. If we compare this volume with that of a cell (Table 1-2) or of an organelle, we see that in one second an enzyme molecule will sweep out a large fraction of the volume of a small cell, mitochondrion, chloroplast, etc.

The "cage effect" and rotation of molecules.

It is of interest to compare the bimolecular rate constant for encounters calculated by the Smoluchowski theory (Eq. 9-29) with the corresponding bimolecular rate constant for molecular collisions given by the kinetic theory of gases (Eq. 9-34).

$$k\,(collision) = \frac{N(r_A + r_B)^2}{1000}\left[8\pi k_B T\left(\frac{1}{m_A} + \frac{1}{m_B}\right)\right]^{1/2} \tag{9-34}$$

Here m_A and m_B are the masses of the two particles. This rate constant is also relatively independent of molecular size and for spheres varies from (4 to 11) $\times$ $10^{11}\,M^{-1}\,s^{-1}$, over an order of magnitude greater than the rate constant for encounters. In a solution, molecules still collide at about the same rate as in a gas so that *100 to 200 collisions occur between two particles for each encounter*. However, during the time of the single encounter, the particles are together in a solvent "cage." While in this solvent cage, both substrate and enzyme molecules undergo random rotational motions. Successive collisions bring them together in different orientations, one of which is likely to lead to a sufficiently close match of complementary surfaces (of substrate and binding site) so that formation of a "productive" ES complex takes place.

Molecular rotation in a solution is described quantitatively by diffusion laws (analogous to Fick's laws) for which a **rotary diffusion constant** θ is defined.[35,36] Consider a group of molecules all oriented the same way initially, then undergoing rotary diffusion until their orientations become random. If we measure the orientation of each molecule by an angle we see that initially the value of cos α is 1 but that when the angles become random the mean value of cos α averaged over all molecules is zero. The rotary relaxation time τ is the time required for the mean value of cos α to fall to

Initial orientation   Later orientation

$1/e$ (which occurs when $\alpha = 68.5°$). For a sphere θ is given by Eq. 9-35.

$$\theta = \tfrac{1}{2}\tau = \frac{k_B T}{8\pi\eta r^3} \qquad (9\text{-}35)$$

Ellipsoidal or rod-shaped molecules have two different rotary diffusion constants while, if the dimensions of the molecules are different along all three axes, three constants must be specified.[36]

From Eq. 9-35 we can calculate (if $\eta = 0.01$ poise) the following values for a small spherical molecule (substrate) of ~1 nm length and for a spherical enzyme of 5 nm diameter:

$r = 0.5$ nm $\theta \approx 1.3 \times 10^9 \text{ s}^{-1}$
$r = 2.5$ nm $\theta \approx 1.0 \times 10^7 \text{ s}^{-1}$

We see that smaller molecules rotate much faster than large ones and that rotational relaxation times for small proteins are of the same order of magnitude as k_D for diffusion-limited encounter. However, for very large molecules, especially long rods, the rotary relaxation time about a short axis may be a large fraction of a second.

The rate of substrate binding. At very low substrate concentrations the Michaelis–Menten equation (Eq. 9-15) simplifies as follows:

$$v = (V_{max}/K_m)[S] \qquad (9\text{-}36)$$

Since $V_{max} = k_{cat}[E]_t$ (Eq. 9-6) and at low [S] most of $[E]_t$ is free E, we obtain the following equation:[2]

$$v = (k_{cat}/K_m)[E][S] \qquad (9\text{-}37)$$

From this we see that k_{cat}/K_m is *the apparent second-order rate constant for the reaction of free enzyme with substrate*. As such it cannot exceed the diffusion controlled limit k_D of Eqs. 9-28 to 9-30 which falls in the range of $10^9 - 10^{11} \text{ M}^{-1} \text{ s}^{-1}$. Experimentally observed

values of k_{cat}/K_m are always less than this limit, indicating that a certain time is required for a substrate molecule to become oriented and seated in the active site.[37] However, for several real enzymes values of k_{cat}/K_m of 10^7 to $3 \times 10^8 \text{ M}^{-1} \text{ s}^{-1}$ have been observed. For triose phosphate isomerase Albery and Knowles obtained a value of $4 \times 10^8 \text{ M}^{-1} \text{ s}^{-1}$, so close to the diffusion controlled limit that these authors regard this enzyme as a *nearly perfect catalyst*, one that could not have evolved further because it is already catalyzing the reaction with substrate at almost the maximum velocity that is possible.[37–39]

The displacement of bound ligands by substrate. Jenkins pointed out that in many instances a substrate must displace another ligand from the active site to form the ES complex.[40] For example, a binding site for an ionic substrate often already contains an ion, either of the product P or some other ion X which Jenkins calls a **substrate surrogate**. For this situation Eq. 9-14 must be replaced by the following set of equations:

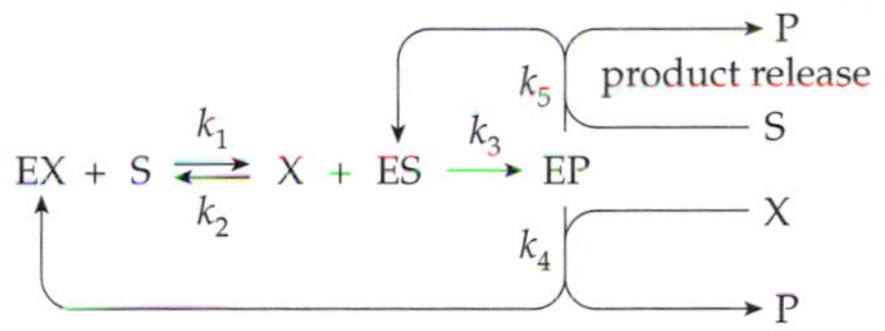

$$\text{EX} + \text{S} \underset{k_2}{\overset{k_1}{\rightleftharpoons}} \text{X} + \text{ES} \xrightarrow{k_3} \text{EP} \qquad (9\text{-}38)$$

Here k_2 and also k_4 and k_5, are second-order rate constants. The release of product, as determined by k_4 and k_5, may be rate-limiting. At zero time the reverse reactions may be ignored, and steady-state analysis shows that the Michaelis–Menten equation (Eq. 9-16b) will be replaced by Eq. 9-39. Here, D is a constant and A is also constant if X is present at a fixed concentration.

$$\frac{V_{max}}{v} = 1 + \frac{k_2[X] + k_3}{k_1[S]} + \frac{k_3(1/k_5 - 1/k_1)}{k_4[X]/k_5 + [S]}$$
$$= 1 + K_m/[S] + D/(A + [S]) \qquad (9\text{-}39)$$

When k_5, the rate constant for displacement of product by substrate, is very small, this equation simplifies to Eq. 9-40.

$$\frac{V_{max}}{v} = 1 + k_3/k_4[X] + (k_2[X] + k_3)/k_1[S]$$
$$= 1 + k_3/k_4[X] + K_m'/[S] \qquad (9\text{-}40)$$

At high concentrations of X the second term becomes negligible and the equation becomes identical to the Michaelis–Menten equation (Eq. 9-16) except that K_m', the apparent Michaelis constant, now increases as [X] increases (Eq. 9-41). This is exactly what is observed for competitive inhibitors where

$$K_m' = \frac{k_2[X] + k_3}{k_1} \tag{9-41}$$

Thus substrate surrogates at high concentrations are often competitive inhibitors. However, at low [X] the second term in Eq. 9-41 may be important and *X then serves as an activator*. Under many circumstances, Eq. 9-39 does not simplify further and a nonlinear relationship between $1/v$ and $1/[S]$ is observed, with the shape of the curves being influenced by the values of constants A and D (see also Section B,4).[40]

4. Reversible Enzymatic Reactions

For some enzyme-catalyzed reactions the equilibrium lies far to one side. However, many other reactions are freely reversible. Since a catalyst promotes reactions in both directions, we must consider the action of an enzyme on the reverse reaction. Let us designate the maximum velocity in the forward direction as V_f and that in the reverse direction as V_r. There will be a Michaelis constant for reaction of enzyme with product K_{mP}, while K_{ms} will refer to the reaction with substrate.

As in any other chemical reaction, there is a relationship between the rate constants for forward and reverse enzyme-catalyzed reactions and the equilibrium constant. This relationship, first derived by the British kineticist J. B. S. Haldane and proposed in his book *Enzymes*[41] in 1930, is known as the **Haldane relationship**. It is obtained by setting $v_f = v_r$ for the condition that product and substrate concentrations are those at equilibrium. For a single substrate–single product system it is given by Eq. 9-42.

$$K_{eq} = V_f K_{mP} / V_r K_{ms} \tag{9-42}$$

Because the Haldane relationship imposes constraints on the values of the velocity constants and Michaelis constants, it is of some value in understanding regulation of metabolism. Consider the case that the maximum forward velocity V_f is high and that K_{ms} has a moderately low value (fairly strong binding of substrate). If the reaction is freely reversible ($K_{eq} \sim 1$) and the velocity of the reverse reaction V_r is about the same as that of the forward reaction, it will necessarily be true (see Eq. 9-42) that the product P will also be fairly tightly bound. If $V_r << V_f$, the value of K_{mP} will have to be much *lower* than that of K_{mA}. In such a situation P will remain tightly bound to the enzyme and since V_r is low, it will tend to clog the enzyme. Such **product inhibition** may sometimes slow down a whole pathway of metabolism. In such a case, the

only way that an enzymatic sequence can keep going in the forward direction is for product P to be removed rapidly by a subsequent reaction with a second enzyme. The first enzyme may be thought of as possessing a kind of "one-way valve" that turns off the flow in a metabolic pathway when the concentration of its product rises.

Reactions of two or more substrates. Enzymes frequently catalyze the reaction of two, three, or even more different molecules to give one, two, three, or more products. Sometimes all of the substrate molecules must be bound to an active site at the same time and are presumably lined up on the enzyme molecule in such a way that they can react in proper sequence. In other cases, the enzyme may transform molecule A to a product, and then cause the product to react with molecule B. The number of variations is enormous.[1,10,12]

The order in which two molecules A and B bind to an enzyme to form a complex EAB may be completely *random* or it may be *obligatorily ordered*. Both situations occur with real enzymes. Cleland introduced a widely used method of depicting the possibilities.[42] For example, Eq. 9-43 shows the reaction of A and B in an *ordered sequence* to form the complex EAB which is then isomerized to EPQ, the complex formed by binding the two products P and Q to the enzyme. The rate constant to the left of each vertical arrow or above each horizontal arrow refers to the reaction in the

$$
\begin{array}{ccccccccc}
& A & & B & & & P & & Q \\
& k_1 \big\downarrow k_2 & & k_3 \big\downarrow k_4 & & & k_7 \big\uparrow k_8 & & k_9 \big\uparrow k_{10} \\
E & \xrightarrow{} EA & \xrightarrow{} EAB & \underset{k_6}{\overset{k_5}{\rightleftarrows}} EPQ & \xrightarrow{} EQ & \xrightarrow{} E
\end{array}
\tag{9-43}
$$

forward direction as indicated by the arrow while the other constants (to the right or below the arrows) refer to the reverse reactions. The velocity in the forward direction for an enzyme with ordered binding is given by Eq. 9-44a,

$$v_f = \frac{V_f[A][B]}{K_{eqA}K_{mB} + K_{mB}[A] + K_{mA}[B] + [A][B]} \tag{9-44a}$$

which may also be written in the reciprocal form (Eq. 9-44b):

$$\frac{1}{v_f} = \frac{1}{V_f}\left(1 + \frac{K_{mA}}{[A]} + \frac{K_{mB}}{[B]} + \frac{K_{eqA}K_{mB}}{[A][B]}\right) \tag{9-44b}$$

An alternative form of this equation (Eq. 9-45), proposed by Dalziel,[43] is sometimes used.

$$\frac{[E]_t}{v_f} = \phi_0 + \frac{\phi_1}{[A]} + \frac{\phi_2}{[B]} + \frac{\phi_{12}}{[A][B]} \qquad (9\text{-}45)$$

The kinetic parameters of Eq. 9-44 are V_f, the maximum velocity in the forward direction, the two Michaelis constants, K_{mB} and K_{mA}, and the equilibrium constant K_{eqA}, for reversible dissociation of the complex EA and which is equal to k_2/k_1. The relationship between the parameters of Eq. 9-44 (K_m's, V's, and K_{eqA}'s) and the rate constants k_1–k_{10} is not obvious. However, remember that the parameters are experimental quantities determined by measurements on the enzyme. Sometimes, but not always, it is possible to deduce some of the values of individual rate constants from the experimental parameters.

An equation similar to 9-44a can be written for the velocity v_r of the reaction of P and Q. Also an equation can be written in similar form for $v_f – v_r$, i.e., the instantaneous velocity of the reaction in any mixture of all four components A, B, P, and Q.

The kinetic parameters of Eq. 9-44b are often obtained from experimental data through the use of reciprocal plots (Fig. 9-6). However, Eq. 9-44b is linear only if the concentration of one or the other of the substrates A and B is kept constant. For this reason a series of experiments is usually performed in which [A] is varied while [B] is held constant. Then [A] is held constant and [B] is varied. Each of these experiments leads to a family of lines (Fig. 9-6A) whose slopes and intercepts are measured. The slopes and intercepts of this family of curves are then plotted against the reciprocal of the second concentration, i.e.,

the one that was held fixed.

From a set of these **secondary plots**, V_f and one of the Michaelis constants can be determined (Fig. 9-6B and C). Using two sets of secondary plots, all of the constants of Eq. 9-44 may be established. Alternatively, a computer can be used to examine all of the data at once and to obtain the best values of the parameters. The latter approach is desirable because estimates of the standard deviations of the parameters can be obtained. However, the user must take care to ensure that the experimental errors are correctly estimated and are not simply estimates of how well the computer has fitted the points on the assumption that they contain no error.[23]

The meaning of the kinetic parameters may be slightly difficult to grasp. V_f is the velocity that would be obtained if both [A] and [B] were high enough to saturate the enzyme. Each K_m corresponds to that for a simple system in which the concentration of second substrate is at a high, saturating value.

For the bimolecular reaction that we have considered, there are two Haldane relationships:

$$K_{eq} = \frac{V_f K_{mP} K_{dQ}}{V_r K_{dA} K_{mB}} = \left(\frac{V_f}{V_r}\right)^2 \frac{K_{dP} K_{mQ}}{K_{mA} K_{dB}} \qquad (9\text{-}46)$$

Of these, only the first is ordinarily used.

"Ping-pong" mechanisms. A common type of mechanism that is especially prevalent for enzymes with tightly bound cofactors has been dubbed **ping-pong** because the enzyme alternates between

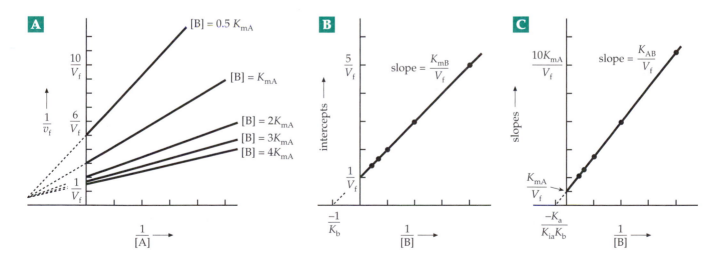

Figure 9-6 Reciprocal plots used to analyze kinetics of two-substrate enzymes. (A) Plot of $1/v_f$ against $1/[A]$ for a series of different concentrations of the second substrate B. (B) A secondary plot in which the intercepts from graph A are plotted against $1/[B]$. (C) Secondary plot in which the slopes from graph A have been plotted against $1/[B]$. The figures have been drawn for the case that $K_{mA} = 10^{-3}$ M, $K_{mB} = 2\,K_{mA}$, and $K_{AB} = K_{eqA} K_{mB}$ (Eq. 9-46) $= K_{mA}/200$ and [A] and [B] are in units of moles per liter. Eadie–Hofstee plots of $v_f/[A]$ vs v_f at constant [B] can also be used as the primary plots. The student can easily convert Eq. 9-44 to the proper form analogous to Eq. 9-21.

two forms E and E′ (Eq. 9-47). Substrate A reacts via complex EA to form E′, a modified enzyme that often

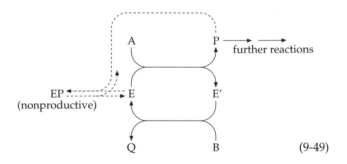

Ping-pong mechanism for reaction
of A and B to form P and Q   (9-47)

contains an altered coenzyme. At the same time A is changed to product P still bound to the enzyme. P dissociates leaving E′ which is then able to react with the second substrate B and to go through the second half of the cycle during which E′ is converted back to E. An example is provided by the aminotransferases (Eq. 14-25) in which the coenzyme pyridoxal phosphate is interconverted with pyridoxamine phosphate.

The rate equations of the ping-pong mechanism resemble that for the ordered bimolecular reaction (Eq. 9-44), but each has one less term (Eq. 9-48):

$$v_f = \frac{V_f[A][B]}{K_{mB}[A] + K_{mA}[B] + [A][B]}$$ (9-48a)

or

$$\frac{1}{v_f} = \frac{1}{V_f}\left(1 + \frac{K_{mA}}{[A]} + \frac{K_{mB}}{[B]}\right)$$ (9-48b)

A diagnostic feature of the ping-pong mechanism is that the families of lines (in double reciprocal plots) obtained when one substrate is held constant while the other is varied no longer intersect as they do in Fig. 9-6A but are parallel. They must be truly parallel and the experimentalist must be aware that nearly parallel lines may sometimes be observed for sequential reactions. Thus, if K_{eqA} of Eq. 9-44b is small enough the last term of that equation will be small compared to the other terms and the equation will be approximated by Eq. 9-48b. The reaction will appear to be ping-pong even though it is sequential and the reaction proceeds through the ternary complex EAB.

One less kinetic parameter can be obtained from an analysis of the data for a ping-pong mechanism than can be obtained for ordered reactions. Nevertheless, in Eq. 9-47, twelve rate constants are indicated. At least this many steps must be considered to describe the behavior of the enzyme. Not all of these constants can be determined from a study of steady-state kinetics, but they may be obtained in other ways.

Isomechanisms. A catalyst functions over and over again without being altered. However, during a single turnover it is likely to undergo a temporary change. For example, if an enzyme assists in removing a proton from a substrate, some functional group of

the enzyme will be protonated during part of the catalytic cycle. This proton must dissociate before another catalytic cycle begins, and in some instances this dissociation may be a rate-limiting step. Northrup and coworkers developed the use of product inhibition to identify such isomechanisms.[44]

Dead-end complexes. In the steady state of action of an enzyme with ping-pong kinetics, part of the enzyme is in form E and part in form E′. Ideally, E would have affinity only for A and Q, while E′ would have affinity only for B and P. However, in many real situations E also binds B and P weakly; similarly, E′ binds A and Q. This is because the products and reactants usually have structural features in common. The propensity of enzymes with ping-pong kinetics to form **dead-end complexes** (also called abortive complexes) may sometimes have a regulatory function. In Eq. 9-49 the reactions of Eq. 9-47 have been rearranged to depict a situation in which product P normally undergoes a sequence of further reactions. However, if P accumulates to a high enough concentration it can react reversibly to form a dead-end complex EP. This is an effective form of product inhibition which can be relieved only by a lowering of the concentration of P through its further metabolism.

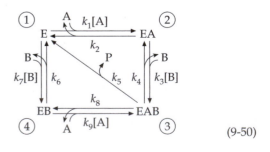

(9-49)

Handling rate equations for complex mechanisms. While steady-state rate equations can be derived easily for the simple cases discussed in the preceding sections, enzymes are often considerably more complex and the derivation of the correct rate equations can be extremely tedious. The **topological theory of graphs**, widely used in analysis of electrical networks, has been applied to both steady-state and nonsteady-state enzyme kinetics.[45–50] The method employs diagrams of the type shown in Eq. 9-50. Here

(9-50)

the reaction of an enzyme with two substrates A and B with a **random** order of binding is depicted. (In contrast, Eq. 9-43 shows the case of ordered binding of two substrates.) When complex EAB is formed, it can decompose to free enzyme and to the single product P. Each one of the **nodes**, which are numbered 1–4 in the diagram, corresponds to a single form of the enzyme. The appropriate first-order rate constant or apparent first-order constant is placed by each arrow. The methods provide easy rules for deriving from such a scheme the steady-state rate equation.

The importance of the simplified schematic methods is apparent when one considers that the steady-state rate equation for Eq. 9-50 would have 6 terms in the numerator and 12 terms in the denominator.[51] In the more complex case in which EAB breaks down to two products P and Q with a random order of release, the rate equation contains 672 terms in the denominator. In such cases it is worthwhile to enlist the help of a computer in deriving the equation.[24,52–54]

The rapid equilibrium assumption.
Rate equations for enzymes are often simplified if a single step, e.g., that of reaction of complex EAB to product in Eq. 9-50, is **rate limiting**.[54a] If it is assumed that all reaction steps preceding or following the rate-limiting step are at equilibrium, the equation for random binding with a two-substrate and two-product reaction simplifies to one whose form is similar to that obtained for ordered binding (Eq. 9-44). In the absence of products P and Q Eq. 9-44 will correctly represent the steady-state rate equation corresponding to Eq. 9-50. However, this simplification may not be valid for a very rapidly acting enzyme.

Isotope exchange at equilibrium.
Consider the reaction of substrates A and B to form P and Q (Eq. 9-51). If both reactants and both products are present with the enzyme and in the ratio found at equilibrium no net reaction will take place. However, the reactants and products will be continually interconverted under the action of the enzyme. Now if a small amount of

$$A + B \rightleftarrows P + Q \qquad (9\text{-}51)$$

highly labeled reactant (A* or B*) is added, the rate at which isotope is transferred from the labeled reactant into one or the other of the products can be measured. In general, a label in one of the substrates will appear in only one of the products.

Figure 9-7A shows the rate of exchange of isotopically labeled glucose (glucose*) with glucose 6-phosphate as catalyzed by the enzyme hexokinase (Chapter 12). The exchange rate is plotted against the concentration of glucose 6-phosphate with the ratio [glucose] / [glucose 6-phosphate] constant at 1/19, such that an equilibrium ratio for reactants and products is always

maintained. As can be seen from the graph, this exchange rate increases monotonically as substrate concentrations are increased. This is also true for the rate of ATP–ADP exchange. The fact that both exchange rates increase continuously indicates random binding of substrates.[55] The inequality of the two maximal exchange rates suggests that release of glucose 6-phosphate may be slower than that of ADP.

Figure 9-7B shows similar plots for lactate dehydrogenase.[53] In this case, after an initial rise (that is not regarded as significant), the pyruvate*–lactate exchange reaches a high constant value as the amount of pyruvate is increased (with a constant [pyruvate] / [lactate] ratio of 1/35). However, the NAD*–NADH exchange increases rapidly at first but then drops abruptly as the pyruvate and lactate concentrations continue to increase. This suggests an ordered mechanism (Eq. 9-43) in which NAD^+ and NADH represent A and Q, respectively, and pyruvate and lactate represent B and P. As the concentrations of B and P become very high, the

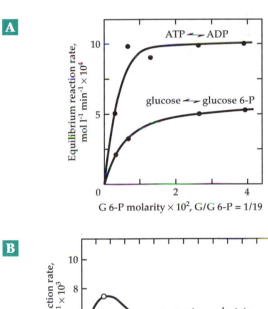

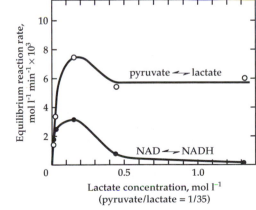

Figure 9-7 (A) Effect of glucose and glucose 6-phosphate concentrations on reaction rate of yeast hexokinase at equilibrium. Reaction mixtures contain 1–2.2 mM ATP, and 25.6 mM ADP at pH 6.5. From Fromm *et al.*[51] (B) Effect of lactate and pyruvate concentrations on equilibrium reaction rates of rabbit muscle lactate dehydrogenase. Reaction mixtures contained 1.7 mM NAD^+, and 30 – 46 μM NADH in Tris-nitrate buffer, pH 7.9, 25°C. From Silverstein and Boyer.[53]

enzyme shuttles back and forth between EA and EQ, but these two complexes rarely dissociate to give free enzyme and A or Q. Hence, the A*–Q exchange rate drops.

In other cases a label may be transferred from A into P or from B into Q. Information on such exchanges has provided a valuable criterion of mechanism which is considered in Chapter 12, Section B,4.

5. Kinetics of Rapid Reactions

The fastest steps in an enzymatic process cannot be observed by conventional steady-state kinetic methods because the latter cannot be applied to reactions with half-times of less than about 10 s. Consequently, a variety of methods have been developed[18,56–59a] to measure rates in the range of 1 to 10^{13} s^{-1}.

Flowing substrates together. One of the first rapid kinetic methods to be devised consists of rapidly mixing two flowing solutions together in a special mixing device and allowing the resulting reaction mixture to move at a rate of several meters per second down a straight tube. At a flow velocity of 10 m s^{-1} a solution will move 1 cm in 10^{-3} s. Observations of the mixture are made at a suitable distance, e.g., 1 cm, and with various flow rates. Using spectrophotometry or other observation techniques, the formation or disappearance of a product or reactant can be followed. The special advantage of this technique is that observation can be made slowly. However, it may require large amounts of precious reactant solutions, e.g., those of purified enzymes.

In the **stopped flow** technique two solutions are mixed rapidly by the flow technique during a period of only 1–2 (or a few) milliseconds. A ram drives the solutions from syringes through a mixing chamber into an observation chamber. After the flow stops light absorption, fluorescence, conductivity, or other property, is measured. A means of rapid observation of changes during the reaction is essential. For example, light absorption may be measured by a photomultiplier with data being collected by a computer. Relaxation times as short as a few milliseconds or less can be observed in this way.[59a,b]

Observing relaxation. Kinetic measurements over periods of tens of microseconds or less can be made by rapidly inducing a small displacement from the equilibrium position of a reaction (or series of reactions) and observing the rate of return (relaxation) of the system to equilibrium. Best known is the **temperature jump** method devised by Eigen and associates. Over a period of about 10^{-6} s a potential difference of ~100 kV is applied across the experimental solution. A rapid electrical discharge from a bank of condensers passes through the solution (without any sparking) raising the temperature 2–10 degrees. All the chemical equilibria for which $\Delta H \neq 0$ are perturbed. If some property, such as the absorbance at a particular wavelength or the conductivity of the solution, is measured, very small relaxation times can be determined.

While it may not be intuitively obvious, if the displacement from equilibrium is small, the rate of return to equilibrium can always be expressed as a first-order process (e.g., see Eq. 9-13). In the event that there is more than one chemical reaction required to reequilibrate the system, each reaction has its own characteristic relaxation time. If these relaxation times are close together, it is difficult to distinguish them; however, they often differ by an order of magnitude or more. Therefore, two or more relaxation times can often be evaluated for a given solution. In favorable circumstances these relaxation times can be related directly to rate constants for particular steps. For example, Eigen measured the conductivity of water following a temperature jump[18] and observed the rate of combination of H^+ and OH^- for which τ at 23°C equals 37 x 10^{-6} s. From this, the rate constant for combination of OH^- and H^+ (Eq. 9-52) was calculated as follows (Eq. 9-53):

$$H^+ + OH^- \longrightarrow H_2O \qquad (9\text{-}52)$$

$$k = 1/\{\tau([OH^-] + [H^+])\} = 1.3 \times 10^{11} \text{ M}^{-1}\text{ s}^{-1} \quad (9\text{-}53)$$

Pressure jump and electric field jump methods have also been used, as have methods depending upon periodic changes in some property. For example, absorption of ultrasonic sound causes a periodic change in the pressure of the system.

Rapid photometric methods. Another useful method has been to discharge a condenser through a flash tube over a period of 10^{-12} to 10^{-4} s, causing a rapid light absorption in a sample in an adjacent parallel tube. Following the flash, changes in absorption spectrum or fluorescence of the sample can be followed. The availability of intense lasers as light sources has made it possible to follow the results of light flashes of 5–10 picosecond duration and to measure extremely short relaxation times (Chapter 23).[58,59]

Some results. Rapid kinetic methods have revealed that enzymes often combine with substrates extremely quickly,[60] with values of k_1 in Eq. 9-14 falling in the range of 10^6 to 10^8 M^{-1} s^{-1}. Helix–coil transitions of polypeptides have relaxation times of about 10^{-8} s, but renaturation of a denatured protein may be much slower. The first detectable structural change in the vitamin A-based chromophore of the light-operated proton pump bacteriorhodopsin occurs in ~5 x 10^{-8} s, while a proton is pumped through the membrane in

~10^{-4} s.[61] Interconversion between chair and boat forms of cyclohexane derivatives may have $\tau \sim 10^{-5}$ s at room temperature, while rotation about a C–N bond in an amide linkage may be very slow with $\tau \sim 0.1$ s. The nonenzymatic hydration of the aldehyde pyridoxal phosphate via Eq. 13-1 occurs with $\tau = 0.01 - 0.1$ s, depending upon the pH.[59]

6. Cryoenzymology

An alternative to studying rapid reactions is to cool enzymes to a subzero temperature (down to $-100°C$) where reactions proceed more slowly.[60,62] The enzyme must be dissolved in a suitable "cryosolvent," often containing 50–80% by volume of an organic solvent or solvents such as methanol, ethanol, dimethyl sulfoxide, dimethyl formamide, or ethylene glycol. Those containing methanol are especially desirable because of their low viscosities. Kinetics can be studied by various spectroscopic methods and stopped-flow, temperature-jump, and other rapid-reaction techniques can be applied. One goal of cryoenzymology is to stabilize otherwise unstable intermediates. X-ray crystallographic measurements can also be made at these low temperatures and it should be possible to observe structures of intermediate ES complexes. A problem is that the forms of the complexes stabilized may be side-products rather than true intermediates. However, as discussed in Chapter 3, a combination of low-temperature Laue X-ray diffraction and a laser-induced temperature jump may be feasible.

7. The Effect of pH on Enzymatic Action

Because proteins contain many acidic and basic groups it is not surprising that the activity of enzymes often varies strongly with pH. However, it is usually found that the state of protonation of only a few groups has a strong effect on activity. This is understandable because most ionized groups are on the outer surface of protein molecules and most are not close to an active site. Protonation or deprotonation of those groups will hardly ever have a major influence on events in the active site. Often only one or two ionizable groups have highly significant effects.

For many enzymes a plot of V_{max} against pH is a bell-shaped curve (Fig. 9-8). The **optimum** rate is observed at some intermediate pH, which is often, but not always, in the range of pH 6–9. This type of curve can be interpreted most simply by assuming the presence in the active site of two ionizable groups and three forms of the enzyme with different degrees of protonation: E, EH, and EH$_2$.

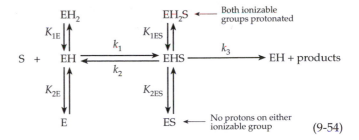

$$(9\text{-}54)$$

Let us designate the acid dissociation constants (K_a) for the two groups in the enzyme as K_{1E} and K_{2E} and those of the ES complex K_{1ES} and K_{2ES}. However, as discussed in Chapter 6, Section E,2 these consecutive K_a values cannot necessarily be assigned to single groups. They may belong to a system of interacting groups. Each may be the sum of more than one microscopic K_a and there may be extensive tautomerism within the active site. The rate constants k_1, k_2, and k_3 define the rates of formation and breakdown of the ES complex.

If it is assumed that only form EHS reacts to give products, the bell-shaped curves of Fig. 9-8 are obtained. The frequent observation of such curves supports the model of Eq. 9-54. It also suggests that the two ionizable groups may be intimately involved in catalysis, one as an acid and the other as a conjugate base (see Section E,5).

For the simple case illustrated in Eq. 9-54, the pH dependence of the initial maximum velocity, the apparent Michaelis constant, and V_{max} / K_m are given by Eqs. 9-55 to 9-57.

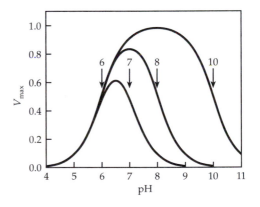

Figure 9-8 Expected dependence of V_{max} on the pH according to Eq. 9-54 with $k_3[E_t] = 1$, $pK_{1ES} = 6$, and pK_{2ES} as given on the graph. After Alberty.[63] Computer-drawn graph courtesy of Carol M. Metzler.

$$V_{max} = \frac{k_3[E]_t}{(1 + [H^+]/K_{1ES} + K_{2ES}/[H^+])} = \frac{k_3[E]_t}{F_{2ES}} \qquad (9\text{-}55)$$

$$K_m = \frac{(k_2 + k_3)}{k_1} \cdot \frac{(1 + [H^+]/K_{1E} + K_{2E}/[H^+])}{(1 + [H^+]/K_{1ES} + K_{2ES}/[H^+])} = K_m F_{2E}/F_{2ES} \qquad (9\text{-}56)$$

$$\frac{V_{max}}{K_m} = \frac{k_1 k_3}{(k_2 + k_3)} \cdot \frac{1}{(1 + [H^+]/K_{1E} + K_{2E}/[H^+])} = \frac{k_3[E]_t}{K_m F_{2E}} \qquad (9\text{-}57)$$

The denominators (which are the Michaelis pH functions given by the first three terms of Eq. 3-6) represent the fraction of enzyme or of ES complex in the monoprotonated state. The pH dependence of enzymatic action is often more complex than that shown in Fig. 9-8 and given by the foregoing equations. However, it is easy to write Michaelis pH functions (see Chapter 3) for enzymes with any number of dissociable groups in both E and ES and to write appropriate equations of the type of

BOX 9-B GROWTH RATES OF CELLS

How do we correctly describe the rate at which a cell grows? Consider bacteria in their rapidly growing "log phase."[a] Each cell divides after a fixed length of time, the **doubling time**. For *E. coli* this may be as short as 17 min in the early stages of growth but becomes somewhat longer as time goes on. A mean value of about 26 min. for *E. coli* at 37°C is typical. In contrast, the doubling time for mammalian cells in tissue culture is often about one day.

If a given volume of culture contains N_0 cells initially, the number N_n after n cell divisions will be:

$$N_n = 2^n N_0$$

From this equation we can calculate that a single bacterium with a generation time of 20 min can produce 2^{144} cells in 48 h of exponential growth. The **exponential growth rate constant k** is equal to *the number of doublings per unit time*. Thus, k is the reciprocal of the doubling time. It is easy to show that the number of bacteria present at time t will be given by the following equation.

$$N_t = 2^{kt} N_0$$

This can be rewritten in a form that can be used to determine k by counting the number of bacteria at zero time and at time t.

$$kt = \log_2(N_t/N_0) = \log_{10}(N_t/N_0)/0.301$$

Another way of expressing the growth is to equate the rate of increase of the number of bacteria with a growth rate constant μ multiplied by the number of bacteria present at that time.

$$dN/dt = \mu N$$

This is a general equation for an autocatalytic reaction and N could be replaced with a concentration, for example, the total content of cellular matter per

liter in the medium. From the two preceding equations the following can be derived.[a]

$$\mu = k \ln 2 = 0.69k$$

When bacteria are transferred to new medium there is usually a lag before exponential growth begins. Exponential growth eventually stops and the culture enters **stationary phase**, which is usually followed by relatively rapid death of cells in the culture. It is often desirable to study cell growth under conditions of **continuous cultivation** in which a constant generation time is maintained but the density of cells in the medium does not increase. This can be done with a simple device known as the **chemostat**.[a,b] A culture vessel containing bacteria is stirred to ensure homogeneity. Fresh culture medium continuously enters the vessel from a reservoir and part of the content of the vessel, suspended bacteria included, is continuously removed through another tube. The bacterial population in the vessel builds up to a constant level and can be maintained at the same level for relatively long periods of time.

Here are some other statistics on cell growth. Bacteria growing exponentially expand their linear dimensions by 1.5 nm and synthesize 1.6×10^7 Da of new cell material in one second. This is equivalent to ~1000 small proteins and includes 23 ribosomes and 3000 base pairs of DNA at each growing point. The much larger HeLa human tumor cell grows by 0.13 nm / s but makes 4.6×10^8 Da/s of material.[c]

[a] Stanier, R. Y., Douderoff, M., and Adelberg, E. A. (1970) *The Microbial World*, 3rd ed., Prentice-Hall, Englewood Cliffs, New Jersey (pp. 298–324)

[b] Smith, H. L., and Waltman, P., eds. (1995) *The Theory of the Chemostat; Dynamics of Microbial Competition*, Cambridge University Press, London and New York

[c] Pollard, E. C. (1973) in *Cell Biology in Medicine* (Bittar, E. E., ed), pp. 357–377, Wiley, New York

Eqs. 9-55 to 9-57. Bear in mind that if the **free substrate** contains groups dissociating in the pH range of interest, a Michaelis pH function for the free substrates will also appear in the numerator of Eq. 9-56. If the pH dependence of the enyzme is regulated by a conformational change in the protein, there may be a cooperative gain or loss of more than one proton and the Michaelis pH function must reflect this fact. This can sometimes be accomplished by addition of a term related to Eq. 7-45. For more information see Dixon and Webb,[64] Cleland,[65,66] or Kyte.[67]

Plots of log V_{max} (or log-specific activity) and $-\log K_m$ or log (V_{max}/K_m) versus pH yield graphs of the type shown in Fig. 9-9. The curved segments of the graphs that extend for about 1.5 pH units on either side of each pK_a are asymptotic to straight lines of slope 1 when a single proton is involved or of a higher slope for multiple cooperative proton dissociation. The straight lines can be extrapolated and intersect at the pK_a values. (However, it is better to fit a complete curve of theoretically correct shape to the points.) Note that the curved line always passes below or above the intersection point at the value of log 2 = 0.30 except in the case of cooperative proton dissociation when it is closer.

Upward turns in the curve of log K_m vs pH correspond to pK_a values in the free enzyme or substrate and downward turns to pK_a values in ES. This approach to analysis of pH dependence has been adopted widely but often incorrectly. For example, many published curves have very sharp bends in which the curved portion covers less than 3 pH units and the curve is much closer than 0.30 to the extrapolated point. This suggests cooperative proton binding and an *apparent* pK_a that is related to $\bar{K}$ of Eq. 7-21. Cooperativity is always a possibility if a conformational change in the protein is involved.[67]

The simple treatment given above is based on the assumption that all proton dissociations are rapid compared to k_{cat}, that enzyme in only one state of protonation binds substrate, and that ES in only one state of protonation yields products. These assumptions are not always valid. It also assumes that both binding and dissociation of substrate are rapid, that is, to use Cleland's terminology the substrate is not "sticky." For a sticky substrate that dissociates more slowly than it reacts to form products ($k_3 > k_2$; Eq. 9-54), the values of pK_{1E} will be lowered and pK_{1E} of Eq. 9-53 will be raised by log $(1 + k_3/k_2)$.[65,66] In addition to the articles by Cleland, other detailed treatments of pH effects have been prepared by Brocklehurst and Dixon[69] and Tipton and Dixon.[70]

Fumarate hydratase (fumarase), which is discussed in Chapter 13, catalyzes the reversible hydration of fumaric acid to malic acid (Eq. 13-11). It was one of the first enzymes whose pH dependence was studied intensively. A bell-shaped pH dependence

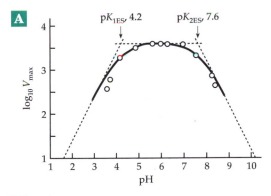

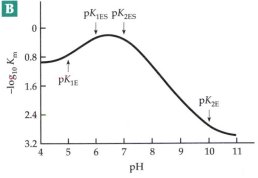

Figure 9-9 (A) Plot of log V_{max} vs pH for a crystalline bacterial α-amylase. From Ono, et al.[68] (B) Theoretical curve of log K_m vs pH for Eq. 9-56 with pK_{1E} = 5, pK_{2E} = 10, pK_{1ES} = 6, and pK_{2ES} = 7. Courtesy of C. Metzler.

for both forward and reverse reactions was observed by Alberty and coworkers[63] and, using Eqs. 9-54 and 9-55, the two apparent pK_a values were measured.

B. Inhibition and Activation of Enzymes

The action of most enzymes is inhibited by many substances. Inhibition is often specific, and studies of the relationship between inhibitor structure and activity have been important to the development of our concepts of active sites and of complementarity of surfaces of biomolecules. Inhibition of enzymes is also the basis of the action of a very large fraction of important drugs. Inhibition may be **reversible** or **irreversible**, the latter leading to permanent inactivation of the enzyme. Often, but not always, irreversible inhibition is preceded by reversible binding of the inhibitor at a complementary site on the enzyme surface.

1. Competitive Inhibitors

Inhibitors with close structural similarities to a substrate tend to bind to the substrate site. In truly competitive inhibition, substrate and inhibitor not only compete for the same site but also their binding is reversible and mutually exclusive. The affinity of the inhibitor for the enzyme is expressed quantitatively through the **inhibition constant** K_i which is the *dissociation constant of the enzyme inhibitor complex EI*:

$$K_i = [E][I] / [EI] \qquad (9\text{-}58)$$

Using the steady state assumption for the mechanism shown in Eq. 9-14, and writing a mass balance equation that includes not only free enzyme and ES but also EI we obtain an equation relating rate to substrate concentration. It is entirely analogous to Eq. 9-15 but K_m is replaced by an apparent Michaelis constant, K'_m:

$$K'_m = K_m \left(1 + \frac{[I]}{K_i} \right) \qquad (9\text{-}59)$$

The relationships between v and [S] and between $1/v$ and $1/[S]$ are as follows:

$$\frac{v}{[S]} = \frac{V_{max}}{K'_m} - \frac{v}{K'_m} \qquad (9\text{-}60)$$

$$\frac{1}{v} = \frac{1}{V_{max}} + \frac{K_m}{V_{max}[S]} \left(1 + \frac{[I]}{K_i} \right) \qquad (9\text{-}61)$$

A commonly used test for competitive inhibition is to plot $1/v$ vs $1/[S]$ (Eq. 9-61), both in the absence of inhibitor and in the presence of one or more fixed concentrations of I. The result, in each case, is a family of lines of varying slope (Fig. 9-10) that converge on one of the axes at the value $1/V_{max}$. We see that the maximum velocity is unchanged by the presence of inhibitor. If sufficient substrate is added, the enzyme will be saturated with substrate and the inhibitor cannot bind. The value of K_i can be calculated using Eq. 9-61 from the change in slope caused by addition of inhibitor.

The effect of a fixed concentration of a competitive inhibitor on a plot of v against log [S] (Fig. 9-11) is to shift the curve to the right, i.e., toward higher values of [S], but without any change in shape (or in the value of V_{max}).

Another plot, introduced by Dixon,[71] is that of $1/v$ versus [I] at two or more fixed substrate concentrations. The student should be able to demonstrate that this plot contains a family of straight lines that intersect at a point to the left of the origin with coordinates $[I] = -K_i$ and $1/v = 1/V_{max}$. This plot may fail to distinguish certain types of inhibition discussed in the next section.[72]

Competitive inhibition is extremely common and has great significance for metabolic control and for the effects of drugs and of poisons. Simple ions are often competitive inhibitors. Since many biochemical sub-

stances carry negative charges, anions such as Cl^-, HCO_3^-, HPO_4^{2-}, and acetate$^-$ frequently act as competitive inhibitors.

Two special classes of competitive inhibitors are characterized by **slow binding** and **slow, tight binding** to active sites.[73–77] Among the very tight-binding inhibitors, which may also be slow to dissociate from active sites, are transition state inhibitors discussed in Section D,1.

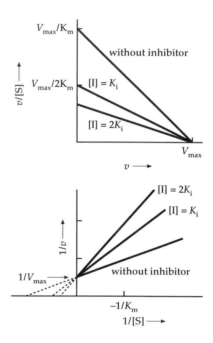

Figure 9-10 Effect of a competitive inhibitor on the Eadie–Hofstee plot (top) and on a double reciprocal plot (bottom). The apparent K_m (Eq. 9-59) is increased by increasing [I], but V_{max} is unchanged.

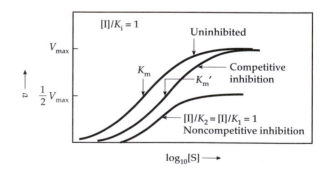

Figure 9-11 Plots of v vs. log [S] for competitive and noncompetitive inhibition.

2. Noncompetitive Inhibition and Activation; Allosteric Sites

If an inhibitor binds not only to free enzyme but also to the enzyme substrate complex ES, inhibition is **noncompetitive**. In this case, S and I do not mutually exclude each other and both can be bound to the enzyme at the same time. Why does such an inhibitor slow an enzymatic reaction? In most instances, the structure of the inhibitor does not show a close similarity to that of substrate, which suggests that the binding of inhibitors is at an **allosteric site**, that is, at a site other than that of the substrate. The inhibition of the enzyme may result from a distortion of the three-dimensional structure of the enzyme which is caused by the binding of the inhibitor. This distortion may be transmitted to the active site even though the inhibitor binds far from that site. In some cases two distinctly different conformers of the protein may exist, one binding substrate well and the other binding inhibitor well (Fig. 9-12). In other instances the bound inhibitor may interfere with the catalytic action by partially overlapping the active site. In either case the ES complex reacts to give product in a normal way, but the ESI complex reacts more slowly or not at all.

Binding of a substance to an allosteric site sometimes has the effect of *increasing* the activity of an enzyme rather than inhibiting it. This may occur because the **activator** stabilizes the conformation that binds substrate best (Fig. 9-12). The quantitative treatment of such activation is similar to that of inhibition; allosteric inhibitors and activators are often considered together and are referred to as **modifiers** or

---

BOX 9-C THE SULFONAMIDES AS ANTIMETABOLITES

The development of the "sulfa drugs,"[a-c] derivatives of **sulfanilamide**, originated with studies of the staining of protozoal parasites by Paul Erhlich. In 1932 it was shown that the red dye 2,4-diamino-azobenzene-4'-sulfonamide (Prontosil) dramatically cured systemic infections by gram-positive bacteria. Subsequent studies revealed that bacteria converted

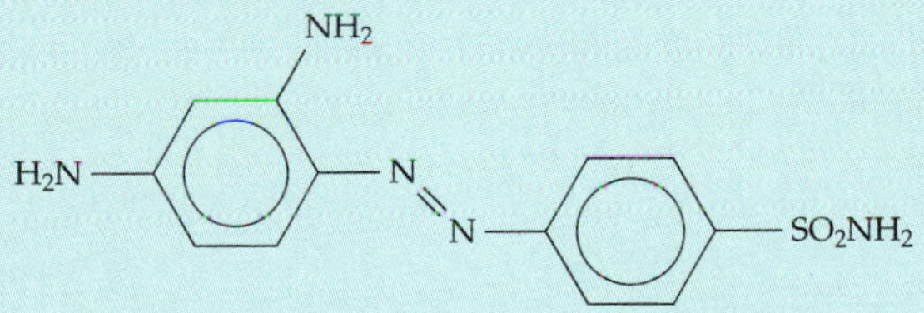

Prontosil

the azo dye to sulfanilamide, a compound with strong bacteriostatic activity; that is, it inhibited bacterial growth without killing the bacteria. Although sulfanilamide had been used in large quantities since 1908 as an intermediate in synthesis of dyes, its potential as an antibacterial agent had not been recognized.

In 1935, D. D. Woods found that the growth inhibition of sulfanilamide was reversed by yeast extract.[d] From this source, in 1940 he isolated **p-aminobenzoic acid** and demonstrated that the inhibitory effect of 3×10^{-4} M sulfanilamide was overcome by 6×10^{-8} M p-aminobenzoate. The relationship between the two compounds was shown to be strictly competitive. If the sulfanilamide concentration was doubled, twice as much p-aminobenzoate was required to reverse the inhibition as before. These facts led to the formulation by Woods and by P. Fildes (in 1940) of the

antimetabolite theory.[d,e] It was proposed that p-aminobenzoate was needed by bacteria and that sulfanilamide competed for a site on an enzyme designed to act on p-aminobenzoate. We now know that the idea was correct and that the enzyme on which the competition occurs catalyzes the synthesis of dihydropteroic acid (Fig. 25-19), a precursor to folic acid.

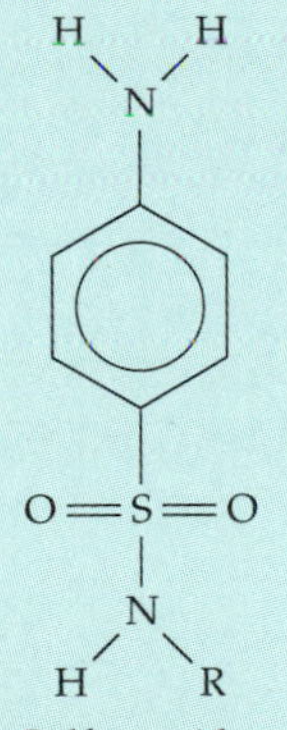

Sulfonamides
R=H in sulfanilamide

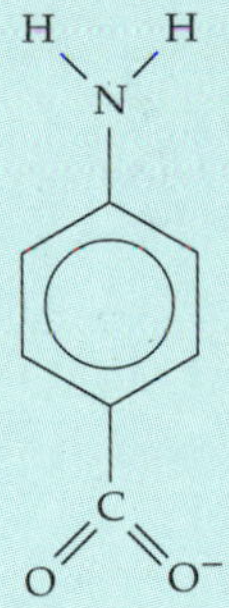

p-Aminobenzoate

While sulfanilamide itself is somewhat toxic, a variety of related drugs of outstanding value have been developed. Over 10,000 sulfonamides and related compounds have been tested for antibacterial action.

[a] Bardos, T. J. (1974) *Top. Curr. Chem.* **52**, 63–98
[b] Gale, E. F., Cundliffe, E., Reynolds, P. E., Richmond, M. H., and Waring, M. J. (1972) *The Molecular Basis of Antibiotic Action*, Wiley, New York
[c] Shepherd, R. G. (1970) in *Medicinal Chemistry*, 3rd ed. (Burger, A., ed), pp. 255–304, Wiley (Interscience), New York
[d] Woods, D. D. (1940) *Brit. J. Exp. Pathol.* **21**, 74–90
[e] Fildes, P. (1940) *Lancet* **1**, 955–957

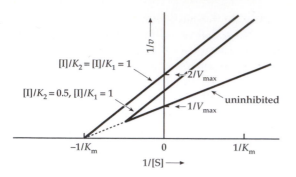

Figure 9-12 Double reciprocal plots for two cases of non-competitive inhibition.

effectors. A general scheme[78,79] is given by Eq. 9-62.

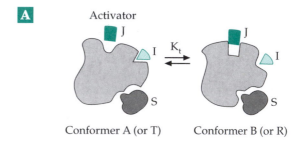

$$\text{(9-62)}$$

Here K_1 and K_2 are equilibrium constants for dissociation of M from EM and ESM, respectively, while K_{ds} and K'_{ds} are the dissociation constant of ES and of ESM (to S and EM), respectively. Notice that, K'_{ds} is not independent of the others (Eq. 9-63):

$$K'_{ds} = K_{ds}K_2/K_1 \qquad \text{(9-63)}$$

We have already considered competitive inhibition which is obtained when $K_2 = 0$ (and therefore $K'_{ds} = 0$). For this case, M is always an inhibitor and no activation is possible. Notice that the inhibition will appear competitive even if M binds at an allosteric site as in Fig. 9-13 or if the inhibited form does not react at all with substrate. Noncompetitive inhibition will be observed if ESM is formed but does not react, i.e., if $k_4 = 0$. Then the rate equation in reciprocal form will be given by Eq. 9-64.

$$\frac{1}{v} = \frac{1}{V_{max}}\left(1 + \frac{[I]}{K_2}\right) + \frac{K_m}{V_{max}[S]}\left(1 + \frac{[I]}{K_1}\right) \qquad \text{(9-64)}$$

We see that $1/V_{max}$ is multiplied by a term containing [I] and K_2. It is a characteristic of noncompetitive inhibition that the maximum velocity is decreased from that observed in the absence of the inhibitor. Also, no matter how high the substrate concentration,

inhibition cannot be reversed completely.

Figure 9-12 shows a plot of $1/v$ against $1/[S]$ at a series of fixed values of [I]. For the case that $K_1 = K_2$ **(classical noncompetitive inhibition)**, a family of reciprocal plots that intersect on the horizontal axis at a value of $-1/K_m$ is obtained. On the other hand, if K_1 and K_2 differ (the general case of noncompetitive inhibition), the family of curves intersect at some other point to the left of the vertical axis and, depending upon the relative values of K_1 and K_2, either above or below the horizontal axis. The example illustrated is for $K_2 = 0.5K_1$; that is, for the binding of M to ES being twice as strong as that to E.

Figure 9-11 shows inhibition data for both the noncompetitive and the competitive cases plotted vs log[S]. The shift of the midpoint to the right in each case reflects the tendency of the inhibitor to exclude the substrate from binding, while the lowered value of the maximum velocity in the case of noncompetitive inhibition results from the failure of the substrate to completely displace the inhibitor from the enzyme

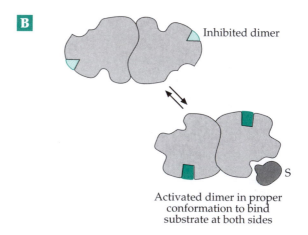

Figure 9-13 (A) An enzyme with binding sites for allosteric inhibitor I and activator J. Conformer A binds inhibitor I strongly but has little affinity for activator J or for substrate S. Conformer B binds S and catalyzes its reaction. It also binds activator J whose presence tends to lock the enzyme in the "on" conformation B. Conformers A and B are designated T and R in the MWC model of Monod, Wyman, and Changeux.[80] (B) Inhibited and activated dimeric enzymes.

even as [S] becomes very high. If an inhibitor binds only to ES and not to E, i.e., $K_1 = 0$, a family of parallel double-reciprocal plots of $1/v$ vs $1/[S]$ will be obtained. This case is referred to as **uncompetitive inhibition**. Multisubstrate enzymes with either ordered sequential or ping-pong mechanisms also often give parallel line plots with inhibitors.

Uncompetitive inhibitors of liver alcohol dehydrogenase (Chapter 15) could be used to treat cases of poisoning by methanol or ethylene glycol.[81–83] The aim is to prevent rapid oxidation to the toxic acids HCOOH and HOCH$_2$COOH, which lower blood pH, while the alcohols are excreted. Uncompetitive inhibitors have an advantage over competitive inhibitors as therapeutic agents in that the inhibition is not overcome when the substrate concentration is saturating.[84]

If rate constant k_4 of Eq. 9-62 is not zero we may observe either inhibition or activation. If $k_3 = k_4$ the effect of a modifier will be to alter the apparent K_m by either increasing it (inhibition) or decreasing it (activation). The maximum velocity will be unchanged. Monod *et al.*[80] referred to enzymes showing such behavior as **K systems**. On the other hand, if $K_1 = K_2$ and k_3 differs from k_4 we have a purely **V system**. In the general case a modifier affects both the apparent K_m and V_{max}. Although in the foregoing discussion it has been assumed that activators bind at allosteric sites, as was pointed out in Section A,11, ions and other small molecules (substrate surrogates) that act as competitive inhibitors at high concentrations may be activators at low concentrations.[40]

Activation of enzymes by specific metallic ions is often observed. In many instances the metallic ion is properly regarded as a *second substrate* which must bind along with the first substrate before reaction can occur. Alternatively, the complex of the organic substrate with the metal ion can be considered the "true substrate." Thus, many enzymes act upon the magnesium complex of ATP (Chapter 12). The enzymes can either be regarded as three-substrate enzymes requiring Mg^{2+} + ATP^{4-} and another substrate or as two-substrate enzymes acting upon ATPMg^{2-} and a second substrate.

3. Inhibitors in the Study of Mechanisms

A substrate analog will frequently inhibit only one of the two forms of a multisubstrate enzyme with a ping-pong mechanism.[1,72] Reciprocal plots made for various inhibitor concentrations consist of a family of parallel lines reminiscent of uncompetitive inhibition. Observation of such parallel line plots can support a ping-pong mechanism for an enzyme but cannot prove it because in some cases parallel lines are observed for inhibition of enzymes acting by an ordered sequential mechanism. The following question arises naturally for any ordered bimolecular reaction (Eq. 9-43): Of the

two substrates required by the enzymes, which one binds to the enzyme first? If the concentration of one substrate is kept constant while varying concentrations of an inhibitory analog of that substrate are added and $1/v$ is plotted against the reciprocal of the concentration of the other substrate, parallel lines are obtained if, and only if, the substrate of fixed concentration is B, the substrate is adding second in the binding sequence, *and* if I is its analog. The substrate binding first (A) is the one whose concentration was varied in the experiment.

Product inhibition (Section A,12) can also provide information about mechanisms. For example, if $1/v$ is plotted against $1/[A]$ in the presence and absence of the product Q, the product will be found to compete with A and to give a typical family of lines for competitive inhibition. On the other hand, a plot of $1/v$ vs $1/[B]$ in the presence and absence of Q will indicate noncompetitive inhibition if the binding of substrates is ordered (Eq. 9-43). In other words, only the A–Q pair of substrates are competitive. Product inhibition is also observed with enzymes having ping-pong kinetics (Eq. 9-47) as a result of formation of nonproductive complexes.

4. Allosteric Effectors in the Regulation of Enzyme Activity

The binding of a substance at an allosteric site with the induction of a conformational change forms the basis for many aspects of regulation. The term **allostery** (allosterism) usually refers to the effects of allosteric modifiers, which may be either inhibitors or activators, on oligomeric enzymes. However, as we have already seen (Eq. 9-62), monomeric enzymes may also be subject to allosteric regulation by modifiers. Consider a monomer that contains binding sites for substrate, inhibitor, and activator and which exists in conformations A and B as in Fig. 9-13. Let us assume (see Eq. 7-31) that conformer B binds both substrate and activator well but that it binds inhibitor poorly or not at all. On the other hand, A binds inhibitor well but binds substrate and activator poorly. This simple combination of two conformers with different binding properties provides a means by which enzymes can be turned "on" or "off" in response to changing conditions.

If an inhibitory substance builds up to a high concentration within a cell, it binds to conformer A; if the inhibitor concentration is high enough, virtually all of the enzyme will be locked in the inactive conformation A. The enzyme will be turned off or at least reduced to a low activity. On the other hand, in the presence of a high concentration of activator the enzyme will be turned on because it is locked in the B conformation. The relative concentrations of inhibitor, activator, and substrate within a cell at any given time will determine what fraction of the enzyme is in active conformation B. It is this interplay of inhibitory and activating effects that provides the basis for much of the regulation of cell chemistry.[80,85]

The effects of inhibitors or activators on the kinetics of the monomeric enzyme of Fig. 9-13 can be described by Eq. 9-62 to 9-64. Separate terms for both inhibition and activation can be included. The equilibrium between the two conformers can also be indicated explicitly according to Eq. 7-30. However, for monomeric enzymes it is usually not profitable to try to separate the two constants K_t and K_{BX} which describe the conformational change and binding of substrate or activator, respectively, in Eq. 7-30.

Most intracellular enzymes are oligomeric, and the binding of allosteric effectors leads to additional interesting effects. Binding constants or dissociation constants must be defined for both inhibitor and activator to both conformers A and B. Since all species must be taken into account in the mass balance, the equations are complex. However, the Monod–Wyman–Changeux (MWC) model (Chapter 7) gives a relatively simple picture. The saturation curve for an oligomeric enzyme following this model may be derived from Eq. 7-39 and is given by the following expression:

$$Y = \frac{Lc\alpha(1 + c\alpha)^{n-1} + \alpha(1 + \alpha)^{n-1}}{L(1 + c\alpha)^n + (1 + \alpha)^n} \qquad (9\text{-}65)$$

Here, L is the **allosteric constant** which is given (for a dimer) by Eq. 7-36. The constant c is the ratio of dissociation constants K_{BS} and K_{AS} for the two conformers:

$$c = K_{BS} / K_{AS} \qquad (9\text{-}66)$$

Notice that in this chapter *dissociation* constants (K_d) of ES complexes are being used, whereas the equations of Chapter 7 are all written in terms of association constant (K_f). The parameter α is defined as follows, where K_{BS} equals $1/K_{BX}$ of Chapter 7.

$$\alpha = [S] / K_{BS}) \qquad (9\text{-}67)$$

To take account of effects of inhibitor and activator the *ratios* of dissociation constants of I from BI and AI and of activator J from BJ and AJ are defined as in Eq. 9-68. Likewise, "normalized concentrations" of I and J are defined (Eq. 9-69) as β and γ, respectively.

$$d = \frac{K_{BI}}{K_{AI}} > 1 \qquad e = \frac{K_{BJ}}{K_{AJ}} < 1 \qquad (9\text{-}68)$$

$$\beta = [I]/K_{AI} \qquad \gamma = [J]K_{BJ} \qquad (9\text{-}69)$$

According to the MWC model, in the presence of inhibitor and activator at normalized concentrations β and γ an enzyme will still follow Eq. 9-65, but the allosteric constant L will be replaced by an apparent allosteric constant L' (Eq. 9-70).[86] Figure 9-14 shows plots of Y vs. log α for two different values of L' for a tetramer with a specific value assumed for c. In both

$$L' = L\left[\frac{(1 + \beta d)(1 + \gamma e)}{(1 + \beta)(1 + \gamma)}\right]^n \qquad (9\text{-}70)$$

cases, Y approaches 1 as log α increases, but since we are dealing with noncompetitive inhibition at high values of L', much of the enzyme will be in the T(A) conformation at saturation.

Noncompetitive inhibition cannot be completely reversed by very high substrate concentrations. Monod *et al.* defined for an allosteric enzyme a **function of state $\overline{R}$** (Eq. 9-71) which is the fraction of total enzyme in the $\overline{R}$ (B) conformation:

$$\overline{R} = \frac{(1 + \alpha)^n}{L(1 + c\alpha)^n + (1 + \alpha)^n}$$
$$= \text{function of state} \qquad (9\text{-}71)$$

In a K system (Section B,2) it is the value of $\overline{R}$ that determines the velocity with which an enzyme reacts. Figure 9-14 also shows $\overline{R}$ as a function of log [α]. Note that when L' is low $\overline{R}$ does not approach zero even when [S] $\rightarrow$ 0. In other words, the enzyme is never completely turned off, just as when L' is high the enzyme is never completely turned on.

Figure 9-14 may be compared with Fig. 9-11 which shows similar curves for noncompetitive inhibition of a monomeric enzyme. The significant difference between the two figures is that saturation of the oligomeric enzyme occurs over a narrower concentration range than does that of the monomer, i.e., saturation of the oligomeric enzyme, especially in the presence of inhibitor, is *cooperative*. Note that cooperative binding of substrate requires that the free enzyme be largely in conformation T (A), as it is in the presence of an inhibitor. Allosteric interactions between two identical molecules, whether of substrate or of effector, are described by Monod *et al.*[80] as **homotropic interactions**. Such interactions lead to cooperativity or anticooperativity in binding. Allosteric interactions between two different molecules, e.g., a substrate and an activator are designated **heterotropic**.

For many enzymes the MWC model is unrealistically simple. The more general treatment of binding equilibria given in Chapter 7 may be applicable. However, in addition to K systems there are V systems in which a conformational change alters the maximum velocity (see Eq. 9-62)[87] and sometimes both substrate

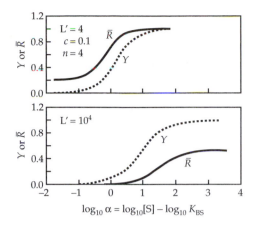

Figure 9-14 Fractional saturation Y and "function of state" $\bar{R}$ for hypothetical tetrameric enzymes following the MWC model. Curves are calculated for two different values of the apparent allosteric constant L' (Eq. 9-70) and for $c = 0.1$ (Eq. 9-66). After Rubin and Changeux.[86]

affinity and maximum velocity.

The fact that data can be fitted to an equation is not proof that a mathematical model is correct; other models may predict the results just as well. For example, Jenkins has shown that the presence of ions that act as substrate surrogates (Section A,11) can produce cooperative or anticooperative binding curves for substrates.[40,88] Rabin[89] suggested the following explanation of cooperativity for a monomomeric enzyme with a single substrate-binding site. In its active conformation E the enzyme reacts with substrate rapidly and the ES complex yields product rapidly as in the upper loop of Eq. 9-72. However, a slow conformational change interconverts E and E', a less active form with much lower affinity for substrate.

$$S \longrightarrow ES \longrightarrow E + P$$
$$S \longrightarrow ES' \quad \text{slow}$$
$$E' \text{ (inactive)} \qquad (9\text{-}72)$$

At the same time, the complex ES', if formed, can equilibriate with ES and thereby alter the conformational state of the protein. At low substrate concentrations E' will predominate and the enzymatic activity will be low. At high substrate concentrations most E will exist as ES and after release of product will tend to remain in the active conformation long enough to bind and act on another substrate molecule. This and other kinetic models of cooperativity are discussed by Newsholme and Start.[90]

The physiological significance of cooperative binding of substrates to enzymes is analogous to that of cooperative binding of oxygen by hemoglobin which provides for more efficient release of oxygen to tissues. However, in the presence of excess activator an enzyme is locked in the *R* (B) conformation and no cooperativity is seen in the binding of substrate. In this case, each binding site behaves independently. On the other hand, *there will be strong cooperativity in the binding of the activator*. The result is that control of the enzyme is sensitive to a higher power than the first of the activator concentration. Likewise, the turning off of the enzyme is more sensitive to inhibitor concentration as a result of cooperative binding of the latter. It seems likely that the evolution of oligomeric enzymes is at least partly a result of the greater efficiency of control mechanisms based on cooperative binding of effectors.

5. Irreversible Inhibition of Enzymes

Some of our most effective drugs, such as penicillin (Box 20-G) are irreversible inhibitors of specific enzymes. Such inhibitors are also of practical importance to biochemists who wish to inhibit specific enzymes such as proteases that might otherwise destroy proteins they are studying. Transition state analogs (Section E,1) and slow-binding inhibitors (Section B,1) may appear to inactivate an enzyme irreversibly because they bind tightly and dissociate slowly. However, true irreversible inhibitors, such as oxidizing agents or alkylating agents, cause a more permanent chemical modification, usually at or near the active site. Two groups of irreversible inhibitors are of special interest: affinity labeling agents and enzyme-activated inhibitors.

Affinity labeling agents (active-site directed inactivating reagents) have two essential properties: *a high affinity for the active site* of the specific enzyme to be inhibited and *the presence of a chemically highly reactive group* which can attack a functional group in a protein. A good example is provided by derivatives of chloromethylketones, which are reactive alkylating agents. One of these, *N*-tosylphenylalanine chloromethyl ketone (TPCK), is a potent inhibitor of chymotrypsin. In addition to the chloromethyl ketone group it contains

L-*N*-tosylphenylalanine chloromethylketone (TPCK)

the phenylalanine side chain, which helps direct it into the substrate binding site of chymotrypsin, and a sulfonamide linkage that mimics the normal amide linkages of the substrate. The group Y, which becomes alkylated in an S_N2-like displacement of a chloride ion (see Eq. 9-76) is thought to be His 57 of the active site (see Fig. 12-10). For additional discussion see Walsh,[91] Kyte,[67] and Plapp.[91a]

Enzyme-activated inhibitors (also called suicide substrates, k_{cat} inhibitors, or mechanism-based inhibitors) are chemically inert until they enter the active site of their target enzymes. Then, by passing through at least some of the normal stages of the catalytic action of the enzyme, they are converted to reactive intermediates that can become irreversibly bound to an enzyme.[92,93] For example, a halogen atom (F, Cl, Br) together with a proton may be eliminated from an intermediate to give an unsaturated compound to which a nucleophilic side chain of the protein adds. Several of these inhibitors are discussed in later chapters. Because of their high specificity many enzyme-activated inhibitors are potential drugs. One of them, α-difluoromethylornithine (Box 14-C), is said to bind covalently to only one protein in the body of a rat, namely, ornithine decarboxylase, the target enzyme for the drug.

C. The Specificity of Enzymatic Action

Enzymes are usually impressively specific in their action. The specificity toward substrate is sometimes almost absolute. For many years urea was believed to be the *only* substrate for the enzyme **urease** and succinate the only substrate for **succinate dehydrogenase**. Even after much searching for other substrates, only

one or two closely related compounds could be found that were acted on at all. In other cases enzymes can use a class of compounds as substrates. For example, the **D-amino acid oxidase** of kidney oxidizes a variety of D-amino acids but does not touch L-amino acids.

Almost as impressive as the substrate specificity of enzymes is the specificity for a given type of reaction. Many substrates are capable of undergoing a variety of different chemical reactions, either unimolecular or with water or some other compound present in the cell. The enzyme catalyzes only one of these reactions.

Although side reactions may occur to a small extent, the most impressive thing in comparing an enzyme-catalyzed reaction with an uncatalyzed organic reaction is that the latter often produces large amounts of side reaction products, but the enzymatic reaction does not.

1. Complementarity of Substrate and Enzyme Surfaces

Impressed by the specificity of enzymatic action, biochemists early adopted a "lock-and-key" theory which stated that for a reaction to occur the substrate must fit into an active site precisely. Modern experiments have amply verified the idea. A vast amount of kinetic data on families of substrates and related competitive inhibitors support the idea and numerous X-ray structures of enzymes with bound inhibitors or with very slow substrates have given visual evidence of the reality of the lock-and-key concept. Directed mutation of genes of many enzymes of known three-dimensional structure has provided additional proof.

As anticipated, hydrophobic groups of substrates or inhibitors usually contact hydrophobic regions of the protein and the fit is tight. For example, the active site of chymotrypsin contains a "hydrophobic pocket" designed to hold a large hydrophobic side chain, thus providing the specificity observed with this enzyme (Table 3-2). Likewise, polar groups of substrates contact polar groups of the enzyme. The interactions are complementary, positive charges fitting against negative and with correctly formed hydrogen bonds. Trypsin, whose structure is similar to that of chymotrypsin, exerts its specificity for a positively charged side chain next to the bond cleaved (Table 3-2) by virtue of the presence of a negatively charged carboxylate at the bottom of the hydrophobic pocket. Several C=O and N–H groups of the peptide linkages in the substrate form hydrogen bonds to the edge of a β sheet in the protein, in effect making the substrate an added β strand in the sheet (Fig. 12-10). Aspartate aminotransferase acts on the dicarboxylic amino acids glutamate and aspartate. A pair of arginine side chains bind the two carboxylates of the substrate while the $-NH_3^+$ of the substrate is attracted to a negatively charged group in the coenzyme pyridoxal phosphate, which is also present in the active site (Eq. 14-39, Fig. 14-10).

2. Stereospecificity and Prochiral Centers

Most enzymes possess an infallible ability to recognize the difference between the right side and the left side of an organic substrate even when the latter has perfect bilateral symmetry. In fact, this ability is limited to **prochiral centers** of molecules and is a natural consequence of their reaction with the chiral enzyme.[94-98] Consider carbon atom number 1 of ethanol. The two attached hydrogen atoms are chemically identical and would react identically with a nonchiral reagent. Nevertheless, these atoms are no more equivalent stereochemically than are your right and your left arms. We say that the molecule has a prochiral center at C-1. A prochiral center on a tetrahedrally bonded carbon atom always contains two identical atoms or groups but a total of three different kinds of atoms or groups. If the priority of this H is elevated, e.g., by substitution of ^{2}H for ^{1}H the configuration would be R (the priority of this ^{2}H would become **c**).

The two hydrogen atoms on C-1 of ethanol are called **enantiotopic** because replacement of one or the other by a fourth different kind of atom or group would produce a pair of enantiomers (page 41). This fact suggests a way of naming the positions occupied by two enantiotopic atoms or groups. We first assign priorities to all of the atoms or groups attached to the central carbon atom according to the RS system. Now, we ask whether the configuration will be R or S when the priority of one of the two identical groups is raised, e.g., by substitution of one of the hydrogen atoms by deuterium. If the configuration becomes R, that group occupies a pro-R position; if the configuration becomes S, it occupies a pro-S position. Referring to the preceding diagram, it is easy to see (by viewing down the bond to the group of lowest priority and applying the usual rule for determining configuration) that if the pro-R hydrogen (H_R) is replaced by deuterium (^{2}H), the configuration will be R. Conversely, replacement of H_s by deuterium will lead to the S configuration.

When ethanol is oxidized by the action of **alcohol dehydrogenase** (Eq. 9-73), only the pro-R hydrogen atom is removed. If the reaction is reversed in such a way that deuterium is introduced into ethanol from the reduced coenzyme the optically active R-2-deuterio-ethanol is formed. The ability of an enzyme to

$$(9\text{-}73)$$

recognize a single hydrogen of a pair of hydrogens on a CH_2 group was at first a surprise to many biochemists.[99] However, it is a natural result of the complementarity of enzyme and substrate surfaces, just as the fit of a shoe is determined by the complementarity of surfaces of foot and shoe. Only a chiral catalyst can have this ability.

Another example is provided by malic acid, a chiral molecule which also contains a prochiral center (see Eq. 9-74). In this case replacement of the pro-R or pro-S hydrogen atom by another atom or group would yield a pair of diastereoisomers rather than enantiomers. Therefore, these hydrogen atoms are **diastereotopic**. When L-malic acid is dehydrated by fumarate hydratase (Chapter 13) the hydrogen in the pro-R position is removed but that in the pro-S position is not touched. This can be demonstrated by allowing the dehydration product, fumarate, to be hydrated to malate in ^2H_2O (Eq. 9-74). The malate formed contains deuterium in the pro-R position. If this malate is now isolated and placed with another portion of enzyme in H_2O, the deuterium is removed cleanly. The fumarate produced contains no deuterium.

BOX 9-D   RECEPTORS, AGONISTS, AND ANTAGONISTS

Inhibition of enzymes provides the basis for many of the effects of antibiotics and other chemotherapeutic substances (e.g., see Box 9-B). However, some drugs act on cell surface **receptors** which have ordinarily not been regarded as enzymes. According to receptor theory, developed around 1937, structurally similar drugs often elicit similar responses because they bind to the same receptor. A receptor may normally bind a hormone, neurotransmitter, or other metabolite whose geometry is partially shared by a drug. Binding of drugs of one class, termed **agonists** in the pharmacological literature, to an appropriate receptor triggers a response in a cell, similar to that of a hormone. On the other hand, compounds of related structure often act as **antagonists**, binding to receptor but failing to elicit a response. Agonist and antagonist often act in a strictly competitive fashion as in competitive inhibition of enzyme action.

We know now that many receptors *are* enzymes, some of which may act quite slowly. The active site may sometimes be far from the receptor binding site and sometimes in a separate subunit. The receptor can be viewed as an allosteric effector which binds at a distant site or as a ligand for a regulatory subunit of the enzyme complex. Alternatively, the active site may be viewed as the site for relaying a signal received from the hormone or other agonist.

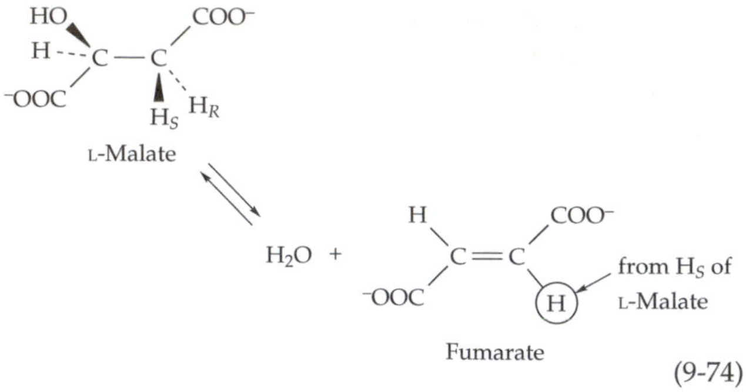

L-Malate

$$H_2O + \quad \text{Fumarate}$$

from H_S of L-Malate

(9-74)

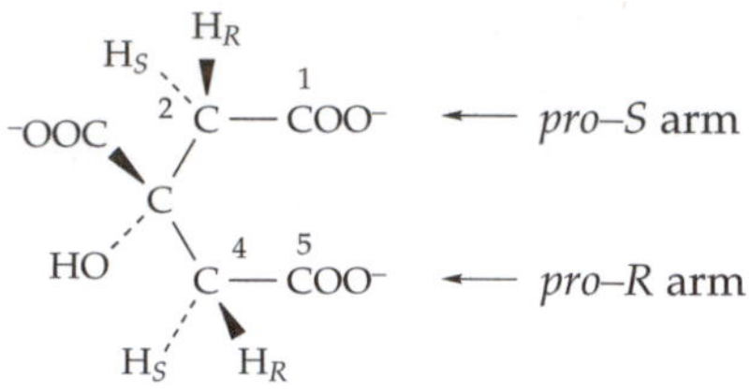

Citrate ion numbered according to the stereochemical system

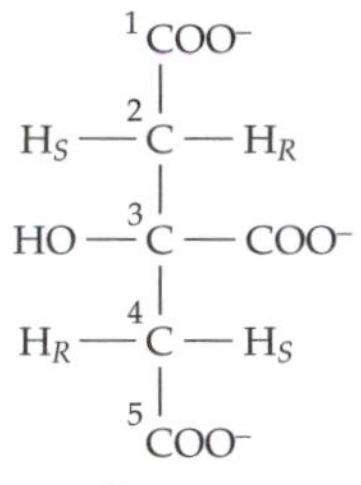

Drawn using the Fisher convention

Citrate

Stereochemical numbering. In some prochiral molecules, such as glycerol, the two ends of the main carbon chain form identical groups. Since the two ends are distinguishable to an enzyme, it is important to decide which should be labeled C-1 and which C-3. Hirschmann proposed a stereochemical numbering system[100] in which the carbons are numbered beginning with the end of the chain that occupies the *pro-S* position.

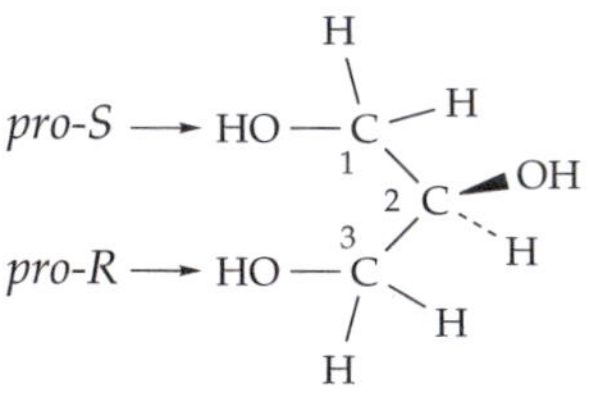

Stereochemical numbering for a symmetric prochiral molecule. Carbons are numbered beginning with the end of the chain that occupies the *pro-S* position.

In this numbering system derivatives of the parent prochiral compound are given the prefix *sn-*. Thus, glycerol phosphate, used by cells to construct phospholipids, usually bears a phosphate group on the $-CH_2OH$ in the *pro-R* position of glycerol and is therefore *sn*-3-glycerol phosphate.

The **citrate ion**, a very important prochiral metabolic intermediate, has three prochiral centers at C-2, C-3, and C-4, respectively. That at C-3 distinguishes the *pro-R* and *pro-S* arms and determines the stereochemical numbering. Citrate containing ^{14}C in the *sn*-1 position is called *sn*-citrate[1-^{14}C] and is the form of labeled citrate that is synthesized in living cells from oxaloacetate and [1-^{14}C]acetyl coenzyme A (see Fig. 10-6). The first step in the further metabolism of citrate is the elimination of the −OH group from C-3 together with the H_R proton from C-4 through the action of the enzyme **aconitate hydratase** (aconitase). In this case the proton at C-4 (in the *pro-R* arm) is selected rather than that at C-2.

Trigonal prochiral centers. Planar trigonal atoms, such as those of aldehydes, ketones, and alkenes, are also prochiral if they are attached to three different kinds of atoms or groups. The two faces are enantiotopic or diastereotopic, if another chiral center is present. Hanson[94,94a] proposed that the faces be named as indicated in Eq. 9-75. The trigonal atom is viewed from one side and the three groups surrounding the carbon atom are given priorities, *a*, *b*, and *c*, as in the *RS* system (Chapter 2). If the sequence *a*, *b*, and *c* of priorities is clockwise, the face toward the reader is *re* (*rectus*); if counterclockwise, it is *si* (*sinister*). Priorities can often be assigned on the basis of the atomic numbers of the first atoms in the three groups (as in the example shown)

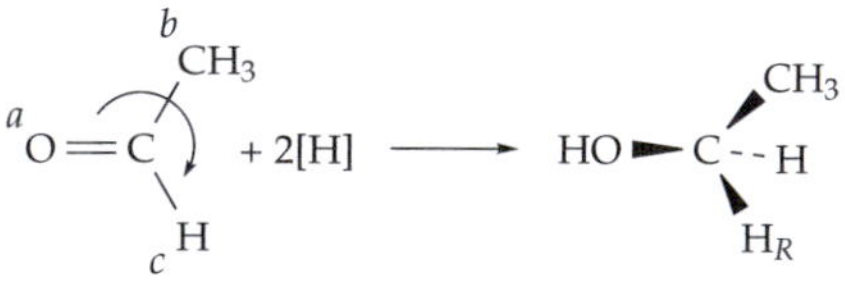

re (or *Re*) face is toward the reader
si (or *Si*) face is behind the page

(9-75)

but, if necessary, replica atoms must be added as described in Chapter 2. Replica atoms attached to an originally trigonal carbon are ignored completely, but those attached to O or N of C=O or C=N may be required to establish the priorities of the groups around the carbon atom. For C=O and C=N the faces of the O and N atoms are taken to be the same as that of the attached trigonal carbon atom. In C = C the two carbons may have their *re* faces either on the same side or on opposite sides of the group.

Notice that addition of an atom of ^2H to the *re* face of acetaldehyde would give *R*-deuteroethanol (Eq. 9-75; reverse of Eq. 9-73). This is the reaction catalyzed by alcohol dehydrogenase. Addition of ^1H to the *re* face places the entering hydrogen in the *pro-R* position. Addition to the *si* face would place it in the *pro-S* position. Fumarate has two trigonal carbon atoms. The *si* faces of both are toward the viewer in the following structure (also shown in Eq. 9-74). Referring to Eq. 9-74, we see that an HO⁻ ion from water adds to the *si* face of one carbon atom of fumarate to give *s*-malate (L-malate). At the same time, a proton combines with the *re* face of the adjacent unsaturated carbon atom to enter the *pro-R* position.

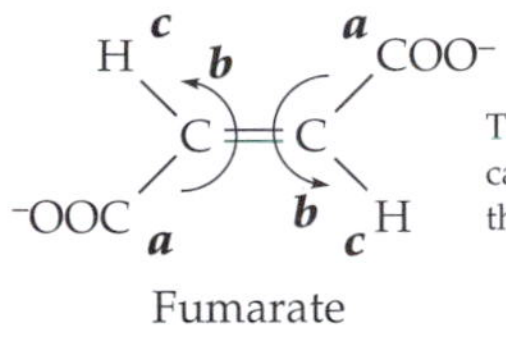

The *si* faces of both trigonal carbon atoms are toward the viewer.

Fumarate

3. Induced Fit and Conformational Changes

Results of many X-ray studies indicate that the lock-and-key picture of enzyme action must be modified. If an enzyme is a "lock" and the substrate the "key," *the entrance of the key into the lock often induces a conformational change in the protein.* Binding of substrates may be imperfect in the initially formed ES complex but may be more nearly perfect a few nanoseconds or microseconds later as the protein readjusts its structure to accommodate the substrate. The substrate has induced a fit. For example, when an amino acid substrate binds to aspartate aminotransferase one whole domain of the enzyme moves inward, packing hydrophobic side chains of the protein against the substrate (Chapter 14). This strengthens the electrostatic interactions between the ion pairs that orient the substrate and align it for reaction. Similar conformational changes have been observed for citrate synthase, glycogen phosphorylase, various kinases (Chapter 12), alcohol dehydrogenase (Chapter 15), and a growing list of other enzymes. Accompanying changes in circular dichroism, ultraviolet spectra, and sedimentation constants are often observed.

For many other enzymes the observed conformational changes are subtle. A single loop of polypeptide chain or even of a side chain moves to cover the bound substrate. With trypsin and other serine proteases a flexible segment of the peptide chain becomes immobilized and forms more hydrogen bonds after substrate binds than before. It is probably quite unusual for a major unfolding and refolding of parts of a protein to take place. The term induced fit usually refers to substrate binding, but as substrates are interconverted within active sites successive changes in geometry and in charge distribution occur. Small conformational changes may be required at several stages of an enzymatic reaction to ensure that complementarity of substrate and enzyme is preserved.[101] In line with this idea are the facts that proteins are less tightly packed at active sites than in other parts of the molecules, and that active sites often lie between domains and usually are formed by several loops of the peptide chain (Chapter 2).[102]

Although conformational changes allow proteins to maintain good complementarity with substrates, it does not follow that substrates are therefore bound very tightly. This is easiest to understand for reversible reactions in which a substrate is the product of the reverse reaction. Not only binding but also the rate of release of product must be rapid. Very tight binding would retard release and cause product inhibition. Furthermore, as we can see from Eq. 9-37, the velocity of catalysis is the product of $[S][E] \times k_{cat} / K_m$, where $[E]$ is, *free enzyme.* Catalysis can be made more efficient in two ways: by increasing k_{cat} or by decreasing K_m, that is, by tighter binding. However, tighter binding will be advantageous only if K_m is not so low that the enzyme is approaching saturation with substrate. Otherwise, $[E]$ will fall as K_m is lowered and little increase in rate will be observed.[2]

In fact, it appears that enzymes have evolved to have high values of k_{cat} and *high* values of K_m, that is, *weak binding* of substrates.[2] Frequently, in one conformational state of a protein the active site is open, with solvent molecules or substrate surrogates present, while in a second state of nearly equal energy it is closed around the substrate. The functional groups of the enzyme's active site can be bound either to external ions or to ionic groups of substrates and either to water or to hydrogen-bonding groups of substrates. In some cases two ionic groups in the protein may pair with each other in the open conformation and with ionic groups of the substrate in the closed conformation. Thus, the energy changes accompanying the conformational changes can be small but very good complementarity can exist in the ES complex, an important factor in establishing specificity.

4. Specificity and k_{cat}

Enzyme specificity is often observed not only in binding but also in the rate at which ES is converted to products. Thus, it is the values of k_{cat} / K_m that determine specificity. Good examples are provided by chymotrypsin and related serine proteases (Chapter 12),[2] for which substrates with the shortest chains are often bound as well as those with longer chains but react more slowly. For example, *N*-acetylphenylalanine amide binds to chymotrypsin about as well as does the longer *N*-acetylphenylalanylalanine amide but reacts only 1/47 as fast.[2] One might anticipate that increasing the length of the substrate would make it bind more tightly because of the greater number of contacts between substrate and enzyme. It has often been suggested that the reason that this does *not* happen is that the binding of the longer, more specific substrates distorts the enzyme and that *the binding energy is now stored in the enzyme. It is as if the enzyme contained an internal spring which would be compressed when the substrate binds.* This would keep the binding weak but the energy in the spring might then be used to increase the velocity.

5. Proofreading

Although enzymes may be very specific they do make mistakes. This is of particular concern for processes such as protein synthesis in which the correct amino acid is placed at each position in the sequence with an error rate that has been estimated for *E. coli* as only 1 in 10^4. It would be impossible for an enzyme designed to attach valine to its specific transfer RNA to avoid attaching the smaller alanine if discrimination between the two were based solely on the Gibbs energy differences of binding.[103] However, it would be easier for an enzyme to exclude *larger* amino acids. This problem may be resolved by use of multistep screening.[104,105] For example, isoleucyl-tRNA synthetase (Chapter 29) does occasionally attach the smaller valine to the specific tRNA,[Ile] but when it does the enzyme in a "proofreading and editing" step hydrolyzes off the incorrect amino acid. The active site for this hydrolyzing activity, whether at a different place on the enzyme surface or created by a conformational change, may be able to exclude sterically the larger isoleucyl residue while acting on the valyl-tRNA. This editing mechanism for isoleucyl-tRNA synthetase was demonstrated directly in 1998 by X-ray crystallography on complexes of the enzyme with L-isoleucine and L-valine. Both substrates fit into the ATP-requiring synthetic site but neither isoleucine nor isoleucyl-tRNA will fit into the editing site which is located in an adjacent β-barrel domain.[104,105] Proofreading steps based on differing chemical properties as well as size can also be visualized.[103,106]

D. Mechanisms of Catalysis

Kinetic studies tell how fast enzymes act but by themselves say nothing about *how* enzymes catalyze reactions. They do not give the **chemical mechanism** of catalysis, the step-by-step process by which a reaction takes place. Most of the individual steps involve the simultaneous breaking of a chemical bond and formation of a new bond. Consider a simple **displacement reaction**, that of a hydroxyl ion reacting with methyl iodide to give the products methanol and iodide ions.

$$HO^- \cdot \underset{H\quad H}{\overset{H}{\underset{|}{C}}} - I \longrightarrow HO-CH_3 + I^-$$

(9-76)

The reaction can be thought of as an "attack" by the OH^- ion on the "back side" of the carbon atom of the methyl groups with a simultaneous displacement of the I^-.

1. The "Transition State"

A reaction such as that of Eq. 9-76 is not instantaneous, and at some time between that at which the reactants exist and that at which products have been formed the C–I bond will be stretched and partially broken and the new C–O bond will be partially formed. The structure at this point is not that of an ordinary compound and is energetically unstable with respect to both the reactant and the products. The intermediate structure of the least stability is known as the **transition state**. Although no one has actually seen a transition state structure, we might represent that for Eq. 9-76 as follows:

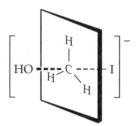

Transition state structure

The negative charge is distributed between the attacking HO^- group and the departing iodide. The bonds to the central carbon atom are no longer tetrahedrally arranged but the C–H bonds lie in a single plane and the partial bonds to the OH and iodine atom lie at right angles to that plane.

It is useful in discussing a reaction mechanism to construct a **transition state diagram** in which Gibbs energy G is plotted against **reaction coordinate** (Fig. 9-15A). Energy E or enthalpy H may be plotted in the same way and authors frequently do not state whether G, E, or H is meant. The reaction coordinate is usually not assigned an exact physical meaning but represents the progress from reactants toward products. It is directly related to the extent to which an existing bond has been stretched and broken or a new one formed. The high energy point is the transition state. A somewhat more detailed idea of a transition state is obtained from a contour diagram such as that of Fig. 9-15B. Here, energy is plotted as a function of two distances, e.g., the lengthening C–I bond distance and the shortening C–O distance for Eq. 9-76. The path of minimum energy across the "saddle point" representing the transition state is indicated by the dashed line.

In a reaction coordinate diagram the difference in value of G between reactants and products is the overall Gibbs energy change ΔG for the reaction, while the difference in G between the transition state and reactants is $\Delta G^{\ddagger}$, the **Gibbs energy of activation**. The magnitude of $\Delta G^{\ddagger}$ represents the "energy barrier" to a reaction and largely determines the rate constant.

The diagrams in Fig. 9-15 are too simple because enzymatic reactions usually occur in several intermediate steps. There will be transition states for each step with valleys in between. The valleys correspond to intermediate species, which are sometimes very unstable. The passing from reactants to products in an enzymatic reaction can be likened to wandering through a series of mountain ranges of various heights and finally reaching the other side.

Quantitative transition state theory.[107–113] In the 1880s Arrhenius observed that the rate of chemical reactions varies with temperature according to Eq. 9-77 in a manner similar to the variation of an equilibrium constant with temperature (integrate Eq. 6-37 and compare). Here, k is a first-order rate constant, the quantity E_a is known as the **Arrhenius activation energy**, and the constant A is referred to as the "pre-exponential factor" or the "frequency factor".

$$k = Ae^{-E_a/RT} \tag{9-77}$$

The Arrhenius equation, together with studies of the effects of salts on reaction rates and observation of quantitative correlations between rates and equilibrium constants, suggested that a rate constant for a reaction might be a product of a constant term which is nearly independent of temperature and a constant $K^{\ddagger}$ which has the properties of an equilibrium constant for formation of the transition state. Eyring made this quantitative in 1935 with his "absolute rate theory"[112] according to which all transition states break down with a rate constant $\kappa k_B T/h$. Eyring reached this conclusion by assuming that *the rate of a chemical reaction is determined by the frequency of stretching of the bond that is being broken in the transition state.* To be more precise, it is the "normal-mode" oscillation of the transition state complex along the reaction coordinate.[109] This frequency ν was deduced by describing the vibrational energy as $h\nu$ (from quantum mechanics) and as k_B/T (from classical mechanics) and setting them equal.

$$h\nu = k_B T \text{ and}$$
$$\nu = k_B T/h$$

$$k_B \text{ is the Boltzmann constant}$$
$$h \text{ is Planck's constant.} \tag{9-78}$$

The right side of Eq. 9-78 is usually multiplied by a **transmission coefficient** κ, which may vary from 1 to 0.1 or even much less. However, for lack of any better value, κ is usually assumed to be 1. From Eq. 9-78, at 25°C $\nu = 6.2 \times 10^{12}$ s^{-1}. This is the maximum rate for a chemical reaction of molecules in the transition state. This is the rate for a single molecule and must be multiplied by the concentration of the reacting substance X in the transition state. This concentration $[X]^{\ddagger}$ is determined by the equilibrium constant $K^{\ddagger} = [X]^{\ddagger}/[X]$. The velocity of the reaction becomes

$$\frac{-d[X]}{dt} = v = \frac{k_B T}{h} \cdot K^{\ddagger}[X] \tag{9-79}$$

The first-order rate constant is

$$k_1 = \frac{-d[X]/dt}{[X]} = K \frac{k_B T}{h} \cdot K^{\ddagger} \text{ s}^{-1} \tag{9-80}$$

Since $\Delta G^{\ddagger} = -RT \ln K^{\ddagger}$, Eq. 9-80 can be rewritten as Eq. 9-81, in which $\Delta G^{\ddagger}$ is the Gibbs energy of activation.

$$k_1 \approx \frac{k_B T}{h} \cdot e^{-\Delta G^{\ddagger}/RT} \text{ s}^{-1} \tag{9-81}$$

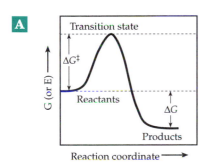

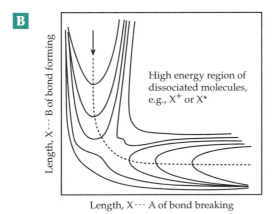

Figure 9-15 (A) Transition state diagram illustrating Gibbs energy vs reaction coordinate for conversion of reactants to products in a chemical reaction. (B) Contour map of Gibbs energy vs interatomic bond distances for reaction B + X – A → B – X + A.

At 25°C, $\Delta G^{\ddagger}$ in kJ / mol, the following "practical" form of Eq. 9-81 can be written.

$$k_1 = 6.2 \times 10^{12} \, e^{-\Delta G^{\ddagger}/2.48} \, s^{-1}$$
$$\log k_1 = 12.79 - \Delta G^{\ddagger}/5.71$$
$$\text{and } \Delta G^{\ddagger} = 73.0 - 5.71 \log k_1$$

Equation 9-81 is approximate and a more correct statistical mechanical treatment is available.[113,114] See also comment on p. 288 about log k (and log k_1) being unitless. Employing Eq. 6-14, we may expand Eq. 9-81 as follows:

$$k \approx \left(\frac{k_B T}{h} e^{\Delta S^{\ddagger}/R} \right) e^{-\Delta H^{\ddagger}/RT} \, s^{-1} \tag{9-82}$$

From this it appears that $\Delta H^{\ddagger} \quad E_a$, the Arrhenius activation energy. A more correct treatment gives $\Delta H^{\ddagger} = E_a - RT$ for reactions in solution. However, since RT at 25°C is only 2.5 kJ mol^{-1}, the approximation that $\Delta H^{\ddagger} = E_a$ is often used. The preexponential term, in parentheses in Eq. 9-82, depends principally on $\Delta S^{\ddagger}$, the entropy change accompanying formation of the transition state. The quantities $\Delta G^{\ddagger}$, $\Delta H^{\ddagger}$, and $\Delta S^{\ddagger}$ are sometimes measured for enzymatic reactions but useful interpretations are difficult.

Equations 9-81 and 9-82 also show that a small decrease in transition state energy will give a large increase in rate; stabilization of the transition state by 5.7 kJ / mol (1.4 kcal / mol) will increase the rate 10-fold. If $\Delta G^{\ddagger} = 400$ kJ/mol, as for a strong covalent single bond, $k_1 = 5 \times 10^{-58} \, s^{-1}$ and the half-life $t_{1/2} = .693 / k_1 = 1.3 \times 10^{57} \, s$ (4 x 10^{49} years, greater than the ~10^{10} years estimated age of the universe). If $\Delta G^{\ddagger} = 100$ kJ / mol, $\log k_1 \approx -4.7 \, s^{-1}$, $k_1 = 1.9 \times 10^{-5}$, and $t_{1/2} \approx 10$ h. This is about the rate of a typical nonenzymatic "model" reaction at 25°C. If $\Delta G^{\ddagger} = 50$ kJ / mol, $\log k_1 \approx 4$, as for a *fast* enzyme.

We can conclude that *enzymes make use of relatively small energy differences in catalyzing reactions*. The energies of numerous van der Waals interactions of hydrogen bonds and electrostatic attraction or repulsion of charges are sufficient. Nevertheless, we can see that anything that stabilizes the transition state (decreases $\Delta G^{\ddagger}$) will increase the rate of reaction. The role of a catalyst is to permit the formation of a transition state of lower energy (higher stability) than that for the uncatalyzed reaction. Stabilization of the transition state of a reaction by an enzyme suggests that the enzyme has a higher affinity for the transition state than it does for substrate or products, an idea that appears to have been expressed first by Haldane[41] and popularized by Pauling.[115]

I think that enzymes are molecules that are complementary in structure to the activated complexes of the reactions that they catalyze. The attraction of the enzyme

molecule for the activated complex would thus lead to a decrease in its energy, and hence to a decrease in energy of activation of the reaction and to an increase in the rate of the reaction.

Transition state inhibitors.

Suppose that a chemical reaction of a compound S takes place with rate constant k_N through transition state T. Let the equilibrium constant for formation of T be $K_N^{\ddagger}$. Assume that an enzyme E can combine either with S with dissociation constant K_{dS} or with the compound in its transition state structure T with dissociation constant K_{dT} (Eq. 9-83).

$$
\begin{array}{ccc}
S & \xrightleftharpoons{K_N^{\ddagger}} & T \longrightarrow \text{product} \\
+ & & + \\
E & & E \\
\Big\updownarrow K_{dS} & & \Big\updownarrow K_{dT} \\
ES & \xrightleftharpoons[K_E^{\ddagger}]{} & ET \longrightarrow \text{product}
\end{array}
\tag{9-83}
$$

If equilibrium is assumed for all four sets of double arrows it is easy to show that Eq. 9-84 holds.

$$K_E^{\ddagger} / K_N^{\ddagger} = K_{dS} / K_{dT} \tag{9-84}$$

According to transition state theory, if the transmission coefficient $\kappa = 1$, T and ET will be transformed to products at the same rate. Thus, if the mechanisms of the nonenzymatic and enzymatic reactions are assumed the same, the ratio of maximum velocities for first-order transformation of ES and S will be given by Eq. 9-85. For some enzymes the ratio

$$k_E / k_N = K_E^{\ddagger} / K_N^{\ddagger} = K_{dS} / K_{dT} \tag{9-85}$$

k_E / k_N may be 10^8 or more. Thus, if $K_{dS} \approx 10^{-3}$ the constant K_{dT} would be $\approx 10^{-11}$. The enzyme would be expected to bind the transition state structure (T) 10^8 times more tightly than it binds S.

The foregoing reasoning suggests that if structural analogs of T could be found for a particular reaction, they too might be very tightly bound—more so than ordinary substrate analogs.[116] Wolfenden[117] listed a series of compounds which may be **transition state inhibitors** of this type and many others have been described.[117a] Nevertheless, very tight, reversible binding is no *proof* that an inhibitor is actually binding to the transition state structure of the enzyme. For example, an inhibitor may bind to a conformational form of the enzyme that lies off to a side rather than directly on the reaction pathway for catalysis.

Describing the transition state.

How can we describe the structure of the transition state when we have no direct knowledge of its structure? We can try

BOX 9-E CHORISMATE MUTASE

Bacteria, fungi, and plants convert products of glucose metabolism into the aromatic amino acids phenylalanine, tyrosine, and tryptophan by a complex sequence of reactions that is described in Chapter 25. One of these reactions, the conversion of **chorismate** into **prephenate,** is unique among enzyme-catalyzed reactions. It is a Claisen rearrangement that occurs readily *without catalysis* in an aqueous solution. However, the rate is increased over one million-fold by the enzyme **chorismate mutase**. The enzyme from *Bacillus subtilis* is a trimer of identical, small 127-residue subunits. Many studies, both of the nonenzymatic and enzymatic reactions, have suggested a "pericyclic" mechanism involving a polar chairlike transition state whose presumed structure is shown in the accompanying equation.[a–f] The transition state analog, whose

structure is also depicted, is a powerful inhibitor. X-ray structural studies show that this inhibitor binds in the conformation shown with a number of interactions with polar side chain groups.[d,g] Some of these interactions are shown for the *E. coli* enzyme in the figure;[g] they are different in enzymes from *B. subtilis*[d] and yeast.[h] However, in no case are there groups in obvious positions to serve as acid–base catalysts and the maximum velocity V_{max} is independent of pH.[b] NMR studies show that the bound product prephenate displays large shifts in some ^{13}C resonances, indicating strong interaction with the polar side chains of the protein.[c] Theoretical calculations also suggested that these interactions, especially those with Arg 90 and Glu 78 of the *B. subtilis* enzymes,[f–l] help to stabilize the partial charge separation in the polar transition state. Studies of mutant enzymes have confirmed that the charged Arg 90 (Lys 39 for *E. coli*) side chain is essential.[k]

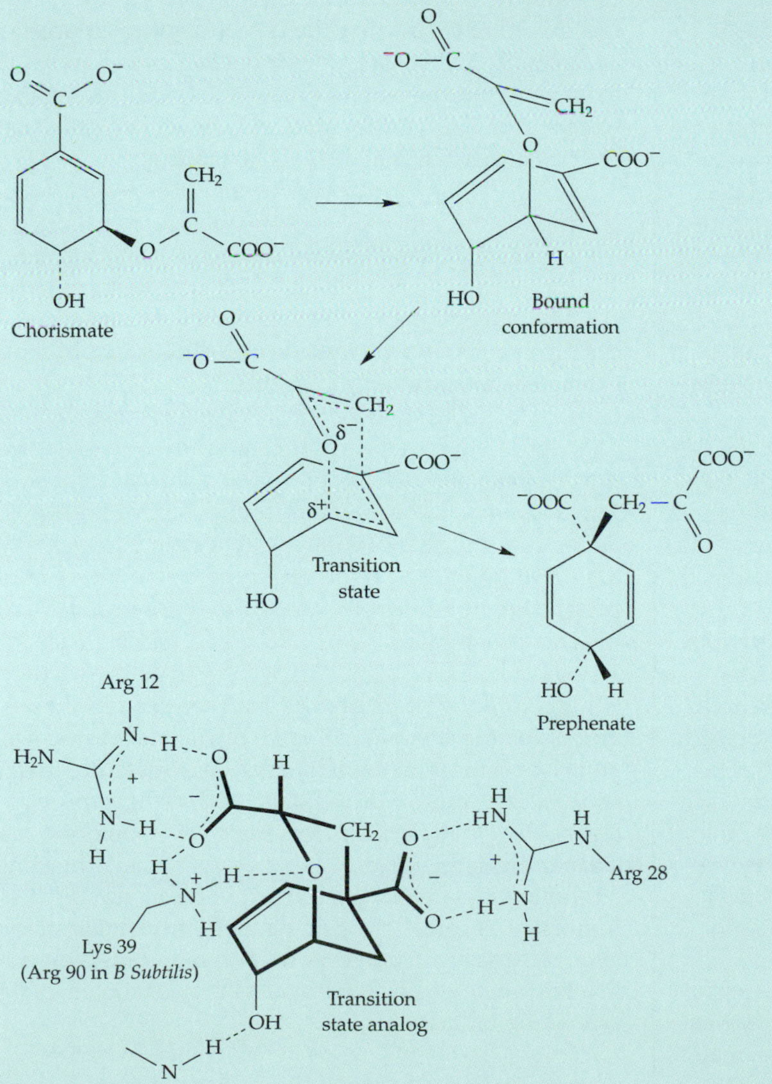

[a] Gray, J. V., Eren, D., and Knowles, J. R. (1990) *Biochemistry* **29**, 8872–8878

[b] Turnbull, J., Cleland, W. W., and Morrison, J. F. (1991) *Biochemistry* **30**, 7777–7782

[c] Rajagopalan, J. S., Taylor, K. M., and Jaffe, E. K. (1993) *Biochemistry* **32**, 3965–3972

[d] Chook, Y. M., Gray, J. V., Ke, H., and Lipscomb, W. N. (1994) *J. Mol. Biol.* **240**, 476–500

[e] Gray, J. V., and Knowles, J. R. (1994) *Biochemistry* **33**, 9953–9959

[f] Lyne, P. D., Mulholland, A. J., and Richards, W. G. (1995) *J. Am. Chem. Soc.* **117**, 11345–11350

[g] Lee, A. Y., Karplus, P. A., Ganem, B., and Clardy, J. (1995) *J. Am. Chem. Soc.* **117**, 3627–3628

[h] Xue, Y., Lipscomb, W. N., Graf, R., Schnappauf, G., and Braus, G. (1994) *Proc. Natl. Acad. Sci. U.S.A.* **91**, 10814–10818

[i] Wiest, O., and Houk, K. N. (1995) *J. Am. Chem. Soc.* **117**, 11628–11639

[j] Kast, P., Hartgerink, J. D., Asif-Ullah, M., and Hilvert, D. (1996) *J. Am. Chem. Soc.* **118**, 3069–3070

[k] Lin, S. L., Xu, D., Li, A., Rosen, M., Wolfson, H. J., and Nussinov, R. (1997) *J. Mol. Biol.* **271**, 838–845

[l] Ma, J., Zheng, X., Schnappauf, G., Braus, G., Karplus, M., and Lipscomb, W. N. (1998) *Proc. Natl. Acad. Sci. U.S.A.* **95**, 14640–14645

to predict the atomic coordinates of all of the atoms of the reacting substrates and of the protein. The transition state structure of the substrate must be between that of the last ES complex and the first EP complex formed. It should be similar to that of a transition state inhibitor. We can hope to obtain suitable X-ray or NMR structures and deduce an appropriate enzyme structure by molecular modeling. Theoretical calculations using quantum chemical methods are of value in this effort.[118,119]

Lengths of the breaking and forming bonds in a transition state are often estimated using **kinetic isotope effects** (KIEs) on the velocity of the reaction.[118–122] This is somewhat indirect. Nevertheless, by measuring these effects for a large number of substrates and several isotopes a picture of the transition state structure that is good enough to use as a model for design of an enzyme inhibitor may be obtained.[117a,121–124] Other physical measurements such as Raman difference spectroscopy of isotopically labeled inhibitors can also be of value.[125]

Getting to the transition state. Since some bonds must lengthen and some shorten and bond angles must also change, it may not be simple for an enzyme to catalyze necessary steps prior to the rate-limiting breakdown of the transition state. Hydrogen bonds within the protein may need to be broken and reformed with new bonding partners. Nonpolar side chains may need to shift to give a packing arrangement that will allow the movement of nuclei that must occur in forming the transition state structure. The enzyme must not only stabilize the transition state but also hold the substrate and escort it into the transition state configuration.

The attacking hydroxyl ion in Eq. 9-76 carries a negative charge which becomes distributed more or less equally between the hydroxyl and the iodine atom in the transition state. In this state the CH_3 group carries a partial positive charge as well. To provide good complementarity between an enzyme and this transition state, a change in the initial charge distribution within the enzyme will also be required. If a positively charged group initially binds the OH^- ion it may have to lose part of its charge and a group next to the iodide atom may have to gain positive charge. In the nonenzymatic reaction of Eq. 9-76 the ionic atmosphere provided by positive counterions in solution can continuously readjust to keep the negative charge effectively balanced at every step along the reaction coordinate and through the transition state. Within enzymes this adjustment may occur via redistribution of electrical charges within the polarizable network of internal hydrogen bonds. The enzyme structure must allow this. Because of the complexity of an enzymatic transition state it may be hard to compare it with the transition state of the corresponding nonenzymatic

reaction. Effects of pressure[125a] and of temperature[125b] also raise doubts about the simple picture of Eq. 9-83. It also follows that tight binding of substrates in the ground state does not necessarily interfere with transition state stabilization.[126–129a]

2. Microscopic Reversibility

The important statistical mechanical principle of microscopic reversibility asserts that *the mechanism of any chemical reaction considered in the reverse direction must be exactly the inverse of the mechanism of the forward reaction.* A consequence of this principle is that if the mechanism of a reaction is known, that of the reverse reaction is also known. Furthermore, it follows that the *forward and reverse reactions catalyzed by an enzyme must occur at the same active site* on the enzyme and the transition state must be the same in both directions. The principle of microscopic reversibility is often useful when the likelihood of a given mechanism is being considered. If a mechanism is proposed for a reversible reaction in one direction the principle of microscopic reversibility will give an unambiguous mechanism for the reverse reaction. Sometimes this reverse mechanism will be chemically untenable and, recognizing this, the enzymologist can search for a better one.

3. Acid and Base Catalysis

Many reactions that are promoted by enzymes can also be catalyzed by acids or bases or by both. An example is **mutarotation**, the reversible interconversion of the α- and β-anomeric forms of sugars (Eqs. 4-1 and 9-86). This reaction is catalyzed by a specific **mutarotase** and also by inorganic acids and bases. The frequently observed bell-shaped curve for the dependence of rate of catalysis on pH (e.g., Eqs. 9-55 to 9-57 and Fig. 9-8) also suggests participation of both protonated and unprotonated acid–base groups present in enzymes.

Acidic and basic groups in enzymes. In this discussion the symbols HB or H^+B will be used for acids and B^- or B for their conjugate bases. Remember that *strong acids* have *low pK_a values* and that the *conjugate bases formed from them are weak bases.* Likewise, *very weak acids have high pK_a values and their conjugate bases are strong.* The following are among the pK_a values that may be important in considering an enzyme mechanism. However, depending upon the environment of the side chain in a protein, these pK_a values may fall substantially outside of the indicated ranges.

−1.74	H_3O^+ in 55.5 M water
~3.6	−COOH, terminal in peptide
4.3–4.7	−COOH in a glutamic or aspartic acid side chain
6.4–7.0	Imidazole (histidine) and phosphate ($-OPO_3H^-$)
7.5–8.0	$-NH_3^+$, terminal in peptide
8.5–10	−SH of cysteine side chains
9.5–11	Phenolic −OH of tyrosine
~10.5	$-NH_3^+$ of lysine side chain
~13.6	$-CH_2OH$ of serine side chain
15.7	free HO^- in 55.5 M H_2O, 25°C

Acid–base catalysis of mutarotation.

The mutarotation of glucose proceeds through the free aldehyde form as an intermediate (Eq. 9-86). A hydrogen atom is removed as H^+ from the oxygen at carbon 1 in step *a* and a proton (probably a different one) is transferred to the oxygen of the ring with cleavage of the O – C bond to the anomeric carbon atom. A similar process in reverse is required for step *b*. Transfers of hydrogen ions between atoms of oxygen, nitrogen, and sulfur atoms are a common feature of biochemical reactions. The bonds between hydrogen and O, N, and S atoms tend to be polarized strongly, leaving a partial positive charge on the hydrogen atoms. Consequently, the groups are weakly acidic and protons can be transferred from them relatively easily. It is reasonable to suppose that acid and base catalysis is related to these proton transfers.

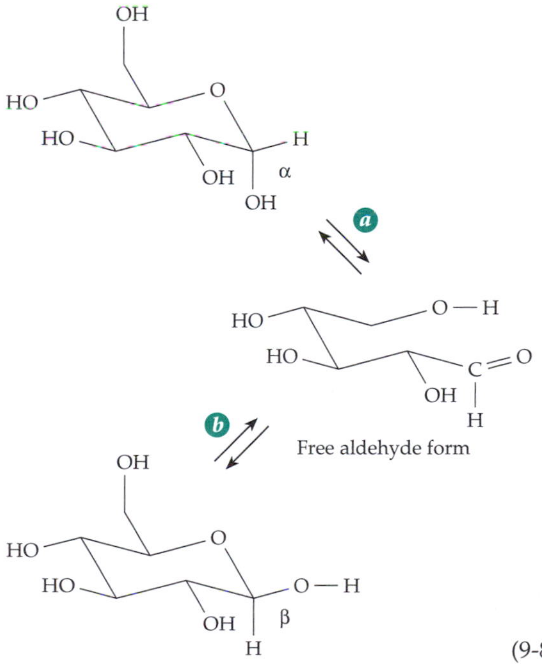

(9-86)

General base and general acid catalysis.

Base-catalyzed mutarotation might be formulated as follows: A hydroxyl ion or some other base attacks the proton on the anomeric −OH group of the sugar, removing it to form an anion and the conjugate acid BH^+ (Eq. 9-87,

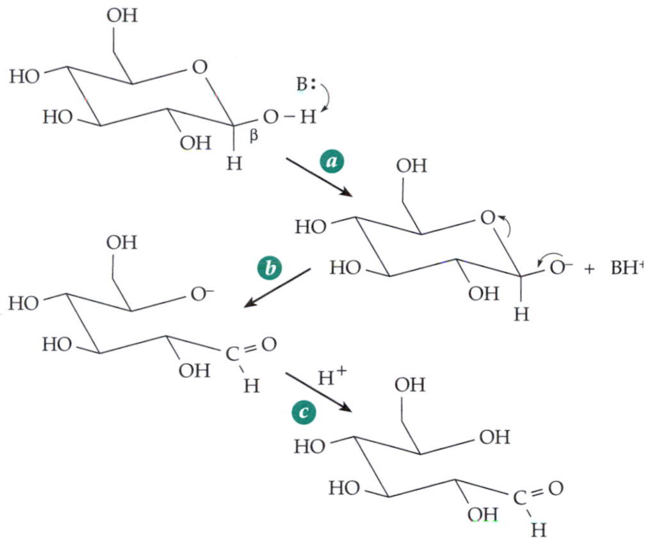

(9-87)

step *a*). The anion is isomerized to a second anion with the ring opened (step *b*). Addition of a proton (transfer of a proton from H_3O^+) produces the free aldehyde form of the sugar (step *c*).

The catalytic base B: might be HO^- or a weaker base such as ammonia or even water. For reactions the rate is proportional only to the concentration of OH^- and the presence of other weaker bases has no effect.[129b] Such catalysis is referred to as **specific hydroxyl ion catalysis**.[19] More commonly, the rate is found to depend both on [OH^-] and on the concentration of other weaker bases. In such cases the apparent first-order rate constant (k_{obs}) for the process is represented by a sum of terms (Eq. 9-88). The term k_{H_2O} is the rate in

$$k_{obs} = k_{H_2O} + K_{OH^-}[OH^-] + k_B[B] \qquad (9-88)$$

pure water and represents catalysis occurring by the action of water alone as either an acid or a base. The last two terms represent the contributions to the catalysis by OH^- and by the other base, respectively. The term $k_B[B]$ represents **general base catalysis**.

Catalysis of mutarotation by acids occurs if an acid donates a proton to the oxygen in the sugar ring as

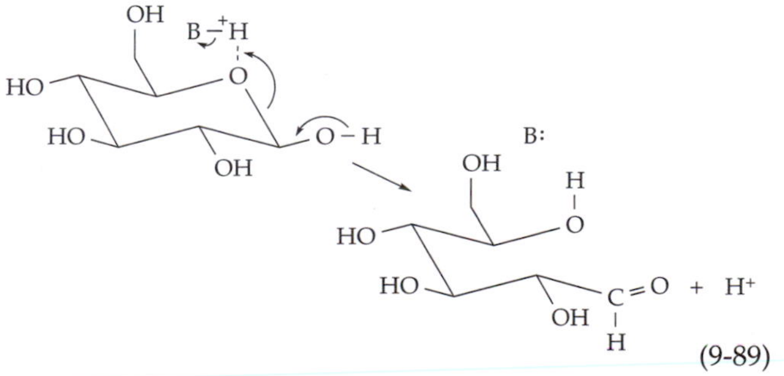

(9-89)

BOX 9-F IMMUNOPHILINS AS ROTAMASES

Since its introduction into clinical use in about 1979 the immunosuppressant **cyclosporin** has been responsible for a revolution in human organ transplantation.[a-c] The exact mechanism of action in suppressing T-lymphocyte-mediated autoimmune responses is still not completely clear, but cyclosporin, a cyclic lipophilic peptide from a fungus, was found to bind to specific proteins that were named **cyclophilins**.[d] Human cyclophilin A is a 165-residue protein which associates, in the crystal form, as a decamer with five-fold rotational and dihedral symmetry.[e] This protein is also found in almost all other organisms and has a highly conserved sequence.[f] Another immunosuppressant, a synthetic molecule known only as **FK506**, has an action similar to that of cyclosporin but binds to distinctly different proteins, the FK506-binding proteins (FKBPs), of which the 107-residue FKBP-12 is the best known.[g-j] These proteins also bind another immunosuppressant, **rapamycin**, which, however, has different biological properties than does FK506. Both FK506 and rapamycin are macrocyclic compounds but with structures very different from that of cyclosporin.

It was a surprise to discover that all of the cyclophilins and FK506-binding proteins are **peptidyl prolyl *cis–trans* isomerases** or **rotamases**. They all catalyze the following simple and reversible reaction of a prolyl peptide linkage:

Cyclosporin A

The transition state energy for the reaction is lowered by 33 kJ mol^{-1} by cyclophilin A[k] and by 27 kJ mol^{-1} by FKBP-12[g] with corresponding rate increases of ~ -6×10^5- and 5×10^4-fold, respectively.

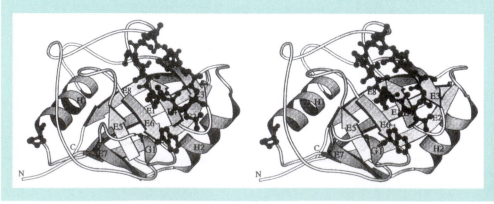

Cyclosporin (black) bound to human cyclophilin. From Pflügl *et al.*[e] Courtesy of J. N. Jansonius.

BOX 9-F (continued)

For both enzymes the maximum velocity is independent of pH over a wide region. For cyclophilin the rate is nearly independent of the nature of residue X in the foregoing structure but FKBP prefers a hydrophobic residue.

How do these enzymes work? One possibility would be for the protein to transfer a proton to the nitrogen atom of the proline ring. This would destroy the partial double-bond character of the amide linkage and allow free rotation about a single bond. The same thing could be accomplished if a nucleophilic group from the enzyme formed a covalent adduct with the C=O of the substrate. However, this seems unlikely because there is no suitably placed nucleophile. A third possibility is that the enzyme distorts the substrate and stabilizes the transition state using only noncovalent interactions.[g,h,l] This would account for the lack of pH dependence and observable solvent isotope effects. However, in human cyclophilin the guanidinium group of arginine 55 hydrogen bonds to the peptide C=O of the substrate's proline residue and could easily shift to place a guanidinium proton against the proline ring nitrogen effectively protonating it and permitting rotation about the peptide linkage.[m]

Do cyclosporin and FK506 act as transition state inhibitors? If so, we might learn something about the mechanism from the three-dimensional structures of the inhibited rotamases.[i,l,n,o] From these structures as well as those with simpler substrate analogs, it is seen that the substrate is desolvated and that many nonbonded (van der Waals) interactions stabilize the binding. Hydrogen bonding is also important.[p,q] The presence of distinct hydrogen bonds to the peptide NH and C=O groups on the N-terminal side of the substrate X-Pro linkage and less well-defined bonding on the terminal side suggests that the C-terminal part may be rotated while the enzyme holds the N-terminal part.[p] It can also be concluded that mechanisms are probably not identical for all rotamases.

Cyclophilins and FKBPs are large families of proteins. In E. coli there are two cyclophilin genes,[r,s] three encoding FK506-binding proteins, and one of

a third family of bacterial rotamases (parvulins) that lack sequence similarity to the other two families.[t,u] Yeast contains at least five cyclophilins[v] and even more eukaryotic FKBPs are known.[j] There is abundant evidence that these proteins play an important role in protein folding in vivo (Chapter 29).

[a] Kahan, B. D. (1989) N. Engl. J. Med. 321, 1725–1738
[b] Schreiber, S. L. (1991) Science 251, 283–287
[c] High, K. P., Joiner, K. A., and Handschumacher, R. E. (1994) J. Biol. Chem. 269, 9105–9112
[d] Fruman, D. A., Burakoff, S. J., and Bierer, B. E. (1994) FASEB J. 8, 391–400
[e] Pflügl, G. M., Kallen, J., Jansonius, J. N., and Walkinshaw, M. D. (1994) J. Mol. Biol. 244, 385–409
[f] Lippuner, V., Chou, I. T., Scott, S. V., Ettinger, W. F., Theg, S. M., and Gasser, C. S. (1994) J. Biol. Chem. 269, 7863–7868
[g] Fischer, S., Michnick, S., and Karplus, M. (1993) Biochemistry 32, 13830–13837
[h] Orozco, M., Tirado-Rives, J., and Jorgensen, W. L. (1993) Biochemistry 32, 12864–12874
[i] Van Duyne, G. D., Standaert, R. F., Karplus, P. A., Schreiber, S. L., and Clardy, J. (1993) J. Mol. Biol. 229, 105–124
[j] Lam, E., Martin, M. M., Timerman, A. P., Sabers, C., Fleischer, S., Lukas, T., Abraham, R. T., O'Keefe, S. J., O'Neill, E. A., and Wiederrecht, G. J. (1995) J. Biol. Chem. 270, 26511–26522
[k] Eberhardt, E. S., Loh, S. N., Hinck, A. P., and Raines, R. T. (1992) J. Am. Chem. Soc. 114, 5437–5439
[l] Kakalis, L. T., and Armitage, I. M. (1994) Biochemistry 33, 1495–1501
[m] Zhao, Y., and Ke, H. (1996) Biochemistry 35, 7362–7368
[n] Thériault, Y., Logan, T. M., Meadows, R., Yu, L., Olejniczak, E. T., Holzman, T. F., Simmer, R. L., and Fesik, S. W. (1993) Nature (London) 361, 88–91
[o] Konno, M., Ito, M., Hayano, T., and Takahashi, N. (1996) J. Mol. Biol. 256, 897–908
[p] Kern, D., Kern, G., Scherer, G., Fischer, G., and Drakenberg, T. (1995) Biochemistry 34, 13594–13602
[q] Göthel, S. F., Herrler, M., and Marahiel, M. A. (1996) Biochemistry 35, 3636–3640
[r] Clubb, R. T., Ferguson, S. B., Walsh, C. T., and Wagner, G. (1994) Biochemistry 33, 2761–2772
[s] Edwards, K. J., Ollis, D. L., and Dixon, N. E. (1997) J. Mol. Biol. 271, 258–265
[t] Rudd, K. E., Sofia, H. J., Koonin, E. V., Plunkett, G., III, Lazar, S., and Rouviere, P. E. (1995) Trends Biochem. Sci. 20, 12–14
[u] Rahfeld, J.-U., Rücknagel, K. P., Stoller, G., Horne, S. M., Schierhorn, A., Young, K. D., and Fischer, G. (1996) J. Biol. Chem. 271, 22130–22138
[v] Matouschek, A., Rospert, S., Schmid, K., Glick, B. S., and Schatz, G. (1995) Proc. Natl. Acad. Sci. U.S.A. 92, 6319–6323

shown in Eq. 9-89. Again, either **specific acid catalysis** (by H_3O^+) or **general acid catalysis** is possible.

Enzymes are not able to concentrate protons or hydroxyl ions to the point that specific base or acid catalysis would be effective. However, either general acid or general base catalysis can be accomplished by groups present in an enzyme in their normal states of

protonation at the pH of the cell. Thus, if one of the enzymes must be dissociated to its conjugate base and a second must be protonated for reaction to take place (as in Eq. 9-54), it is reasonable to suppose that these two groups participate in acid–base catalysis. Of the acidic and basic groups present in proteins, the imidazole group of histidine would appear to be the most ideal both for general base catalysis and, because both forms may exist in nearly equal amounts at pH 7, for general acid catalysis. However, carboxyl, thiol, lysyl, tyrosyl, and N-terminal amino groups are all thought to function in various enzymes.

The mutarotase of *E. coli* has a turnover number of $10^4 \, s^{-1}$. The plot of $-\log K_m$ vs pH indicates two pK_a values in the free enzyme at 5.5 and 7.6, while the plot of $\log V_{max}$ yields a single pK_a of 4.75 for the ES complex.[130] The latter might represent a carboxyl group on an imidazole group in its conjugate base form. Why doesn't the group having pK_a 7.6 in the free enzyme also show a pK_a in the ES complex? Either the group has no catalytic function so that EH_2S of Eq. 9-54 reacts to form products just as fast as does HES or the pK_a is so strongly shifted by substrate binding that it is not detected in the $\log V_{max}$ plot.

The Brönsted relationships. The effectiveness of a specific base as a general base catalyst can usually be related to its basicity (pK_a) via the **Brönsted equation** (Eq. 9-90). Here, k_B is defined by Eq. 9-88 and G_B is a

$$\log k_B = \log G_B + \beta \, (pK_a) \tag{9-90}$$

constant for a particular reaction. A similar equation (9-91) relates the constant k_A for general acid catalysis

$$\log k_{HB} = \log G_A - \alpha (pK_a) \tag{9-91}$$

to pK_a. These equations are linear Gibbs energy relationships similar to the ones discussed in Box 6-C. For the Brönsted equations to hold, the Gibbs energy of activation for the reaction must be directly related to the basicity or acidity of the catalyst.

The exponents β and α of Eqs. 9-90 and 9-91 measure the *sensitivity* of a reaction toward the basicity or acidity of the catalyst. It is easy to show that as β and α approach 1.0 general base or general acid catalysis is lost and that the rate becomes exactly that of specific hydroxyl ion or specific hydrogen ion catalysis.[131] As β or α approach zero basic or acidic catalysis is indetectable. Thus, general base or general acid catalysis is most significant when β or α is in the neighborhood of 0.5. Under these circumstances it is easy to see how a moderately weak basic group, such as the imidazole group of histidine, can be an unusually effective catalyst at pH 7.

To determine α or β experimentally a plot of $\log k_B$ or $\log k_{HB}$ vs pK_a (a **Brönsted plot**) is made and the slope is measured. Statistical corrections (Chapter 7) should be applied for dicarboxylic acids and for ammonium ions from which one of three protons may be lost from the nitrogen atom. General base or general acid catalysis implies an important feature of any mechanism for which it is observed, namely, that removal of a proton or addition of a proton is involved in the

rate-determining step of a reaction.

Concerted acid–base or tautomeric catalysis. A third possible type of catalysis requires that a base and an acid act *synchronously* to effect the breaking and formation of bonds in a single step. Thus, tetramethylglucose mutarotates very slowly in benzene containing either pyridine (a base) or phenol (an acid). However, when both pyridine and phenol are present, mutarotation is rapid. This suggested to Swain and Brown[132] a **concerted mechanism** (Eq. 9-92) in which both an acid and a base participate.

During the reaction shown, the acid BH^+ is converted to its conjugate base B and the base B' to its conjugate acid H^+B'. It might seem that these agents, having been altered by the reaction, are not serving as true catalysts. However, a simple proton exchange will restore the original forms and complete the catalytic cycle. In aqueous solutions, water itself might act as the acid or the base or even both in concerted catalysis.

The original experimental evidence for concerted acid–base catalysis of the mutarotation in benzene is now considered unsound[133,134] and concerted acid–base catalysis has been difficult to prove for nonenzymatic reactions in aqueous solution. However, measurements of kinetic isotope effects seem to support Swain and Brown's interpretation.[135] Concerted acid–base catalysis by acetic acid and acetate ions may have been observed for the enolization of acetone[136] and it may be employed by enzymes.[136a]

Swain and Brown showed that a more effective catalyst for the mutarotation of sugars than a mixture of an acid and a base can be designed by incorporating the acidic and basic groups into *the same molecule*.[132,135] Thus, with 0.1 M tetramethylglucose in benzene solution, 0.001 M α-hydroxypyridine is 7000 times as effective a catalyst as a mixture of 0.001 M pyridine + 0.001 M phenol. Swain and Brown suggested the following completely concerted reaction mechanism for the **polyfunctional catalyst α-hydroxypyridine** in which the hydrogen-bonded complex formed (Eq. 9-93) is analogous to an enzyme–substrate complex. The product of the catalyst is 2-pyridone, a tautomer of

$$\tag{9-92}$$

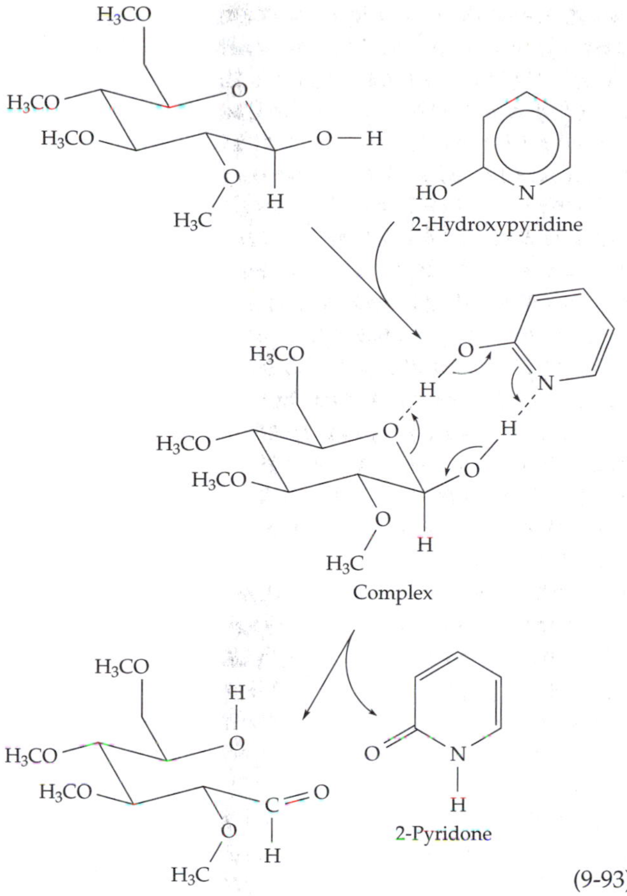

suggested that rapid transfer of protons along rigidly and accurately held hydrogen bonds in the ES complex may be an essential feature of enzymatic catalysis.[137] This conclusion is supported by many more recent observations. A remaining question is whether this proton transfer can take place in such a way that the transition state barrier is lowered.

Ultrafast proton transfer. The diffusion-controlled limit for second-order rate constants (Section A3) is ~ 10^{10} M^{-1} s^{-1}. In 1956, Eigen, who had developed new methods for studying very fast reactions, discovered that protons and hydroxide ions react much more rapidly when present in a lattice of ice than when in solution.[138] He observed second-order rate constants of 10^{13} to 10^{14} M^{-1} s^{-1}. These represent rates almost as great as those of molecular vibration. For example, the frequency of vibration of the OH bond in water is about 10^{14} s^{-1}. The latter can be deduced directly from the frequency of infrared light absorbed in exciting this vibration: Frequency v equals wave number (3710 cm^{-1} for –OH stretching) times c, the velocity of light (3×10^{10} cm s^{-1}).

This ultrafast transfer of protons can be explained as follows: The OH$^-$ ion and the proton, which is combined with a water molecule to form H_3O^+, are both hydrogen-bonded to adjacent water molecules. In ice a chain of hydrogen-bonded water molecules links the hydroxide and the hydrogen ions (Eq. 9-94). By synchronous movement of electron pairs from the OH$^-$ ion and from each of the water molecules in the

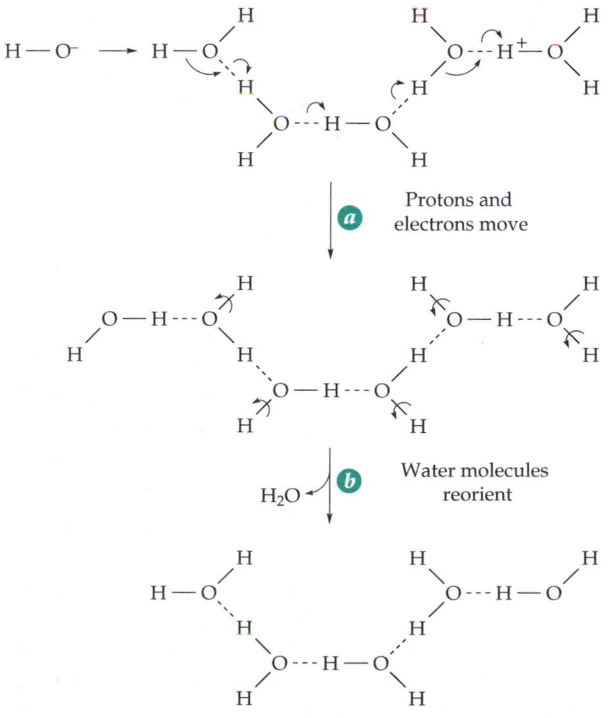

2-hydroxypyridine, with which it is in a rapid reversible equilibrium.

Rony called catalysis of the type illustrated in Eq. 9-93 tautomeric catalysis and suggested that its efficiency lies not simply in the close proximity of acidic and basic groups in the same molecule but also in the ability of the catalyst to repeatedly cycle between the two tautomeric states.[133] For an enzyme the tautomerization of the free catalyst could sometimes be rate determining (see Section A,4 on isomechanisms).

4. Hydrogen Bonding and the Transfer of Protons within Active Sites

An extensive network of hydrogen bonds runs throughout most proteins and may be especially complex within active sites. The network often runs through bound substrate molecules and immobilized water molecules in the active site cavity. This network arises in part because of the frequent occurrence of acidic and basic catalytic groups in active sites and by the fact that many substrates contain polar groups. The structure of an enzyme often seems to be more rigid in a complex with substrates or inhibitors than in the free state. Does this network of linked hydrogen bonds play a role in catalysis? If so, what? Wang

Complex

2-Pyridone

(9-93)

a Protons and electrons move

b Water molecules reorient

A second pair of HO$^-$ and H$^+$ can now use the same pathway (9-94)

chain (as indicated by the little arrows) the neutralization can take place during the time of one molecular vibration (step a, Eq. 9-94). The positions of the oxygen atoms remain unchanged at the end of the reaction but the protons that were engaged in hydrogen bond formation have moved toward the left and are now attached to different oxygen atoms. The central chain of water molecules can be restored to its original state if each molecule rotates around one of its single bonds, swinging a hydrogen from right back to left (Eq. 9-94, step b). This must be a slower process which may occur one molecule at a time. The rotation of the leftmost water molecule would leave an empty space between two O-atoms, a "fault" in the H-bonded chain of the ice structure. This fault can be corrected by

(9-95)

rotation of the second molecule (Eq. 9-95); however, this would create a fault at the right of the second water molecule. This would induce rotation of the third molecule, etc., causing the fault to migrate from left to right across the entire chain of water molecules (Eq. 9-94, step b) and leaving the chain ready to function again in transferring another proton.

Because of the disorder present in ice, an array of water molecules such as that in Eq. 9-94 wouldn't revert to its exact original form in step b. However, active sites of enzymes are highly structured and proton transfers may occur with precision. For example, a synchronous shift of protons in an array of carboxylic acid, imidazolium, and phosphate groups can be envisioned readily (Eq. 9-96). The net effect of the process is to transfer a proton from one end of the chain to the other (as in Eq. 9-94) with facile tauto-

(9-96)

merization reactions providing the pathway. Such a pathway might be constructed by protein side chains to join the two sides of an active center promoting a concerted acid–base catalyzed reaction such as that of Eq. 9-92. Other tautomerization processes are possible within proteins if the existence of less stable minor tautomers of selected amide groups in the peptide backbone is allowed.[139] Nagle and Morowitz suggested a

process similar to that of Eq. 9-94 but involving side chains of serine residues.[140]

Proton transfer rates. Consider the reversible reaction of a proton acceptor B with acid H–A (Eq. 9-97). Eigen pointed out that the reaction will be fastest if the two reactants form a hydrogen-bonded complex (Eq. 9-97, step a).[138] The hydrogen bonding shortens the distance from the proton to B and allows for very rapid transfer of the proton from A to B within the

(9-97)

hydrogen-bonded complex (step b). The activation energy is close to zero. The complex dissociates in step c to form the products. The reactions are reversible, even though they have been indicated by unidirectional arrows.

The hydrogen-bonded complexes A–H⎯B and A⎯H–B can be formed readily if A and B are oxygen or nitrogen bases such as $-COO^-$, $-OH$, $-NH_2$, or imidazole. In such cases, as in ice, the interconversion of the two complexes (step b in Eq. 9-97) is very fast. The overall rate of the proton transfer will then be limited by diffusion of B and H–A together or by diffusion of A and H–B apart. It might seem that these two processes should also both be very fast. However, the rates will be determined by the concentration ratio [A–H⎯B]/[A⎯H–B]. If [A–H⎯B] equals [A⎯H–B] the rate of dissociation of A–H–B will be half what it is if nearly all of the complex is A⎯H–B. If A–H⎯B predominates by 1000 to 1, the rate will be slowed much more.

The equilibrium constant for the reaction of Eq. 9-97 will be:

$$K_{eq}^{AB} = K_a^{HA} / K_a^{HB} = k_f / k_r$$

or $\quad \log K_{eq} = pK_a^{HA} - pK_a^{HB} = \log k_f - \log k_r$

(9-98)

Here k_f and k_r refer to the rates in the forward and reverse directions. If the pK_a's of HA and HB are equal k_f and k_r will be the same, but if they are very far from equal the reaction will be slowed in one direction. If proton transfer is a step in an enzymatic reaction it may be slowed enough to become rate limiting.

The difference in pK_a values between the proton donor and the proton acceptor in Eq. 9-97 can be expressed as the Gibbs energy change which at 25°C is equal to $\pm 5.71 \times \Delta pK_a$. This is often referred to as the **thermodynamic barrier** $\Delta G°$ to a reaction and $\Delta G^{\ddagger}$ can be expressed as the sum of the thermodynamic barrier $\Delta G°$ plus an **intrinsic barrier** $\Delta G^{\ddagger}_{int}$. For the proton transfer of Eq. 9-97 the intrinsic barrier (for step b) is thought to be near zero so that $\Delta G^{\ddagger} \approx 5.71 \Delta pK_a$.

From this we can conclude that two pK_a values can be as much as eight units apart and $\Delta G^{\ddagger}$ will still be less than 50 kJ / mol, low enough to permit rapid enzymatic reactions. However, for transfer of a proton from a C–H bond to a catalytic group, for example, to form an enolate ion for an aldol condensation (Chapter 13), the intrinsic barrier is known to be about 50 kJ / mol.[141] In this case, to allow rapid enzymatic reaction either the thermodynamic barrier must be very low, as a result of closely matching pK_a values, or the enzyme must lower the intrinsic barrier. It may do both.

Marcus theory. Discussion of intrinsic barriers is often approached using a quantitative theory proposed by Marcus.[142–145] It was first applied to electron transfer (Chapter 16) but has been used for a great variety of nonenzymatic and enzymatic reactions. As used by Gerlt and Gassman,[141] the Marcus formalism describes the reaction coordinate (Fig. 9-15A) as an inverted parabola whose shape is determined by the overall Gibbs energy change $\Delta G°$ and the intrinsic barrier $\Delta G^{\ddagger}_{int}$. The value of the Gibbs energy at any point on the curve is designated G and the reaction coordinate x is taken as 0 for the reactants and 1 for the products.

$$G = -4 \, \Delta G^{\ddagger}_{int} \, (x - 0.5)^2 \ + \ \Delta G° \, (x - 0.5)$$
$$\text{Limits: } \Delta G°/4 \leq \Delta G^{\ddagger}_{int} \leq 4 \, \Delta G° \qquad (9\text{-}99)$$

Differentiation of Eq. 9-99 yields the position of the transition state coordinate $x^{\ddagger}$ as follows:

$$x^{\ddagger} = 0.5 + \Delta G°/ 8 \, \Delta G^{\ddagger}_{int} \qquad (9\text{-}100)$$

and it follows that

$$\Delta G^{\ddagger} = \Delta G^{\ddagger}_{int} \, (1 + \Delta G°/ 4 \, \Delta G^{\ddagger}_{int})^2 \qquad (9\text{-}101)$$

Diffusion-controlled dissociation of protons. The direct proton transfer between C-1 and C-2 during the action of sugar isomerases may seem puzzling. How can a highly mobile proton remain attached to a group in the enzyme for a millisecond or more instead of being transferred out to a solvent molecule? This can mean that the enzyme promotes the transfer of a hydride ion or of a hydrogen atom rather than a proton (see Chapter 13). If so, the observed proton exchange with solvent would be an unimportant side reaction. On the other hand, could the group in the enzyme that removes the proton be out of contact with the aqueous medium and thus able to hold onto the proton more tightly? In recent years, it has been recognized that neither of these explanations may be necessary. An imidazole group is a likely proton-carrying group at many active sites and it is thought that a proton cannot be expected to transfer out from an imidazolium group with a rate constant greater than $\sim 10^3$ s^{-1}.

The argument is as follows.[146] The rate of donation of a proton from H_3O^+ to imidazole (reverse of Eq. 9-102) is known to be diffusion controlled with a rate constant of 1.5×10^{10} M^{-1} s^{-1}.

$$ImH^+ + H_2O \rightleftharpoons [ImH^+\text{----}OH_2] \rightleftharpoons Im + OH_3^+$$

Very fast diffusion controlled
$k \sim 1.5 \times 10^{10}$ M^{-1} s^{-1} $\qquad (9\text{-}102)$

The equilibrium constant for Eq. 9-102, calculated from the pK_a of 7.0 for imidazole, is 10^{-7} M. Since K_{eq} is also the *ratio of the overall rate constants for the forward and reverse reactions*, we see that for the forward reaction $k_f = 10^{-7} \times 1.5 \times 10^{10} = 1.5 \times 10^3$ s^{-1}. This slow rate results from the fact that in the intermediate complex (in brackets in Eq. 9-102) the proton is on the imidazole group most of the time. For a small fraction of the time it is on the coordinated molecule H_2O but reverts to being on the imidazole many times before the imidazole and OH_3^+ separate (see also Eqs. 9-97 and 9-98). Because of this unfavorable equilibrium within the complex, the diffusion-controlled rate of proton transfer from a protonated imidazole to water is far less than for proton transfer in the reverse direction.

Coupled proton transfers. Enzymatic reactions often require the transfer of two or more protons. They may move individually, one proton at a time, or as in Eq. 9-93 they may move synchronously in a **coupled** or concerted process. Such coupled movement is generally not possible for heavier nuclei.[147] However, studies of solvent isotope effects using a **proton inventory** technique[111,148,149] have provided evidence favoring coupled proton transfers for a variety of enzymes. Movement of protons along hydrogen-bonded paths, as well as electron transfer, may take place with some participation of **quantum mechanical tunneling**.[150–152] Coupling to vibrational modes of the hydrogen-bonded protons may provide **vibration-assisted tunneling**.[153–154d] These reactions are associated with unusually large kinetic isotope effects.

Unusually strong hydrogen bonds. The strength of a hydrogen bond is thought to be directly related to the length, which is ordinarily taken to be the distance between the two surrounding heavier atoms (see Chapter 2, Section B,3). Hydrogen bonds are sometimes classified on the basis of N–H---O distances[155,156] as:

<0.25 nm	very strong
0.25–0.265 nm	strong
>0.28 nm	weak
>0.37 nm	no van der Waals contact but electrostatic interaction still occurs

N–H---O distances may be a little longer than these.

The 0.276-nm hydrogen bonds in ice are regarded as moderately strong. However, if one of the oxygen atoms in an O–H---O hydrogen bond carries a negative charge, as in the maleate monoanion, it will be

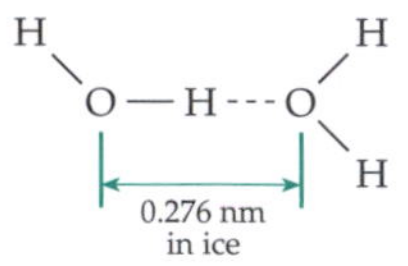

shorter.[156–158] Although the proton will be closer to one oxygen atom than to the other, it will be able to move between them by passing a very low transition state barrier. For this to occur the microscopic pK_a values for the two groups (when they are protonated) must be similar; for the maleate anion they are identical.

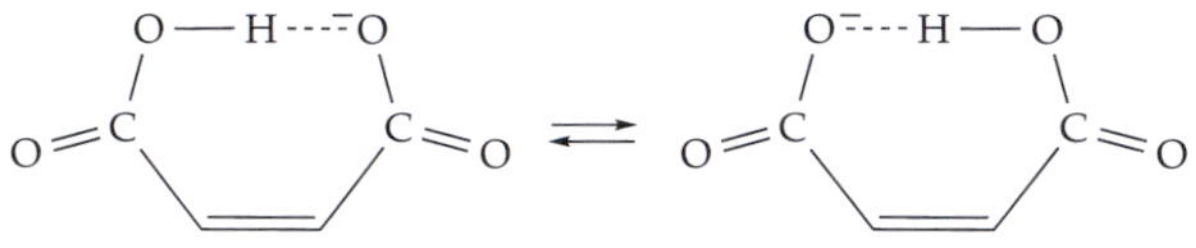

Monohydrogen maleate

The *strength* of a hydrogen bond can be measured for hydrogen-bonded complexes in the gas phase and range from 10–100 kJ/mol and even higher[155] for such complexes as (FHF)⁻. It is more difficult to establish the strength in the liquid state or within the active site of an enzyme, but shorter hydrogen bonds are usually stronger than longer ones. Hydrogen bond distances in crystals of small molecules can be measured precisely by X-ray or neutron diffraction but greater uncertainty is present in distances within proteins. Lengths and probably strengths of hydrogen bonds in proteins can be measured by NMR methods.[158a,b] For example, solid-state NMR measurements on crystals of amino acids and other carboxylic acids have provided a plot of ¹H chemical shift vs hydrogen bond length.[159] For very short (0.24- to 0.25-nm) hydrogen bonds the ¹H chemical shift may be as great as 21 ppm. See also Fig. 3-30 and associated discussion. When dissolved in an ¹H₂O–²H₂O mixture strongly hydrogen-bonded protons within a protein become enriched in ¹H. The ¹H/²H ratio of the hydrogen-bonded protons provides another measure of the hydrogen bond strength.[158c]

A short hydrogen bond is also present in such cations as the following:

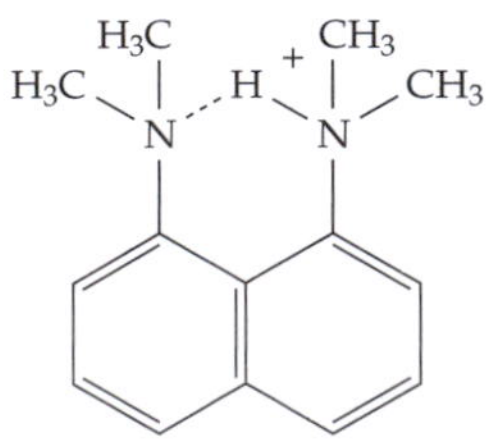

The pK_a for dissociation of its proton is 12.3 and the hydrogen-bonded proton is probably located in the *center* of the bond with both amino groups sharing the charge.[160] Enols can also form unusually strong "resonance-assisted" hydrogen bonds:

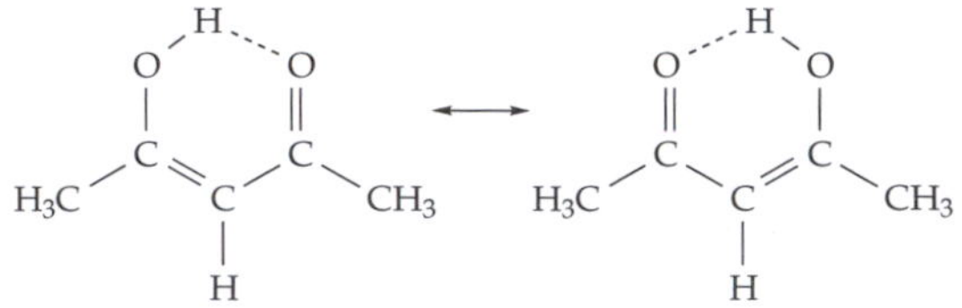

The structure can be thought of as a resonance-stabilized enolate anion with a proton bound between the two oxygen atoms and equidistant from them.[156,157]

Do these short "**low-barrier hydrogen bonds**" have a special significance in enzymology? Proposals that they contribute to stabilization of transition states[161–166] have received some support[154,167] and aroused controversy.[158b,168–173] In later chapters we will examine specific enzymes in which low-barrier hydrogen bonds have been observed.

5. Covalent Catalysis

In addition to participating in acid–base catalysis, some amino acid side chains may enter into covalent bond formation with substrate molecules, a phenomenon that is often referred to as covalent catalysis.[174] When basic groups participate this may be called **nucleophilic catalysis**. Covalent catalysis occurs frequently with enzymes catalyzing nucleophilic displacement reactions and examples will be considered in Chapter 12. They include the formation of an acyl-enzyme intermediate by chymotrypsin (Fig. 12-11). Several of the coenzymes discussed in Chapters 14 and 15 also participate in covalent catalysis. These coenzymes combine with substrates to form reactive intermediate compounds whose structures allow them to be converted rapidly to the final products.

6. Proximity and Orientation

One of the earliest ideas about enzymes was that they simply brought reactants together and bound them side by side for a long enough time that the reactive groups might bump together and finally react. How important is this **proximity factor**? Page and Jencks estimated that rate enhancements by factors of 10^3 or more may be expected solely from the loss in the entropy of two reactants when they are bound in close proximity on an enzyme surface.[107,175,176] In view of the large entropy decrease involved, *the enthalpy of* *binding must be high*, and if this explanation is correct the *binding of the substrates* to the enzyme provides much of the driving force for catalysis. Westheimer described this by stating that enzymes use the substrate-binding force as an **entropy trap**.[108] The losses of translational and rotational entropy, which Page and Jencks estimated as up to -160 to -210 J/deg/mol, overcome the unfavorable entropy of activation that is usual in bimolecular reactions.

How precise must the orientation of substrates be for rapid reaction?[107,177,177a] Compounds such as the acid shown in Eq. 9-103 form an internal ester (lactone) spontaneously with elimination of water.

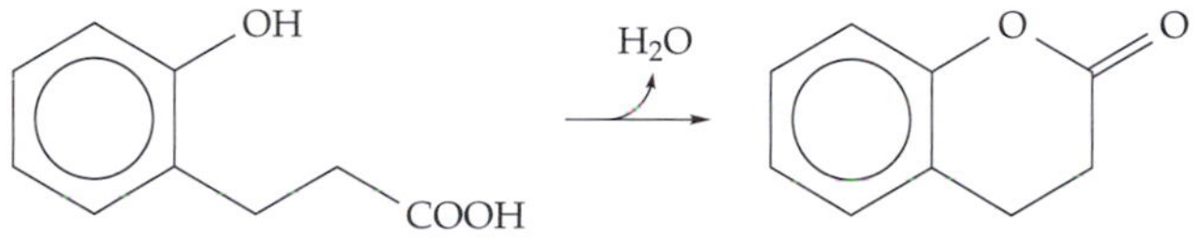

$$(9\text{-}103)$$

However, the following compound reacts at a rate over 10^{11} times that of the acid shown in Eq. 9-102. This is presumably because its conformation is highly restricted and the –COOH is constrained to frequently collide with the – OH group.[178,179] The three methyl

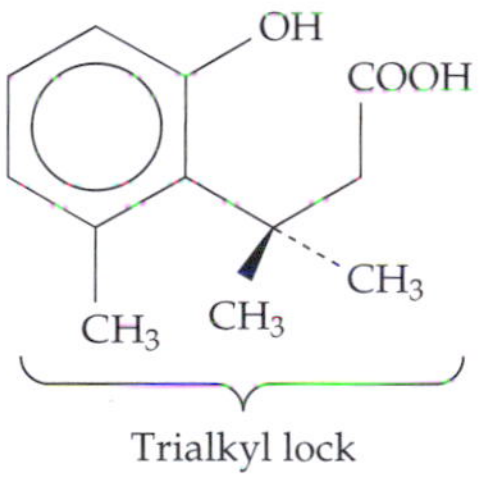

Trialkyl lock

groups interdigitate and form a **trialkyl lock**. Orientation must also play a large role in enzymatic catalysis. As previously emphasized, enzymes often orient substrates precisely by formation of multiple hydrogen bonds. Small distortions by mutation or substitution of an essential metal by a different one can have very great effects. For example, because of an altered coordination pattern isocitrate dehydrogenases with Ca^{2+} in the active site has a maximum velocity of catalysis only 2.5×10^{-3} that with the normal Mg^{2+}-containing enzyme.[180] Bruice and associates concluded that enzymes must bring reacting groups into close proximity with orbitals of the reactants properly aligned in the ground state prior to moving to the transition state. They suggested that enzymes preorganize the enzyme-substrate complex into a **near attack conformation** in which the positions of reacting groups, the arrangements of hydrogen bonds, and the local dielectric constant in the active site, resemble those in the transition state. In this conformation the energy barrier to the transition

state may be very low.[128-128b] Preorganization of the complex also acts to eliminate the slow components of solvent reorganization required for reaction in aqueous solutions.[128c,128d]

The necessity for reacting groups of substrates to collide with an orientation that allows productive interaction of electronic orbitals is often called a **stereoelectronic effect**. An example is the addition reaction of Eq. 9-74. The orbital of an unshared pair of electrons on the HO^- ion must be perpendicular to the plane of the double bond. Furthermore, if the proton becomes attached to the adjacent carbon in a synchronous or concerted manner it must enter from the opposite side, as it does in Eq. 9-74.

In many biochemical reactions an alcohol or amine is eliminated from a tetrahedrally bonded intermediate as in Eq. 9-104. Deslongchamps proposed a stereoelectronic theory[181,182] according to which elimination of either NH—R_2 or –O—R_1 from this intermediate will depend upon the values of the torsion angles θ_1 and θ_2 (Eq. 9-104). The theory asserts that for rapid elimination of OR_1 (Eq. 9-104a), unshared electron pairs on *both* the –O^- and N atoms must be antiperiplanar to the bond being broken (see Figure 2-2). If rotation around the C–O bond (θ_1) occurs an orientation can be found in which an electron pair on each of the two oxygens will be antiperiplanar to the bond to NHR_2. From this orientation the latter will be eliminated (Eq. 9-104b). The theory has been supported by much experimental evidence involving reaction rates and product distribution among competing reactions of small conformationally restricted organic molecules.[181–184] Although more recent experiments[185] suggest that stereoelectronic factors may be of less significance than had been assumed, even a small decrease in transition state energy can be significant in an enzyme-catalyzed reaction. Enzymes may not only orient substrates in accord with stereoelectronic principles but also be able to promote conformational alterations in intermediates that allow them to take advantage of stereoelectronic factors. Examples are considered in Chapter 12.

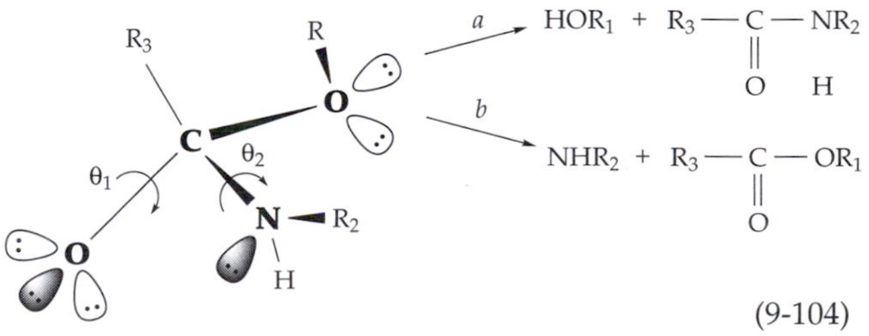

7. The Microenvironment

A substrate bound at an active site may be in an environment quite different than that in an ordinary

$$(9\text{-}104)$$

aqueous solution. In fact, the protein often surrounds the substrate to the extent that *the protein is the solvent*. What kind of **microenvironment** does the protein provide and can it assist in the catalytic process? Charged and dipolar groups of the protein provide an electrostatic field that provides part of the binding energy and which may also assist in catalysis, a concept expressed by Quastel as early as 1926.[183a] The ends of protein helices often seem to point at active centers.[186,187] In many cases more than one N terminus with its positive electrical potential (see Chapter 2) or more than one C terminus (negative potential) of a helix point to an active site. These helix dipoles may be important in stabilizing transition state structures[188] or in altering pK_a values of functional groups.[186] Fluctuations in charge distribution within a protein and in the hydrogen-bonding pattern of the protein with solvent and substrate may also be important.[189]

For some enzymatic reactions a transition state will be favored by a medium of very low dielectric constant and a correctly constructed active site can provide just such a surrounding.[189a] Hydrophobic groups may be packed around a site where an ion pair or other ionic interaction between enzyme and substrate occurs increasing the strength of that interaction. Conformational changes may enhance such effects. The dehydration of polar groups that must often occur upon binding of substrate may make these groups more reactive.[187,190,191] It has been suggested that the substantial volume changes ($\Delta V^{\ddagger}$) that sometimes occur during formation of transition states for enzyme-catalyzed reactions may result largely from changes in hydration of groups on the enzyme surface and that these changes may play an important role in catalysis.

In the past most enzymologists tacitly assumed that the external medium in which enzymes act must be aqueous. However, many enzymes function well in media containing largely hydrocarbons. Enzymes in a dry, powdered form have been suspended directly in organic solvents.[192,193] Under these conditions, enzymes may contain only tightly bound "structural" water, together with less than one equivalent of a monolayer of water outside the protein. The enzymes often remain active and are able to catalyze reactions with an altered substrate specificity as well as different reactions overall.

8. Strain and Distortion

The fact that enzymes appear to bind their substrates in such a way as to surround and immobilize them means that something other than the kinetic energy of the substrate is needed to provide energy for the ES complex to pass over the transition state barrier. What is the source of this activation energy? As with nonenzymatic reactions, it must come ultimately from the translational energy of solvent or solute molecules bombarding the complex. Can enzymes act as "energy funnels" that effectively channel kinetic energy from spots on the enzyme surface to the active site?[194] This could either be through strictly mechanical movement or through induction by fluctuations of the electrical field.[195]

It is often suggested that enzymes assist in catalysis by distorting bond lengths or angles away from their normal values. If the distorted structure were closer to the transition state geometry than the undistorted structure, catalysis would be assisted (see *lysozyme*, Chapter 12). However, Levitt[196] concluded (see also Fersht[2]) that forces provided this way by a protein are small and that the protein would become distorted rather than a substrate. On the other hand, if binding of a substrate distorts the protein, could the resultant "stored energy" be used in some way to assist in catalysis? If binding of a substrate bends a "spring" (Section C,4), can the tension in the spring then be used to distort the enzyme to be more exactly complementary to the transition state? This concept has become popular.[2,113,197]

The amino acid side chains of proteins are usually well packed. However, neither the side chains nor the main chain are rigid and immobile. Some regions of the protein will contain empty spaces – packing defects. Lumry called these **mobile defects** because, as a result of fluctuations in side chain packing, they can move within a protein from one site to another for a considerable distance.[198] Much evidence, including X-ray studies at low temperatures,[199] supports the existence of many **conformational substates** in proteins. Some substates may bind substrates better than others and some may allow conversion to the transition state more readily than do others. Rotation of histidine rings, peptide linkages, – OH groups, or amide side chains[199a] may be required and has, in fact, been observed for some enzymes. Perhaps in one of the substates of an enzyme–substrate complex an especially favorable vibrational mode[200] leads the complex to the transition state. In the transition state the packing of side chains may be especially tight. A mobile defect present in the active site may have moved elsewhere. The binding of substrates to the protein in the transition state will also be tighter because of the conformational alterations that have occurred. The binding energy of the substrates is now being utilized to lower the transition state barrier. The substrate is literally squeezed into the transition state configuration.

Is it possible that the protein domains forming an active site act like **ferroelectric crystals**, which change their dipole moment in response to a change in electric field? The highly polarizable hydrogen-bonded network, the amide linkages, imidazole rings, guanidinium group, etc. of active sites may permit a flip-flop of the dipole moment as in domains of ferroelectric crystals.

In such crystals, e.g., those of the hydrogen-bonded KH_2PO_4, a 180° change of dipole-moment direction results from very small movements of heavy atoms together with larger movements of the hydrogen-bonded protons.[200a,b,c] Could a similar flip-flop in a protein domain be coupled to the passage of a substrate over the transition state barrier? I have not seen any discussion of this possibility, but the structure of protein domains would seem to allow it. One could also imagine that with a small change in the hydrogen-binding arrangement of protein groups an active site could become preorganized to favor a flip-flop along a hydrogen-bonded chain in a different direction for a subsequent step in a reaction sequence.

9. Why Oligomeric Enzymes?

As we have seen (Chapter 7), a large fraction of all proteins exist as dimers, trimers, and higher oligomers. Oligomeric proteins raise the osmotic pressure much less than would the same number of monomeric subunits and this may be crucial to a cell. Another advantage of oligomers may be reflected in the fact that active sites of enzymes are often at interfaces between two or more subunits. This may enhance the ability of enzymes to undergo conformational rearrangements that are required during their action, just as hemoglobin changes its oxygen affinity in concert with a change in intersubunit contacts (Fig. 7-25).

A curious observation is that crystals of the dimeric pig heart malate dehydrogenase bind only one molecule of the substrate NAD^+ per dimer tightly; the second NAD^+ is bound weakly.[201,202] Similar *anticooperativity* has been reported for other crystalline dehydrogenases[203] and various other enzymes. An intriguing idea is that anticooperativity in binding might reflect a cooperative action between subunits during catalysis. Suppose that only conformation A of a dehydrogenase bound reduced substrate and NAD^+, while conformation B bound NADH and oxidized substrate. If reduced substrate and NAD^+ were present in excess and if oxidized substrate were efficiently removed from the scene by further oxidation, the following cooperative events could occur in the mixed AB dimer. The subunit of conformation A would bind substrates, react, and be converted to conformation B. At the same time, because of the strong AB interaction, the subunit that was originally in conformation B would be converted back to A and would be ready to initiate a new round of catalysis. Since conformation A has a low affinity for NADH, dissociation of the reduced coenzyme, which is often the slow step in dehydrogenase action, would be facilitated.[204–207] Such a **reciprocating** or **flip-flop** mechanism, suggested first by Harada and Wolfe,[204] is attractive because it provides a natural basis for the existence of the many known dimeric enzymes that do not exhibit evident allosteric properties. Attempts to verify this idea have largely failed. However, recent crystallographic studies of both glyceraldehyde phosphate dehydrogenase[206,207] and thymidylate synthase[208] are consistent with the proposal. There is still a possibility that coordinated reciprocal changes in distribution of electrical charges in the two subunits may also be important.

10. Summary

It appears that enzymes exert their catalytic powers by **first** bringing together substrates and binding them in proper orientations at the active site. **Second**, they often provide acidic and basic groups in the proper orientations to promote proton transfers within the substrate. **Third**, groups within the enzyme (especially nucleophilic groups) may enter into covalent interaction with the substrates to form structures that are more reactive than those originally present in the substrate. **Fourth**, the protein often closes around the substrate to immobilize it and to hold it in an environment, often lacking water, which could impede catalysis. The enzyme is also probably able to make small readjustments of its structure to provide good complementarity to the substrate at every stage and especially to the transition state structure. **Fifth**, the enzyme may be able to induce strain or distortion in the substrate perhaps accompanying a conformational change within the protein. The following question is often asked: "Why are enzymes such large molecules?"[209] At least part of the answer is that an enzyme usually interacts, sometimes via special domains, with other proteins.[206] Another part of the answer is evident when we consider that the formation of a surface complementary to that of the substrate and possessing reasonable rigidity requires a complex geometry in the peptide backbone. In addition, the enzyme must provide functional groups at the proper places to enter into catalysis. It may require a certain bulk to provide a low dielectric medium. Finally, if essential conformational changes occur during the course of the catalysis, we can only be surprised that nature has succeeded in packing so much machinery into such a small volume.

E. Classification of Enzymes

An official commission of the International Union of Biochemistry (IUB) has classified enzymes in the following six categories:[210]

1. *Oxidoreductases.* Enzymes catalyzing dehydrogenation or other oxidation and reduction reactions.

2. *Transferases*. These catalyze group transfer reactions.
3. *Hydrolases*. Enzymes catalyzing transfer of groups to the HO⁻ ion of H_2O.
4. *Lyases*. Enzymes promoting addition to double bonds or the reverse.
5. *Isomerases*. Enzymes catalyzing rearrangement reactions.
6. *Ligases* (synthetases). Enzymes that catalyze condensation with simultaneous cleavage of ATP and related reactions.

As an example, chymotrypsin is classified EC 3.4.4.5 according to the IUB system.[210] In this book a more mechanistically based classification is used. Because some official names are quite long, traditional trivial names for enzymes often have been retained. Remember that to be precise it is always necessary to mention the species from which an enzyme was isolated and, if possible, the strain. Also remember that almost every significant genetic difference is reflected in some change in some protein. It is possible that the enzyme you are working with is slightly different from the same enzyme prepared in a different laboratory.

References

1. Fromm, H. (1975) *Initial Rate Enzyme Kinetics*, Springer-Verlag, New York
2. Fersht, A. (1999) *Structure and Mechanism in Protein Science: A Guide to Enzyme Catalysis and Protein Folding*, Freeman, New York
3. Dixon, M., and Webb, E. C., eds. (1979) *Enzymes*, 3rd ed., Academic Press, New York
4. Wharton, C. W., and Eisenthal, R. (1981) *Molecular Enzymology*, Wiley, New York
5. Engel, P. C. (1982) *Enzyme Kinetics, the Steady-State Approach*, Chapman and Hall, London
6. Hammes, G. G. (1982) *Enzyme Catalysis and Regulation*, Academic Press, New York
7. Kull, F. J. (1994) *Principles of Biomolecular Kinetics and Binding*, CRC Press, Boca Raton, Florida
8. Suckling, C. J., ed. (1990) *Enzyme Chemistry*, 2nd ed., Chapman and Hall, New York
9. Price, N. C., and Stevens, L., eds. (1989) *Fundamentals of Enzymology*, 2nd ed., Oxford Univ. Press, Oxford, England
10. Kuby, S. A. (1990) *Enzymes: a Comprehensive Study*, CRC Press, Boca Raton, Florida
11. Schulz, A. R. (1994) *Enzyme Kinetics From Diastase to Multi-enzyme Systems*, Cambridge Univ. Press, New York
12. Segel, I. H. (1975) *Enzyme Kinetics:Behavior and Analysis of Rapid Equilibrium and Steady-State Enzyme Systems*, Wiley, New York
13. Cornish-Bowden, A. (1995) *Fundamentals of Enzyme Kinetics*, 2nd ed., Portland Press, Brookfield, Vermont
13a. Purich, D. L., and Allison, R. D. (2000) *Trends Biochem. Sci.* **25**, 455
14. Lowry, O. H., and Passonneau, J. V. (1972) *A Flexible System of Enzymatic Analysis*, Academic Press, New York
15. Passonneau, J., and Lowry, O., eds. (1993) *Enzymatic Analysis*, 1st ed., Humana Press, Totowa, New Jersey
16. Liu, Y.-M., and Sweedler, J. V. (1995) *J. Am. Chem. Soc.* **117**, 8871–8872
17. Liébecq, C., ed. (1992) *Biochemical Nomenclature*, Portland Press, London and Chapel Hill, North Carolina (for the International Union of Biochemistry and Molecular Biology)
18. Caldin, E. F. (1964) *Fast Reactions in Solution*, Wiley, New York
19. Jencks, W. P. (1987) *Catalysis in Chemistry and Enzymology*, Dover, Mineola, New York (p. 393)
20. Glick, N., Landman, A. D., and Roufogalis, B. D. (1979) *Trends Biochem. Sci.* **4**, N82–N83
20a. Wilkinson. (1980) *Biochem. J.* **80**, 324–
21. Orsi, B. A., and Tipton, K. E. (1979) *Methods Enzymol.* **63**, 159–183
22. Wharton, C. W., and Szawelski, R. J. (1982) *Biochem. J.* **203**, 351–360

23. Cleland, W. W. (1979) *Methods Enzymol.* **63**, 103–138
24. Frieden, C. (1994) *Methods Enzymol.* **240**, 311–322
25. Schechter, A. N. (1970) *Science* **170**, 273–280
26. Cha, S. (1970) *J. Biol. Chem.* **245**, 4814–4818
27. Martinez, M. B., Flickinger, M. C., and Nelsestuen, G. L. (1996) *Biochemistry* **35**, 1179–1186
28. Bull, H. B. (1971) *An Introduction to Physical Biochemistry*, 2nd ed., Davis Co., Philadelphia, Pennsylvania
29. Eisenstein, B. I. (1987) *Escherichia coli and Salmonella typhimurium*, Am. Soc. Microbiology, Washington, D.C., FC Niedhardt, ed. (pp.84–90)
30. Smoluchowski, M. (1917) *Z. Phys. Chem* **92**, 129–168
31. Caldin, E. F. (1964) *Fast Reactions in Solution*, Wiley, New York (pp. 10 and 279)
32. Debye, P. (1942) *Trans. Electrochem. Soc.* **82**, 265–272
32a. Elcock, A. H., Huber, G. A., and McCammon, J. A. (1997) *Biochemistry* **36**, 16049–16058
32b. Selzer, T. and Schreiber, G. (1999) *J. Mol. Biol.* **287**, 409–419
33. Mastro, A. M., Babich, M. A., Taylor, W. D., and Keith, A. D. (1984) *Proc. Natl. Acad. Sci. U.S.A.* **81**, 3414–3418
34. French, D. (1957) *Brewers Digest* **32**, 50–56
35. Cohn, E. J., and Edsall, J. T. (1943) *Proteins, Amino Acids and Peptides as Ions and Dipolar Ions*, Van Nostrand-Reinhold, Princeton, New Jersey (pp. 90–93)
36. Koenig, S. H. (1975) *Biopolymers* **14**, 2421–2423
37. Blacklow, S. C., Raines, R. T., Lim, W. A., Zamore, P. D., and Knowles, J. R. (1988) *Biochemistry* **27**, 1158–1167
38. Albery, W. J., and Knowles, J. R. (1976) *Biochemistry* **15**, 5627–5631
39. Albery, W. J., and Knowles, J. R. (1976) *Biochemistry* **15**, 5631–5640
40. Jenkins, W. T. (1982) *Adv. Enzymol.* **53**, 307–344
41. Haldane, J. B. S. (1930) *Enzymes*, Longmans, Green, New York
42. Cleland, W. W. (1970) in *The Enzymes*, 3rd ed., Vol. 2 (Boyer, P. D., ed), Academic Press, New York
43. Dalziel, K. (1957) *Acta Chem. Scand.* **11**, 1706–1723
44. Rebholz, K. L., and Northrop, D. B. (1994) *Arch. Biochem. Biophys.* **312**, 227–233
45. Volkenstein, M. V., and Goldstein, B. N. (1966) *Biokhim.* **31**, 541–547
46. Volkenstein, M. V., and Goldstein, B. N. (1966) *Biochim. Biophys. Acta.* **115**, 471–477
47. Fromm, H. J. (1970) *Biochem. Biophys. Res. Commun.* **40**, 692–697

48. Seshagiri, N. (1972) *J. Theor. Biol.* **34**, 469–486
49. Huang, C. Y. (1979) *Methods Enzymol.* **63**, 54–84
50. Chou, K., and Forsén, S. (1980) *Biochem. J.* **187**, 829–835
51. Fromm, H. J., Silverstein, E., and Boyer, P. D. (1964) *J. Biol. Chem.* **239**, 3645–3652
52. Fromm, S. J., and Fromm, H. J. (1999) *Biochem. Biophys. Res. Commun.* **265**, 448–452
53. Silverstein, E., and Boyer, P. D. (1964) *J. Biol. Chem.* **239**, 3901–3907
54. Hurst, R. O. (1969) *Can. J. Biochem. Physiol.* **47**, 941–944
54a. Cha, S. (1968) *J. Biol. Chem.* **243**, 820–825
55. Srere, P. A. (1967) *Science* **158**, 936–937
56. Hiromi, K., ed. (1979) *Kinetics of Fast Enzyme Reaction-Theory and Practice*, John Wiley, New York
57. Rentzepis, P. M. (1978) *Science* **202**, 174–182
58. Hammes, G. G., and Schimmel, P. R. (1970) in *The Enzymes*, 3rd ed., Vol. 2 (Boyer, P. D., ed), pp. 67–114, Academic Press, New York
59. Ahrens, M.-L., Maass, G., Schuster, P., and Winkler, H. (1970) *J. Am. Chem. Soc.* **92**, 6134–6139
59a. Gutfreund, H. (1999) *Trends Biochem. Sci.* **24**, 457–460
59b. Beechem, J. M. (1998) *Biophys. J.* **74**, 2141
60. Fink, A. L. (1979) *Trends Biochem. Sci.* **4**, 8–10
61. Balashov, S. P., Imasheva, E. S., Ebrey, T. G., Chen, N., Menick, D. R., and Crouch, R. K. (1997) *Biochemistry* **36**, 8671–8676
62. Fink, A. L., and Geeves, M. A. (1979) *Methods Enzymol.* **63**, 336–370
63. Alberty, R. A. (1956) *Adv. Enzymol.* **17**, 1–64
64. Dixon, M., and Webb, E. C., eds. (1979) *Enzymes*, 3rd ed., Academic Press, New York (pp.138–163)
65. Cleland, W. W. (1977) *Adv. Enzymol.* **45**, 273–387
66. Cleland, W. W. (1982) *Methods Enzymol.* **87**, 390–405
67. Kyte, J. (1995) *Mechanism in Protein Chemistry*, Garland Publ., New York
68. Ono, S., Hiromi, K., and Yashikawa, Y. (1958) *Bull. Chem. Soc. Jap.* **31**, 957–962
69. Brocklehurst, K., and Dixon, H. B. F. (1976) *Biochem. J.* **155**, 61–70
70. Tipton, K. F., and Dixon, H. B. F. (1979) *Methods Enzymol.* **63**, 183–234
71. Dixon, M. (1953) *Biochem. J.* **55**, 170–171
72. Purich, L. D., and Fromm, H. J. (1972) *Biochim. Biophys. Acta.* **268**, 1–3
73. Williams, J. W., and Morrison, J. F. (1979) *Methods Enzymol.* **63**, 437–467
74. Morrison, J. F. (1982) *Trends Biochem. Sci.* **7**, 102–105
75. Frieden, C. (1964) *J. Biol. Chem.* **239**, 3522–3531

76. Morrison, J. F., and Walsh, C. T. (1988) *Adv. Enzymol.* **61**, 201–301

77. Gutheil, W. G., and Bachovin, W. W. (1993) *Biochemistry* **32**, 8723–8731

78. Di Cera, E., Hopfner, K.-P., and Dang, Q. D. (1996) *Biophys. J.* **70**, 174–181

79. Botts, J., and Morales, M. (1953) *Trans. Faraday Soc.* **49**, 696–707

80. Monod, J., Wyman, J., and Changeux, J. D. (1965) *J. Mol. Biol.* **12**, 88–118

81. Plapp, B. V., Leidal, K. G., Smith, R. K., and Murch, B. P. (1984) *Arch. Biochem. Biophys.* **230**, 30–38

82. Chadha, V. K., Leidal, K. G., and Plapp, B. V. (1983) *Journal of Medicinal Chemistry* **26**, 916–922

83. Cho, H., Ramaswamy, S., and Plapp, B. V. (1997) *Biochemistry* **36**, 382–389

84. Westley, A. M., and Westley, J. (1996) *J. Biol. Chem.* **271**, 5347–5352

85. Hervé, G., ed. (1989) *Allosteric Enzymes*, CRC Press, Boca Raton, Florida

86. Rubin, M. M., and Changeux, J. D. (1966) *J. Mol. Biol.* **21**, 265–274

87. Grant, G. A., Schuller, D. J., and Banaszak, L. J. (1996) *Protein Sci.* **5**, 34–41

88. Williams, R. O., Young, J. R., and Majiwa, P. A. O. (1979) *Nature (London)* **282**, 847–849

89. Rabin, B. R. (1967) *Biochem. J.* **102**, 22c

90. Newsholme, E. A., and Start, C. (1973) *Regulation in Metabolism*, Wiley, New York

91. Walsh, C. (1979) *Enzymatic Reaction Mechanism*, Freeman, San Francisco, California

91a. Plapp, B. V. (1982) *Methods Enzymol.* **87**, 469–499

92. Sandler, M., ed. (1980) *Enzyme Inhibitors as Drugs*, Univ. Park Press, Baltimore, Maryland

93. Bey, P. (1981) *Chem. Ind. (London)*, 139–144

94. Hanson, K. R. (1966) *J. Am. Chem. Soc.* **88**, 2731–2742

94a. Prelog, V., and Helmchen, G. (1982) *Angew. Chem. Int. Ed. Engl.* **21**, 567–583

95. Bentley, R. (1969) *Molecular Asymmetry in Biology*, Vol. 1, Academic Press, New York (pp. 49–56)

96. Bentley, R. (1970) *Molecular Asymmetry in Biochemistry*, Vol. 2, Academic Press, New York

97. Alworth, W. L. (1972) *Stereochemistry and its Application in Biochemistry*, Wiley-Interscience, New York

98. Bentley, R. (1978) *Nature (London)* **276**, 673–676

99. Barry, J. M. (1997) *Trends Biochem. Sci.* **22**, 228–230

100. Hirschmann, H. (1960) *J. Biol. Chem.* **235**, 2762–2767

101. Post, C. B., and Ray, W. J. J. (1995) *Biochemistry* **34**, 15881–15885

102. Tsou, C.-L. (1986) *Trends Biochem. Sci.* **11**, 427–429

103. Ferscht, A. R. (1980) *Trends Biochem. Sci.* **5**, 262–265

104. Fersht, A. R. (1998) *Science* **280**, 541

105. Nureki, O., Vassylyev, D. G., Tateno, M., Shimada, A., Nakama, T., Fukai, S., Konno, M., Hendrickson, T. L., Schimmel, P., and Yokoyama, S. (1998) *Science* **280**, 578–582

106. Hopfield, J. J. (1980) *Proc. Natl. Acad. Sci. U.S.A.* **77**, 5248–5252

107. Jencks, W. P. (1975) *Adv. Enzymol.* **43**, 219–410

108. Westheimer, F. H. (1962) *Adv. Enzymol.* **24**, 441–482

109. Kraut, J. (1988) *Science* **242**, 533–539

110. Fersht, A. (1985) *Enzyme Structure and Mechanism*, 2nd ed., Freeman, New York

111. Gandour, R. D., and Schowen, R. L., eds. (1978) *Transition States of Biochemical Processes*, Plenum, New York

112. Laidler, K. J. (1969) *Theories of Chemical Reaction Rates*, McGraw-Hill, New York

113. Kyte, J. (1995) *Mechanism in Protein Chemistry*, Garland Publ., New York (pp. 199–213)

114. Marcus, R. A. (1992) *Science* **256**, 1523–1524

115. Pauling, L. (1948) *Nature (London)* **161**, 707–713

116. Linehard, G. E. (1973) *Science* **180**, 149–154

117. Wolfenden, R. (1976) *Annu Rev Biophys Bioeng.* **5**, 271–306

117a. Schramm, V. L. (1998) *Ann. Rev. Biochem.* **67**, 693–720

118. Nielsen, P. A., Glad, S. S., and Jensen, F. (1996) *J. Am. Chem. Soc.* **118**, 10577–10583

119. Zheng, Y.-J., and Bruice, T. C. (1997) *J. Am. Chem. Soc.* **119**, 8137–8145

120. Merkler, D. J., Kline, P. C., Weiss, P., and Schramm, V. L. (1993) *Biochemistry* **32**, 12993–13001

121. Degano, M., Almo, S. C., Sacchettini, J. C., and Schramm, V. L. (1998) *Biochemistry* **37**, 6277–6285

122. Glad, S. S., and Jensen, F. (1997) *J. Am. Chem. Soc.* **119**, 227–232

123. Schramm, V. L., Horenstein, B. A., and Kline, P. C. (1994) *J. Biol. Chem.* **269**, 18259–18262

124. Kline, P. C., and Schramm, V. L. (1994) *J. Biol. Chem.* **269**, 22385–22390

125. Deng, H., Kurz, L. C., Rudolph, F. B., and Callender, R. (1998) *Biochemistry* **37**, 4968–4976

125a. Cho, Y.-K., and Northrop, D. B. (1999) *Biochemistry* **38**, 7470–7475

125b. Wolfenden, R., Snider, M., Ridgway, C., and Miller, B. (1999) *J. Am. Chem. Soc.* **121**, 7419–7420

126. Menger, F. M. (1992) *Biochemistry* **31**, 5368–5373

127. Goldsmith, J. O., and Kuo, L. C. (1993) *J. Biol. Chem.* **268**, 18481–18484

128. Lightstone, F. C., and Bruice, T. C. (1996) *J. Am. Chem. Soc.* **118**, 2595–2605

128a. Bruice, T. C., and Lightstone, F. C. (1999) *Acc. Chem. Res.* **32**, 127–136

128b. Bruice, T. C., and Benkovic, S. J. (2000) *Biochemistry* **39**, 6267–6274

128c. Cannon, W. R., Singleton, S. F., and Benkovic, S. J. (1996) *Nature Struct. Biol.* **3**, 821–833

128d. Cannon, W. R., and Benkovic, S. J. (1998) *J. Biol. Chem.* **273**, 26257–26260

129. Murphy, D. J. (1995) *Biochemistry* **34**, 4507–4510

129a. Warshel, A. (1998) *J. Biol. Chem.* **273**, 27035–27038

129b. Jencks, W. P. (1987) *Catalysis in Chemistry and Enzymology*, Dover, Mineola, New York (pp. 210–213)

130. Hucho, F., and Wallenfels, K. (1971) *Eur. J. Biochem.* **23**, 489–496

131. Jencks, W. P. (1969) *Catalysis in Chemistry and Enzymology*, McGraw-Hill, New York (pp. 170–199)

132. Swain, C. G., and Brown, J. F., Jr. (1952) *J. Am. Chem. Soc.* **74**, 2534–2537 and 2538–2543

133. Rony, P. R. (1969) *J. Am. Chem. Soc.* **91**, 6090–6096

134. Jencks, W. P. (1987) *Catalysis in Chemistry and Enzymology*, Dover, Mineola, New York (p. 199)

135. Engdahl, K.-Å., Bivehed, H., Ahlberg, P., and Saunders, W. H., Jr. (1983) *J. Am. Chem. Soc.* **105**, 4767–4774

136. Hegarty, A. F., and Jencks, W. P. (1975) *J. Am. Chem. Soc.* **97**, 7188–7189

136a. Williams, A. (1999) *Concerted Organic and Bio-Organic Mechanisms*, CRC Press, Boca Raton, Florida

137. Wang, J. H. (1968) *Science* **161**, 328–334

138. Eigen, M. (1964) *Angew. Chem. Int. Ed. Engl.* **3**, 1–19

139. Metzler, D. E. (1979) *Adv. Enzymol.* **50**, 1–40

140. Nagle, J. F., and Morowitz, H. J. (1978) *Proc. Natl. Acad. Sci. U.S.A.* **75**, 298–302

141. Gerlt, J. A., and Gassman, P. G. (1993) *J. Am. Chem. Soc.* **115**, 11552–11568

142. Cohen, A. O., and Marcus, R. A. (1968) *J. Phys. Chem.* **72**, 4249–4256

143. Marcus, R. A. (1969) *J. Am. Chem. Soc.* **91**, 7224–7225

144. Guthrie, J. P. (1996) *J. Am. Chem. Soc.* **118**, 12878–12885

145. Guthrie, J. P. (1996) *J. Am. Chem. Soc.* **118**, 12886–12890

146. Jencks, W. P. (1987) *Catalysis in Chemistry and Enzymology*, Dover, Mineola, New York (pp. 207–213)

147. Dewar, M. J. S. (1984) *J. Am. Chem. Soc.* **106**, 209–219

148. Gandour, R. D., Maggiora, G. M., and Schowen, R. L. (1974) *J. Am. Chem. Soc.* **96**, 6967–6979

149. Quinn, D. M. (1987) *Chem. Rev.* **87**, 955–979

150. Klinman, J. P. (1989) *Trends Biochem. Sci.* **14**, 368–373

151. Rucker, J., Cha, Y., Jonsson, T., Grant, K. L., and Klinman, J. P. (1992) *Biochemistry* **31**, 11489–11499

152. Hwang, J.-K., and Warshel, A. (1996) *J. Am. Chem. Soc.* **118**, 11745–11751

153. Barbara, P. F., Walker, G. C., and Smith, T. P. (1992) *Science* **256**, 975–981

154. Kearley, G. J., Fillaux, F., Baron, M.-H., Bennington, S., and Tomkinson, J. (1994) *Science* **264**, 1285–1289

154a. Antoniou, D., and Schwartz, S. D. (1997) *Proc. Natl. Acad. Sci. U.S.A.* **94**, 12360–12365

154b. Basran, J., Sutcliffe, M. J., and Scrutton, N. S. (1999) *Biochemistry* **38**, 3218–3222

154c. Sutcliffe, M. J., and Scrutton, N. S. (2000) *Trends Biochem. Sci.* **25**, 405–408

154d. Huang, Y., Rettner, C. T., Auerbach, D. J., and Wodtke, A. M. (2000) *Science* **290**, 111–114

155. Hibbert, F., and Emsley, J. (1990) *Adv. Phys. Org. Chem.* **26**, 255–379

156. Gilli, P., Bertolasi, V., Ferretti, V., and Gilli, G. (1994) *J. Am. Chem. Soc.* **116**, 909–915

157. Perrin, C. L. (1994) *Science* **266**, 1665–1668

158. Garcia-Viloca, M., González-Lafont, A., and Lluch, J. M. (1997) *J. Am. Chem. Soc.* **119**, 1081–1086

158a. Bao, D., Huskey, W. P., Kettner, C. A., and Jordan, F. (1999) *J. Am. Chem. Soc.* **121**, 4684–4689

158b. Garcia-Viloca, M., Gelabert, R., González-Lafont, A., Moreno, M., and Lluch, J. M. (1998) *J. Am. Chem. Soc.* **120**, 10203–10209

158c. Bowers, P. M., and Klevit, R. E. (2000) *J. Am. Chem. Soc.* **122**, 1030–1033

159. McDermott, A., and Ridenour, C. F. (1996) in *Encyclopedia of NMR*, (Emsley, J. W., ed.), pp. 3820–3824, Wiley, Sussex, England

160. Frey, P. A. (1995) *Science* **269**, 104–106

161. Cleland, W. W., Frey, P. A., and Gerlt, J. A. (1998) *J. Biol. Chem.* **273**, 25529–25532

162. Cleland, W. W., and Kreevoy, M. M. (1994) *Science* **264**, 1887–1890

163. Frey, P. A., Whitt, S. A., and Tobin, J. B. (1994) *Science* **264**, 1927–1930

164. Cassidy, C. S., Lin, J., and Frey, P. A. (1997) *Biochemistry* **36**, 4576–4584

165. Halkides, C. J., Wu, Y. Q., and Murray, C. J. (1996) *Biochemistry* **35**, 15941–15948

166. Zheng, Y.-J., and Bruice, T. C. (1997) *Proc. Natl. Acad. Sci. U.S.A.* **94**, 4285–4288

167. Richard, J. P. (1998) *Biochemistry* **37**, 4305–4309

168. Scheiner, S., and Kar, T. (1995) *J. Am. Chem. Soc.* **117**, 6970–6975

169. Warshel, A., Papazyan, A., Kollman, P. A., Cleland, W. W., Kreevoy, M. M., and Frey, P. A. (1995) *Science* **269**, 102–106

170. Shan, S.-o, Loh, S., and Herschlag, D. (1996) *Science* **272**, 97–101

171. Ash, E. L., Sudmeier, J. L., De Fabo, E. C., and Bachovchin, W. W. (1997) *Science* **278**, 1128–1132

References

172. Shan, S.-o, and Herschlag, D. (1996) *J. Am. Chem. Soc.* **118**, 5515–5518
173. Warshel, A., and Papazyan, A. (1996) *Proc. Natl. Acad. Sci. U.S.A.* **93**, 13665–13670
174. Spector, L. B. (1982) *Covalent Catalysis by Enzymes*, Springer-Verlag, New York
175. Page, M. I., and Jencks, W. P. (1971) *Proc. Natl. Acad. Sci. U.S.A.* **68**, 1678–1683
176. Kirsch, J. F. (1973) *Ann. Rev. Biochem.* **42**, 205–234
177. Koshland, D. E., Jr., Carraway, K. W., Dafforn, G. A., Goss, J. D., and Storm, D. R. (1971) *Cold Spring Harbor Symp. Quant. Biol.* **36**, 13–20
177a. Villà, J., Strajbl, M., Glennon, T. M., Sham, Y. Y., Chu, Z. T., and Warshel, A. (2000) *Proc. Natl. Acad. Sci. U.S.A.* **97**, 11899–11904
178. Milstien, S., and Cohen, L. A. (1972) *J. Am. Chem. Soc.* **94**, 9158–9165
179. Karle, J. M., and Karle, I. L. (1972) *J. Am. Chem. Soc.* **94**, 9182–9189
180. Mesecar, A. D., Stoddard, B. L., and Koshland, D. E., Jr. (1997) *Science* **277**, 202–206
181. Deslongchamps, P., Lebreux, C., and Taillefer, R. (1973) *Can. J. Chem.* **51**, 1665–1669
182. Deslongchamps, P. (1975) *Tetrahedron* **31**, 2463–2490
183. Perrin, C. L., and Arrhenius, M. L. (1982) *J. Am. Chem. Soc.* **104**, 2839–2842
184. Evans, C. M., Glenn, R., and Kirby, A. J. (1982) *J. Am. Chem. Soc.* **104**, 4706–4707
185. Kuo, L. C., Fukuyama, J. M., and Makinen, M. W. (1983) *J. Mol. Biol.* **163**, 63–105
186. Plogman, J. H., Drenth, G., Kalk, K. H., and Hol, W. G. J. (1979) *J. Mol. Biol.* **127**, 149–162
187. Warshel, A. (1981) *Biochemistry* **20**, 3167–3177
188. Hol, W. G. J., van Duijen, P. T., and Berendsen, H. J. C. (1978) *Nature (London)* **273**, 443–446
189. Welch, G. R., ed. (1986) *The Fluctuating Enzyme*, Wiley, New York
189a. Mertz, E. L., and Krishtalik, L. I. (2000) *Proc. Natl. Acad. Sci. U.S.A.* **97**, 2081–2086
190. Dewar, M. J. S., and Storch, D. M. (1985) *Proc. Natl. Acad. Sci. U.S.A.* **82**, 2225–2229
191. Dewar, M. J. S., and Dieter, K. M. (1988) *Biochemistry* **27**, 3302–3308
192. Yennawar, H. P., Yennawar, N. H., and Farber, G. K. (1995) *J. Am. Chem. Soc.* **117**, 583–585
193. Schmitke, J. L., Wescott, C. R., and Klibanov, A. M. (1996) *J. Am. Chem. Soc.* **118**, 3360–3365
194. Kell, D. B. (1982) *Trends Biochem. Sci.* **7**, 1
195. Robertson, B., and Astumian, R. D. (1992) *Biochemistry* **31**, 138–141
196. Levitt, M. (1974) in *Peptides, Polypeptides and Proteins* (Blout, E. R., Bouey, F. A., Goodman, M., and Lotan, N., eds), p. 99, Wiley, New York
197. Avis, J. M., and Fersht, A. R. (1993) *Biochemistry* **32**, 5321–5326
198. Lumry, R., and Rosenberg, A. (1974) *Colloques Internationaux du C.N.R.S.* **246**, 53–62
199. Rader, S. D., and Agard, D. A. (1997) *Protein Sci.* **6**, 1375–1386
199a. Word, J. M., Lovell, S. C., Richardson, J. S., and Richardson, D. C. (1999) *J. Mol. Biol.* **285**, 1735–1747
200. Bialek, W., and Onuchic, J. N. (1988) *Proc. Natl. Acad. Sci. U.S.A.* **85**, 5908–5912
200a. Bystrov, D. S., and Popova, E. A. (1987) *Ferroelectrics* **72**, 147–155
200b. Emsley, J., Reza, N. M., Dowes, H. M., Hursthouse, M. B., and Kurodo, R. (1988) *Phosphorus and Sulfur* **35**, 141–149
200c. Rakvin, and Dalal. (1992) *Ferroelectrics* **135**, 227–236
201. Banaszak, L. A., and Bradshaw, R. A. (1975) in *The Enzymes*, 3rd ed., Vol. 11 (Boyer, P. D., ed), Academic Press, New York (pp. 369–396)
202. Zimmerle, C. T., and Alter, G. M. (1993) *Biochemistry* **32**, 12743–12748
203. Niefind, K., Hecht, H.-J., and Schomburg, D. (1995) *J. Mol. Biol.* **251**, 256–281
204. Harada, K., and Wolfe, R. G. (1968) *J. Biol. Chem.* **243**, 4131–4137
205. Lazdunski, M., Petitclerc, C., Chappelet, D., and Lazdunski, C. (1971) *Eur. J. Biochem.* **20**, 124–139
206. Skarzynski, T., Moody, P. C. E., and Wonacott, A. J. (1987) *J. Mol. Biol.* **193**, 171–187
207. Yun, M., Park, C.-G., Kim, J.-Y., and Park, H.-W. (2000) *Biochemistry* **39**, 10702–10710
208. Anderson, A. C., O'Neil, R. H., DeLano, W. L., and Stroud, R. M. (1999) *Biochemistry* **38**, 13829–13836
209. Srere, P. A. (1984) *Trends Biochem. Sci.* **9**, 387–390
210. Nomenclature Committee of the International Union of Biochemistry and Molecular Biology. (1992) *Enzyme Nomenclature*, Academic Press, San Diego, California

1. Give two reasons why enzymes are important to living organisms.

2. If an enzyme catalyzes the reaction A→ B + C, will it also catalyze the reaction B + C → A?

3. Can an enzyme use a site to catalyze a reverse reaction different from what it uses to catalyze the reaction in the forward direction? Explain your answer.

4. Define the "steady state" of living cells and contrast with a state of chemical equilibrium.

5. How is the instantaneous initial velocity of an enzymatic reaction measured? What precautions must be taken to ensure that a true instantaneous velocity is being obtained?

6. How is the rate of an enzyme action influenced by changes in:
   a) temperature
   b) enzyme concentration
   c) substrate concentration
   d) pH

7. In what ways are enzymatic reactions typical of ordinary catalytic reactions in organic or inorganic chemistry, and in what ways are they distinct?

8. In general, for a reaction that can take place with or without catalysis by an enzyme, what is the effect, if any, of the enzyme on
   a) standard Gibbs energy change of the reaction
   b) energy of activation of the reaction
   c) initial velocity of the reaction
   d) temperature coefficient of the rate constant

9. What are the dimensions of the rate constant for zero-, first-, and second-order reactions? If a first-order reaction is half completed in 2 min, what is its rate constant?

10. A sample of hemin containing radioactive iron, ^{59}Fe, was assayed with a Geiger-Müller counter at intervals of time with the following results:

Time (days)	Radioactive counts per min
0	3981
2	3864
6	3648
10	3437
14	3238
20	2965

Determine the half-life of ^{59}Fe and the value for the decay constant.

11. The kinetics of the aerobic oxidation of enzymatically reduced nicotinamide adenine dinucleotide (NADH) have been investigated at pH 7.38 at 30°C. The reaction rate was followed spectrophotometrically by measuring the decrease in absorbance at 340 nm over a period of 30 min. The reaction may be represented as

$$\text{NADH} + \text{H}^+ + \text{riboflavin} \rightarrow$$
$$\text{NAD}^+ + \text{dihydroriboflavin}$$

Time (days)	A (at 340 nm)
1	0.347
2	0.339
5	0.327
9	0.302
16.5	0.275
23	0.254
27	0.239
30	0.229

Determine the rate constant and the order of the reaction.

12. You have isolated and purified a new enzyme (E) which converts a single substrate (S) into a single product (P). You have determined M_r by gel filtration as ~ 46,400. However, in SDS gel electrophoresis, a molecular mass of ~ 23 kDa was indicated for the single protein band observed. A solution of the enzyme was analyzed in the following way. The absorbance at 280 nm was found to be 0.512. A 1.00 ml portion of the same solution was subjected to amino acid analysis and was found to contain 71.3 nmol of tryptophan. N-terminal analysis on the same volume of enzyme revealed 23.8 nmol of N-terminal alanine.

   a) What is the approximate molecular mass of the enzyme? Discuss this answer. Be sure to use an appropriate number of significant figures in this and other calculations.
   b) What is the concentration of your enzyme in moles per liter of active sites?
   c) What is the molar extinction coefficient ε at 280 nm where A = εcl; A = absorbance, c = mol / liter, and l = cell width in cm. Assume that all spectrophotometric measurements are made in 1.00 cm cuvettes.

   A second preparation of the enzyme had an absorbance at 280 nm of 0.485. This enzyme was diluted very carefully: 1.00 ml into 250 ml and this diluted enzyme was used for the following experiments (I to III).

Study Questions

Experiment I. A 1.00 ml portion of the diluted enzyme was added to 250 ml of buffered substrate at pH 7.0 and was mixed rapidly. The resulting initial substrate concentration $[S]_o$ was 1.000 mM. This reaction mixture was held at 25.0°C and portions were removed periodically at time t for analysis of the product P formed. The results follow. Plot $[P]$ vs. time.

Time t (s)	$[P]$ (mM)	Time t (s)	Remaining $[S]$ (mM)
200	0.104	2800	0.070
400	0.208	3200	0.040
800	0.392	3600	0.022
1200	0.554	4400	0.0060
1600	0.695	6000	0.00048
2000	0.800		
2400	0.881		

The integrated rate equation corresponding to the Michaelis–Menten equation

$$v = V_{max} [S] / (K_m + [S])$$ is as follows:

$$V_{max} \cdot t = [S]_o - [S] + K_m \ln ([S]_o/[S])$$

d) Plot $([S]_o - [S])/t$ vs. $1/t \times \ln ([S]_o/[S])$. From this plot evaluate K_m and V_{max}. Make this and other plots to appropriate scale on good quality graph paper.

e) What is k_{cat}?

Experiment II. In a second experiment, a series of test tubes were set up, each containing a different amount of buffered substrate at pH 7 but each in a volume of exactly 4.00 ml. The same enzyme solution used in part d (absorbance at 280 nm = 0.485) was diluted 2.00 ml in <u>250</u> ml as in I, then again 2.00 ml in <u>200</u> ml. Portions of 1.00 ml of this twice diluted enzyme were added at $t = 0$ to each of the test tubes of buffered substrate. The reaction was stopped in just 10.0 minutes by adding perchloric acid; a suitable reagent was added to provide for a colorimetric determination of the product. The results were as follows:

$[S]$ (mM)	Amount of product (μmol/tube)	$[S]$ (mM)	Amount of product (μmol/tube)
10.0	2.29	0.600	1.31
5.00	2.18	0.400	1.07
2.50	2.00	0.200	0.686
1.200	1.69	0.100	0.400
0.800	1.48		

f) Plot $1/v$ vs. $1/[S]$ where v is in units of μmol per tube and $[S]$ in millimoles/liter. Evaluate K_m, V_{max}, and k_{cat} from this plot. Fit your data by eye. It is generally agreed that a linear least squares fit is inappropriate unless suitable weighting factors are used.

g) Plot the same data as $v/[S]$ vs. v. Again evaluate K_m and V_{max}.

Experiment III. The preceding experiment was repeated but an inhibitor was present in each tube in a concentration equal to 5.00 mM, 10.0 mM, or 25.0 mM. Two different inhibitors were used, A and B. The following results were obtained.

	\multicolumn{4}{c}{μmol of product formed in 10 min/tube}		
$[S]$ (mM)	$[I] =$ 5.00 mM Inhibitor: A	5.00 mM B	25.0 mM Inhibitor: A	10.0 mM B
5.00	2.00	1.09	1.50	0.727
2.50	0.71	1.00	1.09	0.667
1.20	1.31	0.848	0.686	0.565
0.800	1.07	0.739	0.505	0.492
0.600	0.900	0.654	0.400	0.436
0.400	0.686	0.533	0.282	0.356
0.200	0.400	0.343	0.150	0.229
0.100	0.218	0.200	0.077	0.133

h) Plot $1/v$ vs. $1/[S]$ for each of these sets of data. For each case evaluate V_{max}, apparent K_m, and inhibitor constants $K_I = [I] [E]/[EI]$. There may be two K_I values.

i) For uninhibited enzymes, when $[S] = 0.4$ mM what fraction of the enzyme is ES? Free E?

j) For enzyme in the presence of 10.00 mM inhibitor and 0.8 mM substrate, what fraction of the total enzyme is ES? EI? free E?

k) Recall that $v = [ES] k_{cat} = [E][S] \cdot k_{cat}/K_m$ in many cases. What will be the relative rates of product formation from your substrate S and another competing substrate S′ present at the same concentration which also reacts with your enzyme with $K_m = 0.01$ mM and $k_{cat} = 200/s$ if $[S] = [S′]$?

13. Using the method of graphs, write the initial rate equation for the following system with A, B, P, and Q present.

Odd-numbered rate constants are for forward reactions; even-numbered constants are for reverse.

a) Satisfy yourself that

$$\frac{-d[A]}{dt} = k_1[A][E] - k_2[EA]$$

$$= \frac{-d[B]}{dt} = k_3[B][EA] - k_4[EXY]$$

$$= \frac{d[P]}{dt} = k_5[EXY] - k_6[P][EQ]$$

$$= \frac{d[Q]}{dt} = k_7[EQ] - k_8[E][Q]$$

The determinants (given by the method of graphs) which provide the steady-state concentrations of [E] and of the various exzyme–substrate and enzyme–product complexes are

$[E] = k_2k_7[k_4 + k_5] + k_3k_5k_7[B] + k_2k_4k_6[P]$

$[EA] = k_1k_7[k_4 + k_5][A] + k_1k_4k_6[A][P] + k_4k_6k_8[P][Q]$

$[EXY] = k_1k_3k_7[A][B] + k_2k_6k_8[P][Q] + k_1k_3k_6[A][B][P]$
$\quad\quad\quad + k_3k_6k_8[B][P][Q]$

$[EQ] = k_2k_8[k_4 + k_5][Q] + k_1k_3k_5[A][B] + k_3k_5k_8[B][Q]$

b) To obtain an expression for v_f the expression for $-d[A]/dt$ may be multiplied by $[E]_t$ and divided by $[E] + [EA] + [EXY] + [EQ] = [E]_t$. Then the above determinants may be substituted.

c) What is $V_{max,forward}$ and $V_{max,reverse}$? HINT: $V_{max,forward}$ is obtained when $[P] = [Q] = 0$ and $[A] = [B] = \infty$. What are K_{mA} and K_{mB}? K_{mA} is obtained when $[P] = [Q] = 0$, $[B] = \infty$, and $v = 1/2\, V_{max,forward}$.

d) With a knowledge of the kinetic parameters, indicate how the eight rate constants may be obtained if the total concentration of enzyme, $[E]_t$, is known.

14. Anticooperativity was observed in the plot of velocity vs. substrate concentration for an enzyme. Can this observation be explained by the Monod–Wyman–Changeux model for oligomeric enzymes? By the model of Koshland? Explain.

15. Interpret, for each of the following cases, the curve showing measured initial velocity at constant substrate concentration (*not* maximum velocity) against pH for an enzyme-catalyzed reaction.

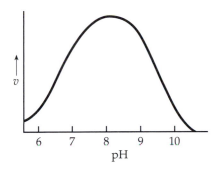

a) The substrate is neutral and contains no acidic or basic groups. The Michaelis constant is found to be independent of pH over the range studied.

b) The substrate is neutral and contains no acidic or basic groups, and the maximum velocity is found to be independent of pH.

c) The substrate is an α-amino acid, and the maximum velocity is found to be independent of pH.

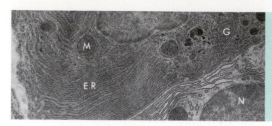

This electron micrograph of a thin section of pancreatic epithelial cells shows parts of two secretory cells that are synthesizing proenzymes (zymogens). Nuclei (N) are seen at the top center and lower right. Numerous ribosomes (barely seen here) line the many membranes of the endoplasmic reticulum (ER) and are most abundant near lateral and basal (left) surfaces of the cells. A few mitochondria (M) are present. The synthesized proteins move toward the apical surfaces (a small piece is seen at upper right), passing through the Golgi region (G) and accumulating in zymogen granules (Z) before secretion. From Porter, K. R. and Bonneville, M. A., *Fine Structure of Cells and Tissues*, Lea and Febiger, Philadelphia 1973. Courtesy of Mary Bonneville.

Contents

An Introduction to Metabolism

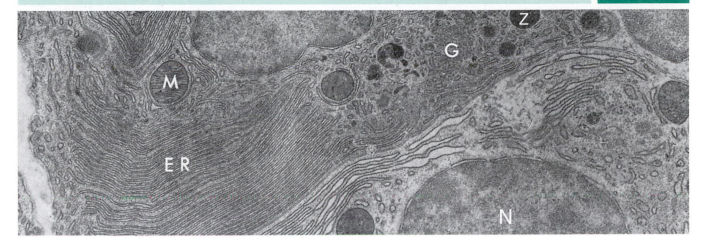

10

The first section of this book has dealt with the basic structures and with many of the complexities of the materials formed by living cells. In the next major section we will look at the chemical reactions that build and maintain a cell and that permit it to grow and to be responsive to external stimuli. These reactions are organized into **metabolic sequences** or **pathways** that form a complex, branched, and interconnected network. It would be pointless to try to memorize all of them. However, at this point it will be worthwhile to consider the significance of a few of the major sequences, which describe central pathways of metabolism.

Beginning in the nineteenth century, the investigation of these pathways and of the associated enzymes has provided much of our knowledge about the chemistry of living things. The protein products that are encoded in the structural genes that direct these pathways make up a large fraction of the material in a cell. It will be well worth the reader's effort to examine the chemistry of these metabolic sequences in detail. Regulatory mechanisms that are applicable to them, and to other pathways, are described in Chapter 11 and chemical details of enzyme action are considered in Chapters 12–16. Then, in Chapter 17, the chemical logic of the reaction sequences is considered in more detail.

A. The Starting Materials

All cells of all organisms take in chemical starting materials and give off chemical products. They all have a source of energy and generate heat during their metabolism. They all synthesize complex organic substances and maintain a high degree of organization, that is, a state of relatively low entropy. The materials taken up by cells are often organic compounds which

not only supply material for the synthesis of cellular constituents but also may be degraded to provide energy. A characteristic of all cells is the ability simultaneously to synthesize thousands of complex proteins and other materials and, at the same time, to break down (catabolize) the same types of compounds. Since cells both synthesize and catabolize most cellular components there is a continuous turnover of the very structural components of which they are composed. Metabolism encompasses all of these processes.

The most rapid catabolic reactions are usually those that provide the cell's energy. Organisms vary greatly in the materials used for food. Human beings, as well as many other organisms, break down carbohydrates, lipids, and proteins to obtain energy and starting materials for biosynthesis. In contrast, some organisms use only one or a few simple organic compounds while autotrophs satisfy their needs entirely with inorganic materials. Looking at all of the species of living things we find an extraordinary range of specialization in metabolism as well as in structure. Nevertheless, there are many common features of metabolism. For example, most cells utilize glucose or a close relative as a source of biosynthetic intermediates. Pathways for synthesis of nucleic acids and proteins are similar in all species. Even the control of growth and development depends upon proteins whose structures are often conserved throughout the living world.

1. Digestion

We humans must digest most foods before we can utilize them. The same is true for most bacteria, which need amino acids generated by the breakdown of

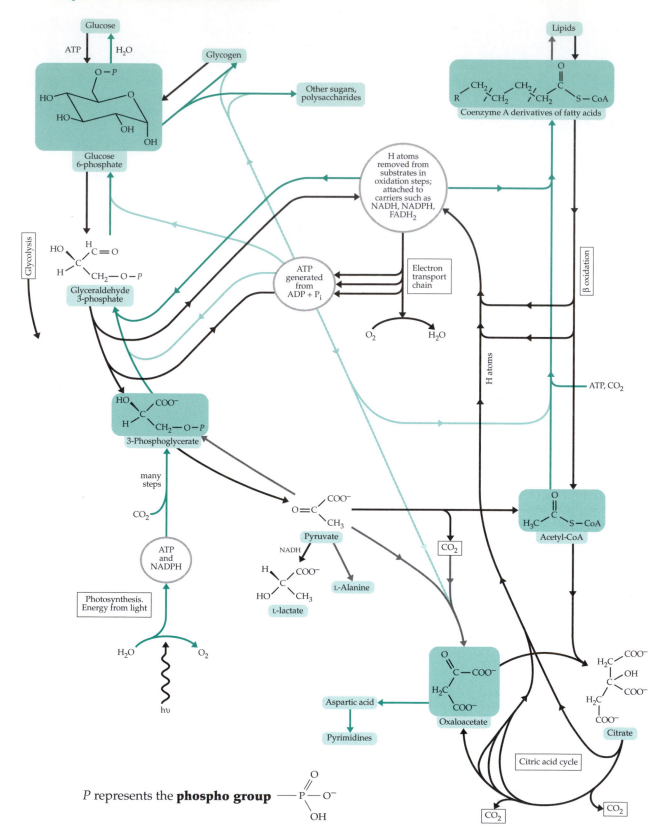

Figure 10-1 An overall view of some metabolic sequences. Several major pathways of catabolism are indicated by heavy lines. The glycolysis pathway, leading to pyruvate and lactate, starts at the top left, while the β oxidation pathway of fatty acids is on the right. Biosynthetic routes are shown in green lines. A few of the points of synthesis and utilization of ATP are indicated by the lighter green lines. Some of the oxidation–reduction reactions that produce or utilize the reduced hydrogen carriers NADH, NADPH, and FADH₂ are also indicated. The citric acid cycle, a major supplier of these molecules, is shown at the bottom right, while photosynthesis, the source of reduced hydrogen carriers in green plants, and the source of nearly all energy for life, is shown at the bottom left.

proteins. Our digestive enzymes, which act on carbohydrates, proteins, lipids, and nucleic acids, are synthesized by cells in the salivary glands, stomach, pancreas, and intestinal lining. The properties of these enzymes are described in Chapter 12. Bacteria secrete digestive enzymes into their surroundings. Although most green plants have no need to digest foods, they too have enzymes closely related to our digestive enzymes. These enzymes break bonds in proteins or carbohydrate polymers to allow expansion of leaves and stems, ripening of fruits, and other physical changes. All cells carry out **processing** or **maturation** of their newly synthesized polymers. This often involves the cutting off of one or more pieces of a protein, polysaccharide, or nucleic acid. The processing enzymes are also relatives of digestive enzymes.

2. Sources of Energy

All cells require a source of chemical energy. This is provided to a large extent by ATP (Chapter 6), whose hydrolysis can be coupled to reactions of synthesis, to transport across membranes, and to other endergonic processes. For this reason, all cells have active pathways for the synthesis of ATP. *Biosynthesis often also involves chemical reduction* of intermediates; therefore, cells must have a means of generating suitable reductants. The reagent of choice is often the hydrogen-carrying coenzyme **NADPH**. It is a phosphorylated form of NADH, a derivative of the vitamin **nicotinamide** (Box 15-A), whose structure follows. Reduction of metabolic intermediates by NADPH provides a second mechanism by which chemical energy is harnessed for biosynthesis. A third source of energy, which is considered in Chapter 18, is the gradient of hydrogen ions that is set up across cell membranes by oxidative or photochemical processes.

The reduced coenzyme NADH

B. Catabolism and the Synthesis of ATP

The catabolic sequences by which cells obtain energy often appear dominant. For animals, fungi, and nonphotosynthetic bacteria these pathways are used to metabolize large amounts of food and to produce large amounts of ATP. Because they are also related to biosynthetic pathways they have a central importance in virtually all organisms.

1. Priming or Activation of Metabolites

After a polymeric nutrient is digested and the monomeric products are absorbed into a cell, an energy-requiring **priming reaction** is usually required. For example, the hydrolysis of fats (whether in the gut or within cells) produces free fatty acids. Before undergoing further metabolism, these fatty acids are combined with the thiol compound **coenzyme A** to form **fatty acyl-CoA** derivatives. This requires a thermodynamically unfavorable reaction that necessitates the "expenditure" of ATP, that is, its hydrolysis to AMP and inorganic pyrophosphate (PP_i). Likewise, glucose, when taken into cells, is converted into the phosphate ester glucose 6-phosphate (glucose-6-*P*), again with an associated cleavage of ATP. *The major metabolic pathways often start with one of these two substances: a fatty acyl-CoA derivative or glucose 6-phosphate.* The structures of both are given at the top of Fig. 10-1, which is designed to provide an overall view of metabolism.

The phospho group of glucose-6-*P* is not very reactive but it provides a "handle" which helps enzymes to recognize and to hold onto this glucose derivative. Likewise, the coenzyme A molecule, whose complete structure is given in Chapter 14, provides a large and complex handle for enzymes. However, from a chemical viewpoint, the formation of the thioester linkage of a fatty acid with the –SH group of coenzyme A in the acyl-CoA is more important. This alters the electronic structure of the fatty acid molecule, "activating" it toward the reactions that it must undergo subsequently. Thus, priming reactions often provide chemical changes essential to the mechanisms of subsequent reactions.

Activation of a carboxylic acid by formation of an acyl-CoA derivative (Eq. 10-1) is of special importance because of its widespread use by all organisms. The basic reaction occurs by two steps. The first (Eq. 10-1, step *a*) is a displacement reaction on the phosphorus atom of ATP to form an **acyl adenylate**, a mixed anhydride of the carboxylic acid and a substituted phosphoric acid. Such mixed anhydride intermediates, or **acyl phosphates**, are central to much of cellular energy metabolism because *they preserve the high group-transfer potential of a phospho group of ATP (Chapter 6) while imparting a high group-transfer potential to the acyl group.* This allows the acyl group to be transferred in step *b*

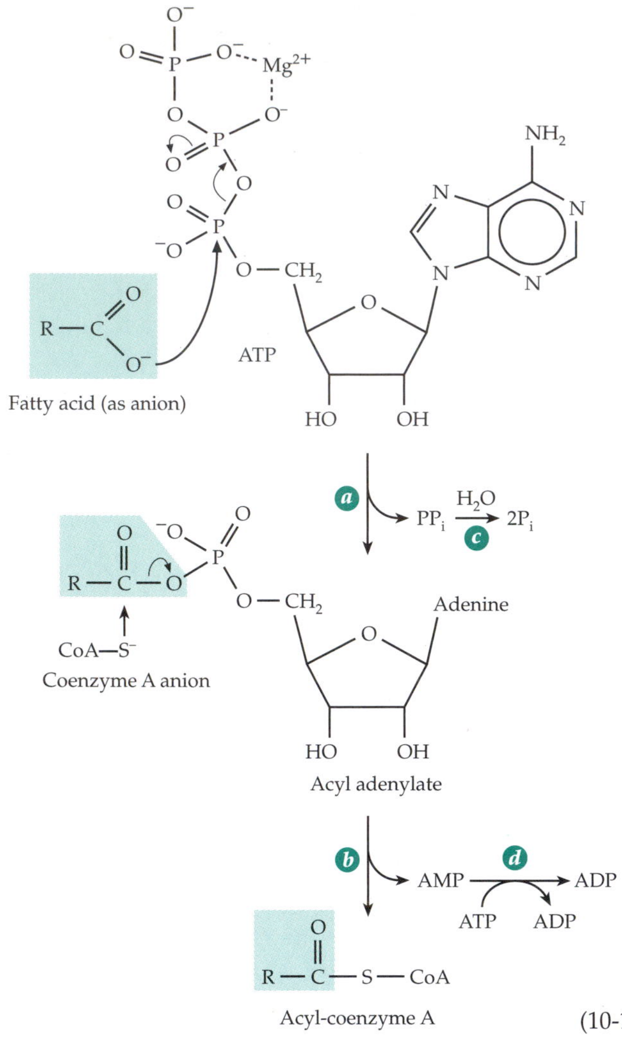

Fatty acid (as anion)

Coenzyme A anion

Acyl adenylate

Acyl-coenzyme A   (10-1)

of Eq. 10-1 onto the sulfur atom of coenzyme A to form an acyl-coenzyme A in which the high group transfer potential of the acyl group is conserved (see Table 6-6). Two additional steps are linked to this sequence. In step c the inorganic pyrophosphate that is displaced in step a is hydrolyzed to two molecules of HPO_4^{2-} (P_i), and in step d the AMP formed is phosphorylated by ATP to form ADP. The overall sequence of Eq. 10-1 leads to hydrolysis of two molecules of ATP to ADP + P_i. In other words, two molecules of ATP are spent in activating a carboxylic acid by conversion to an acyl-coenzyme A.

2. Interconversions of Sugar Phosphates

The strategy employed by most cells in the catabolism of several 6-carbon sugars is to convert them to glucose 6-phosphate and, in the several steps outlined in Fig. 10-2, to cleave this hexose phosphate to two equivalent molecules of **glyceraldehyde 3-phosphate**. This triose phosphate can then be metabolized further. Notice the chemical nature of the reactions involved in

the formation of glyceraldehyde 3-phosphate. The first step (Fig. 10-2, step a) is the transfer of the phospho group from ATP to the 6 position of glucose. Step b is the reversible opening of the sugar ring and step c the isomerization of an aldose to a ketose. Step d is a second transfer of a phospho group from ATP, another priming reaction that provides the second of the phosphate handles for the two molecules of triose phosphate that are formed. Step e is an aldol cleavage which breaks the C–C bond in the center of the ketose chain, while step f is another sugar isomerization that is chemically similar to the one in step c.

We see that in the six steps shown in Fig. 10-2 for conversion of glucose to two molecules of glyceraldehyde 3-phosphate, there are only four kinds of reaction: (1) phospho transfer from ATP; (2) opening of a sugar ring; (3) aldose-ketose isomerization; and (4) aldol cleavage. These types of reactions, which are described in more detail in Chapters 12 and 13, occur in many places in metabolism. As is indicated in Fig. 10-2, glucose 6-phosphate can also be generated from glycogen or starch. The first step (Fig. 10-2, step g) is phosphorolysis, cleavage by an inorganic phosphate ion. This leads directly to glucose 1-phosphate. The latter is isomerized by a phospho transfer process (step h).

3. Glycolysis and Fermentation

A major route of breakdown of carbohydrates is the **Embden–Meyerhof–Parnas** pathway, often referred to simply as **glycolysis**. It is indicated on the left side of Fig. 10-1 and in more detail in Figs. 10-2 and 10-3. The pathway begins with the reactions of Fig. 10-2, with either free glucose or glycogen as starting materials. Its end products may be reduced materials such as **lactic acid** or **ethanol**, which are formed under *anaerobic conditions* (Fig. 10-3). However, under *aerobic conditions* the product is **acetyl-coenzyme A**, whose acetyl group can then be oxidized to carbon dioxide and water in the **citric acid cycle**.

After the sugar chain is cleaved in the glycolysis sequence, the two resultant molecules of glyceraldehyde 3-phosphate are oxidized to 3-phosphoglycerate (Fig. 10-3, steps a,b). The oxidant is the hydrogen carrier NAD^+, the oxidized form of NADH. Cells frequently use NAD^+ to dehydrogenate alcohols to aldehydes or ketones, with one atom of hydrogen and an electron from the alcohol becoming attached to the NAD^+ to give NADH and the other hydrogen being released as H^+ (Eq. 10-2).

The oxidation of glyceraldehyde 3-phosphate is considerably more complex. The oxidation of an aldehyde to a carboxylic acid is a strongly exergonic process and the oxidation of glyceraldehyde 3-phosphate by cells is almost always coupled to the synthesis

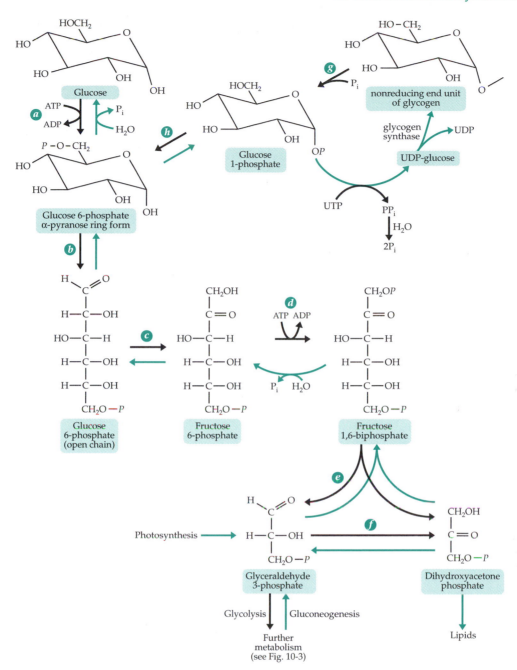

Figure 10-2 The interconversion of glucose, glycogen, and glyceraldehyde-3-phosphate in the pathways of glycolysis, gluconeogenesis, and glycogen synthesis. Pathways of catabolism are indicated with black lines and those of biosynthesis with green lines.

of ATP. The importance of this coupling can be understood by the fact that *for many organisms living by anaerobic fermentation reactions, this one oxidation step provides the sole source of ATP.* It is the mechanism for coupling this oxidation to synthesis of ATP that accounts for the complexity of the glyceraldehyde 3-phosphate dehydrogenase reaction, whose chemistry is presented in detail in Chapter 15. As is indicated in the simplified version of Fig. 10-3 (step *a*) inorganic phosphate (P$_i$) is

a reactant and the product released by the enzyme is 1,3-bisphosphoglycerate. This compound, which has been formed by the oxidative process, is an anhydride of phosphoric acid with 3-phosphoglyceric acid. The phospho group of such **acyl phosphates**, like that of ATP, has a high group transfer potential and can be transferred to ADP to generate ATP (Fig. 10-3, step *b*). Since each glucose molecule is cleaved to two molecules of triose phosphate, two molecules of ATP are generated in this process by the fermentation of one molecule of glucose. This is enough for a bacterium to live on if it ferments enough sugar.

Further metabolism of 3-phosphoglycerate involves an isomerization by means of a phosphotransferase (mutase reaction, step *c*) to form 2-phosphoglycerate which then loses water in an elimination reaction (step *d*). Elimination of a hydroxyl group in a β position relative to a carboxyl group together with an α-proton is another very frequently used type of metabolic reaction. In this case the product, **phosphoenol-pyruvate** (PEP), is a unique and important metabolite. It is a key intermediate in the biosynthesis of aromatic amino acids and in numerous other biosynthetic sequences.

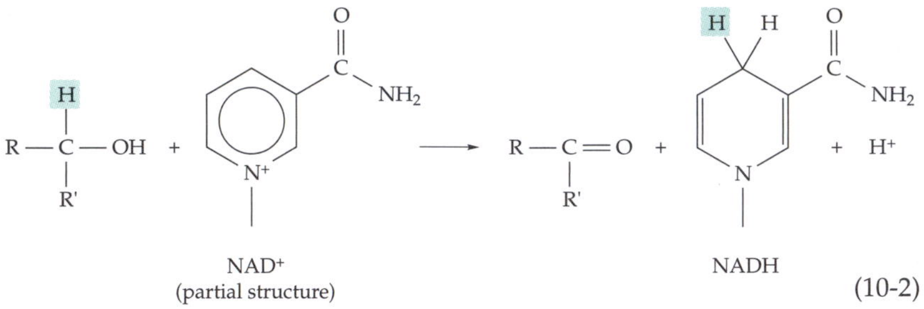

In the Embden–Meyerhof–Parnas pathway PEP transfers its phospho group to ADP (step *e*) to generate ATP and pyruvate. The latter is shown enclosed in brackets in Fig. 10-3 as the enolic form, which, tautomerizes to the oxo form (step *f*). The fact that the oxo form of pyruvate is much more stable than the enol gives the phospho group of phosphoenolpyruvate its high group transfer potential. The metabolic significance of this step is that the phosphate handles of the glycolytic intermediates, which were initially transferred on from ATP, are now returned to ADP with the regeneration of an equivalent amount of ATP.

The product of this metabolic sequence, **pyruvate**, is a metabolite of central importance. Its fate depends upon the conditions within a cell and upon the type of cell. When oxygen is plentiful pyruvate is usually converted to acetyl-coenzyme A, but under anaerobic conditions it may be reduced by NADH + H⁺ to the alcohol **lactic acid** (Fig. 10-3, step *h*). This reduction exactly balances the previous oxidation step, that is, the oxidation of glyceraldehyde 3-phosphate to 3-phosphoglycerate (steps *a* and *b*). With a balanced sequence of an oxidation reaction, followed by a reduction reaction, glucose can be converted to lactate in the absence of oxygen, a **fermentation** process. The lactic acid fermentation occurs not only in certain bacteria but also in our own muscles under conditions of extremely vigorous exercise. It also occurs continuously in some tissues, e.g., the transparent lens and cornea of the eye.

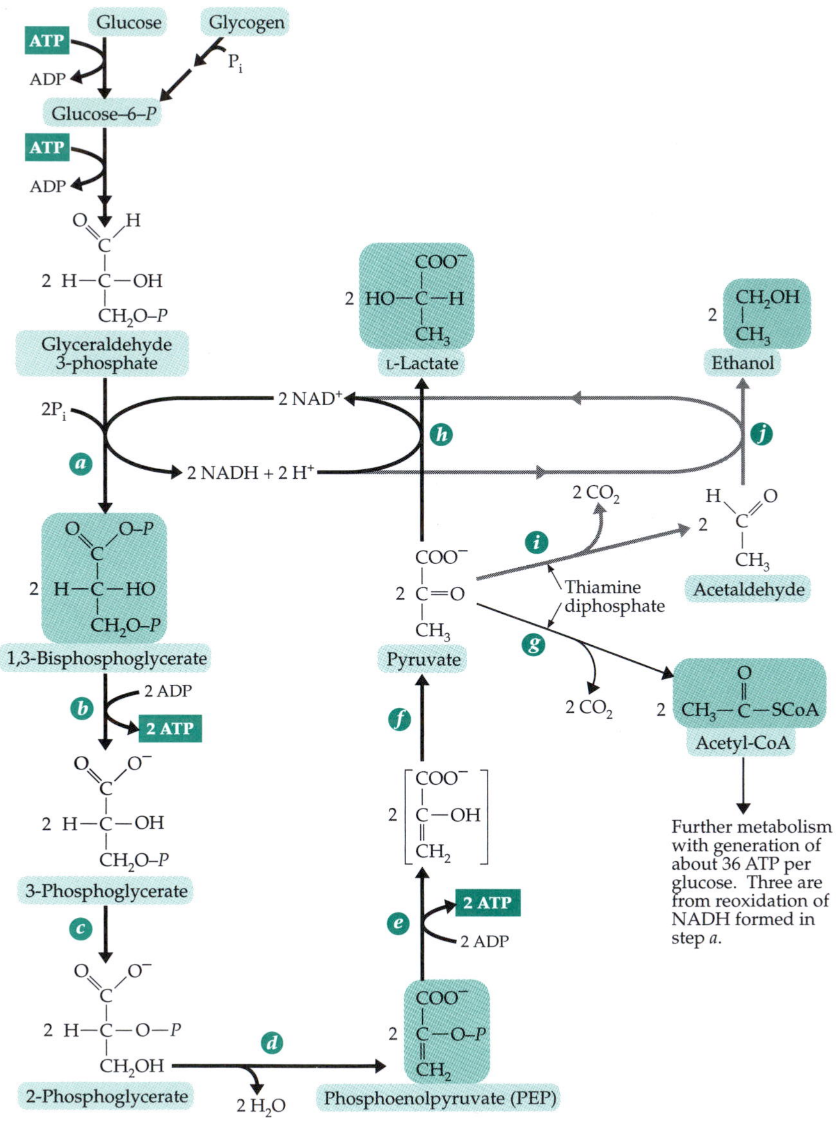

$$(10\text{-}2)$$

Figure 10-3 Coupling of the reactions of glycolysis with formation of lactic acid and ethanol in fermentations. Steps *a* to *g* describe the Embden–Meyerhof–Parnas pathway. Generation of 2 ATP in step *b* can provide all of the cell's energy.

In the well-known fermentation of sugar to ethanol by yeast, the pyruvate generated by glycolysis is first decarboxylated to acetaldehyde (Fig. 10-3, step *i*). This decarboxylation of a 2-oxo acid is chemically difficult and the enzyme catalyzing it makes use of a special reagent known as a **coenzyme**. For this type of reaction, the coenzyme is the diphosphate (pyrophosphate) ester of **thiamin** (**vitamin B$_1$**). Its mode of action is discussed in Chapter 14. It is usually needed when a 2-oxo acid is decarboxylated. The alcoholic fermentation by yeast is completed by reduction of acetaldehyde to ethanol (Fig. 10-3, step *j*), again using the NADH produced in the oxidation of glyceraldehyde 3-phosphate. The conversion of glucose to lactic acid or to ethanol and CO_2 are just two of many fermentation reactions, most of which are carried out by bacteria and which are dealt with further in Chapter 17. An important requirement is that the Gibbs energy change for the overall fermentation reaction be sufficiently negative that ATP synthesis can be coupled to it. Thus, using data from Table 6-4:

$$C_6H_{12}O_6 \text{ (glucose)} \rightarrow 2\ C_3H_5O_3 \text{ (lactate)} + 2H^+$$
$$\Delta G' \text{ (pH 7)} = -196 \text{ kJ/mol} (-46.8 \text{ kcal/mol}) \quad (10\text{-}3)$$

$$2ADP^{3-} + 2P_i^{2-} + H^+ \rightarrow H_2O + ATP^{4-}$$
$$\Delta G' \text{ (pH 7)} = +69 \text{ kJ/mol} \quad (10\text{-}4)$$

$$\text{Sum: } C_6H_{12}O_6 + 2ADP^{3-} + 2P_i^{2-} \rightarrow$$
$$H_2O + 2\ C_3H_5O_3 + 2H^+ + 2\ ATP^{4-}$$
$$\Delta G' \text{ (pH 7)} = -127 \text{ kJ/mol} \quad (10\text{-}5)$$

A requirement for all fermentations is the existence of a mechanism for coupling ATP synthesis to the fermentation reactions. In the lactic acid and ethanol fermentations this coupling mechanism consists of the formation of the intermediate 1,3-bisphosphoglycerate by the glyceraldehyde 3-phosphate dehydrogenase (Fig. 10-3, step *a*). This intermediate contains parts of both the products ATP and lactate or ethanol.

4. Pyruvate Dehydrogenase

In most organisms undergoing aerobic metabolism, pyruvate is oxidized to acetyl-CoA in a complex process involving its decarboxylation (Eq. 10-6). This **oxidative decarboxylation**, like the decarboxylation of pyruvate to acetaldehyde, requires thiamin diphosphate. In addition, an array of other catalysts participate in the process (see Fig. 15-15). Among these are the electron carrier **flavin adenine diphosphate** (**FAD**), which is derived from the vitamin **riboflavin**. Like NAD$^+$, this

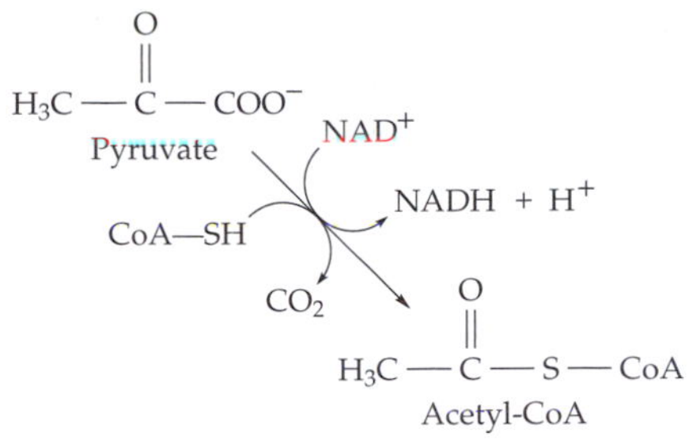

$$(10\text{-}6)$$

compound can accept two electrons (plus two protons) to form FADH$_2$. However, it sometimes serves as a one-electron carrier.

Acetyl-CoA is another major metabolic intermediate.

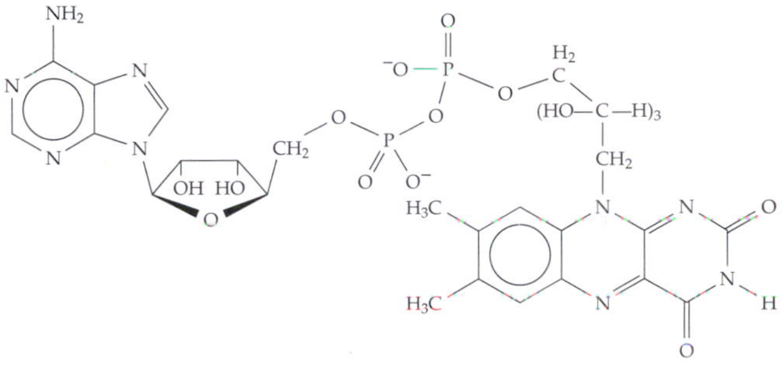

The coenzyme flavin adenine diphosphate (FAD)

It is an acyl-CoA of the type mentioned in Section 1 and can also be formed from acetate, ATP, and coenzyme A. Although the human diet contains some acetic acid, the two major sources of acetyl-CoA in our bodies are the oxidative decarboxylation of pyruvate (Eq. 10-6) and the breakdown of fatty acid chains. Let us consider the latter process before examining the further metabolism of acetyl-CoA.

5. Beta Oxidation

Whether fatty acids are oxidized to obtain energy or are utilized for biosynthesis, they are first converted to their acyl-CoA forms and are then cleaved to the two-carbon units represented by the acetyl groups of acetyl-CoA. The **beta oxidation** sequence, by which this occurs, is represented by the solid vertical arrow on the right side of Fig. 10-1 and is shown in greater detail in Fig. 10-4. We see from the latter figure that there are two dehydrogenation steps. The first (step *b*) removes hydrogen atoms from the α- and β-carbon atoms to produce a *trans* α, β unsaturated fatty acyl-CoA (Enoyl-CoA). The hydrogen acceptor is FAD. It is needed because this reaction requires a more powerful oxidant than does the dehydrogenation of an alcohol, for which NAD$^+$ is adequate. Addition of water to the double bond of the unsaturated acyl-CoA (step *c*) generates an alcohol which is then dehydrogenated (step *d*) by NAD$^+$ to form a β-oxoacyl-CoA. This is cleaved (step *e*) by a reaction (thiolysis) with another molecule of coenzyme A to form acetyl-CoA and a

new acyl-CoA with a shortened chain. The latter is "recycled" by passage through the β oxidation sequence repeatedly until the chain is shortened to a 2- or 3-carbon fragment, acetyl-CoA, or propionyl-CoA. Since most dietary fatty acids contain an even number of carbon atoms, acetyl-CoA is the predominant product.

The beta oxidation of fatty acids, like the dehydrogenation of pyruvate to acetyl-CoA, takes place within the inner **matrix** of the mitochondria in eukaryotic organisms. The reduced hydrogen carriers FADH$_2$ and NADH transfer their electrons to other carriers located within the **inner membrane** of the mitochondria. In bacteria the corresponding reactions occur with electron carriers present in the plasma membrane. Both FAD and NAD$^+$ are regenerated in this way and are able to again accept hydrogen from the beta oxidation reactions. This transfer of hydrogen atoms from substrates to hydrogen carriers is typical of biological oxidation processes.

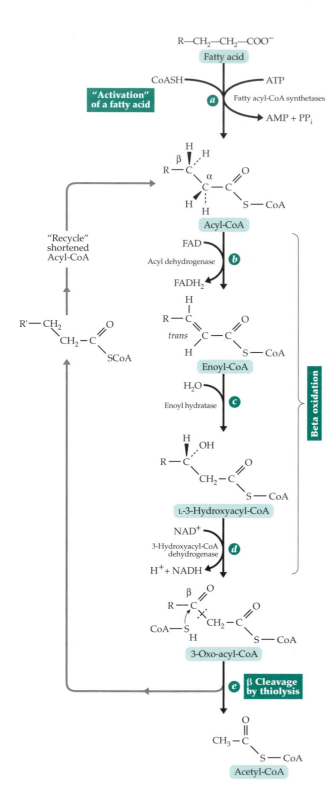

Figure 10-4 Reactions of fatty acid activation and of breakdown by β oxidation.

6. The Electron Transport Chain, Oxidative Phosphorylation

Reoxidation of the reduced carriers NADH and FADH$_2$ actually involves a sequence of electron carriers, the **electron transport chain**, whose function is indicated below the circle near the center of Fig. 10-1. The oxidation of reduced NADH by O$_2$ (Eq. 10-7) is a highly exergonic process and is accompanied by the

$$NADH + H^+ + 1/2\ O_2 \rightarrow NAD^+ + H_2O$$
$$\Delta G'\ (pH\ 7) = -219\ kJ/mol \tag{10-7}$$

generation of about three molecules of ATP (from ADP and inorganic phosphate). This process, termed **oxidative phosphorylation**, is the principal source of usable energy (in the form of ATP) provided by breakdown of both carbohydrates and fats in the human body.

The mechanism of oxidation of NADH in the electron transport chain appears to occur by transfer of a hydrogen atom together with two electrons (a hydride ion H$^-$). Oxidation of FADH$_2$ to FAD might occur by transfer of two hydrogen atoms or by transfer of H$^-$ + H$^+$. However, it is useful to talk about all of these compounds as **electron carriers** with the understanding that movement of one or both of the electrons may be accompanied by transfer of H$^+$. The electron transport complex is pictured in a very simplified form in Fig. 10-5.

The electrons donated from NADH or other reductants, upon entering this complex, travel from one carrier to the next, with each carrier being a somewhat more powerful oxidant than the previous one. The

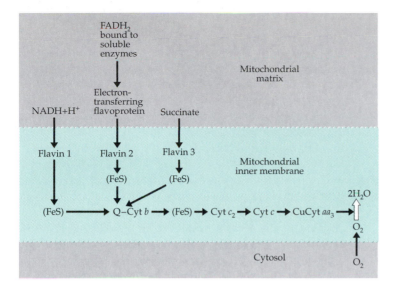

Figure 10-5 An abbreviated version of the electron transport chain of mitochondria. Four electrons reduce O_2 to 2 H_2O. For details see Figs. 18-5 and 18-6.

final carrier, known as cytochrome aa_3 or **cytochrome oxidase**, reacts with molecular oxygen. Each molecule of O_2, together with 4 H^+, is converted to two molecules of water. This stoichiometry requires that two molecules of NADH pass four electrons down the chain for each O_2 reduced.

The chemical structures of the components of the mitochondrial electron transport chain are varied but fall into several distinct classes. Most are proteins but these proteins always contain special coenzymes or **prosthetic groups** able to engage in oxidation–reduction reactions. At the substrate end of the chain NADH passes a hydrogen atom together with its bonding electron pair to a riboflavin-containing coenzyme **riboflavin 5'-phosphate** (FMN), which is tightly bound into a protein (designated Flavin 1 in Fig. 10-5). Similar **flavoproteins**, Flavin 2, and Flavin 3, act as oxidants for $FADH_2$ arising during beta

BOX 10-A   AN EARLY LABELING EXPERIMENT

In 1904, long before the advent of radioactive tracers, Knoop synthesized fatty acids labeled by chemical attachment of a benzene ring at the end opposite the carboxyl group. He prepared these compounds with both odd and even numbers of carbon atoms in straight chains and fed them to dogs. From the dogs' urine he isolated **hippuric acid** and **phenylaceturic acid**, which are the amides of glycine with benzoic acid and phenylacetic acid, respectively. Knoop showed that the phenylacetic acid was produced from those fatty acids with an

even number of carbon atoms, while the benzoic acid was formed from those with an odd number. From these experimental results, Knoop deduced that *fatty acid degradation occurs two carbon atoms at a time* and proposed his famous **β oxidation theory**.

Later experiments using isotopic labeling with ^{13}C verified Knoop's proposals,[a] but study of isolated enzymes was not possible until after the discovery of coenzyme A in 1950. Then, studies of fatty acid oxidation by extracts from isolated mitochondria established the details of the pathway.[b]

[a] Weinhouse, S. (1995) *FASEB J.* **9**, 820–821
[b] Quastel, J. H. (1984) *Trends Biochem. Sci.* **9**, 117–118

oxidation and oxidation of succinate, whose significance to metabolism is discussed in the next section.

A significant difference between NADH and $FADH_2$ is that the former diffuses freely between the **dehydrogenases** that transfer hydrogen to it and the flavoprotein NADH dehydrogenase (Flavin 1) that reoxidizes it. However, FAD, whether in its oxidized state or as $FADH_2$, stays tightly bound to proteins at all times. Only hydrogen atoms or electrons are transferred into or out of these proteins. During beta oxidation the $FADH_2$ generated remains tied to the fatty acyl-CoA dehydrogenase protein. However, the two hydrogen atoms of this $FADH_2$ are transferred to another molecule of FAD, which is bound to an **electron-transferring flavoprotein.** This protein carries electrons one at a time, as FADH, by diffusion to the inner surface of the inner mitochondrial membrane. There it transfers the hydrogen atoms that it carries to the flavoprotein designated Flavin 2 in Fig. 10-5. The electron-transferring flavoprotein can also be viewed as a carrier of single electrons, each accompanied by H^+.

A second group of electron carriers in mitochondrial membranes are the **iron–sulfur [Fe–S] clusters** which are also bound to proteins. Iron–sulfur proteins release Fe^{3+} or Fe^{2+} plus H_2S when acidified. The "inorganic clusters" bound into the proteins have characteristic compositions such as Fe_2S_2 and Fe_4S_4. The sulfur atoms of the clusters can be regarded as sulfide ions bound to the iron ions. The iron atoms are also attached to other sulfur atoms from cysteine side chains from the proteins. The Fe–S proteins are often tightly associated with other components of the electron transport chain. For example, the flavoproteins Flavin 1, Flavin 2, and Flavin 3 shown in Fig. 10-5 all contain Fe–S clusters as does the Q-cytochrome b complex. All of these Fe–S clusters seem to be one-electron carriers.

A third hydrogen carrier of mitochondrial membranes, and the only one that is not unequivocally associated with a specific protein, is the isoprenoid quinone **ubiquinone** or **coenzyme Q** (Q in Fig. 10-5). Ubiquinone apparently serves as a common carrier, collecting electrons from three or more separate input ends of the chain and directing them along a single pathway to O_2.

The final group of mitochondrial redox components are one-electron carriers, small proteins (**cytochromes**) that contain iron in the form of the porphyrin complex known as **heme**. These carriers, which are discussed in Chapter 16, exist as several chemically distinct types: a, b, and c. Two or more components of each type are present in mitochondria. The complex cytochrome aa_3 deserves special comment. Although cytochromes are single-electron carriers, the cytochrome aa_3 complex must deliver four electrons to a single O_2 molecule. This may explain why the monomeric complex contains two hemes and two **copper** atoms which are also able to undergo redox reactions.[1,2]

Although the components of the electron transport chain have been studied intensively, there is still some mystery associated with the process by which ATP synthesis is coupled to electron transport (oxidative phosphorylation). A theory originally proposed by Mitchell[3] and now generally accepted[4,5] is that passage of electrons through the chain "pumps" protons from the inside to the outside of the tight inner mitochondrial membrane. As a result, protons accumulate along the outside of this membrane, as do negative counterions along the inside. The membrane becomes charged like a miniature electrical condenser. The synthesis of ATP takes place in the small knoblike **ATP synthase** which is partially embedded in the same membranes. The Gibbs energy for the formation of ATP from ADP and inorganic phosphate is apparently supplied by the flow of protons back to the inside of the membrane through the ATP synthase. Possible

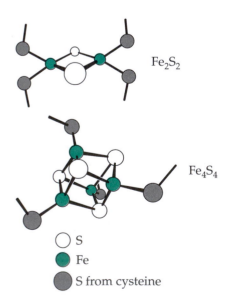

Fe₂S₂

Fe₄S₄

○ S

● Fe

⬤ S from cysteine

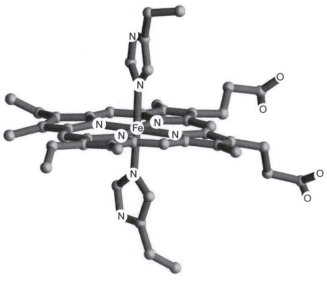

Iron in a heme in a b type cytochrome

ways in which this may occur as well as many other aspects of electron transport are dealt with in Chapter 18.

7. The Citric Acid Cycle

The 2-carbon acetyl units removed from fatty acid chains or generated from sugars by glycolysis and the action of pyruvate dehydrogenase must be completely oxidized to carbon dioxide to provide cells with the maximum amount of energy. The oxidation of an acetyl group is a difficult chemical process, and probably for this reason nature has devised an elegant catalytic cycle, the **citric acid cycle**, which is indicated at the lower right in Fig. 10-1 and is shown in detail in Fig. 10-6. The cycle begins when the 4-carbon **oxaloacetate** (also spelled oxalacetate) is condensed with an acetyl group of acetyl-CoA to form the 6-carbon **citrate**. Then, in the remaining reactions of the cycle, two carbon atoms are removed as CO_2 and oxaloacetate is regenerated. Several oxidation steps are involved, each of which feeds additional **reducing equivalents** (i.e., hydrogen atoms removed from substrates) into the pool of hydrogen carriers and allows for more synthesis of ATP via the electron transport chain. The importance of this pathway to an organism can be understood from the fact that when glucose is oxidized completely to CO_2 and water via the citric acid cycle, about 38 molecules of ATP are formed, 19 times as much as by fermentation.

Although oxaloacetate is regenerated and therefore not consumed during the operation of the citric acid cycle, it also enters other metabolic pathways. To replace losses, oxaloacetate can be synthesized from pyruvate and CO_2 in a reaction that uses ATP as an energy source. This is indicated by the heavy gray line leading downward to the right from pyruvate in Fig. 10-1 and at the top center of Fig. 10-6. This reaction depends upon yet another coenzyme, a bound form of the vitamin **biotin**. Pyruvate is formed from breakdown of carbohydrates such as glucose, and the need for oxaloacetate in the citric acid cycle makes the oxidation of fats in the human body dependent on the concurrent metabolism of carbohydrates.

C. Biosynthesis

At the same time that cells break down foodstuffs to obtain energy, they are continuously creating new materials. The green lines in Fig. 10-1 indicate some pathways by which such biosynthesis takes place. Examining the left side of the figure, we see that either pyruvate or 3-phosphoglycerate can be converted back to glucose 6-phosphate and that the latter can be used to synthesize glycogen or other sugars or polysaccharides.

1. Reversing Catabolic Pathways

Breakdown of carbohydrates is thermodynamically spontaneous. Therefore, *cells cannot simply use catabolic pathways operating in reverse without finding ways of coupling the cleavage of ATP to synthesis.* In the formation of glucose from pyruvate in the liver, a process known as **gluconeogenesis**, there are three distinct points at which the enzymes used differ from those used in catabolism: (1) Pyruvate is converted to phosphoenol-pyruvate by a mechanism utilizing more than one molecule of ATP, a pathway that is discussed in detail in Chapter 17; (2) as is shown in Fig. 10-2, fructose 1,6-bisphosphate is hydrolyzed to fructose 6-phosphate and inorganic phosphate by a **phosphatase** rather than through reversal of step *d*, which would form ATP; and (3) glucose-6-phosphate is hydrolyzed by a phosphatase rather than following the reverse of step *a* in Fig. 10-2. Furthermore, glycogen is synthesized from glucose 6-phosphate, not by reversal of the phosphorylase reaction (Fig. 10-2, step *g*), but via a new intermediate, **uridine diphosphate glucose** (UDPG), whose formation involves cleavage of UTP. The latter is generated by phospho transfer from ATP. Inorganic **pyrophosphate** (PP_i) is a product of UDPG formation and is removed by hydrolytic cleavage to two molecules of inorganic phosphate by the enzyme **pyrophosphatase**. This reaction helps to make the biosynthesis thermodynamically spontaneous by removing one of the reaction products.

Uridine diphosphate glucose (UDPG)

A similar situation exists in the case of fatty acid synthesis, which proceeds from acetyl-CoA and reverses fatty acid breakdown. However, both carbon dioxide and ATP, a source of energy, are needed in the synthetic pathway. Furthermore, while oxidation of fatty acids requires NAD^+ as one of the oxidants, and generates NADH, the biosynthetic process often requires the related NADPH. These patterns seen in biosynthesis of sugars and fatty acids are typical. *Synthetic reactions resemble the catabolic sequences in reverse, but distinct differences are evident.* These can usually be related to the requirement for energy and often also to control mechanisms.

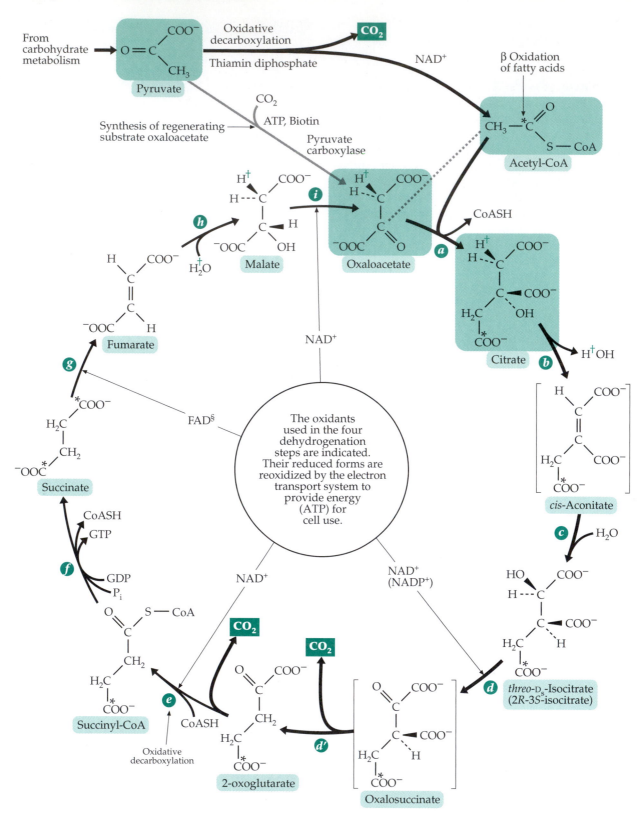

Figure 10-6 Reactions of the citric acid cycle (Krebs' tricarboxylic acid cycle). Asterisks designate positions of isotopic label from entrance of carboxyl-labeled acetate into the cycle. Note that it is *not* the two carbon atoms from acetyl-CoA that are immediately removed as CO_2 but two atoms from oxaloacetate. Only after several turns of the cycle are the carbon atoms of the acetyl-CoA completely converted to CO_2. Nevertheless, the cycle can properly be regarded as a mechanism of oxidation of acetyl groups to CO_2. Green daggers (†) designate the position of ^{2}H introduced into malate as ^{2}H$^+$ from the medium. FAD§ designates covalently bound 8-histidyl-FAD (see Chapter 15).

BOX 10-B DISCOVERY OF THE CITRIC ACID CYCLE (KREBS' TRICARBOXYLIC ACID CYCLE)

One of the first persons to study the oxidation of organic compounds by animal tissues was T. Thunberg, who between 1911 and 1920 discovered about 40 organic compounds that could be oxidized by animal tissues. Salts of succinate, fumarate, malate, and citrate were oxidized the fastest. Well aware of Knoop's β oxidation theory, Thunberg proposed a cyclic mechanism for oxidation of acetate. Two molecules of this two-carbon compound were supposed to condense (with reduction) to succinate, which was then oxidized as in the citric acid cycle to oxaloacetate. The latter was decarboxylated to pyruvate, which was oxidatively decarboxylated to acetate to complete the cycle. *One of the reactions essential for this cycle could not be verified experimentally.* It is left to the reader to recognize which one.

In 1935, A. Szent-Györgyi discovered that all of the carboxylic acids that we now recognize as members of the citric acid cycle stimulated respiration of animal tissues that were oxidizing other substrates such as glucose. Drawing on this knowledge, Krebs and Johnson in 1937 proposed the citric acid cycle.[a,b,c] Krebs provided further confirmation

in 1940 by the observation that malonate, a close structural analog and competitive inhibitor of succinate, in concentrations as low as 0.01 M blocked the respiration of tissues by stopping the oxidation of succinate to fumarate.[c]

$$^-OOC—CH_2—COO^-$$
Malonate

$$^-OOC—CH_2—CH_2—COO^-$$
Succinate

In muscle, 90% of all respiration was inhibited and succinate was shown to accumulate, powerful proof of the importance of the citric acid cycle in the respiration of animal tissues.

[a] Krebs, H. A., and Johnson, W. A. (1937) *Enzymologia* **4**, 148–156
[b] Krebs, H. A. (1981) *Reminiscences and Reflections* Clarendon Press, Oxford
[c] Fruton, J. S. (1999) *Proteins, Enzymes, Genes: the Interplay of Chemistry and Biology* Yale Univ. Press, New Haven, CT
[d] Quastel, J. H. (1978) *Trends Biochem. Sci.* **3**, 68–69

2. Photosynthesis

The principal means of formation of glucose in nature is through photosynthesis in green plants (Fig. 10-1, lower left). Light energy is captured by chlorophyll and is used to transfer electrons from chlorophyll to other electron carriers, the most important of which is NADP⁺. It is reduced to NADPH which is used to reduce carbon dioxide to sugar phosphates in a complex series of reactions known as the **reductive pentose phosphate pathway**, which is described in Chapter 17. ATP is also required for photosynthetic reduction of CO_2. It is generated by allowing some of the electrons to flow back through an electron transport chain in the membranes of the chloroplasts. This chain closely resembles that from Q to cytochrome *c* in mitochondria (Fig. 10-5), and the generation of ATP in this **photosynthetic phosphorylation** occurs in a manner analogous to that in the electron transport chain of mitochondria. In green plants the electrons removed from chlorophyll in one light-requiring reaction are replaced by electrons formed during the cleavage of water in a second light-dependent reaction, a reaction that also releases oxygen, O_2, and generates hydrogen ions (H⁺). The first stable product from reduction of CO_2 in photosynthesis is 3-phosphoglycerate. It can be converted to sugars by pathways analogous to those employed by animals in gluconeogenesis. One inter-

esting difference is the use in chloroplasts of NADPH + H⁺ in reduction of 3-phosphoglycerate (reverse of step *a* in Fig. 10-3).

A small number of other biosynthetic pathways, which are used by both photosynthetic and nonphotosynthetic organisms, are indicated in Fig. 10-1. For example, pyruvate is converted readily to the amino acid **L-alanine** and oxaloacetate to **L-aspartic acid**; the latter, in turn, may be utilized in the biosynthesis of pyrimidines. Other amino acids, purines, and additional compounds needed for construction of cells are formed in pathways, most of which branch from some compound shown in Fig. 10-1 or from a point on one of the pathways shown in the figure. In virtually every instance biosynthesis is dependent upon a supply of energy furnished by the cleavage to ATP. In many cases it also requires one of the hydrogen carriers in a reduced form. While Fig. 10-1 outlines in briefest form a minute fraction of the metabolic pathways known, the ones shown are of central importance.

D. Synthesis and Turnover of Macromolecules

Proteins make up the bulk of the catalytic machinery of cells and together with other macromolecules most of the structure. Therefore, the synthesis and degradation of proteins and the control of those processes are of great importance to cells. Although the emphasis in this section is on proteins, similar considerations apply to nucleic acids, polysaccharides, and other macromolecules.

With the exception of some antibiotics and other short-chain molecules, all polypeptides are formed on ribosomes, which assemble proteins according to the sequences of nucleotides in the messenger RNA (mRNA) molecules. The basic chemistry is simple. The carboxyl groups of the amino acids are converted to reactive acyl adenylates by reaction with ATP, just as in Eq. 10-1. Each "activated" amino acid is carried on a molecule of transfer RNA (tRNA) and is placed in the reactive site of a ribosome when the appropriate codon of the mRNA has moved into the site. The growing peptide chain is then transferred by a displacement reaction onto the amino group of the activated amino acid that is being added to the peptide chain. In this manner, new amino acids are added one at a time to the carboxyl end of the chain, which always remains attached to a tRNA molecule. The process continues until a stop signal in the mRNA ends the process and the completed protein chain is released from the ribosome. Details are given in Chapter 29.

1. Folding and Maturation of Proteins

A newly synthesized peptide chain probably folds quickly. However, the cytosol provides an environment rich in other proteins and other macromolecules that can interact with the new peptide and may catalyze or inhibit folding. Among the most abundant proteins of bacteria or eukaryotes are proteins known as **chaperonins**. They apparently help polypeptide chains to fold correctly, partly by "chaperoning" them through the cytoplasm and across cell membranes, protecting them from becoming entangled with other proteins and macromolecules while they fold.[6–9] There are several classes of chaperonins. Most are oligomers made up of 10- to 90-kDa subunits. They have a variety of names, which may be somewhat confusing. The first chaperonins were identified only as **heat shock proteins** (Hsp). They are produced in large amounts by bacteria or other cells when the temperature is raised quickly, and are designated Hsp70, Hsp90, etc., where the number is the subunit mass in kDa. Other chaperonins were recognized as products of genes needed for replication of the DNA of bacteriophage λ in *E. coli*. Consequently, one major chaperonin of the Hsp70 class is designated **DnaK**. The Hsp70 protein of mitochondria was named "binding protein" or BiP.[10] Other workers have abbreviated chaperonins as Cpn70, Cpn60, etc.

The abbreviations Hsp70, Cpn70, DnaK, and BiP all refer to a group of similar 70-kDa proteins that are apparently found in all organisms. Their role seems to be to stabilize unfolded proteins prior to final folding

in the cytosol or after translocation into the endoplasmic reticulum (ER) or into mitochondria or other organelles. Each of these proteins consists of two functional domains. A 52-kDa domain at the C terminus binds 7- or 8-residue segments of unfolded peptide chains in an elongated conformation.[9] At some point in its reaction cycle the N-terminal 40-kDa domain, which binds ATP tightly, causes the ATP to be hydrolyzed to ADP and inorganic phosphate: $ATP + H_2O \rightarrow ADP + P_i$. Binding and release of a polypeptide by the *E. coli* DnaK protein is coupled tightly to this exergonic **ATPase** reaction.[10a,10b] The reaction is dependent upon potassium ions[11] and is regulated by two **co-chaperones, DnaJ** (Hsp40)[11a-11c] and **GrpE**.[10a] Both ATP and extended polypeptides bind weakly to DnaK, and the ATP in the DnaK•polypeptide•ATP complex is hydrolyzed slowly to ADP and inorganic phosphate.[10a,10b,12] Co-chaperone DnaJ, which shares a largely α-helical structural motif with J-domains in various other proteins, binds to the complex.[11a, 11b, 11c] It probably induces a conformational change that leads to rapid hydrolysis of ATP. In the resulting complex both ADP and the polypeptide are bound tightly to the DnaK protein and dissociate from it very slowly. The *E. coli* co-chaperone GrpE acts as a **nucleotide exchange factor** that catalyzes rapid loss of ADP from the complex. If ATP binds to this DnaK•polypeptide complex the polypeptide is released.[10a] This cycle, which can be repeated, accomplishes the function of DnaK in protecting extended polypeptides and releasing them under appropriate conditions. There is also evidence that DnaK participates in refolding of misfolded proteins.[11d] It cooperates with a ribosome-associated prolylisomerase in bacteria.[11e] The three-dimensional structure of the ATPase domain of Hsp70 is strikingly similar to that of the enzyme **hexokinase** (Chapter 12) and to that of the muscle protein **actin**, Fig. 7-10.[13,14] Archaeal chaperonins lack the Cpn10 ring but have lid-like extensions at the cylinder ends.[15a]

The Cpn60 class of chaperonins are amazing cage-like structures. Each oligomer is composed of two rings, each made up of seven 60-kDa subunits stacked back-to-back. Cpn60 structures from archaeobacteria are similar but may have 8- or 9-subunit rings.[15,15a] The best known member of this group is the *E. coli* protein known as **GroEL** whose three-dimensional structure is depicted in Box 7-A.[16,17] This 14-subunit oligomer of GroEL is a cylinder which is capped by a smaller ring composed of seven 10-kDa subunits of **GroES**, a Cpn10 protein.[17,18] Like Hsp70, GroEL has ATPase activity and an ATP binding site in its large equatorial domain.[16] Archaeal chaperonins lack the Cpn10 ring but have lid-like extensions at the cylinder ends.[15a]

Within the GroEL–GroES cage polypeptide chains can fold without becoming entangled with other proteins or being cleaved by protein-hydrolyzing enzymes of the cytoplasm. An unfolded or partially folded protein may diffuse into the open end of the complex and bind temporarily via noncovalent interactions. The hydrophobic inner surface of the complex may favor the formation of helices within the folding protein. Binding of the polypeptide substrate and capping with GroES causes a major cooperative conformational change with doubling of the internal volume (Box 7-A). The character of the inner wall also changes as a result of exposure of hydrophilic groups.[18a,18b] Perhaps the expansion also stretches segments of the unfolded polypeptide. Subsequent hydrolysis of ATP may be coupled to release of the bound polypeptide, which may then leave the complex or bind again for another chance to fold correctly.[17,19] The two ends of the GroEL·GroES complex may function alternately, with each end in turn receiving the GroES cap.[17]

Hsp90 is one of the most abundant cytosolic proteins in eukaryotic cells, but it seems to chaperone only a few proteins, among which are steroid-hormone receptors.[20,20a] Another group of over 30 different chaperone proteins participate in assembly of external bacterial cell-surface structures such as pili (Fig. 7-9). The chaperone PapD consists of two domains, both having an immunoglobulin-like fold.[20b] PapD binds to the pilin subunits, escorting them to the site of pilus assembly. Folding of integral membrane proteins may be facilitated by a chaperone function of membrane lipids.[20c]

For most proteins the initial synthesis is followed by a sequence of **processing** or **maturation** reactions. These reactions sometimes involve alteration of amino acid side chains and very often include hydrolytic cleavages by which pieces of the peptide chain are cut off. The initially synthesized polypeptides are often not functional and gain biological activity only after one or more pieces have been removed. For example, digestive enzymes are usually secreted as **zymogens** or **proenzymes** which are activated by hydrolytic cleavages only after secretion into the digestive tract. Many proenzymes are components of the extremely complex cascades of activating reactions involved in **blood clotting** (Chapter 12) and in the defensive **complement system** (Chapter 31).

Most peptide hormones are cut out from larger proteins. For example, human insulin is synthesized as a 110-residue **preproinsulin** which is converted in stages to the active two-chain, 51-residue hormone (Eq. 10-8 and Fig. 10-7).[20d,20e]

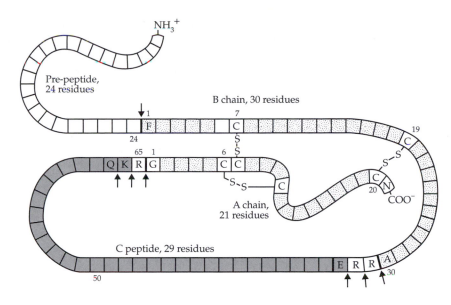

Figure 10-7 Schematic diagram of the structure of human preproinsulin. The 24-residue prepeptide, the 29-residue C-peptide and basic residues 31, 32, 64, and 65 are cut from the peptide upon conversion to insulin as indicated by the small arrows. Some amino acid residues are identified using the one-letter codes. See Fig. 7-17 for details of insulin structure.

$$\text{preproinsulin} \rightarrow \text{proinsulin} \rightarrow \text{insulin}$$

preproinsulin (110 residues)   proinsulin (86 residues, one chain)   insulin (51 residues, two disulfide-linked chains)   (10-8)

The 24-residue prepeptide is cleaved from preproinsulin within a few minutes of synthesis. Then, over a period of about an hour additional cleavages occur to give the final product.[21] This crystallizes as the zinc hexamer (Fig. 7-18) within the "dense cores" of storage vesicles from which it is released into the bloodstream as needed.[22] Other examples are described in Section 3.

2. Transport of Proteins to Their Destinations within a Cell

The synthesis of peproinsulin on ribosomes in the cytosol and the release of mature insulin from secretion vesicles involve not only the chain cleavages but also transport of the polypeptide across membranes of the ER, Golgi compartments, and secretion vesicles.

Signal sequences and translocation. In 1971, Blobel and Sabatini postulated that for some proteins an N-terminal segment of newly synthesized polypeptide chains contains a signal sequence or leader peptide of 15–30 amino acids which carries information concerning the location of the mature protein in the cell.[23–28b] The rather nonpolar signal sequences of proteins destined for secretion from cells would interact with and pass through the membrane of the ER. On the other side of the membrane, a **signal peptidase** would cut off the signal peptide. In many cases, glycosyltransferases would add sugar residues near the N terminus to create a hydrophilic end which would help to pull the rest of the peptide chain through the membrane. A similar situation would hold for secretion of proteins by bacteria.

The signal hypothesis has been proven correct but with considerable added complexity. It now appears that when the N terminus of a new polypeptide chain carrying the proper signal sequence emerges from a ribosome it is intercepted by a **signal recognition particle (SRP)**, a 250-kDa ribonucleoprotein consisting of six polypeptide chains and a small 300-residue 7S RNA chain.[29–33] Binding to the ribosome, the SRP temporarily blocks further elongation of the peptide chain until it bumps against and binds to a 72-kDa membrane protein called the **SRP receptor** or **docking protein**. Then translation begins again and the protein moves into the ER, where it is cleaved by the signal peptidase and is modified further. Although the bacterial SRP is a single protein called **Ffh** (for 54 homolog),[33a] the basic machinery for signal recognition and secretion is remarkably conserved from E. coli to humans.[33b] Bacterial proteins that are destined for secretion or for a function in the periplasmic space or in external membranes also contain signal sequences which are cut off by a signal peptidase embedded in the plasma membrane.[33c]

Signal sequences vary in structure but usually have a net positive charge within the first 5–8 residues at the N terminus. This region is followed by a "hydrophobic core" made up of 8–10 residues with a strong tendency toward α helix formation. This sequence is often followed by one or a few proline, glycine, or serine residues and then a sequence AXA that immediately precedes the cleavage site. Here, A is usually alanine in prokaryotes but may also be glycine, serine, or threonine in eukaryotes. Residue X is any amino acid.[25,27,28a,28b]

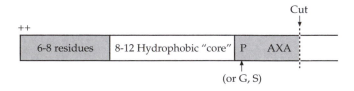

(or G, S)

It had often been assumed that a hydrophobic signal sequence, perhaps folded into a hairpin loop, spontaneously inserts itself into an ER membrane to initiate translocation. However, study of the genetics of protein transport suggests otherwise. Over 50 different genetic loci affect the translocation of proteins in yeast.[31,32,33d] Products of these **secretory genes**,

which are named *SEC* 61, *SEC* 62, *SEC* 63, etc., are proteins with corresponding names, e.g., Sec 61 protein (or Sec 61p). Study of these proteins suggested that the Sec 61, Sec 62, and Sec 63 proteins are all directly involved in transport into the ER.[31,34,35] Additional proteins are needed for movement of vesicles out from the Golgi and for delivery of secretory vesicles to the plasma membrane.[33d] The protein Sec 61p has been identified as homologous to the α subunit of a similar trimeric αβγ Sec 61p complex of mammalian tissues.[32] It also appears to correspond to Sec Y, a protein required for secretion of proteins through the plasma membrane of E. coli into the periplasmic space. Proteins corresponding to the β and γ subunits of mammalian related proteins have been identified in yeast and E. coli[36] and in *Arabidopsis*, where it is necessary to bring proteins into the thylakoid membrane of the chloroplasts.[37]

In E. coli the products of genes *SecY*, *SecE*, and *SecG* are integral membrane proteins that, together with additional proteins, form a proteinaceous pore through which proteins pass.[38–40a] Additional proteins required for translocation of some bacterial proteins are the chaperonin **SecB**, a tetramer of 17.3-kDa subunits,[41] and **SecA**, a soluble ATPase that may participate in docking and serve as an engine for transport throught the pore.[38,40a,42] In eukaryotic cells a membrane-bound Ca^{2+}-dependent chaperonin called **calnexin** assists in bringing the protein into the ER in a properly folded state.[43,44]

Unlike transport across the membranes of the ER, transport across plasma membranes of bacteria often requires both hydrolysis of ATP and energy provided by the membrane electrical potential.[33,38,44–48] Secretion into the periplasmic space has been well characterized but less is known about transport of proteins into the external membranes of E. coli.[48] A 16 kDa periplasmic chaperone may be required.[48a]

Many bacteria have a second complete secretion system.[48b] This multi-gene type II system is present but usually inactive in E. coli. A third (type III) system is present in many pathogenic bacteria, and has evolved for delivery of specialized structural and regulatory proteins into host cells.[48c,48d]

Ticketing destinations. We have seen that proteins that contain suitable signal sequences are exported from the cytoplasm while other proteins remain. Some proteins are secreted while others take up residence as integral or peripheral membrane proteins or as soluble proteins within an organelle. All of the available evidence indicates that it is the sequence of a protein that determines its destination. Proteins targeted to pass through the inner mitochondrial or chloroplast membrane have 20- to 70-residue presequences that are rich in arginine and lysine and which are removed when the protein reaches its destination.[49-53] Proteins meant to go to the intermembrane space or the outer membrane of mitochondria have a sequence containing basic amino acids followed by a long stretch of uncharged residues.[54] Peroxisomal preproteins may have signal sequences such as SKL at the C terminus[54-56] and proteins destined to become attached to phosphatidylinositol glycan anchors (Fig. 8-13)[57] as well as some bacterial surface proteins[57a] have signal peptides at both N and C termini. Some lysosomal membrane glycoproteins have an LE pair in the N-terminal cytoplasmic tail[58] and proteins with suitable dileucine and related pairs are often taken up in lysosomes.[59] Soluble proteins that are resident in the cisternae of the ER often have the C-terminal sequence KDEL or HDEL.[29,60-64] It may serve both as a **retention signal** and as a **retrieval signal** for return of the protein if it passes on into the Golgi vesicles.[65,66]

Transport of proteins into mitochondria is dependent upon both cytosolic chaperonins and mitochondrial chaperonins of the Hsp70 and Hsp60 classes.[67-70] Entrance into mitochondria[71-73] resembles passage through membranes of the ER. However, entry to the nucleus through the nuclear pore complex requires other localization signals[73a] as well as specialized proteins.[74,74a] The sorting of proteins into six different compartments within a chloroplast requires a whole series of recognition signals as well as chaperonins and channel proteins.[75-79] The various signal sequences that determine a protein's interactions with chaperonins, docking proteins, and proteinaceous pore complexes may be complex and overlapping but evolution has selected sequences that allow cells to live and function. Computer programs for detecting sorting signals are being developed.[79a] Not only proteins but also other macromolecules are automatically sorted to their correct destination.

Vesicular transport and the Golgi system.
Movement of secreted proteins through the cytosol from the ER to the external surface occurs through the formation and opening up of small vesicles about 70 nm in diameter. It occurs in steps that involve passage from the ER to the various Golgi membranes. Figure 10-8 provides a sketch of the system of ER, Golgi, secretion vesicles, and lysosomes.[63] Newly formed secretory proteins flow from the ER into an intermediate compartment where vesicles are formed and are carried to the *cis* Golgi network (CGN). They move step-by-step through the Golgi stack (GS) and into the *trans* Golgi network (TGN). New vesicles are formed to carry proteins between each pair of membranous compartments. In each compartment new glycosylation reactions or other modifications may occur. Finally, vesicles carry the mature proteins to the plasma membrane, lysosomes, or vacuoles.[79b] A related process is *uptake* of proteins by endocytosis to form, consecutively, early and late endosomes which fuse with lysosomes

(Chapter 8, Section C,6, Fig. 10-8).

At every step in these processes vesicles are formed and are carried to the next destination where the vesicles fuse with the new membrane and discharge their contents.[80-85a] This remarkable process is complex and highly specific. Rothman proposed that the vesicles to be transported are "docked" on appropriate receptor molecules (called SNARES) on the destination membrane. This is accomplished with the aid of specific soluble marker proteins (called SNAPS) with surfaces complementary to those of the receptors.[80] Proteins are sorted in this way according to their destination signals. The Golgi system is considered further in Chapter 20.

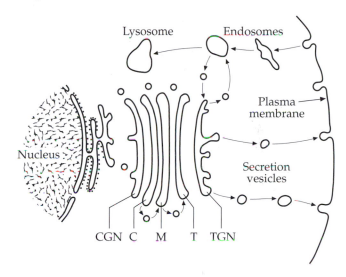

Figure 10-8 Current version of protein synthesis and processing via ER, Golgi, and secretory vesicles. CGN, *cis*-Golgi network: C, T, M are the *cis*, medial, and *trans* compartments of the Golgi stack; TGN, *trans* Golgi network. Arrows indicate some of the movements of transport vesicles.

3. Posttranslational Alterations

Chemical alterations to a protein begins as soon as the peptide chain is formed. Proteins translocated into the periplasmic space of bacteria or into the cisternae of the ER of eukaryotic cells meet several important enzymes. Peptidylprolyl isomerase (discussed in Box 9-F) assists the folding of the peptide chain and protein disulfide isomerases help to form disulfide linkages. The cytosol provides a reducing environment in which many proteins that have cysteine side chains carry free –SH groups. In contrast, the periplasmic space and the ER have more oxidizing environments in which disulfide bridges may form. As shown in Eq. 10-9, the oxidizing agent is often assumed to be the disulfide form of the tripeptide **glutathione** (see Box 11-B),

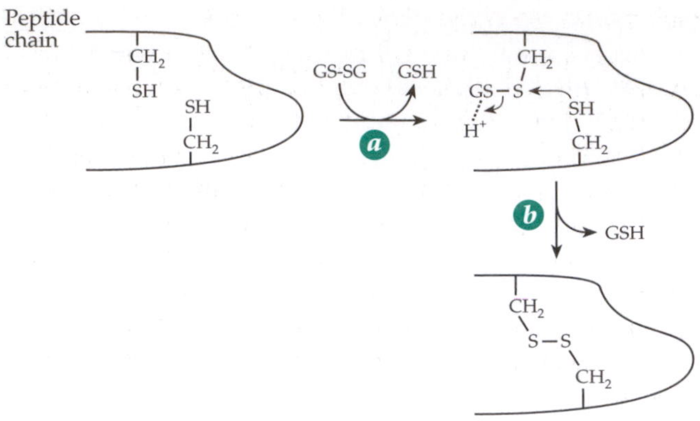

$$(10-9)$$

which is commonly abbreviated GS–SG. It can react with proteins by disulfide exchange reactions to form mixed disulfide groups with release of reduced glutathione, GSH (Eq. 10-9, step *a*). The mixed disulfides can be converted to disulfide bridges by a second exchange reaction (Eq. 10-9, step *b*). However, in the bacterial periplasm the dithiol protein DsbA appears to be the major oxidant (see Box 15-C). It functions together with other proteins, some of which carry electrons from the electron transport chains of the bacteria.[85a,85b] The rates of these reactions are greatly increased by the protein disulfide isomerases.[85a–85e,86–91] These enzymes tend to promote formation of disulfide linkages between the correct pairs of SH groups. If there are three or more –SH groups in a chain some incorrect pairing may, and often does, occur. The protein disulfide isomerases break these bonds and allow new ones to form.[92] The active sites of these isomerases contain pairs of –SH groups which can be oxidized to internal –S–S– bridges by NAD[+]-dependent enzymes. These enzymes and their relatives **thioredoxin** and **glutaredoxin** are discussed further in Box 15-C. Glutathione and oxidation–reduction buffering are considered in Box 11-B.

Proteolytic processing.

For proteins exported from the cytosol, a signal peptidase (or leader peptidase) is usually waiting in the bacterial plasma membrane or in membranes of the ER. The *E. coli* leader peptidase is an integral membrane protein with its catalytic domain in the periplasm.[45,93] For insulin the 24-residue prepiece serves as the signal sequence which is hydrolyzed off in the ER. However, the other cleavages shown in Fig. 10-7 appear to take place in immature secretory granules.[21,22] For many proteins proteolytic processing begins earlier than this and may occur in more than one location. Eukaryotic processing proteases often resemble the bacterial protease subtilisin[20e] but cut peptide chains preferentially after dibasic amino acid pairs, e.g., Arg-Arg, Arg-Lys.[94–96] There are also other processing enzymes with differing specificities.[97]

Although most polypeptides leave ribosomes as single proteins, some are **polyproteins**, which give rise to two or more functional peptides. Polyproteins occur in all cells but are especially prevalent among virally encoded peptides. For example, the polio virus polyprotein is cut into at least ten pieces by proteases, some present in the host cell normally and some encoded by the virus.[98] Many neurohormones arise from polyproteins that undergo processing as they travel down an axon from the cell body before being secreted into a synapse (Chapter 30). Within our own brains the peptide **prepro-opiomelanocortin** undergoes numerous cleavages to give rise to at least seven different neurohormones (see Fig. 30-2).[96,99–101]

Altered ends.

Proteins often contain "blocked" end groups in place of free –NH$_3^+$ or –COO$^-$. For many of the cytosolic proteins, which do not carry leader sequences, the N-terminal methionine that is always used to initiate the ribosomal synthesis of proteins in eukaryotes is removed. If the next residue after the N-terminal methionine is small and uncharged, the methionine is usually hydrolyzed off enzymatically after 30–40 residues have been added to the growing peptide chain.[102–105] An acetyl group is transferred onto the NH$_2$ termini of about 85% of all cytosolic proteins, whether or not the initiator methionine has been removed.[106] In other cases fatty acyl groups may be transferred to the terminal –NH$_2$ group or to side chain SH groups. For example, a **myristoyl** (tetradecanoyl) group is frequently combined in amide linkage at the N terminus of cellular and virally encoded proteins. A palmitoyl group is joined in thioester linkage to a cysteine side chain near the N terminus of the *E. coli* periplasmic lipoprotein (Fig. 8-28).[107] Together with the N-terminal myristoyl group it forms a membrane anchor for this protein. Polyprenyl groups are transferred onto cysteine –SH groups at or near the C termini of many eukaryotic proteins to form a different membrane anchor.[108–110] Some of the enzymes act on the sequence CAAX, where X = M, S, Q, C but not L or I. After transfer of a farnesyl or geranylgeranyl group onto the cysteine the AAX is removed proteolytically and the new terminal cysteinyl carboxyl group is methylated. Surface proteins of gram-positive bacteria are joined by amide linkage from their C termini to a pentaglycine chain of the peptidoglycan layer (Fig. 8-28)[57a]

A pyroglutamyl N terminus is found in the thyrotropin-releasing hormone (Fig. 2-4) and in many other peptide hormones and proteins. It presumably arises by attack of the α-NH$_2$ group of an N-terminal glutamine on the side chain amide group with release of NH$_3$ (Eq. 10-10).[111,112]

Another frequent modification at the C terminus of peptide hormones and of other proteins is **amidation**. In this reaction a C-terminal glycine is oxidatively removed as glyoxylate in an O$_2^-$, copper- and ascorbate

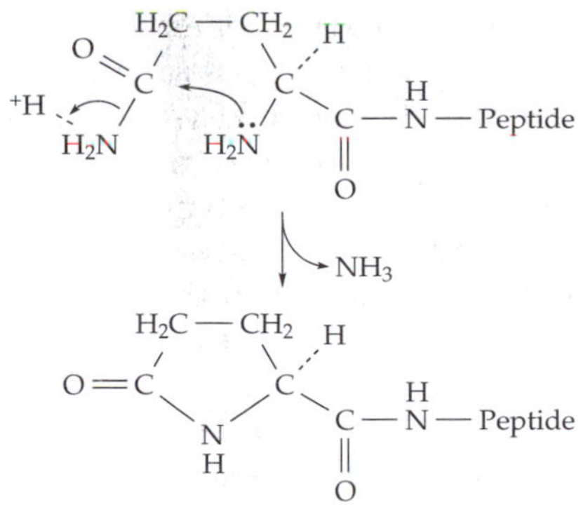

$$(10\text{-}10)$$

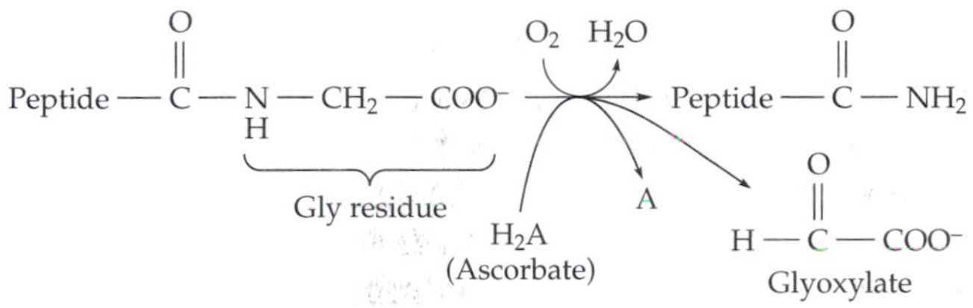

$$(10\text{-}11)$$

(vitamin C)-dependent process (Eq. 10-11).[113–117] See also Chapter 18, Section F,2.

An example of both of these modifications is formation of thyrotropin-releasing hormone shown in Fig. 2-4 from its immediate precursor:

$$\text{Gln–His–Pro–Gly} \rightarrow \text{pyroglutamyl–His–Pro–NH}_2$$

$$(10\text{-}12)$$

Proteins with long C-terminal hydrophobic signal sequences may become attached to phosphatidylinositol-glycan anchors embedded in the plasma membrane (Fig. 8-13). An example is a human alkaline phosphatase in which the α carboxyl of the terminal aspartate residue forms an amide linkage with the ethanolamine part of the anchor. Attachment may occur by a direct attack of the $-NH_2$ group of the ethanolanine on a peptide linkage in a transacylation reaction that releases a 29-residue peptide from the C terminus.[118,119] (See Chapter 29).

Many other covalent modifications of proteins are dealt with in other sections of the book.[120–122] A few are described in Chapter 2, Eqs. 2-14 to 2-22. Reversible alterations used to regulate enzymes are considered in Chapter 11. Of these, the phosphorylation of –OH groups of serine, threonine, and tyrosine is the most important. A large fraction of all cellular proteins appear to be modified in this way. Protein glycosylation, the transfer of glycosyl groups onto –OH side chain groups of serine and threonine (Chapter 4, and Chapter 20) and nonenzymatic glycation (Eq. 4-8) also affect many proteins, often at turns in the peptide chain. Hydroxylation, glycosylation, and other modifications of collagen are described in Chapter 8. Another common

reversible alteration is formation of sulfate esters of tyrosine –OH groups.[99,123] Reactions by which cofactors become attached to proteins[124] are described in Chapters 15–17.

Methylation,[125] hydroxylation, and other irreversible modifications often affect specific residues in a protein. Oxidative alterations occur during aging of proteins (Chapter 18).[126] A few proteins even undergo "splicing" that alters the amino acid sequence (Box 29-E).[127] All of these reactions not only affect the properties of the proteins but also participate in driving the turnover of these macromolecules.

4. Intracellular Degradation of Proteins

Once a protein has reached its correct location and has acquired its proper function, it usually has a limited lifetime, which may average only a few hours or a few days. The protein is then hydrolytically degraded back to its constituent amino acids.[128] Defective and damaged proteins are usually degraded much more rapidly than are intact proteins.[129,130] Under conditions of starvation, proteins are broken down more rapidly than usual to supply the cell with energy. Rapid degradation of proteins is often induced at certain stages of differentiation. For example, spore-forming bacteria contain a protease that becomes activated upon germination of the spore.[131] Within minutes this enzyme digests stored proteins to provide amino acids for the synthesis of new proteins during growth.

Eukaryotic cells degrade proteins within both the cytosol and lysosomes. Lysosomes apparently take up many proteins but have a preference for N-terminal KFERQ[132] and also for particular types of glycosylation (Chapter 20). Lysosomes act on many long-lived proteins.[133,133a] Once within the lysosomes, the proteins are broken down into amino acids with a half-life of ~8 minutes. During nutritional deprivation, the rate of uptake of proteins by lysosomes increases markedly. The same is true during certain developmental changes, for example, when a tadpole loses its tail.

Many short-lived proteins are degraded within the cytosol in ATP-dependent processes. A major process involves the small protein **ubiquitin** (Box 10-C).[134] Once "labeled" by formation of an isopeptide linkage to ubiquitin, a peptide is attacked by proteases in the **proteasome** complexes (Box 7-A, Chapter 12). There it is quickly degraded. Other proteases, most of which do not require ATP, are also present in the cytoplasm (Chapter 12). How do these enzymes as well as those within the lysosomes work together to produce a harmonious turnover of the very substance of our tissues? How is it possible that one protein has a long half life of many days while another lasts only an hour or two in the same cell? The answer seems to be that

BOX 10-C UBIQUITIN

The small 76-residue protein called ubiquitin[a-c] is probably present in all eukaryotic cells. Found in the nucleus, cytoplasm, cell surface membranes, and extracellular fluids, ubiquitin is often joined by isopeptide linkages from its C-terminal glycine to the ε-amino groups of lysine side chains of other proteins. Ubiquitin has one of the most conserved of all known amino acid sequences. No amino acid substitutions have been found among animal species and only three differences distinguish plant ubiquitins from that of humans. In its three-dimensional structure ubiquitin is compact, tightly hydrogen bonded, and roughly spherical. It contains an α helix, a mixed β sheet, and a distinct hydrophobic core.

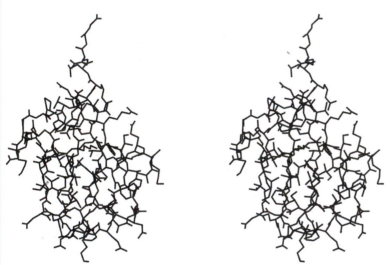

Stereoscopic drawing of human ubiquitin. The C terminus is at the top. From Vijay-Kumar *et al.*[c] Courtesy of William J. Cook

Linkage of ubiquitin to other proteins occurs through the action of a ligase system which catalyzes four sequential reactions as shown in the accompanying equation. In step *a* a **ubiquitin activating enzyme** (E_1) forms a C-terminal acyl adenylate by reaction of ubiquitin (Ub) and ATP.[d,e] In step *b* a sulfhydryl group of the same enzyme then displaces the AMP part to form a thioester linkage to ubiquitin. The chemistry of the reaction is the same as that in Eq. 10-1. The activated ubiquitin is next transferred (step *c*) by transacylation to several **ubiquitin-conjugating enzymes** (E_2), also called ubiquitin-carrier proteins.[f-i] These in turn (step *d*) transfer the ubiquitin to amino groups of lysine side chains of target proteins (Prot-NH$_2$). A third protein, **ubiquitin-protein ligase** (E_3), is sometimes required for this

last step. Most E_2 enzymes, acting without E_3, couple ubiquitin with amines or with small basic proteins whose cellular functions are still unclear. For example, an E_2 which is the product of the yeast gene CDC34 appears to function in the **cell cycle** (Fig. 11-15). Its absence from yeast is lethal[f,i,j] and it is clear that ubiquitin-mediated hydrolysis of the specialized proteins called **cyclins** is essential to operation of the cycle.[k,l]

The best understood function of ubiquitin is in nonlysosomal degradation of proteins.[j,m,n] Protein E_3 appears to select the proteins for degradation, binding them and catalyzing the formation of the isopeptide linkage to substrate. After one ubiquitin molecule has been attached, and while still held by E_3, a second activated ubiquitin is coupled to Lys 48 of the first ubiquitin. This process may continue until several molecules of ubiquitin are joined by Gly-Lys isopeptide linkages to form a **polyubiquitin** chain. Sometimes more than one lysine of the substrate becomes polyubiquitinated to form branched chains. A free α-NH$_3^+$ group on the protein being degraded is essential for rapid conjugation with ubiquitin and certain N-terminal residues such as arginine favor the conjugation and subsequent hydrolytic breakdown. Some proteins become attached to ubiquitin only after arginine is transferred

BOX 10-C (continued)

onto their N-termini from an aminoacyl-tRNA.[o]

Polyubiquitin chains serve as recognition markers that induce rapid hydrolysis of the marked proteins in the 2000 kDa 26S proteasome or **multicatalytic protease**.[p–s] This complex is discussed in Box 7-A and again in Chapter 12. During hydrolytic destruction of the protein the ubiquitin is released for reuse by **ubiquitin carboxyl-terminal hydrolases** or **isopeptidases** which cleave the thioester or isopeptide linkages that tie ubiquitin to proteins.[t,u]

While proteins may be modified to favor rapid ubiquitination, others may be altered to protect them from ubiquitination. For example, calmodulin produced from a cloned gene in bacteria is a good substrate for ubiquitination but within cells it appears to be protected by the posttranslational conversion of Lys 115 to trimethyllysine.[v]

About 10% of the histone H2A present in higher eukaryotes is ubiquitinated at Lys 119.[w] In the slime mold *Physarum* the content of ubiquitinated histones H2A and H2B changes rapidly during the various stages of mitosis. Apparently, ubiquitin must be cleaved from the histones to permit packaging of DNA into metaphase chromosomes and must become attached to the histones in some regions of the chromosomes to allow unfolding of the highly packed nucleosomes. A yeast enzyme that attaches ubiquitin to histones is encoded by the gene *RAD6*, which is required for DNA repair, sporulation, and

other cellular processes. Terminal differentiation of reticulocytes to form erythrocytes involves loss of specific enzymes as well as of entire mitochondria. These processes also depend upon ubiquitin. Ubiquitin is one of the components of the paired helical filaments present in brains of persons with Alzheimer's disease.[x]

In most organisms there are two arrangements for ubiquitin genes. There is a cluster of up to 100 tandomly repeated genes whose transcription gives rise to polyubiquitin, a chain of ubiquitin molecules joined by Gly-Met linkages. These must be cleaved, perhaps by the same ubiquitin C-terminal hydrolase that releases ubiquitin from its conjugates.[y] Other ubiquitin genes are fused to genes encoding ribosomal proteins. The resulting polyproteins have the ribosomal peptides fused to the C termini of the ubiquitin sequences and must be proteolytically cleaved to give mature proteins.[z]

Recently a variety of modifiers of ubiquitin ligases have been discovered[aa,bb] as have ubiquitin-like domains in other proteins.[cc] These findings elucidate the complexity of the sorting of proteins and removal of improperly folded and otherwise defective proteins from the secretory pathway and return to the proteasomes in the cytosol.[dd,ee] They also suggest important roles for ubiquitination in a broad range of metabolic controls.

[a] Vijay-Kumar, S., Bugg, C. E., Wilkinson, K. D., Vierstra, R. D., Hatfield, P. M., and Cook, W. J. (1987) *J. Biol. Chem.* **262**, 6396–6399

[b] Vierstra, R. D., Langan, S. M., and Schaller, G. E. (1986) *Biochemistry* **25**, 3105–3108

[c] Vijay-Kumar, S., Bugg, C. E., and Cook, W. J. (1987) *J. Mol. Biol.* **194**, 531–544

[d] Hershko, A. (1991) *Trends Biochem. Sci.* **16**, 265–268

[e] Pickart, C. M., Kasperek, E. M., Beal, R., and Kim, A. (1994) *J. Biol. Chem.* **269**, 7115–7123

[f] Jentsch, S., Seufert, W., Sommer, T., and Reins, H.-A. (1990) *Trends Biochem. Sci.* **15**, 195–198

[g] Cook, W. J., Jeffrey, L. C., Sullivan, M. L., and Vierstra, R. D. (1992) *J. Biol. Chem.* **267**, 15116–15121

[h] Cook, W. J., Jeffrey, L. C., Xu, Y., and Chau, V. (1993) *Biochemistry* **32**, 13809–13817

[i] Blumenfeld, N., Gonen, H., Mayer, A., Smith, C. E., Siegel, N. R., Schwartz, A. L., and Ciechanover, A. (1994) *J. Biol. Chem.* **269**, 9574–9581

[j] Ciechanover, A., and Schwartz, A. L. (1994) *FASEB J.* **8**, 182–191

[k] Barinaga, M. (1995) *Science* **269**, 631–632

[l] Dorée, M., and Galas, S. (1994) *FASEB J.* **8**, 1114–1121

[m] Pickart, C. M. (2000) *Trends Biochem. Sci.* **25**, 544–548

[n] Johnson, E. S., Ma, P. C. M., Ota, I. M., and Varshavsky, A. (1995) *J. Biol. Chem.* **270**, 17442–17456

[o] Ciechanover, A., Ferber, S., Ganoth, D., Elias, S., Hershko, A., and Arfin, S. (1988) *J. Biol. Chem.* **263**, 11155–11167

[p] Goldberg, A. L. (1995) *Science* **268**, 522–523

[q] Löwe, J., Stock, D., Jap, B., Zwickl, P., Baumeister, W., and Huber, R. (1995) *Science* **268**, 533–539

[r] Thrower, J. S., Hoffman, L., Rechsteiner, M., and Pickart, C. M. (2000) *EMBO J.* **19**, 94–102

[s] Peters, J.-M. (1994) *Trends Biochem. Sci.* **19**, 377–382

[t] Stein, R. L., Chen, Z., and Melandri, F. (1995) *Biochemistry* **34**, 12616–12623

[u] Wilkinson, K. D., Tashayev, V. L., O'Connor, L. B., Larsen, C. N., Kasperek, E., and Pickart, C. M. (1995) *Biochemistry* **34**, 14535–14546

[v] Johnston, S. C., Riddle, S. M., Cohen, R. E., and Hill, C. P. (1999) *EMBO J.* **18**, 3877–3887

[w] Davie, J. R., and Murphy, L. C. (1990) *Biochemistry* **29**, 4752–4757

[x] Mori, H., Kondo, J., and Ihara, Y. (1987) *Science* **235**, 1641–1644

[y] Finley, D., and Varshavsky, A. (1985) *Trends Biochem. Sci.* **10**, 343–347

[z] Baker, R. T., Tobias, J. W., and Varshavsky, A. (1992) *J. Biol. Chem.* **267**, 23364–23375

[aa] Tyers, M., and Willems, A. R. (1999) *Science* **284**, 601–604

[bb] Turner, G. C., Du, F., and Varshavsky, A. (2000) *Nature (London)* **405**, 579–583

[cc] Hochstrasser, M. (2000) *Science* **289**, 563–564

[dd] Ellgaard, L., Molinari, M., and Helenius, A. (1999) *Science* **286**, 1882–1888

[ee] Plemper, R. K., and Wolf, D. H. (1999) *Trends Biochem. Sci.* **24**, 266–270

TABLE 10-1
Types of Biochemical Reactions with Ionic Mechanisms

---

1. **Nucleophilic displacement, often via an addition–elimination sequence**

A. 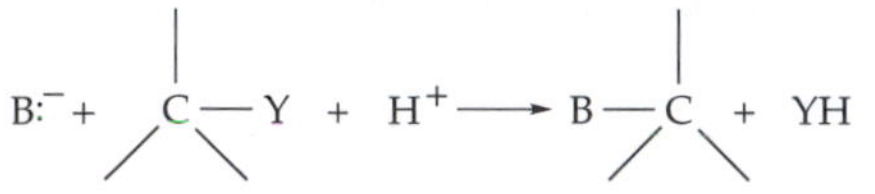 $B:^- + \quad C-Y + H^+ \longrightarrow B-C + YH$

(*Examples: methyl-, glycosyl-transferases*)

B. 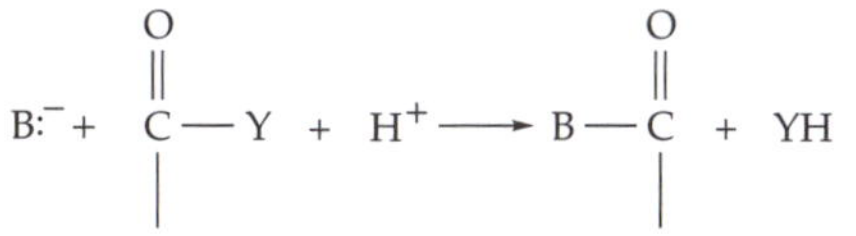 $B:^- + \quad C-Y + H^+ \longrightarrow B-C + YH$

(*Examples: esterases, peptidases*)

C. 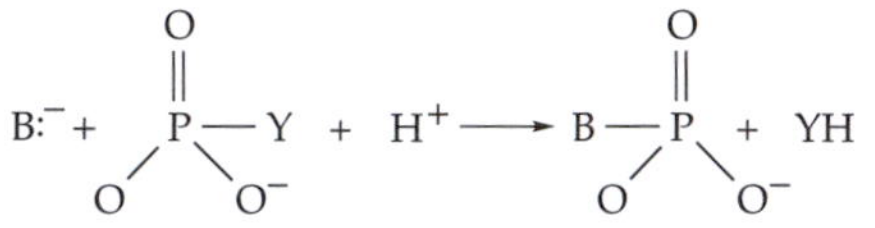 $B:^- + \quad P-Y + H^+ \longrightarrow B-P + YH$

(*Examples: phosphatases, kinases, phosphomutases, ribozymes*)

D. 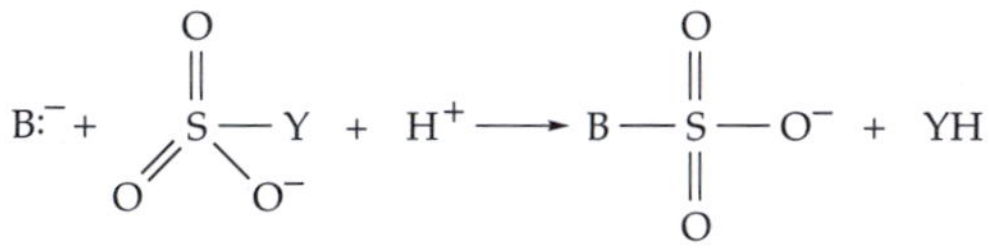 $B:^- + \quad S-Y + H^+ \longrightarrow B-S-O^- + YH$

(*Examples: sulfatases, sulfotransferases*)

---

2. **Addition**

A. To polarized double bonds such as $\diagup C=O$

and $\diagup C=N-$

(*This reaction most often occurs as a step in an enzymatic process e.g., formation of hemiacetals, hemiketals, hemimercaptals, carbinolamines*)

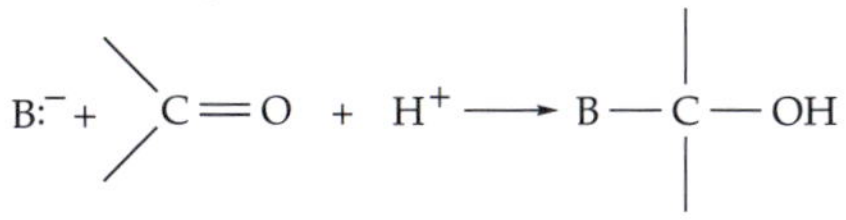

$B:^- + \diagup C=O + H^+ \longrightarrow B-C-OH$

(*Example: mutarotase*)

B. To double bond conjugated with $\diagup C=O$ or

$\diagup C=N-$

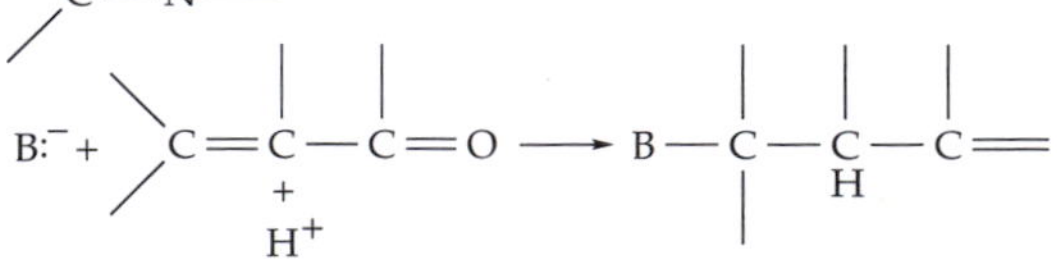

$B:^- + \diagup C=C-C=O \longrightarrow B-C-C-C=$

$+ H^+$

(*Examples: acyl-CoA, fumarate hydratases, aspartase*)

C. To isolated double bonds (*Example: oleate hydratase*)
(*see p. 689*)

---

3. **Elimination**

A. and B. Precisely the opposite of *addition*

(*Eliminations that are the reverse of type 2A are frequent steps in more complex enzyme mechanisms*)

C. Decarboxylative elimination

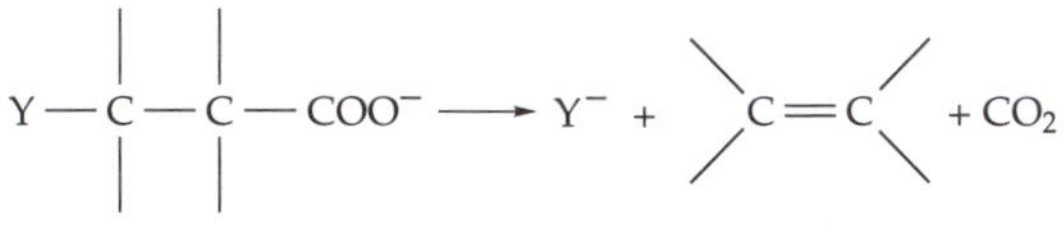

$Y-C-C-COO^- \longrightarrow Y^- + \quad C=C \quad + CO_2$

---

4. **Formation of stabilized enolate anions and enamines**

A.

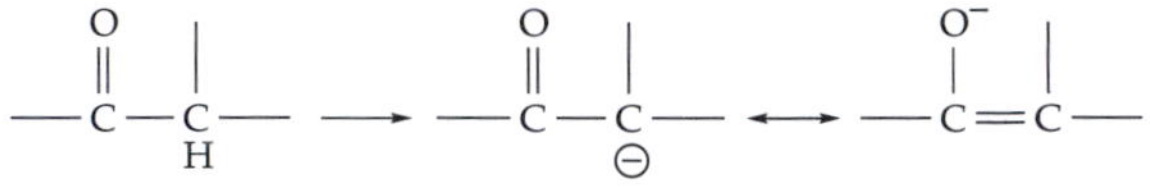

$-C-C- \longrightarrow -C-C- \longleftrightarrow -C=C-$

B.

$-C-C- \longrightarrow -C-C- \longleftrightarrow -C=C-$

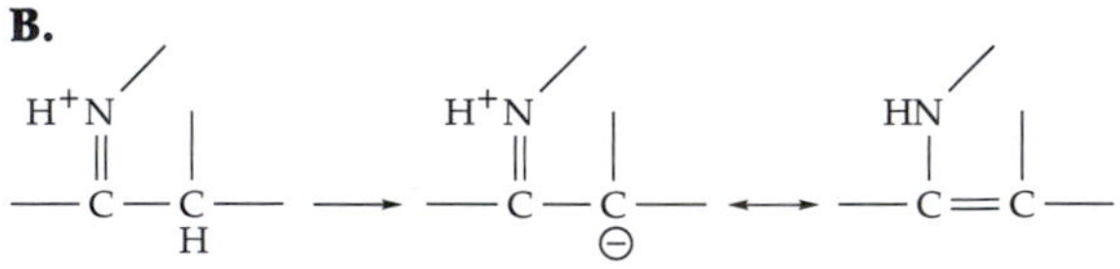

Isomerization reactions

C.

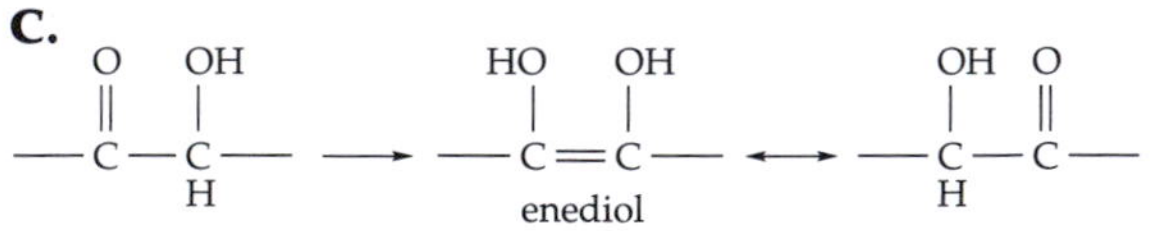

$-C-C- \longrightarrow -C=C- \longleftrightarrow -C-C-$
       enediol

(*Example: sugar isomerases*)

D.

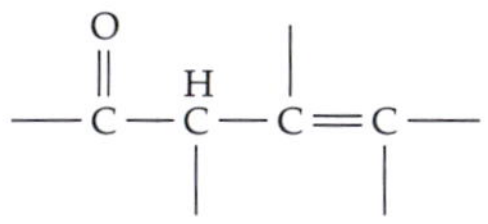

$-C-C-C=C- \quad$ (*Example: oxosteroid isomerases*)

$-C-C-C=C- \longrightarrow -C-C=C-C-$

TABLE 10-1
(continued)

5. Stabilized enolate anions as nucleophiles: formation of carbon–carbon bonds (β condensation)

A. Displacement on a carbonyl group

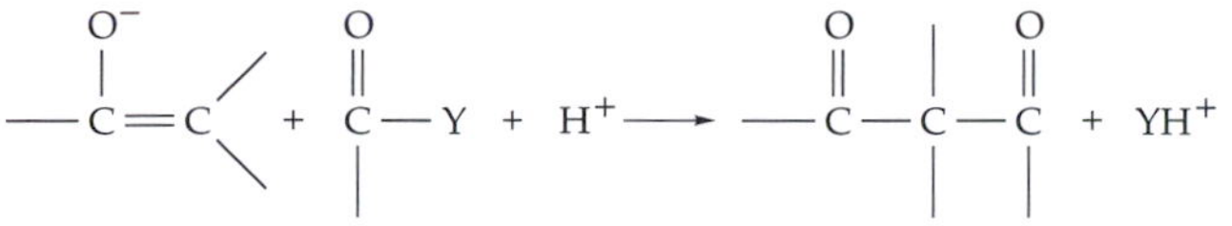

(Example: 3-Oxoacyl–CoA transferase (Y = —S—CoA)

B. Addition to a carbonyl group: aldol condensation

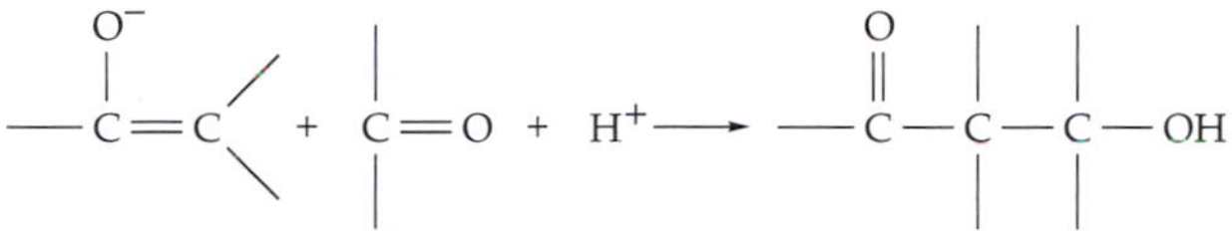

(Examples: aldolases, citrate synthases)

C. Addition to a carbon dioxide (β carboxylation); decarboxylation

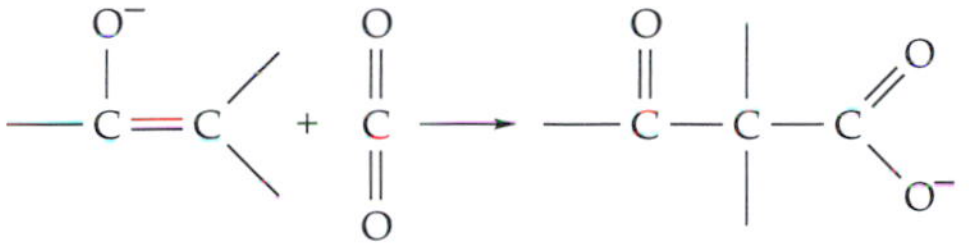

(Examples: phosphoenolpyruvate carboxylase, oxaloacetate decarboxylases)

6. Some isomerization and rearrangement reactions

A. Allylic rearrangement (1,3-proton shift)

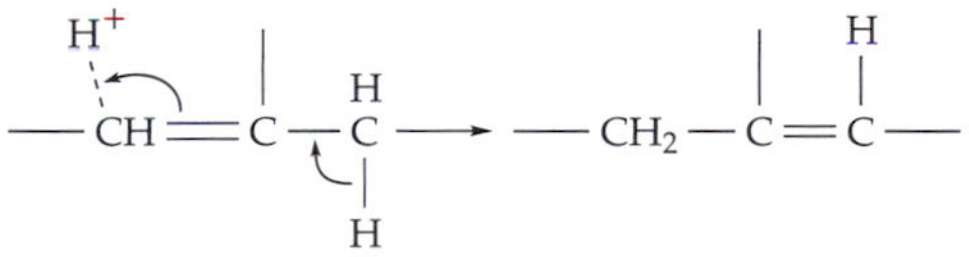

B. Allylic rearrangement with condensation

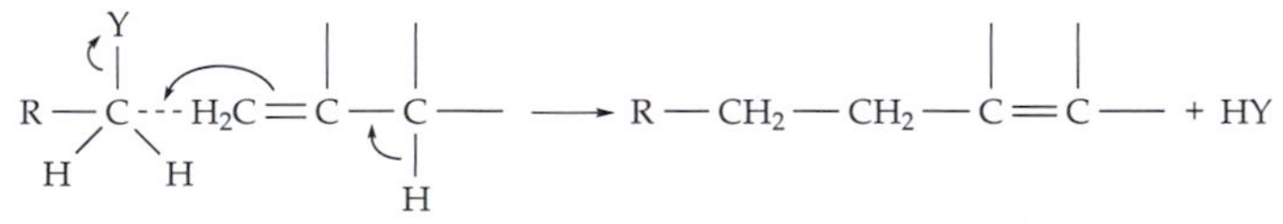

(Example: condensation of dimethylallyl pyrophosphate with isopentenyl pyrophosphate)

C. Rearrangements with alkyl or hydride ion shift

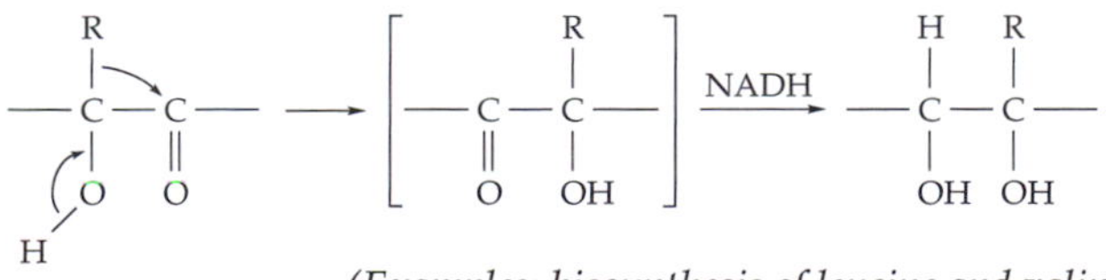

Glyoxalase

(Examples: biosynthesis of leucine and valine, xylose isomerase, glyoxalase)

the turnover rate of a protein is determined in large part by its sequence. Some proteins are tightly folded and have few bends on the outside that have sequences meeting the specificity requirements of intracellular proteolytic enzymes. These have long half lives. Other proteins may have external loops with sequences susceptible to attack or sequences that favor rapid reaction with the ubiquitin system or uptake into lysosomes. Prematurely terminated proteins and peptide fragments from partial degradation of proteins may tend to be unfolded at the N terminus and to be attacked rapidly.[128,135] Proteins that have undergone covalent modifications or oxidative damage also seem to be hydrolyzed rapidly.[136]

Regions rich in proline, glutamate, serine, and threonine (PEST regions) may be good substrates for Ca^{2+}-activated cytosolic proteases.[137] The ubiquitin system appears to act most slowly on a protein when the normal initiation amino acid methionine is present at the N terminus. For example, the half-life of β-galactosidase in yeast is over 20 hours. Replacement of the methionine with S, A, T, V, or G has little effect. However, replacement with other amino acids shortens the half-life as follows: I and E, 30 min.; Y and Q, 10 min.; F, L, D, and K, 3 min.; and R, 2 min.[138–140a]

5. Turnover of Nucleic Acids

Because of its special role in carrying genetic information, DNA is relatively stable. An elaborate system of repair enzymes (Chapter 27) act to correct errors and to help DNA to preserve its genetic information. However, in some specialized cells such as those forming immunoglobulin, the DNA too undergoes major rearrangements (Chapter 31). RNA molecules are subject to extensive processing. This includes the conversion of RNA bases to modified forms, chain cleavages during maturation of ribosomal and transfer RNAs, cutting and splicing of gene transcripts to form mRNAs and finally degradation of the mRNA (Chapter 28). Proteins that serve as **RNA chaperones** assist in folding these molecules.[141]

BOX 10-D DRAWING THOSE LITTLE ARROWS

Organic mechanisms are often indicated by arrows that show the flow of electrons in individual steps of a reaction. Many errors are made by students on exams and even in published research papers. The arrows are often drawn backward, are too numerous, or do not clearly indicate electron flow. Here are some tips.

1 Always write a mechanism step-by-step. Never combine two steps (e.g., *a* and *b* at right) *in which electron flow occurs in opposite directions*. Notice that step *c*, the hydrolysis of a Schiff base, is also a two-step process. The reaction is a familiar one that is commonly indicated as shown here. However, this scheme does not show a detailed mechanism for step *c*.

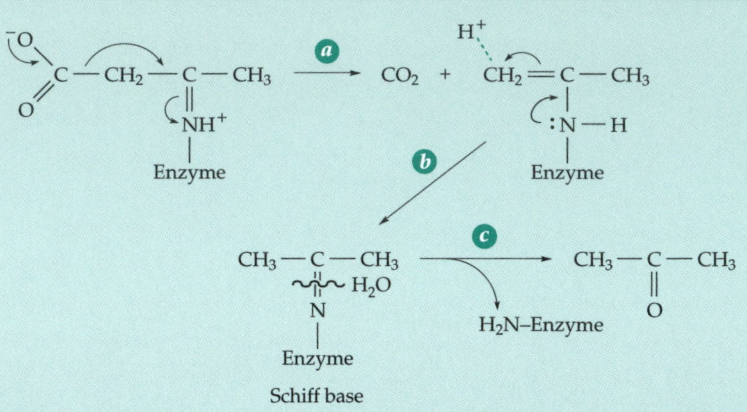

2 Identify the **nucleophilic** and **electrophilic** centers before starting to write a mechanism. Oxygen, sulfur, and nitrogen atoms are usually nucleophilic, e.g., those in —OH, $—NH_2$, $—COO^-$, —SH, $—OPO_3^{2-}$, and enolate anions. The weak nucleophiles —OH and —SH may be converted into strong nucleophiles $—O^-$ $—S^-$ by removal of a proton by a basic group of an enzyme. The nonnucleophilic $—NH_3^+$ becomes the good nucleophile $—NH_2$ by loss of a proton. Nucleophilic centers contain unshared electron pairs. Nucleophiles are basic but the basicity, as indicated by proton binding (by the pK_a), is not necessarily proportional to nucleophilic strength (nucleophilicity). Enolate anions and enamines provide nucleophilic centers on carbon atoms, important in formation and cleavage of C—C bonds.

Electrophilic centers include acidic hydrogen atoms, metal ions, the carbon atoms of carbonyl groups, and the β-carbon atoms of α, β unsaturated acids, ketones, or acyl-CoA derivatives. Highly polarized groups such as carbonyl and enamine generate electrophilic centers as indicated by the positive charges. They also affect more distant positions in conjugated systems, e.g., in α, β-unsaturated acyl-CoA derivatives, and in intermediates formed from thiamin diphosphate and pyridoxal phosphate.

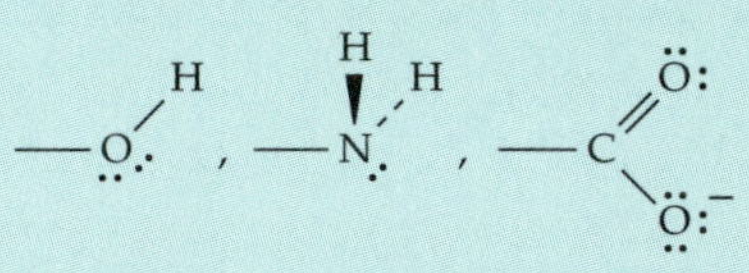

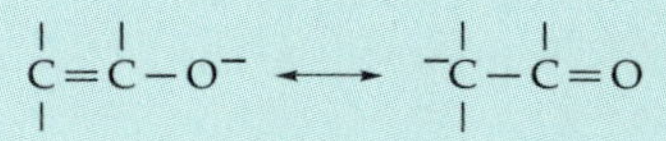

Enolate anion

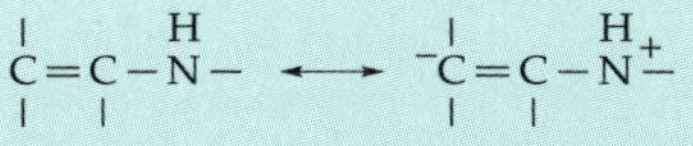

Enamine

Groups with nucleophilic centers indicated by unshared electron pairs and/or negative charge.

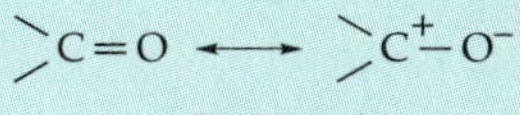

Carbonyl

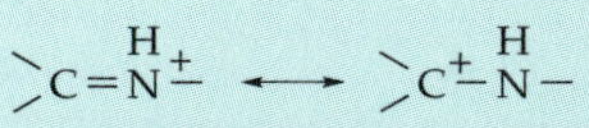

Protonated imine

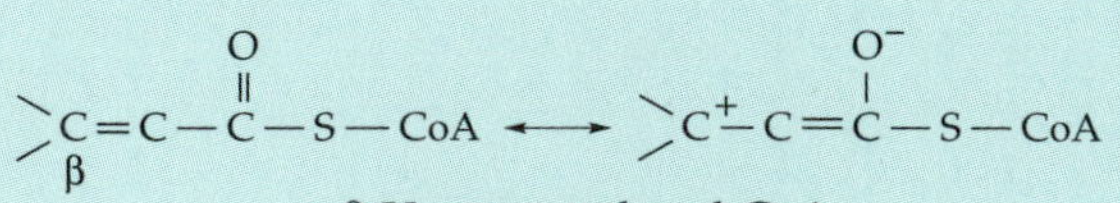

α,β-Unsaturated acyl-CoA

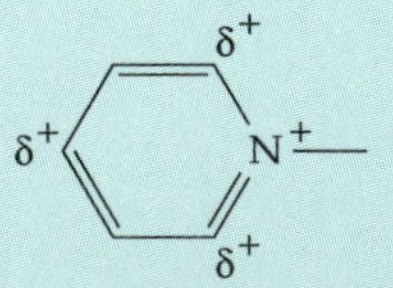

Some electrophilic centers are indicated by + or δ^+

BOX 10-D (continued)

3 When an ionic organic reaction (the kind catalyzed by most enzymes) occurs a nucleophilic center joins with an electrophilic center. We use arrows to show the movement of pairs of electrons. The movement is always *away* from the nucleophile which can be thought of as "attacking" an electrophilic center. Notice the first step in the second example at right. The unsaturated ketone is polarized initially. However, this is not shown as a separate step. Rather, the flow of electrons from the double bond, between the α- and β-carbons into the electron-accepting C=O groups, is coordinated with the attack by the nucleophile. Dotted lines are often used to indicate bonds that will be formed in a reaction step, e.g., in an aldol condensation (right). Dashed or dotted lines are often used to indicate partially formed and partially broken bonds in a transition state, e.g., for the aldol condensation (with prior protonation of the aldehyde). However, *do not put arrows on transition state structures.*

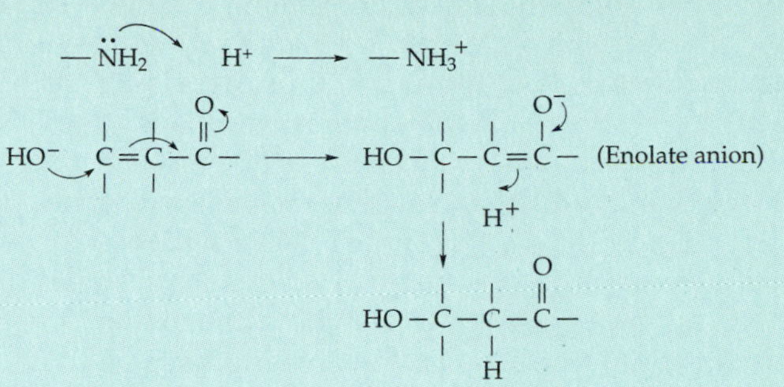

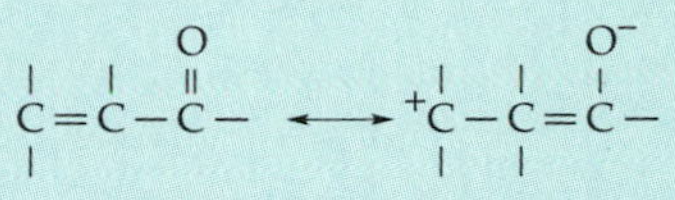

$$-\overset{\cdot\cdot}{N}H_2 \;+\; H^+ \longrightarrow \;-NH_3^+$$

(Enolate anion)

Polarized unsaturated ketone

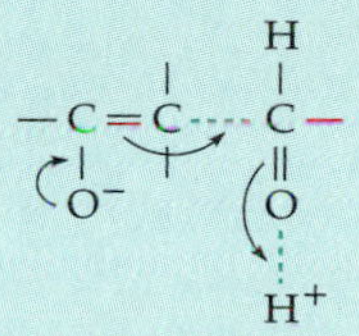

Aldol condensation 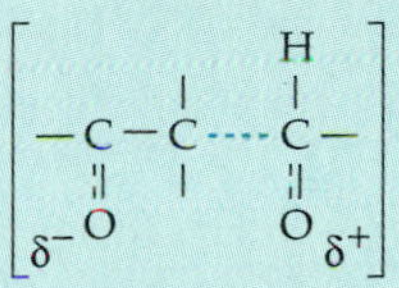   Transition state

4 In a given reaction step all of the arrows must point the same way. *The arrows point into bonds that are forming* or toward atoms that will carry an unshared electron pair in the product. *Arrows originate from unshared electron pairs or from bonds that are breaking.*

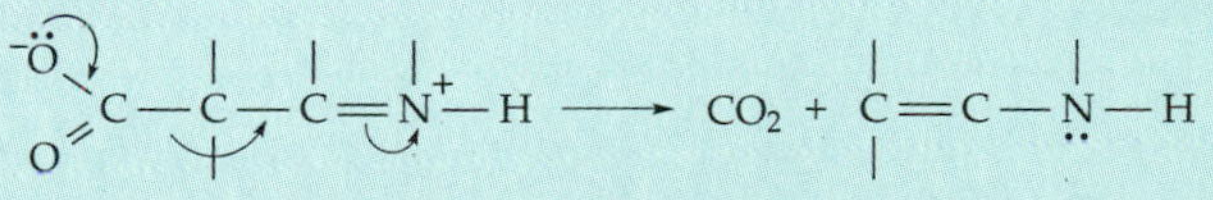

All arrows point the same way

5 Never start an arrow from the same bond that another arrow is forming; i.e., electrons flow out of *alternate* bonds.

6 For reactions of radicals (homolytic reactions) arrows are used to indicate motion of single electrons rather than of electron pairs. It is desirable to use arrows with *half-heads*. For example, reaction of a superoxide radical ($\cdot O_2^-$) with $FADH_2$ could occur as follows:

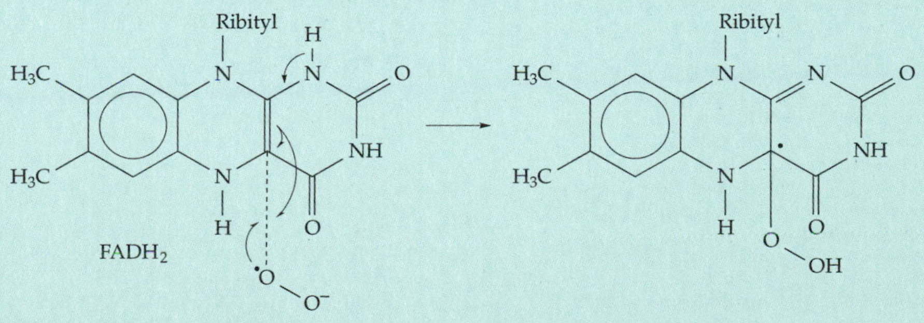

$FADH_2$

Half-heads on arrows indicate one-electron movement

E. Classifying Enzymatic Reactions

The majority of enzymes appear to contain, in their active centers, only the side chains of amino acids. Most of the reactions catalyzed by these enzymes can be classified into a small number of types as is indicated in Table 10-1: (Type 1) **displacement** or **substitution** reactions in which one base or nucleophile replaces another, (Type 2) **addition** reactions in which a reagent adds to a double bond, and (Type 3) **elimination** reactions by which groups are removed from a substrate to create double bonds. Note that the latter are the reverse of addition reactions. Two other groups of reactions depend upon formation of transient **enolic forms**. These include (Type 4) **isomerases** and (Type 5) **lyases**, a large and important group of reactions that form or cleave carbon–carbon bonds. Finally, there is a group (Type 6) of **isomerization** and **rearrangement** reactions that do not appear to fit any of the foregoing categories. Another quite different group of enzyme-catalyzed reactions, which are considered in Chapter 16, function with the participation of free radical intermediates.

Biochemical displacement reactions include all of the hydrolytic reactions by which biopolymers are broken down to monomers as well as most of the reactions by which the monomers are linked together to form polymers. Addition reactions are used to introduce oxygen, nitrogen, and sulfur atoms into biochemical compounds and elimination reactions often provide the driving force for biosynthetic sequences. Complex enzymatic processes are often combinations of several steps involving displacement, addition, or elimination. The reactions involving formation or cleavage of C–C bonds are essential to biosynthesis and degradation of the various carbon skeletons found in biomolecules, while the isomerization reactions provide connecting links between the other kinds of reactions in the establishment of metabolic pathways.

In Chapters 12 and 13 the individual reactions of metabolism are classified into these types and the enzymes that catalyze them are described in some detail. The chemistry of coenzymes and metalloenzymes are presented systematically in Chapters 14 to 16, and in Chapter 17 the logic of the combining of individual reactions into metabolic sequences is considered. It is not necessary to read Chapters 12–16 in their entirety since much of their content is reference material. In the later chapters on metabolism, cross-references point out the discussions of individual enzymes in Chapters 12–16.

The following are topics that may be especially valuable to the student and which might be read initially: in Chapter 12, lysozyme (Section B,5), chymo-trypsin (Section C,1), kinases (Section D,9), multiple displacement, reactions (Section G); in Chapter 13, imines (Section A,2), addition to C=C bonds (Section A, 4,5), beta cleavage and condensation (Section C); in Chapter 14, thiamin diphosphate (Section D), pyridoxal phosphate (Section E); in Chapter 15, NAD (Section A).

References

1. Iwata, S., Ostermeier, C., Ludwig, B., and Michel, H. (1995) *Nature* **376**, 660–669
2. Tsukihara, T., Aoyama, H., Yamashita, E., Tomizaki, T., Yamaguchi, H., Shinzawa-Itoh, K., Nakashima, R., Yaono, R., and Yoshikawa, S. (1995) *Science* **269**, 1069–1074
3. Mitchell, P. (1966) *Biol. Rev. Cambridge Philos. Soc.* **41**, 445–502
4. Mitchell, P. (1979) *Science* **206**, 1148–1159
5. Nicholls, D. G., and Ferguson, S. J. (1992) *Bioenergetics 2*, Academic Press, London
6. Gething, M.-J., and Sambrook, J. (1992) *Science* **355**, 33–45
7. Frydman, J., Nimmesgern, E., Ohtsuka, K., and Hartl, F. U. (1994) *Nature* **370**, 111–117
8. Buchner, J. (1994) *Trends Biochem. Sci.* **19**, 559
9. Hartl, F.-U., Hlodan, R., and Langer, T. (1994) *Trends Biochem. Sci.* **19**, 20–25
10. Flynn, G. C., Pohl, J., Flocco, M. T., and Rothman, J. E. (1991) *Nature* **353**, 726–730
10a. Russell, R., Jordan, R., and McMacken, R. (1998) *Biochemistry* **37**, 596–607
10b. Harrison, C. J., Hayer-Hartl, M., Di Liberto, M., Hartl, F.-U., and Kuriyan, J. (1997) *Science* **276**, 431–435
11. Wilbanks, S. M., and McKay, D. B. (1995) *J. Biol. Chem.* **270**, 2251–2257
11a. Kelley, W. L. (1998) *Trends Biochem. Sci.* **23**, 222–227
11b. Martinez-Yamout, M., Legge, G. B., Zhang, O., Wright, P. E., and Dyson, H. J. (2000) *J. Mol. Biol.* **300**, 805–818
11c. Westermann, B., and Neupert, W. (1997) *J. Mol. Biol.* **272**, 477–483
11d. Mogk, A., Tomoyasu, T., Goloubinoff, P., Rüdiger, S., Röder, D., Langen, H., and Bukau, B. (1999) *EMBO J.* **18**, 6934–6949
11e. Deuerling, E., Schulze-Specking, A., Tomoyasu, T., Mogk, A., and Bukau, B. (1999) *Nature (London)* **400**, 693–696
12. McCarty, J. S., Buchberger, A., Reinstein, J., and Bukau, B. (1995) *J. Mol. Biol.* **249**, 126–137
13. Flaherty, K. M., DeLuca-Flaherty, C., and McKay, D. B. (1990) *Nature* **346**, 623–628
14. Flaherty, K. M., Wilbanks, S. M., DeLuca-Flaherty, C., and McKay, D. B. (1994) *J. Biol. Chem.* **269**, 12899–12907
15. Phipps, B. M., Typke, D., Hegeri, R., Volker, S., Hoffmann, A., Stetter, K. O., and Baumeister, W. (1993) *Nature* **361**, 475–477
15a. Schoehn, G., Hayes, M., Cliff, M., Clarke, A. R., and Saibil, H. R. (2000) *J. Mol. Biol.* **301**, 323–332
16. Braig, D., Otwinowski, Z., Hegde, R., Boisvert, D. C., Joachimiak, A., Horwich, A. L., and Sigler, P. B. (1994) *Nature* **371**, 578–586
17. Hayer-Hartl, M. K., Martin, J., and Hartl, F. U. (1995) *Science* **269**, 836–841
18. Chen, S., Roseman, A. M., Hunter, A. S., Wood, S. P., Burston, S. G., Ranson, N. A., Clarke, A. R., and Saibil, H. R. (1994) *Nature* **371**, 261–264
18a. Betancourt, M. R., and Thirumalai, D. (1999) *J. Mol. Biol.* **287**, 627–644
18b. Wang, Z., Feng, H.-p, Landry, S. J., Maxwell, J., and Gierasch, L. M. (1999) *Biochemistry* **38**, 12537–12546
19. Shtilerman, M., Lorimer, G. H., and Englander, S. W. (1999) *Science* **284**, 822–825
20. Weaver, A. J., Sullivan, W. P., Felts, S. J., Owen, B. A. L., and Toft, D. O. (2000) *J. Biol. Chem.* **275**, 23045–23052
20a. Buchner, J. (1999) *Trends Biochem. Sci.* **24**, 136–141
20b. Sauer, F. G., Fütterer, K., Pinkner, J. S., Dodson, K. W., Hultgren, S. J., and Waksman, G. (1999) *Science* **285**, 1058–1061
20c. Bogdanov, M., and Dowhan, W. (1999) *J. Biol. Chem.* **274**, 36827–36830
20d. Orci, L., Vassalli, J.-D., and Perrelet, A. (1988) *Sci. Am.* **259**(Sep), 85–94
20e. Zhou, A., Webb, G., Zhu, X., and Steiner, D. F. (1999) *J. Biol. Chem.* **274**, 20745–20748
21. Smeekens, S. P., Montag, A. G., Thomas, G., Albiges-Rizo, C., Carroll, R., Benig, M., Phillips, L. A., Martin, S., Ohagi, S., Gardner, P., Swift, H. H., and Steiner, D. F. (1992) *Proc. Natl. Acad. Sci. U.S.A.* **89**, 8822–8826
22. Huang, X. F., and Arvan, P. (1995) *J. Biol. Chem.* **270**, 20417–20423
23. Blobel, G., and Sabatini, D. D. (1971) in *Biomembranes*, Vol. 2 (Manson, L. A., ed), pp. 193–195, Plenum, New York
24. Blobel, G., and Dobberstein, B. (1975) *J. Cell Biol.* **67**, 852

References

25. Landry, S. J., and Gierasch, L. M. (1991) *Trends Biochem. Sci.* **16**, 159–163
26. Gierasch, L. M. (1989) *Biochemistry* **28**, 923–930
27. Jain, R. G., Rusch, S. L., and Kendall, D. A. (1994) *J. Biol. Chem.* **269**, 16305–16310
28. Andersson, H., and von Heijne, G. (1993) *J. Biol. Chem.* **268**, 21389–21393
28a. von Heijne, G. (1998) *Nature (London)* **396**, 1 11–113
28b. Emanuelsson, O., Nielsen, H., Brunak, S., and von Heijne, G. (2000) *J. Mol. Biol.* **300**, 1005–1016
29. Verner, K., and Schatz, G. (1988) *Science* **241**, 1307–1313
30. Rapoport, T. A. (1992) *Science* **258**, 931–935
30a. Batey, R. T., Rambo, R. P., Lucast, L., Rha, B., and Doudna, J. A. (2000) *Science* **287**, 1232–1239
31. Sanders, S. L., and Schekman, R. (1992) *J. Biol. Chem.* **267**, 13791–13794
32. Hartmann, E., Sommer, T., Prehn, S., Görlich, D., Jentsch, S., and Rapoport, T. A. (1994) *Nature* **367**, 654–657
32a. Clemons, W. M., Jr., Gowda, K., Black, S. D., Zwieb, C., and Ramakrishnan, V. (1999) *J. Mol. Biol.* **292**, 697–705
33. Wickner, W. T. (1994) *Science* **266**, 1197–1198
33a. Moser, C., Mol, O., Goody, R. S., and Sinning, I. (1997) *Proc. Natl. Acad. Sci. U.S.A.* **94**, 11339–11344
33b. Samuelson, J. C., Chen, M., Jiang, F., Möller, I., Wiedmann, M., Kuhn, A., Phillips, G. J., and Dalbey, R. E. (2000) *Nature (London)* **406**, 637–640
33c. Paetzel, M., Strynadka, N. C. J., Tschantz, W. R., Casareno, R., Bullinger, P. R., and Dalbey, R. E. (1997) *J. Biol. Chem.* **272**, 9994–10003
33d. Guo, W., Grant, A., and Novick, P. (1999) *J. Biol. Chem.* **274**, 23558–23564
34. Noël, P. J., and Cartwright, I. L. (1994) *EMBO J.* **13**, 5253–5261
35. Brodsky, J. L., Goeckeler, J., and Schekman, R. (1995) *Proc. Natl. Acad. Sci. U.S.A.* **92**, 9643–9646
36. Dobberstein, B. (1994) *Nature* **367**, 599–600
37. Laidler, V., Chaddock, A. M., Knott, T. G., Walker, D., and Robinson, C. (1995) *J. Biol. Chem.* **270**, 17664–17667
38. Dalbey, R. E., and Robinson, C. (1999) *Trends Biochem. Sci.* **24**, 17–22
39. Douville, K., Price, A., Eichler, J., Economou, A., and Wickner, W. (1995) *J. Biol. Chem.* **270**, 20106–20111
39a. Eichler, J., Brunner, J., and Wickner, W. (1997) *EMBO J.* **16**, 2188–2196
39b. Matsumoto, G., Yoshihisa, T., and Ito, K. (1997) *EMBO J.* **16**, 6384–6393
39c. Scotti, P. A., Urbanus, M. L., Brunner, J., de Gier, J.-W. L., von Heijne, G., van der Does, C., Driessen, A. J. M., Oudega, B., and Luirink, J. (2000) *EMBO J.* **19**, 542–549
40. Meyer, T. H., Ménétret, J.-F., Breitling, R., Miller, K. R., Akey, C. W., and Rapoport, T. A. (1999) *J. Mol. Biol.* **285**, 1789–1800
40a. Yahr, T. L., and Wickner, W. T. (2000) *EMBO J.* **19**, 4393–4401
41. Randall, L. L., and Hardy, S. J. S. (1995) *Trends Biochem. Sci.* **20**, 65–69
42. Ulbrandt, N. D., London, E., and Oliver, D. B. (1992) *J. Biol. Chem.* **267**, 15184–15192
43. Bergeron, J. J. M., Brenner, M. B., Thomas, D. Y., and Williams, D. B. (1994) *Trends Biochem. Sci.* **19**, 124–128
44. Jungery, M., Pasvol, G., Newbold, C. I., and Weatherall, D. J. (1983) *Proc. Natl. Acad. Sci. U.S.A.* **80**, 1018–1022
45. Cao, G., Kuhn, A., and Dalbey, R. E. (1995) *EMBO J.* **14**, 866–875
46. Kawasaki, S., Mizushima, S., and Tokuda, H. (1993) *J. Biol. Chem.* **268**, 8193–8198
47. Wickner, W., Driessen, A. J. M., and Hartl, F.-U. (1991) *Ann. Rev. Biochem.* **60**, 101–124

48. Matsuyama, S.-i, Tajima, T., and Tokuda, H. (1995) *EMBO J.* **14**, 3365–3372
48a. Schäfer, U., Beck, K., and Müller, M. (1999) *J. Biol. Chem.* **274**, 24567–24574
48b. Russel, M. (1998) *J. Mol. Biol.* **279**, 485–499
48c. Galán, J. E., and Collmer, A. (1999) *Science* **284**, 1322–1328
48d. Tamano, K., Aizawa, S.-I., Katayama, E., Nonaka, T., Imajoh-Ohmi, S., Kuwae, A., Nagai, S., and Sasakawa, C. (2000) *EMBO J.* **19**, 3876–3887
49. Schatz, G. (1993) *Protein Sci.* **2**, 141–146
50. Pfeffer, S. R., and Rothman, J. E. (1987) *Ann. Rev. Biochem.* **56**, 829–852
51. Lodish, H. F. (1988) *J. Biol. Chem.* **263**, 2107–2110
52. Smeekens, S., Bauerle, C., Hageman, J., Keegstra, K., and Weisbeek, P. (1986) *Cell* **46**, 365–375
53. Roise, D., and Schatz, G. (1988) *J. Biol. Chem.* **263**, 4509–4511
54. Hurt, E. C., and van Loon, A. P. G. M. (1986) *Trends Biochem. Sci.* **11**, 204–207
55. Aitchison, J. D., Murray, W. W., and Rachubinski, R. A. (1991) *J. Biol. Chem.* **266**, 23197–23203
56. Wolins, N. E., and Donaldson, R. P. (1994) *J. Biol. Chem.* **269**, 1149–1153
57. Takeda, J., and Kinoshita, T. (1995) *Trends Biochem. Sci.* **20**, 367–371
57a. Mazmanian, S. K., Liu, G., Ton-That, H., and Schneewind, O. (1999) *Science* **285**, 760–762
58. Ogata, S., and Fukuda, M. (1994) *J. Biol. Chem.* **269**, 5210–5217
59. Pond, L., Kuhn, L. A., Teyton, L., Schutze, M.-P., Tainer, J. A., Jackson, M. R., and Peterson, P. A. (1995) *J. Biol. Chem.* **270**, 19989–19997
60. Pelham, H. R. B. (1990) *Trends Biochem. Sci.* **15**, 483–486
61. Wilson, D. W., Lewis, M. J., and Pelham, H. R. B. (1993) *J. Biol. Chem.* **268**, 7465–7468
62. Mallabiabarrena, A., Jiménez, M. A., Rico, M., and Alarcón, B. (1995) *EMBO J.* **14**, 2257–2268
63. Luzio, J. P., and Banting, G. (1993) *Trends Biochem. Sci.* **18**, 395–398
64. Peter, F., Van, P. N., and Soling, H.-D. (1992) *J. Biol. Chem.* **267**, 10631–10637
65. Beh, C. T., and Rose, M. D. (1995) *Proc. Natl. Acad. Sci. U.S.A.* **92**, 9820–9823
66. Hoe, M. H., Slusarewicz, P., Misteli, T., Watson, R., and Warren, G. (1995) *J. Biol. Chem.* **270**, 25057–25063
67. Schneider, H.-C., Berthold, J., Bauer, M. F., Dietmeier, K., Guiard, B., Brunner, M., and Neupert, W. (1994) *Nature* **371**, 768–773
68. Bhattacharyya, T., Karnezis, A. N., Murphy, S. P., Hoang, T., Freeman, B. C., Phillips, B., and Morimoto, R. I. (1995) *J. Biol. Chem.* **270**, 1705–1710
69. Schmitt, M., Neupert, W., and Langer, T. (1995) *EMBO J.* **14**, 3434–3444
70. Stuart, R. A., Cyr, D. M., Craig, E. A., and Neupert, W. (1994) *Trends Biochem. Sci.* **19**, 87–92
71. Lithgow, T., Glick, B. S., and Schatz, G. (1995) *Trends Biochem. Sci.* **20**, 98–101
72. Pfanner, N., Craig, E. A., and Meijer, M. (1994) *Trends Biochem. Sci.* **19**, 368–372
73. Mayer, A., Nargang, F. E., Neupert, W., and Lill, R. (1995) *EMBO J.* **14**, 4204–4211
73a. Heard, T. S., and Weiner, H. (1998) *J. Biol. Chem.* **273**, 29389–29393
74. Görlich, D., Vogel, F., Mills, A. D., Hartmann, E., and Laskey, R. A. (1995) *Nature* **377**, 246–248
74a. Koehler, C. M., Merchant, S., and Schatz, G. (1999) *Trends Biochem. Sci.* **24**, 428–432
75. Smeekens, S., Weisbeek, P., and Robinson, C. (1990) *Trends Biochem. Sci.* **15**, 73–76
76. Pilon, M., Wienk, H., Sips, W., de Swaaf, M., Talboom, I., van 't Hof, R., de Korte-Kool, G., Demel, R., Weisbeek, P., and de Kruijff, B. (1995) *J. Biol. Chem.* **270**, 3882–3893

77. Schnell, D. J., Kessler, F., and Blobel, G. (1994) *Science* **266**, 1007–1011
78. Voelker, R., and Barkan, A. (1995) *EMBO J.* **14**, 3905–3914
79. Viitanen, P. V., Schmidt, M., Buchner, J., Suzuki, T., Vierling, E., Dickson, R., Lorimer, G. H., Gatenby, A., and Soll, J. (1995) *J. Biol. Chem.* **270**, 18158–18164
79a. Nakai, K., and Horton, P. (1999) *Trends Biochem. Sci.* **24**, 34–35
79b. Klionsky, D. J. (1998) *J. Biol. Chem.* **273**, 10807–10810
80. Rothman, J. E. (1994) *Nature* **372**, 55–63
81. Pryer, N. K., Wuestehube, L. J., and Schekman, R. (1992) *Ann. Rev. Biochem.* **61**, 471–516
82. Bennett, M. K., and Scheller, R. H. (1993) *Proc. Natl. Acad. Sci. U.S.A.* **90**, 2559–2563
83. Duden, R., Hosobuchi, M., Hamamoto, S., Winey, M., Byers, B., and Schekman, R. (1994) *J. Biol. Chem.* **269**, 24486–24495
84. Edelmann, L., Hanson, P. I., Chapman, E. R., and Jahn, R. (1995) *EMBO J.* **14**, 224–231
84a. Dulubova, I., Sugita, S., Hill, S., Hosaka, M., Fernandez, I., Südhof, T. C., and Rizo, J. (1999) *EMBO J.* **18**, 4372–4382
84b. Katz, L., Hanson, P. I., Heuser, J. E., and Brennwald, P. (1998) *EMBO J.* **17**, 6200–6209
84c. Gerona, R. R. L., Larsen, E. C., Kowalchyk, J. A., and Martin, T. F. J. (2000) *J. Biol. Chem.* **275**, 6328–6336
85. Morgan, A., and Burgoyne, R. D. (1995) *EMBO J.* **14**, 232–239
85a. Glockshuber, R. (1999) *Nature (London)* **401**, 30–31
85b. Bolhuis, A., Venema, G., Quax, W. J., Bron, S., and van Dijl, J. M. (1999) *J. Biol. Chem.* **274**, 24531–24538
85c. Ban, N., Nissen, P., Hansen, J., Moore, P. B., and Steitz, T. A. (2000) *Science* **289**, 905–920
85d. Berardi, M. J., and Bushweller, J. H. (1999) *J. Mol. Biol.* **292**, 151–161
85e. van den Berg, B., Chung, E. W., Robinson, C. V., Mateo, P. L., and Dobson, C. M. (1999) *EMBO J.* **18**, 4794–4803
86. Freedman, R. B., Hirst, T. R., and Tuite, M. F. (1994) *Trends Biochem. Sci.* **19**, 331–335
87. Martin, J. L., Bardwell, J. C. A., and Kuriyan, J. (1993) *Nature* **365**, 465–468
88. Kishigami, S., Kanaya, E., Kikuchi, M., and Ito, K. (1995) *J. Biol. Chem.* **270**, 17072–17074
89. Jander, G., Martin, N. L., and Beckwith, J. (1994) *EMBO J.* **13**, 5121–5127
90. Kanaya, E., Anaguchi, H., and Kikuchi, M. (1994) *J. Biol. Chem.* **269**, 4273–4278
91. Frech, C., and Schmid, F. X. (1995) *J. Biol. Chem.* **270**, 5367–5374
92. Chivers, P. T., Laboissiére, M. C. A., and Raines, R. T. (1996) *EMBO J.* **15**, 2659–2667
93. Tschantz, W. R., Paetzel, M., Cao, G., Suciu, D., Inouye, M., and Dalbey, R. E. (1995) *Biochemistry* **34**, 3935–3941
94. Steiner, D. F., Smeekens, S. P., Ohagi, S., and Chan, S. J. (1992) *J. Biol. Chem.* **267**, 23435–23438
95. De Bie, I., Savaria, D., Roebroek, A. J. M., Day, R., Lazure, C., Van de Ven, W. J. M., and Seidah, N. G. (1995) *J. Biol. Chem.* **270**, 1020–1028
96. Fisher, J. M., and Scheller, R. H. (1988) *J. Biol. Chem.* **263**, 16515–16518
97. Rehfeld, J. F., Hansen, C. P., and Johnsen, A. H. (1995) *EMBO J.* **14**, 389–396
98. Ypma-Wong, M. F., Filman, D. J., Hogle, J. M., and Semler, B. L. (1988) *J. Biol. Chem.* **263**, 17846–17856
99. Bateman, A., Solomon, S., and Bennett, H. P. J. (1990) *J. Biol. Chem.* **265**, 22130–22136
100. Richter, D. (1983) *Trends Biochem. Sci.* **8**, 278–280

References

101. Douglass, J., Civelli, O., and Herbert, E. (1984) *Ann. Rev. Biochem.* **53**, 665–715
102. Arfin, S. M., and Bradshaw, R. A. (1988) *Biochemistry* **27**, 7979–7984
103. Tolan, D. R., Amsden, A. B., Putney, S. D., Urdea, M. S., and Penhoet, E. E. (1984) *J. Biol. Chem.* **259**, 1127–1131
104. Rubenstein, P. A., and Martin, D. J. (1983) *J. Biol. Chem.* **258**, 3961–3966
105. Sheff, D. R., and Rubenstein, P. A. (1992) *J. Biol. Chem.* **267**, 20217–20224
106. Kulkarni, M. S., and Sherman, F. (1994) *J. Biol. Chem.* **269**, 13141–13147
107. Gan, K., Gupta, S. D., Sankaran, K., Schmidt, M. B., and Wu, H. C. (1993) *J. Biol. Chem.* **268**, 16544–16550
108. Vogt, A., Sun, J., Qian, Y., Tan-Chiu, E., Hamilton, A. D., and Sebti, S. M. (1995) *Biochemistry* **34**, 12398–12403
109. Parish, C. A., and Rando, R. R. (1994) *Biochemistry* **33**, 9986–9991
110. Pompliano, D. L., Rands, E., Schaber, M. D., Mosser, S. D., Anthony, N. J., and Gibbs, J. B. (1992) *Biochemistry* **31**, 3800–3807
111. Busby, W. H., Jr., Quackenbush, G. E., Humm, J., Youngblood, W. W., and Kizer, J. S. (1987) *J. Biol. Chem.* **262**, 8532–8536
112. Fischer, W. H., and Spiess, J. (1987) *Proc. Natl. Acad. Sci. U.S.A.* **84**, 3628–3632
113. Katopodis, A. G., Ping, D., Smith, C. E., and May, S. W. (1991) *Biochemistry* **30**, 6189–6194
114. Merkler, D. J., Kulathila, R., Consalvo, A. P., Young, S. D., and Ash, D. E. (1992) *Biochemistry* **31**, 7282–7288
115. Ping, D., Katopodis, A. G., and May, S. W. (1992) *J. Am. Chem. Soc.* **114**, 3998–4000
116. Eipper, B. A., Milgram, S. L., Husten, E. J., Yun, H.-Y., and Mains, R. E. (1993) *Protein Sci.* **2**, 489–497
117. Bradbury, A. F., and Smyth, D. G. (1991) *Trends Biochem. Sci.* **16**, 112–115
118. Low, M. G., and Saltiel, A. R. (1988) *Science* **239**, 268–275
119. Lisanti, M. P., and Rodriguez-Boulan, E. (1990) *Trends Biochem. Sci.* **15**, 113–118
120. Graves, D. J., Martin, B. L., and Wang, J. H. (1994) *Co-and post-translational modification of proteins*, Oxford Univ. Press, New York
121. Tuboi, S., Taniguchi, N., and Katunuma, N., eds. (1992) *Post-Translation Modification of Proteins*, CRC Press, Boca Raton, Florida
122. Barrett, G. C., ed. (1985) *Chemistry and Biochemistry of the Amino Acids*, Chapman and Hall, London; New York
123. Niehrs, C., Kraft, M., Lee, R. W. H., and Huttner, W. B. (1990) *J. Biol. Chem.* **265**, 8525–8532
124. Rucker, R. B., and Wold, F. (1988) *FASEB J.* **2**, 2252–2261
125. Klotz, A. V., Leary, J. A., and Glazer, A. N. (1986) *J. Biol. Chem.* **261**, 15891–15894
126. Oliver, C. N., Ahn, B., Moerman, E. J., Goldstein, S., and Stadtman, E. R. (1987) *J. Biol. Chem.* **262**, 5488–5491
127. Cooper, A. A., and Stevens, T. H. (1995) *Trends Biochem. Sci.* **20**, 351–356
128. Rechsteiner, M., Rogers, S., and Rote, K. (1987) *Trends Biochem. Sci.* **12**, 390–394
129. Stadtman, E. R. (1986) *Trends Biochem. Sci.* **11**, 11–12
130. Rivett, A. J. (1985) *J. Biol. Chem.* **260**, 300–305
131. Loshon, C. A., Swerdlow, B. M., and Setlow, P. (1982) *J. Biol. Chem.* **257**, 10838–10845
132. Olden, K., Parent, J. B., and White, S. C. (1982) *Biochim. Biophys. Acta.* **650**, 209–232
133. Chiang, H.-L., and Dice, J. F. (1988) *J. Biol. Chem.* **263**, 6797–6805
133a. Dell'Angelica, E. C., Mullins, C., Caplan, S., and Bonifacino, J. S. (2000) *FASEB J.* **14**, 1265–1278
134. Hershko, A., and Ciechanover, A. (1992) *Ann. Rev. Biochem.* **61**, 761–807
135. Dice, J. F. (1987) *FASEB J.* **1**, 349–357
136. Stadtman, E. R. (1990) *Biochemistry* **29**, 6323–6331
137. Rogers, S., Wells, R., and Rechsteiner, M. (1986) *Science* **234**, 364–368
138. Bachmair, A., Finley, D., and Varshavsky, A. (1986) *Science* **234**, 179–186
139. Gonda, D. K., Bachmair, A., Wünning, I., Tobias, J. W., Lane, W. S., and Varshavsky, A. (1989) *J. Biol. Chem.* **264**, 16700–16712
140. Madura, K., and Varshavsky, A. (1994) *Science* **265**, 1454–1458
140a. Davydov, I. V., and Varchavsky, A. (2000) *J. Biol. Chem.* **275**, 22931–22941
141. Herschlag, D. (1995) *J. Biol. Chem.* **270**, 20871–20874

Study Questions

1. Outline in detail, using structural formulas, the enzyme-catalyzed reactions by which cells in the human body convert glyceraldehyde 3-phosphate into pyruvate.

2. Describe, using chemical structural formulas, the reactions involved in the breakdown of glycogen to glucose 1-phosphate and the synthesis of glycogen from glucose 1-phosphate.

3. Describe the reaction steps in gluconeogenesis by which pyruvate is converted into glyceraldehyde 3-phosphate.

4. Compare the reactions of pyruvate that give rise to the following three compounds. List coenzymes or electron-carriers involved in each case and indicate any intermediate compounds.
   a) Ethanol
   b) Lactic acid
   c) Acetyl-Coenzyme A

5. Mammalian sperm cells metabolize D-fructose preferentially as a source of energy. Fructose is formed in cells of the seminal vesicle from D-glucose via reduction to the sugar alcohol sorbitol using NADPH, followed by oxidation of sorbitol to fructose using NAD^+. The fructose concentration in human semen is about 12 mM, whereas the glucose concentration within cells is usually less than 1 mM. If the ratio $[NADPH]/[NADP^+]$ is 10^4 times higher than the ratio $[NADH]/[NAD^+]$, what is the minimum glucose concentration in cells that could allow formation of 12 mM fructose? The standard Gibbs energies of formation from the elements G°_f in kJ/mol at 25°C are: D-glucose –917, D-fructose –915.

6. Outline the reactions by which glyceraldehyde 3-phosphate is converted to 3-phosphoglycerate with coupled synthesis of ATP in the glycolysis pathway. Show important mechanistic details.

7. Why can't acetyl CoA be converted to glucose in animals?

8. Describe the parallel reaction sequences between the citric acid cycle and the β oxidation pathway.

9. Contrary to legend, camels do not store water in their humps, which actually consist of a large fat deposit. How can these fat deposits serve as a source of water? Calculate the amount of water (liters) that can be produced by the camel from 500 g of fat. Assume for simplicity that the fat consists entirely of tripalmitin.

10. A little-known microorganism carries out the following net reaction in a series of enzymatic steps:

$$\text{heptanoyl CoA} + \text{CoASH} + \text{NAD}^+ + \text{FAD} + H_2O \rightarrow \text{pentanoyl CoA} + \text{acetyl CoA} + \text{NADH} + \text{FADH}_2 + H^+$$

The net result is exactly the same as that of the β oxidation pathway, but for this microorganism, the pathway is demonstrably different in that pentanal, $CH_3(CH_2)_3CHO$, is an intermediate in the sequence. Which enzyme(s) of the β oxidation pathway does this microorganism lack? Propose an enzymatic reaction to account for the formation of this intermediate and a series for its conversion to the acyl CoA. ATP is not required.

11. Most natural fatty acids are even-numbered. What is the product of the *final* thiolase reaction to form an odd-chain fatty acid? Give a brief explanation.

12. It has been calculated that an average man takes in 1.5 kg of solid food and 1.4 kg of water per day. He gives off 3.5 kg of waste and 0.75 kg of sweat. It would thus appear that he should be losing over 1 kg per day through normal activites. How do you account for the fact that his weight remains relatively constant?

13. Suppose that fatty acids, instead of being broken down in two carbon fragments, were metabolized in three carbon units. What product(s) would Dr. Knoop have observed in the urine of his experimental dogs?

14. A renowned pharmacologist has announced the discovery of a new drug that specifically inhibits the fatty acid oxidation pathway within minutes of ingestion. The effects last for several hours only. The drug has no other effects on the subjects. The pharmacologist argues that this will increase athletic performance by shifting oxidative metabolism entirely to the more rapidly mobilized carbohydrate degradation pathway. Assume that the drug does work exactly as he suggests. Explain in a sentence or two how the drug would affect the performance of
    a) a sprinter in the 100 meter dash
    b) a long-distance runner in a marathon

15. Some bacteria use a "dicarboxylic acid cycle" to oxidize glyoxylate $OHC-COO^-$ to CO_2. The regenerating substrate for this cycle is acetyl-CoA. It is synthesized from glyoxylate by a complex pathway that begins with conversion of two mole-cules of glyoxylate to tartronic semialdehyde:

$^-OOC-CHOH-CHO$. The latter is then dehydro-genated to D-glycerate. Write out a detailed scheme for the dicarboxylate cycle. Also indicate how glucose and other cell constituents can be formed from intermediates created in this biosynthetic pathway.

16. Write a balanced fermentation sequence by which glycogen can be converted rapidly to *sn*-glycerol 3-phosphate and pyruvate in insect flight muscle. How many molecules of ATP per glucose unit of glycogen will be formed?

17. Show which parts (if any) of the citric acid cycle are utilized in each of the following reactions and what if any additional enzymes are needed in each case.
    a) Oxidation of acetyl-CoA to CO_2
    b) Catabolism of glutamate to CO_2
    c) Biosynthesis of glutamate from pyruvate
    d) Formation of propionate from pyruvate

18. a) If the Gibbs energy change $\Delta G'$ (pH 7) for the reaction $A \rightarrow B$ is $+25$ kJ/mol at 25°C, what would the ratio of $[B]/[A]$ be at equilibrium?
    b) Suppose that the reaction were coupled to the cleavage of ATP as follows:

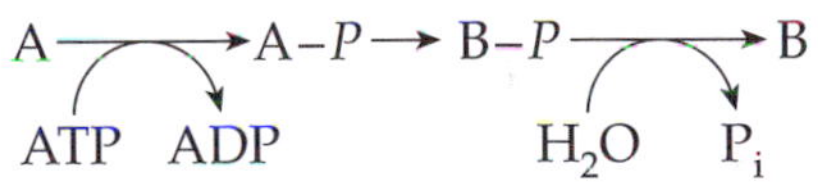

Suppose further that the group transfer potential $(-\Delta G')$ for the phospho group of $A-P$ at 25°C, pH 7 is 12 kJ/mol and that the equilibrium constant for conversion of $A-P$ to $B-P$ is the same as that for conversion of A to B. Calculate the concentrations of A, B, $A-P$ and $B-P$ at equilibrium if the phosphorylation state ratio R_p is 10^4 M^{-1}.

19. This problem refers to the 13 meter climb described in Chapter 6, Study Question 14.
    a) Assuming that muscle accounts for 30% of total body mass, estimate the amounts (mmol/kg, total mmol, and total grams) of each of the following in their coenzyme forms: nicotinamide, pantothenic acid, thiamin. You may be able to obtain rough estimates from the vitamin content of pork.
    b) Calculate how many times, on the average, each molecule of nicotinamide would undergo reduction and re-oxidation (i.e., turn over) during the climb. Do the same for pantothenic acid (cycling between acyl-CoA and free CoA forms) and for thiamin through its catalytic cycles.

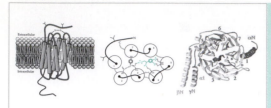

Much of the control of cellular metabolism is accomplished by hormones and other molecules that bind to receptor proteins embedded in the plasma membrane. Left: Many receptors are seven-helix transmembrane proteins, the best known being the light receptors of the rhodopsin family (Chapter 23). Center: A related adrenergic receptor, viewed here from the extracellular side, has a molecule of the hormone adrenaline (green) bound in the center (Fig. 11-6). Right: These receptors interact with GTP-hydrolyzing "G proteins," which pass signals, often for short periods of time, to enzymes and other proteins. One structural domain of a G protein (Fig. 11-7C) shown here contains a β propeller domain which binds to the Ras-like domain of the α subunit. A few residues of the latter are visible in the center of this image.

Contents

The Regulation of Enzymatic Activity and Metabolism

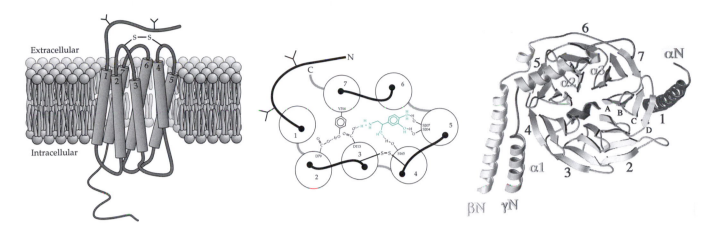

11

Living cells must operate with controls that provide a stable environment and a relatively constant supply of materials needed for biosynthesis and for meeting the energy needs of cells. They must also be responsive to changes in their environment and must be able to undergo mitosis and reproduce when appropriate. The necessary control of metabolism and of growth is accomplished largely through mechanisms that regulate the locations, the amounts, and the catalytic activities of enzymes. The purpose of this chapter is to summarize these control mechanisms and to introduce terminology and shorthand notations that will be used throughout this book. Many of the **control elements** considered are summarized in Fig. 11-1.

A. Pacemakers and the Control of Metabolic Flux

Metabolic control can be understood to some extent by focusing attention on those enzymes that catalyze rate-limiting steps in a reaction sequence. Such **pacemaker enzymes**[1-4] are often involved in reactions that determine the overall respiration rate of a cell, reactions that initiate major metabolic sequences, or reactions that initiate branch pathways in metabolism. Often the first step in a unique biosynthetic pathway for a compound acts as the pacemaker reaction. Such a reaction may be described as the **committed step** of the pathway. It usually proceeds with a large decrease in Gibbs energy and tends to be tightly controlled. Both the rate of synthesis of the enzyme protein and the activity of the enzyme, once it is formed, may be inhibited by **feedback inhibition** which occurs when an end product of a biosynthetic pathway accumulates

and inhibits the enzyme. Enzyme activity may also be turned on or off by the effect of a hormone, by some other external stimulus, or by internal mechanisms that sense the metabolic state of the cell. Enzymes, other than the pacemaker, that catalyze reactions in a pathway may not be regulated and may operate at a steady state close to equilibrium.

If conditions within a cell change, a pacemaker reaction may cease to be rate limiting. A reactant plentiful in one circumstance may, in another, be depleted to the point that the rate of its formation from a preceding reaction determines the overall rate. Thus, as we have seen in Chapter 10, metabolism of glucose in our bodies occurs through the rapidly interconvertible phosphate esters glucose 6-phosphate and fructose 6-phosphate. The pacemaker enzyme in utilization of glucose or of glycogen is often **phosphofructokinase** (Fig. 11-2, step *b*) which catalyzes further metabolism of fructose 6-phosphate. However, if metabolism by this route is sufficiently rapid, the rate of formation of glucose 6-phosphate from glucose catalyzed by **hexokinase** (Fig. 11-2, step *a*) may become rate limiting.

Some catabolic reactions depend upon ADP, but under most conditions its concentration is very low because it is nearly all phosphorylated to ATP. Reactions utilizing ADP may then become the rate-limiting pacemakers in reaction sequences. Depletion of a reactant sometimes has the effect of changing the whole pattern of metabolism. Thus, if oxygen is unavailable to a yeast, the reduced coenzyme NADH accumulates and reduces pyruvate to ethanol plus CO_2 (Fig. 10-3). The result is a shift from oxidative metabolism to fermentation.

Pacemaker enzymes are often identified by the fact that the measured **mass action ratio**, e.g., for the reaction A + B → P + Q ,the ratio [P][Q]/[A][B], is far

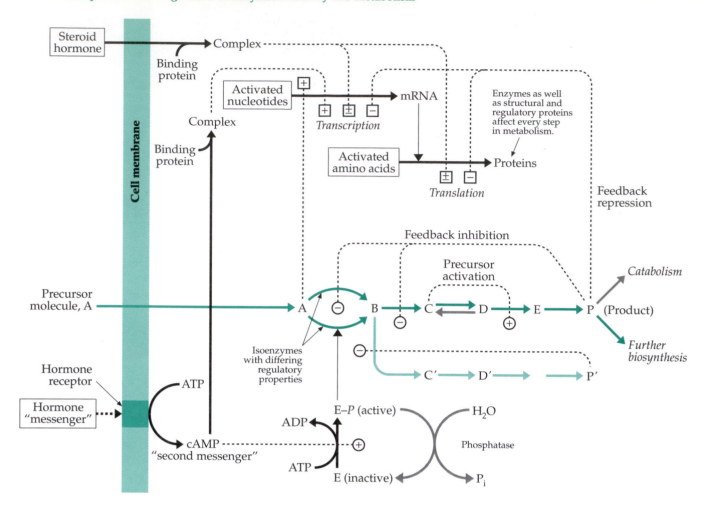

Figure 11-1 Some control elements for metabolic reactions. *Throughout the book modulation of the activity of an enzyme by allosteric effectors or of transcription and translation of genes is indicated by dotted lines from the appropriate metabolite.* The lines terminate in a minus sign for inhibition or repression and in a plus sign for activation or derepression. Circles indicate direct effects on enzymes, while boxes indicate repression or induction of enzyme synthesis.

from that predicted by the known equilibrium constant for the reaction. Another approach to identifying a pacemaker enzyme is to inhibit it and observe resulting changes in steady-state metabolite concentrations.[5] If the pacemaker enzyme is inhibited, its substrate and other compounds preceding the step catalyzed by this enzyme will accumulate. At the same time the concentrations of products of the pacemaker reaction will drop as a result of their relatively more rapid removal by enzymes catalyzing subsequent steps. However, conclusions drawn by this approach may sometimes be erroneous.[6]

In spite of its usefulness, the pacemaker concept is oversimplified. It is often impossible to identify a specific pacemaker enzyme. When both catabolism and biosynthesis occur (e.g., as in the scheme in Fig. 11-2) it may be more useful to model the entire system with a computer than to try to identify pacemakers.[7,8] It is also important to realize that reaction rates may be

determined by the rate of diffusion of a compound through a membrane. Thus, membrane transport processes can serve as pacemakers.

A general approach to analysis of complex metabolic pathways or to cell growth was introduced by Savageau,[9–11] and similar approaches have been followed by others.[3,12–19] They emphasize the **flux** of material through a pathway under steady-state conditions and recognize that every enzyme in a sequence can have some effect on the overall rate. Consider a chain of enzymes E_1 to E_n acting on substrate S_0, which is converted via substrates $S_1, S_2 \ldots S_{n-1}$ into product S_n:

$$S_0 \xrightarrow{E_1} S_1 \xrightarrow{E_2} S_2 \xrightarrow{E_3} S_3 \cdots \xrightarrow{E_{n-1}} S_{n-1} \xrightarrow{E_n} S_n \quad (11\text{-}1)$$

The flux $F = d[S_n]/dt$ is a constant. If the concentration of E_i were to change by an infinitesimal amount a corresponding change in F might be observed. The

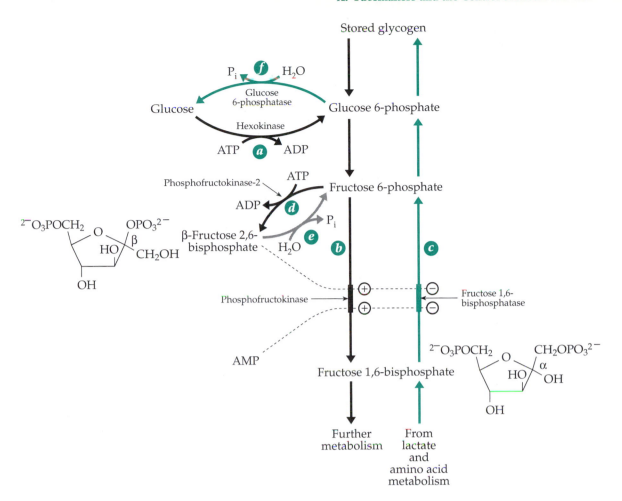

Figure 11-2 Roles of phosphofructose kinase and fructose 1,6-bisphosphatase in the control of the breakdown (➡) and storage (➡) of glycogen in muscle. The uptake of glucose from blood and its release from tissues is also illustrated. The allosteric effector fructose 2,6-bisphosphate (Fru-2,6-P_2) regulates both phosphofructokinase and fructose 2,6-bisphosphatase. These enzymes are also regulated by AMP if it accumulates. The activity of phosphofructokinase-2 (which synthesizes Fru-2,6-P_2) is controlled by a cyclic AMP-dependent kinase and by dephosphorylation by a phosphatase.

ratio of the change in flux to the change in [E_i] is the sensitivity coefficient Z_i:

$$Z_i = \frac{dF/F}{d[E_i]/[E_i]} = \frac{d \ln F}{d \ln[E_i]} \qquad (11\text{-}2)$$

The sensitivity coefficient can range from 1 for a pacemaker enzyme to almost zero for a very active enzyme that does not significantly limit the flux. The sum of Z_1 to Z_n is equal to 1:

$$\sum_{n}^{i=1} Z_i = 1 \qquad (11\text{-}3)$$

Not only can the concentration [E_i] change but also allosteric effectors can alter the activity. Kacser and Burns defined this in terms of a **controllability coefficient** κ,

the ratio of the logarithmic change in velocity to the change in the concentration of an effector:

$$\kappa = d \ln v/d \ln [P] \qquad (11\text{-}4)$$

It follows that the logarithmic change in flux $d \ln F$ can be related to the change in effector concentration [P] in the following way:

$$d \ln F = Z_i \kappa d \ln [P] \qquad (11\text{-}5)$$

From this equation we see that for the flux to be sensitively dependent upon [P], both Z_i and κ must be reasonably large. This is exactly the case for many of the enzymes that have been identified as pacemakers. This approach can be applied to many aspects of metabolic control including cell growth.[20]

B. Genetic Control of Enzyme Synthesis

All cellular regulatory mechanisms depend upon the synthesis of proteins, that is, upon the expression of genetic information. Within a given cell many genes are transcribed continuously but others may remain unexpressed. The rates of both transcription and degradation of mRNA are regulated, as are the rates of synthesis of enzymes on the ribosomes in the cytoplasm and the rates of protein turnover. Although a single copy of a gene in each member of a pair of chromosomes is often adequate, there are situations in which extra copies of part of the DNA of a cell are formed. Such **gene amplification**, which is dealt with further in Chapter 27, provides the possibility for very rapid synthesis of an enzyme or other protein. It often happens in highly specialized cells such as those of the silkworm's silk glands which must make enormous amounts of a small number of proteins.[21]

1. One Gene or Many?

Many enzymes exist within a cell as two or more **isoenzymes**, enzymes that catalyze the same chemical reaction and have similar substrate specificities. They are not isomers but are distinctly different proteins which are usually encoded by different genes.[22,23] An example is provided by aspartate aminotransferase (Fig. 2-6) which occurs in eukaryotes as a pair of cytosolic and mitochondrial isoenzymes with different amino acid sequences and different isoelectric points. Although these isoenzymes share less than 50% sequence identity, their internal structures are nearly identical.[24-27] The two isoenzymes, which also share structural homology with that of *E. coli*,[28] may have evolved separately in the cytosol and mitochondria, respectively, from an ancient common precursor. The differences between them are concentrated on the external surface and may be important to as yet unknown interactions with other protein molecules.

Isoenzymes are designated in various ways. They are often *numbered in the order of decreasing electrophoretic mobility* at pH 7 to 9: Most enzymes are negatively charged in this pH range and the one that migrates most rapidly toward the anode is numbered one. This is the same convention used in the electrophoresis of blood proteins (e.g. see Box 2-A).

Lactate dehydrogenase exists in the cytoplasm of humans and most animals as *five forms* which are easily separable by electrophoresis and are evenly spaced on electropherograms.[8] This enzyme is a tetramer made of two kinds of subunits. Isoenzyme 1, which has the highest electrophoretic mobility, consists of four identical type B subunits. The slowest moving tetramer (isoenzyme 5) consists of four type A subunits, while the other three forms, AB_3, A_2B_2, and A_3B, contain both subunits in different proportions. The two subunits are encoded by separate genes which are expressed to different extents in different tissues. Thus, heart muscle and liver produce mainly subunit B, while skeletal muscle produces principally subunit A. A third subunit type (C) is found only in the testes.[29]

Why do cells produce isoenzymes? One reason may be that enzymes with differing kinetic parameters are needed.[30] Substrate concentrations may vary greatly between different tissues; between different subcellular compartments; and at different developmental stages of an organism. While the need for various isoenzymes of lactate dehydrogenase is not well understood,[31] it is easier to understand the roles of the multiple forms of hexokinase, the enzyme that catalyzes the reaction of step *a* in Fig. 11-2. The brain enzyme has a high affinity for glucose ($K_m = 0.05$ mM). Thus, it is able to phosphorylate glucose and to make that substrate available to the brain for metabolism, even if the glucose concentration in the tissues falls to low values.[31a] On the other hand, **glucokinase**, the hexokinase isozyme found in liver, has a much higher K_m of ~10 mM. It functions to remove excess glucose from blood, whose normal glucose content is ~5.5 mM. Glucokinase reaches its maximal activity only when the glucose concentration becomes much higher.[32] This happens after a meal when the absorbed glucose passes through the portal circulation directly to the liver.

Another important source of variation in enzymes as well as in other proteins is **alternative splicing** of mRNA.[33] For example, transcription of the mouse α-amylase gene in the salivary gland starts at a different site (promoter) than does transcription in the liver. The two common isoforms of the insulin receptor (Fig. 11-11) arise because a 36-nucleotide (12-amino acid) exon is spliced out of the mRNA for the shorter protein. Isoenzymes of aldolase[34] and of many other proteins are formed in a similar manner. **Frame-shifting** during protein synthesis (Chapter 29) and also post-translational alterations may give rise to additional modified forms. They are often synthesized in relatively small amounts but may be essential to the life of the cell. In addition, genetic variants of almost any protein will be found in any population. These often differ in sequence by a single amino acid.

2. Repression, Induction, and Turnover

The synthesis of some enzymes is referred to as **constitutive**, implying that the enzyme is formed no matter what the environmental conditions of the cell. For example, many bacteria synthesize the enzymes required to catabolize glucose under all conditions of growth. Other enzymes, known as **inducible**, are often produced only in small amounts. However, if cells are grown in specific inducing conditions for these enzymes, they are synthesized in larger quantities. For example, when *E. coli* is cultivated in the presence of lactose, several of the enzymes required for the catabolism of that disaccharide are formed. Synthesis of these enzymes is normally **repressed**. The genes which code for them are kept turned off through the action of protein **repressors** which bind to specific sites on the DNA and block transcription of the genes that they control (Fig. 11-1). Repressors have allosteric properties; in one conformation they bind tightly to DNA but in another they do not bind. For example, the free tryptophan repressor binds to DNA only weakly, but if a high concentration of tryptophan develops within the cell the tryptophan binds to an allosteric site on the repressor protein (Fig. 5-35). This changes the conformation to one that binds tightly to the appropriate control sequence in the DNA. In the case of lactose catabolism the free repressor binds to control sequences in the DNA until the **inducer**, allolactose (Chapters 4 and 28), binds at an allosteric site. This decreases the affinity of the repressor for DNA and the controlled genes are **derepressed**. There are also many protein **transcription factors** that have a positive effect, binding to DNA and promoting transcription of specific genes.

Synthesis of many enzymes is repressed most of the time. The appearance of an enzyme at a particular stage in the life of an organism as well as the differing distributions of isoenzymes within differentiated tissue result from derepression. The control of enzyme synthesis may also be exerted during the splicing of transcripts and at the translational level as well. These control mechanisms are often relatively slow, with response times of hours or even days. However, effects on the synthesis of some hormones, such as insulin (Section G), may be observed within a few minutes.

Genetic factors influence the rate of not only synthesis of proteins but also their breakdown, i.e., the rate of turnover. As we have seen in Chapter 10, some enzymes are synthesized as inactive proenzymes which are later modified to active forms, and active enzymes are destroyed, both by accident and via deliberate hydrolytic pathways. Protein **antienzymes** may not only inhibit enzymes but may promote their breakdown.[35] An example is the antienzyme that controls ornithine decarboxylase, a key enzyme in the synthesis of the polyamines that are essential to growth.[36,37] As with all cell constituents, the synthesis of enzymes and other proteins is balanced by degradation.

3. Differences among Species

Catalytic mechanisms of enzymes have usually been conserved throughout evolution, and certain residues in an enzyme may be invariant among many species. However, there are usually many differences in the distribution of amino acid residues on the surface of the proteins. Since changes in the surface shape of a protein molecule may alter the sensitivity to a potential allosteric effector, very different regulatory properties are sometimes found between species.

C. Regulation of the Activity of Enzymes

Some regulation of metabolism is provided by the kinetic properties of the enzymes. Thus, the value (mol/l) of the Michaelis constant K_m for an enzyme is usually low if the substrate normally occurs in very low concentration. It is likely to be higher if the enzyme acts on an abundant substrate. A tightly bound product of an enzymatic reaction will be released slowly from an enzyme if the product concentration within the cell is too high. However, while some regulation of metabolism is provided in such simple ways, the rapid changes that result from stimulation by hormones or by nerve impulses depend upon additional specific regulatory mechanisms that are discussed in the following sections.

1. Allosteric Control

Probably the most common and widespread control mechanisms in cells are **allosteric inhibition** and **allosteric activation**. These mechanisms are incorporated into metabolic pathways in many ways, the most frequent being **feedback inhibition**. This occurs when an end product of a metabolic sequence accumulates and turns off one or more enzymes needed for its own formation. It is often *the first enzyme unique to the specific biosynthetic pathway for the product that is inhibited*. When a cell makes two or more isoenzymes, only one of them may be inhibited by a particular product. For example, in Fig. 11-1 product P inhibits just one of the two isoenzymes that catalyzes conversion of A to B; the other is controlled by an enzyme modification reaction. In bacteria such as *E. coli*, three isoenzymes, which are labeled I, II, and III in Fig. 11-3, convert aspartate to β-aspartyl phosphate, the precursor to the end products threonine, isoleucine, methionine, and lysine. Each product inhibits only one of the isoenzymes as shown in the figure.

Feedback can also be positive. Since AMP is a product of the hydrolysis of ATP, its accumulation is a signal to *activate* key enzymes in metabolic pathways that generate ATP. For example, AMP causes allosteric activation of glycogen phosphorylase, which catalyzes the first step in the catabolism of glycogen. As is shown in Fig. 11-5, the allosteric site for AMP or IMP binding is more than 3 nm away from the active site. Only a

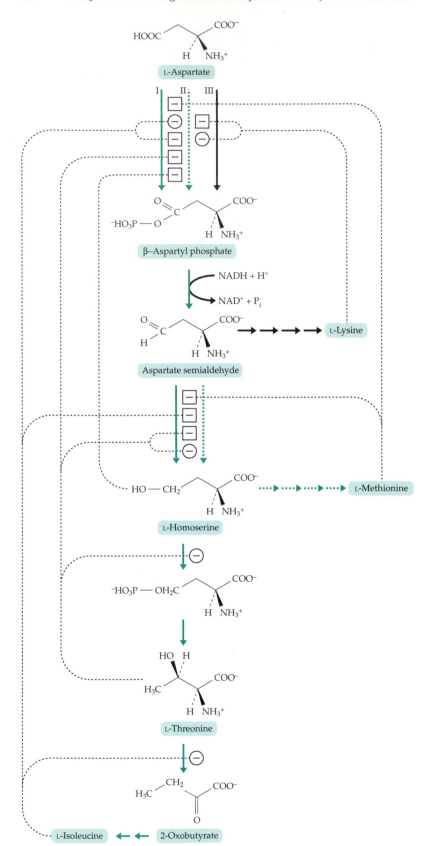

subtle conformational change accompanies binding of IMP.[38] However, ^{31}P NMR studies indicate that binding of the activator may induce a change in the state of protonation of the phosphate group of the coenzyme pyridoxal phosphate at the active site as explained in Chapter 12.

A second pattern of allosteric control may be referred to as **precursor activation** or feed-forward control. A metabolite acting as an allosteric effector turns on an enzyme that either acts directly on that metabolite or acts on a product that lies further ahead in the sequence. For example, in Fig. 11-1, metabolite C activates the enzyme that catalyzes an essentially irreversible reaction of compound D. An actual example is provided by glycogen synthase, whose inactive "dependent" or D form is activated allosterically by the glycogen precursor glucose 6-phosphate.[39] See also phosphorylase kinase (Section 2).

Regulatory subunits. Some enzymes consist not only of catalytic subunits, which contain the active sites, but also of **regulatory subunits**. The latter bind to the catalytic subunits and serve as allosteric modifiers. The binding of inhibitors or activators to specific sites on the regulatory subunits induces conformational changes in these subunits, altering their interaction with the catalytic subunits. A well-known example is aspartate carbamoyltransferase (ACTase) from *E. coli* (Fig. 7-20).[40-42] Its regulatory subunits carry binding sites for cytidine triphosphate (CTP), which acts as a specific allosteric inhibitor of the enzyme. The significance of this fact is that ACTase catalyzes the first reaction specific to the pathway of synthesis of pyrimidine nucleotides (Chapter 25). CTP is an end product of that pathway and exerts feedback inhibition on the enzyme that initiates its synthesis. ATP also binds to the regulatory subunits and causes inhibition of the enzyme.

Figure 11-3 Feedback inhibition of enzymes involved in the biosynthesis of threonine, isoleucine, methionine, and lysine in *E. coli*. These amino acids all arise from L-aspartate, which is formed from oxaloacetate generated by the biosynthetic reactions of the citric acid cycle (Fig. 10-6). ⊖ Allosteric inhibition. ⊟ Repression of transcription of the enzyme or of its synthesis on ribosomes.

In the presence of CTP the binding of the substrates is cooperative, as would be anticipated if there is a two-state conformational equilibrium involving the catalytic subunits. This would be similar to the case depicted in Fig. 9-13 except that trimers rather than dimers are involved and the inhibitor is a part of the regulatory subunit and is controlled by binding of CTP.

Binding of ATP to ACTase decreases the cooperativity in substrate binding, again as predicted for a two-state model. However, as suggested by its structure, this enzyme system is more complex, as also indicated by the observed anticooperative binding of the activator CTP. X-ray studies show that binding of ligands causes a movement along subunit interfaces as well as localized conformational changes within the subunits. These are reminiscent of the changes seen upon binding of oxygen to hemoglobin (Fig. 7-25).

Glycolysis and gluconeogenesis. The highly regulated enzymes phosphofructokinase and fructose 1,6-bisphosphatase catalyze steps *b* and *c* of the reactions in Fig. 11-2, reactions that control glucose metabolism in cells. These enzymes have been studied for many years but the important allosteric effector **β-fructose 2,6-bisphosphate** was not discovered until 1980.[43-46]

β-Fructose
2,6-bisphosphate

This compound, which is formed from fructose 6-phosphate by a new enzyme, **phosphofructokinase-2** (also called fructose 6-phosphate 2-kinase) in step *d*, Fig. 11-2 activates phosphofructokinase allosterically. At the same time it inhibits fructose 1,6-bisphosphatase, an enzyme required for reversal of glycogen breakdown, that is, for the conversion of various metabolites arising from amino acids into glycogen (Fig. 11-2, step *c*).[46a] These same two key regulated enzymes are also affected by many other metabolites. For example, ATP in excess inhibits phosphofructokinase, decreasing the overall rate of glucose metabolism and consequently of ATP production. Citrate, which is exported from mitochondria when carbohydrate metabolism is excessive, inhibits the same enzyme. On the other hand, AMP acts together with fructose 2,6-bisphosphate to activate the pathway for glycogen breakdown and to inhibit that for its synthesis (Fig. 11-2). The concentration of the regulator fructose 2,6-bisphosphate is controlled by mechanisms that are discussed in the following section.

2. Covalent Modification by Phosphorylation and Dephosphorylation

Rapid alteration of the activities of enzymes is often accomplished by **reversible covalent modification**.[39,47] Many different modification reactions are known (Table 11-1) and there are doubtless many more to be discovered. Probably the most widespread and certainly the most studied is **phosphorylation**, the transfer of a phospho group from ATP or other suitable donor to a side chain group on the enzyme. An example is the phosphorylation by ATP of hydroxyl groups of specific serine residues in the two enzymes glycogen phosphorylase and glycogen synthase. These modifications are accomplished through a series or **cascade** of reactions initiated by the binding of hormones to cell surface receptors or by nerve impulses as is shown in Fig. 11-4. In the absence of such a stimulus, glycogen phosphorylase is present in its unphosphorylated or *b* form. Although this form can be activated allosterically by AMP, it is normally nearly inactive. When an appropriate hormone binds to the cell surface a cascade of reactions, as described in Section D, leads to activation of an enzyme called **phosphorylase kinase**.[47a] This enzyme transfers a phospho group from ATP to the −OH group of the side chain of Ser 14 in each subunit (Fig. 11-4, left center), converting the enzyme into the active glycogen-degrading **phosphorylase *a***. This switches the cellular metabolism from that designed to deposit the storage polysaccharide glycogen to one that degrades glycogen to provide the cell with energy. Serine 14 of glycogen phosphorylase is located adjacent to the allosteric AMP binding site (Fig. 11-5) and is surrounded by positively charged arginine and lysine side chains. Phosphorylation induces a rearrangement of hydrogen bonds involving these residues and in some way sends an appropriate signal to the active site. It also increases the affinity for AMP in the allosteric sites.[48,48a] Phosphorylase kinase is allosterically activated by AMP, a product of its action – a feed-forward activator.

The control of glycogen phosphorylase by the phosphorylation–dephosphorylation cycle was discovered in 1955 by Edmond Fischer and Edwin Krebs[50] and was at first regarded as peculiar to glycogen breakdown. However, it is now abundantly clear that similar reactions control most aspects of metabolism.[51] Phosphorylation of proteins is involved in control of carbohydrate, lipid, and amino acid metabolism; in control of muscular contraction, regulation of photosynthesis in plants,[52] transcription of genes,[51] protein syntheses,[53] and cell division; and in mediating most effects of hormones.

Protein kinases and cyclic AMP. Phosphorylase kinase is one of hundreds of different protein kinases which differ in specificity toward their substrates, in

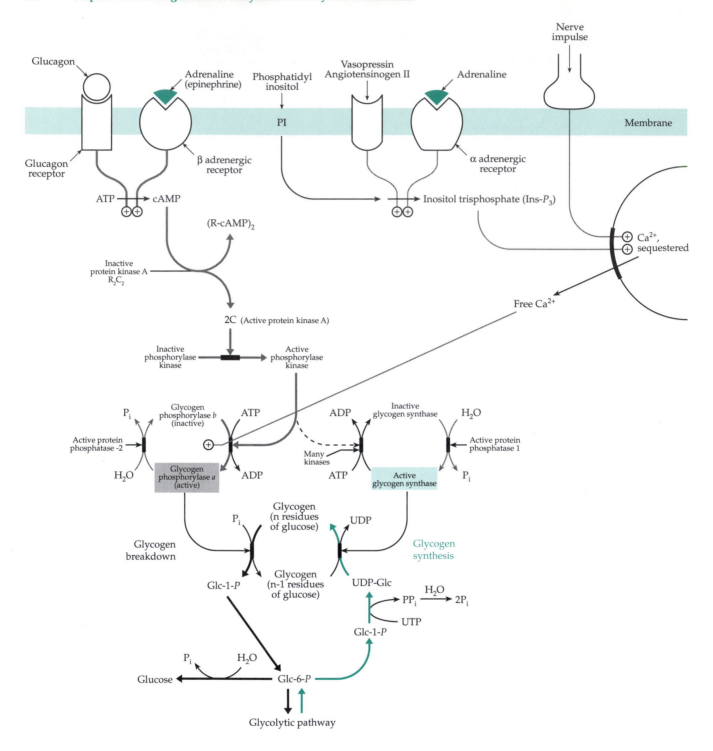

Figure 11-4 Cascades of phosphorylation and dephosphorylation reactions involved in the control of the metabolism of glycogen. Heavy arrows (➡) show pathways by which glucosyl units of glycogen are converted into free glucose or enter the glycolytic pathway. Green arrows (➡) trace the corresponding biosynthetic pathways. Gray arrows (➡) trace the pathway of activation of glycogen phosphorylase by a hormone such as adrenaline (epinephrine) or glucagon and by the action of protein kinases. A few of the related pathways in the network of reactions that affect glycogen metabolism are also shown. This includes protein phosphatases, which remove phospho groups from proteins and allow cells to relax to the state that preceded activation. One of these (protein phosphatase 1) is activated by phosphorylation by an insulin-stimulated protein kinase.[49] However, the significance is uncertain; control of glycogen synthase is complex.[49a]

TABLE 11-1
A Few Covalent Modification Reactions Utilized to Control Metabolism

Reaction	Example	Location of discussion
A. Phosphorylation–dephosphorylation		
Phosphorylation of Ser, Thr	Glycogen phosphorylase	This section
Phosphorylation of Tyr	Insulin receptor	Section G
Adenylylation, Uridylylation	Glutamine synthetase	Chapter 25
ADP-ribosylation		This section
B. Methylation of carboxyl groups	Bacterial glutamyl	Section D,5
	Aspartyl	Box 12-A
	Protein phosphatase 2A	Section C,3
	Ras	Section D,3
C. Formation of carbamino groups		
	In hemoglobin	Eq. 7-23
	In ribulose bisphosphate carboxylase	Chapters 13,23
D. Acylation		
Acetylation	Histones	Chapter 27
Palmitoylation	Ras	Section D,3
Prenylation	Ras	Chapter 22
E. Disulfide formation and cleavage		Chapters 10,15
		Section C,4

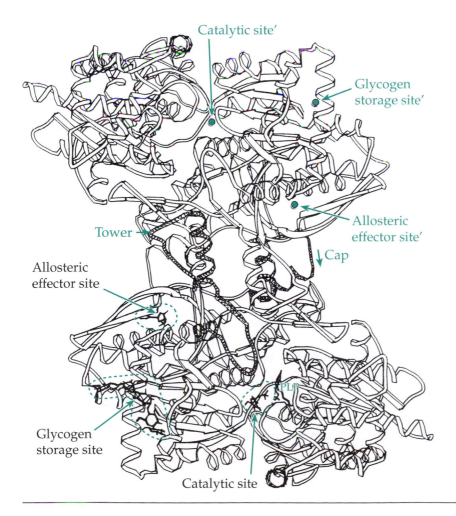

Figure 11-5 Schematic diagram of the glycogen phosphorylase dimer. The view is down the twofold axis of the dimer with the allosteric and Ser-*P* sites toward the viewer. Access to the catalytic site is from the far side of the molecule. The diagram shows the major change in conformation of the amino-terminal residues on phosphorylation. Residues 10–23 of glycogen phosphorylase *b* are shown as a thick solid line. These residues are not well ordered and make intrasubunit contacts. Upon phosphorylation, residues 10–23 change their conformation and are shown dark crosshatched with the position of Ser14-*P* indicated at the intersubunit interface. The fold of residues 24–80 through the α1 helix, the cap, and the α2 helix is shown lightly cross-hatched. The AMP allosteric effector site is located between the α2 helix and the cap region of the other subunit. The glycogen storage site is located on the surface of the subunit and is associated with a long α helix. The catalytic site is at the center of the subunit where the two domains come together. Courtesy of Louise N. Johnson.[48]

the functional groups phosphorylated, and in their allosteric activators.[39,51,54–56] There may be more than 1000 different protein kinase genes in vertebrate animals, accounting for ~2% of the genome.[57] Many cytosolic protein kinases transfer a phospho group from ATP to either a serine or threonine side chain at a β bend or other surface feature of the substrate protein. Some sites of phosphorylation in the substrate proteins contain lysine or arginine residues, separated from the serine or threonine by only one residue but many other sequences may also surround phosphorylation sites.[51,58] In the case of the 750-residue glycogen synthase, seven serine residues in different parts of the chain are phosphorylated by the action of at least five different kinases.[49,49a,59,60] The various kinases phosphorylate groups at different sites and their effects are roughly additive.

Some of the best known protein kinases (designated PKA or cAPK) are those that depend upon **3', 5'-cyclic AMP (cAMP)** as an allosteric effector. They are oligomeric proteins of composition R_2C_2, where R is a regulatory subunit and C is a catalytic subunit. Unless

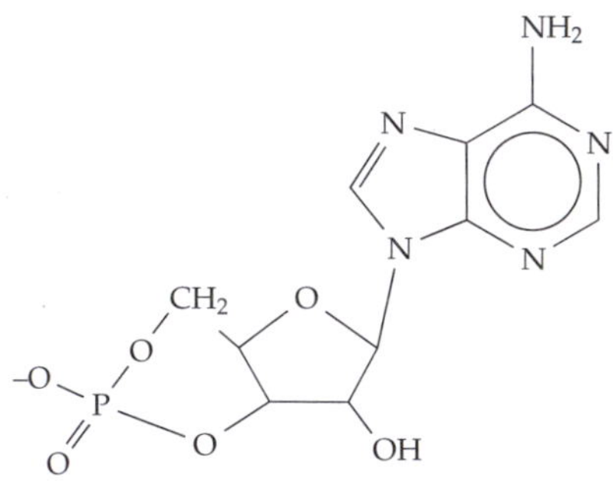

NH₂ structure diagram

3',5'-cyclic AMP

cAMP is present, the regulatory subunits interact with the catalytic subunits, keeping them in an inactive inhibited form. However, when cAMP is present it binds to the regulatory subunit dimer, releasing the two active catalytic units (Eq. 11-6). This reaction is

$$R_2C_2 + 4\ cAMP \rightarrow R_2\ (cAMP)_4 + 2C \qquad (11\text{-}6)$$

reversible and as the concentration of cAMP is reduced by hydrolysis (see Eq. 11-8) the regulatory units recombine with the catalytic subunits and again inhibit them.

There are two prominent types of mammalian cAMP-dependent protein kinases.[51,61] The catalytic subunit is identical for both; the 41-kDa peptide as isolated from beef heart has 350 residues and an N terminus blocked by a myristoyl (tetradecanoyl) group.[62] One phosphoserine and one phosphothreonine are also present.[51] The ~50-kDa regulatory subunits vary in size and may also be subject to additional regulation by phosphorylation.[63] Three-dimensional structures are known for both the catalytic[62,64,65] and the regulatory[66] subunits. A **cyclic GMP** (cGMP)-activated protein

kinase is present in some mammalian tissues[67,68] and is widespread in invertebrates.[51,67] The cyclic nucleotide-binding domains of these kinases have structures similar to that of the *E. coli* catabolite activator protein (Fig. 28-6)[51,68] and the catalytic domains are structurally similar to those of many other kinases.

Among the kinases that phosphorylate glycogen synthase is a **casein kinase**, named for the fact that the milk protein casein is also a good substrate. A family of casein kinases are found in the cytoplasm and nuclei of all eukaryotic cells. They phosphorylate serine and threonine side chains but have structures distinct from those of the cAMP-dependent kinases.[69–72] Casein kinase-2 (CK2) phosphorylates many proteins, including several that function in gene replication, transcription, and cell growth and division.[72a] **Theileriosis**, a parasitic disease of cattle in Africa, is caused by the tick-borne protist *Theileria parva*. The condition is often fatal as a result of a leukemia-like condition resulting from overexpression of the casein kinase-2 gene.[72]

Phosphorylase kinase is one of a large group of *specialized* protein kinases, each of which acts on a small number of proteins. It is regulated both by covalent modification and allosterically by **calcium ions**.[73–78] It contains four different kinds of subunits ranging in size from 17 kDa to about 145 kDa and has the composition $(\alpha\beta\gamma\delta)_4$. Phosphorylation of one serine on each of the 145-kDa α and 120-kDa β regulatory subunits is catalyzed by a cAMP-dependent protein kinase. The δ subunit is the Ca^{2+}-binding protein **calmodulin** (Fig. 6-8) and serves as a regulatory subunit sensitive to Ca^{2+}. The 45-kDa γ subunit contains the catalytic domain as well as a calmodulin-binding domain.[79] Other Ca^{2+}-dependent protein kinases include the **protein kinase C** family, which is discussed further in Section E, Ca^{2+}/calmodulin-dependent protein kinases,[80,81] and a plant kinase with a regulatory domain similiar to calmodulin.[82]

Protein tyrosine kinases (PTKs) place phospho groups on the phenolic oxygen atoms of tyrosyl residues of some proteins.[83–85] The resulting phosphotyrosine accounts for only about 1/3000 of the phospho groups in proteins but has aroused interest for two reasons. First, binding of growth hormones, such as **epidermal growth factor** (EGF), **platelet-derived growth factor** (PDGF), and **insulin**, to their receptors stimulates tyrosine-specific protein kinase activity of the receptor proteins (see Figs. 11-11 to 11-13).[69,83,86] Second, a number of cancer-causing **oncogenes** encode similar kinases (Section H). Tyrosine protein kinases are essential components of the cell division cycle (see Fig. 11-15).

Protein phosphatases. Most regulatory alterations in enzymatic activity are spontaneously reversible. The concentration of the allosteric effector soon drops and the covalent modifications are reversed so that the system relaxes to a state approximating the original one. Among the enzymes required for this reversal are the protein phosphatases that remove the phospho groups placed on amino acid side chains by kinases.[54,86–91] For example, four different phosphatases act on the (inactive) phosphorylated glycogen synthase and on the phosphorylated forms of glycogen phosphorylase and phosphorylase kinase. At least one of these (a protein phosphatase 1; see Fig. 11-4) is regulated by phosphorylation that is stimulated by insulin.[49,49a] The matter is made yet more complex by the presence in tissues of small protein inhibitors which can prevent the action of these phosphatases.[88,92–94] Another naturally occurring inhibitor is the polyether fatty acid **okadaic acid**, a "shellfish poison," actually produced by dinoflagellates but which accumulates in sponges and in mussels and other bivalves.[94] See also Chapter 12, D,4.

Most protein phosphatases are specific toward phosphoserine and phosphothreonine[89,95] or toward phosphotyrosine residues.[90,96–98] Some have a dual specificity.[99] There are Ca^{2+}-activated phosphatases[100] and phosphatases with transmembrane segments connected to receptors on the external cell surface. Some phosphorylation–dephosphorylation regulatory systems utilize bifunctional enzymes consisting of a protein kinase fused to a protein phosphatase. For example, the same *E. coli* enzyme that phosphorylates isocitrate dehydrogenase also hydrolyzes off the phospho group that it has put on.[101] Another bifunctional kinase–phosphatase catalyzes both the formation (reaction *d* of Fig. 11-2) and breakdown (reaction *e*) of fructose 2,6–bisphosphate in eukaryotic tissues. This kinase–phosphatase enzyme is, in turn, controlled by phosphorylation by a cAMP-dependent protein kinase. In the liver, this latter phosphorylation strongly inhibits the kinase activity preventing buildup of fructose 2,6-bisphosphate, the allosteric activator of phosphofructokinase-1, which catalyzes reaction *b* of Fig. 11-2. Consequently, when a pulse of cAMP is generated within a hepatocyte, glycolysis is blocked but **glycogenolysis**, the breakdown of glycogen by phosphorylase, is stimulated. The glucose 6-phosphate that is formed is hydrolyzed by glucose-6-phosphatase releasing glucose into the bloodstream. On the other hand, in heart and probably in most other tissues, a different isoenzyme form of the bifunctional enzyme is present. Phosphorylation by cAMP activates the kinase and inhibits the phosphatase. Consequently, in these tissues cAMP induces the activation of glycogenolysis and glycolysis coordinately.[102–105] In brain a different allosteric activator, **ribose 1, 5-bisphosphate**, may regulate phosphofructokinase-1.[106]

Phosphorylation in bacteria. A bacterial enzyme whose activity is controlled by phosphorylation is **isocitrate dehydrogenase**. Transfer of a phospho group to the –OH of Ser 113 completely inactivates the

E. coli enzyme,[107,108] causing isocitrate to build up in the citric acid cycle (Fig. 10-6) and to be diverted into the glyoxylate pathway, which is depicted in Fig. 17-16. In this instance, it appears likely that the negative charge of the added phospho group causes electrostatic repulsion of the substrate isocitrate. In agreement with this concept, mutation of Ser 113 of this enzyme to Asp (mutant S113 D) also inactivates the enzyme.[109]

In addition to kinase–phosphatase cycles, bacteria use at least two other ATP-dependent regulatory mechanisms.[110] In the **sensor kinase/response regulator** (or "two-component") systems[110-111] a sensor protein, upon being allosterically activated, phosphorylates itself (autophosphorylation) on a specific histidyl residue to form an *N*-phosphohistidine derivative. The phosphohistidine then transfers its phospho group onto a specific aspartyl group in the response regulator causing the regulator to bind to its target protein and to exert its regulatory effect. The best known example involves the control of the motion of bacterial flagella which is discussed briefly in Section D,5 and further in Chapter 19.

The third bacterial regulatory device is the phosphoenolpyruvate:sugar phosphotransferase system (Eq. 8-4). It is involved not only in transport but also in controlling a variety of physiological processes.[110,112,113]

3. Other Modification Reactions Involving Group Transfer

Nucleotidylation, the transfer of an entire nucleotidyl unit, rather than just a phospho group, to a protein is sometimes utilized for regulation. For example, glutamine synthetase of *E. coli* is modified by **adenylylation**, transfer of an adenylyl group from ATP to a tyrosine side chain.[114–116] Relaxation to the unmodified enzyme is catalyzed by a deadenylylating enzyme. A regulatory protein that undergoes reversible uridylylation[117] also functions in this system (see Fig. 24-7). Several mitochondrial and cytoplasmic proteins are modified by attachment of ADP-ribosyl groups to specific guanidino groups of arginine and other side chain groups.[114,118–123] This **ADP–ribosylation** requires the coenzyme NAD^+ as the ADP-ribosyl donor and is also catalyzed by bacterial toxins such as cholera toxin (Box 11-A) and by diphtheria toxin (Box 29-A). A second kind of ADP–ribosylation occurs within nuclei where the enzyme **poly(ADP–ribose) synthase** catalyzes both an initial ADP–ribosylation of an amino acid side chain group and also addition of more ADP-ribosyl units to form a polymer[124] (see Chapter 27).

Like inorganic phosphate, inorganic sulfate can be converted into an activated form in which the sulfate resembles the terminal phospho group of ATP (Eq. 17-38). The resulting **activated sulfo group** can be transferred to other compounds including enzymes. Formation of

BOX 11-A CHOLERA TOXIN AND OTHER DANGEROUS PROTEINS

As recently as 1980 it was estimated that there were ~ 100 million cases of acute diarrhea in Asia, Africa, and Latin America;[a] in 1991 there were four million deaths among children under five years of age.[b] The causative agents are bacteria and one of the most dangerous is *Vibrio cholerae*, which multiplies in the small intestine and secretes an **exotoxin**. Cholera toxin causes such a rapid loss of fluid and salts from the body that death occurs very quickly, even in adults. There is little cellular damage and almost all deaths can be prevented by intravenous administration of water, salts, and the antibiotic tetracycline. Fluids can also be given orally if glucose, which promotes intestinal absorption, is included with the Na^+, K^+, Cl^-, and HCO_3^- salts.[b]

Cholera toxin consists of one 240-residue A subunit and a ring of five 103-residue B subunits.[c-e] The latter is a "targeting complex," a bacterial lectin each subunit of which binds to the galactose and sialic acid termini of a single molecule of ganglioside G_{M1}.

Galβ1→3Galβ1→4Galβ1→4Glcβ1–Ceramide
$$3$$
$$\uparrow$$
NeuNAcβ2

Ganglioside G_{M1} (See also Fig. 20-11)

The nonpolar ceramide chains of the ganglioside are buried in the membrane bilayer. The initial tight binding (K_f ~ 2 x 10^7 M^{-1}) of the toxin is followed by a poorly understood process in which the A subunit enters the cell. The peptide chain of this subunit is

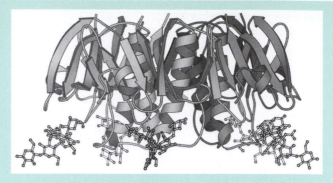

The three-dimensional structure of cholera toxin.[e] Side view of the β subunit pentamer as a ribbon drawing. Bound noncovalently to it are five molecules of the ganglioside G_{M1} (compare with the structure in Fig. 7-5). The diacyl glycerol parts of the gangliosides are buried in the membrane that lies below the toxin molecule. Courtesy of W. G. J. Hol.

cut by bacterial proteases to give A_1 and A_2 fragments joined by an S–S bridge. Within the victim's cells the disulfide is reduced to generate the active toxin, the 192- to 194- residue A_1 fragment.[f,g]

This toxin subunit is an enzyme, an ADP-ribosyltransferase which catalyzes transfer of ADP-ribosyl units from the coenzyme NAD^+ to specific arginine side chains to form *N*-ADP-ribosyl derivatives of various proteins. Of the proteins modified by cholera toxin, the most significant is the guanyl nucleotide regulatory protein G_s of the adenylate cyclase system.[c,f,h] ADP ribosylation of arginine 201 of the α subunit of protein G_s inhibits the GTP hydrolysis that normally allows the protein to relax to an unactivated form.[e] The ADP-ribosylated G_s keeps adenylate cyclase activated continuously and

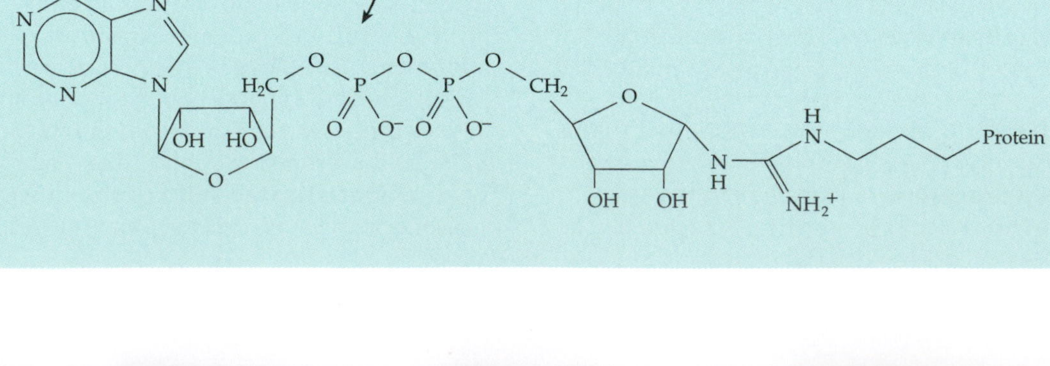

BOX 11-A (continued)

(A) Stereoscopic view of the G_{M1} pentasaccharide binding site, showing both direct and solvent-mediated hydrogen bonding interactions between sugar and protein residues. The viewpoint is from "underneath" the membrane binding surface of the B pentamer. Starred residues are from an adjacent monomer. The terminal galactose residue of the pentasaccharide (upper right) is most deeply inserted into the binding site and is involved in the greatest number of identifiable binding interactions. The sialic acid residue, near the bottom of the figure, is also involved in several hydrogen-bonding interactions. Hydrophobic interactions include the approach of the sialic acid acetyl group to the edge of the Tyr 12 phenyl ring and the positioning of the terminal galactose sugar ring parallel to the indole ring of Trp 88. Figures generated using program MolScript (Kraulis, 1991). From Merritt et al.[e] (B) Side view of the intact AB_5 toxin as an α-carbon trace. The nicked A chain is at the top. The nick can be seen in the upper right corner, as can a disulfide bridge connecting the A_1 and A_2 segments. A single α helix extends into the "pore" in the center of the B pentamer. The side chains of Trp 88 of the B subunits have been added to mark the ganglioside binding sites. These side chains are also seen in (A). (C) View of the AB_5 molecule from the bottom showing the helix from the A_2 fragment surrounded by a tight cage of five long helices from the B subunits. Residues 237–240 of the A_2 fragment have the KDEL sequence and may extend from the pore to contact the membrane. (B) and (C) courtesy of Edwin M. Westbrook. See Zhang et al.[c]

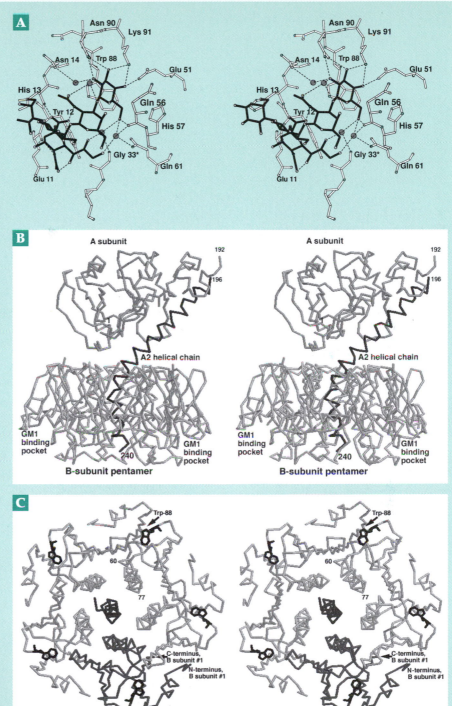

generating abnormally high concentrations of cAMP. It is the effect of the cAMP on proteins of the intestinal mucosal cell membranes that causes the disastrous excessive secretion of water and salts that are characteristic of cholera.

Most strains of *V. cholerae* are relatively harmless but they may suddenly be transformed into a virulent toxin-producing form by infection with a bacterial virus similar to M13 (Chapter 5), which carries the toxin gene. Entrance into the *Vibrio* cell occurs with the help of pili which are present in many strains.[i,j]

Many other bacterial toxins have AB_n structures similar to that of cholera toxin. For example, toxin-producing strains of *E. coli* are also important causes of diarrhea in humans and in domestic animals.[k–m]

BOX 11-A CHOLERA TOXIN AND OTHER DANGEROUS PROTEINS (continued)

The heat-labile *E. coli* enterotoxin, whose gene is carried on a plasmid, is a close relative of cholera toxin[n,o] and also catalyzes ADP ribosylation of arginine 201 of the $G_{s\alpha}$ subunit.[m] *Bordetella pertussis*, which causes whooping cough, forms a similar toxin that attacks the inhibitory regulatory protein G_i[p,q] as well as transducin and inactivates them by ADP ribosylation. Diphtheria toxin (Box 29-A), the exotoxin from *Pseudomonas aeruginosa*, and the toxin from *Clostridium botulinum* also catalyze ADP-ribosylation reactions.[k,r]

Some bacteria produce effects similar to those of cholera toxin in different ways. For example, among a variety of toxic proteins produced by *Bacillus anthracis*, the causative agent of anthrax, is an adenylate cyclase that is able to enter the host's cells.[s] Similarly, *B. pertussis*, in addition to

its ADP-ribosylating toxin, produces a calmodulin-stimulated adenylate cyclase which, when taken up by phagocytic cells, disrupts their function in the body's defense system.[t–w] In addition to the heat-labile enterotoxin, some strains of *E. coli* produce *heat-stable* toxins, small 18-residue peptides related in structure to the intestinal peptide **guanylin**.[l,x] These peptides bind to a membrane-bound guanylate cyclase activating fluid secretion into the intestine.

Why do our cells obligingly provide both initial receptors and means of uptake for these dangerous toxic proteins? Some ADP-ribosylation reactions are a natural part of cell function[y] and some hormones, for example thyrotropin, seem to stimulate their activity. It may be that the toxins use mechanisms designed to respond to normal hormonal stimulation.

[a] Holmgren, J. (1981) *Nature (London)* **292**, 413–417

[b] Hirschhorn, N., and Greenough, W. B., III (1991) *Sci. Am.* **264**(May), 50–56

[c] Zhang, R.-G., Scott, D. L., Westbrook, M. L., Nance, S., Spangler, B. D., Shipley, G. G., and Westbrook, E. M. (1995) *J. Mol. Biol.* **251**, 563–573

[d] Zhang, R.-G., Westbrook, M. L., Westbrook, E. M., Scott, D. L., Otwinowski, Z., Maulik, P. R., Reed, R. A., and Shipley, G. G. (1995) *J. Mol. Biol.* **251**, 550–562

[e] Merritt, E. A., Sarfaty, S., van den Akker, F., L'Hoir, C., Martial, J. A., and Hol, W. G. J. (1994) *Protein Sci.* **3**, 166–175

[f] Moss, J., Stanley, S. J., Morin, J. E., and Dixon, J. E. (1980) *J. Biol. Chem.* **255**, 11085–11087

[g] Janicot, M., Fouque, F., and Desbuquois, B. (1991) *J. Biol. Chem.* **266**, 12858–12865

[h] Tsai, S.-C., Adamik, R., Moss, J., and Vaughan, M. (1991) *Biochemistry* **30**, 3697–3703

[i] Williams, N. (1996) *Science* **272**, 1869–1870

[j] Waldor, M. K., and Mekalanos, J. J. (1996) *Science* **272**, 1910–1914

[k] Moss, J., and Vaugham, M., eds. (1990) *ADP–Ribosylating Toxins and G Proteins: Insights into Signal Transduction*, American Society for Microbiology, Washington, DC

[l] Sato, T., Ozaki, H., Hata, Y., Kitagawa, Y., Katsube, Y., and Shimonishi, Y. (1994) *Biochemistry* **33**, 8641–8650

[m] van den Akker, F., Merritt, E. A., Pizza, M. G., Domenighini,

M., Rappuoli, R., and Hol, W. G. J. (1995) *Biochemistry* **34**, 10996–11004

[n] Sixma, T. K., Pronk, S. E., Kalk, K. H., van Zanten, B. A. M., Berghuis, A. M., and Hol, W. G. J. (1992) *Nature (London)* **355**, 561–564

[o] Sixma, T. K., Kalk, K. H., van Zanten, B. A. M., Dauter, Z., Kingma, J., Witholt, B., and Hol, W. G. J. (1993) *J. Mol. Biol.* **230**, 890–918

[p] Antoine, R., and Locht, C. (1994) *J. Biol. Chem.* **269**, 6450–6457

[q] Goodemote, K. A., Mattie, M. E., Berger, A., and Spiegel, S. (1995) *J. Biol. Chem.* **270**, 10270–10277

[r] Ohtsuka, T., Nagata, K.-i, Iiri, T., Nozawa, Y., Ueno, K., Ui, M., and Katada, T. (1989) *J. Biol. Chem.* **264**, 15000–15005

[s] Arora, N., Klimpel, K. R., Singh, Y., and Leppla, S. H. (1992) *J. Biol. Chem.* **267**, 15542–15548

[t] Benz, R., Maier, E., Ladant, D., Ullmann, A., and Sebo, P. (1994) *J. Biol. Chem.* **269**, 27231–27239

[u] Heveker, N., Bonnaffé, D., and Ullmann, A. (1994) *J. Biol. Chem.* **269**, 32844–32847

[v] Otero, A. S., Yi, X. B., Gray, M. C., Szabo, G., and Hewlett, E. L. (1995) *J. Biol. Chem.* **270**, 9695–9697

[w] Hackett, M., Guo, L., Shabanowitz, J., Hunt, D. F., and Hewlett, E. L. (1994) *Science* **266**, 433–435

[x] Ozaki, H., Sato, T., Kubota, H., Hata, Y., Katsube, Y., and Shimonishi, Y. (1991) *J. Biol. Chem.* **266**, 5934–5941

[y] Moss, J., and Vaughan, M. (1995) *J. Biol. Chem.* **270**, 12327–12330

tyrosine-*O*-sulfate residues may be a widespread regulatory mechanism.[125–127] One of the proteins known to contain a tyrosine sulfate residue is the blood protein fibrinogen. Many polysaccharides and oligosaccharides on glycoproteins exist in part as sulfate esters (Fig. 4-11).[128,129]

Carboxyl groups of certain glutamate side chains in proteins that control bacterial **chemotaxis** are methylated reversibly to form methyl esters.[130] This **carboxylmethylation** occurs as part of a reaction sequence by which the bacteria sense compounds that can serve as food or that indicate the presence of food

(Section D,5, Fig. 11-8, and Chapter 19). Demethylation occurs by hydrolysis, which is catalyzed by esterases. Carboxylmethylation also occurs in eukaryotic cells but is often substoichiometric and part of a mechanism for repair of isomerized or racemized aspartyl residues in aged proteins (Box 12-A). However, the major eukaryotic protein phosphatase 2A is carboxylmethylated at its C terminus,[131] as are the Ras proteins discussed in Section D,3.

It was pointed out in Chapter 4 that many proteins, especially those secreted from cells or taking up residence within membranes, are **glycosylated**. Specific glycosyl groups may be removed or added to a protein. Such alterations may all be regarded as parts of control mechanisms that direct the proteins to their proper locations or determine the length of time that they remain active. The nuclear histones undergo extensive and reversible **acetylation**[132,133] which is thought to be important to replication of DNA and to transcription (Chapters 27 and 28).

4. Thiol-Disulfide Equilibria

The activity of enzymes is sometimes controlled by the formation or the reductive cleavage of disulfide linkages between cysteine residues within the protein. An example is provided by the effects of light on enzymes of chloroplasts. Light absorbed by the photosynthetic reaction centers generates NADPH, which in turn reduces **thioredoxin**. This is a small protein containing a readily accessible S–S bridge that can be reduced to a pair of SH groups (Box 15-C). Within illuminated chloroplasts, these newly formed SH groups reduce disulfide bridges to activate a series of enzymes including fructose 1,6-bisphosphatase that participate in the photosynthetic incorporation of CO_2 into sugars[134,135] (see Chapter 23). Thioredoxins also function within bacteria, fungi, and animals, serving as electron carriers for processes, some of which are involved in metabolic control. A number of known enzymes and other proteins, including insulin, contain reducible S–S bridges within peptide loops as in the thioredoxins. Another possible control mechanism for SH-containing enzymes depends upon formation of **mixed disulfides** with small SH-containing metabolites. For example, **cysteamine disulfide**, a minor constituent of cells, cysteine, or some small disulfide, could be released

$$H_3{}^+N-CH_2-CH_2-S-S-CH_2-CH_2-NH_3{}^+$$

Cysteamine disulfide (cystamine)

following hormonal or other stimulation and could react by disulfide exchange (Eq. 11-7) to either inactivate or activate an enzyme.

$$
\begin{array}{ccc}
E-SH & \longrightarrow & E-S-S-R \\
R-S-S-R & & R-SH \\
\text{Cystamine} & & \text{Cysteamine} \\
\text{(Cystine)} & & \text{(Cysteine)} \\
\end{array}
\qquad (11\text{-}7)
$$

Almost all cells contain a high concentration (3–9 mM) of the thiol-containing tripeptide **glutathione** (G–SH, Box 11-B). In its disulfide form it participates in forming disulfide bridges in secreted extracellular proteins (Eq. 10-9) via intermediate mixed disulfides. Mixed disulfides with glutathione as well as with other thiols can also be formed within cells by oxidative

reactions. Reduction of these disulfides by reduced glutathione will return the enzymes to their reduced states. The small protein **glutaredoxin**, whose eukaryotic forms are also called **thioltransferase**, resembles thioredoxins. It undergoes reduction by glutathione and, in turn, reduces S–S linkages in a different set of proteins than those reduced by thioredoxin.

Low-molecular-mass thiols such as coenzyme A and protein-bound thiol cofactors such as phosphopantetheine are present in all cells. Their SH groups can also be oxidized to disulfides and it is of interest that in resting bacterial spores these compounds exist largely as disulfides or mixed disulfides. Upon germination of the spores special enzymes reduce the disulfides.[136] Some proteins involved in control of protein synthesis contain SH groups that add covalently to C-6 atoms of a uracil ring in specific mRNA molecules. Control of their state of reduction may also be important.[137]

5. Regulatory Effects of H⁺, Ca²⁺, and Other Specific Ions

The pH of the cytoplasm is controlled tightly. Yet transient changes can occur and may affect many aspects of metabolism. For example, rapid glycolysis leads to lactic acid formation with an associated drop in internal pH. Both increases and decreases in pH have been associated with successive stages in embryonic or larval development.[138] Cytoplasmic pH changes may serve as regulatory signals. A well-understood example is the Bohr effect on oxygenation of hemoglobin (Chapter 7). Another is the protein kinase C-stimulated H^+/Na^+ exchange through membranes. Because the Na^+ concentration is high outside cells and low inside, the exchange leads to an increase in cytosolic pH with many resultant effects on metabolism.[139,140] Exchange of external Cl^- for internal HCO_3^- also affects pH.[141]

Uptake of Ca^{2+} into cells, or release of this ion from intracellular stores, is a major regulatory mechanism in many if not all cells (see Section E). Mn^{2+} activates **phosphoenolpyruvate carboxykinase** (Eq. 13-46) and may be a regulator of gluconeogenesis.[142] Iron controls the synthesis of ferritin and of transferrin receptors[137] (Chapter 16). The specific metal ions present in many biological macromolecules are likely to participate in additional regulatory processes.

Phosphate and bicarbonate ions are important substrates for many enzymatic processes and as such have regulatory functions. Bicarbonate controls the key enzyme of photosynthesis, **ribulose bisphosphate carboxylase**, by carbamate formation (Fig. 13-12). Chloride ions activate amylases and may affect the action of "G proteins" that mediate hormone actions. Other observed effects of ions are too numerous to mention.

BOX 11-B GLUTATHIONE, INTRACELLULAR OXIDATION–REDUCTION BUFFER

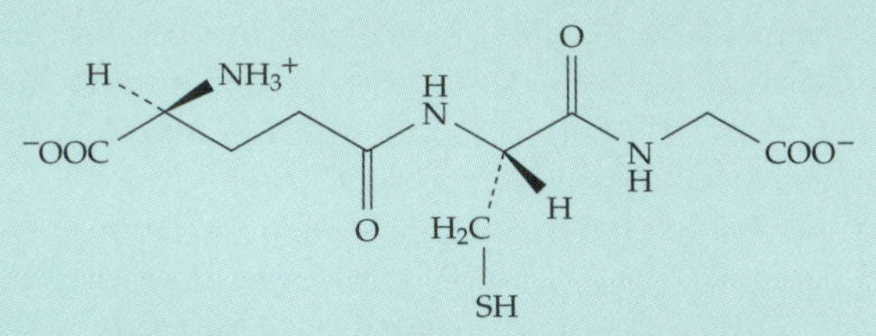

Glutathione (γ-L-Glutamyl-L-cysteinylglycine)

In 1888, de Rey-Pailhade discovered the sulfur-containing tripeptide that we now know as glutathione (G–SH).[a] By 1929 it had been characterized by F. G. Hopkins and others as an unusual tripeptide present in most, if not all, eukaryotic cells. Within animal cells the concentration is typically 1–5 mM. Lower levels are found in many bacteria.[b-g]

The most interesting chemical characteristics of glutathione are the γ-glutamyl linkage and the presence of a free SH group. The latter can be oxidized to form a disulfide bridge linking two glutathione molecules.

$$2G \text{—} SH - 2[H] \rightarrow G \text{—} S \text{—} S \text{—} G$$
$$E°' \text{ (pH 7)} = -0.25 \text{ V}$$

It is this redox reaction that has focused attention on glutathione as an intracellular reducing agent whose primary function may be to keep the SH groups of proteins reduced (see Section C). Glutathione is usually maintained in its reduced form by the flavoprotein glutathione reductase (Fig. 15-12).

Although it is primarily an intracellular compound, glutathione is secreted by epithelial and other cells. It may regulate the redox state of proteins in plasma and other extracellular fluids as well as within cells. In addition, glutathione released from the liver may be an important source of cysteine for other tissues. In the endoplasmic reticulum and the periplasm of bacteria glutathione functions in crosslinking thiol groups in newly formed proteins (Eq. 10-9).

Glutathione also has a series of **protective functions**. It reduces peroxides via the selenium-containing **glutathione peroxidase** (Box 15-H) and is part of a system for trapping and detoxifying harmful free radicals. The importance of this function is suggested by the fact that *Entamoeba histolytica*, an organism that lacks both mitochondria and aerobic respiration, produces no glutathione.[h] It may be that the primary function of glutathione in eukaryotes is to protect cells against oxygen toxicity associated with their mitochondria.[i] The renewal of free radicals and regeneration of protein –SH groups

may involve the cooperation of glutathione, glutaredoxin (Box 15 - C), and ascorbic acid.[j]

Another protective function is fulfilled by the formation of soluble **mercaptides** and other "conjugates" of glutathione with many foreign substances (**xenobiotics**). These conjugates are made by the action of a large family of **glutathione transferases**[e,k,l] (see also Chapter 13) which catalyze addition reactions of the thiolate group of glutathione with epoxides, alkylating compounds, and chlorinated aromatic hydrocarbons. The addition step is often followed by elimination, e.g., of chloride ion, as in the following example. Two steps of hydrolysis and an acetylation by acetyl-CoA form a mercapturic acid:

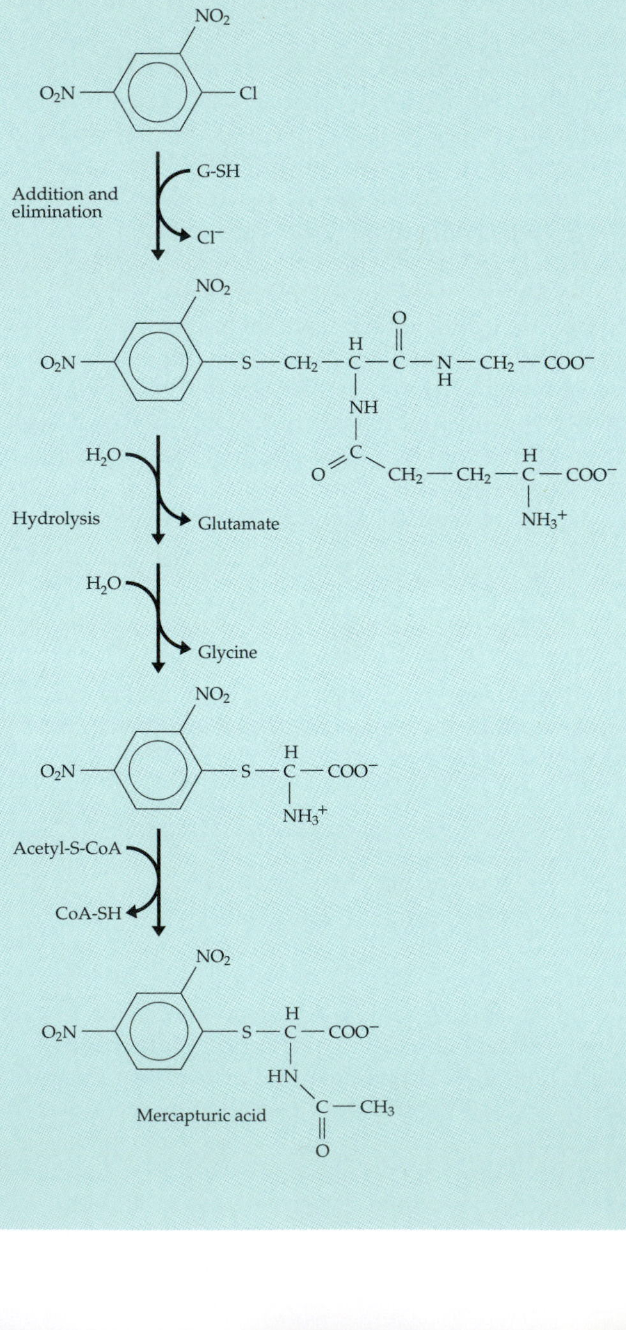

BOX 11-B (continued)

The mercapturic acids and related compounds can then be exported from cells by an ATP-dependent export pump.[m] Glutathione is a coenzyme for **glyoxalase** (Eq. 13-33), **maleylacetoacetate isomerase** (Eq. 13-20), and **DDT dehydrochlorinase**. The latter enzyme catalyzes elimination of HCl from molecules of the insecticide and is especially active in DDT-resistant flies.[a] Glutathione is said to be the specific factor eliciting the feeding reaction of *Hydra*; that is, the release of glutathione from injured cells causes the little animal to engulf food.

Synthesis of glutathione occurs within cells via the ATP-dependent reactions in the following scheme.[n,o] Much more γ-glutamylcysteine is synthesized than is converted to glutathione and the excess is degraded by γ-glutamyl cyclotransferase to form the cyclic amide 5-oxoproline. Cleavage of ATP is required to reopen the ring to form glutamate. Although biosynthesis is exclusively intracellular, most glutathione appears to be secreted and degraded by extracellular enzymes. The membrane-bound γ-glutamyl transferase catalyzes hydrolysis of

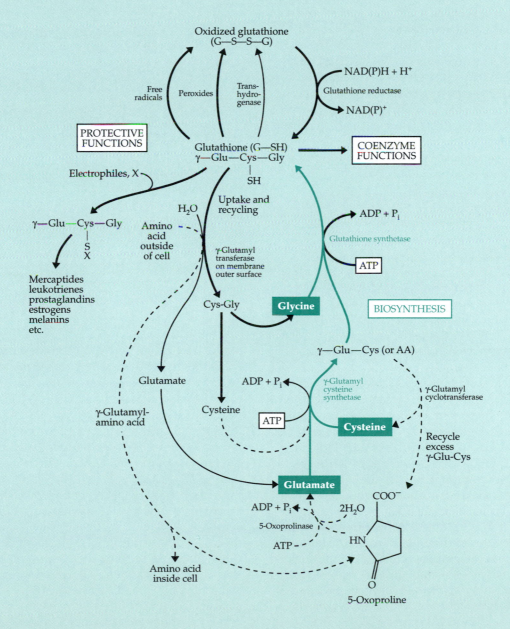

Scheme illustrating interrelationships of the biosynthesis and protective, coenzymatic, and transport functions of glutathione. See also Meister.[c]

BOX 11-B GLUTATHIONE, INTRACELLULAR OXIDATION–REDUCTION BUFFER (continued)

glutathione to cysteinylglycine which is further cleaved by a peptidase. The activity of γ-glutamyl-transferase varies among tissues and is especially high in cells of the kidney tubules. The cysteine and glycine released by the peptidase may reenter the cells by a Na$^+$-dependent process. Meister proposed that the γ-glutamyltransferase, acting on amino acids, forms γ-glutamyl amino acids (see scheme) which are released *within* cells and are cleaved by γ-glutamyl cyclotransferase.[c] The cleavage of gluta-thione would provide the driving force for amino acid uptake. However, this is probably a minor pathway.[p]

Trypanosomes contain little glutathione but a large amount of **trypanothione** [N^1,N^8-bis (gluta-thionyl) spermidine].[q] This diamide of spermidine (Chapter 24) is in equilibrium with its disulfide, a 24-membered macrocyclic structure, and appears to have functions similar to those of glutathione. Trypanothione reductase, which is unique to trypa-nosomes, is a potential target for antitrypanosomal drugs,[q–s] as is trypanothione synthetase.[t] Bacteria that do not synthesize glutathione usually accumulate some other thiol, e.g., α-glutamylcysteine or coen-zyme A, in high concentrations.[u,v]

[a] Meister, A. (1988) *Trends Biochem. Sci.* **13**, 185–188
[b] Bernofsky, C., and Wanda, S.-Y. C. (1982) *J. Biol. Chem.* **257**, 6809–6817
[c] Meister, A. (1988) *J. Biol. Chem.* **263**, 17205–17208
[d] Inoue, M. (1985) in *Renal Biochemistry* (Kinne, R. K. H., ed), pp. 225–269, Elsevier, Amsterdam

[e] Orrenius, S., Ormstad, K., Thor, H., and Jewell, S. A. (1983) *Fed. Proc.* **42**, 3177–3188
[f] Viña, J., ed. (1990) *Glutathione: Metabolism and Physiological Functions*, CRC Press, Inc., Boca Raton, Florida
[g] Dolphin, D., Poulson, R., and Avramovic, O., eds. (1989) *Glu-tathione-Chemical, Biochemical and Medical Aspects (Coenzymes and Cofactors)*, Vol. 3, Wiley, New York (Parts A & B)
[h] Fahey, R. C., Newton, G. L., Arrick, B., Overdank-Bogart, T., and Aley, S. B. (1984) *Science* **224**, 70–72
[i] Garcia de la Asuncion, J., Millan, A., Pla, R., Bruseghini, L., Esteras, A., Pallardo, F. V., Sastre, J., and Viña, J. (1996) *FASEB J.* **10**, 333–338
[j] Meister, A. (1994) *J. Biol. Chem.* **269**, 9397–9400
[k] Ji, X., Johnson, W. W., Sesay, M. A., Dickert, L., Prasad, S. M., Ammon, H. L., Armstrong, R. N., and Gilliland, G. L. (1994) *Biochemistry* **33**, 1043–1052
[l] Hebert, H., Schmidt-Krey, I., and Morgenstern, R. (1995) *EMBO J.* **14**, 3864–3869
[m] Ishikawa, T. (1992) *Trends Biochem. Sci.* **17**, 463–468
[n] Polekhina, G., Board, P. G., Gali, R. R., Rossjohn, J., and Parker, M. W. (1999) *EMBO J.* **18**, 3204–3213
[o] Lu, S. C. (1999) *FASEB J.* **13**, 1169–1183
[p] Lee, W., Hawkins, R., Peterson, D., and Viña, J. (1996) *J. Biol. Chem.* **271**, 19129–19133
[q] Henderson, G. B., Ulrich, P., Fairlamb, A. H., Rosenberg, I., Pereira, M., Sela, M., and Cerami, A. (1988) *Proc. Natl. Acad. Sci. U.S.A.* **85**, 5374–5378
[r] Bollinger, J. M., Jr., Kwon, D. S., Huisman, G. W., Kolter, R., and Walsh, C. T. (1995) *J. Biol. Chem.* **270**, 14031–14041
[s] Sullivan, F. X., Sobolov, S. B., Bradley, M., and Walsh, C. T. (1991) *Biochemistry* **30**, 2761–2767
[t] Henderson, G. B., Yamaguchi, M., Novoa, L., Fairlamb, A. H., and Cerami, A. (1990) *Biochemistry* **29**, 3924–3929
[u] Sundquist, A. R., and Fahey, R. C. (1989) *J. Biol. Chem.* **264**, 719–725
[v] Swerdlow, R. D., Green, C. L., Setlow, B., and Setlow, P. (1979) *J. Biol. Chem.* **254**, 6835–6837

6. Compartments and Organized Assemblies

The geometry of cell construction provides another important aspect of metabolic control. In a bacterium, the periplasmic space (Fig. 8-28) provides a compart-ment that is separate from the cytosol. Some enzymes are localized in this space and do not mix with those within the cell. Other enzymes are fixed within or attached to the membrane. Eukaryotic cells have more compartments: nuclei, mitochondria (containing both matrix and intermembrane spaces), lysosomes, micro-bodies, plastids, and vacuoles. Within the cytosol the tubules and vesicles of the endoplasmic reticulum (ER) separate off other membrane-bounded compartments. The rate of transport of metabolites through the mem-branes between compartments is limited and often is controlled tightly.

While many enzymes appear to be dissolved in the cytosol and have no long-term association with other proteins, enzymes that catalyze a series of con-secutive reactions may form complexes within which substrates are **channeled**.[143–147] Many enzymes are attached to membranes where they may be held close together as organized assemblies.[146] This appears to be the case for oxidative enzymes of mitochondria (Chapter 18) and for the cytoplasmic fatty acid synthe-tases (Chapter 21). In bacterial fatty acid synthesis, the product of the first enzyme is covalently attached to a "carrier" and, while so attached, is subjected to the action of a series of other enzymes. In eukaryotes several enzymes form domains of a single fatty acid synthase. Efficient substrate channeling results. Tryp-tophan synthase (Fig. 25-3) passes indole through a tunnel between subunits.[146a] Both NH$_3$ and carbamate ions pass through tunnels between subunits of *E. coli* carbamoyl phosphate synthetase (Eq. 24-22). The product carbamoyl phosphate may then be passed directly to aspartate carbamoyltransferase for synthesis of carbamoylaspartate in the pyrimidine biosynthetic pathway.[146b] Channeling is sometimes difficult to prove. Geck and Kirsch have provided a generally useful technique for testing. A large amount of a genetically modified, inactive form of an enzyme is added. Unless channeling occurs, this will decrease the rate of a reaction whose rate is limited by diffusion or by instability of an intermediate.[146c]

Proteasomes (Box 7-A) have enzymatic sites within a protected box which limits the escape of long peptide fragments. Membrane anchors (Chapter 8), often consisting of acyl or polyprenyl groups, hold many proteins to cell surfaces and strong protein–protein bonds hold many others.[146d] Enzymes involved in cell signaling, and discussed in the following section, are often anchored close together on membrane surfaces.[148,149]

D. Hormones and Their Receptors

A major element in the control of the metabolism of a cell is provided by chemical messages sent from *other cells* and sensed by receptors on the cell membrane or in the cytoplasm. Hormones such as insulin, adrenaline, and the sex hormones are released from an organ and travel through the blood, affecting tissues throughout the body. There are many such hormones. A large number of other hormones have more local effects, influencing mostly adjacent cells. When released at nerve endings these substances are called **neurotransmitters**. Chemical messages are probably sent from virtually all cells to their immediate neighbors, affecting both their growth and their behavior. In recent years the very reactive and toxic compounds **nitric oxide (NO)** and **carbon monoxide (CO)** have been identified as important hormones. These compounds can diffuse rapidly and react with many compounds within cells. Their actions are more rapid, more rapidly ended, and probably less specific than those of most hormones.

Hormones bind to **receptors** on their "target" cells. The receptors are often integral membrane proteins on cell surfaces. Binding of the hormone often causes a conformational alteration that "activates" the receptor. In some cases the receptor is an enzyme and the hormone an allosteric activator. In others, the activated receptor interacts with an enzyme in the cytoplasm or on the membrane facing the cytoplasm. This enzyme may generate a **second messenger**, a substance that can diffuse throughout the cell and alter metabolism by exerting allosteric effects on various other enzymes. The best known second messenger is cyclic AMP but there are many others, a few of which are listed in Table 11-2.

Some hormone–receptor complexes enter the cell via endocytosis in coated vesicles. Within the cell both the receptor and the bound hormone, if a peptide, may be degraded by proteases. The initial binding of the hormone may induce the release of a second messenger,

while the degradation of the receptor complex at a later time releases peptides that may be additional second messengers. This is one way in which hormones may elicit a rapid response followed by delayed responses.

While many hormones bind to surface receptors the steroid hormones, which are lipid in nature, pass through the cell membrane and bind to receptor proteins in the nucleus. The resulting hormone–protein complexes induce changes in gene expression through regulation of transcription (Fig. 11-1, top). These receptors are considered in Chapter 22 and hormones are considered further in Chapter 30.

1. Beta Adrenergic Receptors and Related Seven-Helix Proteins

Sites that bind **adrenaline** (epinephrine), **noradrenaline** (norepinephrine), and related **catecholamines** (see Chapter 30) to almost all cell surfaces are classified as either **α adrenergic** or **β adrenergic receptors**. The β receptors, which have been studied the most,[150] occur as two major types. The β_1 receptors have approximately equal affinity for adrenaline and noradrenaline, whereas the β_2 receptors, the most common type, are more nearly specific for adrenaline. Binding of the hormone or other agonist such as **isoproterenol** to any of the β receptors stimulates cAMP formation within the cell.[153] The receptors belong to a family of integral membrane proteins related in sequence to the **opsins** of the retina and the light-operated proton pump bacteriorhodopsin (Fig. 23-45) and probably have a very similar three-dimensional structure. Based upon the known three-dimensional structures of rhodopsin and bacteriorhodopsin (Chapter 23), sequence comparisons, and much other evidence, all of these receptors are probably folded into groups of seven hydrophobic membrane-spanning helices arranged as closely packed bundles with folded loops protruding into the cytoplasm and into the extracellular space. The N termini of the proteins are thought to be in the extracellular space and the C termini in the cytoplasm as shown in Fig. 11-6.[153a,b,c] Defects in structure or functioning of human β receptors have been associated with both asthma[154] and heart failure.[155,155a,b]

Among the proteins phosphorylated in response to formation of cAMP are the β adrenergic receptors themselves. Phosphorylation occurs within a cluster of serine and threonine residues near the C terminus as is indicated in Fig. 11-6A by the action of various kinases including a **receptor kinase**[156,156a] which may be anchored nearby.[157] The effect of this C-terminal phosphorylation is to decrease the sensitivity of the receptor so that after a few minutes it conveys a diminished response. Thus, cAMP exerts feedback inhibition of its own synthesis. *Desensitization upon continuous occupancy*

TABLE 11-2
Some Molecules (Second Messengers) That Carry Intracellular Signals

Compound	Metabolic state	Response	Location of discussion
Cyclic AMP	Stimulation of β adrenergic receptors	Increased glycogenolysis, glycolysis	Sections C,2 and D,2;
Cyclic GMP	Visual stimulation	Neuronal signal	Chapter 23, Section D
Ca^{2+}	Ca^{2+} channels open	Muscular contraction, others	Box 6-D, Section E Chapter 19, Section B,4
Inositol-1,4,5-trisphosphate and related compounds	Stimulation of α adrenergic receptors, various other stimuli	Opens Ca^{2+} channels in ER	Section E
Diadenosine 5'-tetraphosphate (Ap_4A)	Oxidative or heat stress		Chapters 28, 29
Guanosine 5'-diphosphate, 3-diphosphate (ppGpp)	Nutritional stress		Chapter 29
Mn^{2+}	Low glucose	Increased PECK activity	Chapter 17, Section L
		Increased arginase activity	Chapter 16, Section E
Cyclic ADP-ribose (cADPR) and 2'-P-cADPR			Chapter 15, Section E

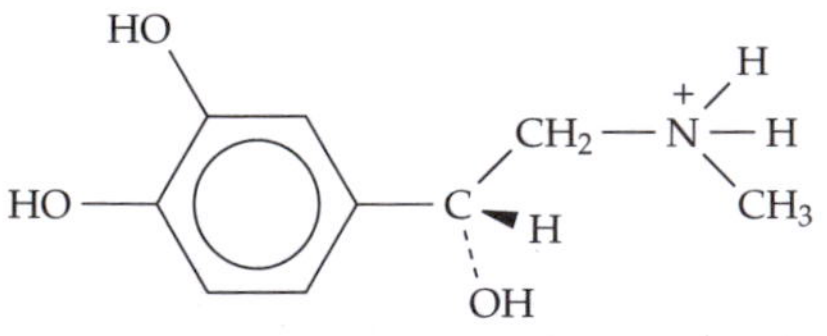

Adrenaline (epinephrine)

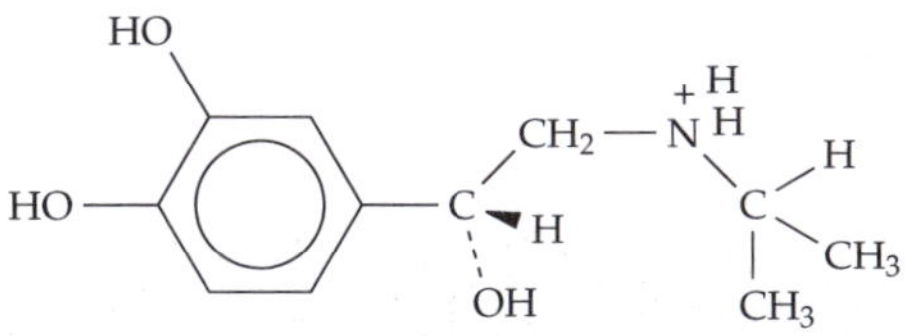

Isoproterenol

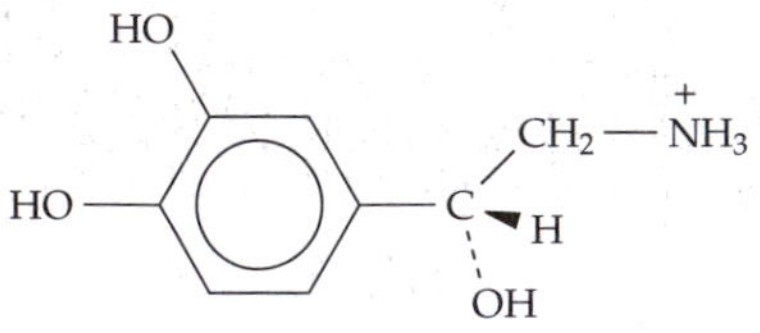

Noradrenaline

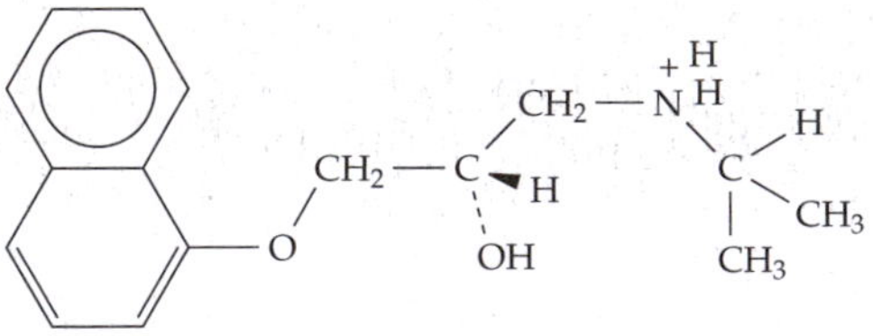

Propranolol

by an agonist is a property of many other receptors as well. When a receptor is no longer occupied, phosphatases remove the phospho groups added by the receptor kinase permitting the sensitivity to rise again. This phosphorylation–dephosphorylation cycle appears to be only one of several mechanisms by which cells regulate receptor sensitivity.[155b,158,158a,158b]

Many other receptors also have a seven-helix structure similar to that of the adrenergic receptors. These include receptors for the following: **glucagon**, one of the pancreatic hormones regulating glucose metabolism;[159,160] **vasopressin** (Fig. 2-4);[161,162] **lutropin**, another pituitary hormone;[163] other gonadotropins;[164] the **thyrotropin-releasing factor** (TRF; Fig. 2-4);[165] a receptor for **KDEL** peptide sequences. The KDEL receptor functions in the return of soluble proteins containing the KDEL motif from the Golgi to the endoplasmic reticulum.[152]

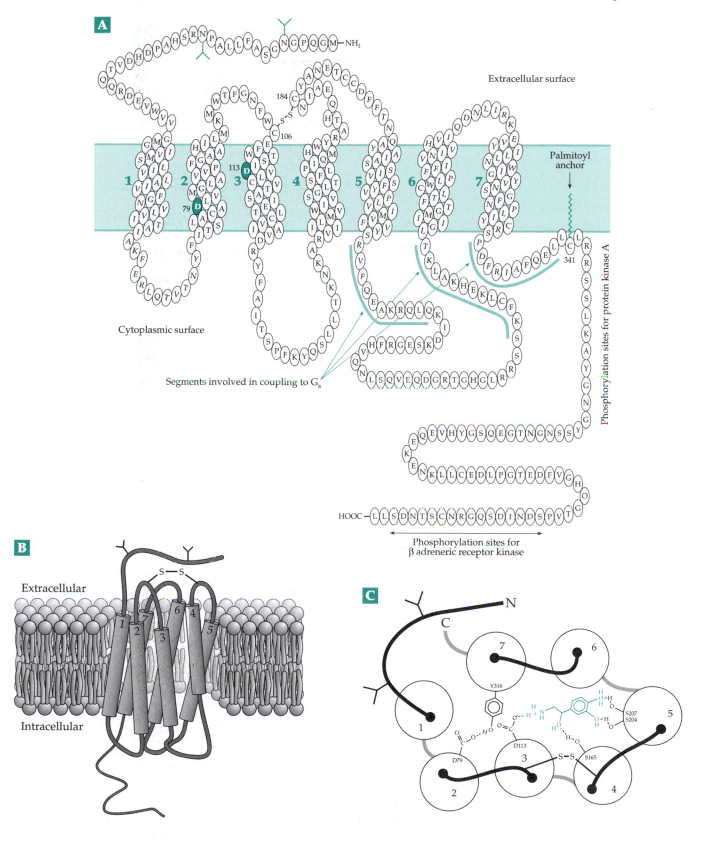

Figure 11-6 (A) Proposed organization of the human β_2 adrenergic receptor. The 413-residue polypeptide chain is arranged according to the model for rhodopsin as seven transmembrane helices. The two N-glycosylation sites, which may carry very large oligosaccharides, are indicated by Y. The palmitoylated cysteine 341 is shown with its alkyl side chain embedded in the membrane. Aspartates 79 and 113, also shown in (C), are shaded. After Strosberg[151] and Scheel and Pelham.[152] (B) Arrangement of the seven helices suggested by the rhodopsin structure. (C) Hypothetical view of the receptor from the external membrane surface showing a molecule of noradrenaline bound to hydrophilic residues deep in the cleft between the helices. From Strosberg.[151]

2. Adenylate Cyclases (Adenylyl Cyclases)

All of the effects of the catecholamines bound to β adrenergic receptors and of **glucagon**, **ACTH**, and many other hormones appear to be mediated by **adenylate cyclase**. This integral membrane protein catalyzes the formation of cAMP from ATP (Eq. 11-8, step *a*). The reaction, whose mechanism is considered in Chapter 12, also produces inorganic pyrophosphate. The released cAMP acts as the second messenger and diffuses rapidly throughout the cell to activate the cAMP-dependent protein kinases and thereby to stimulate phosphorylation of a selected group of proteins (Fig. 11-4). Subsequent relaxation to a low level of cytosolic cAMP is accomplished by hydrolysis of the cAMP by a phosphodiesterase (Eq. 11-8, step *b*).[166,167] In the absence of phosphodiesterase cAMP is extremely stable kinetically. However, it is thermodynamically unstable with respect to hydrolysis.

The existence of cAMP as a compound mediating the action of adrenaline and glucagon on glycogen phosphorylase was first recognized in 1956 by Sutherland.[168,169] However, for many years most biochemists regarded cAMP as a curiosity and the regulatory chemistry of phosphorylase as an unusual specialization. That view was altered drastically when cAMP was found to function as a second messenger in the action of over 20 different hormones. Phosphorylation by cAMP-dependent protein kinases regulates several enzymes concerned with energy metabolism or with the control of cell division. Phosphorylation reactions can lead to different responses in different specialized cells. Proteins of membranes, microtubules, and ribosomes are phosphorylated, as are nuclear histones. Cyclic AMP mediates the action of some neurotransmitters released at synapses and functions as a signal between cells of some slime molds (Box 11-C). Transcription of genes can be either stimulated or inhibited by cAMP.[170] Even in *E. coli* cAMP is generated and acts as a positive effector for transcription of some genes (Chapter 28). Cyclic AMP may function in signaling *stress* in plants.[171]

The isolation and characterization of adenylate cyclases have been difficult because of their location within membranes and because there is so little (usually only 0.001– 0.01%) of the membrane protein.[172] The 1060- to 1250-residue, 120 kDa proteins are also easily denatured. However, the sequences of cloned adenylate cyclase genes and the observed patterns of synthesis of the corresponding mRNAs have been revealing. There are at least eight mammalian adenylate cyclase genes with complex regulatory properties.[173 – 177] The sequences suggest that most of the isoenzymes are integral membrane proteins, *each with 12 transmembrane helices* organized as two sets of six with a large ~40-kDa cytoplasmic domain between the sets. This is similar to the organization of the cystic fibrosis transmembrane conductance regulator (Box 26-A) and some other membrane transporters. However, there is no firm evidence that adenylate cyclases contain ion channels. These enzymes are also discussed in Chapter 12, Section D,9.

A quantitative indication of the importance of the cAMP system within cells can be derived from measurement of the kinetics of the incorporation of ^{18}O from water into the α-phospho groups of AMP, ADP, and ATP. This incorporation will result from hydrolysis of cAMP by the phosphodiesterases that allow relaxation to a low cAMP level (Eq. 11-8). It is thought that in human blood platelets this represents the major pathway of this labeling (Eq. 11-9), which occurs at a rate of about 1.1 μmol of ^{18}O kg^{-1}s^{-1}.

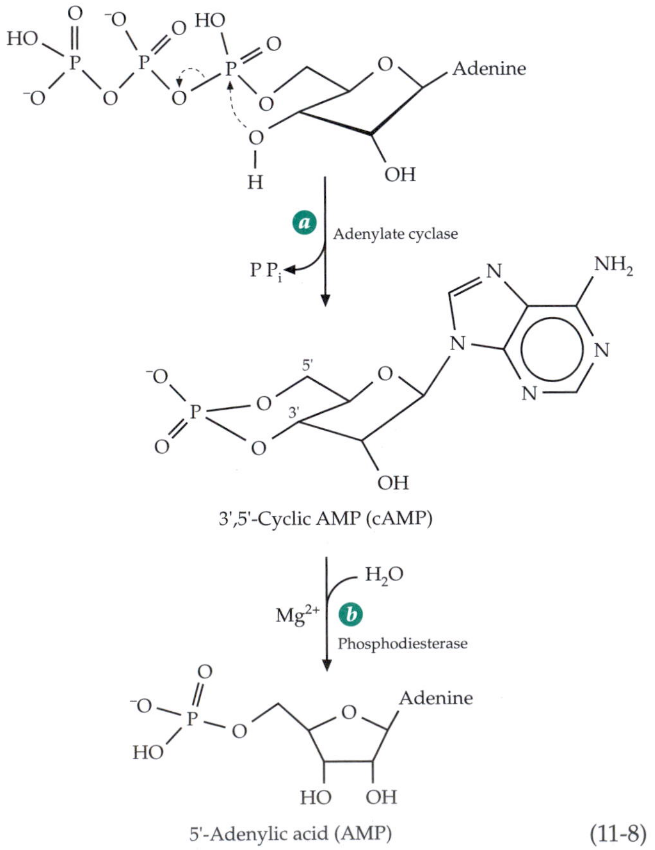

3',5'-Cyclic AMP (cAMP)

5'-Adenylic acid (AMP)   (11-8)

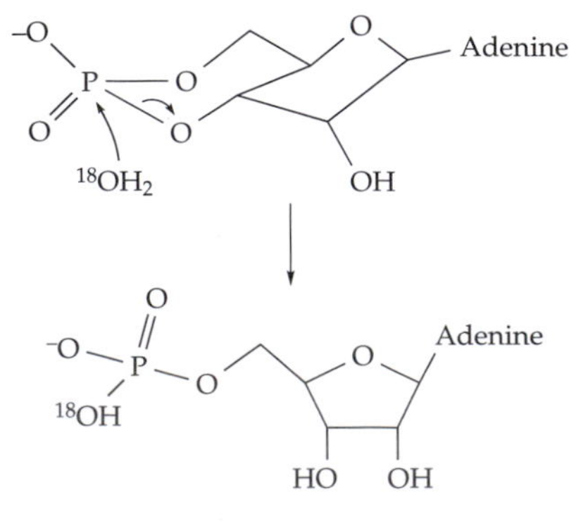

(11-9)

Stimulation of the platelets by prostacyclin (prostaglandin I_2; see Chapter 22) leads to a 10- to 40- fold increase in cAMP concentration and a 4- to 5-fold increase in the rate of ^{18}O incorporation.[178] A quite different *soluble* adenylate cyclase is present in spermatozoa. It is stimulated directly by bicarbonate ions. It may be a **bicarbonate sensor** in sperm cells and in some other bicarbonate-responsive tissues as well as in cyanobacteria.[178a,178b]

3. Guanine Nucleotide-Binding Proteins (G Proteins)

The β adrenergic receptors are not coupled directly to adenylate cyclase but interact through an intermediary stimulatory protein **G_s**, which contains three subunits, α, β, and γ.[179–182] We know that the G_s protein associated with β adrenergic stimulation is only one of a very large number of related **G proteins**, so named because of their property of binding and hydrolyzing GTP. In its unactivated state the α subunit of a $G_{\alpha\beta\gamma}$ heterotrimer carries a molecule of bound GDP. Apparently, the G_s proteins and the hormone receptors,

which are activated by the binding of hormones, diffuse within the membrane until they make contact and form molecules of the G_s•GDP complex (Eq. 11-10, step *b*). The complex then undergoes a rapid exchange of the bound GDP for GTP, after which the hormone and receptor dissociate. The G_α•GTP complex may also dissociate from the complex, perhaps entering the cytosol as a soluble protein (Eq. 11-10, step *c*). The G_α•GTP complex combines with adenylate cyclase (step *d*) and activates it to generate cAMP. However, the activation is transient. G_α also contains GTP-hydrolyzing (GTPase) activity and within a few minutes G_α•GTP is completely converted to G_α•GDP and the adenylate cyclase dissociates (Eq. 11-10, step *e*). The G_α•GDP recombines with the βγ complex which may serve as a membrane anchor, to complete the regulatory cycle. The overall effect is for hormone binding to cause a rapid release of cAMP in a short burst that may last only about 15 s.

It is not the binding of GTP but a slow subsequent conformational alteration that activates the G_α•GTP adenylate cyclase complex. This was deduced by study of analogs of GTP, such as guanosine 5'-(β, γ-imido) triphosphate (GMP-*P*-(NH)-*P* or GppNp), which are

---

BOX 11-C THE ATTRACTION OF *DICTYOSTELIUM* TO CYCLIC AMP

In higher animals cyclic AMP (cAMP) is an intracellular second messenger, but in the cellular slime mold *Dictyostelium discoideum* it serves to convey signals *between* cells.[a–c] As was mentioned on p. 20, the organism exists as individual amebas until the food supply is exhausted. Then the cells begin to signal their lack of food by secreting pulses of cAMP. Because they also secrete a phosphodiesterase, the cAMP is short-lived.[d] However, it is present long enough for any other nearby ameba to sense the gradient of cAMP concentration from one end of the cell to the other. As little as a 2% difference in concentration can induce chemotaxis.[e] The amebas move toward the source of cAMP and emit pulses of the compound. This results in the formation of aggregation centers in which the concentration of cAMP oscillates spontaneously as the cAMP moves outward in waves. The cells move up the concentration gradient until the peak of a wave reaches them. Then they move in a random direction until the next wave reaches them and again orients their motion.

After ~50 movement steps an aggregation center contains ~10^5 cells which now follow a "development program." The amebas adhere to each other to form 1- to 2 mm-long multicellular "slugs" in which all of the cells move forward together. About 30 h after aggregation begins the slugs stop and form stalks with spore-containing fruiting bodies on top. The

lead cells in a slug become the stalk.[a,c]

Two types of G protein-coupled cAMP receptors in the cell membranes of the amebas have been identified. One leads to activation of adenylate cyclase and the release of new pulses of cAMP and the other to activation of guanylate cyclase.[e] This enzyme causes a rapid 7- to 10-fold increase in intracellular cGMP which plays an important role in controlling chemotaxis.

Some species of cellular slime molds use other chemical attractants (**acrasins**). For example, *D. minutum* secretes an analog of folic acid[f] and cells of *Polysphondylium violaceum* are attracted by the ethyl ester of *N*-propionyl-γ-L-glutamyl-L-ornithine-δ-lactam.[g]

[a] Gerisch, G. (1987) *Ann. Rev. Biochem.* **56**, 853–879

[b] Devreotes, P. (1989) *Science* **245**, 1054–1058

[c] Alberts, B., Bray, D., Lewis, J., Raff, M., Roberts, K., and Watson, J. D. (1994) *Molecular Biology of the Cell*, 3rd ed., Garland, New York

[d] Levine, H., Aranson, I., Tsimring, L., and Truong, T. V. (1996) *Proc. Natl. Acad. Sci. U.S.A.* **93**, 6382–6386

[e] Kuwayama, H., and Van Haastert, P. J. M. (1996) *J. Biol. Chem.* **271**, 23718–23724

[f] Schapp, P., Konijn, T. M., and van Haastert, P. J. M. (1984) *Proc. Natl. Acad. Sci. U.S.A.* **81**, 2122–2126

[g] Shimomura, O., Suthers, H. L. B., and Bonner, J. T. (1982) *Proc. Natl. Acad. Sci. U.S.A.* **79**, 7376–7379

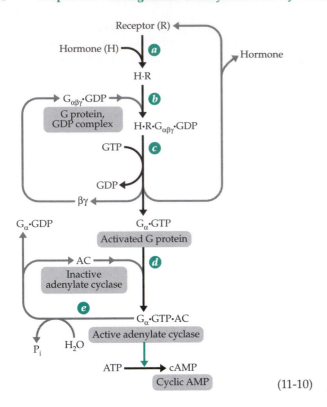

(11-10)

not hydrolyzed to GDP. These compounds may require many minutes to activate the complex, but because the GppNp is not hydrolyzed, the complex remains activated and able to catalyze cAMP formation continuously.

Guanosine 5'-(β,γ-imido) triphosphate (GppNp); in place of the NH, O is present in GTP, S is present is GTPγS

Occupied receptors for adrenaline, glucagon, ACTH, and histamine activate adenylate cyclase via G_s proteins. Other G_s proteins, which contain subunits designated α_{olf} and which exist as a number of subtypes, mediate **olfactory responses**. Subunit α_o is another specialized polypeptide which is located primarily in neural tissues. A variety of additional G proteins have been discovered in organisms ranging from bacteria to mammals.[179,183–186] All have similar structures with 39- to 45-kDa α subunits, 35- to 36-kDa β subunits and 5- to 8-kDa γ subunits. Whereas the α subunits are unique to each G protein, β and γ subunits may be shared among several G proteins. These proteins appear to function with many kinds of hormone receptors and

in nerve transmission, secretion, and endocytosis.[187] Mammals can form at least 23 different α_s subunits, which are coded for by 16 genes. Some of the forms arise by alternative splicing of mRNA transcribed from a single gene.[180] There are at least 6 β and 12 γ subunits.[179,180,182,188,189] Other G proteins, found in heart muscle, induce the opening of K^+ or Ca^{2+} channels in cell membranes.[185,190] It has often been assumed that all of the effects of G proteins are mediated by activated α subunits. However, the $\beta\gamma$ complex, which sticks together very tightly, can also have signaling functions.[191] For example, an auxiliary protein called **phosducin** binds to the $\beta\gamma$ complex of G_s, slowing the reconversion to the $\alpha\beta\gamma$ trimer.[192] Phosducin was discovered in rod cells of the retina but has been found to be distributed broadly in other tissues as well.[192,193] In yeast the $\beta\gamma$ subunit of a G protein linked to the receptor for the yeast mating factor initiates the pheromone response of a cell.[193a]

There is another large class of receptors whose occupancy by an agonist leads to *inhibition* of adenylate cyclase. These include the α_2 adrenergic receptors, receptors for acetylcholine, adenosine, prostaglandin E_2 (Chapter 21), somatostatin, and some receptors for dopamine. Their responses are mediated by inhibitory proteins G_i, which closely resemble G_s in their sizes, amino acid sequences, and heterotrimeric structures, but which inhibit adenylate cyclase when activated.[180] A clear distinction between the G_s and G_i proteins is evident in the fact that G_s is irreversibly activated by the action of cholera toxin, while G_i loses its ability to respond to occupied receptors when modified by the action of *Pertussis* toxin (Box 11-A). A specialized heterotrimeric G_i protein known as **transducin** mediates the light-induced activation of a **cyclic GMP phosphodiesterase** in the retina[194,195] (see Chapter 23). Its α subunit is designated α_t. The related **gustducin** is found in taste buds.[196]

Monomeric G proteins. An entirely different class of G proteins was discovered when it was found that the small 189-residue, 21-kDa protein products of the human oncogenes and proto-oncogenes known as **ras** are monomeric G proteins.[186,197–200] There are over 80 related proteins of this group of nine families.[200a] They include the much larger **elongation factor** EF-Tu, which functions in protein synthesis (Chapter 29). Even though there is a large difference in size, the three-dimensional structures of the GTP-binding domains of EF-Tu[201–204] and of the 21-kDa Ras p21 proteins[205–209] are very similar (Fig. 11-7). They also resemble that of adenylate kinase (Fig. 12-30).[205] The human genome contains four "true" *ras* genes: H-*ras*, N-*ras*, K-*ras*, and K-*rasB*. The occurrence of mutations in these genes in human cancer is discussed in Section H, as are the complex signaling functions of *ras* genes, (see also Fig. 11-13).[210]

Other proteins in the Ras superfamily include some members of the **Rho** family, which function in the cytoskeleton.[200a,211,212,212a,b,c,213] These include at least 14 mammalian proteins, among them RhoA, RhoB, etc., Rac1, Rac2, etc., and Cdc42. The latter is associated with the formation of fingerlike extensions of the cell membrane containing actin bundles.[200a,212d,212e] Genes *rac1* and *rac2* encode proteins involved in the initiation of superoxide radical formation by activated phagocytes (Chapter 18).[214] The **Rab** family of proteins are involved in exocytosis and endocytosis and vesicular transport. There are 24 or more human *rab* genes.[212,215,216] Yeast cells also contain Ras proteins. One *RAS* gene must be present for yeast to form spores.[217] **ADP-ribosylation factors** (ARFs), which stimulate the action of cholera toxin A subunit (Box 11-A) are also members of the Ras superfamily.[120,217a]

Like the trimeric G proteins, the monomeric proteins also serve as "timed switches" that are turned on by GDP–GTP exchange and are turned off by the hydrolysis of the bound GTP. In the absence of other regulatory proteins both the exchange of bound GDP for GTP and the hydrolysis of bound GTP are slow. While the receptor-associated trimeric G proteins are activated by the hormone receptors, the monomeric G proteins are activated toward GDP–GTP exchange by binding to proteins called **guanine nucleotide dissociation stimulators** (GDSs).[217b] They speed the release of GDP from the G protein allowing the active GTP complex to be formed. The velocity of hydrolysis of the bound GTP in the activated G protein is greatly increased by **GTPase activating proteins** (**GAPs**). These auxiliary proteins can be thought of as signaling molecules that pass their messages to the G protein, controlling the extent of its activation and therefore the strength of a signal that it sends on to the processes that it is controlling.[200,210,218] One group of GAP proteins are known as **regulators of G-protein signaling** (RGS).[218-218b] These proteins, which contain a characteristic 12-residue core, were first recognized as negative regulators of signaling in yeast cells called **GDP-dissociation inhibitors** (GDIs).[219] They can also be regarded as a family of GAPs.

Acylation and prenylation. The amino terminus (usually glycine) of the α subunit of any G protein is nearly always converted to an **N-myristoyl group**.[220-223] This modification occurs in a **cotranslational process** after removal of the initiating methionine (Chapter 29) and can be described as an acyl transfer from coenzyme A.[224] The C termini of the γ subunits of heterotrimeric G proteins, and also of the monomeric proteins of the Ras family, also undergo processing. For example, the C-terminal end of an intact Ras protein contains 18 residues which probably assume a largely α-helical conformation. A cysteine side chain near the terminus and having the sequence CAAX is

converted within cells to a **thioether** with all-*trans*-farnesol or longer polyprenyl alcohols.[225] Another nearby cysteine may form a **thioester** with palmitic acid,[221,226,227] for example:

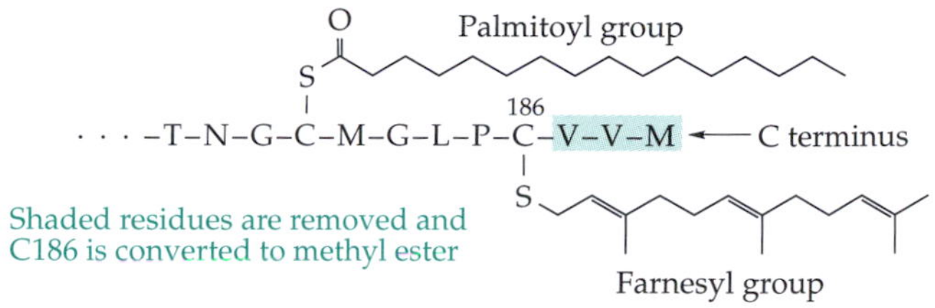

Shaded residues are removed and C186 is converted to methyl ester

N-*ras* p21

The farnesyl (or longer) polyprenyl and palmitoyl groups, together with nearby nonpolar side chains, serve as a membrane anchor. After the prenylation the C-terminal three amino acids (VVM in N-*ras* p21) are removed proteolytically and the new terminal carboxyl group of the farnesylated Cys 186 is converted to a methyl ester.[212c,225,228] N-terminal myristoylation is usually regarded as irreversible. However, the thioester linkages by which palmitoyl groups are attached to proteins are labile and may be cleaved rapidly by hydrolases. It follows that palmitoylation can be a rapid, regulated modification of proteins,[229-230a,b] strongly affecting the location and extent of adherance to membrane surfaces.

Three-dimensional structures. The structure of the GTP-binding domain of elongation factor EF-Tu was determined by Jurnak in 1985[201] and that of the complete three-domain structure later.[202,203] When the structure of the *catalytic domain* of the first Ras protein was determined (Fig. 11-7A) it was clear that it was similar to that of EF-Tu.[205,207] The same was true for the transducin α$_t$,[194,231,232] for the inhibitory G$_{iα1}$,[218b,233,234] and for other G proteins.[235] In every case the differences in structure of the enzyme with GDP or with analogs of GTP were small and limited to a region close to the γ-phospho group of the bound GTP. This group can be seen clearly in Fig. 11-7A adjacent to residue 60 of the protein. See also Fig. 12-36.

The β strands and helices in the GTP-binding domain of the G proteins are connected by a series of loops. The first, second, and fourth loops (residues 10–17, 32–40, and 57–65, respectively) in the Ras structure of Fig. 11-7A form the catalytic site in which the hydrolysis to GDP takes place. The first loop, also called the P loop, is conserved in all GTP-binding proteins. In the Ras proteins it has the following sequence:

10 17	
G–A–G–G–V–G–K–S	Ras
G–X–X–X–X–G–K–S(T)	Consensus

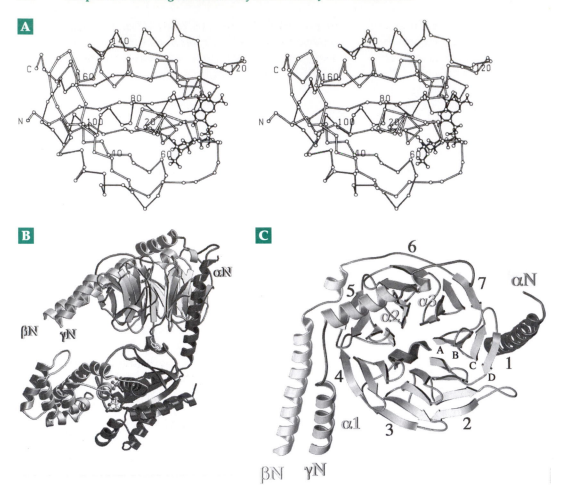

Figure 11-7 (A) Stereoscopic drawing showing α-carbon positions in the 166-residue "catalytic domain" of the human c-H-*ras* gene product and the bound GTP analog GppNp (solid bond lines). The intact protein contains an additional 23-residue C-terminal extension. From Pai *et al.*[241] (B), (C) Ribbon drawings showing two nearly orthogonal views of a hetero-trimeric inhibitory G$_i$ brain protein produced using a cloned bovine gene. (B) The amino termini (N) of the three subunits are seen in the left-to-right order: γ, β, α. A side view of the β propeller domain of the β subunit is seen at top center. The Ras-like domain and the additional large helical domain of the α subunit are marked. (C) View from the flared end of the β propeller looking toward the α subunit. The strands of each propeller blade are labeled A, B, C, and D and the seven blades are numbered around the periphery of the propeller. From Wall *et al.*[242] (B) and (C) courtesy of Stephen Sprang.

Also shown here is the "consensus" sequence for all GTP-binding proteins. Cancer cells frequently have mutations in Ras at Gly12, which may be substituted by Asp, Lys, Val, or Arg, and at Gly 13 and Gln 61. The latter is in loop 4, a part of the sequence that appears especially mobile but is highly conserved. These mutations all activate the protein, i.e., they decrease the GTPase activity, allowing the G protein to remain in its active conformation longer than normal.

Ras proteins fulfill their functions by interacting closely with two or more proteins in signaling pathways as described in Section H. Other G proteins have additional domains. The 405-residue EF-Tu from *Thermus thermophilus* has three domains: the C-terminal nucleotide-binding domain and two β-barrel domains following it. A major difference in conformation is observed between forms of the protein with bound

GTP or bound GDP, the GDP form being opened up by a hinging motion of ~ 90° between domains I and II.[202,203] The conformational change is triggered by the GDP–GTP exchange. An enhanced rate of GTP hydrolysis, which results from binding of a transfer RNA and binding to a ribosome, is apparently accomplished by interaction with loop 2 of the catalytic domain (corresponding to residues 32 – 40 in Fig. 11-7A). The function of this protein is considered further in Chapter 29.

Transducin has a 113-residue domain inserted into loop 2 of the catalytic domain. In this case, too, a large hinging movement is associated with the GDP–GTP exchange. The light-activated receptor rhodopsin induces the conformational change.[194,231] Again, structural changes resulting from the presence or absence of interactions of the protein with the γ-phospho group of GTP are involved. The GTPase activity of hetero-trimeric G proteins can also be activated by aluminum fluoride AlF_4^-.[232,236] From the X-ray structure of the transducin $\alpha \cdot GDP \cdot AlF_4^-$ complex it is seen that hydrated AlF_4^- is covalently bonded in the position of the γ-phospho group and makes hydrogen bonds to active site groups, mimicking a possible transition state for the GTPase reaction[232] (see Chapter 12, Section D).

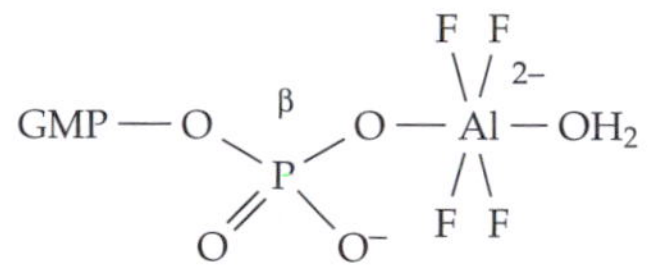

$$GMP - O \underset{\underset{O^-}{\overset{\parallel}{\underset{\|}{P}}}}{\overset{\beta}{}} O - \overset{F \quad F}{\underset{F \quad F}{\overset{\backslash /}{\underset{/ \backslash}{Al}}}}^{2-} - OH_2$$

The β subunit of the $\beta\gamma$ complex of transducin is a **seven-bladed β propeller** (Fig.11-7C). It is composed of seven GH–WD repeat units: $-[GH - X_n - WD]_{4-8}-$ where GH = Gly – His, WD = Trp – Asp, and X_n is a core repeating sequence, usually 32 – 42 residues in length. This motif is also found in at least 40 other eukaryotic proteins.[188,237–240] In the $\beta\gamma$ complex (Fig. 11-7B) the γ subunit assumes an elongated, largely α-helical structure. It is often anchored at its C terminus by a farnesyl or geranylgeranyl chain, while G_α may be myristoylated or palmitoylated.[195,230,243]

The three-dimensional structure of the GDP complex of the intact transducin heterotrimer[195] also shows a tight interaction between α and β subunits. The major interaction is probably disrupted by replacement of the bound GDP by GTP and the conformational change that occurs around the γ- phospho group. This explains the dissociation of the α subunit from $\beta\gamma$ upon activation. An entirely similar picture has been obtained for the action of the inhibitory G protein, G_{i2}, for which structures of the α subunit and of the $\alpha\beta\gamma$ heterotrimer (Fig. 11-7,B,C) have been determined.[188,233,234,242] The structures resemble those of transducin, but differ in details.

4. Guanylate Cyclase (Guanylyl Cyclase), Nitric Oxide, and the Sensing of Light

Formation of the less abundant cyclic guanosine monophosphate (cGMP) is catalyzed by guanylate cyclases found in both soluble and particulate fractions of tissue homogenates.[244–247] However, its significance in metabolism is only now becoming well understood. There are cGMP-dependent protein kinases,[247] but, until recently cGMP could not be regarded as a second messenger for any mammalian hormone. Now we know that cGMP has several essential functions. It mediates the effects of the **atrial natriuretic factor**, a peptide hormone causing dilation of blood vessels (Box 23-D),[245,248–250] and also of a peptide called **guanylin**, which is formed in the intestinal epithelium.[245,251] The receptors for both of these peptides are *transmembrane proteins with cytoplasmic domains that have guanylate cyclase activity*. The released cGMP appears to induce relaxation of smooth muscles of the blood vessel walls. Another mediator of smooth muscle relaxation is the **endothelial cell-derived relaxing factor**, which is evidently **nitric oxide, NO**[252] (see Chapter 18). A soluble guanylate cyclase appears to be an NO receptor and is activated by binding of the NO to a heme group in one of the two subunits of the cyclase.[253–255] The cyclase can also be activated by carbon monoxide, CO, in a similar fashion.[256,257] Carbon monoxide is produced in the body by degradation of heme (Fig. 24-24) and is thought to be a neurohormone.[258]

Cyclic GMP also plays an important role in vision. Specific phosphodiesterases hydrolyze cGMP to GMP. The cGMP phosphodiesterases of the rod and cone cells of the retina are activated by the G protein transducin, which has been activated by light absorbed by the visual pigments.[194] The resulting decrease in the cGMP concentration is thought to cause the closing of cation channels and thereby to initiate a nerve impulse (see Chapter 23). Cyclic nucleotide forms of the pyrimidine nucleotides are present only in very small amounts in cells.

5. Bacterial Chemoreceptors

Bacteria are attracted to foods with the aid of chemoreceptors that bind certain amino acids such as aspartate. The receptors send signals to the mechanisms that ensure that the bacterium is swimming toward the food[113,259, 259a] (more details are given in Chapter 19). The aspartate and serine chemoreceptors found in membranes of *E. coli* and *Salmonella typhimurium* are among the most carefully studied of all receptors. Like the seven-helix receptors, they have an extracellular sensory domain to which the signaling molecule, aspartate or serine, binds, and a transmembrane helical bundle. In these receptors the transmembrane part consists of four helical segments coming from the two subunits of the dimeric protein. For the aspartate receptor there is a high-resolution X-ray structure for the cytoplasmic domain[260,261] including the bound aspartate[262] as is shown in Fig. 11-8A,B.[260,263] The structures of the receptors for aspartate, serine, and many other signaling molecules are evidently very similar. The structure of the cytosolic domain of the serine receptor has been determined and a model for the complete receptor has been proposed.[259a] The molecular mass of the 188-residue extracellular domain is ~18 kDa, while that of the larger cytoplasmic domains is ~36 kDa. The latter includes a linker region as well as a long four-helical bundle domain which can be divided into a methylation domain and a signaling domain as shown in (Fig. 11-8C).

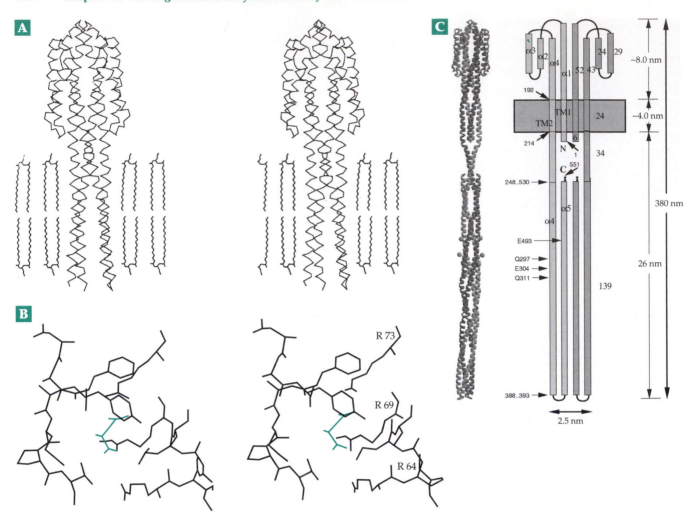

Figure 11-8 (A) Stereoscopic view of the sensory domain of the dimer with modeled transmembrane region. Each monomer contains a four α-helix bundle with two helices continuing through the membrane. From Scott *et al.*[260] (B) Stereoscopic image of aspartate in the major binding site of the receptor. The atomic model of aspartate (green) has been fitted into the observed difference electron density map. From Yeh *et al.*[262] (C) Model of an intact *E. coli* serine chemoreceptor. Left. Ribbon drawing viewed perpendicular to the molecular twofold axis. Methylation sites are represented by the dark balls by the cytoplasmic domain. The bound serine is drawn as a partially hidden green ball at the upper left in the extracellular domain. Right. Diagram of the receptor. The presumed membrane bilayer is represented by the gray band. Positions of some residues are marked on the left side. On the right, the numbers of residues in various peptide segments are indicated. From Kim *et al.*[259a] Courtesy of Sung-Hou Kim.

What happens chemically when aspartate binds into a deep pocket in the center of the sensory domain? What kind of signal can be passed from the aspartate to the "signaling domain"? Several possibilities were discussed by Kim,[264] who suggested that the binding of aspartate causes the rotation of the sensory domain of one subunit relative to that of the other and that this rotation is transmitted through the membrane to the signaling domain. The signaling domain of the receptor forms a complex with a multimeric **protein histidine kinase** called **CheA** (chemotaxis protein A) together with an auxiliary protein **CheW**. A more likely possibility is that some kind of "piston" action occurs between the transmembrane helices of the two subunits.[264a] Binding of aspartate to the receptor, and the associated alterations in the receptor•CheA•CheW complex, causes a strong *decrease* in the catalytic activity of CheA.

In its active form CheA undergoes **autophosphorylation**, that is, the phosphorylation of a histidine imidazole group in one of its subunits by the protein kinase active site of an adjacent subunit. The phospho group is then transferred from phospho-CheA to another protein, **CheY**. Phospho-CheY interacts with the flagellar motor proteins (Chapter 19) periodically causing a reversal of direction of the bacterial flagella. As a result the bacteria tumble and then usually move in a new direction. If the attractant molecule aspartate is present on the receptor the signal is weakened and the bacteria continue to swim in the same direction— toward food. Phospho-CheY is spontaneously hydrolyzed to remove the phospho group, but the dephosphorylation is promoted by an additional protein **CheZ**. At least two other auxiliary proteins also function in this control system. **CheR** is a **methyltransferase** that methylates the glutamate side chain carboxylates in the methylation domain of the coiled coil region and **CheB** is a **methylesterase** that removes the methyl groups. These two proteins control the methylation level in the coiled coil region of the receptor (Fig. 11-8A). Increased methylation increases the strength of the signal sent from the receptor, perhaps because methylation removes negative carboxylate charges that repel each other and weaken the structure, interfering with signaling.[259] See also Fig. 19-3.

E. Calcium, Inositol Polyphosphates, and Diacylglycerols

Calcium ions entering cells from the outside or released from internal stores trigger many biological responses (see Box 6-D). Within cells Ca^{2+} often accumulates in mitochondria, in the ER, or in vesicles called **calciosomes**.[265] Release of the stored Ca^{2+} is induced by hormones or by nerve impulses. For example, impulses flow from the nerve endings into the muscle fibers and along the invaginations of the plasma membrane called transverse tubules (Chapter 19). There they induce release of Ca^{2+} from the ER. The released ions activate enzymes[266] and induce contraction of the muscle fibers. In many cells, Ca^{2+} causes release of secreted materials, for example, neurotransmitters in the brain.[267,268]

1. Alpha Adrenergic Receptors

The release of stored Ca^{2+} is often triggered by α adrenergic receptors.[269] Like β adrenergic receptors, the α receptors are activated by adrenaline. Specific inhibitors distinguish them. For example, the β receptors are inhibited by propranolol, while the α receptors are blocked by phenoxybenzamine. The synthetic agonist **phenylephrine** activates only α receptors and no increases in cAMP or in protein kinase activity are observed.

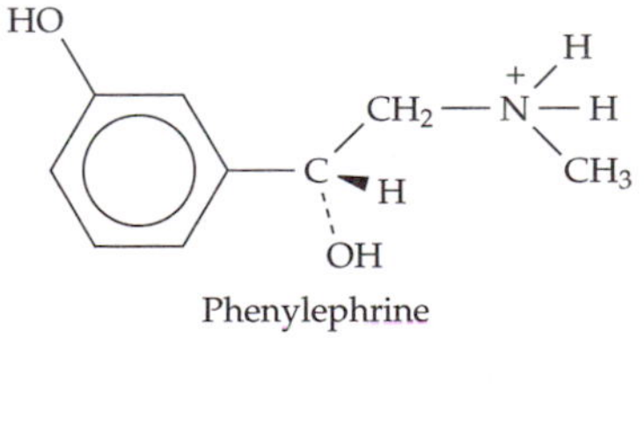

Phenylephrine

There are two major α adrenergic receptor subtypes.[151,269] Activation of the α_2 receptors, which are present in various tissues including blood platelets, causes *inhibition* of adenylate cyclase. This inhibition is evidently mediated by the G_i protein considered in Section D,3. The nucleotide sequences of cloned α_2 receptor genes and other properties suggest close structural similarity to the β receptors.[270–273] Subtle differences in hydrogen bonding to the serine side chains shown in Fig. 11-6C may distinguish α_2 from β_2 receptors.[273] A characteristic of α_2 receptors is that pertussis toxin (Box 11-A) abolishes the inhibition of adenylate cyclase, which they mediate. In contrast, the action of α_1 adrenergic receptors is not affected by pertussis toxin.

The α_1 receptors are activated not only by catecholamines but also by the hormones **vasopressin** and **angiotensin II**. Binding of these hormones to α_1 receptors induces a complex response that involves rapid hydrolysis of **phosphatidylinositol** derivatives and release of Ca^{2+} into the cytoplasm, and of diacylglycerols into the lipid bilayer of the membrane. The response is mediated by another G protein called G_q.[274,275] When this G protein is activated it induces the hydrolysis of **phosphatidylinositol 4,5-bisphosphate** (PtdInsP_2), a normal minor component of the lipid bilayer, by **phospholipase C** (phosphoinositidase C).[265,276–281]

Receptor $\to G_q \to$ phospholipase C $\to$
$$\begin{cases} \text{inositol phosphates} \\ + \text{ diacylglycerol (DAG)} \to \text{increased } [Ca^{2+}] \end{cases} \quad (11\text{-}11)$$

The products, **inositol 1,4,5-trisphosphate** [abbreviated InsP_3 or Ins(1,4,5)P_3], and **diacylglycerol** (DAG) are both regarded as second messengers for the catecholamines acting on α_1 receptors and for about 20 other hormones, neurotransmitters, and growth factors upon binding to their specific receptors.[269,282] In addition to vasopressin, the gonadotropin-releasing hormones, histamine, thrombin (upon binding to platelet surfaces), and acetylcholine (upon binding to its "muscarinic" receptors; Chapter 30) all stimulate inositol phosphate release.

2. Phosphatidylinositol and the Release of Calcium Ions

The phosphoinositides constitute ~2–8% of the lipid of eukaryotic cell membranes but are metabolized more rapidly than are other lipids.[265,278,279,283–285] A simplified picture of this metabolism is presented in Fig. 11-9. Phosphatidylinositol is converted by the consecutive action of two kinases into phosphatidylinositol 4,5-bisphosphate.[286,287] The InsP_3 released from this precursor molecule by receptor-stimulated phospholipase C is thought to mobilize calcium ions by opening "gates" of calcium channels in membranes of the ER or of calciosomes.[282,288–289a] It diffuses across the peripheral cytoplasm to **InsP_3 receptors** which are embedded in the membranes of the ER.[282,290] One of the several isotypes of InsP_3 receptors is a 2749-residue protein thought to contain a calcium ion channel.[290] Similar receptors are also found in inner membranes of the nucleus.[291,292]

Several uncertainties have complicated our understanding of the role of Ca^{2+} in signaling. What is the source of Ca^{2+}? How much of it enters cells from the outside and how much is released from internal stores? Where are the internal stores? What other kinds of ion channels are present and what second messengers regulate them? The sarcoplasmic reticulum of skeletal muscle and also membranes in many other cells contain **ryanodine receptors** as well as InsP_3 receptors.[282,293] Both of these receptors have similar structures and contain Ca^{2+} channels. However, the ryanodine receptors are activated by **cyclic ADP ribose** (cADPR),[294,295] which was first discovered as a compound inducing the release of Ca^{2+} in sea urchin eggs.[296] The 2-phospho derivative of cADPR may also have a similar function.[297]

2'-phospho in 2'-*P*-CADPR

Cyclic ADP-ribose (cADPR)

Phospholipase C, which initiates the release of phosphatidylinositol derivatives, also requires Ca^{2+} for activity. It is difficult to determine whether release of Ca^{2+} is a primary or secondary response. There are three isoenzyme types of phospholipase C–β, γ, and δ– and several subforms of each with a variety of regulatory mechanisms.[298–300a] For example, the γ isoenzymes are activated by binding to the tyrosine kinase domain of receptors such as that for epidermal growth factor (see Fig. 11-13). In contrast, the β forms are often activated by inhibitory G_i proteins and also by G_q, which is specific for inositol phosphate release.

Calcium ions are usually released in distinct pulses or "quanta." The kinetic characteristics of the system of receptors, diffusing InsP_3, calcium buffers, and calcium pumps in the cell membrane, and the membrane potential may account for this behavior.[174,301–306] Since

the phosphoinositols are chelators of Ca^{2+}, the equilibria involved in this control system are complex.[307]

In addition to InsP_3, several other inositol derivatives are released by adrenergic stimulation. **Inositol 1,3,4,5-tetrakisphosphate** (InsP_4) is formed from InsP_3 by action of a soluble kinase.[308,309,309a,b] A controversial suggestion is that InsP_4 may induce opening of Ca^{2+} channels through the outer membrane of the cell[310–312] and may also function to promote storage of Ca^{2+}. It also acts as a transcriptional regulator. Other possible second messengers are inositol 1,2-cyclic-3,4-trisphosphate and related metabolites that arise by the action of phospholipase C on PtdIns 4-*P* followed by additional actions of kinases and phosphatases.[278,309,313–315] A different inositol tetrakisphosphate, Ins(3,4,5,6)P_4, may control chloride ion channels.[316,317] Both inositol 1,3,4,5,6-pentakisphosphate (**InsP_5**) and inositol hexakisphosphate (**InsP_6**) are also found in plants[318] and animals.[319] InsP_5 serves as an allosteric effector regulating hemoglobin in avian erythrocytes (Chapter 7). InsP_6, also known as **phytate**, is present in large amounts in cereals and recently has been found to play a role in regulating the export of mRNA from the nucleus.[319a] Additional phospho groups can be added to InsP_6 to form pyrophosphates[320,320a] and other complex polyphosphates.[321] Both InsP_5 and InsP_6 can also serve as precursors to Ins(1,3,4,5)P_4 and Ins(1,4,5)P_3 by hydrolytic dephosphorylation.[322]

Following stimulation of a cell the induced metabolism of phosphoinositides decays rapidly. The diacylglycerols are converted into phosphatidic acid and resynthesized into phospholipids (Chapter 21). The InsP_4, InsP_3, and other inositol metabolites are hydrolyzed by phosphatases.[278,309,323–327] Two of these phosphatases are inhibited by Li^+ as indicated in Fig. 11-9. They may represent one site of action of **lithium ions** in the brain.[324,327,328] Lithium salts are one of the most important drugs for treatment of **bipolar (manic-depressive) illness**. By blocking the release of free inositol, which can be resynthesized into PtdInsP_2, Li^+ may prevent neuronal receptors from becoming too active. However, the basis for the therapeutic effect of lithium ions remains uncertain.

The diacylglycerols released by phospholipase C diffuse laterally through the bilayer and, together with the incoming Ca^{2+}, activate **protein kinases C**. These kinases also require **phosphatidylserine** for their activity and phosphorylate serine and threonine side chains in a variety of proteins.[329–330b] They are stimulated by the released unsaturated diacylglycerols. In addition protein kinases C can be activated by **phorbol esters**, which are the best known tumor promoters (Box 11-D). The diacylglycerol requirement favors a function for these protein kinases in membranes. They also appear to cooperate with calmodulin to activate the Ca^{2+}-dependent contraction of smooth muscle.[330]

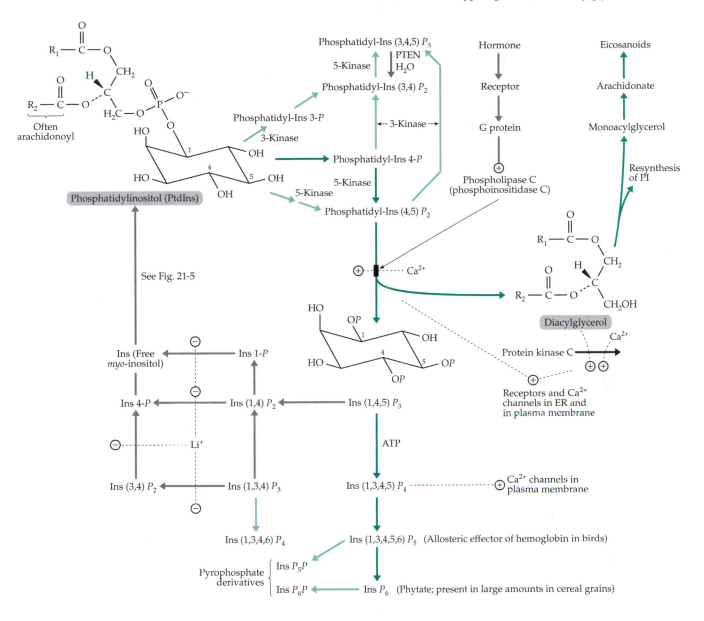

Figure 11-9 Scheme showing synthesis and release of diacylglycerol and inositol phosphates and their regulation of calcium concentration in response to hormonal stimulation.

Diacylglycerols released by phospholipase C usually contain **arachidonic acid** in the 2 position. Hydrolytic cleavage of this linkage[284,331] is a major source of arachidonate for synthesis of **eicosanoids** such as the **prostaglandins** whose functions are discussed in Chapter 21. Arachidonate can also be formed directly by the action of phospholipase A_2 on membrane phospholipids[332,332a] (see the structure on p. 566). It has been suggested that protein G_q, which activates the hydrolysis of phosphoinositides, may also regulate phospholipase A_2 directly. Released diacylglycerols may not only activate protein kinases C but also have a direct role in promoting the membrane fusion required in exocytosis and endocytosis.[332] Other breakdown pathways of phosphatidylinositols

are also indicated in the following structure.[298,333]

Initially most attention was paid to the water-soluble inositol phosphates that are released from phosphoinositides. However, phosphoinositol derivatives that retain the diacylglycerol part of the molecule have regulatory functions while remaining in membranes. A phosphatidylinositol 3-kinase phosphorylates the 3-OH of inositol in PtdIns, PtdIns 4-P and PtdIns $(4,5)P_2$ to give PtdIns 3-P, PtdIns$(3,4)P_2$, and PtdIns$(3,4,5)P_2$, respectively.[335–337d] However, they are ideally suited to function as spatially restricted membrane signals.[337a] They affect protein kinase C as well as the Ser / Thr protein kinase Akt (discussed in Section F,3),[326b] and have several functions in vesicular membrane transport and in regulation of the cytoskeleton.[337e,f]

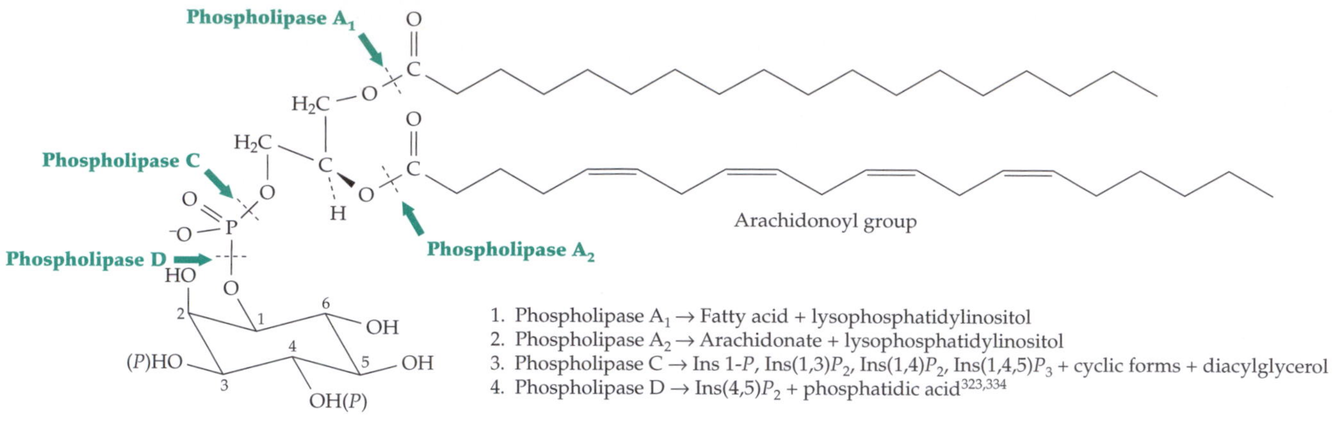

Arachidonoyl group

1. Phospholipase $A_1 \rightarrow$ Fatty acid + lysophosphatidylinositol
2. Phospholipase $A_2 \rightarrow$ Arachidonate + lysophosphatidylinositol
3. Phospholipase $C \rightarrow$ Ins 1-P, Ins(1,3)P_2, Ins(1,4)P_2, Ins(1,4,5)P_3 + cyclic forms + diacylglycerol
4. Phospholipase $D \rightarrow$ Ins(4,5)P_2 + phosphatidic acid[323,334]

Phosphatidylinositol

Characteristic of many of the phosphoinositide-regulated proteins is a proline-rich PH domain (Table 7-3) which can transmit regulatory signals to additional proteins in a cascade. Their importance is emphasized by the finding that a 405-residue phosphatase **PTEN** (named after its gene symbol) which catalyses hydro-lytic removal of one phospho group from PtdIns(3,4,5)P_3 to form PtdIns(3,4) P_2, (Fig. 11-9) is a major human tumor suppressor (see Box 11-D).[338,338a,b] These lipids are also important in insulin action (Section G,3). The steroid-like fungal metabolite wortmannin is a specific inhibitor of the 3-kinase.[339,339a]

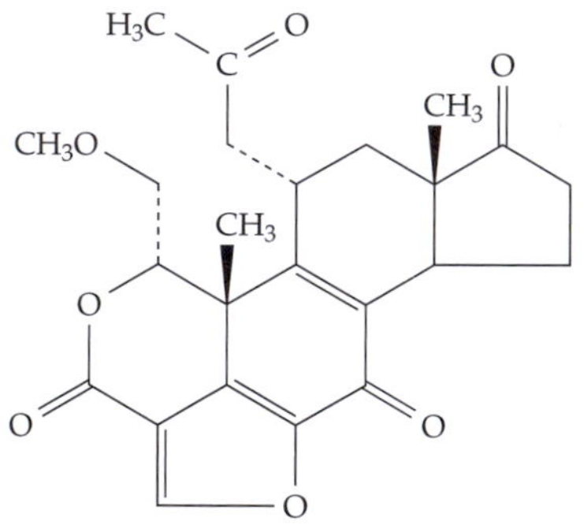

Wortmannin

F. Regulatory Cascades

The effect of a regulated change in the activity of an enzyme is often amplified through a cascade mechanism. The first enzyme acts on a second enzyme, the second on a third, etc. The effect is to rapidly create a large amount of the active form of the last enzyme in the series.

We have already considered regulatory cascades initiated respectively by the β and α_2 adrenergic receptors. The effect of these cascades on glycogen phosphorylase is outlined on the left side of Fig. 11-4. One branch of the cascade sequence begins with release of

adrenaline under control of the autonomic nervous system. In muscle, binding of this hormone to the β receptors on the cell membrane releases cAMP which activates a protein kinase. The kinase then phosphoryl-ates phosphorylase kinase. At this point the muscles are prepared for the rapid breakdown of glycogen. However, an additional initiating signal is the release of Ca^{2+} into the cytoplasm in response to impulses to specific muscles via the motor neurons. Calcium ions can also be released in the liver by α adrenergic stimulation. Phosphorylase kinase is activated by the calcium ions, and in their presence it converts inactive phosphorylase b to the active phosphorylase a. Both protein kinases and phosphorylase kinase also act on glycogen synthase, phosphorylating it and converting it to an inactive form. This turns off the biosynthetic pathway at the same time that glycogenolysis (glycogen breakdown) is turned on. Spontaneous reversion of the enzymes to their resting states occurs through the action of phosphatases that cut off the phospho groups placed on the protein by the kinases. Also essential are phosphodiesterases, which destroy the cAMP, and the calcium ion pump, which reduces the concentration of the activating calcium ion to a low level.

Elaborate cascades initiate the clotting of blood (Chapter 12) and the action of the protective complement system (Chapter 31). Cascades considered later in the book are involved in controlling transcription (Fig. 11-13) and in the regulation of mammalian pyruvate dehydrogenase (Eq. 17-9), 3-hydroxy-3-methyl-glutaryl-CoA reductase and eicosanoids (Chapter 21), and glutamine synthetase (Chapter 24).

1. Advantages of Regulatory Cascades

Computer simulations as well as studies of experimental models have led to the following conclusions.[340,341] Even simple cascade mechanisms, such as the one shown in Fig. 11-10, can provide a more flexible response to allosteric effectors (such as e_1 in the figure) than if the effector acted directly on the enzyme rather than on the protein kinase. The cascade also provides **amplification**. This is especially true if additional cycles are added. A response can result from the binding of only a small number of hormone molecules to a receptor in a cell membrane or from activation of only a few molecules of a protease in the initiation of blood clotting. A striking amplification occurs in visual responses. Under appropriate conditions a single quantum of light falling on a receptor cell in the retina of an eye can initiate a nerve impulse (Chapter 23). The latter requires the flow of a large number of Na^+ ions across the plasma membrane. It would be hard to imagine how absorption of one quantum could initiate a photochemical reaction leading to that much sodium transport without intermediate amplification stages. Another advantage of cascades is that they may provide **ultrasensitive responses**. Not only can a response be sensitive to a higher power than the first of the concentration of a signaling molecule but also the amplification provided by the cascade confers a high sensitivity to the response.[342]

Cascade systems also provide for response to more than one allosteric stimulus in a single pathway. Thus, as shown in Fig. 11-4, glycogen catabolism can be initiated in more than one way. Two pathways are known for initiation of both blood clotting and activation of the complement system. Many pathways activate the MAP kinase pathway shown in Fig. 11-13.

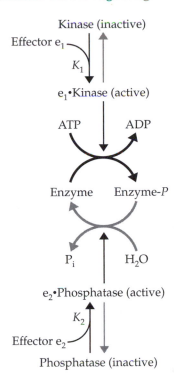

Figure 11-10 A "monocyclic" regulatory cascade involving phosphorylation and dephosphorylation of an enzyme. After Stadtman and Chock.[341]

2. Substrate Cycles

Although cascade systems offer advantage to cells, there is a distinct energy expenditure associated with the controls.[343] As can be seen in Fig. 11-10, in addition to providing for turning on and turning off the regulated enzyme, the kinase and the phosphatase together catalyze hydrolysis of ATP to ADP and inorganic phosphate. Thus, the regulated enzyme is continually cycling between active and inactive forms. The relative amounts of each are determined by the amounts of the effectors e_1 and e_2, the concentrations of modifying enzymes, and the kinetic constants. Such cycles have sometimes been called "futile cycles" because they seemingly waste ATP. This is particularly true of cycles that involve major metabolites, sometimes called **substrate cycles**. An example is the conversion of fructose 6-P to fructose 1,6-bisphosphate by phosphofructokinase and hydrolysis of the bisphosphate back to fructose 6-P by a phosphatase (Fig. 11-2). However, *the cycles are not futile* because the hydrolysis of the ATP provides the energy required to maintain the concentrations of modified (phosphorylated) enzyme or enzymes at steady-state levels that are required for efficient catalysis.[341,344] The energy utilized in this way

in regulation of enzymes is often small.[345] Substrate cycles are not only an unavoidable consequence of the need to regulate enzymes but also provide for improved sensitivity in metabolic control. Consider Fig. 11-2 again. If the enzymes are set for conversion of fructose 1,6-P_2 to glycogen (backwards direction), the flux of materials flowing in the forward direction (glycolysis) will be efficiently curtailed because any fructose 1,6-P_2 formed will be hydrolyzed rapidly by the active phosphatase. Under these circumstances the flux in the forward direction will be ultrasensitive to the activation of phosphofructokinase and to the inhibition of fructose 1,6-bisphosphatase.[346,347]

G. Insulin and Related Growth-Regulating Hormones

Since its isolation in 1921, insulin has been the object of an enormous amount of experimentation aimed at clarifying its mode of action. It is produced by the β cells of the pancreatic islets of Langerhans and released into the bloodstream in response to elevated glucose levels. The absence of insulin or of a normal response to insulin results in the condition of **diabetes mellitus**, which is the most prevalent human metabolic disorder (see Box 17-G).[348]

Insulin or insulin-like material is also produced in ciliated protozoa, vertebrate and invertebrate animals, fungi, green plants,[349,350] and even in *E. coli*.[351]

1. Metabolic Effects of Insulin

Insulin has many effects on metabolism.[348,352,353] (Some of these are listed in Table 17-3). They can be summarized by saying that: (1) In most tissues insulin stimulates synthesis of proteins, glycogen, and lipids. (2) It affects the permeability of membranes, promoting the uptake and utilization of glucose and amino acids and of various ions from the blood. (3) It promotes both synthesis of glycogen and the breakdown of glucose by glycolysis. At the same time, it inhibits synthesis of glucose from amino acids by the gluconeogenesis pathway. The effects are not uniformly the same in all tissues. Some can be observed within a few minutes after administration, presumably as a result of regulation of enzymes. Effects on mRNA metabolism, protein synthesis, and cell growth are seen at later times.

The uptake of glucose by brain, liver, kidneys, erythrocytes, and the islets of Langerhans is unaffected by insulin. However, in muscle and adipose tissues insulin stimulates glucose uptake. Part of this effect results from insulin-induced translocation of molecules of the 509-residue glucose transport protein GLUT4 (Chapter 8) from the cytosol into the plasma membrane where it can function.[354–356a] Insulin apparently also increases the rate of synthesis of the transporters.

Insulin stimulates the phosphorylation of serine side chains of many proteins, including ATP citratelyase, acetyl-CoA carboxylase, and ribosomal subunit S6. At the same time it stimulates the dephosphorylation of other proteins, including acetyl-CoA carboxylase, glycogen synthase in skeletal muscle (Fig. 11-4),[49] pyruvate dehydrogenase, and hormone-sensitive lipase in adipose tissue. Yet another effect of insulin is to alter the amounts of specific messenger RNA molecules. For example, the transcription of the gene for phosphoenolpyruvate carboxykinase (PEPCK; Eq. 13-46) a key enzyme in gluconeogenesis, is inhibited by insulin within seconds after binding. The hexokinase isoenzyme called **glucokinase** phosphorylates glucose to glucose-6-*P* in liver and in pancreatic β cells. Its synthesis in liver is induced by insulin. (However, in the β cells the synthesis of glucokinase is induced by glucose.)

2. Insulin Receptors

All of the effects of insulin appear to result from its binding to insulin receptors, of which ~10^2 to 10^5 are present in the plasma membranes of most animal cells. First isolated in 1972,[357] insulin receptors and their cloned genes have been studied intensively. The receptors are $\alpha_2\beta_2$ disulfide crosslinked oligomers composed of pairs of identical 120- to 135-kDa α subunits and 95-kDa β subunits. An α and a β subunit are cut from a single precursor chain. The human insulin receptor precursor exists as two isoforms, A and B, which arise as a result of a difference in splicing of the mRNA.[358] Twenty-one introns are removed by splicing to form the mRNA for the 1355-residue B precursor.[348,358a] The mRNA for the A form lacks a 36-nucleotide segment (exon 11) that is discarded during splicing. The α chains come from the N-terminal part of the precursor and the β chains from the C-terminal part. The receptor sequence is numbered as in the longer B form precursor. (However, many authors number the chains of the A form receptor as in its precursor, 12 less than the numbers given here for residues 719 or higher[359]). The B form receptor has 731-residue α chains while the A form has 719-residue α chains as a result of the missing sequence from exon 11 (see Fig. 11-11A). Four residues (732–735 in the B form) are cut out and discarded, leaving 620-residue β chains for both isoforms. The chains become linked by three or more disulfide crossbridges.[360]

The two α subunits, which contain the insulin binding sites, are apparently present entirely on the outer surface of the plasma membrane. The β subunits pass through the membrane with their C termini in the cytoplasm.[361–363] Both α and β subunits are glycoproteins. Study of the amino acid sequences suggests that only one 23-residue segment (residues 930–952 of the B isoform) of hydrophobic amino acid residues in each β subunit is likely to exist as an α helix that spans the membrane. This raises several questions. How can a signal be sent from the cell surface into the cytoplasm through a single pair of α helices? Are there additional parts of the receptor that span the bilayer of the membrane? To make it easier to visualize these questions refer to Fig. 11-11B, which is a more realistic, although fanciful, drawing than that in Fig.11-11A. Electron microscopy and crystallography are now providing the first direct images of the receptor.[363a, b, c]

One possible way in which a signal could be sent through the membrane is for binding of insulin to promote aggregation of two or more receptors.[364] If insulin induces the receptors to stick together on the outside of the cell, the parts protruding into the cytoplasm would also tend to aggregate. This could induce a response. A second possibility is that a conformational change in the α subunits pulls or pushes on the β subunit allowing the latter to be exposed less or more on the cytoplasmic side. This difference might be sufficient to cause a conformational change in the cytoplasmic domain of the subunit. A third possibility involves activation by a twisting mechanism as proposed for the aspartate receptor (Fig. 11-8).

A large part of the cytoplasmic domain of the insulin

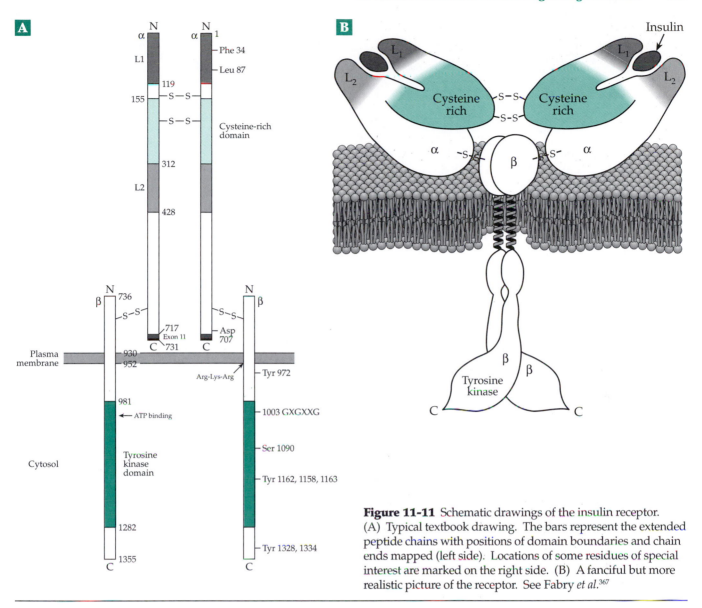

Figure 11-11 Schematic drawings of the insulin receptor. (A) Typical textbook drawing. The bars represent the extended peptide chains with positions of domain boundaries and chain ends mapped (left side). Locations of some residues of special interest are marked on the right side. (B) A fanciful but more realistic picture of the receptor. See Fabry *et al.*[367]

receptor consists of a tyrosine-specific protein kinase whose three-dimensional structure is depicted in Fig. 11-12.[365,365a] Not only can it phosphorylate – OH groups on tyrosine side chains of other proteins[362] but also it catalyzes ATP-dependent **autophosphorylation** of several residues in the C-terminal region. The most readily phosphorylated residue is Tyr 1158.[366] Some diabetic individuals have receptors with impaired tyrosine kinase activity.[362]

Using directed mutation of the cloned receptor gene, Lys 1018 in the ATP-binding part of the tyrosine kinase domain was replaced by alanine. This caused a loss both of kinase activity and of biologic response to insulin.[362] Thus, both the tyrosine kinase activity and autophosphorylation appear essential. If so, aggregation of two or more receptors may increase the extent of autophosphorylation and initiate a response.

Studies of many mutant proteins indicate that insulin binds to the α chains of the receptor between the two domains labeled L1 and L2 in Fig. 11-11.[368] Among essential residues is Phe 39 (marked).[369] Insulin contains three disulfide linkages (Fig. 7-16) and might undergo a thiol–disulfide exchange reaction (Eq. 11-7) with SH groups present in a cytoplasmic cysteine-rich domain of the α subunit of the receptor or with an external thiol compound.[369a,b] Such an exchange may also be essential for activation of the receptor.[370]

3. A Second Messenger for Insulin?

The known regulatory effects of insulin (Table 17-3) often involve phosphorylation of serine or threonine side chains on specific proteins. The tyrosine kinase of the activated insulin receptors does not catalyze such phosphorylation. Therefore, it seems likely that one or more second messengers or mediator substances are needed. Much effort has gone into searching for

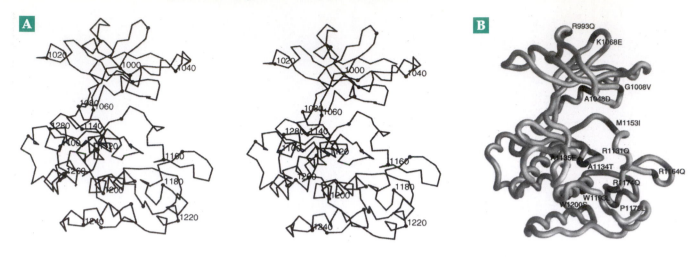

Figure 11-12 (A) Stereoscopic view of an α-carbon trace of the insulin receptor kinase domain. Every tenth residue is marked with a filled circle and every twentieth residue is labeled. (B) Locations of missense mutations in noninsulin-dependent diabetes mellitus patients mapped onto the receptor kinase structure. The mutations are R993Q, G1008V, A1048D, R1164Q, R1174Q, P1178L, W1193L, and W1200S. Here, R993Q is the mutant in which arginine 993 is replaced with glutamine, etc. From Hubbard *et al.*[365]

substrates for the tyrosine kinase,[353,362,371,372] whose activity on external protein substrates reaches a maximum after receptor tyrosines 1158, 1162, and 1163 have been autophosphorylated. Phosphorylation of such **insulin receptor substrates (IRSs)** as well as of some smaller "adapter" proteins is thought to initiate a series of complex cascades that serve to pass the insulin signal on to a variety of sites of action within a cell.[373–376] The large 185-kDA **insulin receptor substrate-1 (IRS-1)**, which is present in most cells, is phosphorylated on several tyrosines, usually within the sequences YXXM or YMXM. Phosphotyrosine within such sequences is known to be a ligand for the recognition domain SH2, which is present in many proteins. Binding of SH2 domains of other proteins to IRS-1 allows the insulin signal to be passed to several different proteins in the branched signaling pathways. Other insulin receptor substrates include **IRS-2**,[376a,b] protein Gab-1[376c] and the smaller protein called **Shc**, which occurs as 46-, 52-, and 66-kDa isoforms and is discussed further on p. 568.[376,377] Shc is an adapter protein which forms a complex that triggers the activation of the G protein Ras and the MAP kinase cascade shown in Fig. 11-13.[377,378] This pathway (see Section H,2) is thought to mediate the mitogenic (growth promoting) effects of insulin and also to promote phosphorylation of serine and threonine side chains of many proteins in the cytoplasm, the cytoskeleton, ribosomes, membranes, and the nucleus.[379,380] One serine kinase specifically phosphorylates the insulin receptor on serine 1078.[381]

In addition to IRS-1, IRS-2, and Shc, there are additional adapter proteins that interact with the phosphorylated tyrosine kinase domain of the insulin receptor.[382] Furthermore, the receptor tyrosine kinase may catalyze *direct* phosphorylation of some proteins, e.g. a cytoplasmic loop of the β2 adrenergic receptor (Fig. 11-6), without intervention of an adapter.[383]

Phosphorylated IRS-1 activates a second signaling pathway by interacting with an 85-kDa SH2-containing protein that is a subunit of phophatidylinositol 3-kinase.[384–386] This activates the 110-kDa catalytic subunit of the 3-kinase, which catalyzes formation of phosphatidylinositol 3-phosphate as well as PtdIns $(3,4)P_2$ and PtdIns $(3,4,5)P_3$.[387,387a] These compounds, which remain within membranes, activate other branches of the signaling cascade, some of which may converge with those of the MAP kinase cascade. However, there appears to be specific activation of a ribosomal Ser/Thr kinase that, among other activities, phosphorylates ribosomal protein S6, a component of the small ribosomal subunit.[388] It also phosphorylates some isoforms of protein kinase C and other enzymes. PtdIns 3-kinase may also activate 6-phosphofructo-2-kinase (Fig. 11-2, step *d*).[384,388]

One of the most important effects of insulin is to increase glucose uptake by cells.[373,389] The mechanism is thought to depend upon the transporter protein GLUT4, which is stored within the membranes of small cytoplasmic vesicles. Binding of insulin to its receptors induces movement of these vesicles to the plasma membrane where fusion with the plasma membrane makes the GLUT4 molecules available for glucose transport.[390] Phosphatidylinositol 3-kinase also plays an important role. The PtdIns$(3,4)P_2$ and PtdIns$(3,4,5)P_3$ generated by this enzyme (Fig. 11-9) remain in the membrane and provide sites to which various proteins, e.g., those containing PH domains (Chapter 7) may bind. Among the proteins "recruited" to the inner membrane surface in this way is a Ser/Thr protein kinase known as PKB/Akt. The Akt abbreviation refers to its relationship to a particular viral oncogene.[390a–390e] and the phosphatidyl-inositol derivatives generated by the 3-kinase may have a direct effect upon the exocytosis of the GLUT-4 containing vesicles as may Gq-coupled receptors.[391]

Although insulin unquestionably stimulates phosphorylation of many proteins, the major metabolic effect appears to result from *dephosphorylation* by phosphatases of phosphorylated forms of several enzymes.[379,392] These include **glycogen synthase** and **pyruvate dehydrogenase**, which are both activated by dephosphorylation, and **glycogen phosphorylase**[393] and **hormone-sensitive lipase**, which are deactivated by dephosphorylation. These changes result in increased synthesis and storage of both glycogen and triacyglycerols. Insulin also stimulates a membrane-bound cAMP phosphodiesterase causing a reduction in cAMP concentrations.[394]

Insulin injected into the human body has a half-life of only 5–10 minutes. Much of the hormone is destroyed by hydrolytic cleavage of the peptide chain by a non-lysosomal protease.[395–397] The receptors together with the hormone are taken into cells by endocytosis from clathrin-coated pits and both may be degraded in the lysosomes.[396,398] Some signaling may arise from insulin receptors in endosomes.[399] Some of the hormone enters the nucleus, perhaps still bound to the receptors,[400] and may have a direct effect upon gene transcription. Some receptors are recycled to the cell surface.[398] This latter process, together with that of new receptor synthesis, controls the number of receptors on the surface and, in turn, the sensitivity of the cell to insulin. This is only one mechanism used for desensitization or **downregulation** of receptor sensitivity and numbers, something that happens normally for any hormone when its concentration is high.[401,402]

H. Growth Factors, Oncogenes, and the Cell Cycle

Insulin is just one of a large number of proteins that are secreted by cells and which influence the growth of nearby cells.[403–405] For example, blood platelets, which aggregate at the site of an injury to a blood vessel, contain granules (α granules) which, when the platelet is "activated" (see Box 8-A) release **platelet-derived growth factor** (PDGF), a 28- to 31-kDa glycosylated peptide which stimulates growth and tissue repair in the injured region.[406,407] There are many other protein growth factors, a few of which are considered here. Others are discussed in Chapters 30–32.

1. Oncogenes and Proto-oncogenes

Studies of cancer-causing cellular oncogenes and **proto-oncogenes** (**c-onc**) identified in the human genome[408–411] have contributed greatly to our understanding of the action of growth factors. Proto-oncogenes are segments of DNA that code for proteins that have a normal function but which may become "activated" by a mutation or by chromosomal rearrangement to become cancer-causing oncogenes (Box 11-D). Oncogenes were discovered first in oncogenic (cancer-causing) retroviruses,[412] where they are designated v-*onc*. The v-*onc* genes are usually very similar but not identical to the corresponding c-*onc* genes and are thought to have arisen from them. The v-*onc* genes are often incomplete or have become fused to other genes by genetic recombination. Considerable excitement has attended the discovery that *many solid human tumors contain activated oncogenes*,[413] several of which are listed in Table 11-3.

In 1983 it was found that the oncogene *sis*, which is carried by the simian sarcoma virus, has a 104-residue sequence nearly identical to that of human PDGF.[404,414] The PDGF receptor has a large extracellular domain consisting of five immunoglobulin-like domains. Like the insulin receptor, it has tyrosine kinase activity which resides in a C-terminal cytoplasmic domain.[406,407,415] This suggests that the malignant transformation of cells by the viral v-*sis* gene leads to an excessive production of normal PDGF and, consequently, to excessive growth. The sequences related to v-*sis* are often found in human tumors and are located on chromosome 22.[416] Rearrangements by which part of this chromosome is moved to another location are well known and sometimes lead to conversion of the proto-oncogene to an active cancer-inducing oncogene.

The avian v-*erbB* oncogene and oncogene *neu*[417-419] both have sequences homologous to that of the gene for the receptor for the 53-residue **epidermal growth factor** EGF.[420–423a] The corresponding cellular c-*erbB* is the gene for the EGF receptor, a large 170-kDa 1186-residue protein. It also resembles the insulin receptor in having an N-terminal domain outside the cell, a single hydrophobic helix that spans the membrane, and a cytoplasmic domain with tyrosine-specific protein kinase activity.[403,419,424,425] It differs from the insulin receptor in being a single peptide chain. However, when EGF binds, the receptor dimerizes and the protein kinase is activated.[426] It phosphorylates its own Tyr 1173 near the C terminus, as well as tyrosines in the **lipocortins**, 36- to 38-kDa calcium-binding proteins located on the cytosolic face of plasma membranes.[427] The activated receptor also stimulates phospholipase C with a resultant increase in concentrations of inositol trisphosphate and Ca^{2+} (see Fig. 11-13).[428] The active oncogene v-*erbB* encodes a major fragment of the EGF receptor. However, most of the N-terminal domain

TABLE 11-3
A Few Oncogenes That Have Interested Biochemists

Oncogene Symbol	Source	Properties
v-*sis*	Simian sarcoma virus	Gene product is closely related to the B-chain of platelet-derived growth factor
v-*erbB*	Avian erythroblastosis virus	Gene product is shortened version of the EGF receptor, a tyrosine kinase
neu	(Her-2) Rat neuroblastomas A similar gene is found in human breast cancers and adenocarcinomas	Homologous to EGF receptor; has tyrosine kinase activity; may control phosphotidylinositol 3-kinase
v-*src*	Rous avian sarcoma virus	Gene product is another tyrosine-specific protein kinase
abl	Chronic myelogenous leukemia	Similar to *src*; characteristic chromosomal translocations yield cancer
ras, has, v-H-*ras* v-K-*ras* N-*ras*	Human bladder, colon, lung carcinomas	Homology with G_s and G_i regulatory proteins; several human proto-oncogenes exist; a single mutation in H-*ras* may lead to cancer
bas, kis	Murine sarcoma viruses	
v-*myc*	Burkitt's lymphoma, mouse plasmacytomas, avian retrovirus MC29	Nuclear location for gene product; which forms complex with protein Jun
mos	Burkitt's lymphoma, mouse plasmacytomas	
v-*fos*	Osteosarcoma virus (mouse)	Nuclear location for gene product, which forms complex with protein Jun
v-*jun*		Protein product is subunit of transcription factor AP-1
*bcl*1		Cyclin D1

including the EGF binding site is missing and the C-terminal end has been shortened. The tyrosine kinase domain is intact but Tyr 1173 is missing.

The rat *neu* oncogene (also called *erbB*-2 and *HER*-2), which apparently is derived from the gene for another growth factor receptor, differs from normal *neu* by a single nucleotide. This change causes valine to be substituted for glutamic acid at position 664 in the membrane-spanning domain of the 185-kDa protein.[418] This evidently gives an overactive and perhaps uncontrolled tyrosine-specific protein kinase. The c-*erbA* proto-oncogene appears to be the nuclear receptor for the thyroid hromone **triiodothyronine** (Chapter 25).

The Rous sarcoma virus oncogene v-*src* and a family of related oncogenes are derived from protein tyrosine kinases that are attached with the aid of a myristoyl anchor to the inner, cytoplasmic surfaces of membranes.[429–431] They may be activated by interaction with an occupied surface receptor. It has been difficult to understand the functioning of the normal Src protein. However, inactivation of c-*src* in mice caused the serious bone disease **osteopetrosis** in which the osteoclasts fail to function properly in resorbing the bone matrix, thereby allowing excessive accumulation of calcium phosphate.[432,433] A c-*src* deficiency also decreases formation of the bone adhesion protein **osteopontin**, an RGD protein.[434] There is actually a *family* of src proteins, some of which have important functions in lymphocytes.[435–438] The gene for one of these is mutated in the β cell disorder **agammaglobulinemia**.[437]

Another oncogene derived from a tyrosine kinase

BOX 11-D CANCER

Although cancer occurs in about 200 clinically distinct types, most cancers can be classified into four categories. In **leukemias**, which account for 3% of the ~700,000 cases of cancer diagnosed per year in the United States, an abnormal number of leukocytes are produced by the bone marrow. **Lymphomas**, such as Hodgkin's disease and Burkitt's lymphoma, arise from lymphocytes. They account for ~5% of human cancers. In these diseases malignant cells are produced in the spleen and lymph nodes and sometimes aggregate in lymphoid tissues. **Sarcomas**, solid tumors of bone or other connective tissue, contribute ~2% to the total of human cancers, while **carcinomas**, cancers of epithelial tissue, account for 85%. Carcinomas may develop from either the external or the internal epithelia, including the glands, lungs, and nerves.[a-d] More than one-third of all cancers in the United States are nonmelanoma carcinomas of the skin;[e] for these the mortality rate is low. Lung, colorectal, breast, and prostate tumors account for 55% of cancer deaths.[c] Carcinomas predominate in humans, but lymphomas, leukemias, and sarcomas are much more prevalent in laboratory animals and fowl.

An important characteristic of cancer cells is their uncontrolled proliferation. They don't respond to the normal signals from adjacent cells that indicate that cell division should stop. Cancer cells also differ dramatically from those present in warts and other benign tumors and in psoriasis. These conditions also result in excessive proliferation of cells and partial derangement of normal regulatory processes.

A second characteristic of cancer cells is that they usually appear less differentiated than the tissues from which they arise and are more like embryonic cells. Many cancers produce **ectopic proteins**, proteins inappropriate to the tissue involved and often identical to proteins synthesized by embryonic or fetal cells. A well-known example is **α-fetoprotein**, a 72-kDa glycoprotein normally present in serum in almost undetectable amounts but present in large amounts when some types of cancer are present.[f] A third property of cancers is the tendency toward **metastasis**, the detachment of cells from the cancer and their development in distant parts of the body.[g-i]

Cancer cells don't grow any faster than normal cells but they continue to divide when normal cells would not. For this reason, a cancer can grow rapidly and its demands for nutrients can literally starve the host. Cancer tends to weaken the immune system, making the host more susceptible to infections. In addition, cancers often interfere directly with the functioning of various organs and may cause death in this way.

What initiates a cancer? We know that cancers can arise from only one or a very small number of cells and that cancer can be induced by carcinogenic chemical compounds, by certain viruses, and by radiation. Use of tobacco appears to be responsible for about 30% of all human tumors.[c,j] Diet also affects the likelihood of developing cancer. For example, diets containing less animal fat and more fruits and vegetables are associated with lower levels of colon cancer.[c,k] Many carcinogenic compounds are naturally present in foods. Genetic factors help to determine susceptibility to cancer[b,l] and characteristic chromosomal aberrations are usually associated with cancer.[m-o] The incidence of cancer increases markedly with age.

A common feature of all the agents that induce cancer is the production of mutations and cancer probably always involves some alteration in the cell's DNA. The long lag between exposure to carcinogenic materials and development of cancers, often 20 years or more, suggested that more than one mutation or chromosomal rearrangement is required for production of a cancer. Recent evidence confirms that several mutations are required.[b,p] Relevant to this conclusion is the fact that carcinogenic compounds can be applied to the skin in amounts sufficient to cause a number of mutations in the epithelial cells but insufficient to actually induce cancer. Then, even many years later, irritant compounds known as **cancer promoters** can be applied and cancer will develop promptly. The promoters apparently induce cell proliferation which leads to more errors in DNA replication, converting an initially mutated cell to a cancerous cell. The most studied promoters are the phorbol esters, which are known to activate the protein kinase C isoenzymes (Section E,2). Any factors that increase rates of cell division such as some hormones, excess calories, or chronic inflammation cause increased cancer.[c] Cell divisions are accompanied by errors in replication of DNA and sometimes by translocation or deletion of parts of chromosomes. Chronic infection by bacteria, viruses, or other organisms may cause cancer as a result of continuing inflammation. Other cancers arise from integration of viral DNA into the host's DNA.

Virally induced cancers are often epidemic among poultry and rodents. When infected with cancer-causing viruses from these animals, cells in culture often become **transformed**. Whereas normal cells tend to respond to **contact inhibition** and grow as a monolayer, transformed cells continue to divide after the monolayer is complete. In laboratory studies, transformation of cells is often taken as the equivalent to an early step in cancer production in an animal. Studies of transformation led to the identification and characterization of several **viral oncogenes** which cause the transformation. These are designated by abbreviations such as v-*src* (the oncogene of Rous sarcoma virus, which induces cancer in chickens) and v-*sis* (the oncogene of simian sarcoma virus, which causes cancers in monkeys). Some other oncogenes are described in the main text. Naturally occurring gene

BOX 11-D CANCER (continued)

sequences closely homologous to those of the viral oncogenes have been found in many solid human tumors. Study of oncogenes has shown that they are related to and derived from **proto-oncogenes**, normal cellular genes that are involved in control of growth and differentiation.

Oncogenes are often "amplified" in tumor cells so that their copy number is greater than that of the corresponding genes in normal cells. For example an oncogene related to the viral oncogene *neu* is amplified in many human breast and ovarian cancers[q] and oncogene *src* in many colon cancers.[r] Mutated *ras* genes have been found in over one-third of human colorectal cancers.[s] Amplified oncogenes *ras*, *myc*, and *myb* have been observed in other cancers.[t] Oncogenes are often overexpressed or are responsible for over-expression of other genes. One idea that developed from these observations is that cancer cells may secrete new or mutated growth factors that stimulate their own receptors (**autocrine** stimulation) in a way that promotes uncontrolled growth.[u]

Cancer develops in stages and in many cases defective proto-oncogenes appear before truly malignant cells appear. The latter must result from additional mutations that often involve *loss* of parts of chromosomes. A major breakthrough in our understanding of cancer and how it is induced by loss of genes has come from studies of some rare cancers that are inherited in a Medelian fashion. One of these is **retinoblastoma**, an intraocular tumor which affects 1 child in 20,000. Homozygotes always develop the disease between the ages of 1 and 5. The hereditary defect has been traced to the absence of a functional retinoblastoma gene *RB1*, which is found in band q14 of chromosome 13.[v] Additional mutational events are required to induce cancer. The *RB1* gene was the first **tumor-suppressor** gene (anti-oncogene) identified. These suppressor genes encode proteins that inhibit growth.[w – z] The retinoblastoma gene encodes a 105-kDa DNA-binding phosphoprotein (Rb-*P*).[aa – bb] The Rb protein is phosphorylated and dephosphorylated in a cyclic fashion that is synchronized with the cell replication cycle (Fig. 11-15). It forms a complex with a transcription factor E2F that functions in transcription of the adenovirus genes and is also involved in control of the cell replication cycle.[cc] Deletion of the Rb gene from mice leads to death of embryos homozygous for the mutation.[dd]

Study of other rare hereditary cancers has led to the location of 20 or more additional probable tumor-suppressor genes. One of these, **p53**, is inactive in over 50% of all human cancers and over 90% of sqamous cell carcinomas of the skin.[ee] In small-cell lung cancers and osteosarcoma *both RB and p53* are inactive.[z] Protein p53 is a stronger tumor suppressor than protein Rb. Results of a variety of experiments have suggested that p53, a DNA-binding protein of known structure,[ff] plays a key role in checking DNA for damage at the G_1 to S-phase checkpoint in the cell cycle. If the DNA has too many defects the cycle is stopped in the G_1 stage and the cell may be killed by the process known as **apoptosis**.[gg – jj] Protein p53 has been called the "guardian of the genome." The mechanisms by which it functions are complex and poorly understood. It may act with the assistance of Rb and many other proteins. Mutations in DNA and their repair are discussed in Chapter 27 and the cell cycle is discussed in this chapter and further in Chapters 26 and 32.

Many other cancer susceptibility genes also encode suppressors. Mutation in genes *BRACA1*[kk] and *BRACA2*[ll] are responsible for early onset ovarian and breast cancer. Gene *DPC4* may be a suppressor of pancreatic cancer.[mm] The gene *ptc* (patched), first studied as a developmental gene in *Drosophila*, may encode a suppressor of basal cell carcinoma, the commonest form of human skin cancer.[nn] Gene *p16* (also called *CDKN2*) may be a major suppressor that is mutated in many cancers including the dangerous skin melanoma.[oo] Mutations in the *NF* gene, which may be a cytoskeletal protein, are associated with **neurofibromatosis**,[pp] a relatively common hereditary disease causing tumors of the nervous system. The tumors are usually not malignant but are numerous and disfiguring. Several cancer susceptibility genes are associated with faulty mismatch repair of DNA. Among these is the *APC* gene, whose malfunction is associated with human **familial adenomatous polyposis** which causes thousands of benign tumors in the lining of the large intestine and often colorectal cancer.[qq,rr,ss] In the much more common nonpolyposis colon cancer a complex pattern of instability in several genes is associated with DNA repair,[tt] a topic dealt with further in Chapter 27. The *ATM* gene defective in **ataxia telangiectasia** (Chapter 27) may encode a phosphatidylinositol kinase that is in some way involved in repair of DNA.[ss,uu,vv] A transcription factor gene *nm23* may be a suppressor gene for metastasis[ww] and the cell–cell adhesion molecule **E-cadherin** may suppress tumor invasion in some kinds of breast cancer.[xx] The **VHL** (van Hippel-Lindau cancer syndrome) suppressor protein is defective in the majority of kidney cancers. It normally binds to the **elongin complex**, a DNA-binding complex that functions in control of transcription.[zz]

In addition to treatment by surgery there are numerous chemical approaches to combating cancer.[xy] They usually exploit the tendency of cancers to grow continuously. For example, a toxic analog of a metabolite needed for growth, such as methotrexate (Chapter 16) and 5-fluorouridine (Box 28-C) may be taken up more rapidly by tumor cells than by normal cells. A variety of DNA-binding compounds are useful in chemotherapy (Box 5-B). Alkylating agents such as the nitrogen mustards (Eq. 5-20) and certain antibiotic compounds are also used widely, as are intercalating compounds

BOX 11-D (continued)

such as adriamycin (14-hydroxydaunomycin; Fig. 5-22) and cisplatin (Box 5-B). A disadvantage to most present-day chemotherapy is that normal proliferation of cells, especially of glandular tissues, intestinal epithelium, hair, etc., is severely damaged. A possibility for circumventing this problem is to put normal growth "on hold" temporarily while cancer growth is being inhibited. Some inhibitors of **topoisomerase I** (see Chapter 27) have low toxicity and could be useful. Another therapeutic approach is based on the fact that one function of the immune system is to destroy cancerous or precancerous cells. The immune system tends to weaken with age, which may be one reason that the incidence of cancers rises rapidly in older age Are there ways of stimulating the immune system into increased activity against cancer cells? Another approach is to find ways of increasing the activity of tumor-suppressor genes.

Can drugs be developed to *prevent* cancer? Such "chemoprevention"[yz] may be appropriate for persons carrying genes that make them highly susceptible to cancer. The antiestrogenic drug tamoxifen (Chapter 22) is currently being tested on women with a high risk for breast cancer. Use of oral contraceptives appears to have cut the risk of endometrial cancer substantially.[yy] Newer approaches to contraception may help prevent breast cancer and chemoprevention may also be possible for prostate cancer.[yy]

[a] Cairns, J. (1978) *Cancer: Science and Society*, Freeman, San Francisco, California

[b] Cavenee, W. K., and White, R. L. (1995) *Sci. Am.* **272**(Mar), 72–79

[c] Ames, B. N., Gold, L. S., and Willett, W. C. (1995) *Proc. Natl. Acad. Sci. U.S.A.* **92**, 5258–5265

[d] Ruddon, R. W. (1987) *Cancer Biology*, 2nd ed., Oxford Univ. Press, London

[e] Preston, D. S., and Stern, R. S. (1992) *N. Engl. J. Med.* **327**, 1649–1662

[f] Zhang, D., Hoyt, P. R., and Papaconstantinou, J. (1990) *J. Biol. Chem.* **265**, 3382–3391

[g] Packard, B. (1986) *Trends Biochem. Sci.* **11**, 490–491

[h] Feldman, M., and Eisenbach, L. (1988) *Sci. Am.* **259**(Nov), 60–85

[i] Marx, J. (1993) *Science* **259**, 626–629

[j] zur Hausen, H. (1991) *Science* **254**, 1167–1173

[k] Willett, W. (1989) *Nature (London)* **338**, 389–394

[l] Dragani, T. A., Canzian, F., and Pierotti, M. A. (1996) *FASEB J.* **10**, 865–870

[m] Solomon, E., Borrow, J., and Goddard, A. D. (1991) *Science* **254**, 1153–1160

[n] Nowell, P. C. (1994) *FASEB J.* **8**, 408–413

[o] Pennisi, E. (1996) *Science* **272**, 649

[p] Marx, J. (1989) *Science* **246**, 1386–1388

[q] Slamon, D. J., Godolphin, W., Jones, L. A., Holt, J. A., Wong, S. G., Keith, D. E., Levin, W. J., Stuart, S. G., Udove, J., Ullrich, A., and Press, M. F. (1989) *Science* **244**, 707–712

[r] Cartwright, C. A., Meisler, A. I., and Eckhart, W. (1990) *Proc. Natl. Acad. Sci. U.S.A.* **87**, 558–562

[s] Bos, J. L., Fearon, E. R., Hamilton, S. R., Verlaan-de Vries, M., van Boom, J. H., van der Eb, A. J., and Vogelstein, B. (1987) *Nature (London)* **327**, 293–297

[t] Yokota, J., Tsunetsugu-Yokota, Y., Battifora, H., Le Fevre, C., and Cline, M. J. (1986) *Science* **231**, 261–265

[u] Sporn, M. B., and Roberts, A. B. (1985) *Nature (London)* **313**, 745–747

[v] Weinberg, R. A. (1990) *Trends Biochem. Sci.* **15**, 199–202

[w] Stanbridge, E. J. (1990) *Science* **247**, 12–13

[x] Weinberg, R. A. (1991) *Science* **254**, 1138–1146

[y] Knudson, A. G. (1993) *Proc. Natl. Acad. Sci. U.S.A.* **90**, 10914–10921

[z] Yokota, J., and Sugimura, T. (1993) *FASEB J.* **7**, 920–925

[aa] Wiman, K. G. (1993) *FASEB J.* **7**, 841–845

[bb] Cobrinik, D., Dowdy, S. F., Hinds, P. W., Mittnacht, S., and Weinberg, R. A. (1992) *Trends Biochem. Sci.* **17**, 312–315

[cc] Nevins, J. R. (1992) *Science* **258**, 424–429

[dd] Jacks, T., Fazeli, A., Schmitt, E. M., Bronson, R. T., Goodell, M. A., and Weinberg, R. A. (1992) *Nature (London)* **359**, 295–300

[ee] Ziegler, A., Jonason, A. S., Leffell, D. J., Simon, J. A., Sharma, H. W., Kimmelman, J., Remington, L., Jacks, T., and Brash, D. E. (1994) *Nature (London)* **372**, 773–776

[ff] Cho, Y., Gorina, S., Jeffrey, P. D., and Pavletich, N. P. (1994) *Science* **265**, 346–355

[gg] Marx, J. (1993) *Science* **262**, 1644–1645

[hh] Enoch, T., and Norbury, C. (1995) *Trends Biochem. Sci.* **20**, 426–430

[ii] Kaufmann, W. K., and Paules, R. S. (1996) *FASEB J.* **10**, 238–247

[jj] Hartwell, L. H., and Kastan, M. B. (1994) *Science* **266**, 1821–1828

[kk] Futreal, P. A., and 26 other authors (1994) *Science* **266**, 120–122

[ll] Wooster, R., and 30 other authors (1994) *Science* **265**, 2088–2090

[mm] Hahn, S. A., Schutte, M., Hoque, A. T. M. S., Moskaluk, C. A., da Costa, L. T., Rozenblum, E., Weinstein, C. L., Fischer, A., Yeo, C. J., Hruban, R. H., and Kern, S. (1996) *Science* **271**, 350–353

[nn] Pennisi, E. (1996) *Science* **272**, 1583–1584

[oo] Marx, J. (1994) *Science* **265**, 1364–1365

[pp] Rouleau, G. A., and 20 other authors (1993) *Nature (London)* **363**, 515–521

[qq] Peltomäki, P., Aaltonen, L. A., Sistonen, P., Pylkkänen, L., Mecklin, J.-P., Järvinen, H., Green, J. S., Jass, J. R., Weber, J. L., Leach, F. S., Petersen, G. M., Hamilton, S. R., de la Chapelle, A., and Vogelstein, B. (1993) *Science* **260**, 810–819

[rr] Huang, J., Papadopoulos, N., McKinley, A. J., Farrington, S. M., Curtis, L. J., Wyllie, A. H., Zheng, S., Willson, J. K. V., Markowitz, S. D., Morin, P., Kinzler, K. W., Vogelstein, B., and Dunlop, M. G. (1996) *Proc. Natl. Acad. Sci. U.S.A.* **93**, 9049–9054

[ss] Kolodner, R. D. (1995) *Trends Biochem. Sci.* **20**, 397–401

[tt] Papadopoulos, N., and 19 other authors (1994) *Science* **18**, 1625–1629

[uu] Keith, C. T., and Schreiber, S. L. (1995) *Science* **270**, 50–51

[vv] Sanchez, Y., Desany, B. A., Jones, W. J., Liu, Q., Wang, B., and Elledge, S. J. (1996) *Science* **271**, 357–360

[ww] Marx, J. (1993) *Science* **261**, 428–429

[xx] Berx, G., Cleton-Jansen, A.-M., Nollet, F., de Leeuw, W. J. F., van de Vijver, M. J., Cornelisse, C., and van Roy, F. (1995) *EMBO J.* **14**, 6107–6115

[xy] Hoffman, E. J. (1999) *Cancer and the Search for Selective Biochemical Inhibitors*, CRC Press, Boca Raton, Florida

[yy] Henderson, B. E., Ross, R. K., and Pike, M. C. (1993) *Science* **259**, 633–638

[yz] Young, R. C. (2000) *Nature (London)* **408**, 141

[zz] Stebbins, C. E., Kaelin, W. G., Jr., and Pavletich, N. P. (1999) *Science* **284**, 455–461

gene is the Abelson murine leukemia virus v-*abl*.[438a] In the mouse the c-*abl* gene is split into at least ten exons. However, v-*abl* contains all of these as a correctly spliced sequence suggesting that v-*abl* was derived from a c-*abl* messenger RNA. Part of one exon of the cellular gene is missing in v-*abl* and there is at least one base substitution mutation as well. Human c-*abl* is located on the long arm q of chromosome 9. This arm has long been known to be translocated to chromosome 22 (the "Philadelphia translocation") in patients with chronic myelogenous leukemia. There the c-*abl* gene is fused with another gene.[439] The human c-*sis* gene is also located in the region of chromosome 22 that is translocated to chromosome 9 in the same patients.

A group of human pituitary tumors have been shown to contain an oncogene that is apparently a mutated gene for the stimulatory protein G_s.[440] This is the protein that activates adenylate cyclase in response to hormonal activation. The oncogenic mutations inhibit the GTPase activity that normally turns off this activation. These tumors secrete growth hormone which binds to receptors on the tumor cells activating the defective G_s proteins and causing excessive synthesis of cAMP. This in turn promotes growth of the tumor.

The ras oncogenes. Activated *ras*-oncogenes have been found in at least 25% of all human tumors. The proto-oncogenes, which are designated c-H-*ras*, c-K-*ras*, and c-N-*ras*, are found on the short arms of chromosomes 11, 12, and 1, respectively. A single base substitution (G → T) at position 35 of any of the genes, resulting in a Ras protein containing valine instead of glycine at position 12 of the 21 kDa protein product (usually designated p21) produces an active oncogene. Substitutions at position 13 or at positions 59 and 61, which are adjacent to Gly 12 in the three-dimensional structure, can also activate the oncogenes.[441] From the drawing in Fig. 11-7A the locations of glycines 12 and 13 and of residues 59/61 are seen to be close to the β phospho group of bound GTP. Proteins encoded by activated *ras* oncogenes are less active in catalyzing GTP hydrolysis than are the corresponding normal proteins. In addition, the GTPase-activating protein (GAP) that binds to normal *ras* proteins and stimulates their GTPase activity does not affect the mutant oncogenic proteins.

A single-base alteration is capable of activating a *ras* gene with respect to cell transformation. However, initiation of a malignant tumor requires the additional presence of at least a second activated oncogene such as *myc*[442,443] or *fos*[444] or previous transformation of a fibroblast into a nonmalignant but "immortalized" form by treatment with carcinogens. The c-*myc* gene, which is found in active form in plasmacytomas (tumors of B lymphocytes) of mice as well as in the human Burkitt's lymphoma, is normally located on human chromosome 8. In most Burkitt's lymphomas a trans-

location has brought the c-*myc* gene into the locus of the immunoglobulin heavy chains on chromosome 14. There its transcription may be subject to different controls than in its original location.[445,446] During the translocation process the c-*myc* gene is often broken within the first intron. Thus, the activated gene lacks the first exon and is placed after a new controlling sequence that may drastically alter its transcription rate. Viral *myc* genes are found in at least one human virus, **cytomegalovirus**, which has been associated with carcinomas.

Transcription factors. The proto-oncogenes c-*myc*[447–451a], c-*myb*,[452–454] c-*fos*, c-*jun* and c-*ets*[455] all encode nuclear proteins involved in regulation of transcription. The 39 kDa protein Jun, which is encoded by c-*jun*, is a major component of the **transcriptional activator** called **AP-1**.[456–459] It binds to palindromic **enhancer** sites (Chapter 28) in DNA promoters to increase the transcription rate for a group of genes. Jun is actually a multigene family whose encoded proteins bind to DNA as complexes formed with the 62 kDa phosphoprotein Fos, the product of the c-*fos* gene.[460] The heterodimeric Fos/Jun complex is held together, at least in part, by interactions between leucine side chains lying along a pair of parallel α-helices in a "leucine zipper" (Fig. 5-36; Fig. 2-21).[461,462]

Regulation of the synthesis of Jun is complex, but growth factors such as Neu, EGF,[463] and PDGF stimulate transcription of c-*jun* in cultured cells. Messenger RNA for synthesis of Fos appears within a few minutes of stimulation of PDGF receptors.[464,465] This is one of the earliest known nuclear reactions to a mitogenic stimulus and suggested that a Ras p21 protein is involved in stimulating transcription of c-*fos* in the signaling pathway from PDGF.[466] Synthesis of Fos is also induced by a variety of other stimuli.[456] Upon translocation into the nucleus Fos combines with pre-existing Jun to form AP-1, which binds to sites on DNA and induces the transcription of a large number of proteins (Chapter 28). Deletion of the c-*fos* gene in mice leads to defects in developing bone, teeth, and blood cells,[467] while excessive synthesis of Fos has been associated with the human bone disease **fibrous dysplasia**.[468]

2. The MAP Kinase Cascade

Insulin, platelet-derived growth factor (PDGF), epidermal growth factor (EGF), and many other proteins have **mitogenic activity**, that is, they induce cells to transcribe genes, to grow, and to divide. How is this accomplished? Study of such oncogenes as *src* and *ras* suggested that the proteins that they encode also participate in the process, as do transcription factors, including those encoded by the proto-oncogenes

Table 11-4
A Few Abbreviations Used in Discussions of Cell Signaling

AP-1	A major transcriptional activator protein	PKC, PKCα, β, γ	Protein kinase C,α, β, γ subforms
βARK	Beta adrenergic receptor kinase	PKR	dsRNA-activated protein kinase
EGF	Epidermal growth factor	PLA	Phospholipase A
EGFR	Epidermal growth factor receptor	PLC	Phospholipase C
ERK-1, ERK-2	Extracellular signal-regulated protein kinases (proline-directed protein kinases)	PP2A	Protein phosphatase, type 2A
		PTK	Protein tyrosine kinase
		PTP (PTPase, PTP 1B, etc.)	Protein tyrosine phosphatases
Fos	Protein encoded by proto-oncogene *fos* (Table 11-3)	pY	Phosphotyrosine residue
Grb2	Adapter protein containing one SH2 and two SH3 domains	Raf-1	A cytoplasmic serine/threonine protein kinase; also called MAP kinase kinase kinase (MAP3K)
IGF-1	Insulin-like growth factor 1	Ras	A monomeric G protein encoded by the proto-oncogene *ras*
InsRTK	Insulin receptor tyrosine kinase		
InsP_3 or IP$_3$	Inositol (1,3,5)-trisphosphate	RSK	Ribosomal protein S6 kinase
Jun	Protein encoded by proto-oncogene *jun*, an AP-1 gene	RTK	Receptor tyrosine kinase
		RTK-*P*	Phosphorylated receptor tyrosine kinase
MAPK	Mitogen-activated protein kinase (also designated ERK)	SH2, SH3	Src homology domains 2 and 3
MAPKK	A kinase acting on MAPK	Sos	"son of sevenless," a GDP – GTP exchange factor named for a similarity to the protein encoded by the *Drosophila* sevenless gene
MEK-1,MEK-2	Mitogen-activated ERK-activating kinases; dual function Ser/Thr and Tyr protein kinases; also called MAP kinase kinases		
c-myc	A cellular proto-oncogene	Src	Protein tyrosine kinase encoded by oncogene *src* (Table 11-3); contains recognition domains SH2 and SH3
PDGF	Platelet-derived growth factor		
PtdIns (or PI) 3-kinase	Phosphatidylinositol 3-kinase	V2R	Type 2 vasopressin receptor
		Shc	A proline-rich adapter protein containing SH2 and PH domains
PKA (cAPK)	cyclic AMP-dependent protein kinase		

myc, fos, and *jun*. The Src and Ras proteins are anchored on the inner surfaces of cytoplasmic membranes. Although the exact functional relationships of all of the components, one to another, has not been completely established, a general picture, usually described as the **mitogen-activated protein kinase** (MAP kinase) **cascade**, has emerged.[69,380,469-470b] This is sketched in simplified form in Fig. 11-13. The many amplification steps provide for ultrasensitive responses.[342]

The binding of a hormone or growth factor (a ligand) to a dimeric receptor activates the protein kinase domain of the receptor which phosphorylates a number of tyrosine hydroxyl groups of the receptor itself. This autophosphorylation is followed by a variety of events, which include phosphorylation of tyrosine side chains of various other proteins.[426] An-

other major event is the *binding of a variety of different protein molecules containing recognition domains to the phosphotyrosyl groups of the activated receptors.*[471] The major recognition motif is the SH2 domain. See Figs. 7-30 and 11-14.[472 – 475] Proteins containing SH2 domains can bind to the phosphotyrosyl groups of the activated receptors and while bound become phosphorylated by the receptor tyrosine kinase action and/or be activated allosterically.

Other proteins interact with the receptors indirectly through adapter molecules which have no catalytic activity. Two well-known adapter proteins are **Grb**2[476] and **Shc**.[477,478] The 25-kDa protein Grb2 consists entirely of recognition domains, one SH2 and two SH3 domains (Fig. 11-14). The larger Shc, which is found in all mammalian tissues, contains a 200-residue phosphotyrosyl-binding PH domain (Chapter 7) at the N terminus, a collagen-like domain that binds to Grb2 and an SH2 domain at the C terminus.[376,377,477]

Adapter Grb2 binds to a phosphotyrosine side chain of an activated receptor, such as that for EGF, and simultaneously binds to the GDP-GTP exchange protein called **Sos** (Table 11-4). This signals Sos to activate the membrane-bound Ras protein by converting it into the GTP form. The second adapter Shc may also participate in formation of the receptor kinase - Grb2–Sos complex,[479] perhaps permitting formation of a more robust complex that may receive signals from more than one kind of receptor. Functions of other members of the Grb adapter family are being discovered.[479a]

Activated Ras binds to and activates the cytoplasmic serine/threonine protein kinase called **Raf-1**.[380,480,481] This kinase becomes transiently activated within 2–3 min of the binding of a mitogen to a receptor. Raf-1 initiates a cascade of other protein kinases by acting on the dual-function Ser/Thr and tyrosine protein kinases called **MEK-1** and **MEK-2**. The phosphorylated, active MEK proteins phosphorylate the mitogen-activated protein kinases **MAPK** which act on a variety of other proteins. Two of the best known MAPK proteins are designated **ERK-1** and **ERK-2**. These are *proline-directed* kinases which phosphorylate serines and threonines that are neighbors to prolines, e.g. in the sequence PLS/TP.[380] The activated ERKs are able to phosphorylate a large number of different proteins including nuclear proteins that control the transcription of such protein transcription factors as AP-1 and Myc. A protein known as the **serum response factor** binds to nucleotide sequences $CC(A/T)_6GC$ in the DNA to locate initiation sites for transcription. Other proteins that have been phosphorylated by the MAP kinase cascade then induce transcription.[482] The induction of c-*fos* mRNA is one of the earliest identified responses to growth factors.[375,456,483] Protein Jun, whose synthesis is induced independently by almost all growth factors, is usually present in excess.

The Fos/Jun complex is transcription factor AP-1, which induces transcription of many genes needed for cell growth. However, transcription of specific genes often depends upon additional nucleotide sequences. For example, the sequence CGGAAA is present in an **insulin response element** found in the promoter sequences of genes encoding such proteins as phosphoenolpyruvate carboxykinase, glyceraldehyde phosphate dehydrogenase, and prolactin–proteins whose synthesis is induced by insulin.[484]

The MAPK cascade also has direct effects upon protein synthesis, i.e., on the translation of mRNA messages. For example, insulin stimulates phosphorylation of proteins that regulate a translation initiation factor, a protein called eIF-4E (see Chapter 29). Phosphorylation of inhibitory proteins allows them to dissociate from the initiation factor so that protein synthesis can proceed.[485,486]

The scheme in Fig. 11-13 is complex, but in reality it is *much* more complex than is shown. Each protein kinase (receptor kinase, Raf-1, MEK, and MAPK) will phosphorylate not only the proteins indicated in this scheme but also any others that meet the specificity requirements of the kinases. Thus, there will be branches diverging from the pathways shown.[486a,b,c] There are isoenzymes that provide further divergence and interaction.[487] Not only do pathways diverge but also others *converge*. Thus, binding of many different ligands to their receptors activates the same MAPK cascade. For example, seven-helix G protein-coupled receptors release their βγ subunits which may also activate Ras as indicated on the right edge of Fig. 11-13A. At the same time the α subunits of the heterotrimer G proteins can affect not only adenylate cyclase but also phospholipases C which can, in some cases, also activate the MAP kinase pathway.[488,489] Sphingosine 1-phosphate may be released from membrane spingo-lipids and activate the same cascade.[490] However, hormones do not all affect cells in the same way. In view of all the converging pathways, how is this possible? Part of the answer lies in the proximity or spatial separation of components of the pathway. The kinases exist in complexes with other signaling proteins and may phosphorylate them, sending a signal back, as well as forward via other protein substrates. There are also unknown kinetic considerations. Hormones, neurotransmitters, and calcium ions are often released in pulses. The signaling system must integrate effects of all the stimuli that arise from different parts of the cell, at different times, and from differing receptors. At the same time all of the phosphorylated proteins are acted upon by phosphatases that may either activate or deactivate the proteins and by proteases that process newly formed peptides and modify or destroy mature proteins. The various modifying enzymes act on cytosolic proteins, proteins of membranes, of the cytoskeleton, and of the nucleus. The regulatory processes that we discuss in such minute detail involve the very substance of living cytoplasm which is ever-changing and responding to its surroundings. The flow of energy, provided by synthesis of ATP and by the use of ATP by kinases, phosphatases, and the protein synthetic machinery, goes along with the flow of information and drives the signaling network. Evolution has shaped this system to allow it to respond appropriately for every species.

3. The Cell Cycle and Control of Growth

When a cell divides it is of utmost importance that the DNA be replicated reliably. This requires that the dividing cell be large enough and contain enough biosynthetic precursor materials to complete the elaborate process of DNA synthesis and of mitosis. The **cell**

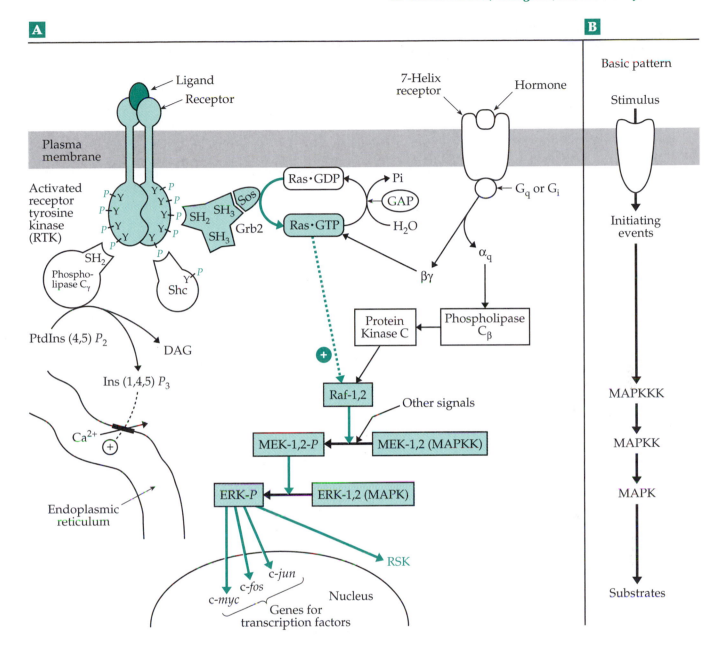

Figure 11-13 (A) A simplified version of the mitogen-activated kinase (MAPK) signaling cascade. At left is shown a hormone receptor, e.g., that for the epidermal growth factor (EGF). The receptor tyrosine kinase undergoes autophosphorylation on numerous tyrosines. The resulting phosphotyrosyl (Y-P) groups bind to SH2 domains of adapters such as Grb2 and Shc. Two pathways from the activated receptor are shown. At the left is activation of phospholipase Cγ and formation, at a membrane-bound site, of inositol trisphosphate and diacylglycerol (DAG). The main pathway, in the center, activates Ras with the aid of the G protein Sos. Activated Ras, in turn, activates Raf and successive components of the MAPK cascade. At the right a seven-helix receptor activates both phospholipase Cβ and Ras via interaction with a βγ subunit. (B) A generalized scheme for the MAP kinase pathway. See Seger and Krebs.[380]

replication cycle (or simply cell cycle) is commonly shown as a circle in which the time from one cell division to the next, in a rapidly growing organism or tissue, is represented by the circumference. The time required for DNA synthesis is the **S-phase** and the time required for mitosis the mitotic or **M-phase** (Fig. 11-15). After metaphase there is a **gap** in time denoted **G_1**. A second

gap **G_2** separates the synthetic S-phase and the M-phase. The total time required for one cycle varies with conditions. It may be as short as 8–60 min in an early embryo but is usually two hours or more. A slowly growing cell may pause before the G_1 phase in a nongrowing **G_0 phase.**[491,492]

What controls the cell replication cycle? Signals

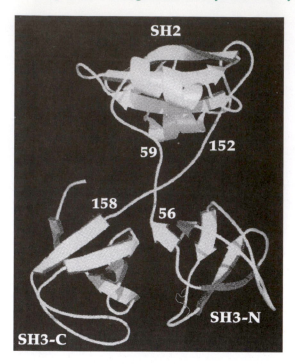

Figure 11-14 Ribbon drawing of the three-dimensional structure of adapter protein Grb2. The two SH3 domains at the N and C termini are labeled, as is the central SH2 domain. Produced with programs MolScript and Raster3D. From Maignan et al.[476] Courtesy of Arnaud Ducruix.

from growth factors can stimulate a cell to leave G_0 and enter G_1. However, to go further the cell depends upon a large number of proteins of which two types are prominent. The **cyclins** are labile 45- to 60-kDa proteins which are degraded by the ubiquitin system as part of the cycle. The cyclins associate with a series of protein Ser/Thr kinases known as **cyclin-dependent kinases** (**CDK**s). It is the cyclin–CDK complexes that signal the start of the next step in the cycle.[493–495]

The number of different cyclins and CDK enzymes needed varies with the organism, often being greater for more complex species. As a rule, there are at least two types of cyclins. The G_1 or **start cyclins** initiate the passage through the start (G_1 checkpoint) into the S-phase, while **mitotic cyclins** initiate the passage from the S-phase into the M-phase. Both vertebrates and *Drosophila* utilize at least four different types of cyclin (A, B, D, and E) and also four or more CDKs (CDK1, CDK2, CDK4, and CDK6).[492] A somewhat different set of these proteins are found in yeast.

Control of cell growth is directly related to the cell cycle. Mitogenic signals from growth factors act to initiate progression through G_1, apparently by stimulating transcription of D-type cyclins.[496] They may affect other steps as well. Oncogenic signals can arise from such oncogenes as *ras* and *abl*.[497,498] As pointed out in Box 11-D, the tumor-suppressor protein Rb becomes phosphorylated in synchrony with the cell

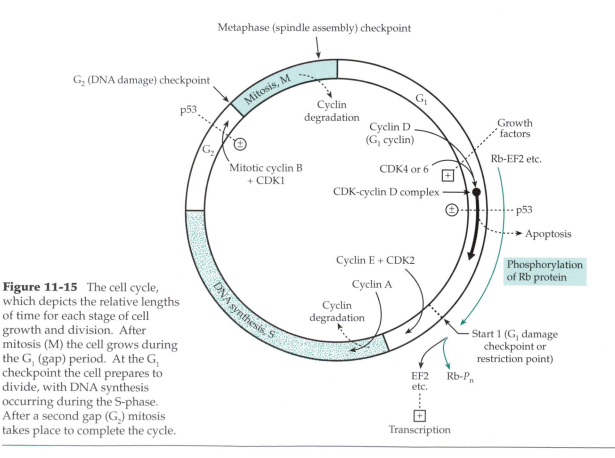

Figure 11-15 The cell cycle, which depicts the relative lengths of time for each stage of cell growth and division. After mitosis (M) the cell grows during the G_1 (gap) period. At the G_1 checkpoint the cell prepares to divide, with DNA synthesis occurring during the S-phase. After a second gap (G_2) mitosis takes place to complete the cycle.

cycle. This seems to be an essential process for passage through the G_1 "checkpoint." The checkpoints in the cycle should be viewed not as points but as essential interlocking processes that must be accomplished before transcription of genes essential for the next step in the cycle can take place.[499] A possibility is that the cyclin-dependent kinase in the cyclin D•CDK4 (or CDK6) complex phosphorylates the Rb protein which in its unphosphorylated state forms a complex with transcriptional regulators of the E2F family. Phosphorylation of Rb allows this complex to dissociate and permits E2F to induce transcription of genes encoding essential proteins for the next step in the cycle.[499a,b,c] Each CDK may also phosphorylate its associated cyclin, a modification that may be essential to the proteolysis of the cyclin and progression to the next stage of the cycle.

The powerful cancer suppressor, p53, which is also described in Box 11-D, in some way senses DNA damage. It prevents passage through the G_1 checkpoint and also through the G_2 checkpoint if the DNA has not been adequately repaired.[496,497] Protein p53, whose three-dimensional structure is known,[500,501] binds to DNA and also induces transcription of genes that cause arrest of the cell cycle. It may also induce cell death (apoptosis), a process that may also require the protein product of protooncogene c-*myc*.[502-505] A variety of protein kinases and phosphatases act on p53 and influence its activity.[506]

References

1. Krebs, H. A. (1957) *Endeavour* **16**, 125–132
2. Newsholme, E. A., and Start, C. (1973) *Regulation in Metabolism*, Wiley, New York
3. Heinrich, R., Rapoport, S. M., and Rapoport, T. A. (1977) *Prog. Biophys. and Mol. Biol.* **32**, 1–82
4. Martin, B. R. (1987) *Metabolic Regulation*, Blackwell Scientific Publ., Oxford
5. Chance, B., and Williams, G. R. (1955) *J. Biol. Chem.* **217**, 477–488
6. Heinrich, R., and Rapoport, T. A. (1974) *Eur. J. Biochem.* **42**, 89–95; 97–105
7. Mendes, P. (1997) *Trends Biochem. Sci.* **22**, 361–363
8. Wright, B. E. (1986) *Trends Biochem. Sci.* **11**, 164–165
9. Savageau, M. A. (1987) *Trends Biochem. Sci.* **12**, 219–220
10. Voit, E. O., and Savageau, M. A. (1987) *Biochemistry* **26**, 6869–6880
11. Hlavacek, W. S., and Savageau, M. A. (1996) *J. Mol. Biol.* **255**, 121–139
12. Voit, E. O. (1987) *Trends Biochem. Sci.* **12**, 221
13. Kacser, H., and Burns, J. A. (1979) *Biochem. Soc. Trans.* **7**, 1149–1166
14. Crabtree, B., and Newsholme, E. A. (1987) *Trends Biochem. Sci.* **12**, 4–12
15. Kholodenko, B. N., and Westerhoff, H. V. (1995) *Trends Biochem. Sci.* **20**, 52–54
16. Fell, D. (1997) *Understanding the Control of Metabolism*, Portland Press, London
17. Quant, P. A. (1993) *Trends Biochem. Sci.* **18**, 26–30
18. Salter, M., Knowles, R. G., and Pogson, C. I. (1994) in *Essays in Biochemistry*, Vol. 28 (Tipton, K. F., ed), Portland Press, London
19. Schulze, E.-D. (1994) *Flux Control in Biological Systems*, Academic Press, San Diego, California
20. Kacser, H., and Burns, J. A. (1973) *Symp. Soc. Exp. Biol* **XXVII**, 65–104
21. Stark, G. R., and Wahl, G. M. (1984) *Ann. Rev. Biochem.* **53**, 447–491
22. Wilkinson, J. H. (1970) *Isoenzymes*, 2nd ed., Lippincott, Philadelphia, Pennsylvania
23. Markert, C. L., ed. (1975) *Isozymes*, Vol. 4, Academic Press, New York
24. Christen, P., and Metzler, D. E., eds. (1985) *Transaminases*, Wiley, New York
25. Salerno, C., Giartosio, A., and Fasella, P. (1986) in *Vitamin B$_6$ Pyridoxal Phosphate*, Vol. 1b (Dolphin, D., Poulson, R., and Avramovic, O., eds), Wiley, New York
26. McPhalen, C. A., Vincent, M. G., and Jansonius, J. N. (1992) *J. Mol. Biol.* **225**, 495–517

27. Mehta, P. K., Hale, T. I., and Christen, P. (1989) *Eur. J. Biochem.* **186**, 249–253
28. Kamitori, S., Okamoto, A., Hirotsu, K., Higuchi, T., Kuramitsu, S., Kagamiyama, H., Matsuura, Y., and Katsube, Y. (1990) *J. Biochem.* **108**, 175–184
29. Hogrefe, H. H., Griffith, J. P., Rossman, M. G., and Goldberg, E. (1987) *J. Biol. Chem.* **262**, 13155–13162
30. Purich, D. L., and Fromm, H. J. (1972) *Curr. Top. Cell. Regul.* **6**, 131–167
31. Huijing, F. (1979) *Trends Biochem. Sci.* **4**, N132–N134
31a. Aleshin, A. E., Kirby, C., Liu, X., Bourenkov, G. P., Bartunik, H. D., Fromm, H. J., and Honzatko, R. B. (2000) *J. Mol. Biol.* **296**, 1001–1015
32. Purich, D. L., Fromm, H. J., and Rudolph, F. R. (1973) *Adv. Enzymol.* **39**, 249–326
33. Breitbar, R. E., Andreadis, A., and Nadal-Ginard, B. (1987) *Ann. Rev. Biochem.* **56**, 467–495
34. Joh, K., Arai, Y., Mukai, T., and Hori, K. (1986) *J. Mol. Biol.* **190**, 401–410
35. Haldane, J. B. S. (1930) *Enzymes*, Longmans, Green, New York
36. Hayashi, S.-i, Murakami, Y., and Matsufuji, S. (1996) *Trends Biochem. Sci.* **21**, 27–30
37. Li, X., Stebbins, B., Hoffman, L., Pratt, G., Rechsteiner, M., and Coffino, P. (1996) *J. Biol. Chem.* **271**, 4441–4446
38. Sprang, S. R., Acharya, K. R., Goldsmith, E. J., Stuart, D. I., Varvill, K., Fletterick, R. J., Madsen, N. B., and Johnson, L. N. (1988) *Nature (London)* **336**, 215–221
39. Cohen, P. (1983) *Control of Enzyme Activity*, 2nd ed., Chapman and Hall, London
40. Kantrowitz, E. R., and Lipscomb, W. N. (1990) *Trends Biochem. Sci.* **15**, 53–59
41. Zhou, B.-B., and Schachman, H. K. (1993) *Protein Sci.* **2**, 103–112
42. Xi, X. G., De Staercke, C., Van Vliet, F., Triniolles, F., Jacobs, A., Stas, P. P., Ladjimi, M. M., Simon, V., Cunin, R., and Hervé, G. (1994) *J. Mol. Biol.* **242**, 139–149
43. Uyeda, K., Furuya, E., and Sherry, A. D. (1981) *J. Biol. Chem.* **256**, 8679–8684
44. Hers, H., Hue, L., and Schaftingen, E. (1982) *Trends Biochem. Sci.* **7**, 329–331
45. Lively, M. O., El-Maghrabi, M. R., Pilkis, J., D'Angelo, G., Colosia, A. D., Ciavola, J.-A., Fraser, B. A., and Pilkis, S. J. (1988) *J. Biol. Chem.* **263**, 839–849
46. Kitamura, K., and Uyeda, K. (1988) *J. Biol. Chem.* **263**, 9027–9033

46a. Choe, J.-Y., Fromm, H. J., and Honzatko, R. B. (2000) *Biochemistry* **39**, 8565–8574
47. Graves, D. J., Martin, B. L., and Wang, J. H. (1994) *Co – and Post – Translational Modification of Proteins*, Oxford Univ. Press, New York
47a. Skamnaki, V. T., Owen, D. J., Noble, M. E. M., Lowe, E. D., Lowe, G., Oikonomakos, N. G., and Johnson, L. N. (1999) *Biochemistry* **38**, 14718–14730
48. Johnson, L. N. (1992) *FASEB J.* **6**, 2274–2282
48a. Cheng, A., Fitzgerald, T. J., Bhatnagar, D., Roskoski, R., Jr., and Carlson, G. M. (1988) *J. Biol. Chem.* **263**, 5534–5542
49. Dent, P., Lavoinne, A., Nakielny, S., Caudwell, F. B., Watt, P., and Cohen, P. (1990) *Nature (London)* **348**, 302–308
49a. Brady, M. J., Bourbonais, F. J., and Saltiel, A. R. (1998) *J. Biol. Chem.* **273**, 14063–14066
50. Krebs, E. G., and Fischer, E. H. (1956) *Biochim. Biophys. Acta.* **20**, 150–157
51. Edelman, A. M., Blumenthal, D. K., and Krebs, E. G. (1987) *Ann. Rev. Biochem.* **56**, 567–613
52. Allen, J. F. (1992) *Trends Biochem. Sci.* **17**, 12–17
53. Hershey, J. W. B. (1989) *J. Biol. Chem.* **264**, 20823–20826
54. Kennelly, P. J., and Krebs, E. G. (1991) *J. Biol. Chem.* **266**, 15555–15558
55. Hanks, S. K., and Hunter, T. (1995) *FASEB J.* **9**, 576–596
56. Woodgett, J. R., ed. (1994) *Protein Kinases*, IRL Press, Oxford
57. Hunter, T. (1994) *Sem. Cell Biol.* **5**, 367–376
58. Kemp, B. E., Parker, M. W., Hu, S., Tiganis, T., and House, C. (1994) *Trends Biochem. Sci.* **19**, 440–444
59. Imazu, M., Stricklund, W. G., Chrisman, T. D., and Exton, J. H. (1984) *J. Biol. Chem.* **259**, 1813–1821
60. He, X., Saint-Jeannet, J.-P., Woodgett, J. R., Varmus, H. E., and David, I. B. (1995) *Nature (London)* **374**, 617–622
61. Hartl, F. T., and Roskoski, R., Jr. (1983) *J. Biol. Chem.* **258**, 3950–3955
62. Zheng, J., Knighton, D. R., Xuong, N.-H., Taylor, S. S., Sowadski, J. M., and Eyck, L. F. T. (1993) *Protein Sci.* **2**, 1559–1573
63. Beebe, S. J., Reimann, E. M., and Schlender, K. K. (1984) *J. Biol. Chem.* **259**, 1415–1422
64. Taylor, S. S., Knighton, D. R., Zheng, J., Sowadski, J. M., Gibbs, C. S., and Zoller, M. J. (1993) *Trends Biochem. Sci.* **18**, 84–89
65. Narayana, N., Cox, S., Shaltiel, S., Taylor, S. S., and Xuong, N.-h. (1997) *Biochemistry* **36**, 4438–4448

References

66. Su, Y., Dostmann, W. R. G., Herberg, F. W., Durick, K., Xuong, N.-h, Eyck, L. T., Taylor, S. S., and Varughese, K. I. (1995) *Science* **269**, 807–813

67. Lincoln, T. M., Thompson, M., and Cornwell, T. L. (1988) *J. Biol. Chem.* **263**, 17632–17637

68. Weber, I. T., Shabb, J. B., and Corbin, J. D. (1989) *Biochemistry* **28**, 6122–6127

69. Fry, M. J., Panayotou, G., Booker, G. W., and Waterfield, M. D. (1993) *Protein Sci.* **2**, 1785–1797

70. Xu, R.-M., Carmel, G., Sweet, R. M., Kuret, J., and Cheng, X. (1995) *EMBO J.* **14**, 1015–1023

71. Allende, J. E., and Allende, C. C. (1995) *FASEB J.* **9**, 313–323

72. Seldin, D. C., and Leder, P. (1995) *Science* **267**, 894–897

72a. Niefind, K., Guerra, B., Pinna, L. A., Issinger, O.-G., and Schomburg, D. (1998) *EMBO J.* **17**, 2451–2462

73. Chan, K.- F., and Graves, D. J. (1984) in *Calcium and Cell Function*, Vol. 5 (Cheung, W. Y., ed), pp. 1–32, Academic Press, New York

74. Heilmeyer, L. M. G., Jr. (1991) *Biochim. Biophys. Acta.* **1094**, 168–174

75. Norcum, M. T., Wilkinson, D. A., Carlson, M. C., Hainfeld, J. F., and Carlson, G. M. (1994) *J. Mol. Biol.* **241**, 94–102

76. Kee, S. M., and Graves, D. J. (1987) *J. Biol. Chem.* **262**, 9448–9453

77. Paudel, H. K., and Carlson, G. M. (1987) *J. Biol. Chem.* **262**, 11912–11915

78. Cheng, A., Fitzgerald, T. J., Bhatnagar, D., Roskoski, R., Jr., and Carlson, G. M. (1988) *J. Biol. Chem.* **263**, 5534–5542

79. Owen, D. J., Papageorgiou, A. C., Garman, E. F., Noble, M. E. M., and Johnson, L. N. (1995) *J. Mol. Biol.* **246**, 374–381

80. Ohmstede, C.-A., Bland, M. M., Merrill, B. M., and Sahyoun, N. (1991) *Proc. Natl. Acad. Sci. U.S.A.* **88**, 5784–5788

81. Hanson, P. I., and Schulman, H. (1992) *Ann. Rev. Biochem.* **61**, 559–601

82. Harper, J. F., Sussman, M. R., Schaller, G. E., Putnam-Evans, C., Charbonneau, H., and Harmon, A. C. (1991) *Science* **252**, 951–954

83. Ullrich, A., Bell, J. R., Chen, E. Y., Herrera, R., Petruzzelli, L. M., Dull, T. J., Gray, A., Coussens, L., Liao, Y.-C., Tsubokawa, M., Mason, A., Seeburg, P. H., Grunfield, C., Roden, O. M., and Ramachandran, J. (1985) *Nature (London)* **313**, 756–761

84. Wang, J. Y. J. (1994) *Trends Biochem. Sci.* **19**, 373–376

85. Clark, E. A., Shattil, S. J., and Brugge, J. S. (1994) *Trends Biochem. Sci.* **19**, 464–469

86. Sun, H., and Tonks, N. K. (1994) *Trends Biochem. Sci.* **19**, 480–485

87. Ingebritsen, T. S., and Cohen, P. (1983) *Science* **221**, 331–338

88. Cohen, P. (1989) *Ann. Rev. Biochem.* **58**, 453–508

89. Goldberg, J., Huang, H., Kwon, Y., Greengard, P., Nairn, A. C., and Kuriyan, J. (1995) *Nature (London)* **376**, 745–753

90. Fischer, E. H., Charbonneau, H., and Tonks, N. K. (1991) *Science* **253**, 401–406

91. Yuvaniyama, J., Denu, J. M., Dixon, J. E., and Saper, M. A. (1996) *Science* **272**, 1328–1331

92. MacKintosh, C., and MacKintosh, R. W. (1994) *Trends Biochem. Sci.* **19**, 444–448

93. Hidaka, H., and Kobayashi, R. (1994) in *Essays in Biochemistry*, Vol. 28 (Tipton, K. F., ed), pp. 73–98, Portland Press, London

94. Cohen, P., Holmes, C. F. B., and Tsukitani, Y. (1990) *Trends Biochem. Sci.* **15**, 98–102

95. Price, N. E., and Mumby, M. C. (2000) *Biochemistry* **39**, 11312–11318

96. Zhang, M., Van Etten, R. L., and Stauffacher, C. V. (1994) *Biochemistry* **33**, 11097–11105

97. Barford, D., Flint, A. J., and Tonks, N. K. (1994) *Science* **263**, 1397–1403

98. Stone, R. L., and Dixon, J. E. (1994) *J. Biol. Chem.* **269**, 31323–31326

99. Chen, W., Wilborn, M., and Rudolph, J. (2000) *Biochemistry* **39**, 10781–10789

100. Kissinger, C. R., Parge, H. E., Knighton, D. R., Lewis, C. T., Pelletier, L. A., Tempczyk, A., Kalish, V. J., Tucker, K. D., Showalter, R. E., Moomaw, E. W., Gastinel, L. N., Habuka, N., Chen, X., Maldonado, F., Barker, J. E., Bacquet, R., and Villafranca, J. E. (1995) *Nature (London)* **378**, 641–644

101. LaPorte, D. C., and Chung, T. (1985) *J. Biol. Chem.* **260**, 15291–15297

102. Pilkis, S. J., Claus, T. H., Kurland, I. J., and Lange, A. J. (1995) *Ann. Rev. Biochem.* **64**, 799–835

103. Istvan, E. S., Hasemann, C. A., Kurumbail, R. G., Uyeda, K., and Deisenhofer, J. (1995) *Protein Sci.* **4**, 2439–2441

104. Vertommen, D., Bertrand, L., Sontag, B., Di Pietro, A., Louckx, M. P., Vidal, H., Hue, L., and Rider, M. H. (1996) *J. Biol. Chem.* **271**, 17875–17880

105. Abe, Y., Minami, Y., Li, Y., Nguyen, C., and Uyeda, K. (1995) *Biochemistry* **34**, 2553–2559

106. Ogushi, S., Lawson, J. W. R., Dobson, G. P., Veech, R. L., and Uyeda, K. (1990) *J. Biol. Chem.* **265**, 10943–10949

107. Nimmo, H. G. (1984) *Trends Biochem. Sci.* **9**, 475–478

108. Dean, A. M., and Koshland, D. E., Jr. (1993) *Biochemistry* **32**, 9302–9309

109. Dean, A. M., and Koshland, D. E., Jr. (1990) *Science* **249**, 1044–1046

110. Saier, M. H., Jr., Wu, L.-F., and Reizer, J. (1990) *Trends Biochem. Sci.* **15**, 391–395

110a. Song, H. K., Lee, J. Y., Lee, M. G., Moon, J., Min, K., Yang, J. K., and Suh, S. W. (1999) *J. Mol. Biol.* **293**, 753–761

111. Swanson, R. V., Alex, L. A., and Simon, M. I. (1994) *Trends Biochem. Sci.* **19**, 485–490

112. Saier, M. H., Jr. (1989) *Microbiol. Rev.* **53**, 109–120

113. Stock, J. B., Stock, A. M., and Mottonen, J. M. (1990) *Nature (London)* **344**, 395–400

114. Boon, P., Chock, P. B., Rhee, S. G., and Stadtman, E. R. (1980) *Ann. Rev. Biochem.* **49**, 813–843

115. Mura, U., Chock, P. B., and Stadtman, E. R. (1981) *J. Biol. Chem.* **256**, 13022–13029

116. Almassy, R. J., Janson, C. A., Hamlin, R., Xuong, N.-H., and Eisenberg, D. (1986) *Nature (London)* **323**, 304–309

117. Garcia, E., and Rhee, S. G. (1983) *J. Biol. Chem.* **258**, 2246–2253

118. Moss, J., and Vaughan, M. (1995) *J. Biol. Chem.* **270**, 12327–12330

119. Amor, J. C., Harrison, D. H., Kahn, R. A., and Ringe, D. (1994) *Nature (London)* **372**, 704–708

120. Boman, A. L., and Kahn, R. A. (1995) *Trends Biochem. Sci.* **20**, 147–150

121. Lee, F.-J. S., Stevens, L. A., Hall, L. M., Murtagh, J. J., Jr., Kao, Y. L., Moss, J., and Vaughan, M. (1994) *J. Biol. Chem.* **269**, 21555–21560

122. Moss, J., Stanley, S. J., Nightingale, M. S., Murtagh, J. J., Jr., Monaco, L., Mishima, K., Chen, H.-C., Williamson, K.-C., and Tsai, S.-C. (1992) *J. Biol. Chem.* **267**, 10481–10488

123. Guse, A. H., da Silva, C. P., Weber, K., Ashamu, G. A., Potter, B. V. L., and Mayr, G. W. (1996) *J. Biol. Chem.* **271**, 23946–23953

124. Gaal, J. C., Smith, K. R., and Pearson, C. K. (1987) *Trends Biochem. Sci.* **12**, 129–130

125. Huttner, W. B. (1987) *Trends Biochem. Sci.* **12**, 361–363

126. Sakakibara, Y., Takami, Y., Zwieb, C., Nakayama, T., Suiko, M., Nakajima, H., and Liu, M.-C. (1995) *J. Biol. Chem.* **270**, 30470–30478

127. Rosenquist, G. L., and Nicholas, H. B., Jr. (1993) *Protein Sci.* **2**, 215–222

128. Sundaram, K. S., and Lev, M. (1992) *J. Biol. Chem.* **267**, 24041–24044

129. Hooper, L. V., Manzella, S. M., and Baenziger, J. U. (1996) *FASEB J.* **10**, 1137–1146

130. Terwilliger, T. C., and Koshland, D. E., Jr. (1984) *J. Biol. Chem.* **259**, 7719–7725

131. Favre, B., Zolnierowicz, S., Turowski, P., and Hemmings, B. A. (1994) *J. Biol. Chem.* **269**, 16311–16317

132. Edmonson, D. G., and Roth, S. Y. (1996) *FASEB J.* **10**, 1173–1182

133. Kuo, M.-H., Brownell, J. E., Sobel, R. E., Ranalli, T. A., Cook, R. G., Edmonson, D. G., Roth, S. Y., and Allis, C. D. (1996) *Nature (London)* **383**, 269–272

134. Brandes, H. K., Larimer, F. W., and Hartman, F. C. (1996) *J. Biol. Chem.* **271**, 3333–3335

135. Brandes, H. K., Hartman, F. C., Lu, T.-Y. S., and Larimer, F. W. (1996) *J. Biol. Chem.* **271**, 6490–6496

136. Swerdlow, R. D., Green, C. L., Setlow, B., and Setlow, D. (1979) *J. Biol. Chem.* **254**, 6835–6837

137. Hentze, M. W., Rouault, T. A., Harford, J. B., and Klausner, R. D. (1989) *Science* **244**, 357–359

138. Wadsworth, W. G., and Riddle, D. L. (1988) *Proc. Natl. Acad. Sci. U.S.A.* **85**, 8435–8438

139. Boron, W. F. (1984) *Nature (London)* **312**, 312

140. Siffert, W., and Akkerman, J. W. N. (1988) *Trends Biochem. Sci.* **13**, 148–151

141. Reinertsen, K. V., Tonnessen, T. I., Jacobsen, J., Sandvig, K., and Olsnes, S. (1988) *J. Biol. Chem.* **263**, 11117–11125

142. Schramm, V. L. (1982) *Trends Biochem. Sci.* **7**, 369–371

143. Srere, P. A. (1987) *Ann. Rev. Biochem.* **56**, 89–124

144. Srivastava, D. K., and Bernhard, S. A. (1987) *Ann. Rev. Biophys. Biophys. Chem.* **16**, 175–204

145. Ovádi, J. (1988) *Trends Biochem. Sci.* **13**, 486–490

146. Kaprelyants, A. S. (1988) *Trends Biochem. Sci.* **13**, 43–46

146a. Miles, E. W., Rhee, S., and Davies, D. R. (1999) *J. Biol. Chem.* **274**, 12193–12196

146b. Serre, V., Guy, H., Penverne, B., Lux, M., Rotgeri, A., Evans, D., and Hervé, G. (1999) *J. Biol. Chem.* **274**, 23794–23801

146c. Geck, M. K., and Kirsch, J. F. (1999) *Biochemistry* **38**, 8032–8037

146d. Dell'Acqua, M. L., and Scott, J. D. (1997) *J. Biol. Chem.* **272**, 12881–12884

147. Ovádi, J., and Srere, P. A. (1992) *Trends Biochem. Sci.* **17**, 445–447

148. Inagaki, N., Ito, M., Nakano, T., and Inagaki, M. (1994) *Trends Biochem. Sci.* **19**, 448–452

149. Faux, M. C., and Scott, J. D. (1996) *Trends Biochem. Sci.* **21**, 312–315

150. Fraser, C. M. (1989) *J. Biol. Chem.* **264**, 9266–9270

151. Strosberg, A. D. (1993) *Protein Sci.* **2**, 1198–1209

152. Scheel, A. A., and Pelham, H. R. B. (1996) *Biochemistry* **35**, 10203–10209

153. Levitzki, A. (1988) *Science* **241**, 800–806

153a. Stacey, M., Lin, H.-H., Gordon, S., and McKnight, A. J. (2000) *Trends Biochem. Sci.* **25**, 284–289

153b. Leurs, R., Smit, M. J., Alewijnse, A. E., and Timmerman, H. (1998) *Trends Biochem. Sci.* **23**, 418–422

153c. Bockaert, J., and Pin, J. P. (1999) *EMBO J.* **18**, 1723–1729

154. Insel, P. A., and Wasserman, S. I. (1990) *FASEB J.* **4**, 2732–2736

155. Turki, J., Lorenz, J. N., Green, S. A., Donnelly, E. T., Jacinto, M., and Liggett, S. B. (1996) *Proc. Natl. Acad. Sci. U.S.A.* **93**, 10483–10488

References

155a. Rohrer, D. K., Chruscinski, A., Schauble, E. H., Bernstein, D., and Kobilka, B. K. (1999) *J. Biol. Chem.* **274**, 16701–16708

155b. Limbird, L. E., and Vaughan, D. E. (1999) *Proc. Natl. Acad. Sci. U.S.A.* **96**, 7125–7127

156. Premont, R. T., Inglese, J., and Lefkowitz, R. J. (1995) *FASEB J.* **9**, 175–182

156a. Pitcher, J. A., Freedman, N. J., and Lefkowitz, R. J. (1998) *Ann. Rev. Biochem.* **67**, 653–692

157. Mochly-Rosen, D. (1995) *Science* **268**, 247–251

158. Benovic, J. L., DeBlasi, A., Stone, W. C., Caron, M. G., and Lefkowitz, R. J. (1989) *Science* **246**, 235–240

158a. Danner, S., Frank, M., and Lohse, M. J. (1998) *J. Biol. Chem.* **273**, 3223–3229

158b. Gabilondo, A. M., Hegler, J., Krasel, C., Boivin-Jahns, V., Hein, L., and Lohse, M. J. (1997) *Proc. Natl. Acad. Sci. U.S.A.* **94**, 12285–12290

159. Rodbell, M., Birnbaumer, L., Pohl, S. L., and Krans, H. M. J. (1971) *J. Biol. Chem.* **246**, 1877–1882

160. Unson, C. G., Cypess, A. M., Wu, C.-R., Goldsmith, P. K., Merrifield, R. B., and Sakmar, T. P. (1996) *Proc. Natl. Acad. Sci. U.S.A.* **93**, 310–315

161. Rosenthal, W., Antaramian, A., Gilbert, S., and Birnbaumer, M. (1993) *J. Biol. Chem.* **268**, 13030–13033

162. Sharif, M., and Hanley, M. R. (1992) *Nature (London)* **357**, 279–280

163. McFarland, K. C., Sprengel, R., Phillips, H. S., Köhler, M., Rosemblit, N., Nikolics, K., Segaloff, D. L., and Seeburg, P. H. (1989) *Science* **245**, 494–499

164. Kudo, M., Osuga, Y., Kobilka, B. K., and Hsueh, A. J. W. (1996) *J. Biol. Chem.* **271**, 22470–22478

165. Perlman, J. H., Laakkonen, L., Osman, R., and Gershengorn, M. C. (1994) *J. Biol. Chem.* **269**, 23383–23386

166. Conti, M., Iona, S., Cuomo, M., Swinnen, J. V., Odeh, J., and Svoboda, M. E. (1995) *Biochemistry* **34**, 7979–7987

167. Smith, K. J., Scotland, G., Beattie, J., Trayer, I. P., and Houslay, M. D. (1996) *J. Biol. Chem.* **271**, 16703–16711

168. Sutherland, E. W. (1972) *Science* **177**, 401–408

169. Pastan, I. (1972) *Sci. Am.* **227**(Aug), 97–105

170. Lalli, E., and Sassone-Corsi, P. (1994) *J. Biol. Chem.* **269**, 17359–17362

171. Bolwell, G. P. (1995) *Trends Biochem. Sci.* **20**, 492–495

172. Taussig, R., and Gilman, A. G. (1995) *J. Biol. Chem.* **270**, 1–4

173. Dessauer, C. W., Tesmer, J. J. G., Sprang, S. R., and Gilman, A. G. (1998) *J. Biol. Chem.* **273**, 25831–25839

174. Cooper, D. M. F., Mons, N., and Karpen, J. W. (1995) *Nature (London)* **374**, 421–424

175. Désaubry, L., Shoshani, I., and Johnson, R. A. (1996) *J. Biol. Chem.* **271**, 14028–14034

176. Hurley, J. H. (1999) *J. Biol. Chem.* **274**, 7599–7602

177. Hellevuo, K., Yoshimura, M., Mons, N., Hoffman, P. L., Cooper, D. M. F., and Tabakoff, B. (1995) *J. Biol. Chem.* **270**, 11581–11589

178. Walseth, T. F., Gander, J. E., Eide, S. J., Krick, T. P., and Goldberg, N. D. (1983) *J. Biol. Chem.* **258**, 1544–1558

178a. Chen, Y., Cann, M. J., Litvin, T. N., Lourgenko, V., Sinclair, M. L., Levin, L. R., and Buck, J. (2000) *Science* **289**, 625–628

178b. Kaupp, U. B., and Weyand, I. (2000) *Science* **289**, 559–560

179. Hamm, H. E. (1998) *J. Biol. Chem.* **273**, 669–672

180. Neer, E. J. (1994) *Protein Sci.* **3**, 3–14

181. Collins, S., Caron, M. G., and Lefkowitz, R. J. (1992) *Trends Biochem. Sci.* **17**, 37–39

182. Hepler, J. R., and Gilman, A. G. (1992) *Trends Biochem. Sci.* **17**, 383–387

183. Levitzki, A. (1988) *Trends Biochem. Sci.* **13**, 298–301

184. Rawls, R. L. (1987) *Chem. Eng. News* **65** **December 21**, 26–39

185. Wang, N., Yan, K., and Rasenick, M. M. (1990) *J. Biol. Chem.* **265**, 1239–1242

186. Wang, Q., Mullah, B. K., and Robishaw, J. D. (1999) *J. Biol. Chem.* **274**, 17365–17371

187. Mayorga, L. S., Diaz, R., and Stahl, P. D. (1989) *Science* **244**, 1475–1477

188. Coleman, D. E., and Sprang, S. R. (1996) *Trends Biochem. Sci.* **21**, 41–44

189. Ray, K., Kunsch, C., Bonner, L. M., and Robishaw, J. D. (1995) *J. Biol. Chem.* **270**, 21765–21771

190. Brown, A. M. (1991) *FASEB J.* **5**, 2175–2179

191. Chen, J., DeVivo, M., Dingus, J., Harry, A., Li, J., Sui, J., Carty, D. J., Blank, J. L., Exton, J. H., Stoffel, R. H., Inglese, J., Lefkowitz, R. J., Logothetis, D. E., Hildebrandt, J. D., and Iyengar, R. (1995) *Science* **268**, 1166–1169

192. Schulz, K., Danner, S., Bauer, P., Schröder, S., and Lohse, M. J. (1996) *J. Biol. Chem.* **271**, 22546–22551

193. Yoshida, T., Willardson, B. M., Wilkins, J. F., Jensen, G. J., Thornton, B. D., and Bitensky, M. W. (1994) *J. Biol. Chem.* **269**, 24050–24057

193a. Whiteway, M., Hougan, L., Dignard, D., Thomas, D. Y., Bell, L., Saari, G. C., Grant, F. J., O'Hara, P., and MacKay, V. L. (1989) *Cell* **56**, 467–477

194. Noel, J. P., Hamm, H. E., and Sigler, P. B. (1993) *Nature (London)* **366**, 654–663

195. Lambright, D. G., Sondek, J., Bohm, A., Skiba, N. P., Hamm, H. E., and Sigler, P. B. (1996) *Nature (London)* **379**, 311–319

196. McLaughlin, S. K., McKinnon, P. J., and Margolskee, R. F. (1992) *Nature (London)* **357**, 563–569

197. Hall, A. (1990) *Science* **249**, 635–640

198. Valencia, A., Chardin, P., Wittinghofer, A., and Sander, C. (1991) *Biochemistry* **30**, 4637–4648

199. Macara, I. G., Lounsbury, K. M., Richards, S. A., McKiernan, C., and Bar-Sagi, D. (1996) *FASEB J.* **10**, 625–630

200. Boguski, M. S., and McCormick, F. (1993) *Nature (London)* **366**, 643–654

200a. Scita, G., Tenca, P., Frittoli, E., Tocchetti, A., Innocenti, M., Giardina, G., and Di Fiore, P. P. (2000) *EMBO J.* **19**, 2393–2398

201. Jurnak, F. (1985) *Science* **230**, 32–36

202. Sprinzl, M. (1994) *Trends Biochem. Sci.* **19**, 245–250

203. Berchtold, H., Reshetnikova, L., Reiser, C. O. A., Schirmer, N. K., Sprinzl, M., and Hilgenfeld, R. (1993) *Nature (London)* **365**, 126–132

204. Kawashima, T., Berthet-Colominas, C., Wulff, M., Cusack, S., and Leberman, R. (1996) *Nature (London)* **379**, 511–518

205. Jurnak, F. (1988) *Trends Biochem. Sci.* **13**, 195–198

206. Downward, J. (1990) *Trends Biochem. Sci.* **15**, 469–472

207. Wittinghofer, A., and Pai, E. F. (1991) *Trends Biochem. Sci.* **16**, 382–387

208. Schlichting, I., Almo, S. C., Rapp, G., Wilson, K., Petratos, K., Lentfer, A., Wittinghofer, A., Kabsch, W., Pai, E. F., Petsko, G. A., and Goody, R. S. (1990) *Nature (London)* **345**, 309–315

209. Kraulis, P. J., Domaille, P. J., Campbell-Burk, S. L., Van Aken, T., and Laue, E. D. (1994) *Biochemistry* **33**, 3515–3531

210. Bokoch, G. M., and Der, C. J. (1993) *FASEB J.* **7**, 750–759

211. Takai, Y., Sasaki, T., Tanaka, K., and Nakanishi, H. (1995) *Trends Biochem. Sci.* **20**, 227–231

212. Fischer von Mollard, G., Stahl, B., Li, C., Südhof, T. C., and Jahn, R. (1994) *Trends Biochem. Sci.* **19**, 164–168

212a. Burridge, K. (1999) *Science* **283**, 2028–2029

212b. Lin, R., Cerione, R. A., and Manor, D. (1999) *J. Biol. Chem.* **274**, 23633–23641

212c. Desrosiers, R. R., Gauthier, F., Lanthier, J., and Béliveau, R. (2000) *J. Biol. Chem.* **275**, 14949–14957

212d. Schwartz, M. A., and Shattil, S. J. (2000) *Trends Biochem. Sci.* **25**, 388–391

212e. Zhang, B., Zhang, Y., Wang, Z.-x, and Zheng, Y. (2000) *J. Biol. Chem.* **275**, 25299–25307

213. Symons, M. (1996) *Trends Biochem. Sci.* **21**, 178–181

214. Ando, S., Kaibuchi, K., Sasaki, T., Hiraoka, K., Nishiyama, T., Mizuno, T., Asada, M., Nunoi, H., Matsuda, I., Matsuura, Y., Polakis, P., McCormick, F., and Takai, Y. (1992) *J. Biol. Chem.* **267**, 25709–25713

215. Nuoffer, C., and Balch, W. E. (1994) *Ann. Rev. Biochem.* **63**, 949–990

216. Rybin, V., Ullrich, O., Rubino, M., Alexandrov, K., Simon, I., Seabra, M. C., Goody, R., and Zerial, M. (1996) *Nature (London)* **383**, 266–269

217. Garrett, M. D., Self, A. J., van Oers, C., and Hall, A. (1989) *J. Biol. Chem.* **264**, 10–13

217a. Moss, J., and Vaughan, M. (1998) *J. Biol. Chem.* **273**, 21431–21434

217b. Hutchinson, J. P., and Eccleston, J. F. (2000) *Biochemistry* **39**, 11348–11359

218. Scheffzek, K., Ahmadian, M. R., and Wittinghofer, A. (1998) *Trends Biochem. Sci.* **23**, 257–262

218a. Berman, D. M., and Gilman, A. G. (1998) *J. Biol. Chem.* **273**, 1269–1272

218b. Coleman, D. E., and Sprang, S. R. (1999) *J. Biol. Chem.* **274**, 16669–16672

219. Schalk, I., Zeng, K., Wu, S.-K., Stura, E. A., Matteson, J., Huang, M., Tandon, A., Wilson, I. A., and Balch, W. E. (1996) *Nature (London)* **381**, 42–48

220. Buss, J. E., Mumby, S. M., Casey, P. J., Gilman, A. G., and Sefton, B. M. (1987) *Proc. Natl. Acad. Sci. U.S.A.* **84**, 7493–7497

221. Wedegaertner, P. B., Wilson, P. T., and Bourne, H. R. (1995) *J. Biol. Chem.* **270**, 503–506

222. Song, J., Hirschman, J., Gunn, K., and Dohlman, H. G. (1996) *J. Biol. Chem.* **271**, 20273–20283

223. Hepler, J. R., Biddlecome, G. H., Kleuss, C., Camp, L. A., Hofmann, S. L., Ross, E. M., and Gilman, A. G. (1996) *J. Biol. Chem.* **271**, 496–504

224. Peseckis, S. M., and Resh, M. D. (1994) *J. Biol. Chem.* **269**, 30888–30892

225. Giner, J.-L., and Rando, R. R. (1994) *Biochemistry* **33**, 15116–15123

226. Liu, L., Dudler, T., and Gelb, M. H. (1996) *J. Biol. Chem.* **271**, 23269–23276

227. Casey, P. J. (1995) *Science* **268**, 221–225

228. Maltese, W. A., Sheridan, K. M., Repko, E. M., and Erdman, R. A. (1990) *J. Biol. Chem.* **265**, 2148–2155

229. Milligan, G., Parenti, M., and Magee, A. I. (1995) *Trends Biochem. Sci.* **20**, 181–186

230. Duncan, J. A., and Gilman, A. G. (1996) *J. Biol. Chem.* **271**, 23594–23600

230a. Iiri, T., Backlund, P. S., Jr., Jones, T. L. Z., Wedegaertner, P. B., and Bourne, H. R. (1996) *Proc. Natl. Acad. Sci. U.S.A.* **93**, 14592–14597

230b. Lee, T. W., Seifert, R., Guan, X., and Kobilka, B. K. (1999) *Biochemistry* **38**, 13801–13809

231. Lambright, D. G., Noel, J. P., Hamm, H. E., and Sigler, P. B. (1994) *Nature (London)* **369**, 621–628

232. Sondek, J., Lambright, D. G., Noel, J. P., Hamm, H. E., and Sigler, P. B. (1994) *Nature (London)* **372**, 276–279

References

233. Coleman, D. E., Berghuis, A. M., Lee, E., Linder, M. E., Gilman, A. G., and Sprang, S. R. (1994) *Science* **265**, 1405–1412

234. Mixon, M. B., Lee, E., Coleman, D. E., Berghuis, A. M., Gilman, A. G., and Sprang, S. R. (1995) *Science* **270**, 954–960

235. Kjeldgaard, M., Nyborg, J., and Clark, B. F. C. (1996) *FASEB J.* **10**, 1347–1368

236. Higashijima, T., Graziano, M. P., Suga, H., Kainosho, M., and Gilman, A. G. (1991) *J. Biol. Chem.* **266**, 3396–3401

237. Sondek, J., Bohm, A., Lambright, D. G., Hamm, H. E., and Sigler, P. B. (1996) *Nature (London)* **379**, 369–374

238. Clapham, D. E. (1996) *Nature (London)* **379**, 297–299

239. Neer, E. J., Schmidt, C. J., Nambudripad, R., and Smith, T. F. (1994) *Nature (London)* **371**, 297–300

240. Faber, X. X. (1994) *Structure* **3**, 551–559

241. Pai, E. F., Kabsch, W., Krengel, U., Holmes, K. C., John, J., and Wittinghofer, A. (1989) *Nature (London)* **341**, 209–214

242. Wall, M. A., Posner, B. A., and Sprang, S. R. (1998) *Structure* **6**, 1169–1183

243. Wilson, P. T., and Bourne, H. R. (1995) *J. Biol. Chem.* **270**, 9667–9675

244. Zhao, Y., Brandish, P. E., DiValentin, M., Schelvis, J. P. M., Babcock, G. T., and Marletta, M. A. (2000) *Biochemistry* **39**, 10848–10854

245. Garbers, D. L., and Lowe, D. G. (1994) *J. Biol. Chem.* **269**, 30741–30744

246. Subbaraya, I., Ruiz, C. C., Helekar, B. S., Zhao, X., Gorczyca, W. A., Pettenati, M. J., Rao, P. N., Palczewski, K., and Baehr, W. (1994) *J. Biol. Chem.* **269**, 31080–31089

247. Lincoln, T. M., and Cornwell, T. L. (1993) *FASEB J.* **7**, 328–338

248. Houslay, M. D. (1985) *Trends Biochem. Sci.* **10**, 465–466

249. Garbers, D. L. (1989) *J. Biol. Chem.* **264**, 9103–9106

250. Lowe, D. G., Chang, M.-S., Hellmiss, R., Chen, E., Singh, S., Garbers, D. L., and Goeddel, D. V. (1989) *EMBO J.* **8**, 1377–1384

251. Hamra, F. K., Forte, L. R., Eber, S. L., Pidhorodeckyj, N. V., Krause, W. J., Freeman, R. H., Chin, D. T., Tompkins, J. A., Fok, K. F., Smith, C. E., Duffin, K. L., Siegel, N. R., and Currie, M. G. (1993) *Proc. Natl. Acad. Sci. U.S.A.* **90**, 10464–10468

252. Stamler, J. S., Singel, D. J., and Loscalzo, J. (1992) *Science* **258**, 1898–1902

253. Yuen, P. S. T., Doolittle, L. K., and Garbers, D. L. (1994) *J. Biol. Chem.* **269**, 791–793

254. Stone, J. R., and Marletta, M. A. (1996) *Biochemistry* **35**, 1093–1099

255. Mayer, B., Schrammel, A., Klatt, P., Koesling, D., and Schmidt, K. (1995) *J. Biol. Chem.* **270**, 17355–17360

256. Stone, J. R., and Marletta, M. A. (1995) *Biochemistry* **34**, 16397–16403

257. Deinum, G., Stone, J. R., Babcock, G. T., and Marletta, M. A. (1996) *Biochemistry* **35**, 1540–1547

258. Verma, A., Hirsch, D. J., Glatt, C. E., Ronnett, G. V., and Snyder, S. H. (1993) *Science* **259**, 381–384

259. Stock, J. B., Surette, M. G., McCleary, W. R., and Stock, A. M. (1992) *J. Biol. Chem.* **267**, 19753–19756

259a. Kim, K. K., Yokota, H., and Kim, S.-H. (1999) *Nature (London)* **400**, 787–792

260. Scott, W. G., Milligan, D. L., Milburn, M. V., Privé, G. G., Yeh, J., Koshland, D. E., Jr., and Kim, S.-H. (1993) *J. Mol. Biol.* **232**, 555–573

261. Danielson, M. A., Biemann, H.-P., Koshland, D. E., Jr., and Falke, J. J. (1994) *Biochemistry* **33**, 6100–6109

262. Yeh, J. I., Biemann, H.-P., Privé, G. G., Pandit, J., Koshland, D. E., Jr., and Kim, S.-H. (1996) *J. Mol. Biol.* **262**, 186–201

263. Danielson, M. A., Bass, R. B., and Falke, J. J. (1997) *J. Biol. Chem.* **272**, 32878–32888

264. Kim, S.-H. (1994) *Protein Sci.* **3**, 159–165

264a. Ottemann, K. M., Xiao, W., Shin, Y.-K., and Koshland, D. E., Jr. (1999) *Science* **285**, 1751–1754

265. Berridge, M. J., and Irvine, R. F. (1989) *Nature (London)* **341**, 197–205

266. Schulman, H., and Lou, L. L. (1989) *Trends Biochem. Sci.* **14**, 62–66

267. Carafoli, E. (1987) *Ann. Rev. Biochem.* **56**, 395–433

268. Rasmussen, H. (1989) *Sci. Am.* **261**(Oct), 66–73

269. Kobilka, B. K., Matsui, H., Kobilka, T. S., Yang-Feng, T. L., Francke, U., Caron, M. G., Lefkowitz, R. J., and Regan, J. W. (1987) *Science* **238**, 650–656

270. Cotecchia, S., Kobilka, B. K., Daniel, K. W., Nolan, R. D., Lapetina, E. Y., Caron, M. G., Lefkowitz, R. J., and Regan, J. W. (1990) *J. Biol. Chem.* **265**, 63–69

271. Ceresa, B. P., and Limbird, L. E. (1994) *J. Biol. Chem.* **269**, 29557–29564

272. Scheer, A., Fanelli, F., Costa, T., De Benedetti, P. G., and Cotecchia, S. (1996) *EMBO J.* **15**, 3566–3578

273. Hwa, J., and Perez, D. M. (1996) *J. Biol. Chem.* **271**, 6322–6327

274. Boyer, J. L., Waldo, G. L., Evans, T., Northup, J. K., Downes, C. P., and Harden, T. K. (1989) *J. Biol. Chem.* **264**, 13917–13922

275. Wu, D., Jiang, H., and Simon, M. I. (1995) *J. Biol. Chem.* **270**, 9828–9832

276. Nishizuka, Y. (1995) *FASEB J.* **9**, 484–496

277. Majerus, P. W. (1992) *Ann. Rev. Biochem.* **61**, 225–250

278. Majerus, P. W., Conolly, T. M., Bansal, V. S., Inhorn, R. C., Ross, T. S., and Lips, D. L. (1988) *J. Biol. Chem.* **263**, 3051–3054

279. Rhee, S. G., Suh, P.-G., Ryu, S.-H., and Lee, S. Y. (1989) *Science* **244**, 546–550

280. Hough, E., Hansen, L. K., Birknes, B., Jynge, K., Hansen, S., Hordvik, A., Little, C., Dodson, E., and Derewenda, Z. (1989) *Nature (London)* **338**, 357–360

281. Essen, L.-O., Perisic, O., Cheung, R., Katan, M., and Williams, R. L. (1996) *Nature (London)* **380**, 595–602

282. Berridge, M. J. (1993) *Nature (London)* **361**, 315–325

283. Parthasarathy, R., and Eisenberg, F. J. (1986) *Biochem. J.* **235**, 313–322

284. Majerus, P. W., Connolly, T. M., Deckmyn, H., Ross, T. S., Bross, T. E., Ishii, H., Bansal, V. S., and Wilson, D. B. (1986) *Science* **234**, 1519–1526

285. Houslay, M. D. (1987) *Trends Biochem. Sci.* **12**, 1–2

286. Nickels, J. T., Jr., Buxeda, R. J., and Carman, G. M. (1994) *J. Biol. Chem.* **269**, 11018–11024

287. Boronenkov, I. V., and Anderson, R. A. (1995) *J. Biol. Chem.* **270**, 2881–2884

288. Hughes, A. R., Takemura, H., and Putney, J. W., Jr. (1988) *J. Biol. Chem.* **263**, 10314–10319

289. Furuichi, T., Yoshikawa, S., Miyawaki, A., Wada, K., Maeda, N., and Mikoshiba, K. (1989) *Nature (London)* **342**, 32–38

289a. Zimmermann, B., and Walz, B. (1999) *EMBO J.* **18**, 3222–3231

290. Yoshikawa, F., Morita, M., Monkawa, T., Michikawa, T., Furuichi, T., and Mikoshiba, K. (1996) *J. Biol. Chem.* **271**, 18277–18284

291. Humbert, J.-P., Matter, N., Artault, J.-C., Köppler, P., and Malviya, A. N. (1996) *J. Biol. Chem.* **271**, 478–485

292. Hennager, D. J., Welsh, M. J., and DeLisle, S. (1995) *J. Biol. Chem.* **270**, 4959–4962

293. Taylor, C. W., and Marshall, I. C. B. (1992) *Trends Biochem. Sci.* **17**, 403–407

294. Gu, Q.-M., and Sih, C. J. (1994) *J. Am. Chem. Soc.* **116**, 7481–7486

295. Berridge, M. J. (1993) *Nature (London)* **365**, 388–389

296. Lee, H. C., Walseth, T. F., Bratt, G. T., Hayes, R. N., and Clapper, D. L. (1989) *J. Biol. Chem.* **264**, 1608–1615

297. Vu, C. Q., Lu, P.-J., Chen, C.-S., and Jacobson, M. K. (1996) *J. Biol. Chem.* **271**, 4747–4754

298. Dennis, E. A., Rhee, S. G., Billah, M. M., and Hannun, Y. A. (1991) *FASEB J.* **5**, 2068–2077

299. Sternweis, P. C., and Smrcka, A. V. (1992) *Trends Biochem. Sci.* **17**, 502–506

300. Rhee, S. G., and Choi, K. D. (1992) *J. Biol. Chem.* **267**, 12393–12396

300a. Kim, M. J., Chang, J.-S., Park, S. K., Hwang, J.-I., Ryu, S. H., and Suh, P.-G. (2000) *Biochemistry* **39**, 8674–8682

301. Jafri, M. S., and Keizer, J. (1995) *Biophys. J.* **69**, 2139–2153

302. Combettes, L., Cheek, T. R., and Taylor, C. W. (1996) *EMBO J.* **15**, 2086–2093

303. van de Put, F. H. M. M., De Pont, J. J. H. H. M., and Willems, P. H. G. M. (1994) *J. Biol. Chem.* **269**, 12438–12443

304. Kukuljan, M., Rojas, E., Catt, K. J., and Stojilkovic, S. S. (1994) *J. Biol. Chem.* **269**, 4860–4865

305. Berridge, M. J., and Galione, A. (1988) *FASEB J.* **2**, 3074–3082

306. Meyer, T., Wensel, T., and Stryer, L. (1990) *Biochemistry* **29**, 32–37

307. Luttrell, B. M. (1993) *J. Biol. Chem.* **268**, 1521–1524

308. Wilson, M. P., and Majerus, P. W. (1996) *J. Biol. Chem.* **271**, 11904–11910

309. Shears, S. B. (1989) *J. Biol. Chem.* **264**, 19879–19886

309a. Zhu, D.-M., Tekle, E., Huang, C. Y., and Chock, P. B. (2000) *J. Biol. Chem.* **275**, 6063–6066

309b. Odom, A. R., Stahlberg, A., Wente, S. R., and York, J. D. (2000) *Science* **287**, 2026–2029

310. Ryu, S. H., Lee, S. Y., Lee, K.-Y., and Rhee, S. G. (1987) *FASEB J.* **1**, 388–393

311. Fukuda, M., and Mikoshiba, K. (1996) *J. Biol. Chem.* **271**, 18838–18842

312. Bird, G. St. J., and Putney, J. W., Jr. (1996) *J. Biol. Chem.* **271**, 6766–6770

313. Dixon, J. F., and Hokin, L. E. (1987) *J. Biol. Chem.* **262**, 13892–13895

314. Heinz, D. W., Ryan, M., Bullock, T. L., and Griffith, O. H. (1995) *EMBO J.* **14**, 3855–3863

315. Ali, N., Craxton, A., and Shears, S. B. (1993) *J. Biol. Chem.* **268**, 6161–6167

316. Xie, W., Kaetzel, M. A., Bruzik, K. S., Dedman, J. R., Shears, S. B., and Nelson, D. J. (1996) *J. Biol. Chem.* **271**, 14092–14097

317. Ismailov, I. I., Fuller, C. M., Berdiev, B. K., Shlyonsky, V. G., Benos, D. J., and Barrett, K. E. (1996) *Proc. Natl. Acad. Sci. U.S.A.* **93**, 10505–10509

318. Dasgupta, S., Dasgupta, D., Sen, M., Biswas, S., and Biswas, B. B. (1996) *Biochemistry* **35**, 4994–5001

319. Voglmaier, S. M., Bembenek, M. E., Kaplin, A. I., Dormán, G., Olszewski, J. D., Prestwich, G. D., and Snyder, S. H. (1996) *Proc. Natl. Acad. Sci. U.S.A.* **93**, 4305–4310

319a. Chi, T. H., and Crabtree, G. R. (2000) *Science* **287**, 1937–1939

320. Menniti, F. S., Oliver, K. G., Putney, J. W., Jr., and Shears, S. B. (1993) *Trends Biochem. Sci.* **18**, 53–56

320a. Yang, X., Safrany, S. T., and Shears, S. B. (1999) *J. Biol. Chem.* **274**, 35434–35440

References

321. Stephens, L., Radenberg, T., Thiel.U, Vogel, G., Khoo, K.-H., Dell, A., Jackson, T. R., Hawkins, P. T., and Mayr, G. W. (1993) *J. Biol. Chem.* **268**, 4009–4015

322. Van Dijken, P., de Haas, J.-R., Craxton, A., Erneux, C., Shears, S. B., and Van Haastert, P. J. M. (1995) *J. Biol. Chem.* **270**, 29724–29731

323. Bansal, V. S., Caldwell, K. K., and Majerus, P. W. (1990) *J. Biol. Chem.* **265**, 1806–1811

324. York, J. D., Ponder, J. W., Chen, Z.-w, Mathews, F. S., and Majerus, P. W. (1994) *Biochemistry* **33**, 13164–13171

325. Norris, F. A., Auethavekiat, V., and Majerus, P. W. (1995) *J. Biol. Chem.* **270**, 16128–16133

326. Jefferson, A. B., and Majerus, P. W. (1995) *J. Biol. Chem.* **270**, 9370–9377

326a. Norris, F. A., Wilson, M. P., Wallis, T. S., Galyov, E. E., and Majerus, P. W. (1998) *Proc. Natl. Acad. Sci. U.S.A.* **95**, 14057–14059

326b. Munday, A. D., Norris, F. A., Caldwell, K. K., Brown, S., Majerus, P. W., and Mitchell, C. A. (1999) *Proc. Natl. Acad. Sci. U.S.A.* **96**, 3640–3645

327. Bone, R., Springer, J. P., and Atack, J. R. (1992) *Proc. Natl. Acad. Sci. U.S.A.* **89**, 10031–10035

328. Ganzhorn, A. J., Lepage, P., Pelton, P. D., Strasser, F., Vincendon, P., and Rondeau, J.-M. (1996) *Biochemistry* **35**, 10957–10966

329. Burgoyne, R. D. (1989) *Trends Biochem. Sci.* **14**, 87–88

330. Huang, K.-P., Huang, F. L., Nakabayashi, H., and Yoshida, Y. (1988) *J. Biol. Chem.* **263**, 14839–14845

330a. Ron, D., and Kazanietz, M. G. (1999) *FASEB J.* **13**, 1658–1676

330b. Ward, N. E., Stewart, J. R., Ionnides, C. G., and O'Brian, C. A. (2000) *Biochemistry* **39**, 10319–10329

331. Pessin, M. S., and Raben, D. M. (1989) *J. Biol. Chem.* **264**, 8729–8738

332. Exton, J. H. (1990) *J. Biol. Chem.* **265**, 1–4

332a. Perisic, O., Paterson, H. F., Mosedale, G., Lara-González, S., and Williams, R. L. (1999) *J. Biol. Chem.* **274**, 14979–14987

333. Roberts, M. F. (1996) *FASEB J.* **10**, 1159–1172

334. Liscovitch, M., Chalifa, V., Pertile, P., Chen, C.-S., and Cantley, L. C. (1994) *J. Biol. Chem.* **269**, 21403–21406

335. Stein, R. L., Melandri, F., and Dick, L. (1935) *Biochemistry* **35**, 3899–3908

336. Ward, S. G., Mills, S. J., Liu, C., Westwick, J., and Potter, B. V. L. (1995) *J. Biol. Chem.* **270**, 12075–12084

337. Woscholski, R., Kodaki, T., Palmer, R. H., Waterfield, M. D., and Parker, P. J. (1995) *Biochemistry* **34**, 11489–11493

337a. Odorizzi, G., Babst, M., and Emr, S. D. (2000) *Trends Biochem. Sci.* **25**, 229–235

337b. Walker, E. H., Perisic, O., Ried, C., Stephens, L., and Williams, R. L. (1999) *Nature (London)* **402**, 313–320

337c. Ching, T.-T., Wang, D.-S., Hsu, A.-L., Lu, P.-J., and Chen, C.-S. (1999) *J. Biol. Chem.* **274**, 8611–8617

337d. Bertsch, U., Deschermeier, C., Fanick, W., Girkontaite, I., Hillemeier, K., Johnen, H., Weglöhner, W., Emmrich, F., and Mayr, G. W. (2000) *J. Biol. Chem.* **275**, 1557–1564

337e. Jackson, T. R., Kearns, B. G., and Theibert, A. B. (2000) *Trends Biochem. Sci.* **25**, 489–495

337f. Gillooly, D. J., Morrow, I. C., Lindsay, M., Gould, R., Bryant, N. J., Gaullier, J.-M., Parton, R. G., and Stenmark, H. (2000) *EMBO J.* **19**, 4577–4588

338. Pennisi, E. (1997) *Science* **275**, 1876–1878

338a. Stambolic, V., Suzuki, A., de la Pompa, J. L., Brothers, G. M., Mirtsos, C., Sasaki, T., Ruland, J., Penninger, J. M., Siderovski, D. P., and Mak, T. W. (1998) *Cell* **95**, 29–39

338b. Toker, A., and Cantley, L. C. (1997) *Nature (London)* **387**, 673–675

339. Ui, M., Okada, T., Hazeki, K., and Hazeki, O. (1995) *Trends Biochem. Sci.* **20**, 303–307

339a. Takasuga, S., Katada, T., Ui, M., and Hazeki, O. (1999) *J. Biol. Chem.* **274**, 19545–19550

340. Chock, P. B., Rhee, S. G., and Stadtman, E. R. (1980) *Ann. Rev. Biochem.* **49**, 813–843

341. Stadtman, E. R., and Chock, P. B. (1977) *Proc. Natl. Acad. Sci. U.S.A.* **74**, 2761–2765

342. Huang, C.-Y. F., and Ferrell, J. E., Jr. (1996) *Proc. Natl. Acad. Sci. U.S.A.* **93**, 10078–10083

343. Goldbeter, A., and Koshland, D. E., Jr. (1987) *J. Biol. Chem.* **262**, 4460–4471

344. Schacter-Noiman, E., Chock, P. B., and Stadtman, E. R. (1983) *Philos. Trans. R. Soc. London B* **302**, 157–166

345. Schacter, E., Chock, P. B., and Stadtman, E. R. (1984) *J. Biol. Chem.* **259**, 12260–12264

346. Newsholme, E. A., Challiss, R. A. J., and Crabtree, B. (1984) *Trends Biochem. Sci.* **9**, 277–280

347. Koshland, D. E., Jr. (1987) *Trends Biochem. Sci.* **12**, 225–228

348. Taylor, S. I. (1995) in *The Metabolic and Molecular Bases of Inherited Disease*, 7th ed., Vol. 1 (Scriver, C. R., Beaudet, A. L., Sly, W. S., and Valle, D., eds), pp. 843–896, McGraw-Hill, New York

349. Müller, G., Rouveyre, N., Crecelius, A., and Bandlow, W. (1998) *Biochemistry* **37**, 8683–8695

350. Smit, A. B., Vreugdenhil, E., Ebberink, R. H. M., Geraerts, W. P. M., Klootwijk, J., and Joosse, J. (1988) *Nature (London)* **331**, 535–538

351. LeRoith, D., Shiloach, J., Roth, J., and Lesniak, M. A. (1983) *J. Biol. Chem.* **256**, 6533–6536

352. Rosen, O. M. (1987) *Science* **237**, 1452–1458

353. White, M. F., and Kahn, C. R. (1994) *J. Biol. Chem.* **269**, 1–4

354. Barnard, R. J., and Youngren, J. F. (1992) *FASEB J.* **6**, 3238–3244

355. Bell, G. I., Burant, C. F., Takeda, J., and Gould, G. W. (1993) *J. Biol. Chem.* **268**, 19161–19164

356. Mueckler, M. (1994) *Eur. J. Biochem.* **219**, 713–725

356a. Li, J., Houseknecht, K. L., Stenbit, A. E., Katz, E. B., and Charron, M. J. (2000) *FASEB J.* **14**, 1117–1125

357. Cuatrecasas, P. (1972) *Proc. Natl. Acad. Sci. U.S.A.* **69**, 1277–1281

358. Pashmforoush, M., Yoshimasa, Y., and Steiner, D. F. (1994) *J. Biol. Chem.* **269**, 32639–32648

359. Ullrich, A., Bell, J. R., Chen, E. Y., Herrera, R., Petruzzelli, L. M., Dull, T. J., Gray, A., Coussens, L., Liao, Y.-C., Tsubokawa, M., Mason, A., Seeburg, P. H., Grunfeld, C., Rosen, O. M., and Ramachandran, J. (1985) *Nature (London)* **313**, 756–761

360. Xu, Q.-Y., Paxton, R. J., and Fujita-Yamaguchi, Y. (1990) *J. Biol. Chem.* **265**, 18673–18681

361. Herrera, R., Petruzzelli, L., Thomas, N., Bramson, H. N., Kaiser, E. T., and Rosen, O. M. (1985) *Proc. Natl. Acad. Sci. U.S.A.* **82**, 7899–7903

362. McClain, D. A., Maegawa, H., Lee, J., Dull, T. J., Ullrich, A., and Olefsky, J. M. (1987) *J. Biol. Chem.* **262**, 14663–14671

363. Perlman, R., Bottaro, D. P., White, M. F., and Kahn, C. R. (1989) *J. Biol. Chem.* **264**, 8946–8950

363a. Rouard, M., Bass, J., Grigorescu, F., Garrett, T. P. J., Ward, C. W., Lipkind, G., Jaffiole, C., Steiner, D. F., and Bell, G. I. (1999) *J. Biol. Chem.* **274**, 18487–18491

363b. Luo, R. Z.-T., Beniac, D. R., Fernandes, A., Yip, C. C., and Ottensmeyer, F. P. (1999) *Science* **285**, 1077–1080

363c. Woldin, C. N., Hing, F. S., Lee, J., Pilch, P. F., and Shipley, G. G. (1999) *J. Biol. Chem.* **274**, 34981–34992

364. Lemmon, M. A., and Schlessinger, J. (1994) *Trends Biochem. Sci.* **19**, 459–563

365. Hubbard, S. R., Wei, L., Ellis, L., and Hendrickson, W. A. (1994) *Nature (London)* **372**, 746–754

365a. Hubbard, S. R. (1997) *EMBO J.* **16**, 5572–5581

366. White, M. F., Shoelson, S. E., Keutmann, H., and Kahn, C. R. (1988) *J. Biol. Chem.* **263**, 2969–2980

367. Fabry, M., Schaefer, E., Ellis, L., Kojro, E., Fahrenholz, F., and Brandenburg, D. (1992) *J. Biol. Chem.* **267**, 8950–8956

368. Williams, P. F., Mynarcik, D. C., Yu, G. Q., and Whittaker, J. (1995) *J. Biol. Chem.* **270**, 3012–3016

369. Kjeldsen, T., Wiberg, F. C., and Andersen, A. S. (1994) *J. Biol. Chem.* **269**, 32942–32946

369a. Sparrow, L. G., McKern, N. M., Gorman, J. J., Strike, P. M., Robinson, C. P., Bentley, J. D., and Ward, C. W. (1997) *J. Biol. Chem.* **272**, 29460–29467

369b. Garant, M. J., Kole, S., Maksimova, E. M., and Bernier, M. (1999) *Biochemistry* **38**, 5896–5904

370. Clark, S., and Harrison, L. C. (1983) *J. Biol. Chem.* **258**, 11434–11437

371. Saltiel, A. R. (1994) *FASEB J.* **8**, 1034–1040

372. Carter, W. G., Asamoah, K. A., and Sale, G. J. (1995) *Biochemistry* **34**, 9488–9499

373. Myers, M. G., Jr., Sun, X. J., and White, M. F. (1994) *Trends Biochem. Sci.* **19**, 289–293

374. Yenush, L., Makati, K. J., Smith-Hall, J., Ishibashi, O., Myers, M. G. J., and White, M. F. (1996) *J. Biol. Chem.* **271**, 24300–24306

375. Kowalski-Chauvel, A., Pradayrol, L., Vaysse, N., and Seva, C. (1996) *J. Biol. Chem.* **271**, 26356–26361

376. Paz, K., Voliovitch, H., Hadari, Y. R., Roberts, C. T., Jr., LeRoith, D., and Zick, Y. (1996) *J. Biol. Chem.* **271**, 6998–7003

376a. Rother, K. I., Imai, Y., Caruso, M., Beguinot, F., Formisano, P., and Accili, D. (1998) *J. Biol. Chem.* **273**, 17491–17497

376b. Clark, S. F., Molero, J.-C., and James, D. E. (2000) *J. Biol. Chem.* **275**, 3819–3826

376c. Lehr, S., Kotzka, J., Herkner, A., Sikmann, A., Meyer, H. E., Krone, W., and Müller-Wieland, D. (2000) *Biochemistry* **39**, 10898–10907

377. Ricketts, W. A., Rose, D. W., Shoelson, S., and Olefsky, J. M. (1996) *J. Biol. Chem.* **271**, 26165–26169

378. Okada, S., and Pessin, J. E. (1996) *J. Biol. Chem.* **271**, 25533–25538

379. Czech, M. P., Klarlund, J. K., Yagaloff, K. A., Bradford, A. P., and Lewis, R. E. (1988) *J. Biol. Chem.* **263**, 11017–11020

380. Seger, R., and Krebs, E. G. (1995) *FASEB J.* **9**, 726–735

381. Carter, W. G., Sullivan, A. C., Asamoah, K. A., and Sale, G. J. (1996) *Biochemistry* **35**, 14340–14351

382. Kasus-Jacobi, A., Perdereau, D., Auzan, C., Clauser, E., Van Obberghen, E., Mauvais-Jarvis, F., Girard, J., and Burnol, A.-F. (1998) *J. Biol. Chem.* **273**, 26026–26035

383. Baltensperger, K., Karoor, V., Paul, H., Ruoho, A., Czech, M. P., and Malbon, C. C. (1996) *J. Biol. Chem.* **271**, 1061–1064

384. Hara, K., Yonezawa, K., Sakaue, H., Ando, A., Kotani, K., Kitamura, T., Kitamura, Y., Ueda, H., Stephens, L., Jackson, T. R., Hawkins, P. T., Dhand, R., Clark, A. E., Holman, G. D., Waterfield, M. D., and Kasuga, M. (1994) *Proc. Natl. Acad. Sci. U.S.A.* **91**, 7415–7419

385. Lam, K., Carpenter, C. L., Ruderman, N. B., Friel, J. C., and Kelly, K. L. (1994) *J. Biol. Chem.* **269**, 20648–20652

386. Sutherland, C., O'Brien, R. M., and Granner, D. K. (1995) *J. Biol. Chem.* **270**, 15501–15506

387. Domin, J., Dhand, R., and Waterfield, M. D. (1996) *J. Biol. Chem.* **271**, 21614–21621

387a. Kosaki, A., Yamada, K., Suga, J., Otaka, A., and Kuzuya, H. (1998) *J. Biol. Chem.* **273**, 940–944

References

388. Lefebvre, V., Méchin, M.-C., Louckx, M. P., Rider, M. H., and Hue, L. (1996) *J. Biol. Chem.* **271**, 22289–22292

389. Quon, M. J., Butte, A. J., Zarnowski, M. J., Sesti, G., Cushman, S. W., and Taylor, S. I. (1994) *J. Biol. Chem.* **269**, 27920–27924

390. Kandror, K. V., and Pilch, P. F. (1996) *J. Biol. Chem.* **271**, 21703–21708

390a. Marte, B. M., and Downward, J. (1997) *Trends Biochem. Sci.* **22**, 355–358

390b. Hemmings, B. A. (1997) *Science* **275**, 628–630

390c. Chan, T. O., Rittenhouse, S. E., and Tsichlis, P. N. (1999) *Ann. Rev. Biochem.* **68**, 965–1014

390d. Aguirre, V., Uchida, T., Yenush, L., Davis, R., and White, M. F. (2000) *J. Biol. Chem.* **275**, 9047–9054

390e. Nakae, J., Park, B.-C., and Accili, D. (1999) *J. Biol. Chem.* **274**, 15982–15985

391. Kishi, K., Hayashi, H., Wang, L., Kamohara, S., Tamaoka, K., Shimizu, T., Ushikubi, F., Narumiya, S., and Ebina, Y. (1996) *J. Biol. Chem.* **271**, 26561–26568

392. Yamauchi, K., Ribon, V., Saltiel, A. R., and Pessin, J. E. (1995) *J. Biol. Chem.* **270**, 17716–17722

393. Zhang, J., Hiken, J., Davis, A. E., and Lawrence, J. C., Jr. (1989) *J. Biol. Chem.* **264**, 17513–17523

394. Robinson, F. W., Smith, C. J., Flanagan, J. E., Shibata, H., and Kono, T. (1989) *J. Biol. Chem.* **264**, 16458–16464

395. Najjar, S. M., Choice, C. V., Soni, P., Whitman, C. M., and Poy, M. N. (1998) *J. Biol. Chem.* **273**, 12923–12928

396. Haft, C. R., Klausner, R. D., and Taylor, S. I. (1994) *J. Biol. Chem.* **269**, 26286–26294

397. Safavi, A., Miller, B. C., Cottam, L., and Hersh, L. B. (1996) *Biochemistry* **35**, 14318–14325

398. Kublaoui, B., Lee, J., and Pilch, P. F. (1995) *J. Biol. Chem.* **270**, 59–65

399. Bevan, A. P., Burgess, J. W., Drake, P. G., Shaver, A., Bergeron, J. J. M., and Posner, B. I. (1995) *J. Biol. Chem.* **270**, 10784–10791

400. Podlecki, D. A., Smith, R. M., Kao, M., Tsai, P., Huecksteadt, T., Brandenburg, D., Lasher, R. S., Jarett, L., and Olefsky, J. M. (1987) *J. Biol. Chem.* **262**, 3362–3368

401. Cherniack, A. D., Klarlund, J. K., Conway, B. R., and Czech, M. P. (1995) *J. Biol. Chem.* **270**, 1485–1488

402. Unoue, G., Cheatham, B., and Kahn, C. R. (1996) *J. Biol. Chem.* **271**, 28206–28211

403. Yarden, Y., and Ullrich, A. (1988) *Biochemistry* **27**, 3113–3119

404. James, R. (1984) *Ann. Rev. Biochem.* **53**, 259–292

405. Nilsen-Hamilton, M., ed. (1994) *Growth Factors and Signal Transduction in Development*, Wiley-Liss, New York

406. Claesson-Welsh, L. (1994) *J. Biol. Chem.* **269**, 32023–32026

407. Williams, L. T. (1989) *Science* **243**, 1564–1570

408. Hesketh, R. (1995) *The Oncogene Facts Book*, Academic Press, San Diego, California

409. Hunter, T. (1984) *Sci. Am.* **251**(Aug), 70–79

410. Marx, J. (1994) *Science* **266**, 1942–1944

411. Cooper, G. M., ed. (1990) *Oncogenes*, Jones & Bartlett, Boston, Massachusetts

412. Bishop, J. M. (1996) *FASEB J.* **10**, 362–364

413. Weinberg, R. A. (1983) *Sci. Am.* **249**(Nov), 126–142

414. Hickman, C. P. (1973) *Biology of the Invertebrates*, Mosby, St. Louis, Missouri

415. Yarden, Y., Escobedo, J. A., Kuang, W. J., Yang-Feng, T. L., Daniel, T. O., Tremble, P. M., Chen, E. Y., Ando, M. E., Harkins, R. N., Francke, U., Fried, V. A., Ullrich, A., and Williams, L. T. (1986) *Nature (London)* **323**, 226–232

416. Robbins, K. C., Antoniades, H. N., Devare, S. G., Hunkapillar, M. W., and Aaronson, S. A. (1983) *Nature (London)* **305**, 605–608

417. Yarden, Y., and Peles, E. (1991) *Biochemistry* **30**, 3543–3550

418. Peles, E., Lamprecht, R., Ben-Levy, R., Tzahar, E., and Yarden, Y. (1992) *J. Biol. Chem.* **267**, 12266–12274

419. Carpenter, G., and Cohen, S. (1990) *J. Biol. Chem.* **265**, 7709–7712

420. Suen, T.-C., and Goss, P. E. (2000) *J. Biol. Chem.* **275**, 6600–6607

421. Stover, D. R., Becker, M., Liebetanz, J., and Lydon, N. B. (1995) *J. Biol. Chem.* **270**, 15591–15597

422. Kauffmann-Zeh, A., Thomas, G. M. H., Ball, A., Prosser, S., Cunningham, E., Cockcroft, S., and Hsuan, J. J. (1995) *Science* **268**, 1188–1190

422a. Jones, J. T., Ballinger, M. D., Pisacane, P. I., Lofgren, J. A., Fitzpatrick, V. D., Fairbrother, W. J., Wells, J. A., and Sliwkowski, M. X. (1998) *J. Biol. Chem.* **273**, 11667–11674

423. Montelione, G. T., Wüthrich, K., and Scheraga, H. A. (1988) *Biochemistry* **27**, 2235–2243

424. Schlessinger, J. (1988) *Biochemistry* **27**, 3119–3123

425. Montelione, G. T., Wüthrich, K., and Scheraga, H. A. (1988) *Biochemistry* **27**, 2235–2243

426. Chantry, A. (1995) *J. Biol. Chem.* **270**, 3068–3073

427. Karasik, A., Pepinsky, R. B., Shoelson, S. E., and Kahn, C. R. (1988) *J. Biol. Chem.* **263**, 11862–11867

428. Kauffmann-Zeh, A., Klinger, R., Endemann, G., Waterfield, M. D., Wetzker, R., and Hsuan, J. J. (1994) *J. Biol. Chem.* **269**, 31243–31251

429. Kaplan, K. B., Bibbins, K. B., Swedlow, J. R., Arnaud, M., Morgan, D. O., and Varmus, H. E. (1994) *EMBO J.* **13**, 4745–4756

429a. Williams, J. C., Weijland, A., Gonfloni, S., Thompson, A., Courtneidge, S. A., Superti-Furga, G., and Wierenga, R. K. (1997) *J. Mol. Biol.* **274**, 757–775

430. Okada, M., and Nakagawa, H. (1989) *J. Biol. Chem.* **264**, 20886–20893

431. Buss, J. E., and Sefton, B. M. (1985) *Journal of Virology* **53**, 7–12

432. Soriano, P., Montgomery, C., Geske, R., and Bradley, A. (1991) *Cell* **64**, 693–702

433. Lowe, C., Yoneda, T., Boyce, B. F., Chen, H., Mundy, G. R., and Soriano, P. (1993) *Proc. Natl. Acad. Sci. U.S.A.* **90**, 4485–4489

434. Tezuka, K.-i, Denhardt, D. T., Rodan, G. A., and Harada, S.-i. (1996) *J. Biol. Chem.* **271**, 22713–22717

435. Bolen, J. B., and Veillette, A. (1989) *Trends Biochem. Sci.* **14**, 404–407

436. Mustelin, T., and Burn, P. (1993) *Trends Biochem. Sci.* **18**, 215–220

437. Vetrie, D., Vorechovsky, I., Sideras, P., Holland, J., Davies, A., Flinter, F., Hammarström, L., Kinnon, C., Levinsky, R., Bobrow, M., Smith, C. I. E., and Bentley, D. R. (1993) *Nature (London)* **361**, 226–233

438. Hatakeyama, M., Kono, T., Kobayashi, N., Kawahara, A., Levin, S. D., Perlmutter, R. M., and Taniguchi, T. (1991) *Science* **252**, 1523–1528

438a. Zou, X., and Calame, K. (1999) *J. Biol. Chem.* **274**, 18141–18144

439. Daley, G. Q., Van Etten, R. A., and Baltimore, D. (1990) *Science* **247**, 824–830

440. Landis, C. A., Masters, S. B., Spada, A., Pace, A. M., Bourne, H. R., and Vallar, L. (1989) *Nature (London)* **340**, 692–696

441. Barbacid, M. (1987) *Ann. Rev. Biochem.* **56**, 779–827

442. Kato, G. J., and Dang, C. V. (1992) *FASEB J.* **6**, 3065–3072

443. Marcu, K. B., Bossone, S. A., and Patel, A. J. (1992) *Ann. Rev. Biochem.* **61**, 809–860

444. Greenhalgh, D. A., Welty, D. J., Player, A., and Yuspa, S. H. (1990) *Proc. Natl. Acad. Sci. U.S.A.* **87**, 643–647

445. Rabbitts, T. H., Forster, A., Hamlyn, P., and Baer, R. (1984) *Nature (London)* **309**, 592–597

446. Shima, E. A., Le Beau, M. M., McKeithan, T. W., Minowada, J., Showe, L. C., Mak, T. W., Minden, M. D., Rowley, J. D., and Diaz, M. O. (1986) *Proc. Natl. Acad. Sci. U.S.A.* **83**, 3439–3443

447. Cramer, C. J., and Truhlar, D. G. (1993) *J. Am. Chem. Soc.* **115**, 5745–5753

448. Perrin, C. L., Armstrong, K. B., and Fabian, M. A. (1994) *J. Am. Chem. Soc.* **116**, 715–722

449. Gabb, H. A., and Harvey, S. C. (1993) *J. Am. Chem. Soc.* **115**, 4218–4227

450. Plavec, J., Tong, W., and Chattopadhyaya, J. (1993) *J. Am. Chem. Soc.* **115**, 9734–9746

451. Ellervik, U., and Magnusson, G. (1994) *J. Am. Chem. Soc.* **116**, 2340–2347

451a. Hermeking, H., Rago, C., Schuhmacher, M., Li, Q., Barrett, J. F., Obaya, A. J., O'Connell, B. C., Mateyak, M. K., Tam, W., Kohlhuber, F., Dang, C. V., Sedivy, J. M., Eick, D., Vogelstein, B., and Kinzler, K. W. (2000) *Proc. Natl. Acad. Sci. U.S.A.* **97**, 2229–2234

452. Eden, A., Simchen, G., and Benvenisty, N. (1996) *J. Biol. Chem.* **271**, 20242–20245

453. Postel, E. H., Berberich, S. J., Flint, S. J., and Ferrone, C. A. (1993) *Science* **261**, 478–480

454. Ying, G.-G., Proost, P., van Damme, J., Bruschi, M., Introna, M., and Golay, J. (2000) *J. Biol. Chem.* **275**, 4152–4158

455. Macleod, K., Leprince, D., and Stehelin, D. (1992) *Trends Biochem. Sci.* **17**, 251–256

456. Karin, M. (1995) *J. Biol. Chem.* **270**, 16483–16486

457. Turner, R., and Tjian, R. (1989) *Science* **243**, 1689–1694

458. Vogt, P. K., and Bos, T. J. (1989) *Trends Biochem. Sci.* **14**, 172–175

459. Neuberg, M., Schuermann, M., Hunter, J. B., and Müller, R. (1989) *Nature (London)* **338**, 589–590

460. Treisman, R. (1995) *EMBO J.* **14**, 4905–4913

461. Abel, T., and Maniatis, T. (1989) *Nature (London)* **341**, 24–25

462. Glover, J. N. M., and Harrison, S. C. (1995) *Nature (London)* **373**, 257–261

463. Quantin, B., and Breathnach, R. (1988) *Nature (London)* **334**, 538–539

464. Greenberg, M. E., and Ziff, E. B. (1984) *Nature (London)* **311**, 433–438

465. Müller, R., Bravo, R., Burckhardt, J., and Curran, T. (1984) *Nature (London)* **312**, 716–720

466. Fukumoto, Y., Kaibuchi, K., Oku, N., Hori, Y., and Takai, Y. (1990) *J. Biol. Chem.* **265**, 774–780

467. Wang, Z.-Q., Ovitt, C., Grigoriadis, A. E., Möhle-Steinlein, U., Rüther, U., and Wagner, E. F. (1992) *Nature (London)* **360**, 741–745

468. Candeliere, G. A., Glorieux, F. H., Prud'homme, J., and St.-Arnaud, R. (1995) *N. Engl. J. Med.* **332**, 1546–1551

469. Egan, S. E., and Weinberg, R. A. (1993) *Nature (London)* **365**, 781–783

470. Davis, R. J. (1994) *Trends Biochem. Sci.* **19**, 470–473

470a. Gutkind, J. S. (1998) *J. Biol. Chem.* **273**, 1839–1842

470b. Whitmarsh, A. J., and Davis, R. J. (1998) *Trends Biochem. Sci.* **23**, 481–485

471. Pawson, T. (1995) *Nature (London)* **373**, 573–580

472. Daly, R. J., Sanderson, G. M., Janes, P. W., and Sutherland, R. L. (1996) *J. Biol. Chem.* **271**, 12502–12510

473. Feller, S. M., Ren, R., Hanafusa, H., and Baltimore, D. (1994) *Trends Biochem. Sci.* **19**, 453–458

474. Pei, D., Lorenz, U., Klingmüller, U., Neel, B. G., and Walsh, C. T. (1994) *Biochemistry* **33**, 15483–15493

References

475. Ladbury, J. E., Hensmann, M., Panayotou, G., and Campbell, I. D. (1996) *Biochemistry* **35**, 11062–11069

476. Maignan, S., Guilloteau, J.-P., Fromage, N., Arnoux, B., Becquart, J., and Ducruix, A. (1995) *Science* **268**, 291–293

477. Mikol, V., Baumann, G., Zurini, M. G. M., and Hommel, U. (1995) *J. Mol. Biol.* **254**, 86–95

478. Zhou, M.-M., Ravichandran, K. S., Olejniczak, E. T., Petros, A. M., Meadows, R. P., Sattler, M., Harlan, J. E., Wade, W. S., Burakoff, S. J., and Fesik, S. W. (1995) *Nature (London)* **378**, 584–592

479. Bonfini, L., Migliaccio, E., Pelicci, G., Lanfrancone, L., and Pelicci, P. G. (1996) *Trends Biochem. Sci.* **21**, 257–261

479a. Nantel, A., Mohammad-Ali, K., Sherk, J., Posner, B. I., and Thomas, D. Y. (1998) *J. Biol. Chem.* **273**, 10475–10484

480. Burgering, B. M. T., and Bos, J. L. (1995) *Trends Biochem. Sci.* **20**, 18–22

481. Pumiglia, K. M., LeVine, H., Haske, T., Habib, T., Jove, R., and Decker, S. J. (1995) *J. Biol. Chem.* **270**, 14251–14254

482. Thompson, M. J., Roe, M. W., Malik, R. K., and Blackshear, P. J. (1994) *J. Biol. Chem.* **269**, 21127–21135

483. Jhun, B. H., Haruta, T., Meinkoth, J. L., Leitner, J. W., Draznin, B., Saltiel, A. R., Pang, L., Sasaoka, T., and Olefsky, J. M. (1995) *Biochemistry* **34**, 7996–8004

484. Jacob, K. K., Ouyang, L., and Stanley, F. M. (1995) *J. Biol. Chem.* **270**, 27773–27779

485. Pause, A., Belsham, G. J., Gingras, A.-C., Donzé, O., Lin, T.-A., Lawrence, J. C., Jr., and Sonenberg, N. (1994) *Nature (London)* **371**, 762–767

486. Proud, C. G. (1994) *Nature (London)* **371**, 747–748

486a. Wartmann, M., Hofer, P., Turowski, P., Saltiel, A. R., and Hynes, N. E. (1997) *J. Biol. Chem.* **272**, 3915–3923

486b. Weng, G., Bhalla, U. S., and Lyengar, R. (1999) *Science* **284**, 92–96

486c. Petosa, C., Masters, S. C., Bankston, L. A., Pohl, J., Wang, B., Fu, H., and Liddington, R. C. (1998) *J. Biol. Chem.* **273**, 16305–16310

487. Cobb, M. H., and Goldsmith, E. J. (1995) *J. Biol. Chem.* **270**, 14843–14846

488. Inglese, J., Koch, W. J., Touhara, K., and Lefkowitz, R. J. (1995) *Trends Biochem. Sci.* **20**, 151–156

489. van Biesen, T., Hawes, B. E., Raymond, J. R., Luttrell, L. M., Koch, W. J., and Lefkowitz, R. J. (1996) *J. Biol. Chem.* **271**, 1266–1269

490. Wu, J., Spiegel, S., and Sturgill, T. W. (1995) *J. Biol. Chem.* **270**, 11484–11488

491. Nurse, P. (2000) *Science* **289**, 1711–1716

492. Edgar, B. A., and Lehner, C. F. (1996) *Science* **274**, 1646–1652

493. Kirschner, M. (1992) *Trends Biochem. Sci.* **17**, 281–285

494. Horne, M. C., Goolsby, G. L., Donaldson, K. L., Tran, D., Neubauer, M., and Wahl, A. F. (1996) *J. Biol. Chem.* **271**, 6050–6061

495. Tyson, J. J., Novak, B., Odell, G. M., Chen, K., and Thron, C. D. (1996) *Trends Biochem. Sci.* **21**, 89–96

496. Sherr, C. J. (1996) *Science* **274**, 1672–1677

497. Hartwell, L. H., and Kastan, M. B. (1994) *Science* **266**, 1821–1828

498. Hennig, M., Jansonius, J. N., Terwisscha van Scheltinga, A. C., Dijkstra, B. W., and Schlesier, B. (1995) *J. Mol. Biol.* **254**, 237–246

499. Elledge, S. J. (1996) *Science* **274**, 1664–1672

499a. Chin, L., Pomerantz, J., and DePinho, R. A. (1998) *Trends Biochem. Sci.* **23**, 291–296

499b. Nead, M. A., Baglia, L. A., Antinore, M. J., Ludlow, J. W., and McCance, D. J. (1998) *EMBO J.* **17**, 2342–2352

499c. Bakiri, L., Lallemand, D., Bossy-Wetzel, E., and Yaniv, M. (2000) *EMBO J.* **19**, 2056–2068

500. Kussie, P. H., Gorina, S., Marechal, V., Elenbaas, B., Moreau, J., Levine, A. J., and Pavletich, N. P. (1996) *Science* **274**, 948–953

501. Gorina, S., and Pavletich, N. P. (1996) *Science* **274**, 1001–1005

502. Galaktionov, K., Chen, X., and Beach, D. (1996) *Nature (London)* **382**, 511–517

503. Agarwal, M. L., Taylor, W. R., Chernov, M. V., Chernova, O. B., and Stark, G. R. (1998) *J. Biol. Chem.* **273**, 1–4

504. Carr, A. M. (2000) *Science* **287**, 1765–1766

505. Hirao, A., Kong, Y.-Y., Matsuoka, S., Wakeham, A., Ruland, J., Yoshida, H., Liu, D., Elledge, S. J., and Mak, T. W. (2000) *Science* **287**, 1824–1827

506. Gatti, A., Li, H.-H., Traugh, J. A., and Liu, X. (2000) *Biochemistry* **39**, 9837–9842

Study Questions

1. Illustrated is a generalized metabolic pathway in which capital letters indicate major metabolites in the pathway, small letters indicate cofactors and numbers indicate enzymes catalyzing the reactions.

$$A \xrightarrow[\substack{m \quad n}]{1} B \underset{2}{\rightleftharpoons} C \underset{3}{\overset{\substack{o \quad p}}{\rightleftharpoons}} D \underset{4}{\rightleftharpoons} E \xrightarrow{5} F$$

List and describe four different ways in which the pathway might be regulated, referring to the specific enzymes, reactants, and cofactors indicated in the diagram. NOTE: Do not just refer to four different reactions as possible sites of regulation, but give four different general methods for regulation.

2. Describe the role of β-fructose-2,6-bisphosphate in the control of the further breakdown and resynthesis of glucose 1-phosphate.

3. It has been proposed that the substrate cycle involving phosphofructokinase and fructose bisphosphatase is used by bumblebees to warm their flight muscles to 30°C before flight begins.

Clark *et al.* (1973) *Biochem. J.* **134**, 589–597 found maximal rates of catalytic activity for both enzymes to be about 44 μ mol / min / g of fresh tissue. In flying bees glycolysis occurred at a rate of about 20 μ mol / min / g of tissue with no substrate cycling. In resting bees at 27°C no cycling was detected, but at 5°C substrate cycling occurred at the rate of 10.4 μ mol / min / g while glycolysis had slowed to 5.8 μ mol / min / g. If the cycling provides heat to warm the insect, estimate how long it would take to reach 30°C if a cold (5°C) bee could carry out cycling at the maximum rate of 40 μ mol / min / g and if no heat were lost to the surroundings.

4. a) Compare the reaction cycle of a small GTPase (G protein) that is regulated by the actions of a GTPase-activating protein (GAP) and a guanine-nucleotide exchange factor (GEF) with that of a G protein linked to a receptor.

   b) Some G proteins have been described as "timed switches" and others as "triggered switches." In what ways might these two groups differ? See Kjeldgaard *et al.* (1996) *FASEB J.* **10**, 1347–1368.

View of a modified bovine fibrinogen molecule. The 45-nm-long disulfide-linked dimer is composed of three nonidentical polypeptide chains. The N termini of the six chains from the two halves come together in the center in a small globular "disulfide knot." The C termini form globular domains at the ends. The 340-kDa molecule has been treated with a lysine-specific protease which has removed portions of two chains to give the ~285-kDa molecule whose crystal structure is shown. Arrows point to attached oligosaccharides. From Brown *et al.* (2000) *Proc. Natl. Acad. Sci. U. S. A.* **97**, 85–90. Courtesy of Carolyn Cohen.

Contents

Transferring Groups by Displacement Reactions

12

The majority of enzymes that are apt to be mentioned in any discussion of metabolism catalyze nucleophilic displacement reactions (Type 1, Table 10-1). These include most of the reactions by which the energy of ATP cleavage is harnessed and by which polymers are assembled from monomers. They include reactions by which pieces, large or small, are transferred onto or off of polymers as well as the reactions by which polymers are cleaved into pieces.[1–3d]

In these reactions a **nucleophilic group** (a base), designated B⁻ in Table 10-1, approaches an **electrophilic center,** often an electron-deficient carbon or phosphorus atom. It forms a bond with this atom, at the same time displacing some other atom, usually O, N, S, or C. The displaced atom leaves with its bonding electron pair and with whatever other chemical group is attached, the entire unit being called the **leaving group**. This is designated YH in Table 10-1. Simultaneous or subsequent donation of a proton from an acidic group of the enzyme, or from water, to the O, N, or S atom of the leaving group is usually required to complete the reaction. The entering base B, which may or may not carry a negative charge, must often be generated by enzyme-catalyzed removal of a proton from the conjugate acid BH. If BH is water, the enzyme is a **hydrolase**; otherwise, it is called a **transferase**.

For purposes of classifying the reactions of metabolism, in this book the nucleophilic displacements are grouped into four subtypes (Table 10-1). These are displacements on (A) a saturated carbon atom, often from a methyl group or a glycosyl group; (B) a carbonyl group of an ester, thioester, or amide; (C) a phospho group; or (D) a sulfur atom. In addition, many enzymes employ in sequence a displacement on a carbon atom followed by a second displacement on a phosphorus atom (or vice versa).

A. Factors Affecting Rates of a Displacement Reaction

In Chapter 9, the displacement of an iodide ion from methyl iodide by a hydroxide ion (Eq. 9-76) was considered. Can we similarly displace a methyl group from ethane, CH_3–CH_3, to break the C–C bond and form CH_3OH? The answer is no. Ethane is perfectly stable in sodium hydroxide and is not cleaved by a simple displacement process within our bodies. Likewise, long hydrocarbon chains such as those in the fatty acids cannot be broken by a corresponding process during metabolism of fatty acids. Not every structure will allow a nucleophilic displacement reaction to occur and not every anion or neutral base can act to displace another group.

At least four factors affect the likelihood of a displacement reaction:[1–3d]

(1) *The position of the equilibrium in the overall reaction.* An example is provided by the hydrolases that catalyze cleavage of amide, ester, and phosphodiester linkages using water as the entering nucleophile. Because enzymes usually act in an environment of high water content, the equilibrium almost always favors hydrolysis rather than the reverse reactions of synthesis. However, in a nonaqueous solvent the same enzyme will catalyze synthetic reactions.

(2) *The reactivity of the entering nucleophile.* Nucleophilic reactivity or **nucleophilicity** is partly determined by basicity. Compounds that are strong bases tend to react more rapidly in nonenzymatic displacement reactions than do weaker bases; the hydroxyl ion, HO⁻, is a better nucleophile than –COO⁻. However, enzymes usually act optimally near a neutral pH. Under these conditions the –COO⁻ group may be

more reactive than strongly basic groups such as $-NH_2$ or HO^- because at pH 7 the latter groups will be almost completely protonated and the active nucleophile will be present in low concentrations.

A second factor affecting nucleophilic reactivity is **polarizability**, which is the ease with which the electronic distribution around an atom or within a chemical group can be distorted. A large atomic radius and the presence of double bonds in a group both tend to increase polarizability. In most cases, the more highly polarizable a group, the more rapidly it will react in a nucleophilic displacement, apparently because a polarizable group is able to form a partial bond at a greater distance than can a nonpolarizable group. Thus, I^- is more reactive than Br^-, which is more reactive than Cl^-. Polarizable bases such as imidazole are often much more reactive than nonpolarizable ones such as $-NH_2$. Sulfur compounds also tend to have a high nucleophilic reactivity. In displacement reactions on carbonyl groups (reaction type 1B), the less polarizable "hard" nucleophiles are more reactive than polarizable ones such as I^-. Attempts have been made to relate quantitatively nucleophilic reactivity to basicity plus polarizability.[1]

Certain chemical groups, e.g., those in which an atom with unpaired electrons is directly bonded to the nucleophilic center undergoing reaction, are more reactive than others of similar basicity. This **α effect** has been invoked to explain the high reactivity of the poisons hydroxylamine (NH_2OH) and cyanide ion[4] and other puzzling results.[1]

(3) *The chemical nature of the leaving group that is displaced.* The chemistry of the leaving group affects both the rate and the equilibrium position in nucleophilic displacements. The leaving group must accommodate a pair of electrons and often must bear a negative charge. A methyl group displaced from ethane or methane as CH_3^- would be an extremely poor leaving group; the pK_a of methane as an acid[5] has been estimated as 47. Iodide ion is a good leaving group, but F^- is over 10^4 times poorer.[5] In an aqueous medium, phosphate is a much better leaving group than OH^-, and pyrophosphate and tripolyphosphate are even better.

(4) *Special structural features present in the substrate.* Enzymes are usually constructed so that they recognize unique features in substrates. As a consequence, they have many ways of lowering the energy of the transition state and increasing the apparent nucleophilic reactivity.

B. Nucleophilic Displacements on Singly Bonded Carbon Atoms

The enzymatic counterpart of Eq. 9-76, the displacement of I^- from a saturated carbon atom by HO^-,

is catalyzed by the **haloacid dehalogenases** of soil pseudomonads (Eq. 12-1).[6–8] Even the poor leaving group F^- can be displaced by OH^- in the active site of these enzymes.

$$HO^- + F-CH_2COO^- \rightarrow HO-CH_2-COO^- + F^-$$
(12-1)

1. Inversion as a Criterion of Mechanism

An interesting result was obtained when a haloacid dehalogenase was tested with a substrate containing a chiral center.[6] Reaction of L-2-chloropropionate with hydroxyl ion gave only D-lactate, a compound with a chirality opposite to that of the reactant. A plausible explanation is that the hydroxyl ion attacks the central carbon from behind the chlorine atom (Eq. 12-2). The resulting five-bonded transition state (center) loses a chloride ion to form the product D-lactate in which inversion of configuration has occurred. *Inversion always accompanies single displacement reactions in which bond breaking and bond formation occur synchronously, as in Eq. 12-2.* This is true for both enzymatic and nonenzymatic reactions. However, the occurrence of inversion does not rule out more complex mechanisms. Indeed, in this case the displacing group is evidently not HO^- but a carboxylate group from the enzyme.

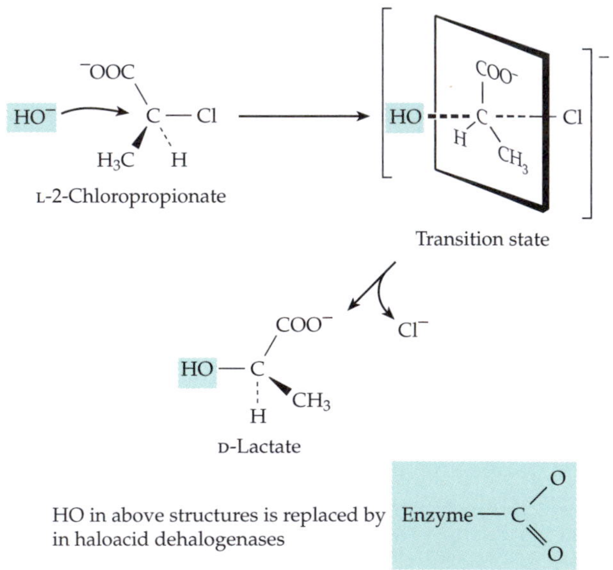

L-2-Chloropropionate

Transition state

D-Lactate

HO in above structures is replaced by $Enzyme-C$ in haloacid dehalogenases

(12-2)

In one of the haloacid dehalogenases, a 232-residue protein for which the three-dimensional structure is known,[8,9] Asp 10 is in a position to carry out the initial attack which would give an enzyme-bound intermediate with an ester linkage:

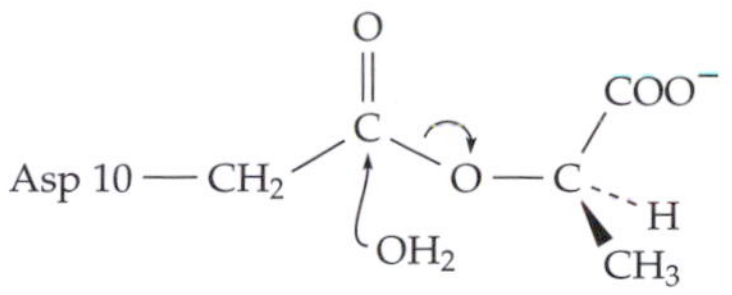

Hydrolytic cleavage, as indicated in the diagram, yields the product D-lactate.[7] Thus, we have a direct displacement with inversion followed by an additional hydrolytic step. This is an example of **covalent catalysis**, the enzyme providing a well-oriented reactive group instead of generating a hydroxyl group from a bound water molecule.

Related **haloalkane dehalogenases** as well as **epoxide hydrolases** and a large superfamily of other enzymes utilize similar mechanisms.[10,11,11a] In the active site of a haloalkane dehalogenase from *Xanthobacter* (Fig. 12-1) the carboxylate of Asp 124 acts as the attacking nucleophile that displaces Cl⁻ from the substrate dichloroethane, which is held in a small cavity with the aid of two tryptophan indole rings.[12–14] However, the histidine–aspartate pair and the bound water molecule shown in Fig. 12-1 are essential for the subsequent hydrolysis of the intermediate ester. (see also Fig. 12-11).[15] The substrate shown in Fig. 12-1 is 1,2-dichloroethane, a widespread environmental pollutant that is not known to occur naturally. An interesting question considered by Pries *et al.*[16] is how this dehalogenase has evolved in the years since 1922 when industrial production of dichloroethane began.

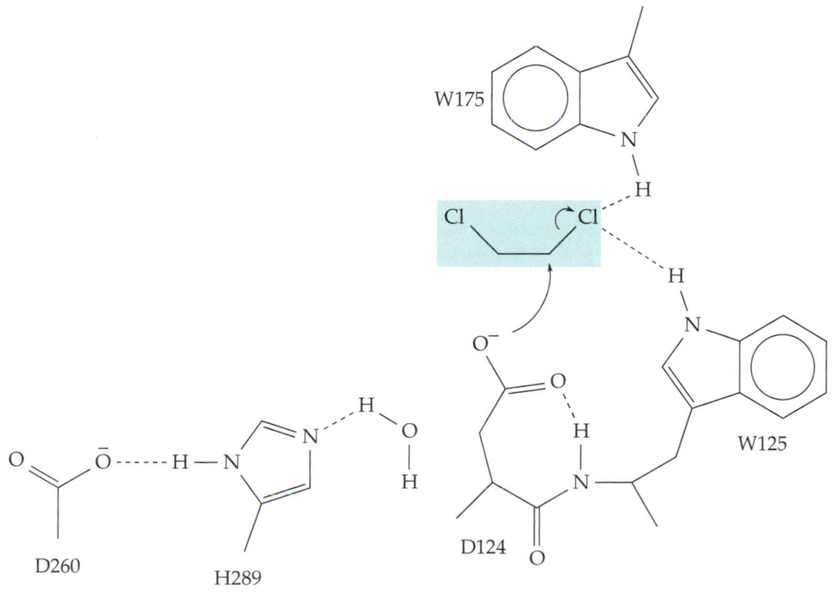

2. Transmethylation

Nucleophilic attack on a methyl group or other alkyl group occurs most readily if the carbon is attached to an atom bearing a positive charge, for example, the sulfur atom of a sulfonium ion such as that present in **S-adenosylmethionine** (abbreviated **AdoMet** or SAM). The ensuing **transmethylation** reaction results in the transfer of the methyl group from the sulfur to the attacking nucleophile (Eq. 12-3). Transmethylation is an important metabolic process by which various oxygen, nitrogen, and other atoms at precise positions in proteins, nucleic acids, phospholipids, and other small molecules are methylated.[17] The methyl group donor is usually S-adenosylmethionine. This compound has two chiral centers, one at the α-carbon of the amino acid and one at the sulfur atom, with an unshared electron pair serving as the fourth group around the S atom. In the naturally occurring AdoMet both centers have the S configuration.[18] The reaction of Eq. 12-3, which is catalyzed by **catechol O-methyltransferase** (COMT), inactivates the neurotransmitters adrenaline, dopamine, and related compounds by methylation. When the substrate contains a chiral methyl group (– C¹H²H³H; see also Chapter 13), the inversion of the methyl group expected for a simple S_N2-like reaction is observed.[19]

Structural features similar to those of COMT[20] are found in glycine N-methyltransferase[21] and guanidinoacetate methyltransferase[22] from liver, and transferases that place methyl groups on N-6 of adenine[23–25a] and N-4 of cytosine or on C-5 of cytosine[26] in nucleic acids.[27] A stereoscopic view of the active site of glycine N-methyltransferase is shown in Fig. 12-2. An acetate ion present in the active site in the crystal has been converted into glycine by computer-assisted modeling. The amino group of the bound glycine is adjacent to the methyl group of AdoMet. A nearby glutamic acid side chain (E15) may have removed a proton from the glycine dipolar ion to create the free –NH₂ group required for the reaction. In COMT a magnesium Mg^{2+} ion binds to the two aromatic hydroxyl groups of the substrate and helps to hold it correctly in the active site.[28–29a]

In the case of the more difficult C-methylation of uracil 54 in transfer RNA, an –SH group from Cys 324 of the methyltransferase adds to the C-6 position of U54 of the substrate, which may be any of the *E. coli* tRNAs (Eq. 12-4, step *a*).[30] In the

Figure 12-1 The active site structure of haloalkane dehalogenase from *Xanthobacter autotrophicus* with a molecule of bound dichloroethane. See Pries *et al.*[13] The arrows illustrate the initial nucleophilic displacement. The D260 – H289 pair is essential for the subsequent hydrolysis of the intermediate ester formed in the initial step.

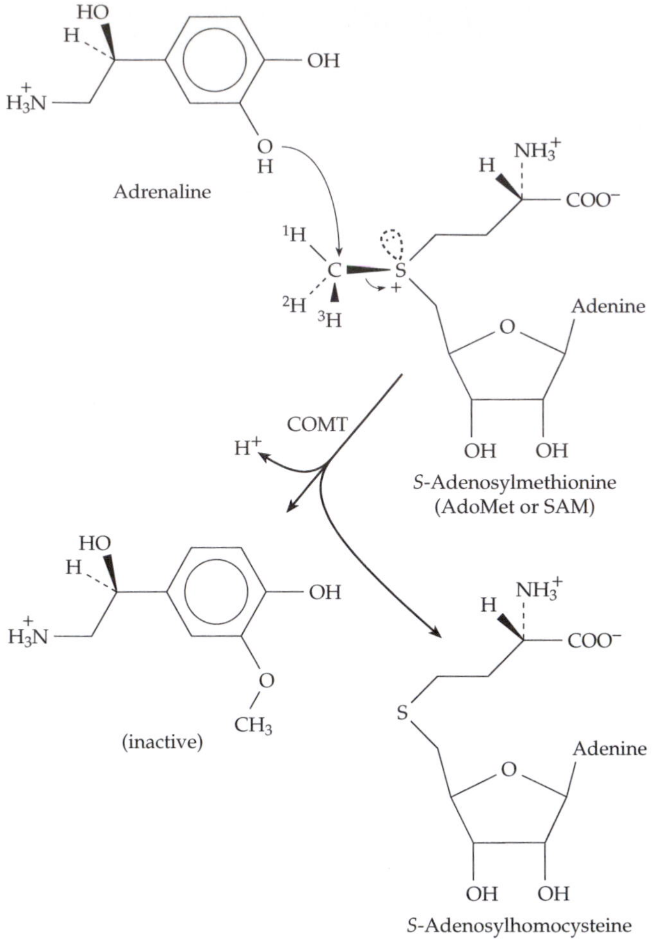

Adrenaline

S-Adenosylmethionine
(AdoMet or SAM)

COMT

(inactive)

S-Adenosylhomocysteine

(12-3)

adduct carbon atom C-5 acquires substantial nucleophilic character which permits the transfer of the methyl group in step *b*. The adduct breaks up (step *c*) and the product is released.

Methylation of nucleic acids is considered further in Chapers 27 and 28. Methylation of carboxyl groups

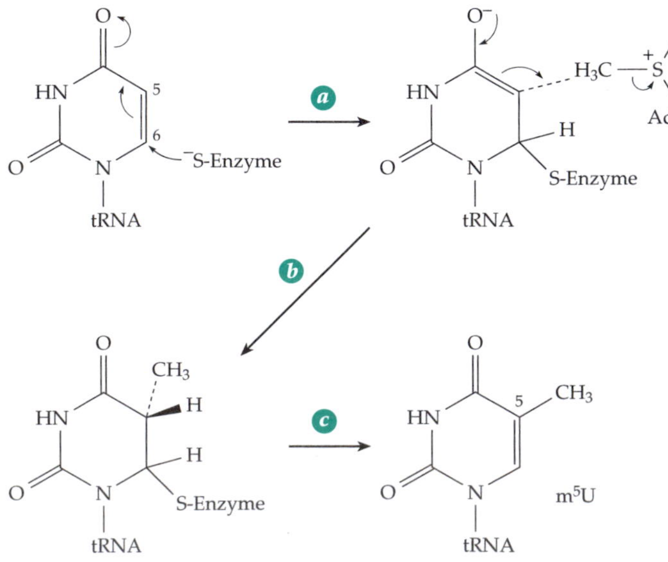

(12-4)

of certain proteins regulates motion of bacterial flagella (Chapter 19) and other aspects of metabolism[31] while methylation of isomerized aspartyl residues is part of a protein repair process (Box 12-A). Methyl groups are usually transferred from S-adenosylmethionine, but sometimes from a folic acid derivative (Chapter 15) or from a cobalt atom of a corrin ring (Chapter 16).

3. Kinetic Isotope Effects

An S–^{12}C bond breaks a little faster than an S–^{13}C bond in a nucleophilic displacement reaction. This **primary kinetic isotope effect** (KIE)[3a,3b,31a–d] is usually discussed first for breakage of a C–H bond. In a linear transition state, in which the C–H bond is being stretched, then cleaved, the difference between the transition state barrier for a C–^1H bond and a C–^2H bond is thought to arise principally from a difference in the energies of the C–H stretching vibration.[3a] This vibrational energy, in the ground state of a molecule (the zero-point energy) is equal to $1/2\,h\nu_0$ where ν_0 may be observed in the infrared absorption spectrum (see Fig. 23-2). For a C–^2H bond, with a stretching vibration at a wave number of ~2900 cm^{-1}, the zero point energy is about + 17.4 kJ mol^{-1}. For a C–^2H bond, with a stretching wave number of ~2200 cm^{-1}, it is about + 17.4 kJ mol^{-1}. This difference is a result of the difference in mass of ^1H (1.67 x 10^{-24} g) and ^2H (2.34 x 10^{-24} g). The isotope effect arises because the vibration occurs along the axis of the bond being broken, and the vibrational energy is converted into translational motion along the reaction coordinate, in effect lowering the transition state barrier by $\Delta G^{0\ddagger}$. The difference in this effect between ^1H and ^2H ($\Delta\Delta G^{0\ddagger}$) gives the ratio of expected rate constants as follows:

$$\frac{k^{C-H}}{k^{C-D}} = e^{-\Delta\Delta G^{0\pm}/RT} = 7$$

For heavier nuclei, such as C, N, O the kinetic isotope effect is much smaller.

When ^{13}C was introduced into the methyl group of S-adenosylmethionine and its rate of reaction with catechol O-methyltransferase was compared with that of the normal ^{12}C-containing substrate, the expected effect on V_{max}, expressed as a first-order rate constant, was seen: $k_{12}/k_{13} = 1.09 \pm 0.05$. This effect is small but it can be measured reliably and establishes that the methyl transfer step rather than substrate binding, product release, or a conformational change in the protein is rate limiting.[32]

Substitution of ^1H by ^2H in the CH$_3$ group has a larger effect. This **secondary kinetic isotope effect** (or α-deuterium

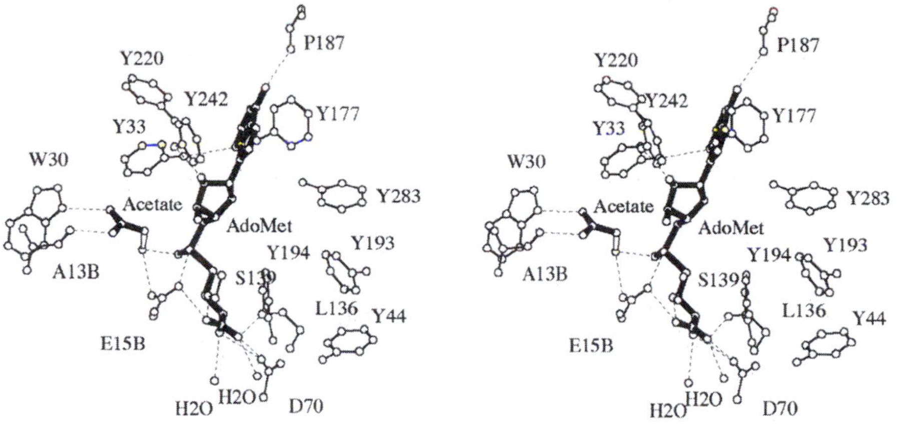

Fig. 12-2 *S*-Adenosylmethionine (AdoMet, solid bonds) bound in the active site of glycine *N*-methyltransferase together with an acetate ion bound in the glycine site. Glycine was built by attaching an amino group (open bond) to the acetate (solid bonds). Possible polar interactions (O–O and O–N < 0.31 nm and O–S < 0.4 nm) are indicated by dotted lines. Tyrosine residues located at the inner surface of the active site are also shown. From Fu *et al.*[21]

effect) arises because of small differences in the vibrational energies of the methyl group resulting from the differences in mass of ^{1}H and ^{2}H. It often leads to a more rapid reaction for the ^{1}H-containing substrates. In a model nonenzymatic displacement reaction of a similar type[32] the ratio of rate constants was $k_{1_H} / k_{2_H} = 1.17 \pm 0.02$. However, for COMT an inverse α-deuterium effect was seen: $k_{(c^1 H_3)} / k_{(c^2 H_3)} = 0.83 \pm 0.05$. Theoretical calculations suggested that such an effect might be observed if the enzyme compresses an S_N2-like transition state, shortening the bonds from the central carbon atom to both the oxygen atom of the entering nucleophile and the sulfur atom of the leaving group.[33,34] However, more recent calculations suggest that it is difficult to draw such conclusions from secondary isotope effects.[35,36] Calculations by Zheng, Kahn, and Bruice predict that in the gas phase the two substrates, upon collision, will react with a very low energy barrier.[29,29a] For this enzyme, as for the haloalkane dehalogenase (Fig. 12-1), it has been concluded that the enzyme excludes water from the active site and binds the two substrates very close together and in a correct orientation for reaction. The computations predicted secondary isotope effects for the enzymatic reaction similar to those measured by Hegazi *et al.*[32] suggesting that the transition states for the enzymatic reactions closely resemble those for the gas phase.

4. Glycosyltransferases

The polarization of a single C–O bond in an ether is quantitatively much less than that of the C–S$^+$ bond of *S*-adenosylmethionine, and simple ethers are not readily cleaved by displacement reactions. However, glycosides, which contain a carbon atom attached to *two* oxygen atoms, do undergo displacement reactions which lead to **hydrolysis**, **phosphorolysis**, or **glycosyl exchange**. The corresponding enzymes are glycosylhydrolases, phosphorylases, and glycosyltransferases. Two characteristics are commonly used to classify glycosylhydrolases that act on polysaccharides: **Endoglycanases** cut at random locations within the chains of sugar units, while **exoglycanases** cut only at one end or another, usually at the nonreducing end. **Inverting enzymes** invert the configuration at the anomeric carbon atom which they attack, while **retaining enzymes** do not. Although simple glycosides undergo acid-catalyzed hydrolysis readily, uncatalyzed hydrolysis is extremely slow, the estimated first order rate constant[37] at 25°C being about 2×10^{-15} s^{-1}. Polysaccharides may be the most stable of all biopolymers.

Glycosyltransferases are numerous. For example, the amino acid sequences for about 500 glucosidases, which hydrolyze linkages between glucose residues, have been determined and this one group of enzymes has been classified into over 60 families[38,39] and eight larger groups.[40] Many three-dimensional structures have been reported and numerous studies of the reaction mechanism for both enzymatic and nonenzymatic hydrolysis of glycosides have been conducted. Most of the experimental results have been carefully verified. Nevertheless, they serve to illustrate how difficult it is to understand how enzymes catalyze this simple type of reaction.[41]

Inversion or retention? Equation 12-5 pictures the reaction of a glycoside (such as a glucose unit at

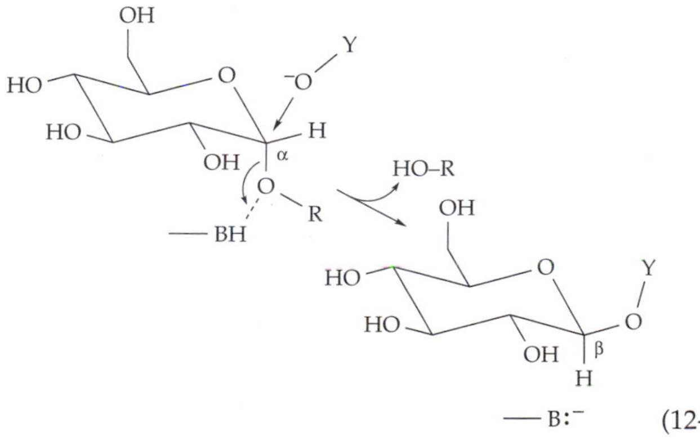

$$(12\text{-}5)$$

BOX 12-A CARBOXYMETHYLATION OF PROTEINS

Methylation and demethylation of the carboxyl groups of the side chains of specific glutamyl residues play a role in bacterial chemotaxis (Chapter 19, Section A). However, similar functions in eukaryotes have not been found. There is a specific methyltransferase that methylates C-terminal carboxyl groups on prenylated and sometimes palmitoylated peptides.[a] Another type of carboxylmethyltransferase acts on only a small percentage of many proteins and forms labile methyl esters. These methyltransferases have a dual specificity, acting on L-**isospartyl** residues and usually also on D-**aspartyl** residues. Both of these amino acids can arise from deamidation of asparagine, especially in Asn-Gly sequences (Eq. 2-24 and steps *a* and *b* of following scheme).[b–g] A similar isomerization of aspartyl residues occurs more slowly. Asn-Gly sequences are present in many proteins and provide weak linkages whose isomerization is an inevitable aspect of aging.[e,h] Furthermore, the α proton of the cyclic imide (green box) is more readily dissociated than in a standard peptide, leading to racemization (step *c*). The isoaspartyl and D-aspartyl residues provide kinks in the peptide chain which may interfere with normal function and turnover.

The L-isoaspartyl / D-aspartyl methyltransferase (step *d*), which is especially abundant in the brain,[i,j] provides for partial repair of these defects.[k–o] The methyl esters of isoaspartyl residues readily undergo demethylation with a return to a cyclic imide (step *e*). The cyclic imide is opened hydrolytically (step *c*) in part to an isoaspartyl residue, but in part to a normal aspartyl form. The combined action of the carboxylmethyltransferase and the demethylation reaction tends to repair the isomerized linkages. Methylation of D- aspartyl residues returns them to the cyclic imide (steps *f*, *e*), allowing them to also return to the normal L configuration. Nevertheless, the protein will have a different net charge than it did originally and must be considered a new modified form.

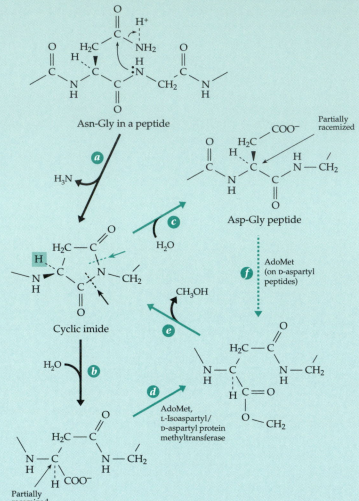

Asn-Gly in a peptide

Asp-Gly peptide

Cyclic imide

Partially racemized

[a] Dai, Q., Choy, E., Chiu, V., Romano, J., Slivka, S. R., Steitz, S. A., Michaelis, S., and Philips, M. R. (1998) *J. Biol. Chem.* **273**, 15030–15034

[b] Geiger, T., and Clarke, S. (1987) *J. Biol. Chem.* **262**, 785–794

[c] Tyler-Cross, R., and Schirch, V. (1991) *J. Biol. Chem.* **266**, 22549–22556

[d] Brennan, T. V., and Clarke, S. (1993) *Protein Sci.* **2**, 331–338

[e] Paranandi, M. V., Guzzetta, A. W., Hancock, W. S., and Aswad, D. W. (1994) *J. Biol. Chem.* **269**, 243–253

[f] Donato, A. D., Ciardiello, M. A., de Nigris, M., Piccoli, R., Mazzarella, L., and D'Alessio, G. (1993) *J. Biol. Chem.* **268**, 4745–4751

[g] Tomizawa, H., Yamada, H., Ueda, T., and Imoto, T. (1994) *Biochemistry* **33**, 8770–8774

[h] Man, E. H., Sandhouse, M. E., Burg, J., and Fisher, G. H. (1983) *Science* **220**, 1407–1408

[i] Orpiszewski, J., and Aswad, D. W. (1996) *J. Biol. Chem.* **271**, 22952–22968

[j] Najbauer, J., Orpiszewski, J., and Aswad, D. W. (1996) *Biochemistry* **35**, 5183–5190

[k] Johnson, B. A., Murray, E. D. J., Clarke, S., Glass, D. B., and Aswad, D. W. (1987) *J. Biol. Chem.* **262**, 5622–5629

[l] Brennan, T. V., Anderson, J. W., Jia, Z., Waygood, E. B., and Clarke, S. (1994) *J. Biol. Chem.* **269**, 24586–24595

[m] Mudgett, M. B., and Clarke, S. (1993) *Biochemistry* **32**, 11100–11111

[n] Mudgett, M. B., and Clarke, S. (1994) *J. Biol. Chem.* **269**, 25605–25612

[o] Aswad, D. W., ed. (1995) *Deamidation and Isoaspartate Formation in Peptides and Proteins*, CRC Press, Boca Raton, Florida

the end of a starch chain) with a nucleophile Y–O⁻ as the displacing group. An enzyme-bound acidic group –BH is shown assisting the process. Inversion of configuration with formation of a product of the β configuration at the anomeric carbon atom would be predicted and is observed for many of these enzymes. However, many others do *not* cause inversion.[41,42] An example is the reaction catalyzed by **sucrose phosphorylase** from *Pseudomonas saccharophila*:

Sucrose (an α-glucoside of fructose) +
inorganic phosphate (P$_i$) →
α-D-glucose 1-phosphate + fructose
(12-6)

Two possible explanations for the lack of inversion during this reaction are that the enzyme acts by a **double-displacement** reaction or through a stabilized **carbocationic intermediate**. Let us consider these possibilities in turn.

Double-displacement mechanisms. In a double-displacement mechanism sucrose phosphorylase would catalyze two consecutive single displacements, each with inversion. A nucleophilic group of the enzyme would react in Eq. 12-7, step *a*. In step *b*, a phosphate would react to regenerate the enzyme with its free nucleophilic group –B⁻.

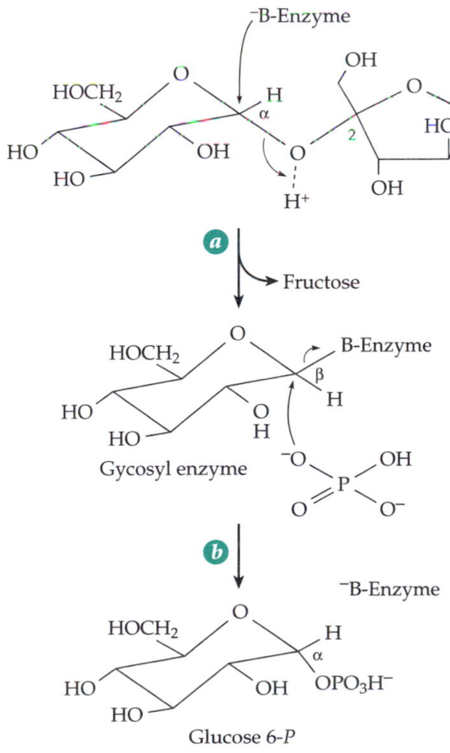

Gycosyl enzyme

Glucose 6-*P*

(12-7)

Four kinds of experiments were used to identify this double-displacement mechanism.

(1) *Kinetics.* In a double-displacement mechanism the enzyme shuttles between free enzyme and the intermediate carrying the substrate fragment (here, the glycosyl enzyme). With sucrose phosphorylase the maximum velocity varies with the concentrations of sucrose and HPO$_4^{2-}$ in the characteristic fashion expected for this "ping-pong" mechanism (Eq. 9-47).[43]

(2) *Exchange reactions.* In a double-displacement mechanism sucrose containing ¹⁴C in the fructose portion of the molecule should react with free enzyme E to form glycosyl enzyme and free radioactive fructose (Eq. 12-8). The ¹⁴C-containing groups are designated here by the asterisks.

D-Glucosyl-fructose* + E ⇄ glucosyl-E + fructose*
(Sucrose)
(12-8)

If a very low molar concentration of enzyme is present, and a large excess of nonradioactive fructose is added, the enzyme will catalyze no net reaction but will change back and forth repeatedly between the free enzyme and glucosyl enzyme. Each time, in the reverse reaction, it will make use primarily of unlabeled fructose. The net effect will be catalysis of a sucrose–fructose exchange:

Sucrose* + fructose ⇄ sucrose + fructose*
(12-9)

This exchange reaction, as well as other predicted exchanges, has been observed.[44] Although the exchange criterion of the mechanism is often applied to enzymatic processes, the observation of exchange reactions does not *prove* the existence of a covalently bound intermediate. Furthermore, enzymes using double-displacement mechanisms may not always catalyze the expected exchanges.

(3) *Arsenolysis.* Sucrose phosphorylase also catalyzes the cleavage of sucrose by arsenate and promotes a rapid cleavage (**arsenolysis**) of glucose 1-phosphate to free glucose. This reaction is evidently a result of a displacement by arsenate on a glycosyl enzyme intermediate. The resulting unstable glucose 1-arsenate (see Box 12-B) is hydrolyzed rapidly. Arsenolysis is a general way of trapping reactive enzyme-bound intermediates that normally react with phosphate groups. Arsenate is one of many substrate analogs that can be used to siphon off reactive enzyme-bound intermediates into nonproductive side paths.

(4) *Identification of glycosyl-enzyme intermediates.* Studies with pure enzymes often make it possible to confirm directly the existence of enzyme-bound intermediates. The intermediates detected are frequently **glycosyl esters** of glutamate or aspartate side chain

BOX 12-B ARSENIC

Arsenate, AsO_4^{3-}, is chemically similar to phosphate in size and geometry and in its ability to enter into biochemical reactions. However, arsenate esters are far less stable than phosphate esters. If formed in the active site of an enzyme, they are quickly hydrolyzed upon dissociation from the enzyme. This fact accounts for some of the toxicity of arsenic compounds.[a]

Arsenate will replace phosphate in all phosphorolytic reactions, e.g., in the cleavage of glycogen by glycogen phosphorylase, of sucrose by sucrose phosphorylase, and in the action of purine nucleoside phosphorylase.[b] Glucose 1-arsenate or ribose-1-arsenate is presumably a transient intermediate which is hydrolyzed to glucose. The overall process is called **arsenolysis**. Another reaction in which arsenate can replace phosphate is the oxidation of glyceraldehyde 3-phosphate in the presence of P_i to form 1,3-bisphosphoglycerate:

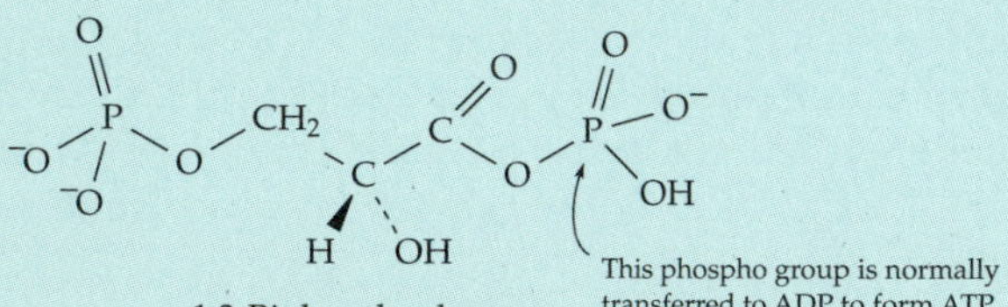

1,3-Biphosphoglycerate

This phospho group is normally transferred to ADP to form ATP

The subsequent transfer of the 1-phospho group to ADP is an important energy-yielding step in metabolism (Chapter 12). When arsenate substitutes for phosphate the acyl arsenate (1-arseno-3-phosphoglycerate) is hydrolyzed to 3-phosphoglycerate. As a consequence, in the presence of arsenate oxidation of 3-phosphoglyceraldehyde continues but ATP synthesis ceases. Arsenate is said to **uncouple** phosphorylation from oxidation. Arsenate can also partially replace phosphate in stimulating the respiration of mitochondria and is an uncoupler of oxidative phosphorylation (Chapter 18). Enzymes that normally act on a phosphorylated substrate will usually catalyze a slow reaction of the corresponding unphosphorylated substrate in the presence of arsenate. Apparently, the arsenate ester of the substrate forms transiently on the enzyme surface, permitting the reaction to occur.

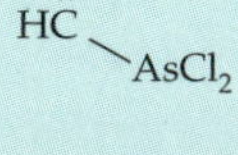

Lewisite

Arsenite is noted for its tendency to react rapidly with thiol groups,[c] especially with pairs of adjacent (vicinal) or closely spaced –SH groups[d] as in lipoic acid. By blocking oxidative enzymes requiring

lipoic acid (Chapter 15), arsenite causes the accumulation of pyruvate and of other 2-oxo acids. Similar chemistry underlies the action of the mustard gas Lewisite, which also attacks lipoic acid.[de]

Many, perhaps all, organisms have enzyme systems for protection against arsenic compounds.

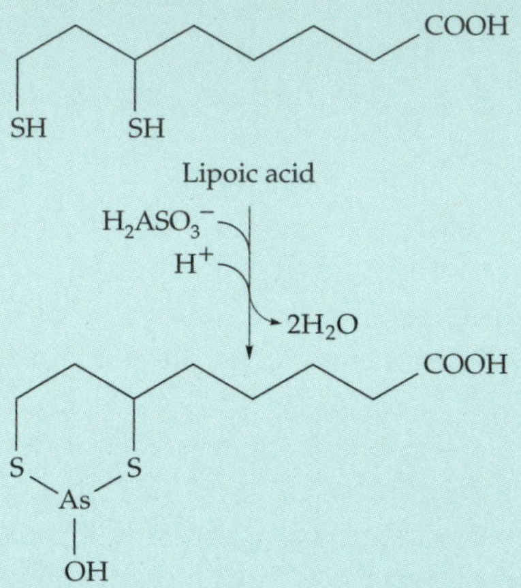

In *E. coli* arsenate is reduced to arsenite by a glutaredoxin- and NADH-dependent system.[e–g] The arsenite as well as antimonite and tellurite are pumped out by an ATP-dependent transporter. The genes for reductase, periplasmic-binding protein, and transporter components are encoded in a conjugative plasmid.[h,i] A quite similar system functions in yeast.[j]

Marine algae as well as some higher aquatic plants detoxify and excrete arsenate by conversion to various water-soluble organic forms such as trimethyarsonium lactic acid (Chapter 8) and the following ribofuranoside.[a]

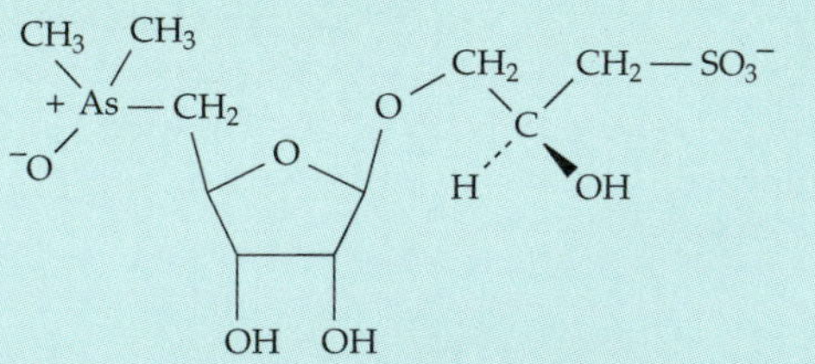

2-Hydroxy-3-sulfopropyl-5-deoxy-5-
(dimethylarseno)-ribofuranoside

Arsenic-containing phospholipids are also formed (Chapter 8).[k]

Arsenic is present at high levels in some soils and contamination of drinking water with arsenic is a major problem is some areas of the world. In West Bengal, India, millions of people drink contaminated well water. As a result hundreds of thousands have developed debilitating nodular keratoses on their feet.[l,m] The problem is made worse by the increasing

BOX 12-B (continued)

use of fresh water for irrigation and the difficulty of removing arsenic contamination.

Although it is most known for its toxicity, arsenic may be an essential nutrient. Data from feeding of chicks, goats, rats, and miniature pigs suggest a probable human need for arsenic of ~12 µg / day.[n]

Compounds of arsenic have been used in medicine for over 2000 years, but only in the past century

have specific arsenicals been created as drugs. In 1905, it was discovered that sodium arsanilate is toxic to trypanosomes. The development by P. Ehrlich of **arsenicals** for the treatment of syphilis (in 1909) first focused attention on the possibility of effective chemotherapy against bacterial infections.

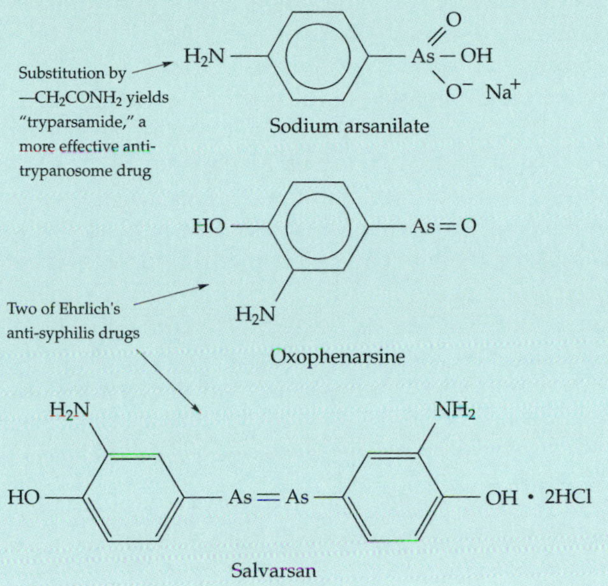

Substitution by —CH$_2$CONH$_2$ yields "tryparsamide," a more effective anti-trypanosome drug

Sodium arsanilate

Two of Ehrlich's anti-syphilis drugs

Oxophenarsine

Salvarsan

[a] Knowles, F. C., and Benson, A. A. (1983) *Trends Biochem. Sci.* **8**, 178–180
[b] Kline, P. C., and Schramm, V. L. (1995) *Biochemistry* **34**, 1153–1162
[c] Lam, W.-C., Tsao, D. H. H., Maki, A. H., Maegley, K. A., and Reich, N. O. (1992) *Biochemistry* **31**, 10438–10442
[d] Li, J., and Pickart, C. M. (1995) *Biochemistry* **34**, 15829–15837
[de] Ord, M. G., and Stocken, L. A. (2000) *Trends Biochem. Sci.* **25**, 253–256
[e] Silver, S., Nucifora, G., Chu, L., and Misra, T. K. (1989) *Trends Biochem. Sci.* **14**, 76–80
[f] Ji, G., Garber, E. A. E., Armes, L. G., Chen, C.-M., Fuchs, J. A., and Silver, S. (1994) *Biochemistry* **33**, 7294–7299
[g] Rosen, B. P., Weigel, U., Karkaria, C., and Gangola, P. (1988) *J. Biol. Chem.* **263**, 3067–3070
[h] Gladysheva, T. B., Oden, K. L., and Rosen, B. P. (1994) *Biochemistry* **33**, 7288–7293
[i] Chen, Y., and Rosen, B. P. (1997) *J. Biol. Chem.* **272**, 14257–14262
[j] Wysocki, R., Bobrowicz, P., and Ulaszewski, S. (1997) *J. Biol. Chem.* **272**, 30061–30066
[k] Cooney, R. V., Mumma, R. O., and Benson, A. A. (1978) *Proc. Natl. Acad. Sci. U.S.A.* **75**, 4262–4264
[l] Bagla, P., and Kaiser, J. (1996) *Science* **274**, 174–175
[m] Saha, D. P., and Subramanian, K. S. (1996) *Science* **274**, 1287–1288
[n] Nielsen, F. H. (1991) *FASEB J.* **5**, 2661–2667

carboxyl groups. Such an intermediate, fructofuranosyl-enzyme, was identified tentatively for sucrose phosphorylase and also for a related levan sucrase.[45] Recently, identification of glycosyl-enzyme intermediates has been accomplished for many other glycosyltransferases. Among these are human pancreatic and salivary α-amylases,[42,46] α-glucosidases, and some cellulases and xylanases.

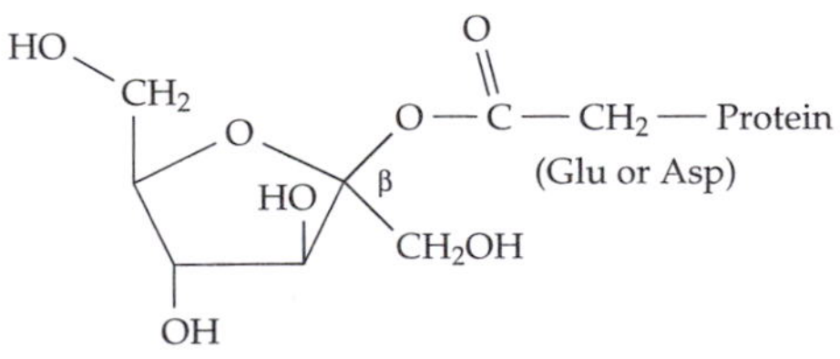

Fructofuranosyl - enzyme

For these hydrolytic enzymes the glycosyl-enzyme would be attacked by a hydroxyl ion derived from H$_2$O, whose deprotonation would presumably be assisted by the conjugate base (—B: in Eq. 12-5) of the catalytic acidic group.

A convenient way to form and identify covalently linked glucosyl-enzyme intermediates, developed by Withers and coworkers, employs enzyme-activated inhibitors such as 2-deoxy-2-fluoroglycosyl fluorides (Eq. 12-10).[42,47,48] The 2-F substituent greatly decreases both the rate of formation of a glycosyl-enzyme and its rate of breakdown by hydrolysis or transglycosylation. This may be in part because these compounds lack the 2–OH group which helps to stabilize, by hydrogen bonding, the complexes of normal substrates. A second factor is the high electronegativity of –F versus –OH, which leads to significant loss of stability for a transition state in which the anomeric carbon atom carries a significant amount of positive charge (see next section). Having a good leaving group, such as fluorine or 2,4-dinitrophenyl, the compounds react rapidly to give stable glucosyl-enzymes which can be characterized. In the example shown in Eq. 12-10, the mannosyl-enzyme was identified by the chemical shifts and linewidths of the ^{19}F resonances of the intermediates and the anomeric configuration was established. More recently, mass spectrometry has been used.[49] For example, a maltotriosyl-enzyme intermediate in the action of glycogen-debranching enzyme was identified by separation of an active site peptide by HPLC followed by mass spectrometry (Chapter 3).[50]

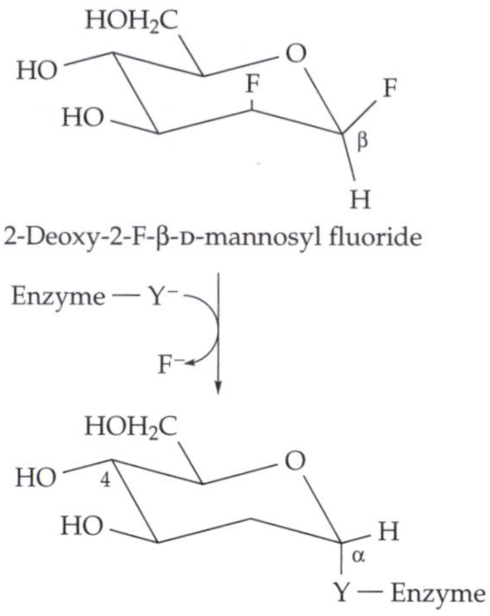

2-Deoxy-2-F-β-D-mannosyl fluoride

(12-10)

Carbocationic intermediates. In a second mechanism of nucleophilic displacement the leaving group departs (often in a protonated form) before the entering nucleophile reacts.

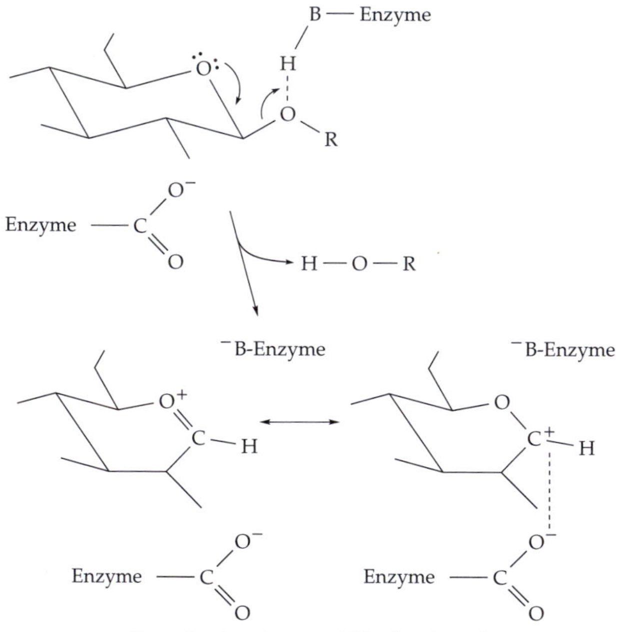

Oxocarbenium ion, a stabilized carbocation

(12-11)

A carbocation is formed as shown in Eq. 12-11, which represents just one-half of the overall displacement reaction. In the common terminology of physical organic chemistry this is an S_N1 reaction rather than an S_N2 reaction of the kind shown in Eqs. 12-3 and 12-5. This terminology is not quite appropriate for enzymes because the breakdown of ES complexes to product is usually a zero-order process and the numbers 1 and 2

in the symbols S_N1 and S_N2 refer to the order or molecularity of the reaction. It is better to speak of S_N1-like or S_N2-like reactions.

The carbocation in Eq. 12-11 is depicted as a resonance hybrid of two forms. One of these is an **oxocarbenium** (or oxonium) ion which contains a double bond between carbon and oxygen.[50a] This double-bonded structure can be visualized as arising from the original structure by an internal displacement or elimination by the unshared electrons on oxygen, as indicated by the small arrows. In the oxocarbenium ion, the ring atoms C-2, C-1, and C-5 and the oxygen atom are almost coplanar and the ring conformation is described as **half-chair** or **sofa**. As pointed out in Chapter 9, the theory of stereoelectronic control predicts that elimination of the group −OR of Eq. 12-11 should occur most easily if a lone pair of electrons on the ring oxygen atom are antiperiplanar to the O−R bond. This is impossible for a β-glycoside with a ring in the chair conformation shown in Eq. 12-11. This fact suggested that enzymes cleaving β-glycosides may preferentially bind substrate with the appropriate sugar ring in a less stable half-chair or flexible boat conformation prior to bond cleavage (Eq. 12-12).[51,52] This would allow an unshared pair of electrons on oxygen antiperiplanar to the C-1 to O−R bond to participate in the elimination reaction as is indicated in the following diagrams.

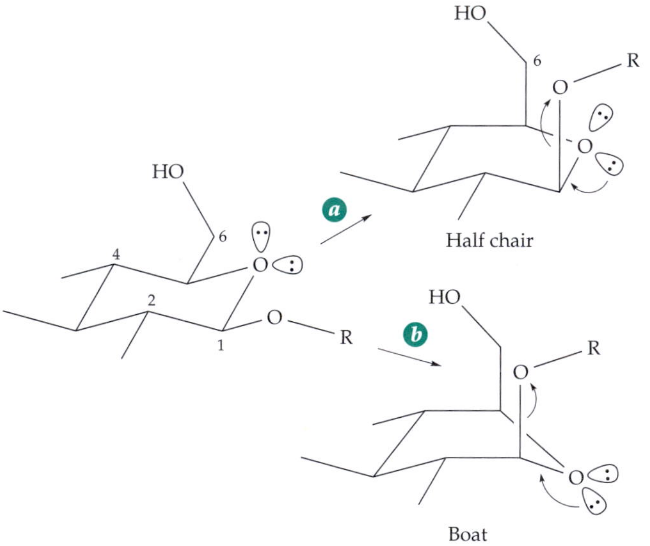

(12-12)

On the other hand, Bennet and Sinnott provided evidence that an antiperiplanar lone electron pair is *not* needed in acid-catalyzed cleavage of glycosides via a carbocationic intermediate.[53,54] Theories of stereoelectronic control must be applied to enzymes with caution!

5. Lysozymes and Chitinases

Polysaccharide chains in the peptidoglycan layer (Fig. 8-29) of the cell walls of bacteria are attacked and cleaved by lysozymes,[55] enzymes that occur in tears and other body secretions and in large amounts in egg white. Some bacteria and fungi, and even viruses, contain lysozymes.[56] Their function is usually to protect against bacteria, but lysozyme of phage T4 is a component of the baseplate of the virus tail (Box 7-C). Its role is to cut a hole in the bacterial cell wall to permit injection of the virus' own DNA. Egg white lysozyme, the first enzyme for which a complete three-dimensional structure was determined by X-ray diffraction,[55] is a 129-residue protein. The active site is in a cleft between a large domain with a nonpolar core and a smaller β-sheet domain that contains many hydrogen-bonded polar side chains (Figs. 12-3, 12-4). Human lysozyme has a similar structure and properties.[57–59] The T4 lysozyme has an additional C-terminal domain whose function may be to bind the crosslinking peptide of the *E. coli* peptidoglycan. Goose lysozyme is similar in part to both hen lysozyme and T4 lysozyme. All three enzymes, as well as that of our own tears, may have evolved from a common ancestral protein.[60] On the other hand, *Streptomyces erythaeus* has developed its own lysozyme with a completely different structure.[61] An extensive series of T4 lysozyme mutants have been studied in efforts to understand protein folding and stability.[61–63]

Catalytic side chain groups. Six *N*-acetylglucosamine (GlcNAc) or *N*-acetylmuramic acid (MurNAc) rings of the polysaccharide substrate are able to fit precisely into six subsites (designated A to F) in a groove between the two structural domains of egg white lysozyme (Fig. 12-4). The bond between the fourth and fifth rings (subsites D and E) is then cleaved. At the active site, the side chain carboxyl group of Glu 35 is positioned correctly to serve as the proton donor BH of Eq. 12-11, while the carboxyl of Asp 52 lies on the opposite side of the groove. Both Glu 35 and Asp 52 have relatively high pK_a values for carboxyl (microscopic pK_a's are ~5.3 and 4.6, respectively, in the fully protonated active site when the ionic strength is ~0.2)[64] as a result of the hydrophobic environment and hydrogen bonding to other groups. If the pH is raised from an initially low value, a proton from Asp 52 usually dissociates first and the electrostatic field of the resulting anion keeps Glu 35 largely protonated until the pH

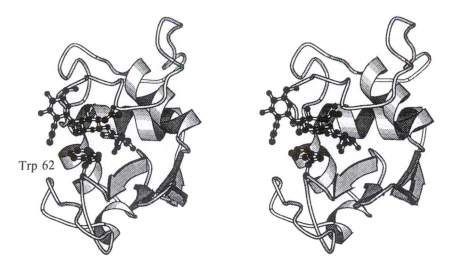

Trp 62

Figure 12-3 Stereoscopic MolScript view of hen egg white lysozyme complexed with a trisaccharide of *N*-acetylglucosamine (GlcNAc)$_3$ in binding subsites A, B, and C. The side chain of Trp 62 and the trisaccharide above it are shown in ball-and-stick form. From Maenaka *et al.*[65] Courtesy of Izumi Kumagai.

approaches 6. Positively charged basic groups nearby affect the pK_a values; hence, the behavior of the enzyme is sensitive to the ionic strength of the medium.[63] The Asp 52 anion lies only ~0.3 nm from the center of positive charge expected for an oxocarbenium ion intermediate and presumably stabilizes it. Replacement of Glu 35 by Gln destroys all catalytic activity and replacement of Asp 52 by Ala or Asn decreases activity to ~4–5% of the original.[66–68] Less than 1% activity remained for the D52S mutant.[69]

Nevertheless, Asp 52 is not absolutely essential for lysozyme activity. Goose lysozyme lacks this catalytic aspartate.[70] Matsumura and Kirsch suggested that carboxyl groups of glycine residues covalently attached to *N*-acetylmuramic acid rings in the natural substrates may participate in catalysis.[66] Many mutant forms of lysozyme have been studied. Of special interest is the D52E mutant. In this enzyme the carboxylate of the longer glutamate side chain reacts with the oxocarbenium ion intermediate to form a covalent adduct and apparently alter the basic mechanism.[71] Replacement of asparagine 46, which can hydrogen bond to Asp 52, also decreases k_{cat} greatly, suggesting a role in catalysis.[72]

The lysozyme-catalyzed reaction is completed by stereospecific addition of a hydroxyl ion to the oxocarbenium ion with the original β configuration being retained in the product. Such stereospecificity for reactions of enzyme-bound carbocations is not surprising because the enzyme probably assists in generation, on the appropriate side of the sugar ring, of the attacking hydroxyl ion.

Kinetic isotope effect for lysozyme. A secondary kinetic isotope effect is expected because a molecule with ^1H in the number 1 position can be converted to the corresponding oxocarbenium ion somewhat more easily than the molecule with ^2H in the same position (Eq. 12-13). For example, in the nonenzymatic acid-catalyzed hydrolysis of a methyl- glucoside, a reaction also believed to proceed through a carbocation inter-mediate,[41,75] the ratio k_{1_H}/k_{2_H} is 1.14 for the α anomer and 1.09 for the β anomer.[53] In the base-catalyzed hydrolysis of the same compound, which is believed to occur by a double-displacement reaction involving participation of the neighboring OH group on C-2, the ratio k_{1_H}/k_{2_H} is 1.03. The corresponding ratio measured

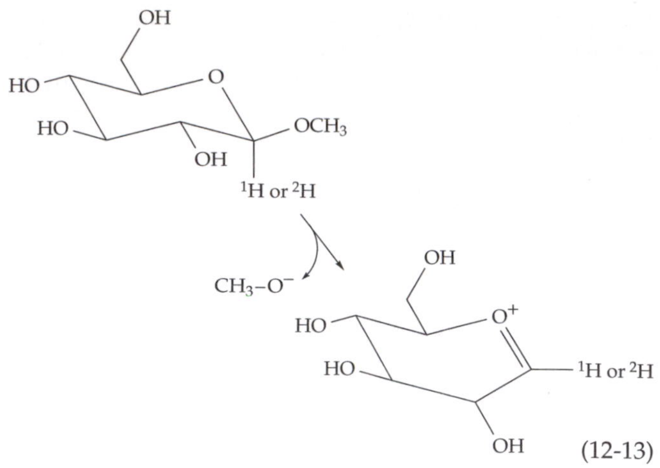

$$(12\text{-}13)$$

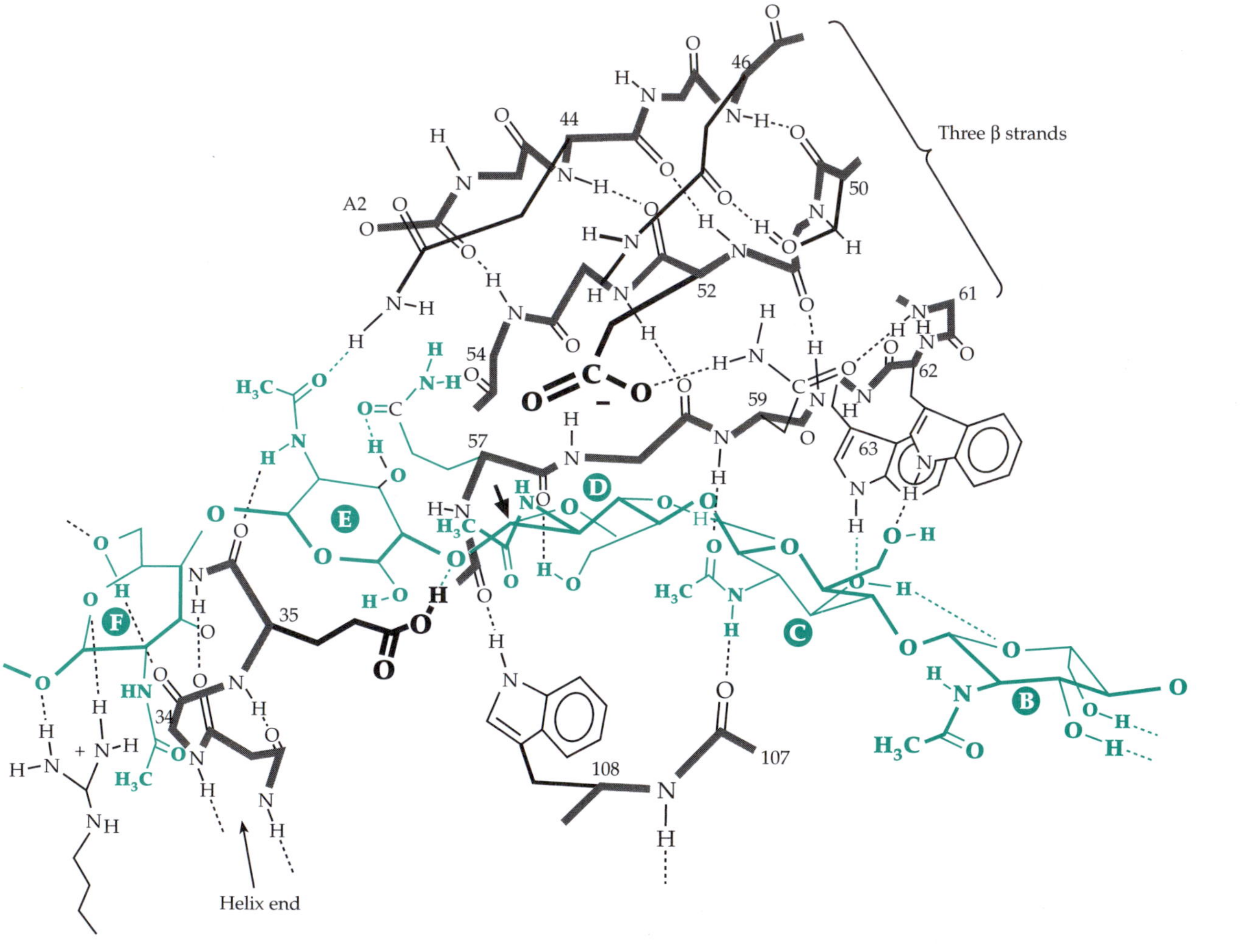

Figure 12-4 Schematic drawing of the active site of egg white lysozyme with substrate in place and about to be cleaved. Three strands of the small β-sheet domain, which contains an extensive hydrogen bond network, are seen at the top. Notice that this view into the active site is different from that in Fig. 12-3. A 180° rotation of either figure will make the views more similar. The three-strand β sheet can be seen (in stereo) at the lower right in Fig. 12-3 and the helix end carrying Glu 35 in the right center foreground. A segment of a chitin oligosaccharide (green), whose reducing end is to the left, is bound into subsites B –F. Cleavage occurs between rings D and E as indicated by the heavy arrow which points to the anomeric carbon atom of ring D. The side chain of Glu 35 is shown protonating the bridge oxygen. Ring D has been distorted into a twist conformation to facilitate cleavage. A larger domain, which contains one long α helix, is at the bottom of the drawing. The active site lies in a cleft between the domains. Notice the chain of hydrogen bonds between the carbonyl of residue 107 in the large domain and the peptide NH of residue 59 in the small domain. Also notice the aromatic side chains which are usually present in carbohydrase active sites. Drawing is based in part on those of Irving Geis[73] and Levitt[74] and on a sketch by author from a three-dimensional model.

for the action of lysozyme is 1.11, much closer to that of the oxocarbenium ion mechanism than to that of the double-displacement mechanism. Similar observations have been made with amylases.[76] Kinetic isotope effects have also been measured for ^{12}C vs ^{13}C in the anomeric position, ^{16}O vs ^{18}O in the leaving group (−OCH_3) of the methyl glucosides, and for other locations[53] as well as for hydrolysis of glucosyl fluorides.[77–79] The results are generally supportive of the oxocarbenium ion mechanism for lysozyme. However, as mentioned in Section B,3, the interpretation of secondary isotope effects is difficult. Such effects cannot reliably identify a carbohydrase mechanism.[80,81]

For acid-catalyzed hydrolysis of methyl glucosides[53] the kinetic isotope effect observed for the oxygen of the leaving group was $k_{16_O} / k_{18_O} = 1.024 – 1.026$. Observation of similar effects for enzymes supports the participation of an acidic group of the protein (Glu 35 of lysozyme) in catalysis but does not eliminate the possibility of concerted involvement of a nucleophilic group, e.g., Asp 52 in lysozyme.[81,82]

Does lysozyme distort its substrate?

An early study of models indicated that for six sugar rings of a substrate to bind tightly into the active site of lysozyme, the ring in subsite D, which contains the carbon atom on which the displacement occurs, had to be distorted from its normal chair conformation into the half-chair conformation.[83] This is illustrated in Fig. 12-4. It was suggested that by binding the substrate chain at six different sites the enzyme provides leverage to distort the ring in subsite D into a conformation similar to that of the transition state. This idea was criticized on the basis that an enzyme would be too flexible to act in this manner.[74] Furthermore, the non-hydrolyzed trisaccharide MurNAc–GlcNac–MurNAc was shown to fit into subsites B, C, and D of the active site groove without distortion.[84] However, it is bound weakly.

Can electrical forces acting within the active site help to distort the substrate and assist in formation of the carbocation intermediate? Levitt and Warshel suggested this possibility and proposed that the necessary electrostatic force arises from the arrangement of dipoles within the peptide backbone of the protein and in the amino acid side chains.[74] As can be seen from Fig. 12-4, the enzyme forms many hydrogen bonds with the substrates. Of special interest is a chain of hydrogen bonds that passes from the backbone carbonyl of Ala 107 through the 2-acetamido group of the substrate in subsite C and into the edge of the β sheet at the backbone NH of Asn 59. This interaction provides specificity toward GlcNac-containing substrates. Perhaps oscillation of charge within such polarizable chains of hydrogen bonds can also help a substrate to move toward its transition state structure.[85,86]

From simulations of lysozyme action by the methods of molecular dynamics, Post and Karplus observed

that Glu 35 tended not to hydrogen bond to the exocyclic bridge oxygen, even though that would be a logical step in the protonation (by −BH) shown in Eq. 12-11. They suggested that Glu 35 may protonate the ring oxygen.[87] The ring could then open to form an exocyclic oxonium ion, which could hydrolyze and cyclize to the final product. The initial elimination could receive stereoelectronic assistance from the antiperiplanar lone pair of electrons on the bridge oxygen. This proposal has been criticized.[81,88] For example, it is hard to explain the rapid cleavage of glycosyl fluorides by this exocyclic oxonium mechanism.

As attractive as the arguments for a oxocarbenium mechanism in lysozyme are, is it still possible to explain the experimental observations by a double displacement in which Asp 52 serves as the nucleophile to form a glycosyl enzyme?[38,82,82a] Sucrose phosphorylase and other glucosyl transferases, like lysozyme, apparently have a carboxylate ion at the active site. In some enzymes, the carboxylate ion forms a covalent glucosyl enzyme, whereas in lysozyme it apparently only stabilizes a carbocation. Is there really a difference in mechanism? Do glucosyl enzymes form only with certain substrates or upon denaturation of the enzymes? Nature hides her secrets well. The difficulty in pinning down the fine mechanistic details of enzymatic action makes it essential to be skeptical, to examine data critically, and to try to imagine all possible alternatives −even when things seem to be proven beautifully.

Help from a neighboring group.

In the 1960s it was suggested that the acetamido group of N-acetylglucosamine residues might participate as a nucleophile, either stabilizing an oxocarbenium ion or forming an oxazoline intermediate.[89–91] The proposal received little support, as applied to lysozyme, until sequences and structures of many larger carbohydrases were determined. Among these are **chitinases**, enzymes that act on the same substrates as the small lysozymes. One group of plant chitinases have a structure similar to that of egg white lysozyme. The 243-residue enzyme from barley seeds apparently has Glu 67 as a proton donor and Glu 89 as a possible stabilizing nucleophile.[92] Another group of chitinases from both plants and bacteria have active sites at ends of $(\alpha\beta)_6$ barrels.[93–95] They all have a proton-donating glutamate in a conserved position but *no aspartate* that could serve as a nucleophile. Study of complexes with substrates and inhibitors with these enzymes has provided direct evidence of ring distortion and of the probable role of the acetamido group as indicated in Eq. 12-14.[94,95] The distortion is beyond that in a sofa conformation and allows for maximum stereoelectronic assistance as well as participation of the acetamido group.

In these relatively large enzymes the substrate is deeply buried and cannot reach a conformation

Oxazoline

(12-14)

approaching that of the transition state without having a strained ring conformation. These enzymes may use binding forces exerted on many parts of the substrate to stabilize the transition state structure.

6. Cellulases and Other β-Glycosidases

Cellulose, the most abundant of all biopolymers, is extremely stable but is attacked by a host of bacterial and fungal β-glycanases.[96] Animals do not ordinarily produce cellulases but some termites do.[97] Cellulase structures are varied, being represented by 10 of 57 different glycosylhydrolase families.[98] Most, like lysozyme, retain the β configuration in their products but some invert.[98–100]

Among the cellulases are **endoglycanases**, which cleave chains at random positions, and **exoglycanases** (also called **cellobiohydrolases**) that cleave cellobiose units from ends of chains. Some act on the nonreducing ends and some on reducing ends; a mixture of enzymes is most effective.[99,101,102] In some bacteria a whole series of different cellulases, together with a large (197-kDa) organizing protein, form a **cellusome**, a complex with high catalytic activity for crystalline cellulose.[103–105] Cellulases usually have tightly packed catalytic domains which may vary in size from ~200 residues[98] to over 400.[106] In most cases the catalytic domain is connected by a flexible linker to one or more small, globular **cellulose-binding domains**. These vary in size from 36 to 200 residues and often have a β-barrel fold.[104,107–108a] Their function is to hold the enzymes to the cellulose surfaces. They may also facilitate disruption of the tightly hydrogen bonded cellulose structure (Fig. 4-5). As with other carbohydrases, carboxyl groups of amino acid side chains provide the major catalytic groups (Table 12-1). An

extensive hydrogen-bond network which often includes imidazole groups may influence activity.[109,110]

A striking feature of a 411-residue endoglucanase from *Fusarium* was revealed by the binding of a nonhydrolyzable thiooligosaccharide substrate analog. The pyranose ring at the cleavage site was distorted in an identical manner to that mentioned in the preceding section for a chitinase thought to use the substrate's acetamido group as a nucleophile (Eq. 12-14).[106] The distortion observed is beyond that required for a sofa conformation and allows for the maximum stereoelectronic assistance (Eq. 12-12b). An oxygen atom of the E197 carboxylate, the catalytic nucleophile, occupies a position in the complex that is coincident with that of the C2 acetamido oxygen in the catalytic site of the chitinase discussed in the previous section.

The *inverting* β-glucanases also have two catalytic acid base groups but they are ~0.9–1.0 nm apart rather than ~0.6 nm for retaining enzymes. This allows space for a water molecule whose $^-$OH is the nucleophilic reactant ($^-$OY in Eq. 12-5) and in which a carboxylate group assists in dissociating the water molecule.[98] (This mechanism is illustrated for glucoamylase in Fig. 12-7).

Structural features met in some cellulases include an α,α barrel[111] similar to that of glucoamylase (Fig. 2-29) and, in a cellobiohydrolase,[101] a 5-nm-long tunnel into which the cellulose chains must enter. Ten well-defined subsites for glycosyl units are present in the tunnel.[101] A feature associated with this tunnel is **processive** action, movement of the enzyme along the chain without dissociation,[105] a phenomenon observed long ago for amylases (see Section 9) and often observed for enzymes acting on nucleic acids.

Another group of **β-glucanases**, found in plants and their seeds, hydrolyze β-1,3-linked glucans[112] and, in some cases, also mixed 1,3- and 1,4-linked polysaccharides. A characteristic enzyme is found in barley (Table 12-1).[113,114] Similar enzymes are produced by some bacteria[113] and also by molluscs.[115]

Xylanases act on the β-1,4-linked xylan, the most abundant of the **hemicelluloses** that constitutes over 30% of the dry weight of terrestrial plants.[116] They resemble cellulases and cooperate with cellulases and **xylosidases**[117] in digestion of plant cell walls.[110,116,118–121] **Galactanase** digests the β-1,4-linked component of pectins.[122]

The final step in degradation of cellulose is hydrolysis of cellobiose to glucose. This is accomplished by **β-glucosidases**, enzymes that also hydrolyze lactose, phosphorylated disaccharides, and cyanogenic glycosides.[123–126] Lactose is also cleaved by **β-galactosidases**.[127] The large 1023-residue β-galactosidase from *E. coli*[128] is famous in the history of molecular biology as a component of the *lac* operon.[129–132] Its properties are also employed to assist in the cloning of genes (Chapter 26).

TABLE 12-1
Acidic and Basic Catalytic Groups in a Few Glycosyltransferases[a]

Enzyme	Number of residues	Inverting?	Glycosyl-enzyme identified?	Nucleophile (–COO⁻)	Electrophile (proton donor) * Assisting group
Lysozyme					
Human, hen[b]	130	No		E35	D53(52)
Bacteriophage T4[c]		No		E11	D20
Chitinase					
Rubber plant[d]	273	No			E127
Cellulases					
E1 endocellulase,					
Acidothermus[e]	521				
catalytic domain	358	No		E282	E162, D252*
Endoglucanase I					
Fusarium[f]	411	No		E197	E202
Endoglucanase Cen A					
Cellulomonas[g]	>351	Yes		D392	D78
1,3-β-D-Glucanase					
Barley[h]		No		E231	E288
Xylanase					
B. circulans[i]		No	Yes	E78	E172
β-Glucosidase					
Agrobacterium[j]	458	No	Yes	E358	E170, Y298*
β-Galactosidase					
E. coli[k]	1023	No	Yes	E537	E461, Y503*
α-Amylase					
Human and pig[l]	496	No		D197	E233, D300*
Barley[m]		No		D179	E204, D289*
Aspergillus[n]		No		D206	E230
Cyclodextrin glucosyltransferase					
B. circulans[o]				D229	E257, D328
α-Glucosidase					
Saccharomyces[p]	No	Yes		D214	E233, D300
Glycogen debranching enzyme, rabbit[s]	No	Yes		D549	
Glucoamylase					
Aspergillus[q]	616	Yes		H₂O, E400	E179
β-Amylase					
Soybean[r]		Yes		E186	E380
Glucocerebrosidase					
Human[t]				E340	

[a] For classification of glycosyl hydrolases into families, see Henrissat, B., Callebaut, I., Fabrega, S., Lehn, P., Mornon, J.-P., and Davies, G. (1995) *Proc. Natl. Acad. Sci. U.S.A.* **92**, 7090–7094

[b] Matsumura, I., and Kirsch, J. F. (1996) *Biochemistry* **35**, 1881–1889; Harata, K., Muraki, M., Hayashi, Y., and Jigami, Y. (1992) *Protein Sci.* **1**, 1447–1453

[c] Brzozowski, A. M., and Davies, G. J. (1997) *Biochemistry* **36**, 10837–10845; Hardy, L. W., and Poteete, A. R. (1991) *Biochemistry* **30**, 9457–9463 Kuroki, R., Weaver, L. H., and Matthews, B. W. (1993) *Science* **262**, 2030–2033

[d] Tews, I., Terwissscha van Scheltinga, A. C., Perrakis, A., Wilson, K. S., and Dijkstra, B. W. (1997) *J. Am. Chem. Soc.* **119**, 7954–7959

[e] Sakon, J., Adney, W. S., Himmel, M. E., Thomas, S. R., and Karplus, P. A. (1996) *Biochemistry* **35**, 10648–10660; Ghidoni, R., Sonnino, S., Tettamanti, G., Baumann, N., Reuter, G., and Schauer, R. (1980) *J. Biol. Chem.* **255**, 6990–6995

[f] Sulzenbacher, G., Schülein, M., and Davies, G. J. (1997) *Biochemistry* **36**, 5902–5911; Sulzenbacher, G., Driguez, H., Henrissat, B., Schülein, M., and Davies, G. J. (1996) *Biochemistry* **35**, 15280–15287; Mackenzie, L. F., Davies, G. J., Schülein, M., and Withers, S. G. (1997) *Biochemistry* **36**, 5893–5901

[g] Damude, H. G., Withers, S. G., Kilburn, D. G., Miller, R. C., Jr., and Warren, R. A. J. (1995) *Biochemistry* **34**, 2220–2224

[h] Chen, L., Garrett, T. P. J., Fincher, G. B., and Hoj, P. B. (1995) *J. Biol. Chem.* **270**, 8093–8101

[i] Wakarchuk, W. W., Campbell, R. L., Sung, W. L., Davoodi, J., and Yaguchi, M. (1994) *Protein Sci.* **3**, 467–475; Lawson, S. L., Wakarchuk, W. W., and Withers, S. G. (1997) *Biochemistry* **36**, 2257–2265; Sidhu, G., Withers, S. G., Nguyen, N. T., McIntosh, L. P., Ziser, L., and Brayer, G. D. (1999) *Biochemistry* **38**, 5346–5354

[j] Wang, Q., Trimbur, D., Graham, R., Warren, R. A. J., and Withers, S. G. (1995) *Biochemistry* **34**, 14554–14562

[k] Gebler, J. C., Aebersold, R., and Withers, S. G. (1992) *J. Biol. Chem.* **267**, 11126–11130; Jacobson, R. H., Zhang, X.-J., DuBose, R. F., and Matthews, B. W. (1994) *Nature (London)* **369**, 761–766; Richard, J. P., Huber, R. E., Lin, S., Heo, C., and Amyes, T. L. (1996) *Biochemistry* **35**, 12377–12386

[l] Brayer, G. D., Luo, Y., and Withers, S. G. (1995) *Protein Sci.* **4**, 1730–1742; Qian, M., Haser, R., Buisson, G., Duée, E., and Payan, F. (1994) *Biochemistry* **33**, 6284–6294

[m] Kadziola, A., Sogaard, M., Svensson, B., and Haser, R. (1998) *J. Mol. Biol.* **278**, 205–217

[n] Brzozowski, A. M., and Davies, G. J. (1997) *Biochemistry* **36**, 10837–10845; Matsuura, Y., Kusunoki, M., Harada, W., and Kakudo, M. (1984) *J. Biochem.* **95**, 697–702

[o] Knegtel, R. M. A., Strokopytov, B., Penninga, D., Faber, O. G., Rozeboom, H. J., Kalk, K. H., Dijkhuizen, L., and Dijkstra, B. W. (1995) *J. Biol. Chem.* **270**, 29256–29264

[p] McCarter, J. D., and Withers, S. G. (1996) *J. Biol. Chem.* **271**, 6889–6894

[q] Christensen, U., Olsen, K., Stoffer, B. B., and Svensson, B. (1996) *Biochemistry* **35**, 15009–15018

[r] Mikami, B., Degano, M., Hehre, E. J., and Sacchettini, J. C. (1994) *Biochemistry* **33**, 7779–7787; Adachi, M., Mikami, B., Katsube, T., and Utsumi, S. (1998) *J. Biol. Chem.* **273**, 19859–19865

[s] Braun, C., Lindhorst, T., Madsen, N. B., and Withers, S. G. (1996) *Biochemistry* **35**, 5458–5463

[t] Withers, S. G., and Aebersold, R. (1995) *Protein Sci.* **4**, 361–372

7. Glycogen Phosphorylase

A large number of different glycosyltransferases act on the α1,4 linkages of glycogen, starch, and related polysaccharides. Among these, one of the most studied is glycogen phosphorylase. It is not a hydrolase, but it catalyzes cleavage of α1,4 linkages at the nonreducing ends of glycogen molecules by displacement with inorganic phosphate to give α-D-glucose-1-phosphate. It is a very large enzyme (841 residues in rabbit muscle) whose structure is shown in Fig. 11-5. Its complex regulatory mechanisms were discussed briefly in Chapter 11.

Neither partial exchange reactions nor inversion of configuration occur when glycogen phosphorylase acts on its substrates. The enzyme apparently does nothing until both substrates are present. These are glycogen + inorganic phosphate or, for the reverse reaction, glycogen (shifted over by one sugar binding subsite) + glucose-1-phosphate. The active site is in a deep groove in the enzyme. Evidently, the protein must close and fold around the substrates before it becomes active.[133] An oxocarbenium ion mechanism has also been proposed for this enzyme, partly on the basis of strong inhibition of phosphorylase by **5-gluconolactone**,[134,135] a compound having a half-chair conformation and perhaps acting as a transition state inhibitor (Chapter 9). This gluconolactone also inhibits many other carbohydrases.[123,136]

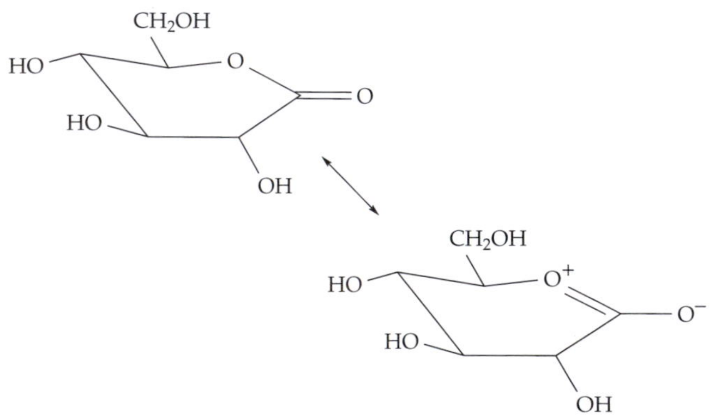

Glucose-1-phosphate

(12-15)

5-Gluconolactone

An unexpected discovery was that glycogen phosphorylase contains a molecule of the coenzyme pyridoxal 5'-phosphate (PLP) bound into the center of the protein behind the active site[134,135,137–139] with its phosphate group adjacent to the binding site of the phosphate of glucose-1-phosphate. It probably serves as a general base catalyst, e.g., assisting the attack of a phosphate ion on the oxocarbenium formed by cleavage of the glycogen chain (Eq. 12-15).[137,140–142] A key observation by Graves and associates was that pyridoxal alone does not activate apo-phosphorylase but that pyridoxal

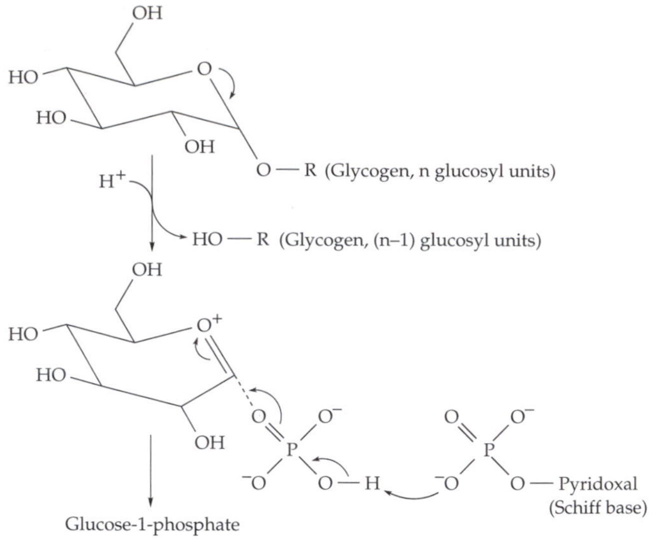

plus phosphite, phosphate, or fluorophosphate does provide up to 20% of full activity.[143] X-ray crystallography confirmed that these activating anions are bound into the active site at the approximate position of the phosphate group of PLP.[139,143a] Glycogen phosphorylase is being studied by X-ray diffraction techniques that allow observation of structural changes in as short a time as a few milliseconds (see Chapter 3, Section H).[144]

The regulation of glycogen phosphorylase, like that of many other allosteric proteins, depends upon the existence of two distinct conformational states, whose structures have been established by crystallography.[137] It is not immediately evident how they can affect the active site. In the R-state the enzyme has a low affinity for both substrates and activators such as AMP. In the T-state the affinities are much higher. For example, that of inorganic phosphate is raised by a factor of fifteen.[137] As we have already seen (Figs. 11-4, 11-5) the relatively inactive phosphorylase *b* is converted to the active phosphorylase *a* by phosphorylation of the side chain of serine 14, a structural change which favors the R-state. In the inactive T-state of phosphorylase *b*, an 18-residue N-terminal segment of the polypeptide is not seen by X-ray diffraction, presumably because it does not assume a fixed conformation but projects into a solvent channel in the crystal and moves freely within it. However, in the active R-state of phosphorylase *a*, in which Ser 14 has been phosphorylated, the N-terminal segment is rigid. The phospho group on Ser 14 binds to Arg 69 and other arginine, lysine, and histidine side chains from both subunits.[137,145] The phosphorylation of Ser 14 occurs 1.5 nm from the active sites. The conformational changes induced by phosphorylation of Ser 14 cause a rotation of about 10° between subunits, somewhat reminiscent of the changes accompanying oxygenation of hemoglobin (Fig. 7-25).

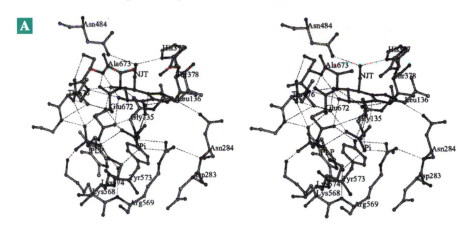

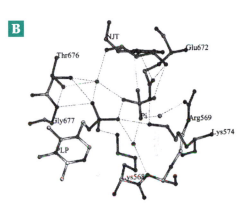

Figure 12-5 (A) Stereoscopic view of the structure of the catalytic site of phosphorylase *b* in the inhibited T-state with the inhibitor nojirimycin tetrazole bound into the active site. Inorganic phosphate (P$_i$) as well as the coenzyme pyridoxal 5'-phosphate (PLP) are also shown. (B) Details of interactions of the inhibitor, P$_i$, and PLP with the protein and with water molecules (small circles). This is a weak-binding state but the P$_i$ has displaced the negatively charged side chain carboxylate of Asp 283 (visible at the lower right in A). This carboxylate blocks access to the P$_i$ site in the free T-state enzyme. Here, the positively charged Arg 569 guanidinium group has swung in to interact with the P$_i$. From Mitchell *et al.*[133] Courtesy of Louise N. Johnson.

A related structural change accompanies binding of caffeine, adenosine, or AMP at high concentrations. These substances inhibit the enzyme by binding at a site next to the catalytic site (Fig. 11-5), stabilizing the T-state and causing a loop of protein to move into the catalytic site and to block it.[137] The catalytic site lies deeply buried in the protein between the two large structural domains (Fig. 11-5). The substrate is held by a network of hydrogen bonds, some of which are shown in Fig. 12-5, in which the active site contains an inhibitory substrate analog **nojirimycin tetrazole**, which is viewed toward the edges of the two rings.

The enzyme is in the weak-binding T-state but inorganic phosphate (P$_i$) is bound below it and next to the phosphate group of PLP, as required by the mechanism of Eq. 12-15.

The difference in binding affinities for the T- and R-states lies in a flexible loop of residues 280–288, which in the T-state blocks access to the substrate-binding cleft. The universally conserved Asp 238 behaves as a substrate mimic, occupying the P$_i$ site.[137,146,147] In the R-state this residue moves, allowing P$_i$ to enter and bind (Fig. 12-5).

Most of the structure of mammalian phosphorylases is conserved in species as diverse as *E. coli* and the potato.[148] However, *E. coli* maltodextrin phosphorylase[143a,149–151] and potato phosphorylases have less sophisticated regulatory mechanisms than do the animal enzymes. Another feature of glycogen phosphorylase is a "glycogen storage site" about 2.5 nm from the active site (Fig. 11-5).[152] This provides a means for the enzyme to hold onto the giant glycogen molecule while "nibbling off" the outside ends of nearby branches.

8. Starch-Hydrolyzing Enzymes

Among the hydrolases are the widely distributed **α-amylases**, *endo*-glycosidases which hydrolyze

Nojirimycin tetrazole

starch chains by random attack at points far from chain ends to form short polysaccharide chains known as **dextrins** as well as simpler sugars.[153] The catalyzed reactions proceed with retention of the original α configuration. Alpha-amylases are found in fungi, plants, and animals. One powerful enzyme of this class is present in the saliva of most humans and other isoenzymes are formed by the pancreas.[154] They are encoded by a family of genes, a fact that accounts for the existence of some healthy individuals completely lacking salivary amylase.[155]

Structures are known for human[154] and porcine[156–159] α-amylases as well as for corresponding enzymes from barley,[160] mealworms,[161] fungi, and bacteria.[162–164] The α-amylase from *Aspergillus oryzae* (Taka-amylase), widely used in laboratory work, was the first for which a structure was determined.[165,166] The α-amylases fold into three domains (Fig. 12-6), with the active site in the center of an $(\alpha/\beta)_8$ barrel. All of the α-amylases contain one or more bound Ca^{2+} ions. Some, including the human α-amylases, also require a **chloride ion**. The Cl^- is held by a pair of arginine guanidinium groups[154] and interacts with adjacent carboxyl groups, inducing pK_a shifts and allosteric activation.[163]

The sequences of α-amylases vary widely but a conserved cluster of one glutamate and two aspartates is usually present (Table 12-1).

Studies of the **action patterns**, i.e., the distribution of products formed by α-amylases when acting on a variety of α 1,4-linked oligosaccharides, suggested that the substrate binding region of the porcine pancreatic enzyme has five subsites, each binding one glucose residue.[153] The α-glucan chain is cleaved between the residues bound at the second and third subsites (numbered from the reducing end of the substrate)

by a lysozyme-like mechanism. Endolytic enzymes, which cleave biopolymer chains internally, are usually thought to carry out random attack. However, after the initial catalytic reaction one of the polysaccharide products of porcine pancreatic amylase action often does not leave the enzyme. The polysaccharide simply "slides over" until it fully occupies all of the subsites of the substrate binding site and a second "attack" occurs. An average of seven catalytic events occur each time it forms a complex with amylase.[167] Is there a mechanism by which the enzyme deliberately promotes the "sliding" of the substrate in this **multiple attack** or **processive** mechanism[168,169] or does the dissociated product simply diffuse a short distance while enclosed in a solvent cage? The latter explanation may be adequate for some enzymes. In contrast to pancreatic α-amylase a bacterial "maltogenic" α-amylase produces principally maltose as a product.[169a,b]

Digestion of starch and glycogen by α-amylases produces a mixture of glucose, oligosaccharides, and dextrins, which in the human body are further degraded by **α-glucosidases** of the brush border membrane of the small intestine.[154] A lysosomal form of the enzyme is missing in Pompe's disease (Box 20-D).[170,171] The α-glucosidases are also members of the α-amylase family.[46,172] For digestion of the branched amylopectin and glycogen a **debranching enzyme** and **α-1,6-glucosidase** activity are required. In mammals these activities are found in a single polypeptide chain with separate but adjacent active sites.[50,173,174] The debranching enzyme catalyzes transfer of oligosaccharide chains from α1,6-linked branch positions to new locations at ends of chains with α1,4 linkages. A bacterial oligo-1,6-glucosidase has a catalytic site formed from

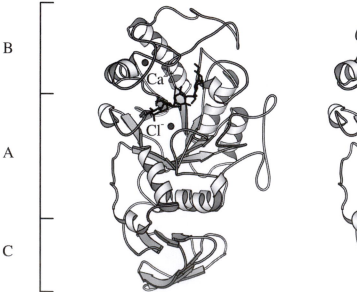
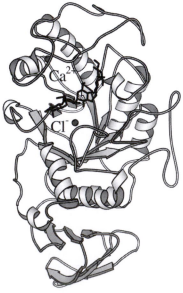

Figure 12-6 Drawing showing the overall polypeptide chain fold and relative positioning of the three structural domains of human pancreatic α-amylase. Also drawn are the locations of the calcium and chloride binding sites. Overlaid is the placement of a modified form of the inhibitor acarbose (p. 607) that binds in the active site cleft. MolScript drawing courtesy of G. Sidhu and G. Brayer.

D199, E255, and D329 similar to that of α-amylases.[175]

Another member of the α-amylase family has principally a glycosyltransferase activity. **Cyclodextrin glucanotransferase** forms cyclodextrins (Box 4-A) by a transferase reaction in which a 6- to 8- member oligo-saccharyl group is transferred from a straight amylase chain onto a protein side chain (Glu 257) and then joins the ends of the oligosaccharide to form the cyclo-dextrin rings (an overall double-displacement process).[176–177b] However, large circular dextrins are produced initially and are then converted into the smaller cyclodextrins.[178]

A starch-digesting enzyme of great industrial importance is **glucoamylase**, whose $(\alpha,\alpha)_6$-barrel structure is shown in Fig. 2-29. That figure also shows the tetrasaccharide inhibitor **acarbose** in the active site. The ring at the nonreducing end is deeply embedded in the protein and, as shown in Fig. 12-7 for the related D-*gluco*-dihydroacarbose, is held by many hydrogen bonds. This slow enzyme ($k_{cat} \sim 50$ s^{-1} at 45°C)[179] cuts off a single glucose unit, then releases

inverting enzyme.

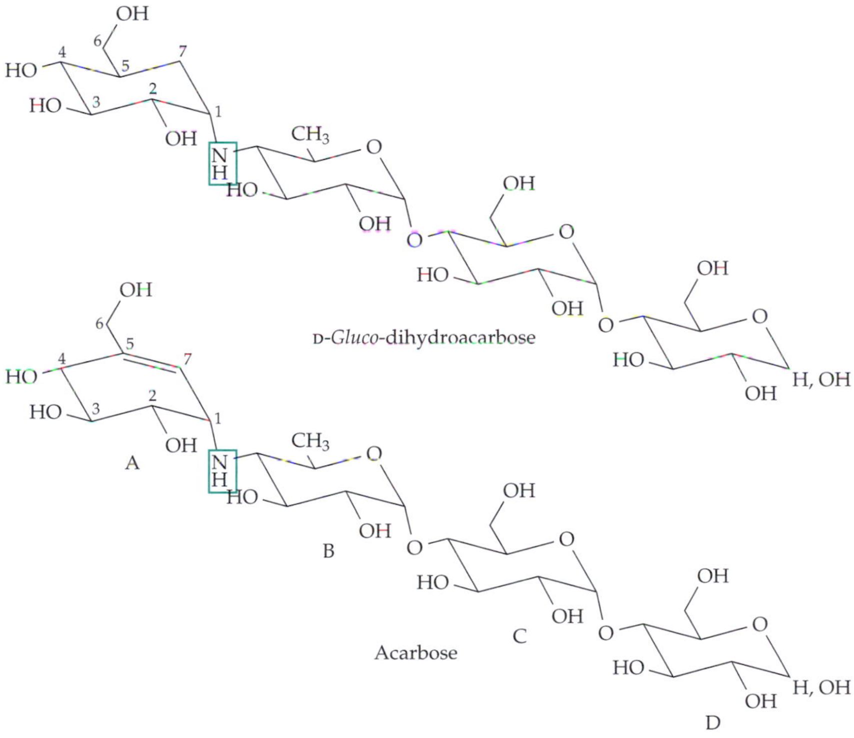

Notice in Fig. 12-7 the abundance of polar groups, many of which are charged. Four carboxyl groups, from D55, E179, E180, and E400, are present in the active site region and participate in the hydrogen bonded network. Active sites of other carbohydrases differ from that of glucoamylase. However, the presence of many hydrogen-bonded polar groups, including two or more carboxyl groups, is characteristic. As illustrated in Figs. 9-8 and 9-9, the pH dependence of the maximum velocity v_{max} is often determined by these carboxyl groups. At least one of them must be protonated for maximum catalytic activity and if two are protonated activity may fall again. Under these circumstances a single proton bound to a molecule of enzyme may spend a fraction of its time on each of several different carboxylate, imidazole, or other groups.[181] In the case of glucoamylase it was concluded from the X-ray structure that the catalytic acid group, E179, is probably *unprotonated* most of the time.[181] However, it can still bind to starch. Then, after a fraction of a second a proton may be transferred onto the Glu 179 carboxylate and from there to the bridge oxygen of the substrate, inducing reaction according to Eq. 12-16. It may even be necessary to have enzyme protonated initially on a group other than E179 to allow small conformational changes to occur prior to formation of the final activated complex. Such essential conformational changes have often been invoked for glucoamylase.[182]

Just as most cellulases have special cellulose-binding domains, glucoamylase has a compact C-terminal starch-binding domain (residues 509–616) similar to the

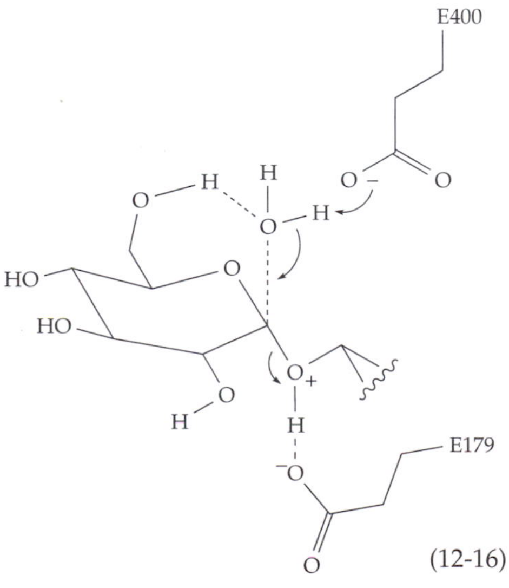

D-*Gluco*-dihydroacarbose

Acarbose

itself from the starch before releasing the glucose and rebinding to the starch. The catalytic acid has been identified as the carboxyl group of Glu 179. In Fig. 12-7 it is seen, presumably as a carboxylate group, tightly hydrogen bonded to what is doubtless the bridge –NH$_2$$^+$– between rings A and B of the inhibitor D-*gluco*-dihydroacarbose. Thus, the complex mimics that of a true substrate protonated on the bridge oxygen, a possible first step in normal catalysis. In accord with this mechanism (Eq. 12-16), glucoamylase is an

(12-16)

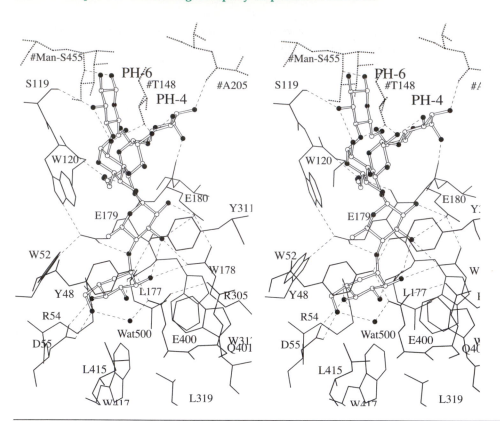

Figure 12-7 Stereoscopic view of the inhibitor D-*gluco*-dihydroacarbose in its complex with glucoamylase of *Aspergillus*. Residues from symmetry-related molecules of the enzyme are shown as dotted lines. Ring A at the nonreducing end of an amylase chain is thought to bind in a similar way (with ring A at the bottom in this figure). Cleavage is between the A and B rings E178 acting as a proton donor to the bridge oxygen (or NH for the inhibitor). The attacking nucleophile is thought to be a water molecule, which is labeled Wat 500 and is held by the assisting carboxylate of E400.[180] Courtesy of Alexander Aleshin.

corresponding domain of cellulases.[183] It is connected by a glycosylated linker to the 470-residue catalytic subunit. Cyclodextrin glucanotransferase also has a starch-binding domain.[184]

Beta-amylases, characteristic plant and bacterial enzymes[184a], have an *exo* action, cutting off chain ends two sugar units at a time as *maltose*. The original α linkage is inverted, the product being β-maltose. The (α/β)₈-barrel structure is unlike that of glucoamylase, but the spacing of active site carboxyl groups suggests that a water molecule is held and activated as in glucoamylase.[185,186] The multiple subsite structure of the active site may permit the substrate amylase to slip forward after a maltose product molecule leaves the site. This may account for the observed processive action.[186]

Many specialized glycosyl transferases synthesize glycogen, starch, cellulose, and other polysaccharides and add glycosyl groups to glycoproteins and glycolipids.[187,188] Often the glycosyl group is transferred from a carrier such as uridine diphosphate (UDP). An example is glycogen synthase (Fig. 11-4), which transfers glucosyl groups of UDP-glucose to the 4'–OH groups at the nonreducing ends of the bushlike glycogen molecules. Other similar synthetic reactions are considered in Chapter 20.

C. Displacement Reactions on Carbonyl Groups

A second major class of nucleophilic displacement reactions (Type 1B in Table 10-1) involve replacement of a group Y attached to a carbonyl carbon:

$$B{:}^- + R-\overset{\overset{\textstyle O}{\|}}{C}-Y + H^+ \longrightarrow R-\overset{\overset{\textstyle O}{\|}}{C}-B + YH$$

$$(12\text{-}17)$$

Group Y may be – OR (esters), – SR (thioesters) or – NHR (amides or peptides). If B⁻ is a hydroxyl ion formed from H_2O, the reaction is **hydrolysis**. If B⁻ is the anion of an alcohol, thiol, or amine, the reaction is **transacylation**. Transacylation is an essential process in biosynthesis of proteins and lipids, but it is the digestive enzymes, which catalyze hydrolysis, that have been studied most intensively.

Uncatalyzed hydrolysis of a peptide linkage is very slow with $t_{1/2}$ at neutral pH and 25°C of ~300–600 years.[189] Both acids and bases catalyze hydrolysis, but enzymes are needed for rapid digestion. The carbonyl group C=O is highly polarized, with the resonance form C^+-O^- contributing substantially to its structure. An attack by a base will take place readily on the electrophilic carbon atom. While the reactivity of the carbonyl group in esters and amides is relatively low because of the resonance stabilization of these groups, the carbon atom still maintains an electrophilic character and may combine with basic groups. Thus, in the base-catalyzed hydrolysis of esters a hydroxyl ion adds to the carbonyl group to form a transient single-bonded "tetrahedral" intermediate (Eq. 12-18). Similar intermediates are believed to form during the action of many enzymes. However, for purposes of classification, we can regard them as simple displacement reactions on a carbon atom with the understanding that there are probably transient single-bonded intermediates.

$$R - \overset{\overset{O}{\|}}{C} - OR' \longrightarrow R - \overset{\overset{O^-}{|}}{\underset{OH}{C}} - OR'$$

"Tetrahedral" intermediate

$$H^+ \Big\downarrow \quad \longrightarrow HOR'$$

$$R - \overset{O}{\underset{OH}{C}}$$

$$(12\text{-}18)$$

A large number of hydrolytic enzymes, the **proteases**, and **peptidases** act on peptide linkages of proteins.[190] At present the traditional name protease, which implies proteolysis, is most often used. However, the IUB encourages use of the name **proteinase**. Although this seems less specific in meaning, its use will probably increase. Some proteases trim newly formed peptide chains, others convert proteins from precursor forms into biologically active molecules, and others digest proteins. In addition to endopeptidases, such as trypsin and chymotrypsin, which cleave at positions *within* a long peptide chain, there are many enzymes that cleave amino acids from the *ends* of chains. These are usually called peptidases and are designated **aminopeptidases** if they cleave from the N terminus and **carboxypeptidases** if they cleave from the C terminus. Most of these enzymes can be classified into **serine** proteases, **cysteine** proteases, **aspartic** proteases, **metallo** proteases, or **N-terminal nucleophile** hydrolases, depending upon the chemical nature of the active site. These groups are further divided into families or "clans."[190]

1. The Serine Proteases

The digestive enzymes **trypsin**, **chymotrypsin**, **elastase**, and **proteinase E** are related serine proteases. All three are synthesized in the pancreas which secretes 5–10 g per day of proteins, mostly the inactive proenzymes (zymogens) of digestive enzymes.[191,192]

These proenzymes are synthesized and "packaged" as **zymogen granules** which travel to the surfaces of the secretory cells. The contents of the granules are secreted into the extracellular medium and are discharged via the pancreatic duct into the small intestine. At their sites of action, the zymogens are converted into active enzymes by the cutting out of one or more pieces from the precursor. This occurs in a cascade-type process triggered by **enteropeptidase** (historically enterokinase), another serine protease which is secreted by the intestinal lining.[192a] Human enteropeptidase consists of a 235-residue catalytic subunit bonded through a disulfide bridge to a larger 784-residue membrane-anchoring subunit.[193–195] It attacks specifically **trypsinogen**, converting it to active trypsin.[196,197] Trypsin in turn activates the other zymogens, as is indicated in Fig. 12-8. Trypsin can also activate its own zymogen, trypsinogen, in an autocatalytic process.

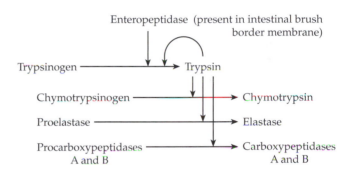

Figure 12-8 Cascade of reactions that activate pancreatic proteases. Enteropeptidase, or trypsin, cleaves the proenzyme (zymogen) at specific sites.

Chymotrypsinogen consists of a single 245-residue chain. The amino acid residues in chymotrypsin, trypsin, and elastase are usually all numbered according to their position in this zymogen. Inactive proenzymes are formed as precursors to enzymes of many different classes and are activated in a variety of ways. A part of the polypeptide chain of the proenzymes is often folded over the active site, interacting in a nonsubstrate-like fashion and blocking the site.[197a]

Serine as a nucleophile. An early clue to the mechanism of action of chymotrypsin came from investigation of the related **acetylcholinesterase**. This key enzyme of the nervous system is inactivated irreversibly by powerful phosphorus-containing poisons that had been developed as insecticides and as war gases (nerve gases, Box 12-C). Around 1949, the nerve gas diisopropylfluorophosphate (DFP) was shown also to inactivate chymotrypsin. When radioactive ^{32}P-containing DFP was allowed to react the ^{32}P became covalently attached to the enzyme. When the labeled enzyme was denatured and subjected to acid hydrolysis the phosphorus stuck tightly; the radioactive fragment was identified as *O*-phosphoserine. It was evident that this product could be formed by an attack of the hydroxyl group of the serine side chain on the phosphorus with displacment of the fluoride ion. This is a nucleophilic displacement on phosphorus, occurring on an enzyme that normally catalyzes displacement on C=O. The DFP molecule acts as a pseudosubstrate which reacts with the enzyme in a manner analogous to that of a true substrate but which does not complete the reaction sequence normally.

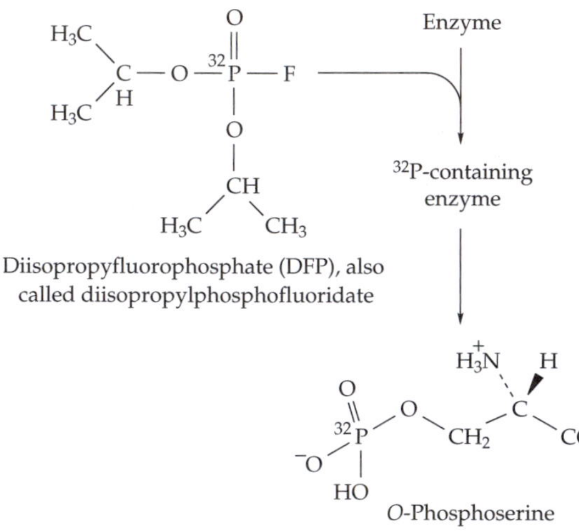

Diisopropylfluorophosphate (DFP), also called diisopropylphosphofluoridate

O-Phosphoserine

(12-19)

From study of peptides formed by partial hydrolysis of the ^{32}P-labeled chymotrypsin, the sequence of amino acids surrounding the reactive serine was established and serine 195 was identified as the residue whose side chain hydroxyl group became phosphorylated. The same sequence Gly-Asp-Ser-Gly was soon discovered around reactive serine residues in trypsin, thrombin, elastase, and in the trypsin-like **cocoonase** used by silkmoths to escape from their cocoons.[198] We know now that these are only a few of the enzymes in a very large family of serine proteases, most of which have related active site sequences.[199,200] Among these are **thrombin** and other enzymes of the blood-clotting cascade (Fig. 12-17), proteases of lysosomes, and secreted proteases.

Numerous serine proteases, including trypsin-like enzymes called **tryptases**[201–204a] and chymotrypsin-like chymases,[205–207] are found within tissues in which they are stored in granules of mast cells,[208] neutrophils, lymphocytes, and cytotoxic T cells.[205] Secretory granules of mast cells present in skin and other tissues contain high concentrations of tryptase and chymase precursors[202,206] which may be released as part of an inflammatory response. Tryptase may be involved in asthma and other allergic resposes.[201] **Cathepsin G**[209]

(proteinase II), **neutrophil elastase**, and **proteinase III**[210] are found in granules of neutrophils and monocytes as well as in mast cells.[209] Cytoplasmic granules of cytotoxic T cells contain at least seven proteases called **granzymes** that can be released to attack target cells.[211–213a]

Many secreted proteins, as well as smaller peptide hormones, are acted upon in the endoplasmic reticulum by tryptases and other serine proteases. They often cut between pairs of basic residues such as KK, KR, or RR.[214–216] A substilisin-like protease cleaves adjacent to methionine.[217] Other classes of proteases (e.g., zinc-dependent carboxypeptidases) also participate in this processing. **Serine carboxypeptidases** are involved in processing human prohormones.[218] Among the serine carboxypeptidases of known structure is one from wheat[219] and **carboxypeptidase Y**, a vacuolar enzyme from yeast.[220] Like the pancreatic metallocarboxypeptidases discussed in Section 4, these enzymes remove one amino acid at a time, a property that has made carboxypeptidases valuable reagents for determination of amino acid sequences. Carboxypeptidases may also be used for modification of proteins by removal of one or a few amino acids from the ends.

The variety of bacterial serine proteases known include the 198-residue **α-lytic protease** of *Myxobacter*,[221] a family of at least 80 **subtilisins** which are produced by various species of *Bacillus*[222–225a] as well as by many other organisms,[226] and a trypsinlike enzyme from *Streptomyces griseus*.[227,228] **Tripeptidyl peptidases**, subtilisin-like enzymes, cut tripeptides from the N-termini of proteins.[228a,b,c] One participates in lysosomal protein degradation and the other, an oligomer of 138 kDa subunits, cuts precursor proteins to form neuropeptides and other hormones (Chapter 30).

Acyl-enzyme intermediates. Serine proteases are probably the most studied of any group of enzymes.[229] Early work was focused on the digestive enzymes. The pseudosubstrate, *p*-nitrophenyl acetate, reacts with chymotrypsin at pH 4 (far below the optimum pH for hydrolysis) with rapid release of *p*-nitrophenol and formation of acetyl derivative of the enzyme.

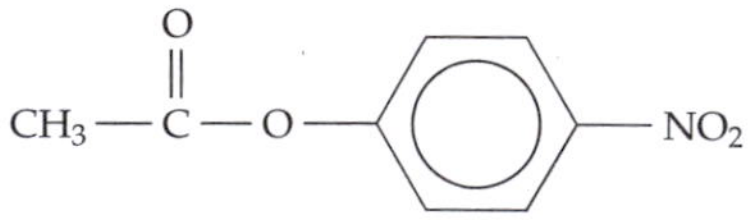

p-Nitrophenyl acetate

This acetyl enzyme hydrolyzes very slowly at pH 4 but rapidly at higher pH. These experiments suggested a double displacement mechanism:

$$ (12\text{-}20) $$

Although the experiments with DFP suggested that the $-O^-$ group from Ser 195 might be the base $-B^-$ in this equation, there was reluctance to accept this deduction because of the very weak acidity of the $-CH_2OH$ group. Furthermore, the pH dependence of catalysis suggested an imidazole group of a histidine side chain as the attacking nucleophile B. Indeed, imidazoles catalyze the nonenzymatic hydrolysis

of *p*-nitrophenyl acetate with formation of unstable *N*-acetyl imidazoles as intermediates. Thus, while the stable end products of reactions with pseudosubstrates were unquestionably derivatives of serine, the possibility remained that these were side products and that histidine was involved in transient, rapidly forming, and reacting intermediates. It remained for the results of the then newly developed science of X-ray crystallography to clarify this question.[230]

Three-dimensional structures. The structures of chymotrypsin,[199,230,231] trypsin,[232,233] elastase,[234,235] thrombin,[236] kallikrein,[237,238] and many other enzymes are similar, with the basic fold shown in Fig. 12-9.[199,228] Both Ser 195 and His 57 (or corresponding residues) are present in the active site (Figs. 12-9, 12-10). From the observed positions of competitive inhibitors occupying the active site, the modes of binding depicted in Fig. 12-10A for the chymotrypsin family and in Fig. 12-10B for the subtilisin family have been deduced. Bear in mind that the X-ray diffraction results do not

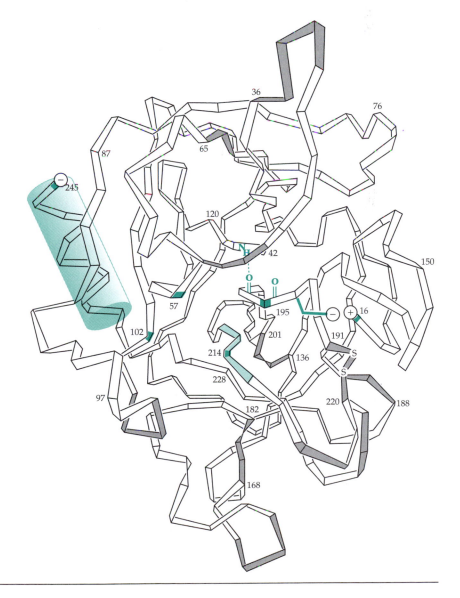

Figure 12-9 Alpha-carbon diagram of the three-dimensional structure of pancreatic elastase. A principal structural feature is a pair of β cylinders. One of these, at the top of the figure, is viewed end-on, while the other, at the bottom of the figure, is viewed from a side. The prominent interface between them is seen in the center. The α-carbon positions of the catalytic triad serine 195, histidine 57, and aspartate 102 are marked. As shown in Fig. 12-10 the catalytic triad is located across the interface between β cylinders, which may allow for easier conformational alterations during the action of the enzyme. The four disulfide bridges and the long C-terminal helix are also emphasized. The cross-hatched loop regions are residues present in elastase but not present in chymotrypsin. The green shaded strand (residues 214–216) is a segment that joins residues P_1 and P_2 of the substrate to the edge of the lower β cylinder in an extended β structure. The ion pair formed by aspartate 194 and isoleucine 16 during zymogen activation is also shown, as are two peptide carbonyl groups that protrude into the interface area, one forming a hydrogen bond across the interface. Modified from a drawing of Sawyer *et al.*[234] which was in turn based on the original chymotrypsin drawing of Annette Snazle.[230]

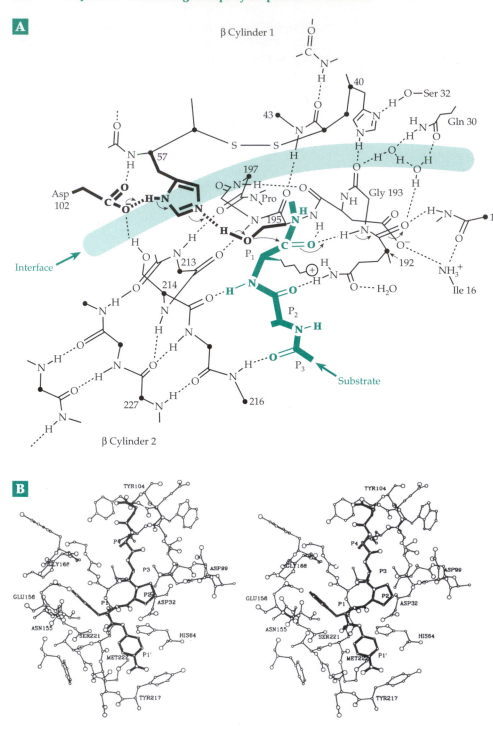

A

β Cylinder 1

Interface

β Cylinder 2

B

Figure 12-10 (A) Part of the hydrogen-bonding network of trypsin and other serine proteases with a bound trypsin substrate (green). Residues P_1 to P_3, as defined in Fig. 12-14, have been marked. The view is similar to that in Fig. 12-9 but some background lines have been omitted. Note the two competing hydrogen-bonded chains passing through C=O of residue 227 in β cylinder 2. One of these chains passes through the backbone of the substrate and the other through residues 214 and 195 across the interface between domains into β cylinder 2. Asp 102, His 57, and Ser 195 of the catalytic triad are emphasized. Arrows indicate probable movement of electrons from the negative charge of Asp 102 into the "oxyanion hole." After Metzler.[85] Based on papers of Sawyer *et al.*[234] and Huber and Bode.[240] (B) Stereoscopic view showing the model substrate *N*-succinyl-L-Ala-L-Ala-L-Pro-L-Phe-*p*-nitroanilide bound into the active site of subtilisin BPN'. Residues of the catalytic triad (Ser 221, His 64, Asp 32) and some others are labeled. Subsites P_1', P_1, P_2, P_3, and P_4 (see Fig. 12-14) are also labeled. Notice that site P_3, which is near the top in this drawing, is at the bottom in (A). Based on X-ray data of R. Bott and M. Ultsch at 0.2 nm resolution. From Wells and Estell.[223]

show where the hydrogen atoms are and that these have been added in Fig. 12-10. The imidazole group of His 57 is located next to the side chain –OH group of Ser 195 and is able to form an N---H–O hydrogen bond to it. The obvious conclusion is that His 57 acts as a general base catalyst that assists in removing the proton from the –OH of Ser 195, making that hydroxyl group more nucleophilic than it would be otherwise. This may happen after the –OH group has started to add to the substrate carbonyl.

The second nitrogen atom of His 57 is hydrogen bonded to the carboxylate group of Asp 102, which is in turn hydrogen bonded to two other groups. Aspartate 102 has one of the few carboxylate side chains that is buried inside the protein. To Blow,[230,231,239] the structure suggested a **charge-relay system** by which negative charges might move synchronously from Asp 102 to the imidazole which could then deprotonate the hydroxyl group of Ser 195, allowing the serine oxygen to add to the substrate carbonyl to form the tetrahedral or **oxyanion** intermediate, which is depicted in step *b* of Fig. 12-11. The small arrows in Fig. 12-10 also indicate the movement of charge. Blow suggested that in the extreme case a *proton* might be transferred also from the His 57 imidazole to the Asp 102 carboxylate. However, a variety of experiments, including studies by ^{15}N NMR[241,242] and by neutron diffraction,[243] suggest that the imidazole does not transfer its proton to the carboxylate of Asp 102,[244,245] unless it does so transiently, e.g., in a transition state complex. Instead, the carboxylate and the imidazolium ions formed by protonation of His 57 probably exist as a tight ion pair:

Support for this concept is provided by ^{1}H NMR studies which have identified a downfield resonance of the hydrogen-bonded proton in this pair at ~18 ppm in chymotrypsinogen and chymotrypsin at low pH and at ~14.9–15.5 ppm at high pH values.[246,247] Similar resonances are seen in the α-lytic protease,[248] in subtilisin,[249] in adducts of serine proteases with boronic acids[250,251] or peptidyl trifluoromethyl ketones,[252] in alkylated derivative of the active site histidine,[253] and in molecular complexes that mimic the Asp-His pair in the active sites of serine proteases.[254]

The catalytic cycle. Figure 12-11 depicts the generally accepted sequence of reactions for a serine protease. If we consider both the formation and the subsequent hydrolysis of the acyl-enzyme intermediate with appropriate oxyanion intermediates, there are at least seven distinct steps. As indicated in this figure, His 57 not only accepts a proton from the hydroxyl

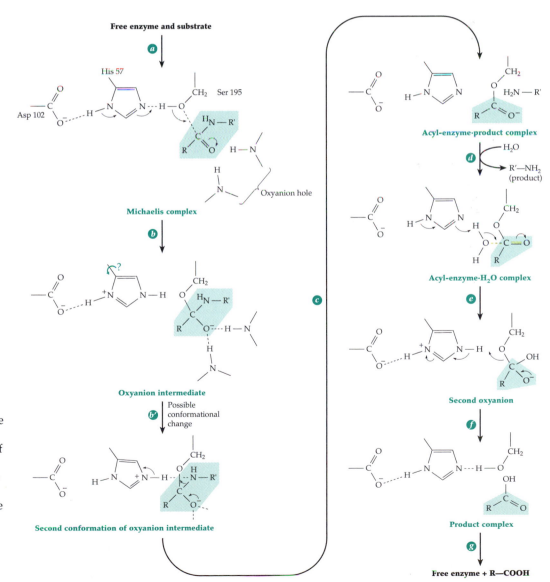

Figure 12-11 Sequence of chemical reactions involved in the action of a serine protease. The oxyanion hole structure has been omitted in the right-hand column. The imidazole ring may rotate, as indicated by the green arrow, to provide two different conformations.

group of Ser 195 (step *b*) but probably also functions in protonation of the –NH–R' leaving group (step *c*). An unprotonated –NH–R' would be such a poor leaving group that the oxyanion intermediate would not go on to acyl-enzyme. However, donation of a proton to that leaving group from the protonated His 57 (general acid catalysis) permits elimination of $H_2N–R'$ (step *c*). The product of this step is the acyl-enzyme intermediate which must be hydrolyzed to complete the catalytic cycle. This is accomplished through steps *d–f* of Fig. 12-11. The product R'–NH_2 is replaced by H_2O, which, in steps paralleling steps *b* and *c*, is converted to an HO^- ion that serves as the attacking nucleophile to form (in step *e*) a second oxyanion intermediate which is cleaved to the second product R–COOH. The water molecule that enters in step *d*, and which participates in hydrolysis of the acyl enzyme, has apparently been observed directly by time-resolved Laue crystallography at low temperature.[255]

The catalytic triad. The significance of the charge-relay effect may not be fully understood but the importance of the Ser•His•Asp cluster, which has become known as the catalytic triad, cannot be doubted. It has evolved independently in several subfamilies of bacterial and plant proteases, a well-known example of "convergent evolution." This triad is also found throughout a broad range of many different kinds of enzymes and other proteins. In pancreatic enzymes the catalytic triad consists of Ser 195•His 57•Asp 102 but in the bacterial subtilisin it consists of Ser 221•His 64•Asp 32 and in a wheat serine carboxypeptidase it is Ser 146•His 397•Asp 338.[219] In all three cases the folding patterns of the polypeptides are entirely different but the geometry of the triad is the same. Another folding pattern is seen in a protease encoded by cytomegalovirus, which contains an active site Ser 132•His 63 pair.[256,257]

Investigation of a host of mutant proteins also demonstrates the importance of the catalytic triad. For example, if either the histidine or the serine of the triad of subtilisin was replaced by alanine the catalytic activity decreased by a factor of 2×10^6 and replacement of the aspartate of the triad by alanine decreased activity by a factor of 3×10^4.[229,258] When Asp 102 of trypsin is replaced by asparagine the catalytic activity falls by four orders of magnitude.[259] This may be in part because the histidine in this mutant is hydrogen bonded to Asn 102 as the tautomer with a proton on N^ε, the nitrogen that should serve as the catalytic base in step *b* (Fig. 12-11).[260] A mutant in which Ser 214 (see Fig. 12-10) was replaced with alanine is fully active but charged residues in this position interfere with catalysis.[261]

Does the "low-barrier hydrogen bond" in the catalytic triad play any special role in catalysis? Blow's suggestion of a charge relay from Asp 102 to

Ser 195 of chymotrypsin is probably correct. Some theoretical calculations have indicated the possibility of synchronous movement of the two protons in the system during step *a* of the sequence shown in Fig. 12-11, with the proton in the strong hydrogen bond to Asp 102 moving away from His 57 and toward the midpoint distance of the hydrogen bond. Could the presumed high energy of the short hydrogen bond be harnessed to lower the transition state energy? A realistic possibility is that the hydrogen bond increases the polarizability of the catalytic triad, facilitating movement of the substrate to the transition state. Cassidy *et al.* suggested that the strong hydrogen bond may be formed by compression of the triad resulting from binding of substrate in the S_1 and S_2 subsites.[252] They suggested that this would raise the pK_a of His 57 from ~7 in the free enzyme to 10–12, high enough to enable it to remove the proton from the Ser 195 –OH group and low enough to allow the protonated form to be the proton donor to the leaving group (step *c*, Fig. 12-11).

The "oxyanion hole." A third mechanism by which an enzyme can assist in a displacement reaction on a carbonyl group is through protonation of the carbonyl oxygen atom by an acidic group of the enzyme (Eq. 12-21). This will greatly increase the positive charge on the carbon atom making attack by a nucleophile easier and will also stabilize the tetrahedral

$$(12\text{-}21)$$

intermediate. Although the carbonyl oxygen is very weakly basic, it can interact with a suitably oriented acidic group of the enzyme. In many serine proteases this acidic function is apparently fulfilled by NH groups of two amide linkages. In chymotrypsin these are the backbone NH groups of Ser 195 and Gly 193 (Figs. 12-10,12-12). No actual transfer of a proton to the carbonyl oxygen of the substrate is expected. However, the NH groups are positive ends of amide dipoles and can interact electrostatically with the negative charge that develops on the oxyanion. The fit of substrate into the **oxyanion hole** between the two

NH groups is apparently good only for the tetrahedrally bonded oxyanion intermediate,[262] a structure thought to be close to that of the transition state. The importance of the oxyanion hole for catalysis has been supported by theoretical calculations[245] and by the study of mutant enzymes.[263] In subtilisin, in which a side chain of asparagine forms part of the oxyanion hole, replacement of Asn with the isosteric Leu causes k_{cat} to fall by a factor of about 200 while K_m is unaffected.[264] It is also of interest that thiono ester substrates, in which the C=O of an oxygen ester has been replaced by C=S, bind with normal affinity to chymotrypsin but are not hydrolyzed at significant rates.[265]

The formation of the oxyanion intermediate during serine protease action is also supported by the existence of tetrahedral forms of enzymes inhibited by substrate-like aldehydes. The –OH group of Ser 195 can add to the carbonyl group to form hemiacetals. For example, a ^{13}C-enriched aldehyde whose carbonyl carbon had a chemical shift of 204 ppm gave a 94 ppm resonance as it formed the tetrahedral hemiacetal with one of the inhibitory aldehydes, *N*-acetyl-L-Leu-L-Leu-L-arginal

(**leupeptin**; Box 12-C). A natural product from a species of *Streptomyces*, this aldehyde inhibits trypsin and several other enzymes strongly.[266] Adducts of wheat serine carboxypeptidases with aldehyde inhibitors have also been observed.[267] While the carbonyl group of the substrate amide linkage that is to be cleaved apparently can't form strong hydrogen bonds to the NH groups of the oxyanion hole, that of the acyl-enzyme intermediate can, as judged by resonance Raman spectroscopy.[268,269] The strength of the hydrogen bonding, as judged by the stretching frequency of the C=O group, is correlated with the reactivity of the acyl group.[269]

Chymotrypsinogen and related proenzymes have extremely low catalytic activity even though a major part of the substrate binding site as well as the catalytic triad system are already in place. However, the oxyanion hole is created during activation of the proenzyme by a subtle conformational change[197,262,271] that involves the chain segment containing Gly 193 (Fig. 12-12). This is further evidence of the importance of this part of the active site structure.

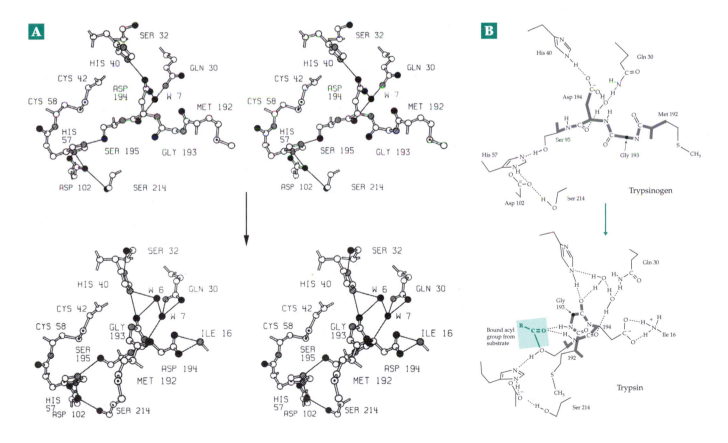

Figure 12-12 Formation of the oxyanion hole following cleavage of trypsinogen between Lys 15 and Ile 16. (A) Stereoscopic view. (B) Schematic representation. The newly created terminal –NH₃⁺ of Ile 16 forms a hydrogen-bonded ion pair with the carboxylate of Asp 194. This breaks the hydrogen bond between Asp 194 and His 40 in trypsinogen, inducing the peptide segment 192–194 to shift from an extended conformation to a helical form in which the NH groups of Gly 193 and Ser 195 form the oxyanion hole. Notice that the positions and interactions of Asp 102, His 57, and Ser 195, the catalytic triad, are very little changed. From Birktoft *et al.*[270]

Stereoelectronic considerations. The amide group that is cleaved by a protease is a resonance hybrid of structures A and B of Fig. 12-13. The unshared pair of electrons on the nitrogen atom of structure A and the third unshared pair on the oxygen atoms of structure B (shaded orbitals) have been drawn in this figure in such a way that they are anti-periplanar to the entering serine oxygen. This is required in the transition state according to stereoelectronic theory. The newly created (shaded) electron pair on oxygen is one of those that hydrogen bonds to an NH group of the oxyanion hole. The tetrahedral intermediate has another unshared pair, which is not hydrogen bonded, and is antiperiplanar to the HN-R leaving group. Thus, it appears to be set up for easy elimination. Nevertheless, there is some doubt about the need for adherence to the stereoelectronic "rule" that two antiperiplanar lone pairs are necessary for elimination.

There is another complication. The leaving group –NH–R' cannot be eliminated from the oxyanion until it is protonated, presumably by the imidazolium group of His 57. However, the proton on His 57 will be adjacent to the proton that is already on this nitrogen rather than to the unshared pair of electrons on the same nitrogen atom. A conformational change in which the tilt of the catalytic imidazole ring is altered (step *b'*, Fig. 12-11) or in which the ring rotates, may have to precede the protonation of the leaving group (step *c*, Fig. 12-11).[272–274b] This change may be assisted by the presence of positive charge on the protonated imidazole, but it still does not solve the problem.

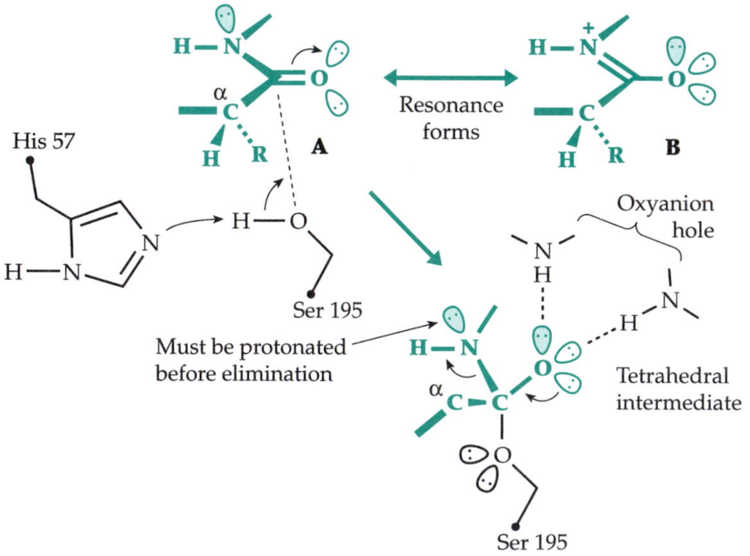

Bizzozero and Butler suggested that a rapid inversion of this chiral center on the –NH–R' group may be required prior to protonation and elimination.[273] This would explain their observation that N-alkylated peptide linkages of otherwise fast substrates are not cleaved by chymotrypsin. A high-resolution structure determination demonstrated that a very slow elastase substrate that contains N-methylleucine at the cleavage site forms a normal ES complex.[275] However, inversion would be hindered by the N-methyl group. An alternative in inversion would be a torsional rearrangement of the substrate during binding.[276] The structure of an acyl-enzyme intermediate with elastase has been determined by crystallographic cryoenzymology. The intermediate was allowed to accumulate at –26°C, after which the temperature was lowered to –50°C and the structure determined. The structure shows the carbonyl group in the oxyanion hole as anticipated.[277]

pH dependence. A plot of k_{cat}/K_m for chymotrypsin is bell shaped with a maximum around pH 7.8 and pK_a values of 6.8 and 8.8. These represent pK_as of the free enzyme (Eq. 9-57). That of 6.8 has been shown to represent His 57. As the pH is lowered from the optimum the affinity for substrate falls off as His 57 becomes protonated. The high pK_a of 8.8 is thought to belong to the N-terminal amino group of Ile 16, a group that is generated during the conversion of the proenzyme to active enzyme.[278] The Ile 16 amino group forms an ion pair with Asp 194 (Fig. 12-12B) which is next to the serine at the active center. This salt linkage helps to hold the enzyme in the required conformation for reaction and its deprotonation at high pH causes a decrease in substrate affinity.[279]

The value of k_{cat} is also pH dependent and falls off at low pH around a pK_a value of from 6–7 depending upon the substrate. However, no higher pK_a affects k_{cat} in the experimentally accessible range. These results provide the basis for believing that an unprotonated His 57 is needed in the ES complex for catalysis to occur. Similar conclusions have been reached for other serine proteases.

Substrate specificity. Like most other enzymes, proteases display distinct preferences for certain substrates. These are often discussed using the nomenclature of Fig. 12-14. The substrate residue contributing the carbonyl of the amide group to be cleaved is designated P_1 and residues toward the N terminus as P_2, P_3, etc., as is shown in Fig. 12-14. Residues toward the C terminus from the peptide linkage to be cleaved are designated P_1', P_2', etc. Chymotrypsin acts

Figure 12-13 The stereochemistry of formation of a tetrahedral intermediate by a serine protease. The most probable orientation of groups as deduced by model building is shown. The shaded orbitals in A and B are antiperiplanar to the entering oxygen of Ser 195. See Polgár and Halász.[272]

most rapidly if the P_1 residue is one of the aromatic amino acids.[280] Thus, the S_1 part of the substrate binding site must bind preferentially to large, flat aromatic rings. The crystal structure of chymotrypsin showed this site to be composed of nonpolar side chain groups. On the other hand, in trypsin the specificity portion of the S_1 site is a deep "specificity pocket" containing a fixed negative charge provided by the carboxylate side chain group of Asp 189. This explains why trypsin acts only upon peptide linkages containing the positively charged arginine or lysine residues in the P_1 position. In elastase the specificity pocket is partly filled by nonpolar side chains and the enzyme can accommodate only small P_1 side chains such as the methyl group of alanine. Replacement of Asp 189 of trypsin by lysine led to the predicted loss of specificity for basic side chains. However, the mutant enzyme did not become specific for negatively charged side chains.[281] This is presumably because the lysine $-NH_3^+$ group was not located in the same position as the Asp 189 carboxylate as a result of a different packing of side chains between native and mutant enzymes. Residues P_1' and P_2' also have major effects on substrate binding by serine proteases.[282–284] Now many mutant forms of subtilisin[225,285] and other serine proteases are being made and are yielding a more sophisticated understanding of the basis of the specificity of these enzymes. An important factor that has emerged is flexibility of surface loops in allowing an enzyme to adjust its structure to give a better fit to some substrates.[286,287] The specificity of the serine proteases is also being exploited in the design of specific inhibitors (Box 12-D).

Many serine proteases react with p-nitrophenylacetate to give acetyl enzymes. However, its rate of hydrolysis to give acetate is orders of magnitude slower than that of acyl-enzymes derived from small substrates such as the chymotrypsin substrate N-acetyltyrosine amide.

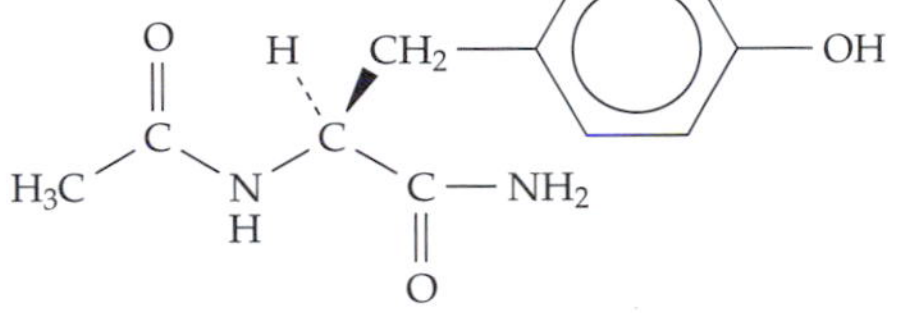

N-Acetyltyrosine amide

In addition to the specificity-determining P_1 aromatic side chain, the amide groups of this substrate can form specific hydrogen bonds to the protein (Fig. 12-10). These hydrogen bonds presumably help the enzyme to recognize the compound, which is bound with K_m of ~0.03 M and is hydrolyzed (with liberation of NH_3)[288,289] with k_{cat} ~0.17 s^{-1}.

What happens when the length of the substrate is extended in the direction of the N terminus? The tyrosyl residue in the foregoing compound may be designated P_1. For extended substrates which contain P_2, P_3, and additional residues (this includes most natural substrates) the K_m values decrease very little from that of short substrates despite the larger number of "subsites" to which the substrate is bound. However, the maximum velocity is often much greater for the extended substrates than for short ones. Thus, for N-acetyltyrosyl-glycine amide K_m is 0.017 M, only a little less than for N-acetyltyrosine amide, but k_{cat} is 7.5 s^{-1}, 440 times greater than for the shorter substrate.[229,288,289] Other examples have been tabulated by Fersht.[279]

These observations suggest that *the binding energy that would be expected to increase the tightness of binding is, instead, causing an increase in V_{max}*[279,290] that is, it is reducing the Gibbs energy of activation. How can this be? Imagine that as the extended substrate binds, for example, into the subsite S_2 (which binds residue P_2), it must compress a spring in the enzyme. Could not the compressed spring now provide a source of energy for assisting in peptide bond cleavage? If this is the case, we must ask "what are the springs?" Are there amide linkages of the peptide backbone that are distorted when the substrate hydrogen bonds into the site? How is the distortion transmitted to the active site and how does it stabilize the transition

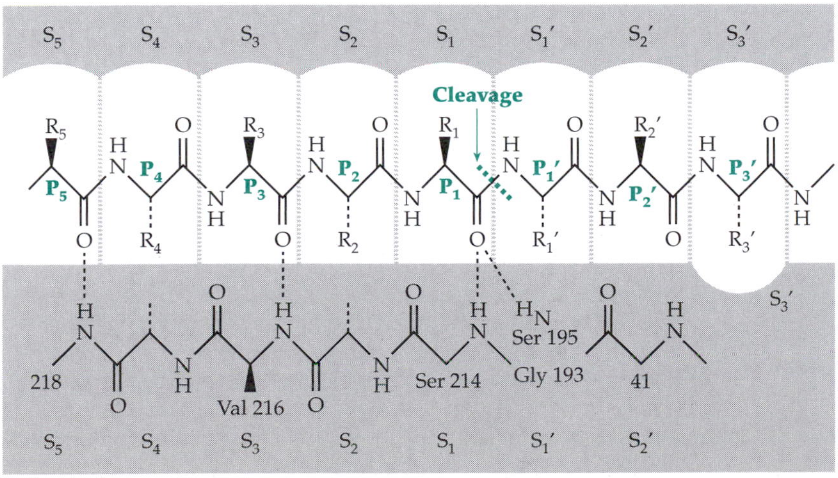

Figure 12-14 Standard nomenclature used to define the residues P_1, P_2 ... toward the N terminus and P_1', P_2' ... toward the C terminus of a peptide substrate for a protease. The corresponding subsites of the protease are designated S_1, S_2 ... S_1', S_2'.

state? We are far from understanding the answers to these questions. However, it is of interest that one chain of H-bonds that passes through the amide between P_1 and P_2 also passes through a carbonyl group of a β bulge (Fig. 12-10). A second chain of H-bonds that connects to the same β bulge passes through the backbone amide of the active site Ser 195, across the interface between the two domains, and through the backbone of the second β cylinder.[85] Perhaps when substrate binds, the strengthening of the first H-bond chain weakens the second through competition at the β bulge, allowing some subtle rearrangement in the active site structure. We are talking here about the P_1–P_2 amide. What happens at P_2–P_3, and further along the chain in both directions? It is important to understand these effects; the phenomenon of increased reaction rate for longer, more specific substrates is observed with many proteases and other enzymes as well.

Another difference between small substrates and longer, more specific substrates has been found in studies of the effects of changing the solvent from pure H_2O to mixtures containing an increasing mole fraction of ^2H_2O. This is called a **proton inventory**.[291] For the serine proteases the rates are decreased as the ^2H content of the solvent increases, a fact that suggests that some step involving a proton transfer, for example, the deprotonation of Ser 195, is rate limiting. For simple substrates, the effect is directly proportional to the mole fraction of ^2H. However, for extended substrates, a quadratic dependence on the mole fraction of ^2H is observed. This suggests that a process involving synchronous transfer of two protons, as in the postulated charge relay system, may be more important in extended substrates than in simple ones and may account for the more rapid action on these substrates.[290,292] However, other interpretations of the data are possible, leaving this conclusion uncertain.[229,291]

2. The Cysteine Proteases (Thiol Proteases)

Papain from the papaya is one of a family of enzymes that includes **bromelain** of the pineapple, **ficin** of the fig, and **actinidin** of the kiwifruit.[293] Additional cysteine proteases[294] from the latex of the papaya tree *Carica papaya* are known as **caricain**, **chymopapain**,[294] and **glycyl endopeptidase**.[295] All are members of a large superfamily which includes at least 12 mammalian enzymes and many others from both eukaryotic and prokaryotic organisms.[296] All share with papain a characteristic structure which was determined by Drenth and coworkers in 1968.[297] The participating nucleophile in the active sites of these enzymes is an SH group, that of Cys 25 in papain. An adjacent imidazole from His 159 removes the proton from the SH group to form a thiolate-imidazolium pair.[298–300]

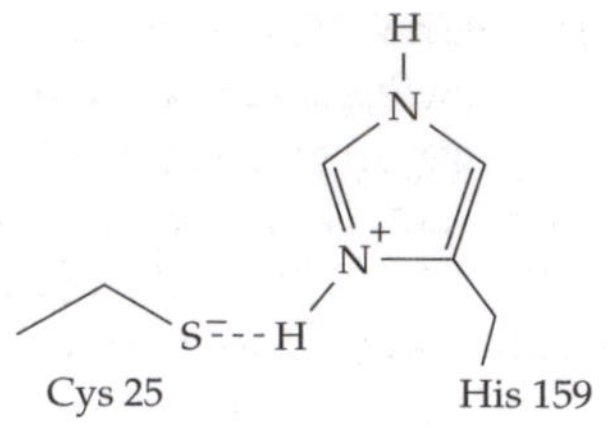

The close proximity of the imidazolium group greatly lowers the microscopic pK_a of the Cys 25 thiol group and the proximity of the resulting hydrogen-bonded S^- group greatly raises the microscopic pK_a of the imidazolium group.

Studies of the pH dependence of V_{max}/k_{cat} (Eq. 9-57) reveal a bell-shaped dependence on pH with pK_a values[301,302] of ~4 and ~9. However, the ion pair is formed at a pH below four with apparent pK_a values of 2.5, 2.9, and 3.3 for ficin, caricain, and papain, respectively.[302] These low values can be assigned principally to Cys 25 with only very small contributions from His 159 (see Eq. 6-75). A third pK_a, perhaps of a nearby carboxylate from Glu 50, affects the rate. For caricain the nearby Asp 158 (Fig. 12-15) has been implicated.[303,304]

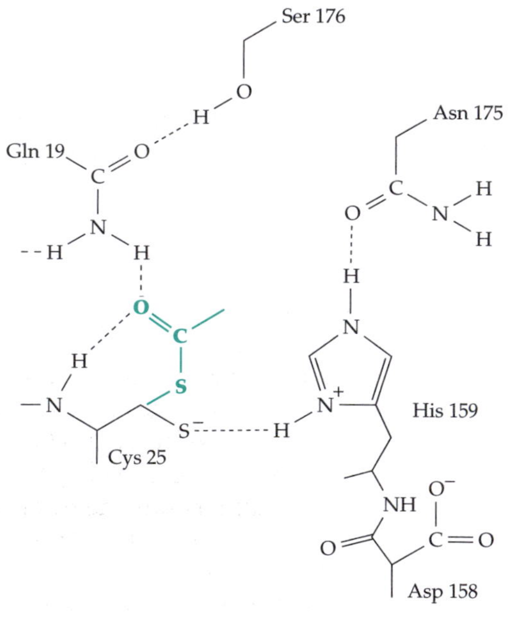

Figure 12-15 Schematic drawing of the active site of a cysteine protease of the papain family with a partial structure of an acyl-enzyme intermediate in green. The thiolate-imidazolium pair of Cys 25•His 159 lies deep in the substrate-binding cleft and bridges an interface between two major structural domains, just as the Ser•His pair does in serine proteases (Fig. 12-10). This may facilitate small conformational changes during the catalytic cycle. Asn 175 provides a polarizable acceptor for positive charge, helping to stabilize the preformed ion pair, and allows easy transfer of an imidazolium proton to the product of substrate cleavage. The peptide NH of Cys 25 and the side chain of Gln 19 form an oxyanion hole.

As shown in Fig. 12-15, the side chain of Asn 175 provides papain with a third member of a catalytic triad analogous to that of serine proteases.[305] Glutamine 19, together with the peptide backbone NH of Cys 25, provides an oxyanion hole.[306–308] Many studies, including structure determinations on bound aldehyde and other inhibitors, and observation by ^{13}C NMR[309] indicate that thiol proteases act by addition of the thiolate anion to the peptide carbonyl of the P_1 residue, just as in step b of Fig. 12-11 (see also Fig. 12-12) for serine proteases.[308,310] However, alternative sequences of proton transfer have been suggested.[311] A possible role for a strong hydrogen bond has also been proposed.[310] For a detailed discussion see Brocklehurst et $al.$[312]

Most of the lysosomal proteases called **cathepsins** are small 20- to 40-kDa glycoproteins found in all animal tissues.[313] Most are cysteine proteases which function best and are most stable in the low pH reducing environment of lysosomes. They resemble papain in size, amino acid sequence, and active site structures. Papain is nonspecific but most cathepsins have definite substrate preferences. Cathepsin B is the most abundant. There are smaller amounts of related cathepsins H (an aminopeptidase)[314] and L[315] and still less of cathepsins C, K, and others. Cathepsin B is both an endopeptidase and an exopeptidase.[316] It acts on peptides with arginine at either P_1 or P_2 but also accepts bulky hydrophobic residues in P_1 and prefers tyrosine at P_3.[317] Cathepsin S is less stable at higher pH than other cathepsins and has a more limited tissue distribution, being especially active in the immune system.[318,319]

Cathepsin K is especially abundant in the bone resorbing osteoclasts (Chapter 8). It is essential to normal bone structure and its absence is associated with the rare hereditary disease pycnodysostosis (pycno) which causes short stature, fragile bones, and skull and skeletal deformities.[320] It may also play a role in the very common bone condition osteoporosis. Cathepsin C is also called **dipeptidyl peptidase**. It removes N-terminal dipeptides from many intracellular proteins activating many enzymes, including some other cathepsins.[321,322] A **prohormone thiol protease** cleaves peptide chains between pairs of basic residues, e.g., in the brain peptide precursor proenkephalin (Chapter 30), and also on the N-terminal side of arginine residues.[323] **Pyroglutamate aminopeptidase** removes pyroglutamyl (5-oxoprolyl) groups from amino termini of some peptides and proteins (see Fig. 2-4).[324] Another cysteine protease cleaves **isopeptide** linkages such as those formed by transglutaminase or those involving ubiquitin (Box 10-C).[325] Another cysteine protease present in animal tissues was recognized by its ability to hydrolyze the anticancer drug bleomycin. This **bleomycin hydrolase** is a hexamer with a central channel lined with papainlike active sites as in the proteasome structure (Box 7-A)[326] The enzyme also binds to DNA.[327] Unfortunately, cancer tissues often contain high levels of the enzyme, whereas it is low in skin and lung tissues, which are damaged by bleomycin.

The Ca^{2+}-dependent neutral proteases called **calpains** are found within the cells of higher animals. The 705-residue multidomain peptide chain of a chicken calpain contains a papain-like domain as well as a calmodulin-like domain.[328] It presumably arose from fusion of the genes of these proteins. At least six calpains with similar properties are known.[329] Some have a preference for myofibrillar proteins or neurofilaments.[330] They presumably function in normal turnover of these proteins and may play a role in numerous calcium-activated cellular processes.[331–332a]

A group of **cysteinyl aspartate-specific proteases (caspases)** play an essential role in programmed cell death (apoptosis).[333–335] Recall that nematodes are cell-constant organisms. For maturation of $Caenorhabditis$ $elegans$ 131 programmed cell deaths must occur at specific stages of development. An essential gene for this process was identified and named **CED-3**. Deletion of this gene completely blocked the death of these cells. CED-3 encodes a cysteine protease that is highly homologous to the mammalian **interleukin-1β-converting enzyme** (**ICE** or caspase 1) which cleaves the 31-kDa **pro-interleukin-1β** to form the active 175 kDa species of this cytokine (Chapter 30). At least ten caspases are known and many observations have confirmed their role in apoptosis (see also Chapter 32). In caspases 1 and 2 the side chains of Cys 285 and His 237 form the catalytic dyad and peptide NH groups of Cys 285 and Gly 238 form the oxyanion hole.[333]

Parasites often use proteases in attacks on their hosts. The cysteine protease **cruzain** is secreted by the trypanosomes that cause Chagas' disease and is essential to their survival within the human body.[336,337] Cruzain is consequently an attractive target for development of drugs for treating this major disease.

Among cysteine proteases of bacteria is a papain-like enzyme from $Clostridium$ $histolyticum$ with a specificity similar to that of trypsin.[338] The anaerobic $Porphyromonas$ $gingivalis$, which is implicated in periodental disease, produces both arginine- and lysine-specific cysteine proteases designated **gingipains**.[339,339a] Some virally encoded cysteine proteases, including one from the polio virus, have trypsin-like sequences with the serine of the catalytic triad replaced by cysteine.[340,341] A human adenovirus protease also has a Cys•His•Glu triad but a totally different protein fold.[342]

Zymogens of cysteine proteases usually have a long terminal extension which is removed, sometimes by autoactivation. Propapain has a 107-residue extension.[343] The 322-residue cathepsin B carries an unusually short 62-residue extension in its proenzyme form.[315,343,344] In every case the N-terminal extension folds into a domain, one of whose functions is to block the active site cleft.

3. N-terminal Nucleophile Hydrolases and Related Enzymes

The most recently discovered group of proteases are the N-terminal threonine hydrolases of the **multicatalytic protease complex** (MPC) of proteasomes. The enzymes are arranged in a regular array inside proteasomal compartments as shown in Box 7-A. The active site is a catalytic dyad formed from the amino group at the N terminus of the β subunits.[345–346a]

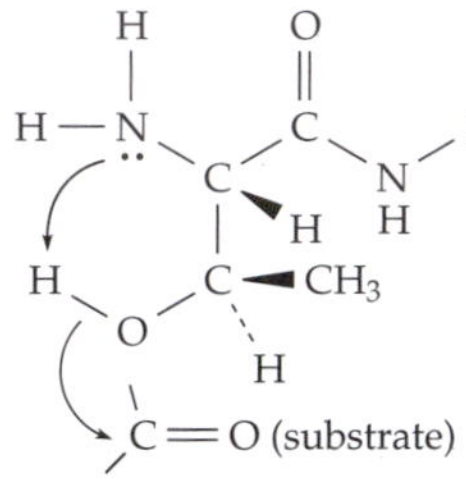

Hydroxyl · α-amine catalytic dyad

Proteasomes of *Thermoplasma* contain a single type of β subunit but eukaryotic proteasomes contain subunits with at least three distinct substrate preferences.[347–349c] They all appear to use the same hydrolytic mechanism but in their substrate specificities they are chymotrypsin-like, peptidylglutamyl-peptide hydrolyzing, branched chain amino acid preferring, and small neutral amino acid preferring based on the P_1 amino acid residue. In the spleen some of the β subunits of the proteasomes appear to have been replaced by proteins encoded by the major histocompatibility complex of the immune system (Chapter 31).[347] This may alter the properties of the proteasome to favor their function in antigen processing. Proteasomes are also ATP- and ubiquitin-dependent, as discussed in Section 6.

The enzyme glucosylasparaginase (aspartylglucosaminidase) is one of a group of other enzymes that use N-terminal threonyl groups as catalytic dyads.[346,350–353] It removes N-linked glycosyl groups from asparagine side chains of proteolytically degraded proteins and, as indicated on the accompanying structural formula, releases free aspartate and a 1-amino-*N*-acetylglucosamine-containing oligosaccharide. The amino group is then released as NH_3 from the product by acid catalysis in the lysosome.[350]

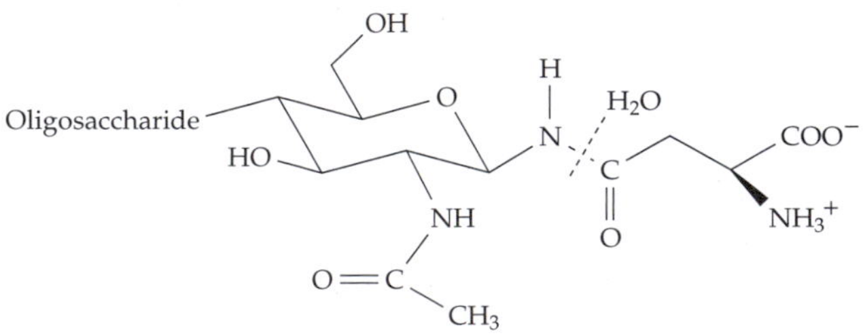

Glutamine PRPP amidotransferase (Fig. 25-15) and a penicillin acylase have similar active sites and overall

structures.[354] The $-NH_2$ group is basic enough in the environment of the protein to remove the proton from the threonine $-OH$ group, activating it as indicated below. Several serine proteases use an $-NH_2$ group of a lysine to form a serine·lysine dyad.[346]

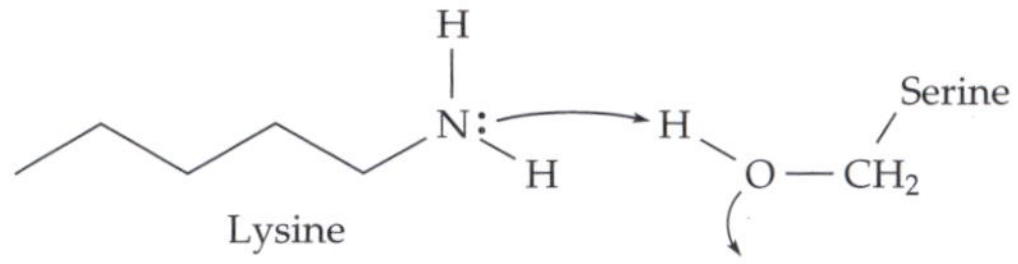

Among these are the well-known *E. coli* **leader peptidase**[355,356] and other signal peptidases.[357] These are integral membrane proteins that cleave N-terminal signal sequences from proteins incorporated into plasma membranes. Another enzyme of this class is the **lexA** repressor and protease discussed in Chapter 28.

A specific inhibitor of the major proteasomal activities is **lactacystin**, a compound formed by *Streptomyces*. Lactacystin is converted reversibly, by loss of *N*-acetyl-cysteine, into a β-lactone known as *clasto*-lactacystin. The N-terminal amino group attacks the reactive four-membered ring of the lactone (Eq. 12-22).[358,359]

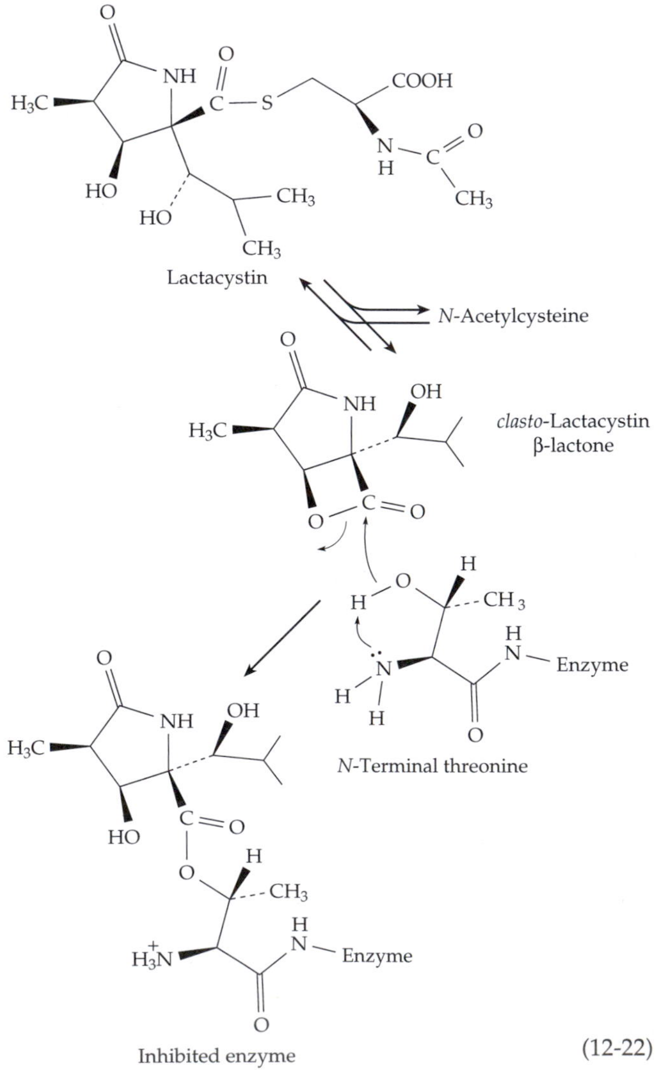

$$(12\text{-}22)$$

The active sites of the N-terminal nucleophile hydrolases are generated *autocatalytically*.[360–362] A single peptide chain is cleaved to form α and β chains as in the following diagram. An activating nucleophile such as histidine removes a proton from an adjacent threonine – OH and the resulting alkoxide ion attacks the adjacent peptide linkage, presumably via a tetrahedral intermediate, to form an ester linkage. Compare this sequence with reactions in Box 12-A and Eq. 14-41. Hydrolysis generates the N-terminal threonine as indicated:

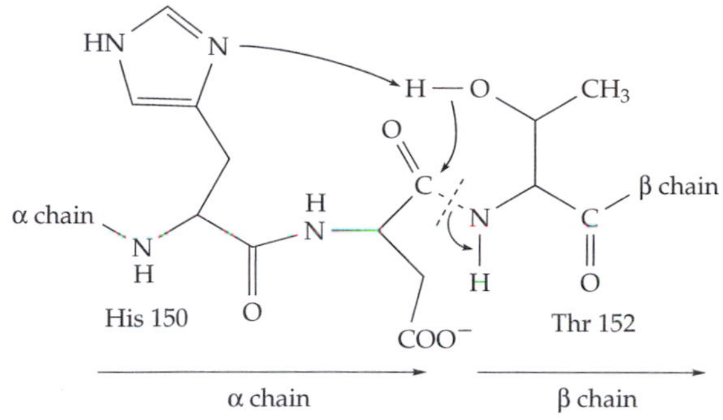

4. The Aspartic Proteases

A fourth large group of protein-hydrolyzing enzymes consists of **pepsin** of the stomach[363] and related enzymes.[364] Each of these ~320-residue proteins is folded into two domains which associate with a pseudo-twofold axis of symmetry that passes through the active site.[365] A second human gastric aspartic proteinase is **gastricsin**.[365] The related **chymosin** (rennin),[366] which is obtained from the fourth stomach of the calf, causes a rapid clotting of milk and is widely used in manufacture of cheese. Pepsin has a broad specificity but cleaves preferentially between pairs of hydrophobic residues, converting proteins into soluble fragments. It is unusual in being able to cleave X-Pro peptide bonds.[367] Chymosin has a more restricted specificity, cutting the κ-casein of milk between a Phe–Met bond. This decreases the stability of the milk micelles, inducing clotting.[366] The serum protein **renin** (distinct from rennin), the lysosomal **cathepsins D** and **E**,[368–370] an **aspartyl aminopeptidase**,[371] and various fungal proteases are also closely related.[372–375] Renin,[376,377] which is synthesized largely in the kidneys, is involved in blood pressure regulation (Box 22-D). More distantly related aspartic proteases are encoded by retroviruses.

The pepsin family is most active in the low pH range 1–5. All of the enzymes contain two especially reactive aspartate carboxyl groups.[378] One of them (Asp 215 in pepsin) reacts with site-directed diazonium compounds and the other (Asp 32) with site-directed epoxides.[379] It is attractive to think that one of these carboxyl groups might be the nucleophile in a double displacement mechanism. The second carboxyl could then be the proton donor to the cleaving group.

The acyl-enzyme would be an acid anhydride. However, X-ray studies on pepsin[363,380] and on related fungal enzymes such as **penicillopepsin**[381] and others[373,375] suggested a different possibility: A water molecule, hydrogen bonded to one of the active site Asp carboxylates or bridging between them, becomes the nucleophile. A proton, held by the carboxylate pair, protonates the substrate carbonyl to facilitate nucleophilic attack.[381] This is illustrated in Eq. 12-23 but without detail. Several intermediate sequences of reaction steps are possible.[382] At the beginning of the sequence one of the two symmetrically placed carboxylates is protonated while the other is not. Also notice that although the active site is symmetric, the substrate is bound asymmetrically as determined by its hydrogen bonding into an extended binding site.

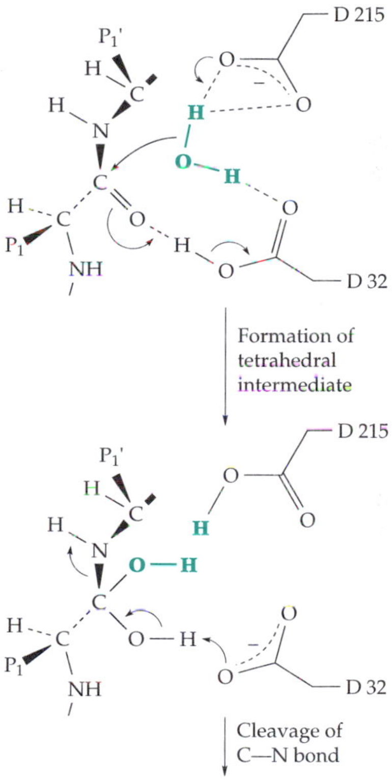

$$(12\text{-}23)$$

A characteristic feature of catalysis by the aspartic proteases is a tendency, with certain substrates, to catalyze transpeptidation reactions of the following type.

$$\text{Acetyl-Phe-Tyr} + \text{Acetyl-Phe*} \rightarrow \text{Acetyl-Phe*-Tyr}$$

$$(12\text{-}24)$$

Here Phe* is an isotopically labeled residue. Although such reactions suggested the possibility of some kind of activated amino group on the tyrosine that is cut off in the initial cleavage of the unlabeled substrate, it is more likely that the released tyrosine stays in the active site,[383] while the acetyl-Phe fragment exchanges with acetyl*-Phe.

BOX 12-C SYNTHETIC PROTEASE INHIBITORS

One of the goals of synthetic medicinal chemistry is to design potent inhibitors of clinically important proteases. Elastase inhibitors may be useful for treatment of emphysema, pancreatitis, and arthritis,[a,b] while inhibitors of the angiotensinogen-converting enzyme or of renin (Box 22-D) can help control blood pressure. Inhibition of thrombin, factor Xa, or other blood clotting factors (Fig. 12-17) may prevent blood clots and inhibition of the cytosolic tryptase may provide a new treatment for asthma. Inhibition of the cysteine protease cathepsin K may help combat osteoporosis and inhibition of cysteine proteases of corona viruses may fight the common cold. Cysteine proteases of schistosomes are also targets for protease inhibitors.[c]

Many chemical approaches are used in designing inhibitors. Often, a naturally occurring inhibitor provides a starting point. The availability of high-resolution structures of the target enzymes and of various enzyme-inhibitor complexes assists in the rational design of tight-binding inhibitors. Use of combinatorial chemistry (Chapter 3)[d] is another source of potential inhibitors. To be of practical use in medicine many criteria of stability, solubility, and low toxicity must be met. While most inhibitors are disappointing as drugs, their use in laboratory experimentation has clarified a great deal of biochemistry.

A straightforward approach is to hunt for short polypeptides that meet the specificity requirement of an enzyme but which, because of peculiarities of the sequence, are acted upon very slowly. Such a peptide may contain unusual or chemically modified amino acids. For example, the peptide **Thr-Pro-nVal-NMeLeu-Tyr-Thr** (nVal = norvaline; NMeLeu = N-methylleucine) is a very slow elastase substrate whose binding can be studied by X-ray diffraction and NMR spectroscopy.[e] Thiol proteases are inhibited by **succinyl-Gln-Val-Val-Ala-Ala-p-nitroanilide**, which includes a sequence common to a number of naturally occurring peptide inhibitors called **cystatins**.[f] They are found in various animal tissues where they inhibit cysteine proteases.

A group of inhibitors such as **leupeptin** have C-terminal aldehyde groups. Small oligopeptides with this structure and with appropriate specificity-determining side chains form tetrahedral hemiacetals, which may mimic transition state structures, at the active sites of the target enzymes.[g,h]

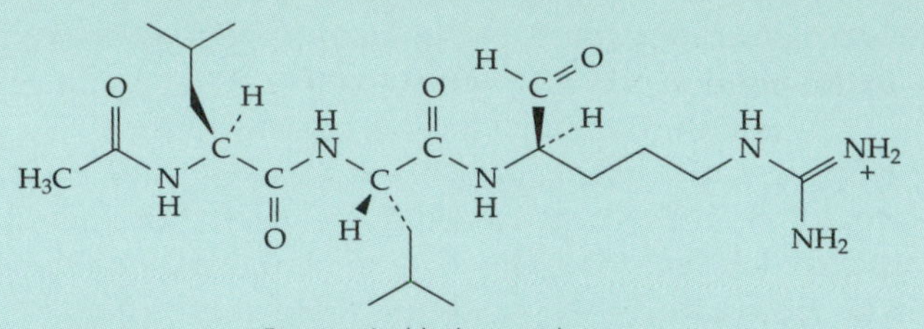

Leupeptin (*Actinomyces*)

Leupeptin is a slow, tight-binding inhibitor of trypsin. Some peptide aldehydes are potent, reversible inhibitors of cysteine proteases forming hemithioacetals with the active site cysteine.[i] Similarly, peptide nitriles form thioimidate adducts.[h]

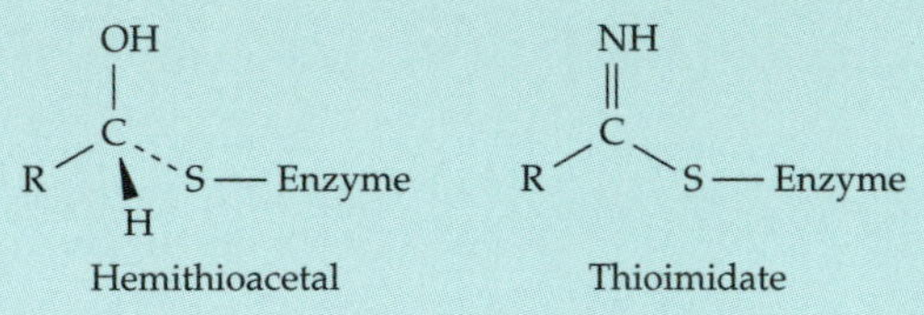

Hemithioacetal   Thioimidate

The peptide **boronic acids** form adducts with the active site serine of serine proteases.[j–l]

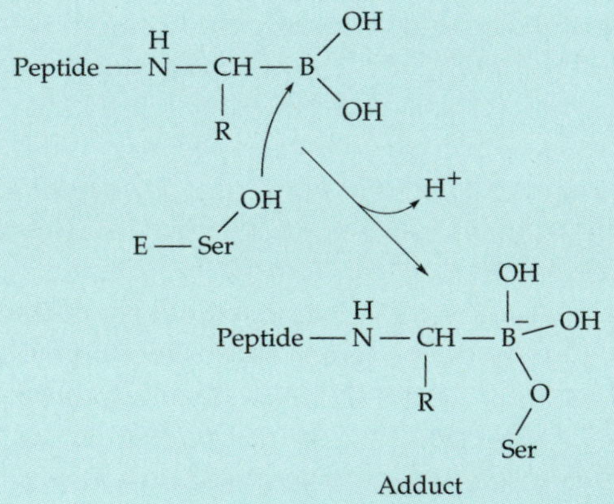

Adduct

However, both X-ray crystallography[m] and ^{11}B NMR[n] have shown that imidazole of the catalytic triad may also add to the boronic acid and that a tetrahedral adduct with both the serine oxygen and histidine nitrogen covalently bonded to boron can also be formed.[m] Thus, in reversibly inhibited enzymes a mixture of different chemical species may exist. Inhibitors can be designed to bind more tightly by providing additional bonding opportunities. For example, a suitably placed cyano group on a phenylalanine ring in the P_1 position of thrombin

BOX 12-C (continued)

or other serine proteases can form a hydrogen bond to the peptide NH of Gly 219 (see Fig. 12-9).[o]

Numerous synthetic active-site directed or enzyme-activated irreversible inhibitors have been designed.[p] For example, the following chloroketone inhibits chymotrypsin but does not act on trypsin. The corresponding structure with a lysine side chain (TLCK) inhibits only trypsin. These **affinity labeling compounds** initially bind noncovalently at the active site.

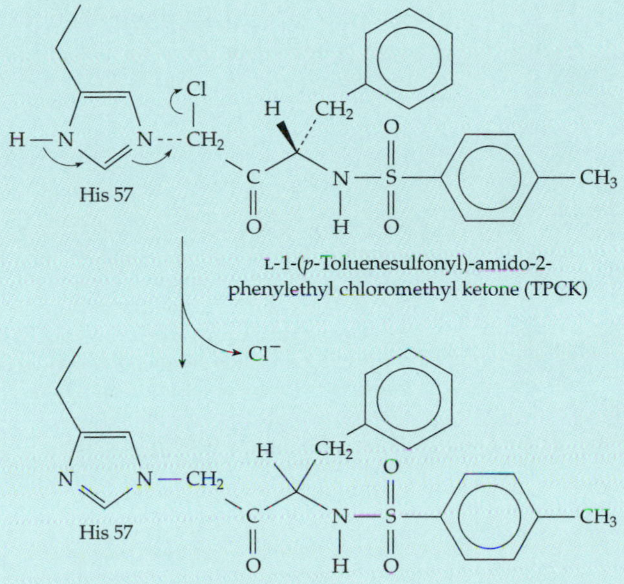

L-1-(*p*-Toluenesulfonyl)-amido-2-phenylethyl chloromethyl ketone (TPCK)

However, α-chloroketones are powerful alkylating agents and the bound inhibitor attacks His 57 of the catalytic triad system. The reaction is probably more complex than is indicated in the foregoing equation and may involve an epoxy ether intermediate.[q] Many other peptide chloromethyl ketone inhibitors have been devised.[q,r]

Isocoumarins inactivate many serine proteases. For example, 7-amino-4-chloro-3-methoxyisocoumarin acylates serine 195 of elastases as follows.[s]

Many other enzyme-activated inhibitors are being developed.[c,d,t]

Epoxy groups, such as that of E-64, a compound isolated from the culture medium of a species of *Aspergillus*, react irreversibly with the active site thiolate group of cysteine proteases.[i,u,v] Related epoxides may become useful medications against abnormal cathepsin levels.

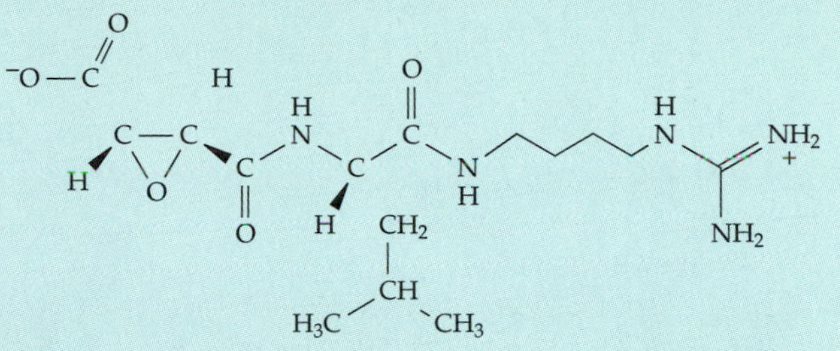

E-64, from *Aspergillus japonicum*

All of the aspartic proteases are inhibited by **pepstatin**, a peptide produced by some species of *Actinomyces* and which contains two residues of the unusual amino acid **statine (sta)**.[w] Pepstatin has the sequence Isovaleryl-L-Val-L-Val-Sta-L-Ala-Sta.

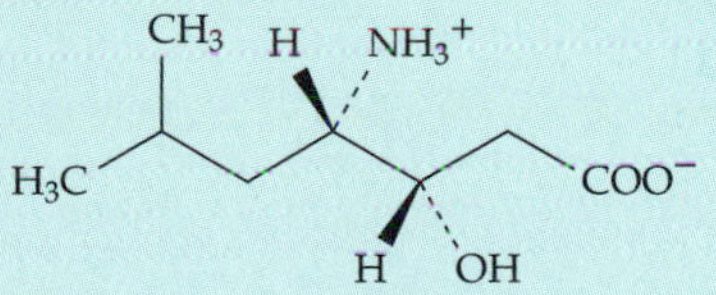

Statine (Sta)

The statine residue mimics the noncovalently bonded tetrahedral intermediate, permitting formation of a very tight complex. Pepstatin is a poor inhibitor of human renin but its existence has inspired the synthesis of numerous related compounds, some of which are effective renin inhibitors.[x,y] Some of these inhibitors use the human angiotensinogen sequence with a secondary alcohol group mimicking the tetrahedral intermediate.[x,z]

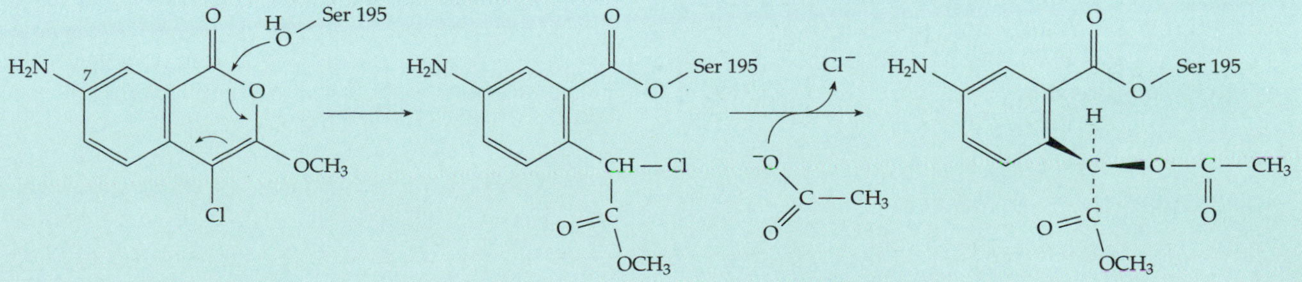

BOX 12-C SYNTHETIC PROTEASE INHIBITORS (continued)

These aspartic protease inhibitors are also "lead compounds" in the development of inhibitors of HIV protease.[aa–cc] As in statine-based inhibitors, the site of occupancy by the catalytic H_2O (green in Eq. 12-23) is occupied in the inhibitor by something that mimics a tetrahedral intermediate with $-CHOH-$, $-PO_2H-$, etc.[bb] Tremendous efforts are being expended in designing these inhibitors and considerable

success has been achieved. However, the rapid development of mutant strains of the virus with drug-resistant proteases presents a major challenge.[cc,dd]

Mercaptans of suitable structure bind tightly to Zn^{2+} in the active sites of metalloproteases. For example, **captopril**[ee] is a tight-binding competitive inhibitor of the angiotensinogen-converting enzyme that is effective in lowering blood pressure and the first of many related inhibitors.[ff] As mentioned in the text, a variety of inhibitors that mimic the geometry of a tetrahedral intermediate or transition state are also potent inhibitors of metalloproteases.

Stereoscopic ribbon structure of the HIV-1 protease with the synthetic inhibitor Sequinivir[cc] bound in the active site. One of the two identical subunits (top) is shaded darker than the second (bottom). When mutated, the amino acid side chains shown in ball-and-stick form with residue numbers shown for the top subunit led to drug-resistant viruses. Courtesy of Alex Wlodawer, National Cancer Institute.[cc]

[a] Powers, J. C., Oleksyszyn, J., Narasimhan, S. L., Kam, C.-M., Radhakrishnan, R., and Meyer, E. F., Jr. (1990) *Biochemistry* **29**, 3108–3118

[b] Mattos, C., Giammona, D. A., Petsko, G. A., and Ringe, D. (1995) *Biochemistry* **34**, 3193–3203

[c] Seife, C. (1997) *Science* **277**, 1602–1603

[d] Peisach, E., Casebier, D., Gallion, S. L., Furth, P., Petsko, G. A., Hogan, J. C., Jr., and Ringe, D. (1995) *Science* **269**, 66–69

[e] Meyer, E. F., Jr., Clore, G. M., Gronenborn, A. M., and Hansen, H. A. S. (1988) *Biochemistry* **27**, 725–730

[f] Yamamoto, A., Tomoo, K., Doi, M., Ohishi, H., Inoue, M., Ishida, T., Yamamoto, D., Tsuboi, S., Okamoto, H., and Okada, Y. (1992) *Biochemistry* **31**, 11305–11309

[g] Ortiz, C., Tellier, C., Williams, H., Stolowich, N. J., and Scott, A. I. (1991) *Biochemistry* **30**, 10026–10034

[h] Dufour, E., Storer, A. C., and Ménard, R. (1995) *Biochemistry* **34**, 9136–9143

[i] Mehdi, S. (1991) *Trends Biochem. Sci.* **16**, 150–153

[j] Bone, R., Shenvi, A. B., Kettner, C. A., and Agard, D. A. (1987) *Biochemistry* **26**, 7609–7614

[k] Takahashi, L. H., Radhakrishnan, R., Rosenfield, R. E., Jr., and Meyer, E. F., Jr. (1989) *Biochemistry* **28**, 7610–7617

[l] Nienaber, V. L., Mersinger, L. J., and Kettner, C. A. (1996) *Biochemistry* **35**, 9690–9699

[m] Stoll, V. S., Eger, B. T., Hynes, R. C., Martichonok, V., Jones, J. B., and Pai, E. F. (1998) *Biochemistry* **37**, 451–462

[n] Zhong, S., Jordan, F., Kettner, C., and Polgar, L. (1991) *J. Am. Chem. Soc.* **113**, 9429–9435

[o] Lee, S.-L., Alexander, R. S., Smallwood, A., Trievel, R., Mersinger, L., Weber, P. C., and Kettner, C. (1997) *Biochemistry* **36**, 13180–13186

[p] Bode, W., Meyer, E., Jr., and Powers, J. C. (1989) *Biochemistry* **28**, 1951–1963

[q] Kreutter, K., Steinmetz, A. C. U., Liang, T.-C., Prorok, M., Abeles, R. H., and Ringe, D. (1994) *Biochemistry* **33**, 13792–13800

[r] Wolf, W. M., Bajorath, J., Müller, A., Raghunathan, S., Singh, T. P., Hinrichs, W., and Saenger, W. (1991) *J. Biol. Chem.* **266**, 17695–17699

[s] Meyer, E. F., Jr., Presta, L. G., and Radhakrishnan, R. (1985) *J. Am. Chem. Soc.* **107**, 4091–4094

[t] Groutas, W. C., Kuang, R., Venkataraman, R., Epp, J. B., Ruan, S., and Prakash, O. (1997) *Biochemistry* **36**, 4739–4750

[u] Yamamoto, D., Matsumoto, K., Ohishi, H., Ishida, T., Inoue, M., Kitamura, K., and Mizuno, H. (1991) *J. Biol. Chem.* **266**, 14771–14777

[v] Varughese, K. I., Su, Y., Cromwell, D., Hasnain, S., and Xuong, N.-h. (1992) *Biochemistry* **31**, 5172–5176

[w] Gómez, J., and Freire, E. (1995) *J. Mol. Biol.* **252**, 337–350

[x] Cooper, J., Quail, W., Frazao, C., Foundling, S. I., Blundell, T. L., Humblet, C., Lunney, E. A., Lowther, W. T., and Dunn, B. M. (1992) *Biochemistry* **31**, 8142–8150

[y] Tong, L., Pav, S., Lamarre, D., Pilote, L., LaPlante, S., Anderson, P. C., and Jung, G. (1995) *J. Mol. Biol.* **250**, 211–222

[z] Cooper, J. B., Foundling, S. I., Blundell, T. L., Boger, J., Jupp, R. A., and Kay, J. (1989) *Biochemistry* **28**, 8596–8603

[aa] Hui, K. Y., Manetta, J. V., Gygi, T., Bowdon, B. J., Keith, K. A., Shannon, W. M., and Lai, M.-H. T. (1991) *FASEB J.* **5**, 2606–2610

[bb] Abdel-Meguid, S. S., Zhao, B., Murthy, K. H. M., Winborne, E., Choi, J.-K., DesJarlais, R. L., Minnich, M. D., Culp, J. S., Debouck, C., Tomaszek, T. A., Meek, T. D., and Dreyer, G. B. (1993) *Biochemistry* **32**, 7972–7980

[cc] Ridky, T., and Leis, J. (1995) *J. Biol. Chem.* **270**, 29621–29623

[dd] Chen, Z., Li, Y., Schock, H. B., Hall, D., Chen, E., and Kuo, L. C. (1995) *J. Biol. Chem.* **270**, 21433–21436

[ee] Vidt, D. G., Bravo, E. L., and Fouad, F. M. (1982) *N. Engl. J. Med.* **306**, 214–219

[ff] Gros, C., Noël, N., Souque, A., Schwartz, J.-C., Danvy, D., Plaquevent, J.-C., Duhamel, L., Duhamel, P., Lecomte, J.-M., and Bralet, J. (1991) *Proc. Natl. Acad. Sci. U.S.A.* **88**, 4210–4214

Pepsin is secreted as the inactive pepsinogen, which is activated by H$^+$ ions at a pH below 5. Determination of its crystal structure revealed that in the proenzyme the N-terminal 44-residue peptide segment lies across the active site, blocking it.[384] At low pH the salt bridges that stabilize the proenzyme are disrupted and the active site is opened up to substrates.

While the cellular aspartate proteases are over 300 residues in length, the retroviral proteases are less than one-half this size.[385–388] That of the human HIV-I protease contains only 99 residues. These enzymes are cut from a polyprotein (encoded by the viral *gag* and *pol* genes) (Fig. 28-26). The *pol* gene encodes four other essential enzymes as well, and these are cut apart at eight different sites by action of the protease.[388] Despite its small size, it displays sequence homologies with the larger cellular aspartate proteases and has a related three-dimensional structure. Each chain has only one active site aspartate; the functional enzymes are dimers and the catalytic mechanism appears to be similar to that of pepsin.[388–392] A great deal of effort is being devoted to designing synthetic HIV protease inhibitors, which are used in the treatment of AIDS (Box 12-C).

5. Metalloproteases

The pancreatic **carboxypeptidases** are characterized by the presence of one firmly bound **zinc ion** in each molecule. The Zn^{2+} can be removed and can be replaced by other metal ions such as Co^{2+} and Ni^{2+}, in some cases with reconstitution of catalytic activity. The human pancreas synthesizes and secretes proenzymes for two forms of carboxypeptidase A, with a preference for C-terminal hydrophobic residues, as well as carboxypeptidases B, which prefer C-terminal basic residues. Additional A and B forms[393] as well as more specialized nondigestive carboxypeptidases are also known. In eukaryotic cells the latter participate in processing of proteins. Following removal of N-terminal signal sequences, processing often continues with removal of basic residues from the C termini by **carboxypeptidases N**, **H** (also called E or enkephalin convertase),[394] and **M**, all of which are metalloenzymes.[395,396] Carboxypeptidase N removes C-terminal arginines from many biologically important peptides. It also circulates in the plasma and protects the body by inactivating such potent inflammatory peptides as the kinins and anaphylotoxins.[397] Carboxypeptidase H is located in secretory granules, while carboxypeptidase M is membrane associated.[396] Dipeptidyl carboxypeptidase (**angiotensin-converting enzyme**) removes the C-terminal Pro–Phe dipeptide from angiotensinogen to generate the potent pressor agent angiotensin I (Box 22-D) and cleaves dipeptides from many other substrates as well.[398] A **D-alanyl-D-alanyl carboxypeptidase** cleaves D-alanine from the ends of cell wall peptides (Chapter 20).[399]

Another well-known zinc-containing enzyme is **thermolysin**, a nonspecific *endopeptidase* widely used in laboratories. Produced by a themophilic bacterium, it is unusually resistant to heat. It contains four bound calcium ions in addition to the active site zinc.[400–402] The active site structure resembles that of pancreatic carboxypeptidase A and the two enzymes appear to act by similar mechanisms.[401,403,404] The mammalian zinc endopeptidase **neprilysin**, an integral membrane protein involved in inactivation of enkephalins and other signaling peptides, also resembles thermolysin.[405] A related neutral endopeptidase is the product of a gene called *PEX* (phosphate-regulating gene with homologies to endopeptidases on the X chromosome). The absence of the *PEX* gene product causes **X-linked hypophosphatemic rickets** which leads to excessive loss of phosphate from the body with defective mineralization of bone.[406] The **mitochondrial processing peptidase**, which removes signal sequences from the N termini of mitochondrial proteins, also contains Zn^{2+} at its active site.[407,408] However, it is an $\alpha\beta$ heterodimer and a member of an additional family of enzymes, one of which includes human insulin-degrading enzyme.

In both carboxypeptidase A and thermolysin the active site Zn^{2+} is chelated by two imidazole groups and a glutamate side chain (Fig. 12-16). In carboxypeptidase A, Arg 145, Tyr 248, and perhaps Arg 127 form hydrogen bonds to the substrate. A water molecule is also bound to the Zn^{2+} ion. The presence of the positively charged side chain of Arg 145 and of a hydrophobic pocket accounts for the preference of the enzyme for C-terminal amino acids with bulky, nonpolar side chains. The Zn^{2+} in thermolysin is also bound to two imidazole groups and that in D-alanyl-D-alanyl carboxypeptidase to three.

The presence of a zinc ion in the metalloproteases immediately suggested a role in catalysis. Unlike protons, which have a weak affinity for the oxygen of an amide carbonyl group, a metal ion can form a strong complex. If held in position by other ligands from the protein, a properly placed zinc ion might be expected to greatly enhance the electrophilic nature of the carbon atom of the C=O group. However, it has been difficult to establish the exact mechanism of action.[410] Carboxypeptidase A cleaves both peptides and ester substrates. For peptides, K_m is the same for various metals while k_{cat} changes, but the converse is true for ester substrates.[411] From its position Glu 270 (Glu 143 in thermolysin) seems to be the logical nucleophile to attack the substrate to form an acyl-enzyme intermediate, an anhydride. Using the following specific *ester* substrate, Makinen *et al.* showed that at very low temperatures of −40° to −60°C, in solvents such as 50:50 ethylene glycol:water, an acyl-enzyme intermediate can be detected spectroscopically. It could even be separated from free enzyme[412] by gel

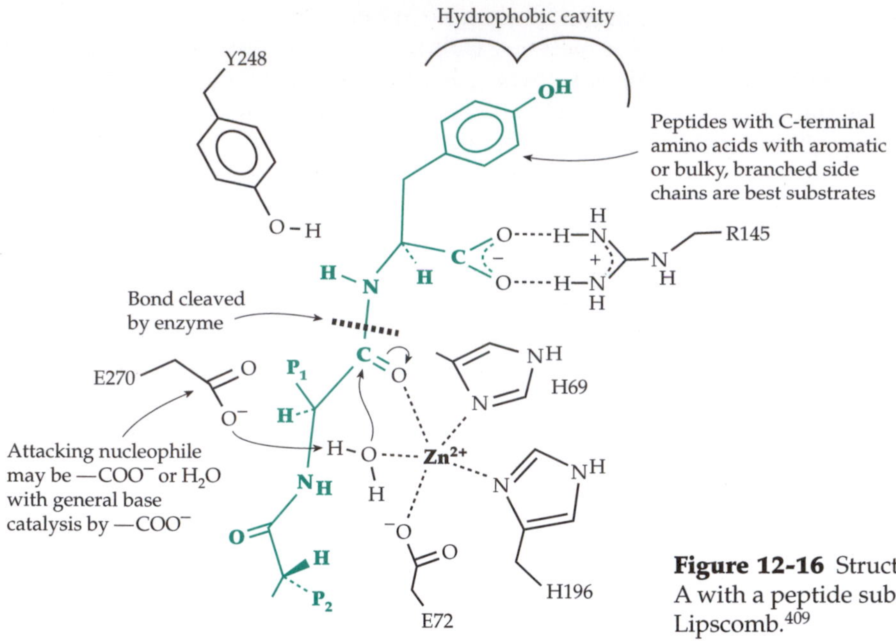

Figure 12-16 Structure of the active site of carboxypeptidase A with a peptide substrate present. See Christianson and Lipscomb.[409]

filtration at −60°C and its cyanoborohydride reduction product was characterized.[413]

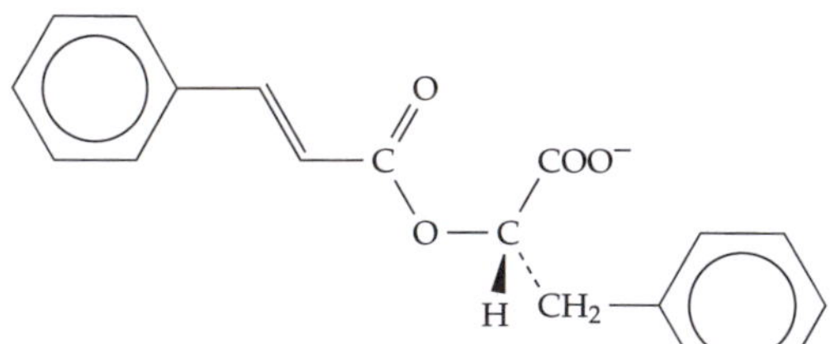

O-(*trans-p*-cinnamoyl)-L-β-phenyllactate

The intermediate appears to be the acid anhydride formed by Glu-270.[414,415] The conformation of the intermediate was deduced by ENDOR spectroscopy and its formation and reaction interpreted according to stereoelectronic principles.[416]

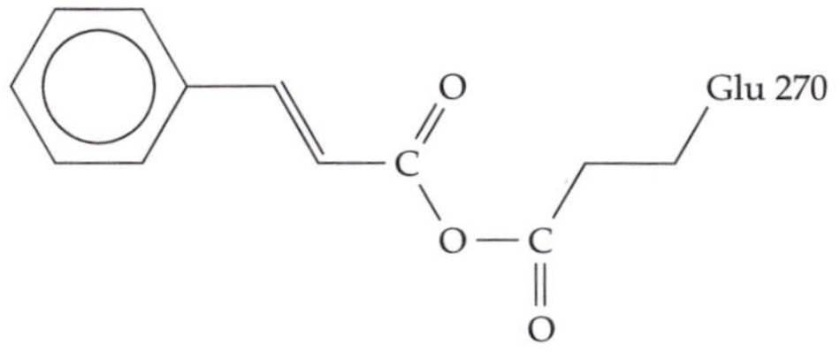

If the anhydride mechanism is correct, the water molecule bound to the Zn^{2+} probably provides an HO⁻ ion necessary for the hydrolysis of the intermediate anhydride.[417]

Although these results seem convincing there are objections.[410] The mechanism deduced for hydrolysis of an ester may be different than that for a peptide.[418]

Furthermore, many observations favor an alternative mechanism. A hydroxide ion derived from a water molecule bound to the zinc ion may be the initial attacking nucleophile.[404,419–422] Both X-ray crystallographic studies and EPR investigations[403,414] show that the zinc in carboxypeptidases can coordinate at least five surrounding atoms. As shown in Fig. 12-16, the Zn^{2+} could hold the attacking water molecule and, at the same time, provide the positive charge for stabilizing the resulting tetrahedral intermediate. Glu 270 (or in thermolysin His 231) acts as a base to deprotonate the bound H_2O. Mock and Stanford argue that the H_2O is probably displaced when the peptide carbonyl binds and that the H_2O is then deprotonated and adds to the polarized carbonyl.[404]

The pH dependence of the action of carboxypeptidase A is determined by pK_a values of ~6 and ~9.5 for the free enzyme[422] and of ~6.4 and ~9 for k_{cat}.[420] For thermolysin the values for k_{cat} are ~5 and ~8.[404] Assignment of pK_a values has been controversial. They may all be composites of two or more microscopic constants but probably, at least for carboxypeptidase, the low pK_a is largely that of Glu 270 while the high one represents largely the dissociation of a proton from the zinc-bound H_2O.

Earlier studies of carboxypeptidase had indicated that Tyr 248 moves its position dramatically upon substrate binding, and it was suggested that its phenolic –OH group protonates the leaving group in the acylation step. However, a mutant in which Tyr 248 was replaced by phenylalanine still functions well.[423]

Phosphonamidates,[424,425] phosphonate esters, and oxo-methylene substrate analogs, which presumably mimic the geometry of the tetrahedral intermediate or transition state of the intermediate or transition state of the catalytic cycle, are effective inhibitors of zinc proteases. X-ray studies of complexes of such inhibitors with both thermolysin and carboxypeptidase A support the suggestion that the Zn^{2+} binds the carbonyl oxygen of the amide bond that is to be cleaved in a

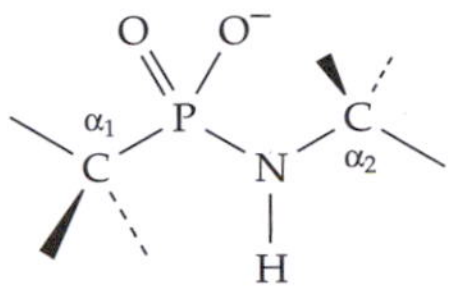

Phosphonamidate

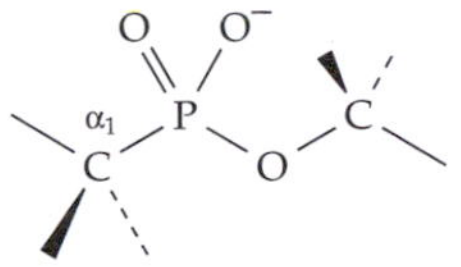

Phosphonate ester

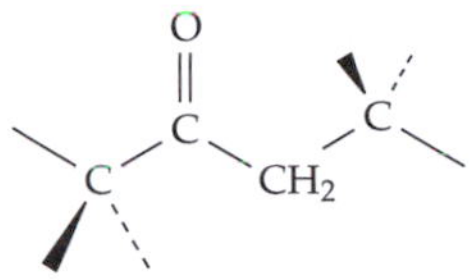

Oxo-methylene analog

substrate and that a glutamate side chain could activate a Zn^{2+}-bound water to form an attacking nucleophile. A similar conclusion was reached from the structure of a thermolysin–product complex.[426]

Another large group of zinc proteases are the **matrix metalloproteases** which act on proteins of the extracellular matrix, such as collagen, proteoglycans, and fibronectin.[427–429c] These enzymes have been classified as **collagenases**, **gelatinases** (which act on denatured collagen), **stromelysins** (which are activated in response to inflammatory stimuli),[430–432] and a membrane type. The group also includes the digestive enzyme **astacin**, from the crayfish,[433] and metalloproteases from snake venoms.[429,434] The matrix metalloproteases are essential to the remodeling of the extracellular matrix that occurs during wound healing, tissue growth, differentiation, and cell death. An example of tissue remodeling is the development of dental enamel (Box 8-G). The proteinaceous matrix formed initially must be digested away, perhaps by the metalloprotease **enamelysin**, and replaced by the dense mineral of enamel.[435] Excessive secretion of collagenase by fibroblasts is observed in **rheumatoid arthritis** and other inflammatory conditions.

At least 64 different matrix metalloproteins are known.[427] Each enzyme consists of three domains. An 80- to 90-residue N-terminal propeptide domain contains a cysteine whose –S⁻ group binds to the active site zinc, screening it from potential substrates. The central catalytic domain is followed by a hinge region and a C-terminal domain that resembles the serum iron binding and transporting **hemopexin**.[427,436] The mechanism of action is probably similar to that of thermolysin.[430]

Many aminopeptidases are metalloenzymes.[437] Most studied is the cytosolic **leucine aminopeptidase** which acts rapidly on N-terminal leucine and removes other amino acids more slowly. Each of the subunits of the hexameric enzyme contains *two* divalent metal ions, one of which must be Zn^{2+} or Co^{2+}.[438,439] A methionine aminopeptidase from *E. coli* contains two Co^{2+} ions[440,441] and a proline-specific aminopeptidase from the same bacterium two Mn^{2+}.[442] In all of these enzymes the metal ions are present as dimetal pairs similar to those observed in phosphatases and discussed in Section D,4 and to the Fe–Fe pairs of hemerythrin and other diiron proteins (Fig. 16-20). A hydroxide ion that bridges the metal ions may serve as the nucleophile in the aminopeptidases.[438] A bound bicarbonate ion may assist.[438a]

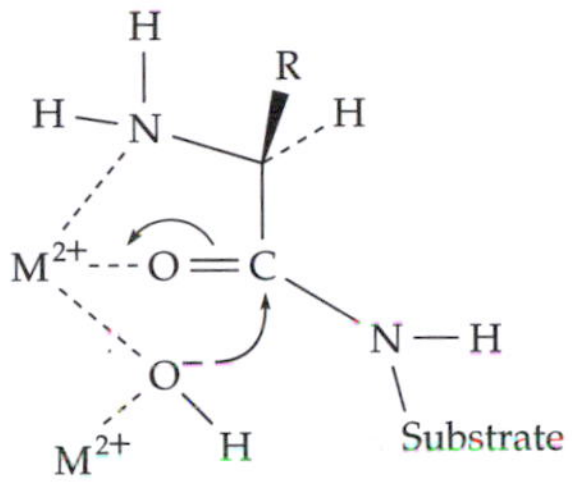

A metalloenzyme **peptide deformylase** removes the formyl groups from the N termini of bacterial proteins. Although the active site is similar to that of thermolysin,[443] the Zn^{2+} form of peptide deformylase is unstable. Both Ni^{2+} and Fe^{2+} form active, stable enzymes.[444,445]

6. ATP-Dependent Proteases

Much of metabolism is driven by the Gibbs energy of hydrolysis of ATP so it shouldn't be surprising that ATP is sometimes rather directly involved in hydrolytic degradation of polypeptide chains. Much of protein processing occurs in the endoplasmic reticulum,[446] which also assists in sorting unneeded and defective proteins for degradation in the ubiquitin–proteasome system. Proteasomal degradation occurs in the cytosol, in the nucleus, and along the ER. However, the predominant location is the *centrosome*.[446a]

Polyubiquitination of proteins requires ATP (Box 10-C) and additional ATP is utilized in the proteasomes (Box 7-A) during the selection of polyubiquitinated proteins for hydrolysis.[447,448] With 28 subunits the 26S proteasome is complex and not fully understood.[448] The ATP-hydrolyzing subunits appear to all be in the cap regions. Is the ATP used to open and close the entry pores? To induce conformational changes in all subunits as part of a catalytic cycle? Or to unfold folded proteins to help them enter the proteasomes?[449]

Some answers may be obtained from smaller bacterial, mitochondrial, and chloroplast ATP-dependent proteases. Cells of *E. coli* contain at least nine proteases, which have been named after the musical scale as Do, Re, Mi, Fa, So, La, Ti, Di, and Ci.[450,451] Most are serine proteases but two, Ci and Pi, are metalloproteins. Protease **La** (**Lon protease**, encoded by gene *lon*) has attracted particular attention because the hydrolysis of two molecules of ATP occurs synchronously with cleavage of a peptide linkage in the protein chain.[452] This enzyme, as well as protease **Ti** (more often called **Clp**, for caseinolytic protease),[451,453,454] is ATP-dependent.[451]

The Lon protease of *E. coli* is a large 88-kDa serine protease with the catalytic domain, containing active site Ser 679, in the C-terminal half. The N-terminal portion contains two ATP-binding motifs and a linker region.[455,456] A homologous protein known either as Lon or as PIM1 is present in mitochondria.[457,458] The *E. coli* proteases Clp (Ti) include ClpAP and ClpXP, which are heterodimers, each containing the catalytic subunit ClpP and either ATPase ClpA or ClpX. The active enzyme may be designated **ClpAP**.[459–461] A homolog of ClpP has been found in human mitochondria.[461a] The ATPase Clp forms seven-subunit rings resembling the rings of proteasomes (Box 7-A), while the catalytic subunit ClpA forms six-membered rings.[451,454] Despite the mismatch in symmetry and the fact that they are serine proteases, these enzymes appear similar to proteasomes. However, each catalytic subunit has an ATPase neighbor. Why is it needed? The related **ClpB** is both an ATPase and a chaperonin. It is essential for survival of *E. coli* at high temperatures.[461b]

Another ATP-dependent protease identified among heat shock proteins of *E. coli* is known as **Hs1V–Hs1U** or (ClpQ–ClpY). It has a threonine protease active site and is even more closely reminiscent of eukaryotic proteasomes.[462–463a] Also active in *E. coli* is another ring-like protease, a membrane-bound zinc endopeptidase **FtsH** (or HflB).[463b] Similar eukaryotic proteases also exist.[460]

7. The Many Functions of Proteases

Many of the enzymes considered in the preceding sections function within the *digestive* tract. Others function in the *processing* of newly formed peptide chains, while others act in the *intracellular degradation* of proteins. Yet others are secreted from cells and function in the external surroundings. Both processing and intracellular degradation can be viewed as parts of *biosynthetic loops* (Chapter 17) that synthesize mature proteins, and then degrade them when they have served their function or have become damaged. The pathways involved are varied and complex, a natural result of thousands of enzymes and other compounds

acting on accessible and chemically appropriate parts of the proteins. Evolutionary selection has evidently led to the particular set of pathways that we observe for any organisms.

Because they must often cleave large **polyproteins**, many viruses encode processing proteases.[464–466b] For example, the entire RNA genome of the poliovirus encodes a large polyprotein which is cut by two virally encoded chymotrypsin-like cysteine proteases within Tyr-Gly and Gln-Gly sequences;[341,465] Asn-Ser sequences are also cut, apparently autocatalytically. As we have already seen, retroviruses encode their own aspartic protease. Most cellular and secreted proteins of bacteria or eukaryotes also undergo processing. This ranges from removal of *N*-formyl groups to cutting off signal peptides, addition or formation of prosthetic groups, internal cleavages, and modifications at the C termini. A variety of peptidases are required.

Degradation of proteins, which converts them back to amino acids as well as other products, takes place in part in the cytosol via the ubiquitin-protease system. Proteolysis also occurs in ER and external cell spaces through the action of membrane-bound and secreted proteases. Loosely folded proteins, which can arise in various ways, are subject to rapid degradation. For example, synthesis of a polypeptide chain on a ribosome may be accidentally disrupted with formation of a protein with a shorted chain. Mature proteins may become damaged. For example, certain histidine residues are readily attacked by oxidizing reagents. Oxidative damage may mark proteins for rapid degeneration.[450] In *E. coli*, proteases So and Re attack oxidized glutamine synthetase much more rapidly than the intact native enzyme.[450,467] In mammalian cells the proteasomes degrade oxidized as well as poorly folded proteins.[468] However, proteasomal digestion is not complete and peptide fragments may be secreted. Peptide fragments of proteins are also formed in the ER and may be secreted. Thus, a variety of small peptide hormones and other biologically active peptides are both generated and inactivated in or around cells.[469–471] Some receptors are activated by proteases.[471a]

Circulating proenzymes of the blood clotting factors, of the complement system (Chapter 31), represent a specialized group of secreted **signaling proteins** that are able to initiate important defensive cascades. Proteases also act more directly in defense systems of the body. For example, serine proteases cause lysis of the target cells of cytotoxic T lymphocytes[472] (Chapter 31) and activated neutrophils[473] (Chapter 18). At the same time, pathogenic bacteria often secrete proteases that assist in attack on their hosts[474] and schistosomes secrete an elastase that helps them penetrate skin and invade their hosts.[475]

8. Protease Inhibitors of Animals and Plants

Premature conversion of proenzymes such as trypsinogen into active proteases in the pancreas would be disastrous. To prevent this, the pancreas also produces inhibitors. The complex cascades that control blood clotting would also be unstable were it not for the presence in blood of numerous inhibitory proteins. Indeed, inhibitors of proteases are found everywhere in animals, plants, and microorganisms.[476-478] They are usually proteins but small antibiotic protease inhibitors are also produced by microorganisms. Protein inhibitors are usually specific for a given type of enzyme: The **serpins** inhibit serine proteases, and the **cystatins** inhibit cysteine proteases.[479-483] Other inhibitors block the action of metalloproteases[484-486] and aspartic proteases. Inhibitors help not only host organisms but also parasites. For example, *Ascaris suum*, a very large nematode that is thought to infect 1/4 of the human population of the earth and nearly all of the pigs, secretes a pepsin inhibitor.[487] However, the large 720-kDa serum **α-macroglobulins** (Box 12-D) inhibit all classes of proteases.

About 20 families of protein inhibitors of proteases have been described.[488] The egg white **ovomucoids** comprise one family. Turkey ovomucoid is a three-domain protein whose 56-residue third domain is a potent inhibitor of most serine proteases.[488,489] The 58-residue **pancreatic trypsin inhibitor**[490] is a member of another family of small proteins. A 36-residue insect (locust) protease inhibitor is even smaller.[491]

Inhibitors that block the action of trypsin and other proteases are found in many plants. The inhibitor activity is usually highest in seeds and tubers, but synthesis of inhibitors can be induced in other parts of plants e.g. tomato and potato by wounding.[492] See also Chapter 31. Legumes contain small 60- to 76-residue inhibitors (Bowman–Birk inhibitors) each containing seven disulfide linkages, and they are relatively stable toward cooking and toward acid denaturation in the stomach.[493,494] Although they interfere with protein digestion they seem to have an anticarcinogen effect.[493] Soybeans also contain a larger 181-residue trypsin inhibitor (Kunitz type).[495]

Several inhibitor–protease complexes have been crystallized and details of their interactions are known. For example, the pancreatic trypsin inhibitor binds at the active site of trypsin with $K_f > 10^{13}$ M^{-1} at neutral pH.[496] The two molecules fit snugly together,[490,497] the inhibitor being bound as if it were a peptide substrate with one edge of the inhibitor molecule forming an antiparallel β structure with a peptide chain in the enzyme. Lysine 15, which forms part of this β structure, enters the specific P_1 binding site for a basic amino acid in a substrate. Thus, the protease inhibitor is a modified substrate which may actually undergo attack at the active site. However, the fit between the two

molecules is so tight that it is hard for a water molecule to enter and complete the catalytic act. The complex reacts very slowly, keeping the enzyme inactive. There is not enough inhibitor to interfere with the large amounts of trypsin formed from trypsinogen in the small intestine so that trypsin can function there.

Most protease inhibitors act by mechanisms similar to that of the pancreatic trypsin inhibitor. They are very slow substrates with a **reactive loop** that carries suitable P_1, P_2, and P_1' residues that meet the specificity requirements of the enzyme. Additional noncovalent interactions prevent dissociation and make the energy barrier for hydrolysis so high that the reaction is extremely slow.[488,494,495]

The serpins are larger ~400-residue proteins.[498-500] They also form complexes in which hydrolysis and release of the serpin occurs very slowly. However, structural analysis before reaction and after release showed that a major rearrangement occurs in the serpin structure. The P_1 through ~P_{15} residues of the cleaved reactive loop are inserted into the center of the main β sheet of the serpin, leaving the P_1 and P_1' residues ~7 nm apart.[499,500] It seems likely that the rearrangement begins during formation of the tight inhibited complex, which cannot be dissociated by boiling in a sodium dodecyl sulfate (SDS) solution and which may be an acyl-enzyme.[501,502] A serpin molecule can act only once.

Blood contains several serpins. They are abundant, accounting for about 10% of the total protein of human plasma.[478,500] The most abundant is the **α1-protease inhibitor** or α_1-antitrypsin, a 394-residue glycoprotein component of the α-globulin fraction of blood serum.[500a] There is no trypsin in tissues, but this inhibitor blocks the action of other serine proteases, including cathepsin G and **leukocyte elastase**.[503] Hereditary absence of α_1-protease inhibitor often leads to severe **pulmonary emphysema** at an early age. Elastase released by neutrophils at sites of inflammation degrades many components of connective tissue including elastin, collagen, and proteoglycans. Without the presence of protease inhibitor too much damage is done to surrounding tissue. A lack of this inhibitor is one of the commonest genetic defects among persons of European ancestry, affecting 1 in 750 persons born.[504-506] The most serious known mutation is a replacement of guanine by adenine at a specific point in the DNA and a resultant replacement of Glu 342 with lysine in the protease inhibitor. This in some manner adversely affected the processing and secretion of the protein.

At its reactive site the α_1-protease inhibitor has the sequence

Ala-Ile-Pro-Met*-Ser-Ile-Pro-Pro,

the Met-Ser pair marked by the asterisk fitting into the P_1-P_1' sites (Fig. 12-14). The methionine (Met 358) in

BOX 12-D MOLECULAR MOUSETRAPS

The large 720-kD α_2-macroglobulin of human serum, as well as related proteins of vertebrate and invertebrate circulatory systems and of egg whites of birds and reptiles, is a trap for proteases.[a–f] Human macroglobulin is a homotetramer consisting of two pairs of identical 180-kDa subunits, each pair being held together in an antiparallel configuration by two disulfide bridges. Each subunit contains a "bait region" with cleavage sites appropriate for nearly all known endoproteases[f] and also a thioester linkage as explained later. Electron microscopic reconstructions of the native protein and its complexes with proteases show that a major structural transformation occurs.[e,f] The macroglobulin traps two protease molecules of the size of trypsin, or one larger one such as plasmin, in an internal cavity. The internal thioesters, which are formed between Cys 949 and Gln 952 (with loss of NH_3) in each subunit, become reactive[g] and form covalent bonds with ε-amino groups of various lysine side chains of the trapped proteases.

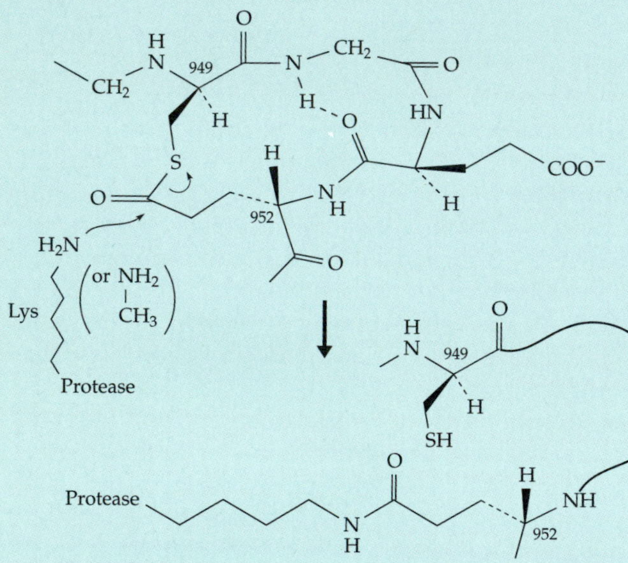

The serum proteins C3 and C4, members of the complement system (Chapter 31), are converted into their active forms, C3b and C4b, by proteolytic removal of short N-terminal peptide fragments. Both C3b and C4b bind tightly to cell surfaces, a feature that helps to direct the complement system's attack to the surfaces of invading organisms. This tight binding also involves covalent attachment of macromolecules by reaction with a preformed thioester just as with α_2-macroglobulin.[h–j] In fact, the thioester linkage was first discovered in the complement proteins. Both C3 and C4 contain the thioester

within the following sequence, which is the same as that in α_2-macroglobulin:

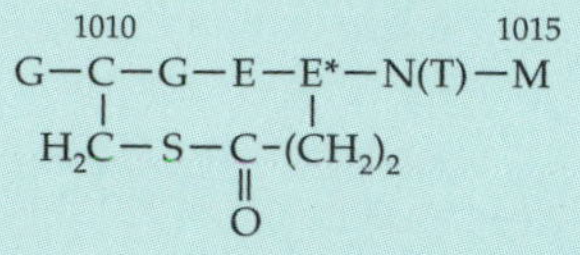

Here the side chains have been added for the thioester-forming cysteine and glutamate and the sequence numbers are for C3. The thioester-forming glutamate is labeled E* because it is not encoded as glutamate but as glutamine, suggesting a mechanism of thioester formation. Protein C4 exists as two subforms, C4A and C4B. Both C3 and C4A react predominately with lysine amino groups, but C4B reacts with –OH groups of cell surface polysaccharides.[h] It has a histidine at position 1106. There is good evidence that it is adjacent to the C=O group of the thioester and reacts to form an acyl-imidazole which is more reactive with hydroxyl groups than is a thioester:

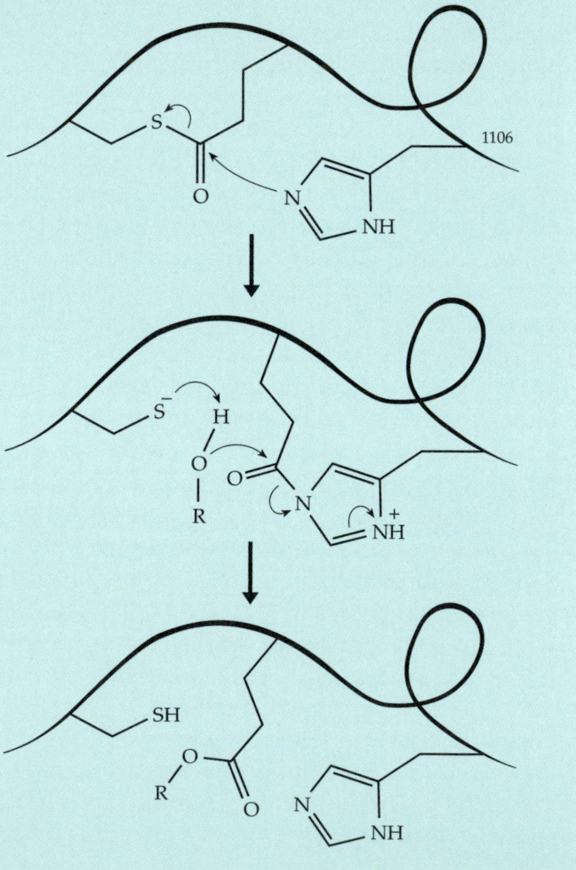

Activation of C3 and C4 apparently allows the preformed thioester, which is buried in the interior

BOX 12-D (continued)

of these proteins, to become exposed on the external surface where it can react. The high group transfer potential of the thioester ensures that the reaction will go to completion.

---

[a] Sottrup-Jensen, L. (1989) *J. Biol. Chem.* **264**, 11539–11542
[b] Fothergill, J. (1982) *Nature (London)* **298**, 705–706
[c] Jacobsen, L., and Sottrup-Jensen, L. (1993) *Biochemistry* **32**, 120–126
[d] Andersen, G. R., Koch, T. J., Dolmer, K., Sottrup-Jensen, L., and Nyborg, J. (1995) *J. Biol. Chem.* **270**, 25133–25141

[e] Boisset, N., Taveau, J.-C., Pochon, F., and Lamy, J. (1996) *J. Biol. Chem.* **271**, 25762–25769
[f] Qazi, U., Gettins, P. G. W., Strickland, D. K., and Stoops, J. K. (1999) *J. Biol. Chem.* **274**, 8137–8142
[g] Gettins, P. G. W. (1995) *Biochemistry* **34**, 12233–12240
[h] Law, S. K. A., and Dodds, A. W. (1997) *Protein Sci.* **6**, 263–274
[i] Khan, S. A., Sekulski, J. M., and Erickson, B. W. (1986) *Biochemistry* **25**, 5165–5171
[j] Dodds, A. W., Ren, X.-D., Willis, A. C., and Law, S. K. A. (1996) *Nature (London)* **379**, 177–179

this sequence, as well as another nearby methionine residue, is very susceptible to oxidation to sulfoxides:

Methionine sulfoxide

The oxidation, which may be caused by such agents as myeloperoxidase (Chapter 16) released from leukocytes,[507] inactivates the inhibitor. This may be physiologically important in permitting the proteases to be *uninhibited* in the immediate vicinity of the leukocyte. Cigarette smoke also inactivates α_1-protease inhibitor by oxidation of the same methionine residues and the lungs of smokers contain the oxidized inhibitor.[508] However, the major cause of emphysema among smokers appears to be an increase in released neutrophil elastase.[509] One approach to the treatment of emphysema involves weekly intravenous injection of α_1-antitrypsin.[510] This treatment may be improved by use of genetically engineered oxidation-resistant variants of the antitrypsin such as Met 385→Val.[504,511] Efforts are also being made to introduce an α_1-antitrypsin gene into lung epithelial cells.[510,512]

Blood plasma also contains at least nine other protease inhibitors. One of these, the thrombin inhibitor **antithrombin III** (Section 9), contains the sequence Arg-Ser-Leu at the P_1, P_1', and P_2' sites. A tragic case of a person born with a Met 385 → Arg mutation in α_1-antitrypsin has been reported.[513] This converted the antitrypsin to an antithrombin causing a bleeding disorder that was eventually fatal.

9. Coagulation of Blood

The clotting of blood following injury and the subsequent dissolving of the clot are familiar phenomena that involve several cascades of proteolytic enzymes together with a number of accessory **cofactors**.[514–516] The first step in clotting results from "activation" of blood platelets which aggregate to form a platelet plug that slows bleeding.[514,517] The clot, which is formed by the insoluble **fibrin**, grows on the platelet surfaces and strengthens the plug. The initial rapid formation of a clot occurs via the **tissue factor pathway** (or extrinsic pathway; right side of Fig. 12-17) which is triggered by the exposure in injured tissues of **tissue factor** (TF), a transmembrane glycoprotein[518–522] and a member of the cytokine receptor superfamily.[518] Human TF is a 263-residue protein with a single membrane-spanning region and a small 20-residue C-terminal cytoplasmic domain. The 219-residue extracellular domain consists largely of two IgG-like domains (Fig. 12-18).[519,523] This protein stimulates the conversion of fibrinogen[524–526] into the insoluble fibrin through the action of three proteases—factor VIIa, factor Xa, and thrombin. These enzymes are generated from proenzymes VII, X, and prothrombin, respectively in a cascade mechanism.

Factor VII binds tightly to TF,[527,527a,b] which also binds Ca^{2+} and phospholipid of the cell membranes. Within this complex a plasma protease, such as thrombin or factor VIIa or Xa, cleaves a single Arg-Ile bond in VII to form active VIIa.[528–530] The TF·VIIa complex is a very active protease which cleaves a specific peptide bond in factor X to form Xa[531–533] which continues the cascade. Notice that there are autocatalytic features: VII can be converted to the active VIIa by the action of Xa and the accessory factor Va is generated from the precursor, factor V, in part through the action of thrombin.[514,534–535] Factors Xa and Va together with Ca^{2+} and phospholipid form the active **prothrombinase** complex which attacks prothrombin to form the active enzyme thrombin.[536,537] The roles of factor Va and Ca^{2+} appear to be to hold the prothrombin and the activated protease Xa together on the phospholipid surface.[538] This localizes the clotting. Factor Xa is unusually specific, cleaving only after arginine in the sequence. Its activation of prothrombin results from cleavage of two

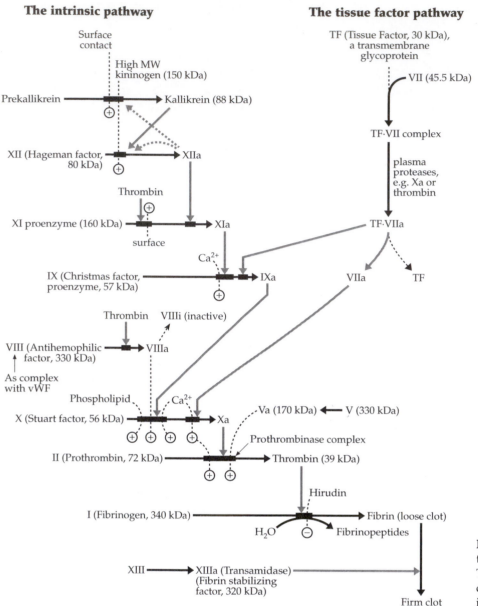

The intrinsic pathway

The tissue factor pathway

Figure 12-17 Major components of the human blood clotting cascades. The site of action of the leech anticoagulant protein hirudin is also indicated.

peptide bonds, which releases the 39-kDa thrombin from the much larger, immobilized 72-kDa proenzyme.[236,531]

Thrombin, like most other clotting factors, is also a serine protease. However, the clotting factors are multidomain proteins that are more elaborate and more specific[539] than the digestive enzymes. Prothrombin, as well as factors VII, IX, and X and the anticoagulant protein C, contain at their N-terminal ends several residues of γ-carboxyglutamate (Gla), an amino acid generated in a posttranslational modification that depends upon vitamin K (Chapter 15). In human prothrombin there are ten of these in the following N-terminal sequence, where E* represents Gla. Since many of these enzymes are also dependent upon activation by calcium ions (see Fig. 12-17), it is thought that the function of Gla is to assist in the binding of

$$
\begin{array}{ccc}
10 & 20 & 30 \\
\bullet & \bullet & \bullet
\end{array}
$$
ANTFLE*E*VRKGNLE*RE*CVE*E*TCSYE*E*AAFE*ALE*SS

Ca^{2+} which helps to tie these proteins to the phospholipids of platelet surfaces. In factors VII, IX, X, and protein C this Ca^{2+}-binding domain is followed by two epidermal growth factor (EGF)-like domains, each containing one residue of *erythro*-β-hydroxyaspartate or hydroxyasparagine formed by hydroxylation of an aspartate or asparagine residue in the first EGF-like domain.[540,540a,b] The C-terminal catalytic domain of each enzyme contains the protease active site.

Fibrinogen is an elongated molecule with an $(\alpha\beta\gamma)_2$ structure.[524,541,541a] Thrombin cleaves specific Arg-Gly bonds in the α and β chains releasing short (14- to16-residue) "fibrinopeptides" from the N termini of the peptide chains. This leaves Gly-Pro-Arg "knobs" at

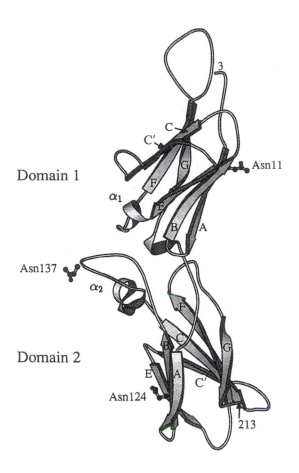

Figure 12-18 A ribbon drawing of the three-dimensional structure of the extracellular region (residues 3 to 213) of human tissue factor. Strands of domain 1 and 2 are labeled A to G. Two helices are labeled α_1, and α_2 and the three asparagines that provide the glycosylation sites are also marked. MolScript drawing from Harlos et al.[520]

the N termini of the α-chains and these fit into complementary "holes" in the γ chains to form noncovalently linked aggregates.[542–544] The clot that forms is unstable, but it is soon crosslinked by the action of the **transamidase** (transglutaminase or **factor XIIIa**; Eq. 2-23).[545,546] The fibrin monomers, von Willebrand factor (discussed in following paragraphs), fibronectin, collagen, and other proteins all become crosslinked.

The slower **intrinsic mechanism** first described in 1964[547,548] consists of a cascade involving six proteases (Fig. 12-17, left side). Again, autocatalytic cycles are present in the activation by XIIa of both prekallikrein[549] and XII and in activation by thrombin of factors XI and VIII. This intrinsic pathway is initiated by the serine protease proenzymes prekallikrein and factor XII together with the accessory protein high-molecular-mass-H-kininigen.[549] Activation occurs when blood contacts surfaces such as glass or kaolin (a clay).[547,548,550] Factor XI can also be activated by thrombin. Heredity absence of factor XI leads to bleeding problems, especially

after surgery, but absence of XII does not. This fact suggests that direct activation of XI by thrombin is important and that the kallikrein–factor XII pathway is usually less important.

Factor IX (Christmas factor) is next in the intrinsic mechanism cascade. It can be activated either by XIa or by VIIa of the tissue factor pathway. The absence of a functional factor IX leads to the inherited X-linked bleeding disorder **hemophilia B** which affects 1 in 30,000 males. The condition can be mild or very serious[551,552] and may be caused by a variety of mutations or by incorrect splicing of the messenger RNA for the 416-residue factor IX. The level of factor IX in blood increases with age, almost doubling by old age.[552a]

Factor IXa causes a rapid activation of factor X only if Ca^{2+}, phospholipid,[553,554] and the accessory factor VIIIa[555] are present. The IXa•VIIIa complex acts on X about 2×10^5 times faster than does IXa alone. This complex cleaves the same bonds in X as does the VIIa•Va complex formed in the tissue factor pathway.[514] The 2332-residue factor VIII and factor V have similar structures that include three repeats of a domain homologous to the blue copper-containing plasma protein **ceruloplasmin** (Chapter 16).[556–559] Tyrosine 1680 of VIII apparently must be converted to a sulfate ester for full activity.[560]

The absence of factor VIII in about 1 / 10,000 males born, leads to the severe X-linked bleeding disorder **hemophilia A**. Human factor VIII is encoded by a 186-kb gene containing 26 coding exons. Severe cases of hemophilia are usually a result of point mutations that produce stop codons in this gene and therefore a shortened protein. Milder cases may result from an amino acid substitution.[561] Factor VIII circulates in the plasma as a complex with the **von Willebrand factor**, (vWF), a large multimeric protein derived by proteolytic cleavage, glycosylation, and sulfation of a 2813-residue precursor.[562–566] The mature 2050-residue vWF stabilizes factor VIII in the blood. Another important function of vWF is to bind platelets to damaged endothelial surfaces.[567,568] Like fibrinogen, vWF contains RGD sequences specific for binding to adhesion receptors.[569,570] It also binds to collagen. A serious inherited bleeding disorder caused by deficiency of vWF was first discovered among inhabitants of islands in the Gulf of Bothnia, Finland, in 1926 by von Willebrand.[563,571] If mild forms of the disease are included, vWD deficiency is the commonest bleeding problem. However, abnormalities have been identified in almost every one of the proteins in the coagulation cascades.[514]

Until recently, these bleeding diseases were treated by regular injections of the appropriate factors isolated from human plasma. This was both costly and carried a high risk of infection by HIV, hepatitis, or other contaminating viruses. Now, cloned genes are used for commercial production of the factors.[572,573] Experiments in animals, designed to lead to eventual gene replacement therapy, have been somewhat successful.[573,574] It is of interest to compare the relative concentrations of several of the clotting factors in plasma and to compare these with their positions in the cascade of the tissue factor (extrinsic) pathway.[515]

	mg / L	Mass kDa	μM
VII	0.5	45.5	0.01
V	10.0	330	0.03
Prothrombin	150	72	2.1
Fibrinogen	3,000	340	8.8

Why do we have the intrinsic pathway when the tissue factor pathway provides rapid clot formation? The answer seems to be that the tissue factor pathway is needed immediately after injury but that it is turned off quickly by the **anticoagulation systems** of the body. As a result the protease **plasmin** begins to dissolve (lyse) the clot within a few hours. The intrinsic pathway is apparently needed to maintain the clot for a longer period.[514]

What prevents the clotting mechanism, with its autocatalytic cycles, from running out of control? In part the answer lies in the localization of the enzymatic activation to tissue surfaces near a wound. The flow of blood rapidly dilutes components that escape into the general circulation and liver cells take up and destroy the active proteases. A variety of circulating antiproteases including the **tissue factor pathway inhibitor**[575] act on these escaped enzymes. Two anticoagulation systems that are localized on vessel walls[576] also come into action very quickly. The circulating polysaccharide heparin (Chapter 4) forms a complex with the serpin **antithrombin**. Antithrombin traps thrombin as an inactive complex or compound[577,578a] and heparin greatly accelerates the inactivation.[578] Kallikrein and factors IXa, Xa, XIa, and XIIa are also inhibited.

Thrombin is an allosteric protein which exists as a mixture, in nearly equal amounts, of a fast-acting form that cleaves fibrinogen and is stabilized by Na$^+$, and a slow-acting form that initiates the second anticoagulant cascade.[579] The slow acting form, bound to **thrombomodulin**,[527a,540b,580,580a,b,581] an endothelial cell surface protein, attacks the proenzyme of another serum protease called **protein C**.[582,582a] Activated protein C (APC) inactivates the accessory clotting factors Va and VIIIa. The accessory factor **protein S**[583] is also needed for rapid inactivation. A blood clot is temporary and its dissolution begins as soon as it is formed, largely through the action of plasmin, a protease derived from the circulating 791-residue **plasminogen**[584–588] through the action of yet other proteases. Plasminogen often becomes crosslinked to cell surface proteins by transglutaminase.[586] Plasminogen activators include **urokinase**,[589] a

protease present in kidney tissue and urine, and **tissue plasminogen activator** (tPA),[590,591] a 527-residue protease which is now produced by recombinant DNA technology. It is sometimes used to dissolve blood clots in emergency situations, such as myocardial infarction and pulmonary embolism,[592] but can also cause serious bleeding problems. Plasmin is inhibited by the plasma **α$_2$ antiplasmin** and is inactivated by action of clotting factor XIIa. Likewise, tPA is inhibited by several protease inhibitors present in tissues and in plasma. One is a fast-acting serpin called plasminogen activator inhibitor.[593] Another anticoagulant compound of medical interest is **hirudin**, a 65-residue peptide from the leech. It binds very tightly to thrombin (K_d = 1 pM) preventing its action.[594–596] Insects also produce antithrombin.[597,598] Ticks[599,600] and some insects[598] inject proteins that inhibit Xa selectively.[599] Anticoagulants are of great medical importance and much effort is being devoted to the design of better inhibitors of thrombin,[601–603] factor Xa,[604] and other components of the blood coagulation cascade.

Inherited deficiencies of the anticoagulant pathways with associated problems of thrombosis are known. These include problems with protein C,[576,605] plasminogen,[606] and antithrombin.[607,608]

10. Esterases and Lipases

A group of esterases hydrolyze simple oxygen esters. Some of these are designed to hydrolyze a particular ester or small group of esters, while others have a more nonspecific action. **Acetylcholinesterase**[609–611a] is specific for acetylcholine (Eq. 12-25), a neurotransmitter that is released at many nerve synapses and neuromuscular junctions (Chapter 30). The acetylcholine, which is very toxic in excess, must be destroyed rapidly to prepare the synapse for transmission of another impulse:

$$(12\text{-}25)$$

The more widely distributed **butyrylcholinesterase**[612] is less specific but prefers butyrylcholine. Acetylcholinesterase is a very efficient catalyst:[613–615] $k_{cat} = 1.6 \times 10^4 \text{ s}^{-1}$

and $k_{cat} / K_m = 2 \times 10^8$ M^{-1}s^{-1} at 25°C. It exists as a series of molecular forms containing varying numbers of 68-kDa catalytic subunits, 100-kDa structural subunits and subunits with triple-helical ~120-kDa collagen-like "tails."[609,616] The subunits are joined by disulfide bridges to give aggregates that range from simple dimers of catalytic subunits to tailed forms containing 8–12 catalytic subunits and non-collagenous structural subunits as well. The tailed forms are secreted from cells and may be designed to take up residence in the basal lamina of synapses, whereas the dimers are apparently attached to phosphatidylinositol anchors in the membranes (Fig. 8-13).[617] Human liver **carboxylesterases** are relatively nonspecific enzymes that hydrolyze ester groups of various drugs and toxins including cocaine and heroin. Products are often excreted in the urine.[618] **Thioesterases** function in biosynthesis of fatty acids, polyketides (Chapter 21), and many other substances.[619]

Lipases[620,621] hydrolyze triacylglycerols. The pancreatic digestive lipase[622,623] acts faster on emulsified fats than on glycerol esters in true solution but requires the cooperation of a small 10-kDa **colipase**.[624] **Cholesterol esterase** hydrolyzes not only cholesterol esters but also esters of fat-soluble vitamins, phospholipids, and triacylglycerols.[625,626] Other lipases include gastric[627] and hepatic lipases and a **lysosomal acid lipase**[628] which also attacks neutral lipids. Plasma **lipoprotein lipase**[629,630] digests fats in the chylomicrons and from the very low-density lipoproteins of blood. **Hormone-sensitive lipase** hydrolyzes stored triacylglycerols in the cytosol in response to catecholamines, ACTH, and other hormones.[631] The **phospholipases** attack phospholipids, while **cutinase**,[632] produced by some fungi, cleaves the ester linkages in the cutin (Chapter 21) of plant surfaces. Fungal lipases are important industrial commodities. Numerous structural and mechanistic studies have been made with them.[620,629,633–635] The gene for a lipase from *Candida rugosa* has been synthesized using codons that maximize its expression in *Saccharomyces cerevisiae* and which allow for further genetic engineering of the lipase.[636]

All of these esterases appear to act by mechanisms closely related to those of proteases. Acetylcholinesterase contains an active site serine that reacts with organophosphorus compounds (Box 12-E) and is part of an Asp-His-Ser catalytic triad which lies in a deep "gorge" as well as an oxanion hole.[637] A surprise is the absence of an essential carboxylate group that might bind the positively charged trimethylammonium group of acetylcholine. Instead, the lining of the gourge is rich in aromatic side chains which may interact with the methyl groups of the substrate and by their polarizability stabilize the charge.[611,638] Most lipases, including cutinase, also have an Asp-His-Ser or Glu-His-Ser triad as well as some form of oxanion hole.[620,632,639] Like the serine proteases, the lipases have bell-shaped pH optima.

Phospholipase A$_2$ cleaves the ester linkage at the 2 position in phospholipids.[640–642] One isoenzyme form is secreted by the pancreas as a proenzyme whose N-terminal seven residues are removed by trypsin to give an active 125-residue enzyme. Phospholipase A$_2$ of a similar type is abundant in venoms of reptiles and bees. The venom and pancreatic enzymes have closely similar three-dimensional structures.[643–645] Although the folding pattern is different from that of chymotrypsin, imidazole (His 48) and carboxylate (Asp 99) groups are present in the active site in an orientation resembling that of the catalytic triad of serine protease. The enzyme requires calcium ions, one of which binds at an appropriate point for complexation with the substrate carbonyl as it is converted to an oxanion intermediate. The backbone NH of Gly 30 may also serve as an oxanion ligand. However, in most phospholipases there is no active site serine. Instead, a water molecule is positioned to serve as the attacking nucleophile in formation of the oxanion as is indicated in the following scheme, which shows a truncated phosphatidylethanolamine as the substrate.[643] Phospholipase A$_2$ is up to 1000 or more times as active on phospholipids in micelles as on dissolved substrates.[621,646,647] Apparently, the

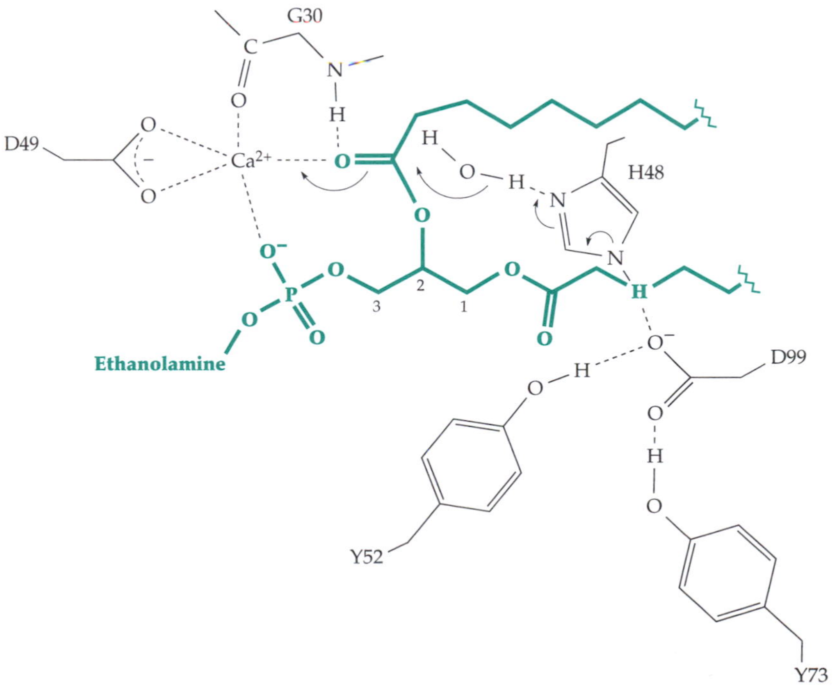

BOX 12-E INSECTICIDES

Over 200 organic insecticides, designed to kill insects without excessive danger to humans and animals, are presently in use.[a-e] Many of these compounds act by inhibiting cell respiration; others uncouple ATP synthesis from electron transport. The chlorinated hydrocarbons such as DDT act on nerves in a manner that is still not fully understood. One of the largest classes of insecticides acts on the enzyme acetylcholinesterase of nerve synapses. Like chymotrypsin, it contains an active site serine residue that reacts with organophosphorus compounds. The extreme toxicity of esters of pyrophosphate and of dialkylphosphonofluoridates was recognized in the 1930s and led to their development in Germany and in England as insecticides and as nerve gases. Among the most notorious is diisopropylphosphonofluoridate (diisopropylfluorophosphate; DFP), for which the LD_{50} (dose lethal to 50% of the animals tested) is only 0.5 mg kg^{-1} intravenously. This exceedingly dangerous compound can cause rapid death by absorption through the skin.

The following structures are a few of the organophosphorus compounds and other acetylcholinesterase inhibitors that are selectively toxic to insects.

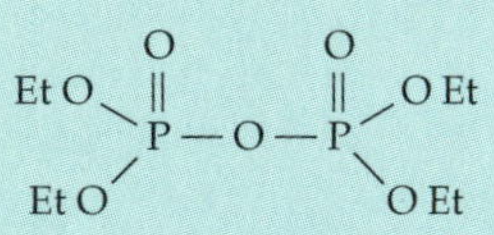

Tetraethyl pyrophosphate (TEPP)

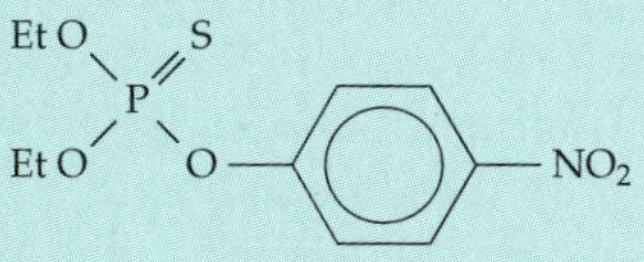

Parathion

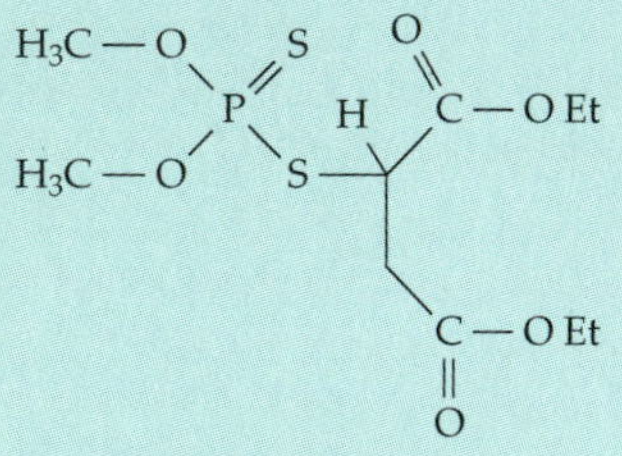

Malathion

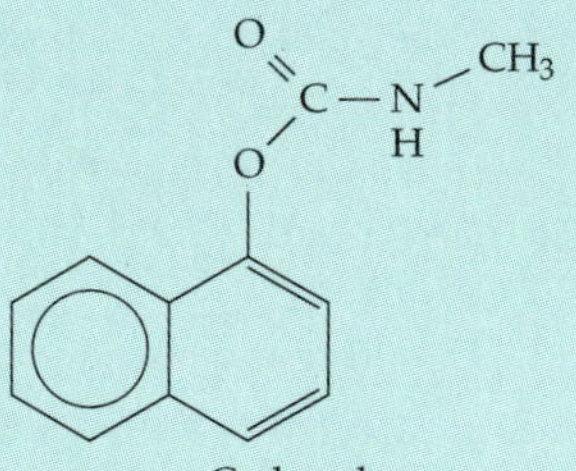

Carbaryl

The characteristic high group transfer potential of a phospho group in pyrophosphate linkage, which makes ATP so useful in cells, also permits tetraethyl pyrophosphate to phosphorylate active sites of acetylcholinesterases. While TEPP is very toxic, it is rapidly hydrolyzed; all harmful residues are gone within a few hours after use.

Two insecticides that have been used widely are **parathion** and **malathion**. They are less toxic than DFP or TEPP and do not become effective insecticides until they undergo bioactivation during which conversion from a P = S to a P = O compound occurs:

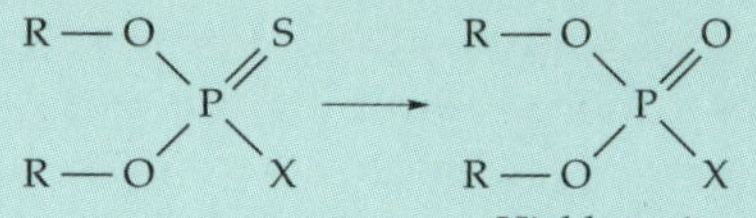

Highly toxic

The desulfuration reaction involves microsomal oxidases of the liver, the sulfur being oxidized ultimately to sulfate.[f] The reactivity of parathion with cholinesterases depends upon the high group transfer potential imparted by the presence of the excellent leaving group, the p-nitrophenolate anion. If the P– O linkage to this group is hydrolyzed before the desulfuration takes place, the phosphorus compound is rendered harmless. Thus, the design of an effective insecticide involves finding a compound which insects activate rapidly but which is quickly degraded by higher animals. Other factors, such as rate of penetration of the insect cuticle and rate of excretion from the organism, are also important.

The phosphorylated esterases formed by the action of organophosphorus inhibitors are very stable, but some antidotes can reverse the inhibition. The oxime of 2-formyl-1-methylpyridinium ion (pralidoxime) is very effective.[g] Its positive charge permits it to bind to the site normally occupied by the quarternary nitrogen of acetylcholine and to displace the dialkylphospho group:

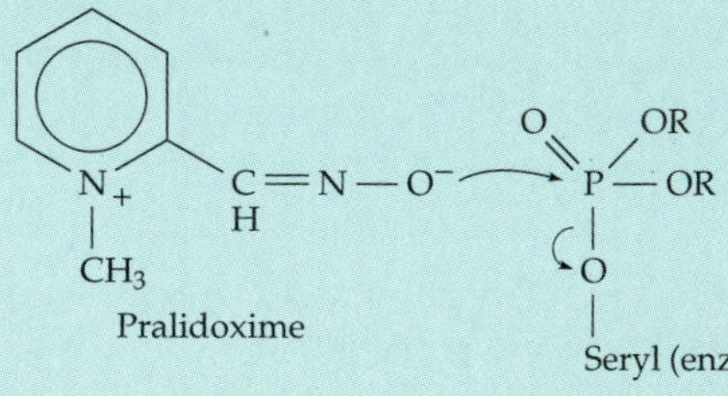

Pralidoxime

Seryl (enzyme)

Carbaryl, a widely used methyl carbamate, is a pseudosubstrate of acetylcholinesterase that reacts 10^5 to 10^6 times more slowly than do normal substrates. The carbamoylated enzyme formed is not as stable as the phosphorylated enzymes and the inhibition is reversible.

A basic problem is that most insecticides are designed to attack the central nervous system of the insect, the system that depends heavily upon acetylcholine. However, in human beings the readily accessible peripheral nervous system also depends upon acetylcholine, e.g., in neuronuscular junctions. The danger of poisoning is great. Another approach is to attack the glutamate-dependent peripheral system in insects, e.g., with inhibitors of glutamate decarboxylase. Glutamate functions as a neurotransmitter in the human body but only in the well-

BOX 12-E (continued)

protected central nervous system.

   Another important problem is the development of insects resistant to insecticides. This often arises as a result of increased levels of carboxylesterases which hydrolyze both organophosphates and carbaryl.[h,i] A mutation that changed a single active site glycine to aspartate in a carboxylesterase of a blowfly changed the esterase to an organophosphorus hydrolase which protected the fly against insecticides.[j]

---

[a] Büchel, K. H., ed. (1983) *Chemistry of Pesticides*, Wiley, New York
[b] Hassall, K. A. (1990) *The Biochemistry and Uses of Pesticides*, 2nd ed., VCH Publ., Weinheim

[c] Kamrin, M. A., ed. (1997) *Pesticide Profiles*, CRC Press, Boca Raton, Florida
[d] Casida, J. E. (1973) *Ann. Rev. Biochem.* **42**, 259–278
[e] Wilkinson, C. F., ed. (1976) *Insect Biochemistry and Physiology*, Plenum, New York
[f] Nakatsugawa, T., Tolman, N. M., and Dahm, P. A. (1969) *Biochem. Pharmacol.* **18**, 1103–1114
[g] Wilson, I. B., and Ginsburg, S. (1955) *Biochim. Biophys. Acta.* **18**, 168–170
[h] Raymond, M., Callaghan, A., Fort, P., and Pasteur, N. (1991) *Nature (London)* **350**, 151–153
[i] Karunaratne, S. H. P. P., Hemingway, J., Jayawardena, K. G. I., Dassanayaka, V., and Vaughan, A. (1995) *J. Biol. Chem.* **270**, 31124–31128
[j] Newcomb, R. D., Campbell, P. M., Ollis, D. L., Cheah, E., Russell, R. J., and Oakeshott, J. G. (1997) *Proc. Natl. Acad. Sci. U.S.A.* **94**, 7464–7468

substrate-binding cavity of the protein is designed to accommodate phospholipid molecules in the preferred conformation found in the micelles.[643] Most of the other lipases have lids which close over the active sites and impede access of substrates. Binding to a phospholipid surface apparently induces a conformational change that opens the lid. This allows substrate to enter from the lipid surface.[620,621,646] Cutinases, which do not display interfacial activation, do not have lids. Phospholipase A$_2$ activity is stimulated by an applied electrical field, a result that suggests that its activity *in vivo* may be regulated in part by the membrane potential.[648]

11. Other Acyltransferases

   Acyl groups are frequently transferred from amides or esters to various acceptors in biosynthetic reactions. Among the many known acyltransferases are the ribosomal **peptidyl transferases** (Chapter 29), a transacylase involved in bacterial peptidoglycan synthesis (Chapter 20), transglutaminase (Eq. 2-23),[649,650] γ-glutamylcyclotransferase (Box 11-B), and transacylation reactions involving acyl-CoA derivatives. Examples of the latter are *N*-acetylation[651] or myristoylation (Chapter 8) of proteins, the formation of acetylcholine from choline[652] and of acetylcarnitine from carnitine (Eq. 17-4), and acetylation of the antibiotic chloramphenicol. The high group transfer potential of thioesters ensures that these reactions proceed to completion. **Chloramphenicol acetyltransferase (CAT)**,[653] the enzyme that catalyzes acetylation and inactivation of the antibiotic by bacteria, is much used in studies of gene expression (Chapter 28). A catalytic histidine removes the proton from the 3-OH group and a serine hydroxyl provides an oxyanion hole to accommodate the anticipated tetrahedral intermediate.[652,654] The

steroidal antibiotic fusidic acid (Chapter 22, Section G) is a competitive inhibitor.[655] A related transferase is aspartate carbamoyltransferase (Fig. 7-20; Chapter 24).

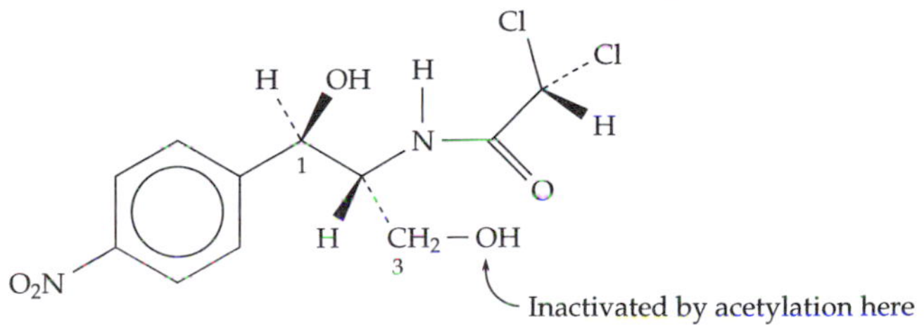

Chloramphenicol, a broad-spectrum *Streptomyces* antibiotic

Inactivated by acetylation here

   Penicillin and related antibiotics are inactivated by β-lactamases (Box 20-G), some of which resemble serine proteases in forming acyl enzymes with active site serine side chains.[656,657] Others are zinc metallogenzymes.[658,659] **Amidohydrolases** such as **asparaginase** and **glutaminase**,[660,661] deacetylases,[662] and many other hydrolases can also be described as acyltransferases.

D. Displacement on a Phosphorus Atom

   Nucleophilic displacements on phosphorus (Table 10-1, reaction type 1C) are involved in virtually every aspect of cellular energetics and in many aspects of biosynthesis. One large group of such enzymes are **phosphotransferases**, which transfer **phospho** (also called **phosphono** or, traditionally in biochemistry, **phosphoryl**) groups from one nucleophilic center to another. When transfer is to water the enzymes are called **phosphatases**, and when from one group in a molecule to another in the same molecule, **mutases**.

Enzymes that transfer a phospho group from ATP to water are **ATPases** and those that transfer the phospho group from ATP to some other nucleophile are **kinases**. Substituted phospho groups can also be transferred. Thus, **nucleases**, members of a larger class of **phosphodiesterases**, hydrolyze nucleic acids by transfer of nucleotidyl groups to a hydroxyl group of water. **Polynucleotide polymerases** transfer nucleotidyl groups to growing polynucleotide chains. An intramolecular nucleotidyl transferase is adenylate cyclase. **Topoisomerases** carry out a sequence of phosphotransferase reactions.

1. Questions about Mechanisms

Consider the following general equation for transfer to nucleophile Y^- of a phospho group attached in an ester or an anhydride linkage to form ROH, which could be an alcohol, carboxylic acid, or a phosphoric acid such as ATP.

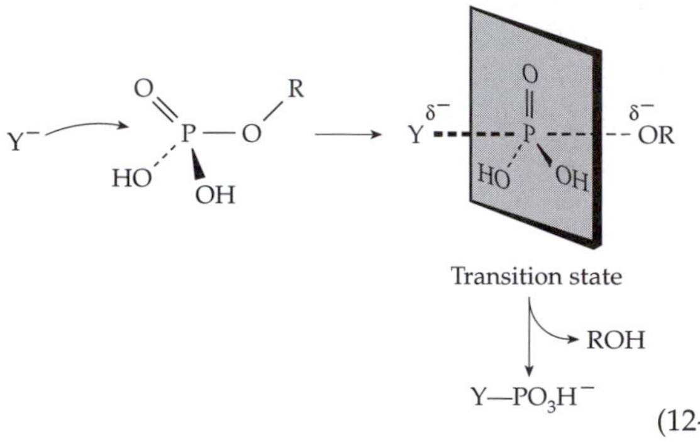

Transition state

$$Y—PO_3H^-$$

(12-26)

This equation could also represent a half-reaction in a double-displacement process. As with displacements on saturated carbon atoms, two basic mechanisms can be imagined. The first is S_N2-like or **associative**.[663–667] The transition state might be represented where the bonds from the phosphorous atom to Y and to O are approximately equally formed. In an **in-line** displacement, where Y, P, and −OR are colinear, this mechanism leads to inversion if a chiral phospho group is used. There is another possibility for the associative mechanism. Whereas a carbon atom can form only four stable covalent bonds, phosphorus is able to form five. While nucleophilic attack on carbon leads to a *transient* five-bonded transition state (Eq. 12-2), attack on phosphorus could produce a relatively long-lived pentacovalent intermediate (Eq. 12-27). Notice that two transition states are involved. Remember that our conventional way of drawing phosphate esters with a double bond from phosphorus to one of the oxygens is misleading. All of the P–O bonds share some of the double-bond character and the phosphate group has many characteristics of a completely single-bonded structure:

Likewise, in the transition state structures of Eq. 12-27, step *b*, the P–O bonds, except for that to −OR, are equivalent and all have partial double-bond character.

Geometric complexities. The geometry of the pentacovalent intermediate in Eq. 12-27 is that of a **trigonal bipyramid**. In this structure the bond angles in the **equatorial** plane are 120°, whereas all of the angles between any of those bonds and the two that attach to the groups in the **apical** positions are 90°. This disparity arises naturally from the fact that it is impossible to place five points on the surface of a sphere all equidistant one from the other. The attack of Y^- from the side opposite O–R (an in-line attack) leads to a trigonal bipyramid in which −O–R and −Y occupy the two apical positions. However, if Y^- attacks a face opposite one of the other oxygens (**adjacent** attack), −O–R will take an equatorial position.

The chemical reactivities of groups in the apical and equatorial positions of pentacovalent intermediates are different.[664] In particular, elimination of a nucleophilic group to form a tetrahedral phosphate is easier from an apical position than from an equatorial position. For the in-line displacement of Eq. 12-27 elimination of RO^- should be easy. However an adjacent attack would leave −OR in an equatorial position. Before it could be eliminated, the intermediate would probably have to undergo a **permutational rearrangement** by which −OR would be transferred from an equatorial to an apical position.

One type of permutational rearrangement, known as **pseudorotation**, can be visualized as in Eq. 12-28.[668,669] The axial groups a and b move back while equatorial groups 2 and 3 move forward, still in the same equato-

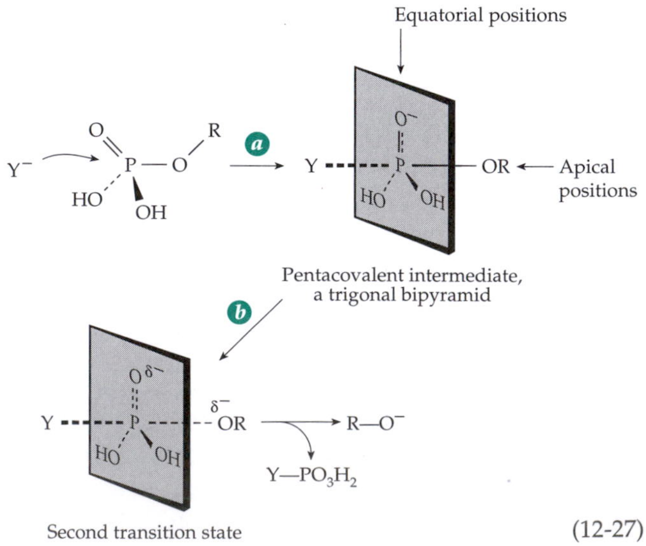

Pentacovalent intermediate, a trigonal bipyramid

Second transition state

$$Y—PO_3H_2$$

(12-27)

rial plane. Equatorial group 1 does not move. This decreases the 120° bond angles between the equatorial groups and increases the bond angles between group 1 and the axial groups until all four bond angles to group 1 are equivalent. The resulting **square pyramid** is a high-energy transition state structure in the pseudo-rotation process and can either revert to the original structure or, by continued motion of the groups in the same directions as before, to the structures shown at

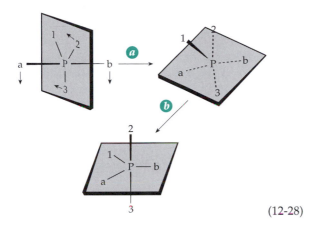

(12-28)

the bottom of Eq. 12-28. In the final structure, groups 2 and 3 are axial and the original axial groups a and b are equatorial. Pseudorotation is slow enough that it could be rate limiting in enzymatic reactions. Stereo-electronic effects could also affect these rates.[670,671] However, much evidence indicates that enzymes almost always avoid these complexities by using in-line mechanisms.

Metaphosphate ions. An alternative to an associative mechanism is an S_N1-like or **dissociative** mechanism which can occur by elimination of a **metaphosphate ion** (Eq. 12-29, step *a*). A nucleophilic reagent can then add to the eliminated metaphosphate in step *b*. The formation of metaphosphate is analogous to formation of carbocation during the action of lysozyme (Eq. 12-11). This dissociative mechanism could lead either to racemization or to inversion of a chiral phosphate. Metaphosphate ions have been shown to exist. They are generated in certain nonenzymatic elimination reactions in aprotic solvents[672–674] and they are reactive electrophiles able to react as in Eq. 12-29. However, there is some doubt that they can exist free in aqueous media.[675] Jencks and associates concluded from studies of linear Gibbs energy relationships that in the transition state for nonenzymatic phospho transfer reactions there is a large amount of bond breaking and a small amount of new bond formation in the transition state[676] but no free metaphosphate. There is a large dependence of the rate on the pK_a of the leaving group, but there is still a small dependence on the pK_a of the entering nucleophile. Thus, we have a

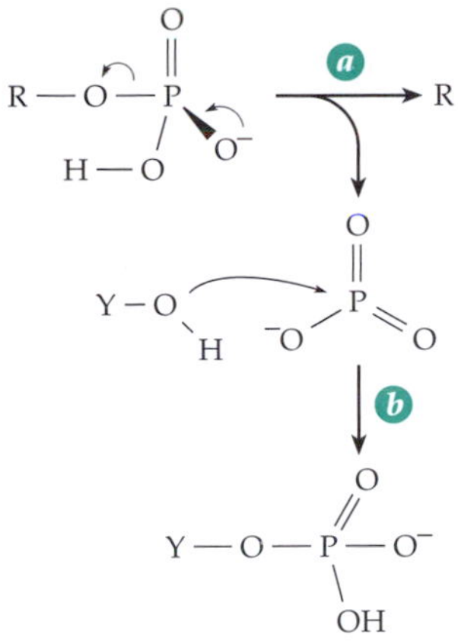

(12-29)

dissociative mechanism but without free metaphosphate. The concept is supported by studies of kinetic isotope effects.[677,678]

Coping with negative charges. We visualize phospho group transfers as involving attack by a nucleophile bearing at least a partial negative charge. However, phospho groups also carry one or two negative charges and that in ATP even more. Therefore, it does not seem surprising that many phosphotransferases are metalloenzymes, sometimes containing bound Zn^{2+} which neutralizes some of the negative charge. Furthermore, enzymes usually accept ATP as a substrate only when it is accompanied by a divalent metal ion,[679] usually Mg^{2+}. Another way in which enzymes deal with the negative charges on phospho groups is to have arginine side chains in appropriate positions to interact by forming strong ion paired-hydrogen bonds. It is often assumed that it is essential to neutralize the charge on the phospho group to avoid electrostatic repulsion from a partial or complete negative charge on the attacking nucleophilic center. However, with a dissociative mechanism and a meta-phosphate-like intermediate the transition state may be reached without charge neutralization.[679,680] Interactions of Mg^{2+} with phosphorus-containing substrates, like those of fixed positive charges in the protein may also be essential for binding the substrate correctly to an enzyme.

2. Magnetic Resonance Studies

There have been many investigations of phosphotransferases by NMR and EPR methods.[681,682] One approach is to use paramagnetic ions such as Mn^{2+}, Cu^{2+}, or Cr^{3+} to induce nuclear relaxation in substrate and coenzyme molecules at active sites of enzymes. Flavin radicals and specifically introduced nitroxide spin labels can serve as well. Paramagnetic ions greatly increase the rate of magnetic relaxation of nearby nuclei (Chapter 3, Section I). Thus, small amounts of

Mn^{2+} in a sample lead to broadening of lines in ordinary ^1H, ^{13}C, and ^{31}P NMR spectra.

Useful information about enzymes can sometimes be obtained by observing effects of paramagnetic ions on the NMR signal of protons in the solvent water. The relaxation time of solvent protons is usually greater than 1 s. However, in the ion $Mn(H_2O)_6^{2+}$ the protons of the coordinated water molecules relax much more rapidly, both T_1 and T_2 being $\sim 10^{-1}$ s. Since the coordinated water molecules usually exchange rapidly with the bulk solvent, a small number of manganese ions can cause a significant increase in relaxation rate for all of the water protons. Broadening of the proton band is observed and differences in T_1 and T_2 can be measured by appropriate methods. Paramagnetic relaxation effects usually increase as the inverse sixth power of the internuclear distance. Knowing the $Mn^{2+}-H$ distance to be 0.287 ± 0.005 nm for hydrated Mn^{2+}, it has been possible to relate the effects on T_1 and T_2 to the number of H_2O molecules coordinated at any one time to a protein-bound metal ion and to their rate of exchange with the bulk solvent.

Relaxation effects on ^1H, ^{13}C, and ^{31}P, while more difficult to observe, can provide geometric information about active sites. The theory is complex, but under some conditions the paramagnetically induced relaxation can be described adequately by Eq. 12-30.

$$r = C[T_{1M}f(\tau_c)]^{1/6} \qquad (12\text{-}30)$$

Here r is the internuclear distance, C is a combination of physical constants, and T_{1M} is the longitudinal relaxation time. The complex function $f(\tau_c)$ depends upon the correlation time τ_c, the resonance frequency of the nucleus being observed, and the frequency of precession of the electron spins at the paramagnetic centers. The value of τ_c can be estimated (Chapter 3) and, in turn, the distance r according to Eq. 12-30.

Such studies on creatine kinase (Eq. 12-31) utilized both a bound Mn^{2+} ion and a nitroxide spin label to estimate distances of various protons from the nitroxide.[683] Together with EPR measurements (Box 8-C), which gave the Mn^{2+}–nitroxide distance, a model of the ATP•Mn^{2+} complex in the active site was constructed. Additional EPR experiments on Mn^{2+} complexes with ATP and ADP containing ^{17}O in the α, β, or γ phospho groups showed that in the enzyme•ATP• creatine complex the metal ion is bound to all three phospho groups of ATP. It remained coordinated with the two phospho groups of ADP and also that of the phosphocreatine product in the enzyme•ADP•creatine-P complex as well as in the transition state, which is pictured occurring via a metaphosphate ion.[684]

$$\text{Creatine-}P + \text{ADP} \underset{k_r}{\overset{k_f}{\rightleftarrows}} \text{ATP} + \text{creatine} \qquad (12\text{-}31)$$

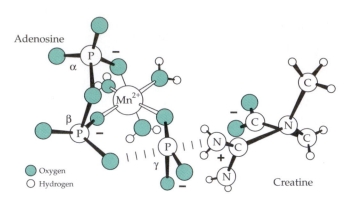

Figure 12-19 Proposed transition state structure formed from Mn^{2+}, ATP, and creatine bound in the active site of muscle creatine kinase. Based on EPR spectroscopy with regiospecifically ^{17}O-labeled substrates. The electrical charges have been added in one possible constellation. However, hydrogen atoms bound to phospho groups are not shown. After Leyh et al.[684]

Phosphorus-31 NMR has been used to measure **internal equilibrium constants** within enzyme-substrate (ES) complexes.[663,685–687] By having both substrate and product concentrations high enough to saturate the enzyme, all of the enzyme exists as ES and enzyme–product (EP) complexes in equilibrium with each other. For a phosphotransferase at least one substrate and one product contain phosphorus. Although the NMR resonances are broadened by binding to the large, slowly tumbling protein, their areas can be measured satisfactorily and can be used to calculate an equilibrium constant such as that for Eq. 12-32:

$$\text{ES} \underset{k_r}{\overset{k_f}{\rightleftarrows}} \text{EP}$$

$$K' = [\text{EP}]/[\text{ES}] \qquad (12\text{-}32)$$

An example is illustrated in Fig. 12-20. In this experiment[685] the relative areas of the ^{31}P signals of ADP (one for free ADP and one, slightly more intense, for MgADP) and of the signal for phosphoenolpyruvate (PEP) were measured in the absence of enzyme and in the presence of a catalytic amount of pyruvate kinase (Fig. 12-20A). The results verified that the equilibrium constant for the overall reaction (Eq. 12-33) is very high (3300).

Phosphoenolpyruvate (PEP) + ADP → ATP + pyruvate

$$(12\text{-}33)$$

However, with an excess of enzyme (Fig. 12-20B) the internal constant was estimated as

$$K' = \frac{[\text{E} \cdot \text{MgATP} \cdot \text{pyruvate}]}{[\text{E} \cdot \text{MgADP} \cdot \text{PEP}]} = 0.5 - 1.0$$

$$(12\text{-}34)$$

This is a difficult measurement and reinvestigation by another group[686] indicated that the amount of the PEP-containing complex had been overestimated and that $K'=10$.

If the rates of the forward and backward reactions in Eq. 12-34 are of the same order as the spin-lattice relaxation times (T_1) of the ^{31}P in the bound substrate and product, the rate constants k_f and k_r can be evaluated by **saturation-transfer NMR**.[688] This is done by irradiating one resonance, e.g., that of the γ-P of ATP, and observing whether this causes a loss of intensity of the resonance for the product which is receiving its phospho group from ATP. This technique was used to observe the creatine kinase reaction (Eq. 12-31) in living muscle in both relaxed and contracting states. For resting muscle the observed forward flux was 1.7×10^{-3} M s^{-1} and the backward flux 1.2×10^{-3} M s^{-1}. Thus, this reaction, which supplies ATP for contraction of the muscle from stored phosphocreatine, appears to be operating at or near equilibrium. This had been assumed but had previously been difficult to prove. Two-dimensional NMR techniques can now be used for this kind of measurement.[689]

When ^{31}P is bonded to ^{18}O the chemical shift of the ^{31}P is altered by 0.0206 ppm from that when the phosphorus is bonded to ^{16}O. This allows ^{18}O labels introduced into phospho groups to serve as tracers which can be followed continuously during reactions.[683] The technique is useful in studies of stereochemistry (see Section 2) and for examination of **positional isotope exchange**.[690] This latter technique is often used with ATP containing ^{18}O in the β,γ-bridge position. If an enzyme transfers the terminal (γ) phospho group to an acceptor via a phosphoenzyme but without loss of the ADP, we may expect positional isomerization. The ^{18}O will move between the β,γ-bridge position and a nonbonding position as the phospho group is repeatedly transferred back and forth between ATP and the acceptor and as the phospho group rotates.[682,690]

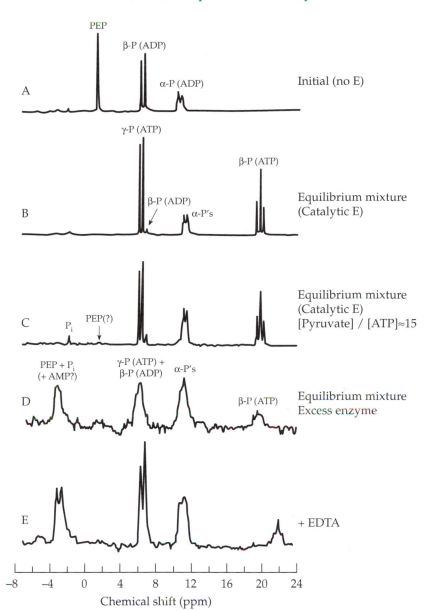

Figure 12-20 Equilibria in pyruvate kinase reaction as studied by ^{31}P NMR at 40.3 MHz, pH 8.0, 15°C. (A–C) Equilibria with low enzyme in levels ~15% ^{2}H$_2$O. (A) ^{31}P NMR spectrum of 1.5 ml of reaction mixture; PEP, 13.3 mM; ADP, 14.1 mM; MgCl$_2$, 20 mM; potassium Hepes buffer, 100 mM; KC, 50 mM without enzyme. (B) Equilibrium mixture after the addition of ~1 mg of pyruvate kinase to the reaction mixture. (C) Equilibrium after the addition of potassium pyruvate (final concentration of 200 mM) to the sample of the spectrum in (B). (D,E) Equilibrium with enzyme concentrations in excess of the substrates. Sample volumes ~1.1 ml with 10% ^{2}H$_2$O. (D) Equilibrium mixture set up with enzyme (2.8 mM active sites); 2.8 mM PEP; 2.4 mM ADP; 5.7 mM MgCl$_2$; 100 mM potassium Hepes; 100 mM KCl. (E) Spectrum after the addition of 50 μl of 400 mM EDTA (pH readjusted to 8.0) to the sample of spectrum D. The EDTA removes metal ions, stopping the catalytic reactions and sharpening the resonances. From Nageswara Rao *et al.*[685]

Equation 12-35 shows one part of this isomerization. The negative oxygens have been omitted here to avoid implying a known state of protonation or a localization of charge.

$$ \text{(Eq. 12-35 scheme)} $$

Enzyme-product complex

(12-35)

Figure 12-21 illustrates the use of the technique in investigating the possible participation of metaphosphate in a nonenzymatic reaction.

3. Stereochemistry

Evidence for an in-line S_N2-like mechanism of most enzymatic phospho group transfer reactions comes largely from study of chiral phospho groups.[663,692–695] A chiral phosphate can be introduced at either the α or β phosphorus of ATP by substitution of one of the oxygen atoms by sulfur. A chiral phospho group in the β position can be formed by substituting one oxygen by S and a second by ^{18}O.

ATP αS (R configuration at P)

Notice that the negative charge is largely localized on sulfur in these phosphorothioate compounds.[696] More general is the use of ^{17}O and ^{18}O to form a chiral phospho group:

ATP βS (S configuration at P)

An ester chiral at the phosphorus atom; R configuration

Considerable ingenuity was required in both the synthesis of these chiral compounds[695,697] and the stereochemical analysis of the products formed from them by enzymes.[698–700] In one experiment the phospho group was transferred from chiral phenyl phosphate to a diol acceptor using *E. coli* alkaline phosphatase as a catalyst (Eq. 12-36). In this reaction transfer of the phospho group occurred without inversion. The chirality of the product was determined as follows. It was cyclized by a nonenzymatic in-line displacement to give equimolar ratios of three isomeric cyclic diesters. These were methylated with diazomethane to a mixture of three pairs of diastereoisomers triesters. These diastereoisomers were separated and the chirality was determined by a sophisticated mass spectrometric analysis.[692] A simpler analysis employs ^{31}P NMR spectroscopy and is illustrated in Fig. 12-22. Since alkaline phosphatase is relatively nonspecific, most phosphate esters produced by the action of phosphotransferases can have their phospho groups transferred without inversion to 1,2-propanediol and the chirality can be determined by this method.

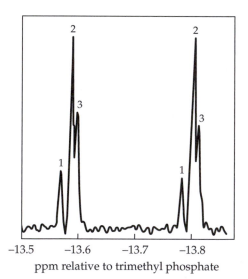

ppm relative to trimethyl phosphate

Figure 12-21 The ^{31}P NMR spectrum at 101.2 MHz of P_α of isotopically labeled ADP. This was recovered from an experiment in which ADP containing 87 atom % of ^{18}O in all four oxygens around P_β was allowed to undergo partial (~20%) nonenzymatic hydrolysis to AMP and P_i. Peaks 1 represent the species containing no ^{18}O bonded to P_α. Peaks 2 represent all species with ^{18}O in the P_α–O–P_β bridge, and peaks 3 represent species with ^{18}O in nonbridging positions at P_α. These have undergone positional isotope exchange. From Lowe and Tuck.[691]

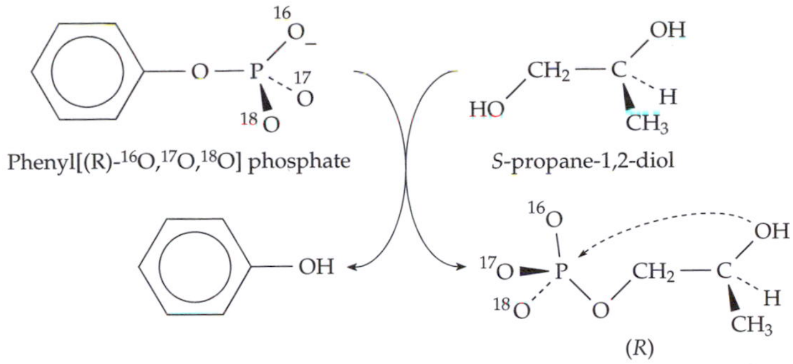

Phenyl[(R)-$^{16}O,^{17}O,^{18}O$] phosphate   S-propane-1,2-diol

(12-36)

Although inversion was not observed with the *E. coli* alkaline phosphatase, it has been observed for ribonucleases and many other hydrolytic enzymes and for most kinases transferring phospho groups from ATP. The difference lies in the existence of a phospho-enzyme intermediate in the action of alkaline phosphatase (see Eq. 12-38). Each of the two phosphotransferase steps in the phosphatase action apparently occurs with inversion. The simplest interpretation of all the experimental results is that *phosphotransferases usually act by in-line S_N2-like mechanisms which may involve metaphosphate-ion-like transition states that are constrained to react with an incoming nucleophile to give inversion.* An adjacent attack with pseudorotation would probably retain the original configuration and is therefore excluded.

The substrate for many phosphotransferases is MgATP. Which of the possible isomers of this chelate complex is utilized by these enzymes? Since Mg^{2+} associates and dissociates rapidly from the complexes there are several possibilities: a tridentate complex with oxygens from α, β, and γ phospho groups coordinated with the metal ion, an α,β-bidentate, a β,γ-bidentate, or a monodentate complex. Most evidence suggests that β,γ-bidentate complexes of the following types are the true substrates.

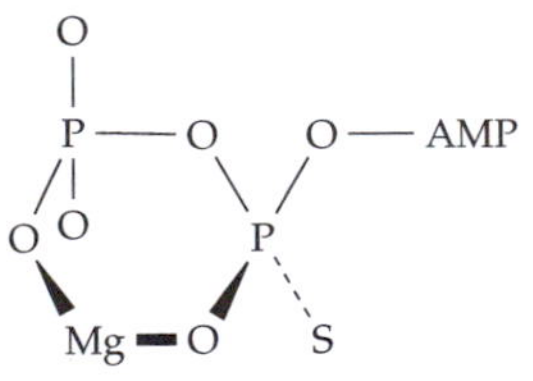

Mg–ATP βS complex with Λ screw sense, R configuration at P

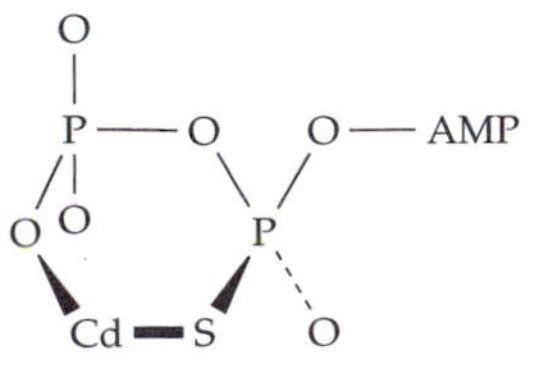

Cd–AT βS complex with Λ screw sense, S configuration at P

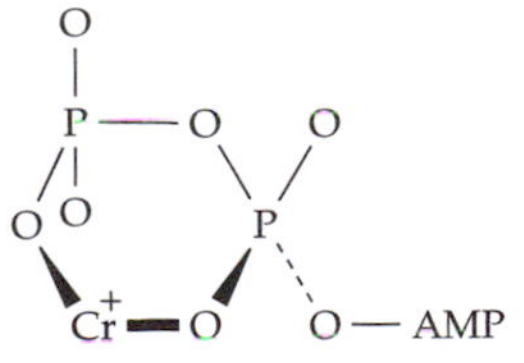

Cr–ATP complex with Δ screw sense

The first structure drawn is for a Mg–ATP *S* complex with a chiral β-phosphothioate group. The Mg^{2+} is expected to bond to oxygen. However, in the second complex, in which Cd^{2+} has been substituted for Mg^{2+}, it is expected that the Cd^{2+} will bond to sulfur. Therefore, the stereochemically equivalent structure will be obtained only when the compound has the *S* configuration at the phosphorus. In the third of the foregoing complexes, Mg^{2+} has been replaced by Cr^{3+} to give an **exchange-inert** complex of ATP in which the Cr^{3+} will remain attached firmly to the same two oxygen atoms under most experimental conditions. Notice also that in this complex, the AMP portion occupies a different position than in the first two complexes. It is called the Δ **screw-sense isomer**,[702] while the upper two complexes are Λ screw-sense isomers. We may reasonably expect that an enzyme, which may provide additional ligands to the metal ions, will prefer one or the other of these screw-sense isomers. For α,β-bidentate ATP complexes both α and β phosphorous atoms become chiral centers and even more isomers are possible.

The use of exchange inert Cr^{3+} and Co^{3+} complexes of ATP has been developed by Cleland and associates.[702–705] The β,γ-bidentate chromium complexes were separated into the Λ and Δ isomers.[704] Each was separated further into a pair of "ring-puckering isomers". These metal complexes are all competitive inhibitors of MgATP and both ring puckering isomers of the Δ screw sense are very slow substrates of various kinases. The Λ isomers serve as very slow substrates for pyruvate kinase, adenylate kinase, and fructose-6-phosphate kinase.[703] The Δ exo isomer is the strongest inhibitor of creatine kinase, suggesting this same conformation for the MgATP substrate. The disastereoisomers of ATPαS and ATPβS have also been tested with kinases in the presence of either Mg^{2+} or Cd^{2+}. With creatine kinase[695,706] the isomers with the R configuration at phosphorus are preferred in the presence of Mg^{2+} but those with the S configuration are preferred in the presence of Cd^{2+}. This might be related to the previously mentioned preferences of the metals for O vs. S ligands, and would suggest that this enzyme prefers the Λ isomer (see foregoing structures). However, EPR studies with the corresponding Mn^{2+} complexes suggest that the Δ isomer is preferred by the enzyme.[693] There are 12 isomers of monoammine Cr(III)ATP. Their use has provided information about the location of water molecules in metal complexes of kinases.[707]

In the following sections we will consider several individual phosphotransferases.

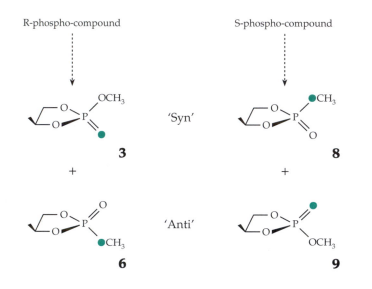

The product 1-phospho-(S)-propane-1,2-diol **1** (here shown as R at phosphorus) is converted by in-line ring closure to an equimolar mixture of three cyclic diesters. These are methylated to give six cyclic triesters. Of these, only **3** and **6** give sharp ³¹P resonance because the ¹⁷O in the others broadens the lines.

O = ¹⁶O
● = ¹⁷O
● = ¹⁸O

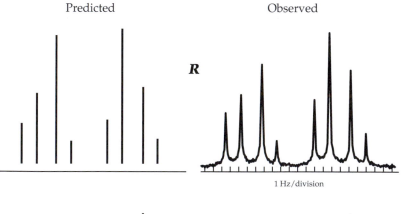

Two cyclic triesters **3** and **6** from an *R*-phospho compound and two others **8** and **9** from an *S*-phospho compound will give sharp ³¹P NMR resonances.

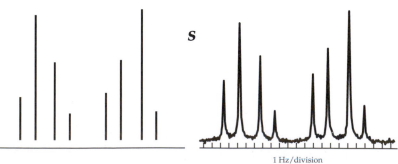

Predicted and observed ³¹P NMR spectra of the mixtures of syn and anti cyclic triesters derived from labeled samples of 1 phospho-(S)-propane-1,2-diols that are R and S at phosphorus.

Figure 12-22 Method for determining chirality of phospho groups containing ¹⁶O, ¹⁷O, and ¹⁸O. From Buchwald and Knowles.[701]

4. Phosphatases

The phosphatases catalyze hydrolysis of phosphate esters to produce inorganic phosphate:[667]

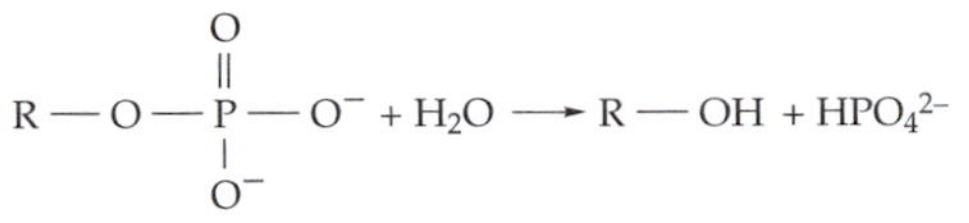

$$R—O—\overset{\overset{O}{\|}}{\underset{\underset{O^-}{|}}{P}}—O^- + H_2O \longrightarrow R—OH + HPO_4^{2-}$$

$$(12\text{-}37)$$

The **acid phosphatases** and **alkaline phosphatases** are nonspecific and cleave many different phosphate esters, whereas **glucose-6-phosphatase, fructose-1,6-bisphosphatase**, and many others are specific for single substrates. The nonspecific phosphatases may provide inorganic phosphate ions in places where they are needed, e.g., in mineralizing bone. All phosphatases help to drive metabolic cycles (Chapter 17).

The alkaline phosphatases are found in bacteria, fungi, and higher animals but not in higher plants. In E. coli alkaline phosphatase is concentrated in the periplasmic space. In animals it is found in the brush border of kidney cells, in cells of the intestinal mucosa, and in the osteocytes and osteoblasts of bone. It is almost absent from red blood cells, muscle, and other tissues which are not involved extensively in transport of nutrients.

The alkaline phosphatase of E. coli is a dimer of 449-residue subunits which requires Zn^{2+}, is allosterically activated by Mg^{2+}, and has a pH optimum above 8.[667,708–711] At a pH of ~4, incubation of the enzyme with inorganic phosphate leads to formation of a phosphoenzyme. Using ^{32}P-labeled phosphate, it was established that the phosphate becomes attached in ester linkages to serine 102. The same active site sequence Asp-Ser-Ala is found in mammalian alkaline phosphatases. These results, as well as the stereochemical arguments given in Section 2, suggest a double-displacement mechanism of Eq. 12-38:

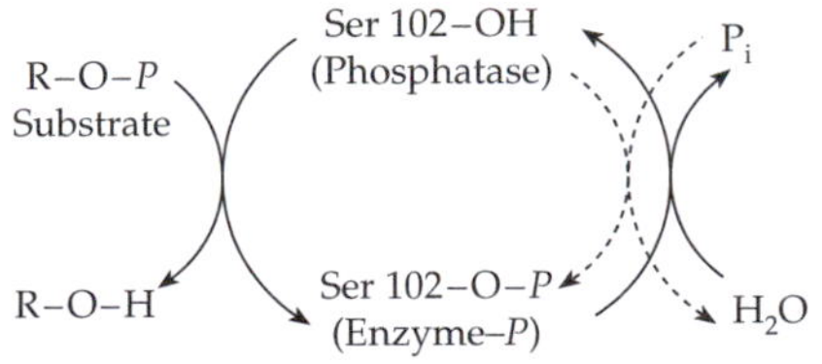

$$(12\text{-}38)$$

The active site contains two Zn^{2+} ions and one Mg^{2+} ion which are held by imidazole and carboxylate groups. The inorganic phosphate in an enzyme–product complex is bound to both zinc ions (Fig. 12-23). The Ser 102 side chain is above one Zn. In the enzyme-P intermediate it would be linked to the phospho group as an ester which would then be hydrolyzed, reversibly, by a water molecule bound to Zn.[712–713a] This water presumably dissociates to Zn^+–OH and its bound hydroxyl ion carries out the displacement. This reaction may be preceded by a proton transfer to an oxygen atom of the phospho group.[714]

Acid phosphatases, which have pH optima of ~5 and are inhibited by fluoride ion, occur in bacteria, fungi, plants, and animals. In bone, the acid phosphatase content is high in the osteoclasts which function in the resorption of calcium from bone. The highest content of acid phosphatase in humans is in the prostate and an elevated serum level has long been used as a diagnostic indicator of prostatic cancer.[715] Acid phosphatase is a periplasmic enzyme in E. coli.[716] Phosphoenzymes have been trapped from both plant and animal acid phosphatases.[717] For example, a brief incubation with ^{32}P-labeled p-nitrophenyl phosphate followed by rapid denaturation in an alkaline medium gave a covalently labeled protein from which ^{32}P-containing N δ-phosphohistidine was isolated. In agreement with kinetic evidence that a phosphoenzyme is a true intermediate, there is no inversion of chiral phospho groups by liver or prostate acid phosphatases.[718]

Some acid phosphatases from animals and plants are violet in color and contain iron (Chapter 16) and an Mn^{3+}-containing acid phosphatase has been isolated from sweet potatoes.[720] These enzymes have dimetal centers, often containing one Zn^{2+} and one Fe^{3+} with bridging carboxylate and hydroxide ions between the metals. Imidazole, tyrosinate, and carboxylate side chains hold the metals as in Fig. 16-20. A water molecule bound to the Fe^{3+} is thought to dissociate with a low pK_a of 4.8 to give an $Fe^{3+} \cdot {}^-OH$ complex. The hydroxyl ion can then attack the phospho groups, one

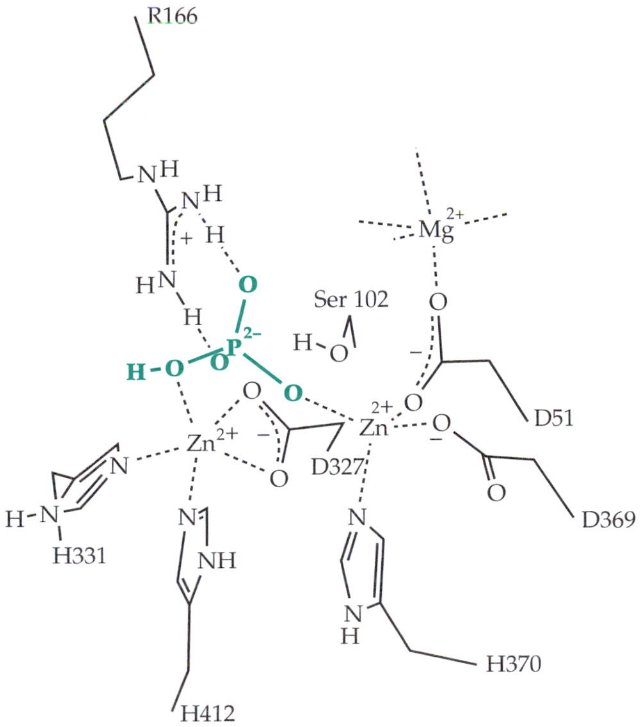

Figure 12-23 Schematic drawing of the product inorganic phosphate bound in the active site of E. coli alkaline phosphatase. See Ma and Kantrowitz.[719]

of whose oxygen atoms is coordinated with the Zn^{2+}.

The mechanism resembles that proposed for a **phosphotriesterase** (Fig. 12-24). The triesterase catalyzes detoxification of organophosphorus toxins such as parathion (Box 12-E) and seems to have evolved rapidly from a homologous protein of unknown function.[721] The phosphotriesterase contains two Zn^{2+} ions in a dimetal center. An unusual structural feature is a carbamate group, formed from Lys 169 and CO_2, which provides a bridging ligand for the metal pair.[721–725] A carbamylated lysine also functions in ribulose bisphosphate carboxylase (Fig. 13-11).

Phosphatases specific for such substrates as glucose-6-phosphate, fructose-1,6-bisphosphate, and phosphoglycolate help to drive metabolic cycles (Chapter 17). The 335-residue **fructose-1,6-bisphosphatase** associates to form a tetramer with D_2 symmetry.[726–730] The allosteric enzyme exists in two conformational states (see Chapter 11). Activity is dependent upon Mg^{2+} or other suitable divalent cation, e.g., Mn^{2+} or Zn^{2+}, and is further enhanced by K^+ or NH_3^+. While the dimetal sites depicted in Figs. 12-23 and 12-24 are quite rigid and undergo little change upon formation of complexes with substrates or products, the active site of fructose-1,6-bisphosphatase is more flexible. There are three metal-binding sites but they contain no histidine side chains and have been seen clearly only in a product complex.[727,728] Perhaps because of the need for

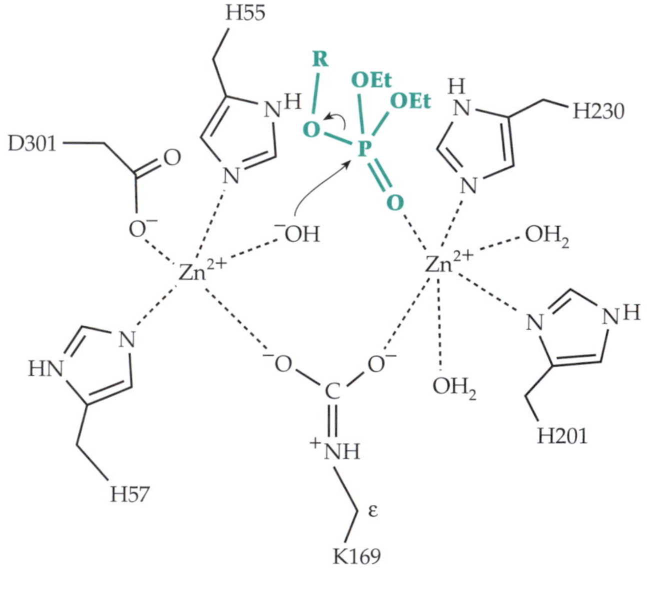

Figure 12-24 Hypothetical event in the action of a phosphotriesterase. A carbamylated lysine (lower center), as well as a water molecule, bridge the two Zn^{2+} ions, which are held by imidazole and aspartate carboxylate groups. The bound H_2O can be deprotonated to give the HO^- complex shown. The substrate may displace the HO^- ion from the right-hand zinc and thereby move close to the bound HO^- which attacks as indicated. Based on Cd^{2+}-containing structure and discussion by Benning *et al.*[722]

flexibility involved in allosteric changes, the active site is not fully formed until the substrate binds. **Fructose-2,6-bisphosphatase** forms one domain of a bifunctional kinase-phosphatase (Chapter 11). It has two imidazole rings, as well as side chains from a glutamate and two arginine residues at the catalytic and substrate-binding site.[728a]

The 357-residue mammalian glucose-6-phosphatase plays an important role in metabolism (Chapter 17). Defects in the enzyme cause a glycogen storage disease (Box 20-D) and severe disruption of metabolism.[731] However, the molecular basis of its action is not well-known. Furthermore, the active site of the enzyme is located in the lumen of the endoplasmic reticulum[732] and glucose-6-phosphate must pass in through the plasma membrane. An additional glucose-6-phosphate transporter subunit may be required to allow the substrate to leave the cytoplasm.[73]

Pyrophosphatases, which are present in all cells, and catalyze hydrolysis of inorganic pyrophosphate (PP_1) to orthophosphate (P_i) (see Chapter 6, Section D), also drive metabolic sequences. The very active pyrophosphatase of *E. coli* has a turnover number of over 2×10^4 s^{-1} at 37°C. The 1000 molecules per cell are sufficient to immediately hydrolyze any pyrophosphate produced by bacterial metabolism.[733] The much studied soluble pyrophosphatases of *E. coli*,[734,735] yeast,[736] and other organisms[736a,b] are metalloenzymes that are most active with Mg^{2+}. Two Mg^{2+} ions are held, mostly by carboxylate side chains, while a third apparently enters the active site as magnesium pyrophosphate, perhaps $MgP_2O_7^-$. As with other metallohydrolases, a metal-bound hydroxyl ion may serve as the attacking nucleophile.

At least three distinct families of **protein phosphatases** remove phosphate groups from serine, threonine, and tyrosine side chains in proteins.[737] Their role in control of numerous biochemical processes has been discussed in Chapter 11, Section C,2. The catalytic domains or subunits of the protein phosphatases act together with regulatory domains or separate regulatory subunits to control thousands of reactions. For example, protein phosphatase 1 (PP1) together with a glycogen-targeting subunit dephosphorylates inactive glycogen kinase (see Fig. 11-4). Belonging to the same family is **calcineurin** (PP2B), a phosphatase activated by Ca^{2+} through binding to calmodulin (Box 6-D). There are two families of Ser/Thr phosphatases. Their polypeptide folding patterns differ, but the active sites have similar dimetal centers resembling those in Figs. 12-24 and 16-20 with $Mn^{2+} + Fe^{2+}$, $Zn^{2+} + Fe^{3+}$, and probably other pairs of metals.[737–740] The family containing PP1 has weak sequence homology with the purple acid phosphatases.[740] Another common feature of these enzymes is a conserved His-Asp dyad (His 125 and Asp 95 in PP1) which is thought to be a proton donor, protonating the leaving group ($-O^-$) in a manner reminiscent of the serine proteases. The reaction is evidently initiated by attack of an $^-$OH ion held by the dimetal center. This would resemble the mechanism pictured in Fig. 12-24 except that the phospho group would carry two negative charges.

The protein tyrosine phosphatases also exist as several families with numerous functions in control of transcription, growth, differentiation, and metabolism.[741–743] These enzymes function by a double-displacement mechanism, as in Eq. 12-38, but with a cysteine side chain rather than serine. The cysteine is present in the conserved sequence (H/V)**C**X$_5$**R**(S/T). The arginine binds the phospho group and helps to stabilize the transition state, which probably is metaphosphate-like.[742]

5. Ribonucleases (RNases)

Many hydrolases act on phosphodiester linkages, which abound in nature.[725,744] Some are digestive enzymes but others serve more specific metabolic functions. **Ribonuclease A** (**RNase A**), the pancreatic digestive enzyme responsible for breakdown of RNA, was one of the first enzymes for which a structure was deduced. By 1963 Moore and Stein and their associates, who had earlier developed ion exchange methods for separating amino acids and peptides (Fig. 3-6), had determined the sequence of the 124-residue bovine enzyme.[745] They observed that Lys 41 was unusually reactive with dinitrofluorobenzene and that photo-oxidation of His 12 and His 119, which are almost at opposite ends of the peptide chain, inactivates the enzyme. They concluded that both histidines are at the active site, a conclusion that was later substantiated by X-ray crystallography.[746–748] A segment 12 nucleotides in length can fit into the cleft in the enzyme that contains the active site. The negatively charged phosphates of the RNA backbone form 8 – 9 electrostatic bonds to lysine and arginine side chains of the enzyme.[749,750] However, the only close interactions of the nucleic acid bases with the enzyme occur at the site of cleavage as shown in Fig. 12-25. The four residues His 12, Lys 41, Thr 45, and His 119 are strictly conserved in the RNAse A superfamily.[749] The carboxylate of Asp 121 apparently helps to orient the proper tautomer of His 119 for catalysis.[751]

Ribonuclease A was the first enzyme to be synthesized in the laboratory. Fully active ribonuclease has been synthesized,[752] as have new modified enzymes. For example a 63-residue peptide made up of five segments of the native RNase sequence retained measurable catalytic activity.[753] Using total synthesis, unnatural amino acids, such as 4-fluorohistidine, have been incorporated at specific positions in RNAse.[752]

Cleavage of a phosphodiester linkage in the substrate chain occurs in two steps. In the first or *trans-*

esterification step, the hydroxyl group on the 2' position of the ribose ring is thought to be deprotonated by attack of either the imidazole of His 12, as shown in Fig. 12-25, or by the adjacent amino group of Lys 41. (In the latter case His 12 would have to remove a proton from the $-NH_3^+$ group of Lys 41 before it could attack.) In either case the positive charge of Lys 41 would help to neutralize the negative charge on the phosphate. The deprotonation of the 2'–OH may occur synchronously with its attack on the adjacent 3' phospho group. An in-line displacement of the oxygen attached to the 5' carbon of the next nucleotide unit is thought to be assisted by His 119. Its protonated imidazolium group may transfer a proton to a phosphate oxygen atom prior to or synchronously with formation of the new P–O bond in step *a* (Fig. 12-25).[754] The intermediate formed in step *a* is a cyclic 2', 3'-diphosphate which then undergoes hydrolysis by attack of a water molecule in step *b* to give the free nucleoside 3'-phosphate. The overall reaction is a two-step double-displacement, analogous to that with chymotrypsin, except that a neighboring group in the substrate rather than an amino acid side chain is the nucleophilic catalyst. The pH dependence of the enzyme is in agreement with this mechanism because there are two pK_a values of ~5.4 and ~6.4 which regulate the catalytic activity. Microscopic pK_a values of His 12 and His 119 have been measured by NMR spectroscopy as ~ 6.1 and ~ 6.3 and are shifted somewhat by binding of nucleotides.

A bacterial peptidase splits a 20-residue fragment containing His 12 from the N-terminal end of RNase A. This "S-peptide" can be recombined with the rest of the molecule, which is inactive, to give a functional enzyme called ribonuclease S. In a similar way, residues 119–124 of RNase A can be removed by digestion with carboxypeptidase to give an inactive protein which lacks His 119. In this case, a synthetic peptide with the sequence of residues 111 – 124 of RNase A forms a complex with the shortened enzyme restoring full activity.[755]

Stereochemical studies support in-line mechanisms for both the transesterification and hydrolysis steps of ribonuclease catalysis. For example, chiral uridine 2',3'-cyclic phosphorothioates are hydrolyzed with inversion of configuration, with the diastereoisomer shown yielding a 2'-monophosphothioate of the *R* configuration at phosphorus.

Uridine 2',3'-cyclic phosphorothioate

Transesterification step is also in-line[757] as it is shown in Fig. 12-25. Study of kinetic isotope effects in H_2O–D_2O mixtures suggested that two protons may move synchronously as the enzyme–substrate complex passes through the transition state.[758] Although RNase A is one of the most studied of all enzymes, there are still uncertainties about the mechanism. Is a proton removed first by His 12 (Fig. 12-25), as has long been assumed, or is a proton transferred first from His 119 to the oxygen of the phospho group?[759,760] Is the reaction concerted, as suggested by kinetic isotope effects,[761] or is there a pentacovalent intermediate?

The specificity of RNase A for a pyrimidine on the 3' side of the phosphodiester bond that is cleaved is evidently ensured by the pair of hydrogen bonds from O-2' of the pyrimidine to the backbone NH of Thr 45 and a second from the N-4' proton to the side chain OH of the same threonine (Fig. 12-25). Other nucleases, such as ribonuclease T_2,[762] with different specificities also make use of hydrogen bonding of the base at the 3' side of the cleavage point with backbone amide groupings.

Various bacterial ribonucleases as well as the fungal ribonucleases T_1, U_1, and U_2 (see also Fig. 5-43) have amino acid sequences related to that of RNase A[763,764,764a] but with distinctly different three-dimensional structures. The active sites contain Glu, His, and Arg side chains. For RNase T_1, Glu 58 and His 92 appear to provide acid–base catalysis with assistance from Tyr 38, Arg 77, and His 40.[763,765] A glutamate carboxylate also appears to be the catalytic base in the related RNase, called **barnase**, from *Bacillus amyloliquefaciens*.[766]

In addition to extracellular digestive enzymes, the RNase family contains many intracellular enzymes that are involved in turnover of RNA.[767,768] RNase H digests away RNA primers during DNA synthesis (Chapter 27). RNase H activity is also present in a domain of viral reverse transcriptases and is absolutely essential for the replication of HIV and other retroviruses.[769] The structures of the reverse transcriptase RNase H domain and of the *E. coli* enzyme are similar.[769-771] Unlike RNase A, the RNases H are metalloenzymes which apparently contain two Mg^{2+} ions held by carboxylate groups and utilize a metal bound HO^- ion as in previously discussed phosphatases. Secreted RNases sometimes have specific functions. For example, the 123-residue **angiogenin** is homologous to pancreatic RNases but acts to induce formation of new blood vessels (angiogenesis).[772-774] This is essential to growth of solid cancers as well as for normal growth. The enzyme is a very poor catalyst but its RNase activity appears essential for its biological function. Mutation of any of the catalytic residues His 13, Lys 40, or His 114 abolishes all angiogenic activity. A neurotoxin secreted by eosinophils[775] is one of a group of selectively toxic RNases.[776,777] Intracellular RNases are often found as complexes of specific inhibitor proteins.[778]

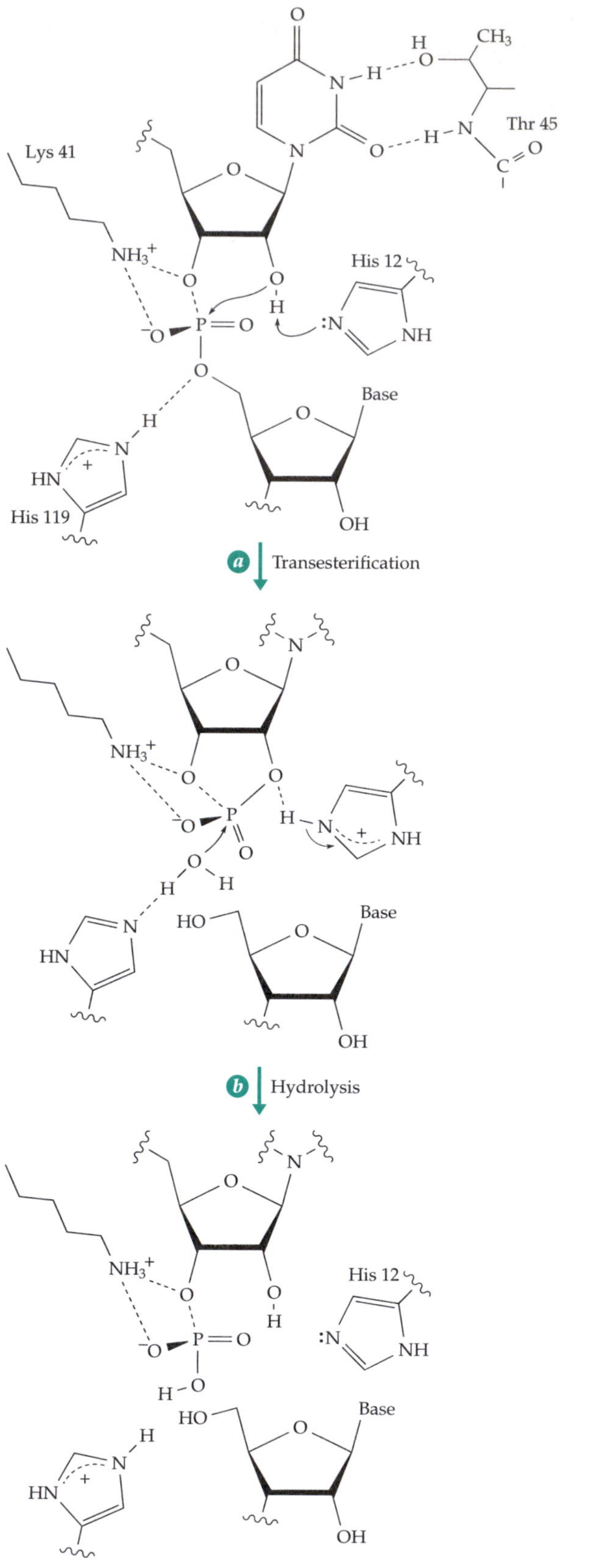

Figure 12-25 Proposed two-step in-line reaction mechanism for ribonuclease A. The hydrogen bonding that provides recognition of the pyrimidine base at the 3' end created by the cleavage is also shown. See Wladkowski *et al.*[756]

6. Ribonuclease P, Ribozymes, and Peptidyl Transferase

A very different ribonuclease participates in the biosynthesis of all of the transfer RNAs of *E. coli*. **Ribonuclease P** cuts a 5' leader sequence from precursor RNAs to form the final 5' termini of the tRNAs. Sidney Altman and coworkers in 1980 showed that the enzyme consists of a 13.7-kDa protein together with a specific 377-nucleotide RNA component (designated M1 RNA) that is about five times more massive than the protein.[779] Amazingly, the M1 RNA alone is able to catalyze the ribonuclease reaction with the proper substrate specificity.[780–782a] The protein apparently accelerates the reaction only about twofold for some substrates but much more for certain natural substrates. The catalytic center is in the RNA, which functions well only in a high salt concentration. A major role of the small protein subunit may be to provide counterions to screen the negative charges on the RNA and permit rapid binding of substrate and release of products.[783] Eukaryotes, as well as other prokaryotes, have enzymes similar to the *E. coli* RNase P. However, the eukaryotic enzymes require the protein part as well as the RNA for activity.[784]

Thomas Cech and associates independently discovered another class of catalytic RNA molecules. These are **self-splicing RNAs** that cut out intervening sequences from themselves to generate ribosomal RNA precursors (see Chapter 28).[785–787] They act only once and are therefore not enzymes. However, the introns that are cut out during self-splicing are **ribozymes** which, like the RNA from ribonuclease P, can act catalytically and have properties similar to those of protein enzymes. They exhibit the kinetic properties of enzymes and are denatured by heat. The RNAs are folded into compact structures resembling those of globular proteins. Like tRNA, they contain loops and hydrogen-bonded stems. Phylogenetic comparisons (Chapter 29) of the M1 RNA of ribonuclease P isolated from various species have allowed prediction of precise secondary structures.[783,788,789] A simplified M1 RNA consisting of 263 nucleotides from conserved regions of the molecule is catalytically effective.[783] Tetrahymena ribozyme also has a complex structure with a 247-nucleotide catalytic core formed by two structural domains (Fig. 12-26).[790,791] The crystal structure of a third ribozyme, one found in the RNA of the human pathogen hepatitis delta virus (HDV), has also been determined.[792] It is a smaller 72-nucleotide self-cleaved molecule with a very different structure from that in Fig. 12-26. It makes use of a double pseudoknot (see Fig. 5-29) to bind the RNA into a compact, tightly hydrogen-bonded structure with a deep active site cleft. It is the fastest known naturally occurring self-cleaving RNA and is able to react at a rate of more than $1 \, s^{-1}$ at its optimum temperature of 65°C.

Smaller self-cleaving RNAs have been found among plant viruses and viroids. Many of them have a common catalytic core which can be converted into 30- to 40-nucleotide ribozymes. Only 17 nucleotides and three hydrogen-bonded helical stems are required to form the self-cleaving "hammerhead" domain, which has a structural similarity to the catalytic core of the *Tetrahymena* ribozyme. The **hammerhead ribozymes** (Fig. 12-27) represent one form of small ribozyme.[793–797] Another is the **hairpin ribozyme** shown in Fig. 12-28,[798,801] which also shows the even smaller lead-dependent "**leadzyme**," a ribozyme that doesn't occur in nature.

In an intact viral self-cleaving RNA the entire catalytic center is formed from a single strand. Stems II and III of the hammerhead ribozyme (Fig. 12-27C) are closed by large loops. In the ribozyme shown, loop III has been cut off and stem II has been closed by a tight loop to form a compact catalytic RNA that will cut a substrate having a suitable nucleotide sequence for binding to the ribozyme. Only 12 bases in this ribozyme are highly conserved. By varying the sequences in the ribozyme half of stems I and III, catalysts that cleave after any sequence GUX, where X=A, C, or V, can be designed. Such catalysts are useful in the laboratory and potentially also in medicine.

What groups of a ribozyme bind to substrates and what groups participate in catalysis? Like peptides, RNAs have amide groups that can hydrogen bond to substrates. Adenine and cytosine can supply protonated amino groups which could participate in acid–base catalysis. This is evidently the case in the **peptidyl transferase** centers of ribosomes. The RNA in these centers catalyzes a transesterification in which an aminoacyl group is transferred from an aminoacyl-tRNA anto the growing polypeptide chain attached to a second tRNA molecule. The reaction is evidently catalyzed by a universally conserved adenine ring located at position 2486 in the *Haloarcula marismortii* 23S RNA (position 2451 in *E. coli*). There are no protein groups within 1.8 nm of the location of peptide bond synthesis.[798a] The active site adenine appears to be much more basic than normal. A high pK_a of 7.6 controls the peptidyl transferase, and also controls the methylation of the active site adenine by dimethylsulfate.[798b] The site of protonation is thought to be largely N3 of adenine 2486, which is probably the basic center involved in catalysis. The peptidyltransferase reaction may be initiated as follows.[798a]

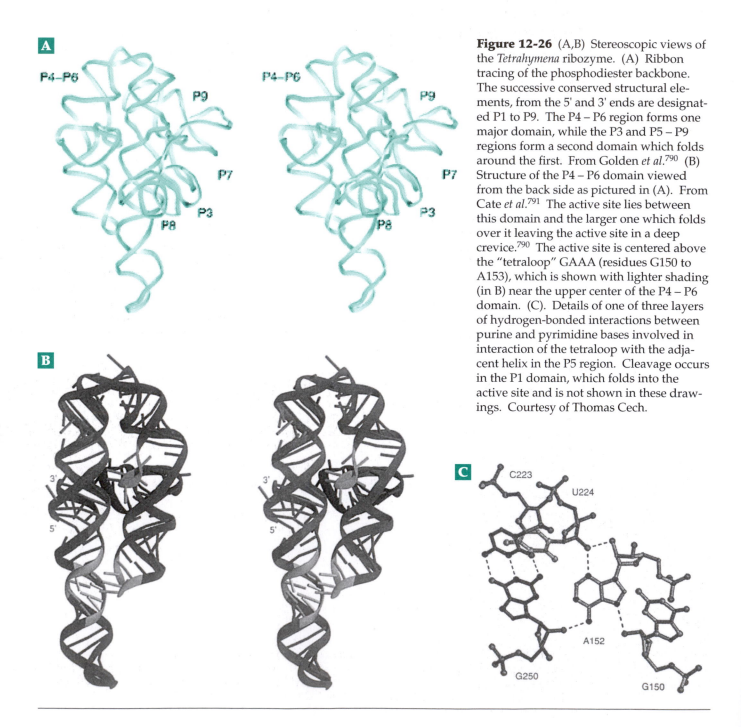

A similar catalytic mechanism is probably used by the small 85-nucleotide hepatitis delta virus ribozyme whose catalytic base is thought to be N3 of cytosine 75, which is associated with a pK_a of ~6.1.[798c] Both this HDV ribozyme and the ribosomal RNA resemble serine proteases with histidine as the catalytic base. However, the self-splicing RNA of *Tetrahymena* initiates a nucleophilic attack with the 3'–OH group of a guanosine molecule that is bound to a site in the P7 region (Fig. 12-16A) and which acts as a cofactor (see Fig. 28-18). Ribonuclease P and all group I and II self-cleaving introns also use an external nucleophile such as a guanosine –OH and form 3'-OH and *O*-phosphate or

Figure 12-26 (A,B) Stereoscopic views of the *Tetrahymena* ribozyme. (A) Ribbon tracing of the phosphodiester backbone. The successive conserved structural elements, from the 5' and 3' ends are designated P1 to P9. The P4 – P6 region forms one major domain, while the P3 and P5 – P9 regions form a second domain which folds around the first. From Golden *et al.*[790] (B) Structure of the P4 – P6 domain viewed from the back side as pictured in (A). From Cate *et al.*[791] The active site lies between this domain and the larger one which folds over it leaving the active site in a deep crevice.[790] The active site is centered above the "tetraloop" GAAA (residues G150 to A153), which is shown with lighter shading (in B) near the upper center of the P4 – P6 domain. (C). Details of one of three layers of hydrogen-bonded interactions between purine and pyrimidine bases involved in interaction of the tetraloop with the adjacent helix in the P5 region. Cleavage occurs in the P1 domain, which folds into the active site and is not shown in these drawings. Courtesy of Thomas Cech.

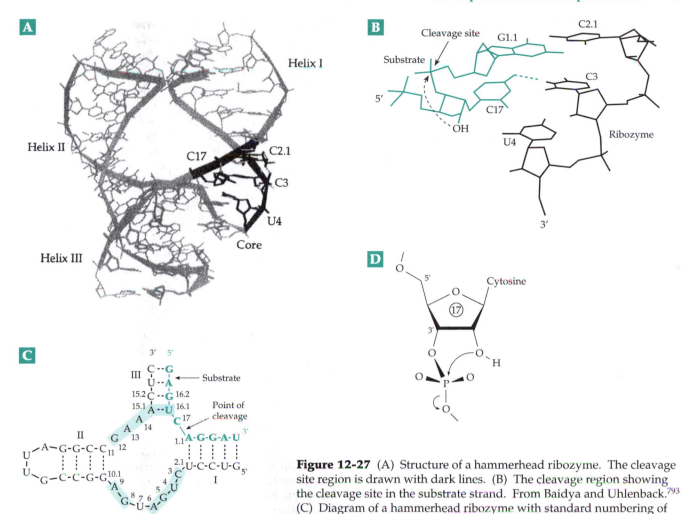

Figure 12-27 (A) Structure of a hammerhead ribozyme. The cleavage site region is drawn with dark lines. (B) The cleavage region showing the cleavage site in the substrate strand. From Baidya and Uhlenback.[793] (C) Diagram of a hammerhead ribozyme with standard numbering of nucleotides. The three helical stems are labeled I, II, and III. From Bevers et al.[796] (D) Simplified cleavage mechanism which resembles step *a* of Fig. 12-25 and ignores the known participation of metal ions.

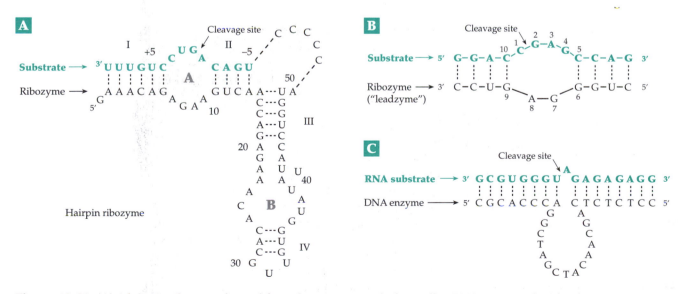

Figure 12-28 (A) A hairpin ribozyme formed from the minus strand of a satellite RNA associated with tobacco ringspot virus. On the basis of hydroxyl radical footprinting (see Fig. 5-50), to identify protected areas a folding pattern that brings domains A and B together to form a compact catalytic core has been proposed.[798] (B) A "leadzyme," a ribozyme dependent upon Pb^{2+} for cleavage of RNA.[802,803] (C) An RNA-cleaving DNA enzyme.[804]

phosphodiester ends at the cleavage points. Hammerhead ribozymes utilize the 2'-OH of the ribose at the cleavage site to form a 2',3'-cyclic phosphate ester as in step *a* of the ribonuclease A reaction (Fig. 12-25) and as indicated in Fig. 12-27D. Ribozymes act by in-line mechanisms, causing inversion of the configuration at the phosphorus.[799,800]

Most ribozymes, as well as RNase P,[805,806] require one or two metal ions for activity.[807] Magnesium ions predominate and many Mg^{2+} ions are bound at distinct sites in crystalline ribozymes. Hammerhead and hairpin ribozymes work reasonably well with monovalent ions. One proposed mechanism is for an HO^- ion bound to the Mg^{2+} to remove H^+ from the 2'– OH of the ribose ring as follows:

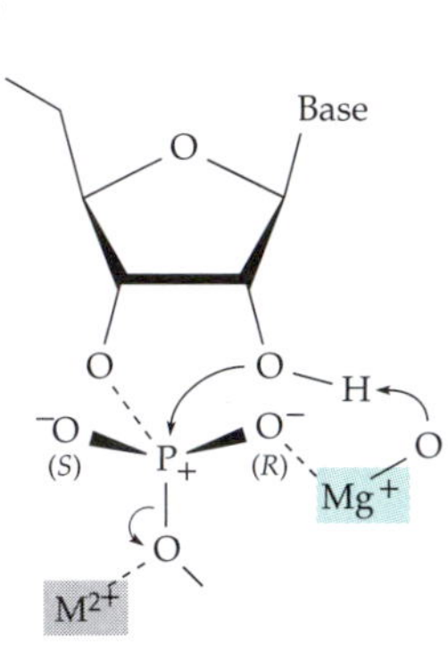

When the *pro-R* oxygen of the phospho group (labeled *R* in the accompanying structure) was replaced by sulfur the rate of cleavage decreased 10^3-fold. However, the rate was restored fully when Cd^{2+} was added, presumably because of the high affinity of cadmium ions for sulfur ligands.[808] This supports the possibility of a dual role for Mg^{2+} in activating a water molecule to provide HO^- and in stabilizing negative charges on the phospho group by interaction with the *pro-R* oxygen in the transition state. Several investigators have suggested that *two* metal ions may be needed. One possibility is for a mechanism similar to that proposed for alkaline phosphatase (Fig. 12-25) and other phosphotransferases.[809,810] The *Tetrahymena* ribozyme functions best if both Mg^{2+} and Mn^{2+} are present.[811] A second metal may act as a Lewis acid facilitating loss of the 5'-OH as indicated by M^{2+} (in gray) in the preceding diagram. Metal ions may also participate in conformational changes as well as have structural functions in ribozymes.[812,813] Studies of the kinetics of action of hammerhead ribozymes have suggested that the intrinsic ribozyme-substrate binding energy is utilized for catalysis.[814] This may be possible because the ribozyme is only partially folded in the ground state but it folds into a tighter conformation in the transition state.[815]

Methods have been devised for generating enormous numbers of RNA molecules with random sequences and for selecting those with unusual catalytic activities.[816–819] Among the new catalysts produced in this way are very small ribozymes that cleave RNA specifically in the presence of Pb^{2+} (Fig. 12-28B).[802,803] The leadzyme is more active with neodymium (Nd^{3+}) + Pb^{2+} than with lead alone, suggesting a two-metal mechanism.[820] Other artificial ribozymes include RNA ligases,[817] acyltransferases,[821] and DNA hydrolases.[822,823] Is it possible to find a DNA enzyme? Without the 2'–OH of ribose to form hydrogen bonds it seemed doubtful, but an RNA-cleaving DNA enzyme has been selected from a population of ~10^{14} different small DNA molecules. The DNA enzyme (Fig. 12-28C) will cleave RNA, whose sequence fulfills the base pairing requirements of two 8-deoxynucleotide recognition domains. Cleavage occurs between an unpaired purine and a paired pyrimidine using a metal-dependent mechanism that gives a 2',3'-cyclic phosphate as in ribonuclease A cleavage.[804]

7. Deoxyribonucleases (DNases)

A multitude of nucleases cleave DNA, single- or double-stranded. They range from the pancreatic digestive enzyme DNase I through specialized nucleases that function during DNA repair and the hundreds of restriction endonucleases that have become so valuable in modern laboratory work. Some nucleases leave a 3'-phosphate ester at a cut end in a DNA chain, while others leave a 5'-phosphate end.[824] Many nucleases are dealt with in later chapters. Only a few will be mentioned here.

One of the most studied enzymes of this group is the 149-residue micrococcal (staphylcoccal) nuclease from *Micrococcus* which cleaves either RNA or single- or double-stranded DNA. The relatively nonspecific enzyme cuts nearly randomly at the 5' side of the phosphodiester linkages, leaving 3'-phosphate groups. It enhances the uncatalyzed hydrolysis rate at least 10^{16}-fold.[825–827] The crystal structure showed that the majority of the acidic and basic side chains of the protein interact with each other through clusters of hydrogen bonds. At the active site the side chains of both Arg 35 and Arg 87 form pairs of hydrogen bonds to the 5'-phosphate group of the specific inhibitor deoxythymidine 3',5'-diphosphate (Fig. 12-29). While Arg 87 appears to be in a position to protonate the leaving group $-O^-$, ^{13}C NMR experiments showed that all of the arginine side chains had pK_a values above 11.6. However, Tyr 85 has a pK_a of 9.5, which appears to control k_{cat} / K_m.[828] A Y85F mutant lacks this pK_a. The X-ray structure also suggests that Glu 43 may be the attacking nucleophile and that it may deprotonate a water molecule bound to the Glu 43 carboxylate. The resulting HO^- probably carries out a direct in-line attack as shown in Fig. 12-29. Mutants such as E43D, E43Q, and E43S have greatly decreased activity,[829] in

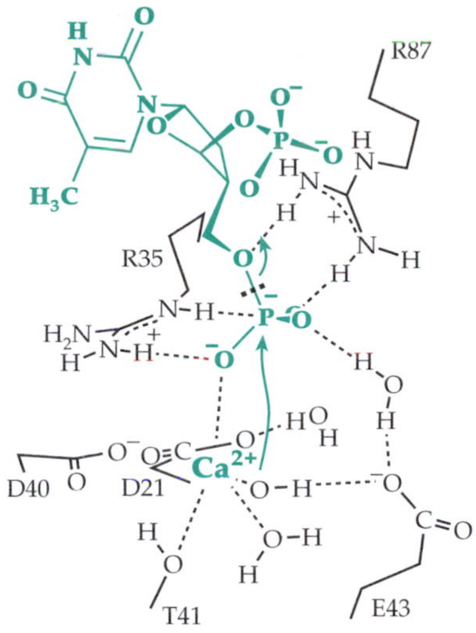

Figure 12-29 Drawing showing the hydrogen-bonding interactions between the guanidinium ions of arginines 35 and 87 of the micrococcal (staphylococcal) nuclease with the 5'-phosphate of the inhibitor thymidine 3',5'-diphosphate in the complex of E + I + Ca^{2+}. A possible mechanism is illustrated. A hydroxyl ion bound to Ca^{2+} carries out an in-line attack on the phosphorus. See Libson *et al*.[826]

agreement with this mechanism. The nearby Ca^{2+} is essential. For this nonspecific nuclease there is no hydrogen bonding of a purine or pyrimidine base of the substrate to the enzyme.

The digestive enzyme, pancreatic DNAse I, makes single-stranded cuts in double-stranded (ds) DNA. An exposed strand of peptide chain from the enzyme binds into the minor groove of B-type DNA.[830] Because this groove becomes too narrow in long (A+T) rich sequences, they are cleaved slowly. Certain hypersensitive sites are cleaved very rapidly, perhaps because the DNA at these regions is bent or is able to bend to give a very good fit to the enzyme active site. A histidine which is hydrogen bonded to a nearby carboxylate of a glutamate side chain appears to be a catalytic base that acts upon a water molecule as in phospholipase A (Section D,10), displacing the 3' oxygen of the phosphodiester linkage. An imidazolium group from a second histidine is hydrogen bonded to an aspartate carboxylate and a tyrosine –OH to form a catalytic triad that can protonate the 3' –O⁻ as it is displaced.[831] Two Mg^{2+} ions are also required. Both are held by different carboxylate side chains and may also interact with oxygen atoms of the phospho group to neutralize charge and stabilize the transition state.

In contrast to DNAse I, the **restriction endonucleases**, which are discussed in Chapter 5, Section H,2 and in Chapter 26, have precise substrate sequence specificities. Three of the best known restriction endonucleases are called *Eco*RI,[832–834] an enzyme which binds

to and cuts both strands of the palindromic sequence 5'-GAATTC; *Eco*RV,[835] which cuts both strands in the center of the sequence 5'-GATATC; and *Bam*H1, which binds to the sequence 5'-GGATCC and cleaves after the 5' G on each strand.[836] A high-resolution structure is also known for *Cfr*10I, which recognizes the less strict sequence 5'-PuCCGGPy and cleaves both strands after the 5' Pu.[837] All of these enzymes require Mg^{2+} and have active sites containing carboxylate groups. Two-metal mechanisms have been suggested.

Restriction endonuclease *Eco*RI is able to cut a chain in dsDNA which has a chiral phophorothioate group at the specific cleavage site.[838] The reaction occurs with inversion of configuration at phosphorus, suggesting direct in-line attack by a hydroxyl ion generated from H_2O.

Attempts are being made to design semisynthetic restriction endonucleases specific for single-stranded DNA or RNA. For example, an oligonucleotide with a sequence complementary to a sequence adjacent the linkage that is to be cut can be covalently linked to a relatively nonspecific nuclease. Such an enzyme derived from micrococal nuclease cuts a single-stranded chain of either DNA or RNA adjacent to the double-stranded region of the ES complex.[839]

8. Mutases

Phosphotransferases that shift phospho groups from one position within a substrate to another are often called **mutases**. For example, **phosphoglucomutase** catalyzes the interconversion of glucose 1-phosphate and glucose 6-phosphate, an important reaction that bridges glycogen metabolism and glycolysis (Fig. 11-2). This 561-residue protein operates through formation of an intermediate **phosphoenzyme**.[840–842] The phospho group becomes attached to the OH of Ser 116 and can be transferred either to the 6 or the 1 position of a glucose phosphate (step *a* and reverse of step *b* in Eq. 12-39).

The two-step reaction accomplishes the reversible isomerization of glucose 1-phosphate and glucose 6-phosphate via **glucose 1,6-bisphosphate**. Evidently, the glucose bisphosphate, without leaving the enzyme is reoriented to allow transfer of the phospho group to either the 1- or 6-position.[841] The phospho enzyme is relatively unstable and can undergo hydrolysis to free enzyme and P_i. To prevent loss of active enzyme in this way, a separate reaction (catalyzed by a kinase; Eq. 12-39, step *c*)[843] generates glucose 1,6-bisphosphate, which rephosphorylates any free enzyme formed by hydrolysis of the phospho enzyme. The glucose 1,6-bisphosphate can be regarded as a cofactor or **cosubstrate** for the reaction.

Phosphoglycerate mutase, which interconverts 2-phosphoglycerate and 3-phosphoglycerate in glycolysis (Fig. 10-3, step *c*), functions by a similar mecha-

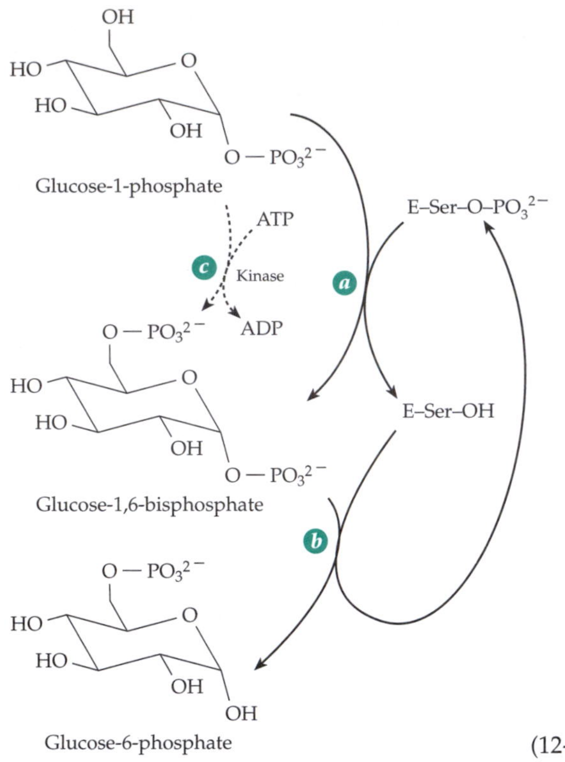

Glucose-1-phosphate

Glucose-1,6-bisphosphate

Glucose-6-phosphate

$$(12\text{-}39)$$

atoms of a second substrate.[848] Examples include **hexokinase**, the enzyme responsible for synthesis of glucose 6-phosphate from free glucose and ATP (Fig. 11-2, step a); **phosphofructokinase**, which forms fructose 1,6 bisphosphate in the glycolysis pathway (Fig. 11-2, step b); and **phosphoglycerate kinase**, and **pyruvate kinase**, both of which form ATP from ADP in the glycolysis pathway (Fig. 10-3, steps b, c, and f). There are many others. Kinases vary greatly in size and in three-dimensional structure. For example, a small **adenylate kinase**, which phosphorylates AMP to ADP (Eq. 6-65), is a 22-kDa monomer of 194 residues. Pyruvate kinase is a tetramer of 60-kDa subunits and muscle phosphofructokinases are tetramers of 75- to 85-kDa subunits. The three-dimensional structures also vary. While all kinases consist of two domains built around central β sheets (Fig. 12-30), there are several different folding patterns.[849] The two-domain structures all have deep clefts which contain the active sites. Both adenylate kinase (Fig. 12-30) and hexokinase crystallize in two or more forms with differing conformations.[850] This and other evidence suggests that as a kinase binds and recognizes its two correct substrates, the active site cleft closes by a hinging action that brings together the reacting molecules in the correct orientation.[851–853] In the crystal structure shown in Fig. 12-30 both ADP and AMP are bound in a nonproductive complex. If the ADP were replaced with ATP (or the AMP with a second ADP) to form a productive complex the two reacting phospho groups would be ~0.8 nm apart. A reaction could not occur without further closing of the active site cleft.[854] Evidence for domain closure has been obtained for many other kinases. For example, substrate complexes of phosphoglycerate kinase have been crystallized in both "open" forms and "closed" forms in which a 30° hinge-bending movement has brought the ligands together for an in-line phospho group transfer.[855]

For many enzymes an ATP binding site has been revealed by study of nonhydrolyzable analogs of ATP such as "AMP-PNP" whose structure is shown in Fig. 12-31.[856] AMP-PNP has been used in thousands of investigations of ATP-dependent processes.[857] For example, the structure of a phosphoglycerate • Mg^{2+} • AMP-PNP complex in the active site of phosphoglycerate kinase has been determined.[858] Modeling of a transition state complex indicates that all three negatively charged oxygens of the ATP portion of structure are stabilized by hydrogen bonding.[852] Related analogs such as AMP-PCH$_2$P (Fig. 12-3) have been used in similar ways.[859] Another analog Mg^{2+} • Ap$_5$A (Fig. 12-31) is a bisubstrate inhibitor which binds to adenylate kinases, fixing the enzymes in a closed conformation that is thought to resemble the transition state.[859,860]

The Mg^{2+} complex of ATP is regarded as the true substrate for kinases. The metal usually also binds both to the phospho groups of ATP and to groups on

nism.[843–845] However, the enzyme-bound phospho group is carried on an imidazole group. The essential cosubstrate required by some phosphoglycerate mutases is 2,3-bisphosphoglycerate.[845] It is formed from the glycolytic intermediate 1,3-bisphosphoglycerate (Fig. 10-3) by action of another mutase, **bisphosphoglycerate mutase**.[843,846]

$$\text{1,3-bisphosphoglycerate} \rightarrow \text{2,3-bisphosphoglycerate}$$
$$(12\text{-}40)$$

This is also the pathway for synthesis of 2,3-biphosphoglycerate in red blood cells where it serves as an important allosteric regulator (Chapter 7). Two human phosphoglycerate mutase isoenzymes are known. One is found in muscle and the other in brain and other tissues.[844] A hereditary lack of the muscle type enzyme is one of the known types of glycogen storage diseases (Box 20-D).

Human **phosphomannomutase**, which catalyzes the interconversion of mannose 1- and 6- phosphates, appears to carry the phospho group on an aspartate side chain in the sequence **D**XDX (T/V), which is conserved in a family of phosphomutases and phosphatases.[847] The first aspartate in the sequence is phosphorylated during the enzymatic reaction.

9. Molecular Properties of Kinases

Kinases transfer phospho groups from polyphosphates such as ATP to oxygen, nitrogen, or sulfur

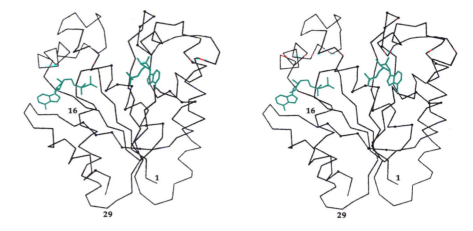

Figure 12-30 Stereoscopic α-carbon plots of a 194-residue subunit of adenylate kinase from the archaebacterium *Sulfolobus acidocaldarius* with ADP (left side) and AMP (right side) bound into the active site. From Vonrheim *et al.*[854] Courtesy of G. E. Schulz.

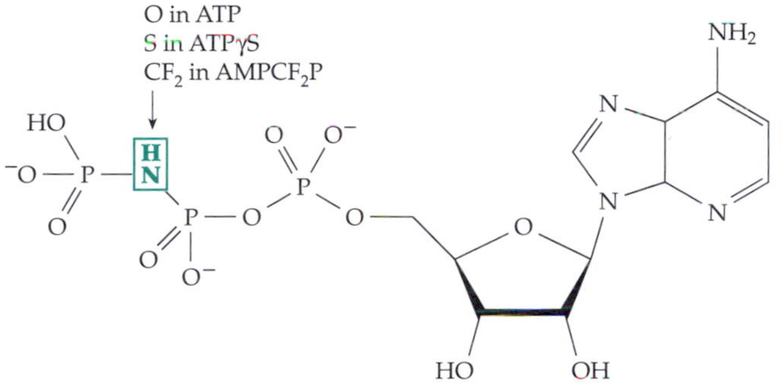

O in ATP
S in ATPγS
CF₂ in AMPCF₂P

AMP-PNP

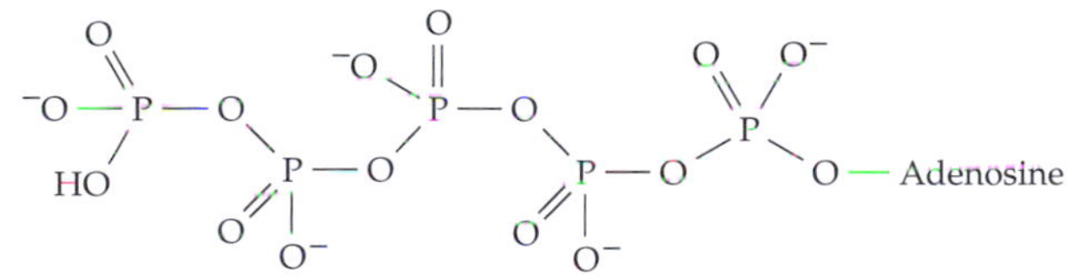

Ap₅A, a bisubstrate inhibitor of adenylate kinase

Figure 12-31 Some useful analogs of ATP.

the enzyme. Crystallographic investigations as well as studies with exchange-inert ATP complexes (Section 2) have suggested that the metal ion is bound initially to the terminal (γ) and adjacent (β) phospho groups of ATP. However, this metal bridge could prevent rather than assist the reaction. For some phosphotransferase reactions with exchange-inert complexes the product appears to be the α,β-bidentate complex of ADP (see Section 2) suggesting that movement of the metal ion from the β,γ- to the α,β-bidentate ATP complex may occur either prior to or during transfer of the phospho group. Theoretical calculations also suggest a movement in the metal complexation as the reaction progresses.[861] EPR studies that involved observation of hyperfine coupling between ^{17}O in substrates and Mn^{2+} in the active site led Reed and Leyh to conclude that the activating metal ion is bound to all three phospho groups in the transition state.[849]

Adenylate kinase performs the essential function of recovering AMP formed by many enzymatic processes and converting it to ADP (Eq. 6-65) which can be reconverted to ATP by oxidative or substrate level phosphorylation. The enzyme is present in all organisms. In vertebrates different isoenzymes function in the cytosol, mitochondrial intermembrane space, and mitochondrial matrix.[862,863] A group of other **nucleotide** and **deoxynucleotide kinases** convert nucleoside monophosphates into diphosphates.[864,865] Some of them, e.g., **uridylate kinase** are similar in structure and properties to adenylate kinase.[866,867] Another member of the adenylate kinase family is phosphoribulokinase, an important photosynthetic enzyme (see Fig. 17-14, step *a*).[868]

Most kinases transfer chiral phospho groups with inversion and fail to catalyze partial exchange reactions that would indicate phosphoenzyme intermediates. However, **nucleoside diphosphate kinase** contains an active site histidine which is phosphorylated to form a phosphoenzyme.[869] The enzyme catalyzes phosphorylation of nucleoside diphosphates other than ADP by a nucleotide triphosphate, usually ATP.

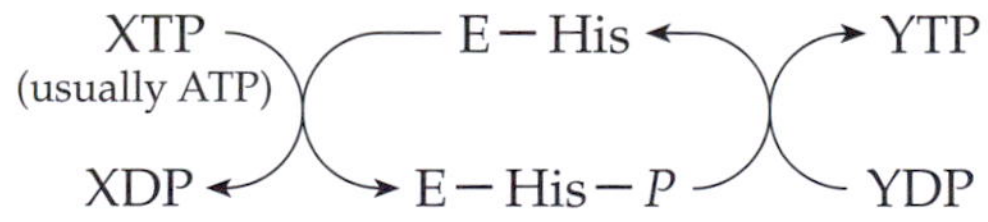

$$\begin{array}{ccc} XTP & E-His & YTP \\ \text{(usually ATP)} & & \\ XDP & E-His-P & YDP \end{array}$$

(12-41)

Here, X is usually adenosine and Y is any ribonucleoside or deoxyribonucleoside. This enzyme supplies all of the nucleotide triphosphates except ATP for use in the many cellular processes that require them.[870–872] The enzyme aligns the substrate, holding the phospho group with a pair of arginines and a magnesium ion. The phospho group is aligned for an in-line displacement by N^δ of His 122, part of a hydrogen-bonded His–Glu dyad. Formation of the phosphoenzyme occurs in less than 1 ms without significant conformational change other than a 30° rotation of the histidine ring.[870,873]

Hexokinase, the enzyme that phosphorylates glucose to glucose 6-phosphate, exists as four isoenzymes in mammals. Hexokinases I, II, and III are large ~ 100-kDa monomers with similar amino acid sequences, a single active site, and complex allosteric regulatory properties. The three-dimensional structure is known for hexokinase I, which is the pacemaker of glycolysis in the brain.[874,875] Hexokinase IV (glucokinase) is a 50-kDa protein that is found in liver and in the β cells of the pancreas. It has a low K_m and is not inhibited by glucose 6-phosphate, properties that allow for rapid uptake of glucose after a meal.[876] Its properties are similar to those of a major isoenzyme of yeast,[877] a dimer of identical 50-kDa subunits whose structure is also known.[878] The glucose-binding residues are conserved in yeast hexokinase, in glucokinase, and in brain hexokinase. The sequence of the latter suggests that it may have arisen by a doubling of a shorter hexokinase gene.[879]

Fructose 6-phosphate kinase (phosphofructokinase) has attracted much attention because of its regulatory properties (Chapter 11).[880–882] Prokaryotic forms are somewhat simpler.[883] The related fructose 6-phosphate 2-kinase is a component of a bifunctional kinase-phosphatase (Fig. 11-2, steps *d* and *e*) and has a structure similar to that of adenylate kinase.[884]

Phosphoglycerate kinase is encoded by the mammalian X chromosome. Several mutant forms are associated with hemolytic anemia and mental disorders.[885,886] Mutation of Glu 190, which is in the hinge region far from the active site, to Gln or Asn markedly reduces enzymatic activity.[887]

Creatine kinase transfers a phospho group to a nitrogen atom of the guanidinium group of creatine (Eq. 12-31 and Fig. 12-19). Several isoenzymes participate in its function of buffering the ATP level in tissues such as muscle fibers, neurons, photoreceptors, and spermatozoa which experience high and fluctuating energy needs.[888] A form from the mitochondrial intermembrane space of chicken heart is an octomer of 380-residue subunits.[889] The structurally and mechanistically similar **arginine kinase** has an analogous function in many invertebrates, e.g., in the horseshoe crab, which provided enzyme for a structure determination.[890]

The ~ 500-residue subunits of **pyruvate kinase** consist of four domains,[891] the largest of which contains an 8-stranded barrel similar to that present in triose phosphate isomerase (Fig. 2-28). Although these two enzymes catalyze different types of reactions, a common feature is an enolic intermediate. One could imagine that pyruvate kinase protonates its substrate phosphoenolpyruvate (PEP) synchronously with the phospho group transfer (Eq. 12-42). However, the enzyme catalyzes the rapid conversion of the enolic form of pyruvate to the oxo form (Eq. 12-43) adding the proton sterospecifically to the *si* face. This and other evidence favors the enol as a true intermediate

(12-42)

(12-43)

and a product of the phosphotransfer step.[891] Pyruvate kinase requires not only two equivalents of a divalent cation such as Mg^{2+} or Mn^{2+} but also a monovalent cation, usually K^+. However, Li^+, Na^+, NH_4^+, Rb^+, Tl^+, and others can substitute. The monovalent cation induces an essential conformational change. Using ^{205}Tl NMR it was found that the thallium ion binds about 0.6 nm from Mn^{2+} that is also present in the active site.[892] All three metals interact directly with the γ phospho group of ATP.[891] Pyruvate kinase is a regulated allosteric enzyme present in four isoenzymic forms in mammals.[893–894a]

Protein kinases, which were discussed in Chapter 11, phosphorylate selected – OH groups of serine, threonine, and tyrosine side chains in proteins. Examination of the sequences in the complete genome of yeast (*Saccharomyces cerevisiae*) indicates the presence of at least 113 protein kinase genes, which account for ~ 2% of the total DNA.[895] Higher eukaryotes have more. While structures of these enzymes vary widely, they share a common two-domain catalytic core structure.[896,897] The best known, and one of the simplest of them, is the catalytic subunit of cyclic AMP-dependent protein kinase.[897,898] The substrate MgATP binds into the active site cleft with the γ-phospho group protruding to meet the appropriate site of a bound protein substrate (Fig. 12-32). Most protein kinases are regulated by an "activation loop" that must be correctly placed before the ES complex can be formed. As discussed in Chapter 11, regulation of the cAMP-dependent kinase depends upon inhibition by a regulatory subunit. In the cAMP-dependent kinases and many other protein kinases the activation loop (not shown in Fig. 12-32), which helps to form the substrate site, contains a phosphothreonine residue which is essential for activity. It is a stable feature of the cAMP-activated kinase, incorporated at position 197, but for some tyrosine kinases it is generated by autophosphorylation.[899]

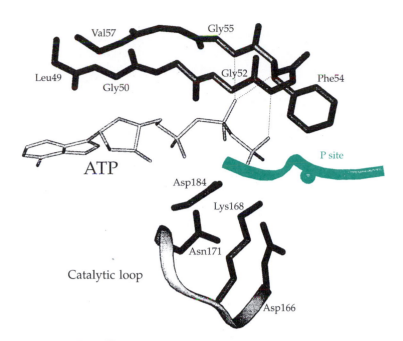

Figure 12-32 The active site of cyclic AMP-dependent protein kinase with bound ATP and a segment of an inhibitory peptide (green) blocking the substrate site. The small ball is an alanine side chain; it would be Ser or Thr in a substrate. The ATP is clamped by the glycine-rich loop at the top of the figure. Hydrogen bonds from peptide NH groups bind the β and γ phospho groups of ATP. A magnesium ion (not shown) also binds to the β and γ phosphate oxygens and to the invariant Asp 184. The Asp 166 carboxylate is probably the catalytic base for deprotonation of the substrate –OH. From Grant *et al.*[896] Courtesy of Susan S. Taylor.

For the enzyme to be activated the phosphothreonine must form a hydrogen-bonded ion pair with Lys 189 and be hydrogen bonded to His 87, tying together critical regions of the catalytic domain. The regulatory subunits, unless occupied by cAMP (Chapter 11) are competitive inhibitors of the substrates.[900] Cyclic GMP-activated kinases also have distinct functions.[901] In **tryosine kinases**[902–904] C-terminal src-homology domains (Fig. 7-30 and Chapter 11) fold over and interact in an inhibitory fashion until an appropriate activating signal is received (see Fig. 11-13).

A relative of the kinases is **adenylate cyclase**, whose role in forming the allosteric effector 3',5'-cyclic AMP (cAMP) was considered in Chapter 11. This enzyme catalyzes a displacement on P_α of ATP by the 3'-hydroxyl group of its ribose ring (see Eq. 11-8, step *a*). The structure of the active site is known.[905] Studies with ATPαS suggest an in-line mechanism resembling that of ribonuclease (step *a*, Eq. 12-25). However, it is Mg^{2+} dependent, does not utilize the two-histidine mechanism of ribonuclease A, and involves an aspartate carboxylate as catalytic base.[906] All isoforms of adenylate cyclase are activated by the α subunits of some G proteins (Chapter 11). The structures[907] of $G_{s\alpha}$ and of its complex with adenylate kinase[905] have been determined. The $G_{s\alpha}$ activator appears to serve as an allosteric effector.

Guanylate cyclases, which form cyclic GMP, occur in particulate and soluble forms.[908] The latter have been of great interest because they are activated by nitric oxide (NO). The soluble guanylate cyclases are αβ heterodimers. The C-terminal regions of both α and β subunits are homologous to the catalytic domain of adenylate cyclase. The N-terminal domain of the α subunits contains heme whose Fe atom is coordinated by a histidine imidazole.[908,908a] This iron atom is apparently the receptor for NO, a major gaseous hormone, which is discussed in Chapter 18.

10. Nucleotidyl Transferases

An important group of enzymes transfer substituted phospho groups, most often nucleotidyl groups. The nucleases, ATPases, and GTPases, which have already been discussed, belong to this group as do the nucleic acid synthesizing enzymes, the **DNA** and **RNA polymerases**,[909–911] **reverse transcriptase**,[912] and **topoisomerases**. As with other phosphotransferases, the nucleotidyl transfers occur with inversion[913,913a] and crystallographic investigations also support in-line mechanisms as illustrated in the following scheme. Two metal ions assist.

The inorganic pyrophosphate formed is hydrolyzed to inorganic phosphate by pyrophosphatase. Specific information about the polymerases and topoisomerases is given in Chapters 27 and 28.

E. The Adenylate Kinase Fold, the P Loop, and ATPases and GTPases

A magnesium–ATP-binding fragment consisting of ~40 residues at the N terminus of adenylate kinase contains sequences homologous to those in the GTP-binding "G proteins" such as the protooncogenes *ras* (Fig. 11-7A) and also to sequences in myosin and in mitochondrial ATP synthase.[914] This includes the glycine-rich "P loop" which extends from Gly 15 to Gly 22 in the porcine cytosolic enzyme and contains a highly conserved lysine [Lys 21 of porcine adenylate

---

BOX 12-F THE TOXICITY OF ALUMINUM

Aluminum, in the form of oxides and silicate minerals, is the most abundant element in the earth's crust. Yet it appears to be actively excluded from most living organisms. It seems surprising that a naturally occuring Al^{3+}-dependent enzyme hasn't been found, but there is no evidence that Al is an essential element. Until recently, it was usually regarded as harmless. The aluminum salts known as alums are used in baking powders and have been added to pickles and other foods. Aluminum sulfate is often used as a coagulant to clarify turbid drinking water. Because of the insolubility of $Al(OH)_3$ and of aluminum phosphate the concentration of Al^{3+} is very low at neutral pH. However, small amounts of $AlOH^{2+}$, $Al(OH)_2{}^+$, and $Al(OH)_4{}^-$ are present in water. Fluoride complexes such as AlF^{2+} may also be present. Soluble complexes, such as that with citrate, may permit some Al^{3+} to be absorbed by the body.[a,b]

The toxicity of aluminum has been recognized most clearly by the development of bone disease caused by deposition of Al in bones of patients on hemodialysis[a–c] and in infants on intravenous therapy.[d,e] Excessive Al in the water used for dialysis may also cause brain damage. Dietary aluminum may be one cause of Alzheimer's disease,[f–h] but this is controversial as is a possible role of aluminum in vaccines in causing inflammation in muscle.[i,j] Solubilization of soil aluminum by acid rain has been blamed for the decline of forests in Europe and North America,[j] for the death of fish in acid waters,[k] and for very large reductions in yield for many crops.[l,m] An aluminum-resistant strain of buckwheat makes and secretes from its roots large amounts of oxalate which binds and detoxifies the Al^{3+} ions.[m]

Al^{3+} has a radius somewhat less than that of Fe^{3+} (Table 6-10) and it may sometimes occupy empty Fe^{3+} binding sites. Thus, the transferrin[n] in blood carries some Al^{3+}, although citrate is probably a more important carrier.[o] Al^{3+} binds preferentially to oxygen ligands and can compete with Mg^{2+}. However, the slower rate of ligand exchange reactions with Al^{3+} may interfere with the proper functioning of the metal. Brain hexokinase is strongly inhibited[p] by Al^{3+} and the binding of Al^{3+} to tubulin decreases the rates of GTP hydrolysis and of Ca^{2+}-induced depolymerization of microtubules.[q] Aluminofluoride ions, such as $AlF_4{}^-$, react with phosphates to form ions such as,

$$R-O-\overset{\displaystyle O}{\underset{\displaystyle O}{\overset{\|}{\underset{\|}{P}}}}-O-AlF_3{}^-$$

which may be potent competitive inhibitors of enzymes acting on ATP, GTP, or other phosphate-containing substrate.[r,s] However, Fe^{3+} can be replaced by Al^{3+} in a purple acid phosphatase (Chapter 16) with retention of good catalytic activity.[t]

[a] Martin, R. B. (1986) *Clinical Chemistry* **32**, 1797–1806
[b] Macdonald, T. L., and Martin, R. B. (1988) *Trends Biochem. Sci.* **13**, 15–19
[c] Andress, D. L., Kopp, J. B., Maloney, N. A., Coburn, J. W., and Sherrard, D. J. (1987) *N. Engl. J. Med.* **316**, 292–296
[d] Sedman, A. B., Klein, G. L., Merritt, R. J., Miller, N. L., Weber, K. O., Gill, W. L., Anand, H., and Alfrey, A. C. (1985) *N. Engl. J. Med.* **312**, 1337–1342
[e] Bishop, N. J., Morley, R., Chir, B., Day, J. P., and Lucas, A. (1997) *N. Engl. J. Med.* **336**, 1557–1561
[f] Good, P. F., and Perl, D. P. (1993) *Nature (London)* **362**, 418
[g] Shen, Z. M., Perczel, A., Hollósi, M., Nagypál, I., and Fasman, G. D. (1994) *Biochemistry* **33**, 9627–9636
[h] Walker, P. R., LeBlanc, J., and Sikorska, M. (1989) *Biochemistry* **28**, 3911–3915
[i] Landsberg, J. P., McDonald, B., and Watt, F. (1992) *Nature (London)* **360**, 65–68
[ij] Malakoff, D. (2000) *Science* **288**, 1323–1324
[j] Godbold, D. L., Fritz, E., and Hüttermann, A. (1988) *Proc. Natl. Acad. Sci. U.S.A.* **85**, 3888–3892
[k] Birchall, J. D., Exley, C., Chappell, J. S., and Phillips, M. J. (1989) *Nature (London)* **338**, 146–148
[l] Barinaga, M. (1997) *Science* **276**, 1497
[m] Ma, J. F., Zheng, S. J., Matsumoto, H., and Hiradate, S. (1997) *Nature (London)* **390**, 569–570
[n] Roskams, A. J., and Connor, J. R. (1990) *Proc. Natl. Acad. Sci. U.S.A.* **87**, 9024–9027
[o] Martin, R. B., Savory, J., Brown, S., Bertholf, R. L., and Wills, M. R. (1987) *Clinical Chemistry* **33**, 405–407
[p] Viola, R. E., Morrison, J. F., and Cleland, W. W. (1987) *Biochemistry* **19**, 3131–3137
[q] Macdonald, T. L., Humphreys, W. G., and Martin, R. B. (1987) *Science* **236**, 183–186
[r] Troullier, A., Girardet, J.-L., and Dupont, Y. (1992) *J. Biol. Chem.* **267**, 22821–22829
[s] Chabre, M. (1990) *Trends Biochem. Sci.* **15**, 6–10
[t] Merkx, M., and Averill, B. A. (1999) *J. Am. Chem. Soc.* **121**, 6683–6689

kinase and Lys 17 of the archaeal enzyme (Fig. 12-30)] and of the *ras* oncogene product).[915] The lysine side chain appears to interact with the β and / or γ phospho group of ATP or GTP. The peptide chain of the P loop wraps around the β phospho group as the chain turns from the first β strand into the first helix. This can be seen in Fig. 12-30, in which the P loop surrounds the β phospho group of ADP. Two peptide NH groups also bind to the β phospho group on a side opposite to that occupied by Mg^{2+}. A similar loop in the dinucleotide-binding domain of a dehydrogenase can be seen hydrogen bonded to the P_α and P_β phospho groups of NAD^+ in Fig. 2-13. Consensus sequences for three groups of glycine-rich loops are:[896]

Dinucleotide-binding	G X G X X G
P loop	G X X X X G K (S/T)
Protein kinase	G X G X X G X V

The protein kinase loop is seen at the top of Fig. 12-32 and extends from Gly 50 to Val 57. These conserved loops help to hold the ATP in place and to orient it correctly. Do they have any other significance? The answer is not clear. These glycine-rich loops fold across the β–γ diphosphate linkage that is broken when ATP or GTP is hydrolyzed. Cleavage of both of these molecules is associated with movement. Kinases close around ATP, and parts of G proteins (Fig. 11-7) move when these regulatory devices function.[915,916] Cleavage of ATP causes movement in the ATPase heads of muscle myosin, providing the force for muscular contraction (Chapter 19). Somewhat the opposite occurs in mitochondrial ATP synthase (Chapter 18) when movement in the synthase heads snaps ADP and inorganic phosphate together to form ATP. Common to all of these processes is a movement of charge within the ATP (or its cleavage products) from the attack nucleophile into the neighboring phospho group, as is indicated by the arrows in the following diagram.

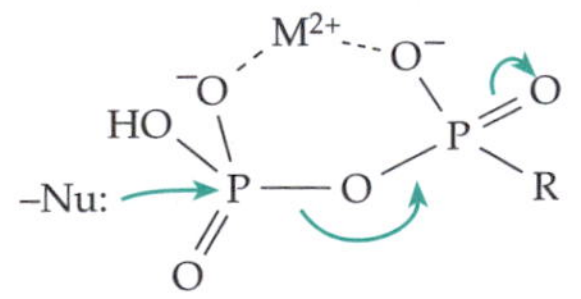

Not shown in this diagram is an accompanying flow of positive charge which may include movement of a metal ion, addition and loss of protons, and which may induce conformational changes.[914] The latter are essential not only to muscle contraction and ATP synthesis but also to many other processes that depend upon the Gibbs energy of cleavage of ATP and related compounds. This includes the pumping of ions against concentration gradients (Chapter 8), the action of **topoisomerases** (Chapter 27) which function to alter the supercoiling in DNA, and the functioning of the phosphotransferase system by which sugars and other compounds are brought into bacterial cells (Chapter 8).

F. Displacements on Sulfur Atoms

Nucleophilic displacement reactions occur on sulfur atoms in various oxidation states. A common reaction is **thiol–disulfide** exchange (Eqs. 10-9, 11-7), a reaction in which a nucleophilic thiolate anion attacks one of the sulfur atoms of a disulfide. Proteins such as **thioredoxin** of *E. coli* and **thioltransferases** (Box 15-C), which contain internal disulfide bridges, can be reduced by disulfide exchange with thiols such as glutathione (Box 11-B). The reduced proteins may then undergo similar exchange reactions that cleave disulfide linkages in other molecules. An example is glutathione reductase (Fig. 15-12). Thioltransferases may also serve as **protein disulfide isomerases** (Chapter 10, Section D,3). Nucleophilic displacements on sulfur or on selenium atoms are steps in a variety of enzymatic reactions. Among these are glutathione peroxidase (Eq. 15-59) and thiosulfate: cyanide sulfo-transferase (Eq. 24-45).

While esters of sulfuric acid do not play as central a role in metabolism as do phosphate esters, they occur widely. Both oxygen esters ($R-O-SO_3^-$, often referred to as **O-sulfates**) and derivatives of sulfamic acid ($R-NH-SO_3^-$, **N-sulfates**) are found, the latter occurring in mucopolysaccharides such as heparin. Sulfate esters of mucopolysaccharides and of steroids are ubiquitous and sulfation is the most abundant known modification of tyrosine side chains. Choline sulfate and ascorbic acid 2-sulfate are also found in cells. Sulfate esters of phenols and many other organic sulfates are present in urine.

Sulfotransferases[917–920a] transfer sulfo groups to O and N atoms of suitable acceptors (reaction type 1D, Table 10-1). Usually, transfer is from the "active sulfate," **3'-phosphoadenosine 5'-phosphosulfate** (**PAPS**),[921] whose formation is depicted in Eq. 17-38. **Sulfatases** catalyze hydrolysis of sulfate esters. The importance of such enzymes is demonstrated by the genetic **mucopolysaccharidoses**. In four of these disease-specific sulfatases that act on iduronate sulfate, heparan *N*-sulfate, galactose-6-sulfate, or *N*-acetylglucosamine-4-sulfate are absent. Some of these, such as heparan *N*-sulfatase deficiency, lead to severe mental retardation, some cause serious skeletal abnormalities, while others are mild in their effects.[922]

Sulfatases are unusual in having a residue of **formylglycine** at the active site. This is generated oxidatively from cysteine in human enzymes[923,923a] and from serine in some bacterial sulfatases.[924,924a] Absence of this modification results in a multiple sulfatase deficiency disease. A probable mechanism of sulfatase based on a crystal structure determination is given by Eq. 12-44.[923]

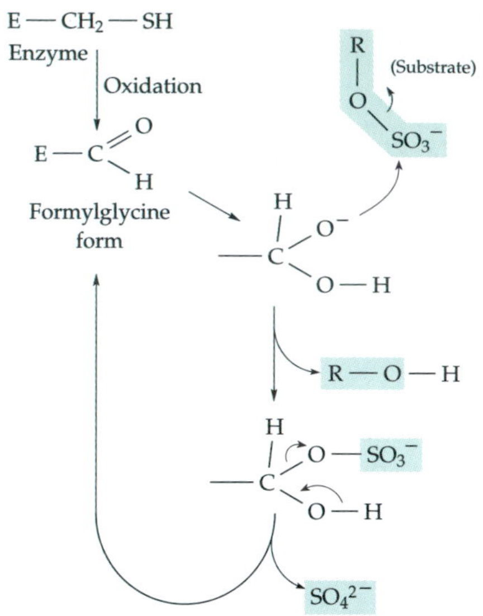

(12-44)

G. Multiple Displacement Reactions and the Coupling of ATP Cleavage to Endergonic Processes

A combination of successive displacement reactions of two types is required in many enzymatic reactions, including most of those by which the cleavage of ATP is coupled to biosynthesis. To harness the group transfer potential of ATP to drive an endergonic metabolic process there must be a mechanism of **coupling**. Otherwise, hydrolysis of ATP within a cell would simply generate heat. *An essential part of the coupling mechanism usually consists of a nucleophilic displacement on phosphorus followed by displacement on carbon.* Likewise, the synthesis of ATP and related compounds often begins with a displacement on carbon followed by one on phosphorus.

1. Transfer of Phospho, Pyrophospho, and Adenylyl Groups from ATP

The first step in coupling ATP cleavage to any process is transfer of part of the ATP molecule to a nucleophile Y, usually by displacement on one of the three phosphorus atoms. The nucleophilic attack may be (*a*) on the terminal phosphorus (P_γ) with displacement of ADP or (*b*) on the internal phosphorus (P_α) with displacement of inorganic pyrophosphate. In the first case, $Y-PO_3H^-$ is formed; in the latter, **Y-adenylyl** (sometimes shortened to Y-adenyl) is formed. More rarely, displacement occurs (*c*) on the central phosphorus (P_β) with transfer of a pyrophospho group to the

nucleophile. Still less frequent (*d*) is a displacement on C-5' as shown in Eq. 17-37. If the nucleophile Y in any of these displacement reactions is H_2O, the resulting hydrolysis tends to go to completion, i.e., the phospho, adenylyl, and pyrophospho groups of ATP all have high group transfer potentials (Table 6-6). If Y is an –OH group in an ordinary alcohol the transfer reaction also tends to go to completion because the group transfer potential of a simple phosphate ester is relatively low. Consequently, phosphorylation by ATP is often used as a means of introducing an essentially irreversible step in a metabolic pathway.

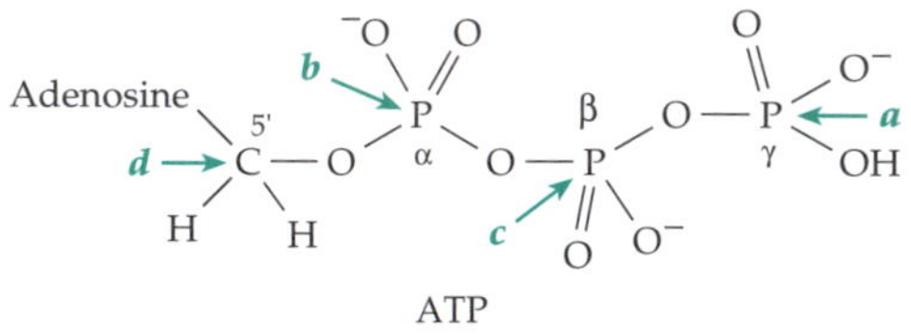

ATP

2. Acyl Phosphates

Transfer of a phospho or adenylyl group from ATP to the oxygen atom of a carboxylate group yields an **acyl phosphate**, a type of metabolic intermediate of special significance. Acyl phosphates are mixed anhydrides of carboxylic and phosphoric acids in which *both the acyl group and the phospho group have high group*

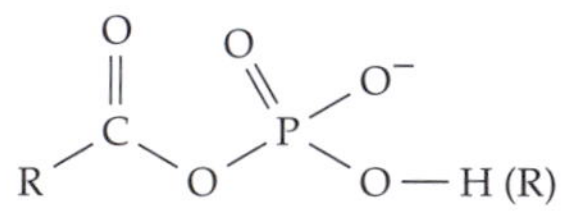

Acyl phosphate

transfer potentials. As a consequence, acyl phosphates can serve as metabolic intermediates through which the group transfer potential of ATP is transferred into other molecules and is harnessed to do chemical work.

A typical example is the synthesis of acetyl coenzyme A (Eq. 12-45). See Fig. 14-1 for the complete structure of –SH group-containing coenzyme A.

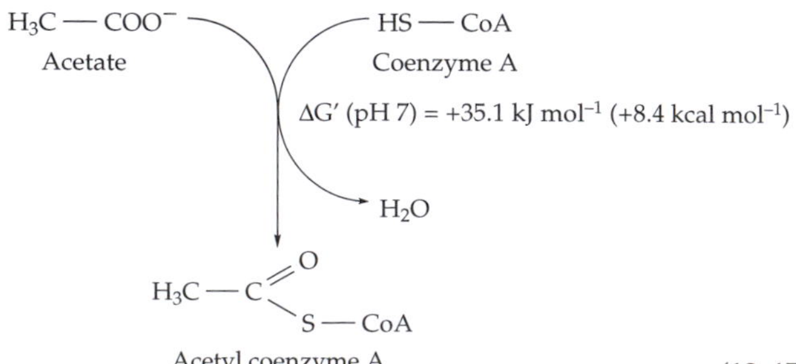

(12-45)

Because the acetyl group in the product also has a high group transfer potential, $\Delta G'$ for this reaction is highly positive and the formation of acetyl-CoA will not occur spontaneously. However, the sum of $\Delta G'$ for the reaction in Eq.12-45 plus that for the hydrolysis of ATP (Eq. 12-46) is nearly zero ($+0.6$ kJ mol^{-1}).

$$ATP^{4-} + H_2O \rightarrow ADP^{3-} + HPO_4^{2-} + H^+$$
$$\Delta G' = -34.5 \text{ kJ mol}^{-1} \ (-8.3 \text{ kcal mol}^{-1}) \quad (12\text{-}46)$$

Coupling of the two reactions is accomplished by first letting an oxygen atom of the nucleophilic carboxylate group attack P_γ of ATP to form acetyl phosphate (Eq. 12-47, step a). In the second step (step b) the sulfur atom of the –SH group of coenzyme A (often abbreviated CoA–SH) attacks the carbon atom of the acetyl phosphate with displacement of the good leaving group P_i. While $\Delta G'$ for Eq. 12-47, step a, is moderately positive (meaning that a relatively low concentration of acetyl phosphate will accumulate unless the [ATP]/[ADP] ratio is high), the equilibrium in Eq. 12-47, step b, favors the products.

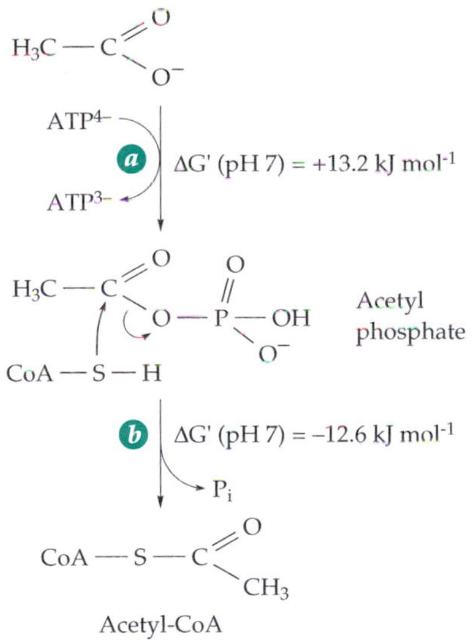

$$ (12\text{-}47) $$

The two reactions of Eq. 12-47 are catalyzed by **acetate kinase**[925] and an **S-acetyltransferase**, respectively. The sequence represents an essential first stage in bacterial utilization of acetate for growth. It is also used in a few bacteria in reverse as a way of *generating* ATP in fermentation reactions. On the other hand, most eukaryotic cells make acetyl-CoA from acetate by coupling the synthesis to cleavage of ATP to AMP and P_i. A single enzyme **acetyl-CoA synthetase** (acetate thiokinase) catalyzes both steps in the reaction (Eq. 10-1). The sequence parallels that of Eq. 12-47, but the initial displacement is on P_α of ATP to form **acetyl adenylate**. This intermediate remains tightly bound

to the enzyme until the second step in the sequence takes place. When ^{18}O is present in the acetate (designated by the asterisks in Eq. 12-48) it appears in the phospho group of AMP as expected for the indicated mechanism.

3. General Mechanism of Formation of Thioesters, Esters, and Amides

The sequences of Eqs. 12-47 and 12-48 are general ones used by cells for linking carboxylic acids to –OH, –SH, and –NH$_2$ groups to form oxygen esters, thioesters, or amides, respectively. ATP can be cleaved at either P_α or P_γ. If cleavage is at P_α (Eq. 12-48) the hydrolysis of inorganic pyrophosphate (PP$_i$) to P$_i$ provides an additional coupling of ATP cleavage to biosynthesis as is discussed in Chapter 17, Section H.

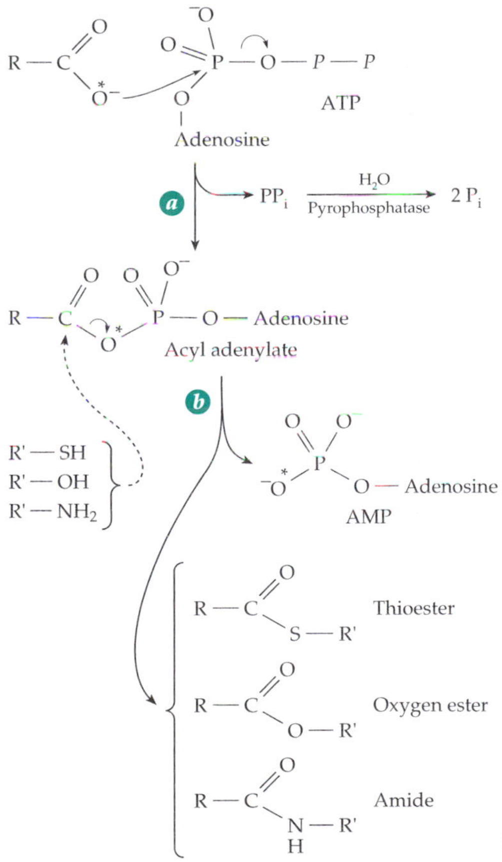

$$ (12\text{-}48) $$

An enzyme catalyzing a reaction similar to that of acetyl-CoA synthetase is **succinyl-CoA synthetase** (succinate thiokinase).[926,927] The enzyme from *E. coli* has been studied most. The first step is formation of a phosphoenzyme by transfer of the γ phospho group from ATP to N$^\epsilon$ of histidine 246 in the α subunit of the 140-kDa $\alpha_2\beta_2$ tetramer.[926] The phospho group is then transferred to succinate to form succinyl phosphate, which reacts with coenzyme A, as in step b, Eq. 12-48, to form succinyl-CoA. Crystallographic studies suggest that for this step the succinyl phosphate and the coenzyme A may be bound to opposite α subunits in the tetramer.[926] However, in some bacteria and in eukaryotes the enzyme appears to operate as an αβ heterodimer.[927]

Glutamine synthetase,[928,929] a large enzyme containing 12 identical 468-residue subunits with 622 symmetry (as in Fig. 7-12), has a major regulatory function in nitrogen metabolism, which is discussed in Chapter 24. Apparently, the intermediate acyl phosphate (γ-glutamyl phosphate) has a transient existence and all three reactants–glutamate, NH_4^+, and ATP–must be bound to the enzyme concurrently before the active site becomes functional. Early evidence for the acyl phosphate[930,931] included reduction by sodium borohydride to an alcohol (Eq. 12-49a) and isolation of the internal amide of glutamic acid **5-oxoproline** (Eq. 12-49b).

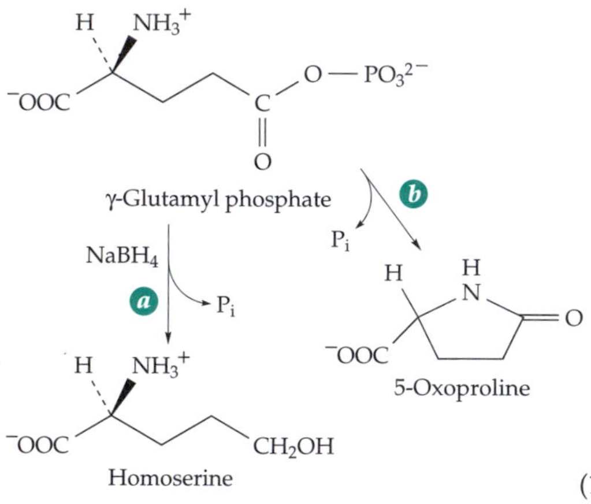

γ-Glutamyl phosphate

NaBH₄ → P_i

5-Oxoproline

Homoserine

(12-49)

Ammonium ions appear to be bound at a specific site and to be deprotonated by the Asp 50 carboxylate.[929]

The **aminoacyl-tRNA synthetases** join amino acids to their appropriate tRNA molecules for protein synthesis. They have the very important task of selecting both a specific amino acid and a specific tRNA and joining them. The enzymes differ in size and other properties. However, they all appear to function by a common basic chemistry that makes use of cleavage of ATP at P_α (Eq. 12-48) via an intermediate aminoacyl adenylate and that is outlined also in Eq. 17-36. These enzymes are discussed in Chapter 29 .

4. Coenzyme A Transferases

The following problem in energy transfer arises occasionally: A thioester, such as succinyl-CoA, is available to a cell and the energy available in its unstable linkage is needed for synthesis of a different thioester. It would be possible for a cell to first form ATP or GTP,

using a synthetase reaction in reverse; then the ATP or GTP formed could be used to make the new linkage by the action of another acyl-CoA synthetase. However, special enzymes, the **CoA transferases**, function more directly (Eq. 12-50). The mechanism is not obvious.

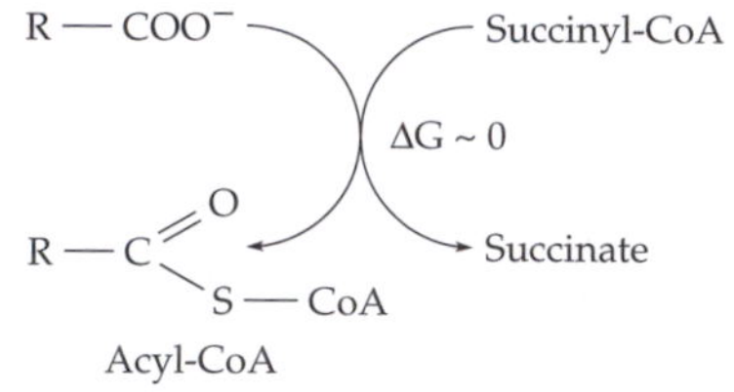

(12-50)

How can the CoA be transferred from one acyl group to another while still retaining the high group transfer potential of the acyl group?

The following experiments shed some light. Kinetic studies of **succinyl-CoA–acetoacetate CoA transferase** indicate a ping-pong mechanism. The enzyme alternates between two distinct forms, one of which has been shown to contain bound CoA.[932–934] The E-CoA intermediate formed from enzyme plus acetoacetyl-CoA was reduced with ³H-containing sodium borohydride and the protein was completely hydrolyzed with HCl. Tritium-containing α-amino-δ-hydroxyvaleric acid was isolated. Since thioesters (as well as oxygen esters) are cleaved in a two-step process

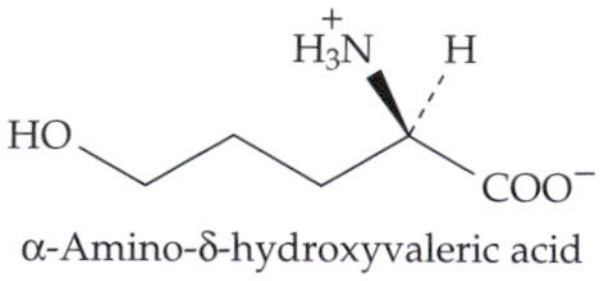

α-Amino-δ-hydroxyvaleric acid

to alcohols by reduction with borohydride, it was concluded that the intermediate E-CoA is a thioester of the Glu 344 side chain. In exchange reactions ¹⁸O from labeled succinate entered both the E-CoA intermediate and the carboxyl group of acetoacetate.

A mechanism involving formation of a transient anhydride is similar to reactions discussed in preceding sections.[717,720] The student should be able to write out the step-by-step detail. Does this mechanism explain the ¹⁸O exchange data? A second possibility is a **4-center reaction** (Eq. 12-51). However, mechanisms of this type have not been demonstrated for enzymatic reactions.

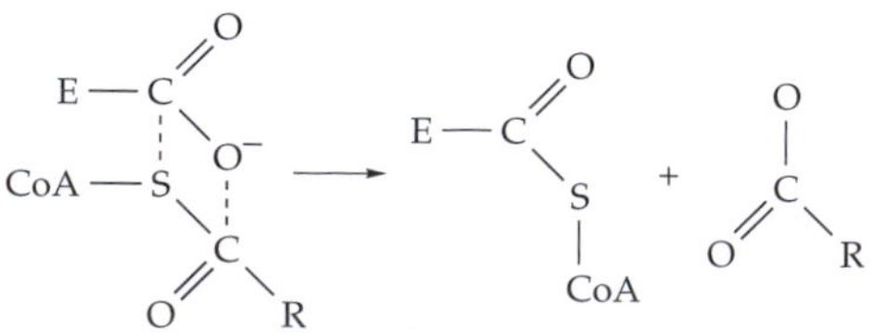

(12-51)

References

1. Jencks, W. P. (1987) *Catalysis in Chemistry and Enzymology*, Dover, Mineola, New York (pp. 78–110)

2. Lowe, J. N., and Ingraham, L. L. (1974) *An Introduction to Biochemical Reaction Mechanisms*, Prentice-Hall, Englewood Cliffs, New Jersey

3. Bruice, T. C., and Benkovic, S. J. (1966) *Bioorganic Mechanisms*, Benjamin, New York (2 vols.)

3a. Kyte, J. (1995) *Mechanism in Protein Chemistry*, Garland Publ., New York

3b. Fersht, A. (1999) *Structure and Mechanism in Protein Science: A Guide to Enzyme Catalysis and Protein Folding*, Freeman, New York

3c. Silverman, R. B. (1999) *The Organic Chemistry of Enzyme-Catalyzed Reactions*, Academic Press, San Diego, California

3d. Sinnott, M., ed. (1998) *Comprehensive Biological Catalysis. A Mechanistic Reference*, Vol. I, Academic Press, San Diego, California

4. Dixon, J. E., and Bruice, T. C. (1971) *J. Am. Chem. Soc.* **93**, 6592–6597

5. Kosower, E. M. (1968) *An Introduction to Physical Organic Chemistry*, Wiley, New York (pp. 28 and 81)

6. Goldman, P., Milne, G. W. A., and Keister, D. B. (1968) *J. Biol. Chem.* **243**, 428–434

7. Liu, J.-Q., Kurihara, T., Miyagi, M., Esaki, N., and Soda, K. (1995) *J. Biol. Chem.* **270**, 18309–18312

8. Hisano, T., Hata, Y., Fujii, T., Liu, J.-Q., Kurihara, T., Esaki, N., and Soda, K. (1996) *J. Biol. Chem.* **271**, 20322–20330

9. Li, Y.-F., Hata, Y., Fujii, T., Hisano, T., Nishihara, M., Kurihara, T., and Esaki, N. (1998) *J. Biol. Chem.* **273**, 15035–15044

10. Grimmelikhuijzen, C. J. P., and Schaller, H. C. (1979) *Trends Biochem. Sci.* **4**, 265–267

11. Koonin, E. V., and Tatusov, R. L. (1994) *J. Mol. Biol.* **244**, 125–132

11a. Nardini, M., Ridder, I. S., Rozeboom, H. J., Kalk, K. H., Rink, R., Janssen, D. B., and Dijkstra, B. W. (1999) *J. Biol. Chem.* **274**, 14579–14586

12. Verschueren, K. H. G., Seljée, F., Rozeboom, H. J., Kalk, K. H., and Dijkstra, B. W. (1993) *Nature (London)* **363**, 693–698

12a. Ridder, I. S., Rozeboom, H. J., Kalk, K. H., and Dijkstra, B. W. (1999) *J. Biol. Chem.* **274**, 30672–30678

13. Pries, F., Kingma, J., Krooshof, G. H., Jeronimus-Stratingh, C. M., Bruins, A. P., and Janssen, D. B. (1995) *J. Biol. Chem.* **270**, 10405–10411

13a. Schindler, J. F., Naranjo, P. A., Honaberger, D. A., Chang, C.-H., Brainard, J. R., Vanderberg, L. A., and Unkefer, C. J. (1999) *Biochemistry* **38**, 5772–5778

14. Lightstone, F. C., Zheng, Y.-J., Maulitz, A. H., and Bruice, T. C. (1997) *Proc. Natl. Acad. Sci. U.S.A.* **94**, 8417–8420

15. Lightstone, F. C., Zheng, Y.-J., and Bruice, T. C. (1998) *J. Am. Chem. Soc.* **120**, 5611–5621

16. Pries, F., van den Wijngaard, A. J., Bos, R., Pentenga, M., and Janssen, D. B. (1994) *J. Biol. Chem.* **269**, 17490–17494

17. Chiang, P. K., Gordon, R. K., Tal, J., Zeng, G. C., Doctor, B. P., Pardhasaradhi, K., and McCann, P. P. (1996) *FASEB J.* **10**, 471–480

18. Cornforth, J. W., Reichard, S. A., Talalay, P., Carrell, H. L., and Glusker, J. P. (1977) *J. Am. Chem. Soc.* **99**, 7292–7301

19. Woodard, R. W., Tsai, M.-D., Floss, H. G., Crooks, P. A., and Coward, J. K. (1980) *J. Biol. Chem.* **255**, 9124–9127

20. Vidgren, J., Svensson, L. A., and Liljas, A. (1994) *Nature (London)* **368**, 354–357

21. Fu, Z., Hu, Y., Konishi, K., Takata, Y., Ogawa, H., Gomi, T., Fujioka, M., and Takusagawa, F. (1996) *Biochemistry* **35**, 11985–11993

22. Takata, Y., Konishi, K., Gomi, T., and Fujioka, M. (1994) *J. Biol. Chem.* **269**, 5537–5542

23. Schluckebier, G., O'Gara, M., Saenger, W., and Cheng, X. (1995) *J. Mol. Biol.* **247**, 16–20

24. Szczelkun, M. D., Jones, H., and Connolly, B. A. (1995) *Biochemistry* **34**, 10734–10743

25. O'Gara, M., Roberts, R. J., and Cheng, X. (1996) *J. Mol. Biol.* **263**, 597–606

25a. Reddy, Y. V. R., and Rao, D. N. (2000) *J. Mol. Biol.* **298**, 597–610

26. Patel, D. J. (1994) *Nature (London)* **367**, 688–690

27. Jeltsch, A., Christ, F., Fatemi, M., and Roth, M. (1999) *J. Biol. Chem.* **274**, 19538–19544

28. Lotta, T., Vidgren, J., Tilgmann, C., Ulmanen, I., Melén, K., Julkunen, I., and Taskinen, J. (1995) *Biochemistry* **34**, 4202–4210

29. Zheng, Y.-J., and Bruice, T. C. (1997) *J. Am. Chem. Soc.* **119**, 8137–8145

29a. Kahn, K., and Bruice, T. C. (2000) *J. Am. Chem. Soc.* **122**, 46–51

30. Kealey, J. T., and Santi, D. V. (1995) *Biochemistry* **34**, 2441–2446

31. Hrycyna, C. A., Yang, M. C., and Clarke, S. (1994) *Biochemistry* **33**, 9806–9812

31a. Cleland, W. W., O'Leary, M. H., and Northrop, D. B., eds. (1977) *Isotope Effects on Enzyme-Catalyzed Reactions*, Univ. Park Press, Baltimore, Maryland

31b. Klinman, J. P. (1978) in *Transition States of Biochemical Processes* (Gandour, R. D., and Schowen, R. L., eds), pp. 165–200, Plenum, New York

31c. Gandour, R. D., and Schowen, R. L., eds. (1978) *Transition States of Biochemical Processes*, Plenum, New York

31d. Schramm, V. L., Horenstein, B. A., and Kline, P. C. (1994) *J. Biol. Chem.* **269**, 18259–18262

32. Hegazi, M. F., Borchardt, R. T., and Schowen, R. L. (1979) *J. Am. Chem. Soc.* **101**, 4359–4365

33. Mihel, I., Knipe, J. O., Coward, J. K., and Schowen, R. L. (1979) *J. Am. Chem. Soc.* **101**, 4349–4351

34. Rodgers, J., Femec, D. A., and Schowen, R. L. (1982) *J. Am. Chem. Soc.* **104**, 3263–3268

35. Boyd, R. J., Kim, C.-K., Shi, Z., Weinberg, N., and Wolfe, S. (1993) *J. Am. Chem. Soc.* **115**, 10147–10152

36. Glad, S. S., and Jensen, F. (1997) *J. Am. Chem. Soc.* **119**, 227–232

37. Wolfenden, R., Lu, X., and Young, G. (1998) *J. Am. Chem. Soc.* **120**, 6814–6815

38. Henrissat, B., and Bairoch, A. (1993) *Biochem. J.* **293**, 781–788

39. Henrissat, B., and Bairoch, A. (1996) *Biochem. J.* **316**, 695–696

40. Henrissat, B., Callebaut, I., Fabrega, S., Lehn, P., Mornon, J.-P., and Davies, G. (1995) *Proc. Natl. Acad. Sci. U.S.A.* **92**, 7090–7094

41. Sinnot, M. L. (1990) *Chem. Rev.* **90**, 1171–1202

42. Withers, S. G., and Aebersold, R. (1995) *Protein Sci.* **4**, 361–372

43. Silverstein, R., Voet, J., Reed, D., and Abeles, R. H. (1967) *J. Biol. Chem.* **242**, 1338–1346

44. Doudoroff, M., Barker, H. A., and Hassid, W. Z. (1947) *J. Biol. Chem.* **168**, 725–732 and 733–746

45. LeBrun, E., and van Rapenbusch, R. (1980) *J. Biol. Chem.* **255**, 12034–12036

46. McCarter, J. D., and Withers, S. G. (1996) *J. Biol. Chem.* **271**, 6889–6894

47. Withers, S. G., Rupitz, K., and Street, I. P. (1988) *J. Biol. Chem.* **17**, 7929–7932

48. Withers, S. G., and Street, I. P. (1988) *J. Am. Chem. Soc.* **110**, 8551–8553

49. Miao, S., Ziser, L., Aebersold, R., and Withers, S. G. (1994) *Biochemistry* **33**, 7027–7032

50. Braun, C., Lindhorst, T., Madsen, N. B., and Withers, S. G. (1996) *Biochemistry* **35**, 5458–5463

50a. Namchuk, M. N., McCarter, J. D., Becalski, A., Andrews, T., and Withers, S. G. (2000) *J. Am. Chem. Soc.* **122**, 1270–1277

51. Gorenstein, D. G., Findlay, J. B., Luxon, B. A., and Kar, D. (1977) *J. Am. Chem. Soc.* **99**, 3473–3479

52. Kirby, A. J. (1983) *The Anomeric Effect and Related Stereoelectronic Effects at Oxygen*, Springer, Berlin

53. Bennet, A. J., and Sinnott, M. L. (1986) *J. Am. Chem. Soc.* **108**, 7287–7294

54. Sinnott, M. L. (1986) in *Mechanisms of Enzymatic Reactions: Stereochemistry A. Steenbock Symposium* (Frey, P. A., ed), pp. 293–305, Elsevier, New York

55. Phillips, D. C. (1966) *Sci. Am.* **215** (Nov), 78–90

56. Sanz, J. M., García, P., and García, J. L. (1992) *Biochemistry* **31**, 8495–8499

57. Artymicek, P. J., and Blake, C. C. F. (1981) *J. Mol. Biol.* **152**, 737–762

58. Harata, K., Muraki, M., and Jigami, Y. (1993) *J. Mol. Biol.* **233**, 524–535

59. Harata, K., Muraki, M., Hayashi, Y., and Jigami, Y. (1992) *Protein Sci.* **1**, 1447–1453

60. Grütter, M. G., Weaver, L. H., and Matthews, B. W. (1983) *Nature (London)* **303**, 828–831

61. Zhang, X.-J., Baase, W. A., and Matthews, B. W. (1991) *Biochemistry* **30**, 2012–2017

62. Dao-pin, S., Sauer, U., Nicholson, H., and Matthews, B. W. (1991) *Biochemistry* **30**, 7142–7153

63. Matsumura, M., Becktel, W. J., Levitt, M., and Matthews, B. W. (1989) *Proc. Natl. Acad. Sci. U.S.A.* **86**, 6562–6566

64. Parsons, S. M., and Raftery, M. A. (1972) *Biochemistry* **11**, 1623–1628

65. Maenaka, K., Matsushima, M., Song, H., Sunada, F., Watanabe, K., and Kumagai, I. (1995) *J. Mol. Biol.* **247**, 281–293

66. Matsumura, I., and Kirsch, J. F. (1996) *Biochemistry* **35**, 1881–1889

67. Kuroki, R., Yamada, H., Moriyama, T., and Imoto, T. (1986) *J. Biol. Chem.* **261**, 13571–13574

68. Malcolm, B. A., Rosenberg, S., Corey, M. J., Allen, J. S., de Baetselier, A., and Kirsch, J. F. (1989) *Proc. Natl. Acad. Sci. U.S.A.* **86**, 133–137

69. Hadfield, A. T., Harvey, D. J., Archer, D. B., MacKenzie, da, Jeenes, D. J., Radford, S. E., Lowe, G., Dobson, C. M., and Jonhson, L. N. (1994) *J. Mol. Biol.* **243**, 856–872

70. Weaver, L. H., Grütter, M. G., and Matthews, B. W. (1995) *J. Mol. Biol.* **245**, 54–68

71. Kuroki, R., Ito, Y., Kato, Y., and Imoto, T. (1997) *J. Biol. Chem.* **272**, 19976–19981

72. Matsumura, I., and Kirsch, J. F. (1996) *Biochemistry* **35**, 1890–1896

73. Dickerson, R. E., and Geis, I. (1969) *The Structure and Action of Proteins*, Harper and Row, New York (p. 71)

74. Levitt, M. (1974) in *Peptides, Polypeptides and Proteins* (Blout, E. R., Bouey, F. A., Goodman, M., and Lotan, N., eds), pp. 99–113, Wiley, New York

75. Huang, X., Surry, C., Hiebert, T., and Bennet, A. J. (1995) *J. Am. Chem. Soc.* **117**, 10614–10621

76. Matsui, H., Blanchard, J. S., Brewer, C. F., and Hehre, E. J. (1989) *J. Biol. Chem.* **264**, 8714–8716

77. Banait, N. S., and Jencks, W. P. (1991) *J. Am. Chem. Soc.* **113**, 7951–7958

78. Zhang, Y., Bommuswamy, J., and Sinnott, M. L. (1994) *J. Am. Chem. Soc.* **116**, 7557–7563

79. Tanaka, Y., Tao, W., Blanchard, J. S., and Hehre, E. J. (1994) *J. Biol. Chem.* **269**, 32306–32312

80. Knier, B. L., and Jencks, W. P. (1980) *J. Am. Chem. Soc.* **102**, 6789–6798

81. Huang, X., Tanaka, K. S. E., and Bennet, A. J. (1997) *J. Am. Chem. Soc.* **119**, 11147–11154

References

82. Hardy, L. W., and Poteete, A. R. (1991) *Biochemistry* **30**, 9457–9463

82a. Davies, G. J., Mackenzie, L., Varrot, A., Dauter, M., Brzozowski, A. M., Schülein, M., and Withers, S. G. (1998) *Biochemistry* **37**, 11707–11713

82b. Vocadlo, D. J., Mayer, C., He, S., and Withers, S. G. (2000) *Biochemistry* **39**, 117–126

83. Banerjee, S. K., and Rupley, J. A. (1975) *J. Biol. Chem.* **250**, 8267–8274

84. Kelly, J. A., Sielecki, A. R., Sykes, B. D., James, M. N. G., and Phillips, D. C. (1979) *Nature (London)* **282**, 875–878

85. Metzler, D. E. (1979) *Adv. Enzymol.* **50**, 1–40

86. Dao-pin, S., Liao, D.-I., and Remington, S. J. (1989) *Proc. Natl. Acad. Sci. U.S.A.* **86**, 5361–5365

87. Post, C. B., and Karplus, M. (1986) *J. Am. Chem. Soc.* **108**, 1317–1319

88. Fife, T. H., Jaffe, S. H., and Natarajan, R. (1991) *J. Am. Chem. Soc.* **113**, 7646–7653

89. Piszkiewicz, D., and Bruice, T. C. (1968) *J. Am. Chem. Soc.* **90**, 2156–2163

90. Lowe, G., and Sheppard, G. (1968) *J. Chem. Soc. Chem. Commun.*, 529–530

91. Capon, B. (1969) *Chem. Rev.* **69**, 407–498

92. Hart, P. J., Pfluger, H. D., Monzingo, A. F., Hollis, T., and Robertus, J. D. (1995) *J. Mol. Biol.* **248**, 402–413

93. Terwissscha van Scheltinga, A. C., Armand, S., Kalk, K. H., Isogai, A., Henrissat, B., and Dijkstra, B. W. (1995) *Biochemistry* **34**, 15619–15623

94. Tews, I., Terwissscha van Scheltinga, A. C., Perrakis, A., Wilson, K. S., and Dijkstra, B. W. (1997) *J. Am. Chem. Soc.* **119**, 7954–7959

95. Brameld, K. A., and Goddard, W. A., III (1998) *J. Am. Chem. Soc.* **120**, 3571–3580

96. Klyosov, A. A. (1990) *Biochemistry* **29**, 10577–10585

97. Watanabe, H., Noda, H., Tokuda, G., and Lo, N. (1998) *Nature (London)* **394**, 330–331

98. Davies, G. J., Tolley, S. P., Henrissat, B., Hjort, C., and Schülein, M. (1995) *Biochemistry* **34**, 16210–16220

99. Damude, H. G., Withers, S. G., Kilburn, D. G., Miller, R. C., Jr., and Warren, R. A. J. (1995) *Biochemistry* **34**, 2220–2224

100. Barr, B. K., Wolfgang, D. E., Piens, K., Claeyssens, M., and Wilson, D. B. (1998) *Biochemistry* **37**, 9220–9229

101. Divne, C., Ståhlberg, J., Teeri, T. T., and Jones, T. A. (1998) *J. Mol. Biol.* **275**, 309–325

102. Barr, B. K., Hsieh, Y.-L., Ganem, B., and Wilson, D. B. (1996) *Biochemistry* **35**, 586–592

103. Choi, S. K., and Ljungdahl, L. G. (1996) *Biochemistry* **35**, 4906–4910

104. Tormo, J., Lamed, R., Chirino, A. J., Morag, E., Bayer, E. A., Shoham, Y., and Steitz, T. A. (1996) *EMBO J.* **15**, 5739–5751

105. Sulzenbacher, G., Schülein, M., and Davies, G. J. (1997) *Biochemistry* **36**, 5902–5911

106. Sulzenbacher, G., Driguez, H., Henrissat, B., Schülein, M., and Davies, G. J. (1996) *Biochemistry* **35**, 15280–15287

107. Johnson, P. E., Joshi, M. D., Tomme, P., Kilburn, D. G., and McIntosh, L. P. (1996) *Biochemistry* **35**, 14381–14394

108. Fukuda, M., Spooncer, E., Oates, J. E., Dell, A., and Klock, J. C. (1984) *J. Biol. Chem.* **259**, 10925–10935

108a. Ponyi, T., Szabó, L., Nagy, T., Orosz, L., Simpson, P. J., Williamson, M. P., and Gilbert, H. J. (2000) *Biochemistry* **39**, 985–991

109. Domínguez, R., Souchon, H., Lascombe, M.-B., and Alzari, P. M. (1996) *J. Mol. Biol.* **257**, 1042–1051

110. Roberge, M., Shareck, F., Morosoli, R., Kluepfel, D., and Dupont, C. (1997) *Biochemistry* **36**, 7769–7775

111. Juy, M., Amit, A. G., Alzari, P. M., Poljak, R. J., Claeyssens, M., Béguin, P., and Aubert, J.-P. (1992) *Nature (London)* **357**, 89–91

112. Chen, L., Fincher, G. B., and Hoj, P. B. (1993) *J. Biol. Chem.* **268**, 13318–13326

113. Chen, L., Garrett, T. P. J., Fincher, G. B., and Hoj, P. B. (1995) *J. Biol. Chem.* **270**, 8093–8101

114. Müller, J. J., Thomsen, K. K., and Heinemann, U. (1998) *J. Biol. Chem.* **273**, 3438–3446

115. Lindley, M. G., Shallenberger, R. S., and Herbert, S. M. (1976) *Food Chemistry* **1**, 149–159

116. Krengel, U., and Dijkstra, B. W. (1996) *J. Mol. Biol.* **263**, 70–78

117. Kasumi, T., Tsumuraya, Y., Brewer, C. F., Kersters-Hilderson, H., Claeyssens, M., and Hehre, E. J. (1987) *Biochemistry* **26**, 3010–3016

118. Lawson, S. L., Wakarchuk, W. W., and Withers, S. G. (1996) *Biochemistry* **35**, 10110–10118

119. Derewenda, U., Swenson, L., Green, R., Wei, Y., Morosoli, R., Shareck, F., Kluepfel, D., and Derewenda, Z. S. (1994) *J. Biol. Chem.* **269**, 20811–20814

120. McIntosh, L. P., Hand, G., Johnson, P. E., Joshi, M. D., Körner, M., Plesniak, L. A., Ziser, L., Wakarchuk, W. W., and Withers, S. G. (1996) *Biochemistry* **35**, 9958–9966

121. He, X. M., and Carter, D. C. (1992) *Nature (London)* **358**, 209–215

122. Braithwaite, K. L., Barna, T., Spurway, T. D., Charnock, S. J., Black, G. W., Hughes, N., Lakey, J. H., Virden, R., Hazlewood, G. P., Henrissat, B., and Gilbert, H. J. (1997) *Biochemistry* **36**, 15489–15500

123. Sanz-Aparicio, J., Hermoso, J. A., Martínez-Ripoll, M., Lequerica, J. L., and Polaina, J. (1998) *J. Mol. Biol.* **275**, 491–502

124. Wang, Q., Trimbur, D., Graham, R., Warren, R. A. J., and Withers, S. G. (1995) *Biochemistry* **34**, 14554–14562

125. Namchuk, M. N., and Withers, S. G. (1995) *Biochemistry* **34**, 16194–16202

126. Hrmova, M., MacGregor, E. A., Biely, P., Stewart, R. J., and Fincher, G. B. (1998) *J. Biol. Chem.* **273**, 11134–11143

127. Febbraio, F., Barone, R., D'Auria, S., Rossi, M., Nucci, R., Piccialli, G., De Napoli, L., Orrù, S., and Pucci, P. (1997) *Biochemistry* **36**, 3068–3075

128. Jacobson, R. H., Zhang, X.-J., DuBose, R. F., and Matthews, B. W. (1994) *Nature (London)* **369**, 761–766

129. Gebler, J. C., Aebersold, R., and Withers, S. G. (1992) *J. Biol. Chem.* **267**, 11126–11130

130. Richard, J. P., Westerfeld, J. G., Lin, S., and Beard, J. (1995) *Biochemistry* **34**, 11713–11724

131. Richard, J. P., Huber, R. E., Lin, S., Heo, C., and Amyes, T. L. (1996) *Biochemistry* **35**, 12377–12386

132. Roth, N. J., Rob, B., and Huber, R. E. (1998) *Biochemistry* **37**, 10099–10107

133. Mitchell, E. P., Withers, S. G., Ermert, P., Vasella, A. T., Garman, E. F., Oikonomakos, N. G., and Johnson, L. N. (1996) *Biochemistry* **35**, 7341–7355

134. Tu, J.-I., Jacobson, G. R., and Graves, D. J. (1971) *Biochemistry* **10**, 1229–1236

135. Gold, A. M., Legrand, E., and Sanchez, G. R. (1971) *J. Biol. Chem.* **246**, 5700–5706

136. Levvy, G. A., and Snaith, S. M. (1972) *Adv. Enzymol.* **36**, 151–181

137. Barford, D., and Johnson, L. N. (1989) *Nature (London)* **340**, 609–616

138. Oikonomakos, N. G., Johnson, L. N., Acharya, K. R., Stuart, D. I., Barford, D., Hajdu, J., Varvill, K. M., Melpidou, A. E., Papageorgiou, T., Graves, D. J., and Palm, D. (1987) *Biochemistry* **26**, 8381–8389

139. Oikonomakos, N. G., Zographos, S. E., Tsitsanou, K. E., Johnson, L. N., and Acharya, K. R. (1996) *Protein Sci.* **5**, 2416–2428

140. Helmreich, E. J. M., and Klein, H. W. (1980) *Angew. Chem. Int. Ed. Engl.* **19**, 441–455

141. Klein, H. W., Im, M. J., Palm, D., and Helmreich, E. J. M. (1984) *Biochemistry* **23**, 5853–5861

142. Street, I. P., Rupitz, K., and Withers, S. G. (1989) *Biochemistry* **28**, 1581–1587

143. Parrish, R. F., Uhing, R. J., and Graves, D. J. (1977) *Biochemistry* **16**, 4824–4831

143a. Watson, K. A., McCleverty, C., Geremia, S., Cottaz, S., Driguez, H., and Johnson, L. N. (1999) *EMBO J.* **18**, 4619–4632

144. Duke, E. M. H., Wakatsuki, S., Hadfield, A., and Johnson, L. N. (1994) *Protein Sci.* **3**, 1178–1196

145. Buchbinder, J. L., Luong, C. B. H., Browner, M. F., and Fletterick, R. J. (1997) *Biochemistry* **36**, 8039–8044

146. Buchbinder, J. L., and Fletterick, R. J. (1996) *J. Biol. Chem.* **271**, 22305–22309

147. Gregoriou, M., Noble, M. E. M., Watson, K. A., Garman, E. F., Krulle, T. M., De La Fuente, C., Fleet, G. W. J., Oikonomakos, N. G., and Johnson, L. N. (1998) *Protein Sci.* **7**, 915–927

148. Nakano, K., and Fukui, T. (1986) *J. Biol. Chem.* **261**, 8230–8236

149. Watson, K. A., Schinzel, R., Palm, D., and Johnson, L. N. (1997) *EMBO J.* **16**, 1–14

150. Becker, S., Palm, D., and Schinzel, R. (1994) *J. Biol. Chem.* **269**, 2485–2490

151. Lin, K., Hwang, P. K., and Fletterick, R. J. (1995) *J. Biol. Chem.* **270**, 26833–26839

152. Goldsmith, E. J., Fletterick, R. J., and Withers, S. G. (1987) *J. Biol. Chem.* **262**, 1449–1455

153. Robyt, J. F., and French, D. (1970) *J. Biol. Chem.* **245**, 3917–3927

154. Brayer, G. D., Luo, Y., and Withers, S. G. (1995) *Protein Sci.* **4**, 1730–1742

155. MacDonald, R. J., Crerar, M. M., Swain, W. F., Pictet, R. L., Thomas, G., and Rutter, W. J. (1980) *Nature (London)* **287**, 117–122

156. Qian, M., Haser, R., Buisson, G., Duée, E., and Payan, F. (1994) *Biochemistry* **33**, 6284–6294

157. Qian, M., Haser, R., and Payan, F. (1995) *Protein Sci.* **4**, 747–755

158. Larson, S. B., Greenwood, A., Cascio, D., Day, J., and McPherson, A. (1994) *J. Mol. Biol.* **235**, 1560–1584

159. Machius, M., Vértesy, L., Huber, R., and Wiegand, G. (1996) *J. Mol. Biol.* **260**, 409–421

160. Kadziola, A., Sogaard, M., Svensson, B., and Haser, R. (1998) *J. Mol. Biol.* **278**, 205–217

161. Strobl, S., Maskos, K., Betz, M., Wiegand, G., Huber, R., Gomis-Rüth, F. X., and Glockshuber, R. (1998) *J. Mol. Biol.* **278**, 617–628

162. Fujimoto, Z., Takase, K., Doui, N., Momma, M., Matsumoto, T., and Mizuno, H. (1998) *J. Mol. Biol.* **277**, 393–407

163. Feller, G., le Bussy, O., Houssier, C., and Gerday, C. (1996) *J. Biol. Chem.* **271**, 23836–23841

164. Aghajari, N., Feller, G., Gerday, C., and Haser, R. (1998) *Protein Sci.* **7**, 564–572

165. Matsuura, Y., Kusunoki, M., Harada, W., and Kakudo, M. (1984) *J. Biochem.* **95**, 697–702

166. Brzozowski, A. M., and Davies, G. J. (1997) *Biochemistry* **36**, 10837–10845

167. Robyt, J. F., and French, D. (1970) *Arch. Biochem. Biophys.* **138**, 662–670

168. Robyt, J. F., and French, D. (1967) *Arch. Biochem. Biophys.* **122**, 8–16

169. delCardayré, S. B., and Raines, R. T. (1994) *Biochemistry* **33**, 6031–6037

169a. Dauter, Z., Dauter, M., Brzozowski, A. M., Christensen, S., Borchert, T. V., Beier, L., Wilson, K. S., and Davies, G. J. (1999) *Biochemistry* **38**, 8385–8392

169b. Park, K. H., Kim, M. J., Lee, H. S., Han, N. S., Kim, D., and Robyt, J. F. (1998) *Carbohydr. Res.* **313**, 235–246

170. Hermans, M. M. P., Kroos, M. A., van Beeuman, J., Oostra, B. A., and Reuser, A. J. J. (1991) *J. Biol. Chem.* **266**, 13507–13512

171. Raben, N., Nagaraju, K., Lee, E., Kessler, P., Byrne, B., Lee, L., LaMarca, M., King, C., Ward, J., Sauer, B., and Plotz, P. (1998) *J. Biol. Chem.* **273**, 19086–19092

172. Howard, S., and Withers, S. G. (1998) *Biochemistry* **37**, 3858–3864

173. Jespersen, H. M., MacGregor, E. A., Henrissat, B., Sierks, M. R., and Svensson, B. (1993) *Journal of Protein Chemistry* **12**, 791–805

174. Liu, W., Madsen, N. B., Fan, B., Zucker, K. A., Glew, R. H., and Fry, D. E. (1995) *Biochemistry* **34**, 7056–7061

175. Watanabe, K., Hata, Y., Kizaki, H., Katsube, Y., and Suzuki, Y. (1997) *J. Mol. Biol.* **269**, 142–153

176. Knegtel, R. M. A., Strokopytov, B., Penninga, D., Faber, O. G., Rozeboom, H. J., Kalk, K. H., Dijkhuizen, L., and Dijkstra, B. W. (1995) *J. Biol. Chem.* **270**, 29256–29264

177. Schmidt, A. K., Cottaz, S., Driguez, H., and Schulz, G. E. (1998) *Biochemistry* **37**, 5909–5915

177a. Uitdehaag, J. C. M., Mosi, R., Kalk, K. H., van der Veen, B. A., Dijkhuizen, L., Withers, S. G., and Dijkstra, B. W. (1999) *Nature Struct. Biol.* **6**, 432–436

177b. van der Veen, B. A., Uitdehaag, J. C. M., Penninga, D., van Alebeek, G.-J. W. M., Smith, L. M., Dijkstra, B. W., and Dijkhuizen, L. (2000) *J. Mol. Biol.* **296**, 1027–1038

178. Terada, Y., Yanase, M., Takata, H., Takaha, T., and Okada, S. (1997) *J. Biol. Chem.* **272**, 15729–15733

179. Fierobe, H.-P., Clarke, A. J., Tull, D., and Svensson, B. (1998) *Biochemistry* **37**, 3753–3759

180. Aleshin, A. E., Stoffer, B., Firsov, L. M., Svensson, B., and Honzatko, R. B. (1996) *Biochemistry* **35**, 8319–8328

181. Firsov, L. M., Neustroev, K. N., Aleshin, A. E., Metzler, C. M., Metzler, D. E., Scott, R. D., Stoffer, B., Christensen, T., and Svensson, B. (1994) *Eur. J. Biochem.* **223**, 293–302

182. Natarajan, S. K., and Sierks, M. R. (1997) *Biochemistry* **36**, 14946–14955

183. Williamson, M. P., Le Gal-Coëffet, M.-F., Sorimachi, K., Furniss, C. S. M., Archer, D. B., and Williamson, G. (1997) *Biochemistry* **36**, 7535–7539

184. Penninga, D., van der Veen, B. A., Knegtel, R. M. A., van Hijum, S. A. F. T., Rozeboom, H. J., Kalk, K. H., Dijkstra, B. W., and Dijkhuizen, L. (1996) *J. Biol. Chem.* **271**, 32777–32784

184a. Mikami, B., Adachi, M., Kage, T., Sarikaya, E., Nanmori, T., Shinke, R., and Utsumi, S. (1999) *Biochemistry* **38**, 7050–7061

185. Mikami, B., Degano, M., Hehre, E. J., and Sacchettini, J. C. (1994) *Biochemistry* **33**, 7779–7787

186. Adachi, M., Mikami, B., Katsube, T., and Utsumi, S. (1998) *J. Biol. Chem.* **273**, 19859–19865

187. Amado, M., Almeida, R., Carneiro, F., Levery, S. B., Holmes, E. H., Nomoto, M., Hollingsworth, M. A., Hassan, H., Schwientek, T., Nielsen, P. A., Bennett, E. P., and Clausen, H. (1998) *J. Biol. Chem.* **273**, 12770–12778

188. Baenziger, J. U. (1994) *FASEB J.* **8**, 1019–1025

189. Radzicka, A., and Wolfenden, R. (1996) *J. Am. Chem. Soc.* **118**, 6105–6109

190. Barrett, A. J., Rawlings, N. D., and Woessner, J. F., eds. (1998) *Handbook of Proteolytic Enzymes*, Academic Press, San Diego, California

191. Fushiki, T., and Iwai, K. (1989) *FASEB J.* **3**, 121–126

192. Pignol, D., Granon, S., Chapus, C., and Carlos, J. (1995) *J. Mol. Biol.* **252**, 20–24

192a. Lu, D., Fütterer, K., Korolev, S., Zheng, X., Tan, K., Waksman, G., and Sadler, J. E. (1999) *J. Mol. Biol.* **292**, 361–373

193. Light, A., and Janska, H. (1989) *Trends Biochem. Sci.* **14**, 110–112

194. Matsushima, M., Ichinose, M., Yahagi, N., Kakei, N., Tsukada, S., Miki, K., Kurokawa, K., Tashiro, K., Shiokawa, K., Shinomiya, K., Umeyama, H., Inoue, H., Takahashi, T., and Takahashi, K. (1994) *J. Biol. Chem.* **269**, 19976–19982

195. Kitamoto, Y., Veile, R. A., Donis–Keller, H., and Sadler, J. E. (1995) *Biochemistry* **34**, 4562–4568

196. Khan, A. R., and James, M. N. G. (1998) *Protein Sci.* **7**, 815–836

197. Hedstrom, L., Lin, T.-Y., and Fast, W. (1996) *Biochemistry* **35**, 4515–4523

197a. Khan, A. R., Khazanovich-Bernstein, N., Bergman, E. M., and James, M. N. G. (1999) *Proc. Natl. Acad. Sci. U.S.A.* **96**, 10968–10975

198. Kramer, K. J., Felsted, R. L., and Law, J. H. (1973) *J. Biol. Chem.* **248**, 3021–3028

199. Lesk, A. M., and Fordham, W. D. (1996) *J. Mol. Biol.* **258**, 501–537

200. Cygler, M., Schrag, J. D., Sussman, J. L., Harel, M., Silman, I., Gentry, M. K., and Doctor, B. P. (1993) *Protein Sci.* **2**, 366–382

201. Pereira, P. J. B., Bergner, A., Macedo, R., S, Huber, R., Matschiner, G., Fritz, H., Sommerhoff, C. P., and Bode, W. (1998) *Nature (London)* **392**, 306–311

201a. Sommerhoff, C. P., Bode, W., Pereira, P. J. B., Stubbs, M. T., Stürzebecher, J., Piechottka, G. P., Matschiner, G., and Bergner, A. (1999) *Proc. Natl. Acad. Sci. U.S.A.* **96**, 10984–10991

202. Schechter, N. M., Eng, G. Y., Selwood, T., and McCaslin, D. R. (1995) *Biochemistry* **34**, 10628–10638

203. Johnson, D. A., and Barton, G. J. (1992) *Protein Sci.* **1**, 370–377

204. Kido, H., Yokogoshi, Y., Sakai, K., Tashiro, M., Kishino, Y., Fukutomi, A., and Katunuma, N. (1992) *J. Biol. Chem.* **267**, 13573–13579

204a. Wong, G. W., Tang, Y., Feyfant, E., Sali, A., Li, L., Li, Y., Huang, C., Friend, D. S., Krilis, S. A., and Stevens, R. L. (1999) *J. Biol. Chem.* **274**, 30784–30793

205. Remington, S. J., Woodbury, R. G., Reynolds, R. A., Matthews, B. W., and Neurath, H. (1988) *Biochemistry* **27**, 8097–8105

206. McGrath, M. E., Mirzadegan, T., and Schmidt, B. F. (1997) *Biochemistry* **36**, 14318–14324

207. Sali, A., Matsumoto, R., McNeil, H. P., Karplus, M., and Stevens, R. L. (1993) *J. Biol. Chem.* **268**, 9023–9034

208. Springman, E. B., Dikov, M. M., and Serafin, W. E. (1995) *J. Biol. Chem.* **270**, 1300–1307

209. Hof, P., Mayr, I., Huber, R., Korzus, E., Potempa, J., Travis, J., Powers, J. C., and Bode, W. (1996) *EMBO J.* **15**, 5481–5491

210. Rao, N. V., Wehner, N. G., Marshall, B. C., Gray, W. R., Gray, B. H., and Hoidal, J. R. (1991) *J. Biol. Chem.* **266**, 9540–9548

211. Hershberger, R. J., Gershenfeld, H. K., Weissman, I. L., and Su, L. (1992) *J. Biol. Chem.* **267**, 25488–25493

212. Beresford, P. J., Kam, C.-M., Powers, J. C., and Lieberman, J. (1997) *Proc. Natl. Acad. Sci. U.S.A.* **94**, 9285–9290

213. Caputo, A., Garner, R. S., Winkler, U., Hudig, D., and Bleackley, R. C. (1993) *J. Biol. Chem.* **268**, 17672–17675

213a. Harris, J. L., Peterson, E. P., Hudig, D., Thornberry, N. A., and Craik, C. S. (1998) *J. Biol. Chem.* **273**, 27364–27373

214. Fuller, R. S., Brake, A. J., and Thorner, J. (1989) *Science* **246**, 482–486

215. Cromlish, J. A., Seidah, N. G., Marcinkiewicz, M., Hamelin, J., Johnson, D. A., and Chrétien, M. (1987) *J. Biol. Chem.* **262**, 1363–1373

216. Schaner, P., Todd, R. B., Seidah, N. G., and Nillni, E. A. (1997) *J. Biol. Chem.* **272**, 19958–19968

217. Rose, C., Vargas, F., Facchinetti, P., Bourgeat, P., Bambal, R. B., Bishop, P. B., Chan, S. M. T., Moore, A. N. J., Ganellin, C. R., and Schwartz, J.-C. (1996) *Nature (London)* **380**, 403–409

218. Shilton, B. H., Thomas, D. Y., and Cygler, M. (1997) *Biochemistry* **36**, 9002–9012

219. Bullock, T. L., Branchaud, B., and Remington, S. J. (1994) *Biochemistry* **33**, 11127–11134

220. Mortensen, U. H., Remington, S. J., and Breddam, K. (1994) *Biochemistry* **33**, 508–517

221. Brayer, G. D., Delbaere, L. T. J., and James, M. N. G. (1979) *J. Mol. Biol.* **131**, 743–775

222. Plou, F. J., Kowlessur, D., Malthouse, J. P. G., Mellor, G. W., Hartshorn, M. J., Pinitglang, S., Patel, H., Topham, C. M., Thomas, E. W., Verma, C., and Brocklehurst, K. (1996) *J. Mol. Biol.* **257**, 1088–1111

223. Wells, J. A., and Estell, D. A. (1988) *Trends Biochem. Sci.* **13**, 291–297

224. Kossiakoff, A. A., Ultsch, M., White, S., and Eigenbrot, C. (1991) *Biochemistry* **30**, 1211–1221

225. Sorensen, S. B., Bech, L. M., Meldal, M., and Breddam, K. (1993) *Biochemistry* **32**, 8994–8999

225a. Smith, C. A., Toogood, H. S., Baker, H. M., Daniel, R. M., and Baker, E. N. (1999) *J. Mol. Biol.* **294**, 1027–1040

226. Koszelak, S., Ng, J. D., Day, J., Ko, T. P., Greenwood, A., and McPherson, A. (1997) *Biochemistry* **36**, 6597–6604

227. Delbaere, L. T. J., and Brayer, G. D. (1980) *J. Mol. Biol.* **139**, 45–51

228. Blanchard, H., and James, M. N. G. (1994) *J. Mol. Biol.* **241**, 574–587

228a. Tomkinson, B. (1999) *Trends Biochem. Sci.* **24**, 355–359

228b. Rose, C., Vargas, F., Facchinetti, P., Bourgeat, P., Bambal, R. B., Bishop, P. B., Chan, S. M. T., Moore, A. N. J., Ganellin, C. R., and Schwartz, J.-C. (1996) *Nature (London)* **380**, 403–409

228c. Renn, S. C. P., Tomkinson, B., and Taghert, P. H. (1998) *J. Biol. Chem.* **273**, 19173–19182

229. Wharton, C. W. (1998) in *Comprehensive Biological Catalysis. A Mechanistic Reference*, Vol. I (Sinnott, M., ed), pp. 345–379, Academic Press, San Diego, California

230. Blow, D. M. (1997) *Trends Biochem. Sci.* **22**, 405–408

231. Tsukada, H., and Blow, D. M. (1985) *J. Mol. Biol.* **184**, 703–711

232. Bode, W., and Schwager, P. (1975) *J. Mol. Biol.* **98**, 693–717

233. Gaboriaud, C., Serre, L., Guy-Crotte, O., Forest, E., and Fontecilla-Camps, J.-C. (1996) *J. Mol. Biol.* **259**, 995–1010

234. Sawyer, L., Shotten, D. M., Campbell, J. W., Wendell, P. L., Muirhead, H., and Watson, H. C. (1978) *J. Mol. Biol.* **118**, 137–208

235. Bode, W., Meyer, E., Jr., and Powers, J. C. (1989) *Biochemistry* **28**, 1951–1963

236. Stubbs, M. T., and Bode, W. (1995) *Trends Biochem. Sci.* **20**, 23–28

237. Katz, B. A., Liu, B., Barnes, M., and Springman, E. B. (1998) *Protein Sci.* **7**, 875–885

238. Timm, D. E. (1997) *Protein Sci.* **6**, 1418–1425

239. Blow, D. M. (1971) in *The Enzymes*, 3rd ed., Vol. 3 (Boyer, P. D., ed), pp. 185–212, Academic Press, New York

240. Huber, R., and Bode, W. (1977) in *NMR in Biology* (Dwek, R. A., Campbell, I. D., Richards, R. E., and Williams, R. J. P., eds), pp. 1–31 , Academic Press, New York

241. Bachovchin, W. W., and Roberts, J. D. (1978) *J. Am. Chem. Soc.* **100**, 8041–8047

References

242. Smith, S. O., Farr-Jones, S., Griffin, R. G., and Bachovchin, W. W. (1989) *Science* **244**, 961–964

243. Kossiakoff, A. A., and Spencer, S. A. (1981) *Biochemistry* **20**, 6462–6474

244. Polgár, L. (1989) *Mechanisms of Protease Action*, CRC Press, Boca Raton, Florida

245. Warshel, A., Naray-Szabo, G., Sussman, F., and Hwang, J.-K. (1989) *Biochemistry* **28**, 3630–3637

246. Robillard, G., and Schulman, R. G. (1972) *J. Mol. Biol.* **71**, 507–511

247. Markley, J. L., and Westler, W. M. (1996) *Biochemistry* **35**, 11092–11097

248. Ash, E. L., Sudmeier, J. L., De Fabo, E. C., and Bachovchin, W. W. (1997) *Science* **278**, 1128–1132

249. Halkides, C. J., Wu, Y. Q., and Murray, C. J. (1996) *Biochemistry* **35**, 15941–15948

250. Zhong, S., Haghjoo, K., Kettner, C., and Jordan, F. (1995) *J. Am. Chem. Soc.* **117**, 7048–7055

251. Bachovchin, W. W., Wong, W. Y. L., Farr-Jones, S., Shenvi, A. B., and Kettner, C. A. (1988) *Biochemistry* **27**, 7689–7697

252. Cassidy, C. S., Lin, J., and Frey, P. A. (1997) *Biochemistry* **36**, 4576–4584

253. Tsilikounas, E., Rao, T., Gutheil, W. G., and Bachovchin, W. W. (1996) *Biochemistry* **35**, 2437–2444

254. Tobin, J. B., Whitt, S. A., Cassidy, C. S., and Frey, P. A. (1995) *Biochemistry* **34**, 6919–6924

255. Singer, P. T., Smalås, A., Carty, R. P., Mangel, W. F., and Sweet, R. M. (1993) *Science* **259**, 669–673

256. Shieh, H.-S., Kurumbail, R. G., Stevens, A. M., Stegeman, R. A., Sturman, E. J., Pak, J. Y., Wittwer, A. J., Palmier, M. O., Wiegand, R. C., Holwerda, B. C., and Stallings, W. C. (1996) *Nature (London)* **383**, 279–282

257. Tong, L., Quin, C., Massariol, M.-J., Bonneau, P. R., Cordingley, M. G., and Lagacé, L. (1996) *Nature (London)* **383**, 272–275

258. Carter, P., and Wells, J. A. (1988) *Nature (London)* **332**, 564–568

259. Corey, D. R., and Craik, C. S. (1992) *J. Am. Chem. Soc.* **114**, 1784–1790

260. Sprang, S., Standing, T., Fletterick, R. J., Stroud, R. M., Finer-Moore, J., Xuong, N.-H., Hamlin, R., Rutter, W. J., and Craik, C. S. (1987) *Science* **237**, 905–909

261. McGrath, M. E., Vásquez, J. R., Craik, C. S., Yang, A. S., Honig, B., and Fletterick, R. J. (1992) *Biochemistry* **31**, 3059–3064

262. Robertus, J. D., Kraut, J., Alden, R. A., and Birktoft, J. J. (1972) *Biochemistry* **11**, 4293–4303

263. Whiting, A. K., and Peticolas, W. L. (1994) *Biochemistry* **33**, 552–561

264. Bryan, P., Pantoliano, M. W., Quill, S. G., Hsiao, H.-Y., and Poulos, T. (1986) *Proc. Natl. Acad. Sci. U.S.A.* **83**, 3743–3745

265. Asbóth, B., and Polgár, L. (1983) *Biochemistry* **22**, 117–122

266. Ortiz, C., Tellier, C., Williams, H., Stolowich, N. J., and Scott, A. I. (1991) *Biochemistry* **30**, 10026–10034

267. Bullock, T. L., Breddam, K., and Remington, S. J. (1996) *J. Mol. Biol.* **255**, 714–725

268. Tonge, P. J., and Carey, P. R. (1989) *Biochemistry* **28**, 6701–6709

269. Tonge, P. J., and Carey, P. R. (1992) *Biochemistry* **31**, 9122–9125

270. Birktoft, J. J., Kraut, J., and Freer, S. T. (1976) *Biochemistry* **15**, 4481–4485

271. Brünger, A. T., Huber, R., and Karplus, M. (1987) *Biochemistry* **26**, 5153 –5162

272. Polgár, L., and Halász, R. (1982) *Biochem. J.* **207**, 1–10

273. Bizzozero, S. A., and Dutler, H. (1981) *Bioorg. Chem.* **10**, 46–62

274. Cruickshank, W. H., and Kaplan, H. (1974) *J. Mol. Biol.* **83**, 267–274

274a. Ash, E. L., Sudmeier, J. L., Day, R. M., Vincent, M., Torchilin, E. V., Haddad, K. C., Bradshaw, E. M., Sanford, D. G., and Bachovchin, W. W. (2000) *Proc. Natl. Acad. Sci. U.S.A.* **97**, 10371–10376

274b. Kidd, R. D., Sears, P., Huang, D.-H., Witte, K., Wong, C.-H., and Farber, G. K. (1999) *Protein Sci.* **8**, 410–417

275. Meyer, E. F., Jr., Clore, G. M., Gronenborn, A. M., and Hansen, H. A. S. (1988) *Biochemistry* **27**, 725–730

276. Wells, G. B., Mustafi, D., and Makinen, M. W. (1994) *J. Biol. Chem.* **269**, 4577–4586

277. Ding, X., Rasmussen, B. F., Petsko, G. A., and Ringe, D. (1994) *Biochemistry* **33**, 9285–9293

278. Oppenheimer, H. L., Labouesse, B., and Hess, G. P. (1966) *J. Biol. Chem.* **241**, 2720–2730

279. Fersht, A. (1985) *Enzyme Structure and Mechanism*, 2nd ed., Freeman, San Francisco, California

280. Perona, J. J., Hedstrom, L., Rutter, W. J., and Fletterick, R. J. (1995) *Biochemistry* **34**, 1489–1499

281. Graf, L., Craik, C. S., Patthy, A., Roczniak, S., Fletterick, R. J., and Rutter, W. J. (1987) *Biochemistry* **26**, 2616–2623

282. Schellenberger, V., Turck, C. W., and Rutter, W. J. (1994) *Biochemistry* **33**, 4251–4257

283. Kurth, T., Ullmann, D., Jakubke, H.-D., and Hedstrom, L. (1997) *Biochemistry* **36**, 10098–10104

284. Le Bonniec, B. F., Myles, T., Johnson, T., Knight, C. G., Tapparelli, C., and Stone, S. R. (1996) *Biochemistry* **35**, 7114–7122

285. Wangikar, P. P., Rich, J. O., Clark, D. S., and Dordick, J. S. (1995) *Biochemistry* **34**, 12302–12310

286. Perona, J. J., and Craik, C. S. (1995) *Protein Sci.* **4**, 337–360

287. Davis, J. H., and Agard, D. A. (1998) *Biochemistry* **37**, 7696–7707

288. Baumann, W. K., Bizzozero, S. A., and Dutler, H. (1970) *FEBS Letters* **8**, 257–260

289. Baumann, W. K., Bizzozero, S. A., and Dutler, H. (1973) *Eur. J. Biochem.* **39**, 381–387

290. Stein, R. L., Elrod, J. P., and Schowen, R. L. (1983) *J. Am. Chem. Soc.* **105**, 2446–2452

291. Chang, T. K., Chiang, Y., Guo, H.-X., Kresge, A. J., Mathew, L., Powell, M. F., and Wells, J. A. (1996) *J. Am. Chem. Soc.* **118**, 8802–8807

292. Stein, R. L., and Stimpler, A. M. (1987) *J. Am. Chem. Soc.* **109**, 4387–4390

293. Baker, E. N. (1980) *J. Mol. Biol.* **141**, 441–484

294. Maes, D., Bouckaert, J., Poortmans, F., Wyns, L., and Looze, Y. (1996) *Biochemistry* **35**, 16292–16298

295. O'Hara, B. P., Hemmings, A. M., Buttle, D. J., and Pearl, L. H. (1995) *Biochemistry* **34**, 13190–13195

296. Berti, P. J., and Storer, A. C. (1995) *J. Mol. Biol.* **246**, 273–283

297. Drenth, J., Jansonius, J. N., Koekoek, R., and Wolthers, B. G. (1971) *Adv. Prot. Chem.* **25**, 79–115

298. Lewis, S. D., Johnson, F. A., and Shafer, J. A. (1981) *Biochemistry* **20**, 48–51

299. Roberts, D. D., Lewis, S. D., Ballou, D. P., Olson, S. T., and Shafer, J. A. (1986) *Biochemistry* **25**, 5595–5601

300. Johnson, F. A., Lewis, S. D., and Shafer, J. A. (1981) *Biochemistry* **20**, 44–48

301. Keillor, J. W., and Brown, R. S. (1992) *J. Am. Chem. Soc.* **114**, 7983–7989

302. Pinitglang, S., Watts, A. B., Patel, M., Reid, J. D., Noble, M. A., Gul, S., Bokth, A., Naeem, A., Patel, H., Thomas, E. W., Sreedharan, S. K., Verma, C., and Brocklehurst, K. (1997) *Biochemistry* **36**, 9968–9982

303. Ménard, R., Khouri, H. E., Plouffe, C., Laflamme, P., Dupras, R., Vernet, T., Tessier, D. C., Thomas, D. Y., and Storer, A. C. (1991) *Biochemistry* **30**, 5531–5538

304. Katerelos, N. A., and Goodenough, P. W. (1996) *Biochemistry* **35**, 14763–14772

305. Vernet, T., Tessier, D. C., Chatellier, J., Plouffe, C., Lee, T. S., Thomas, D. Y., Storer, A. C., and Ménard, R. (1995) *J. Biol. Chem.* **270**, 16645–16652

306. Ménard, R., Carrière, J., Laflamme, P., Plouffe, C., Khouri, H. E., Vernet, T., Tessier, D. C., Thomas, D. Y., and Storer, A. C. (1991) *Biochemistry* **30**, 8924–8928

307. Ménard, R., Laflamme, P., Plouffe, C., Vernet, T., Tessier, D. C., Thomas, D. Y., and Storer, A. C. (1995) *Biochemistry* **34**, 464–471

308. Doran, J. D., Tonge, P. J., Mort, J. S., and Carey, P. R. (1996) *Biochemistry* **35**, 12487–12494

309. Gamcsik, M. P., Malthouse, J. P. G., Primrose, W. U., Mackenzie, N. E., Boyd, A. S. F., Russell, R. A., and Scott, A. I. (1983) *J. Am. Chem. Soc.* **105**, 6324–6325

310. Zheng, Y.-J., and Bruice, T. C. (1997) *Proc. Natl. Acad. Sci. U.S.A.* **94**, 4285–4288

311. Arad, D., Langridge, R., and Kollman, P. A. (1990) *J. Am. Chem. Soc.* **112**, 491–502

312. Brocklehurst, K., Watts, A. B., Patel, M., Verma, C., and Thomas, E. W. (1998) in *Comprehensive Biological Catalysis. A Mechanistic Reference*, Vol. I (Sinnott, M., ed), Academic Press, San Diego, California

313. Bond, J. S., and Butler, P. E. (1987) *Ann. Rev. Biochem.* **56**, 333–364

314. Takahashi, T., Dehdarani, A. H., and Tang, J. (1988) *J. Biol. Chem.* **263**, 10952–10957

315. Coulombe, R., Grochulski, P., Sivaraman, J., Ménard, R., Mort, J. S., and Cygler, M. (1996) *EMBO J.* **15**, 5492–5503

316. Jia, Z., Hasnain, S., Hirama, T., Lee, X., Mort, J. S., To, R., and Huber, C. P. (1995) *J. Biol. Chem.* **270**, 5527–5533

317. Taralp, A., Kaplan, H., Sytwu, I.-I., Vlattas, I., Bohacek, R., Knap, A. K., Hirama, T., Huber, C. P., and Hasnain, S. (1995) *J. Biol. Chem.* **270**, 18036–18043

318. McGrath, M. E., Palmer, J. T., Brömme, D., and Somoza, J. R. (1998) *Protein Sci.* **7**, 1294–1302

319. Shi, G.-P., Webb, A. C., Foster, K. E., Knoll, J. H. M., Lemere, C. A., Munger, J. S., and Chapman, H. A. (1994) *J. Biol. Chem.* **269**, 11530–11536

320. Gelb, B. D., Shi, G.-P., Chapman, H. A., and Desnick, R. J. (1996) *Science* **273**, 1236–1238

321. Ishidoh, K., Muno, D., Sato, N., and Kominami, E. (1991) *J. Biol. Chem.* **266**, 16312–16317

322. Wolters, P. J., Raymond, W. W., Blount, J. L., and Caughey, G. H. (1998) *J. Biol. Chem.* **273**, 15514–15520

323. Schiller, M. R., Mende-Mueller, L., Moran, K., Meng, M., Miller, K. W., and Hook, V. Y. H. (1995) *Biochemistry* **34**, 7988–7995

324. Friedman, T. C., Kline, T. B., and Wilk, S. (1985) *Biochemistry* **24**, 3907–3913

325. Matsui, S.-I., Sandberg, A. A., Negoro, S., Seon, B. K., and Goldstein, G. (1982) *Proc. Natl. Acad. Sci. U.S.A.* **79**, 1535–1539

326. Joshua-Tor, L., Xu, H. E., Johnston, S. A., and Rees, D. C. (1995) *Science* **269**, 945–950

327. Richards, F. M. (1991) *Sci. Am.* **264** (Jan), 54–63

328. Ohno, S., Emori, Y., Imajoh, S., Kawasaki, H., Kisaragi, M., and Suzuki, K. (1984) *Nature (London)* **312**, 566–570

329. Saido, T. C., Sorimachi, H., and Suzuki, K. (1994) *FASEB J.* **8**, 814–822

330. Chian, H.-L., and Dice, J. F. (1988) *J. Biol. Chem.* **263**, 6797–6805

331. Tompa, P., Baki, A., Schád, É., and Friedrich, P. (1996) *J. Biol. Chem.* **271**, 33161–33164

References

332. Stabach, P. R., Cianci, C. D., Glantz, S. B., Zhang, Z., and Morrow, J. S. (1997) *Biochemistry* **36**, 57–65

332a. Hosfield, C. M., Elce, J. S., Davies, P. L., and Jia, Z. (1999) *EMBO J.* **18**, 6880–6889

333. Nicholson, D. W., and Thornberry, N. A. (1997) *Trends Biochem. Sci.* **22**, 299–306

334. Villa, P., Kaufmann, S. H., and Earnshaw, W. C. (1997) *Trends Biochem. Sci.* **22**, 388–393

335. Munday, N. A., Vaillancourt, J. P., Ali, A., Casano, F. J., Miller, D. K., Molineaux, S. M., Yamin, T.-T., Yu, V. L., and Nicholson, D. W. (1995) *J. Biol. Chem.* **270**, 15870–15876

336. McGrath, M. E., Eakin, A. E., Engel, J. C., McKerrow, J. H., Craik, C. S., and Fletterick, R. J. (1995) *J. Mol. Biol.* **247**, 251–259

337. Gillmor, S. A., Craik, C. S., and Fletterick, R. J. (1997) *Protein Sci.* **6**, 1603–1611

338. Porter, W. H., Cunningham, L. W., and Mitchell, W. M. (1971) *J. Biol. Chem.* **246**, 7675–7682

339. Pavloff, N., Pemberton, P. A., Potempa, J., Chen, W.-C. A., Pike, R. N., Prochazka, V., Kiefer, M. C., Travis, J., and Barr, P. J. (1997) *J. Biol. Chem.* **272**, 1595–1600

339a. Nelson, D., Potempa, J., Kordula, T., and Travis, J. (1999) *J. Biol. Chem.* **274**, 12245–12251

340. Malcolm, B. A. (1995) *Protein Sci.* **4**, 1439–1445

341. Mosimann, S. C., Cherney, M. M., Sia, S., Plotch, S., and James, M. N. G. (1997) *J. Mol. Biol.* **273**, 1032–1047

342. Ding, J., McGrath, W. J., Sweet, R. M., and Mangel, W. F. (1996) *EMBO J.* **15**, 1778–1783

343. Fox, T., de Miguel, E., Mort, J. S., and Storer, A. C. (1992) *Biochemistry* **31**, 12571–12576

344. Podobnik, M., Kuhelj, R., Turk, V., and Turk, D. (1997) *J. Mol. Biol.* **271**, 774–788

345. Seemüller, E., Lupas, A., Stock, D., Löwe, J., Huber, R., and Baumeister, W. (1995) *Science* **268**, 579–582

346. Paetzel, M., and Dalbey, R. E. (1997) *Trends Biochem. Sci.* **22**, 28–31

346a. Kisselev, A. F., Songyang, Z., and Goldberg, A. L. (2000) *J. Biol. Chem.* **275**, 14831–14837

347. Eleuteri, A. M., Kohanski, R. A., Cardozo, C., and Orlowski, M. (1997) *J. Biol. Chem.* **272**, 11824–11831

348. Kopp, F., Kristensen, P., Hendil, K. B., Johnsen, A., Sobek, A., and Dahlmann, B. (1995) *J. Mol. Biol.* **248**, 264–272

349. Cardozo, C., Michaud, C., and Orlowski, M. (1999) *Biochemistry* **38**, 9768–9777

349a. Groll, M., Heinemeyer, W., Jäger, S., Ullrich, T., Bochtler, M., Wolf, D. H., and Huber, R. (1999) *Proc. Natl. Acad. Sci. U.S.A.* **96**, 10976–10983

349b. Witt, E., Zantopf, D., Schmidt, M., Kraft, R., Kloetzel, P.-M., and Krüger, E. (2000) *J. Mol. Biol.* **301**, 1–9

349c. Orlowski, M., and Wilk, S. (2000) *Arch. Biochem. Biophys.* **383**, 1–16

350. Tikkanen, R., Riikonen, A., Oinonen, C., Rouvinen, J., and Peltonen, L. (1996) *EMBO J.* **15**, 2954–2960

351. Xuan, J., Tarentino, A. L., Grimwood, B. G., Plummer, T. H., Jr., Cui, T., Guan, C., and Van Roey, P. (1998) *Protein Sci.* **7**, 774–781

352. Liu, Y., Guan, C., and Aronson, N. N., Jr. (1998) *J. Biol. Chem.* **273**, 9688–9694

353. Guan, C., Liu, Y., Shao, Y., Cui, T., Liao, W., Ewel, A., and Whitaker, R. (1998) *J. Biol. Chem.* **273**, 9695–9702

354. Brannigan, J. A., Dodson, G., Duggleby, H. J., Moody, P. C. E., Smith, J. L., Tomchick, D. R., and Murzin, A. G. (1995) *Nature (London)* **378**, 416–419

355. Dalbey, R. E., and Wickner, W. (1988) *J. Biol. Chem.* **263**, 404–408

356. Tschantz, W. R., Sung, M., Delgado-Partin, V. M., and Dalbey, R. E. (1993) *J. Biol. Chem.* **268**, 27349–27354

357. Dalbey, R. E., Lively, M. O., Bron, S., and Van Dijl, J. M. (1997) *Protein Sci.* **6**, 1129–1138

358. Fenteany, G., and Schreiber, S. L. (1998) *J. Biol. Chem.* **273**, 8545–8548

359. Bogyo, M., McMaster, J. S., Gaczynska, M., Tortorella, D., Goldberg, A. L., and Ploegh, H. (1997) *Proc. Natl. Acad. Sci. U.S.A.* **94**, 6629–6634

360. Schmidtke, G., Kraft, R., Kostka, S., Henklein, P., Frömmel, C., Löwe, J., Huber, R., Kloetzel, P. M., and Schmidt, M. (1996) *EMBO J.* **15**, 6887–6898

361. Heinemeyer, W., Fischer, M., Krimmer, T., Stachon, U., and Wolf, D. H. (1997) *J. Biol. Chem.* **272**, 25200–25209

362. Ditzel, L., Huber, R., Mann, K., Heinemeyer, W., Wolf, D. H., and Groll, M. (1998) *J. Mol. Biol.* **279**, 1187–1191

363. Andreeva, N. S., Zdanov, A. S., Gustchina, A. E., and Fedorov, A. A. (1984) *J. Biol. Chem.* **259**, 11353–11365

364. Fusek, M., and Vetvicka, V., eds. (1995) *Aspartic Proteinases: Physiology and Pathology*, CRC Press, Boca Raton, Florida

365. Moore, S. A., Sielecki, A. R., Chernaia, M. M., Tarasova, N. I., and James, M. N. G. (1995) *J. Mol. Biol.* **247**, 466–485

366. Strop, P., Sedlacek, J., Stys, J., Kaderabkova, Z., Blaha, I., Pavlickova, L., Pohl, J., Fabry, M., Kostka, V., Newman, M., Frazao, C., Shearer, A., Tickle, I. J., and Blundell, T. L. (1990) *Biochemistry* **29**, 9863–9871

367. Vance, J. E., LeBlanc, D. A., and London, R. E. (1997) *Biochemistry* **36**, 13232–13240

368. Majer, P., Collins, J. R., Gulnik, S. V., and Erickson, J. W. (1997) *Protein Sci.* **6**, 1458–1466

369. Krieger, T. J., and Hook, V. Y. H. (1992) *Biochemistry* **31**, 4223–4231

370. Fineschi, B., and Miller, J. (1997) *Trends Biochem. Sci.* **22**, 377–382

371. Wilk, S., Wilk, E., and Magnusson, R. P. (1998) *J. Biol. Chem.* **273**, 15961–15970

372. Tong, L., Pav, S., Lamarre, D., Pilote, L., LaPlante, S., Anderson, P. C., and Jung, G. (1995) *J. Mol. Biol.* **250**, 211–222

373. Yang, J., Teplyakov, A., and Quail, J. W. (1997) *J. Mol. Biol.* **268**, 449–459

374. Gómez, J., and Freire, E. (1995) *J. Mol. Biol.* **252**, 337–350

375. Symersky, J., Monod, M., and Foundling, S. I. (1997) *Biochemistry* **36**, 12700–12710

376. Green, D. W., Aykent, S., Gierse, J. K., and Zupec, M. E. (1990) *Biochemistry* **29**, 3126–3133

377. Baldwin, E. T., Bhat, T. N., Gulnik, S., Hosur, M. V., Sowder, R. C., Cachau, R. E., Collins, J., Silva, A. M., and Erickson, J. W. (1993) *Proc. Natl. Acad. Sci. U.S.A.* **90**, 6796–6800

378. Goldblum, A. (1988) *Biochemistry* **27**, 1653–1658

379. Subramanian, E. (1978) *Trends Biochem. Sci.* **3**, 1–3

380. Andreeva, N. S., Zdanov, A. S., Gustchina, A. E., and Fedorov, A. A. (1984) *J. Biol. Chem.* **259**, 11353–11365

381. James, M. N. G., Sielecki, A. R., Hayakawa, K., and Gelb, M. H. (1992) *Biochemistry* **31**, 3872–3886

382. Meek, T. D. (1998) in *Comprehensive Biological Catalysis. A Mechanistic Reference*, Vol. I (Sinnott, M., ed), pp. 327–344, Academic Press, San Diego, California

383. Blum, M., Cunningham, A., Pang, H., and Hofmann, T. (1991) *J. Biol. Chem.* **266**, 9501–9507

384. James, M. N. G., and Sielecki, A. R. (1986) *Nature (London)* **319**, 33–38

385. Blundell, T. L., Lapatto, R., Wilderspin, A. F., Hemmings, A. M., Hobart, P. M., Danley, D. E., and Whittle, P. J. (1990) *Trends Biochem. Sci.* **15**, 425–430

386. Katz, R. A., and Skalka, A. M. (1994) *Ann. Rev. Biochem.* **63**, 133–173

387. Wlodawer, A., Miller, M., Jaskólski, M., Sathyanarayana, B. K., Baldwin, E., Weber, I. T., Selk, L. M., Clawson, L., Schneider, J., and Kent, S. B. H. (1989) *Science* **245**, 616–620

388. Rose, R. B., Craik, C. S., Douglas, N. L., and Stroud, R. M. (1996) *Biochemistry* **35**, 12933–12944

389. Chatfield, D. C., and Brooks, B. R. (1995) *J. Am. Chem. Soc.* **117**, 5561–5572

390. Rodriguez, E. J., Angeles, T. S., and Meek, T. D. (1993) *Biochemistry* **32**, 12380–12385

391. Liu, H., Müller-Plathe, F., and van Gunsteren, W. F. (1996) *J. Mol. Biol.* **261**, 454–469

392. Silva, A. M., Cachau, R. E., Sham, H. L., and Erickson, J. W. (1996) *J. Mol. Biol.* **255**, 321–346

393. Reverter, D., Ventura, S., Villegas, V., Vendrell, J., and Avilés, F. X. (1998) *J. Biol. Chem.* **273**, 3535–3541

394. Varlamov, O., Leiter, E. H., and Fricker, L. (1996) *J. Biol. Chem.* **271**, 13981–13986

395. Rodríguez, C., Brayton, K. A., Brownstein, M., and Dixon, J. E. (1989) *J. Biol. Chem.* **264**, 5988–5995

396. Tan, F., Chan, S. J., Steiner, D. F., Schilling, J. W., and Skidgel, R. A. (1989) *J. Biol. Chem.* **264**, 13165–13170

397. Tan, F., Weerasinghe, D. K., Skidgel, R. A., Tamei, H., Kaul, R. K., Roninson, I. B., Schilling, J. W., and Erdös, E. G. (1990) *J. Biol. Chem.* **265**, 13–19

398. Ehlers, M. R. W., and Riordan, J. F. (1989) *Biochemistry* **28**, 5311–5318

399. Dideberg, O., Charlier, P., Dive, G., Joris, B., Frere, J. M., and Ghuysen, J. M. (1982) *Nature (London)* **299**, 469–470

400. Weaver, L. H., Kester, W. R., and Matthews, B. W. (1977) *J. Mol. Biol.* **114**, 119–132

401. Hausrath, A. C., and Matthews, B. W. (1994) *J. Biol. Chem.* **269**, 18839–18842

402. Holland, D. R., Hausrath, A. C., Juers, D., and Matthews, B. W. (1995) *Protein Sci.* **4**, 1955–1965

403. Holden, H. M., Tronrud, D. E., Monzingo, A. F., Weaver, L. H., and Matthews, B. W. (1987) *Biochemistry* **26**, 8542–8553

404. Mock, W. L., and Stanford, D. J. (1996) *Biochemistry* **35**, 7369–7377

405. Marie-Claire, C., Ruffet, E., Antonczak, S., Beaumont, A., O'Donohue, M., Roques, B. P., and Fournié-Zaluski, M.-C. (1997) *Biochemistry* **36**, 13938–13945

406. Lipman, M. L., Panda, D., Bennett, H. P. J., Henderson, J. E., Shane, E., Shen, Y., Goltzman, D., and Karaplis, A. C. (1998) *J. Biol. Chem.* **273**, 13729–13737

407. Ogishima, T., Niidome, T., Shimokata, K., Kitada, S., and Ito, A. (1995) *J. Biol. Chem.* **270**, 30322–30326

408. Luciano, P., Tokatlidis, K., Chambre, I., Germanique, J.-C., and Géli, V. (1998) *J. Mol. Biol.* **280**, 193–199

409. Christianson, D. W., and Lipscomb, W. N. (1986) *Proc. Natl. Acad. Sci. U.S.A.* **83**, 7568–7572

410. Mock, W. L. (1998) in *Comprehensive Biological Catalysis. A Mechanistic Reference*, Vol. I (Sinnott, M., ed), pp. 425–453, Academic Press, San Diego, California

411. Campbell, P., and Nashed, N. T. (1982) *J. Am. Chem. Soc.* **104**, 5221–5226

412. Makinen, M. W., Fukuyama, J. M., and Kuo, L. C. (1982) *J. Am. Chem. Soc.* **104**, 2667–2669

413. Sander, M. E., and Witzel, H. (1985) *Biochem. Biophys. Res. Commun.* **132**, 681–687

414. Kuo, L. C., Fukuyama, J. M., and Makinen, M. W. (1983) *J. Mol. Biol.* **163**, 63–105

References

415. Suh, J., Hong, S.-B., and Chung, S. (1986) *J. Biol. Chem.* **261**, 7112–7114
416. Mustafi, D., and Makinen, M. W. (1994) *J. Biol. Chem.* **269**, 4587–4595
417. Kuo, L. C., and Makinen, M. W. (1985) *J. Am. Chem. Soc.* **107**, 5255–5261
418. Cleland, W. W. (1977) *Adv. Enzymol.* **45**, 273–286
419. Kim, H., and Lipscomb, W. N. (1990) *Biochemistry* **29**, 5546–5555
420. Osumi, A., Rahmo, A., King, S. W., Przystas, T. J., and Fife, T. H. (1994) *Biochemistry* **33**, 14750–14757
421. Mock, W. L., and Zhang, J. Z. (1991) *J. Biol. Chem.* **266**, 6393–6400
422. Zhang, K., and Auld, D. S. (1995) *Biochemistry* **34**, 16306–16312
423. Hilver, D., Gardell, S. J., Rutter, W. J., and Kaiser, E. T. (1986) *J. Am. Chem. Soc.* **108**, 5298–5304
424. Kam, C.-M., Nishino, N., and Powers, J. C. (1979) *Biochemistry* **18**, 3032–3038
425. Bartlett, P. A., and Marlowe, C. K. (1983) *Biochemistry* **22**, 4618–4624
426. Holden, H. M., and Matthews, B. W. (1988) *J. Biol. Chem.* **263**, 3256–3260
427. Massova, I., Kotra, L. P., Fridman, R., and Mobashery, S. (1998) *FASEB J.* **12**, 1075–1095
428. Bond, J. S., and Beynon, R. J. (1995) *Protein Sci.* **4**, 1247–1261
429. Stöcker, W., Grams, F., Baumann, U., Reinemer, P., Gomis-Rüth, F.-X., McKay, D. B., and Bode, W. (1995) *Protein Sci.* **4**, 823–840
429a. Morgunova, E., Tuuttila, A., Bergmann, U., Isupov, M., Lindqvist, Y., Schneider, G., and Tryggvason, K. (1999) *Science* **284**, 1667–1670
429b. Tortorella, M. D., and 27 other authors. (1999) *Science* **284**, 1664–1666
429c. Nagase, H., and Woessner, J. F., Jr. (1999) *J. Biol. Chem.* **274**, 21491–21494
430. Becker, J. W., Marcy, A. I., Rokosz, L. L., Axel, M. G., Burbaum, J. J., Fitzgerald, P. M. D., Cameron, P. M., Esser, C. K., Hagmann, W. K., Hermes, J. D., and Springer, J. P. (1995) *Protein Sci.* **4**, 1966–1976
431. Wetmore, D. R., and Hardman, K. D. (1996) *Biochemistry* **35**, 6549–6558
432. Arumugam, S., Hemme, C. L., Yoshida, N., Suzuki, K., Nagase, H., Berjanskii, M., Wu, B., and Van Doren, S. R. (1998) *Biochemistry* **37**, 9650–9657
433. Gomis-Rüth, F. X., Stöcker, W., Huber, R., Zwilling, R., and Bode, W. (1993) *J. Mol. Biol.* **229**, 945–968
434. Gomis-Rüth, F. X., Kress, L. F., Kellermann, J., Mayr, I., Lee, X., Huber, R., and Bode, W. (1994) *J. Mol. Biol.* **239**, 513–544
435. Llano, E., Pendás, A. M., Knäuper, V., Sorsa, T., Salo, T., Salido, E., Murphy, G., Simmer, J. P., Bartlett, J. D., and López-Otín, C. (1997) *Biochemistry* **36**, 15101–15108
436. Gomis-Rüth, F. X., Gohlke, U., Betz, M., Knäuper, V., Murphy, G., López-Otín, C., and Bode, W. (1996) *J. Mol. Biol.* **264**, 556–566
437. Taylor, A. (1993) *FASEB J.* **7**, 290–298
438. Sträter, N., and Lipscomb, W. N. (1995) *Biochemistry* **34**, 14792–14800
438a. Sträter, N., Sun, L., Kantrowitz, E. R., and Lipscomb, W. N. (1999) *Proc. Natl. Acad. Sci. U.S.A.* **96**, 11151–11155
439. Chen, G., Edwards, T., D'souza, V. M., and Holz, R. C. (1997) *Biochemistry* **36**, 4278–4286
440. Lowther, W. T., Orville, A. M., Madden, D. T., Lim, S., Rich, D. H., and Matthews, B. W. (1999) *Biochemistry* **38**, 7678–7688
441. Bradshaw, R. A., Brickey, W. W., and Walker, K. W. (1998) *Trends Biochem. Sci.* **23**, 263–267

442. Wilce, M. C. J., Bond, C. S., Dixon, N. E., Freeman, H. C., Guss, J. M., Lilley, P. E., and Wilce, J. A. (1998) *Proc. Natl. Acad. Sci. U.S.A.* **95**, 3472–3477
443. Chan, M. K., Gong, W., Rajagopalan, P. T. R., Hao, B., Tsai, C. M., and Pei, D. (1997) *Biochemistry* **36**, 13904–13909
444. Dardel, F., Ragusa, S., Lazennec, C., Blanquet, S., and Meinnel, T. (1998) *J. Mol. Biol.* **280**, 501–513
445. Becker, A., Schlichting, I., Kabsch, W., Schultz, S., and Wagner, A. F. V. (1998) *J. Biol. Chem.* **273**, 11413–11416
446. Andrews, D. W., and Johnson, A. E. (1996) *Trends Biochem. Sci.* **21**, 365–369
446a. Fabunmi, R. P., Wigley, W. C., Thomas, P. J., and DeMartino, G. N. (2000) *J. Biol. Chem.* **275**, 409–413
447. Sommer, T., and Wolf, D. H. (1997) *FASEB J.* **11**, 1227–1233
448. Ciechanover, A., and Schwartz, A. L. (1998) *Proc. Natl. Acad. Sci. U.S.A.* **95**, 2727–2730
449. Rubin, D. M., Glickman, M. H., Larsen, C. N., Dhruvakumar, S., and Finley, D. (1998) *EMBO J.* **17**, 4909–4919
450. Lee, Y. S., Park, S. C., Goldberg, A. L., and Chung, C. H. (1988) *J. Biol. Chem.* **263**, 6643–6646
451. Shin, D. H., Lee, C. S., Chung, C. H., and Suh, S. W. (1996) *J. Mol. Biol.* **262**, 71–76
452. Chin, D. T., Goff, S. A., Webster, T., Smith, T., and Goldberg, A. L. (1988) *J. Biol. Chem.* **263**, 11718–11728
453. Flanagan, J. M., Wall, J. S., Capel, M. S., Schneider, D. K., and Shanklin, J. (1995) *Biochemistry* **34**, 10910–10917
454. Kessel, M., Maurizi, M. R., Kim, B., Kocsis, E., Trus, B. L., Singh, S. K., and Steven, A. C. (1995) *J. Mol. Biol.* **250**, 587–594
455. Suzuki, C. K., Rep, M., Maarten van Dijl, J., Suda, K., Grivell, L. A., and Schatz, G. (1997) *Trends Biochem. Sci.* **22**, 118–123
456. Roudiak, S. G., and Shrader, T. E. (1998) *Biochemistry* **37**, 11255–11263
457. Wang, N., Maurizi, M. R., Emmert-Buck, L., and Gottesman, M. M. (1994) *J. Biol. Chem.* **269**, 29308–29313
458. Savel'ev, A. S., Novikova, L. A., Kovaleva, I. E., Luzikov, V. N., Neupert, W., and Langer, T. (1998) *J. Biol. Chem.* **273**, 20596–20602
459. Singh, S. K., Guo, F., and Maurizi, M. R. (1999) *Biochemistry* **38**, 14906–14915
460. Kihara, A., Akiyama, Y., and Ito, K. (1998) *J. Mol. Biol.* **279**, 175–188
461. Lupas, A., Flanagan, J. M., Tamura, T., and Baumeister, W. (1997) *Trends Biochem. Sci.* **22**, 399–404
461a. de Sagarra, M. R., Mayo, I., Marco, S., Rodríguez-Vilarino, S., Oliva, J., Carrascosa, J. L., and Castano, J. G. (1999) *J. Mol. Biol.* **292**, 819–825
461b. Kim, K. I., Cheong, G.-W., Park, S.-C., Ha, J.-S., Woo, K. M., Choi, S. J., and Chung, C. H. (2000) *J. Mol. Biol.* **303**, 655–666
462. Rohrwild, M., Coux, O., Huang, H.-C., Moerschell, R. P., Yoo, S. J., Seol, J. H., Chung, C. H., and Goldberg, A. L. (1996) *Proc. Natl. Acad. Sci. U.S.A.* **93**, 5808–5813
463. Missiakas, D., Schwager, F., Betton, J.-M., Georgopoulos, C., and Raina, S. (1996) *EMBO J.* **15**, 6899–6909
463a. Bochtler, M., Hartmann, C., Song, H. K., Bourenkov, G. P., Bartunik, H. D., and Huber, R. (2000) *Nature (London)* **403**, 800–805
463b. Shotland, Y., Teff, D., Koby, S., Kobiler, O., and Oppenheim, A. B. (2000) *J. Mol. Biol.* **299**, 953–964
464. Kräusslich, H.-G., and Wimmer, E. (1988) *Ann. Rev. Biochem.* **57**, 701–754

465. Ypma-Wong, M. F., Filman, D. J., Hogle, J. M., and Semler, B. L. (1988) *J. Biol. Chem.* **263**, 17846–17856
466. López-Otín, C., Simón-Mateo, C., Martínez, L., and Vinuela, E. (1989) *J. Biol. Chem.* **264**, 9107–9110
466a. Wu, Z., Yao, N., Le, H. V., and Weber, P. C. (1998) *Trends Biochem. Sci.* **23**, 92–93
466b. Barbato, G., Cicero, D. O., Cordier, F., Narjes, F., Gerlach, B., Sambucini, S., Grzesiek, S., Matassa, V. G., De Francesco, R., and Bazzo, R. (2000) *EMBO J.* **19**, 1195–1206
467. Roseman, J. E., and Levine, R. L. (1987) *J. Biol. Chem.* **262**, 2101–2110
468. Grune, T., Reinheckel, T., and Davies, K. J. A. (1997) *FASEB J.* **11**, 526–534
469. Erdös, E. G., and Skidgel, R. A. (1989) *FASEB J.* **3**, 145–151
470. Hui, K.-S. (1988) *J. Biol. Chem.* **263**, 6613–6618
471. Zisfein, J. B., Graham, R. M., Dreskin, S. V., Wildey, G. M., Fischman, A. J., and Homey, C. J. (1987) *Biochemistry* **26**, 8690–8697
471a. Loew, D., Perrault, C., Morales, M., Moog, S., Ravanat, C., Schuhler, S., Arcone, R., Pietrapaolo, C., Cazenave, J.-P., van Dorsselaer, A., and Lanza, F. (2000) *Biochemistry* **39**, 10812–10822
472. Lobe, C. G., Finlay, B. B., Paranchych, W., Paetkau, V. H., and Bleackley, R. C. (1986) *Science* **232**, 858–861
473. Melloni, E., Pontremoki, S., Salamino, F., Sparatore, B., Michetti, M., Sacco, O., and Horecker, B. L. (1986) *J. Biol. Chem.* **261**, 11437–11439
474. Pohlner, J., Halter, R., Beyreuther, K., and Meyer, T. F. (1987) *Nature (London)* **325**, 458–462
475. Newport, G. R., McKerrow, J. H., Hedstrom, R., Petitt, M., McGarrigle, L., Barr, D. J., and Agabian, N. (1988) *J. Biol. Chem.* **263**, 13179–13184
476. Folk, J. E. (1980) *Ann. Rev. Biochem.* **49**, 517–531
477. McGrath, M. E., Gillmor, S. A., and Fletterick, R. J. (1995) *Protein Sci.* **4**, 141–148
478. Salier, J.-P. (1990) *Trends Biochem. Sci.* **15**, 435–439
479. Laskowski, M., Jr., and Kato, I. (1980) *Ann. Rev. Biochem.* **49**, 593–626
480. Barrett, A. J. (1987) *Trends Biochem. Sci.* **12**, 193–196
481. Freije, J. P., Balbín, M., Abrahamson, M., Velasco, G., Dalboge, H., Grubb, A., and López-Otín, C. (1993) *J. Biol. Chem.* **268**, 15737–15744
482. Martin, J. R., Craven, C. J., Jerala, R., Kroon-Zitko, L., Zerovnik, E., Turk, V., and Waltho, J. P. (1995) *J. Mol. Biol.* **246**, 331–343
483. Brown, W. M., and Dziegielewska, K. M. (1997) *Protein Sci.* **6**, 5–12
484. Williamson, R. A., Carr, M. D., Frenkiel, T. A., Feeney, J., and Freedman, R. B. (1997) *Biochemistry* **36**, 13882–13889
485. Apte, S. S., Olsen, B. R., and Murphy, G. (1995) *J. Biol. Chem.* **270**, 14313–14318
486. Baumann, U., Bauer, M., Létoffé, S., Delepelaire, P., and Wandersman, C. (1995) *J. Mol. Biol.* **248**, 653–661
487. Martzen, M. R., McMullen, B. A., Smith, N. E., Fujikawa, K., and Peanasky, R. J. (1990) *Biochemistry* **29**, 7366–7372
488. Huang, K., Lu, W., Anderson, S., Laskowski, M., Jr., and James, M. N. G. (1995) *Protein Sci.* **4**, 1985–1997
489. Lu, W., Qasim, M. A., and Kent, S. B. H. (1996) *J. Am. Chem. Soc.* **118**, 8518–8523
490. Huber, R., Kukla, D., Bode, W., Schwager, P., Bartels, K., Diesenhofer, J., and Steigemann, W. (1974) *J. Mol. Biol.* **89**, 73–101

References

491. Mer, G., Hietter, H., Kellenberger, C., Renatus, M., Luu, B., and Lefèvre, J.-F. (1996) *J. Mol. Biol.* **258**, 158–171

492. Conconi, A., and Ryan, C. A. (1993) *J. Biol. Chem.* **268**, 430–435

493. Werner, M. H., and Wemmer, D. E. (1991) *Biochemistry* **30**, 3356–3364

494. McBride, J. D., Brauer, A. B. E., Nievo, M., and Leatherbarrow, R. J. (1998) *J. Mol. Biol.* **282**, 447–457

495. Song, H. K., and Suh, S. W. (1998) *J. Mol. Biol.* **275**, 347–363

496. Brandt, P., and Woodward, C. (1987) *Biochemistry* **26**, 3156–3162

497. Huber, R., and Bode, W. (1978) *Acc. Chem. Res.* **11**, 114–122

498. Potempa, J., Korzus, E., and Travis, J. (1994) *J. Biol. Chem.* **269**, 15957–15960

499. Whisstock, J., Skinner, R., and Lesk, A. M. (1998) *Trends Biochem. Sci.* **23**, 63–67

500. Lukacs, C. M., Rubin, H., and Christianson, D. W. (1998) *Biochemistry* **37**, 3297–3304

500a. Whisstock, J. C., Skinner, R., Carrell, R. W., and Lesk, A. M. (2000) *J. Mol. Biol.* **296**, 685–699

501. Lawrence, D. A., Ginsburg, D., Day, D. E., Berkenpas, M. B., Verhamme, I. M., Kvassman, J.-O., and Shore, J. D. (1995) *J. Biol. Chem.* **270**, 25309–25312

502. Kaslik, G., Kardos, J., Szabó, E., Szilágyi, L., Závodszky, P., Westler, W. M., Markley, J. L., and Gráf, L. (1997) *Biochemistry* **36**, 5455–5464

503. Bode, W., Wei, A.-Z., Huber, R., Meyer, E., Travis, J., and Neumann, S. (1986) *EMBO J.* **5**, 2453–2458

504. Boswell, D. R., and Carrell, R. (1986) *Trends Biochem. Sci.* **11**, 102–103

505. Curiel, D. T., Holmes, M. D., Okayama, H., Brantly, M. L., Vogelmeier, C., Travis, W. D., Stier, L. E., Perks, W. H., and Crystal, R. G. (1989) *J. Biol. Chem.* **264**, 13938–13945

506. Lomas, D. A., Finch, J. T., Seyama, K., Nukiwa, T., and Carrell, R. W. (1993) *J. Biol. Chem.* **268**, 15333–15335

507. Carrell, R. W., Jeppsson, J.-O., Laurell, C.-B., Brennan, S. O., Owen, M. C., Vaughan, L., and Boswell, D. R. (1982) *Nature (London)* **298**, 329–334

508. Carp, H., Miller, F., Hoidal, J. R., and Janoff, A. (1982) *Proc. Natl. Acad. Sci. U.S.A.* **79**, 2041–2045

509. Radinsky, L. (1983) *Science* **221**, 1187–1191

510. Culliton, B. J. (1989) *Science* **246**, 750–751

511. Matheson, N. R., Gibson, H. E., Hallewell, R. A., Barr, P. J., and Travis, J. (1986) *J. Biol. Chem.* **261**, 10404–10409

512. Rosenfeld, M. A., Siegfried, W., Yoshimura, K., Yoneyama, K., Fukayama, M., Stier, L. E., Pääkkö, P. K., Gilardi, P., Stratford-Perricaudet, L. D., Perricaudet, M., Jallat, S., Pavirani, A., Lecocq, J.-P., and Crystal, R. G. (1991) *Science* **252**, 431–434

513. Owen, M. C., Brennan, S. O., Lewis, J. H., and Carrell, R. W. (1983) *N. Engl. J. Med.* **309**, 694–698

514. Davie, E. W., Fujikawa, K., and Kisiel, W. (1991) *Biochemistry* **30**, 10364–10370

515. Smith, E. L., Hill, R. L., Lehman, I. R., Lefkowitz, R. J., Handler, P., and White, A. (1983) in *Principles of Biochemistry, Mammalian Biochemistry*, 7th ed., pp. 17–37, McGraw-Hill, New York

516. Mann, K. G., Jenny, R. J., and Krishnaswamy, S. (1988) *Ann. Rev. Biochem.* **57**, 915–956

517. Zucker, M. B. (1980) *Sci. Am.* **242** (Jun), 86–103

518. Gibbs, C. S., McCurdy, S. N., Leung, L. L. K., and Paborsky, L. R. (1994) *Biochemistry* **33**, 14003–14010

519. Muller, Y. A., Ultsch, M. H., Kelley, R. F., and de Vos, A. M. (1994) *Biochemistry* **33**, 10864–10870

520. Harlos, K., Martin, D. M. A., O'Brien, D. P., Jones, E. Y., Stuart, D. I., Polikarpov, I., Miller, A., Tuddenham, E. G. D., and Boys, C. W. G. (1994) *Nature (London)* **370**, 662–666

521. Martin, D. M. A., Boys, C. W. G., and Ruf, W. (1995) *FASEB J.* **9**, 852–859

522. Ruf, W., Kelly, C. R., Schullek, J. R., Martin, D. M. A., Polikarpov, I., Boys, C. W. G., Tuddenham, E. G. D., and Edgington, T. S. (1995) *Biochemistry* **34**, 6310–6315

523. Muller, Y. A., Ultsch, M. H., and de Vos, A. M. (1996) *J. Mol. Biol.* **256**, 144–159

524. Doolittle, R. F. (1981) *Sci. Am.* **245** (Dec), 126–135

525. Fu, Y., Weissbach, L., Plant, P. W., Oddoux, C., Cao, Y., Liang, T. J., Roy, S. N., Redman, C. M., and Grieninger, G. (1992) *Biochemistry* **31**, 11968–11972

526. Hunziker, E. B., Straub, P. W., and Haeberli, A. (1990) *J. Biol. Chem.* **265**, 7455–7463

527. Banner, D. W., D'Arcy, A., Chène, C., Winkler, F. K., Guha, A., Konigsberg, W. H., Nemerson, Y., and Kirchhofer, D. (1996) *Nature (London)* **380**, 41–46

527a. Banner, D. W. (2000) *Nature (London)* **404**, 449–450

527b. Chang, Y.-J., Hamaguchi, N., Chang, S.-C., Ruf, W., Shen, M.-C., and Lin, S.-W. (1999) *Biochemistry* **38**, 10940–10948

528. Muranyi, A., Finn, B. E., Gippert, G. P., Forsén, S., Stenflo, J., and Drakenberg, T. (1998) *Biochemistry* **37**, 10605–10615

529. Andrews, B. S. (1991) *Trends Biochem. Sci.* **16**, 31–36

530. Neuenschwander, P. F., and Morrissey, J. H. (1992) *J. Biol. Chem.* **267**, 14477–14482

531. Altieri, D. C. (1995) *FASEB J.* **9**, 860–865

532. Brandstetter, H., Kühne, A., Bode, W., Huber, R., von der Saal, W., Wirthensohn, K., and Engh, R. A. (1996) *J. Biol. Chem.* **271**, 29988–29992

533. Sabharwal, A. K., Padmanabhan, K., Tulinsky, A., Mathur, A., Gorka, J., and Bajaj, S. P. (1997) *J. Biol. Chem.* **272**, 22037–22045

534. Dharmawardana, K. R., and Bock, P. E. (1998) *Biochemistry* **37**, 13143–13152

534a. Macedo-Ribeiro, S., Bode, W., Huber, R., Quinn-Allen, M. A., Kim, S. W., Ortel, T. L., Bourenkov, G. P., Bartunik, H. D., Stubbs, M. T., Kane, W. H., and Fuentes-Prior, P. (1999) *Nature (London)* **402**, 434–439

535. Xue, J., Kalafatis, M., Silveira, J. R., Kung, C., and Mann, K. G. (1994) *Biochemistry* **33**, 13109–13116

536. Comfurius, P., Smeets, E. F., Willems, G. M., Bevers, E. M., and Zwaal, R. F. A. (1994) *Biochemistry* **33**, 10319–10324

537. Betz, A., and Krishnaswamy, S. (1998) *J. Biol. Chem.* **273**, 10709–10718

538. Bottenus, R. E., Ichinose, A., and Davie, E. W. (1990) *Biochemistry* **29**, 11195–11209

539. Slon-Usakiewicz, J. J., Purisima, E., Tsuda, Y., Sulea, T., Pedyczak, A., Féthière, J., Cygler, M., and Konishi, Y. (1997) *Biochemistry* **36**, 13494–13502

540. Hughes, P. E., Morgan, G., Rooney, E. K., Brownlee, G. G., and Handford, P. (1993) *J. Biol. Chem.* **268**, 17727–17733

540a. Mathur, A., and Bajaj, S. P. (1999) *J. Biol. Chem.* **274**, 18477–18486

540b. Tolkatchev, D., Ng, A., Zhu, B., and Ni, F. (2000) *Biochemistry* **39**, 10365–10372

541. Xu, W.-f, Chung, D. W., and Davie, E. W. (1996) *J. Biol. Chem.* **271**, 27948–27953

541a. Brown, J. H., Volkmann, N., Jun, G., Henschen-Edman, A. H., and Cohen, C. (2000) *Proc. Natl. Acad. Sci. U.S.A.* **97**, 85–90

542. Doolittle, R. F., Everse, S. J., and Spraggon, G. (1996) *FASEB J.* **10**, 1464–1470

543. Everse, S. J., Spraggon, G., Veerapandian, L., Riley, M., and Doolittle, R. F. (1998) *Biochemistry* **37**, 8637–8642

544. Spraggon, G., Everse, S. J., and Doolittle, R. F. (1997) *Nature (London)* **389**, 455–462

545. Murthy, S. N. P., Wilson, J. H., Lukas, T. J., Veklich, Y., Weisel, J. W., and Lorand, L. (2000) *Proc. Natl. Acad. Sci. U.S.A.* **97**, 44–48

546. Pedersen, L. C., Yee, V. C., Bishop, P. D., Le Trong, I., Teller, D. C., and Stenkamp, R. E. (1994) *Protein Sci.* **3**, 1131–1135

547. Bernardo, M. M., Day, D. E., Olson, S. T., and Shore, J. D. (1993) *J. Biol. Chem.* **268**, 12468–12476

548. Beaubien, G., Rosinski-Chupin, I., Mattei, M. G., Mbikay, M., Chrétien, M., and Seidah, N. G. (1991) *Biochemistry* **30**, 1628–1635

549. Herwald, H., Renné, T., Meijers, J. C. M., Chung, D. W., Page, J. D., Colman, R. W., and Müller-Esterl, W. (1996) *J. Biol. Chem.* **271**, 13061–13067

550. Baglia, F. A., Jameson, B. A., and Walsh, P. N. (1993) *J. Biol. Chem.* **268**, 3838–3844

551. Hamaguchi, N., Charifson, P. S., Pedersen, L. G., Brayer, G. D., Smith, K. J., and Stafford, D. W. (1991) *J. Biol. Chem.* **266**, 15213–15220

552. Kurachi, S., Hitomi, Y., Furukawa, M., and Kurachi, K. (1995) *J. Biol. Chem.* **270**, 5276–5281

552a. Kurachi, S., Deyashiki, Y., Takeshita, J., and Kurachi, K. (1999) *Science* **285**, 739–743

553. Freedman, S. J., Furie, B. C., Furie, B., and Baleja, J. D. (1995) *Biochemistry* **34**, 12126–12137

554. Freedman, S. J., Blostein, M. D., Baleja, J. D., Jacobs, M., Furie, B. C., and Furie, B. (1996) *J. Biol. Chem.* **271**, 16227–16236

555. Lenting, P. J., Christophe, O. D., ter Maat, H., Rees, D. J. G., and Mertens, K. (1996) *J. Biol. Chem.* **271**, 25332–25337

556. Tagliavacca, L., Moon, N., Dunham, W. R., and Kaufman, R. J. (1997) *J. Biol. Chem.* **272**, 27428–27434

557. Pittman, D. D., Wang, J. H., and Kaufman, R. J. (1992) *Biochemistry* **31**, 3315–3325

558. Gilbert, G. E., and Baleja, J. D. (1995) *Biochemistry* **34**, 3022–3031

559. Gilbert, G. E., and Drinkwater, D. (1993) *Biochemistry* **32**, 9577–9585

560. Leyte, A., van Schijndel, H. B., Niehrs, C., Huttner, W. B., Verbeet, M. P., Mertens, K., and van Mourik, J. A. (1991) *J. Biol. Chem.* **266**, 740–746

561. Gitschier, J., Wood, W. I., Shuman, M. A., and Lawn, R. M. (1986) *Science* **232**, 1415–1416

562. Titani, K., and Walsh, K. A. (1988) *Trends Biochem. Sci.* **13**, 94–97

563. Sadler, J. E. (1998) *Ann. Rev. Biochem.* **67**, 395–424

564. Ruggeri, Z. M., and Ware, J. (1993) *FASEB J.* **7**, 308–316

565. Huizinga, E. G., van der Plas, R. M., Kroon, J., Sixma, J. J., and Gros, P. (1997) *Structure* **5**, 1147–1156

566. Emsley, J., Cruz, M., Handin, R., and Liddington, R. (1998) *J. Biol. Chem.* **273**, 10396–10401

567. Sugimoto, M., Dent, J., McClintock, R., Ware, J., and Ruggeri, Z. M. (1993) *J. Biol. Chem.* **268**, 12185–12192

568. George, J. N., Nurden, A. T., and Phillips, D. R. (1984) *N. Engl. J. Med.* **311**, 1084–1098

569. Cruz, M. A., Yuan, H., Lee, J. R., Wise, R. J., and Handin, R. I. (1995) *J. Biol. Chem.* **270**, 10822–10827

570. Beacham, D. A., Wise, R. J., Turci, S. M., and Handin, R. I. (1992) *J. Biol. Chem.* **267**, 3409–3415

571. Zhang, Z. P., Blombäck, M., Nyman, D., and Anvret, M. (1993) *Proc. Natl. Acad. Sci. U.S.A.* **90**, 7937–7940

References

572. Kaufman, R. J. (1989) *Nature (London)* **342**, 207–208

573. Dwarki, V. J., Belloni, P., Nijjar, T., Smith, J., Couto, L., Rabier, M., Clift, S., Berns, A., and Cohen, L. K. (1995) *Proc. Natl. Acad. Sci. U.S.A.* **92**, 1023–1027

574. Kay, M. A., Rothenberg, S., Landen, C. N., Bellinger, D. A., Leland, F., Toman, C., Finegold, M., Thompson, A. R., Read, M. S., Brinkhous, K. M., and Woo, S. L. C. (1993) *Science* **262**, 117–119

575. Jesty, J., Wun, T.-C., and Lorenz, A. (1994) *Biochemistry* **33**, 12686–12694

576. Hajjar, K. A. (1994) *N. Engl. J. Med.* **331**, 1585–1587

577. Björk, I., Nordling, K., and Olson, S. T. (1993) *Biochemistry* **32**, 6501–6505

578. Streusand, V. J., Björk, I., Gettins, P. G. W., Petitou, M., and Olson, S. T. (1995) *J. Biol. Chem.* **270**, 9043–9051

578a. Whisstock, J. C., Pike, R. N., Jin, L., Skinner, R., Pei, X. Y., Carrell, R. W., and Lesk, A. M. (2000) *J. Mol. Biol.* **301**, 1287–1305

579. Di Cera, E., Guinto, E. R., Vindigni, A., Dang, Q. D., Ayala, Y. M., Wuyi, M., and Tulinsky, A. (1995) *J. Biol. Chem.* **270**, 22089–22092

580. Weisel, J. W., Nagaswami, C., Young, T. A., and Light, D. R. (1996) *J. Biol. Chem.* **271**, 31485–31490

580a. Fuentes-Prior, P., Iwanaga, Y., Huber, R., Pagila, R., Rumennik, G., Seto, M., Morser, J., Light, D. R., and Bode, W. (2000) *Nature (London)* **404**, 518–525

580b. Baerga-Ortiz, A., Rezaie, A. R., and Komives, E. A. (2000) *J. Mol. Biol.* **296**, 651–658

581. Esmon, C. T. (1995) *FASEB J.* **9**, 946–955

582. Colpitts, T. L., Prorok, M., and Castellino, F. J. (1995) *Biochemistry* **34**, 2424–2430

582a. Shen, L., Dahlbäck, B., and Villoutreix, B. O. (2000) *Biochemistry* **39**, 2853–2860

583. Öhlin, A., Landes, G., Bourden, P., Oppenheimer, C., Wydro, R., and Stenflo, J. (1988) *J. Biol. Chem.* **263**, 19240–19248

584. Plow, E. F., Herren, T., Redlitz, A., Miles, L. A., and Hoover-Plow, J. L. (1995) *FASEB J.* **9**, 939–945

585. Menhart, N., Hoover, G. J., McCance, S. G., and Castellino, F. J. (1995) *Biochemistry* **34**, 1482–1488

586. Bendixen, E., Harpel, P. C., and Sottrup-Jensen, L. (1995) *J. Biol. Chem.* **270**, 17929–17933

587. Pirie-Shepherd, S. R., Jett, E. A., Andon, N. L., and Pizzo, S. V. (1995) *J. Biol. Chem.* **270**, 5877–5881

588. Marti, D. N., Hu, C.-K., An, S. S. A., von Haller, P., Schaller, J., and Llinás, M. (1997) *Biochemistry* **36**, 11591–11604

589. Bogusky, M. J., Dobson, C. M., and Smith, R. A. G. (1989) *Biochemistry* **28**, 6728–6735

590. Byeon, I.-J. L., Kelley, R. F., Mulkerrin, M. G., An, S. S. A., and Llinás, M. (1995) *Biochemistry* **34**, 2739–2750

591. Parry, M. A. A., Zhang, X. C., and Bode, W. (2000) *Trends Biochem. Sci.* **25**, 53–59

592. Oates, J. A., Wood, A. J. J., Loscalzo, J., and Braunwald, E. (1988) *N. Engl. J. Med.* **319**, 925–931

593. Mottonen, J., Strand, A., Symersky, J., Sweet, R. M., Danley, D. E., Geoghegan, K. F., Gerard, R. D., and Goldsmith, E. J. (1992) *Nature (London)* **355**, 270–273

594. Szyperski, T., Antuch, W., Schick, M., Betz, A., Stone, S. R., and Wüthrich, K. (1994) *Biochemistry* **33**, 9303–9310

595. Betz, A., Hofsteenge, J., and Stone, S. R. (1992) *Biochemistry* **31**, 4557–4562

596. Vitali, J., Martin, P. D., Malkowski, M. G., Robertson, W. D., Lazar, J. B., Winant, R. C., Johnson, P. H., and Edwards, B. F. P. (1992) *J. Biol. Chem.* **267**, 17670–17678

597. van de Locht, A., Lamba, D., Bauer, M., Huber, R., Friedrich, T., Kröger, B., Höffken, W., and Bode, W. (1995) *EMBO J.* **14**, 5149–5157

598. Stark, K. R., and James, A. A. (1998) *J. Biol. Chem.* **273**, 20802–20809

599. Lim-Wilby, M. S. L., Hallenga, K., De Maeyer, M., Lasters, I., Vlasuk, G. P., and Brunck, T. K. (1995) *Protein Sci.* **4**, 178–186

600. van de Locht, A., Stubbs, M. T., Bode, W., Friedrich, T., Bollschweiler, C., Höffken, W., and Huber, R. (1996) *EMBO J.* **15**, 6011–6017

601. Tabernero, L., Chang, C. Y. Y., Ohringer, S. L., Lau, W. F., Iwanowicz, E. J., Han, W.-C., Wang, T. C., Seiler, S. M., Roberts, D. G. M., and Sack, J. S. (1995) *J. Mol. Biol.* **246**, 14–20

602. Cheng, Y., Slon-Usakiewicz, J. J., Wang, J., Purisima, E. O., and Konishi, Y. (1996) *Biochemistry* **35**, 13021–13029

603. Krishnan, R., Tulinsky, A., Vlasuk, G. P., Pearson, D., Vallar, P., Bergum, P., Brunck, T. K., and Ripka, W. C. (1996) *Protein Sci.* **5**, 422–433

604. Kamata, K., Kawamoto, H., Honma, T., Iwama, T., and Kim, S.-H. (1998) *Proc. Natl. Acad. Sci. U.S.A.* **95**, 6630–6635

605. Lu, D., Bovill, E. G., and Long, G. L. (1994) *J. Biol. Chem.* **269**, 29032–29038

606. Ichinose, A. (1992) *Biochemistry* **31**, 3113–3118

607. Olds, R. J., Lane, D. A., Chowdhury, V., De Stefano, V., Leone, G., and Thein, S. L. (1993) *Biochemistry* **32**, 4216–4224

608. Watton, J., Longstaff, C., Lane, D. A., and Barrowcliffe, T. W. (1993) *Biochemistry* **32**, 7286–7293

609. Taylor, P. (1991) *J. Biol. Chem.* **266**, 4025–4028

610. Sussman, J. L., Harel, M., Frolow, F., Oefner, C., Goldman, A., Toker, L., and Silman, I. (1991) *Science* **253**, 872–879

611. Maelicke, A. (1991) *Trends Biochem. Sci.* **16**, 355–356

611a. Koellner, G., Kryger, G., Millard, C. B., Silman, I., Sussman, J. L., and Steiner, T. (2000) *J. Mol. Biol.* **296**, 713–735

612. Vellom, D. C., Radic', Z., Li, Y., Pickering, N. A., Camp, S., and Taylor, P. (1993) *Biochemistry* **32**, 12–17

613. Kovach, I. M., Huber, J. H.-A., and Schowen, R. L. (1988) *J. Am. Chem. Soc.* **110**, 590–593

614. Radic, Z., Kirchhoff, P. D., Quinn, D. M., McCammon, J. A., and Taylor, P. (1997) *J. Biol. Chem.* **272**, 23265–23277

615. Wlodek, S. T., Antosiewicz, J., and Briggs, J. M. (1997) *J. Am. Chem. Soc.* **119**, 8159–8165

616. Haas, R., Marshall, T. L., and Rosenberry, T. L. (1988) *Biochemistry* **27**, 6453–6457

617. Krejci, E., Thomine, S., Boschetti, N., Legay, C., Sketelj, J., and Massoulié, J. (1997) *J. Biol. Chem.* **272**, 22840–22847

618. Pindel, E. V., Kedishvili, N. Y., Abraham, T. L., Brzezinski, M. R., Zhang, J., Dean, R. A., and Bosron, W. F. (1997) *J. Biol. Chem.* **272**, 14769–14775

619. Li, J., Szittner, R., Derewenda, Z. S., and Meighen, E. A. (1996) *Biochemistry* **35**, 9967–9973

620. Derewenda, Z. S., and Sharp, A. M. (1993) *Trends Biochem. Sci.* **18**, 20–25

621. Quinn, D. M., and Feaster, S. R. (1998) in *Comprehensive Biological Catalysis. A Mechanistic Reference*, Vol. I (Sinnott, M., ed), pp. 455–482, Academic Press, San Diego, California

622. Bourne, Y., Martinez, C., Kerfelec, B., Lombardo, D., Chapus, C., and Cambillau, C. (1994) *J. Mol. Biol.* **238**, 709–732

623. Egloff, M.-P., Marguet, F., Buono, G., Verger, R., Cambillau, C., and van Tilbeurgh, H. (1995) *Biochemistry* **34**, 2751–2762

624. Dahim, M., and Brockman, H. (1998) *Biochemistry* **37**, 8369–8377

625. Jentoft, N. (1990) *Trends Biochem. Sci.* **15**, 291–294

626. Chen, J. C.-H., Miercke, L. J. W., Krucinski, J., Starr, J. R., Saenz, G., Wang, X., Spilburg, C. A., Lange, L. G., Ellsworth, J. L., and Stroud, R. M. (1998) *Biochemistry* **37**, 5107–5117

627. Gargouri, Y., Moreau, H., Pieroni, G., and Verger, R. (1988) *J. Biol. Chem.* **263**, 2159–2162

628. Warner, T. G., Dambach, L. M., Shin, J. H., and O'Brien, J. S. (1981) *J. Biol. Chem.* **256**, 2952–2957

629. Emmerich, J., Beg, O. U., Peterson, J., Previato, L., Brunzell, J. D., Brewer, J., HB, and Santamarina-Fojo, S. (1992) *J. Biol. Chem.* **267**, 4161–4165

630. Kobayashi, J., Applebaum-Bowden, D., Dugi, K. A., Brown, D. R., Kashyap, V. S., Parrott, C., Duarte, C., Maeda, N., and Santamarina-Fojo, S. (1996) *J. Biol. Chem.* **271**, 26296–26301

631. Shen, W.-J., Patel, S., Natu, V., and Kraemer, F. B. (1998) *Biochemistry* **37**, 8973–8979

632. Nicolas, A., Egmond, M., Verrips, C. T., de Vlieg, J., Longhi, S., Cambillau, C., and Martinez, C. (1996) *Biochemistry* **35**, 398–410

633. Grochulski, P., Bouthillier, F., Kazlauskas, R. J., Serreqi, A. N., Schrag, J. D., Ziomek, E., and Cygler, M. (1994) *Biochemistry* **33**, 3494–3500

634. Lang, D., Hofmann, B., Haalck, L., Hecht, H.-J., Spener, F., Schmid, R. D., and Schomburg, D. (1996) *J. Mol. Biol.* **259**, 704–717

635. Uppenberg, J., Öhrner, N., Norin, M., Hult, K., Kleywegt, G. J., Patkar, S., Waagen, V., Anthonsen, T., and Jones, T. A. (1995) *Biochemistry* **34**, 16838–16851

636. Brocca, S., Schmidt-Dannert, C., Lotti, M., Alberghina, L., and Schmid, R. D. (1998) *Protein Sci.* **7**, 1415–1422

637. Ordentlich, A., Barak, D., Kronman, C., Ariel, N., Segall, Y., Velan, B., and Shafferman, A. (1998) *J. Biol. Chem.* **273**, 19509–19517

638. Axelsen, P. H., Harel, M., Silman, I., and Sussman, J. L. (1994) *Protein Sci.* **3**, 188–197

639. Brady, L., Brzozowski, A. M., Derewenda, Z. S., Dodson, E., Dodson, G., Tolley, S., Turkenburg, J. P., Christiansen, L., Huge-Jensen, B., Norskov, L., Thim, L., and Menge, U. (1990) *Nature (London)* **343**, 767–770

640. Dennis, E. A. (1997) *Trends Biochem. Sci.* **22**, 1–2

641. Balsinde, J., and Dennis, E. A. (1997) *J. Biol. Chem.* **272**, 16069–16072

642. Leslie, C. C. (1997) *J. Biol. Chem.* **272**, 16709–16712

643. Scott, D. L., White, S. P., Otwinowski, Z., Yuan, W., Gelb, M. H., and Sigler, P. B. (1990) *Science* **250**, 1541–1546

644. Brunie, S., Bolin, J., Gewirth, D., and Sigler, P. B. (1985) *J. Biol. Chem.* **260**, 9742–9749

645. Scott, D. L., Achari, A., Vidal, J. C., and Sigler, P. B. (1992) *J. Biol. Chem.* **267**, 22645–22657

646. van den Berg, B., Tessari, M., de Haas, G. H., Verheij, H. M., Boelens, R., and Kaptein, R. (1995) *EMBO J.* **14**, 4123–4131

647. Gelb, M. H., Jain, M. K., Hanel, A. M., and Berg, O. G. (1995) *Ann. Rev. Biochem.* **64**, 653–688

648. Thuren, T., Tulkki, A.-P., Virtanen, J. A., and Kinnunen, P. K. J. (1987) *Biochemistry* **26**, 4907–4910

649. Huber, M., Yee, V. C., Burri, N., Vikerfors, E., Lavrijsen, A. P. M., Paller, A. S., and Hohl, D. (1997) *J. Biol. Chem.* **272**, 21018–21026

650. Yee, V. C., Pedersen, L. C., Le Trong, I., Bishop, P. D., Stenkamp, R. E., and Teller, D. C. (1994) *Proc. Natl. Acad. Sci. U.S.A.* **91**, 7296–7300

References

651. Kulkarni, M. S., and Sherman, F. (1994) *J. Biol. Chem.* **269**, 13141–13147

652. Wu, D., and Hersh, L. B. (1995) *J. Biol. Chem.* **270**, 29111–29116

653. Schwartz, B., and Drueckhammer, D. G. (1996) *J. Am. Chem. Soc.* **118**, 9826–9830

654. Ellis, J., Bagshaw, C. R., and Shaw, W. V. (1995) *Biochemistry* **34**, 16852–16859

655. Murray, I. A., Cann, P. A., Day, P. J., Derrick, J. P., Sutcliffe, M. J., Shaw, W. V., and Leslie, A. G. W. (1995) *J. Mol. Biol.* **254**, 993–1005

656. Guillaume, G., Vanhove, M., Lamotte-Brasseur, J., Ledent, P., Jamin, M., Joris, B., and Frère, J.-M. (1997) *J. Biol. Chem.* **272**, 5438–5444

657. Maveyraud, L., Pratt, R. F., and Samama, J.-P. (1998) *Biochemistry* **37**, 2622–2628

658. Crowder, M. W., Wang, Z., Franklin, S. L., Zovinka, E. P., and Benkovic, S. J. (1996) *Biochemistry* **35**, 12126–12132

659. Orellano, E. G., Girardini, J. E., Cricco, J. A., Ceccarelli, E. A., and Vila, A. J. (1998) *Biochemistry* **37**, 10173–10180

660. Lubkowski, J., Wlodawer, A., Ammon, H. L., Copeland, T. D., and Swain, A. L. (1994) *Biochemistry* **33**, 10257–10265

661. Stewart, A. E., Arfin, S. M., and Bradshaw, R. A. (1995) *J. Biol. Chem.* **270**, 25–28

662. Nakamura, N., Inoue, N., Watanabe, R., Takahashi, M., Takeda, J., Stevens, V. L., and Kinoshita, T. (1997) *J. Biol. Chem.* **272**, 15834–15840

663. Knowles, J. R. (1980) *Ann. Rev. Biochem.* **49**, 877–919

664. Westheimer, F. H. (1968) *Acc. Chem. Res.* **1**, 70–78

665. Westheimer, F. H. (1987) *Science* **235**, 1173–1178

666. Hengge, A. C. (1998) in *Comprehensive Biological Catalysis. A Mechanistic Reference,* Vol. I (Sinnott, M., ed), pp. 517–542, Academic Press, San Diego, California

667. Vincent, J. B., Crowder, M. W., and Averill, B. A. (1992) *Trends Biochem. Sci.* **17**, 105–110

668. Mislow, K. (1970) *Acc. Chem. Res.* **3**, 321–331

669. Bunton, C. A. (1970) *Acc. Chem. Res.* **3**, 257–265

670. Gorenstein, D. G., Luxon, B. A., and Findlay, J. B. (1979) *J. Am. Chem. Soc.* **101**, 5869–5875

671. Taira, K., Mock, W. L., and Gorenstein, D. G. (1984) *J. Am. Chem. Soc.* **106**, 7831–7835

672. Calvo, K. C., and Westheimer, F. H. (1984) *J. Am. Chem. Soc.* **106**, 4205–4210

673. Friedman, J. M., Freeman, S., and Knowles, J. R. (1988) *J. Am. Chem. Soc.* **110**, 1268–1275

674. Wu, Y.-D., and Houk, K. N. (1993) *J. Am. Chem. Soc.* **115**, 11997–12002

675. Buchwald, S. L., Friedman, J. M., and Knowles, J. R. (1984) *J. Am. Chem. Soc.* **106**, 4911–4916

676. Herschlag, D., and Jencks, W. P. (1989) *J. Am. Chem. Soc.* **111**, 7579–7586

677. Hengge, A. C., Edens, W. A., and Elsing, H. (1994) *J. Am. Chem. Soc.* **116**, 5045–5049

678. Jankowski, S., Quin, L. D., Paneth, P., and O'Leary, M. H. (1994) *J. Am. Chem. Soc.* **116**, 11675–11677

679. Herschlag, D., and Jencks, W. P. (1990) *Biochemistry* **29**, 5172–5179

680. Admiraal, S. J., and Herschlag, D. (1999) *J. Am. Chem. Soc.* **121**, 5837–5845

681. Cohn, M., and Reed, G. H. (1982) *Ann. Rev. Biochem.* **51**, 365–394

682. Villafranca, J. J., and Raushel, F. M. (1980) *Annu Rev Biophys Bioeng.* **9**, 363–392

683. McLaughlin, A. C., Leigh, J. S., Jr., and Cohn, M. (1976) *J. Biol. Chem.* **251**, 2777–2787

684. Leyh, T. S., Goodhart, P. J., Nguyen, A. C., Kenyon, G. L., and Reed, G. H. (1985) *Biochemistry* **24**, 308–316

685. Nageswara Rao, B. D., Kayne, F. J., and Cohn, M. (1979) *J. Biol. Chem.* **254**, 2689–2696

686. Stackhouse, J., Nambiar, K. P., Burbaum, J. J., Stauffer, D. M., and Benner, S. A. (1985) *J. Am. Chem. Soc.* **107**, 2757–2763

687. Rao, B. D. N., and Cohn, M. (1981) *J. Biol. Chem.* **256**, 1716–1721

688. Gadian, D. G., Radda, G. K., Brown, T. R., Chance, E. M., Dawson, M. J., and Wilke, D. R. (1981) *Biochem. J.* **194**, 215–228

689. Mendz, G. L., Robinson, G., and Kuchel, P. W. (1986) *J. Am. Chem. Soc.* **108**, 169–173

690. Midelfort, C. F., and Rose, I. A. (1976) *J. Biol. Chem.* **251**, 5881–5887

691. Lowe, G., and Tuck, S. P. (1986) *J. Am. Chem. Soc.* **108**, 1300–1301

692. Abbott, S. J., Jones, S. R., Weinman, S. A., Bockhoff, F. M., McLafferty, F. W., and Knowles, J. R. (1979) *J. Am. Chem. Soc.* **101**, 4323–4332

693. Pliura, D. H., Schomburg, D., Richard, J. P., Frey, P. A., and Knowles, J. R. (1980) *Biochemistry* **19**, 325–329

694. Sammons, R. D., and Frey, P. A. (1982) *J. Biol. Chem.* **257**, 1138–1141

695. Eckstein, F. (1980) *Trends Biochem. Sci.* **5**, 157–159

696. Baraniak, J., and Frey, P. A. (1988) *J. Am. Chem. Soc.* **110**, 4059–4060

697. Richard, J. P., and Frey, P. A. (1982) *J. Am. Chem. Soc.* **104**, 3476–3481

698. Hassett, A., Blättler, W., and Knowles, J. R. (1982) *Biochemistry* **21**, 6335–6340

699. Mehdi, S., and Gerlt, J. A. (1984) *Biochemistry* **23**, 4844–4852

700. Sammons, R. D., Frey, P. A., Bruzik, K., and Tsai, M.-D. (1983) *J. Am. Chem. Soc.* **105**, 5455–5461

701. Buchwald, S. L., and Knowles, J. R. (1980) *J. Am. Chem. Soc.* **102**, 6602–6604

702. Cornelius, R. D., and Cleland, W. W. (1978) *Biochemistry* **17**, 3279–3286

703. Dunaway-Mariano, D., and Cleland, W. W. (1980) *Biochemistry* **19**, 1506–1515

704. Speckhard, D. C., Pecoraro, V. L., Knight, W. B., and Cleland, W. W. (1986) *J. Am. Chem. Soc.* **108**, 4167–4171

705. Lin, I., and Dunaway-Mariano, D. (1988) *J. Am. Chem. Soc.* **110**, 950–956

706. Burgers, P. M. J., and Eckstein, F. (1980) *J. Biol. Chem.* **255**, 8229–8233

707. Lester, L. M., Rusch, L. A., Robinson, G. J., and Speckhard, D. C. (1998) *Biochemistry* **37**, 5349–5355

708. Matlin, A. R., Kendall, D. A., Carano, K. S., Banzon, J. A., Klecka, S. B., and Solomon, N. M. (1992) *Biochemistry* **31**, 8196–8200

709. Simopoulos, T. T., and Jencks, W. P. (1994) *Biochemistry* **33**, 10375–10380

710. Han, R., and Coleman, J. E. (1995) *Biochemistry* **34**, 4238–4245

711. Craig, D. B., Arriaga, E. A., Wong, J. C. Y., Lu, H., and Dovichi, N. J. (1996) *J. Am. Chem. Soc.* **118**, 5245–5253

712. Kimura, E., Kodama, Y., Koike, T., and Shiro, M. (1995) *J. Am. Chem. Soc.* **117**, 8304–8311

713. Stec, B., Hehir, M. J., Brennan, C., Nolte, M., and Kantrowitz, E. R. (1998) *J. Mol. Biol.* **277**, 647–662

713a. Stec, B., Holtz, K. M., and Kantrowitz, E. R. (2000) *J. Mol. Biol.* **299**, 1303–1311

714. Florián, J., and Warshel, A. (1997) *J. Am. Chem. Soc.* **119**, 5473–5474

715. Herschman, H. R. (1980) *Trends Biochem. Sci.* **5**, 82–84

716. Ostanin, K., and Van Etten, R. L. (1993) *J. Biol. Chem.* **268**, 20778–20784

717. VanEtten, R. L., Waymack, P. P., and Rehkop, D. M. (1974) *J. Am. Chem. Soc.* **96**, 6783–6785

718. Buchwald, S. L., Saini, M. S., Knowles, J. R., and Van Etten, R. L. (1984) *J. Biol. Chem.* **259**, 2208–2213

719. Ma, L., and Kantrowitz, E. R. (1996) *Biochemistry* **35**, 2394–2402

720. Sugiura, Y., Kawabe, H., Tanaka, H., Fujimoto, S., and Ohara, A. (1981) *J. Biol. Chem.* **256**, 10664–10670

721. Hong, S.-B., Kuo, J. M., Mullins, L. S., and Raushel, F. M. (1995) *J. Am. Chem. Soc.* **117**, 7580–7581

722. Benning, M. M., Kuo, J. M., Raushel, F. M., and Holden, H. M. (1995) *Biochemistry* **34**, 7973–7978

723. Vanhooke, J. L., Benning, M. M., Raushel, F. M., and Holden, H. M. (1996) *Biochemistry* **35**, 6020–6025

723a. Benning, M. M., Hong, S.-B., Raushel, F. M., and Holden, H. M. (2000) *J. Biol. Chem.* **275**, 30556–30560

724. Kuo, J. M., Chae, M. Y., and Raushel, F. M. (1997) *Biochemistry* **36**, 1982–1988

724a. Shim, H., and Raushel, F. M. (2000) *Biochemistry* **39**, 7357–7364

725. Williams, N. H. (1998) in *Comprehensive Biological Catalysis. A Mechanistic Reference,* Vol. I (Sinnott, M., ed), pp. 543–561, Academic Press, San Diego, California

726. Villeret, V., Huang, S., Zhang, Y., and Lipscomb, W. N. (1995) *Biochemistry* **34**, 4307–4315

727. Kurbanov, F. T., Choe, J.-y, Honzatko, R. B., and Fromm, H. J. (1998) *J. Biol. Chem.* **273**, 17511–17516

728. Choe, J.-Y., Fromm, H. J., and Honzatko, R. B. (2000) *Biochemistry* **39**, 8565–8574

728a. Lee, Y.-H., Ogata, C., Pflugrath, J. W., Levitt, D. G., Sarma, R., Banaszak, L. J., and Pilkis, S. J. (1996) *Biochemistry* **35**, 6010–6019

729. Lu, G., Giroux, E. L., and Kantrowitz, E. R. (1997) *J. Biol. Chem.* **272**, 5076–5081

730. Villeret, V., Huang, S., Zhang, Y., Xue, Y., and Lipscomb, W. N. (1995) *Biochemistry* **34**, 4299–4306

731. Lei, K.-J., Pan, C.-J., Liu, J.-L., Shelly, L. L., and Chou, J. Y. (1995) *J. Biol. Chem.* **270**, 11882–11886

732. Clottes, E., and Burchell, A. (1998) *J. Biol. Chem.* **273**, 19391–19397

733. Josse, J., and Wong, S. C. K. (1971) in *The Enzymes,* 3rd ed., Vol. 4 (Boyer, P. D., ed), pp. 499–541, Academic Press, New York

734. Harutyunyan, E. H., Oganessyan, V. Y., Oganessyan, N. N., Avaeva, S. M., Nazarova, T. I., Vorobyeva, N. N., Kurilova, S. A., Huber, R., and Mather, T. (1997) *Biochemistry* **36**, 7754–7760

735. Salminen, A., Efimova, I. S., Parfenyev, A. N., Magretova, N. N., Mikalahti, K., Goldman, A., Baykov, A. A., and Lahti, R. (1999) *J. Biol. Chem.* **274**, 33898–33904

736. Pohjanjoki, P., Lahti, R., Goldman, A., and Cooperman, B. S. (1998) *Biochemistry* **37**, 1754–1761

736a. Lappänen, V.-M., Nummelin, H., Hansen, T., Lahti, R., Schäfer, G., and Goldman, A. (1999) *Protein Sci.* **8**, 1218–1231

736b. Baykov, A. A., Fabrichniy, I. P., Pohjanjoki, P., Zyryanov, A. B., and Lahti, R. (2000) *Biochemistry* **39**, 11939–11947

737. Barford, D. (1996) *Trends Biochem. Sci.* **21**, 407–412

738. Goldberg, J., Huang, H., Kwon, Y., Greengard, P., Nairn, A. C., and Kuriyan, J. (1995) *Nature (London)* **376**, 745–753

739. Egloff, M.-P., Cohen, P. T. W., Reinemer, P., and Barford, D. (1995) *J. Mol. Biol.* **254**, 942–959

740. Zhang, J., Zhang, Z., Brew, K., and Lee, E. Y. C. (1996) *Biochemistry* **35**, 6276–6282

References

741. Zhang, M., Zhou, M., Van Etten, R. L., and Stauffacher, C. V. (1997) *Biochemistry* **36**, 15–23

742. Hengge, A. C., Sowa, G. A., Wu, L., and Zhang, Z.-Y. (1995) *Biochemistry* **34**, 13982–13987

743. Yuvaniyama, J., Denu, J. M., Dixon, J. E., and Saper, M. A. (1996) *Science* **272**, 1328–1331

744. D'Alessio, G., and Riordan, J. F., eds. (1997) *Ribonuclease Structure and Functions*, Academic Press, San Diego, California

745. Moore, S., and Stein, W. H. (1973) *Science* **180**, 458–464

746. Wlodawer, A., Svensson, L. A., Sjölin, L., and Gilliland, G. L. (1988) *Biochemistry* **27**, 2705–2717

747. Santoro, J., González, C., Bruix, M., Neira, J. L., Nieto, J. L., Herranz, J., and Rico, M. (1993) *J. Mol. Biol.* **229**, 722–734

748. Tilton, R. F., Jr., Dewan, J. C., and Petsko, G. A. (1992) *Biochemistry* **31**, 2469–2481

749. Fedorov, A. A., Joseph-McCarthy, D., Fedorov, E., Sirakova, D., Graf, I., and Almo, S. C. (1996) *Biochemistry* **35**, 15962–15979

750. Boix, E., Nogués, M. V., Schein, C. H., Benner, S. A., and Cuchillo, C. M. (1994) *J. Biol. Chem.* **269**, 2529–2534

751. Schultz, L. W., Quirk, D. J., and Raines, R. T. (1998) *Biochemistry* **37**, 8886–8898

752. Jackson, D. Y., Burnier, J., Quan, C., Stanley, M., Tom, J., and Wells, J. A. (1994) *Science* **266**, 243–247

753. Gutte, B. (1977) *J. Biol. Chem.* **252**, 663–670

754. Breslow, R., Dong, S. D., Webb, Y., and Xu, R. (1996) *J. Am. Chem. Soc.* **118**, 6588–6600

755. Martin, P. D., Coscha, M. S., and Edwards, B. F. P. (1987) *J. Biol. Chem.* **262**, 15930–15938

756. Wladkowski, B. D., Svensson, L. A., Sjolin, L., Ladner, J. E., and Gilliland, G. L. (1998) *J. Am. Chem. Soc.* **120**, 5488–5498

757. Eckstein, F. (1979) *Acc. Chem. Res.* **12**, 204–210

758. Matta, M. S., and Vo, D. T. (1986) *J. Am. Chem. Soc.* **108**, 5316–5318

759. Herschlag, D. (1994) *J. Am. Chem. Soc.* **116**, 11631–11635

760. Breslow, R., and Chapman, W. H., Jr. (1996) *Proc. Natl. Acad. Sci. U.S.A.* **93**, 10018–10021

761. Sowa, G. A., Hengge, A. C., and Cleland, W. W. (1997) *J. Am. Chem. Soc.* **119**, 2319–2320

762. Kawata, Y., Sakiyama, F., Hayashi, F., and Kyogoku, Y. (1990) *Eur. J. Biochem.* **187**, 255–262

763. Pletinckx, J., Steyaert, J., Zegers, I., Choe, H.-W., Heinemann, U., and Wyns, L. (1994) *Biochemistry* **33**, 1654–1662

764. Kurihara, H., Nonaka, T., Mitsui, Y., Ohgi, K., Irie, M., and Nakamura, K. T. (1996) *J. Mol. Biol.* **255**, 310–320

764a. Noguchi, S., Satow, Y., Uchida, T., Sasaki, C., and Matsuzaki, T. (1995) *Biochemistry* **34**, 15583–15591

765. Cordes, F., Starikov, E. B., and Saenger, W. (1995) *J. Am. Chem. Soc.* **117**, 10365–10372

766. Buckle, A. M., and Fersht, A. R. (1994) *Biochemistry* **33**, 1644–1653

767. Deutscher, M. P. (1988) *Trends Biochem. Sci.* **13**, 136–139

768. Benner, S. A., and Alleman, R. K. (1989) *Trends Biochem. Sci.* **14**, 396–397

769. Davies, J. F., II, Hostomska, Z., Hostomsky, Z., Jordan, S. R., and Matthews, D. A. (1991) *Science* **252**, 88–95

770. Katayanagi, K., Miyagawa, M., Matsushima, M., Ishikawa, M., Kanaya, S., Ikehara, M., Matsuzaki, T., and Morikawa, K. (1990) *Nature (London)* **347**, 306–309

771. Yang, W., Hendrickson, W. A., Crouch, R. J., and Satow, Y. (1990) *Science* **249**, 1398–1405

772. Acharya, K. R., Shapiro, R., Riordan, J. F., and Vallee, B. L. (1995) *Proc. Natl. Acad. Sci. U.S.A.* **92**, 2949–2953

773. Lequin, O., Albaret, C., Bontems, F., Spik, G., and Lallemand, J.-Y. (1996) *Biochemistry* **35**, 8870–8880

774. Shapiro, R. (1998) *Biochemistry* **37**, 6847–6856

775. Rosenberg, H. F., Tenen, D. G., and Ackerman, S. J. (1989) *Proc. Natl. Acad. Sci. U.S.A.* **86**, 4460–4464

776. Leland, P. A., Schultz, L. W., Kim, B.-M., and Raines, R. T. (1998) *Proc. Natl. Acad. Sci. U.S.A.* **95**, 10407–10412

777. D'Alessio, G., Di Donato, A., Parente, A., and Piccoli, R. (1991) *Trends Biochem. Sci.* **16**, 104–106

778. Hofsteenge, J., Kieffer, B., Matthies, R., Hemmings, B. A., and Stone, S. R. (1988) *Biochemistry* **27**, 8537–8544

779. Kole, R., and Altman, S. (1981) *Biochemistry* **20**, 1902–1906

780. Altman, S., Baer, M., Guerrier-Takada, C., and Vioque, A. (1986) *Trends Biochem. Sci.* **11**, 515–518

781. Altman, S., Kirsebom, L., and Talbot, S. (1993) *FASEB J.* **7**, 7–14

782. Yuan, Y., and Altman, S. (1995) *EMBO J.* **14**, 159–168

782a. Massire, C., Jaeger, L., and Westhof, E. (1998) *J. Mol. Biol.* **279**, 773–793

783. Waugh, D. S., Green, C. J., and Pace, N. R. (1989) *Science* **244**, 1569–1571

784. True, H. L., and Celander, D. W. (1998) *J. Biol. Chem.* **273**, 7193–7196

785. Cech, T. R. (1987) *Science* **236**, 1532–1539

786. Zaug, A. J., and Cech, T. R. (1986) *Science* **231**, 470–475

787. Cech, T. R. (1986) *Sci. Am.* **255** (Nov), 64–75

788. Brown, J. W., Haas, E. S., Gilbert, D. G., and Pace, N. R. (1994) *Nucleic Acids Res.* **22**, 3660–3662

789. Pan, T. (1995) *Biochemistry* **34**, 902–909

790. Golden, B. L., Gooding, A. R., Podell, E. R., and Cech, T. R. (1998) *Science* **282**, 259–264

791. Cate, J. H., Gooding, A. R., Podell, E., Zhou, K., Golden, B. L., Kundrot, C. E., Cech, T. R., and Doudna, J. A. (1996) *Science* **273**, 1678–1685

792. Ferré-D'Amaré, A. R., Zhou, K., and Doudna, J. A. (1998) *Nature (London)* **395**, 567–574

793. Baidya, N., and Uhlenbeck, O. C. (1997) *Biochemistry* **36**, 1108–1114

794. Pley, H. W., Flaherty, K. M., and McKay, D. B. (1994) *Nature (London)* **372**, 68–74

795. Scott, W. G., Murray, J. B., Arnold, J. R. P., Stoddard, B. L., and Klug, A. (1996) *Science* **274**, 2065–2069

796. Bevers, S., Xiang, G., and McLaughlin, L. W. (1996) *Biochemistry* **35**, 6483–6490

796a. Bevers, S., Ha, S. B., and McLaughlin, L. W. (1999) *Biochemistry* **38**, 7710–7718

796b. Kore, A. R., and Eckstein, F. (1999) *Biochemistry* **38**, 10915–10918

796c. Lyne, P. D., and Karplus, M. (2000) *J. Am. Chem. Soc.* **122**, 166–167

796d. Torres, R. A., and Bruice, T. C. (2000) *J. Am. Chem. Soc.* **122**, 781–791

797. Simorre, J.-P., Legault, P., Baidya, N., Uhlenbeck, O. C., Maloney, L., Wincott, F., Usman, N., Beigelman, L., and Pardi, A. (1998) *Biochemistry* **37**, 4034–4044

798. Sargueil, B., Pecchia, D. B., and Burke, J. M. (1995) *Biochemistry* **34**, 7739–7748

798a. Nissen, P., Hansen, J., Ban, N., Moore, P. B., and Steitz, T. A. (2000) *Science* **289**, 920–930

798b. Muth, G. W., Ortoleva-Donnelly, L., and Strobel, S. A. (2000) *Science* **289**, 947–950

798c. Nakano, S.-i, Chadalavada, D. M., and Bevilacqua, P. C. (2000) *Science* **287**, 1493–1497

799. Scott, W. G., and Klug, A. (1996) *Trends Biochem. Sci.* **21**, 220–224

800. Grasby, J. A. (1998) in *Comprehensive Biological Catalysis. A Mechanistic Reference*, Vol. I (Sinnott, M., ed), pp. 563–571, Academic Press, San Diego, California

801. Hampel, K. J., Walter, N. G., and Burke, J. M. (1998) *Biochemistry* **37**, 14672–14682

802. Pan, T., and Uhlenbeck, O. C. (1992) *Nature (London)* **358**, 560–563

803. Chartrand, P., Usman, N., and Cedergren, R. (1997) *Biochemistry* **36**, 3145–3150

804. Santoro, S. W., and Joyce, G. F. (1998) *Biochemistry* **37**, 13330–13342

805. Beebe, J. A., Kurz, J. C., and Fierke, C. A. (1996) *Biochemistry* **35**, 10493–10505

806. Torres, R. A., and Bruice, T. C. (1998) *Proc. Natl. Acad. Sci. U.S.A.* **95**, 11077–11082

807. Pyle, A. M. (1993) *Science* **261**, 709–714

808. Peracchi, A., Beigelman, L., Scott, E. C., Uhlenbeck, O. C., and Herschlag, D. (1997) *J. Biol. Chem.* **272**, 26822–26826

809. Pontius, B. W., Lott, W. B., and von Hippel, P. H. (1997) *Proc. Natl. Acad. Sci. U.S.A.* **94**, 2290–2294

810. Bruice, T. C., Tsubouchi, A., Dempcy, R. O., and Olson, L. P. (1996) *J. Am. Chem. Soc.* **118**, 9867–9875

811. Weinstein, L. B., Jones, B. C. N. M., Cosstick, R., and Cech, T. R. (1997) *Nature (London)* **388**, 805–808

812. Christian, E. L., and Yarus, M. (1993) *Biochemistry* **32**, 4475–4480

813. Orita, M., Vinayak, R., Andrus, A., Warashina, M., Chiba, A., Kaniwa, H., Nishikawa, F., Nishikawa, S., and Taira, K. (1996) *J. Biol. Chem.* **271**, 9447–9454

814. Hertel, K. J., Peracchi, A., Uhlenbeck, O. C., and Herschlag, D. (1997) *Proc. Natl. Acad. Sci. U.S.A.* **94**, 8497–8502

815. Peracchi, A., Karpeisky, A., Maloney, L., Beigelman, L., and Herschlag, D. (1998) *Biochemistry* **37**, 14765–14775

816. Joyce, G. F. (1992) *Sci. Am.* **267** (Dec), 90–97

817. Ekland, E. H., Szostak, J. W., and Bartel, D. P. (1995) *Science* **269**, 364–370

818. Wilson, C., and Szostak, J. W. (1995) *Nature (London)* **374**, 777–782

819. Wright, M. C., and Joyce, G. F. (1997) *Science* **276**, 614–617

820. Ohmichi, T., and Sugimoto, N. (1997) *Biochemistry* **36**, 3514–3521

821. Suga, H., Cowan, J. A., and Szostak, J. W. (1998) *Biochemistry* **37**, 10118–10125

822. Tsang, J., and Joyce, G. F. (1996) *J. Mol. Biol.* **262**, 31–42

823. Geyer, C. R., and Sen, D. (1998) *J. Mol. Biol.* **275**, 483–489

824. Ceska, T. A., and Sayers, J. R. (1998) *Trends Biochem. Sci.* **23**, 331–336

825. Weber, D. J., Libson, A. M., Gittis, A. G., Lebowitz, M. S., and Mildvan, A. S. (1994) *Biochemistry* **33**, 8017–8028

826. Libson, A. M., Gittis, A. G., and Lattman, E. E. (1994) *Biochemistry* **33**, 8007–8016

827. Hale, S. P., Poole, L. B., and Gerlt, J. A. (1993) *Biochemistry* **32**, 7479–7487

828. Grissom, C. B., and Markley, J. L. (1989) *Biochemistry* **28**, 2116–2124

829. Serpersu, E. H., Hibler, D. W., Gerlt, J. A., and Mildvan, A. S. (1989) *Biochemistry* **28**, 1539–1548

830. Suck, D., Lahm, A., and Oefner, C. (1988) *Nature (London)* **332**, 464–468

831. Jones, S. J., Worrall, A. F., and Connolly, B. A. (1996) *J. Mol. Biol.* **264**, 1154–1163

832. Frederick, C. A., Grable, J., Melia, M., Samudzi, C., Jen-Jacobson, L., Wang, B.-C., Greene, P., Boyer, H. W., and Rosenberg, J. M. (1984) *Nature (London)* **309**, 327–331

References

833. McClarin, J. A., Frederick, C. A., Wang, B.-C., Greene, P., Boyer, H. W., Grable, J., and Rosenberg, J. M. (1986) *Science* **234**, 1526–1541

834. McCarthy, A. D., and Hardie, D. G. (1984) *Trends Biochem. Sci.* **9**, 60–63

835. Kostrewa, D., and Winkler, F. K. (1995) *Biochemistry* **34**, 683–696

836. Newman, M., Strzelecka, T., Dorner, L. F., Schildkraut, I., and Aggarwal, A. K. (1995) *Science* **269**, 656–663

837. Bozic, D., Grazulis, S., Siksnys, V., and Huber, R. (1996) *J. Mol. Biol.* **255**, 176–186

838. Connolly, B. A., Echstein, F., and Pingoud, A. (1984) *J. Biol. Chem.* **259**, 10760–10763

839. Corey, D. R., Pei, D., and Schultz, P. G. (1989) *Biochemistry* **28**, 8277–8786

840. Ray, W. J., Jr., Hermodsen, M. A., Puvathingal, J. M., and Mahoney, W. C. (1983) *J. Biol. Chem.* **258**, 9166–9174

841. Percival, M. D., and Withers, S. G. (1992) *Biochemistry* **31**, 505–512

842. Dai, J.-B., Liu, Y., Ray, W. J., Jr., and Konno, M. (1992) *J. Biol. Chem.* **267**, 6322–6337

843. Rose, Z. B. (1986) *Trends Biochem. Sci.* **11**, 253–255

844. Shanske, S., Sakoda, S., Hermodson, M. A., DiMauro, S., and Schon, E. A. (1987) *J. Biol. Chem.* **262**, 14612–14617

845. Rigden, D. J., Alexeev, D., Phillips, S. E. V., and Fothergill-Gilmore, L. A. (1998) *J. Mol. Biol.* **276**, 449–459

846. Ravel, P., Craescu, C. T., Arous, N., Rosa, J., and Garel, M. C. (1997) *J. Biol. Chem.* **272**, 14045–14050

847. Collet, J.-F., Stroobant, V., Pirard, M., Delpierre, G., and Van Schaftingen, E. (1998) *J. Biol. Chem.* **273**, 14107–14112

848. Kalckar, H. M. (1985) *Trends Biochem. Sci.* **10**, 291–293

849. Reed, G. H., and Leyh, T. S. (1980) *Biochemistry* **19**, 5472–5480

850. Schulz, G. E., and Schirmer, R. H. (1979) *Principles of Protein Structure*, Springer-Verlag, New York (pp. 222–226)

851. Gerstein, M., Schulz, G., and Chothia, C. (1993) *J. Mol. Biol.* **229**, 494–501

852. Bernstein, B. E., and Hol, W. G. J. (1998) *Biochemistry* **37**, 4429–4436

853. Matte, A., Tari, L. W., and Delbaere, L. T. J. (1998) *Structure* **6**, 413–419

854. Vonrhein, C., Bönisch, H., Schäfer, G., and Schulz, G. E. *J. Mol. Biol.* **282**, 167–179

855. Bernstein, B. E., Michels, P. A. M., and Hol, W. G. J. (1997) *Nature (London)* **385**, 275–278

856. Yount, R. G., Babcock, D., Ballantyne, W., and Ojala, D. (1971) *Biochemistry* **10**, 2484–2489

857. Yount, R. G. (1975) *Adv. Enzymol.* **43**, 1–56

858. McPhillips, T. M., Hsu, B. T., Sherman, M. A., Mas, M. T., and Rees, D. C. (1996) *Biochemistry* **35**, 4118–4127

859. Schlauderer, G. J., Proba, K., and Schulz, G. E. (1996) *J. Mol. Biol.* **256**, 223–227

860. Byeon, I.-J. L., Shi, Z., and Tsai, M.-D. (1995) *Biochemistry* **34**, 3172–3182

861. Ray, B. D., Chau, M. H., Fife, W. K., Jarori, G. K., and Nageswara Rao, B. D. (1996) *Biochemistry* **35**, 7239–7246

862. Schlauderer, G. J., and Schulz, G. E. (1996) *Protein Sci.* **5**, 434–441

863. Schricker, R., Magdolen, V., Strobel, G., Bogengruber, E., Breitenbach, M., and Bandlow, W. (1995) *J. Biol. Chem.* **270**, 31103–31110

864. Teplyakov, A., Sebastiao, P., Obmolova, G., Perrakis, A., Brush, G. S., Bessman, M. J., and Wilson, K. S. (1996) *EMBO J.* **15**, 3487–3497

865. Zhang, Y., Li, Y., Wu, Y., and Yan, H. (1997) *J. Biol. Chem.* **272**, 29343–29350

866. Müller-Dieckmann, H.-J., and Schulz, G. E. (1995) *J. Mol. Biol.* **246**, 522–530

867. Schlichting, I., and Reinstein, J. (1997) *Biochemistry* **36**, 9290–9296

868. Runquist, J. A., Harrison, D. H. T., and Miziorko, H. M. (1998) *Biochemistry* **37**, 1221–1226

869. Moréra, S., Chiadmi, M., LeBras, G., Lascu, I., and Janin, J. (1995) *Biochemistry* **34**, 11062–11070

870. Webb, P. A., Perisic, O., Mendola, C. E., Backer, J. M., and Williams, R. L. (1995) *J. Mol. Biol.* **251**, 574–587

871. Abdulaev, N. G., Karaschuk, G. N., Ladner, J. E., Kakuev, D. L., Yakhyaev, A. V., Tordova, M., Gaidarov, I. O., Popov, V. I., Fujiwara, J. H., Chinchilla, D., Eisenstrein, E., Gilliland, G. L., and Ridge, K. D. (1998) *Biochemistry* **37**, 13958–13967

872. Mesnildrey, S., Agou, F., Karlsson, A., Bonne, D. D., and Véron, M. (1998) *J. Biol. Chem.* **273**, 4436–4442

873. Turano, A., Furey, W., Pletcher, J., Sax, M., Pike, D., and Kluger, R. (1982) *J. Am. Chem. Soc.* **104**, 3089–3095

874. Aleshin, A. E., Zeng, C., Bourenkov, G. P., Bartunik, H. D., Fromm, H. J., and Honzatko, R. B. (1998) *Structure* **6**, 39–50

875. Aleshin, A. E., Zeng, C., Bartunik, H. D., Fromm, H. J., and Honzatko, R. B. (1998) *J. Mol. Biol.* **282**, 345–357

876. Lundblad, V. (1998) *Proc. Natl. Acad. Sci. U.S.A.* **95**, 8415–8416

877. Behlke, J., Heidrich, K., Naumann, M., Müller, E.-C., Otto, A., Reuter, R., and Kriegel, T. (1998) *Biochemistry* **37**, 11989–11995

878. Anderson, C. M., Stenkamp, R. E., and Steitz, T. A. (1978) *J. Mol. Biol.* **123**, 15–23

879. Schwab, D. A., and Wilson, J. E. (1989) *Proc. Natl. Acad. Sci. U.S.A.* **86**, 2563–2567

880. Schirmer, T., and Evans, P. R. (1990) *Nature (London)* **343**, 140–145

881. Auzat, I., Le Bras, G., and Garel, J.-R. (1995) *J. Mol. Biol.* **246**, 248–253

882. Blake, C. C. F., and Evans, P. R. (1974) *J. Mol. Biol.* **484**, 585–601

883. Berger, S. A., and Evans, P. R. (1992) *Biochemistry* **31**, 9237–9242

884. Mizuguchi, H., Cook, P. F., Hasemann, C. A., and Uyeda, K. (1997) *Biochemistry* **36**, 8775–8784

885. Fujii, H., Krietsch, W. K. G., and Yoshida, A. (1980) *J. Biol. Chem.* **255**, 6421–6423

886. Tanaka, K. R., and Paglia, D. E. (1995) in *The Metabolic and Molecular Bases of Inherited Disease* (Scriver, C. R., Beaudet, A. L., Sly, W. S., and Valle, D., eds), pp. 3485–3511, McGraw-Hill, New York

887. Mas, M. T., Bailey, J. M., and Resplandor, Z. E. (1988) *Biochemistry* **27**, 1168–1172

888. Forstner, M., Müller, A., Stolz, M., and Wallimann, T. (1997) *Protein Sci.* **6**, 331–339

889. Fritz-Wolf, K., Schnyder, T., Wallimann, T., and Kabsch, W. (1996) *Nature (London)* **381**, 341–345

890. Zhou, G., Somasundaram, T., Blanc, E., Parthasarathy, G., Ellington, W. R., and Chapman, M. S. (1998) *Proc. Natl. Acad. Sci. U.S.A.* **95**, 8449–8454

891. Larsen, T. M., Benning, M. M., Rayment, I., and Reed, G. H. (1998) *Biochemistry* **37**, 6247–6255

892. Loria, J. P., and Nowak, T. (1998) *Biochemistry* **37**, 6967–6974

893. Lovell, S. C., Mullick, A. H., and Muirhead, H. (1998) *J. Mol. Biol.* **276**, 839–851

894. Cheng, X., Friesen, R. H. E., and Lee, J. C. (1996) *J. Biol. Chem.* **271**, 6313–6321

894a. Valentini, G., Chiarelli, L., Fortin, R., Speranza, M. L., Galizzi, A., and Mattevi, A. (2000) *J. Biol. Chem.* **275**, 18145–18152

895. Hunter, T., and Plowman, G. D. (1997) *Trends Biochem. Sci.* **22**, 18–22

896. Grant, B. D., Hemmer, W., Tsigelny, I., Adams, J. A., and Taylor, S. S. (1998) *Biochemistry* **37**, 7708–7715

897. Cheng, X., Shaltiel, S., and Taylor, S. S. (1998) *Biochemistry* **37**, 14005–14013

898. Engh, R. A., Girod, A., Kinzel, V., Huber, R., and Bossemeyer, D. (1996) *J. Biol. Chem.* **271**, 26157–26164

899. Herberg, F. W., Zimmermann, B., McGlone, M., and Taylor, S. S. (1997) *Protein Sci.* **6**, 569–579

900. Su, Y., Dostmann, W. R. G., Herberg, F. W., Durick, K., Xuong, N.-h, Eyck, L. T., Taylor, S. S., and Varughese, K. I. (1995) *Science* **269**, 807–813

901. Lohmann, S. M., Vaandrager, A. B., Smolenski, A., Walter, U., and DeJonge, H. R. (1997) *Trends Biochem. Sci.* **22**, 307–312

902. Xu, W., Harrison, S. C., and Eck, M. J. (1997) *Nature (London)* **385**, 595–602

903. Sicheri, F., Moarefi, I., and Kuriyan, J. (1997) *Nature (London)* **385**, 602–609

904. Adams, J. A. (1996) *Biochemistry* **35**, 10949–10956

905. Tesmer, J. J. G., Sunahara, R. K., Gilman, A. G., and Sprang, S. R. (1997) *Science* **278**, 1907–1916

906. Tesmer, J. J. G., Sunahara, R. K., Johnson, R. A., Gosselin, G., Gilman, A. G., and Sprang, S. R. (1999) *Science* **285**, 756–760

907. Sunahara, R. K., Tesmer, J. J. G., Gilman, A. G., and Sprang, S. R. (1997) *Science* **278**, 1943–1947

908. Zhao, Y., Schelvis, J. P. M., Babcock, G. T., and Marletta, M. A. (1998) *Biochemistry* **37**, 4502–4509

908a. Zhao, Y., Brandish, P. E., DiValentin, M., Schelvis, J. P. M., Babcock, G. T., and Marletta, M. A. (2000) *Biochemistry* **39**, 10848–10854

909. Joyce, C. M., and Steitz, T. A. (1994) *Ann. Rev. Biochem.* **63**, 777–822

910. Kim, Y., Eom, S. H., Wang, J., Lee, D.-S., Suh, S. W., and Steitz, T. A. (1995) *Nature (London)* **376**, 612–616

911. Doublié, S., Tabor, S., Long, A. M., Richardson, C. C., and Ellenberger, T. (1998) *Nature (London)* **391**, 251–258

912. Patel, P. H., Jacobo-Molina, A., Ding, J., Tantillo, C., Clark, A. D., Jr., Raag, R., Nanni, R. G., Hughes, S. H., and Arnold, E. (1995) *Biochemistry* **34**, 5351–5363

913. Bartlett, P. A., and Eckstein, F. (1988) *J. Biol. Chem.* **257**, 8879–8884

913a. Stivers, J. T., Nawrot, B., Jagadeesh, G. J., Stec, W. J., and Shuman, S. (2000) *Biochemistry* **39**, 5561–5572

914. Smith, C. A., and Rayment, I. (1996) *Biophys. J.* **70**, 1590–1602

915. Wittinghofer, A., and Pai, E. F. (1991) *Trends Biochem. Sci.* **16**, 382–387

916. Coleman, D. E., and Sprang, S. R. (1998) *Biochemistry* **37**, 14376–14385

917. Falany, C. N. (1997) *FASEB J.* **11**, 206–216

918. Varin, L., Marsolais, F., Richard, M., and Rouleau, M. (1997) *FASEB J.* **11**, 517–525

919. Kakuta, Y., Petrotchenko, E. V., Pedersen, L. C., and Negishi, M. (1998) *J. Biol. Chem.* **273**, 27325–27330

920. Zhang, H., Varmalova, O., Vargas, F. M., Falany, C. N., and Leyh, T. S. (1998) *J. Biol. Chem.* **273**, 10888–10892

920a. Hiraoka, N., Nakagawa, H., Ong, E., Akama, T. O., Fukuda, M. N., and Fukuda, M. (2000) *J. Biol. Chem.* **275**, 20188–20196

921. Klaassen, C. D., and Boles, J. W. (1997) *FASEB J.* **11**, 404–418

References

922. Neufeld, E. F., and Muenzer, J. (1995) in *The Metabolic and Molecular Bases of Inherited Disease*, 7th ed., Vol. 2 (Scriver, C. R., Beaudet, A. L., Sly, W. S., and Valle, D., eds), pp. 2465–2494, McGraw-Hill, New York

923. Lukatela, G., Krauss, N., Theis, K., Selmer, T., Gieselmann, V., von Figura, K., and Saenger, W. (1998) *Biochemistry* **37**, 3654–3664

923a. Dierks, T., Lecca, M. R., Schlotterhose, P., Schmidt, B., and von Figura, K. (1999) *EMBO J.* **18**, 2084–2091

924. Waldow, A., Schmidt, B., Dierks, T., von Bülow, R., and von Figura, K. (1999) *J. Biol. Chem.* **274**, 12284–12288

924a. Szameit, C., Miech, C., Balleininger, M., Schmidt, B., von Figura, K., and Dierks, T. (1999) *J. Biol. Chem.* **274**, 15375–15381

925. Fox, D. K., and Roseman, S. (1986) *J. Biol. Chem.* **261**, 13487–13497

926. Joyce, M. A., Fraser, M. E., James, M. N. G., Bridger, W. A., and Wolodko, W. T. (2000) *Biochemistry* **39**, 17–25

927. Johnson, J. D., Muhonen, W. W., and Lambeth, D. O. (1998) *J. Biol. Chem.* **273**, 27573–27579

928. Liaw, S.-H., and Eisenberg, D. (1994) *Biochemistry* **33**, 675–681

929. Alibhai, M., and Villafranca, J. J. (1994) *Biochemistry* **33**, 682–686

930. Meister, A. (1968) *Adv. Enzymol.* **31**, 183–218

931. Meister, A. (1974) in *The Enzymes*, 3rd ed., Vol. 10 (Boyer, P. D., ed), pp. 669–754, Academic Press, New York

932. Whitty, A., Fierke, C. A., and Jencks, W. P. (1995) *Biochemistry* **34**, 11678–11689

933. Rochet, J.-C., and Bridger, W. A. (1994) *Protein Sci.* **3**, 975–981

934. Selmer, T., and Buckel, W. (1999) *J. Biol. Chem.* **274**, 20772–20778

Study Questions

1. Outline the reactions by which glyceraldehyde 3-phosphate is converted to 3-phosphoglycerate with coupled synthesis of ATP in the glycolysis pathway. Show important mechanistic details.

2. Papain is a protein-hydrolyzing (proteolytic) enzyme with an –SH group and an imidazole group at the active site. Write a reasonable structure for a "tetrahedral intermediate" that would be expected to arise during formation of an acyl enzyme intermediate.

3. Adenylate kinase catalyzes the interconversion of ATP, AMP, and ADP.
   a) Draw a reasonable structure for a penta-covalent intermediate derived from ATP and AMP.
   b) Draw a reasonable structure for the transition state leading from ATP + AMP to two molecules of ADP in an S_N2-like reaction.

4. Penicillin inhibits a D-alanyl-D-alanine transpeptidase that catalyzes the reaction

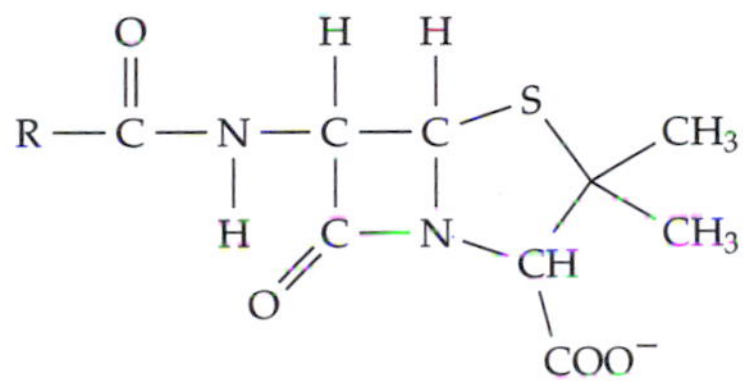

Penicillin

$$R'\text{-D-alanyl-D-alanine} + H_2N\text{-R} \rightarrow$$
$$R'\text{-D-alanyl-NH-R} + \text{D-alanine}$$

where R and R' are different parts of a bacterial peptidoglycan. Write a step-by-step mechanism for this reaction and indicate how penicillin may inhibit the enzyme by combining with it irreversibly. See Strynodka et al., Nature **359**, 700–705, 1992, for related reaction of penicillin with a penicillinase.

5. Write out step-by-step chemical mechanisms for the following enzymatic reaction. Use small arrows to indicate directions of electron flow. Remember to have all electrons move in the same direction in any single structure.

$$\text{Amino acid} + \text{tRNA} + \text{ATP} \rightarrow$$
$$\text{Aminoacyl-tRNA} + \text{AMP} + \text{PP}_i$$

6. Trypsin in which Asp 102 has been replaced by Asn has 10^4 times less catalytic activity than natural trypsin at neutral pH. From the crystal structure of the mutant enzyme it appears that the imidazole group of His 57 is held by the Asn side chain in the wrong tautomeric form for catalysis. Explain. Compare this incorrect tautomeric form with that in the initial structure shown in Fig. 12-11.

7. A recent discovery in biochemistry is that RNA can act as an enzyme in chemical reactions, usually reactions involving RNA hydrolysis. Discuss the features of RNA structure that might favor evolution of enzymes composed entirely of a single polyribonucleotide chain, and describe a proposed mechanism for RNA-catalyzed hydrolysis of RNA molecules.

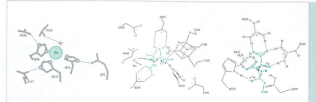

One of the simplest biochemical addition reactions is the hydration of carbon dioxide to form carbonic acid, which is released from the zinc-containing **carbonic anhydrase** (left, Fig. 13-1) as HCO_3^-. **Aconitase** (center, Fig. 13-4) is shown here removing a water molecule from isocitrate, an intermediate compound in the citric acid cycle. The H_2O that is removed will become bonded to an iron atom of the Fe_4S_4 cluster at the active site as indicated by the black H_2O. An enolate anion derived from acetyl-CoA adds to the carbonyl group of oxaloacetate to form citrate in the active site of **citrate synthase** (right, Fig. 13-9) to initiate the citric acid cycle.

Contents

Enzymatic Addition, Elimination, Condensation, and Isomerization: Roles for Enolate and Carbocation Intermediates

13

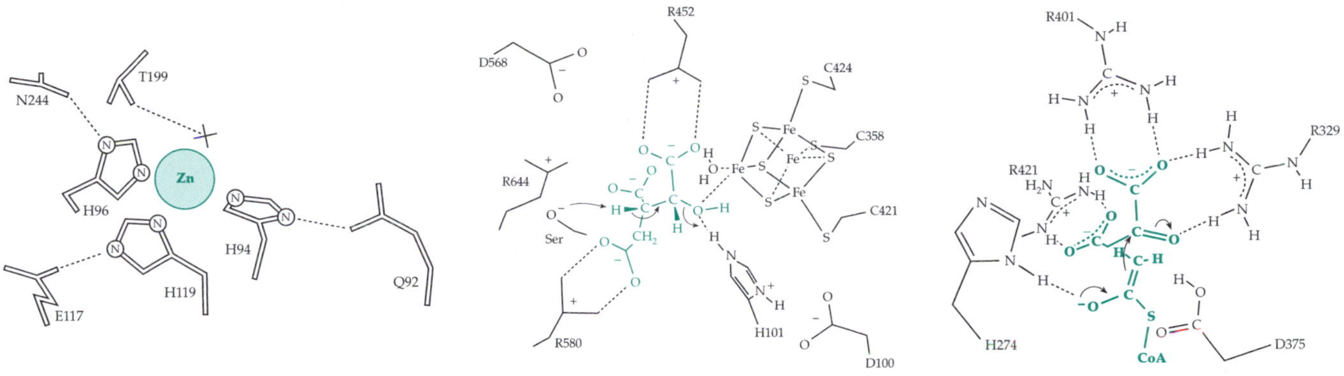

In Chapter 12 we considered reactions by which living cells are able to transfer groups from one molecule to another using nucleophilic displacements. We also showed how transfer reactions can be utilized by ligases to join two molecules together with the Gibbs energy of cleavage of ATP or of a related molecule driving the reaction. In this chapter we will examine addition reactions, which provide a simple way of joining two molecules by means of C–O, C–N, C–S, or C–C bonds. Among these are the aldol and Claisen-type condensations by which C–C bonds are formed. We will also consider elimination reactions and decarboxylations, which are the reverse of addition and condensation reactions, as well as mechanistically related isomerizations. Many reactions of these types occur in the major pathways of metabolism.

A. Addition of R–OH, R–NH$_2$, and R–SH to Polarized Double Bonds

Next to nucleophilic displacement, the commonest mechanistic processes in enzymatic catalysis are addition to double bonds and elimination to form double bonds. These often involve addition of a nucleophile together with a proton to a highly polarized double bond such as C=O or C=N–. In other reactions, which are discussed in Section C,2, the nucleophile attacks one end of a C=C bond that is polarized by conjugation with C=O or C=N.

Alcohols, amines, and thiols add readily to the electrophilic carbon of the carbonyl group to form **hemiacetals, carbinolamines, hemiketals,** and **hemimercaptals**. An example is the formation of ring structures of sugars (Eq. 4-1). Water can also add to carbonyl groups and most aliphatic carbonyl compounds

exist in water as an equilibrium mixture with a covalent hydrate (Eq. 13-1). For example, acetaldehyde in aqueous solution consists of a mixture of about 50% free aldehyde and 50% hydrate in rapidly reversible equilibrium[1,2] and formaldehyde is over 99.9% hydrated.[3]

$$ CH_3 - \overset{\displaystyle O}{\underset{\displaystyle H}{C}} \; + \; H_2O \; \rightleftharpoons \; CH_3 - \overset{\displaystyle OH}{\underset{\displaystyle H}{\underset{|}{\overset{|}{C}}}} - OH $$

(13-1)

Addition reactions often occur as parts of more complex reactions. For example, a thiol group of glyceraldehyde-3-phosphate dehydrogenase reacts with the aldehyde substrate to form a hemimercaptal, which is subsequently oxidized to a thioester (see Fig. 15-6).

1. Carbonic Anhydrase

Another simple addition reaction is the hydration of CO$_2$ to form the bicarbonate ion. Without catalysis the reaction may require several seconds,[4,5] the apparent first-order rate constant being ~0.03 s^{-1} at 25°C. Cells must often hasten the process. The specific catalyst carbonic anhydrase is widespread in its distribution

$$ O=C=O \; \xrightarrow{\;H_2O\;} \; HO-\overset{\displaystyle O}{\underset{\displaystyle OH}{C}} \; \xrightarrow{\;H^+\;} \; HO-\overset{\displaystyle O}{\underset{\displaystyle O^-}{C}} $$

(13-2)

and is especially active in tissues (e.g., red blood cells, and lungs) that are involved in respiration. One liter of mammalian blood contains 1–2 g of this enzyme, a monomeric 30-kDa protein containing ~260 amino acids and one tightly bound ion of Zn^{2+}. Erythrocytes contain two isoenzymes (I and II) of carbonic anhydrase and the human body contains at least eight distinct isoenzymes (I – VIII).[5–8] They are found wherever there is a high demand for CO_2 or bicarbonate. Isoenzyme I, II, III, and VII are cytosolic. Carbonic anhydrase I is specific to erythrocytes, while isoenzyme II is present in most cells. Hereditary lack of carbonic anhydrase II is associated with **osteopetrosis** (marble bone disease), a condition involving failure of bone resorption and the calcification of other tissues.[9] The generation of acidity according to Eq. 13-2 is presumably required for dissolution of bone by the osteoclasts.

Isoenzymes III and VII have a more specialized distribution. Carbonic anhydrase III is abundant in adipocytes which use bicarbonate in fatty acid synthesis.[7] Isoenzyme V is present in the mitochondrial matrix and is also abundant in both adipocytes and liver.[7,8] Isoenzyme IV is a larger membrane-associated form, while VI is secreted into the saliva.[10] Carbonic anhydrase has also been identified in *E. coli.*,[11] in a methanobacterium,[12] and in green plants.[13,13a] A 60-kDa carbonic anhydrase called **nacrein** is found in the organic matrix of the nacreous layer of the pearl oyster, the layer that forms aragonite (orthorhombic calcium carbonate) in the shell and in pearls.[14]

X-ray studies of carbonic anhydrases I and II, from human blood, revealed that both have an ellipsoidal shape of dimensions ~4.1 x 4.1 x 4.7 nm.[15,16] The zinc atom in each molecule lies in a deep pocket ~1.2 nm from the surface and is surrounded by three histidine side chains and one H_2O or OH^- ion, the four ligands forming a distorted tetrahedron (Fig. 13-1). The coordinating imidazole group from His 119 is hydrogen bonded to a carboxylate group of Glu 117, a feature reminiscent of the charge-relay system of serine protease. This carboxylate group is also bound into a more extended hydrogen-bonded network, part of which is indicated in Fig. 13-1. The other imidazole groups also form hydrogen bonds to protein groups and the zinc-bound H_2O is involved in an extensive hydrogen-bonded network with several other bound water molecules and protein side chains.[17a] Most of these structural features are conserved in the other mammalian isoenzymes.[5] However, from X-ray absorption spectroscopy (EXAFS) it appears that spinach carbonic anhydrase contains one or more sulfur ligands to the zinc,[13] while the enzymes from the archaeon *Methanosarcina thermophila* have a left-handed β helix structure (see Figs. 2-17 and 13-3).[12] Nevertheless, this enzyme has the same three-histidine Zn – OH structure found in the mammalian enzymes.

Carbonic anhydrase II is among the most rapid

enzymes known, with the turnover number at 25°C for hydration of CO_2 being ~10^6 s^{-1}. The same enzyme catalyzes hydration of acetaldehyde (Eq. 13-1) but at a 1000-fold slower rate. A pK_a of ~7 controls the activity. This appears to represent the loss of H$^+$ from the Zn^{2+}–OH_2 complex[18] to give Zn^+–OH. The latter is in effect a stabilized hydroxide ion existing at a pH at which OH^- is normally not present in quantity. It is this hydroxide ion that adds to the CO_2 or to the aldehyde substrate (Eq. 13-3, step a).[18–20] In step b a water molecule replaces the departing bicarbonate.[17] A variety of data

(13-3)

indicate that proton transfers mediated by the enzyme are essential parts of the carbonic anhydrase mechanism.[18–19b,21–23] One proposal is that the nearby imidazole group of His 64 (not shown in Fig. 13-1) deprotonates the bound H_2O via a hydrogen-bonded network of bound water molecules (Eq. 13-3, step c). The side chain of Thr 199 may function in a cyclic proton transfer in step c.[19] The proton generated in step c is released to the solvent in a process that is catalyzed by buffer anions or by amines such as histamine. The latter binds at the edge of the active site and forms an additional hydrogen-bonded pathway to the zinc-bound H_2O.[24] A different proton shuttle pathway has been proposed for the slower carbonic anhydrase III.[25]

Related to the reaction catalyzed by carbonic anhydrase is the addition of an amino group to CO_2 (Eq. 2-21) to form a carbamino group (–NH–COO$^-$). This reaction is essential to the functioning of hemoglobin, which must carry large amounts of CO_2, as carbamino groups, from tissues to the lungs, (Eq. 7-47) and to some enzymes such as ribulose bisphosphate carboxylase (Figs. 13-10, 13-11).

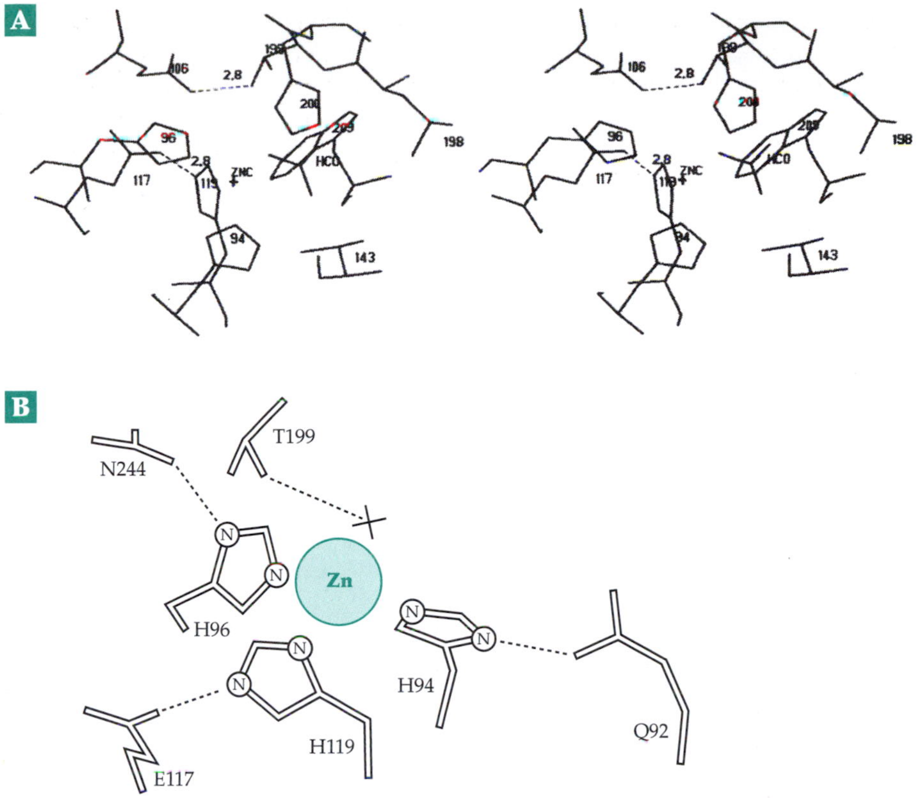

Figure 13-1 (A) Stereoscopic view showing the binding pocket for HCO_3^- in the active site of human carbonic acid anhydrase I. Also shown are the hydrogen-bonded pairs E117–H119 and E106–T199. ZNC, zinc ion; HCO, bicarbonate. From Kumar and Kannan.[16] (B) View of the active site of carbonic anhydrase II. The orientation is a little different than in (A). The location of H_2O or ^-OH bound to the zinc ion is marked X. Hydrogen bonds from the three histidines that form coordinate bonds to the zinc, Q92, E117, the backbone carbonyl of N244, and the hydrogen bond from the zinc-bound hydroxyl to the T199 side chain are shown as dashed lines. From Kiefer *et al.*[17]

2. Imines (Schiff Bases)

As we have seen already, many enzymatic reactions depend upon formation of **imines**, which are commonly called Schiff bases. The two-step formation of Schiff bases consists of addition of an amino group to a carbonyl group to form a carbinolamine followed by elimination of water (Eq. 13-4).[26] One group of **aldolases** (Section D) have, at their active centers,

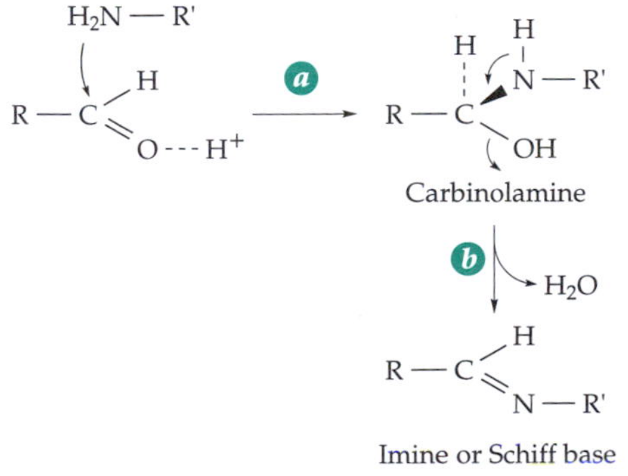

$$(13\text{-}4)$$

lysine side chains which form Schiff bases with the ketone substrates prior to the principal reaction of breaking or forming a C – C bond. Similarly, the initial reaction of the aldehyde coenzyme **pyridoxal phosphate** with amino acid substrates is the formation of Schiff bases (Fig. 14-4). Indeed, *the groups C=O and H_2N– are inherently complementary* and their interaction through imine formation is extremely common.

Schiff bases often form within a fraction of a second, but one or both steps of Eq. 13-4 may require catalysis to achieve enzymatic velocities.[27] The reaction is usually completely reversible and formation constants are often low enough that a carbonyl compound present in small amounts will not react extensively with an amine unless the two are brought together on an enzyme surface. If the amino group is hydrogen bonded to the enzyme, the proton may remain on the Schiff base nitrogen, enhancing the electron-accepting properties of the C=N group. The pK_a values of Schiff bases with aliphatic aldehydes are usually 1– 2 units lower than those of the corresponding primary amines. However, the local environment may sometimes cause the pK_a to be increased. Most Schiff bases are reduced readily by sodium borohydride or sodium cyanoborohydride to form secondary amines in which the original aldehydes are bound covalently to the original amino groups (Eq. 3-34). This provides a method for locating sites of Schiff base formation in enzymes. An isotopically substituted aldehyde or amine can be employed, or an isotopic label can be introduced, by using ^2H- or ^3H-containing sodium borohydride in the reduction.

3. Stereochemistry of Addition to Trigonal Carbon Atoms

Adducts formed by enzymatic addition of nucleophiles to carbonyl groups are usually chiral. For example, addition of an amino group to the "front" side of the carbonyl carbon in Eq. 13-5 (the *si* face as defined in Chapter 9) creates a carbinolamine of the *S* configuration. Addition of the amino group from behind the plane of the paper (the *re* face) would lead to a carbinolamine of the *R* configuration.

BOX 13-A ZINC

The average human ingests 10–15 mg of zinc a day.[a] Although it is poorly absorbed, the concentrations of zinc in tissues are relatively high and the metal plays an essential role in a multitude of enzymes. The total zinc content of a 70-kg person is 1.4–2.3 g. A typical tissue concentration of Zn^{2+} is 0.3–0.5 mM; an unusually high content of ~15 mM is found in the prostate gland.

Zinc ion is much more tightly bound to most organic ligands than is Mg^{2+} (Table 6-9). It has a filled $3d$ shell and tends to form four ionic bonds with a tetrahedral geometry, often with nitrogen- or sulfur-containing ligands.[b] Unlike Mg^{2+}, which interacts rapidly and reversibly with enzymes, Zn^{2+} tends to be tightly bound within over 300 **metallo-enzymes**.[c–e] A common feature is the surrounding of the Zn^{2+} at the active center by three imidazole groups, the fourth coordination position being free for interaction with substrate. The second nitrogen of the imidazole ring in many instances is hydrogen bonded to a main chain carbonyl group of the peptide, a feature that is also shared by histidines in other metalloproteins.

The most important chemical function of Zn^{2+} in enzymes is probably that of a Lewis acid providing a concentrated center of positive charge at a nucleophilic site on the substrate.[f] This role for Zn^{2+} is discussed for carboxypeptidases (Fig.12-16) and thermolysin,[g] alkaline phosphatase (Fig. 12-23),[h] RNA polymerases, DNA polymerases, carbonic anhydrase (Fig. 13-1),[i] class II aldolases (Fig. 13-7), some alcohol dehydrogenases (Fig. 15-5), and superoxide dismutases (Fig.16-22). Zinc ions in enzymes can often be replaced by Mn^{2+}, Co^{2+}, and other ions with substantial retention of catalytic activity.[f,j]

In addition to its function in catalysis, zinc often plays an important structural role, e.g., in the **zinc finger** transcriptional regulators (Fig. 5-38).[k] Zinc ions bind to insulin and stabilize its hexameric structure (Fig. 7-18).[l] Six Zn^{2+} ions are present in the hexagonal tail plate of the T-even bacteriophage (Box 7-C) and appear to be essential for invasion of bacteria.[m] In carnivores, the **tapetum**, the reflecting layer behind the retina of the eye of many animals, contains crystals of the Zn^{2+}–cysteine complex.

Since zinc ions have no color their presence has often been overlooked. Zinc ions will doubtless be found in many more places within cells. Zinc is usually the major component of the bound metals in the **metallothioneins** (Box 6-E). These small 6.6-kDa proteins which contain ~33% cysteine and bind as many as six ions of Cd^{2+}, Hg^{2+}, Cu^{2+}, or Zn^{2+} per molecule are present in all animal tissues as well as in plants and some bacteria.[j]

From a nutritional viewpoint, Cu^{2+} competes with zinc ion, as does the very toxic Cd^{2+}. The latter accumulates in the cortex of the kidney. Dietary cadmium in concentrations less than those found in human kidneys shortens the lives of rats and mice. However, some marine diatoms contain a cadmium-dependent carbonic anhydrase.[n] Although zinc deficiency was once regarded as unlikely in humans, it is now recognized as occurring under a variety of circumstances[o,p] and is well-known in domestic animals.[q] Consumption of excessive amounts of protein as well as alcoholism, malabsorption, sickle cell anemia, and chronic kidney disease can all be accompanied by zinc deficiency.

[a] O'Dell, B. L., and Campbell, B. J. (1971) *Comprehensive Biochemistry* **21**, 179–216
[b] Bock, C. W., Katz, A. K., and Glusker, J. P. (1995) *J. Am. Chem. Soc.* **117**, 3754–3765
[c] Berg, J. M., and Shi, Y. (1996) *Science* **271**, 1081–1085
[d] Vallee, B. L., and Auld, D. S. (1993) *Biochemistry* **32**, 6493–6500
[e] Coleman, J. E. (1992) *Ann. Rev. Biochem.* **61**, 897–946
[f] Mildvan, A. S. (1974) *Ann. Rev. Biochem.* **43**, 357–399
[g] Holland, D. R., Hausrath, A. C., Juers, D., and Matthews, B. W. (1995) *Protein Sci.* **4**, 1955–1965
[h] Kimura, E., Kodama, Y., Koike, T., and Shiro, M. (1995) *J. Am. Chem. Soc.* **117**, 8304–8311
[i] Lesburg, C. A., and Christianson, D. W. (1995) *J. Am. Chem. Soc.* **117**, 6838–6844
[j] Kagi, J. H. R., Himmelhoch, S. R., Whanger, P. D., Bethune, J. L., and Vallee, B. L. (1974) *J. Biol. Chem.* **249**, 3537–3542
[k] Berg, J. M. (1990) *J. Biol. Chem.* **265**, 6513–6516
[l] Hill, C. P., Dauter, Z., Dodson, E. J., Dodson, G. G., and Dunn, M. F. (1991) *Biochemistry* **30**, 917–924
[m] Kozloff, L. M., and Lute, M. (1977) *J. Biol. Chem.* **252**, 7715–7724
[n] Lane, T. W., and Morel, F. M. M. (2000) *Proc. Natl. Acad. Sci. U.S.A.* **97**, 4627–4631
[o] Prasad, A. S. (1984) *Fed. Proc.* **43**, 2829–2834
[p] Day, H. G. (1991) *FASEB J.* **5**, 2315–2316
[q] Luecke, R. W. (1984) *Fed. Proc.* **43**, 2823–2828

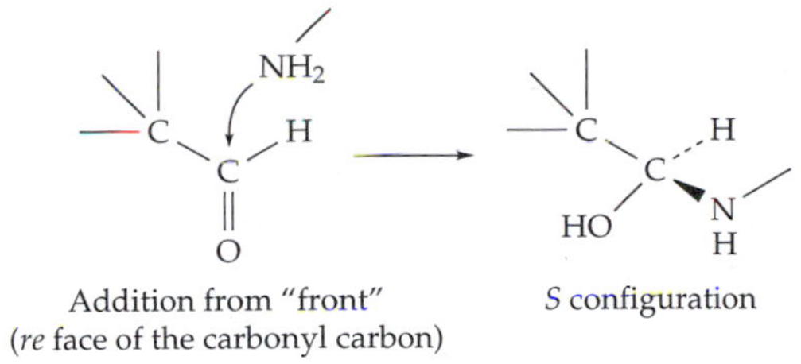

Addition from "front"
(*re* face of the carbonyl carbon)

S configuration

(13-5)

Enolate anion

(13-6)

4. Addition to Carbon–Carbon Double Bonds, Often Reversible Reactions

Most of the reactions that we will consider in this chapter involve addition of a proton to a carbon atom or removal of a proton attached to a carbon atom. A frequent metabolic reaction is addition of water to a carbon–carbon double bond that is conjugated with a carbonyl group. This transmits the polarization of the carbonyl group to a position located two carbon atoms further along the chain.

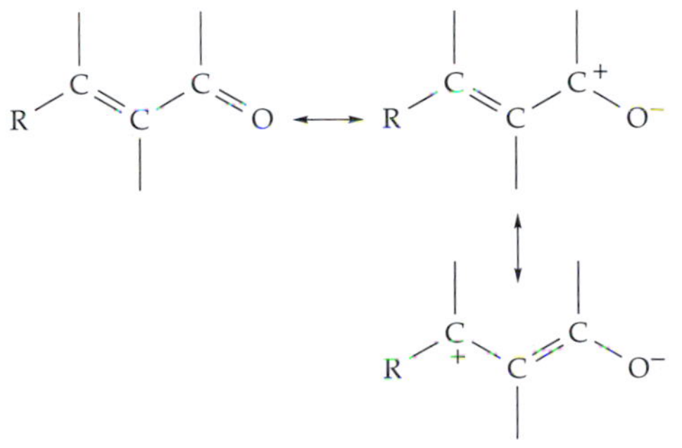

Because of this effect bases can add to a carbon–carbon double bond in a position β to the carbonyl group as in Eq. 13-6, step *a*. The product of addition of HO⁻ is the anion of an **enol** in which the negative charge is distributed by resonance between the oxygen of the carbonyl group and the carbon adjacent to the carbonyl. A stable product results if a proton adds to the latter position (Eq. 13-6, step *b*). Many nucleophilic groups (Nu in Eq. 13-6) other than HO⁻, may add, the reaction being known as a **Michael addition**. If a neutral nucleophile Nu adds, the product may lose a proton, or transfer it to the adjacent carbon in Eq. 13-6, step *b*, to give a neutral end product. The reverse of an addition reaction of this type is also known as an **elimination reaction**.

Enzymatic reactions involving addition to a C=C bond adjacent to a carbonyl group (or in which elimination occurs α,β to a carbonyl) are numerous. Except for some enzymes acting by free radical mechanisms, the nucleophilic group always adds at the β position suggesting that the mechanism portrayed by Eq. 13-7 is probable. It is noteworthy that *frequently in a metabolic sequence a carbonyl group is deliberately introduced to*

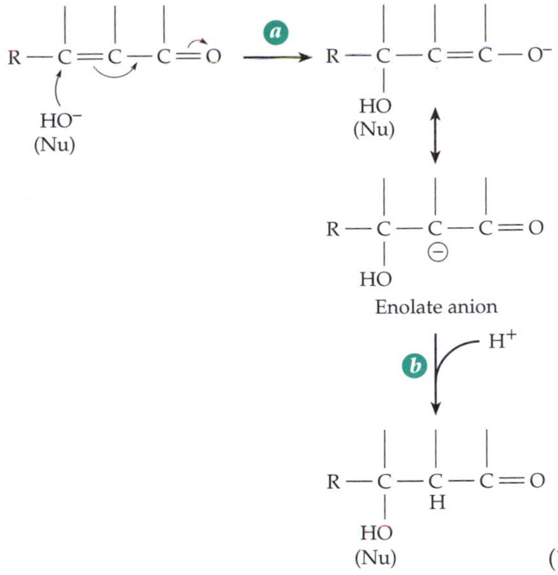

facilitate elimination or addition at adjacent carbon atoms. The carbonyl may be formed by oxidation of a hydroxyl group or it may be provided by a thioester formed with coenzyme A or with an **acyl carrier protein** (Chapter 14).

Enoyl-CoA hydratase. A specific example of the reaction in Eq. 13-6 is the addition of water to *trans*-α,β-unsaturated CoA derivatives (Eq. 13-7). It is catalyzed by enoyl–CoA hydratase (crotonase) from mitochondria and is a step in the β oxidation of fatty acids (Fig. 10-4).

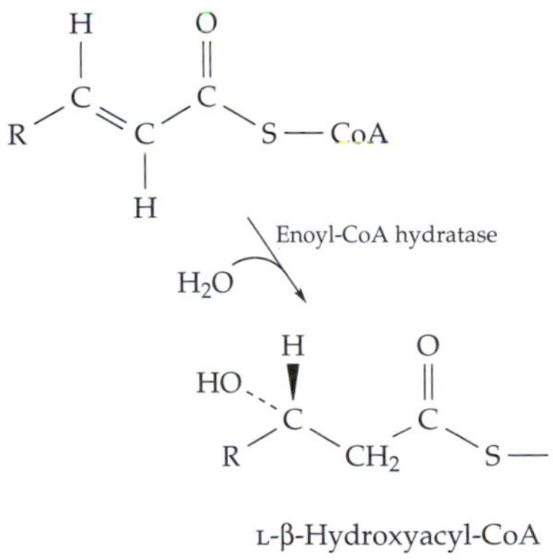

L-β-Hydroxyacyl-CoA

(13-7)

The enzyme is a hexamer, actually a dimer of trimers made up of 291-residue polypeptide chains.[28] Acetoacetyl-CoA is a competitive inhibitor which binds into the active site and locates it. From the X-ray structure of the enzyme–inhibitor complex it can be deduced that the carboxylate group of E144 abstracts a proton from a water molecule to provide the hydroxyl ion that binds to the β position (Eq. 13-6, step *a*) and that the E164 carboxyl group donates a proton to the intermediate enolate anion in step *b*.[28] The hydroxyl group and the proton enter from the same side of the double bond in a *syn* addition. The X-ray structure shows that the E144 and E164 side chains are on the same side of the double bond, accounting for this stereochemistry.[28,28a]

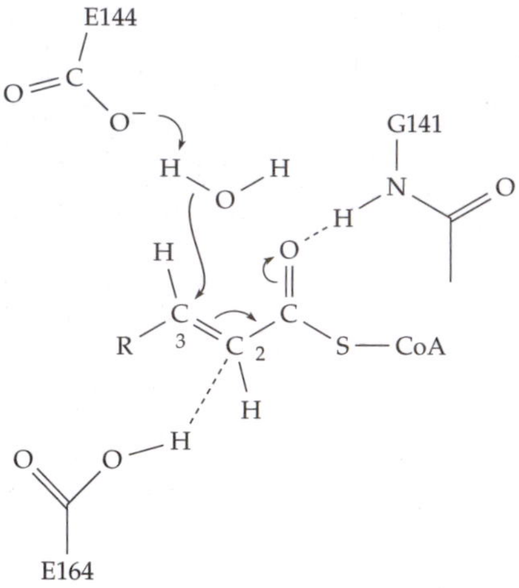

(13-8)

Cinnamoyl-CoA thiol esters (R = phenyl in the foregoing structure) contain a good light-absorbing group (chromophore). Binding to the protein induces distinct shifts in the ultraviolet absorption spectrum and also in ^{13}C NMR and Raman spectra. These shifts suggest that binding induces an enhanced positive charge at C-3 and formation of a strong hydrogen bond to the carbonyl group.[29] The X-ray studies indicate that this bond is to the G141 peptide nitrogen as shown. These results favor an enolate anion intermediate as in Eq. 13-6. However, kinetic isotope effects[30] as well as studies of proton exchange have suggested a concerted mechanism with water adding at the same time that a proton binds at C-2.[30-31a]

A closely related *E. coli* protein is a 79-kDa multifunctional enzyme that catalyzes four different reactions of fatty acid oxidation (Chapter 17). The amino-terminal region contains the enoyl hydratase activity.[32] A quite different enzyme catalyzes dehydration of thioesters of β-hydroxyacids such as 3-hydroxydecanoyl-acyl carrier protein (see Eq. 21-2) to both form and isomerize enoyl-ACP derivatives during synthesis of unsaturated fatty acids by *E. coli*. Again, a glutamate side chain is the catalytic base but an imidazole group of histidine has also been implicated.[33] This enzyme is inhibited irreversibly by the *N*-acetylcysteamine thioester of 3-decynoic acids (Eq. 13-8). This was one of the first enzyme-activated inhibitors to be studied.[34]

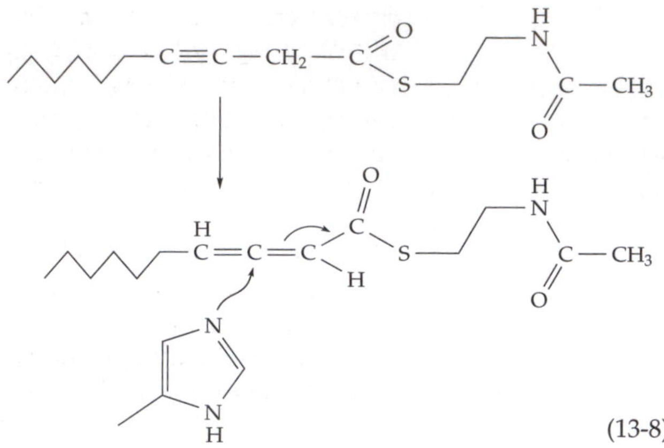

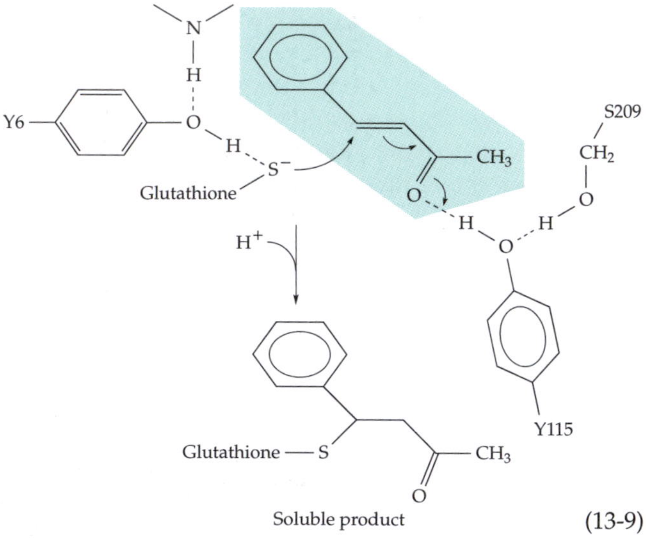

Glutathione S-transferases. Addition of glutathione (Box 11-B) to a large variety of different substrates containing electrophilic centers, such as that at the β position in an α,β-unsaturated ketone (Eq. 13-9), is catalyzed by the ubiquitous group of enzymes called glutathione S-transferases. There are six classes of

(13-9)

eukayotic glutathione S-transferases—one membrane-associated microsomal[35] and five cytosolic.[36-40] They play an important role in detoxifying many xenobiotics and other relatively hydrophobic compounds by converting them to more soluble compounds that can be degraded and excreted easily (see Box 11-B). A pair of tyrosines appear to participate in catalysis, as is indicated in Eq. 13-9.[38,41] Glutathione S-transferases have attracted attention because of their role in detoxifying anticancer drugs. Cancer cells can become resistant to drugs as a result of excessive synthesis of these detoxifying enzymes. At the same time glutathione transferases protect human patients from drugs. In plants these enzymes may provide protection from insecticides and herbicides.[41a] Glutathione transferases of pathogenic organisms such as schistosomes are appropriate targets for new drugs as well as for vaccines.[42] Nonenzymatic Michael addition of glutathione to such compounds as **4-hydroxynonenal**, a product of peroxidation of the polyunsaturated arachidonate (see Eq. 21-15) are also biochemically important.

Chlorobenzoyl-CoA dehalogenase. The enzymatic release of chloride ion from 4-chloroxybenzoyl-CoA would be hard to explain by a simple nucleophilic displacement by HO⁻. However, *addition* of HO⁻, followed by elimination of Cl⁻, would be chemically reasonable. Nevertheless, single-turnover studies revealed a more complex mechanism involving formation of a covalent adduct with the enzyme.[43,44] A carboxylate group adds and Cl⁻ is eliminated to form of an oxygen ester, which is then hydrolyzed to the final product (Eq. 13-10). Studies of mutant enzymes[45] together with X-ray crystallography[46] support this mechanism. As with enoyl-CoA hydratase, binding of substrate to chlorobenzoyl-CoA dehalogenase causes alterations in ultraviolet, NMR, and Raman spectra that can be interpreted as indicating enhanced polarization of the benzoyl group.[47,48] For example, changes in the C=O stretching frequency suggest that the bond is elongated, presumably as a result of hydrogen bonding to the N–H of G114, which lies at the N terminus of an α helix and experiences the additional polarizing effect of the helix dipole (Fig. 2-20).[48,49] This effect can be compared with that mentioned previously for enoyl-CoA hydratase.

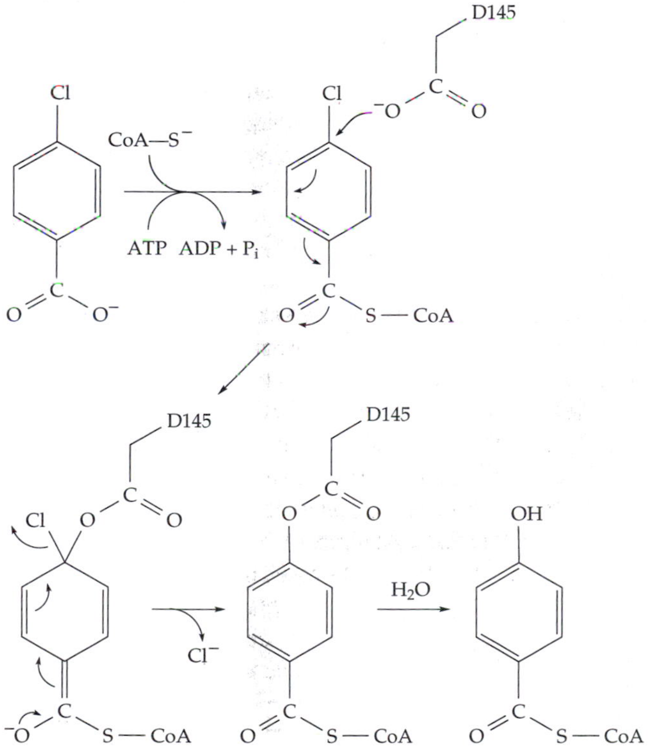

(13-10)

In fact, close sequence and structural homologies show that chlorobenzoyl-CoA dehalogenase, enoyl-CoA hydratase, and a variety of other hydratases, isomerases, synthases, lyases, and hydrolases belong to a large family of related proteins.[49a,49b,49c]

5. Addition to Double Bonds Adjacent to Carboxylate Groups

Biochemical reactions often involve addition to C=C bonds that are not conjugated with a true carbonyl group but with the poorer electron acceptor –COO⁻. While held on an enzyme a carboxylate group may be protonated, making it a better electron acceptor. Nevertheless, there has been some doubt as to whether the carbanion mechanism of Eq. 13-6 holds for these enzymes. Some experimental data suggested a quite different mechanism, one that has been established for the nonenzymatic hydration of alkenes. An example is the hydration of ethylene by hot water with dilute sulfuric acid as a catalyst (Eq. 13-11), an industrial method of preparation of ethanol. The electrons of the double bond form the point of attack by a proton, and the resulting carbocation readily abstracts a hydroxyl

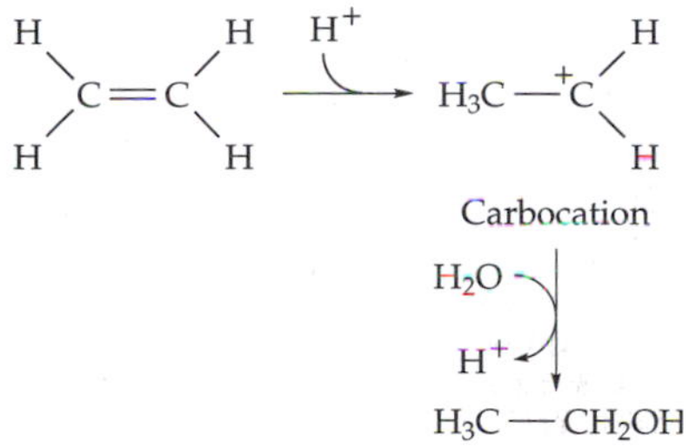

(13-11)

ion from water. Direct addition of OH⁻ to ethylene to form a carbanion is not favored, because there is no adjacent carbonyl group to stabilize the negative charge.

Fumarate hydratase. The most studied enzyme of this group is probably the porcine mitochondrial fumarate hydratase (fumarase; see also Chapter 9), a tetramer of 48.5-kDa subunits[50] with a turnover number of ~2 x 10³ s⁻¹. It accelerates the hydration reaction more than 10¹⁵-fold.[51] A similar enzyme, the 467-residue **fumarase C** whose three-dimensional structure is known,[52,53] is found in cells of *E. coli* when grown aerobically. The product of the fumarate hydratase reaction is L-malate (*S*-malate). The stereospecificity is extremely high. If the reaction is carried out in ²H₂O an atom of ²H is incorporated into the *pro-R* position, i.e., the proton is added strictly from the *re* face of the trigonal carbon (Eq. 13-12). To obtain L-malate the hydroxyl must have been added from the opposite side of the double bond. Such *anti* (*trans*) addition is much more common in both nonenzymatic and enzymatic reactions than is addition of both H and OH (or –Y) from the same side (*syn*, *cis*, or adjacent addition).[54] For concerted addition it is a natural result of stereoelectronic control. Almost *all* enzymatic addition and elimination reactions involving free carboxylic acids are anti with the proton entering from the *re* face,

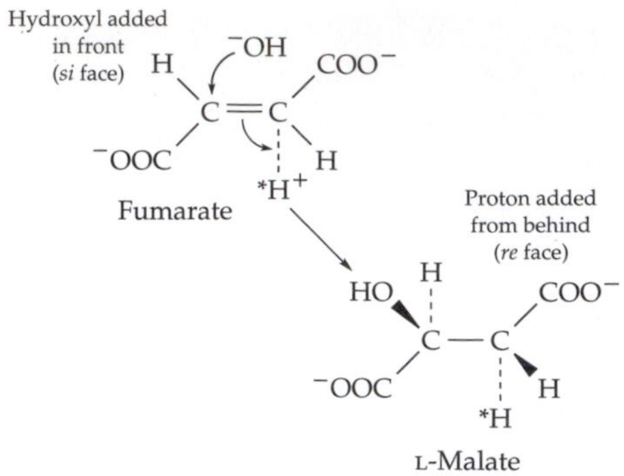

Hydroxyl added in front (si face)

Fumarate

Proton added from behind (re face)

L-Malate   (13-12)

suggesting that there has been conservation through-out evolution of a single basically similar mechanism.[54,55]

The pH dependence of the action of fumarate hydratase indicates participation of both an acidic and a basic group with pK_a values of 5.8 and 7.1.[56] See Chapter 9 for additional information. However, either anion or carbocation mechanisms might be possible. That the cleavage of the C–H bond is not rate limiting is suggested by the observation that malate containing ^2H in the *pro-R* position is dehydrated at the same rate as ordinary malate. If the anion mechanism (Eq. 13-13) is correct, the ^2H from the *pro-R* position of specifically labeled malate might be removed rapidly, while the loss of OH$^-$ could be slower. If so, the ^2H would be "washed out" of L-malate faster than could happen by conversion to fumarate followed by rehydration to malate. In fact, the opposite was observed.

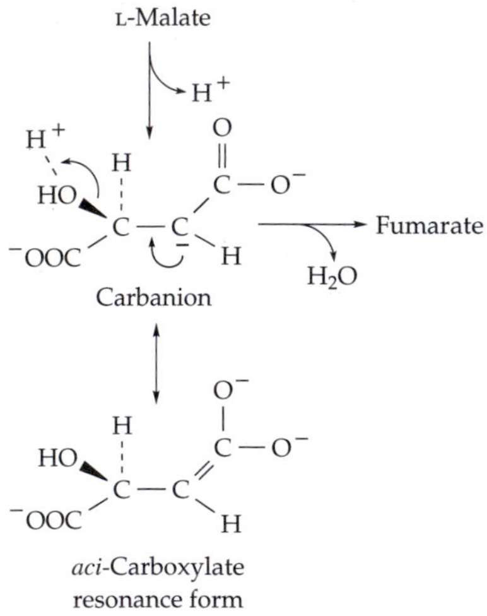

L-Malate

Carbanion

aci-Carboxylate resonance form   (13-13)

The hydroxyl group was lost rapidly and the ^2H more slowly. This result suggested the carbocation ion mechanism of Eq. 13-14.

While no *primary* isotope rate effect was observed, when the hydrogen at C-2 or the *pro-S* hydrogen at C-3 of malate was replaced by ^2H or ^3H distinct secondary isotope effects were seen. Thus, $k(^1H) / k(^2H) = 1.09$ for both the *pro-S* and the C-3 hydrogen atoms.[57] These findings appeared to support the carbocation

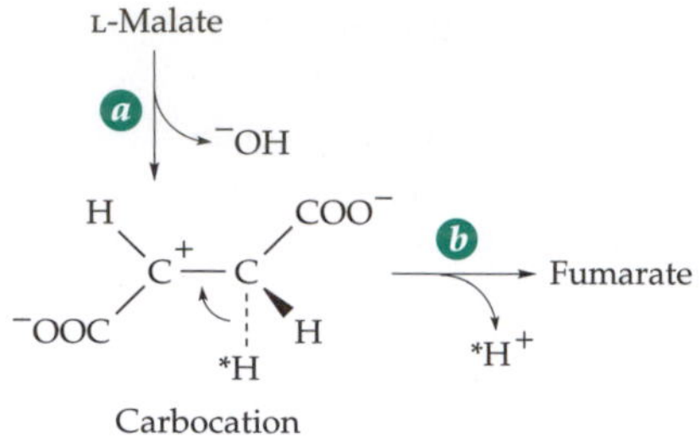

L-Malate

Carbocation   (13-14)

mechanism for the reasons considered in Chapter 12 and suggested that step *b* of Eq. 13-14 is rate limiting. The relative values of V_{max} for hydration of fumarate, fluorofumarate, and difluorofumarate (104, 410, and 86 mol ml^{-1} min^{-1} mg^{-1}) also seemed to support the carbocation mechanism.[58]

However, it has become clear that protons removed from a substrate to a basic group in a protein need not exchange rapidly with solvent (see Eq. 9-102). In fact, the proton removed by fumarate hydratase from malate is held by the enzyme for relatively long periods of time. Its rate of exchange between malate and solvent is slower than the exchange of a bound fumarate ion on the enzyme surface with another substrate molecule from the medium.[59] Thus, the overall rate is determined by the speed of dissociation of products from the enzyme and we cannot yet decide whether removal of a proton precedes or follows loss of OH$^-$.

Two new lines of evidence suggest that proton abstraction comes first. A careful study of both ^{18}O and ^2H isotope effects[60] supports the carbanion intermediate, as does the strong inhibition by anions of 3-nitropropionate and 3-nitro-2-hydroxypropionate.[61] To provide a good electron sink the carboxylate group adjacent to the proton that is removed by fumarate hydratase must either be actually protonated in the enzyme–substrate (ES) complex or paired with and hydrogen bonded to a positively charged group.

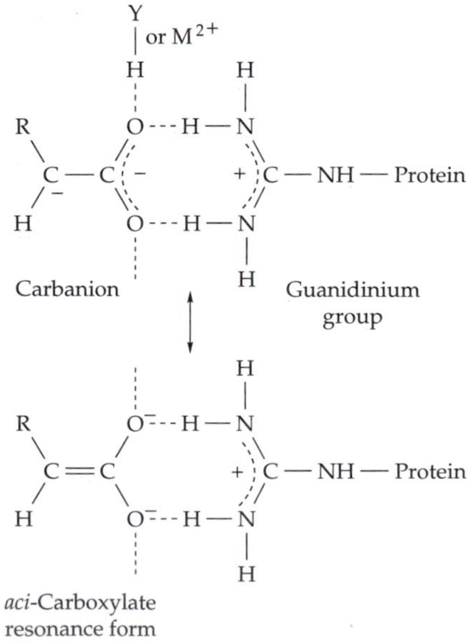

Carbanion   Guanidinium group

aci-Carboxylate resonance form

The anion formed by removal of the 3-H is analogous to the enolate anion of Eq. 13-6 and has a strong structural similarity to the readily formed anions of organic nitro compounds. The nitronate anions may, perhaps, be regarded as transition state inhibitors.

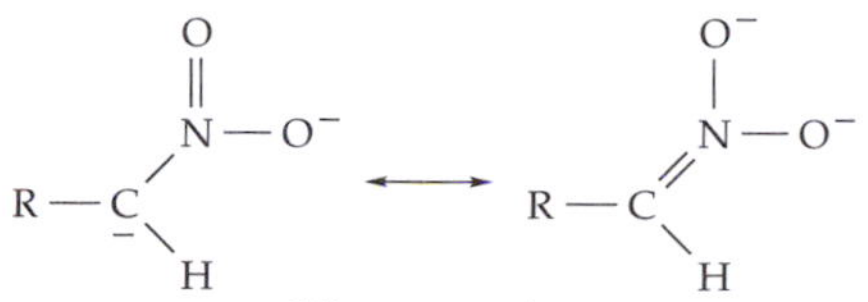

Nitronate anion

There is still a third possible mechanism for the fumarate hydratase reaction. The proton and hydroxyl groups may be added *simultaneously* in a concerted reaction.[62] However, observed kinetic isotope effects are not consistent with this mechanism.[60] In 1997 the structure of fumarase C of *E. coli* was reported.[52,53] Each active site of the tetrameric enzyme is formed using side chains from three different subunits. The H188 imidazole is hydrogen bonded to an active site water molecule and is backed up by the E331 carboxylate which forms a familiar catalytic pair. However, these results have not clarified the exact mode of substrate binding nor the details of the catalytic mechanism. Structural studies of fumarate hydratase from yeast[53a] and the pig[53b] are also in progress.

Some other fumarate-forming reactions.

Other enzymes that catalyze elimination reactions that produce fumarate are **aspartate ammonia-lyase** (aspartase),[63] **argininosuccinate lyase** (Fig. 24-10, reaction *g*),[64,65] and **adenylosuccinate lyase** (Fig. 25-15). In every case it is NH₃ or an amine, rather than an OH group, that is eliminated. However, the mechanisms probably resemble that of fumarate hydratase. Sequence analysis indicated that all of these enzymes belong to a single **fumarase-aspartase family**.[64,65] The three-dimensional structure of aspartate ammonia-lyase resembles that of fumarate hydratase, but the catalytic site lacks the essential H188 of fumarate hydratase. However, the pK_a values deduced from the pH dependence of V_{max} are similar to those for fumarase.[64] **3-Methylaspartate lyase** catalyzes the same kind of reaction to produce ammonia plus *cis*-mesaconate.[63] Its sequence is not related to that of fumarase and it may contain a dehydroalanine residue (Chapter 14).[66]

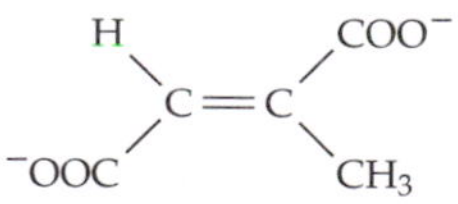

cis-Mesaconate

Enolase. A key reaction in the metabolism of sugars is the dehydration of 2-phosphoglycerate to form **phosphoenolpyruvate** (PEP), the phospho derivative of the enolic form of pyruvic acid:

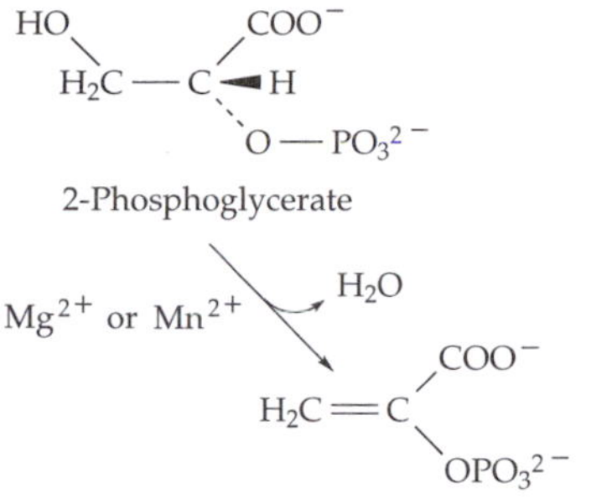

2-Phosphoglycerate

Phosphoenolpyruvate (13-15)

A carbanionic intermediate has often been suggested for this enzyme.[67,68] However, despite measurements of kinetic isotope effects and many other experiments it has been difficult to establish a detailed mechanism.[68]

Enolase has a complex metal ion requirement,[68a,69] usually met by Mg²⁺ and Mn²⁺. From NMR studies of the relaxation of water protons, it was concluded that a Mn²⁺ ion coordinates two rapidly exchangeable water molecules in the free enzyme. When substrate binds, one of these water molecules may be immobilized and may participate in an addition reaction that forms phosphoenolpyruvate (reverse of reaction 13-15). A tightly bound "conformational" metal ion is located in the known three-dimensional structure in such a

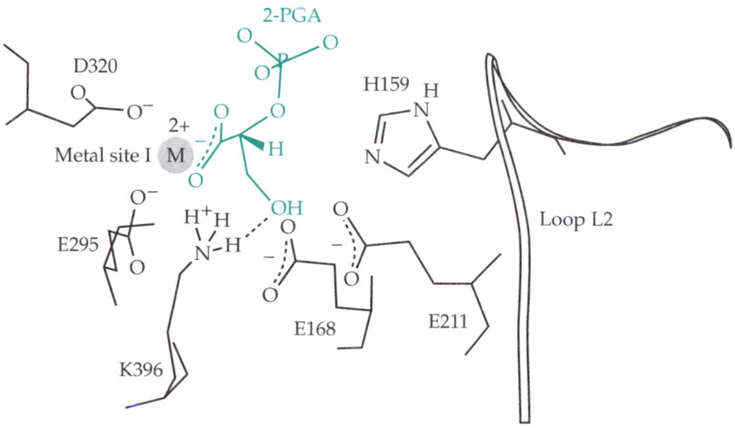

Figure 13-2 View of the active site of yeast enolase containing a bound molecule of 2-phospho-D-glycerate. The catalytic magnesium ion is at the left but the "conformational" metal is not visible here. The imidazole group of His 159 serves as the catalytic base and the —NH₃⁺ of Lys 396 or Lys 345[73b] as the catalytic acid. From Vinarov and Nowak.[69]

way that it might function in this manner. A more loosely bound "catalytic" metal ion is also essential.[69–72]

Figure 13-2 shows a view of the active site of yeast enolase occupied by a molecule of bound 2-phospho-glycerate. Histidine 159 is probably the catalytic base that removes the α-proton to form an *aci*-anion which is stabilized by interaction with the catalytic Mg^{2+} ion. A protonated lysine 396 amino group may be the catalytic acid.[68a,69] The active site is surrounded by a complex hydrogen-bonded network.[73] As is discussed in Section B, a large number of other enzymes belong to an **enolase superfamily** of enzymes. Among them is **glucarate dehydratase**, which initiates a pathway for catabolism of D-glucarate and galactarate in *E. coli*.[73a]

Pectate lyase and related enzymes. A group of polysaccharide lyases cleave the chains of polymers of uronic acids with 1,4 linkages such as pectins, hyaluronan, heparin,[74,75] and dermatan sulfate (Fig. 4-11). These bacterial enzymes also employ an elimination mechanism.[76] The geometry of the β-linked galacturonic acid units of pectin is favorable for *anti* elimination of the 5-H and the *O*-glycosyl group in the 4 position (Eq. 13-16). However, the corresponding **hyaluronate lyase** (hyaluronidase) acting on glucuronic acid residues causes a *syn* elimination. These results suggest the formation of anionic intermediates which can eliminate a substituent from either the equatorial or axial position of the sugar ring. Hyaluronate lyase from the pathogenic *Streptococcus pneumoniae* apparently utilizes an imidazole group as the catalytic base and a tyrosine side chain as the proton donor in the reaction.[76a,76b]

Pectate lyase C from the plant pathogen *Erwinia*, which causes soft-rot in many different plants, has a parallel β barrel structure (Fig. 13-3)[77] which is similar to that of the tailspike protein shown in Fig. 2-17 and represents what may be a very large structural family of proteins.[78] The location of the active site is not

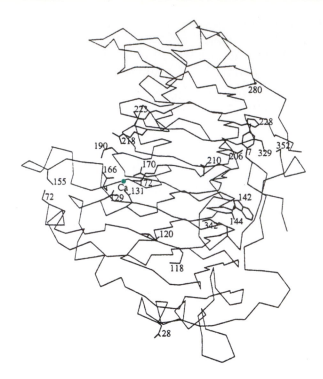

Figure 13-3 The three-dimensional structure of pectate lyase C showing locations of amino acids substituted by oligonucleotide-directed mutation of the cloned gene. The green dot labeled Ca is the Ca^{2+}-binding site.[79] Courtesy of Frances Jurnak.

obvious from the structure but on the basis of the effects of a large number of mutations, it is thought to be near the Ca^{2+}- binding site labeled in Fig. 13-3. Replacement of any of three Asp or Glu and three Lys residues in this region leads to loss of or reduction in catalytic activity.[79]

6. Aconitase and Related Iron – Sulfur Proteins

Two consecutive reactions of the citric acid cycle (Fig. 10-6), the dehydration of citrate to form *cis*-aconitate and the rehydration in a different way to form isocitrate (Eq. 13-17), are catalyzed by aconitase (aconitate hydratase). Both reactions are completely stereospecific. In the first (Eq. 13-17, step *a*), the *pro-R* proton from C-4 (stereochemical numbering) of citrate is removed and in step *c* isocitrate is formed. Proton addition is to the *re* face in both cases.

As with fumarate hydratase, the enzyme holds the abstracted proton (for up to 7×10^{-5} s) long enough so that a *cis*-aconitate molecule sometimes diffuses from the enzyme and (if excess *cis*-aconitate is present) is replaced by another. The result is that the new *cis*-aconitate molecule sometimes receives the proton (intermolecular proton transfer). The proton removed

Galacturonic acid unit of pectin

(13-16)

BOX 13-B EPSP SYNTHASE AND THE HERBICIDE GLYPHOSATE

The reversible reaction of phosphoenolpyruvate (PEP) with shikimate 3-phosphate is a step in the synthesis of the aromatic amino acids (see Fig. 25-1). The chemical mechanism indicated

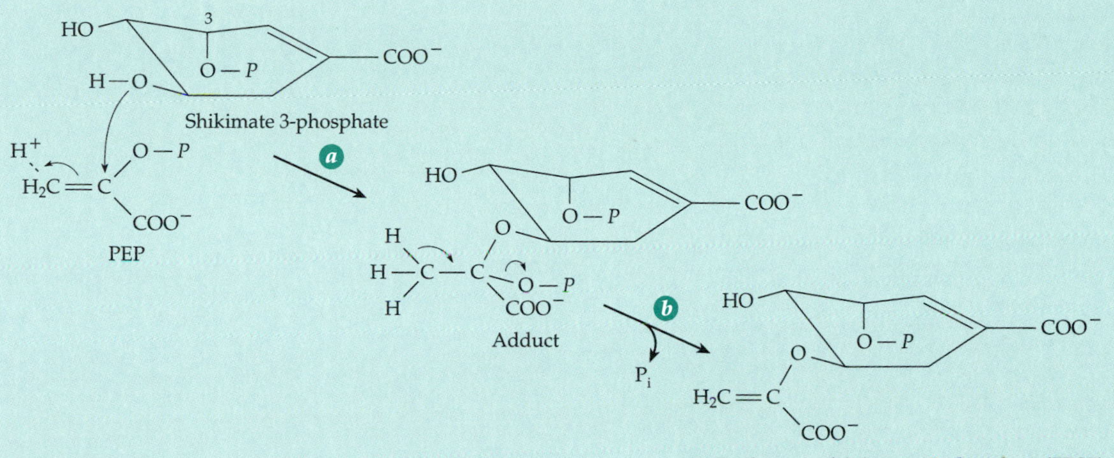

product.[f–h] Glyphosate was for many years viewed as a transition-state analog but more recently has been shown to be a tight-binding noncompetitive inhibitor.[i]

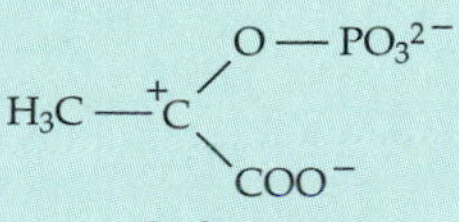

5-Enolpyruvoylshikimate-3-phosphate (EPSP)

was proposed by Leaven and Sprinson in 1964.[a] Step a is unusual[b] because it involves protonation on the methylene carbon of PEP and addition of a nucleophile at C-2, the opposite of the addition in the enolase reaction (Eq. 13-15, reverse). It is likely that formation of a cationic intermediate precedes that of the adduct shown. However, the structure of the adduct has been confirmed by isolation from the active site and by synthesis[c] and it has been observed in the active site by NMR spectroscopy.[d] The three-dimensional structure of EPSP synthase is known.[e,m] It is inhibited by the commercial herbicide **glyphosate** [N-(phosphonomethyl) glycine] whose structure is somewhat similar to that of the proposed carbocation that arises from PEP. The inhibitor (Z)-3-fluorophosphoenolpyruvate is a pseudosubstrate that reacts in step a to give a stable adduct unable to go through step b to form a

A related mechanism is utilized in the biosynthesis of UDP-muramic acid (Eq. 20-6).[j] There is an enolpyruvoyl adduct analogous to that of EPSP synthase; a proposed enolpyruvoyl-enzyme adduct with Cys 115 is not on the major path.[k,l] However, this enzyme is not inhibited by glyphosate.[i]

[a] Levin, J. G., and Sprinson, D. B. (1964) *J. Biol. Chem.* **239**, 1142–1150

[b] Barlow, P. N., Appleyard, R. J., Wilson, B. J. O., and Evans, J. N. S. (1989) *Biochemistry* **28**, 7985–7991

[c] Anderson, K. S., Sikorski, J. A., Benesi, A. J., and Johnson, K. A. (1988) *J. Am. Chem. Soc.* **110**, 6577–6579

[d] Appleyard, R. J., Shuttleworth, W. A., and Evans, J. N. S. (1994) *Biochemistry* **33**, 6812–6821

[e] Stallings, W. C., Abdel-Meguid, S. S., Lim, L. W., Shieh, H.-S., Dayringer, H. E., Leimgruber, N. K., Stegeman, R. A., Anderson, K. S., Sikorski, J. A., Padgette, S. R., and Kishore, G. M. (1991) *Proc. Natl. Acad. Sci. U.S.A.* **88**, 5046–5050

[f] Alberg, D. G., Lauhon, C. T., Nyfeler, R., Fässler, A., and Bartlett, P. A. (1992) *J. Am. Chem. Soc.* **114**, 3535–3546

[g] Walker, M. C., Jones, C. R., Somerville, R. L., and Sikorski, J. A. (1992) *J. Am. Chem. Soc.* **114**, 7601–7603

[h] Ream, J. E., Yuen, H. K., Frazier, R. B., and Sikorski, J. A. (1992) *Biochemistry* **31**, 5528–5534

[i] Sammons, R. D., Gruys, K. J., Anderson, K. S., Johnson, K. A., and Sikorski, J. A. (1995) *Biochemistry* **34**, 6433–6440

[j] Samland, A. K., Amrhein, N., and Macheroux, P. (1999) *Biochemistry* **38**, 13162–13169

[k] Skarzynski, T., Kim, D. H., Lees, W. J., Walsh, C. T., and Duncan, K. (1998) *Biochemistry* **37**, 2572–2577

[l] Jia, Y., Lu, Z., Huang, K., Herzberg, O., and Dunaway-Mariano, D. (1999) *Biochemistry* **38**, 14165–14173

[m] Lewis, J., Johnson, K. A., and Anderson, K. S. (1999) *Biochemistry* **38**, 7372–7379

Carbocation   Glyphosate

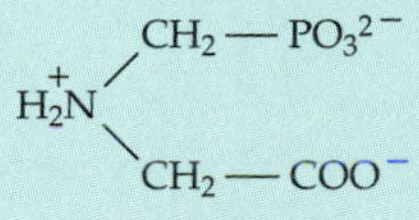

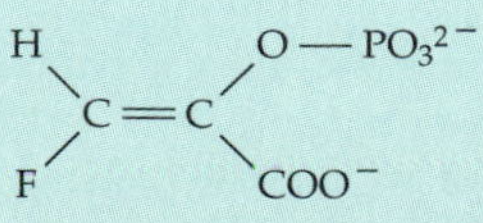

(Z)-3-Fluorophosphonoenolypyruvate

from citrate is often returned to the molecule in Eq. 13-17, step *b*, but the position of reentry is different from that of removal. Apparently, after the initial proton removal, the *cis*-aconitate that is formed "flips over" so that it can be rehydrated (Eq. 13-17, step *c*) with participation of the same groups involved in dehydration but with formation of the new product.[80,80a]

Aconitase contains iron in the form of an Fe_4S_4 iron-sulfur cluster (Fig. 13-4).[81–83] However, the enzyme is usually isolated in a form that does not show its maximum activity until it has been incubated with ferrous iron (Fe^{2+}). The inactive form of the enzyme is thought to contain an Fe_3S_4 cluster (Chapter 16) which is converted back to the Fe_4S_4 cluster by the incubation

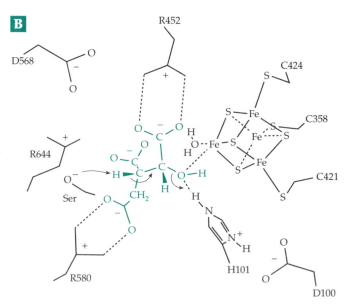

$$(13\text{-}17)$$

A

B

Figure 13-4 (A) Stereoscopic view of the active site of mitochondrial aconitase with a molecule of L-isocitrate placed by modeling next to the Fe_4S_4 cluster. This cluster still has a water molecule bound to one iron atom as in the free enzyme.[83] Courtesy of C. D. Stout. (B) Interpretive diagram.

with Fe^{2+}.[84–87] From Mössbauer spectroscopy it was deduced that the iron atom that is lost upon inactivation is the binding site for the –OH group of citrate or isocitrate and of an adjacent carboxylate.[87] Since iron can engage in oxidation–reduction processes, aconitase could act by a mechanism different from others discussed here. However, an Fe–OH group could act much as does the Zn–OH of carbonic anhydrase (Eq. 13-4). Any redox chemistry may be involved in control of the enzyme.

X-ray studies have confirmed the proximity of the Fe_4S_4 cluster to bound substrate analogs.[83,88] The isocitrate shown in Fig. 13-4 was fitted into the active site by modeling. It shows the water molecule bound to the Fe_4S_4 cluster as observed for the free enzyme.[83] Notice that this cluster is held by three cysteine side chains (C358, C421, and C424) rather than the usual four (Chapter 16). The fourth iron atom is free to bind to water. When the enzyme acts on isocitrate this water must be displaced by the one generated from the substrate. The Fe probably acts as a polarizing electrophile that assists in the elimination but it is apparently the histidine–aspartate pair (H101–D100) that serves as the catalytic acid in generating the H_2O. What is the catalytic base? The only candidate seen in the structure is the Ser 642 –OH, dissociated to –O⁻. Can this be correct? Mutational analysis supports the essential role of this side chain.[80a] The peptide NH and guanidinium groups of R644 appear to provide an "oxyanion hole" (Chapter 12) that stabilizes the negative charge.[86] Mutations also support the role of the His-Asp pair.[88,89] Another His–Asp pair (H167–D165) is located directly behind the bound substrate and Fe–OH₂ in Fig. 13-4A and is apparently also essential.[88]

Aconitase exists as both mitochondrial and cytosolic isoenzyme forms of similar structure. However, the cytosolic isoenzyme has a second function. In its apoenzyme form, which lacks the iron–sulfur cluster, it acts as the much-studied **iron regulatory factor**, or iron-responsive element binding protein (IRE-BP). This protein binds to a specific stem-loop structure in the messenger RNA for proteins involved in iron transport and storage (Chapter 28).[86,90]

Other enzymes in the aconitase family include **isopropylmalate isomerase** and **homoaconitase** enzymes functioning in the chain elongation pathways to leucine and lysine, both of which are pictured in Fig. 17-18.[90] There are also iron–sulfur dehydratases, some of which may function by a mechanism similar to that of aconitase. Among these are the two fumarate hydratases, **fumarases A and B,** which are formed in place of fumarase C by cells of *E. coli* growing anaerobically.[91,92] Also related may be bacterial L-serine and L-threonine dehydratases. These function without the coenzyme pyridoxal phosphate (Chapter 14) but contain iron–sulfur centers.[93–95] A **lactyl-CoA**

dehydratase and some related iron-sulfur enzymes (pp. 861-862) may act via a mechanism related to that of vitamin B_{12}.

7. Addition to or Formation of Isolated Double Bonds

Only a few examples are known in which an enzyme induces addition to a double bond that is *not conjugated* with a carbonyl or carboxyl group. Pseudomonads have been observed to catalyze stereospecific hydration of oleic acid to D-10-hydroxystearate.[96] The addition is *anti* and the proton enters from the *re* face.

8. Conjugative and Decarboxylative Elimination Reactions

Elimination can occur if the electrophilic and nucleophilic groups to be removed are located not on adjacent carbon atoms but rather are separated from each other by a pair of atoms joined by a double bond. Such a conjugative elimination of phosphate is the last step in the biosynthesis of **chorismate** (Eq. 13-18, step *a*). Chorismate is converted to **prephenate** (see Box 9-E) which undergoes a conjugative and decarboxylative elimination (Eq. 13-18, step *c*) with loss of both water and CO_2 to form **phenylpyruvate**, the immediate biosynthetic precursor of phenylalanine. These reactions provide a good example of how *elimination reactions can be used to generate aromatic groups*. In fact, this is the usual method of synthesis of aromatic rings in nature.

Notice the stereochemistry of Eq. 13-18, step *a*. Orbital interaction rules predict that if the elimination is a concerted process it should be syn. The observed anti elimination suggests a more complex mechanism involving participation of a nucleophilic group of the enzyme.[97]

Elimination usually involves loss of a proton together with a nucleophilic group such as –OH, –NH₃⁺, phosphate, or pyrophosphate. However, as in Eq. 13-18, step *c*, electrophilic groups such as –COO⁻ can replace the proton. Another example is the conversion of **mevalonic acid-5-pyrophosphate** to **isopentenyl pyrophosphate** (Eq. 13-19): This is a key reaction in the biosynthesis of isoprenoid compounds such as cholesterol and vitamin A (Chapter 22). The phosphate ester formed in step *a* is a probable intermediate and the reaction probably involves a carbocationic intermediate generated by the loss of phosphate prior to the decarboxylation.

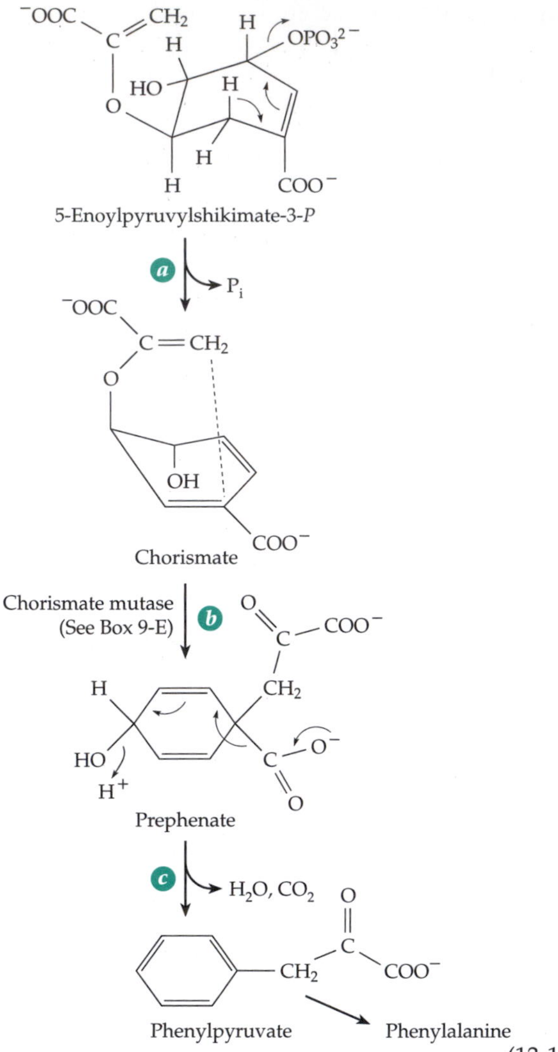

5-Enoylpyruvylshikimate-3-*P*

Chorismate

Chorismate mutase
(See Box 9-E)

Prephenate

Phenylpyruvate → Phenylalanine

(13-18)

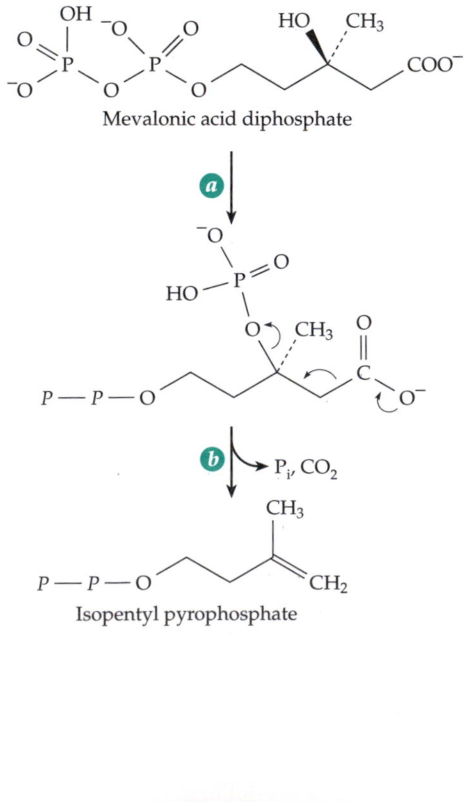

Mevalonic acid diphosphate

Isopentyl pyrophosphate

(13-19)

9. Isomerization Assisted by Addition

An interesting use is made of addition to a double bond by glutathione-dependent *cis–trans* **isomerases**.[76] One of them converts **maleate** to fumarate with a turnover number of 300 s^{-1}. Similar enzymes, which participate in bacterial breakdown of aromatic compounds (Fig. 25-7), isomerize **maleylacetoacetate** and **maleylpyruvate** to the corresponding fumaryl derivatives (Eq. 13-20). The –SH group of bound glutathione is thought to add to the double bond. Rotation can then occur in the enolic intermediate. Thiocyanate ion catalyzes the isomerization of maleic acid nonenzymatically, presumably by a similar mechanism.

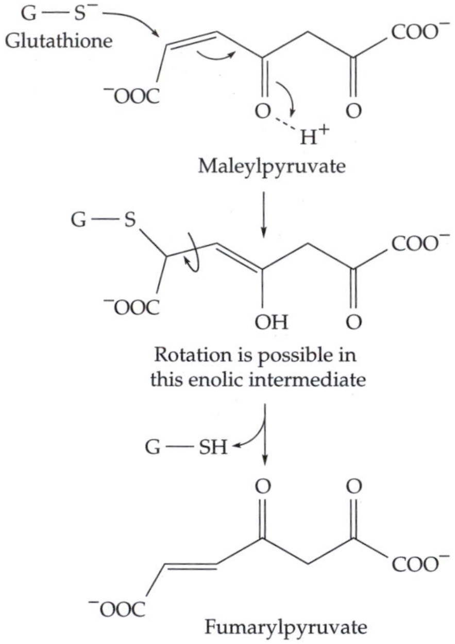

Maleylpyruvate

Rotation is possible in this enolic intermediate

Fumarylpyruvate

(13-20)

10. Reversibility of Addition and Elimination Reactions

Many addition and elimination reactions, e.g., the hydration of aldehydes and ketones, and reactions catalyzed by lyases such as fumarate hydratase are strictly reversible. However, biosynthetic sequences are often nearly irreversible because of the elimination of inorganic phosphate or pyrophosphate ions. Both of these ions occur in low concentrations within cells so that the reverse reaction does not tend to take place. In decarboxylative eliminations, carbon dioxide is produced and reversal becomes unlikely because of the high stability of CO_2. Further irreversibility is introduced when the major product is an aromatic ring, as in the formation of phenylpyruvate.

B. Enolic Intermediates in Enzymatic Reactions

Enzymologists have freely proposed enolate anions, enols, and enamines as intermediates for many years. Such intermediates have been demonstrated for some nonenzymatic acid- or base-catalyzed reactions, but how can enzymes form enolates at pH 7 without the use of strong acids or bases? The microscopic pK_a value of an α-hydrogen in a ketone or aldehyde is about 17–20.[72,98,99]

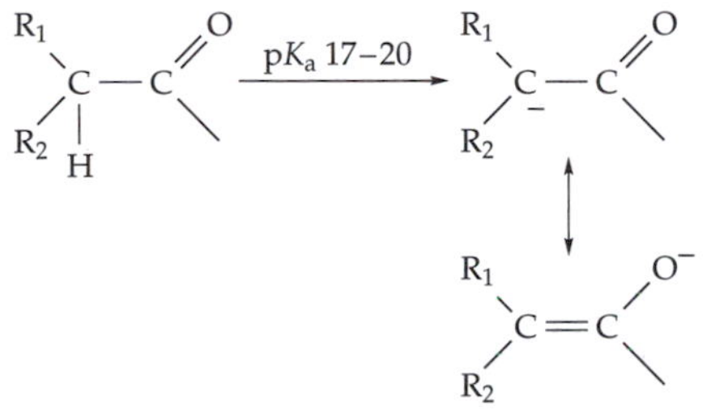

(13-21)

It will be similar for an acyl-CoA. However, for a carboxylic acid, protonated on oxygen, the pK_a will be much higher (~22–25) and for carboxylate anions even higher (~29–32).[100]

1. Mandelate Racemase and Related Enzymes

The degradation of mandelic acid by the bacterium *Pseudomonas putida* (Chapter 25) is initiated by mandelate racemase, another $(\alpha/\beta)_8$-barrel protein.[101] X-ray structures of bound inhibitors together with modeling suggest that the side chain of Lys 264 is the catalytic base that abstracts the α-H from *S*-mandelate (Fig. 13-5) and that the catalytic pair of His 297 and Asp 270 acts as proton donor, or, in the reverse direction, as catalytic base for deprotonation of *R*-mandelate.[102–104] The enzyme is structurally a member of the enolase superfamily[105] and requires Mg^{2+} for activity. The pH dependence of the reaction velocity k_{cat} reveals two pK_a values in the ES complex; ~6.4 and ~10.0. These are the same for *S*- and *R*- isomers. How can the pK_a values be assigned? Do they each belong in part to His 297 and in part to Lys 166? Do other adjacent groups also share? In the K166R mutant the lower pK_a is raised to 8.0. The D270N mutant has lost all but 0.01% of its catalytic activity.[104]

The carboxylate group of the mandelate interacts with side chains of E317, K264, and a Mg^{2+} ion (Fig. 13-5). These may serve both to protonate the carboxylate and also to help stabilize an *aci* anion formed upon dissociation of the α-hydrogen (Eq. 13-22). Both pK_a values for mandelic acid and its enolic form and the equilibrium constants for enolization in water are known[107] and are given beside the arrows in Eq. 13-22. It is difficult to imagine how a base with a pK_a near neutrality could remove an α-proton with a pK_a of 22.0, which is 14 pK units away from the pK_a of the catalytic base. This corresponds to a thermodynamic barrier (Eq. 9-97) of 14 x 5.7 = 80 kJ/ mol, making the reaction impossibly slow. In addition to this thermodynamic barrier the *rates* of dissociation of carbon acids are known to be slow, presumably because of the lack of hydrogen bond formation between the C–H proton and the catalytic base (Eq. 9-97). This *intrinsic* barrier (Chapter 9) for simple ketones has been estimated as 45 kJ/ mol for a total barrier of ~125 kJ. However, the observed $\Delta G^{\ddagger}$ as estimated from Eq. 9-81 is only ~57 kJ/ mol. The enzyme must catalyze the reaction by lowering the very high thermodynamic barrier and perhaps also by lowering the intrinsic barrier.[72,108]

Protonation of the carboxylate greatly decreases the microscopic pK_a for loss of the α-hydrogen as a proton to form the enolic *aci* acid (Eq. 13-22). Double protonation, although depending upon a pK_a of –8,

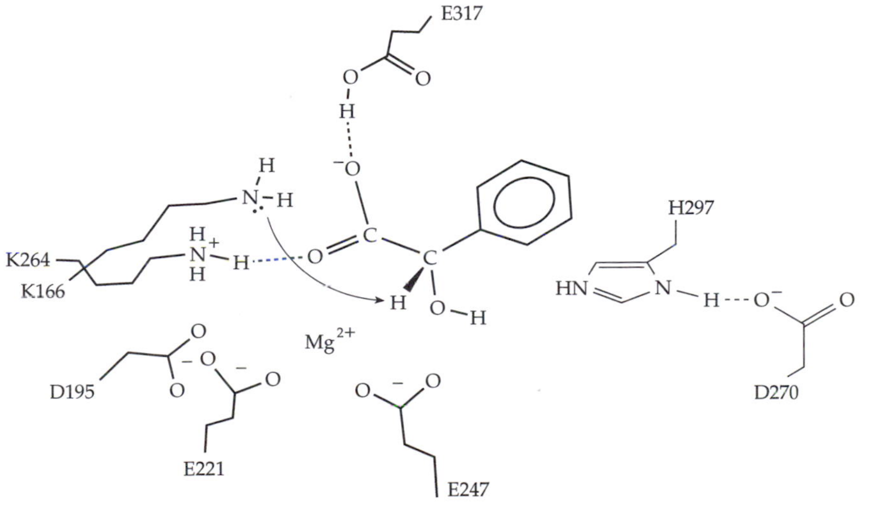

Figure 13-5 An *S*-mandelate ion in the active site of mandelate racemase. Only some of the polar groups surrounding the active site are shown. The enzyme has two catalytic acid–base groups. Lysine 166 is thought to deprotonate *S*-mandelate to form the *aci* anion, while His 297 deprotonates *R*-mandelate to form the same anion.[106]

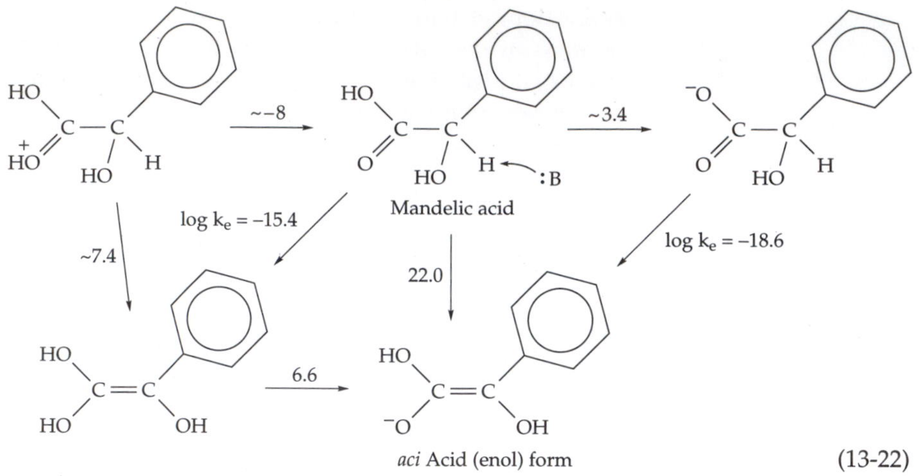

(13-22)

would reduce the pK_a of the α-hydrogen to ~ 7.4, making it very easy to enolize[109] and solving the thermodynamic problem. However, what forces could keep the substrate molecule in this unlikely state of protonation? Gerlt and Gassman proposed formation of a strong, short hydrogen bond to a carboxylate oxygen.[110] Formation of such a bond would lead to an increased positive charge on the alpha carbon, lowering the pK_a and therefore the thermodynamic barrier. It could also lower the intrinsic barrier,[108] for example, by permitting, to some extent, the formation of a C—H---N hydrogen bond in the transition state.[111] Polarization by the Mg^{2+} ion may also be involved. The very strong electrostatic forces within the active site may be sufficient to explain the formation of enolate anions or enols as intermediates in many different enzymes.[108,112]

A number of other racemases and epimerases may function by similar mechanisms. While some amino acid racemases depend upon pyridoxal phosphate (Chapter 14), several others function without this coenzyme. These include racemases for aspartate,[113] glutamate,[114–115a] proline, phenylalanine,[116] and diaminopimelate epimerase.[117] Some spiders are able to interconvert D and L forms of amino acid residues in intact polypeptide chains.[118,119]

Another enzyme of the mandelate pathway of degradation of aromatic rings (Fig. 25-8) is the **cis,cis-muconate lactonizing enzyme** which catalyzes the reaction of Eq. 13-23. It has a three-dimensional struc-

(13-23)

ture almost identical to that of mandelate racemase but has incorporated an additional feature that allows formation, by addition, of the intermediate *aci*-acid.[101,105]

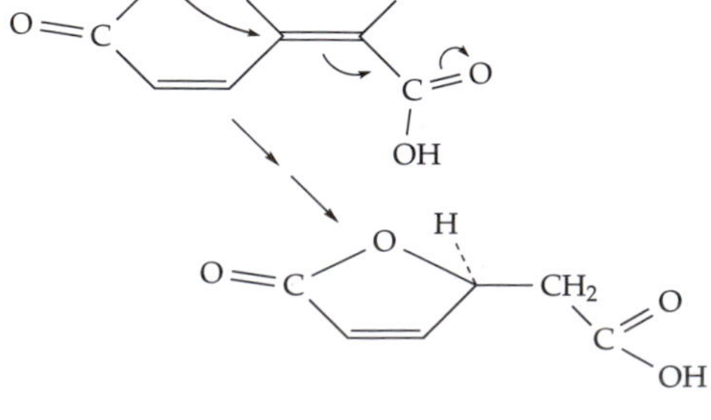

2. Isomerases

The isomerases that catalyze the simplest reactions are **tautomerases** that promote the oxo-enol (keto-enol) transformation. The widely distributed oxaloacetate tautomerase (Eq. 13-24) is especially active in animal tissues.[97,120] Oxaloacetate exists to a substantial extent in the enolic form: at 38°, ~ 6% enol, 13% oxo, and ~ 81% covalent hydrate.[120,121] A mammalian **phenylpyruvate tautomerase** has also been investigated.[122]

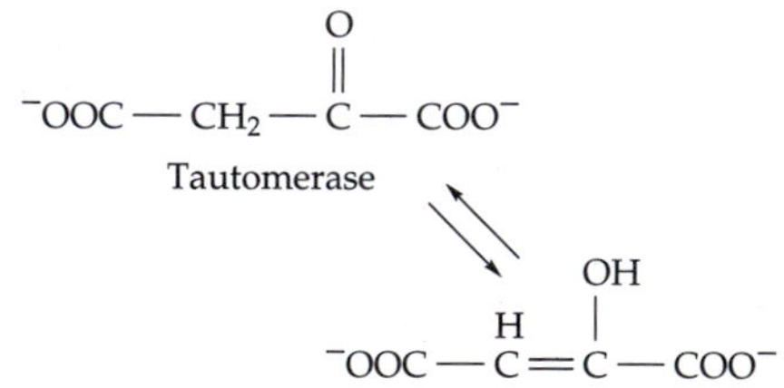

(13-24)

The oxidation of one functional group of a molecule by an adjacent group in the same molecule is a feature of many metabolic sequences. In most cases an enolic intermediate, formed either from a ketone as in Eq. 13-25 or by a dehydration reaction (see Eq. 13-32), is postulated.

Aldose–ketose interconversions. A metabolically important group of enzymes catalyze the interconversion of aldose sugars with the corresponding 2-ketoses (Table 10-1, reaction 4C). Several sugar phosphates undergo rapid isomerization. **Glucose 6-phosphate isomerase** (Eq. 13-25) appears to function in all cells with a high efficiency.[123–125a] The 132 kDa dimeric protein from muscle converts glucose 6-phosphate to

Glucose 6-phosphate   cis-Enediol

Glucose
6-phosphate
isomerase

Fructose 6-phosphate

$$(13\text{-}25)$$

fructose 6-phosphate with a turnover number of $\sim 10^3\ s^{-1}$. Hereditary defects in this enzyme cause a variety of problems that range from mild to very severe.[126] **Mannose 6-phosphate isomerase** also forms fructose 6-phosphate, while **ribose 5-phosphate isomerase**[127] interconverts the 5-phosphates of D-ribose and D-xylulose in the pentose phosphate pathways (Chapter 17). Other enzymes, most often metalloenzymes, catalyze the isomerization of free sugars.

These enzymes vary widely in secondary and tertiary structure.[127a] Mannose-6-phosphate isomerase is a 45 kDa Zn^{2+}-containing monomer. The larger 65 kDa **L-fucose isomerase**, which also acts on D-arabinose, is a hexameric Mn^{2+}-dependent enzyme.[127a] **L-Arabinose isomerase** of *E. coli*, which interconverts arabinose and L-ribulose, is a hexamer of 60-kDa subunits[128] while the **D-xylose isomerase** of *Streptomyces* is a tetramer of 43-kDa subunits.[129] The nonenzymatic counterpart of the isomerization catalyzed by the enzyme is the base-catalyzed **Lobry deBruyn–Alberda van Ekenstein transformation** (Eq. 13-25).[130]

In 1895, Emil Ficher proposed an enediol intermediate for this isomerization. As would be expected, the enzyme-catalyzed isomerization of glucose-6-phosphate in ^2H_2O is accompanied by incorporation of deuterium into the product fructose 6-phosphate at C-1. In the reverse reaction ^2H-containing fructose 6-phosphate was found to react at only 45% of the rate of the ^1H-containing compound. Thus, the primary deuterium isotope effect expected for a rate-limiting cleavage of the C–H bond was observed (see Chapter 12, Section B,3).

When fructose 6-phosphate containing both ^2H and ^{14}C in the 1 position was isomerized in the presence of a large amount of nonlabeled fructose 6-phosphate, the product glucose 6-phosphate contained not only ^{14}C but also ^2H, and the distribution indicated that the ^2H had been transferred from the C-1 position into the C-2 position.[123] It was concluded that in over half the

turnovers of the enzyme, ^2H removed from C-1 is put back on the same molecule at C-2. This intramolecular transfer of a proton suggests a syn transfer, the proton being removed and put back on the same side of the molecule. The carrier of the proton may be a histidine side chain.[131] This result, together with the known configuration of glucose at C-2, indicates that the intermediate is the *cis*-enediol and that addition of a proton at either C-1 or C-2 of the enediol is to the *re* face. However, mannose-6-phosphate isomerase catalyzes addition to the *si* face.

Catalysis of ring opening by isomerases.
Glucose-6-phosphate isomerase catalyzes a second reaction, namely, the opening of the ring of the α-anomer of glucose 6-phosphate (one-half of the mutarotation reaction; Eq. 10-88). Noltmann suggested a concerted acid–base catalysis by two side chain groups as is indicated in Eq. 9-90. An NMR study showed that the isomerization of the α anomer occurs at least ten times faster than that of the β. Thus, β-glucose 6-phosphate is first converted to α-glucose 6-phosphate before it can be isomerized to fructose 6-phosphate.[132]

Triose phosphate isomerase.
This dimeric 53-kDa enzyme interconverts the 3-phosphate esters of glyceraldehyde and dihydroxyacetone and is the fastest enzyme participating in glycolysis. Its molecular activity at 25° is $\sim 2800\ s^{-1}$ in the direction shown in Eq. 13-26 and $\sim 250\ s^{-1}$ in the reverse direction (the predominant direction in metabolism)[133] and is thought to operate at the diffusion-controlled limit (Chapter 9).[133] Each of the identical subunits consists of a striking $(\alpha/\beta)_8$ barrel (Fig. 13-6) with an active site at the carboxyl ends of the β strands.[134–137] Structures of the enzyme containing bound inhibitors such as phosphoglycolohydroxamate have also been determined (Fig. 13-6).[138] Triose phosphate isomerase is also one of the most investigated of all enzymes. Not only are its catalytic properties unusual but also there are known defects in the human enzyme[137] and it is also a potential target for antitrypanosomal drugs.[139]

Phosphoglycolohydroxamate (PGH), a tight-binding inhibitor ($K_d = 15\mu M$) that resembles enediolate 1 of Eq. 13-26

Although its high catalytic activity might favor intramolecular transfer of the proton removed by triose phosphate isomerase, little such transfer has

Dihydroxyacetone phosphate

Enediolate 1

a

b

E-165

Enediolate 2

c

Glyceraldehyde
3-phosphate

(13-26)

trum of dihydroxyacetone bound to the enzyme revealed a shift of the carbonyl bands at 1733 cm^{-1} by about 20 cm^{-1} to ~1713 cm^{-1}. This might indicate a polarization and stretching of the carbonyl group by a positively charged histidine or lysine side chain of the protein.[142] The His 95 side chain is appropriately placed (Fig. 13-6) to function in this way[143] and its replacement by glutamine decreases catalytic activity by a factor of 400.[144] The –NH$_3^+$ group of lysine 12, another essential residue, is apparently needed for substrate binding.[136,145] After product is released, an isomerization within the enzyme is usually required to prepare it for acceptance of a new substrate. This may involve movement of protons between side chain groups or a conformational change or both. In the case of triose phosphate isomerase, a very rapid isomerization requiring about ~1 ms has been detected.[146]

To investigate further the function of His 95, Lodi and Knowles recorded the ^{13}C and ^{15}N NMR spectra of the three histidine rings[148] both in unligated enzyme and in the phosphoglycolohydroxamate (PGH) complex. The results were a surprise. The His 95 resonances

PGH

H95

213.5 → 204.2   138.1 → 137.1   121.3 → 122.6

been observed.[140] This suggested that a relatively weak base such as a carboxylate group might serve as the proton acceptor at the C-3 position of dihydroxyacetone phosphate. Covalent labeling experiments and later X-ray studies have implicated Glu 165, whose carboxylate group is thought to remove the *pro-R* hydrogen atom from the hydroxymethyl group of dihydroxyacetone phosphate as indicated in Eq. 13-26. When this residue is replaced by Asp, most activity is lost.[141] Kinetic studies suggest that this carboxyl group has a pK_a of ~3.9 in the free enzyme. However, k_{cat} is independent of pH up to pH 10.[140] The pK_a of the phosphate also affects the binding (K_m), with only the phosphate dianion being a substrate. Study of the infrared spec-

did not change at all when the pH was changed from below 5 to 9.9, and the chemical shift values (shown above) indicate clearly that the ring is **unprotonated** in both free enzyme and in the complex. The key chemical shift values are shown on the diagram to the

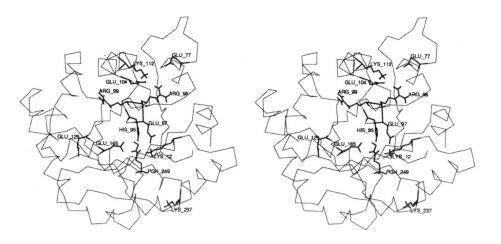

Figure 13-6 Stereoscopic view into the active site of triose phosphate isomerase showing side chains of some charged residues; PGH, a molecule of bound phosphoglycolohydroxamate, an analog of the substrate enolate.[138] The peptide backbone, as an alpha-carbon plot, is shown in light lines.[147] The $(\alpha/\beta)_8$-barrel structure is often called a TIM barrel because of its discovery in this enzyme. Courtesy of M. Karplus.

left; the arrows indicate the changes upon formation of the PGH complex. These results suggest that H95 is never protonated and that if it acts as the catalytic acid it does so by dissociating to an imidazolate ion (Eq. 13-27).[148]

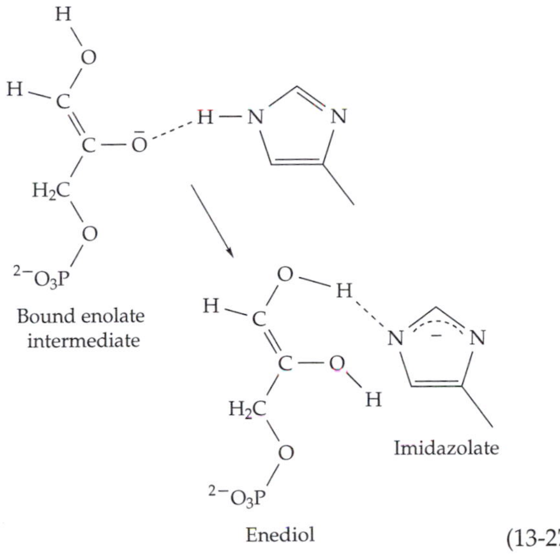

Bound enolate intermediate

Enediol   (13-27)

Imidazolate

The imidazole would donate a proton to the enolate ion and then remove a proton from the resulting enediol. Although the proposed chemistry is unusual, it is argued that the high pK_a of the neutral imidazole acting as a proton donor would be matched with the high pK_a of the enol, permitting rapid reaction.[148] Nevertheless, two other possibilities exist. Theoretical calculations support the idea that the proton transfer between the two oxygens in the enediolate may occur *without catalysis by a proton donor*[149] as indicated in step *b* of Eq. 13-26. Another possibility is that a transient protonation of H95 by the adjacent E97 carboxyl group occurs and allows the histidine to participate in the proton transfer. Study of kinetic isotope effects has indicated coupled motion of protons and proton tunneling.[143]

Another detail should be mentioned. The active site of triose phosphate isomerase is formed by a series of loops connecting the α helices and β strands of the barrel. One of those loops, consisting of residues 167–176, folds over the active site after the substrate is bound to form a hinged lid that helps to hold the substrate in the correct orientation for reaction.[150–152] When the lid, which can be seen in Fig. 13-6, closes, the peptide NH of G171 forms a hydrogen bond to a phosphate oxygen atom of the substrate. This is only one of many known enzymes with deeply buried active sites that close in some similar fashion before a rapid reaction occurs.

Although enzymes tend to be extremely specific they are not always completely able to avoid side reactions. Triose phosphate isomerase releases small amounts of **methylglyoxal** (Eq. 13-28), presumably as

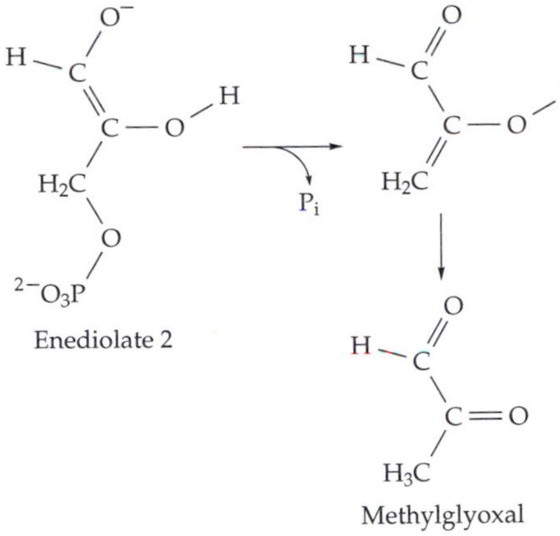

Enediolate 2

Methylglyoxal   (13-28)

a result of elimination of phosphate from enediolate 2 of Eq. 13-26. Deletion of four residues from the hinged lid produced an enzyme in which this side reaction was increased 5.5-fold.[146]

Xylose isomerase and the hydride shift mechanism.

This bacterial enzyme, which isomerizes D-xylose to D-xylulose, has an $(\alpha/\beta)_8$ barrel similar to that of triose phosphate isomerase. It is of industrial importance because it also catalyzes isomerization of D-glucose to the sweeter sugar D-fructose, a reaction used in preparation of high-fructose syrup. For this purpose the isomerase is often immobilized on an insoluble matrix such as diethylaminoethyl cellulose.[153] This enzyme, as well as other isomerases that act on free sugars, requires a metal ion such as Mg^{2+}, Co^{2+}, or Mn^{2+}. The three-dimensional structures, solved in several different laboratories,[129,154–158] show that there are two metal ions (ordinarily Mg^{2+}) about 0.5 nm apart and held by an array of glutamate and aspartate side chains. One glutamate carboxylate forms short ionic bonds to both metal ions and an essential histidine is also coordinated with metal ion 2. The three-dimensional structure is superficially similar to that of triose phosphate isomerase but there are major differences in properties. Xylose isomerase requires metal ions, acts on unphosphorylated non-ionic substrates, and does not catalyze detectable exchange of protons with the solvent. Furthermore, *X-ray structures do not show the presence of any catalytic base that could initiate formation of an intermediate enolate ion*. These facts suggested an alternative isomerization mechanism, one involving a hydride ion shift and well-known from studies of the Cannizarro reaction.[159] Because the non-ionic substrates and inhibitors bind weakly it has been difficult to obtain a clear picture of events in the active site. The substrate is the α-anomer of D-xylopyranose[155] and the enzyme catalyzes the opening of the sugar ring (step *a*, Eq. 13-29).[157] The details of this process are not clear but acid-base catalysis as in Eq. 9-87 is probable.

There is a consensus[156–158] that the open form of

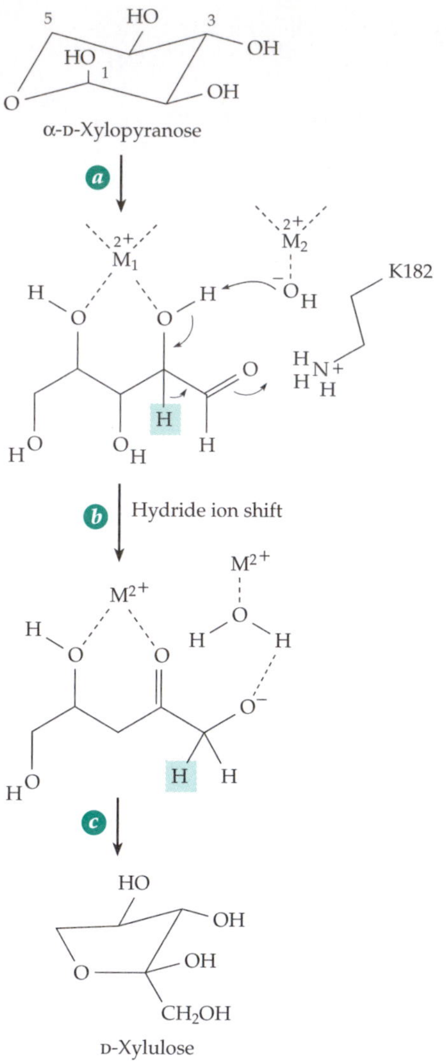

α-D-Xylopyranose

a ↓

Hydride ion shift

b ↓

c ↓

D-Xylulose

(13-29)

associate as dimers or higher oligomers. The enzyme has a high content of nonpolar amino acid residues and is soluble in high concentrations of ethanol. This property is compatible with the location of the eukaryotic enzyme in the endoplasmic reticulum. The substrate binding site is a hydrophobic cavity.

Oxosteroid isomerase has a remarkably high molecular activity ($\sim 0.75 \times 10^5$ s^{-1}). Substrates containing ^{2}H in the 4 position react only one-fourth as rapidly as normal substrates. The large isotope effect suggested that cleavage of the C−H bond to form an enzyme-stabilized enolate anion is rate limiting (Eq. 13-30, step *a*).[160] The proton in the axial position at C-4 is removed preferentially but without complete stereo-specificty.[165] The abstracted proton must be carried by a group on the enzyme and returned to the 6 position of the substrate, again in an axial orientation (Eq. 13-30, step *b*), a suprafacial transfer. Studies of the binding of spin-labeled substrate analogs, molecular modeling, and directed mutation indicate that Asp 38 is this proton carrier.[164,166] Little exchange of the proton with solvent is observed, presumably because of the extreme rapidity of the isomerase action. However, competitive inhibitors such as nortestosterone, whose double-bond arrangement is that of the product in Eq. 13-30, undergo exchange of one of the hydrogens at C-4 with the medium. The ultraviolet absorption band of the inhibitor at 248 nm is shifted to 258 nm upon combination with the enzyme, presumably as a result of formation of the

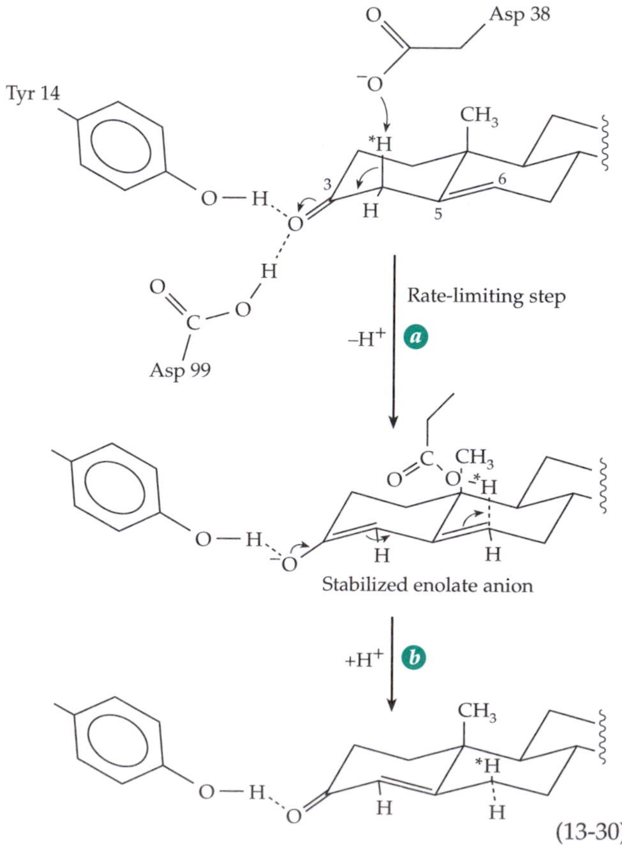

the sugar is bound to one of the two metal ions as shown in Eq. 13-29 and that a bound hydroxide ion on the second metal provides the catalytic base to remove a proton from the −OH at position 2. The isomerization occurs with the shaded hydrogen atom being shifted as H$^-$ to the carbon atom of the carbonyl group. During the reaction metal ion 2 moves apart from metal ion 1 by ~ 0.09 nm,[129,157,158] another fact that has made analysis of X-ray data difficult. Protonation of the alkoxide ion at C-2 is followed by ring closure (step *c*, Eq. 13-29).

3-Oxosteroid isomerases.

Cholesterol serves in the animal body as a precursor of all of the steroid hormones, including the 3-oxosteroids progesterone and testosterone (Chapter 22). The 3-hydroxyl group of cholesterol is first oxidized to an oxo group. This is followed by an essentially irreversible migration of the double bond in the 5,6-position into conjugation with the carbonyl group (Eq. 13-30) catalyzed by a 3-oxo-Δ^5-steroid isomerase. The small 125-residue enzyme from the bacterium *Pseudomonas testosteroni* has been studied by a great variety of methods.[160–164] The subunits

Rate-limiting step

−H$^+$ *a*

Stabilized enolate anion

+H$^+$ *b*

(13-30)

enolate anion. Tyrosine 14 has been identified as the catalytic acid, able to stabilize the enolate anion as depicted in Eq. 13-30.[161,162] A high-resolution NMR structure[167] and X-ray crystal structures[167a,b] have revealed an adjacent Asp 99 side chain that may assist.[167c]

Enzymatic isomerization of *cis*-aconitate to *trans*-aconitate apparently also involves proton abstraction,[165] with resonance in the anion extending into the carboxylic acid group. Its mechanism may be directly related to that of the oxosteroid isomerase. However, there are other 1,3-proton shifts in which neither a carbonyl nor a carboxyl group is present in the substrate (Eqs. 13-55, 13-56).

4-Oxalocrotonate tautomerase. This bacterial enzyme, which functions in the degradation of toluene (Chapter 25), is actually an isomerase. It catalyzes rapid interconversion of an unconjugated unsaturated α-oxoacid such as 4-oxalocrotonate with an intermediate enol (which may leave the enzyme) and the isomeric conjugated oxoacid (Eq. 13-31).[168–170] A related 5-carboxymethyl-2-hydroxymuconate isomerase

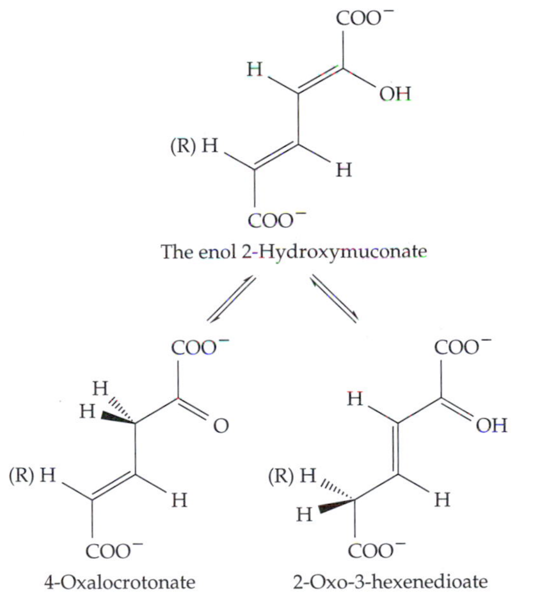

The enol 2-Hydroxymuconate

4-Oxalocrotonate   2-Oxo-3-hexenedioate

(13-31)

catalyzes the same reaction when R = –CH$_2$COO$^-$ in Eq. 13-31. It is an unusually small enzyme consisting of only 62 amino acid residues. The pH dependence reveals pK_a values of 6.2 and 9.0 in the free enzyme and 7.7 and 8.5 in the ES complex.[171]

The pK_a of 6.2 has been associated with the amino-terminal proline 1.[172] 4-Oxalocrotonate tautomerase is one of a small group of enzymes that have been synthesized nonenzymatically with both L amino acids and as a mirror image constructed with D amino acids.[173]

3. Internal Oxidation–Reduction by Dehydration of Dihydroxyacids

When a carboxylic acid contains hydroxyl groups in both the α and β positions, dehydration leads to formation of an enol that can tautomerize to 2-oxo-3-deoxy derivatives of the original acid (Eq. 13-32). Thus, phosphogluconate dehydratase yields 2-oxo-3-deoxyphosphogluconate as the product. When the reaction is carried out in ^2H_2O the ^2H is incorporated with a random configuration at C-3, indicating that the enzyme catalyzes only the dehydration and that the tautomerization of the enol to the ketone is nonenzymatic.

The reaction of Eq. 13-32 initiates a unique pathway of sugar breakdown, the Entner–Douderoff pathway (Eq. 17-18), in certain organisms. The 6-phosphogluconate is formed by oxidation of the aldehyde group of glucose 6-phosphate. This pattern of oxidation of a sugar to an aldonic acid followed by dehydration according to Eq. 13-32 occurs frequently in metabolism. A related reaction is the dehydration of 2-phosphoglycerate by enolase (Eq. 13-15). In this case the product is phosphoenolpyruvate, a stabilized form of the enolic intermediate of Eq. 13-32 (when R = H).

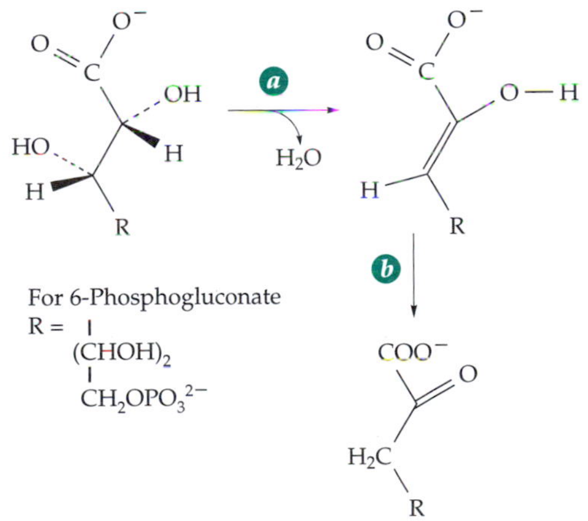

For 6-Phosphogluconate
R = (CHOH)$_2$ CH$_2$OPO$_3^{2-}$

(13-32)

4. Formation and Metabolism of Methylglyoxal (Pyruvaldehyde)

The rather toxic methylglyoxal is formed in many organisms and within human tissues.[174] It arises in part as a side reaction of triose phosphate isomerase (Eq. 13-28) and also from oxidation of acetone (Eq. 17-7) or aminoacetone, a metabolite of threonine (Chapter 24).[175] In addition, yeast and some bacteria, including *E. coli*, have a **methylglyoxal synthase** that converts dihydroxyacetone to methylglyoxal, apparently using a mechanism similar to that of triose phosphate isomerase. It presumably forms enediolate 2 of Eq. 13-26, which eliminates inorganic phosphate to yield methylglyoxal as in Eq. 13-28.[176,176a] Methylglyoxal is converted to D-lactate by the two-enzyme **glyoxalase** system (Eq. 13-33).[177] The combined action of methylglyoxal synthase and the glyoxalases provides a bypass to the usual glycolysis pathway (Fig. 10-3). Although it does not provide energy to the cell, it releases inorganic phosphate from sugar phosphates that may accumulate under conditions of low phosphate because the free P_i concentration is too low to support the glyceraldehyde 3-phosphate dehydrogenase reaction (step a of Fig. 10-3).

Glyoxalase acts not only on methylglyoxal but also on other α-oxo-aldehydes. It is thought to be an important enzyme system that protects cells against these potentially dangerous metabolites. Glyoxalase consists of a pair of enzymes, **glyoxalase I** and **glyoxalase II**, which catalyze the reactions of Eq. 13-33. Each subunit of the 183-residue human glyoxalase I contains one tightly bound Zn^{2+} ion.[178] However, the *E.coli* enzyme[178a] is inactive with Zn^{2+} and maximally active with Ni^{2+}. The enzyme requires glutathione as a cofactor, and in step a of the reaction (Eq. 13-33) the glutathione adds to form a thiohemiacetal[179] which is then isomerized in step b. During this step retention of the abstracted proton is so complete that it was earlier thought to function by an intramolecular hydride ion shift as in xylose isomerase. More recent evidence favors an enolate anion intermediate. However, the three-dimensional structure is not related to that of other isomerases and the exact mechanism remains

uncertain. Glyoxylase II, which catalyzes step c of Eq. 13-33, is an esterase.[175,180]

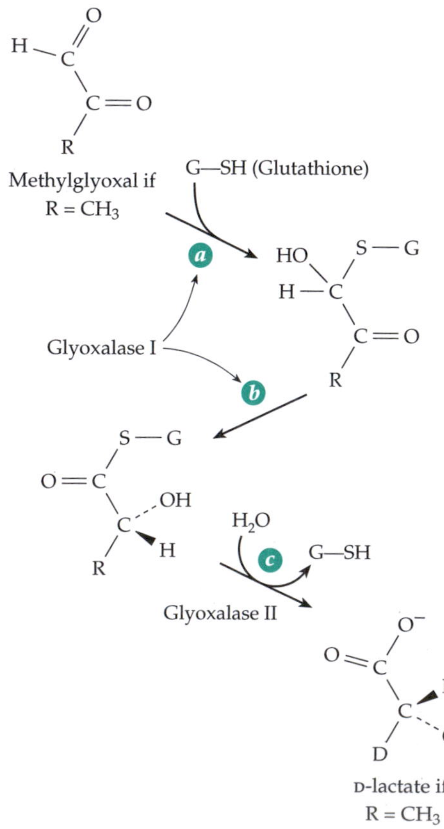

C. Beta Cleavage and Condensation

In Chapter 12 and in the preceding sections of this chapter we examined displacement and addition reactions involving nucleophilic centers on O, N, or S. Bonds from carbon to these atoms can usually be broken easily by acidic or basic catalysis. The breaking and making of C–C bonds does not occur as readily and the "carbon skeletons" of organic molecules often stick together tenaciously. Yet living cells must both form and destroy the many complex, branched carbon compounds found within them.

A major mechanistic problem in cleavage or formation of carbon–carbon bonds is the creation of a nucleophilic center on a carbon atom. The problem is most often solved by using the *activating influence of a carbonyl group to generate a resonance-stabilized enolate anion*. Just as the presence of a carbonyl group facilitates cleavage of an adjacent C–H bond, so it can also assist the cleavage of a C–C bond. The best known reactions of this type are the **aldol cleavage** and the

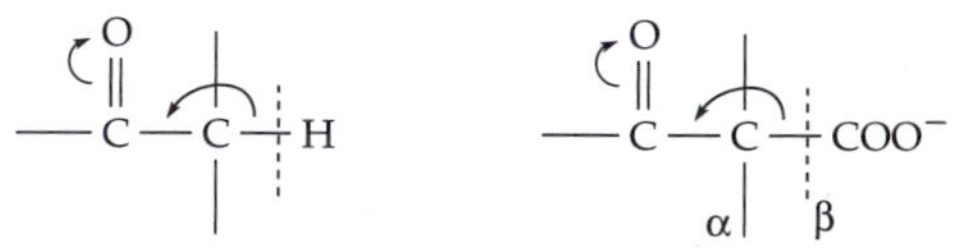

Cleavage of C—H bond
facilitated by a carbonyl group

β Cleavage with
release of CO_2

decarboxylation of **β-oxo acids.** The latter has been referred to as β decarboxylation and its reverse as β carboxylation. In this book these terms have been extended to include other reactions by which bonds between the α and β carbon atoms of a carbonyl compound are broken or formed, and these will be referred to as **β cleavage and β condensation**.

The β *condensation* reactions consist of displacement or addition reactions in which an enzyme-bound enolate anion acts as the nucleophile. We can group these condensation reactions into three categories as indicated by reaction types 5A, 5B, and 5C of Table 10-1.

1. Displacement on a Carbonyl Group

A β-oxo acid, with proper catalysis, is susceptible to hydrolysis by attack of water on the carbonyl group. An example is the reaction catalyzed by **oxaloacetate acetylhydrolase** which has been isolated from *Aspergillus niger* (Eq. 13-34). A related cleavage is catalyzed by ribulose bisphosphate carboxylase (see Eq. 13-48).

Methylglyoxal if
R = CH₃

G—SH (Glutathione)

Glyoxalase I

Glyoxalase II

D-lactate if
R = CH₃   (13-33)

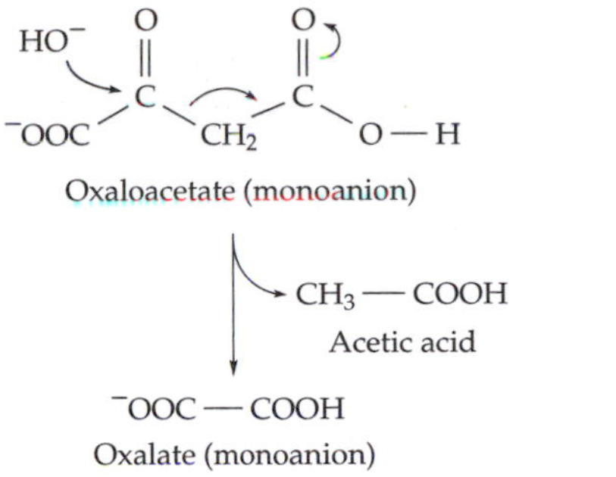

Oxaloacetate (monoanion)

CH₃ — COOH
Acetic acid

⁻OOC — COOH
Oxalate (monoanion) (13-34)

The **thiolases**[181] are lyases that cleave β-oxoacyl derivatives of CoA by displacement with a thiol group of another CoA molecule (Eq. 13-35). This is the chain cleavage step in the β oxidation sequence by which fatty acid chains are degraded (Fig. 10-4). Biosynthetic thiolases catalyze the condensation of two molecules of acetyl-CoA to form acetoacetyl-CoA, (see Eq. 17-5), a precursor to cholesterol and related compounds and to poly-β-hydroxybutyrate (Box 21-D). Because acetyl-CoA is a thioester, the reaction is usually described as a **Claisen condensation**. A related reaction, involving decarboxylation of malonyl-CoA, is a step in fatty acid synthesis. Since the thiolases are inhibited by –SH reagents it has been suggested that a thiol group in the enzyme reacts initially with the β-carbonyl group as in Eq. 13-35 to give an enzyme-bound *S*-acyl intermediate. The acyl group is then transferred to CoA in a second step.

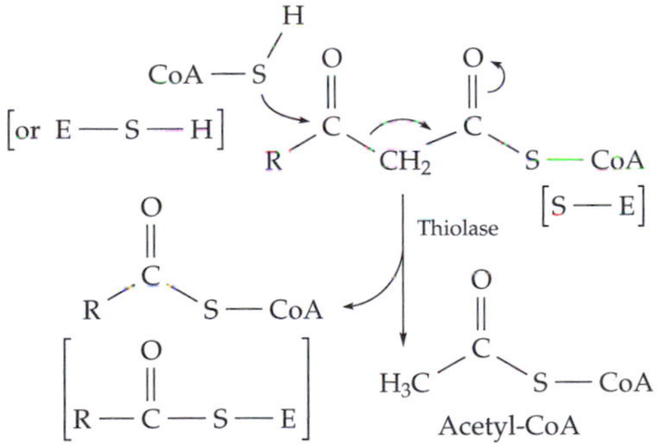

(13-35)

A very similar reaction is catalyzed by 3-hydroxy-3-methylglutaryl-CoA lyase (HMG-CoA lyase), which functions in the formation of acetoacetate in the human body (Eq. 17-5, step c) and also in the catabolism of leucine (Fig. 24-18)[182,183] and in the synthesis of **3-hydroxy-3-methylglutaryl-CoA**, the presursor of cholesterol (Eq. 17-5, step b)[183a]

2. Addition of an Enolate Anion to a Carbonyl Group or an Imine; Aldolases

The **aldol condensation** (Eq. 13-36), which is also illustrated in Box 10-D, is one of the more common

reactions by which C–C bonds are formed[183b] and, in the reverse reaction, cleaved in metabolism. Aldolases are classified into two major types. The type II aldolases are metal-ion dependent, the metal ion stabilizing the intermediate enolate ion (Eq. 13-36, steps a and b). Type I aldolases, which include the most studied mammalian enzymes, have a more complex mechanism involving intermediate Schiff base forms (Eq. 13-36, steps a', b', c', d').[184] The best known members of this group are the **fructose bisphosphate aldolases** (often referred to simply as aldolases), which cleave fructose-1,6-P_2 during glycolysis (Fig. 10-2, step e).

These enzymes have been found in all plant and animal tissues examined and are absent only from a few specialized bacteria. Three closely related isoenzymes are found in vertebrates.[185,186] The much studied rabbit muscle aldolase A is a 158-kDa protein tetramer of identical peptide chains.[186,187] Aldolase B, which is lacking in hereditary fructose intolerance, predominates in liver and isoenzyme C in brain.[185]

Treatment with sodium borohydride of the enzyme-substrate complex of aldolase A and dihydroxyacetone phosphate leads to formation of a covalent linkage between the protein and substrate. This and other evidence suggested a Schiff base intermediate (Eq. 13-36). When ¹⁴C-containing substrate was used, the borohydride reduction (Eq. 3-34) labeled a lysine side chain in the active site. The radioactive label was followed through the sequence determination and was found on Lys 229 in the chain of 363 amino acids.[186,188–188b] The enzyme is another $(\alpha/\beta)_8$-barrel protein and the side chain of Lys 229 projects into the interior of the barrel which opens at the C-terminal ends of the strands. The conjugate base form of another lysine, Lys 146, may represent the basic group B in Eq. 13-36,

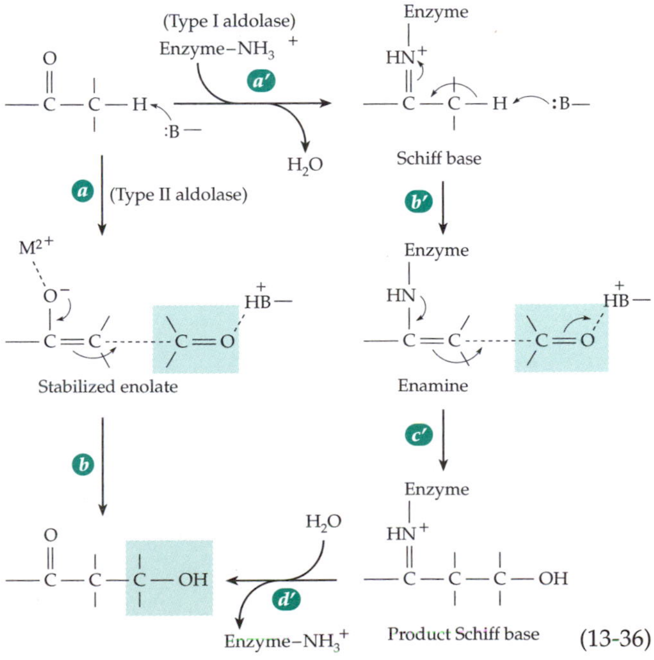

(13-36)

step *b*. Another possibility is that the adjacent phosphate group of the substrate acts as the acid–base catalyst for this step.[189] Aldolase A has been altered by mutations into a monomeric form that retains high catalytic activity,[190] something that has not often been accomplished for oligomeric enzymes.

The **type II aldolases** are not inactivated by sodium borohydride in the presence of substrate. A probable function of the essential metal ion is to polarize the carbonyl group as indicated in Eq. 13-36. In both yeast and *E. coli* aldolase the Zn^{2+} is held by 3 imidazole groups.[191,191a] An arginine side chain is a conserved residue involved in substrate binding in several class II aldolases.[192] Some blue-green algae contain both types of aldolase, as do the flagellates *Euglena* and *Chlamydomonas*.

The catabolism of L-fucose by *E. coli* requires cleavage of L-fuculose-1-phosphate to form dihydroxyacetone phosphate and D-lactaldehyde by a class II aldolase.[193]

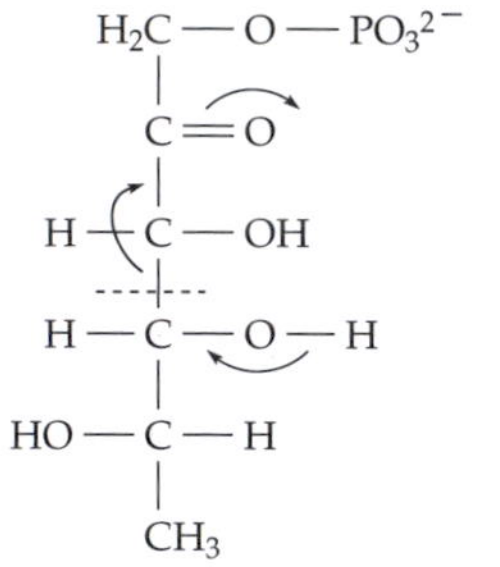

Special aldolases of both classes cleave and form C–C bonds throughout metabolism. Several of them act on 2-oxo-3-deoxy substrates forming pyruvate as one product (Eq. 13-37). The aldehyde product varies. In the Entner–Doudoroff pathway of carbohydrate metabolism (Chapter 21) 3-deoxy-2-oxo-6-phosphogluconate (KDPG) is cleaved to pyruvate and 3-phosphoglyceraldehyde.[194] The same products arise from the corresponding phosphogalactonate derivative.[195] The subunits of the trimeric KDPG aldolase have an $(\alpha / \beta)_8$-barrel structure similar to that of eukaryotic fructose 1,6-bisphosphate aldolase.[194] The 8-carbon sugar acid "KDO" of bacterial cell walls (Fig. 4-26) is cleaved by another aldolase. The catabolism of hydroxyproline leads to 4-hydroxy-2-oxoglutarate, which is cleaved to pyruvate and glyoxylate.[196] In the catabolism of deoxynucleotides, another aldolase converts 2-deoxyribose 5-phosphate to acetaldehyde and glyceraldehyde 3-phosphate.[197]

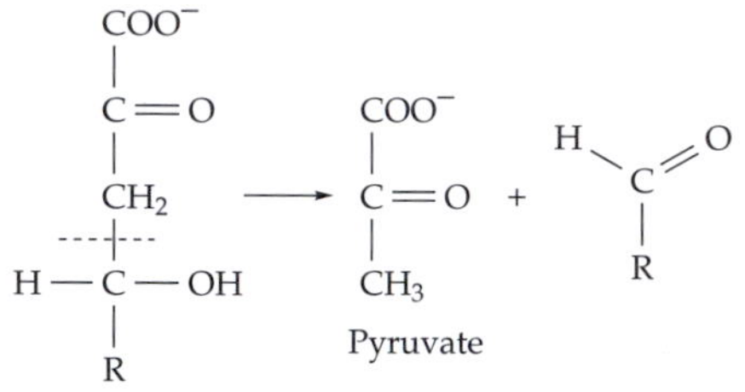

L-Fuculose-1-phosphate

$$(13\text{-}37)$$

The mechanism of chain cleavage proposed on the basis of the structure and modeling[193] is illustrated in Fig. 13-7.

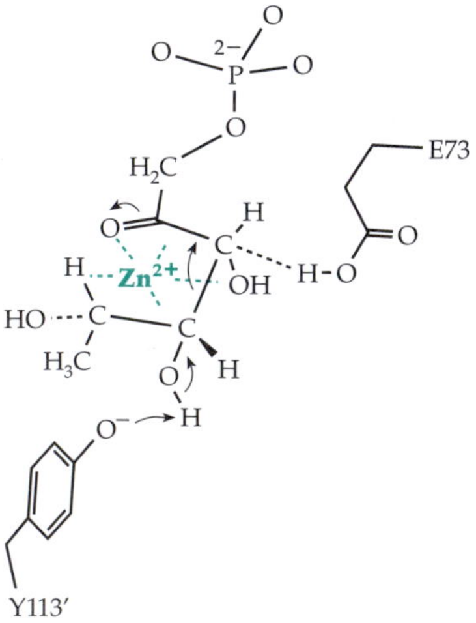

Figure 13-7 Interaction of the bound zinc ion of L-fuculose-1-phosphate aldolase and catalytic side chains with the substrate in the active site of the enzyme as revealed by X-ray crystallography and modeling. See Dreyer and Schulz.[193]

Closely related to aldolases is **transaldolase**, an important enzyme in the pentose phosphate pathways of sugar metabolism and in photosynthesis. The mechanism of the transaldolase reaction (Eq. 17-15) is similar to that used by fructose-1,6-bisphosphate aldolase with a lysine side chain forming a Schiff base and catalytic aspartate and glutamate side chains.[198]

Polycarboxylic acid synthases. Several enzymes, including **citrate synthase,** the key enzyme which catalyzes the first step of the citric acid cycle, promote condensations of acetyl-CoA with ketones (Eq. 13-38). An α-oxo acid is most often the second substrate, and a thioester intermediate (Eq. 13-38) undergoes hydrolysis to release coenzyme A.[199] Because the substrate acetyl-CoA is a thioester, the reaction is often described as a Claisen condensation. The same enzyme that catalyzes the condensation of acetyl-CoA with a ketone also catalyzes the second step, the hydrolysis of the CoA thioester. These polycarboxylic acid synthases are important in biosynthesis. They carry out the initial steps in a *general chain elongation process* (Fig. 17-18). While one function of the thioester group in acetyl-CoA is to activate the methyl hydrogens toward the aldol condensation, the subsequent hydrolysis of the thioester linkage provides for overall irreversibility and "drives" the synthetic reaction.

TABLE 13-1
Products Arising from Reactions of Acetyl-CoA with a Second Substrate with Catalysis by a Polycarboxylate Synthase

Ketone substrate	Product	Further metabolites
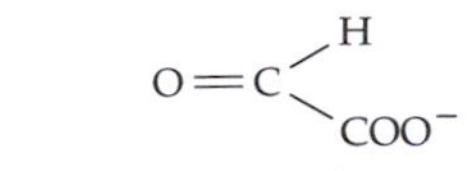 Glyoxylate	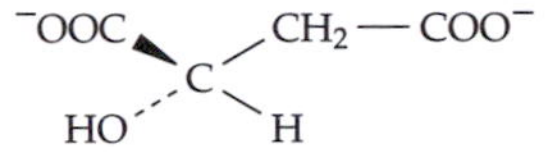 L-Malate	Carbohydrates, etc., via glyoxylate pathway
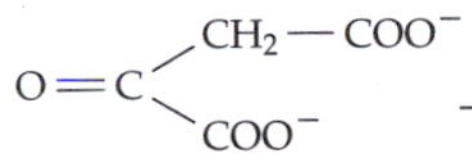 Oxaloacetate (Si-Citrate synthase)	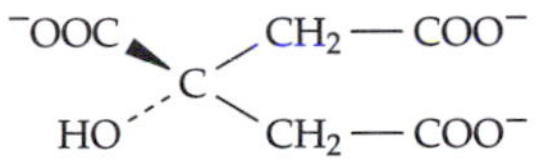 Citrate	2-Carbon unit from acetyl-CoA occupies pro-S position
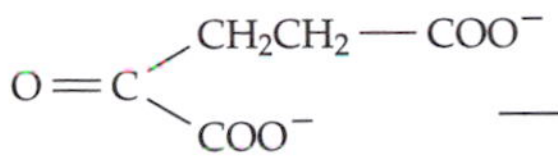 2-Oxoglutarate	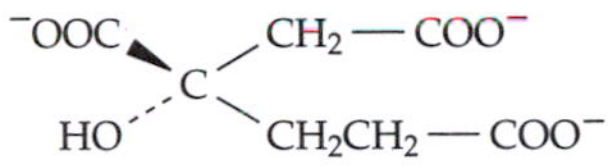 Homocitrate	Lysine via α-aminoadipic acid
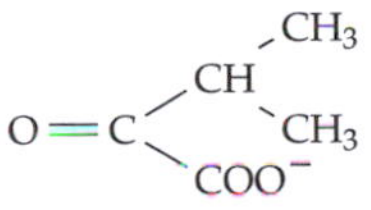 2-Oxoisovalerate	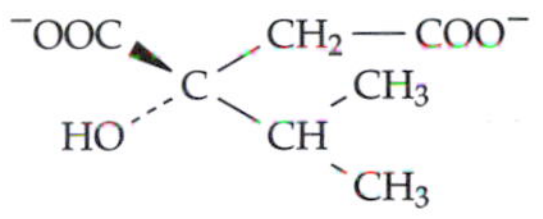 α-Isopropylmalate	Leucine
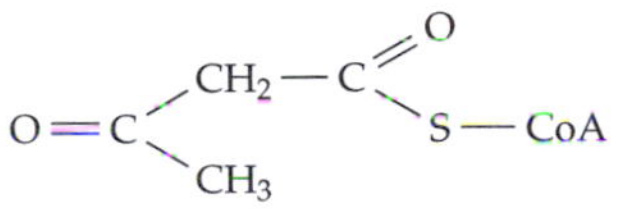 Acetoacetyl-CoA	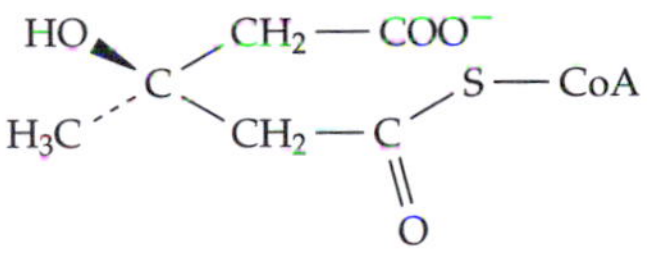 S-3-Hydroxy-3-methylglutaryl-CoA (HMG-CoA)	Free acetoacetate / Isoprenoid compounds

The stereochemistry of the reaction is also illustrated in Eq. 13-38. These enzymes may be classified by designating the face of the carbonyl group to which the enolate anion adds. The *si* face is up in Eq. 13-38. The common citrate synthase of animal tissues[200] and that of *E. coli*[201] condense with the *si* face and are designated (*si*)-citrate synthases. A few anaerobic bacteria use citrate (*re*)-synthase having the opposite stereochemistry.[202] Many citrate synthases are ~100-kDa dimers[203,204] but some are hexamers[203,205] and are allosterically inhibited by NADH. The second substrates and products of several related reactions of acetyl-CoA are summarized in Table 13-1.

The *si*-citrate synthase of pigs is a dimer of 437-residue chains, each of which is organized into a large rigid domain and a smaller

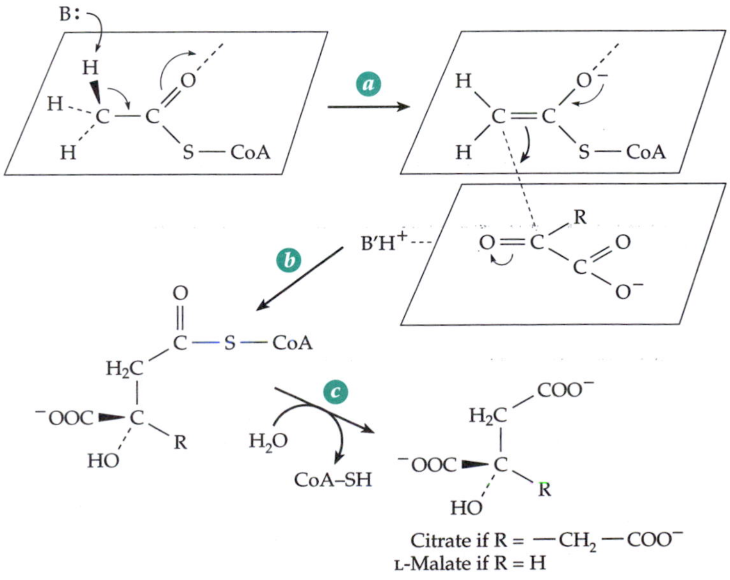

Citrate if R = —CH₂—COO⁻
L-Malate if R = H

(13-38)

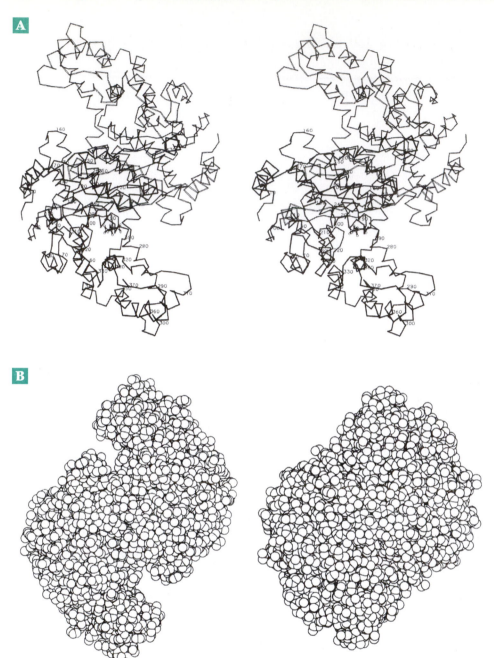

Figure 13-8 Three-dimensional structure of citrate synthase. (A) Stereoscopic alpha carbon trace of the dimeric pig enzyme in its open conformation.[209] A molecule of citrate is shown in the lower subunit. The view is down the two-fold axis. Courtesy of Robert Huber. (B) Space-filling representation of the open (left) and closed (right) forms of the same enzyme.[200] Courtesy of Stephen J. Remington.

flexible domain. The active site lies between the domains.[206] When oxaloacetate binds into the cleft the smaller domain undergoes a complex motion that closes the enzyme tightly around the substrate (Fig. 13-8).[200,207,208] The bound oxaloacetate is shown in Fig. 13-9. It is surrounded by a large number of polar side chains, including several from histidine and arginine residues. Of these, Arg 401 and Arg 421 of the second subunit bind the substrates' two carboxylate groups. In this tight complex the imidazole of His 320 is in the correct position to protonate the carbonyl oxygen of the oxaloacetate. The domain movement has also brought the groups that bind the acetyl-CoA into their proper position creating, by "induced fit,"

the acetyl-CoA binding site. This conformational change also accounts for the observation of an ordered kinetic mechanism with oxaloacetate binding before acetyl-CoA. The imidazole of His 274 is correctly oriented to abstract a proton from the methyl group of acetyl-CoA to generate the intermediate enolate anion. When oxaloacetate alone is bound into the active site the carbonyl stretching frequency, observed by infrared spectroscopy, is shifted from 1718 cm^{-1} for free oxaloacetate to 1697 cm^{-1}. This decrease of 21 cm^{-1} suggests a strong polarization of the C = O bond by its interaction with His 320 in the ground state.[211] This interaction is seen also in the ^{13}C NMR spectrum, the carbonyl resonance being shifted downfield by 6.8 ppm upon binding to the protein.[212] The enolate anion mechanism is also supported by kinetic isotope effects.[213]

The methyl protons of dithia-acetyl-CoA are much more reactive than those of acetyl-CoA ($pK_a \sim 12.5$ vs ~ 20). When citrate acts on this acetyl-CoA analog together with oxaloacetate, the expected enolate anion

$$\begin{array}{c} S \\ \parallel \\ H_3C - C - S - CoA \end{array}$$

Dithia-acetyl-CoA

forms rapidly as is indicated by the appearance of a 306 nm absorption band. However, it condenses very slowly with the oxaloacetate.[214] The binding equilibria, kinetics, and X-ray structures of complexes for a variety of other analogs of acetyl-CoA have also been studied.[210,215-216a] It appears that Asp 375 and His 274 work together to generate the enolate anion (Fig. 13-9). NMR measurements indicate formation of an unusually short hydrogen bond, but its significance is uncertain.[216a,216b] Malate synthase (Table 13-1) operates with a very similar mechanism but has an entirely different amino acid sequence and protein fold.[216c]

Other Claisen condensations are involved in synthesis of fatty acids and polyketides[217] (Chapter 21) and in formation of 3-hydroxy-3-methylglutaryl-CoA, the precursor to the polyprenyl family of compounds (Chapter 22). In these cases the acetyl group of acetyl-CoA is transferred by a simple displacement mechanism onto an –SH group at the active site of the synthase to form an acetyl-enzyme.[218,219] The acetyl-enzyme is the actual reactant in step *b* of Eq. 17-5 where this reaction, as well as that of HMG-CoA lyase, is illustrated.

Citrate cleaving enzymes. In eukaryotic organisms the synthesis of citrate takes place within the mitochondria, but under some circumstances citrate is exported into the cytoplasm. There it is cleaved by **ATP-citrate lyase**. To ensure that the reaction goes to completion, cleavage is coupled to the hydrolysis of ATP to ADP and inorganic phosphate (Eq. 13-39). The value of *G'* given here is extremely dependent upon the concentration of Mg^{2+} as a result of strong chelation of Mg^{2+} by citrate.[220] The reaction sequence is complex but can be understood in terms of an initial ATP-dependent synthesis of

$$\text{Citrate} \quad \text{CoA} \diagdown \diagup \text{ATP}$$
$$\searrow \text{ADP} + P_i$$
$$\text{Acetyl-CoA} \longleftarrow \diagup \diagdown \longrightarrow \text{Oxaloacetate}$$
$$\Delta G' \text{ (pH 7)}^* = +0.9 \text{ kJ mol}^{-1} \qquad (13\text{-}39)$$

citryl-CoA using a mechanism similar to that in Eq. 12-47. There is evidence for both phosphoenzyme and citryl enzyme intermediates (Eq. 13-40).[221] Native ATP-citrate lyase is a tetramer of 110-kDa subunits. It usually contains some phosphoserine and phosphothreonine residues but they apparently have little effect on activity.[222] Phosphorylation is catalyzed by cAMP-dependent and by insulin-dependent protein kinases.[223,224] A related reaction is the ATP-dependent cleavage of malate to acetyl-CoA and glyoxylate. It requires two enzymes, malyl-CoA being an intermediate.[225]

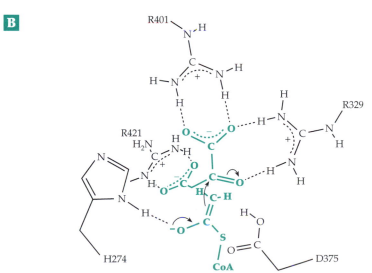

Figure 13-9 Active site of pig citrate synthase. (A) Stereoscopic view with a molecule of citrate in the active site.[200] Courtesy of Stephen J. Remington. (B) Interpretive view of the enolate anion of acetyl-CoA and oxaloacetate bound in the active site. Based on work by Kurz *et al.*[210]

A substrate-induced citrate lyase found in bacteria such as *E. coli* and *Klebsiella* promotes the anaerobic dissimilation of citrate splitting it to oxaloacetate and acetate.[226] The large ~585-kDa protein from *Klebsiella*[225] has the composition $(\alpha\beta\gamma)_6$, where the α, β, and γ sub-units have masses of ~55-, 30-, and 10-kDa, respectively. The subunit carries an unusual covalently bound deriv-ative of coenzyme A[227] (see also Chapter 14). The 10-kDa γ subunit serves as an **acyl-carrier protein**, the –SH of its prosthetic group being acetylated by a separate ATP-dependent ligase. The resulting acetyl-enzyme undergoes an acyl exchange (analogous to a CoA-transferase reaction; Eq. 12-50) to form a citryl enzyme before the aldol cleavage takes place (Eq. 13-40).

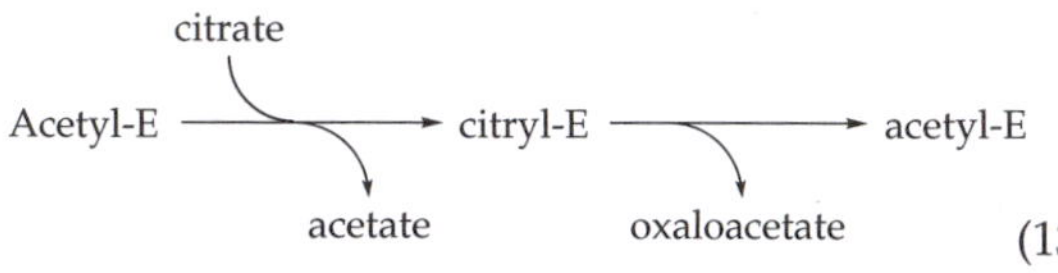

$$\text{Acetyl-E} \xrightarrow[\text{acetate}]{\text{citrate}} \text{citryl-E} \xrightarrow[\text{oxaloacetate}]{} \text{acetyl-E} \qquad (13\text{-}40)$$

The first unique enzyme of the important glyoxy-late pathway (Chapter 17), **isocitrate lyase**, cleaves isocitrate to succinate and glyoxylate (Eq. 13-41).[228] The carboxylate group that acts as electron acceptor would presumably be protonated by the enzyme.

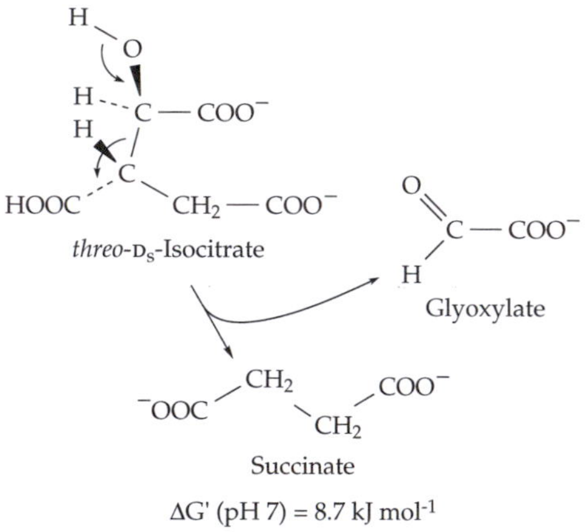

$$\Delta G' \text{ (pH 7)} = 8.7 \text{ kJ mol}^{-1} \qquad (13\text{-}41)$$

3. Chiral Acetates and Their Use in Stereochemical Studies

Consider the series of enzyme-catalyzed reactions shown in Eq. 13-42. Fumarate is hydrated in ^{3}H-con-taining water to malate which is oxidized to oxaloace-tate. Hydrolysis of the latter with oxaloacetate acetyl-hydrolase (Eq. 13-34) in ^{2}H$_2$O gives oxalate and chiral (R) acetate. The identical product can be obtained by condensing oxaloacetate with acetyl-CoA using citrate (*re*)-synthetase. The resulting citrate is cleaved in ^{2}H$_2$O using a citrate lyase having the *si* specificity.[229,230] Acetate of the opposite chirality can be formed enzy-matically beginning with [2,3-^{3}H]fumarate hydrated

by fumarate hydratase in ordinary water. Chiral acetates have also been prepared nonenzymatically,[229] and their configuration has been established unequivocally.

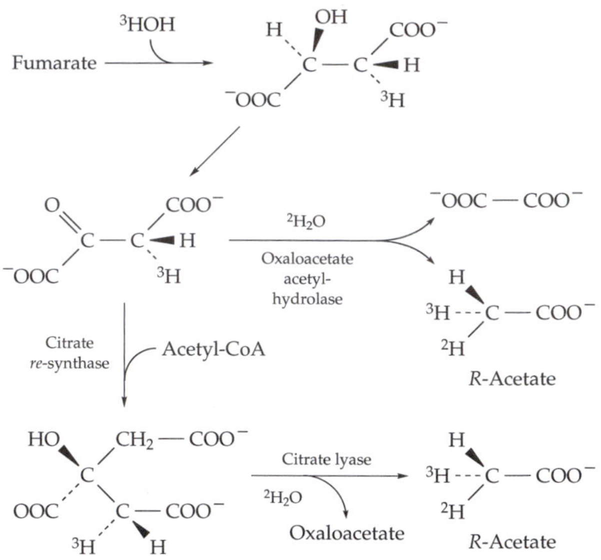

$$(13\text{-}42)$$

During the action of both oxaloacetate acetyl-hydrolase and citrate lyase (Eq. 13-42) inversion of configuration occurs about the carbon atom that carries the negative charge in the departing enolate anion. That inversion also occurs during catalysis by citrate (*re*)-synthase and other related enzymes has been demonstrated through the use of chiral acetates.[229,230] The findings with malate synthase[199] are illustrated in Eq. 13-38. Presumably, a basic group B of the enzyme removes a proton to form the planar enolate anion. The second substrate glyoxylate approaches from the other side of the molecule and condenses as is shown. Since any one of the three protons in either *R* or *S* chiral acetyl-CoA might have been abstracted by base B, several possible combinations of isotopes are possible in the L-malate formed. One of the results of the experi-ment using chiral (*R*) acetyl-CoA is illustrated in Eq. 13-43. The reader can easily tabulate the results of removal of the ^{2}H or ^{3}H. However, notice that if the base –B: removes ^{2}H (D) or ^{3}H (T) the reaction will be much slower because of the kinetic isotope effects which are expected to be $^H k/^D k \approx 7$ and $^H k/^T k \approx 16$. A second important fact is that the *pro-R* hydrogen at C-3 in malate is specifically exchanged out into water by the action of fumarate hydratase. From the distribution of tritium in the malate and fumarate formed using the two chiral acetates, the inversion by malate synthase was established. See Kyte[231] for a detailed discussion.

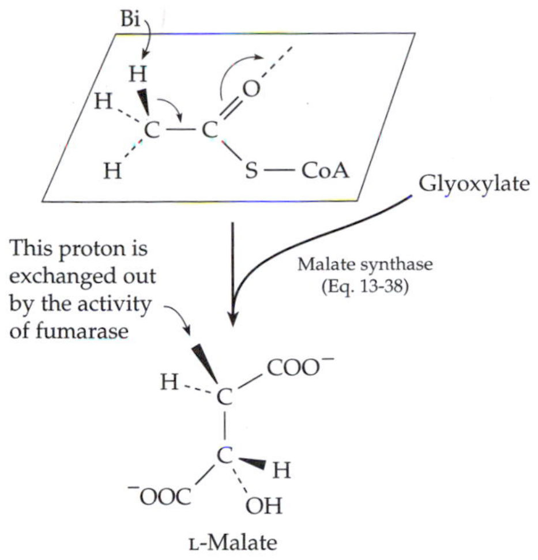

This proton is exchanged out by the activity of fumarase

Glyoxylate

Malate synthase (Eq. 13-38)

L-Malate

$$(13\text{-}43)$$

4. Addition of an Enolate Ion to Carbon Dioxide and Decarboxylation

The addition of an enolate anion to CO_2 to form a β-oxoacid represents one of the commonest means of incorporation of CO_2 into organic compounds. The reverse reaction of **decarboxylation** is a major mechanism of biochemical formation of CO_2. The equilibrium constants usually favor decarboxylation but the cleavage of ATP can be coupled to drive carboxylation when it is needed, e.g., in photosynthesis.

Decarboxylation of β-oxoacids. Beta-oxoacids such as oxaloacetic acid and acetoacetic acid are unstable, their decarboxylation being catalyzed by amines, metal ions, and other substances. Catalysis by amines depends upon Schiff base formation,[232] while metal ions form chelates in which the metal assists in electron withdrawal to form an enolate anion.[233–235]

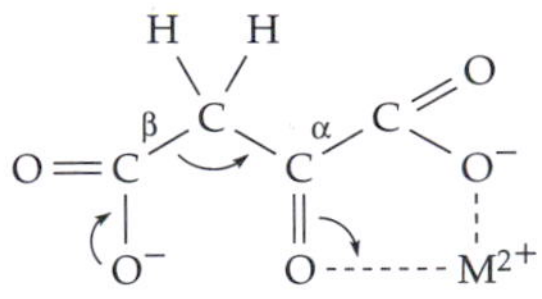

Can we apply any of this information from non-enzymatic catalysis to decarboxylating enzymes? Some decarboxylases do form Schiff bases with their substrates, and some are dependent on metal ions.[235] The acetone-forming fermentation of *Clostridium acetobutylicum* requires large amounts of acetoacetate decarboxylase (Eq. 13-44).

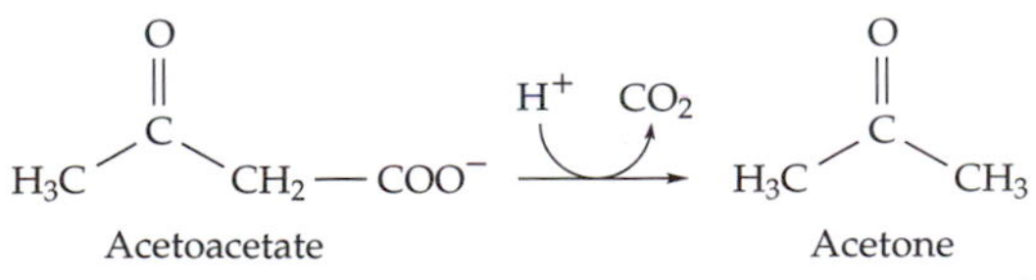

Acetoacetate

Acetone

$$\cdot \ (13\text{-}44)$$

The enzyme is inactivated by borohydride in the presence of substrate, and acid hydrolysis of the inactivated enzyme yielded ε-N-isopropyllysine. Decarboxylation occurs from a Schiff base by a mechanism analogous to that of the aldol cleavage shown in Eq. 13-36.[236] Mechanistically related is 4-oxalocrotonate decarboxylase.[236a]

Linked oxidation and decarboxylation. Metabolic pathways often make use of oxidation of a β-hydroxy acid to a β-oxoacid followed by decarboxylation in the active site of the same enzyme. An example is conversion of L-malate to pyruvate (Eq. 13-45). The Mg^{2+} or Mn^{2+}-dependent decarboxylating malic dehydrogenase that catalyzes the reaction is usually called **the malic enzyme**. It is found in most organisms.[237–240] While a concerted decarboxylation and dehydrogenation may sometimes occur,[241–242] the enzymes of this group appear usually to operate with bound oxoacid intermediates as in Eq. 13-45.

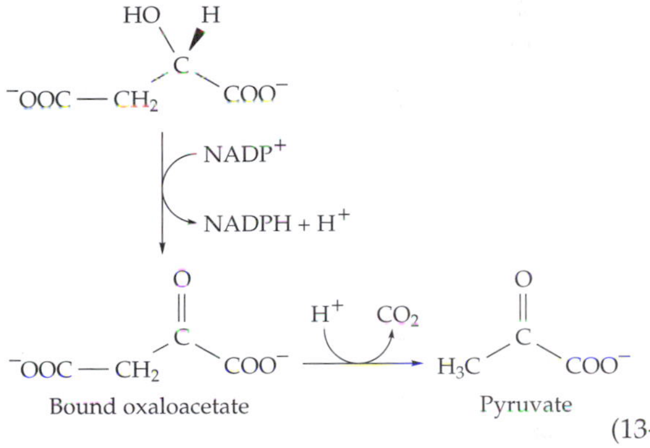

Bound oxaloacetate

Pyruvate

$$(13\text{-}45)$$

Other reactions of this type are the oxidation of isocitrate to 2-oxoglutarate in the citric acid cycle (Fig. 17-4, steps *d* and *e*),[243] oxidation of 6-phosphogluconate to ribulose 5-phosphate (Eq. 17-12),[244] and corresponding reactions of isopropylmalate dehydrogenase[245,246] and tartrate dehydrogenase.[247,248] Crystallographic studies of isocitrate dehydrogenase using both photolabile "caged" isocitrate[247] and slow mutant forms[243] with polychromatic Laue crystallography (Chapter 3) have demonstrated the rapid formation of the anticipated intermediate **oxalosuccinate**.

Phosphoenolpyruvate, a key metabolic intermediate. A compound of central importance in metabolism is the phosphate ester of the enol form of pyruvate, commonly known simply as phosphoenolpyruvate (PEP).[249] It is formed in the glycolysis pathway by dehydration of 2-phosphoglycerate (Eq. 13-15) or by decarboxylation of oxaloacetate. Serving as a preformed enol from which a reactive enolate anion can be released for condensation reactions,[250,251] PEP is utilized in metabolism in many ways.

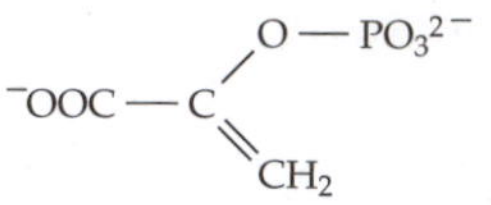

Phosphoenolpyruvate (PEP)

In animals and in many bacteria, PEP is formed by decarboxylation of oxaloacetate. In this reaction, which is catalyzed by **PEP carboxykinase** (PEPCK), a molecule of GTP, ATP, or inosine triphosphate captures and phosphorylates the enolate anion generated by the decarboxylation (Eq. 13-46).[252] The stereochemistry is such that CO_2 departs from the *si* face of the forming enol.[253] The phospho group is transferred from GTP with inversion at the phosphorus atom.[254] The enzyme requires a divalent metal ion, preferably Mn^{2+}. In fact, kinetic studies of the GTP-dependent avian mitochondrial enzyme indicate two metal-binding sites, one on the polyphosphate group of the bound GTP and one on carboxylate side chains of the protein.[252,255] The three-dimensional structure of the ATP-dependent *E. coli* enzyme reveals a nucleotide binding site similar to the ATP site of adenylate kinase (Fig. 12-30).[256] A definite binding site for CO_2 is also present in the enzyme.[257]

PEPCK is also activated by low concentrations of Fe^{2+} and this activation depends upon a protein that has been identified as glutathione peroxidase (Eq. 15-58). By destroying H_2O_2 the latter may allow the Fe^{2+} to prevent oxidation of an SH group on PEP carboxykinase.[258] Synthesis of PEPCK is stimulated by glucagon, evidently through a direct action of cAMP on transcription of the structural gene.[259] Transcription is also stimulated by glucocorticoids and thyroid hormone and is inhibited by insulin.

In some organisms, such as the parasitic *Ascaris suum*, PEPCK functions principally as a means of synthesis of oxaloacetate by reaction of PEP with CO_2

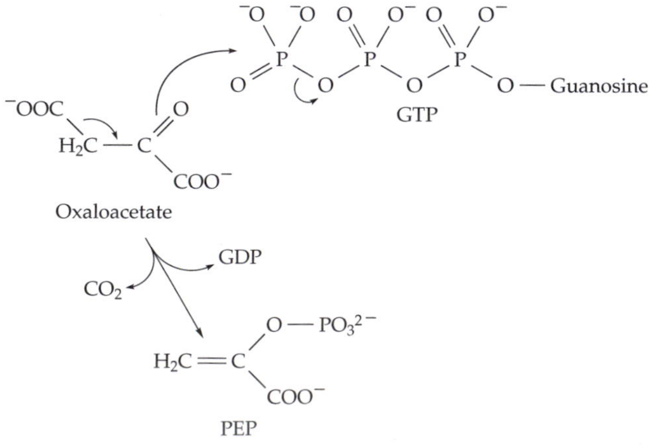

re Face is toward reader   (13-46)

and GDP. These organisms lack pyruvate kinase, which allows for buildup of the high PEP concentration needed to drive this reaction.[260] In a similar way PEP can be converted to oxaloacetate by **PEP carboxytransphosphorylase**, an enzyme found only in propionic acid bacteria and in *Entamoeba*. The reaction (Eq. 13-47) is accompanied by synthesis of inorganic pyrophosphate which may be cleaved to "pull" the reaction in the indicated direction.

$$PEP + CO_2 + P_i \rightarrow PP_i + \text{oxaloacetate} \qquad (13\text{-}47)$$

Oxaloacetate is also decarboxylated without phosphorylation of the enolate anion formed but with release of free pyruvate. Both pyruvate kinase and PEPCK can act as oxaloacetate decarboxylases.[261]

In the important reactions discussed in the following sections enolate ions are intermediates in carbon-carbon bond formation. Other examples are given in Eqs. 20-7 and 20-8, and Fig. 25-1, in which C–C bonds are formed by action of PEP as a carbon nucleophile,

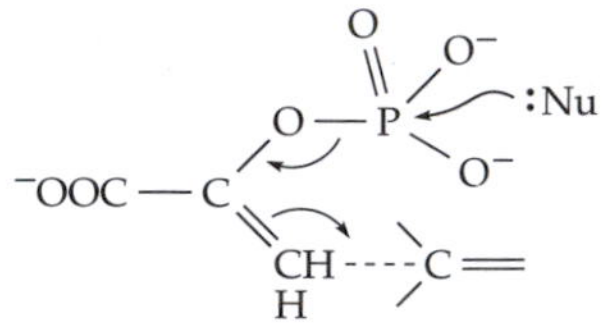

e.g., in aldol-like condensations.[242a] Yet another unusual type of reaction, involving formation of an enol ether linkage is illustrated in Box 13-B and in Eq. 25-3.

Ribulose bisphosphate carboxylase. The major route of incorporation of CO_2 into organic compounds is via photosynthesis. When ^{14}C-labeled CO_2 enters chloroplasts of green plants, the first organic ^{14}C-containing compound detected is 3-phosphoglycerate. Two molecules of this compound are formed through the action of **ribulose-1,5-bisphosphate carboxylase** (abbreviated **rubisco**), an enzyme present in chloroplasts and making up 16% of the protein of spinach leaves. This enzyme, whose structure is illustrated in Fig. 13-10, is thought to be the most abundant protein on earth. Because O_2 competes with CO_2 as a substrate, the enzyme also catalyzes an "oxygenase" reaction. It is therefore often called **ribulose-1,5-bisphosphate carboxylase/oxygenase**.[262] The rubisco from most plants is a 500- to 560-kDa L_8S_8 oligomer as shown in Fig. 13-10. The large subunit is encoded in the chloroplast DNA. However, a family of nuclear genes encode the small subunits, which are synthesized as larger precursors, with N-terminal extensions being removed to give the mature subunits.[263,264] Two types of small subunits occupy different positions in the quasi-symmetric spinach rubisco.[264] In *Euglena* the small subunit is synthesized as a polyprotein precursor containing eight copies of the subunit.[265] Rubisco from the hydrogen-oxidizing bacterium *Alcaligenes eutrophus* has a similar quaternary structure,[266] but the enzyme from *Rhodospirillum rubrum* is a simple dimer.[267] In dinoflagellates the rubisco gene is present in nuclear DNA rather than in the chloroplasts.[268]

The carboxylation reaction catalyzed by rubisco differs from others that we have considered in that the carboxylated product is split by the same enzyme (Eq. 13-48). The mechanism shown in Eq. 13-48 was suggested by Bassham *et al.*[271] Ribulose bisphosphate, for which the enzyme is absolutely specific, is first converted to its enolic form, a 2,3-enediol. Loss of a proton from the 3-OH group forms the endiolate anion needed for the carboxylation in step *b*. The product of this step is a β-oxoacid which undergoes enzyme-catalyzed hydrolytic cleavage (Eq. 13-48, step *c*; see also Eq. 13-34). Support for this mechanism came from the observation that 2-carboxyarabinitol bisphosphate (Fig. 13-11B) is a potent inhibitor, possibly a transition state analog.[272] This inhibitor, bound into the active site of the enzyme from spinach, is seen in both Figs. 13-10B and 13-11A. An expanded version of Eq. 13-48 is given in Fig. 13-12, which is based upon modeling together with a variety of X-ray structures including that shown in Fig. 13-11.

An essential Mg^{2+} ion is held by carboxylates of D203 and E204 and by modified K201. It also coordinates three molecules of H_2O in the free enzyme.[272a] Catalytic roles for the H294 and H327 imidazole groups are still being elucidated.[272b] Lysine 175 protonates C2 of the aci anion generated by C–C bond cleavage (step *e*, Fig. 13-12).[272c,276] Like many enzymes, rubisco exists in two major conformational states: open and closed.[272d]

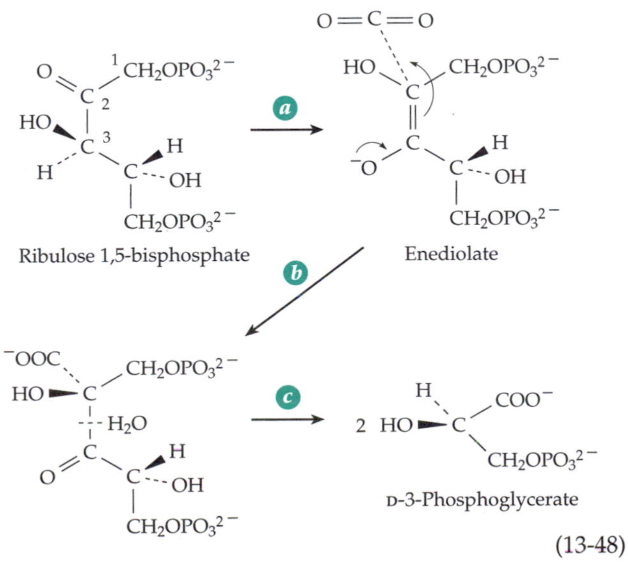

Chemical studies also support the indicated mechanism. For example, the β-oxoacid intermediate formed in step *b* of Eq. 13-48 or Fig. 13-12 has been identified as a product released from the enzyme by acid denaturation during steady-state turnover.[273,274] Isotopic exchange with ^3H in the solvent[275] and measurement of ^{13}C isotope effects[277] have provided additional verification of the mechanism. The catalytic activity of the enzyme is determined by ionizable groups with pK_a values of 7.1 and 8.3 in the ES complex.[278]

The apparent value of K_m for total CO_2 (CO_2 + HCO_3^-) is high, 11–30 mM, but for the true substrate CO_2 it is only 0.45 mM. In intact chloroplasts the affinity for substrate is distinctly higher, with the K_m for total CO_2 dropping to ~0.6 mM. The difference appears to result largely from a regulatory reaction of CO_2 in which the side chain amino group of lysine 201 of the large subunit forms a carbamate (Eq. 13-49). Although carbamylation is spontanous, it is enhanced by an ATP-dependent process catalyzed by **rubisco activase.**[279] The carbamylation converts the side chain of Lys 201 into a negatively charged group that binds to an essential divalent metal, usually Mg^{2+}, in the active center[282] as is shown in Fig. 13-11, A and B. The nature of the reaction is uncertain. One possibility is that the activase is a chaperonin.[280] It appears to assist the enzyme in removing inhibitory sugars that arise by side reactions in the active site.[281] Rubisco is also regulated by the level of a natural inhibitor which has been identified as 2-carboxyarabinitol 1-phosphate. This is the same as the inhibitor shown in Fig. 13-11 but with one less phosphate group and consequent weaker binding.[273a]

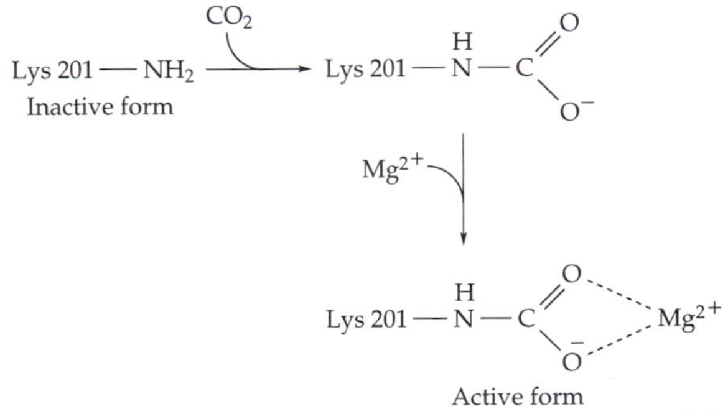

(13-49)

In most plants photosynthesis is also strongly inhibited by O_2. This observation led to the discovery that O_2 competes directly for CO_2 at the active site of rubisco in a process called **photorespiration**. Chloroplasts inhibited by oxygen produce **glycolate** in large amounts[282a] as a result of the reaction of the intermediate enediolate ion formed in step *b* of Eq. 13-48 with O_2

(13-48)

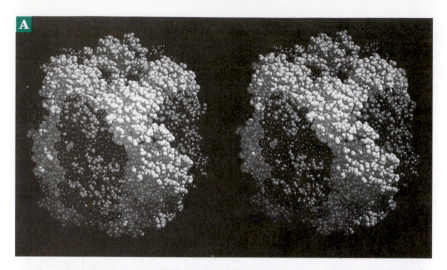

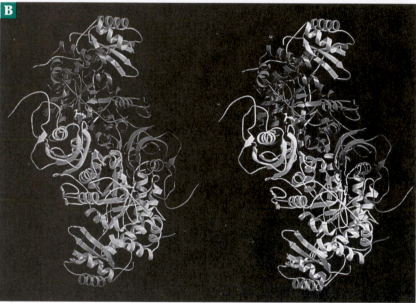

Figure 13-10 Stereoscopic view of ribulose bisphosphate carboxylase (rubisco) from spinach. (A) The symmetric L_8S_8 molecule contains eight 475-residue large subunits (in two shades of gray) and eight 123-residue small subunits (lighter gray in upper half of image). (B) One L_2S_2 substructure containing two active sites shared between adjacent large subunits with the bound inhibitor 2-carboxy-D-arabinitol 1,5-bisphosphate.[269] In the upper LS unit the S subunit (top) is light and the L subunit is dark. Courtesy of Inger Andersson. Similar structures have been determined for enzymes from tobacco[269a] and from the cyanobacterium *Synechoccus*.[270]

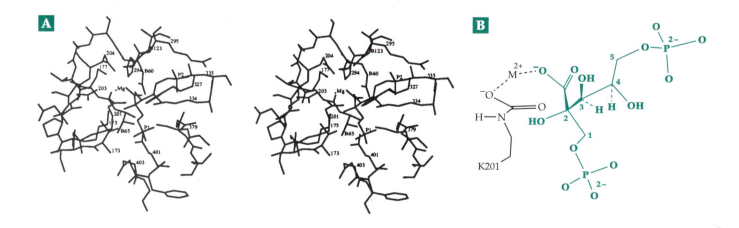

Figure 13-11 (A) Overview of the active site of spinach rubisco showing bound 2-carboxy-D-arabinitol 1,5-bisphosphate and Mg^{2+} and residues within hydrogen-bonding distance of these ligands. The hydroxyl groups at C2 and C3 of the inhibitor are in *cis* conformation.[269] Courtesy of Inger Andersson. (B) Structure of the inhibitor 2-carboxy-D-arabinitol 1,5-bisphosphate. A part of the carbamylated lysine 201 and the essential metal ion are also shown.

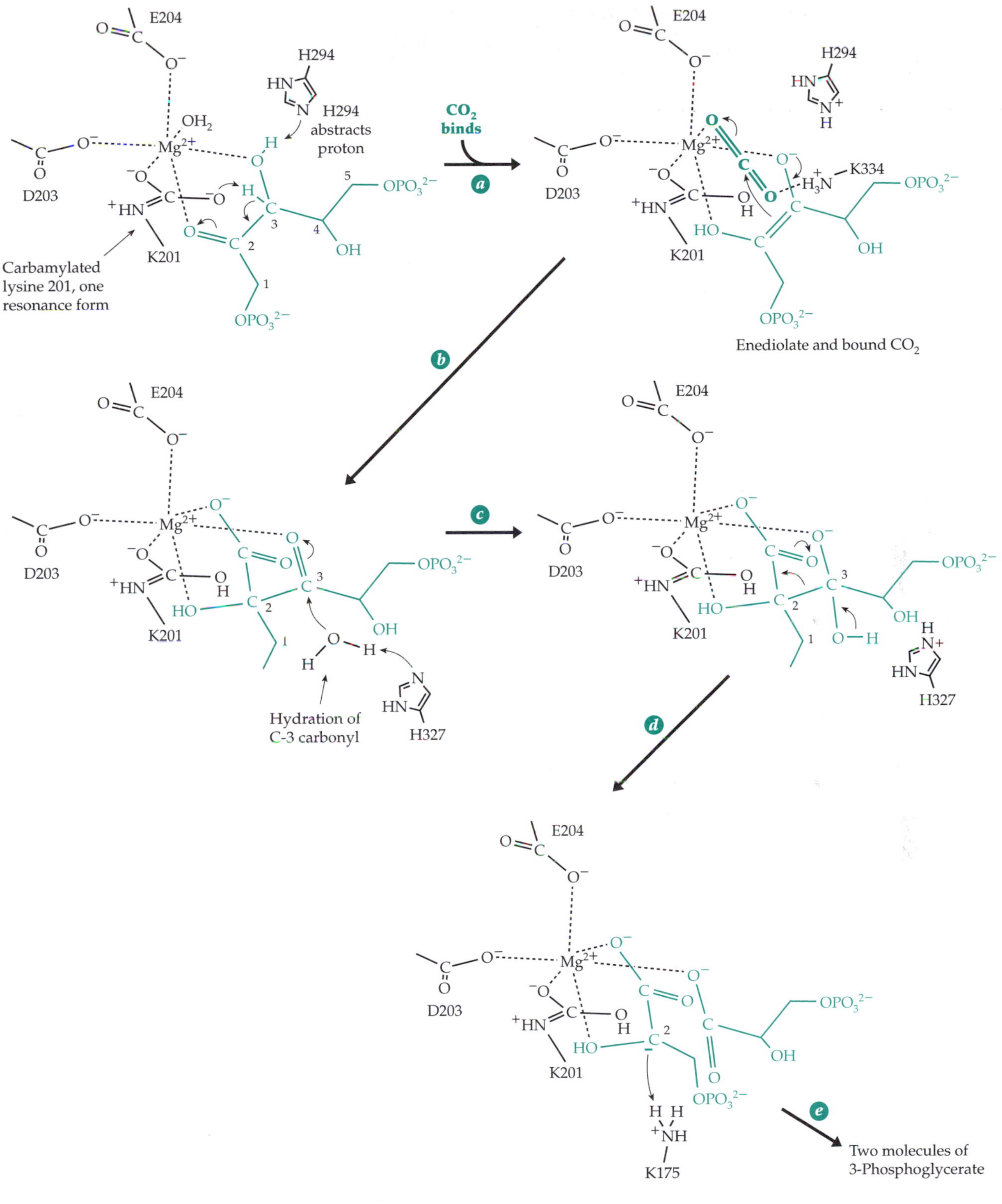

Figure 13-12 Proposed mechanism of action of ribulose bisphosphate carboxylase (rubisco). This is an abbreviated version of the mechanism as presented by Taylor and Andersson.[276] The binding of ribulose 1,5-bisphosphate occurs after carbamylation of lysine 201 and binding of a magnesium ion. Formation of an enediolate intermediate in step a is probably catalyzed by the carbamate group as indicated. Removal of a proton from the 3-OH, perhaps by H294, and addition of a proton to form an –OH group at C2 are also necessary. The CO_2 may bind to the Mg^{2+} and be polarized by interaction with other side chains prior to reaction. Carboxylation occurs by addition of the enediolate to CO_2 in step b. The hydration of the resulting 3-oxo group (step c) is necessary for cleavage of the C–C bond in step d. The participation of the newly formed carboxylate as an *aci* anion coordinated to Mg^{2+} is presumably involved. Protonation, with stereochemical inversion, is thought to involve K175, as shown. The two product molecules dissociate in step e.

(Eq. 13-50, step *a*). The peroxide formed in this way breaks up under the hydrolytic action of the enzyme to form phosphoglycolate and 3-phosphoglycerate (Eq. 13-50, step *b*).

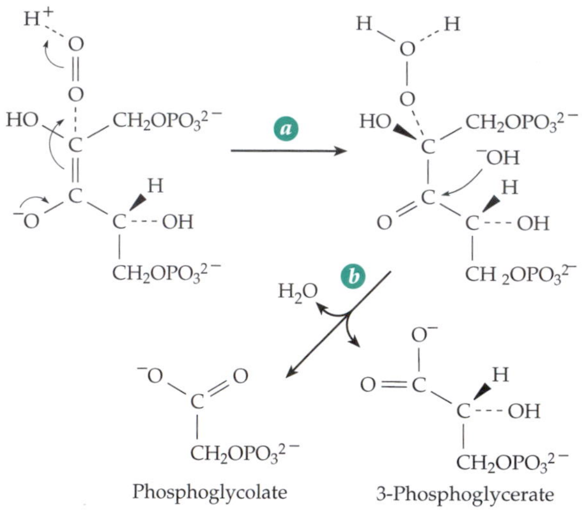

Phosphoglycolate   3-Phosphoglycerate

$$(13\text{-}50)$$

Molecular oxygen usually reacts rapidly with only those organic substrates, such as dihydroflavins, that are able to form stable free radicals. However, the endiolate anion of Eq. 13-50 may be able to donate a single electron to O_2 to form a superoxide–organic radical pair prior to formation of the peroxide (see also Eq. 15-30). Similar oxygenase side reactions have been observed for a variety of other enzymes that utilize carbanion mechanisms.[283] The reaction of rubisco with O_2 is of both theoretical and practical interest, the latter because of its significance in lowering the yield in photosynthesis (Chapter 23).

The simpler dimeric rubisco from *Rhodospirillum* is very inefficient in carboxylation and catalyzes much more oxygenation than do rubiscos of higher plants.[283a] Mutant enzymes that have impaired carboxylase and enhanced oxygenase are also known.[284,284a]

The small subunits of rubisco may help suppress undesirable side reactions.[285] For example, the following deoxypentodiulose phosphate can be formed by β elimination from the second intermediate of Eq. 13-48.

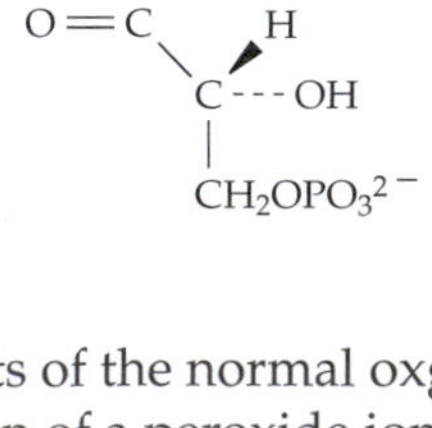

Some side products of the normal oxgenase reaction arise by elimination of a peroxide ion from the first

intermediate of Eq. 13-50. The resulting dicarbonyl bisphosphate can rearrange to give a carboxytetritol bisphosphate (Eq. 13-51).[284]

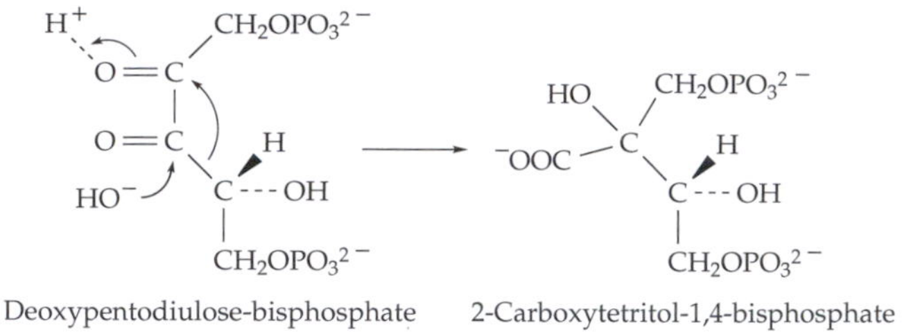

Deoxypentodiulose-bisphosphate   2-Carboxytetritol-1,4-bisphosphate

$$(13\text{-}51)$$

Carbon dioxide or bicarbonate ion? An important question in the consideration of carboxylation and decarboxylation reactions is whether the reactant or the product is CO_2 or HCO_3^-. An approach to answering the question was suggested by Krebs and Roughton,[286]

$$CO_2 + H_2O \xrightarrow{\quad a \quad} H_2CO_3 \xrightarrow{\quad b \quad} H^+ + HCO_3^-$$

$$(13\text{-}52)$$

who pointed out that the attainment of equilibrium between free CO_2 and HCO_3^- (Eq. 13-52) may require several seconds. If an enzyme produces CO_2 as a product and progress of the reaction is followed manometrically, the pressure will rise higher than the equilibrium value as the CO_2 is evolved. Later when substrate is exhausted and the CO_2 equilibrates with bicarbonate, the pressure will fall again. The addition of carbonic anhydrase, which catalyzes Eq. 13-52, step *a*, abolishes the "overshoot." If bicarbonate is the primary product of a decarboxylation, there is a lag in the appearance of free CO_2.

A second approach is to use [^{18}O]bicarbonate and to follow the incorporation of ^{18}O into a carboxylated substrate. If CO_2 is the primary substrate only two labeled oxygen atoms enter the compound, whereas if HCO_3^- is the reactant three are incorporated.[287] A third technique is measurement of the rate of incorporation of CO_2 or bicarbonate in the carboxylated product. Over a short interval of time, e.g., 1 min, different kinetics will be observed for the incorporation of CO_2 and of bicarbonate.[288] Using these methods, it was established that the product formed in Eq. 13-46 and the reactant in Eq. 13-47 is CO_2. However, the carboxylation enzymes considered in the next section use bicarbonate as the substrate.

5. Incorporation of Bicarbonate into Carboxyl Groups

An important enzyme with a biosynthetic function in many bacteria and in all higher plants is **PEP carboxylase**,[289] which catalyzes the reaction of Eq. 13-53. This enzyme, in effect, accomplishes the reverse of Eq. 13-46 by converting the three-carbon PEP, by reaction with bicarbonate, into the four-carbon oxaloacetate. The latter is needed for "priming" of the citric acid cycle and for biosynthesis of such amino acids as aspartate and glutamate. That the enzyme functions in this way is indicated by the fact that mutants of *Salmonella* defective in the enzyme do not grow unless oxaloacetate or some other intermediate in the citric acid cycle is added to the medium. The enzyme from *S. typhimurium* is a 400-kDa tetramer with complex regulatory properties. The corresponding enzyme from spinach has 12 subunits and 12 bound Mn^{2+} ions. The enzyme also has a special function in the C_4 plants,[290] in which it is a component of a carbon dioxide concentrating system (Chapter 23).

When [^{18}O]bicarbonate is a substrate, two labeled oxygen atoms enter the oxaloacetate, while the third appears in P_i. A concerted, cyclic mechanism could explain these results. However, study of kinetic isotope effects,[291] use of a substrate with a chiral thiophospho group,[292] and additional ^{18}O exchange studies[293] have ruled out this possibility. A transient carboxyl phosphate (Eq. 13-53) is evidently an intermediate.[294,295] The incorporation of the ^{18}O from bicarbonate into phosphate is indicated by the asterisks. The carboxyl group enters on the *si* face of PEP. However, there is another possibility.[295,296] The carboxyl phosphate, while in the active site adjacent to the enolate anion, may eliminate phosphate, the enolate ion adding to the resulting CO_2 to form the final product. According to this mechanism the group transfer potential of the phosopho group in PEP is

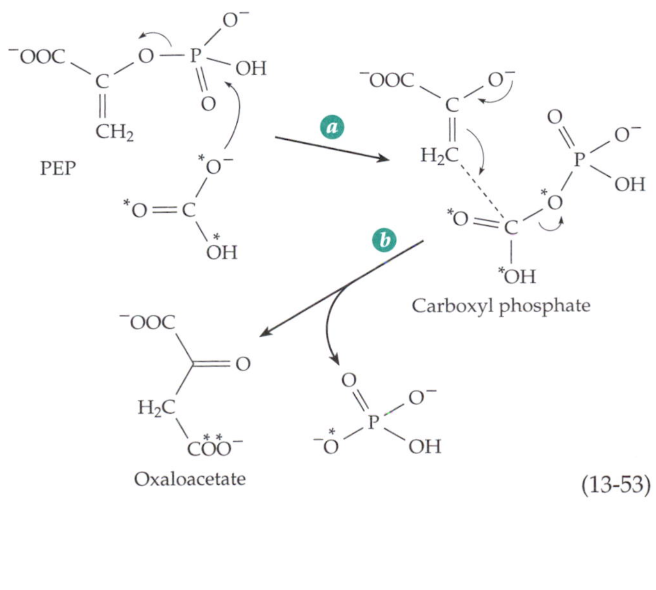

utilized to concentrate the CO_2 and to localize it next to the enolate anion generated by the same process.

PEP carboxylase is lacking from animal tissues and fungi. In these creatures PEP is converted to pyruvate, which is then carboxylated to oxaloacetate with coupled cleavage of ATP by the action of **pyruvate carboxylase** (Eq. 14-3), an enzyme that not only utilizes bicarbonate ion but also contains **biotin**. However, there are mechanistic similarities between its action and that of PEP carboxylase.

PEP mutase and the synthesis of phosphonates. The lipids of some organisms, such as *Tetrahymena*, contain aminoethylphosphonate, a compound with a C–P bond (Chapter 8). There are also many other naturally occurring phosphono compounds and huge quantities of synthetic phosphonates, present in detergents, herbicides, and insecticides, are metabolized by bacteria.[297] Here we will consider only one step in the biosynthesis of phosphonates, the conversion of PEP into phosphonopyruvate (Eq. 13-54), a reaction catalyzed by **PEP mutase**. The phospho group is moved

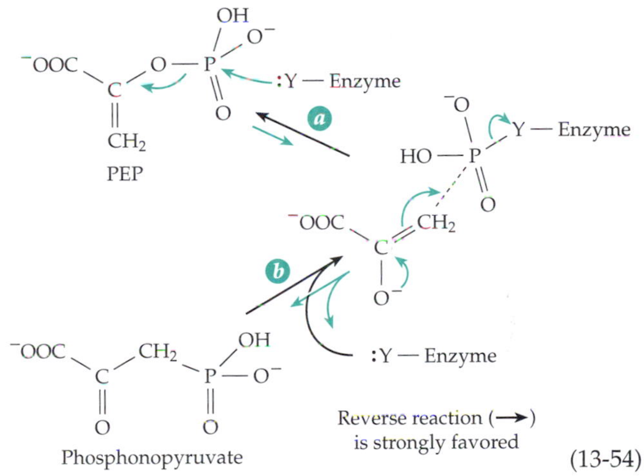

(13-54)

from the oxygen atom of PEP to the methylene carbon atom. When attached to carbon the phospho group is designated **phosphono**. The reaction occurs with retention of the configuration around phosphorus and a phosphoenzyme intermediate seems to be involved[298,299] as shown in Eq. 13-54. The equilibrium in this reaction strongly favors PEP. One further product formed from phosphonopyruvate is phosphonoethylamine, a component of phosphonolipids.[300] It is easy to imagine a synthetic route involving transamination followed by decarboxylation. Another mutase, similar to PEP mutase, shifts the carboxyphospho group $^-OOC–PO_2^-$ exactly as in Eq. 13-54 as part of the biosynthetic pathway of a natural herbicide formed by *Streptomyces hygroscopicus*.[301,302] Soil bacteria contain specialized enzymes that catalyze the hydrolytic cleavage of P–C bonds.[302a]

(13-53)

D. Some Isomerization and Rearrangement Reactions

A few metabolic reactions do not fit into any of the categories discussed so far and apparently do not depend upon a coenzyme. Some of these involve transfer of alkyl groups or of hydrogen atoms from one carbon to another. The hydrogen atoms move by direct transfer without exchange with the medium. All of the reactions could involve carbocations but there is often more than one mechanistic possibility.

A simple 1,3-proton shift is shown in Table 10-1 as reaction type 6A. An example is the isomerization of oleic acid to $trans$-Δ^{10}-octadecenoic acid (Eq. 13-55) catalyzed by a soluble enzyme from a pseudomonad.[303]

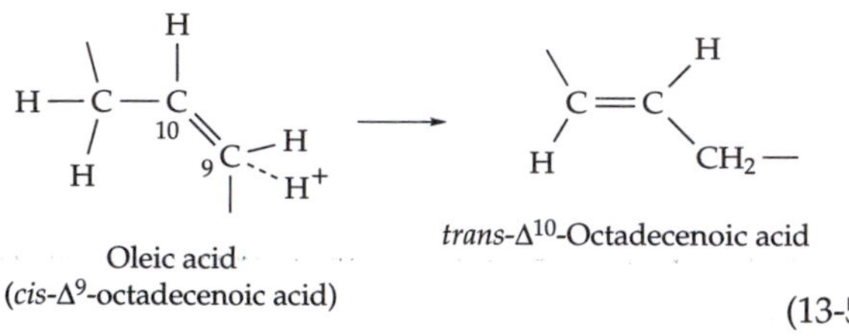

Oleic acid
(cis-Δ^9-octadecenoic acid)

$trans$-Δ^{10}-Octadecenoic acid

(13-55)

A second example is isomerization of **isopentenyl diphosphate** to **dimethylallyl diphosphate** (Eq. 13-56).[304–307] The stereochemistry has been investigated using the ^{3}H-labeled compound shown in Eq. 13-56. The pro-R proton is lost from C-2 and a proton is added to the re face at C-4. When the reaction was carried out in ^2H_2O a chiral methyl group was produced as shown.[304] A concerted proton addition and abstraction is also possible, the observed $trans$ stereochemistry being expected for such a mechanism. However, the

Isopentenyl diphosphate

Dimethylallyl diphosphate   (13-56)

fact that the enzyme is strongly inhibited by the cation of 2-(diethylamino) ethyl pyrophosphate, an analog of a probable carbocation intermediate. Cysteine and glutamate side chains are essential.[307]

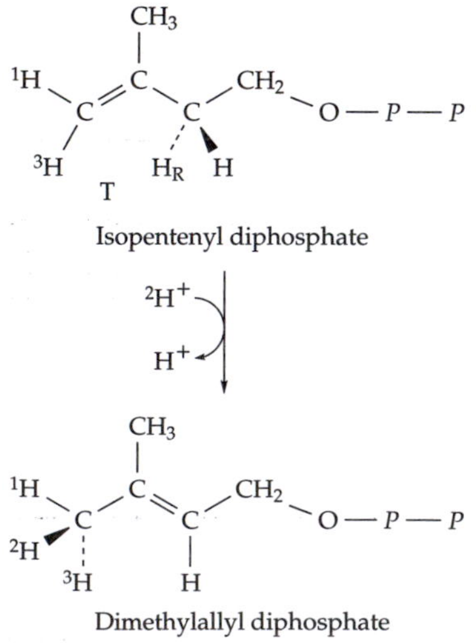

Reaction type 6B of Table 10-1 is allylic rearrangement with simultaneous condensation with another molecule. The reaction, which is catalyzed by **prenyltransferases**,[307a] occurs during the polymerization of polyprenyl compounds (Fig. 22-1, Eqs. 22-2, 22-3). Experimental evidence favors a carbocation mechanism for all of these reactions.[308,309] See Chapter 22.

Reaction type 6C (Table 10-1) occurs during the biosynthesis of leucine and valine (Fig. 24-17). The rearrangement is often compared with the nonenzymatic acid-catalyzed pinacol–pinacolone rearrangement in which a similar shift of an alkyl group takes place (Eq. 13-57). The enzyme-catalyzed rearrangement

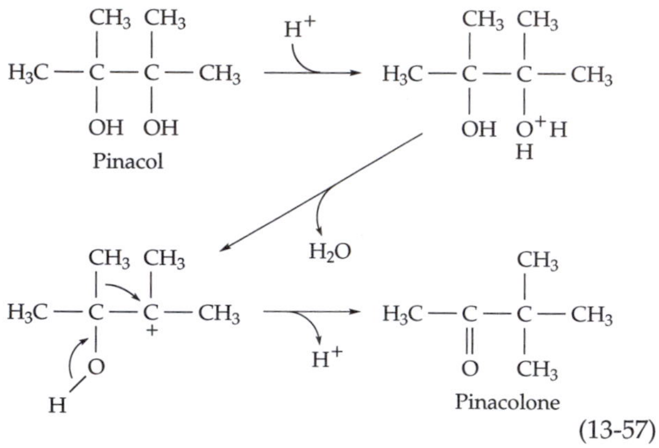

Pinacol

Pinacolone

(13-57)

presumably gives the structure drawn in brackets in Table 10-1. The same enzyme always catalyzes reduction with NADH to the diol, the Mg^{2+}-dependent enzyme being called **acetohydroxy acid isomeroreductase**. Rearrangement has never been observed without the accompanying reduction.[310–313] More complex rearrangements that occur during biosynthesis of sterols are described in Chapter 22.

References

1. Jencks, W. P. (1987) *Catalysis in Chemistry and Enzymology*, Dover, Mineola, N. Y. (p. 465)
2. Sorensen, P. E., and Jencks, W. P. (1987) *J. Am. Chem. Soc.* **109**, 4675–4690
3. Kallen, R. G., and Jencks, W. P. (1966) *J. Biol. Chem.* **241**, 5845–5850, 5851–5863
4. Edsall, J. T., and Wyman, J. (1958) *Biophysical Chemistry*, Vol. I, Academic Press, New York (p. 550ff)
5. Fernley, R. T. (1988) *Trends Biochem. Sci.* **13**, 356–359
6. Sly, W. S., and Hu, P. Y. (1995) *Ann. Rev. Biochem.* **64**, 375–401
7. Hazen, S. A., Waheed, A., Sly, W. S., LaNoue, K. F., and Lynch, C. J. (1996) *FASEB J.* **10**, 481–490
8. Heck, R. W., Boriack-Sjodin, P. A., Qian, M., Tu, C., Christianson, D. W., Laipis, P. J., and Silverman, D. N. (1996) *Biochemistry* **35**, 11605–11611
9. Maren, T. H. (1985) *N. Engl. J. Med.* **313**, 179–181
10. Stams, T., Nair, S. K., Okuyama, T., Waheed, A., Sly, W. S., and Christianson, D. W. (1996) *Proc. Natl. Acad. Sci. U.S.A.* **93**, 13589–13594
11. Guilloton, M. B., Korte, J. J., Lamblin, A. F., Fuchs, J. A., and Anderson, P. M. (1992) *J. Biol. Chem.* **267**, 3731–3734
12. Kisker, C., Schindelin, H., Alber, B. E., Ferry, J. G., and Rees, D. C. (1996) *EMBO J.* **15**, 2323–2330
13. Rowlett, R. S., Chance, M. R., Wirt, M. D., Sidelinger, D. E., Royal, J. R., Woodroffe, M., Wang, Y.-F. A., Saha, R. P., and Lam, M. G. (1994) *Biochemistry* **33**, 13967–13976
13a. Kimber, M. S., and Pai, E. F. (2000) *EMBO J.* **19**, 1407–418
14. Miyamoto, H., Miyashita, T., Okushima, M., Nakano, S., Morita, T., and Matsushiro, A. (1996) *Proc. Natl. Acad. Sci. U.S.A.* **93**, 9657–9660
15. Håkansson, K., Carlsson, M., Svensson, L. A., and Liljas, A. (1992) *J. Mol. Biol.* **227**, 1192–1204
16. Kumar, V., and Kannan, K. K. (1994) *J. Mol. Biol.* **241**, 226–232
17. Kiefer, L. L., Paterno, S. A., and Fierke, C. A. (1995) *J. Am. Chem. Soc.* **117**, 6831–6837
17a. Lesburg, C. A., Huang, C.-c, Christianson, D. W., and Fierke, C. A. (1997) *Biochemistry* **36**, 15780–15791
18. Liang, J.-Y., and Lipscomb, W. N. (1987) *Biochemistry* **26**, 5293–5301
19. Pocker, Y., and Janjic, N. (1989) *J. Am. Chem. Soc.* **111**, 731–733
19a. Earnhardt, J. N., Qian, M., Tu, C., Lakkis, M. M., Bergenhem, N. C. H., Laipis, P. J., Tashian, R. E., and Silverman, D. N. (1998) *Biochemistry* **37**, 10837–10845
19b. Denisov, V. P., Jonsson, B.-H., and Halle, B. (1999) *J. Am. Chem. Soc.* **121**, 2327–2328
20. Paneth, P., and O'Leary, M. H. (1985) *J. Am. Chem. Soc.* **107**, 7381–7384
21. Scolnick, L. R., and Christianson, D. W. (1996) *Biochemistry* **35**, 16429–16434
22. Toba, S., Colombo, G., and Merz, K. M., Jr. (1999) *J. Am. Chem. Soc.* **121**, 2290–2302
23. Merz, K. M., Jr., and Banci, L. (1997) *J. Am. Chem. Soc.* **119**, 863–871
24. Briganti, F., Mangani, S., Orioli, P., Scozzafava, A., Vernaglione, G., and Supuran, C. T. (1997) *Biochemistry* **36**, 10384–10392
25. Ren, X., Tu, C., Laipis, P. J., and Silverman, D. N. (1995) *Biochemistry* **34**, 8492–8498
26. Malatesta, V., and Cocivera, M. (1978) *Journal of Organic Chemistry* **43**, 1737–1742
27. Jencks, W. P. (1987) *Catalysis in Chemistry and Enzymology*, Dover, Mineola, New York
28. Engel, C. K., Mathieu, M., Zeelen, J. P., Hiltunen, J. K., and Wierenga, R. K. (1996) *EMBO J.* **15**, 5135–5145

28a. Engel, C. K., Kiema, T. R., Hiltunen, J. K., and Wierenga, R. K. (1998) *J. Mol. Biol.* **275**, 847–859
29. D'Ordine, R. L., Tonge, P. J., Carey, P. R., and Anderson, V. E. (1994) *Biochemistry* **33**, 12635–12643
30. Bahnson, B. J., and Anderson, V. E. (1991) *Biochemistry* **30**, 5894–5906
31. D'Ordine, R. L., Bahnson, B. J., Tonge, P. J., and Anderson, V. E. (1994) *Biochemistry* **33**, 14733–14742
31a. Hofstein, H. A., Feng, Y., Anderson, V. E., and Tonge, P. J. (1999) *Biochemistry* **38**, 9508–9516
32. Yang, S.-Y., He, X.-Y., and Schulz, H. (1995) *Biochemistry* **34**, 6441–6447
33. Annand, R. R., Kozlowski, J. F., Davisson, V. J., and Schwab, J. (1993) *J. Am. Chem. Soc.* **115**, 1088–1094
34. Brock, D. J. H., Kass, L. R., and Bloch, K. (1967) *J. Biol. Chem.* **242**, 4432–4440
35. Hebert, H., Schmidt-Krey, I., Morgenstern, R., Murata, K., Hirai, T., Mitsuoka, K., and Fujiyoshi, Y. (1997) *J. Mol. Biol.* **271**, 751–758
36. Hu, L., Borleske, B. L., and Colman, R. F. (1997) *Protein Sci.* **6**, 43–52
37. Rushmore, T. H., and Pickett, C. B. (1993) *J. Biol. Chem.* **268**, 11475–11478
38. Ji, X., Johnson, W. W., Sesay, M. A., Dickert, L., Prasad, S. M., Ammon, H. L., Armstrong, R. N., and Gilliland, G. L. (1994) *Biochemistry* **33**, 1043–1052
39. Oakley, A. J., Rossjohn, J., Lo Bello, M., Caccuri, A. M., Federici, G., and Parker, M. W. (1997) *Biochemistry* **36**, 576–585
40. Reinemer, P., Prade, L., Hof, P., Neuefeind, T., Huber, R., Zettl, R., Palme, K., Schell, J., Koelln, I., Bartunik, H. D., and Bieseler, B. (1996) *J. Mol. Biol.* **255**, 289–309
41. Xiao, G., Liu, S., Ji, X., Johnson, W. W., Chen, J., Parsons, J. F., Stevens, W. J., Gilliland, G. L., and Armstrong, R. N. (1996) *Biochemistry* **35**, 4753–4765
41a. Neuefeind, T., Huber, R., Reinemer, P., Knäblein, J., Prade, L., Mann, K., and Bieseler, B. (1997) *J. Mol. Biol.* **274**, 577–587
42. McTigue, M. A., Williams, D. R., and Tainer, J. A. (1995) *J. Mol. Biol.* **246**, 21–27
43. Yang, G., Liang, P.-H., and Dunaway-Mariano, D. (1994) *Biochemistry* **33**, 8327–8331
44. Crooks, G. P., Xu, L., Barkley, R. M., and Copley, S. D. (1995) *J. Am. Chem. Soc.* **117**, 10791–10798
45. Yang, G., Liu, R.-Q., Taylor, K. L., Xiang, H., Price, J., and Dunaway-Mariano, D. (1996) *Biochemistry* **36**, 10879–10885
46. Benning, M. M., Taylor, K. L., Liu, R.-Q., Yang, G., Xiang, H., Wesenberg, G., Dunaway-Mariano, D., and Holden, H. M. (1996) *Biochemistry* **35**, 8103–8109
47. Taylor, K. L., Liu, R.-Q., Liang, P.-H., Price, J., Dunaway-Mariano, D., Tonge, P. J., Clarkson, J., and Carey, P. R. (1995) *Biochemistry* **34**, 13881–13888
48. Clarkson, J., Tonge, P. J., Taylor, K. L., Dunaway-Mariano, D., and Carey, P. R. (1997) *Biochemistry* **36**, 10192–10199
49. Taylor, K. L., Xiang, H., Liu, R.-Q., Yang, G., and Dunaway-Mariano, D. (1997) *Biochemistry* **36**, 1349–1361
49a. Benning, M. M., Haller, T., Gerlt, J. A., and Holden, H. M. (2000) *Biochemistry* **39**, 4630–4639
49b. Xiang, H., Luo, L., Taylor, K. L., and Dunaway-Mariano, D. (1999) *Biochemistry* **38**, 7638–7652
49c. Babbitt, P. C., and Gerlt, J. A. (1997) *J. Biol. Chem.* **272**, 30591–30594

50. Sacchettini, J. C., Meininger, T., Rodrick, S., and Banaszak, L. J. (1986) *J. Biol. Chem.* **261**, 15183–15185
51. Bearne, S. L., and Wolfenden, R. (1995) *J. Am. Chem. Soc.* **117**, 9588–9589
52. Weaver, T., and Banaszak, L. (1996) *Biochemistry* **35**, 13955–13965
53. Weaver, T., Lees, M., and Banaszak, L. (1997) *Protein Sci.* **6**, 834–842
53a. Weaver, T., Lees, M., Zaitsev, V., Zaitseva, I., Duke, E., Lindley, P., McSweeny, S., Svensson, A., Keruchenko, J., Keruchenko, I., Gladilin, K., and Banaszak, L. (1998) *J. Mol. Biol.* **280**, 431–442
53b. Beeckmans, S., and Van Driessche, E. (1998) *J. Biol. Chem.* **273**, 31661–31669
54. Mohrig, J. R., Moerke, K. A., Cloutier, D. L., Lane, B. D., Person, E. C., and Onasch, T. B. (1995) *Science* **269**, 527–529
55. Rose, I. A. (1972) *CRC Critical Review of Biochemistry* **1**, 33–57
56. Brant, D. A., Barnett, L. B., and Alberty, R. A. (1963) *J. Am. Chem. Soc.* **85**, 2204–2209
57. Schmidt, D. E., Jr., Nigh, W. G., Tanzer, C., and Richards, J. H. (1969) *J. Am. Chem. Soc.* **91**, 5849–5854
58. Nigh, W. G., and Richards, J. H. (1969) *J. Am. Chem. Soc.* **91**, 5847–5848
59. Hansen, J. N., Dinovo, E. C., and Boyer, P. D. (1969) *J. Biol. Chem.* **244**, 6270–6279
60. Blanchard, J. S., and Cleland, W. W. (1980) *Biochemistry* **19**, 4506–4513
61. Porter, D. J. T., and Bright, H. J. (1980) *J. Biol. Chem.* **255**, 4772–4780
62. Jones, V. T., Lowe, G., and Potter, B. V. L. (1980) *Eur. J. Biochem.* **108**, 433–437
63. Botting, N. P., and Gani, D. (1992) *Biochemistry* **31**, 1509–1520
64. Shi, W., Dunbar, J., Jayasekera, M. M. K., Viola, R. E., and Farber, G. K. (1997) *Biochemistry* **36**, 9136–9144
65. Jayasekera, M. M. K., Shi, W., Farber, G. K., and Viola, R. E. (1997) *Biochemistry* **36**, 9145–9150
66. Goda, S. K., Minton, N. P., Botting, N. P., and Gani, D. (1992) *Biochemistry* **31**, 10747–10756
67. Weiss, P. M., Boerner, R. J., and Cleland, W. W. (1987) *J. Am. Chem. Soc.* **109**, 7201–7202
68. Anderson, S. R., Anderson, V. E., and Knowles, J. R. (1994) *Biochemistry* **33**, 10545–10555
68a. Vinarov, D. A., and Nowak, T. (1998) *Biochemistry* **37**, 15238–15246
69. Vinarov, D. A., and Nowak, T. (1999) *Biochemistry* **38**, 12138–12149
70. Zhang, E., Hatada, M., Brewer, J. M., and Lebioda, L. (1994) *Biochemistry* **33**, 6295–6300
71. Wedekind, J. E., Reed, G. H., and Rayment, I. (1995) *Biochemistry* **34**, 4325–4330
72. Duquerroy, S., Camus, C., and Janin, J. (1995) *Biochemistry* **34**, 12513–12523
73. Larsen, T. M., Wedekind, J. E., Rayment, I., and Reed, G. H. (1996) *Biochemistry* **35**, 4349–4358
73a. Hubbard, B. K., Koch, M., Palmer, D. R. J., Babbitt, P. C., and Gerlt, J. A. (1998) *Biochemistry* **37**, 14369–14375
73b. Thompson, T. B., Garrett, J. B., Taylor, E. A., Meganathan, R., Gerlt, J. A., and Rayment, I. (2000) *Biochemistry* **39**, 10662–10676
74. Linhardt, R. J., Turnbull, J. E., Wang, H. M., Loganathan, D., and Gallagher, J. T. (1990) *Biochemistry* **29**, 2611–2617
75. Desai, U. R., Wang, H.-M., and Linhardt, R. J. (1993) *Biochemistry* **32**, 8140–8145
76. Kiss, J. (1974) *Adv. Carbohydrate Chem. Biochem.* **29**, 229–303
76a. Li, S., Kelly, S. J., Lamani, E., Ferraroni, M., and Jedrzejas, M. J. (2000) *EMBO J.* **19**, 1228–1240

References

76b. Ponnuraj, K., and Jedrzejas, M. J. (2000) *J. Mol. Biol.* **299**, 885–895

77. Yoder, M. D., Keen, N. T., and Jurnak, F. (1993) *Science* **260**, 1503–1507

78. Cohen, F. E. (1993) *Science* **260**, 1444–1445

79. Kita, N., Boyd, C. M., Garrett, M. R., Jurnak, F., and Keen, N. T. (1996) *J. Biol. Chem.* **271**, 26529–26535

80. Kuo, D. J., and Rose, I. A. (1987) *Biochemistry* **26**, 7589–7596

80a. Lloyd, S. J., Lauble, H., Prasad, G. S., and Stout, C. D. (1999) *Protein Sci.* **8**, 2655–2662

81. Werst, M. M., Kennedy, M. C., Houseman, A. L. P., Beinert, H., and Hoffman, B. M. (1990) *Biochemistry* **29**, 10533–10540

82. Kilpatrick, L. K., Kennedy, M. C., Beinert, H., Czernuszewicz, R. S., Qiu, D., and Spiro, T. G. (1994) *J. Am. Chem. Soc.* **116**, 4053–4061

83. Lauble, H., Kennedy, M. C., Beinert, H., and Stout, C. D. (1992) *Biochemistry* **31**, 2735–2748

84. Kennedy, M. C., Emptage, M. H., and Beinert, H. (1984) *J. Biol. Chem.* **259**, 3145–3151

85. Kennedy, M. C., and Beinert, H. (1988) *J. Biol. Chem.* **263**, 8194–8198

86. Beinert, H., and Kennedy, M. C. (1993) *FASEB J.* **7**, 1442–1448

87. Kent, T. A., Emptage, M. H., Merkle, H., Kennedy, M. C., Beinert, H., and Münck, E. (1985) *J. Biol. Chem.* **260**, 6871–6881

88. Lauble, H., Kennedy, M. C., Beinert, H., and Stout, C. D. (1994) *J. Mol. Biol.* **237**, 437–451

89. Zheng, L., Kennedy, M. C., Beinert, H., and Zalkin, H. (1992) *J. Biol. Chem.* **267**, 7895–7903

90. Gruer, M. J., Artymiuk, P. J., and Guest, J. R. (1997) *Trends Biochem. Sci.* **22**, 3–6

91. Flint, D. H. (1993) *Biochemistry* **32**, 799–805

92. Flint, D. H., and McKay, R. G. (1994) *J. Am. Chem. Soc.* **116**, 5534–5539

93. Hofmeister, A. E. M., Grabowski, R., Linder, D., and Buckel, W. (1993) *Eur. J. Biochem.* **215**, 341–349

94. Grabowski, R., Hofmeister, A. E. M., and Buckel, W. (1993) *Trends Biochem. Sci.* **18**, 297–300

95. Hofmeister, A. E. M., Berger, S., and Buckel, W. (1992) *Eur. J. Biochem.* **205**, 743–749

96. Schroepfer, G. J., Jr. (1966) *J. Biol. Chem.* **241**, 5441–5447

97. Hill, R. K., and Newkome, G. R. (1969) *J. Am. Chem. Soc.* **91**, 5893–5894

98. Chiang, Y., and Kresge, A. J. (1991) *Science* **253**, 395–400

99. Vellom, D. C., Radic', Z., Li, Y., Pickering, N. A., Camp, S., and Taylor, P. (1993) *Biochemistry* **32**, 12–17

100. Gerlt, J. A., and Gassman, P. G. (1993) *Biochemistry* **32**, 11943–11952

101. Petsko, G. A., Kenyon, G. L., Gerlt, J. A., Ringe, D., and Kozarich, J. W. (1993) *Trends Biochem. Sci.* **18**, 372–376

102. Neidhart, D. J., Howell, P. L., Petsko, G. A., Powers, V. M., Li, R., Kenyon, G. L., and Gerlt, J. A. (1991) *Biochemistry* **30**, 9264–9273

103. St. Maurice, M., and Bearne, S. L. (2000) *Biochemistry* **39**, 13324–13335

104. Schafer, S. L., Barrett, W. C., Kallarakal, A. T., Mitra, B., Kozarich, J. W., Gerlt, J. A., Clifton, J. G., Petsko, G. A., and Kenyon, G. L. (1996) *Biochemistry* **35**, 5662–5669

105. Babbitt, P. C., Hasson, M. S., Wedekind, J. E., Palmer, D. R. J., Barrett, W. C., Reed, G. H., Rayment, I., Ringe, D., Kenyon, G. L., and Gerlt, J. A. (1996) *Biochemistry* **35**, 16489–16501

106. Babbitt, P. C., Mrachko, G. T., Hasson, M. S., Huisman, G. W., Kolter, R., Ringe, D., Petsko, G. A., Kenyon, G. L., and Gerlt, J. A. (1995) *Science* **267**, 1159–1161

107. Chiang, Y., Kresge, A. J., Pruszynski, P., Schepp, N. P., and Wirz, J. (1990) *Angew.*

Chem. Int. Ed. Engl. **29**, 792–794

108. Guthrie, J. P., and Kluger, R. (1993) *J. Am. Chem. Soc.* **115**, 11569–11572

109. Gerlt, J. A., Kozarich, J. W., Kenyon, G. L., and Gassman, P. G. (1991) *J. Am. Chem. Soc.* **113**, 9667–9669

110. Gerlt, J. A., and Gassman, P. G. (1993) *J. Am. Chem. Soc.* **115**, 11552–11568

111. Guthrie, J. P. (1997) *J. Am. Chem. Soc.* **119**, 1151–1152

112. Tobin, J. B., and Frey, P. A. (1996) *J. Am. Chem. Soc.* **118**, 12253–12260

113. Yamauchi, T., Choi, S.-Y., Okada, H., Yohda, M., Kumagai, H., Esaki, N., and Soda, K. (1992) *J. Biol. Chem.* **267**, 18361–18364

114. Glavas, S., and Tanner, M. E. (1999) *Biochemistry* **38**, 4106–4113

115. Ho, H.-T., Falk, P. J., Ervin, K. M., Krishnan, B. S., Discotto, L. F., Dougherty, T. J., and Pucci, M. J. (1995) *Biochemistry* **34**, 2464–2470

115a. Hwang, K. Y., Cho, C.-S., Kim, S. S., Sung, H.-C., Yu, Y. G., and Cho, Y. (1999) *Nature Struct. Biol.* **6**, 422–426

116. Stein, T., Kluge, B., Vater, J., Franke, P., Otto, A., and Wittmann-Liebold, B. (1995) *Biochemistry* **34**, 4633–4642

117. Wiseman, J. S., and Nichols, J. S. (1984) *J. Biol. Chem.* **259**, 8907–8914

118. Shikata, Y., Watanabe, T., Teramoto, T., Inoue, A., Kawakami, Y., Nishizawa, Y., Katayama, K., and Kuwada, M. (1995) *J. Biol. Chem.* **270**, 16719–16723

119. Heck, S. D., Faraci, W. S., Kelbaugh, P. R., Saccomano, N. A., Thadeio, P. F., and Volkmann, R. A. (1996) *Proc. Natl. Acad. Sci. U.S.A.* **93**, 4036–4039

120. Johnson, J. D., Creighton, D. J., and Lambert, M. R. (1986) *J. Biol. Chem.* **261**, 4535–4541

121. Cooper, A. J. L., Gines, J. Z., and Meister, A. (1983) *Chem. Rev.* **83**, 321–358

122. Pirrung, M. C., Chen, J., Rowley, E. G., and McPhail, A. T. (1993) *J. Am. Chem. Soc.* **115**, 7103–7110

123. Noltmann, E. A. (1972) in *The Enzymes*, 3rd ed., Vol. 6 (Boyer, P. D., ed), pp. 271–354, Academic Press, New York

124. McGee, D. M., Hathaway, G. H., Palmieri, R. H., and Noltmann, E. A. (1980) *J. Mol. Biol.* **142**, 29–42

125. Mushegian, A. R., and Koonin, E. V. (1996) *Proc. Natl. Acad. Sci. U.S.A.* **93**, 10268–10273

125a. Jeffery, C. J., Bahnson, B. J., Chien, W., Ringe, D., and Petsko, G. A. (2000) *Biochemistry* **39**, 955–964

126. Tanaka, K. R., and Paglia, D. E. (1995) in *The Metabolic and Molecular Bases of Inherited Disease* (Scriver, C. R., Beaudet, A. L., Sly, W. S., and Valle, D., eds), pp. 3485–3511, McGraw-Hill, New York

127. Woodruff, W. W., III, and Wolfenden, R. (1979) *J. Biol. Chem.* **254**, 5866–5867

127a. Seemann, J. E., and Schulz, G. E. (1997) *J. Mol. Biol.* **273**, 256–268

128. Wallace, L. J., Eiserling, F. A., and Wilcox, G. (1978) *J. Biol. Chem.* **253**, 3717–3720

129. Bogumil, R., Kappl, R., Hüttermann, J., and Witzel, H. (1997) *Biochemistry* **36**, 2345–2352

130. Speck, J. C., Jr. (1958) *Adv. Carbohydrate Chem.* **13**, 63–103

131. Gibson, D. R., Gracy, R. W., and Hartman, F. C. (1980) *J. Biol. Chem.* **255**, 9369–9374

132. Balaban, R. S., and Ferretti, J. A. (1983) *Proc. Natl. Acad. Sci. U.S.A.* **80**, 1241–1245

133. Blacklow, S. C., Raines, R. T., Lim, W. A., Zamore, P. D., and Knowles, J. R. (1988) *Biochemistry* **27**, 1158–1167

134. Muirhead, H. (1983) *Trends Biochem. Sci.* **8**, 326–330

135. Lolis, E., Alber, T., Davenport, R. C., Rose, D., Hartman, F. C., and Petsko, G. A. (1990) *Biochemistry* **29**, 6609–6618

136. Joseph-McCarthy, D., Lolis, E., Komives, E. A., and Petsko, G. A. (1994) *Biochemistry* **33**, 2815–2823

137. Mande, S. C., Mainfroid, V., Kalk, K. H., Goraj, K., Martial, J. A., and Hol, W. G. J. (1994) *Protein Sci.* **3**, 810–821

138. Davenport, R. C., Bash, P. A., Seaton, B. A., Karplus, M., Petsko, G. A., and Ringe, D. (1991) *Biochemistry* **30**, 5821–5826

139. Verlinde, C. L. M. J., Witmans, C. J., Pijning, T., Kalk, K. H., Hol, W. G. J., Callens, M., and Opperdoes, F. R. (1992) *Protein Sci.* **1**, 1578–1584

140. Rose, I. A., Fung, W.-J., and Warms, J. V. B. (1990) *Biochemistry* **29**, 4312–4317

140a. Harris, T. K., Cole, R. N., Comer, F. I., and Mildvan, A. S. (1998) *Biochemistry* **37**, 16828–16838

141. Joseph-McCarthy, D., Rost, L. E., Komives, E. A., and Petsko, G. A. (1994) *Biochemistry* **33**, 2824–2829

142. Belasco, J. G., and Knowles, J. R. (1980) *Biochemistry* **19**, 472–477

143. Alston, W. C., II, Kanska, M., and Murray, C. J. (1996) *Biochemistry* **35**, 12873–12881

144. Nickbarg, E. B., Davenport, R. C., Petsko, G. A., and Knowles, J. R. (1988) *Biochemistry* **27**, 5948–5960

145. Lodi, P. J., Chang, L. C., Knowles, J. R., and Komives, E. A. (1994) *Biochemistry* **33**, 2809–2814

146. Raines, R. T., and Knowles, J. R. (1987) *Biochemistry* **26**, 7014–7020

147. Bash, P. A., Field, M. J., Davenport, R. C., Petsko, G. A., Ringe, D., and Karplus, M. (1991) *Biochemistry* **30**, 5826–5832

148. Lodi, P. J., and Knowles, J. R. (1991) *Biochemistry* **30**, 6948–6956

149. Alagona, G., Ghio, C., and Kollman, P. A. (1995) *J. Am. Chem. Soc.* **117**, 9855–9862

150. Pompliano, D. L., Peyman, A., and Knowles, J. R. (1990) *Biochemistry* **29**, 3186–3194

151. Yüksel, K. Ü., Sun, A.-Q., Gracy, R. W., and Schnackerz, K. D. (1994) *J. Biol. Chem.* **269**, 5005–5008

152. Williams, J. C., and McDermott, A. E. (1995) *Biochemistry* **34**, 8309–8319

153. Carrell, H. L., Rubin, B. H., Hurley, T. J., and Glusker, J. P. (1984) *J. Biol. Chem.* **259**, 3230–3236

154. Farber, G. K., Glasfeld, A., Tiraby, G., Ringe, D., and Petsko, G. A. (1989) *Biochemistry* **28**, 7289–7297

155. Collyer, C. A., Goldberg, J. D., Viehmann, H., Blow, D. M., Ramsden, N. G., Fleet, G. W. J., Montgomery, F. J., and Grice, P. (1992) *Biochemistry* **31**, 12211–12218

156. Jenkins, J., Janin, J., Rey, F., Chiadmi, M., van Tilbeurgh, H., Lasters, I., De Maeyer, M., Van Belle, D., Wodak, S. J., Lauwereys, M., Stanssens, P., Mrabet, N. T., Snauwaert, J., Matthyssens, G., and Lambeir, A.-M. (1992) *Biochemistry* **31**, 5449–5458

157. Whitaker, R. D., Cho, Y., Cha, J., Carrell, H. L., Glusker, J. P., Karplus, P. A., and Batt, C. A. (1995) *J. Biol. Chem.* **270**, 22895–22906

158. Allen, K. N., Lavie, A., Petsko, G. A., and Ringe, D. (1995) *Biochemistry* **34**, 3742–3749

159. Hall, S. S., Doweyko, A. M., and Jordan, F. (1978) *J. Am. Chem. Soc.* **100**, 5934

160. Xue, L., Talalay, P., and Mildvan, A. S. (1991) *Biochemistry* **30**, 10858–10865

161. Brooks, B., and Benisek, W. F. (1994) *Biochemistry* **33**, 2682–2687

162. Austin, J. C., Zhao, Q., Jordan, T., Talalay, P., Mildvan, A. S., and Spiro, T. G. (1995) *Biochemistry* **34**, 4441–4447

163. Zhao, Q., Li, Y.-K., Mildvan, A. S., and Talalay, P. (1995) *Biochemistry* **34**, 6562–6572

164. Hawkinson, D. C., Pollack, R. M., and Ambulos, N. P., Jr. (1994) *Biochemistry* **33**, 12172–12183

165. Viger, A., Coustal, S., and Marquet, A. (1983) *J. Am. Chem. Soc.* **103**, 451–458

166. Holman, C. M., and Benisek, W. F. (1994) *Biochemistry* **33**, 2672–2681

167. Wu, Z. R., Ebrahimian, S., Zawrotny, M. E., Thornburg, L. D., Perez-Alvarado, G. C., Brothers, P., Pollack, R. M., and Summers, M. F. (1997) *Science* **276**, 415–418

167a. Kim, S. W., Cha, S.-S., Cho, H.-S., Kim, J.-S., Ha, N.-C., Cho, M.-J., Joo, S., Kim, K. K., Choi, K. Y., and Oh, B.-H. (1997) *Biochemistry* **36**, 14030–14036

167b. Choi, G., Ha, N.-C., Kim, S. W., Kim, D.-H., Park, S., Oh, B.-H., and Choi, K. Y. (2000) *Biochemistry* **39**, 903–909

167c. Thornburg, L. D., Hénot, F., Bash, D. P., Hawkinson, D. C., Bartel, S. D., and Pollack, R. M. (1998) *Biochemistry* **37**, 10499–10506

168. Chen, L. H., Kenyon, G. L., Curtin, F., Harayama, S., Bembenek, M. E., Hajipour, G., and Whitman, C. P. (1992) *J. Biol. Chem.* **267**, 17716–17721

169. Subramanya, H. S., Roper, D. I., Dauter, Z., Dodson, E. J., Davies, G. J., Wilson, K. S., and Wigley, D. B. (1996) *Biochemistry* **35**, 792–802

170. Taylor, A. B., Czerwinski, R. M., Johnson, W. H., Jr., Whitman, C. P., and Hackert, M. L. (1998) *Biochemistry* **37**, 14692–14700

171. Stivers, J. T., Abeygunawardana, C., Mildvan, A. S., Hajipour, G., and Whitman, C. P. (1996) *Biochemistry* **35**, 814–823

172. Czerwinski, R. M., Harris, T. K., Johnson, W. H., Jr., Legler, P. M., Stivers, J. T., Mildvan, A. S., and Whitman, C. P. (1999) *Biochemistry* **38**, 12358–12366

173. Fitzgerald, M. C., Chernushevich, I., Standing, K. G., Kent, S. B. H., and Whitman, C. P. (1995) *J. Am. Chem. Soc.* **117**, 11075–11080

174. Lo, T. W. C., Westwood, M. E., McLellan, A. C., Selwood, T., and Thornalley, P. J. (1994) *J. Biol. Chem.* **269**, 32299–32305

175. Ridderström, M., Saccucci, F., Hellman, U., Bergman, T., Principato, G., and Mannervik, B. (1996) *J. Biol. Chem.* **271**, 319–323

176. Richard, J. P. (1991) *Biochemistry* **30**, 4581–4585

176a. Saadat, D., and Harrison, D. H. T. (2000) *Biochemistry* **39**, 2950–2960

177. Lan, Y., Lu, T., Lovett, P. S., and Creighton, D. J. (1995) *J. Biol. Chem.* **270**, 12957–12960

178. Cameron, A. D., Ridderström, M., Olin, B., Kavarana, M. J., Creighton, D. J., and Mannervik, B. (1999) *Biochemistry* **38**, 13480–13490

178a. He, M. M., Clugston, S. L., Honek, J. F., and Matthews, B. W. (2000) *Biochemistry* **39**, 8719–8727

179. Rae, C., O'Donoghue, S. I., Bubb, W. A., and Kuchel, P. W. (1994) *Biochemistry* **33**, 3548–3559

180. Bito, A., Haider, M., Hadler, I., and Breitenbach, M. (1997) *J. Biol. Chem.* **272**, 21509–21519

181. Mathieu, M., Modis, Y., Zeelen, J. P., Engel, C. K., Abagyan, R. A., Ahlberg, A., Rasmussen, B., Lamzin, V. S., Kunau, W. H., and Wierenga, R. K. (1997) *J. Mol. Biol.* **273**, 714–728

181a. Modis, Y., and Wierenga, R. K. (2000) *J. Mol. Biol.* **297**, 1171–1182

182. Roberts, J. R., Narasimhan, C., and Miziorko, H. M. (1995) *J. Biol. Chem.* **270**, 17311–17316

183. Narasimhan, C., Roberts, J. R., and Miziorko, H. M. (1995) *Biochemistry* **34**, 9930–9935

183a. Vinarov, D. A., and Miziorko, H. M. (2000) *Biochemistry* **39**, 3360–3368

183b. Richard, J. P., and Nagorski, R. W. (1999) *J. Am. Chem. Soc.* **121**, 4763–4770

184. Marsh, J. J., and Lebherz, H. G. (1992) *Trends Biochem. Sci.* **17**, 110–113

185. Cox, T. M. (1994) *FASEB J.* **8**, 62–71

186. Morris, A. J., and Tolan, D. R. (1994) *Biochemistry* **33**, 12291–12297

187. Sygusch, J., Beaudry, D., and Allaire, M. (1987) *Proc. Natl. Acad. Sci. U.S.A.* **84**, 7846–7850

188. Lai, C. Y., and Oshima, T. (1971) *Arch. Biochem. Biophys.* **144**, 363–

188a. Choi, K. H., Mazurkie, A. S., Morris, A. J., Utheza, D., Tolan, D. R., and Allen, K. N. (1999) *Biochemistry* **38**, 12655–12664

188b. Dalby, A., Dauter, Z., and Littlechild, J. A. (1999) *Protein Sci.* **8**, 291–297

189. Periana, R. A., Motiu-DeGrood, R., Chiang, Y., and Hupe, D. J. (1980) *J. Am. Chem. Soc.* **102**, 3923–3927

190. Beernink, P. T., and Tolan, D. R. (1996) *Proc. Natl. Acad. Sci. U.S.A.* **93**, 5374–5379

191. Kadonaga, J. T., and Knowles, J. R. (1983) *Biochemistry* **22**, 130–136

191a. Hall, D. R., Leonard, G. A., Reed, C. D., Watt, C. I., Berry, A., and Hunter, W. N. (1999) *J. Mol. Biol.* **287**, 383–394

192. Qamar, S., Marsh, K., and Berry, A. (1996) *Protein Sci.* **5**, 154–161

193. Dreyer, M. K., and Schulz, G. E. (1996) *J. Mol. Biol.* **259**, 458–466

194. Lebioda, L., Hatada, M. H., Tulinsky, A., and Mavridis, I. M. (1982) *J. Mol. Biol.* **162**, 445–458

195. Meloche, H. P. (1981) *Trends Biochem. Sci.* **6**, 38–41

196. Dekker, E. E., and Kitson, R. P. (1992) *J. Biol. Chem.* **267**, 10507–10514

197. Wong, C.-H., Garcia-Junceda, E., Chen, L., Blanco, O., Gijsen, H. J. M., and Steensma, D. H. (1995) *J. Am. Chem. Soc.* **117**, 3333–3339

198. Jia, J., Schörken, U., Lindqvist, Y., Sprenger, G. A., and Schneider, G. (1997) *Protein Sci.* **6**, 119–124

199. Higgins, M. J. P., Kornblatt, J. A., and Rudney, H. (1972) in *The Enzymes*, 3rd ed., Vol. 7 (Boyer, P. D., ed), pp. 407–434, Academic Press, New York

200. Wiegand, G., and Remington, S. J. (1986) *Ann. Rev. Biophys. Biophys. Chem.* **15**, 97–117

201. Anderson, D. H., and Duckworth, H. W. (1988) *J. Biol. Chem.* **263**, 2163–2169

202. Wiegand, G., Remington, S., Deisenhofer, J., and Huber, R. (1984) *J. Mol. Biol.* **174**, 205–219

203. Rault-Leonardon, M., Atkinson, M. A. L., Slaughter, C. A., Moomaw, C. R., and Srere, P. A. (1995) *Biochemistry* **34**, 257–263

204. Russell, R. J. M., Ferguson, J. M. C., Hough, D. W., Danson, M. J., and Taylor, G. L. (1997) *Biochemistry* **36**, 9983–9994

205. Pereira, F. D., Donald, L. J., Hosfield, D. J., and Duckworth, H. W. (1994) *J. Biol. Chem.* **269**, 412–417

206. Karpusas, M., Holland, D., and Remington, S. J. (1991) *Biochemistry* **30**, 6024–6031

207. Chothia, C., and Lesk, A. M. (1985) *Trends Biochem. Sci.* **10**, 116–118

208. Evans, C. T., Kurz, L. C., Remington, S. J., and Srere, P. A. (1996) *Biochemistry* **35**, 10661–10672

209. Remington, S., Wiegand, G., and Huber, R. (1982) *J. Mol. Biol.* **158**, 111–152

210. Kurz, L. C., Roble, J. H., Nakra, T., Drysdale, G. R., Buzan, J. M., Schwartz, B., and Drueckhammer, D. G. (1997) *Biochemistry* **36**, 3981–3990

211. Kurz, L. C., and Drysdale, G. R. (1987) *Biochemistry* **26**, 2623–2627

212. Kurz, L. C., Ackerman, J. J. H., and Drysdale, G. R. (1985) *Biochemistry* **24**, 452–457

213. Clark, J. D., O'Keefe, S. J., and Knowles, J. R. (1988) *Biochemistry* **27**, 5961–5971

214. Wlassics, I. D., and Anderson, V. E. (1989) *Biochemistry* **28**, 1627–1633

215. Usher, K. C., Remington, S. J., Martin, D. P., and Drueckhammer, D. G. (1994) *Biochemistry* **33**, 7753–7759

215a. Kurz, L. C., Nakra, T., Stein, R., Plungkhen, W., Riley, M., Hsu, F., and Drysdale, G. R. (1998) *Biochemistry* **37**, 9724–9737

216. Schwartz, B., Drueckhammer, D. G., Usher, K. C., and Remington, S. J. (1995) *Biochemistry* **34**, 15459–15466

216a. Gu, Z., Drueckhammer, D. G., Kurz, L., Liu, K., Martin, D. P., and McDermott, A. (1999) *Biochemistry* **38**, 8022–8031

216b. Mulholland, A. J., Lyne, P. D., and Karplus, M. (2000) *J. Am. Chem. Soc.* **122**, 534–535

216c. Howard, B. R., Endrizzi, J. A., and Remington, S. J. (2000) *Biochemistry* **39**, 3156–3168

217. Smith, S. (1994) *FASEB J.* **8**, 1248–1259

218. Misra, I., Narasimhan, C., and Miziorko, H. M. (1993) *J. Biol. Chem.* **268**, 12129–12135

219. Misra, I., and Miziorko, H. M. (1996) *Biochemistry* **35**, 9610–9616

220. Guynn, R. W., and Veech, R. L. (1979) *J. Biol. Chem.* **254**, 1691–1698

221. Linn, T. C., and Srere, P. A. (1979) *J. Biol. Chem.* **254**, 1691–1698

222. Ranganathan, N. S., Linn, T. C., and Srere, P. A. (1982) *J. Biol. Chem.* **257**, 698–702

223. Elshourbagy, N. A., Near, J. C., Kmetz, P. J., Sathe, G. M., Southan, C., Strickler, J. E., Gross, M., Young, J. F., Wells, T. N. C., and Groot, P. H. E. (1990) *J. Biol. Chem.* **265**, 1430–1435

224. Pentyala, S. N., and Benjamin, W. B. (1995) *Biochemistry* **34**, 10961–10969

225. Hersch, L. B. (1973) *J. Biol. Chem.* **248**, 7295–7303

226. Nilekani, S., and SivaRaman, C. (1983) *Biochemistry* **22**, 4657–4663

227. Oppenheimer, N. J., Singh, M., Sweeley, C. C., Sung, S.-J., and Srere, P. A. (1979) *J. Biol. Chem.* **254**, 1000–1002

228. Ko, Y. H., Vanni, P., Munske, G. R., and McFadden, B. A. (1991) *Biochemistry* **30**, 7451–7456

229. Lenz, H., Buckel, W., Wunderwald, P., Biedermann, G., Buschmeier, V., Eggerer, H., Cornforth, J. W., Redmond, J. W., and Mallaby, R. (1971) *Eur. J. Biochem.* **24**, 207–215

230. Retey, J., Luthy, J., and Arigoni, D. (1970) *Nature (London)* **226**, 519–521

231. Kyte, J. (1995) *Mechanism in Protein Chemistry*, Garland Publ., New York (pp. 293–313)

232. Leussing, D. L., and Raghavan, N. V. (1980) *J. Am. Chem. Soc.* **102**, 5635–5643

233. Steinberger, R., and Westheimer, F. H. (1951) *J. Am. Chem. Soc.* **73**, 429–435

234. Kubala, G., and Martell, A. E. (1982) *J. Am. Chem. Soc.* **104**, 6602–6609

235. Waldrop, G. L., Braxton, B. F., Urbauer, J. L., Cleland, W. W., and Kiick, D. M. (1994) *Biochemistry* **33**, 5262–5267

236. Highbarger, L. A., Gerlt, J. A., and Kenyon, G. L. (1996) *Biochemistry* **35**, 41–46

236a. Stanley, T. M., Johnson, W. H., Jr., Burks, E. A., Whitman, C. P., Hwang, C.-C., and Cook, P. F. (2000) *Biochemistry* **39**, 718–726

237. Loeber, G., Infante, A. A., Maurer-Fogy, I., Krystek, E., and Dworkin, M. B. (1991) *J. Biol. Chem.* **266**, 3016–3021

238. Chou, W.-Y., Liu, M.-Y., Huang, S.-M., and Chang, G.-G. (1996) *Biochemistry* **35**, 9873–9879

239. Winning, B. M., Bourguignon, J., and Leaver, C. J. (1994) *J. Biol. Chem.* **269**, 4780–4786

240. Wei, C.-H., Chou, W.-Y., and Chang, G.-G. (1995) *Biochemistry* **34**, 7949–7954

241. Karsten, W. E., and Cook, P. F. (1994) *Biochemistry* **33**, 2096–2103

References

241a. Liu, D., Karsten, W. E., and Cook, P. F. (2000) *Biochemistry* **39**, 11955–11960

242. Edens, W. A., Urbauer, J. L., and Cleland, W. W. (1997) *Biochemistry* **36**, 1141–1147

242a. Gruys, K. J., and Sikorski, J. A. (1998) in *Comprehensive Biological Catalysis. A Mechanistic Reference*, Vol. I (Sinnott, M., ed), pp. 273–291, Academic Press, San Diego, California

243. Bolduc, J. M., Dyer, D. H., Scott, W. G., Singer, P., Sweet, R. M., Koshland, D. E., Jr., and Stoddard, B. L. (1995) *Science* **268**, 1312–1318

244. Berdis, A. J., and Cook, P. F. (1993) *Biochemistry* **32**, 2041–2046

245. Dean, A. M., and Dvorak, L. (1995) *Protein Sci.* **4**, 2156–2167

246. Wallon, G., Kryger, G., Lovett, S. T., Oshima, T., Ringe, D., and Petsko, G. A. (1997) *J. Mol. Biol.* **266**, 1016–1031

247. Brubaker, M. J., Dyer, D. H., Stoddard, B., and Koshland, D. E., Jr. (1996) *Biochemistry* **35**, 2854–2864

248. Tipton, P. A. (1993) *Biochemistry* **32**, 2822–2827

249. Kalckar, H. M. (1985) *Trends Biochem. Sci.* **10**, 132–133

250. Peliska, J. A., and O'Leary, M. H. (1991) *J. Am. Chem. Soc.* **113**, 1841–1842

251. Seeholzer, S. H., Jaworowski, A., and Rose, I. A. (1991) *Biochemistry* **30**, 727–732

252. Matte, A., Tari, L. W., Goldie, H., and Delbaere, L. T. J. (1997) *J. Biol. Chem.* **272**, 8105–8108

253. Hwang, S. H., and Nowak, T. (1986) *Biochemistry* **25**, 5590–5595

254. Konopka, J. M., Lardy, H. A., and Frey, P. A. (1986) *Biochemistry* **25**, 5571–5575

255. Hlavaty, J. J., and Nowak, T. (1997) *Biochemistry* **36**, 3389–3403

256. Matte, A., Goldie, H., Sweet, R. M., and Delbaere, L. T. J. (1996) *J. Mol. Biol.* **256**, 126–143

257. Arnelle, D. R., and O'Leary, M. H. (1992) *Biochemistry* **31**, 4363–4368

258. Punekar, N. S., and Lardy, H. A. (1987) *J. Biol. Chem.* **262**, 6714–6719

259. Roseler, W. J., Vandenbark, G. R., and Hanson, R. H. (1989) *J. Biol. Chem.* **264**, 9657–9664

260. Rohrer, S. D., Saz, H. J., and Nowak, T. (1986) *J. Biol. Chem.* **261**, 13049–13055

261. Hebdo, C. A., and Nowak, T. (1982) *J. Biol. Chem.* **257**, 5503–5514

262. Hartman, F. C., and Harpel, M. R. (1994) *Ann. Rev. Biochem.* **63**, 197–234

263. Wasmann, C. C., Reiss, B., and Bohnert, H. J. (1988) *J. Biol. Chem.* **263**, 617–619

264. Shibata, N., Inoue, T., Fukuhara, K., Nagara, Y., Kitagawa, R., Harada, S., Kasai, N., Uemura, K., Kato, K., Yokota, A., and Kai, Y. (1996) *J. Biol. Chem.* **271**, 26449–26452

265. Tessier, L. H., Paulus, F., Keller, M., Vial, C., and Imbault, P. (1995) *J. Mol. Biol.* **245**, 22–33

266. Holzenburg, A., Mayer, F., Harauz, G., van Heel, M., Tokuoka, R., Ishida, T., Harata, K., Pal, G. P., and Saenger, W. (1987) *Nature (London)* **325**, 730–732

267. Erijman, L., Lorimer, G. H., and Weber, G. (1993) *Biochemistry* **32**, 5187–5195

268. Morse, D., Salois, P., Markovic, P., and Hastings, J. W. (1995) *Science* **268**, 1622–1624

269. Andersson, I. (1996) *J. Mol. Biol.* **259**, 160–174

269a. Schreuder, H. A., Knight, S., Curmi, P. M. G., Andersson, I., Cascio, D., Sweet, R. M., Brändén, C.-I., and Eisenberg, D. (1993) *Protein Sci.* **2**, 1136–1146

270. Newman, J., and Gutteridge, S. (1993) *J. Biol. Chem.* **268**, 25876–25886

271. Bassham, J. A., Benson, A. A., Kay, L. D., Harris, A. Z., Wilson, A. T., and Calvin, M. (1954) *J. Am. Chem. Soc.* **76**, 1760–1770

272. Siegel, M. I., and Lane, M. D. (1973) *J. Biol. Chem.* **248**, 5486–5498

272a. Taylor, T. C., and Andersson, I. (1996) *Nature Struct. Biol.* **3**, 95–101

272b. Harpel, M. R., Larimer, F. W., and Hartman, F. C. (1998) *Protein Sci.* **7**, 730–738

272c. Harpel, M. R., and Hartman, F. C. (1996) *Biochemistry* **35**, 13865–13870

272d. Duff, A. P., Andrews, T. J., and Curmi, R. M. G. (2000) *J. Mol. Biol.* **298**, 903–916

273. Pierce, J., Andrews, T. J., and Lorimer, G. H. (1986) *J. Biol. Chem.* **261**, 10248–10256

273a. Berry, J. A., Lorimer, G. H., Pierce, J., Seeman, J. R., Meeks, J., and Freas, S. (1987) *Proc. Natl. Acad. Sci. U.S.A.* **84**, 734–738

274. Jaworowski, A., Hartman, F. C., and Rose, I. A. (1984) *J. Biol. Chem.* **259**, 6783–6789

275. Saver, B. G., and Knowles, J. R. (1982) *Biochemistry* **21**, 5398–5403

276. Taylor, T. C., and Andersson, I. (1997) *J. Mol. Biol.* **265**, 432–444

277. Roeske, C. A., and O'Leary, M. H. (1984) *Biochemistry* **23**, 6275–6284

278. Van Dyke, D. E., and Schloss, J. V. (1986) *Biochemistry* **25**, 5145–5156

279. van de Loo, F. J., and Salvucci, M. E. (1996) *Biochemistry* **35**, 8143–8148

280. de Jiménez, E. S., Medrano, L., and Martínez-Barajas, E. (1995) *Biochemistry* **34**, 2826–2831

281. Larson, E. M., O'Brien, C. M., Zhu, G., Spreitzer, R. J., and Portis, A. R., Jr. (1997) *J. Biol. Chem.* **272**, 17033–17037

282. Mueller, D. D., Schmidt, A., Pappan, K. L., McKay, R. A., and Schaefer, J. (1995) *Biochemistry* **34**, 5597–5603

282a. Ogren, W. L. (1984) *Ann. Rev. Plant Physiol.* **35**, 415–442

283. Tse, J. M.-T., and Schloss, J. V. (1993) *Biochemistry* **32**, 10398–10403

283a. Sugawara, H., Yamamoto, H., Shibata, N., Inoue, T., Okada, S., Miyake, C., Yokota, A., and Kai, Y. (1999) *J. Biol. Chem.* **274**, 15655–15661

284. Harpel, M. R., Serpersu, E. H., Lamerdin, J. A., Huang, Z.-H., Gage, D. A., and Hartman, F. C. (1995) *Biochemistry* **34**, 11296–11306

284a. Gutteridge, S., Rhoades, D. F., and Herrmann, C. (1993) *J. Biol. Chem.* **268**, 7818–7824

285. Morell, M. K., Wilkin, J.-M., Kane, H. J., and Andrews, T. J. (1997) *J. Biol. Chem.* **272**, 5445–5451

286. Krebs, H. A., and Roughton, F. J. W. (1948) *Biochem. J.* **43**, 550–555

287. Kaziro, Y., Hass, L. F., Boyer, P. D., and Ochoa, S. (1962) *J. Biol. Chem.* **237**, 1460–1468

288. Cooper, T. G., Tchen, T. T., Wood, H. G., and Benedict, C. R. (1968) *J. Biol. Chem.* **243**, 3857–3863

289. González, D. H., and Andreo, C. S. (1988) *Biochemistry* **27**, 177–183

290. Harpster, M. H., and Taylor, W. C. (1986) *J. Biol. Chem.* **261**, 6132–6136

291. O'Leary, M. H., Rife, J. E., and Slater, J. D. (1981) *Biochemistry* **20**, 7308–7314

292. Hansen, D. E., and Knowles, J. R. (1982) *J. Biol. Chem.* **257**, 14795–14798

293. Fujita, N., Izui, K., Nishino, I., and Katsuki, H. (1984) *Biochemistry* **23**, 1774–1779

294. González, D. H., and Andreo, C. S. (1989) *Trends Biochem. Sci.* **14**, 24–27

295. Knowles, J. R. (1989) *Ann. Rev. Biochem.* **58**, 195–221

296. Janc, J. W., Urbauer, J. L., O'Leary, M. H., and Cleland, W. W. (1992) *Biochemistry* **31**, 6432–6440

297. Lee, S.-L., Hepburn, T. W., Swartz, W. H., Ammon, H. L., Mariano, P. S., and Dunaway-Matiano, D. (1992) *J. Am. Chem. Soc.* **114**, 7346–7354

298. McQueney, M. S., Lee, S.-l, Bowman, E., Mariano, P. S., and Dunaway-Mariano, D. (1989) *J. Am. Chem. Soc.* **111**, 6885–6887

299. Seidel, H. M., and Knowles, J. R. (1994) *Biochemistry* **33**, 5641–5646

300. Bowman, E., McQueney, M., Barry, R. J., and Dunaway-Mariano, D. (1988) *J. Am. Chem. Soc.* **110**, 5575–5576

301. Hidaka, T., and Seto, H. (1989) *J. Am. Chem. Soc.* **111**, 8012–8013

302. Freeman, S., Pollack, S. J., and Knowles, J. R. (1992) *J. Am. Chem. Soc.* **114**, 377–378

302a. Morais, M. C., Zhang, W., Baker, A. S., Zhang, G., Dunaway-Mariano, D., and Allen, K. N. (2000) *Biochemistry* **39**, 10385–10396

303. Mortimer, C. E., and Niehaus, W. G., Jr. (1974) *J. Biol. Chem.* **249**, 2833–2842

304. Clifford, K., Cornforth, J. W., Mallaby, R., and Phillips, G. T. (1971) *J. Chem. Soc. Chem. Commun.*, 1599–1600

305. Reardan, J. E., and Abeles, R. H. (1986) *Biochemistry* **25**, 5609–5616

306. Poulter, C. D., Muehlbacker, M., and Davis, D. R. (1989) *J. Am. Chem. Soc.* **111**, 3740–3742

307. Street, I. P., Coffman, H. R., Baker, J. A., and Poulter, C. D. (1994) *Biochemistry* **33**, 4212–4217

307a. Oh, S. K., Han, K. H., Ryu, S. B., and Kang, H. (2000) *J. Biol. Chem.* **275**, 18482–18488

308. Cane, D. E., Abell, C., Harrison, P. H., Hubbard, B. R., Kane, C. T., Lattman, R., Oliver, J. S., and Weiner, S. W. (1991) *Philos Trans R Soc Lond B Biol Sci* **332**, 123–129

309. Cane, D. E., Pawlak, J. L., and Horak, R. M. (1990) *Biochemistry* **29**, 5476–5490

310. Aulabaugh, A., and Schloss, J. V. (1990) *Biochemistry* **29**, 2824–2830

311. Biou, V., Dumas, R., Cohen-Addad, C., Douce, R., Job, D., and Pebay-Peyroula, E. (1997) *EMBO J.* **16**, 3405–3415

312. Halgand, F., Dumas, R., Biou, V., Andrieu, J.-P., Thomazeau, K., Gagnon, J., Douce, R., and Forest, E. (1999) *Biochemistry* **38**, 6025–6034

313. Proust-De Martin, F., Dumas, R., and Field, M. J. (2000) *J. Am. Chem. Soc.* **122**, 7688–7697

Study Questions

1. Discuss the role of carbonyl groups in facilitating reactions of metabolism.

2. Write a step-by-step sequence showing the chemical mechanisms involved in the action of a type I aldolase that catalyzes cleavage of fructose 1,6-bisphosphate. The enzyme is inactivated by sodium borohydride in the presence of the substrate. Explain this inactivation.

3. Some methylotrophic bacteria dehydrogenate methanol to formaldehyde. The latter undergoes an aldol condensation to form a hexulose-6-phosphate:

$$
\begin{array}{c}
CH_2OH \\
| \\
HC-OH \\
| \\
C=O \\
| \\
(HC-OH)_2 \\
| \\
CH_2OP
\end{array}
$$

Write a reasonable cycle of biochemical reactions (utilizing this compound) by which three molecules of methanol can yield one molecule of glyceraldehyde 3-phosphate, a compound that can be either catabolized for energy or utilized for biosynthesis.

4. Malate is formed in the glyoxylate pathway by reaction of glyoxylate with Acetyl-CoA. Indicate the chemical mechanism of the reactions involved and structure of an intermediate species.

5. Acetyl-CoA is condensed with acetoacetyl-CoA to give an intermediate which is eventually converted to mevalonate.
   a) Show the sequence of reactions leading from acetyl-CoA to mevalonate with the structures of important intermediates.
   b) Mevalonate is an important intermediate on the pathway leading to what *class* of compounds?

6. a) Show the structures of the reactants for the 3-hydroxy-3-methylglutaryl-CoA synthase reaction.
   b) Free coenzyme A is liberated in the above reaction. From which of the two reactants did it come? Explain the metabolic significance of the liberation of free CoA.

7. Phosphoenolpyruvate carboxykinase catalyzes the following reaction:

   Oxaloacetate + GTP→ CO_2 + phosphoenolpyruvate + GDP

   Illustrate the probable mechanism of the reaction.

8. Write a complete step-by-step mechanism for the action of ATP-citrate lyase which catalyzes the following reaction (Eq. 13-39):

   ATP + citrate + CoA – SH → Acetyl – CoA + oxaloacetate + ADP + P_i

9. Malonyl-CoA synthetase forms its product from free malonate and MgATP. Both phospho-enzyme and malonyl-enzyme intermediates have been detected. Suggest a sequential mechanism of action. See Kim, Y. S., and Lee, J. K. (1986) *J. Biol. Chem.* **261**, 16295–16297.

10. The carboxylation of acetone with HCO_3^- to form acetoacetate (the reverse of Eq. 13-44) is not a thermodynamically spontaneous process ($\Delta G^{\circ\prime}$ ~17 kJ mol^{-1}). An anaerobic strain of *Xanthobacter* couples this reaction to the cleavage of ATP to AMP + 2 P_i. Suggest possible mechanisms. See Sluis, M. K., and Ensign, S. A. (1997) *Proc. Natl. Acad. Sci. U.S.A.* **94**, 8456–8461.

11. The reaction of Eq. 10-11 begins with a hydroxylation, which can be viewed as insertion of an oxygen atom into a C–H bond. Draw the probable intermediate formed in this reaction and indicate how it can be converted to the final products.

12. Cyanase from *E. coli* decomposes cyanate (NCO^-) to ammonia and CO_2. Isotopic labeling studies show that the second oxygen atom in the product CO_2 comes not from H_2O but from a second substrate, bicarbonate ion. The correct equation for the enzymatic process is:

   $NCO^- + H^+ + HCO_3^- \rightarrow NH_2 – COO^- + CO_2$

   Propose a sequence of steps for this enzymatic reaction, which has been investigated by Anderson and coworkers. Johnson, W. V., and Anderson, P. M. (1987) *J. Biol. Chem.* **262**, 9021–9025 and Anderson, P. M., Korte, J. J., Holcomb, T. A., Cho, Y.-g, Son, C.-m, and Sung, Y.-c. (1994) *J. Biol. Chem.* **269**, 15036–15045.

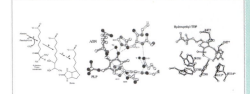

The vitamin **biotin** is formed in nature (left) by condensation of L-alanine with pimeloyl-CoA to form 8-amino-7-oxononanoate (AON). This compound is seen at the upper left of the center structure joined as a Schiff base with the coenzyme **pyridoxal phosphate** (PLP). This is a product complex of the enzyme AON synthase (see Webster *et al.*, *Biochemistry* **39**, 516-528, 2000) Courtesy of D. Alexeev, R. L. Baxter, and L. Sawyer. Biotin synthesis requires three other enzymes (steps *b, c, d*). Step *b* is catalyzed by a PLP-dependent transaminase. At the left is **thiamin diphosphate**, in the form of its 2-(1-hydroxyethyl) derivative, an intermediate in the enzyme pyruvate decarboxylase (Dobritzsch *et al.*, *J. Biol. Chem.* **273**, 20196-20204, 1998). Courtesy of Guoguang Lu. Thiamin diphosphate functions in all living organisms to cleave C–C bonds adjacent to C=O groups.

Contents

Coenzymes: Nature's Special Reagents

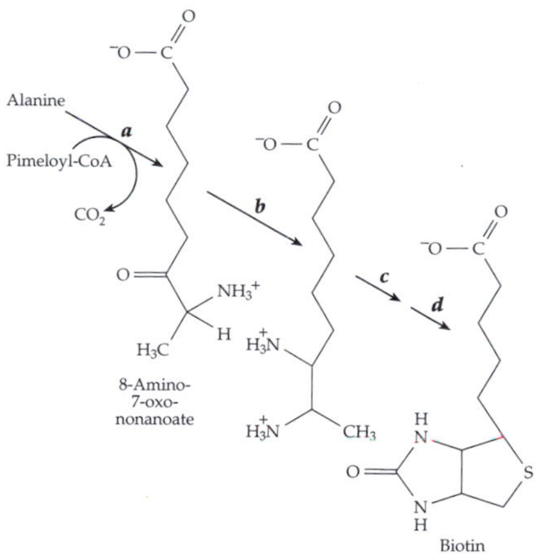

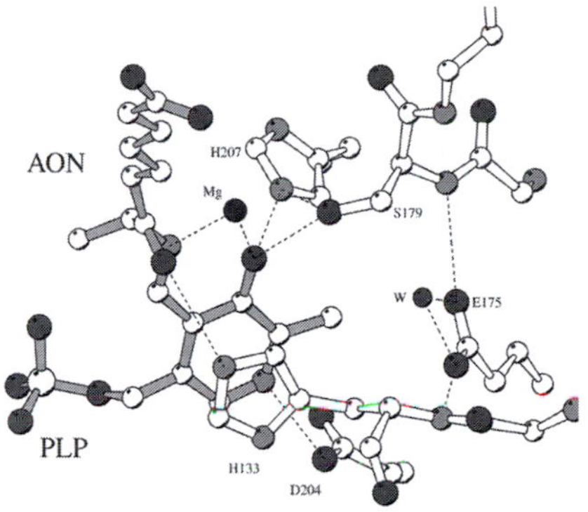

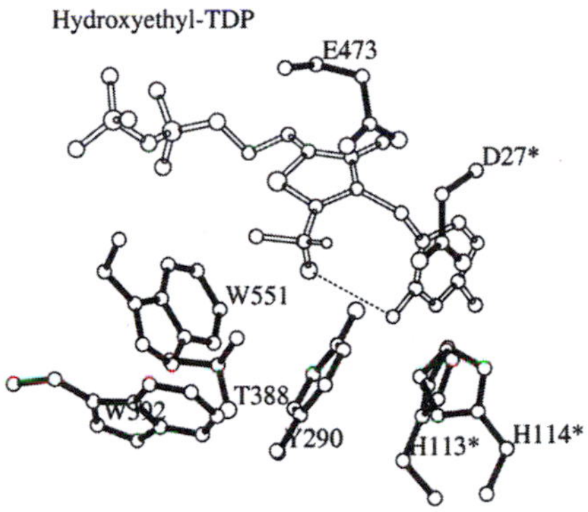

14

Most of the reactions discussed in Chapters 12 and 13 are catalyzed by enzymes that contain only those functional groups found in the side chains of the constituent amino acids. **Coenzymes** are nonprotein molecules that function as essential parts of enzymes. Coenzymes often serve as "special reagents" needed for reactions that would be difficult or impossible using only simple acid–base catalysis. In many instances, they also serve as **carriers**, alternating catalysts that accept and donate chemical groups, hydrogen atoms, or electrons. Coenzymes will be considered here in three groups:

1. Compounds of high group transfer potential such as ATP and GTP that function in energy coupling within cells. Because it is cleaved and then dissociates from the enzyme to which it is bound, ATP may be regarded as a substrate rather than a coenzyme. However, as a phosphorylated form of ADP it may also be viewed as a carrier of high-energy phospho groups.

2. Compounds, often derivatives of **vitamins** that, while in the active site of the enzyme, alter the structure of a substrate in a way that permits it to react more readily. Coenzyme A, pyridoxal phosphate, thiamin diphosphate, and vitamin B_{12} coenzymes fall into this group.

3. Oxidative coenzymes with structures of precisely determined oxidation–reduction potential. Examples are NAD^+, $NADP^+$, FAD, and lipoic acid. They serve as carriers of hydrogen atoms or of electrons. Some of these coenzymes, such as NAD^+ and $NADP^+$, can usually dissociate rapidly and reversibly from the enzymes with which they function. Others, including FAD, are much more tightly bound and rarely if ever dissociate from the protein catalyst. Heme groups are covalently linked to proteins such as cytochrome c and cannot be dissociated without destroying the enzyme. Very tightly bound coenzyme groups are often called **prosthetic groups**, but there is no sharp line that divides prosthetic groups from the loosely bound coenzymes. For example, NAD^+ is bound weakly to some proteins but tightly to others. Oxidative coenzymes are discussed in Chapter 15.

A. ATP and the Nucleotide "Handles"

The role of ATP in "driving" biosynthetic reactions has been considered in Chapter 12, where attention was focused on the polyphosphate group which undergoes cleavage. What about the adenosine end? Here is a shapely structure borrowed from the nucleic acids. What is it doing as a carrier of phospho groups? At least part of the answer seems to be that the adenosine monophosphate (AMP) portion of the molecule is a "handle" which can be "grasped" by catalytic proteins. For some enzymes, such as acetyl-CoA synthetase (Eq. 12-45), the handle is important because the intermediate acyl adenylate must remain bound to the protein. Without the large adenosine group, there would be little for the protein to hold onto.

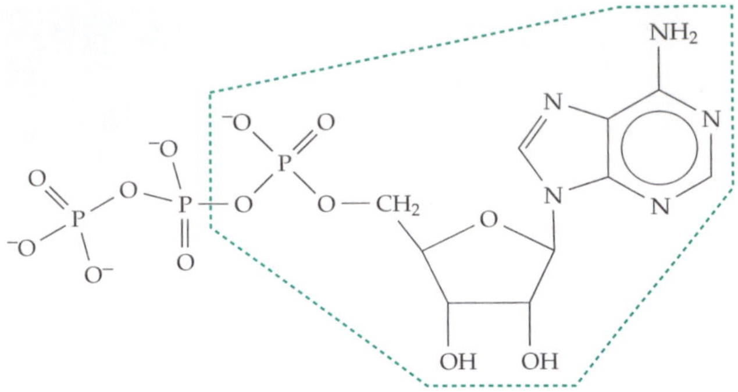

The AMP "handle" in ATP

AMP is only one of several handles to which nature attaches phospho groups to form di- and triphosphate derivatives. Like AMP, the other handles are nucleotides, the monomer units of nucleic acids. Thus, one enzyme requiring a polyphosphate as an energy source selects ATP, another CTP, or GTP. The nucleotide handles not only carry polyphosphate groups but also are present in other coenzymes, such as CoA, NAD[+], NADP[+], and FAD. In addition, they serve as carriers for small organic molecules. For example, **uridine diphosphate glucose** (UDP-Glc) Chapter 10, is a carrier of active glucosyl groups important in sugar metabolism and **cytidine diphosphate choline** is an intermediate in synthesis of phospholipids.

Recalling that acetyl adenylate (acetyl-AMP) is an intermediate in synthesis of acetyl-CoA, and comparing the biosynthesis of sugars, phospholipids, and acetyl-CoA, we see that in each case the enzyme involved requires a different nucleotide handle. The handle may provide a means of recognition which can help an enzyme to pick the right bit of raw material out of the sea of molecules surrounding it. Figure 5-6 shows the shapes of the four purine and pyrimidine bases forming the most common nucleotide handles. The distinctive differences both in shape and in hydrogen bond patterns are obvious. In binding to proteins, the hydrogen bond-forming groups in the purine and pyrimidine bases sometimes interact with precisely positioned groups in the protein. However, in some enzymes the "handle" is not precisely bound. This seems to be the case for adenine, which makes surprisingly few hydrogen bonds to proteins.[1] Hydroxyl groups of the ribose or deoxyribose ring also often form hydrogen bonds and the negatively charged oxygen atoms of the 5'-phosphate may interact with positively charged protein side chains.

B. Coenzyme A and Phosphopantetheine

The existence of a special coenzyme required in biological acetylation was recognized by Fritz Lipmann in 1945.[2–3a] The joining of acetyl groups to other

molecules is a commonplace reaction within living cells, one example being the formation of the neurotransmitter **acetylcholine**. In the laboratory acetylation is carried out with reactive compounds such as acetic anhydride or acetyl chloride.

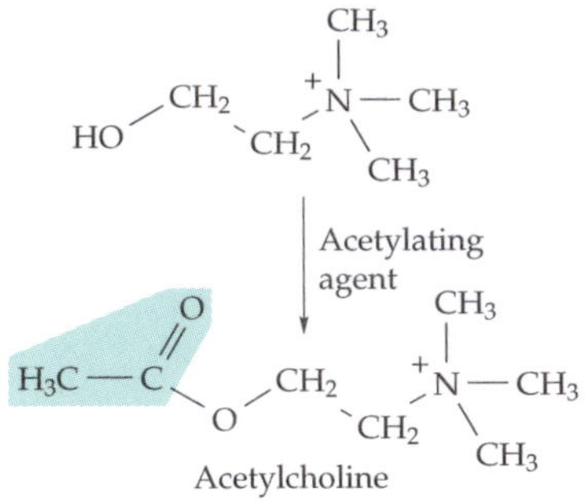

Acetylcholine   (14-1)

Lipmann wondered what nature used in their place. His approach in seeking the biological "active acetate" is one that has been used successfully in solving many biochemical problems. He first set up a test system to examine the ability of extracts prepared from fresh liver tissue to catalyze the acetylation of sulfanilamide (Eq. 14-2). A specific color test was available for quantitative determination of very small amounts of the

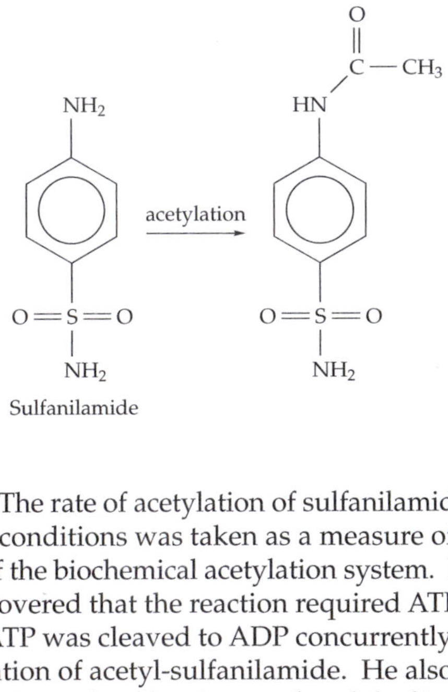

Sulfanilamide   (14-2)

product. The rate of acetylation of sulfanilamide under standard conditions was taken as a measure of the activity of the biochemical acetylation system. Lipmann soon discovered that the reaction required ATP and that the ATP was cleaved to ADP concurrently with the formation of acetyl-sulfanilamide. He also found that dialysis or ultrafiltration rendered the liver extract almost inactive in acetylation. Apparently some essential material passed out through the semipermeable dialysis membrane. When the dialysate or ultrafiltrate was concentrated and added back, acetylation activity was restored. The unknown material was not destroyed by boiling, and Lipmann postulated that it was a new coenzyme which he called **coenzyme A** (CoA). Now the test system was used to estimate the amount of the

BOX 14-A DISCOVERY OF THE VITAMINS

Several mysterious and often fatal diseases which resulted from vitamin deficiencies were prevalent until the past century. Sailors on long sea voyages were often the victims. In the Orient, the disease **beriberi** was rampant and millions died of its strange paralysis called "polyneuritis." In 1840, George Budd predicted that the disease was caused by the lack of some organic compound that would be discovered in "a not too distant future."[a] In 1893, C. Eijkman, a Dutch physician working in Indonesia, observed paralysis in chicks fed white rice consumed by the local populace. He found that the paralysis could be relieved promptly by feeding an extract of rice polishings. This was one of the pieces of evidence that led the Polish biochemist Casimir Funk to formulate the "vitamine theory" in about 1912. Funk suggested that the diseases **beriberi**, **pellagra**, **rickets**, and **scurvy** resulted from lack in the diet of four different vital nutrients. He imagined them all to be amines, hence the name **vitamine**.

In the same year in England, F. G. Hopkins announced that he had fed rats on purified diets and discovered that amazingly small amounts of **accessory growth factors**, which could be obtained from milk, were necessary for normal growth.[b] By 1915, E. V. McCollum and M. Davis at the University of Wisconsin had recognized that rat growth depended on not one but at least two accessory factors. The first, soluble in fatty solvents, they called **A**, and the other, soluble in water, they designated **B**. Factor B cured beriberi in chicks. Later, when it was shown that vitamine A was not an amine, the "e" was dropped and **vitamin** became a general term.

Progress in isolation of the vitamins was slow, principally because of a lack of interest. According to R. R. Williams, when he started his work on isolation of the antiberiberi factor in 1910 most people were convinced that his efforts were doomed to failure, so ingrained was Pasteur's idea that diseases were caused only by bacteria. In 1926, Jansen isolated a small amount of thiamin, but it was not until 1933 that Williams, working almost without financial support, succeeded in preparing a large amount of the crystalline compound from rice polishings. Characterization and synthesis followed rapidly.[c,d]

It was soon apparent that the new vitamin alone would not satisfy the dietary need of rats for the B factor. A second thermostable factor (B_2) was required in addition to thiamin (B_1), which was labile and easily destroyed by heating. When it became clear that factor B_2 contained more than one component, it was called **vitamin B complex**. There was some confusion until relatively specific animal tests for each one of the members had been devised.

Riboflavin was found to be most responsible for the stimulation of rat growth, while **vitamin B_6** was needed to prevent a facial dermatitis or "rat pellagra." **Pantothenic acid** was especially effective in curing a chick dermatitis, while **nicotinamide** was required to cure human pellagra. **Biotin** was required for growth of yeast.

The antiscurvy (antiscorbutic) activity was called **vitamin C**, and when its structure became known it was called **ascorbic acid**. The fat-soluble factor preventing rickets was designated **vitamin D**. By 1922, it was recognized that another fat-soluble factor, **vitamin E**, is essential for full-term pregnancy in the rat. In the early 1930s **vitamin K** and the **essential fatty acids** were added to the list of fat-soluble vitamins. Study of the human blood disorders "tropical macrocytic anemia" and "pernicious anemia" led to recognition of two more water-soluble vitamins, **folic acid** and **vitamin B_{12}**. The latter is required in minute amounts and was not isolated until 1948. Have all the vitamins been discovered? Rats can be reared on an almost completely synthetic diet. However, there is the possibility that for good health humans require some as yet undiscovered compounds in our diet. Furthermore, it is quite likely that we receive some essential nutrients that we cannot synthesize from bacteria in our intestinal tracts. An example may be the pyrroloquinoline quinone (PQQ).[e]

Why do we need vitamins? Early clues came in 1935 when nicotinamide was found in NAD^+ by H. von Euler and associates and in $NADP^+$ by Warburg and Christian. Two years later, K. Lohman and P. Schuster isolated pure **cocarboxylase**, a dialyzable material required for decarboxylation of pyruvate by an enzyme from yeast. It was shown to be thiamin diphosphate (Fig. 15-3). Most of the water-soluble vitamins are converted into coenzymes or are covalently bound into active sites of enzymes. Some lipid-soluble vitamins have similar functions but others, such as vitamin D and some metabolites of vitamin A, act more like hormones, binding to receptors that control gene expression or other aspects of metabolism.

[a] Hughes, R. E. (1973) *Medical History* **17**, 127–134
[b] Harris, L. D. (1937) *Vitamins in Theory and Practice*, Cambridge Univ. Press, London and New York
[c] Karlson, P. (1984) *Trends Biochem. Sci.* **9**, 536–537
[d] Williams, R. R., and Spies, T. D. (1938) *Vitamin B₁ and Its Use in Medicine*, Macmillan, New York
[e] Killgore, J., Smidt, C., Duich, L., Romero-Chapman, N., Tinker, D., Reiser, K., Melko, M., Hyde, D., and Rucker, R. B. (1989) *Science* **245**, 850–852

coenzyme in a given volume of dialysate or in any other sample. When small amounts of CoA were supplied to the test system, only partial restoration of the acetylation activity was observed and the amount of restoration was proportional to the amount of CoA. With test system in hand to monitor various fractionation methods, Lipmann soon isolated the new coenzyme in pure form from yeast and liver.

Coenzyme A (Fig. 14-1) is a surprisingly complex molecule. The handle is AMP with an extra phospho group on its 3'-hydroxyl. The phosphate of the 5'-carbon is linked in anhydride (pyrophosphate) linkage to another phosphoric acid, which is in turn esterified with **pantoic acid**. Pantoic acid is linked to **β-alanine** and the latter to **β-mercaptoethylamine** through amide linkages, the reactive SH group being attached to a long (1.9-nm) semi-flexible chain. Coenzyme A can be cleaved by hydrolysis to **pantetheine** (Fig. 14-1), **pantetheine 4'-phosphate**, and **pantothenic acid**. These three compounds are all **growth factors**.

Pantothenic acid is a vitamin, which is essential to human life. Its name is derived from a Greek root that reflects its universal occurrence in living things. The bacterium *Lactobacillus bulgaricus*, which converts milk

to yogurt, needs the more complex pantetheine for growth. It finds a ready supply of pantetheine in milk and has lost its ability to synthesize this compound. However, it can convert pantetheine to CoA. Pantetheine 4'-phosphate is required for growth of *Acetobacter suboxydans*. While it is not a dietary essential for most organisms, it is found in a covalently bound form in several enzymes.

While CoA was discovered as the "acetylation coenzyme," it has a far more general function. It is required, in the form of acetyl-CoA, to catalyze the synthesis of citrate in the citric acid cycle. It is essential to the β oxidation of fatty acids and carries propionyl and other acyl groups in a great variety of other metabolic reactions. About 4% of all known enzymes require CoA or one of its esters as a substrate.[4]

Coenzyme A has two distinctly different biochemical functions, which have already been considered briefly in Chapters 12 and 13 and can be summarized as follows:

1. Activation of acyl group, $R-C-$, toward transfer by nucleophilic displacment

2. Activation of a hydrogen atom adjacent to the carbonyl of a thioester

These functions depend, to a considerable extent, on the fact that the properties of the carbonyl group of a thioester are closely similar to those of an isolated carbonyl group in a ketone.[5]

Synthesis of fatty acids in bacteria requires a small **acyl carrier protein (ACP)** whose functions are similar to those of CoA. However, it contains pante-

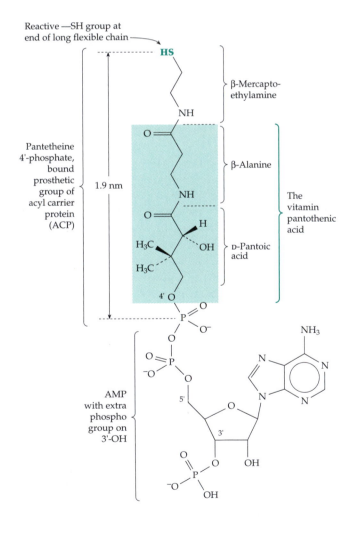

Figure 14-1 Coenzyme A, an acyl-activating coenzyme containing the vitamin pantothenic acid.

theine 4'-phosphate covalently bonded through phosphodiester linkage to a serine side chain. In *E. coli* this is at position 36 in the 77-residue protein.[6,7] Here the nucleotide handle of CoA has been replaced with a larger and more complex protein which can interact in specific ways with the multiprotein fatty acid synthase complex described in Chapter 21. In higher organisms ACP is usually not a separate protein but a domain in a large synthase molecule. Bound phosphopantetheine is also found in enzymes involved in synthesis of peptide antibiotics (Chapter 29)[8,9] and polyketides (Chapter 21).[10,11] It is also present in subunits of cytochrome oxidase and of ATP synthase of *Neurospora* but appears to play only a structural role, being needed for proper assembly of these multimeric proteins.[12] In a citrate-cleaving enzyme, and in a bacterial malonate decarboxylase, phosphopantethine is attached to a serine side chain as 2'-(5''-phosphoribosyl)-3'-dephospho-CoA.[13–14a]

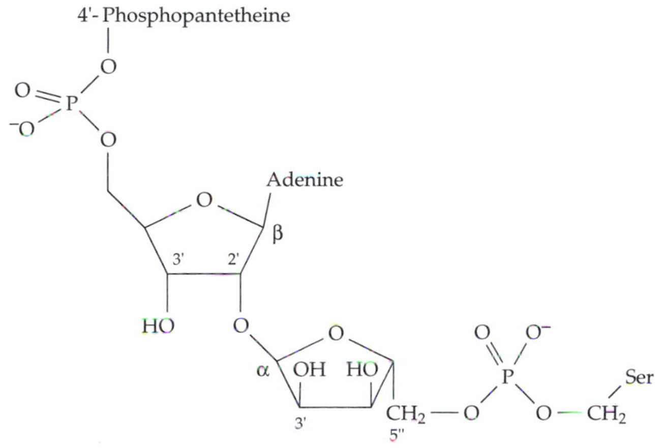

protein through hydrophobic interactions and hydrogen-bond recognition interactions. Additional specific hydrophobic interactions allow the enzyme to hold the acetyl group in a precise position where it can be acted upon by the catalytic groups. In the case of CoA transferase (Eq. 12-50) smaller thiols can replace CoA but with much lower catalytic rates. The acyl-CoA derivatives are weakly bound, the expected intrinsic binding energy of the pantetheine portion of the molecule apparently being used to increase k_{cat}.[17] From a study of kinetics and equilibria it was concluded that the binding energy of the interaction of the nucleotide portion of coenzyme A with the enzyme is utilized to increase the rate of formation and to stabilize the covalently linked E-CoA (see Eq. 12-50). On the other hand, binding of the pantoic acid part of the molecule decreases the stability of the transition state for break-up of the complex by ~40 kJ/mol, which corresponds to a 10^7-fold increase in the reaction rate.[16]

4'-Phosphopantetheine

2'-(5''-Phophoribosyl)-3'-dephospho-CoA

Attachment of phosphopantetheine to proteins is catalyzed by a phosphotransferase that utilizes CoA as the donor. A phosphodiesterase removes the phosphopantetheine, providing a turnover cycle.[15–15b] A variety of synthetic analogs have been made.[4,16] The reactive center of CoA and phosphopantetheine is the SH group, which is carried on a flexible arm that consists in part of the β-alanine portion of pantothenic acid. A mystery is why pantoic acid, a small odd-shaped molecule that the human body cannot make, is so essential for life. The hydroxyl group is a potential reactive site and the two methyl groups may enter into formation of a "trialkyl lock" (p. 485), part of a sophisticated "elbow" or shoulder for the SH-bearing arm.

When it binds to citrate synthase acetyl-CoA appears to bind only after the enzyme has undergone a conformational change that closes the enzyme around its other substrate oxaloacetate (Fig. 13-8). The adenine ring of the long CoA handle is tightly bonded to the

C. Biotin and the Formation of Carboxyl Groups from Bicarbonate

By 1901 it was recognized that yeast required for its growth an unknown material which was called **bios**. This was eventually found to be a mixture of pantothenic acid, inositol, and a third component which was named **biotin**. Biotin was also recognized as a factor promoting growth and respiration of the clover root nodule organism *Rhizobium trifolii* and as vitamin H, a material that prevented dermatitis and paralysis in rats that were fed large amounts of uncooked egg white. Isolation of the pure vitamin was a heroic task accomplished by Kögl in 1935. In one preparation 250 kg of dried egg yolk yielded only 1.1 mg of crystalline biotin.[18]

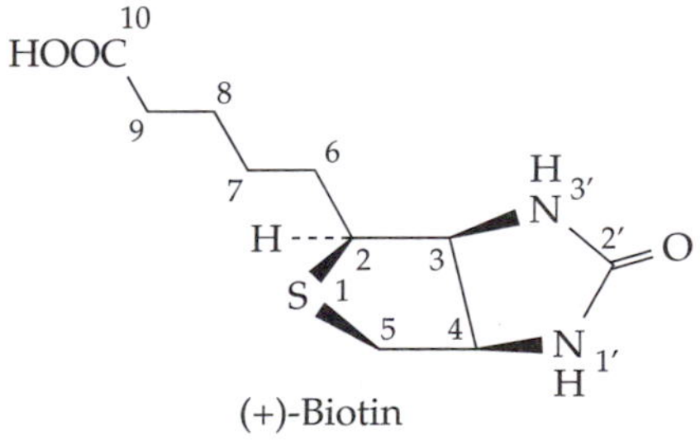

(+)-Biotin

Biotin contains three chiral centers and therefore has eight stereoisomers.[18,19] Of these, only one, the dextrorotatory (+)-biotin, is biologically active.[19,20] The vitamin is readily oxidized to the sulfoxide and sulfone. The sulfoxide can be reduced back to biotin by a molybdenum-containing reductase in some bacteria (see also Chapter 16, Section H).[20a] Biotin is synthesized from pimeloyl-CoA (see chapter banner, p. 719 and Eq. 24-39). Four enzymes are required. Two of them, a synthase that catalyzes step *a* (banner) and a transaminase that catalyzes step *b*, contain the coenzyme pyridoxal phosphate (PLP).[20b] The final step is insertion of a sulfur atom from an iron-sulfur center.

1. Biotin-Containing Enzymes

Within cells the biotin is covalently bonded to proteins, its double ring being attached to a 1.6-nm flexible arm. The first clue to this fact was obtained from isolation of a biotin-containing material **biocytin**, ε-N-biotinyl-L-lysine, from autolysates (self-digests) of rapidly growing yeast.[21] It was subsequently shown that the lysine residue of the biocytin was originally present in proteins at the active sites of biotin-containing enzymes usually within the sequence AMKM[22] or VMKM.[23] Other conserved features also mark the attachment site.[24]

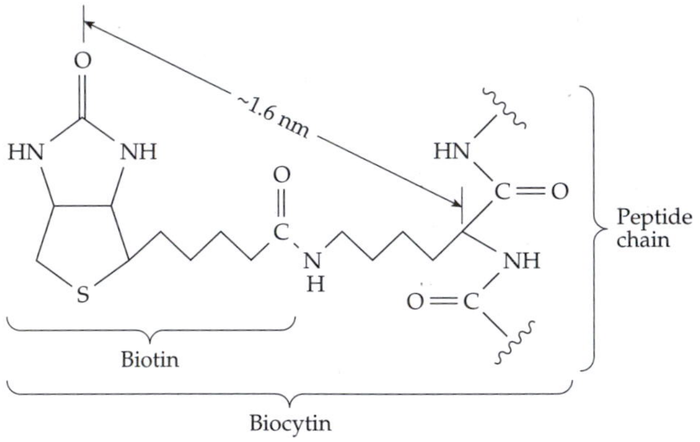

Biotin acts as a **carboxyl group carrier** in a series of carboxylation reactions, a function originally suggested by the fact that aspartate partially replaces biotin in promoting the growth of the yeast *Torula cremonis*. Aspartate was known to arise by transamination from oxaloacetate, which in turn could be formed by carboxylation of pyruvate. Subsequent studies showed that biotin was needed for an enzymatic ATP-dependent reaction of pyruvate with bicarbonate ion to form oxaloacetate (Eq. 14-3). This is a β carboxylation coupled to the hydrolysis of ATP.

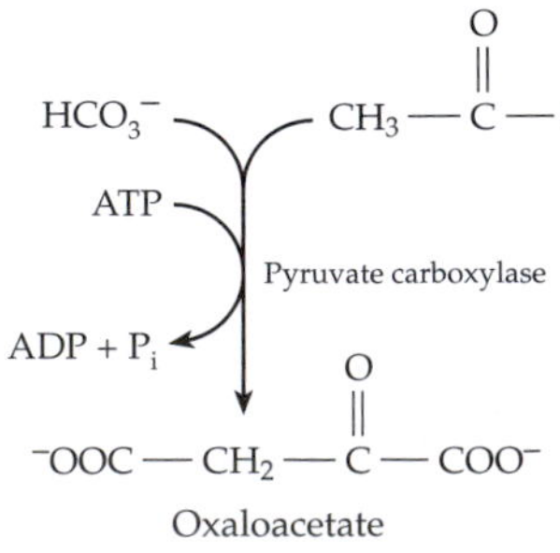

In addition to pyruvate carboxylase, other biotin-requiring enzymes act on **acetyl-CoA**, **propionyl-CoA**, and **β-methylcrotonyl-CoA**, using HCO_3^- to add carboxyl groups at the sites indicated by the arrows in the accompanying structures. Because of the presence of the C=C double bond conjugated with the carbonyl group, the carboxylation of β-methylcrotonyl-CoA is electronically analogous to β carboxylation.

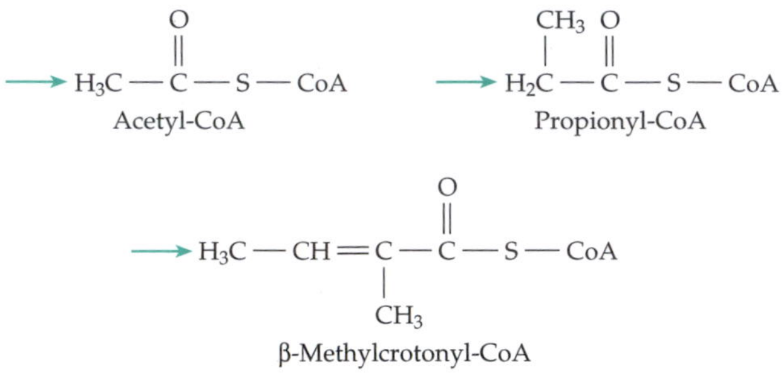

Human cells, as well as those of other higher eukaryotes, carry genes for all four of these enzymes.[25–29] Acetyl-CoA carboxylase is a cytosolic enzyme needed for synthesis of fatty acids but the other three enzymes enter mitochondria when they function. The biotin-dependent carboxylases, which are listed in Table 14-1, have a variety of molecular sizes and subunits but show much evolutionary conservation in their sequences and chemical mechanisms.[22,25] In higher eukaryotes all of the catalytic apparatus of the enzymes is present in single large 190- to 200-kDa subunits. The 251-kDa subunit of yeast acetyl-CoA carboxylase consists of 2337 amino acid residues.[30] That of rats contains 2345[31] and that of the alga *Cyclotella* 2089.[32] Human cytosolic acetyl-Co carboxylase has 2347 residues while a mitochondria-associated form has an extra 136 residues, most of them in a hydrophobic N-terminal extension.[32a,b] Plants have two forms of the enzyme, one cytosolic and one located in plastids. In wheat, they have 2260 and 2311 residues, respectively.[32c,32d] Animal and fungal pyruvate carboxylases are also large ~500-kDa tetramers.[32e] The yeast enzyme consists of 1178-residue monomers.[33] In contrast, the 560-kDa human propionyl-CoA carboxylase is an $\alpha_4\beta_4$ tetramer.[34,35]

In bacteria and in at least some plant chloroplasts,[36,37] acetyl-CoA carboxylase consists of three different kinds of subunit and four different peptide chains. The much studied *E. coli* enzyme is composed of a 156-residue **biotin carboxyl carrier protein**,[38] a 449-residue **biotin carboxylase**, whose three-dimensional structure in known,[39,39a] and a **carboxyltransferase** subunit consisting of 304 (α)- and 319 (β)- residue chains. These all associate as a dimer of the three subunits (eight peptide chains).[40–42]

Biotin becomes attached to the proper ε-amino groups at the active centers of biotin enzymes by the action of **biotin holoenzyme synthetase** (biotinyl protein ligase), which utilizes ATP to form an intermediate biotinyl-AMP.[43–47a] Hereditary deficiency of this enzyme has been observed in a few children and has been treated by administration of extra biotin.[48] The *E. coli* biotin holoenzyme synthetase, whose three-dimensional structure is known, has a dual function. It is also a repressor of transcription of the biotin biosynthetic operon.[45] Intracellular degradation of biotin-containing proteins yields biotin-containing oligopeptides as well as biocytin. These are acted on by **biotinidase** to release free biotin.[49,50] The action of this enzyme in recycling biotin may be a controlling factor in the rate of formation of new biotin-dependent enzymes.

2. The Mechanism of Biotin Action

It may seem surprising that a coenzyme is needed for these carboxylation reactions. However, unless the cleavage of ATP were coupled to the reactions, the equilibria would lie far in the direction of decarboxylation. For example, the measured apparent equilibrium constant K' for conversion of propionyl-CoA to S methylmalonyl-CoA at pH 8.1 and 28°C[51] is given by Eq. 14-4.

$$K' = \frac{[ADP][P_i][\text{methylmalonyl–CoA}]}{[ATP][HCO_3^-][\text{propionyl–CoA}]} = 5.7$$

$$\Delta G' = -4.36 \text{ kJ mol}^{-1} \qquad (14\text{-}4)$$

TABLE 14-1
Enzymes Containing Bound Biotin

1. Catalyzing beta carboxylation using HCO_3^- with coupled cleavage of ATP to ADP + P_i

    Acetyl-CoA carboxylase
    Propionyl-CoA carboxylase
    Pyruvate carboxylase
    β-Methylcrotonyl-CoA carboxylase
    (δ carboxylation)

2. Carboxyl group transfer without cleavage of ATP

    Carboxyltransferase of *Propionobacterium*

3. Biotin-dependent Na⁺ pumps

    Oxaloacetate decarboxylase
    Methylmalonyl-CoA decarboxylase
    Glutaconyl-CoA decarboxylase

4. Other
    Malonate decarboxylase
    Urea carboxylase

The function of biotin is to mediate the coupling of ATP cleavage to the carboxylation, making the overall reaction exergonic. This is accomplished by a two-stage process in which a **carboxybiotin** intermediate is formed (Eq. 14-5). There is *one known biotin-containing enzyme that does not utilize ATP*. Propionic acid bacteria contain a **carboxyltransferase** which transfers a carboxyl group reversibly from methylmalonyl-CoA to pyruvate to form oxaloacetate and propionyl-CoA (see Fig. 17-10). This huge enzyme consists of a central hexameric core of large 12S subunits to which six 5S dimeric subunits and twelve 123-residue biotinylated peptides are attached.[25,52] No ATP is needed because free HCO_3^- is not a substrate. However, biotin serves as the carboxyl group carrier in this enzyme too.

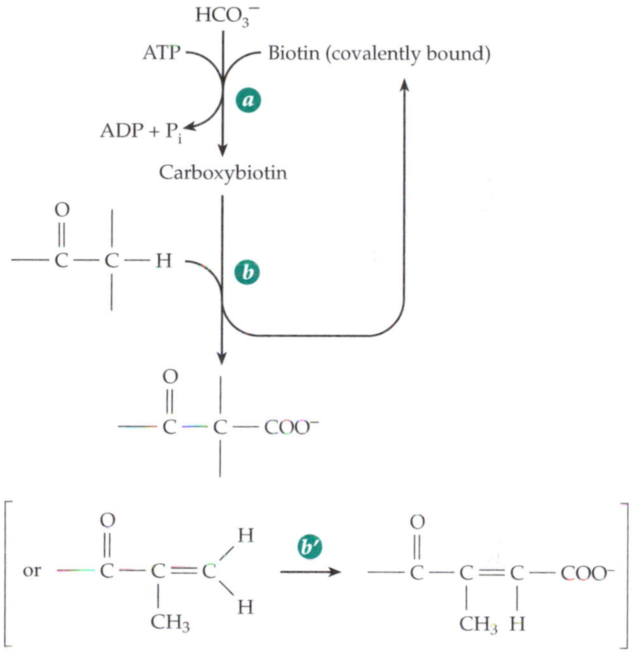

$$(14\text{-}5)$$

Carboxybiotin. The structure of biotin suggested that bicarbonate might be incorporated reversibly into its position 2'. However, this proved not to be true and it remained for F. Lynen and associates to obtain a clue from a "model reaction." They showed that purified β-methylcrotonyl-CoA carboxylase promoted the carboxylation of *free* biotin with bicarbonate ($H^{14}CO_3^-$) and ATP. While the carboxylated biotin was labile, treatment with diazomethane (Eq. 14-6) gave a stable dimethyl ester of **N-1'-carboxybiotin**.[53,54] The covalently bound biotin at active sites of enzymes was also successfully labeled with $^{14}CO_2$. Treatment of the labeled enzymes with diazomethane followed by hydrolysis with trypsin and pepsin gave authentic N-1'-carboxybiocytin. It was now clear that the cleavage of ATP is required to couple the CO_2 from HCO_3^- to the biotin to form carboxybiotin. The enzyme must

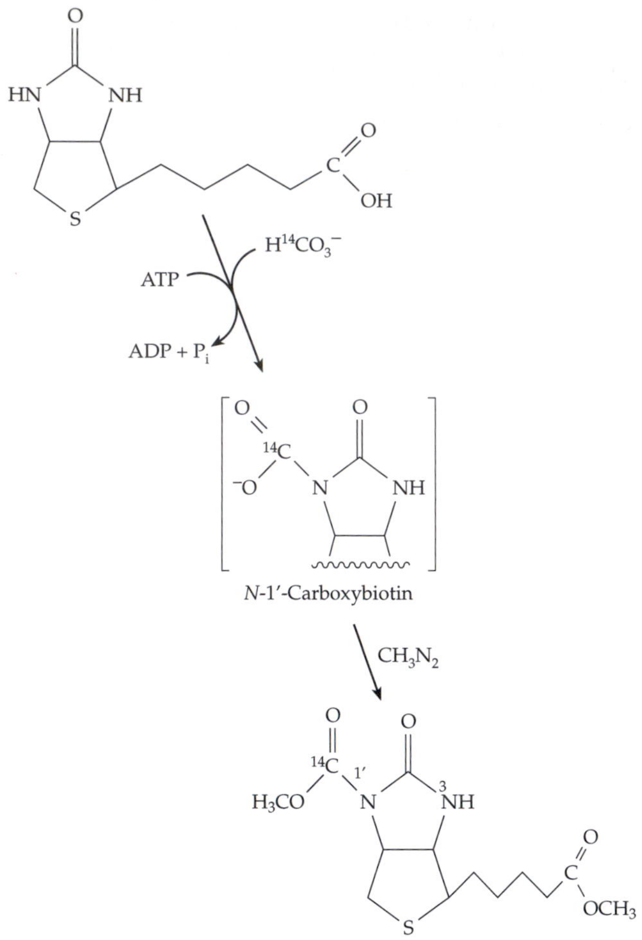

N-1'-Carboxybiotin

(14-6)

then transfer the carboxyl group from carboxybiotin to the substrate that is to be carboxylated. Enzymatic transfer of a carboxyl group from chemically synthesized carboxybiotin onto specific substrates confirmed the proposed mechanism.[55]

The biotin carboxyl carrier subunit of *E. coli* acetyl-CoA carboxylase contains the covalently bound biotin.[55a,b] The larger biotin carboxylase subunit catalyzes the ATP-dependent attachment of CO_2 to the biotin and the carboxyltransferase subunit catalyzes the final transcarboxylation step (Eq. 14-5, step *b*) by which acetyl-CoA is converted into malonyl-CoA. The biotin, which is attached to the carrier protein, is presumably able to move by means of its flexible arm from a site on the carboxylase to a site on the transcarboxylase.

Carboxyphosphate. During the initial carboxylation step ^{18}O from labeled bicarbonate enters the P_i that is split from ATP. This suggested transient formation of **carboxyphosphate** by nucleophilic attack of HCO_3^- on ATP (Eq. 14-7). The carboxyl group of this reactive mixed anhydride[56] could then be transferred to biotin. This mechanism is supported by the fact that biotin carboxylase catalyzes the transfer of a

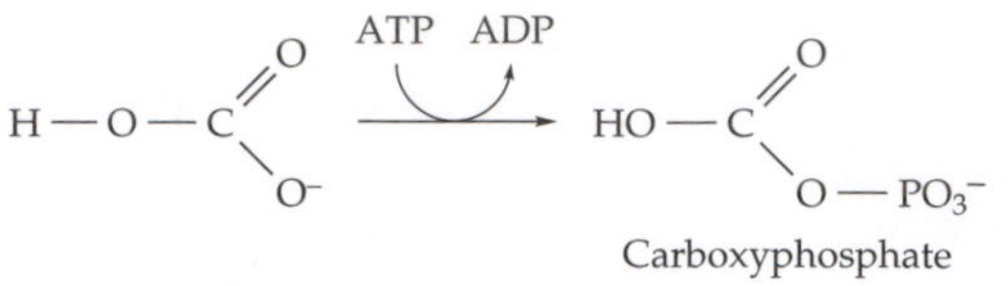

Carboxyphosphate

(14-7)

phospho group to ADP from carbamoyl phosphate, an analog of carboxyphosphate in a reaction that is analogous to the reverse of that in Eq. 14-7 and also by a slow bicarbonate-dependent ATPase activity that does not depend upon biotin.[57,58,58a]

The simplest mechanism for transfer of the carboxyl group of carboxyphosphate to biotin would appear to be nucleophilic displacement of the phosphate leaving group by N1' of biotin. The enzyme could presumably first catalyze removal of the N1' hydrogen to form a ureido anion.[59] Another reasonable possibility would be for the terminal phospho group of ATP to be transferred to biotin to form an *O*-phosphate[60,61]

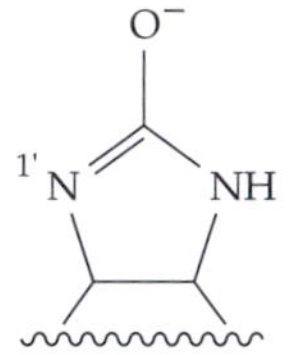

Ureido anion

(Eq. 14-8, step *a*) which could react with bicarbonate as in Eq. 14-8, step *b*. Cleavage of the enol phosphate by attack of HCO_3^- would simultaneously create a nucleophilic center at N1' and carboxyphosphate ready to react with N1'. However, this could not easily explain the ATPase activity in the absence of biotin.

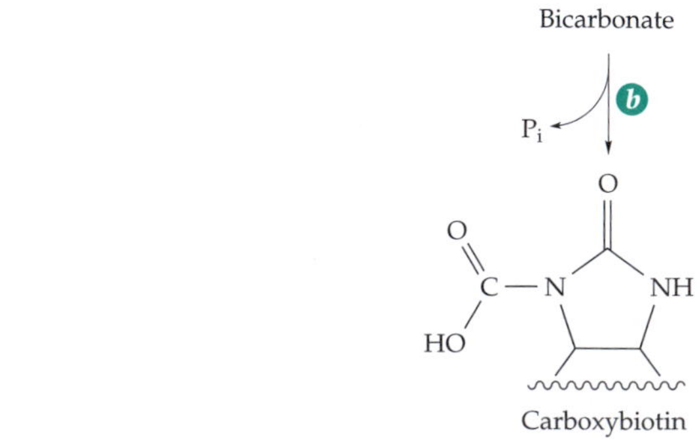

Bicarbonate

Carboxybiotin

(14-8)

Either of the foregoing mechanisms requires that the ureido anion of biotin attack the rather unreactive carbon atom of carboxyphosphate. Another alternative, which is analogous to that suggested for PEP carboxylase (Eq. 13-53) is for carboxyphosphate to eliminate inorganic phosphate to give the more electrophilic CO_2 (Eq. 14-9, step a). The very basic inorganic phosphate trianion PO_4^{3-} that is eliminated could remove the proton from N1' of biotin to create the biotin ureido anion (step b), which could then add to CO_2 (step c).[22]

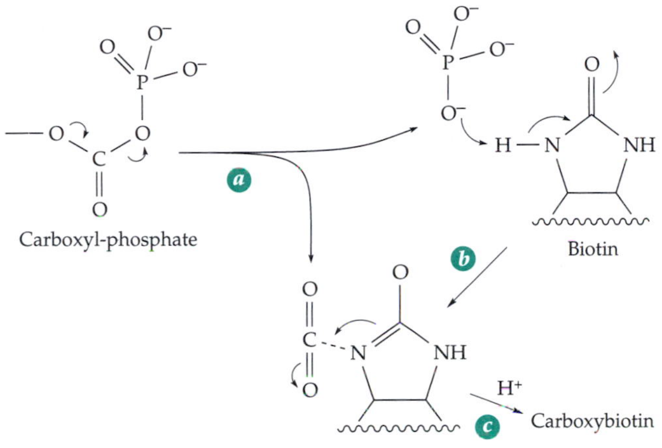

(step c), driving the reaction to completion. The observed transfer of isotope from ^{18}O-containing bicarbonate into ADP would be observed.

The β carboxylation step. Once formed, the carboxybiotin "head group" could swing to the carboxyltransferase site where transfer of the carboxyl group into the final product takes place. This might occur either by nucleophilic attack of an enolate anion on the carbonyl carbon (Eq. 14-11) or on CO_2 generated by reversal of the reactions of Eq. 14-9, steps b and c.

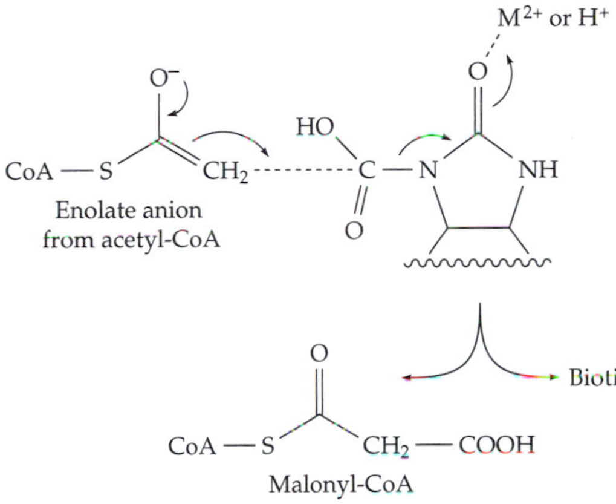

(14-9)

(14-11)

Have we checked all of the possibilities for the mechanism of biotin carboxylation? Kruger and associates[62,63] suggested that biotin, as a ureido anion, might add to bicarbonate to form a highly unstable intermediate which, however, could be phosphorylated by ATP (Eq. 14-10, steps a and b). This intermediate could undergo elimination of inorganic phosphate

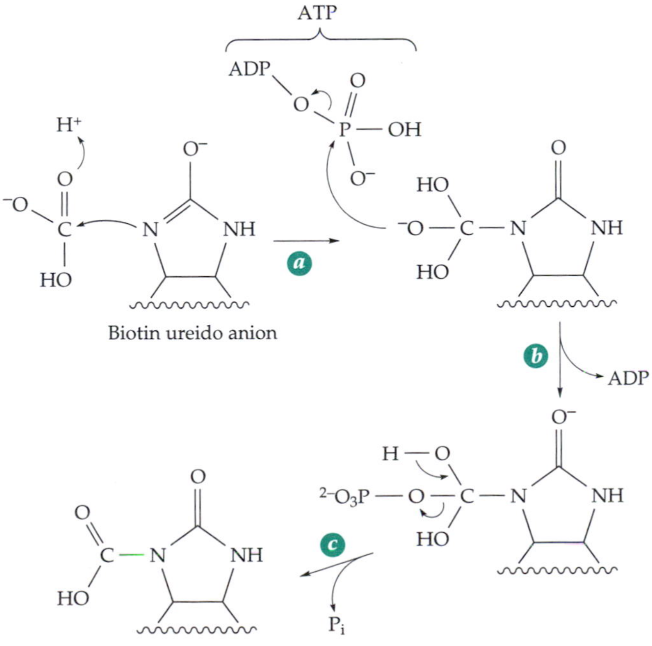

When pyruvate with a chiral methyl group is carboxylated by pyruvate carboxylase the configuration at C-3 is retained. The carboxyl enters from the 2-*si* side, the same side from which the proton (marked H*) was removed to form the enolate anion (Eq. 14-12). Comparable stereochemistry has been established for other biotin-dependent enzymes.[64,65]

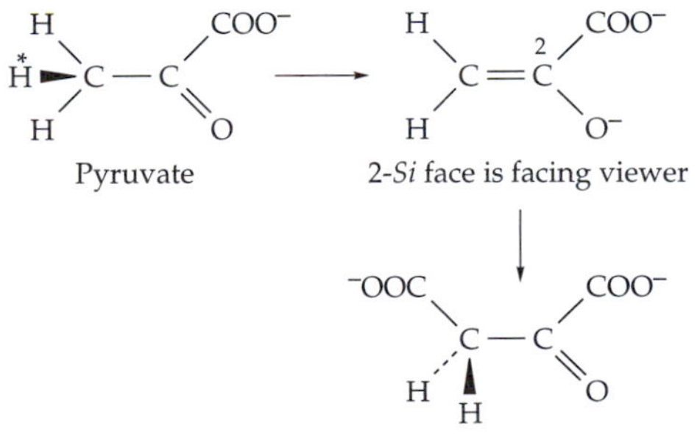

Biotin ureido anion

Carboxybiotin

(14-10)

(14-12)

These enzymes do not catalyze any proton exchange at C-3 of pyruvate or at C-2 of an acyl-CoA unless the biotin is first carboxylated. This suggested that removal of the proton to the biotin oxygen and carboxylation might be synchronous. However, ^{13}C and ^2H kinetic isotope effects and studies of ^3H exchange[66] support the existence of a discrete enolate anion intermediate as shown in Eq. 14-11.[65,67] This mechanism is also consistent with the observation that propionyl-CoA

BOX 14-B THE BIOTIN-BINDING PROTEINS AVIDIN AND STREPTAVIDIN

A biochemical curiosity is the presence in egg white of the glycoprotein **avidin**.[a,b] Each 68-kDa subunit of this tetrameric protein binds one molecule of biotin tenaciously with $K_f \sim 10^{15}$ M^{-1}. Nature's purpose in placing this unusual protein in egg white is uncertain. Perhaps it is a storage form of biotin, but it is more likely an antibiotic that depletes the environment of biotin. A closely similar protein **streptavidin** is secreted into the culture medium by *Streptomyces avidinii*.[c] Its sequence is homologous to that of avidin. It has a similar binding constant for biotin and the two proteins have similar three-dimensional structures.[a,d–j] Biotin binds at one end of a β barrel formed from antiparallel strands and is held by multiple hydrogen bonds and a conformational alteration that allows a peptide loop to close over the bound vitamin.

Historically, avidin was important to the discovery of biotin. The bonding between avidin and biotin is so tight that inclusion of raw egg white in the diet of animals is sufficient to cause a severe biotin deficiency. Avidin has also been an important tool to enzymologists interested in biotin-containing enzymes. Avidin invariably inhibits these enzymes and inhibition by avidin is diagnostic of a biotin-containing protein. Recently, avidin and streptavidin have found widespread application in affinity chromatography, in immunoassays, and in the staining of cells and tissues.[d,k–p] These uses are all based on the ability to attach biotin covalently to side chain groups of proteins, polysaccharides, and other substances. The carboxyl group of the biotin "arm," which lies at the surface of the complex with avidin or streptovidin, can be converted to any of a series of reactive derivatives. For example, p-nitrophenyl or N-hydroxysuccinimide esters of biotin can be used to attach biotin to amino groups of proteins to form biotinylated proteins.

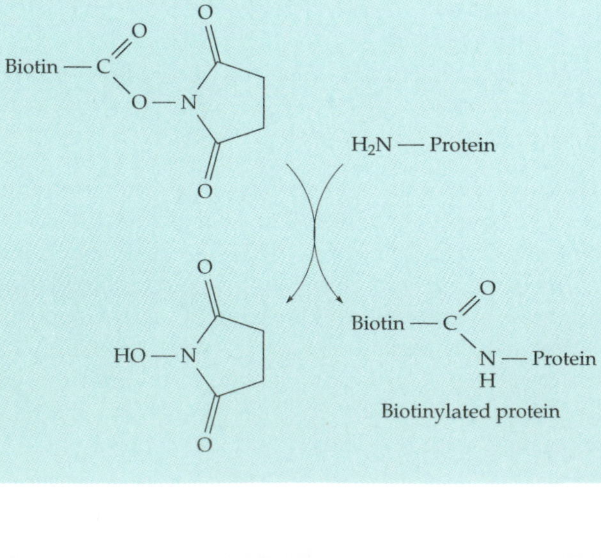

Other reactive derivatives can be used to attach biotin to phenolic, thiol, or carbonyl groups.

The affinity of avidin or streptavidin for the resulting biotinylated materials is still very high. This fact has permitted the application of this "biotin–avidin technology" to numerous aspects of research and diagnostic medicine. For example, a specific antibody (IgG) can be utilized for immunoassay of a hormone or other ligand. A second antibody, specific for the IgG–ligand complex, can be produced in a biotinylated form. A steptavidin complex of a biotinylated enzyme such as alkaline phosphatase, β-galactosidase, or horseradish peroxidase, for which a sensitive colorimetric assay is available, is allowed to react with the biotin·anti-IgG·IgG·ligand complex. The streptavidin now releases enzyme in proportion to the amount of ligand originally present. Since the biotinylated anti IgG and streptavidin·biotinylated enzyme can be stored as a stable mixture, the assay is simple and fast.

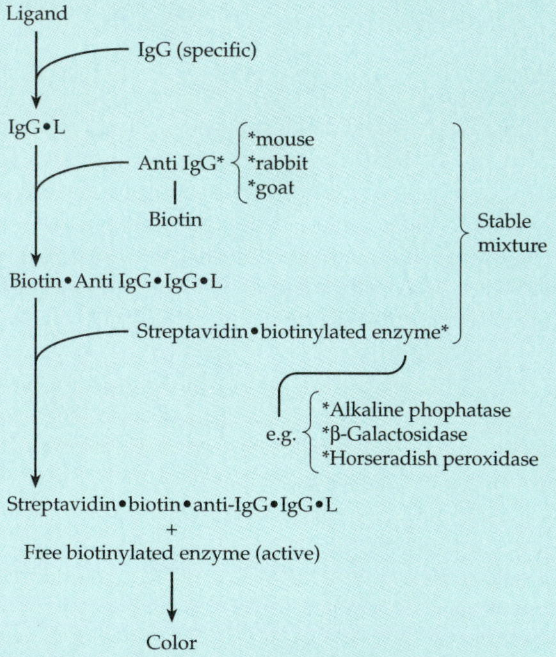

Avidin technology can also be applied to the isolation of proteins and other materials from cells. Because the irreversibility of the binding of biotin may be a problem, photocleavable biotin derivatives have been developed.[q] In the following structure, the biotin derivative has been joined to a protein (as in the first equation in this box) and is ready for separation, perhaps on a column containing immobilized avidin or streptavidin. After separation the biotin together with the linker and photocleavable group are cut off by a short irradiation with ultra-

Biotinylated protein

BOX 14-B (continued)

violet light, leaving the protein in a free form.

Photocleavable group

a Livnah, O., Bayer, E. A., Wilshek, M., and Sussman, J. L. (1993) Proc. Natl. Acad. Sci. U.S.A. 90, 5076–5080

b Pugliese, L., Coda, A., Malcovati, M., and Bolognesi, M. (1993) J. Mol. Biol. 231, 698–710

c Meslar, H. W., Camper, S. A., and White, H. B., III. (1978) J. Biol. Chem. 253, 6979–6982

d Wilchek, M., and Bayer, E. A. (1989) Trends Biochem. Sci. 14, 408–412

e Punekar, N. S., and Lardy, H. A. (1987) J. Biol. Chem. 262, 6714–6719

f Henderson, W. A., Pähler, A., Smith, J. L., Satow, Y., Merritt, E. A., and Phizackerley, R. P. (1989) Proc. Natl. Acad. Sci. U.S.A. 86, 2190–2194

g Weber, P. C., Pantoliano, M. W., and Thompson, L. D. (1992) Biochemistry 31, 9350–9354

h Schmidt, T. G. M., Koepke, J., Frank, R., and Skerra, A. (1996) J. Mol. Biol. 255, 753–766

i Weber, P. C., Ohlendorf, D. H., Wendoloski, J. J., and Salemme, F. R. (1989) Science 243, 85–88

j Sano, T., and Cantor, C. R. (1995) Proc. Natl. Acad. Sci. U.S.A. 92, 3180–3184

k Bayer, E. A., and Wilchek, M. (1980) Meth. Biochem. Anal. 26, 1–46

l Childs, G. V., Naor, Z., Hazum, F., Tibolt, R., Westlund, K. N., and Hancock, M. B. (1983) J. Histochem. Cytochem. 31, 1422–1425

m Wilchek, M., and Bayer, E. A., eds. (1990) Methods in Enzymology, Vol. 184, Academic Press, San Diego, California

n Savage, M. D., Mattson, G., Desai, S., Nielander, G. W., Morgensen, S., and Conklin, E. J. (1992) Avidin-Biotin Chemistry: A Handbook, Pierce, Rockford, Illinois

o Donnelson, J. E., and Wu, R. (1972) J. Biol. Chem. 247, 4661–4668

p Laundon, C. H., and Griffith, J. D. (1987) Biochemistry 26, 3759–3762

q Olejnik, J., Sonar, S., Krzymańska-Olejnik, E., and Rothschild, K. J. (1995) Proc. Natl. Acad. Sci. U.S.A. 92, 7590–7594

carboxylase and transcarboxylase both catalyze elimination of HF from β-fluoropropionyl-CoA to form the unsaturated acrylyl-CoA.[67] The elimination presumably occurs via an enolate anion intermediate as in Eq. 13-28.

A bound divalent metal ion, usually Mn^{2+}, is required in the transcarboxylation step. A possible function is to assist in enolization of the carboxyl acceptor. However, measurement of the effect of the bound Mn^{2+} on ^{13}C relaxation times in the substrate for pyruvate carboxylase indicated a distance of ~0.7 nm between the carbonyl carbon and the Mn^{2+}, too great for direct coordination of the metal to the carbonyl oxygen.[68] Another possibility is that the metal binds to the carbonyl of biotin as indicated in Eq. 14-11. Pyruvate carboxylase utilizes two divalent metal ions and at least one monovalent cation.[68a]

What is the role of the sulfur atom in biotin? Perhaps it interacts with CO_2, helping to hold it in a correct orientation for reaction. Perhaps it helps to keep the ureido ring of biotin planar, or perhaps it has no special function.[69]

3. Control Mechanisms

Most pyruvate carboxylases of animal and of yeast are allosterically activated by acetyl-CoA, but those of bacteria are usually not. The enzyme from chicken liver has almost no activity in the absence of acetyl-CoA, which appears to increase greatly the rate of formation of carboxyphosphate and to slow the side reaction by which carboxyphosphate is hydrolyzed to bicarbonate and phosphate.[70] The acetyl-CoA carboxylases of rat or chicken liver aggregate in the presence of citrate to form ~8000-kDa rods. Citrate is an allosteric activator for this enzyme but it acts only on a phosphorylated form and the primary control mechanism.[71,72] This enzyme in plants is a target for a group of herbicides that are selectively toxic to grasses.[73]

4. Pumping Ions with the Help of Biotin

Biotin-dependent **decarboxylases** act as sodium ion pumps in *Klebsiella*[74] and in various anaerobes.[22,75] For example, oxaloacetate is converted to pyruvate and bound carboxybiotin.[74,74a] The latter is decarboxylated to CO_2 at the same time that two Na^+ ions are transported from the inside to the outside of the cell. The function of this pump, like that of the Na^+, K^+-ATPase (Fig. 8-25) is to provide an electrochemical gradient that drives the transport of other ions and molecules through the membrane. Similar ion pumps are operated by decarboxylation of methylmalonyl-CoA[76] and glutaconyl-CoA.[77] Yeast (*Saccharomyces cerevisiae*) cannot make biotin and requires an unusually large amount of the vitamin when urea, allantoin, allantoic acid, and certain other compounds are supplied as the sole source of nitrogen for growth. The reason is that in this organism urea must first be carboxylated by the biotin-containing **urea carboxylase**[78] (see Eq. 24-25) before it can be hydrolyzed to NH_3 and CO_2.

Thiamin diphosphate

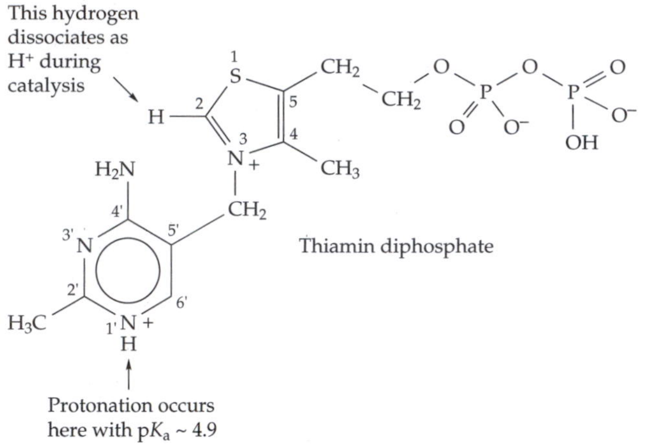

This hydrogen dissociates as H^+ during catalysis

Protonation occurs here with $pK_a \sim 4.9$

D. Thiamin Diphosphate

In Chapter 13 we considered the breaking of a bond between two carbon atoms, one of which is also bonded to a carbonyl group. These β cleavages are catalyzed by simple acidic and basic groups of the protein side chains. On the other hand, the decarboxylation of 2-oxo acids (Eq. 14-13) and the cleavage and formation of α-hydroxyketones (Eq. 14-14) depend upon thiamin diphosphate (**TDP**).[79-85] These reactions represent a second important method of making and breaking carbon–carbon bonds which we will designate **α condensation and α cleavage**. The common feature of all thiamin-catalyzed reactions is that the bond broken (or formed) is *immediately adjacent to the carbonyl group*, not one carbon removed, as in β cleavage reactions. No simple acid–base catalyzed mechanisms can be written; hence the need for a coenzyme.

$$R-\overset{\overset{\displaystyle O}{\|}}{C}-COO^- + H^+ \longrightarrow R-\overset{\overset{\displaystyle O}{\|}}{C}H + CO_2 \tag{14-13}$$

$$R-\overset{\overset{\displaystyle HO}{|}}{\underset{\underset{\displaystyle H}{|}}{C}}-\overset{\overset{\displaystyle O}{\|}}{C}-R' \longrightarrow R-\overset{\overset{\displaystyle O}{\|}}{C}-H + H-\overset{\overset{\displaystyle O}{\|}}{C}-R' \tag{14-14}$$

1. Chemical Properties of Thiamin

The weakly basic portion of thiamin or of its coenzyme forms is protonated at low pH, largely on N-1 of the pyrimidine ring.[86-88] The pK_a value is ~ 4.9. In basic solution, thiamin reacts in two steps with an opening of the thiazole ring (Eq. 14-15) to give the anion of a thiol form which may be crystallized as the sodium salt.[79,84] This reaction, like the competing reaction described in Eq. 7-19, and which leads to a yellow

unstable form of the thiamin anion, is an example of a cooperative two-proton dissociation with linked structural changes. A very low concentration of the intermediate "pseudobase" is present during the titration. This property, which is unusual among small molecules, was instrumental in leading Williams *et al.* to the correct structure for the vitamin.[89] A still unanswered question is, What biological significance is associated with these reactions? Perhaps the thiol form depicted in Eq. 14-15 or the "yellow form" (Eq. 7-19) becomes attached to active sites of some proteins through disulfide linkages.

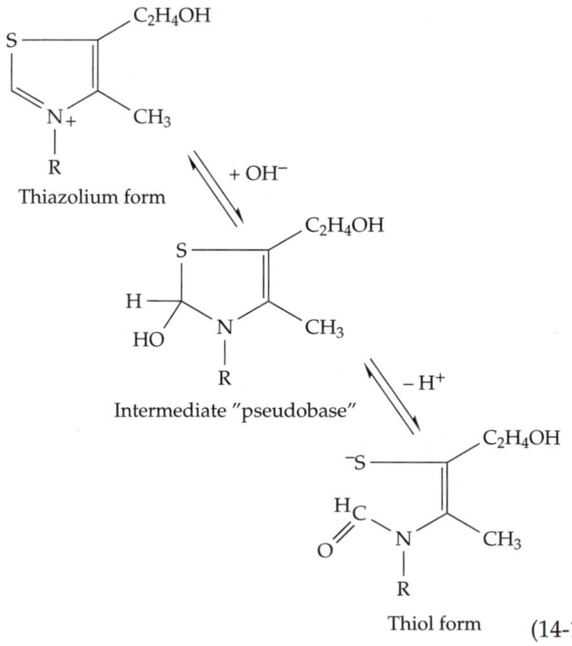

Thiazolium form

$+ OH^-$

Intermediate "pseudobase"

$-H^+$

Thiol form   (14-15)

Thiamin is unstable at high pH[90,91] and is destroyed by the cooking of foods under mildly basic conditions. The thiol form undergoes hydrolysis and oxidation by air to a disulfide. The tricyclic form (Eq. 7-19) is oxidized to **thiochrome**, a fluorescent compound whose formation from thiamin by treatment with alkaline hexacyanoferrate (III) (ferricyanide; Eq. 14-16) is the basis of a much used fluorimetric assay.

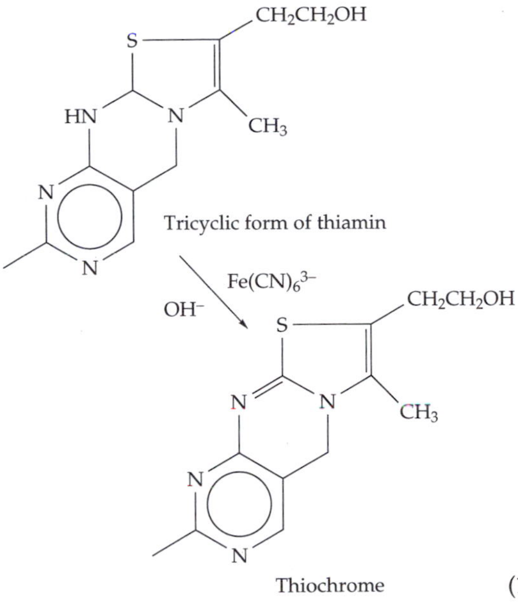

Tricyclic form of thiamin

Thiochrome   (14-16)

Treatment of thiamin with boiling 5N HCl deaminates it to the hydroxy analogue **oxythiamin**, a potent antagonist. **Pyrithiamin**, another competitor containing

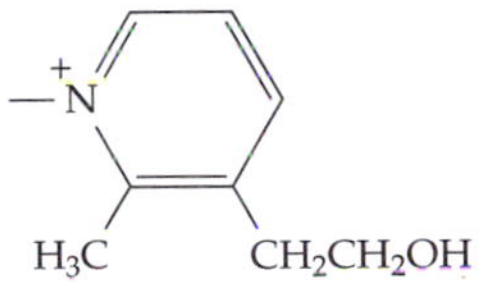

in place of the thiazolium ring, is very toxic especially to the nervous system.

In a solution of sodium sulfite at pH 5, thiamin is cleaved by what appears to be a nucleophilic displacement reaction on the methylene group to give the free thiazole and a sulfonic acid.

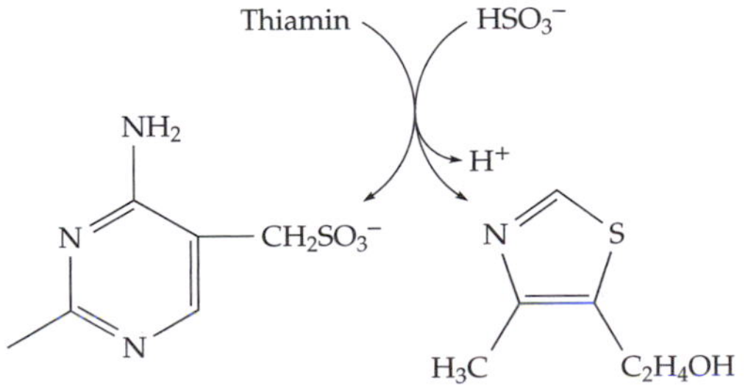

Thiamin   HSO$_3^-$

(14-17)

In fact the mechanism of the reaction is more complex and is evidently initiated by addition of a nucleophile Y, such as $^-$OH or bisulfite, followed by elimination of the thiazole (Eq. 14-18).

A similar cleavage is catalyzed by thiamin-degrading enzymes known as thiaminases which are found in a number of bacteria, marine organisms, and plants. In a bacterial thiaminase, group Y of Eq. 14-18 is a cysteine –SH.[92,92a]

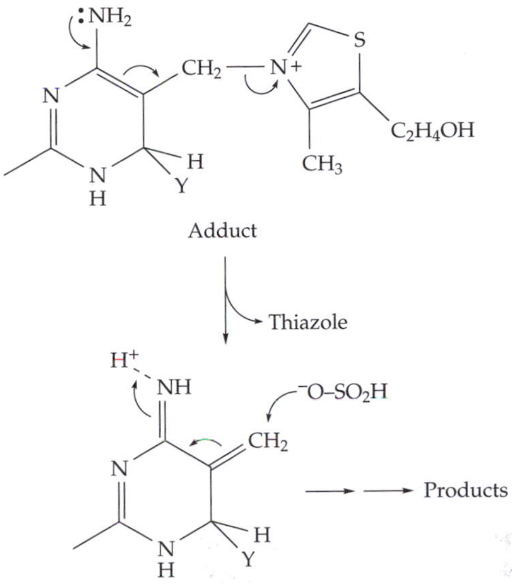

Adduct

Thiazole

Products   (14-18)

Thiamin is synthesized in bacteria, fungi, and plants from 1-deoxyxylulose 5-phosphate (Eq. 25-21), which is also an intermediate in the nonmevalonate pathway of polyprenyl synthesis. However, thiamin diphosphate is a coenzyme for synthesis of this intermediate (p. 736), suggesting that an alternative pathway must also exist. Each of the two rings of thiamin is formed separately as the esters 4-amino-5-hydroxymethylpyrimidine diphosphate and 4-methyl-5-(β-hydroxyethyl) thiazole monophosphate. These precursors are joined with displacement of pyrophosphate to form thiamin monophosphate.[92b] In eukaryotes this is hydrolyzed to thiamin, then converted to thiamin diphosphate by transfer of a diphospho group from ATP.[92b,c] In bacteria thiamin monophosphate is converted to the diphosphate by ATP and thiamin monophosphate kinase.[92b]

2. Catalytic Mechanisms

The first real clue to the mechanism of thiamin-dependent cleavage came in about 1950 when Mizuhara showed that at pH 8.4 thiamin catalyzes the nonenzymatic conversion of pyruvate into acetoin (Eq. 14-19).[93] Following Mizuhara's lead, Breslow investigated the same reaction using the then new NMR method.[94] He made the surprising discovery that the hydrogen atom in the 2 position of the thiazolium ring, between the sulfur and the nitrogen atoms, exchanged easily with deuterium of ^2H_2O. The pK$_a$ of this proton has been estimated as ~ 18, low enough to permit rapid

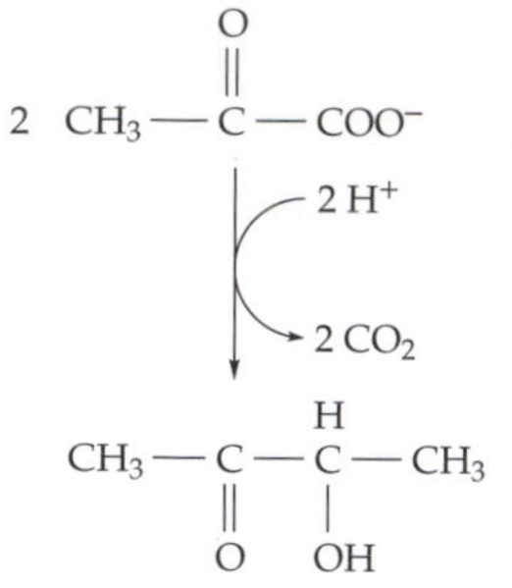

$$2 \quad CH_3 - \overset{\overset{\displaystyle O}{\|}}{C} - COO^-$$

(14-19)

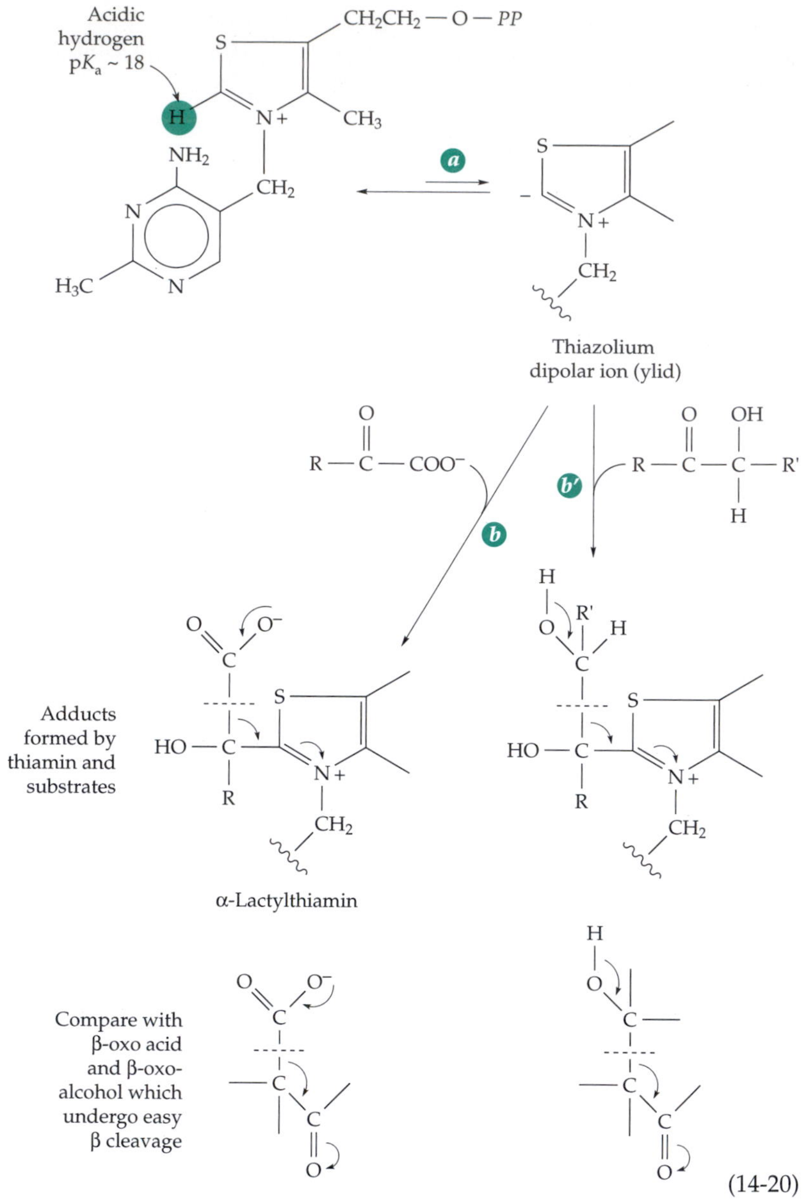

$$CH_3 - \overset{\overset{\displaystyle H}{|}}{\underset{\overset{\|}{O}}{C}} - \overset{}{\underset{OH}{C}} - CH_3$$

dissociation and replacement with ^2H.[84,95,96] The resulting **thiazolium dipolar ion** (or **ylid**) formed by this dissociation (Eq. 14-20, step *a*) is stabilized by the electrostatic interaction of the adjacent positive and negative charges. Breslow suggested that this dipolar ion is the key intermediate in reactions of thiamin-dependent enzymes.[94,97] The anionic center of the dipolar ion can react with a substrate such as an 2-oxo acid or 2-oxo alcohol by addition to the carbonyl group (Eq. 14-20, step *b* or 14-20, step *b'*). The resulting adducts are able to undergo cleavage readily, as indicated by the arrows showing the electron flow toward the =N⁺– group.

Below the structures of the adducts in Eq. 14-20 are those of a 2-oxo acid and a β-ketol with arrows indicating the electron flow in decarboxylation and in the aldol cleavage. The similarities to the thiamin-dependent cleavage reaction are especially striking if one remembers that in some aldolases and decarboxylases the substrate carbonyl group is first converted to an N-protonated Schiff base before the bond cleavage.

We see that *the essence of the action of thiamin diphosphate as a coenzyme is to convert the substrate into a form in which electron flow can occur from the bond to be broken into the structure of the coenzyme.* Because of this alteration in structure, a bond breaking reaction that would not otherwise have been possible occurs readily. To complete the catalytic cycle, the electron flow has to be reversed again. The thiamin-bound cleavage product (an enamine) from either of the adducts in Eq. 14-20 can be reconverted to the thiazolium dipolar ion and an aldehyde as shown in step *b* of Eq. 14-21 for decarboxylation of pyruvate to acetaldehyde.

The adducts α-lactylthiamin and α-lactylthiamin diphosphate have both been synthesized.[84,98–100] As long as α-lactylthiamin is kept as a dry solid or at low pH, it is stable. However, it decarboxylates readily in neutral solution (Eq. 14-21). Decarboxylation is much

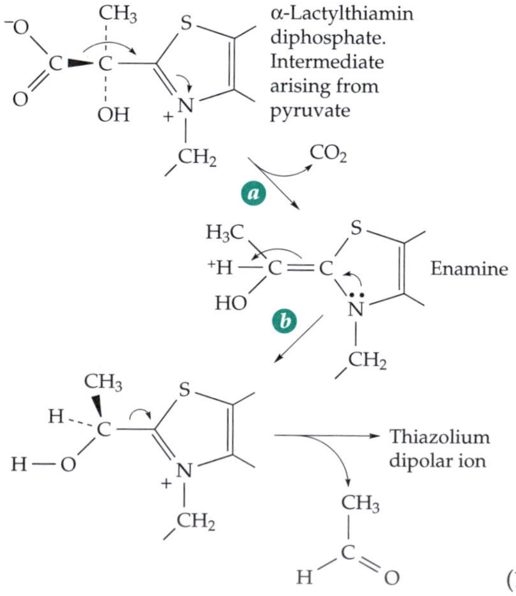

(14-20)

(14-21)

more rapid in methanol, a fact that was predicted by Lienhard and associates.[101] They suggested that decarboxylation is easier in a solvent of low polarity because the transition state has a lower polarity than does lactylthiamin. An enzyme could assist the reaction by providing a relatively nonpolar environment.

The crystal structures of thiamin-dependent enzymes (see next section) as well as modeling[102,103] suggest that lactylthiamin pyrophosphate has the conformation shown in Eq. 14-21. If so, it would be formed by the addition of the ylid to the carbonyl of pyruvate in accord with stereoelectronic principles, and the carboxylate group would also be in the correct orientation for elimination to form the enamine in Eq. 14-21, step b.[82–83a] A transient 380- to 440-nm absorption band arising during the action of pyruvate decarboxylase has been attributed to the enamine.

What is the role of the pyrimidine portion of the coenzyme in these reactions? The pyrimidine ring has a large inductive effect on the basicity of the thiazolium nitrogen and may increase the rate of dissociation of the C-2 proton somewhat.[104] More significant is the fact that the $-NH_2$ group is properly placed to function as a basic catalyst in the generation of the thiazolium dipolar ion. However, the amino group of thiamin is not very basic ($pK_a \sim 4.9$) and the site of protonation at low pH is largely N-1 of the pyrimidine ring. In the protonated form the $-NH_2$ group is even less basic because of electron withdrawal into the ring. Studies of thiamin analogs suggested another possibility. Schellenberger[105] found pyruvate decarboxylase inactive when TDP was substituted by analogs with modified aminopyrimidine rings, e.g., with methylated or dimethylated amino groups or with N1 of the ring replaced by carbon (an aminopyridyl analog). More recently the experiment has been repeated with additional enzymes[106] and X-ray studies have shown that the analogs bind into the active site of transketolase in a normal way.[107] Of the compounds studied *only an aminopyridyl analog of TDP having a nitrogen atom at 1' (but CH at 3') had substantial catalytic activity*. Jordan and Mariam showed that N-1'-methylthiamin is a superior catalyst in non-enzymatic catalysis.[108] These results are consistent with the speculative scheme illustrated in the following drawing from the first edition of this book.[109] The $-NH_2$ group of the N1-protonated aminopyrimidine has lost a proton to form a normally *minor tautomer* in which the resulting imino group would be quite basic. Assisted by a basic group from the protein, it could abstract the proton from the thiazolium ring to form the ylid.

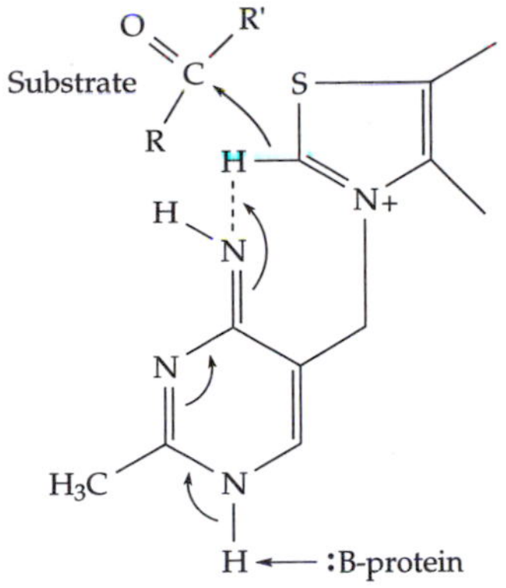

Crystallographic studies show that the catalytic base (:B-protein) is the carboxylate group of a conserved glutamate side chain (E59' in Fig. 14-2). Kern *et al.* used NMR spectroscopy of thiamin diphosphate present in native and mutant pyruvate decarboxylase and transketolase to monitor the exchange rates of the C2-H proton of the thiazolium ring.[109a] The results confirmed the importance of the conserved glutamate side chain for dissociation of the C2-H proton. Participation of other catalytic groups from the enzyme may also be important. However, these groups are not conserved in the whole family of enzymes. For example, glutamine 122, which is within hydrogen-bonding distance of both the substrate and thiamin amino group, is replaced by histidine in transketolase.[110] A variety of kinetic studies involving mutants,[111,111a] alternative substrates,[112,113] and isotope effects in substrates[114–116] and solvent[117,118] have not yet resolved the details of the proton transfers that occur within the active site.

3. Structures of Thiamin-Dependent Enzymes

By 1998, X-ray structures had been determined for four thiamin diphosphate-dependent enzymes: (1) a bacterial pyruvate oxidase,[119,120] (2) yeast and bacterial pyruvate decarboxylases,[121–122c] (3) transketolase,[110,123,124] and (4) benzoylformate decarboxylase.[124a] The reactions catalyzed by these enzymes are all quite different, as are the sequences of the proteins. However, the thiamin diphosphate is bound in a similar way in all of them. A conserved pattern of hydrogen bonds holds the diphosphate group to the protein and also provides ligands to a metal ion. This is normally Mg^{2+}, which is held in nearly perfect octahedral coordination by two phosphate oxygen atoms, a conserved aspartate carboxylate, a conserved asparagine amide, and a water molecule. The thiamin rings are in a less polar region. The amino group of the pyrimidine is adjacent to the 2–CH of the thiazole and N1' of the pyrimidine is apparently protonated and hydrogen bonded to the carboxylate group of a conserved glutamate side chain. This is shown in Fig. 14-2. Substitution of the corresponding glutamate 51 of yeast pyruvate decarboxylase by glutamine or alanine greatly reduced or eliminated catalytic activity.[125]

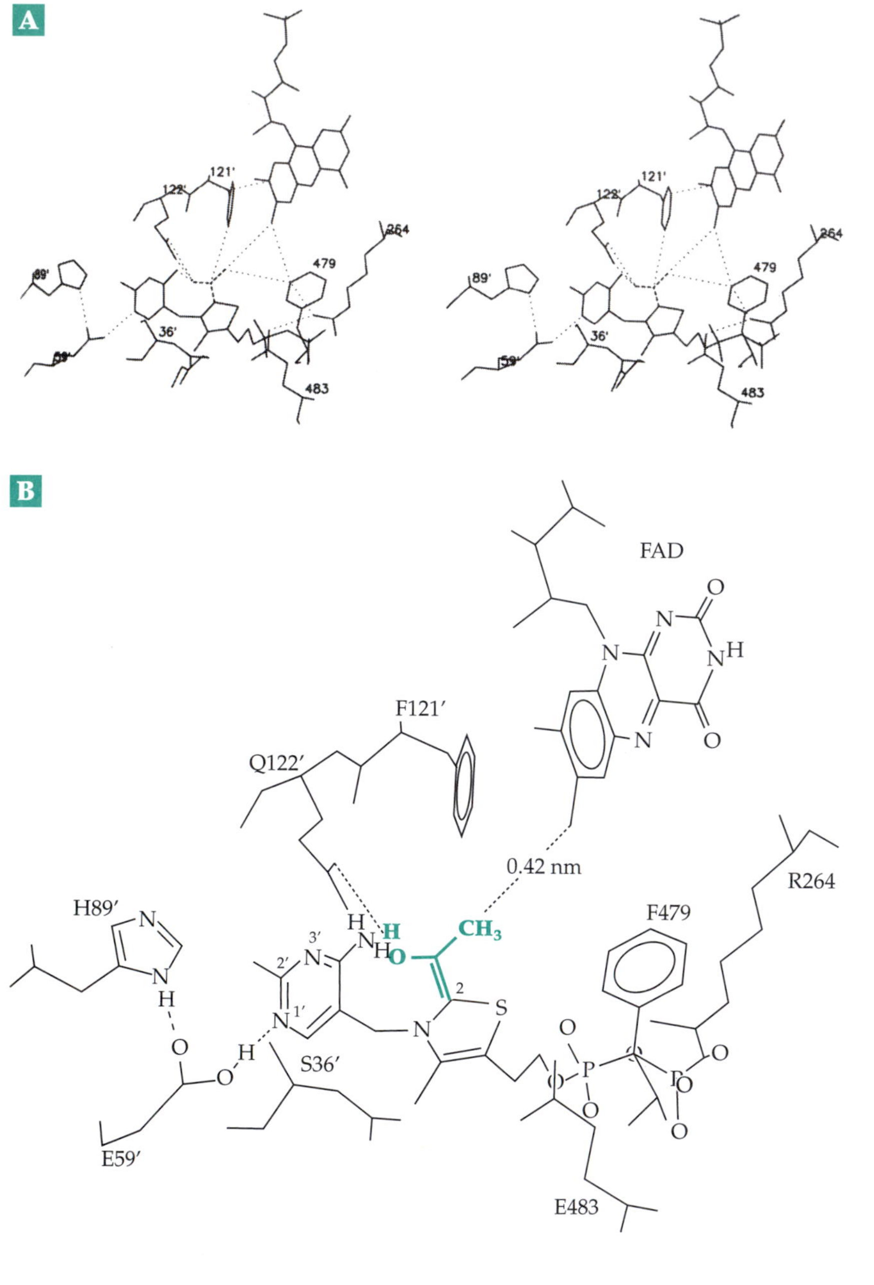

4. The Variety of Enzymatic Reactions Involving Thiamin

Most known thiamin diphosphate-dependent reactions (Table 14-2) can be derived from the five half-reactions, *a* through *e*, shown in Fig. 14-3. Each half-reaction is an α cleavage which leads to a thiamin- bound enamine (center, Fig. 14-3) The decarboxylation of an α-oxo acid to an aldehyde is represented by step *b* followed by *a* in reverse. The most studied enzyme catalyzing a reaction of this type is yeast **pyruvate decarboxylase,** an enzyme essential to alcoholic fermentation (Fig. 10-3). There are two ~250-kDa isoenzyme forms, one an α$_4$ tetramer and one with an (αβ)$_2$ quaternary structure. The isolation of α-hydroxyethylthiamin diphosphate from reaction mixtures of this enzyme with pyruvate[52] provided important verification of the mechanisms of Eqs. 14-14, 14-15. Other decarboxylases produce aldehydes in specialized metabolic pathways: indolepyruvate decarboxylase[126] in the biosynthesis of the plant hormone **indole-3-acetate** and benzoylformate decarboxylase in the mandelate pathway of bacterial metabolism (Chapter 25).[124a,127]

Formation of α-ketols from α-oxo acids also starts with step *b* of Fig. 14-3 but is followed by condensation with another carbonyl compound in step *c*, in reverse. An example is decarboxylation of pyruvate and condensation of the resulting active acetaldehyde with a second pyruvate molecule to give *R*-α-acetolactate, a reaction catalyzed by **acetohydroxy acid synthase** (acetolactate synthase).[128] Acetolactate is the precursor to valine and leucine. A similar ketol condensation, which is catalyzed by the same synthase, is

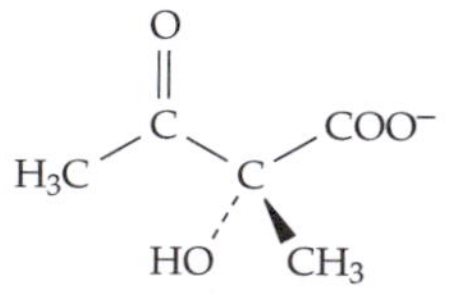

Figure 14-2 (A) Stereoscopic view of the active site of pyruvate oxidase from the bacterium *Lactobacillus plantarium* showing the thiamin diphosphate as well as the flavin part of the bound FAD. The planar structure of the part of the intermediate enamine that arises from pyruvate is shown by dotted lines. Only some residues that may be important for catalysis are displayed: G35', S36', E59', H89', F121', Q122', R264, F479, and E483. Courtesy of Georg E. Schulz.[119] (B) Simplified view with some atoms labeled and some side chains omitted. The atoms of the hypothetical enamine that are formed from pyruvate, by decarboxylation, are shown in green.

required in the biosynthesis of isoleucine (Fig. 24-17). Since this

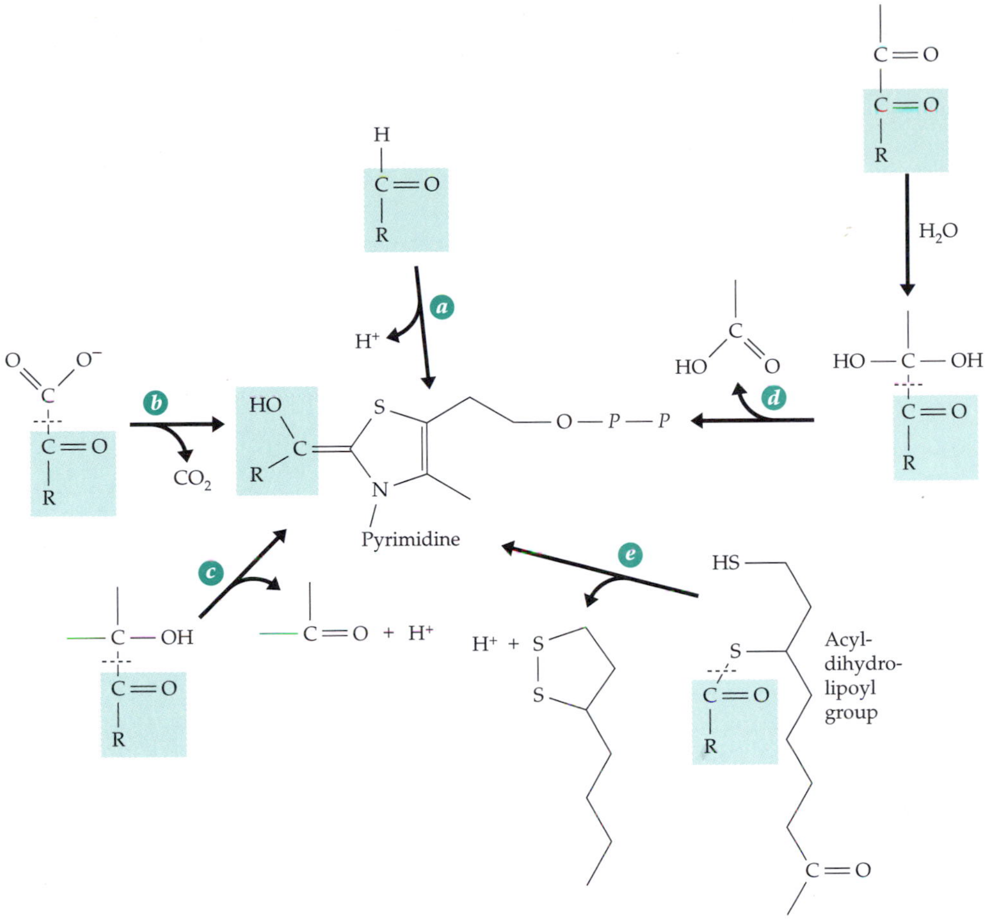

Figure 14-3 Half-reactions making up the thiamin-dependent α cleavage and α condensation reactions.

TABLE 14-2
Enzymes Dependent upon Thiamin Diphosphate as a Coenzyme

1. Nonoxidative
    Pyruvate decarboxylase*
    Indolepyruvate decarboxylase
    Benzoylformate decarboxylase*
    Glyoxylate carboligase
    Acetohydroxy acid synthase (acetolactate synthase)
    1-Deoxy-D-xylulose 5-phosphate synthase
    Transketolase*
    Phosphoketolase

2. Oxidative decarboxylase
    Pyruvate oxidase (FAD)*
    Pyruvate dehydrogenase (Lipoyl, FAD, NAD+)
       multienzyme complex
    Pyruvate:ferredoxin oxidoreductase
    Indolepyruvate:ferredoxin oxidoreductase

---

* Three-dimensional structures for these enzymes had been determined by 1998.

synthase is not present in mammals it is a popular target for herbicides.[129,130] It is inhibited by many of the most widely used herbicides including sulfometuron methyl, whose structure is shown here.

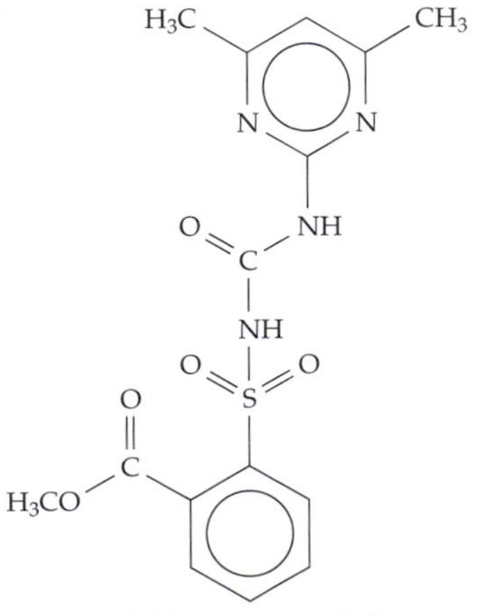

Sulfometuron methyl

Acetolactate is a β-oxo acid and is readily decarboxylated to acetoin, a reaction of importance in bacterial fermentations (Eq. 17-26). Acetoin, of both R and S

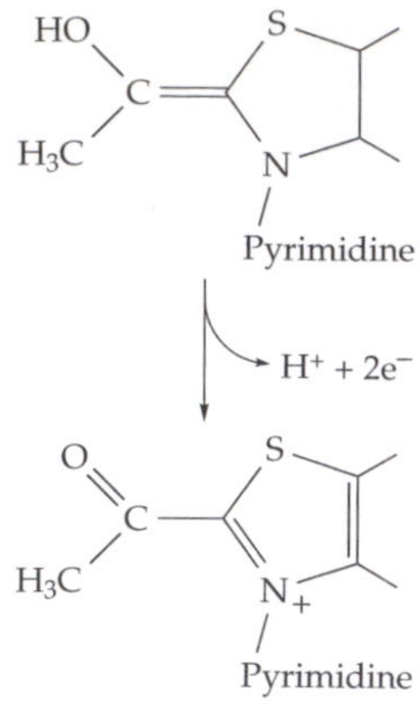

2-Acetylthiamin diphosphate   (14-22)

organisms. This reaction is usually formulated as the reverse of step e of Fig. 14-3, which shows the cleavage of an **acyl-dihydrolipoyl** derivative. However, there is a possibility that the lipoyl group functions not as shown in Fig. 14-3 but as an oxidant that converts the TDP enamine to 2-acetylthiamin diphosphate (Eq. 14-22)[134,135] and only after that as an acyl group carrier. A related reaction that is known to proceed through acetyl-TDP is the previously mentioned bacterial pyruvate oxidase. As seen in Fig. 14-2, this enzyme has its own oxidant, FAD, which is ready to accept the two electrons of Eq. 14-22 to produce bound acetyl-TDP. The electrons may be able to jump directly to the FAD, with thiamin and flavin radicals being formed at an intermediate stage.[135a] The electron transfers as well as other aspects of oxidative decarboxylation are discussed in Chapter 15, Section C.

A reaction that is related to that of transketolase but is likely to function via acetyl-TDP is **phosphoketolase**, whose action is required in the energy metabolism of some bacteria (Eq. 14-23). A product of phosphoketolase is acetyl phosphate, whose cleavage can be coupled to synthesis of ATP. Phosphoketolase presumably catalyzes an α cleavage to the thiamin-containing enamine shown in Fig. 14-3. A possible mechanism of formation of acetyl phosphate is elimination of H_2O from this enamine, tautomerization to 2-acetylthiamin, and reaction of the latter with inorganic phosphate.

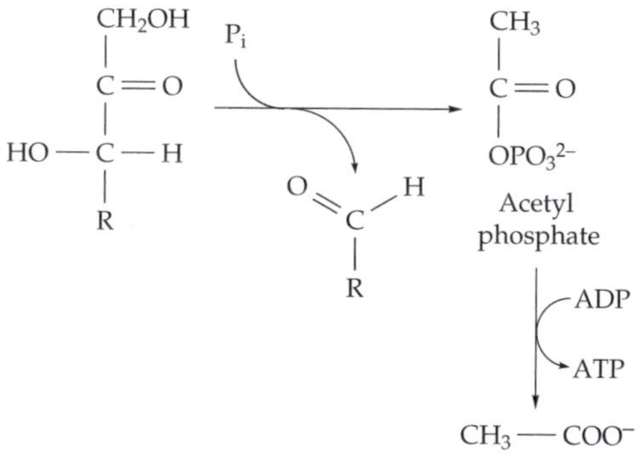

configurations, is also formed by pyruvate decarboxylases acting on acetaldehyde.[103,131] The ketol condensation of two molecules of glyoxylate with decarboxylation to form tartronic semialdehyde (see Fig. 17-6) is an important reaction in bacterial metabolism. It is catalyzed by **glyoxylate carboligase,**[132] another thiamin diphosphate-dependent enzyme. Formation of 1-deoxy-D-xylulose 5-phosphate, an intermediate in the nonmevalonate pathway of isoprenoid synthesis, is formed in a thiamin diphosphate-catalyzed condensation of pyruvate with glyceraldehyde 3-phosophate (Fig. 22-2)[132a] However, there is an unresolved problem. As previously mentioned, the same intermediate is thought to be a precursor to thiamin diphosphate (Eq. 25-21). This suggests the presence of an alternative pathway.

Ketols can also be formed enzymatically by cleavage of an aldehyde (step a, Fig. 14-3) followed by condensation with a second aldehyde (step c, in reverse). An enzyme utilizing these steps is **transketolase** (Eq. 17-15),[132b] which is essential in the pentose phosphate pathways of metabolism and in photosynthesis. α-Diketones can be cleaved (step d) to a carboxylic acid plus active aldehyde, which can react either via a or c in reverse. These and other combinations of steps are often observed as side reactions of such enzymes as pyruvate decarboxylase. A related thiamin-dependent reaction is that of pyruvate and acetyl-CoA to give the α-diketone, **diacetyl**, $CH_3COCOCH_3$.[133] The reaction can be viewed as a displacement of the CoA anion from acetyl-CoA by attack of thiamin-bound active acetaldehyde derived from pyruvate (reverse of step d, Fig. 14-3 with release of CoA).

(14-23)

5. Oxidative Decarboxylation and 2-Acetylthiamin Diphosphate

The oxidative decarboxylation of pyruvate to form acetyl-CoA or acetyl phosphate plays a central role in the metabolism of our bodies and of most other

6. Thiamin Coenzymes in Nerve Action

The striking paralysis caused by thiamin deficiency together with studies of thiamin analogs as metabolites suggested a special action for this vitamin in nerves.[136] The thiamin analog **pyrithiamin** (p. 731) both induces paralytic symptoms and displaces thiamin from nerve preparations. The nerve poison **tetrodotoxin** (Chapter 30) blocks nerve conduction by inhibiting inward diffusion of sodium, but it also promotes release of thiamin from nerve membranes. Evidence for a metabolic significance of thiamin triphosphate comes from identification of soluble and membrane-associated thiamin triphosphatases[137] as well as a kinase that forms protein-bound thiamin triphosphate in the brain.[138] Mono-, tri-, and tetraphosphates also occur naturally in smaller amounts. One might speculate about a possible role for the rapid interconversion of cationic and yellow anionic forms of thiamin via the tricyclic form (Eq. 7-19) in some aspect of nerve conduction.

E. Pyridoxal Phosphate

The phosphate ester of the aldehyde form of vitamin B$_6$, **pyridoxal phosphate** (pyridoxal-P or PLP), is required by many enzymes catalyzing reactions of amino acids and amines. The reactions are numerous, and pyridoxal phosphate is surely one of nature's most versatile catalysts. The story begins with biochemical **transamination**, a process of central importance in nitrogen metabolism. In 1937, Alexander Braunstein and Maria Kritzmann, in Moscow, described the transamination reaction by which amino groups can be transferred from one carbon skeleton to another.[139,140] For example, the amino group of glutamate can be transferred to the carbon skeleton of oxaloacetate to form aspartate and 2-oxoglutarate (Eq. 14-24).

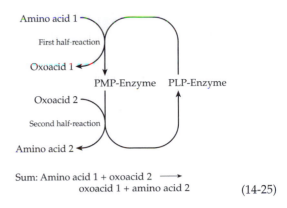

$$\text{Sum: Amino acid 1 + oxoacid 2} \longrightarrow \text{oxoacid 1 + amino acid 2} \qquad (14\text{-}25)$$

At about the same time, Gunsalus and coworkers noticed that the activity of **tyrosine decarboxylase** produced by lactic acid bacteria was unusually low when the medium was deficient in pyridoxine. Addition of pyridoxal plus ATP increased the decarboxylase activity of cell extracts.[146] PLP was synthesized and was found to be the essential coenzyme for this and a variety of other enzymes.[147]

1. Nonenzymatic Models

Pyridoxal or PLP, in the complete absence of enzymes, not only undergoes slow transamination with amino acids but also catalyzes many other reactions of amino acids that are identical to those catalyzed by PLP-dependent enzymes. Thus, *the coenzyme itself can be regarded as the active site of the enzymes* and can be studied in nonenzymatic reactions. The latter can be thought of as *models* for corresponding enzymatic reactions. From such studies Snell and associates drew the following conclusions.[148]

a. The aldehyde group of PLP reacts readily and reversibly with amino acids to form Schiff bases (Fig. 14-4) which react further to give products.

b. For an aldehyde to be a catalyst, a strong electron-attracting group, e.g., the ring nitrogen of pyridine (as in PLP), must be *ortho* or *para* to the –CHO group. A nitro group, also strongly electron attracting, can replace the pyridine nitrogen in model reactions.

This transamination reaction is a widespread process of importance in many aspects of the nitrogen metabolism of organisms. A large series of **transaminases** (**aminotransferases**), for which glutamate is most often one of the reactants, have been shown to catalyze the reactions of other oxoacids and amino acids.[141–143]

In 1944, Esmond Snell reported the nonenzymatic conversion of pyridoxal into pyridoxamine (Box 14-C) by heating with glutamate. He recognized that this was also transamination and proposed that pyridoxal might be a part of a coenzyme needed for aminotransferases and that these enzymes might act via two half-reactions that interconverted pyridoxal and pyridoxamine (Eq. 14-25). The hypothesis was soon verified and the coenzyme was identified as pyridoxal 5'-phosphate or pyridoxamine 5'-phosphate (Fig. 14-5).[144,145]

BOX 14-C THE VITAMIN B₆ FAMILY: PYRIDOXINE, PYRIDOXAL, AND PYRIDOXAMINE

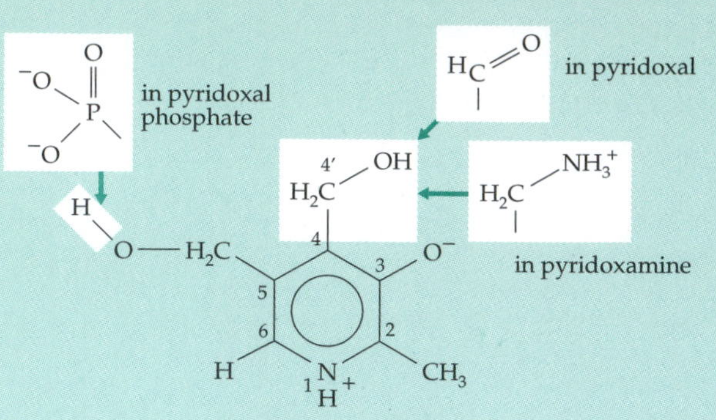

Pyridoxine (pyridoxol)
Vitamin B₆ alcohol

Pyridoxine, the usual commercial form of vitamin B₆, was isolated and synthesized in 1938. However, studies of bacterial nutrition soon indicated the existence of other naturally occurring forms of the new vitamin which were more active than pyridoxine in promoting growth of certain lactic acid bacteria. The amine **pyridoxamine** and the aldehyde **pyridoxal** were identified by Snell, who found that pyridoxal could be formed from pyridoxine by mild oxidation and that pyridoxamine could be formed from pyridoxal by heating in a solution with glutamic acid via a transamination reaction. These simple experiments also suggested the correct structures of the new forms of vitamin B₆. Animal tissues contain largely pyridoxal, pyridoxamine, and their phosphate esters. The lability of the aldehyde explains the ease of destruction of the vitamin by excessive heat or by light. On the other hand, plant tissues contain mostly pyridoxine, which is more stable. Kinases use ATP to form the phosphate esters, which are interconvertible within cells.[a–e] Pyridoxine 5'-phosphate can be oxidized to PLP[d,e,f] and the latter may undergo transamination to PMP. The acid–base chemistry and tautomerism of pyridoxine were discussed in Chapter 6, Section E,2.

Many poisonous substances as well as useful drugs react with PLP-requiring enzymes. Thus, much of the toxic effect of the "carbonyl reagents" hydroxylamine, hydrazine, and semicarbazide stems from their formation of stable derivatives analogous to Schiff bases with PLP.

H — O — NH₂, R — O — NH₂
Hydroxylamines

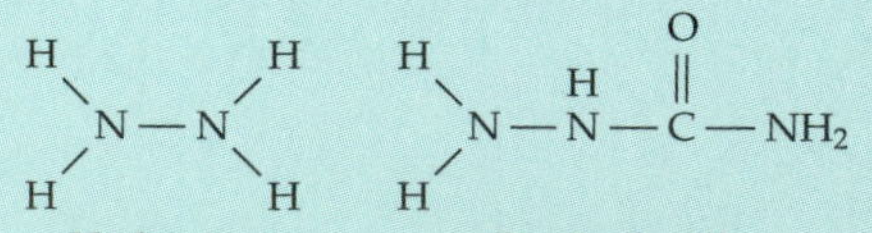

Hydrazine   Semicarbazide

Isonicotinyl hydrazide (INH), one of the most effective drugs against tuberculosis, is inhibitory to pyridoxal kinase, the enzyme that converts pyridoxal to PLP.[c] Apparently, the drug reacts with pyridoxal to form a hydrazone which blocks the enzyme. Pyridoxal kinase is not the primary target of INH in mycobacteria. However, patients on long-term isonicotinyl hydrazide therapy sometimes suffer symptoms of vitamin B₆ deficiency.[g]

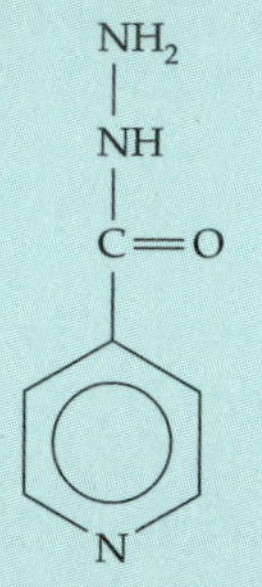

PLP-dependent enzymes are inhibited by a great variety of enzyme-activated inhibitors that react by several distinctly different chemical mechanisms.[h] Here are a few. The naturally occurring **gabaculline** mimics γ-aminobutyrate (Gaba) and inhibits γ-aminobutyrate aminotransferase as well as other PLP-dependent enzymes. The inhibitor follows the normal catalytic pathway as far as the ketimine. There, a proton is lost from the inhibitor permitting formation of a stable benzene ring and leaving the inhibitor stuck in the active site:

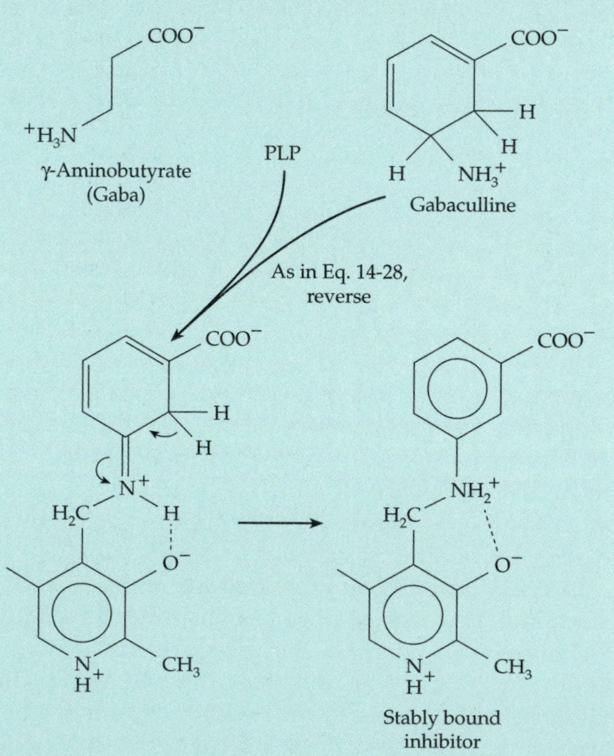

Isonicotinyl hydrazide (isoniazid)

γ-Aminobutyrate (Gaba)

Gabaculline

As in Eq. 14-28, reverse

Stably bound inhibitor

BOX 14-C (continued)

Beta-chloroalanine and serine O-sulfate can undergo β elimination (as in Eq. 14-29) in active sites of glutamate decarboxylase or aspartate aminotransferase. The enzymes then form free aminoacrylate, a reactive molecule that can undergo an aldol-type condensation with the external aldimine to give the following product.[i]

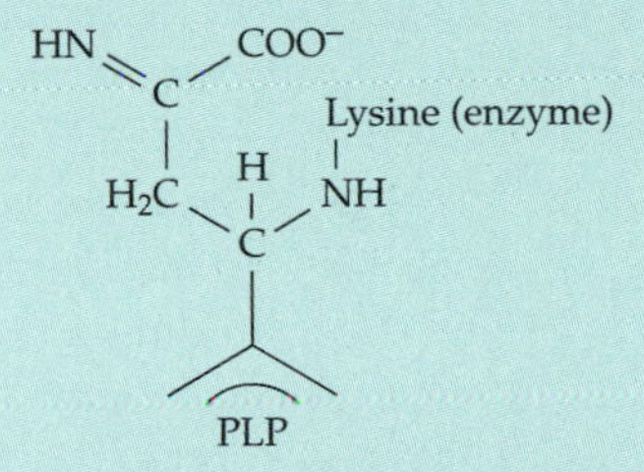

Nucleophilic groups from enzymes can add to double bonds, e.g., in an aminoacrylate Schiff base, or to multiple bonds present in the inhibitor. An example is γ-vinyl γ-aminobutyrate (4-amino-5-hexenoic acid), another inhibitor of brain γ-aminobutyrate aminotransferase which is a useful anticonvulsant drug.

Another enzyme-activated inhibitor is the streptomyces antibiotic **D-cycloserine** (oxamycin), an antitubercular drug that resembles D-alanine in structure. A potent inhibitor of alanine racemase, it also inhibits the non-PLP, ATP-dependent, **D-alanyl-D-alanine synthetase** which is needed in the biosynthesis of the peptidoglycan of bacterial cell walls.

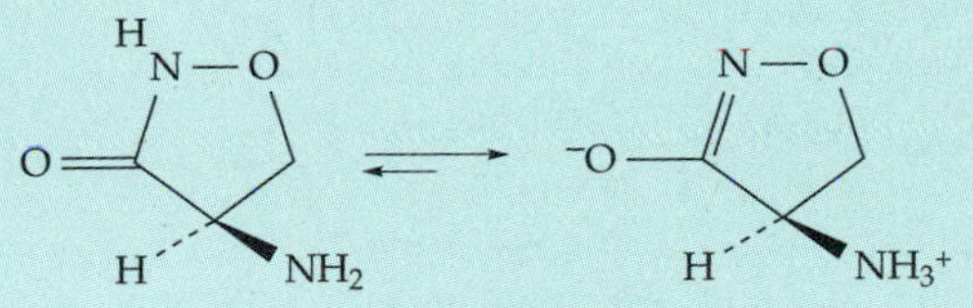

D-Cycloserine

L-Cycloserine inhibits many PLP enzymes and is toxic to humans. This observation led Khomutov *et al.* to synthesize the following more specific "cycloglutamates," structural analogs of glutamic acid with fixed conformations.[j,k] Nature apparently anticipated the synthetic chemist, because it has been reported that the mushroom *Tricholoma muscar-*

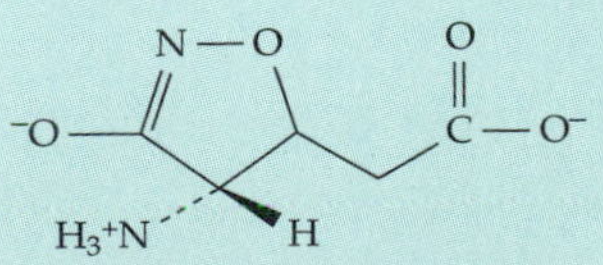

A cycloglutamate that inhibits aspartate aminotransferase

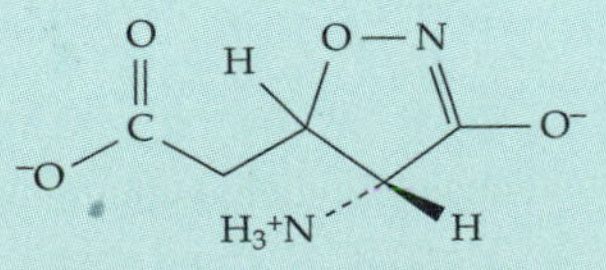

An isomeric cycloglutamate tricholomic acid found in certain mushrooms

ium contains one of the same compounds. It is said to impart two interesting properties to the mushroom: a delicious flavor and a lethal action on flies that alight on the mushroom's surface![l]

The substituted cysteine derivative L-penicillamine causes convulsions and low glutamate decarboxylase levels in the brain, presumably because the Schiff base formed with PLP can then undergo cyclization, the SH group adding to the C=N to form a stable thiazolidine ring.

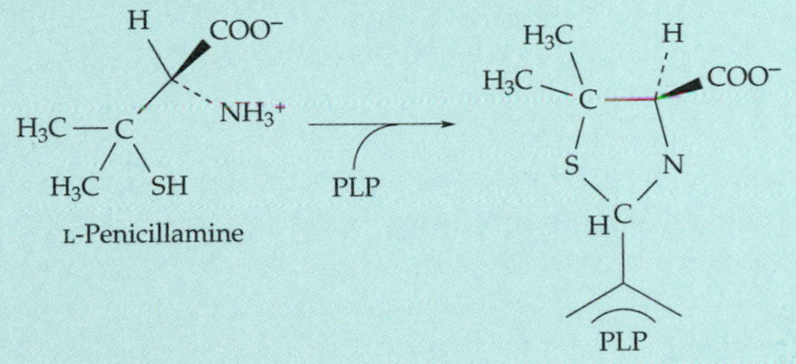

L-Penicillamine

Toxopyrimidine, the alcohol derived from the pyrimidine portion of the thiamin molecule, is a structural analog of pyridoxal. When fed to rats or mice it induces running fits which can be stopped by administration of vitamin B_6. Phophorylation of toxopyrimidine by pyridoxal kinase may produce an antagonistic analog of PLP. In a similar fashion, 4-deoxypyridoxine, which was tested as a possible anticancer drug, caused convulsions and other symptons of vitamin B_6 deficiency in humans. A host of synthetic PLP derivatives have been made, some of which are effective in blocking PLP enzymes.[h]

[a] Lepkovsky, S. (1979) *Fed. Proc.* **38**, 2699–2700

[b] McCormick, D. B., Gregory, M. E., and Snell, E. E. (1961) *J. Biol. Chem.* **236**, 2076–2084

[c] Snell, E. E., and Haskell, B. E. (1970) *Comprehensive Biochemistry* **21**, 47–71

[d] McCormick, D. B. and Chen, H. (1999) *J. Nutr.* **129**, 325–327

[e] Hanna, M. L., Turner, A. J., and Kirkness, E. F. (1997) *J. Biol. Chem.* **272**, 10756–10760

[f] Ngo, E. O., LePage, G. R., Thanassi, J. W., Meisler, N., and Nutter, L. M. (1998) *Biochemistry* **37**, 7741–7748

[g] Lui, A., and Lumeng, L. (1986) in *Vitamin B_6, Pyridoxal Phosphate: Chemical, Biochemical and Medical Aspects*, Vol. 1B (Dolphin, D., Poulson, R., and Avramovíc, O., eds), pp. 601–674, Wiley, New York

[h] Walsh, C. T. (1986) in *Vitamin B_6, Pyridoxal Phosphate: Chemical, Biochemical and Medical Aspects*, Vol. 1B (Dolphin, D., Poulson, R., and Avramovíc, O., eds), pp. 43–70, Wiley, New York

[i] Likos, J. J., Ueno, H., Fedhaus, R. W., and Metzler, D. E. (1982) *Biochemistry* **21**, 4377–4386

[j] Khomutov, R. M., Koveleva, G. K., Severin, E. S., and Vdovina, L. V. (1967) *Biokhim.* **32**, 900–907

[k] Sastchenko, L. P., Severin, E. S., Metzler, D. E., and Khomutov, R. M. (1971) *Biochemistry* **10**, 4888–4894

[l] Iwasaki, H., Kamiya, T., Oka, O., and Veyanagi, J. (1969) *Chem. Pharm. Bull.* **17**, 866–872

c. The presence of an – OH group adjacent to the – CHO group greatly enhances the catalytic activity. Since certain metal ions, such as Cu^{2+} and Al^{3+}, increase the rates in model systems and are known to chelate with Schiff bases of the type formed with PLP, it was concluded that either a metal ion or a proton formed a chelate ring and helped to hold the Schiff base in a planar conformation (Fig. 14-6). *However, such a function for metal ions has not been found in PLP-dependent enzymes.*

d. In model systems the 5-hydroxymethyl and 2-methyl groups are not needed for catalysis. However, in enzymes the 5-CH_2OH group is essential for attachment of the phosphate handle. The 2-CH_3 group is usually not necessary for coenzymatic activity.

Many investigations of nonenzymatic reactions of PLP and related compounds have been and are still being conducted[149,150]

2. A General Mechanism of Action of PLP

Based upon consideration of the various known PLP-dependent enzymes of amino acid metabolism, Braunstein and Shemyakin in 1952 proposed a general mechanism of PLP action[151,152] which, in most details, was the same as the one proposed independently by Snell and associates on the basis of the nonenzymatic reactions.[148] The general mechanism, which has been verified by studies of many enzymes, can be stated as follows: *Pyridoxal phosphate reacts to convert the amino group of a substrate into a Schiff base that is electronically the equivalent of an adjacent carbonyl* (Fig. 14-4). However, a Schiff base of an amino acid with a simple aldehyde (for example, acetaldehyde) has the opposite polarity from that of C=O (see the following structures). Such an imine could not substitute for a carbonyl group in activating an α-hydrogen nor in facilitating C–C bond cleavage in the amino acid. It is necessary to have the strongly electron-attracting pyridine group conjugated with the C=N group in such a way that electrons can flow from the substrate into the coenzyme.

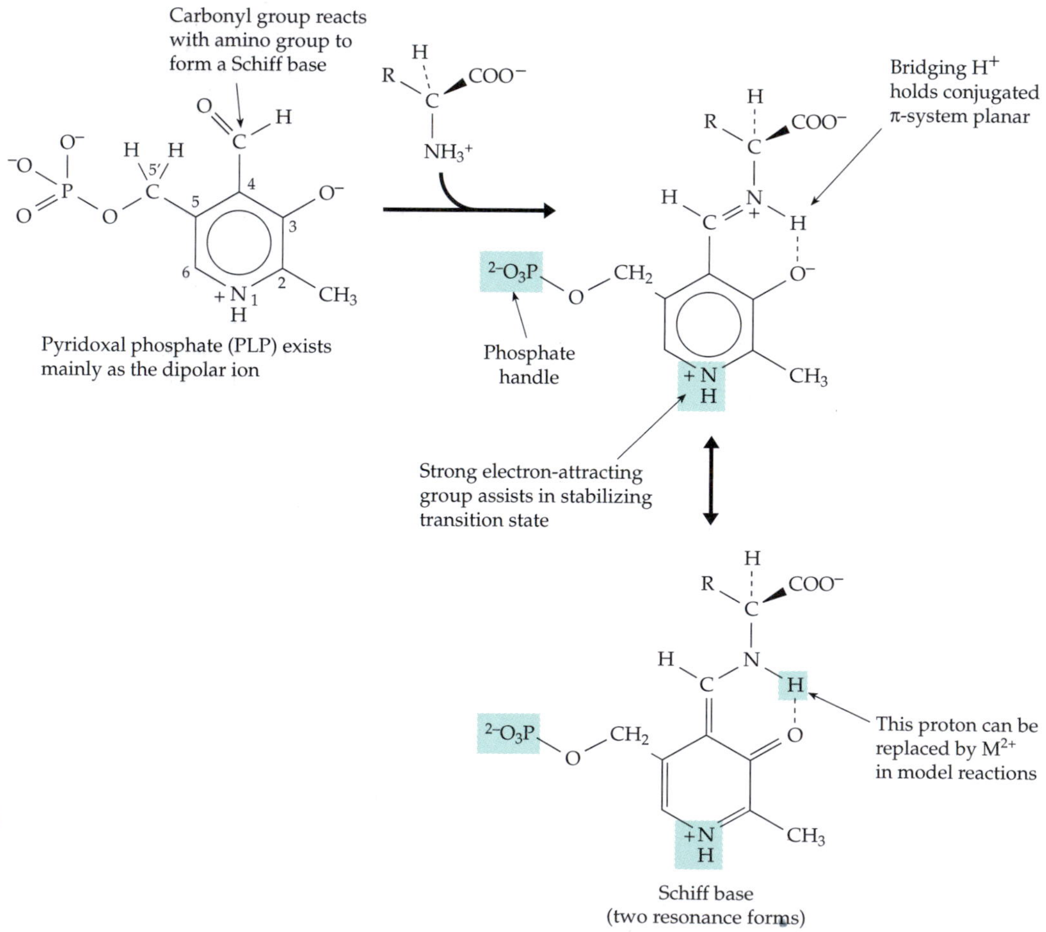

Figure 14-4 Pyridoxal 5'-phosphate (PLP), a special coenzyme for reactions of amino acids.

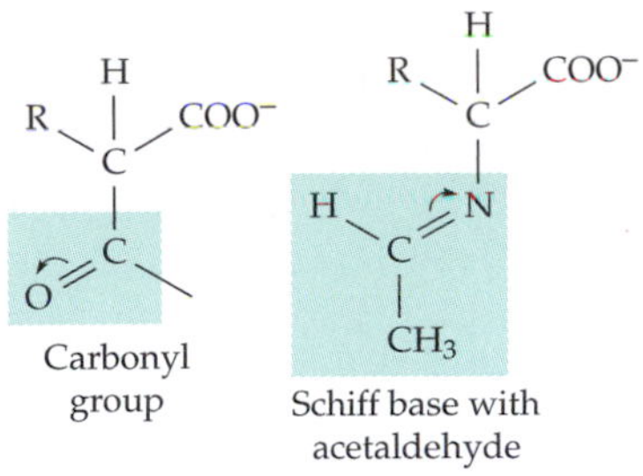

Carbonyl group

Schiff base with acetaldehyde

3. The Variety of PLP-Dependent Reactions

In Fig. 14-5 the reactions of PLP-amino acid Schiff bases are compared with those of β-oxo-acids. Beta-hydroxy-α-oxo acids and Schiff bases of PLP with β-hydroxy-α-amino acids can react in similar ways. The reactions fall naturally into three groups (*a,b,c*) depending upon whether the bond cleaved is from the α-carbon of the substrate to the hydrogen atom, to the carboxyl group, or to the side chain. A fourth group of reactions of PLP-dependent enzymes (*d*) also involve removal of the α-hydrogen but are mechanistically more complex. Some of the many reactions catalyzed by these enzymes are listed in Table 14-3.

Loss of the α-hydrogen (Group a). Dissociation of the α-hydrogen from the Schiff base leads to a **quinonoid–carbanionic intermediate** whose structure in depicted in Fig. 14-5. The name reflects the characteristics of the two resonance forms drawn. Like an enolate anion, this intermediate can react in several ways (1-4).

(1) **Racemization.** A proton can be added back to the original alpha position but without stereospecificity. A racemase which does this is important to bacteria. They must synthesize D-alanine and D-glutamic acid from the corresponding L-isomers for use in formation of their peptidoglycan envelopes.[153–154a] The combined actions of alanine racemase plus D-alanine aminotransferase, which produces D-glutamate as a product, provide bacteria with both D amino acids. A fungal alanine racemase is necessary for synthesis of the immunosuppresant cyclosporin (Box 9-F).[155,155a] High concentrations of free D-alanine are found in certain regions of the brain and also in various glands.[156]

The carboxyl group of an amino acid can also activate the α-hydrogen. This may be the basis for an aspartate racemase and other racemases that are *not* dependent upon PLP.[156–158] See also Chapter 13, Section B,4.

(2) **Cyclization.** A second kind of reaction is represented by the conversion of *S*-adenosylmethionine to **aminocyclopropanecarboxylic acid**, a precursor to the plant hormone **ethylene** (see Chapter 24).[159] The quinonoid intermediate cyclizes with elimination of methylthioadenosine to give a Schiff base of the product (Eq. 14-27).[160–161a] The cyclization step appears to be a simple S_N2-like reaction.[162]

(3) **Transamination.** A proton can add to the carbon attached to the 4 position of the PLP ring (Fig. 14-5) to form a second Schiff base, often referred to as a **ketimine** (Eq. 14-28). The latter can readily undergo hydrolysis to **pyridoxamine phosphate** (PMP) and an α-oxo acid. This sequence represents one of the

Before discussing the reactions of Schiff bases of PLP we should consider one fact that was not known in 1952. PLP is bound into an enzyme's active site as a Schiff base with a specific lysine side chain before a substate binds . This is often called the **internal aldimine**. When the substrate binds it reacts with the internal Schiff base by a two-step process called **transimination** (Eq. 14-26) to form the substrate Schiff base, which is also called the **external aldimine**.

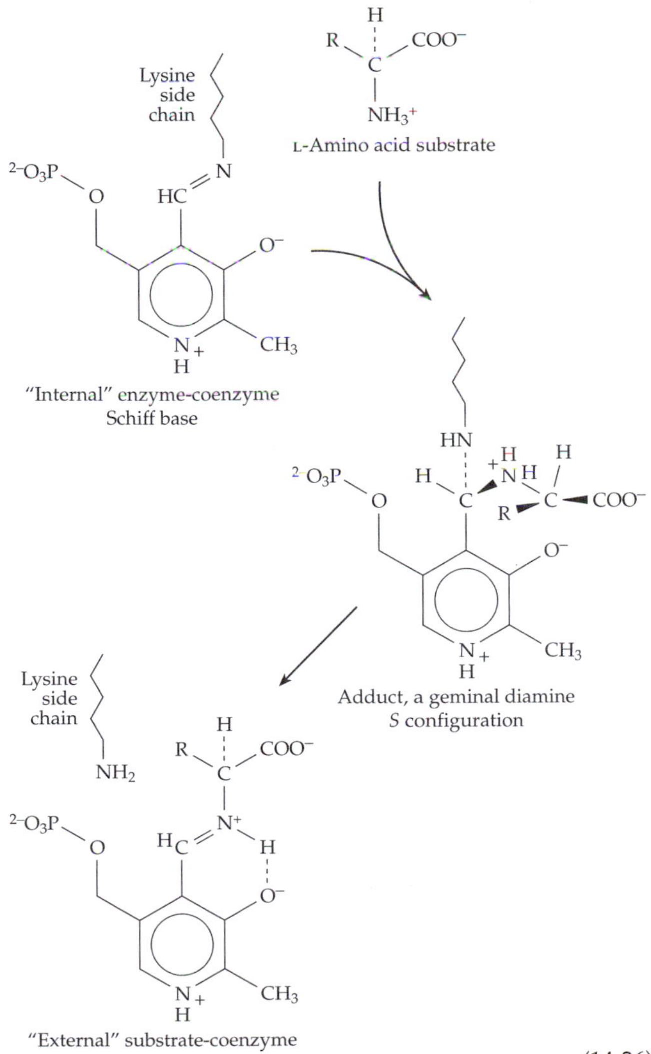

(14-26)

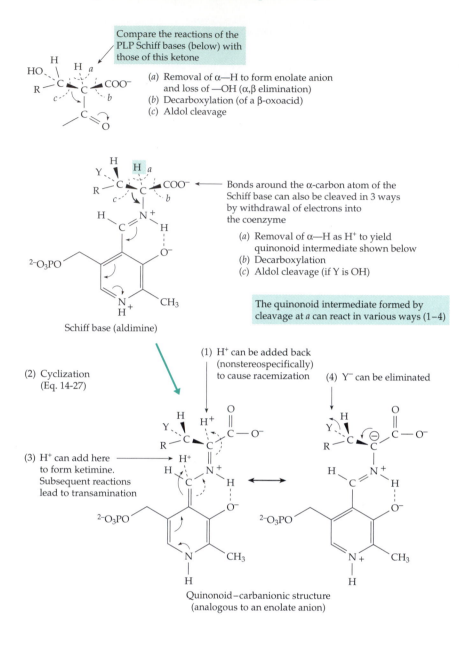

(Chapter 18). The sequences of the two proteins differ greatly, with only 50% of the residues being the same in both isoenzymes. However, these differences are largely on the outside surface, the folding pattern (Fig. 2-6) and internal structure (Fig. 14-6) are almost identical.[142,163–167a] Three-dimensional structures of aspartate aminotransferases of *E. coli*, yeast, chickens, and mammals are extremely similar, even though sequence identity may be as low as 20%.[167a] Most other transaminases also use the L-gutamate–oxoglutarate pair as one of the product–reactant pairs but a few prefer smaller substrates with uncharged side chains. An example is serine: pyruvate (or alanine:glyoxylate) aminotransferase, an important mitochondrial and peroxisomal enzyme in both animals and plants.[168] Other specialized aminotransferases act on aromatic amino acids,[168a] the branched chain amino acids valine, leucine, and isoleucine,[168b,c] and D-amino acids.[168d] Many of them are highly specific for individual amino acids such as phosphoserine,[168e] ornithine,[168f,g] N-acetylornithine,[168h,i] and 8-amino-7-oxononanoate (see banner, p. 719).[168j] An apparently internal transamination, which requires PMP and PLP, converts glutamate-1-semialdehyde into δ-aminolevulinate (see Eq. 24-44) in the pathway of porphyrin biosynthesis used by bacteria and plants.[168k]

(4) Elimination and β replacement. When a good leaving group is present in the β position of the amino acid it can be eliminated (Fig. 14-5, Eq. 14-29).[170] A large number of enzymes catalyze such reactions. Among them are **serine** and **threonine dehydratases,** which eliminate OH⁻ as H_2O;[171–173a] **tryptophan indole-lyase** (tryptophanase) of bacteria, which eliminates indole;[174–176c] **tyrosine phenol-lyase** (elimination of phenol);[177–178a] and **alliinase** of garlic (elimination of 1-propenylsulfenic acid).[179,180] Cystathionine, a precursor to methionine, eliminates L-homocysteine through the action of **cystathionine β lyase** (cystathionase).[181,182] Ammonia is eliminated from the β position of 2,3-diaminopropionate by a bacterial lyase.[183]

Figure 14-5 Some reactions of Schiff bases of pyridoxal phosphate. (*a*) Formation of the quinonoid intermediate, (*b*) elimination of a β substituent, and (*c*) transamination. The quinonoid–carbanionic intermediate can react in four ways (1–4) if enzyme specificity and substrate structure allow.

two half-reactions (Eqs. 14-24 and 14-25) required for enzymatic transamination.

Transaminases participate in metabolism of most of the amino acids, over 60 different enzymes have been identified.[142,163] Best studied are the **aspartate aminotransferases**, a pair of cytosolic and mitochondrial isoenzymes which can be isolated readily from animal hearts. Their presence in heart muscle and brain in high concentration is thought to be a result of their functioning in the malate–aspartate shuttle

TABLE 14-3
Some Enzymes That Require Pyridoxal Phosphate as a Coenzyme

(a) Removing alpha hydrogen as H^+
   (1) Racemization
      Alanine racemase*
   (2) Cyclization
      Aminocyclopropane carboxylate synthase
   (3) Amino group transfer
      Aspartate aminotransferase*
      Alanine aminotransferase
      D-Amino acid aminotransferase*
      Branched chain aminotransferase
      Gamma-aminobutyrate aminotransferase
      ω-Amino acid:pyruvate aminotransferase*
      Tyrosine aminotransferase
      Serine:pyruvate aminotransferase
   (4) Beta elimination or replacement
      D- and L- Serine dehydratases (deaminases)
      Tryptophan indole-lyase (tryptophanase)*
      Tyrosine phenol-lyase*
      Alliinase
      Cystathionine β-lyase (cystathionase)*
      O-Acetylserine sulfhydrylase (cysteine
        synthase)
      Cystathionine β-synthase
      Tryptophan synthase*
(b) Removal of alpha carboxylate as CO_2
      Diaminopimelate decarboxylase
      Glycine decarboxylase (requires lipoyl group)
      Glutamate decarboxylase
      Histidine decarboxylase
      Dopa decarboxylase
      Ornithine decarboxylase*
      Tyrosine decarboxylase
      Dialkylglycine decarboxylase (a decarboxylating
        transaminase)*
(c) Removal or replacement of side chain (or –H) by aldol
   cleavage
      Serine hydroxymethyltransferase
      Threonine aldolase
      δ-Aminolevulinate synthase
      Serine palmitoyltransferase
      2-Amino-3-oxobutyrate-CoA ligase
(d) Reactions of ketimine intermediates
      Aspartate γ-decarboxylase
      Selenocysteine lyase
      Nif S protein of nitrogenase
      Gamma elimination and replacement
        Cystathionine γ-synthase
        Cystathionine γ-lyase
        Threonine synthase
(e) Other enzymes
      Lysine 2,3-aminomutase
      Glycogen phosphorylase*
      Pyridoxamine phosphate (PMP) in synthesis of
        3,6-dideoxy hexoses

---

* The three-dimensional structures of these and other PLP-dependent enzymes were determined by 2000.

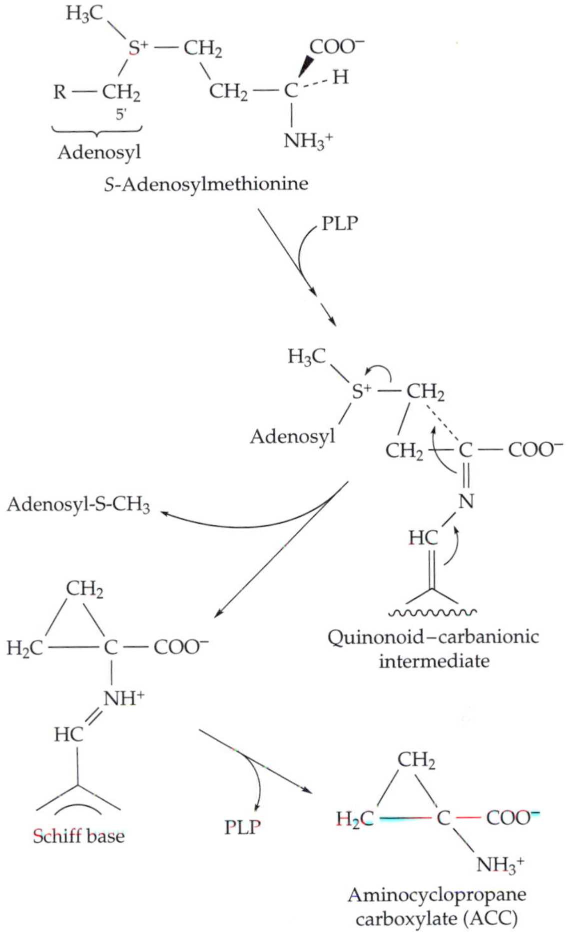

(14-27)

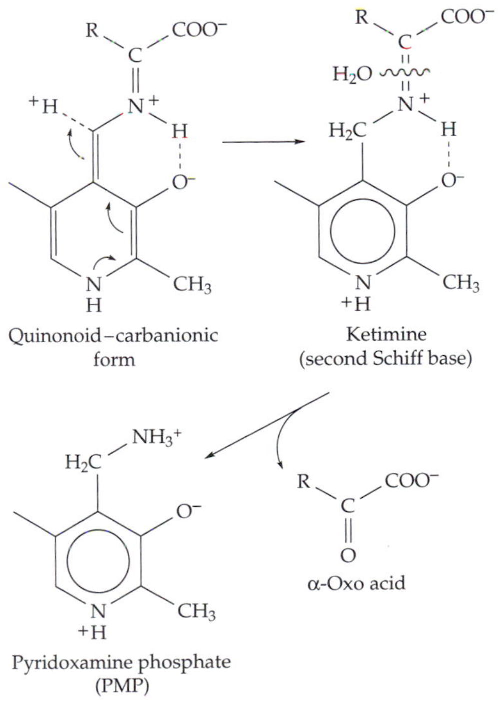

(14-28)

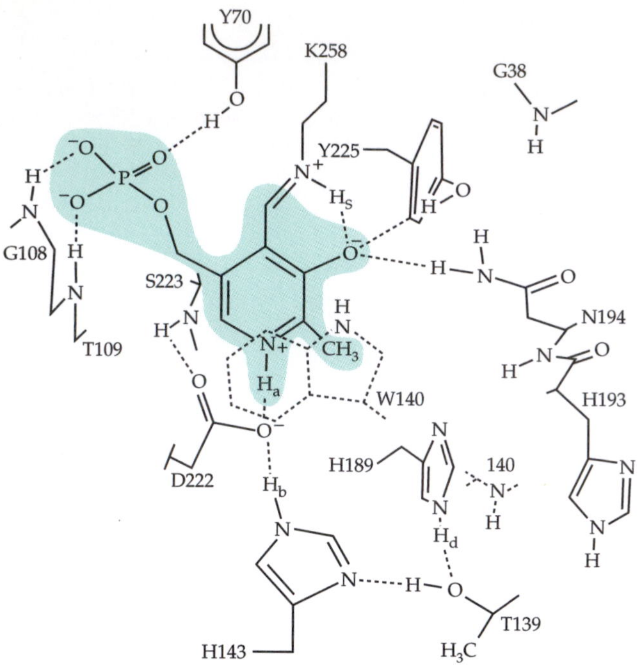

Figure 14-6 Drawing showing pyridoxal phosphate (shaded) and some surrounding protein structure in the active site of cytosolic aspartate aminotransferase. This is the low pH form of the enzyme with an N-protonated Schiff base linkage of lysine 258 to the PLP. The tryptophan 140 ring lies in front of the coenzyme. Several protons, labeled H_a, H_b, and H_d, are represented in ^1H NMR spectra by distinct resonances whose chemical shifts are sensitive to changes in the active site.[169]

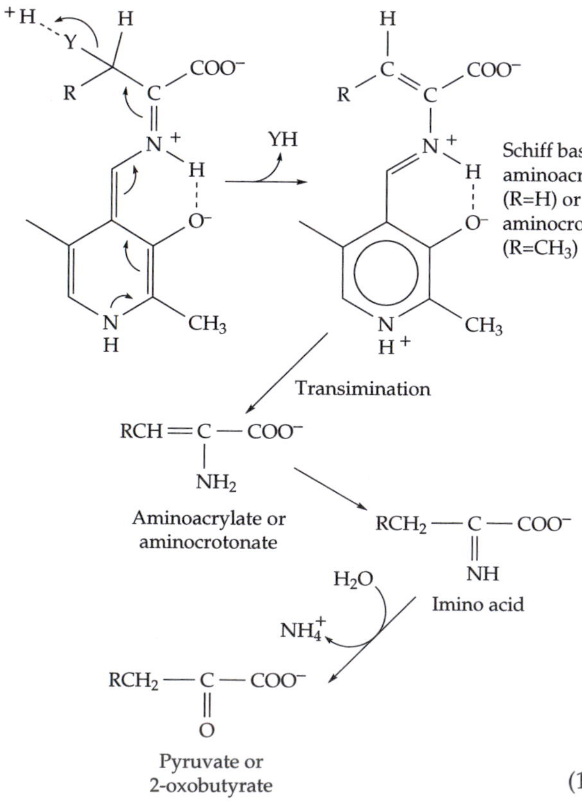

Beta replacement is catalyzed by such enzymes of amino acid biosynthesis as **tryptophan synthase** (Chapter 25),[184] **O-acetylserine sulfhydrylase** (cysteine synthase),[185–186a] and **cystathionine β-synthase** (Chapter 24).[187–188c] In both elimination and β replacement an unsaturated Schiff base, usually of aminoacrylate or aminocrotonate, is a probable intermediate (Eq. 14-29). Conversion to the final products is usually assumed to be via hydrolysis to free aminoacrylate, tautomerization to an imino acid, and hydrolysis of the latter, e.g., to pyruvate and ammonium ion (Eq. 14-29). However, the observed stereospecific addition of a

Quinonoid–carbanionic intermediate with tautomerized indole ring

proton at the β-C atom of 2-oxobutyrate[189] suggests that these steps may occur with the participation of groups from the enzyme. Before indole can be eliminated by tryptophan indolelyase the indole ring must presumably be tautomerized to the following form of the quinonoid intermediate. The same species may be created by tryptophan synthase upon addition of indole to the enzyme-bound aminoacrylate. The green arrows on the structure indicate the tautomerization that would occur to convert the indole ring to the structure found in tryptophan. The three-dimensional structure of tryptophan synthase is shown in Fig. 25-3. It is a complex of two enzymes with a remarkable tunnel through which the intermediate indole can pass.[184,190] An unusual PLP-dependent β replacement is used to synthesize a transfer RNA ester of **selenocysteine** prior to its insertion into special locations in a few proteins (Chapter 15, Section G, and Chapter 29).

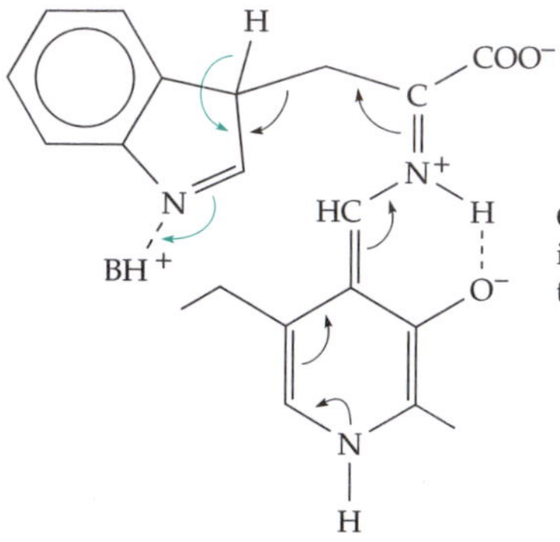

Decarboxylation (Group b). The bond to the carboxyl group of an amino acid substrate is broken in reactions catalyzed by **amino acid decarboxylases**.[191,192] These also presumably lead to a transient quinonoid-carbanionic intermediate. Addition of a proton at the original site of decarboxylation followed by breakup of the Schiff base completes the sequence. Decarboxylation of amino acids is nearly irreversible and frequently appears as a final step in synthesis of amino compounds. For example, in the brain glutamic acid is decarboxylated to **γ-aminobutyric acid** (Gaba),[193–196b] while 3,4-dihydroxyphenylalanine (dopa) and 5-hydroxy-

Schiff base of aminoacrylate (R=H) or aminocrotonate (R=CH₃)

Transimination

Aminoacrylate or aminocrotonate

Imino acid

Pyruvate or 2-oxobutyrate   (14-29)

tryptophan are acted upon by an **aromatic amino acid decarboxylase** to form, respectively, the neurotransmitters **dopamine** and **serotonin**.[197–199b] Histidine is decarboxylated to **histamine**.[200–202] However, not all histidine decarboxylases use PLP as a coenzyme (Section F).

Arginine is converted by a PLP-dependent decarboxylase to agmatine (Fig. 24-12) which is hydrolyzed to **1,4-diaminopropane**.[191] This important cell constituent is also formed by hydrolysis of arginine to **ornithine** (Fig 24-10) and decarboxylation of the latter.[203–206c] **Lysine** is formed in bacteria by decarboxylation of *meso*-diamino-pimelic acid (Fig. 24-14). **Glycine** is decarboxylated oxidatively in mitochondria in a sequence requiring lipoic acid and tetrahydrofolate as well as PLP (Fig. 15-20).[207–209b] A **methionine** decarboxylase has been isolated in pure form from a fern.[210] The bacterial **dialkylglycine** decarboxylase is both a decarboxylase and an aminotransferase which uses pyruvate as its second substrate forming a ketone and L-alanine as products (See Eq. 14-37)[210a, 210b]

Side chain cleavage (Group c).

In a third type of reaction the side chain of the Schiff base of Fig. 14-5 undergoes aldol cleavage. Conversely, a side chain can be added by β condensation. The best known enzyme of this group is **serine hydroxymethyltransferase**, which converts serine to glycine and formaldehyde.[211–213b] The latter is not released in a free form but is transferred by the same enzyme specifically to **tetrahydrofolic acid** (Eq. 14-30), with which it forms a cyclic adduct.

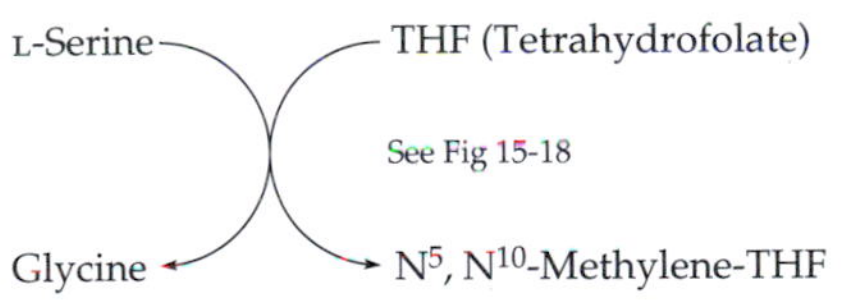

$$\text{L-Serine} \longrightarrow \text{THF (Tetrahydrofolate)}$$
$$\text{See Fig 15-18}$$
$$\text{Glycine} \longleftarrow N^5, N^{10}\text{-Methylene-THF} \qquad (14\text{-}30)$$

Threonine is cleaved to acetaldehyde by the same enzyme. A related reaction is indicated in Fig. 24-27 (top). In a more important pathway of degradation of threonine the hydroxyl group of its side chain is dehydrogenated to form 2-amino-3-oxobutyrate which is cleaved by a PLP-dependent enzyme to glycine and acetyl-CoA (Eq. 14-31).[214,215]

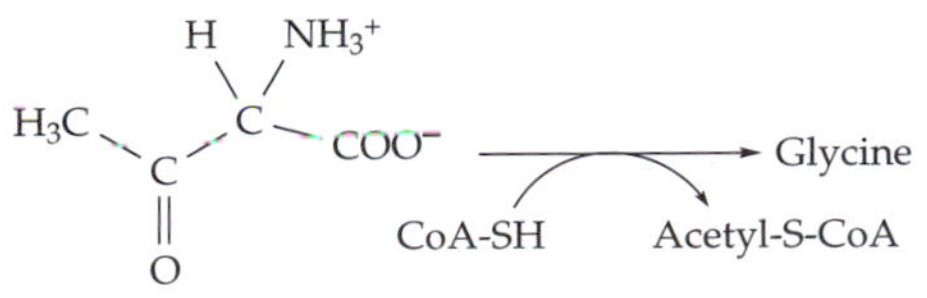

$$(14\text{-}31)$$

Conversely, ester condensation reactions join acyl groups from CoA derivatives to Schiff bases derived from glycine or serine. Succinyl-CoA is the acyl donor

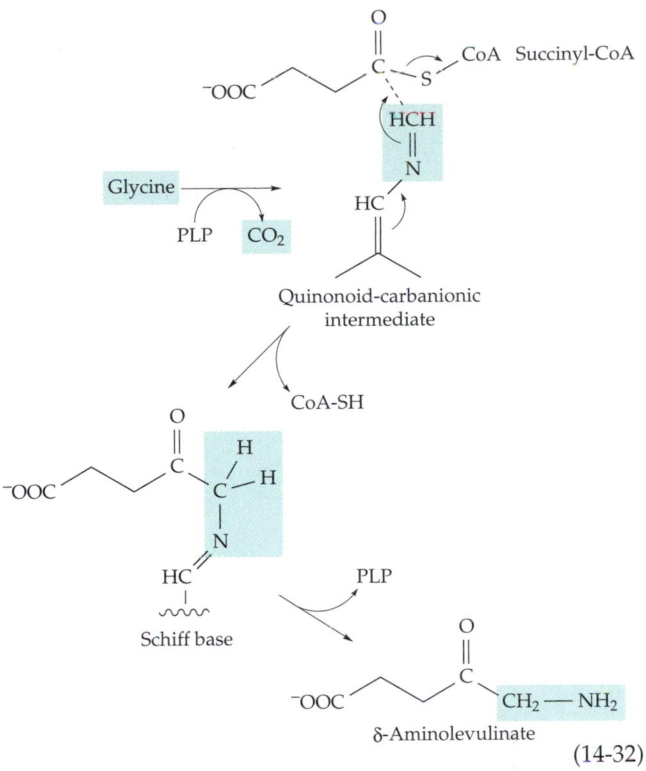

$$(14\text{-}32)$$

in Eq. 14-32 for the second known pathway for biosynthesis of **δ-aminolevulinic acid**, an intermediate in porphyrin synthesis (Chapter 24).[216–218b] The enzyme does not catalyze decarboxylation of glycine in the absence of succinyl-CoA, and the decarboxylation probably follows the condensation as indicated in Eq. 14-32.[219] In a similar reaction in the biosynthesis of **sphingosine** serine is condensed with palmitoyl-CoA and decarboxylated to form an aminoketone intermediate (Fig. 21-6).[219a] 8-Amino-7-oxonanoate synthase forms a precursor of biotin (see banner, p. 719).[220]

Ketimine intermediate as electron acceptor (Group d).

The fourth group of PLP-dependent reactions are thought to depend upon formation of the ketimine intermediate of Eq. 14-28. In this form the original α-hydrogen of the amino acid has been removed and the $C=NH^+$ bond of the ketimine is polarized in a direction that favors electron withdrawal from the amino acid into the imine group. This permits another series of enzymatic reactions analogous to those of the β-oxo acid shown at the top of Fig. 14-5. Both elimination and C–C bond cleavage α,β to the $C=N$ group of the ketimine can occur.

Enzymes of this group catalyze elimination of γ substituents from amino acids as illustrated in Fig. 14-7. Eliminated groups may be replaced by other substituents, either in the α or the β positions. The ketimine formed initially by such an enzyme (step *a*) undergoes elimination of the γ substituent (β with respect to the $C=N$ group) along with a proton from the β position of the original amino acid to form an unsaturated intermediate which can react in one of three ways, depending upon the enzyme. Addition of HY' leads to γ replacement (step *c*), while addition of a proton at the α position leads, via reaction step *d*, to an α,β-unsaturated Schiff base. The latter can react by addition of HY' (β replacement, step *e*) or it can break down to an α-oxo acid and ammonium ion (step *f*), just as in the β elimination reactions of Eq. 14-29. An important γ replacement reaction is conversion of *O*-acetyl-, *O*-succinyl-, or *O*-phosphohomoserine to cystathionine (Eq. 14-33). This **cystathionine γ-synthase** reaction[220a] lies on the pathway of biosynthesis of methionine by bacteria, fungi, and

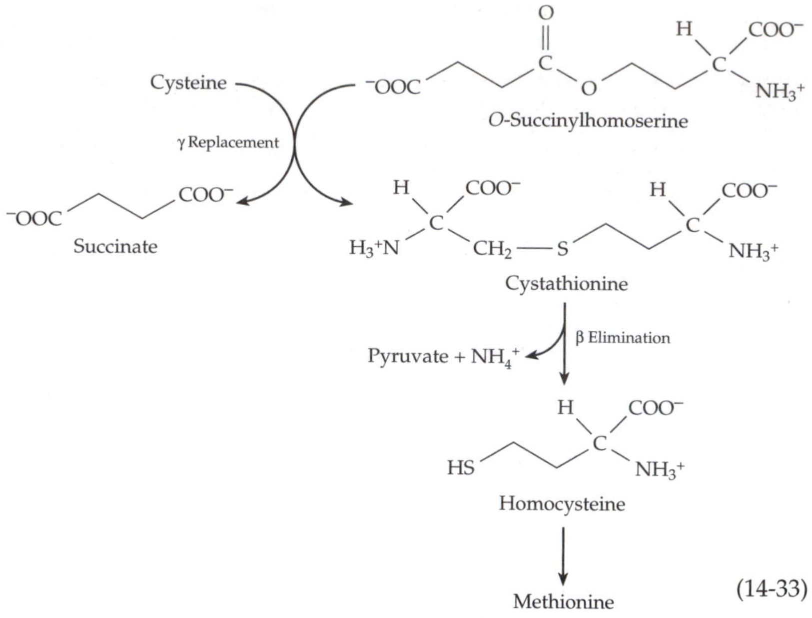

(14-33)

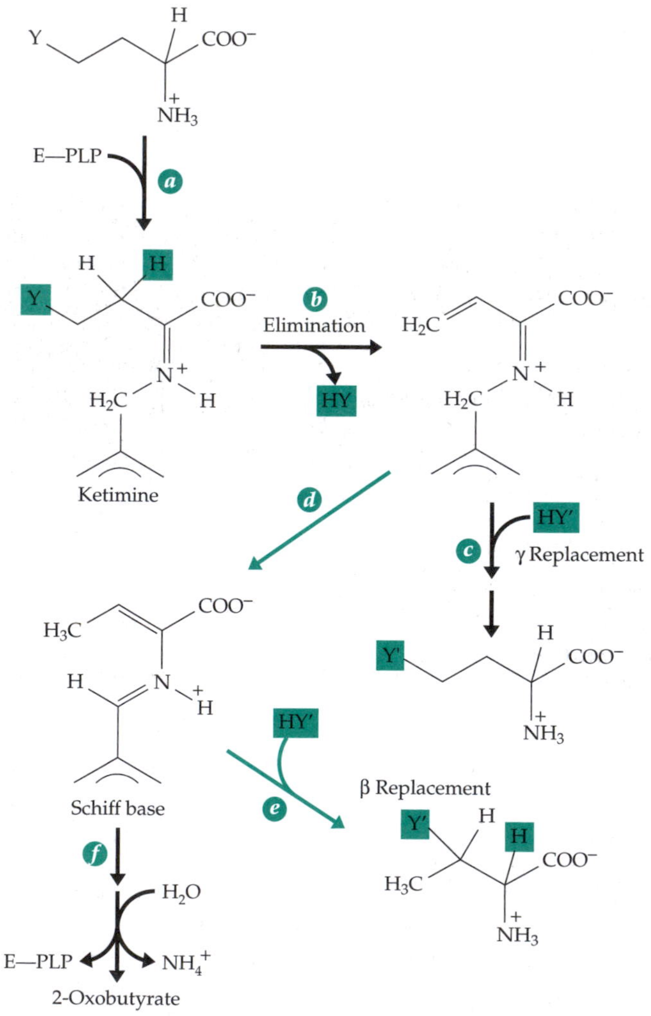

Figure 14-7 Some PLP-dependent reactions involving elimination of a γ substituent. Replacement by another γ substituent or by a substituent in the β position is possible, as is deamination to an α-oxo acid.

higher plants. Subsequent reactions include β elimination from cystathionine of **homocysteine**[220b] which is then converted to methionine (Eq. 14-33). Threonine is formed from **O-phosphohomoserine** via γ elimination followed by β replacement with HO⁻, a reaction catalyzed by **threonine synthase** (Fig. 24-13).[220c]

The loss of a β-carboxyl group as CO_2 can also occur through a ketimine or quinonoid intermediate. For example, the bacterial **aspartate β-decarboxylase**[221] converts aspartate to alanine and CO_2. **Selenocysteine** is utilized to create the active sites of several enzymes (Chapter 15). Excess selenocysteine is degraded by the PLP-dependent **selenocysteine lyase**,[222–223b] which evidently eliminates elemental selenium from a ketimine or quinonoid state of an intermediate Schiff base (Eq. 14-34). A similar reaction may occur in the biosynthesis of iron–sulfur clusters. The **Nif S** protein is essential for formation of Fe_4S_4 clusters in the nitrogen-fixing enzyme nitrogenase. This enzyme is in some way involved in transferring the sulfur atom of cysteine into an iron–sulfur cluster.[224] Alanine is the other product suggesting transfer of S^0 into the cluster using the sequence of Eq. 14-34.

Another related reaction that goes through a ketimine is the conversion of the amino acid **kynurenine** to alanine and anthranilic acid.[225] It presumably depends upon hydration of the carbonyl group prior to β cleavage (Eq. 14-35). An analogous thiolytic cleavage utilizes CoA to convert 2-amino-4-ketopentanoate to acetyl-CoA and alanine.[226]

Glycogen phosphorylase. While PLP is ideally designed to catalyze reactions of amino compounds it was surprising to find it as an essential cofactor for glycogen phosphorylase (Fig. 11-5). The PLP is linked as a Schiff base in the same way as in other PLP-depen-

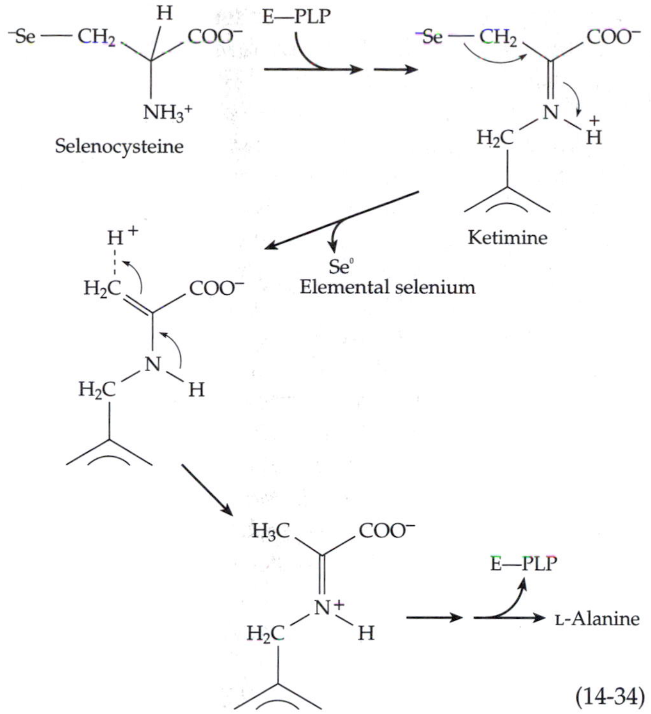

(14-34)

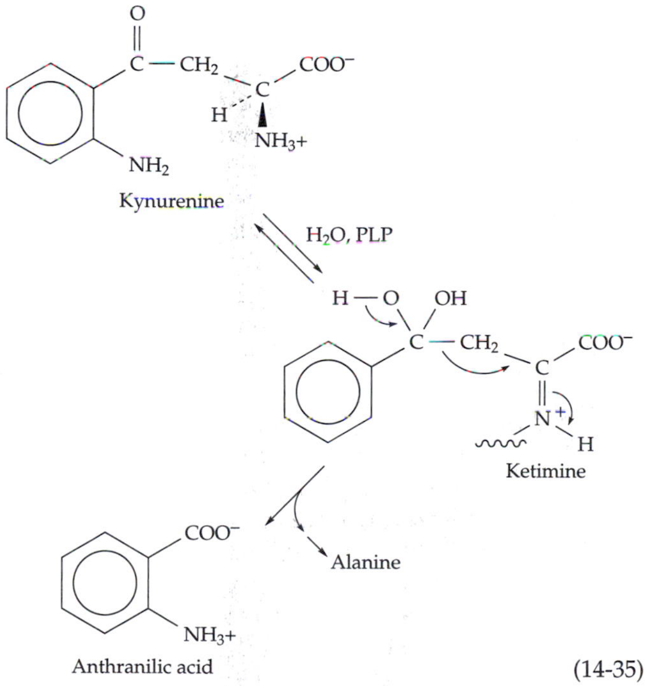

(14-35)

4. Pyridoxamine Phosphate as a Coenzyme

If PLP is a cofactor designed to react with amino groups of substrates, might not pyridoxamine phosphate (PMP) act as a coenzyme for reactions of carbonyl compounds? An example of this kind of function has been found[228-230] in the formation of 3,6-dideoxyhexoses needed for bacterial cell surface antigens (Fig. 4-15). Glucose (as cytidine diphosphate glucose; CDP-glucose) is first converted to 4-oxo-6-deoxy-CDP-glucose. The conversion of the latter to 3,6-dideoxy-CDP-glucose (Eq. 14-36) requires PMP as well as NADH or NADPH.

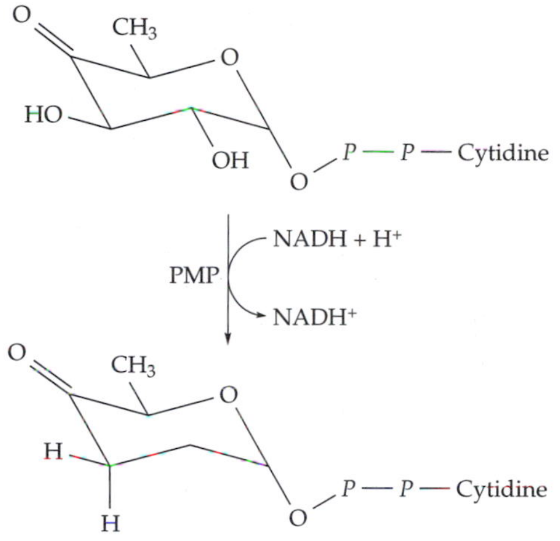

(14-36)

The student may find it of interest to propose a mechanism for this reaction, taking into account the expected direct transfer of a hydrogen from NADH as described in Chapter 15, before consulting published papers. Part of the reaction cycle appears to involve a free radical derived from the PMP. This is discussed further in Chapter 20 together with free radical-forming PLP enzymes

5. Stereochemistry of PLP-Requiring Enzymes

According to stereoelectronic principles, the bond in the substrate amino acid that is to be broken by a PLP-dependent enzyme should lie in a plane perpendicular to the plane of the cofactor–imine π system (Fig. 14-8). This would minimize the energy of the transition state by allowing maximum σ–π overlap between the breaking bond and the ring–imine π system. It also would provide the geometry closest to that of the planar quinonoid intermediate to be formed, thus minimizing molecular motion in the approach to the transition state. Figure 14-8 shows three orientations of an amino acid in which the α-hydrogen, the carboxyl group, and the side chain, respectively, are positioned for cleavage. For each orientation shown, another geometry suitable for cleavage of the same bond is

dent enzymes, but there is no obvious function for the coenzyme ring. As suggested in Chapter 12, the phosphate group probably acts as an acid–base catalyst. It has been estimated that 50% of the vitamin B_6 in our body is present as PLP in muscle phosphorylase.[227] Studies of vitamin B_6-deficient rats suggest that PLP in phosphorylase serves as a reserve supply, much of which can be taken for other purposes during times of deficiency.

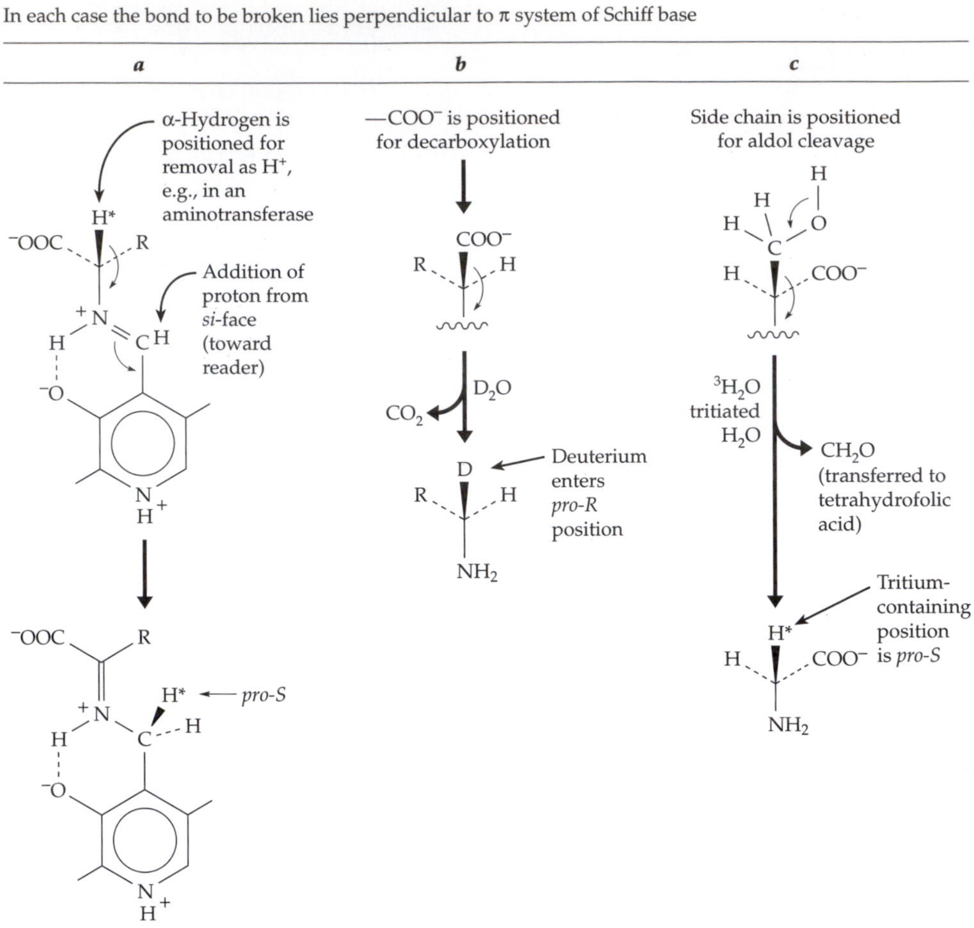

In each case the bond to be broken lies perpendicular to π system of Schiff base

Figure 14-8 Some stereochemical aspects of catalysis of PLP-requiring enzymes.

obtained by rotating the amino acid through 180°.

Dunathan suggested that this stereoelectronic requirement explains certain side reactions observed with PLP-requiring enzymes. The idea also received support from experiments with a bacterial α-dialkyl-glycinedecarboxylase.[231,232] The enzyme ordinarily catalyzes, as one half-reaction, the combination decarboxylation–transamination reaction shown in Eq. 14-37. It also acts on both D- and L-alanine, decarboxylating the former but catalyzing only removal of the α-H

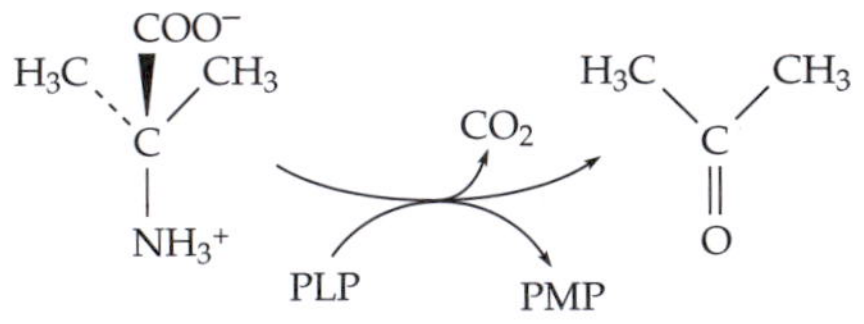

from L-alanine. The results can be rationalized by assuming that the enzyme possesses a definite site for one alkyl group but that the position of the second alkyl group can be occupied by –H or –COO⁻ and that the group labilized lies perpendicular to the π system:

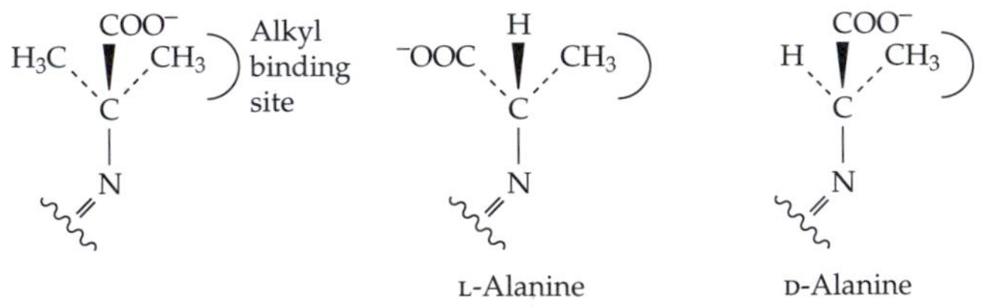

Glycine is unreactive, suggesting that occupation of the alkyl binding site is required for catalysis.

According to Dunathan's postulate, there are only two possible orientations of the amino acid substrate in an aminotransferase. One is shown in Fig. 14-8. In the other, the amino acid is rotated 180° so that the α-hydrogen protrudes *behind* the plane of the paper. Dunathan studied **pyridoxamine:pyruvate aminotransferase**, an enzyme closely related to PLP-requiring aminotransferases and which catalyzes the transamination of pyridoxal with L-alanine to form pyridoxamine and pyruvate. The same reaction is catalyzed by the apoenzyme of aspartate aminotransferase. In both cases, when the alanine contained ²H in the α position the ²H was transferred stereospecifically into the *pro-S* position at C-4' of the pyridoxam-ine (indicated by asterisks in Fig. 14-8).[233] The results suggested that a group from the protein abstracts a proton from the α position and transfers it on the same side of the π system (**suprafacial transfer**), adding it to the *si* face of the C=N group as shown in Fig. 14-8. Later, the same stereospecific proton transfer was demon-strated for the PLP present in the holoen-zyme.[234] Not surprisingly, the D-amino acid aminotransferase adds the proton to the *re* face of the C=N group.[235]

When a decarboxylase acts on an amino acid in ^2H_2O, an atom of ^2H is incorporated in the *pro-R* position, the position originally occupied by the carboxyl group (Fig. 14-8). Cleavage of serine by serine hydroxymethyltransferase in ^3H-containing water leads to incorporation of ^3H in the *pro-S* position. Stereospecific introduction of ^2H or ^3H has been ob-served in the β position of 2-oxobutyrate formed in β or γ elimination reactions. Conversion of serine to tryptophan by tryptophan synthetase occurs without inversion at C-3.[236] These and many other observations on PLP-dependent enzymes[189,237,238] can be generalized by saying that enzymatic reactions of PLP Schiff bases usually take place on only one face of the relatively planar structure. This is the *si* face at C-4' of the coen-zyme (see Fig. 14-8). This result is expected if a single acid–base group serves as proton acceptor in one step and as proton donor in a later step. This leads naturally to the observed retention of configuration in steps involving replacement and the suprafacial transfer of protons from one position on that face to another.

6. Seeing Changes in the Optical Properties of the Coenzyme

The absorption of light in the ultraviolet and visible regions is a striking characteristic of many coenzymes. It can be measured accurately and displayed as an absorption spectrum and may also give rise to circular dichroism and to fluorescence (Chapter 23). The optical properties of the vitamin B_6 coenzymes are sensitive to changes both in environment and in the state of proto-nation of groups in the molecule. For example, PMP in the neutral dipolar ionic form, which exists at pH 7, has three strong light absorption bands centered at 327, 253, and 217 nm.[239] The other ionic forms of PMP and other derivatives of vitamin B_6 also each have three absorption bands spaced at roughly similar intervals, but with varying positions and intensities. The minor tautomer of PMP containing an uncharged ring (Eq. 14-38) has its low-energy (long-wavelength) band at 283 nm. When both the ring nitrogen and phenolic oxygen are protonated, the band shifts again

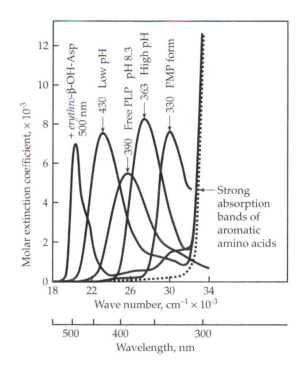

Dipolar ionic form
I   327 nm
II   253 nm
III 217 nm

Minor non-dipolar ionic tautomer
I   283 nm

(14-38)

to 294 nm and if both groups are deprotonated the resulting anion absorbs at 312 nm. Thus, observation of the absorption spectrum of the coenzyme bound to an enzyme surface can tell us whether particular groups are protonated or unprotonated. The peak of bound PMP at 330 nm in aspartate aminotransferase (Fig. 14-9) is indicative of the dipolar ionic form (Eq. 14-38). How-ever, the 5-nm shift from the position of free PMP sug-gests a distinct change in environment.

Figure 14-9 Absorption spectra of various forms of aspar-tate aminotransferase compared with that of free pyridoxal phosphate. The low pH form of the enzyme observed at pH <5 is converted to the high pH form with pK_a ~6.3. Addi-tion of *erythro*-3-hydroxyaspartate produces a quinonoid form whose spectrum here is shown only 1/3 its true height. The spectrum of free PLP at pH 8.3 is also shown . The spectrum of the apoenzyme (---) contains a small amount of residual absorption of uncertain origin in the 300- to 400-nm region.

Pyridoxal phosphate exists in an equilibrium between the aldehyde and its covalent hydrate (as in Eq. 13-1). The aldehyde has a yellow color and absorbs at 390 nm (Fig. 14-9), while the hydrate absorbs at nearly the same position as does PMP. The absorption bands of Schiff bases of PLP are shifted even further to longer wavelengths, with N-protonated forms absorbing at 415–430 nm. Forms with an unprotonated $C=N$ group absorb at shorter wavelengths.[149,240]

Imine groups in free enzymes. When, in 1957, W. T. Jenkins examined one of the first highly purified aspartate aminotransferase preparations he noted a surprising fact: The bound coenzyme, at pH 5, absorbed not at 390 nm, as does PLP, but at 430 nm, like a Schiff base (Fig. 14-9). When the pH was raised, the absorption band shifted to 363 nm. The result suggests dissociation of a proton (with a pK_a of ~ 6.3) from the hydrogen-bonded position in a Schiff base of the type shown in Fig. 14-6. It was quickly demonstrated for this enzyme and for many other PLP-dependent enzymes that reduction with sodium borohydride caused the spectrum to revert to one similar to that of PMP and fixed the coenzyme to the protein. After complete HCl digestion of such borohydride-reduced proteins a fluorescent amino acid containing the reduced pyridoxyl group was obtained and in every case was identified as ε-pyridoxyl-lysine. Thus, PLP-containing enzymes in the absence of substrates usually exist as Schiff bases with lysine side chains of the proteins. Even the PLP in glycogen phosphorylase is joined in this way. However, its absorption maximum at 330 nm shows

ε-Pyridoxyllysine

that in phosphorylase it is present as the nondipolar ionic tautomer with a 3-OH group on the ring as in Eq. 14-38.

Absorption bands at 500 nm. With many PLP enzymes certain substrates and inhibitors cause the appearance of intense and unusually narrow bands at ~500 nm. Such a band is observed with aspartate aminotransferases acting on *erythro*-3-hydroxyaspartate (Fig. 14-9). This substrate undergoes transamination very slowly, and the 500-nm absorbing form which accumulates is probably an intermediate in the normal reaction sequence. A similar spectrum is produced by tryptophan indole-lyase acting on the competitive inhibitor L-alanine. Under the same conditions the

enzyme promotes a rapid exchange of the α-hydrogen of the alanine with ^2H of ^2H_2O. Serine hydroxymethyl-transferase gives a 495- to 500-nm band with both D-alanine and the normal product glycine.[241] Similar spectra have been produced in nonenzymatic model reactions[242] and probably represent the postulated quinonoid–carbanionic intermediates.

7. Atomic Structures

The three-dimensional structures of aspartate aminotransferases from *E. coli* to humans are very similar.[163–167a] The folding pattern (Fig. 2-6) and active site structure (Figs. 14-6 and 14-10) are completely conserved. The major domain of the protein contains a central β sheet surrounded by helices with coenzyme attached to a lysine at the C terminus of one of the β strands. The protein is a dimer with the two major domains held together by both polar and nonpolar interactions. The two active sites are located at the interface between the subunits and residues from both subunits participate in forming the active site (Fig. 14-10). The internal Schiff base is formed with Lys 258. The protonated ring nitrogen of the dipolar ionic PLP forms an ion pair with the carboxylate of Asp 222 which protrudes from a central seven-stranded β sheet. The phenolic $-O^-$ forms a hydrogen bond with the $-OH$ of Tyr 225. The interactions of Asp 222 and Tyr 225 fix the ring as the dipolar ionic tautomer.

The phosphate group of the coenzyme, which ^{31}P NMR shows to be predominantly dianionic,[243] forms an ion pair with the side chain of Arg 266 and hydrogen bonds to a backbone $N-H$ at the N terminus of a long helix where it can interact with the positive end of the helix dipole. In addition, the phosphate forms hydrogen bonds to four OH groups of Ser, Thr, and Tyr side chains. In front of the coenzyme ring are two guanidinium groups from the side chains of Arg 386 and of Arg 292* (from the second subunit). These have been shown by X-ray crystallography to bind the two carboxylate groups of a substrate such as glutamate, 2-oxoglutarate, aspartate, oxaloacetate, or cysteine sulfinate; of quasi-substrates such as 2-methylaspartate and *erythro*-β-hydroxyaspartate; or of dicarboxylic inhibitors (Fig. 14-10).

Comparison of amino acid sequences suggests that many other PLP-dependent enzymes have folding patterns similar to those of aspartate aminotransferase but that there are four or more additional different folding patterns.[205,243a,b] Among the enzymes resembling aspartate aminotransferase are ω-amino acid: pyruvate aminotransferase,[244] 2,3-dialkylglycine decarboxylase,[232] tyrosine phenol-lyase,[177] a bacterial ornithine decarboxylase,[182,204] and cystathionine β-lyase.[182] The tryptophan synthase β subunit has a second folding pattern,[184] while alanine racemase[153] and eukaryotic ornithine decarboxylases[205] have $(\alpha\beta)_8$-barrel structures resembling that in Fig. 2-28. A fourth structural pattern is that of D-amino acid aminotransferase.[235] It is anticipated that the branched-chain aminotransferase[245] will have a similar stucture. Glycogen phosphorylase (Fig 11-5) has a fifth folding pattern.

8. Constructing a Detailed Picture of the Action of a PLP Enzyme

Consider the number of different steps that must occur in about one-thousanths of a second during the action of an aminotransferase. First, the substrate binds to form the "Michaelis complex." Then the transimination (Eq. 14-26) takes place in two steps and is followed by the removal of the α-hydrogen to form the quinonoid intermediate. An additional four steps are needed to form the ketimine, to hydrolyze it, and to release the oxoacid product to give the PMP form of the enzyme. The reaction sequences in some of the other enzymes are even more complex. How can one enzyme do all this?

The first step in the sequence is the binding of the substrate to form the "Michaelis complex." The positive charges on Arg 386 and Arg 292 doubtless attract the carboxylate groups of the substrate and aid in guiding it toward a correct fit. In a similar manner the $-O^-$ of the coenzyme, which is distributed by resonance into the $-C=N$ of the Schiff base linkage, attracts the $-NH_3^+$ of the substrate. When a substrate or inhibitor binds to the two guanidinium groups a small structural domain of the enzyme moves and closes around the substrate which now has very little contact with the external solvent. In the initial "Michaelis complex" the $-NH_3^+$ group of the substrate lies directly in front of the C-4' carbon of the coenzyme (Fig. 14-10), where it can initiate the transimination reaction of Eq. 14-26. However, before this can happen a proton must be removed to convert the $-NH_3^+$ to $-NH_2$.

Long before the three-dimensional structures were known, Ivanov and Karpeisky[246] suggested that in the free enzyme the positively charged group (Arg 386) that binds the α-carboxylate of the substrate interacts electrostatically with the $-O^-$ of the coenzyme. This is one of the factors that keeps the pK_a of the $-CH=N^+H-$ that is conjugated with this $-O^-$ at a low value of ~6.3 (at 0.1 M anion concentration). However, in the Michaelis complex this interaction of the + charge of Arg 386 with the imine group must be weakened because of the pairing of the $\alpha-COO^-$ of the substrate with the + charge. This will increase the basicity of the imine nitrogen and will also cause a decrease in the pK_a of the substrate $-NH_3^+$, making is easier for a proton to jump from the $-NH_3^+$ to the imine group. This proton transfer is shown in Eq. 14-39, step a. Thus, the nucleophilic $-NH_2$ group is generated by a

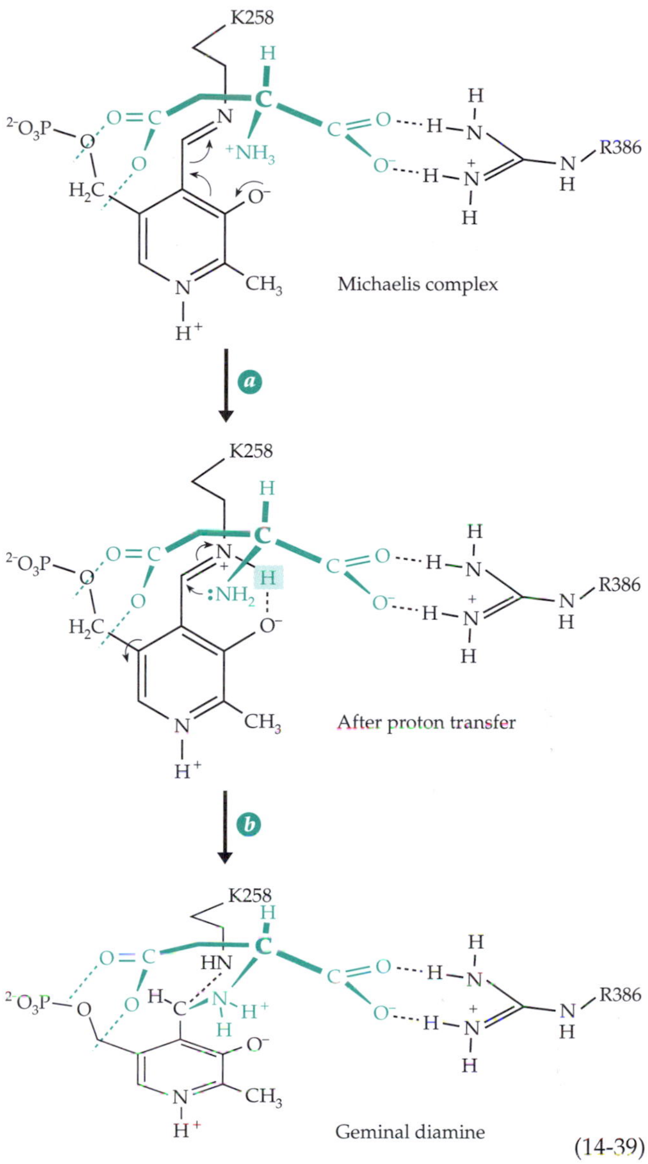

K258

Michaelis complex

a

K258

After proton transfer

b

K258

Geminal diamine   (14-39)

process that at the same time increases the electrophilic properties of the carbon atom of the imine group. This favors the immediate addition of $-NH_2$ to $-C=N^+H-$ to give the adduct a **geminal diamine** (Eq. 14-39, step b), which is shown in three dimensions in Fig. 14-10B.

Notice that each step in the overall sequence changes the electronic or steric characteristics of the complex in a way that facilitates the next step.[246] This is an important principle that is applicable throughout enzymology: *For an enzyme to be an efficient catalyst each step must lead to a change that sets the stage for the next.* These consecutive steps often require proton transfers, and each such transfer will influence the subsequent step in the sequence. Some steps also require alterations in the conformation of substrate, coenzyme, and enzyme. One of these is the transimination sequence (Eqs. 14-26, 14-39). On the basis of the observed loss of circular dichroism in the external aldimine, Ivanov and Karpeisky suggested that a

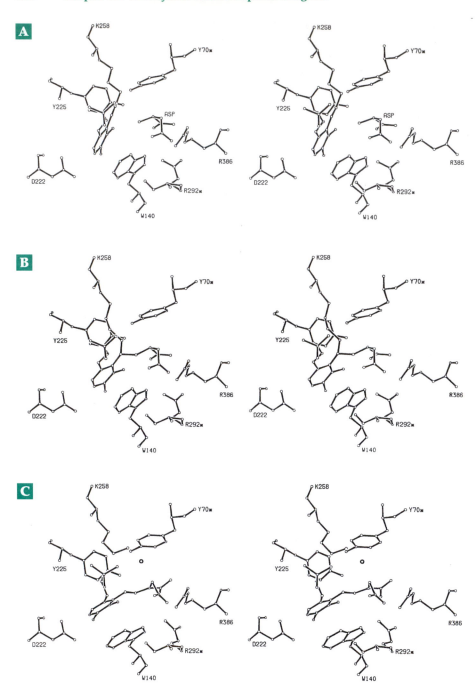

Figure 14-10 Models of catalytic intermediates for aspartate aminotransferase in a half-transamination reaction from aspartate to oxalocetate. For clarity, only a selection of the active site groups are shown. (A) Michaelis complex of PLP enzyme with aspartate. (B) Geminal diamine. (C) Ketimine intermediate. The circle indicates a bound water molecule. See Jansonius and Vincent in Jurnak and McPherson.[163] Courtesy of J.N. Jansonius.

groups of the substrate remain bound in their initial positions.

That the coenzyme really does change its orientation was suggested by a dramatic change in the absorption spectrum of a crystal recorded with plane polarized light (linear dichroism) when 2-methylaspartate was soaked into a crystal (Fig. 23-9).[247] X-ray crystallography confirmed rotation of the ring by ~30°.[163,242,247] To complete the transimination sequence, which is shown only partially by Eq. 14-39, another proton transfer is needed to move the positive charge on the substrate $-N^+H_2-$ to that of lysine 258, whose amino group is then eliminated. This requires additional tilting of the ring. The crystal structure of the external aldimine with α-methylaspartate has been determined[248] as have those of ketimines with glutamate and aspartate,[249,250] a carbinolamine,[251] and quinonoid complexes of related enzymes.[212a]

The ε-amino group, which is eliminated in Eq. 14-26, is basic and functions in the next several steps of catalysis, including the abstraction of the α-hydrogen and in its transfer to the 4'-carbon (Eq. 14-28). This amino group can be seen in Fig. 14-10C, where it is positioned beside a water molecule that is needed to hydrolyze the ketimine. From this figure it can be seen that the group is able to move from site to site on one side of the planar external aldimine, quinonoid–carbanionic, and nearly planar ketimine forms. If the Lys 258 amino group is the catalytic base for these reactions it must be present at pH 7. What is its pK_a? And why doesn't this pK_a show up in the plot of V_{max} vs pH as in Fig 9-9? In fact, the maximum

velocity of aspartate aminotransferase and many other PLP-dependent enzymes is independent of pH over a broad range. The answer is probably that the pK_a's of both the $-NH_2$ groups of Lys 258 and the amino acid in the Michalis complex are high when there is no positive charge on the adjacent Schiff base $-C=N-$ but

rotation of the coenzyme occurs as the $-NH_2$ of the substrate adds to the C = N bond during transimination (Eq. 14-39, step *b*). This accomplishes the essential shortening by ~0.15 nm of the distance between C-4' and the N atom from a van der Waals contact distance to a covalent bond distance while the carboxylate

very low when the Schiff base is protonated (–C = NH$^+$–). This is a result of the large electrostatic effect of a closely adjacent charge in a medium of low dielectric constant (see Chapter 7, Section A). When the Schiff base is unprotonated, as in the first structure of Eq. 14-39, the adjacent amino group has a *high* pK_a and is mostly protonated. However, the proton can jump as in step *a* of that equation to give a protonated Schiff base. Now the unprotonated –NH$_2$ group has a very *low* pK_a, but it is still strongly nucleophilic and can readily add to the Schiff base double bond in step *b* of Eq. 14-39. In a similar fashion the amino group of Lys 258 will alternately be unprotonated and then protonated, its microscopic pK_a alternating between low and high, as it catalyzes the steps of abstracting the α-hydrogen and forming and hydrolyzing the ketimine. Only two microscopic pK_a values, one very high and one very low appear in the V_{max} vs pH profile. This alternation of microscopic pK_a values may be a common characteristic of enzymes that bind ionized substrates and can make good use of the strong electrostatic effects that arise in the active sites to facilitate essential proton transfers.

Aspartate aminotransferases are distinguished from most other PLP-dependent enzymes including transaminases by the relatively low pK_a of ~6.3 for the *free* enzyme. The unprotonated Schiff base in the free enzyme can then react with the protonated amino group of the substrate. How can other PLP enzymes react with amino acids when they have protonated Schiff bases even at relatively high pH? A logical answer is that some basic group with a low pK_a is close to the Schiff base and acts to deprotonate the substrate –NH$_3^+$ so that transimination can occur. Clausen *et al.* suggested that in cystathionine β-lyase this is probably tyrosine (Y111), which is adjacent to the Schiff base –C=NH$^+$– of the external aldimine and is thought to be ionized in the active enzyme.[182] NMR evidence suggests that protonation of an adjacent catalytic base occurs upon substrate binding in D-serine dehydratase as well.[252] Many other variations in active site environments are seen among PLP-dependent enzymes. As with aspartate aminotransferases, there is often an essential carboxylate group that holds a proton onto the pyridine ring of the coenzyme. However, in alanine racemase a guanidinium group from arginine is hydrogen bonded to the coenzyme ring.[153]

Below the active site of aspartate aminotransferase, as shown in Fig. 14-6, is a cluster of three buried histidine side chains in close contact with each other. The imidazole of H143 is hydrogen bonded to the D222 carboxylate, the same carboxylate that forms an ion pair with the coenzyme. This system looks somewhat like the catalytic triad of the serine proteases in reverse. As with the serine proteases, the proton-labeled H$_b$ in Fig. 14-6 can be "seen" by NMR spectroscopy (Fig. 3-30). So can the proton H$_a$ on the PLP ring. These protons

act as built-in sensors able to detect small changes in the electronic environment. For example, when the Schiff base proton dissociates around the pK_a of ~6.2 the NMR resonance of H$_a$ shifts upfield from 17.2 to 15.2 ppm as a result of donation of electons into the ring from the –O$^-$ of the coenzyme.[169] This shift illustrates the reality of the strong electrostatic forces that operate across heterocyclic aromatic rings within active sites of proteins. In alanine racemase a different histidine cluster is present beneath the active site and constitutes part of the "solvent" in which the catalyzed reaction takes place. At least in the case of aspartate aminotransferase, none of the histidines are absolutely essential for activity but the hydrogen-bonded network, which can be altered in mutant forms, may be important.

The detailed description of a reaction sequence given here has to be altered for each specific enzyme. A vast amount of work, only a little of which is cited here, has been done on PLP enzymes.[253–254a] These studies involve calorimetry,[255] kinetics,[210,256–259] crystallography, optical spectroscopy,[188,260,261] NMR,[243,262–264] and genetic engineering and chemical modification.[178,185,186,265]

F. Pyruvoyl Groups and Other Unusual Electrophilic Centers

A few enzymes that might be expected to have PLP at their active sites have instead a prosthetic group consisting of pyruvic acid bound by an amide linkage, a **pyruvoyl group** (Table 14-4). These and several apparently related enzymes are the subject of this section.

TABLE 14-4
Some Pyruvoyl Enzymes

Decarboxylases	Product
Histidine (bacterial)	
S-Adenosylmethionine	
Aspartate α- decarboxylase	β-Alanine
Phosphatidylserine	Phosphatidyl-ethanolamine
4' - Phosphopantothenylcysteine	4' - Phosphopan-tetheine

Reductases (clostridial)
Proline
Glycine

Adapted from van Peolje and Snell.[267]

1. Decarboxylases

Mammalian **histidine decarboxylase** contains PLP but the enzyme from many bacteria contains a pyruvoyl group, as do a few other decarboxylases both of bacterial and eukaryotic origin.[266,267] These enzymes are inhibited by carbonyl reagents and by borohydride. When ^{3}H-containing borohydride was used to reduce the histidine decarboxylase of *Lactobacillus*, ^{3}H was incorporated and was recovered in lactic acid following hydrolysis. This suggested the presence of a pyruvoyl group attached by an amide linkage and undergoing the chemical reactions that are shown in Eq. 14-40. Reduction in the presence of the substrate histidine resulted in covalent binding of the histidine to the bound pyruvate. Thus, as with the PLP-containing decarboxylases, a Schiff base is formed with the substrate. Decarboxylation is presumably accomplished by using the electron-attracting properties of the carbonyl group of the amide:

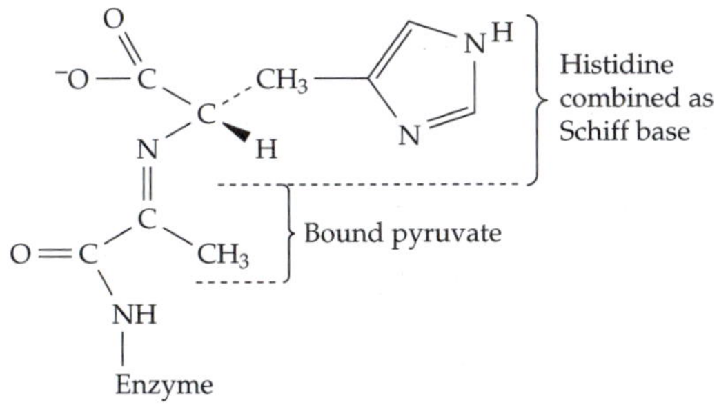

When ^{14}C-labeled serine was fed to organisms producing histidine decarboxylase, ^{14}C was incorporated into the bound pyruvoyl group (Fig. 14-11). Thus, serine is a precursor of the bound pyruvate. The enzyme is manufactured in the cell as a longer 307-residue proenzyme which associates as hexamers (designated π_6). The active enzyme was found to be formed by cleavage of the π chains between Ser 81 and Ser 82 to form 226-residue α chains and 81-residue β chains which associate as $(\alpha\beta)_6$.[270,271] The α chains

carry the N-terminal pyruvoyl group, which was formed from Ser 82. The activation occurs spontaneously by incubations of the proenzyme for 24–48 h at pH ~7 in the presence of divalent metal ions.

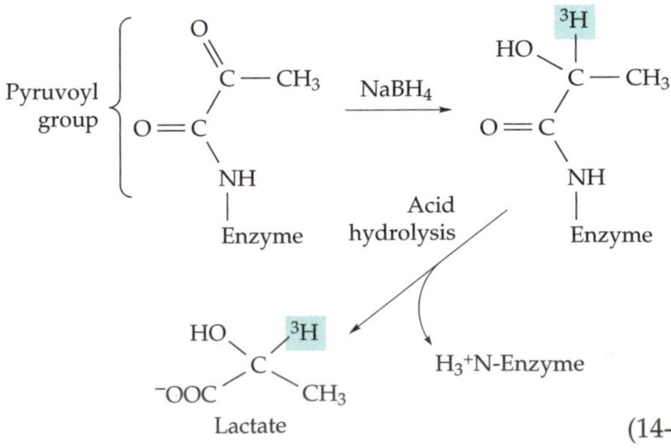

Substituted serine residues under mildly alkaline conditions readily undergo α,β elimination to form **dehydroalanine** residues. When prohistidine decarboxylase containing [18]O in its serine side chains was activated [18]O was found in the carboxylate group of Ser 81 of the β chains. It was shown that it had been transferred from the side chain of Ser 82. This suggested the formation of an intermediate oxygen ester of Ser 81 during formation of the pyruvoyl group (Eq. 14-41).[272]

S-Adenosylmethionine decarboxylase is the first enzyme in the biosynthetic pathway to spermidine (Chapter 24). Whether isolated from bacteria, yeast, animals, or other eukaryotes, this enzyme always contains a bound pyruvoyl group.[273–274b] Both the

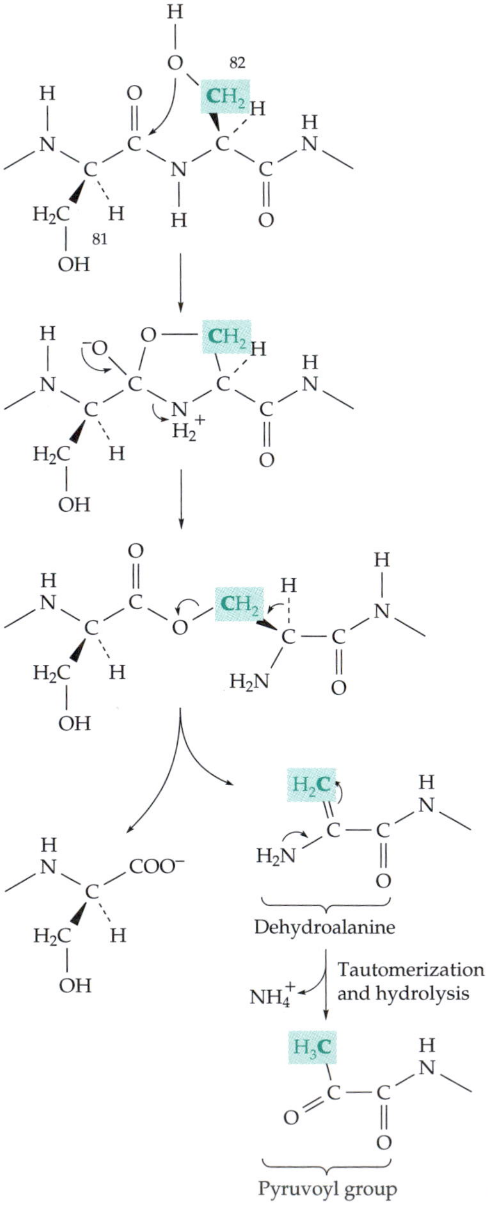

(14-40)

(14-41)

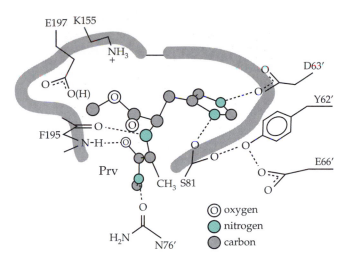

Figure 14-11 Schematic diagram of the active site of the pyruvoyl enzyme histidine decarboxylase showing key polar interactions between the pyruvoyl group and groups of the inhibitor *O*-methylhistidine and surrounding enzyme groups. Aspartate 63 appears to form an ion pair with the imidazolium group of the substrate.[268] Hydrogen bonds are indicated by dotted lines. See Gallagher *et al.*[269]

mammalian enzyme and that from *E. coli* are $(\alpha\beta)_4$ tetramers formed in a manner similar to that of bacterial histidine decarboxylase. Other pyruvoyl decarboxylases are **phosphatidylserine decarboxylase**,[275,276] an intrinsic membrane protein used to form phosphatidylethanolamine, **aspartate α-decarboxylase**,[277] which forms β alanine needed for biosynthesis of coenzyme A, and **4'-phosphopantethenoylcysteine decarboxylase**,[278] the second of two decarboxylases required in the synthesis of coenzyme A. Because of the lack of a primary amino group, its mechanism must be somewhat different from that of other enzymes in this group.[267]

2. Proline and Glycine Reductases

An enzyme required in the anaerobic breakdown of proline by clostridia utilizes a dithiol-containing protein to reductively open the ring (Eq. 14-42).[279,280]

(14-42)

This enzyme also contains an N-terminal pyruvoyl residue as does one subunit of a selenium-containing glycine reductase which utilizes a dithiol to convert glycine into acetate with coupled formation of ATP:

$$\text{Glycine} + \text{ADP} + P_i + 2\,e^- \rightarrow$$
$$\text{acetate}^- + NH_4^+ + \text{ATP} \qquad (14\text{-}43)$$

In both cases it has been proposed that the pyruvoyl group forms a positively charged Schiff base with the substrate. The bond-breaking mechanisms are not obvious but some ideas have been proposed.[281–283a] The ATP may arise from reaction with an intermediate acetyl phosphate.[282] See Eq. 15-61.

3. Dehydroalanine and Histidine and Phenylalanine Ammonia-Lyases

Catabolism of histidine in most organisms proceeds via an initial elimination of NH_3 to form **urocanic acid** (Eq. 14-44). The absence of the enzyme **L-histidine ammonia-lyase** (histidase) causes the genetic disease histidinemia.[284,285] A similar reaction is catalyzed by the important plant enzyme **L-phenylalanine ammonia-lyase**. It eliminates $-NH_3^+$ along with the *pro-S* hydrogen in the β position of phenylalanine to form *trans*-cinnamate (Eq. 14-45). Tyrosine is converted to *p*-coumarate by the same enzyme. Cinnamate and coumarate are formed in higher plants and are converted into a vast array of derivatives (Box 21-E, Fig. 25-8).

The reactions of Eqs. 14-44 and 14-45 are unusual because the nucleophilic substituent eliminated is on the α-carbon atom rather than the β. There is nothing in the structures of the substrates that would permit an

(14-44)

(14-45)

easy elimination of the α-amino group. Thus, it should not be surprising that both enzymes contain a special active center. When ^2H is introduced into the β position of phenylalanine, no isotope effect on the rate is observed. Rather, the rate-limiting step appears to be the release of ammonia from the coenzyme group. It appears that the enzyme must in some way make the amino group a much better leaving group than it would be otherwise.

Both enzymes are inhibited by sodium borohydride and also by nitromethane. After reduction with NaB^3H_4 and hydrolysis, ^3H-containing alanine was isolated. This suggested that they contain **dehydroalanine**, which could arise by dehydration of a specific serine residue.[286,287] For phenylalanine ammonia-lyase from *Pseudomonas putida* this active site residue has been identified as S143. Replacement by cysteine in the S143C mutant also gave active enzyme while S143A

BOX 14-D DIETARY REQUIREMENTS FOR B VITAMINS

It is difficult to determine the amounts of vitamins needed for good health and they may differ considerably from one individual to another. The quantities listed below are probably adequate for most young adults but must be increased during pregnancy and lactation and after very strenuous exercise.

Pantothenic acid: 10–15 mg/day. Deficiency causes apathy, depression, impaired adrenal function, and muscular weakness. ω -Methylpantothenic acid is a specific antagonist. The calcium salt, calcium pantothenate, is the usual commercial form.

Biotin: 0.15–0.3 mg/day. The discovery that biotin deficiency in young chickens can lead to sudden death resulted in a recommendation to supplement infant formulations with biotin.[a] **Desthiobiotin**, in which the sulfur has been removed and replaced by two hydrogen atoms, can replace biotin in some organisms and appears to lie on one pathway of biosynthesis.[b,c] **Oxybiotin,** in which the sulfur has been replaced by oxygen, is active for many organisms and partially active for others. No evidence for conversion to biotin itself has been reported, and oxybiotin may function satisfactorily in at least some enzymes.

Thiamin: 0.23 mg or more per 1000 kcal of food consumed and a minimum total of 0.8 mg/day. Replacement of the methyl group on the pyrimidine ring by ethyl, propyl, or isopropyl gives compounds with some vitamin activity, but replacement by hydrogen cuts activity to 5% of the original. The butyl analog is a competitive inhibitor.

Vitamin B$_6$: 1.5–2 mg/day; 0.4 mg/day for infants. Vitamin B$_6$ is widely distributed in foods, and symptoms of severe deficiency are seldom observed. However, a number of cases of convulsions have been attributed to partial destruction of vitamin B$_6$ in infant liquid milk formulas. Convulsions occurred when the vitamin B$_6$ content was reduced to about one-half that normally present in human milk.

Several cases of children with an abnormally high vitamin B$_6$ requirement (2–10 mg/day) have

been reported, and rare metabolic diseases are known[d-f] in which specific enzymes, such as cystathionine synthase, have a reduced affinity for PLP. Patients with these diseases also benefit from a higher than normal intake of the vitamin. Excessive excretion of the vitamin may also occur, an example being provided by a strain of laboratory mice that require twice the normal amount of vitamin B$_6$ and which die in convulsions after a brief period of vitamin B$_6$ depletion.[d] Dietary supplementation with large amounts of vitamin B$_6$ for treatment of medical conditions such as carpal tunnel syndrome has been controversial.[g] Amounts of pyridoxine over 50 mg per day may damage peripheral nerves, probably because of the chemical reactivity of pyridoxal and PLP.

Nicotinamide: About 7.5 mg/day. Tryptophan can substitute to some extent. See also Box 15-A.

Riboflavin: About 1.5 mg/day. See also Box 15-B.

Folic acid: About 0.2–0.4 mg/day. See also Box 15-D.

Vitamin B$_{12}$: About 2 µg/day. See also Box 16-B.

Vitamin C (ascorbic acid): 50–200 mg/day. See also Box 18-D.

---

[a] Parry, R. J., and Kunitni, M. G. (1976) *J. Am. Chem. Soc.* **98**, 4024–4025

[b] Gibson, K. J., Lorimer, G. H., Rendina, A. R., Taylor, W. S., Cohen, G., Gatenby, A. A., Payne, W. G., Roe, D. C., Lockett, B. A., Nudelman, A., Marcovici, D., Nachum, A., Wexler, B. A., Marsilii, E. L., Turner, S., IM, Howe, L. D., Kalbach, C. E., and Chi, H. (1995) *Biochemistry* **34**, 10976–10984

[c] Huang, W., Jia, J., Gibson, K. J., Taylor, W. S., Rendina, A. R., Schneider, G., and Lindqvist, Y. (1995) *Biochemistry* **34**, 10985–10995

[d] Mudd, S. H., Levy, H. L., and Skovby, F. (1995) in *The Metabolic and Molecular Bases of Inherited Disease*, 7th ed., Vol. 1 (Scriver, C. R., Beaudet, A. L., Sly, W. S., and Valle, D., eds), pp. 1279–1327, McGraw-Hill, New York

[e] Bell, R. R., and Haskell, B. E. (1971) *Arch. Biochem. Biophys.* **147**, 588–601

[f] Pascal, T. A., Gaull, G. E., Beratis, N. G., Gillam, B. M., Tallan, H. H., and Hirschhorn, K. (1975) *Science* **190**, 1209–1211

[g] Bender, D. A. (1999) *Br. J. Nutr.* **81**, 7–20

was inactive.[288,289] These results support the formation of dehydroalanine. One proposed mechanism of action of the enzymes involved addition of the substrate amino group to the C = C bond of the dehydroalanine followed by elimination (Eq. 14-46). Dissociation of the C – H bond in this step would be assisted by the electron-accepting properties of the adjacent aromatic ring. However, the proposed chemistry was not convincing. Noting that a 5-nitro substituent on the histidine greatly enhances the rate of reaction, Langer

et al. proposed that the histidine reacts with the electrophilic carbon in the pyruvoyl center as in Eq. 14-47.[290,291] However, this chemistry, too, was unprecedented in enzymology. Determination of the three-dimensional structure of histidine ammonia-lyase[292] led to the discovery of a new prosthetic group and a solution to the problem. Two dehydration steps, (Eq. 14-47) convert an Ala-Gly-Ser sequence within the protein into **4-methylidine-imidazole-5-one (MIO)**, a modified dehydroalanine with enhanced electron-accepting properties. The proposed mechanism of action is portrayed in Eq. 14-48.

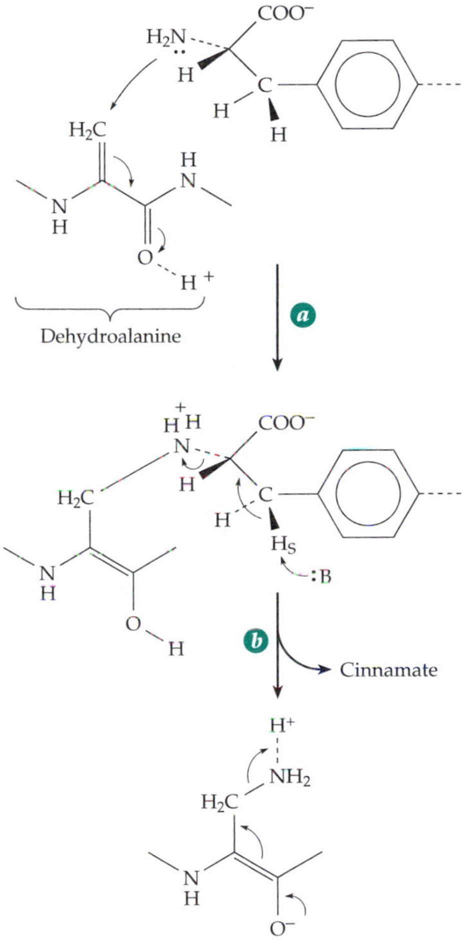

Dehydroalanine

Cinnamate

(14-46)

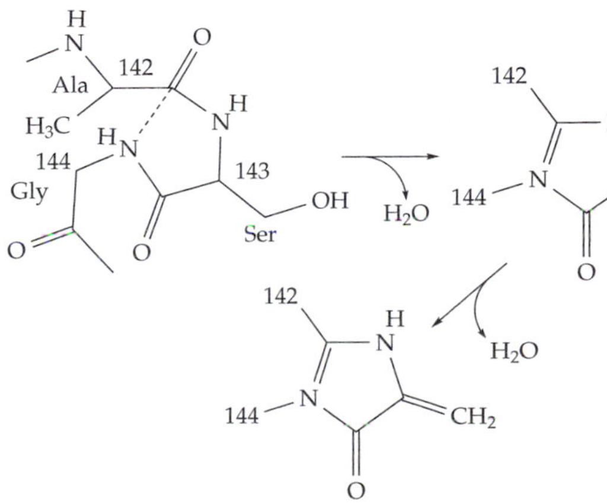

4-Methylidene-imidazole-5-one (MIO)   (14-47)

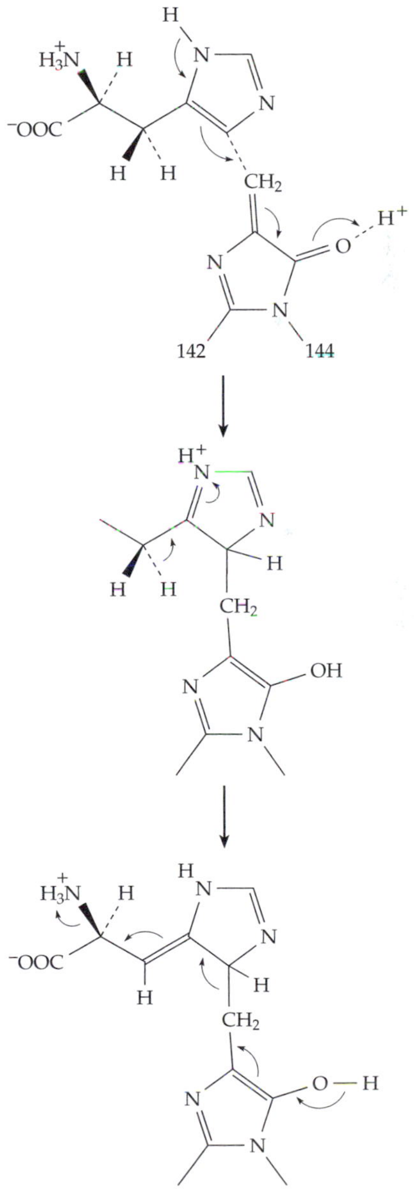

(14-48)

References

1. Moodie, S. L., Mitchell, J. B. O., and Thornton, J. M. (1996) *J. Mol. Biol.* **263**, 486–500
2. Lipmann, F. (1945) *J. Biol. Chem.* **160**, 173–190
3. Roskoski, R. N., Jr. (1987) *Trends Biochem. Sci.* **12**, 136–138
3a. Bentley, R. (2000) *Trends Biochem. Sci.* **25**, 302–305
4. Martin, D. P., and Drueckhammer, D. G. (1992) *J. Am. Chem. Soc.* **114**, 7287–7288
5. Baraniak, J., and Frey, P. A. (1988) *J. Am. Chem. Soc.* **110**, 4059–4060
6. Richter, D., and Hilz, H. (1979) *Trends Biochem. Sci.* **1**, N123–N124
7. Ohlrogge, J. B., Jaworski, J. G., and Post-Beittenmiller, D. (1993) in *Lipid Metabolism in Plants* (Moore, T. S., Jr., ed), pp. 3–32, CRC Press, Boca Raton, Florida
8. Pavela-Vrancic, M., Pfeifer, E., Schröder, W., von Döhren, H., and Kleinkauf, H. (1994) *J. Biol. Chem.* **269**, 14962–14966
9. Haese, A., Pieper, R., von Ostrowski, T., and Zocher, R. (1994) *J. Mol. Biol.* **243**, 116–122
10. Pieper, R., Ebert-Khosla, S., Cane, D., and Khosla, C. (1996) *Biochemistry* **35**, 2054 –2060
11. Cortes, J., Wiesmann, K. E. H., Roberts, G. A., Brown, M. J. B., Staunton, J., and Leadlay, P. F. (1995) *Science* **268**, 1487–1489
12. Brambl, R., and Plesofsky-Vig, N. (1986) *Proc. Natl. Acad. Sci.* **83**, 3644–3648
13. Berg, M., Hilbi, H., and Dimroth, P. (1996) *Biochemistry* **35**, 4689–4696
14. Dimroth, P. (1987) *Microbiol. Rev.* **51**, 320–340
14a. Hoenke, S., Wild, M. R., and Dimroth, P. (2000) *Biochemistry* **39**, 13223–13232
15. Jackowski, S., and Rock, C. O. (1984) *J. Biol. Chem.* **259**, 1891–1895
15a. McAllister, K. A., Peery, R. B., Meier, T. I., Fischl, A. S., and Zhao, G. (2000) *J. Biol. Chem.* **275**, 30864–30872
15b. Chirgadze, N. Y., Briggs, S. L., McAllister, K. A., Fischl, A. S., and Zhao, G. (2000) *EMBO J.* **19**, 5281–5287
16. Whitty, A., Fierke, C. A., and Jencks, W. P. (1995) *Biochemistry* **34**, 11678–11689
17. Fierke, C. A., and Jencks, W. P. (1986) *J. Biol. Chem.* **261**, 7603–7606
18. Langer, B. W., Jr., and Gyorgy, P. (1968) in *The Vitamins*, 2nd ed., Vol. 2 (Sebrell, W. H., Jr., and Harris, R. S., eds), pp. 294–322, Academic Press, New York
19. Bentley, R. (1985) *Trends Biochem. Sci.* **10**, 51–56
20. DeTitta, G. T., Edmonds, J. W., Stallings, W., and Donohue, J. (1976) *J. Am. Chem. Soc.* **98**, 1920–1926
20a. Garton, S. D., Temple, C. A., Dhawan, I. K., Barber, M. J., and Rajagopalan, K. V. (2000) *J. Biol. Chem.* **275**, 6798–6805
20b. McIver, L., Baxter, R. L., and Campopiano, D. J. (2000) *J. Biol. Chem.* **275**, 13888–13894
21. Wright, L. D., Cresson, E. L., Skeggs, H. R., Peck, R. L., Wolf, D. E., Wood, T. R., Valent, J., and Folkers, K. (1951) *Science* **114**, 635–636
22. Knowles, J. R. (1989) *Ann. Rev. Biochem.* **58**, 195–221
23. López-Casillas, F., Bai, D.-H., Luo, X., Kong, I. S., Hermodson, M. A., and Kim, K.-H. (1988) *Proc. Natl. Acad. Sci. U.S.A.* **85**, 5784–5788
24. Browner, M. F., Taroni, F., Sztul, E., and Rosenberg, L. E. (1989) *J. Biol. Chem.* **264**, 12680–12685
25. Samols, D., Thornton, C. G., Murtif, V. L., Kumar, G. K., Haase, F. C., and Wood, H. G. (1988) *J. Biol. Chem.* **263**, 6461–6464
26. León-Del-Rio, A., Leclerc, D., Akerman, B., Wakamatsu, N., and Gravel, R. A. (1995) *Proc. Natl. Acad. Sci. U.S.A.* **92**, 4626–4630
27. Ahmad, P. M., and Ahmad, F. (1991) *FASEB J.* **5**, 2482–2485

28. Wang, X., Wurtele, E. S., Keller, G., McKean, A. L., and Nikolau, B. J. (1994) *J. Biol. Chem.* **269**, 11760–11769
28a. McKean, A. L., Ke, J., Song, J., Che, P., Achenbach, S., Nikolau, B. J., and Wurtele, E. S. (2000) *J. Biol. Chem.* **275**, 5582–5590
29. Weaver, L. M., Lebrun, L., Wurtele, E. S., and Nikolau, B. J. (1995) *Plant Physiol.* **107**, 1013–1014
30. Al-Feel, W., Chirala, S. S., and Wakil, S. J. (1992) *Proc. Natl. Acad. Sci. U.S.A.* **89**, 4534–4538
31. Lamzin, V. S., Dauter, Z., Popov, V. O., Harutyunyan, E. H., and Wilson, K. S. (1994) *J. Mol. Biol.* **236**, 759–785
32. Roessler, P. G., and Ohlrogge, J. B. (1993) *J. Biol. Chem.* **268**, 19254–19259
32a. Abu-Elheiga, L., Almarza-Ortega, D. B., Baldini, A., and Wakil, S. J. (1997) *J. Biol. Chem.* **272**, 10669–10677
32b. Abu-Elheiga, L., Brinkley, W. R., Zhong, L., Chirala, S. S., Woldegiorgis, G., and Wakil, S. J. (2000) *Proc. Natl. Acad. Sci. U.S.A.* **97**, 1444–1449
32c. Nikolskaya, T., Zagnitko, O., Tevzadze, G., Haselkorn, R., and Gornicki, P. (1999) *Proc. Natl. Acad. Sci. U.S.A.* **96**, 14647–14651
32d. Schulte, W., Töpfer, R., Stracke, R., Schell, J., and Martini, N. (1997) *Proc. Natl. Acad. Sci. U.S.A.* **94**, 3465–3470
32e. Jitrapakdee, S., Booker, G. W., Cassady, A. I., and Wallace, J. C. (1997) *J. Biol. Chem.* **272**, 20522–20530
33. Lim, F., Morris, C. P., Occhiodoro, F., and Wallace, J. C. (1988) *J. Biol. Chem.* **263**, 11493–11497
34. Kalousek, F., Darigo, M. D., and Rosenberg, L. E. (1980) *J. Biol. Chem.* **255**, 60–65
35. Leon–Del-Rio, A., and Gravel, R. A. (1994) *J. Biol. Chem.* **269**, 22964–22968
36. Sasaki, Y., Hakamada, K., Suama, Y., Nagano, Y., Furusawa, I., and Matsuno, R. (1993) *J. Biol. Chem.* **268**, 25118–25123
37. Gornicki, P., Podkowinski, J., Scappino, L. A., DiMaio, J., Ward, E., and Haselkorn, R. (1994) *Proc. Natl. Acad. Sci. U.S.A.* **91**, 6860–6864
38. Nenortas, E., and Beckett, D. (1996) *J. Biol. Chem.* **271**, 7559–7567
39. Waldrop, G. L., Rayment, I., and Holden, H. M. (1994) *Biochemistry* **33**, 10249–10256
39a. Blanchard, C. Z., Lee, Y. M., Frantom, P. A., and Waldrop, G. L. (1999) *Biochemistry* **38**, 3393–3400
40. Li, S.-J., and Cronan, J. E., Jr. (1992) *J. Biol. Chem.* **267**, 855–863
41. Li, S.-J., and Cronan, J. E. J. (1992) *J. Biol. Chem.* **267**, 16841–16847
42. Ha, J., Daniel, S., Broyles, S. S., and Kim, K.-H. (1994) *J. Biol. Chem.* **269**, 22162–22168
43. Wood, H. G., Harmon, F. R., Wuhr, B., Hubner, K., and Lynen, F. (1980) *J. Biol. Chem.* **255**, 7397–7409
44. Shenoy, B. C., and Wood, H. G. (1988) *FASEB J.* **2**, 2396–2401
45. Wilson, K. P., Shewchuk, L. M., Brennan, R. G., Otsuka, A. J., and Matthews, B. W. (1992) *Proc. Natl. Acad. Sci. U.S.A.* **89**, 9257–9261
46. Wolf, B. (1995) in *The Metabolic and Molecular Bases of Inherited Disease*, 7th ed., Vol. 2 (Scriver, C. R., Beaudet, A. L., Sly, W. S., and Valle, D., eds), pp. 3151–3177, McGraw-Hill, New York
47. Taroni, F., and Rosenberg, L. E. (1991) *J. Biol. Chem.* **266**, 13267–13271
47a. Reche, P., and Perham, R. N. (1999) *EMBO J.* **18**, 2673–2682

48. Tanaka, K. (1981) *N. Engl. J. Med.* **304**, 839–840
49. Craft, D. V., Goss, N. H., Chandramouli, N., and Wood, H. G. (1985) *Biochemistry* **24**, 2471–2476
50. Chauhan, J., and Dakshinamurti, K. (1986) *J. Biol. Chem.* **261**, 4268–4275
51. Alberts, A. W., and Vagelos, P. R. (1972) in *The Enzymes*, 3rd ed., Vol. 6 (Boyer, P. D., ed), pp. 37–82, Academic Press, New York
52. Kumar, G. K., Beegen, H., and Wood, H. G. (1988) *Biochemistry* **27**, 5972–5978
53. Lynen, F., Knappe, J., Lorch, E., Jutting, G., and Ringelmann, E. (1959) *Angew. Chem. Int. Ed. Engl.* **71**, 481–486
54. Wood, H. G. (1979) *Trends Biochem. Sci.* **4**, N300–N302
55. Polakis, S. E., Guchhait, R. B., and Lane, M. D. (1972) *J. Biol. Chem.* **247**, 1235–1337
55a. Chapman-Smith, A., Forbes, B. E., Wallace, J. C., and Cronan, J. E., Jr. (1997) *J. Biol. Chem.* **272**, 26017–26022
55b. Reddy, D. V., Shenoy, B. C., Carey, P. R., and Sönnichsen, F. D. (2000) *Biochemistry* **39**, 2509–2516
56. Tipton, P. A., and Cleland, W. W. (1988) *J. Am. Chem. Soc.* **110**, 5866–5869
57. Tipton, P. A., and Cleland, W. W. (1988) *Biochemistry* **27**, 4317– 4325
58. Ogita, T., and Knowles, J. R. (1988) *Biochemistry* **27**, 8028–8033
58a. Branson, J. P., and Attwood, P. V. (2000) *Biochemistry* **39**, 7480–7491
59. Perrin, C. L., and Dwyer, T. J. (1987) *J. Am. Chem. Soc.* **109**, 5163–5167
60. Kluger, R., Davis, P. P., and Adawadkar, P. D. (1979) *J. Am. Chem. Soc.* **101**, 5995–6000
61. King, S. W., Natarajan, R., Bembi, R., and Fife, T. H. (1992) *J. Am. Chem. Soc.* **114**, 10715–10721
62. Kluger, R., and Taylor, S. D. (1991) *J. Am. Chem. Soc.* **113**, 996–1001
63. Taylor, S. D., and Kluger, R. (1993) *J. Am. Chem. Soc.* **115**, 867 – 871
64. Hoving, H., Crysell, B., and Leadbay, P. F. (1985) *Biochemistry* **24**, 6163–6169
65. O'Keefe, S. J., and Knowles, J. R. (1986) *Biochemistry* **25**, 6077– 6084
66. Kuo, D. J., and Rose, I. A. (1993) *J. Am. Chem. Soc.* **115**, 387–390
67. Stubbe, J., Fish, S., and Abeles, R. H. (1980) *J. Biol. Chem.* **255**, 236–242
68. Reed, G. H., and Scrutton, M. G. (1974) *J. Biol. Chem.* **249**, 6156–6162
68a. Werneburg, B. G., and Ash, D. E. (1997) *Biochemistry* **36**, 14392–14402
69. DeTitta, G. T., Blessing, R. H., Moss, G. R., King, H. F., Sukumaran, D. K., and Roskwitalski, R. L. (1994) *J. Am. Chem. Soc.* **116**, 6485–6493
70. Legge, G. B., Branson, J. P., and Attwood, P. V. (1996) *Biochemistry* **35**, 3849–3856
71. Mabrouk, G. M., Helmy, I. M., Thampy, K. G., and Wakil, S. J. (1990) *J. Biol. Chem.* **265**, 6330–6338
72. Kim, K.-H., López-Casillas, F., Bai, D. H., Luo, X., and Pape, M. E. (1989) *FASEB J.* **3**, 2250–2256
73. Harwood, J. L. (1988) *Trends Biochem. Sci.* **13**, 330 – 331
74. LauBermair, E., Schwarz, E., Oesterhelt, D., Reinke, H., Beyreuther, K., and Dimroth, P. (1989) *J. Biol. Chem.* **264**, 14710–14715
74a. Jockel, P., Schmid, M., Choinowski, T., and Dimroth, P. (2000) *Biochemistry* **39**, 4320–4326
75. Laussermair, E., Schwarz, E., Oesterhelt, D., Reinke, H., Beyreuther, K., and Dimroth, P. (1989) *J. Biol. Chem.* **264**, 14710–14715
76. Huder, J. B., and Dimroth, P. (1993) *J. Biol. Chem.* **268**, 24564–24571

References

77. Bendrat, K., and Buckel, W. (1993) *Eur. J. Biochem.* **211**, 697–702

78. Ahmad, F., Lygre, D. G., Jacobson, B. E., and Wood, H. G. (1972) *J. Biol. Chem.* **247**, 6299–6305

79. Metzler, D. E. (1960) in *The Enzymes*, 2nd ed., Vol. 2 (Boyer, P. D., Lardy, H., and Myrback, K., eds), pp. 295–337, Academic Press, New York

80. Schellenberger, A., and Schowen, R. L., eds. (1988) *Thiamin Pyrophosphate Biochemistry*, CRC Press, Boca Raton, Florida (2 Vols.)

81. Krampitz, L. O. (1970) *Thiamin Diphosphate and Its Catalytic Functions*, Dekker, New York

82. Gubler, C. J., ed. (1976) *Thiamine*, Wiley, New York

83. Gallo, A., Mieyal, J. J., and Sable, H. Z. (1978) *Bioorg. Chem.* **4**, 147–177

83a. Jordan, F., Li, H., and Brown, A. (1999) *Biochemistry* **38**, 6369–6373

84. Kluger, R. (1987) *Chem. Rev.* **87**, 863–876

85. Sable, H. Z., ed., and Gubler, C. J., ed. (1982) *Ann. N.Y. Acad. Sci.* **378**

86. Karlson, P. (1984) *Trends Biochem. Sci.* **9**, 536–537

87. Panijpan, B. (1979) *J. Chem. Educ.* **56**, 805–806

88. Suchy, J., Mieyal, J. J., Bantle, G., and Sable, H. Z. (1972) *J. Biol. Chem.* **247**, 5905–5912

89. Williams, R. R., and Spies, T. D. (1938) *Vitamin B₁ and Its Use in Medicine*, Macmillan, New York

90. Kluger, R., Lam, J. F., Pezacki, J. P., and Yang, C.-M. (1995) *J. Am. Chem. Soc.* **117**, 11383–11389

91. Windheuser, J. J., and Higuchi, T. (1962) *J. Pharm. Sci.* **51**, 354–364

92. Costello, C. A., Kelleher, N. L., Abe, M., McLafferty, F. W., and Begley, T. P. (1996) *J. Biol. Chem.* **271**, 3445–3452

92a. Campobasso, N., Costello, C. A., Kinsland, C., Begley, T. P., and Ealick, S. E. (1998) *Biochemistry* **37**, 15981–15989

92b. Webb, E., and Downs, D. (1997) *J. Biol. Chem.* **272**, 15702–15707

92c. Nosaka, K., Onozuka, M., Nishino, H., Nishimura, H., Kawasaki, Y., and Ueyama, H. (1999) *J. Biol. Chem.* **274**, 34129–34133

93. Mizuhara, S., and Handler, P. (1954) *J. Am. Chem. Soc.* **76**, 571–573

94. Breslow, R. (1958) *J. Am. Chem. Soc.* **80**, 3719–3726

95. Hopmann, R. F. W., Brugnoni, G. P., and Fol, B. (1982) *J. Am. Chem. Soc.* **104**, 1341–1344

96. Washabaugh, M. W., and Jencks, W. P. (1989) *J. Am. Chem. Soc.* **111**, 683–692

97. Breslow, R., and McNelis, E. (1959) *J. Am. Chem. Soc.* **81**, 3080–3082

98. Kluger, R., Chin, J., and Smyth, T. (1981) *J. Am. Chem. Soc.* **103**, 884–888

99. Kluger, R., and Smyth, T. (1981) *J. Am. Chem. Soc.* **103**, 1216–1218

100. Kluger, R., Karimian, K., Gish, G., Pangborn, W. A., and DeTitta, G. T. (1987) *J. Am. Chem. Soc.* **109**, 618–620

101. Crosby, J., Stone, R., and Lienhard, G. E. (1970) *J. Am. Chem. Soc.* **92**, 2891–2900

102. Shin, W., Oh, D.-G., Chae, C.-H., and Yoon, T.-S. (1993) *J. Am. Chem. Soc.* **115**, 12238–12250

103. Lobell, M., and Crout, D. H. G. (1996) *J. Am. Chem. Soc.* **118**, 1867–1873

104. Gallo, A. A., and Sable, H. Z. (1974) *J. Biol. Chem.* **249**, 1382–1389

105. Schellenberger, A. (1967) *Angew. Chem. Int. Ed. Engl.* **6**, 1024–1035

106. Golbik, R., Neef, H., Hübner, G., König, S., Seliger, B., Meshalkina, L., Kochetov, G. A., and Schellenberger, A. (1991) *Bioorg. Chem.* **19**, 10–17

107. König, S., Schellenberger, A., Neef, H., and Schneider, G. (1994) *J. Biol. Chem.* **269**, 10879–10882

108. Jordan, F., and Mariam, Y. H. (1978) *J. Am. Chem. Soc.* **100**, 2534–2541

109. Metzler, D. E. (1977) *Biochemistry; The Chemical Reactions of Living Cells*, Academic Press, New York (440–441)

109a. Kern, D., Kern, G., Neef, H., Tittmann, K., Killenberg-Jabs, M., Wikner, C., Schneider, G., and Hübner, G. (1997) *Science* **275**, 67–70

110. Singleton, C. K., Wang, J. J.-L., Shan, L., and Martin, P. R. (1996) *Biochemistry* **35**, 15865–15869

111. Guo, F., Zhang, D., Kahyaoglu, A., Farid, R. S., and Jordan, F. (1998) *Biochemistry* **37**, 13379–13391

111a. Li, H., Furey, W., and Jordan, F. (1999) *Biochemistry* **38**, 9992–10003

112. Gish, G., Smyth, T., and Kluger, R. (1988) *J. Am. Chem. Soc.* **110**, 6230–6234

113. Annan, N., Paris, R., and Jordan, F. (1989) *J. Am. Chem. Soc.* **111**, 8895–8901

114. Alvarez, F. J., Ermer, J., Hübner, G., Schellenberger, A., and Schowen, R. L. (1991) *J. Am. Chem. Soc.* **113**, 8402–8409

115. Alvarez, F. J., Ermer, J., Hübner, G., Schellenberger, A., and Schowen, R. L. (1995) *J. Am. Chem. Soc.* **117**, 1678–1683

116. Sun, S., Duggleby, R. G., and Schowen, R. L. (1995) *J. Am. Chem. Soc.* **117**, 7317–7322

117. Harris, T. K., and Washabaugh, M. W. (1995) *Biochemistry* **34**, 13994–14000

118. Harris, T. K., and Washabaugh, M. W. (1995) *Biochemistry* **34**, 14001–14011

119. Muller, Y. A., and Schulz, G. E. (1993) *Science* **259**, 965–967

120. Muller, Y. A., Schumacher, G., Rudolph, R., and Schulz, G. E. (1994) *J. Mol. Biol.* **237**, 315–335

121. Dyda, F., Furey, W., Swaminathan, S., Sax, M., Farrenkopf, B., and Jordan, F. (1993) *Biochemistry* **32**, 6165–6170

122. Arjunan, P., Umland, T., Dyda, R., Swaminathan, S., Furey, W., Sax, M., Farrenkopf, B., Gao, Y., Zhang, D., and Jordan, F. (1996) *J. Mol. Biol.* **256**, 590–600

122a. Dobritzsch, D., König, S., Schneider, G., and Lu, G. (1998) *J. Biol. Chem.* **273**, 20196–20204

122b. Chang, A. K., Nixon, P. F., and Duggleby, R. G. (2000) *Biochemistry* **39**, 9430–9437

122c. Li, H., and Jordan, F. (1999) *Biochemistry* **38**, 10004–10012

123. Wikner, C., Meshalkina, L., Nilsson, U., Nikkola, M., Lindqvist, Y., Sundström, M., and Schneider, G. (1994) *J. Biol. Chem.* **269**, 32144–32150

124. Nikkola, M., Lindqvist, Y., and Schneider, G. (1994) *J. Mol. Biol.* **238**, 387–404

124a. Hasson, M. S., Muscate, A., McLeish, M. J., Polovnikova, L. S., Gerlt, J. A., Kenyon, G. L., Petsko, G. A., and Ringe, D. (1998) *Biochemistry* **37**, 9918–9930

125. Killenberg-Jabs, M., König, S., Eberhardt, I., Hohmann, S., and Hübner, G. (1997) *Biochemistry* **36**, 1900–1905

126. Koga, J., Adachi, T., and Hidaka, H. (1992) *J. Biol. Chem.* **267**, 15823–15828

127. Hasson, M. S., Muscate, A., Henehan, G. T. M., Guidinger, P. F., Petsko, G. A., Ringe, D., and Kenyon, G. L. (1995) *Protein Sci.* **4**, 955–959

128. Ibdah, M., Bar-Ilan, A., Livnah, O., Schloss, J. V., Barak, Z., and Chipman, D. M. (1996) *Biochemistry* **35**, 16282–16291

129. Bernasconi, P., Woodworth, A. R., Rosen, B. A., Subramanian, M. V., and Siehl, D. L. (1995) *J. Biol. Chem.* **270**, 17381–17385

130. Ott, K.-H., Kwagh, J.-G., Stockton, G. W., Sidorov, V., and Kakefuda, G. (1996) *J. Mol. Biol.* **263**, 359–368

131. Stivers, J. T., and Washabaugh, M. W. (1993) *Biochemistry* **32**, 13472–13482

132. Chang, Y.-Y., Wang, A.-Y., and Cronan, J. E., Jr. (1993) *J. Biol. Chem.* **268**, 3911–3919

132a. Kuzuyama, T., Takahashi, S., Takagi, M., and Seto, H. (2000) *J. Biol. Chem.* **275**, 19928–19932

132b. Wikner, C., Nilsson, U., Meshalkina, L., Udekwu, C., Lindqvist, Y., and Schneider, G. (1997) *Biochemistry* **36**, 15643–15649

133. Chuang, L. F., and Collins, E. B. (1968) *J. Bacteriol.* **95**, 2083–2089

134. Gruys, K. J., Datta, A., and Frey, P. A. (1989) *Biochemistry* **28**, 9071–9080

135. Flournoy, D. S., and Frey, P. A. (1989) *Biochemistry* **28**, 9594–9602

135a. Tittmann, K., Proske, D., Spinka, M., Ghisla, S., Rudolph, R., Hübner, G., and Kern, G. (1998) *J. Biol. Chem.* **273**, 12929–12934

136. Matsuda, T., and Cooper, J. R. (1983) *Biochemistry* **22**, 2209–2213

137. Barchi, R. L., and Viale, R. O. (1976) *J. Biol. Chem.* **251**, 193–197

138. Nishino, K., Itokawa, Y., Nishino, N., Piros, K., and Cooper, J. R. (1983) *J. Biol. Chem.* **258**, 1871–11878

139. Braunstein, A. E., and Kritzmann, M. G. (1937) *Enzymologia* **2**, 751–752

140. Braunstein, A. E., and Kritzmann, M. G. (1937) *Biokhim.* **2**, 242–262,859–874

141. Braunstein, A. E. (1973) in *The Enzymes*, 3rd ed., Vol. 9 (Boyer, P. D., ed), pp. 379–481, Academic Press, New York

142. Christen, P., and Metzler, D. E., eds. (1984) *Transaminases*, Wiley, New York

143. Torchinsky, Y. (1986) in *Vitamin B₆, Pyridoxal Phosphate: Chemical, Biochemical and Medical Aspects*, Vol. 1 (Dolphin, R., Poulson, R., and Avramovic, O., eds), pp. 169–221, Wiley, New York

144. Snell, E. E. (1944) *J. Biol. Chem.* **154**, 313–314

145. Snell, E. E. (1986) in *Vitamin B₆, Pyridoxal Phosphate: Chemical, Biochemical and Medical Aspects*, Vol. 1A (Dolphin, D., Poulson, R., and Avramovic, O., eds), pp. 1–12, Wiley, New York

146. Gunsalus, I. C., Bellamy, W. D., and Umbreit, W. W. (1944) *J. Biol. Chem.* **155**, 685–686

147. Snell, E. E. (1981) in *Methods in Vitamin B₆ Nutrition* (Leklem, J. E., and Reynolds, R. D., eds), pp. 1–19, Plenum, New York

148. Metzler, D. E., Ikawa, M., and Snell, E. E. (1954) *J. Am. Chem. Soc.* **76**, 648–652

149. Kallen, R. G., Korpela, T., Martell, A. E., Matsushima, Y., Metzler, C. M., Metzler, D. E., Morozov, Y. V., Ralston, I. M., Savin, F. A., Torchinsky, Y. M., and Ueno, H. (1984) in *Transaminases* (Christen, P., and Metzler, D. E., eds), pp. 37–106, Wiley, New York

150. Leussing, D. L. (1986) in *Vitamin B₆, Pyridoxal Phosphate: Chemical, Biochemical and Medical Aspects*, Vol. 1A (Dolphin, D., Poulson, R., and Avramovic, O., eds), pp. 69–115, Wiley, New York

151. Braunstein, A. E., and Kritzmann, M. G. (1952) *Dokl. Akad. Nauk. SSSR* **85**, 1115–1118

152. Braunstein, A. E., and Kritzmann, M. G. (1952) *Biokhim.* **18**, 393–411

153. Shaw, J. P., Petsko, G. A., and Ringe, D. (1997) *Biochemistry* **36**, 1329–1342

153a. Stamper, C. G. F., Morollo, A. A., and Ringe, D. (1998) *Biochemistry* **37**, 10438–10445

153b. Morollo, A. A., Petsko, G. A., and Ringe, D. (1999) *Biochemistry* **38**, 3293–3301

153c. Sun, S., and Toney, M. D. (1999) *Biochemistry* **38**, 4058–4065

154. Aswad, D. W., and Johnson, B. A. (1987) *Trends Biochem. Sci.* **12**, 155–158

154a. Watababe, A., Kurokawa, Y., Yoshimura, T., Kurihara, T., Soda, K., and Esaki, N. (1999) *J. Biol. Chem.* **274**, 4189–4194

References

155. Hoffmann, K., Schneider-Scherzer, E., Kleinkauf, H., and Zocher, R. (1994) *J. Biol. Chem.* **269**, 12710–12714

155a. Cheng, Y.-Q., and Walton, J. D. (2000) *J. Biol. Chem.* **275**, 4906–4911

156. Schnell, M. J., Cooper, O. B., and Snyder, S. H. (1997) *Proc. Natl. Acad. Sci. U.S.A.* **94**, 2013–2018

157. Yamauchi, T., Choi, S.-Y., Okada, H., Yohda, M., Kumagai, H., Esaki, N., and Soda, K. (1992) *J. Biol. Chem.* **267**, 18361–18364

158. Yohda, M., Endo, I., Abe, Y., Ohta, T., Iida, T., Maruyama, T., and Kagawa, Y. (1996) *J. Biol. Chem.* **271**, 22017–22021

159. Li, N., and Mattoo, A. K. (1994) *J. Biol. Chem.* **269**, 6908– 6917

160. Li, N., Jiang, X. N., Cai, G. P., and Yang, S. F. (1996) *J. Biol. Chem.* **271**, 25738–25741

161. Hohenester, E., White, M. F., Kirsch, J. F., and Jansonius, J. N. (1994) *J. Mol. Biol.* **243**, 947–949

161a. Capitani, G., Hohenester, E., Feng, L., Storici, P., Kirsch, J. F., and Jansonius, J. N. (1999) *J. Mol. Biol.* **294**, 745–756

162. Ramalingam, K., Lee, K.-M., Woodard, R. W., Bleecker, A. B., and Kende, H. (1985) *Proc. Natl. Acad. Sci. U.S.A.* **82**, 7820–7824

163. Rhee, S., Silva, M. M., Hyde, C. C., Rogers, P. H., Metzler, C. M., Metzler, D. E., and Arnone, A. (1997) *J. Biol. Chem.* **272**, 17293–17302

164. Malashkevich, V. N., Strokopytov, B. V., Borisov, V. V., Dauter, Z., Wilson, K. S., and Torchinsky, Y. M. (1995) *J. Mol. Biol.* **247**, 111–124

165. McPhalen, C. A., Vincent, M. G., and Jansonius, J. N. (1992) *J. Mol. Biol.* **225**, 495–517

166. Kamitori, S., Okamoto, A., Hirotsu, K., Higuchi, T., Kuramitsu, S., Kagamiyama, H., Matsuura, Y., and Katsube, Y. (1990) *J. Biochem.* **108**, 175–184

167. Okamoto, A., Higuchi, T., Hirotsu, K., Kuramitsu, S., and Kagamiyama, H. (1994) *J. Biochem.* **116**, 95–107

167a. Jeffery, C. J., Barry, T., Doonan, S., Petsko, G. A., and Ringe, D. (1998) *Protein Sci.* **7**, 1380–1387

168. Uchida, C., Funai, T., Oda, T., Ohbayashi, K., and Ichiyama, A. (1994) *J. Biol. Chem.* **269**, 8846–8856

168a. Matsui, I., Matsui, E., Sakai, Y., Kikuchi, H., Kawarabayasi, Y., Ura, H., Kawaguchi, S.-i, Kuramitsu, S., and Harata, K. (2000) *J. Biol. Chem.* **275**, 4871–4879

168b. Okada, K., Hirotsu, K., Sato, M., Hayashi, H., and Kagamiyama, H. (1997) *J. Biochem.* **121**, 637–641

168c. Davoodi, J., Drown, P. M., Bledsoe, R. K., Wallin, R., Reinhart, G. D., and Hutson, S. M. (1998) *J. Biol. Chem.* **273**, 4982–4989

168d. van Ophem, P. W., Peisach, D., Erickson, S. D., Soda, K., Ringe, D., and Manning, J. M. (1999) *Biochemistry* **38**, 1323–1331

168e. Hester, G., Stark, W., Moser, M., Kallen, J., Markovic'-Housley, Z., and Jansonius, J. N. (1999) *J. Mol. Biol.* **286**, 829–850

168f. Storici, P., Capitani, G., Müller, R., Schirmer, T., and Jansonius, J. N. (1999) *J. Mol. Biol.* **285**, 297–309

168g. Shah, S. A., Shen, B. W., and Brünger, A. T. (1997) *Structure* **5**, 1067–1075

168h. Ledwidge, R., and Blanchard, J. S. (1999) *Biochemistry* **38**, 3019–3024

168i. Smith, M. A., King, P. J., and Grimm, B. (1998) *Biochemistry* **37**, 319–329

168j. Käck, H., Sandmark, J., Gibson, K., Schneider, G., and Lindqvist, Y. (1999) *J. Mol. Biol.* **291**, 857–876

168k. Contestabile, R., Angelaccio, S., Maytum, R., Bossa, F., and John, R. A. (2000) *J. Biol. Chem.* **275**, 3879–3886

169. Kintanar, A., Metzler, C. M., Metzler, D. E., and Scott, R. D. (1991) *J. Biol. Chem.* **266**, 17222–17229

170. Miles, E. W. (1986) in *Vitamin B₆, Pyridoxal Phosphate: Chemical, Biochemical and Medical Aspects*, Vol. 1B (Dolphin, D., Poulson, R., and Avramovíc, O., eds), pp. 253–310, Wiley, New York

171. Marceau, M., Lewis, S. D., Kojiro, C. L., and Shafer, J. A. (1989) *J. Biol. Chem.* **264**, 2753–2757

172. Ogawa, H., Gomi, T., Konishi, K., Date, T., Nakashima, H., Nose, K., Matsuda, Y., Peraino, C., Pilot, H. C., and Fujioka, M. (1989) *J. Biol. Chem.* **264**, 15818–15823

173. Obmolova, G., Tepliakov, A., Harutyunyan, E., Wahler, G., and Schnackerz, K. D. (1990) *J. Mol. Biol.* **214**, 641–642

173a. Ogawa, J., Takusagawa, F., Wakaki, K., Kishi, H., Eskandarian, M. R., Kobayashi, M., Date, T., Huh, N.-H., and Pitot, H. C. (1999) *J. Biol. Chem.* **274**, 12855–12860

174. Snell, E. E. (1975) *Adv. Enzymol.* **42**, 287–333

175. Metzler, C. M., Viswanath, R., and Metzler, D. E. (1991) *J. Biol. Chem.* **266**, 9374–9381

176. Sloan, M. J., and Phillips, R. S. (1996) *Biochemistry* **35**, 16165–16173

176a. Isupov, M. N., Antson, A. A., Dodson, E. J., Dodson, G. G., Dementieva, I. S., Zakomirdina, L. N., Wilson, K. S., Dauter, Z., Lebedev, A. A., and Harutyunyan, E. H. (1998) *J. Mol. Biol.* **276**, 603–623

176b. Ikushiro, H., Hayashi, H., Kawata, Y., and Kagamiyama, H. (1998) *Biochemistry* **37**, 3043–3052

176c. Phillips, R. S., Sundararaju, B., and Faleev, N. G. (2000) *J. Am. Chem. Soc.* **122**, 1008–1014

177. Antson, A. A., Demidkina, T. V., Gollnick, P., Dauter, Z., Von Tersch, R. L., Long, J., Berezhnoy, S. N., Phillips, R. S., Harutyunyan, E. H., and Wilson, K. S. (1993) *Biochemistry* **32**, 4195–4206

177a. Sundararaju, B., Antson, A. A., Phillips, R. S., Demidkina, T. V., Barbolina, M. V., Gollnick, P., Dodson, G. G., and Wilson, K. S. (1997) *Biochemistry* **36**, 6502–6510

178. Chen, H. Y., Demidkina, T. V., and Phillips, R. S. (1995) *Biochemistry* **34**, 12276–12283

178a. Sundararaju, B., Chen, H., Shilcutt, S., and Phillips, R. S. (2000) *Biochemistry* **39**, 8546–8555

179. Block, E. (1985) *Sci. Am.* **252**(Mar), 114–119

180. Block, E., Gillies, J. Z., Gillies, C. W., Bazzi, A. A., Putman, D., Revelle, L. K., Wang, D., and Zhang, X. (1996) *J. Am. Chem. Soc.* **118**, 7492–7501

181. Gentry-Weeks, C. R., Spokes, J., and Thompson, J. (1995) *J. Biol. Chem.* **270**, 7695–7702

182. Clausen, T., Huber, R., Laber, B., Pohlenz, H.-D., and Messerschmidt, A. (1996) *J. Mol. Biol.* **262**, 202–224

183. Nagasawa, T., Tanizawa, K., Satoda, T., and Yamada, H. (1988) *J. Biol. Chem.* **263**, 958–964

184. Hyde, C. C., Ahmed, S. A., Padlan, E. A., Miles, E. W., and Davies, D. R. (1988) *J. Biol. Chem.* **263**, 17857–17871

184a. Rhee, S., Miles, E. W., and Davies, D. R. (1998) *J. Biol. Chem.* **273**, 8553–8555

184b. Weyand, M., and Schlichting, I. (1999) *Biochemistry* **38**, 16469–16480

184c. Sachpatzidis, A., Dealwis, C., Lubetsky, J. B., Liang, P.-H., Anderson, K. S., and Lolis, E. (1999) *Biochemistry* **38**, 12665–12674

184d. Rhee, S., Miles, E. W., Mozzarelli, A., and Davies, D. R. (1998) *Biochemistry* **37**, 10653–10659

185. Rege, V. D., Kredich, N. M., Tai, C.-H., Karsten, W. E., Schnackerz, K. D., and Cook, P. F. (1996) *Biochemistry* **35**, 13485–13493

185a. Burkhard, P., Tai, C.-H., Jansonius, J. N., and Cook, P. F. (2000) *J. Mol. Biol.* **303**, 279–286

186. Cook, P. F., Tai, C.-H., Hwang, C.-C., Woehl, E. U., Dunn, M. F., and Schnackerz, K. D. (1996) *J. Biol. Chem.* **271**, 25842–25849

186a. Benci, S., Vaccari, S., Mozzarelli, A., and Cook, P. F. (1997) *Biochemistry* **36**, 15419–15427

187. Kery, V., Bukovska, G., and Kraus, J. P. (1994) *J. Biol. Chem.* **269**, 25283–25288

188. Peracchi, A., Bettati, S., Mozzarelli, A., Rossi, G. L., Miles, E. W., and Dunn, M. F. (1996) *Biochemistry* **35**, 1872–1880

188a. Jhee, K.-H., McPhie, P., and Miles, E. W. (2000) *J. Biol. Chem.* **275**, 11541–11544

188b. Taoka, S., Widjaja, L., and Banerjee, R. (1999) *Biochemistry* **38**, 13155–13161

188c. Ojha, S., Hwang, J., Kabil, O., Penner-Hahn, J. E., and Banerjee, R. (2000) *Biochemistry* **39**, 10542–10547

189. Palcic, M. M., and Floss, H. G. (1986) in *Vitamin B₆, Pyridoxal Phosphate: Chemical, Biochemical and Medical Aspects*, Vol. 1A (Dolphin, D., Poulson, R., and Avramovíc, O., eds), pp. 25–68, Wiley, New York

190. Pan, P., Woehl, E., and Dunn, M. F. (1997) *Trends Biochem. Sci.* **22**, 22–27

191. Sukhareva, B. S. (1986) in *Vitamin B₆, Pyridoxal Phosphate: Chemical, Biochemical and Medical Aspects*, Vol. 1B (Dolphin, D., Poulson, R., and Avramovíc, O., eds), pp. 325–353, Wiley, New York

192. Sandmeier, E., Hale, T. I., and Christen, P. (1994) *Eur. J. Biochem.* **221**, 997–1002

193. Nathan, B., Hsu, C.-C., Bao, J., Wu, R., and Wu, J.-Y. (1994) *J. Biol. Chem.* **269**, 7249–7254

194. Porter, T. G., and Martin, D. L. (1988) *Biochim. Biophys. Acta.* **874**, 235–244

195. Bu, D.-F., Erlander, M. G., Hitz, B. C., Tillakaratne, N. J. K., Kaufman, D. L., Wagner-McPherson, C. B., Evans, G. A., and Tobin, A. J. (1992) *Proc. Natl. Acad. Sci. U.S.A.* **89**, 2115–2119

196. Dirkx, R., Jr., Thomas, A., Li, L., Lernmark, Å., Sherwin, R. S., DeCamilli, P., and Solimena, M. (1995) *J. Biol. Chem.* **270**, 2241–2246

196a. Qu, K., Martin, D. L., and Lawrence, C. E. (1998) *Protein Sci.* **7**, 1092–1105

196b. Kanaani, J., Lissin, D., Kash, S. F., and Baekkeskov, S. (1999) *J. Biol. Chem.* **274**, 37200–37209

197. Malashkevich, V. N., Filipponi, P., Sauder, U., Dominici, P., Jansonius, J. N., and Voltattorni, C. B. (1992) *J. Mol. Biol.* **224**, 1167–1170

198. Dominici, P., Filipponi, P., Schinninà, M. E., Barra, D., and Voltattorni, C. B. (1990) *Ann. N.Y. Acad. Sci.* **585**, 162–171

199. Reith, J., Benkelfat, C., Sherwin, A., Yasuhara, Y., Kuwabara, H., Andermann, F., Bachneff, S., Cumming, P., Diksic, M., Dyve, S. E., Etinne, P., Evans, A. C., Lal, S., Shevell, M., Savard, G., Wong, D. F., Chouinard, G., and Gjedde, A. (1994) *Proc. Natl. Acad. Sci. U.S.A.* **91**, 11651–11654

199a. Hayashi, H., Tsukiyama, F., Ishii, S., Mizuguchi, H., and Kagamiyama, H. (1999) *Biochemistry* **38**, 15615–15622

199b. Nishino, J., Hayashi, H., Ishii, S., and Kagamiyama, H. (1997) *J. Biochem.* **121**, 604–611

200. Bhattacharjee, M. K., and Snell, E. E. (1990) *J. Biol. Chem.* **265**, 6664–6668

201. Chudomelka, P. J., Ramaley, R. F., and Murrin, L. C. (1990) *Neurochemical Research* **15**, 17–24

202. Yatsunami, K., Tsuchikawa, M., Kamada, M., Hori, K., and Higuchi, T. (1995) *J. Biol. Chem.* **270**, 30813–30817

References

203. Osterman, A. L., Kinch, L. N., Grishin, N. V., and Phillips, M. A. (1995) *J. Biol. Chem.* **270**, 11797–11802

204. Momany, C., Ernst, S., Ghosh, R., Chang, N.-L., and Hackert, M. L. (1995) *J. Mol. Biol.* **252**, 643–655

205. Osterman, A. L., Brooks, H. B., Jackson, L., Abbott, J. J., and Phillips, M. A. (1999) *Biochemistry* **38**, 11814–11826

206. Viguera, E., Trelles, O., Urdiales, J. L., Matés, J. M., and Sánchez-Jiménez, F. (1994) *Trends Biochem. Sci.* **19**, 318–319

206a. Almrud, J. J., Oliveira, M. A., Kern, A. D., Grishin, N. V., Phillips, M. A., and Hackert, M. L. (2000) *J. Mol. Biol.* **295**, 7–16

206b. Jackson, L. K., Brooks, H. B., Osterman, A. L., Goldsmith, E. J., and Phillips, M. A. (2000) *Biochemistry* **39**, 11247–11257

206c. Grishin, N. V., Osterman, A. L., Brooks, H. B., Phillips, M. A., and Goldsmith, E. J. (1999) *Biochemistry* **38**, 15174–15184

207. Kume, A., Koyata, H., Sakakibara, T., Ishiguro, Y., Kure, S., and Hiraga, K. (1991) *J. Biol. Chem.* **266**, 3323–3329

208. Pasternack, L. B., Laude, D. A., Jr., and Appling, D. R. (1992) *Biochemistry* **31**, 8713–8719

209. Kennard, O., and Salisbury, S. A. (1993) *J. Biol. Chem.* **268**, 10701–10704

209a. Okamura-Ikeda, K., Fujiwara, K., and Motokawa, Y. (1999) *J. Biol. Chem.* **274**, 17471–17477

209b. Guilhaudis, L., Simorre, J.-P., Blackledge, M., Marion, D., Gans, P., Neuburger, M., and Douce, R. (2000) *Biochemistry* **39**, 4259–4266

210. Akhtar, M., Stevenson, D. E., and Gani, D. (1990) *Biochemistry* **29**, 7648–7660

210a. Zhou, X., and Toney, M. D. (1999) *Biochemistry* **38**, 311–320

210b. Malashkevich, V. N., Strop, P., Keller, J. W., Jansonius, J. N., and Toney, M. D. (1999) *J. Mol. Biol.* **294**, 193–200

211. Scarsdale, J. N., Kazanina, G., Radaev, S., Schirch, V., and Wright, H. T. (1999) *Biochemistry* **38**, 8347–8358

212. Stover, P. J., Chen, L. H., Suh, J. R., Stover, D. M., Keyomarsi, K., and Shane, B. (1997) *J. Biol. Chem.* **272**, 1842–1848

212a. Szebenyi, D. M. E., Liu, X., Kriksunov, I. A., Stover, P. J., and Thiel, D. J. (2000) *Biochemistry* **39**, 13313–13323

213. Webb, H. K., and Matthews, R. G. (1995) *J. Biol. Chem.* **270**, 17204–17209

213a. Jagath, J. R., Sharma, B., Appaji Rao, N., and Savithri, H. S. (1997) *J. Biol. Chem.* **272**, 24355–24362

213b. Scarsdale, J. N., Radaev, S., Kazanina, G., Schirch, V., and Wright, H. T. (2000) *J. Mol. Biol.* **296**, 155–168

214. Tong, H., and Davis, L. (1994) *J. Biol. Chem.* **269**, 4057–4064

215. Tong, H., and Davis, L. (1995) *Biochemistry* **34**, 3362–3367

216. Cotter, P. D., Baumann, M., and Bishop, D. F. (1992) *Proc. Natl. Acad. Sci. U.S.A.* **89**, 4028–4032

217. Ferreira, G. C., Neame, P. J., and Dailey, H. A. (1993) *Protein Sci.* **2**, 1959–1965

218. Cox, T. C., Bottomley, S. S., Wiley, J. S., Bawden, M. J., Matthews, C. S., and May, B. K. (1994) *N. Engl. J. Med.* **330**, 675–679

218a. Hunter, G. A., and Ferreira, G. C. (1999) *J. Biol. Chem.* **274**, 12222–12228

218b. Hunter, G. A., and Ferreira, G. C. (1999) *Biochemistry* **38**, 3711–3718

219. Williams, R. D., Nixon, D. W., and Merrill, A. H., Jr. (1984) *Cancer Research* **44**, 1918–1923

220. Webster, S. P., Alexeev, D., Campopiano, D. J., Watt, R. M., Alexeeva, M., Sawyer, L., and Baxter, R. L. (2000) *Biochemistry* **39**, 516–528

220a. Steegborn, C., Messerschmidt, A., Laber, B., Streber, W., Huber, R., and Clausen, T. (1999) *J. Mol. Biol.* **290**, 983–996

220b. Steegborn, C., Clausen, T., Sondermann, P., Jacob, U., Worbs, M., Marinkovic, S., Huber, R., and Wahl, M. C. (1999) *J. Biol. Chem.* **274**, 12675–12684

220c. Laber, B., Gerbling, K.-P., Harde, C., Neff, K.-H., Nordhoff, E., and Pohlenz, H.-D. (1994) *Biochemistry* **33**, 3413–3423

221. Tate, S. E., and Meister, A. (1971) *Adv. Enzymol.* **35**, 503–543

222. Mihara, H., Kurihara, T., Watanabe, T., Yoshimura, T., and Esaki, N. (2000) *J. Biol. Chem.* **275**, 6195–6200

223. Esaki, N., Seraneeprakarn, V., Tanaka, H., and Soda, K. (1988) *J. Bacteriol.* **170**, 751–756

223a. Mihara, H., Maeda, M., Fujii, T., Kurihara, T., Hata, Y., and Esaki, N. (1999) *J. Biol. Chem.* **274**, 14768–14772

223b. Fujii, T., Maeda, M., Mihara, H., Kurihara, T., Esaki, N., and Hata, Y. (2000) *Biochemistry* **39**, 1263–1273

224. Zheng, L., and Dean, D. R. (1994) *J. Biol. Chem.* **269**, 18723–18726

225. Phillips, R. S., and Dua, R. K. (1991) *J. Am. Chem. Soc.* **113**, 7385–7388

226. Dua, R. K., Taylor, E. W., and Phillips, R. S. (1993) *J. Am. Chem. Soc.* **115**, 1264–1270

227. Krebs, E. G., and Fischer, E. H. (1964) *Vitam. Horm.(N. Y.)* **22**, 399–410

228. Rubenstein, P. A., and Strominger, J. L. (1974) *J. Biol. Chem.* **249**, 3776–3781

229. Johnson, D. A., Gassner, G. T., Bandarian, V., Ruzicka, F. J., Ballou, D. P., Reed, G. H., and Liu, H.-w. (1996) *Biochemistry* **35**, 15846–15856

230. Gonzalez-Porqué, P. (1986) in *Vitamin B₆, Pyridoxal Phosphate: Chemical, Biochemical and Medical Aspects*, Vol. 1B (Dolphin, D., Poulson, R., and Avramovic, O., eds), pp. 391–419, Wiley, New York

231. Bailey, G. B., Chotamangsa, O., and Vuttivej, K. (1970) *Biochemistry* **9**, 3243–3248

232. Toney, M. D., Hohenester, E., Cowan, S. W., and Jansonius, J. N. (1993) *Science* **261**, 756–759

233. Dunathan, H. C. (1971) *Adv. Enzymol.* **35**, 79–134

234. Tobler, H. P., Gehring, H., and Christen, P. (1987) *J. Biol. Chem.* **262**, 8985–8989

235. Sugio, S., Petsko, G. A., Manning, J. M., Soda, K., and Ringe, D. (1995) *Biochemistry* **34**, 9661–9669

236. Syke, G. E., Potts, R., and Floss, H. G. (1974) *J. Am. Chem. Soc.* **96**, 1593–1595

237. Floss, H. G., and Vederas, J. C. (1982) in *New Comprehensive Biochemistry Stereochemistry*, Vol. 3 (Tamm, C., ed), pp. 161–199, Elsevier, Amsterdam

238. Palcic, M. M., and Floss, H. G. (1986) in *Vitamin B₆, Pyridoxal Phosphate: Chemical, Biochemical and Medical Aspects*, Vol. 1A (Dolphin, D., Poulson, R., and Avramovic, O., eds), pp. 25–68, Wiley, New York

239. Metzler, D. E., Harris, C. M., Johnson, R. J., Siano, D. B., and Thomson, J. A. (1973) *Biochemistry* **12**, 5377–5392

240. Metzler, C. M., Cahill, A. E., and Metzler, D. E. (1980) *J. Am. Chem. Soc.* **102**, 6075–6082

241. Schirch, L., and Jenkins, W. T. (1964) *J. Biol. Chem.* **239**, 3801–3807

242. Metzler, C. M., Harris, A. G., and Metzler, D. E. (1988) *Biochemistry* **27**, 4923–4933

243. Schnackerz, K. D. (1986) in *Vitamin B₆, Pyridoxal Phosphate: Chemical, Biochemical and Medical Aspects*, Vol. 1A (Dolphin, D., Poulson, R., and Avramovic, O., eds), pp. 245–264, Wiley, New York

243a. Mehta, P. K., and Christen, P. (1998) in *Advances in Enzymology and Related Areas of Molecular Biology*, Vol. 74 (Purich, D. L., ed), Wiley, New York (pp. 129–184)

243b. Jansonius, J. N. (1998) *Current Opinion in Structural Biology* **8**, 759–769

244. Watanabe, N., Yonaha, K., Sakabe, K., Sakabe, N., Aibara, S., and Morita, Y. (1991) in *Enzymes Dependent on Pyridoxal Phosphate and Other Carbonyl Compounds as Cofactors* (Fukui, T., Kagamiyama, H., Soda, K., and Wada, H., eds), pp. 121–124, Pergamon Press, Oxford

245. Huston, S. M., Bledsoe, R. K., Hall, T. R., and Dawson, P. A. (1995) *J. Biol. Chem.* **270**, 30344–30352

246. Ivanov, V. I., and Karpeisky, M. Y. (1969) *Adv. Enzymol.* **32**, 21–53

247. Arnone, A., Christen, P., Jansonius, J. N., and Metzler, D. E. (1985) in *Transaminases* (Christen, P., and Metzler, D. E., eds), pp. 326–370, Wiley, New York

248. Rhee, S., Silva, M. M., Hyde, C. C., Rogers, P. H., Metzler, C. M., Metzler, D. E., and Arnone, A. (1997) *J. Biol. Chem.* **272**, 17293–17302

249. Metzler, C. M., Mitra, J., Metzler, D. E., Makinen, M. W., Hyde, C. C., Rogers, P., and Arnone, A. (1988) *J. Mol. Biol.* **203**, 197–220

250. Malashkevich, V. N., Toney, M. D., and Jansonius, J. N. (1993) *Biochemistry* **32**, 13451–13462

251. von Stosch, A. G. (1996) *Biochemistry* **35**, 15260–15268

252. Metzler, C. M., Metzler, D. E., Kintanar, A., Scott, R. D., and Marceau, M. (1991) *Biochem. Biophys. Res. Commun.* **178**, 385–392

253. Fukui, T., Kagamiyama, H., Soda, K., and Wada, H., eds. (1991) *Enzymes Dependent on Pyridoxal Phosphate and Other Carbonyl Compounds as Cofactors*, Pergamon Press, Oxford

254. Marino, G., Sannia, G., and Bossa, F., eds. (1994) *Biochemistry of Vitamin B₆ and PQQ*, Birkhäuser, Basel

254a. Iriarte, A., Kagan, H. M., and Martinez-Carrion, M., eds. (2000) *Biochemistry and Molecular Biology of Vitamin B6 and PQQ-dependent Proteins*, Birkhäuser Verlag, Basel

255. Relimpio, A., Iriarte, A., Chlebowski, J. F., and Martinez-Carrion, M. (1981) *J. Biol. Chem.* **256**, 4478–4488

256. Gloss, L. M., and Kirsch, J. F. (1995) *Biochemistry* **34**, 12323–12332

257. Hwang, C.-C., Woehl, E. U., Minter, D. E., Dunn, M. F., and Cook, P. F. (1996) *Biochemistry* **35**, 6358–6365

258. Pan, P., and Dunn, M. F. (1996) *Biochemistry* **35**, 5002–5013

259. Anderson, K. S., Kim, A. Y., Quillen, J. M., Sayers, E., Yang, X.-J., and Miles, E. W. (1995) *J. Biol. Chem.* **270**, 29936–29944

260. Schnackerz, K. D., Tai, C.-H., Simmons, J. W., III, Jacobson, T. M., Rao, G. S. J., and Cook, P. F. (1995) *Biochemistry* **34**, 12152–12160

261. Hayashi, H., and Kagamiyama, H. (1995) *Biochemistry* **34**, 9413–9423

262. Mattingly, M. E., Mattingly, J. R., Jr., and Martinez-Carrion, M. (1982) *J. Biol. Chem.* **257**, 8872–8878

263. Metzler, D. E. (1997) *Methods Enzymol.* **280**, 30–40

264. Higaki, T., Tanase, S., Nagashima, F., Morino, Y., Scott, A. I., Williams, H. J., and Stolowich, N. J. (1991) *Biochemistry* **30**, 2519–2526

265. Goldberg, J. M., and Kirsch, J. F. (1996) *Biochemistry* **35**, 5280–5291

266. Recsei, P. A., and Snell, E. E. (1984) *Ann. Rev. Biochem.* **53**, 357–387

267. van Peolje, P. D., and Snell, E. E. (1990) *Ann. Rev. Biochem.* **59**, 29–59

References

268. Pishko, E. J., and Robertus, J. D. (1993) *Biochemistry* **32**, 4943–4948

269. Gallagher, T., Snell, E. E., and Hackert, M. L. (1989) *J. Biol. Chem.* **264**, 12737–12743

270. Recsei, P. A., Huynh, Q. K., and Snell, E. E. (1983) *Proc. Natl. Acad. Sci. U.S.A.* **80**, 973–977

271. Van Poelje, P. D., and Snell, E. E. (1988) *Proc. Natl. Acad. Sci. U.S.A.* **85**, 8449–8453

272. Huynh, Q. H., and Snell, E. E. (1986) *J. Biol. Chem.* **261**, 1521–1524

273. Shirahata, A., and Pegg, A. E. (1986) *J. Biol. Chem.* **261**, 13833–13837

274. Stanley, B. A., Shantz, L. M., and Pegg, A. E. (1994) *J. Biol. Chem.* **269**, 7901–7907

274a. Xiong, H., and Pegg, A. E. (1999) *J. Biol. Chem.* **274**, 35059–35066

275. Clancey, C. J., Chang, S.-C., and Dowhan, W. (1993) *J. Biol. Chem.* **268**, 24580–24590

276. Satre, M., and Kennedy, E. P. (1978) *J. Biol. Chem.* **253**, 479–483

277. Williamson, J. M., and Brown, G. M. (1979) *J. Biol. Chem.* **254**, 8074–8082

278. Yang, H., and Abeles, R. H. (1987) *Biochemistry* **26**, 4076–4081

279. Seto, B., and Stadtman, T. C. (1976) *J. Biol. Chem.* **251**, 2435–2439

280. Seto, B. (1978) *J. Biol. Chem.* **253**, 4525–4529

281. Hodgins, D. S., and Abeles, R. H. (1969) *Arch. Biochem. Biophys.* **130**, 274–285

282. Arkowitz, R. A., and Abeles, R. H. (1989) *Biochemistry* **28**, 4639–4644

283. Stadtman, T. C., and Davis, J. N. (1991) *J. Biol. Chem.* **266**, 22147–22153

283a. Kabisch, U. C., Gräntzdörffer, A., Schierhorn, A., Rücknagel, K. P., Andreesen, J. R., and Pich, A. (1999) *J. Biol. Chem.* **274**, 8445–8454

284. Furuta, T., Takahashi, H., Shibasaki, H., and Kasuya, Y. (1992) *J. Biol. Chem.* **267**, 12600–12605

285. Levy, H. L., Taylor, R. G., and McInnes, R. R. (1995) in *The Metabolic and Molecular Bases of Inherited Disease*, 7th ed., Vol. 1 (Scriver, C. R., Beaudet, A. L., Sly, W. S., and Valle, D., eds), pp. 1107–1123, McGraw-Hill, New York

286. Havir, E. A., and Hanson, K. R. (1985) *Biochemistry* **24**, 2959–2967

287. Consevage, M. W., and Phillips, A. T. (1985) *Biochemistry* **24**, 301–308

288. Langer, M., Reck, G., Reed, J., and Rétey, J. (1994) *Biochemistry* **33**, 6462–6467

289. Langer, M., Lieber, A., and Rétey, J. (1994) *Biochemistry* **33**, 14034–14038

290. Langer, M., Pauling, A., and Rétey, J. (1995) *Angew. Chem. Int. Ed. Engl.* **34**, 1464–1465

291. Schuster, B., and Rétey, J. (1995) *Proc. Natl. Acad. Sci. U.S.A.* **92**, 8433–8437

292. Schwede, T. F., Rétey, J., and Schulz, G. E. (1999) *Biochemistry* **38**, 5355–5361

1. Discuss the role of biotin in metabolism and the chemical mechanism of its action. Illustrate with examples.

2. Biotin has been shown to be an essential component of a bacterial oxaloacetate *decarboxylase* that pumps two sodium ions out of a cell for each oxaloacetate molecule decarboxylated. Propose a chemical mechanism for the functioning of biotin and also any ideas that you may have for the operation of the sodium pump.

3. Write out the equation by which **acetyl-CoA** and **pyruvate** can be converted into L-**glutamate**, which can be observed in living animals by ^{13}C NMR. If [2-^{13}C] sodium acetate were injected into the animal what labeling pattern could be anticipated in the L-glutamate?

4. Compare mechanisms and coenzyme requirements for biological decarboxylation of the following three groups of compounds:

    β-oxoacids
    α-oxoacids
    α-amino acids

5. Illustrate, using structural equations, the chemical mechanisms of the following biochemical reactions.

    a) Pyruvate → acetaldehyde + CO_2
    b) 2-Pyruvate → α-acetolactate + CO_2
    c) Pyruvate + NAD^+ + CoA-SH → acetyl-CoA + NADH + H^+ + CO_2
    d) Fructose 6-P + glyceraldehyde-3-P → xylulose-5-P + erythrose-4-P

6. 3-Fluoropyruvate is converted quantitatively by pyruvate decarboxylase from wheat germ into acetate, fluoride (F^-), and carbon dioxide. Propose a reaction mechanism. See Gish, G., Smyth, T., and Kluger, R. (1988) *J. Am. Chem. Soc.* **110**, 6230-6234.

7. In *E. coli* L-**cysteine** is formed from L-serine and the sulfide ion S^{2-} in a reaction that also requires acetyl-CoA and is catalyzed by the consecutive action of an acyl transferase and cysteine synthase. Outline the mechanism of this conversion indicating participation of any essential coenzymes.

8. Illustrate, using structural equations, the chemical mechanisms of the following biochemical reactions.

    a) L-Glutamate + oxaloacetate → 2-oxoglutarate + L-aspartate
    b) L-Serine → pyruvate + NH_4^+
    c) L-Serine + indole (from cleavage of indole-3-glycerol phosphate) → L-tryptophan
    d) L-Serine + tetrahydrofolate → glycine + N^5, N^{10}-methylene-tetrahydrofolate
    e) L-Selenocysteine → L-alanine + Se^0

9. Threonine is formed by *E. coli* from homoserine via the intermediate γ-phosphohomoserine. Write out an abbreviated reaction sequence for its conversion to L-threonine by the action of threonine synthase.

10. Write out a plausible step-by-step mechanism by which 1-aminocyclopropane-1-carboxylate synthase (ACC synthase) of plant tissues can form ACC from *S*-adenosylmethionine. This reaction requires a specific cofactor

11. Tissues of the mammalian central nervous system contain a pyridoxal phosphate-dependent glutamate decarboxylase that catalyzes conversion of Glu to γ-aminobutyrate (GABA), an inhibitory synaptic transmitter. GABA is degraded by transimination with α-oxoglutarate as the acceptor to yield succinic semialdehyde, which then is oxidized to succinate by an NAD-linked dehydrogenase.

    a) Show how these reactions can operate as a shunt pathway that allows the citric acid cycle to function without the enzymes α-oxoglutarate dehydrogenase and succinate thiokinase.
    b) Is the shunt more or less efficient than the normal cycle from the standpoint of energy recovery? Explain.

Biological oxidation-reduction reactions rely upon many organic coenzymes and transition metal ions. Left, a molecule of the hydrogen carrier pyrroloquinoline quinone (**PQQ**, Fig. 15-23) is seen at the bottom of a 7-bladed β propeller of a bacterial methanol dehydrogenase. In adrenodoxin reductase (right) the reducing power of **NADPH** is passed to FAD and then to the small redox protein adrenodoxin. In this stereoscopic view the pyridine ring of the oxidized coenzyme NADP⁺ and the tricyclic flavin ring of FAD are seen stacked against each other in the center; the adenylate ends stretching toward the top and bottom of the complex. From Ziegler and Schulz, (2000) *Biochemistry* **39**, 10986-10995.

Contents

Coenzymes of Oxidation–Reduction Reactions

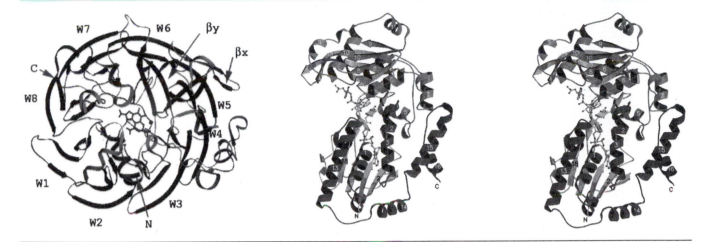

15

The **dehydrogenation** of an alcohol to a ketone or aldehyde (Eq. 15-1) is one of the most frequent biological oxidation reactions. Although the hydrogen atoms removed from the substrate are often indicated simply as 2[H], it was recognized early in the twentieth century that they are actually transferred to hydrogen-carrying coenzymes such as NAD$^+$, NADP$^+$, FAD, and **riboflavin**

$$H-C-OH \longrightarrow C=O + 2[H]$$

(15-1)

5'-phosphate (FMN). This chapter deals with these coenzymes and also with a number of other organic oxidation–reduction coenzymes. They may be considered either as carriers of hydrogen or as carriers of electrons (H = H$^+$ + e$^-$) in metabolic reactions.

When NAD$^+$ becomes reduced by dehydrogenation of an alcohol, one of the hydrogen atoms removed from the alcohol becomes firmly attached to the NAD$^+$, converting it to NADH. The other is released as a proton (Eq. 15-2). Study of ^{2}H-labeled alcohols and their oxidation by NAD$^+$ has shown that *dehydrogenases catalyze direct transfer to NAD$^+$ of the hydrogen that is attached to carbon in the alcohol.* There is never any exchange of this hydrogen atom with protons of the medium. At the same time, the hydrogen attached to the oxygen of the alcohol is released into the medium as H$^+$:

$$NAD^+ + H-C-O-H \longrightarrow C=O + NADH + H^+$$

This H is transferred directly to NAD$^+$

This hydrogen is released as a proton, H$^+$

(15-2)

The foregoing observations suggested that these biological dehydrogenations may be viewed as removal of a hydride ion (H$^-$) together with a proton (H$^+$) rather than as removal of two hydrogen atoms. NAD$^+$ and NADP$^+$ are regarded as hydride ion-accepting coenzymes. However, it has been impossible to establish conclusively that the hydrogen atom and electron are transferred simultaneously as H$^-$. Transfer of the hydrogen atom to or from these coenzymes may conceivably be followed by or preceded by transfer of an electron. The situation is even less clear for the flavin coenzymes FAD and riboflavin phosphate for which intermediate **free radical** oxidation states are known to exist. However, regardless of the actual mechanism, it is convenient to classify most hydrogen transfer reactions of metabolism as if they occurred by transfer of a hydride ion. *The hydride ion can be regarded as a nucleophile which can add to double bonds or can be eliminated from substrates in reactions of types that we have already considered.* Some of these reactions are listed in Table 15-1.

Why are there *four* major hydrogen transfer coenzymes, NAD$^+$, NADP$^+$, FAD, and riboflavin phosphate (FMN), instead of just one? Part of the answer is that the **reduced pyridine nucleotides** NADPH and NADH are more powerful reducing agents than are reduced **flavins** (Table 6-7). Conversely, flavin coenzymes are more powerful oxidizing agents than are

NAD$^+$ and NADP$^+$. This difference reflects the chemical difference between the vitamins **riboflavin** and **nicotinamide** which form the oxidation–reduction centers of the coenzymes. Another difference is that NAD$^+$ and NADP$^+$ tend to be present in free forms within cells, diffusing from a site on one enzyme to a site on another. These coenzymes are *sometimes* tightly bound but flavin coenzymes are *usually* firmly bound to proteins, fixed, and unable to move. Thus, they tend to accept hydrogen atoms from one substrate and to pass them to a second substrate while attached to a single enzyme.

The oxidation–reduction potential of a pyridine nucleotide coenzyme system is determined by the standard redox potential for the free coenzyme (Table 6-8) together with the ratio of concentrations of oxidized to reduced coenzyme ([NAD$^+$] / [NADH], Eq. 6-64). If these concentrations are known, a redox

TABLE 15-1
Some Biochemical Hydrogen Transfer Reactions[a]

Reaction	Example (oxidant)
A Dehydrogenation of an alcohol 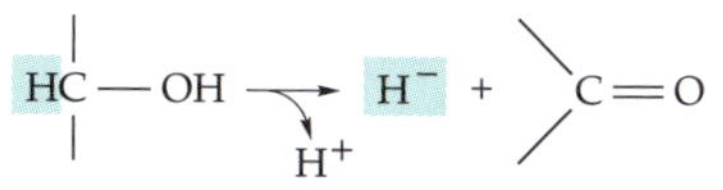	Alcohol dehydrogenase (NAD$^+$)
B Dehydrogenation of an amine 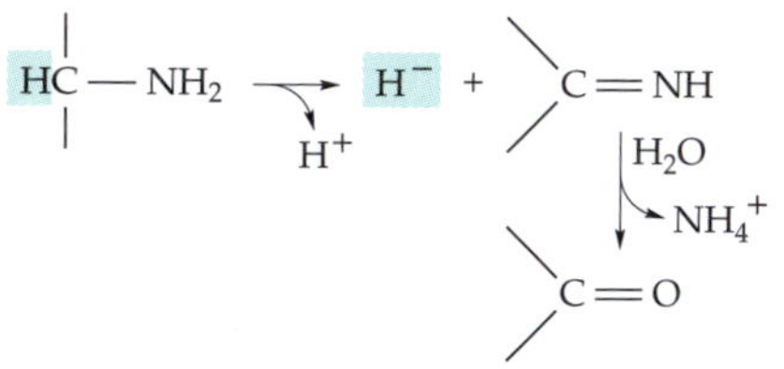	Amino acid dehydrogenases, amine oxidases (NAD$^+$ or flavin)
C Dehydrogenation of adduct of thiol and aldehyde 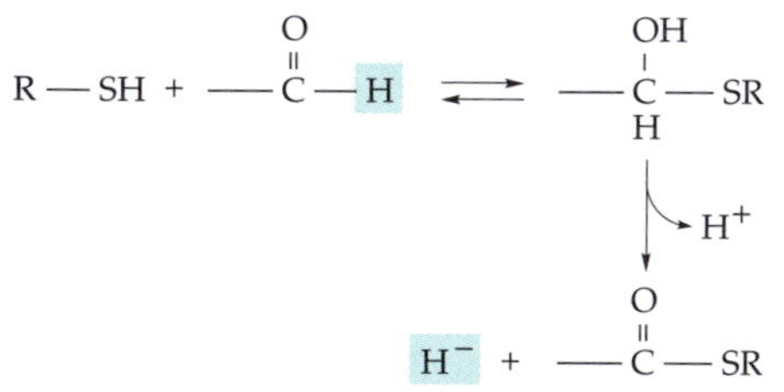	Glyceraldehyde 3-phosphate dehydrogenase (NAD$^+$ or NADP$^+$)
D Dehydrogenation of acyl-CoA, acyl-ACP, or carboxylic acid	Acyl-CoA dehydrogenases (Flavin)
	Succinic dehydrogenase (Flavin)
(Y = —S—CoA or —OH)	Opposite: enoyl reductase (NADPH)
E Reduction of desmosterol to cholesterol by NADPH 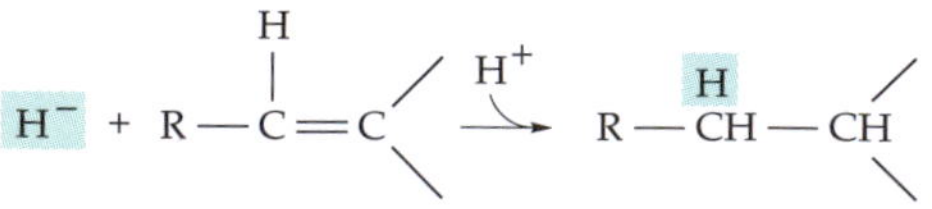	Reduction of desmosterol to cholesterol by NADPH

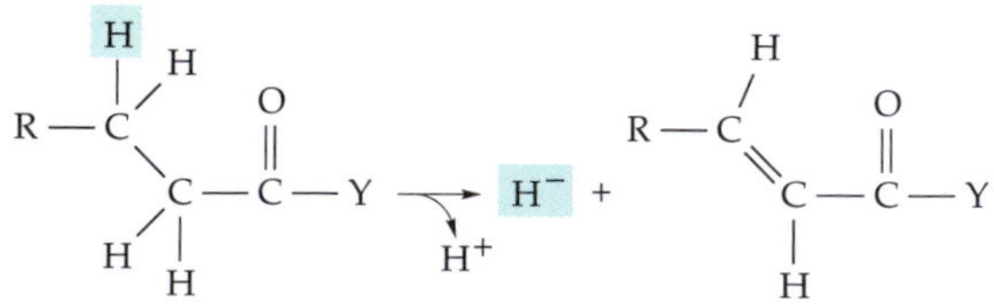

[a] Reaction type 9 A,B of Table on the inside cover at the end of the book: A hypothetical hydride ion H$^-$ is transferred from the substrate to a coenzyme of suitable reduction potential such as NAD$^+$, NADP$^+$, FAD, or riboflavin 5′-phosphate. The reverse of hydrogenation is shown for E. Many of the reactions are reversible and often go spontaneously in the reverse direction from that shown here.

potential can be defined for the NAD^+ system within a cell. This potential may vary in different parts of the cell because of differences in the $[NAD^+] / [NADH]$ ratio, but within a given region of the cell it is constant. On the other hand, the redox potentials of flavoproteins vary. Since the flavin coenzymes are not dissociable, two flavoproteins may operate at very different potentials even when they are physically close together.

Why are there two pyridine nucleotides, NAD^+ and $NADP^+$, differing only in the presence or absence of an extra phosphate group? One important answer is that they are members of two different oxidation–reduction systems, both based on nicotinamide but functionally independent. The experimentally measured ratio $[NAD^+] / [NADH]$ is much higher than the ratio $[NADP^+] / [NADPH]$. Thus, these two coenzyme systems also can operate within a cell at different redox potentials. A related generalization that holds much of the time is that *NAD$^+$ is usually involved in pathways of catabolism, where it functions as an oxidant, while NADPH is more often used as a reducing agent in biosynthetic processes.* See Chapter 17, Section I for further discussion.

A. Pyridine Nucleotide Coenzymes and Dehydrogenases

In 1897, Buchner discovered that "yeast juice," prepared by grinding yeast with sand and filtering, catalyzed fermentation of sugar. This was a major discovery which excited the interest of many other biochemists.[1] Among them were Harden and Young, who, in 1904, showed that Buchner's cell-free yeast juice lost its ability to ferment glucose to alcohol and carbon dioxide when it was dialyzed. Apparently, fermentation depended upon a low-molecular-weight substance that passed out through the pores of the dialysis membrane. Fermentation could be restored by adding back to the yeast juice either the concentrated dialysate or boiled yeast juice (in which the enzyme proteins had been destroyed). The heat-stable material, which Harden and Young called **cozymase**, was eventually found to be a mixture of inorganic phosphate ions, thiamin diphosphate, and NAD^+. However, characterization of NAD^+ was not accomplished until 1935.

Pure $NADP^+$ was isolated from red blood cells in 1934 by Otto Warburg and W. Christian, who had been studying the oxidation of glucose 6-phosphate by erythrocytes.[1a] They demonstrated a requirement for a dialyzable coenzyme which they characterized and named **triphosphopyridine nucleotide** (TPN$^+$, but now officially $NADP^+$; Fig. 15-1). Thus, even before its recognition as an important vitamin in human nutrition, nicotinamide was identified as a component of $NADP^+$.

Warburg and Christian recognized the relationship of $NADP^+$ and NAD^+ (then called DPN$^+$) and proposed that both of these compounds act as hydrogen carriers through alternate reduction and oxidation of the pyridine ring. They showed that the coenzymes could be reduced either enzymatically or with sodium dithionite $Na_2S_2O_4$.

$$S_2O_4^{2-} + NAD^+ + H^+ \rightarrow NADH + 2 SO_2 \qquad (15\text{-}3)$$

The reduced coenzymes NADH and NADPH were characterized by a new light-absorption band at 340 nm. This is not present in the oxidized forms, which absorb maximally at 260 nm (Fig. 15-2). The reduced forms are stable in air, but their reoxidation was found

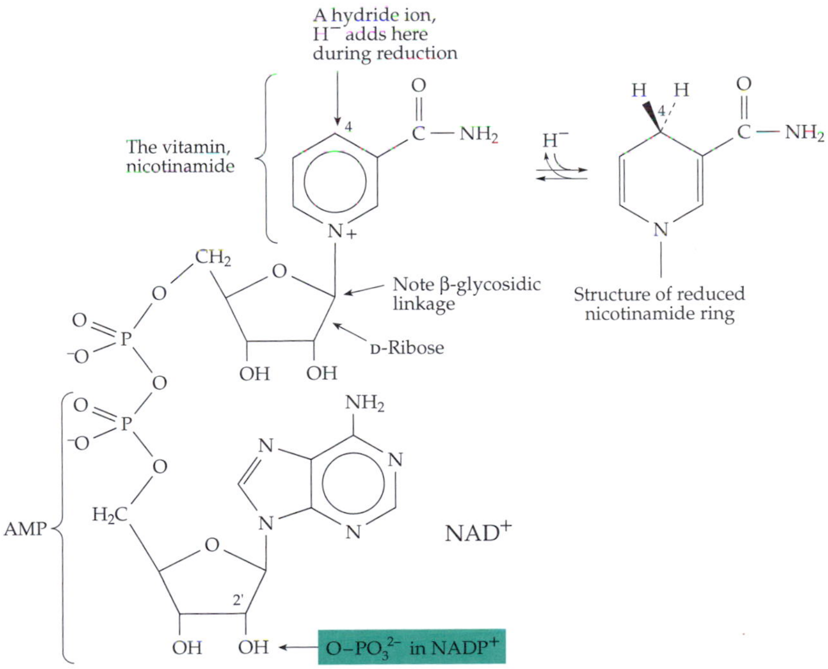

Figure 15-1 The hydrogen-carrying coenzymes NAD^+ (nicotinanide adenine dinucleotide) and $NADP^+$ (nicotinamide adenine dinucleotide phosphate). We use the abbreviations NAD^+ and $NADP^+$, even though the net charge on the entire molecule at pH 7 is negative because of the charges on oxygen atoms of the phospho groups.

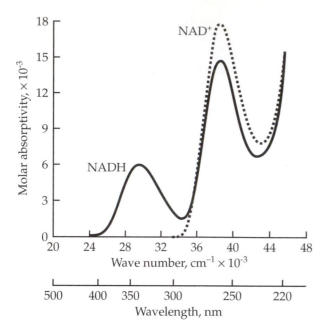

Figure 15-2 Absorption spectra of NAD⁺ and NADH. Spectra of NADP⁺ and NADPH are nearly the same as these. The difference in absorbance between oxidized and reduced forms at 340 nm is the basis for what is probably the single most often used spectral measurement in biochemistry. Reduction of NAD⁺ or NADP⁺ or oxidation of NADH or NADPH is measured by changes in absorbance at 340 nm in many methods of enzyme assay. If a pyridine nucleotide is not a reactant for the enzyme being studied, a **coupled assay** is often possible. For example, the rate of enzymatic formation of ATP in a process can be measured by adding to the reaction mixture the following enzymes and substrates: hexokinase + glucose + glucose-6-phosphate dehydrogenase + NADP⁺. As ATP is formed, it phosphorylates glucose via the action of hexokinase. NADP⁺ then oxidizes the glucose 6-phosphate that is formed with production of NADPH, whose rate of appearance is monitored at 340 nm.

to be catalyzed by certain yellow enzymes which were later identified as flavoproteins.

1. Three-Dimensional Structures of Dehydrogenases

Most NAD⁺- or NADP⁺- dependent dehydrogenases are dimers or trimers of 20- to 40-kDa subunits. Among them are some of the first enzymes for which complete structures were determined by X-ray diffraction methods. The structure of the 329-residue per subunit muscle (M_4) isoenzyme of **lactate dehydrogenases** (see Chapter 11) from the dogfish was determined to 0.25 nm resolution by Rossmann and associates in 1971.[2–4] More recently, structures have been determined for mammalian muscle and heart type (H_4) isoenzymes[5] for the testicular (C_4) isoenzyme from the

mouse,[6] and for bacterial lactate dehydrogenases.[6–8] In all of these the polypeptide is folded nearly identically. The structures of the homologous cytosolic and mitochondrial isoenzymes of **malate dehydrogenase** are also similar,[9–11] as are those of the bacterial enzyme.[12,13] All of these proteins consist of two structural domains and the NAD⁺ is bound to the nucleotide-binding domain in a similar manner as is shown in Fig. 2-13 for glyceraldehyde phosphate dehydrogenase. The coenzyme-binding domains of the dehydrogenases of known structures all have this nearly constant structural feature (often called the Rossmann fold) consisting of a six-stranded parallel sheet together with several α helical coils[14] (Figs. 2-13 and 2-27).

The coenzyme molecule curls around one end of the nucleotide-binding domains in a "C" conformation with the nicotinamide ring lying in a pocket (Figs. 2-13, 15-3). Even before the crystal structure of lactate dehydrogenase was known, the lack of pH dependence of coenzyme binding from pH 5 to 10 together with observed inactivation by butanedione suggested that the pyrophosphate group of NAD⁺ binds to a guanidinium group of an arginine residue. This was identified by X-ray diffraction studies as Arg 101. This ion pairing interaction, as well as the hydrogen bond between Asp 53 and the 2' oxygen atom of a ribose ring (Fig. 15-3), is present in all of the lactate and malate dehydrogenases.

The adenine ring of the coenzyme is bound in a hydrophobic pocket with its amino group pointed out into the solvent. A second structural domain holds additional catalytic groups needed to form the active site.

2. Stereospecificity and Mechanism

When NAD⁺ is reduced in ^2H_2O by dithionite (Eq. 15-3) an atom of ^2H is introduced into the reduced pyridine. Chemical degradation showed the ^2H to be present at the 4 position of the ring para to the nitrogen atom[15] (see Fig. 15-1). As shown by Westheimer and associates, during enzymatic reduction of NAD⁺ by deuterium-containing ethanol, $CH_3–C^2H_2OH$, one of the ^2H atoms is transferred into the NADH formed, thus establishing the *direct transfer of a hydrogen atom*.[16,17] When the NAD^2H formed in this way is reoxidized enzymatically with acetaldehyde, with regeneration of NAD⁺ and ethanol, the ^2H is completely removed.

This was one of the first recognized examples of the ability of an enzyme to choose between two identical atoms at a *pro*-chiral center (Chapter 9). The two sides of the nicotinamide ring of NAD were designated A and B and the two hydrogen atoms at the 4 position of NADH as H_A (now known as *pro-R*) and H_B (*pro-S*). Alcohol dehydrogenase always removes the *pro-R* hydrogen. Malate, isocitrate, lactate, and D-glycerate dehydrogenases select the same hydrogen. However,

BOX 15-A NICOTINIC ACID AND NICOTINAMIDE

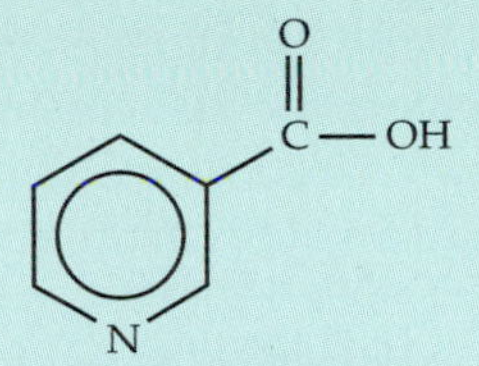

Nicotinic acid

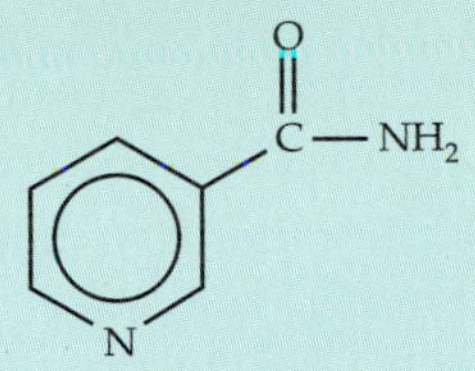

Nicotinamide

Nicotinic acid was prepared in 1867 by oxidation of nicotine. Although it was later isolated by Funk and independently by Suzuki in 1911–1912 from yeast and rice polishings, it was not recognized as a vitamin. Its biological significance was established in 1935 when nicotinamide was identified as a component of NAD^+ by von Euler and associates and of $NADP^+$ by Warburg and Christian.[a] Both forms of the vitamin are stable, colorless compounds highly soluble in water.

In 1937, Elvehjem and coworkers demonstrated that nicotinic acid cured canine "blacktongue." In the same year it was found to cure human **pellagra**, a terrible disease characterized by weakness, indigestion, and loss of appetite followed by dermatitis, diarrhea, mental disorders, and eventual death. At that time pellagra was common in the United States, especially in the south. The U.S. Public Health Service estimated that during 1912–1916 there were 100,000 victims and 10,000 deaths a year.[b,c]

The daily requirement for an adult is about 7.5 mg. The amount is decreased by the presence in the diet of tryptophan, which can be converted partially to nicotinic acid (Chapter 25).[d,e] Tryptophan is about 1/60 as active as nicotinic acid itself. The one-time prevalence of pellagra in the southern United States was a direct consequence of a diet high in maize whose proteins have an unusually low tryptophan content.

[a] Schlenk, F. (1984) *Trends Biochem. Sci.* **9**, 286–288
[b] Wagner, A. F., and Folkers, K. (1964) *Vitamins and Coenzymes*, Wiley (Interscience), New York (p. 73)
[c] Rosenkrantz, B. G. (1974) *Science* **183**, 949–950
[d] Teply, L. J. (1993) *FASEB J.* **7**, 1300
[e] Sauberlich, H. E. (1987) in *Pyridine Nucleotide Coenzymes: Chemical, Biochemical and Medical Aspects*, Vol. B (Dolphin, D., Avramović, O., and Poulson, R., eds), pp. 599–626, Wiley (Interscience), New York

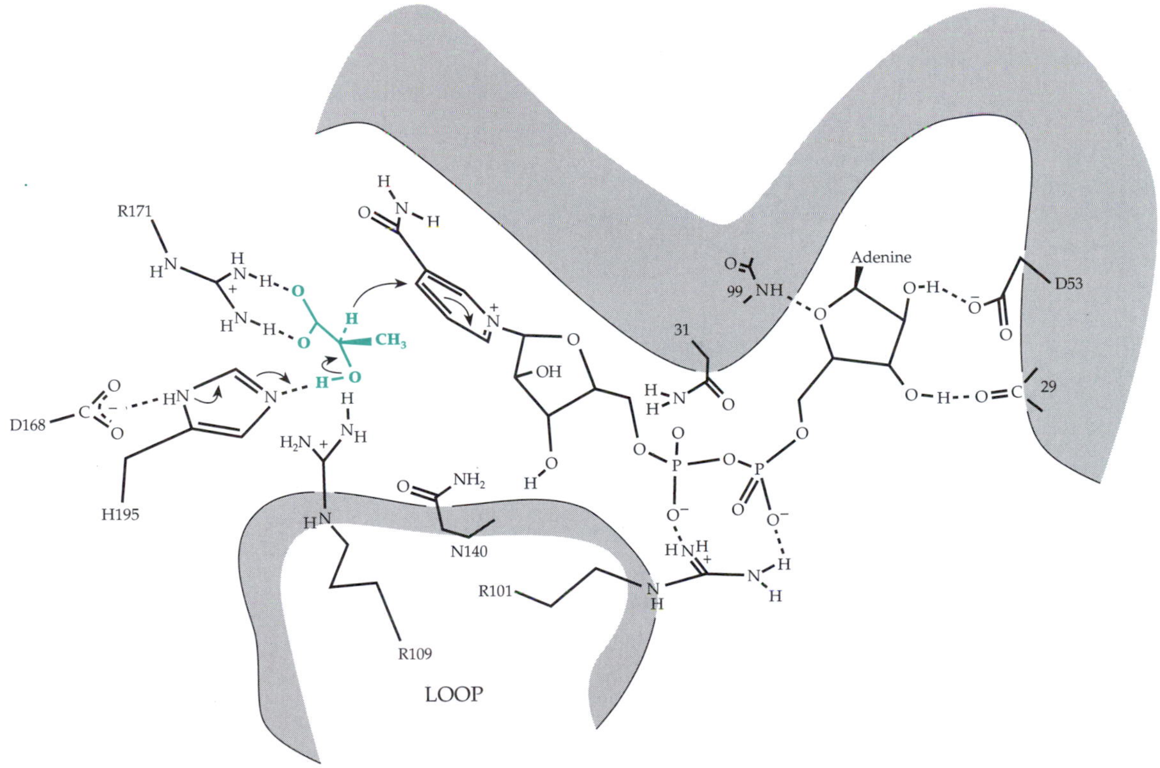

Figure 15-3 Diagrammatic structure of NAD^+ and L-lactate bound into the active site of lactate dehydrogenase. See Eventhoff *et al.*[5]

dehydrogenases acting on glucose 6-phosphate, glutamate, 6-phosphogluconate, and 3-phosphoglyceralde-
hyde remove the *pro-S* hydrogen. By 1979 this stereospecificity had been determined for 127 dehydrogenases,
about half of which were found A-specific and half
B-specific.[18] Isotopically labeled substrates

$$pro\text{-}R \longrightarrow H_A \quad H_B \longleftarrow pro\text{-}S$$

have usually been used for this purpose but a simple
NMR method has been devised.[19,20]

How complete is the stereospecificity? Does the
enzyme sometimes make a mistake? *Lactate dehydrogenase displays nearly absolute specificity*, transferring a
proton into the "incorrect" *pro-S* position no more than
once in 5×10^7 catalytic cycles.[21] This suggests a difference in transition-state energies of about 40 kJ / mol for
the two isomers. X-ray structural studies suggested
that a major factor determining this stereoselectivity
might be the location of hydrogen-bonding groups
that hold the $-CONH_2$ group of the nicotinamide. If
these are located in such a way that the nicotinamide
ring has a *syn* orientation with respect to the ribose
ring to which it is attached (Fig. 15-4), the A face of the
coenzyme will be against the substrate undergoing
oxidation or reduction. If the ring has an *anti* conformation, the B side will be against the substrate.[18,22]
However, the explanation cannot be this simple. High
specificity is still retained when NADH is modified,
for example, by replacement of the $-CONH_2$ on the
pyridine ring with $-COCH_3$ or $-CHO$ or by use of
α-NADH, in which the ribose–nicotinamide linkage
is α instead of β as in normal NADH.[23]

An intriguing idea is that the enzyme selectively
stabilizes one of the boat conformations of NADH or
NADPH. According to stereoelectronic principles the
axially oriented hydrogen in such a boat structure
(*pro-R* in Fig. 15-4) will be the most readily transferred.
On the basis of this principle, together with the idea
that enzymes adjust the Gibbs energies of intermediate
states to achieve optimum catalytic rates, the following
prediction was made by Benner and associates:[24] *The
thermodynamically most easily reduced carbonyl compounds
will react by enzymatic transfer of the hydride ion from the
pro-R position of NADH.* Conversely, the most difficultly
reduced carbonyl compounds will receive the hydride
ion from the *pro-S* position of NADH. The proposal
has been controversial and it has been argued that
evolutionary relationships play a dominant role in
determining stereoselectivity.[23] Theoretical computations suggest that the boatlike puckering of the ring is

flexible, raising some doubts about the rigidity of the
reactants in the active site.[12,25] Nevertheless, it is likely
that the coenzyme as well as bound substrates must
assume very specific conformations before the enzyme
is able to move to the transition state for the reaction.
A great deal of effort has been expended in trying to
understand other factors that may explain the high
stereospecificity of dehydrogenases.[12,22,26–28]

When a hydrogen atom is transferred by an enzyme
from the 4 position of NADH or NADPH to an aldehyde
or ketone to form àn alcohol, the placement of the
hydrogen atom on the alcohol is also stereospecific.
Thus, alcohol dehydrogenase acting on NAD^2H converts acetaldehyde to (*R*)-mono-[^2H]ethanol (Eq. 9-73).
Pyruvate is reduced by lactate dehydrogenase to
L-lactate, and so on. Even mutations that disrupt the
binding of the carboxylate of lactate, e.g., substitution
of arginine 171 (see Fig. 15-3) by tryptophan or phenylalanine and introduction of a new arginine on the other
side of the active site, do not alter the L-stereospecificity.
However, the specificity for lactate is lost.[29]

One step or two-step transfer? Another major
question about dehydrogenases is whether the hydrogen atom that is transferred moves as a hydride ion,
as is generally accepted, or as a hydrogen atom with
separate transfer of an electron and with an intermediate NAD or NADPH free radical. In one study para-
substituted benzaldehydes were reduced with NADH
and NAD^2H using yeast alcohol dehydrogenase as a
catalyst.[30] This permitted the application of the Hammett equation (Box 6-C) to the rate data. For a series
of benzaldehydes for which σ+ varied widely, a value

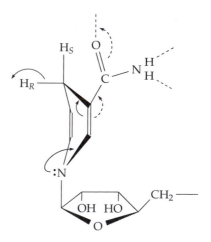

Figure 15-4 The nicotinamide ring of NADH in a *syn* boat
conformation suitable for transfer of an axially oriented *pro-R*
hydrogen atom from its A face as H⁻. The flow of electrons
is shown by the solid arrows. The dashed arrows indicate
competing resonance which favors planarity of the ring and
opposes the H⁻ transfer. Hydrogen bonds from the protein
to the carboxamide group (dashed lines) affect both this
tendency and the conformation of the nucleotide.

of $\rho = +2.2$ was observed for the rate constant with both NADH and NAD^{2}H. Thus, electron-accepting substituents in the para position hasten the reaction. While the significance of this observation is not immediately obvious, the relatively low value of ρ is probably incompatible with a mechanism that requires complete transfer of a single electron from NADH to acetaldehyde in the first step. A primary isotope effect on the rates was $k_H/k_D = 3.6$, indicating that the C–H bond in NADH is broken in the rate-limiting step. The fact that the isotope effect is the same for all of the substituted benzaldehydes also argued in favor of a hydride ion transfer.

Studies of the kinetics of nonenzymatic model reactions of NADH with quinones[31,32] have also been interpreted to favor a single-step hydride ion transfer. Application of Marcus theory (Chapter 9, Section D,4) to data from model systems also supports the hydride transfer mechanism.[33,34] Quantum mechanical tunneling may be involved in enzymatic transfer of protons, hydride ions, and electrons.[35–36] Tunneling is often recognized by unusually large primary or secondary kinetic isotope effects. According to semi-classical theory (Chapter 12), the maximum effects for deuterium and tritium are given by the following ratios:

$$(k_{2_H}/k_{1_H}) = (k_{3_H}/k_{2_H})^{3.3} \approx 7.$$

Higher ratios, which suggest tunneling, are frequently observed for dehydrogenases.[36a,b] Tunneling is apparently coupled to fluctuations in motion within the enzyme-substrate complex. Study of effects of pressure on reactions provides a new approach that can aid interpretation.[36c,36d]

Coenzyme and substrate analogs. The structures of enzyme•NAD$^+$•substrate complexes (Fig. 15-3) may be studied by X-ray crystallography under certain conditions or can be inferred from those of various stable enzyme-inhibitor complexes or from enzyme reconstituted with NAD$^+$ that has been covalently

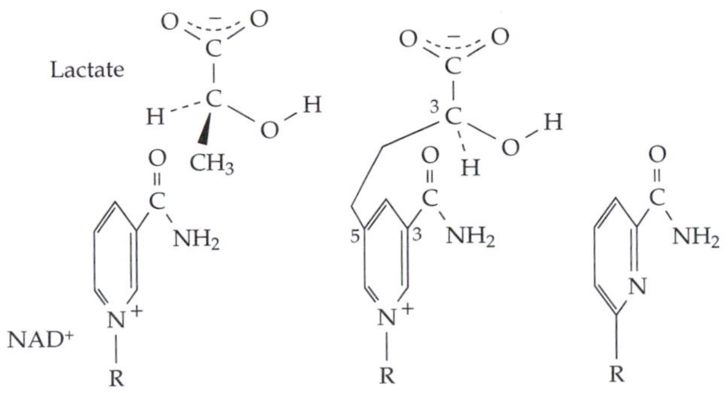

linked to the substrate. In the following diagram,[27] NAD$^+$ (I) is shown with L-lactate lying next to its A face, ready to transfer a hydride ion to the *pro-R* position in NADPH. Also shown (II) is 5-(S)-lac-NAD$^+$, with the covalently linked lactate portion in nearly the same position as in diagram I. This NAD derivative was used to obtain the first 0.27-nm structure with a bound substrate-like molecule in lactate dehydrogenase.[37] Since then the structures of many complexes with a variety of dehydrogenases have been studied. In coenzyme analog III the ring is bound to ribose with a C–C bond and it lacks the positive charge on the nicotinamide ring in NAD$^+$. It does not react with substrate. However, it binds to the coenzyme site of alcohol dehydrogenase and forms with ethanol a ternary complex whose structure has been solved.[38]

A related approach is to study complexes formed with normal NAD$^+$ but with an unreactive second substrate. An example is oxamate, which binds well to lactate dehydrogenase to form stable ternary complexes for which equilibrium isotope effects have been studied.[39]

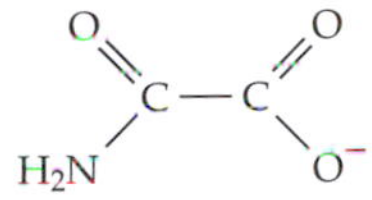

Oxamate

In the structure of the lactate dehydrogenase active site shown in Fig. 15-3, the lactate carboxylate ion is held and neutralized by the guanidinium group of Arg 171, and the imidazole group of His 195 is in position to serve as a general base catalyst to abstract a proton from the hydroxyl group of the substrate (Eq. 15-4). This imidazole group is also hydrogen bonded to the carboxylate of Asp 168 as in the "charge-relay" system of the serine proteases (Fig. 12-10). The same features are present in the active site of malate dehydrogenases and have been shown essential by study of various mutant forms.[11,40–42] The His:Asp pair of the dehydrogenases is not part of the nucleotide-binding domain but is present in the second structural domain, the "catalytic domain." This is another feature reminiscent of the serine proteases. A bacterial D-glycerate dehydrogenase also has a similar structure and the same catalytic groups. However, the placement of the catalytic histidine and the arginine that binds the α-carboxylate group of the substrate is reversed, allowing the enzyme to act on the D-isomer.[43]

Conformational changes during dehydrogenase action. Dehydrogenases bind coenzyme and substrate in an ordered sequence. The coenzyme binds first, then the oxidizable or reducible substrate. The binding of the coenzyme to lactate dehydrogenase is accompanied by a conformational change by which a loop, involving residues 98–120 and including one helix,

R = Ribose-P-P-adenosine

I   II   III

5-(S)-lac-NAD$^+$,
(3S)-5-(3-carboxy-3
hydroxypropyl)-NAD$^+$

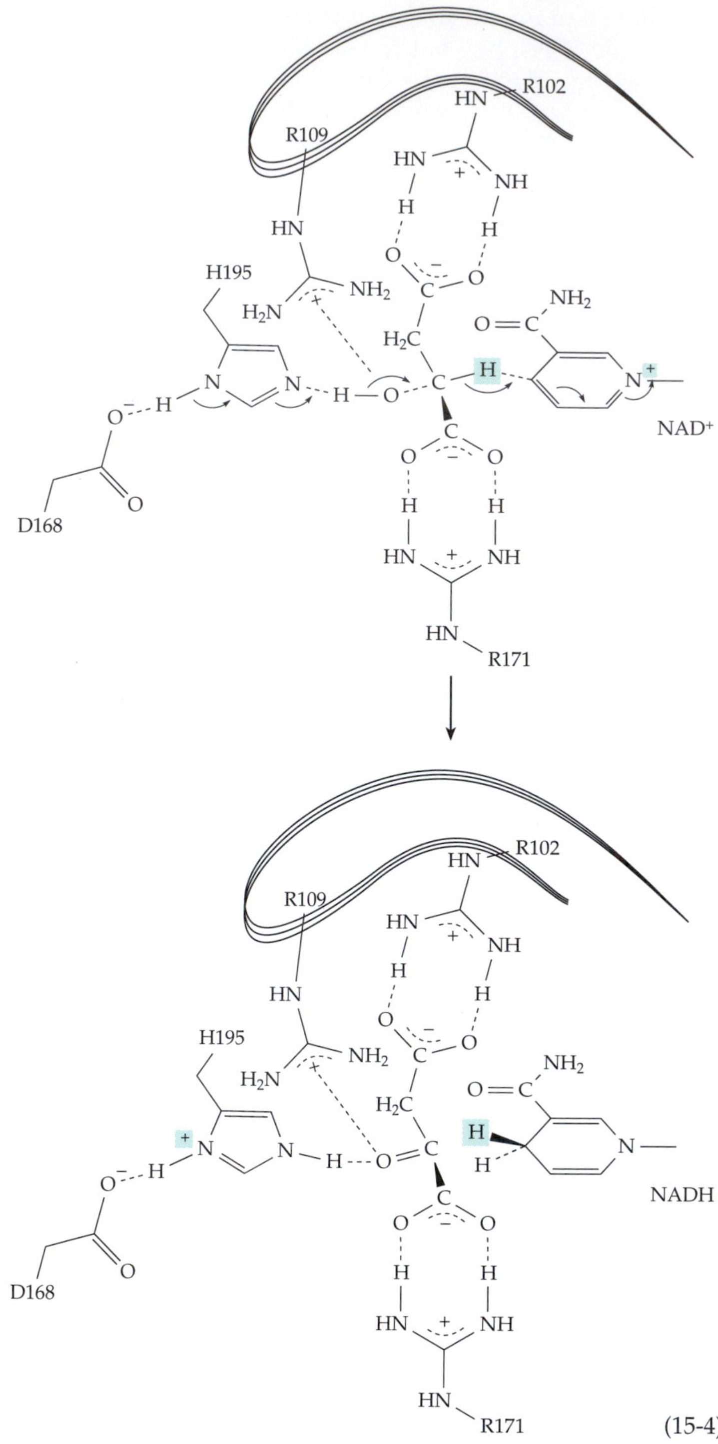

(15-4)

ment of the positive charge toward the 4' carbon of the ring, assisting in the transfer of the hydride ion (Eq. 15-4).[41] At the end of this transfer both Arg 109 and His 196 are positively charged and electrostatic repulsion between them may help to move the loop and release the products. If the reaction proceeds in the opposite direction the presence of two positive charges will assist in the hydride ion transfer. The importance of Arg 109 is demonstrated by the fact that a mutant with glutamine in place of Arg 109 has a value of k_{cat} only 1/400 that of native enzyme.[45]

The loop closure also causes significant changes in the Raman spectra of the bound NAD+, especially in vibrational modes that involve the carboxamide group of the nicotinamide ring. These have been interpreted as indicating an increased ridigity of binding of the coenzyme in the closed conformation.[46] As with many other enzymes acting on polar substrates, a characteristic of the pretransition state complex appears to be formation of a complex with a network of hydrogen bonds extending into the protein, exclusion of most water molecules, and tight packing of protein groups around the substrate. It is also significant that an overall net electrical charge on the ES complexes must be correct for tight binding of substrates to occur for lactate or malate dehydrogenase.[44] Positive and negative charges are balanced except for one excess positive charge which may be needed for catalysis.

Zinc-containing alcohol dehydrogenases.

Liver alcohol dehydrogenase is a relatively nonspecific enzyme that oxidizes ethanol and many other alcohols. The much studied horse liver enzyme[47–60] is a dimer of 374-residue subunits, each of which contains a "catalytic" Zn^{2+} ion deeply buried in a crevice between the nucleotide-binding and catalytic domains. The enzyme also contains a "structural" Zn^{2+} ion[48] that is bound by four sulfur atoms from cysteine side chains but does not represent a conserved feature of all alcohol dehydrogenases. The catalytic Zn^{2+} is ligated by sulfur atoms from cysteines 46 and 174 and by a nitrogen atom of the imidazole group of His 67 (Fig. 15-5). In the free enzyme a water molecule is thought to occupy the 4th coordination position and its dissociation to form the Zn^{+}–HO^{-} complex (Eq. 15-5) may account for a pK_a of 9.2 in the free enzyme and of 7.6 in the NAD^{+} complex.[48] The apparent pK_a drops further to about 6.4 in the presence of the substrate

folds over the coenzyme like a lid.[41,44] This conformational change must occur during each catalytic cycle, just as in the previously discussed cases of citrate synthase (Chapter 13) and aspartate aminotransferase (Chapter 14). One effect of folding of the loop is to bring the side chain of Arg 109 into close proximity to His 195 and to the OH group of the bound lactate (Eq. 15-4). The closing of the loop also forces the positively charged nicotinamide ring more deeply into a relatively nonpolar pocket. This may induce a move-

$$Zn^{2+}-OH_2 \rightarrow Zn^{+}-OH + H^{+} \qquad (15-5)$$

ethanol. The assignment of this pK_a value has been controversial.[48,49] Histidine 51 (Fig. 15-5), Glu 68 and Asp 49 have side chains close to the zinc and, as we have seen (Chapter 7), macroscopic pK_a values of proteins can be *shared* by two or more closely placed groups.

Substrate binding also induces a conformational change in this enzyme. When both coenzymes and substrate bind the "closed" conformation of the enzyme is formed by a rotation of the catalytic domains of the two subunits relative to the coenzyme-binding domains.[50–51a] Structures of ternary complexes with inhibitors and with substrates have also been established. For example, liver alcohol dehydrogenase was crystallized as the enzyme•NAD⁺• *p*-bromobenzyl alcohol complex with saturating concentrations of substrates in an equilibrium mixture[51b] and studied at low resolution. Transient kinetic studies or direct spectroscopic determinations led to the conclusion that the internal equilibrium (E•NAD⁺•alcohol = E•NADH•aldehyde) favors the NAD⁺•alcohol complex.[52] Subsequently, the complex was studied at higher resolution, and the basic structural features were confirmed with a

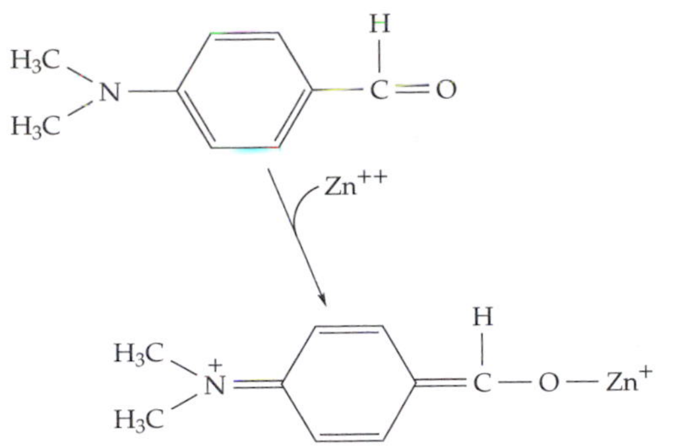

$$(15\text{-}6)$$

structure of the enzyme complexed with NAD⁺ and 2,3,4,5,6-pentafluorobenzyl alcohol.[53] From the crystal structures of the NAD⁺ • *p*-bromobenzyl alcohol complex and the previously mentioned complex with ethanol and analog III it appears that the oxygen of the alcohol substrate coordinates with the Zn^{2+}, displacing the bound water. Binding of the chromophoric aldehyde 4-*trans*-(N, N-dimethylamino) cinnamaldehyde shifts the wavelength of maximum absorbance by 66 nm, suggesting that Zn^{2+} binds directly to the oxygen of this ligand (Eq. 15-6).[54] Resonance-enhanced Raman spectroscopy (Chapter 23) of the complex of dimethylamino-·benzaldehyde with alcohol dehydrogenase also supports an intermediate in which the Zn^{2+} becomes coordinated directly with the substrate oxygen (Eq. 15-6).[55] Rapid scanning spectrophotometry of complexes with another chromophoric substrate, 3-hydroxy-4-nitrobenzyl alcohol, also suggested that the alcohol first formed a complex with an undissociated alcoholic – OH group (Eq. 15-7, step *a*) and then lost a proton to form a zinc alcoholate complex (step *b*)[56] which could react by hydride ion transfer (step *c*).

Eklund *et al.* suggested that the side chains of Ser 48 and His 51 act as a proton relay system to remove the proton from the alcohol, in step *b* of Eq. 15-7, leaving the transient zinc-bound alcoholate ion, which can then transfer a hydride ion to NAD⁺, in step *c*.[52] The shaded hydrogen atom leaves as H⁺. The role of His 51 as a base is supported by studies of the inactivation of the horse liver enzyme by diethyl pyrocarbonate[57] and by directed mutation of yeast and liver enzymes. When His 51 was substituted by Gln the pK_a of 7 was abolished and the activity was decreased ten-fold.[58]

The functioning of zinc ions in enzymes has been controversial and other mechanisms have been proposed. Makinen *et al.* suggested a transient pentacoordinate Zn^{2+} complex on the basis of EPR measurements on en-

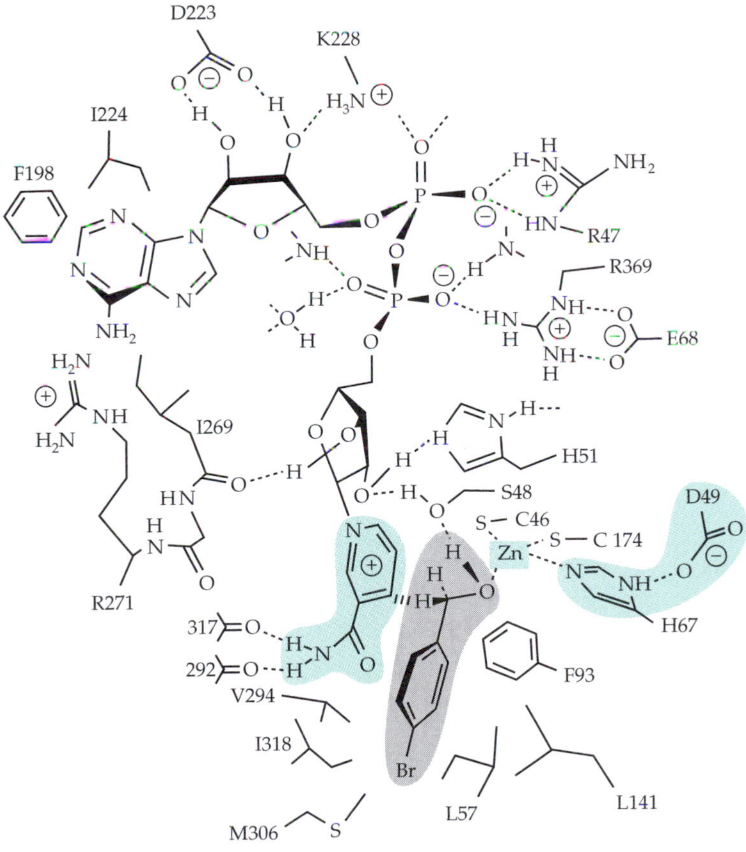

Figure 15-5 Structure of the complex of horse liver alcohol dehydrogenase with NAD⁺ and the slow substrate *p*-bromobenzyl alcohol. The zinc atom and the nicotinamide ring of the bound NAD⁺ are shaded. Adjacent to them is the bound substrate. Courtesy of Bryce Plapp. Based on Ramaswamy *et al.*[53]

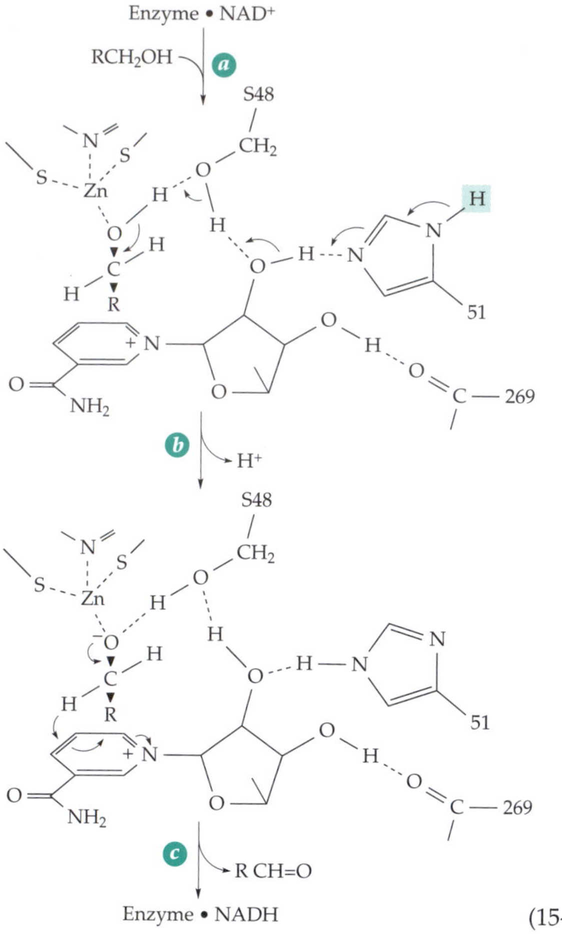

Enzyme • NAD$^+$

RCH_2OH (a)

S48

... S---Zn ... (mechanism diagram) ... 51 ... C — 269

(b) H$^+$

S48

... S---Zn ... 51 ... C — 269

(c) R CH=O

Enzyme • NADH   (15-7)

zyme containing Co^{2+}. Such a complex, in which the side chain of nearby Glu 68 would participate, would allow the coordinated water molecule to act as the base in deprotonation of the alcohol.[59] Molecular dynamics calculations indicate that Glu 68 can coordinate the zinc ion in this fashion, but Ryde suggests that its function may be to assist the exchange of ligands, i.e., the release of an alcohol or aldehyde product.[60] A variety of kinetic and spectroscopic studies have provided additional information that makes alcohol dehydrogenase one of the most investigated of all enzymes.

Liver alcohol dehydrogenase is important to the metabolism of ethanol by drinkers. Human beings exhibit small individual differences in their rates of alcohol metabolism which may reflect the fact that there are several isoenzymes and a number of genetic variants whose distribution differs from one person to the next as well as among tissues.[61–64] Inhibition of these isoenzymes by *uncompetitive* inhibitors, discussed in Chapter 9, is a goal in treatment of poisoning by methanol or ethylene glycol. Inhibition of the dehydrogenases slows the two-step oxidation of these substrates to toxic carboxylic acids.[65]

Yeast contains two cytosolic alcohol dehydrogenase isoenzymes.[66] Alcohol dehydrogenase I, present in large amounts in cells undergoing fermentation,

functions to reduce acetaldehyde in the fermentation process. Alcohol dehydrogenase II is synthesized by cells growing on such carbon sources as ethanol itself and needing to oxidize it to obtain energy. A third isoenzyme is present in mitochondria. An alcohol dehydrogenase isoenzyme of green plants is induced by anaerobic stress such as flooding. It permits ethanolic fermentation to provide energy temporarily to submerged roots and other tissues.[67] Some bacteria contain an NADP$^+$-dependent, Zn^{2+}-containing alcohol dehydrogenase.[67a]

Other alcohol dehydrogenases and aldo-keto reductases. The oxidation of an alcohol to a carbonyl compound and the reverse reaction of reduction of a carbonyl group are found in so many metabolic pathways that numerous specialized dehydrogenases exist. A large group of "short-chain" dehydrogenases and reductases had at least 57 known members by 1995.[68–69a] Their structures and functions are variable. Most appear to be single-domain proteins with a large β sheet, a nucleotide-binding pocket, and a conserved pair of residues: Tyr 152 and Lys 156. These may function in a manner similar to that of the His-Asp pair in lactate dehydrogenase. A possible role of a cysteine side chain has been suggested for another member of this group, 3-hydroxyisobutyrate dehydrogenase, an enzyme of valine catabolism.[70]

Dehydrogenases often act primarily to reduce a carbonyl compound rather than to dehydrogenate an alcohol. These enzymes may still be called dehydrogenases. For example, in the lactic acid fermentation lactate is formed by reduction of pyruvate but we still call the enzyme lactate dehydrogenase. In our bodies this enzyme functions in both directions. However, some enzymes that act mainly in the direction of reduction are called reductases. An example is **aldose reductase**, a member of a family of **aldo-keto reductases**[71–73] which have $(\alpha / \beta)_8$-barrel structures.[74–76]

The normal physiological function of aldose reductase is uncertain but it can cause a problem in diabetic persons by reducing glucose to sorbitol (glucitol), ribose to ribitol, etc. The resulting sugar alcohols accumulate in the lens and are thought to promote cataract formation and may also be involved in the severe damage to retinas and kidneys that occurs in diabetes mellitus. Inhibitors of aldose reductase delay development of these complications in animals but the compounds tested are too toxic for human use.[76]

The active sites of aldo-keto reductases contain an essential tyrosine (Tyr 48) with a low pK_a value. The nearby His 110, Asp 43, and Lys 77 may also participate in catalysis.[76,77] The kinetics are unusual. Both NAD$^+$ and NADH are bound tightly and the overall rate of reduction of a substrate is limited by the rate of dissociation of NAD$^+$.[78] Citrate is a natural uncompetitive inhibitor of aldose reductase.[79]

3. Dehydrogenation of Amino Acids and Amines

The dehydrogenation of an amine or the reverse reaction, the reduction of a Schiff base (Table 15-1, reaction type B), is another important pyridine nucleotide-dependent process. **Glutamate dehydrogenase,** a large oligomeric protein whose subunits contain 450 or more residues, is the best known enzyme catalyzing this reaction. An intermediate Schiff base of 2-oxoglutarate and NH_3 is a presumed intermediate.[80,81] Similar reactions are catalyzed by dehydrogenases for alanine,[82,83] leucine,[84] phenylalanine,[85] and other amino acids.[83,86,87]

4. Glyceraldehyde–3–Phosphate Dehydrogenase and the Generation of ATP in Fermentation Reactions

The oxidation of an aldehyde to a carboxylic acid, a highly exergonic process, often proceeds through a thioester intermediate whose cleavage can then be coupled to synthesis of ATP. This sequence is of central importance to the energy metabolism of cells (Chapters 10 and 17) and is shown in Fig. 15-6.

The best known enzyme catalyzing the first step of this reaction sequence is glyceraldehyde 3-phosphate dehydrogenase which functions in the glycolytic sequence (steps *a* and *b* of Fig. 10-3). It is present in both prokaryotes and eukaryotes as a tetramer of identical 36- to 37-kDa subunits. Three-dimensional structures have been determined for enzyme from several species, including lobster,[47,88] *E. coli*,[89] the thermophilic bacterium *Bacillus stearothermophilus* (Fig. 2-13),[90–90a] and trypanosomes.[91] Recall that aldehydes are in equilibrium with their covalent hydrates (Eq. 15-8, step *a*). Dehydrogenation of such a hydrate yields an acid (Eq. 15-8, step *b*) but such a mechanism offers no possibility of conserving the energy available from the reaction. However, during catalysis by glyceraldehyde-phosphate dehydrogenase the SH group of Cys 149, in the first step (step *a*, Fig. 15-6), adds to the substrate carbonyl group to form an adduct, a thiohemiacetal. This adduct is oxidized by NAD^+ to a thioester, an *S*-acyl enzyme (step *b*), which is then cleaved by the same enzyme through a displacement on carbon by an oxygen atom of P_i (phosphorolysis; step *c*). The sulfhydryl group of the enzyme is released simultaneously and the product, the acyl phosphate **1,3-bisphosphoglycerate**, is formed. The imidazole group of His 176 may catalyze both steps *a* and *b*.[89,92] A separate enzyme then transfers the phospho group from the 1 position of 1,3-bisphosphoglycerate to ADP to form ATP and 3-phosphoglycerate (step *d*). The overall sequence of Fig. 15-6 is the synthesis of one mole of ATP coupled to the oxidation of an aldehyde

to a carboxylic acid and the conversion of NAD^+ to NADH.

In green plants and in some bacteria an $NADP^+$-dependent cytoplasmic glyceraldehyde 3-phosphate dehydrogenase *does not* use inorganic phosphate to form an acyl phosphate intermediate but gives 3-phosphoglycerate with a free carboxylate as in Eq. 15-8.[93,93a] Because it doesn't couple ATP cleavage to the dehydrogenation, it drives the [NADPH] / [$NADP^+$] ratio to a high value favorable to biosynthetic processes (see Chapter 17).

$$(15\text{-}8)$$

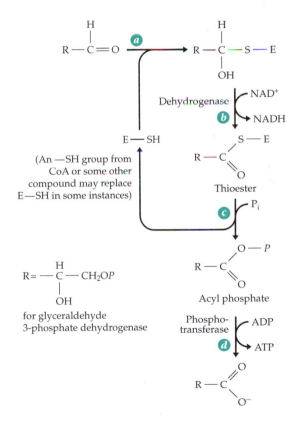

Figure 15-6 Generation of ATP coupled to oxidation of an aldehyde to a carboxylic acid. The most important known example of this sequence is the oxidation of glyceraldehyde 3-phosphate to 3-phosphoglycerate (Fig. 10-3, steps *a* and *b*). Other important sequences for "substrate-level" phosphorylation are shown in Eq. 14-23 and in Fig. 15-16.

Animal tissues also contain aldehyde dehydrogenases of a nonspecific type which are thought to act to remove toxic aldehydes from tissues.[94,95] Like glyceraldehyde 3-phosphate, these enzymes form thioester intermediates as in Fig. 15-6 but which are hydrolyzed rather than being converted to acyl phosphates. A mutation (E487K) in the mitochondrial enzyme occurs in about 50% of the Asian population. Although the structural alteration is not at the active site, the enzyme activity is low. Individuals carrying the mutation are healthy but have an aversion to alcohol, whose consumption causes an elevated blood level of acetaldehyde, facial flushing, dizziness, and other symptoms. A similar effect is exerted by the drug disulfiram (Antabuse), which has been used to discourage drinking and whose metabolites are thought to inhibit aldehyde dehydrogenase.[95]

Alcohol dehydrogenases also oxidize aldehydes, probably most often as the geminal diol forms, according to Eq. 15-8. No ATP is formed. The same enzymes can catalyze the dismutation of aldehydes, with equal numbers of aldehyde molecules going to carboxylic acid and to the alcohol.[96–98]

5. Reduction of Carboxyl Groups

The last two reaction steps in Fig. 15-6, steps c and d, are in essence the reverse of the sequence used for synthesis of a thioester such as a fatty acyl coenzyme A. Thus, the chemistry by which ATP is generated during glycolysis and that by which it is utilized in biosynthesis is nearly the same. Furthermore, a standard biochemical method for reduction of carboxyl groups to aldehyde groups is conversion, in an ATP-requiring process, to a thioester followed by reduction of the thioester (Table 15-1, reaction type C). For example, the sequence of Fig. 15-6 is reversed during gluconeogenesis (see Fig. 17-17). The carboxyl group of the side chain of aspartate can be reduced in two steps to form the alcohol homoserine (Figs. 11-3; 24-13).

The aldehyde generated by reduction of a thioester is not always released from the enzyme but may be converted on to the alcohol in a second reduction step.[99] This is the case for **3-hydroxy-3-methylglutaryl-CoA reductase** (HMG-CoA reductase), a large 887-residue protein that synthesizes mevalonate (Eq. 15-9).[100–102] This highly regulated enzyme controls the rate of synthesis of cholesterol and is a major target of drugs designed to block cholesterol synthesis. The structure of a smaller 428-residue bacterial enzyme is known.[103,104] Aspartate 766 is a probable proton donor in both reduction steps and Glu 558 and His 865 may act as a catalytic pair that protonates the sulfur of coenzyme A.[100]

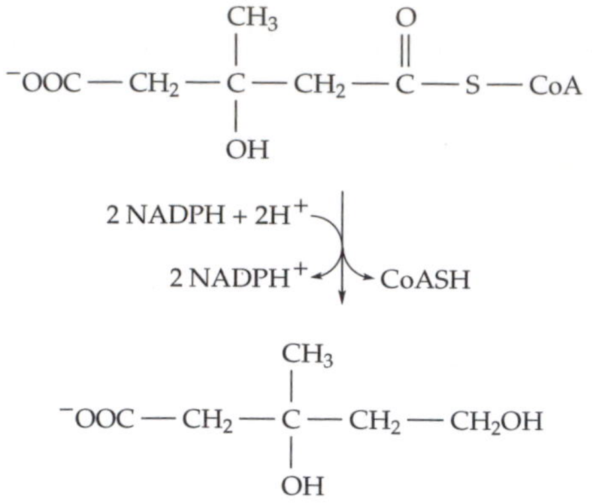

$$(15\text{-}9)$$

A related oxidation reaction is catalyzed by **glucose-6-phosphate dehydrogenase**, the enzyme that originally attracted Warburg's attention and led to the discovery of NADP$^+$. The substrate, the hemiacetal ring form of glucose, is oxidized to a lactone which is then hydrolyzed to 6-phosphogluconate (Eq. 15-10).[104a] This oxidation of an aldehyde to a carboxylic acid is not linked directly to ATP synthesis as in Fig. 15-6. The ring-opening step ensures that the reaction goes to completion. This reaction is a major supplier of NADPH for reductive biosynthesis and the large Gibbs energy decrease for the overall reaction ensures that the ratio [NADPH] / [NADP$^+$] is kept high within cells. This is the only source of NADPH for mature erythrocytes and a deficiency of glucose 6-phosphate dehydrogenase is a common cause of drug- and food-

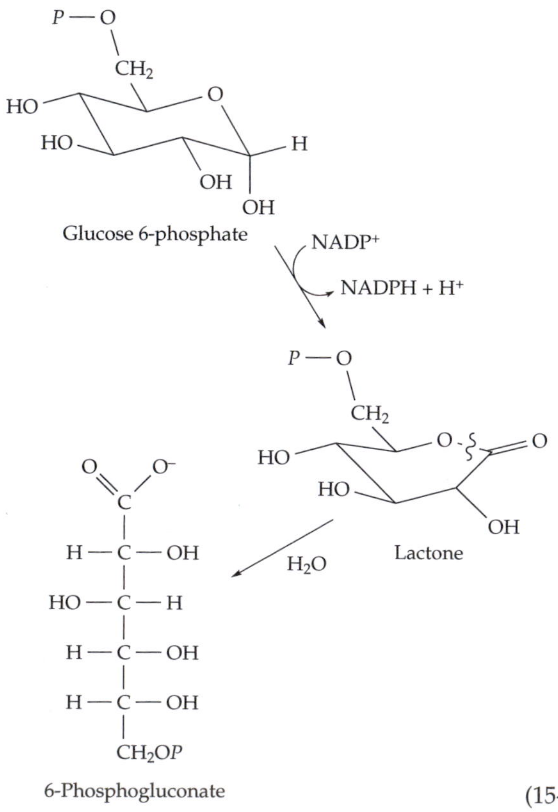

Glucose 6-phosphate

Lactone

6-Phosphogluconate

$$(15\text{-}10)$$

induced hemolytic anemia in human beings. About 400 variant forms of this enzyme are known.[105] Like the sickle cell trait (Box 7-B) some mutant forms of glucose-6-phosphate dehydrogenase appear to confer resistance to malaria.

6. Reduction of Carbon–Carbon Double Bonds

Neither NADP⁺ nor NAD⁺ is a strong enough oxidant to carry out the dehydrogenation of an acyl-CoA (reaction type D of Table 15-1). However, NADPH or NADH can participate in the opposite reaction. Thus, NADPH transfers a hydride ion to the β-carbon of an unsaturated acyl group during the biosynthesis of fatty acids (Chapter 17) and during elongation of shorter fatty acids (Chapter 21).[106–107b] A discovery of medical importance is that isonicotinyl hydrazide (INH), the most widely used antituberculosis drug, forms an adduct (of an INH anion or radical) with NAD⁺ of long-chain **enoyl-acyl carrier protein reductase** (enoyl-ACP reductase).[133a,b] This enzyme utilizes NADH in reduction of a C=C double bond during synthesis of mycolic acids. The same enzyme is blocked by **triclosan**, an antibacterial compound used widely in household products such as antiseptic soaps, toothpastes, cosmetics, fabrics, and toys.[133c–e]

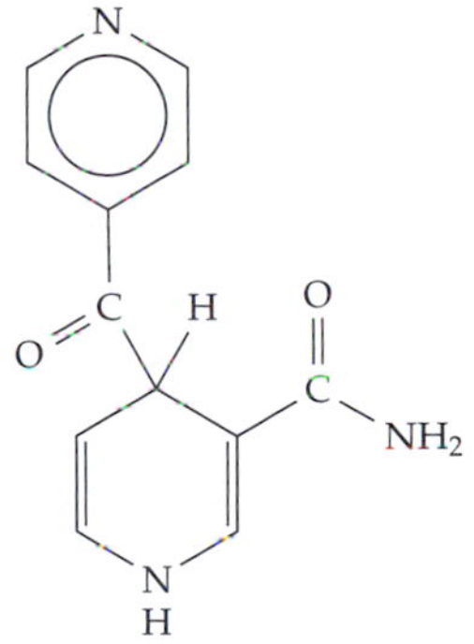

Adduct of isonicotinyl hydrazide
anion with NAD⁺

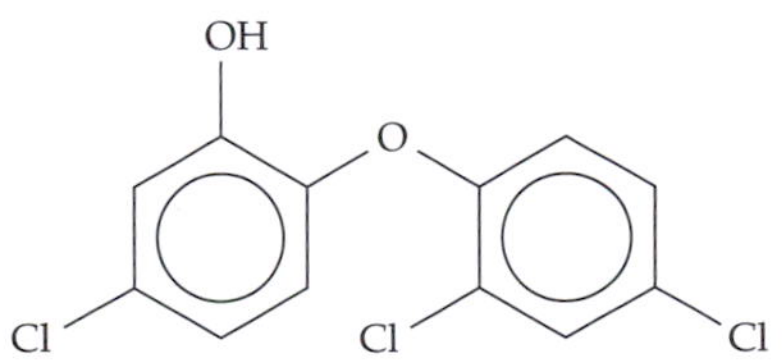

Triclosan

Less frequently NADPH is used to reduce an *isolated double bond*. An example is the hydrogenation of **desmosterol** by NADPH (Eq. 15-11), the final step in one of the pathways of biosynthesis of cholesterol (Fig. 22-7). In this and in other reactions of the same

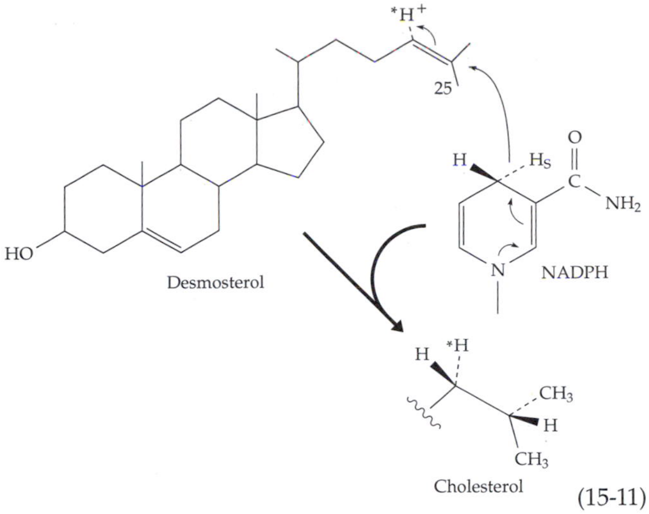

(15-11)

type, hydrogen transfer has been shown to be from the *pro-S* position in NADPH directly to C-25 of the sterol. The proton introduced from the medium (designated by the asterisks in Eq. 15-11) enters *trans* to the H⁻ ion from NADPH. The proton always adds to the more electron-rich terminus of the double bond, i.e., it follows the Markovnikov rule. This result suggests that protonation of the double bond may precede H⁻ transfer.[108]

Additional pyridine nucleotide-dependent dehydrogenases include **glutathione reductase** (Figs. 15-10, 15-12), **dihydrofolate reductase** (Fig. 15-19), isocitrate dehydrogenase, *sn*-**glycerol-3-phosphate dehydrogenase** (Chapter 21), L-3-hydroxyacyl-CoA dehydrogenase (Chapter 21), **retinol dehydrogenase** (Chapter 23), and a bacterial quinone oxidoreductase.[109] Some of these also contain a flavin coenzyme.

7. Transient Carbonyl Groups in Catalysis

Some enzymes contain bound NAD⁺ which oxidizes a substrate alcohol to facilitate a reaction step and is then regenerated. For example, the **malolactic enzyme** found in some lactic acid bacteria and also in *Ascaris* decarboxylates L-malate to lactate (Eq. 15-12). This reaction is similar to those of isocitrate dehydrogenase,[110–112] 6-phosphogluconate dehydrogenase,[113] and the malic enzyme (Eq. 13-45)[114] which utilize free NAD⁺ to first dehydrogenate the substrate to a bound oxoacid whose β carbonyl group facilitates decarboxylation. Likewise, the bound NAD⁺ of the malolactic

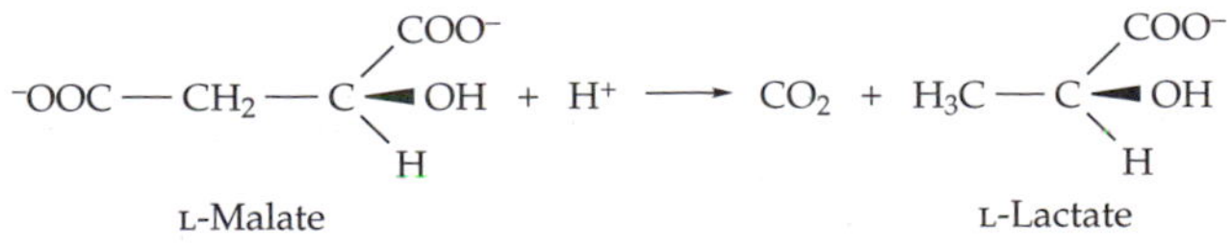

L-Malate

L-Lactate

(15-12)

enzyme apparently dehydrogenates the malate to a bound oxaloacetate which is decarboxylated to pyruvate.[115] The latter remains in the active site and is reduced by the bound NADH to lactate which is released from the enzyme.

Another reason for introducing a carbonyl group is to form a symmetric intermediate in a reaction that inverts the configuration about a chiral center. An example is **UDP-galactose 4-epimerase**, an enzyme that converts UDP-galactose to UDP-glucose (Eq. 15-13; Chapter 20) and is essential in the metabolism of galactose in our bodies. The enzyme contains bound NAD⁺ and forms a transient 4-oxo intermediate and bound NADH. Rotation of the intermediate allows nonstereospecific reduction by the NADH, leading to epimerization.[116–117a]

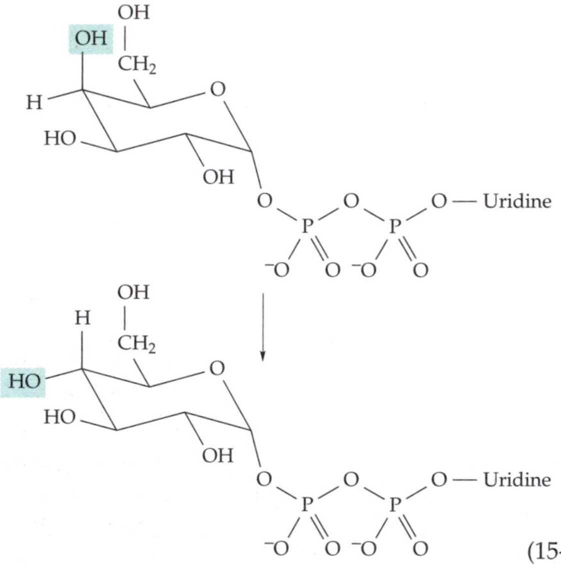

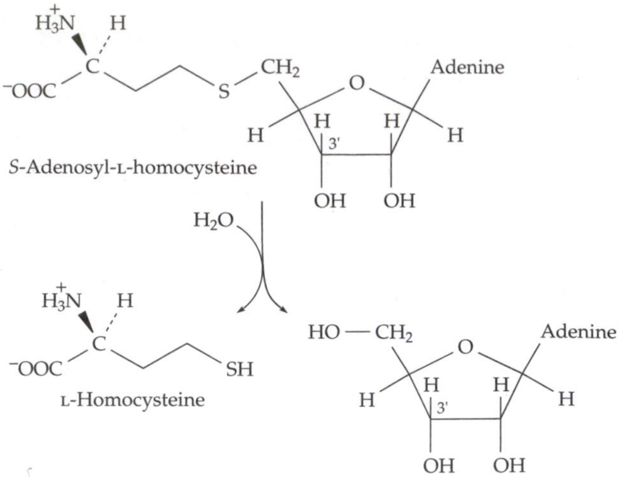

S-Adenosyl-L-homocysteine

L-Homocysteine

Adenosine   (15-14)

The enzyme is a member of the short-chain dehydrogenase group with a catalytic Tyr–Lys pair in the active site.[118] Another way that formation of an oxo group can assist in epimerization of a sugar is through enolization with nonstereospecific return of a proton to the intermediate enediol. A third possible mechanism of epimerization is through aldol cleavage followed by aldol condensation, with inversion of configuration. In each case the initial creation of an oxo group by dehydrogenation is essential.

Bound NAD⁺ is also present in **S-adenosylhomocysteine hydrolase**,[119,120] which catalyzes the irreversible reaction of Eq. 15-14. Transient oxidation at the 3 position of the ribose ring facilitates the reaction. The reader can doubtless deduce the function that has been established for the bound NAD⁺ in this enzyme. However, the role of NAD in the **urocanase** reaction (Eq. 15-15) is puzzling. This reaction, which is the second step in the catabolism of histidine, following Eq. 14-44, appears simple. However, there is no obvious

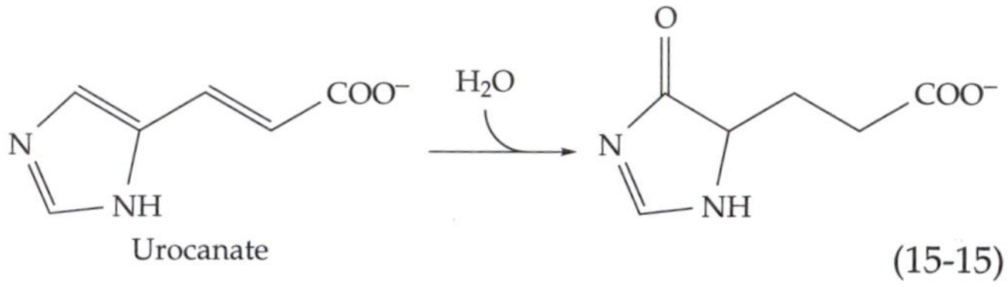

Urocanate   (15-15)

mechanism and no obvious role for NAD⁺. See Frey for a discussion.[117]

8. ADP Ribosylation and Related Reactions of NAD⁺ and NADP⁺

The linkage of nicotinamide to ribose in NAD⁺ and NADP⁺ is easily broken by nucleophilic attack on C-1 of ribose. In Chapter 11 enzyme-catalyzed ADP ribosylation, which can be shown as in Eq. 15-16, is discussed briefly. The nucleophilic group –Y from an ADP-ribosyltransferase carries the ADP-ribosyl group which can then be transferred by a second displacement onto a suitable nucleophilic acceptor group.[121,122] Hydrolysis (Eq. 15-16, step *a*) gives free ADP-ribose.[123] Other known products of enzymatic action are indicated in steps *c–f*. Poly-(ADP ribosylation) is discussed in Chapter 27. The structure of cyclic ADP-ribose (cADPR)[124] is shown in Chapter 11, Section E,2. The acceptor nucleophile is N-1 of the adenine ring which is made more nucleophilic by electron donation from the amino group. A similar reaction with ADP ribose (Eq. 15-16, step *e*) produces a dimeric ADP-ribose (ADPR)₂[125] while reaction with free nicotinic acid (step *f*) yields, in an overall base exchange, nicotinic acid adenine dinucleotide (NAADP⁺).[126,127] Some of these compounds, e.g., cADPR, NAADP⁺, and (ADPR)₂, are involved in signaling with calcium ions.[124]

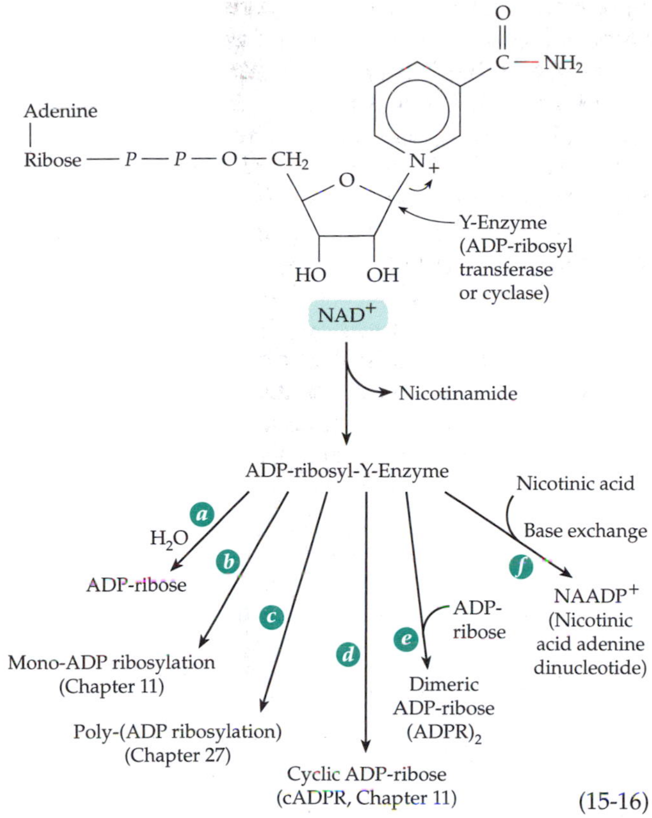

$$(15\text{-}16)$$

9. The Varied Chemistry of the Pyridine Nucleotides

Despite the apparent simplicity of their structures, the chemistry of the nicotinamide ring in NAD$^+$ and NADP$^+$ is surprisingly complex.[128,129] NAD$^+$ is extremely unstable in basic solutions, whereas NADH is just as unstable in slightly acidic media. These properties, together with the ability of NAD$^+$ to undergo condensation reactions with other compounds, have sometimes caused serious errors in interpretation of experiments and may be of significance to biological function.

Addition to NAD$^+$ and NADP$^+$. Many nucleophilic reagents add reversibly at the para (or 4) position (Eq. 15-17) to form adducts having structures resembling those of the reduced coenzymes. Formation of

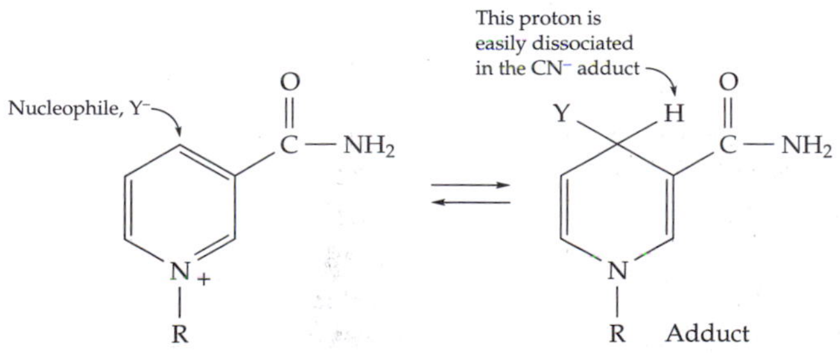

the cyanide adduct, whose absorption maximum is at 327 nm, has been used to introduce deuterium into the para position of the pyridine nucleotides. In the adduct, the hydrogen adjacent to the highly polarized C – N bond is easily dissociated as a proton. Thiolate anions and bisulfite also add. Dithionite ion, $S_2O_4^{2-}$, can lose SO_2 and acquire a proton to form the sulfoxylate ion HSO_2^- which also adds to the 4 position of the NAD$^+$ ring.[130] The resulting adduct is unstable and loses SO_2 to give NADH + H$^+$. Addition can also occur at the two ortho positions. The adducts of HO$^-$ to the 4 position of NAD$^+$ are stable but those to the 2 position undergo ring opening (Eq. 15-18) in base-catalyzed reactions which are followed by further degradation.[128,131]

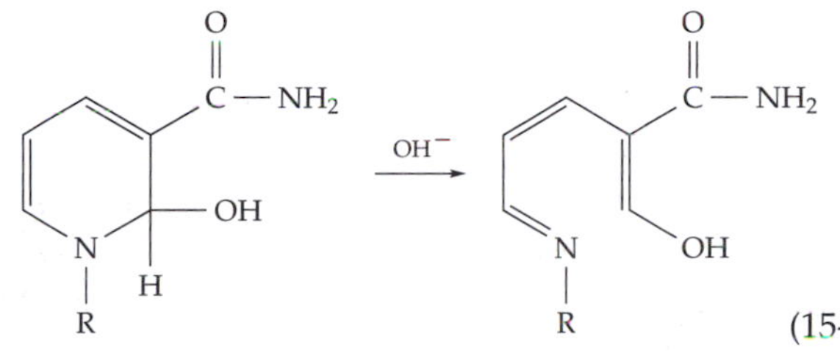

$$(15\text{-}18)$$

Another base-catalyzed reaction is the addition of enolate anions derived from ketones to the 4 position of the pyridine nucleotides (Eq. 15-19). The adducts undergo ring closure and in the presence of oxygen are converted slowly to fluorescent materials. While forming the basis for a useful analytical method for determination of NAD$^+$ (using 2-butanone), these reactions also have created a troublesome enzyme inhibitor from traces of acetone present in commercial NADH.[132]

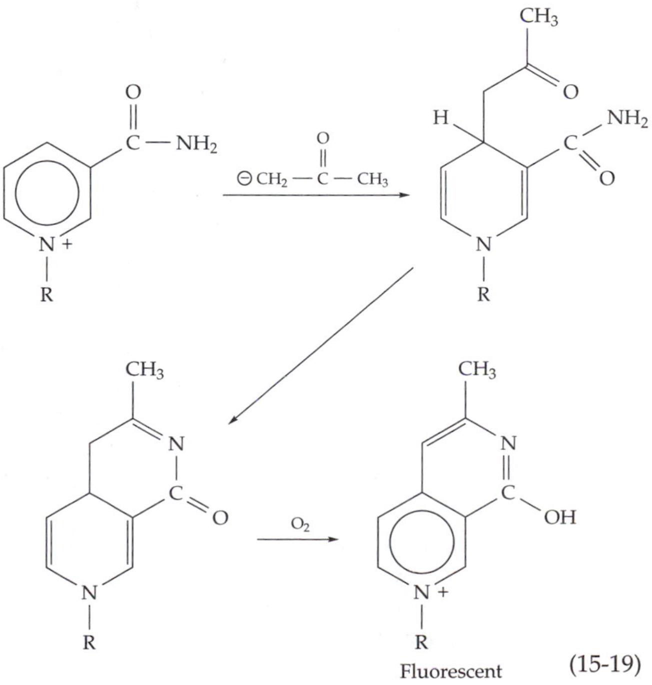

$$(15\text{-}17)$$

$$(15\text{-}19)$$

The reactions of Eq. 15-19 occur nonenzymatically only under the influence of strong base but dehydrogenases often catalyze similar condensations relatively rapidly and reversibly. Pyruvate inhibits lactate dehydrogenase, 2-oxoglutarate inhibits glutamate dehydrogenase, and ketones inhibit a short-chain alcohol dehydrogenase in this manner.[133,69a]

Modification of NADH in acid.

Reduced pyridine nucleotides are destroyed rapidly in dilute HCl and more slowly at pH 7 in reactions catalyzed by buffer acids.[128,131,134] Apparently the reduced nicotinamide ring is first protonated at C-5, after which a nucleophile Y⁻ adds at the 6 position (Eq. 15-20). The nucleophile may be OH⁻, and the adduct may undergo further reactions. For example, water may add to the other double bond and the ring may open on either side of the nitrogen. The glycosidic linkage can be isomerized from β to α or can be hydrolytically cleaved. The early steps in the modification reaction are partially reversible, but the overall sequence is irreversible. One of the products, which has been characterized by crystal structure determination, is shown in Eq. 15-20.[128,135] It can arise if the group Y of Eq. 15-20 is the C-2' hydroxyl of the ribose ring and if the configuration of the glycosidic linkage is inverted (anomerized).

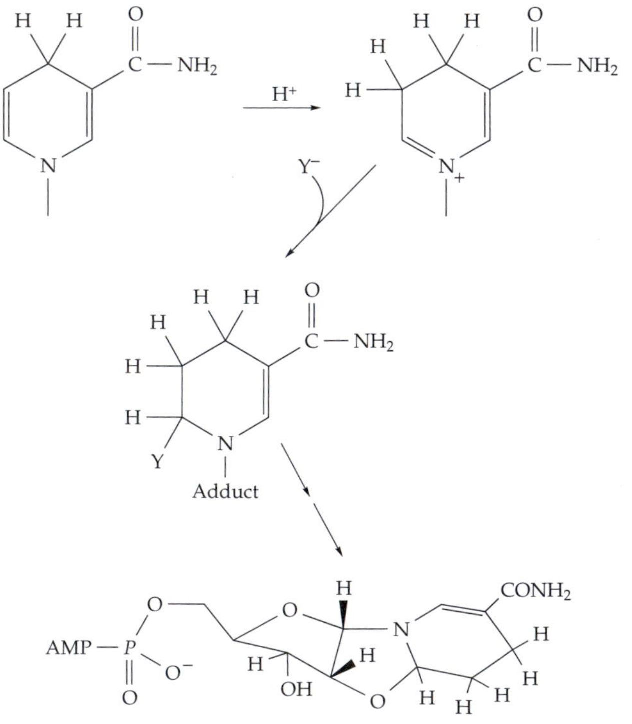

An acid modification product from NADH

(15-20)

The foregoing reactions have attracted interest because glyceraldehyde-3-phosphate dehydrogenase, in a side reaction, converts NADH to a substance referred to as NADH-X which has been shown to be the 6(R) adduct of Eq. 15-20, where Y is –OH. In an ATP-dependent reaction an enzyme from yeast reconverts NADH-X to NADH.[136]

Mercury (II) ions can add in place of H⁺ in the first step of Eq. 15-20 and subsequent reactions similar to those promoted by acid can occur.[137]

Other reactions of pyridine nucleotides.

Alkaline hexacyanoferrate (III) oxidizes NAD⁺ and NADP⁺ to 2-,4-, and 6-pyridones. The 6-pyridone of N-methylnicotinamide is a well-known excretion product of nicotinic acid in mammals. Reoxidation of NADH and NADPH to NAD⁺ and NADP⁺ can be accomplished with hexacyanoferrate (III), quinones, and riboflavin

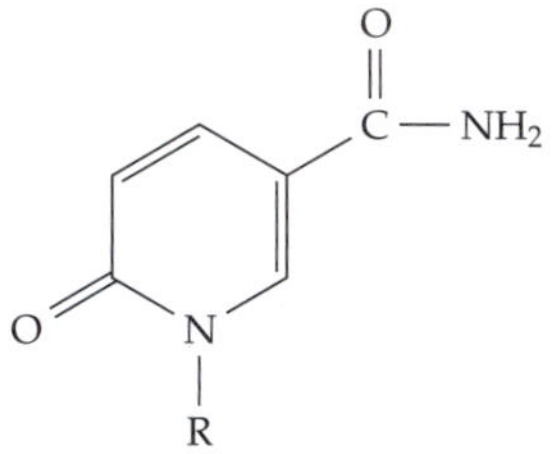

6-Pyridone

but not by H_2O_2 or O_2. However, O_2 does react at neutral pH with uptake of a proton to form a peroxide derivative of NADH.[138] When heated in 0.1 N alkali at 100°C for 5 min, NAD⁺ is hydrolyzed to nicotinamide and adenosine-diphosphate-ribose.

Treatment of NAD⁺ with nitrous acid deaminates the adenine ring. The resulting deamino NAD⁺ as well as synthetic analogs containing the following groups in place of the carboxyamide have been used

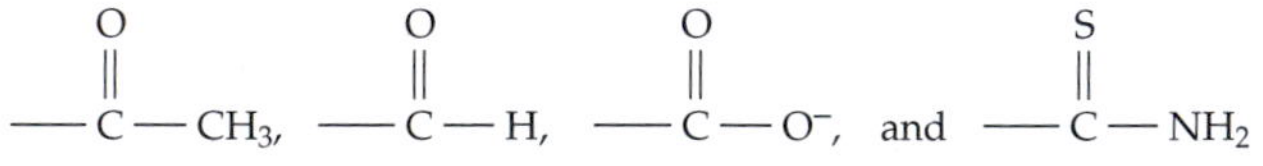

widely in enzyme studies. In fact, almost every part of the coenzyme molecule has been varied systematically and the effects on the chemical and enzymatic properties have been investigated.[139–141] "Caged" NAD⁺ and NADP⁺ have also been made.[142] These compounds do not react as substrates until they are released ("uncaged") by photolytic action of a laser beam (see Chapter 23).

B. The Flavin Coenzymes

Flavin adenine diphosphate (FAD, flavin adenine dinucleotide) and **riboflavin 5'-monophosphate (FMN**, flavin mononucleotide), whose structures are shown in Fig. 15-7, are perhaps the most versatile of all

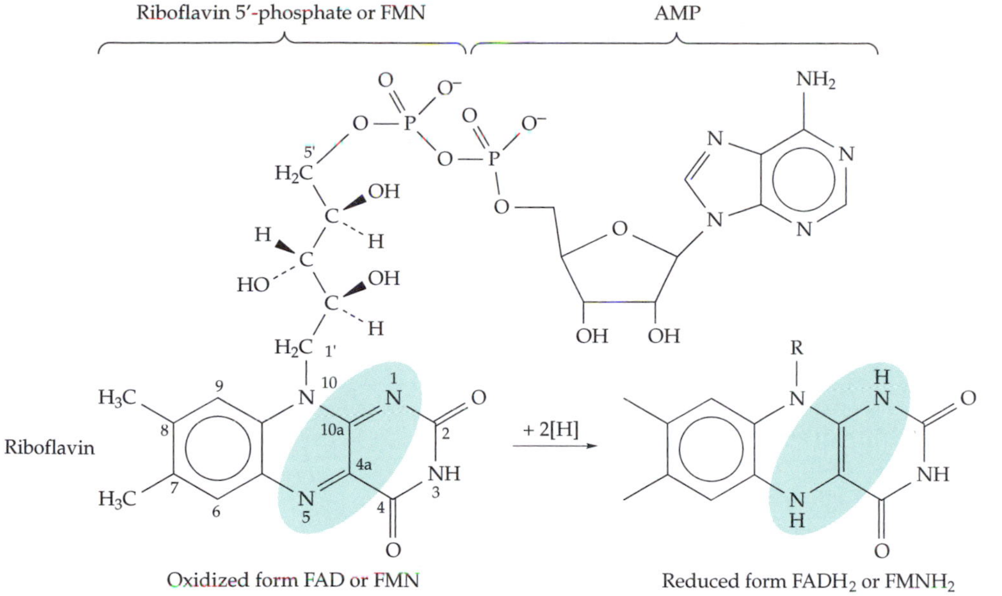

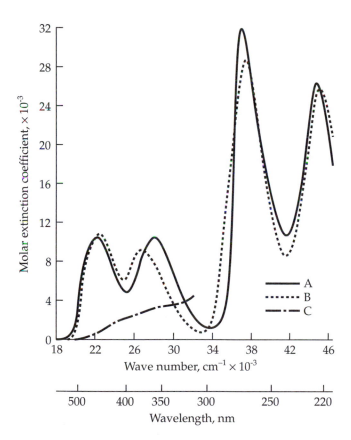

Figure 15-7 The flavin coenzymes flavin adenine dinucleotide (FAD) and riboflavin 5'-phosphate (FMN). Dotted lines enclose the region that is altered upon reduction.

Figure 15-8 Absorption spectrum of neutral, uncharged riboflavin (A), the riboflavin anion (B), and reduced to the dihydro form (Fig. 15-7) by the action of light in the presence of EDTA (C). A solution of 1.1×10^{-4} M riboflavin containing 0.01 M EDTA was placed 11.5 cm from a 40-W incandescent lamp for 30 min.

the oxidation coenzymes. The name flavin adenine dinucleotide is not entirely appropriate because the D-ribityl group is not linked to the riboflavin in a glycosidic linkage. Hemmerich suggested that FAD be called flavin adenine diphosphate.[143]

The attention of biochemists was first attracted to flavins as a result of their color and fluorescence. The study of spectral properties of flavins (Fig. 15-8) has been of importance in understanding these coenzymes. The biochemical role of the flavin coenzymes was first recognized through studies of the "old yellow enzyme"[144,145] which was shown by Theorell to contain riboflavin 5'-phosphate. By 1938, FAD was recognized as the coenzyme of a different yellow protein, **D-amino acid oxidase** of kidney tissue. Like the pyridine nucleotides, the new flavin coenzymes were reduced by dithionite to nearly colorless dihydro forms (Figs. 15-7 and 15-8) revealing the chemical basis for their function as hydrogen carriers.

Flavins are also among the natural light receptors and display an interesting and much studied photochemistry (Fig. 3-5).[143] Flavins may function in some photoresponses of plants, and they serve as light emitters in bacterial bioluminescence (Chapter 23).[146,147]

Three facts account for the need of cells for both the flavin and pyridine nucleotide coenzymes: (1) Flavins are usually stronger oxidizing agents than is NAD+. This property fits them for a role in the electron transport chains of mitochondria where a sequence of increasingly more powerful oxidants is needed and makes them ideal oxidants in a variety of other dehydrogenations. (2) Flavins can be reduced either by one- or two-electron processes. This enables them to participate in oxidation reactions involving free radicals and in reactions with metal ions. (3) Reduced flavins are "autooxidizable," i.e., they can be reoxidized directly and rapidly by O_2, a property shared with relatively few other organic substances. For example, NADH and NADPH are not spontaneously reoxidized by oxygen. Autooxidizability allows flavins of some enzymes to pass electrons directly to O_2 and also provides a basis for the functioning of flavins in hydroxylation reactions.

1. Flavoproteins and Their Reduction Potentials

Flavin coenzymes are usually bound tightly to proteins and cycle between reduced and oxidized states while attached to the same protein molecule. In a free unbound coenzyme the redox potential is determined by the structures of the oxidized and reduced forms of the couple. Both riboflavin and the pyridine nucleotides contain aromatic ring systems that are stabilized by resonance. Part of this resonance stabilization is lost upon reduction. The value of $E^{\circ\prime}$ depends in part upon the varying amounts of resonance in the oxidized and reduced forms. The structures of the coenzymes have apparently evolved to provide values of $E^{\circ\prime}$ appropriate for their biological functions.

The relative strengths of binding of oxidized and reduced flavin coenzymes to a protein also have strong effects upon the reduction potential of the coenzyme.[148] If the oxidized form is bound weakly, but the reduced form is bound tightly, a bound flavin will have a greater tendency to stay in the reduced form than it did when free. The reduction potential $E^{\circ\prime}$ will be less negative than it is for the free flavin-dihydroflavin couple. On the other hand, if the oxidized form of the flavin is bound more tightly by the protein than is the reduced form, $E^{\circ\prime}$ will be more negative and the flavoenzyme will be a less powerful oxidizing agent. In fact, the values of $E^{\circ\prime}$ at pH 7 for flavoproteins span a remarkably wide range from – 0.49 to +0.19 V. The state of protonation of the reduced flavin when bound to the enzyme will also have a major effect on the oxidation–reduction potential. For example, acyl-CoA dehydrogenases are thought to form an anionic species of reduced FAD (FADH⁻) which is tightly bound to the protein.[149]

Every flavoprotein accepts electrons from the substrate that it oxidizes and passes these electrons on to another substrate, an oxidant. In the following sections we will consider for several enzymes how the electrons may get into the flavin from the oxidizable substrate and how they may flow out of the flavin into the final electron acceptor.

2. Typical Dehydrogenation Reactions Catalyzed by Flavoproteins

The functions of flavoprotein enzymes are numerous and diversified.[151–153a] A few of them are shown in Table 15-2 and are classified there as follows: (A) oxidation of hemiacetals to lactones, (B) oxidation of alcohols to aldehydes or ketones, (C) oxidation of amines to imines, (D) oxidation of carbonyl compounds or carboxylic acids to α,β-unsaturated compounds, (E) oxidation of NADH and NADPH in electron transport chains, and (F) oxidation of dithiols to disulfides or the reverse reaction. Three-dimensional structures are known for enzymes of each of these types.

Reactions of types A–C could equally well be catalyzed by pyridine nucleotide-requiring dehydrogenases. Recall that D-glucose-6-phosphate dehydrogenase uses NADP⁺ as the oxidant (Eq. 15-10). The first product is the lactone which is hydrolyzed to 6-phosphogluconic acid. The similar reaction of free glucose (Table 15-2, reaction type A) is catalyzed by fungal **glucose oxidase**, a 580-residue FAD-containing enzyme.[154–155] A bacterial cholesterol oxidase has a similar structure.[156] The important plant enzyme **glycolate oxidase** is a dimer containing riboflavin 5'-phosphate.[157–159] It catalyzes a reaction of type B, which plays an important role in photorespiration (Chapter 23). **Amino acid oxidases** (reaction type C) are well-known. The peroxisomal D-amino acid oxidase from kidney was the source from which Warburg first isolated FAD and has been the subject of much investigation of mechanism and structure.[153,160–163] Many snake venoms contain an active 140-kDa L-amino-acid oxidase which contains FAD. Flavin-dependent **amine oxidases**, important in the human body, catalyze the related reaction with primary, secondary, or tertiary amines and in which a carboxyl group need not be present.[164–166a] Reduced flavin produced by all of these oxidases is reoxidized with molecular oxygen and hydrogen peroxide is the product. Nature has chosen to forego the use of an electron transport chain (Fig. 10-5), giving up the possible gain of ATP in favor of simplicity and a more direct reaction with oxygen. In some cases there is specific value to the organism in forming H_2O_2 (see Chapter 18).

In contrast to the flavin oxidases, flavin dehydrogenases pass electrons to carriers within electron transport chains and the flavin does *not* react with O_2. Examples include a bacterial **trimethylamine dehydrogenase** (Fig. 15-9) which contains an iron–sulfur cluster that serves as the immediate electron acceptor[167–169] and yeast **flavocytochrome b_2**, a lactate dehydrogenase that passes electrons to a built-in heme group which can then pass the electrons to an external acceptor, another heme in cytochrome c.[170–173] Like glycolate oxidase, these enzymes bind their flavin coenzyme at the ends of 8-stranded αβ barrels similar

BOX 15-B RIBOFLAVIN

The bright orange-yellow color and brilliant greenish fluorescence of riboflavin first attracted the attention of chemists. Blyth isolated the vitamin from whey in 1879 and others later obtained the same fluorescent, yellow compound from eggs, muscle, and urine. All of these substances, referred to as **flavins** because of their yellow color, were eventually recognized as identical. The structure of riboflavin was established in 1933 by R. Kuhn and associates, who had isolated 30 mg of the pure material from 30 kg of dried albumin from 10,000 eggs. The intense fluorescence assisted in the final stages of purification. The vitamin was synthesized in 1935 by P. Karrer.[a]

Riboflavin, a yellow solid, has a low solubility of ~ 100 mg / l at 25°C. Three crystalline forms are known. One of these, the "readily soluble form," is ten times more soluble than the others and can be used to prepare metastable solutions of higher concentration. One crystalline form is platelike and occurs naturally in the tapetum (Box 13-C) of the nocturnal lemur.

Discovery of the role of riboflavin in biological oxidation was an outgrowth of biochemists' interest in respiration. In the 1920s Warburg provided evidence that oxygen reacted with an iron-containing respiration catalyst and it was shown that the dye **methylene blue** could often substitute for oxygen as an oxidant (Box 18-A). Oxidation of glucose 6-phosphate by methylene blue within red blood cells required both a "ferment" (enzyme) and a "coferment," later identified as NADP+. A yellow protein, isolated from yeast, was found to have the remarkable property of being decolorized by the reducing system of glucose 6-phosphate plus the protein and coferment from red blood cells.

Warburg and Christian showed that the color of this **old yellow enzyme** came from a flavin and proposed that its cyclic reduction and reoxidation played a role in cellular oxidation. When NADP+ was isolated the proposal was extended to encompass a **respiratory chain**. The two hydrogen carriers NADP+ and flavin would work in sequence to link dehydrogenation of glucose to the iron-containing catalyst that interacted with oxygen. While we still do not know the physiological function of the old yellow enzyme,[b] the concept of respiratory chain was correct.

Human beings require about 1.5 mg of riboflavin per day. Because of its wide distribution in food, a deficiency, which affects skin and eyes, is rarely seen. Riboflavin is produced commercially in large quantities by fungi such as *Eremothecium asbyii* which, apparently because of some metabolic anomaly, produce the vitamin in such copious amounts that it crystallizes in the culture medium.

When taken up by the body, riboflavin is converted into its coenzyme forms (Chapter 25) and any excess is quickly excreted in the urine. Urine also contains smaller amounts of metabolites. The ribityl group may be cut by the action of intestinal bacteria acting on riboflavin before it is absorbed. The resulting 10-hydroxyethyl flavin may sometimes be a major urinary product.[c,d] The related 10-formylmethyl flavin is also excreted,[c] as are small amounts of 7α- and 8α- hydroxyriboflavins, apparently formed in the body by hydroxylation. These may be degraded farther to the 7α- and 8α- carboxylic acids of lumichrome (riboflavin from which the ribityl side chain is totally missing).[e] A riboflavin glucoside has also been found in rat urine.[f]

The choroid layer of the eye (behind the retina) in many animals contains a high concentration of free riboflavin. Cats' eyes also contain a large amount of 7 α-hydroxyriboflavin (nekoflavin), as do their livers.[g] Nekoflavin is also present in human blood.[h] Hen egg white contains a 219-residue riboflavin-binding protein whose functions are thought to be storage of the vitamin and delivery to the developing embryo.[i-k] Most of the riboflavin in human blood is bound to proteins such as albumin and immunoglobulins. However, during pregnancy a riboflavin-binding protein similar to that of the chicken appears, apparently to carry riboflavin to the fetus.[j]

Riboflavin is stable to heat but is extremely sensitive to light, a fact of some nutritional significance. Do not leave bottled milk in the sunshine (see Fig. 15-8)! Many products of photolysis are formed (Fig. 3-5). Among them is lumichrome.

[a] Yagi, K. (1990) in *Flavins and Flavoproteins* (Curti, B., Ronchi, S., and Zanetti, G., eds), pp. 3 –16, Walter de Gruyter, Berlin

[b] Kohli, R. M., and Massey, V. (1998) *J. Biol. Chem.* **273**, 32763–32770

[c] Owen, E. C., West, D. W., and Coates, M. E. (1970) *Br. J. Nutr.* **24**, 259–267

[d] Roughead, Z. K., and McCormick, D. B. (1991) *European Journal of Clinical Nutrition* **45**, 299–307

[e] Ohkawa, H., Ohishi, N., and Yagi, K. (1983) *J. Biol. Chem.* **258**, 5623–5628

[f] Ohkawa, H., Ohishi, N., and Yagi, K. (1983) *J. Nutr. Sci. Vitaminol.* **29**, 515–522

[g] Matsui, K., and Kasai, S. (1996) *J. Biochem.* **119**, 441–447

[h] Zempleni, J., Galloway, J. R., and McCormick, D. B. (1995) *Int. J. Vitamins Nutr. Res.* **66**, 151–157

[i] Matsui, K., Sugimoto, K., and Kasai, S. (1982) *J. Biochem.* **91**, 469–475

[j] Miura, R., Tojo, H., Fujii, S., Yamano, T., Miyake, Y., (1984) *J. Biochem.* **96**, 197–206

[k] Monaco, H. L. (1997) *EMBO J.* **16**, 1475–1483

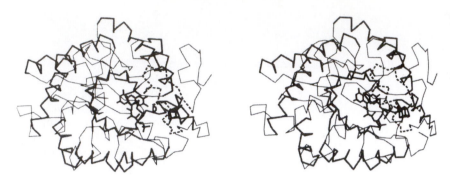

Figure 15-9 Stereoscopic view of the large domain (residues 1–383) of trimethylamine dehydrogenase from a methylotrophic bacterium. The helices and β strands of the $(\alpha\beta)_8$ barrel are drawn in heavy lines as are the FMN (center) and the Fe_4S_4 iron–sulfur cluster at the lower right edge. The α/β loop to which it is bound is drawn with dashed lines. The 733-residue protein also contains two other structural domains. From Lim *et al.*[150] Courtesy of F. S. Mathews.

to that of triose phosphate isomerase (Fig. 2-28). Flavocytochrome b_2 has an additional domain which carries the bound heme. Two additional domains of trimethylamine dehydrogenase have a topology resembling that of the FAD- and NADH-binding domains of glutathione reductase shown in Fig. 15-10. A bacterial **mandelate dehydrogenase** is structurally and mechanistically closely related to the glycolate oxidase family.[173a]

Reaction type D of Table 15-2, the dehydrogenation of an acyl-coenzyme A (CoA), could not be accomplished by a pyridine nucleotide system because the reduction potential ($E°'$, pH 7 = –0.32 V) is inappropriate. The more powerfully oxidizing flavin system is needed. (However, the reverse reaction, hydrogenation of a C=C bond, is often carried out biologically with a reduced pyridine nucleotide.) Dehydrogenation reactions of this type are important in the energy metabolism of aerobic cells. For example, the first oxidative step in the β oxidation of fatty acids (Fig. 17-1) is the α,β dehydrogenation of fatty acyl-CoA derivatives. The *pro-R* hydrogen atoms are removed from both the α- and β-carbon atoms to create the double bond (Table 15-2, type C).

TABLE 15-2
Some Dehydrogenation Reactions Catalyzed by Flavoproteins[a]

A   D-Glucose $\xrightarrow[\text{Glucose oxidase}]{-2[H]}$ Gluconolactone $\xrightarrow{H_2O}$ D-Gluconic acid

(See Eq. 17-12 for structures in a closely related reaction.)

B $HOOC-CH_2-OH \xrightarrow[\text{Glycolate oxidase}]{-2[H]} HOOC-CH=O$

C $R-\overset{\overset{\displaystyle H}{|}}{\underset{\underset{\displaystyle NH_2}{|}}{C}}-COOH \xrightarrow{-2[H]} R-\overset{\displaystyle ||}{\underset{\displaystyle NH}{C}}-COOH \xrightarrow{H_2O} R-\overset{\displaystyle ||}{\underset{\displaystyle O}{C}}-COOH + NH_3$

Amino acid oxidases

D $R-CH_2-CH_2-\overset{\overset{\displaystyle O}{||}}{C}-S-CoA \xrightarrow{-2[H]} R-CH=CH-\overset{\overset{\displaystyle O}{||}}{C}-S-CoA$

Acyl-CoA dehydrogenases

E $NADH \text{ (or NADPH)} + H^+ \xrightarrow{-2[H]} NAD^+ \text{ (NADP}^+\text{)}$

F   [Dihydrolipoic acid amide structure] $\xrightarrow[\text{GSH}]{-2[H]}$ [oxidized structure]

Dihydrolipoic acid amide

[a] These are shown as removal of two H atoms [H] and may occur by transfer of H^-, $H^+ + e^-$, or $2H^+ + 2e^-$. They represent reaction type 9C of the table inside the back cover.

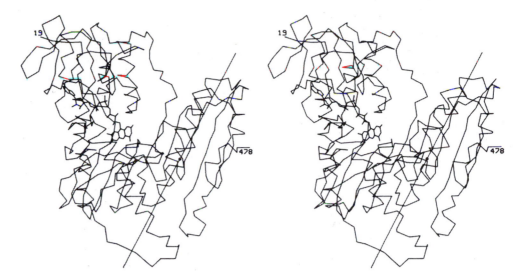

Figure 15-10 The three-dimensional structure of glutathione reductase. Bound FAD is shown. NAD$^+$ binds to a separate domain below the FAD. The two cysteine residues forming the reducible disulfide loop are indicated by dots. From Thieme *et al.*[182]

Animal mitochondria contain several different **acyl-CoA dehydrogenases** with differing preferences for chain length or branching pattern.[174–177] A related reaction that occurs in the citric acid cycle is dehydrogenation of succinate to fumarate by **succinate dehydrogenase** (Eq. 15-21).[177a] The dehydrogenation also involves *trans* removal of one of the two hydrogens, one *pro-S* hydrogen and one *pro-R*.[178] The enzyme has a large 621-residue flavoprotein subunit and a smaller 27-kDa iron–sulfur protein subunit.[179–181]

$$\text{(15-21)}$$

Neither the acyl-CoA dehydrogenases nor succinate dehydrogenase react with O$_2$. Acyl-CoA dehydrogenases pass the electrons removed from substrates to another flavoprotein, a soluble electron transferring flavoprotein (p. 794), which carries the electrons to an iron–sulfur protein embedded in the inner mitochondrial membrane where they enter the electron-transport chain. Succinate dehydrogenase as well as NADH dehydrogenase (Table 15-2, reaction E)[183] are embedded in the same membrane and also pass their electrons to iron–sulfur clusters and eventually to oxygen through the electron transport chain of the mitochondria (Chapter 18). **Fumarate reductase**[184] has properties similar to those of succinate dehydrogenase but catalyzes the opposite reaction in "anaerobic respiration" (Chapter 18),[184] as do similar reductases of bacteria[185] and of some eukaryotes.[186]

Dihydrolipoyl dehydrogenase (lipoamide dehydrogenase), **glutathione reductase**, and human **thioredoxin reductase**[187–190] belong to a subclass of flavoproteins that act on dithiols or disulfides. The reaction catalyzed by the first of these is illustrated in Table 15-2 (reaction type F). The other two enzymes usually promote the reverse type of reaction, the reduction of a disulfide to two SH groups by NADPH (Eq. 15-22). Glutathione reductase splits its substrate into two halves while reduction of the small 12-kDa protein **thioredoxin** (Box 15-C) simply opens a loop in its peptide chain. The reduction of lipoic acid opens the small disulfide-containing 5-membered ring in that molecule. Each of these flavoproteins also contains within its structure a reducible disulfide group that participates in catalysis.

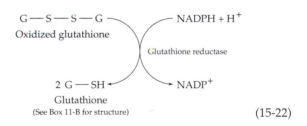

$$\text{(15-22)}$$

Each 50-kDa chain of the dimeric glutathione reductase is organized into three structural domains (Fig. 15-10).[191–193] Two of the domains each contain a nucleotide-binding motif resembling those of the NAD$^+$-dependent dehydrogenases. One of these domains binds NADPH and the other FAD. The latter domain also contains the reducible disulfide which is formed from Cys 58 and Cys 63. It serves as an intermediate hydrogen carrier which can in turn reduce oxidized glutathione. A **trypanothione reductase** from trypanosomes and related flagellated protozoa has a similar structure and acts on trypanothione, which replaces glutathione in these organisms.[188,194-196] Because it is unique to trypanosomes, this enzyme is a target for design of drugs against these organisms

BOX 15-C THIOREDOXIN AND GLUTAREDOXIN

The small proteins thioredoxin and glutaredoxin are present in relatively high concentrations in bacteria, plants, and animals. For example, thioredoxin has a concentration of 15 μM in *E. coli*. Both proteins were discovered by their role as reducing agents in conversion of the ribonucleotides AMP, GMP, CMP, and UMP to the corresponding 2-deoxyribonucleotides which are needed for synthesis of DNA:[a,b]

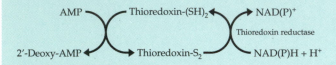

The redox group in thioredoxin is a **disulfide loop** located on a protrusion at one end of the molecule. In the 108-residue *E. coli* thioredoxin, as well as in the 105-residue human thioredoxin,[c] it is formed by cysteines 32 and 35 which are present in the conserved sequence CGPC. The –SH groups of these two interacting cysteines have pK_a values of ~6.9 and 7.5, the former belonging predominantly to the more exposed Cys 32.[d] The buried Asp 26 carboxyl group, which may be a proton donor to Cys 35 during reduction of the disulfide form,[e] forms a salt bridge with the Lys 57 $-NH_3^+$ and has a high pK_a of ~7.4.[e–g]

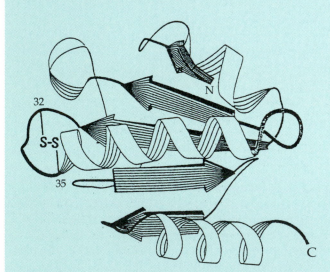

Oxidized *E. coli* thioredoxin. From Langsetmo *et al.*[f]

The pK_a assignments, which have been controversial, are discussed in Chapter 7. This disulfide loop is reduced by NADPH through the action of the flavoprotein enzyme **thioredoxin reductase**.

The resulting thiol pair of the reduced thioredoxin is the reductant used for ribonucleotide reductase (Chapter 16). The standard redox potential $E°'$ of *E. coli* thioredoxin is −0.27 V, appropriately low for coupling to the NADPH / NADP$^+$ system.

Reduced thioredoxin has a variety of functions.[h] It is the reductant for conversion of methionine sulfoxide to methionine in bacteria, for reduction of sulfate in yeast, and for additional specific enzymatic reactions.[b,h] However, its major function may be to reduce disulfide linkages in various proteins.[i] Several photosynthetic enzymes are activated by reduction of disulfide linkages via photosynthetically generated reduced ferredoxin and thioredoxin[j,k] (Chapter 23). Reduced thioredoxin may also play a similar role in nonphotosynthetic cells. It may reduce mixed disulfides such as those formed between glutathione and proteins (Box 11-B).[l–n] Thioredoxin may participate in regulation of the level of nitric oxide (NO) in tissues[m] and it is needed in the assembly of filamentous bacteriophages.[b,o] For reasons that are not clear, thioredoxin is also an essential subunit for a virus-induced DNA polymerase formed in *E. coli* following infection by bacteriophage T7 (see Chapter 27).

It was a surprise to discover that a mutant of *E. coli* lacking thioredoxin can still reduce ribonucleotides. In the mutant cells thioredoxin is replaced by glutaredoxin, whose active site disulfide linkage is reduced by glutathione rather than directly by NADPH. Oxidized glutathione is, in turn, reduced by NADPH and glutathione reductase. Thus, the end result is the same with respect to ribonucleotide reduction.

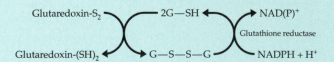

However, the two proteins have significantly different specificities and functions. The disulfide loop in glutaredoxin, whose eukaryotic forms are often called **thioltransferases**,[p] has the sequence CPYC. Although glutaredoxins are weaker reductants of mixed disulfides of proteins with glutathione than are thioredoxins,[q–s] they are more specific.

Both thioredoxin and glutaredoxin are members of a larger group of thiol:disulfide oxidoreductases which are found in all known organisms. In *E. coli* there are one thioredoxin, three different glutaredoxins,[h,t] and the periplasmic protein disulfide

BOX 15-C (continued)

isomerases DsbA and DsbC, which were discussed in Chapter 10.[u-w] Similar enzymes are found in the endoplasmic reticulum of eukaryotic cells.[u,x] Like thioredoxins, they contain disulfide loops which contain the sequences CGHC and CGYC as well as CPHC. Thioredoxin reductase itself also can keep cytoplasmic bacterial proteins reduced.[y] Redox potentials vary within this family of proteins:[n,s,z,aa]

	$E^{o'}$, pH 7
Thioredoxins	–.27 V
Glutaredoxins	–.20 to –.23 V
Protein disulfide isomerase	–.18 V
DsbA	–.09 to –.12 V

These differences are correlated with differing functions. The intracellular reduced thioredoxins are thermodynamically the best reductants of disulfide linkages in proteins and they help keep intracellular proteins reduced. Glutaredoxin can be reduced efficiently by reduced glutathione or by NADPH and glutathione reductase and can, in turn, reduce cysteine and the oxidized form of vitamin C, **dehydroascorbic acid** (Box 18-D).[bb,cc] The periplasmic bacterial proteins DsbA and DsbC have the highest redox potentials and an unusually low first pK_a for the dithiol pair in their thioredoxinlike domains.[aa] The basis for these properties has been hard to understand but is consistent with their role in assisting formation of disulfide bridges in extracellular proteins and with the role of the related protein disulfide isomerase in the ER.[u,dd-ff] These disulfide exchanges are nucleophilic displacement reactions (Chapter 12).

[a] Holmgren, A. (1981) Trends Biochem. Sci. 6, 26–29

[b] Holmgren, A. (1989) J. Biol. Chem. 264, 13963–13966

[c] Forman-Kay, J. D., Clore, G. M., Wingfield, P. T., and Gronenborn, A. M. (1991) Biochemistry 30, 2685–2698

[d] Dyson, H. J., Jeng, M.-F., Tennant, L. L., Slaby, I., Lindell, M., Cui, D.-S., Kuprin, S., and Holmgren, A. (1997) Biochemistry 36, 2622–2636

[e] Jeng, M.-F., and Dyson, H. J. (1996) Biochemistry 35, 1–6

[f] Langsetmo, K., Fuchs, J. A., and Woodward, C. (1991) Biochemistry 30, 7603–7609

[g] Ladbury, J. E., Wynn, R., Hellinga, H. W., and Sturtevant, J. M. (1993) Biochemistry 32, 7526–7530

[h] Gvakharia, B. O., Hanson, E., Koonin, E. K., and Mathews, C. K. (1996) J. Biol. Chem. 271, 15307–15310

[i] Thomas, J. A., Poland, B., and Honzatko, R. (1995) Arch. Biochem. Biophys. 319, 1–9

[j] Capitani, G., Markovic-Housley, Z., DelVal, G., Morris, M., Jansonius, J. N., and Schürmann, P. (2000) J. Mol. Biol. 302, 135–154

[k] Buchanan, B. B., Schürmann, P., Decottignies, P., and Lozano, R. M. (1994) Arch. Biochem. Biophys. 314, 257–260

[l] Wynn, R., Cocco, M. J., and Richrds, F. M. (1995) Biochemistry 34, 11807–11813

[m] Nikitovic, D., and Holmgren, A. (1996) J. Biol. Chem. 271, 19180–19185

[n] Prinz, W. A., Åslund, F., Holmgren, A., and Beckwith, J. (1997) J. Biol. Chem. 272, 15661–15667

[o] Russel, M., and Model, P. (1985) Proc. Natl. Acad. Sci. U.S.A. 82, 29–33

[p] Srinivasan, U., Mieyal, P. A., and Mieyal, J. J. (1997) Biochemistry 36, 3199–3206

[q] Katti, S. K., Robbins, A. H., Yang, Y., and Wells, W. W. (1995) Protein Sci. 4, 1998–2005

[r] Gravina, S. A., and Mieyal, J. J. (1993) Biochemistry 32, 336–3376

[s] Nikkola, M., Gleason, F. K., and Eklund, H. (1993) J. Biol. Chem. 268, 3845–3849

[t] Åslund, F., Nordstrand, K., Berndt, K. D., Nikkola, M., Bergman, T., Ponstingl, H., Jörnvall, H., Otting, G., and Holmgren, A. (1996) J. Biol. Chem. 271, 6736–6745

[u] Freedman, R. B., Hirst, T. R., and Tuite, M. F. (1994) Trends Biochem. Sci. 19, 331–335

[v] Rietsch, A., Belin, D., Martin, N., and Beckwith, J. (1996) Proc. Natl. Acad. Sci. U.S.A. 93, 13048–13053

[w] Darby, N. J., and Creighton, T. E. (1995) Biochemistry 34, 16770–16780

[x] Kanaya, E., Anaguchi, H., and Kikuchi, M. (1994) J. Biol. Chem. 269, 4273–4278

[y] Derman, A. I., Prinz, W. A., Belin, D., and Beckwith, J. (1993) Science 262, 1744–1747

[z] Chivers, P. T., Prehoda, K. E., and Raines, R. T. (1997) Biochemistry 36, 4061–4066

[aa] Jacobi, A., Huber-Wunderlich, M., Hennecke, J., and Glockshuber, R. (1997) J. Biol. Chem. 272, 21692–21699

[bb] Wells, W. W., Xu, D. P., Yang, Y., and Rocque, P. A. (1990) J. Biol. Chem. 265, 15361–15364

[cc] Bischoff, R., Lepage, P., Jaquinod, M., Cauet, G., Acker-Klein, M., Clesse, D., Laporte, M., Bayol, A., Van Dorsselaer, A., and Roitsch, C. (1993) Biochemistry 32, 725–734

[dd] Hwang, C., Sinskey, A. J., and Lodish, H. F. (1992) Science 257, 1496–1502

[ee] Ruoppolo, M., Freedman, R. B., Pucci, P., and Marino, G. (1996) Biochemistry 35, 13636–13646

[ff] Couprie, J., Vinci, F., Dugave, C., Quéméneur, E., and Mourtiez, M. (2000) Biochemistry 39, 6732–6742

which cause such terrible diseases as African sleeping sickness and Chagas disease.[195,197]

Another flavoprotein constructed on the glutathione reductase pattern is the bacterial plasmid-encoded **mercuric reductase** which reduces the highly toxic Hg^{2+} to volatile elemental mercury, Hg^0. A reducible disulfide loop corresponding to that in glutathione reductase is present in this enzyme but there is also a second pair of cysteines nearby. All of these may participate in binding and reduction of Hg^{2+}.[198–199a]

3. More Flavoproteins

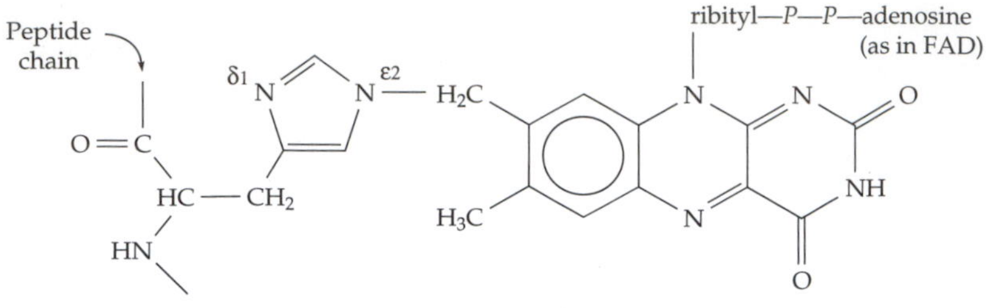

Flavoproteins function in virtually every area of metabolism and we have considered only a small fraction of the total number. Here are a few more. Flavin-dependent reductases use hydrogen atoms from NADH or NADPH to reduce many specific substances or classes of compounds. The FAD-containing ferredoxin: NADP$^+$ oxidoreductase catalyzes the reduction of free NADP$^+$ by reduced ferredoxin generated in the chloroplasts of green leaves.[200,201] Similar enzymes, some of which utilize reduced flavodoxins, are found in bacteria.[202,203] The FMN-containing subunit of NADH: ubiquinone oxidoreductase is an essential link in the mitochondrial electron transport chain for oxidation of NADH in plants and animals[183,204,205] and for related processes in bacteria. **Glutamate synthase**, a key enzyme in the nitrogen metabolism of plants and microorganisms, uses electrons from NADPH to reduce 2-oxoglutarate to glutamate in a complex glutamine-dependent process (see Fig. 24-5). The enzyme contains both FMN and FAD and three different iron–sulfur clusters.[205a] Flavin reductases use NADH or NADPH to reduce free riboflavin, FMN, or FAD needed for various purposes[206,206a] including emission of light by luminous bacteria.[207] They provide electrons to many enzymes that react with O_2 such as the cytochromes P450[208,209] and nitric oxide synthase (Chapter 18). An example is adrenodoxin reductase (see chapter banner, p. 764), which passes electrons from NADPH to cytochrome P450 via the small redox protein adrenodoxin. This system functions in steroid biosynthesis as is indicated in Fig. 22-7.[209a,b] Other flavin-dependent reductases have protective functions catalyzing the reduction of ascorbic acid radicals,[210,211] toxic quinones,[212–214] and peroxides.[215–218]

Covalently bound modified FAD
of succinate dehydrogenase

has been isolated from a bacterial electron-transferring flavoprotein.[225] Commercial FAD may contain some riboflavin 5'-pyrophosphate which activates some flavoproteins and inhibits others.[226]

Methanogenic bacteria contain a series of unique coenzymes (Section F) among which is **coenzyme F$_{420}$**, a 5-deazaflavin substituted by H at position 7 and –OH at position 8 (8-hydroxy-7,8-didemethyl-5-deazariboflavin).[227,228]

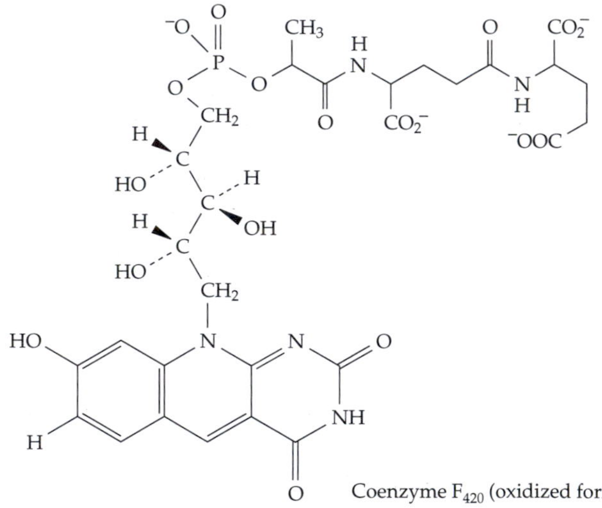

Coenzyme F$_{420}$ (oxidized form)

4. Modified Flavin Coenzymes

Mitochondrial succinate dehydrogenase, which catalyzes the reaction of Eq. 15-21, contains a flavin prosthetic group that is covalently attached to a histidine side chain. This modified FAD was isolated and identified as 8α-($N^{\epsilon 2}$-histidyl)-FAD.[219] The same prosthetic group has also been found in several other dehydrogenases.[220] It was the first identified member of a series of modified FAD or riboflavin 5'-phosphate derivatives that are attached by covalent bonds to the active sites of more than 20 different enzymes.[219]

These include 8α-($N^{\epsilon 2}$-histidyl)-FMN,[221] 8α-($N^{\delta 1}$-histidyl)-FAD,[222] 8α-(O-tyrosyl-FAD),[223] and 6-(S-cysteinyl)-riboflavin 5'-phosphate, which is found in trimethylamine dehydrogenase (Fig. 15-9).[224] An 8-hydroxy analog of FAD (–OH in place of the 8-CH$_3$)

This unique redox catalyst links the oxidation of H_2 or of formate to the reduction of NADP^{+}[229] and also serves as the reductant in the final step of methane biosynthesis (see Section E).[228] It resembles NAD$^+$ in having a redox potential of about –0.345 volts and the tendency to be only a two-electron donor. More recently free 8-hydroxy-7,8-didemethyl-5-deazaribo-flavin has been identified as an essential light-absorbing chromophore in DNA photolyase of *Methanobacterium*, other bacteria, and eukaryotic algae.[230] **Roseoflavin** is not a coenzyme but an antibiotic from *Streptomyces davawensis*.[231] Many synthetic flavins have been used in studies of mechanisms and for NMR[232] and other forms of spectroscopy.

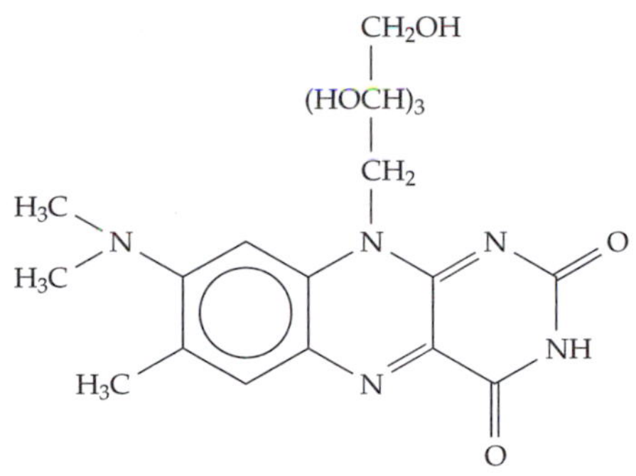

Roseoflavin

5. Mechanisms of Flavin Dehydrogenase Action

The chemistry of flavins is complex, a fact that is reflected in the uncertainty that has accompanied efforts to understand mechanisms. For flavoproteins at least four mechanistic possibilities must be considered.[153a,233] (a) A reasonable **hydride-transfer** mechanism can be written for flavoprotein dehydrogenases (Eq. 15-23). The hydride ion is donated at N-5 and a proton is accepted at N-1. The oxidation of alcohols, amines, ketones, and reduced pyridine nucleotides can all be visualized in this way. Support for such a mechanism came from study of the nonenzymatic oxidation of NADH by flavins, a reaction that occurs at moderate speed in water at room temperature. A variety of flavins and dihydropyridine derivatives have been studied, and the electronic effects observed for the reaction are compatible with the hydride ion mechanism.[234–236]

According to the mechanism of Eq. 15-23, a hydride ion is transferred directly from a carbon atom in a substrate to the flavin. However, a labeled hydrogen atom transferred to N-5 or N-1 would immediately exchange with the medium, rapid exchange being characteristic of hydrogens attached to nitrogen.

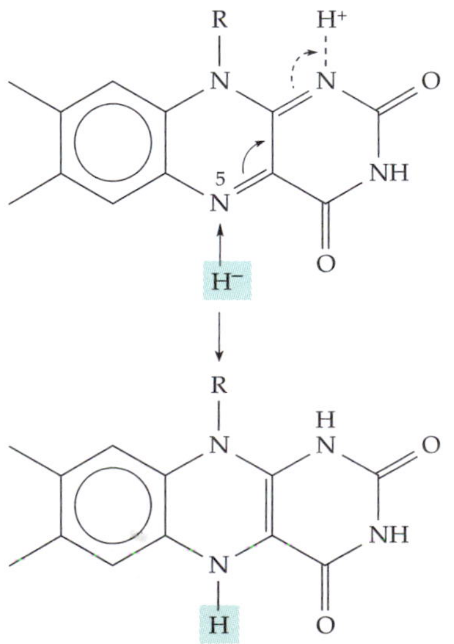

(15-23)

To avoid this problem, Brustlein and Bruice used a **5-deazaflavin** to oxidize NADH nonenzymatically.[237] When this reaction was carried out in ^2H_2O, no ^2H entered the product at C-5, indicating that a hydrogen atom (circled in Eq. 15-24) had been transferred directly from NADH to the C-5 position. Similar direct transfer of hydrogen to C-5 of 5-deazariboflavin 5'-phosphate is catalyzed by flavoproteins such as N-methylglutamate synthase[238] and acyl-CoA dehydrogenase.[237–239]

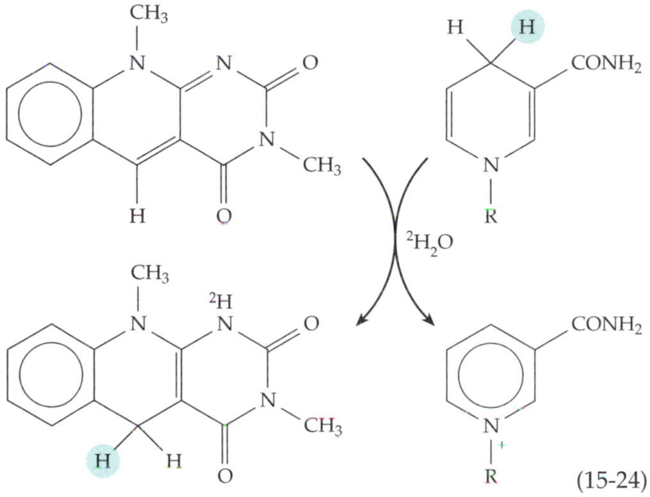

(15-24)

However, these experiments may not have established a mechanism for natural flavoprotein catalysis because the properties of 5-deazaflavins resemble those of NAD$^+$ more than of flavins.[239] Their oxidation–reduction potentials are low, they do not form stable free radicals, and their reduced forms don't react readily with O_2. Nevertheless, for an acyl-CoA dehydrogenase the rate of reaction of the deazaflavin is almost as fast as that of natural FAD.[238] For these enzymes a hydride ion transfer from the β CH (reaction type D of Table 15-1) is made easy by removal of the α-H of the acyl-CoA to form an enolate anion intermediate.

The three-dimensional structure of the medium chain acyl-CoA dehydrogenases with bound substrates and inhibitors is known.[174,175,240] A conserved glutamate side chain is positioned to pull the *pro-R* proton from the α carbon to create the initial enolate anion.[174,175,241] The *pro-R* β C–H lies by N-5 of the flavin ring seemingly ready to donate a hydride ion as in Eq. 15-23. NMR spectroscopy has been carried out with ^{13}C or ^{15}N in each of the atoms of the redox active part of the FAD. The results show directly the effects of strong hydrogen bonding to the protein at N-1, N-3, and N-5 and also suggest that the bound $FADH_2$ is really FADH$^-$ with the negative charge localized on N-1 by strong hydrogen bonding.[149] Many mutants have been made,[242] substrate analogs have been tested,[176,243] kinetic isotope effects have been measured,[242a] and potentiometric titrations have been done.[243] All of the results are compatible with the enolate anion hydride-transfer mechanism. Questions about the acidity of the α-H and the mechanism of its removal to form the enolate anion[242a] are similar to those discussed in Chapter 13, Section B.

A peculiarity of several acyl-CoA dehydrogenases is a bright green color with an absorption maximum at 710 nm. This was found to result from tightly bound coenzyme A persulfide (CoA–S–S⁻).[244,245]

(b) A second possible mechanism of flavin reduction is suggested by the occurrence of addition reactions involving the isoalloxazine ring of flavins. Sulfite adds to flavins by forming an N–S bond at the 5 position and nitroethane, which is readily dissociated to the carbanion $H_3C-CH^--NO_2$, acts as a substrate for D-amino acid oxidase.[246,247] This fact suggested a **carbanion mechanism** according to which normal D-amino acid substrates form carbanions by dissociation of the αH (Eq. 15-25). Ionization would be facilitated by binding of the substrate carboxylate to an adjacent arginine side chain and the carbanion could react at N-5 of the flavin as in Eq. 15-25. Similar mechanisms have been suggested for other flavin enzymes.[248,249]

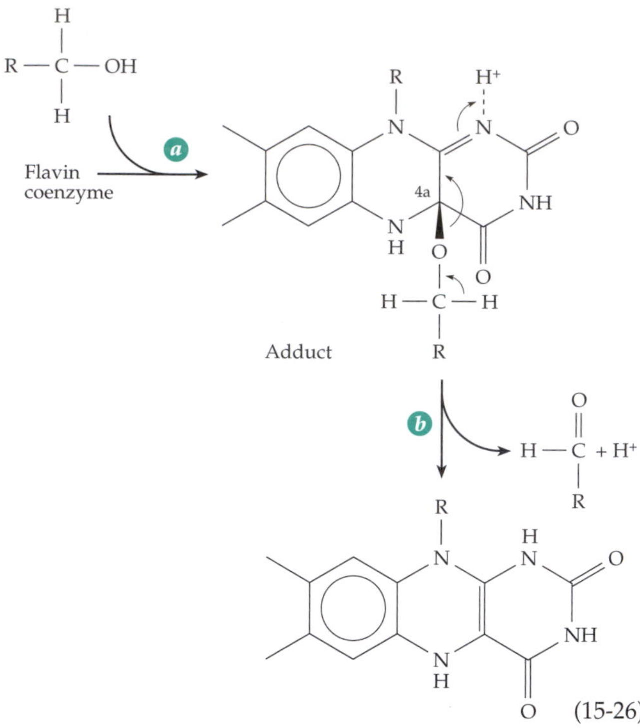

(15-26)

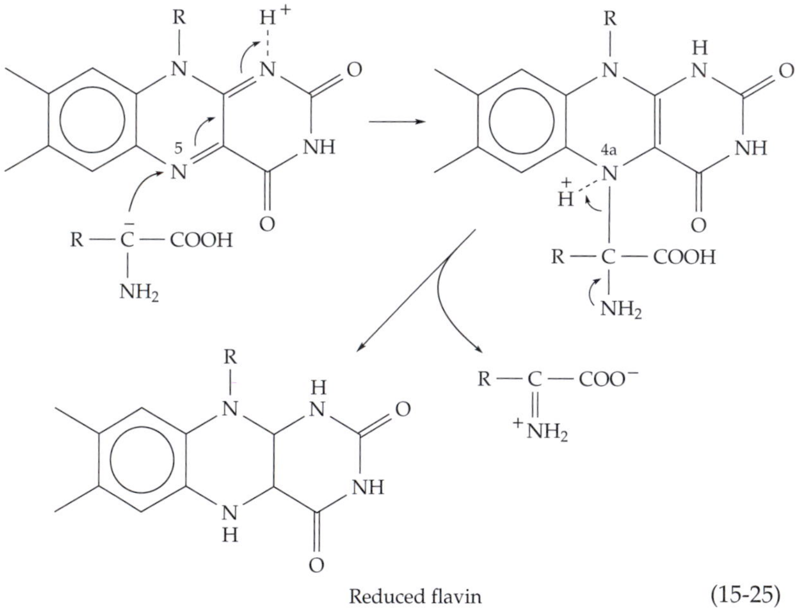

4a adduct mechanism are the reduced flavin and an aldehyde, the same as would be obtained by the hydride ion mechanism. However, in Eq. 15-26, both hydrogens in the original substrate (that on oxygen and that on carbon) have dissociated as protons, the electrons having moved as a pair during the cleavage of the adduct. Hamilton argued that an isolated hydride ion has a large diameter while a proton is small and mobile; for this reason dehydrogenation may often take place by proton transfer mechanisms.

Experimental support for the mechanism of Eq. 15-26 has been obtained using D-chloroalanine as a substrate for D-amino acid oxidase.[252–254] Chloro-pyruvate is the expected product, but under anaerobic conditions pyruvate was formed. Kinetic data obtained with α-²H and α-³H substrates suggested a common intermediate for formation of both pyruvate and chloro-pyruvate. This intermediate could be an anion formed by loss of H⁺ either from alanine or from a C-4a adduct. The anion could eliminate chloride ion as indicated by the dashed arrows in the following structure. This would lead to formation of pyruvate without reduction of the flavin. Alternatively, the electrons from the carbanion could flow into the flavin (green arrows), reducing it as in Eq. 15-26. A similar mechanism has been suggested for other flavoenzymes.[249,255] Objections to the carbanion mechanism are the expected

Reduced flavin   (15-25)

(c) The adducts with nitroethane and other compounds[250] pointed to reaction at N-5, but Hamilton[251] suggested that a better position for addition of nucleophiles is carbon 4a, which together with N-5 forms a cyclic Schiff base. He argued that other electrophilic centers in the flavin molecule, such as carbons 2, 4, and 10a, would be unreactive because of their involvement in amide or amidine-type resonance but an amine, alcohol, or other substrate could add to a flavin at position 4a (Eq. 15-26, step a). Cleavage of the newly formed C–O bond could then occur by movement of electrons from the alcohol part of the adduct into the flavin as indicated in step b. The products of this

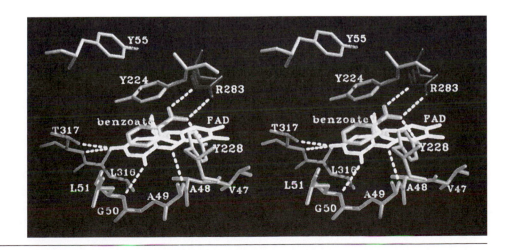

This H removed as H+ → H
in rate-limiting step

very high pK_a for loss of the α-H to form the carbanion[256] and the observed formation of only chloroalanine and no pyruvate in the reverse reaction of chloropyruvate, ammonia, and reduced flavoprotein.

A long-known characteristic of D-amino acid oxidase is its tendency to form charge-transfer complexes with amines, complexes in which a nonbonding electron has been transferred partially to the flavin. Complete electron transfer would yield a flavin radical and a substrate radical which could be intermediates in a **free radical mechanism**, as discussed in the next section.[256]

The three-dimensional structure of the complex of D-amino acid oxidase with the substrate analog benzoate has been determined. The carboxyl group of the inhibitor is bound by an arginine side chain (Fig. 15-11) that probably also holds the amino acid substrate. There is no basic group nearby in the enzyme that could serve to remove the α-H atom in Eq. 15-26 but the position is appropriate for a direct transfer of the hydrogen to the flavin as a hydride ion as in Eq. 15-23.[161,162,257] In spite of all arguments to the contrary the hydride ion mechanism could be correct! However, an adduct mechanism is still possible.

Experimental evidence supports a 4a adduct mechanism for glutathione reductase and related enzymes[191,258,258a] (Fig. 15-12). In this figure the reac-

tion sequence is opposite to that in Eq. 15-26. The enzyme presumably functions as follows. NADPH binds next to the bound FAD and reduces it by transfer of the 4-pro-S hydrogen of the NADPH (Fig. 15-12, step a). The sulfur atom of Cys 63 is in van der Waals contact with the bound FAD at or near carbon atom 4a. In step b the nucleophilic center on atom C-4a of FADH$_2$ attacks a sulfur atom of the disulfide loop between cysteines 58 and 63 in the protein to create a C-4a adduct of a thiolate ion with oxidized FAD and to cleave the −S−S− linkage in the loop. In step c the thiol group of cysteine 63 is eliminated, after which the thiol of Cys 58 attacks the nearer sulfur atom of the oxidized glutathione in a nucleophilic displacement (step d) to give one reduced glutathione (GSH) and a mixed disulfide of glutathione and the enzyme (G-S-S-Cys 58). The thiolate anion of Cys 63, which is stabilized by interaction with the adjacent flavin ring, then attacks this disulfide (step e) to regenerate the internal disulfide and to free the second molecule of reduced glutathione. The imidazole group of the nearby His 467 of the second subunit presumably participates in catalysis as may some other side chains.[191] The disulfide exchange reactions are similar to those discussed in Chapter 12.

A variation is observed for E. coli thioredoxin reductase. The reducible disulfide and the NADPH binding site are both on the same side of the flavin rather than on opposite sides as in Fig. 15-12.[190,259] Mercuric reductase also uses NADPH as the reductant transferring the 4S hydrogen. The Hg^{2+} presumably binds to a sulfur atom of the reduced disulfide loop and there undergoes reduction. The observed geometry of the active site is correct for this mechanism.

6. Half-Reduced Flavins

A possible mechanism of flavin dehydrogenation consists of consecutive transfer of a hydrogen atom and of an electron with intermediate radicals being formed both on the flavin and on the substrate. Such

Figure 15-11 Stereoscopic view of the benzoate ion in its complex with D-amino acid oxidase. A pair of hydrogen bonds binds the carboxylate of the ligand to the guanidinium group of R283. Several hydrogen bonds to the flavin ring of the FAD are also indicated. Courtesy of Retsu Miura.[161]

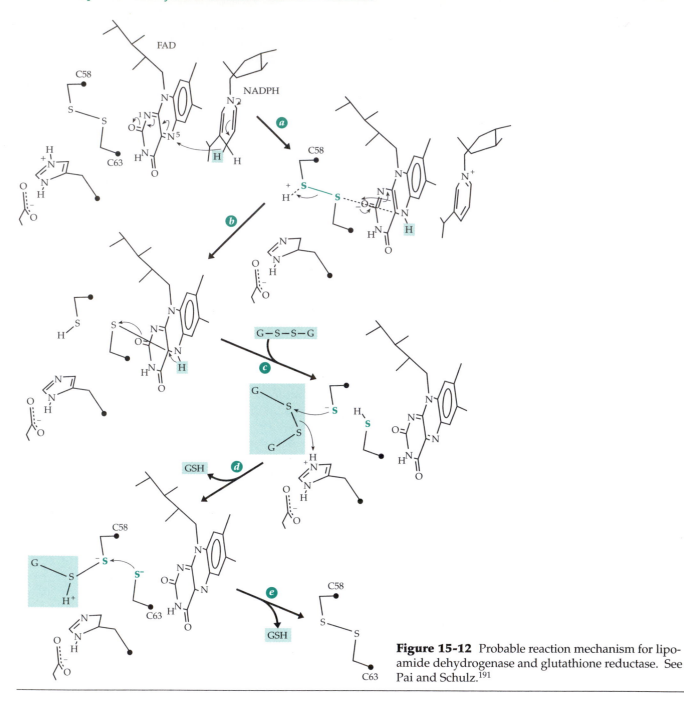

Figure 15-12 Probable reaction mechanism for lipo-amide dehydrogenase and glutathione reductase. See Pai and Schulz.[191]

a mechanism takes full advantage of one of the most characteristic properties of flavins, their ability to accept single electrons to form **semiquinone** radicals. If the oxidized form Fl of a flavin is mixed with the reduced form FlH$_2$, a single hydrogen atom is transferred from FlH$_2$ to Fl to form two ·FlH radicals (Eq. 15-27).

$$Fl + FlH_2 \xrightleftharpoons{K_f} 2 \,^{\cdot}FlH \qquad (15\text{-}27)$$

The equilibrium represented by this equation is independent of pH, but because all three forms of the flavin have different pK_a values (Fig. 15-13) the appar-

ent equilibrium constants relating total concentrations of oxidized, reduced, and radical forms vary with pH.[143,260–262] The fraction of radicals present is greater at low pH and at high pH than at neutrality. For a 3-alkylated flavin the formation constant K_f has been estimated as 2.3×10^{-2} and for riboflavin[260] as 1.5×10^{-2}. From these values and the pK_a values in Fig. 15-13, it is possible to estimate the amount of radical present at any pH.

Neutral flavin radicals have a blue color (the wavelength of the absorption maximum, λ_{max}, is ~560 nm) but either protonation at N-1 or dissociation of a proton from N-5 leads to red cation or anion radicals with λ_{max} at ~477 nm. Both blue and red radicals are

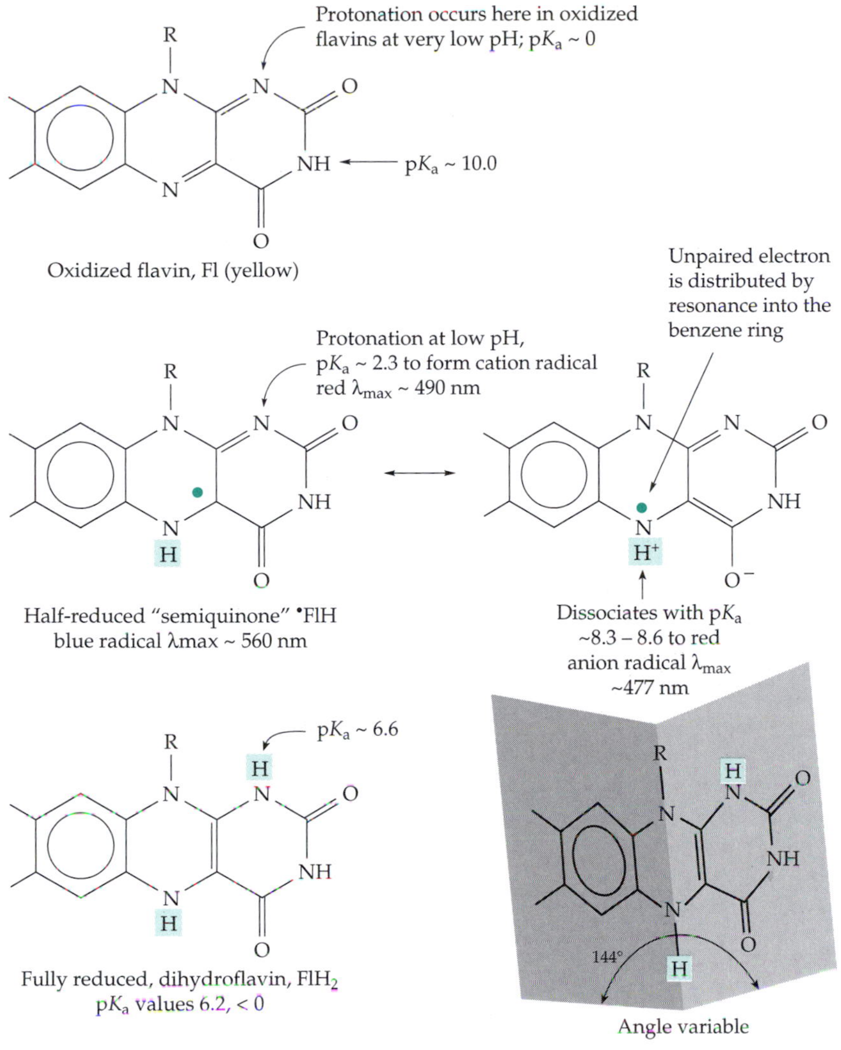

Figure 15-13 Properties of oxidized, half-reduced, and fully reduced flavins. See Müller et al.[263,264]

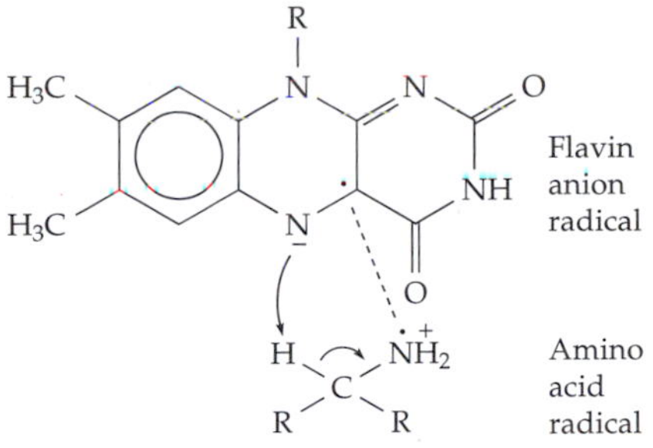

observed in enzymes, with some enzymes favoring one and some the other. Hemmerich suggested that enzymes forming red radicals make a strong hydrogen bond to the proton in the 5 position of the flavin. This increases the basicity of N-1 leading to its protonation and formation of the red cation radicals.

It is possible that many flavoprotein oxygenases and dehydrogenases react via free radicals. For example, instead of the mechanism of Eq. 15-26, an electron could be transferred to the flavin, leaving a radical pair (at right). The crystallographic structure and modeling of the substrate complex support this possibility.[265,265a] In this pair the flavin radical would be more basic than in the fully oxidized form and the amino acid radical would be more acidic than in the uncharged form. A proton transfer as indicated together with coupling of the radical pair would yield the same product as the mechanism of Eq. 15-26.

Another alternative to a radical pair is a hydrogen atom transfer followed by a second electron transfer.[265]

If an enzyme binds a flavin radical much more tightly than the fully oxidized or reduced forms, reduction of the flavoprotein will take place in two one-electron steps. In such proteins the values of $E°'$ for the two steps may be widely separated. The best known examples are the small, low-potential electron-carrying proteins known as **flavodoxins**.[266–269a] These proteins, which carry electrons between pairs of other redox proteins, have a variety of functions in anaerobic and photosynthetic bacteria, cyanobacteria, and green algae. Their functions are similar to those of the **ferredoxins**, iron–sulfur proteins that are considered in Chapter 16. In some bacteria ferredoxin and flavodoxin are interchangeable and the synthesis of flavodoxin is induced if the bacteria become deficient in iron. Flavodoxins all contain riboflavin monophosphate, which functions by cycling between the fully reduced anionic form and a blue semiquinone radical.[270] The two reduction steps, from oxidized flavin to semiquinone and from semiquinone to dihydroflavin, are well separated. For example, the values of $E°'$ (pH 7) for the flavodoxin from *Megasphaera elsdenii* are –0.115 and –0.373 V, while those of the *Azotobacter vinlandii* flavodoxin (azotoflavin) are + 0.050 and – 0.495 V. The latter is the lowest known for any flavoprotein.

Flavodoxins are small proteins with an α/β structure resembling that of the nucleotide binding domain of dehydrogenases. According to ^{31}P NMR data, the phosphate group of the coenzyme bound to flavodoxin is completely ionized,[271] even though it is deeply buried in the protein and is not bound to any positively charged side chain but to the N terminus of an α helix and to four –OH groups of serine and threonine side chains. The flavin ring is partially buried near the surface of the 138-residue protein. An aromatic side chain, from tryptophan or tyrosine, lies against the flavin on the outside of the molecule. Flavodoxins can be crystallized in all three forms: oxidized, semiquinone, and fully reduced. In the crystals the flavin semiquinone, like the oxidized flavin, is nearly planar.

The **DNA photolyase** of *E. coli*, an enzyme that participates in the photochemical repair of damaged DNA (Chapter 23), contains a blue neutral FAD radical with a 580-nm absorption band and an appropriate ESR signal.[230,272] In contrast, the mitochondrial **electron-transferring flavoprotein** (ETF), a 57-kDa αβ dimer containing one molecule of FAD, functions as a single electron carrier cycling between oxidized FAD and the red anionic semiquinone.[273,274] The reduced forms of the acyl-CoA dehydrogenases transfer their electrons one at a time from their FAD to the FAD of two molecules of electron-transferring flavoprotein. Therefore, an intermediate enzyme-bound radical must be present in the FAD of acyl-CoA dehydrogenase at one stage of its catalytic cycle. A related ETF from a methylotrophic bacterium accepts single electrons from reduced trimethylamine dehydrogenase (Fig. 15-9).[275]

Another flavoprotein that makes use of both one- and two-electron transfer reactions is **ferredoxin-NADP⁺ oxidoreductase** (Eq. 15-28). Its bound FAD accepts electrons one at a time from each of the two

$$2Fd_{red} + NADP^+ + H^+ \rightarrow 2\ Fd_{ox}^+ + NADPH \tag{15-28}$$

reduced ferredoxins (Fd_{red}) in chloroplasts and then presumably transfers a hydride ion to the $NADP^+$. The enzyme is organized into two structural domains,[201] one of which binds FAD and the other $NADP^+$. Similar single-electron transfers through flavoproteins also occur in many other enzymes. Chorismate mutase, an important enzyme in biosynthesis of aromatic rings (Chapter 25), contains bound FMN. Its function is unclear but involves formation of a neutral flavin radical.[276,277]

7. Metal Complexes of Flavins and Metalloflavoproteins

The presence of metal ions in many flavoproteins suggested a direct association of metal ions and flavins. Although oxidized flavins do not readily bind most metal ions, they form red complexes with Ag^+ and Cu^+ with a loss of a proton from N-3.[278] Flavin semiquinone radicals also form strong red complexes with many metals.[264] If the complexed metal ion can exist in more than one oxidation state, electron transfer between the flavin and a substrate could take place through the metal atom. However, *chelation by flavins in nature has not been observed*. Metalloflavoproteins probably function by having the metal centers close enough to the

flavin for electron transfer to occur but not in direct contact. This is the case for a bacterial trimethylamine dehydrogenase in which the FeS cluster is bound about 0.4 nm from the alloxazine ring of riboflavin 5′-phosphate as shown in Fig. 15-9.

Some metalloflavoproteins contain heme groups. The previously mentioned **flavocytochrome b_2** of yeast is a 230-kDa tetramer, one domain of which carries riboflavin phosphate and another heme. A flavocytochrome from the photosynthetic sulfur bacterium *Chromatium* (cytochrome c-552)[279] is a complex of a 21-kDa cytochrome c and a 46-kDa flavoprotein containing 8α-(S-cysteinyl)-FAD. The 670-kDa **sulfite reductase** of *E. coli* has an $\alpha_8\beta_4$ subunit structure. The eight α chains bind four molecules of FAD and four of riboflavin phosphate, while the β chains bind three or four molecules of **siroheme** (Fig. 16-6) and also contain Fe_4S_4 clusters.[280,281] Many nitrate and some nitrite reductases are flavoproteins which also contain Mo or

Fe prosthetic groups.[282,283] A group of **aldehyde oxidases** and **xanthine dehydrogenases** also contain molybdenum as well as iron (Chapter 16). In every case the metal ions are bound independently of the flavin.[283a]

8. Reactions of Reduced Flavins with Oxygen

Free dihydroriboflavin reacts nonenzymatically in seconds and reduced flavin oxygenases react even faster with molecular oxygen to form hydrogen peroxide (Eq. 15-29).

$$FlH_2 + O_2 \rightarrow Fl + H_2O_2 \tag{15-29}$$

The reaction is more complex than it appears. As soon as a small amount of oxidized flavin is formed, it reacts with reduced flavin to generate flavin radicals $^\bullet FlH$ (Eq. 15-27). The latter react rapidly with O_2 each donating an electron to form superoxide anion radicals $^\bullet O_2^-$ (Eq. 15-30a) which can then combine with flavin radicals (Eq. 15-30b).[284]

$$^\bullet FlH + O_2 \rightarrow {}^\bullet O_2^- + H^+ + Fl \tag{15-30a}$$

$$^\bullet O_2^- + {}^\bullet FlH + H^+ \rightarrow Fl + H_2O_2 \tag{15-30b}$$

During the corresponding reactions of reduced flavoproteins with O_2, intermediates have been detected. For example, spectrophotometric studies of the FAD-containing bacterial *p*-hydroxybenzoate hydroxylase (Chapter 18) revealed the consecutive appearance of three intermediate forms.[285–287] The first, whose absorption maximum is at 380–390 nm, is thought to be an adduct at position 4a (Eq. 15-31). That such a 4a peroxide really forms with the riboflavin phosphate of the light-emitting bacterial **luciferase** (Chapter 23) was demonstrated using coenzyme enriched with ^{13}C at position 4a.[147] A large shift to lower frequency (from 104 to 83 ppm) accompanied formation of the transient adduct. Comparison with reference compounds showed that this change agreed with that predicted.

Other structures for O_2 adducts have also been considered, as has the possibility of rearrangements among these structures.[288] Nevertheless, the products observed from many different flavoprotein reactions can be explained on the basis of a 4a peroxide.[289]

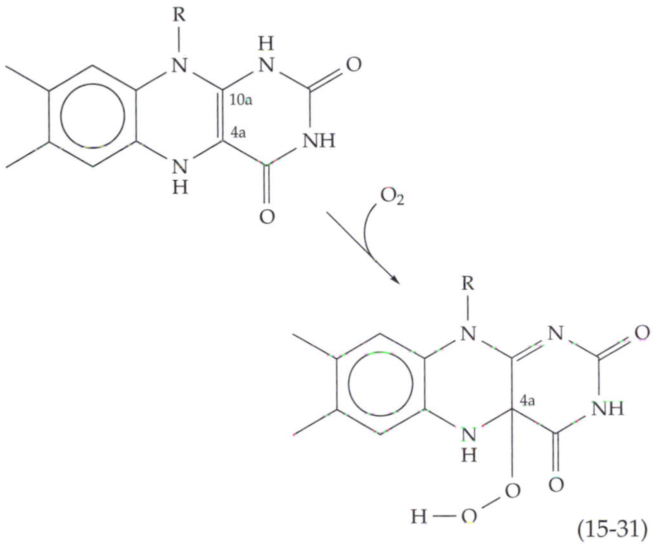

(15-31)

Formation of H_2O_2 by flavin oxidases can occur via elimination of a peroxide anion HOO^- from the adduct of Eq. 15-31 with regeneration of the oxidized flavin. In the active site of a hydroxylase, an OH group can be transferred from the peroxide to a suitable substrate (Eq. 18-42). Although radical mechanisms are likely to be involved, such hydroxylation reactions can also be viewed as transfer of OH^+ to the substrate together with protonation on the inner oxygen atom of the original peroxide to give a 4a –OH adduct. The latter is a covalent hydrate which can be converted to the oxidized flavin by elimination of H_2O. This hydrate is believed to be the third spectral intermediate identified during the action of *p*-hydroxybenzoate hydroxylase.[286,287,290]

C. Lipoic Acid and the Oxidative Decarboxylation of α-Oxoacids

The isolation of lipoic acid in 1951 followed an earlier discovery that the ciliate protozoan *Tetrahymena geleii* required an unknown factor for growth. In independent experiments acetic acid was observed to promote rapid growth of *Lactobacillus casei*, but it could be replaced by an unknown "acetate replacing factor." Another lactic acid bacterium *Streptococcus faecalis* was unable to oxidize pyruvate without addition of "pyruvate oxidation factor." By 1949, all three unknown substances were recognized as identical.[291,291a] After working up the equivalent of 10 tons of water-soluble residue from liver, Lester Reed and his collaborators isolated 30 mg of a fat-soluble acidic material which was named **lipoic acid** (or 6-thioctic acid).[292–294]

While *Tetrahymena* must have lipoic acid in its diet, we humans can make our own, and it is not considered a vitamin. Lipoic acid is present in tissues in extraordinarily small amounts. Its major function is to participate in the oxidative decarboxylation of α-oxoacids but it also plays an essential role in glycine catabolism in the human body as well as in plants.[295,296] The structure is simple, and the functional group is clearly the cyclic disulfide which swings on the end of a long arm. Like biotin, which is also present in tissues in very small amounts, lipoic acid is bound in covalent amide linkage to lysine side chains in active sites of enzymes:[296a]

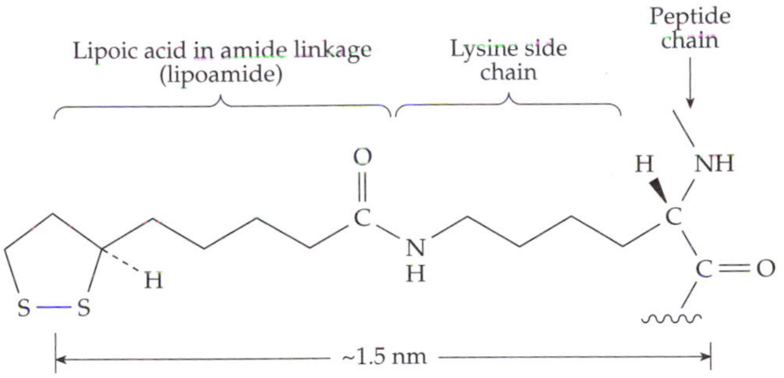

1. Chemical Reactions of Lipoic Acid

The most striking chemical property of lipoic acid is the presence of ring strain of ~17–25 kJ mol^{-1} in the cyclic disulfide. Because of this, thiol groups and cyanide ions react readily with oxidized lipoic acid to give mixed disulfides (Eq. 15-32a) and isothiocyanates (Eq. 15-32b), respectively.

Another result of the ring strain is that the reduction potential $E^{\circ\prime}$ (pH 7, 25°C), is –0.30 V, almost the same as that of reduced NAD (–0.32V). Thus, reoxidation of reduced lipoic acid amide by NAD^+ is thermodynamically feasible. Yet another property attributed to the ring strain in lipoic acid is the presence of an absorption maximum at 333 nm.

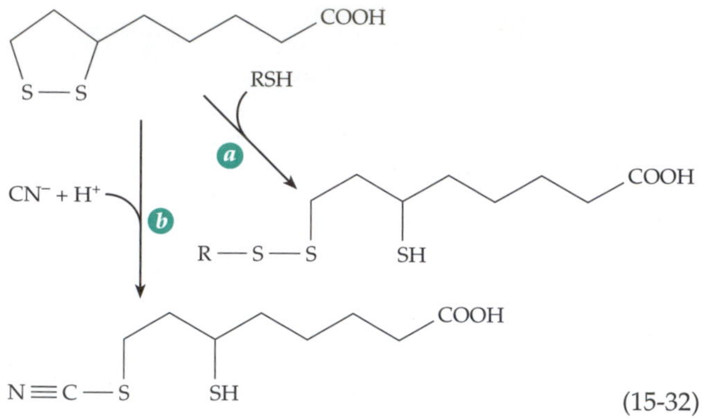

$$\tag{15-32}$$

2. Oxidative Decarboxylation of Alpha-Oxoacids

The oxidative cleavage of an α-oxoacid is a major step in the metabolism of carbohydrates and of amino acids and is also a step in the citric acid cycle. In many bacteria and in eukaryotes the process depends upon both thiamin diphosphate and lipoic acid. The oxoacid anion is cleaved to form CO_2 and the remaining acyl group is combined with coenzyme A (Eq. 15-33). NAD^+ serves as the oxidant. The reaction is catalyzed by a complex of enzymes whose molecular mass varies from ~4 to 10×10^6, depending on the species and exact substrate.[297] Separate oxoacid dehydrogenase systems are known for pyruvate,[298–300] 2-oxoglutarate,[301] and the 2-oxoacids with branched side chains derived metabolically from leucine, isoleucine, and

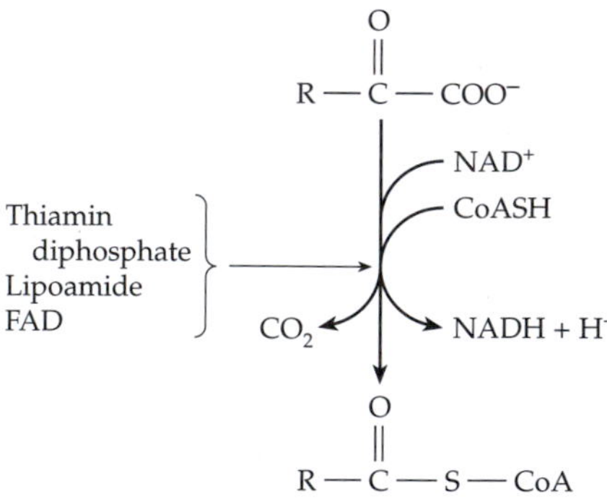

$$\tag{15-33}$$

valine.[302,302a] In eukaryotes these enzymes are located in the mitochondria. The **pyruvate** and **2-oxoglutarate dehydrogenase** complexes of E. coli and Azotobacter vinelandii have been studied most. In both cases there are three major protein components. The first (E_1) is a **decarboxylase** (also referred to as a dehydrogenase) for which the thiamin diphosphate is the dissociable cofactor. The second (E_2) is a lipoic acid amide-containing "core" enzyme which is a **dihydrolipoyl transacylase.** The third (E_3) is the flavoprotein **dihydrolipoyl dehydrogenase**, a member of the glutathione reductase family with a three-dimensional structure and

catalytic mechanism similar to those of glutathione reductase (Figs. 15-10, 15-12).[303–305]

Electron microscopy of the core dihydrolipoyl transacylase from E. coli reveals a striking octahedral symmetry which has been confirmed by X-ray diffraction.[306–307a] The core from pyruvate dehydrogenase has a mass of ~2390 kDa and contains 24 identical 99.5-kDa E_2 subunits. The 2-oxoglutarate dehydrogenase from E. coli has a similar but slightly less symmetric structure. Each core subunit is composed of three domains. A lipoyl group is bound in amide linkage to lysine 42 and protrudes from one end of the domain. A second domain is necessary for binding to subunits E_1 and E_3, while the third major 250-residue domain contains the catalytic acyltransferase center.[308,309] This center closely resembles that of chloramphenicol acetyltransferase (Chapter 12).[310,311] The lipoyl[301,309,312] and catalytic[307] domains of the dihydrolipoyl succinyltransferase from 2-oxoglutarate dehydrogenase resemble those of pyruvate dehydrogenase and also of the branched chain oxoacid dehydrogenase. The three domains of the proteins are joined by long 25- to 30-residue segments rich in alanine, proline, and ionized hydrophilic side chains.[309] This presumably provides flexibility for the lipoyl groups which must move from site to site. The presence of unexpectedly sharp lines in the proton NMR spectrum of the core protein may be a result of this flexibility.[309]

To obtain the X-ray structure of the core protein it was necessary to delete the lipoyl- and $E_1(E_2)$- binding domains. The resulting 24-subunit structure is shown in Fig 15-14A,B.[306] It has been assumed for many years that 12 of the dimeric decarboxylase units (E_1) are bound to the 12 edges of the transacetylase cube, while six (50.6×2 kDa) flavoprotein (E_3) dimers bind on the six faces of the cube. The active centers of all three types of subunits are thought to come close together in the regions where the subunits touch, permitting the sequence of catalytic reactions indicated in Fig. 15-15 to take place. Eukaryotic as well as some bacterial pyruvate decarboxylases have a core of 60 subunits in an icosahedral array with 532 symmetry. This can be seen in the image reconstructions of the enzyme from Saccharomyces cerevisiae shown in Fig. 15-14C and D. A surprising discovery is that in this yeast enzyme the E_3 units are not on the outside of the 5-fold symmetric faces but protrude into the inner cavity.[299] Each of the 12 E_3 subunits is assisted in binding correctly to the E_2 core by a molecule of the 47-kDa **E_3-binding protein** (BP), also known as protein X.[313,314] Absence of this protein is associated with congenital lactic acidosis.

The unique function of lipoic acid is in the oxidation of the thiamin-bound active aldehyde (Fig. 15-15) in such a way that when the complex with thiamin breaks up, the acyl group formed by the oxidative decarboxylation of the oxoacid is attached to the

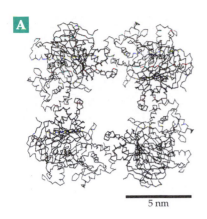

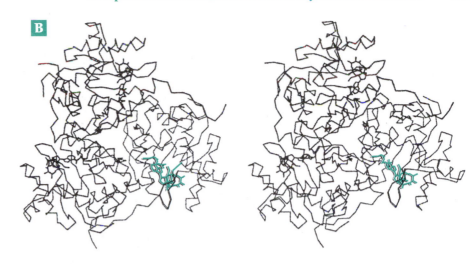

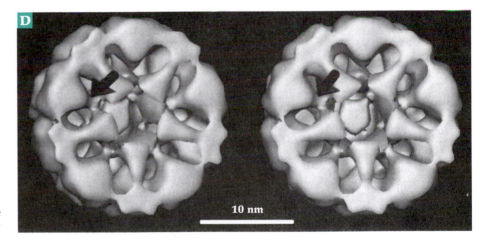

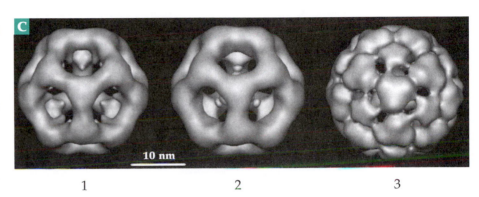

Figure 15-14 (A) View of the complex down a fourfold axis through the face of the cube. (B) Stereoscopic view of three subunits of the 24-subunit cubic core of the *E. coli* pyruvate dehydrogenase complex. The view is down the threefold axis through opposite vertices of the cube. Bound lipoate and CoA molecules are visible in each subunit. (A) and (B) from Knapp *et al.*[307a] (C) Image of the 60-subunit core of yeast pyruvate dehydrogenase reconstructed from electron micrographs of frozen-hydrated complexes. The five-, three-, and twofold symmetries can be seen. The view is along the threefold axis. 1. Truncated E_2; 2. Truncated E_2 plus binding proteins; 3. Truncated E_2 + binding protein + E_3. The reconstructions show that the E_3 and BP components are inside the truncated E_2 scaffold. The morphologies of E_3 and BP are not correctly represented since they are not bound with icosahedral symmetry. (D) Stereoscopic image of complex shown in B,3 with the front half cut away to show the interior. The conical objects protruding in are the E_3 molecules. The arrow points to a binding site for attachment of the binding protein to the core. Courtesy of J. K. Stoops *et al.*[299]

dihydrolipoyl group at the S-8 position.[315] The lipoic acid, which is attached to the flexible lipoyl domains of the core enzyme on a 1.5-nm-long arm, apparently first contacts the thiamin diphosphate site on one of the decarboxylase subunits. Bearing the acyl group, it now swings to the catalytic site on the core enzyme where CoA is bound. The acyl group is transferred to CoA producing a dihydrolipoyl group which then swings to the third subunit where the disulfide loop and a bound FAD of dihydrolipoamide dehydrogenase

reoxidize the lipoyl group. The reduced flavin-disulfide enzyme is then oxidized by NAD^+ (Fig. 15-15) by the reverse of the mechanism depicted in Fig. 15-12.

Although the direct reaction of a lipoyl group with the thiamin-bound enamine (active aldehyde) is generally accepted, and is supported by recent studies,[315a] an alternative must be considered.[315] Hexacyanoferrate (III) can replace NAD^+ as an oxidant for pyruvate dehydrogenase and is also able to oxidize nonenzymatically thiamin-bound active acetaldehyde

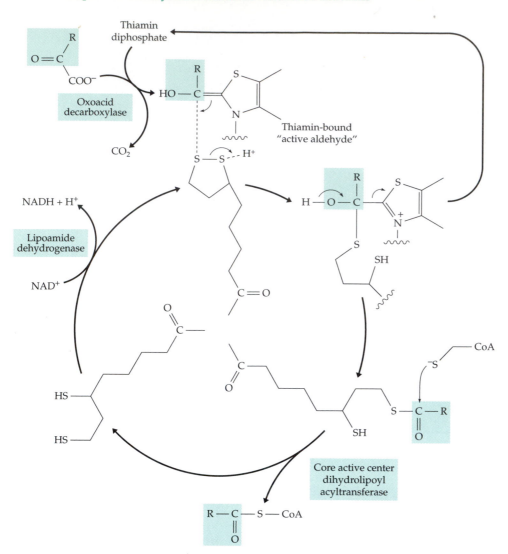

Figure 15-15 Sequence of reactions catalyzed by α-oxoacid dehydrogenases. The substrate and product are shown in boxes, and the path of the oxidized oxoacid is traced by the heavy arrows. The lipoic acid "head" is shown rotating about the point of attachment to a core subunit. However, a whole flexible domain of the core is also thought to move.

to 2-acetylthiamin, a compound in which the acetyl group has a high group transfer potential (Eq. 14-22). Thus, the lipoyl group could first oxidize the active aldehyde to a thiamin-bound acyl derivative, and then in the second step accept the acyl group by a nucleophilic displacement reaction. This mechanism fails to explain the unique role of lipoic acid in oxidative decarboxylation. However, as we will see in the next section, oxidative decarboxylation does not always require lipoic acid and acetylthiamin is probably an intermediate whenever lipoamide is not utilized.

Within many tissues the enzymatic activities of the pyruvate and branched chain oxoacid dehydrogenases complexes are controlled in part by a phosphorylation–dephosphorylation mechanism (see Eq. 17-9). Phosphorylation of the decarboxylase subunit by an ATP-dependent kinase produces an inactive phosphoenzyme. A phosphatase reactivates the dehydrogenase to complete the regulatory cycle (see Eq. 17-9 and associated discussion). The regulation is apparently accomplished, in part, by controlling the affinity of the protein for

thiamin diphosphate.[315b] The lipoamide dehydrogenase component of all three dehydrogenase complexes appears to be the same.

3. Other Functions of Lipoic Acid

In addition to its role in the oxidative decarboxylation of 2-oxoacids, lipoic acid functions in the human body as part of the essential mitochondrial glycine-cleavage system described in Section E. It may also participate in bacterial glycine reductase (Eq. 15-61) and other enzyme systems. Dihydrolipoamide dehydrogenase binds to G4-DNA structures of telomeres and may have a biological role in DNA-binding.[315c] Lipoic acid is being utilized as a nutritional supplement and appears to help in maintaining cellular levels of glutathione.[316]

4. Additional Mechanisms of Oxidative Decarboxylation

Pyruvate is a metabolite of central importance and a variety of mechanisms exist for its cleavage. The pyruvate dehydrogenase complex is adequate for strict aerobes and is very important in aerobic bacteria and in facultative anaerobes such as *E. coli*. Lactic acid bacteria, which lack cytochromes and other heme proteins, are able to carry out limited oxidation with flavoproteins such as **pyruvate oxidase**.[317] However, *strict* anaerobes have to avoid accumulation of reduced pyridine nucleotides and use a nonoxidative cleavage of pyruvate by **pyruvate formate-lyase**. The pyruvate dehydrogenase complex of *our* bodies generates NADH which can be oxidized in mitochondria to provide energy to our cells. However, NADH is a weak reductant. Some cells, such as those of nitrogen-fixing bacteria, require more powerful reductants such as reduced ferredoxin and utilize **pyruvate: ferredoxin oxidoreductase**. *Escherichia coli*, a facultative anaerobe, is adaptable and makes use of all of these types of pyruvate cleavage.[318]

Pyruvate oxidase. The soluble flavoprotein pyruvate oxidase, which was discussed briefly in Chapter 14 (Fig. 14-2, Eq. 14-22), acts together with a membrane-bound electron transport system to convert pyruvate to acetyl phosphate and CO_2.[319] Thiamin diphosphate is needed by this enzyme but lipoic acid is not. The flavin probably dehydrogenates the thiamin-bound intermediate to 2-acetylthiamin as shown in Eq. 15-34. The electron acceptor is the bound FAD and the reaction may occur in two steps as shown with a thiamin diphosphate radical intermediate.[319a] Reaction with inorganic phosphate generates the energy storage metabolite **acetyl phosphate**.

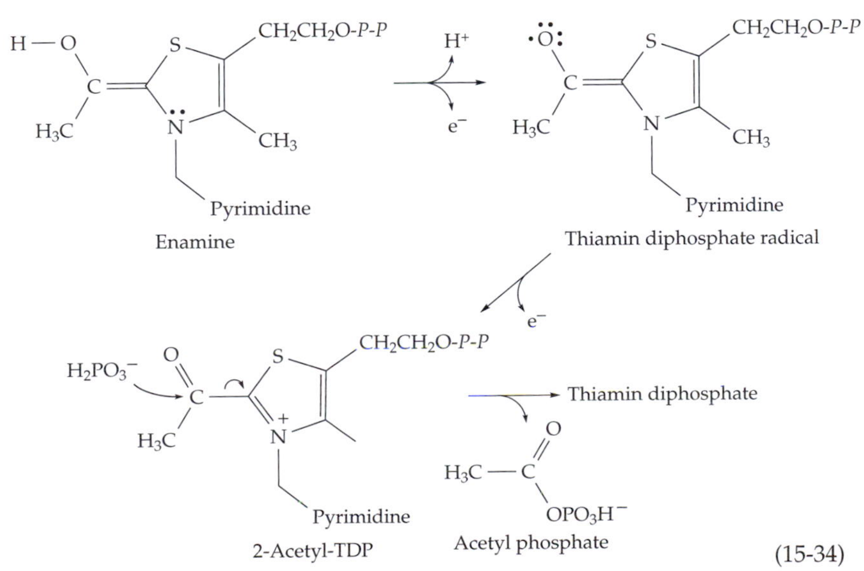

Pyruvate:ferredoxin oxidoreductase. Within clostridia and other strict anaerobes this enzyme catalyzes *reversible* decarboxylation of pyruvate (Eq. 15-35). The oxidant used by clostridia is the low-potential iron–sulfur ferredoxin.[320,320a] Clostridial ferredoxins contain two Fe-S clusters and are therefore two-electron oxidants. Ferredoxin substitutes for NAD^+ in Eq. 15-33 but the Gibbs energy decrease is much less (−16.9 vs − 34.9 kJ / mol. for oxidation by NAD^+).

$$\text{Pyruvate}^- + \text{CoA} \longrightarrow \text{Acetyl-CoA} + CO_2$$
$$\text{Ferredoxin (ox)} \quad \text{Ferredoxin (red)}$$
$$\Delta G' \text{ (pH 7)} = -16.9 \text{ kJ mol}^{-1}$$

(15-35)

The enzyme does not require lipoic acid. It seems likely that a thiamin-bound enamine is oxidized by an iron–sulfide center in the oxidoreductase to 2-acetylthiamin which then reacts with CoA. A free radical intermediate has been detected[318,321] and the proposed sequence for oxidation of the enamine intermediate is that in Eq. 15-34 but with the Fe−S center as the electron acceptor. Like pyruvate oxidase, this enzyme transfers the acetyl group from acetylthiamin to coenzyme A. Cleavage of the resulting acetyl-CoA is used to generate ATP. An indolepyruvate: ferredoxin oxidoreductase has similar properties.[322]

In methanogenic bacteria[320] the low-potential 5-deazaflavin coenzyme F_{420} serves as the reductant in a reversal of Eq. 15-35. A similar enzyme, **2-oxoglutarate synthase**, apparently functions in synthesis of 2-oxoglutarate from succinyl-CoA and CO_2 by photosynthetic bacteria.[323] Either reduced ferredoxin or, in *Azotobacter*, reduced flavodoxin is generated in nitrogen-fixing bacteria (Chapter 24) by cleavage of pyruvate and is used in the N_2 fixation process.

Oxidative decarboxylation by hydrogen peroxide. The nonenzymatic oxidative decarboxylation of α-oxoacids by H_2O_2 is well-known. The first step is the formation of an adduct, an organic peroxide, which breaks up as indicated in Eq. 15-36. An enzyme-catalyzed version of this reaction is promoted by **lactate monooxygenase**, a 360-kDa octameric flavoprotein obtained from *Mycobacterium smegmatis*[324] and a member of the glycolate oxidase family. Under anaerobic conditions, the enzyme produces pyruvate by a simple dehydrogenation. However, the pyruvate dissociates slowly and in the presence of

(15-34)

oxygen it forms acetic acid, with one of the oxygen atoms of the carboxyl group coming from O_2.[324,325] Hydrogen peroxide is the usual product formed from oxygen by flavoprotein oxidases, and it seems likely that with lactate monooxygenase the hydrogen peroxide formed immediately oxidizes the pyruvate according to Eq. 15-36. The α-oxoacids formed by amino acid oxidases *in vitro* are also oxidized by accumulating hydrogen peroxide. However, if catalase (Chapter 16) is present it destroys the H_2O_2 and allows the oxoacid to accumulate.

$$R-C\overset{O}{\underset{COO^-}{=}} \xrightarrow{H_2O_2} \left[R-\overset{OH}{\underset{\underset{HO}{O}}{\overset{|}{C}}-C\overset{O}{\underset{O^-}{=}} \right] \xrightarrow{H_2O + CO_2} R-C\overset{O}{\underset{O^-}{=}}$$

(15-36)

Pyruvate formate-lyase reaction. Anaerobic cleavage of pyruvate to acetyl-CoA and formate (Eq. 15-37) is essential to the energy economy of many cells, including those of *E. coli*. No external oxidant is needed, and the reaction does not require lipoic acid.

The reaction has sometimes been called the "phosphoroclastic reaction" because the product acetyl-CoA usually reacts further with inorganic phosphate to form acetyl phosphate (see Fig. 15-16). The latter can then transfer its phospho group to ADP to form ATP.

$$CH_3-C\overset{O}{\underset{COO^-}{=}} \xrightarrow{CoA-SH} \quad CH_3-C\overset{O}{\underset{\|}{}}-S-CoA \quad + \quad H-C\overset{O}{\underset{O^-}{=}}$$

ΔG' (pH 7) = –12.9 kJ mol⁻¹   (15-37)

The mechanism of the cleavage of the pyruvate in Eq. 15-37 is not obvious. Thiamin diphosphate is not involved, and free CO_2 is not formed. The first identified intermediate is an acetyl-enzyme containing a thioester linkage to a cysteine side chain. This is cleaved by reaction with CoA-SH to give the final product. A clue came when it was found by Knappe and coworkers that the active enzyme, which is rapidly inactivated by oxygen, contains a long-lived free radical.[326] Under anaerobic conditions cells convert the inactive form E_i to the active form E_a by an enzymatic reaction with *S*-adenosylmethionine and reduced flavodoxin Fd(red) as shown in Eq. 15-38.[327–329] A deactivase reverses the process.[330]

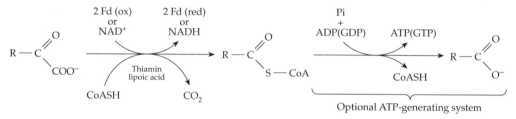

A Oxidative decarboxylation of an α-oxoacid with thiamin diphosphate

1. with lipoic acid and NAD⁺ as oxidants   ΔG' (pH 7, for pyruvate) = –35.5 kJ/mol overall
2. with ferredoxin as oxidant   ΔG' (pH 7, for pyruvate) = –13.9 kJ/mol overall

B The pyruvate formate-lyase reaction

ΔG' (pH 7) = –13.5 kJ/mol overall

Figure 15-16 Two systems for oxidative decarboxylation of α-oxoacids and for "substrate-level" phosphorylation. The value of ΔG' = +34.5 kJ mol⁻¹ (Table 6-6) was used for the synthesis of ATP⁴⁻ from ADP³⁻ and HPO₄²⁻ in computing the values of ΔG' given.

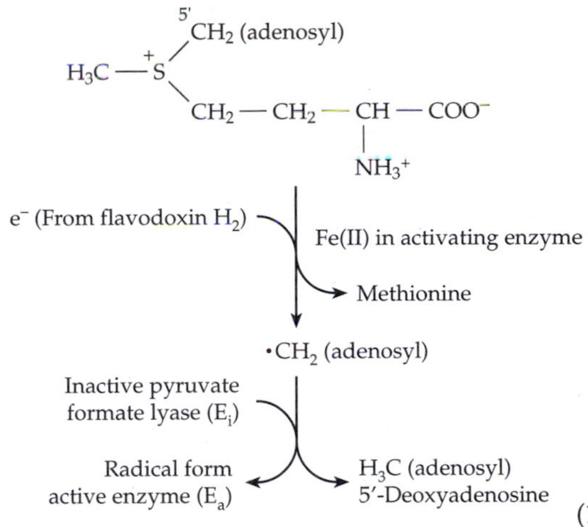

(15-38)

(15-40)

The activating enzyme, which is allosterically activated by pyruvate, is an iron–sulfur (Fe_4S_4) protein (Chapter 16).[331,331a] Formation of the observed radical may proceed via a 5'-deoxyadenosyl radical as has been proposed for lysine 2,3-aminomutase (Eq. 16-42).[332] The activation reaction also resembles the free radical-dependent reactions of vitamin B_{12} which are discussed in Chapter 16. When subjected to O_2 of air at 25°C pyruvate formate-lyase is destroyed with a half-life of ~ 10 s. The peptide chain is cleaved and sequence analysis and mass spectrometry of the resulting fragments show that the specific sequence Ser-Gly-Tyr at positions 733–755 is cut with formation of an oxalyl group on the N-terminal tyrosine of one fragment (Eq. 15-39). Various ^{13}C-containing amino acids were supplied to growing cells of *E. coli* and were incorporated into the proteins of the bacteria. Pyruvate formate-lyase containing ^{13}C in carbon-2 of glycine gave an EPR spectrum with hyperfine splitting arising from coupling of the unpaired electron of the radical with the adjacent ^{13}C nucleus.[333,333a] This experiment, together with the results described by Eq. 15-39, suggested that the radical is derived from Gly 734.

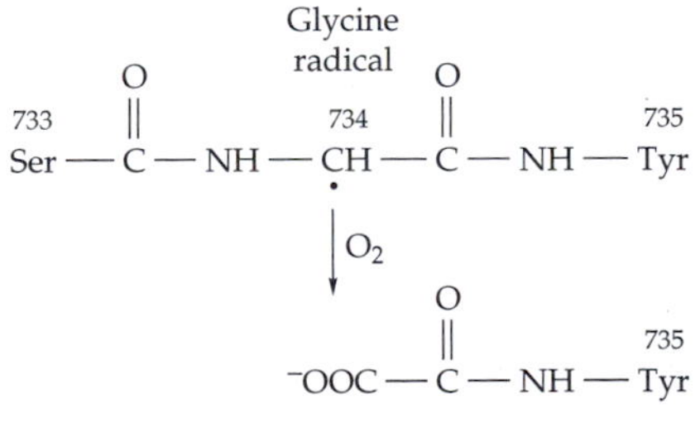

(15-39)

The activating enzyme will also generate radicals from short peptides such as Arg-Val-Ser-Gly-Tyr-Ala-Val, which corresponds to residues 731–737 of the pyruvate formate-lyase active site. If Gly 734 is replaced by L-alanine, no radical is formed, but radical *is* formed if D-alanine is in this position. This suggests that the *pro-S* proton of Gly 734 is removed by the activating

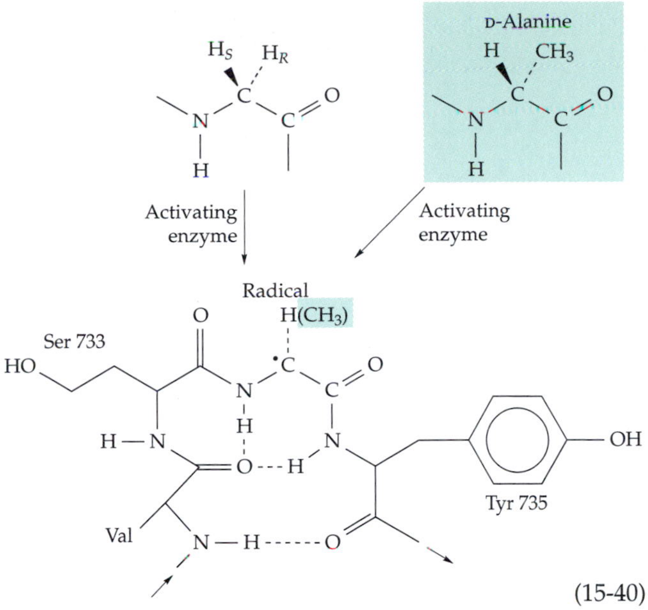

enzyme[329] as illustrated in Eq. 15-40. It has been suggested that the peptide exists in the β-bend conformation shown.

How does the enzyme work? The α-CH proton of the glycyl radical of the original *pro-R* proton undergoes an unexpected exchange with the deuterium of ^2H_2O. The exchange is catalyzed by the thiol group of Cys 419, suggesting that it is close to Gly 734 in the three-dimensional structure.[334] The mutants C419S and C418S are inactive but still allow formation of the Gly 734 radical.[334–335a] The mechanism has been proposed in Eq. 15-41.

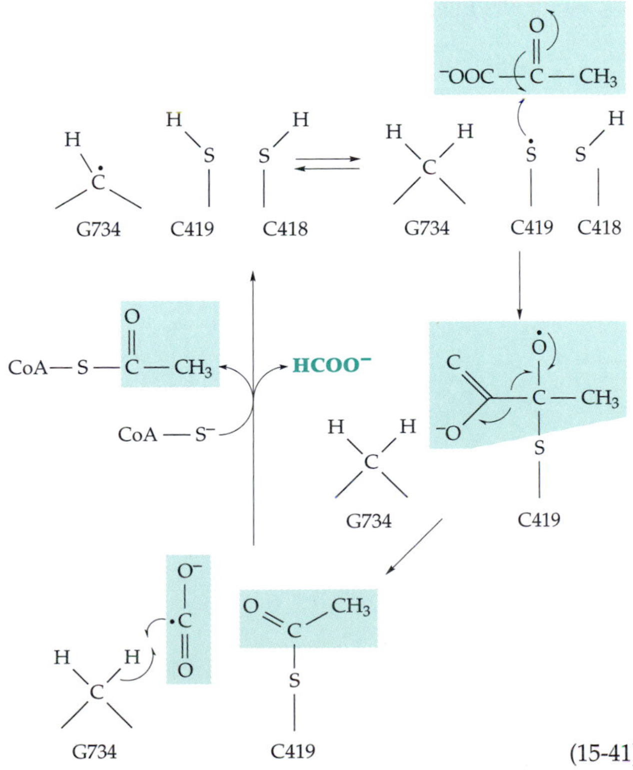

(15-41)

5. Cleavage of α-Oxoacids and Substrate-Level Phosphorylation

The α-oxoacid dehydrogenases yield CoA derivatives which may enter biosynthetic reactions. Alternatively, the acyl-CoA compounds may be cleaved with generation of ATP. The pyruvate formate-lyase system also operates as part of an ATP-generating system for anaerobic organisms, for example, in the "mixed acid fermentation" of enterobacteria such as *E. coli* (Chapter 17). These two reactions, which are compared in Fig. 15-16, constitute an important pair of processes both of which accomplish substrate-level phosphorylation. They should be compared with the previously considered examples of substrate level phosphorylation depicted in Eq. 14-23 and Fig. 15-16.

D. Tetrahydrofolic Acid and Other Pterin Coenzymes

In most organisms reduced forms of the vitamin **folic acid** serve as *carriers for one-carbon groups* at three

different oxidation levels corresponding to **formic acid**, **formaldehyde**, and the **methyl group** and also facilitate their interconversion.[336,337] Moreover, folic acid is just one of the derivatives of the **pteridine** ring system that enjoy a widespread natural distribution.[338] One of these derivatives functions in the hydroxylation of aromatic amino acids and in nitric oxide synthase and yet another within several molybdenum-containing enzymes. Pteridines also provide coloring to insect wings and eyes and to the skins of amphibians and fish. They appear to act as protective filters in insect eyes and may function as light receptors. An example is a folic acid derivative found in some forms of the DNA photorepair enzyme **DNA photolyase** (Chapter 23).

1. Structure of Pterins

Because of the prevalence of its derivatives, 2-amino-4-hydroxypteridine has been given the trivial name **pterin**. Its structure resembles closely that of guanine, the 5-membered ring of the latter having been expanded to a 6-membered ring. In fact, pterins are derived

---

BOX 15-D FOLIC ACID (PTEROYLGLUTAMIC ACID)

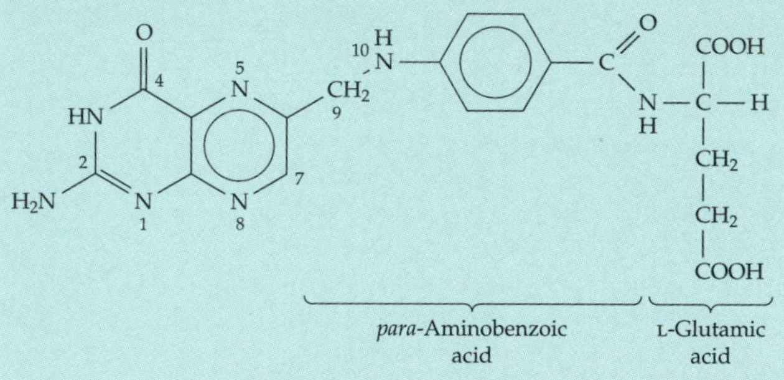

para-Aminobenzoic acid   L-Glutamic acid

In 1931, Lucy Wills, working in India, observed that patients with **tropical macrocytic anemia**, a disease in which the erythrocytes are enlarged but reduced in numbers, were cured by extracts of yeast or liver. The disease could be mimicked in monkeys fed the local diet, and a similar anemia could be induced in chicks. By 1938 it had also been shown that a factor present in yeast, alfalfa, and other materials was required for the growth of chicks. Isolation of the new vitamin came rapidly after it was recognized that it was also an essential nutrient for *Lactobacillus casei* and *Streptococcus faecalis* R.[a,b] Spinach was a rich source of the new compound, and it was named folic acid (from the same root as the word foliage).

The microbiological activity attributed to folic acid in extracts of natural materials was largely that of di- and triglutamyl derivatives, one of the facts

that has led to the description of the history of folic acid as "the most complicated chapter in the story of the vitamin B complex."[a] Metabolic functions for folic acid were suggested by the observations that the requirement for *Streptococcus faecalis* could be replaced by thymine plus serine plus a purine base. Folic acid is required for the biosynthesis of all of these substances. A function in the interconversion of serine and glycine was suggested by the observation that certain mutants of *E. coli* required either serine or glycine for growth. Isotopic labeling experiments established that in the rat as well as in the yeast *Torulopsis* serine and glycine could be interconverted. It was also shown that the amount of interconversion decreased in the folate-deficient rat.

Deficiency of folic acid is a common nutritional problem of worldwide importance.[b] A recommended daily intake is 0.2 mg, but because of the association between low folic acid intake and neural tube defects in infants, women of child-bearing age should have 0.4 mg / day.[c–e]

[a] Wagner, A. F., and Folkers, E. (1964) *Vitamins and Coenzymes*, Wiley (Interscience), New York
[b] Jukes, T. H. (1980) *Trends Biochem. Sci.* **5**, 112–113
[c] Cziezel, A. E., and Dudás, I. (1992) *N. Engl. J. Med.* **327**, 1832–1835
[d] Jukes, T. H. (1997) *Protein Sci.* **6**, 254–256
[e] Rosenquist, T. H., Ratashak, S. A., and Selhub, J. (1996) *Proc. Natl. Acad. Sci. U.S.A.* **93**, 15227–15232

biosynthetically from guanine. The two-ring system of pterin is also related structurally and biosynthetically to that of riboflavin (Box 15-B).

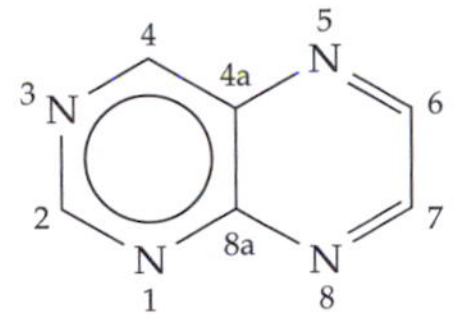

Pteridine

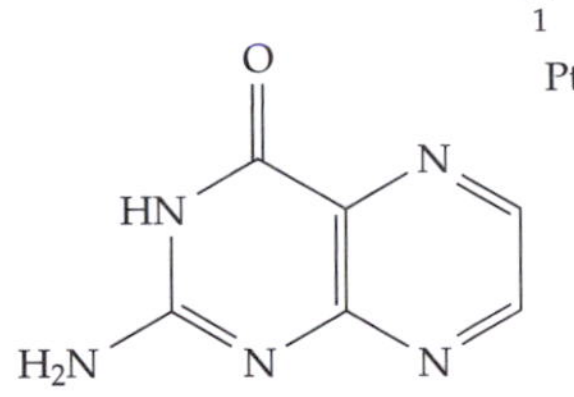

Pterin: 2-amino-4-hydroxypteridine 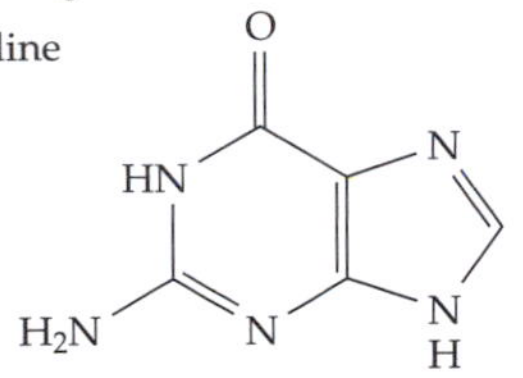   Guanine

Interest in pteridines began with Frederick G. Hopkins, who in 1891 started his investigation of the yellow and white pigments of butterflies. Almost 50 years and a million butterflies later, the structures of the two pigments, **xanthopterin** and **leucopterin** (Fig. 15-17), were established.[339] These pigments are produced in such quantities as to suggest that their synthesis may be a means of deposition of nitrogenous wastes in dry form.

Among the simple pterins isolated from the eyes of *Drosophila*[340] is **sepiapterin** (Fig. 15-17), in which the pyrazine ring has been reduced in the 7,8 position and a short side chain is present at position 6. Reduction of the carbonyl group of sepiapterin with $NaBH_4$ followed by air oxidation produces **biopterin**, the most widely distributed of the pterin compounds. First isolated from human urine, biopterin (Fig. 15-17) is present in liver and other tissues where it functions in a reduced form as a **hydroxylation coenzyme** (see Chapter 18).[338] It is also present in nitric oxide synthase (Chapter 18).[341,342] Other functions in oxidative reactions, in regulation of electron transport, and in photosynthesis have been proposed.[343] **Neopterin**, found in honeybee larvae, resembles biopterin but has a D-*erythro* configuration in the side chain. The red eye pigments of *Drosophila*, called **drosopterins**, are complex dimeric pterins containing fused 7-membered rings (Fig. 15-17).[344,345]

In the pineal gland, as well as in the retina of the eye, light-sensitive pterins may be photochemically cleaved to generate such products as **6-formylpterin**, a compound that could serve as a metabolic regulator.[346] Another pterin acts as a chemical attractant for aggregation of the ameboid cells of *Dictyostelium lacteum*[347] (Box 11-C). **Molybdopterin** (Fig. 15-17) is a component of several

Mo-containing enzymes discussed in Chapter 16,[348,349] and **methanopterin** is utilized by methanogenic bacteria.[350–353]

2. Folate Coenzymes

The coenzymes responsible for carrying single-carbon units in most organisms are derivatives of **5,6,7,8-tetrahydrofolic acid** (abbreviated H_4PteGlu, H_4folate or THF). However, in methanogenic bacteria the tetrahydro derivative of the structurally unique methanopterin (Fig. 15-17) is the corresponding single-carbon carrier.[353] Naturally occurring tetrahydrofolates contain a chiral center of the S configuration.[354,355] They exhibit negative optical rotation at 589 nm. The folate coenzymes are present in extremely low concentrations and the reduced ring is readily oxidized by air.

In addition to the single L-glutamate unit present in tetrahydrofolic acid, the coenzymes occur to the greatest extent as conjugates called **folyl polyglutamates** in which one to eight or more additional molecules of L-glutamic acid have been combined via amide linkages.[356–357a] The first two of the extra glutamates are always joined through the γ (side chain) carboxyl groups but in *E. coli* the rest are joined through their α carboxyls.[358] The distribution of the polyglutamates varies from one organism to the next. Some bacteria contain exclusively the triglutamate derivatives, while in others almost exclusively the tetraglutamate or octaglutamate derivatives predominate.[359] The serum

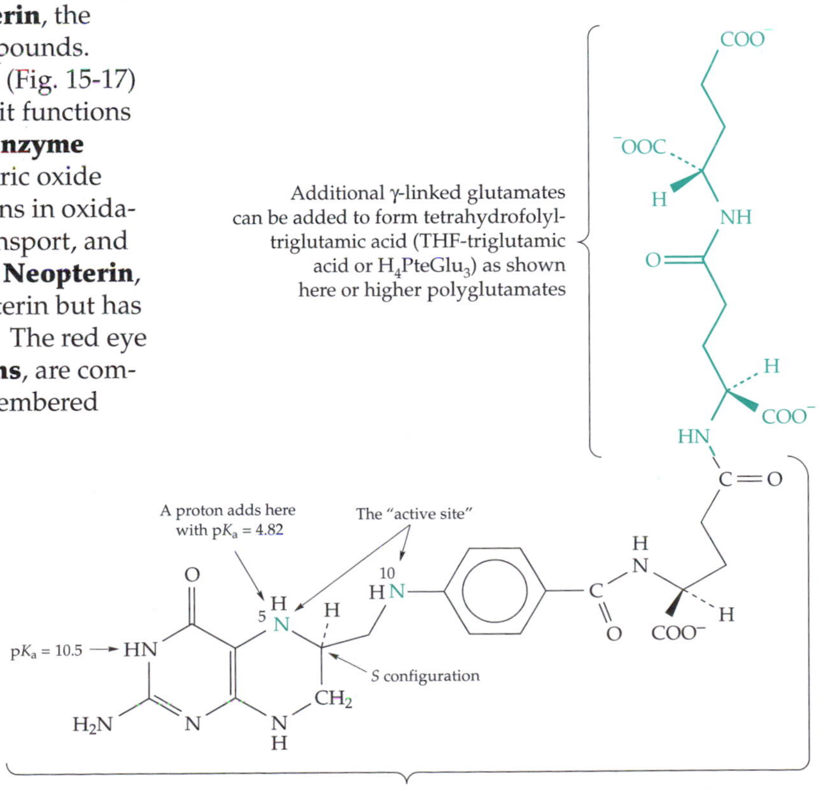

Additional γ-linked glutamates can be added to form tetrahydrofolyl-triglutamic acid (THF-triglutamic acid or H_4PteGlu₃) as shown here or higher polyglutamates

A proton adds here with pK_a = 4.82   The "active site"

pK_a = 10.5   S configuration

Tetrahydrofolic acid (THF) or tetrahydropteroylglutamic acid (H_4PteGlu)

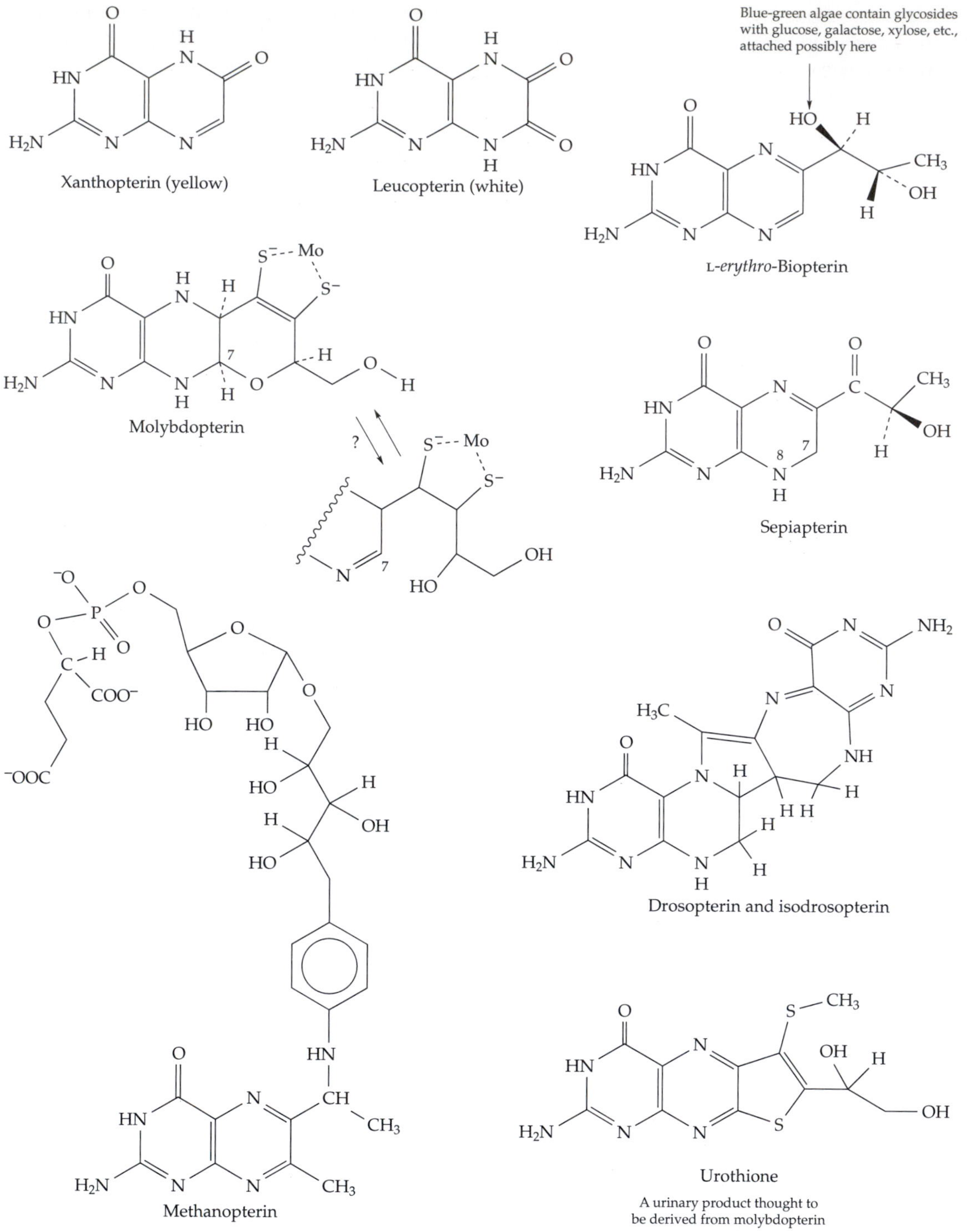

Figure 15-17 Structures of several biologically important pterins.

of many species contains only derivatives of folic acid itself but pteroylpentaglutamate is the major folate derivative present in rodent livers.[360,361] A large fraction of these pentaglutamates consists of the 5-methyl-THF derivative, while at the heptaglutamate level most consists of the free THF derivative. Folyl polyglutamate in its oxidized form is a component of some DNA photolyases (Chapter 23).[362]

3. Dihydrofolate Reductase

Folic acid and its polyglutamyl derivatives can be reduced to the THF coenzymes in two stages: the first step is a slow reduction with NADPH to 7,8-dihydro-folate (step a, Fig. 15-18). The same enzyme that catalyzes this reaction rapidly reduces the dihydrofolates to tetrahydrofolates (step *b*, Fig. 15-18). Again, NADPH is the reducing agent and the enzyme has been given the name **dihydrofolate reductase**. An unusual fact is that bacteriophage T4 not only carries a gene for its own dihydrofolate reductase but also the enzyme is a structural unit in the phage baseplate.[363]

Inhibitors. Aside from its role in providing reduced folate coenzymes for cells, this enzyme has attracted a great deal of attention because it appears to be a site of action of the important anticancer drugs **methotrexate** (amethopterin) and aminopterin.[293,364,365] These compounds inhibit dihydrofolate reductase in concentrations as low as 10^{-8} to 10^{-9} M. Methotrexate is also widely used as an **immunosuppresant** drug and in the treatment of parasitic infections.

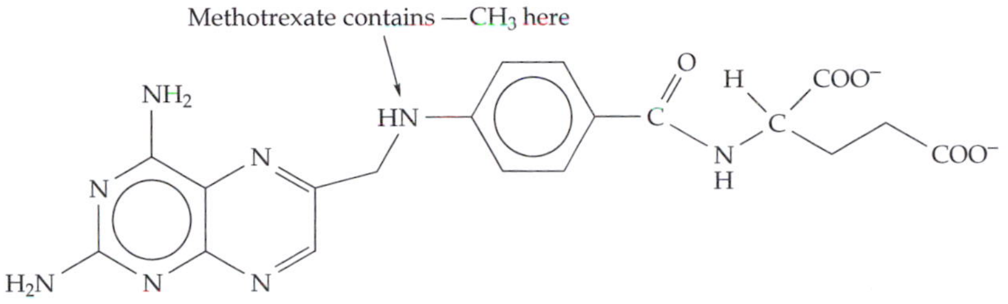

Methotrexate contains —CH$_3$ here

Aminopterin (4-amino-4-deoxyfolic acid)

Another dihydrofolate inhibitor **trimethoprim** is an important antibacterial drug, usually given together with a sulfonamide. Although it is not as close a structural analog of folic acid as is methotrexate, it is

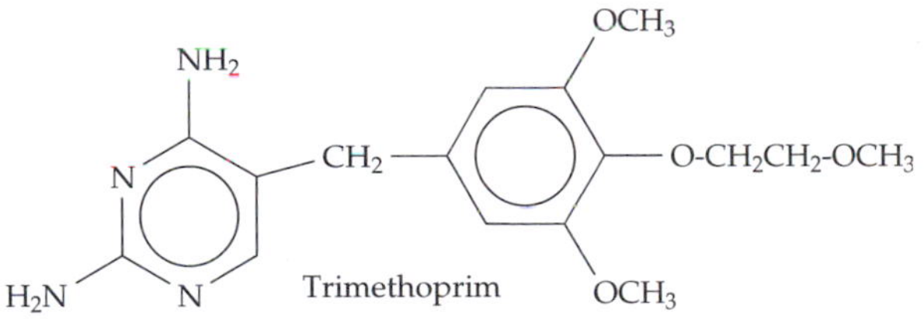

Trimethoprim

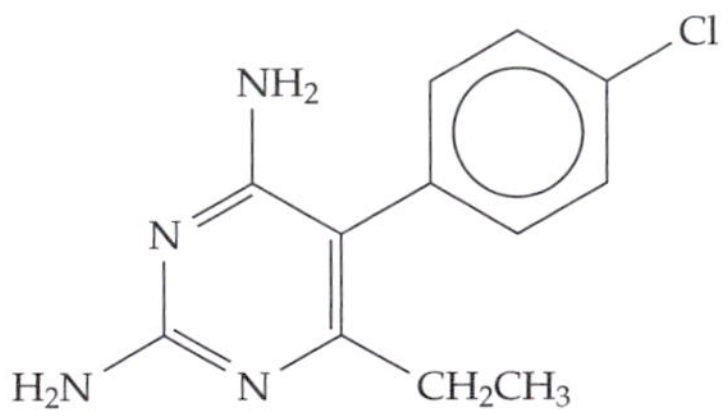

Pyrimethamine

a potent inhibitor which binds far more tightly to the enzyme from bacteria than to that from humans.[366–368] The inhibitors **pyrimethamine** and **cycloguanil** are effective antimalarial drugs.[369,370]

An enormous number of synthetic compounds have been prepared in the hope of finding still more effective inhibitors. By 1984, over 1700 different inhibitors

of dihydrofolate reductase had been studied.[367] However, methotrexate and trimethoprim remain outstanding. Methotrexate has been in clinical use for nearly 50 years and is very effective against leukemia and some other cancers.[371] Before 1960 persons with acute leukemia lived no more than 3–6 months. However, with antifolate treatment some have lived for 5 years or more and complete cures of the relatively rare choriocarcinoma have been achieved. New methods of chemotherapy use the antifolates in combination with other drugs.

Folate coenzymes are required in the biosynthesis of both purines and thymine. Consequently, rapidly growing cancer cells have a high requirement for activity of this enzyme. However, since all cells require the enzyme the antifolates are toxic and cannot be used for prolonged therapy. An even more serious problem is the development of resistance to the drug by cancer cells, often through "amplification" of the dihydrofolate reductase gene[365,372–374] as discussed in Chapter 27. Cells may also become resistant to methotrexate and other antifolates as a result of mutations that prevent efficient uptake of the drug,[375] cause increased action of effux pumps in the cell,[376] reduce the affinity of dihydrofolate reductase for the drug,[377] or interfere with conversion of the drug to polyglutamate derivatives which are better inhibitors than free methotrexate.[378,379] Cell surface **folate receptors**, present in large numbers in some tumor cells, can be utilized to bring suitably designed antifolate drugs or even unrelated cytotoxic compounds into tumor cells.[379a] Some tumor cells are more active than normal cells in generating folyl polyglutamates, contributing to the effectiveness of methotrexate.[379] Resistance of *E. coli* cells to trimethoprin may result from acquisition by the bacteria of a new form of dihydrofolate reductase carried by a plasmid.[380]

Structure and mechanism. Dihydrofolate reductase from *E. coli* is a small 159-residue protein with a central parallel stranded sheet,[381–383] while that from higher animals is 20% larger. The three-dimensional structures of the enzymes from *Lactobacillus*,[384] chicken, mouse, and human are closely similar. The NADP binds at the C-terminal ends of β strands as in other dehydrogenases. In Fig. 15-19 the reduced nicotinamide end of NADP$^+$ is seen next to a molecule of bound dihydrofolate.[381] The side chain carboxylate of Asp 27 makes a pair of hydrogen bonds to the pterin ring as shown on the right-hand side of Eq. 15-42. As can be seen from Fig. 15-19, the nicotinamide ring of NADP$^+$ is correctly positioned to have donated a hydride ion to C-6 to form THF with the 6*S* configuration. Notice that in Eq. 15-42 the NH hydrogen atom at the 3' position

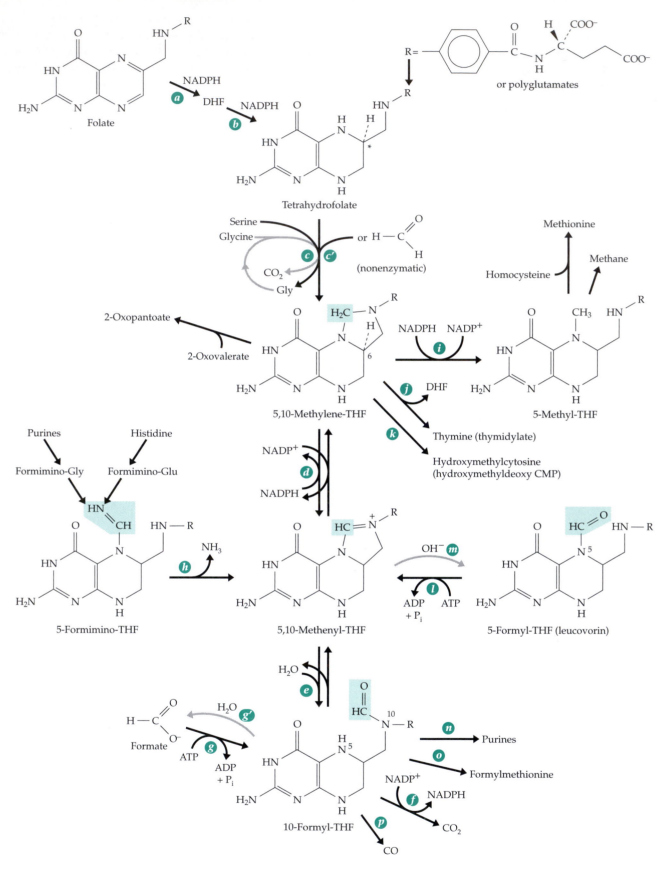

Figure 15-18 Tetrahydrofolic acid and its one-carbon derivatives.

of the pterin makes a hydrogen bond to Asp 27. However, X-ray studies have shown that the pteridine ring of methotrexate is turned 180° so that its 4-NH_2 group forms hydrogen bonds to backbone carbonyls of Ala 97 and Leu 4 at the edge of the central pleated sheet while a protonated N-1 interacts with the side chain carboxylate of Asp 27.

Because of its significance in cancer therapy and its small size, dihydrofolate reductase is one of the most studied of all enzymes. Numerous NMR studies[385–387] and investigations of catalytic mechanism and of other properties[388] have been conducted. Many mutant forms have been created.[389–391] For example, substitution of Asp 27 (Asp 26 in *L. casei*) of the *E. coli*

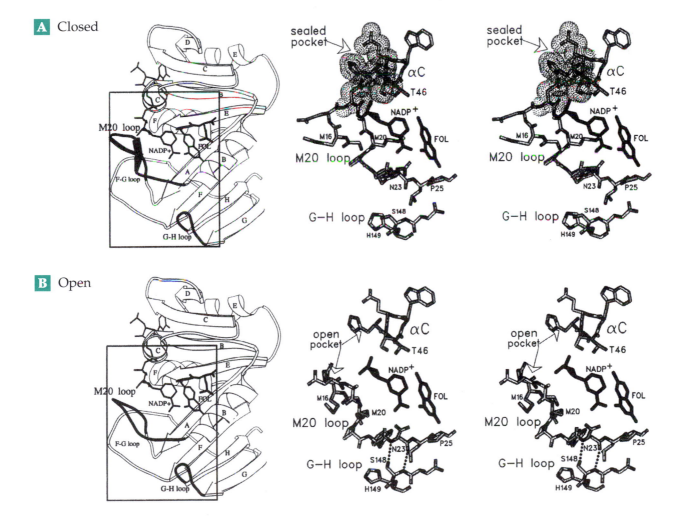

Figure 15-19 Drawings of the active site of *E. coli* dihydrofolate reductase showing the bound ligands NADP+ and tetrahydrofolate. Several key amino acid side chains are shown in the stereoscopic views on the right. The complete ribbon structures are on the left. (A) Closed form. (B) Open form into which substrates can enter and products can escape. From Sawaya and Kraut.[381] Courtesy of Joseph Kraut. Molscript drawings (Kraulis, 1991).

enzyme by Asn reduced the specific catalytic activity to 1/300 that of native enzyme indicating that this residue is important for catalysis.[390] The value of k_{cat} for reduction of dihydrofolate is highest at low pH and varies around a pK_a of ~6.5.[392–394] One interpretation is that this pK_a belongs to the Asp 27 carboxyl group and another is that it belongs to an N-5 protonated species of the coenzyme. As we have seen (Chapter 6) it is often impossible to assign a pK_a to a single group because there may be a mixture of interacting tautomeric species. Despite the enormous amount of study, we still don't know quite how the proton gets to N-5 in this reaction. Only one possibility is illustrated in Eq. 15-42. Asp 27 is protonated, the pterin ring is enolized, and a buried water molecule serves to relay a proton to N-5. The X-ray crystallographic studies on the *E. coli* enzyme show that after the binding of substrates a conformational change closes a lid over the active site (Fig. 15-19). This excludes water from N-5 but it may permit intermittent access, allowing transfer of a proton from a water molecule bound to O-4 as shown in Eq. 15-42.[381] During the reduction of folate to dihydrofolate a different mechanism of proton donation must be followed to allow protonation at C-7.

TABLE 15-3
Single-Carbon Compounds in Order of Oxidation State of Carbon

CH_4	$H_3C—Y$	H—C(=O)—H	H—C(=O)—H	CO_2
Methane	Methyl groups bound to O, N, S	Formaldehyde	Formic acid	H_2CO_3
		—CH_2OH	—C(=O)—H	
		Hydroxymethyl group	Formyl group	
		—CH_2—	—C(=)—H	
		Methylene group	Methenyl group	
			—C(=NH)—H	
			Formimino group	
			C=O	
			Carbon monoxide	

4. Single-Carbon Compounds and Groups in Metabolism

Some single-carbon compounds and groups important in metabolism are shown in Table 15-3 in order (from left to right) of increasing oxidation state of the carbon atom. Groups at three different oxidation levels, corresponding to **formic acid**, **formaldehyde**, and the **methyl group**, are carried by tetrahydrofolic acid coenzymes. While the most completely reduced compound, methane, cannot exist in a combined form, its biosynthesis depends upon reduced methanopterin, as does that of carbon monoxide. Figure 15-18 summarizes the metabolic interrelationships of the compounds and groups in this table.

Serine as a C1 donor. For many organisms, from *E. coli* to higher animals, **serine** is a major precursor of C-1 units.[395–397] The β-carbon of serine is removed as formaldehyde through direct transfer to tetrahydrofolate with formation of methylene-THF and glycine (Eq. 14-30, Fig. 15-18, step *c*). This is a stereospecific transfer in which the *pro-S* hydrogen on C-3 of serine enters the *pro-S* position also in methylene-THF.[398] The glycine formed in this reaction can, in turn, yield another single-carbon unit by loss of CO_2 under the influence of the THF and the PLP-requiring glycine cleavage system which is discussed in the next paragraph. Free formaldehyde in a low concentration can also combine with THF to form methylene-THF (Fig. 15-18, step *c′* and Eq. 15-43).[399]

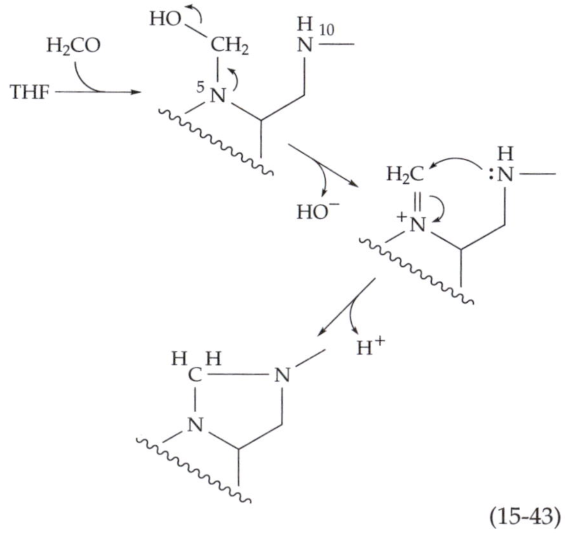

(15-43)

The glycine decarboxylase–synthetase system. Glycine is cleaved reversibly within mitochondria of

plants and animals and also by bacteria to CO_2, NH_3, and a methylene group which is carried by tetrahydrofolic acid[296,400-403] (Fig. 15-20). Four proteins are required. The P-protein consists of two identical 100-kDa subunits, each containing a molecule of PLP. This protein is a **glycine decarboxylase** which, however, replaces the lost CO_2 by an electrophilic sulfur of lipoate rather than by a proton. Serine hydroxymethyltransferase can also catalyze this step of the sequence.[404] The lipoate is bound to a second protein, the H-protein. A third protein, the T-protein, carries bound tetrahydrofolate which displaces the aminomethyl group from the dihydrolipoate and converts it to N^5,N^{10}-methylene tetrahydrofolate with release of ammonia. The dihydrolipoate is then reoxidized by NAD^+ and dihydrolipamide dehydrogenase (Fig. 15-20). Glycine can be oxidized completely in liver mitochondria with the methylene group of methylene-THF being converted to CO_2 through reaction steps d, e, and f of Fig. 15-18. The glycine cleavage system is reversible and is used by some organisms to synthesize glycine.

Whether it arises from the hydroxymethyl group of serine or from glycine, the single-carbon unit of methylene-THF (which is at the formaldehyde level of oxidation) can either be oxidized further to 5,10-methenyl-THF and 10-formyl-THF (steps d and e, Fig. 15-18)

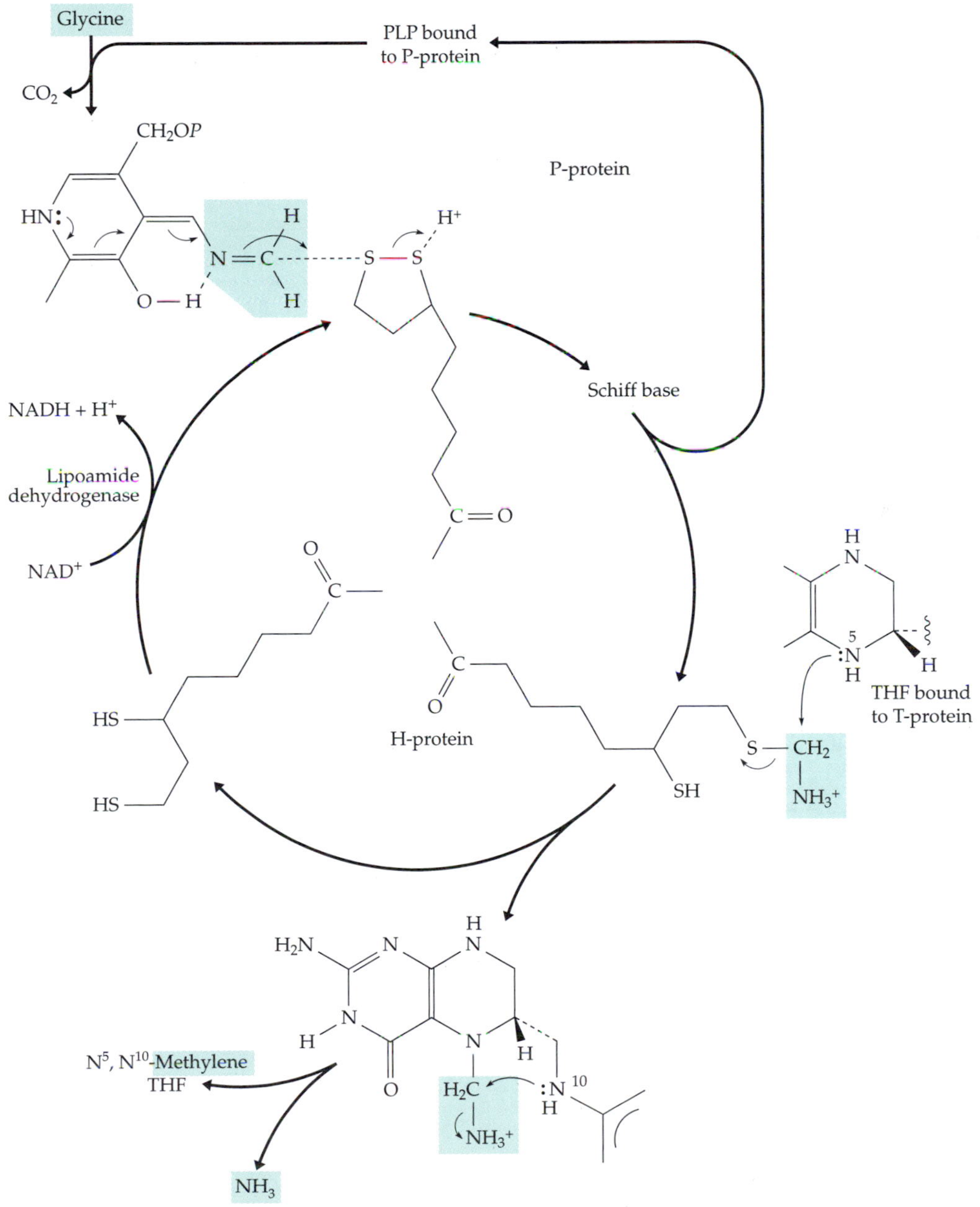

Figure 15-20 The reversible glycine cleavage system of mitochondria. Compare with Fig. 15-15.

or it can be reduced to methyl-THF (step *i*). The reactions of folate metabolism occurs in both cytoplasm and mitochondria at rates that are affected by the length of the polyglutamate chain. The many complexities of these pathways are not fully understood.[395,397]

Starting with formate or carbon dioxide.

Most organisms can, to some extent, also utilize formate as a source of single-carbon units. Human beings have a very limited ability to metabolize formate and the accumulation of formic acid in the body following ingestion of methanol is often fatal. However, many bacteria are able to subsist on formate as a sole carbon source. Archaea may generate the formate by reduction of CO_2. Utilization of formate begins with formation of 10-formyl-THF (Fig. 15-18, step *g*, lower left corner). 10-Formyl-THF can be reduced to methylene-THF and the single-carbon unit can be transferred to glycine to form serine. In some bacteria three separate enzymes are required to convert formate to methylene-THF: 10-formyl-THF synthetase (Fig. 15-18, step *g*, presumably via formyl phosphate),[405] methenyl-THF cyclohydrolase (step *e*, reverse), and methylene-THF dehydrogenase (step *d*, reverse). NADH is used by the acetogens but in *E. coli* and in most higher organisms NADPH is the reductant. In some bacteria and in some tumor cells, two of the three enzymes are present as a bifunctional enzyme.[406–407a] In most eukaryotes all three of the enzymatic activities needed for converting formate to methylene-THF are present in a single large (200-kDa) dimeric trifunctional protein called **formyl-THF synthetase**.[407,408]

N^{10}-Formyl-THF serves as *a biological formylating agent* needed for two steps in the synthesis of purines[409–410b] (Chapter 25) and, in bacteria and in mitochondria and chloroplasts, for synthesis of formylmethionyl-tRNA[411] which initiates synthesis of all polypeptide chains in bacteria and in these organelles (Chapter 29).

N^5-Formyl-THF (leucovorin).

The N^5-formyl derivative of THF (5-formyl-THF) is a growth factor for *Leuconostoc citrovorum*. Following this discovery in 1949 it was called the "citrovorum factor" or **leucovorin**. Its significance in metabolism is not clear. Perhaps it serves as a storage form of folate in cells that have a dormant stage, e.g., of seeds or spores.[412] It may also have a regulatory function.[413] In some ants and in certain beetles, it is stored and hydrolyzed to formic acid. The carabid beetle *Galerita lecontei* ejects a defensive spray that contains 80% formic acid.[414]

5-Formyl-THF can arise by transfer of a formyl group from formylglutamate and there is an enzyme that converts 5-formyl-THF to 5,10-methenyl-THF with concurrent cleavage of ATP.[412] 5-Formyl-THF is sometimes used in a remarkable way to treat certain highly malignant cancers. Following surgical removal of the tumor the patient is periodically given what would normally be a lethal dose of methotrexate, then, about 36 h later, the patient is rescued by injection of 5-formyl-THF. The mechanism of rescue is thought to depend upon the compound's rapid conversion to 10-formyl-THF within cells. Tumor cells are not rescued, perhaps because they have a higher capacity than normal cells for synthesis of polyglutamates of methotrexate.[379] This observation also suggests that the major anti-cancer effect of methotrexate may not be on dihydrofolate reductase but on enzymes of formyl group transfer that are inhibited by polyglutamate derivatives of methotrexate.

Catabolism of histidine and purines.

Another source of single-carbon units in metabolism is the degradation of histidine which occurs both in bacteria and in animals via **formiminoglutamate**. The latter transfers the $-CH=NH$ group to THF forming 5-formimino-THF, which is in turn converted (step *h*, Fig. 15-18) to 5,10-methenyl-THF and ammonia. In bacteria that ferment purines, **formiminoglycine** is an intermediate. Again, the formimino group is transferred to THF and deaminated, the eventual product being 10-formyl-THF. In these organisms, the enzyme 10-formyl-THF synthase also has a very high activity.[415] It probably operates in reverse as a mechanism for synthesis of ATP in this type of fermentation. In other organisms excess 10-formyl-THF may be oxidized to CO_2 via an $NADP^+$-dependent dehydrogenase (step *f*, Fig. 15-16), providing a mechanism for detoxifying formic acid. In some organisms excess 10-formyl-THF may simply be hydrolyzed with release of formate (step *q*).[415a]

Thymidylate synthase.

Methylene-THF serves as the direct precursor of the 5-methyl group of thymine as well as of the hydroxymethyl groups of **hydroxymethylcytosine**[416] and of **2-oxopantoate**, an intermediate in the formation of pantothenate and coenzyme A.[417] The latter is a simple hydroxymethyl transfer reaction (Eq. 15-44) that is related to an aldol condensation and which may proceed through an imine of the kind shown in Eq. 15-43.

$$N^{5,10}\text{-Methylene-THF} \qquad (15\text{-}44)$$

During thymine formation the coenzyme is oxidized to dihydrofolate, which must be reduced by dihydrofolate reductase to complete the catalytic cycle. A possible mechanistic sequence for **thymidylate synthase**, an enzyme of known three-dimensional structure,[354,418–421a] is given in Fig. 15-21. In the first step (a) a thiolate anion, from the side chain of Cys 198 of the 316-residue *Lactobacillus* enzyme, adds to the 5 position of the substrate 2'-deoxyuridine monophosphate

(dUMP). As a consequence the 6 position becomes a nucleophilic center which can combine with methylene-THF as shown in step b of Fig. 15-21. After tautomerization (step c) this adduct eliminates THF (step d) to give a 5-methylene derivative of the dUMP. The latter immediately oxidizes the THF by a hydride ion transfer to form thymidylate and dihydrofolate.[422,422a]

In protozoa thymidylate synthase and dihydrofolate reductase exist as a single bifunctional protein.

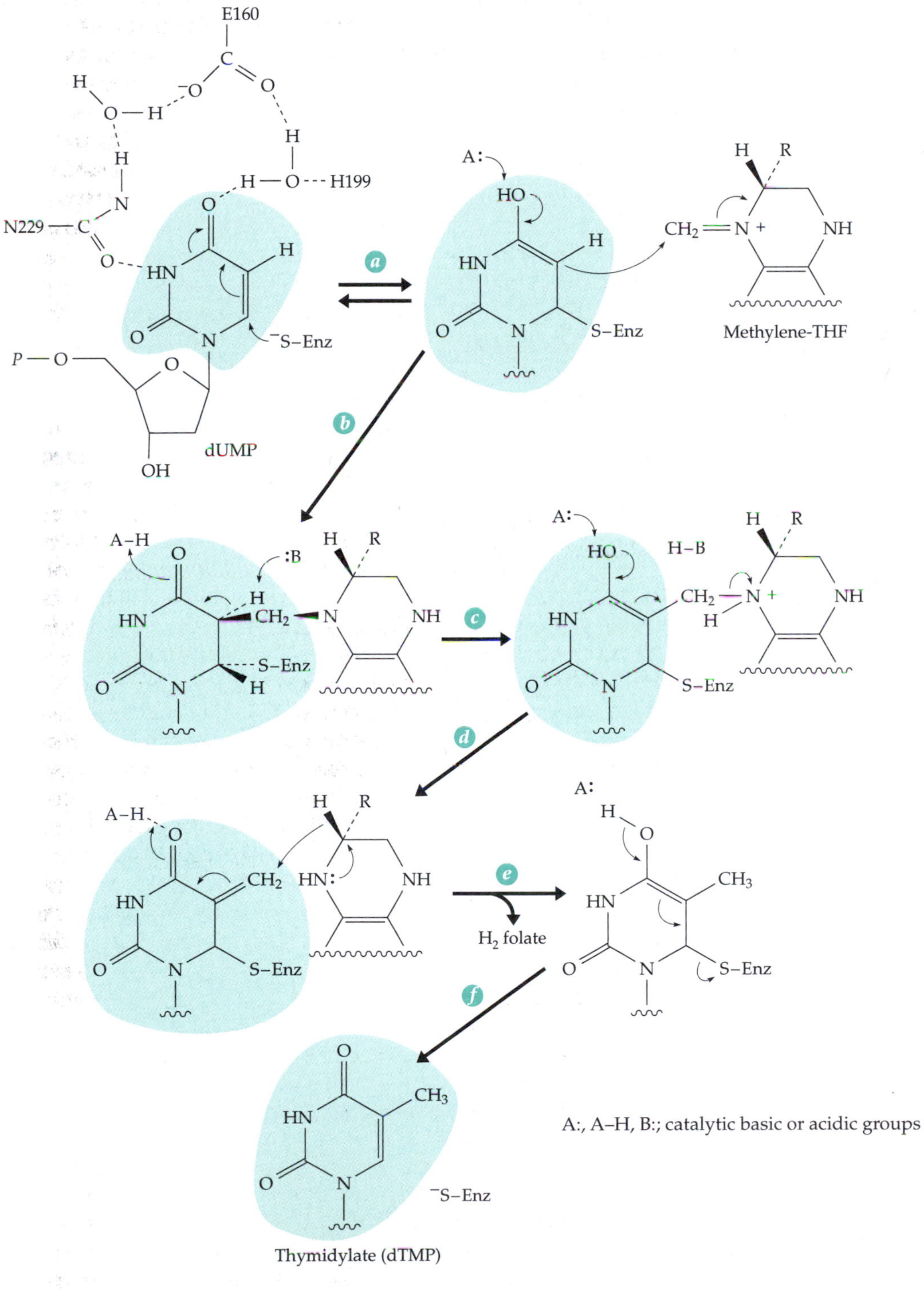

Figure 15-21 Probable mechanism of action of thymidylate synthase. After Huang and Santi[422] and Hyatt *et al.*[354]

Consequently, in methotrexate-resistant strains of *Leishmania* (and also in cancer cells) both enzyme activities are increased equally by gene amplification.[423] This presents a serious problem in the treatment of protozoal parasitic diseases for which few suitable drugs are known.

Cells of *E. coli*, when infected with T-even bacterio-phage, convert dCMP to 5-hydroxymethyl-dCMP[423a]

BOX 15-E THYMIDYLATE SYNTHASE, A TARGET ENZYME IN CANCER CHEMOTHERAPY[a]

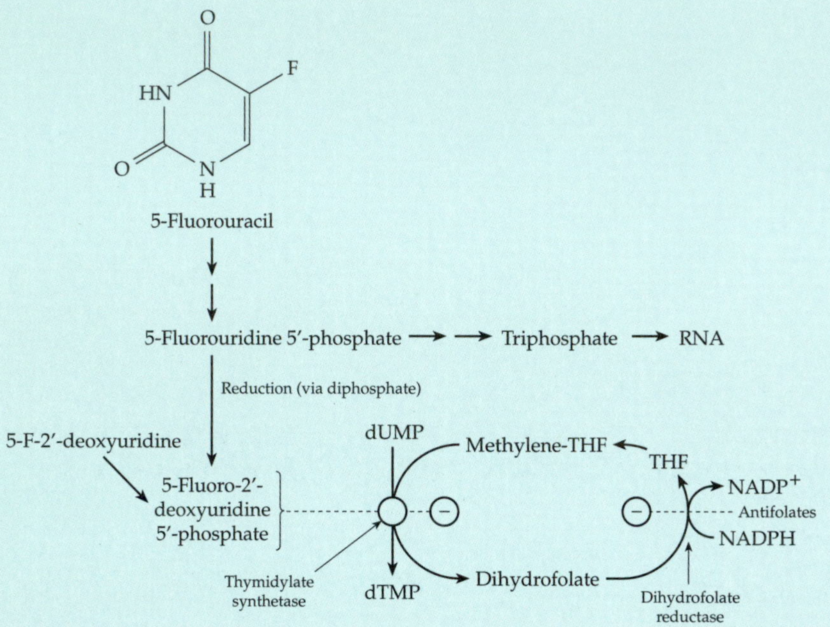

If an animal or bacterial cells are deprived of thymine they can no longer make DNA. However, synthesis of proteins and of RNA continues for some time. This can be demonstrated experimentally with thymine-requiring mutants. However, sooner or later such cells lose their vitality and die. The cause of this **thymineless death**[b] is not entirely clear. Perhaps thymine is needed to repair damage to DNA and if it is not available transcription eventually becomes faulty. Chromosome breakage is also observed.[c] Whatever the cause of death, the phenomenon provides the basis for some of the most effective chemotherapeutic attacks on cancer. Rapidly metabolizing cancer cells are especially vulnerable to thymineless death. Consequently, thymidylate synthase is an important target enzyme for inhibition. One powerful inhibitor is the mono-phosphate of **5-fluoro-2'-deoxyuridine**. The inhibition was originally discovered when 5-fluoro-uracil was recognized as a useful cancer chemo-therapeutic agent.

Fluorouracil has many effects in cells, including incorporation into RNA,[b] but the inhibition of thymidylate synthase by the reduction product may be the most useful effect in chemotherapy. In fact, 5-fluoro-2'-deoxyuridine is a much less toxic and more potent drug than 5-fluoro-uracil. It binds into the active site of thymidylate synthase and reacts in the initial steps of catalysis. However, the 5-H of 2'-deoxyuridine, the normal substrate, must be removed as H⁺ (step *c* of Fig. 15-21) in order for the reaction to continue. The 5-F atom cannot be removed in this way and a stable adduct is formed.[d,e] A crystal structure containing both the bound 5-fluorodeoxyuridine and methylene-THF supports this conclusion.[e]

Thymidylate synthase requires methylene tetrahydro-folate as a reductant and the reduction of dihydrofolate is also an important part of the process. In protozoa dihydrofolate reduc-tase and thymidylate synthase occur as a single-chain bifunctional enzyme.[f] As has been pointed out in the main text, such folic acid analogs as methotrexate are among the most useful anticancer drugs. By inhibiting dihydrofolate reductase they deprive thymidylate synthase of an essential substrate.

Because 5-fluorouracil acts on normal cells as well as cancer cells, its usefulness is limited. Knowledge of the chemistry and three-dimensional structure of thymidylate synthase complexes is being used in an attempt to discover more specific and effective drugs that attack this enzyme.[g,h]

---

[a] Friedkin, M. (1973) *Adv. Enzymol.* **38**, 235–292

[b] Sahasrabudhe, P. V., and Gmeiner, W. H. (1997) *Biochemistry* **36**, 5981–5991

[c] Ayusawa, D., Shimizu, K., Koyama, H., Takeishi, K., and Seno, T. (1983) *J. Biol. Chem.* **258**, 12448–12454

[d] Huang, X. F., and Arvan, P. (1995) *J. Biol. Chem.* **270**, 20417–20423

[e] Hyatt, D. C., Maley, F., and Montfort, W. R. (1997) *Biochemistry* **36**, 4585–4594

[f] Ivanetich, K. M., and Santi, D. V. (1990) *FASEB J.* **4**, 1591–1597

[g] Shoichet, B. K., Stroud, R. M., Santi, D. V., Kuntz, I. D., and Perry, K. M. (1993) *Science* **259**, 1445–1449

[h] Weichsel, A., Montfort, W. R., Ciesla, J., and Maley, F. (1995) *Proc. Natl. Acad. Sci. U.S.A.* **92**, 3493–3497

and suitably infected cells of *Bacillus subtilis* form 5-hydroxymethyl-dUMP.[424,425] These products can be formed by attack of OH^- on quinonoid intermediates of the type postulated for thymidylate synthetase (formed in step *d* of Fig. 15-21).[416] A related reaction is the posttranscriptional conversion of a single uracil residue in tRNA molecules of some bacteria to a thymine ring (a **ribothymidylic acid** residue). In this case, $FADH_2$ is used as a reducing agent to convert the initial adduct of Fig. 15-21 to the ribothymidylic acid and THF.[426]

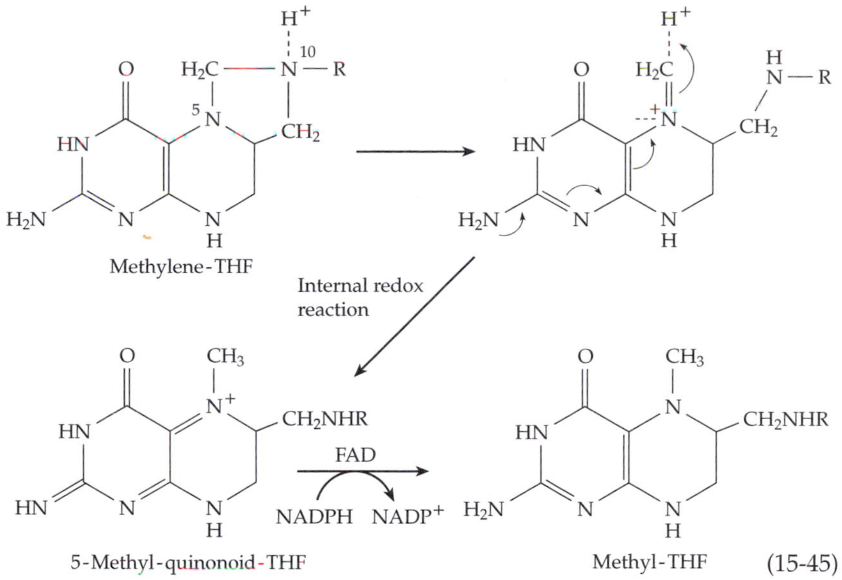

Synthesis of methyl groups.

The reduction of methylene-THF to 5-methyl-THF within all living organisms from bacteria to higher plants and animals provides the methyl groups needed in biosynthesis.[426a] These are required for formation of methionine and, from it, *S*-adenosylmethionine. The latter is used to modify proteins, nucleic acids, and other biochemicals through methylation of specific groups. In the methanogens, reduction of a corresponding methylene derivative of methanopterin gives rise to methane (Section F).

Mammalian methylene-THF reductase is a FAD-containing flavoprotein that utilizes NADPH for the reduction to 5-methyl-THF.[427,428] Matthews[429] suggested that the mechanism of this reaction involves an internal oxidation–reduction reaction that generates a 5-methyl-quinonoid dihydro-THF (Eq. 15-45). Methylene-THF reductase of acetogenic bacteria is also a flavoprotein but it contains Fe–S centers as well. The 237-kDa $\alpha_4\beta_4$ oligomer contains two molecules of FAD and four to six of both Fe and S^{2-} ions.[430,431]

The methyl group of methyl-THF is incorporated into methionine by the vitamin B_{12}-dependent methionine synthase which is discussed in Chapter 16. Matthews suggested that methionine synthase may also make use of the 5-methyl-quinonoid-THF of Eq. 15-45. An initial reduction step would precede transfer of the methyl group. Methyl-THF is a precursor to acetate in

Methylene-THF → (Internal redox reaction) → 5-Methyl-quinonoid-THF → (FAD, NADPH → NADP⁺) → Methyl-THF (15-45)

the acetogenic bacteria.[432] This corrinoid coenzyme-dependent process is also considered in Chapter 16.

E. Specialized Coenzymes of Methanogenic Bacteria

Methane-producing bacteria[433,434] obtain energy by reducing CO_2 with molecular hydrogen:

$$CO_2 + 4H_2 \rightarrow CH_4 + 2H_2O$$
$$\Delta G° = -131 \text{ kJ / mol} \qquad (15\text{-}46)$$

Some species are also able to utilize formate or formaldehyde as reducing agents.[435] These compounds are oxidized to CO_2, the reducing equivalents formed being used to reduce CO_2 to methane. Carbon monoxide can also be converted to CO_2.

$$CO + H_2O \rightarrow CO_2 + H_2$$
$$\Delta G° = -20 \text{ kJ / mol} \qquad (15\text{-}47)$$

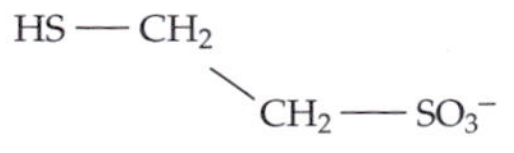

Coenzyme M

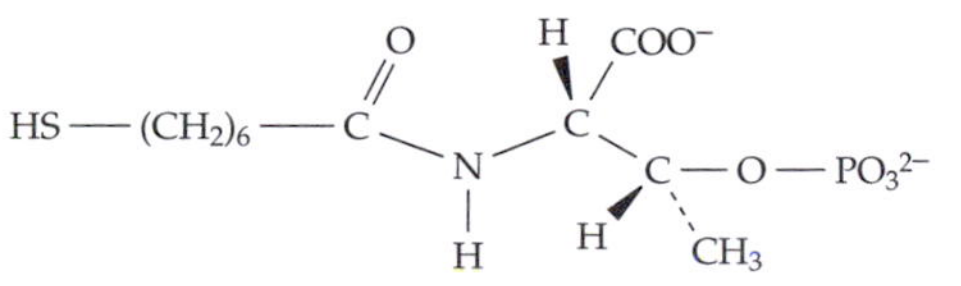

7-Mercaptoheptanoylthreonine phosphate

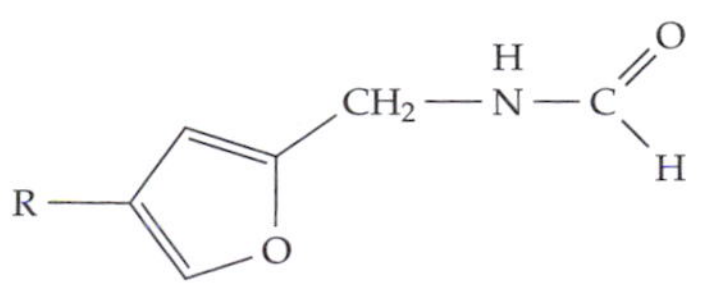

Formylmethanofuran

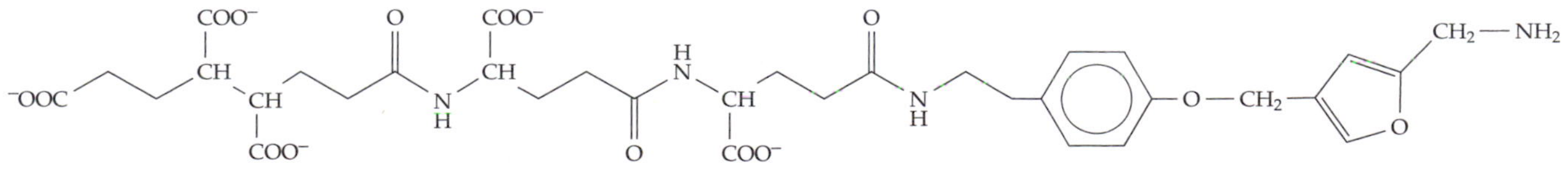

Methanofuran

By using the combined reactions of Eqs. 15-46 and 15-47 the bacteria can subsist on CO alone (Eq. 15-48):

$$4CO + 2\,H_2O \rightarrow 3CO_2 + CH_4$$
$$\Delta G° = -211 \text{ kJ / mol of}$$
$$\text{methane formed} \qquad (15\text{-}48)$$

Some species reduce methanol to methane via Eqs. 15-49 or 15-50.

$$CH_3OH + H_2 \rightarrow CH_4 + H_2O$$
$$\Delta G° = -113 \text{ kJ / mol} \qquad (15\text{-}49)$$

$$4CH_3OH \rightarrow 3CH_4 + CO_2 + 2\,H_2O$$
$$\Delta G° = -107 \text{ kJ / mol of}$$
$$\text{methane formed} \qquad (15\text{-}50)$$

To accomplish these reactions a surprising variety of specialized cofactors are needed.[351,352,434] The first of these, **coenzyme M**, 2-mercaptoethane sulfonate, was discovered in 1974.[436] It is the simplest known coenzyme. Later, the previously described **5-deazaflavin F_{420}** (Section B), a **nickel tetrapyrrole F_{430}** (Chapter 16), **methanopterin** (Fig. 15-17),[437] the "carbon dioxide reduction factor" called **methanofuran**,[352,438] and **7-mercaptoheptanoylthreonine phosphate**[439,440] were also identified.

A sketch of the metabolic pathways followed in methane formation is given in Fig. 15-22.[352,435] In the first step (*a*) the amino group of methanofuran is thought to add to CO_2 to form a carbamate which is reduced to formylmethanofuran by H_2 and an intermediate carrier H_2X in step *b*. The formyl group is then transferred to tetrahydromethanopterin (H_4MPT) (step *c*)[440a] and is cyclized and reduced in two stages in steps *d*, *e*, and *f*. The reductant is the deazaflavin F_{420} and the reactions parallel those for conversion of formyl-THF to methyl-THF (Fig. 15-18).[431,440b,441] The methyl group of methyl-H_4MPT is then transferred to the sulfur atom of the thiolate anion of coenzyme M, from which it is reduced off as CH_4. This is a complex process requiring the nickel-containing F_{430}, FAD, and 7-mercaptoheptanoylthreonine phosphate (HS-HTP).

The HS-HTP may be the 2-electron donor for the reduction. The mixed-disulfide CoM-S-S-HTP appears to be an intermediate and also an allosteric effector for the first step, the reduction of

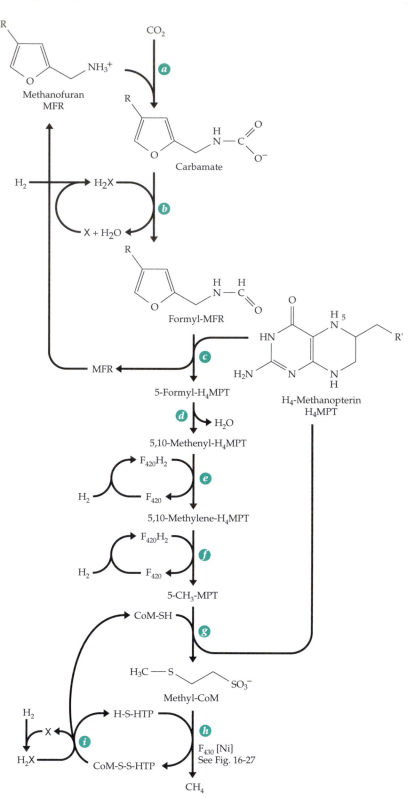

Figure 15-22 Tentative scheme for reduction of carbon dioxide to methane by methanogens. After Rouvière *et al.*[352] and Thauer *et al.*[435]

CO_2 to formylmethanofuran. A complex of proteins that is unstable in oxygen is also needed. The principal component of the methyl reductase binds two molecules of the nickel-containing F_{430}. It also binds two moles of CoM-SH noncovalently. The binding process requires ATP as well as the presence of other proteins, notably those of the hydrogenase system.

Coenzyme M, previously found only in methanogenic archaeobacteria, has recently been discovered in both gram-negative and gram-positive alkene-oxidizing eubacteria. It seems to function in cleavage of epoxy rings and as a carrier of hydroxyalkyl groups.[441a] See also Chapter 17.

F. Quinones, Hydroquinones, and Tocopherols

1. Pyrroloquinoline Quinone (PQQ)

Bacteria that oxidize methane or methanol (**methylotrophs**) employ a periplasmic methanol dehydrogenase that contains as a bound coenzyme, the pyrroloquinoline quinone designated **PQQ** or methoxatin (Eq. 15-51).[442–444] This fluorescent *ortho*-quinone

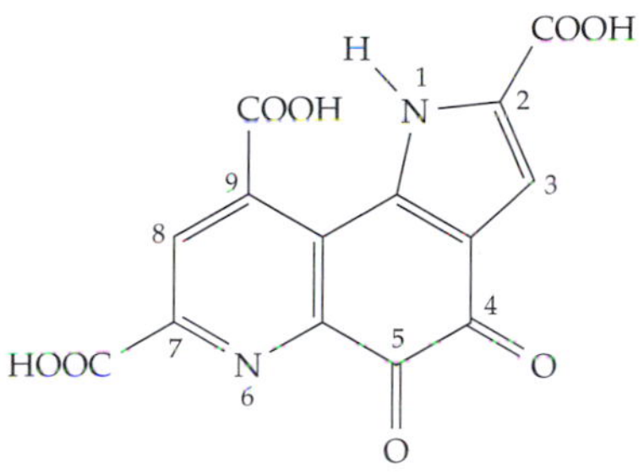

PQQ (oxidized)

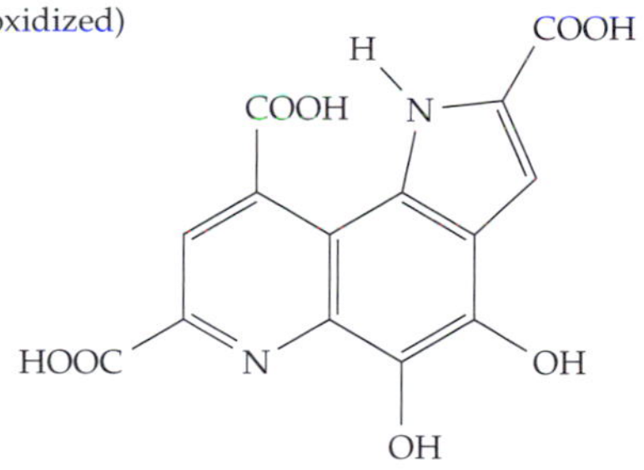

PQQH$_2$ (reduced)

is released when the protein is denatured. Some other bacterial alcohol dehydrogenases[445,445a] and glucose dehydrogenases[446-446d] contain the same cofactor.

PQQ-containing methanol dehydrogenase from *Methylophilus* is a dimer of a 640-residue protein consisting of two disulfide-linked peptide chains with one noncovalently bound PQQ.[447] The large subunit contains a β propellor (Fig. 15-23), which is similar to that in the protein G_i shown in Fig. 11-7. The PQQ binding site lies above the propellor and is formed by a series of loops. The coenzyme interacts with several polar protein side chains and with a bound calcium ion.[448] Bacterial PQQ-dependent glucose dehydrogenases

have also been studied intensively. The 450-residue soluble enzyme from *Acinetobacter calcoaceticus* is partially homologous to the methanol dehydrogenase.[446a,446b]

PQQ and the other quinone prosthetic groups described here all function in reactions that would be possible for pyridine nucleotide or flavin coenzymes. All of them, like the flavins, can exist in oxidized, half-reduced semiquinone and fully reduced dihydro forms. The questions to be asked are the same as we asked for flavins. How do the substrates react? How is the reduced cofactor reoxidized? In nonenzymatic reactions alcohols, amines, and enolate anions all add at C-5 of PQQ to give adducts such as that shown for methanol in Eq. 15-51, step *a*.[444,449,449a] Although many additional reactions are possible, this addition is a reasonable first step in the mechanism shown in Eq. 15-51. An enzymatic base could remove a proton as is indicated in step *b* to give PQQH$_2$. The pathway for reoxidation (step *c*) might involve a cytochrome *b*, cytochrome *c*, or bound ubiquinone.[445,446]

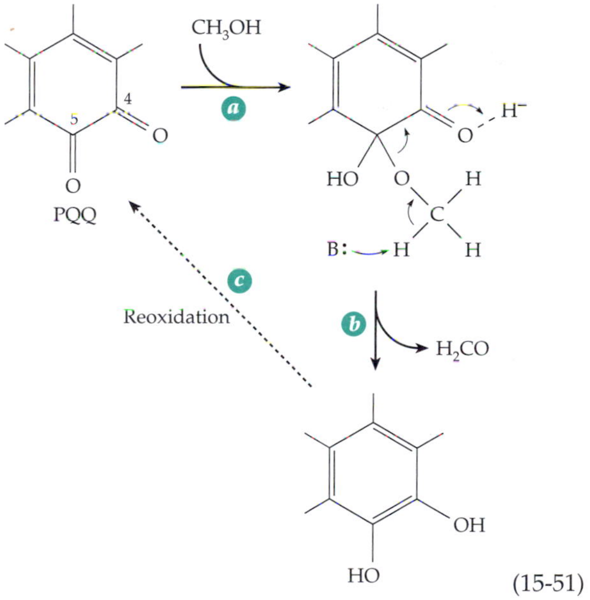

(15-51)

Although the soluble PQQ-dependent glucose dehydrogenase forms, with methylhydrazine, an adduct similar to that depicted in Eq. 15-51,[449b] the structure of a glucose complex of the PPQH$_2$- and Ca^{2+}-containing enzyme at a resolution of 0.2 nm suggests a direct hydride ion transfer. The only base close to glucose C1 is His 144. The C1-H proton lies directly above the PQQH$_2$ C5 atom. Oubrie *et al.*[446a] propose a deprotonation of the C1-OH by His 144 and H$^-$ transfer to C5 of PQQ followed by tautomerization (Eq. 15-52). Theoretical calculations by Zheng and Bruice[449c] favor a simpler mechanism with hydride transfer to the oxygen atom of the C4 carbonyl (green arrows). In either case, the Ca^{2+} would assist by polarizing the C5 carbonyl group.

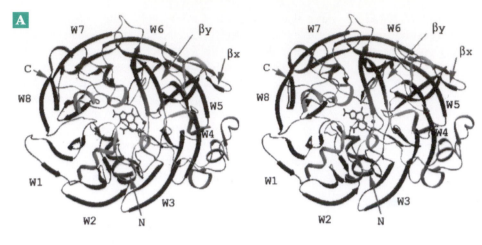

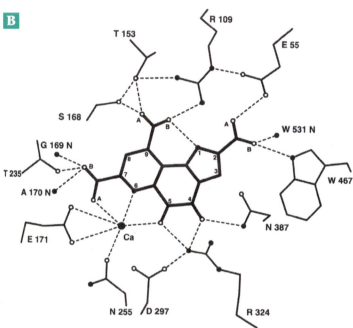

Figure 15-23 (A) Stereoscopic view of the H subunit of methanol dehydrogenase. Eight four-stranded antiparallel β sheets, labeled W1–W8, form the base of the subunit. Several helices and two additional β-sheet structures (labeled βx and βy) form a cap over the base. The PQQ is located in a funnel within the cap approximately on an eight-fold axis of pseudosymmetry. Courtesy of Xia *et al.*[447] (B) Schematic view of the active site. W467 is parallel to the plane of PQQ. All hydrogen-bond interactions between PQQ and its surrounding atoms, except for the three water molecules, are indicated. Courtesy of White *et al.*[448]

(15-52)

2. Copper Amine Oxidases

At first it appeared that PQQ had a broad distribution in enzymes, including eukaryotic amine oxidases. However, it was discovered, after considerable effort, that there are additional quinone cofactors that function in oxidation of amines. These are derivatives of tyrosyl groups of specific enzyme proteins. Together with enzymes containing bound PQQ they are often called **quinoproteins**.[450–454]

Topaquinone (TPQ). Both bacteria and eukaryotes contain amine oxidases that utilize bound copper ions and O_2 as electron acceptors and form an aldehyde, NH_3, and H_2O_2. The presence of an organic cofactor was suggested by the absorption spectra which was variously attributed to pyridoxal phosphate or PQQ. However, isolation from the active site of bovine serum amine oxidase established the structure of the reduced form of the cofactor as a trihydroxyphenylalanyl group covalently linked in the peptide chain.[455,456] Its oxidized form is called topaquinone:

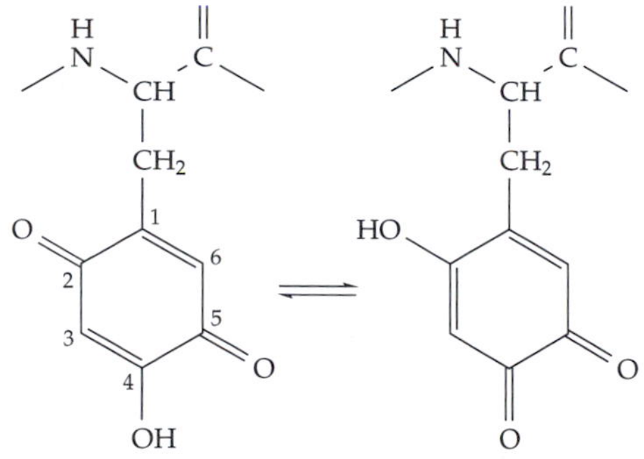

Topaquinone (TPQ)

The same cofactor is present in an amine oxidase from *E. coli*[457–460] and in other bacterial[461–463] fungal,[463a] plant,[464] and mammalian[464a] amine oxidases.

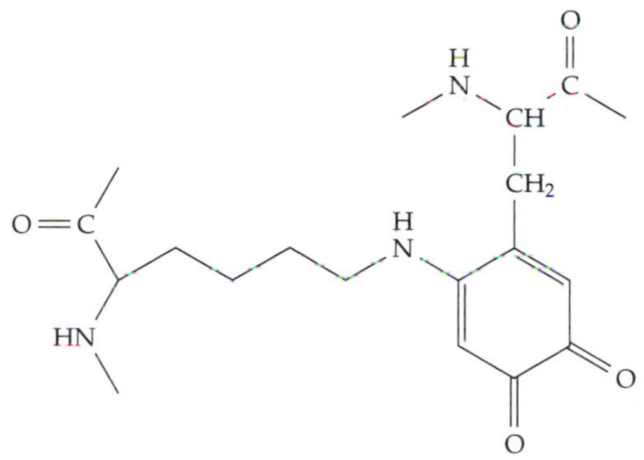

Lysine tyrosylquinone (LTQ)

Lysine tyrosylquinone (LTQ).

Another copper amine oxidase, **lysyl oxidase**, which oxidizes side chains of lysine in collagen and elastin (Eq. 8-8) contains a cofactor that has been identified as having a lysyl group of a different segment of the protein in place of the –OH in the 2 position of topaquinone.[465] Lysyl oxidase plays an essential role in the crosslinking of collagen and elastin.

Tryptophan tryptophanylquinone (TTQ).

This recently discovered quinone cofactor is similar to the lysyl tyrosylquinone but is formed from two tryptophanyl side chains.[466] It has been found in **methyl-amine dehydrogenase** from methylotrophic gram-negative bacteria[467–469] and also in a bacterial aromatic amine dehydrogenase.[470]

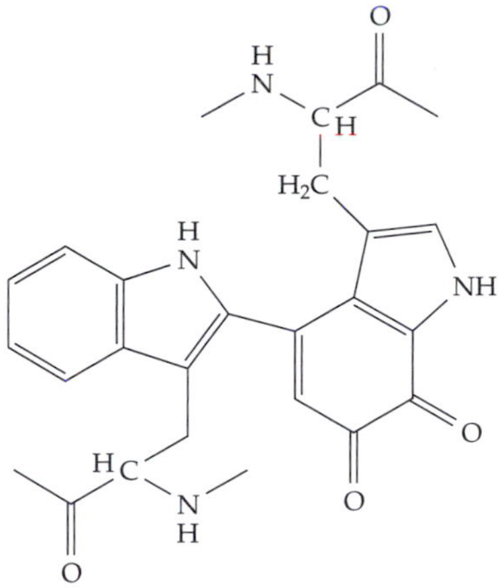

Tryptophan tryptophylquinone (TTQ)

Three-dimensional structures.

The TPQ-containing amine oxidase from *E. coli* is a dimer of 727-residue subunits with one molecule of TPQ at position 402 in each subunit.[457,458] Methylamine dehydrogenase is also a large dimeric protein of two large 46.7-kDa subunits and two small 15.5-kDa subunits. Each large subunit contains a TTQ cofactor group.[468,471,471a] Reduced TTQ is reoxidized by the 12.5-kDa blue copper protein **amicyanin**. Crystal structures have been determined for complexes of methylamine dehydrogenase with amicyanin[471] and of these two proteins with a third protein, a small bacterial cytochrome *c*.[472,472a]

Mechanisms.

Studies of model reactions[473–476] and of electronic, Raman,[466,477,478] ESR,[479,480] and NMR spectra and kinetics[481] have contributed to an understanding of these enzymes.[459,461,464,482,483] For these copper amine oxidases the experimental evidence suggests an aminotransferase mechanism.[450,453,474,474a-d] The structure of the *E.coli* oxidase shows that a single copper ion is bound by three histidine imidazoles and is located adjacent to the TPQ (Eq. 15-53). Asp 383 is a conserved residue that may be the catalytic base in Eq. 15-53.[474b] A similar mechanism can be invoked for LTQ and TTQ.

How is the reduced cofactor reoxidized? Presumably the copper ion adjacent to the TPQ functions in this process, passing electrons one at a time to the next carrier in a chain. There is no copper in the TTQ-containing subunits. Electrons apparently must jump about 1.6 nm to the copper ion of amicyanin, then another 2.5 nm to the iron ion of the cytochrome *c*.[472] Reoxidation of the aminoquinol formed in Eq. 15-53, step *d*, yields a Schiff base whose hydrolysis will release ammonia and regenerate the TTQ. Intermediate states with Cu[+] and a TTQ semiquinone radical have been observed.[483a]

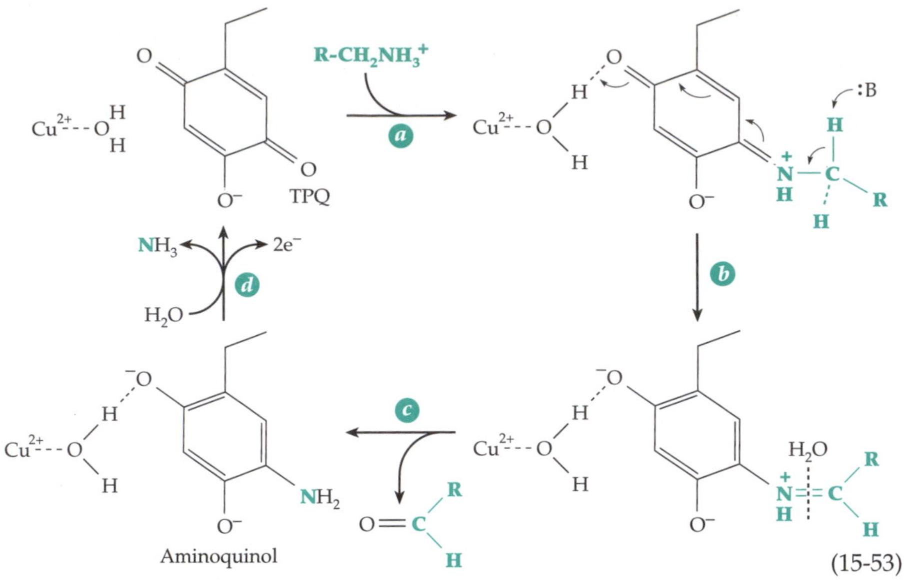

TPQ

Aminoquinol

(15-53)

3. Ubiquinones, Plastoquinones, Tocopherols, and Vitamin K

In 1955, R. A. Morton and associates in Liverpool announced the isolation of a quinone which they named ubiquinone for its ubiquitous occurrence.[484,485] It was characterized as a derivative of benzoquinone attached to an unsaturated polyprenyl (isoprenoid) side chain (Fig. 15-24). In fact, there is a family of ubiquinones: that from bacteria typically contains six prenyl units in its side chain, while most ubiquinones from mammalian mitochondria contain ten. Ubiquinone was also isolated by F. L. Crane and associates using isooctane extraction of mitochondria. These workers proposed that the new quinone, which they called **coenzyme Q**, might participate in electron transport. As is described in Chapter 18, this function has been fully established. Both the name ubiquinone and the abbreviation Q are in general use. A subscript indicates the number of prenyl units, e.g., Q_{10}. Ubiquinones can be reversibly reduced to the hydroquinone forms (Fig. 15-24), providing a basis for their function in electron transport within mitochondria and chloroplasts.[486–490]

While vitamin E (Box 15-G) and vitamin K (Box 15-F) are dietary essentials for humans, ubiquinone is apparently not. Animals are able to make ubiquinones in quantities adequate to meet the need for this essential component of mitochondria. However, an extra dietary supplement may sometimes be of value.[491] A reduced level of ubiquinone has been reported in gum tissues of patients with periodontal disease.[492] Ubiquinones are being tested as possible protectants against heart damage caused by lack of adequate oxygen or by drugs.[491]

A closely related series of **plastoquinones** occur in chloroplasts (Chapter 23). In these compounds the two methoxyl groups of ubiquinone are replaced by methyl groups (Fig. 15-24). The most abundant of these compounds, plastoquinone A, contains nine prenyl units.[493,494]

Addition of the hydroxyl group of the reduced ubiquinone or plastoquinone to the adjacent double bond leads to a chroman-6-ol structure. The compounds derived from ubiquinone in this way are called **ubichromanols** (Fig. 15-24). The corresponding ubichromenol (Fig. 15-24) has been isolated from human kidney. **Plastochromanols** are derived from plastoquinone. That from plastoquinone A, first isolated from tobacco, is also known as solanochromene.

A closely related and important family of chromanols are the **tocopherols** or vitamins E (Fig. 15-24, Box 15-G). Tocopherols are plant products found primarily in plant oils and are essential to proper nutrition of humans and other animals. α-Tocopherol is the most abundant form of the vitamin E family; smaller amounts of the β, δ, and γ forms occur, as do a series of **tocotrienols** which contain unsaturated isoprenoid units.[495] The configuration of α-tocopherol is 2R,4'R,8'R as indicated in Fig. 15-24. When α-tocopherol is oxidized, e.g., with ferric chloride, the ring can be opened by hydrolysis to give **tocopherolquinones** (Fig. 15-24), which can in turn be reduced to tocopherolhydroquinones. Large amounts of the tocopherolquinones have been found in chloroplasts.

Another important family of quinones, related in structure to those already discussed, are the **vitamins K** (Fig. 15-24, Box 15-F). These occur naturally as two families. The vitamins K_1 (**phylloquinones**) have only one double bond in the side chain and that is in the prenyl unit closest to the ring. This suggests again the possibility of chromanol formation. In the vitamin K_2 (**menaquinone**) series, a double bond is present in each of the prenyl units. A synthetic compound **menadione** completely lacks the polyprenyl side chain and bears a hydrogen in the corresponding position on the ring. Nevertheless, menadione serves as a synthetic vitamin K, apparently because it can be converted in the body to forms containing polyprenyl side chains.

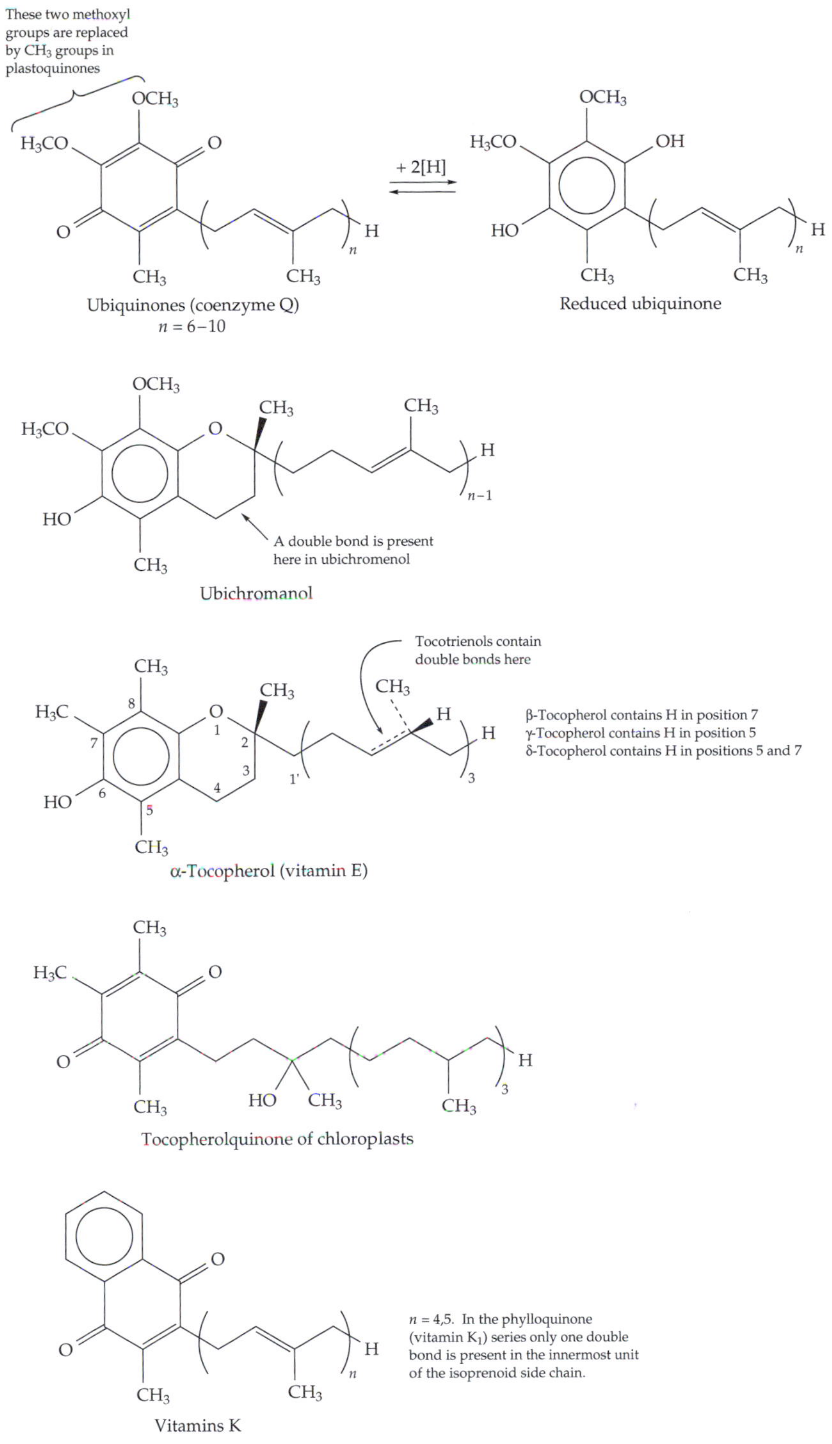

These two methoxyl groups are replaced by CH_3 groups in plastoquinones

Ubiquinones (coenzyme Q)
$n = 6-10$

$+ 2[H]$

Reduced ubiquinone

A double bond is present here in ubichromenol

Ubichromanol

Tocotrienols contain double bonds here

β-Tocopherol contains H in position 7
γ-Tocopherol contains H in position 5
δ-Tocopherol contains H in positions 5 and 7

α-Tocopherol (vitamin E)

Tocopherolquinone of chloroplasts

$n = 4,5$. In the phylloquinone (vitamin K_1) series only one double bond is present in the innermost unit of the isoprenoid side chain.

Vitamins K

Figure 15-24 Structures of the isoprenoid quinones and vitamin E.

4. Quinones as Electron Carriers

Ubiquinones function as electron transport agents within the inner mitochondrial membranes[496] and also within the reaction centers of the photosynthetic membranes of bacteria (Eq. 23-32).[484,488,494] The plastoquinones also function in electron transport within these membranes. Within some mycobacteria vitamin K apparently participates in electron transport chains in the same way (see Chapter 18). Some bacteria contain both menaquinones and ubiquinones.

The vitamin E derivative **α-tocopherolquinone** (Fig. 15-24) can also serve as an electron carrier, being reversibly reduced to the hydroquinone form α-tocopherolquinol. Such a function has been proposed for the anaerobic rumen bacterium *Butyrovibrio fibrisolvens*.[497]

When a single hydrogen atom is removed from a hydroquinone or from a chromanol such as a tocopherol, a free radical is formed (Eq. 15-54). Phenols substituted in the 2, 4, and 6 positions give especially stable radicals.

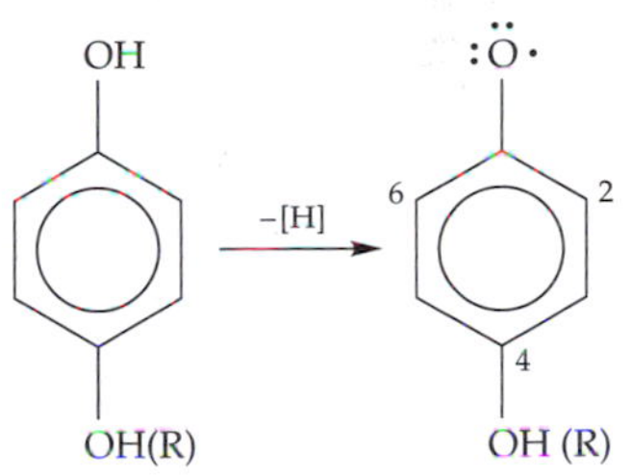

(15-54)

Both the presence of methyl substituents in the tocopherols and their chromanol structures increase the ability of these compounds to form relatively stable radicals.[498,499] This ability is doubtless probably important also in the function of ubiquinones and plastoquinones. Ubiquinone radicals (semiquinones) are probably intermediates in mitochondrial electron transport (Chapter 18) and radicals amounting to as much as 40% of the total ubiquinone in the NADH-ubiquinone reductase of heart mitochondria have been detected by EPR measurements.[500,501]

The equilibria governing semiquinone formation from quinones are similar to those for the flavin semiquinones which were discussed in Section B,6. Two consecutive one-electron redox steps can be defined. Their redox potentials will vary with pH because of a pK_a for the semiquinone in the pH 4.5 −6.5 region. For ubiquinone this pK_a is about 4.9 in water and 6.45 in methanol. A pK_a of over 13 in the hydroquinone form[502] will have little effect on redox potentials near pH 7. The potential for the one-electron reaction $Q + e^- \rightarrow Q^-$ is evaluated most readily. For this reaction $E^{\circ\prime}$ (pH 7) is -0.074, -0.13, -0.17, and -0.23 for 2,3-dimethylbenzoquinone, plastoquinone, ubiquinone, and phylloquinone, respectively.

Why does this entire family of compounds have the long polyprenyl side chains? A simple answer is that they serve to anchor the compounds in the lipid portion of the cell membranes where they function. In the case of ubiquinones both the oxidized and the reduced forms may move freely through the lipid phase shuttling electrons between carriers.

5. Vitamin K and γ-Carboxyglutamate Formation

In higher animals the only known function of vitamin K is in the synthesis of γ-carboxyglutamate (Gla)-containing proteins, several of which are needed in blood clotting (Box 15-F, Chapter 12). Following the discovery of γ-carboxyglutamate, it was shown that liver microsomes were able to incorporate ^{14}C-containing bicarbonate or CO_2 into the Gla of prothrombin and could also generate Gla in certain simple peptides such as Phe-Leu-Glu-Val. Three enzymes are required. All are probably bound to the microsomal membranes.[503–507a] An NADPH-dependent reductase reduces vitamin K quinone to its hydroquinone form. Conversion of Glu residues to Gla residues requires this reduced vitamin K as well as O_2 and CO_2. During the carboxylation reaction the reduced vitamin K is converted into vitamin K 2,3-epoxide (Eq. 15-55).[508] The mechanism is uncertain but a peroxide intermediate such as that shown in Eq. 15-56 is probably involved. This could be used to generate a hydroxide ion adjacent to the pro-S -H of the glutamate side chain of the substrate. This hydrogen could be abstracted by the OH⁻ to form

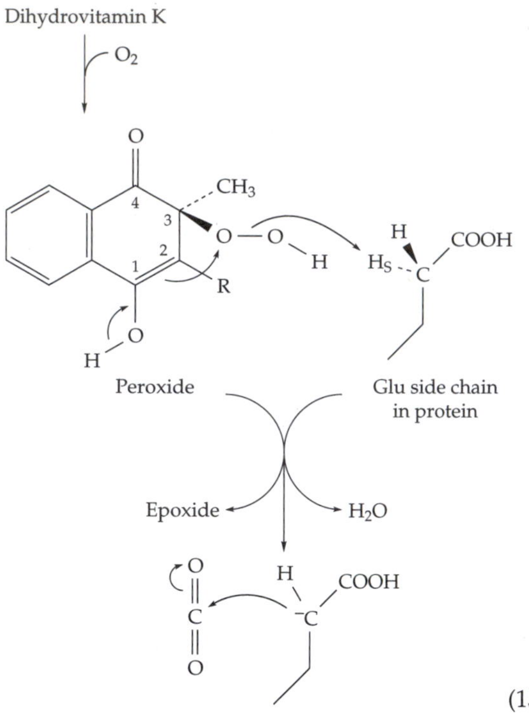

Dihydrovitamin K

$$(15\text{-}56)$$

H_2O and a carbanion which would be stabilized by the adjacent carboxyl group[508–511] (Eq. 15-56).

Dowd and coworkers raised doubts that a hydroxide ion released in the active site in this manner is a strong enough base to generate the anion shown in Eq. 15-52.[512,513] They hypothesized a *"base strength amplification"* mechanism that begins with a peroxide formed at C-4 followed by ring closure to form a dioxetane and rearrangement to the following hypothetical strong base.[507,512,514]

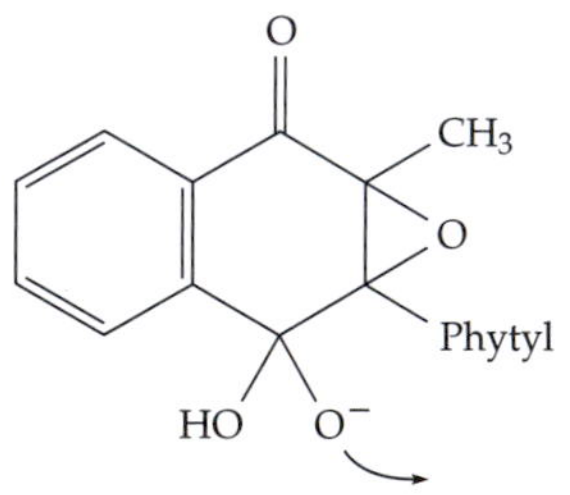

Hypothetical strong base

The proposal was supported by model experiments and also by observation of some incorporation of both atoms of $^{18}O_2$ into vitamin K epoxide. However, theoretical calculations support the simpler mechanism of Eq. 15-56.

The fact that L-*threo*-γ-fluoroglutamate residues, in which the fluorine atom is in the position corresponding to the pro-S hydrogen, are not carboxylated but that an *erythro*-γ-fluoroglutamate is carboxylated

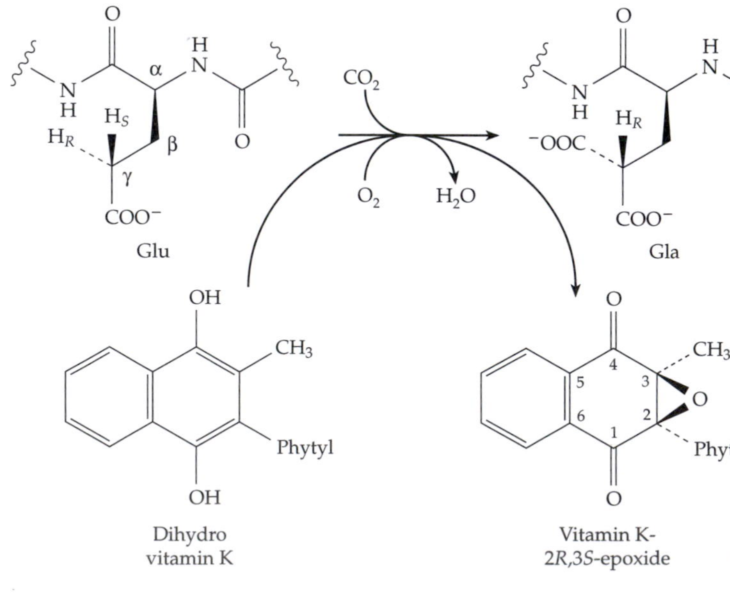

$$(15\text{-}55)$$

Glu → Gla

Dihydro vitamin K → Vitamin K-2R,3S-epoxide

BOX 15-F THE VITAMIN K FAMILY

The existence of an "antihemorrhagic factor" required in the diet of chicks to ensure rapid clotting of blood was reported in 1929 by Henrik Dam at the University of Copenhagen.[a,b] The fat-soluble material, later designated vitamin K, causes a prompt (2–6 h) decrease in the clotting time when administered to deficient animals and birds. The clotting time for a vitamin K-deficient chick may be greater than 240 s, but 6 h after injection of 2 μg of vitamin K_3 it falls to 50–100 s.[c] Pure vitamin K (Fig. 15-24), a 1,4-naphthoquinone, was isolated from alfalfa in 1939. Within a short time two series, the phylloquinones (vitamin K_1) and the menaquinones (vitamin K_2), were recognized. The most prominant phylloquinone contains the phytyl group, which is also present in the chlorophylls. For a human being a dietary intake of about 30 μg per day is recommended.[d] Additional vitamin K is normally supplied by intestinal bacteria.

The most obvious effect of a deficiency in vitamin K in animals is delayed blood clotting, which has been traced to a decrease in the activity of **prothrombin** and of clotting factors VII, IX, and X (Chapter 12, Fig. 12-17). Prothrombin formed by the liver in the absence of vitamin K lacks the ability to chelate calcium ions essential for the binding of prothrombin to phospholipids and to its activation to thrombin. The structural differences between this abnormal protein and the normal prothrombin have been pinpointed at the N terminus of the ~560 residue glycoprotein.[e,f] Tryptic peptides from the N termini differed in electrophoretic mobility. As detailed in Chapter 12, ten residues within the first 33, which were identified as glutamate residues by the sequence analysis on normal prothrombin, are actually **γ-carboxyglutamate** (Gla). The same amino acid is present near the N termini of clotting factors VII, IX, and X.

γ-Carboxyglutamate (Gla)

The fact that γ-carboxyglutamate had not been identified previously as a protein substituent is explained by its easy decarboxylation to glutamic acid during treatment with strong acid. The function of vitamin K is to assist in the incorporation of the additional carboxyl group into the glutamate residues of preformed prothrombin and other blood-clotting factors[g,h] with a resulting increase in calcium ion affinity.

Four other plasma proteins designated C, S, M, and Z contain γ-carboxyglutamate. The functions of proteins M and Z are unknown but protein C is a serine protease involved in regulation of blood coagulation and protein S is a cofactor that assists the action of protein C. Other proteins that require vitamin K for synthesis include the 49-residue **bone Gla protein** (or **osteocalcin**) and the 79-residue **matrix Gla protein** found in bone and cartilage.[i,j] These proteins contain three and five residues of Gla, respectively. Their possible functions in mineralization are considered in Chapter 8. At least two additional small human proline-rich Gla proteins of unknown function are synthesized in many tissues.[k] Gamma-carboxyglutamate also occurs in an invertebrate peptide from fish-hunting cone snails. This "sleeper peptide," which induces sleep in mice after intracerebral injection, has the sequence GEE*E*LQE*NQE*LIRE*KSN. Here E* designates the 5 residues of Gla.[l]

An interesting facet of vitamin K nutrition and metabolism was revealed by the observation that cattle fed on spoiled sweet clover develop a fatal hemorrhagic disease. The causative agent is **dicoumarol**, a compound arising from coumarin, a natural constituent of clover. Dicourmarol and the closely related synthetic **warfarin** are both potent vitamin K antagonists. Warfarin is used both as a rat poison and in the treatment of thromboembolic disease. As rodenticides hydroxycourmarin derivatives are usually safe because a single accidental ingestion by a child or pet does little harm, whereas regular ingestion by rodents is fatal.

Dicoumarol

[a] Wasserman, R. H. (1972) *Ann. Rev. Biochem.* **41**, 179–202
[b] Tim Kim, X. (1979) *Trends Biochem. Sci.* **4**, 118–119
[c] Olson, R. E. (1964) *Science* **45**, 926–928
[d] Shils, M. E., Olson, J. A., and Shike, M., eds. (1994) *Modern Nutrition in Health and Disease*, 8th ed., Vol. 1, Lea & Febiger, Philadelphia, Pennsylvania (pp. 353–355)
[e] Friedman, P. A. (1984) *N. Engl. J. Med.* **310**, 1458–1460
[f] Stenflo, J. (1976) *J. Biol. Chem.* **251**, 355–363
[g] Wood, G. M., and Suttie, J. W. (1988) *J. Biol. Chem.* **263**, 3234–3239
[h] Wu, S.-M., Mutucumarana, V. P., Geromanos, S., and Stafford, D. W. (1997) *J. Biol. Chem.* **272**, 11718–11722
[i] Price, P. A., and Williamson, M. K. (1985) *J. Biol. Chem.* **260**, 14971–14975
[j] Price, P. A., Rice, J. S., and Williamson, M. K. (1994) *Protein Sci.* **3**, 822–830
[k] Kulman, J. D., Harris, J. E., Haldeman, B. A., Davie, E. W. (1997) *Proc. Natl. Acad. Sci., USA* **94**, 9058–9062
[l] Prorok, M., Warder, S. E., Blandl, T., and Castellino, F. J. (1996) *Biochemistry* **35**, 16528–16534

suggested the indicated stereospecificity.[515,516] This was confirmed by observation of a kinetic isotope effect when ^2H is present in the *pro-S* position.[517] Addition of the carbanion to CO_2 would generate the Gla residue. The two glutamates in the following sequence, in which X may be various amino acids, are carboxylated if the protein also carries a suitable N-terminal signal sequence: EXXXEXC. If a suitable glutamyl peptide is not available for carboxylation the postulated peroxide intermediate (Eq. 15-55) is still converted slowly to the 2,3-epoxide of vitamin K.

A third enzyme is required to reduce the epoxide to vitamin K (Eq. 15-57). The biological reductant is uncertain but dithiols such as dithiothreitol serve in the laboratory.[518] See also Eq. 18-47. Protonation of an intermediate enolate anion would give 3-hydroxy-2, 3-dihydrovitamin K, an observed side reaction product.

$$\text{Vitamin K epoxide} \ + 2[H] \rightarrow H_2O + \text{Vitamin K}$$
$$(15\text{-}57)$$

This reaction is of interest because of its specific inhibition by such coumarin derivatives as Warfarin:

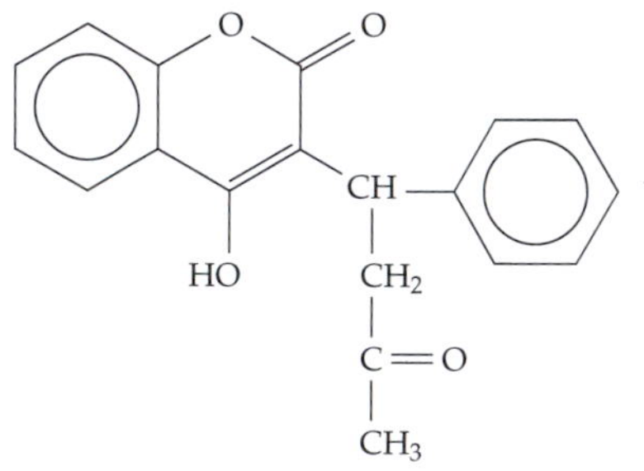

This synthetic compound, as well as natural coumarin anticoagulants (Box 15-F), inhibits both the vitamin K reductase and the epoxide reductase.[518,519] The matter is of considerable practical importance because of the spread of warfarin-resistant rats in Europe and the United States. One resistance mutation has altered the vitamin K epoxide reductase so that it is much less susceptible to inhibition by warfarin.[519,520]

While glutamate residues in peptides of appropriate sequence are carboxylated by the vitamin K-dependent system, aspartate peptides scarcely react.[503] Beta-carboxyaspartate is present in protein C of the blood anticoagulant system (Fig. 12-17)[521] and in various other proteins containing EGF homology domains (Table 7-3),[522] but the mechanism of its formation is unknown.

6. Tocopherols (Vitamin E) as Antioxidants

The major function of the tocopherols is thought to be the protection of phospholipids of cell membranes against oxidative attack by free radicals and organic peroxides. Peroxidation of lipids, which is described in Chapter 21, can lead to rapid development of rancidity in fats and oils. However, the presence of a small amount of tocopherol inhibits this decomposition, presumably by trapping the intermediate radicals in the form of the more stable tocopherol radicals (Eq. 15-54) which may dimerize or react with other radicals to terminate the chain. Even though only one molecule of tocopherol is present for a thousand molecules of phospholipid, it is enough to protect membranes.[523] That vitamin E does function in this way is supported by the observation that much of the tocopherol requirement of some species can be replaced by *N,N'*-diphenyl-*p*-phenylenediamine, a synthetic antioxidant (see Table 18-5 for the structure of a related substance). Three generations of rats have been raised on a tocopherol-free diet containing this synthetic antioxidant. However, not all of the deficiency symptoms are prevented.

The antioxidant role of α-tocopherol in membranes is generally accepted.[524–526] It is thought to be critical to defense against oxidative injury and to help the body combat the development of tumors and to slow aging. Gamma-tocopherol may be more reactive than α-tocopherol in removing radicals created by NO and other nitrogen oxides.[526] Its actions are strongly linked to those of ascorbic acid (Box 18-D) and selenium. Ascorbate may reduce tocopherol semiquinone radicals, while selenium acts to enhance breakdown of peroxides as described in the next section.

G. Selenium-Containing Enzymes

In 1957, Schwartz and associates showed that the toxic element selenium was also a nutritional factor essential for prevention of the death of liver cells in rats.[527] Liver necrosis would be prevented by as little as 0.1 ppm of selenium in the diet. Similar amounts of selenium were shown to prevent a muscular dystrophy called "white muscle disease" in cattle and sheep grazing on selenium-deficient soil. Sodium selenite and other inorganic selenium compounds were more effective than organic compounds in which Se had replaced sulfur. **Keshan disease**, an often fatal heart condition that is prevalent among childen in Se-deficient regions of China, can be prevented by supplementation of the diet with $NaSeO_3$.[528] Even the little crustacean "water flea" *Daphnia* needs 0.1 part per billion of Se in its water.[529]

Selenium has long been known to enhance the antioxidant activity of vitamin E. Recent work suggests that vitamin E acts as a radical scavenger, preventing

BOX 15-G VITAMIN E: THE TOCOPHEROLS

Vitamin E was recognized in 1926 as a factor preventing sterility in rats that had been fed rancid lipids.[a–e] The curative factor, present in high concentration in wheat germ and lettuce seed oils, is a family of vitamin E compounds, the tocopherols (Fig. 15-24). The first of these was isolated by Evans and associates in 1936. Vitamin E deficiency in the rabbit or rat is accompanied by muscular degeneration (**nutritional muscular dystrophy**; see also Box 15-A) and a variety of other symptoms that vary from one species to another. Animals deficient in vitamin E display obvious physical deterioration followed by sudden death. Muscles of deficient rats show abnormally high rates of oxygen uptake, and abnormalities appear in the membranes of the endoplasmic reticulum as viewed with the electron microscope. It is thought that deterioration of lysosomal membranes may be the immediate cause of death.

The tocopherol requirement of humans is not known with certainty, but about 5 mg (7.5 IU) / day plus an additional 0.6 mg for each gram of polyunsaturated fatty acid consumed may be adequate. It is estimated that the average daily intake is about 14 mg, but the increasing use of highly refined foods may lead to dangerously low consumption. Recent interest[d–f] has been aroused by studies that show that much larger amounts of vitamin E (e.g., 100–400 mg/day) substantially reduce the risk of coronary disease and stroke in both women[g] and men[h] and also decrease oxidative modification of brain proteins.[i] The decrease in heart attacks and stroke may be in part an indirect effect of the anticlotting action of vitamin E quinone.[f] Plant oils are usually the richest sources of tocopherols, while animal products contain lower quantities.

To some extent the vitamin E requirement may be lessened by the presence in the diet of synthetic antioxidants and by selenium. Much evidence supports a relationship between the nutritional need for selenium and that for vitamin E. Lack of either causes muscular dystrophy in many animals as well as severe edema (exudative diathesis) in chicks. Since vitamin E-deficient rats have a low selenide (Se^{2-}) content, it has been suggested that vitamin E protects reduced selenium from oxidation.[j] Vitamin C (ascorbic acid), in turn, protects vitamin E.

[a] Sebrell, W. H., Jr., and Harris, R. S., eds. (1972) *The Vitamins*, Vol. 5, Academic Press, New York
[b] DeLuca, H. F., and Suttie, J. W., eds. (1970) *The Fat-Soluble Vitamins*, Univ. of Wisconsin Press, Madison, Wisconsin
[c] Machlin, L. J., ed. (1980) *Vitamin E*, Dekker, New York
[d] Mino, M., Nakamura, H., Diplock, A. T., and Kayden, H. J., eds. (1993) *Vitamin E: Its Usefulness in Health and in Curing Diseases*, Japan Scientific Socities Press, Tokyo
[e] Packer, L., and Fuchs, J., eds. (1993) *Vitamin E in Health and Disease*, Dekker, New York
[f] Dowd, P., and Zhend, Z. B. (1995) *Proc. Natl. Acad. Sci. U.S.A.* **92**, 8171–8175
[g] Stampfer, M. J., Hennekens, C. H., Manson, J. E., Colditz, G. A., Rosner, B., and Willett, W. C. (1993) *N. Engl. J. Med.* **328**, 1444–1449
[h] Rimm, E. B., Stampfer, M. J., Ascherio, A., Giovannucci, E., Colditz, G. A., and Willett, W. C. (1993) *N. Engl. J. Med.* **328**, 1450–1456
[i] Poulin, J. E., Cover, C., Gustafson, M. R., and Kay, M. M. B. (1996) *Proc. Natl. Acad. Sci. U.S.A.* **93**, 5600–5603
[j] Diplock, A. T., and Lucy, J. A. (1973) *FEBS Lett.* **29**, 205–210

excessive peroxidation of membrane lipids, while selenium, in the enzyme **glutathione peroxidase**, acts to destroy the small amounts of peroxides that do form. This was the first established function of selenium in human beings, but there are others. If we include proteins from animals and bacteria, at least ten selenoproteins are known (Table 15-4).[530–534] Seven of these are enzymes and most catalyze redox processes. The active sites most often contain **selenocysteine**, whose selenol side chain is more acidic (pK_a ~5.2) than that of cysteine and exists as $-CH_2-Se^-$ at neutral pH.[530]

Glutathione peroxidases catalyze the reductive decomposition of H_2O_2 or of organic peroxides by glutathione (G–SH) according to Eq. 15-58. At least three isoenzyme forms have been identified in mammals: a cellular form,[531,535–537] a plasma form, and a

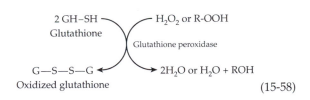

$$(15\text{-}58)$$

form with a preference for organic peroxides derived from phospholipids.[327,538–540] A related selenoprotein has been found in a human poxvirus.[540a] Selenocysteine is present at a position (residues 41–47) near the N terminus of an α helix, in the ~ 180-residue polypeptides. A possible reaction mechanism involves attack by the selenol on the peroxide to give a selenic acid intermediate which is reduced by glutathione in two nucleophilic displacement steps (Eq. 15-59).

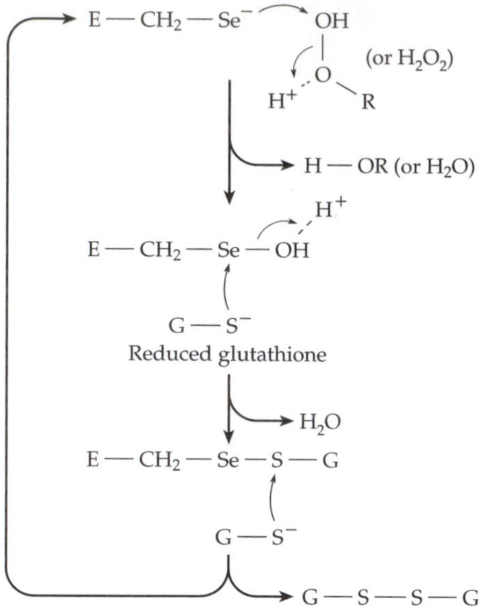

$$(15-59)$$

Three types of **iodothyronine deiodinase** remove iodine atoms from thyroxine to form the active thyroid hormone triiodothyronine and also to inactivate the hormone by removing additional iodine[531,541–546] (see also Chapter 25). In this case the $-CH_2-Se^-$ may attach the iodine atom, removing it as I^+ to form $-CH_2-Se-I$. The process could be assisted by the phenolic $-OH$ group if it were first tautomerized (Eq. 15-60).

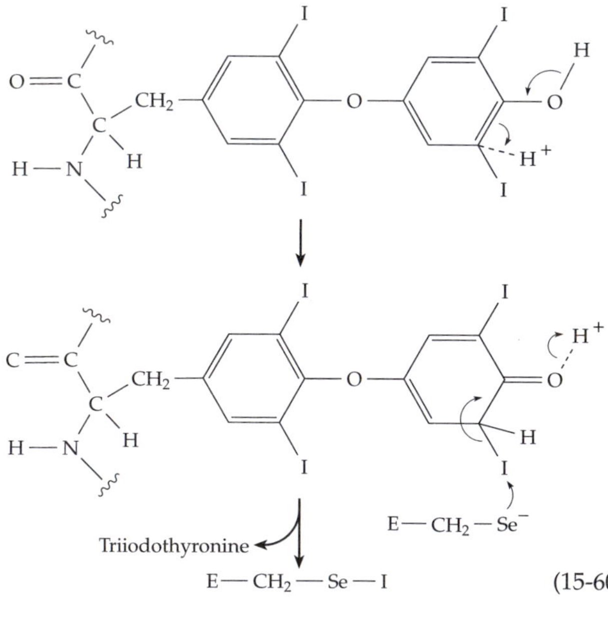

$$(15-60)$$

A recently discovered human selenoprotein is a **thioredoxin reductase** which is present in the T cells of the immune system as well as in placenta and other tissues.[189,547–549] The 55-kDa protein has one seleno-cysteine as the penultimate C-terminal residue. Another mammalian selenoprotein, of uncertain function, is the 57-kDa **selenoprotein P**. It contains over 60% of the

selenium in rat plasma and is also present in the human body. Selenoprotein P contains ten selenocysteine residues.[550–552a] Some of these may be replaced by serine in a fraction of the molecules.[553] A smaller 9.6-kDa skeletal muscle protein, **selenoprotein W**, contains a single selenocysteine.[554–556] Another seleno-protein has been found in sperm cells, both in the tail and in a keratin-rich capsule that surrounds the mito-chondria in the sperm midpiece.[557] Lack of this protein may be the cause of the abnormal immotile spermatozoa observed in Se-deficient rats and of reproductive diffi-culties among farm animals in Se-deficient regions.[558]

Several selenoproteins have been found in certain bacteria and archaea. A **hydrogenase** from *Methano-coccus vannielii* contains selenocysteine.[559,560] This enzyme transfers electrons from H_2 to the C-5 *si* face of the 8-hydroxy-5-deazaflavin cofactor F_{420} (Section B,4). The same bacterium synthesizes two **formate dehydrogenases** (see Fig 15-23), one of which con-tains Se. Two Se-containing formate dehydrogenases are made by *E. coli*. One of them, which is coupled to a hydrogenase in the formate hydrogen-lyase system (see Eq. 15-37), is a 715-residue protein containing selenocysteine at position 140.[561–563] The second has selenocysteine at position 196 and functions with a nitrate reductase in anaerobic nitrate respiration.[561]

Glycine reductase is a complex enzyme[530,564–566] that catalyzes the reductive cleavage of glycine to acetyl phosphate and ammonia (Eq. 15-61) with the

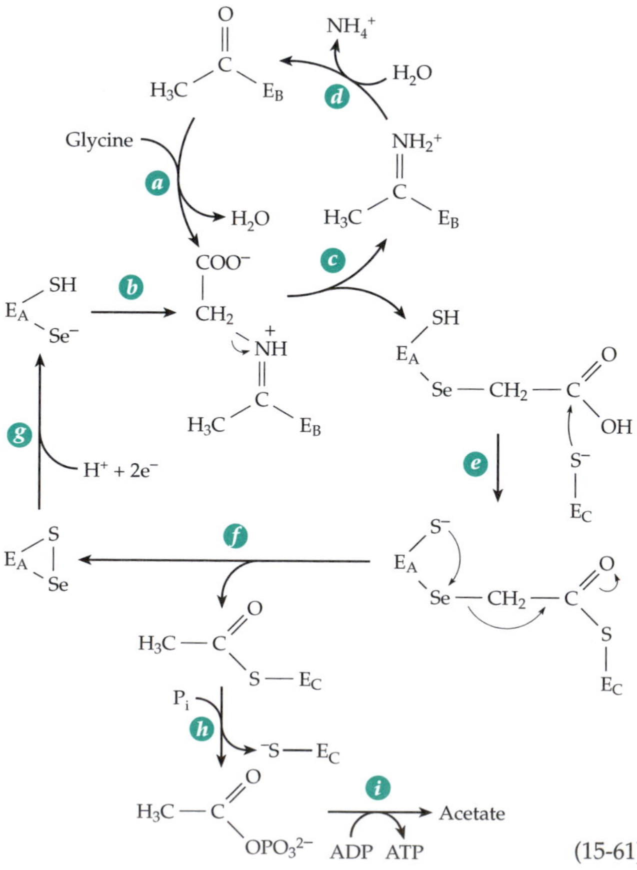

$$(15-61)$$

subsequent synthesis of ATP (Eq. 14-43). Electrons for reduction of the disulfide that is formed are provided by NADH. A single selenocysteine residue is present in the small 12-kDa subunits. The enzyme contains a dehydroalanine residue (Chapter 14) in subunit B and a thiol group in subunit C.[566] An acetyl-enzyme derivative of subunit C, perhaps of its –SH group, has been identified.[546] The mechanism of action is uncertain but the steps in Eq. 15-61 have been suggested.[567] The subunits are designated E_A, E_B, and E_C. Step *e* is particularly hard to understand because formation of a thioester in this manner is not expected to occur spontaneously and must be linked in some way to other steps.

A selenium-containing **xanthine dehydrogenase** is present in purine-fermenting clostridia. Like other xanthine dehydrogenases (Chapter 16), it converts xanthine to uric acid and contains nonheme iron, molybdenum, FAD, and an Fe–S center. The selenium is probably present as Se^{2-} bound to Mo as is S^{2-} in xanthine oxidase (Fig. 16-32).[567a] A related reaction is catalyzed by the Fe–S protein **nicotinic acid hydroxylase** (Eq. 15-62) found in some clostridia.[568] Splitting of the Mo(V) EPR signal when [77]Se is present in the enzyme shows that the selenium is present as a ligand of molybdenum. Another member of the family is a purine hydroxylase that converts purine 2-hydroxypurine, or hypoxanthine to xanthine.[567a]

TABLE 15-4
Selenium-Containing Proteins

Enzyme	Source	Mass (kDa)	Subunit composition	Other cofactors
Glutathione peroxidases				None
Cellular	Mammals	21 × 4		
Plasma	Humans			
Phospholipid hydroperoxide	Pig, rat	18		
Iodothyronine deiodinases	Vertebrates			
Thioredoxin reductase	Humans	55 × 2		
Selenoprotein *P*	Mammals	57		None
Selenoprotein *W*	Rat	19.6		
Formate dehydrogenase	Bacteria - *E. coli*	600	$\alpha_2\beta_4\gamma_{2-4}$	Heme *b*
	Archaea			Mo; molybdopterin
Hydrogenase	Bacteria		$\alpha_2\beta_4\gamma_2$	FAD; NiFeSe
	Archaea			
Glycine reductase	Some clostridia		ABC	
Selenoprotein A		12		
Selenoprotein B		200		
Carbonyl protein C		250		Fe?
Nicotinic acid hydroxylase	Clostridia	300		FAD, FeS, Mo
Purine hydroxylase	Clostridia			FAD, FeS, Mo
Thiolase (contains				
selenomethionine)	*Clostridium kluyverii*	39 × 4		None
Carbon monoxide dehydrogenase	*Oligotropha carboxidovorans*	137 × 2	$\alpha_2\beta_2\gamma_2$	FAD; Mo; molybdopterin

BOX 15-H GLUTATHIONE PEROXIDASE AND ABNORMALITIES OF RED BLOOD CELLS

The processes by which hemoglobin is kept in the Fe(II) state and functioning normally within intact erythrocytes is vital to our health. Numerous hereditary defects leading to a tendency toward anemia have helped to unravel the biochemistry indicated in the accompanying scheme.[a]

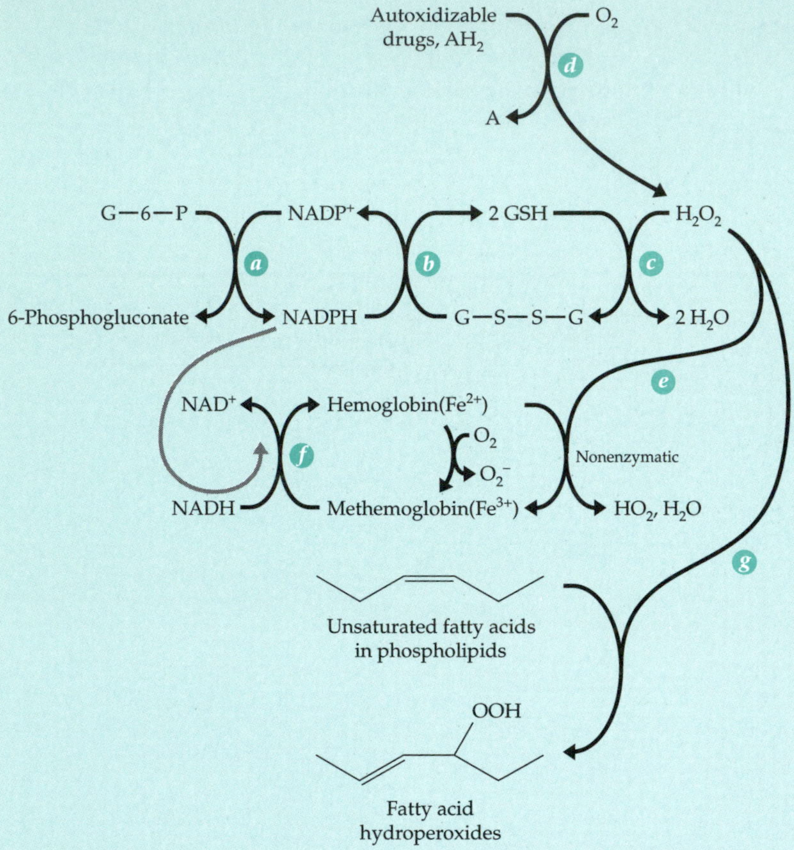

About 90% of the glucose utilized by erythrocytes is converted by glycolysis to lactate, but about 10% is oxidized (via glucose 6-phosphate) to 6-phosphogluconate. The oxidation (reaction a) is catalyzed by glucose-6-phosphate dehydrogenase (Eq.15-10) using $NADP^+$. This is the principal reaction providing the red cell with NADPH for reduction of glutathione (Box 11-B) according to reaction b. Despite the important function of glucose-6-P dehydrogenase, ~400 million persons, principally in tropical and Mediterranean areas, have a hereditary deficiency of this enzyme. The genetic variations are numerous, with about 400 different ones having been identified. Although most individuals with this dificiency have no symptoms, the lack of the enzyme is truly detrimental and sometimes leads to excessive destruction of red cells and anemia during some sicknesses and in response to administration of various drugs.[b] The survival of the defective genes, like that for sickle cell hemoglobin (Box 7-B) is

thought to result from increased resistance to malaria parasites.

Other erythrocyte defects that lead to drug sensitivity include a deficiency of glutathione (resulting from a decrease in its synthesis) and a deficiency of glutathione reductase (reaction b). The effects of drugs have been traced to the production of H_2O_2 (reaction c) in red blood cells; catalase, which converts H_2O_2 into H_2O and O_2, is thought to function in a similar way. Both enzymes are probably necessary for optimal health.

An excess of H_2O_2 can damage erythrocytes in two ways. It can cause excessive oxidation of functioning hemoglobin to the Fe(III)-containing methemoglobin. (Methemoglobin is also formed spontaneously during the course of the oxygen-carrying function of hemoglobin. It is estimated that normally as much as 3% of the hemoglobin may be oxidized to methemoglobin daily.) The methemoglobin formed is reduced back to hemoglobin through the action of **NADH-methemoglobin reductase** (reaction f). A smaller fraction of the methemoglobin is reduced by a similar enzyme requiring NADPH (as indicated by the colored arrow). A hereditary lack of the NADH-methemoglobin reductase is also known.

A second destructive function of H_2O_2 is attack on double bonds of unsaturated fatty acids of the phospholipids in cell membranes. The resulting fatty acid hydroperoxides can react further with C–C chain cleavage and disruption of the membrane. This is thought to be the principal cause of the hemolytic anemia induced by drugs in susceptible individuals. The selenium-containing glutathione peroxidase is thought to decompose these fatty acid hydroperoxides. Vitamin E (Box 15-G), acting as an antioxidant within membranes, is also needed for good health of erythrocytes.[c,d]

[a] Chanarin, I. (1970) in *Biochemical Disorders in Human Disease*, 3rd ed. (Thompson, R. H. S., and Wooton, I. D. P., eds), pp. 163–173, Academic Press, New York

[b] Luzzatto, L., and Mehta, A. (1995) in *The Metabolic and Molecular Bases of Inherited Disease*, 7th ed., Vol. 1 (Scriver, C. R., Beaudet, A. L., Sly, W. S., and Valle, D., eds), pp. 3367–3398, McGraw-Hill, New York

[c] Constantinescu, A., Han, D., and Packer, L. (1993) *J. Biol. Chem.* **268**, 10906–10913

[d] Liebler, D. C., and Burr, J. A. (1992) *Biochemistry* **31**, 8278–8284

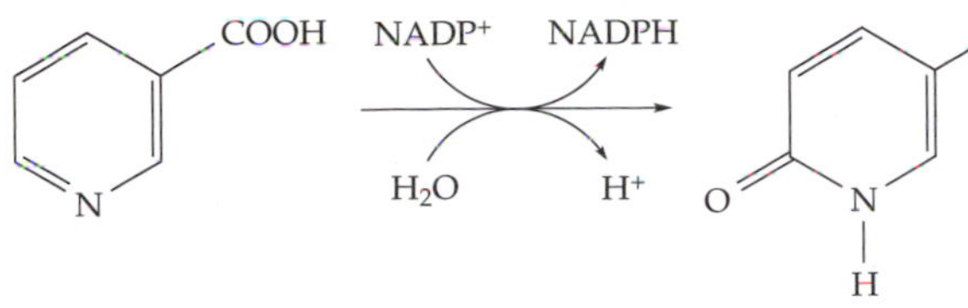

$$\text{(15-62)}$$

A **thiolase** (Eq. 13-35) from *Clostridium kluyveri* is one of only two known selenoproteins that contain selenomethionine.[569] However, the selenomethionine is incorporated randomly in place of methionine. This occurs in all proteins of all organisms to some extent and the toxicity of selenium may result in part from excessive incorporation of selenomethionine into various proteins.

Selenium is found to a minor extent wherever sulfur exists in nature. This includes the sulfur-containing modified bases of tRNA molecules. In addition to a small amount of nonspecific incorporation of Se into all S-containing bases there are, at least in bacteria, specific Se-containing tRNAs. In *E. coli* one of these is specific for lysine and one for glutamate. One of the modified bases has been identified as 5-methyl-amino-methyl-2-selenouridine.[570] It is present at the first position of the anticodon, the "wobble" position.[571]

Selenium has its own metabolism. Through the use of ^{75}Se as a tracer, normal rat liver has been shown to contain Se^{2-}, SeO_3^{2-}, and selenium in a higher oxidation state.[572] Glutathione may be involved in reduction of selenite to selenide.[573] The nonenzymatic reduction of selenite by glutathione yields a selenotrisulfide derivative (Eq. 15-63). The latter is spontaneously decomposed to oxidized glutathione and elemental selenium or by the action of glutathione reductase to glutathione and selenium. Selenocysteine can be converted to alanine + elemental selenium (Eq. 14-34). Some bacteria are able to oxidize elemental Se back to selenite.[574] Selenium undergoes biological methylation readily in bacteria, fungi, plants, and animals (Chapter 16).[575,576] This may in some way be related to the reported effect of selenium in protecting animals against the toxicity of mercury. Excess selenium may appear in the urine as trimethylselenonium ions.[577]

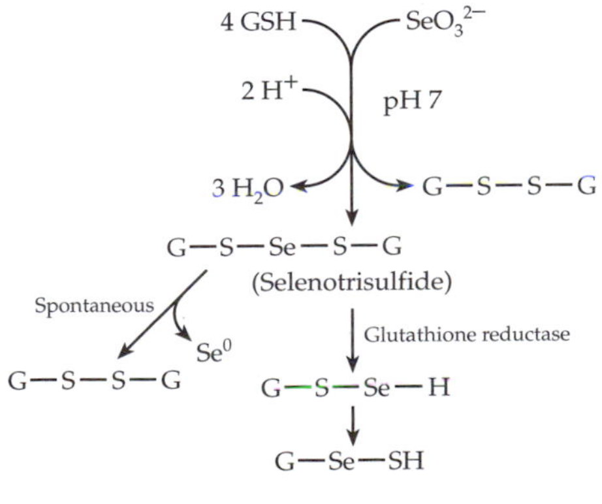

How is selenium incorporated into selenocysteine-containing proteins? This element does enter amino acids to a limited extent via the standard synthetic pathways for cysteine and methionine. However, the placement of selenocysteine into specific positions in selenoproteins occurs by the use of a minor serine-specific tRNA that acts as a suppressor of chain termination during protein synthesis.[532,533,578] (This topic is dealt with further in Chapter 29.) The genes for these and presumably for other selenocysteine-containing proteins have the "stop" codon TGA at the selenocysteine positions. However, when present in a suitable "context" the minor tRNA, carrying selenocysteine in place of serine, is utilized to place selenocysteine into the growing peptide chain. In bacteria, and presumably also in eukaryotes, the selenocysteinyl-tRNA is formed from the corresponding seryl-tRNA by a PLP-catalyzed β-replacement reaction. The selenium donor is not Se^{2-} but selenophosphate $Se-PO_3^{2-}$ in which the Se–P bond is quite weak.[579–581] After addition to the aminoacrylate intermediate in the PLP enzyme the Se–P bond may be hydrolytically cleaved to HPO_4^{2-} and selenocysteyl-tRNA.

$$\text{(15-63)}$$

References

1. Cornish-Bowden, A., ed. (1997) *New Beer in an Old Bottle (Eduard Buchner and the Growth of Biochemical Knowledge)*, Valencia, Universitat de València

1a. Kalckar, H. M. (1969) *Biological Phosphorylations*, Prentice-Hall, Englewood Cliffs, New Jersey (pp. 86–97)

2. Rossmann, M. G., Adams, M. J., Buehner, M., Ford, G. C., Hackert, M. L., Lentz, P. J., Jr., McPherson, A., Jr., Schevitz, R. W., and Smiley, I. E. (1971) *Cold Spring Harbor Symposia on Quant. Biol.* **36**, 179–191

3. Abad-Zapatero, C., Griffith, J. P., Sussman, J. L., and Rossmann, M. G. (1987) *J. Mol. Biol.* **198**, 445–467

4. Rossmann, M. (1974) *New Scientist* **61**, 266–268

5. Eventoff, W., Rossmann, M. G., Taylor, S. S., Torff, H.-J., Meyer, H., Keil, W., and Kiltz, H.-H. (1977) *Proc. Natl. Acad. Sci. U.S.A.* **74**, 2677–2681

6. Hogrefe, H. H., Griffith, J. P., Rossmann, M. G., and Goldberg, E. (1987) *J. Biol. Chem.* **262**, 13155–13162

7. Iwata, S., and Ohta, T. (1993) *J. Mol. Biol.* **230**, 21–27

8. Ostendorp, R., Auerbach, G., and Jaenicke, R. (1996) *Protein Sci.* **5**, 862–873

9. Birktoft, J. J., Rhodes, G., and Banaszak, L. J. (1989) *Biochemistry* **28**, 6065–6081

10. Gleason, W. B., Fu, Z., Birktoft, J., and Banaszak, L. (1994) *Biochemistry* **33**, 2078–2088

11. Goward, C. R., and Nicholls, D. J. (1994) *Protein Sci.* **3**, 1883–1888

12. Hall, M. D., Levitt, D. G., and Banaszak, L. J. (1992) *J. Mol. Biol.* **226**, 867–882

13. Kelly, C. A., Nishiyama, M., Ohnishi, Y., Beppu, T., and Birktoft, J. J. (1993) *Biochemistry* **32**, 3913–3922

14. Bellamacina, C. R. (1996) *FASEB J.* **10**, 1257–1269

15. Pullman, M. E., and Colowick, S. P. (1954) *J. Biol. Chem.* **206**, 129–141

16. Fisher, H. F., Conn, E. E., Vennesland, B., and Westheimer, F. H. (1953) *J. Biol. Chem.* **202**, 687–697

17. Westheiner, F. H. (1987) in *Pyridine Nucleotide Coenzymes: Chemical, Biochemical and Medical Aspects*, Vol. A (Dolphin, D., Avramovic, O., and Poulson, R., eds), pp. 253–322, Wiley (Interscience), New York

18. You, K.-s, Arnold, L. J., Jr., Allison, W. S., and Kaplan, N. O. (1978) *Trends Biochem. Sci.* **3**, 265–268

19. Nakajima, N., Nakamura, K., Esaki, N., Tanaka, H., and Soda, K. (1989) *J. Biochem.* **106**, 515–517

20. Esaki, N., Shimoi, H., Nakajima, N., Ohshima, T., Tanaka, H., and Soda, K. (1989) *J. Biol. Chem.* **264**, 9750–9752

21. Anderson, V. E., and LaReau, R. D. (1988) *J. Am. Chem. Soc.* **110**, 3695–3697

22. Almarsson, Ö., and Bruice, T. C. (1993) *J. Am. Chem. Soc.* **115**, 2125–2138

23. Oppenheimer, N. J. (1986) *J. Biol. Chem.* **261**, 12209–12212

24. Benner, S. A., Nambiar, K. P., and Chambers, G. K. (1985) *J. Am. Chem. Soc.* **107**, 5513–5517

25. Young, L., and Post, C. B. (1993) *J. Am. Chem. Soc.* **115**, 1964–1970

26. Almarsson, Ö, Karaman, R., and Bruice, T. C. (1992) *J. Am. Chem. Soc.* **114**, 8702–8704

27. LaReau, R. D., and Anderson, V. E. (1992) *Biochemistry* **31**, 4174–4180

28. Ohno, A., Tsutsumi, A., Kawai, Y., Yamazaki, N., Mikata, Y., and Okamura, M. (1994) *J. Am. Chem. Soc.* **116**, 8133–8137

29. Kallwass, H. K. W., Hogan, J. K., Macfarlane, E. L. A., Martichonok, V., Parris, W., Kay, C. M., Gold, M., and Jones, J. B. (1992) *J. Am. Chem. Soc.* **114**, 10704–10710

30. Klinman, J. P. (1972) *J. Biol. Chem.* **247**, 7977–7987

31. Ostovic, D., Roberts, R. M. G., and Kreevoy, M. M. (1983) *J. Am. Chem. Soc.* **105**, 7629–7631

32. Coleman, C. A., Rose, J. G., and Murray, C. J. (1992) *J. Am. Chem. Soc.* **114**, 9755–9762

33. Kong, Y. S., and Warshel, A. (1995) *J. Am. Chem. Soc.* **117**, 6234–6242

34. Lee, I.-S. H., Jeoung, E. H., and Kreevoy, M. M. (1997) *J. Am. Chem. Soc.* **119**, 2722–2728

35. Rucker, J., Cha, Y., Jonsson, T., Grant, K. L., and Klinman, J. P. (1992) *Biochemistry* **31**, 11489–11499

36. Huskey, W. P., and Schowen, R. L. (1983) *J. Am. Chem. Soc.* **105**, 5704–5706

36a. Kohen, A., Cannio, R., Bartolucci, S., and Klinman, J. P. (1999) *Nature (London)* **399**, 496–499

36b. Chin, J. K., and Klinman, J. P. (2000) *Biochemistry* **39**, 1278–1284

36c. Northrop, D. B. (1999) *J. Am. Chem. Soc.* **121**, 3521–3524

36d. Karsten, W. E., Hwang, C.-C., and Cook, P. F. (1999) *Biochemistry* **38**, 4398–4402

37. Grau, U. M., Trommer, W. E., and Rossmann, M. G. (1981) *J. Mol. Biol.* **151**, 289–307

38. Li, H., Hallows, W. H., Punzi, J. S., Pankiewicz, K. W., Watanabe, K. A., and Goldstein, B. M. (1994) *Biochemistry* **33**, 11734–11744

39. Gawlita, E., Paneth, P., and Anderson, V. E. (1995) *Biochemistry* **34**, 6050–6058

40. Clarke, A. R., Wilks, H. M., Barstow, D. A., Atkinson, T., Chia, W. N., and Holbrook, J. J. (1988) *Biochemistry* **27**, 1617–1622

41. Clarke, A. R., Atkinson, T., and Holbrook, J. J. (1989) *Trends Biochem. Sci.* **14**, 101–105

42. Clarke, A. R., Atkinson, T., and Holbrook, J. J. (1989) *Trends Biochem. Sci.* **14**, 145–148

43. Goldberg, J. D., Yoshida, T., and Brick, P. (1994) *J. Mol. Biol.* **236**, 1123–1140

44. Cortes, A., Emery, D. C., Halsall, D. J., Jackson, R. M., Clarke, A. R., and Holbrook, J. J. (1992) *Protein Sci.* **1**, 892–901

45. Clarke, A. R., Wigley, D. B., Chia, W. N., Barstow, D. A., Atkinson, T., and Holbrook, J. J. (1986) *Nature (London)* **324**, 699–702

46. Deng, H., Burgner, J., and Callender, R. (1992) *J. Am. Chem. Soc.* **114**, 7997–8003

47. Grau, W. M., Trommer, W. E., and Rossmann, M. G. (1981) *J. Mol. Biol.* **151**, 289–307

48. Pettersson, G. (1987) *CRC Critical Review of Biochemistry* **21**, 349–389

49. Maret, W., and Makinen, M. W. (1991) *J. Biol. Chem.* **266**, 20636–20644

50. Eklund, H., and Brändén, C.-I. (1987) in *Pyridine Nucleotide Coenzyme: Chemical, Biochemical and Medical Aspects*, Vol. 2, Part A (Dolphin, D., Poulson, R., and Avramovic, O., eds), pp. 51–98, Wiley, New York

51. Adolph, H. W., Kiefer, M., and Cedergren-Zeppezauer, E. (1997) *Biochemistry* **36**, 8743–8754

51a. Colby, T. D., Bahnson, B. J., Chin, J. K., Klinman, J. P., and Goldstein, B. M. (1998) *Biochemistry* **37**, 9295–9304

51b. Plapp, B. V., Eklund, H., and Brändén, C.-I. (1978) *J. Mol. Biol.* **122**, 23–32

52. Shearer, G. L., Kim, K., Lee, K. M., Wang, C. K., and Plapp, B. V. (1993) *Biochemistry* **32**, 11186–11194

53. Ramaswamy, S., Eklund, H., and Plapp, B. V. (1994) *Biochemistry* **33**, 5230–5237

54. Dunn, M. F., Dietrich, H., MacGibbon, A. K. H., Koerber, S. C., and Zeppezauer, M. (1982) *Biochemistry* **21**, 354–363

55. Jagodzinski, P. W., and Peticolas, W. L. (1981) *J. Am. Chem. Soc.* **103**, 234–236

56. MacGibbon, A. K. H., Koerber, S. C., Pease, K., and Dunn, M. F. (1987) *Biochemistry* **26**, 3058–3067

57. Hennecke, M., and Plapp, B. V. (1983) *Biochemistry* **22**, 3721–3728

58. Plapp, B. V. (1995) *Methods Enzymol.* **249**, 91–119

59. Makinen, M. W., Maret, W., and Yim, M. B. (1983) *Proc. Natl. Acad. Sci. U.S.A.* **80**, 2584–2588

60. Ryde, U. (1995) *Protein Sci.* **4**, 1124–1132

61. Ehrig, T., Muhoberac, B. B., Brems, D., and Bosron, W. F. (1993) *J. Biol. Chem.* **268**, 11721–11726

62. Hurley, T. D., Bosron, W. F., Stone, C. L., and Amzel, L. M. (1994) *J. Mol. Biol.* **239**, 415–429

63. Charlier, H. A., Jr., and Plapp, B. V. (2000) *J. Biol. Chem.* **275**, 11569–11575

64. Xie, P., Parsons, S. H., Speckhard, D. C., Bosron, W. F., and Hurley, T. D. (1997) *J. Biol. Chem.* **272**, 18558–18563

65. Cho, H., Ramaswamy, S., and Plapp, B. V. (1997) *Biochemistry* **36**, 382–389

66. Ganzhorn, A. J., Green, D. W., Hershey, A. D., Gould, R. M., and Plapp, B. V. (1987) *J. Biol. Chem.* **262**, 3754–3761

67. Hake, S., Kelley, P. M., Taylor, W. C., and Freeling, M. (1985) *J. Biol. Chem.* **260**, 5050–5054

67a. Korkhin, Y., Kalb-Gilboa, A. J., Peretz, M., Bogin, O., Burstein, Y., and Frolow, F. (1998) *J. Mol. Biol.* **278**, 967–981

68. Jörnvall, H., Persson, B., Krook, M., Atrian, S., Gonzàlez-Duarte, R., Jeffery, J., and Ghosh, D. (1995) *Biochemistry* **34**, 6003–6013

69. Ribas de Pouplana, L., and Fothergill-Gilmore, L. A. (1994) *Biochemistry* **33**, 7047–7055

69a. Benach, J., Atrian, S., Gonzàlez-Duarte, R., and Ladenstein, R. (1999) *J. Mol. Biol.* **289**, 335–355

70. Hawes, J. W., Crabb, D. W., Chan, R. M., Rougraff, P. M., and Harris, R. A. (1995) *Biochemistry* **34**, 4231–4237

71. Pawlowski, J. E., and Penning, T. M. (1994) *J. Biol. Chem.* **269**, 13502–13510

72. Wilson, D. K., Nakano, T., Petrash, J. M., and Quiocho, F. A. (1995) *Biochemistry* **34**, 14323–14330

73. Hoog, S. S., Pawlowski, J. E., Alzari, P. M., Penning, T. M., and Lewis, M. (1994) *Proc. Natl. Acad. Sci. U.S.A.* **91**, 2517–2521

74. Wilson, D. K., Bohren, K. M., Gabbay, K. H., and Quiocho, F. A. (1992) *Science* **257**, 81–84

75. Rondeau, J.-M., Tête-Favier, F., Podjarny, A., Reymann, J.-M., Barth, P., Biellmann, J.-F., and Moras, D. (1992) *Nature (London)* **355**, 469–471

76. Tarle, I., Borhani, D. W., Wilson, D. K., Quiocho, F. A., and Petrash, J. M. (1993) *J. Biol. Chem.* **268**, 25687–25693

77. Grimshaw, C. E., Bohren, K. M., Lai, C.-J., and Gabbay, K. H. (1995) *Biochemistry* **34**, 14374–14384

78. Grimshaw, C. E. (1992) *Biochemistry* **31**, 10139–10145

79. Harrison, D. H., Bohren, K. M., Ringe, D., Petsko, G. A., and Gabbay, K. H. (1994) *Biochemistry* **33**, 2011–2020

80. Maniscalco, S. J., Saha, S. K., Vicedomine, P., and Fisher, H. F. (1996) *Biochemistry* **35**, 89–94

81. Saha, S. K., Maniscalco, S. J., and Fisher, H. F. (1996) *Biochemistry* **35**, 16483–16488

82. Smith, M. T., and Emerich, D. W. (1993) *J. Biol. Chem.* **268**, 10746–10753

83. Delforge, D., Devreese, B., Dieu, M., Delaive, E., Van Beeumen, J., and Remacle, J. (1997) *J. Biol. Chem.* **272**, 2276–2284

84. Sekimoto, T., Fukui, T., and Tanizawa, K. (1994) *J. Biol. Chem.* **269**, 7262–7266

85. Brunhuber, N. M. W., Banerjee, A., Jacobs, W. R., Jr., and Blanchard, J. S. (1994) *J. Biol. Chem.* **269**, 16203–16211

86. Britton, K. L., Baker, P. J., Engel, P. C., Rice, D. W., and Stillman, T. J. (1993) *J. Mol. Biol.* **234**, 938–945

87. Scapin, G., Blanchard, J. S., and Sacchettini, J. C. (1995) *Biochemistry* **34**, 3502–3512

88. Moras, D., Olsen, K. W., Sabesan, M. N., Buehner, M., Ford, G. C., and Rossmann, M. G. (1975) *J. Biol. Chem.* **250**, 9137–9162

89. Duée, E., Olivier-Deyris, L., Fanchon, E., Corbier, C., Branlant, G., and Dideberg, O. (1996) *J. Mol. Biol.* **257**, 814–838

90. Skarzynski, T., Moody, P. C. E., and Wonacott, A. J. (1987) *J. Mol. Biol.* **193**, 171–187

90a. Roitel, O., Sergienko, E., and Branlant, G. (1999) *Biochemistry* **38**, 16084–16091

91. Kim, H., and Hol, W. G. J. (1998) *J. Mol. Biol.* **278**, 5–11

92. Soukri, A., Mougin, A., Corbier, C., Wonacott, A., Branlant, C., and Branlant, G. (1989) *Biochemistry* **28**, 2568–2592

93. Habenicht, A., Hellman, U., and Cerff, R. (1994) *J. Mol. Biol.* **237**, 165–171

93a. Marchal, S., Rahuel-Clermont, S., and Branlant, G. (2000) *Biochemistry* **39**, 3327–3335

94. Sheikh, S., Ni, L., Hurley, T. D., and Weiner, H. (1997) *J. Biol. Chem.* **272**, 18817–18822

95. Steinmetz, C. G., Xie, P., Weiner, H., and Hurley, T. D. (1997) *Structure* **5**, 701–711

96. Henehan, G. T. M., and Oppenheimer, N. J. (1993) *Biochemistry* **32**, 735–738

97. Henehan, G. T. M., Chang, S. H., and Oppenheimer, N. J. (1995) *Biochemistry* **34**, 12294–12301

98. Shearer, G. L., Kim, K., Lee, K. M., Wang, C. K., and Plapp, B. V. (1993) *Biochemistry* **32**, 11186–11194

99. Feingold, D. S., and Franzen, J. S. (1981) *Trends Biochem. Sci.* **6**, 103–105

100. Frimpong, K., and Rodwell, V. W. (1994) *J. Biol. Chem.* **269**, 11478–11483

101. Omkumar, R. V., and Rodwell, V. W. (1994) *J. Biol. Chem.* **269**, 16862–16866

102. Friesen, J. A., and Rodwell, V. W. (1997) *Biochemistry* **36**, 1157–1162

103. Lawrence, C. M., Rodwell, V. W., and Stauffacher, C. V. (1995) *Science* **268**, 1758–1762

104. Friesen, J. A., Lawrence, C. M., Stauffacher, C. V., and Rodwell, V. W. (1996) *Biochemistry* **35**, 11945–11950

104a. Cosgrove, M. S., Naylor, C., Paludan, S., Adams, M. J., and Levy, H. R. (1998) *Biochemistry* **37**, 2759–2767

105. Luzzatto, L., and Mehta, A. (1995) in *The Metabolic and Molecular Bases of Inherited Disease*, 7th ed., Vol. 1 (Scriver, C. R., Beaudet, A. L., Sly, W. S., and Valle, D., eds), pp. 3367–3398, McGraw-Hill, New York

106. Wakil, S. J. (1989) *Biochemistry* **28**, 4523–4530

107. Smith, S. (1994) *FASEB J.* **8**, 1248–1259

107a. Barycki, J. J., O'Brien, L. K., Bratt, J. M., Zhang, R., Sanishvili, R., Strauss, A. W., and Banaszak, L. J. (1999) *Biochemistry* **38**, 5786–5798

107b. Fillgrove, K. L., and Anderson, V. E. (2000) *Biochemistry* **39**, 7001–7011

108. Watkinson, I. A., Wilton, D. L., Rahimtula, A. D., and Akhtar, M. M. (1971) *Eur. J. Biochem.* **23**, 1–6

109. Thorn, J. M., Barton, J. D., Dixon, N. E., Ollis, D. L., and Edwards, K. J. (1995) *J. Mol. Biol.* **249**, 785–799

110. Bolduc, J. M., Dyer, D. H., Scott, W. G., Singer, P., Sweet, R. M., Koshland, D. E., Jr., and Stoddard, B. L. (1995) *Science* **268**, 1312–1318

111. Lee, M. E., Dyer, D. H., Klein, O. D., Bolduc, J. M., Stoddard, B. L., and Koshland, D. E., Jr. (1995) *Biochemistry* **34**, 378–384

112. Schütz, M., and Radler, F. (1973) *Arch. Mikrobiol.* **91**, 183

113. Hwang, C.-C., and Cook, P. F. (1998) *Biochemistry* **37**, 15698–15702

114. Karsten, W. E., Gavva, S. R., Park, S.-H., and Cook, P. F. (1995) *Biochemistry* **34**, 3253–3260

115. Caspritz, G., and Radler, F. (1983) *J. Biol. Chem.* **258**, 4907–4910

116. Thoden, J. B., Frey, P. A., and Holden, H. M. (1996) *Biochemistry* **35**, 5137–5144

117. Frey, P. A. (1987) in *Pyridine Nucleotide Coenzymes: Chemical, Biochemical and Medical Aspects*, Vol. B (Dolphin, D., Avramovic, O., and Poulson, R., eds), pp. 461–512, Wiley (Interscience), New York

117a. Thoden, J. B., Wohlers, T. M., Fridovich-Keil, J. L., and Holden, H. M. (2000) *Biochemistry* **39**, 5691–5701

118. Thoden, J. B., Hegeman, A. D., Wesenberg, G., Chapeau, M. C., Frey, P. A., and Holden, H. M. (1997) *Biochemistry* **36**, 6294–6304

119. Palmer, J. L., and Abeles, R. H. (1979) *J. Biol. Chem.* **254**, 1217–1226

120. Yuan, C.-S., Ault-Riché, D. B., and Borchardt, R. T. (1996) *J. Biol. Chem.* **271**, 28009–28016

121. Koch-Nolte, F., Petersen, D., Balasubramanian, S., Haag, F., Kahlke, D., Willer, T., Kastelein, R., Bazan, F., and Thiele, H.-G. (1996) *J. Biol. Chem.* **271**, 7686–7693

122. Takada, T., Iida, K., and Moss, J. (1995) *J. Biol. Chem.* **270**, 541–544

123. Rising, K. A., and Schramm, V. L. (1997) *J. Am. Chem. Soc.* **119**, 27–37

124. Muller-Steffner, H. M., Augustin, A., and Schuber, F. (1996) *J. Biol. Chem.* **271**, 23967–23972

125. De Flora, A., Guida, L., Franco, L., Zocchi, E., Bruzzone, S., Benatti, U., Damonte, G., and Lee, H. C. (1997) *J. Biol. Chem.* **272**, 12945–12951

126. Aarhus, R., Graeff, R. M., Dickey, D. M., Walseth, T. F., and Lee, H. C. (1995) *J. Biol. Chem.* **270**, 30327–30333

127. Chini, E. N., Beers, K. W., and Dousa, T. P. (1995) *J. Biol. Chem.* **270**, 3216–3223

128. Oppenheimer, N. J. (1987) in *Pyridine Nucleotide Coenzymes: Chemical, Biochemical and Medical Aspects*, Vol. A (Dolphin, D., Avramovic, O., and Poulson, R., eds), pp. 323–365, Wiley (Interscience), New York

129. Everse, J., Anderson, B., and You, K.-S., eds. (1982) *The Pyridine Nucleotide Coenzymes*, Academic Press, New York

130. Blankenhorn, G., and Moore, E. G. (1980) *J. Am. Chem. Soc.* **102**, 1092–1098

131. Bernofsky, C. (1987) in *Pyridine Nucleotide Coenzymes: Chemical, Biochemical and Medical Aspects*, Vol. B (Dolphin, D., Avramovic, O., and Poulson, R., eds), pp. 105–172, Wiley (Interscience), New York

132. Chaykin, S. (1967) *Ann. Rev. Biochem.* **36**, (I), 149–170

133. Everse, J., and Kaplan, N. O. (1973) *Adv. Enzymol.* **37**, 61–133

133a. Rozwarski, D. A., Grant, G. A., Barton, D. H. R., Jacobs, W. R. J., and Sacchettini, J. C. (1998) *Science* **279**, 98–102

133b. Parikh, S., Moynihan, D. P., Xiao, G., and Tonge, P. J. (1999) *Biochemistry* **38**, 13623–13634

133c. McMurry, L. M., Oethinger, M., and Levy, S. B. (1998) *Nature (London)* **394**, 531–532

133d. Roujeinikova, A., and 14 other authors. (1999) *J. Mol. Biol.* **294**, 527–535

133e. Parikh, S. L., Xiao, G., and Tonge, P. J. (2000) *Biochemistry* **39**, 7645–7650

134. Johnson, R. W., Marschner, T. M., and Oppenheimer, N. J. (1988) *J. Am. Chem. Soc.* **110**, 2257–2263

135. Oppenheimer, N. J. (1973) *Biochem. Biophys. Res. Commun.* **50**, 683–690

136. Acheson, S. A., Kirkman, H. N., and Wolfenden, R. (1988) *Biochemistry* **27**, 7371–7375

137. Marshall, J. L., Booth, J. E., and Williams, J. W. (1984) *J. Biol. Chem.* **259**, 3033–3036

138. Bernofsky, C., and Wanda, S.-Y. C. (1982) *J. Biol. Chem.* **257**, 6809–6817

139. Woenckhaus, C. H. (1974) in *Topics in Current Chemistry (Fortschritte der chemischen Forschung,* Vol. 52 (Boschke, F. L., ed), pp. 209–233, Springer-Verlag, Berlin

140. Anderson, B. M. (1982) in *The Pyridine Nucleotide Coenzymes* (Everse, J., Anderson, B., and You, K.-S., eds), Academic Press, New York

141. Woenckhaus, C., and Jeck, R. (1987) in *Pyridine Nucleotide Coenzymes: Chemical, Biochemical and Medical Aspects*, Vol. A (Dolphin, D., Avramovic, O., and Poulson, R., eds), pp. 449–568, Wiley (Interscience), New York

142. Cohen, B. E., Stoddard, B. L., and Koshland, D. E., Jr. (1997) *Biochemistry* **36**, 9035–9044

143. Hemmerich, P. (1976) in *Progress in the Chemistry of Organic Natural Products*, Vol. 33 (Herz, W., Grisebach, H., and Kirby, G. W., eds), pp. 451–527, Springer-Verlag, New York

144. Fox, K. M., and Karplus, P. A. (1999) *J. Biol. Chem.* **274**, 9357–9362

145. Niino, Y. S., Chakraborty, S., Brown, B. J., and Massey, V. (1995) *J. Biol. Chem.* **270**, 1983–1991

146. Meighen, E. A. (1993) *FASEB J.* **7**, 1016–1022

147. Fisher, A. J., Raushel, F. M., Baldwin, T. O., and Rayment, I. (1995) *Biochemistry* **34**, 6581–6586

148. Van den Berghe-Snorek, S., and Stankovich, M. T. (1984) *J. Am. Chem. Soc.* **106**, 3685–3687

149. Miura, R., Nishina, Y., Sato, K., Fujii, S., Kuroda, K., and Shiga, K. (1993) *J. Biochem.* **113**, 106–113

150. Lim, L. W., Shamala, N., Mathews, F. S., Steenkamp, D. J., Hamlin, R., and Xuong, N. (1986) *J. Biol. Chem.* **261**, 15140–15146

151. Massey, V. (1995) *FASEB J.* **9**, 473–475

152. Müller, F. (1991) *Chemistry and Biochemistry of Flavoenzymes*, Vol. I, CRC Press, Boca Raton, Florida

153. Curti, B., Ronchi, S., and Zanetti, G., eds. (1990) *Flavins and Flavoproteins*, Walter de Gruyter, Berlin

153a. Fraaije, M. W., and Mattevi, A. (2000) *Trends Biochem. Sci.* **25**, 126–132

154. Kohen, A., Jonsson, T., and Klinman, J. P. (1997) *Biochemistry* **36**, 2603–2611

154a. Su, Q., and Klinman, J. P. (1999) *Biochemistry* **38**, 8572–8581

155. Bourdillon, C., Demaille, C., Moiroux, J., and Savéant, J.-M. (1993) *J. Am. Chem. Soc.* **115**, 2–10

156. Li, J., Vrielink, A., Brick, P., and Blow, D. M. (1993) *Biochemistry* **32**, 11507–11515

157. Lindqvist, Y., and Brändén, C.-I. (1985) *Proc. Natl. Acad. Sci. U.S.A.* **82**, 6855–6859

158. Lindqvist, Y., and Brändén, C.-I. (1989) *J. Biol. Chem.* **264**, 3624–3628

159. Stenberg, K., and Lindqvist, Y. (1997) *Protein Sci.* **6**, 1009–1015

160. Yagi, K., and Ozawa, T. (1989) *Biochim. Biophys. Acta.* **1000**, 203–206

160a. Curti, B., Ronchi, S., and Simonetta, M. P. (1990) in *Chemistry and Biochemistry of Flavoenzymes*, Vol. III (Müller, F., ed), pp. 69–94, CRC Press, Boca Raton, Florida

References

161. Mizutani, H., Miyahara, I., Hirotsu, K., Nishina, Y., Shiga, K., Setoyama, C., and Miura, R. (1996) *J. Biochem.* **120**, 14–17

162. Mattevi, A., Vanoni, M. A., Todone, F., Rizzi, M., Teplyakov, A., Coda, A., Bolognesi, M., and Curti, B. (1996) *Proc. Natl. Acad. Sci. U.S.A.* **93**, 7496–7501

163. Vanoni, M. A., Cosma, A., Mazzeo, D., Mattevi, A., Todone, F., and Curti, B. (1997) *Biochemistry* **36**, 5624–5632

164. Tan, A. K., and Ramsay, R. R. (1993) *Biochemistry* **32**, 2137–2143

165. Woo, J. C. G., and Silverman, R. B. (1995) *J. Am. Chem. Soc.* **117**, 1663–1664

166. Walker, M. C., and Edmondson, D. E. (1994) *Biochemistry* **33**, 7088–7098

166a. Pawelek, P. D., Cheah, J., Coulombe, R., Macheroux, P., Ghisla, S., and Vrielink, A. (2000) *EMBO J.* **19**, 4204–4215

167. Barber, M. J., Neame, P. J., Lim, L. W., White, S., and Mathews, F. S. (1992) *J. Biol. Chem.* **267**, 6611–6619

168. Trickey, P., Basran, J., Lian, L.-Y., Chen, Z.-w, Barton, J. D., Sutcliffe, M. J., Scrutton, N. S., and Mathews, F. S. (2000) *Biochemistry* **39**, 7678–7688

168a. Jang, M.-H., Basran, J., Scrutton, N. S., and Hille, R. (1999) *J. Biol. Chem.* **274**, 13147–13154

169. Falzon, L., and Davidson, V. L. (1996) *Biochemistry* **35**, 12111–12118

170. Lindqvist, Y., Brändén, C.-I., Mathews, F. S., and Lederer, F. (1991) *J. Biol. Chem.* **266**, 3198–3207

171. Tegoni, M., and Cambillau, C. (1994) *Protein Sci.* **3**, 303–313

172. Gondry, M., and Lederer, F. (1996) *Biochemistry* **35**, 8587–8594

173. Tegoni, M., Gervais, M., and Desbois, A. (1997) *Biochemistry* **36**, 8932–8946

173a. Lehoux, I. E., and Mitra, B. (2000) *Biochemistry* **39**, 10055–10065

174. Schaller, R. A., Mohsen, A.-W. A., Vockley, J., and Thorpe, C. (1997) *Biochemistry* **36**, 7761–7768

175. Rudik, I., Ghisla, S., and Thorpe, C. (1998) *Biochemistry* **37**, 8437–8445

176. Mancini-Samuelson, G. J., Kieweg, V., Sabaj, K. M., Ghisla, S., and Stankovich, M. T. (1998) *Biochemistry* **37**, 14605–14612

176a. Srivastava, D. K., and Peterson, K. L. (1998) *Biochemistry* **37**, 8446–8456

177. Mohsen, A.-W. A., and Vockley, J. (1995) *Biochemistry* **34**, 10146–10152

177a. Ackrell, B., McIntire, B., and Vessey, D. (2000) *Trends Biochem. Sci.* **25**, 9–10

178. Rétey, J., Seibl, J., Arigoni, D., Cornforth, J. W., Ryback, G., Zeylemaker, W. P., and Veeger, C. (1970) *Eur. J. Biochem.* **14**, 232–242

179. Birch-Machin, M. A., Farnsworth, L., Ackrell, B. A. C., Cochran, B., Jackson, S., Bindoff, L. A., Aitken, A., Diamond, A. G., and Turnbull, D. M. (1992) *J. Biol. Chem.* **267**, 11553–11558

180. Schmidt, D. M., Saghbini, M., and Scheffler, I. E. (1992) *Biochemistry* **31**, 8442–8448

181. Sucheta, A., Ackrell, B. A. C., Cochran, B., and Armstrong, F. A. (1992) *Nature (London)* **356**, 361–362

182. Thieme, R., Pai, E. F., Schirmer, R. H., and Schulz, G. E. (1981) *J. Mol. Biol.* **152**, 763–782

183. Sled, V. D., Rudnitzky, N. I., Hatefi, Y., and Ohnishi, T. (1994) *Biochemistry* **33**, 10069–10075

184. Westenberg, D. J., Gunsalus, R. P., Ackrell, B. A. C., Sices, H., and Cecchini, G. (1993) *J. Biol. Chem.* **268**, 815–822

185. Hägerhäll, C., Fridén, H., Aasa, R., and Hederstedt, L. (1995) *Biochemistry* **34**, 11080–11089

186. Van Hellemond, J. J., Klockiewicz, M., Gaasenbeek, C. P. H., Roos, M. H., and Tielens, A. G. M. (1995) *J. Biol. Chem.* **270**, 31065–31070

187. Waksman, G., Krishna, T. S. R., Williams, J., CH, and Kuriyan, J. (1994) *J. Mol. Biol.* **236**, 800–816

188. Mulrooney, S. B., and Williams, C. H. (1994) *Biochemistry* **33**, 3148–3154

189. Arscott, L. D., Gromer, S., Schirmer, R. H., Becker, K., and Williams, C. H., Jr. (1997) *Proc. Natl. Acad. Sci. U.S.A.* **94**, 3621–3626

190. Lennon, B. W., and Williams, C. H., Jr. (1995) *Biochemistry* **34**, 3670–3677

191. Pai, E. F., and Schulz, G. E. (1983) *J. Biol. Chem.* **258**, 1752–1757

192. Karplus, P. A., and Schulz, G. E. (1987) *J. Mol. Biol.* **195**, 701–729

193. Mittl, P. R. E., and Schulz, G. E. (1994) *Protein Sci.* **3**, 799–809

194. Zhang, Y., Bond, C. S., Bailey, S., Cunningham, M. L., Fairlamb, A. H., and Hunter, W. N. (1996) *Protein Sci.* **5**, 52–61

195. Walsh, C., Bradley, M., and Nadeau, K. (1991) *Trends Biochem. Sci.* **16**, 305–309

196. Zheng, R., Cenas, N., and Blanchard, J. S. (1995) *Biochemistry* **34**, 12697–12703

197. Krauth-Siegel, R. L., and Schöneck, R. (1995) *FASEB J.* **9**, 1138–1146

198. Brown, N. L. (1985) *Trends Biochem. Sci.* **10**, 400–403

199. Schiering, N., Kabsch, W., Moore, M. J., Distefano, M. D., Walsh, C. T., and Pai, E. F. (1991) *Nature (London)* **352**, 168–172

199a. Engst, S., and Miller, S. M. (1999) *Biochemistry* **38**, 3519–3529

200. Serre, L., Vellieux, F. M. D., Gomez-Moreno, M. C., Fontecilla-Camps, J. C., and Frey, M. (1996) *J. Mol. Biol.* **263**, 20–39

201. Aliverti, A., Bruns, C. M., Pandini, V. E., Karplus, P. A., Vanoni, M. A., Curti, B., and Zanetti, G. (1995) *Biochemistry* **34**, 8371–8379

202. Ermler, U., Siddiqui, R. A., Cramm, R., and Friedrich, B. (1995) *EMBO J.* **14**, 6067–6077

203. Ingelman, M., Bianchi, V., and Eklund, H. (1997) *J. Mol. Biol.* **268**, 147–157

204. Menz, R. I., and Day, D. A. (1996) *J. Biol. Chem.* **271**, 23117–23120

205. Takano, S., Yano, T., and Yagi, T. (1996) *Biochemistry* **35**, 9120–9127

205a. Morandi, P., Valzasina, B., Colombo, C., Curti, B., and Vanoni, M. A. (2000) *Biochemistry* **39**, 727–735

206. Fieschi, F., Nivière, V., Frier, C., Décout, J.-L., and Fontecave, M. (1995) *J. Biol. Chem.* **270**, 30392–30400

206a. Ingelman, M., Ramaswamy, S., Nivière, V., Fontecave, M., and Eklund, H. (1999) *Biochemistry* **38**, 7040–7049

207. Tanner, J. J., Lei, B., Tu, S.-C., and Krause, K. L. (1996) *Biochemistry* **35**, 13531–13539

208. Porter, T. D. (1991) *Trends Biochem. Sci.* **16**, 154–158

209. Vang, M., Roberts, D. L., Paschke, R., Shea, T. M., Masters, B. S. S., and Kin, J.-J. P. (1997) *Proc. Natl. Acad. Sci. U.S.A.* **94**, 8411–8416

209a. Uhlmann, H., and Bernhardt, R. (1995) *J. Biol. Chem.* **270**, 29959–29966

209b. Ziegler, G. A., and Schulz, G. E. (2000) *Biochemistry* **39**, 10986–10995

210. Kobayashi, K., Tagawa, S., Sano, S., and Asada, K. (1995) *J. Biol. Chem.* **270**, 27551–27554

211. Sano, S., Miyake, C., Mikami, B., and Asada, K. (1995) *J. Biol. Chem.* **270**, 21354–21361

212. Tedeschi, G., Chen, S., and Massey, V. (1995) *J. Biol. Chem.* **270**, 1198–1204

213. Li, R., Bianchet, M. A., Talalay, P., and Amzel, L. M. (1995) *Proc. Natl. Acad. Sci. U.S.A.* **92**, 8846–8850

214. Chen, S., Clarke, P. E., Martino, P. A., Dang, P. S. K., Yeh, C.-H., Lee, T. D., Prochaska, H. J., and Talalay, P. (1994) *Protein Sci.* **3**, 1296–1304

215. Claiborne, A., Ross, R. P., and Parsonage, D. (1992) *Trends Biochem. Sci.* **17**, 183–186

216. Parsonage, D., and Claiborne, A. (1995) *Biochemistry* **34**, 435–441

217. Yeh, J. I., Claiborne, A., and Hol, W. G. J. (1996) *Biochemistry* **35**, 9951–9957

218. Poole, L. B. (1996) *Biochemistry* **35**, 65–75

219. Mewies, M., McIntire, W. S., and Scrutton, N. S. (1998) *Protein Sci.* **7**, 7–20

220. Brandsch, R., and Bichler, V. (1991) *J. Biol. Chem.* **266**, 19056–19062

221. Willie, A., Edmondson, D. E., and Jorns, M. S. (1996) *Biochemistry* **35**, 5292–5299

222. Kenney, W. C., Singer, T. P., Fukuyama, M., and Miyake, Y. (1979) *J. Biol. Chem.* **254**, 4689–4690

223. Kim, J., Fuller, J. H., Kuusk, V., Cunane, L., Chen, Z.-w, Mathews, F. S., and McIntire, W. S. (1995) *J. Biol. Chem.* **270**, 31202–31209

224. Mewies, M., Basran, J., Packman, L. C., Hille, R., and Scrutton, N. S. (1997) *Biochemistry* **36**, 7162–7168

225. Ghisla, S., and Mayhew, S. G. (1976) *Eur. J. Biochem.* **63**, 373–390

226. Hartman, H. A., Edmondson, D. E., and McCormick, D. B. (1992) *Anal. Biochem.* **202**, 348–355

227. Eirich, L. D., Vogels, G. D., and Wolfe, R. S. (1979) *J. Bacterial.* **140**, 20–27

228. Jacobson, F., and Walsh, C. (1984) *Biochemistry* **23**, 979–988

229. Jones, J. B., and Stadtman, T. C. (1980) *J. Biol. Chem.* **255**, 1049–1053

230. Sancar, A. (1994) *Biochemistry* **33**, 2–9

231. Otani, S., Takatsu, M., Nakano, M., Kasai, S., Miura, R., and Matsui, K. (1974) *The Journal of Antibiotics* **27**, 88–89

232. Murthy, Y. V. S. N., and Massey, V. (1995) *J. Biol. Chem.* **270**, 28586–28594

233. Ghisla, S., and Massey, V. (1989) *Eur. J. Biochem.* **2**, 243–289

234. Powell, M. F., and Bruice, T. C. (1983) *J. Am. Chem. Soc.* **105**, 1014–1021

235. Fox, J. L., and Tolin, G. (1966) *Biochemistry* **5**, 3865–3872

236. Lee, I.-S. H., Ostović, D., and Kreevoy, M. (1988) *J. Am. Chem. Soc.* **110**, 3989–3993

237. Brustlein, M., and Bruice, T. C. (1972) *J. Am. Chem. Soc.* **94**, 6548–6549

238. Jorns, M. S., and Hersch, L. B. (1975) *J. Biol. Chem.* **250**, 3620–3628

239. Ghisla, S., Thorpe, C., and Massey, V. (1984) *Biochemistry* **23**, 3154–3161

240. Kim, J.-J. P., Wang, M., and Paschke, R. (1993) *Proc. Natl. Acad. Sci. U.S.A.* **90**, 7523–7527

241. Dakoji, S., Shin, I., Becker, D. F., Stankovich, M. T., and Liu, H.-w. (1996) *J. Am. Chem. Soc.* **118**, 10971–10979

242. Tompkins, L. S., and Falkow, S. (1995) *Science* **267**, 1621–1622

242a. Vock, P., Engst, S., Eder, M., and Ghisla, S. (1998) *Biochemistry* **37**, 1848–1860

243. Johnson, B. D., Mancini-Samuelson, G. J., and Stankovich, M. T. (1995) *Biochemistry* **34**, 7047–7055

244. Thorpe, C. (1989) *Trends Biochem. Sci.* **14**, 148–151

245. Djordjevic, S., Pace, C. P., Stankovich, M. T., and Kim, J.-J. P. (1995) *Biochemistry* **34**, 2163–2171

246. Porter, D. J. T., Voet, J. G., and Bright, H. J. (1973) *J. Biol. Chem.* **248**, 4400–4416

247. Alston, T. A., Porter, D. J. T., and Bright, H. J. (1983) *J. Biol. Chem.* **258**, 1136–1141

248. Gadda, G., Edmondson, R. D., Russell, D. H., and Fitzpatrick, P. F. (1997) *J. Biol. Chem.* **272**, 5563–5570

249. Lehoux, I., and Mitra, B. (1997) *FASEB J.* **11**, A889

250. Ghisla, S., and Massey, V. (1980) *J. Biol. Chem.* **255**, 5688–5696

251. Hamilton, G. H. (1971) *Prog. Bioorg. Chem.* **1**, 83–157

252. Walsh, C. T., Krodel, E., Massey, V., and Abeles, R. H. (1973) *J. Biol. Chem.* **248**, 1946–1951

253. Ghisla, S., and Massey, V. (1989) *Eur. J. Biochem.* **181**, 1–17

254. Curti, B., Ronchi, S., and Simonetta, M. P. (1992) in *Chemistry and Biochemistry of Flavoenzymes* (Müller, F., ed), pp. 69–94, CRC Press, Boca Raton, Florida

255. Lederer, F. (1992) *Protein Sci.* **1**, 540–548

256. Miura, R., and Miyake, Y. (1988) *Bioorg. Chem.* **16**, 97–110

257. Todone, F., Vanoni, M. A., Mozzarelli, A., Bolognesi, M., Coda, A., Curti, B., and Mattevi, A. (1997) *Biochemistry* **36**, 5853–5860

258. Rietveld, P., Arscott, L. D., Berry, A., Scrutton, N. S., Deonarain, M. P., Perham, R. N., and Williams, C. H. (1994) *Biochemistry* **33**, 13888–13895

258a. Krauth-Siegel, R. L., Arscott, L. D., Schönleben-Janas, A., Schirmer, R. H., and Williams, C. H., Jr. (1998) *Biochemistry* **37**, 13968–13977

259. Williams, C. H., Jr. (1995) *FASEB J.* **9**, 1267–1276

260. Barman, B. G., and Tollin, G. (1972) *Biochemistry* **11**, 4760–4765

261. Su, Y., and Tripathi, G. N. R. (1994) *J. Am. Chem. Soc.* **116**, 4405–4407

262. Zheng, Y.-J., and Ornstein, R. L. (1996) *J. Am. Chem. Soc.* **118**, 9402–9408

263. Müller, F., Hemmerich, P., and Ehrenberg, A. (1971) in *Flavins and Flavoproteins* (Kamin, H., ed), pp. 107–180, Univ. Park Press, Baltimore, Maryland

264. Müller, F., Eriksson, L. E. G., and Ehrenberg, A. (1970) *Eur. J. Biochem.* **12**, 93–103

265. Miura, R., Setoyama, C., Nishina, Y., Shiga, K., Mizutani, H., Miyahara, I., and Hirotsu, K. (1997) *FASEB J.* **11**, A1306

265a. Miura, R., Setoyama, C., Nishina, Y., Shiga, K., Mizutani, H., Miyahara, I., and Hirotsu, K. (1977) *J. Biochem.* **122**, 825–833

266. Smith, W. W., Burnett, R. M., Darling, G. D., and Ludwig, M. L. (1977) *J. Mol. Biol.* **117**, 195–225

267. Chang, F.-C., and Swenson, R. P. (1999) *Biochemistry* **38**, 7168–7176

268. Zhou, Z., and Swenson, R. P. (1996) *Biochemistry* **35**, 15980–15988

269. Ludwig, M. L., Pattridge, K. A., Metzger, A. L., and Dixon, M. M. (1997) *Biochemistry* **36**, 1259–1280

269a. Lostao, A., Gómez-Moreno, C., Mayhew, S. G., and Sancho, J. (1997) *Biochemistry* **36**, 14334–14344

270. Hoover, D. M., Drennan, C. L., Metzger, A. L., Osborne, C., Weber, C. H., Pattridge, K. A., and Ludwig, M. L. (1999) *J. Mol. Biol.* **294**, 725–743

271. Moonen, C. T. W., and Müller, F. (1982) *Biochemistry* **21**, 408–414

272. Jorns, M. S., Sancar, G. B., and Sancar, A. (1984) *Biochemistry* **23**, 2673–2679

273. McKean, M. C., Beckman, J. D., and Frerman, F. E. (1983) *J. Biol. Chem.* **258**, 1866–1870

274. Roberts, D. L., Salazar, D., Fulmer, J. P., Frerman, F. E., and Kim, J.-J. P. (1999) *Biochemistry* **38**, 1977–1989

275. Huang, L., Rohlfs, R. J., and Hille, R. (1995) *J. Biol. Chem.* **270**, 23958–23965

276. Balasubramanian, S., Coggins, J. R., and Abell, C. (1995) *Biochemistry* **34**, 341–348

277. Macheroux, P., Petersen, J., Bornemann, S., Lowe, D. J., and Thorneley, R. N. F. (1996) *Biochemistry* **35**, 1643–1652

278. Yu, M. W., and Fritchie, C. J. J. (1975) *J. Biol. Chem.* **250**, 946–951

279. Kenney, W. C., and Singer, T. P. (1977) *J. Biol. Chem.* **252**, 4767–4772

280. Ostrowski, J., Barber, M. J., Rueger, D. C., Miller, B. E., Siegel, L. M., and Kredich, N. M. (1989) *J. Biol. Chem.* **264**, 15796–15808

281. Covès, J., Zeghouf, M., Macherel, D., Guigliarelli, B., Asso, M., and Fontecave, M. (1997) *Biochemistry* **36**, 5921–5928

282. Campbell, W. H., and Kinghorn, J. R. (1990) *Trends Biochem. Sci.* **15**, 315–319

283. Hyde, G. E., Crawford, N. M., and Campbell, W. H. (1991) *J. Biol. Chem.* **266**, 23542–23547

283a. Terao, M., Kurosaki, M., Saltini, G., Demontis, S., Marini, M., Salmona, M., and Garattini, E. (2000) *J. Biol. Chem.* **275**, 30690–30700

284. Bruice, T. C. (1980) *Acc. Chem. Res.* **13**, 256–262

285. Entsch, B., Palfey, B. A., Ballou, D. P., and Massey, V. (1991) *J. Biol. Chem.* **266**, 17341–17349

286. Gatti, D. L., Entsch, B., Ballou, D. P., and Ludwig, M. L. (1996) *Biochemistry* **35**, 567–578

287. Schreuder, H. A., Hol, W. G. J., and Drenth, J. (1990) *Biochemistry* **29**, 3101–3108

288. Yamasaki, M., and Yamano, T. (1973) *Biochem. Biophys. Res. Commun.* **51**, 612–619

289. Merényi, G., and Lind, J. (1991) *J. Am. Chem. Soc.* **113**, 3146–3153

290. Entsch, B., and van Berkel, W. J. H. (1995) *FASEB J.* **9**, 476–483

291. Snell, E. E., and Broquist, H. P. (1949) *Arch. Biochem. Biophys.* **23**, 326–328

291a. Reed, L. J. (1998) *Protein Sci.* **7**, 220–224

292. Reed, L. J., DeBusk, B. G., Gunsalus, I. C., and Hornberger, C. S. (1951) *Science* **114**, 93–94

293. Jukes, T. H. (1997) *Protein Sci.* **6**, 254–256

294. Schmidt, U., Grafen, P., Altland, K., and Goedde, H. W. (1969) *Adv. Enzymol.* **32**, 432–469

295. Reed, K. E., Morris, T. W., and Cronan, J. E., Jr. (1994) *Proc. Natl. Acad. Sci. U.S.A.* **91**, 3720–3724

296. Pares, S., Cohen-Addad, C., Sieker, L., Neuburger, M., and Douce, R. (1994) *Proc. Natl. Acad. Sci. U.S.A.* **91**, 4850–4853

296a. Morris, T. W., Reed, K. E., and Cronan, J. E., Jr. (1994) *J. Biol. Chem.* **269**, 16091–16100

297. Roche, T. E., and Patel, M. E., eds. (1990) *Alpha-Keto Acid Dehydrogenase Complexes Organization, Regulation, and Biomedical Ramifications*, Vol. 573, New York Academy of Sciences, New York

298. Patel, M. S., and Roche, T. E. (1990) *FASEB J.* **4**, 3224–3233

299. Stoops, J. K., Cheng, R. H., Yazdi, M. A., Maeng, C.-Y., Schroeter, J. P., Klueppelberg, U., Kolodziej, S. J., Baker, T. S., and Reed, L. J. (1997) *J. Biol. Chem.* **272**, 5757–5764

300. Yi, J., Nemeria, N., McNally, A., Jordan, F., Machado, R. S., and Guest, J. R. (1996) *J. Biol. Chem.* **271**, 33192–33200

301. Berg, A., Vervoort, J., and de Kok, A. (1996) *J. Mol. Biol.* **261**, 432–442

302. Wynn, R. M., Davie, J. R., Zhi, W., Cox, R. P., and Chuang, D. T. (1994) *Biochemistry* **33**, 8962–8968

302a. Wynn, R. M., Davie, J. R., Chuang, J. L., Cote, C. D., and Chuang, D. T. (1998) *J. Biol. Chem.* **273**, 13110–13118

303. Mattevi, A., Obmolova, G., Kalk, K. H., van Berkel, W. J. H., and Hol, W. G. J. (1993) *J. Mol. Biol.* **230**, 1200–1215

304. Liu, T.-C., Korotchkina, L. G., Hyatt, S. L., Vettakkorumakankav, N. N., and Patel, M. S. (1995) *J. Biol. Chem.* **270**, 15545–15550

305. Guan, Y., Rawsthorne, S., Scofield, G., Shaw, P., and Doonan, J. (1995) *J. Biol. Chem.* **270**, 5412–5417

306. Mattevi, A., Obmolova, G., Schulze, E., Kalk, K. H., Westphal, A. H., de Kok, A., and Hol, W. G. J. (1992) *Science* **255**, 1544–1550

306a. Mattevi, A., Obmolova, G., Kalk, K. H., Westphal, A. H., de Kok, A., and Hol, W. G. J. (1993) *J. Mol. Biol.* **230**, 1183–1199

306b. Izard, T., AEvarsson, A., Allen, M. D., Westphal, A. H., Perhan, R. N., de Kok, A., and Hol, W. G. J. (1999) *Proc. Natl. Acad. Sci. U.S.A.* **96**, 1240–1245

307. Reed, L. J., and Hackert, M. L. (1990) *J. Biol. Chem.* **265**, 8971–8974

307a. Knapp, J. E., Mitchell, D. T., Yazdi, M. A., Ernst, S. R., Reed, L. J., and Hackert, M. L. (1998) *J. Mol. Biol.* **280**, 655–668

308. Dardel, F., Davis, A. L., Laue, E. D., and Perham, R. N. (1993) *J. Mol. Biol.* **229**, 1037–1048

309. Green, J. D. F., Laue, E. D., Perham, R. N., Ali, S. T., and Guest, J. R. (1995) *J. Mol. Biol.* **248**, 328–343

310. Mattevi, A., Obmolova, G., Kalk, K. H., Teplyakov, A., and Hol, W. G. J. (1993) *Biochemistry* **32**, 3887–3901

311. Hendle, J., Mattevi, A., Westphal, A. H., Spee, J., de Kok, A., Teplyakov, A., and Hol, W. G. J. (1995) *Biochemistry* **34**, 4287–4298

312. Ricaud, P. M., Howard, M. J., Roberts, E. L., Broadhurst, R. W., and Perham, R. N. (1996) *J. Mol. Biol.* **264**, 179–190

313. Maeng, C.-Y., Yazdi, M. A., Niu, X.-D., Lee, H. Y., and Reed, L. J. (1994) *Biochemistry* **33**, 13801–13807

314. McCartney, R. G., Sanderson, S. J., and Lindsay, J. G. (1997) *Biochemistry* **36**, 6819–6826

315. Yang, Y.-S., and Frey, P. A. (1986) *Biochemistry* **25**, 8173–8178

315a. Pan, K., and Jordan, F. (1998) *Biochemistry* **37**, 1357–1364

315b. Hennig, J., Kern, G., Neef, H., Spinka, M., Bisswanger, H., and Hübner, G. (1997) *Biochemistry* **36**, 15772–15779

315c. Kee, K., Niu, L., and Henderson, E. (1998) *Biochemistry* **37**, 4224–4234

316. Han, D., Handelman, G., Marcocci, L., Sen, C. K., Roy, S., Kobuchi, H., Tritschler, H. J., Flohé, L., and Packer, L. (1997) *BioFactors* **6**, 321–338

317. Muller, Y. A., Schumacher, G., Rudolph, R., and Schulz, G. E. (1994) *J. Mol. Biol.* **237**, 315–335

318. Kerscher, L., and Oesterhelt, D. (1982) *Trends Biochem. Sci.* **7**, 371–374

319. Muller, Y. A., and Schulz, G. E. (1993) *Science* **259**, 965–967

319a. Tittmann, K., Golbik, R., Ghisla, S., and Hübner, G. (2000) *Biochemistry* **39**, 10747–10754

320. Menon, S., and Ragsdale, S. W. (1997) *Biochemistry* **36**, 8484–8494

320a. Bouchev, V. F., Furdui, C. M., Menon, S., Muthukumaran, R. B., Ragsdale, S. W., and McCracken, J. (1999) *J. Am. Chem. Soc.* **121**, 3724–3729

321. Smith, E. T., Blamey, J. M., and Adams, M. W. W. (1994) *Biochemistry* **33**, 1008–1016

322. Mai, X., and Adams, M. W. W. (1994) *J. Biol. Chem.* **269**, 16726–16732

323. Gehring, V., and Arnon, D. I. (1972) *J. Biol. Chem.* **247**, 6963–6969

324. Sun, W., Williams, C. H., Jr., and Massey, V. (1996) *J. Biol. Chem.* **271**, 17226–17233

325. Müh, U., Williams, C. H., and Massey, V. (1994) *J. Biol. Chem.* **269**, 7994–8000

326. Unkrig, V., Neugebauer, F. A., and Knappe, J. (1989) *Eur. J. Biochem.* **184**, 723–728

327. Maiorino, M., Chu, F. F., Ursini, F., Davies, K. J. A., Doroshow, J. H., and Esworthy, R. S. (1991) *J. Biol. Chem.* **266**, 7728–7732

328. Wong, K. K., Murray, B. W., Lewisch, S. A., Baxter, M. K., Ridky, T. W., Ulissi-DeMario, L., and Kozarich, J. W. (1993) *Biochemistry* **32**, 14102–14110

References

329. Frey, M., Rothe, M., Wagner, A. F. V., and Knappe, J. (1994) *J. Biol. Chem.* **269**, 12432–12437

330. Kraus, R. J., Foster, S. J., and Ganther, H. E. (1983) *Biochemistry* **22**, 5853–5858

331. Broderick, J. B., Duderstadt, R. E., Fernandez, D. C., Wojtuszewski, K., Henshaw, T. F., and Johnson, M. K. (1997) *J. Am. Chem. Soc.* **119**, 7390–7391

331a. Külzer, R., Pils, T., Kappl, R., Hüttermann, J., and Knappe, J. (1998) *J. Biol. Chem.* **273**, 4897–4903

332. Wu, W., Lieder, K. W., Reed, G. H., and Frey, P. A. (1995) *Biochemistry* **34**, 10532–10537

333. Wagner, A. F. V., Frey, M., Neugebauer, F. A., Schäfer, W., and Knappe, J. (1992) *Proc. Natl. Acad. Sci. U.S.A.* **89**, 996–1000

333a. Gauld, J. W., and Eriksson, L. A. (2000) *J. Am. Chem. Soc.* **122**, 2035–2040

334. Parast, C. V., Wong, K. K., Lewisch, S. A., Kozarich, J. W., Peisach, J., and Magliozzo, R. S. (1995) *Biochemistry* **34**, 2393–2399

335. Parast, C. V., Wong, K. K., Kozarich, J. W., Peisach, J., and Magliozzo, R. S. (1995) *Biochemistry* **34**, 5712–5717

335a. Reddy, S. G., Wong, K. K., Parast, C. V., Peisach, J., Magliozzo, R. S., and Kozarich, J. W. (1998) *Biochemistry* **37**, 558–563

336. Mastropaolo, D., Camerman, A., and Camerman, N. (1980) *Science* **210**, 334–336

337. Bailey, L. B., ed. (1995) *Folate in Health and Disease*, Dekker, New York

338. Curtius, H.-C., Matasovic, A., Schoedon, G., Kuster, T., Guibaud, P., Giudici, T., and Blau, N. (1990) *J. Biol. Chem.* **265**, 3923–3930

339. Blakley, R. L., and Benkovic, S. J., eds. (1984–1986) *Folates and Pterins*, Wiley, New York (3 Vols.)

340. Hadorn, E. (1962) *Sci. Am.* **206**(April), 101–110

341. Ghosh, D. K., and Stuehr, D. J. (1995) *Biochemistry* **34**, 801–807

342. Witteveen, C. F. B., Giovanelli, J., and Kaufman, S. (1996) *J. Biol. Chem.* **271**, 4143–4147

343. Rembold, H., and Gyure, W. L. (1972) *Angew. Chem. Int. Ed. Engl.* **11**, 1061–1072

344. Jacobson, K. B., Dorsett, D., Pfleiderer, W., McCloskey, J. A., Sethi, S. K., Buchanan, M. V., and Rubin, I. B. (1982) *Biochemistry* **21**, 5700–5706

345. Wiederrecht, G. J., Paton, D. R., and Brown, G. M. (1984) *J. Biol. Chem.* **259**, 2195–2200

346. Cremer-Bartels, G., and Ebels, I. (1980) *Proc. Natl. Acad. Sci. U.S.A.* **77**, 2415–2418

347. Van Haastert, P. J. M., DeWitt, R. J. W., Grijpma, J., and Konijn, T. M. (1982) *Proc. Natl. Acad. Sci. U.S.A.* **79**, 6270–6274

348. Irby, R. B., and Adair, J., WL. (1994) *J. Biol. Chem.* **269**, 23981–23987

349. Stiefel, E. I. (1996) *Science* **272**, 1599–1600

350. Escalante-Semerena, J. C., Rinehart, K. L., Jr., and Wolfe, R. S. (1984) *J. Biol. Chem.* **259**, 9447–9455

351. Wolfe, R. S. (1985) *Trends Biochem. Sci.* **10**, 396–399

352. Rouvière, P. E., and Wolfe, R. S. (1988) *J. Biol. Chem.* **263**, 7913–7916

353. White, R. H. (1993) *Biochemistry* **32**, 745–753

354. Hyatt, D. C., Maley, F., and Montfort, W. R. (1997) *Biochemistry* **36**, 4585–4594

355. Blakeley, R. L. (1984) in *Folates and Pterins*, Vol. I (Blakeley, P. L., and Bencovic, S. J., eds), pp. 191–253, Wiley, New York

356. Stover, P., and Schirch, V. (1993) *Trends Biochem. Sci.* **18**, 102–106

357. Lin, B.-F., and Shane, B. (1994) *J. Biol. Chem.* **269**, 9705–9713

357a. Cherest, H., Thomas, D., and Surdin-Kerjan, Y. (2000) *J. Biol. Chem.* **275**, 14056–14063

358. Keshavjee, K., Pyne, C., and Bognar, A. L. (1991) *J. Biol. Chem.* **266**, 19925–19929

359. Scott, J. M. (1976) *Biochem. Soc. Trans.* **4**, 845–850

360. Brody, T., Watson, J. E., and Stokstad, E. L. R. (1982) *Biochemistry* **21**, 276–282

361. Kim, D. W., Huang, T., Schirch, D., and Schirch, V. (1996) *Biochemistry* **35**, 15772–15783

362. Lipman, R. S. A., Bailey, S. W., Jarrett, J. T., Matthews, R. G., and Jorns, M. S. (1995) *Biochemistry* **34**, 11217–11220

363. Mathews, C. K. (1971) *Bacteriophage Biochemistry*, Van Nostrand-Reinhold, Princeton, New Jersey

364. Roth, B. (1986) *Fed. Proc.* **45**, 2765–2772

365. Schweitzer, B. I., Dicker, A. P., and Bertino, J. R. (1990) *FASEB J.* **4**, 2441–2452

366. Groom, C. R., Thillet, J., North, A. C. T., Pictet, R., and Geddes, A. J. (1991) *J. Biol. Chem.* **266**, 19890–19893

367. Blaney, J. M., Hansch, C., Silipo, C., and Vittoria, A. (1984) *Chem. Rev.* **84**, 333–407

368. Dale, G. E., Broger, C., D'Arcy, A., Hartman, P. G., DeHoogt, R., Jolidon, S., Kompis, I., Labhardt, A. M., Langen, H., Locher, H., Page, M. G. P., Stüber, D., Then, R. L., Wipf, B., and Oefner, C. (1997) *J. Mol. Biol.* **266**, 23–30

369. Sirawaraporn, W., Sathitkul, T., Sirawaraporn, R., Yuthavong, Y., and Santi, D. V. (1997) *Proc. Natl. Acad. Sci. U.S.A.* **94**, 1124–1129

370. Birdsall, B., Tendler, S. J. B., Arnold, J. R. P., Feeney, J., Griffin, R. J., Carr, M. D., Thomas, J. A., Roberts, G. C. K., and Stevens, M. F. G. (1990) *Biochemistry* **29**, 9660–9667

371. Jolivet, J., Cowan, K. H., Curt, G. A., Clendeninn, N. J., and Chabner, B. A. (1983) *N. Engl. J. Med.* **309**, 1094–1104

372. Federspiel, N. A., Beverley, S. M., Schilling, J. W., and Schimke, R. T. (1984) *J. Biol. Chem.* **259**, 9127–9140

373. Chu, E., Takimoto, C. H., Voeller, D., Grem, J. L., and Allegra, C. J. (1993) *Biochemistry* **32**, 4756–4760

374. Friesheim, J. H., and Matthews, D. A. (1984) in *Folate Antagonists as Therapeutic Agents*, Vol. 1 (Sirotnak, F. M., Burchall, J. J., Ensminger, W. D., and Montgomery, J. A., eds), pp. 69–131, Academic Press, Orlando, Florida

375. Fan, J., Vitols, K. S., and Huennekens, F. M. (1991) *J. Biol. Chem.* **266**, 14862–14865

376. Henderson, G. B., Hughes, T. R., and Saxena, M. (1994) *J. Biol. Chem.* **269**, 13382–13389

377. Thompson, P. D., and Freisheim, J. H. (1991) *Biochemistry* **30**, 8124–8130

378. Chen, L., Qi, H., Korenberg, J., Garrow, T. A., Choi, Y.-J., and Shane, B. (1996) *J. Biol. Chem.* **271**, 13077–13087

379. Roy, K., Mitsugi, K., and Sirotnak, F. M. (1997) *J. Biol. Chem.* **272**, 5587–5593

379a. Maziarz, K. M., Monaco, H. L., Shen, F., and Ratnam, M. (1999) *J. Biol. Chem.* **274**, 11086–11091

380. Park, H., Zhuang, P., Nichols, R., and Howell, E. E. (1997) *J. Biol. Chem.* **272**, 2252–2258

381. Sawaya, M. R., and Kraut, J. (1997) *Biochemistry* **36**, 586–603

382. Reyes, V. M., Sawaya, M. R., Brown, K. A., and Kraut, J. (1995) *Biochemistry* **34**, 2710–2723

383. Lee, H., Reyes, V. M., and Kraut, J. (1996) *Biochemistry* **35**, 7012–7020

384. Verma, C. S., Caves, L. S. D., Hubbard, R. E., and Roberts, G. C. K. (1997) *J. Mol. Biol.* **266**, 776–796

385. Cheung, H. T. A., Birdsall, B., Frenkiel, T. A., Chau, D. D., and Feeney, J. (1993) *Biochemistry* **32**, 6846–6854

386. Epstein, D. M., Benkovic, S. J., and Wright, P. E. (1995) *Biochemistry* **34**, 11037–11048

387. Meiering, E. M., and Wagner, G. (1995) *J. Mol. Biol.* **247**, 294–308

388. Cannon, W. R., Garrison, B. J., and Benkovic, S. J. (1997) *J. Am. Chem. Soc.* **119**, 2386–2395

389. Nakano, T., Spencer, H. T., Appleman, J. R., and Blakley, R. L. (1994) *Biochemistry* **33**, 9945–9952

390. Howell, E. E., Villafranca, J. E., Warren, M. S., Oatley, S. J., and Kraut, J. (1986) *Science* **231**, 1123–1128

391. Benkovic, S. J., Fierke, C. A., and Naylor, A. M. (1988) *Science* **239**, 1105–1110

392. Chen, Y.-Q., Kraut, J., Blakley, R. L., and Callender, R. (1994) *Biochemistry* **33**, 7023–7032

393. Basran, J., Casarotto, M. G., Barsukov, I. L., and Roberts, G. C. K. (1995) *Biochemistry* **34**, 2872–2882

394. Jeong, S.-S., and Gready, J. E. (1995) *Biochemistry* **34**, 3734–3741

395. Appling, D. R. (1991) *FASEB J.* **5**, 2645–2651

396. Stover, P. J., Chen, L. H., Suh, J. R., Stover, D. M., Keyomarsi, K., and Shane, B. (1997) *J. Biol. Chem.* **272**, 1842–1848

397. Rosenblatt, D. S. (1995) in *The Metabolic and Molecular Bases of Inherited Disease*, 7th ed., Vol. II (Scriver, C. R., Beaudet, A. L., Sly, W. S., and Valle, D., eds), pp. 3111–3128, McGraw-Hill, New York

398. Slieker, L. J., and Bencovic, S. J. (1984) *J. Am. Chem. Soc.* **106**, 1833–1838

399. Kallen, R. G., and Jencks, W. P. (1966) *J. Biol. Chem.* **241**, 5845–5850,5851–5863

400. Walker, J. L., and Oliver, D. J. (1986) *J. Biol. Chem.* **261**, 2214–2221

401. Kume, A., Koyata, H., Sakakibara, T., Ishiguro, Y., Kure, S., and Hiraga, K. (1991) *J. Biol. Chem.* **266**, 3323–3329

402. Pasternack, L. B., Laude, D. A., Jr., and Appling, D. R. (1992) *Biochemistry* **31**, 8713–8719

403. Kennard, O., and Salisbury, S. A. (1993) *J. Biol. Chem.* **268**, 10701–10704

404. Zieske, L. R., and Davis, L. (1983) *J. Biol. Chem.* **258**, 10355–10359

404a. Barber, R. D., and Donohue, T. J. (1998) *J. Mol. Biol.* **280**, 775–784

404b. Fernández, M. R., Biosca, J. A., Torres, D., Crosas, B., and Parés, X. (1999) *J. Biol. Chem.* **274**, 37869–37875

405. Mejillano, M. R., Jahansouz, H., Matsunaga, T. O., Kenyon, G. L., and Himes, R. H. (1989) *Biochemistry* **28**, 5136–5145

406. D'Ari, L., and Rabinowitz, J. C. (1991) *J. Biol. Chem.* **266**, 23953–23958

406a. Shen, B. W., Dyer, D. H., Huang, J.-Y., D'Ari, L., Rabinowitz, J., and Stoddard, B. L. (1999) *Protein Sci.* **8**, 1342–1349

407. Schmidt, A., Wu, H., MacKenzie, R. E., Chen, V. J., Bewly, J. R., Ray, J. E., Toth, J. E., and Cygler, M. (2000) *Biochemistry* **39**, 6325–6335

407a. Pawelek, P. D., and MacKenzie, R. E. (1998) *Biochemistry* **37**, 1109–1115

408. Wahls, W. P., Song, J. M., and Smith, G. R. (1993) *J. Biol. Chem.* **268**, 23792–23798

409. Klein, C., Chen, P., Arevalo, J. H., Stura, E. A., Marolewski, A., Warren, M. S., Benkovic, S. J., and Wilson, I. A. (1995) *J. Mol. Biol.* **249**, 153–175

409a. Greasley, S. E., Yamashita, M. M., Cai, H., Benkovic, S. J., Boger, D. L., and Wilson, I. A. (1999) *Biochemistry* **38**, 16783–16793

409b. Thoden, J. B., Firestine, S., Nixon, A., Benkovic, S. J., and Holden, H. M. (2000) *Biochemistry* **39**, 8791–8802

410. Caperelli, C. A., and Giroux, E. L. (1997) *Arch. Biochem. Biophys.* **341**, 98–103

411. Schmitt, E., Blanquet, S., and Mechulam, Y. (1996) *EMBO J.* **15**, 4749–4758

412. Kruschwitz, H. L., McDonald, D., Cossins, E. A., and Schirch, V. (1994) *J. Biol. Chem.* **269**, 28757–28763

413. Lutsenko, S., Daoud, S., and Kaplan, J. H. (1997) *J. Biol. Chem.* **272**, 5249–5255

414. Rossini, C., Attygalle, A. B., González, A., Smedley, S. R., Eisner, M., Meinwald, J., and Eisner, T. (1997) *Proc. Natl. Acad. Sci. U.S.A.* **94**, 6792–6797

415. Curthoys, N. P., and Rabinowitz, J. C. (1972) *J. Biol. Chem.* **247**, 1965–1971

415a. Krupenko, S. A., and Wagner, C. (1999) *J. Biol. Chem.* **274**, 35777–35784

416. Hardy, L. W., Graves, K. L., and Nalivaika, E. (1995) *Biochemistry* **34**, 8422–8432

417. Powers, S. G., and Snell, E. E. (1976) *J. Biol. Chem.* **251**, 3786–3793

418. Carreras, C. W., and Santi, D. V. (1995) *Ann. Rev. Biochem.* **64**, 721–762

419. Schiffer, C. A., Clifton, I. J., Davisson, V. J., Santi, D. V., and Stroud, R. M. (1995) *Biochemistry* **34**, 16279–16287

420. Birdsall, D. L., Finer-Moore, J., and Stroud, R. M. (1996) *J. Mol. Biol.* **255**, 522–535

421. Sage, C. R., Rutenber, E. E., Stout, T. J., and Stroud, R. M. (1996) *Biochemistry* **35**, 16270–16281

421a. Anderson, A. C., O'Neil, R. H., DeLano, W. L., and Stroud, R. M. (1999) *Biochemistry* **38**, 13829–13836

422. Huang, W., and Santi, D. V. (1997) *Biochemistry* **36**, 1869–1873

422a. Strop, P., Changchien, L., Maley, F., and Montfort, W. R. (1997) *Protein Sci.* **6**, 2504–2511

423. Knighton, E. R., Kan, C.-C., Howland, E., Janson, C. A., Hostomska, Z., Welsh, K. M., and Matthews, D. A. (1994) *Nature Struct. Biol.* **1**, 186–194

423a. Song, H. K., Sohn, S. H., and Suh, S. W. (1999) *EMBO J.* **18**, 1104–1113

424. Butler, M. M., Graves, K. L., and Hardy, L. W. (1994) *Biochemistry* **33**, 10521–10526

425. Graves, K. L., and Hardy, L. W. (1994) *Biochemistry* **33**, 13049–13056

426. Santi, D. V., and Hardy, L. W. (1987) *Biochemistry* **26**, 8599–8606

426a. Roje, S., Wang, H., McNeil, S. D., Raymond, R. K., Appling, D. R., Shachar-Hill, Y., Bohnert, H. J., and Hanson, A. D. (1999) *J. Biol. Chem.* **274**, 36089–36096

427. Green, J. M., Ballou, D. P., and Matthews, R. G. (1988) *FASEB J.* **2**, 42–47

428. Vanoni, M. A., Lee, S., Floss, H. G., and Matthews, R. G. (1990) *J. Am. Chem. Soc.* **112**, 3987–3992

429. Matthews, R. G. (1982) *Fed. Proc.* **41**, 2600–2604

430. te Brömmelstroet, B. W., Hensgens, C. M. H., Keltjens, J. T., van der Drift, C., and Vogels, G. D. (1990) *J. Biol. Chem.* **265**, 1852–1857

431. Clark, J. E., and Ljungdahl, L. G. (1984) *J. Biol. Chem.* **259**, 10845–10849

432. Zhao, S., Roberts, D. L., and Ragsdale, S. W. (1995) *Biochemistry* **34**, 15075–15083

433. Müller, V., Blaut, M., and Gottschalk, G. (1993) in *Methanogenesis: Ecology, Physiology, Biochemistry and Genetics* (Ferry, J. G., ed), pp. 360–406, Chapman and Hall, New York

434. Ferry, J. G., ed. (1993) *Methanogenesis: Ecology, Physiology, Biochemistry and Genetics*, Chapman & Hall, New York

435. Thauer, R. K., Hedderich, R., and Fischer, R. (1993) in *Methanogenesis: Ecology, Physiology, Biochemistry and Genetics* (Ferry, J. G., ed), pp. 209–252, Chapman and Hall, New York

436. Taylor, G. D., and Wolfe, R. S. (1974) *J. Biol. Chem.* **249**, 4879–4885

437. Keltjens, J. T., Raemakers-Franken, P. C., and Vogels, G. D. (1993) in *Microbial Growth on C1 Compounds* (Murrell, J. C., and Kelly, D. P., eds), pp. 135–150, Intercept Ltd., Andover, UK

438. Leigh, J. A., Rinehart, K. L., Jr., and Wolfe, R. S. (1985) *Biochemistry* **24**, 995–999

439. Leigh, J. A., Rinehart, K. L., Jr., and Wolfe, R. S. (1984) *J. Am. Chem. Soc.* **106**, 3636–3640

440. White, R. H. (1989) *Biochemistry* **28**, 860–865

440a. Ermler, U., Merckel, M. C., Thauer, R. K., and Shima, S. (1997) *Structure* **5**, 635–646

440b. Shima, S., Warkentin, E., Grabarse, W., Sordel, M., Wicke, M., Thauer, R. K., and Ermler, U. (2000) *J. Mol. Biol.* **300**, 935–950

441. Mukhopadhyay, B., Purwantini, E., Pihl, T. D., Reeve, J. N., and Daniels, L. (1995) *J. Biol. Chem.* **270**, 2827–2832

441a. Allen, J. R., Clark, D. D., Krum, J. G., and Ensign, S. A. (1999) *Proc. Natl. Acad. Sci. U.S.A.* **96**, 8432–8437

442. Duine, J. A., and Frank, J. (1981) *Trends Biochem. Sci.* **6**, 278–280

443. Ohta, S., Fujita, T., and Tobari, J. (1981) *J. Biochem.* **90**, 205–213

444. Itoh, S., Ogino, M., Fukui, Y., Murao, H., Komatsu, M., Ohshiro, Y., Inoue, T., Kai, Y., and Kasai, N. (1993) *J. Am. Chem. Soc.* **115**, 9960–9967

445. de Jong, G. A. H., Caldeira, J., Sun, J., Jongejan, J. A., de Vries, S., Loehr, T. M., Moura, I., Moura, J. J. G., and Duine, J. A. (1995) *Biochemistry* **34**, 9451–9458

445a. Keitel, T., Diehl, A., Knaute, T., Stezowski, J. J., Höhne, W., and Görisch, H. (2000) *J. Mol. Biol.* **297**, 961–974

446. Matsushita, K., Shinagawa, E., Adachi, O., and Ameyama, M. (1989) *Biochemistry* **28**, 6276–6280

446a. Oubrie, A., Rozeboom, H. J., Kalk, K. H., Olsthoorn, A. J. J., Duine, J. A., and Dijkstra, B. W. (1999) *EMBO J.* **18**, 5187–5194

446b. Oubrie, A., Rozeboom, H. J., Kalk, K. H., Duine, J. A., and Dijkstra, B. W. (1999) *J. Mol. Biol.* **289**, 319–333

446c. Elias, M. D., Tanaka, M., Izu, H., Matsushita, K., Adachi, O., and Yamada, M. (2000) *J. Biol. Chem.* **275**, 7321–7326

446d. Dewanti, A. R., and Duine, J. A. (2000) *Biochemistry* **39**, 9384–9392

447. Xia, Z.-x, Dai, W.-w, Zhang, Y.-f, White, S. A., Boyd, G. D., and Mathews, F. S. (1996) *J. Mol. Biol.* **259**, 480–501

448. White, S., Boyd, G., Mathews, F. S., Xia, Z., Dai, W., Zhang, Y., and Davidson, V. L. (1993) *Biochemistry* **32**, 12955–12958

449. Ishida, T., Kawamoto, E., In, Y., Amano, T., Kanayama, J., Doi, M., Iwashita, T., and Nomoto, K. (1995) *J. Am. Chem. Soc.* **117**, 3278–3279

449a. Itoh, S., Kawakami, H., and Fukuzumi, S. (1998) *Biochemistry* **37**, 6562–6571

449b. Oubrie, A., Rozeboom, H. J., and Dijkstra, B. W. (1999) *Proc. Natl. Acad. Sci. U.S.A.* **96**, 11787–11791

449c. Zheng, Y.-J., and Bruice, T. C. (1997) *Proc. Natl. Acad. Sci. U.S.A.* **94**, 11881–11886

450. McIntire, W. S. (1994) *FASEB J.* **8**, 513–521

451. Anthony, C. (1996) *Biochem. J.* **320**, 697–711

452. Klinman, J. P., and Mu, D. (1994) *Ann. Rev. Biochem.* **63**, 299–344

453. Klinman, J. P. (1996) *J. Biol. Chem.* **271**, 27189–27192

454. Davidson, V. L. (1992) *Principles and Applications of Quinoproteins*, Dekker, New York

455. Janes, S. M., Mu, D., Wemmer, D., Smith, A. J., Kaur, S., Maltby, D., Burlingame, A. L., and Klinman, J. P. (1990) *Science* **248**, 981–987

456. Mu, D., Medzihradszky, K. F., Adams, G. W., Mayer, P., Hines, W. M., Burlingame, A. L., Smith, A. J., Cai, D., and Klinman, J. P. (1994) *J. Biol. Chem.* **269**, 9926–9932

457. Parsons, M. R., Convey, M. A., Wilmot, C. M., Yadav, K. D. S., Blakeley, V., Corner, A. S., Phillips, S. E. V., McPherson, M. J., and Knowles, P. F. (1995) *Structure* **3**, 1171–1184

458. Wilmot, C. M., Murray, J. M., Alton, G., Parsons, M. R., Convery, M. A., Blakeley, V., Corner, A. S., Palcic, M. M., Knowles, P. F., McPherson, M. J., and Phillips, S. E. V. (1997) *Biochemistry* **36**, 1608–1620

459. Moënne-Loccoz, P., Nakamura, N., Steinebach, V., Duine, J. A., Mure, M., Klinman, J. P., and Sanders-Loehr, J. (1995) *Biochemistry* **34**, 7020–7026

460. Parsons, M. R., Convery, M. A., Wilmot, C. M., Yadav, K. D., Blakley, V., Corner, A. S., Phillips, S. E. V., McPherson, J. J., and Knowles, P. F. (1995) *Structure* **3**, 1171–1184

461. Warncke, K., Babcock, G. T., Dooley, D. M., McGuirl, M. A., and McCracken, J. (1994) *J. Am. Chem. Soc.* **116**, 4028–4037

462. Choi, Y.-H., Matsuzaki, R., Fukui, T., Shimizu, E., Yorifuji, T., Sato, H., Ozaki, Y., and Tanizawa, K. (1995) *J. Biol. Chem.* **270**, 4712–4720

463. Choi, Y.-H., Matsuzaki, R., Suzuki, S., and Tanizawa, K. (1996) *J. Biol. Chem.* **271**, 22598–22603

463a. Schwartz, B., Green, E. L., Sanders-Loehr, J., and Klinman, J. P. (1998) *Biochemistry* **37**, 16591–16600

464. Medda, R., Padiglia, A., Pedersen, J. Z., Rotilio, G., Agrò, A. F., and Floris, G. (1995) *Biochemistry* **34**, 16375–16381

464a. Holt, A., Alton, G., Scaman, C. H., Loppnow, G. R., Szpacenko, A., Svendsen, I., and Palcic, M. M. (1998) *Biochemistry* **37**, 4946–4957

465. Wang, S. X., Mure, M., Medzihradszky, K. F., Burlingame, A. L., Brown, D. E., Dooley, D. M., Smith, A. J., Kagan, H. M., and Klinman, J. P. (1996) *Science* **273**, 1078–1084

465a. Wang, S. X., Nakamura, N., Mure, M., Klinman, J. P., and Sanders-Loehr, J. (1997) *J. Biol. Chem.* **272**, 28841–28844

466. Backes, G., Davidson, V. L., Huitema, F., Duine, J. A., and Sanders-Loehr, J. (1991) *Biochemistry* **30**, 9201–9210

466a. Zhu, Z., and Davidson, V. L. (1998) *J. Biol. Chem.* **273**, 14254–14260

467. McIntire, W. S., Wemmer, D. E., Chistoserdov, A., and Lidstrom, M. E. (1991) *Science* **252**, 817–824

468. Huizinga, E. G., van Zanten, B. A. M., Duine, J. A., Jongejan, J. A., Huitema, F., Wilson, K. S., and Hol, W. G. J. (1992) *Biochemistry* **31**, 9789–9795

469. Kuusk, V., and McIntire, W. S. (1994) *J. Biol. Chem.* **269**, 26136–26143

470. Hyun, Y.-L., and Davidson, V. L. (1995) *Biochemistry* **34**, 816–823

471. Chen, L., Durley, R., Poliks, B. J., Hamada, K., Chen, Z., Mathews, F. S., Davidson, V. L., Satow, Y., Huizinga, E., Vellieux, F. M. D., and Hol, W. G. J. (1992) *Biochemistry* **31**, 4959–4964

471a. Chen, L., Doi, M., Durley, R. C. E., Chistoserdov, A. Y., Lidstrom, M. E., Davidson, V. L., and Mathews, F. S. (1998) *J. Mol. Biol.* **276**, 131–149

472. Chen, L., Durley, R. C. E., Mathews, F. S., and Davidson, V. L. (1994) *Science* **264**, 86–90

472a. Zhu, Z., and Davidson, V. L. (1999) *Biochemistry* **38**, 4862–4867

473. Mure, M., and Klinman, J. P. (1995) *J. Am. Chem. Soc.* **117**, 8707–8718

474. Lee, Y., and Sayre, L. M. (1995) *J. Am. Chem. Soc.* **117**, 11823–11828

References

474a. Su, Q., and Klinman, J. P. (1998) *Biochemistry* **37**, 12513–12525

474b. Wilmot, C. M., Hajdu, J., McPherson, M. J., Knowles, P. F., and Phillips, S. E. V. (1999) *Science* **286**, 1724–1728

474c. Plastino, J., Green, E. L., Sanders-Loehr, J., and Klinman, J. P. (1999) *Biochemistry* **38**, 8204–8216

474d. Singh, V., Zhu, Z., Davidson, V. L., and McCracken, J. (2000) *J. Am. Chem. Soc.* **122**, 931–938

475. Itoh, S., Ogino, M., Haranou, S., Terasaka, T., Ando, T., Komatsu, M., Ohshiro, Y., Fukuzumi, S., Kano, K., Takagi, K., and Ikeda, T. (1995) *J. Am. Chem. Soc.* **117**, 1485–1493

476. Itoh, S., Kawakami, H., and Fukuzumi, S. (1997) *J. Am. Chem. Soc.* **119**, 439–440

477. Gorren, A. C. F., Moenne-Loccoz, P., Backes, G., de Vries, S., Sanders-Loehr, J., and Duine, J. A. (1995) *Biochemistry* **34**, 12926–12931

478. Moënne-Loccoz, P., Nakamura, N., Itoh, S., Fukuzumi, S., Gorren, A. C. F., Duine, J. A., and Sanders-Loehr, J. (1996) *Biochemistry* **35**, 4713–4720

479. Pedersen, J. Z., EL-Sherbini, S., Finazzi-Agrò, A., and Rotilio, G. (1992) *Biochemistry* **31**, 8–12

480. Warncke, K., Brooks, H. B., Lee, H.-i, McCracken, J., Davidson, V. L., and Babcock, G. T. (1995) *J. Am. Chem. Soc.* **117**, 10063–10075

481. Gorren, A. C. F., and Duine, J. A. (1994) *Biochemistry* **33**, 12202–12209

482. Hartmann, C., Brzovic, P., and Klinman, J. P. (1993) *Biochemistry* **32**, 2234–2241

483. Steinebach, V., de Vries, S., and Duine, J. A. (1996) *J. Biol. Chem.* **271**, 5580–5588

484. Morton, R. A. (1971) *Biol. Rev. Cambridge Philos. Soc.* **46**, 47–96

485. Morton, R. A. (1977) *Biol. Rev. Cambridge Philos. Soc.* **46**, 47–96

486. Morton, R. A. (1972) *Vitamins* **5**, 355–391

487. Suzuki, H., and King, T. E. (1983) *J. Biol. Chem.* **258**, 352–358

488. Matsuma, K., Bowyer, J. R., Ohnishi, T., and Dutton, L. P. (1983) *J. Biol. Chem.* **258**, 1571–1579

489. He, D.-Y., Yu, L., and Yu, C.-A. (1994) *J. Biol. Chem.* **269**, 27885–27888

490. He, D.-Y., Gu, L.-Q., Yu, L., and Yu, C.-A. (1994) *Biochemistry* **33**, 880–884

491. Folkers, K., Yamamuro, Y., and Ito, K., eds. (1980) *Biomedical and Clinical Aspects of Coenzyme Q*, Elsevier, Amsterdam

492. Morton, R. A. (1972) *Biochem. Soc. Symp.* **35**, 203–217

493. Gibbs, M. (1971) *Structure and Function of Chloroplasts*, Springer-Verlag, Berlin and New York

494. Trumpower, B. L., ed. (1982) *Function of Quinones in Energy Conserving Systems*, Academic Press, New York

495. Suzuki, Y. J., Tsuchiya, M., Wassall, S. R., Choo, Y. M., Govil, G., Kagan, V. E., and Packer, L. (1993) *Biochemistry* **32**, 10692–10699

496. Sun, I. L., Sun, E. E., Crane, F. L., Morré, D. J., Lindgren, A., and Löw, H. (1992) *Proc. Natl. Acad. Sci. U.S.A.* **89**, 11126–11130

497. Hughes, P. E., and Tove, S. B. (1980) *J. Biol. Chem.* **255**, 7095–7097

498. Valgimigli, L., Banks, J. T., Ingold, K. U., and Lusztyk, J. (1995) *J. Am. Chem. Soc.* **117**, 9966–9971

499. Nagaoka, S.-i, and Ishihara, K. (1996) *J. Am. Chem. Soc.* **118**, 7361–7366

500. van Belzen, R., Kotlyar, A. B., Moon, N., Dunham, R. D., and Albracht, S. P. J. (1997) *Biochemistry* **36**, 886–893

501. Salerno, J. C., Osgood, M., Liu, Y., Taylor, H., and Scholes, C. P. (1990) *Biochemistry* **29**, 6987–6993

502. Swallow, A. J. (1982) in *Function of Quinones in Energy Conserving Systems* (Trumpower, B. L., ed), pp. 59–71, Academic Press, New York

503. McTigue, J. J., Dhaon, M. K., Rich, D. H., and Suttie, J. W. (1984) *J. Biol. Chem.* **259**, 4272–4278

504. Suttie, J. W. (1993) *FASEB J.* **7**, 445–452

505. Morris, D. P., Soute, B. A. M., Vermeer, C., and Stafford, D. W. (1993) *J. Biol. Chem.* **268**, 8735–8742

506. Lingenfelter, S. E., and Berkner, K. L. (1996) *Biochemistry* **35**, 8234–8243

507. Kuliopulos, A., Hubbard, B. R., Lam, Z., Koski, I. J., Furie, B., Furie, B. C., and Walsh, C. T. (1992) *Biochemistry* **31**, 7722–7728

507a. Wu, S.-M., Mutucumarana, V. P., Geromanos, S., and Stafford, D. W. (1997) *J. Biol. Chem.* **272**, 11718–11722

508. Wood, G. M., and Suttie, J. W. (1988) *J. Biol. Chem.* **263**, 3234–3239

509. Metzler, D. E. (1977) *Biochemistry. The Chemical Reactions of Living Cells*, Academic Press, New York (p. 581)

510. Sadowski, J. A., Esmon, C. T., and Suttie, J. W. (1976) *J. Biol. Chem.* **251**, 2770–2776

511. Anton, D. L., and Friedman, P. A. (1983) *J. Biol. Chem.* **258**, 14084–14087

512. Naganathan, S., Hershline, R., Ham, S. W., and Dowd, P. (1994) *J. Am. Chem. Soc.* **116**, 9831–9839

513. Dowd, P., Hershline, R., Ham, S. W., and Naganathan, S. (1995) *Science* **269**, 1684–1691

514. Bouchard, B. A., Furie, B., and Furie, B. C. (1999) *Biochemistry* **38**, 9517–9523

514a. Zheng, Y.-J., and Bruice, T. C. (1998) *J. Am. Chem. Soc.* **120**, 1623–1624

515. Dubois, J., Gaudry, M., Bory, S., Azerad, R., and Marquet, A. (1983) *J. Biol. Chem.* **258**, 7897–7899

516. Dubois, J., Dugave, C., Fourès, C., Kaminsky, M., Tabet, J.-C., Bory, S., Gaudry, M., and Marquet, A. (1991) *Biochemistry* **30**, 10506–10512

517. Decottignies-Le Maréchal, P., Ducrocq, C., Marquet, A., and Azerad, R. (1984) *J. Biol. Chem.* **259**, 15010–15012

518. Lee, J. J., Principe, L. M., and Fasco, M. J. (1985) *Biochemistry* **24**, 7063–7070

519. Hildebrandt, E. F., and Suttie, J. W. (1982) *Biochemistry* **21**, 2406–2411

520. Fasco, M. J., Preusch, P. C., Hildebrandt, E., and Suttie, J. W. (1983) *J. Biol. Chem.* **258**, 4372–4380

521. Esmon, C. T. (1989) *J. Biol. Chem.* **264**, 4743–4746

522. Stenflo, J., Öhlin, A.-K., Owen, W. G., and Schneider, W. J. (1988) *J. Biol. Chem.* **263**, 21–24

523. Pryor, W. A. (1982) *Ann. N.Y. Acad. Sci.* **393**, 1–22

524. Ham, A.-J. L., and Liebler, D. C. (1995) *Biochemistry* **34**, 5754–5761

525. Valgimigli, L., Ingold, K. U., and Lusztyk, J. (1996) *J. Am. Chem. Soc.* **118**, 3545–3549

526. Christen, S., Woodall, A. A., Shigenaga, M. K., Southwell-Keely, P. T., Duncan, M. W., and Ames, B. N. (1997) *Proc. Natl. Acad. Sci. U.S.A.* **94**, 3217–3222

527. Schwarz, K., and Foltz, C. M. (1957) *J. Am. Chem. Soc.* **79**, 3292–3293

528. Shamberger, R. J. (1983) *Biochemistry of Selenium*, Plenum, New York

529. Keating, K. I., and Dagbusan, B. C. (1984) *Proc. Natl. Acad. Sci. U.S.A.* **81**, 3433–3437

530. Stadtman, T. C. (1991) *J. Biol. Chem.* **266**, 16257–16260

531. Stadtman, T. C. (1996) *Ann. Rev. Biochem.* **65**, 83–100

532. Burk, R. F. (1991) *FASEB J.* **5**, 2274–2279

533. Low, S. C., and Berry, M. J. (1996) *Trends Biochem. Sci.* **21**, 203–208

534. Ren, B., Huang, W., Åkesson, B., and Ladenstein, R. (1997) *J. Mol. Biol.* **268**, 869–885

535. Ladenstein, R., and Wendel, A. (1983) *Eur. J. Biochem.* **133**, 51–69

536. Shen, Q., Chu, F.-F., and Newburger, P. E. (1993) *J. Biol. Chem.* **268**, 11463–11469

537. Chu, F.-F., Doroshow, J. H., and Esworthy, R. S. (1993) *J. Biol. Chem.* **268**, 2571–2576

538. Duan, Y.-J., Komura, S., Fiszer-Szafarz, B., Szafarz, D., and Yagi, K. (1988) *J. Biol. Chem.* **263**, 19003–19008

539. Schnurr, K., Belkner, J., Ursini, F., Schewe, T., and Kühn, H. (1996) *J. Biol. Chem.* **271**, 4653–4658

540. Friedman, P. A. (1984) *N. Engl. J. Med.* **310**, 1458–1460

540a. Shisler, J. L., Senkevich, T. G., Berry, M. J., and Moss, B. (1998) *Science* **279**, 102–105

541. Visser, T. J. (1980) *Trends Biochem. Sci.* **5**, 222–224

542. Toyoda, N., Harney, J. W., Berry, M. J., and Larsen, P. R. (1994) *J. Biol. Chem.* **269**, 20329–20334

543. DePalo, D., Kinlaw, W. B., Zhao, C., Engelberg-Kulka, H., and St. Germain, D. L. (1994) *J. Biol. Chem.* **269**, 16223–16228

544. Davey, J. C., Becker, K. B., Schneider, M. J., St. Germain, D. L., and Galton, V. A. (1995) *J. Biol. Chem.* **270**, 26786–26789

545. Croteau, W., Whittemore, S. L., Schneider, M. J., and St. Germain, D. L. (1995) *J. Biol. Chem.* **270**, 16569–16575

546. St. Germain, D. L., Schwartzman, R. A., Groteau, W., Kanamori, A., Wang, Z., Brown, D. D., and Galton, V. A. (1994) *Proc. Natl. Acad. Sci. U.S.A.* **91**, 7767–7771

547. Tamura, T., and Stadtman, T. C. (1996) *Proc. Natl. Acad. Sci. U.S.A.* **93**, 1006–1011

548. Gladyshev, V. N., Jeang, K.-T., and Stadtman, T. C. (1996) *Proc. Natl. Acad. Sci. U.S.A.* **93**, 6146–6151

548a. Sun, Q.-A., Wu, Y., Zappacosta, F., Jeang, K.-T., Lee, B. J., Hatfield, D. L., and Gladyshev, V. N. (1999) *J. Biol. Chem.* **274**, 24522–24530

549. Marcocci, L., Flohé, L., and Packer, L. (1997) *BioFactors* **6**, 351–358

550. Hill, K. E., Lloyd, R. S., Yang, J.-G., Read, R., and Burk, R. F. (1991) *J. Biol. Chem.* **266**, 10050–10053

551. Himeno, S., Chittum, H. S., and Burk, R. F. (1996) *J. Biol. Chem.* **271**, 15769–15775

552. Steinert, P., Ahrens, M., Gross, G., and Flohé, L. (1997) *BioFactors* **6**, 311–319

552a. Saito, Y., Hayashi, T., Tanaka, A., Watanabe, Y., Suzuki, M., Saito, E., and Takahashi, K. (1999) *J. Biol. Chem.* **274**, 2866–2871

553. Read, R., Bellew, T., Yang, J.-G., Hill, K. E., Palmer, I. S., and Burk, R. F. (1990) *J. Biol. Chem.* **265**, 17899–17905

554. Vendeland, S. C., Beilstein, M. A., Chen, C. L., Jensen, O. N., Barofsky, E., and Whanger, P. D. (1993) *J. Biol. Chem.* **268**, 17103–17107

555. Vendeland, S. C., Beilstein, M. A., Yeh, J.-Y., Ream, W., and Whanger, P. D. (1995) *Proc. Natl. Acad. Sci. U.S.A.* **92**, 8749–8753

556. Yeh, J.-Y., Beilstein, M. A., Andrews, J. S., and Whanger, P. D. (1995) *FASEB J.* **9**, 392–396

557. Calvin, H. I., Cooper, G. W., and Wallace, E. (1981) *Gamete Research* **4**, 139–149

558. Shamberger, R. J. (1983) *Biochemistry of Selenium*, Plenum, New York (pp.47–49)

559. Yamazaki, S. (1982) *J. Biol. Chem.* **257**, 7926–7929

560. Yamazaki, S., Tsai, L., Stadtman, T. C., Teshima, T., and Nakaji, A. (1985) *Proc. Natl. Acad. Sci. U.S.A.* **82**, 1364–1366

561. Berg, B. L., Baron, C., and Stewart, V. (1991) *J. Biol. Chem.* **266**, 22386–22391

References

562. Gladyshev, V. N., Boyington, J. C., Khangulov, S. V., Grahame, D. A., Stadtman, T. C., and Sun, P. D. (1996) *J. Biol. Chem.* **271**, 8095–8100

563. Boyington, J. C., Gladyshev, V. N., Khangulov, S. V., Stadtman, T. C., and Sun, P. D. (1997) *Science* **275**, 1305–1306

564. Arkowitz, R. A., and Abeles, R. H. (1990) *J. Am. Chem. Soc.* **112**, 870–872

565. Arkowitz, R. A., and Abeles, R. H. (1991) *Biochemistry* **30**, 4090–4097

566. Stadtman, T. C., and Davis, J. N. (1991) *J. Biol. Chem.* **266**, 22147–22153

567. Arkowitz, R. A., and Abeles, R. H. (1989) *Biochemistry* **28**, 4639–4644

567a. Self, W. T., and Stadtman, T. C. (2000) *Proc. Natl. Acad. Sci. U.S.A.* **97**, 7208–7213

568. Gladyshev, V. N., Khangulov, S. V., and Stadtman, T. C. (1994) *Proc. Natl. Acad. Sci. U.S.A.* **91**, 232–236

569. Sliwkowski, M. X., and Stadtman, T. C. (1985) *J. Biol. Chem.* **260**, 3140–3144

570. Veres, Z., Tsai, L., Scholz, T. D., Politino, M., Balaban, R. S., and Stadtman, T. C. (1992) *Proc. Natl. Acad. Sci. U.S.A.* **89**, 2975–2979

571. Diamond, A. M., Choi, I. S., Crain, P. F., Hashizume, T., Pomerantz, S. C., Cruz, R., Steer, C. J., Hill, K. E., Burk, R. F., McCloskey, J. A., and Hatfield, D. L. (1993) *J. Biol. Chem.* **268**, 14215–14223

572. Diplock, A. T., and Lucy, J. A. (1973) *FEBS Lett.* **29**, 205–210

573. Sandholm, M., and Sipponen, P. (1973) *Arch. Biochem. Biophys.* **155**, 120–124

574. Sarathchandra, S. V., and Watkinson, J. H. (1981) *Science* **211**, 600–601

575. Reamer, D. C., and Zoller, W. H. (1980) *Science* **208**, 500–502

576. Mozier, N. M., McConnell, K. P., and Hoffman, J. L. (1988) *J. Biol. Chem.* **263**, 4527–4531

577. Harrison, D. J., Fluri, K., Seiler, K., Fan, Z., Effenhauser, C. S., and Manz, A. (1993) *Science* **261**, 895–897

578. Kromayer, M., Wilting, R., Tormay, P., and Böck, A. (1996) *J. Mol. Biol.* **262**, 413–420

579. Glass, R. S., Singh, W. P., Jung, W., Veres, Z., Scholz, T. D., and Stadtman, T. C. (1993) *Biochemistry* **32**, 12555–12559

580. Veres, Z., Kim, I. Y., Scholz, T. D., and Stadtman, T. C. (1994) *J. Biol. Chem.* **269**, 10597–10603

581. Mullins, L. S., Hong, S.-B., Gibson, G. E., Walker, H., Stadtman, T. C., and Raushel, F. M. (1997) *J. Am. Chem. Soc.* **119**, 6684–6685

Study Questions

1. *S*-adenosylmethionine is also a biological methyl group donor. The product of its methyl transferase reactions is *S*-adenosylhomocysteine. This product is further degraded by *S*-adenosylhomocysteine hydrolase, an enzyme that contains tightly bound NAD^+, to form homocysteine and adenosine.

   Write a step-by-step mechanism for the action of this hydrolase.

2. Compare the chemical mechanisms of enzyme-catalyzed decarboxylation of the following:
   a) a β-oxo-acid such as acetoacetate or oxalo-acetate
   b) an α-oxo-acid such as pyruvate
   c) an amino acid such as L-glutamate

3. Describe the subunit structure of the enzyme pyruvate dehydrogenase. Discuss the functioning of each of the coenzymes that are associated with these subunits and write detailed mechanisms for each step in the pyruvate dehydrogenase reaction.

4. Free **formate** can be assimilated by cells via the intermediate **10-formyl-tetrahydrofolate** (10-formyl-THF).
   a) Describe the mechanism of synthesis of this compound from formate and tetrahydrofolate.
   b) Diagram a hypothetical transition state for the first step of this reaction sequence.
   c) Describe two or more uses that the human body makes of 10-formyl-THF.

5. Using partial structural formulas, describe the reactions by which serine and methionine react to form *N*-formylmethionine needed for protein synthesis.

6. Write the equations for each of the reactions shown below. Using the $E^{0'}$ values below, calculate approximate Gibbs energies for each reaction, and show by the relative length of the arrows on which side of the reaction the equilibrium lies.
   a) The oxidation of malate by NAD^+
   b) The oxidation of succinate by NAD^+
   c) The oxidation of succinate by enzyme-bound FAD
   d) What can you say about the cofactor required for oxidation of succinate from your calculations?

   The values of $E^{0'}$ for several half reactions are given below. Everything has been rounded to one significant figure so that a calculator is unnecessary.

Reaction	$E^{0'}$ (volts)
$NAD^+ + H^+ + 2\,e^- \rightarrow NADH$	−0.3
enzyme bound FAD + 2 H^+ + 2 $e^- \rightarrow$ enzyme bound $FADH_2$	0.0
fumarate + 2 H^+ + 2 $e^- \rightarrow$ succinate	0.0
oxalacetate + 2 H^+ + 2 $e^- \rightarrow$ malate	−0.2

7. Some acetogenic bacteria, which convert CO_2 to acetic acid, form pyruvate for synthesis of carbohydrates, etc., by formation of formaldehyde and conversion of the latter to glycine by reversal of the PLP and lipoic acid-dependent glycine decarboxylase, a 4-protein system. The glycine is then converted to serine, pyruvate, oxaloacetate, etc. Propose a detailed pathway for this sequence.

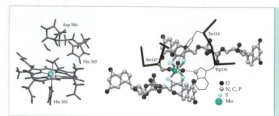

Transition metal ions function in many biological oxidation-reduction processes. (Left) A heme ring in **nitrite reductase** chelates an iron ion in its center, holding the Fe^{2+} or Fe^{3+} with bonds from four nitrogen atoms and a fifth bond from an imidazole ring below. Above the iron is a nitrite ion, NO_2^-, awaiting reduction to nitric oxide NO in a denitrifying bacterium. From Ranghino *et al.* (2000) *Biochemistry* **39**, 10958–10966. (Right) The active site of a bacterial **dimethylsulfoxide reductase** has an atom of molybdenum or tungsten at its center. The metal is held by four sulfur atoms from two molybdopterin molecules and an oxygen atom of a serine side chain. Two other oxygen atoms are bound as oxo groups and may participate in catalysis. From Stewart *et al.* (2000) *J. Mol. Biol.* **299**, 593–600.

Contents

Transition Metals in Catalysis and Electron Transport

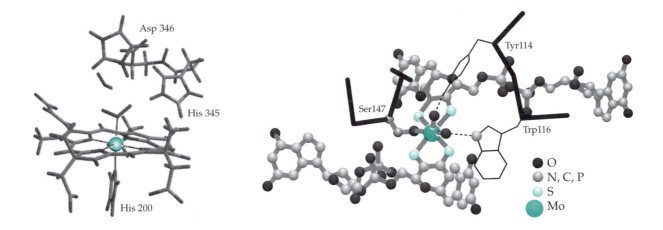

16

Asp 346

His 345

His 200

Tyr114

Ser147

Trp116

○ O
○ N, C, P
○ S
● Mo

Although the amounts present within living cells are very small, the ions of the transition metals Fe, Co, Ni, Cu, and Mn are extremely active centers for catalysis, especially of reactions that take advantage of the ability of these metals to exist in more than one oxidation state.[1–4] Iron, copper, and nickel are also components of the electron carrier proteins that function as oxidants or reductants in many biochemical processes. These metals are all nutritionally essential, as are chromium and vanadium. Among the heavier transition elements molybdenum is a constituent of an important group of enzymes that includes the sulfite oxidase of human liver and nitrogenase of nitrogen-fixing bacteria. Tungsten occasionally substitutes for molybdenum.

A. Iron

Iron is one of the most abundant elements in the earth's crust, being present to the extent of ~4% in a typical soil. Its functions in living cells are numerous and diverse.[2,5–8] The average overall iron content of both bacteria and fungi is ~1 mmol / kg, but that of animal tissues is usually less. Seventy percent of the 3–5 g of iron present in the human body is located in the blood's erythrocytes, whose overall iron content is ~20 mM. In other tissues the total iron averages closer to 0.3 mM and consists principally of storage forms. The total concentration of iron in all of the iron-containing *enzymes* of tissues amounts to only about 0.01 mM. Although these concentrations are low, the iron is concentrated in oxidative enzymes of membranes and may attain much higher concentrations locally. Only a few parasitic or anaerobic bacteria, e.g. the lactic acid

bacteria, possess no oxygen-requiring enzymes and are almost devoid of both iron and copper. All other organisms appear to require iron for life.

1. Uptake by Living Cells

A major problem for cells is posed by the relative insolubility of ferric hydroxide and other compounds from which iron must be extracted by the organism. A consequence is that iron is often taken up in a chelated form and is transferred from one organic ligand, often a protein, to another with little or no existence as free Fe^{3+} or Fe^{2+}. As can be calculated from the estimated solubility product of $Fe(OH)_3$ (Eq. 16-1),[7] the equilibrium concentration of Fe^{3+} at pH 7 is only 10^{-17} M.

$$K_{sp} = [Fe^{3+}][OH^-]^3 < 10^{-38} \text{ M}^4 \qquad (16\text{-}1a)$$

$$\text{or } [Fe^{3+}] / [H^+]^3 < 10^4 \text{ M}^4 \text{ at } 25° \text{ C} \qquad (16\text{-}1b)$$

For a 2 μm³ bacterial cell this amounts to just one free Fe^{3+} ion in almost 100 million cells at any single moment. The importance of chelated forms of iron becomes obvious. It is also evident from Eq. 16-1 that, in addition to chelation, a low external pH can also facilitate uptake of Fe^{3+} by organisms.

The values of the formation constants for chelates of Fe^{2+} typically lie between those of Mn^{2+} and Co^{2+} (Fig. 6-6, Table 6-9). For example, $K_1 = 10^{14.3}$ M⁻¹ for formation of the Fe^{2+} chelate of EDTA. The smaller and more highly charged Fe^{3+} is bonded more strongly ($K_1 = 10^{25}$ M⁻¹). These binding constants are independent of pH. However, the binding of any metal ion is affected by pH, as discussed in Chapter 6. A fact of considerable biochemical significance is the stronger binding of Fe^{3+} to oxygen-containing ligands than to nitrogen atoms, while Fe^{2+} tends to bind preferentially to nitrogen. It is also significant that Fe^{3+} bound to oxygen ligands tends to exchange readily with other ferric ions in the medium, whereas Fe^{3+} bound to nitrogen-containing ligands such as heme exchanges slowly. This fact is important for both iron-transport compounds and enzymes.

Siderophores.

If a suitably high content of iron (e.g., 50 µM or more for *E. coli*) is maintained in the external medium, bacteria and other microorganisms have little problem with uptake of iron. However, when the external iron concentration is low, special compounds called siderophores are utilized to render the iron more soluble.[7–11c] For example, at iron concentrations below 2 µM, *E. coli* and other enterobacteria secrete large amounts of **enterobactin** (Fig. 16-1). The stable Fe^{3+}–enterobactin complex is taken up by a transport system that involves receptors on the outer bacterial membrane.[9,12,13] Siderophores from many bacteria have in common with enterobactin the presence of **catechol** (*ortho*-dihydroxybenzene) groups that chelate the iron.

The three catechol groups of enterobactin are carried on a cyclic serine triester structure. A variety of both cyclic and linear structures are found among other catechol siderophores.[14–19] For example, **parabactin** and **agrobactin** (Fig. 16-1) contain a backbone of spermidine[20] (Chapter 24). After the Fe^{3+}–enterobactin complex enters a bacterial cell the ester linkages of a siderophore are cleaved by an esterase. Because of the extremely high formation constant of ~10^{52} M^{-1} for the complex[11] the only way for a cell to release the Fe^{3+} is through this irreversible destruction of the iron carrier.[11d] Reduction to Fe^{2+} may be involved in release of iron from some siderophores.[11e]

The first known siderophore, isolated in 1952 by Neilands,[22] is **ferrichrome** (Fig. 16-1), a cyclic hexapeptide containing **hydroxamate** groups at the iron-binding centers. Oxygen atoms form the bonds to iron

Hydroxamate group

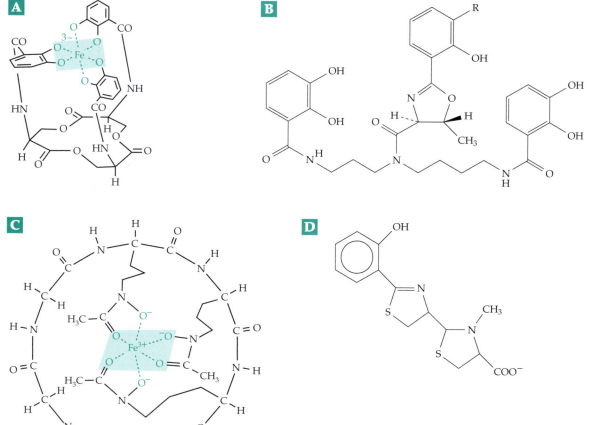

Figure 16-1 Structures of several siderophores and of their metal complexes. (A) Enterobactin of *E. coli* and other enteric bacteria;[12] (B) parabactin (R = H) from *Paracoccus denitrificans* and agrobactin (R = OH) from *Agrobacterium tumefaciens*;[20] (C) ferrichrome;[9] (D) pyochelin from *Pseudomonas aeruginosa*.[21]

in this compound also. Ferrichrome binds Fe^{3+} with a formation constant of $\sim 10^{29}$ M^{-1}. The binding is not as tight as with enterobactin and the iron can be released by enzymatic reduction to Fe^{2+} which is much less tightly bound than is Fe^{3+}. The released ferrichrome can be secreted and used repeatedly to bring in more iron. Ferrichrome is produced by various fungi and bacilli and is only one of a series of known hydroxamate siderophores.[16] Since iron is essential to virtually all parasitic organisms the ability to obtain iron is often the limiting factor in establishing an infection.[23,24]

In *E. coli* there are seven different outer **membrane receptors** for siderophores.[25] One of these, the gated porin **FepA**, is specific for ferric enterochelin. With the assistance of another protein, **TonB**, it allows the ferric siderophore to penetrate the outer membrane.[26] A different receptor, **FhuA**, binds ferrichrome. Both FepA and FhaA are large 22-strand porins resembling the 16-strand porin shown in Fig. 8-20. However, they are nearly 7 nm long with internal diameters three times those of the 16-strand porins. In addition, loops of polypeptide chain on the outer edges can close while an N-terminal domain forms a "cork" that remains in place until the Fe^{3+}-siderophore complex enter the channel and binds. Like the hatches in an air lock on a spacecraft, the outer loops then close, after which the inner cork unwinds to allow the siderophore complex to enter the periplasmic space. The apparatus requires an energy supply, which apparently is provided by an additional complex consisting of proteins TonB, ExbA, and ExbD. They evidently couple the electrochemical gradient across the cell membrane (Chapter 8, B1 and C5 and Chapter 18) with the operation of the channel gates. Some bacteria use a different strategy for passage across the outer membrane.[11c] The channel of a receptor protein contains a molecule of an iron-free siderophore. When a molecule of Fe^{3+}-siderophore binds in the outside part of the channel the Fe^{3+} jumps to the inner siderophore, which then dissociates from the receptor, carrying the Fe^{3+}-siderophore complex into the periplasmic space. This mechanism also seems to be available in *E. coli*. The **ferric uptake regulation** (Fur) protein binds excess free Fe^{2+}, the resulting complex acting as a repressor of all of the iron uptake genes in *E. coli*.[11e]

Additional proteins are required for passage through the membrane.[11b,23,25,27] These are ABC transporters, which utilize hydrolysis of ATP as an energy source (Chapter 8, Section C,4). For uptake of the Fe^{3+}-ferrichrome complex protein **FhuD** is the periplasmic binding protein, **FhuC** is an integral membrane component, and the cytosolic **FhuC** contains the ATPase center.[11b] Another ABC transporter, found in many bacteria, carries unchelated Fe^{3+} across the inner membrane. The binding protein component for *Hemophilus influenzae*, designated Hit, resembles one lobe of mammalian transferrin (Fig. 16-2).[11b] The siderophore

receptors of bacteria have been "parasitized" by various bacteriophages and toxic proteins. For example, FepA is also a receptor for the toxic colicins B and D (Box 8-D) and tonB is a receptor for bacteriophage T1.[9,13]

Some bacteria do not form siderophores but take up Fe^{2+}. Even *E. coli*, when grown anaerobically, synthesizes an uptake system for Fe^{2+}. It utilizes a 75-residue peptide encoded by gene *feoA* and a 773-residue protein encoded by *feoB*.[28,11b]

Uptake of iron by eukaryotic cells. The yeast *Saccharomyces cerevisiae* utilizes two systems for uptake of iron.[29,30] A low-affinity system transports Fe^{2+} with an apparent K_m of ~ 30 μM, while a high-affinity system has a K_m of ~ 0.15 μM. Study of these systems has been greatly assisted by the use of genetic methods developed for both bacteria and yeast (Chapter 26). The low-affinity iron uptake depends upon a protein transporter encoded by gene *FET4* and on a reductase encoded by genes *FRE1* and *FRE2*, proteins that are embedded in the cytoplasmic membrane.[29,31-33] It might seem reasonable that the *FET3* copper oxidoreductase should keep Fe^{2+} *reduced* while it is transported. However, it appears to *oxidize* Fe^{2+} to Fe^{3+}. The high-affinity uptake system is more puzzling. It requires a permease encoded by *FTR1* and an additional protein encoded by *FET3*.[30,33-35] The Fet3 protein is a copper oxidoreductase related to **ceruloplasmin** (Section D). The protein **Fre1p** (encoded by *FRE1*) is a metalloreductase that reduces Cu^{2+} to Cu^+, as well as Fe^{3+} to Fe^{2+}. It is essential for copper uptake (Section D).[33] It has long been known that ceruloplasmin is required for mobilization of iron from mammalian tissues.[30] Hereditary ceruloplasmin deficiency causes accumulation of iron in tissues.[36] Yeast also contains both *mitochondrial* and *vacuolar* iron transporters.[37,37a,b]

The uptake of iron by animals is not as well understood[38-40] but it resembles that of yeast.[41] A general divalent cation transporter that is coupled to the membrane proton gradient is involved in intestinal iron uptake.[42,60] Ascorbic acid promotes the uptake of iron, presumably by reducing it to Fe(II), which is more readily absorbed than Fe(III), and also by promoting ferritin synthesis.[43] Uptake is also promoted by meat in the diet.[44] Within the body iron is probably transferred from one protein to another with only a transient existence as free Fe^{2+}. An average daily human diet contains ~ 15 mg of iron, of which ~ 1 mg is absorbed. This is usually enough to compensate for the small losses of the metal from the body, principally through the bile. Once it enters the body, iron is carefully conserved. The 9 billion red blood cells destroyed daily yield 20–25 mg of iron which is almost all reused or stored. The body apparently has no mechanism for excretion of large amounts of iron; a person's iron content is regulated almost entirely by the rate of uptake. This rate is increased during pregnancy and, in young women, to compensate for iron lost in menstrual bleeding. Nevertheless, control of iron uptake is imperfect and perhaps 500 million people around the world suffer from iron deficiency.[44,45] For others, an excessive intake of iron or a genetic defect lead to accumulation of iron to toxic levels, a condition called **hemochromatosis**.[44,46,46a,b] This condition may also arise in any disease that leads to excessive destruction of hemoglobin or accumulation of damaged erythrocytes. Examples are β-thalassemia (Chapter 28) and cerebral malaria.[47] Treatment with chelating agents designed to remove iron is often employed.[47,48]

Transferrins. Within the body iron is moved from one location to another while bound as Fe^{3+} to transferrins, a family of related 680- to 700-residue 80-kDa proteins.[38,49–53] Each transferrin molecule contains two Fe^{+3} binding sites, one located in each of two similar domains of the folded peptide chain. A dianion, usually CO_3^{2-}, is bound together with each Fe^{3+}. Milk transferrin (**lactoferrin**[51a,b,c] also found in leukocytes), hen egg transferrin (**ovotransferrin**),[52,52a] and rabbit and human serum transferrin[54,54a] all have similar structures. Each Fe^{3+} is bonded to oxygen anions from two tyrosine side chains, an aspartate carboxylate, an imidazole group, and the bound carbonate ion (Fig. 16-2B). Transferrin of blood plasma is encoded by a separate gene but has a similar structure. Transferrin of chickens appears to be identical to conalbumin of egg whites. The iron-binding proteins of body fluids are sometimes given the group name **siderophilins**. Transferrins may function not only in transport of iron throughout the body but also as iron buffers that provide a relatively constant iron concentration within tissues.

The entrance of iron into the body through the intestinal mucosal cells may involve the transferrin present in those cells[44] and the influx of iron may also be regulated by blood plasma transferrin. There is also a nontransferrin pathway.[42,55]

Transferrins bind Fe^{2+} weakly and it is likely that a transferrin–Fe^{2+}–HCO_3^- complex formed initially undergoes oxidation to the Fe^{3+}–CO_3^{2-} complex within cells and within the bloodstream. A conformational change closes the protein around the iron ions.[56] In yeast the previously mentioned copper oxidoreductase encoded by the *FET3* gene appears to not only oxidize Fe^{2+} but also transfer the resulting Fe^{3+} to transferrin. Ceruloplasmin may play a similar role in mammals.[33]

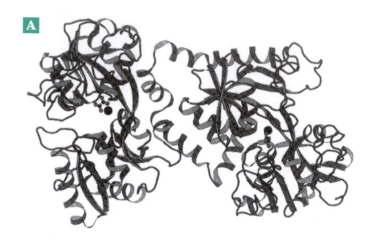

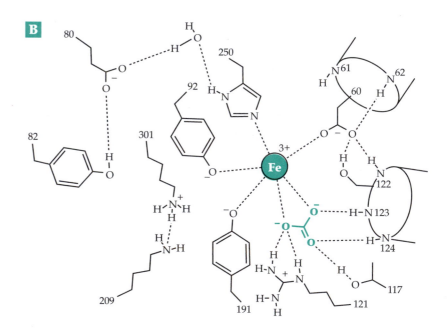

Figure 16-2 (A) Ribbon drawing of the polypeptide chain of a transferrin, human lactoferrin. The N lobe is to the left and the C lobe to the right. Each active site contains bound Fe^{3+} and a molecule of oxalate dianion which replaces the physiological CO_3^{2-}. From Baker *et al.*[51] Courtesy of Edward Baker. (B) Schematic diagram showing part of the hydrogen-bond network involved in binding the Fe^{3+} in the N lobe of hen ovotransferrin. Some side chain groups and water molecules have been omitted. The positions of hydrogen atoms and the charge state of acid–base groups are uncertain. Most of the hydrogen-bond distances (O - - O, O - - N, N - - N) indicated by dashed lines are between 0.27 and 0.3 nm. Release of the bound Fe^{3+} may be accomplished in part by protonation of the bound CO_3^{2-} to form HCO_3^-. See Kurokawa *et al.*[52]

Iron is transferred from the plasma transferrin into cells of the body following binding of the Fe^{3+}–transferrin complex to specific receptors. The surface of an immature red blood cell (reticulocyte) may contain 300,000 transferrin receptors, each capable of catalyzing the entry of ~36 iron ions per hour.[38] The receptor is a 180-kDa dimeric glycoprotein. When the Fe^{3+}–transferrin complex is bound, the receptors aggregate in coated pits and are internalized. The mechanism of release of the Fe^{3+} may occur by different mechanisms in the two lobes.[57] The pH of the endocytic vesicles containing the receptor complex is probably lowered to ~5.6 as in lysosomes. This may protonate the bound CO_3^{2-} in the complex[51c,54a,58] and assists in the release of the Fe^{3+}, possibly after reduction to Fe^{2+}. Both the apo-transferrin and its receptor are returned to the cell surface for reuse, the apotransferrin being released into the blood. Chelating agents such as pyrophosphate, ATP, and citrate as well as simple anions[59] may also assist in removal of iron from transferrin. The same trans-membrane transporter that is involved in intestinal iron uptake[42] is needed to remove iron from the endosome after release.[60]

2. Storage of Iron

Within tissues of animals, plants, and fungi much of the iron is packaged into the red-brown water-soluble protein **ferritin**, which stores Fe(III) in a soluble, nontoxic, and readily available form.[61–64] Although bacteria store very little iron,[65] some of them also contain a type of ferritin.[66–67a] On the other hand, the yeast *S. cerevisieae* stores iron in polyphosphate-rich granules, even though a ferritin is also present.[65] Ferritin contains 17–23% iron as a dense core of hydrated ferric oxide ~7 nm in diameter surrounded by a protein coat made up of twenty-four subunits of mo-

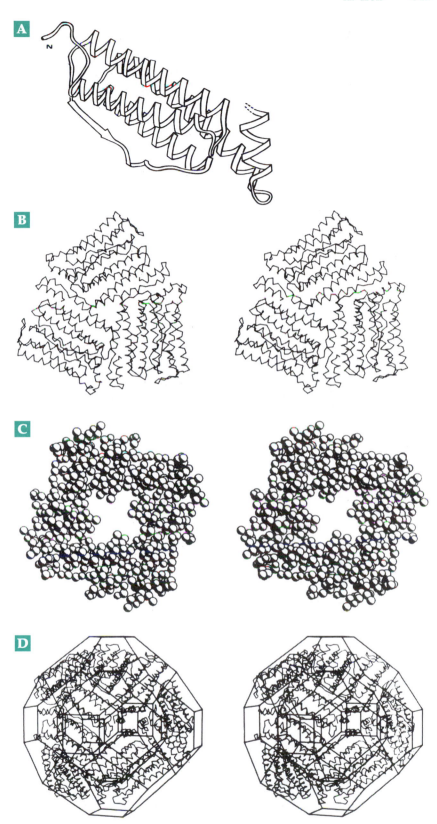

Figure 16-3 Structure of the protein shell of ferritin (apoferritin). (A) Ribbon drawing of the 163-residue monomer. From Crichton.[62] (B) Stereo drawing of a hexamer composed of three dimers. (C) A tetrad of four subunits drawn as a space-filling diagram and viewed down the four-fold axis from the exterior of the molecule. (D) A half molecule composed of 12 subunits inscribed within a truncated rhombic dodecahedron. B–D from Bourne *et al.*[74]

lecular mass 17- to 21-kDa. Each subunit is folded as a four-helix bundle (Fig. 16-3). Mammalian ferritins consist of combinations of subunits of two or more types. For example, human ferritins contain similarly folded 19-kDa L (light) and 21-kDa H (heavy) subunits.[68] The twenty-four subunits are arranged in a cubic array (Fig. 7-13, Fig. 16-3). The outer diameter of the 444-kDa apoferritin is ~12 nm. The completely filled ferritin molecule contains 23% Fe and over 2000 atoms of iron in a crystalline lattice. Larger ferritins may contain as many as 4500 atoms of iron with the approximate composition $[Fe(O)OH]_8FeOPO_3H_2$.[69] Phosphate ions are sometimes bound into surface layers of the ferritin cores.[69a] Ferritin cores are readily visible in the electron microscope, and ferritin is often used as a labeling reagent in microscopy. Another

storage form of iron, **hemosiderin**, seems to consist of ferritin partially degraded by lysosomes and containing a higher iron content than does ferritin. Depositions of hemosiderin in the liver can rise to toxic levels if excessive amounts of iron are absorbed.

Iron can be deposited in ferritin by allowing apoferritin to stand with an Fe(II) salt and a suitable oxidant, which may be O_2. Physiological transfer of Fe(III) from transferrin to ferritin is thought to require prior reduction to Fe(II). The reoxidation by O_2 to Fe(III) for deposition in the ferritin core (Eq. 16-2) is catalyzed by **ferrooxidase sites** located in the centers of the helical bundles of the H-chains.[70–73]

$$2\,Fe^{2+} + O_2 + 4\,H_2O \;\rightarrow$$
$$2\,Fe(O)OH_{core} + H_2O_2 + 4\,H^+ \qquad (16\text{-}2)$$

BOX 16-A MAGNETIC IRON OXIDE IN ORGANISMS

An unusual form of stored iron is the magnetic iron oxide **magnetite** (Fe_3O_4). Honeybees,[a,b] monarch butterflies,[b–c] homing pigeons,[c–e] migrating birds, and even magnetotactic bacteria[f] contain deposits of Fe_3O_4 that are suspected of being used in nagivation.[c,g] Some bacteria have magnetic iron sulfide particles.[h–j] Human beings have magnetic bones in their sinuses[k] and in their brains[l] and may be able to sense direction magnetically. A set of possible magnetoreceptor cells, as well as associated nerve pathways, have been identified in trout.[j] In the magnetotactic bacteria found in the Northern Hemisphere the magnetic domains are oriented parallel with the axis of motility of the bacteria which tend to swim toward the geomagnetic North and downward into sediments. Similar bacteria from the Southern Hemisphere prefer to swim south and downward. The magnetic polarity of the bacterial magnetite crystals can be reversed by strong magnetic pulses, after which the bacteria swim in the direction opposite to their natural one.[m] Magnetic ferritin can be produced artificially in the laboratory.[n] The resulting particles may have practical uses, for

example, in medical magnetic imaging. Magnetic materials in the human body are of interest not only in terms of a possible sensory function but also because of possible effects of electromagnetic fields on human health and behavior.[o]

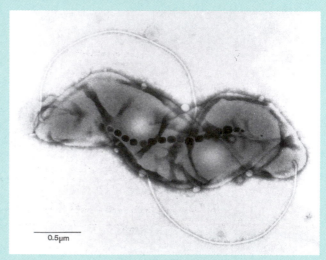

Magnetotactic soil bacterium containing 36 magnetite-containing magnetosomes. Courtesy of Dennis Bazylinski.

---

[a] Hsu, C.-Y., and Li, C.-W. (1994) *Science* **265**, 95–97

[b] Nichol, H., and Locke, M. (1995) *Science* **269**, 1888–1889

[bc] Etheredge, J. A., Perez, S. M., Taylor, O. R., and Jander, R. (1999) *Proc. Natl. Acad. Sci. U.S.A.* **96**, 13845–13846

[c] Gould, J. L. (1982) *Nature (London)* **296**, 205–211

[d] Guilford, T. (1993) *Nature (London)* **363**, 112–113

[e] Moore, B. R. (1980) *Nature (London)* **285**, 69–70

[f] Blakemore, R. P., and Frankel, R. B. (1981) *Sci. Am.* **245** (Dec), 58–65

[g] Maugh, T. H., II,(1982) *Science* **215**, 1492–1493

[h] Dunin-Borkowski, R. E., McCartney, M. R., Frankel, R. B., Bazylinski, D. A., Pósfai, M., and Buseck, P. R. (1998) *Science* **282**, 1868–1870

[i] Pósfai, M., Buseck, P. R., Bazylinski, D. A., and Frankel, R. B. (1998) *Science* **280**, 880–883

[j] Walker, M. M., Diebel, C. E., Haugh, C. V., Pankhurst, P. M., Montgomery, J. C., and Green, C. R. (1997) *Nature (London)* **390**, 371–376

[k] Baker, R. R., Mather, J. G., and Kennaugh, J. H. (1983) *Nature (London)* **301**, 78–80

[l] Kirschvink, J. L., Kobayashi-Kirschvink, A., and Woodford, B. J. (1992) *Proc. Natl. Acad. Sci. U.S.A.* **89**, 7683–7687

[m] Blakemore, R. P., Frankel, R. B., and Kalmijn, A. J. (1980) *Nature (London)* **286**, 384–385

[n] Meldrum, F. C., Heywood, B. R., and Mann, S. (1992) *Science* **257**, 522–523

[o] Barinaga, M. (1992) *Science* **256**, 967

The ferrooxidase site is a dinuclear iron center (see Section 8) in which two iron ions (probably Fe^{2+}) are bound as in Fig. 16-4. They are then converted to Fe^{3+} ions by O_2, which may bind initially to the Fe^{2+}, forming a transient blue intermediate that is thought to have a peroxodiferric structure, perhaps of the following type.[71–73] Reaction of this intermediate with H_2O

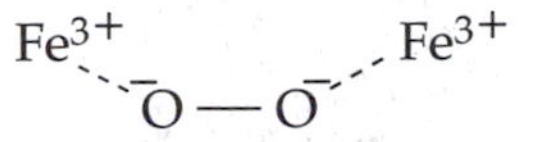

$$Fe^{3+} \underset{O-O}{} Fe^{3+}$$

may yield H_2O_2 plus a biomineral precursor $Fe^{2+} - O - Fe^{2+}$, which is incorporated into the core.[72a] Ferritin H subunits predominate in tissues with high oxygen levels, e.g., heart and blood cells, while the L subunits predominate in tissues with slower turnover of iron, e.g., liver.[72] The L subunits lack ferrooxidase activity but, in the centers of their helical bundles, contain polar side chains that may help to initiate growth of the mineral core.[64]

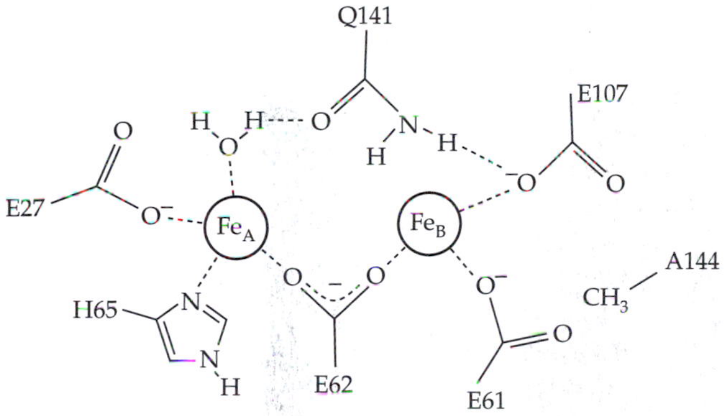

Figure 16-4 The dinuclear iron center or ferroxidase center of human ferritin based on the structure of a terbium(III) derivative.[73] Courtesy of Pauline Harrison.

Removal of Fe(III) from storage in ferritin cores may require reduction to Fe(II) again, possibly by ascorbic acid[75] or glutathione. Some bacterial ferritins contain a bound cytochrome *b* which may assist in reduction.[67,67a] Released iron in the Fe^{2+} state can be incorporated into iron-containing proteins or into heme. The enzyme **ferrochelatase**[76–76b] catalyzes the transfer of free Fe^{2+} into protoporphyrin IX (Section 3) to form protoheme (Fig. 16-5). Iron in the Fe(II) state may also be oxidized to Fe^{3+} through action of the copper-containing ceruloplasmin (Section D) and be incorporated into heme by direct transfer from ferritin.[53]

3. Heme Proteins

In 1879, German physiological chemist Hoppe-Seyler showed that two of the most striking pigments of nature are related. The red iron-containing **heme** from blood and the green magnesium complex **chlorophyll a** of leaves have similar ring structures. Later, H. Fischer proved their structures and provided them with the names and numbering systems that are used today. This information is summarized in the following section.

Some names to remember. **Porphins** are planar molecules which contain large rings made by joining four pyrrole rings with methine bridges. In the **chlorins**, found in the chlorophylls, one of the rings (ring D in chlorophyll, Fig. 23-20) is reduced. The specific class of porphins known as **porphyrins** have eight substituents around the periphery of the large ring. Like the chlorins and the **corrins** of vitamin B_{12} (Section B), the porphyrins are all formed biosynthetically from **porphobilinogen**. This compound is polymerized in two ways (see Fig. 24-21) to give porphyrins of types I and III (Fig. 16-5). In type I porphyrins, polymerization of porphobilinogen has taken place in a regular way so that the sequences of the carboxymethyl and carboxyethyl side chains (often referred to as acetic acid and propionic acid side chains, respectively) are the same all the way around the outside of the molecule. However, most biologically important porphyrins belong to type III, in which the first three rings A, B, and C have the same sequence of carboxymethyl and carboxyethyl side chains, but in which ring D has been incorporated in a reverse fashion. Thus, the carboxyethyl side chains of rings C and D are adjacent to each other (Fig. 16-5). Porphyrins containing all four carboxymethyl and four carboxyethyl side chains intact are known as **uroporphyrins**.

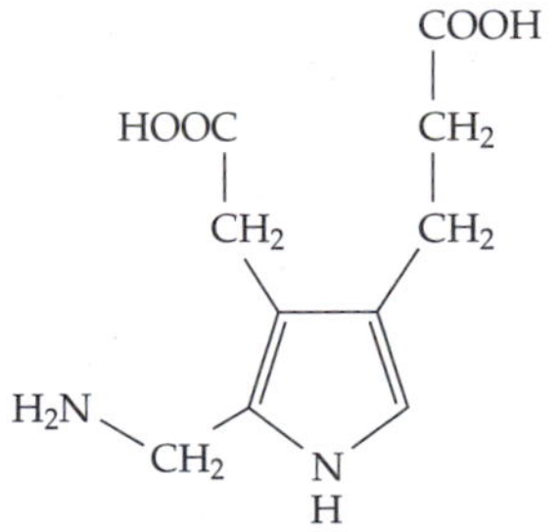

Porphobilinogen

Uroporphyrins I and III are both excreted in small amounts in the urine. Another excretion product is **coproporphyrin** III, in which all of the carboxymethyl side chains have been decarboxylated to methyl groups. The feathers of the tropical touraco are colored with copper(II) complex of coproporphyrin III and this porphyrin as well as others are commonly found in birds' eggs. The heme proteins are all derived from protoporphyrin IX, which is formed by decarboxylation and dehydrogenation of two of the carboxyethyl side chains of uroporphyrin III to vinyl groups (Fig. 16-5).

Hemes and heme proteins. Protoporphyrin IX contains a completely conjugated system of double bonds. In the center two hydrogen atoms are attached, one each to two of the nitrogens; they are free to move to other nitrogens in the center with rearrangement of the double bonds. Thus, there is tautomerism as well as resonance within the heme ring.[77–78a] The two central hydrogens can be replaced by many metal ions to form stable chelates. The complexes with Fe^{2+} are known as **hemes** and the Fe^{2+} complex with protoporphyrin IX as **protoheme**. Heme complexes of Fe^{2+} may be designated as **ferrohemes** and the Fe^{3+} compounds as **ferrihemes**. The Fe^{3+} protoporphyrin IX

is also called **hemin**, and may be crystallized as a chloride salt.[78b] Iron tends to have a coordination number of six, and other ligands can attach to the iron from the two axial positions on opposite sides of the planar heme. If these are nitrogen ligands, such as pyridine or imidazole, the resulting compounds, called **hemochromes**, have characteristic absorption spectra. An example is cytochrome b_5, which contains two axial imidazole groups.

Several modifications of protoheme are indicated in Fig. 16-5. To determine which type of heme exists in a particular protein, it is customary to split off the heme by treatment with acetone and hydrochloric acid and to convert it by addition of pyridine to the pyridine hemochrome for spectral analysis. By this means, protoheme was shown to occur in hemoglobin, myoglobin, cytochromes of the b and P450 types, and catalases and many peroxidases. Cytochromes a and a_3 contain **heme *a***, while one of the terminal oxidase

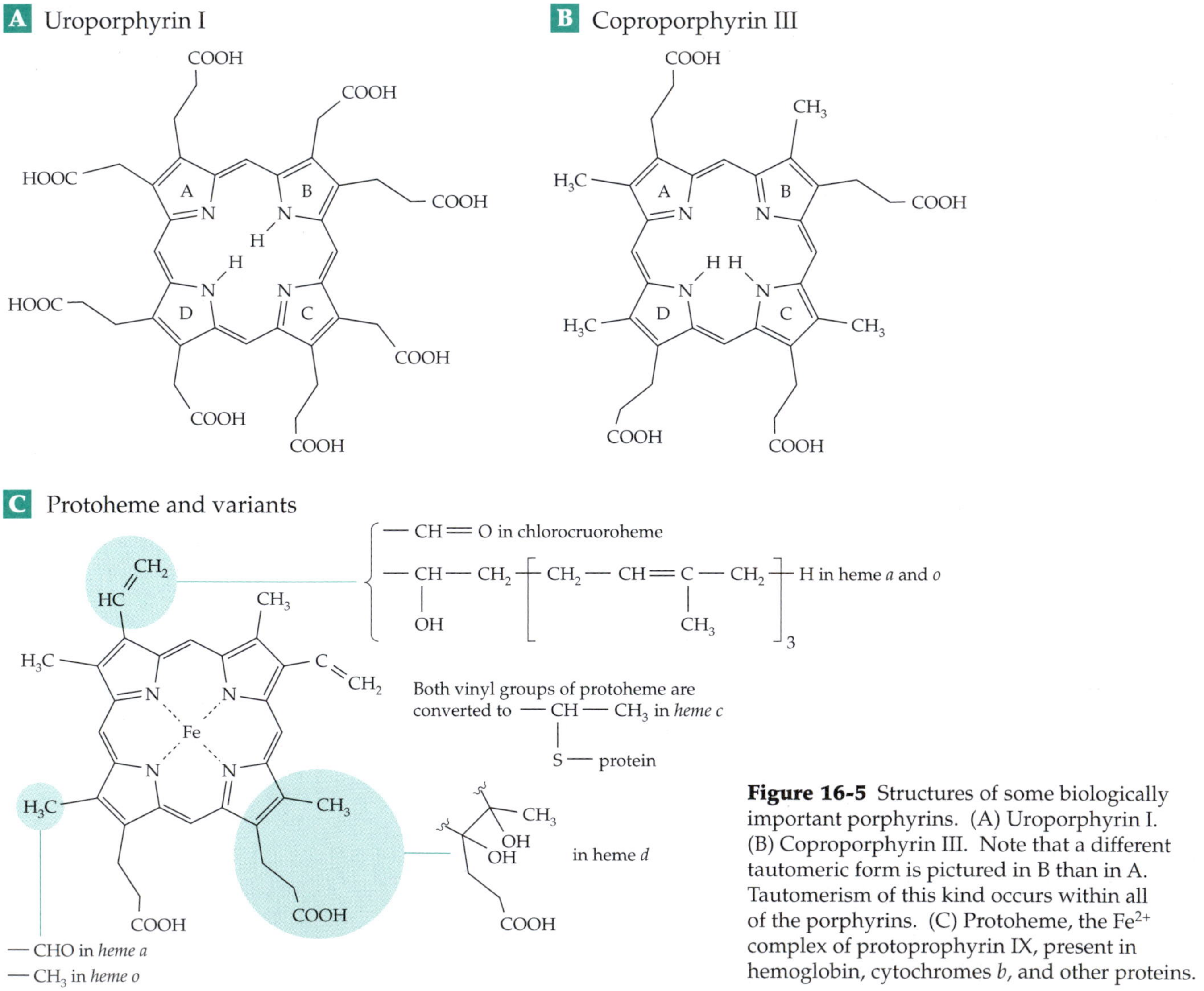

A Uroporphyrin I

B Coproporphyrin III

C Protoheme and variants

— CHO in *heme a*
— CH_3 in *heme o*

Both vinyl groups of protoheme are converted to — CH — CH_3 in *heme c*

S — protein

in *heme d*

Figure 16-5 Structures of some biologically important porphyrins. (A) Uroporphyrin I. (B) Coproporphyrin III. Note that a different tautomeric form is pictured in B than in A. Tautomerism of this kind occurs within all of the porphyrins. (C) Protoheme, the Fe^{2+} complex of protoprophyrin IX, present in hemoglobin, cytochromes b, and other proteins.

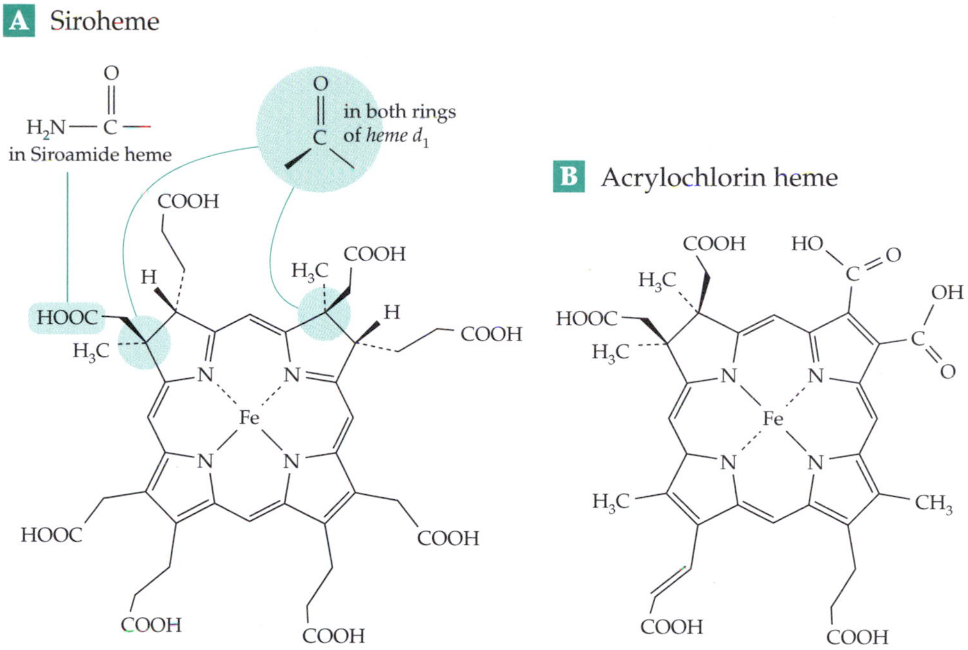

A Siroheme

B Acrylochlorin heme

Figure 16-6 Structures of isobacteriochlorin prosthetic groups. (A) Siroheme from nitrite and sulfite reductases; (B) acrylochlorin heme from dissimilatory nitrite reductases of *Pseudomonas* and *Paracoccus*.

systems of enteric bacteria contains the closely related heme *o* (Fig. 16-5).[79,80] A second terminal oxidase of those same bacteria contains **heme d** (formerly a_2). **Heme c** (present in cytochromes *c* and *f*) is a variation in which two SH groups of the protein have added to the vinyl groups of protoheme to form two thioether linkages (Fig. 16-5). A few cytochromes *c* have only one such linkage. In myeloperoxidase (Section 6) three covalent linkages, different than those in cytochrome *c*, join the heme to the protein.[81,82] There is a possibility that heme *a* may sometimes form a Schiff base with a lysyl amino group through its formyl group.[83,84]

Heme *d* is a chlorin,[85] as is **acrylochlorin heme** from certain bacterial nitrite reductases (Fig. 16-6).[86,87] **Siroheme** (Fig. 16-6), which is found in both nitrite and sulfite reductases of bacteria (Chapter 24),[88,89] is an isobacteriochlorin in which both the A and B rings are reduced. It apparently occurs as an amide **siroamide** (Fig. 16-6) in *Desulfovibrio*.[90] Heme d_1 of nitrite reductases of denitrifying bacteria is a dioxo-bacteriochlorin derivative (Fig. 16-6).[91,92]

As in myoglobin, hemoglobin (Fig. 7-23), and cytochrome *c* (see Fig 16-8), one axial coordination position on the iron of most heme proteins (customarily called the *proximal* position) is occupied by an imidazole group of a histidine side chain. However, in cytochrome P450 and chloroperoxidase a thiolate (–S⁻) group from a cysteinyl side chain, and in catalase a phenolate anion from a tyrosyl side chain, occupies the proximal position. The sixth or *distal* coordination position is occupied by the sulfur atom of methionine in cytochrome *c* and most other cytochromes with low-spin iron but cytochromes b_5 and c_3 have histidine. The high-spin heme proteins, such as cytochromes *c'*,

globins, peroxidase, and catalase, usually have no ligand other than weakly bound H_2O in the distal position.[93]

Hemes are found in all organisms except the anaerobic clostridia and lactic acid bacteria. Heme proteins of blood carry oxygen reversibly, whereas those of **terminal oxidase systems**, **hydroxylases**, and **oxygenases** "activate" oxygen, catalyzing reactions with hydrogen ions and electrons or with carbon compounds. The heme-containing **peroxidases** and **catalases** catalyze reactions not with O_2 but with H_2O_2. Another group of heme proteins includes most of the cytochromes, which are purely electron-transferring compounds.

4. The Cytochromes

The iron in the small proteins known as cytochromes acts as an electron carrier, undergoing alternate reduction to the +2 state and oxidation to the +3 state. The cytochromes, discovered in 1884 by McMunn,[94] were first studied systematically in the 1920s by Keilin (Chapter 18) and have been isolated from many sources.[95–97] The classification into groups *a, b,* and *c* according to the position of the longest wavelength light absorption band (the α band; Fig. 16-7) follows a practice introduced by Keilin. However, it is now customary to designate a new cytochrome by giving the heme type (*a, b, c* or *d*) together with the wavelength of the α band, e.g., cytochrome c_{552} or cyt $b_{557.5}$.

Cytochromes of the *b* type including bacterial cytochrome *o* contain protoheme. Because the sixth position is ligated, most cytochromes *b* do not react with O_2. However, cytochromes *o* and *d* serve as terminal electron acceptors (cytochrome oxidases) and are oxidizable by O_2. Another protoheme-containing cytochrome, involved in hydroxylation (Chapter 18), is **cytochrome P450**. Here the 450 refers to the position of the intense "Soret band" (also called the γ band) of the spectrum (Fig. 16-7) in a difference spectrum run in the presence and absence of CO. Other properties are also used in arriving at designations for cytochromes. For example, cytochrome a_3 has a spectrum similar to that of cytochrome *a* but it reacts readily with both CO and O_2.

Another property that distinguishes various cytochromes is the redox potential $E^{\circ\prime}$ (Table 6-8), which in this discussion is given for pH 7.0. Cytochromes carry electrons between other oxidoreductase proteins of widely varying values of E^0. Because of the various heme environments cytochromes have greatly differing values of E^0, allowing them to function in many different biochemical systems.[97a,97b] For mitochondrial cytochrome *c* the value of $E^{\circ\prime}$ is ~ + 0.265 V but for the closely related cytochrome *f* of chloroplasts it is ~ +0.365 V and for cytochrome c_3 of *Desulfovibrio* about −0.330 V. There is more than an 0.6-volt difference between $E^{\circ\prime}$

of cytochromes *f* and c_3. Cytochromes *b* tend to have lower $E^{\circ\prime}$ values, close to zero, than most cytochromes *c*, while cytochrome a_3 has $E^{\circ\prime}$ ~ +0.385 V.

The c-type cytochromes. Mitochondrial cytochrome *c* is one of the few intracellular heme pigments that is soluble in water and that can be removed easily from membranes. A small 13-kDa protein typically containing about 104 amino acid residues, cytochrome *c* has been isolated from plants, animals, and eukaryotic microorganisms.[95–97,99,100] Complete amino acid sequences have been determined for over 100 species. Within the peptide chain 28 positions are invariant and a number of other positions contain only conservative substitutions. Cytochrome *c* was one of the first proteins to be used in attempting to trace evolutionary relationships between species by observing differences in sequence. Humans and chimpanzees have identical cytochrome *c*, but 12 differences in amino acid sequence occur between humans and the horses and 44 between human and *Neurospora*.[96] The related cytochrome c_2 of the photosynthetic bacterium *Rhodospillum rubrum* is thought to have diverged in evolution 2×10^9 years ago from the precursor of mammalian cytochrome *c*. Even so, 15 residues remain invariant.[101]

Structural studies[95–97,101–103] on cytochromes of the *c* and c_2 types show that the heme group provides a core around which the peptide chain is wound. The 104 residues of mitochondrial cytochrome *c* are enough to do little more than envelope the heme. In both the oxidized and reduced forms of the protein, methionine 80 (to the left in Fig. 16-8A) and histidine 18 (to the right) fill the axial coordination positions of the iron. The heme is nearly "buried" and inaccessible to the surrounding solvent.

The shorter chains of the 82- to 86-residue cytochromes c_{550} (from *Pseudomonas*[102]), c_{553},[102a] and c_{555} (from *Chlorobium*[105]) as well as the longer 112-residue polypeptide of cytochrome c_2 from *Rhodospirillum rubrum*[93] have nearly the same folding pattern as that in mitochondrial cytochrome *c*. However, the 128-residue chain of the dimeric cytochrome *c'* from *Rhodospirillum molischianum* forms an antiparallel four-helix bundle (Fig. 16-8).[106–109] This is the same folding pattern present in the ferritin monomer (Fig. 16-3), hemerythrin (Fig. 2-22), and many other proteins including cytochrome b_{562} of *E. coli*.[110] Cytochrome *f*, which functions in photosynthetic electron transport, is also a *c*-type cytochrome but with a unique protein fold.[111,112]

Most cytochromes have only one heme group per polypeptide chain,[112] but the 115-residue cytochrome c_3 from the sulfate-reducing bacterium *Desulfovibrio* binds four hemes (Fig. 16-8C).[104,113–115] Each one seems to have a different redox potential in the −0.20 to −0.38 V range.[114] Another *c*-type cytochrome, also from *Desulfovibrio*, contains six hemes in a much larger 66-kDa protein and functions as a nitrite reductase.[116]

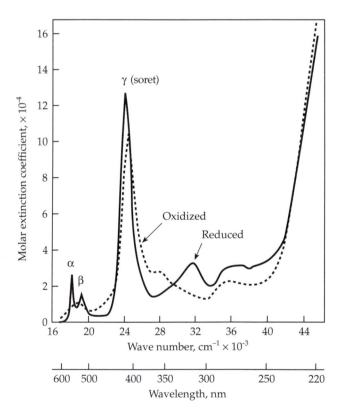

Figure 16-7 Absorption spectra of oxidized and reduced horse heart cytochrome *c* at pH 6.8. From data of Margoliash and Frohwirt.[98]

Figure 16-8 Structures of three cytochromes of the *c* type. (A) Horse mitochondrial cytochrome *c*; (B) a subunit of the dimeric cytochrome *c'* from *Rhodospirillum molischianum*; (C) cytochrome c_3 from *Desulfovibrio desulfuricans*. (A) and (B) courtesy of Salemme;[101] (C) from Simões *et al.*[104] Courtesy of Maria A. Carrondo.

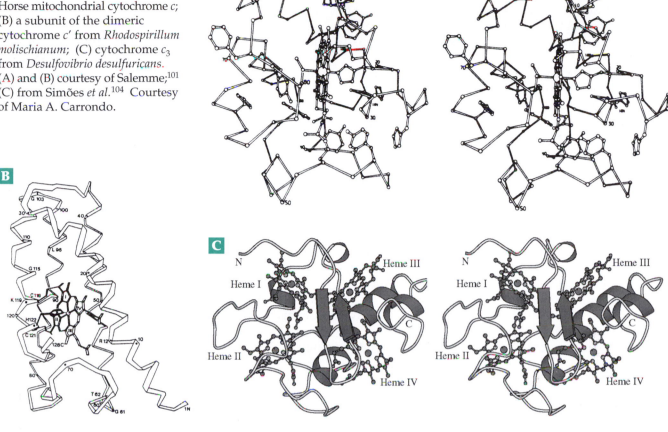

Triheme and octaheme proteins are also known.[117]

Many cytochromes *c* are soluble but others are bound to membranes or to other proteins. A well-studied tetraheme protein binds to the reaction centers of many purple and green bacteria and transfers electrons to those photosynthetic centers.[118–120] Cytochrome c_2 plays a similar role in *Rhodobacter*, forming a complex of known three-dimensional structure.[121] Additional cytochromes participate in both cyclic and noncyclic electron transport in photosynthetic bacteria and algae (see Chapter 23).[120,122–124] Some bacterial membranes as well as those of mitochondria contain a **cytochrome bc_1 complex** whose structure is shown in Fig. 18-8.[125,126]

Cytochromes b, a, and o. Protoheme-containing cytochromes *b* are widely distributed.[127,128] There are at least five of them in *E. coli*. Whether in bacteria, mitochondria, or chloroplasts, the cytochromes *b* function within electron transport chains, often gathering electrons from dehydrogenases and passing them on to *c*-type cytochromes or to iron–sulfur proteins. Most cytochromes *b* are bound to or embedded within membranes of bacteria, mitochondria, chloroplasts, or endoplasmic reticulum (ER). For example, cyto-

chrome b_5[129,129a] delivers electrons to a fatty acid desaturating system located in the ER of liver cells and to many other reductive biosynthetic enzymes.[130–132] The protein contains 132 amino acid residues plus another 85 largely hydrophobic N-terminal residues that provide a nonpolar tail which is thought to be buried in the ER membranes.[130] Solubilization of the protein causes loss of this N-terminal sequence. The heme in cytochrome b_5 is not covalently bonded to the protein but is held tightly between two histidine side chains. The polypeptide chain is folded differently than in either cytochrome *c* or myoglobin.

The folding pattern of cytochrome b_5 is also found in the complex heme protein **flavocytochrome b_2** from yeast (Chapter 15)[133] and probably also in liver **sulfite oxidase**.[134,135] Both are 58-kDa peptides which can be cleaved by trypsin to 11-kDa fragments that have spectroscopic similarities and sequence homology with cytochrome b_5. Sulfite oxidase also has a molybdenum center (Section H). The 100-residue N-terminal portion of flavocytochrome b_2 has the cytochrome b_5 folding pattern but the next 386 residues form an eight-stranded $(\alpha / \beta)_8$ barrel that binds a molecule of FMN.[133,136] All of these proteins pass electrons to cytochrome *c*. In contrast, the folding of **cytochrome**

b_{562} of *E. coli* resembles that of cytochrome *c'* (Fig. 16-8).[110,137] However, it has methionine side chains as both the fifth and sixth iron ligands.

Cytochromes *b* of mitochondrial membranes are involved in passing electrons from succinate to ubiquinone in complex II[138] and also from reduced ubiquinone to cytochrome c_1 in the 248-kDa complex III (Fig. 18-8). A similar complex is present in photosynthetic purple bacteria.[123,139] Cytochrome b_{560} functions in the transport of electrons from succinate dehydrogenase to ubiquinone,[138] and cytochrome b_{561} of secretory vesicle membranes has a specific role in reducing ascorbic acid radicals.[140]

In bacteria some cytochromes *b* and d_1 serve as terminal electron carriers able to react with O_2, nitrite, or nitrate, while others act as carriers between redox systems.[141–143a] The aldehyde heme *a* is utilized by animals and by some bacteria in **cytochrome c oxidase**, a complex enzyme whose three-dimensional structure is known (see Fig. 18-10) and which is discussed further in Chapter 18.

5. Mechanisms of Biological Electron Transfer

The heme groups of the cytochromes as well as many other transition metal centers act as carriers of electrons. For example, cytochrome *c* may accept an electron from reduced cytochrome c_1 and pass it to cytochrome oxidase or cytochrome *c* peroxidase. The electron moves from one heme group to another over distances as great as 2 nm. Similar electron-transfer reactions between defined redox sites are met in photosynthetic reaction centers (Fig. 23-31), in metalloflavoproteins (Fig. 15-9), and in mitochondrial membranes.

What are the factors that determine the probability of an electron transfer reaction and the rate at which it may occur? They include: (1) The distance from the electron donor to the acceptor. (2) The thermodynamic driving force $\Delta G°$ for the reaction. This can be approximated using the difference in standard electrode potentials (as in Table 6-1) between donor and acceptor. $\Delta G° = -96.5\, \Delta E°$ kJ/mol at 25° C. (3) The chemical makeup of the material through which the electron transfer takes place. (4) Any changes in the geometry or charge state of the donor or acceptor that accompany the transfer. (5) The orientation of the acceptor and donor groups.[144–146] It is usually assumed that the Franck–Condon principle is obeyed, i.e., that the electron jump occurs so rapidly ($<10^{-12}$ s) that there is no change in the positions of atomic nuclei (see also Chapter 23). Subsequent rearrangement of nuclear positions may occur at rates that allow a rapid overall reaction.

The various factors that affect the rate of electron transfer were incorporated by Marcus into a quantitative theory. Electron transfer is often discussed in terms of this classic Marcus theory together with effects of quantum mechanical tunneling.[144,147–150] According to Marcus the electron transfer rate from a donor to an acceptor at a fixed separation depends upon $\Delta G°$, a nuclear reorganization parameter (λ), and the electronic coupling strength $|H_{AB}|$ between reactant and product in the transition state (Eq. 16-3):

$$k_{ET} = (4\pi^3 / h^2 \lambda k_B T)|H_{AB}|^2 \exp[-(\Delta G° + \lambda)^2 / 4\lambda k_B T]$$
(16-3)

Here $|H_{AB}|$ is a quantum mechanical matrix whose strength decreases exponentially with the distance of separation *R* as $e^{-\beta R}$ where β is a coefficient of the order of 9–14 nm^{-1}. At the closest contact (R = 0) the rate k_{ET}, by extrapolation from experimental data on small synthetic compounds, is close to the molecular vibration frequency of 10^{13} s^{-1}.[151,152] At distances greater than 2 nm the rate would be negligible were it not for other factors.

Using mutant proteins as well as a variety of redox pairs and electron-transfer distances the validity of the Marcus equation with respect to the thermodynamic driving force and distance dependence has been verified.[153] This is even true for cytochrome *c* mutants functioning in living yeast cells.[146]

A huge amount of experimental work with proteins has been done to test and refine the theories of electron transport. For example, electron donor groups with various reduction potentials have been attached to various sites on the surface of a protein containing a heme or other electron accepting group. Ruthenium complexes such as Ru(III) $(NH_3)_5^{3+}$ form tight covalent linkages to imidazole nitrogens[154,155] such as that of His 33 of horse heart cytochrome *c*. This metal can be reduced rapidly to Ru(II) by an external reagent, after which the transfer of an electron from the Ru(II) across a distance of 1.2 nm to the heme Fe(III) can be followed spectroscopically. The reduction potentials $E°$ for the Ru(III)/Ru(II) and Fe(III)/Fe(II) couples at pH 7 in these compounds are 0.16 and 0.27 V, respectively. Thus, an electron will jump spontaneously from the Ru(II) to the Fe(III) with $\Delta G° = -15.4$ kJ/mol. A rate constant of ~5 s^{-1}, which was nearly independent of temperature, was observed. Since the structures of Fe(II) and Fe(III) forms of cytochrome *c* differ only slightly,[156] the electron transfer apparently occurs with only a small amount of geometric rearrangement. The distribution of charges and dipoles within the protein may be such that the Fe^{2+} and Fe^{3+} complexes have almost equal thermodynamic stability.

Electron-transfer pathways? In spite of the success of the Marcus theory, rates of electron-transfer from the iron of cytochrome *c* have been found to vary for different pathways.[150,153,155] For example, transfer of an electron from Fe(II) in reduced cytochrome *c* to an Ru(III) complex on His 33 was fast (~440 s^{-1})[157] but

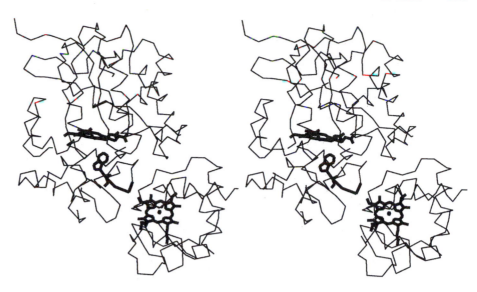

Figure 16-9 Stereoscopic α-carbon plot of yeast cytochrome *c* peroxidase (top) and yeast cytochrome *c* (below) as determined from a cocrystal by Pelletier and Kraut.[164] The heme rings of the two proteins appear in bold lines, as does the ring of tryptophan 191 and the backbone of residues 191–193 of the cytochrome *c* peroxidase. Drawing from Miller *et al.*[165]

the rate of transfer to an equidistant Fe(III) ion on Met 65 was at most 0.6 s$^{-1.}$ These results suggested that distinct electron-transfer pathways exist. One suggestion was that the sulfur atom of Met 80 donates an electron to Fe^{3+} leaving an electron-deficient radical. The "hole" so created could be filled by an electron jumping in from the –OH group of the adjacent Tyr 67, which might then accept an electron from an external acceptor via Tyr 74 at the protein surface. Do electrons flow singly or as pairs from the surface through hydrogen-bonded paths? Use of both semisynthesis[155] and directed mutation[153] of cytochromes *c* is permitting a detailed study of these effects. A striking result is that substitution of the conserved residue phenylalanine 82 in a yeast cytochrome *c* with leucine or isoleucine retards electron transfer by a factor of ~10^4.

"Docking." It is now recognized that there are distinct "docking sites" on the surface of electron-transport proteins. For rapid electron transfer to occur the two electron carriers must be properly oriented and docked by formation of correct polar and nonpolar interactions. Early indications of the importance of docking came from study of modified cytochromes *c*. Each one of the 19 lysine side chains was individually altered by acylation or alkylation to remove the positive charge or to replace it with a negative charge. The rate of electron transfer into cytochrome *c* from hexacyanoferrate was decreased by a factor of 1.3–2.0 when any one lysine at positions 8, 13, 27, 72, or 79, which are clustered around the heme edge, was modified.[158] Modification of Lys 22, 55, 99, or 100, distant from this edge, had no effect. Electron transfer *into* cytochrome *c* from its natural electron donor ubiquinol:cytochrome *c* reductase was also strongly inhibited by modification of lysines that surround the heme edge.[159] Modification of these lysines also inhibited electron transfer *out of*

cytochrome *c* into its natural acceptors, cytochrome *c* oxidase and cytochrome *c* peroxidase (Fig. 16-9).[160] A major factor in these effects is probably the large dipole moment in the cytochrome *c* that arises from the unequal distribution of surface charges. This charge distribution must assist in the proper docking of the cytochrome with its natural electron donors and acceptors. The positive surface charges presumably also facilitate the reaction with hexacyanoferrate (II) or ascorbate, both of which are negatively charged reductants that react rapidly with cytochrome *c*.

Measurements of many kinds have been made between natural donor–acceptor pairs such as cytochrome *c*–cytochrome b_5,[161,162] cytochrome *c*–cytochrome *c* peroxidase (Fig. 16-9),[153,163–166] trimethylamine dehydrogenase–FMN to Fe$_4$S$_4$ center (Fig. 15-9),[167] and methylamine dehydrogenase (TTQ radical)–amicyanin (Cu^{2+}).[168] Designed metalloproteins are being studied as well.[169] Femtosecond laser spectroscopy is providing a new approach.[169a]

Coupling and gating of electron transfer.
Electrons are thought to be transferred into or out of cytochrome *c* through the exposed edge of the heme. The rate depends upon effective coupling, which in turn may depend upon orientation as well as the structure and dynamics of the protein.[170] Proteins with much β structure appear to provide stronger coupling than those that are largely composed of α helices.[144,171] The high nuclear reorganization energy λ of α helices may block electron transfer along some pathways.[153] A conformational change,[167,169] transfer of a proton, or binding of some other specific ion[172] before electron transfer occurs can be the "gating" process that determines the rate of electron transfer.[162,167,173] Electron transfer can also be "coupled" to an unfavorable, but fast, equilibrium.

Effects of ionic equilibria on electron transfer.

The charge on an ion of Fe^{2+} in a heme is exactly balanced by two negative charges on the porphyrin ring. However, when the Fe^{2+} loses an electron to become Fe^{3+} an extra positive charge is suddenly present in the center of the protein. This change in charge will have a powerful electrostatic effect on charged groups in the immediate vicinity of the iron and even at the outer surface of the molecule. For example, an anion from the medium or from a neighboring protein molecule might become bound to the heme protein (Eq. 16-4).

$$X^- + [Fe(II)(heme)] \rightarrow [Fe(III)(heme)]^+ X^- + e^- \tag{16-4}$$

In this case the presence of a high concentration of X^- in the medium would favor the oxidation of Fe(II) to Fe(III). The reduced heme would be a better reducing agent and the oxidized form a weaker oxidant than in the absence of X^-. If –YH were a group in the protein the loss of an electron could cause –YH to dissociate so that Y^- and the Fe(III) would interact more tightly (Eq. 16-5).

$$HY - [Fe(II)(heme)] \rightarrow$$
$$Y^- - [Fe(III)(heme)]^+ + H^+ + e^- \tag{16-5}$$

We see that electron transfer can be accompanied by loss of a proton and that $E^{\circ\prime}$ may become pH dependent. (See also Eq. 16-18.) Even with cytochrome c, although there is little structural change upon electron transfer, there is an increased structural mobility in the oxidized form.[156] This may be important for coupling and could also facilitate associated proton-transfer reactions. For example, it is possible that in some cytochromes the imidazole ring in the fifth coordination position may become deprotonated upon oxidation. This possibility is of special interest because cytochromes are components of proton pumps in mitochondrial membranes (Chapter 18).

6. Reactions of Heme Proteins with Oxygen or Hydrogen Peroxide

As Ingraham remarked,[174] "Living in a bath of 20% oxygen, we tend to forget how reactive it is." From a thermodynamic viewpoint, all living matter is extremely unstable with respect to combustion by oxygen. Ordinarily, a high temperature is required and if we are careful with fire, we can expect to escape a catastrophe. However, one mole of properly chelated copper could catalyze consumption of all of the air in an average room within one second.[174] Biochemists are interested in both the fact that O_2 is kinetically stable and unreactive and also that oxidative enzymes such as cytochrome c oxidase are able to promote rapid reactions. Two oxygen atoms, each with six valence electrons, might reasonably be expected to form dioxygen, O_2, as a double-bonded structure with one σ and one π bond as follows (left):

$$:\!\ddot{O}\!=\!\ddot{O}\!: \qquad :\!\ddot{O}\!-\!\dot{O}\!:$$

Dioxygen

However, O_2 is paramagnetic and contains two unpaired electrons.[175] From this evidence O_2 might be assigned the structure on the right.[175] The oxygen molecule is very stable, and it is relatively difficult to add an electron to form the reactive **superoxide anion radical** O_2^-.

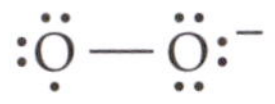

$$:\!\ddot{O}\!-\!\ddot{O}\!:^-$$

Superoxide anion radical

For this reason, oxidative attack by O_2 tends to be slow. However, once an electron has been acquired, it is easy for additional electrons to be added to the structure and further reduction occurs more easily. The biochemical question suggested is, "How can some heme proteins carry O_2 reversibly without any oxidation of the iron contained in them while others *activate* oxygen toward reaction with substrates?" Among this latter group, cytochrome c oxidase transfers electrons to both oxygen atoms so that only H_2O is a product, whereas the hydroxylases and oxygenases, which are discussed in Chapter 18, incorporate either one or two of the atoms of O_2, respectively, into an organic substrate. Before examining these reactions let us reconsider the heme oxygen carriers.

Oxygen-carrying proteins.

In Chapter 7, we examined the behavior of hemoglobin in the cooperative binding of four molecules of O_2 and studied its structural relationship to the monomeric muscle protein myoglobin.[176] The iron in functional hemoglobin and myoglobin is always Fe(II) and is only very slowly converted by O_2 into the Fe(III) forms methemoglobin or metmyoglobin.[177,178] Erythrocytes contain an enzyme system for immediately reducing methemoglobin back to the Fe(II) state (see Box 15-H).

Binding of O_2 to the iron in the heme is usually considered not to cause a change in the oxidation state of the metal. However, oxygenated heme has some of the electronic characteristics of an $Fe^{3+}-OO^-$ peroxide anion. Bonding of the heme iron to oxygen is thought to occur by donation of a pair of electrons by the oxygen to the metal. In deoxyhemoglobin the Fe(II) ion is in the "high-spin" state; four of the five $3d$ orbitals in the valence shell of the iron contain one unpaired electron and the fifth orbital contains two paired electrons. The binding of oxygen causes the iron to revert to the "low-spin" state in which all of the electrons are paired and the paramagnetism of hemoglobin is lost. The stability of heme–oxygen complexes is thought to be enhanced by "back-bonding," i.e., the donation of an electron pair from one of the filled d orbitals of the iron atom to form a bond with the adjacent oxygen.[179] This can be indicated symbolically as follows:

These structures, which have been formulated by assuming that one of the unshared electron pairs on O_2 forms the initial bond to the metal, are expected to lead to an angular geometry which has been observed in X-ray structures of model compounds,[180] in oxy-myoglobin (Fig. 16-10),[181,182] and in oxyhemoglobin.[183] Neutron diffraction studies have shown that the outermost oxygen atom of the bound O_2 is hydrogen bonded to the H atom on the N^ε atom of the distal imidazole ring of His E7 (Fig. 16-10). Carbon monoxide binds with the C≡O axis perpendicular to the heme plane and unable to form a corresponding hydrogen bond.[184] This decreases the affinity for CO and helps to protect us from carbon monoxide poisoning.

All oxygen-carrying heme proteins have another imidazole group that binds to iron on the side opposite the oxygen site. Without this proximal imidazole group, heme does not combine with oxygen. Coordination with heterocyclic nitrogen compounds favors formation of low-spin iron complexes and simple synthetic compounds that closely mimic the behavior of myoglobin have been prepared by attaching an imidazole group by a chain of appropriate length to the edge of a heme ring.[179,185] Similar compounds bearing a pyridine ring in the fifth coordination position have a low affinity for oxygen. Thus, the polarizable imidazole ring itself seems to play a role in promoting oxygen binding. The π electrons of the imidazole ring may also participate in bonding to the iron as is indicated in the following structures.[179] The π bonding to the iron would allow the iron to back-bond more strongly to an O_2 atom entering the sixth coordination position. These diagrams illustrate another feature found frequently in heme proteins: The N–H group of the imidazole is hydrogen bonded to a peptide backbone carbonyl group.

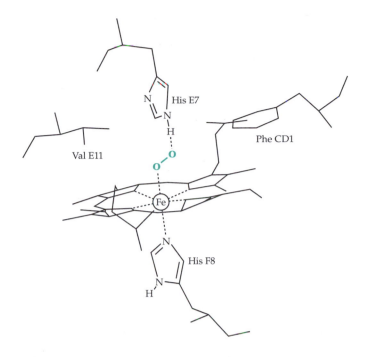

Figure 16-10 Geometry of bonding of O_2 to myoglobin and position of hydrogen bond to N^ε of the distal histidine E7 side chain. After Perutz.[182]

The coordination of the heme iron to histidine also appears to provide the basis for the cooperativity in binding of oxygen by hemoglobin.[186] The radius of high-spin iron, whether Fe(II) or Fe(III), is so large that the iron cannot fit into the center of the porphyrin ring but is displaced toward the coordinated imidazole group by a distance of ~0.04 nm for Fe(II).[187] Thus, in deoxyhemoglobin both iron and the imidazole group lie further from the center of the ring than they do in oxyhemoglobin. In the latter, the iron lies in the center of the porphyrin ring because the change to the low-spin state is accompanied by a decrease in ionic radius.[186,188] The change in protein conformation induced by this small shift in the position of the iron ion was described in Chapter 7. However, the exact nature of the linkage between the Fe position and the conformational changes is not clear.

The mechanical response to the movement of the iron and proximal histidine, described in Chapter 7, may explain this linkage. However, oxygenation may also induce a change in the charge distribution within the hydrogen-bond network of the protein. The carbonyl group shown in the foregoing structure is attached to the F helix (see Fig. 7-23) and is also hydrogen bonded to other amide groups. Electron withdrawal into the heme–oxygen complex would tend to strengthen the hydrogen bond as indicated by the resonance forms shown and also to weaken competing hydrogen bonds.[189,190] This could affect the charge distribution in the upper end of the F helix and could conceivably induce a momentary conformational change that could facilitate the rearrangement of structure that was discussed in Chapter 7 (Fig. 7-25). The $\alpha_1\beta_2$ contact in which a change of hydrogen bonding takes place is located nearby behind the F and G helices. In any event, it is remarkable that nature has so effectively made use of the subtle differences in the properties of iron induced by changes in the electron distribution within the d orbitals of this transition metal.

A few groups of invertebrates, e.g., the sipunculid worms, use a nonheme iron-containing protein, **hemerythrin**, as an oxygen carrier.[191,192] Its 113-residue subunits are often associated as octamers of C_4 symmetry, each peptide chain having a four-helix bundle structure (Fig. 2-22). Instead of a heme group, each monomer contains two atoms of high-spin Fe(II) held by a cluster of histidine and carboxylate side chains (see Fig. 16-20).[193,194] Hemerythrin is a member of a group of such diiron oxoproteins which are considered further in Section 8. The copper oxygen carrier **hemocyanin** is discussed in Section D.

Catalases and peroxidases. Many iron and copper proteins do not bind O_2 reversibly but "activate" it for further reaction. We will look at such metalloprotein oxidases in Chapter 18. Here we will consider heme enzymes that react not with O_2 but with peroxides. The peroxidases,[194a] which occur in plants, animals, and fungi, catalyze the following reactions (Eq. 16-6, 16-7):

$$H_2O_2 + AH_2 \rightarrow 2\ H_2O + A \qquad (16\text{-}6)$$

$$ROOH + AH_2 \rightarrow R\text{—}OH + H_2O + A \qquad (16\text{-}7)$$

Here AH_2 is an oxidizable organic compound such as an alcohol or a pair of one-electron donor molecules. Catalases, which are found in almost all aerobic cells,[194b] may sometimes account for as much as 1% of the dry weight of bacteria. The enzyme catalyzes the breakdown of H_2O_2 to water and oxygen by a mechanism similar to that employed by peroxidases. If Eq. 16-7 is rewritten with H_2O_2 for AH_2 and O_2 for A, we have the following equation:

$$2\ H_2O_2 \xrightarrow{\text{Catalase}} 2\ H_2O + O_2 \qquad (16\text{-}8)$$

The action of catalase is very fast, almost 10^4 times faster than that of peroxidases. The molecular activity per catalytic center is about $2 \times 10^5\ s^{-1}$.

Catalase exerts a protective function by preventing the accumulation of H_2O_2 which might be harmful to cell constituents. The complete intolerance of obligate anaerobes to oxygen may result from their lack of this

enzyme. Support for this protective function comes from the existence of the human hereditary condition **acatalasemia**.[195,196] Persons with extremely low catalase activity are found worldwide but are especially numerous in Korea. In Japan it is estimated that there are 1800 persons lacking catalase. Because about half of them have no symptoms, catalase might be judged unessential. However, many of the individuals affected develop ulcers around their teeth. Apparently, hydrogen peroxide produced by bacteria accumulates and oxidizes hemoglobin to methemoglobin (Box 15-H) depriving the tissues of oxygen.

Catalase from most eukaryotic species is tetrameric.[197] The protein from beef liver consists of 506-residue subunits.[198] Human catalase is similar.[198a] The proximal ligand to the heme Fe^{3+} is a tyrosinate anion (Tyr 358), while side chains of His 75 and Asn 148 lie close to the heme on the distal H_2O_2-binding side (Fig. 16-11). Larger ~650-residue fungal and bacterial catalases have a similar folding pattern but an extra C-terminal domain with a flavodoxin-like structure.[197,199] Catalase is gradually inactivated by its very reactive substrate. As isolated, beef liver catalase usually contains about two subunits in which the heme ring has been oxidatively cleaved to **biliverdin**[200] (Fig. 24-24) and various other alterations have been found.[197] Each subunit of mammalian catalases normally contains a bound molecule of NADPH which helps to protect against inactivation by H_2O_2.[201,202] Catalases from *Neurospora* and from *E. coli* contain heme d rather than protoporphyrin.[197,203] Some lactobacilli, lacking heme altogether, form a manganese-containing pseudocatalase.[204,205]

Of the plant peroxidases, which are found in abundance in the peroxisomes, the 40-kDa monomeric **horseradish peroxidase** has been studied the most.[206–208a] It occurs in over 30 isoforms and has an extracellular role in generating free radical intermediates for polymerization and crosslinking of plant cell wall components.[209] Secreted fungal peroxidases, e.g., such as those from *Coprinus*[210] and *Arthromyces*,[211] form a second class of peroxidases with related structues.[212] A third class is represented by **ascorbate peroxidase** from the cytosol of the pea[212–214] and by the small 34-kDa **cytochrome c peroxidase** from yeast mitochondria[215] (Fig. 16-11). The latter has a strong preference for reduced cytochrome c as a substrate (Eq. 16-9).[216,217]

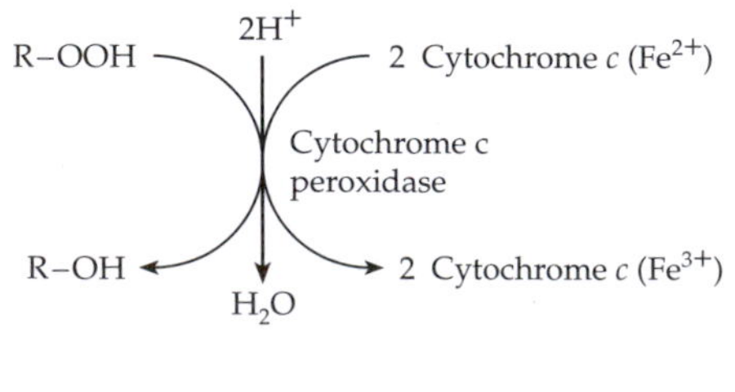

$$(16\text{-}9)$$

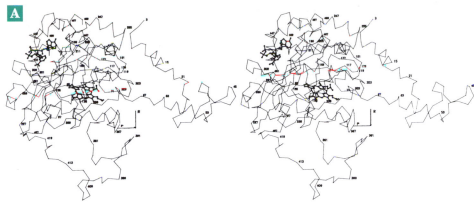

A

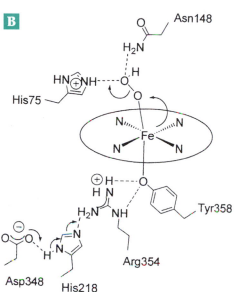

B

Figure 16-11 (A) Stereo drawing showing folding pattern for beef liver catalase and the positions of the NADPH (upper left) and heme (center). From Fita and Rossmann.[198] (B) Diagram of proposed structure of an Fe(III)–OOH ferric peroxide complex of human catalase (see also Fig. 16-14). A possible mechanism by which the peroxide is cleaved (step *b* in Fig. 16-14) is indicated by the arrows. His 75 and Asn 148 are directly involved, and a charge relay system below the ring may also participate. From Putnam *et al.*[198a]

Because the three-dimensional structures of the peroxidase, its reductant cytochrome *c*, and the complex of the two (Fig. 16-9) are known, cytochrome *c* peroxidase is the subject of much experimental study. Other fungal peroxidases, some of which contain manganese rather than iron, act to degrade lignin (Chapter 25).[218] A lignin peroxidase from the white wood-rot fungus *Phanerochaete chrysosporium* has a surface tryptophan with a specifically hydroxylated Cβ carbon atom which may have a functional role in catalysis.[218a,b]

The human body contains **lactoperoxidase**, a product of exocrine secretion into milk, saliva, tears, etc., and peroxidases with specialized functions in **saliva**, the **thyroid, eosinophils**,[219] and **neutrophils**.[220] The functions are largely protective but the enzymes also participate in biosynthesis. Mammalian peroxidases have heme covalently linked to the proteins, as indicated in Fig. 16-12.[220–222a]

The active site structure of peroxidases (Fig. 16-13) is quite highly conserved. As in myoglobin, an imidazole group is the proximal heme ligand, but it is usually hydrogen bonded to an aspartate carboxylate as a catalytic diad (Fig. 16-13).[223] In cytochrome *c* peroxidase

there is also a buried tryptophan, which has already been highlighted in Fig. 16-9. A conserved and essential feature on the distal side is another histidine, which is hydrogen bonded to an asparagine[224] and which can also hydrogen bond to the substrate H_2O_2. Fungal peroxidases also have a conserved arginine on the distal side. However, even an octapeptide with a bound heme cut from cytochrome *c* acts as a "microperoxidase" with properties similar to those of natural peroxidases.[225]

Peroxidases and catalases contain high-spin Fe(III) and resemble metmyoglobin in properties. The enzymes are reducible to the Fe(II) state in which form they are able to combine (irreversibly) with O_2. We see that the same active center found in myoglobin and hemoglobin is present but its chemistry has been modified by the proteins. The affinity for O_2 has been altered drastically and a new group of catalytic activities for ferriheme-containing proteins has emerged.

Mechanisms of catalase and peroxidase catalysis. Attention has been focused on a series of strikingly colored intermediates formed in the presence of substrates. When a slight excess of H_2O_2 is added to a solution of horseradish peroxidase, the dark brown enzyme first turns olive green as **compound I** is formed, and then pale red as it turns into **compound II**. The latter reacts slowly with substrate AH_2 or with another H_2O_2 molecule to regenerate the original enzyme. This sequence of reactions is indicated by the colored arrows in Fig. 16-14, steps *a–d*.

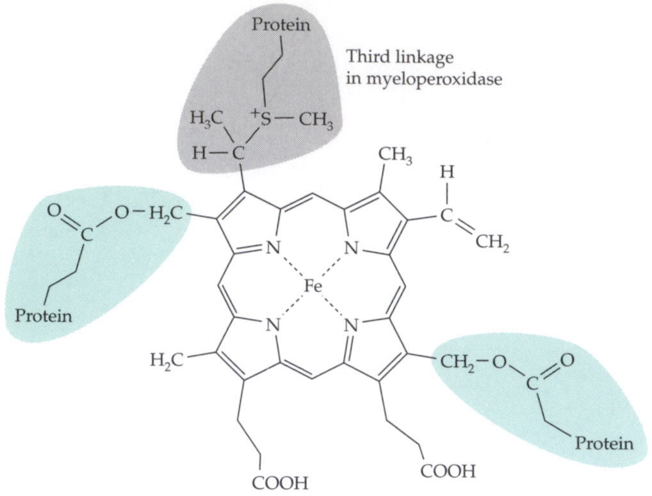

Figure 16-12 Linkage of heme to mammalian peroxidases. There are two ester linkages to carboxylate side chains from the protein.[220,221] Myeloperoxidase contains a third linkage.[222,222a]

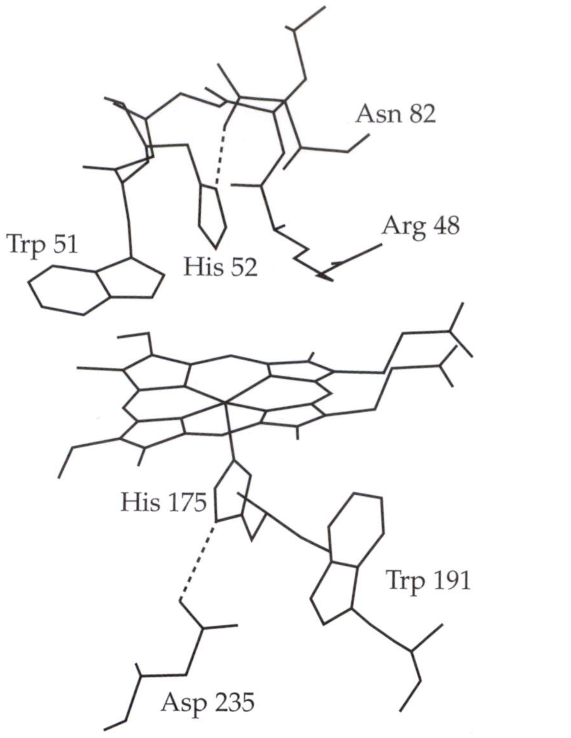

Titrations with such reducing agents as ferrocyanide or K_2IrBr_6 have established that compound I is converted into compound II by a one-electron reduction and compound II to free peroxidase by another one-electron reduction. Thus, the iron in compound I may formally be designated Fe(V) and that in II as Fe(IV). However, this does not tell us whether or not the oxygen atoms of H_2O_2 are present in compounds I and II. The enzyme in the Fe^{3+} form can be reduced to Fe^{2+} (Fig. 16-14, step e), as previously mentioned, and when the Fe^{2+} enzyme reacts with H_2O_2 it is apparently converted into compound II (Fig. 16-14, step f). This suggests that the latter is an Fe^{2+} complex of the peroxide anion. Here, P^{2-} represents the porphyrin ring:

$$P^{2-}—Fe^+(II)—OOH$$

However, spectroscopic evidence suggests that compound II is a **ferryl iron** complex which could be derived from the preceding structure by addition of a proton and loss of water.[226,227]

$$P^{2-}—Fe^{2+}(IV)=O$$
Compound II

High concentrations of H_2O_2 convert II into compound III, which is thought to be the same as the **oxyperoxidase** that is formed upon addition of O_2 to the Fe(II) form of the free enzyme (Fig. 16-14, step g) and corresponds in structure to oxyhemoglobin.[228]

Compound I was at one time thought to be a complex of H_2O_2 or its anion with Fe(III), but its magnetic and spectral properties are inconsistent with this structure. Rather, it too appears to contain ferryl iron

Figure 16-13 The active site of yeast cytochrome c peroxidase. Access for substrates is through a channel above the front edge of the heme ring as viewed by the reader. A pathway for entrance of electrons may be via Trp 191 and His 175. From Holzbaur *et al.*[206] Based on coordinates of Finzel *et al.*[215]

bound to an electron-deficient porphyrin π-cation radical.[229] The reaction with peroxide probably involves initial formation of a peroxide anion complex (Fig. 16-14, step a) which is cleaved with release of water (step b).[215,230] The resulting $Fe(V)^+=O$ compound is converted to compound I by transfer of a single elec-

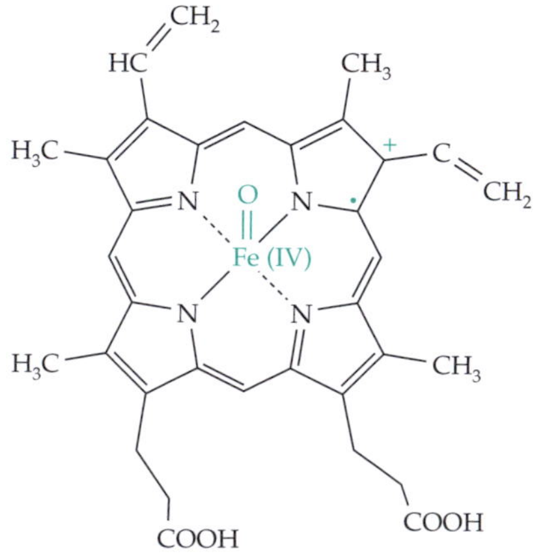

Compound I. The unpaired electron and the positive charge are delocalized over the porphyrin ring and perhaps into the proximal histidine ring

tron from the porphyrin to the iron. In cytochrome *c* peroxidase compound I contains a free radical on the nearby Trp 191 ring instead of on the porphyrin radical.[216] Consistent with this is the fact that horseradish peroxidase contains phenylalanine in place of Trp 191.

If we consider the fate of substrate AH_2 during the action of a peroxidase, we see that donation of an electron to compound I to convert it into II (Fig. 16-14, step *c*) will generate a free radical •AH as well as a proton. The radical may then donate a second electron to II to form the free enzyme. Alternatively, a second molecule of AH_2 may react (Fig. 16-14, step *c*) to form a second radical •AH. The two •AH radicals may then disproportionate to form A and AH_2 or they may leave the enzyme and react with other molecules in their environment. Compound II of horse radish peroxidase is able to exchange the oxygen atom of its Fe(IV)=O center with water rapidly at pH 7, presumably by donation of a proton from the nearby histidine side chain (corresponding to His 52 of Fig. 16-13).[227,230a,b] This histidine presumably also functions in proton transfer during reactions with substrates (see Fig. 16-11B).[224]

The catalase compound I appears to be converted in a two-electron reduction by H_2O_2 directly to free ferricatalase without intervention of compound II (Fig. 16-14, step *c'*). The catalytic histidine probably donates a proton to help form water from one of the oxygen

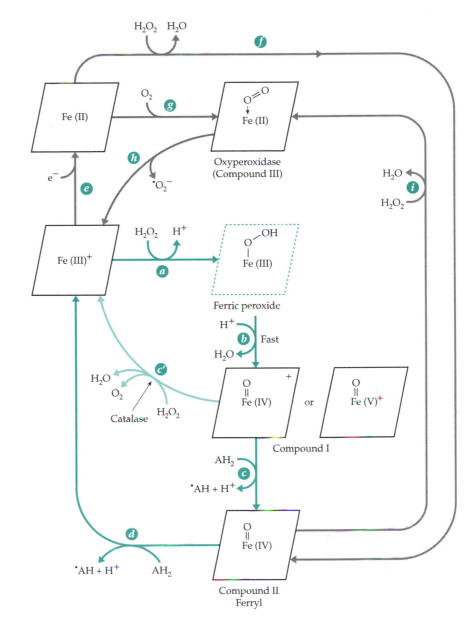

Figure 16-14 The catalytic cycles and other reactions of peroxidases and catalases. The principal cycle for peroxidases is given by the colored arrows. That of catalases is smaller, making use of step *a*, *b*, and *c'*, which is marked by a light green line.

$$•P^+—Fe(IV)=O + H_2O_2 \rightarrow H_2O + O_2 + P—Fe(III)^+$$
Compound I

(16-10)

atoms of the H_2O_2. Nevertheless, compound II does form slowly, especially if a slow substrate such as ethanol is present. The previously mentioned bound NADPH apparently reduces compound II formed in this way, converting the inactivated enzyme back to active catalase.[201,231] This may involve unusual one-electron oxidation steps for the NADPH.[231] Under some circumstances compound II of peroxidases reacts with substrates in a two-electron process[232–235] with transfer of an oxygen atom to the substrate, a characteristic also of reactions catalyzed by cytochromes P450 (see Eq. 18-57).

Haloperoxidases. Many specialized peroxidases are active in halogenation reactions. **Chloroperoxidases** from fungi[236,237] catalyze chlorination reactions like that of Eq. 16-11 using H_2O_2 and Cl^- as well as the usual peroxidase reaction.

Chloroperoxidase is isolated in a low-spin Fe(III) state. The reduced Fe(II) enzyme is a high-spin form

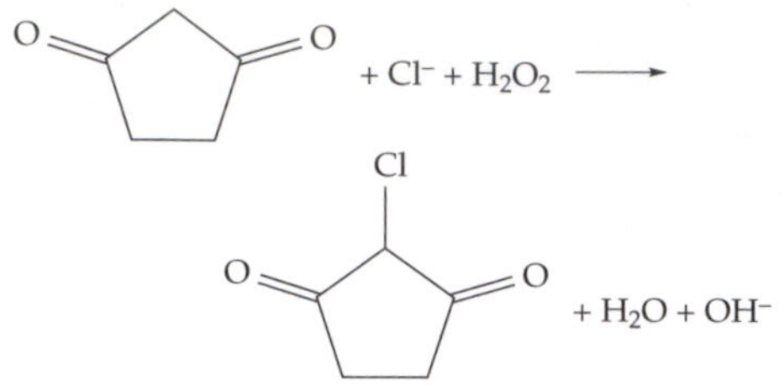

$$+ \, Cl^- + H_2O_2 \longrightarrow$$

$$+ \, H_2O + OH^-$$

(16-11)

with spectroscopic properties similar to those of the oxygenases of the cytochrome P450 family, which are discussed in Chapter 18.[238,239] Like the cytochromes P450 chloroperoxidase contains a thiolate group of a cysteine side chain as the fifth iron ligand.[235,239,240] Chloroperoxidase forms compounds I and II, as do other peroxidases. A chloride ion may combine with compound I to form a complex of hypochlorite with the Fe(III) heme.

$$^\bullet P^+ - Fe(IV) = O + Cl^- \rightarrow P - Fe(III)^+ - OCl$$

(16-12)

This intermediate could then halogenate substrate AH (Eq. 16-13, step a), lose HOCl (Eq. 16-13, step b), or generate Cl_2 by reaction with Cl^- (Eq. 16-13, step c). These are all well-established reactions for the enzyme. In each case the chlorine in the peroxidase complex can be viewed as an electrophile which is transferred to an attacking nucleophile. A fourth reaction that can go

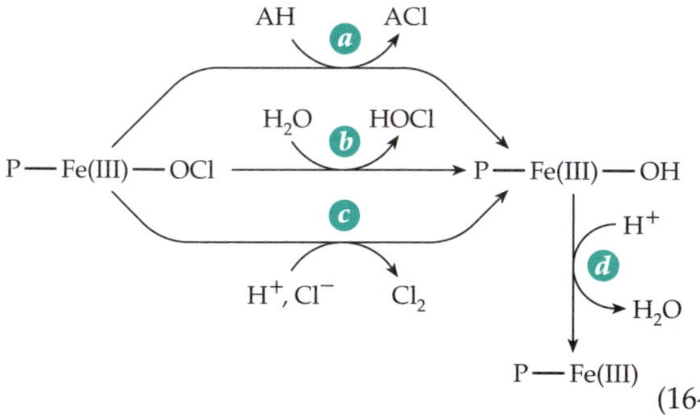

(16-13)

through the same intermediate is conversion of alkenes to α,β-halohydrins (Eq. 16-14).[241] **Lactoperoxidase** of milk, reacting with $I^- + H_2O_2$, promotes an analogous iodination of tyrosine and histidine residues of proteins. With radioactive $^{125}I^-$ or $^{131}I^-$ it provides a convenient and much used method for labeling of proteins in

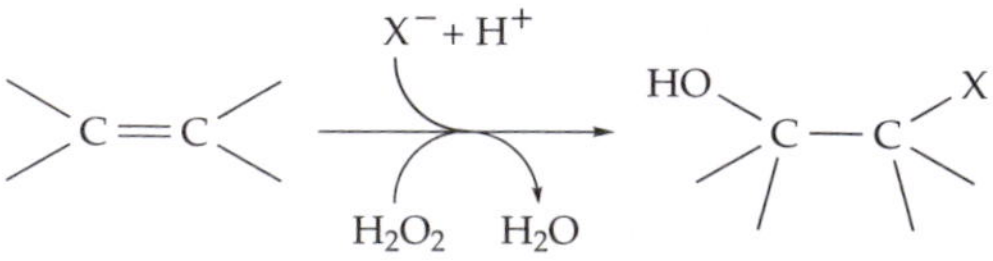

(16-14)

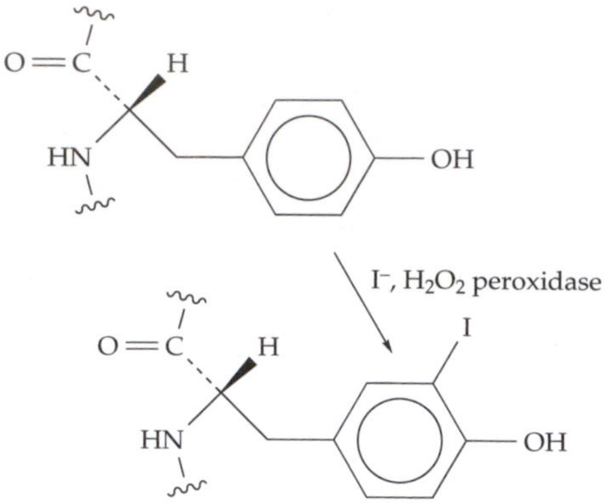

(16-15)

exposed surfaces of membranes (Eq. 16-15).[242] Iodinated tyrosine derivatives are formed in the thyroid gland by a similar reaction catalyzed by **thyroid peroxidase**.[232] Even horseradish peroxidase can oxidize iodide ions but neither it nor lactoperoxidase will carry out chlorination or bromination reactions.

Myeloperoxidase, present in specialized lysosomes of polymorphonuclear leukocytes (neutrophils),[243] utilizes H_2O_2 and a halide ion to kill ingested bacteria.[244,245] Phagocytosis induces increased respiration by the leukocyte and generation of H_2O_2, partly by the membrane-bound NADPH oxidase described in Chapter 18. Some of the H_2O_2 is used by myeloperoxidase to attack the bacteria, apparently through generation of HOCl by peroxidation of Cl^-. Human myeloperoxidase is a tetramer of two 466-residue chains and two 108-residue chains, which carry the covalently linked heme.[222a,246] Another oxygen-dependent killing mechanism that may also be used by neutrophils is the generation of the reactive **singlet oxygen**.[247] This can occur by reaction of hypochlorite with H_2O_2 (Eq. 16-16) or from an enzyme-bound hypochlorite intermediate such as that shown in Eq. 16-13. Hereditary deficiency of myeloperoxidase is relatively common.[245]

$$OCl^- + H_2O_2 \rightarrow O_2(^1\Delta g) + H_2O + Cl^-$$

singlet O_2

(16-16)

Lactoperoxidase[248] and chloroperoxidase[249] also generate singlet oxygen. The possible biological significance is discussed in Chapter 18. Eosinophil peroxidase appears to promote formation of **hydroxyl radicals**.[250] **Bromoperoxidases** are found in many red and green marine algae.[251] Many of them contain **vanadium** and function by a mechanism different than that used by heme peroxidases (see Section G).[252,253]

Another related nonheme enzyme is the selenoprotein glutathione peroxidase. It reacts by a mechanism very different (Eq. 15-59) from those discussed here, as does **NADPH peroxidase**, a flavoprotein with a cysteine sulfinate side chain in the active site.[254-255a] A lignin-degrading peroxidase from the white wood rot fungus *Phanerochaete chrysosporium* is a simple heme protein,[256] while other peroxidases secreted by this organism contain Mn.[257,258]

7. The Iron–Sulfur Proteins

Not all of the iron within cells is chelated by porphyrin groups. Hemerythrin (see Fig. 16-20) has been known for many years, but the general significance of nonheme iron proteins was not appreciated until large-scale preparation of mitochondria was developed by Crane in about 1945. The iron content of mitochondria was found to far exceed that of the heme proteins present. In 1960, Beinert, who was studying the mitochondrial dehydrogenase systems for succinate and for NADH, observed that when the electron transport chain was partially reduced by these substrates and the solutions were frozen at low temperature and examined, a strong EPR signal was observed at $g = 1.94$. The signal was obtained only upon reduction by substrate, and fractionation pointed to the nonheme iron proteins. Six or more proteins of this type are involved in the mitochondrial electron transport chain (Eq. 18-1), and numerous others have become recognized as members of the same large family of **iron–sulfur proteins** (Fe–S proteins).[259,260]

Ferredoxins, high-potential iron proteins, and rubredoxins. The presence of nonheme iron proteins is most evident in the anaerobic clostridia, which contain no heme. It was from these bacteria that the first Fe–S protein was isolated and named **ferredoxin**. This protein has a very low reduction potential of $E°'(pH 7) = -0.41$ V. It participates in the pyruvate– ferredoxin oxidoreductase reaction (Eq. 15-35), in nitrogen fixation in some species, and in formation of H_2. A small green-brown protein, the ferredoxin of *Peptococcus aerogenes* contains only 54 amino acids but complexes eight atoms of iron. If the pH is lowered to ~1, eight molecules of H_2S are released. Thus, the protein contains eight "labile sulfur" atoms in an iron sulfide linkage. There are also eight iron atoms.

Another group of related electron carriers, the **high-potential iron proteins** (HIPIP) contain four labile sulfur and four iron atoms per peptide chain.[261-266] X-ray studies showed that the 86-residue polypeptide chain of the HIPIP of *Chromatium* is wrapped around a single **iron–sulfur cluster** which contains the side chains of four cysteine residues plus the four iron and four sulfur atoms (Fig. 16-15D).[261] This kind of cluster is referred to as [4Fe–4S], or as Fe_4S_4. Each cysteine sulfur is attached to one atom of Fe, with the four iron atoms forming an irregular tetrahedron with an Fe–Fe

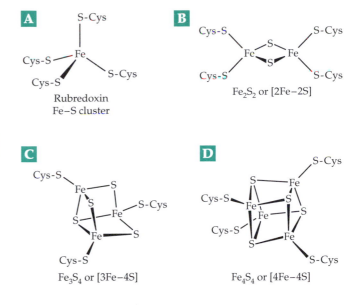

Figure 16-15 Four different iron–sulfur clusters of a type found in many proteins. From Beinert[259] with permission.

distance of ~0.28 nm. The four labile sulfur atoms (S^{2-}) form an interpenetrating tetrahedron 0.35 nm on a side with each of the sulfur atoms bonded to three iron atoms. The cluster is ordinarily able to accept only a single electron. The iron–sulfur cluster structure was a surprise, but after its discovery it was found that ions such as $[Fe_4S_4(S-CH_2CH_2COO^-)_4]^{6-}$ assemble spontaneously from their components and have a similar cluster structure.[259,266a] Thus, living things have simply improved upon a natural bonding arrangement.

The bacterial ferredoxins from *Peptococcus*, *Clostridium* (Fig. 16-16B),[267,268] *Desulfovibrio*, and other anaerobes each contain two Fe_4S_4 clusters with essentially the same structure as that of the *Chromatium* HIPIP.[267,269] Each cluster can accept one electron. Much of the amino acid sequence in the first half of the ferredoxin chain is repeated in the second half, suggesting that the chain may have originated as a result of gene duplication. Many invariant positions are present in the sequence, including those of the cysteine residues forming the Fe–S cluster. Ferredoxins with single Fe_4S_4 clusters are also known.[270]

The ferredoxins have reduction potentials $E°'$ (pH 7) from about -0.4 V to as low as -0.6 V. However, the corresponding values for HIPIP proteins range from $+0.05$ to $+0.50$ V at pH 7.[271] This wide range of potentials initially seemed strange because the structures of the active centers of both the clostridial ferredoxins and the *Chromatium* HIPIP appear virtually identical.[272] Part of the explanation lies in the fact that Fe_4S_4 clusters can exist in three oxidation states (Eq. 16-17) that differ, one from another, by a single electron.[273]

$$[Fe_4S_4]^+ \xrightarrow{\quad} [Fe_4S_4]^{2+} \xrightarrow{\quad} [Fe_4S_4]^{3+}$$

$e-$ $e-$

Reduced Fd   Oxidized Fd;   Oxidized high
              reduced high   potential iron
              potential iron protein; Super-
              protein   oxidized Fd   (16-17)

Here the charges shown are those on the cluster. The cysteine ligands from the protein each add an additional negative charge. The *Chromatium* HIPIP and the clostridial ferredoxins have the middle oxidation state in common. The cluster is a little smaller in the more oxidized states; in the *Chromatium* HIPIP the Fe–Fe distance changes from 0.281 to 0.272 nm upon oxidation.

Rubredoxins. The simplest of the Fe–S proteins are the rubredoxins. These proteins contain iron but no labile sulfur. The rubredoxin of *Clostridium pasteurianum* is a 54-residue peptide containing four cysteines whose side chains form a distorted tetrahedron about a single iron atom (Fig. 16-16A).[274] Not shown for any

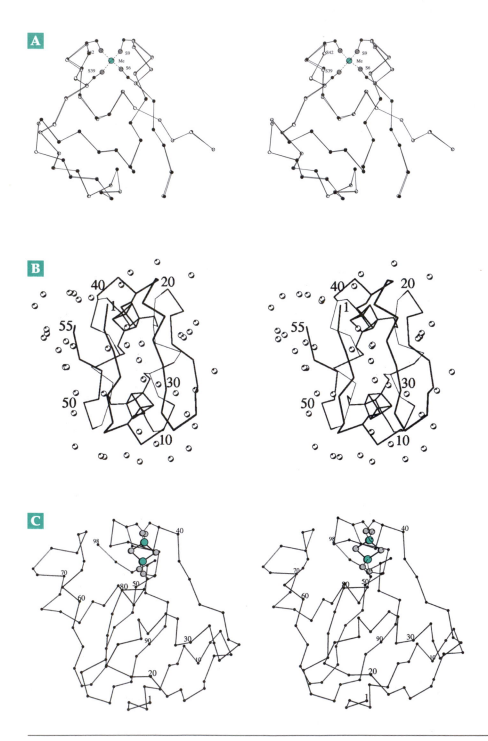

Figure 16-16 (A) Superimposed stereoscopic α-carbon traces of the peptide chain of rubredoxin from *Clostridium pasteurianum* with either Fe^{3+} (solid circles) or Zn^{2+} (open circles) bound by four cysteine side chains. From Dauter *et al.*[274] (B) Alpha-carbon trace for ferredoxin from *Clostridium acidurici*. The two Fe_4S_4 clusters attached to eight cysteine side chains are also shown. The open circles are water molecules. Based on a high-resolution X-ray structure by Duée *et al.*[267] Courtesy of E. D. Duée. (C) Polypeptide chain of a chloroplast-type ferredoxin from the cyanobacterium *Spirulina platensis*. The Fe_2S_2 cluster is visible at the top of the molecule. From Fukuyama *et al.*[276] Courtesy of K. Fukuyama.

of the structures in Fig. 16-16 are NH—S hydrogen bonds that connect backbone NH groups of the peptide chain to the sulfur atoms of the cysteine groups, forming the clusters.[275] These bonds may have important effects on properties of the cluster. Rubredoxins also participate in electron transport and can substitute for ferredoxins in some reactions. Larger 14-kDa and 18-kDa rubredoxins able to bind two iron ions participate in electron transport in a hydroxylase system of *Pseudomonas* (Chapter 18).[277] A smaller 7.9-kDa 2-Fe **desulforedoxin** functions in sulfate-reducing bacteria.[278]

Chloroplast-type ferredoxins.

Members of a large class of [2Fe–2S] or Fe_2S_2 ferredoxins each contain two iron atoms and two labile sulfur atoms with the linear structure of Fig. 16-15B.[279,279a] Best known are the chloroplast ferredoxins, which transfer electrons from photosynthetic centers of chloroplasts to the flavoprotein reductase that reduces $NADP^+$ to NADPH.[280,280a] The structure of a cyanobacterial protein of this type is shown in Fig. 16-16C.[276] A second group of Fe_2S_2 ferredoxins are found in bacteria including *E. coli*[281] and in human mitochondria.[282] For example, in steroid hormone-forming tissues the ferredoxin **adrenodoxin**[282a] carries electrons to cytochromes P450. Its Fe_2S_2 center receives electrons from adrenodoxin reductase (Chapter 15 banner).[282b] The nitrogen-fixing *Clostridium pasteurianum* also contains a ferredoxin of this class.[283] In *Pseudomonas putida* (Chapter 18) the related 106-residue **putidaredoxin** transfers electrons to cytochromes P450.[284]

The 3Fe–4S clusters.

A 106-residue ferredoxin from *Azotobacter vinelandii* contains seven iron atoms in two Fe–S clusters that operate at very different redox potentials of −0.42 and +0.32 V.[285] Other similar seven-iron proteins are known.[286] From EPR measurements it appeared that both clusters function between the 2+ and 3+ (oxidized and superoxidized) states of Eq. 16-17, despite the widely differing potentials. A super-reduced all-Fe(II) form with $E^{\circ'}$ (pH 7) = −0.70 V can also be formed.[286] X-ray crystallographic studies have revealed that the protein contains one Fe_4S_4 cluster and one Fe_3S_4 cluster (Fig. 16-15C).[286–288] The structures and environments of the Fe_3S_4 clusters are similar to those of Fe_4S_4 clusters but they lack one iron and one cysteine side chain. An Fe_4S_4 cluster may sometimes lose S^{2-} to form an Fe_3S_4 cluster such as the one in the *A. vinelandii* ferredoxin. A less likely possibility is isomerization to a linear Fe_3S_4 structure.[289]

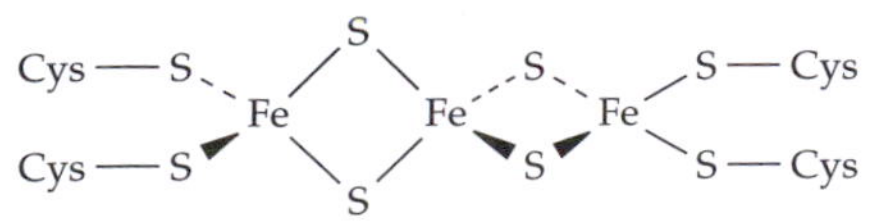

Aconitase (Eq. 13-17) isolated under aerobic conditions contains an Fe_3S_4 cluster and is catalytically inactive. Incubation with Fe^{2+} activates the enzyme and reconverts the Fe_3S_4 to an Fe_4S_4 cluster (Fig. 13-4).[260,289,290]

Properties of iron–sulfur clusters.

These clusters were viewed for many years as unstable and unable to exist outside of a protein. However, if protected from oxygen and manipulated in the presence of soluble organic thiols they are stable and "cofactor-like."[260] Intact clusters can be "extruded" from proteins by treatment with thiols in nonaqueous media. Both Fe_4S_4 and Fe_2S_2 clusters as well as more complex forms have been synthesized[291] and nonenzymatic cluster interconversions have been demonstrated. Binding to proteins stabilizes the clusters further, but some (Fe–S) proteins are labile and difficult to study. This is evidently because of partial exposure of the cluster to the surrounding solvent. Not only can O_2 cause oxidation of exposed clusters[291a] but also superoxide,[292] nitric oxide, and peroxynitrite can react with the iron. Aconitase has only three cysteine side chains available for coordination with Fe and the protein is unstable. Apparently, a superoxidized $[Fe_4S_4]^{3+}$ cluster is formed in the presence of O_2 but loses Fe^{2+} to give an $[Fe_3S_4]^+$ cluster.[293]

Another interesting cluster conversion is the joining of two Fe_2S_2 clusters in a protein to form a single Fe_4S_4 cluster at the interface between a dimeric protein. Such a cluster is present in the nitrogenase iron protein (Fig. 24-2) and probably also in biotin synthase.[294] The clusters in such proteins can also be split to release the monomers.

Synthetic iron–sulfur clusters have weakly basic properties[273] and accept protons with a pK_a of from 3.9 to 7.4. Similarly, one clostridial ferredoxin, in the oxidized form, has a pK_a of 7.4; it is shifted to 8.9 in the reduced form.[295] If we designate the low-pH oxidized form of such a protein as HOx^+ and the reduced form as HRed, we can depict the reduction of each Fe_4S_4 cluster as follows.

$$HOx^+ + e^- \rightarrow HRed \qquad (16\text{-}18)$$

Comparing with Eq. 6-64 and using the Michaelis pH functions (first two terms of Eq. 7-13) for HOx^+ and HRed, it is easy to show that the value of $E^{\circ'}$ ($E_{1/2}$) at which equal amounts of oxidized protein (HOx^+ + Ox) and reduced protein (HRed + Red^-) are present is given by Eq. 16-19, in which K_{ox} and K_{red} are the K_a values for dissociation of the protonated oxidized and reduced forms, respectively.

$$E_{1/2} = E^0 \,(\text{low pH}) + 0.0592 \log\left[\frac{(1 + K_{red}/[H^+])}{(1 + K_{ox}/[H^+])}\right] \text{V}$$

$$(16\text{-}19)$$

At the high pH limit this becomes $E_{1/2} = E°$ (low pH) + 0.0592 (pK_{ox}–pK_{red}) and V = –0.371 + 0.0592 (7.4–8.9) V = –0.431 V. Thus, the value of $E_{1/2}$ changes from –0.371 to –0.460 V as the pH is increased. In the pH range between the pK_a values of 7.4 and 8.9 reduction of the protein will lead to binding of a proton from the medium and oxidation to loss of a proton. Human and other vertebrate ferredoxins also show pH-dependent redox potentials.[282] This suggests, as with the cytochromes, a possible role of Fe–S centers in the operation of proton pumps in membranes. Nevertheless, many ferredoxins, such as that of C. *pasteurianum,* show a constant value of $E°'$ from pH 6.3 to 10[296] and appear to be purely electron carriers.

Both the iron and labile sulfur can be removed from Fe–S proteins and the active proteins can often be reconstituted by adding sulfide and Fe^{2+} ions. Using this approach, the natural isotope ^{56}Fe (nuclear spin zero) has been exchanged with ^{57}Fe, which has a magnetic nucleus[297] and ^{32}S has been replaced by ^{77}Se. The resulting proteins appear to function naturally and give EPR spectra containing hyperfine lines that result from interaction of these nuclei with unpaired electrons in the Fe_4S_4 clusters (Fig. 16-17). These observations suggest that electrons accepted by Fe_4S_4 clusters are not localized on a single type of atom but interact with nuclei of both Fe and S. The native proteins as well as many mutant forms are being studied by

NMR[299,300] and other spectroscopic techniques (Fig. 16-18),[260] by theoretical computations,[301–304] and by protein engineering and "rational design."[305,306]

Functions of iron–sulfur enzymes. Numerous iron–sulfur clusters are present within the membrane-bound electron transport chains discussed in Chapter 18. Of special interest is the Fe_2S_2 cluster present in a protein isolated from the cytochrome *bc* complex (complex III) of mitochondria. First purified by Rieske *et al.,*[307] this protein is often called the **Rieske iron–sulfur protein**.[308] Similar proteins are found in cytochrome *bc* complexes of chloroplasts.[125,300,309,310] In

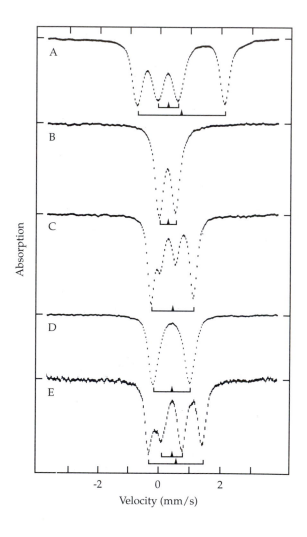

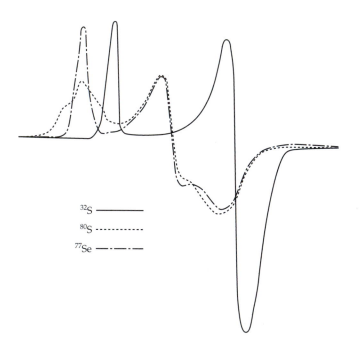

Figure 16-17 Electron paramagnetic resonance spectrum of the Fe–S protein putidaredoxin in the natural form (^{32}S) and with labile sulfur replaced by selenium isotopes. Well-developed shoulders are seen in the low-field end of the spectrum of the ^{77}Se (spin = 1/2)-containing protein. From Orme-Johnson *et al.*[298] Courtesy of W. H. Orme-Johnson.

Figure 16-18 Mössbauer X-ray absorption spectra of iron–sulfur clusters. (See Chapter 23 for a brief description of the method.) Quadrupole doublets are indicated by brackets and isomer shifts are marked by triangles. (A) $[Fe_2S_2]^{1+}$ cluster of the Rieske protein from *Pseudomonas mendocina*, at temperature $T = 200$ K. (B) $[Fe_3S_4]^{1+}$ state of *D. gigas* ferredoxin II, $T = 90$ K. (C) $[Fe_3S_4]^0$ state of *D. gigas* ferredoxin II, $T = 15$ K. (D) $[Fe_4S_4]^{2+}$ cluster of *E. coli* FNR protein, $T = 4.2$ K. (E) $[Fe_4S_4]^{1+}$ cluster of *E. coli* sulfite reductase, $T = 110$ K. From Beinert *et al.*[260]

some bacteria Rieske-type proteins deliver electrons to oxygenases.[300] The 196-residue mitochondrial protein has an unusually high midpoint potential, E_m of ~ 0.30 V. The Fe_2S_2 cluster, which is visible in the atomic structure of complex III shown in Fig. 18-8, is coordinated by two cysteine thiolates and two histidine side chains. In a *Rhodobacter* protein they occur in the following conserved sequences:[311] C133-T-H-L-G-C138 and C153-P-C-H-G-S158. One iron is bound by C133 and C153 and the other by H135 and H156 as follows:

These proteins may also have an ionizable group with a pK_a of ~8.0, perhaps from one of the histidines that is linked to the oxidation–reduction reaction.

Iron–sulfur clusters are found in flavoproteins such as NADH dehydrogenase (Chapter 18) and trimethylamine dehydrogenase (Fig. 15-9) and in the siroheme-containing **sulfite reductases** and **nitrite reductases**.[312] These two reductases are found both in bacteria and in green plants. Spinach nitrite reductase,[313] which is considered further in Chapter 24, utilizes reduced ferredoxin to carry out a six-electron reduction of NO_2^- to NH_3 or of SO_3^{2-} to S^{2-}. The 61-kDa monomeric enzyme contains one siroheme and one Fe_4S_4 cluster. A sulfite reductase from *E. coli* utilizes NADPH as the reductant. It is a large $\beta_8\alpha_4$ oligomer.[312] The 66-kDa α chains contain bound flavin

(4 FAD + 4 FMN),[312,314,315] while the 64-kDa β subunits contain both siroheme and a neighboring Fe_4S_4 cluster (Fig. 16-19).[89,304,312,316] The iron of the siroheme and the closest iron atom of the cluster are bridged by a single sulfur atom of a cysteine side chain.

A somewhat similar double cluster is present in **all-Fe hydrogenases** from *Clostridium pasteurianum*.[316a,b,c] These enzymes also contain two or three Fe_4S_4 clusters and, in one case, an Fe_2S_2 cluster.[316a] At the presumed active site a special **H cluster** consists of an Fe_3S_4 cluster with one cysteine sulfur atom shared by an adjoining Fe_2S_2 cluster: Because many hydrogenases contain nickel, their chemistry and functions are discussed in Section C,2.

The role of the iron–sulfur clusters in many of the proteins that we have just considered is primarily one of single-electron transfer. The Fe–S cluster is a place for an electron to rest while waiting for a chance to react. There may sometimes be an associated proton pumping action. In a second group of enzymes, exemplified by aconitase (Fig. 13-4), an iron atom of a cluster functions as a **Lewis acid** in facilitating removal of an –OH group in an α,β dehydration of a carboxylic acid (Chapter 13). A substantial number of other bacterial dehydratases as well as an important plant dihydroxyacid dehydratase also apparently use Fe–S clusters in a catalytic fashion.[317] Fumarases A and B from *E. coli*,[317] L-serine dehydratase of a *Peptostreptococcus* species,[317–319] and the dihydroxyacid

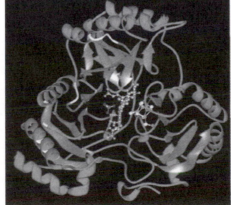
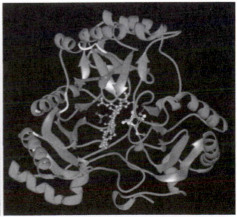

The hydrogenase Fe–S H cluster

Figure 16-19 Stereoscopic view of *E. coli* assimilatory sulfite reductase. The siroheme (Fig. 16-6) is in the center with one edge toward the viewer and the Fe_4S_4 cluster is visible on its right side. A single S atom from a cysteine side chain bridges between the Fe of the siroheme and the Fe_4S_4 cluster. A phosphate ion is visible in the sulfite-binding pocket at the left center of the siroheme. From Crane et al.[312] Courtesy of E. D. Getzoff.

dehydratase[320,321] may all use their Fe_4S_4 clusters in a manner similar to that of aconitase (Eq. 13-17). However, the Fe−S enzymes that dehydrate R-lactyl-CoA to crotonyl-CoA,[322] 4-hydroxybutyryl-CoA to crotonyl-CoA,[323,324] and R-2-hydroxyglutaryl-CoA to E-glutaconyl-CoA[325] must act by quite different mechanisms, perhaps similar to those utilized in vitamin B_{12}-dependent reactions. In these enzymes, as in pyruvate formate lyase (Eq. 15-38), the Fe−S center may act as a radical generator.

The molybdenum-containing enzymes considered in Section F also contain Fe−S clusters. Nitrogenases (Chapter 24) contain a more complex Fe−S−Mo cluster. Carbon monoxide dehydrogenase (Section C) contains 2 Ni, ~ 11 Fe, and 14 S^{2-} as well as Zn in a dimeric structure. In these enzymes the Fe−S clusters appear to participate in catalysis by undergoing alternate reduction and oxidation.

For a few enzymes such as aconitase and amido-transferases,[326] an Fe−S cluster plays a **regulatory role** in addition to or instead of a catalytic function.[327] Cytosolic aconitase is identical to **iron regulatory proteins 1** (IRP1), which binds to iron responsive elements in RNA to inhibit translation of genes asso-ciated with iron uptake. A high iron con-centration promotes assembly of the Fe_4S_4 cluster (see Chapter 28, Section C,6). Another example is provided by the *E. coli* transcription factor SOXR. This protein, which controls a cellular defense system against oxygen-derived superoxide radicals, contains two Fe_2S_2 centers. Oxidation of these centers by superoxide radicals appears to induce the transcription of genes encoding superoxide dismutase and other proteins involved in protecting cells against oxidative injury (Chapter 18).[328–330]

8. The (μ-oxo) Diiron Proteins

Both the ferroxidase center of ferritin (Fig. 16-4) and the oxygen-carrying hemerythrin are members of a family of diiron proteins with similar active site structures.[331,332] The pair of iron atoms, with the bridging (μ) ligands such as O_2, HO^-, HOO^-, O_2^{2-}, is often held between four α helices, as can be seen in Fig. 2-22 and in Fig. 16-20C. In most cases each iron is ligated by at least one histidine, one glutamate side chain, and frequently a tyrosinate side chain. As many as three side chains may bind to each iron. In addition, one or two carboxylate groups from glutamate or aspartate side chains bridge to both irons, as does the μ-oxo group, which is typically H_2O or ^-OH. Examples of this structure are illustrated in Figs. 16-4 and 16-20. Although the active sites of all of the diiron proteins appear similar, the chemistry of the catalyzed reactions is varied.

Hemerythrin. When both iron atoms are in the Fe(II) state, hemerythrin, like hemoglobin, functions as a carrier of O_2. In the oxidized Fe(III) **methemerythrin** form the iron atoms are only 0.32 nm apart. Three bridging (μ) groups lie between them: two carboxylate groups and a single oxygen atom which may be either O^{2-} or OH^-. One coordination position on one of the hexacoordinate iron atoms is open and appears to be the site of binding of oxygen. The O_2 is thought to accept two electrons, oxidizing the two iron atoms to Fe(III) and itself becoming a peroxide dianion O_2^{2-}. The process is completely reversible. The conversion of the oxygen to a bound peroxide ion is supported by studies of resonance Raman spectra (see Chapter 23) which also suggest that the peroxide group is proto-nated. In the diferrous protein the μ-oxo bridge is thought to be an ^-OH group. Upon oxygenation the proton could be shared with or donated to the peroxo group (Eq. 16-20).[334] A similar binding of O_2 as a peroxide dianion appears to occur in the copper-containing hemocyanins (Section C,3).

Deoxyhemerythrin

Oxyhemerythrin   (16-20)

Purple acid phosphatases. Diiron-tyrosinate proteins with acid phosphatase activity occur in mam-mals, plants, and bacteria. Most are basic glycoproteins with an intense 510- to 550-nm light absorption band. Well-studied members come from beef spleen, from the uterine fluid of pregnant sows (**uteroferrin**),[335] and from human macrophages and osteoclasts.[336–336b] One of the two iron atoms is usually in the Fe(III) oxidation state, but the second can be reduced to Fe(II) by mild reductants such as ascorbate. This half-reduced form is enzymatically active and has a pink color and a characteristic EPR signal. Treatment with oxidants such as H_2O_2 or hexacyannoferrate (III)

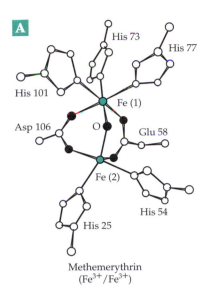

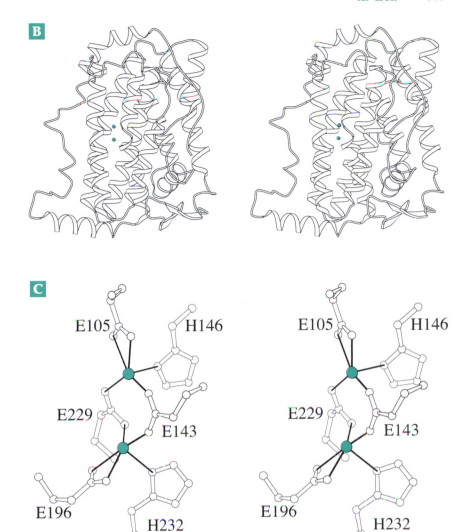

Figure 16-20 (A) The active site of hemerythrin showing the two iron atoms (green) and their ligands which include the μ oxo bridge and two bridging carboxylate groups. From Lukat et al.[193] The active site is between four parallel helices as shown in Fig. 2-22. (B) Stereoscopic view of the backbone structure of a Δ⁹ stearoyl-acyl carrier protein desaturase which also contains a diiron center. Notice the nine-antiparallel-helix bundle. The diiron center is between four helices as in hemerythrin. (C) Stereoscopic view of the diiron center of the desaturase.[333] (B) and (C) courtesy of Ylva Lindqvist.

generates purple inactive forms which lack a detectable EPR signal.[337,338] The Fe^{3+} of the active enzyme can be replaced[339] with Ga^{3+} and the Fe^{2+} with Zn^{2+} with retention of activity; also, some plants contain phosphatases with Fe–Zn centers.[340,341] The catalytic mechanism resembles those of other metallophosphatases (Chapter 12) and the change of oxidation state of the Fe may play a regulatory role. On the other hand, a principal function of uteroferrin may be in transplacental transport of iron to the fetus.[342]

Diiron oxygenases and desaturases. In the (μ-oxo) diiron oxygenases O_2 is initially bound in a manner similar to that in hemerythrin but one atom of the bound O_2 is reduced to H_2O using electrons supplied by a cosubstrate such as NADPH. The other oxygen atom enters the substrate. This is illustrated by Eq. 16-21 for methane monooxygenase.[332,343–345] A toluene monooxygenase has similar properties.[346]

$$CH_4 + NADH + H^+ + O_2 \rightarrow CH_3OH + NAD^+ + H_2O$$

$$(16\text{-}21)$$

In green plants a soluble $Δ^9$ stearoyl-acyl carrier protein desaturase uses O_2 and NADH or NADPH to introduce a double bond into fatty acids. The structure of this protein (Fig. 16-20B,C) is related to those of methane oxygenase and ribonucleotide reductase.[333,347] The desaturase mechanism is discussed in Chapter 21.

9. Ribonucleotide Reductases

Ribonucleotides are reduced to the 2'-deoxyribonucleotides (Eq. 16-21) that are needed for DNA synthesis by enzymes that act on either the di- or triphosphates of the purine and pyrimidine nucleosides[348–351] (Chapter 25). These ribonucleotide reductases utilize either thioredoxin or glutaredoxin (Box 15-C) as the immediate hydrogen donors (Eq. 16-22). The pair of closely spaced –SH groups in the reduced thioredoxin or glutaredoxin are converted into a disulfide bridge at the same time that the 2'-OH of the ribonucleotide di- (or tri-) phosphate is converted to H_2O. While some organisms employ a vitamin B_{12}-

dependent enzyme for this purpose, most utilize iron–tyrosinate enzymes (Class I ribonucleotide reductases). These are two-protein complexes of composition $\alpha_2\beta_2$. The enzyme from *E. coli* contains 761-residue α chains and 375-residue β chains. That from *Salmonella typhimurium*[351a] is similar, as are corresponding mammalian enzymes and a virus-encoded ribonucleotide reductase formed in *E. coli* following infection by T4 bacteriophage.[352] In every case the larger α_2 dimer, which is usually called the **R1 protein**, contains the substrate binding sites, allosteric effector sites, and redox-active SH groups. Each α chain is folded into an unusual $(\alpha/\beta)_{10}$ barrel.[353,354] As in the more familiar $(\alpha/\beta)_8$ barrels, the active site is at the N termini of the β strands.

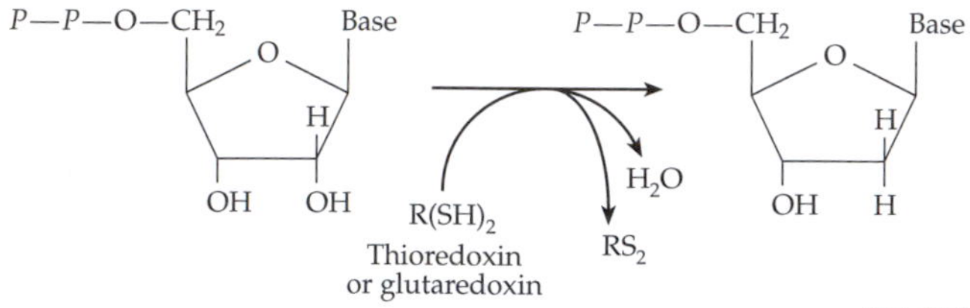

(16-22)

Each polypeptide chain of the β_2 dimer or **R2 protein** contains a diiron center which serves as a free radical generator.[354a,b,c] A few bacteria utilize a dimanganese center.[355] Oxygenation of this center is linked to the uptake of both a proton and an electron and to the removal of a hydrogen atom from the ring of tyrosine 166 to form H_2O and an organic radical (Eq. 16-23):[356–360]

$$2\ Fe(II)^{2+} + O_2 + H^+ + e^- + \text{Y122-OH} \rightarrow$$
$$Fe(III)^{3+} - O^{2-} - Fe(III)^{3+} - \text{Y122} - \text{O}\bullet + H_2O$$

(16-23)

The tyrosyl radical is used to initiate the ribonucleotide reduction at the active site in the R1 protein ~3.5 nm away. The tyrosyl radical is very stable and was discovered by a characteristic EPR spectrum of isolated enzyme. Alteration of this spectrum when bacteria were grown in deuterated tyrosine indicated that the radical is located on a tyrosyl side chain and that the spin density is delocalized over the tyrosyl ring.[361,362] Using protein engineering techniques the ring was located as Tyr 122 of the *E. coli* enzyme. A few of the resonance structures that can be used to depict the radical are the following:

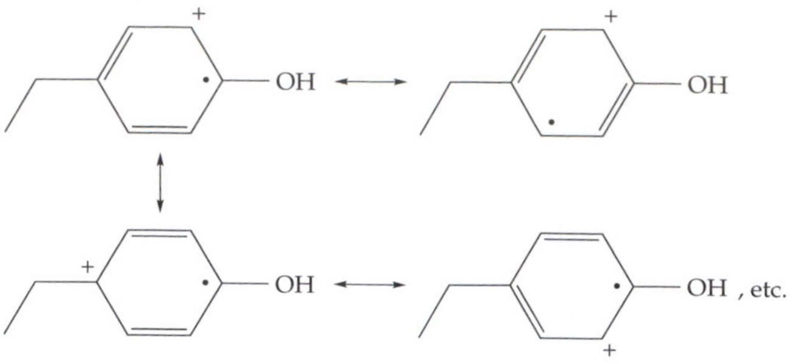

A chain of hydrogen-bonded side chains apparently provides a pathway for transfer of an unpaired electron from the active site to the Tyr 122 radical and from there to the radical generating center.[363] The tyrosyl radical can be destroyed by removal of the iron by exposure to O_2 or by treatment of ribonucleotide reductases with hydroxyurea, which reduces the radical and also destroys catalytic activity:

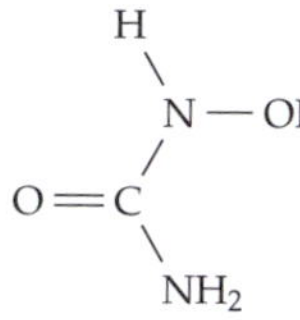

A second group of ribonucleotide reductases (Class II), found in many bacteria, depend upon the cobalt-containing **vitamin B$_{12}$ coenzyme** which is discussed in Section B. These enzymes are monomeric or homodimeric proteins of about the size of the larger α subunits of the Class I enzymes. The radical generating center is the 5'-deoxyadenosyl coenzyme.[350,364,365]

Class III or **anaerobic nucleotide reductases** are used by various anaerobic bacteria including *E. coli* when grown anaerobically[350,366–369a] and also by some bacteriophages.[370] Like the Class I reductases, they have an $\alpha_2\beta_2$ structure but each β subunit contains an Fe_4S_4 cluster which serves as the free radical generator,[369] that forms a stable glycyl radical at G580.[369a] In this respect the enzyme resembles pyruvate formate-lyase (Eq. 15-40). As with other enzymes using Fe_4S_4 clusters as radical generators, *S*-adenosylmethionine is also required. All ribonucleotide reductases may operate by similar radical mechanisms.[350,351]

When a 2'-Cl or -F analog of UDP was used in place of the substrate an irreversible side reaction occurred by which Cl$^-$ or F$^-$, inorganic pyrophosphate, and uracil were released.[349] When one of these enzyme-activated inhibitors containing ^{3}H in the 3' position was tested, the tritium was shifted to the 2' position with loss of Cl$^-$ and formation of a reactive 3'-carbonyl compound (Eq. 16-24) that can undergo β elimination at each end to give an unsaturated ketone which inactivates the enzyme. This suggested that the Fe-tyrosyl radical abstracts an electron (through a

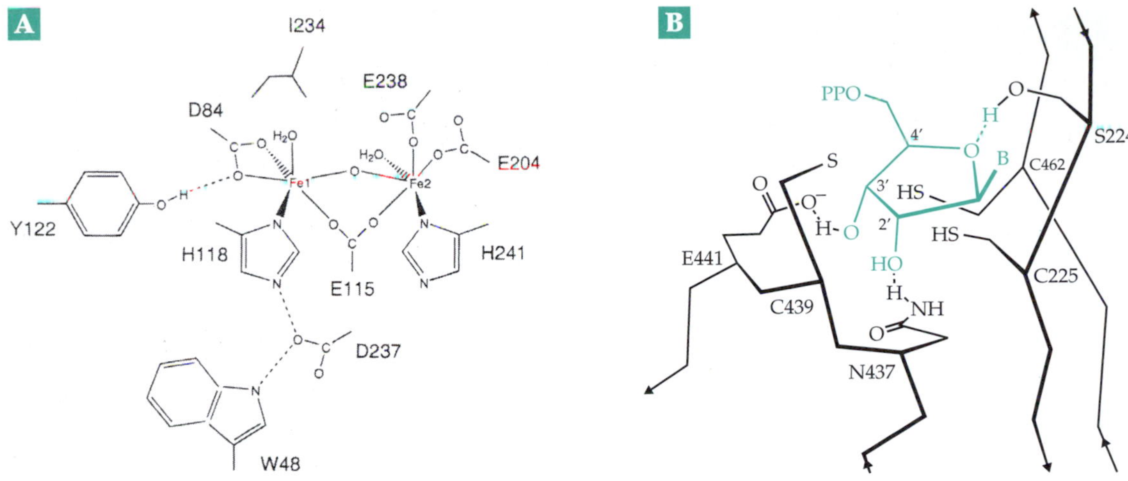

Figure 16-21 (A) Scheme showing the diiron center of the R2 subunit of *E. coli* ribonucleotide reductase. Included are the side chains of tyrosine 122, which loses an electron to form a radical, and of histidine 118, aspartate 237, and tryptophan 48. These side chains provide a pathway for radical transfer to the R1 subunit where the chain continues to tyrosines 738 and 737 and cysteine 429.[354a–c] From Andersson et al.[354c] (B) Schematic drawing of the active site region of the *E. coli* class III ribonucleotide reductase with a plausible position for a model-built substrate molecule. Redrawn from Lenz and Giese[373] with permission.

chain of intermediate groups) from an –SH group, now identified as C439 in the *E. coli* enzyme[371–372] (Fig. 16-21). The resulting thiyl radical is thought to abstract a hydrogen atom from C′-3 of a true substrate to form a substrate radical (Eq. 16-25, step *a*) which, with the help of the C462 –SH group, facilitates the loss of ⁻OH from C-2 in step *b*. The resulting C2 radical would be reduced by the nearby redox-active thiol pair C462 and C225 (Fig. 16-21 and Eq. 16-25, step *c*). In Eq. 16-25 the reaction is shown as a hydride transfer with an associated one-electron shift but the mechanism is uncertain. In step *d* of Eq. 16-25 the thiyl radical is regenerated and continues to function in subsequent rounds of catalysis. In

the final step (step *e*) the redox active pair is reduced by reduced thioredoxin or glutaredoxin. The active site must open to release the product and to permit this reduction, which may involve participation of still other –SH groups in the protein.

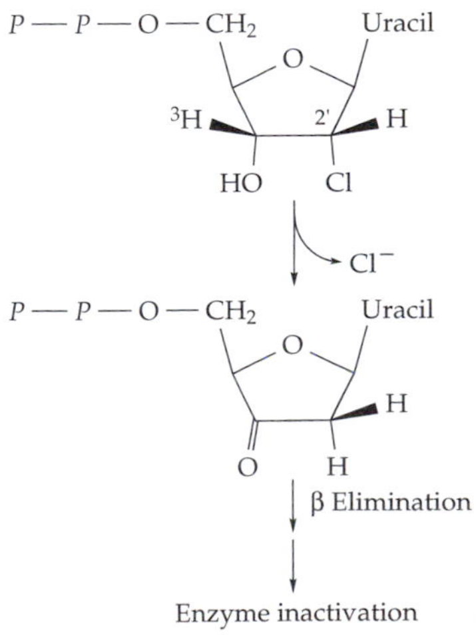

$$(16\text{-}24)$$

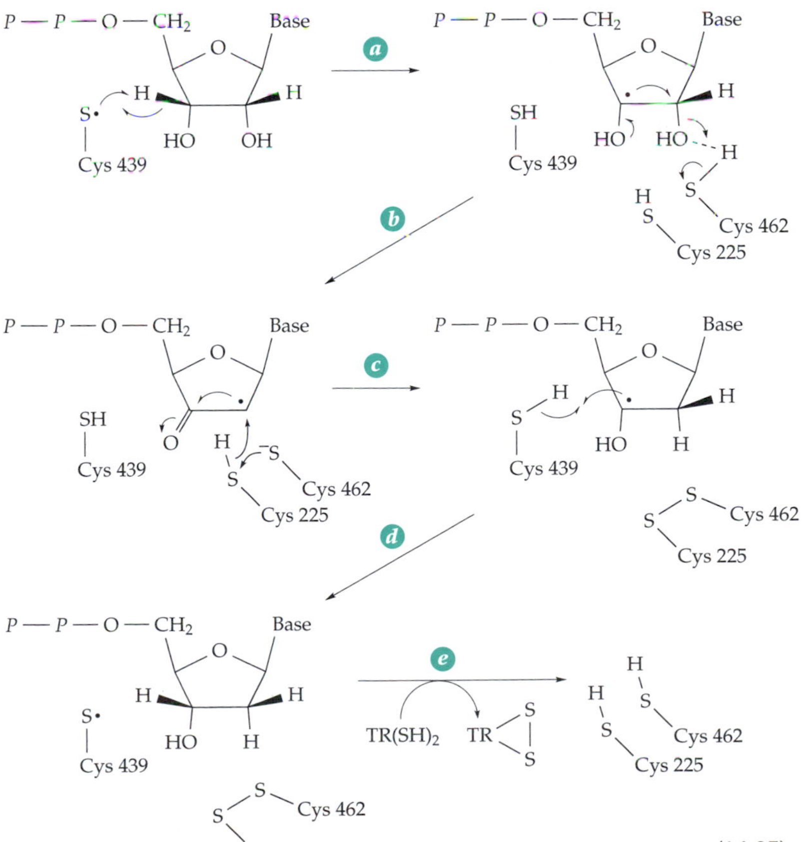

$$(16\text{-}25)$$

10. Superoxide Dismutases

Metalloenzymes of at least three different types catalyze the destruction of superoxide radicals that arise from reactions of oxygen with heme proteins, reduced flavoproteins, and other metalloenzymes. These superoxide dismutases (SODs) convert super-oxide anion radicals $\cdot O_2^-$ into H_2O_2 and O_2 (Eq. 16-26). The H_2O_2 can then be destroyed by catalase (Eq. 16-8).

$$2\,O_2^- + 2\,H^+ \longrightarrow H_2O_2 + O_2 \qquad (16\text{-}26)$$

The much studied Cu / Zn superoxide dismutase of eukaryotic cytoplasm is described in Section D. However, eukaryotic mitochondria contain manganese SOD and some eukaryotes also synthesize an iron-containing SOD. For example, the protozoan *Leishmania tropica*, which takes up residence in the phagolyso-somes of a victim's macrophages, synthesizes an iron-containing SOD[374] to protect itself against superoxide generated by the macrophages. *Mycobacterium tuberculosis* secretes an iron SOD which assists its survival in living tissues and is also a target for the immune response of human hosts.[375]

Iron and manganese SODs have ~ 20-kDa subunits in each of which a single ion of Fe or Mn is bound by three imidazole groups and a carboxylate group.[376,377] The metal ion undergoes a cyclic change in oxidation state as illustrated by Eq. 16-27. Notice that two protons must be taken up for formation of H_2O_2. In Cu / Zn SOD the copper cycles between Cu^{2+} and Cu^+. The structure of the active site of an Fe SOD is shown in Fig. 16-22. That of Mn SOD is almost identical.[376] In addition to the histidine and carboxylate ligands, the metal binds a hydroxyl ion ^-OH or H_2O and has a site

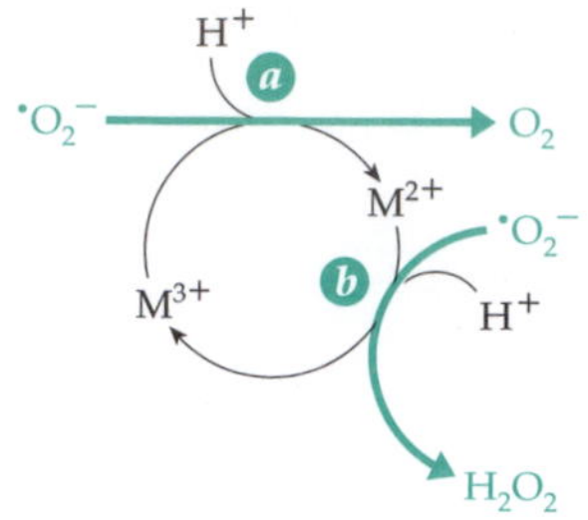

$$(16\text{-}27)$$

open for binding of $\cdot O_2^-$. As indicated in Eq. 16-27, uptake of one proton is associated with each reaction step. As illustrated in Fig 16-22 for step *a* of Eq. 16-27, the first proton may be taken up to convert the bound ^-OH to H_2O. The enzyme in the Fe^{2+} form has a pK_a of 8.5 that has been associated with tyrosine 34. Perhaps this residue is involved in the proton uptake process.[378] A similarly located Tyr 41 from *Sulfolobus* is covalently modified, perhaps by phosphorylation.[378a]

B. Cobalt and Vitamin B_{12}

The human body contains only about 1.5 mg of cobalt, almost all of it is in the form of **cobalamin**, vitamin B_{12}. Ruminant animals, such as cattle and sheep, have a relatively high nutritional need for cobalt and in regions with a low soil cobalt content, such as Australia, cobalt deficiency in these animals is a serious problem. This need for cobalt largely reflects the high requirement of the microorganisms of the rumen (paunch) for vitamin B_{12}. All bacteria require vitamin B_{12} but not all are able to synthesize it. For example, *E. coli* lacks one enzyme in the biosynthetic

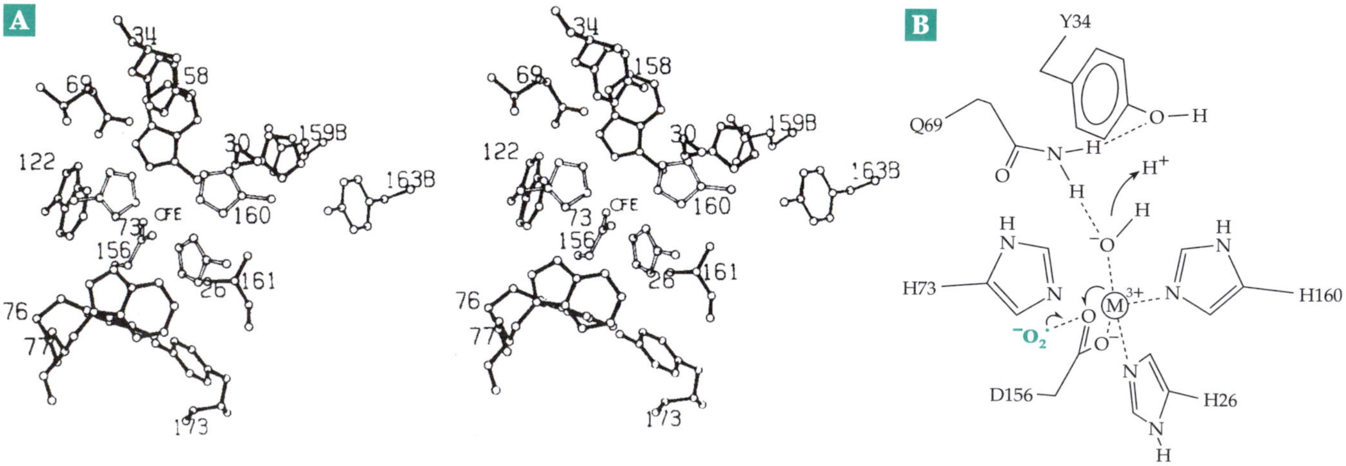

Figure 16-22 (A) Structure of the active site of iron superoxide dismutase from *E. coli*. From Carlioz *et al.*[379] Courtesy of M. Ludwig. (B) Interpretive drawing illustrating the single-electron transfer from a superoxide molecule to the Fe^{3+} of super-oxide dismutase and associated proton uptake. Based on Lah *et al.*[376]

pathway and must depend upon other bacteria to complete the synthesis.

1. Coenzyme Forms

For several years after the discovery of cobalamin its biochemical function remained a mystery, a major reason being the extreme sensitivity of the coenzymes to decomposition by light. Progress came after Barker and associates discovered that the initial step in the anaerobic fermentation of glutamate by *Clostridium tetanomorphum* is rearrangement to β-methylaspartate[380,381] (Eq. 16-28).

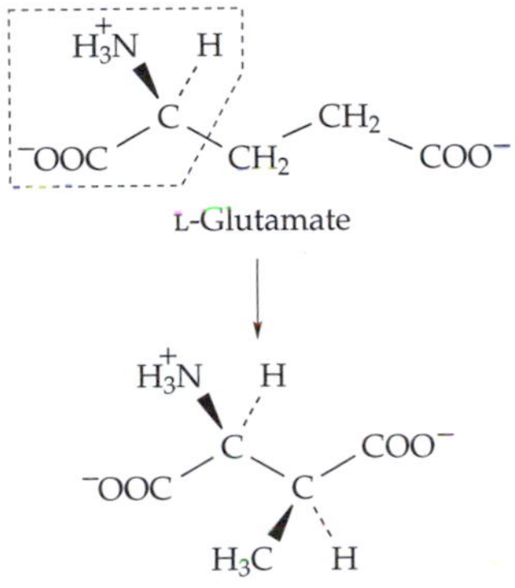

L-Glutamate

threo-β-Methylaspartate　　　　(16-28)

The latter compound can be catabolized by reactions that cannot be used on glutamate itself. Thus, the initial rearrangement is an indispensable step in the energy metabolism of the bacterium. A new coenzyme required for this reaction was isolated in 1958 after it was found that protection from light during the preparation was necessary. The coenzyme was characterized in 1961 by X-ray diffraction[382] as **5'-deoxyadenosylcobalamin**. It is related to cyanocobalamin (Box 16-B) by replacement of the CN group by a 5'-deoxyadenosyl group as indicated in the following abbreviated formulas.[383–385] Here the planes represent the corrin ring system and Bz the dimethylbenzimidazole that is coordinated with the cobalt from below the ring.

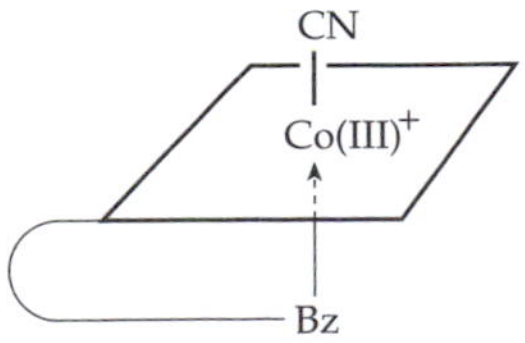

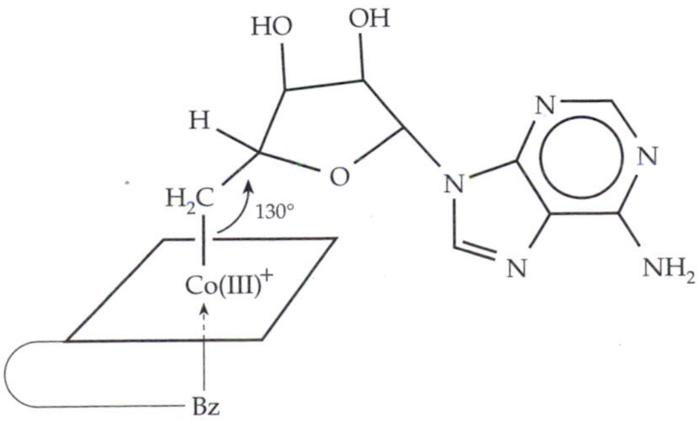

The most surprising structural feature is the Co–C single bond of length 0.205 nm. Thus, the coenzyme is an alkyl cobalt, the first such compound found in nature. In fact, alkyl cobalts were previously thought to be unstable. Vitamin B$_{12}$ contains Co(III), and cyanocobalamin can be imagined as arising by replacement of the single hydrogen on the inside of the corrin ring by Co^{3+} plus CN$^-$. However, bear in mind that three other nitrogens of the corrin ring and a nitrogen of dimethylbenzimidazole also bind to the cobalt. Each nitrogen atom donates an electron pair to form coordinate covalent linkages. Because of resonance in the conjugated double-bond system of the corrin, all four of the Co–N bonds in the ring are nearly equivalent and the positive charge is distributed over the nitrogen atoms surrounding the cobalt.

The strength of the axial Co–C bond is directly influenced by the strength of bonding of the dimethylimidazole whose conjugate base has a microscopic pK_a of 5.5.[386] Protonation of this base breaks its bond to cobalt and may thereby strengthen the Co–C bond.[387] Steric factors are also important in determining the strength of this bond. NMR techniques are now playing an important role in investigation of these factors.[386–388]

In both bacteria and liver, the 5'-deoxyadenosyl coenzyme is the most abundant form of vitamin B$_{12}$, while lesser amounts of **methylcobalamin** are present. Other naturally occurring analogs of the coenzymes include **pseudo vitamin B$_{12}$** which contains adenine in place of the dimethylbenzimidazole.

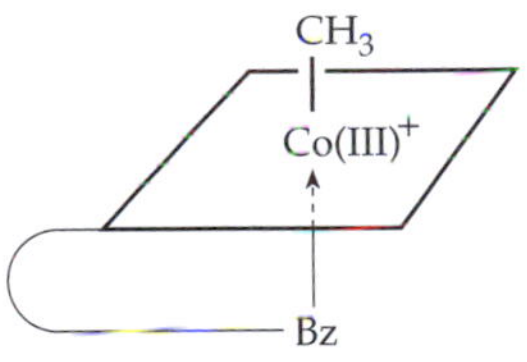

Methylcobalamin

Like dimethylbenzimidazole, it is combined with ribose in the unusual α linkage. A compound called factor A is the vitamin B$_{12}$ analog with 2-methyladenine. Related compounds have been isolated from such sources as sewage sludge which abounds in anaerobic bacteria. It has been suggested that plants may contain vitamin B$_{12}$-like materials which do not support growth of bacteria. Thus, we may not have discovered all of the alkyl cobalt coenzymes.

Cyanocobalamin　　　　5'-Deoxyadenosylcobalamin

BOX 16-B COBALAMIN (VITAMIN B$_{12}$)

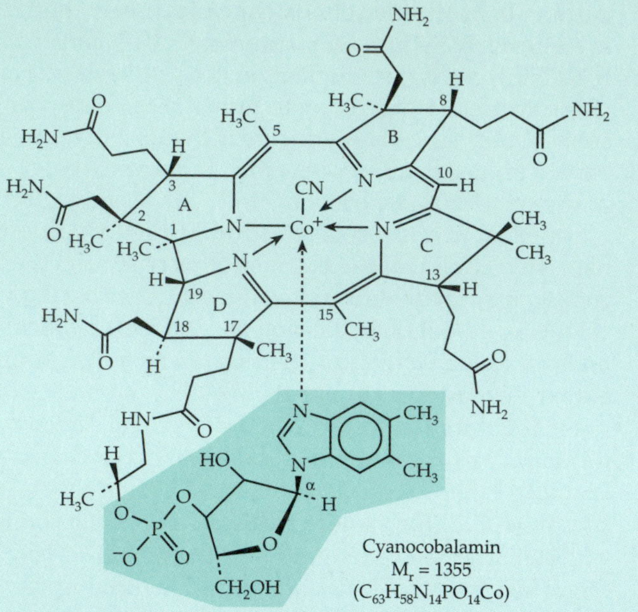

Cyanocobalamin
$M_r = 1355$
$(C_{63}H_{58}N_{14}PO_{14}Co)$

The story of vitamin B$_{12}$ began with pernicious anemia, a disease that usually affects only persons of age 60 or more but which occasionally strikes children.[a] Before 1926 the disease was incurable and usually fatal. Abnormally large, immature, and fragile red blood cells are produced but the total number of erythrocytes is much reduced from 4–6 x 10^6 mm^{-3} to 1– 3 x 10^6 mm^{-3}. Within the bone marrow mitosis appears to be blocked and DNA synthesis is suppressed. The disease also affects other rapidly growing tissues such as the gastric mucous membranes (which stop secreting HCl) and nervous tissues. Demyelination of the central nervous system with loss of muscular coordination (ataxia) and psychotic symptoms is often observed.

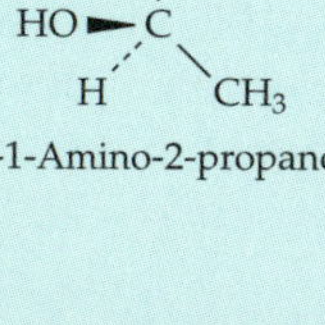

R-1-Amino-2-propanol 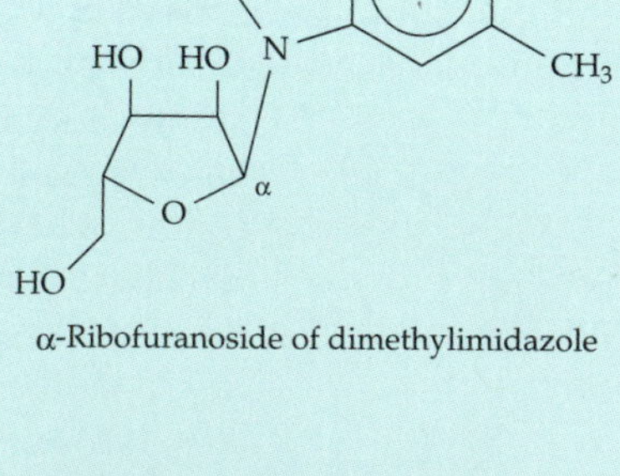   α-Ribofuranoside of dimethylimidazole

In 1926, Minot and Murphy discovered that pernicious anemia could be controlled by eating one-half pound of raw or lightly cooked liver per day, a treatment which not all patients accepted

with enthusiasm. Twenty-two years later vitamin B$_{12}$ was isolated (as the crystalline derivative cyanocobalamin) and was shown to be the curative agent. It is present in liver to the extent of 1 mg kg^{-1} or ~10^{-6} M. Although much effort was expended in preparation of concentrated liver extracts for the treatment of pernicious anemia, the lack of an assay other than treatment of human patients made progress slow.

In the early 1940s nutritional studies of young animals raised on diets lacking animal proteins and maintained out of contact with their own excreta (which contained vitamin B$_{12}$) demonstrated the need for "animal protein factor" which was soon shown to be the same as vitamin B$_{12}$. The animal feeding experiments also demonstrated that waste liquors from streptomyces fermentations used in production of antibiotics were extremely rich in vitamin B$_{12}$. Later this vitamin was recognized as a growth factor for a strain of *Lactobacillus lactis* which responded with half-maximum growth to as little as 0.013 µg / l (10^{-11} M).

In 1948, red cobalt-containing crystals of vitamin B$_{12}$ were obtained almost simultaneously by two pharmaceutical firms. Charcoal adsorption from liver extracts was followed by elution with alcohol and numerous other separation steps. Later fermentation broths provided a richer source. Chemical studies revealed that the new vitamin had an enormous molecular weight, that it contained one atom of phosphorus which could be released as P$_i$, a molecule of aminopropanol, and a ribofuranoside of dimethyl benzimidazole with the unusual α configuration.

Note the relationship of the dimethylbenzimidazole to the ring system of riboflavin (Box 15-B). Several molecules of ammonia could be released from amide linkages by hydrolysis, but all attempts to remove the cobalt reversibly from the ring system were unsuccessful. The structure was determined in 1956 by Dorothy C. Hodgkin and coworkers using X-ray diffraction.[b] At that time, it was the largest organic structure determined by X-ray diffraction. The complete laboratory synthesis was accomplished in 1972.[c]

The ring system of vitamin B$_{12}$, like that of porphyrins (Fig. 16-5), is made up of four pyrrole rings whose biosynthetic relationship to the corresponding rings in porphyrins is obvious from the structures. In addition, a number of "extra" methyl groups are present. A less extensive conjugated system of double bonds is present in the **corrin** ring of vitamin B$_{12}$ than in porphyrins, and as a result, many chiral centers are found around the periphery

BOX 16-B (continued)

of the somewhat nonplanar rings.

Cyanocobalamin, the form of vitamin B$_{12}$ isolated initially, contains cyanide attached to cobalt. It occurs only in minor amounts, if at all, in nature but is generated through the addition of cyanide during the isolation. **Hydroxocobalamin** (vitamin B$_{12a}$) containing OH$^-$ in place of CN$^-$ does occur in nature. However, the predominant forms of the vitamin are the coenzymes in which an alkyl group replaces the CN$^-$ of cyanocobalamin.

Intramuscular injection of as little as 3–6 µg of crystalline vitamin B$_{12}$ is sufficient to bring about a remission of pernicious anemia and 1 µg daily provides a suitable maintenance dose (often administered as hydroxocobalamin injected once every 2 weeks). For a normal person a dietary intake of 2–5 µg / day is adequate. There is rarely any difficulty in meeting this requirement from ordinary diets. Vitamin B$_{12}$ has the distinction of being synthesized only by bacteria, and plants apparently contain none. Consequently, strict vegetarians sometimes have symptoms of vitamin B$_{12}$ deficiency.

Pernicious anemia is usually caused by poor absorption of the vitamin. Absorption depends upon the **intrinsic factor**, a mucoprotein (or mucoproteins) synthesized by the stomach lining.[a,d-f] Pernicious anemia patients often have a genetic predilection toward decreased synthesis of the intrinsic factor. Gastrectomy, which decreases synthesis of the intrinsic factor, or infection with fish tapeworms, which compete for available vitamin B$_{12}$ and interfere with absorption, can also induce the disease. Also essential are a plasma membrane receptor[g,h] and two blood transport proteins

transcobalamin[d,i] and **cobalophilin**. The latter is a glycoprotein found in virtually every human biological fluid and which may protect the vitamin from photodegradation by light that penetrates tissues.[j] A variety of genetic defects involving uptake, transport, and conversion to vitamin B$_{12}$ coenzyme forms are known.[f,k]

Normal blood levels of vitamin B$_{12}$ are ~2 x 10^{-10} M or a little more, but in vegetarians the level may drop to less than one-half this value. A deficiency of folic acid can also cause megaloblastic anemia, and a large excess of folic acid can, to some extent, reverse the anemia of pernicious anemia and mask the disease.

[a] Karlson, P. (1979) *Trends Biochem. Sci.* **4**, 286
[b] Hodgkin, D. C. (1965) *Science* **150**, 979–988
[c] Maugh, T. H., II (1973) *Science* **179**, 266–267
[d] Gräsbeck, R., and Kouvonen, I. (1983) *Trends Biochem. Sci.* **8**, 203–205
[e] Allen, R. H., Stabler, S. P., Savage, D. G., and Lindenbaum, J. (1993) *FASEB J.* **7**, 1344–1353
[f] Fenton, W. A., and Rosenberg, L. E. (1995) in *The Metabolic and Molecular Bases of Inherited Disease*, 7th ed., Vol. 2 (Scriver, C. R., Beaudet, A. L., Sly, W. S., and Valle, D., eds), pp. 3129–3149, McGraw-Hill, New York
[g] Seetharam, S., Ramanujam, K. S., and Seetharam, B. (1992) *J. Biol. Chem.* **267**, 7421–7427
[h] Birn, H., Verroust, P. J., Nexo, E., Hager, H., Jacobsen, C., Christensen, E. I., and Moestrup, S. K. (1997) *J. Biol. Chem.* **272**, 26497–26504
[i] Fedosov, S. N., Berglund, L., Nexo, E., and Petersen, T. E. (1999) *J. Biol. Chem.* **274**, 26015–26020
[j] Frisbie, S. M., and Chance, M. R. (1993) *Biochemistry* **32**, 13886–13892
[k] Rosenblatt, D. S., Hosack, A., Matiaszuk, N. V., Cooper, B. A., and Laframboise, R. (1985) *Science* **228**, 1319–1321

2. Reduction of Cyanocobalamin and Synthesis of Alkyl Cobalamins

Cyanocobalamin can be reduced in two one-electron steps (Eq. 16-29).[385,388] The cyanide ion is lost in the first step (Eq. 16-29, step *a*), which may be accomplished with chromous acetate at pH 5 or by catalytic hydrogenation. The product is the brown paramagnetic compound B$_{12r}$, a tetragonal low-spin cobalt(II) complex. In the second step (Eq. 16-29, step *b*), an additional electron is added, e.g., from sodium borohydride or from chromous acetate at pH 9.5, to give the gray-green exceedingly reactive B$_{12s}$. The latter is thought to be in equilibrium with cobalt(III) hydride, as shown in Eq. 16-30, step *a*. The hydride is unstable and breaks down slowly to H$_2$ and B$_{12r}$ (Eq. 16-30, step *b*).[389]

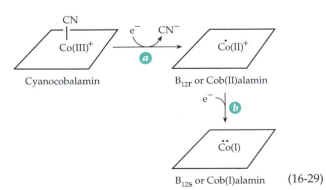

Cyanocobalamin   B$_{12r}$ or Cob(II)alamin   B$_{12s}$ or Cob(I)alamin   (16-29)

$$B_{12s} \xrightleftharpoons[]{H^+ \ \ a} \ \overset{H}{Co(III)^+} \xrightarrow[Slow]{b} B_{12r}$$
$$\frac{1}{2}\,H_2$$

(16-30)

Vitamin B_{12s} reacts rapidly with alkyl iodides (e.g., methyl iodide or a 5'-chloro derivative of adenosine) via nucleophilic displacement to form the alkyl cobalt forms of vitamin B_{12} (Eq. 16-31). These reactions provide a convenient way of preparing isotopically labeled alkyl cobalamins, including those selectively

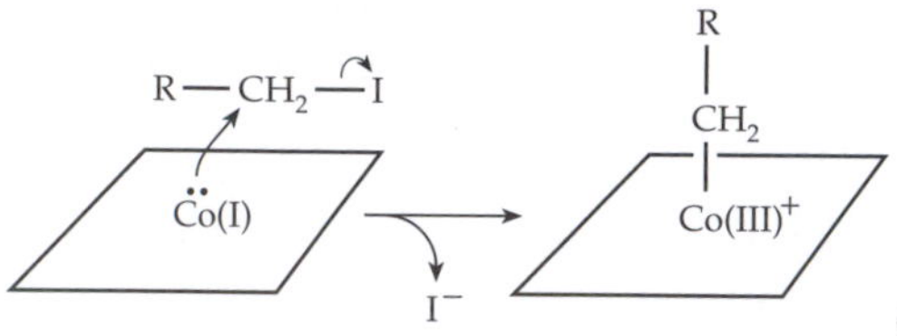

(16-31)

enriched in ^{13}C for use in NMR studies.[390] The biosynthesis of 5'-deoxyadenosylcobalamin utilizes the same type of reaction with ATP as a substrate.[391] A **B_{12s} adenosyltransferase** catalyzes nucleophilic displacement on the 5' carbon of ATP with formation of the coenzyme and displacement of inorganic tripolyphosphate PPP_i.

3. Three Nonenzymatic Cleavage Reactions of Vitamin B$_{12}$ Coenzymes

The 5'-deoxyadenosyl coenzyme is easily decomposed by a variety of agents. Anaerobic irradiation with visible light yields principally vitamin B_{12r} and a cyclic 5',8-deoxyadenosine which is probably formed through an intermediate radical[391a] (Eq. 16-32):

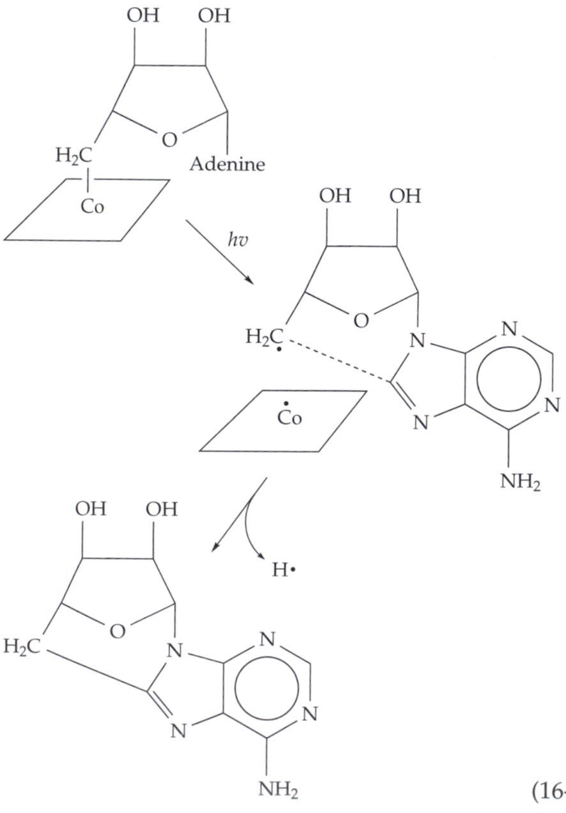

(16-32)

Irradiation in the presence of air gives a variety of products.

Hydrolysis of deoxyadenosylcobalamin by acid (1 M HCl, 100°, 90 min) yields hydroxycobalamin, adenine, and an unsaturated sugar (Eq. 16-33). The initial reaction step is thought to be protonation of the oxygen of the ribose ring.

A related cleavage by alkaline cyanide can be viewed as a nucleophilic displacement of the deoxyadenosyl anion by cyanide. The end product is **dicyanocobalamin**, in which the loosely bound nucleotide containing dimethyl benzimidazole is replaced by a second cyanide ion. Methyl and other simple alkyl cobalamins are stable to alkaline cyanide. A number of other cleavage reactions of alkyl cobalamins are known.[392,393]

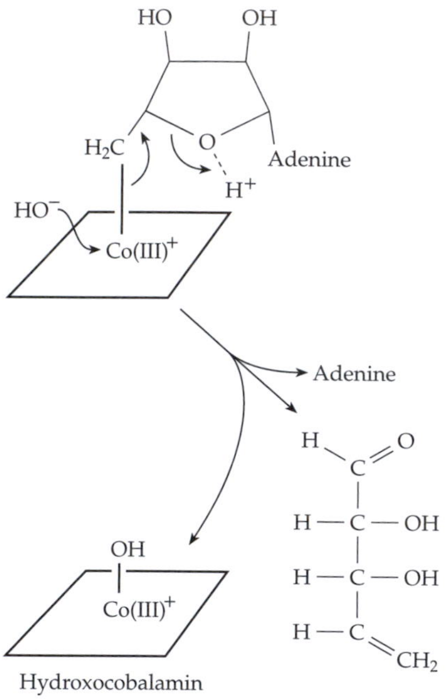

Hydroxocobalamin

(16-33)

4. Enzymatic Functions of B$_{12}$ Coenzymes

Three types of enzymatic reactions depend upon alkyl corrin coenzymes. The first is the reduction of ribonucleotide triphosphates by cobalamin-dependent ribonucleotide reductase, a process involving *intermolecular* hydrogen transfer (Eq. 16-21). The second type of reaction encompasses the series of isomerizations shown in Table 16-1. These can all be depicted as in Eq. 16-34. Some group X, which may be attached by a C–C, C–O, or C–N bond, is transferred to an adjacent carbon atom bearing a hydrogen. At the same time,

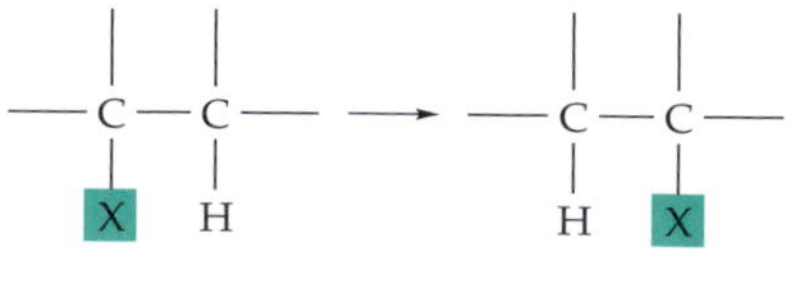

(16-34)

the hydrogen is transferred to the carbon to which X was originally attached. The third type of reaction is the transfer of methyl groups via methylcobalamin and some related bacterial metabolic reactions.

Cobalamin-dependent ribonucleotide reductase.

Lactic acid bacteria such as *Lactobacillus leichmanni* and many other bacteria utilize a 5'-deoxyadenosylcobal-amin-containing enzyme to reduce nucleoside triphos-phates according to Eq. 16-21. Thioredoxin or dihydrolipoic acid can serve as the hydrogen donor. Early experiments showed that protons from water are reversibly incorporated at C-2' of the reduced

nucleotide with retention of configuration. A more important finding was a large kinetic isotope effect of 1.8 when 3'-^{3}H-containing UTP was reduced by the enzyme.[394]

Reaction of the reductase with dihydrolipoic acid in the presence of deoxy-GTP, which apparently serves as an allosteric activator, leads to formation, within a few milliseconds, of a radical with a characteristic EPR spectrum that can be studied when the reaction mixture is rapidly cooled to 130°K. When GTP (a true substrate) is used instead of dGTP, the radical signal reaches a maximum in about 20 ms and then decays. Of the various oxidation states of cobalt (3+, 2+, and 1+)

TABLE 16-1
Isomerization Reactions Involving Hydrogen Transfer and Dependent upon a Vitamin B$_{12}$ Coenzyme

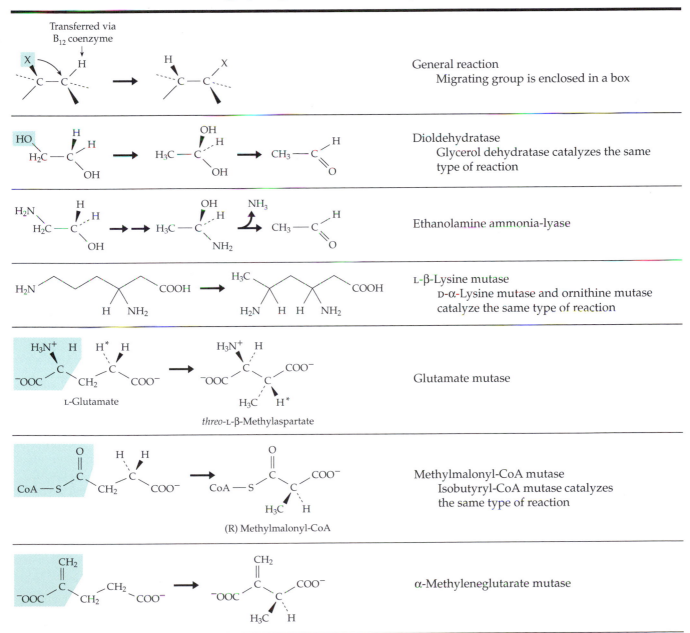

only the 2+ state of vitamin B_{12r} is paramagnetic and gives rise to an EPR signal. The electronic absorption spectrum of the coenzyme of ribonucleotide reductase is also changed rapidly by substrate in a way that suggests formation of B_{12r}. Thus, it was proposed that a *homolytic* cleavage occurs to form B_{12r} and a stabilized 5'-deoxyadenosyl radical (Eq. 16-35).[395–396a] However, H^3 is not transferred from the 3' position of the substrate into the deoxyadenosyl part of the coenzyme.[394] The enzyme has many properties in common with the previously discussed iron–tyrosinate ribonucleotide reductases (Fig. 16-21) including limited peptide sequence homology.[350] Stubbe and coworkers suggested that the deoxyadenosyl radical is formed as a radical chain initiator[394,396b] and that the mechanism of ribonucleotide reduction is as shown in Eq. 16-25. Studies of enzyme-activated inhibitors support this mechanism.[364]

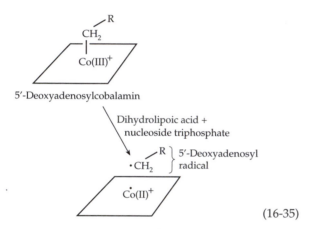

5'-Deoxyadenosylcobalamin

Dihydrolipoic acid + nucleoside triphosphate

5'-Deoxyadenosyl radical

(16-35)

The isomerization reactions. At least 10 reactions of the type described by Eq. 16-34 are known[397] (Table 16-1). They can be subdivided into three groups. First, X = OH or NH_2 in Eq. 16-34; isomerization gives a *geminal*-diol or aminoalcohol that can eliminate H_2O or NH_3 to give an aldehyde. All of these enzymes, which are called **hydro-lyases** or **ammonia-lyases**, specifically require K^+ as well as the vitamin B_{12} coenzyme. Second, X = NH_2 in Eq. 16-34; For this group of **aminomutases** PLP is required as a second coenzyme. Third, X is attached via a carbon atom; the enzymes are called **mutases**. **Methylmalonyl-Co mutase** is required for catabolism of propionate in the human body, and is one of only two known vitamin B_{12}-

dependent enzymes. The related isobutyryl-CoA mutase participates in the microbial synthesis of such polyether antibiotics as monensin A.[397a,b] Other mutases are involved in anaerobic bacterial fermentations.

In these reactions the hydrogen is always transferred via the B_{12} coenzyme. No exchange with the medium takes place. Since X may be an electronegative group such as OH, the reactions could all be treated formally as hydride ion transfers but it is more likely that they occur via homolytic cleavages. Such cleavage is indicated by the observation of EPR signals for several of the enzymes in the presence of their substrates.[398]

Abeles and associates showed that when **dioldehydratase** (Table 16-1) catalyzes the conversion of 1,2-[1-³H]propanediol to propionaldehyde, tritium appears in the coenzyme as well as in the final product. When ³H-containing coenzyme is incubated with unlabeled propanediol, the product also contains ³H, which was shown by chemical degradation to be exclusively on C-5'. Synthetic 5'-deoxyadenosyl coenzyme containing ³H in the 5' position transferred ³H to product. Most important, using a mixture of propanediol and ethylene glycol, a small amount of *intermolecular* transfer was demonstrated; that is, ³H was transferred into acetaldehyde, the product of dehydration of ethylene glycol. Similar results were also obtained with ethanolamine ammonia-lyase.[399]

Another important experiment[398] showed that ¹⁸O from [2-¹⁸O]propanediol was transferred into the 1 position without exchange with solvent. Furthermore, ¹⁸O from (S)-[1-¹⁸O]propanediol was retained in the product while that from the (R) isomer was not. Thus, it appears that the enzyme stereospecifically dehydrates the final intermediate. From these and other experiments, it was concluded that initially a 5'-deoxyadenosyl radical is formed via Eq. 16-35. This radical then abstracts the hydrogen atom marked by a shaded box in Eq. 16-36 to form a substrate radical and 5'-deoxyadenosine. One proposal, illustrated in Eq. 16-36, is that the substrate radical immediately recombines with

(16-36)

the Co(II) of the coenzyme to form an organo-cobalt substrate compound (step *a* of Eq. 16-36). The 5'-deoxyadenosine now contains hydrogen from the substrate; because of rotation of the methyl group this hydrogen becomes equivalent to the two already present in the coenzyme. The substrate–cobalamin compound formed in this step then undergoes isomerization, which, in the case of dioldehydratase, leads to intramolecular transfer of the OH group (step *b*). In step *c* the hydrogen atom is transferred back from the 5'-deoxyadenosine to its new location in the product and in step *d* the resulting *gem*-diol is dehydrated to form the aldehyde product.[401a]

Carbocation, carboanion, and free radical intermediates have all been proposed for the isomerization step in the reaction. A carbocation would presumably be cyclized to an epoxide which could react with the B$_{12s}$ (Eq. 16-37, step *b*) to complete the isomerization.

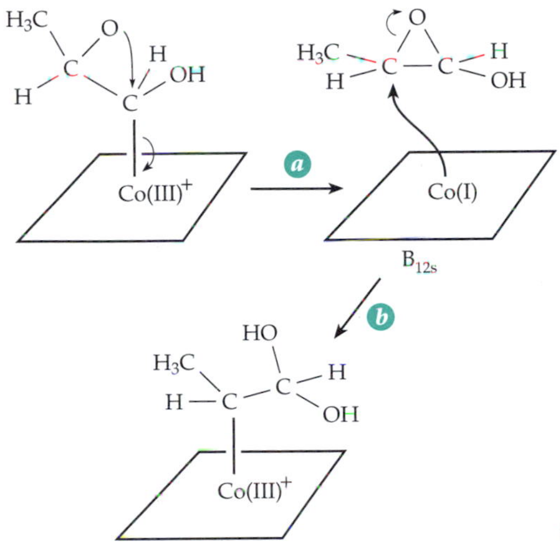

Recently, EPR spectroscopy with ^{2}H- and ^{13}C-labeled glutamates as substrates for glutamate mutase permitted identification of a 4-glutamyl radical as a probable intermediate for that enzyme.[402,402a,b]

On the other hand, for methylmalonyl-CoA mutase Ingraham suggested cleavage of the Co–C bond of an organocobalt intermediate to form a carbanion, the substrate–cobalamin compound serving as a sort of "biological Grignard reagent." The carbanion would be stabilized by the carbonyl group of the thioester forming a "homoenolate anion" (Eq. 16-39). The latter could break up in either of two ways reforming a C–Co bond and causing the isomerization.[403] Some experimental results also favor an ionic or organocobalt pathway.[404]

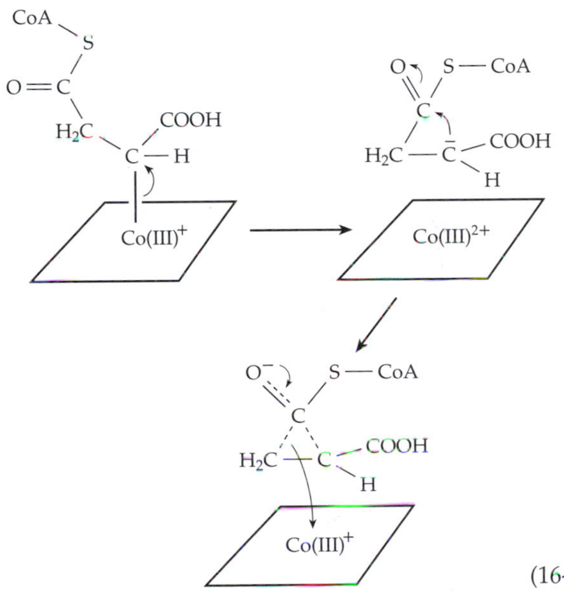

(16-37)

(16-39)

At present most evidence favors, for the isomerization reactions, an enzyme-catalyzed rearrangement of the substrate radical produced initially during formation of the 5'-deoxyadenosine (Eq. 16-38).[400–401c]

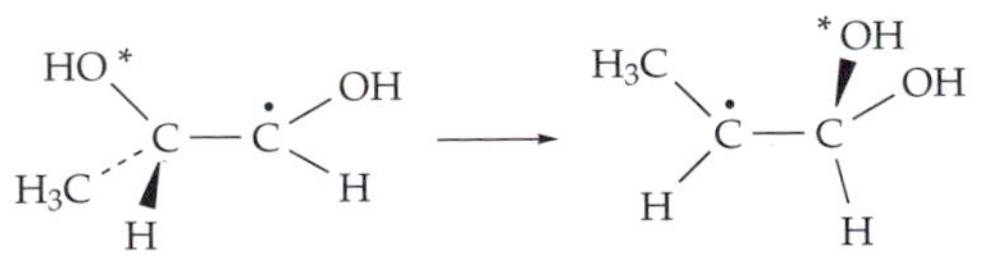

(16-38)

According to this mechanism the Co(II) of the B$_{12r}$ formed in Eq. 16-35 has no active role in the isomerization and does not form an organocobalt intermediate as in Eq. 16-36. Its only role is to be available to recombine with the 5'-deoxyadenosyl radical at the end of the reaction sequence. Support for this interpretation has been obtained from study of model reactions and of organic radicals generated in other ways.[400]

The three-dimensional structure of methylmalonyl-CoA mutase from *Propionibacterium shermanii* shows that the vitamin B$_{12}$ is bound in a base-off conformation with the dimethylbenzimidazole group bound to the protein far from the corrin ring (Fig. 16-23; see also Fig. 16-24).[405,405a] A histidine of the protein coordinates the cobalt, as also in methionine synthase. The entrance to the deeply buried active site is blocked by the coenzyme part of the substrate. The buried active site may be favorable for free radical rearrangement reactions. The structure of substrate complexes shows that the coenzyme is in the cob(II)alamin (B$_{12r}$) form with the 5'-deoxyadenosyl group detached from the cobalt, rotated, shifted, and weakly bound to the protein.[405a] Side chains from neighboring Y89, R207, and H244 all hydrogen bond to the substrate. The R207 guanidinium group makes an ion pair with the substrate carboxylate, and the phenolic and imidazole groups may have catalytic functions.[405b,c,d,406] Studies are also in progress on crystalline glutamate mutase.[406,406a]

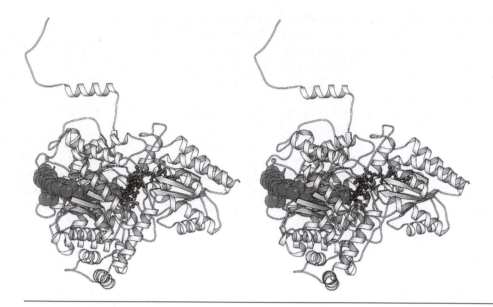

Figure 16-23 Three-dimensional structure of methylmalonyl-CoA mutase from *Propionobacterium shermanii*. The B$_{12}$ coenzyme is deeply buried, as is the active site. A molecule of bound desulfo-coenzyme A, a substrate analog, blocks the active site entrance on the left side. From Mancia *et al.*[405] Courtesy of Philip R. Evans.

According to all of these mechanisms, 5'-deoxyadenosine is freed from its bond to cobalt during the action of the enzymes. Why then does the deoxyadenosine not escape from the coenzyme entirely, leading to its inactivation? Substrate-induced inactivation is not ordinarily observed with coenzyme B$_{12}$-dependent reactions, but some quasi-substrates do inactivate their enzymes. Thus, glycolaldehyde converts the coenzyme of dioldehydratase to 5'-deoxyadenosine and ethylene glycol does the same with ethanolamine deaminase. When 5'-deoxyinosine replaces the 5'-deoxyadenosine of the normal coenzyme in dioldehydratase, 5'-deoxyinosine is released quantitatively by the substrate. This suggests that the dehydratase may normally hold the adenine group of 5'-deoxyadenosine through hydrogen bonding to the amino group. Because the OH group of inosine tautomerizes to C=O, inosine may not be held as tightly. The deeply buried active site (Fig. 16-23) may also prevent escape of the deoxyadenosine.

Despite the evidence in its favor, there was initially some reluctance to accept 5'-deoxyadenosine as an intermediate in vitamin B$_{12}$-dependent isomerization reactions. It was hard to believe that a methyl group could exchange hydrogen atoms so rapidly. It was suggested that protonation of the oxygen of the ribose ring as in Eq. 16-33 might facilitate release of a hydrogen atom. However, substitution of the ring oxygen by CH$_2$ in a synthetic analog did not destroy the coenzymatic activity.[407] Another possibility is that the methyl group has an unusual reactivity if the cobalt is reduced to Co(II).

Stereochemistry of the isomerization reactions.
Dioldehydratase acts on either the (*R*) or (*S*) isomers of 1,2-propanediol (Eqs. 16-40 and 16-41; asterisks and daggers mark positions of labeled atoms in specific experiments).

$$(S)\text{-1,2-Propanediol} \qquad (16\text{-}40)$$

$$(R)\text{-1,2-Propanediol} \qquad (16\text{-}41)$$

In both cases the reaction proceeds with retention of configuration at C-2 and with stereochemical specificity[408] for one of the two hydrogens at C-1. The reaction catalyzed by methylmalonyl-CoA mutase likewise proceeds with retention of configuration at C-2 (Table 16-1)[409] but the glutamate mutase reaction is accompanied by inversion (Eq. 16-28).

Aminomutases.
The enzymes L-β-lysine mutase (which is also D-α-lysine mutase) and D-ornithine mutase catalyze the transfer of an ω-amino group to an adjacent carbon atom[410] (Table 16-1). Two proteins are needed for the reaction; pyridoxal phosphate is required and is apparently directly involved in the amino group migration. In the β-lysine mutase the 6-amino group of L-β-lysine replaces the pro-*S* hydrogen at C-5 but with inversion at C-5 to yield (3*S*, 5*S*)-3,5-diaminohexanoic acid.[411] A bacterial D-lysine 5,6-aminomutase interconverts D-lysine with 2,5-diaminohexanoic acid.[411a] Another related enzyme

is **L-leucine 2,3-aminomutase**, which catalyzes the reversible interconversion of L-leucine and β-leucine.[410,412] It was reported to be present in plants and also in the human body, but the latter could not be confirmed.[413]

The interconversion of L-α-lysine and L-β-lysine is catalyzed by a lysine 2,3-aminomutase found in certain clostridia.[414-415a] This enzyme also requires pyridoxal phosphate and catalyzes a reaction with the same stereochemistry as that of β-lysine mutase. However, it does not contain vitamin B$_{12}$ but depends upon *S*-adenosylmethionine (AdoMet) and an iron–sulfur cluster. The adenosyl group of AdoMet may function in the same manner as does the deoxyadenosyl group of the adenosylcobalamin coenzyme. In these mutases a 5'-deoxyadenosyl radical may abstract a hydrogen atom from the β position of a Schiff base of PLP with the amino acid substrate. The radical isomerizes (Eq. 16-42) and then accepts a hydrogen atom back from the 5'-deoxyadenosine to complete the reaction. Its

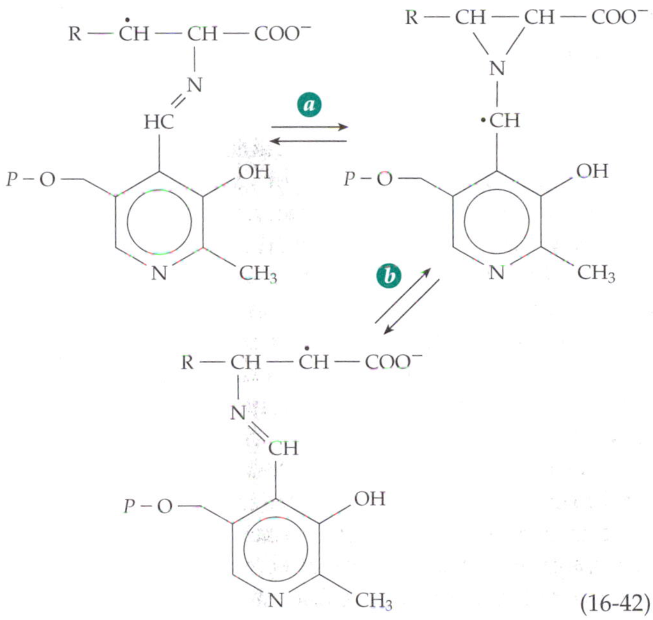

properties suggest that lysine 2,3-aminomutase is related to the pyruvate formate-lyase of *E. coli* (Eq. 15-37),[410] class I ribonucleotide reductases, and other enzymes that act by homolytic mechanisms.

Transfer reactions of methyl groups. The generation and utilization of methyl groups is a quantitatively important aspect of the metabolism of all cells. As we have seen (Fig. 15-18), methyl groups can be created by the reduction of one-carbon compounds attached to tetrahydrofolic acid. Methyl groups of methyltetrahydrofolic acid (N^5-CH$_3$THF) can then be transferred to the sulfur atom of homocysteine to form methionine (Eq. 16-43). The latter is converted to *S*-adenosylmethionine, the nearly universal methyl group donor for transmethylation reactions (Eq. 12-3). In some bacteria, fungi, and higher plants, the methyl-THF-homocysteine transmethylase does not depend upon vitamin B$_{12}$ but is a metalloenzyme with zinc at the active center.[416] However, human beings share with certain strains of *E. coli* and other bacteria the need for methylcobalamin.

The structure of the *E. coli* enzyme (Fig. 16-24) shows methylcobalamin bound in a base-off conformation, with histidine 759 of the protein replacing dimethylbenzimidazole in the distal coordination position on the cobalt. This histidine is part of a sequence Asp-X-His-X-X-Gly that is found not only in methionine synthase but also in methylmalonyl-CoA mutase, glutamate mutase, and 2-methyleneglutarate mutase. However, diol dehydratase lacks this sequence and binds adenosylcobalamin with the dimethylbenzimidazole–cobalt bond intact.[417]

The coenzyme evidently functions in a cyclic process. The cobalt alternates between the +1 and +3 oxidation states as shown in Eq. 16-43. The first indication of such a cyclic process was the report by Weissbach that ^{14}C-labeled methylcobalamin could be isolated following treatment of the enzyme with such methyl donors as AdoMet and methyl iodide

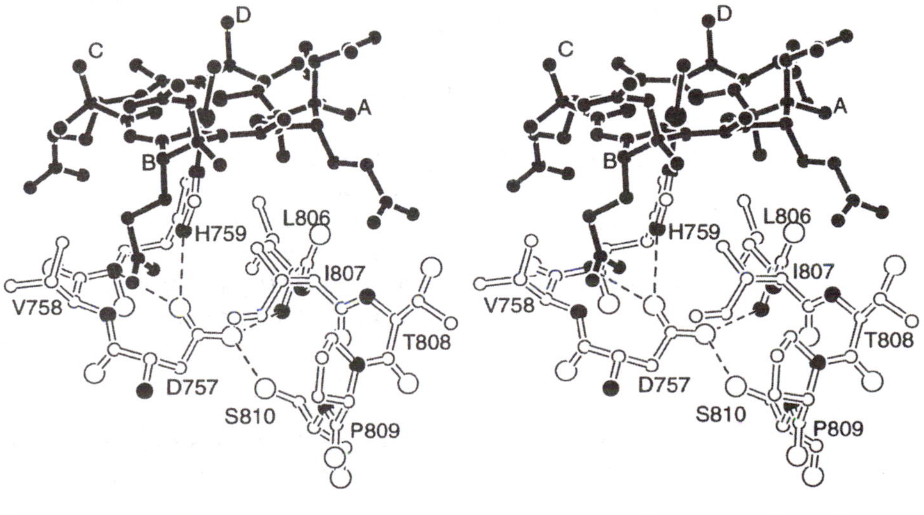

(16-42)

Figure 16-24 Stereoscopic views of the active site of methionine synthase from *E. coli*. Methylcobalamin (black) is in the active site with His 759 of the protein in the distal position of the coenzyme in a base-off conformation. The dimethylbenzimidole nucleotide has been omitted for clarity. Notice the hydrogen-bonded His 759 – D757 – S810 triad. From Jarret *et al.*[418a] Courtesy of R. G. Matthews.

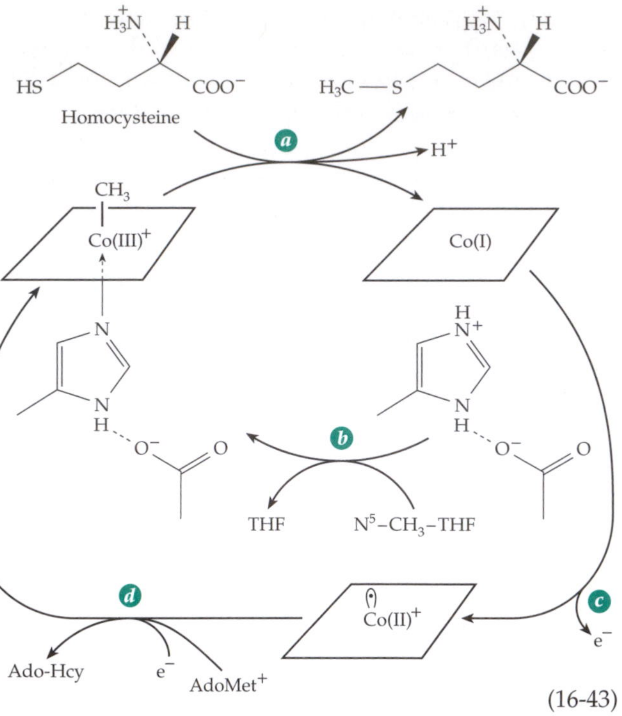

(16-43)

after reduction (e.g., with reduced riboflavin phosphate). The sequence parallels that of Eq. 16-31 for the laboratory synthesis of methylcobalamin. Nevertheless, the transmethylase demonstrates some complexities. Initially, it must be "activated" by AdoMet or methyl iodide, after which it cycles according to Eq. 16-43, steps *a* and *b*, but is gradually inactivated. This apparently happens by oxidation to a Co(II) form of the enzyme (Eq. 16-43, step *c*) that must be reductively methylated with AdoMet and reduced flavodoxin (step *d*) to regenerate the active form.[418] It is also possible that the methyl group is not transferred as a formal CH_3^+ as pictured in Eq. 16-43 but as a $^{\bullet}CH_3$ radical generated by homolytic cleavage of methylcobalamin to cob(II)alamin. Whatever the mechanism, a chiral methyl group is transferred from 5-methyl-THF to homocysteine with overall retention of configuration.[419]

Another important group of methyl transfer reactions are those from methyl corrinoids to mercury, tin, arsenic, selenium, and tellurium. For example, Eq. 16-44 describes the methylation of Hg^{2+}. These reactions are of special interest because of the generation of toxic methyl and dimethyl mercury and dimethylarsine.

Notice that whereas in Eq. 16-43 the methyl group is transferred as CH_3^+ by nucleophilic displacement on a carbon atom, the transfer to Hg^{2+} in Eq. 16-44 is that of a carbanion, CH_3^-, with no valence change occurring in the cobalt. However, it is also possible that transfer occurs as a methyl radical.[420] Methyl corrinoids are able to undergo this type of reaction nonenzymatically, and the ability to transfer a methyl anion is a property of methyl corrinoids not shared by other transmethy-

lating agents such as AdoMet. At the conclusion of the reaction in Eq. 16-44 the cobalt is in the +3 state. To be remethylated, it must presumably be reduced to Co(II). A second methyl group can be transferred by the same type of reaction to form $(CH_3)_2Hg$.

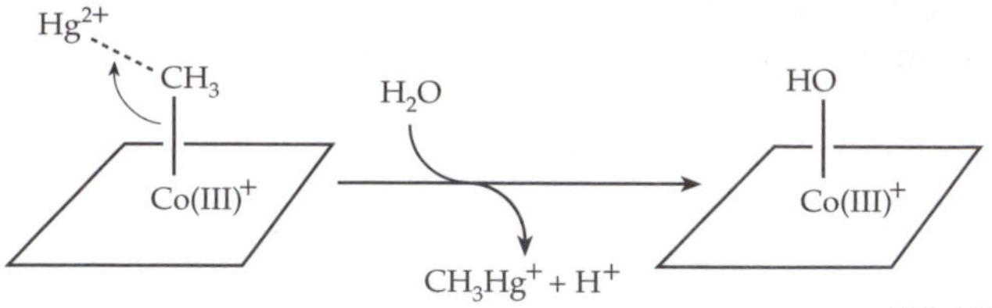

(16-44)

Methylation of arsenic is an important pollution problem because of the widespread use of arsenic compounds in insecticides and because of the presence of arsenate in the phosphate used in household detergents.[421,422] After reduction to arsenite, methylation occurs in two steps (Eq. 16-45). Additional reduction steps result in the formation of **dimethylarsine**, one of the principal products of action of methanogenic bacteria on arsenate. The methyl transfer is shown as occurring through CH_3^+, with an accompanying loss of a proton from the substrate. However, a CH_3 radical may be transferred with formation of a cobalt(II) corrinoid.[423]

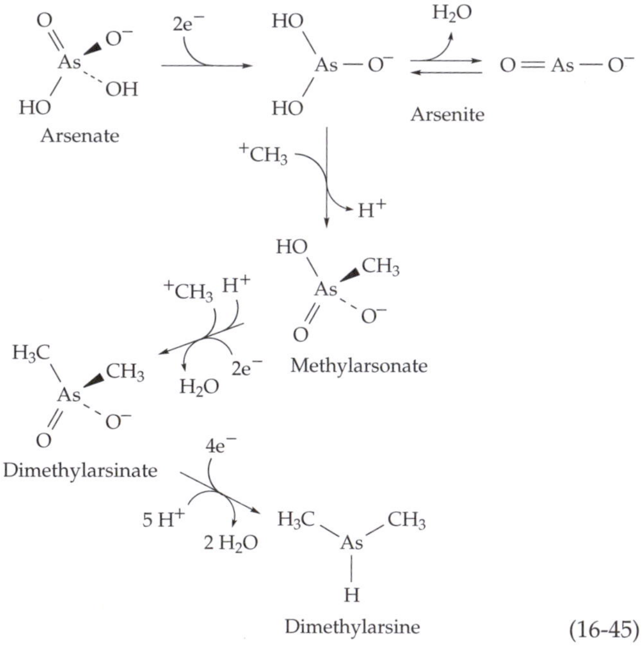

(16-45)

Corrinoid-dependent synthesis of acetyl-CoA.

The anaerobic bacterium *Clostridium thermoaceticum* obtains its energy for growth by reduction of CO_2 with hydrogen (Eq. 16-46). One of the CO_2 molecules is reduced to formate which is converted via 5-methyl-THF to the methyl corrinoid 5-methoxybenzimidazolylcobamide. The methyl group of the latter

$$4\,H_2 + 2\,CO_2 \longrightarrow CH_3COO^- + H^+ + 2\,H_2O$$
$$\Delta G'\,(\text{pH 7}) = -94.9\ \text{kJ mol}^{-1} \qquad (16\text{-}46)$$

combines with another CO_2 to form acetyl-CoA. The process, which requires the nickel-containing carbon monoxide dehydrogenase, is discussed in Section C.

C. Nickel

It was not until the 1970s that nickel was first recognized as a dietary essential for animals.[424–426] Nickel-deficient chicks grew poorly, had thickened legs, and developed dermatitis. Tissues of deficient animals contained swollen mitochondria and swollen perinuclear space suggesting a function for Ni in membranes. However, doubts have been raised about the conclusions based on these experiments.[427] Within tissues the nickel content ranges from 1 to 5 µg / l. Some of the metal in serum is present as complexes of low molecular mass and some is bound to serum albumin[428] and to a specific nickel-containing protein of the macroglobulin class, known as **nickeloplasmin**.[429] Nickel is also present in plants; in some, e.g., *Allysum*, it accumulates to high concentrations.[430] It is essential for legumes and possibly for all plants.[431] Nickel uptake proteins have been identified in bacteria and fungi.[432,432a] Because of its ubiquitous occurrence it is difficult to prepare a totally Ni-free diet. The acute toxicity of orally ingested Ni(II) is low, and homeostatic mechanisms exist in the animal body for regulating its concentration. However, the volatile nickel carbonyl $Ni(CO)_4$ is very toxic[426] and Ni from jewelry is a common cause of dermatitis.[428,433]

In its compounds nickel usually has the +2 oxidation state but the +3 and +4 states occur rarely in complexes. The Ni^{2+} ion contains eight 3d electrons, a configuration that favors square-planar coordination of four ligands. However, the ion is also able to form a complex with six ligands and an octahedral geometry. It has been suggested that this "ambivalence" may be of biochemical significance.

Nickel is found in at least four enzymes: **urease**, certain **hydrogenases, methyl-CoM reductase** (in its **cofactor F₄₃₀**) of methanogenic bacteria, and **carbon monoxide dehydrogenase** of acetogenic and methanogenic bacteria.[434]

1. Urease

Urease, which was first isolated from the jack bean has a special place in biochemical history as the first enzyme to be crystallized. This was accomplished by J. B. Sumner in 1926, and although Sumner eventually

obtained the Nobel Prize, his first reports were greeted with skepticism and outright disbelief. The presence of two atoms of nickel in each molecule of urease[435] was not discovered until 1975. The metal ions had been overlooked previously, despite the fact that the absorption spectrum of the purified enzyme contains an absorption "tail" extending into the visible region with a shoulder at 425 nm and weak maxima at 725 and 1060 nm. Urease catalyzes the hydrolytic cleavage of urea to two molecules of ammonia and one of bicarbonate and is useful in the analytical determination of urea.

Jack bean urease is a trimer or hexamer of identical 91-kDa subunits while that of the bacterium *Klebsiella* has an $(\alpha\beta_2\gamma_2)_2$ stoichiometry. Nevertheless, the enzymes are homologous and both contain the same binickel catalytic center (Fig. 16-25).[435–437a] The three-dimensional structure of the *Klebsiella* enzyme revealed that the two nickel ions are bridged by a carbamyl group of a carbamylated lysine. Like ribulose bisphosphate carboxylase (Fig. 13-10), urease also requires CO_2 for formation of the active enzyme.[438] Formation of the metallocenter also requires four additional proteins, including a chaperonin and a nickel-binding protein.[438,439]

The mechanism of urease action is probably related to those of metalloproteases such as carboxypeptidase A (Fig. 12-16) and of the zinc-dependent carbonic

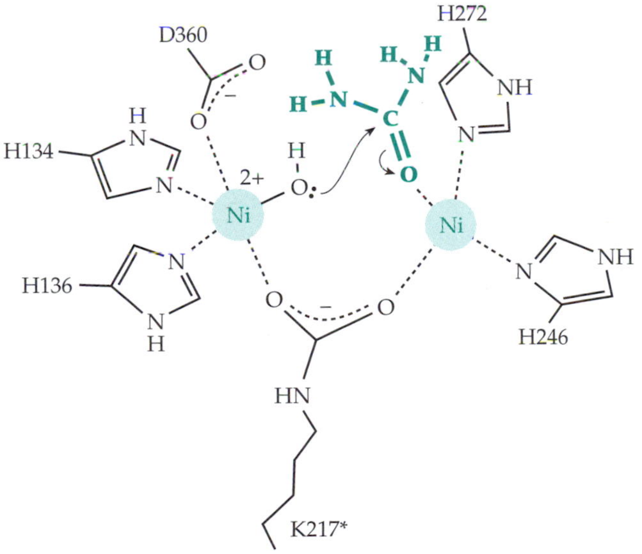

Figure 16-25 The active site of urease showing the two Ni^+ ions held by histidine side chains and bridged by a carbamylated lysine (K217*). A bound urea molecule is shown in green. It has been placed in an open coordination position on one nickel and is shown being attacked for hydrolytic cleavage by a hydroxyl group bound to the other nickel. Based on a structure by Jabri *et al.*[436] and drawing by Lippard.[437]

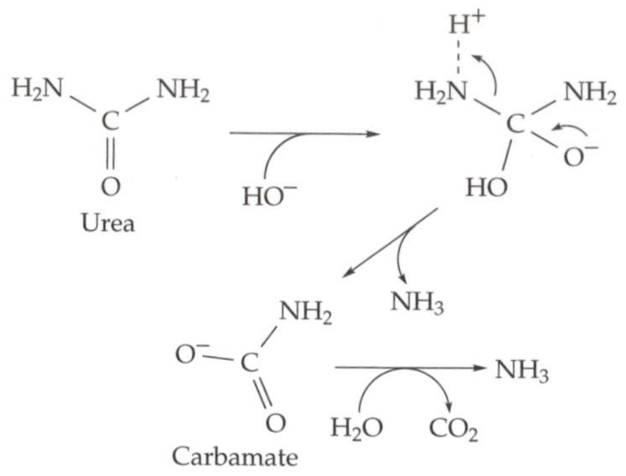

Urea

Carbamate

$$(16\text{-}47)$$

anhydrase (Fig. 13-1). In Fig. 16-25 one nickel ion is shown polarizing the carbonyl group, while the second provides a bound hydroxyl ion that serves as the attacking nucleophile. A probable intermediate product is a carbamate ion (Eq. 16-47).

Urease is an essential enzyme for bacteria and other organisms that use urea as a primary source of nitrogen. The peptic ulcer bacterium *Helicobacter pylori* uses urease to hydrolyze urea in order to defend against the high acidity of the stomach.[439a] The enzyme is also present in plant leaves and may play a necessary role in nitrogen metabolism.[431] In nitrogen-fixing legumes urea derivatives, the **ureides**, have an important function (Chapter 25) but urease may not be involved in the catabolism of these compounds.[440]

2. Hydrogenases

Many plants, animals, and microorganisms are able to evolve H_2 by reduction of hydrogen ions (Eq. 16-48) or to oxidize H_2 by the reverse of this reaction.[441,442]

$$2H^+ + 2e^- \rightarrow H_2 \qquad (16\text{-}48)$$

Hydrogenases have been classified into two main types: **Fe-hydrogenases**, which contain iron as the only metal,[443] and **Ni-hydrogenases**, which contain both iron and nickel.[444] In a few Ni-hydrogenases a selenocysteine residue replaces a conserved cysteine

side chain.[445,446] Fe-hydrogenases are often extremely active and are utilized to rid organisms of an excess of electrons by evolution of H_2. Since they may also be used to acquire electrons by oxidation of H_2, they are often described as *bidirectional*. Ni-hydrogenases, as well as some Fe-hydrogenases, are involved primarily in *uptake* of H_2.[447] All hydrogenases contain one or more Fe–S centers in addition to the H_2-forming catalytic center.[441] Some hydrogenases are membrane bound and are often coupled through unidentified carriers to formate dehydrogenase (Chapter 17). In the strict anaerobes such as clostridia, hydrogenases are linked to ferredoxins. Hydrogenases are inactivated readily by O_2, which oxidizes the catalytic centers but can sometimes be reactivated by treatment with reducing agents.[448]

The 60-kDa all-iron monomeric hydrogenase I of *Clostridium pasteurianum* (mentioned on p. 861) contains ferredoxin-like Fe_4S_4 clusters plus additional Fe and sulfur atoms organized as a special H cluster. The EPR spectrum of the catalytic center, recognized because the spectrum is altered by the binding of carbon monoxide, is unusual. Its g values of 2.00, 2.04, and 2.10 are similar to those of oxidized high-potential iron proteins.[449] The *C. pasteurianum* hydrogenase II is a 53-kDa monomer containing eight Fe and eight S^{2-} ions. These are organized into one ferredoxin-like Fe_4S_4 cluster plus a three-Fe cluster and one iron ion in a unique environment.[449]

In contrast to these iron-only hydrogenases, the large periplasmic hydrogenase of *Desulfovibrio gigas* consists of one 28-kDa subunit and one 60-kDa subunit and contains two Fe_4S_4 clusters, one Fe_3S_4 cluster, and another dimetal center containing a single atom of Ni.[450,451] These are seen clearly in the three-dimensional structure depicted in Fig. 16-26. The three Fe–S clusters, at 5- to 6-nm intervals, form a chain from the external surface to the deeply buried nickel–iron center.[450–453a] A plausible pathway for transport of protons to the active center can also be seen.[450,451] The nickel center also contains an atom of Fe. Four cysteine side chains participate in forming the Ni–Fe cluster, two of them provide sulfur atoms that bridge between the metals while two others are ligands to Ni (Fig. 16-26).

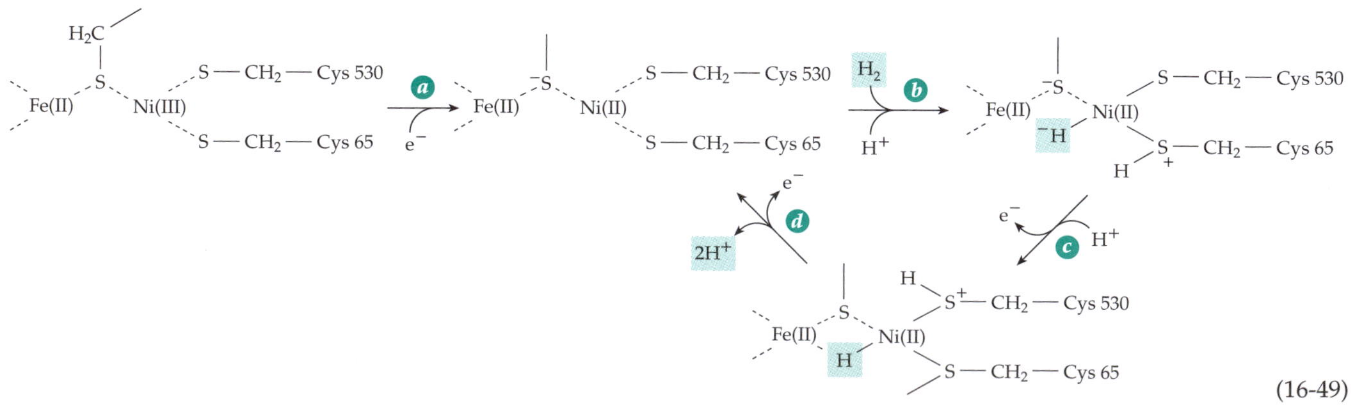

$$(16\text{-}49)$$

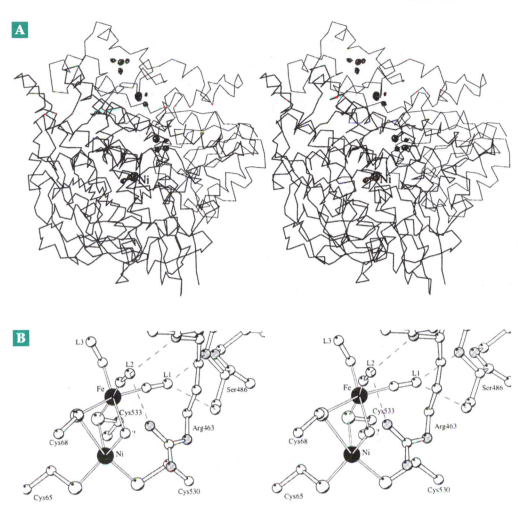

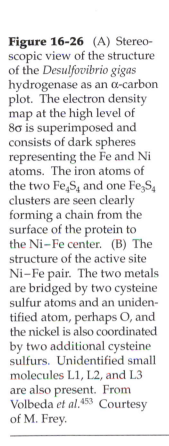

Figure 16-26 (A) Stereoscopic view of the structure of the *Desulfovibrio gigas* hydrogenase as an α-carbon plot. The electron density map at the high level of 8σ is superimposed and consists of dark spheres representing the Fe and Ni atoms. The iron atoms of the two Fe_4S_4 and one Fe_3S_4 clusters are seen clearly forming a chain from the surface of the protein to the Ni−Fe center. (B) The structure of the active site Ni−Fe pair. The two metals are bridged by two cysteine sulfur atoms and an unidentified atom, perhaps O, and the nickel is also coordinated by two additional cysteine sulfurs. Unidentified small molecules L1, L2, and L3 are also present. From Volbeda *et al.*[453] Courtesy of M. Frey.

A hydrogenase from *chromatium vinosum* has a similar structure.[453b] There are still uncertainties about other nonprotein ligands such as H_2O.[452–453a]

All of the Ni-hydrogenases display an EPR signal that can be assigned to Ni(III).[452] However, the active enzyme from *D. gigas* contains Ni(II). A proposed mechanism[452] is indicated in Eq. 16-49. Step *a* of this equation is a reductive activation. In step *b* a molecule of H_2 is bound as a hydride ion on Ni and a proton on a nearby sulfur. Protonation of a second sulfur ligand to Ni is needed to promote the cleavage of H_2 prior to the two-step oxidation of the bound H^-. One of two Ni-containing hydrogenases of *Methanobacterium thermoautotrophicum* contains FAD as well as Fe−S clusters.[454] It specifically reduces the 5-deazaflavin cofactor F_{420} (Chapter 15). A major function of this deazaflavin is reduction of the nickel-containing cofactor F_{430}.

3. Cofactor F_{430} and Methyl-Coenzyme M Reductase

Cofactor F_{430} is a nickel tetrapyrrole with a structure

(Fig. 16-27)[455,456] similar to those of vitamin B_{12} and of siroheme. The tetrapyrrole ring is the most highly reduced in cofactor F_{430}, which functions in reduction of methyl-CoM to methane in methanogens (Fig. 16-28). The methyl CoM reductase of *Methanobacterium thermoautotrophicum* is a large 300-kDa protein with subunit composition $\alpha_2\beta_2\gamma_2$ and containing two molecules of bound F_{430}. The nickel in F_{430} is first thought to be reduced, in an activation step, to Ni(I),[456a] which may attack the methyl group of methyl-CoM homolytically to yield a methyl nickel complex and a sulfur radical.[457,458] Alkyl nickel compounds react with protons, and in this case they would yield methane and would regenerate the Ni(II) form of the cofactor. The CoM radical could be reduced back to free CoM.

High-resolution crystal structures of the enzyme in two inactive Ni(II) forms[458] show the two F_{430} molecules. Each is bound in an identical channel about 3 nm in length and extending from the surface deep into the interior of the protein. The F_{430} lies at the bottom of this channel with its nickel atom coordinated with the oxygen atom of a glutamine side chain. In one form CoM lies directly above the nickel, with its

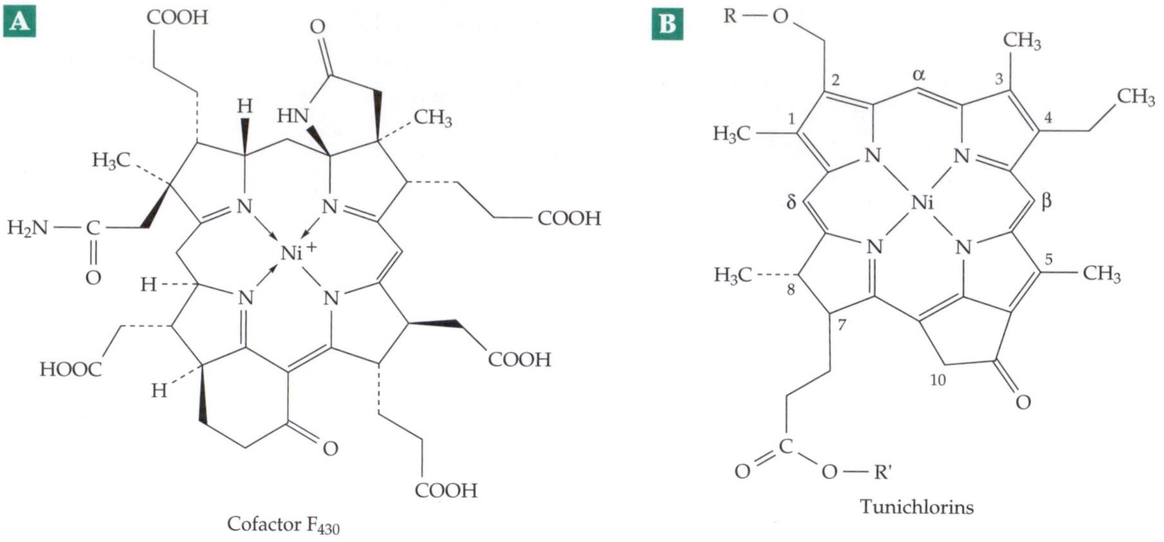

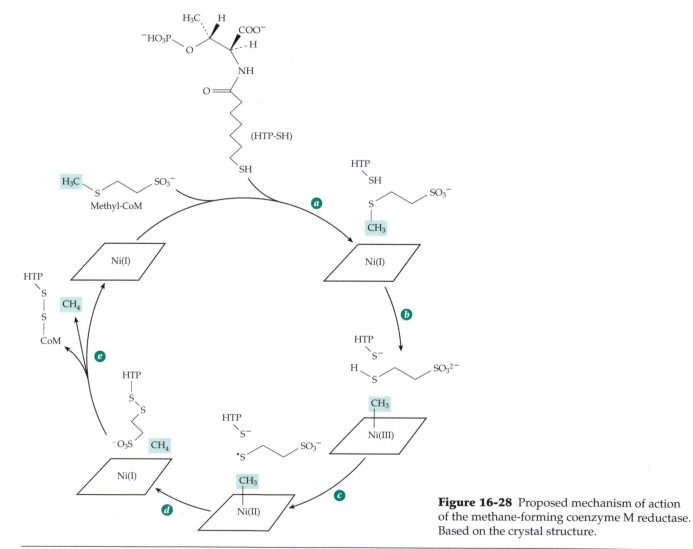

A   Cofactor F$_{430}$

B   Tunichlorins

Figure 16-27 (A) Structure of the nickel-containing prosthetic group F$_{430}$ as isolated in the esterified (methylated) form. From Pfaltz *et al.*[459] The "front" face, which reacts with methyl-coenzyme M, is toward the reader.[458]   (B) Structure of a representative member of a family of tunichlorins isolated from marine tunicates.[460] For tunichlorin R = R' = H. Related compounds have R' = CH$_3$ and/or R = an alkyl group with 13–21 carbon atoms and up to six double bonds.

Figure 16-28 Proposed mechanism of action of the methane-forming coenzyme M reductase. Based on the crystal structure.

thiolate sulfur providing the sixth ligand for the nickel. The long –SH-containing side chain of **heptanoyl threonine phosphate** (HTP; also called coenzyme B; Chapter 15, Section E) also lies within the channel with its amino acid head group blocking the entrance. In a second crystal form the mixed disulfide HTP–S–S–CoM, an expected product of the reaction (Fig. 16-28), is present in the channel.[458] Because of the distance from the –SH group of HTP and the nickel atom it is clear that there must be some motion of the methyl-CoM and that the methane formed may stay trapped in the active site until the HTP–S–S–CoA product leaves.

A proposed mechanism for the catalytic cycle based on the X-ray results as well as previous chemical studies and EPR spectroscopy is shown in Fig. 16-28. The substrates enter in step *a*. The position of the HTP, with its extended side chain, is probably the same as that seen in the X-ray structures of the Ni(II) complexes but the conformation of the methyl-CoM is different. The methyl transfer in step *b* is reminiscent of that of methionine synthase (Eq. 16-43). Although the distance from the CoM sulfur and the HTP–SH is too great for direct proton transfer between the two, as indicated for step *b*, there are two tyrosine hydroxyls that could provide a pathway for proton transfer. The region around the surface of the nickel coenzyme is largely hydrophobic and could facilitate formation of the thiyl radical in step *c*. In the structure with the bound HTP–S–S–CoM heterodisulfide an oxygen atom of the sulfonate group of CoM is bonded tightly to the Ni(II). However, in the active Ni(I) form the nickel is nucleophilic and would probably repel the sulfonate, perhaps assisting the product release.[458]

The enzyme contains five posttranslationally modified amino acids near the active site: *N*-methylhistidine, 5-methylarginine, 2-methylglutamine, 2-methylcysteine, and thioglycine in a thiopeptide bond. The latter may be the site of radical formation.[458a,b]

4. Tunichlorins

Nickel is found in various marine invertebrates. In the tunicates (sea squirts and their relatives) it occurs in a fixed ratio with cobalt, suggesting a metabolic role.[460] A new class of nickel chelates called tunichlorins have been isolated. An example is shown in Fig. 16-27B. The function of tunichlorins is unknown but their existence suggests the possibility of unidentified biochemical roles for nickel.

5. Carbon Monoxide Dehydrogenases and Carbon Monoxide Dehydrogenase/Acetyl-CoA Synthase

There are several bacterial carbon monoxide dehydrogenases that catalyze the reversible oxidation of CO to CO_2:

$$CO + H_2O \rightarrow CO_2 + H^+ + 2\ e^- \qquad (16\text{-}50a)$$

$$CO + H_2O \rightarrow CO_2 + H_2$$
$$\Delta G° = -38.1\ kJ/mol \qquad (16\text{-}50b)$$

Some bacteria use CO as both a source of energy and for synthesis of carbon compounds. The purple photosynthetic bacterium *Rhodospirillum rubrum* employs a relatively simple monomeric Ni-containing CO dehydrogenase containing one atom of Ni and seven or eight iron atoms, apparently arranged in Fe_4S_4 clusters. These bacteria can grow anaerobically with CO as the sole source of both energy and carbon.[461] Some aerobic bacteria oxidize CO using a molybdenum-containing enzyme (Section H). However, the most studied CO dehydrogenase is a complex enzyme that also synthesizes, reversibly, acetyl-CoA from CO and a methyl corrin. Employed by methanogens, acetogens, and sulfate-reducing bacteria, it is at the heart of the **Wood–Ljundahl pathway** of autotrophic metabolism, which is discussed further in Chapter 17.

Both oxidation of CO to CO_2 and reduction of CO_2 to CO are important activities of CO dehydrogenase / acetyl-CoA synthase. During growth on CO, some CO must be oxidized to CO_2 and then reduced by the pathways of Fig. 15-22 to form a methyl-tetrahydropterin which can be used to form the methyl group of acetyl-CoA. During growth on any other carbon compound CO_2 must be reduced to CO to form the carbonyl group of acetyl-CoA which can serve as a precursor to all other carbon compounds. Native CO dehydrogenase/acetyl CoA synthase was isolated from cells of *Clostridium thermoaceticum* grown in the presence of radioactive ^{63}Ni. The protein is a 310-kDa $\alpha_2\beta_2$ oligomer. Each $\alpha\beta$ dimer contains 2 atoms of Ni, 1 of Zn, ~12 of Fe and ~12 sulfide ions,[462–464] which are organized into three metal clusters referred to as A, B, and C. Each cluster contains 4 Fe atoms and clusters A and C also contain 1 Ni each. Oxidation of CO occurs in the β subunits, each of which contains both cluster B, an Fe_4S_4 ferredoxin-type cluster, and cluster C, where the oxidation of CO is thought to occur. Cluster C contains 1 nickel ion as well as an Fe_4S_4 cluster that resembles that of aconitase (Fig. 13-4). Cluster A, which is in the α subunit, also contains 1 atom of Ni and 4 Fe ions and is probably the site of synthesis of acetyl-CoA.

Oxidation of CO may require cooperation of the nickel ion and the Fe_4S_4 group within the C cluster.

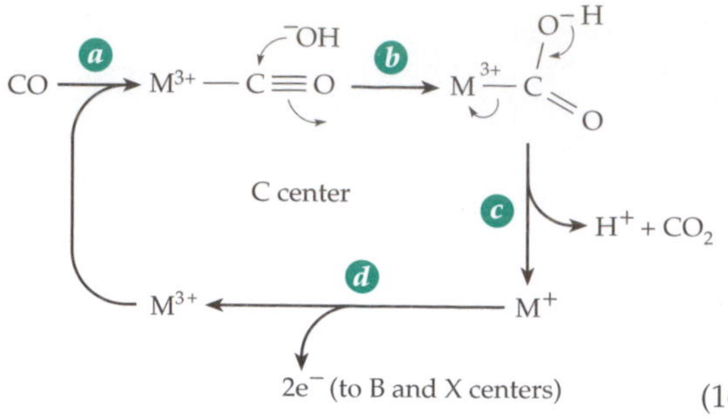

$$2e^- \text{ (to B and X centers)} \qquad (16\text{-}51)$$

CO probably binds to one of these metals and is attacked by a hydroxyl ion (Eq. 16-51, step *b*) which may be donated by the other metal of the pair.[465,465a] CO_2 and a proton are released rapidly (step *c*), after which the reduced metal center is reoxidized (step *d*). One electron is thought to be transferred directly to the Fe_4S_4 cluster B and the second by an alternative route.[465] A multienzyme complex isolated from the methanogen *Methanosarcina* has an $(\alpha\beta\gamma\delta\epsilon)_6$ structure, with the subunits having masses of 89, 60, 50, 48, and 20 kDa, respectively.[466] In this complex the CO dehydrogenase/acetyl-CoA synthase activity appears to reside in the $\alpha_2\epsilon_2$ complex, the $\gamma\delta$ complex has a tetrahydropteridine:cob(I)amide–protein transferase, and the β subunit has an acetyltransferase that binds acetyl-CoA and transfers the acetyl group to a group on the β subunit.[466]

Although the *Clostridium* and *Methanosarcina* systems are not identical, similar mechanisms are presumably involved.[467,467a] To generate acetyl-CoA a methyl group is first transferred from a tetrahydro-

pterin such as tetrahydrofolate or, in methanogens, tetrahydromethanopterin or tetrahydrosarcinapterin (Fig. 15-17) to form a methylcorrinoid. At the A center of the CO dehydrogenase a molecule of CO, which may be bound to the Ni, equilibrates with the methyl group, and with acetyl-CoA. As depicted in Eq. 16-52, acetyl-Ni may be an intermediate. Other details shown here are hypothetical. It is possible that the methyl group is transferred to the Ni atom in the M cluster before reaction with the CO, which might be bound to either Ni of cluster C or to Fe. This reaction of two transition-metal-bound ligands parallels a proposed industrial process for synthesis of acetic acid from methanol and CO and involving catalysis by rhodium metal and methyl iodide. It is thought that rhodium-bound CO is inserted into bound $Rh-CH_3$ to form an intermediate $Rh-\overset{O}{\overset{\|}{C}}-CH_3$. The acetyl group is released as acetyl iodide which is hydrolyzed to acetic acid. An acetyl-nickel intermediate may be involved in the corresponding biological reaction of Eq. 16-52. The stereochemistry of the sequence has been investigated using methyl-THF containing a chiral methyl group. Overall retention of the configuration of the methyl group in acetyl-CoA was observed.[468,469]

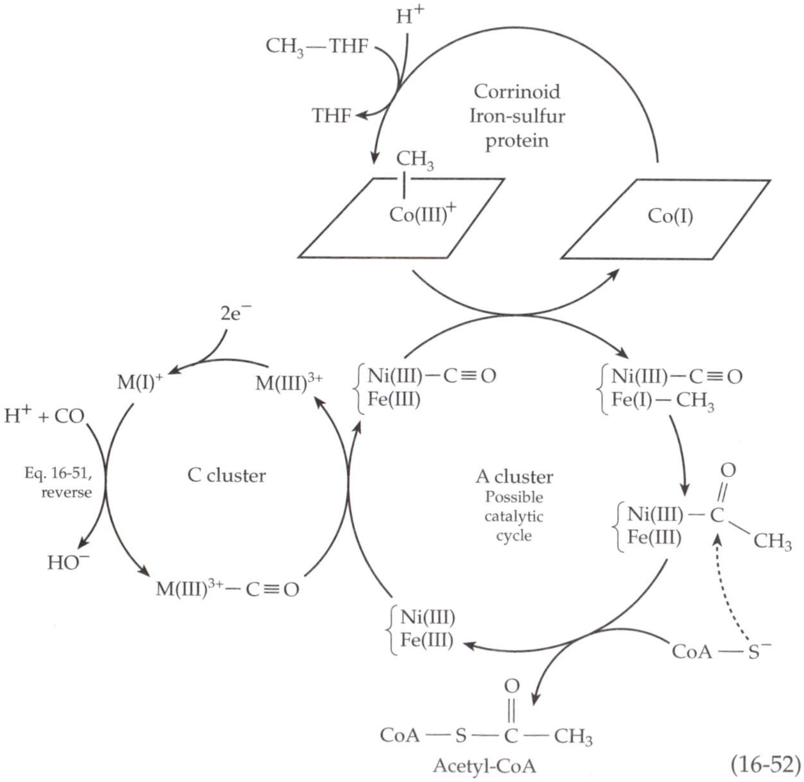

D. Copper

Copper was recognized as nutritionally essential by 1924 and has since been found to function in many cellular proteins.[470–474] Copper is so broadly distributed in foods that a deficiency has only rarely been observed in humans.[474a] However, animals may sometimes receive inadequate amounts because absorption of Cu^{2+} is antagonized by Zn^{2+} and because copper may be tied up by molybdate as an inert complex. There are copper-deficient desert areas of Australia where neither plants nor animals survive. Copper-deficient animals have bone defects, hair color is lacking, and hemoglobin synthesis is impaired. Cytochrome oxidase activity is low. The protein elastin of arterial walls is poorly crosslinked and the arteries are weak. Genetic defects in copper metabolism can have similar effects.

An adult human ingests ~2–5 mg of copper per day, about 30% of which is absorbed. The total body content of copper is ~100 mg ($\sim 2 \times 10^{-4}$ mol / kg), and both uptake and excretion (via the bile) are regulated. Since an excess of copper is toxic, regulation is important.

$$\text{Acetyl-CoA} \qquad (16\text{-}52)$$

Because Cu^{2+} is the most tightly bound metal ion in most chelating centers (Table 6-9), almost all of the copper present in living cells is complexed with proteins. Copper is transported in the blood by a 132-kDa, 1046-residue sky-blue glycoprotein called **ceruloplasmin**.[471,475–477] This one protein contains 3% of the total body copper.

Regulation of copper uptake has been studied in most detail in the yeast *Saccharomyces cerevisiae*. Uptake of Cu^{2+} is similar to that of Fe^{3+}. The same plasma membrane reductase system, consisting of proteins Fre1p and Fre2p (encoded by genes *FRE*1 and *FRE*2), acts to reduce both Fe^{3+} and Cu^{2+}.[478–481] These two genes are controlled in part by a transcriptional activator that responds to the internal copper concentration.[410,482,483] Similar regulation is thought to occur in both plants[484] and animals.

The human hereditary disorders **Wilson's disease** and **Menkes' disease** have provided further insight into copper metabolism.[485,486] In Wilson's disease the ceruloplasmin content is low and copper gradually accumulates to high levels in the liver and brain. In Menkes syndrome, there is also a low ceruloplasmin level and an accumulation of copper in the form of copper metallothionein.[487,488] Persons with this disease have abnormalities of hair, arteries, and bones and die in childhood of cerebral degeneration.[489,490] Similar symptoms are seen in some patients with Ehlers–Danlos syndrome (Box 8-E).[491] Genes for the proteins that are defective in both Wilson's and Menkes' diseases have been cloned and both proteins have been identified as P-type ATPase cation transporters (Chapter 8).[492–495c] The two proteins must be similar in structure as indicated by a 55% sequence identity.[496] Homologous genes involved in copper homeostasis have been located in both yeast[497] and the cyanobacterium *Synechococcus*.[498] The transporter encoded by this yeast gene, designated *Ccc*2, apparently functions to export copper from the cytosol into an extracytosolic compartment. In a similar way the Wilson and Menkes disease proteins, which reside in the *trans*-Golgi network, are thought to export copper or to provide copper for incorporation into essential proteins.[493,499] The Wilson disease protein is also found in a shortened form in mitochondrial membranes.[492] Other proteins associated with intracellular copper metabolism seem to be chaperones for Cu(I).[500–501a]

The ability of copper ions to undergo reversible changes in oxidation state permits them to function in a variety of oxidation–reduction processes. Like iron, copper also provides sites for reaction with O_2, with superoxide radicals, and with nitrite ions.

1. Electron-Transferring Copper Proteins

A large group of small, intensely blue copper proteins function as single-electron carriers within bacteria and plants. Best known is **plastocyanin**, which is ubiquitous in green plants and functions in the electron transport chain between the light-absorbing photosynthetic centers I and II of chloroplasts (Chapter 23). The bacterial **azurins**[502] are thought to carry electrons between cytochrome c_{441} and cytochrome oxidase. **Amicyanin** accepts electrons from the coenzyme TTQ of methylamine dehydrogenase of methylotrophic bacteria and passes them to a cytochrome c (Chapter 15).[168,503,504] A basic blue copper protein **phytocyanin** of uncertain function occurs in cucumber seeds.[505]

The 10.5-kDa peptide chain of plastocyanin is folded into an eight-stranded β barrel (Fig. 2-16), which contains a single copper atom. In poplar plastocyanin, the Cu is coordinated by the side chains of His 37, His 87, Met 92, and Cys 84 in a tetrahedral but distorted toward a trigonal bipyramidal geometry. Since copper-free apoplastocyanin has essentially the same structure, this geometry may be imposed by the protein onto the Cu^{2+}, which usually prefers square-planar or tetrahedral coordination (Chapter 7).[506] Calculations suggest that there is little or no strain and that the "Franck–Condon barrier" to electron transfer is low.[507,508] The three-dimensional structure and copper environment of azurin are similar to those of plastocyanin.[509–511] Messerschmidt *et al*.[512] suggested that the copper site in these proteins is perfectly adapted to its function because its geometry is a compromise between the optimal geometries of the Cu(I) and Cu(II) states between which it alternates. **Stellacyanin**, present in the Japanese lac tree and some other plants, is a mucoprotein; the 108-residue protein is over 40% carbohydrate.[513] While its spectrum resembles that of plastocyanins and azurins, stellacyanin contains no methionine and this amino acid cannot be a ligand to copper.[514] **Rusticyanin** functions in the periplasmic space of some chemolithotropic sulfur bacteria to transfer electrons from Fe^{2+} to cytochrome c as part of the energy-providing reaction for these organisms (Eq. 18-23).[515–517] **Halocyanin** functions in membranes of the archaeobacterium *Natronobacterium*,[518] and **aurocyanin** functions in green photosynthetic bacteria.[519]

The blue color of these "type 1" copper proteins is much more intense than are the well known colors of the hydrated ion $Cu(H_2O)_4^{2+}$ or of the more strongly absorbing $Cu(NH_3)_4^{2+}$. The blue color of these simple complexes arises from a transition of an electron from one d orbital to another within the copper atom. The absorption is somewhat more intense in copper peptide chelates of the type shown in Eq. 6-85. However, the ~ 600 nm absorption bands of the blue proteins are an order of magnitude more intense, as is illustrated by the absorption spectrum of azurin (Fig. 23-8). The intense blue is thought to arise as a result of transfer of electronic charge from the cysteine thiolate to the Cu^{2+} ion.[520,521]

A third type of copper center, first recognized in cytochrome c oxidase (see Fig. 18-10) is called Cu_A or **purple CuA**. Each copper ion is bonded to an imidazole and two cysteines serve as bridging ligands. The two copper ions are about 0.24 nm apart, and the two Cu^{2+} ions together can accept a single electron from an external donor such as cytochrome c or azurin to give a half-reduced form.[521a,b]

2. Copper, Zinc-Superoxide Dismutase

Although of similar topology to the blue electron-transferring proteins, Cu, Zn-superoxide dismutase, has a different function. This dimeric 153-residue protein has been demonstrated in the cytoplasm of virtually all eukaryotic cells[522] and in the periplasmic space of some bacteria[523] where it converts superoxide ions $\cdot O_2^-$ to O_2 and H_2O_2. The enzyme, which has a major protective role against oxidative damage to cells, presumably functions in a manner similar to that indicated in Eq. 16-27 for iron or manganese. However, copper cycles between Cu^{2+} and Cu^+, alternately accepting and donating electrons.

The active site of cytosolic superoxide dismutase (SOD) contains both Cu^{2+} and Zn^{2+}. The copper ion is of "type 2": nonblue and paramagnetic. It is surrounded by four imidazole groups with an irregular square planar geometry.[524-527] One of these imidazole groups (that of His 61) is shared with the Zn^{2+}, which is also bonded to two additional imidazole groups and a side chain carboxylate. The metal ions have evidently replaced the hydrogen atom that would otherwise be present on the imidazole of His 61 (see the following diagram). It has been suggested, as is also indicated in the diagram, that when the bound superoxide

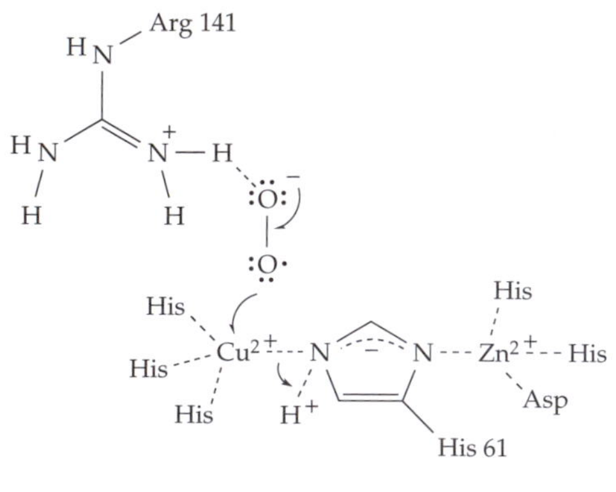

donates an electron to the Cu(II) to become O_2 (first step of Eq. 16-27), a proton becomes attached to the bridging imidazole with breakage of its linkage to the Cu(I). The structure of a new crystalline form of reduced yeast SOD shows that the Cu(I) has moved 0.1 nm away from the bridging imidazole in agreement with this possibility.[528] In the second half reaction the

imidazole proton, together with a second proton from the medium and an electron from the Cu(I), would react to convert the second O_2^- into H_2O_2. The role of the Zn^{2+} may be in part structural but it may also serve to ensure that His 61 is protonated on the correct nitrogen atom. Arg 141 may assist in binding the O_2^- as is shown in the diagram. However, the fact that a mutant containing leucine in place of the active site arginine has over 10% of the activity of the native enzyme shows that the arginine is not absolutely essential.[529] Additional nearby positively charged arginine and lysine side chains may provide "electrostatic guidance" that increases the velocity of reaction of superoxide ions.[530,531] Cu, Zn-SOD is one of the fastest enzymes known.

In addition to the cytosolic SOD there is a longer ~222-residue extracellular form that binds to the proteoglycans found on cell surfaces.[522,527] Manganese SODs are found in mitochondria and in bacteria and iron SODs in plants and bacteria. They all appear to be important in protecting cells from superoxide radicals.[522,532,533] This importance was dramatically emphasized when it was found that a defective SOD is present in persons (about 1 in 100,000) with a hereditary form of **amylotrophic lateral sclerosis** (ALS), which is also called Lou Gehrig's disease after the baseball hero who was stricken with this terrible disease of motor neurons in 1939 at the age of 36.[534-536]

3. Nitrite and Nitrous Oxide Reductases

Copper enzymes participate in two important reactions catalyzed by denitrifying bacteria. Nitrite reductases from species of *Achromobacter*[537,538] and *Alcaligenes*[539-542] are trimeric proteins[543] made up of 37-kDa subunits, each of which contains one type 1 (blue) copper and one type 2 (nonblue) copper. The first copper serves as an electron acceptor from a small blue **pseudoazurin**.[544,544a] The second copper, which is in the active site, is thought to bind to nitrite throught its nitrogen atom and to reduce it to NO.

$$NO_2^- + 2H^+ + e^- \rightarrow NO + H_2O \qquad (16\text{-}53)$$

Crystallographic studies on the 343-residue *Alcaligenes* enzyme reveal two β barrel domains with the type 1 copper embedded in one of them and the type 2 copper in an interface between the domains. Studies of EPR[541] and ENDOR[545] spectra and of various mutant forms have shown that, as for other copper enzymes, the type 1 copper is an electron-transferring center, accepting electrons from the pseudoazurin and passing them to the type 2 copper which binds and reduces nitrite.[540,545a]

The immediate product of nitrite reductase is NO, which is reduced in two one-electron steps to N_2O, and then to N_2. The second of these steps is catalyzed by another copper enzyme, nitrous oxide reductase.[546]

$$N_2O + 2H^+ + 2e^- \rightarrow N_2 + H_2O \qquad (16\text{-}54)$$

The biological significance of these reactions is considered further in Chapters 18 and 24. The 132-kDa dimeric N_2O reductase from *Pseudomonas stutzeri* contains four copper atoms per subunit.[546] One of its copper centers resembles the Cu_A centers of cytochrome *c* oxidase. A second copper center consists of four copper ions, held by seven histidine side chains in a roughly tetrahedral array around one sulfide (S^{2-}) ion. Rasmussen *et al.* speculate that this copper-sulfide cluster may be an acceptor of the oxygen atoms of N_2O in the formation of N_2.[546a] There is also a cytochrome cd_1 type of nitrite reductase.[143a]

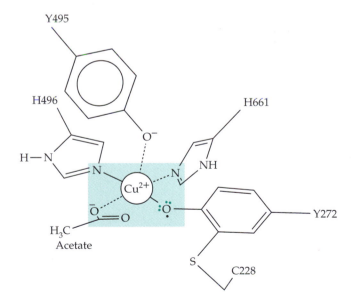

4. Hemocyanins

While many copper proteins are catalysts for oxidative reactions of O_2, hemocyanin reacts with O_2 reversibly. This water-soluble O_2 carrier is found in the blue blood of many molluscs and arthropods, including snails, crabs, spiders, and scorpions. Hemocyanins are large oligomers ranging in molecular mass from 450 to 13,000 kDa. Molluscan hemocyanins are cylindrical oligomers which have a striking appearance under the electron microscope. Simpler hemocyanins, found in arthropods, are hexamers of 660-residue 75-kDa subunits. Each subunit of the hemocyanin from the spiny lobster is folded into three distinct domains, one of which contains a pair of Cu(I) atoms which bind the O_2. Each copper ion is held by three imidazole groups without any bridging groups between them, the Cu–Cu distance being 0.36–0.46 nm.[547] *Octopus* hemocyanin has a different fold and forms oligomers of ten subunits. However, the active sites are very similar.[548] The O_2 is thought to bind between the two copper atoms. An allosteric mechanism may involve changes in the distance between the copper atoms.[549] The oxygenated compound is distinctly blue with a molar extinction coefficient 5–10 times greater than that of cupric complexes. This fact suggests that the Cu(I) has been oxidized to Cu(II) and that the O_2 has been reduced to the peroxide dianion O_2^{2-} in the complex.[550–552] Further support for this idea comes from the observation that treatment of oxygenated hemocyanin with glacial acetic acid leads to the formation of

equal amounts of Cu^{2+} and Cu^+ and protonated superoxide.

$$(CuO_2Cu)^{2+} + H^+ \longrightarrow Cu^{2+} + Cu^+ + HO_2 \qquad (16\text{-}55)$$

5. Copper Oxidases

A large group of copper-containing proteins activate oxygen toward chemical reactions of dehydrogenation, hydroxylation, or oxygenation. **Galactose oxidase** (Fig. 16-29), from the mushroom *Polyporus*, is a dehydrogenase which converts the 6-hydroxymethyl group of galactose to an aldehyde while O_2 is reduced to H_2O_2.

$$R - CH_2OH \longrightarrow R - CHO$$
$$O_2 \quad H_2O_2 \qquad (16\text{-}56)$$

Galactose oxidase has been used frequently to label glycoproteins of external cell membrane surfaces. Exposed terminal galactosyl or *N*-acetylgalactosaminyl residues are oxidized to the corresponding C-6 aldehydes and the latter are reduced under mild conditions with tritiated sodium borohydride.[553]

The single 639-residue polypeptide chain contains one type 2 copper ion.[554] Neither oxygen nor galactose affects the absorption spectrum of the light murky green enzyme, but the combination of the two does, suggesting that both substrates bind to the enzyme before a reaction takes place. A side reaction releases

Figure 16-29 Drawing of the active site of galactose oxidase showing both the Cu(II) atom and the neighboring free radical on tyrosine 272, which has been modified by addition of the thiol of cysteine 228 and oxidation. See Halfen *et al.*[557] Based on a crystal structure of Ito *et al.*[558]

superoxide ion and leaves the enzyme in the inactive Cu(II) state. EPR spectroscopic observations on the enzyme were puzzling. The active enzyme shows no EPR signal but a one-electron reduction gives an inactive form with an EPR signal that arises from Cu(II). Experimental studies eventually pointed to the presence of a second reducible center which contains an organic free radical. In the active form this radical is **antiferromagnetically coupled** (spin-coupled) giving an "EPR-silent" enzyme able to accept two electrons.[555–557]

Another surprise was the discovery, from the X-ray structure,[558–559] that a tyrosine side chain at the active site has been modified by addition of a thiolate group

from a cysteine residue. This structure, which is also the site of the organic free radical, is shown in Fig. 16-29. A possible mechanism of action is portrayed in Fig. 16-30. The substrate binds in step a and its $-OH$ group is deprotonated, perhaps by transfer of H[+] to the phenolate oxygen of tyrosine 495 (Fig. 16-29) in step b of Fig. 16-30. In step c a free radical hydrogen transfer occurs to form a ketyl radical which is immediately oxidized by Cu(II) in step d. In steps c and f the oxidant O_2 is converted to H_2O and the aldehyde product is also released. A fungal glyoxal oxidase has similar characteristics.[560] The galactose oxidase proenzyme is self-processing. The Tyr-Cys cofactor arises as a result of copper-catalyzed oxidative modification via a tyrosine free radical.[560a]

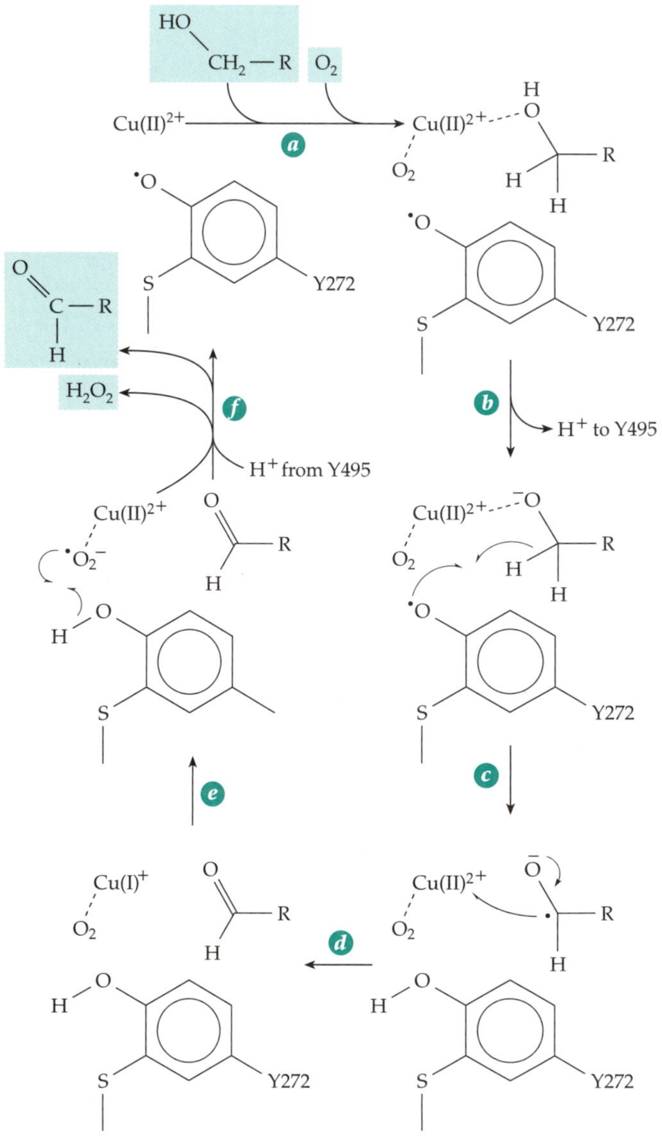

Several copper-containing amine oxidases[561,561a] convert amines to aldehydes and H_2O_2. They also contain one of the organic quinone cofactors discussed in Chapter 15. The dimeric **plasma amine oxidase** contains a molecule of the coenzyme TPQ and one Cu^{2+} per 90-kDa subunit.[562–563a] Whether the O_2 binds only to the single atom Cu(I) or also interacts directly with a cofactor radical in step d of Eq. 15-53 is uncertain.[562] **Lysyl oxidase**, which is responsible for conversion of ε-amino groups of side chains of lysine into aldehyde groups in collagen and elastin (Chapter 8) contains coenzyme LTQ as well as Cu.[564] The enzyme is specifically inhibited by β-aminopropionitrile (Box 8-G) and its activity is decreased in genetic diseases of copper metabolism. The **glycerol oxidase** of *Aspergillus* is a large 400 kDa protein containing one heme and two atoms of Cu.[565] It converts glycerol + O_2 into glyceraldehyde and H_2O_2. Also containing copper is **urate oxidase**, whose action is indicated in Fig. 25-18.

Tyrosinase catalyzes hydroxylation followed by dehydrogenation (Eq. 16-57). First identified in mushrooms, the enzyme has a widespread distribution in

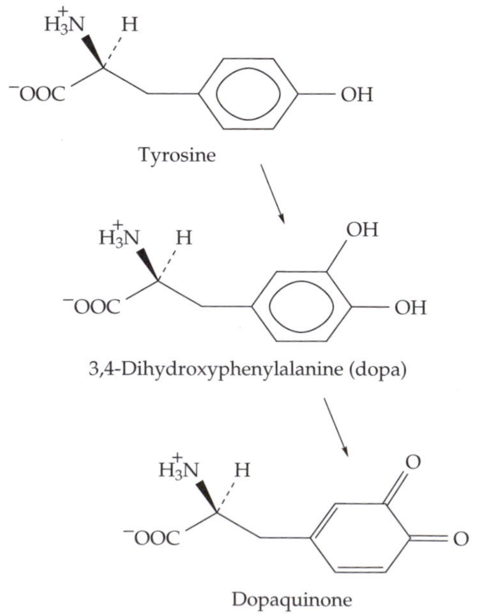

Figure 16-30 Possible reaction cycle for catalysis of the 6–OH of D-galactose or other suitable alcohol substrate by galactose oxidase. See Wachter *et al.*[566] and Whittaker *et al.*[567]

Tyrosine

3,4-Dihydroxyphenylalanine (dopa)

Dopaquinone   (16-57)

nature. It is present in large amounts in plant tissues and is responsible for the darkening of cut fruits. In animals tyrosinase participates in the synthesis of **dihydroxyphenylalanine** (dopa) and in the formation of the black **melanin** pigment of skin and hair. Either a lack of or inhibition of this enzyme in the melanin-producing melanocytes causes **albinism** (Chapter 25).

The 46-kDa monomeric tyrosinase of *Neurospora* contains a pair of spin-coupled Cu(II) ions.[568,569] The structure of this copper pair (**type 3 copper**) has many properties in common with the copper pair in hemocyanin.[569a] For example, in the absence of other substrates, tyrosinase binds O_2 to form "oxytyrosinase," a compound with properties resembling those of oxyhemocyanin and containing a bound peroxide dianion.[569]

Tyrosinase is both an oxidase and a hydroxylase. Some other copper enzymes have only a hydroxylase function. One of the best understood of these is the **peptidylglycine α-hydroxylating monoxygenase**, which catalyzes the first step of the reaction of Eq. 10-11. The enzyme is a colorless two-copper protein but the copper atoms are 1.1 nm apart and do not form a binuclear center.[570] Ascorbate is an essential cosubstrate, with two molecules being oxidized to the semidehydroascorbate radical as both coppers are reduced to Cu(I). A ternary complex of reduced enzyme, peptide, and O_2 is formed and reacts to give the hydroxylated product.[570] A related two-copper enzyme is **dopamine β-monooxygenase**, which utilizes O_2 and ascorbate to hydroxylate dopamine to noradrenaline (Chapter 25).[571,572] These and other types of hydroxylases are compared in Chapter 18.

The **blue multicopper oxidases** couple the oxidation of substrates to the four-electron reduction of molecular oxygen to H_2O.[573] In this respect they resemble cytochrome *c* oxidase, which also contains copper. However, they do not contain iron. The best known member of this group is the plant enzyme **ascorbate oxidase**, which dehydrogenates ascorbic acid to dehydroascorbic acid (See Box 18-D). It is a dimeric blue copper with identical 70-kDa subunits. The three-dimensional structure revealed one type 1 copper ion held in a typical blue copper environment as in plastocyanin or azurin and also a **three-copper center**. In this center a pair of copper ions, each held by three imidazole groups, and bridged by a μ-oxo group as in hemerythrin (Fig. 16-20), lie 0.51 nm apart and 0.41–0.44 nm away from the third copper, which is held by two other imidazoles.[574] The type 1 copper shows typical intense 600-nm absorption and characteristic EPR signal, while the pair with the oxo bridge are antiferromagnetically coupled and EPR silent but with strong near ultraviolet light absorption (type 3 copper). The additional metal ion in the trinuclear center is a type 2 copper which lacks characteristic spectroscopic features.[575,576] Reduction of O_2 to 2 H_2O

is thought to procede via superoxide radical intermediates. When substrate is added to the enzyme, the blue color fades and it can be shown that the copper is reduced to the +1 state. The reduced enzyme then reacts with O_2, converting it into two molecules of H_2O. Similar to ascorbate oxidase in structure and properties are **laccase**, found in the latex of the Japanese lac tree and in the mushroom *Polyporus*, and the previously discussed **ceruloplasmin.**[477] Laccase is a catalyst for oxidation of phenolic compounds by a free radical mechanism involving the trinuclear copper center.[577] Studied by Gabriel Bertrand in the 1890s, it was one of the first oxidative enzymes investigated.[578]

In addition to its previously mentioned role in copper transport, ceruloplasmin is an amine oxidase, a superoxide dismutase, and a ferrooxidase able to catalyze the oxidation of Fe^{2+} to Fe^{3+}. Ceruloplasmin contains three consecutive homologous 350-residue sequences which may have originated from an ancestral copper oxidase gene. Like ascorbate oxidase, this blue protein contains copper of the three different types. Blood clotting factors V and VIII (Fig. 12-17), and the iron uptake protein Fet3 (Section A,1) are also closely related.

6. Cytochrome *c* Oxidase

The most studied of all copper-containing oxidases is cytochrome *c* oxidase of mitochondria. This multi-subunit membrane-embedded enzyme accepts four electrons from cytochrome *c* and uses them to reduce O_2 to 2 H_2O. It is also a proton pump. Its structure and functions are considered in Chapter 18. However, it is appropriate to mention here that the essential catalytic centers consist of two molecules of heme *a* (*a* and a_3) and three Cu+ ions. In the fully oxidized enzyme two metal centers, one Cu^{2+} (of the two-copper center Cu_A) and one Fe^{3+} (heme *a*), can be detected by EPR spectroscopy. The other Cu^{2+} (Cu_B) and heme a_3 exist as an EPR-silent exchange-coupled pair just as do the two copper ions of hemocyanin and of other type 3 binuclear copper centers.

E. Manganese

Tissues usually contain less than one part per million of manganese on a dry weight basis, less than 0.01 mM in fresh tissues. This compares with a total content in animal tissues of the more abundant Mg^{2+} of 10 mM. A somewhat higher Mn content (3.5 ppm) is found in bone. Nevertheless, manganese is nutritionally essential[579,580] and its deficiency leads to well-defined symptoms. These include ovarian and testicular degeneration, shortening and bowing of legs, and other skeletal abnormalities such as the "slipped tendon disease" of chicks. In Mn deficiency the organic matrix of bones and cartilage develops poorly. The galactosamine, hexuronic acids, and chondroitin sulfates content of cartilage is decreased. Manganese is also essential for plant growth and plays a unique and essential role in the photosynthetic reaction centers of chloroplasts. Two magnetically coupled pairs of manganese ions bound in a protein act as the O_2 evolving center in photosynthetic system II. This function is considered in Chapter 23. An ABC transporter for managanese uptake has been identified in the cyanobacterium *Synechocystis*.[581]

Manganese lies in the center of the first transition series of elements. The stable Mn^{2+} (manganous ion) contains five $3d$ electrons in a high-spin configuration. The less stable Mn^{3+} (manganic ion) appears to be of importance in some enzymes and is essential to the photosynthetic evolution of oxygen. Many enzymes specifically require or prefer Mn^{2+}. These include galactosyl and *N*-acetylgalactosaminyltransferases[582] needed for synthesis of mucopolysaccharides (Chapter 20), lactose synthetase (Eq. 20-15), and a muconate-lactonizing enzyme (Eq. 13-23).[583]

Arginase, essential to the production of urea in the human body (Fig. 24-11), specifically requires Mn^{2+} which exists as a spin-coupled dimetal center with a bridging water or ^-OH ion. The Mn^{2+} may act much as does the Ni^{2+} ions of urease (Fig. 16-25).[584,585] Pyruvate carboxylase (Eq. 14-3) contains four atoms of tightly bound Mn^{2+}, one for each biotin molecule present. This manganese is essential for the transcarboxylation step in the action of this enzyme. Either Mn^{2+} or Mg^{2+} is also needed in the initial step of carboxylation of biotin (Eq. 14-5). Another Mn^{2+}-containing protein is the lectin concanavalin A (Chapter 4). The joining of *O*-linked oligosaccharides to secreted glycoproteins also seems to require manganese.[586]

Manganese is a component of a "pseudocatalase" of *Lactobacillus*,[204] of lignin-degrading peroxidases,[257,258] and of the wine-red superoxide dismutases found in bacteria and in the mitochondria of eukaryotes.[376,587] The dimeric dismutase from *E. coli* has a structure nearly identical to that of bacterial iron SOD (Fig. 16-22). The manganese ions are presumed to alternate between the +3 and +2 states during catalysis (Eq. 16-27). "Knockout" mice with inactivated Mn SOD genes live no more than three weeks, indicating that this enzyme is essential to life. However, mice lacking CuZn SOD appear normal in most circumstances.[588] Some dioxygenases contain manganese.[589] Many enzymes that require Mg^{2+} can utilize Mn^{2+} in its place, a fact that has been exploited in study of the active sites of enzymes.[590] The highly paramagnetic Mn^{2+} is the most useful ion for EPR studies (Box 8-C) and for investigations of paramagnetic relaxation of NMR signals. Manganese can also replace Zn^{2+} in some enzymes and may alter catalytic properties.

Manganese may function in the regulation of some enzymes. For example, glutamine synthetase (Fig. 24-7) in one form requires Mg^{2+} for activity but upon adenylylation binds Mn^{2+} tightly.[591] Nucleases and DNA polymerases often show altered specificity when Mn^{2+} substitutes for Mg^{2+}. However, the significance of these differences *in vivo* is uncertain. Manganese is mutagenic in living organisms, apparently because it diminishes the fidelity of DNA replication.[592]

A striking accumulation of Mn^{2+} often occurs within bacterial spores (Chapter 32). *Bacillus subtilus* absolutely requires Mn^{2+} for initiation of sporulation. During logarithmic growth the bacteria can concentrate Mn^{2+} from 1 μM in the external medium to 0.2 mM internally; during sporulation the concentrations become much higher.[593]

F. Chromium

Animals deficient in chromium grow poorly and have a reduced life span.[594–596] They also have decreased "glucose tolerance," i.e., glucose injected into the blood stream is removed only half as fast as it is normally.[597,598] This is similar to the effect of a deficiency of insulin. Fractionation of yeast led to the isolation of a chromium-containing **glucose tolerance factor** which appeared to be a complex of Cr^{3+}, nicotinic acid, and amino acids.[597] The chromium in this material is apparently well absorbed by the body but is probably not an essential cofactor.[596] Nevertheless, dietary supplementation with chromium appears to improve glucose utilization, apparently by enhancing the action of insulin.[596] Ingestion of glucose not only increases insulin levels in blood but also causes increased urinary loss of chromium,[599] perhaps as a result of insulin-induced mobilization of stored chromium.[596] It has been suggested that a specific **chromium-binding oligopeptide** isolated from mammalian liver[596,600] may be released in response to insulin and may activate a membrane phosphotyrosine phosphatase.[596]

Chromium concentrations in animal tissues are usually less than 2 μM but tend to be much higher in the caudate nucleus of the brain. High concentrations of Cr^{3+} have also been found in RNA–protein complexes.[601] While several oxidation states, including +2, +3, and +6, are known for chromium, only Cr(III) is found to a significant extent in tissues. The Cr(VI) complex ions, chromate and dichromate, are toxic and chronic exposure to chromate-containing dust can lead to lung cancer. Ascorbate is a principal biological reductant of chromate and can create mutagenic Cr(V) compounds that include a Cr(V)–ascorbate–peroxo complex.[602] However, Cr(III) compounds administered orally are not significantly toxic. Evidently, the Cr(VI) compounds can cross cell membranes and be reduced to Cr(III), which forms stable complexes with many constituents of cells including DNA.[603] The use of such "exchange-inert" Cr(III) complexes of ATP in enzymology was considered in Chapter 12.

Most forms of Cr(III) are not absorbed and utilized by the body. For this reason, and because of the increased use of sucrose and other refined foods, a marginal human chromium deficiency may be widespread.[604,605] This may result not only in poor utilization of glucose but also in other effects on lipid and protein metabolism.[597] However, questions have been raised about the use of chromium picolinate as a dietary supplement. High concentrations have been reported to cause chromosome damage[606] and there may be danger of excessive accumulation of chromium in the body.[607]

G. Vanadium

Vanadium is a dietary essential for goats and presumably also for human beings,[427] who typically consume ~2 mg / day. However, because vanadium compounds have powerful pharmacological effects it has been difficult to establish the nutritional requirement for animals.[427,604,607a] The adult body contains only about 0.1 mg of vanadium. Typical tissue concentrations are 0.1–0.7 μM[608,609] and serum concentration may be 10 nM or less. Vanadium can assume oxidation states ranging from +2 to +5, the vanadate ion VO_4^{3-} being the predominant form of V(V) in basic solution and in dilute solutions at pH 7. However, at millimolar concentrations, $V_3O_9^{3-}$, $V_4O_{12}^{4-}$, and other polynuclear forms predominate.[610–612] In plasma most exists as metavanadate, VO_2^-, but within cells it is reduced to the vanadyl cation VO^{2+} which is an especially stable double-bonded unit in compounds of V(IV).[609] Only at very low pH is V(III) stable.

The first suggestion of a possible biochemical function for vanadium came from the discovery that **vanadocytes**, the green blood cells of tunicates (sea squirts), contain ~1.0 M V(III) and 1.5–2 M H_2SO_4 .[613] It was proposed that a V-containing protein is an oxygen carrier. However, the V^{3+} appears not to be associated with proteins[612] and it does not carry O_2. It may be there to poison predators.[614] The vanadium-accumulating species also synthesize several complex, yellow catechol-type chelating agents (somewhat similar to enterobactin; Fig. 16-1) which presumably complex V(V) and perhaps also reduce it to V(III).[615] Vanadium is also accumulated by other marine organisms and by the mushroom *Amanita muscaria*.

Vanadoproteins are found in most marine algae and seaweed and in some lichens.[616] Among these are **haloperoxidases**,[252,253,617–618b] enzymes that are quite different from the corresponding heme peroxidases discussed in Section A,6. The vanadium is bound as

hydrogen vanadate, HVO_4^{2-}, in trigonal bipyrimidal coordination with the three oxygens in equatorial positions and a histidine in one axial position. In the crystal structure an azide (N_3^-) ion occupies the other axial position, but it is presumably the site of interaction with peroxide.[619] The structure is similar to that of acid phosphatases inhibited by vanadate.[620,621] Many nitrogen-fixing bacteria contain genes for a vanadium-dependent nitrogenase that is formed only if molybdenum is not available.[622] The nitrogenases are discussed in Chapter 24.

Much of current interest in vanadium stems from the discovery that vanadate (HVO_4^{2-} at pH 7) is a powerful inhibitor of ATPases such as the sodium pump protein ($Na^+ + K^+$)ATPase (Chapter 8), of phosphatases,[623] and of kinases.[624] This can be readily understood from comparison of the structure of phosphate and vanadate ions.

Phosphate   Vanadate

Other enzymes such as the cyclic AMP-dependent protein kinase are *stimulated* by vanadium.[624] Vanadate seems to inhibit most strongly those enzymes that form a phosphoenzyme intermediate. This inhibition may be diminished within cells because vanadate is readily reduced by glutathione and other intracellular reductants. The resulting vanadyl ion is a much weaker inhibitor and also stimulates several metabolic processes.[608]

Also of great interest is an insulin-like action of vanadium[607a,625] and evidence that vanadium may be essential to proper cardiac function.[626] A role in lipid metabolism was suggested by the observation that in high doses vanadium inhibits cholesterol synthesis and lowers the phospholipid and cholesterol content of blood. Vanadium is reported to inhibit development of caries by stimulating mineralization of teeth. Unlike tungsten, vanadium does not compete with molybdenum in the animal body.[627] The sometimes dramatic effects of vanadate as an inhibitor, activator, and metabolic regulator are shared also by molybdate and tungstate.[628,629] Even greater effects are observed with vanadate, molybdate, or tungstate plus H_2O_2.[630] The resulting **pervanadate**, **permolybdate**, and **pertungstate** are often assumed to be monoperoxo compounds, e.g., vanodyl hydroperoxide. However, there is some uncertainty.[631]

H. Molybdenum

Long recognized as an essential element for the growth of plants, molybdenum has never been directly demonstrated as a necessary animal nutrient. Nevertheless, it is found in several enzymes of the human body, as well as in 30 or more additional enzymes of bacteria and plants.[632] **Aldehyde oxidases,**[633] **xanthine oxidase** of liver and the related **xanthine dehydrogenase**, catalyze the reactions of Eqs. 16-58 and 16-59 and contain molybdenum that is essential for catalytic activity. Xanthine oxidase also contains two Fe_2S_2 clusters and bound FAD. The enzymes can also

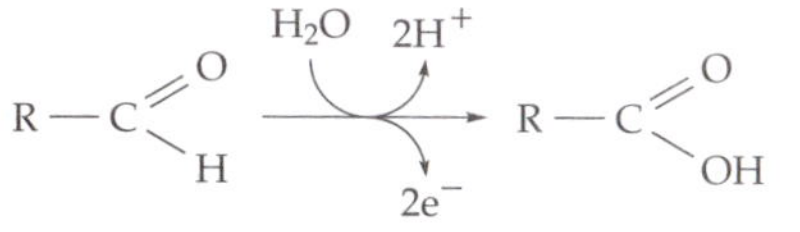

$$(16-58)$$

oxidize xanthine further (Eq. 16-59, step *b*) by a repetition of the same type of oxidation process at positions 8 and 9 to form **uric acid**. The much studied xanthine dehydrogenase has been isolated from milk,[634,635] liver, fungi,[636] and some bacteria.[637] In the dehydrogenase NAD^+ is the electron acceptor that oxidizes the bound $FADH_2$ formed in Eq. 16-59. Xanthine dehydrogenase, in the absence of thiol compounds, is converted spontaneously into xanthine oxidase, probably as a result of a conformational change and formation of a disulfide bridge within the protein. Treatment with thiol compounds such as dithiothreitol reconverts the enzyme to the dehydrogenase. Evidently in the oxidase form the NAD^+ binding site has moved away from the FAD, permitting oxidation of $FADH_2$ by O_2 with formation of hydrogen peroxide.[635,638]

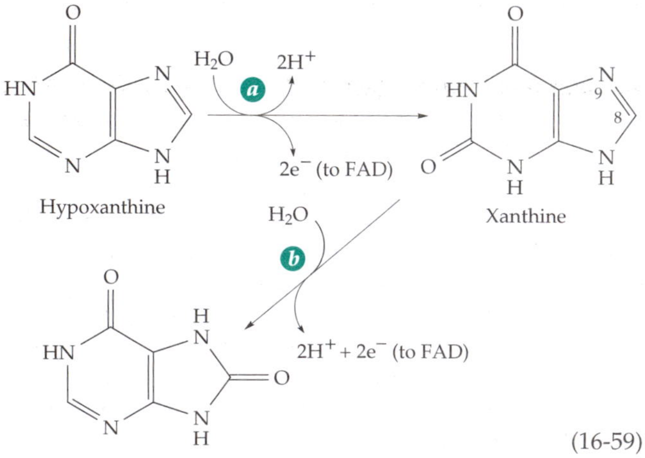

$$(16-59)$$

A purine hydroxylase from fungi,[639] bacterial quinoline and isoquinoline oxidoreductases,[640,641] and a selenium-containing nicotinic acid hydroxylase from *Clostridium barberei*[642] are members of the

xanthine oxidase family (or molybdenum hydroxylase family).[632,641,643] Also included in the family are aldehyde oxidoreductases from the sulfate-reducing *Desulfovibrio gigas*[633] and from the tomato.[644]

Two other families of molybdoenzymes are the **sulfite oxidase family**[646a,b] and the **dimethylsulfoxide reductase family**.[632,641] **Nitrogenase** (Chapter 24) constitutes a fourth family. Sulfite oxidase (Eq. 16-60) is an essential human liver enzyme (see also Chapter 24).[645,646]

$$SO_3^{2-} + H_2O \rightarrow SO_4^{2-} + 2\ e^- + 2\ H^+ \quad (16-60)$$

The **assimilatory nitrate reductase** (Eq. 16-61) of fungi and green plants (Chapter 24) also belongs to the sulfite oxidase family.

$$NO_3^2 + 2\ e^- + 2\ H^+ \rightarrow NO_2^- + H_2O \quad (16-61)$$

DMSO reductase reduces dimethylsulfoxide to dimethylsulfide (Eq. 16-62) as part of the biological sulfur cycle.[647–648d]

$$(H_3C)_2S{=}O + 2\ H^+ + 2\ e^- \rightarrow (CH_3)_2S \quad (16-62)$$

A number of other reductases and dehydrogenases, including **dissimilatory nitrate reductases** of *E. coli* and of denitrifying bacteria (Chapter 18), belong to the DMSO reductase family. Other members are reductases for biotin *S*-oxide,[649] trimethylamine *N*-oxide, and polysulfides as well as **formate dehydrogenases** (Eq. 16-63), formylmethanofuran dehydrogenase (Fig. 15-22,

$$HCOO^- + 2\ e^- + 3\ H^+ \rightarrow CO_2 + 2\ H_2O \quad (16-63)$$

step *b*), and arsenite oxidase.[632] Several other molybdoenzymes, such as pyridoxal oxidase, had not been classified by 1996.[632]

1. Molybdenum Ions and Coenzyme Forms

Molybdenum is a metal of the second transition series, one of the few heavy elements known to be essential to life. Its most stable oxidation state, Mo(VI), has $4d$ orbitals available for coordination with anionic ligands. Coordination numbers of 4 and 6 are preferred, but molybdenum can accommodate up to eight ligands. Most of the complexes are formed from the oxycation $Mo(VI)O_2^{2+}$. If two molecules of water are coordinated with this ion, the protons are so acidic that they dissociate completely to give $Mo(VI)O_4^{2-}$, the molybdate ion. Other oxidation states vary from Mo(III) to Mo(V). In these lower oxidation states, the tendency for protons to dissociate from coordinated ligands is less, e.g., $Mo(III)(H_2O)_6^{3+}$ does not lose protons even in a very basic medium. Molybdenum tends to form dimeric

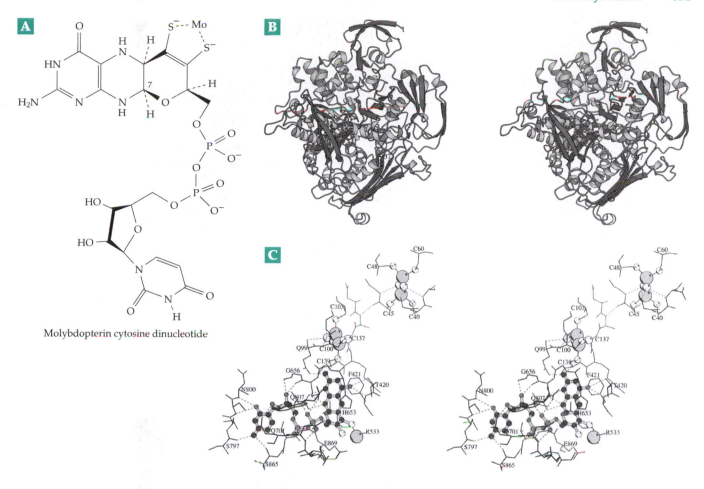

A

Molybdopterin cytosine dinucleotide

Figure 16-31 (A) Structure of molybdopterin cytosine dinucleotide complexed with an atom of molybdenum. (B) Stereoscopic ribbon drawing of the structure of one subunit of the xanthine oxidase-related aldehyde oxidoreductase from *Desulfovibrio gigas*. Each 907-residue subunit of the homodimeric protein contains two Fe_2S_2 clusters visible at the top and the molybdenum–molybdopterin coenzyme buried in the center. (C) Alpha-carbon plot of portions of the protein surrounding the molybdenum–molybdopterin cytosine dinucleotide and (at the top) the two plant-ferredoxin-like Fe_2S_2 clusters. Each of these is held by a separate structural domain of the protein. Two additional domains bind the molybdopterin coenzyme and there is also an intermediate connecting domain. In xanthine oxidase the latter presumably has the FAD binding site which is lacking in the *D. gigas* enzyme. From Romão *et al.*[633] Courtesy of R. Huber.

or polymeric oxygen-bridged ions. However, within the enzymes it exists as the unique **molybdenum coenzymes.** The Mo-containing enzymes usually also contain additional bound cofactors, including Fe – S clusters and flavin coenzymes or heme.

The recognition that the Mo in the molybdoproteins exists in organic cofactor forms came from studies of mutants of *Aspergillus* and *Neurospora*.[650] In 1964, Pateman and associates discovered mutants that lacked both nitrate reductase and xanthine dehydrogenase. Later, it was shown that acid-treated molybdoenzymes released a material that would restore activity to the inactived nitrate reductase from the mutant organisms. This new coenzyme, a phosphate ester of molybdopterin (Fig. 15-17), was characterized by Rajagopalan and coworkers.[650,651] A more complex form of the coenzyme, **molybdopterin cytosine dinucleotide**

(Fig. 16-31), is found in the *D. gigas* aldehyde oxidoreductase. Related coenzyme forms include nucleotides of adenine, guanine (see chapter banner, p. 837), and hypoxanthine.[651a,651b] The structure of molybdopterin is related to that of **urothione** (Fig. 15-17), a normal urinary constituent. The relationship to urothione was strengthened by the fact that several children with severe neurological and other symptoms were found to lack both sulfite oxidase and xanthine dehydrogenase as well as the molybdenum cofactor and urinary urothione.[646,646a,646b]

Study by X-ray absorption spectroscopy of the extended **X-ray absorption fine structure** (EXAFS) has provided estimates of both the nature and the number of the nearest neighboring atoms around the Mo. The EXAFS spectra of xanthine dehydrogenase and of nitrate reductase from **Chlorella** confirmed the presence of both the $Mo(VI)O_2$ unit with Mo–O distances of 0.17 nm and two or three sulfur atoms at distances of 0.24 nm.[652,653] The two sulfur atoms were presumed to come from the molybdopterin. A peculiarity of the xanthine oxidase family is the presence on the molybdenum of a "cyanolyzable" sulfur.[654] This is a sulfide attached to the molybdenum, which is present as Mo(VI)OS rather than $Mo(VI)O_2$. Reaction with cyanide produces thiocyanate (Eq. 16-64).

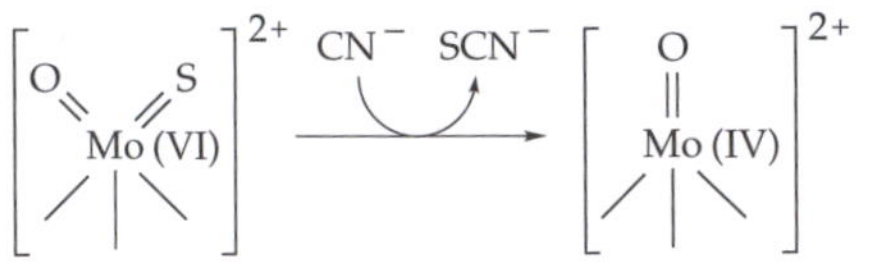

$$\left[\begin{array}{c} O \diagdown \quad \diagup S \\ Mo\,(VI) \end{array} \right]^{2+} \xrightarrow[\quad]{CN^- \quad SCN^-} \left[\begin{array}{c} O \\ \| \\ Mo\,(IV) \end{array} \right]^{2+}$$

(16-64)

The active site structures of the three classes of molybdenum-containing enzymes are compared in Fig. 16-32. In the DMSO reductase family there are two identical molybdopterin dinucleotide coenzymes complexed with one molybdenum. However, only one of these appears to be functionally linked to the Fe_2S_2 center.

Nitrogenase, which catalyzes the reduction of N_2 to two molecules of NH_3, has a different **molybdenum –iron cofactor (FeMo-co)**. It can be obtained by acid denaturation of the very oxygen-labile iron–molybdenum protein of nitrogenase followed by extraction with dimethylformamide.[655,656] The coenzyme is a complex Fe–S–Mo cluster also containing **homo-citrate** with a composition $MoFe_7S_9$–homocitrate (see Fig. 24-3). Nitrogenase and this coenzyme are considered further in Chapter 24.

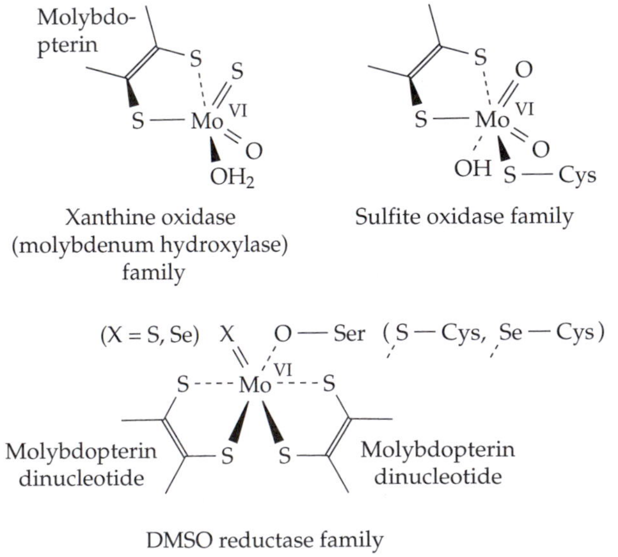

Molybdo-pterin

Xanthine oxidase
(molybdenum hydroxylase)
family

Sulfite oxidase family

(X = S, Se)

Molybdopterin dinucleotide

Molybdopterin dinucleotide

DMSO reductase family

Figure 16-32 Structures surrounding molybdenum in three families of molybdoenzymes. See Hille.[632]

2. Enzymatic Mechanisms

Although several of the reactions catalyzed by molybdoenzymes are classified as dehydrogenases, all of them except nitrogenase involve H_2O as either a reactant or a product. The EXAFS spectra suggest that the $Mo(VI)O_2$ unit is converted to $Mo(IV)O$ during reaction with a substrate Sub (Eq. 16-65, step a). Reaction of the $Mo(IV)O$ with water (step b) completes the catalysis.

$$a$$
$$[Mo(VI)O_2]^{2+} + Sub \rightarrow [Mo(IV)O]^{2+} + Sub\text{–}O$$

$$b$$
$$[Mo(IV)O]^{2+} + H_2O \rightarrow 2H^+ + 2e^- + [Mo(VI)]^{2+}O_2$$
$$SumSub + H_2O \rightarrow Sub\text{–}O + 2H^+ + 2e^- \quad (16\text{-}65)$$

Step a of all of these reactions can be regarded as an **oxo-transfer**.[653]

To complete the reaction, two electrons must be passed from Mo(IV) to a suitable acceptor, usually an Fe–S cluster or a bound heme group. FAD is also often present. Xanthine oxidase[634,635,643,657,657a] contains two Fe_2S_2 clusters and a FAD for each of the two atoms of Mo in the dimer. Since this enzyme acts like a typical flavin oxidase that generates H_2O_2 from O_2, it may be that electrons pass from Mo to the Fe–S center and then to the flavin. Since the EPR signal of the paramagnetic Mo(V), with its characteristic six-line hyperfine structure, is seen during the action of xanthine oxidase and other molybdenum-containing enzymes, single-electron transfers are probably involved.

In bacteria such as *E. coli* a dissimilatory nitrate reductase allows nitrate to serve as an oxidant in place of O_2. An oxygen atom is removed from the nitrate to form nitrite as two electrons are accepted from a membrane-bound cytochrome b. The nitrate reductase consists of a 139-kDa Mo-containing catalytic subunit, a 58-kDa electron-transferring subunit that contains both Fe_3S_4 and Fe_4S_4 centers, and a 26-kDa heme-containing membrane anchor subunit.[658–660] The assimilatory nitrate reductase of fungi, green algae, and higher plants contains both a b-type cytochrome and FAD and a molybdenum coenzyme in a large oligomeric complex.[661–663a]

Formate dehydrogenases from many bacteria contain molybdopterin and also often selenium (Table 15-4).[664,665] A membrane-bound Mo-containing formate dehydrogenase is produced by *E. coli* grown anaerobically in the presence of nitrate. Under these circumstances it is coupled to nitrate reductase via an electron-transport chain in the membranes which permits oxidation of formate by nitrate (Eq. 18-26). This enzyme is also a multisubunit protein.[665,666] Two other Mo- and Se- containing formate dehydrogenases are produced by *E. coli*.[667,668] The three-dimensional structure is known for one of them, **formate dehydrogenase H**, a component of the anaerobic formate hydrogen lyase complex (Eq. 17-25).[669,670] The structure shows Mo held by the sulfur atoms of two molybdopterin molecules, as in DMSO reductase. The Se atom of SeCys 140 is also coordinated with the Mo atom, and the imidazole of His 141 is in close proximity. When ^{13}C-labeled formate was oxidized in ^{18}O-enriched water no ^{18}O was found in the released product, CO_2.[671] This suggested that formate may be bound to Mo and dehydrogenated, with Mo(VI) being reduced to Mo(IV). The formate hydrogen might be transferred as H^+ to the His 140 side chain. Mo(IV) could then be reoxidized by electron transfer in two one-electron steps.[670] However, recent X-ray absorption spectra suggest the presence in the enzyme of a selenosulfide ligand to Mo.[672] Mechanistic uncertainties remain!

A flavin-dependent formate dehydrogenase system found in *Methanobacterium* passes electrons from dehydrogenation of formate to FAD and then to the deazaflavin coenzyme F_{420}.[673] In contrast to these Mo-containing enzymes, the formate dehydrogenase from *Pseudomonas oxalaticus*, which oxidizes formate with NAD^+ (Eq. 16-66), contains neither Mo or Se.[674]

$$HCOO^- + NAD^+ \longrightarrow CO_2 + NADH \qquad (16\text{-}66)$$

It is a large 315-kDa oligomer containing 2 FMN and ~20 Fe / S. Formate dehydrogenases of green plants and yeasts are smaller 70- to 80-kDa proteins lacking bound prosthetic groups.[674] A key enzyme in the metabolism of carbon monoxide-oxidizing bacteria is CO oxidase, another membrane-bound molybdoenzyme.[675–676c] It also contains selenium, which is attached to a cysteine side chain as **S-selanylcysteine.** A proposed reaction requence[676a] is shown in Eq. 16-67.

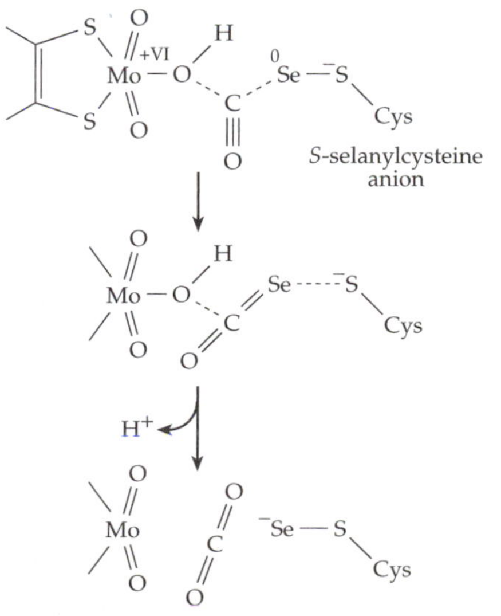

S-selanylcysteine anion

(16-67)

3. Nutritional Need for Mo

The first hint of an essential role of molybdenum in metabolism came from the discovery that animals raised on a diet deficient in molybdenum had decreased liver xanthine oxidase activity. There is no evidence that xanthine oxidase is essential for all life, but a human genetic deficiency of sulfite oxidase or of its molybdopterin coenzyme can be lethal.[646,646a,b] The conversion of molybdate into the molybdopterin cofactor in *E. coli* depends upon at least five genes.[677] In *Drosophila* the addition of the cyanolyzable sulfur (Eq. 16-64) is the final step in formation of xanthine dehydrogenase.[678] It is of interest that sulfur (S^0) can be transferred from rhodanese (see Eq. 24-45), or from a related mercaptopyruvate sulfurtransferase[679] into the desulfo form of xanthine oxidase to generate an active enzyme.[680]

Uptake of molybdate by cells of *E. coli* is accomplished by an ABC-type transport system.[681] In some bacteria, e.g., the nitrogen-fixing *Azotobacter*, molybdenum can be stored in protein-bound forms.[682]

I. Tungsten

For many years tungsten was considered only as a potential antagonist for molybdenum. However, in 1970 growth stimulation by tungsten compounds was observed for some acetogens, some methanogens, and a few hyperthermophilic bacteria. Since then over a dozen tungstoenzymes have been isolated.[683,683a] These can be classified into three categories: aldehyde oxidoreductases, formaldehyde oxidoreductases, and the single enzyme acetylene hydratase. In most cases the tungstoenzymes resemble the corresponding molybdoenzymes and in most instances organisms containing a tungsten-requiring enzyme also contain the corresponding molybdenum enzyme. However, a few hyperthermophilic archaea appear to require W and are unable to use Mo.

The aldehyde ferredoxin oxidoreductase from the hyperthermophile *Pyrococcus furiosus* was the first molybdopterin-dependent enzyme for which a three-dimensional structure became available.[683,684] The tungstoenzyme resembles that of the related molybdoenzyme (Fig. 16-31). A similar ferredoxin-dependent enzyme reduces glyceraldehyde-3-phosphate.[685] Another member of the tungstoenzyme aldehyde oxidoreductase family is **carboxylic acid reductase**, an enzyme found in certain acetogenic clostridia. It is able to use reduced ferredoxin to convert unactivated carboxylic acids into aldehydes, even though $E^{o'}$ for the acetaldehyde / acetate couple is −0.58 V.[686]

Tungsten- and sometimes Se-containing formate dehydrogenases together with *N*-formylmethanofuran dehydrogenases (Fig. 15-22, step *b*) form a second family.

Again, these appear to resemble the corresponding Mo-dependent enzymes. The unique **acetylene hydratase** from the acetylene-utilizing *Pelobacter acetylenicus* catalyzes the hydration of acetylene to acetaldehyde.[687]

$$H-C \equiv C-H + H_2O \rightarrow H_3C-CHO \qquad (16\text{-}68)$$

In *Thermotoga maritima*, the most thermophilic organism known, tungsten promotes synthesis of an Fe-containing hydrogenase as well as some other enzymes but seems to have a regulatory rather than a structural role.[688]

References

1. Ochiai, E. I. (1977) *Bioinorganic Chemistry. An Introduction*, Allyn and Bacon, Boston
2. Fraústo da Silva, J. J. R., and Williams, R. J. P. (1991) *The Biological Chemistry of the Elements: The Inorganic Chemistry of Life*, Clarendon Press, Oxford
3. Lippard, S. J., and Berg, J. M. (1994) *Principles of Bioinorganic Chemistry*, Univ. Science Books, Mill Valley, California
4. Holm, R. H., Kennepohl, P., and Solomon, E. I. (1996) *Chem. Rev.* **96**, 2239–2314
5. Neilands, J. B., ed. (1974) *Microbial Iron Metabolism*, Academic Press, New York
6. Jacobs, A., and Woodwood, M., eds. (1974) *Iron in Biochemistry and Medicine*, Academic Press, New York
7. Neilands, J. B. (1973) in *Inorganic Biochemistry*, Vol. 1 (Eichhorn, G. L., ed), pp. 167–202, Elsevier, Amsterdam
8. Bergeron, R. J. (1986) *Trends Biochem. Sci.* **11**, 133–136
9. Neilands, J. B. (1995) *J. Biol. Chem.* **270**, 26723–26726
10. Sakaitani, M., Rusnak, F., Quinn, N. R., Tu, C., Frigo, T. B., Berchtold, G. A., and Walsh, C. T. (1990) *Biochemistry* **29**, 6789–6798
11. Harris, W. R., Carrano, C. J., Cooper, S. R., Sofen, S. R., Avdeef, A. E., McArdle, J. V., and Raymond, K. N. (1979) *J. Am. Chem. Soc.* **101**, 6097–6104
11a. Ferguson, A. D., Hofmann, E., Coulton, J. W., Diederichs, K., and Welte, W. (1998) *Science* **282**, 2215–2220
11b. Braun, V., and Killmann, H. (1999) *Trends Biochem. Sci.* **24**, 104–109
11c. Stintzi, A., Barnes, C., Xu, J., and Raymond, K. N. (2000) *Proc. Natl. Acad. Sci. U.S.A.* **97**, 10691–10696
11d. Cohen, S. M., Meyer, M., and Raymond, K. N. (1998) *J. Am. Chem. Soc.* **120**, 6277–6286
11e. Adrait, A., Jacquamet, L., Le Pape, L., Gonzalez de Peredo, A., Aberdam, D., Hazemann, J.-L., Latour, J.-M., and Michaud-Soret, I. (1999) *Biochemistry* **38**, 6248–6260
12. Neilands, J. B., Erickson, T. J., and Rastetter, W. H. (1981) *J. Biol. Chem.* **256**, 3831–3832

13. Newton, S. M. C., Allen, J. S., Cao, Z., Qi, Z., Jiang, X., Sprencel, C., Igo, J. D., Foster, S. B., Payne, M. A., and Klebba, P. E. (1997) *Proc. Natl. Acad. Sci. U.S.A.* **94**, 4560–4565
14. Persmark, M., Expert, D., and Neilands, J. B. (1989) *J. Biol. Chem.* **264**, 3187–3193
15. Khalil-Rizvi, S., Toth, S. I., van der Helm, D., Vidavsky, I., and Gross, M. L. (1997) *Biochemistry* **36**, 4163–4171
16. Wong, G. B., Kappel, M. J., Raymond, K. N., Matzanke, B., and Winkelmann, G. (1983) *J. Am. Chem. Soc.* **105**, 810–815
17. Shanzer, A., Libman, J., Lifson, S., and Felder, C. E. (1986) *J. Am. Chem. Soc.* **108**, 7609–7619
18. Reid, R. T., Live, D. H., Faulkner, D. J., and Butler, A. (1993) *Nature (London)* **366**, 455–458
19. Llinás, M., Wilson, D. M., and Neilands, J. B. (1973) *Biochemistry* **12**, 3836–3843
20. Eng-Wilmot, D. L., and van der Helm, D. (1980) *J. Am. Chem. Soc.* **102**, 7719–7725
21. Cox, C. D., Rinehart, K. L., Jr., Moore, M. L., and Cook, J. C., Jr. (1981) *Proc. Natl. Acad. Sci. U.S.A.* **78**, 4256–4260
22. Neilands, J. B. (1952) *J. Am. Chem. Soc.* **74**, 4846–4847
23. Braun, V. (1985) *Trends Biochem. Sci.* **10**, 75–78
24. Letendre, E. D. (1985) *Trends Biochem. Sci.* **10**, 166–168
25. Köster, W., and Braun, V. (1990) *J. Biol. Chem.* **265**, 21407–21410
26. Jiang, X., Payne, M. A., Cao, Z., Foster, S. B., Feix, J. B., Newton, S. M. C., and Klebba, P. E. (1997) *Science* **276**, 1261–1264
27. Coy, M., and Neilands, J. B. (1991) *Biochemistry* **30**, 8201–8210
28. Kammler, M., Schoin, C., and Hantke, D. (1993) *J. Bacteriol.* **175**, 6212–6219
29. Dix, D. R., Bridgham, J. T., Broderius, M. A., Byersdorfer, C. A., and Eide, D. J. (1994) *J. Biol. Chem.* **269**, 26092–26099
30. Kaplan, J., and O'Halloran, T. V. (1996) *Science* **271**, 1510–1512
31. Dix, D., Bridgham, J., Broderius, M., and Eide, D. (1997) *J. Biol. Chem.* **272**, 11770–11777
32. de Silva, D., Davis-Kaplan, S., Fergestad, J., and Kaplan, J. (1997) *J. Biol. Chem.* **272**, 14208–14213

33. Radisky, D., and Kaplan, J. (1999) *J. Biol. Chem.* **274**, 4481–4484
34. Stearman, R., Yuan, D. S., Yamaguchi-Iwai, Y., Klausner, R. D., and Dancis, A. (1996) *Science* **271**, 1552–1557
35. Hassett, R. F., Romeo, A. M., and Kosman, D. J. (1998) *J. Biol. Chem.* **273**, 7628–7636
36. Mukhopadhyay, C., Attieh, Z. K., and Fox, P. L. (1998) *Science* **279**, 714–717
37. Li, L., and Kaplan, J. (1997) *J. Biol. Chem.* **272**, 28485–28493
37a. Lange, H., Kispal, G., and Lill, R. (1999) *J. Biol. Chem.* **274**, 18989–18996
37b. Urbanowski, J. L., and Piper, R. C. (1999) *J. Biol. Chem.* **274**, 38061–38070
38. Octave, J.-N., Schneider, Y.-J., Trouet, A., and Crichton, R. R. (1983) *Trends Biochem. Sci.* **8**, 217–222
39. Conrad, M. E., Umbreit, J. N., Moore, E. G., Peterson, R. D. A., and Jones, M. B. (1990) *J. Biol. Chem.* **265**, 5273–5279
40. Ponka, P., Schulman, H. M., Woodworth, R. C., and Richter, G. W., eds. (1990) *Iron Transport and Storage*, CRC Press, Boca Raton, Florida
41. Yu, J., and Wessling-Resnick, M. (1998) *J. Biol. Chem.* **273**, 6909–6915
42. Gunshin, H., Mackenzie, B., Berger, U. V., Gunshin, Y., Romero, M. F., Boron, W. F., Nussberger, S., Gollan, J. L., and Hediger, M. A. (1997) *Nature (London)* **388**, 482–488
43. Toth, I., Rogers, J. T., McPhee, J. A., Elliott, S. M., Abramson, S. L., and Bridges, K. R. (1995) *J. Biol. Chem.* **270**, 2846–2852
44. Finch, C. A., and Huebers, H. (1982) *N. Engl. J. Med.* **306**, 1520–1528
45. Scrimshaw, N. S. (1991) *Sci. Am.* **265**(Oct), 46–52
46. Kaplan, J., and Kushner, J. P. (2000) *Nature (London)* **403**, 711,713
46a. Kühn, L. C. (1999) *Trends Biochem. Sci.* **24**, 164–166
46b. Attieh, Z. K., Mukhopadhyay, C. K., Seshadri, V., Tripoulas, N. A., and Fox, P. L. (1999) *J. Biol. Chem.* **274**, 1116–1123
47. Gordeuk, V., Thuma, P., Brittenham, G., McLaren, C., Parry, D., Backenstose, A., Biemba, G., Msiska, R., Holmes, L., McKinley, E., Vargas, L., Gilkeson, R., and Poltera, A. A. (1992) *N. Engl. J. Med.* **327**, 1518–1521

48. Faller, B., and Nick, H. (1994) *J. Am. Chem. Soc.* **116**, 3860–3865
49. Baker, E. N., Rumball, S. V., and Anderson, B. F. (1987) *Trends Biochem. Sci.* **12**, 350–353
50. Welch, S. (1992) *Transferrin: The Iron Carrier,* CRC Press, Boca Raton, Florida
51. Baker, H. M., Anderson, B. F., Brodie, A. M., Shongwe, M. S., Smith, C. A., and Baker, E. N. (1996) *Biochemistry* **35**, 9007–9013
51a. Sharma, A. K., Paramasivam, M., Srinivasan, A., Yadav, M. P., and Singh, T. P. (1998) *J. Mol. Biol.* **289**, 303–317
51b. Peterson, N. A., Anderson, B. F., Jameson, G. B., Tweedie, J. W., and Baker, E. N. (2000) *Biochemistry* **39**, 6625–6633
51c. Bou Abdallah, F., and El Hage Chahine, J.-M. (2000) *J. Mol. Biol.* **303**, 255–266
52. Kurokawa, H., Mikami, B., and Hirose, M. (1995) *J. Mol. Biol.* **254**, 196–207
52a. Mizutani, K., Yamashita, H., Mikami, B., and Hirose, M. (2000) *Biochemistry* **39**, 3258–3265
53. Moore, S. A., Anderson, B. F., Groom, C. R., Haridas, M., and Baker, E. N. (1997) *J. Mol. Biol.* **274**, 222–236
54. Bailey, S., Evans, R. W., Garratt, R. C., Gorinsky, B., Hasnain, S., Horsburgh, C., Jhoti, H., Lindley, P. F., Mydin, A., Sarra, R., and Watson, J. L. (1988) *Biochemistry* **27**, 5804–5812
54a. He, Q.-Y., Mason, A. B., Tam, B. M., MacGillivray, R. T. A., and Woodworth, R. C. (1999) *Biochemistry* **38**, 9704–9711
55. Randell, E. W., Parkes, J. G., Olivieri, N. F., and Templeton, D. M. (1994) *J. Biol. Chem.* **269**, 16046–16053
56. Grossmann, J. G., Mason, A. B., Woodworth, R. C., Neu, M., Lindley, P. F., and Hasnain, S. S. (1993) *J. Mol. Biol.* **231**, 554–558
57. Zak, O., Tam, B., MacGillivray, R. T. A., and Aisen, P. (1997) *Biochemistry* **36**, 11036–11043
58. Zak, O., Aisen, P., Crawley, J. B., Joannou, C. L., Patel, K. J., Rafiq, M., and Evans, R. W. (1995) *Biochemistry* **34**, 14428–14434
59. Mecklenburg, S. L., Donohoe, R. J., and Olah, G. A. (1997) *J. Mol. Biol.* **270**, 739–750
60. Fleming, M. D., Romano, M. A., Su, M. A., Garrick, L. M., Garrick, M. D., and Andrews, N. C. (1998) *Proc. Natl. Acad. Sci. U.S.A.* **95**, 1148–1153
61. Theil, E. C. (1987) *Ann. Rev. Biochem.* **56**, 289–315
62. Crichton, R. R. (1984) *Trends Biochem. Sci.* **9**, 283–286
63. Trikha, J., Theil, E. C., and Allewell, N. M. (1995) *J. Mol. Biol.* **248**, 949–967
64. Hempstead, P. D., Yewdall, S. J., Fernie, A. R., Lawson, D. M., Artymiuk, P. J., Rice, D. W., Ford, G. C., and Harrison, P. M. (1997) *J. Mol. Biol.* **268**, 424–448
65. Crichton, R. R., Soruco, J.-A., Roland, F., Michaux, M.-A., Gallois, B., Précigoux, G., Mahy, J.-P., and Mansuy, D. (1997) *Biochemistry* **36**, 15049–15054
66. Watt, G. D., Frankel, R. B., Papaefthymiou, G. C., Spartalian, K., and Stiefel, E. I. (1986) *Biochemistry* **25**, 4330–4336
67. Garg, R. P., Vargo, C. J., Cui, X., and Kurtz, D. M., Jr. (1996) *Biochemistry* **35**, 6297–6301
67a. Romão, C. V., Regalla, M., Xavier, A. V., Teixeira, M., Liu, M.-Y., and Le Gall, J. (2000) *Biochemistry* **39**, 6841–6849
68. Dickey, L. F., Sreedharan, S., Theil, E. C., Didsbury, J. R., Wang, H.-H., and Kaufman, R. E. (1987) *J. Biol. Chem.* **262**, 7901–7907
69. Sayers, D. E., Theil, E. C., and Rennick, F. J. (1983) *J. Biol. Chem.* **258**, 14076–14079
69a. Johnson, J. L., Cannon, M., Watt, R. K., Frankel, R. B., and Watt, G. D. (1999) *Biochemistry* **38**, 6706–6713

70. Bauminger, E. R., Treffry, A., Quail, M. A., Zhao, Z., Nowik, I., and Harrison, P. M. (1999) *Biochemistry* **38**, 7791–7802
71. Yang, X., Chen-Barrett, Y., Arosio, P., and Chasteen, N. D. (1998) *Biochemistry* **37**, 9743–9750
72. Pereira, A. S., Small, W., Krebs, C., Tavares, P., Edmondson, D. E., Theil, E. C., and Huynh, B. H. (1998) *Biochemistry* **37**, 9871–9876
72a. Hwang, J., Krebs, C., Huynh, B. H., Edmondson, D. E., Theil, E. C., and Penner-Hahn, J. E. (2000) *Science* **287**, 122–125
73. Treffry, A., Zhao, Z., Quail, M. A., Guest, J. R., and Harrison, P. M. (1997) *Biochemistry* **36**, 432–441
74. Bourne, P. E., Harrison, P. M., Rice, D. W., Smith, J. M. A., and Stansfield, R. F. D. (1982) in *The Biochemistry and Physiology of Iron* (Saltman, P., and Hegenauer, J., eds), pp. 427–, Elsevier North Holland, Amsterdam
75. Bridges, K. R., and Hoffman, K. E. (1986) *J. Biol. Chem.* **261**, 14273–14277
76. Medlock, A. E., and Dailey, H. A. (2000) *Biochemistry* **39**, 7461–7467
76a. Franco, R., Ma, J.-G., Lu, Y., Ferreira, G. C., and Shelnutt, J. A. (2000) *Biochemistry* **39**, 2517–2529
76b. Lecerof, D., Fodje, M., Hansson, A., Hansson, M., and Al-Karadaghi, S. (2000) *J. Mol. Biol.* **297**, 221–232
77. Eaton, S. S., and Eaton, G. R. (1977) *J. Am. Chem. Soc.* **99**, 1601–1604
78. Crossley, M. J., Harding, M. M., and Sternhell, S. (1986) *J. Am. Chem. Soc.* **108**, 3608–3613
78a. Braun, J., Schwesinger, R., Williams, P. G., Morimoto, H., Wemmer, D. E., and Limbach, H.-H. (1996) *J. Am. Chem. Soc.* **118**, 11101–11110
78b. Treibs, A. (1979) *Trends Biochem. Sci.* **4**, 71–72
79. Wu, W., Chang, C. K., Varotsis, C., Babcock, G. T., Puustinen, A., and Wikström, M. (1992) *J. Am. Chem. Soc.* **114**, 1182–1187
80. Sone, N., Ogura, T., Noguchi, S., and Kitagawa, T. (1994) *Biochemistry* **33**, 849–855
81. Fenna, R., Zeng, J., and Davey, C. (1995) *Arch. Biochem. Biophys.* **316**, 653–656
82. Taylor, K. L., Strobel, F., Yue, K. T., Ram, P., Pohl, J., Woods, A. S., and Kinkade, J. M., Jr. (1995) *Arch. Biochem. Biophys.* **316**, 635–642
83. Callahan, P. M., and Babcock, G. T. (1983) *Biochemistry* **22**, 452–461
84. Petke, J. D., and Maggiora, G. M. (1984) *J. Am. Chem. Soc.* **106**, 3129–3133
85. Anraku, Y., and Gennis, R. B. (1987) *Trends Biochem. Sci.* **12**, 262–266
86. Margoliash, E., and Schejter, H. (1984) *Trends Biochem. Sci.* **9**, 364–367
87. Sotiriou, C., and Chang, C. K. (1988) *J. Am. Chem. Soc.* **110**, 2264–2270
88. Timkovich, R., Cork, M. S., and Taylor, P. V. (1984) *J. Biol. Chem.* **259**, 15089–15093
89. Kaufman, J., Spicer, L. D., and Siegel, L. M. (1993) *Biochemistry* **32**, 2853–2867
90. Matthews, J. C., Timkovich, R., Liu, M.-Y., and Le Gall, J. (1995) *Biochemistry* **34**, 5248–5251
91. Chang, C. K., Timkovich, R., and Wu, W. (1986) *Biochemistry* **25**, 8447–8453
92. Mylrajan, M., Andersson, L. A., Loehr, T. M., Wu, W., and Chang, C. K. (1991) *J. Am. Chem. Soc.* **113**, 5000–5005
93. Meyer, T. E., and Kamen, M. D. (1982) *Adv. Prot. Chem.* **35**, 105–212
94. Margoliash, E., and Schejter, A. (1984) *Trends Biochem. Sci.* **9**, 364–367
95. Dickerson, R. E., and Timkovich, R. (1975) in *The Enzymes*, 3rd ed., Vol. 11 (Boyer, P. D., ed), pp. 397–547, Academic Press, New York
96. Dickerson, R. E. (1972) *Sci. Am.* **226**(Apr), 58–72

97. Takano, T., Kallei, O. B., Swanson, R., and Dickerson, R. E. (1973) *J. Biol. Chem.* **248**, 5234–5255
97a. Sebban-Kreuzer, C., Blackledge, M., Dolla, A., Marion, D., and Guerlesquin, F. (1998) *Biochemistry* **37**, 8331–8340
97b. Dolla, A., Arnoux, P., Protasevich, I., Lobachov, V., Brugna, M., Giudici-Orticoni, M. T., Haser, R., Czjzek, M., Makarov, A., and Bruschi, M. (1999) *Biochemistry* **38**, 33–41
98. Margoliash, E., and Frohwirt, N. (1959) *Biochem. J.* **71**, 570–572
99. Margoliash, E., and Bosshard, H. A. (1983) *Trends Biochem. Sci.* **8**, 316–320
100. Poerio, E., Parr, G. R., and Taniuchi, H. (1986) *J. Biol. Chem.* **261**, 10976–10989
101. Salemme, F. R., Kraut, J., and Kamen, M. D. (1973) *J. Biol. Chem.* **248**, 7701–7716
102. Almassy, R. J., and Dickerson, R. E. (1978) *Proc. Natl. Acad. Sci. U.S.A.* **75**, 2674–2678
102a. Benini, S., González, A., Rypniewski, W. R., Wilson, K. S., Van Beeumen, J. J., and Ciurli, S. (2000) *Biochemistry* **39**, 13115–13126
103. Sogabe, S., and Miki, K. (1995) *J. Mol. Biol.* **252**, 235–247
104. Simões, P., Matias, P. M., Morais, J., Wilson, K., Dauter, Z., Carrondo, M. A., (1998) *Inorganica Chimica Acta* **273**, 213–224
105. Sczekan, S. R., and Joshi, J. G. (1987) *J. Biol. Chem.* **262**, 13780–13788
106. Weber, P. C. (1982) *Biochemistry* **21**, 5116–5119
107. La Mar, G. N., Jackson, J. T., Dugad, L. B., Cusanovich, M. A., and Bartsch, R. G. (1990) *J. Biol. Chem.* **265**, 16173–16180
108. Ren, Z., Meyer, T., and McRee, D. E. (1993) *J. Mol. Biol.* **234**, 433–445
109. Weber, P. C., Salemme, F. R., Mathews, F. S., and Bethge, P. H. (1981) *J. Biol. Chem.* **256**, 7702–7704
110. Hamada, K., Bethge, P. H., and Mathews, F. S. (1995) *J. Mol. Biol.* **247**, 947–962
111. Huang, D., Everly, R. M., Cheng, R. H., Heymann, J. B., Schägger, H., Sled, V., Ohnishi, T., Baker, T. S., and Cramer, W. A. (1994) *Biochemistry* **33**, 4401–4409
112. Prince, R. C., and George, G. N. (1995) *Trends Biochem. Sci.* **20**, 217–218
113. Higuchi, Y., Bando, S., Kusunoki, M., Matsurura, Y., Yasuska, N., Kakuda, M., Yamanaka, T., Yagi, T., and Inokuchi, H. (1981) *J. Biochem.* **89**, 1659–1662
114. Bruschi, M., Woudstra, M., Guigliarelli, B., Asso, M., Lojou, E., Retillot, Y., and Abergel, C. (1997) *Biochemistry* **36**, 10601–10608
115. Czjzek, M., Payan, F., Guerlesquin, F., Bruschi, M., and Haser, R. (1994) *J. Mol. Biol.* **243**, 653–667
116. Liu, M.-C., and Peck, H. D., Jr. (1981) *J. Biol. Chem.* **256**, 13159–13164
117. Banci, L., Bertini, I., Bruschi, M., Sompornpisut, P., and Turano, P. (1996) *Proc. Natl. Acad. Sci. U.S.A.* **93**, 14396–14400
118. Jenney, F. E., Jr., Prince, R. C., and Daldal, F. (1994) *Biochemistry* **33**, 2496–2502
119. Menin, L., Schoepp, B., Garcia, D., Parot, P., and Verméglio, A. (1997) *Biochemistry* **36**, 12175–12182
120. Menin, L., Schoepp, B., Parot, P., and Verméglio, A. (1997) *Biochemistry* **36**, 12183–12188
121. Adir, N., Axelrod, H. L., Beroza, P., Isaacson, R. A., Rongey, S. H., Okamura, M. Y., and Feher, G. (1996) *Biochemistry* **35**, 2535–2547
122. Kerfeld, C. A., Anwar, H. P., Interrante, R., Merchant, S., and Yeates, T. O. (1995) *J. Mol. Biol.* **250**, 627–647
123. Gao, H., Qin, H., Simpson, M. C., Shelnutt, J. A., Knaff, D. B., and Ondrias, M. R. (1996) *Biochemistry* **35**, 12812–12819

References

124. Okkels, J. S., Kjaer, B., Hansson, Ö., Svendsen, I., Møller, B. L., and Scheller, H. V. (1992) *J. Biol. Chem.* **267**, 21139–21145

125. Link, T. A., Hatzfeld, O. M., Unalkat, P., Shergill, J. K., Cammack, R., and Mason, J. R. (1996) *Biochemistry* **35**, 7546–7552

126. Xia, D., Yu, C.-A., Kim, H., Xia, J.-Z., Kachurin, A. M., Zhang, L., Yu, L., and Deisenhofer, J. (1997) *Science* **277**, 60–66

127. von Jagow, G., and Sebald, W. (1980) *Ann. Rev. Biochem.* **49**, 281–314

128. Kiel, J. L. (1995) *Type-B Cytochromes: Sensors and Switches*, CRC Press, Boca Raton, Florida

129. Mathews, F. S., Levine, M., and Argos, P. (1972) *J. Mol. Biol.* **64**, 449–464

129a. Kostanjevecki, V., Leys, D., Van Driessche, G., Meyer, T. E., Cusanovich, M. A., Fischer, U., Guisez, Y., and Van Beeumen, J. (1999) *J. Biol. Chem.* **274**, 35614–35620

130. Vergères, G., Ramsden, J., and Waskell, L. (1995) *J. Biol. Chem.* **270**, 3414–3422

131. Nagi, M., Cook, L., Prasad, M. R., and Cinti, D. L. (1983) *J. Biol. Chem.* **258**, 14823–14828

132. Nishida, H., Inaka, K., Yamanaka, M., Kaida, S., Kobayashi, K., and Miki, K. (1995) *Biochemistry* **34**, 2763–2767

133. Lindqvist, Y., Brändén, C.-I., Mathews, F. S., and Lederer, F. (1991) *J. Biol. Chem.* **266**, 3198–3207

134. Guiard, B., and Lederer, F. (1979) *J. Mol. Biol.* **135**, 639–650

135. Garrett, R. M., and Rajagopalan, K. V. (1996) *J. Biol. Chem.* **271**, 7387–7391

136. Tegoni, M., Begotti, S., and Cambillau, C. (1995) *Biochemistry* **34**, 9840–9850

137. Weber, P. C., Salemme, F. R., Mathews, F. C., and Bethge, P. H. (1981) *J. Biol. Chem.* **257**, 7702–7704

138. Yu, L., Wei, Y.-Y., Usui, S., and Yu, C.-A. (1992) *J. Biol. Chem.* **267**, 24508–24515

139. Knaff, D. B. (1990) *Trends Biochem. Sci.* **15**, 289–291

140. Jalukar, V., Kelley, P. M., and Njus, D. (1991) *J. Biol. Chem.* **266**, 6878–6882

141. Kita, K., Konishi, K., and Anraku, Y. (1984) *J. Biol. Chem.* **259**, 3368–3374

142. Nakamura, K., Yamaki, M., Sarada, M., Nakayama, S., Vibat, C. R. T., Gennis, R. B., Nakayashiki, T., Inokuchi, H., Kojima, S., and Kita, K. (1996) *J. Biol. Chem.* **271**, 521–527

143. Yang, X., Yu, L., and Yu, C.-A. (1997) *J. Biol. Chem.* **272**, 9683–9689

143a. Ranghino, G., Scorza, E., Sjögren, T., Williams, P. A., Ricci, M., and Hajdu, J. (2000) *Biochemistry* **39**, 10958–10966

144. Gray, H. B., and Winkler, J. R. (1996) *Ann. Rev. Biochem.* **65**, 537–561

145. Farver, O., and Pecht, I. (1991) *FASEB J.* **5**, 2554–2559

146. Komar-Panicucci, S., Sherman, F., and McLendon, G. (1996) *Biochemistry* **35**, 4878–4885

147. Page, C. C., Moser, C. C., Chen, X., and Dutton, P. L. (1999) *Nature (London)* **402**, 47–52

148. DeVault, D. (1984) *Quantum-Mechanical Tunnelling in Biological Systems*, Cambridge Univ. Press, Cambridge, UK

149. Closs, G. L., and Miller, J. R. (1988) *Science* **240**, 440–447

150. Evenson, J. W., and Karplus, M. (1993) *Science* **262**, 1247–1249

151. Moser, C. C., Keske, J. M., Warncke, K., Farid, R. S., and Dutton, P. L. (1992) *Nature (London)* **355**, 796–802

152. Williams, R. J. P. (1992) *Nature (London)* **355**, 770–771

153. Pappa, H. S., and Poulos, T. L. (1995) *Biochemistry* **34**, 6573–6580

154. Mayo, S. L., Ellis, W. R., Jr., Crutchley, R. J., and Gray, H. B. (1986) *Science* **233**, 948–952

155. Wuttke, D. S., Bjerrum, M. J., Winkler, J. R., and Gray, H. B. (1992) *Science* **256**, 1007–1009

156. Banci, L., Bertini, I., De la Rosa, M. A., Koulougliotis, D., Navarro, J. A., and Walter, O. (1998) *Biochemistry* **37**, 4831–4843

157. Moreira, I., Sun, J., Cho, M. O.-K., Wishart, J. F., and Isied, S. S. (1994) *J. Am. Chem. Soc.* **116**, 8396–8397

158. Ahmed, A. J., and Millett, F. (1981) *J. Biol. Chem.* **256**, 1611–1615

159. Koppenol, W. H., and Margoliash, E. (1982) *J. Biol. Chem.* **257**, 4426–4437

160. Hahm, S., Miller, M. A., Geren, L., Kraut, J., Durham, B., and Millett, F. (1994) *Biochemistry* **33**, 1473–1480

161. Willie, A., Stayton, P. S., Sligar, S. G., Durham, B., and Millett, F. (1992) *Biochemistry* **31**, 7237–7242

162. Qin, L., and Kostic, N. M. (1994) *Biochemistry* **33**, 12592–12599

163. Liang, N., Mauk, A. G., Pielak, G. J., Johnson, J. A., Smith, M., and Hoffman, B. M. (1988) *Science* **240**, 311–314

164. Pelletier, H., and Kraut, J. (1992) *Science* **258**, 1748–1755

165. Miller, M. A., Vitello, L., and Erman, J. E. (1995) *Biochemistry* **34**, 12048–12058

166. Mei, H., Wang, K., Peffer, N., Weatherly, G., Cohen, D. S., Miller, M., Pielak, G., Durham, B., and Millett, F. (1999) *Biochemistry* **38**, 6846–6854

167. Falzon, L., and Davidson, V. L. (1996) *Biochemistry* **35**, 12111–12118

168. Merli, A., Brodersen, D. E., Morini, B., Chen, Z.-w, Durley, R. C. E., Mathews, F. S., Davidson, V. L., and Rossi, G. L. (1996) *J. Biol. Chem.* **271**, 9177–9180

169. Mutz, M. W., McLendon, G. L., Wishart, J. F., Gaillard, E. R., and Corin, A. F. (1996) *Proc. Natl. Acad. Sci. U.S.A.* **93**, 9521–9526

169a. Castleman, A. W., Jr., Zhong, Q., and Hurley, S. M. (1999) *Proc. Natl. Acad. Sci. U.S.A.* **96**, 4219–4227

170. Daizadeh, I., Medvedev, E. S., and Stuchebrukhov, A. A. (1997) *Proc. Natl. Acad. Sci. U.S.A.* **94**, 3703–3708

171. Langen, R., Chang, I.-J., Germanas, J. P., Richards, J. H., Winkler, J. R., and Gray, H. B. (1995) *Science* **268**, 1733–1735

172. Bishop, G. R., and Davidson, V. L. (1997) *Biochemistry* **36**, 13586–13592

173. Ivkovic-Jensen, M. M., and Kostic, N. M. (1997) *Biochemistry* **36**, 8135–8144

174. Ingraham, L. L. (1966) *Comprehensive Biochemistry* **14**, 424–446

175. Pauling, L. (1948) *The Nature of the Chemical Bond*, 2nd ed., Cornell Univ. Press, Ithaca, New York

176. Smulevich, G., Mantini, A. R., Paoli, M., Coletta, M., and Geraci, G. (1995) *Biochemistry* **34**, 7507–7516

177. Van Dyke, B. R., Saltman, P., and Armstrong, F. A. (1996) *J. Am. Chem. Soc.* **118**, 3490–3492

178. Sugawara, Y., Matsuoka, A., Kaino, A., and Shikama, K. (1995) *Biophys. J.* **69**, 583–592

179. Geibel, J., Chang, C. K., and Traylor, T. G. (1975) *J. Am. Chem. Soc.* **97**, 5924–5926

180. Gerothanassis, I. P., and Momenteau, M. (1987) *J. Am. Chem. Soc.* **109**, 6944–6947

181. Phillips, S. E. V., and Schoenborn, B. P. (1981) *Nature (London)* **292**, 81–84

182. Perutz, M. F. (1989) *Trends Biochem. Sci.* **14**, 42–44

183. Brzozowski, A., Derewenda, Z., Dodson, E., Dodson, G., Grabowski, M., Liddington, R., Skarzynski, T., and Vallely, D. (1984) *Nature (London)* **307**, 74–76

184. McMahon, M. T., deDios, A. C., Godbout, N., Salzmann, R., Laws, D. D., Le, H., Havlin, R. H., and Oldfield, E. (1998) *J. Am. Chem. Soc.* **120**, 4784–4797

185. Collman, J. P., Gagne, R. R., Reed, C. A., Halbert, T. R., Lang, G., and Robinson, W. T. (1975) *J. Am. Chem. Soc.* **97**, 1427–1439

186. Hoard, J. L. (1971) *Science* **174**, 1295–1302

187. Fermi, G., Perutz, M. F., and Shulman, R. G. (1987) *Proc. Natl. Acad. Sci. U.S.A.* **84**, 6167–6168

188. Klotz, I. M., Klippenstein, G. L., and Hendrickson, W. A. (1976) *Science* **192**, 335–344

189. Dou, T., Admiraal, S. J., Ikeda-Saito, M., Krzywda, S., Wilkinson, A. J., Li, T., Olson, J. S., Prince, R. C., Pickering, I. J., and George, G. N. (1995) *J. Biol. Chem.* **270**, 15993–16001

190. Shiro, Y., Iizuka, T., Marubayashi, K., Ogura, T., Kitagawa, T., Balasubramanian, S., and Boxer, S. G. (1994) *Biochemistry* **33**, 14986–14992

191. Kaminaka, S., Takizawa, H., Handa, T., Kihara, H., and Kitagawa, T. (1992) *Biochemistry* **31**, 6997–7002

192. Zhang, J.-H., and Kurtz, D. M., Jr. (1991) *Biochemistry* **30**, 9121–9125

193. Lukat, G. S., Kurtz, D. M., Jr., Shiemke, A. K., Loehr, T. M., and Sanders-Loehr, J. (1984) *Biochemistry* **23**, 6416–6422

194. Stenkamp, R. E., Sieker, L. C., Jensen, L. H., McCallum, J. D., and Sanders-Loehr, J. (1985) *Proc. Natl. Acad. Sci. U.S.A.* **82**, 713–716

194a. Dunford, H. B. (2000) *Heme Peroxidases*, Wiley-VCH, New York

194b. Maté, M. J., Zamocky, M., Nykyri, L. M., Herzog, C., Alzari, P. M., Betzel, C., Koller, F., and Fita, I. (1999) *J. Mol. Biol.* **268**, 135–149

195. Shaffer, J. B., Sutton, R. B., and Bewley, G. C. (1987) *J. Biol. Chem.* **262**, 12908–12911

196. Eaton, J. W., and Ma, M. (1995) in *The Metabolic and Molecular Bases of Inherited Disease*, 7th ed., Vol. 2 (Scriver, C. R., Beaudet, A. L., Sly, W. S., and Valle, D., eds), pp. 2371–2383, McGraw-Hill, New York

197. Bravo, J., Fita, I., Ferrer, J. C., Ens, W., Hillar, A., Switala, J., and Loewen, P. C. (1997) *Protein Sci.* **6**, 1016–1023

198. Fita, I., and Rossmann, M. G. (1985) *J. Mol. Biol.* **185**, 21–37

198a. Putnam, C. D., Arvai, A. S., Bourne, Y., and Tainer, J. A. (2000) *J. Mol. Biol.* **296**, 295–309

199. Vainshtein, B. K., Melik-Adamyan, W. R., Barynin, V. V., Vagin, A. A., and Grebenko, A. I. (1981) *Nature (London)* **293**, 411–412

200. Fita, I., and Rossmann, M. G. (1985) *Proc. Natl. Acad. Sci. U.S.A.* **82**, 1604–1608

201. Kirkman, H. N., Rolfo, M., Ferraris, A. M., and Gaetani, G. F. (1999) *J. Biol. Chem.* **274**, 13908–13914

202. Gouet, P., Jouve, H.-M., and Dideberg, O. (1995) *J. Mol. Biol.* **249**, 933–954

203. Jacob, G. S., and Orme-Johnson, W. H. (1979) *Biochemistry* **18**, 2975–2980

204. Kono, Y., and Fridovich, I. (1983) *J. Biol. Chem.* **258**, 6015–6019

205. Khangulov, S., Sivaraja, M., Barynin, V. V., and Dismukes, G. C. (1993) *Biochemistry* **32**, 4912–4924

206. Holzbaur, I. E., English, A. M., and Ismail, A. A. (1996) *J. Am. Chem. Soc.* **118**, 3354–3359

207. Ryan, O., Smyth, M. R., and Fágáin, C. O. (1994) *Essays in Biochemistry* **28**, 129–146

208. Newmyer, S. L., and Ortiz de Montellano, P. R. (1995) *J. Biol. Chem.* **270**, 19430–19438

208a. Rodríguez-López, J. N., Gilabert, M. A., Tudela, J., Thorneley, R. N. F., and García-Cánovas, F. (2000) *Biochemistry* **39**, 13201–13209

References

209. Rodriguez-Lopez, J. N., Hernández-Ruiz, J., Garcia-Cánovas, F., Thorneley, R. N. F., Acosta, M., and Arnano, M. B. (1997) *J. Biol. Chem.* **272**, 5469–5476

210. Abelskov, A. K., Smith, A. T., Rasmussen, C. B., Dunford, H. B., and Welinder, K. G. (1997) *Biochemistry* **36**, 9453–9463

211. Fukuyama, K., Sato, K., Itakura, H., Takahashi, S., and Hosoya, T. (1997) *J. Biol. Chem.* **272**, 5752–5756

212. Nissum, M., Neri, F., Mandelman, D., Poulos, T. L., and Smulevich, G. (1998) *Biochemistry* **37**, 8080–8087

213. Patterson, W. R., and Poulos, T. L. (1995) *Biochemistry* **34**, 4331–4341

214. Patterson, W. R., Poulos, T. L., and Goodin, D. B. (1995) *Biochemistry* **34**, 4342–4345

215. Finzel, B. C., Poulos, T. L., and Kraut, J. (1984) *J. Biol. Chem.* **259**, 13027–13036

216. Bonagura, C. A., Sundaramoorthy, M., Pappa, H. S., Patterson, W. R., and Poulos, T. L. (1996) *Biochemistry* **35**, 6107–6115

217. Wang, J., Larsen, R. W., Moench, S. J., Satterlee, J. D., Rousseau, D. L., and Ondrias, M. R. (1996) *Biochemistry* **35**, 453–463

218. Sinclair, R., Copeland, B., Yamazaki, I., and Powers, L. (1995) *Biochemistry* **34**, 13176–13182

218a. Doyle, W. A., Blodig, W., Veitch, N. C., Piontek, K., and Smith, A. T. (1998) *Biochemistry* **37**, 15097–15105

218b. Choinowski, T., Blodig, W., Winterhalter, K. H., and Piontek, K. (1999) *J. Mol. Biol.* **286**, 809–827

219. Yamaguchi, Y., Zhang, D.-E., Sun, Z., Albee, E. A., Nagata, S., Tenen, D. G., and Ackerman, S. J. (1994) *J. Biol. Chem.* **269**, 19410–19419

220. Andersson, L. A., Bylkas, S. A., and Wilson, A. E. (1996) *J. Biol. Chem.* **271**, 3406–3412

221. Rae, T. D., and Goff, H. M. (1996) *J. Am. Chem. Soc.* **118**, 2103–2104

222. Kooter, I. M., Moguilevsky, N., Bollen, A., van der Veen, L. A., Otto, C., Dekker, H. L., and Wever, R. (1999) *J. Biol. Chem.* **274**, 26794–26802

222a. Fiedler, T. J., Davey, C. A., and Fenna, R. E. (2000) *J. Biol. Chem.* **275**, 11964–11971

223. Nagano, S., Tanaka, M., Ishimori, K., Watanabe, Y., and Morishima, I. (1996) *Biochemistry* **35**, 14251–14258

224. Tanaka, M., Ishimori, K., and Morishima, I. (1998) *Biochemistry* **37**, 2629–2638

225. Low, D. W., Gray, H. B., and Duus, J. ø. (1997) *J. Am. Chem. Soc.* **119**, 1–5

226. Makino, R., Uno, T., Nishimura, Y., Iuzuka, T., Tsuboi, M., and Ishimura, Y. (1986) *J. Biol. Chem.* **261**, 8376–8382

227. Hashimoto, S., Tatsuno, Y., and Kitagawa, T. (1986) *Proc. Natl. Acad. Sci. U.S.A.* **83**, 2417–2421

228. Nakajima, R., and Yamazaki, I. (1987) *J. Biol. Chem.* **262**, 2576–2581

229. Morishima, I., Takamuki, Y., and Shiro, Y. (1984) *J. Am. Chem. Soc.* **106**, 7666–7672

230. Loew, G., and Dupuis, M. (1996) *J. Am. Chem. Soc.* **118**, 10584–10587

230a. Filizola, M., and Loew, G. H. (2000) *J. Am. Chem. Soc.* **122**, 18–25

230b. Roach, M. P., Ozaki, S.-i, and Watanabe, Y. (2000) *Biochemistry* **39**, 1446–1454

231. Olson, L. P., and Bruice, T. C. (1995) *Biochemistry* **34**, 7335–7347

232. Nakamura, M., Yamazaki, I., Kotani, T., and Ohtaki, S. (1985) *J. Biol. Chem.* **260**, 13546–13552

233. Kobayashi, S., Nakano, M., Kimura, T., and Schaap, A. P. (1987) *Biochemistry* **26**, 5019–5022

234. Ortiz de Montellano, P. R., Choe, Y. S., DePillis, G., and Catalano, C. E. (1987) *J. Biol. Chem.* **262**, 11641–11646

235. Dawson, J. H. (1988) *Science* **240**, 433–439

236. Libby, R. D., Thomas, J. A., Kaiser, L. W., and Hager, L. P. (1982) *J. Biol. Chem.* **257**, 5030–5037

237. Hosten, C. M., Sullivan, A. M., Palaniappan, V., Fitzgerald, M. M., and Terner, J. (1994) *J. Biol. Chem.* **269**, 13966–13978

238. Sono, M., Eble, K. S., Dawson, J. H., and Hager, L. P. (1985) *J. Biol. Chem.* **260**, 15530–15535

239. Sono, M., Dawson, J. H., and Hager, L. P. (1984) *J. Biol. Chem.* **259**, 13209–13216

240. Dawson, J. H., and Sono, M. (1987) *Chem. Rev.* **87**, 1255–1276

241. Geigert, J., Neidleman, S. L., and Dalietos, D. J. (1983) *J. Biol. Chem.* **258**, 2273–2277

242. Mueller, T. J., and Morrison, M. (1974) *J. Biol. Chem.* **259**, 7568–7573

243. Andersson, E., Hellman, L., Gullberg, U., and Olsson, I. (1998) *J. Biol. Chem.* **273**, 4747–4753

244. van Dalen, C. J., Winterbourn, C. C., Senthilmohan, R., and Kettle, A. J. (2000) *J. Biol. Chem.* **275**, 11638–11644

245. Nauseef, W. M., Brigham, S., and Cogley, M. (1994) *J. Biol. Chem.* **269**, 1212–1216

246. Davey, C. A., and Fenna, R. E. (1996) *Biochemistry* **35**, 10967–10973

247. Thomas, E. L., Bozeman, P. M., Jefferson, M. M., and King, C. C. (1995) *J. Biol. Chem.* **270**, 2906–2913

248. Kanofsky, J. R. (1983) *J. Biol. Chem.* **258**, 5991–5993

249. Khan, A. U., Gebauer, P., and Hager, L. P. (1983) *Proc. Natl. Acad. Sci. U.S.A.* **80**, 5195–5197

250. McCormick, M. L., Roeder, T. L., Railsback, M. A., and Britigan, B. E. (1994) *J. Biol. Chem.* **269**, 27914–27919

251. Manthey, J. A., and Hager, L. P. (1981) *J. Biol. Chem.* **256**, 11232–11238

252. Colpas, G. J., Hamstra, B. J., Kampf, J. W., and Pecoraro, V. L. (1996) *J. Am. Chem. Soc.* **118**, 3469–3478

253. Messerschmidt, A., and Wever, R. (1996) *Proc. Natl. Acad. Sci. U.S.A.* **93**, 392–396

254. Mande, S. S., Parsonage, D., Claiborne, A., and Hol, W. G. J. (1995) *Biochemistry* **34**, 6985–6992

255. Crane, E. J., III, Parsonage, D., Poole, L. B., and Claiborne, A. (1995) *Biochemistry* **34**, 14114–14124

255a. Claiborne, A., Yeh, J. I., Mallett, T. C., Luba, J., Crane, E. J., III, Charrier, V., and Parsonage, D. (1999) *Biochemistry* **38**, 15407–15416

256. Miki, K., Renganathan, V., and Gold, M. H. (1986) *Biochemistry* **25**, 4790–4796

257. Sundaramoorthy, M., Kishi, K., Gold, M. H., and Poulos, T. L. (1994) *J. Biol. Chem.* **269**, 32759–32767

258. Mauk, M. R., Kishi, K., Gold, M. H., and Mauk, A. G. (1998) *Biochemistry* **37**, 6767–6771

259. Beinert, H. (1990) *FASEB J.* **4**, 2483–2491

260. Beinert, H., Holm, R. H., and Münck, E. (1997) *Science* **277**, 653–659

261. Carter, C. W., Jr., Kraut, J., Freer, S. T., Xuong, N., Alden, R. A., and Bartsch, R. G. (1974) *J. Biol. Chem.* **249**, 4212–4225

262. Breiter, D. R., Meyer, T. E., Rayment, I., and Holden, H. M. (1991) *J. Biol. Chem.* **266**, 18660–18667

263. Agarwal, A., Li, D., and Cowan, J. A. (1996) *J. Am. Chem. Soc.* **118**, 927–928

264. Banci, L., Bertini, I., Ciurli, S., Ferretti, S., Luchinat, C., and Piccioli, M. (1993) *Biochemistry* **32**, 9387–9397

265. Benning, M. M., Meyer, T. E., Rayment, I., and Holden, H. M. (1994) *Biochemistry* **33**, 2476–2483

266. Soriano, A., Li, D., Bian, S., Agarwal, A., and Cowan, J. A. (1996) *Biochemistry* **35**, 12479–12486

266a. Mulholland, S. E., Gibney, B. R., Rabanal, F., and Dutton, P. L. (1999) *Biochemistry* **38**, 10442–10448

267. Duée, E. D., Fanchon, E., Vicat, J., Sieker, L. C., Meyer, J., and Moulis, J.-M. (1994) *J. Mol. Biol.* **243**, 683–695

268. Dauter, Z., Wilson, K. S., Sieker, L. C., Meyer, J., and Moulis, J.-M. (1997) *Biochemistry* **36**, 16065–16073

269. Jensen, L. H. (1974) *Ann. Rev. Biochem.* **43**, 461–474

270. Séry, A., Housset, D., Serre, L., Bonicel, J., Hatchikian, C., Frey, M., and Roth, M. (1994) *Biochemistry* **33**, 15408–15417

270a. Kyritsis, P., Kümmerle, R., Huber, J. G., Gaillard, J., Guigliarelli, B., Popescu, C., Münck, E., and Moulis, J.-M. (1999) *Biochemistry* **38**, 6335–6345

271. Brereton, P. S., Verhagen, M. F. J. M., Zhou, Z. H., and Adams, M. W. W. (1998) *Biochemistry* **37**, 7351–7362

272. Bruschi, M. H., Guerlesquin, F. A., Bovier-Lapience, G. E., Bonicel, J. J., and Couchoud, P. M. (1985) *J. Biol. Chem.* **260**, 8292–8296

273. Bruice, T. C., Maskiewicz, R., and Job, R. (1975) *Proc. Natl. Acad. Sci. U.S.A.* **72**, 231–234

274. Dauter, Z., Wilson, K. S., Sieker, L. C., Moulis, J.-M., and Meyer, J. (1996) *Proc. Natl. Acad. Sci. U.S.A.* **93**, 8836–8840

275. Xiao, Z., Lavery, M. J., Ayhan, M., Scrofani, S. D. B., Wilce, M. C. J., Guss, J. M., Tregloan, P. A., George, G. N., and Wedd, A. G. (1998) *J. Am. Chem. Soc.* **120**, 4135–4150

276. Fukuyama, K., Hase, T., Matsumoto, S., Tsukihara, T., Katsube, Y., Tanaka, N., Kakudo, M., Wada, K., and Matsubara, H. (1980) *Nature (London)* **286**, 522–524

277. Kok, M., Oldenhuis, R., van der Linden, M. P. G., Meulenberg, C. H. C., Kingma, J., and Witholt, B. (1989) *J. Biol. Chem.* **264**, 5442–5451

278. Archer, M., Huber, R., Tavares, P., Moura, I., Moura, J. J. G., Carrondo, M. A., Sieker, L. C., LeGall, J., and Romao, M. J. (1995) *J. Mol. Biol.* **251**, 690–702

279. Hurley, J. K., Weber-Main, A. M., Stankovich, M. T., Benning, M. M., Thoden, J. B., Vanhooke, J. L., Holden, H. M., Chae, Y. K., Xia, B., Cheng, H., Markley, J. L., Martinez-Júlvez, M., Gómez-Moreno, C., Schmeits, J. L., and Tollin, G. (1997) *Biochemistry* **36**, 11100–11117

279a. Müller, J. J., Müller, A., Rottmann, M., Bernhardt, R., and Heinemann, U. (1999) *J. Mol. Biol.* **294**, 501–513

280. Hurley, J. K., Weber-Main, A. M., Hodges, A. E., Stankovich, M. T., Benning, M. M., Holden, H. M., Cheng, H., Xia, B., Markley, J. L., Genzor, C., Gomez-Moreno, C., Hafezi, R., and Tollin, G. (1997) *Biochemistry* **36**, 15109–15117

280a. Morales, R., Chron, M.-H., Hudry-Clergeon, G., Pétillot, Y., Norager, S., Medina, M., and Frey, M. (1999) *Biochemistry* **38**, 15764–15773

281. Ta, D. T., and Vickery, L. E. (1992) *J. Biol. Chem.* **267**, 11120–11125

282. Xia, B., Cheng, H., Skjeldal, L., Coghlan, V. M., Vickery, L. E., and Markley, J. L. (1995) *Biochemistry* **34**, 180–187

282a. Uhlmann, H., and Bernhardt, R. (1995) *J. Biol. Chem.* **270**, 29959–29966

282b. Ziegler, G. A., and Schulz, G. E. (2000) *Biochemistry* **39**, 10986–10995

283. Meyer, J., Fujinaga, J., Gaillard, J., and Lutz, M. (1994) *Biochemistry* **33**, 13642–13650

284. Pochapsky, T. C., and Ye, X. M. (1991) *Biochemistry* **30**, 3850–3856

285. Stout, C. D. (1988) *J. Biol. Chem.* **263**, 9256–9260

286. Duff, J. L. C., Breton, J. L. J., Butt, J. N., Armstrong, F. A., and Thomson, A. J. (1996) *J. Am. Chem. Soc.* **118**, 8593–8603

References

287. Schipke, C. G., Goodin, D. B., McRee, D. E., and Stout, C. D. (1999) *Biochemistry* **38**, 8228–8239

288. George, G. N., and George, S. J. (1988) *Trends Biochem. Sci.* **13**, 369–370

289. Kent, T. A., Emptage, M. H., Merkle, H., Kennedy, M. C., Beinert, H., and Münck, E. (1985) *J. Biol. Chem.* **260**, 6871–6881

290. Tong, J., and Feinberg, B. A. (1994) *J. Biol. Chem.* **269**, 24920–24927

291. Lane, R. W., Ibers, J. A., Frankel, R. B., and Holm, R. H. (1975) *Proc. Natl. Acad. Sci. U.S.A.* **72**, 2868–2872

291a. Camba, R., and Armstrong, F. A. (2000) *Biochemistry* **39**, 10587–10598

292. Gardner, P. R., and Fridovich, I. (1991) *J. Biol. Chem.* **266**, 19328–19333

293. Kennedy, M. C., Antholine, W. E., and Beinert, H. (1997) *J. Biol. Chem.* **272**, 20340–20347

294. Duin, E. C., Lafferty, M. E., Crouse, B. R., Allen, R. M., Sanyal, I., Flint, D. H., and Johnson, M. K. (1997) *Biochemistry* **36**, 11811–11820

295. Magliozzo, R. S., McIntosh, B. A., and Sweeney, W. V. (1982) *J. Biol. Chem.* **257**, 3506–3509

296. Prince, R. C., and Adams, M. W. W. (1987) *J. Biol. Chem.* **262**, 5125–5128

297. Orme-Johnson, W. H. (1972) *Ann. Rev. Biochem.* **42**, 159–204

298. Orme-Johnson, W. H., Hansen, R. E., Beinert, H., Tsibris, J. C. M., Bartholomaus, R. C., and Gunsalus, I. C. (1968) *Proc. Natl. Acad. Sci. U.S.A.* **60**, 368–372

299. Scrofani, S. D. B., Brereton, P. S., Hamer, A. M., Lavery, M. J., McDowall, S. G., Vincent, G. A., Brownlee, R. T. C., Hoogenraad, N. J., Sadek, M., and Wedd, A. G. (1994) *Biochemistry* **33**, 14486–14495

300. Holz, R. C., Small, F. J., and Ensign, S. A. (1997) *J. Biol. Chem.* **36**, 14690–14696

301. Mouesca, J.-M., Chen, J. L., Noodleman, L., Bashford, D., and Case, D. A. (1994) *J. Am. Chem. Soc.* **116**, 11898–11914

302. Swartz, P. D., Beck, B. W., and Ichiye, T. (1996) *Biophys. J.* **71**, 2958–2969

303. Kemper, M. A., Stout, C. D., Lloyd, S. E. J., Prasad, G. S., Fawcett, S., Armstrong, F. A., Shen, B., and Burgess, B. K. (1997) *J. Biol. Chem.* **272**, 15620–15627

304. Bominaar, E. L., Hu, Z., Münck, E., Girerd, J.-J., and Borshch, S. A. (1995) *J. Am. Chem. Soc.* **117**, 6976–6989

305. Scott, M. P., and Biggins, J. (1997) *Protein Sci.* **6**, 340–346

306. Coldren, C. D., Hellinga, H. W., and Caradonna, J. P. (1997) *Proc. Natl. Acad. Sci. U.S.A.* **94**, 6635–6640

307. Rieske, J. S., MacLennan, D. H., and Coleman, R. (1964) *Biochem. Biophys. Res. Commun.* **15**, 338–344

308. Beckmann, J. D., Ljungdahl, P. O., Lopez, J. L., and Trumpower, B. L. (1987) *J. Biol. Chem.* **262**, 8901–8909

309. Brasseur, G., Sled, V., Liebl, U., Ohnishi, T., and Daldal, F. (1997) *Biochemistry* **36**, 11685–11696

310. Mosser, G., Breyton, C., Olofsson, A., Popot, J.-L., and Rigaud, J.-L. (1997) *J. Biol. Chem.* **272**, 20263–20268

311. Liebl, U., Sled, V., Brasseur, G., Ohnishi, T., and Daldal, F. (1997) *Biochemistry* **36**, 11675–11684

312. Crane, B. R., Siegel, L. M., and Getzoff, E. D. (1995) *Science* **270**, 59–67

313. Wilkerson, J. O., Janick, P. A., and Siegel, L. M. (1983) *Biochemistry* **22**, 5048–5054

314. Crane, B. R., Siegel, L. M., and Getzoff, E. D. (1997) *Biochemistry* **36**, 12120–12137

315. Covès, J., Zeghouf, M., Macherel, D., Guigliarelli, B., Asso, M., and Fontecave, M. (1997) *Biochemistry* **36**, 5921–5928

316. Crane, B. R., Siegel, L. M., and Getzoff, E. D. (1997) *Biochemistry* **36**, 12101–12119

316a. Peters, J. W., Lanzilotta, W. N., Lemon, B. J., and Seefeldt, L. C. (1998) *Science* **282**, 1853–1858

316b. Popescu, C. V., and Münck, E. (1999) *J. Am. Chem. Soc.* **121**, 7877–7884

316c. Nicolet, Y., Lemon, B. J., Fontecilla-Camps, J. C., and Peters, J. W. (2000) *Trends Biochem. Sci.* **25**, 138–143

317. Grabowski, R., Hofmeister, A. E. M., and Buckel, W. (1993) *Trends Biochem. Sci.* **18**, 297–300

318. Hofmeister, A. E. M., Grabowski, R., Linder, D., and Buckel, W. (1993) *Eur. J. Biochem.* **215**, 341–349

319. Hofmeister, A. E. M., Berger, S., and Buckel, W. (1992) *Eur. J. Biochem.* **205**, 743–749

320. Flint, D. H., and Emptage, M. H. (1988) *J. Biol. Chem.* **263**, 3558–3564

321. Flint, D. H., Tuminello, J. F., and Miller, T. J. (1996) *J. Biol. Chem.* **271**, 16053–16067

322. Hofmeister, A. E. M., and Buckel, W. (1992) *Eur. J. Biochem.* **206**, 547–552

323. Müh, U., Cinkaya, I., Albracht, S. P. J., and Buckel, W. (1996) *Biochemistry* **35**, 11710–11718

324. Scherf, U., Söhling, B., Gottschalk, G., Linder, D., and Buckel, W. (1994) *Arch. Microbiol.* **161**, 239–245

325. Klees, A.-G., Linder, D., and Buckel, W. (1992) *Arch. Microbiol.* **158**, 294–301

326. Vollmer, S. J., Switzer, R. L., and Debrunner, P. G. (1983) *J. Biol. Chem.* **258**, 14284–14293

327. Ramsay, R. R., Dreyer, J.-L., Schloss, J. V., Jackson, R. H., Coles, C. J., Beinert, H., Cleland, W. W., and Singer, T. P. (1981) *Biochemistry* **20**, 7476–7482

328. Gaudu, P., and Weiss, B. (1996) *Proc. R. Soc. (London)* **93**, 10094–10098

329. Ding, H., Hidalgo, E., and Demple, B. (1996) *J. Biol. Chem.* **271**, 33173–33175

330. Hentze, M. W. (1996) *Trends Biochem. Sci.* **21**, 282–283

331. Feig, A. L., Masschelein, A., Bakac, A., and Lippard, S. J. (1997) *J. Am. Chem. Soc.* **119**, 334–342

332. Pulver, S. C., Froland, W. A., Lipscomb, J. D., and Solomon, E. I. (1997) *J. Am. Chem. Soc.* **119**, 387–395

333. Lindqvist, Y., Huang, W., Schneider, G., and Shanklin, J. (1996) *EMBO J.* **15**, 4081–4092

334. Que, L., Jr. (1991) *Science* **253**, 273–274

335. Doi, K., Gupta, R., and Aisen, P. (1987) *J. Biol. Chem.* **262**, 6982–6985

336. Hayman, A. R., and Cox, T. M. (1994) *J. Biol. Chem.* **269**, 1294–1300

336a. Lindqvist, Y., Johansson, E., Kaija, H., Vihko, P., and Schneider, G. (1999) *J. Mol. Biol.* **291**, 135–147

336b. Uppenberg, J., Lindqvist, F., Svensson, C., Ek-Rylander, B., and Andersson, G. (1999) *J. Mol. Biol.* **290**, 201–211

337. Wang, D. L., Holz, R. C., David, S. S., Que, L., Jr., and Stankovich, M. T. (1991) *Biochemistry* **30**, 8187–8194

338. Cohen, S. S. (1984) *Trends Biochem. Sci.* **9**, 334–336

339. Merkx, M., and Averill, B. A. (1998) *Biochemistry* **37**, 8490–8497

340. Klabunde, T., Sträter, N., Fröhlich, R., Witzel, H., and Krebs, B. (1996) *J. Mol. Biol.* **259**, 737–748

341. Sträter, N., Klabunde, T., Tucker, P., Witzel, H., and Krebs, B. (1995) *Science* **268**, 1489–1492

342. Baumbach, G. A., Ketcham, C. M., Richardson, D. E., Bazer, F. W., and Roberts, R. M. (1986) *J. Biol. Chem.* **261**, 12869–12878

343. Bender, C. J., Rosenzweig, A. C., Lippard, S. J., and Peisach, J. (1994) *J. Biol. Chem.* **269**, 15993–15998

344. Nesheim, J. C., and Lipscomb, J. D. (1996) *Biochemistry* **35**, 10240–10247

345. Shu, L., Nesheim, J. C., Kauffmann, K., Münck, E., Lipscomb, J. D., and Que, L., Jr. (1997) *Science* **275**, 515–518

346. Pikus, J. D., Studts, J. M., McClay, K., Steffan, R. J., and Fox, B. G. (1997) *Biochemistry* **36**, 9283–9289

347. Herold, S., and Lippard, S. J. (1997) *J. Am. Chem. Soc.* **119**, 145–156

348. Jordan, A., and Reichard, P. (1998) *Ann. Rev. Biochem.* **67**, 71–98

349. Stubbe, J. (1990) *J. Biol. Chem.* **265**, 5329–5332

350. Reichard, P. (1997) *Trends Biochem. Sci.* **22**, 81–85

351. Stubbe, J. (1998) *Proc. Natl. Acad. Sci. U.S.A.* **95**, 2723–2724

351a. Eriksson, M., Jordan, A., and Eklund, H. (1998) *Biochemistry* **37**, 13359–13369

352. Sahlin, M., Petersson, L., Gräslund, A., Ehrenberg, A., Sjöberg, B.-M., and Thelander, L. (1987) *Biochemistry* **26**, 5541–5548

353. Uhlin, U., and Eklund, H. (1994) *Nature (London)* **370**, 533–539

354. Uhlin, U., and Eklund, H. (1996) *J. Mol. Biol.* **262**, 358–369

354a. Parkin, S. E., Chen, S., Ley, B. A., Mangravite, L., Edmondson, D. E., Huynh, B. H., and Bollinger, J. M., Jr. (1998) *Biochemistry* **37**, 1124–1130

354b. Rova, U., Adrait, A., Pötsch, S., Gräslund, A., and Thelander, L. (1999) *J. Biol. Chem.* **274**, 23746–23751

354c. Andersson, M. E., Högbom, M., Rinaldo-Matthis, A., Andersson, K. K., Sjöberg, B.-M., and Nordlund, P. (1999) *J. Am. Chem. Soc.* **121**, 2346–2352

355. Fieschi, F., Torrents, E., Toulokhonova, L., Jordan, A., Hellman, U., Barbe, J., Gibert, I., Karlsson, M., and Sjöberg, B.-M. (1998) *J. Biol. Chem.* **273**, 4329–4337

356. Ling, J., Sahlin, M., Sjöberg, B.-M., Loehr, T. M., and Sanders-Loehr, J. (1994) *J. Biol. Chem.* **269**, 5595–5601

357. Silva, K. E., Elgren, T. E., Que, L., Jr., and Stankovich, M. T. (1995) *Biochemistry* **34**, 14093–14103

358. Sturgeon, B. E., Burdi, D., Chen, S., Huynh, B.-H., Edmondson, D. E., Stubbe, J., and Hoffman, B. M. (1996) *J. Am. Chem. Soc.* **118**, 7551–7557

359. Kauppi, B., Nielsen, B. B., Ramaswamy, S., Larson, I. K., Thelander, M., Thelander, L., and Eklund, H. (1996) *J. Mol. Biol.* **262**, 706–720

360. Katterle, B., Sahlin, M., Schmidt, P. P., Pötsch, S., Logan, D. T., Gräslund, A., and Sjöberg, B.-M. (1997) *J. Biol. Chem.* **272**, 10414–10421

361. Sealy, R. C., Harman, L., West, P. R., and Mason, R. P. (1985) *J. Am. Chem. Soc.* **107**, 3401–3406

362. Sjöberg, B.-M., Karlsson, M., and Jörnvall, H. (1987) *J. Biol. Chem.* **262**, 9736–9743

363. Ekberg, M., Sahlin, M., Eriksson, M., and Sjöberg, B.-M. (1996) *J. Biol. Chem.* **271**, 20655–20659

364. Ong, S. P., McFarlan, S. C., and Hogenkamp, H. P. C. (1993) *Biochemistry* **32**, 11397–11404

365. Gerfen, G. J., Licht, S., Willems, J.-P., Hoffman, B. M., and Stubbe, J. (1996) *J. Am. Chem. Soc.* **118**, 8192–8197

366. Reichard, P. (1993) *J. Biol. Chem.* **268**, 8383–8386

367. Sun, X., Eliasson, R., Pontis, E., Andersson, J., Buist, G., Sjöberg, B.-M., and Reichard, P. (1995) *J. Biol. Chem.* **270**, 2443–2446

368. Sun, X., Ollagnier, S., Schmidt, P. P., Atta, M., Mulliez, E., Lepape, L., Eliasson, R., Gräslund, A., Fontecave, M., Reichard, P., and Sjöberg, B.-M. (1996) *J. Biol. Chem.* **271**, 6827–6831

369. Ollagnier, S., Mulliez, E., Schmidt, P. P., Eliasson, R., Gaillard, J., Deronzier, C., Bergman, T., Gräslund, A., Reichard, P., and Fontecave, M. (1997) *J. Biol. Chem.* **272**, 24216–24223

369a. Logan, D. T., Andersson, J., Sjöberg, B.-M., and Nordlund, P. (1999) *Science* **283**, 1499–1504

370. Young, P., Andersson, J., Sahlin, M., and Sjöberg, B.-M. (1996) *J. Biol. Chem.* **271**, 20770–20775

371. van der Donk, W. A., Stubbe, J., Gerfen, G. J., Bellew, B. F., and Griffin, R. G. (1995) *J. Am. Chem. Soc.* **117**, 8908–8916

371a. Lawrence, C. C., Bennati, M., Obias, H. V., Bar, G., Griffin, R. G., and Stubbe, J. (1999) *Proc. Natl. Acad. Sci. U.S.A.* **96**, 8979–8984

372. Covès, J., Le Hir de Fallois, L., Le Pape, L., Décout, J.-L., and Fontecave, M. (1996) *Biochemistry* **35**, 8595–8602

373. Lenz, R., and Giese, B. (1997) *J. Am. Chem. Soc.* **119**, 2784–2794

374. Trant, N. L., Meshnick, S. R., Kitchener, K., Eaton, J. W., and Cerami, A. (1983) *J. Biol. Chem.* **258**, 125–130

375. Cooper, J. B., McIntyre, K., Badasso, M. O., Wood, S. P., Zhang, Y., Garbe, T. R., and Young, D. (1995) *J. Mol. Biol.* **246**, 531–544

376. Lah, M. S., Dixon, M. M., Pattridge, K. A., Stallings, W. C., Fee, J. A., and Ludwig, M. L. (1995) *Biochemistry* **34**, 1646–1660

377. Stallings, W. C., Pattridge, K. A., Strong, R. K., and Ludwig, M. L. (1985) *J. Biol. Chem.* **260**, 16424–16432

378. Sorkin, D. L., Duong, D. K., and Miller, A.-F. (1997) *Biochemistry* **36**, 8202–8208

378a. Ursby, T., Adinolfi, B. S., Al-Karadaghi, S., De Vendittis, E., and Bocchini, V. (1999) *J. Mol. Biol.* **286**, 189–205

379. Carlioz, A., Ludwig, M. L., Stallings, W. C., Fee, J. A., Steinman, H. M., and Touati, D. (1988) *J. Biol. Chem.* **263**, 1555–1562

380. Barker, H. A. (1972) in *The Enzymes*, 3rd ed., Vol. 6 (Boyer, P. D., ed), pp. 509–537, Academic Press, New York ·

381. Barker, H. A. (1976) in *Relections on Biochemistry* (Kornberg, A., ed), pp. 95–, Pergamon Press, New York

382. Lenhert, P. G., and Hodgkin, D. C. (1961) *Nature (London)* **192**, 937–938

383. Lenhert, P. G., and Hodgkin, D. C. (1961) *Nature (London)* **192**, 937–938

384. Zagalak, B., and Friedrich, W., eds. (1979) *Vitamin B12*, de Gruyter, Berlin

385. Dolphin, D., ed. (1982) *B12*, Wiley, New York (2 vols.)

386. Brown, K. L. (1987) *J. Am. Chem. Soc.* **109**, 2277–2284

387. Anton, D. L., Hogenkamp, H. P. C., Walker, T. E., and Matwiyoff, N. A. (1982) *Biochemistry* **21**, 2372–2378

388. Wagner, F. (1966) *Ann. Rev. Biochem.* **35**, 405–428

389. Schrauzer, G. N., Deutsch, E., and Windgassen, R. J. (1968) *J. Am. Chem. Soc.* **90**, 2441–2442

390. Needham, T. E., Matwiyoff, N. A., Walker, T. E., and Hogenkamp, H. P. C. (1973) *J. Am. Chem. Soc.* **95**, 5019–5024

391. Parry, R. J., Ostrander, J. M., and Arzu, I. Y. (1985) *J. Am. Chem. Soc.* **107**, 2190–2191

391a. Walker, L. A., II, Shiang, J. J., Anderson, N. A., Pullen, S. H., and Sension, R. J. (1998) *J. Am. Chem. Soc.* **120**, 7286–7292

392. Hogenkamp, H. P. C. (1982) in *B₁₂*, Vol. I (Dolphin, D., ed), pp. 295–323, Wiley, New York

393. Hay, B. P., and Finke, R. G. (1986) *J. Am. Chem. Soc.* **108**, 4820–4829

394. Ashley, G. W., Harris, G., and Stubbe, J. (1986) *J. Biol. Chem.* **261**, 3958–3964

395. Sando, G. N., Blakely, R. L., Hogenkamp, H. P. C., and Hoffman, P. J. (1975) *J. Biol. Chem.* **250**, 8774–8779

396. Thelander, L., and Reichard, P. (1979) *Ann. Rev. Biochem.* **48**, 133–158

396a. Brown, K. L., and Li, J. (1998) *J. Am. Chem. Soc.* **120**, 9466–9474

396b. Licht, S. S., Lawrence, C. C., and Stubbe, J. (1999) *J. Am. Chem. Soc.* **121**, 7463–7468

397. Pratt, J. M. (1982) in *B₁₂*, Vol. I (Dolphin, D., ed), pp. 325–392, Wiley, New York

397a. Zerbe-Burkhardt, K., Ratnatilleke, A., Philippon, N., Birch, A., Leiser, A., Vrijbloed, J. W., Hess, D., Hunziker, P., and Robinson, J. A. (1998) *J. Biol. Chem.* **273**, 6508–6517

397b. Ratnatilleke, A., Vrijbloed, J. W., and Robinson, J. A. (1999) *J. Biol. Chem.* **274**, 31679–31685

398. Retéy, J., Umani-Ronchi, A., Seibl, J., and Arigoni, D. (1966) *Experientia* **22**, 502–503

399. Sato, K., Orr, J. C., Babior, B. M., and Abeles, R. H. (1976) *J. Biol. Chem.* **251**, 3734–3737

400. Wollowitz, S., and Halpern, J. (1984) *J. Am. Chem. Soc.* **106**, 8319–8321

401. Graves, S. W., Krouwer, J. S., and Babior, B. M. (1980) *J. Biol. Chem.* **255**, 7444–7448

401a. Smith, D. M., Golding, B. T., and Radom, L. (1999) *J. Am. Chem. Soc.* **121**, 5700–5704

401b. Ke, S.-C., and Warncke, K. (1999) *J. Am. Chem. Soc.* **121**, 9922–9927

401c. Warncke, K., Schmidt, J. C., and Ke, S.-C. (1999) *J. Am. Chem. Soc.* **121**, 10522–10528

402. Bothe, H., Darley, D. J., Albracht, S. P. J., Gerfen, G. J., Golding, B. T., and Buckel, W. (1998) *Biochemistry* **37**, 4105–4113

402a. Chih, H.-W., and Marsh, E. N. G. (1999) *Biochemistry* **38**, 13684–13691

402b. Roymoulik, I., Chen, H.-P., and Marsh, E. N. G. (1999) *J. Biol. Chem.* **274**, 11619–11622

403. Lowe, J. N., and Ingraham, L. L. (1971) *J. Am. Chem. Soc.* **93**, 3801–3802

404. He, M., and Dowd, P. (1998) *J. Am. Chem. Soc.* **120**, 1133–1137

405. Mancia, F., Keep, N. H., Nakagawa, A., Leadlay, P. F., McSweeney, S., Rasmussen, B., Bösecke, P., Diat, O., and Evans, P. R. (1996) *Structure* **4**, 339–350

405a. Mancia, F., Smith, G. A., and Evans, P. R. (1999) *Biochemistry* **38**, 7999–8005

405b. Thomä, N. H., Meier, T. W., Evans, P. R., and Leadlay, P. F. (1998) *Biochemistry* **37**, 14386–14393

405c. Smith, D. M., Golding, B. T., and Radom, L. (1999) *J. Am. Chem. Soc.* **121**, 9388–9399

405d. Maiti, N., Widjaja, L., and Banerjee, R. (1999) *J. Biol. Chem.* **274**, 32733–32737

406. Reitzer, R., Gruber, K., Jogl, G., Wagner, V. G., Bothe, H., Buckel, W., and Kratky, C. (1999) *Structure* **7**, 891–902

406a. Champloy, F., Jogl, G., Reitzer, R., Buckel, W., Bothe, H., Beatrix, B., Broeker, G., Michalowicz, A., Meyer-Klaucke, W., and Kratky, C. (1999) *J. Am. Chem. Soc.* **121**, 11780–11789

407. Stadtman, T. C. (1971) *Science* **171**, 859–867

408. Zagalak, B., Frey, P. A., Karabatsos, G. L., and Abeles, R. H. (1966) *J. Biol. Chem.* **241**, 3028–3035

409. Sprecher, M., Clark, M. J., and Sprinson, D. B. (1966) *J. Biol. Chem.* **241**, 872–877

410. Baker, J. R., and Stadtman, T. C. (1982) in *B₁₂*, Vol. II (Dolphin, D., ed), pp. 203–232, Wiley, New York

411. Kunz, F., Retéy, J., Arigoni, D., Tsai, L., and Stadtman, T. C. (1978) *Helv. Chim. Acta* **61**, 1139–1145

411a. Chang, C. H., and Frey, P. A. (2000) *J. Biol. Chem.* **275**, 106–114

412. Poston, J. M. (1977) *Science* **195**, 301–302

413. Stabler, S. P., Lindenbaum, J., and Allen, R. H. (1988) *J. Biol. Chem.* **263**, 5581–5588

414. Ballinger, M. D., Frey, P. A., Reed, G. H., and LoBrutto, R. (1995) *Biochemistry* **34**, 10086–10093

415. Wu, W., Lieder, K. W., Reed, G. H., and Frey, P. A. (1995) *Biochemistry* **34**, 10532–10537

415a. Wu, W., Booker, S., Lieder, K. W., Bandarian, V., Reed, G. H., and Frey, P. A. (2000) *Biochemistry* **39**, 9561–9570

416. Zhou, Z. S., Peariso, K., Penner-Hahn, J. E., and Matthews, R. G. (1999) *Biochemistry* **38**, 15915–15926

417. Yamanishi, M., Yamada, S., Muguruma, H., Murakami, Y., Tobimatsu, T., Ishida, A., Yamauchi, J., and Toraya, T. (1998) *Biochemistry* **37**, 4799–4803

418. Hall, D. A., Jordan-Starck, T. C., Loo, R. O., Ludwig, M. L., and Matthews, R. G. (2000) *Biochemistry* **39**, 10711–10719

418a. Jarrett, J. T., Amaratunga, M., Drennan, C. L., Scholten, J. D., Sands, R. H., Ludwig, M. L., and Matthews, R. G. (1996) *Biochemistry* **35**, 2464–2475

419. Zydowsky, T. M., Courtney, L. F., Frasca, V., Kobayashi, K., Shimizu, H., Yuen, L., Matthews, R. G., Benkovic, S. J., and Floss, H. G. (1986) *J. Am. Chem. Soc.* **108**, 3152–3153

420. Wood, J. M. (1982) in *B₁₂*, Vol. II (Dolphin, D., ed), pp. 151–164, Wiley, New York

421. Wood, J. M. (1974) *Science* **183**, 1049–1052

422. McBride, B. C., and Wolfe, R. S. (1971) *Biochemistry* **10**, 4312–4317

423. Parker, D. J., Wood, H. G., Ghambeer, R. K., and Ljungdahl, L. G. (1972) *Biochemistry* **11**, 3074–3080

424. Thauer, R. K., Diekert, G., and Schonheit, P. (1980) *Trends Biochem. Sci.* **5**, 304–306

425. Nielson, F. H. (1974) in *Trace Element Metabolism in Animals-2* (Hoekstra, W. G., Suttie, J. W., Ganther, H. E., and Mertz, W., eds), pp. 381–395, Univ. Park Press, Baltimore, Maryland

426. Schnegg, A., and Kirchgessner, M. (1976) *Int. J. Vitamins Nutr. Res.* **46**, 96–99

427. Nielsen, F. H. (1991) *FASEB J.* **5**, 2661–2667

428. Patel, S. U., Sadler, P. J., Tucker, A., and Viles, J. H. (1993) *J. Am. Chem. Soc.* **115**, 9285–9286

429. Nomoto, S., McNeely, M. D., and Sunderman, F. W., Jr. (1971) *Biochemistry* **10**, 1647–1651

430. Severne, B. C. (1974) *Nature (London)* **248**, 807–808

431. Eskew, D. L., Welch, R. M., and Cary, E. E. (1983) *Science* **222**, 621–623

432. Eitinger, T., and Friedrich, B. (1991) *J. Biol. Chem.* **266**, 3222–3227

432a. Eitinger, T., Degen, O., Böhnke, U., and Müller, M. (2000) *J. Biol. Chem.* **275**, 18029–18033

433. Frieden, E., ed. (1984) *Biochemistry of the Essential Ultratrace Elements*, Plenum, New York

434. Walsh, C. T., and Orme-Johnson, W. H. (1987) *Biochemistry* **26**, 4901–4906

435. Todd, M. J., and Hausinger, R. P. (1987) *J. Biol. Chem.* **262**, 5963–5967

436. Jabri, E., Carr, M. B., Hausinger, R. P., and Karplus, P. A. (1995) *Science* **268**, 998–1003

437. Lippard, S. J. (1995) *Science* **268**, 996–997

437a. Barrios, A. M., and Lippard, S. J. (1999) *J. Am. Chem. Soc.* **121**, 11751–11757

438. Park, I.-S., and Hausinger, R. P. (1995) *Science* **267**, 1156–1158

References

439. Lee, M. H., Pankratz, H. S., Wang, S., Scott, R. A., Finnegan, M. G., Johnson, M. K., Ippolito, J. A., Christianson, D. W., and Hausinger, R. P. (1993) *Protein Sci.* **2**, 1042–1052

439a. Doolittle, R. F. (1997) *Nature (London)* **388**, 515–516

440. Winkler, R. G., Blevins, D. G., Polacco, J. C., and Randall, D. D. (1988) *Trends Biochem. Sci.* **13**, 97–100

441. Mayhew, S. G., and O'Connor, M. E. (1982) *Trends Biochem. Sci.* **7**, 18–21

442. Elsden, S. R. (1981) *Trends Biochem. Sci.* **6**, 252–253

443. Fu, W., Drozdzewski, P. M., Morgan, T. V., Mortenson, L. E., Juszczak, A., Adams, M. W. W., He, S.-H., Peck, H. D., Jr., DerVartanian, D. V., LeGall, J., and Johnson, M. K. (1993) *Biochemistry* **32**, 4813–4819

444. Bagley, K. A., Van Garderen, C. J., Chen, M., Duin, E. C., Albracht, S. P. J., and Woodruff, W. H. (1994) *Biochemistry* **33**, 9229–9236

445. He, S. H., Teixeira, M., LeGall, J., Patil, D. S., Moura, I., Moura, J. J. G., DerVartanian, D. V., Huynh, B. H., and Peck, H. D., Jr. (1989) *J. Biol. Chem.* **264**, 2678–2682

446. Sorgenfrei, O., Duin, E. C., Klein, A., and Albracht, S. P. J. (1996) *J. Biol. Chem.* **271**, 23799–23806

447. Arp, D. J., and Burris, R. H. (1981) *Biochemistry* **20**, 2234–2240

448. Seefeldt, L. C., and Arp, D. J. (1989) *Biochemistry* **28**, 1588–1596

449. Telser, J., Benecky, M. J., Adams, M. W. W., Mortensen, L. E., and Hoffman, B. M. (1986) *J. Biol. Chem.* **261**, 13536–13541

450. Volbeda, A., Charon, M.-H., Piras, C., Hatchikian, E. C., Frey, M., and Fontecilla-Camps, J. C. (1995) *Nature (London)* **373**, 580–587

451. Cammack, R. (1995) *Nature (London)* **373**, 556–557

452. Dole, F., Fournel, A., Magro, V., Hatchikian, E. C., Bertrand, P., and Guigliarelli, B. (1997) *Biochemistry* **36**, 7847–7854

453. Volbeda, A., Garcin, E., Piras, C., de Lacey, A. L., Fernandez, V. M., Hatchikian, E. C., Frey, M., and Fontecilla-Camps, J. C. (1996) *J. Am. Chem. Soc.* **118**, 12989–12996

453b. Davidson, G., Choudhury, S. B., Gu, Z., Bose, K., Roseboom, W., Albracht, S. P. J., and Maroney, M. J. (2000) *Biochemistry* **39**, 7468–7479

453a. Amara, P., Volbeda, A., Fontecilla-Camps, J. C., and Field, M. J. (1999) *J. Am. Chem. Soc.* **121**, 4468–4477

454. Tau, S. L., Fox, J. A., Kojima, N., Walsh, C. T., and Orme-Johnson, W. H. (1984) *J. Am. Chem. Soc.* **106**, 3064–3066

455. Won, H., Olson, K. D., Wolfe, R. S., and Summers, M. F. (1990) *J. Am. Chem. Soc.* **112**, 2178–2184

456. Telser, J., Fann, Y.-C., Renner, M. W., Fajer, J., Wang, S., Zhang, H., Scott, R. A., and Hoffman, B. M. (1997) *J. Am. Chem. Soc.* **119**, 733–743

456a. Telser, J., Horng, Y.-C., Becker, D. F., Hoffman, B. M., and Ragsdale, S. W. (2000) *J. Am. Chem. Soc.* **122**, 182–183

457. Varadarajan, R., and Richards, F. M. (1992) *Biochemistry* **31**, 12315–12327

458. Ermler, U., Grabarse, W., Shima, S., Goubeaud, M., and Thauer, R. K. (1997) *Science* **278**, 1457–1462

458a. Selmer, T., Kahnt, J., Goubeaud, M., Shima, S., Grabarse, W., Ermler, U., and Thauer, R. K. (2000) *J. Biol. Chem.* **275**, 3755–3760

458b. Grabarse, W., Mahlert, F., Shima, S., Thauer, R. K., and Ermler, U. (2000) *J. Mol. Biol.* **303**, 329–344

459. Pfaltz, A., Jaun, B., Fässler, A., Eschenmoser, A., Jaenchen, R., Gilles, H. H., Diekert, G., and Thauer, R. K. (1982) *Helv. Chim. Acta* **65**, 828–865

460. Sings, H. L., Bible, K. C., and Rinehart, K. L. (1996) *Proc. Natl. Acad. Sci. U.S.A.* **93**, 10560–10565

461. Ensign, S. A. (1995) *Biochemistry* **34**, 5372–5381

462. Xia, J., Sinclair, J. F., Baldwin, T. O., and Lindahl, P. A. (1996) *Biochemistry* **35**, 1965–1971

463. Seravalli, J., Kumar, M., Lu, W.-P., and Ragsdale, S. W. (1997) *Biochemistry* **36**, 11241–11251

464. Xia, J., Hu, Z., Popescu, C. V., Lindahl, P. A., and Münck, E. (1997) *J. Am. Chem. Soc.* **119**, 8301–8312

465. Anderson, M. E., and Lindahl, P. A. (1996) *Biochemistry* **35**, 8371–8380

465a. Fraser, D. M., and Lindahl, P. A. (1999) *Biochemistry* **38**, 15706–15711

466. Grahame, D. A., and DeMoll, E. (1996) *J. Biol. Chem.* **271**, 8352–8358

467. Menon, S., and Ragsdale, S. W. (1998) *Biochemistry* **37**, 5689–5698

467a. Murakami, E., and Ragsdale, S. W. (2000) *J. Biol. Chem.* **275**, 4699–4707

468. Lebertz, H., Simon, H., Courtney, L. F., Benkovic, S. J., Zydowsky, L. D., Lee, K., and Floss, H. G. (1987) *J. Am. Chem. Soc.* **109**, 3173–3174

469. Raybuck, S. A., Bastian, N. R., Zydowsky, L. D., Kobayashi, K., Floss, H. G., Orme-Johnson, W. H., and Walsh, C. T. (1987) *J. Am. Chem. Soc.* **109**, 3171–3173

470. Howell, J. M., and Gawthorne, J. M., eds. (1987) *Copper in Animals and Man*, Vols. I and II, CRC Press, Boca, Raton, Florida

471. Frieden, E. (1968) *Sci. Am.* **218**(May), 103–114

472. Spiro, T. G., ed. (1981) *Copper Proteins*, Wiley, New York

473. Sigel, H., ed. (1981) *Metal Ions in Biological Systems*, Vol. 13, (Copper Proteins) Dekker, New York

474. Linder, M. C. (1991) *Biochemistry of Copper*, Plenum, New York

474a. Turnland, J. R. (1999) in *Modern Nutrition in Health and Disease*, 9th ed. (Shils, M. E., Olson, J. A., Shike, M., and Ross, A. C., eds), pp. 241–252, Williams & Wilkins, Baltimore, Maryland

475. Calabrese, L., Carbonaro, M., and Musci, G. (1988) *J. Biol. Chem.* **263**, 6480–6483

476. Yang, F., Friedrichs, W. E., Cupples, R. L., Bonifacio, M. J., Sanford, J. A., Horton, W. A., and Bowman, B. H. (1990) *J. Biol. Chem.* **265**, 10780–10785

477. Mukhopadhyay, C. K., Mazumder, B., Lindley, P. F., and Fox, P. L. (1997) *Proc. Natl. Acad. Sci. U.S.A.* **94**, 11546–11551

478. Hassett, R., and Kosman, D. J. (1995) *J. Biol. Chem.* **270**, 128–134

479. Georgatsou, E., Mavrogiannis, L. A., and Fragiadakis, G. S. (1997) *J. Biol. Chem.* **272**, 13786–13792

480. Glerum, D. M., Shtanko, A., and Tzagoloff, A. (1996) *J. Biol. Chem.* **271**, 20531–20535

481. Askwith, C., and Kaplan, J. (1998) *Trends Biochem. Sci.* **23**, 135–138

482. Yamaguchi-Iwai, Y., Serpe, M., Haile, D., Yang, W., Kosman, D. J., Klausner, R. D., and Dancis, A. (1997) *J. Biol. Chem.* **272**, 17711–17718

483. Zhu, Z., Labbé, S., Pena, M. M. O., and Thiele, D. J. (1998) *J. Biol. Chem.* **273**, 1277–1280

484. Kampfenkel, K., Kushnir, S., Babiychuk, E., Inzé, D., and Van Montagu, M. (1995) *J. Biol. Chem.* **270**, 28479–28486

485. Danks, D. M. (1995) in *The Metabolic and Molecular Bases of Inherited Disease*, 7th ed., Vol. 2 (Scriver, C. R., Beaudet, A. L., Sly, W. S., and Valle, D., eds), pp. 2211–2235, McGraw-Hill, New York

486. Lutsenko, S., and Cooper, M. J. (1998) *Proc. Natl. Acad. Sci. U.S.A.* **95**, 6004–6009

487. Nielson, K. B., and Winge, D. R. (1984) *J. Biol. Chem.* **259**, 4941–4946

488. Hamer, D. H., Thiele, D. J., and Lemontt, J. E. (1985) *Science* **228**, 685–690

489. Holtzman, N. A. (1976) *Fed. Proc.* **35**, 2276–2280

490. Riordan, J. R., and Jolicoeur-Paquet, L. (1982) *J. Biol. Chem.* **257**, 4639–4645

491. Peltonen, L., Kuivaniemi, H., Palotie, A., Horn, N., Kaitila, I., and Kivirikko, K. I. (1983) *Biochemistry* **22**, 6156–6163

492. Hung, I. H., Suzuki, M., Yamaguchi, Y., Yuan, D. S., Klausner, R. D., and Gitlin, J. D. (1997) *J. Biol. Chem.* **272**, 21461–21466

493. Payne, A. S., and Gitlin, J. D. (1998) *J. Biol. Chem.* **273**, 3765–3770

494. Zhou, B., and Gitschier, J. (1997) *Proc. Natl. Acad. Sci. U.S.A.* **94**, 7481–7486

495. Lutsenko, S., Petrukhin, K., Cooper, M. J., Gilliam, C. T., and Kaplan, J. H. (1997) *J. Biol. Chem.* **272**, 18939–18944

495a. La Fontaine, S., Firth, S. D., Camakaris, J., Englezou, A., Theophilos, M. B., Petris, M. J., Howie, M., Lockhart, P. J., Greenough, M., Brooks, H., Reddel, R. R., and Mercer, J. F. B. (1998) *J. Biol. Chem.* **273**, 31375–31380

495b. Forbes, J. R., Hsi, G., and Cox, D. W. (1999) *J. Biol. Chem.* **274**, 12408–12413

495c. Cobine, P. A., George, G. N., Winzor, D. J., Harrison, M. D., Mogahaddas, S., and Dameron, C. T. (2000) *Biochemistry* **39**, 6857–6863

496. Yamaguchi, Y., Heiny, M. E., Suzuki, M., and Gitlin, J. D. (1996) *Proc. Natl. Acad. Sci. U.S.A.* **93**, 14030–14035

497. Yuan, D. S., Stearman, R., Dancis, A., Dunn, T., Beeler, T., and Klausner, R. D. (1995) *Proc. Natl. Acad. Sci. U.S.A.* **92**, 2632–2636

498. Phung, L. T., Ajlani, G., and Haselkorn, R. (1994) *Proc. Natl. Acad. Sci. U.S.A.* **91**, 9651–9654

499. Klomp, L. W. J., Lin, S.-J., Yuan, D. S., Klausner, R. D., Culotta, V. C., and Gitlin, J. D. (1997) *J. Biol. Chem.* **272**, 9221–9226

500. Rae, T. D., Schmidt, P. J., Pufahl, R. A., Culotta, V. C., and O'Halloran, T. V. (1999) *Science* **284**, 805–808

500a. O'Halloran, T. V., and Culotta, V. C. (2000) *J. Biol. Chem.* **275**, 25057–25060

501. Valentine, J. S., and Gralla, E. B. (1997) *Science* **278**, 817–818

501a. Harrison, M. D., Jones, C. E., Solioz, M., and Dameron, C. T. (2000) *Trends Biochem. Sci.* **25**, 29–32

502. Hoitnik, C. W. G., Driscoll, P. C., Hill, H. A. O., and Canters, G. W. (1994) *Biochemistry* **33**, 3560–3571

503. Romero, A., Nar, H., Huber, R., Messerschmidt, A., Kalverda, A. P., Canters, G. W., Durley, R., and Mathews, F. S. (1994) *J. Mol. Biol.* **236**, 1196–1211

504. Kalverda, A. P., Wymenga, S. S., Lomman, A., van de Ven, F. J. M., Hilbers, C. W., and Canters, G. W. (1994) *J. Mol. Biol.* **240**, 358–371

505. Guss, J. M., Merritt, E. A., Phizackerley, R. P., and Freeman, H. C. (1996) *J. Mol. Biol.* **262**, 686–705

506. Garrett, T. P. J., Clingeleffer, D. J., Guss, J. M., Rogers, S. J., and Freeman, H. C. (1984) *J. Biol. Chem.* **259**, 2822–2825

507. Ryde, U., Olsson, M. H. M., Pierloot, K., and Roos, B. O. (1996) *J. Mol. Biol.* **261**, 586–596

508. Guckert, J. A., Lowery, M. D., and Solomon, E. I. (1995) *J. Am. Chem. Soc.* **117**, 2817–2844

References

509. Groeneveld, C. M., and Canters, G. W. (1988) *J. Biol. Chem.* **263**, 167–173

510. Karlsson, B. G., Tsai, L.-C., Nar, H., Sanders-Loehr, J., Bonander, N., Langer, V., and Sjölin, L. (1997) *Biochemistry* **36**, 4089–4095

511. van de Kamp, M., Canters, G. W., Wijmenga, S. S., Lommen, A., Hilbers, C. W., Nar, H., Messerschmidt, A., and Huber, R. (1992) *Biochemistry* **31**, 10194–10207

512. Messerschmidt, A., Prade, L., Kroes, S. J., Sanders-Loehr, J., Huber, R., and Canters, G. W. (1998) *Proc. Natl. Acad. Sci. U.S.A.* **95**, 3443–3448

513. Farver, O., Licht, A., and Pecht, I. (1987) *Biochemistry* **26**, 7317–7321

514. Vila, A. J., and Fernández, C. O. (1996) *J. Am. Chem. Soc.* **118**, 7291–7298

515. Casimiro, D. R., Toy-Palmer, A., Blake, R. C., II, and Dyson, H. J. (1995) *Biochemistry* **34**, 6640–6648

516. Grossmann, J. G., Engledew, W. J., Harvey, I., Strange, R. W., and Hasnain, S. S. (1995) *Biochemistry* **34**, 8406–8414

517. Botuyan, M. V., Toy-Palmer, A., Chung, J., Blake, R. C., II, Beroza, P., Case, D. A., and Dyson, H. J. (1996) *J. Mol. Biol.* **263**, 752–767

518. Hildebrandt, P., Matysik, J., Schrader, B., Scharf, B., and Engelhard, M. (1994) *Biochemistry* **33**, 11426–11431

519. McManus, J. D., Brune, D. C., Han, J., Sanders-Loehr, J., Meyer, T. E., Cusanovich, M. A., Tollin, G., and Blankenship, R. E. (1992) *J. Biol. Chem.* **267**, 6531–6540

520. Penfield, K. W., Gerwith, A. A., and Solomon, E. I. (1985) *J. Am. Chem. Soc.* **107**, 4519–4529

521. Blair, D. F., Campbell, G. W., Schoonover, J. R., Chan, S. I., Gray, H. B., Malmstrom, B. G., Pecht, I., Swanson, B. I., Woodruff, W. H., Cho, W. K., English, A. M., Fry, H. A., Lum, V., and Norton, K. A. (1985) *J. Am. Chem. Soc.* **107**, 5755–5766

521a. Robinson, H., Ang, M. C., Gao, Y.-G., Hay, M. T., Lu, Y., and Wang, A. H.-J. (1999) *Biochemistry* **38**, 5677–5683

521b. Holz, R. C., Bennett, B., Chen, G., and Ming, L.-J. (1998) *J. Am. Chem. Soc.* **120**, 6329–6335

522. Fridovich, I. (1995) *Ann. Rev. Biochem.* **64**, 97–112

523. Beck, B. L., Tabatabai, L. B., and Mayfield, J. E. (1990) *Biochemistry* **29**, 372–376

524. Tainer, J. A., Getzoff, E. D., Richardson, J. S., and Richardson, D. C. (1983) *Nature (London)* **306**, 284–287

525. Pesce, A., Capasso, C., Battistoni, A., Folcarelli, S., Rotilio, G., Desideri, A., and Bolognesi, M. (1997) *J. Mol. Biol.* **274**, 408–420

526. Rypniewski, W. R., Mangani, S., Bruni, B., Orioli, P. L., Casati, M., and Wilson, K. S. (1995) *J. Mol. Biol.* **251**, 282–296

527. Stenlund, P., Andersson, D., and Tibell, L. A. E. (1997) *Protein Sci.* **6**, 2350–2358

528. Ogihara, N. L., Parge, H. E., Hart, P. J., Weiss, M. S., Goto, J. J., Crane, B. R., Tsang, J., Slater, K., Roe, J. A., Valentine, J. S., Eisenberg, D., and Tainer, J. A. (1996) *Biochemistry* **35**, 2316–2321

529. Beyer, W. F., Jr., Fridovich, I., Mullenbach, G. T., and Hallewell, R. (1987) *J. Biol. Chem.* **262**, 11182–11187

530. Banci, L., Bertini, I., Bauer, D., Hallewell, R. A., and Viezzoli, M. S. (1993) *Biochemistry* **32**, 4384–4388

531. Getzoff, E. D., Cabelli, D. E., Fisher, C. L., Parge, H. E., Viezzoli, M. S., Banci, L., and Hallewell, R. A. (1992) *Nature (London)* **358**, 347–351

532. Afanas'ev, I. B. (1989,1991) *Superoxide Ion: Chemistry and Biological Implications*, Vol. 2, CRC Press, Boca Raton, Florida

533. Hiraishi, H., Terano, A., Razandi, M., Sugimoto, T., Harada, T., and Ivey, K. J. (1992) *J. Biol. Chem.* **267**, 14812–14817

534. Deng, H.-X., Hentati, A., Tainer, J. A., Iqbal, Z., Cayabyab, A., Hung, W.-Y., Getzoff, E. D., Hu, P., Herzfeldt, B., Roos, R. P., Warner, C., Deng, G., Soriano, E., Smyth, C., Parge, H. E., Ahmed, A., Roses, A. D., Hallewell, R. A., Pericak-Vance, M. A., and Siddique, T. (1993) *Science* **261**, 1047–1051

535. McNamara, J. O., and Fridovich, I. (1993) *Nature (London)* **362**, 20–21

536. Wiedau-Pazos, M., Goto, J. J., Rabizadeh, S., Gralla, E. B., Roe, J. A., Lee, M. K., Valentine, J. S., and Bredesen, D. E. (1996) *Science* **271**, 515–518

537. Adman, E. T., Godden, J. W., and Turley, S. (1995) *J. Biol. Chem.* **270**, 27458–27474

538. Suzuki, S., Kohzuma, T., Deligeer, Yamaguchi, K., Nakamura, N., Shidara, S., Kobayashi, K., and Tagawa, S. (1994) *J. Am. Chem. Soc.* **116**, 11145–11146

539. Murphy, M. E. P., Turley, S., and Adman, E. T. (1997) *J. Biol. Chem.* **272**, 28455–28460

540. Kukimoto, M., Nishiyama, M., Murphy, M. E. P., Turley, S., Adman, E. T., Horinouchi, S., and Beppu, T. (1994) *Biochemistry* **33**, 5246–5252

540a. Strange, R. W., Murphy, L. M., Dodd, F. E., Abraham, Z. H. L., Eady, R. R., Smith, B. E., and Hasnain, S. S. (1999) *J. Mol. Biol.* **287**, 1001–1009

541. Howes, B. D., Abraham, Z. H. L., Lowe, D. J., Brüser, T., Eady, R. R., and Smith, B. E. (1994) *Biochemistry* **33**, 3171–3177

542. Murphy, M. E. P., Turley, S., Kukimoto, M., Nishiyama, M., Horinouchi, S., Sasaki, H., Tanokura, M., and Adman, E. T. (1995) *Biochemistry* **34**, 12107–12117

543. Grossmann, J. G., Abraham, Z. H. L., Adman, E. T., Neu, M., Eady, R. R., Smith, B. E., and Hasnain, S. S. (1993) *Biochemistry* **32**, 7360–7366

544. Peters Libeu, C. A., Kukimoto, M., Nishiyama, M., Horinouchi, S., and Adman, E. T. (1997) *Biochemistry* **36**, 13160–13179

544a. Inoue, T., Nishio, N., Suzuki, S., Kataoka, K., Kohzuma, T., and Kai, Y. (1999) *J. Biol. Chem.* **274**, 17845–17852

545. Veselov, A., Olesen, K., Sienkiewicz, A., Shapleigh, J. P., and Scholes, C. P. (1998) *Biochemistry* **37**, 6095–6105

546. Neese, F., Zumft, W. G., Antholine, W. E., and Kroneck, P. M. H. (1996) *J. Am. Chem. Soc.* **118**, 8692–8699

546a. Rasmussen, T., Berks, B. C., Sanders-Loehr, J., Dooley, D. M., Zumft, W. G., and Thomson, A. J. (2000) *Biochemistry* **39**, 12753–12756

547. Gaykema, W. P. J., Hol, W. G. J., Vereijken, J. M., Soeter, N. M., Bak, H. J., and Beintema, J. J. (1984) *Nature (London)* **309**, 23–29

548. Cuff, M. E., Miller, K. I., van Holde, K. E., and Hendrickson, W. A. (1998) *J. Mol. Biol.* **278**, 855–870

549. Hazes, B., Magnus, K. A., Bonaventura, C., Bonaventura, J., Dauter, Z., Kalk, K. H., and Hol, W. G. J. (1993) *Protein Sci.* **2**, 597–619

550. Solomon, E. I., and Lowery, M. D. (1993) *Science* **259**, 1575–1581

551. Kitajima, N., Fujisawa, K., Fujimoto, C., Moro-oka, Y., Hashimoto, S., Kitagawa, T., Toriumi, K., Tatsumi, K., and Nakamura, A. (1992) *J. Am. Chem. Soc.* **114**, 1277–1291

552. Tian, G., and Klinman, J. P. (1993) *J. Am. Chem. Soc.* **115**, 8891–8897

553. Gahmberg, C. G. (1976) *J. Biol. Chem.* **251**, 510–515

554. Driscoll, J. J., and Kosman, D. J. (1987) *Biochemistry* **26**, 3429–3436

555. Clark, K., Penner-Hahn, J. E., Whittaker, M., and Whittaker, J. W. (1994) *Biochemistry* **33**, 12553–12557

556. Wachter, R. M., and Branchaud, B. P. (1996) *Biochemistry* **35**, 14425–14435

557. Halfen, J. A., Jazdzewski, B. A., Mahapatra, S., Berreau, L. M., Wilkinson, E. C., Que, L., Jr., and Tolman, W. B. (1997) *J. Am. Chem. Soc.* **119**, 8217–8227

558. Ito, N., Phillips, S. E. V., Yadav, K. D. S., and Knowles, P. F. (1994) *J. Mol. Biol.* **238**, 794–814

559. Baron, A. J., Stevens, C., Wilmot, C., Seneviratne, K. D., Blakeley, V., Dooley, D. M., Phillips, S. E. V., Knowles, P. F., and McPherson, M. J. (1994) *J. Biol. Chem.* **269**, 25095–25105

560. Whittaker, M. M., Kersten, P. J., Nakamura, N., Sanders-Loehr, J., Schweizer, E. S., and Whittaker, J. W. (1996) *J. Biol. Chem.* **271**, 681–687

560a. Rogers, M. S., Baron, A. J., McPherson, M. J., Knowles, P. F., and Dooley, D. M. (2000) *J. Am. Chem. Soc.* **122**, 990–991

561. Dooley, D. M., Scott, R. A., Knowles, P. F., Colangelo, C. M., McGuirl, M. A., and Brown, D. E. (1998) *J. Am. Chem. Soc.* **120**, 2599–2605

561a. Itoh, S., Taniguchi, M., Takada, N., Nagatomo, S., Kitagawa, T., and Fukuzumi, S. (2000) *J. Am. Chem. Soc.* **122**, 12087–12097

562. Scott, R. A., and Dooley, D. M. (1985) *J. Am. Chem. Soc.* **107**, 4348–4350

563. Suzuki, S., Sakurai, T., and Nakahara, A. (1986) *Biochemistry* **25**, 338–341

563a. Dove, J. E., Schwartz, B., Williams, N. K., and Klinman, J. P. (2000) *Biochemistry* **39**, 3690–3698

564. Williamson, P. R., and Kagan, H. M. (1987) *J. Biol. Chem.* **262**, 8196–8201

565. Uwajima, T., Shimizu, Y., and Terada, O. (1984) *J. Biol. Chem.* **259**, 2748–2753

566. Wachter, R. M., Montague-Smith, M. P., and Branchaud, B. P. (1997) *J. Am. Chem. Soc.* **119**, 7743–7749

567. Whittaker, M. M., Ballou, D. P., and Whittaker, J. W. (1998) *Biochemistry* **37**, 8426–8436

568. Toussaint, O., and Lerch, K. (1987) *Biochemistry* **26**, 8567–8571

569. Wilcox, D. E., Porras, A. G., Hwang, Y. T., Lerch, K., Winkler, M. E., and Solomon, E. I. (1985) *J. Am. Chem. Soc.* **107**, 4015–4027

569a. Decker, H., and Tuczek, F. (2000) *Trends Biochem. Sci.* **25**, 392–397

570. Prigge, S. T., Kolhekar, A. S., Eipper, B. A., Mains, R. E., and Amzel, L. M. (1997) *Science* **278**, 1300–1305

571. Menniti, F. S., Knoth, J., Peterson, D. S., and Diliberto, E. J., Jr. (1987) *J. Biol. Chem.* **262**, 7651–7657

572. Tian, G., Berry, J. A., and Klinman, J. P. (1994) *Biochemistry* **33**, 226–234

573. Meyer, T. E., Marchesini, A., Cusanovich, M. A., and Tollin, G. (1991) *Biochemistry* **30**, 4619–4623

574. Messerschmidt, A., Luecke, H., and Huber, R. (1993) *J. Mol. Biol.* **230**, 997–1014

575. Woolery, G. L., Powers, L., Peisach, J., and Spiro, T. G. (1984) *Biochemistry* **23**, 3428–3434

576. Farver, O., Wherland, S., and Pecht, I. (1994) *J. Biol. Chem.* **269**, 22933–22936

577. Palmer, A. E., Randall, D. W., Xu, F., and Solomon, E. I. (1999) *J. Am. Chem. Soc.* **121**, 7138–7149

578. Lehn, J.-M., Malmström, B. G., Selin, E., and Öblad, M. (1986) *Trends Biochem. Sci.* **11**, 228–230

579. Keen, C. L., Lonnerdal, B., and Hurley, L. S. (1984) in *Biochemistry of the Essential Ultratrace Elements* (Frieden, E., ed), pp. 89–132, Plenum, New York

References

580. Schramm, V. L., and Wedler, F. C., eds. (1986) *Manganese in Metabolism and Enzyme Function,* Academic Press, Orlando, Florida

581. Bartsevich, V. V., and Pakrasi, H. B. (1996) *J. Biol. Chem.* **271**, 26057–26061

582. Baker, A. P., Griggs, L. J., Munro, J. R., and Finkelstein, J. A. (1973) *J. Biol. Chem.* **248**, 800–883

583. Goldman, A., Ollis, D. L., and Steitz, T. A. (1987) *J. Mol. Biol.* **194**, 147–153

584. Kanyo, Z. F., Scolnick, L. R., Ash, D. E., and Christianson, D. W. (1996) *Nature (London)* **383**, 554–557

585. Khangulov, S. V., Sossong, T. M., Jr., Ash, D. E., and Dismukes, G. C. (1998) *Biochemistry* **37**, 8539–8550

586. Zhou, G. W., Guo, J., Huang, W., Fletterick, R. J., and Scanlan, T. S. (1994) *Science* **265**, 1059–1064

587. Hirose, K., Longo, D. L., Oppenheim, J. J., and Matsushima, K. (1993) *FASEB J.* **7**, 361–368

588. MacMillan-Crow, L. A., Crow, J. P., and Thompson, J. A. (1998) *Biochemistry* **37**, 1613–1622

589. Boldt, Y. R., Whiting, A. K., Wagner, M. L., Sadowsky, M. J., Que, L., Jr., and Wackett, L. P. (1997) *Biochemistry* **36**, 2147–2153

590. Mildvan, A. S. (1974) *Ann. Rev. Biochem.* **43**, 357–399

591. Villafranca, J. J., and Wedler, F. C. (1974) *Biochemistry* **13**, 3286–3291

592. Goodman, M. F., Keener, S., Guidotti, S., and Branscomb, E. W. (1983) *J. Biol. Chem.* **258**, 3469–3475

593. Deuel, T. F., and Prusiner, S. (1974) *J. Biol. Chem.* **249**, 257–264

594. Schwartz, K., and Mertz, W. (1959) *Arch. Biochem. Biophys.* **85**, 292–295

595. Mertz, W. (1993) *J. Nutr. Sci. Vitaminol.* **123**, 626–633

596. Davis, C. M., Sumrall, K. H., and Vincent, J. B. (1996) *Biochemistry* **35**, 12963–12969

597. Saner, G. (1980) *Chromium in Nutrition and Disease,* Liss, New York

598. Frieden, E. (1972) *Sci. Am.* **227**(Jul), 52–60

599. Anderson, R. A., Bryden, N. A., and Canary, J. J. (1991) *Am. J. Clin. Nutr.* **51**, 864–868

600. Davis, C. M., and Vincent, J. B. (1997) *Arch. Biochem. Biophys.* **339**, 335–343

601. O'Dell, B. L., and Campbell, B. J. (1970) *Comprehensive Biochemistry* **21**, 179–226

602. Zhang, L., and Lay, P. A. (1996) *J. Am. Chem. Soc.* **118**, 12624–12637

602a. Thompson, K. H. (1999) *BioFactors* **10**, 43–51

603. Zhitkovich, A., Voitkun, V., and Costa, M. (1996) *Biochemistry* **35**, 7275–7282

604. Anderson, R. A. (1994) in *Risk Assessment of Essential Elements* (Merz, W., Abernathy, C. O., and Olin, S. S., eds), pp. 187–196, ISLI Press, Washington, D.C.

605. Conconi, A., Smerdon, M. J., Howe, G. A., and Ryan, C. A. (1996) *Nature (London)* **383**, 826–829

606. Stearns, D. M., Wise, J. P., Sr., Patierno, S. R., and Wetterhahn, K. E. (1995) *FASEB J.* **9**, 1643–1648

607. Stearns, D. M., BelBruno, J. J., and Wetterhahn, K. E. (1995) *FASEB J.* **9**, 1650–1657

608. Macara, I. G. (1980) *Trends Biochem. Sci.* **5**, 92–94

609. Nechay, B. R., Nanninga, L. B., Nechay, P. S. E., Post, R. L., Grantham, J. J., Macara, I. G., Kubena, L. F., Phillips, T. D., and Nielsen, F. H. (1986) *Fed. Proc.* **45**, 123–132

610. Chasteen, N. D. (1983) *Struct. Bonding* **53**, 104–138

611. Boyd, D. W., and Kustin, K. (1985) *Adv. Inorg. Biochem.* **6**, 311–365

612. Tullius, T. D., Gillum, W. O., Carlson, R. M. K., and Hodgson, K. O. (1980) *J. Am. Chem. Soc.* **102**, 5670–5676

613. Ryan, D. E., Grant, K. B., Nakanishi, K., Frank, P., and Hodgson, K. O. (1996) *Biochemistry* **35**, 8651–8661

614. Orgel, L. E. (1980) *Trends Biochem. Sci.* **5**, X (Aug.)

615. Ryan, D. E., Grant, K. B., and Nakanishi, K. (1996) *Biochemistry* **35**, 8640–8650

616. Chasteen, N. D., ed. (1990) *Vanadium in Biological Systems,* Kluwer Acad. Publ., Dordrecht, Germany

617. Winter, G. E. M., and Butler, A. (1996) *Biochemistry* **35**, 11805–11811

618. Wever, R., and Krenn, B. E. (1990) in *Vanadium in Biological Systems* (Chasteen, N. D., ed), pp. 81–97, Kluwer Acad. Publ., Dordrecht, Germany

618a. Isupov, M. N., Dalby, A. R., Brindley, A. A., Izumi, Y., Tanabe, T., Murshudov, G. N., and Littlechild, J. A. (2000) *J. Mol. Biol.* **299**, 1035–1049

618b. Weyand, M., Hecht, H.-J., Kiess, M., Liaud, M.-F., Vilter, H., and Schomburg, D. (1999) *J. Mol. Biol.* **293**, 595–611

619. Haupts, U., Tittor, J., Bamberg, E., and Oesterhelt, D. (1997) *Biochemistry* **36**, 2–7

620. Lindquist, Y., Schneider, G., and Vihko, P. (1994) *Eur. J. Biochem.* **221**, 139–142

621. Hemrika, W., Renirie, R., Dekker, H. L., Barnett, P., and Wever, R. (1997) *Proc. Natl. Acad. Sci. U.S.A.* **94**, 2145–2149

622. Chatterjee, R., Allen, R. M., Ludden, P. W., and Shah, V. K. (1997) *J. Biol. Chem.* **272**, 21604–21608

623. Zhang, M., Zhou, M., Van Etten, R. L., and Stauffacher, C. V. (1997) *Biochemistry* **36**, 15–23

624. Elberg, G., Li, J., and Shechter, Y. (1994) *J. Biol. Chem.* **269**, 9521–9527

625. Li, J., Elberg, G., Crans, D. C., and Shechter, Y. (1996) *Biochemistry* **35**, 8314–8318

626. Heyliger, C. E., Tahiliani, A. G., and McNeill, J. H. (1985) *Science* **227**, 1474–1476

627. Johnson, J. L., Rajagopalan, K. V., and Cohen, H. J. (1974) *J. Biol. Chem.* **249**, 859–866

628. Barberà, A., Rodríguez-Gil, J. E., and Guinovart, J. J. (1994) *J. Biol. Chem.* **269**, 20047–20053

629. Huyer, G., Liu, S., Kelly, J., Moffat, J., Payette, P., Kennedy, B., Tsaprailis, G., Gresser, M. J., and Ramachandran, C. (1997) *J. Biol. Chem.* **272**, 843–851

630. Li, J., Elberg, G., Gefel, D., and Shechter, Y. (1995) *Biochemistry* **34**, 6218–6225

631. Mikalsen, S.-O., and Kaalhus, O. (1998) *J. Biol. Chem.* **273**, 10036–10045

632. Hille, R. (1996) *Chem. Rev.* **96**, 2757–2816

633. Romão, M. J., Archer, M., Moura, I., Moura, J. J. G., LeGall, J., Engh, R., Schneider, M., Hof, P., and Huber, R. (1995) *Science* **270**, 1170–1176

634. Hille, R., and Nishino, T. (1995) *FASEB J.* **9**, 995–1003

635. Harris, C. M., and Massey, V. (1997) *J. Biol. Chem.* **272**, 8370–8379

636. Glatigny, A., Hof, P., Romao, M. J., Huber, R., and Scazzocchio, C. (1998) *J. Mol. Biol.* **278**, 431–438

637. Xiang, Q., and Edmondson, D. E. (1996) *Biochemistry* **35**, 5441–5450

638. Harris, C. M., and Massey, V. (1997) *J. Biol. Chem.* **272**, 28335–28341

639. Glatigny, A., and Scazzocchio, C. (1995) *J. Biol. Chem.* **270**, 3534–3550

640. Bläse, M., Bruntner, C., Tshisuaka, B., Fetzner, S., and Lingens, F. (1996) *J. Biol. Chem.* **271**, 23068–23079

641. Canne, C., Stephan, I., Finsterbusch, J., Lingens, F., Kappl, R., Fetzner, S., and Hüttermann, J. (1997) *Biochemistry* **36**, 9780–9790

642. Gladyshev, V. N., Khangulov, S. V., and Stadtman, T. C. (1996) *Biochemistry* **35**, 212–223

643. Huber, R., Hof, P., Duarte, R. O., Moura, J. J. G., Moura, I., Liu, M.-Y., LeGall, J., Hille, R., Archer, M., and Romao, M. J. (1996) *Proc. Natl. Acad. Sci. U.S.A.* **93**, 8846–8851

644. Ori, N., Eshed, Y., Pinto, P., Paran, I., Zamir, D., and Fluhr, R. (1997) *J. Biol. Chem.* **272**, 1019–1025

645. Garrett, R. M., Johnson, J. L., Graf, T. N., Feigenbaum, A., and Rajagopalan, K. V. (1998) *Proc. Natl. Acad. Sci. U.S.A.* **95**, 6394–6398

646. Johnson, J. L., and Wadman, S. K. (1995) in *The Metabolic and Molecular Bases of Inherited Disease,* 7th ed., Vol. 2 (Scriver, C. R., Beaudet, A. L., Sly, W. S., and Valle, D., eds), pp. 2271–2283, McGraw-Hill, New York

646a. Raitsimring, A. M., Pacheco, A., and Enemark, J. H. (1998) *J. Am. Chem. Soc.* **120**, 11263–11278

646b. Brody, M. S., and Hille, R. (1999) *Biochemistry* **38**, 6668–6677

647. Schindelin, H., Kisker, C., Hilton, J., Rajagopalan, K. V., and Rees, D. C. (1996) *Science* **272**, 1615–1621

648. Schneider, F., Löwe, J., Huber, R., Schindelin, H., Kisker, C., and Knäblein, J. (1996) *J. Mol. Biol.* **263**, 53–69

648a. Stewart, L. J., Bailey, S., Bennett, B., Charnock, J. M., Garner, C. D., and McAlpine, A. S. (2000) *J. Mol. Biol.* **299**, 593–600

648b. Rothery, R. A., Trieber, C. A., and Weiner, J. H. (1999) *J. Biol. Chem.* **274**, 13002–13009

648c. Temple, C. A., George, G. N., Hilton, J. C., George, M. J., Prince, R. C., Barber, M. J., and Rajagopalan, K. V. (2000) *Biochemistry* **39**, 4046–4052

648d. Li, H.-K., Temple, C., Rajagopalan, K. V., and Schindelin, H. (2000) *J. Am. Chem. Soc.* **122**, 7673–7680

649. Pollock, V. V., and Barber, M. J. (1997) *J. Biol. Chem.* **272**, 3355–3362

650. Rajagopalan, K. V. (1984) in *Biochemistry of the Essential Ultratrace Elements* (Freiden, E., ed), Plenum, New York

651. Kramer, S. P., Johnson, J. L., Ribeiro, A. A., Millington, D. S., and Rajagopalan, K. V. (1987) *J. Biol. Chem.* **262**, 16357–16363

651a. Rajagopalan, K. V., and Johnson, J. L. (1992) *J. Biol. Chem.* **267**, 10199–10202

651b. Luykx, D. M. A. M., Duine, J. A., and de Vries, S. (1998) *Biochemistry* **37**, 11366–11375

651c. Liu, M. T. W., Wuebbens, M. M., Rajagopalan, K. V., and Schindelin, H. (2000) *J. Biol. Chem.* **275**, 1814–1822

651d. Feng, G., Tintrup, H., Kirsch, J., Nichol, M. C., Kuhse, J., Betz, H., and Sanes, J. R. (1998) *Science* **282**, 1321–1324

652. Cramer, S. P., Solomonson, L. P., Adams, M. W. W., and Mortenson, L. E. (1984) *J. Am. Chem. Soc.* **106**, 1467–1471

653. Berg, J. M., and Holm, R. H. (1985) *J. Am. Chem. Soc.* **107**, 917–925

654. Cramer, S. P., Wahl, R., and Rajagopalan, K. V. (1981) *J. Am. Chem. Soc.* **103**, 7721–7727

655. Shah, V. K., Ugalde, R. A., Imperial, J., and Brill, W. J. (1984) *Ann. Rev. Biochem.* **53**, 231–257

656. Orme-Johnson, W. H. (1985) *Ann. Rev. Biophys. Biophys. Chem.* **14**, 419–459

657. Howes, B. D., Bray, R. C., Richards, R. L., Turner, N. A., Bennett, B., and Lowe, D. J. (1996) *Biochemistry* **35**, 1432–1443

657a. Caldeira, J., Belle, V., Asso, M., Guigliarelli, B., Moura, I., Moura, J. J. G., and Bertrand, P. (2000) *Biochemistry* **39**, 2700–2707

658. Rothery, R. A., Magalon, A., Giordano, G., Guigliarelli, B., Blasco, F., and Weiner, J. H. (1998) *J. Biol. Chem.* **273**, 7462–7469

References

659. Augier, V., Guigliarelli, B., Asso, M., Bertrand, P., Frixon, C., Giordano, G., Chippaux, M., and Blasco, F. (1993) *Biochemistry* **32**, 2013–2023

660. Magalon, A., Asso, M., Guigliarelli, B., Rothery, R. A., Bertrand, P., Giordano, G., and Blasco, F. (1998) *Biochemistry* **37**, 7363–7370

661. Campbell, W. H., and Kinghorn, J. R. (1990) *Trends Biochem. Sci.* **15**, 315–319

662. Garde, J., Kinghorn, J. R., and Tomsett, A. B. (1995) *J. Biol. Chem.* **270**, 6644–6650

663. Ratnam, K., Shiraishi, N., Campbell, W. H., and Hille, R. (1995) *J. Biol. Chem.* **270**, 24067–24072

663a. George, G. N., Mertens, J. A., and Campbell, W. H. (1999) *J. Am. Chem. Soc.* **121**, 9730–9731

664. Stadtman, T. C. (1991) *J. Biol. Chem.* **266**, 16257–16260

665. Berg, B. L., Baron, C., and Stewart, V. (1991) *J. Biol. Chem.* **266**, 22386–22391

666. Berg, B. L., Li, J., Heider, J., and Stewart, V. (1991) *J. Biol. Chem.* **266**, 22380–22385

667. Stewart, V. (1988) *Microbiol. Rev.* **52**, 190–232

668. Sawers, G., Heider, J., Zehelein, E., and Böck, A. (1991) *J. Bacteriol.* **173**, 4983–4993

669. Gladyshev, V. N., Boyington, J. C., Khangulov, S. V., Grahame, D. A., Stadtman, T. C., and Sun, P. D. (1996) *J. Biol. Chem.* **271**, 8095–8100

670. Boyington, J. C., Gladyshev, V. N., Khangulov, S. V., Stadtman, T. C., and Sun, P. D. (1997) *Science* **275**, 1305–1306

671. Khangulov, S. V., Gladyshev, V. N., Dismukes, G. C., and Stadtman, T. C. (1998) *Biochemistry* **37**, 3518–3528

672. George, G. N., Colangelo, C. M., Dong, J., Scott, R. A., Khangulov, S. V., Gladyshev, V. N., and Stadtman, T. C. (1998) *J. Am. Chem. Soc.* **120**, 1267–1273

673. Shuber, A. P., Orr, E. C., Recny, M. A., Schendel, P. F., May, H. D., Schauer, N. L., and Ferry, J. G. (1986) *J. Biol. Chem.* **261**, 12942–12947

674. Muller, U., Willnow, P., Ruschig, U., and Höpner, T. (1978) *Eur. J. Biochem.* **83**, 485–498

675. Rohde, M., Mayer, F., and Meyer, O. (1984) *J. Biol. Chem.* **259**, 14788–14792

676. Ribbe, M., Gadkari, D., and Meyer, O. (1997) *J. Biol. Chem.* **272**, 26627–26633

676a. Dobbek, H., Gremer, L., Meyer, O., and Huber, R. (1999) *Proc. Natl. Acad. Sci. U.S.A.* **96**, 8884–8889

676b. Gremer, L., Kellner, S., Dobbek, H., Huber, R., and Meyer, O. (2000) *J. Biol. Chem.* **275**, 1864–1872

676c. Hänzelmann, P., Dobbek, H., Gremer, L., Huber, R., and Meyer, O. (2000) *J. Mol. Biol.* **301**, 1221–1235

677. Rajagopalan, K. V., and Johnson, J. L. (1992) *J. Biol. Chem.* **267**, 10199–10202

678. Wahl, R. C., Warner, C. K., Finnerty, V., and Rajagopalan, K. V. (1982) *J. Biol. Chem.* **257**, 3958–3962

679. Nagahara, N., Okazaki, T., and Nishino, T. (1995) *J. Biol. Chem.* **270**, 16230–16235

680. Nishino, T., Usami, C., and Tsushima, K. (1983) *Proc. Natl. Acad. Sci. U.S.A.* **80**, 1826–1829

681. Rech, S., Wolin, C., and Gunsalus, R. P. (1996) *J. Biol. Chem.* **271**, 2557–2562

682. Blanchard, C. Z., and Hales, B. J. (1996) *Biochemistry* **35**, 472–478

683. Chan, M. K., Mukund, S., Kletzin, A., Adams, M. W. W., and Rees, D. C. (1995) *Science* **267**, 1463–1469

683a. Hu, Y., Faham, S., Roy, R., Adams, M. W. W., and Rees, D. C. (1999) *J. Mol. Biol.* **286**, 899–914

684. Koehler, B. P., Mukund, S., Conover, R. C., Dhawan, I. K., Roy, R., Adams, M. W. W., and Johnson, M. K. (1996) *J. Am. Chem. Soc.* **118**, 12391–12405

685. Mukund, S., and Adams, M. W. W. (1995) *J. Biol. Chem.* **270**, 8389–8392

686. Johnson, M. K., Rees, D. C., and Adams, M. W. W. (1996) *Chem. Rev.* **96**, 2817–2839

687. Yadav, J., Das, S. K., and Sarkar, S. (1997) *J. Am. Chem. Soc.* **119**, 4315–4316

688. Juszczak, A., Aono, S., and Adams, M. W. W. (1991) *J. Biol. Chem.* **266**, 13834–13841

Study Questions

1. Describe one or more metabolic functions of ions or chelate complexes of ions derived from each of the following metallic elements: Ca, Mg, Fe, Cu, Ni, Co.

2. If the concentration of Cu, Zn-superoxide dismutase (SOD) in a yeast cell is 10 μM, the total copper (bound and free) is 70 μM, and the dissociation constant for loss of Cu^+ from SOD is 6 fM, what will be the concentration of free Cu^+ within the cell? If the cell volume is 10^{-14} liter, how many copper ions will be present in a single cell? See Roe *et al.* (1999) *Science* **284**, 805–808.

3. Outline the metabolic pathways that are utilized by acetic acid-producing bacteria (acetogens) in the stoichiometric conversion of one molecule of glucose into three molecules of acetic acid. Indicate briefly the nature of any unusual coenzymes or metalloproteins that are required.

4. Suppose that you could add a solution containing micromolar concentrations of Cu^{2+}, Mn^{2+}, Fe^{3+}, Co^{2+}, Zn^{2+}, and MoO_4^{2-} and millimolar amounts of Mg^{2+}, Ca^{2+}, and K^+ to a solution that contains a large excess of a mixture of many cellular proteins. What would be the characteristics of the sites that would become occupied by each of these metal ions? How tightly do you think they would be bound?

5. What factors affect the rate of electron transfer from an electron **donor** (atom or molecule) to an **acceptor**?

6. List some mechanism that cells can use to combat the toxicity of metal ions?

7. NADH peroxidase (p. 857, Eq. 15-59) does not contain a transition metal ion. Propose a reasonable detailed mechanism for its action and compare it with mechanisms of action of heme peroxidases.

Index to Volume 1

Page numbers set in **boldface** refer to major discussions.
The symbol *s* after a page number refers to a chemical structure.

I

O

Q

Quanine quartet 208s
Quantum mechanical tunneling 494, 771, 848
Quasi-equivalence
        in oligomers 344
        in virus coats 346
Queosine (nucleoside Q) 234s
Quinone(s) 815 – 822
        as electron carriers 819
Quinonoid-carbanionic intermediate 741, 744s
        from pyridoxal phosphate 743, 744s
Quinoproteins 815 – 818
        copper 815

R

Rab protein 559
Rabies 247
Racemase(s) 284
        amino acid 692
        aspartate 741
Racemization 741
        of amino acids 284
        of lens proteins 85
        PLP-dependent enzymes 741
Rac protein 559
Radicals, free. See Free radicals
Radioautogram (radioautograph, autoradio-
        graph) 111
Radioautography 110
Radioimmunoassays 110
Radioisotope 258
Radiolaria 18
Radius
        collision 462
        covalent 40, 41
        van der Waals 40
Raf 578, 579
Raffinose 169
Ramachandran plot 60, 61
Random order of binding 464
Random coil conformation 69
Rapamycin 488
Rapid equilibrium assumption 467
Rapid photometric methods 468
Ras 577 – 579
        human c-H ras 560s
        P loop 559
Ras oncogenes (ras) 572, 576
Ras protein superfamily 558
Rat liver, composition of 31
Rates
        of approach to equilibrium 458
        of displacement reactions 589, 590
        of enzymatic reactions 455 – 497. See
                also Kinetics
        of substrate binding 463
Rate constants 455
        apparent first order 458
        bimolecular 458, 462
        prediction by Hammett equation 309
Rate equation for enzymes 455
Reaction coordinate 482
Reaction rates, diffusion controlled limit 462
Reaction types in metabolism, 530
        table of 526, 527

Reading frames in nucleic acids 236, 237
Rearrangement reactions 527, 530, 712
        recA protein 219
Receptor(s) 1, 553 – 563
        acetylcholine 422
        alpha adrenergic. See α-Adrenergic
                receptors
        autophosphorylation of 562
        bacterial, for aspartate 561
        beta adrenergic. See β-Adrenergic
                receptors
        on cell surfaces 479
        down regulation of 571
        for hormones 553 – 563
        human adrenergic 555
        for surface proteins 407
        seven-helix structures 555
Receptor theory 479
Reciprocal plots
        for kinetics of enzymes 460, 465, 472
Recognition domains in proteins 367
Rectus configuration 42
Red blood cell. See Erythrocytes
Redox potential
        within cells 767
        of flavoproteins 767
Reduced flavins, reactions with oxygen 794
Reducing equivalents from citric acid cycle
        515
Reducing powers of redox couples 300
Reduction potential(s)
        definition of 300
        table, 301
Reference electrode 300
Refinement of X-ray structure 136
Refolding of proteins 82
Regulation
        of activity of enzymes 539 – 553
        of aspartate carbamoyltransferase 540
        control elements in 535 – 537
        of enzymatic activity 535 – 581
        pacemaker enzymes 535 – 537
Regulators of G-protein signaling (RGS) 559
Regulatory cascade(s) 566, 567
        monocyclic (scheme) 567
Regulatory subunits 348, 540, 541
Relative molecular mass (M_r)
        determination of 108 – 115
        estimation by gel filtration 109
        estimation by PAGE 109
Relaxation methods 468
Relaxation time
        approach to equilibrium 458, 468
        first order reactions 457
Renin 621
Reoviruses 248. See also front cover
Repair systems 16
Replication cycle. See Cell cycle
Repression of enzyme synthesis 536, 538, 539
Repressor(s) 76, 239
        of gene expression 539
Resolvase 229
        Tn3 219
Resonance 45, 46, 46
        energy values, table 299
        in esters and amides 608
Respiration 300
Respiratory chain 783
Respiratory distress syndrome 386
Resting potential of cell membrane 400
Restriction endonuclease(s) 249, 653
Restriction maps 260

Resurrection plant 168
Retention signal 521
Retinoblastoma protein (Rb) 574
Retinol-binding protein 58
Retrieval signal KDEL 521
Retroviruses 248
Reverse transcriptase 248, 257, 657
Reversed phase columns 103
Reversible chemical reactions 284
        kinetics of 458
Rhabdoviruses 247
Rhamnogalacturonan 177
Rhamnose (Rha) 165s, 180
Rheumatoid arthritis 627
Rhinovirus 247
        icosahedral 344s
Rhizopoda 18
Rho protein 558
Rhodophyta 22
Ribbon drawing 64, 65, 71, 240, 243, 336, 338,
        343, 347, 348, 372, 409, 412, 413, 419, 488,
        560, 580, 599, 606, 614, 633, 708, 786, 807,
        840, 841
Ribisco activase 707
Ribitol, conformation 44
Ribitolteichoic acids 431s
Riboflavin 721, 766, 783s
        absorption spectrum 781
        fluorescence 783
        light sensitivity 783
        nutritional requirement 756, 783
        properties 783
Riboflavin 5'-phosphate (flavin mononucleo-
        tide, FMN) 513, 765, 780, 781s
Ribonuclease 264, 647, 648
        fungal 648
        pancreatic, diffusion constant of 461
        refolding of 82
Ribonuclease A
        mechanism of catalysis 647
        NMR spectra of 647
        pK_a values of 647
        reaction mechanism 648
        sequence of 647
        X-ray structural studies 647
        in-line mechanisms 647
Ribonuclease P 649 – 652
Ribonuclease S 647
Ribonucleoprotein domain 244
Ribonucleotide reductase 863 – 865
        active site 865
        cobalamin dependent 871
        enzyme-activated inhibitors 864
Ribose 1, 5-bisphosphate
        in brain 545
Ribose 5-phosphate isomerase 693
D-Ribose (Rib) 162, 163s, 200s
Ribosome(s) 3, 5, 11, 233 – 235
        structure of 233, 234
Ribosylthymidine (Thd) 203
Ribosylthymidine 5'-phosphate (Thd-5'P) 203
Ribothymidine 234
        of tRNA 231
Ribothymidylic acid residue 813
Ribozyme(s) 649 – 652, 239
        artificial 652
        hairpin 649, 651s
        hammerhead 649, 651
        leadzyme, 651s
Ribulose 164s
Ribulose bisphosphate 707s
Ribulose bisphosphate carboxylase (Rubisco)

W

X

Y

Z

Save $10 with this coupon!

Backorder *Volume 2* of Metzler's *Biochemistry, Second Edition* for only $69.95 —
ten dollars off the retail price —when you use this coupon. Reserve your copy **before**
publication* and save money!

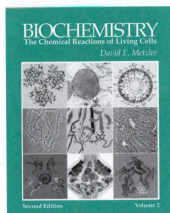

Retail Price ~~$79.95~~ *$69.95*

Hardbound, approx. 900 pages

Available Fall 2001

Table of Contents for *Volume 2*

17. The Organization of Metabolism
18. Electron Transport, Oxidative Phosphorylation, and Hydroxylation
19. The Chemistry of Movement
20. Some Pathways of Carbohydrate Metabolism
21. Specific Aspects of Lipid Metabolism
22. Polyprenyl (Isoprenoid) Compounds
23. Light and Life
24. The Metabolism of Nitrogen and Amino Acids
25. Metabolism of Aromatic Compounds and Nucleic Acid Bases
26. Biochemical Genetics
27. Organization, Replication, Transposition, and Repair of DNA
28. The Transcription of Genes
29. Ribosomes and the Synthesis of Proteins
30. Chemical Communication between Cells
31. Biochemical Defense Mechanisms
32. Growth and Development

**Backordering your copy of Volume 2 of Metzler's *Biochemistry, Second Edition*
(ISBN 0-12-492541-3) is as easy as 1, 2, 3.**

Complete this order form and get it to us one of three ways:

1. **By Fax:** (800) 874-6418
2. **By Phone:** Call our customer service line: 800-321-5068. Please have your credit card available.
3. **By Mail:** Send this form to Marketing & Sales Dept., Harcourt/Academic Press, 525 B Street, Suite 1900, San Diego, CA 92101.

☐ YES, I want to backorder Volume 2 of Metzler's *Biochemistry, Second Edition*
(ISBN 0-12-492541-3) for only $69.95.

☐ MasterCard ☐ Visa ☐ American Express
Account # _____ Expiration Date: _____
Signature _____ (Month/Year)

Name _____
Title _____
Department _____
University _____
Street Address _____
City/State/Zip Code _____

Other courses I teach: _____
Course Title/Number _____
Text in use _____
Course Title/Number _____
Text in use _____

E-mail _____
Phone _____
Course Title/Number _____
Enrollment (per year/semester) _____

How do you prefer to be contacted?
☐ Mail
☐ Phone
☐ E-mail

*Note: This offer is valid only for single-copy purchases of Volume 2 placed as backorders **before** publication. Offer expires upon publication of Volume 2.